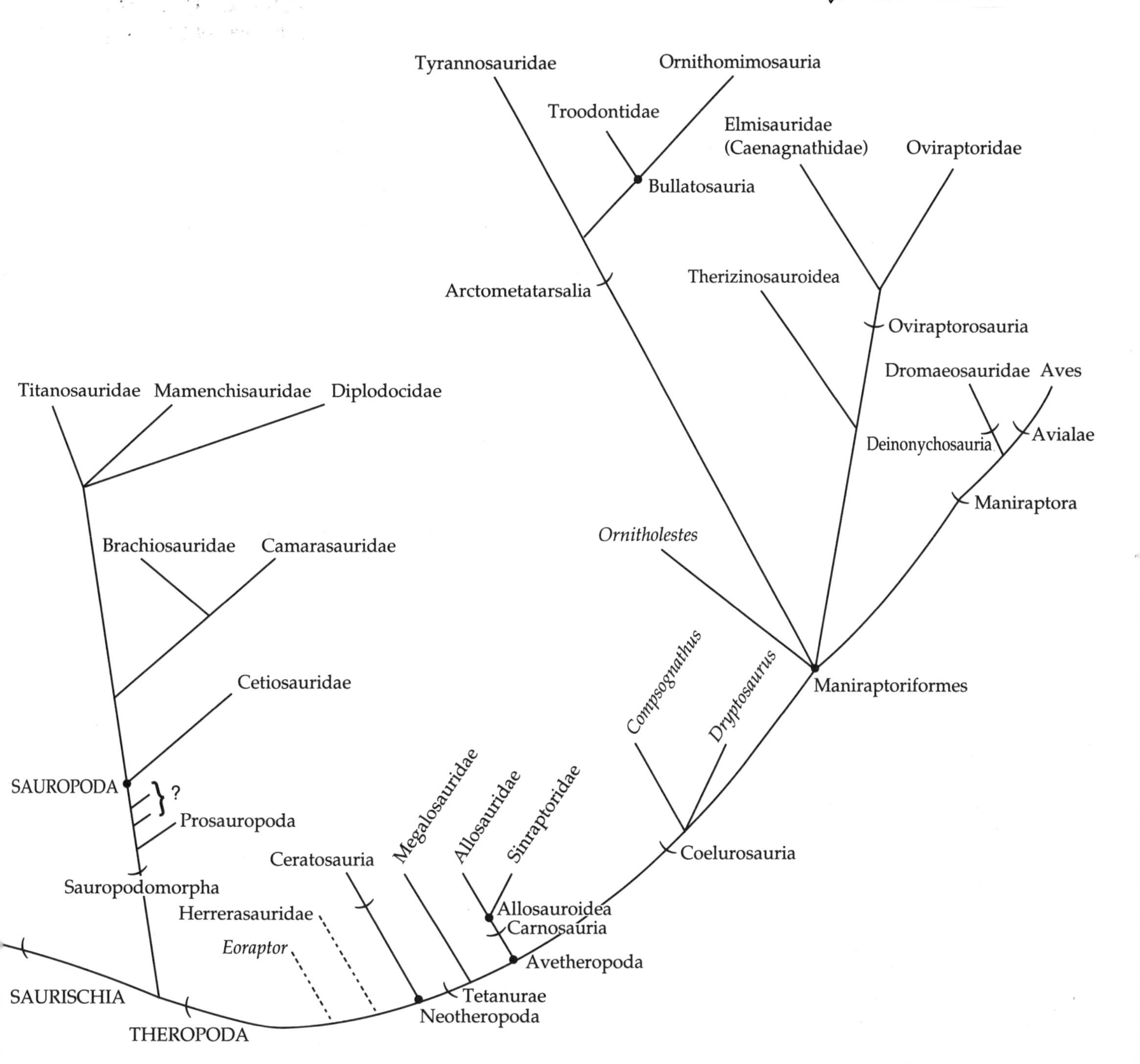
Tyrannosauridae
Ornithomimosauria
Troodontidae
Bullatosauria
Elmisauridae
(Caenagnathidae)
Oviraptoridae
Arctometatarsalia
Therizinosauroidea
Oviraptorosauria
Dromaeosauridae
Aves
Deinonychosauria
Avialae
Maniraptora
Titanosauridae
Mamenchisauridae
Diplodocidae
Ornitholestes
Brachiosauridae
Camarasauridae
Compsognathus
Dryptosaurus
Maniraptoriformes
Cetiosauridae
SAUROPODA
?
Prosauropoda
Megalosauridae
Allosauridae
Sinraptoridae
Coelurosauria
Ceratosauria
Sauropodomorpha
Allosauroidea
Carnosauria
Herrerasauridae
Eoraptor
Avetheropoda
SAURISCHIA
Tetanurae
Neotheropoda
THEROPODA

Encyclopedia of

DINOSAURS

Encyclopedia of DINOSAURS

EDITED BY

Philip J. Currie

Royal Tyrrell Museum of Palaeontology
Drumheller, Alberta, Canada

Kevin Padian

Museum of Paleontology
University of California, Berkeley

ACADEMIC PRESS

San Diego London Boston New York Sydney Tokyo Toronto

Cover Art: *Sinosauropteryx prima*, the feathered dinosaur recently discovered in China, lends additional credence to the hypothesis of a close relationship between dinosaurs and birds. Illustration by Michael Skrepnick.

Frontispiece: Research workers examine Jurassic dinosaur fossils that are exposed on the rock face of the Carnegie Quarry. This site is part of the Dinosaur National Monument, near Jensen, Utah. Photo by Louis Psihoyos.

Letter Openers: A group of Troodontids, in silhouette, pursue a potential prey item. Illustration by Donna Braginetz.

This book is printed on acid-free paper. ∞

Academic Press
a division of Harcourt Brace & Company
525 B Street, Suite 1900, San Diego, California 92101-4495, USA
http://www.apnet.com

Academic Press Limited
24-28 Oval Road, London NW1 7DX, UK
http://www.hbuk.co.uk/ap/

Library of Congress Cataloging-in-Publication Data

Currie, Philip J.
Encyclopedia of dinosaurs / edited by Philip J. Currie, Kevin Padian
p. cm.
Includes index.
ISBN 0-12-226810-5 (alk. paper)
1. Dinosaurs--Encyclopedias. I. Padian, Kevin. II. Title.
QE862.D5C862 1997
567.9'03--dc21 97-23430
CIP

PRINTED IN THE UNITED STATES OF AMERICA
97 98 99 00 01 02 MM 9 8 7 6 5 4 3 2 1

CONTENTS

C

D

E

F

G

H

P

Q

R

Thematic Table of Contents

The Biology of Dinosaurs

CONTRIBUTORS

William L. Abler
Oak Park, Illinois, USA

R. McNeill Alexander
University of Leeds
Leeds, United Kingdom

J. David Archibald
San Diego State University
San Diego, California, USA

Andrea B. Arcucci
Museo de Paleontologica
Universidad Nacional de La Rioja
La Rioja, Argentina

Yoichi Azuma
Fukui Prefectural Museum
Fukui City, Japan

Rinchen Barsbold
Geology Institute
Academy of Sciences of Mongolia
Ulaan Baatar, Mongolia

James F. Basinger
University of Saskatchewan
Saskatoon, Saskatchewan, Canada

Michael J. Benton
University of Bristol
Bristol, United Kingdom

Hervé Bocherens
Université Pierre et Marie Curie
Paris, France

Brent H. Breithaupt
University of Wyoming
Laramie, Wyoming, USA

Michael K. Brett-Surman
George Washington University
Washington, DC, USA

Brooks B. Britt
Museum of Western Colorado
Grand Junction, Colorado, USA

Emily A. Buchholtz
Wellesley College
Wellesley, Massachusetts, USA

Eric Buffetaut
University of Paris VI
Paris, France

Kenneth Carpenter
Denver Museum of Natural History
Denver, Colorado, USA

Ralph E. Chapman
National Museum of Natural History
Smithsonian Institution
Washington, DC, USA

Luis M. Chiappe
American Museum of Natural History
New York, New York, USA

Karen Chin
University of California, Santa Barbara
Santa Barbara, California, USA

Anusuya Chinsamy
South African Museum
Cape Town, South Africa

Per Christiansen
University of Copenhagen
Copenhagen, Denmark

Daniel J. Chure
Dinosaur National Monument
Jensen, Utah, USA

Leon Claessens
Utrecht University
Utrecht, The Netherlands

James M. Clark
George Washington University
Washington, DC, USA

Edwin H. Colbert
Museum of Northern Arizona
Flagstaff, Arizona, USA

Clive Coy
Royal Tyrrell Museum of Palaeontology
Drumheller, Alberta, Canada

Philip J. Currie
Royal Tyrrell Museum of Palaeontology
Drumheller, Alberta, Canada

Brian D. Curtice
Earth Science Museum
Brigham Young University
Provo, Utah, USA

Stephen A. Czerkas
The Dinosaur Museum
Monticello, Utah, USA

Sylvia J. Czerkas
The Dinosaur Museum
Monticello, Utah, USA

Paul G. Davis
National Science Museum
Tokyo, Japan

Lowell Dingus
American Museum of Natural History
New York, New York, USA

Peter Dodson
University of Pennsylvania
Philadelphia, Pennsylvania, USA

Dong Zhiming
Institute of Vertebrate Paleontology and Paleoanthropology
Beijing, China

David A. Eberth
Royal Tyrrell Museum of Paleaontology
Drumheller, Alberta, Canada

Gregory M. Erickson
University of California, Berkeley
Berkeley, California, USA

James O. Farlow
Indiana University at Fort Wayne
Fort Wayne, Indiana, USA

Anthony R. Fiorillo
Dallas Museum of Natural History
Dallas, Texas, USA

Catherine A. Forster
State University of New York at Stony Brook
Stony Brook, Long Island, New York, USA

Eberhard Frey
Staatliches Museum für Naturkunde
Karlsruhe, Germany

Peter M. Galton
University of Bridgeport
Bridgeport, Connecticut, USA

Roland A. Gangloff
University of Alaska Museum
Fairbanks, Alaska, USA

Donald F. Glut
The Dinosaur Society
Burbank, California, USA

William R. Hammer
Augustana College
Rock Island, Illinois, USA

Hartmut Haubold
Martin Luther Universität
Halle, Germany

Rene Hernandez-Rivera
Instituto de Geología UNAM
Del. Coyoacan, Mexico

Ella Hoch
Geological Museum
University of Copenhagen
Copenhagen, Denmark

John R. Horner
Museum of the Rockies
Montana State University
Bozeman, Montana, USA

John R. Hutchinson
Museum of Paleontology
University of California, Berkeley
Berkeley, California, USA

Louis L. Jacobs
Shuler Museum of Paleontology
Southern Methodist University
Dallas, Texas, USA

Aase R. Jacobsen
Royal Tyrrell Museum of Palaeontology
Drumheller, Alberta, Canada

Tom Jerzykiewicz
Geological Survey of Canada
Calgary, Alberta, Canada

Kirk R. Johnson
Denver Museum of Natural History
Denver, Colorado, USA

James I. Kirkland
Dinamation International Society
Fruita, Colorado, USA

Eva B. Koppelhus
Geological Survey of Denmark and Greenland
Copenhagen, Denmark

Jean Le Loeuff
Director, Musée des Dinosaures
Espéraza, Aude, France

J. F. Lerbekmo
University of Alberta
Edmonton, Alberta, Canada

Don Lessem
Co-founder, The Dinosaur Society
Waban, Massachusetts, USA

Martin Lockley
University of Colorado
Denver, Colorado, USA

Donald F. Lofgren
Raymond M. Alf Museum
Claremont, California, USA

John A. Long
Western Australian Museum
Perth, Australia

Robert A. Long
Pleasanton, California, USA

Spencer G. Lucas
New Mexico Museum of Natural History
Albuquerque, New Mexico, USA

Gerhard Maier
Calgary, Alberta, Canada

Peter Makovicky
University of Copenhagen
Copenhagen, Denmark

John Martin
Leicestershire Museum
Leicestershire, United Kingdom

Niall J. Mateer
University of California, Berkeley
Berkeley, California, USA

W. Desmond Maxwell
New York College of Osteopathic Medicine
Old Westbury, New York, USA

John S. McIntosh
Wesleyan University
Middletown, Connecticut, USA

Kenneth J. McNamara
Western Australian Museum
Perth, Australia

Konstantin E. Mikhailov
Russian Academy of Sciences
Moscow, Russia

Angela Milner
Natural History Museum
London, United Kingdom

Ralph E. Molnar
Queensland Museum
Fortitude Valley, Queensland, Australia

José J. Moratalla
Universidad Autónoma de Madrid
Madrid, Spain

Philip A. Murry
Tarleton State University
Stephenville, Texas, USA

Bruce G. Naylor
Royal Tyrrell Museum of Palaeontology
Drumheller, Alberta, Canada

Mark A. Norell
American Museum of Natural History
New York, New York, USA

David B. Norman
Sedgwick Museum
Cambridge University
Cambridge, England, UK

Fernando E. Novas
Museo Argentino de Ciencias Naturales
Buenos Aires, Argentina

Halszka Osmólska
Institute of Paleobiology
Warsaw, Poland

Kevin Padian
Museum of Paleontology
University of California, Berkeley
Berkeley, California, USA

J. Michael Parrish
Northern Illinois University
DeKalb, Illinois, USA

Gregory S. Paul
Baltimore, Maryland, USA

Peng Guangzhao
Zigong Dinosaur Museum
Zigong, Sichuan, China

Bernardino P. Pérez-Moreno
Universidad Autónoma de Madrid
Madrid, Spain

Louie Psihoyos
Boulder, Colorado, USA

Diego Rasskin-Gutman
National Museum of Natural History
Smithsonian Institution
Washington, DC, USA

Robin E. H. Reid
Chesham Bois
Amersham, United Kingdom

Thomas H. Rich
Museum of Victoria
Victoria, Australia

Raymond R. Rogers
Cornell College
Mount Vernon, Iowa, USA

Bruce M. Rothschild
Arthritis Center of Northeast Ohio
Youngstown, Ohio, USA

Timothy Rowe
University of Texas
Austin, Texas, USA

Anthony P. Russell
University of Calgary
Calgary, Alberta, Canada

Dale A. Russell
North Carolina State University
Raleigh, North Carolina, USA

Michael J. Ryan
Royal Tyrrell Museum of Palaeontology
Drumheller, Alberta, Canada

Ashok Sahni
Punjab University
Chandigarh, India

Scott D. Sampson
New York College of Osteopathic Medicine
Old Westbury, New York, USA

P. Martin Sander
Institute of Paleontology
Universität Bonn
Bonn, Germany

Vincent L. Santucci
Slippery Rock University
Slippery Rock, Pennsylvania, USA

José L. Sanz
Universidad Autónoma de Madrid
Madrid, Spain

William A. S. Sarjeant
University of Saskatchewan
Saskatoon, Saskatchewan, Canada

Judith A. Schiebout
Museum of Natural Science
Louisiana State University
Baton Rouge, Louisiana, USA

Mary Schweitzer
Museum of the Rockies
Montana State University
Bozeman, Montana, USA

Paul Sereno
University of Chicago
Chicago, Illinois, USA

David K. Smith
Pima Community College
Tucson, Arizona, USA

Joshua B. Smith
University of Pennsylvania
Philadelphia, Pennsylvania, USA

Kenneth L. Stadtman
Earth Science Museum
Brigham Young University
Provo, Utah, USA

Hans-Dieter Sues
Royal Ontario Museum and
University of Toronto
Toronto, Ontario, Canada

Carl C. Swisher
Berkeley Geochronological Institute
Berkeley, California, USA

Darren H. Tanke
Royal Tyrrell Museum of Palaeontology
Drumheller, Alberta, Canada

Leonid P. Tatarinov
Russian Academy of Sciences
Moscow, Russia

Bruce H. Tiffney
University of California, Santa Barbara
Santa Barbara, California, USA

Yukimitsu Tomida
National Science Museum
Tokyo, Japan

Clive Trueman
University of Bristol
Bristol, United Kingdom

Mary Ann Turner
Peabody Museum of Natural History
Yale University
New Haven, Connecticut, USA

Ron Tykoski
University of Texas
Austin, Texas, USA

Paul Upchurch
Cambridge University
Cambridge, United Kingdom

David J. Varricchio
Old Trail Museum
Choteau, Montana, USA

Matthew K. Vickaryous
Royal Tyrrell Museum of Palaeontology
Drumheller, Alberta, Canada

David B. Weishampel
Johns Hopkins University School of Medicine
Baltimore, Maryland, USA

Dale A. Winkler
Shuler Museum of Paleontology
Southern Methodist University
Dallas, Texas, USA

Lawrence M. Witmer
Ohio University
Athens, Ohio, USA

Joanna Wright
University of Bristol
Bristol, United Kingdom

Xiao-Chun Wu
University of Calgary
Calgary, Alberta, Canada

A Guide to Using the Encyclopedia

The *Encyclopedia of Dinosaurs* is a complete source of information on the subject of dinosaurs, contained within the covers of a single volume. Each article in the Encyclopedia provides an overview of the selected topic to inform a broad spectrum of readers, from researchers to the interested general public.

In order that you, the reader, will derive the maximum benefit from the *Encyclopedia of Dinosaurs*, we have provided this Guide. It explains how the book is organized and how the information within it can be located.

Subject Areas

The *Encyclopedia of Dinosaurs* presents 275 separate articles on the whole range of dinosaur study. It includes information not only on the organisms themselves, of course, but also on all other aspects of this field.

The articles in the *Encyclopedia of Dinosaurs* fall within nine general subject areas, as follows:

- Kinds of Dinosaurs Around the World
- Groups of Dinosaurs and Related Taxa
- The Biology of Dinosaurs
- Environments of the Past
- Important Dinosaur Localities
- Geology and Dinosaurs
- Institutions of Dinosaur Study
- Dinosaur Expeditions
- Dinosaur Research and Techniques

A Thematic Table of Contents appears in the introductory section of the Encyclopedia on page xv. It has a complete list of all the articles in the book, placed according to subject area and in relation to other topics.

Organization

The *Encyclopedia of Dinosaurs* is organized to provide the maximum ease of use for its readers. All of the articles are arranged in a single alphabetical sequence by title, from "A" (Abelisauridae, African Dinosaurs, etc.) to "Z" (Zigong Museum). An alphabetical Table of Contents for the articles can be found on p. v of this introductory section.

As a reader of the Encyclopedia, you can use this alphabetical Table of Contents by itself to locate a topic. Or you can first identify the topic in the Thematic Table of Contents and then go to the alphabetical Table to find the page location.

So that they can be more easily identified, article titles begin with the key word or phrase indicating the topic, with any descriptive terms following this. For example, "Pelvis, Comparative Anatomy" is the title assigned to this article rather than "Comparative Anatomy of the Pelvis," because the specific term *Pelvis* is the key word.

Article Format

Articles in the *Encyclopedia of Dinosaurs* are divided into three general categories. The first category includes concise entries that deal with highly focused topics, such as "Albany Museum," "Cañon City," "Dinoturbation," "Gastralia," and "Tooth Marks." These entries vary in length from one to several paragraphs. The second category, the bulk of the text, includes entries of the standard length of 1 to 4 pages, such as "African Dinosaurs," "Djadokhta Formation," "Dry Mesa Quarry," "Egg Mountain," "Musculature," "Paleoclimatology," and "Trace Fossils."

Major articles constitute the third category. They deal with broad areas of dinosaur study, such as "Behavior," "Bird Origins," "Evolution," and "Systematics." These articles are of extended length, typically 5 to 15 pages, and are identified as to general theme by the following system of symbols:

Kinds of Dinosaurs

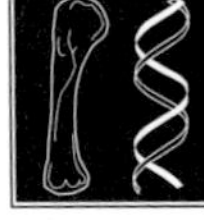
Dinosaur Biology

Geology & Sites

Research Methods

Dinosaur Institutions

Articles in all three categories have been written by individual contributors, as indicated in the article heading. The only exception is that some shorter entries are unsigned. These have been prepared collectively by the general editors, Drs. Currie and Padian.

Cross-References

The *Encyclopedia of Dinosaurs* has an extensive system of cross-referencing. Cross-references to other articles appear in three forms: as marginal headings within the A-to-Z article sequence; as designations within the running text of an article; and as indications of related topics at the end of an article.

As an example of the first type of reference cited above, the following marginal entry appears in the A-to-Z article list between the entries "Archosauria" and "Arctometatarsalia:"

> ***Arctic Dinosaurs*** *see* POLAR DINOSAURS

This reference indicates that the topic of dinosaurs of the Arctic region is discussed elsewhere, under the article title "Polar Dinosaurs."

An example of the second type, a cross-reference within the running text of an article, is this excerpt from the entry "Central Asiatic Expeditions":

> Discovery of non-avian dinosaurs in Central Asia was pioneered by the AMERICAN MUSEUM OF NATURAL HISTORY (AMNH) during a series of swashbuckling Central Asiatic Expeditions in the 1920s.

This indicates that the item "American Museum of Natural History," which is set off in the text by small capital letters, appears as a separate article within the Encyclopedia. (This reference appears here because the Museum was the sponsor of these expeditions.)

An example of the third type, a cross-reference at the end of the article, can be found in the entry "Distribution and Diversity." This article concludes with the statement:

> See also the following related entries:
> BIOGEOGRAPHY • MIGRATION • PLATE TECTONICS

This reference indicates that these three articles all provide some additional information about the distribution and diversity of dinosaurs.

Bibliography

The bibliography section appears as the last element in an article, under the heading "References." This section lists recent secondary sources that will aid the reader in locating more detailed or technical information. Review articles and research papers that are important to a more detailed understanding of the topic are also listed here.

The bibliographic entries in this Encyclopedia are for the benefit of the reader, to provide references for further reading or research on the given topic. Thus they typically consist of a limited number of entries. They are not intended to represent a complete listing of all materials consulted by the author or authors in preparing the article.

Index

The Subject Index for the *Encyclopedia of Dinosaurs* contains more than 5,840 entries. Within the entry for a given topic, references to general coverage of the topic appear first, such as a complete article on the subject. References to more specific aspects of the topic then appear below this in an indented list.

Encyclopedia Website

The *Encyclopedia of Dinosaurs* maintains an editorial Web Page on the Internet at:

http://www.apnet.com/dinosaur/

The *Encyclopedia of Dinosaurs* site provides information about this project and links to many related sites that feature dinosaurs. The site will continue to evolve as more information becomes available.

FOREWORD

We human beings are fascinated by dinosaurs. The first reports of the giant bones of extinct creatures caused a worldwide sensation, and in the century and a half since then, our interest has never diminished. In every country of the world, children and adults are entranced by dinosaurs.

It is often imagined that the current dinosaur mania is a recent phenomenon; in fact it is not. When I first started writing a novel about dinosaurs, back in 1981, I put the project aside because at that time, Americans seemed to be in the grip of an unprecedented dinosaur mania. There were dinosaur cups and saucers, dinosaur toys, dinosaur bedspreads; museums were having dinosaur shows; it seemed there were dinosaurs everywhere you looked. I did not want to write a book that exploited a fashion of the moment. So I waited. But year after year, the fashion never went away. Finally I realized that our fascination with dinosaurs is a permanent phenomenon. It is always there.

Children, of course, have always been captivated by dinosaurs. To go to a museum and see young children, barely able to walk and talk, shrieking "Stegosaurus" and "Tyrannosaurus" as they view the creatures is a very striking thing. Why does it happen? What is going on in their minds as they shout out those complex Latin names? How do we explain the fact that dinosaurs excite the imagination of adults and children throughout the world?

Over the years, I have entertained many theories. For a while, I thought the phenomenon might be characteristic of those countries, like the United States, where many fossils have been found—a kind of nationalistic interest, if you will. But dinosaurs are just as popular in countries such as Japan and Italy, where few remains have been found.

For a while, I thought it was primarily a childish interest. But in museums, you'll notice that adults are equally fascinated. To be honest, it often seems that children are only an excuse for adults to visit the dinosaurs.

Later on, I suspected the interest in dinosaurs might be something that children passed on to each other, a trait of a children's subculture. But my own daughter showed a marked interest in dinosaurs long before she went to preschool—indeed, before she was even very verbal.

Still later, I thought this enthusiasm was provoked by the great size of these creatures. But smaller dinosaurs excite just as much interest among children. Baby dinosaurs are very appealing. And in any case, the dinosaur toys are all small. . . .

For a time, I wondered whether the interest had something to do with the fact that the dinosaurs had become extinct. But children are not clear about this. When my daughter was two years old, she asked to see dinosaurs at the zoo. She had been to the zoo several times, and apparently believed the dinosaurs were housed in some section we hadn't visited yet. When she was told that she could not see dinosaurs, she gave a resigned shrug—parents never do what you want them to do!

Perhaps, I thought, that was a clue. Children spend much of their lives powerless and frustrated. I began to entertain a Freudian notion that being able to pronounce the complex names of huge creatures afforded children a sense of control. In a child's world, after all, everything is big—parents, cars, everything. And naming things is a classic human procedure to reduce anxiety. (Patients are always relieved to hear that they have "idiopathic hypertension," even though the term is literally meaningless.) But once again, careful observation cast doubt on this idea. When my daughter was four, I took her and two friends to Stan Winston's workshop to see the dinosaurs being constructed for Jurassic Park. I thought they'd enjoy it, but they didn't. Although the dinosaurs were then only sculpted in clay, the girls were distressed by what they saw. The animals were simply too big, and too real-looking. It is one thing to play with little dinosaur toys. It is quite another to walk beneath the enormous scaly legs of a towering tyrannosaur, or to touch the big claw of a Velociraptor. The kids were very uneasy. They wanted to leave.

So in the end, I decided I just don't understand why children are fascinated by dinosaurs. And I don't believe anybody understands. In the end, it is a mystery.

And it may be that the mystery is part of the fascination. Certainly for adults, dinosaurs present an intriguing puzzle, in which fantasies are inevitably provoked. Although we know far more about dinosaurs than we did a few decades ago, the truth is that we still know very little. We don't really know what these creatures looked like, or how they behaved. We have some bones, impressions of skin, some trackways, and many fascinating speculations about their biology and social organization. But what hard evidence remains of their long-vanished world is tantalizing and incomplete.

And so they still provoke our dreams. And, probably, they always will.

Michael Crichton

PREFACE

Is it possible that the scientific understanding of dinosaurs has now come so far that a volume such as this, as big as the Manhattan telephone directory, is necessary to compile even a synopsis of what we know about them?

Could Dr. Gideon Mantell and Dean William Buckland, when they described the first remains of what would become known as dinosaurs, have ever imagined the scientific attention that would be paid to them more than 170 years later?

Could Richard Owen, who gave Dinosauria its name in 1842, have intended that his "fearfully great lizards" would be used as metaphors for both evolutionary success and obsolescence?

As we sit today on three decades of the most awesome explosion of knowledge about dinosaurs in history, can we imagine how much more will be learned before most of us now active in the field reach retirement age?

The answers to these questions are probably yes, no, yes, and no, perhaps in no particular order. This is not the first phonebook-sized compendium of dinosaur information, and we wish at the outset to acknowledge both our debt to and admiration for the already classic work, *The Dinosauria* (D. B. Weishampel, P. Dodson, and H. Osmólska, eds.; University of California Press, 1990), which organizes so much of what is known in such an accessible way. This volume should be seen in many respects as a companion to that one, which was dedicated to exploring the individual dinosaurian groups in considerable depth.

Our focus in this *Encyclopedia* is to provide background and a point of reference to the recent literature on dinosaurian subjects in general. The two books can be read in similar and completely different ways; in the organization of the *Encyclopedia* we have tried to foresee the uses that it may serve for its audience. It is unlikely, of course, that Mantell and Buckland could have foreseen much of what has transpired around their *Iguanodon* and *Megalosaurus,* and hundreds of their reptilian kin, so far into the future; indeed, in the 1820s these men had themselves only the scarcest idea of the nature and importance of the fossil bones they were describing. As for Richard Owen, there was little that escaped his eye, and he was always looking to posterity. In his own time he regarded his creations as approaching the great mammals in physiological sophistication; yet, like all reptilian forms of the Secondary Era (Mesozoic), they were doomed to extinction and replacement. But even he, like us toiling in the fields today, can have no idea of what will come next. The possibilities seem almost limitless.

A few words about what this book is and is not. (This part seems to be read only infrequently by reviewers, but we hope it will help the general reader.)

An encyclopedia is designed to be a concise summary of knowledge, ideas, historical background, and current thinking on a general topic. The information for a given entry is not exhaustive; rather, through cross-references and citations to other literature, readers are invited to learn more and explore wider resources. Consequently, the name of every dinosaur, geological formation, quarry, museum, and idea about how dinosaurs lived and died will not be found under its own entry. Many of these names may be found in the indexes at the end of the book, subsumed under other entries, and still others will be gleaned by surfing the cross-references as one would on the Internet. We have assembled a most knowledgeable contingent of experts from all over the globe on various subjects, and we hope that they, and you, enjoy and learn as much from perusing this book as we did in editing it.

For more exhaustive listings of dinosaurian names and histories, a good start can be made in *The Dinosauria* or in Don Glut's *Dinosaurs, the Encyclopedia* (McFarland & Company, Inc., 1997). For an entry into the primary literature about dinosaurs and all extinct backboned things, there is no finer source than the many volumes of the *Bibliography of Fossil Vertebrates*, produced over the years by A. S. Romer, Charles L. Camp, Joseph T. Gregory, Judy Bacskai, George Shkurkin, and a host of colleagues, under the most recent auspices of the American Geological Institute and the Society of Vertebrate Paleontology.

Recent textbook-style works about dinosaurs include D. B. Norman's *The Illustrated Encyclopedia of Dinosaurs* (Salamander, 1985), Spencer G. Lucas's *Dinosaurs: The Textbook* (William C. Brown, 1996), and D. E. Fastovsky and D. B. Weishampel's *The Evolution and Extinction of the Dinosaurs* (Cambridge University Press, 1996). Finally, for younger dinosaur enthusiasts, The Dinosaur Society (East Islip, New York) works with both professional paleontologists and educators to provide the best of what is new and what is known to the many children and adults who want to learn more.

The taxonomic conventions of this book do not follow the venerable Linnean System, in place since 1763 (well before evolution was a mainstream scientific concept), but the newer Phylogenetic System, based on phylogenetic systematics or cladistics. The principles of this system are explained in this work and in others referenced herein, and need not be detailed at this point. Cladistic conventions hold that all taxa (named groups of organisms) must include a common ancestor and all its descendants in theory (i.e., monophyletic groups). All taxa must have both a definition of their ancestry and membership and a diagnosis of the uniquely shared evolutionary features by which they may be recognized. For the sake of stability, we have restricted definitions to node-based and stem-based kinds, as explained in the entry "Phylogenetic System." We have tried to provide definitions for every taxon (stem- and node-based) and diagnoses for every node-based taxon in this book, though many will change with future research. A data matrix of the characteristics and taxa used in phylogenetic analyses is a *sine qua non* of formal systematic research, and it was our initial hope to include matrices in this work; however, the constant revision and expansion of these matrices would soon outdate any printed effort, and we are now more hopeful that these may be made available and updated on CD-ROM or in World Wide Web format in the future.

Controversies are the business of science, which thrives by expanding, testing, or overturning what we think we already know. And dinosaurs are no strangers to debate; many questions from phylogeny to physiology have strong cases for divergent conclusions. In this book we present these controversies as they are seen today. Many are new, some are old; some may well be resolved with further work, and some may never be. The viewpoints of the authors of individual entries may not always coincide; the authors were not recruited because they agreed with each other's conclusions, but because they had something intelligent to say. Of course, not all apparent controversies are real; some have been settled, at least to the satisfaction of the paleontological community's consensus, and excessive attention is not paid to these here. Many apparent controversies regarding dinosaurs live more in the minds of the representatives of the press than in those of the scientists.

Finally, we express our appreciation to our editorial crew at Academic Press, including Gail Rice, Chris Morris, and especially Chuck Crumly, who remained the driving force behind this volume's realization; to Eva Koppelhus, John Hutchinson, and Leakena Au, for pragmatic assistance beyond the call of duty; to the staff of the Tyrrell Museum of Palaeontology, particularly Pat Bobra, and the support of the University of California Museum of Paleontology; to the various individuals, journal editors, publishers, and copyright holders who allowed us to reprint many of the wonderful illustrations in this book; and especially to all our authors, who met deadlines, extraordinary requests, and last-minute pleas with patience and cooperation. To everyone, our best thanks.

Philip Currie and Kevin Padian
Drumheller, Alberta, and Berkeley, California

DEDICATION

To

John H. Ostrom

In a career spanning some forty years in vertebrate paleontology, John Ostrom has worked on so many groups of dinosaurs, and on so many problems relating to the paleobiology of dinosaurs, that he has become *the* central figure in dinosaur research since the mid-1960s. Originally headed for a career in medicine, John became sidetracked under the spell of George Gaylord Simpson and Ned Colbert, from whom he learned vertebrate paleontology at the American Museum of Natural History while a student at Columbia University in the 1950s. He produced several studies of living and extinct amphibians and reptiles, and finally settled on a Ph.D. thesis project studying the skulls of North American hadrosaurs, mainly using the spectacular collections of the AMNH. It was not long before he challenged prevailing ideas about the paleoecology of hadrosaurs, too, suggesting that they were not aquatic but terrestrial. This approach of using anatomy and functional morphology to ask broader questions about paleobiology and behavior would become a hallmark of John Ostrom's work.

After a brief stint teaching at Beloit College, John joined the faculty at Yale, where he has spent the rest of his career. Undoubtedly his most important contributions to the collections of the Yale Peabody Museum were those from the Cloverly Formation of Wyoming, which included various remains of ornithopods, ankylosaurs, and particularly an "unusual" theropod dinosaur that John christened *Deinonychus*, or "terrible claw," in 1969. This beast was to change our concept of dinosaurs in more ways than one. John brought to life an animal that stalked its prey with jaws full of large teeth, long arms with prehensile hands and sharp talons, and a foot that bore a huge curved claw on its second toe, which could not have been used in walking. *Deinonychus* captured the imagination of dinosaur fans everywhere, notably in Michael Crichton's *Jurassic Park.* But that was only a small part of the maelstrom of interest in dinosaurs that grew from John's work.

In 1970 John attended the First North American Paleontological Convention and presented an innocent-sounding paper called "Terrestrial Vertebrates as Indicators of Mesozoic Climates." In it, he maintained that the zoogeography and behavior of living reptiles were inappropriately stereotyped, and that Mesozoic dinosaurs were at least as widespread and ecologically varied. His ideas on the subject were amplified and studied further by his student Robert Bakker, who was largely responsible for popularizing the "renaissance" of ideas about dinosaurian warm-bloodedness. These debates thrived for a decade, and still survive today in modified form.

Perhaps John's greatest contribution to dinosaurs was to recognize that a whole group of them was largely unrecognized: namely, birds. In 1973 John first advanced a synopsis of the evidence that birds had descended from small coelurosaurian dinosaurs. He had gotten this idea (which can be traced back to T. H. Huxley, but was long abandoned) while in Germany studying pterosaurs in the company of his old friend Peter Wellnhofer. John had traveled to Haarlem, in The Netherlands, to look at their relatively small collection of Solnhofen pterosaurs, when he found that one specimen, known only from a hindlimb, had a very dinosaur-like foot. When he looked more closely, he also saw impressions of feathers. This was obviously not a pterosaur, but an *Archaeopteryx*.

As John tells the story, he was faced with a dilemma. Should he ask to borrow the specimen, bring it home, and then "suddenly" discover its true identity? Or should he come clean from the start, risking the loss of ever seeing it again once its value became apparent? He took the latter course. The old, white-haired curator was nonplussed; he gasped and wiped his brow. "You have made our museum famous," he managed to say gratefully, and walked away with the newly precious relic. John forlornly began to gather his things, and was about to leave when the curator reappeared—with the world's newest *Archaeopteryx* specimen wrapped in a shoebox, tied with string. He handed John the box, and history took its course.

The *Archaeopteryx* specimen at Teyler, like the others, had reminded John not only of dinosaurs, but of theropod dinosaurs—particularly of the ones he had been recently studying at Yale, such as *Deinonychus* and *Ornitholestes*, and *Compsognathus* in Munich. Detailed comparisons of these animals, as well as living birds and more distantly related archosaurs, resulted in a series of papers in which John established that birds evolved from small coelurosaurs, probably sometime in the Middle or Late Jurassic. Controversy has flared around this issue intermittently for nearly 25 years, yet it is one of the more firmly established hypotheses in vertebrate history. But John was not content with the origin of birds; he wanted to explore the origin of their flight. To him, the terrestrial, predatory habits of the theropod relatives of birds suggested their origin from the ground up, running and flapping, perhaps initially after small insects. John's heuristic model of *Archaeopteryx,* using its wings as flyswatters, attracted support, intrigue, and brickbats, but stimulated a look into this question from many disciplines and a great deal of further research on the origin of major evolutionary adaptations.

Perhaps one of the reasons for John's pervasive influence, both in the professional field and among interested laymen, is the clarity and simplicity of his writing style. Had he chosen to obfuscate his ideas in a mountain of impenetrable scientific prose, they would not have gotten much of the attention they have. But he has always written directly, modestly, and accessibly, avoiding hyperbole and dogma. (How many paleontologists could have confined themselves to the understatement of calling *Deinonychus* "an unusual theropod"?) He has been a model for his students and colleagues alike, and many entries in this book testify to the endurance and importance of his work and his thought. On behalf of all the authors, it is our great pleasure to dedicate the *Encyclopedia of Dinosaurs* to John Ostrom.

The Editors

Abelisauridae

FERNANDO E. NOVAS
Museo Argentino de Ciencias Naturales
Buenos Aires, Argentina

Abelisauridae constitutes a clade of CRETACEOUS theropods widely documented in the Gondwanan continents (South America, Madagascar, and India). As far as Argentina is concerned, abelisaurids are the best represented group of predatory dinosaurs from the Cretaceous deposits in both number of specimens and species diversity. Abelisaurids underwent a significant evolutionary radiation during the Cretaceous in South America, becoming active predators of large size (see SOUTH AMERICAN DINOSAURS).

Abelisauridae was originally established by Bonaparte and Novas (1985) to include the Late Cretaceous Patagonian dinosaur *Abelisaurus comahuensis* as a mean of emphasizing the distinctness of this taxon with respect to the remaining theropods. The list of abelisaurids, however, rapidly increased: *Carnotaurus sastrei* (Bonaparte *et al.*, 1990), *Xenotarsosaurus bonapartei* (Martínez *et al.*, 1986), *Indosaurus matleyi* and *Indosuchus raptorius* (both taxa from the Lameta Group, Late Cretaceous of India; von Huene and Matley, 1933; Chatterjee, 1978; Bonaparte and Novas, 1985; Bonaparte, 1991), yet undescribed *Carnotaurus*-like creatures (Coria and Salgado, 1993), and *Majungasaurus crenatissimus* (Maevarano Formation, Campanian; Madagascar; Lavocat, 1955; Bonaparte; 1991; Sampson *et al.*, 1996) were added to the family.

Abelisauridae is defined to include the previously listed taxa and all the descendants of their common ancestor. Diagnostic characters of Abelisauridae include craniocaudally short and deep premaxilla; dorsoventrally deep snout at the level of the narial openings; frontals dorsoventrally thickened resulting in a dorsal bulking (*Abelisaurus*), paired horn-like structures (*Carnotaurus*) or a dome-like prominence (*Majungasaurus*; Sampson *et al.*, 1996); posterior surface of basioccipital wide and smooth below occipital condyle; dentary short, with convex ventral margin; and loose contacts among dentary, splenial, and postdentary bones.

Abelisauridae seems to be related to the small theropod *Noasaurus leali* (Bonaparte and Powell, 1980) because both taxa share maxillae with subvertical ascending rami and cervical vertebrae with hypertrophied epipophyses and reduced neural spines (Bonaparte, 1991; Bonaparte *et al.*, 1990; Novas, 1991, 1992). Novas (1991) coined the name Abelisauria to encompass Abelisauridae, *Noasaurus*, and their most recent common ancestor. *Ligabueino andesi* resembles abelisaurs in the morphology of the cervical vertebrae (Bonaparte, 1996).

Interestingly, abelisaurids show close resemblances with the Cenomanian carcharodontosaurids *Giganotosaurus carolinii* and *Carcharodontosaurus saharicus* (Coria and Salgado, 1995; Sereno *et al.*, 1996). These taxa exhibit many derived traits in common (e.g., antorbital fossae reduced, similar patterns of rugosities on external surfaces of nasals and maxillae, preorbital openings anteroposteriorly expanded, wide contacts between lacrimals and postorbitals forming thick "brows" above orbits, and eyes enclosed by sinuous orbital margins of lacrimals and postorbitals), suggesting that they are more closely related than previously thought (Coria and Salgado, 1995; Rauhut, 1994; Sereno *et al.*, 1996). The resemblances noted among the different South American, Malagasian, and Indian taxa suggest Gondwanan origins for abelisaurs plus carcharodontosaurids, contrary to some recent proposals (e.g., Sereno *et al.*, 1996).

The phylogenetic relationships of Abelisauridae need to be studied in depth. *Abelisaurus* and *Carnotaurus*, at least, exhibit several apomorphic resemblances in the morphology of the dorsal and sacral vertebrae with the JURASSIC *Ceratosaurus nasicornis*, and the taxon Neoceratosauria has been erected for this clade (Novas, 1991, 1992).

References

Bonaparte, J. F. (1991). The Gondwanan theropod families Abelisauridae and Noasauridae. *Historical Biol.* **5,** 1–25.

Bonaparte, J. F. (1996). Cretaceous tetrapods of Argentina. *Münchner Geowissenchaftliche Abhandlungen (A)* **30,** 73–130.

Bonaparte, J. F., and Novas, F. E. (1985). *Abelisaurus comahuensis*, n.g. et n.sp., Carnosauria del Cretácico Tardío de Patagonia. *Ameghiniana* **21**(2–4), 259–265.

Bonaparte, J. F., and Powell, J. E. (1980). A continental assemblage of tetrapods from the Upper Cretaceous beds of El Brete, northwestern Argentina. *Mem. Soc. Geol. France N. Ser.* **139,** 19–28.

Bonaparte, J. F., Novas, F. E., and Coria, R. A. (1990). *Carnotaurus sastrei* Bonaparte, the horned, lightly built carnosaur from the middle Cretaceous of Patagonia. *Contrib. Sci. Nat. History Museum Los Angeles County* **416,** 1–42.

Chatterjee, S. (1978). *Indosuchus* and *Indosaurus*, Cretaceous carnosaurs from India. *Journal of Paleontology* **52,** 570–580.

Coria, R. A., and Salgado, L. (1993). Un probable nuevo neoceratosauria Novas, 1989 (Theropoda) del Miembro Huincul, Formación Río Limay (Cretácico presenoniano) de Neuquen. *Ameghiniana* **30**(3), 327.

Coria, R. A., and Salgado, L. (1995). A new giant carnivorous dinosaur from the Cretaceous of Patagonia. *Nature* **377,** 224–226.

Lavocat, R. (1955). Sur une portion de mandibule de théropode provenant du Crétace supérieur de Madagascar. *Bull. Musee Natl. d'Histoire Nat. Ser. 2* **27,** 256–259.

Martínez, R., Ginmenez, O., Rodriguez, J., and Bochatey, G. (1986). *Xenotarsosaurus bonapartei* gen. et. sp. nov. (Carnosauria, Abelisauridae), un nuevo Theropoda de la Formación Bajo Barreal, Chubut, Argentina. *IV Congr. Argentino Paleontol. Bioestratigr. Actas* **2,** 23–31.

Novas, F. E. (1991). Relaciones filogenéticas de los dinosaurios terópodos ceratosaurios. *Ameghiniana* **28**(3/4), 410.

Novas, F. E. (1992). La evolución de los dinosaurios carnívoros. In *Los dinosaurios y su entorno biótico. Actas II Curso de Paleontología en Cuenca* (J. L. Sanz and A. Buscalioni, Eds.), pp. 125–163. Instituto "Juan de Valdés," Ayuntamiento de Cuenca, España.

Rauhut, O. (1994). Zur systematischen Stellung der afrikanischen Theropoden *Carcharodontosaurus* Stromer 1931 und *Bahariasaurus* Stromer 1934. *Berliner Geowissenchaften Abhandlungen* **16,** 357–375.

Sampson, S., Krause, D., Forster, C., and Dodson, P. (1996). Non-avian theropod dinosaurs from the Late Cretaceous of Madagascar and their paleobiogeographic implications. *J. Vertebr. Paleontol.* **16**(3). [Abstract]

Sereno, P., Dutheil, D., Iarochene, M., Larsson, H., Lyon, G., Magwene, P., Sidor, C., Varrichio, D., and Wilson, J. (1996). Predatory dinosaurs from the Sahara and Late Cretaceous faunal differentiation. *Science* **272,** 986–991.

von Huene, F., and Matley, C. (1933). The Cretaceous Saurischia and Ornithischia of the central provinces of India. *Mem. Geol. Survey India* **21,** 1–74.

Academy of Natural Sciences, Philadelphia, Pennsylvania, USA

Specimens collected by Joseph Leidy are a part of this collection, but specimens that were the basis of many of E. D. Cope's publications are now at the American Museum of Natural History.

see MUSEUMS AND DISPLAYS

African Dinosaurs

LOUIS L. JACOBS
Southern Methodist University
Dallas, Texas, USA

The fossil record of dinosaurs in Africa extends from the Late TRIASSIC, over 200 million years ago, until the Late CRETACEOUS, presumably 65 million years ago, although the extinction event that ended the reign of dinosaurs has yet to be documented in Africa. Throughout this length of time, Africa remained relatively stable geologically, changing position only slightly by drifting and rotating northward. By contrast, Africa's neighboring continents moved greatly, resulting in ocean barriers between what were once contiguous land masses. The changing geography of Africa and its neighbors throughout the MESOZOIC is fundamental to understanding the dinosaurs found there.

During the Late Triassic through the Early JURASSIC, major continental land masses were united into the supercontinent of Pangaea. Because the land was not divided into separate continents, dinosaurs and other animals were more or less free to expand across the entire area, not constrained by ocean barriers but rather by environmental and ecological differentiation of this large land area. Thus, the dinosaur fauna of the Late Triassic and Early Jurassic is generally similar across the globe because the globe had only one continent rather than several continents acting as separate theaters of evolution.

Late Triassic dinosaur sites are found extensively in southern Africa (particularly South Africa, Lesotho, and Zimbabwe) and to a lesser extent in northern Africa (Morocco). Herbivorous prosauropods (*Azendohsaurus*, *Blikanasaurus*, *Euskelosaurus*, and *Melanorosaurus*) are the best known of African Triassic dinosaurs. Footprints and incomplete remains indicate the presence of small THEROPODS and ORNITHISCHIANS. The Triassic–Jurassic boundary is marked by extinctions globally, but the boundary has not been studied in detail in Africa.

Early Jurassic localities, like those of the Late Triassic, are concentrated in southern (Lesotho, Namibia, South Africa, and Zimbabwe) and northern Africa (Algeria and Morocco). The northern record is predominantly footprints, although tracks are also found in the south. PROSAUROPODS, represented by *Massospondylus*, appear to be relatively abundant. *Massospondylus* and the ceratosaur *Syntarsus* are also known from North America. SAUROPODS are represented by *Vulcanodon*, a primitive genus known from Zimbabwe. ORNITHOPODS are small and primitive but apparently diverse, being represented by *Abrictosaurus*, *Heterodontosaurus*, *Lanasaurus*, *Lesothosaurus*, and *Lycorhinus*.

The Middle Jurassic is poorly represented and poorly studied in Africa. Large sauropods, usually referred to *Cetiosaurus*, are known from Morocco and Algeria. There are few Late Jurassic localities. Theropod, sauropod, and ornithopod footprints are reported from Morocco and Niger. However, the most impressive concentration of Late Jurassic dinosaurs in Africa is TENDAGURU, Tanzania. This collection of sites was worked first by Germans, until they were disrupted by World War I, then by British. Although theropods, including the ornithomimosaur *Elaphrosaurus*, are present, by far the bulk of the material pertains to large sauropods (*Barosaurus*, *Brachiosaurus*, *Dicraeosaurus*, and *Janenschia*). ORNITHISCHIANS are represented by the ornithopod *Dryosaurus* and the stegosaur *Kentrosaurus*. Perhaps most surprising about the Tendaguru fauna is the similarity it shows to that of the Morrison Formation of North America.

Madagascar has a history separate from that of Africa for the latter half of the Mesozoic Era. *Bothriospondylus* and *Lapparentosaurus*, both sauropods, are reported from the Jurassic. Those records are particularly important because Madagascar separated from Africa approximately 160–150 million years ago, at approximately the same time or slightly postdating the Jurassic fossils. They are perhaps among the last dinosaurs that could have inhabited a Madagascar connected to the African mainland. Although separated from Africa, in the Jurassic and Early Cretaceous Madagascar remained conjoined with India, and through India, to the land masses now known as Antarctica and Australia, and through Antarctica to South America. In the Early Cretaceous, Madagascar plus India separated from Australia and Antarctica. Then, in the Late Cretaceous, they separated from each other to drift to the configuration they have now achieved. This sequence of geographic events is important because it means that the biogeographic affinities of Late Cretaceous Madagascan dinosaurs may lie elsewhere than Africa.

Recent work in the Late Cretaceous of Madagascar is greatly improving our knowledge of that island. Theropods are best represented by the probable abelisaurid *Majungasaurus*. Sauropods are best represented by a derived titanosaurid that was referred to as *Titanosaurus* in earlier literature, a genus first named from India. In addition to the biogeographic implications, titanosaurid remains in Madagascar include the first documented bony dermal armor from a sauropod. Of particular interest among recent finds are birds found in the quarries with the dinosaurs. Earlier studies indicated the presence of a pachycephalosaur, *Majungatholus*, but this animal is an ABELISAURID.

Localities are more widespread across the continent during the Cretaceous Period, but with the Cretaceous lasting from 144 to 65 million years, with few radiometric dates in Africa, and with the density of localities sparse relative to the area and the time involved, it is not surprising that chronological resolution is poor. During the Early Cretaceous Africa remained connected to South America. By the end of the Early Cretaceous or in the early portion of the Late Cretaceous, Africa and South America split apart with the completion of the South Atlantic. This was an important event because, with the completion of the Atlantic, new ocean current patterns were established, distributing heat across the globe and affecting climates. Besides the ecological changes this would bring about, the growing Atlantic formed a widening barrier, allowing the prediction that the similarity of South American and African dinosaur faunas decreases after the Early Cretaceous. Recent work suggests this may be the case.

Most Early Cretaceous localities have yielded fragmentary theropod and sauropod material lacking detailed contextual data. Notable exceptions are the tetanuran *Afrovenator* from Niger, the primitive titanosaurid sauropod *Malawisaurus* from Malawi, the high-spined ornithopod *Ouranosaurus* from Niger, and the stegosaur *Paranthodon* from South Africa. The Late Cretaceous is equally in need of more field work and discovery, although numerous localities are scattered through northern Africa in particular. Notable taxa from the Late Cretaceous of Africa include the coelurosaur *Deltadromeus*, the tetanuran *Spinosaurus*, and the allosauroid *Carcharodontosaurus*, all from northern Africa, and *Kangnasaurus*, an ornithopod from southern Africa. In terms of collected and adequately described taxa, this is clearly unbalanced for both the Early and Late Cretaceous, indicating the fertile ground that Africa is for discovery.

In the current geography of the earth, the Middle East is distinct from Africa. In the Mesozoic it was not. Therefore, indeterminate sauropod remains from Late Jurassic coastal deposits in Yemen and Late Cretaceous theropod footprints from Israel must be considered African. In addition, Croatian localities from a Mesozoic carbonate platform yield fragmentary bones and dinosaur footprints that may have been made on land that was originally a broad, intermittently submerged promontory of Africa or possibly a microplate that drifted northward to join Europe. Either way, the Croatian sites have great implications for the biogeography of African dinosaurs in the Cretaceous.

See also the following related entries:

Biogeography • Extinction, Triassic • Footprints and Trackways • Indian Dinosaurs • Plate Tectonics • South American Dinosaurs

References

Attridge, J., Crompton, A. W., and Jenkins, F. A. (1985). The southern African Liassic prosauropod *Massospondylus* discovered in North America. *J. Vertebr. Paleontol.* **5**(2), 128–132.

Dalla Vecchia, F. M. (1994). Jurassic and Cretaceous sauropod evidence in the Mesozoic carbonate platforms of the southern Alps and Dinarids. *Gaia* **10,** 65–73.

Jacobs, L. L., Winkler, D. A., and Gomani, E. M. (1997). Cretaceous dinosaurs of Africa: Examples from Cameroon and Malawi. In *Gondwana Dinosaurs* (R. Molnar and F. Novas, Eds.), in press. Memoirs of the Queensland Museum, Brisbane.

Krause, D. W., Hartman, J. W., and Wells, N. A. (1997). Late Cretaceous vertebrates from Madagascar: Implications for biotic change in deep time. In *Natural and Human-Induced Change in Madagascar* (B. Patterson and S. Goodman, Eds.), in press. Smithsonian Institution Press, Washington, DC.

Padian, K. (Ed.) (1986). *The Beginning of the Age of Dinosaurs: Faunal Change across the Triassic–Jurassic Boundary,* pp. 378. Cambridge Univ. Press, Cambridge, UK.

Weishampel, D. B., Dodson, P., and Osmólska, H. (1990). *The Dinosauria,* pp. 733. Univ. of California Press, Berkeley.

Age Determination of Dinosaurs

Gregory M. Erickson
University of California
Berkeley, California, USA

Early attempts to estimate the longevity of dinosaurs used allometric scaling principles. Ages were determined by dividing individual mass estimates by rates of growth for extant taxa. For very large species, growth rates were extrapolated to dinosaur proportions using regression analysis. The results of these investigations have been extremely variable because they depend on mass estimates and growth rates that are highly disparate. For example, longevity estimates for the sauropod *Hypselosaurus priscus* range from a few decades to several hundred years (Case, 1978). Recently it has been shown that most dinosaur bones have growth lines that are visible in thin sectioned material viewed under polarized light (e.g., Reid, 1990; Fig. 1). Two types of growth lines exist: annuli and lines of arrested growth (Francillon-Vieillot *et al.,* 1990). Histological examinations have re-

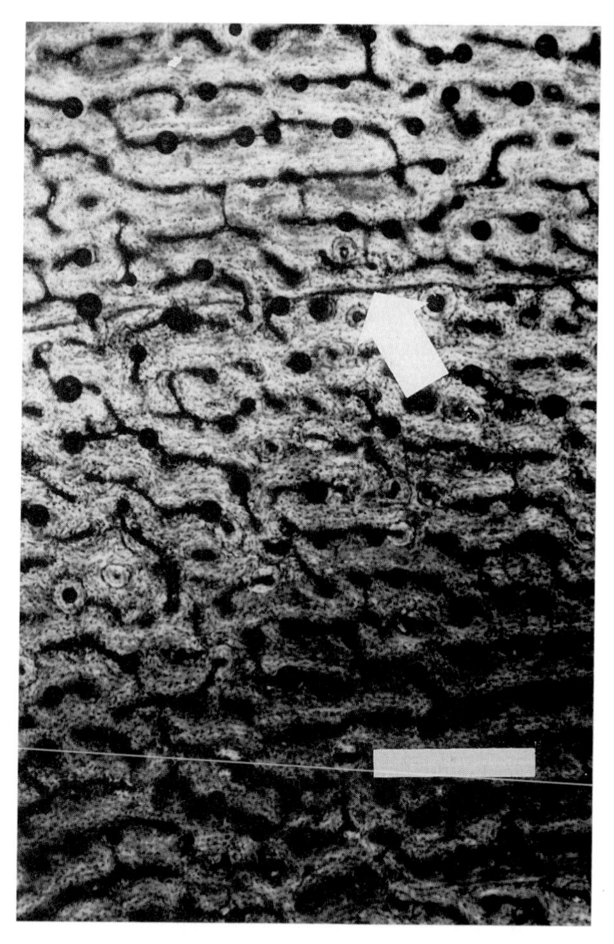

FIGURE 1 Thin-sectioned tibia of the tyrannosaur *Albertosaurus lancensis* (LACM 23845) exhibiting a line of arrested growth (arrow) between zones of highly vascularized fibrolamellar bone. Scale = 1 mm.

vealed that annuli are composed of thin layers of avascular bone with parallel-aligned bone fibers. The growth line annuli are sandwiched between broad vascularized regions of bone with more randomly oriented fibrillar patterns known as zones (Fig. 1). Lines of arrested growth, like annuli, are found between zones and are avascular. They are, however, thinner and have relatively fewer bone fibers by volume (Fig. 1). Studies on extant vertebrates indicate that the vascularized zones form during moderate to rapid skeletogenesis, and that abrupt metabolic disruptions of bone formation trigger growth line deposition. Interruptions that significantly reduce bone growth cause the genesis of annuli, whereas lines of arrested growth form in response to near or complete cessations in bone formation (Francillon-Viellot *et al.*, 1990).

Both types of growth lines may be deposited in synchrony with endogenous biorhythms. For example, captive crocodilians exposed to constant temperature, diet, and photoperiod still exhibit the periodic and cyclical skeletal growth banding of their wild counterparts. In many extant vertebrates, including most actinopterygian fish, amphibians, lepidosaurian reptiles, and crocodilians, the growth lines have an annual periodicity of deposition (Castanet *et al.*, 1993). Consequently, it is assumed by many paleontologists that the growth lines of dinosaurs reflect annual rhythms and that they can be used to determine individual ages. However, in the long bones of many taxa, resorption of internal and external bone proceeds even as new external cortical bone continues to be deposited, so growth lines deposited early in development may need to be inferred. The results of pioneering efforts to age dinosaurs using growth ring counts suggest that the longevity of the basal ceratopsian *Psittacosaurus mongoliensis* was 10 or 11 years (G. Erickson and T. Tumanova, unpublished data), the prosauropod *Massospondylus carinatus* 15 years (Chinsamy, 1994), the sauropod *Bothriospondylus madagascariensis* 43 years (Ricqlès, 1983), the ceratosaur *Syntarsus rhodesiensis* 7 years (Chinsamy, 1994), and the maniraptor *Troodon formosus* 3–5 years (Varricchio, 1993). These data are being used in conjunction with mass estimates to infer the metabolic status and growth rates of dinosaurs and to reconstruct the trophic dynamics of Mesozoic ecosystems.

See also the following related entry:

GROWTH LINES

References

Case, T. J. (1978). Speculations on the growth rate and reproduction of some dinosaurs. *Paleobiology* **4,** 320–328.

Castanet, J., Francillon-Viellot, H., Meunier, F. J., and Ricqlès, A. de (1993). Bone and individual aging. In *Bone* (B. K. Hall, Ed.), pp. 245–283. CRC Press, Boca Raton, FL.

Chinsamy, A. (1994). Dinosaur bone histology: Implications and inferences. In *Dino Fest* (G. D. Rosenberg and D. L. Wolberg, Eds.), pp. 213–227. The Paleontological Society, Department of Geological Sciences, Univ. of Tennessee, Knoxville.

Francillon-Viellot, H., de Buffrénil, V., Castanet, J., Géraudie, J., Meunier, F. J., Sire, J. Y., Zylberberg, L., and Ricqlés, A. de (1990). Microstructure and mineralization of vertebrate skeletal tissues. In *Skeletal Biomineralization: Patterns, Processes and Evolutionary Trends* (J. G. Carter, Ed.). Vol. 1, pp. 471–530. Van Nostrand–Reinhold, New York.

Reid, R. E. H. (1990). Zonal "growth rings" in dinosaurs. *Mod. Geol.* **15,** 19–48.

Ricqlès, A. de (1983). Cyclical growth in the long limb bones of a sauropod dinosaur. *Acta Palaeontol. Polonica* **28,** 225–232.

Varricchio, D. V. (1993). Bone microstructure of the Upper Cretaceous theropod dinosaur *Troodon formosus*. *J. Vertebr. Paleontol.* **13,** 99–104.

Albany Museum, Grahamstown, South Africa

Anusuya Chinsamy
South African Museum
Cape Town, South Africa

The Albany Museum, established in 1855, has postcranial material of *Euskelosaurus* and *Massospondylus*. It also houses probable *Paranthodon africanus* fossils, including several bone fragments. In a recent expedition to the Kirkwood beds of the Algoa Basin, soon-to-be described hypsilophodontid skeletal elements recovered included a partial jaw. Titanosaurid-like, brachiosaurid-like, and theropod teeth from the Kirkwood formation are represented in the collections. The paleontology section of the Albany Museum is currently being renovated. Dinosaurs from the Eastern Cape will be represented by recently collected brachiosaurid material and by full-scale models of *Paranthodon africanus* (the first dinosaur discovered in South Africa) and *Syntarsus rhodesiensis*.

See also the following related entries:

African Dinosaurs • Bernard Price Institute for Palaeontological Research • National Museum, Bloemfontein • South African Museum

Algerian Dinosaurs

Fragmentary remains of dinosaurs have been recovered from the Late Cretaceous of Algeria.

see African Dinosaurs

Allosauroidea

Kevin Padian
John R. Hutchinson
University of California
Berkeley, California, USA

Allosauroidea (Fig. 1) subsumes by content the concepts Allosauria, Allosauridae, and *Allosaurus*. As recently constituted, Allosauroidea includes the Allosauridae and Sinraptoridae (Currie and Zhao, 1993; Sereno *et al.*, 1994; Holtz, 1994, 1996). Because Carnosauria is defined as all Tetanurae closer to *Allosaurus* than to birds, Allosauroidea are by definition carnosaurs, and any potential members of Carnosauria must be evaluated against *Allosaurus*.

Resolution of allosauroid phylogeny beyond this statement is a difficult matter, partly because many taxa that are clearly allied to the group are incompletely known or have had their systematic characters interpreted differently by different workers. Consequently, the membership of Allosauridae and Sinraptoridae is currently not agreed upon apart from their eponymous genera. Sereno *et al.* (1994), for example, place *Sinraptor* and *Yangchuanosaurus* in Sinraptoridae and *Acrocanthosaurus, Allosaurus, Cryolophosaurus,* and *Monolophosaurus* in Allosauridae. Holtz (1994) placed only *Allosaurus* and *Acrocanthosaurus* in Allosauridae. Holtz (1995) regarded *Monolophosaurus* as a tetanurine outside Avetheropoda, but in 1996 found it to be the sister taxon to Allosauroidea,

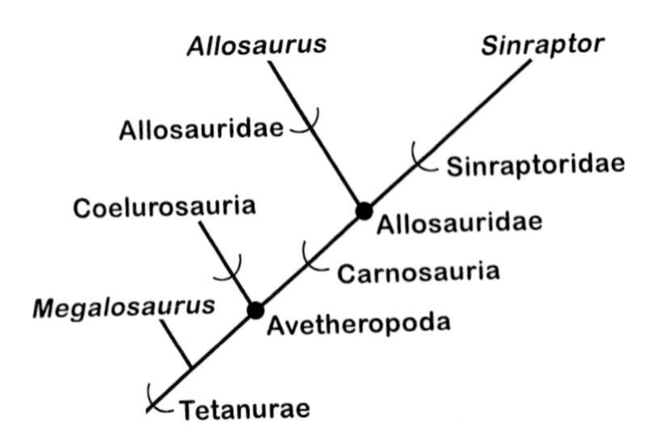

FIGURE 1 Phylogeny of Allosauroidea. For synapomorphies see text. Taxa of debated placement, not listed here, include *Acrocanthosaurus, Monolophosaurus, Cryolophosaurus, Chilantaisaurus, Piatnitzkysaurus, Carcharodontosaurus,* and *Giganotosaurus*. These are all recognized as Carnosauria and a possible phylogeny is given under that entry.

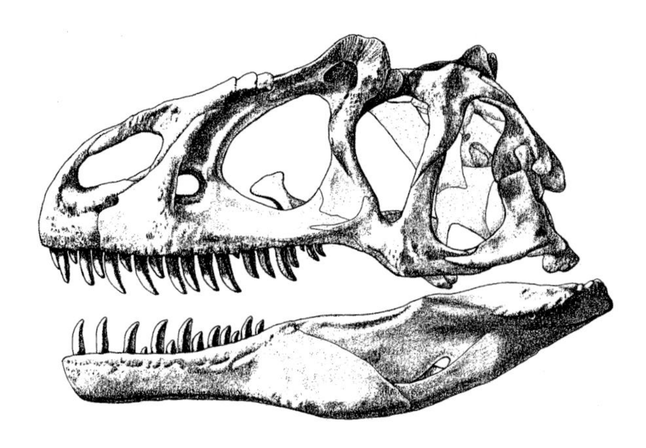

FIGURE 2 Skull of *Allosaurus* (after Madsen, 1976).

with *Cryolophosaurus* as a more basal carnosaur. Holtz (1996) also found *Giganotosaurus* and the Carcharodontosauridae to be allosauroids closer to Allosauridae than to Sinraptoridae. Meanwhile, Sereno *et al.* (1996) also found Carcharodontosauridae to be in Allosauroidea but suggested that *Acrocanthosaurus* and *Giganotosaurus* might belong in Carcharodontosauridae instead of in Allosauridae. *Acrocanthosaurus* was known until recently only from incomplete specimens, but more material has been discovered in the past few years and awaits formal description. Other poorly known taxa, such as *Pianitzkysaurus* and *Chilantaisaurus*, may be allosauroids as well, but their exact relationships have not yet been conclusively established.

Sereno *et al.* (1994) provided four synapomorphies of Allosauroidea, including the participation of the nasal in the antorbital fossa, a flange-shaped lacrimal process on the palatine, the basioccipital excluded from the basal tubera, and the articular with a pendant medial process (Fig. 2). The synapomorphies of Allosauridae include a short quadrate with the head level with the middle of the orbit, a deep anterior ramus of the surangular, and the small diameter of the external mandibular fenestra. Sinraptoridae (Currie and Zhao, 1993) is characterized by two accessory pneumatic excavations on the maxilla, an external nares with a marked inset of the posterior margin, a bulbous, anteriorly projecting rugosity on the postorbital, and a flange on the squamosal that covers the quadrate head in lateral view (Sereno *et al.*, 1994). Holtz (1994) diagnosed Allosauridae by the possession of a pubic "foot" that is longer anteriorly than posteriorly and triangular in ventral view. Because the synapomorphies diagnosing other nodes in his 1995 and 1996 works have not yet been published, differences in phylogenetic conclusions cannot currently be evaluated.

Given the current instability in diagnosing the content and hence the synapomorphies of the allosauroid groups, Allosauridae and Sinraptoridae can be defined only with reference to their eponymous genera. Hence, Allosauridae comprises *Allosaurus* and all Allosauroidea closer to it than to *Sinraptor;* Sinraptoridae comprises *Sinraptor* (Fig. 3) and all Allosauroidea closer to it than to *Allosaurus.* Allosauroidea is a node-based taxon, diagnosed by the synapomorphies previously discussed, that includes *Allosaurus* and *Sinraptor* and all descendants of their most recent common ancestor. It will be noted that Allosauridae and Sinraptoridae are herein defined as stem-based taxa; neither Sereno *et al.* (1994) nor Holtz (1994) indicated node- or stem-based definitions but rather used characters and included taxa (see PHYLOGENETIC SYS-

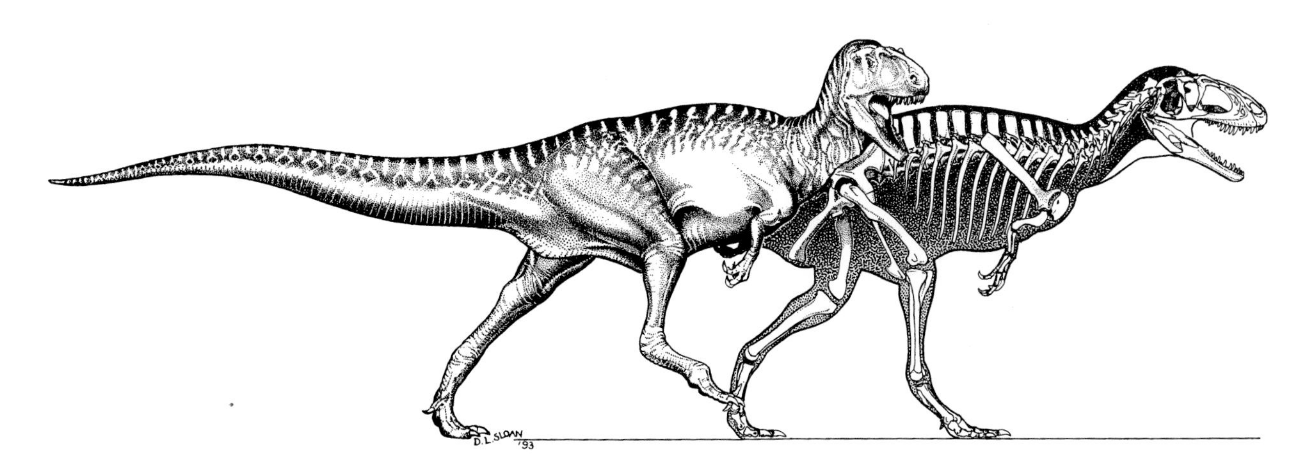

FIGURE 3 *Sinraptor* (after Currie and Zhao, 1993).

TEM). Stem-based definitions are included here because currently the taxa are minimally monotypic: No consensus exists on the membership of more than one genus per taxon.

The genus *Allosaurus* has had a slightly confusing history. Leidy (1870) assigned a partial caudal vertebra, collected from Colorado by the Ferdinand Hayden expedition, to *Poicilopleuron* [sic] *valens*, noting its putative similarity to the European genus *Poikilopleuron* (which has had its own tortured history and has usually been synonymized with, or allied to, MEGALOSAURUS). However, Leidy also provided the generic name *Antrodemus*, should his specimen eventually prove different from *Poikilopleuron*. Marsh (1877) described a tooth, two dorsal vertebrae, and a phalanx, collected by Benjamin Mudge from the Late Jurassic MORRISON FORMATION of Fremont County, Colorado ("Garden Park Quarry"), as *Allosaurus fragilis*, and Marsh eventually described some of the further remains from the same quarry, excavated by M. P. Felch, which included an almost complete skeleton, several partial skeletons, and other bones. Hence, *Allosaurus* came to be well known. Gilmore (1920), however, in describing the full material from Garden Park, decided that Leidy's caudal half-centrum of *Antrodemus valens* had diagnostic characters also seen in the *Allosaurus* material, so he regarded all the material as belonging properly to *Antrodemus*. Gilmore's judgment, however logical at the time, has not been sustained by character analysis, and the name *Allosaurus* is accepted today (Madsen, 1976).

The greatest collection of *Allosaurus* material has been made at the CLEVELAND–LLOYD QUARRY, discovered in 1927 and worked intermittently by crews from the University of Utah, Princeton University, and the Earth Science Museum, Brigham Young University, where most of the material is now jointly stored (see Miller *et al.*, 1996). Several thousand bones now provide a very full picture of this animal, including its osteology and ontogeny, which preserve bones of individuals (unfortunately, none articulated) ranging from approximately 3 to 12 m in length. A detailed study and map of the Cleveland–Lloyd Quarry (Miller *et al.*, 1996) and recent Ph.D. work by David K. Smith at Brigham Young University on the morphometrics of the *Allosaurus* material testify to the continuing importance of this bonanza of skeletal material.

Allosaurus was the largest well-known carnivore in the Morrison Formation, presumably feeding on SAUROPODS, HYPSILOPHODONTIDS, STEGOSAURS, ANKYLOSAURS, and probably other animals; *Ceratosaurus*, another large carnivore, is present but rarer, and an even larger, possibly allosauroid theropod, *Saurophaganax* (which apparently reached the size of some adult *Tyrannosaurus* specimens) has not yet been well described. *Torvosaurus* is a basal tetanurine theropod also from the Morrison Formation (see DRY MESA QUARRY) and it reached comparable size, plus a few other taxa have been reported from fragmentary material that may or may not be diagnostic; therefore, at least four large theropods are known from the Morrison Formation, and there may well have been a considerable diversity of large carnivores feeding on the other components of the Morrison fauna and perhaps on each other. Fragmentary specimens referable to *Allosaurus*, or at least to Allosauridae, are reported from the TENDAGURU Beds of Tanzania and the Strzelecki Group of Victoria, Australia, extending the survival of the allosaurid lineage into the Early CRETACEOUS Period (Molnar *et al.*, 1981, 1985).

The upper jaw of *Allosaurus* bears 20 or more trenchant, laterally compressed teeth; the dentary bears up to 13, but the lower tooth row does not extend as far posteriorly as the upper row, as in most theropods. An extensive system of pneumatic spaces characterizes the orbital "brow ridge" and the skull roof bones behind the orbit (see CRANIOFACIAL AIR SINUS SYSTEMS). The brow ridge in *Allosaurus, Sinraptor*, and apparently *Yangchuanosaurus* is centrally excavated in a particular way, but the function is unknown.

One of the most important recent discoveries concerning *Allosaurus* is that it has a furcula (Chure and Madsen, 1996). This was first discovered during the excavation of a still undescribed, *Allosaurus*-like theropod from DINOSAUR NATIONAL MONUMENT. Comparison of this specimen's furcula with *Allosaurus* material from the Cleveland–Lloyd Quarry revealed that such elements were common at Cleveland–Lloyd Quarry but had been taken for median bones of the ventral cuirass, or gastralia (Madsen, 1976). Gilmore (1920) had figured them as proceeding down the length of the abdomen, when in fact there was only one per individual, situated properly between the pectoral girdles. Their anatomy is easily distinguished from that of the true median gastral elements (Chure and Madsen, 1996). As Holtz (1996) noted, the possession of this distinctive, boomerang-shaped furcular morphology is a synapomorphy of AVETHEROPODA (see also PECTORAL GIRDLE).

References

Chure, D. J., and Madsen, J. H. (1996). On the presence of furculae in some non-maniraptoran theropods. *J. Vertebr. Paleontol.* **16,** 573–577.

Currie, P. J., and Zhao, X. J. (1993). A new carnosaur (Dinosauria: Theropoda) from the Jurassic of Xinjiang, People's Republic of China. *Can. J. Earth Sci.* **30,** 2037–2081.

Gilmore, C. W. (1920). Osteology of the carnivorous Dinosauria in the United States National Museum, with special reference to the genera *Antrodemus (Allosaurus)* and *Ceratosaurus.* Smithsonian Institution, United States National Museum Bulletin No. 110, pp. 1–159.

Holtz, T. R., Jr. (1994). The phylogenetic position of the Tyrannosauridae: Implications for theropod systematics. *J. Paleontol.* **68,** 1100–1117.

Holtz, T. R., Jr. (1995). A new phylogeny of the Theropoda. *J. Vertebr. Paleontol.* **15**(Suppl. to No. 3), 35A.

Holtz, T. R., Jr. (1996). Phylogenetic analysis of the nonavian tetanurine dinosaurs (Saurischia: Theropoda). *J. Vertebr. Paleontol.* **16**(Suppl. to No. 3), 42A.

Leidy, J. (1870). [Remarks on *Poicilopleuron valens*, etc.] *Proc. Acad. Nat. Sci. Philadelphia* **1870,** 3–5.

Madsen, J. H. (1976). *Allosaurus fragilis*: A revised osteology. *Bull. Utah Geol. Miner. Survey* **109,** 1–163.

Marsh, O. C. (1877). Notice of new dinosaurian reptiles from the Jurassic formation. *Am. J. Sci.* **3**(14), 513–514.

Miller, W. E., Horrocks, R. D., and McIntosh, J. H. (1996). The Cleveland–Lloyd Quarry, Emery County, Utah: A U.S. national landmark (including history and quarry map). *Brigham Young Univ. Geol. Stud.* **41,** 3–24, 4 maps.

Molnar, R. E., Flannery, T. F., and Rich, T. H. V. (1981). An allosaurid theropod dinosaur from the Early Cretaceous of Victoria, Australia. *Alcheringa* **5,** 141–146.

Molnar, R. E., Flannery, T. F., and Rich, T. H. V. (1985). Aussie *Allosaurus* after all. *J. Paleontol.* **59,** 1511–1513.

Sereno, P. C., Wilson, J. A., Larsson, H. C. E., Dutheil, D. B., and Sues, H.-D. (1994). Early Cretaceous dinosaurs from the Sahara. *Science* **265,** 267–271.

A diorama of *Albertosaurus* over a prey item, *Centrosaurus*, perhaps a common scene from the Cretaceous of Alberta, Canada. (Photo by François Gohier.)

American Dinosaurs

Peter Dodson
University of Pennsylvania
Philadelphia, Pennsylvania, USA

The United States has a great diversity of dinosaurs spanning a wide stratigraphic range. Although the concept of dinosaur was born in England, it found fertile ground in the United States. The United States has more different kinds of dinosaurs than any other country by a wide margin. A recent tabulation based on data as of 1988 shows that the United States has 64 known genera of dinosaurs compared with 40 for Mongolia and 36 for China. Such figures rapidly become dated as new kinds from around the world are described. In 1993, for instance, four new dinosaurs were described from the United States: *Shuvosaurus* from Texas, *Utahraptor* from Utah, and *Naashoibitosaurus* and *Anasazisaurus* from New Mexico. In 1994, *Mymoorapelta* from Utah was added, and in 1995 the ceratopsids *Einiosaurus* and *Achelousaurus* from Montana were formally described (see Variation). Forthcoming are a theropod from the Early Cretaceous of Utah, an ankylosaurian and an ornithopod from Texas, and a basal ornithischian from New Mexico. Thus, growth of knowledge of new kinds of dinosaurs continues at least as rapidly in the United States as in China and at a greater rate than in Argentina or in Mongolia.

There are four fundamental reasons why the United States has so many different kinds of dinosaurs: stratigraphy, climate and geography, human resources, and history. Like Argentina and China, and unlike Canada and Mongolia, the United States has dinosaur-bearing continental strata that span most of the stratigraphic interval in which dinosaurs may be expected from the Carnian stage of the Late Triassic to the Maastrichtian stage of the Late Cretaceous. The United States has large areas of outcrop in semiarid climates, principally in the west, where erosion is relatively unencumbered by vegetation, unlike Canada, England, or the eastern United States, for example. There is also a large corps of professional, commercial, and amateur dinosaur collectors in this country, all of whom contribute to ongoing discoveries.

The explicit history of dinosaur paleontology in the United States extends back to 1856, when Joseph Leidy applied names to a collection of teeth from the Judith River beds along the Missouri River in Montana that was sent to Philadelphia by Ferdinand Hayden. The four names are *Deinodon, Trachodon, Paleoscincus,* and *Troodon*. Unfortunately, the first three names are *nomina dubia,* as these teeth are diagnostic only at the family level (this being the rule for dinosaur teeth, making it generally unwise to name dinosaurs on that basis). Before this time, an interesting bone had turned up in Cretaceous deposits from Woodbury, New Jersey. Such material had been discussed as early as 1787 at the American Philosophical Society in Philadelphia, but dinosaurs had not yet been recognized scientifically, and the report was forgotten. (Donald Baird has proposed that a hadrosaur metatarsal in the collection of the Academy of Natural Sciences of Philadelphia is this specimen.)

The discovery and description of *Hadrosaurus foulkii* from Haddonfield, New Jersey, by Leidy in 1858 marks the first time that a major portion of a dinosaur skeleton, including fore- and hind-limbs, had been found. This allowed Leidy to reconstruct *Hadrosaurus* as a biped, showing that the Owen–Hawkins reconstruction of *Iguanodon,* exhibited at the Crystal Palace since 1854, was incorrect. The reconstruction and exhibition of *Hadrosaurus* at the Academy of Natural Sciences in 1868 marked the first time that a dinosaur skeleton had ever been exhibited anywhere in the world. Casts of this specimen were exhibited at Princeton University Geology Museum, the Smithsonian, and the Field Columbian Museum in Chicago, but it was not until the first decade of the 20th century that other dinosaur skeletons were exhibited at the American Museum of Natural History, Yale Peabody Museum, and the Smithsonian. E. D. Cope described a partial skeleton of the enigmatic theropod *Laelaps* (preoccupied; renamed *Dryptosaurus* Marsh 1877) from New Jersey in 1868. Cope named the ceratopsids *Agathaumas* in 1872 and *Polyonax* in 1874 from Wyoming and Colorado, respectively, but these are *nomina dubia* based on fragmentary material. In 1876, he collected and named *Monoclonius* from the Judith River Formation of Montana, the first valid ceratopsid. Up to this point, dinosaur finds had been geographically widespread and generally of poor quality. Montana had produced the most dinosaurs up to this time, but most finds were not memorable.

In 1877, dinosaurs were discovered in abundance for the first time anywhere in the world at three separate localities: CAÑON CITY and Morrison, both in Colorado, and COMO BLUFF, Wyoming. The beds proved to be of Late Jurassic age and have produced a remarkable fauna dominated by large sauropods, with stegosaurs also important; theropods and ornithopods were less abundant; recently an ankylosaur was reported. Intensive examination of the Morrison fauna waned after 1885. Renewed interest in the Morrison at the turn of the century, after Marsh and Cope had died, produced further sauropods (*Brachiosaurus* and *Haplocanthosaurus*) and the small theropod *Ornitholestes*. Beds of Triassic age were documented with the description of *Coelophysis bauri* by Cope in 1889. The study of beds of latest Cretaceous age began with the description of *Triceratops* Marsh 1889, followed by *Torosaurus* Marsh 1891, and then *Tyrannosaurus* (1905) and *Ankylosaurus* (1908) early in this century. Lancian hadrosaurine species were described by Marsh in 1890 and 1892, but the proper generic assignments (to *Anatotitan* and *Edmontosaurus*) were not recognized until much more recently. The United States lacks a major dinosaur fauna correlative with the Horseshoe Canyon Formation (early Maastrichtian) of Alberta, Canada, although the TWO MEDICINE FORMATION of Montana, first studied by C. W. Gilmore beginning in 1914, contains an antecedent fauna, as do the FRUITLAND/KIRTLAND FORMATIONS of New Mexico and the Aguja Formation of Texas. A major fauna of Early Cretaceous age, very broadly correlative with the British WEALDEN fauna, was unknown in the United States until John Ostrom described the fauna of the CLOVERLY FORMATION of Wyoming and Montana in 1970. Lateral equivalents of the Cloverly (CEDAR MOUNTAIN FORMATION of Utah is partially equivalent; Trinity Group, TX) are now producing important specimens (*Utahraptor*; Proctor Lake ornithopod). Late Cretaceous dinosaurs from New Mexico began to be described in 1910. The Late Triassic is sparsely productive of dinosaurs, the rich deposits of *Coelophysis* being a conspicuous exception. There are Early Jurassic dinosaurs in Connecticut and the southwest; the Middle Jurassic is essentially unknown.

Primitive theropods are well represented, the most prominent being *Coelophysis* (known from scores of skeletons from the mass death assemblage at GHOST RANCH, NM), *Dilophosaurus*, and *Ceratosaurus*. Large theropods are represented by two principal taxa, the Allosauridae (*Allosaurus*) and the TYRANNOSAURIDAE. *Tyrannosaurus* now appears to be one of the most common large theropods. Good specimens of *Albertosaurus* are common in Canada but are very rare in the United States. The fossil record of maniraptorans in the United States is rather sparse, apart from the imperfect material of *Ornitholestes* and *Coelurus*. *Deinonychus* is the most important American maniraptoran, and recently the larger *Utahraptor* has been described. Ornithomimids are poorly represented but were surely present. Many MANIRAPTORAN taxa are documented principally by teeth (e.g., *Aublysodon*, *Paronychodon*, and *Ricardoestesia*) and thus are in perilous condition taxonomically. No segnosaurs have been confirmed.

"Prosauropods" (basal SAUROPODOMORPHS) are somewhat sparse in the American fossil record. *Anchisaurus* and *Ammosaurus* are the principal taxa, although *Massospondylus* has been reported from the Early Jurassic of Arizona. No sauropods of Early and Middle Jurassic age are known, but the Late Jurassic Morrison Formation contains a sauropod assemblage that is rivaled in quality, quantity, and diversity only by the correlative assemblages from China. For nearly a century, these sauropods presented the basis for understanding sauropods everywhere in the world. The important taxa Camarasauridae (*Camarasaurus*), Brachiosauridae (*Brachiosaurus*), and Diplodocidae (*Diplodocus*, *Apatosaurus*, and *Barosaurus*) were established on Morrison sauropods. The taxa Cetiosauridae (*Haplocanthosaurus*) and Titanosauridae (*Alamosaurus*) are known but are much less important here.

Basal ornithischians are poorly represented at present, but *Technosaurus* from Texas seems representative of such basal taxa. In addition, teeth of basal ornithischians have been documented in Late Triassic sediments from Pennsylvania to Arizona. *Scutellosaurus* and *Scelidosaurus* are good basal thyreophorans. The Stegosauridae are magnificently characterized by *Stegosaurus*, but there is otherwise very low diversity of this family, in contrast to China. There are few basal ankylosaurians, but there are good representatives of the Nodosauridae (*Sauropelta*) and of the Ankylosauridae (*Ankylosaurus*), both taxa being established on American taxa. A very important recent discovery is that of the ankylosaur *Mymoorapelta* from the MORRISON FORMATION of Colorado. The nodosaur *Edmontonia* is now reported from Alaska. For both families, there are more skulls than skeletons, with no complete skeletons in either taxa having yet been collected. ORNITHOPODS are well represented in the United States. Hypsilophodontids are somewhat fragmentary (*Othnielia* and *Orodromeus*), although there are several good specimens of the enigmatic *Thescelosaurus*. Basal iguanodontians are also well represented by *Dryosaurus* and *Tenontosaurus*, the latter of which is particularly abundant and widespread with specimens being reported from Montana, Wyoming, Utah, Oklahoma, and Texas (possibly Maryland as well). *Camptosaurus* is an abundant American iguanodontian, and a fine skull of *Iguanodon* itself, named *I. lakotensis*, has been described. HADROSAURS are abundant in the United States, including both lambeosaurines and hadrosaurines. The former are represented only by *Parasaurolophus* from New Mexico and Utah and *Hypacrosaurus* from northern Montana. Hadrosaurines come from New Jersey (*Hadrosaurus*), Alabama (*Lophorothon*), New Mexico (*Kritosaurus*), and extensively from Wyoming, Montana, North Dakota, and South Dakota (*Anatotitan* and especially *Edmontosaurus*, which is one of the most abundant dinosaurs both in the United States and in the world). *Edmontosaurus* is also reported from the North Slope of Alaska. Pachycephalosaurs are principally represented by crania, particularly of *Pachycephalosaurus* itself. Protoceratopsids are documented by a few incomplete specimens of *Leptoceratops* and by a specimen of *Montanoceratops*. Ceratopsids include both centrosaurines (*Monoclonius*) from Montana and chasmosaurines, especially *Triceratops*, from Wyoming, Montana, North Dakota, South Dakota, and Colorado; *Chasmosaurus* from Texas; *Torosaurus* from Montana and South Dakota; and *Pentaceratops* from New Mexico. Ceratopsids are endemic to North America. *Triceratops* is among the most abundant of all dinosaurs. There is a fragmentary occurrence of *Pachyrhinosaurus* from the North Slope of Alaska.

Because dinosaurs are so diverse in the United States, it is tempting to think of this country as a center of evolution for worldwide faunas. This may not be so. In the Late Triassic, plateosaurids, common in Europe, Asia, and South America, are rare in the United States. Rare Early Jurassic sauropodomorphs have been found in Arizona and Connecticut. There are significant resemblances between Late Triassic *Coelophysis* of New Mexico and Early Jurassic *Syntarsus* of Zimbabwe and South Africa, but the resemblances between these relatively primitive theropods include few derived characters. There are essentially no Middle Jurassic beds in the United States to document the antecedents of the marvelous Late Jurassic sauropods, ornithopods, and stegosaurs of the Morrison Formation. *Haplocanthosaurus* may be presumed to be representative of the basal cetiosaurid radiation better documented in England, Europe, and South America. *Brachiosaurus* from Colorado has affinities with congeneric fossils from Tanzania. Other faunal elements having congeners in East Africa are *Dryosaurus*, probably *Barosaurus*, and less certainly *Ceratosaurus* and *Allosaurus*. It is significant that stegosaurs are much less diverse in the United States than they are in China, although *Stegosaurus* itself may be the most highly derived stegosaur. *Camptosaurus* is an important basal iguanodontian in the United States, with a sister species in the Middle Jurassic of England. In the Early Cretaceous, *Tenontosaurus* is an endemic ornithopod more basal than *Camptosaurus*. Although *Iguanodon* appears to have reached North America, it seems to have been uncommon there. Important new evidence suggests that *Polacanthus* from the WEALDEN of England also lived in Utah. In the Late Cretaceous, there is scant evidence for the titanosaurid sauropod fauna that dominated much of the world. It is postulated that *Alamosaurus* was a late migrant from South America, reintroducing sauropods which had been absent since the Early Cretaceous. There is evidence of faunal interchange with Asia based on similarities at the level of family and genus. A close relationship, possibly at the species level, of *Tyrannosaurus* with the Asiatic *Tarbosaurus* is recognized. Other evidence for exchange is better documented by Canadian dinosaurs, notably the hadrosaurine *Saurolophus*. Due to the relatively impoverished faunas of the Judith River Formation and dearth of early Maastrichtian dinosaurs in the United States, coupled with the relatively restricted area of late Maastrichtian strata in Alberta and Saskatchewan, faunal overlap between the United States and Canada is not as great as expected, and the greater diversity and completeness of specimens favors Canada. Although ceratopsids range from Alaska to Mexico, the only identifiable specimens of this family in Asia are teeth and horn core fragments from Uzbekistan. Mid-Cretaceous dinosaurs are found in Maryland, and Late Cretaceous dinosaurs are known from the eastern seaboard of the United States, from New Jersey to North Carolina, and also along the Gulf Coast and Mississippi embayment from Alabama to western Tennessee and Missouri. Few skeletons have been described (*Hadrosaurus* and *Dryptosaurus* from New Jersey), and the faunal relationship to dinosaurs in the West, across the Inland Sea, is not evident. *Hadrosaurus* appears to be a sister group of *Kritosaurus* from New Mexico and / or *Gryposaurus* from Alberta. *Dryptosaurus* is a nonarctometatarsalian of unclear relationship to any other theropod. It is claimed that there is a specimen of *Albertosaurus* from Alabama, but it is undescribed. This would present the same biogeographic challenge that *Hadrosaurus* presents. The mechanism of faunal exchange across a 1000- to 1500-km inland sea has yet to be elucidated.

See also the following related entries:

CANADIAN DINOSAURS • HADROSAURIDAE • MEXICAN DINOSAURS • POLAR DINOSAURS • SOUTH AMERICAN DINOSAURS

References

Dodson, P. (1990). Counting dinosaurs: How many kinds were there? *Proc. Natl. Acad. Sci. USA* **87,** 7608–7612.

Russell, D. A. (1993). The role of Central Asia in dinosaurian biogeography. *Can. J. Earth Sci.* **30,** 2002–2012.

Weishampel, D. B., Dodson, P., and Osmolska, H. (Eds.) (1990). *The Dinosauria,* pp. 733. Univ. of California Press, Los Angeles.

Perhaps the largest dinosaur of all time, *Seismosaurus* dwarfs most other large sauropods. (Illustration by Donna Braginetz.)

American Museum of Natural History

Lowell Dingus
American Museum of Natural History
New York, New York, USA

The American Museum of Natural History (AMNH) in New York City is the largest private museum in the world, and it has made a similarly large contribution to vertebrate paleontology. The museum attracts millions of visitors each year to its famous dinosaur displays. Some of the most notable dinosaur discoveries have been made by field expeditions sponsored by the AMNH, including the excavations by Barnum Brown and others at Como Bluff and Hell Creek in the 1890s, the Roy Chapman Andrews expeditions in the Gobi Desert in the 1920s, and the recent trips back to Gobi by AMNH curators Michael Novacek and Mark Norell.

The AMNH was incorporated in 1869 in New York City. In 1877 the first permanent building opened at the current site, in midtown Manhattan west of Central Park. Albert S. Bickmore, a zoologist who studied under Louis Agassiz at Harvard, is regarded as the founder of the museum. He envisioned a natural science museum in the center of the metropolis of New York, comparable to the Museum of Comparative Zoology founded in Cambridge by Agassiz in 1859. Bickmore brought together a group of prominent New Yorkers who raised the money for the museum, including the financier J. P. Morgan.

In its earliest years the AMNH had virtually no vertebrate fossils, a fact that museum president Morris K. Jesup set out to rectify in 1891 by his hiring of Henry Fairfield Osborn, a noted paleontologist and a faculty member at Princeton University. Osborn founded the museum's Department of Vertebrate Paleontology and staffed it with an outstanding group of paleontologists, including Barnum Brown, William Diller Matthew, Jacob Wortman, Walter Granger, and Albert "Bill" Thomson. They were supplemented by preparators such as Peter Kaisen, Otto Falkenbach, and Adam Hermann.

Field Expeditions of the Museum

During the Jesup–Osborn era, the museum initiated a series of highly productive field expeditions. Beginning in 1897, Wortman, Brown, Granger, and others explored the Jurassic fossil beds in the Como Bluff area of Wyoming. Although this area had already been extensively worked by the expeditions of Othniel Charles Marsh, the AMNH group made many additional important finds, including the sauropods *Apatosaurus, Diplodocus, Allosaurus,* and *Ornitholestes.* In 1898, museum paleontologists in this area located the distinctive Bone Cabin site.

In 1902 Barnum Brown led an AMNH expedition to the Cretaceous beds of the Hell Creek region of Montana. This resulted in the first known specimen of *Tyrannosaurus rex,* in 1902, and a second, more highly preserved specimen in approximately 1908. This second specimen is generally regarded as the most famous dinosaur fossil in the world and has long been a centerpiece of the AMNH. Brown went on to lead museum-sponsored expeditions in 1910–1915 to the Red Deer River region of Alberta, Canada. These also yielded rich discoveries, especially of hadrosaurs such as *Saurolophus* and *Corythosaurus.* In the 1930s another AMNH expedition led by Brown excavated a large collection of Jurassic fossils from the Howe Ranch site in Wyoming.

What has been termed the golden age of the museum's field expeditions extended from 1890 to 1930. In addition to Brown's highly publicized efforts, various other paleontologists from the AMNH staff also made important finds in this period, including Matthew, Granger, Thomson, and Henry Fairfield Osborn himself. Osborn also obtained many specimens from independent collectors, such as Charles H. Sternberg.

Expeditions of Roy Chapman Andrews

The expeditions described previously all were situated in western North America, but the most renowned expeditions sponsored by the AMNH were carried out in central Asia. These took place in Mongolia in the period 1920–1930 under the leadership of the legendary Roy Chapman Andrews.

Andrews originally went to Mongolia's remote Gobi Desert region at the invitation of Henry Fairfield Osborn, who had succeeded Morris K. Jesup as president of the museum. Osborn believed that an investigation of the region could substantiate his theory that Asia, not Africa, was the original site of human habitation.

Andrews found no human fossil evidence to support Osborn's theory but did find many other significant vertebrate fossils, including the first dinosaur bone discovered in eastern Asia. The Flaming Cliffs region in particular produced important remains of dinosaurs such as *Protoceratops, Oviraptor, Saurornithoides, Pinacosaurus,* and *Velociraptor,* as well as Cretaceous mammals. The most noted single discovery was that of the predatory dinosaur *Oviraptor* lying on a clutch of supposed *Protoceratops* eggs.

Later Expeditions

Both Barnum Brown and Roy Chapman Andrews achieved celebrity status as dinosaur hunters, and Andrews is now often cited as the model for the Hollywood character Indiana Jones. However, after 1930 a combination of reduced museum funds and the historic circumstances of the Depression and World War II meant that high-profile expeditions such as theirs were no longer feasible.

The AMNH nevertheless continued to sponsor important expeditions such as Roland T. Bird's examination of the GLEN ROSE, TEXAS, dinosaur trackway site in the 1930s and Edwin Colbert's discovery of *Coelophysis* skeletons at GHOST RANCH, New Mexico, in the late 1940s.

The AMNH has maintained this tradition of field explorations to the present day. The most notable recent example was a program to revisit the sites explored by Roy Chapman Andrews. Since 1990, field crews from the museum have participated in annual expeditions to the Gobi Desert in conjunction with colleagues from the Mongolian Academy of Sciences. These efforts have resulted in the discovery of a new Late Cretaceous flightless bird, *Mononykus,* and the first known embryo of a theropod dinosaur.

Exhibits

In the new Halls of Ornithischian and Saurischian Dinosaurs, the AMNH exhibits the largest collection of real dinosaur fossils anywhere in the world. More than 100 specimens are on display, and approximately 85% of them include real fossil material. Many new specimens have been added, and several of the older mounts have been modified, including those of *Tyrannosaurus* and *Apatosaurus.*

In contrast to most exhibitions, the primary organizing framework is based on systematics rather than geologic time. Labels have intentionally been developed at different levels of technical difficulty to address the needs of a diverse audience. A main path down the center of each hall represents the trunk of an evolutionary tree. By walking down this path, one can see the most spectacular specimens and encounter labels addressing the major themes. Collection alcoves along the sides of the halls represent branches that contain fossil representatives of the principal dinosaurian clades. One system of computer interactives located in each alcove is utilized to present curatorial views about the evolutionary relationships of the dinosaurs on that branch. A second system of computer interactives is used to present the "walk-through time" approach utilized in most exhibitions.

The main point of the presentation is to illustrate for the visitor what we really know about these extinct animals. Many controversial issues are addressed by simply presenting the evidence for different ideas and letting the visitor make up his or her own mind about what to think. This is intentionally done in order to provide visitors with some insight into how the scientific process works.

The two dinosaur halls are part of a loop of six halls on the fourth floor of the museum that are designed to tell the story of vertebrate evolution. In all, these halls contain 57,000 square feet of exhibition space.

In emphasizing evolutionary relationships based on cladistic methods, the dinosaur exhibits also help make visitors aware of the kind of scientific research conducted in the museum's Department of Vertebrate Paleontology. In terms of dinosaurs, the curatorial staff and associates actively pursue research and fieldwork, including topics such as theropod evolution, the origin or birds, and the extinction of non-avian dinosaurs.

Along with its prominence in field exploration and dinosaur displays as described previously, the museum is also a noted research center. In addition to research by the museum staff and students, each year scientists from around the world come to New York City to work in the museum's dinosaur collections.

See also the following related entries:

CENTRAL ASIATIC EXPEDITIONS • HISTORY OF DINOSAUR DISCOVERIES • UKHAA TOLGOD

References

Colbert, Ed. (1992). *William Diller Matthew, Paleontologist.* Columbia Univ. Press, New York.

Dingus, L. (1996). *Next of Kin: Great Fossils at the American Museum of Natural History.* Rizzoli International, New York.

Hellman, G. (1969). *Bankers, Bones, and Beetles: The First Century of the American Museum of Natural History.* Simon & Schuster, New York.

Novacek, M. (1996). *Dinosaurs of the Flaming Cliffs.* Doubleday, New York.

Preston, D. J. (1986). *Dinosaurs in the Attic: An Excursion into the American Museum of Natural History.* St. Martin's Press, New York.

Rogers, K. (1991). *The Sternberg Fossil Hunters: A Dinosaur Dynasty.* Mountain Press, LaHonda, CA.

Wallace, J. (1994). *American Museum of Natural History's Book of Dinosaurs and Other Ancient Creatures.* Simon & Schuster, New York.

Ankylosauria

KENNETH CARPENTER
Denver Museum of Natural History
Denver, Colorado, USA

Ankylosaurs are four-legged, armor-plated ornithischians that first appeared in the Middle Jurassic. Specimens range in size from 1-m-long juveniles of *Pinacosaurus* to 10-m-long adult *Ankylosaurus*. Ankylosaurs are united by several synapomorphic characters, among which are a low, wide skull; cheek teeth deeply inset from the sides of the jaws; fusion of armor to the skull masking the cranial sutures and covering the supratemporal fenestra; fusion of the last three or four dorsals with the sacrum into a rod of vertebrae or synsacrum; horizontal rotation of the ilium so the ilium faces upwards, not outwards; secondary closure of the acetabulum; reduction in size of the pubis; and a body encased in armor plates (Coombs and Maryanska, 1990). Ankylosauria may be defined as all thyreophoran ornithischians closer to *Ankylosaurus* than to *Stegosaurus*.

The ankylosaurs have been placed into one of two families (Coombs, 1978) based primarily on the presence or absence of a bone club on the end of the tail (Fig. 1). Those with a club are placed in the family Ankylosauridae (or ankylosaurids in the vernacular), and those without in the Nodosauridae (or nodosaurids). Actually, there are many other differences between the two families.

The skull of ankylosaurids seen in top view is triangular, with small "horns" at the upper and lower corners of the skull (Fig. 2). These horns are actually triangular armor plates. The entire surface of the skull is covered in a mosaic of small, irregularly shaped armor, except in *Pinacosaurus,* in which much of the cranial armor is secondarily lost. The front of the ankylosaurid skull is usually broad, with a wide beak suggesting nonselective cropping of low vegetation. The exception to this is the primitive ankylosaurid *Shamosaurus* from the Lower Cretaceous of Mongolia. It has a narrow, pointed muzzle. Ankylosaurid teeth are small for the size of the skull and have a swollen base or cingulum. In at least one form (*Euoplocephalus*) a bony eyelid was also present (Coombs and Maryanska, 1990).

The external nares face forward in many ankylosaurids, with *Ankylosaurus* and *Shamosaurus* being the few exceptions (Tumanova, 1987). During respiration, the air moves through a complex air passage within the ankylosaurid skull, with at least one loop in it. The purpose for the complexity is puzzling, but it may have increased the surface area of the olfactory tissue or acted as a resonating chamber.

In nodosaurids, the skull is elongated with rounded corners. A single large armor plate is present on the top center of the skull and pairs of regular shaped plates in front of this over the snout. The external nares often face laterally at the front of the snout and the air passage is more direct. The beak is narrow for selectively cropping vegetation, perhaps

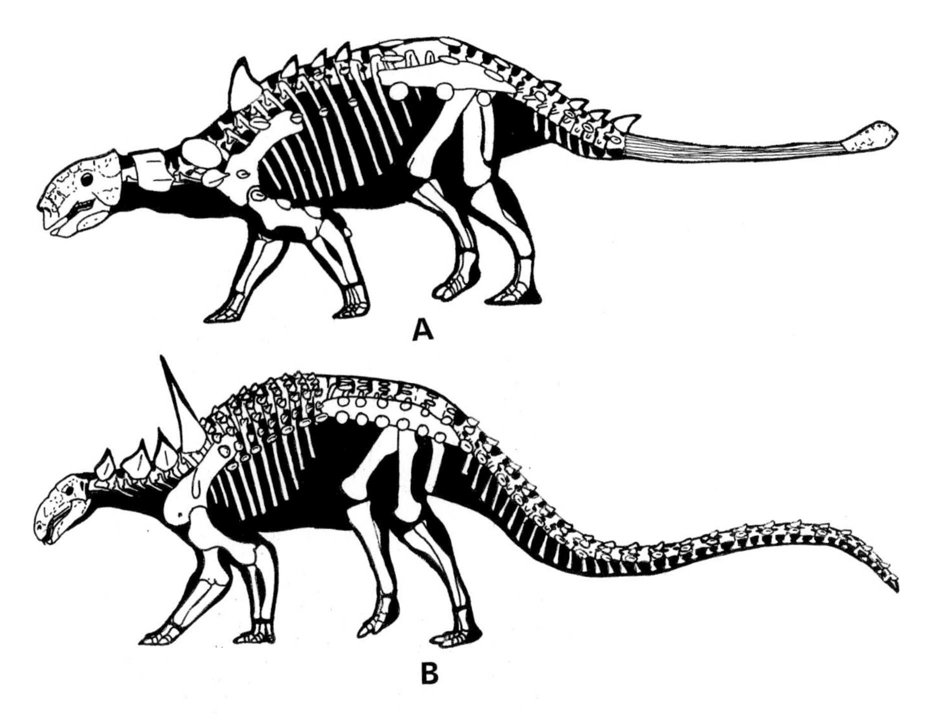

FIGURE 1 The skeleton of an ankylosaurid, *Dyoplosaurus* (A), and nodosaurid, *Sauropelta* (B).

leaves. In some primitive forms, such as *Silvisaurus*, conical premaxillary teeth are present. The cheek teeth are larger than those in ankylosaurids and have a shelf-like cingulum at their base. In *Edmontonia*, an oval cheek plate is present.

The vertebral column of ankylosaurs differs from that of many dinosaurs in that the neural spines are low. In many dinosaurs, the neural spines are often tallest on the posterior dorsals, sacrals, and anterior-most caudals, possibly to increase surface areas for ligaments holding the tail horizontally. In ankylosaurs, the armor may have functioned to hold the tail aloft as the dorsal armor does in crocodilians (Carpenter, 1997).

Another major difference in the vertebral column between ankylosaurs and most dinosaurs is the co-ossification of the last three or four dorsals with the sacrals into a rod called the synsacrum. The ribs from these dorsals arch out to fuse to the underside of the ilium. Sometimes the first and second caudals may also fuse to the rear of the synsacrum. The ribs of the mid-dorsals may co-ossify with the vertebrae as well.

In the tail, the caudal ribs are fused to the centra. These ribs in the anterior portion of the tail are long and slender and often curve downwards. In ankylosaurids, the pre- and postzygapophyses and chevrons are elongated in the posterior portion of the tail. The result is a rigid "handle" to the bone club that terminates the tail. This club is actually formed from enlarged bone plates that fuse together and to the vertebrae. In nodosaurids, there is no such modification of the tail vertebrae and no club.

The shoulder girdle of ankylosaurs is massive, especially the scapula. This is probably due to the enlarged muscles associated with quadrupedality because the scapula of bipedal dinosaurs is proportionally much more slender. In both nodosaurids and ankylosaurids there is a knob near the scapula–coracoid suture for the scapulohumeralis muscle. This knob is large and occurs on a ridge, or pseudoacromion process, in nodosaurids. This ridge may have functioned much like the acromion process in the mammalian scapula—to divide muscle masses. The position of the pseudoacromion process relative to the glenoid varies among the different genera of nodosaurids, making the scapula a taxonomically important bone.

The forelimb is short and stocky in many ankylosaurs and superficially resembles that of ceratopsians and stegosaurs. These similarities are unquestionably

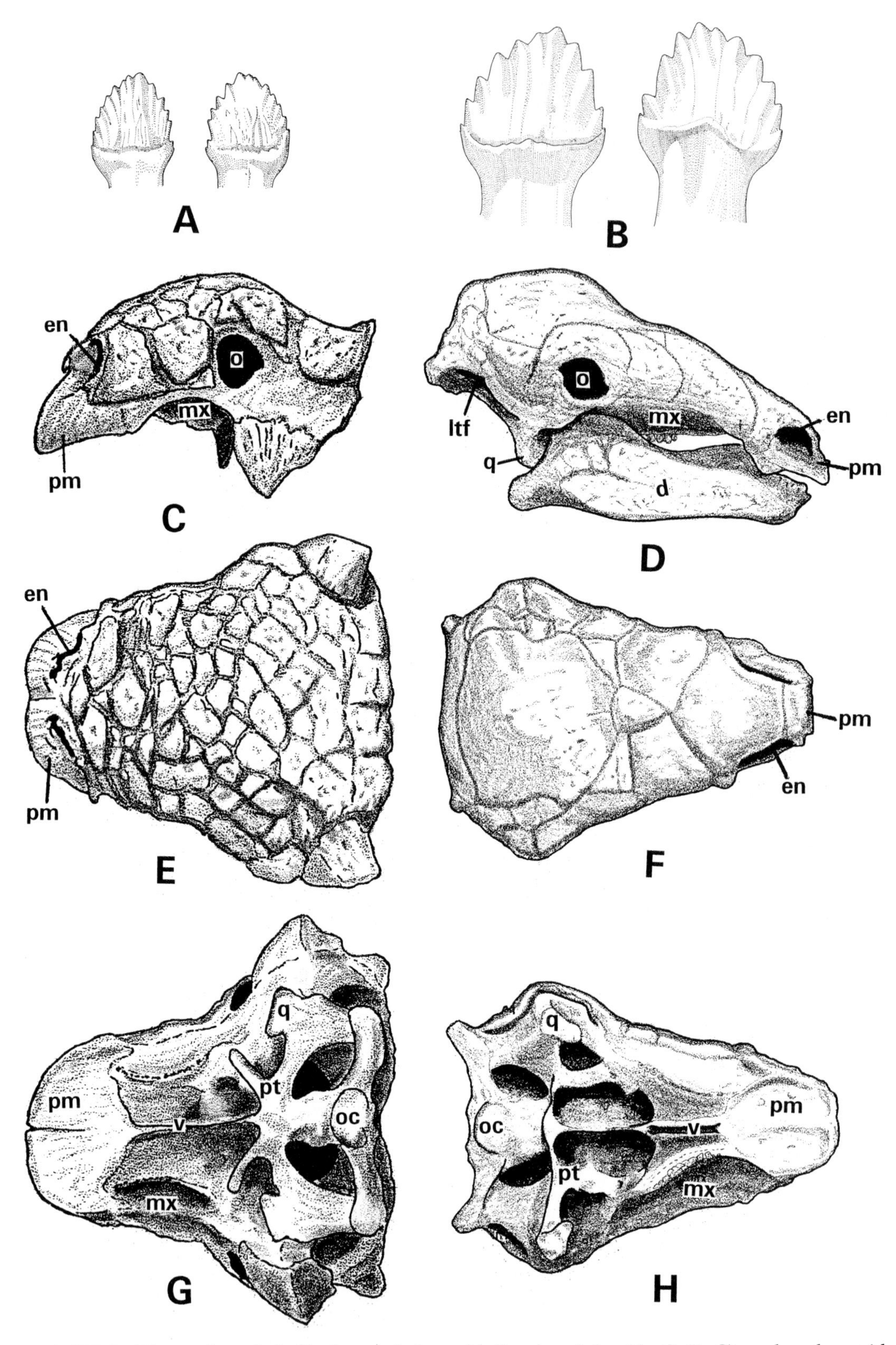

FIGURE 2 The teeth and skull of an ankylosaurid, *Euoplocephalus* (A, C, E, G), and nodosaurid, *Sauropelta* (B, D, F, H).

due to the quadrupedal stance of the three groups. The humerus of most ankylosaurs is short and stocky, although less so in some nodosaurids such as *Sauropelta* (Ostrom, 1970). The ulna has a very prominent olecranon process that can occupy more than a third of the length of the ulna. Such a well-developed olecranon indicates that the elbow was flexed and probably held near the body. The bones of the manus are short for bearing the weight of the animal.

The pelvic girdle of ankylosaurs is considerably modified, with the ilium horizontal, the pubis reduced in size, and the ischium a nearly vertical rod. Furthermore, the neural spines are fused into a vertical sheet of bone, and the acetabulum is closed off by the ilium and ischium. Although both ceratopsians and stegosaurs show partial modification of the ilium into the horizontal plane, this is considerably less than seen in ankylosaurs. The preacetabular portion of the ilium is also expanded to provide a large surface area for the iliotibialis muscles (Coombs, 1979). It is puzzling, however, why this protractor muscle for the leg needed to be so large. Perhaps we have misinterpreted the muscle scars.

The pubis is a small rectangular bone with a very short postpubic process in both ankylosaurids and nodosaurids. The ischium is a long, heavy bar of bone that projects downward, as in the saurischian pelvis, instead of horizontally as in the typical ornithischian pelvis.

As with the forelimb, the hindlimb is adapted for carrying considerable weight. Not surprisingly, the ankylosaur hindleg also resembles that of stegosaurs and ceratopsians. The femur is robust, having a larger midshaft circumference compared to that in a bipedal ornithischian. The fourth trochanter is a large scar on the femur and does not project outwards like the flange seen in hadrosaurs. The tibia and fibula are short and robust. The distal end of the tibia has an enormous fibular process, thus preventing movement between the two bones. The pes is short for bearing weight. Primitively, in ankylosaurs there are four functional toes, but only three in advanced forms, such as the ankylosaurid *Euoplocephalus* (Carpenter, 1982).

The armor of ankylosaurs is perhaps their most distinguishing character, separating them from all other dinosaurs. This armor consists of keeled plates of bone, short spines, and tall spikes. Ankylosaurids typically have plates arranged in transverse bands along the neck, back, and tail. This armor may be supplemented by thin-walled, conical spines on the back, such as in *Euoplocephalus* (Carpenter, 1982). In addition, the neck armor of ankylosaurids is often fused to an underlying band of bone. In *Pinacosaurus*, and probably other ankylosaurids, this neck armor is the first to ossify in juveniles, the rest being cartilaginous. As stated previously, the characteristic tail club of ankylosaurids is made from the fusion of large terminal plates (Coombs, 1978).

In nodosaurids, the keeled plates are co-ossified into a solid shield over the pelvis in some genera, such as *Polacanthus*. Nodosaurids also supplement their armor with outward-projecting spines along the sides of the body, except in *Panoplosaurus*. In *Edmontonia*, a pair of these spines is enlarged into forward-projecting spikes on each shoulder, whereas in *Sauropelta*, four pairs of spikes project upwards from the neck. (Carpenter, 1984, 1990). The spines and spikes in nodosaurids probably played a dual role of defense and display. In *Polacanthus* a small club is present on the end of the tail, formed by a pair of enlarged plates [J. Kirkland (personal communication) suggests this character may indicate a third lineage of ankylosaurs].

Ankylosaurs are known from almost every continent, including Antarctica. Surprisingly, they have not been identified from South America. Keeled armor plates, once identified as ankylosaurian, are now known to belong to a titanosaur sauropod. Globally, ankylosaurs are rare, making up only a small percentage of any dinosaur fauna. Still, there are a few places where they occur in greater numbers, such as the sand dune deposits of the DJADOKHTA FORMATION of Mongolia and China (Tumanova, 1987). Nodosauridae and Ankylosauridae co-occur only in North America. Elsewhere, nodosaurids are known only from Europe and Antarctica and ankylosaurids from Asia. Another lineage of Ankylosauria is hinted by *Minmi* from Australia, which apparently shows a mixture of both ankylosaurid and nodosaurid features (R. Molnar, personal communication).

References

Carpenter, K. (1982). Skeletal and dermal armor reconstruction of *Euoplocephalus tutus* (Ornithischia: Ankylosauridae) from the Late Cretaceous Oldman Formation of Alberta. *Can. J. Earth Sci.* **19,** 689–697.

Carpenter, K. (1984). Skeletal reconstruction and life restoration of *Sauropelta* (Ankylosauria: Nodosauridae) from the Cretaceous of North America. *Can. J. Earth Sci.* **21,** 1491–1498.

Carpenter, K. (1990). Ankylosaur systematics: Example using *Panoplosaurus* and *Edmontonia* (Ankylosauria: Nodosauridae). In *Dinosaur Systematics: Approaches and Perspectives* (K. Carpenter and P. Currie, Eds.), pp. 281–298.

Coombs, W. (1978). The families of the ornithischian dinosaur order Ankylosauria. *Palaeontology* **21,** 143–170.

Coombs, W. (1979). Osteology and myology of the hindlimb in the Ankylosauria (Reptilia, Ornithischia). *J. Paleontol.* **53,** 666–684.

Coombs, W., and Maryanska, T. (1990). Ankylosauria. In *The Dinosauria* (D. Weishampel, P. Dodson, and H. Osmólska, Eds.). pp. 456–483. Univ. of California Press, Berkeley.

Ostrom, J. (1970). Stratigraphy and paleontology of the Cloverly Formation (Lower Cretaceous) of the Bighorn Basin area, Wyoming and Montana. *Peabody Museum Nat. History Bull.* **35,** 1–234.

Tumanova, T. (1987). Pantsirnye dinozavry Mongolii. *Sovmestnaya Sovetsko–Mongolskaya Paleontologischeskaya Ekspeditsiya.* **32,** 1–76. [In Russian]

Ankylosauridae

A member of the Ankylosauria along with Nodosauridae.

see ANKYLOSAURIA

Annie Riggs Museum, Texas, USA

see MUSEUMS AND DISPLAYS

Anniston Museum of Natural History, Alabama, USA

see MUSEUMS AND DISPLAYS

Antarctic Dinosaurs

see POLAR DINOSAURS

Antorbital Fenestra

This opening is between the orbit and the nostril on the side of the skull and is a characteristic of dinosaurs and other archosaurs. It probably housed an air-filled sinus.

see CRANIOFACIAL AIR SINUS SYSTEMS

Archaeopteryx

The first known bird (Aves), from the Late Jurassic (Solnhofen Formation) of Germany.

see AVES; BIRD ORIGINS; SOLNHOFEN FORMATION

Archosauria

J. MICHAEL PARRISH
Northern Illinois University
DeKalb, Illinois, USA

The Archosauria comprises one of the major radiations of terrestrial vertebrates, including such assemblages as the DINOSAURIA (including birds), the CROCODYLIA, and the PTEROSAURIA, along with a number of less familiar, extinct groups. The name Archosauria ("ruling reptile") was erected by Cope in 1869 to include a somewhat different group of living and extinct reptiles. In recent decades, the Subclass Archosauria was considered to include the dinosaurs, crocodilians, pterosaurs, their common ancestors, and a number of other closely related TRIASSIC diapsid groups such as the Proterosuchia and Erythrosuchia. With the advent of wide use of phylogenetic systematics as a basis of classification in the 1980s, three significant changes occurred in the constitution of the

Archosauria. First, the birds were included within the Archosauria (with the subclass designation dropped, along with all other nomenclature designating rank) because they evolved from dinosaur ancestors and thus form part of the monophyletic (single clade and continuous lineage) group including the dinosaurs and all their descendants.

Second, the PERMIAN and Triassic archosaurs excluding dinosaurs, pterosaurs, and crocodilians had traditionally been placed into their own order, the THECODONTIA. One of the goals of phylogenetic systematics is to have classification reflect complete lineages of organisms (see SYSTEMATICS). The Thecodontia, by definition, consisted of only the basal parts of the archosaur lineage through excluding groups such as the dinosaurs and crocodilians that were derived from the basal archosaurs. The smallest monophyletic group that included all the 'thecodonts' was the Archosauria, so that name was retained and the Thecodontia discarded (see THECODONTIA).

Finally, the Archosauria has been redefined as a crown group consisting of the last common ancestor of the two extant groups of archosaurs (birds and crocodiles) and all of its descendants. By this convention, animals that were formerly considered archosaurs but that appeared prior to the split of the crocodile and bird lineages are considered to be part of a newly erected monophyletic group, the Archosauriformes (Gauthier, 1984) but are excluded from the Archosauria (Fig. 1A).

Looking at the broad pattern of evolution within the Amniota (the monophyletic group including birds, reptiles, and mammals and everything descended from their last common ancestor), three major lines diverged within that group (Fig. 1B). The first to split off, the Synapsida, include the mammals and a variety of other forms often informally called the "mammal-like reptiles" (an incorrect designation because they were never reptiles in the contemporary sense). The Anapsida include turtles and a number of fossil groups. The third group, the Diapsida, again consists of two main branches. The Lepidosauromorpha includes lizards, snakes, and the Sphenodontida (a mostly extinct group with one modern representative, the tuatara), along with a number of extinct clades. The Archosauromorpha includes the Archosauria, most of the marine reptile groups, and several extinct groups such as the beaked rhynchosaurs and the long-necked tanystropheids.

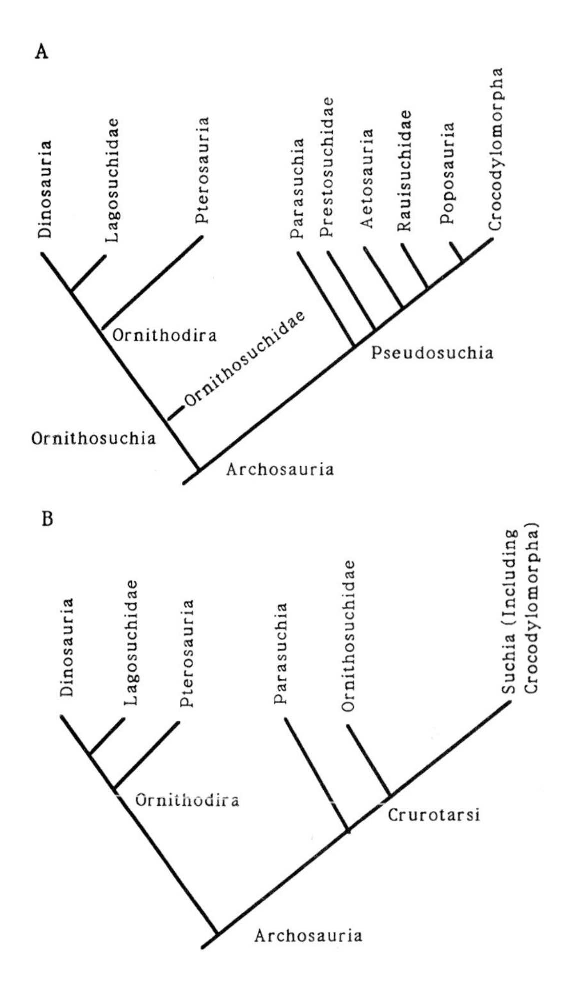

FIGURE 1 (A) Cladogram summarizing the current views of relationships among the Diapsida. (B) Cladogram summarizing the current views about the relationships among the major groups of amniotes with extant members.

The Archosauria proper appear to have originated near the end of the Early Triassic. The small archosauriform *Euparkeria* has often been cited as the best fossil example of what the common archosaurian ancestor might have looked like, although it lacks several key derived characters that diagnose the archosaurs proper. Known from a handful of specimens from a single locality in the Aliwal North Region of South Africa, *Euparkeria* was a small, agile terrestrial carnivore with a tall, mediolaterally compressed skull of a type that is common within several archosaur lineages. *Euparkeria* appears to be the first known archosauriform taxon with dermal armor, a feature that became widespread in the Archosauria. Another group that branched off near the base of the Archo-

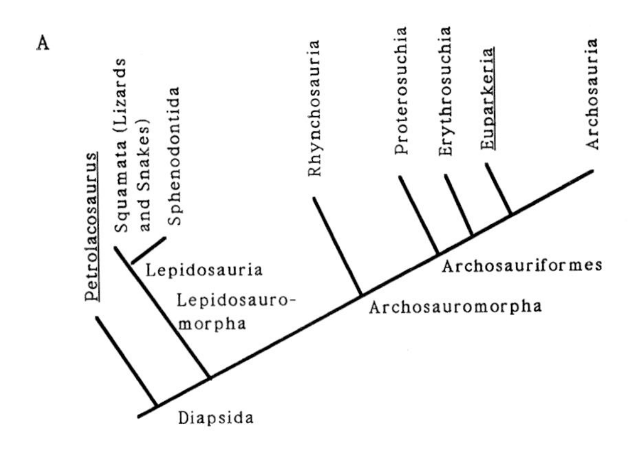

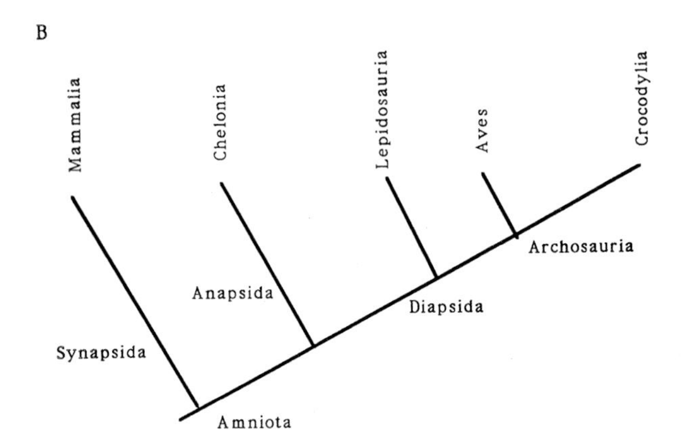

FIGURE 2 Cladograms of archosaur relationships. (A) Cladogram combining Gauthier's (1994) arrangement for the Ornithosuchia with Parrish's (1993) phylogeny of the Crocodylotarsi (Pseudosuchia here). (B) Cladogram depicting Sereno's (1991) views of archosaur relationships. Sereno did not present a phylogeny of relationships among the group he called Suchia.

sauria was the Proterochampsia, a group of mostly amphibious, crocodile-like quadrupeds that seem to be restricted to the Middle and Upper Triassic of South America.

By its current definition, the Archosauria consists of two diverging lineages: one that leads to dinosaurs and ultimately to birds and another that leads to crocodilians. Several different names have been applied to these groups in recent years (Fig. 2). Gauthier (1986) called the archosaurs that are more closely related to birds than crocodilians, the ORNITHOSUCHIA, and those more closely related to crocodilians than birds, the PSEUDOSUCHIA. Benton and Clark (1988) and Parrish (1993) used Crocodylotarsi instead of Pseudosuchia, largely because of the philosophical paradox of putting true crocodilians into a group with a name that means 'false crocodile' in Greek (Fig. 2A). Sereno (1991) used neither of these names, instead defining a group including most of the Crocodylotarsi and the Ornithosuchidae as the Crurotarsi (Fig. 2B). However, priority dictates that Pseudosuchia and Ornithosuchia are sister stem taxa of Archosauria; Crurotarsi is a node-defined taxon containing phytosaurs and crocodiles and all descendants of their most recent common ancestor Crocodylotarsi is a redundant junior synonym.

A key issue in archosaurian systematics involves the interpretation of the structure of the proximal tarsus (ankle) in Archosauriformes. Crocodilians, and a number of related archosaur groups, have a distinctive tarsal pattern in which a ball on the astragalus articulates with a socket on the calcaneum, with the functional result that the calcaneum, structurally part of the foot, rotates on the astragalus, which is structurally united with the lower leg. One group of extinct archosaurs, the Ornithosuchidae, had a second mobile tarsal pattern, with a ball on the calcaneum that articulates with a socket on the astragalus. In basal archosauriforms and several other closely related groups, the proximal tarsals are united by a pair of facets that essentially prevent mobility between them. Instead, the main movement of the ankle takes place between the proximal tarsus and the distal tarsus, which is functionally connected to the foot.

Most of the recent phylogenies of the Archosauria (Benton and Clark, 1988; Gauthier, 1994; Parrish, 1993; Sereno, 1991) agree about the constitution of the crown group Archosauria, although there are considerable differences of opinion about the composition of its constituent groups. A commonly recognized group is the Ornithodira, which comprises the Lagosuchidae, Pterosauria, and Dinosauria.

The pterosaurs, the familiar clade of flying archosaurs, are discussed under their own entry. The Lagosuchidae consists of three small, long-legged archosaurs known from the Middle Triassic Chañares Formation of Argentina. The best represented of these three taxa was originally placed into *Lagosuchus* by Romer (1971) but was later transferred to a new genus, *Marasuchus*, by Sereno and Arcucci (1994) (see DINOSAUROMORPHA).

Gauthier (1986) united the ORNITHODIRA with the Ornithosuchidae, a group of Upper Triassic, quadrupedal carnivores, in a larger group, the ORNITHOSUCHIA. Sereno (1991) considered the Ornithosuchia an invalid grouping. He instead erected a group he called the Crurotarsi that united the Ornithosuchia with Gauthier's Pseudosuchia, with the most important character linking these groups being the presence of mobility in the proximal tarsus as opposed to the mesotarsal ankles in basal archosauromorphs and ornithodirans. However, regardless of the position of the Ornithosuchidae, Pseudosuchia cannot be redefined, and Crurotarsi is now recognized as a node group within Pseudosuchia (see above) (Padian and May, 1993).

The rest of the non-dinosaurian archosaurs are now generally put into the Pseudosuchia, as noted previously. The position of one group, the Late Triassic Parasuchia, is also controversial. The parasuchians, or phytosaurs, were an abundant group of large, long-snouted archosaurs that appear to have been rough ecological equivalents of modern crocodilians. In most archosaurian phylogenies (e.g., Benton and Clark, 1988; Gauthier, 1986; Parrish, 1993), the phytosaurs occupy a basal position within the Pseudosuchia. Sereno (1993) places the phytosaurs as the sister group of his Crurotarsi because he interprets their tarsus as being immobile (but see Parrish, 1986).

Several other taxa of pseudosuchians are recognized. The Stagonolepididae (aetosaurs) consist of a Late Triassic group of apparently herbivorous archosaurs that had a complete carapace of dermal armor and ranged in size from less than a meter to nearly 10 meters in length. They are distinctive in that they are the first archosaurian herbivores to appear in the fossil record. There are also several groups of tall-snouted, terrestrial pseudosuchians that have often been grouped together as the Rauisuchidae or Rauisuchia. Parrish (1993) divided members of this ecomorph into one group, the Prestosuchidae, that occurs near the base of the Pseudosuchia and a second clade, the Rauisuchia proper, that comprises the other members of the ecomorph plus the Crocodylomorpha. Within the Rauisuchia, two lineages were recognized, the Rauisuchidae (which includes several carnivores along with the enigmatic, beaked, amphibious *Lotosaurus*) and the Poposauria, a relatively poorly known group including carnivores, some of which may have been capable of running on their hind limbs. Poposaurs are recognized by most of the recent phylogenies (e.g., Clark in Benton and Clark, 1988; Parrish, 1987, 1993; Sereno, 1991) as the closest relatives of the Crocodylomorpha.

Benton in Benton and Clark (1988) restricted the Pseudosuchia to a monophyletic clade that he recognized comprising the Stagonolepididae and the Rauisuchidae, which he united on the basis of having ventrally facing hip sockets; however, given priority in the phylogenetic system, this constitution of Pseudosuchia is invalid.

The Crocodylomorpha in its modern constitution was recognized by Walker (1970) as a clade including the Crocodylia and a group of long-legged, Triassic–Jurassic pseudosuchians that are often combined within a group, the Sphenosuchia, which may or may not be a monophyletic lineage (e.g., Clark in Benton and Clark, 1988; Walker, 1990). The Crocodylomorpha were initially terrestrial carnivores; they took on their current amphibious/aquatic habitus later in the Mesozoic (Parrish, 1987; Walker, 1970).

During the Early Triassic, archosauriforms were still a relatively unimportant element in most terrestrial ecosystems. The Archosauria appeared early in the Triassic and became much more prominent, in terms of both abundance and diversity, during the Middle Triassic. The skeletal record of dinosaurs appeared near the beginning of the Late Triassic, although they did not become abundant parts of terrestrial ecosystems until well into the Norian (latest Triassic). By the end of the Triassic, all the archosaurs other than crocodylomorphs, pterosaurs, and dinosaurs became extinct.

See also the following related entry:

THECODONTIA

References

Benton, M. J., and Clark, J. M. (1988). Archosaur phylogeny and the relationships of the Crocodylia. in *Phylogeny and Classification of Amniotes* (M. J. Benton, Ed.), Systematics Association Special Vol. 35A, pp. 295–338. Clarendon Press, Oxford.

Bonaparte, J. F. (1975). The family Ornithosuchidae (Archosauria: Thecodontia). *Colloq. Int. CNRS* **218,** 485–501.

Brinkman, D. L. (1981). The origin of the crocodiloid tarsi and the interrelationships of thecodontian reptiles. *Breviora* **464,** 1–23.

Cruickshank, A. R. I. (1979). The ankle joint in some early archosaurs. *South African J. Sci.* **75,** 168–178.

Gauthier, J. A. (1984). A cladistic analysis of the higher systematic categories of the Diapsida, pp. 564. PhD dissertation, University of California, Berkeley.

Gauthier, J. A. (1986). Saurischian monophyly and the origin of birds. *Mem. California Acad. Sci.* **8,**155.

Gauthier, J. A. (1994). The diversification of the amniotes. In *Major Features in Vertebrate Evolution. Short Courses in Paleontology 7* (D. R. Prothero and R. M. Schoch, Eds.), pp. 129–159. Paleontological Society, Knoxville, TN.

Padian, K., and May, C. L. (1993). The earliest dinosaurs. *New Mexico Museum Natl. History Bull.* **3,** 379–382.

Parrish, J. M. (1986). Structure and function of the tarsus in the phytosaurs (Reptilia: Archosauria). In *The Beginning of the Age of Dinosaurs* (K. Padian, Ed.), pp. 35–43. Cambridge Univ. Press, New York.

Parrish, J. M. (1987). The origin of crocodilian locomotion. *Paleobiology* **13,** 396–414.

Parrish, J. M. (1993). Phylogeny of the Crocodylotarsi and a consideration of archosaurian and crurotarsan monophyly. *J. Vertebr. Paleontol.* **13,** 287–308.

Romer, A. S. (1971). The Chañares (Argentina) Triassic reptile fauna. X. Two new but incompletely known long-limbed pseudosuchians. *Breviora* **378,** 1–10.

Sereno, P. C. (1991). Basal archosaurs: Phylogenetic relationships and functional implications. Society of Vertebrate Paleontology Memoir 2. *J. Vertebr. Paleontol.* **11**(Suppl. No. 4), 1–53.

Sereno, P. C. (1994). Dinosaurian precursors from the Middle Triassic of Argentina: *Marasuchus lilloensis*, gen. nov. *J. Vertebr. Paleontol.* **14,** 53–73.

Sereno, P. C., and Arcucci, A. B. (1990). The monophyly of crurotarsal archosaurs and the origin of bird and crocodile ankle joints. *Neues Jahrbuch Geol. Paläontol. Abhandlungen* **180,** 21–52.

Walker, A. D. (1970). A revision of the Jurassic reptile *Hallopus victor* (Marsh). *Philos. Trans. R. Soc. London Ser. B* **257,** 323–372.

Walker, A. D. (1990). A revision of *Sphenosuchus acutus* Haughton, a crocodylomorph reptile from the Elliot Formation (Late Triassic or Early Jurassic) of South Africa. *Philos. Trans. R. Soc. London Ser. B* **330,** 1–120.

Arctic Dinosaurs

see POLAR DINOSAURS

Arctometatarsalia

JOHN R. HUTCHINSON
KEVIN PADIAN
University of California
Berkeley, California, USA

Arctometatarsalia (Fig. 1) was established by Holtz (1994b) to encompass all coelurosaurian theropods that shared the "arctometatarsalian" condition, a name given to the proximally pinched third metatarsal by Holtz (1994b; from the Latin *arctus*, meaning compressed) (see HINDLIMBS and FEET). He defined Arctometarsalia as the first theropod with this condition and all of its descendants, which included by his formulation Ornithomimosauria, Troodontidae, Tyrannosauridae, Elmisauridae, and *Avimimus.* However, Holtz (1996b) revised this definition because he recognized that an apomorphy-based definition was potentially unstable, and the condition had also been found in *Mononykus* (Perle *et al.*, 1994), an early bird (see AVIALAE). Moreover, Elmisauridae may be closer to Oviraptoridae than to the other Arctometatarsalia, and *"Avimimus"* may be composed of several different taxa (Holtz, 1996a,b). Consequently, Holtz (1996b) amended the definition of Arctometatarsalia from an apomorphy-based to a stem-based taxon: the clade comprising *Ornithomimus* and all theropods closer to *Ornithomimus* than to birds. Currently, the Arctometatarsalia principally comprise Ornithomimosauria, Troodontidae (which together form BULLATOSAURIA), and Tyrannosauridae.

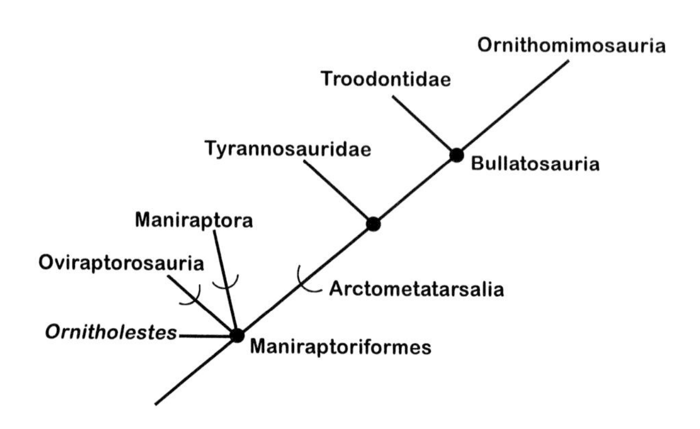

FIGURE 1 Phylogeny of Arctometatarsalia, after Holtz and other sources.

Tyrannosaurs are united to bullatosaurs in Arctometatarsalia by the contact of the iliac blades along most of their dorsal surface, a semicircular scar on the anterior of the ischium, the large surangular foramen, an elongate tibia and metatarsus, metatarsals that are deeper anteroposteriorly than mediolaterally, the loss of flexed cervical zygapophyses, and the pinched "arctometatarsalian" condition.

In the derived arctometatarsalian pes, the proximal articular surfaces of metatarsals II and IV are mostly to entirely dominant at the proximal end of the metatarsal; metatarsal III is only expressed on the proximal end of the metatarsus in those taxa with a less derived arctometatarsus (e.g., *Allosaurus*). The solid, compressed shaft of the third metatarsal forms a rigid structure with metatarsals II and IV proximally, and in life must have been bound together with strong ligaments so that the whole metatarsus acted as a cohesive functional unit. In the distal portion of the metatarsus, metatarsal III is hollow, like metatarsals II and IV, and is expanded, often forming the bulk of the distal metatarsus. Metatarsals II and IV are heavily buttressed in this region for their articulation with metatarsal III, which completes the functional integration of the arctometatarsus (see Fig. 2; Holtz 1994a). Some arctometatarsalian-grade theropods, such as *Elmisaurus*, fuse some of the tarsal elements, including the metatarsals, which would add further rigidity to the pes. This condition is convergent with the condition in some other theropods, such as some ceratosaurs (e.g., *Syntarsus*) and avialians (e.g., enantiornithines and other birds).

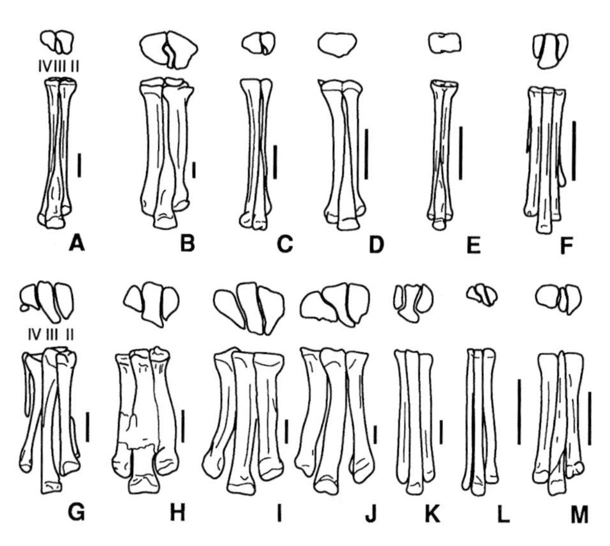

FIGURE 2 Representative theropod right metatarsals II–IV (metatarsal V illustrated in F and G; metatarsal I not included); dorsal (above) and anterior (below) views, with scale bar = 50 mm. From Holtz (1994a, Fig. 1, p. 481). Metatarsi A–E demonstrate the arctometatarsus morphology, whereas F–M exhibit less-derived morphology. Taxa pictured: A, *Struthiomimus* (Ornithomimidae); B, *Albertosaurus* (Tyrannosauridae); C, *Tochisaurus* (Troodontidae); D, *Elmisaurus* (Elmisauridae); E, "*Avimimus*" (validity questionable); F, *Coelophysis* (Coelophysidae); G, *Dilophosaurus* (Coelophysoidea); H, *Ceratosaurus* (Neoceratosauria); I, *Allosaurus* (Allosauridae); J, *Chilantaisaurus* (Tetanurae *incertae sedis*); K, *Elaphrosaurus* (Abelisauroidea); L, *Ornitholestes* (Coelurosauria *incertae sedis*); M, *Deinonychus* (Dromaeosauridae).

The arctometatarsalian pes has certain functional implications (Holtz, 1994a) (see Functional Morphology). Arctometatarsalian-grade theropods in general seem to have a more elongate and gracile pes (and hindlimb) than in other theropods. The biological significance of this is uncertain but it may indicate a degree of increased cursorial ability (Coombs, 1978; Holtz, 1994a). The precise biological action of the arctometatarsus has been a subject of some debate, but Holtz (1994a) and Wilson and Currie (1985) have convincingly argued that it is best interpreted as a force-transducing structure, channeling and evenly distributing ground-reaction forces during locomotion proximally from the pes across the mesotarsal joint. There does not seem to be strong support for any alternate hypotheses, such as a pistoning action of metatarsal III, "snap ligaments" (Coombs, 1978), or rotation of metatarsal III (Wilson and Currie, 1985; see Fig. 3 for further discussion). Holtz (1994a) has provided the only detailed functional analysis of the arctometatarsus, but further mysteries regarding the functional morphology of the pes during locomotion remain unresolved, such as the range of possible motion (joint angles) of the intrapedal and tarsal joints, the relationship between metapodial joint mobility and overall theropod hindlimb kinematics, and similarities or differences between nonavian and avian theropod locomotion. Further studies of theropod hindlimb functional morphology combined with trackway studies offer hope of further clarifying the matter.

Holtz's (1996a,b) revisions of the Arctometatarsalia have revealed that the characteristic arctometatarsal-

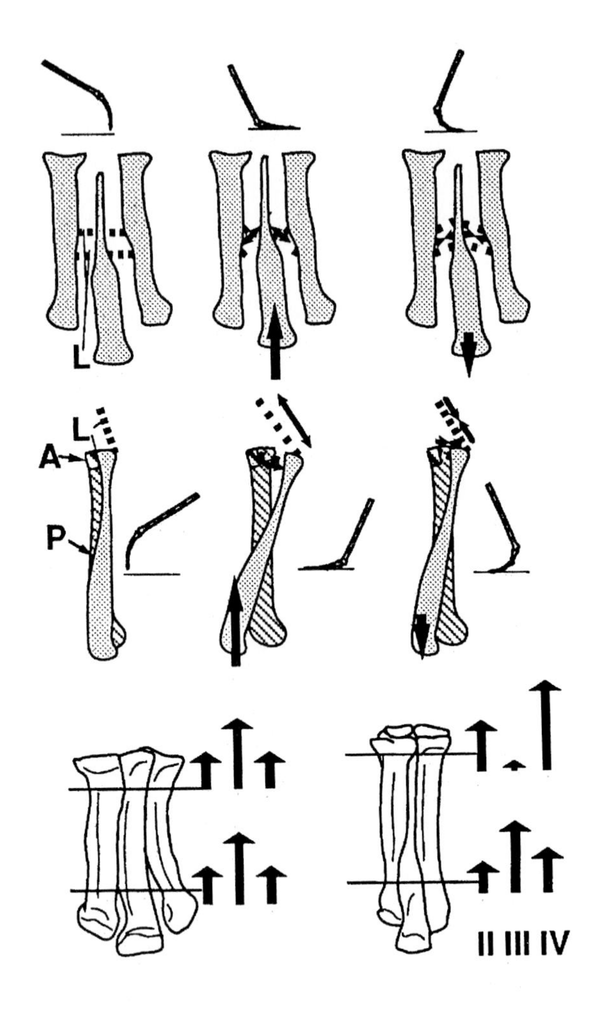

FIGURE 3 Three hypotheses of arctometatarsalian pes function during locomotion. From Holtz (1994a, Fig. 10, p. 499). Ranges of motion are exaggerated for clarity; force vectors are approximated. (A) "Pistoning" or "snap ligament" model from Coombs (1978). From left to right: (left) metatarsal III is suspended between metatarsals II and IV by elastic ligaments (L), which (middle) store potential energy from the ground-reaction force at footfall as metatarsal III moves proximally, and (right) release the stored potential energy at takeoff, adding thrust to the hindlimb. (B) "Rotational" model from Wilson and Currie (1985). From left to right: (left) elastic ligaments (L) running from the epipodium and the anterior projection (A) of the proximal end of metatarsals II and IV attach to the anterior surface of the proximal end of metatarsal III (shaded). The ligaments are stretched by the weight of the animal at footfall (middle) as metatarsal III rotates around the pivot point (P) and (right) release elastically stored potential energy at takeoff, returning metatarsal III to its relaxed position and adding thrust to the hindlimb. (C) "Force transmission" model from Wilson and Currie (1985) and Holtz (1994a). (Left) Underived theropod metatarsus example: *Allosaurus* metatarsus showing independent ground-reaction force transmission along each metatarsal, at proximal and distal sections. (Right) Arctometatarsus example: *Albertosaurus* metatarsus showing the transmission of forces from metatarsal III (distal metapodium; bottom section) to metatarsals II and IV (proximal metapodium; top section) via the wedge and buttress system (metatarsal III forms the "wedge"; the surfaces of metatarsals II and IV facing metatarsal III along the distal half of the metatarsus form the "buttresses").

ian-grade pes has evolved several times within Theropoda—most likely independently in the Elmisauridae (including Caenagnathidae), AVIALAE (alvarezsaurids), Arctometatarsalia, and perhaps *Avimimus* as well. Examples of intermediate character states between the derived arctometatarsus and the primitive, noncompressed theropod metatarsus are few, although the third metatarsals of *Alvarezsaurus* (Alvarezsauridae), *Chilantaisaurus* (TETANURAE *incertae sedis*), *Ornitholestes*, and some basal Arctometatarsalia [e.g., *Harpymimus* (Ornithomimosauria)] seem to be less compressed than in more derived taxa within their respective clades (see Holtz, 1994a, pp. 494–496 for discussion). The third metatarsal in nonalvarezsaurid MANIRAPTORA, such as *Deinonychus* and *Archaeopteryx*, is not technically arctometatarsalian, but it is reduced compared to the third metatarsals of noncoelurosaurian theropods (cf. Fig. 2).

See also the following related entries:

COELUROSAURIA • ORNITHOMIMOSAURIA • THEROPODA

References

Coombs, W. P., Jr. (1978). Theoretical aspects of cursorial adaptations in dinosaurs. *Q. Rev. Biol.* **53,** 393–418.

Holtz, T. R., Jr. (1994a). The arctometatarsalian pes, an unusual structure of the metatarsus of Cretaceous Theropoda (Dinosauria: Saurischia). *J. Vertebr. Paleontol.* **14,** 480–519.

Holtz, T. R., Jr. (1994b). The phylogenetic position of the Tyrannosauridae: Implications for theropod systematics. *J. Paleontol.* **68,** 1100–1117.

Holtz, T. R., Jr. (1995). A new phylogeny of the Theropoda. *J. Vertebr. Paleontol.* **15**(Suppl. to No. 3), 45A.

Holtz, T. R., Jr. (1996a). Phylogenetic analysis of the non-avian Tetanurine dinosaurs (Saurischia, Theropoda). *J. Vertebr. Paleontol.* **16**(Suppl. to No. 3), 42A.

Holtz, T. R., Jr. (1996b). Phylogenetic taxonomy of the Coelurosauria (Dinosauria: Theropoda). *J. Paleontol.* **70,** 536–538.

Karhu, A. A., and Rautian, A. S. (1996). A new family of Maniraptora (Dinosauria: Saurischia) from the Late Cretaceous of Mongolia. *Paleontol. J.* **30,** 583–592.

Perle A., Chiappe, L., Barsbold, R., Clark, J. M., and Norell, M. A. (1994). Skeletal morphology of *Mononykus olecranus* (Theropoda: Avialae) from the Late Cretaceous of Mongolia. *Am. Museum Novitates* **3105,** 1–29.

Wilson, M. C., and Currie, P. J. (1985). *Stenonychosaurus inequalis* (Saurischia: Theropoda) from the Judith River Formation of Alberta: New findings on metatarsal structure. *Can. J. Earth Sci.* **22,** 1813–1817.

Argentinean Dinosaurs

Argentinean dinosaurs are found in Upper Triassic to Upper Cretaceous sediments and include prosauropods, theropods, herrerasaurids, sauropods, hadrosaurs, etc.

see SOUTH AMERICAN DINOSAURS

Arizona Museum of Science and Technology, Arizona, USA

see MUSEUMS AND DISPLAYS

Arkansas Geological Commission, Arkansas, USA

see MUSEUMS AND DISPLAYS

Arkansas Museum of Science and History, Arkansas, USA

see MUSEUMS AND DISPLAYS

Armor

Armor is a characteristic of ankylosaurs, some sauropods, and stegosaurs.

see ORNAMENTATION

Asian Dinosaurs

Dinosaurs are known from most of the continent of Asia. The best known sites are in Mongolia and China, but there are also well-known sites in middle Asia, Japan, and India.

see AUSTRALASIAN DINOSAURS; CHINESE DINOSAURS; INDIAN DINOSAURS; JAPANESE DINOSAURS; MIDDLE ASIAN DINOSAURS; MONGOLIAN DINOSAURS

Australasian Dinosaurs

RALPH E. MOLNAR
Queensland Museum
Queensland, Australia

The unique modern mammals of Australia originated and evolved on a continent separated from all others since the beginning of the Cenozoic. During the Mesozoic Era, Australia and New Zealand were parts of the southern supercontinent, and so their dinosaurs might be expected to be similar to those from elsewhere. However, at least in part, there seems to have been an endemic fauna that shared some of the unusual features of the modern Australasian faunas.

Dinosaurs of Australia

Australia is not only the smallest but also the flattest of the continents, having the lowest proportion of land with topographic relief. Few fossil-bearing rocks are well exposed. In the Mesozoic, terrestrial body fossils (and the rocks that yield them) are known only from the Early Triassic (Scythian), Late Liassic through Bajocian, and Late 'Neocomian' through Cenomanian. This represents about 20% of the Mesozoic but only about 10% can be considered well-known. Many of these rocks are marine, deposited in the epeiric sea that periodically covered the north-central part of the continent during the Early Cretaceous; only in Victoria are there continental deposits (Valanginian–Albian). Therefore, dinosaur bones are few and far between. However, the trackway record is better, extending (intermittently) from the Late Triassic through the Cenomanian and representing about 45% of the Mesozoic. Late Cretaceous terrestrial tetrapods and continental rocks are almost unknown in Australia.

The best known dinosaurs and dinosaurian faunas come from Queensland, which has the greatest area of outcrop, and Victoria, due to the exceptional diligence and perseverance of Tom and Pat Rich. Other specimens are known from South Australia, Western Australia, and the Northern Territory, but only from New South Wales are there more than isolated bones of single taxa.

The oldest reported Australian dinosaur, the prosauropod *Agrosaurus,* was described by Seeley in 1891. The specimen was reportedly collected during the voyage of *H. M. S. Fly* up the eastern coast of Australia, but extraordinarily there is no record of the event in the log. The specimen may have come from anywhere along the *Fly's* route from England and back. However, tracks (probably *Eubrontes* sp.) show that dinosaurs were present in Late Triassic Queensland.

Jurassic bones of any kind are rare, and at no locality is more than one taxon of terrestrial tetrapod known. A partial skeleton of the cetiosaurid *Rhoetosaurus* was found in southeastern Queensland in Bajocian rocks otherwise yielding only fossil plants. However, a contemporaneous caudal from Western Australia indicates that sauropods were widespread in Australia at this time (Long, 1992). The only other nonmarine tetrapods known are Liassic temnospondyls from southeastern Queensland. Tracks are more informative. They show that ornithopods were present already in the Liassic, and theropods not only were present but also had become quite large (to 10 m long) by Middle Jurassic times. A quadrupedal form, tentatively but probably incorrectly reported as stegosaur (Hill *et al.*, 1966), was also present. *Rhoetosaurus* seems similar to the later *Shunosaurus* from Sichuan, China, where Jurassic temnospondyls have also been found.

Four regions yield Cretaceous faunas: the southern coast of Victoria, where both a Valanginian–Aptian and an Aptian–Albian fauna are known; northern New South Wales (Albian); north-central Queensland (Albian and Cenomanian); and the northwestern coast of Western Australia ('Neocomian').

Sauropod, theropod, and ornithopod tracks have recently been found in the Broome Sandstone of Western Australia (Long, 1990). Other tracks may represent stegosaurs, otherwise unknown from Australasia. New discoveries by Tony Thulborn indicate that the trackways are associated with a variety of habitats ranging from lagoons to swamp and forest (Thulborn *et al.*, 1997).

The two faunas from Victoria are basically similar in composition, as far as dinosaurs are concerned, but the earlier fauna also has temnospondyls, which are absent from the later (Rich and Rich, 1989). The most common dinosaurs are small hypsilophodontians. A moderately large theropod has been referred to *Allosaurus.* Unexpectedly, a ceratopsian and ornithomimosaur have been described (Rich and Vickers-Rich, 1994), and a caenagnathid has been reported. The specimens, almost all isolated single bones, were deposited in the braided channels of a large river system flowing westward into the nascent rift valley opening between Australia and Antarctica. At least some of the dinosaurs probably inhabited the closed forests prevalent in the region (Dettmann *et al.*, 1992). All seem to have been relatively small forms, the ornithopods about 1 or 2 m in length and even the *Allosaurus* only about 6 m long. This is consistent with a closed forest habitat, although the apparent small size may result from fluvial sorting of the isolated bones.

Hypsilophodontids are also the most common dinosaurs from the opal mining district at Lightning Ridge, New South Wales (Molnar and Galton, 1986). Theropods, too, were present and miners report having observed what may be sauropod tracks. Other trackways are probably from large ornithopods (Molnar, 1991). The tracks indicate the presence of large dinosaurs, whereas only small forms are represented by the bony fossils, presumably the result of sorting. Their habitat appears to have been an estuary opening into the inland sea (Dettmann *et al.*, 1992).

Two kinds of sauropods, *Austrosaurus* and an unidentified species, are known from the Albian of Queensland, but the most common dinosaurs seem to have been ankylosaurs (*Minmi*) and large ornithopods (*Muttaburrasaurus*). Both forms are quite distinct and probably represent endemic lineages. *Muttaburrasaurus* has several unusual features relating to the feeding apparatus. The postorbital region of the skull is very broad and anteroposteriorly lengthened. All teeth in each toothrow are erupted to the same degree, suggesting that they were replaced simultaneously (Bartholomai and Molnar, 1981). Other character states were previously known only in pachycephalosaurs (transversely broad postorbital bar and contact between pterygoid and BRAINCASE). *Minmi* had but a single supraorbital and a broad dorsal ossification connecting the ilium to the sacral centra, seemingly in addition to the sacral ribs. Its pelvis is unexpectedly plesiomorphic, with a long postacetabular process on the ilium and a fabrosaurian-like pubis (see PELVIS, COMPARATIVE ANATOMY). These animals presumably lived in the open woodlands thought to have covered the coastal regions to the north and east of the epeiric sea (Dettmann *et al.*, 1992).

Sauropod bones, referred to the genus *Austrosaurus*, are the only Cenomanian body fossils of dinosaurs from Australia (Coombs and Molnar, 1981). However, large and small theropod and ornithopod tracks are known from the trackways at Lark Quarry, near Winton, central Queensland (Thulborn and Wade, 1984). None of the Cenomanian taxa represented by tracks have been found as bones and vice versa. The affinities of the Cretaceous sauropods are not known. They may yet prove to be primitive relatives of the titanosaurids.

The dinosaurs of the Queensland Cretaceous seem quite distinct from those elsewhere, whereas those from Victoria appear more similar to other dinosaurs, especially if the identifications of a ceratopsian and ornithomimosaur are correct. On the other hand, these are based on single elements; therefore, maybe distinct but convergent lineages are represented.

A single incomplete bone from marine deposits in Western Australia may represent a Late Cretaceous (Maastrichtian) theropod (Long, 1992).

Dinosaurs of New Zealand

New Zealand has recently produced isolated, but identifiable, dinosaur bones (Molnar and Wiffen, 1997) serendipitously discovered during a research program carried out by amateur workers on the marine saurians of North Island. They were deposited on the bed of a Late Cretaceous estuary. A large but incomplete rib almost certainly represents a sauropod. Pedal phalanges of large and small theropods have been found, as has an incomplete ilium of a small dryosaur-like ornithopod (Wiffen and Molnar, 1989). A rib and caudals indicate a small, probably nodosaurid, ankylosaur. Although none of these taxa can confidently be identified to genus, they indicate that a fauna consisting of at least four and probably five taxa inhabited New Zealand during Late Campanian or Early Maastrichtian times, and thus provide information on a period not represented in Australia. Because New Zealand rifted away from Antarctica, these dinosaurs probably represent components of the Antarctic dinosaur fauna.

Biogeographic and Physiological Implications of Australasian Dinosaurs

Australasia today is known for its unusual large tetrapods and prominent relict taxa. In New Zealand—until the coming of humans—the largest land-dwelling animals were not mammals but birds: This is a place where the descendants of Mesozoic dinosaurs still reigned supreme. The unusual nature of the Australian ornithischians *Minmi and Muttaburrasaurus* suggests that during the Early Cretaceous there was sufficient isolation between Australia and the other continents for these endemic lineages to develop.

This is unexpected in view of the connection of Australia with Antarctica until the Eocene and the Patagonian aspect of the Eocene land mammal faunas from West Antarctica (Marenssi *et al.*, 1994). Thulborn (1986) has argued that already in the Scythian, tetrapod faunas of Australia were significantly different—not in composition, but in proportions—from those elsewhere. These anomalies suggest some barrier to dispersal to (and from) Australia during the Mesozoic. This is supported by the occurrence of what seem to have been relict forms, *Allosaurus* among the dinosaurs, and more notably the temnospondyls (Rich *et al.*, 1992).

During the Mesozoic, Australia was at 'the end of the earth,' situated at the end of what was basically a long peninsula made up of Africa, South America, and Antarctica. It was the region furthest from the Laurasian lands whose dinosaurs are well known. Therefore, it should be no surprise that some Australasian dinosaurs would be different from those of the Northern Hemisphere and that this isolation permitted the survival of forms that had become extinct elsewhere (Molnar, 1989, 1992).

The Late Cretaceous New Zealand fauna lived there after it had rifted north from Antarctica to become insular. This adds another to the short list of insular dinosaurian faunas. The small size of the ankylosaur is consistent with insular ankylosaurs known from elsewhere (e.g., Late Cretaceous Europe), but the sauropod rib fragment shows that large animals did live on these islands.

Furthermore, during Campanian–Maastrichtian times New Zealand was still near Antarctica. The fauna lived near the south polar circle, so an insular, near-polar fauna is represented. Marine temperatures suggest that these dinosaurs lived in a climate not significantly different from that of North Island today (mean annual temperature of ~14°C. Although it has been suggested that Alaskan dinosaurs migrated south during the winter, this option was not open to the dinosaurs of New Zealand; it is unlikely that like giant lemmings they swam to Australia for the winter. No large reptiles today live in such climates.

This point is made even more strongly by the Early Cretaceous dinosaurs of Victoria that lived south of the Antarctic circle.

Paleotemperature measurements indicate a mean annual temperature of between −1 and 9°C (Rich and Rich, 1989) (see PALEOCLIMATOLOGY). Although most Australian workers feel the latter is the more reliable of the temperatures, even this is significantly lower than that for New Zealand. These clearly indicate that dinosaurs were capable of inhabiting regions and climates not accessible to large, land-dwelling lepidosaurs, crocodilians, and chelonians.

The Cretaceous dinosaurs of Australasia show both regional endemism and the ability to survive under climatic conditions that no later reptiles have been able to tolerate.

See also the following related entries:

ASIAN DINOSAURS • POLAR DINOSAURS

References

Bartholomai, A., and Molnar, R. E. (1981). *Muttaburrasaurus*, a new iguanodontid (Ornithischia: Ornithopoda) dinosaur from the Lower Cretaceous of Queensland. *Mem. Queensland Museum* **20,** 319–349.

Coombs, W. P., Jr., and Molnar, R. E. (1981). Sauropoda (Reptilia, Saurischia) from the Cretaceous of Queensland. *Mem. Queensland Museum* **20,** 351–373.

Dettmann, M. E., Molnar, R. E., Douglas, J. G., Burger, D., Fielding, C., Clifford, H. T., Francis, J., Jell, P., Rich, T., Wade, M., Rich, P. V., Pledge, N., Kemp, A., and Rozefelds, A. (1992). Australian Cretaceous terrestrial faunas and floras: Biostratigraphic and biogeographic implications. *Cretaceous Res.* **13,** 207–262.

Hill, D., Playford, G., and Woods, J. T. (1966). *Jurassic Fossils of Queensland*, pp. 32. Queensland Palaeontographical Society, Brisbane.

Long, J. A. (1990). *Dinosaurs of Australia*, pp. 87. Reed Books, Sydney.

Long, J. A. (1992). First dinosaur bones from Western Australia. *Beagle* **9,** 21–27.

Marenssi, S. A., Reguero, M. A., Santillana, S. N., and Vizcaino, S. F. (1994). Eocene land mammals from Seymour Island, Antarctica: Palaeobiogeographical implications. *Antarctic Sci.* **6,** 3–15.

Molnar, R. E. (1989). Terrestrial tetrapods in Cretaceous Antarctica. *Special publication, Geological Soc. Am.* **47,** 131–140.

Molnar, R. E. (1991). Fossil reptiles in Australia. In *Vertebrate Paleontology of Australasia* (P. V. Rich, J. M. Monaghan, R. F. Baird, T. K. Rich, E. M. Thompson, and C. Williams, Eds.), pp. 605–702. Pioneer Design Studio, Melbourne.

Molnar, R. E. (1992). Paleozoogeographic relationships of Australian Mesozoic tetrapods. In *New Concepts in Global Tectonics* (S. Chatterjee and N. Hotton, III, Eds.), pp. 259–266. Texas Tech Univ. Press, Lubbock.

Molnar, R. E., and Galton, P. M. (1986). Hypsilophodontid dinosaurs from Lightning Ridge, New South Wales, Australia. *Geobios* **19,** 231–239.

Molnar, R. E., and Wiffen, J. (1997). A Late Cretaceous polar dinosaur fauna from New Zealand Cretaceous. *Cretaceous Res.*, in press.

Rich, T. H. V., and Rich, P. V. (1989). Polar dinosaurs and biotas of the Early Cretaceous of southeastern Australia. *Natl. Geogr. Res.* **5,** 15–53.

Rich, T. H. V., Rich, P. V., Wagstaff, B., Mason, J. R. C. McE., Flannery, T. F., Archer, M., Molnar, R. E., and Long, J. A. (1992). Two possible chronological anomalies in the Early Cretaceous tetrapod assemblages of southeastern Australia. In *Aspects of Nonmarine Cretaceous Geology* (N. J. Mateer and P.-J. Chen, Eds.), pp. 165–176. China Ocean Press, Beijing.

Rich, T. H., and Vickers-Rich, P. (1994). Neoceratopsians and ornithomimosaurs: Dinosaurs of Gondwana origin. *Res. Exploration* **10,** 129–131.

Seeley, H. G. (1891). 0n *Agrosaurus macgillivrayi* (Seeley) a saurischian reptile from the N. E. coast of Australia. *Q. J. Geological Soc. London* **47,** 164–165.

Thulborn. R. A. (1986). Early Triassic tetrapod faunas of southeastern Gondwana. *Alcheringa* **10,** 297–313.

Thulborn. R. A., Hamley, T., and Foulkes, P. (1997). Preliminary report on sauropod dinosaur tracks in the Broome Sandstone (Lower Cretaceous) of Western Australia. *Gaia,* in press.

Thulborn, R. A., and Wade, M. (1984). Dinosaur trackways in the Winton Formation (mid-Cretaceous) of Queensland. *Mem. Queensland Museum* **21,** 413–517.

Wiffen, J., and Molnar, R. E. (1989). An Upper Cretaceous ornithopod from New Zealand. *Geobios* **22,** 531–536.

Australian Museum

see Museums and Displays

Aves

Luis M. Chiappe
American Museum of Natural History
New York, New York, USA

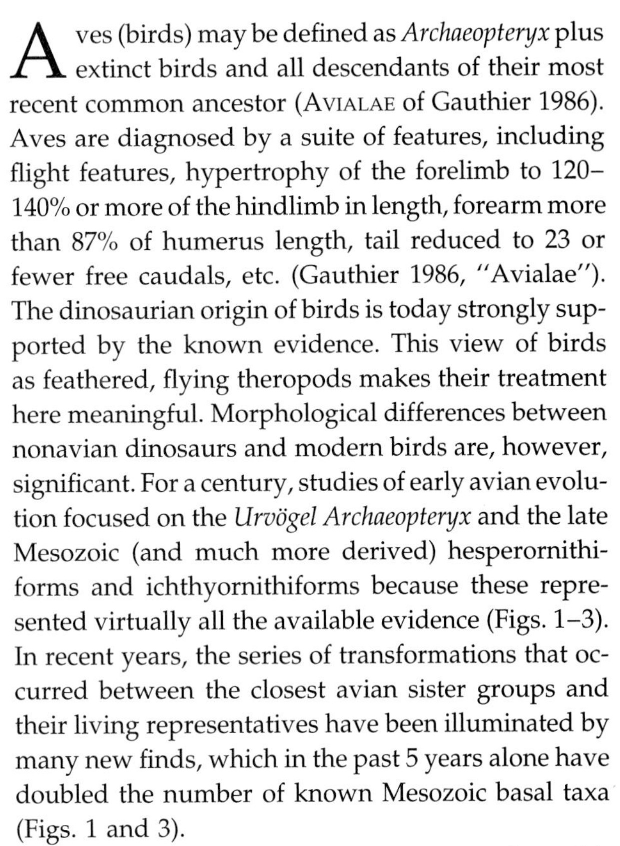

Aves (birds) may be defined as *Archaeopteryx* plus extinct birds and all descendants of their most recent common ancestor (Avialae of Gauthier 1986). Aves are diagnosed by a suite of features, including flight features, hypertrophy of the forelimb to 120–140% or more of the hindlimb in length, forearm more than 87% of humerus length, tail reduced to 23 or fewer free caudals, etc. (Gauthier 1986, "Avialae"). The dinosaurian origin of birds is today strongly supported by the known evidence. This view of birds as feathered, flying theropods makes their treatment here meaningful. Morphological differences between nonavian dinosaurs and modern birds are, however, significant. For a century, studies of early avian evolution focused on the *Urvögel Archaeopteryx* and the late Mesozoic (and much more derived) hesperornithiforms and ichthyornithiforms because these represented virtually all the available evidence (Figs. 1–3). In recent years, the series of transformations that occurred between the closest avian sister groups and their living representatives have been illuminated by many new finds, which in the past 5 years alone have doubled the number of known Mesozoic basal taxa (Figs. 1 and 3).

The diversity of living dinosaurs is very large (almost 10,000 species of birds are usually recognized, and recent estimates are much greater), and their Cenozoic evolutionary history was complex. The current discussion, however, is restricted to their diversity and patterns of evolution in the Mesozoic. This is mainly due to limitations in space and because it is more closely related to other topics treated in this volume (for a discussion of Cenozoic diversity see Olson, 1985).

Mesozoic Avian Diversity

The actual fossil record of birds starts in the Late Jurassic (Fig. 1). Claims for an older record are not supported by reliable evidence. The recent identification of *Protoavis* from the Late Triassic of Texas (Chatterjee, 1991) as a bird is not based on substantiated evidence. The available material of *Protoavis* is fragmentary and its association in the two specimens alleged by Chatterjee is not clear. In addition, some of the elements regarded as avian (e.g., furcula and carinate sternum) can be alternatively interpreted as something else. Likewise, the suggestion of bird-like footprints in Early Jurassic deposits, though interesting, is far from persuasive (Chiappe, 1995a).

Discovered over a span of almost 140 years, the eight specimens (including an isolated feather) of the Bavarian *Archaeopteryx* constitute the most informative evidence of Late Jurassic birds. Many aspects of the biology of *Archaeopteryx*, however, are surrounded by controversy. Several anatomical characteristics (in particular, the braincase and temporal region of the skull) are matters of intense debate. Its mode of life and flying capability are also hotly debated. It is even unclear whether these specimens belong to a single taxon, *Archaeopteryx lithographica*, or to several closely related species.

Until very recently, with the exception of an isolated feather from Kazakhstan (which some workers have regarded as a leaf), the specimens of *Archaeopteryx* were the only known Late Jurassic birds. This singular position of the spectacular specimens of *Archaeopteryx* has been challenged recently by new birds found in beds alleged to be of Late Jurassic age from northeastern China and North Korea. The former report includes several specimens of *Confuciusornis* from the Yixian Formation—a startling toothless bird with relatively short, clawed wings (Hou *et al.*, 1995, 1996). Although Hou *et al.* (1995, 1996) have regarded these birds as Late Jurassic, palynological studies suggest a lowermost Cretaceous age for the Yixian Formation (Li and Liu, 1994) and recent ^{40}Ar–^{39}Ar dates (~121–122 million years) from the base and top of this formation indicate an even younger, Hauterivian to Aptian age, depending on the selected geological time scale (Smith *et al.*, 1995). The specimen from North Korea appears to preserve portions of the skull, neck, and wing associated with feathers. Although

TAXON	AGE	DISTRIBUTION	YEAR
Archaeopteryx	Late Jurassic	Germany	1861
Nanantius	Early Cretaceous	Australia	1986
Sinornis	Early Cretaceous	China	1992
Cathayornis	Early Cretaceous	China	1992
Otogornis	Early Cretaceous	China	1994
Boluochia	Early Cretaceous	China	1995
Concornis	Early Cretaceous	Spain	1992
Eoalulavis	Early Cretaceous	Spain	1996
Gobipteryx	Late Cretaceous	Mongolia	1974
Alexornis	Late Cretaceous	Mexico	1976
Enantiornis	Late Cretaceous	Argentina	1981
Avisaurus	Late Cretaceous	USA	1985
Soroavisaurus	Late Cretaceous	Argentina	1993
Yungavolucris	Late Cretaceous	Argentina	1993
Lectavis	Late Cretaceous	Argentina	1993
Neuquenornis	Late Cretaceous	Argentina	1994
Alvarezsaurus	Late Cretaceous	Argentina	1991
Patagonykus	Late Cretaceous	Argentina	1996
Mononykus	Late Cretaceous	Mongolia	1993
Enaliornis	Early Cretaceous	England	1876
Ambiortus	Early Cretaceous	Mongolia	1982
Gansus	Early Cretaceous	China	1984
Hesperornis	Late Cretaceous	USA-Canada	1872
Ichthyornis	Late Cretaceous	USA-Canada	1873
Apatornis	Late Cretaceous	USA	1873
Baptornis	Late Cretaceous	USA-Canada	1877
Parahesperornis	Late Cretaceous	USA	1984
Noguerornis	Early Cretaceous	Spain	1989
Iberomesornis	Early Cretaceous	Spain	1992
Chaoyangia	Early Cretaceous	China	1993
Confuciusornis	Early Cretaceous	China	1995
Liaoningornis	Early Cretaceous	China	1996
Patagopteryx	Late Cretaceous	Argentina	1992
Vorona	Late Cretaceous	Madagascar	1996

FIGURE 1 Nonneornithine genera of the Mesozoic. Very fragmentary, nondiagnosable taxa have been excluded. The year of publication is listed on the right; note that nearly 50% of these taxa have been described after 1990. Patterns on the right refer to those in Fig. 2.

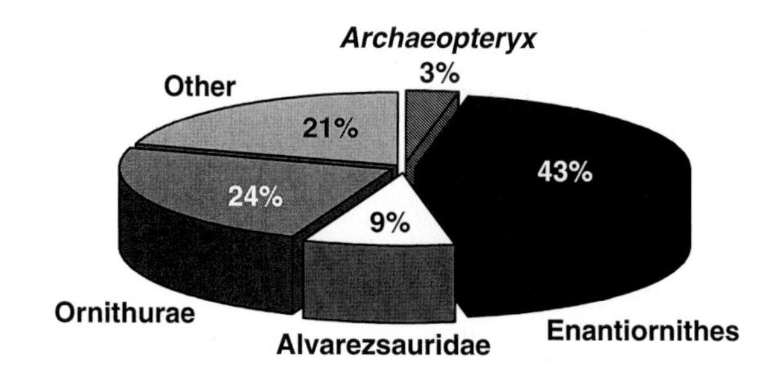

FIGURE 2 Proportions represented by the genera listed in Fig. 1. Patterns correspond to those given in Fig. 1.

dubbed the "North Korean *Archaeopteryx*," the manual proportions of this specimen are different from those of *Archaeopteryx*, in which the digits are proportionally shorter than the metacarpals. None of these new specimens have been fully described in the literature and their chronological significance has yet to be evaluated in light of clarification of their stratigraphic position.

The record of birds is far more abundant in the Cretaceous (Fig. 1). Unquestionable osseous remains of birds have been found in all continents with the exception of Africa, where only footprints have been

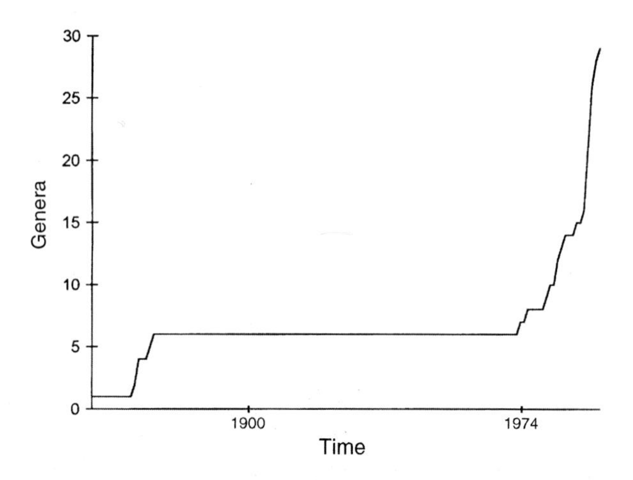

FIGURE 3 Rate of discoveries of Mesozoic birds. Data based on the year of publication of the genera listed in Fig. 1.

recovered. The most primitive and bizarre looking are *Mononykus* (Fig. 4) and its allies (e.g., *Alvarezsaurus*; Fig. 1) from the Late Cretaceous of central Asia and southern Argentina. Bearing a short and robust forelimb that at first glance little resembles an avian wing, the general anatomy of the flightless *Mononykus* supports its avian affinity. The functional meaning of the forelimb of *Mononykus* is puzzling. Although its overall appearance is suggestive of the morphology of digging mammals (e.g., moles), the long, gracile hindlimbs do not support such an idea. In fact, in contrast to digging mammals, in *Mononykus* the forelimbs were not used in locomotory activities. *Mononykus* differs from other flightless birds in that its forelimbs, instead of having simplified structures, have robust muscular attachments suggesting a particular function.

The next branch of the cladogram illustrated in Fig. 4 is *Iberomesornis*. This sparrow-sized bird comes from the Early Cretaceous of Spain. In contrast to *Archaeopteryx* and *Mononykus*, *Iberomesornis* shows characters that strongly suggest an enhanced flying capability. Among these characters are the distal caudal vertebrae fused into a pygostyle; an elongate, strut-like coracoid; and a U-shaped furcula. Also from the Cretaceous of Spain but from significantly older deposits is *Noguerornis*. Although unquestionably basal within avian phylogeny, the fragmentary nature of the only known specimen prevents a precise

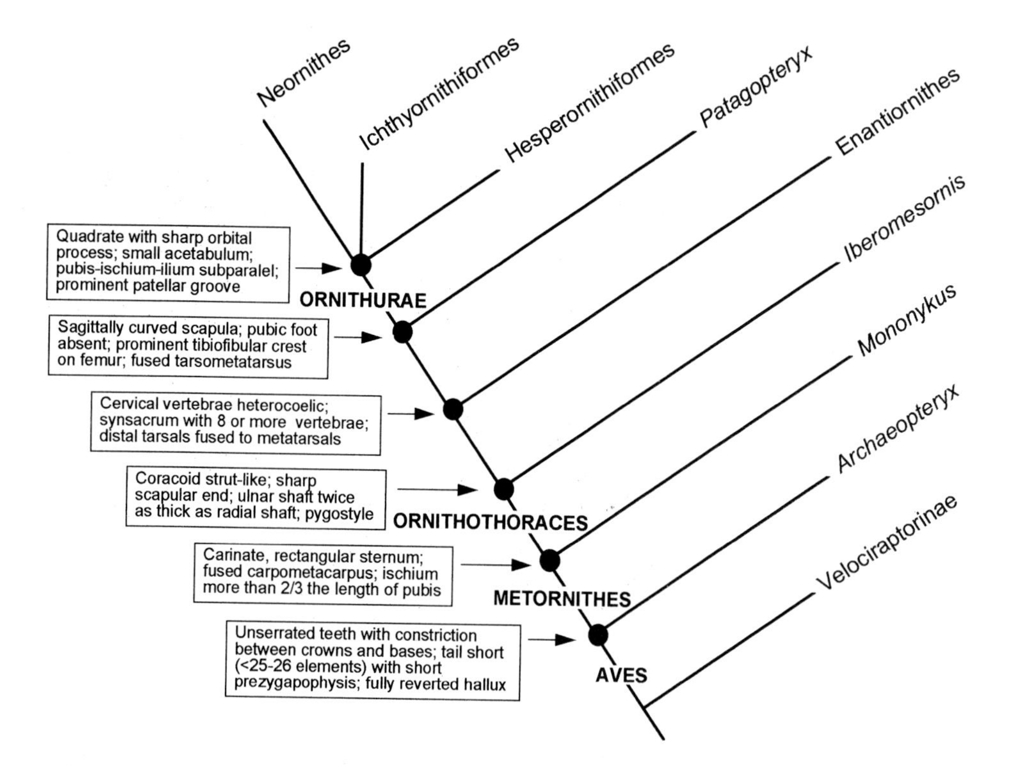

FIGURE 4 Cladogram of best known avian taxa. Synapomorphies diagnosing each node are listed on the left. Derived from Perle *et al.* (1993) and Chiappe (1995a,b).

determination of the phylogenetic relationship of *Noguerornis*. Nevertheless, this taxon is important in that it documents the presence of a fused carpometacarpus, a U-shaped furcula with an enormous hypocleideum, and a well-developed propatagium as early as 10 million years after *Archaeopteryx*.

The most diversified birds of the Mesozoic were the ENANTIORNITHES, of which more than a dozen valid species have been named (Figs. 1 and 2). First recognized in 1981 (although members of this group were collected as early as the 19th century), the Enantiornithes are known from the very Early Cretaceous to the end of this period and have been recorded in South America, North America, Europe, Asia, and Australia (Chiappe, 1995a,b). The Early Cretaceous enantiornithines include taxa such as *Concornis* from Spain and *Sinornis* and *Cathayornis* from China. These early enantiornithines were small toothed birds, showing characteristics indicating an enhanced flying ability and perching capabilities. More derived enantiornithines, such as the Mongolian *Gobipteryx* and other Late Cretaceous forms, were toothless and significantly larger. Some of the Argentine forms of the terminal Cretaceous, such as *Enantiornis*, had a wingspan of over a meter. Likewise, at the end of the Cretaceous, the Enantiornithes present different specializations. At El Brete, in northwestern Argentina, in addition to forms suggesting arboreal and perching specialization, taxa of wading and aquatic habits have been recorded.

The Enantiornithes shared a common ancestor with the clade formed by *Patagopteryx* and the Ornithurae (Fig. 4), the group encompassing hesperornithiforms, ichthyornithiforms, and neornithines (so-called "modern birds"). *Patagopteryx* was a flightless cursorial bird, of the size of a hen, known thus far only from the Late Cretaceous of southern Argentina. Although originally thought to be related to ratites (flightless birds such as the ostrich and its allies), the cladogram in Fig. 4 shows that flightlessness evolved independently in this taxon.

The flightless foot-propelled divers, hesperornithiforms, are principally known from Late Cretaceous marine deposits of the North American western interior (Fig. 1), where excellent specimens were collected as early as the 1870s. Fragmentary remains of hesperornithiforms have been reported from the Late Cretaceous of eastern Europe and western Asia. Whether this group was present in the Early Cretaceous is not clear because the affinity of *Enaliornis*, an alleged hesperornithiform from the Albian of England, is still controversial. The presence of hesperornithiform remains in estuarine deposits indicates that these birds were not exclusively oceanic.

Historically (and in some ways both chronologically and geographically) linked to hesperornithiforms are the ichthyornithiforms, also known from the marine Late Cretaceous of North America and Asia. In contrast to hesperornithiforms, ichthyornithiforms were flying birds of the size of a tern. One of their most remarkable features is the notably large head. The early Cretaceous *Ambiortus*, from the Mongolian Gobi Desert, may be closely related to ichthyornithiforms.

The question of which groups of modern birds were already differentiated in the Mesozoic and when they had their first occurrence in the fossil record is puzzling because their putative records (in most cases isolated bones) are very fragmentary. It is quite clear, though, that forms related to shorebirds (Charadriiformes), loons (Gaviiformes), ducks (Anseriformes), and possibly petrels (Procellariiformes) were present at the end of the Cretaceous. Several fragmentary specimens from older deposits, such as the Early Cretaceous *Gansus* and *Paleocursornis*, have also been related to modern birds, although these need to be studied further.

Major Patterns of Avian Evolution

The information provided by the numerous new Mesozoic findings have substantially overcome our ignorance about the "intermediate" evolutionary steps between *Archaeopteryx* and the much more advanced ichthyornithyforms and hesperornithiforms (Fig. 4). Clarification of the pattern of historical relationships of early birds has been coupled with a remarkable interest in physiological, morphofunctional, and developmental studies of nonavian theropods and extant birds.

Bone Histology, Rates of Growth, and Physiology

Extant birds are known to grow rapidly. Rates of growth in modern birds vary depending on their mode of development (altricial or precocial), but they normally reach adult size within the first year. The bone tissue of extant birds is a fast-deposited, uninterrupted woven-like tissue, known as fibrolamellar bone. The bone microstructure of Mesozoic birds has been studied only in a few taxa. In *Hesperornis*, the bone microstructure resembles that of modern birds (Houde, 1987), consistent with a rapid rate of growth. Enantiornithines and *Patagopteryx*, however, have shown important differences with respect to their living counterparts (Chinsamy *et al.*, 1994, 1995). In these basal birds, the bone deposition was cyclically interrupted by nondepositional pauses (lines of arrested growth; LAGs), a pattern of deposition typically absent in extant birds although known to occur in nonavian theropods (Chinsamy and Dodson, 1995). The bone tissue of *Patagopteryx* was highly vascularized and of fibrolamellar type, as in modern birds and *Hesperornis*. In contrast, that of the Enantiornithes was completely formed of lamellated bone, a slow-deposited, parallel-fibered bone tissue. The presence of LAGs in Enantiornithes and *Patagopteryx*, if annual, indicates cyclical pauses during postnatal growth. This pattern of bone deposition is typical of extant vertebrates in which each LAG is formed annually. Thus, the occurrence of LAGs in Enantiornithes and *Patagopteryx* suggests that their growth was not continuous throughout the year. These Cretaceous birds took more than a single year to acquire adult size. For example, the presence of four and five LAGs in the examined enantiornithine birds indicates that these birds were still growing 4 and 5 years, respectively, after hatching (Chinsamy *et al*, 1994, 1995) (see GROWTH LINES).

Extant birds, along with mammals, are unique among tetrapods in being able to maintain steady rates of temperature-constrained physiological processes under variable climatic conditions. The presence of pauses in bone deposition and the associated slower rate of growth in both *Patagopteryx* and Enantiornithes may suggest physiological differences with respect to their modern counterparts. In the Enantiornithes, the presence of a compacta formed of only lamellated bone suggests that physiological differences between these and extant birds were even greater. As suggested by Chinsamy *et al.* (1994, 1995), if rates of growth are related to metabolic rates (see Ruben, 1995, for a skeptical view), these birds may not have been fully endothermic homeotherms in the sense that we think of extant birds. However, this is not to say that they were "cold-blooded" ectotherms. They might have had an intermediate thermal physiology within the spectrum of ectothermy–endothermy, as has been suggested for some nonavian dinosaurs (Chinsamy and Dodson, 1995). Moreover, mammals, which are considered endothermic homeotherms, show a broad range of thermal strategies and abilities to regulate body temperature.

A "warm-blooded" metabolism has been proposed for basal birds such as *Archaeopteryx* and the enantiornithine *Sinornis* on the basis of allometric estimates, theoretical predictions, and aerodynamic capabilities, although neither of these birds has been studied histologically. In light of the new inferences discussed previously, the physiological interpretations previously made for *Archaeopteryx* and *Sinornis* should be revisited (see PHYSIOLOGY).

Ontogeny and Developmental Modes Extant neornithine birds are known to have a broad range of hatchling appearance and conduct, ranging from one end to the other of the precocial–altricial spectrum. Precocial hatchlings are covered with down, they are capable of locomotion, and they leave the nest soon after hatching. In contrast, altricial hatchlings are usually naked and blind, they are incapable of locomotion, and they are fed by their parents in the nest. Several studies have shown that precociality—in particular the stages such as Precocial 1 and 2, in which the active and downy hatchling has a low rate of growth and follows its parents in search of food—is the ancestral condition for modern birds (Starck, 1993). Basal neornithines, such as paleognaths, anseriforms, and galliforms, typically have this type of early postnatal ontogeny (see BEHAVIOR).

Mesozoic avian embryos are currently known only from the Late Cretaceous of Mongolia. Enantiornithine embryos regarded as *Gobipteryx* were likely precocial (Elzanowski, 1985). This interpretation is congruent with the low rates of growth inferred for the Enantiornithes because living precocial birds have slower rates of growth than altricial birds, which generally grow rapidly (Starck, 1993). The hypothesis that precociality is ancestral for modern birds is supported by the presence of this developmental mode in Enantiornithes (and likely in *Patagopteryx*) and also by recent findings suggesting that nonavian theropod bones were fully ossified (and likely precocial) at the time of hatching (Norell *et al.*, 1994) (see also HETEROCHRONY).

Locomotion and Habits Birds are unique among living vertebrates in that forelimbs and hindlimbs are involved in decoupled locomotory systems (see BIPEDALITY). The series of structural transformations from obligatory bipedal nonavian dinosaurs to the cursorial–aerial locomotion of modern birds has evolved differentially during avian evolution. Namely, the development of a modern flight system preceded the evolution of a modern system of cursorial locomotion (Chiappe, 1991, 1995b). Debates surround the aerial capabilities of *Archaeopteryx*, but soon after the occurrence of this taxon in the fossil record basal taxa such as *Noguerornis* and *Iberomesornis* show a series of structural transformations indicative of an enhanced flying capacity. Furthermore, the pedal morphology of *Iberomesornis*, *Concornis*, and *Sinornis* clearly suggests that these early birds were able to perch, indicating that an arboreal type of lifestyle was acquired very early in the evolution of birds.

Basal birds inherited the bipedal capabilities of their theropodan ancestors, and a terrestrial habitat was probably ancestral for birds as shown by the anatomy of *Archaeopteryx* as well as *Mononykus* and its kin. Gatesy (1995) has shown that the pattern of hindlimb kinematics in basal theropods—retaining mainly the hip-extension mechanism characteristic of extant crocodiles and lizards—was significantly different from that of modern birds in which the hindlimb is moved through a knee-flexion mechanism. Gatesy interpreted the shortening and general reduction of the tail and associated musculature (*M. caudifemoralis longus*) from basal theropods to modern birds as the primary factor involved in the shift between these mechanisms. The transformation of this complex would have been correlated with the decoupling of the tail from hindlimb kinematics and its final linkage with the flight locomotor mechanism, and with the forward migration of the center of mass and the acquisition of the typical modern avian hindlimb posture. Interestingly, these differences in hindlimb kinematics appear to have had little effect on the actual gait because footprint evidence suggests that the bipedal gait of nonavian theropods was passed almost unaltered to their modern counterparts (Padian and Olsen, 1989). Most important for the current discussion is that the final stages toward the typical modern pattern of hindlimb kinematics clearly occur rather late in avian history. Basal birds such as *Archaeopteryx* or *Mononykus* have relatively long tails and the pattern of limb kinematics was certainly more similar to that of nonavian maniraptoran theropods. Although basal ornithothoracine birds have a short tail with a distal pygostyle, they have retained primitive characters that might have played a role in their hindlimb kinematics. For example, ischiadic and pubic symphyses are preserved in *Noguerornis* and the enantiornithine *Concornis*, respectively. Furthermore, in *Patagopteryx* the tail was probably quite long, as suggested by the long neural spines of the mid-caudals. It is not until the rise of the Ornithurae that the modern morphology of the hindlimb and pelvis is ultimately acquired. Although the tail morphology of several basal birds is still not well known, it appears that a modern avian mechanism of hindlimb movement was not fully developed until the differentiation of the Ornithurae (Chiappe, 1991, 1995b).

Minimal Ages and Taxonomic Dynamics The oldest fossils of a lineage provide evidence only for the minimal age of that particular taxon. Thus, a more precise picture of the temporal pattern of clade origination (and character evolution) emerges when the fossil record is calibrated with a phylogenetic hypothesis (Fig. 4), because sister groups originated at the same time. The minimal age for ornithurine birds is Early Cretaceous; although the phylogenetic relationships of *Enaliornis* and *Ambiortus* are not yet clear, they certainly belong to the Ornithurae. This implies that the lineage leading to *Patagopteryx*, the sister group of Ornithurae (Fig. 4), was already differentiated by the Early Cretaceous. Likewise, the earliest Enantiornithes and *Iberomesornis* are also of Early Cretaceous age. This predicts that early relatives of *Mononykus* had already arisen by this age as well. Therefore, calibration of the record of these Mesozoic avian clades with the information provided by their interrelationships indicates that all of them were already present at the beginning of the Cretaceous. Likewise, the fact that several neognaths (e.g., charadriiforms and anseriiforms) have reliable occurrences at the end of the Cretaceous implies that their extant outgroups (ratites and tinamous), so far unknown from the Mesozoic, were differentiated in this time. In fact, if the Early Cretaceous *Gansus* is truly a neornithine, the origin of this clade (including ratites as well) is pushed back into the Early Cretaceous.

Numerous new findings have shown that the Cretaceous was a time of active diversification for birds. Several basal lineages, in particular the Enantiornithes, evolved diverse lifestyles, and several modern bird lineages have their earliest occurrences. This period, however, was also one of dramatic extinction. The end of the Cretaceous (not necessarily the K–T boundary) documents the extinction of all the basal diversity, including *Mononykus* and its kin, Enantiornithes, *Patagopteryx*, hesperornithiforms, and ichthyornithiforms (Chiappe, 1995a).

See also the following related entries:

AVIALAE • BIPEDALITY • BIRD ORIGINS • HETEROCHRONY

References

Chatterjee, S. (1991). Cranial anatomy and relationships of a new Triassic bird from Texas. *Philos. Trans. R. Soc. (Biol. Sci.)* **332**(1265), 277–346.

Chiappe L. M. (1991). Cretaceous avialian remains from Patagonia shed new light on the early radiation of birds. *Alcheringa* **15**(3–4), 333–338.

Chiappe, L. M. (1995a). The first 85 million years of avian evolution. *Nature* **378,** 349–355.

Chiappe, L. M. (1995b). Phylogenetic position of the Cretaceous birds of Argentina: Enantiornithes and *Patagopteryx deferrariisi*. In *3rd Symposium of the Society of Avian Paleontology and Evolution.* Courier Forschungsinstitut, Senckenberg, Germany.

Chinsamy, A., and Dodson, P. (1995). Inside a dinosaur bone. *Am. Sci.* **83,** 174–180.

Chinsamy, A., Chiappe, L. M., and Dodson, P. (1994) Growth rings in Mesozoic avian bones: Physiological implications for basal birds. *Nature* **368,** 196–197.

Chinsamy, A., Chiappe, L. M., and Dodson, P. (1995). Mesozoic avian bone microstructure: Physiological implications. *Paleobiology* **21**(4), 561–574.

Elzanowski, A. (1985). The evolution of parental care in birds with reference to fossil embryos. *Acta XVIII Congr. Int. Ornithol.* **1,** 178–183.

Elzanowski, A., and Galton, P. M. (1991). Braincase of *Enaliornis*, an Early Cretaceous bird from England. *J. Vertebr. Paleontol.* **11**(1), 90–107.

Gatesy, S. M. (1995). Functional evolution of the hindlimb and tail from basal theropods to birds. In *Functional Morphology in Vertebrate Paleontology* (J. Thomason, Ed.), pp. 219–234. Cambridge Univ. Press, Cambridge, UK.

Hou, L., Zhou, Z., Martin, L. D., and Feduccia, A. (1995). A beaked bird from the Jurassic of China. *Nature* **377,** 616–618.

Hou, L., Martin, L. D., Zhou, Z., and Feduccia, A. (1996). Early adaptive radiation of birds: Evidence from fossils from northeastern China. *Science* **274,** 1164–1167.

Houde, P. (1987). Histological evidence for the systematic position of *Hesperornis* (Odontornithes: Hesperornithiformes). *Auk* **104,** 125–129.

Li, W., and Liu, Z. (1994). The Cretaceous palynofloras and their bearing on stratigraphic correlation in China. *Cretaceous Res.* **15,** 333–365.

Norell, M. A., Clark, J. M., Dashzeveg, D., Barsbold, R., Chiappe, L. M., Davidson, A. R., McKenna, M. C., and Novacek, M. J. (1994). A theropod dinosaur embryo, and the affinities of the Flaming Cliffs dinosaur eggs. *Science* **266,** 779–782.

Olson, S. L. (1985). The fossil record of birds. In *Avian Biology* (D. S. Farner, J. King, and K. C. Parkes, Eds.), Vol. 8, pp. 79–238. Academic Press, New York.

Padian, K., and Olsen, P. E. (1989). Ratite footprints and the stance and gait of Mesozoic theropods. In *Dinosaur Tracks and Traces* (D. D. Gillette and M. G. Lockley, Eds.), pp. 231–241. Cambridge Univ. Press, Cambridge, UK.

Perle, A., Norell, M. A., Chiappe, L. M., and Clark, J. M. (1993). Flightless bird from the Cretaceous of Mongolia. *Nature* **362,** 623–626.

Ruben, J. (1995). The evolution of endothermy in mammals and birds: From physiology to fossils. *Annu. Rev. Physiol.* **57,** 69–95.

Smith, P. E., Evensen, N. M., York, D., Chang, M., Jin, F., Li, J., Cumbaa, S., and Russell, D. (1995). Dates and rates in ancient lakes: ^{40}Ar–^{39}Ar evidence for an Early Cretaceous age for the Jehol Group, northeast China. *Can. J. Earth Sci.* **32,** 1426–1431.

Starck, J. M. (1993). Evolution of avian ontogenies. In *Current Ornithology* (D. M. Power, Ed.), Vol. 10, pp. 275–366. Plenum Press, New York.

Avetheropoda

The theropod dinosaur taxon Avetheropoda (Fig. 1) was formally defined in the phylogenetic system by Holtz (1994; following use of the name by Paul, 1988) as the node within TETANURAE comprising the stem groups COELUROSAURIA and CARNOSAURIA, as redefined by Gauthier (1986). The node is diagnosed by loss of the obturator foramen, the proximally placed lesser trochanter of the femur, the basal half of metacarpal I closely appressed to metacarpal II, the cnemial process arising out of the lateral surface of the tibial shaft, the pronounced pubic "foot" or "boot," the posterior tapering of the coracoid, the U-shaped premaxillary symphysis, and the asymmetrical tooth crowns of the premaxilla in cross section (Holtz, 1994). Its members ranged from as early as the late Middle or Late Jurassic through the latest Cretaceous and, in the case of Aves, to the present day. This node appears to be the same as NEOTETANURAE (Sereno *et al.*, 1994), which is a junior synonym of Avetheropoda by priority of publication.

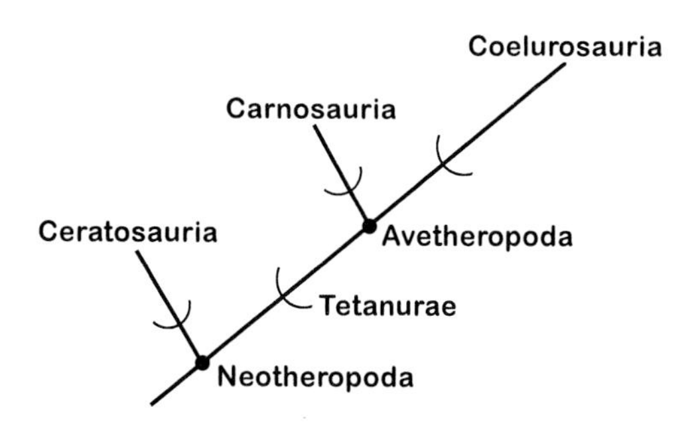

FIGURE 1 Phylogenetic relations of the major groups of Avetheropoda and its out-groups. For details see CARNOSAURIA; COELUROSAURIA.

References

Gauthier, J. A. (1986). Saurischian monophyly and the origin of birds. *Mem. California Acad. Sci.* **8,** 1–55.

Holtz, T. R., Jr. (1994). The phylogenetic position of the Tyrannosauridae: Implications for theropod systematics. *J. Paleontol.* **68,** 1100–1117.

Paul, G. S. (1988). *Predatory Dinosaurs of the World.* Simon & Schuster, New York.

Sereno, P. C., Wilson, J. A., Larsson, H. C. E., Dutheil, D. B., and Sues, H.-D. (1994). Early Cretaceous dinosaurs from the Sahara. *Science* **265,** 267–271.

Avialae

KEVIN PADIAN
University of California
Berkeley, California, USA

Gauthier (1986) established the term Avialae ("bird-wings") to encompass *Archaeopteryx* plus ornithurine birds. Gauthier used the term "ornithurine" birds in a somewhat different sense than other workers, defining it as living birds and all taxa closer to them than to *Archaeopteryx.* In this work, Gauthier also restricted the use of the term "Aves" to crown group birds, that is, extant taxa of birds and all the descendants of their most recent common ancestor (see PHYLOGENETIC SYSTEM; SYSTEMATICS). His purpose in doing so was to maximize the information to taxonomists of soft parts and other structures not usually available in fossils. Moreover, Linneaus's original concept of Aves did not include fossil forms, inasmuch as Linneaus did not know of them.

Acceptance of this redefinition of Aves has been problematic. *Archaeopteryx* was recognized as a primitive but true bird by the 1880s (because it had feathers, a short tail, and other avian features) following several decades of dispute after its initial description in 1860. In standard textbooks in both ornithology and paleontology of the 20th century, *Archaeopteryx* has been treated unexceptionally as a bird, though usually in its own Subclass Archaeornithes, Order Archaeopterygiformes, and Family Archaeopterygidae to recognize its distinctness from other birds. Moreover, to millions of Spanish-speaking people around the world, the word Aves means "bird." Hence, there seems to be strong reason as well as convention for retaining the term Aves to encompass *Archaeopteryx,* extant birds, and all the descendants of their most recent common ancestor (see BIRD ORIGINS). The term Neornithes is normally applied to the node defining

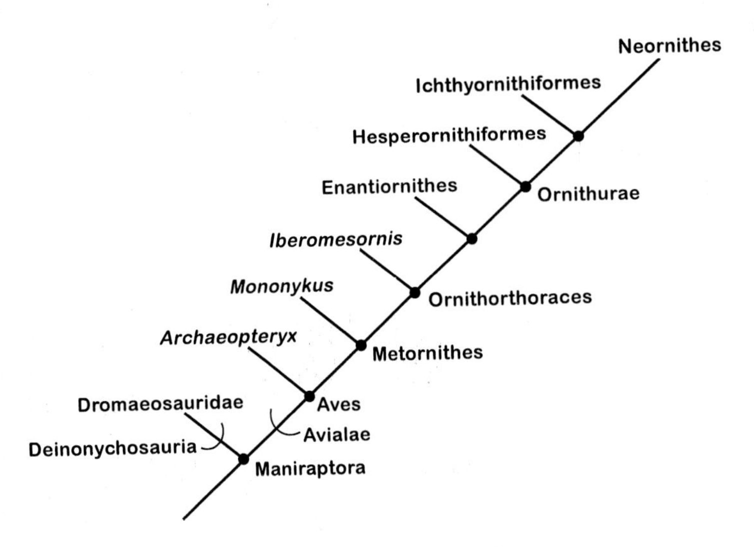

FIGURE 1 Phylogeny of Avialae.

crown group birds. Avialae (Fig. 1), then, is available to define the stem group consisting of Neornithes and all MANIRAPTORANS closer to them than to *Deinonychus;* the name DEINONYCHOSAURIA, used by Gauthier (1986) to include Dromaeosauridae plus Troodontidae (the latter since removed to propinquity with Ornithomimidae), can now be used as the stem-based sister taxon to Avialae, defined as *Deinonychus* and all maniraptorans closer to it than to birds. This is consistent with Gauthier's original formulation of Deinonychosauria and Avialae as sister stem taxa within Maniraptora. Despite the removal of Troodontidae from the former taxon, the meanings of the two groups remain the same because they can be interpreted as stem-based taxa (see Gauthier, 1986, Fig. 9).

Reference

Gauthier, J. A. (1986). Saurischian monophyly and the origin of birds. *Mem. California Acad. Sci.* **8,** 1–55.

Barun Goyot Formation

Halszka Osmólska
Polska Akademia Nauk
Warsaw, Poland

The Barun Goyot Formation (previously referred to as the "Lower Nemegt Beds") was identified in 1948 as the "unfossiliferous lacustrine sandstones" or "barren deposits" in the Nemegt Basin, Mongolia, by the Palaeontological Expedition of the USSR Academy of Sciences. The Khulsan locality was subsequently designated as the type locality of the Barun Goyot Formation. The Barun Goyot Formation is directly overlaid by a layer of the intraformational conglomerate and the Nemegt Formation. The lower boundary of the formation is covered, and the physical contact with the presumably older Djadokhta Formation has not yet been discovered. The total thickness of the Barun Goyot Formation cannot be precisely determined, but it is not less than 110 m. The estimated age of this formation is not precisely determined. It has been considered as ?middle Campanian, Campanian, or even Maastrichtian by various authors; the last referral seems doubtful however.

Red and brownish-red, fine-grained, poorly cemented sandstones dominate the Barun Goyot Formation more than the interstratifying sandy mudstones and sandy claystones. The sandstones are either not stratified or have characteristic large-scale cross-stratification. The dominant lithology of the Barun Goyot Formation resembles that of the underlying Djadokhta Formation but lacks the mature caliche paleosol horizons characteristic of the latter and has thicker, more common claystone layers. Sedimentation of the Barun Goyot Formation is interpreted to have occurred among eolian dunes and interdune deposits but also in small intermittent lakes and ephemeral streams. Thus, the climatic conditions were more humid than those of the Djadokhta.

The Barun Goyot Formation is widely distributed in the Gobi Basin and occurs in the Pre-Altai (locality Udan Sayr), Trans-Altai (localities: Khulsan, southeast Nemegt, Khermeen Tsav, and Ingeni Tsav) and in the Eastern Gobi (localities: Shara Tsav and Khara Khutul).

The Barun Goyot Formation yields a vertebrate assemblage similar to that of the Djadokhta Formation, and several species of lizards are shared by these formations. However, lizards are much more diverse in the Barun Goyot Formation. There are also aquatic (fish remains) and amphibious vertebrates (frogs), which are absent in the Djadokhta Formation. Uncommon but relatively diverse turtles and solitary representatives of two small terrestrial crocodiles (a gobiosuchid and a notosuchian) have also been reported from Barun Goyot deposits. The formation has yielded a small bird (*Gobipteryx minuta*) and numerous vertebrate eggs and eggshells.

Except for one sauropod species, dinosaurs are represented by small- and medium-sized species, of which only one species (?*Velociraptor mongoliensis*) is shared with the Djadokhta. Dinosaur species found in the Barun Goyot Formation include the theropods *Avimimus portentosus*, *Conchoraptor gracilis*, *Hulsanpes perlei*, *Ingenia yanshini*, and ?*Velociraptor mongoliensis*; the sauropod *Quesitosaurus orientalis*; the pachycephalosaurid *Tylocephale gilmorei*; the ceratopsians *Bagaceratops rozhdestvenskyi*, *Breviceratops kozlowskii*, and *Udanoceratops tschizhovi*, and the ankylosaurs *Saichania chulsanensis* and *Tarchia kielanae*. Dinosaur eggs and eggshells include smooth and ornamented protoceratopsid eggs, and dendroolithid eggshells.

See also the following related entries:

Bayn Dzak • Djadokhta Formation • Mongolian Dinosaurs • Nemegt Formation

References

Fox, R. C. (1987). Upper Cretaceous terrestrial vertebrate stratigraphy of the Gobi Desert (Mongolian People's Republic) and western North America *Special paper Geological Assoc. Can.* **18,** 571–594.

Gradzinski, R., and Jerzykiewicz, T. (1974). Sedimentation of the Barun Goyot Formation. *Palaeontol. Polonica* **30,** 11–146.

Gradzinski, R., Kielan-Jaworowska, Z., and Maryanska, T. (1977). Upper Cretaceous Djadokhta, Barun Goyot and Nemegt formations of Mongolia, including remarks on previous subdivisions. *Acta Geol. Polonica* **27,** 281–318.

Jerzykiewicz, T., and Russell, D. A. (1991). Late Mesozoic stratigraphy and vertebrates of the Gobi Basin. *Cretaceous Res.* **12,** 345–377.

Osmolska, H. (1980). The Late Cretaceous vertebrate assemblages of the Gobi Desert, Mongolia. *Mém. Soc. Géol. France* **139,** 145–150.

Bastús Nesting Site

José L. Sanz
José J. Moratalla
Universidad Autónoma de Madrid
Madrid, Spain

The Bastús dinosaur nesting site is found in the Arenisca de Arén Formation (Maastrichtian, Upper Cretaceous) in the province of Lérida (northeastern Spain) (Sanz *et al.*, 1995). The sediments are red sandstones laid down in a shoreline environment. The outcrop has a volume of about 12,000 m^3 and contains a huge number of eggshell fragments. The eggshell material represents about 0.5% of the whole deposit. Using the volume of the sandstone body, an average egg diameter of 20 cm, and eggshell that is 1.45 mm thick, the number of dinosaur eggs at the Bastús site can be estimated at 300,000. The outcrop contains the remains of 24 nesting structures, most of them having two or three eggs, with a maximum of seven. The eggs are subspherical in shape (Fig. 1), with spherulitic shell type.

The dinosaurs were nesting in the exposed sandy sediments of a beach-ridge plain. Because the nesting structures are generally well preserved, postdepositional transport can be excluded as the primary cause for the high level of fragmentation. This is better explained by the trampling and nesting activities of dinosaurs and also by subsequent paedogenesis. Some factor, possibly territorial behavior of the dinosaurs, prevented the destruction of the reasonably well-preserved nests.

FIGURE 1: Dinosaur nest remains at the Bastús site (Lérida province, Spain). Sections of the eggs are visible in the reddish sandstone surface.

Two main conclusions may be inferred from evidence provided by the Bastús nesting site: (i) Remains of the huge number of eggs in the sandstone body indicate some kind of nesting fidelity, as has been reported in other dinosaur nesting localities (Horner, 1982); and (ii) Bastús represents unambiguous evidence of nesting behavior at a seashore locality.

See also the following related entries:

BEHAVIOR • EGG MOUNTAIN • EGGS, EGGSHELLS, AND NESTS

References

Horner, J. R. (1982). Evidence of colonial nesting and "site fidelity" among ornithischian dinosaurs. *Nature* **297,** 675–676.

Sanz, J. L., Moratalla, J. J., Díaz-Molina, M., López-Martínez, N., Kälin, O., and Vianey-Liaud, M. (1995). Dinosaur nests at the sea shore. *Nature* **376,** 731–732.

Bavarian State Collection for Paleontology and Historical Geology, Munich

The Bavarian State Collections, formerly the royal collections of the king of Bavaria, are housed in association with the University of Munich, Germany. There is a fine collection of material, including many important specimens from the Jurassic Solnhofen limestones and other Mesozoic formations of Germany. The Solnhofen collection is perhaps the finest in Europe and includes many pterosaurs, lizards, sphenodontidans, fishes, and invertebrates, as well as casts of all the *Archaeopteryx* specimens and the type specimen of the small coelurosaur *Compsognathus*. There is also a well-prepared collection of tetrapods from the Permian–Triassic deposits of South Africa and a fine specimen of *Triceratops* acquired from Maastrichtian beds of Montana. Displays highlight these fossils and many others, notably a fine pareiasaur skull and skeleton, some South American archosaurs recovered by von Huene in the 1930s, and restorations of a complete *Pteranodon* skeleton and a flock of *Rhamphorhynchus*. Recent acquisitions include fishes, pterosaurs, and other material from the Late Cretaceous Santana Formation of Brazil.

See also the following related entry:

MUSEUMS AND DISPLAYS

Bayan Mandahu

PHILIP J. CURRIE
Royal Tyrrell Museum of Palaeontology
Drumheller, Alberta, Canada

Bayan Mandahu was made famous by the Sino–Canadian Dinosaur Project. Found in Inner Mongolia (China) approximately 300 km from Bayn Dzak in Mongolia, this locality has produced most of the dinosaurs that are characteristic of the Djadokhta Formation (Jerykiewicz et al. 1993). Evidence from this site suggests that *Pinacosaurus* and *Protoceratops* were gregarious animals that sometimes died en masse in sandstorms. An *Oviraptor* found at Bayan Mandahu had died while laying its eggs in a nest (Dong and Currie 1996).

See also the following related entries:

BAYN DZAK • BEHAVIOR • CHINESE DINOSAURS • DJADOKHTA FORMATION • SINO–CANADIAN DINOSAUR PROJECT

References

Dong, Z. M. and P. J. Currie (1996). On the discovery of an oviraptorid skeleton on a nest of eggs at Bayan Mandahu, Inner Mongolia, People's Republic of China. *Canadian Journal of Earth Sciences* **33**:631–636.

Jerzykiewicz, T., P. J. Currie, D. A. Eberth, P. A. Johnston, E. H. Koster and Zheng J. J. (1993). Djadokhta Formation correlative strata in Chinese Inner Mongolia: an overview of the stratigraphy, sedimentary geology, and paleontology and comparisons with the type locality in the pre-Altai Gobi. *Canadian Journal of Earth Sciences* **30**:2180–2195.

Bayn Dzak

HALSZKA OSMÓLSKA
Polska Akademia Nauk
Warsaw, Poland

Bayn Dzak lies in the pre-Altai Gobi Desert, Mongolia. The Upper Cretaceous sediments crop out here in a 10-km-long escarpment, the general direction of which is WNW–ESE. Along its course, the escarpment is cut by numerous small canyons. Bayn Dzak is the type locality of the Djadokhta Formation, and it is famous for yielding Late Cretaceous fossil vertebrates, especially primitive mammals, lizards, dinosaurs, and dinosaur eggs.

Bayn Dzak was discovered in 1922 by the Central Asiatic Expedition of the American Museum of Natural History and was called "Shabarakh Usu." The American group also explored Bayn Dzak in 1923 and 1925, and they gave the name "The Flaming Cliffs" to the highest (up to 50 m) group of the cliffs. The majority of the Bayn Dzak fossil vertebrates were found at the base of these. Subsequently, Bayn Dzak was visited in 1948 by the Palaeontological Expedition of the USSR Academy of Sciences, and in 1963–1971 by the Polish–Mongolian Expeditions.

Although other localities that are richer in dinosaur remains were discovered in Asia later, Bayn Dzak was the first locality where representatives of two new dinosaur groups (the protoceratopsids and oviraptorids) were detected as well as the first Asian representatives of dinosaur groups known previously from other continents (ankylosaurids, troodontids, and dromaeosaurids) and the first known dinosaur hatchlings (of *Protoceratops andrewsi*). Additionally, the first known dinosaurian (and other vertebrate) eggs, the (then) oldest placental mammals and lizards, also came from Bayn Dzak. Subsequent exploration of the Upper Cretaceous sediments at Bayn Dzak by the Polish–Mongolian Expedition increased the number of fossil vertebrate species, mainly of mammals, lizards, and crocodiles.

The fossils found at Bayn Dzak include theropods (*Velociraptor mongoliensis*, *Saurornithoides mongoliensis*, and *Oviraptor philoceratops*), ceratopsians (*Protoceratops andrewsi*), ankylosaurs (*Pinacosaurus grangeri*), and at least four types of dinosaur eggs.

See also the following related entries:

Barun Goyot Formation • Djadokhta Formation • Mongolian Dinosaurs • Nemegt Formation

References

Andrews, R. C. (1932). *The New Conquest of Central Asia*, Vol. 1.

Gradzinski, R., Kazmierczak, J., and Lefeld, J. (1969). Geographical and geological data from the Polish–Mongolian Palaeontological Expeditions. *Palaeontol. Polonica* **19,** 33–82.

Granger, W., and Gregory, W. K. (1923). *Protoceratops andrewsi*, a preceratopsian dinosaur from Mongolia. *Am. Museum Novitates* **42,** 1–9.

Novacek, M. J. (1996). *Dinosaurs of the Flaming Cliffs*. Doubleday / Anchor Books, New York.

Behavior

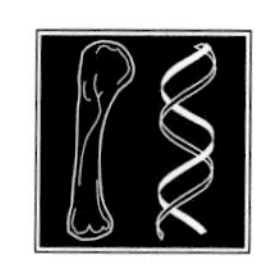

John R. Horner
Montana State University
Bozeman, Montana, USA

Interpretation of the social behaviors of dinosaurs ranges from factual to highly speculative, depending on the completeness and specificity of the data and their explanation. We know, for example, that the dinosaur *Oviraptor* sat on its eggs in at least one instance (Norell *et al.*, 1996) and that a *Troodon* individual constructed a rim around at least one clutch of eggs (Varricchio *et al.*, 1997). We also know that the hadrosaurs *Maiasaura* and *Hypacrosaurus* and the theropod *Troodon* nested in colonies and used particular nesting areas for more than a single year (Horner, 1982; Horner and Weishampel, 1988, 1996; Currie and Horner, 1988; Horner and Currie, 1994). However, we only know with any degree of certainty that these specific animals accomplished these specific behaviors. We can only speculate that other individuals of the same species behaved similarly, or that other related taxa behaved similarly. Additional behaviors, including nesting, gregariousness and its derivatives, (such as herding, migrating, or pack hunting), parental care and its derivatives (such as protection and feeding of young), and display behaviors, are speculative and can only be hypothesized based on geological and paleontological evidence and comparisons to related living taxa (Coombs, 1989, 1990). In modern comparative biology, such hypotheses can be tested not only by analogy to living forms with structures of presumably similar form and function but also by phylogenetic analysis. Phylogenies can test and compare hypotheses about the presence, absence, and sequence of correlated progression of features related to particular behavior (Padian, 1987; Brooks and McLennan, 1991; Weishampel, 1995).

Nesting

There is no evidence to suggest that any dinosaurs were born outside the confines of eggs, and yet very few eggs can be demonstrated to have been laid by particular dinosaur species (see Eggs, Eggshells, and Nests). *In situ*, identifiable embryonic remains are the only conclusive evidence that can be used to determine the identity of particular eggs or egg clutches. Those that are known and have been described include *Troodon* cf. *formosus* (Horner and Weishampel, 1988, 1996), *Oviraptor philoceratops* (Norell *et al.*, 1994), *Maiasaura peeblesorum* (Horner and Makela, 1979; Hirsch and Quinn, 1990), and *Hypacrosaurus stebingeri* (Horner and Currie, 1994). Among these four taxa, complete egg clutches are known only from *Troodon*, which laid approximately 24 eggs in a clutch (Horner, 1987; Varricchio *et al.*, 1997), and *Oviraptor*, which laid an average clutch of 22 eggs (Sabath, 1991; Norell *et al.*, 1994).

Individual eggs of both *Troodon* and *Oviraptor* are elongated, and these elongated eggs are found in circular clutches with the tops of each egg pointing inward toward the clutch centers. The clutches of both taxa have average diameters of 50 cm. Also common to both *Oviraptor* and *Troodon* are indications of parental attention to the eggs. A skeleton of *Oviraptor* was found in a brooding position, sitting directly on a clutch of eggs from Mongolia (Norell *et al.*, 1996), and a similar association between an adult *Troodon* skeleton and a clutch of eggs was discovered in Montana (Varricchio *et al.*, 1997). A very well-preserved clutch of 24 *Troodon* eggs, discovered on Egg Mountain, also revealed a 12-cm-high, built-up sediment rim extending around the periphery of the clutch, about 12 cm out from the eggs (Varricchio *et al*, 1997). The rim indicates that the adult *Troodon* actually invested time and energy into the act of nest construction.

Oviraptor clutches are very common in the Upper Cretaceous Djadokhta and Barun Goyot Formations of Mongolia (Sabath, 1991; Mikhailov, 1991; Norell *et al.*, 1994), but because the eggs are found in eolian sandstones it has not been possible to determine whether these were actual *Oviraptor* nesting horizons or nesting grounds. In contrast, the Egg Mountain and Egg Island sites from the Upper Cretaceous Willow Creek Anticline of Montana reveal extensive nesting horizons of *Troodon* (Horner, 1987; Horner and Weishampel, 1988, 1996).

The two hadrosaurs, *Maiasaura peeblesorum* and *Hypacrosaurus stebingeri,* have less complete clutches of eggs, and none can be demonstrated to have had any particular clutch arrangement. Only one clutch of *Maiasaura* eggs has been prepared and it contains only 11 eggs, of which the maximum egg diameter does not exceed 12 cm. The eggs appear to have been either spherical or bluntly ovoid in shape, with a maximum volume of 1250 ml. The eggs of *Hypacrosaurus,* on the other hand, have an average volume of about 3900 ml and are very nearly spherical in shape (Horner and Currie, 1994). The eggs of a related taxon, an as yet unidentified lambeosaurine from the Judith River Formation of central Montana, produced 22 spherical eggs, each with a volume of 4000 ml. The clutch of this lambeosaur is oblong in shape, measuring 150 × 90 cm. These figures show that there was clearly a difference in egg size among very closely related species. Hadrosaur clutches are found in associated groups, indicating colonial nesting, and through relatively thick units of sediment suggesting prolonged use of the nesting areas.

Egg clutches hypothesized to be dinosaurian in origin are found throughout the world (Carpenter and Alf, 1994; Currie, 1996) and are most commonly found on nesting horizons in association with other clutches of similar morphology, suggesting that numerous dinosaurian taxa probably nested in colonies. In addition, most clutches of eggs attributable to dinosaurs have geometric arrangements indicating that the adult invested time in nest construction.

Gregariousness

Gregarious behaviors include those in which individuals of particular species congregate in groups, including nesting creches, herding, and pack hunting. In each of these situations interpretations are based on studies of TAPHONOMY, the context in which fossils have accumulated.

Nesting crêches are groups of juveniles that remain in their respective nesting areas for a period of time preceding hatching. Among living birds these crêches are actually groups of juveniles protected by a number of adults. Abundances of juvenile bones, representing individuals between embryo and neonate size (2.0–2.5 times the linear dimensions of full-term embryos), are commonly found on the nesting horizons of *M. peeblesorum, H. stebingeri* (Horner and Makela, 1979; Horner, 1994), and the unidentified Upper Cretaceous lambeosaurine from Montana. Comparative studies of the nesting grounds of colonial birds have shown that the greatest majority of bones found on a nesting ground are derived from babies that died during their nesting periods or are the remains of animals brought to the nesting area by parents feeding their young (Horner, 1994).

The occurrences of abundant baby hadrosaur skeletal remains on horizons that yield hadrosaur eggs strongly suggest that the young posthatching hadrosaurs remained in their nesting areas for some time after eclosion (hatching). A group of 15 *Maiasaura* juveniles, all of equal size, found in a nest-like structure on a nesting horizon, has been used as one line of evidence to hypothesize that maiasaurs were born altricial and that the siblings remained together, attended to by one or both adults (Horner and Makela, 1979; Horner, 1984). This is elaborated on under Parental Care. Other groups of juveniles of similar size have been discovered (Gilmore, 1917, 1929; Dodson, 1971; Horner and Makela, 1979; Forster, 1990) and suggest that juvenile dinosaurs representing several species may have engaged in some fashion of congregational behavior (see NEOCERATOPSIA).

Another kind of gregarious behavior in which some dinosaurs appear to have engaged is aggregation or herding, as evidenced by nearly monospecific bone beds of a variety of taxa (see Varricchio and Horner, 1993, for details). Interestingly, the largest bone beds, composed primarily of one dinosaur species, are those representing taxa of ceratopsians and hadrosaurians (Currie and Dodson, 1984; Nelms, 1989; Rogers, 1990; Varricchio and Horner, 1993). Most of these monospecific groups are represented by tens of individuals to several hundred individuals, and there is a rare instance in which the numbers of individuals appear to be in the thousands (Hooker, 1987). Most of these ceratopsian and hadrosaurian groups appear to represent animals that died in catastrophic events such as floods, droughts, or volcanic-related events. Also, because the animals died in these catastrophes, it stands to reason that they were actually in groups or herds (Currie, 1981; Currie and Dodson, 1984) before their catastrophic deaths. Because both the hadrosaurs and the ceratopsians were plant eaters, it is reasonable to hypothesize that these animals migrated from one area to another seeking food resources (Currie, 1989). Other nearly monotaxic groups of herbivorous dinosaurs that have been hypothesized to represent herds or aggregations include the *Iguanodon* assemblage of Bernissart, Belgium (Dupont, 1878) and the *Plateosaurus* assemblage of Trossingen, Germany (Huene, 1928). Both of these have recently been determined not to represent catastrophic events but rather time-averaged accumulations or some other more complex taphonomic situation (Norman, 1985; Weishampel, 1984).

Additional evidence of herding and aggregating can be postulated from dinosaur trackways (Bird, 1944; Bakker, 1968; Ostrom, 1972, 1985; Thulborn, 1979; Lockley *et al.*, 1983; Lockley and Hunt, 1995), although it is rare to be able to demonstrate that a group of animals actually traveled together (Currie, 1983) rather than in the same area over an extended period of time. Animals walking along the shores of lakes, rivers, or marine shores are most likely to travel in one of two directions, parallel to the body of water, regardless of whether they are together in a group (Lockley and Hunt, 1995).

Packing or pack hunting is another form of gregariousness, and probably the most speculative. Pack hunting has been postulated, based on the discovery of a group of five *Deinonychus* skeletons in association with a partial *Tenontosaurus* skeleton (Ostrom, 1969), and more reliably on the basis of several associations of multiple shed *Deinonychus* teeth in aggregation with skeletons of *Tenontosaurus* that appear to have been preyed upon by the theropods (Maxwell and Ostrom, 1995). In one instance a *Tenontosaurus* skeleton is missing portions of its legs and its rib cage has obviously been pulled open. Eleven *Deinonychus* teeth found in close association with the herbivore skeleton imply that the animal was eaten by more than one carnivore. Because the tenontosaur shows no sign of having been scavenged, it can be hypothesized that more than one individual was involved in the actual kill.

In addition to these occurrences are the nearly monospecific assemblages of *Coelophysis* from the GHOST RANCH QUARRY and *Allosaurus* from the CLEVELAND–LLOYD QUARRY. Both of these unusual assemblages contain numerous individuals representing a variety of ontogenetic stages. Farlow (1987) suggested that these assemblages might represent habitat preferences, and that the animals may have died during a particular part of their breeding season. Coombs (1990) hypothesized that a catastrophic event might have driven the animals together. Additionally, a group of four associated specimens of *Troodon* from Montana, representing two juveniles, a subadult, and an adult, may represent a family unit that perished together on the shores of a freshwater lake (Varricchio, 1995).

Parental Care

Parental care includes any investment that the parent or parents of a particular taxon make toward their offspring. Among dinosaurs it appears that several taxa made investments, from brooding (Norell *et al.*, 1996; Varricchio *et al.*, 1997) of eggs to protecting and possibly feeding helpless young (Horner and Weishampel, 1988; Horner *et al.*, 1997). The only cases of direct evidence of parental care, however, are the association of the adult *Oviraptor* on the clutch of eggs discovered in Mongolia (Norell *et al.*, 1996) and a partial *Troodon* skeleton found atop a clutch of *Troodon* eggs (Varricchio *et al.*, 1997).

The association of juvenile dinosaurs on the various hadrosaur nesting grounds, described previously, strongly suggests that the juveniles were being protected by adults (Horner, 1994), as occurs among living archosaurs. The group of 15 maiasaurs of equal size found in a bowl-shaped, nest-like structure may suggest that these dinosaurs were nest-bound, and therefore in need of parental care and feeding. Additional evidence for this behavior is suggested by studies of the internal structure of the leg bones. A recent quantitative study (Horner *et al.*, 1997) determined that the leg bones of a full-term embryonic hadrosaurs contained very little ossified bone, and that the percentages of ossified bone and cartilaginous tissues in these dinosaurs are equivalent to similar percentages found in extant altricial birds. These data contradict a previous qualitative study (Geist and Jones, 1996) suggesting that *Maiasaura* and other hadrosaurian taxa were born precocial.

Visual Display

Another controversial aspect of behavior pertains to behaviors elicited by physical releasing mechanisms. These include morphological features that command behavioral responses through visualization. Examples include characteristics such as horns, frills, spikes, crests, bosses, plates, thickened skulls, long teeth, big eyes, and so on. (see ORNAMENTATION)

Some of the most obvious display features are the horns and crests found on the skulls of many dinosaurs and, in particular, dinosaurs hypothesized to have traveled in groups, such as the ceratopsians and hadrosaurs (Farlow and Dodson, 1975; Molnar, 1977; Spassov, 1979; Sampson, 1995a,b; Dodson, 1996). Farlow and Dodson (1975) and Sampson (1995b) have postulated that the horned dinosaurs used their horns and frills in frontal engagements, shoving and wrestling to settle their hierarchical disputes, much like horned mammals that use their horns to attract mates and determine hierarchical rank (Geist, 1966). Molnar (1977) extended this argument to include all dinosaurian taxa that have exceptional cranial characteristics, including caniniform teeth characteristic of some heterodontosaurs and jugal bosses characteristic of ceratopsians. The thickened skulls of pachycephalosaurs have also been suggested to be linked to hierarchical combat and display (Galton, 1971), and it is likely that the enlarged crest-like structures on the skulls of the oviraptors were also some kind of display feature. Other dinosaurs, such as *Ceratosaurus* and *Monolophosaurus,* have horns on the tops of their noses, and *Allosaurus* had horns over the upper front part of its orbits. *Carnotarsus* and the tyrannosaurids had horns over their orbits and apparently some kind of nasal ornamentation as well.

The broad, flat dorsal plates or dorsal spikes of *Stegosaurus* and other STEGOSAURS could have served a dual purpose—in active defense mechanism and also in lateral display that would have made the animals look larger or more formidable. This could have acted as a passive defense mechanism or an attractant for mates.

Some neotenous or paedomorphic morphological characters (see HETEROCHRONY) have been suggested as mechanisms that may have stimulated adults to respond to their juveniles in particular ways, such as care and feeding (Lorenz, 1971). Interestingly, baby hadrosaurs do not have the cranial characteristics of their adult counterparts, but instead have features, including large eyes, rounded heads, and shortened snouts, that are very similar to the features found in babies of extant birds and mammals that are cared for by parents. Although highly speculative, it may be reasonable to hypothesize that these neotenous features in the baby hadrosaurs triggered some form of parental care.

Obviously, many other behaviors can be hypothesized as related to social interactions, including vocalization (Weishampel, 1981), but such features or characters require considerable hypothetical reasoning and are based primarily on the notion that all related living forms possess the characters.

See also the following related entries:

EGGS, EGGSHELLS, AND NESTS • FOOTPRINTS AND TRACKWAYS • MIGRATION • ORNAMENTATION • PHYSIOLOGY

References

Bakker, R. T. (1968). The superiority of dinosaurs. *Discovery* **3,** 11–22.

Bird, R. T. (1944). Did "brontosaurs" ever walk on land? *Nat. History* **53,** 61–67.

Brooks, D. R., and McLennan, D. A. (1991). *Phylogeny, Ecology, and Behavior.* Univ. of Chicago Press, Chicago.

Carpenter, K., and Alf, K. (1994). Global distribution of dinosaur eggs, nests and babies. In *Dinosaur Eggs and Babies* (K. Carpenter, K. F. Hirsch, and J. R. Horner, Eds.), pp. 15–30. Cambridge Univ. Press, Cambridge, UK.

Coombs, W. P., Jr. (1989). Modern analogues for dinosaur nesting and parental behavior. In *Paleobiology of Dinosaurs* (J. O. Farlow, Ed.), Spec. Pap. Geol. Soc. Amer. No. 238, pp. 21–53.

Coombs, W. P., Jr. (1990). Behavior patterns of dinosaurs. In *The Dinosauria* (D. B. Weishampel, P. Dodson, and H. Osmólska, Eds.), pp. 32–42. Univ. of California Press, Berkeley.

Currie, P. J. (1981). Hunting dinosaurs in Alberta's huge bonebed. *Can. Geogr. J.* **101**(4), 32–39.

Currie, P. J. (1983). Hadrosaur trackways from the lower Cretaceous of Canada. *Acta Palaeontol. Polonica* **28,** 63–73.

Currie, P. J. (1989). Long distance dinosaurs. *Nat. History* **1989**(6), 61–65.

Currie, P. J. (1996). The Great Dinosaur Egg Hunt. *Natl. Geogr.* **189**(5), 96–111.

Currie, P. J., and Dodson, P. (1984). Mass death of a herd of ceratopsian dinosaurs. In *Third Symposium on Mesozoic Terrestrial Ecosystems* (W.-E. Reif and F. Westphal, Eds.), pp. 61–66. Attempto Verlag, Tubingen, Germany.

Currie, P. J., and Horner, J. R. (1988). Lambeosaurine hadrosaur embryos (Reptilia: Ornithischia). *J. Vertebr. Paleontol.* **8,** 13A.

Dodson, P. (1971). Sedimentology and taphonomy of the Oldman Formation (Campanian), Dinosaur Provincial Park, Alberta (Canada). *Palaeogeogr. Palaeoclimatol. Palaeoecol.* **10,** 21–74.

Dodson, P. (1996). *The Horned Dinosaurs*, pp. 346. Princeton Univ. Press, Princeton, NJ.

Dupont, E. (1878). Sur la découverte d'ossements d'*Iguanodon*, de poissons et de végétaux dans la fosse Sainte Barbe du Charbonnage de Bernissart. *Bull. Acad. R. Belgique Sér. 2* **46,** 387–408.

Farlow, J. O. (1987). Speculations on the diet and digestive physiology of herbivorous dinosaurs. *Paleobiology* **13,** 60–72.

Farlow, J. O., and Dodson, P. (1975). The behavioral significance of frill and horn morphology in ceratopsian dinosaurs. *Evolution* **29,** 353–361.

Forster, C. A. (1990). Evidence for juvenile groups in the ornithopod dinosaur *Tenontosaurus tilletti* Ostrom. *J. Paleontol.* **64**(1), 164–165.

Galton, P. M. (1971). A primitive dome-headed dinosaur (Ornithischia: Pachycephalosauridae) from the Lower Cretaceous of England, and the function of the dome in pachycephalosaurids. *J. Paleontol.* **45,** 40–47.

Geist, N. R., and Jones, T. D. (1996). Juvenile skeletal structure and the reproductive habits of dinosaurs. *Science* **272,** 712–714.

Geist, V. (1966). The evolution of horn-like organs. *Behaviour* **27,** 175–214.

Gilmore, C. W. (1917). *Brachyceratops*, a ceratopsian dinosaur from the Two Medicine Formation of Montana. United States Geological Survey Professional Paper No. 103, pp. 1–45.

Gilmore, C. W. (1929). Hunting dinosaurs in Montana. *Explorations Field Work Smithsonian Inst.* **1928,** 7–12.

Greene, H. W. (1986). Diet and arboreality in the emerald monitor, *Varanus prasinus,* with comments on the study of adaptation. *Fieldiana* **1370,** 1–12.

Hirsch, K. F., and Quinn, B. (1990). Eggs and eggshell fragments from the Upper Cretaceous Two Medicine Formation of Montana. *J. Vertebr. Paleontol.* **10,** 491–511.

Hooker, J. S. (1987). Late Cretaceous ashfall and the demise of a hadrosaurian "herd." *Geol. Soc. Am.* (Rocky Mountain Section), **19,** 284. [Abstracts with Programs]

Horner, J. R. (1982). Evidence of colonial nesting and "site fidelity" among ornithischian dinosaurs. *Nature* **297,** 675–676.

Horner, J. R. (1987). Ecologic and behavioral implications derived from a dinosaur nesting site. In *Dinosaurs Past and Present, Vol. II* (S. J. Czerkas and E. C. Olson, Eds.), pp. 51–63. Univ. Washington Press, Seattle.

Horner, J. R. (1994). Comparative taphonomy of some dinosaur and extant bird colonial nesting grounds. In *Dinosaur Eggs and Babies* (K. Carpenter, K. F. Hirsch, and J. R. Horner, Eds.), pp. 116–123. Cambridge Univ. Press, Cambridge, UK.

Horner, J. R., and Currie, P. J. (1994). Embryonic and neonatal morphology and ontogeny of a new species of *Hypacrosaurus* (Ornithischia, Lambeosauridae) from Montana and Alberta. In *Dinosaur Eggs and Babies* (K. Carpenter, K. F. Hirsch, and J. R. Horner, Eds.), pp. 312–336. Cambridge Univ. Press, Cambridge, UK.

Horner, J. R., and Makela, R. (1979). Nest of juveniles provides evidence of family structure among dinosaurs. *Nature* **282,** 296–298.

Horner, J. R., and Weishampel, D. B. (1988). A comparative embryological study of two ornithischian dinosaurs. *Nature* **332,** 256–257.

Horner, J. R., and Weishampel, D. B. (1996). Correction to: A comparative embryological study of two ornithischian dinosaurs (1988). *Nature* **383,** 103.

Horner, J. R., Horner, C. C., and Eschberger, B. (1997). Altricial dinosaurs: Evidence from embryonic and perinatal bone histology. *Paleobiology*, in press.

Huene, F. von (1928). Lebensbild des Saurischier-Vorkommens imobersten Keuper von Trossingen in Wurttemberg. *Palaeobiologica* **1,** 103–116.

Lockley, M., and Hunt, A. P. (1995). *Dinosaur Tracks, and Other Fossil Footprints of the Western United States*, pp. 338. Columbia Univ. Press, New York.

Lockley, M. G., Yang, S. Y., and Carpenter, K. (1983). Hadrosaur locomotion and herding behavior: evidence from footprints in the Mesaverde Formation, Grand Mesa Coal Field, Colorado. *Mountain Geologist* **20**(1), 5–14.

Lorenz, K. (1971). *Studies in Human and Animal Behavior, Vol. 2*, pp. 366. Harvard Univ. Press, Cambridge, MA.

Maxwell, W. D., and Ostrom, J. H. (1995). Taphonomy and paleobiological implications of *Tenontosaurus–Deinonychus* associations. *J. Vertebr. Paleontol.* **15**(4), 707–712.

Mikhailov, K. E. (1991). Classification of fossil eggshells of amniotic vertebrates. *Acta Palaeontol. Polonica* **36,** 193–238.

Molnar, R. E. (1977). Analogies in the evolution of display structures in ornithopods and ungulates. *Evolutionary Theor.* **3,** 165–190.

Nelms, L. G. (1989). Late Cretaceous dinosaurs from the North Slope of Alaska. *J. Vertebr. Paleontol,* **9,** 34A.

Norell, M. A., Clark, J. M., Demberelyin, D., Rhinchen, B., Chiappe, L. M., Davidson, A. R., McKenna, M. C., Altangerel, P., and Novacek, M. J. (1994). A theropod dinosaur embryo and the affinities of the Flaming Cliffs dinosaur eggs. *Science* **266,** 779–782.

Norell, M. A., Clark, J. M., Chiappe, L. M., and Dashzeveg, D. (1996). A nesting dinosaur. *Science* **378,** 774–776.

Norman, D. B. (1985). *The Illustrated Encyclopedia of Dinosaurs*, pp. 208. Crecent, New York.

Ostrom, J. H. (1969). Osteology of *Deinonychus antirrhopus*, an unusual theropod from the Lower Cretaceous of Montana. *Peabody Museum Nat. History Bull.* **30,** 1–165.

Ostrom, J. H. (1972). Were some dinosaurs gregarious? *Palaeogeogr. Palaeoclimatol. Palaeoecol.* **11,** 287–301.

Ostrom, J. H. (1985). Social and unsocial behavior in dinosaurs. *Field Museum Nat. History* **55**(9), 10–21.

Padian, K. (1987). A comparative phylogenetic and functional approach to the origin of vertebrate flight. In *Re-*

cent *Advances in the Study of Bats* (B. Fenton, P. A. Pacey, and J. M. V. Rayner, Eds.), pp. 3–22. Cambridge Univ. Press, Cambridge, UK.

Rogers, R. R. (1990). Taphonomy of three dinosaur bonebeds in the Upper Cretaceous Two Medicine Formation of northwestern Montana: Evidence for drought-related mortality. *Palaios* **5,** 394–413.

Sabath, K. (1991). Upper Cretaceous amniote eggs from the Gobi Desert. *Acta Palaeontol. Polonica* **36,** 151–192.

Sampson, S. D. (1995a). Two new horned dinosaurs from the Upper Cretaceous Two Medicine Formation of Montana; with a phylogenetic analysis of the Centrosaurinae (Ornithischia: Ceratopsidae). *J. Vertebr. Paleontol.* **15**(4), 743–760.

Sampson, S. D. (1995b). Horns, herds, and heirarchies. *Nat. History* **104**(6), 36–40.

Spassov, N. B. (1979). Sexual selection and the evolution of horn-like structures in ceratopsian dinosaurs. *Paleontol. Stratigr. Lithol.* **11,** 37–48.

Thulborn, R. A. (1979). Dinosaur stampede in the Cretaceous of Queensland. *Lethaia* **12,** 275–279.

Varricchio, D. J. (1995). Taphonomy of Jack's Birthday Site, a diverse dinosaur bonebed from the Upper Cretaceous Two Medicine Formation of Montana. *Palaeogeogr, Palaeoclimatol, Palaeoecol.* **114,** 297–323.

Varricchio, D. J., and Horner, J. R. (1993). Hadrosaurid and lambeosaurid bone beds from the Upper Cretaceous Two Medicine Formation of Montana: taphonomic and biologic implications. *Can. J. Earth Sci.* **30,** 997–1006.

Varricchio, D. J., Jackson, F., Borlowski, J., and Horner, J. R. (1997). Nest and egg clutches for the theropod dinosaurs *Troodon formosus* and evolution of avian reproductive traits. *Nature,* in press.

Weishampel, D. B. (1981). Acoustic analysis of potential vocalization in lambeosaurine dinosaurs (Reptilia: Ornithischia). *Paleobiology* **7**(2), 252–261.

Weishampel, D. B. (1984). Trossingen: E. Fraas, F. von Huene, R. Seemann and the "Schwäbische Lindwurm" *Plateosaurus*. In *3rd Symposium Mesozoic Terrestrial Ecosystems, Short Papers* (W.-E. Reif and F. Westphal, Eds.), pp. 249–253. Attempto Verlag, Tübingen, Germany.

Weishampel, D. B. (1995). Fossils, function, and phylogeny. In *Functional Morphology in Vertebrate Paleontology* (J. J. Thomason, Ed.), pp. 34–54. Cambridge Univ. Press, Cambridge, UK.

Beijing Natural History Museum, People's Republic of China

see MUSEUMS AND DISPLAYS

Bernard Price Institute for Palaeontological Research, University of the Witwatersrand, Johannesburg, South Africa

ANUSUYA CHINSAMY
South African Museum
Cape Town, South Africa

The Bernard Price Institute for Palaeontological Research, established in 1945, houses a fairly extensive collection of cranial and postcranial elements of the prosauropod *Massospondylus* and fewer postcranial elements of *Euskelosaurus*. Of the heterodontosaurids, specimens of *Heterodontosaurus tucki* and *Lycorhinus angustidens* (=*Lanasaurus scalpridens*) can be found in the Johannesburg collections. Theropods are represented by skull and postcranial elements of *Syntarsus rhodesiensis*, as are eggs and eggshell fragments from the Early Jurassic of South Africa. The recently completed museum includes a mounted, partial skeleton of *Euskelosaurus* and an articulated skull and postcranial skeleton of *Massospondylus* that is still embedded in its original matrix.

See also the following related entries:

AFRICAN DINOSAURS • ALBANY MUSEUM • NATIONAL MUSEUM, BLOEMFONTEIN • SOUTH AFRICAN MUSEUM

Bernissart Museum, Belgium

see MUSEUMS AND DISPLAYS

Biogeochemistry

see CHEMICAL COMPOSITION OF DINOSAUR FOSSILS

Biogeography

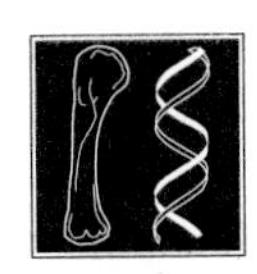

JEAN LE LOEUFF
Musée des Dinosaures
Espéraza, France

As land animals incapable of crossing wide water barriers, dinosaurs yield important information about the ancient distribution of land masses and seas (paleogeography) during the Mesozoic Era. When previously continuous dinosaur populations were isolated from each other, they evolved on their own and produced so-called endemic forms. As a modern example of an endemic terrestrial fauna, Australia shows a completely original assemblage of marsupial mammals because it became isolated from other continents at the end of the Mesozoic Era before placental mammals became dominant in the rest of the world. The ancestors of modern Australian marsupials had no contact with placentals and evolved on their own while their relatives on other lands were largely replaced by placentals. To the contrary, when two endemic dinosaur faunas were reassociated because of the regression of a sea, they mixed together and many taxa became extinct. A review of the dinosaur distribution patterns from the Late Triassic to the Late Cretaceous adds new information to the paleogeographic reconstructions of the world established by geologists.

The distribution of extinct land animals and its relationship to former geographies is not a new subject. It was discussed as early as 1750 by the French naturalist Buffon, who considered that Europe and America had been separated recently because fossil elephants are found on both continents. In Buffon's mind, it was clear that elephants had colonized a large area that was later divided by water. The first contribution to dinosaur biogeography itself is probably that of the American paleontologist Richard Swann Lull in 1910. Of course, Lull's contribution predates the establishment of the theory of plate tectonics, and it reflects a now old-fashioned view. Like most geologists of his time, Lull believed the continents had always occupied their present position, and that variations in sea level resulted in geographies only slightly different from those of today. Although claiming that "the significance of terrestrial vertebrates . . . in throwing light upon the isolation and connection of the continents is becoming more and more appreciated" and that "the dinosaurs, with their known geological range throughout the entire Mesozoic period, and of almost world wide distribution, are the most significant vertebrates of Secondary times," Lull did not understand the importance of his data for changing concepts of geology. This vision of fixed paleogeography became the model for later reflections on dinosaur biogeography and culminated in the 1960s with the paleogeographic atlases of the French paleontologists H. and G. Termier, who drew "intercontinental bridges" of a mysterious nature reaching several thousand kilometers in length and allowing transoceanic migrations of various animals.

With the establishment of a universally accepted model of plate tectonics in the 1960s, these kinds of explanations became untenable, and later papers were written in a completely different way. A single and major exception to the stability dogma of the first half of the century was the brilliant Hungarian paleontologist Franz Nopcsa's posthumous paper, published in 1934, that incorporated Wegener's contemporary theory of continental drift with his incomparable knowledge of the fossil record outside North America. The mechanisms of the observed migrations ("The formation of an Equatorial belt of folding, helping the migration of reptiles from one block of sial to another") and observed differences ("The gradual cleavage of old masses of sial, beginning in the Permian and lasting until the Tertiary period. This cleavage accentuated more and more the separation of the different faunas. . . . Transitory inundations of old masses (transgressions) which isolated some part of the firm land and, when receding, opened again new connections, thereby disturbing the 'economic equilibrium of Nature") were also analyzed by Nopcsa in a very modern way. The differences between Nopcsa's work and the hypotheses advanced below rely mainly on a better fossil record and new advances in plate tectonics rather than on a fundamental change in the concepts.

With increasing worldwide information about dinosaurs, new patterns of dinosaur biogeography are emerging (Bonaparte and Kielan-Jaworovska, 1987; Rage, 1981; Buffetaut *et al.*, 1988; Le Loeuff, 1991; Russell, 1993). A paleobiogeographic study relies at first on the identification of the different faunal provinces (paleobioprovinces) at a given time, characterized by their own land animals. Initially, one can consider that each continental mass is occupied by a particular biota. However, other nonoceanic barriers may prove to be uncrossable for terrestrial vertebrates because of latitudinal, altitudinal, ecological, climatic, or marine (epicontinental seas) barriers. Although latitudinal variations seem to have had little influence on the global distribution of dinosaurs, epicontinental seas on the different land masses have provoked dramatic long-time separations between the terrestrial faunas separated by the barriers. An epicontinental sea (called the Uralian Sea) separated Europe from Asia during the Jurassic and the Cretaceous, preventing any exchange between the two provinces and leading to an important endemism in central Asia during the Jurassic and Early Cretaceous. As a result, paleobioprovinces do not fit exactly with the continental plates recognized by geologists. On the other hand, when two plates are converging, land communication may be possible before the collision across island arcs; in this case, plates are still unique in a geographic sense until complete suture, but a single paleobioprovince exists.

Late Paleozoic and Early Mesozoic paleogeography is characterized by the progressive unification of the main land masses into a single continent called Pangaea (Fig. 1). This enormous continent was not divided by epicontinental seas, and the Early Triassic fauna is rather uniform throughout the world. Pangaea then began to break into different continents, and dinosaur biogeography reflects both this dislocation of Pangaea and the later adventures of its different pieces, which first diverged and then sometimes touched each other again.

In the Late Triassic, the supercontinent Pangaea was indented to the east by an equatorial ocean, called Tethys, which existed until the Tertiary (its closure led to the formation of alpine mountains from Central Europe to the Himalayas). This triangular ocean extended from southern Europe and Africa to the west and joined the megaocean Panthalassa between Australia and Asia.

The most abundant dinosaurs of Pangaean times (Late Triassic to Middle Jurassic; 230–165 Ma) are the herbivorous prosauropods, of which six families are now recognized. Several families had a large distribution, such as the Plateosauridae (North and South America, Europe, and China) and the Melanorosauridae (South America, Europe, and Africa). Others appear to be endemic, such as the North American Anchisauridae, the Chinese Yunnanosauridae, and the South African Blikanasauridae. The theropod *Syntarsus* is known from South Africa and North America. It is difficult during these early times of the dinosaur era to recognize paleobioprovinces. Using other land

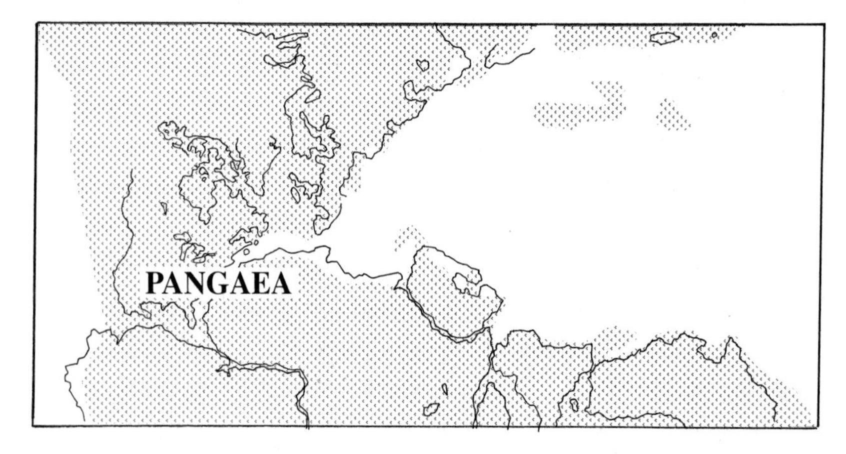

FIGURE 1 Late Triassic reconstruction of the paleobioprovinces (after Dercourt *et al.*, 1993; drawing by Guy Le Roux).

vertebrates, Martin (1981) and Rage (1988) defined a "Peritethysian" province including Europe, North Africa, North America, India, Madagascar, and perhaps Southeast Asia. The Peritethysian province is characterized by phytosaurs (crocodile-like reptiles) and metoposaurian amphibians. However, despite regional differences that are difficult to analyze, Gondwanan and Laurasiatic faunas were not different in the Late Triassic, although one can suggest that a kind of latitudinal variation existed.

Russell (1993) has demonstrated that two paleobioprovinces might be defined during the Jurassic: a central Asian province with marked endemicity and a "Neopangean" one corresponding to Pangaea without central Asia. The isolation of central Asia from the rest of the world is possibly linked with the transgression of an epicontinental sea between Europe and Asia, called the Uralian Sea. From the Bathonian (165 Ma) to the Tertiary, it seems that this Uralian Sea acted as an uncrossable barrier for land vertebrates. Central Asia was isolated from Neopangaea (165–145 Ma) and later from Euramerica (145–110 Ma) until the Middle Cretaceous, when a land route opened between North America and Asia across the Bering Strait.

During the Jurassic, the Tethys ocean continued opening to the west, between North America to the north and South America and Africa to the south (Fig. 2); the separation became effective during Callovian or Kimmeridgian times (155–141 Ma), when seas invaded the rifting corridor south of North America. The Neopangaean province ceased to exist at this time.

In the Kimmeridgian Stage (146–141 Ma), similarities still existed between Africa and North America; the rich Tanzanian locality of Tendaguru has yielded several dinosaur genera that are also recorded in the North American Morrison Formation (*Brachiosaurus*, *Barosaurus*, and *Dryosaurus*). This confirms that land communications were available between Africa and North America until the Late Jurassic. The absence in Africa of the Euro-American Tithonian (141–135 Ma) families Camarasauridae and Camptosauridae, as well as the absence in the north of titanosaurid and dicraeosaurid sauropods, may reflect the beginning of separate evolution in the different parts of breaking Neopangaea, i.e., Euramerica (Europe and North America) and Gondwana (South America, Africa, India, Australia, and Antarctica).

To the east, Chinese dinosaurs had no close relationships with their western counterparts, as was shown by Valérie Martin (1995): The well-known Chinese sauropods *Mamenchisaurus*, *Euhelopus*, *Shunosaurus*, and so on, which were usually referred to western families, in fact constitute an endemic Chinese family (Euhelopodidae) with no clear relationships to other sauropods. Russell (1993) reached nearly the same conclusions about sauropods, adding that theropods (*Yangchuanosaurus* and *Sinraptor*, which constitute the new family Sinraptoridae), stegosaurs (*Chialingosaurus*, *Chungkingosaurus*, and *Tuojiangosaurus*), and even mammals were endemic. This

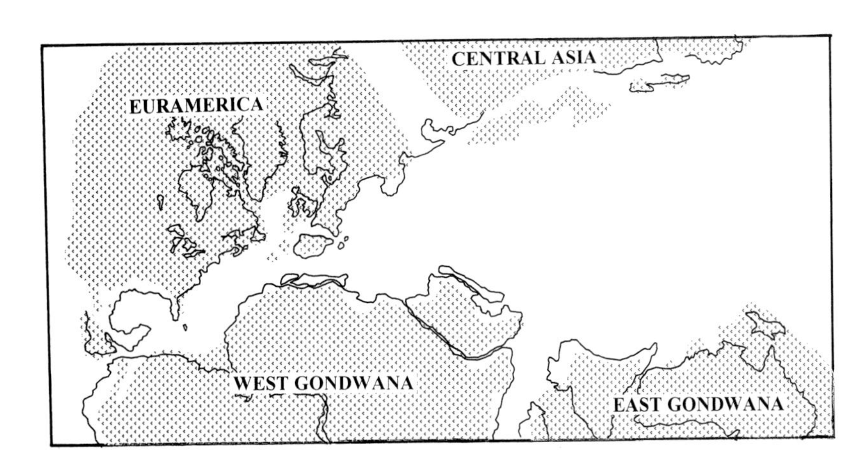

FIGURE 2 Late Jurassic reconstruction of the paleobioprovinces (after Dercourt *et al.*, 1993; drawing by Guy Le Roux).

endemicity indicates that central Asia was isolated from the rest of the world.

The distribution pattern of Early Cretaceous dinosaurs is more clear than that of the Jurassic. With the opening of the central Atlantic Ocean (approximately 141 Ma), communications became impossible between North America and the southern continents, leading to the differentiation of Gondwanan (South America, Africa, India, Australia, and Antarctica) and Euramerican (Europe and North America) faunas. The transgression of epicontinental seas occurred 110 Ma ago as the North Atlantic Ocean broke the faunal continuity between North America and Europe (end of the Euramerican province), while the establishment of a land bridge across the Bering Strait allowed communications between Asia and America (birth of the Asiamerican province).

A Gondwanan fauna was well defined in South America and Africa by titanosaurid and dicraeosaurid sauropods and abelisaurid theropods. The Argentinian Early Cretaceous fauna of La Amarga is close to the Late Jurassic Tendaguru fauna, with the dicraeosaurid *Amargasaurus*. During this period Gondwana began to break into different plates corresponding to the current continents. The eastern Gondwanan continents (India, Australia, and Madagascar) had begun to separate from Africa and South America as early as the Late Jurassic. In the Early Cretaceous, India and Australia separated from each other. These geological interpretations are problematic because they imply a long separation between India and west Gondwana (Africa–South America) that is not at all confirmed by terrestrial vertebrate distribution. The Late Cretaceous Indian fauna looks like the South American one, with abelisaurid and titanosaurid dinosaurs and madtsoiid snakes. During the Middle Cretaceous, rifting began between Africa and South America, leading to the opening of the South Atlantic Ocean and the isolation of the continental faunas (end of West Gondwana).

North American and European dinosaurs were very similar during the earliest Cretaceous. Several examples of intercontinental genera (the ornithopods *Iguanodon* and *Hypsilophodon* and the nodosaurid ankylosaur *Polacanthus*) are recorded from the Lakota Formation of Dakota and the Wealden of England. These indicate that continuous dinosaur populations lived on both continents 120 million years ago. Other Euramerican dinosaurs include ornithomimids (*Pelecanimimus* from Spain and an unnamed ornithomimid from the Cloverly Formation) and troodontids (unnamed species from the Isle of Wight, England). It is likely that this Euramerican paleobioprovince no longer existed when North American and Asian faunas merged at the end of the Early Cretaceous. Euramerica was "broken" by the invasion of the future North Atlantic Ocean by seas, possibly before Aptian times (113 Ma).

The European province also shows Gondwanan components such as the titanosaurid *Iuticosaurus* (Isle of Wight, England) and the spinosaurid *Baryonyx*. The ornithopod *Valdosaurus* is recorded from Africa (Niger) and Europe (England). These similarities indicate that Europe had land connections with both North America and Africa during the Early Cretaceous. Connections with Africa existed across the Mediterranean.

To the east, central Asia came into contact with North America by Aptian–Albian times, forming a new Asiamerican province. It seems that Euramerican taxa, such as iguanodontids, dromaeosaurids, and troodontids, invaded central Asia at this time (Russell, 1993). Before this important event, the Early Cretaceous Asian vertebrates were still provincial, with psittacosaurids and peculiar sauropods such as *Phuwiangosaurus* from Thailand.

With the dislocation of Gondwana in the Cretaceous, and high marine levels leading to the inundation of lands by epicontinental seas, the Late Cretaceous is the most complex period for dinosaur biogeography (Fig. 3). Unfortunately, little is known of the Late Cretaceous "Gondwanan" faunas, which theoretically should have undergone many independent evolutionary events.

Western North America, with its well-known dinosaurs (including *Tyrannosaurus*, *Triceratops*, and *Edmontosaurus*), was isolated from eastern North America by the "Western Interior Sea" from the Cenomanian until the uppermost Cretaceous. Possibly because of the northern position of the Bering "bridge," which may have acted as a climatic or ecological filter, exchanges with Asia were limited because there were few common genera but many common families (such as the Tyrannosauridae with *Tyrannosaurus* in North America and *Tarbosaurus* in Asia). Russell (1993) suggested that most of the North American Late Cretaceous families, including Tyrannosauridae, Ceratopsidae, Protoceratopsidae, and Troodontidae, originated in Asia during the Middle Cretaceous.

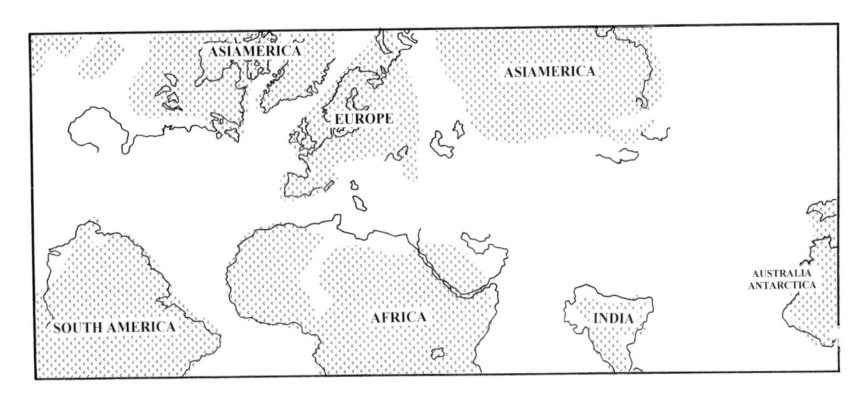

FIGURE 3 Late Cretaceous reconstruction of the paleobioprovinces (after Dercourt *et al.*, 1993; drawing by Guy Le Roux).

Despite the advanced stage of dislocation of Gondwana in the Late Cretaceous, the known faunas from its different pieces show strong affinities until the Late Cretaceous. However, this is possibly due to a poor fossil record outside South America.

South American dinosaurs of the Late Cretaceous include abelisaurids, titanosaurids, noasaurids, and basal hadrosaurids. Despite the fragmentation of Gondwana, titanosaurids and abelisaurids are found at this time in Africa, India, and Madagascar.

Europe, as Nopcsa showed 60 years ago, supported both conservative Euramerican taxa (*Telmatosaurus, Rhabdodon*, Dromaeosauridae, and *Struthiosaurus*) and Gondwanan elements (the abelisaurid *Tarascosaurus* and the titanosaurids *Ampelosaurus* and *Magyarosaurus*) (Fig. 4). The Asiamerican taxa show many affinities to Albian North American dinosaurs: *Rhabdodon* is close to *Tenontosaurus*, and the dromaeosaurid looks like *Deinonychus* (Fig. 5). One can postulate that their Early Cretaceous ancestors (unknown in Europe) belonged to the same intercontinental populations.

During the very Late Cretaceous, collisions between North America and South America to the west, and India and Asia to the east, led to faunal exchanges well documented in America with the arrival of southern taxa such as the titanosaurid *Alamosaurus*. Due to a poor fossil record, it is not possible to know

FIGURE 4 A Late Cretaceous "Gondwanan" element of the European fauna: the titanosaurid sauropod *Ampelosaurus atacis* (drawing by Guy Le Roux).

FIGURE 5 Rhabdodon priscus, an "Euramerican" Late Cretaceous ornithopod from Europe (drawing by Guy Le Roux).

about the extinctions that probably followed these collisions, a few million years before the final extinction of dinosaurs.

From this review of dinosaur biogeography, one can conclude that it fits rather well with modern paleogeographic reconstitutions (which, admittedly, are also based on paleontological data). Of course, the main source of error in dinosaur biogeography is still linked to the imperfection of the fossil record. As suggested by Russell (1993), "it is conceivable that dinosaurian diversity in Gondwana will one day be found to have surpassed that in Paleolaurasia" (Asiamerica plus Europe); from a paleobiogeographical point of view, it is not only conceivable but predictable. Overall, however, a study of dinosaur paleobiogeography shows the essential influence of geological and tectonic factors in the evolution of continental ecosystems.

See also the following related entries:

DISTRIBUTION AND DIVERSITY • PLATE TECTONICS

References

Bonaparte, J. F., and Kielan-Jaworowska, Z. (1987). Late Cretaceous dinosaur and mammal faunas of Laurasia and Gondwana. In *Fourth Symposium on Mesozoic Terrestrial Ecosystems, Short Papers* (P. J. Currie and E. H. Foster, Eds.), pp. 24–29. Tyrrel Museum of Paleontology, Drumheller, Alberta, Canada.

Buffetaut, E., Méchin, P., and Méchin-Salessy, A. (1988). Un dinosaure théropode d'affinités gondwaniennes dans le Crétacé supérieur de Provence. *Comptes Rendus Acad. Sci. Paris Sér. 2* **306,** 153–158.

Dercourt, J., Ricou, L. E., and Vrielynck, B. (1993). Atlas Tethys palaeoenvironmental maps, 14 maps, 1 plate, pp. 307. Gauthier-Villars, Paris.

Le Loeuff, J. (1991). The Campano–Maastrichtian vertebrate faunas from Southern Europe and their relationships with other faunas in the world: Palaeobiogeographical implications. *Cretaceous Res.* **12,** 93–114.

Lull, R. S. (1910). Dinosaurian distribution. *Am. J. Sci.* **29,** 1–39.

Martin, M. (1981). Les dipneustes mésozoïques malgaches, leurs affinités et leur intérêt paléobiogéographique. *Bull. Soc. Géol. France* **23,** 579–585.

Martin, V. (1995).

Nopcsa, F. (1934). The influence of geological and climatological factors on the distribution of non-marine fossil reptiles and Stegocephalia. *Q. J. Geol. Soc. London* **90,** 76–140.

Rage, J. C. (1981). Les continents péri-atlantiques au Crétacé supérieur: Migration des faunes continentales et problèmes paléobiogéographiques. *Cretaceous Res.* **2,** 65–84.

Russell, D. A. (1993). The role of Central Asia in dinosaurian biogeography. *Can. J. Earth Sci.* **30,** 2002–2012.

Biomechanics

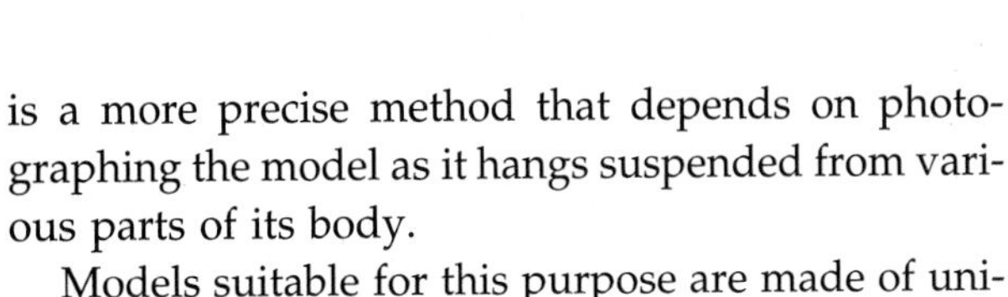

R. McNeill Alexander
University of Leeds
Leeds, United Kingdom

Biomechanics is the application of engineering methods to the study of animals and plants. It deals with the strength and elasticity of skeletons, with mechanisms of movement, and with the energy that is needed. One of the landmarks in the subject's history is a book, *On Growth and Form*, published in 1917 by D'Arcy Thompson, who was professor of natural history at St. Andrew's University, Scotland. In that book he compared the skeleton of a sauropod dinosaur to a nearby bridge that was at the time the most remarkable feat of engineering in his country. He compared the dinosaur's legs to the bridge's piers, the muscles and ligaments of the back to the bridge's tension members, and so on. Since then biomechanics has become immensely more sophisticated and has been applied in far more detail to dinosaurs, which present such obvious problems in engineering.

Some of the problems are discussed in another article, SIZE AND SCALING. It shows that larger animals need relatively thicker bones if they are to be strong enough to be as athletic as small animals. It uses a biomechanical approach to assess the athleticism of large dinosaurs. It concludes that large sauropods were amply strong enough to support their weight on land and could probably have moved much like elephants, which can run at moderate speeds but cannot gallop or jump.

Many biochemical problems require estimates of the mass of the body or of parts of the body. Methods for obtaining these are described in the article titled SIZE AND SCALING. Many problems also require an estimate of the position of the body's center of gravity. This is generally most easily obtained by experiments with scale models of uniform composition—for example, the solid plastic models that can be bought at many museums.

A rough-and-ready way to find a model's center of gravity is to hold it between two pin points, one on each side of the body. If the model balances with the body horizontally the center of mass must be vertically below the line joining the two points. There is a more precise method that depends on photographing the model as it hangs suspended from various parts of its body.

Models suitable for this purpose are made of uniform material throughout, but real dinosaurs consisted partly of muscle and guts, which are slightly denser than water; partly of bone, which is more than twice as dense; and partly of air-filled lungs. The heavy bones are distributed throughout the body and probably do not greatly alter the position of the body's center of gravity, but the light lungs are all in the chest. Because the lungs are placed well forward in the body, and have such low density, the center of gravity of a living dinosaur would be slightly behind the position corresponding to the center of gravity of a solid plastic model. The effect is easily calculated and is small, moving the center of gravity of a 25-m *Diplodocus* only about 20 cm.

Figure 1 shows the positions of the centers of gravity of two sauropods determined in this way. *Diplodocus*, with its very long tail, has its center of gravity far back in the trunk, close to the hip joints. In contrast, *Brachiosaurus*, with a shorter tail, large forelegs, and heavy neck, has its center of gravity much further forward. This tells us that the hindlegs must have supported most of *Diplodocus*' weight but that the weight of *Brachiosaurus* was more evenly shared between fore- and hindlegs. We can be more precise. In Fig. 1, *Diplodocus* has its left hindfoot well forward, directly under the center of gravity, and its right hindfoot well back, about 1.4 m behind the center of gravity. At the mid-point of its step, each foot would be about 0.7 m behind the center of gravity. Each forefoot, at the mid-point of its step, would be about 2.5 m in front of the center of gravity. This tells us that the load supported by the hindfeet must have been 2.5/0.7 times the load supported by the forefeet: The hindlegs supported 78% of body weight and the forelegs only 22%. Similarly, in the case of *Brachiosaurus*, the hindlegs supported 52% and the forelegs 48% of body weight.

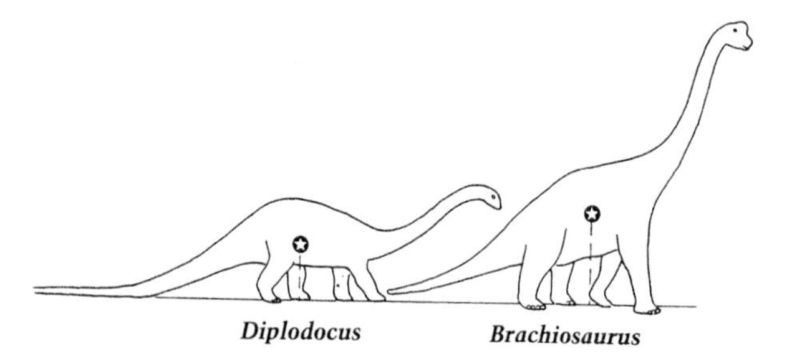

FIGURE 1 The stars show the positions of these dinosaurs' centers of gravity as estimated by experiments with models. From Dynamics of Dinosaurs and Other Extinct Giants by R. McNeill Alexander. Copyright © 1989 by Columbia University Press. Reprinted with permission of the publisher.

It cannot be claimed that these (or any other calculations in dinosaur biomechanics) are very accurate estimates—the results would be modified by changes in the shape and posture of the models—but they seem unlikely to be too inaccurate. They enable us to assess the strengths of leg bones in relation to the loads they have to bear. In particular, they were used in the calculations of "strength indicators" described in the article SIZE AND SCALING.

Bone is a fairly uniform material about equally strong whether it comes from a large animal or from a small one or from a bird or a mammal. (I have no data for the strength of reptile bone.) It seems reasonable to assume that dinosaur bone was about as strong as bone from modern animals.

Calculations of the loads that leg bones can bear must take account of the angles at which they are held. A bone held erect like a pillar will support a much greater load than if it is held horizontally, with a vertical load on its end. Modern lizards stand and run with their feet well out to either side of the body, with the humerus (upper arm bone) and femur (thigh bone) horizontal: The forelimbs are positioned like the arms of a person doing press-ups. We know from fossil footprints, however, that dinosaurs kept their feet directly under the body, and we believe that the large ones walked with their legs nearly straight, like elephants. Simple calculations tell us that the leg bones of, for example, *Apatosaurus* were amply strong enough for walking like this, but that the femur would have broken if it had tried to walk like a lizard.

The necks of sauropods raise another biomechanical question: How were they supported? Figure 2 shows the neck of *Diplodocus* in the horizontal position in which it is usually restored. There are deep notches in the neural spines of the vertebrae that seem likely to have housed a ligament, similar to the large ligament in the necks of cattle. It seems likely that (as in cattle) the weight of the head and neck was supported largely by tension in this ligament. In cattle, the ligament is made mainly of elastin, a rubber-like protein that stretches to allow the animal to lower its head to graze or drink and recoils elastically when the head is raised. Could *Diplodocus* also have had an elastin ligament?

The mass of the head and neck of a *Diplodocus* was estimated as 1.3 tons from measurements of a model's neck. (There is room for doubt here because of uncertainty about the mass of muscle in the neck). Think of the neck as pivoting about the joint in the backbone where the joint reaction force is shown acting in Fig. 2. The weight acts 2.2 m in front of that joint and the ligament is 0.42 m above it; hence, the ligament force is 2.2/0.42 times the neck's weight or 7 tons force (70 kN). The dimensions of the ligament can be estimated from the size of the notches in the vertebrae, and it has been calculated that (if it was elastin) it was just strong enough to withstand this force. At least it was strong enough to take a large part of the load, relieving the neck muscles.

It has been suggested that sauropods such as *Diplodocus* were not limited to feeding like giraffes, keeping all their feet on the ground, but could rear up on their hindlegs to reach leaves on high branches. Is this biomechanically feasible? The strength of the hindlegs presents no problem: The bones were amply strong enough to carry the whole weight of the body and in any case (as we have seen) they carried most of the body's weight even in four-footed standing. However, could the animals have reared up? To do that, they must get their hindlegs directly under the

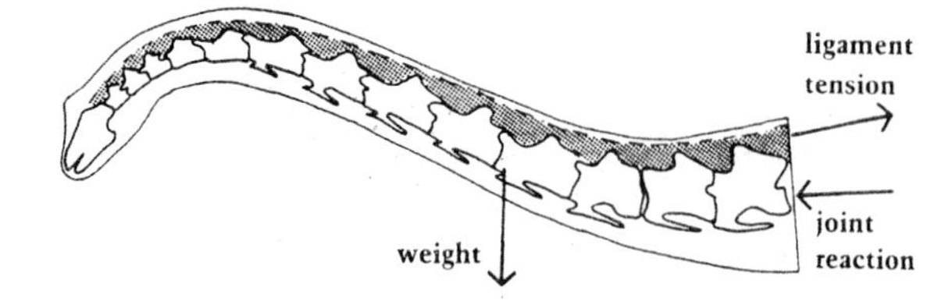

FIGURE 2 A diagram of the neck of *Diplodocus* showing how it may have been supported by tension in a large ligament. From Dynamics of Dinosaurs and Other Extinct Giants by R. McNeill Alexander. Copyright © 1989 by Columbia University Press. Reprinted with permission of the publisher.

center of gravity so that no load would remain on the forelegs. That, too, presented no difficulty. In Fig. 1, the left hindfoot is already under the center of gravity. If the right one were brought alongside it, the animal would be able to rear up.

That does not, of course, prove that *Diplodocus* did rear up, and in any case there is another biomechanical problem to be considered: Is the heart likely to have been strong enough to keep blood flowing to the brain if the head were raised 6 m above the heart (as it might be when the dinosaur reared up). The blood pressures of sauropod dinosaurs are discussed in the article titled PHYSIOLOGY.

There are a wide variety of other problems concerning dinosaurs that have, or could be, tackled by biomechanical methods. However, we will finish with just one more example.

The thick skull roofs of dome-headed dinosaurs have been interpreted as adaptations for fighting by head-butting, similar to the contests between bighorn rams. The interpretation is controversial but let us consider its implications. Imagine two 20-kg males colliding at 3 m per second (a jogging speed for a human). Each would have 90 J kinetic energy ($1/2 \times 20 \times 32$). The force needed to bring the males to a halt would be (kinetic energy)/(stopping distance). If both kept their necks rigid, they would be halted in a few centimeters, the force would be enormous, and they would surely break their necks. If, on the other hand, they allowed their necks to bend in the impact, using their neck muscles to absorb the energy, the stopping distance would be much longer, perhaps 0.3 m. If it were that, the force required would be only $90/0.3 = 300$ N, or 30 kg force, which could probably be tolerated. As often happens in biomechanics, a very simple calculation can be illuminating.

See also the following related entries:

BIPEDALITY • CONSTRUCTIONAL MORPHOLOGY • FUNCTIONAL MORPHOLOGY • SIZE AND SCALING

References

Alexander, R. McN. (1989). *Dynamics of Dinosaurs and Other Extinct Giants.* Columbia Univ. Press, New York.

McGowan, C. (1994). *Diatoms to Dinosaurs. The Size and Scale of Living Things.* Island Press, Washington, DC.

Biometrics

RALPH E. CHAPMAN
National Museum of Natural History
Smithsonian Institution
Washington, DC, USA

DAVID B. WEISHAMPEL
Johns Hopkins University School of Medicine
Baltimore, Maryland, USA

Biometrics or biometry is the measurement of life. In their classic textbook, *Biometry: The Principles and Practice of Statistics in Biological Research*, Sokal and Rohlf (1995) define biometry as the application of statistical methods to the solution of biological problems. *Webster's Ninth New Collegiate Dictionary* similarly defines biometry as the statistical analysis of biological observations or phenomena. We would define biometry more broadly to include other forms of measurement and quantification from simple counting to advanced areas within subjects such as geometry, mathematics, and computer science. As such, we would modify Sokal and Rohlf's (1995) definition as follows:

> Biometry is the application of statistical, geometrical, mathematical, and/or algorithmic approaches to the solution of biological problems.

As thus defined, biometry is not only at the heart of all theoretical studies of dinosaurs but also an important contributor to more descriptive studies, and its importance increases with time. Classically, biometry is associated with studies of shape and size within morphometric and allometric contexts. Following the definition given previously, however, there are also a number of other research areas in which biometrical methods not only are used but also are major factors in the research. These include phylogenetic analysis, distributional analyses, and functional morphology.

Morphometrics and Allometry

Morphometrics is the quantitative measurement of shape and includes a broad range of techniques from the simple measurement of bone or footprint dimensions to the application of very sophisticated statistical and geometrical methods for comparing the shapes of these structures (see Chapman, 1990, for a detailed discussion). Allometry is the study of size and its consequences and is usually carried out using bivariate (two-variable) and multivariate (many variable) morphometric analyses comparing the relative rate of growth of biological structures.

Through time, morphometric analyses have progressed from relatively simple methods to the very complex approaches used today. Early studies consisted of taking simple measurements and presenting tables of them. This progressed to presenting comparative tables among specimens and taxa and noting differences. Two-dimensional analyses, in which two different sets of measurements are used for the same specimens, were introduced early on as well, mostly as unquantified observations (e.g., noting that one vertebrae was relatively broader than another) that later were quantified as ratios of these variables. Differences in ratios typically were used as evidence for differences among specimens and as evidence for the erection of new taxa. Morphometric analyses became more sophisticated as these analyses of two variables progressed to bivariate allometric analyses in which the investigator was able to note relative growth rates and correlations between variables.

From two-dimensional analyses the next step was to multivariate analyses, those utilizing many variables simultaneously (e.g., principal components analysis), and to early techniques of shape analysis such as D'Arcy Thompson's (1942) transformation grids. These two approaches then converged toward techniques of modern shape analysis (Chapman, 1990).

The morphometric study of dinosaurs followed the same progression as the general field of morphometrics, although typically with some lag time. This progression can be seen in the analyses of many different dinosaur groups, such as pachycephalosaurs and hadrosaurs, and even in footprints. Studies of saurischians are less common or typically less advanced due to the relatively small number of specimens available for all but a few taxa. As an example, here we will concentrate on studies done on the ceratopsians.

Morphometric analyses of dinosaur material began early. These began as very simple analyses amounting to measuring and comparing the simple dimensions of bones (e.g., vertebrae). In one of the first papers on dinosaurs, even predating the term dinosaur, Mantell (1833) measured the dimensions of bones of the fossil reptiles found from southeast England, including various marine reptiles and dinosaur material referred to *Megalosaurus*, *Iguanodon*, and the ankylosaur *Hylaeosaurus*. He presented these dimensions and even made note of differences among the different specimens and in comparison with modern analogs such as crocodiles and iguanas. One of his most interesting uses of this data was an attempt to estimate the size of *Iguanodon* by noting the ratio of the size of its various skeletal parts to those of *Iguana* and using their average to estimate the total size of the dinosaur.

The level of sophistication of the analysis of dinosaurs increased at a slow pace during the 19th century and into the first part of this century, and then increased rapidly. In early studies of ceratopsians, typified by Marsh (1890), most of the biometrical data were presented as selected measurements (e.g., skull length) and observations that some proportions supposedly separated new taxa from others closely related to it. For example, Marsh (1890, p. 82), in his description of *Triceratops prorsus*, suggests it is different from *T. horridus* in various ways including a more compressed rostral bone (a ratio of length to width), a less broad parietal crest (another ratio), and a larger frontal horn core (a volume estimate compared with an implied overall size for the specimens). These differences were noted but, as was typical for the time, no real data were presented. Hatcher *et. al*, (1907) went a little further by providing tables of measurements of specimens of the same taxon and, later still, Lull (1933) provided more comparative data, including comparisons of limb proportions between fore- and hindlimbs of a single taxon.

Brown and Schlaikjer (1940) increased the level of sophistication slightly with their detailed analysis of the growth of *Protoceratops*. They provided tables of measurements as well as long lists of differences in shapes between juveniles and adults. They even published an interesting summary diagram showing the proportionate changes that take place in the skull of *Protoceratops* from juvenile to adult. It is surprising, however, that they never attempted any bivariate (two-variable) allometric analyses, which were very popular at that time because of the publication of Sir Julian Huxley's (1932) *Problems of Relative Growth.* This was finally done for known ceratopsian dinosaurs in a pioneering paper by Gray (1946) who looked at growth trajectories, within both phylogenetic and ontogenetic contexts, and noted that they seemed to support Lull's phylogenetic analysis of the Ceratopsia. Later, Lull and Gray (1949) used Thompson's (1942) grids to analyze shape differences among ceratopsians. In this method, a rectangular grid is superimposed on a starting specimen and the shape change is shown by distorting the grid to show the second specimen.

The two papers by Gray (1946) and Lull and Gray (1949) were the most sophisticated morphometric analyses ever done on dinosaurs until the work by Dodson on the growth and taxonomic structure of lambeosaurine hadrosaurs (Dodson, 1975) and, again, *Protoceratops* (Dodson, 1976). In these studies, Dodson used detailed bivariate allometric analyses in combination with multivariate procedures, those that analyze many variables at the same time, to analyze the growth patterns and shape variations in these groups and note the implications of them for the analysis of taxonomic structure and sexual dimorphism. This work was developed further by Dodson (1990a), who combined allometric and shape analysis with character analysis within a phylogenetic context to explore the relationships between *Monoclonius* and *Centrosaurus*. At the same time Lehman (1990) used detailed allometric analyses combined with some basic shape analysis to study systematics and sexual dimorphism within the chasmosaurine ceratopsians. Finally, Dodson (1993) has applied a high-level morphometric approach, RFTRA (see Chapman, 1990, for a detailed discussion), to analyze changes in the shapes of ceratopsian heads in great detail by noting changes in the position of anatomical landmarks on the skulls. Additional studies of ceratopsians using other advanced shape analysis methods are also in development (C. A. Forster, 1994, personal communication).

Biometry is also an important part of many other types of analyses; space limitations allow us to mention only three here.

In phylogenetic analyses various techniques are used to reconstruct the relationships among various organisms. Biometrical procedures are used to document many of the characters used in reconstructing these phylogenies. One dinosaurian example among the many possible is Russell and Zheng's (1993) study of sauropodomorphs. In their phylogenetic analysis they use counts, ratios, and overall dimensions to define their character states. These include the number of vertebrae of certain types, the relative lengths of different parts of the vertebral column, the shapes of teeth defined by ratios, and many other biometrically defined characters. These data are assembled and run through various algorithmic procedures using computer programs such as PAUP (Swofford and Begle, 1993) to provide the resulting phylogenetic trees showing these relationships.

Distributional analyses attempt to analyze the distribution of organisms through space and time. Doing this incorporates biometrical procedures ranging from simple counting to detailed multivariate analyses. Studies of dinosaurs through time and space are rather rare because of the small number of specimens, but initial studies have been published (see Dodson, 1990b; Dodson and Dawson, 1991; Weishampel, 1990; Weishampel and Norman, 1989) documenting the number of species found and documenting species and higher level taxonomic diversity, as well as literature citations, through time. More sophisticated analyses are currently in progress (R. Chapman and D. Weishampel, manuscript in preparation).

Studies of functional morphology attempt to determine the function of structures given their form and context within the organism. Here again, biometrical procedures are very important and many important studies have been published and are in development. For example, in Weishampel's (1981) acoustical analysis of the nasal system in lambeosaurines, analyzing and reconstructing the systems depends strongly on detailed measurements of the heads and nasal systems of the lambeosaurines. Similarly, Alexander's (1989) detailed analyses of the movements of large dinosaurs utilize limb proportions, estimates of dinosaur mass and bone strength, and mechanical equations to assess the abilities of different taxa to move around at various speeds, especially with an eye toward determining maximum running speeds.

See also the following related entries:

BIOMECHANICS • COMPUTERS AND RELATED TECHNOLOGY

References

Alexander, R. McN. (1989). *Dynamics of Dinosaurs and Other Extinct Giants,* pp. 167. Columbia Univ. Press, New York.

Brown, B., and Schlaikjer, E. M. (1940). The structure and relationships of *Protoceratops. Ann. N. Y. Acad. Sci.* **40**(3), 133–266.

Chapman, R. E. (1990). Shape analysis in the study of dinosaur morphology. In *Dinosaur Systematics—Perspectives and Approaches* (K. Carpenter and P. J. Currie, Eds.), pp. 21–42. Cambridge Univ. Press, New York.

Dodson, P. (1975). Taxonomic implications of relative growth in lambeosaurine hadrosaurs. *Systematic Zool.* **24,** 37–54.

Dodson, P. (1976). Quantitative aspects of relative growth and sexual dimorphism in *Protoceratops. J. Paleontol.* **50,** 929–940.

Dodson, P. (1990a). On the status of the ceratopsids *Monoclonius* and *Centrosaurus.* In *Dinosaur Systematics—Perspectives and Approaches* (K. Carpenter and P. J. Currie, Eds.), 231–244. Cambridge Univ. Press, New York.

Dodson, P. (1990b). Counting dinosaurs: How many kinds were there? *Proc. Natl. Acad. Sci. USA* **87,** 7608–7612.

Dodson, P. (1993). Comparative craniology of the Ceratopsia. *Am. J. Sci.* **293A,** 200–234.

Dodson, P., and Dawson, S. D. (1991). Making the fossil record of dinosaurs. *Mod. Geol.* **16**(1/2), 3–15.

Gray, S. W. (1946). Relative growth in a phylogenetic series and in an ontogenetic series of one of its members. *Am. J. Sci.* **244**(11), 792–807.

Hatcher, J. B., Marsh, O. C., and Lull, R. S. (1907). *The Ceratopsia,* Monograph No. 49, pp. 300. U. S. Geological Survey, Washington, DC.

Huxley, J. (1932). *Problems of Relative Growth,* pp. 276. Methuen, London.

Lehman, T. M. (1990). The ceratopsian subfamily Chasmosaurinae: Sexual dimorphism and systematics. In *Dinosaur Systematics—Perspectives and Approaches* (K. Carpenter and P. J. Currie, Eds.), pp. 211–230. Cambridge Univ. Press, New York.

Lull, R. S. (1933). A revision of the ceratopsia or horned dinosaurs. *Mem. Peabody Museum Nat. History* **3**(3), 1–175.

Lull, R. S., and Gray, S. W. (1949). Growth patterns in the Ceratopsia. *Am. J. Sci.* **247**(7), 492–503.

Mantell, G. (1833). *The Geology of the South-East of England,* pp. 415. Longman, London.

Marsh, O. C. (1890). Description of new dinosaurian reptiles. *Am. J. Sci. Third Ser.* **39**(229), 81–86.

Russell, D. A., and Zheng, Z. (1993). A large mamenchisaurid from the Junggar Basin, Xinjiang, People's Republic of China. *Can. J. Earth Sci.* **30**(10/11), 2082–2095.

Sokal, R. R., and Rohlf, F. J. (1995). *Biometry: The Principles and Practice of Statistics in Biological Research, Third Edition.* Freeman, San Francisco.

Swofford, D. L., and Begle, D. P. (1993, March). *User's Manual for PAUP: Phylogenetic Analysis Using Parsimony,* Version 3.1. Laboratory of Molecular Systematics, Smithsonian Institution, Washington, DC.

Thompson, D'A. W. (1942). *On Growth and Form: Revised and Enlarged Edition,* pp. 1116. Cambridge Univ. Press, New York.

Webster's Ninth New Collegiate Dictionary. Merriam-Webster, Springfield, MA.

Weishampel, D. B. (1981). Acoustic analysis of potential vocalization in lambeosaurine dinosaurs (Reptilia: Ornithischia). *Paleobiology* **7,** 252–261.

Weishampel, D. B. (1990). Dinosaur distribution. In *The Dinosauria* (D. B. Weishampel, P. Dodson, and H. Osmolska, Eds.), pp. 63–139. Univ. California Press, Berkeley.

Weishampel, D. B., and Norman, D. B. (1989). Vertebrate herbivory in the Mesozoic; Jaws, plants, and evolutionary metrics [Special paper]. *Geological Soc. Am.* **238,** 87–100.

Biomineralization

Clive Trueman
University of Bristol
Bristol, United Kingdom

Biomineralization is the process by which organisms produce mineral or inorganic tissues. Bone is the most common mineralized tissue in vertebrates. It is used as a structural support and as a reservoir for physiologically important ions such as calcium. It must therefore be both strong and soluble in body fluids.

Growth and Mineralization of Bone

The general sequence of bone mineralization is synthesis and extracellular assembly of the organic matrix framework, followed by mineralization (nucleation and growth of bone mineral) within the framework (Lowenstam and Weiner, 1989).

On a macroscopic level, new bone forms either on a scaffolding of preexisting cartilage (endochondral ossification) or directly onto outer periosteal bone surfaces (membranous ossification).

Bone Mineral

The inorganic or mineral component of bone is usually considered to be some form of hydroxyapatite (HAP), although most sites within the apatite lattice may be subject to substitutions; therefore, the formula may be better represented as

$$(Ca,Sr,Mg,Na,H_2O,REE,[\ ])_2\,(PO_4,HPO_4,CO_3,P_2O_7)_6\,(OH,F,CO_3,Cl,H_2O,O,[\ ])_2,$$

where ([ ]) represents unfilled lattice positions.

Bone is formed of extremely small crystallites, giving bone a very high surface area. A large amount (3–5%) of structural carbonate substituting for phosphate is also found in bone. The carbonate substitution causes lattice defects and increases the reactivity of bone. These two features of bone apatite ensure that it has a relatively high solubility and can be resorbed fairly easily. This is essential to bone growth and repair; in addition, the vertebrate skeleton serves as an important store for calcium and many trace elements.

Bone–Collagen Relationships

There is a well-documented intimate relationship between bone apatite crystallites and collagen fibers. The apatite crystallites are oriented with their long (c) axes parallel to the collagen fibers. This orientation can be seen in polarized light, XRD, or TEM studies, and has been seen in dinosaur bone fragments from *Seismosaurus* (Gillette, 1994) and *Allosaurus* (Zocco and Schwartz, 1994).

Bone collagen is a type 1 collagen, and the fibers are linked to form a triple-helix structure. This crosslinking occurs in a stepped fashion between the molecules, and space considerations mean that a gap of 35 nm exists between the head and tail of individual collagen molecules. This forms a regular spacing of holes within collagen fibers. The intimate linking of bone apatite and collagen, the small size of bone apatite crystallites, and the presence of gaps in the collagen structure in which bone crystals may occur lead to many suggestions that primary bone nucleation occurs within the collagen molecule. Direct evidence of this has not been found, however, and there is still much controversy surrounding the nucleation of apatite crystallites in bone.

Nucleation of HAP

When hydroxyapatite is synthesized from supersaturated solutions, an amorphous precursor phase is found with lower Ca:P ratio. This spontaneously converts slowly to hexagonal hydroxyapatite. This has led to the theory that an amorphous calcium phosphate may form as a precursor to true bone mineral. However, other precursor phases have been suggested, such as brushite ($CaHPO_4{\cdot}2H_2O$), octacalcium phosphate [$Ca_8H_2(PO_4)_6{\cdot}5H_2O$], and also direct precipitation of hydroxyapatite (Posner, 1985). The identification of the precursor phase has great implications for the conditions of initial biomineralization of bone because each phase suggests different conditions of biomineralization, particularly in terms of pH and saturation of Ca and PO_4 ions.

The processes leading to nucleation of HAP are poorly understood, with many unresolved questions. A number of problems can be highlighted (e.g., Simkiss and Wilbur, 1989);

1. To create a sound structural unit, the bulk of mineralization must occur extracellularly, but must be under cellular control.
2. Extracellular fluid levels of Ca^{2+} and P must be kept only slightly supersaturated with respect to precursor phases to avoid unintentional ectopic calcification.
3. Even though Ca and P levels must be slightly supersaturated, some method of overcoming the nucleation barrier (i.e., activation energy) must be found to initiate ectopic calcification.
4. Nucleation must be confined to specific sites and therefore some inhibitory mechanism must be present on the extracellular fluid.
5. Mineralization must be at least as rapid as growth to maintain structural integrity. Stability of apatite appears to grow at the expense of speed—enamel forms much more slowly than bone—and produces very large, well-ordered stable crystals.

Inhibitor Molecules

Many body fluids are supersaturated with respect to Ca and P but do not precipitate any form of calcium phosphate. This has lead to the suggestion that certain inhibitory molecules may be present in body fluids, which may be selectively removed at sites of bone formation, thus allowing a site-specific mechanism of precipitating calcium phosphates. The first of these inhibitor molecules to be recognized was pyrophosphate (Fleisch and Neuman, 1961). This molecule is found in plasma, urine, and saliva and is a successful inhibitor of most forms of calcification. In addition, it may be hydrolyzed by a number of enzymes to orthophosphate, thus providing a biological "on–off switch" for calcium phosphate precipitation. Since this discovery was made, several alternative inhibitors have been suggested, many of which may work in conjunction or separately under differing conditions.

Overcoming the Energy Barrier

Slightly supersaturated solutions may precipitate if certain materials are present. The ion concentration of the solution will determine the sensitivity of the precipitation to the exact catalyst. As the solution becomes more supersaturated, more substances may act as catalysts. Assuming condition 2 (see previous list) to be true, then condition 3 may be met by a variety of mediating substances found in mineralizing bone.

Initially, collagen was inferred to mediate the nucleation process, principally due to its presence in abundance in bone, the intimate association of apatite and collagen, and the apparent association of newly formed apatite crystallites with collagen molecule periodicity. Regular spacing of bone crystallites within mineralizing collagen fibers has been found, and it is postulated that hydroxyapatite is nucleated at a specific site within this gap.

Initial positive studies were later found to be misleading because the association between primary apatite crystallites and collagen molecule periodicity was not upheld. Many theories of bone apatite mineralization now suggest that initial primary nucleation of apatite crystallites is not directly associated with the collagen molecule, but subsequent growth and organization of the apatite crystallites is controlled or at least influenced by the regular spacing in the collagen molecule.

Most other noncollagenous proteins found in bone, such as osteonectin and osteocalcin, have also been implicated as nucleation triggers.

Matrix Vesicles

A second potential mechanism of apatite nucleation involves small extracellular microstructures known as matrix vesicles. These were initially found within mineralizing growth cartilage but subsequently discovered in primary mineralizing dentine and bone. These vesicles contain high concentrations of enzymes as well as high Ca and P concentrations. It has been inferred that the matrix vesicles could act to raise the ion concentrations to the point where direct precipitation of HAP or a precursor phase occurs or could release enzymes that act to destroy inhibitor molecules, and hence allow precipitation of apatite. This mechanism is attractive because it would allow precipitation of bone mineral extracellularly but within cellular control.

Summary

- Bone mineral is some form of HAP. There is still debate as to whether there is a precursor phase and, if so, what that phase is.
- Bone matrix is composed largely of collagen fibers, and it is likely that after initial nucleation the collagen spaces play an important role in biomineralization, forming sites for collagen–protein–apatite bonds.
- A number of substances may act as inhibitors to nucleation and must be removed from the sites of biomineralization.
- The major roles of bone cells are known, but cellular influences on formation of extracellular material are still poorly understood.

Paleoecological Information from Conditions of Biomineralization

The physical and chemical environment in which bone mineral forms can be recorded by a number of chemical signatures. For instance, the temperature at which bone mineral formed may be recorded by the ratio of oxygen isotopes contained within the phosphate group in HAP. If fossil bone phosphate can be shown to be unaffected by diagnoses, then information concerning the body temperature of dinosaurs may be retrieved. This has been attempted by Barrick and Showers (1995). Comparing the temperatures recorded from oxygen isotope signatures of internal bones and bones from extremities, they produced an approximation of the homogeneity of the body temperature of different dinosaurs and compared this to that recorded by a varanid lizard.

The trace element signature of fossil bone may also record paleobiological signals. Differing diets (e.g., shellfish vs meat vs plant) have different trace element contents, and the levels of these elements within bones may reflect the diet of that animal. It is more difficult to retrieve dietary signals from fossil bones because most of the diagnostic trace elements reside in the Ca site of bone apatite, and it is this site that is most susceptible to diagenetic ion exchange.

Eggshell Formation

The biomineralization of eggshell has been well documented in a series of papers by K. Simkiss. This summary is principally taken from that in Simkiss and Wilbur (1989).

In the oviduct of reptiles and birds, the epithelial layer is used as the basis for biomineralization. In birds, the oocyte passes down the oviduct and is held at the end for 15-30 hr (depending on the species) while calcite is deposited on the egg membrane. In birds, eggshell formation is extremely rapid. In domestic fowl, 5 g of calcite is deposited in 20 hr (Simkiss and Wilbur, 1989). This requires a huge reserve of Ca, which is held in bone.

$CaCO_3$ production [($Ca^{2+} + CO_2 + H_2O - CaCO_3 + H_2^+$)] produces protons that can induce an acidosis in the bird. The urine of laying hens acidifies during eggshell formation, and respiratory activity is often increased by panting.

Within the eggshell itself, calcite crystals form from a number of discrete nucleation sites and grow outward from the membrane in a progressively more uniform direction because lateral growth is inhibited by abutting with neighboring crystals. The directional growth of eggshell can be studied by XRD (Silyn-Roberts and Sharp, 1986) and different growth patterns determined between turtles and ostrich related to the distribution and nature of organics within the eggshell. This is in turn related to the strength of the eggshell.

See also the following related entries:

Chemical Composition of Dinosaur Fossils • Permineralization

References

Particular attention is drawn to the excellent reviews by Simkiss and Wilbur, and Lowenstam and Weiner, from which much of this summary was drawn.

Barrick, R. E., and Showers, W. J. (1995). Oxygen isotope variability in juvenile dinosaurs (*Hypacrosaurus*): Evidence for thermoregulation. *Paleobiology* **21**(4), 552–560.

Fleisch, H., and Neuman, W. F. (1961). Mechanisms of classification: Role of collagen, polyphosphates and phosphatase. *Am. J. Physiol.* **200,** 1296–1300.

Gillette, D. G. (1994). *Seismosaurus—The Earth Shaker*. Columbia Univ. Press, New York.

Lowenstam, H. A., and Weiner, S. (1989). *On Biomineralization*. Oxford Univ. Press, New York.

Posner, A. S. (1985). The mineral of bone. *Clin. Orthopaedics* **200,** 87–99.

Silyn-Roberts, H., and Sharp, R. M. (1986). Crystal growth and the role of the organic network in eggshells. *Proc. R. Soc. London Ser. B* **227,** 303–324.

Simkiss, K., and Wilbur, K. M. (1989). *Biomineralization*. Academic Press, San Diego.

Zocco, T. G., and Schwartz, H. L. (1994). Microstructural analysis of bone of the sauropod dinosaur *Seismosaurus* by transmission electron microscopy. *Palaeontology* **37**(3), 493–503.

Biostratigraphy

Spencer G. Lucas
New Mexico Museum of Natural History and Science
Albuquerque, New Mexico, USA

Biostratigraphy identifies and distinguishes strata (layers of sedimentary rock) by their fossil content (Salvador, 1994). Strata with distinctive fossils are the units of biostratigraphy. The most basic unit is the biostratigraphic zone, called biozone, or zone for short. Extensive collecting of fossils over the vertical and lateral extent of a stratum, or strata, allows biozones to be defined.

Historical Development

Biostratigraphy was born in the early 1800s in western Europe. William Smith, a British civil engineer, realized that a given stratum usually contains distinctive fossils. Smith's recognition that many of these distinctive fossils could be traced over large areas, even when the nature of the enclosing rocks changed, allowed him to publish the first geological map of England.

Parallel to Smith's work, French comparative anatomist Georges Cuvier and geologist Alexander Brongniart independently discovered that distinctive kinds of fossils are often associated with specific strata. Whereas Smith did not attach geologic time significance to his observations, Cuvier and Brongniart, and other French paleontologists of the early 18th century (especially A. d'Orbigny), did. They argued that each stratum with its distinctive fossils represents a particular "stage" in the history of life. This conclusion forms the basis for biochronology.

Biostratigraphy vs Biochronology

Biozones, by themselves, have no necessary time significance. They are simply bodies of rock characterized by their fossil content. As such, zones can be mapped, and their thicknesses can be measured. The time equivalent to a biozone is a biochron. Biochronology is thus the use of fossils to delineate intervals of geological time. Many American geologists and paleontologists do not distinguish biostratigraphy from biochronology; they use the term biostratigraphy to encompass the delineation both of rock units and of intervals of geologic time by fossils. The distinction between biostratigraphy and biochronology, however, has long been made by Canadian and many European scientists. They do so because it is conceptually useful to distinguish the biozone, which may be of varying ages over its lateral extent, from the biochron, which theoretically is of the same time value everywhere (Fig. 1).

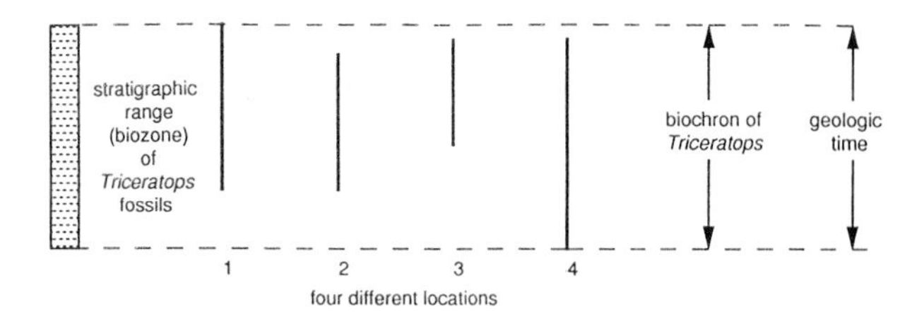

FIGURE 1 A biostratigraphic zone may represent different time intervals at four different locations, but the biochron based on the zone is of one duration everywhere.

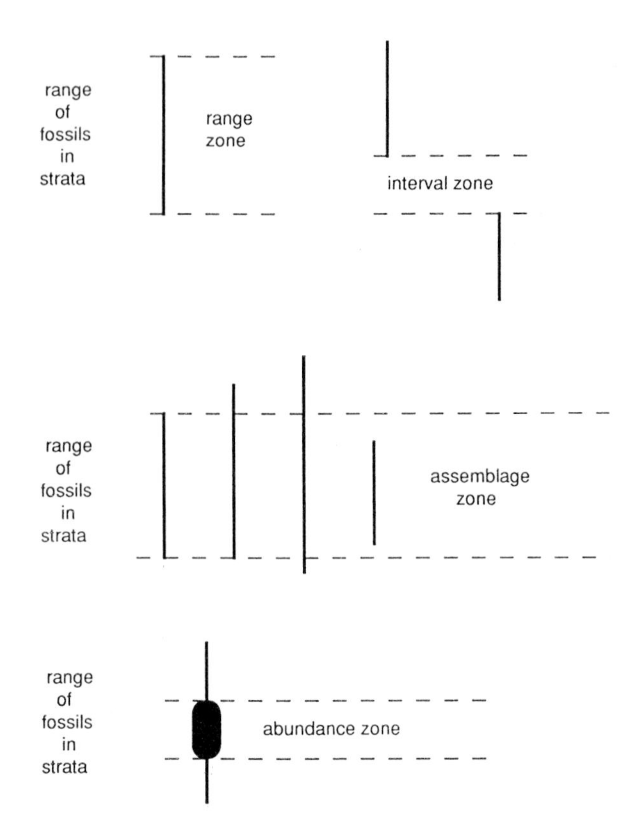

FIGURE 2 Biostratigraphers recognize four principal kinds of zones: range zones, interval zones, assemblage zones, and abundance zones.

Different Kinds of Biozones

The four most commonly identified kinds of biozones are range, interval, assemblage, and abundance (acme) zones (Fig. 2). Recognition of one or more kinds of zones depends largely on the nature of the fossil record being studied and the purpose of the investigation.

Range zones are strata that encompass the range (vertical distribution) of a particular kind of fossil. For example, in the western United States, the stratigraphic range of fossils of the Late Cretaceous dinosaur *Triceratops* defines a *Triceratops* range zone.

Interval zones are the strata between two biostratigraphically significant horizons. Often they are the intervals between the last record of one kind of fossil and the first record of another kind. For example, the last record of the thyreophoran dinosaur *Scelidosaurus* well predates the first record of *Stegosaurus*, and an interval zone could be defined between these two records.

Assemblage zones are strata with a characteristic assemblage (association) of fossils. Many kinds of fossils define an assemblage zone. For example, the Late Cretaceous dinosaurian and other vertebrate fossils from the Judith River Formation of Montana could be considered to define an assemblage zone. Indeed, this assemblage zone has time significance as a biochronological unit, the Judithian land-vertebrate "age."

Abundance zones are strata recognized by the abundance (acme) of a kind (or kinds) of fossil, regardless of range or association. In western North America, hadrosaurs reached their maximum abundance between about 75 and 68 million years ago. This abundance of hadrosaur fossils could thus define a hadrosaur abundance zone.

Correlation

The principal goal of biostratigraphy and biochronology is to correlate rock strata and the physical and biological events that the strata record. To correlate is to establish the equivalence of age or stratigraphic position of strata in separate areas. Broadly speaking, to correlate is to establish the contemporaneity of events in the geological histories of separate regions. Understanding geological and biological history is central to all geological and paleontological investigations, and stratigraphic correlation is critical to this understanding.

The Geological Timescale

The Phanerozoic (last 570 million years) geological timescale is based largely on biostratigraphic data and biochronological correlations (Berry, 1987). This is because the intervals of the timescale are rooted in biozones that form the basis of the biochrons used in global correlation.

Biostratigraphic events thus are critical to the geological timescale. For example, the Mesozoic Era, during which the dinosaurs lived, was originally defined by two biostratigraphic events—the global mass extinctions at the end of the Paleozoic and at the end of the Mesozoic. Today, the boundaries of the Mesozoic are generally recognized by smaller scale biostratigraphic events. Thus, the beginning of the Mesozoic corresponds to the first record of the ammonoid (extinct cephalopod) *Otoceras*, a biostratigraphic event.

Index Fossils and Facies Fossils

Fossils are fundamental to biostratigraphy and biochronology; therefore, they are essential to most stratigraphic correlation. Fossils valuable to correlation are called index fossils because they identify and determine the age of the strata in which they are found. A good index fossil has a short stratigraphic range, is geographically widespread, and is easy to identify.

In contrast, fossils of animals and plants that were very sensitive to environmental conditions are termed facies fossils (facies refers to a specific kind of environment). Facies fossils are usually specific to a certain kind of rock that represents a particular ancient environment. Therefore, facies fossils usually have much longer stratigraphic ranges and more restricted geographic ranges than do index fossils.

In reality, the dichotomy between index and facies fossils is somewhat artificial, and both terms should be seen as endpoints of a spectrum. All organisms of the past had a definite stratigraphic range; those with the shortest ranges make the best index fossils. Also, all organisms of the past lived in particular environments; those most restricted environmentally produce the best facies fossils.

Dinosaurs and Biostratigraphy

Dinosaur fossils are often distributed very unevenly in strata. They rarely are extremely numerous and dense through a stratigraphic interval. For this reason, dinosaurs have been little used by biostratigraphers until recently.

Extensive collecting and a desire to apply the dinosaur fossil record to solving new problems has led to new efforts in the areas of dinosaur biostratigraphy and biochronology. Detailed documentation of dinosaurian range zones has been undertaken to examine problems of dinosaur microevolution (Horner *et al.*, 1992) and extinction (Sullivan, 1987). Dinosaur-dominated assemblage zones have formed the basis for new biochronological units (Jerzykiewicz and Russell, 1991). Extensive regional and intercontinental correlations based on dinosaurs are being advocated (Lucas, 1993). Dinosaur footprints have also proven biochronological significance (Lockley, 1991).

As Lucas (1991) argued, many kinds of dinosaurs make excellent index fossils. Further biostratigraphic organization of the dinosaurian fossil record is needed so they can assume their rightful place as important fossils for Mesozoioc biochronology.

See also the following related entries:

PALEOMAGNETIC CORRELATION • RADIOMETRIC DATING

References

Berry, W. B. N. (1987). *Growth of a Prehistoric Time Scale* (rev. ed.), pp. 202. Blackwell, Palo Alto, CA.

Horner, J. R., Varrichio, D. J., and Goodwin, M. B. (1992). Marine transgressions and the evolution of Cretaceous dinosaurs. *Nature* **358,** 59–61.

Jerzykiewicz, T., and Russell, D. A. (1991). Late Mesozoic stratigraphy and vertebrates of the Gobi basin. *Cretaceous Res.* **12,** 345–377.

Lockley, M. (1991). *Tracking Dinosaurs: A New Look at an Ancient World*, pp. 238. Cambridge Univ. Press, Cambridge, UK.

Lucas, S. G. (1991). Dinosaurs and Mesozoic biochronology. *Mod. Geol.* **16,** 127–138.

Lucas, S. G. (1993). Vertebrate biochronology of the Jurassic–Cretaceous boundary, North American Western Interior. *Mod. Geol.* **18,** 371–390.

Salvador, A. (Ed.) (1994). *International Stratigraphic Guide*, 2nd ed., pp. 214. International Union of Geological Sciences and Geological Society of America, Boulder, CO.

Sullivan, R. M. (1987). A reassessment of reptilian diversity across the Cretaceous–Tertiary boundary. *Nat. History Museum Los Angeles County Contrib. Sci.* **391,** 1–26.

Bipedality

KEVIN PADIAN
University of California
Berkeley, California, USA

Bipedality, or the habit of walking on two legs, was a characteristic of basal dinosaurs, which were small (up to 2 m in length) and lightly built (see ORNITHODIRA; PTEROSAURIA; DINOSAUROMORPHA). Their hindlimbs were considerably longer than their forelimbs, so much so that it is doubtful that the forelimbs in most early dinosaurs could have been used for walking. Their femora (thigh bones) were shorter than their tibiae (shin bones), though this eventually reversed when large forms evolved in several lineages. Their metatarsals (sole bones) are longer than those in typical reptiles and nearly equal in length, with the middle toe the longest (Fig. 1). These features are considered typical of bipedal animals and are clearly seen in birds (Coombs, 1978). It is generally accepted that dinosaurs were digitigrade; that is, they walked on their toes, holding their sole bones off the ground almost without exception, even in the largest sauropods (Alexander, 1985).

The bones of the dinosaurian ankle are also modified (see HINDLIMBS AND FEET). Typically in tetrapods, especially amniotes, the ankle consists of two rows of tarsal (ankle) bones: proximal and distal (Fig. 2). In dinosaurs, the proximal tarsals (the astragalus, which is distal to the tibia, and the calcaneum, which is distal to the fibula) do not rotate against each other or against the tibia and fibula during locomotion. In adult forms these bones are often fused to each other and to the leg bones. The distal tarsals, meanwhile, have a similar functional relationship to each other and to the metatarsals: They tend to cap, as opposed to rotate against, the sole bones. The ankle joint flexes

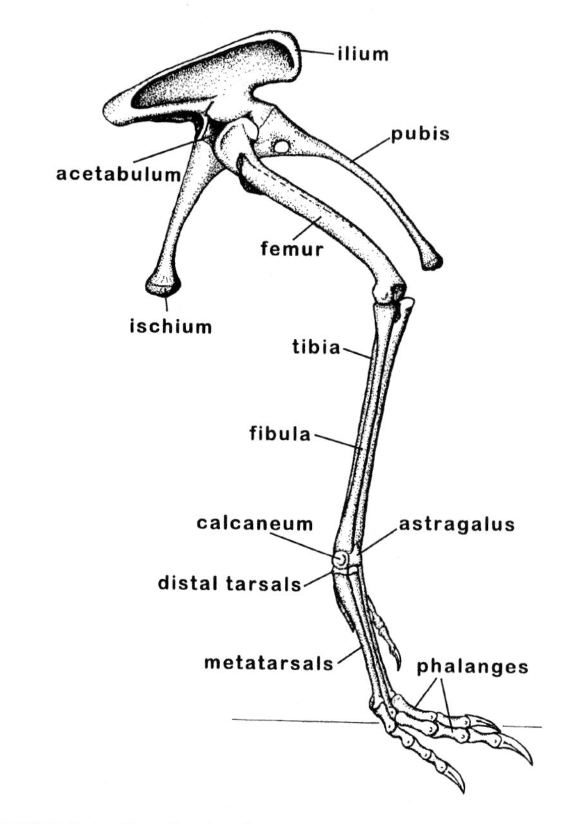

FIGURE 1 Hindlimb of a generalized theropod dinosaur (after *Coelophysis*).

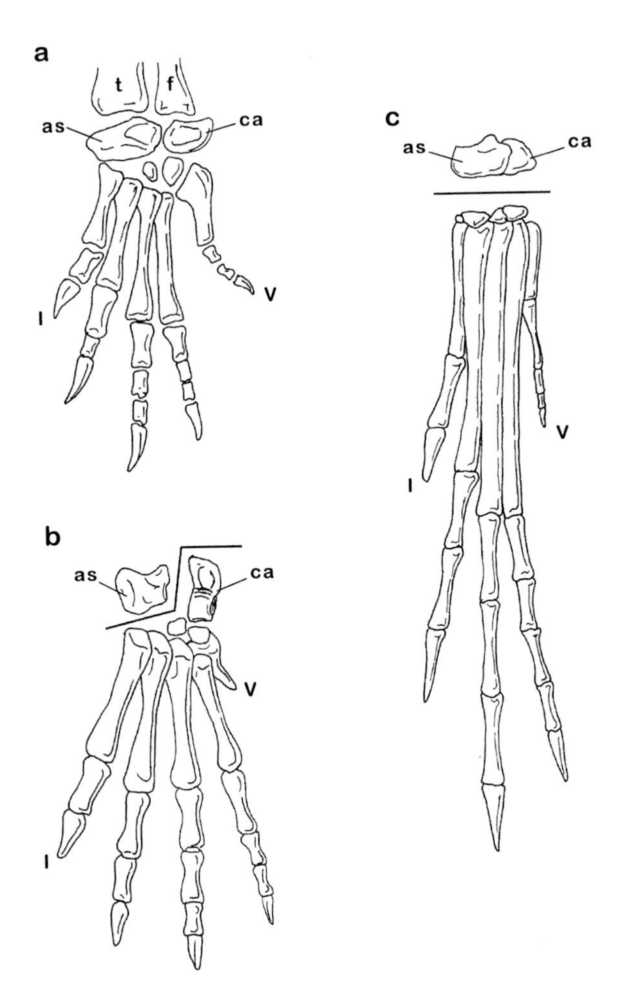

FIGURE 2 Crurotarsal and mesotarsal ankle joints, simplified. (a) Generalized archosauriform ankle *(Euparkeria);* (b) crurotarsal ankle (the crocodylomorph *Notochampsa);* (c) mesotarsal ankle *(Lagosuchus).* Bold line represents the general plane of flexion between leg and foot: The crurotarsal joint flexes obliquely, whereas the mesotarsal joint flexes as a hinge.

between the two rows of tarsals in a hinge joint; this configuration is called mesotarsal ("mid-ankle").

Changes in the form of the hindlimb bones also accompany bipedality (Fig. 1). The femoral head becomes offset from the shaft; the head is inturned to some degree, ranging from approximately 60 to 90°. The shaft is not sigmoid (curved in both dorsoventral and mediolateral planes) but rather is straight (most ornithischians) or only dorsally bowed (all theropods except the largest ones and some ornithischians), reflecting the restriction of movement to the dorsoventral plane. The distal condyles are more strongly developed ventrally and are relatively similar in size, whereas in sprawling reptiles the medial condyle is appreciably larger. The tibia and fibula are no longer of similar size because they are not needed as opposable columns against which the limb muscles rotate the lower leg. Instead, the fibula is reduced (sometimes to a splint) and virtually no rotation of the lower leg occurs at the knee. The positions of trochanters and other muscle attachments are adjusted accordingly. In particular, the calcaneum lacks the prominent heel of crocodilians: The M. gastrocnemius no longer needs to rotate the foot against the leg because the mesotarsal ankle acts as a hinge, as noted previously (Gatesy and Dial, 1996).

All the features discussed so far are considered typical not only of dinosaurs but also of their closest relatives, including other dinosauromorphs, *Pseudolagosuchus, Lagosuchus (Marasuchus), Lagerpeton,* and pterosaurs. This branch of archosaurs is collectively called the ORNITHOSUCHIA, and the distribution of these features among its members suggests that the common ancestor of all these animals had bipedal abilities. In contrast, the PSEUDOSUCHIAN branch of the archosaurs, those closer to crocodiles, flexed their ankles between the proximal tarsals: The astragalus is functionally connected to the tibia, whereas the calcaneum flexes with the foot (Parrish, 1986). This accounts for the sprawling walk of crocodiles, in which the feet splay slightly to the side. This functional arrangement is called crurotarsal ("cross-ankle"). However, when crocodiles execute the "high walk" (in which, instead of sprawling, they tuck their hindlimbs under their bodies and move the legs more parasagittally), their ankles are considerably stiffer and crurotarsal motion is less emphasized (Brinkman, 1980; Parrish, 1987; Gatesy, 1991; see ARCHOSAURIA; FUNCTIONAL MORPHOLOGY; PSEUDOSUCHIA).

It should be noted that in dinosaurs bipedal stance and parasagittal gait were both primitive and obligatory: Dinosaurs could not sprawl. Some living lizards can run bipedally but this is primarily a consequence of high rates of the step cycle coupled with a high disparity between the forelimb and hindlimb lengths. That is, the forelimb is lifted off the ground at high speeds because it is so much shorter than the hindlimb that it would have to move at a prohibitively high rate in order to keep up with the hindlimb. Also, lizards never approach the true parasagittal step cycle of birds and other ornithodirans (Christian *et al.*, 1994).

Footprints and trackways of dinosaurs reveal patterns in their locomotion. THEROPODS, apparently like other bipeds, typically placed one foot virtually in front of the other, and usually the axis of the foot pointed inward slightly; this "pigeon-toed" feature is retained in most avian descendants of these dinosaurs, despite the loss of the fleshy tail and the evolution of flight (Padian and Olsen, 1989). Even quadrupedal dinosaurs retained a narrow lateral distance between left and right feet.

QUADRUPEDALITY in dinosaurs is secondary and evolved at least four times. All theropods were bipedal. Within SAUROPODOMORPHS, basal forms such as *Anchisaurus* were bipedal and this may have been true of juveniles of larger forms. Larger "PROSAUROPODS" such as *Plateosaurus* were at least facultatively quadrupedal, and the larger melanorosaurids were mostly or entirely quadrupedal. SAUROPODS were all quadrupedal. On the view that sauropods evolved from melanorosaurid or closely related basal sauropodomorphs, this quadrupedality was a continuation of an inherited condition. However, if sauropods did not evolve from prosauropods, they must have become quadrupedal independently from unknown relatives because their saurischian outgroups (prosauropods and theropods) were bipedal, at least originally (Sereno, 1991). Some groups appear to have been facultatively quadrupedal, depending on the situation.

Within ORNITHISCHIA, the basal forms *(Fabrosaurus, Lesothosaurus,* and related taxa) were bipeds, and this is true for basal members of the major ornithischian branches. In the THYREOPHORA, *Scutellosaurus,* a lightly built armored biped, is the outgroup to *Scelidosaurus* and the ankylosaurs and stegosaurs, which were habitually quadrupedal. *Heterodontosaurus,* in turn, is the corresponding basal taxon to *Scutellosaurus* for the Ornithischia, and it is clearly a bipedal, small and long-legged form. Conventionally, all the typical ORNITHOPODA are considered to have been bipedal, although some larger forms, notably the hadrosaurids and some iguanodontids, were facultatively quadrupedal, especially while foraging. Some recent analyses have considered these larger forms quadrupedal most of the time, given the hoof-like unguals that are borne on the hand as well as the foot. The MARGINOCEPHALIA (ceratopsians and pachycephalosaurs) are sometimes considered a third branch of Ornithischia and sometimes an offshoot from Ornithopoda, but in either case the common marginocephalian ancestor was almost certainly bipedal: All pachycephalosaurs were and the basal ceratopsians such as *Psittacosaurus* were as well.

See also the following related entries:

BIOMECHANICS • FOOTPRINTS AND TRACKWAYS • FORELIMBS AND HANDS • FUNCTIONAL MORPHOLOGY • HINDLIMBS AND FEET • PELVIS, COMPARATIVE ANATOMY • QUADRUPEDALITY

References

Alexander, R. McN. (1985). Mechanics of posture and gait of some large dinosaurs. *Zool. J. Linnean Soc.* **83,** 1–25.

Brinkman, D. (1980). The hind limb step cycle of *Caiman sclerops* and the mechanics of the crocodile tarsus and metatarsus. *Can. J. Zool.* **58,** 2187–2200.

Christian, A., Horn, H.-G., and Preuschoft, H. (1994). Biomechanical reasons for bipedalism in reptiles. *Amphibia-Reptilia* **15,** 275–284.

Coombs, W. P. (1978). Theoretical aspects of cursorial adaptations in dinosaurs. *Q. Rev. Biol.* **53,** 393–418.

Gatesy, S. M. (1991). Hind limb movements of the American alligator (*Alligator mississipiensis*) and postural grades. *J. Zool. (London)* **224,** 577–588.

Gatesy, S. M., and Dial, K. P. (1996). Locomotor modules and the evolution of avian flight. *Evolution* **50,** 331–340.

Padian, K., and Olsen, P. E. (1989). Ratite footprints and the stance and gait of Mesozoic theropods. In *Dinosaur Tracks and Traces* (D. D. Gillette and M. Lockley, Eds.), pp. 231–241. Cambridge Univ. Press, New York.

Parrish, J. M. (1986). Locomotor adaptations in the hindlimb and pelvis of the thecodontians. *Hunteria* **1**(2), 1–35.

Parrish, J. M. (1987). The origin of crocodilian locomotion. *Paleobiology* **13,** 396–414.

Sereno, P. C. (1991). Basal archosaurs: Phylogenetic relationships and functional implications. *Soc. Vertebr. Paeontol. Mem.* **2,** 1–53.

Bird Origins

Kevin Padian
University of California
Berkeley, California, USA

Luis M. Chiappe
American Museum of Natural History
New York, New York, USA

Since the 1970s it has come to be nearly universally accepted that birds evolved from small carnivorous dinosaurs most closely related to DROMAEOSAURIDS, probably sometime in the Middle to early Late Jurassic. *Archaeopteryx* (Fig. 1) is the first known bird, now represented by seven skeletons and a feather from the Late Jurassic SOLNHOFEN limestones of Germany. Other records of Late Jurassic birds so far have been questionable or apocryphal, although research in the past decade continues to unearth new Early Cretaceous birds that are only slightly more derived than *Archaeopteryx* (Chiappe, 1995; Padian and Chiappe, 1997; see AVES). These new finds help to fill the stratigraphic and morphological gaps between *Archaeopteryx* and more derived, Late Cretaceous birds such as *Hesperornis* and *Ichthyornis,* which have been known for well over a century.

Hypotheses of Bird Origins

As reviewed by Gauthier (1986) and Witmer (1991), there are three major hypotheses of bird origins. One is that they evolved from an unspecified group of basal archosaurs characterized by the disused wastebasket term "THECODONTS." A second is that they share an immediate common ancestor with crocodylomorphs. A third is that they evolved from small THEROPOD dinosaurs. Other suggestions have been made, including common ancestry with lizards, pterosaurs, or mammals, but these ideas were based on only superficial resemblances in a few features and were discredited long ago (Gauthier, 1986).

"Thecodont" Hypothesis This can be traced to the early 1900s but reached its most detailed statement in 1926 with the English-language publication of Gerhard Heilmann's classic *The Origin of Birds* (an earlier version, *Fuglenes Afstamning,* was published in Danish in 1916). Heilmann's book was an exceptionally thorough consideration of avian biology, including skeletal anatomy, embryology, musculature, pterylosis, paleontology, and many other subjects. Heilmann found that theropod dinosaurs were most similar of all fossil groups to *Archaeopteryx* and other birds, but he rejected a theropod ancestry because theropods lacked clavicles; hence, under his interpretation of Dollo's law of evolutionary irreversibility, the clavicles (furcula) of birds could not have reevolved from a theropod precursor. He concluded that the origin of birds must have been among more archaic ARCHOSAURS (see also ORNITHOSUCHIA; PSEUDOSUCHIA), perhaps forms related to *Ornithosuchus* or *Euparkeria,* which had clavicles. Clavicles have since been found in the basal ceratosaurian theropods *Coelophysis* and *Segisaurus,* and a fully formed furcula has been recently discovered in tetanuran theropods ranging from *Allosaurus* and tyrannosaurs to *Velociraptor, Oviraptor,* and *Ingenia* (Fig. 2; see PECTORAL GIRDLE). Some critics contend that the avian furcula is a neomorph not homologous to the reptilian clavicles, partly because the latter are apparently lost in orni-

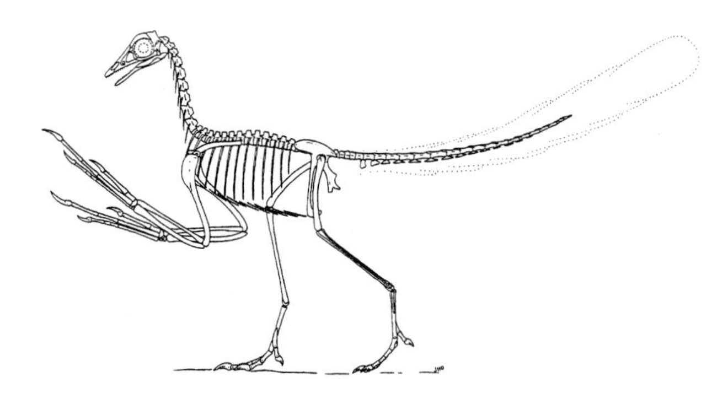

FIGURE 1 *Archaeopteryx,* the first known bird, restored by J. H. Ostrom.

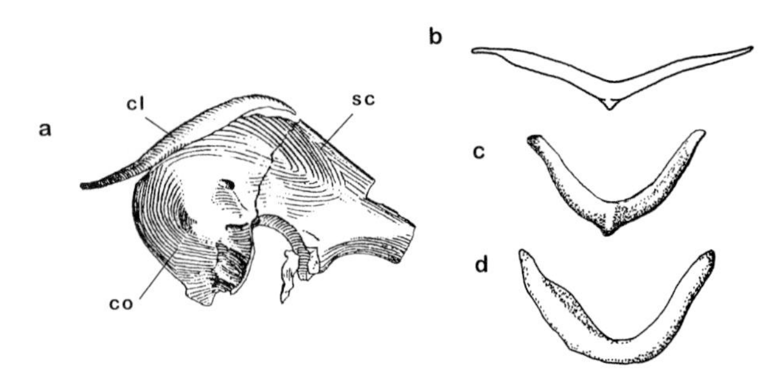

FIGURE 2 The clavicles in several theropods, including (a) the ceratosaur *Segisaurus*, (b) a new allosauroid, the maniraptoran coelurosaurs, (c) *Ingenia*, and (d) *Archaeopteryx*.

thodirans (they are absent in pterosaurs and not known in any nontheropodan dinosauromorph) and partly because in some recent birds the furcula seems to be composed of both dermal and endochondral bone. Regardless of these facts, there is no doubt about theropod monophyly, so the homology of the ceratosaurian clavicles and the tetanuran furcula would not seem to be in question; also, there is no mistaking the identity in shape and position of the boomerang-shaped furcula in nonavian tetanurans, *Archaeopteryx*, and other birds.

Heilmann's retreat to a thecodont hypothesis was a default argument; as he recognized, no features that linked any particular basal archosaur to birds are also not found in theropods, usually with greater similarity. Since Heilmann's work, other authors have advocated a thecodont hypothesis, and the approach is very much the same. No specific candidate among basal archosaurs has been presented as the direct ancestor or the closest known animal to birds; rather, a range of forms with one or two supposedly bird-like characters is advanced, even though most of their character states are far more primitive than those in theropods (see Witmer, 1991, for a thoughtful review). Tarsitano (1991) has advanced most forcefully the idea that the origin of birds is to be found among such "avimorph thecodonts," but cladistic analyses have found that all these animals are more closely related to other forms quite distant from birds. For example, *Cosesaurus* and *Megalancosaurus* are aquatic prolacertiform archosauromorphs, *Scleromochlus* is the closest known sister group to pterosaurs, and *Lagosuchus* and *Lagerpeton* are closest to basal dinosaurs (Gauthier, 1986; Sereno, 1991; Benton, 1988; Padian and Chiappe, 1997). *Protoavis* (Chatterjee, 1991, 1995, 1997) has been advanced as a Triassic bird but has been met with substantial skepticism (reviewed in Padian and Chiappe, 1997); there are apparently even more differences of opinion about the interpretation of its morphology than there are about *Archaeopteryx*. The question has been made more difficult by the circumstance that, in the more than two decades since this controversy was renewed, no advocate of a thecodont ancestry has produced a cladogram incorporating all, or even any, of the available evidence that would support such a case. Cladistic analyses (see SYSTEMATICS; PHYLOGENETIC SYSTEM) are not infallible but at least they are explicit: If the weight of evidence supports a different interpretation than the several independent analyses that have placed birds within the theropods, then this result would be very interesting to see expressed in cladistic terms.

Crocodilian Hypothesis This should more properly be termed the "crocodylomorph" hypothesis because its advocates regard the ancestry of birds as complete before true CROCODILES evolved. Indeed, it is within the sphenosuchian crocodylomorphs, outside Crocodylia, that bird-like characters appear to have been most pronounced. A. D. Walker (e.g., 1977) has been the chief advocate of this view, based on his detailed studies of the braincase, quadrate, ear region, and other features of *Sphenosuchus*, an Early Jurassic crocodylomorph from South Africa. His view has been generally supported by L. D. Martin and his students (e.g., Martin, 1983), although they have tended to draw similarities to birds from true crocodilians as much as from crocodylomorphs. Many of these similarities are valid but have been shown to apply either to a more general level among archosaurs or to have evolved convergently in certain crocodilians and early birds (but not present in the hypothesized common ancestor of both groups) (Gauthier, 1986). Again, no cladogram incorporating all the available evidence has to date supported a crocodylomorph origin of birds.

Theropod Hypothesis This had its roots in the 1860s with T. H. Huxley, who noted in a series of papers (e.g., 1870) a suite of 35 characters shared uniquely by birds and theropod dinosaurs (reviewed by Gauthier, 1986, pp. 4–6). Many of these are still considered valid today, whereas others have turned out to be more general to dinosaurian or other archosaurian groups. Desmond (1982) and Gauthier (1986) have both noted that Huxley's hypothesis, presented to the Geological Society of London in 1870, was contested by Harry Govier Seeley, who "thought it possible that the peculiar structure of the hinder limbs of the Dinosauria was due to the functions they performed, rather than to any actual affinity with birds" (See DINOSAURIA: DEFINITION). This shadow of potential convergence, though not explicitly tested either by Seeley or anyone since, nonetheless not only frustrated the acceptance of Huxley's views at the time but also continues to be contended by opponents of the theropod hypothesis (e.g., Feduccia, 1996).

In the 1970s John Ostrom, in a series of papers (e.g., 1975a,b, 1976b), demonstrated the detailed similarities of *Archaeopteryx* to theropod dinosaurs. Although he did not specify a taxon within Theropoda to which birds might be directly connected, his comparisons tended to run to the dromaeosaurid *Deinonychus*, which he had recently described in a monograph (Ostrom, 1969). As it turned out, dromaeosaurids were found to be the closest sister group to birds in several independent cladistic analyses beginning in the early 1980s by Padian, Gauthier, Benton, Sereno, Holtz, Perle *et al.*, and others (see Padian and Chiappe, 1997). Synapomorphies that link dromaeosaurids and *Archaeopteryx* include the presence of dorsal, caudal, and rostral tympanic recesses, the semilunate carpal, thin metacarpal III, longer pubic peduncle, posteroventrally directed pubis with only a posteriorly projecting foot, shortened ischium, and other features of the skull, pectoral girdle, and hindlimb (Gauthier, 1986; Padian and Chiappe, 1997).

A great many characters classically considered "avian" apply to more general levels within Theropoda. Basal theropods have lightly built bones and a foot reduced to three main toes, with the first usually held off the ground and the fifth lost. Closer to birds, the fifth and fourth digits of the hand are progressively reduced and lost, the skeleton (especially the vertebrae) becomes lighter, and the tail becomes shorter as its vertebrae partially interlock through the elongation of zygapophyses to reinforce its stiffness. In COELUROSAURS (*sensu* Gauthier, 1986, the sister taxon to CARNOSAURS), contrary to the picture suggested by opponents of the theropod hypothesis, the forelimbs become progressively longer until they are nearly as long as the hindlimbs in some dromaeosaurs (Fig. 3); the first toe (hallux) begins to rotate behind the metatarsus, although it does not descend to the point seen in perching birds; the metatarsals become longer; and the scapular blade becomes longer and more strap-like. The presence of the furcula may turn out to be a general tetanuran character, and it is not yet clear how general the calcified sternum in adults may be (it is known, for example, in some oviraptorids, dromaeosaurs, tyrannosaurs, and sinraptorids).

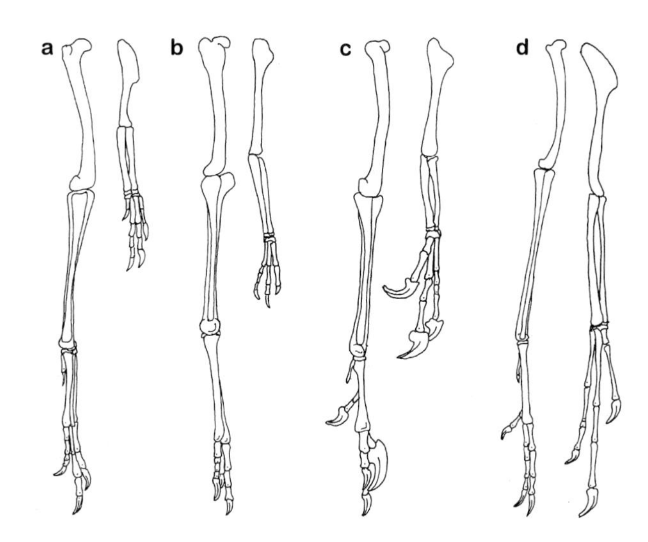

FIGURE 3 The forelimbs compared to the hindlimbs in (a) the ceratosaur *Coelophysis*, (b) the ostrich dinosaur *Struthiomimus*, (c) the maniraptoran coelurosaur *Deinonychus*, and (d) *Archaeopteryx*. Compare these to the phylogeny in Fig. 4.

Holtz (1994, 1996) reevaluated the phylogenetic relationships of Theropoda, and his conclusions uphold Gauthier's (1986) comprehensive analysis, with some adjustments that do not affect the position of birds within Theropoda (Fig. 4).

Clavicles, Digits, Feathers, and Stratigraphy

Individual characters are sometimes advanced as conclusive evidence that birds could not have descended from theropod dinosaurs. For example, it is claimed that the digits of the bird hand are II–III–IV, whereas they are I–II–III in theropods; that the semilunate carpal bones of maniraptorans and other theropods are different from those of *Archaeopteryx* and the birds; and that the ascending process of the astragalus is not the same in theropods and birds. [Feduccia (1996) reviews these claims favorably, and to obviate

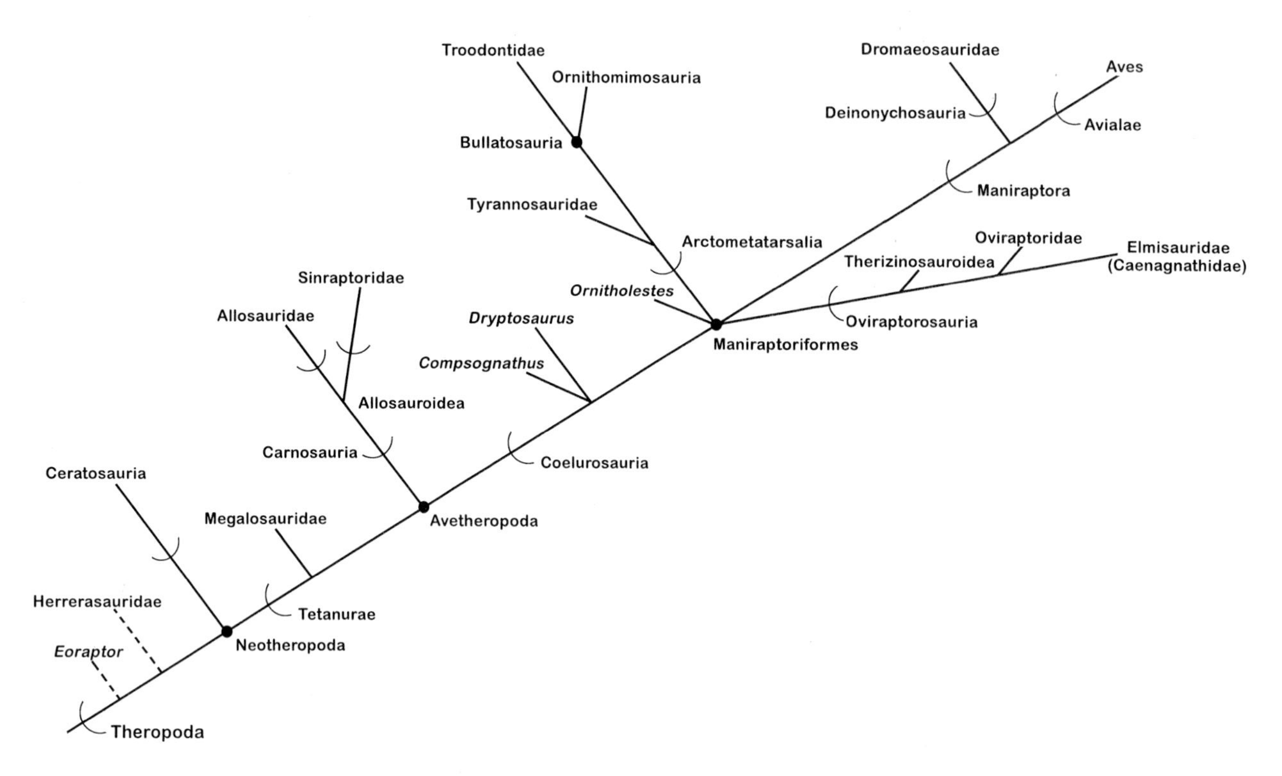

FIGURE 4 Phylogeny of Theropoda, after Holtz (1994, 1996).

extensive literature citations readers are referred to his work for historical background and strong advocacy.] These statements amount to hypotheses that the characters are not homologs but homoplasies. Since Darwin's day, homologies have been recognized as features inherited from common ancestors; even to non-Darwinians, such as Richard Owen, homologies were established by similarity of morphology, position, development, and histological structure. Roth (1988) proposed that, like monophyletic taxa, homologies are defined by ancestry but diagnosed by features such as the criteria just listed. In comparative biology, the recognition of homologous structures in two or more organisms can be tested using the phylogenetic distributions of other, presumably independent characters and by including more taxa in the analysis (see SYSTEMATICS).

Proceeding according to this method, the hypothesis that the furcula of birds is not homologous to the clavicles of theropod dinosaurs is weak both because undoubted nonavian tetanuran theropods have boomerang-shaped clavicles in the same position as those of *Archaeopteryx* and other basal birds and because basal theropods have structures that are manifestly similar in morphology and position to clavicles in other tetrapods.

Digits of the Bird Hand The three remaining digits of the bird hand are sometimes regarded as I, II, and III and sometimes as II, III, and IV. The theropod hand is unquestionably I, II, and III, and opponents of the theropod hypothesis of avian descent are unanimous in their contention that the bird hand is II, III, and IV (therefore, the digits could not be homologous between the two groups). This is often inaccurately portrayed as a difference between paleontologists and ornithologists: Feduccia (1996, p. 2), for example, reproduced a figure of the avian skeleton by Lucas and Stettenheim with the digits numbered II, III, and IV, but reproduced on page 7 of the same work two figures, one by Van Tyne and Berger and another by Burton and Milne, in which the digits are numbered I, II, and III. These illustrations are all by ornithologists. Classical ornithologists from W. K. Parker to Proctor and Lynch (1993) have agreed, and Heilmann (1926) examined the problem at length and reached the same conclusion: I, II, and III.

The evidence for II, III, and IV comes entirely from

some interpretations of the ontogeny of the hand in some living birds. As Heilmann (1926) noted, interpretations on this basis have varied. Some have been based on an assumption of Morse's "law" that digits must be lost from the both sides inward (as in the bird foot: 5, then 1; and in horses' feet: 5, then 1, 4, and 2), although this pattern must surely be related to the weight-bearing function of the locomotory structures. Some authors have even claimed to find *Anlagen* of four digits in the bird hand. Hinchliffe and Hecht (see Hecht and Hecht, 1994) have strongly advocated the modern developmental case for II, III, and IV based on the presence of an ephemeral "element X" medial and palmar to the wrist in early ontogeny that is taken for a remnant of digit I. However, as Shubin (1994) and Padian and Chiappe (1997) note, *Anlagen* do not appear with labels on them but have to be interpreted. To accept the II–III–IV view, the first digit, including its carpal and metacarpal, has to be lost, and digits II, III, and IV have to assume the precise forms, articulations, and proportions of the original digits I, II, and III. If this is possible, developmental biology has so far not provided examples or mechanisms from living tetrapods that support this process. Unfortunately, we do not have similar embryological stages for *Archaeopteryx* or any other Mesozoic birds or dinosaurs, so we cannot examine phylogenetically the hypothesis of element X outside living birds.

In favor of the I–II–III hypothesis, the theropod hand follows a consistent pattern of reduction and loss of elements from the lateral side medially (Shubin, 1994; Padian and Chiappe, 1997) (Fig. 5). Crocodilians and other ornithodiran outgroups have a five-fingered manus in which the third digit is longest. In Dinosauria the fourth and fifth digits are reduced in size, and the phalangeal formula is reduced to 2-3-4-3-2. In Saurischia the second digit becomes the longest, and the fourth digit's phalanges are reduced. In Theropoda the fifth digit is reduced to a nubbin of the metacarpal or lost altogether, and the fourth digit is quite small. In tetanurans all trace of the outer two digits is lost. In tyrannosaurs [which von Huene (1914, 1920, 1926) and Novas (1992) showed are not carnosaurs but actually gigantic coelurosaurs], the third digit is lost, thus proving the invalidity of Morse's law at least in the case of dinosaurs. Given the suite of dozens of synapomorphies from other parts of the skeleton that also support the placement of birds within theropods, it is difficult not to accept the manifest similarities of other features of the hand of *Archaeopteryx* as homologous to those of theropods, including 14 characters related to the form and proportion of hand and wrist elements in theropods (Gauthier, 1986: characters 21–26, 43–46, 61, 62, 75, and 76).

Semilunate Carpal In maniraptorans, and perhaps at a more general level within tetanurans, a half-moon-shaped wrist element overlaps the bases of

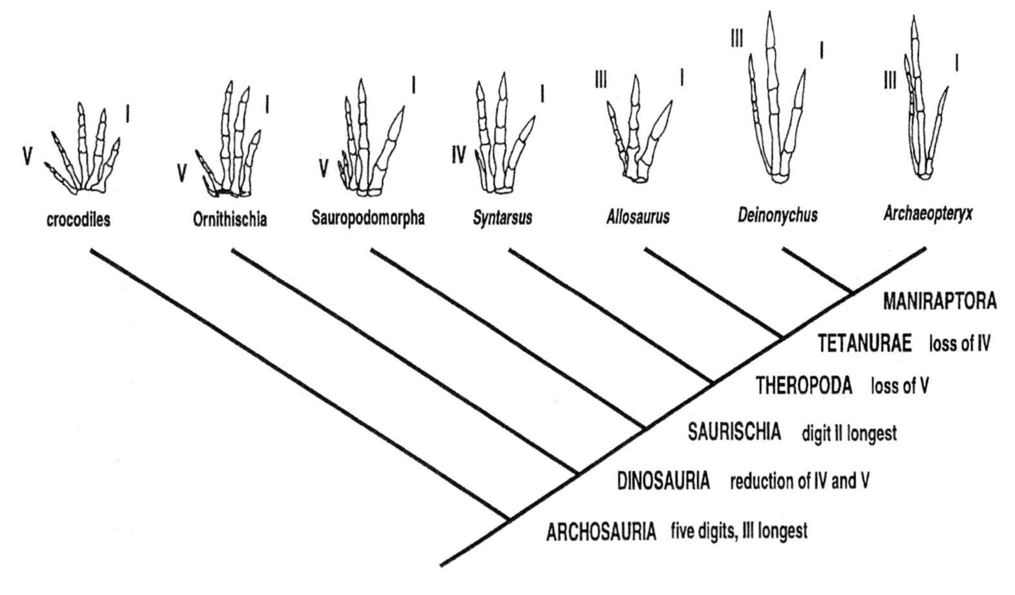

FIGURE 5 The reduction of the hand in theropods, including birds, from Padian and Chiappe (1997).

metacarpals I and II (Ostrom, 1969, 1975a,b, 1976b; Gauthier, 1986). The rounded proximal surface allows the hand to swivel sideways, a feature that appears to have been related to predation in basal maniraptorans and exapted for the flight stroke in birds (Gauthier and Padian, 1985). There appears to be little question about the morphological similarity of this bone in *Archaeopteryx* and troodontids, oviraptors, and dromaeosaurs such as *Deinonychus* and *Velociraptor*; the question surrounds its homology. Carpal elements are often incomplete or unknown for Mesozoic theropods: They may not have been preserved because they were not ossified, perhaps because the specimen in question was not adult; or they may have been removed or destroyed by taphonomic processes, preserved out of position and so unrecognized and not collected, or collected and not correctly identified. Furthermore, as in other amniotes, during ontogeny many of these elements attained better definition and frequently fused, and phylogenetically fusion appears to have increased (Gauthier, 1986). There is no simple answer to the identity of some elements of the wrist in theropods (Feduccia, 1996), but some points are clear.

Contrary to some inferences (Feduccia, 1996), theropods have not "lost" a row of carpals, but sometimes they are incompletely ossified or preserved. Ostrom (1969 *et passim*) identified the semilunate carpal in *Deinonychus* and *Archaeopteryx* as the radiale and the element next to it as the ulnare, but this cannot be because the radiale contacts the radius and does not contact the metacarpals in any tetrapod. Comparison to basal theropods, such as *Coelophysis* and *Syntarsus*, in which the manus is preserved, demonstrates a row of carpals between those contacting the radius and ulna (*de facto* the radiale and ulnare) and the metacarpals (Fig. 6). A single distal carpal, identified by both Colbert and Raath as a fusion of distal carpals 1 and 2, overlaps metacarpal II, and this element is exactly in the position of the semilunate carpal of tetanurans (Gauthier, 1986). In birds, the radiale and ulnare have become more tightly associated with the radius and ulna as the sideways flexion of the wrist has evolved between the proximal and distal carpal rows (Padian and Chiappe, 1997); the distal carpals have become associated immovably with the metacarpals (Ostrom, 1976a).

Hence, the semilunate carpal of birds and other tetanurans is the result of the fusion of distal carpals

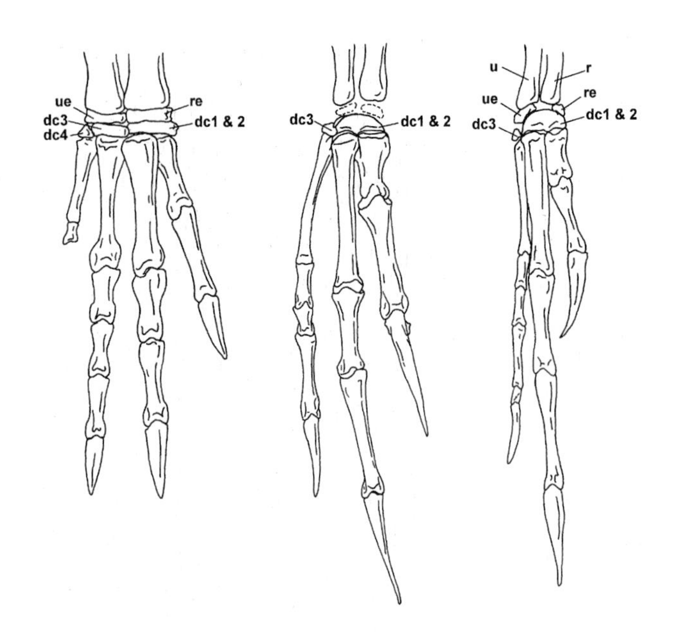

FIGURE 6 The wrists of *Coelophysis*, *Deinonychus*, and *Archaeopteryx* (modified from Colbert and Ostrom).

1 and 2; the radiale and ulnare are not lost but are not always preserved and should be sought in association with the radius and ulna; and a third distal carpal is usually present, if ossified, in tetanurans, but the fourth distal carpal, like the fifth, has been lost in tetanurans along with these digits.

Pubic Foot The end of the pubis is unexpanded in basal dinosaur groups, including all ornithischians, sauropodomorphs, and basal theropods (ceratosaurs, including *Coelophysis*, *Syntarsus*, *Ceratosaurus*, etc.). In tetanurans it is expanded fore and aft: This can be seen both in carnosaurs such as *Allosaurus* and in coelurosaurs such as ornithomimids and tyrannosaurs (Fig. 7). In maniraptorans, such as the dromaeosaurs *Deinonychus*, *Velociraptor*, and *Adasaurus*, the anterior projection of this foot is severely reduced or lost, as it is in *Archaeopteryx* and the birds. Moreover, in dromaeosaurids, as in birds, the pubis itself is retroverted (and convergently in therizinosaurids), although apparently not to the extent seen in living birds. Herrerasaurids, seen variously as basal theropods, basal saurischians, or a sister taxon to dinosaurs, also have a development of the distal pubis similar in some respects to the theropod pubic foot, but this is expected to be a convergence because they are not otherwise closer to birds than are dromaeo-

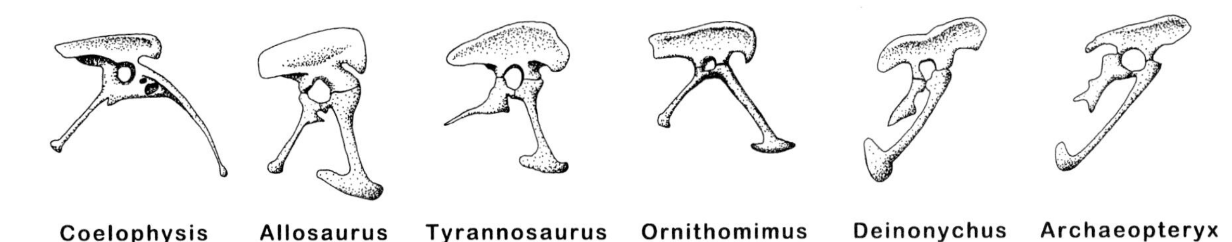

FIGURE 7 The anatomy of the pubic foot varies and can be used to establish affinities among and between lineages of dinosaurs.

saurids. Likewise, the Triassic archosaur *Postosuchus* (like other poposaurids) appears to have a similar expansion of the distal pubis, but this is clearly a convergence because *Postosuchus* is closely allied not to birds or dinosaurs but rather to crocodylomorphs (see PSEUDOSUCHIA).

Ascending Process of the Astragalus In ornithodiran ornithosuchians, posture and gait have changed from the general sprawled or semierect reptilian condition to a more upright stance and a parasagittal gait (Fig. 8). The ankle flexes mesotarsally so that the proximal ankle elements (astragalus and calcaneum) are associated with the lower leg and the distal tarsals with the metatarsals and phalanges. Because parasagittal gait virtually eliminates rotation at the knee, and skeletal mass is concentrated medially, the tibia and fibula have the same action. The tibia becomes the dominant element and the fibula is reduced, and the same happens to their corresponding proximal tarsals. Astragalus and calcaneum frequently fuse in adults, and in no cases do they rotate against each other, as in crurotarsal archosaurs (see PSEUDOSUCHIA). The astragalus expands transversely in basal ornithodirans such as pterosaurs (in which the proximal tarsals are always fused to each other and to the tibia and the fibula is reduced to a splint), *Marasuchus/Lagosuchus*, and *Lagerpeton*. Hence, for the first time, the astragalus articulates with both the tibia and fibula. A dorsal process of the astragalus separating these two articulations begins in basal ornithodirans and is known in all taxa in which the elements are preserved. It is continuous with an ascending process that is posterior to the lower leg bones in *Lagerpeton*, medial in *Marasuchus/Lagosuchus*, but anterior in dinosaurs and much more expanded in theropods (especially tetanurans) than in any other taxa. Based on congruence with other characters, it would appear parsimonious to conclude that birds carry on this basal ornithodiran feature because it is seen to trend through Theropoda.

The identity of this ossification, however, is at issue; it is held by some to be different from the avian "pretibial" bone (summarized in Feduccia, 1996), which is a separate ossification. Its shape and position on the tibia also varies, but generally it is centered anterolaterally, as is the pretibial bone. Feduccia (1996, p. 75) misinterpreted the variation in the ascending process described by Welles and Long (1974): They identified five morphological types, but these

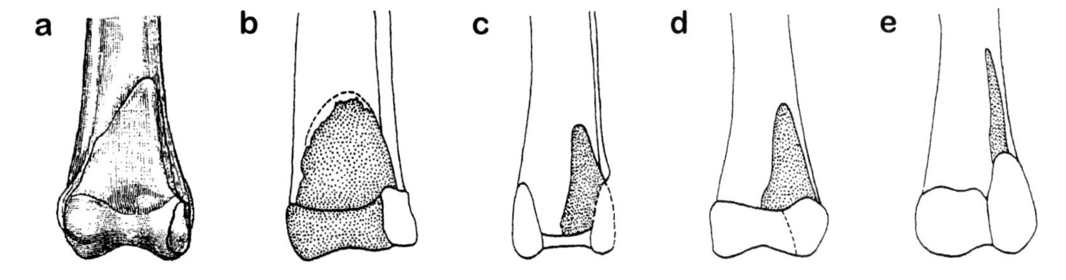

FIGURE 8 The ankle region in (a) *Coelophysis*, (b) *Deinonychus*, (c) *Archaeopteryx*, (d) the Cretaceous bird *Baptornis* (after Martin, 1983, with permission), and (e) the hoatzin (after Martin, Stewart, and Whetstone, *The Auk* **97,** 86–93).

are not of independent phylogenetic origin, and in fact the types they identified are mostly not characteristic of any natural taxa within theropods; they are simply morphological types that vary for reasons apparently connected with relative size or functional features. As Gauthier (1986, p. 29) noted, the ascending process may be a separate ossification in *Dilophosaurus* (as S. P. Welles first discovered) and other theropods, as it is in birds. Moreover, although the pretibial bone fuses with the precociously developed calcaneum in neognath birds, this is not the case in ratites and tinami in which, as in other theropods, it fuses with the astragalus, and in all cases it is located on the anterolateral side of the tibia. Hence, the pretibial bone of birds appears to be homologous to the ascending process of the astragalus in theropods.

Stratigraphic Disjunction A difficulty regarded as insurmountable by opponents of the theropod origin of birds is the presumption that the taxa identified as closest to *Archaeopteryx* among theropods—the dromaeosaurids—do not appear in the fossil record until Albian–Aptian times (perhaps 110 mya: *Deinonychus*, Cloverly Formation, Wyoming), whereas *Archaeopteryx* comes from Late Jurassic (Tithonian) times (approximately 150 mya). The apparent absence of earlier records of dromaeosaurids, although puzzling, is not unusual in the Mesozoic fossil record: For example, although stegosaurs and ankylosaurs are regarded as sister taxa that must have diverged by the late Early Jurassic, stegosaurs are not known before the Bathonian–Callovian (approximately 170 mya), whereas before the 1980s, ankylosaurs were not known before the Aptian–Albian (approximately 110 mya; Weishampel *et al.*, 1990). The situation is not unique to dinosaurs. No one doubts today that marsupials and placentals are sister taxa within mammals, and monotremes are their sister taxon. Hence, the split between therians (marsupials + placentals) and monotremes must have taken place before the first recognizable marsupials and placentals evolved. However, the first marsupials and placentals are known from Early Cretaceous times (approximately 100 mya), whereas until recently, monotremes were not known until the Oligocene (approximately 20 mya), a disjunction of 80 million years—over twice that between *Archaeopteryx* and *Deinonychus* (Carroll, 1988)! Moreover, small maniraptorans are not at all absent from Late Jurassic sediments: Jensen and Padian (1989) described a collection of bones pertaining to small maniraptorans from the Dry Mesa Quarry (Late Jurassic: ?Tithonian; Morrison Formation, Colorado). These bones unfortunately could not be identified to the generic level but nonetheless indicated that if they are not bones of birds, then they are certainly those of their sister taxon, the dromaeosaurids. These arguments would appear to dispose of the fatality of the stratigraphic argument to the theropod hypothesis.

In summary, birds, as Gauthier (1986) pointed out, must be considered dinosaurs because phylogenetic analysis clearly indicates that they evolved from dinosaurs. They are not only dinosaurs but also saurischian, theropodan, tetanuran, and maniraptoran dinosaurs. Arguments to the contrary have been proposed for 20 years since the theropod hypothesis was advanced by Ostrom, but these can no longer be considered matters of evidence. Rather, it is a question of whether one uses the methods of modern comparative biology (see SYSTEMATICS). Issues related to the origin of flight in birds and other topics with starkly contrasting viewpoints are discussed at length in Hecht *et al.* (1985), Schultze and Trueb (1991), Feduccia (1996), and Padian and Chiappe (1997).

See also the following related entries:

AVES • AVIALAE • COELUROSAURIA • DROMAEOSAURIDAE

References

Benton, M. J. (Ed.) (1988). *The Phylogeny and Classification of the Tetrapods, Volume 1.* Clarendon, Oxford.

Carroll, R. L. (1988). *Vertebrate Paleontology and Evolution.* Freeman, New York.

Chatterjee, S. (1991). Cranial anatomy and relationships of a new Triassic bird from Texas. *Philos. Trans. R. Soc. London B* **332,** 277–346.

Chatterjee, S. (1995). The Triassic bird *Protoavis. Archaeopteryx* **13,** 15–31.

Chatterjee, S. (1997). *Protoavis* and the early evolution of birds. *Palaeontographica,* in press. [Abstract A].

Chiappe, L. M. (1995). The first 85 million years of avian evolution. *Nature* **378,** 349–355.

Desmond, A. J. (1982). *Archetypes and Ancestors: Palaeontology in Victorian London, 1850–1875.* Muller, London.

Feduccia, A. (1996). *The Origin and Evolution of Birds*. Yale Univ. Press, New Haven, CT.

Gauthier, J. (1986). Saurischian monophyly and the origin of birds. *Mem. California Acad. Sci.* **8,** 1–55.

Gauthier, J., and Padian, K. (1985). Phylogenetic, functional, and aerodynamic analyses of the origin of birds and their flight. In *The Beginnings of Birds* (M. K. Hecht, J. H. Ostrom, G. Viohl, and P. Wellnhofer, Eds.), pp. 185–197. Freunde des Jura-Museums, Eichstatt.

Hecht, M. K., and Hecht, B. M. (1994). Conflicting developmental and paleontological data: The case of the bird manus. *Acta Paleontol. Polonica* **38**(3 / 4), 329–338.

Hecht, M. K., Ostrom, J. H., Viohl, G., and Wellnhofer, P. (1985). *The Beginnings of Birds*. Freunde des Jura-Museums, Eichstatt.

Heilmann, G. (1926). *The Origin of Birds*. pp. 210. Appleton, New York.

Holtz, T. R., Jr. (1994). The phylogenetic position of the Tyrannosauridae: Implications for theropod systematics. *J. Paleontol.* **68**(5), 1100–1117.

Holtz, T. R., Jr. (1996). Phylogenetic taxonomy of the Coelurosauria (Dinosauria: Theropoda). *J. Paleontol.* **70,** 536–538.

Huxley, T. H. (1870). Further evidence of the affinities between the dinosaurian reptiles and birds. *Q. J. Geol. Soc. London* **26,** 12–31.

Jensen, J. A., and Padian, K. (1989). Small pterosaurs and dinosaurs from the Uncompahgre Fauna (Brushy Basin Member, Morrison Formation: ?Tithonian), Late Jurassic, Western Colorado. *J. Paleontol.* **63,** 364–373.

Martin, L. D. (1983). The origin of birds and of avian flight. In *Current Ornithology* (R. F. Johnston, Ed.), Vol. 1, pp. 106–129. Plenum, New York.

Novas, F. E. (1992). La evolución de los dinosaurios carnívoros. In *Los Dinosaurios y su entorno biótico* (J. L. Sanz and A. D. Buscalione, Eds.), pp. 125–163. Instituto "Juan Valdes," Cuenca, Spain.

Ostrom, J. H. (1969). Osteology of *Deinonychus antirrhopus*, an unusual theropod from the Lower Cretaceous of Montana. *Bull. Peabody Museum Nat. History Yale Univ.* **30,** 1–165.

Ostrom, J. H. (1974). *Archaeopteryx* and the origin of flight. *Q. Rev. Biol.* **49,** 27–47.

Ostrom, J. H. (1975a). The origin of birds. *Annu. Rev. Earth Planet. Sci.* **3,** 35–57.

Ostrom, J. H. (1975b). On the origin of *Archaeopteryx* and the ancestry of birds. *Proc. CNRS Colloq. Int. Prob. Act. Paleontol.—Evol. Vertébr.* **218,** 519–532.

Ostrom, J. H. (1976a). Some hypothetical anatomical stages in the evolution of avian flight. *Smithsonian Contrib. Paleobiol.* **27,** 1–27.

Ostrom, J. H. (1976b). *Archaeopteryx* and the origin of birds. *Biol. J. Linnean Soc.* **8,** 91–182.

Padian, K., and Chiappe, L. M. (1997). The early evolution of birds. *Biol. Rev.,* in press.

Proctor, N. S., and Lynch, P. J. (1993). *Manual of Ornithology: Avian Structure and Function*. Yale Univ. Press, New Haven, CT.

Roth, V. L. (1988). The biological basis of homology. In *Ontogeny and Systematics* (C. J. Humphries, Ed.), pp. 1–26. Columbia Univ. Press, New York.

Schultze, H.-P., and Trueb, L. (Eds.) (1991). *Origins of the Higher Groups of Tetrapods*. Cornell Univ. Press, Ithaca, NY.

Sereno, P. (1991). Basal archosaurs: phylogenetic relationships and functional implications. *J. Vertebr. Paleontol.* **11**(Suppl. 4), 1–53.

Shubin, N. (1994). History, ontogeny, and evolution of the archetype. In *Homology: The Hierarchical Basis of Comparative Biology* (B. K. Hall, Ed.), pp. 248–271. Academic Press, New York.

Tarsitano, S. (1991). *Archaeopteryx:* Quo Vadis? In *Origins of the Higher Groups of Tetrapods* (H.-P. Schultze and L. Trueb, Eds.), pp. 541–576. Cornell Univ. Press, Ithaca, NY.

von Huene, F. (1914). Das naturliche System der Saurischia. *Zentralblatt Mineral. Geol. Paläontol. B* **1914,** 154–158.

von Huene, F. (1920). Bemerkungen zur Systematik und Stammesgeschichte einiger Reptilien. *Zeitschrift Indukt. Abstammungslehre Vererbungslehre* **24,** 162–166.

von Huene, F. (1926). The carnivorous Saurischia in the Jura and Cretaceous formations, principally in Europe. *Revista Museo de La Plata* **29,** 35–167.

Walker, A. D. (1977). Evolution of the pelvis in birds and dinosaurs. In *Problems in Vertebrate Evolution* (S. M. Andrews, R. S. Miles, and A. D. Walker, Eds.), pp. 319–357. Academic Press, New York.

Walker, A. D. (1985). The braincase of *Archaeopteryx*. In *The Beginnings of Birds* (M. K. Hecht, J. H. Ostrom, G. Viohl, and P. Wellnhofer, Eds.), pp. 123–134. Freunde des Jura-Museums, Eichstatt.

Weishampel, D. B., Dodson, P., and Osmólska, H. (Eds.) (1990). *The Dinosauria*. Univ. of California Press, Berkeley.

Welles, S. P., and Long, R. A. (1974). The tarsus of theropod dinosaurs. *Ann. South African Museum* **64,** 191–218.

Witmer, L. (1991). Perspectives on avian origins. In *Origins of the Higher Groups of Tetrapods* (H.-P. Schultze and L. Trueb, Eds.), pp. 427–466. Cornell Univ. Press, Ithaca, NY.

Birmingham Museum, United Kingdom

see Museums and Displays

Black Hills Museum of Natural History, South Dakota, USA

see Museums and Displays

Bloemfontein National Museum

see National Museum, Bloemfontein

Bolivian Dinosaurs

Late Cretaceous dinosaur footprints have been discovered in Bolivia.

see South American Dinosaurs

Bone Beds

see Behavior; Migration; Taphonomy; Variation

Bone Cabin Quarry

Brent H. Breithaupt
University of Wyoming
Laramie, Wyoming, USA

Although Cope and Marsh carried their bitter feud to their deaths in 1897 and 1899, respectively, by the 1890s the "bone wars" were over in the West and other institutions had begun collecting there. In 1897 Henry Fairfield Osborn at the American Museum of Natural History sent collecting crews to the area of Wyoming where Marsh and Cope's crews had collected dinosaurs years earlier. The American Museum crews, under the direction of Jacob Wortman, found that much of Como Bluff was barren of vertebrate fossils. However, they did discover and collect partial skeletons of *Diplodocus* and *Apatosaurus* the first year (see History of Discovery: Early Years).

Patience and persistence resulted in American Museum of Natural History crews finding several new dinosaur quarries in the region. One of the most spectacular of these was the famous Bone Cabin Quarry found in 1898 on a hill approximately 15 km north of Como Bluff on the Little Medicine Bow Anticline. Taking its name from the remains of a sheepherder's cabin foundation made entirely of dinosaur bone fragments, Bone Cabin Quarry preserved more than 50 partial dinosaur skeletons (e.g., *Diplodocus, Camarasaurus, Apatosaurus, Allosaurus, Ornitholestes, Camptosaurus, Dryosaurus,* and *Stegosaurus*) in a total area of only 1,529 m^2. Major collections were made for the American Museum from this and other nearby Morrison Formation quarries (e.g., Nine Mile Quarry and Quarry R) between 1898 and 1905 (McIntosh, 1990), primarily under the direction of Walter Granger after Wortman took a position with the Carnegie Museum in 1899. A large number of complete articulated sauropod fore- and hindlimbs, in some cases with complete feet, were found (Osborn, 1899). Many long tail segments were also found, in addition to several skulls.

Famed dinosaur collector William Harlow Reed wrote that he and Frank Williston had come across this site decades earlier but because of the deteriorated condition of many of the bones (called "head cheese") they decided to ignore it (Breithaupt, 1990). Reed was independently working in this region in 1901 when American Museum crews were collecting fossils there. Because Reed had recently resigned from the field crew of the Carnegie Museum, he proposed working for the American Museum of Natural History. Reed's Quarry R (located to the southeast of Bone Cabin Quarry on the Prager Anticline) and the Nine Mile Quarry (located south of Bone Cabin Quarry on the Little Medicine Bow Anticline) proved to be quite rich in dinosaur remains. *Camarasaurus* and *Allosaurus* bones were found at Quarry R and a partial skeleton of *Apatosaurus* was located at Nine Mile Quarry.

During the major field seasons in Wyoming (1898–1903), crews from the American Museum of Natural History collected approximately 275 boxes weighing more than 68 metric tons (Colbert, 1968). More than 500 specimens representing approximately 69 animals (i.e., 44 sauropods, 3 stegosaurs, 4 ornithopods, 9 theropods, 4 crocodiles, and 5 turtles) were found (McIntosh, 1990; Osborn, 1904). American Museum of Natural History exhibition specimens from this region of Wyoming collected between 1897 and 1905 included *Apatosaurus* (Nine Mile Quarry and Bone Cabin Quarry; 1898), *Stegosaurus* (Como Bluff; 1901), *Ornitholestes* (Bone Cabin Quarry; 1901), and *Camptosaurus* (Bone Cabin Quarry; 1905) (Norell *et al.*, 1995). Another exhibit sauropod skeleton was found by American Museum crews in Bone Cabin Quarry. This specimen of *Diplodocus* was subsequently sent to Germany (McIntosh, 1990). In 1905 the *Apatosaurus* remains from the Nine Mile and Bone Cabin quarries were used by the American Museum of Natural History to mount the first sauropod skeleton in the world (Norell *et al.*, 1995). In the 1990s Bone Cabin Quarry was reopened and important remains of *Allosaurus, Diplodocus, Camarasaurus, Stegosaurus*, and an ankylosaur were recovered.

See also the following related entries:

BEHAVIOR • MIGRATION • MORRISON FORMATION • TAPHONOMY

References

Breithaupt, B. H. (1990). Biography of William Harlow Reed: The story of a frontier fossil collector. *Earth Sci. History* **9**, 6–13.

Colbert, E. H. (1968). *Men and Dinosaurs*, pp. 283. Dutton, New York.

McIntosh, J. S. (1990). The second Jurassic dinosaur rush. *Earth Sci. History* **9**, 22–27.

Norell, M. A., Gaffney, E. S., and Dingus, L. (1995). *Discovering Dinosaurs in the American Museum of Natural History*, pp. 204. Knopf, New York.

Osborn, H. F. (1899). Fore and hind limbs of carnivorous and herbivorous dinosaurs from the Jurassic of Wyoming. *Am. Museum Nat. History Bull.* **5**, 161–172.

Osborn, H. F. (1904). Fossil wonders of the West. The dinosaurs of Bone Cabin Quarry. *Century Magazine* **68**, 680–694.

Bone Chemistry

In addition to information on the chemical composition of bone and other preservable material, there are also issues relevant to preservation of interest to paleontologists.

see CHEMICAL COMPOSITION OF DINOSAUR FOSSILS; PERMINERALIZATION

Brain

Studies of the brains of dinosaurs are done by making computerized axial tomography scans of well-preserved, relatively complete, and undeformed skulls, by making endocranial casts, and by examining natural endocranial casts.

see BRAINCASE ANATOMY; PALEONEUROLOGY; SKULL, COMPARATIVE ANATOMY

Braincase Anatomy

PHILIP J. CURRIE
Royal Tyrrell Museum of Palaeontology
Drumheller, Alberta, Canada

The braincase is generally one of the most poorly understood regions of the dinosaur skeleton. It is often partially or completely obscured by other skull bones. Parts of it do not ossify or are fragile and easily destroyed, and it is a complex of numerous bones pierced by nerves, blood vessels, and pneumatic diverticula. Nevertheless, braincases have been described for each of the major dinosaurian lineages. Increased use of noninvasive computerized tomography (CT) scanning (see COMPUTERS AND RELATED TECHNOLOGY) has also revealed details that could previously only have been visible through serial sectioning, which would invariably result in the destruction of most of the specimen.

Because the braincase is not directly subject to the same selective pressures as parts of the skeleton involved in the acquisition and processing of food (teeth, limb proportions, etc.), in the protection of the animal from predators (i.e., defensive horns, spikes, and armor), or in sexual or other display structures, its morphology tends to be conservative within a lineage. Comparison of braincases among taxa can therefore provide clues to relationships that may otherwise be obscured in more rapidly evolving parts of the body.

The braincase houses the brain; study of the endocranial cavity can approximate the overall size of the brain (see INTELLIGENCE) and show the relative development of different parts of the braincase. Because most cranial nerves and blood vessels pass through foramina and canals in the braincase, and the positions of these openings are conservative in all tetrapods, most of the openings can be identified in fossil skulls. The positions and sizes of these openings can provide information on interrelationships and can give clues about sensory abilities. Because the braincase also forms the inner walls of the middle ear, adjacent bones are often invaded by pneumatic diverticula from the middle ear air sac. These openings are less regular than those of the nerves and blood vessels and can be asymmetrical.

The ossified braincase can develop from as many as 23 separate centers of ossification, from bones of both dermal and endochondral origin. Some of the braincase components are paired, some are medial and singular, and others are complexes of several bones that are co-ossified in even the youngest animals. Braincases usually fuse up completely in mature animals, and most of the sutures are difficult to see. Dermal bones include the frontals and parietals on the skull roof and the parasphenoid, which fuses with the basisphenoid. Specific endochondral bones ossify in particular regions of the chondrocranium and can be identified by their consistent relationships to the cranial nerves and to the inferred positions of the cartilages.

The braincase is roofed by the frontals and parietals in a consistent manner. The ventral surface of the frontal usually has well-defined impressions of the olfactory tract and the cerebral hemispheres, and depressions in the ventral surface of the paired parietals show what the top of the back part of the brain looked like. In dinosaurs such as troodontids (Russell, 1969), the enlarged cerebral hemispheres mostly covered the midbrain so that it left no impression in the frontals. Hadrosaurs were relatively large-brained animals (Hopson, 1979), which is reflected by the doming of the frontal bones.

The occiput is normally formed by four bones (supraoccipital, basioccipital, and a pair of exoccipitals), all of which form the margins of the foramen magnum. The supraoccipital generally makes a contribution to the dorsal margin of the foramen magnum, the sides are formed by the paired exoccipitals, and the basioccipital makes a small contribution to the ventral margin. The supraoccipitals of *Lesothosaurus* (Weishampel and Witmer, 1990a), heterodontosaurids (Weishampel and Witmer, 1990b), hypsilophodonts (Galton, 1989), stegosaurs (Gilmore, 1914), and protoceratopsians (Brown and Schlaikjer, 1940) are larger than normal and contribute to the entire dorsal margin of the foramen magnum. In iguanodonts (Taquet, 1976; Norman, 1986; Galton, 1989), hadrosaurs (Langston, 1960; Weishampel and Horner, 1990), ceratopsids (Hatcher *et al.*, 1907), and possibly *Pachycephalosaurus* (Maryanska and Osmólska, 1974), the supraoccipital is excluded from the margin of the foramen magnum by the exoccipitals, which meet on the midline. The supraoccipital normally makes only a narrow contribution to the dorsal margin of the foramen magnum in the vast majority of dinosaurs, including *Camptosaurus* (Gilmore, 1909), sauropods (Madsen *et al.*, 1995), and most pachycephalosaurids (Maryanska and Osmólska, 1974). An anterior extension of the supraoccipital contacts the prootic and laterosphenoid in all neoceratopsians.

The two epiotic bones probably form from separate centers of ossification than the supraoccipital. They can make a small contribution to the occiput, but in most specimens they have fused indistinguishably to the supraoccipital (Currie and Zhao, 1993a).

In all known dinosaurs the exoccipital is fused without a trace of sutures to the opisthotic, and together they form a conspicuous paroccipital process. The paroccipital process meets the squamosal, parietal, and quadrate in a loose butt joint in most dinosaurs but fuses to the quadrate and squamosal in nodosaurid ankylosaurs. In neoceratopsians, the distal end of the paroccipital process is expanded and is embedded in a slot in the squamosal (Brown and Schlaikjer, 1940). The passages of the 12th cranial nerves are completely enclosed within the exoccipital, which also forms the posteromedial margin of the metotic fissure, through which passed the internal jugular vein and cranial nerves X and XI. The jugular and associated nerves can be diverted posteriorly to exit on the occiput in some dinosaurs (Currie and Zhao, 1993b). The skull has a prominent knob-like process, known as the occipital condyle, that articulates with the first cervical vertebra (the atlas). It is formed by the basioccipital and both exoccipitals. Usually more than two-thirds of the condyle is formed by the basioccipital, which separates the dorsolateral contributions from the exoccipitals. In neoceratopsians (Hatcher *et al.*, 1907; Brown and Schlaikjer, 1940), the basioccipital is excluded from the foramen magnum by the exoccipitals, which can form up to two-thirds of the ball-like occipital condyle in ceratopsids.

The otic capsule is formed from three centers of ossification, although only the outer two—the prootic and opisthotic—can generally be distinguished. As previously mentioned, the epiotic ossifies with the supraoccipital in the tectum synoticum of the chondrocranium. In protoceratopsians and probably ceratopsids (Brown and Schlaikjer, 1940), part of the epiotic is exposed on the lateral surface of the braincase between the prootic, opisthotic, laterosphenoid, and parietal. The facial nerve (VII) and the anterior (vestibular) and posterior (cochlear) branches of the eighth cranial nerve pass through the prootic, which also forms the posterior margin of the exit for cranial nerve V. The prootic sends a tongue-like process posteriorly to extensively overlap the opisthotic above the stapedial recess. The crista prootica often extends ventrolaterally into a wing-like process that forms the anterior wall of a pneumatic cavity in the side of the basisphenoid. This process, referred to by S. Welles (personal communication, 1996) as the preotic pendant, is usually formed in part by the basisphenoid, but in some cases can be formed almost entirely by the basisphenoid (in which case it is called the *ala basisphenoidalis*). The opisthotic co-ossifies with the exoccipital within the occipital arch of the chondrocranium. It forms the anteroventral borders of the opening for the ninth and tenth cranial nerves and for the stapedial recess. The otic capsule encloses the inner ear and semicircular canals and forms a conspicuous bulge on the inner wall of the endocranial cavity. The floccular recess invades the capsule anteromedially and tends to be relatively large in small animals. These and other bones adjacent to the middle ear sac are invaded by pneumatic diverticula in theropods such as ornithomimids (Osmólska *et al.*, 1972), troodontids (Currie, 1985; Currie and Zhao, 1993b), and tyrannosaurs (Russell, 1970) (Fig. 1).

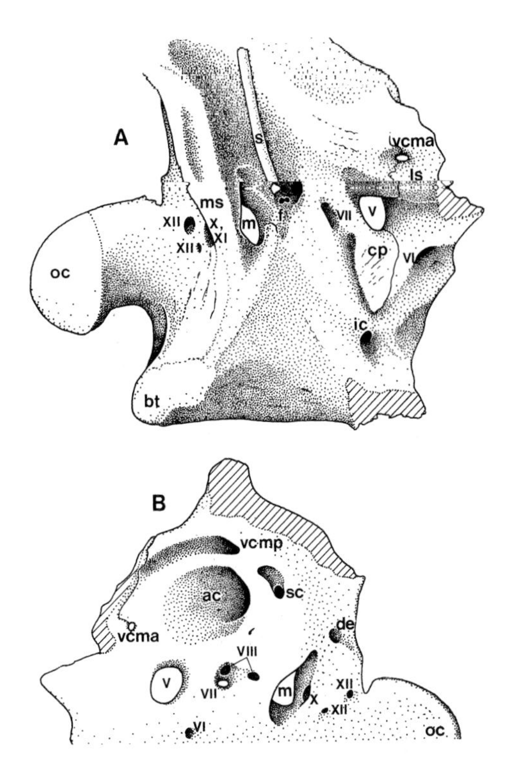

FIGURE 1 Braincase of *Dromaeosaurus albertensis* from right posteroventral (A) and medial (B) views. ac, fossa auriculae cerebelli; bt, basal tubera; de, ductus endolymphaticus; f, fenestra ovalis and fenestra pseudorotunda; ic, internal carotid; ls, laterosphenoid; m, metotic fissure; ms, metotic strut; oc, occipital condyle; pn, pneumatic space, s, stapes; vcma, anterior canal of middle cerebral vein. Roman numerals represent cranial nerves. (From Currie (1995) with permission.)

The floor of the braincase ossifies within the basal plate of the chondrocranium into the basioccipital posteriorly and the basisphenoid anteriorly. The basioccipital forms the floor of the metotic fissure and effectively makes up the lower margins of the exit foramina for cranial nerves IX to XI. Beneath the occipital condyle are a pair of processes called the basal tubera, formed primarily by the basioccipital but supported anteriorly by the basisphenoid. Primitively, a large pneumatic sinus (known by many names, including the basisphenoidal recess and Rathke's Pouch) opens ventrally. Although bounded mostly by the basisphenoid, the posterior wall is formed primarily by the basioccipital. The basicranial complex is pierced on the midline by a eustachian opening in most neoceratopsians (Dodson and Currie, 1990), except in *Bagaceratops* in which the opening passes between the basioccipital and basisphenoid. The sixth cranial nerve passes anteroventrally through the basisphenoid behind the dorsum sellae. In coelurosaurian dinosaurs (Currie and Zhao, 1993b), including birds, cranial nerve VI exits the basisphenoid lateral to the pituitary, whereas in most dinosaurs (including allosaurids, sauropods, hadrosaurs, and pachycephalosaurids) it enters the hypophyseal fossa in primitive fashion. The pituitary is nested in the fossa (which is also called the hypophyseal recess or pituitary fossa) anterior to the dorsum sellae.

The basisphenoid and parasphenoid are indistinguishably fused in all dinosaurs, with the possible exception of pachycephalosaurids (Maryanska and Osmólska, 1974). Anteriorly they taper into the distinctive cultriform process (parasphenoid rostrum), which supports the interorbital septum. In ornithomimids, therizinosaurids, and troodontids (Barsbold, 1983), the parasphenoid is expanded into a pneumatic, balloon-like structure referred to as the bulbous parasphenoid. Ventrally the same bones form the paired basipterygoid processes, which articulate with the pterygoids. Posterodorsal to the basipterygoid process, the lateral wall of the basisphenoid is invaded by the internal carotid. This artery passes anteromedially to meet its counterpart on the midline in the hypophyseal recess.

Like other archosaurs, including crocodiles and birds, the laterosphenoid (pleurosphenoid of some authors) of dinosaurs is ossified in the pila antotica. It extends dorsolaterally in a wing-like process that contacts the frontal, parietal, and postorbital. Ventrally it forms the margin of the foramen for the various branches of the fifth cranial nerve. One of these branches—the ophthalmic—is usually expressed on the laterosphenoid either by a groove on the lateral surface or by a canal enclosed within the bone. The ophthalmic branch of the trigeminal is separate in *Allosaurus* (Hopson, 1979), protoceratopsids (Brown and Schlaikjer, 1940), ceratopsids (Brown, 1914), troodontids (Currie and Zhao, 1993b), and tyrannosaurids (Bakker *et al.*, 1988), but the branches exit from a single opening in all other dinosaurs. Anteriorly, the laterosphenoid forms the posterior border of the foramen for the third cranial nerve, and the fourth passes through its upper regions.

Anterior to the laterosphenoid is a series of ossifications within the interorbital cartilages that show great ontogenetic and taxonomic variability. They are either absent or unknown in most smaller and primitive dinosaurs, including ceratosaurs (Welles, 1984; Raath, 1985), protoceratopsids (Brown and Schlaikjer, 1940), and psittacosaurids (Sereno, 1987). There is also considerable confusion concerning what each of these ossifications should be called. The orbitosphenoids, which develop late in the pila metotica of the chondrocranium, are a pair of small ossifications that form around the common (carcharodontosaurids, troodontids, and tyrannosaurids) or separate openings (allosaurids) on the midline for the second (optic) cranial nerves. The orbitosphenoid also borders the opening for the third cranial nerve and probably the fourth in at least some cases. Other than theropods, the orbitosphenoid is also known in sauropods (Madsen *et al.*, 1995), hadrosaurs (Lull and Wright, 1942), iguanodontids (Norman, 1986), and pachycephalosaurids (Maryanska and Osmólska, 1974). Dorsally, the sphenethmoid can ossify to form an elongate tube or pair of tubes beneath the frontal for the olfactory tracts and bulbs. According to S. Welles (personal communication, 1996), there is a separate ossification, which he calls the septosphenoid, in some theropods beneath the parietal and behind the sphenethmoid. The interorbital septum can also ossify into a thin, vertical sheet of bone between the sphenethmoid and the cultriform process (Maryanska, 1977; Madsen *et al.* 1995; Coria and Currie, 1997).

Despite the complexity of braincases and lack of much comparative information about them, they provide much information about individual dinosaurs. One research area where they have been particularly useful in elucidating relationships has been in the origin of birds debate. Differences in terminology are slowly being resolved, and with the advent of CT scanning, braincases have become more accessible. It is therefore highly likely that the number of publications describing braincases will steadily continue to increase in coming decades.

See also the following related entries:

PALEONEUROLOGY • SKULL, COMPARATIVE ANATOMY

References

Bakker, R. T., Williams, M., and Currie, P. J. (1988). *Nanotyrannus*, a new genus of pygmy tyrannosaur from the latest Cretaceous of Montana. *Hunteria* **1**(5), 1–30.

Barsbold, R. (1983). Carnivorous dinosaurs from the Cretaceous of Mongolia. *Joint Soviet–Mongolian Paleontol. Expedition Trans.* **19,** 5–120. [In Russian]

Brown, B. (1914). *Anchiceratops*, a new genus of horned dinosaur from the Edmonton Cretaceous of Alberta. With discussion of the origin of the ceratopsian crest and the brain casts of *Anchiceratops* and *Trachodon*. *Am. Museum Nat. History Bull.* **33,** 539–548.

Brown, B., and Schlaikjer, E. M. (1940). The structure and relationships of *Protoceratops*. *Proc. N. Y. Acad. Sci.* **40,** 133–266.

Coria, R. A., and Currie, P. J. (1997). The braincase of *Giganotosaurus carolinii* (Dinosauria: Theropoda) from the Upper Cretaceous of Argentina. Manuscript in preparation.

Currie, P. J. (1985). Cranial anatomy of *Stenonychosaurus inequalis* (Saurischia, Theropoda) and its bearing on the origin of birds. *Can. J. Earth Sci.* **22,** 1643–1658.

Currie, P. J. (1995). New information on the anatomy and relationships of *Dromaeosaurus albertensis* (Dinosauria: Theropoda). *J. Vertebr. Paleontol.* **15,** 576–591.

Currie, P. J., and Zhao, X. J. (1993a). A new large theropod (Dinosauria, Theropoda) from the Jurassic of Xinjiang, People's Republic of China. *Can. J. Earth Sci.* **30,** 2037–2081.

Currie, P. J., and Zhao, X. J. (1993b). A new troodontid (Dinosauria, Theropoda) braincase from the Dinosaur Park Formation (Campanian) of Alberta. *Can. J. Earth Sci.* **30,** 2231–2247.

Dodson, P., and Currie, P. J. (1990). Neoceratopsians. In *The Dinosauria* (D. B. Weishampel, P. Dodson, and H. Osmólska, Eds.), pp. 593–618. Univ. of California Press, Berkeley.

Galton, P. M. (1989). Crania and endocranial casts from ornithopod dinosaurs of the families Dryosauridae and Hypsilophodontidae (Reptilia: Ornithischia). *Geol. Palaeontol.* **23,** 217–239.

Galton, P. M. (1990). Stegosauria. In *The Dinosauria* (D. B. Weishampel, P. Dodson, and H. Osmólska, Eds.), pp. 435–455. Univ. of California Press, Berkeley.

Gilmore, C. W. (1909). Osteology of the Jurassic reptile *Camptosaurus*, with a revision of the species of the genus, and descriptions of two new species. *U.S. Natl. Museum Proc.* **36,** 197–332.

Gilmore, C. W. (1914): Osteology of the armoured Dinosauria in the United States National Museum, with special reference to the genus *Stegosaurus*. *Bull. U.S. Natl. Museum* **89,** 1–136.

Hatcher, J. B., Marsh, O. C., and Lull, R. S. (1907). The Ceratopsia. *U.S. Geol. Survey Monogr.* **49,** 1–300.

Hopson, J. A. (1979). Paleoneurology. In *Biology of the Reptilia, Volume 9* (C. Gans, Ed.), pp. 39–146. Academic Press, London.

Langston, W., Jr. (1960). The vertebrate fauna of the Selma Formation of Alabama. Part VI, the Dinosaurs. *Fieldiana Geol. Mem.* **3,** 313–360.

Lull, R. S., and Wright, N. F. (1942). Hadrosaurian dinosaurs of North America. *Geol. Soc. Am. Spec. Papers* **40,** pp. 200.

Madsen, J. H., Jr., McIntosh, J. S., and Berman, D. S. (1995). Skull and atlas–axis complex of the Upper Jurassic sauropod *Camarasaurus* Cope (Reptilia: Saurischia). *Carnegie Museum Nat. History Bull.* **31,** 1–115.

Maryanska, T. (1977). Ankylosauridae (Dinosauria) from Mongolia. *Palaeontol. Polonica* **37,** 85–151.

Maryanska, T., and Osmólska, H. (1974). Pachycephalosauria, a new suborder of ornithischian dinosaurs. *Palaeontol. Polonica* **30,** 45–102.

Norman, D. B. (1986). On the anatomy of *Iguanodon atherfieldensis* (Ornithischia: Ornithopoda). *Inst. R. Sci. Nat. Belgique Bull.* **56,** 281–372.

Osmólska, H., Roniewicz, E., and Barsbold, R. (1972). A new dinosaur, *Gallimimus bullatus*, n. gen. n. sp. (Ornithomimidae) from the Upper Cretaceous of Mongolia. *Palaeontol. Polonica* **27,** 103–143.

Raath, M. A. (1985). The theropod *Syntarsus* and its bearing on the origin of birds. In *The beginnings of Birds* (M. K. Hecht, J. H. Ostrom, G. Viohl, and P. Wellnhofer, Eds.), pp. 219–227. Freunde des Jura-Museums Eichstätt, Willibaldsburg, Eichstätt.

Russell, D. A. (1969). A new specimen of *Stenonychosaurus* from the Oldman Formation of Alberta. *Can. J. Earth Sci.* **6,** 595–612.

Russell, D. A. (1970). Tyrannosaurs from the Late Cretaceous of western Canada. National Museum of Natural Sciences, Publ. Palaeontology No. 1, pp. 34.

Sereno, P. C. (1987). The ornithischian dinosaur *Psittacosaurus* from the Lower Cretaceous of Asia and the relationships of the Ceratopsia. Ph.D. thesis, Columbia University, New York.

Taquet, P. (1976). *Géologie et Paléontologie du gisement de Gadoufaoua (Aptien du Niger)*, pp. 180. Cahiers de Paléontologie, CNRS, Paris.

Weishampel, D. B., and Horner, J. R. (1990). Hadrosauridae. In *The Dinosauria* (D. B. Weishampel, P. Dodson, and H. Osmólska, Eds.), pp. 534–561. Univ. of California Press, Berkeley.

Weishampel, D. B., and Witmer, L. M. (1990a). *Lesothosaurus, Pisanosaurus* and *Technosaurus*. In *The Dinosauria* (D. B. Weishampel, P. Dodson, and H. Osmólska, Eds.), pp. 416–425. Univ. of California Press, Berkeley.

Weishampel, D. B., and Witmer, L. M. (1990b). Heterodontosauridae. In *The Dinosauria* (D. B. Weishampel, P. Dodson, and H. Osmólska, Eds.), pp. 486–497. Univ. of California Press, Berkeley.

Welles, S. P. (1984). *Dilophosaurus wetherilli* (Dinosauria, Theropoda), osteology and comparisons. *Palaeontographica A* **185,** 85–180.

Brazilian Dinosaurs

Theropods and sauropods have been found mostly in Campanian–Maastrichian sediments.

see South American Dinosaurs

British Dinosaurs

The first scientific reports of dinosaurs were based on specimens from Britain. New specimens and taxa continue to be described and analyzed.

see European Dinosaurs; History of Dinosaur Discoveries

British Museum of Natural History

see Natural History Museum, London

Bullatosauria

Kevin Padian
John R. Hutchinson
University of California
Berkeley, California, USA

Bullatosauria (Fig. 1) refers to the clade of arctometatarsalian theropods recognized originally by Kurzanov (1976), then by Currie (1985), and formalized by Holtz (1994), who later (Holtz, 1996) clarified it to represent a node-based clade including *Ornithomimus* and *Troodon* (= *Stenonychosaurus*) and all descendants of their most recent common ancestor. The monophyly of the taxon Bullatosauria (composed of Troodontidae and Ornithomimosauria) has been well supported by recent analyses (e.g., Holtz 1995) and seems to have its origin in the Early Cretaceous (Holtz, 1994). Tyrannosaurids appear to be their closest relatives.

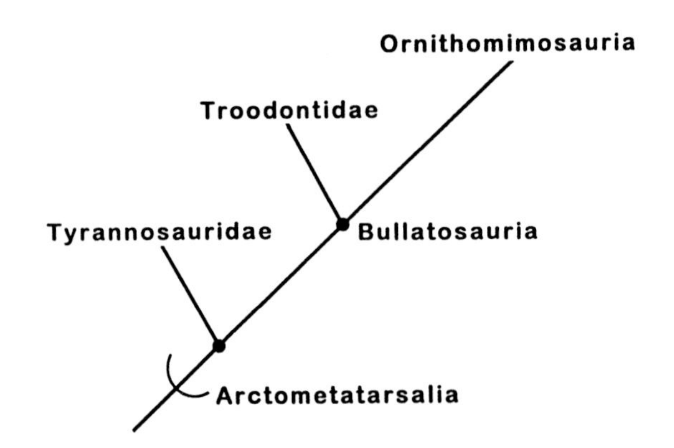

FIGURE 1 Phylogeny of Bullatosauria.

The term Bullatosauria (Latin *bullatus*, meaning inflated) refers to the bulbous parasphenoid capsule beneath and behind the orbit (see Braincase Anatomy), which is considered a valid synapomorphy that unites the clade. Other diagnostic characteristics of Bullatosauria include an enlarged braincase and eyes and a ventrally deflected occipital region. Braincase size in these animals is four to seven times higher than expected in a crocodile of the same size ("encephalization quotient": see Intelligence). These characteristics suggest but do not necessarily indicate some degree of heightened cerebral function compared to most other nonavian theropods. Both ornithomimosaurs and troodontids have brain:body size ratios among the highest for nonavian Reptilia, that is, encephalization comparable to that of an ostrich or an early mammal.

See also the following related entries:

Arctometatarsalia • Coelurosauria

References

Currie, P. C. (1985). Cranial anatomy of *Stenonychosaurus inequalis* (Saurischia, Theropoda) and its bearing on the origin of birds. *Can. J. Earth Sci.* **22,** 1643–1658.

Holtz, T. R., Jr. (1994). The phylogenetic position of the Tyrannosauridae: Implications for theropod systematics. *J. Paleontol.* **68,** 1100–1117.

Holtz, T. R., Jr. (1995). A new phylogeny of the Theropoda. *J. Vertebr. Paleontol.* **15**(Suppl. to No. 3), 35A.

Holtz, T. R., Jr. (1996). Phylogenetic taxonomy of the Coelurosauria (Dinosauria: Theropoda). *J. Paleontol.* **70,** 536–538.

Kurzanov, S. M. (1976). Braincase structure in the carnosaur *Itemirus* n. gen. and some aspects of the cranial anatomy of dinosaurs. *Paleontol. Zhurnal* **1976,** 127–137. [In Russian]

Cabo Espichel

Martin Lockley
University of Colorado at Denver
Denver, Colorado, USA

Situated southwest of Lisbon, Portugal, in the cliffs below the famous monastery of Cabo Espichel, are a series of about 10 Late Jurassic, track-bearing layers. Most of the trackways are attributable to sauropods, but one was made by a limping theropod (Lockley *et al.*, 1994). One layer reveals evidence of a small herd of seven juvenile sauropods heading southwest. Not only is this the richest area for sauropod trackways in all of Europe but also local folklore reveals that the tracks have been known since the 13th century.

See also the following related entries:

European Dinosaurs • Footprints and Trackways

Reference

Lockley, M. G., Santos, V. F., Meyer, C. A., and Hunt, A. P. (Eds.) (1994). *Aspects of Sauropod Biology*. Gaia: Geoscience Magazine of the National Natural History Museum, Lisbon, Portugal.

California Academy of Sciences

see Museums and Displays

Cameroon Dinosaurs

see African Dinosaurs

Cameros Basin Megatracksite

José J. Moratalla
José L. Sanz
Universidad Autónoma de Madrid
Madrid, Spain

The Cameros Basin is located in the north branch of the Iberian Range, including part of the Spanish provinces of Burgos, Soria, and La Rioja. The basin is about 8000 km^2 wide and the sediments are approximately 9 km thick. The sediments of the Cameros Basin are Upper Jurassic (Tithonian)–Lower Cretaceous (Aptian) in age. Most of the dinosaur tracks are located in the "Huérteles" Alloformation (Middle Berriasian) and the "Enciso" group (Berriasian–Lower Aptian) (Moratalla, 1993). The deposits of the Huérteles Alloformation come from a fluvial channel system discharging into a shallow saline lake. During the dryness periods playa-like environments appeared. The sediments of the Enciso group indicate fluvial channels and lacustrine environments.

The scarce sauropod tracks at Cameros Basin have been identified as *Brontopodus* and *Parabrontopodus*. The hindlimb prints are up to 50–60 cm in length.

Four theropod track morphotypes can be found. Among the nonavian theropod prints the ichnogenus *Buckeburgichnus* is one of the most abundant. This ichnogenus ranges up to 70 cm in length. The digits are broad and robust, and the length/width ratio ranges between 0.95 and 1.2. *Buckeburgichnus* is well represented at the "Los Cayos" tracksite (La Rioja) (Fig. 1) in which 425 tracks, including 36 trackways, have been identified. Some footprints from Los Cayos occasionally present a clear impression of the hallux.

The second theropod track morphotype ranges between 15 and 30 cm in length. The digits are robust. The heel is occasionally elongated and there is a strong medial indentation. The length/width ratio

FIGURE 1 Los Cayos tracksite (La Rioja province, Spain): several theropod trackways. The site has been protected, fenced, and roofed by Iberdrola Co.

ranges between 1.3 and 1.5. Similar dinosaur tracks have been reported from Moab, Utah (Lockley, 1991).

The most slender nonavian theropod tracks from Cameros have a length range between 12 and 30 cm. Digits are long and thin and the length/width ratio is 1–1.2.

Finally, some tracks from the Los Cayos (Moratalla and Sanz, 1992) and Serrantes tracksite (Soria) (Fuentes, 1997) have been interpreted as avian in origin.

Ornithopod tracks from the Cameros Basin are represented by four morphotypes. The first one comprises broad footprints of large size, between 40 and 60 cm in length. The digits are short and robust and two indentations (lateral and medial) delimit the heel area. Most of these tracks belong to bipedal trackways, but occasional quadrupedal ones have been reported. This is the case for the sites known as Cabezón de Cameros (La Rioja) (Fig. 2) (Moratalla *et al.*, 1992) and Regumiel de la Sierra (Burgos) (Moratalla *et al.*, 1994).

The largest ornithopod footprints have a broad heel surface and shorter and more robust digits than the first morphotype referred to previously (Fig. 3). The length/width ratio ranges from 0.9 to 1.12. The trackway gauge is wide and the stride length is short. All the available evidence seems to indicate that the trackmaker is a stout, graviportal iguanodontid with a foot skeleton close to that of hadrosaurs (Moratalla *et al.*, 1988).

At Valdeté tracksite (La Rioja), another ornithopod track morphotype has been found. The length/width ratio is larger than 1.2, and the trackway is narrow. The trackmaker has been tentatively identified as being close to camptosaurs.

Finally, the smallest ornithopod tracks, 10–15 cm in length, have been found at Valdevajes site (La Rioja). The trackmaker has been tentatively assigned to hypsilophodontids.

Most of the Cameros Basin trackway ichnorecord has been produced by theropods (80%). Ornithopods are represented by 16% and, finally, sauropod tracks are scarce, representing about 4% (Moratalla, 1993). These percentages are similar to those of the Glen Rose Formation (Texas), and, to a lesser degree, to those of the Sousa Formation (Brazil) (Lockley and Conrad, 1989). The great differences between the Cameros Basin and the English Wealden are striking, taking into account the facies similarity and the synchronicity of both areas. These differences are difficult to explain, but it is possible that many Wealden tracks, supposedly ornithopod in origin, can actually be identified as theropod footprints.

The Cameros Basin is characterized by an evident change in the dinosaur ichnofauna from the Middle Berriasian to the Upper Berriasian–Lower Aptian. Two main features can be distinguished: (i) The oldest ichnofauna is more clearly dominated by theropods than the youngest one—the ichnogenus *Buckeburgichnus* is predominantly recorded in this youngest ichnofauna; and (ii) both the record and the diversity of ornithopod and sauropod trackways clearly increase from the Mid-Berriasian to the Upper Berriasian–Lower Aptian.

The Cameros Basin ichnorecord suggests a gregarious behavior within theropods, iguanodontids, and

FIGURE 2 Cabezón de Cameros tracksite (La Rioja province, Spain): quadrupedal ornithopod trackway.

FIGURE 3 La Magdalena tracksite (La Rioja province, Spain): graviportal ornithopod trackway.

small ornithopods (hypsilophodontids?). The relative speeds in all the recorded trackways have the following ratio: $\lambda / h < 2$ (Alexander, 1976). The values range from 0.40 to 1.84. The average ratio for theropods is 1.18, suggesting a larger biodynamic index than that of ornithopods. The lowest value is that of the iguanodontid trackway of Regumiel de la Sierra. The Cameros Basin track evidence indicates an inversely proportional relationship between the λh ratio and the size of the trackmaker. Thus, if the smallest theropod footprints of the Los Cayos (and other sites) actually belong to the same species of those of the largest ones, juvenile individuals had a greater biodynamic index. This evidence suggests that immature individuals would be more agile and mobile than the adults. This conclusion could also be suggested for other evidence involving structurally similar (but distinct) species of different sizes.

See also the following related entries:

European Dinosaurs • Footprints and Trackways

References

Alexander, R. McN. (1976). Estimates of speeds of dinosaurs. *Nature* **261,** 129–130.

Fuentes, C. (1997). Primeras huellas de aves en el Jurásico de Soria (Wealdense, Formación Cameros, Grupo

Oncala). Descripción del nuevo icnogénero "Archaeornithipus." *Estudios Geológicos*, in press.

Lockley, M. G. (1991). *Tracking Dinosaurs*, pp. 250. Cambridge Univ. Press, Cambridge, UK.

Lockley, M. G., and Conrad, K. (1989). The paleoenvironmental context, preservation and paleoecological significance of dinosaur tracksites in the western USA. In *Dinosaur Tracks and Traces* (D. D. Gillette and M. G. Lockley, Eds.), pp. 121–134. Cambridge Univ. Press, Cambridge, UK.

Moratalla, J. J. (1993). Restos indirectos de dinosaurios del registro español: Paleoicnología de la Cuenca de Cameros (Jurásico superior–Cretácico inferior) y Paleoología del Cretácico superior, pp. 729. Unpublished PhD thesis, Univerisidad Autónoma de Madrid. Spain.

Moratalla, J. J., and Sanz, J. L. (1992). Icnitas aviformes en el yacimiento del Cretácico inferior de Los Cayos (Cornago, La Rioja, España). *Zubía* **10,** 153–160.

Moratalla, J. J., Sanz, J. L., and Jiménez, S. (1988). Nueva evidencia icnológica de dinosaurios en el Cretácico inferior de La Rioja (España). *Estudios geológicos* **44,** 119–131.

Moratalla, J. J., Sanz, J. L., Jiménez, S., and Lockley, M. G. (1992). A quadrupedal ornithopod trackway from the Lower Cretaceous of La Rioja (Spain): Inferences on gait and hand structure. *J. Vertebr. Paleontol.* **12**(2), 150–157.

Moratalla, J. J., Sanz, J. L., and Jiménez, S. (1994). Dinosaur tracks from the Lower Cretaceous of Regumiel de la Sierra (province of Burgos, Spain): Inferences on a new quadrupedal ornithopod trackway. *Ichnos* **3,** 89–97.

Canada–China Dinosaur Project

see SINO–CANADIAN DINOSAUR PROJECT

Canadian Dinosaurs

CLIVE COY
Royal Tyrrell Museum of Palaeontology
Drumheller, Alberta, Canada

Canada has a well-deserved reputation as a productive field for dinosaur discoveries. The western prairie provinces and their extensive modern river valleys have produced thousands of dinosaur specimens that are housed in institutions around the world. The first dinosaur bones found in Canada were discovered by G. M. Dawson of the Canadian Geological Survey near Morgan Creek, Saskatchewan, in 1874. In 1884, Joseph Tyrrell discovered dinosaurs along the Red Deer River valley near the present city of Drumheller, Alberta. Between 1910 and 1917 extensive, valuable collections of dinosaur skeletons were made along the Red Deer River. Since this period of intense collection on the prairies, dinosaurs have been discovered from Canada's chilling Arctic to the pounding surf of the Bay of Fundy in Nova Scotia.

Late Triassic vertebrates have been recovered from the Heilberg Formation of the Northwest Territories and from several formations on the east coast. *Arctosaurus osborni* from the Carnian–Rhaetian is sometimes thought to be an indeterminate theropod. The Lower Wolfville Formation (Carnian) has produced indeterminate prosauropod and ornithischian remains, whereas the Upper Wolfville Formation (Norian) has ornithischian footprints. The Blomidon Formation (Norian) has yielded theropod footprints.

Early Jurassic dinosaurs have been recovered from the McCoy Brook Formation (Hettangian) of Nova Scotia and include saurischian footprints referred to the dinosaurs cf. *Anchisaurus sp.* and cf. *Ammosaurus sp.* and ornithischian footprints referred to the osteotaxon *Scutellosaurus sp.*

Early Cretaceous dinosaurs from the Gething and equivalent formations (Barremian–Aptian) of British Columbia and Alberta include theropod, ankylosaur, and ornithopod footprints.

By far the majority of dinosaurs that have been recovered from Canada lived during Late Cretaceous times. In the north, hadrosaur bones have been found in the Maastrichtian Bonnet Plume Formation of the Yukon and the Kangguk Formation of the Northwest Territories. Theropod bones have also been recovered from the latter, whereas the Summit Creek Formation (Late Maastrichtian) has produced indeterminate ceratopsid bones. Cenomanian footprints (theropods, ankylosaurs, and ornithopods) are known from the Dunvegan Formation of Alberta and British Columbia, along with some poorly preserved bones and teeth.

Early Campanian dromaeosaurids, tyrannosaurids, hadrosaurs, ceratopsids, and nodosaurids are represented mostly by teeth in the Milk River Formation. Some of the richest dinosaur deposits in the world are located in DINOSAUR PROVINCIAL PARK and nearby sites in southern Alberta. These middle to late Campanian beds (Judithian LAND MAMMAL AGE) have produced a huge range of dinosaurs including indeterminate theropods (*Richardoestesia*), dromaeosaurids (*Dromaeosaurus* and *Saurornitholestes*), caenagnathids (*Chirostenotes*), elmisaurids (*Elmisaurus*), avimimids, tyrannosaurids (*Aublysodon, Gorgosaurus,* and *Daspletosaurus*), ornithomimids (*Struthiomimus, Dromicieomimus,* and *Ornithomimus*), troodontids (*Troodon*), a possible therizinosaur (cf. *Erlikosaurus*), many hadrosaurs (*Brachylophosaurus, Gryposaurus, Prosaurolophus, ?Maiasaura sp., Corythosaurus, Hypacrosaurus, Lambeosaurus,* and *Parasaurolophus*), hypsilophodonts (*Orodromeus*), pachycephalosaurids (*Stegoceras, Gravitholus, Pachycephalosaurus,* and *Ornatotholus*), protoceratopsids (*Leptoceratops* and *Montanoceratops*), ceratopsids (*Anchiceratops, Chasmosaurus, Centrosaurus,* and *Styracosaurus*), and ankylosaurs (*Edmontonia, Panoplosaurus,* and *Euoplocephalus*).

The late Campanian–early Maastrichtian beds of the Horseshoe Canyon Formation near Drumheller are also rich with skeletons of indeterminate theropods (*Richardoestesia*), dromaeosaurids (*Dromaeosaurus* and *Saurornitholestes*), caenagnathids (*Chirostenotes*), tyrannosaurids (*Aublysodon, Albertosaurus,* and *Daspletosaurus*), ornithomimids (*Struthiomimus, Dromicieomimus,* and *Ornithomimus*), troodontids (*Troodon*), hypsilophodontids (*Parksosaurus*), hadrosaurs (*Edmontosaurus, Saurolophus,* and *Hypacrosaurus*), pachycephalosaurids (*Stegoceras*), ceratopsids (*Anchiceratops, Arrhinoceratops,* and *Pachyrhinosaurus*), and ankylosaurs (*Edmontonia* and *Euoplocephalus*).

Latest Maastrichtian dinosaurs of the Scollard Formation include *Tyrannosaurus rex, Thescelosaurus, Edmontosaurus, Leptoceratops, Triceratops, Torosaurus,* and *Ankylosaurus.*

In Saskatchewan, Judithian beds have yielded the remains of tyrannosaurids, ornithomimids, caenagnathids (*Chirostenotes*), dromaeosaurids (*Saurornitholestes* and *Dromaeosaurus*), hadrosaurs, and ceratopsians. *Dromaeosaurus, Saurornitholestes, Chirostenotes, Tyrannosaurus, Thescelosaurus, Edmontosaurus, Triceratops, Torosaurus,* and *Ankylosaurus* have all come from the higher Lancian age beds.

This is not a complete listing of the localities, formations, and genera of dinosaurs found in Canada but serves to show the richness of these resources. Traditional sites such as DINOSAUR PROVINCIAL PARK (see DINOSAUR PARK FORMATION) and Drumheller (see HORSESHOE CANYON FORMATION; ROYAL TYRRELL MUSEUM OF PALAEONTOLOGY) continue to produce new specimens every year, while new sites such as DEVIL'S COULEE continue to be found.

See also the following related entries:

AMERICAN DINOSAURS • MEXICAN DINOSAURS • POLAR DINOSAURS

Cañon City

KENNETH CARPENTER
Denver Museum of Natural History
Denver, Colorado, USA

The Upper Jurassic dinosaur beds of the Morrison Formation near Cañon City, Colorado, were first excavated in 1877 by crews working for O. C. Marsh and E. D. Cope (e.g., Cope, 1877; Marsh, 1877). Some of the best known dinosaurs from the Morrison Formation, such as *Camarasaurus supremus, Diplodocus longus, Allosaurus fragilis, Ceratosaurus nasicornis,* and *Stegosaurus stenops,* were first named from specimens collected from there. Many other genera and species were also named but most of these are no longer considered valid taxa. One exception is the little enigmatic *Nanosaurus agilis*, the first dinosaur named from Cañon City.

The most prolific site, Marsh Quarry 1, has produced more than 12 species of dinosaurs, including the sauropods "*Morosaurus*" *agilis, Diplodocus longus, Haplocanthosaurus priscus,* and *Brachiosaurus*; the theropods *Allosaurus fragilis, Ceratosaurus nasicornis, Coelurus agilis,* and *Elaphrosaurus* sp.; the ornithopods *Dryosaurus altus* and *Othnelia rex*; and the stegosaurs *Stegosaurus armatus* and *Stegosaurus stenops* (Carpenter, 1997a).

From Cope's sites, at a stratigraphically higher level, the sauropods *Camarasaurus supremus, Amphicoelias altus,* and the giant *Amphicoelias fragillium* were named. *Amphicoelias fragillium* is based on a partial vertebra that, whole, would have been more than 2.4 m tall! The entire animal would have been about 45 m long, making it the biggest dinosaur yet discovered (Carpenter, 1997a).

Recently, another species of *Haplocanthosaurus*, *H. delfsi*, has been named from a level lower than the Marsh Quarry (Carpenter, 1997a; McIntosh and Williams, 1988). Also at this level, the oldest dinosaur eggs from North America, called *Prismatoolithus coloradensis*, have recently been found (Hirsch, 1994).

Another recent discovery is that of a nearly complete skeleton of *Stegosaurus stenops* with most of the armor preserved in place (Carpenter, 1997b). This specimen showed the plates on the back in two alternating rows on both sides of the neural spines. In addition, the tail spikes were found projecting posterolaterally, numerous small, keeled disks of armor were found over the pelvis, and a wattle of small ossicles was found under the neck.

Recent work has shown that the Morrison Formation at Cañon City can be divided into a lower *Haplocanthosaurus* zone and an upper *Camarasaurus* zone (Carpenter, 1997a). The replacement of *Haplocanthosaurus* by *Camarasaurus* corresponds to an environmentally linked facies change. With dinosaur specimens distributed throughout the formation, the Cañon City dinosaur beds offer a unique place to study the evolution of dinosaur faunas during the Upper Jurassic.

See also the following related entry:

MORRISON FORMATION

References

Carpenter, K. (1997a). Vertebrate biostratigraphy of the Morrison Formation near Cañon City, Colorado. *Modern Geology*, in press.

Carpenter, K. (1997b). Armor of *Stegosaurus stenops*, and the taphonomic history of a new specimen from Garden Park, Colorado. *Modern Geology*, in press.

Cope, E. D. (1877). On reptilian remains from the Dakota beds of Colorado. *Proc. Am. Philos. Soc.* **17,** 193–196.

Hirsch, K. F. (1994). Upper Jurassic eggshells from the Western Interior of North America. In *Dinosaur Eggs and Babies* (K. Carpenter, K. F. Hirsch, and J. Horner, Eds.), pp. 137–150. Cambridge Univ. Press, New York.

Marsh, O. C. (1877). Notice of some new dinosaurian reptiles from the Jurassic formation. *Am. J. Sci.* **14,** 514–516.

McIntosh, J. S., and Williams, M. (1988), A new species of sauropod dinosaur, *Haplocanthosaurus delfsi* sp. nov., from the Upper Jurassic Morrison Formation of Colorado. *Kirtlandia* **43,** 3–26.

Canterbury Museum, New Zealand

see MUSEUMS AND DISPLAYS

Carenque

MARTIN LOCKLEY
University of Colorado at Denver
Denver, Colorado, USA

The Carenque dinosaur tracksite is situated in the suburb of Lisbon that bears that name. It reveals what, at the time of discovery in 1992, was the longest trackway in the world (127 m), which was made by a large bipedal dinosaur in Cretaceous (Cenomanian) sediments. As detailed in an entire book on the subject (Galopim, 1994), the site was saved from destruction by diverting a freeway through a tunnel beneath the tracks. During the course of this project the trackway was further excavated to a length of 141 m. (see FATIMA and KHODJA-PILATA entries for further information on the world's longest trackways).

See also the following related entries:

EUROPEAN DINOSAURS • FOOTPRINTS AND TRACKWAYS

Reference

Galopim, A. M. de C. (1994). *Dinosaurios e a batalha de Carenque*, pp. 291. Editorial Noticias, Lisbon, Spain.

Carnegie Museum of Natural History

JOHN S. MCINTOSH
Wesleyan University
Middletown, Connecticut, USA

The Carnegie Museum of Natural History in Pittsburgh, Pennsylvania, was one of the first institutions in the United States to amass a major collection of dinosaur fossils. Its collection of SAUROPOD dinosaurs remains today the greatest in the United States. On the basis of a newspaper article reporting the discovery of a large dinosaur in Wyoming, Andrew Carnegie, steel magnate and philanthropist, charged Dr. William J. Holland, director of his new museum, to purchase the animal. Holland hired experienced paleontologists such as Jacob Wortman and later John Bell Hatcher, collectors including the legendary William H. Reed, and preparators such as Arthur Coggeshall to carry out major expeditions to the West, beginning in 1898 and continuing unabated until 1923. The first of these succeeded in collecting the greater portion of a skeleton of the JURASSIC sauropod *Diplodocus carnegii* from the MORRISON FORMATION of Sheep Creek, Wyoming. A second skeleton of that animal was collected from the same quarry in the following year, and together they formed the basis of a composite mounted skeleton, only the second such of a sauropod dinosaur. From 1899 until his untimely death in 1904, Hatcher broadened the scope of dinosaur exploration, dispatching parties to a number of quarries in Wyoming, on Sheep Creek and also to the Freezeout Hills and the Red Fork of the Powder River. He also reopened the famous quarry in Garden Park, Colorado, that had provided Professor O. C. Marsh of Yale with most of the Jurassic dinosaur skulls collected before 1900, including those of *Diplodocus*, *Allosaurus*, *Ceratosaurus*, and *Stegosaurus*. Nearly complete skeletons of the latter three of these ended up in the National Museum of Natural History in Washington, DC, where they are now mounted and on exhibition. This quarry also yielded Hatcher two partial skeletons of the new and very rare sauropod *Haplocanthosaurus*. From a second quarry on Sheep Creek, a party led by C. W. Gilmore collected a fine skeleton of *Apatosaurus* (*Brontosaurus*), which Holland planned to mount next to the *Diplodocus*. A partial skeleton of a very young individual of the same animal (originally called *Elosaurus*) was found with it. These plans were disrupted by the discovery in August 1909 of the greatest of all Jurassic dinosaur quarries by Earl Douglass in what is now the DINOSAUR NATIONAL MONUMENT, north of Jensen, Utah. The first specimen taken from this quarry proved to be the most complete skeleton of *Apatosaurus* ever found. Named *A. louisae* by Holland for Andrew Carnegie's wife, this skeleton was mounted next to the *Diplodocus* in 1913, and the Wyoming *Apatosaurus* was eventually sent to the University of Wyoming in Laramie, where it is now on exhibition. The Carnegie Quarry at Dinosaur National Monument continued to yield skeleton after skeleton, many of which, including those of *Allosaurus*, *Camptosaurus*, *Dryosaurus*, and *Stegosaurus*, were eventually mounted in the Carnegie Museum. Perhaps the finest specimen was a nearly complete and articulated skeleton of a juvenile *Camarasaurus*, which formed the basis of a panel mount just as it lay in the quarry. For the first time many sauropod skulls were recovered, including three of *Diplodocus*, several of *Camarasaurus*, and, until very recently, the only known skull of *Apatosaurus*. The Carnegie collection from the quarry includes a large number of skeletal elements of all three of these animals as well as the neck and part of the thorax of the huge, long-necked, and very rare sauropod *Barosaurus*. Also among the non-sauropods were a partial skeleton of a very young *Dryosaurus* and much *Stegosaurus* material.

Although not nearly as numerous, important dinosaur specimens from the CRETACEOUS period are also present in the Carnegie collections. Among these are mounted skeletons of *Tyrannosaurus rex* (the original or type specimen obtained from the American Museum of Natural History), the duck-billed *Corythosaurus*, and the primitive Mongolian *Protoceratops*. Also on exhibit is a fine skull of *Triceratops*. The collections further contain a skeleton of *Edmontosaurus*, casts of many bones of the primitive European ankylosaur *Struthiosaurus*, as well as many scattered specimens of tyrannosaurids, hadrosaurids, and ceratopsids. The only TRIASSIC dinosaur material in the collection consists of specimens of *Coelophysis* from the famous GHOST RANCH QUARRY in New Mexico and the cast of a complete hindfoot of the prosauropod *Plateosaurus* from Trossingen, Germany.

See also the following related entries:

DINOSAUR NATIONAL MONUMENT • MUSEUMS AND DISPLAYS

Carnosauria

John R. Hutchinson
Kevin Padian
University of California
Berkeley, California, USA

Carnosauria (Figs. 1 and 2) was a name coined by F. von Huene (1914, 1920, 1926) to include a variety of large theropod dinosaurs with great skulls and enormous teeth. It was distinguished from Coelurosauria, which were generally the smaller, lightly built theropods. Since that time both terms have been used in variously formal and informal senses, but it has often been thought that the large forms comprised a more or less natural group, the lineage beginning with Early Jurassic forms such as *Dilophosaurus* and extending through the Late Jurassic *Ceratosaurus, Megalosaurus,* and *Allosaurus* to the Late Cretaceous tyrannosaurs (though von Huene did not regard tyrannosaurs or *Ceratosaurus* as members of this group).

Gauthier (1986, p. 26) redefined Coelurosauria in cladistic terms as a stem-based taxon comprising birds and all theropods closer to birds than to Carnosauria. Hence, Coelurosauria and Carnosauria were sister taxa within what Gauthier (1986) called Tetanurae, a stem-based taxon. Holtz (1994) formalized Avetheropoda as the node-based taxon within Tetanurae comprising Coelurosauria and Carnosauria. Gauthier, however, did not formally define Carnosauria apart from listing included taxa: the genera *Allosaurus, Acrocanthosaurus, Indosaurus, Alectrosaurus, Dryptosaurus, Albertosaurus, Alioramus, Daspletosaurus, Indosuchus, Tarbosaurus,* and *Tyrannosaurus*. By implication, Carnosauria would be defined as these taxa and all others closer to them than to birds within Tetanurae.

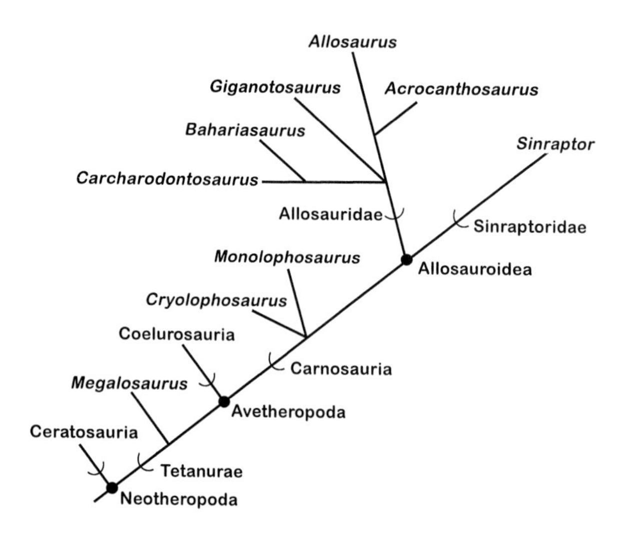

FIGURE 1 Phylogeny of Carnosauria, drawn mostly from Holtz (1994, 1995, 1996). Sereno *et al.* (1996), in contrast, find a monophyletic group composed of *Giganotosaurus, Acrocanthosaurus,* and *Carcharodontosaurus,* with the other taxa in this diagram collapsed into a polytomy together comprising Allosauridae.

Gauthier had formally defined Carnosauria with some misgivings, acknowledging that many characters that appeared to distinguish the group were probably size-related convergences, as they were with large ceratosaurs and "megalosaurs." He also noted that tyrannosaurids were further derived within the group, approaching in some respects the features of elmisaurids, ornithomimids, and *Hulsanpes*. These misgivings turned out to be prophetic. Novas (1992) realized that tyrannosaurs did not belong in Carnosauria, and an analysis by Holtz (1994) indicated that they were the sister group to Bullatosauria (Troodontidae + Ornithomimidae) within Coelurosauria. Hence tyrannosaurs, despite their great size, are not carnosaurs but rather coelurosaurs, as von Huene (1914, 1920, 1926) first recognized. In short, the implication of this taxonomic shift is that large body size evolved many times within Theropoda rather than only once within the traditional "Carnosauria."

The reassignment of tyrannosaurids to the Coelurosauria removes *Albertosaurus, Alectrosaurus, Alioramus, Daspletosaurus, Tarbosaurus,* and *Tyrannosaurus* from Carnosauria. As Molnar (1990) summarized and Bonaparte (1991) elaborated on, some large forms often considered carnosaurs in a more or less formal sense, such as *Carnotaurus, Indosaurus, Indosuchus, Majungasaurus,* and *Xenotarsosaurus,* are related to *Abelisaurus* as members of the Abelisauridae (Bonaparte and Novas, 1985), neoceratosaurian theropods, and many others are represented by fragmentary remains that defy precise taxonomic assignment. This leaves *Allosaurus, Acrocanthosaurus,* and *Dryptosaurus. Dryptosaurus* may actually be a basal coelurosaur rather than a carnosaur (Denton, 1990; Holtz, 1995), so the taxonomic composition of Gauthier's cladistically defined Carnosauria reduces to two genera, *Allosaurus*

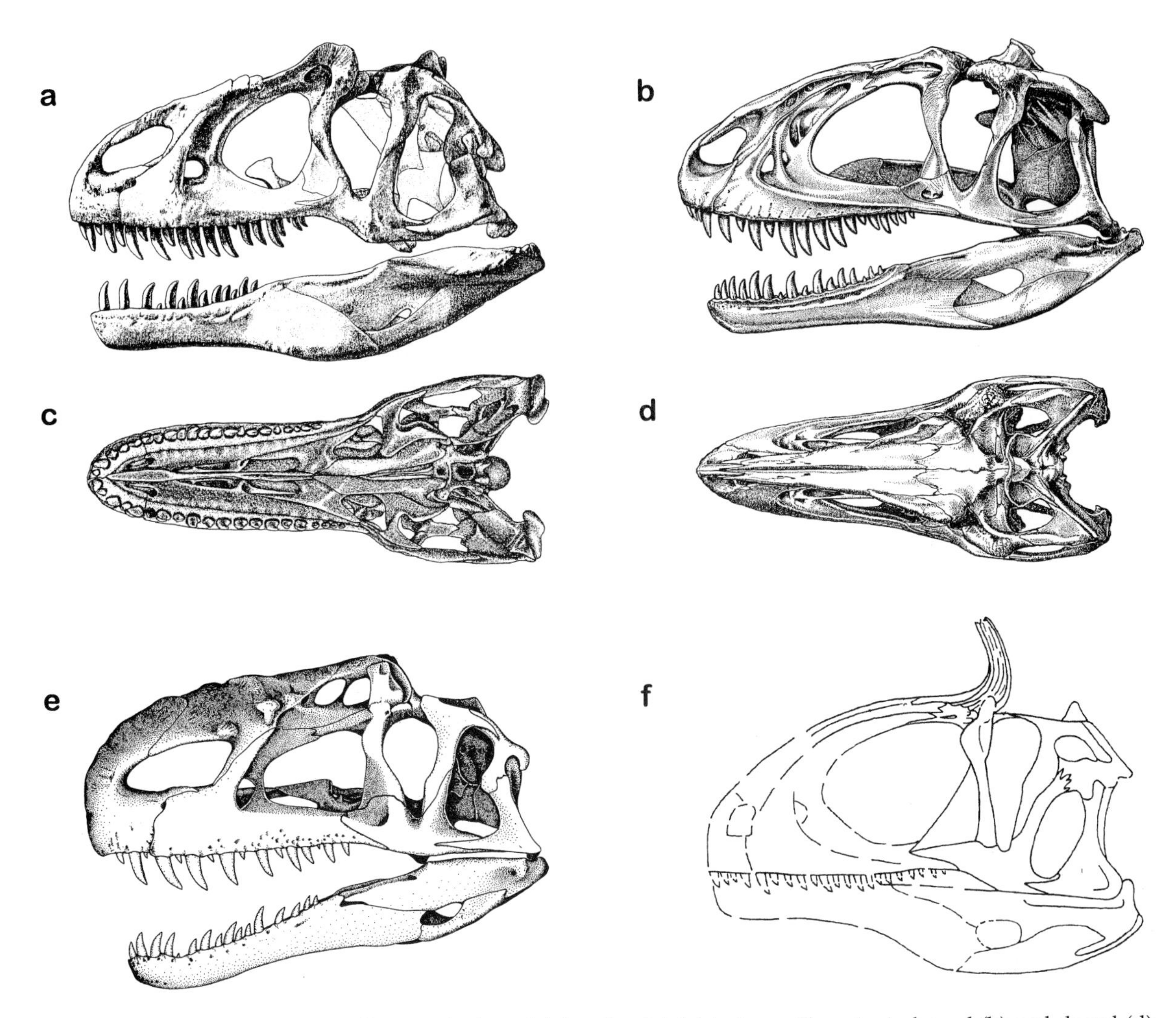

FIGURE 2 Carnosaur skulls. *Allosaurus* in lateral (a) and palatal (c) views; *Sinraptor* in lateral (b) and dorsal (d) views; (e) *Monolophosaurus;* (f) *Cryolophosaurus.*

and *Acrocanthosaurus,* that are already regarded as closely related. Holtz (1994) included both in the Allosauridae, distinguished by the form of the pubic foot (longer anteriorly than posteriorly, triangular in ventral view). This rather weak unity is by default of incomplete or incompletely described material in other taxa, including *Acrocanthosaurus, Chilantaisaurus, Piatnitzkysaurus,* and *Szechuanosaurus* (Molnar *et al.*, 1990). Holtz (1995) regarded the last two genera as possible members of this branch but outside Allosauridae.

Accordingly, the definition of Carnosauria can be formally amended to include *Allosaurus* and all Avetheropoda closer to *Allosaurus* than to birds. Recent finds, such as *Giganotosaurus* in Argentina (Coria and Salgado, 1995) and new *Carcharodontosaurus* material from Morocco (Sereno *et al.*, 1996), have added to the known diversity of the Carnosauria. They have also provided evidence for new biogeographical hypotheses (Currie, 1996) and revealed the huge sizes that some carnosaurs attained (the latter two taxa may have reached larger body sizes than any known tyrannosaurs). Currie (1996) and Rauhut (1995) pointed out that, during the Late Cretaceous Period, different continental regions were dominated by different clades of large theropods (abelisaurids in South America, tyrannosaurids in North America and Asia, and carcharodontosaurs in Africa), although there were obviously numerous regions of overlap where these predators must have come into contact.

Current versions of the Carnosauria eloquently demonstrate the changing views of theropod phylogeny. Holtz (1995) included many recently described taxa in his analysis and found that many of them were grouped as a monophyletic, globally distributed Carnosauria, including taxa from Antarctica (*Cryolophosaurus*), China (*Monolophosaurus* and the Sinraptoridae: *Sinraptor* and *Yangchuanosaurus*), Africa (the Carcharodontosauridae: *Carcharodontosaurus* and *Bahariasaurus*), South America (*Giganotosaurus*, also apparently a carcharodontosaurid), and North America [the Allosauridae: *Allosaurus*, *Acrocanthosaurus* (possibly a carcharodontosaurid; Sereno *et al.*, 1996), and related taxa]. The stem-based ALLOSAUROIDEA includes the latter three node-defined clades (Sinraptoridae, Carcharodontosauridae, and Allosauridae).

The tremendous sizes reached by some carnosaurs (estimated at up to 8 tons; Coria and Salgado, 1995) raise many interesting biological questions (Molnar and Farlow, 1990). How did such gigantic bipeds manage to support their own weight? Scaling constraints must have limited their locomotory capabilities somewhat, but exactly how much is a difficult question to answer, as Molnar and Farlow (1990) have noted. It is, however, unlikely that giant carnosaurs were multiton speedsters, as some authors have depicted them (e.g., Paul, 1988); known trackways suggest that their gaits were moderate, though speeds more rapid than those recorded or calculated on the basis of available evidence are certainly possible (e.g., Alexander, 1985). At least one study concludes that if a large theropod fell at even a moderate speed, it would have risked breaking its leg, with the resultant disability soon proving fatal (Farlow *et al.*, 1995).

Biomechanical problems are not the only issues; ecological (especially trophic) considerations have been the focus of some amount of controversy. Could gigantic bipeds such as *Carcharodontosaurus* have been efficient predators, or were their primary ecological roles more as scavengers? The behavior and ecological interactions of extinct organisms cannot, of course, be monitored directly. Studies of large extant predators suggest that most carnivorous terrestrial vertebrates are opportunistic feeders, taking live or dead prey as they need it to sustain their metabolic demands. This does not preclude the existence of giant scavenging theropods, but it does cast some doubt on the ecological restriction of such animals to a wholly scavenging lifestyle. All known carnosaurs are from localities also inhabited by large herbivores that would have been adequate fodder for them, and in many cases these faunas curiously lack other potential predators. It may be more fruitful to attempt to ask what kind and how much of an impact their predation had on large herbivore populations rather than whether they hunted them or not. Most predators do not engage the most robust individuals of prey populations but rather feed on the very young, old, diseased, isolated, or injured. Nor do they usually risk their own safety more than necessary, which is why prolonged combative engagements are generally avoided in favor of quick strikes that produce locomotor disability, bleeding, shock, and lacerations that encourage disease (note that the TOOTH SERRATIONS of theropods have pockets that might have harbored infective bacteria, as in the living Komodo monitor lizard). Injuries both to the skeletons of allosaurids and to those of their likely prey suggested to earlier investigators such as Gilmore and Lambe that carnosaurs focused their aggression both on the animals they hunted and on each other (see Molnar and Farlow, 1990). In carnosaurs and other large theropods, the forelimbs are short compared to the hindlimbs—perhaps as low as one-third the hindlimb length in allosaurids and independently as low as one-fifth in tyrannosaurids. Part of the hindlimb disparity is a carryover from the condition in basal tetrapods, as well as the ancestral bipedal habit of dinosaurs and other ornithodirans, which put less emphasis on the forelimb (see BIPEDALITY). It has often been assumed that reduction of the forelimb was a characteristic of theropods in general, which has been claimed as evidence against the possibility of birds, with their long forelimbs, evolving from theropods (see BIRD ORIGINS). However, this trend did not hold for coelurosaurs, particularly the maniraptorans, which include birds. Forelimb reduction has also been considered an inherent developmental consequence of large size, in which heterochronic processes favored the robusticity of the jaws and hindlimbs at the expense of the forelimbs, which had a reduced predatory function (see HETEROCHRONY). Developmental processes alone, however, cannot have driven this trend for large theropods in general, as witnessed by the 2.5-m-long arms of the giant ornithomimid *Deinocheirus*. The negative allometry that describes the small forelimbs of large theropods may or may not have been the work of selective forces. It is possible that the immediate forerunners of these animals used their forelimbs less in predation than their great jaws and tearing teeth. With increased body size, the predatory function of the forelimb would have been lessened further or eliminated, hence reducing its selective value as well as its unnecessary mass. Phylogenetic analysis of the basal members of the clades of large theropods could test this prediction, but currently these animals are not sufficiently known.

See also the following related entries:

COELUROSAURIA • ORNITHOMIMOSAURIA • TETANURAE • THEROPODA

References

Alexander, R. McN. (1985). Mechanics of posture and gait of some large dinosaurs. *Zool. J. Linnean Soc. London* **83,** 1–25.

Bonaparte, J. F. (1991). The Gondwanan theropod families Abelisauridae and Noasauridae. *Historical Biol.* **5,** 1–25.

Bonaparte, J. F., and Novas, F. E. (1985). *Abelisaurus comahuensis*, n.g., n. sp., Carnosauria del Cretacico tardio de Patagonia. *Ameghiniana* **21,** 259–265.

Coria, R. A., and Salgado, L. (1995). A new giant carnivorous dinosaur from the Cretaceous of Patagonia. *Nature* **377,** 224–226.

Currie, P. J. (1996). Out of Africa: Meat-eating dinosaurs that challenge *Tyrannosaurus rex*. *Science* **272,** 971–972.

Denton, R. K., Jr. (1990). A revision of the theropod *Dryptosaurus (Laelaps) aquilunguis* (Cope 1869). *J. Vertebr. Paleontol.* **9**(Suppl. to No. 3), 20A.

Farlow, J. O., Smith, M. B., and Robinson, J. M. (1995). Body mass, bone "strength indicator," and cursorial potential of *Tyrannosaurus rex*. *J. Vertebr. Paleontol.* **15,** 713–725.

Gauthier, J. A. (1986). Saurischian monophyly and the origin of birds. *Mem. California Acad. Sci.* **8,** 1–55.

Holtz, T. R., Jr. (1994). The phylogenetic position of the Tyrannosauridae: Implications for theropod systematics. *J. Paleontol.* **68,** 1100–1117.

Holtz, T. R., Jr. (1995). A new phylogeny of the Theropoda. *J. Vertebr. Paleontol.* **15**(Suppl. to No. 3), 35A.

Holtz, T. R., Jr. (1996). Phylogenetic taxonomy of the Coelurosauria (Dinosauria: Theropoda). *J. Paleontol.* **70,** 536–538.

Molnar, R. E. (1990). Problematic Theropoda: "Carnosaurs." In *The Dinosauria* (D. B. Weishampel, P. Dodson, and H. Osmólska, Eds.), pp. 280–305. Univ. of California Press, Berkeley.

Molnar, R. E., and Farlow, J. O. (1990). Carnosaurian paleobiology. In *The Dinosauria* (D. B. Weishampel, P. Dodson, and H. Osmólska, Eds.), pp. 210–224. Univ. of California Press, Berkeley.

Molnar, R. E., Kurzanov, S. M., and Dong, Z. (1990). Carnosauria. In *The Dinosauria* (D. B. Weishampel, P. Dodson, and H. Osmólska, Eds.), pp. 169–209. Univ. of California Press, Berkeley.

Novas, F. E. (1992). La evolución de los dinosaurios carnivóros. In *Los Dinosaurios y su Entorno Biotico. Actas II, Curso de Paleontológia en Cuenca* (J. L. Sanz and A. Buscalione, Eds.), pp. 123–163. Instituto "Juan de Valdes," Cuenca, Spain.

Paul, G. S. (1988). *Predatory Dinosaurs of the World*. Simon & Schuster, New York.

Rauhut, O. W. M. (1995). Zur systematischen Stellung der afrikanischen Theropoden *Carcharodontosaurus* Stromer 1931 und *Bahariasaurus* Stromer 1934. *Berliner Geowissenschaftlichen Abhandlungen* **E16,** 357–375.

Sereno, P. C., Dutheil, D. B., Iarochene, M., Larsson, H. C. E., Lyon, G. H., Magwene, P. M., Sidor, C. A., Varrichio, D. J., and Wilson, J. A. (1996). Predatory dinosaurs from the Sahara and Late Cretaceous faunal differentiation. *Science* **272,** 986–991.

von Huene, F. (1914). Das naturliche System der Saurischia. *Zentralblatt Mineral. Geol. Paläontol. B* **1914,** 154–158.

von Huene, F. (1920). Bemerkungen zur Systematik und Stammesgeschichte einiger Reptilien. *Zeitschrift Indukt. Abstammungslehre Vererbungslehre* **24,** 162–166.

von Huene, F. (1926). The carnivorous Saurischia in the Jura and Cretaceous formations, principally in Europe. *Revista Museo de La Plata* **29,** 35–167.

Carter County Museum, Montana, USA

see MUSEUMS AND DISPLAYS

Cedar Mountain Formation

JAMES I. KIRKLAND
Dinamation International Society
Fruita, Colorado, USA

The Lower Cretaceous of the Colorado Plateau is represented by the Cedar Mountain Formation in Utah and the largely correlative Burro Canyon Formation east of the Colorado River. When first named, the Cedar Mountain Formation was characterized as differing from the underlying Upper Jurassic MORRISON FORMATION in having an abundance of dinosaur gizzard stones (see GASTROLITHS) and in lacking preserved dinosaur bone. Its age was based on palynology and it was assumed to be largely correlative to the Aptian to middle Albian Cloverly Formation of Wyoming and Montana. In recent years, research has shown that there are considerable dinosaur remains in the Cedar Mountain Formation and that instead of preserving only one dinosaur fauna correlative with the Cloverly Formation, it preserves three successive distinct dinosaur faunas (Table I).

The basal fauna ranges upward from a regionally persistent calcrete, which marks the contact between the underlying Morrison Formation and the Jurassic–Cretaceous unconformity, upward to a regionally persistent ledge forming sandstone. This fauna is best developed in the region around Arches National Park. The fauna is characterized by a dominance of nodosaurid ankylosaurs such as "*Gastonia burgei*"; two species of iguanodont, *Iguanodon ottingeri* and a large undescribed sail-backed species; two undescribed species of sauropod, one with large spatulate teeth (cf. *Ornithopsis*); a small undescribed theropod; and the large dromaeosaur *Utahraptor* as the dominant meat-eating dinosaur. This fauna has close ties with the Upper Wealden fauna of the Isle of Wight in southern England and with a poorly known fauna from the Lakota Formation of the Black Hills region of South Dakota, which suggests a Barremian Age for the fauna. Apparently this is the oldest faunal level in the Cretaceous of North America and indicates that at least 25 million years of the earliest Cretaceous terrestrial record is missing in North America. This fauna also helps to establish the final time when terrestrial dispersal routes between Europe and Utah were open during the Cretaceous. The fauna predates the appearance of common flowering plants. Carbonate soil nodules indicate that the climate was dry and seasonal, but a greater abundance of fossil fishes, turtles, and crocodilians suggest that it was somewhat wetter than during the Late Jurassic.

The cliff-forming sandstone separating this fauna from the overlying middle fauna may represent as much as 10 million years. Within this sandstone interval a very large undescribed nodosaurid ankylosaur was excavated by Jim Jensen during the mid-1960s.

Although the middle Cedar Mountain fauna is the most widespread, being recognized in both the

TABLE I
Dinosaurs Unearthed from the Cedar Mountain Formation of Utah

Cedar Mountain Formation (Bodily 1969; Jensen 1970; Galton and Jensen 1979*b*; Nelson and Crooks 1987; Madsen pers. comm.; Stadtman pers. comm.; Nelson pers. comm.; Britt pers. comm.)
- Theropoda
  - Theropoda indet.
  - Troodontid indet.
  - Dromaeosauridae
  - cf. *Deinonychus* sp.
    - Dromaeosaurid indet.
- Sauropoda indet.
- Ornithischia indet.
- Ornithopoda
  - Iguanodontia
    - *Tenontosaurus tilletti*
  - ?Iguanodontidae
    - Undescribed ?iguanodontid (?2 species; Britt pers. comm.)
    - ?Iguanodontid indet. (= *Iguanodon ottingeri*)
  - ?Hadrosaurid indet.
  - Ankylosauria
    - Nodosauridae
      - *Sauropelta edwardsi* (= *Hoplitosaurus* sp.)
    - Ankylosaurid indet.
  - Dinosaur eggshell
  - Age: Albian (Tschudy et al. 1984)

?Dakota Formation (Marsh 1899; Carpenter pers. comm.; not figured)
- Sauropoda
  - Diplodocidae
  - ?*Barosaurus lentus*
- Ornithopoda
  - Hypsilophodontid indet.
  - Iguanodontid indet.
- Age: late Aptian-early Cenomanian (Tschudy et al. 1984)

NOTE. From Weishampel (1990).

Arches region and in the area of the San Rafael Swell to the west, it also is the most poorly known. It occurs from the top of the cliff-forming sandstone up to a thin discontinuous sandstone horizon that marks a sharp break between sediments containing abundant calcareous soil nodules below with sediments completely lacking carbonate soil nodules above. To date, the meager dinosaur remains excavated indicate the presence of the generalized ornithopod *Tenontosaurus*, the basal nodosaurine ankylosaur *Sauropelta*, the sauropod *Pleurocoelus*, the dromaeosaur *Deinonychus*, and the giant sail-backed theropod *Acrocanthosaurus*. This fauna was apparently unique to North America. Flowering plants were beginning to come into their own, signaling one of the greatest floristic changes in the earth's history. Although the climate had not changed appreciably, sea levels were rising globally and tectonic activity increased worldwide.

The upper fauna ranges to the top of the Cedar Mountain Formation and is associated with lignitic intervals that preserve a diversity of flowering plants and highly smectitic sediments, indicating increased volcanic activity. A diversity of freshwater actinopterygians, elasmobranchs, turtles, and crocodilians indicate considerably wetter climatic conditions. This upper interval is best developed in the San Rafael Swell region toward the Sevier Thrustbelt but may extend eastward as far as the east side of Arches. Radiometric dates of 98.5 million years old indicate that this fauna lived during the earliest Cenomanian on the western margin of the Mowry Sea. It is dominated by an undescribed species of basal hadrosaur, "*Eohadrosaurus caroljonesi*." The fauna also includes a small iguanodontid, hypsilophodontids, nodosaurs, cf. *Alectrosaurus*, and a diversity of small theropods including cf. *Troodon*, cf. *Paranycodon*, cf. *Richardoestei*, and dromaeosaurs. This fauna appears to have Asian affinities, where the probable ancestors of many of these dinosaurs resided. With the draining of the Dawson Strait across northwestern Canada at the end of the Lower Cretaceous, migration of dinosaurs from the contiguous Siberian–Alaskan land area could have led to the replacement of much of the preexisting Cloverly fauna. What is truly remarkable is that there is less change in the western North American dinosaur fauna over the following 35 million years of the Cretaceous than there had been between each of these three faunas in the Cedar Mountain Formation and that of the underlying Upper Jurassic Morrison Formation. This apparently reflects the changes in North America's ecosystems with the rapid diversification and spread of flowering plants and the transition from faunas with European affinities to those with Asian affinities.

Central Asian Dinosaurs

see CHINESE DINOSAURS; MONGOLIAN DINOSAURS

Central Asiatic Expeditions

Mark A. Norell
American Museum of Natural History
New York, New York, USA

The use of powdered fossil bones as a key ingredient in traditional Asian medicine implies that the "dragon bones" of central Asia have been known for centuries. They have been known to Western science for only about 75 years. Discovery of non-avian dinosaurs in central Asia was pioneered by the American Museum of Natural History (AMNH) during a series of swashbuckling Central Asiatic Expeditions (CAE) in the 1920's (Fig. 1) (Andrews, 1932). Conceived by the AMNH's Roy Chapman Andrews, the work was pitched by Director H. F. Osborn as a search for human ancestors, based on AMNH President Henry Fairfield Osborn's conviction that Asia, not Africa, was the cradle of humanity. On this account the expedition failed, but its successes highlighted the region as one of the most important dinosaur hunting grounds on the planet. The CAE spent field seasons in 1922, 1923, and 1925 in Mongolia and several more in China. More expeditions were planned, but the volatile political climate of the times impeded further exploration.

Incidents concerning the accidental discovery of the Flaming Cliffs at the end of the 1922 field season have been recounted several times (Andrews, 1932). Lost, the CAE caravan stopped near a couple of *gers* (the Mongolian noun for the Russian yurt—the familiar dwelling of central Asian nomads) to ask directions. While Andrews went to the ger, J. B. Schackelford (the expedition photographer) went in the other direction to check out some promising red rocks. As he closed in on the outcrop, he found himself standing on the edge of a large cliff, bordering an extensive area of badlands. Within minutes, Schackelford began to find fossils and alerted the rest of the party to his discovery. By the end of the afternoon a wealth of fossil material had been collected, including a *Protoceratops* skull and a fossil egg. Later this exposure was named the Flaming Cliffs, after the intense red

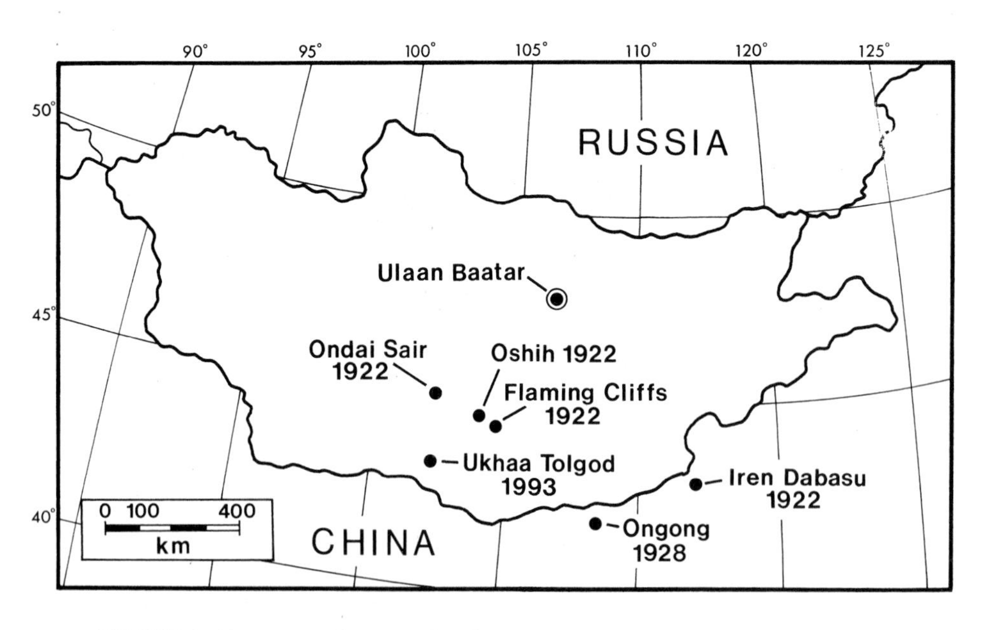

FIGURE 1 Important dinosaur localities discovered by the CAE and MAE.

color that they "take on" during sunset at Shabarakh Usu (*shabarak* = muddy; *usu* = water: referring to the spring at the base of the cliffs). In recent times the locality has become known by its traditional Mongolian name Bayn Dzak (*bayn* = many; *dzak* = a small tree). The size of the badlands and the large amount of bone made it a priority for continued work. However, because the season was running short, the expedition needed to push on immediately toward Kalgan and Beijing.

The following year they hot-footed it toward the locality and spent nearly the entire summer collecting fossils at this locality, returning again in 1925 (Fig. 2). Their total haul included the first definitive dinosaur eggs and nests ever discovered, complete skeletons of *Protoceratops* (Granger and Gregory, 1923), *Pinacosaurus* (Gilmore, 1933a), and remains of the three major Djadokhta theropods, *Oviraptor*, *Velociraptor*, and *Saurornithoides* (Osborn, 1923). The expedition also collected the first skulls of Cretaceous mammals (Gregory and Simpson, 1926) and lizards (Gilmore, 1943), discovered the nearby Paleocene locality of Gashato (khashat = corral) (Matthew and Granger, 1925), and collected archeological materials (including beads made from dinosaur egg-shell) from stabilized sand dunes at the cliff base (Berkey and Nelson, 1926).

In addition to the finds at the Flaming Cliffs, CAE paleontologists discovered several other sites that provided important specimens. Ondai Sair and Oshih are located in what is known as "Valley of the Lakes" in central Mongolia. These Early Cretaceous sites produced the type specimens of *Psittacosaurus* and its synonym *Protiguanodon* (Osborn, 1924). Although fossils are sparse at these localities, the CAE succeeded in collecting juvenile remains of *Psittacosaurus* (Andrews, 1932), dinosaur eggs, and fragmentary sauropod and theropod remains (Osborn, 1924).

In Inner Mongolia, under Chinese control, the CAE also made important dinosaur finds. In 1922 the first dinosaur fossils collected by the CAE were found at the Early Cretaceous locality of Iren Dabasu. At this locality remains of the primitive ornithomimid *Archeornithomimus*, the tyrannosaurid *Alectrosaurus*, and

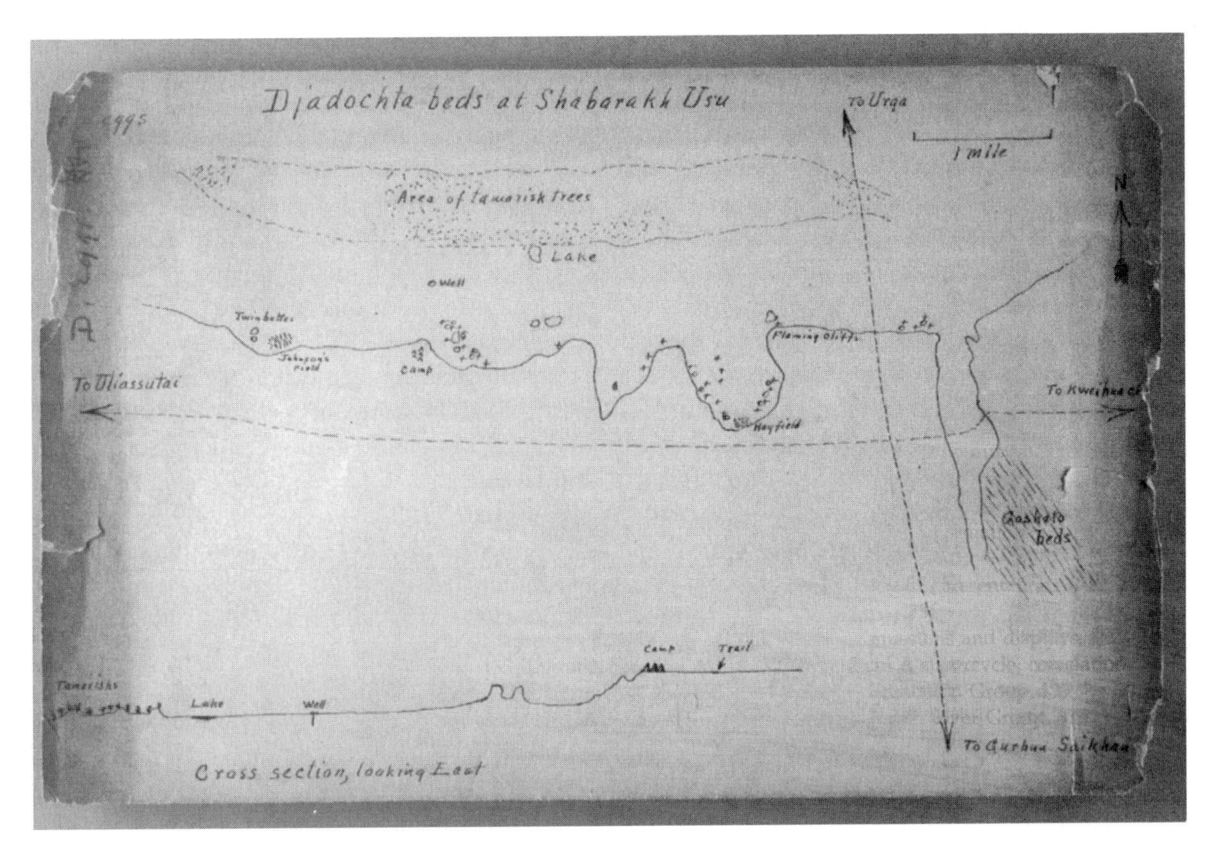

FIGURE 2 Sketch map of the Flaming Cliffs (Bayn Dzak) made after the 1925 expedition. X, fossil occurrence, O, egg or nest occurrence.

the hadrosaurs *Gilmoreosaurus* and *Bactrosaurus* were collected (Gilmore, 1933b). Sauropod material was collected in 1928 at other Inner Mongolian localities (Gilmore, 1933a). Because of political realignment, American Museum dinosaur collecting expeditions retreated from work in Asia at the end of the 1920s. However, this was only the end of the first chapter in Mongolian dinosaur collecting. In subsequent years important Mongolian, Russian (see Lavas, 1993), and Polish expeditions (Kielan-Jaworowska, 1969; see also Lavas, 1993) picked up where AMNH paleontologists left off and made important discoveries.

When the Cold War ended, AMNH expeditions were invited to resume field-work in 1990. These expeditions are organized in collaboration with the Mongolian Academy of Sciences and have been informally called the Mongolian American Museum Expedition (MAE). They differ substantially from the CAE in that they are truly collaborative and include a number of Mongolian scientists including Demberilyin Dashzeveg, Altangarel Perle, and Rinchen Barsbold. Another significant difference is that specimens collected during the MAE remain the property of the Mongolian Academy of Sciences and will be returned to Mongolia. Like the CAE, the MAE concerns itself with all aspects of the fossil fauna and is not restricted to dinosaurs. The MAE expeditions are still in progress; however, it is not too early to propose that results from the Mesozoic part of the work have equaled or exceeded those of the CAE (Novacek, 1996).

In 1990, AMNH paleontologists Michael Novacek, Malcolm McKenna, and Mark Norell traveled to Mongolia to negotiate plans for this new set of expeditions. During this visit they signed an agreement initiating the MAE. A preliminary field trip was taken in 1990, and full-scale expeditions commenced in 1991 and have continued through 1995. At this writing additional field excursions and visits to New York by our Mongolian colleagues have been approved through several years.

Most of the work has concentrated on localities near the Flaming Cliffs and in the Nemegt Basin. During the early years, classic localities discovered by American, Russian, Polish, and Mongolian scientists were visited. Several excellent specimens were recovered, especially from Khulsan (Norell *et al.*, 1992a) and Tugrugeen Shireh (Novacek and McKenna, 1993; Norell *et al.*, 1992b).

A major achievement of these reconnoiterings was the description of the primitive bird *Mononykus* (Perle *et al.*, 1993). The original description was based on two specimens, one collected during the Mongolian–Russian expedition of 1987 at Bugin Tsav and a second by the MAE at Tugrugeen Shireh. Since then, *Mononykus* specimens have been collected by MAE expeditions at several localities (Chiappe *et al.*, 1997). A *Mononykus* specimen collected at Bayn Dzak during the 1923 expedition, labeled only as "bird-like dinosaur," was identified in the AMNH collection (Norell *et al.*, 1993) (see Aves).

Thus far, the most significant result of the MAE is the discovery of Ukhaa Tolgod in July 1993 (Novacek *et al.*, 1994 Dashzeveg *et al.* 1995); (Fig. 3). Like most of the other Djadokhta (Bayn Dzak and Bayan Mandahu) and Djadokhta-like (red beds of Kheerman Tsav and Khulsan variously considered to be referable to the Barun Goyot Formation) localities, Ukhaa Tolgod is composed of predominantly red sandstones. The major unit at these localities is a massive cross-bedded eolian sandstone, sandwiched between two fluvial units (Eberth, 1993; Fastovsky *et al.*, 1994; Dashzeveg *et al.*, 1995). At Ukhaa Tolgod, most of the fossil specimens occur in the thick eolian unit. The site is unparalleled in the abundance of Cretaceous vertebrate fossil remains in all of Asia and perhaps the world (Novacek *et al.*, 1994; Dashzeveg *et al.*, 1995).

The most significant dinosaur discovery yet from Ukhaa Tolgod is a dinosaur nest (Norell *et al.*, 1994; Clark, 1995). Among the fragmented eggs, one contained a nearly complete embryo of a near-hatchling oviraptorid (Fig. 4). These are the first definitive remains of a theropod embryo. Curiously, in the same nest lay skulls of two neonate dromaeosaurids.

The oviraptorid eggs are identical to eggs collected by the CAE in 1923 at the Flaming Cliffs. The CAE eggs were referred to *Protoceratops* based on the abundance of this taxon at the site, although no direct association between these eggs and this ornithischian taxon was ever demonstrated. Since then, the protoceratopsian affinities of these eggs have been challenged (Sabath, 1991); however, the myth that these were the eggs of the small ceratopsian dinosaur continued to be propagated by museum displays, popular books, and the scientific literature. The egg and embryo discovered by the MAE provide definitive evidence of this misidentification.

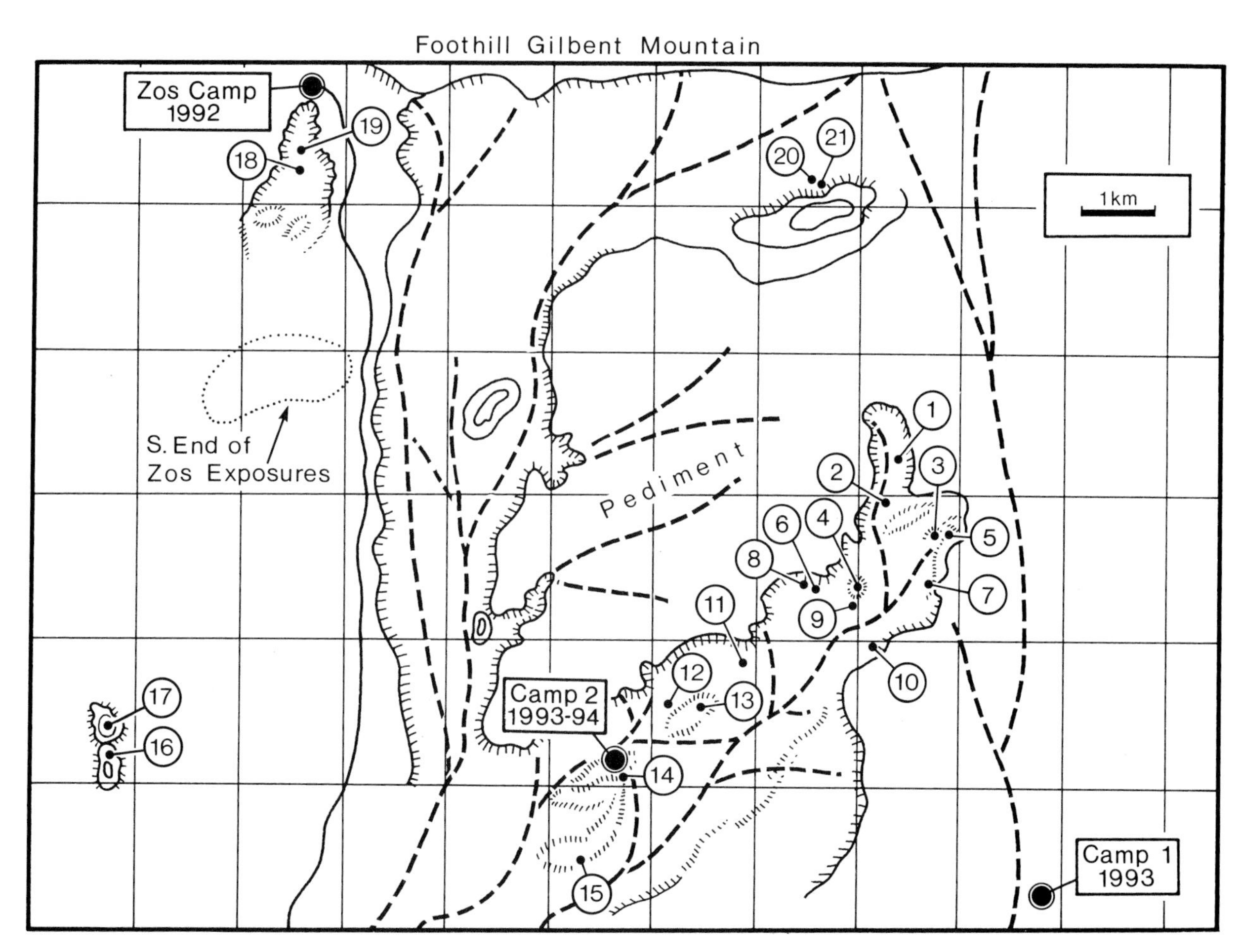

FIGURE 3 Map of the Ukhaa Tolgod basin with sublocalities indicated. Numbers correspond to section in Dashzeveg *et al.* (1995). The oviraptorid embryo (Norell *et al.*, 1993) was found at sublocality 7 (Xanadu).

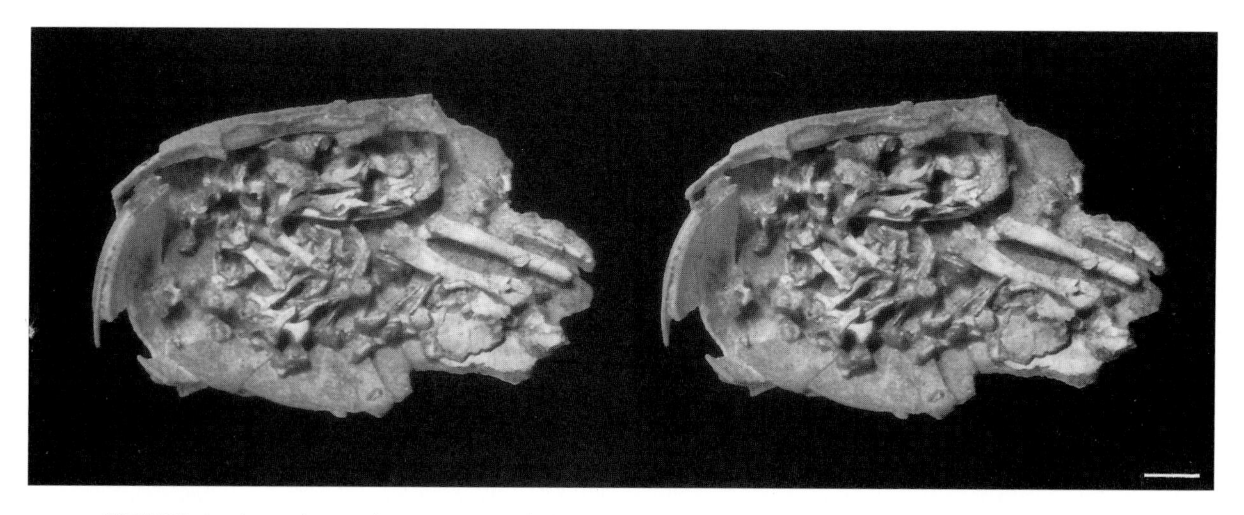

FIGURE 4 An embryo of an oviraptorid theropod collected from Ukhaa Tolgod. Scale bar = 1 cm.

In conclusion, the American Museum of Natural History has had a major impact on our knowledge of Asian dinosaurs. These expeditions discovered many of the classic dinosaur sites and paved the way for Mongolian, Russian (see Lavas, 1993), Polish (Kielan-Jaworowska, 1969; see also Lavas, 1993), Chinese, Canadian (Currie, 1994), Japanese (Watabe, 1994), and later generations of AMNH paleontologists (Novacek *et al.*, 1994; Norell *et al.*, 1995). Some of these, like Bayn Dzak and Ondai Sair, are still producing important dinosaur remains. AMNH expeditions through the early 1990's have been extremely successful in collecting excellent specimens from many of the classic Mongolian localities and discovering important new localities such as Ukhaa Tolgod. This work is still in its infancy, but important new information on dinosaurs is still being produced from old and new AMNH collecting efforts.

See also the following related entries:

American Museum of Natural History • Bayn Dzak • Polish–Mongolian Paleontological Expeditions • Sino-Canadian Dinosaur Project • Sino-Soviet Expeditions • Ukhaa Tolgod

References

Andrews, R. C. (1932). *The New Conquest of Central Asia. A Narrative of the Central Asiatic Expeditions in Mongolia and China, 1921–1930*, pp. 678. American Museum of Natural History, New York.

Berkey, C. P., and Nelson, N. C. (1926). Geology and prehistoric archeology of the Gobi Desert. *Amer. Mus. Novitates* **222,** 1–16.

Chiappe, L., Norell, M. A., and Clark, J. M. (1997). Phylogenetic position of *Mononykus olecranus* from the Upper Cretaceous of the Gobi Desert. Submitted for publication.

Clark, J. M. (1995). The egg thief exonerated. *Nat. History* **June,** 56–57.

Currie, P. J. (1994). Hunting ancient dragons in China and Canada. In *Dino Fest* (R. S. Spencer, Ser. Ed.; G. D. Rosenberg, and D. L. Wolberg, Eds.), Spec. Publ. No. 7, pp. 387–396. Paleontolgical Society.

Dashzeveg, D., Novacek, M. J., Norell, M. A., Clark, J. M., Chiappe, L. M., Davidson, A., McKenna, M. C., Dingus, L., Swisher, C., and Perle, A. (1995). Unusual preservation in a new vertebrate assemblage from the Late Cretaceous of Mongolia. *Nature* **374,** 446–449.

Eberth, D. A. (1993). Depositional environments and facies transitions of dinosaur-bearing Upper Cretaceous redbeds of Bayan Mandahu (Inner Mongolia, People's Republic of China). *Can. J. Earth Sci.* **30**(10/11), 2196–2213.

Fastovsky, D. E., Badamgarav, D., Ishimoto, H., and Watabe, M. (1994). Paleoenvironments of Tugrikin-Shire (Late Cretaceous: Mongolia) and *Protoceratops* (Dinosauria: Ornithischia). *J. Vertebr. Paleontol.* **14**(3), Suppl. 24A.

Gilmore, C. W. (1933a). Two new dinosaurian reptiles from Mongolia with notes on some fragmentary specimens. *Amer. Mus. Novitates* **679,** 1–20.

Gilmore, C. W. (1933b). On the dinosaurian fauna of the Iren Dabasu Formation. *Bull. Am. Museum Nat. History* **67,** 23–78.

Gilmore, C. W. (1943). Fossil lizards of Mongolia. *Bull. Am. Museum Nat. History* **81,** 361–384.

Granger, W., and Gregory, W. K. (1923). *Protoceratops andrewsi*, a pre-ceratopsian dinosaur from Mongolia. *Amer. Mus. Novitates* **72,** 1–9.

Gregory, W. K., and Simpson, G. G. (1926). Cretaceous mammal skulls from Mongolia. *Amer. Mus. Novitates* **22,** 1–20.

Kielan-Jaworowska, Z. (1969). *Hunting for Dinosaurs*, pp. 177. MIT Press, Cambridge, MA.

Lavas, J. R. (1993). *Dragons from the Dunes*, pp. 137. Published by the author.

Matthew, W. D., and Granger, W. (1925). Fauna and correlation of the Gashato Formation of Mongolia. *Amer. Mus. Novitates* **189,** 1–12.

Norell, M. A., McKenna, M. C., and Novacek, M. J. (1992a). *Estesia mongoliensis*, a new fossil varanoid from the Cretaceous Barun Goyot Formation of Mongolia. *Amer. Mus. Novitates* **3045,** 1–24.

Norell, M. A., Clark, J. M., and Perle, A. (1992b). New dromaeosaur material from the Late Cretaceous of Mongolia. *J. Vertebr. Paleontol.* **12**(3), Suppl. 45A.

Norell, M. A., Chiappe, L. M., and Clark, J. M. (1993, September). New limb on the avian family tree. *Nat. History*, 38–43.

Norell, M. A., Clark, J. M., Dashzeveg, D., Barsbold, R., Chiappe, L. M., Davidson, A. R., McKenna, M. C., Perle, A., and Novacek, M. J. (1994). A Theropod dinosaur embryo and the affinities of the Flaming Cliffs dinosaur eggs. *Science* **266,** 779–782.

Norell, M. A., Dingus, L., and Gaffney, E. S. (1995). *Discovering Dinosaurs*, pp. 225. Knopf, New York.

Novacek, M. J. (1996). *Dinosaurs of the Flaming Cliffs*. Anchor/Doubleday, New York.

Novacek, M. J., and McKenna, M. C. (1993). Therian mammals from the Late Cretaceous of Mongolia. *J. Vertebr. Paleontol.* **13**(3), Suppl. 51A.

Novacek, M. J., Norell, M. A., McKenna, M. C., and Clark J. (1994). Fossils of the Flaming Cliffs. *Sci. Am.* **December,** 60–69.

Osborn, H. F. (1923). Two Lower Cretaceous dinosaurs from Mongolia. *Amer. Mus. Novitates* **95,** 1–10.

Osborn, H. F. (1924). Three new Theropoda, *Protoceratops* zone, central Mongolia. *Amer. Mus. Novitates* **144,** 1–12.

Perle, A., Norell, M. A., Chiappe, L. M., and Clark, J. M. (1993). Flightless bird from the Cretaceous of Mongolia. *Nature* **362,** 623–626.

Sabath, K. (1991). Upper Cretaceous amniotic eggs from the Gobi Desert. *Acta Paleontol. Polonica* **36**(2), 151–192.

Watabe, M. (1994). Results of the Hayashibara Museum of Natural Sciences–Institute of geology, Academy of Sciences of Mongolia Joint Paleontological Expedition to the Gobi Desert in 1993. *J. Vertebr. Paleontol.* **14**(3), Suppl. 51A.

Central Geological and Prospecting Museum, Russia

see MUSEUMS AND DISPLAYS

Central State Museum, Mongolia

see MUSEUMS AND DISPLAYS;
PALEONTOLOGICAL MUSEUM, ULAAN BAATAR

Cerapoda

Cerapoda (Fig. 1) was established by Sereno (1986) to represent EUORNITHOPODA (ORNITHOPODA of other authors; see PHYLOGENY OF DINOSAURS) + MARGINOCEPHALIA, a major branch of ornithischia. The node is supported by several synapomorphies, including the substantial diastema between the premaxillary and maxillary teeth, the asymmetrical enamel on upper and lower teeth, reduction to five or fewer premaxillary teeth, and other features. An alternate view of

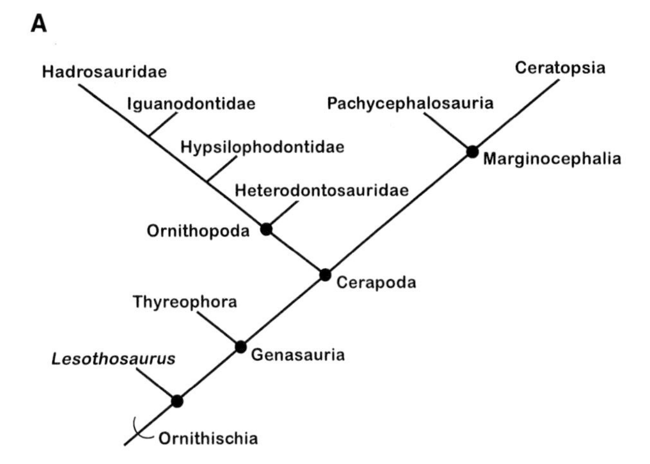

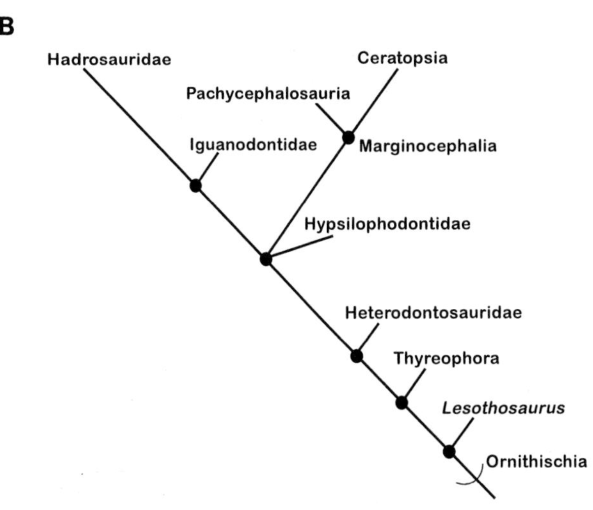

FIGURE 1 Phylogeny of Cerapoda, after (a) Sereno 1986, (b) Norman 1984.

ornithischian interrelationships (Fig. 1b; see NEOCERATOPSIA and PHYLOGENY OF DINOSAURS) is that the origin of Marginocephalia is within Ornithopoda, perhaps among Hypsilophodontidae (Norman 1984; See Benton 1990 for review); if so, Cerapoda would be a problematic taxon.

See also the following related entries:

MARGINOCEPHALIA • ORNITHISCHIA • ORNITHOPODA • PHYLOGENY OF DINOSAURS

References

Benton, M. J. (1990). Origin and interrelationships of dinosaurs. pp. 11–29 *in* D. B. Weishampel, P. Dodson, and H. Osmolska (eds.), *The Dinosauria.* Berkeley: University of California Press.

Norman, D. B. (1984). A systematic appraisal of the reptile order Ornithischia. pp. 157–162 in W.-E. Reif and F. Westphal (eds.), Third Symposium on Mesozoic Terrestrial Ecosystems: Short Papers. ATTEMPTO Verlag, Tübingen.

Sereno, P. C. (1986). Phylogeny of the bird-hipped dinosaurs (Order Ornithischia). *Natl. Geogr. Res.* **2,** 234–256.

Ceratopsia

PETER DODSON
University of Pennsylvania
Philadelphia, Pennsylvania, USA

The Ceratopsia are the horned dinosaurs, important herbivores of the Late Cretaceous. They are defined as all Marginocephalia closer to Ceratopsidae than to Pachycephalosauria. Diagnostic features include one rostral bone, a maxilla at least two-thirds as tall as its length, a broad immobile mandibular symphysis, and a tall snout with relatively broad premaxilla (Sereno, 1990). The name Ceratopsia, meaning "horn faced," was coined by O. C. Marsh in 1890, who understood them only to include the large, quadrupedal, frilled horn-bearers of western North America. With the description of *Leptoceratops* from Alberta in 1914, and of *Protoceratops* from Mongolia in 1923, the concept of Ceratopsia was extended to include not only Marsh's Ceratopsidae, but also the PROTOCERATOPSIDAE. Protoceratopsians have flaring jugals and at least rudimentary parieto-squamosal frills; most lack true horn cores, but all share a number of derived characters with ceratopsids, first among which is the rostral bone in front of the premaxilla, a character found in no other dinosaur. PSITTACOSAURUS from the Early Cretaceous of Mongolia was a small, ornithopod-like biped that lacked both a frill and horns of any kind, but whose face was otherwise remarkably ceratopsian in appearance, including a toothless beak and flaring jugals. In 1975, the existence of a rostral in *Psittacosaurus* was definitively demonstrated by Teresa Maryánska and Halszka Osmólska, requiring the admission of *Psittacosaurus* into the Ceratopsia. In 1986, Paul Sereno created the NEOCERATOPSIA, comprising the Protoceratopsidae and the Ceratopsidae, to stand as a sister group to the Psittacosauria as monophyletic clades within the Ceratopsia. Sereno also created the MARGINOCEPHALIA, comprising the Ceratopsia plus its sister group the PACHYCEPHALOSAURIA, both characterized by an incipient parieto-squamosal frill overhanging the back of the skull.

See also the following related entries:

MARGINOCEPHALIA • NEOCERATOPSIA • PSITTACOSAURIDAE

References

Dodson, P. (1990a). Ceratopsia. In *The Dinosauria* (D. B. Weishampel, P. Dodson, and H. Osmolska, Eds.), p. 578. Univ. of California Press, Berkeley.

Dodson, P. (1990b). Marginocephalia. In *The Dinosauria* (D. B. Weishampel, P. Dodson, and H. Osmolska, Eds.), pp. 562–563. Univ. of California Press, Berkeley.

Dodson, P., and Currie, P. J. (1990). Neoceratopsia. In *The Dinosauria* (D. B. Weishampel, P. Dodson, and H. Osmolska, Eds.), pp. 593–618. Univ. of California Press, Berkeley.

Sereno, P. C. (1990). Psittacosauridae. In *The Dinosauria* (D. B. Weishampel, P. Dodson, and H. Osmolska, Eds.), pp. 579–592. Univ. of California Press, Berkeley.

Ceratosauria

TIMOTHY ROWE
RON TYKOSKI
University of Texas
Austin, Texas, USA

JOHN HUTCHINSON
University of California
Berkeley, California, USA

Ceratosauria is a stem lineage of dinosaurs that comprises the sister lineage of Tetanurae within Theropoda. Many of the taxa now recognized as members of Ceratosauria have a long history of controversial taxonomic assignment. Before the advent of phylogenetic systematics and the recognition of lineages based on shared derived characters, size alone was the principal criterion used by systematists to assign theropods to major taxonomic categories (see CARNOSAURIA; COELUROSAURIA). As a result the larger theropod ceratosaurs, such as *Ceratosaurus nasicornis* and *Dilophosaurus wetherilli,* were grouped as "carnosaurs" or "megalosaurs," whereas the smaller taxa, such as *Coelophysis bauri,* were grouped as "coelurosaurs" or "podokesaurs." Recently, the recognition

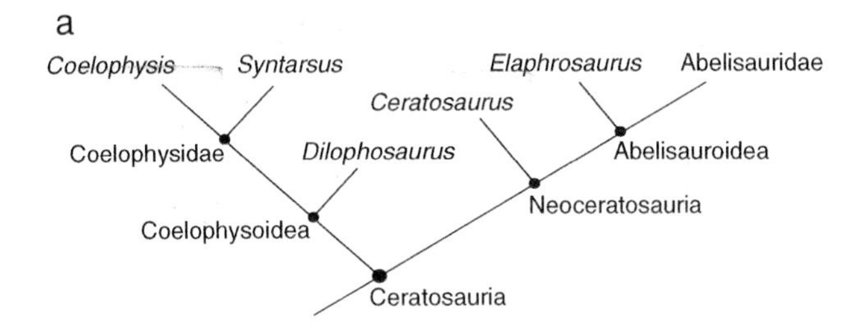

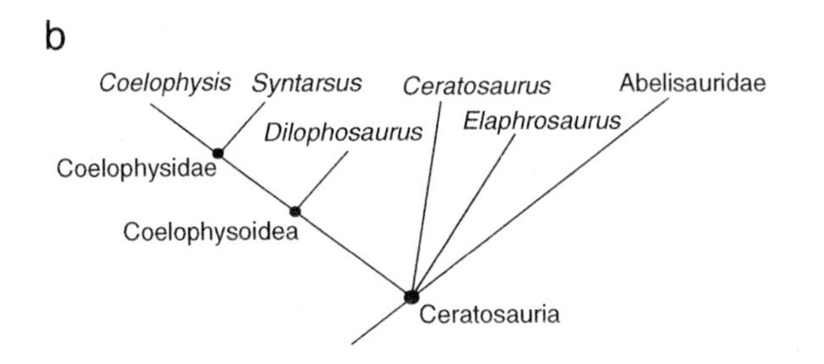

FIGURE 1 (a) A tree showing proposed names and relationships among the better known ceratosaurs (from Holtz, 1994) and (b) a consensus tree that may be a more accurate representation of current knowledge (based on Rowe and Gauthier, 1990; Holtz, 1994; Sereno *et al.*, 1994).

of uniquely derived characters has led to the recognition of a monophyletic Ceratosauria (Figs. 1a and 1b) that, like Tetanurae, includes a highly diversified assemblage of large and small theropods.

Ceratosaurs first appear in the Late Triassic fossil record with *Coelophysis* (Figs. 2 and 3) in North America and the poorly known *Liliensternus* in Europe (Rowe and Gauthier, 1990). Currently, only approximately 20 named theropod species have been assigned to Ceratosauria; several are incomplete, leaving doubts about their assignments. Collectively, these 20 or so taxa may document a lineage extending from the Triassic into the Late Cretaceous, with a global distribution. The known fossil record of ceratosaurs spans approximately 170 million years. The most extensive records of ceratosaurs are from Late Triassic and Early Jurassic rocks, where they are the most common theropods in all currently known faunas. Several ceratosaur taxa are known from burials that preserve multiple individuals, and some offer rare opportunities to study posthatching development in early dinosaurs. One of the richest Mesozoic dinosaur burials ever discovered is the *Coelophysis* quarry at Ghost Ranch, in northwestern New Mexico. Hundreds of *Coelophysis* individuals were buried together *en masse* in sediments of the Triassic Chinle Formation (Colbert, 1995; Schwartz and Gillette, 1994).

Despite the spectacular preservation in the case of *Coelophysis*, the ceratosaur lineage as a whole is only very poorly represented in the fossil record. O. C. Marsh (1884) described the first known member and namesake of the lineage, *C. nasicornis*, based on a nearly complete skeleton from the Late Jurassic Morrison Formation. A few years later, E. D. Cope (1889) described the first fragments of *C. bauri*, but it was another 60 years before a complete skeleton was found. Relatively complete skeletons are also known for the Early Jurassic *D. wetherilli* (Welles, 1954, 1970), two species of *Syntarsus* (Raath, 1969; Rowe, 1989), and the Late Cretaceous *Carnotaurus sastrei* (Bonaparte, 1985). The remaining ceratosaur taxa are known from only fragmentary specimens.

Coelophysis is the oldest known ceratosaur, but it already exhibits many skeletal peculiarities of its own, which evolved following its divergence from the an-

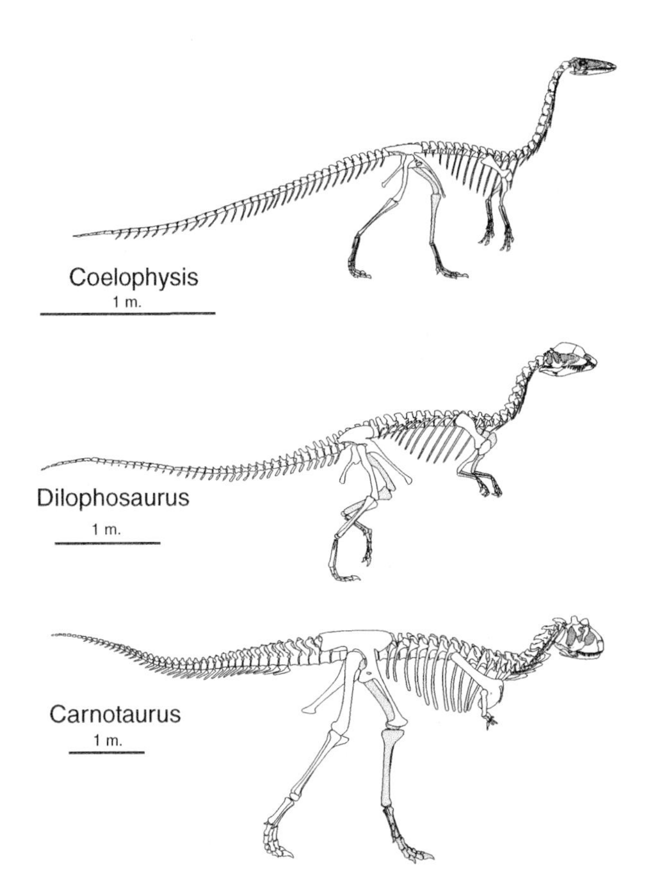

FIGURE 2 Skeletons of three representative ceratosaurs: (a) *Coelophysis* (after Colbert, 1995), (b) *Dilophosaurus* (after Welles, 1984), and (c) *Carnotaurus* (after Bonaparte, 1985).

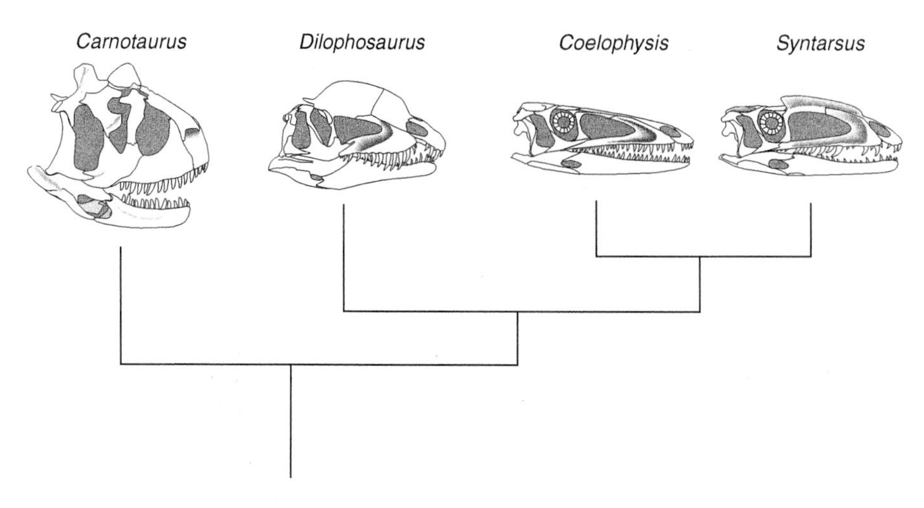

FIGURE 3 Skulls of some of the best known ceratosaurs.

cestral ceratosaur. This implies that the earliest stages of ceratosaur history remain as yet undiscovered. The Late Jurassic *C. nasicornis* is the last known member of the lineage in the Northern Hemisphere, but in Argentina, Madagascar, and India ceratosaurs evidently persisted until the Late Cretaceous. Because documentation of the entire ceratosaur lineage rests on the scant remains of only 20 or so taxa, our knowledge of it is obviously very incomplete. We expect many features of the ceratosaur phylogeny, and especially the diagnoses presented below, to change with new discoveries and we urge caution in making sweeping generalizations with so little evidence.

The first studies to recognize a monophyletic Ceratosauria defined its name in reference to a stem-based lineage that includes *C. nasicornis* (Marsh, 1884; Gilmore, 1920) and its closest relatives among Theropoda (Gauthier, 1984; Rowe and Gauthier, 1990). To date, the relationships among ceratosaurs have not been studied rigorously. Initial studies included within Ceratosauria the taxa *C. nasicornis, Sarcosaurus woodi, Segisaurus halli, D. wetherilli, Liliensternus lilliensterni, C. bauri, Syntarsus rhodesiensis,* and *Syntarsus kayentakatae* (Gauthier, 1984; Rowe and Gauthier, 1990). Later authors have generally supported a monophyletic Ceratosauria composed of two clades, Coelophysoidea and Neoceratosauria (Novas, 1991; Holtz, 1994; Sereno *et al.*, 1994, 1996). Coelophysoidea is composed of *Dilophosaurus* and Coelophysidae (= *Coelophysis* + *Syntarsus;* see Figs. 1a and 1b). Neoceratosauria (Novas, 1991) is made up of *Ceratosaurus,* Abelisauridae (Bonaparte, 1985), and *Elaphrosaurus bambergi* (Holtz, 1994). These authors reject a conflicting suggestion that Ceratosauria is paraphyletic, and that *C. nasicornis* is more closely related to Tetanurae than to coelophysoids (Bakker *et al.*, 1988; Currie, 1995).

Diagnosing Ceratosauria is hindered by incompleteness of the record as well as our state of understanding among basal theropods. Owing to recent discoveries of Triassic dinosauromorphs, several characters previously thought to be diagnostic of the lineage now have equivocal distribution and might or might not be diagnostic of Ceratosauria (Novas, 1996). In addition, some diagnostic ceratosaur features, such as those involving fusion among skeletal elements, are only expressed in mature individuals, and some workers have been inadvertently misled by their absence in subadult specimens. With those caveats, Ceratosauria is diagnosed by the presence of two pairs of pleurocoels in the cervical vertebrae, perforation of the pubic plate by two fenestrae, fusion of sacral vertebrae and ribs in adults, fusion of the astragalus and calcaneum in adults, and a flange of the distal end of the fibula that flares medially to overlap the ascending process of the astragalus anteriorly. The characters involving fusion are convergent upon conditions seen in ornithurine theropods. Additionally, ceratosaurs are distinguished from most other theropods by the retention of a number of plesiomorphic features, including a lack of maxillary fenestration, and the retention of four fingers in the hand. Due to their fragmentary nature, most referrals of taxa to Ceratosauria are only based on a few characters and will warrant reevaluation as more complete specimens are discovered.

The most distinctive and strongly supported lineage within Ceratosauria is Coelophysoidea, which includes small to medium-sized, lightly built species. These are the most common theropod remains from Upper Triassic and Lower Jurassic deposits of North America, Europe, and southern Africa. They lack axial pleurocoels and the transverse processes of their dorsal vertebrae are roughly triangular in dorsal view. The premaxilla and maxilla meet along a loose contact that creates a gap or incisure in the tooth row termed the "subnarial gap" (Welles, 1984). The premaxillary teeth are subcircular in cross section and are not serrated. Two species, *D. wetherilli* and *S. kayentakatae,* sport thin, paired crests on the skull (Welles, 1984; Rowe, 1989). These crests are far too fragile to have served in combat or any other mechanically demanding function and were likely used for visual display. Whether the crests were dimorphic or possessed by only one sex is unknown.

Preserved sclerotic rings in *Syntarsus* (Fig. 3) indicate a large eyeball that filled the orbit. The skulls in coelophysoids are delicately built and set on long necks. The arms are of moderate length, compared to *Tyrannosaurus* at one extreme and *Deinonychus* at the other. The hands are equipped with large claws on digits I–III. Digit IV consists of a reduced metacarpal and one or two tiny phalanges and was so reduced in overall size that it was nonfunctional as a separate digit. The ankle of coelophysoids was primitive in being equipped with only a short ascending process of the astragalus and in the retention of participation by all five metatarsals in the ankle joint.

Since its first recognition (Bonaparte *et al.*, 1990), both the diagnosis and composition of Neoceratosauria have been debated. A thorough reanalysis of basal theropods will be needed to resolve the question, but in lieu of such a study the popular opinion among most experts is that Abelisauridae and *C. nasicornis* both belong to this lineage. Some authors have also included *E. bambergi* (Holtz, 1994). These are all medium- to large-sized theropods known from the Late Jurassic of western North America and East Africa and the Late Cretaceous of South America, India, and Madagascar. One purported synapomorphy is that the premaxilla is as tall or taller than it is long, with nearly vertical anterior and posterior borders. In addition, the quadrate is tall and posteroventrally angled, which places the quadrate/articular joint posteroventrally as well, yielding a large infratemporal fenestra. The effects of size alone may explain some of these features. Another purported synapomorphy is that the femoral head is directed anteromedially, in contrast to the medially directed femoral head of tetanurine theropods. However, all ceratosaurs have anteromedially directed femoral heads, so the monophyly of Neoceratosauria remains at best only weakly defended.

The name Abelisauroidea was suggested (Bonaparte, 1991) for the most inclusive clade within Neoceratosauria. The name was coined in reference to a stem-based lineage that includes Abelisauridae and all taxa closer to abelisaurids than to the North American *C. nasicornis.* Currently, Abelisauroidea includes only *Elaphrosaurus and* Abelisauridae. Purported synapomorphies include pleurocoelous dorsal vertebrae, a cnemial process arising from the lateral surface of the tibia shaft, a pronounced pubic boot, an aliform anterior (= lesser) trochanter, and more than five sacral vertebrae (Holtz, 1994). Each of these characters is homoplastic with many tetanurine theropods, so further work is needed to defend the monophyly of Abelisauroidea.

ABELISAURIDAE (Bonaparte, 1991) is generally taken to be a stem-based name that refers to all taxa closer to *Carnotaurus* than to *Elaphrosaurus.* Currently, it includes *Abelisaurus, Carnotaurus, Xenotarsosaurus, Indosuchus, Indosaurus, Majungasaurus* (Sampson *et al.*, 1996), and possibly some other fragmentary taxa such as *Noasaurus.* Only *Carnotaurus* is known from a relatively complete skeleton. To date, only phenetic resemblances have been offered in support of the recognition of Abelisauridae. Geographic (Gondwanan) and stratigraphic distribution of its members are consistent with abelisaurid monophyly, but the record is so incomplete as to be consistent with many other scenarios of theropod phylogeny. Because stem-based names such as Abelisauridae are predicated on a more general node, in this case the weakly supported Abelisauroidea, and because no rigorous phylogenetic analysis of these poorly known taxa has been conducted, the monophyly of Abelisauridae should be viewed with caution.

Carnotaurus (Figs. 2 and 3) exhibits unusual morphology in which the frontal bones bear blunt, laterally facing bony outgrowths over the orbits. The premaxilla, and the snout in general, is blunt and deep, as has been observed in other abelisaurid taxa (Bonaparte *et al.*, 1990; Sampson *et al.*, 1996). The forelimbs of *Carnotaurus* are profoundly reduced, like the famous condition in *Tyrannosaurus,* but the pectoral girdle remains well developed. The scapula is strap-like, as is the case in tetanurine theropods (Gauthier, 1984). The hand has four stubby fingers, including an incongruously elongated fourth metacarpal. Both the pubis and the ischium are distally expanded. Skin impressions from the neck, shoulder, torso, and tail were found with the original skeleton. The texture of the skin in these regions was rough and pebbly, with larger, intermittently spaced conical thickenings of the hide.

The skulls of several ceratosaurs display bizarre ornamentation. *Ceratosaurus* has a nasal horn and brow hornlets. *Carnotaurus sastrei* lacks a nasal horn, but instead sports massive brow horns and a rugose surface on top of its snout. *Majungasaurus crenatissimus* (Sampson *et al.*, 1996) appears to have a large knob of bone on its skull roof and rugose nasal bones. These structures may have functioned in visual displays, although the robust construction has led to speculation about their use in intraspecific bouts.

The lack of a rigorous phylogenetic analysis, plus discordance in the distribution of purported synapomorphies, leaves doubt regarding the monophyly of Abelisauridae, Abelisauroidea, and Neoceratosauria. Nevertheless, all taxa currently referred to these names seem clearly to lie outside Tetanurae, and all are extinct. The relationships among these taxa, and among basal theropods generally, remain vexing problems (Fig. 1b).

See also the following related entries:

TETANURAE • THEROPODA

References

Bakker, R. T., Williams, M., and Currie, P. (1988). *Nanotyrannus,* a new genus of pygmy tyrannosaur, from the latest Cretaceous of Montana. *Hunteria* **1,** 1–30.

Bonaparte, J. F. (1985). A horned Cretaceous dinosaur from Patagonia. *Natl. Geogr. Res.* **1,** 149–151.

Bonaparte, J. F. (1991). The Gondwanan theropod families Abelisauridae and Noasauridae. *Historical Biol.* **5,** 1–25.

Bonaparte, J. F., Novas, F. E., and Coria, R. A. (1990). *Carnotaurus sastrei* Bonaparte, the horned, lightly built carnosaur from the middle Cretaceous of Patagonia. *Contrib. Sci. Nat. History Museum Los Angeles County* **416,** 1–42.

Colbert, N. E. (1995). *The Little Dinosaurs of Ghost Ranch,* pp. 250. Columbia Univ. Press, New York.

Cope, E. D. (1889). On a new genus of Triassic Dinosauria. *Am. Nat.* **23,** 626.

Currie, P. J. (1995). Phylogeny and systematics of theropods (Dinosauria). *J. Vertebr. Paleontol.* **15**(Suppl. to No. 3), 25A.

Gauthier, J. A. (1984). A cladistic analysis of the higher systematic categories of the Diapsida. Ph.D. dissertation, University of California, Berkeley.

Gilmore, C. W. (1920). Osteology of the carnivorous Dinosauria in the United States National Museum, with special reference to the genera *Antrodemus (Allosaurus)* and *Ceratosaurus. Bull. U.S. Natl. Museum* **110,** 1–154.

Holtz, T. R., Jr. (1994). The phylogenetic position of the Tyrannosauridae: Implications for theropod systematics. *J. Paleontol.* **86,** 1100–1117.

Marsh, O. C. (1884). Principal characters of American Jurassic dinosaurs. Part VIII: The order Theropoda. *Am. J. Sci.* **27**(38), 329–341.

Novas, F. E. (1991). Relaciones filogeneticas de los dinosaurios teropodos ceratosaurios. *Ameghiniana* **28,** 401–414.

Novas, F. E. (1996). Dinosaur monophyly. *J. Vertebr. Paleontol.* **16,** 723–741.

Raath, M. A. (1969). A new coelurosaurian dinosaur from the Forest Sandstone of Rhodesia. *Arnoldia* **4,** 1–25.

Rowe, T. (1989). A new species of the theropod dinosaur *Syntarsus* from the Early Jurassic Kayenta Formation of Arizona. *J. Vertebr. Paleontol.* **9,** 125–136.

Rowe, T., and Gauthier, J. A. (1990). Ceratosauria. In *The Dinosauria* (D. B. Weishampel, P. Dodson, and H. Osmólska, Eds.), pp. 151–168. Univ. of California Press, Berkeley.

Sampson, S. D., Krause, D. W., Dodson, P., and Forster, C. A. (1996). The premaxilla of *Majungasaurus* (Dinosauria: Theropoda), with implications for Gondwanan paleobiogeography. *J. Vertebr. Paleontol.* **16,** 601–605.

Schwartz, H. L., and Gillette, D. D. (1994). Geology and taphonomy of the *Coelophysis* quarry, Upper Triassic Chinle Formation, Ghost Ranch, New Mexico. *J. Paleontol.* **68,** 1118–1130.

Sereno, P. C., Wilson, J. A., Larsson, H. C. E., Dutheil, D. B., and Sues, H.-D. (1994). Early Cretaceous dinosaurs from the Sahara. *Science* **266,** 267–271.

Sereno, P. C., Dutheil, D. B., Iarochene, M., Larsson, H. C. E. *et al.* (1996). Predatory dinosaurs from the Sahara and Late Cretaceous faunal differentiation. *Science* **272,** 986–991.

Welles, S. P. (1954). New Jurassic dinosaur from the Kayenta Formation of Arizona. *Bull. Geol. Soc. Am.* **65,** 591–598.

Welles, S. P. (1970). *Dilophosaurus* (Reptilia: Saurischia), a new name for a dinosaur. *J. Paleontol.* **44,** 989.

Welles, S. P. (1984). *Dilophosaurus wetherilli* (Dinosauria, Theropoda), osteology and comparisons. *Palaeontographica A* **185,** 85–180.

Chemical Composition of Dinosaur Fossils

Hervé Bocherens
Université Pierre et Marie Curie
Paris, France

The chemical composition of fossilized vertebrate hard tissues is the result of the uptake, exchange, and loss of chemical elements, in two different sets of circumstances. First, during the life of the animal, chemical elements are taken from the surrounding environment through food and drinking water, and they are incorporated into the living tissues under physiological conditions. Second, during the diagenetic evolution of the mineralized tissues (i.e., fossilization), this original organization of the chemical elements is altered under the physical, chemical, and microbiological conditions existing when the dead tissues are buried in the sediment. The organization of the chemical elements can be linked to some aspects of the biology of the living animal, such as its genetic pool, its physiology, its diet, and the climatic conditions of the ecosystem. Knowing these relationships in modern animals, it is possible to retrieve some information about the paleobiology of extinct animals from the chemical composition of their fossilized remains, but only if the modifications occurring after death are minimal and do not totally overprint the biological signals.

Bones, teeth, and eggshells are the only preserved fragments of dinosaurs that could eventually provide paleobiological information from their chemical composition. Whenever remnants of dinosaur soft tissues are exceptionally preserved, they are only permineralized pseudomorphs or prints in the sediment, and thus they do not bear any chemical relationship to the original living tissue. In contrast, most fossilized bones, teeth, and eggshells present the same global chemical composition as the living original tissue. The inorganic part is still formed of calcium phosphate in fossil bones and teeth, and of calcium carbonate in fossil eggshells. Although the organic part of these tissues is extensively degraded, some recognizable fragments may still be recovered. The question for the "chemical" paleontologist is whether these minerals and these fragments of organic matter still bear any useful information about the paleobiology of the animal when it was alive. This possibility has been explored sporadically for dinosaurs and other extinct vertebrate groups since the beginning of this century, but this field of research became more firmly based during the past decades (for a historical perspective on this research, see Fig. 1). In "classical" paleontology, information is retrieved at the morphological and at the histological levels of the fossil hard tissues. In chemical paleontology, information is tentatively retrieved at lower levels of organization—the molecular, chemical, and isotopic levels.

We will examine for each level of organization the kind of information recorded in the living tissue and then the possibility of preservation of this information in the fossilized dinosaur remains.

Molecules

Two major groups of molecules can be distinguished in vertebrate mineralized tissues: mineral and organic molecules. The mineral molecules are crystals—calcium phosphate (apatite), with hydroxyl (OH^-) and numerous ionic substitutions including carbonate (CO_3^{2-}) ions (this mineral is thus called carbonate hydroxylapatite; CHA), in bones and teeth; and calcium carbonate in eggshells. The crystallographic properties of CHA, such as its size and crystallinity (i.e., the perfection of the crystals), are mostly determined by the type of tissue (bone and dentine have very small crystals, whereas the crystals are much bigger in enamel) and the genetics (the size of the bioapatite crystals in dentine differs according to the species). Although the great majority of dinosaur fossil bones are still formed of calcium phosphate, the crystallographic properties of the CHA crystals of bone are clearly changed during diagenesis, with an increase in the perfection of the crystals, as shown by X-ray diffractometry of fossil bone powder (Pflug

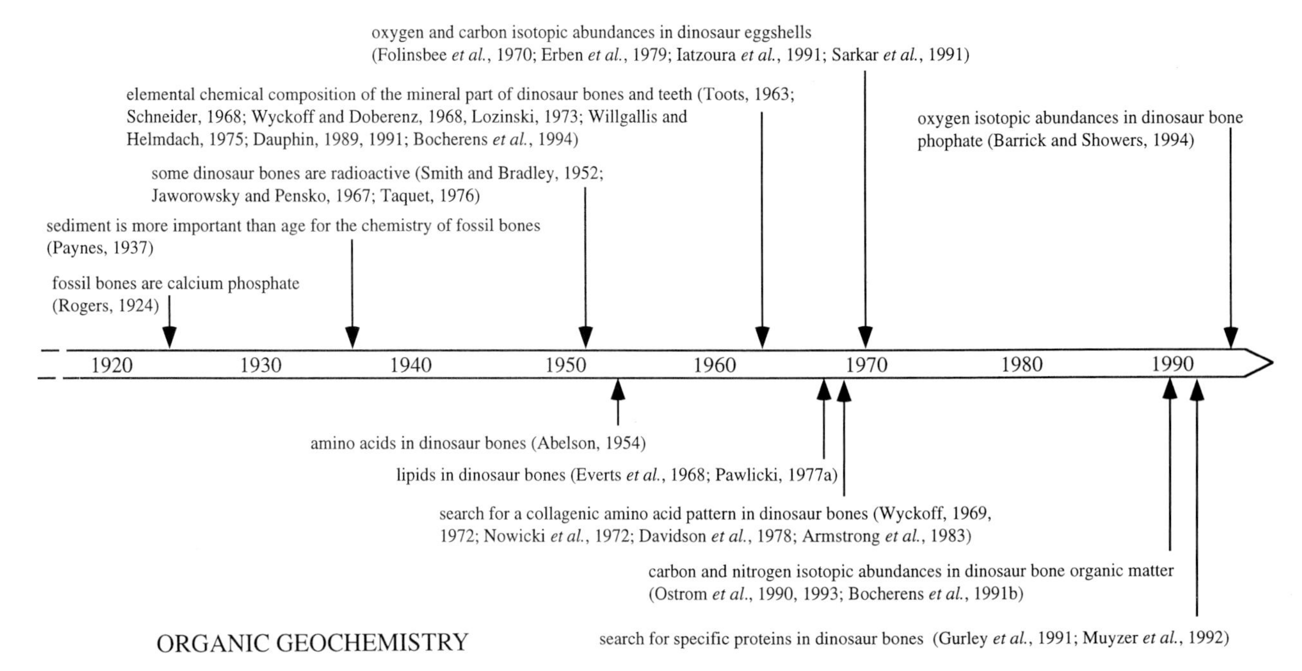

FIGURE 1 Historical perspective on the studies of dinosaur fossils' chemical composition. The arrows indicate the date of the first publication about a topic, and references to relevant work on the same topic are also indicated.

and Strübel, 1967; Person *et al.*, 1995). However, these changes do not alter the histological features of the fossil bones in most cases. The enamel crystals seem less affected by the diagenetic changes than the bone and dentine crystals. Eggshell calcium carbonate crystals have a specific mineralogy (aragonite for turtles, and calcite for other reptiles and birds).

The occurrence and composition of the organic molecules in mineralized tissues are mostly determined by the genetic program. They do not depend on environmental parameters. If organic molecules are retrieved intact from fossil tissues, they may provide phylogenetic information by comparison of homologous molecules in modern species or other fossil species. They can also be used as an uncontaminated support of biogenic chemical or isotopic signals bearing paleoenvironmental information.

Collagen and Other Proteins

Fresh bone and dentine contain about 20% organic matter, 90% of which is collagen. This fibrillar protein is easily recognizable by its specific amino acid composition, with 30% glycine, about 10% proline and 10% hydroxyproline, and a few percent hydroxylysine. These last two amino acids are found almost exclusively in collagen. This protein is not ideal for phylogenetic studies because it does not change very much from one group to another. Collagen is rather insoluble when intact, but it is quickly degraded after death by hydrolysis and generates smaller peptides as degradation products. These peptides are soluble and are likely to be leached out of the bone or tooth during fossilization (Hare, 1980). Amino acid patterns very similar to those of collagen have been reported for organic matter extracted from dinosaur bones (Wyckoff, 1969; Nowicki *et al.*, 1972; Davidson *et al.*, 1978; Bocherens *et al.*, 1991b), and even for Devonian material 300 million years old (Davidson *et al.*, 1978). However, actual peptide remnants of collagen, that would indicate the preservation of this protein at the molecular level have been reported only once, and even so, the amino acid composition of the peptides was not identical to collagen, suggesting a significant amount of contamination (Gurley *et al.*, 1991).

The non-collagenic proteinic material present in bone and dentine, as well as in enamel and eggshells, may have a better potential for preservation at the molecular level than collagen. In bone and dentine, this material is composed of sialoproteins, osteocalcin, phosphoproteins, and serum-derived proteins. Some non-collagenous proteins have been detected immunologically in dinosaur bones (Muyzer *et al.*, 1992), but they have not been extracted and purified yet. Phosphorylated amino acids suggesting the preservation of phosphoproteins have been extracted from Lower Cretaceous crocodile enamel from Niger (Glimcher *et al.*, 1990), but no investigation has been performed on the contemporaneous dinosaur material. Proteinic material has also been reported from dinosaur eggshells (Voss-Foucard, 1968; Kolesnikov and Sochava, 1972; Krampitz *et al.*, 1977; Marin and Dauphin, 1991).

DNA

DNA has been successfully retrieved from mammal fossil bones as old as 150,000 years (Hagelberg *et al.*, 1994). This ancient DNA allowed phylogenetic studies of extinct species such as moa, mammoth, saber-toothed cat, and cave bear (Cooper *et al.*, 1992; Janczewski *et al.*, 1992; Hagelberg *et al.*, 1994; Hänni *et al.*, 1994). The discovery of ancient DNA in insects preserved in amber of the same age as dinosaurs (Cano *et al.*, 1993) raised the hope to extract DNA from dinosaur bones as well. No DNA has yet been retrieved from dinosaur bones, but the research is under way (Morell, 1993).

Other Organic Molecules

Carbohydrates such as mucopolysaccharides and lipids are also contained in fresh bones. Lipids (Everts *et al.*, 1968; Pawlicki, 1977a) and mucopolysaccharides (Pawlicki, 1977b) have been identified in dinosaur bones but with no clear paleobiological implication. However, a reassessment of these results is necessary with improved technologies.

Chemical Elements

The amount of the major chemical elements in bone, such as carbon (C), hydrogen (H), nitrogen (N), oxygen (O), and sulfur (S) in organic matter and calcium (Ca), phosphorus (P), oxygen, and carbon in CHA, is determined mostly genetically and partly by the health status. However, for some elements present in very small amounts (less than 1%) in CHA, also called trace elements, there is a control through the external environment. Indeed, the calcium phosphate of bone, dentine, and enamel includes many other chemical elements than just calcium and phosphate. These atoms or ions substitute into the CHA structure to form a very complex chemical formula (Fig. 2). For instance, there is more fluorine (F) in bones and teeth of marine vertebrates than in those of non-marine vertebrates (Klement, 1938; Schmitz *et al.*, 1991). Strontium (Sr) and barium (Ba) may replace calcium in the bioapatite, but because the organisms discriminate against Sr and Ba during digestion, there is a decrease in the amount of these elements from herbivores to carnivores (Elias *et al.*, 1982; Ezzo, 1994). Similar phenomena have been reported for eggshells, but the details are less well known (Dauphin, 1988).

The amount of such trace elements in archaeological bones has been tentatively used to investigate the diet of ancient people and of extinct animals. However, the problem of diagenetic alteration of trace element amounts in permineralized fossil bones precludes their use in dinosaur bones. The deposition and formation of authigenic minerals in the bone cavities, such as calcite, quartz, sulfides, iron, and manganese oxides, dramatically changes the chemical composition of fossil bone relative to fresh bone. Careful study of these authigenic minerals can yield valuable information about the diagenetic evolution of a fossil bone (Clarke and Barker, 1993) and thus

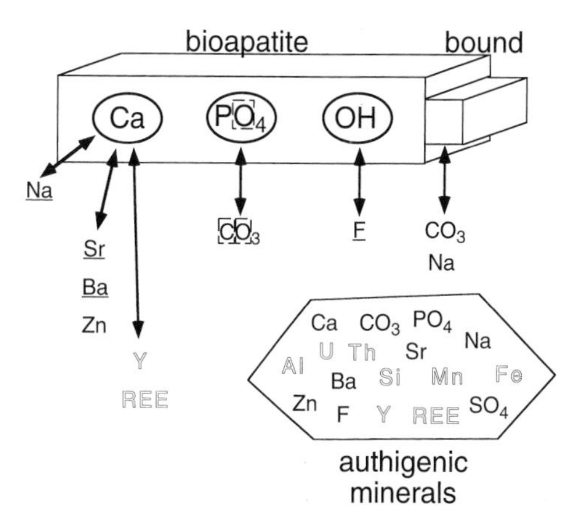

FIGURE 2 Summary of the possible substitution in bioapatite, their bearing for biological studies, and the evolution of this chemical composition during fossilization. Arrows represent the possible substitutions. The underlined elements have a chemical composition that depends on the environmental conditions; elements in relief are not present in living bioapatite but incorporated during diagenesis. Authigenic minerals are formed during fossilization in the porosity of bone.

indirectly help to solve some alteration problems. Microprobe techniques allow the analysis of well-defined spots in a fossil tissue and thus focus on the apatite part of the bone, which is the only one in which the preservation of a biogenic signature is possible. However, these techniques are not as accurate as the techniques for analyzing the bulk of the fossil bone. Microprobe studies have demonstrated the diagenetic alteration of phosphate in dentine (Bocherens *et al.*, 1994). A few studies attempted to link the amounts of different chemical elements in bones, such as calcium, phosphorus, lead, iron, and magnesium, to the health status of a Late Cretaceous ornithomimid from Mongolia (Pawlicki and Bolechala, 1987, 1991). Unfortunately, the conclusions are impaired by the diagenetic alteration of these chemical compositions, and they remain very dubious. The chemical composition of enamel may be less altered than in bone and dentine, due to the greater compactness and the larger size of the enamel crystals (Dauphin, 1989; Bocherens *et al.*, 1994). Applications to the paleobiology of dinosaurs have not yet been achieved. An additional problem when dealing with ancient reptiles is the poor knowledge of the variability of the amount of these trace elements in modern reptile bones. Thus, the interpretation of the values measured in the fossil specimens is difficult (Bocherens *et al.*, 1994). Chemical contents of dinosaur eggshells have also been used (Iatzoura *et al.*, 1991), but, in that case, the control for diagenetic alteration was not satisfactory.

Isotopes

Chemical elements usually exist in the natural environment under different forms called isotopes. The different isotopes of a given chemical element exhibit identical gross chemical properties, but their atomic masses are slightly different. For each element, one isotope is in much greater abundance than the other(s), and very accurate measurements can distinguish between substances of different origins. The isotopes relevant for dinosaur studies are carbon ($^{13}C/^{12}C$), nitrogen ($^{15}N/^{14}N$), and oxygen ($^{18}O/^{16}O$). For these light elements, the isotopic ratios in the mineralized tissues reflect the ratios in the source food or drinking water, and the fractionations in the organism during metabolism, which depends on the environmental conditions and the physiology of the animal.

Carbon is the major constituent of organic matter and is also present in the mineral phase of bone and tooth as carbonate (CO_3^{2-}) substituted for phosphate (Fig. 3), and in the carbonate of eggshells. The source of carbon is the diet, and the isotopic abundance of carbon varies according to the diet. For instance, in modern environments, the ^{13}C amount is lower in terrestrial plants than in marine plants, and many tropical grasses have higher ^{13}C amounts than other terrestrial plants and marine plants.

Nitrogen is a major constituent of proteins. The amount of ^{15}N is higher in an animal's tissues than in its diet. Thus, a carnivore has a higher ^{15}N amount than its herbivorous prey. Moreover, ^{15}N amount increases in a given animal when the food becomes scarcer, which is usually linked to arid conditions of the environment.

Oxygen is present in the phosphate and carbonate of bones, teeth, and eggshells. Its isotopic composition depends primarily on those of its drinking water and food and also on the temperature of crystallization of the minerals. The amount of ^{18}O decreases when drinking and food water have a lower temperature and also when crystallization occurs at a lower temperature. In an animal with constant body temperature, the oxygen isotopic abundances provide information about the oxygen isotopic abundances of water in the environment and thus indirectly about the temperature of the environment. In an animal whose temperature changes according to that of the environment, the oxygen isotopic abundances also reflect the temperature of formation of the crystals, which may differ among different locations within the body—for example, if the organ is close to the source of body heat (trunk) or at the periphery of the body (limbs and tail: Barrick and Showers, 1994).

Stable isotopes have great potential for providing information about the paleobiology of fossil animals. Diet, aridity, temperature, and possibly thermal physiology are reflected in carbon, nitrogen, or oxygen isotopic ratios of bones, teeth, and eggshells of a living animal. However, the biggest problem in using stable isotopes in fossil bones is the possible alteration of these values due to fossilization. One must find a stable support for isotopic values and show that the biogenic isotopic values are preserved. Organic molecules such as collagen can be used as a support for stable isotopic information. It has been shown on Pleistocene bones and teeth up to around 60,000 years

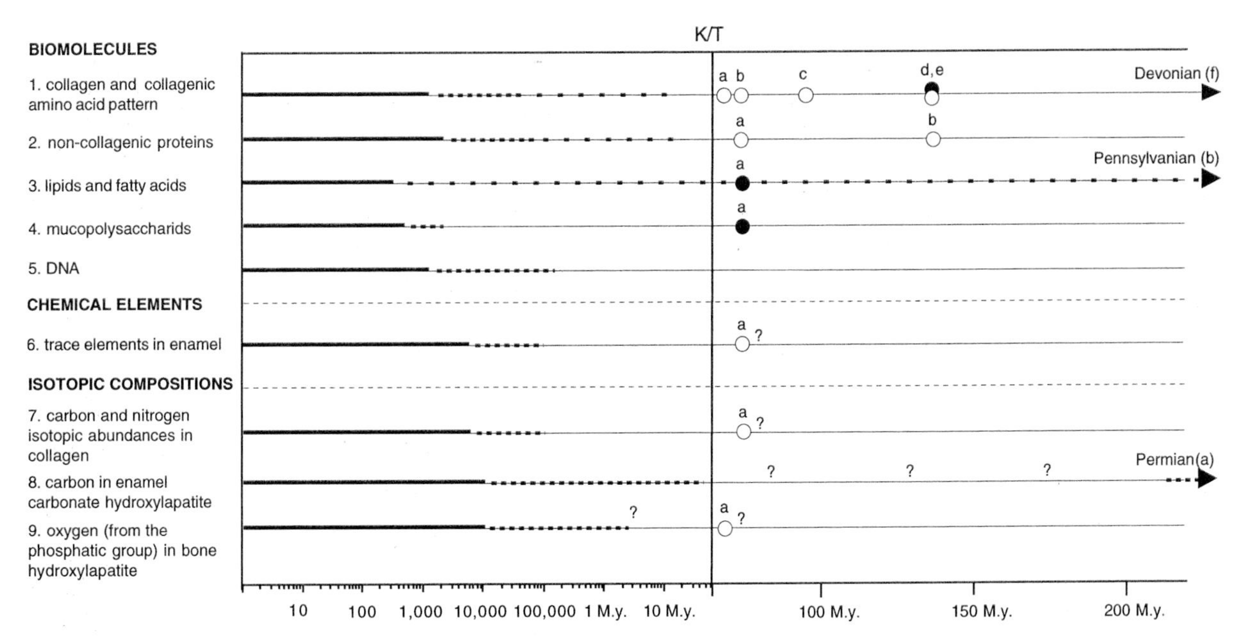

FIGURE 3 Preservation of biomolecules and chemical and isotopic compositions in fossil bones and teeth. The time-scale is logarithmic from the present to 65 million years and is linear for the Mesozoic period (65–220 million years). Solid bold lines indicate the range of collagenic amino acid pattern in a dinosaur bone from the Maastrichtian of Laño, Spain (Bocherens *et al.*, 1991b). (1) Collagen and collagenic amino acid pattern: a, b, collagenic amino acid pattern in a dinosaur toe bone from the Campanian Nemegt Beds, Mongolia (Nowicki *et al.*, 1972); c, collagenic amino acid pattern in an atlantosaurid bone from the Lower Cretaceous, Gadoufaoua, Niger (Davidson *et al.*, 1978); d, fragments of proteins with some collagenic affinity in a *Seismosaurus* bone from the Upper Jurassic Morrison Formation, New Mexico (Gurley *et al.*, 1991); e, collagenic amino acid pattern in a dinosaur bone from the Upper Jurassic Morrison Formation, Wyoming (Wyckoff, 1969); f, collagenic amino acid pattern in a fish dermal plate from the upper Devonian of Niger (Davidson *et al.*, 1978). (2) Non-collagenic proteins: a, immunological evidence for osteocalcin in a *Lambeosaurus* bone from the Campanian Judith River Formation, Alberta, Canada (Muyzer *et al.*, 1992); b, immunological evidence for osteocalcin in a sauropod bone from the Upper Jurassic Morrison Formation (Muyzer *et al.*, 1992). (3) Lipids and fatty acids: a, lipids (staining thin sections with Sudan B) in a *Tarbosaurus* bone from the Campanian, Gobi Desert, Mongolia (Pawlicki, 1977b). (4) Mucopolysaccharids: a, mucopolysaccharids (histochemical reaction) in a *Tarbosaurus* bone from the Campanian, Gobi Desert, Mongolia (Pawlicki, 1977a). (5) DNA: preservation of chemical compositions. (6) Trace elements in enamel: a, trace element composition in dinosaur enamel from the Judith River Formation, Alberta (Bocherens *et al.*, 1994); preservation of isotopic compositions. (7) Carbon and nitrogen isotopic abundances in collagen: a, isotopic abundances in "collagen" extracted from bones and teeth of dinosaur and other reptiles from the Campanian Judith River Formation, Alberta, are claimed to reflect the trophic structure (Ostrom *et al.*, 1993). These values are likely to be diagenetically altered. (8) Carbon in enamel carbonate hydroxylapatite: a, carbon isotopic abundances in enamel of synapsids from the Late Permian from South Africa are claimed to reflect the isotopic changes in the contemporaneous marine carbonates (Thackeray *et al.*, 1993). (9) Oxygen (from the phosphatic group) in bone hydroxylapatite: a, oxygen isotopic abundances in phosphate of *Tyrannosaurus* bone are claimed to reflect thermal physiology (Barrick and Showers, 1994). These values are probably diagenetically altered (Morell, 1993).

old that when collagen with its specific amino acid composition can be extracted from a bone, the carbon and nitrogen isotopic values are preserved. Checking the preservation of isotopic values in collagen-like organic matter extracted from fossil bones is performed by measuring isotopic abundances in bones from species of known dietary habits found in the fossil locality. For instance, horse, reindeer, wolf, and hyena are found in abundance in European Pleistocene localities, and their isotopic abundances can be predicted according to their respective diets (Bocherens *et al.*, 1991a, 1995). A similar attempt was made by Ostrom *et al.* (1990, 1993) on organic matter extracted from dinosaur and other vertebrate bones and teeth from the Late Cretaceous Dinosaur Park Formation in Alberta, Canada. In this study, the amino acid composition of the extracted organic matter was not identical to the amino acid composition of collagen. Moreover, the carbon isotopic values were different from what is expected for such an ecosystem, and the nitrogen isotopic values, although showing some tendencies that could be related to the trophic level of the specimens, presented a far too large range of variation within a given species when compared to modern ecosystems. It is thus likely that these values have been at least partially overprinted by diagenesis and cannot be used for paleobiological reconstructions. Other organic molecules could be better preserved than collagen in dinosaur fossils (Muyzer *et al.*, 1992) and could be a good support for stable isotopic signature (Ajie *et al.*, 1991). However, the quantities that could be recovered would be very low. Another possibility to short-cut the contamination problem of proteinic residues in dinosaur bones would be to use new techniques of analysis, such as a gas chromatograph system, to separate the different amino acids, connected to a combustion oven generating CO_2 from each separated amino acid separately, and finally an isotopic mass spectrometer to analyze the isotopic abundances of carbon in the different amino acids. By carefully choosing the relevant amino acids, it is conceivable to track back the isotopic value of the diet, and the quantities required for such an analysis are very small.

The carbon isotopic abundances of a dinosaur diet have also been recorded in the carbonate of its CHA and eggshells. In bones, the isotopic values of CHA are very quickly altered (Koch *et al.*, 1990) but they can be retained in enamel for millions of years (Lee-Thorp and van der Merwe, 1987). Carbon isotopic abundances are possibly preserved in Paleocene mammal tooth enamel (Koch *et al.*, 1992) and in Permian therapsid tooth enamel (Thackeray *et al.*, 1990). These specimens are just younger and older, respectively, than typical (Mesozoic) dinosaurs. The problem with dinosaurs is the thinness of enamel, less than .5 mm thick, and no attempt has been made to date to measure carbon isotopic abundances on dinosaur enamel CHA. Carbon isotopic abundances have been measured on dinosaur eggshell carbonates (Folinsbee *et al.*, 1970; Erben *et al.*, 1979; Iatzoura *et al.*, 1991; Sarkar *et al.*, 1991), but due to diagenetic alterations these values could well be those of recrystallized carbonates and thus be very different from the biogenic values.

Oxygen isotopic abundances are also likely to be diagenetically altered in eggshell carbonates. On the other hand, oxygen isotopic abundances in bone phosphate have been claimed to be preserved, in some cases at least, and to bring new light on dinosaur thermal physiology (Barrick and Showers, 1994). In this study, oxygen isotopic abundances in different bones from one *Tyrannosaurus rex* skeleton were compared, assuming that a rather uniform distribution of these values would indicate a homeothermic physiology. Unfortunately, the test for diagenetic alteration, i.e.. trying to find the expected isotopic variations on an ectotherm vertebrate, such as a lizard or a crocodile, has not been made. Also, preliminary data from Kolodny *et al.* (1997) indicate that the oxygen isotopic values in Late Cretaceous vertebrate bones from North America are related to latitude and not to the thermal physiology of the specimens within one locality, suggesting an isotopic equilibration of the oxygen phosphate with the circulating groundwater. Due to the possibility of diagenetic alteration, the conclusions about *Tyrannosaurus* thermal physiology cannot be taken for granted until the control for diagenetical alteration is made.

Conclusions

Biogeochemistry of fossil biomineralizations is an expanding new field of vertebrate paleontology. This approach has already brought valuable new lines of information for relatively young specimens, dozens of thousands of years old. Until now, biogeochemical investigations of dinosaur fossils have provided no unambiguous paleobiological information, but many clues indicate the preservation of biogenic material in some dinosaur specimens—bones, teeth, and possibly eggshells. The continuous improvement of technology, using lower quantities of a purer material, and the search for different kinds of preservations, will most probably lead in the near future to spectacular results from this approach in the case of dinosaurs. More complete studies of the fossilization process itself, at the chemical level, are also required to understand the preservation of geochemical signals in fossil material.

See also the following related entries:

BIOMINERALIZATION • GENETICS • JURASSIC PARK • PERMINERALIZATION

References

Barrick, R. E., and Showers, W. J. (1994). Thermophysiology of *Tyrannosaurus rex*: Evidence from oxygen isotopes. *Science* **265,** 222–224.

Bocherens, H., Fizet, M., Mariotti, A., Lange-Badré, B., Vandermeersch, B., Borel, J. P., and Bellon, G. (1991). Isotopic biogeochemistry (^{13}C, ^{15}N) of fossil vertebrate collagen: Implications for the study of fossil food web including Neandertal Man. *J. Hum. Evol.* **20,** 481–492.

Bocherens, H., Brinkman, D. B., Dauphin, Y., and Mariotti, A.. (1994). Microstructural and geochemical investigations on Late Cretaceous archosaur teeth from Alberta (Canada). *Can. J. Earth Sci.* **31,** 783–792.

Clarke, J. B., and Barker, M. J. (1993). Diagenesis in *Iguanodon* bones from the Wealden Group, Isle of Wright, Southern England. *Kaupia, Darmstädter Beitrage zur Naturgeschichte* **2,** 57–65.

Dauphin, Y. (1991). Chemical composition of reptilian teeth. 2. Implications for paleodiets. *Palaeontographica, A* **219**(4–6), 97–105.

Davidson, F. D., Lehman, J.-P., Taquet, P., and Wyckoff, R. W. G. (1978). Analyse des protéines de Vertébrés fossiles dévoniens et crétacés du Sahara. *Comptes Rendus l'Académie Sci. Paris* **287,** 919–922.

Erben, H. K., Hoefs, J., and Wedepohl, K. H. (1979). Paleobiological and isotopic studies of eggshells from a declining dinosaur species. *Paleobiology* **5,** 380–414.

Folinsbee, R. E., Fritz, P., Krouse, H. R., and Robblee, A. R. (1970). Carbon-13 and oxygen-18 in dinosaur, crocodile, and bird eggshells indicate environmental conditions. *Science* **168,** 1353–1356.

Gurley, L. R., Valdez, J. G., Spall, W. D., Smith, B. F., and Gillette, D. D. (1991). Proteins in the fossil bone of the dinosaur, *Seismosaurus*. *J. Prot. Chem.* **10,** 75–90.

Marin, F., and Dauphin, Y. (1991). Composition de la phase protéique soluble des coquilles d'œufs de dinosaures du Rognacien (Crétacé) du Sud-Est de la France. Neues *Jahrbuch Geol. Paläontol.* **4,** 243–255.

Morell, V. (1993). Dino DNA: The hunt and the hype. *Science* **261,** 160–162.

Muyzer, G., Sandberg, P., Knapen, M. H. J., Vermeer, C., Collins, M., and Westbroek, P. (1992). Preservation of the bone protein osteocalcin in dinosaurs. *Geology* **20,** 811–814.

Ostrom, P. H., Macko, S. A., Engel, M. H., Silfer, J. A., and Russell, D. A. (1990). Geochemical characterization of high molecular weight material isolated from Late Cretaceous fossils. *Organic Geochem.* **16,** 1139–1144.

Ostrom, P. H., Macko, S. A., Engel, M. H., and Russell, D. A. (1993). Assessment of trophic structure of Cretaceous communities based on stable nitrogen isotope analyses. *Geology* **21,** 491–494.

Person, A., Bocherens, H., Saliège, J.-Fr., Paris, F., Zeitoun, V., and Gérard, M. (1995). Early diagenetic evolution of bone phosphate: A X-ray diffractometry analysis. *J. Archaeol. Sci.* **22,** 211–221.

Wyckoff, R. W. G. (1972). *The Biochemistry of Animal Fossils,* pp. 152. Scientechnica, Baltimore, MD.

Chengdu College of Geology Museum, People's Republic of China

see MUSEUMS AND DISPLAYS

Children's Museum of Indianapolis, USA

see MUSEUMS AND DISPLAYS

Chilean Dinosaurs

see SOUTH AMERICAN DINOSAURS

Chinese Dinosaurs

Dong Zhiming
Institute of Vertebrate Paleontology and Paleoanthropology, Academia Sinica
Beijing, People's Republic of China

In the past two decades, great progress has been made in the knowledge of Chinese dinosaurs. A huge number of dinosaur skeletons, eggs, and footprints have been collected, studied, and displayed, with more than 100 species of dinosaurs, including eggs and footprints, described and named so far (Dong, 1992). The study of dinosaurs in China not only has greatly extended the list of dinosaurs over the world but also has made great contributions to the understanding of general theoretical problems concerning these fascinating animals (Russell, 1993).

A Brief History of the Study of Dinosaurs in China

The Chinese people call themselves "descendants of the dragon." The fossilized vertebrate bones are known as "dragon bones." For instance, as early as the Jin Dynasty (265–317 A.D.), a book titled *Hua Yang Guo Zhi* already recorded the discovery of dragon bones in Wuchen, which covered the present Santain County, Sichuan Province. Because most of the exposed strata belong to Jurassic deposits, it is highly probable that the bones discovered were actually dinosaur bones.

The earliest scientific discoveries of dinosaurs in China were made in the 1910s by Russians in the southern banks of the Heilongjiang (Amur) River. The finds were referred to a hadrosaur, *Mandschurosaurus*. It is the first named dinosaur in China (Riabinin, 1925, 1930).

In 1913–1915 an American geologist, George Louderback of the University of California at Berkeley, reported the first dinosaur fossils to be found in the Sichuan Basin (Louderback, 1935; Camp, 1935). A German mining engineer, Berhagel, found several fossils of dinosaurs in Mengyin, Shandong, in 1916. This site was excavated by an Austrian paleontologist, Zdansky, in 1922 and 1923. This was the well-known *Euhelopus* (Wiman, 1929).

Reports on these early discoveries of dinosaur fossils aroused the attention and interest of many Western paleontologists. A series of multinational expeditions came to China, such as the Central Asiatic Expeditions of the American Museum of Natural History (1921–1930), the Sino–Swedish Expeditions (1927–1935), and Sino–French Scientific Expeditions (1930). The Central Asiatic Expedition was the largest of the three and proved to be the most fruitful (Osborn, 1924; Andrews, 1932).

The years 1933 to 1949 represent the initial stage in the study of dinosaur fossils in China. The progress accomplished during this period was mainly undertaken by Dr. C. C. Young. After pursuing his studies in Germany, he returned to China in 1928 and devoted himself to the study of paleovertebrates. From 1933 onwards, he began to focus his attention on the study of reptiles and conducted a series of excavations for dinosaur fossils in Sichuan, Yunnan, Xinjiang, and Gansu. He undertook the famous excavation at Lufeng Basin, Yunnan, in 1938 (Young, 1951).

In 1951 Dr. Young led an excavation of Late Cretaceous dinosaurs at Laiyang, Shandong, and collected a complete skeleton of *Tsintaosaurus*. A large sauropod skeleton, *Mamenchisaurus*, was studied by Young in 1954. In 1957, a new nearly complete skeleton of *Mamenchisaurus* was unearthed from Hechuan, Sichuan. This giant dinosaur was named *M. hechuanensis* and is the longest dinosaur ever discovered in Asia (Fig. 1).

From 1959 to 1960, Sino–Soviet paleontological expeditions conducted large-scale excavations in the Erlen and Alxa Gobi areas of Inner Mongolia and obtained a sizable quantity of Early Cretaceous dinosaur fossils, which included *Probactrosaurus* (Iguanodontidae), *Chilantaisaurus* (Theropoda), and an ankylosaurian dinosaur. For political reasons, the expeditions were suspended in 1960.

From 1963 to 1966, the Institute of Vertebrate Paleontology and Paleoanthropology of Chinese Academy of Sciences organized expeditions that explored

FIGURE 1 *Tsintaosaurus*, a hadrosaur from China.

the Junggar and Turpan basins of Xinjiang for 3 years and discovered the pterosaur fauna of Urhe (Dong, 1973). Nearly simultaneously, a large hadrosaur skeleton was found in Shandong. It was named *Shantungosaurus*, measuring 15 m in length. In 1974, 106 crates of dinosaur fossils weighing more than 10 tons were collected by the Chongqing Museum from the Late Jurassic of Zigong area. These finds include two skeletons of *Omeisaurus*, one skeleton of *Szechuanosaurus*, and *Tuojiangosaurus* (Dong *et al.*, 1983). In an attempt to fill in the gaps in the evolution of dinosaurs, Chinese paleontologists engaged in the study of dinosaurs have focused on the Early and Middle Jurassic strata. They began by exploring Yunnan, Guizhou, Sichuan, and the eastern part of Xizang (Tibet) from 1976 to 1979. They found Middle Jurassic sauropods, theropods, and stegosaurs (Dong, 1992).

In 1976, a large, incomplete skeleton of a sauropod was collected from the Xiashaximiao (Lower Shaximiao) Formation of the Middle Jurassic in Dashanpu of Zigong City. This specimen was named *Shunosaurus lii* (Dong *et al.*, 1983). Now this site has become the well-known Dashanpu Dinosaur Quarry.

From May 1979 to July 1981, the author led the dinosaur excavations in the Dashanpu dinosaur quarry. More than 40 tons of dinosaur fossils were uncovered by the Institute of Vertebrate Paleontology and Paleoanthropology, the Chongqing Municipal Museum, and the Zigong Salt Industry History Museum. Approximately 8000 bones have been excavated, many of them large and some articulated as skeletons. The fossils include complete skeletons of sauropods (*Shunosaurus, Omeisaurus,* and *Datousaurus*), carnosaurs (*Gasosaurus*), stegosaurs (*Huayangosaurus*), ornithopods (*Agilisaurus* and *Xiaosaurus*), pterosaurs, a plesiosaur, amphibians, and fishes. They represent more than 100 individual animals and include at least 12 reptiles, including 6 different kinds of dinosaurs. From July 1981 to May 1982, this quarry was worked by a Sichuan expedition.

Dashanpu Quarry has proved to be one of the richest and most rewarding localities for Middle Jurassic dinosaurs in the world (see ZIGONG DINOSAUR MUSEUM). The Zigong Dinosaur Museum was built on the site and opened in the spring of 1987. An outstanding feature of this magnificent museum is that it was erected over the stratum containing the dinosaur skeletons.

The newly reformed, open politics of China welcome foreign scholars to cooperate with Chinese colleagues. Some projects were in cooperation with the British Museum of Natural History (1982), Texas Tech University (1985), and Canada (1986–1990). The most exciting finds were made by the China–Canadian Dinosaur Project (CCDP) from 1986 to 1990.

In 1986, a dinosaur project, the CCDP was organized. This is the first joint paleontological expedition into the northwestern interior of China since the 1940s (Dong *et al.*, 1988, 1989; Currie, 1991). The main aim of CCDP is to study Mesozoic continental strata in the north part of China, to find superb dinosaur fossils, and to trace the relationships among various groups of dinosaurs of Asia and North America. Fieldwork has been carried out in the Junggar Basin of Xinjiang, the Gobi Desert of Inner Mongolia, and the badlands of Alberta and the Arctic islands of Canada. Two special issues of the *Canadian Journal of Earth Sciences* were published as the scientific results of CCDP. Eight new genera and 11 new species of turtles and dinosaurs were described in these special issues. A traveling exhibition, "Dinosaur World Tour," was displayed in several countries.

From 1992 to 1993, an expedition from the Institute of Vertebrate Paleontology and Paleoanthropology (IVPP) explored for dinosaurs along an ancient silk road. A new dinosaur site from the Early Cretaceous was found in the Mazunshan area, Gansu. Many dinosaur fossils, including iguanodontids, sauropods, and primitive protoceratopsids, were collected.

Distribution and Biostratigraphy of Chinese Dinosaurs

China has excellent outcrops yielding dinosaur remains. Rocks deposited during the lifetime of the dinosaurs were laid down on low plains and in basins by rivers and lakes. This was important for the preservation of the remains of land vertebrates. The current climate and topography is such that the rocks at many of the best dinosaur sites are not deeply covered with soil or concealed under thick vegetation. This allows the discovery and excavation of dinosaur bones.

Chinese dinosaur fossils can be divided into five dinosaur faunas and provide an almost unbroken record from the Late Triassic to the Late Cretaceous periods (Dong, 1980, 1992) (Fig. 2):

1. Early Jurassic, Prosauropod–*Lufengosaurus* fauna
2. Middle Jurassic (Bathonian–Callovian), Sauropod–*Shunosaurus* fauna
3. Late Jurassic, Sauropod–*Mamenchisaurus* fauna
4. Early Cretaceous, *Psittacosaur*–Pterosaur fauna
5. Late Cretaceous, Hadrosaurid–Titanosaurid fauna

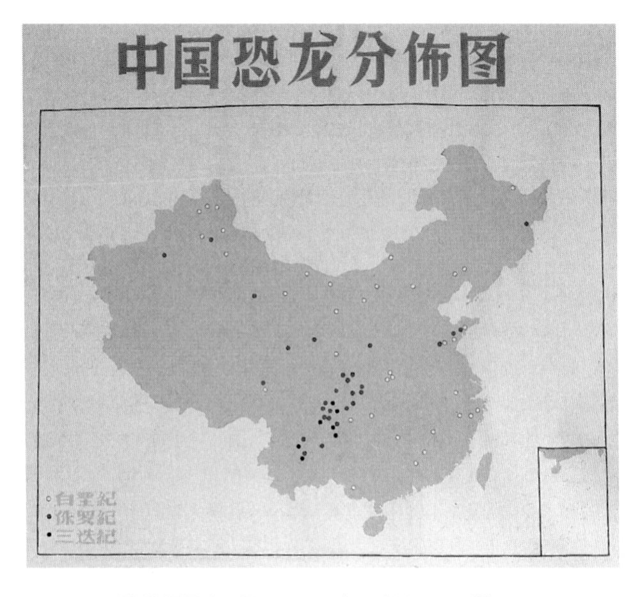

FIGURE 2 Dinosaur localities in China.

The Early Jurassic *Lufengosaurus* Fauna

This period is a significant time in dinosaur evolution. Dinosaurs were extensively distributed throughout all the continents. This cosmopolitan fauna could be named the Circum–Tethys Dinosaur Fauna (Dong, 1983). In China, this dinosaur assemblage comes mainly from southern China, such as the Lufeng and Yimen basins of Yunnan; the Weiyuan Basin of Sichuan, and the Dafang Basin of Guinzhou. So far, 23 genera of vertebrates have been recorded from the Lower Lufeng Formation. It represents what is here termed the prosauropod–*Lufengosaurus* faunal complex (Dong, 1980, 1992; Zhen *et al.*, 1985). Representative dinosaurian taxa include three genera of prosauropods, *Lufengosaurus, Anchisaurus (Gyposaurus),* and *Yunnanosaurus;* theropods (*Lukosaurus, Sinosaurus,* and *Dilophosaurus*); a stegosaur (*Tatisaurus*); and small ornithopods (*Dianchungosaurus*) (Young, 1941, 1942, 1951).

Kunmingosaurus wudingensis was a primitive sauropod from the Early Jurassic of Wuding, Yunnan. Its head and lower jaw are rather deep, with spoon-shaped teeth; the sacrum consists of six vertebrae. Its pelvic girdle shows sauropod-like features, modified in the form of a plate. This dinosaur is 7.5 m long and has a relatively short neck. Its heavy body is supported by four massive and straight legs (Dong, 1992).

The discovery of *Dilophosaurus* in the Lower Lufeng Formation was an important find in China made by the Kunming Museum in 1986 (Wu, 1992). It provided evidence that the age of the Lower Lufeng Formation was Early Jurassic (see Glen Canyon Group). Young (1951) argued for a Rhaetic age of the prosauropod–*Lufengosaurus* fauna of the Lower Lufeng Formation. It is now considered Early Jurassic in age by most paleontologists (Chen *et al.*, 1982; Wu, 1991; Wu *et al.*, 1993) (Fig. 3).

The Middle Jurassic *Shunosaurus* Fauna

The Middle Jurassic is a period when many major dinosaur taxa seemed to make their first appearances. They became the dominant members through a rapid radiation. Middle Jurassic dinosaurs in China come from two major areas: the Sichuan and the Junggar basins. In Sichuan, the Dashanpu is a well-known Middle Jurassic (Bathonian–Callovian) dinosaur site. It produced a primitive sauropod–*Shunosaurus* fauna

FIGURE 3 *Lufengosaurus*, appropriately on display in its namesake, the Lufeng Dinosaur Museum.

from the Lower Shaximiao (Xiashaximiao) Formation. The *Shunosaurus* fauna contained sauropods (*Protognasaurus, Shunosaurus, Datousaurus*), theropods (*Gasosaurus* and *Szechuanosaurus*), a stegosaur (*Huayangosaurus*), and ornithopods (*Xiaosaurus* and *Agilisaurus*).

Shunosaurus, a short-necked sauropod, is the best known dinosaur, with 12 or more complete skeletons and three well-preserved skulls discovered. *Omeisaurus* is a large sauropod. *Omeisaurus tianfuensis* has a bony club at the end of the tail for defense.

Agilisaurus louderbacki was a small fabrosaurid dinosaur (Peng, 1990). The material includes a nearly complete skeleton, with a complete skull. Many remains ranging from juveniles to adults were found at the Dashanpu site.

Huayangosaurus was a rather primitive stegosaur, with six or seven small teeth on the premaxilla. The bony plates are variable in shape and are arranged symmetrically along its back from the neck to the end of the tail. A pair of large bony plates also lie on the shoulders. Twelve individuals were found at Dashanpu Quarry.

Middle Jurassic deposits are also distributed extensively in the Junggar Basin, Xinjiang. They are called the Wuciawan Formation and are composed of light gray, fine to medium-grained feldspathic quartzitic sandstones, sandy mudstones, and siltstones. The environments of deposits were fluvial to deltaic. *Bellusaurus sui* was a small sauropod (4.8 m). Seventeen individuals have been found from the single quarry of Konglonggou (Dinosaur ravine), Kelamaili region. Evidently a herd of these animals had been overwhelmed in a flash flood. Morphological features suggested that they could be a group of juveniles.

Monolophosaurus is an allosaurid with a well-developed ridge on the top of the head. The material is a nearly complete skeleton with a complete skull and was collected from the Wucaiwan Formation in the Jiangjunmiao site in 1984 (Zhao and Currie, 1993).

The Late Jurassic *Mamenchisaurus* Fauna

The Late Jurassic represents a golden age of dinosaurs. Sauropods were flourishing and became the most abundant taxon, evolving giantism during the Late Jurassic. They reached their maximum size and greatest diversity, with a nearly global distribution. Chinese records of the Late Jurassic dinosaur-bearing strata to date are mainly from the Shishugou Formation of the Junggar Basin, the Xiangtang Formation of Gansu Province, the Mengyin Formation of Shandong Province, and the Upper Shaximiao (Shangshaximiao) Formation of the Sichuan Basin.

The main dinosaur fossils yielded by the Upper Shaximiao Formation include sauropods (*Mamenchisaurus* and *Omeisaurus*), theropods (*Szechuanosaurus, Yangchuanosaurus*, and *Sinraptor*), ornithopods

FIGURE 4 *Tuojiangosaurus*, a stegosaur from the Late Jurassic of Sichuan Basin.

(*Gongbusaurus* and *Yandusaurus*), and stegosaurs (*Chialingosaurus*, *Tuojiangosaurus*, and *Chungkingosaurus*) (Fig. 4).

Mamenchisaurus hechuanensis is the most famous sauropod of China, known from a fairly complete skeleton lacking only the skull and forelimbs. *Mamenchisaurus* was placed in a family of its own. C. C. Young pointed out that the Mamenchisauridae is similar to the Diplodocidae. Recently, however, an incomplete skeleton with a nearly complete skull of Mamenchisaurus was found in Raxian County by the Municipal Museum of Chongqing and a complete lower jaw was collected by CCDP. The skull of *Mamenchisaurus* is tiny when compared with the enormous size of the animal. It has the spatula-like teeth instead of the pencil-like teeth of the Diplodocidae, so Mamenchisauridae is valid.

Yangchuanosaurus was an allosauroid. It is known from an almost complete skeleton that lacks only forelimbs and some caudal vertebrae. This genus has three species, *Y. shangyouensis*, *Y. magnus*, and *Y. hepingensis*. The latter is a large form and was collected from the same beds as the former. The material consists of a complete skull, vertebrae, pelvic girdle, and hindlimbs. Recently it was referred to *Sinraptor* as a new species, *S. hepingensis* (Currie and Zhao, 1993), in a new taxon, Sinraptoridae.

Stegosaurs are the most bizarre dinosaurs found in the Upper Shaximiao Formation. Since the first skeleton of stegosaurs was found, the bony back plates of these dinosaurs have vexed paleontologists; both their function and their arrangement are still being argued. *Tuojiangosaurus's* bony plates are symmetrically arranged in pairs. This is similar to those of *Kentrosaurus* of eastern Africa. A pair of large and symmetrical bony plates lying on the shoulders is preserved in a new specimen collected from Heping, near Zigong, with a piece of skin also found from this specimen.

In the summer of 1987, the CCDP worked in the Jiangjunmiao region and found a lower jaw and a series of cervical vertebrae at the Shishugou Formation. It was identified as *Mamenchisaurus sinocandorum*, the largest sauropod in Asia (Russell and Zheng, 1993). A nearly complete allosauroid, *Sinraptor*, was collected from the same horizon. It was described and named *S. dongi* by Currie and Zhao in 1993. This is the most complete skeleton of a theropod from China (Fig. 5).

The Early Cretaceous *Psittacosaurus* Fauna

In Late Mesozoic times, East Asia comprised three major blocks: Siberia, north China, and south China (Lee *et al.*, 1987). Paleomagnetic data appear to point out that all these parts of East Asia occupied the same relative latitude in the Lower Cretaceous. The landscape in the northeastern part of Asia was dominated by a plateau, including north China, Mongolia, and southern Siberia (North China Block). It was covered with alluvial plains or some large lake basins. A river, the ancient Heilongjiang, flowed west to east on this highland, where vegetation flourished (Chen, 1977). This area formed a special ecological province where there lived an endemic dinosaur assemblage (Dong, 1992; Russell, 1993). This unique group of dinosaurs (the *Psittacosaurus* fauna) had evolved by the early Cretaceous in northeast Asia. Several fossil birds were found in the Early Cretaceous deposits of northeast China (Sereno *et al.*, 1990; Hou *et al.*, 1993). These animals are collectively known as the *Psittacosaurus* fauna and are found in the Qingshan Formation of Shandong, the Tugulu Group of the Junggar of Xinjiang, the Zhidan Group, the Ejinhoro Formation of Ordos, and so on. This fauna contains *Psittacosaurus*, theropods (*Kelmayisaurus*, *Phaedrolosaurus*, and *Tugulusaurus*), an iguanodontid (*Probactrosaurus*), a stegosaur (*Wuerhosaurus*), a protoceratopsid (*Microceratops*), and a pterosaur (*Dsungaripterus*).

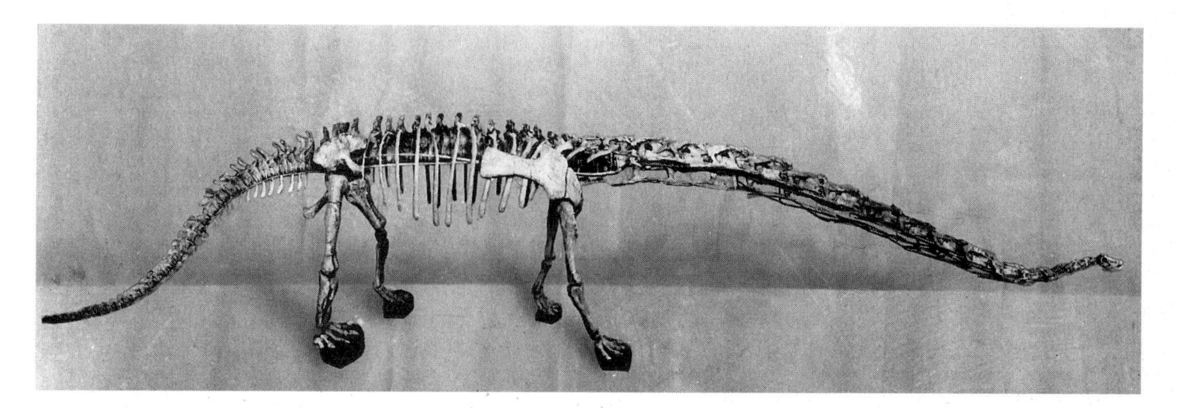

FIGURE 5 *Mamenchisaurus*, in China, is the analog of the sauropod *Diplodocus* from North America.

During the Early Cretaceous Period, there were two separate dinosaur faunas in the nonmarine deposits of north and south China, which are regarded as a separate district of the biogeographic province. The Psittacosaur–Pterosaur fauna is mainly found in the northern part of China (Dong, 1993).

A dinosaur fauna from the Early Cretaceous was reported at Tebch by Bohlin in 1950. Tebch means the "black plate" in Mongolian because the black lava (basalt) that may be of the mid-Aptian age (110 + 0.52 Ma) lies on top of the hill, with a thickness of 1.5–3 m (Eberth *et al.*, 1993). The dinosaur fauna consists of *Prodeinodon* sp. and *Psittacosaurus mongoliensis*. In the summer of 1990, this locality was reexamined by the CCDP.

Nanshiungosaurus is a most interesting dinosaur from the Nanxiong Basin and is known from an incomplete skeleton. This animal has a special pelvis that identifies it as a therizinosaur. The ilium is low, and the anterior apophysis is well developed and extends outward. The pubis is straight, and the exterior edge is thick. The ischium is thin and plate-like and the distal end expands and is fused. The age of the Nanxiong Formation was suggested as Maastrichtian.

Wannanosaurus is a small dome-headed dinosaur (pachycephalosaur), with a large supratemporal fenestra and a completely flattened cranial roof. The frontoparietal region is thick, and the external surface of the skull roof cranial bone is ornamented by small and densely distributed bony processes; the ornamentation on the temporal region is well developed.

Protoceratops was the most common dinosaur discovered from Bayan Mandahu, with 66 specimens of all sizes ranging from skull lengths of 2 cm to more than 1 m. The most important and interesting discovery at this locality was two mass graves of the ankylosaurid *Pinacosaurus* including 12 individual juveniles the size of small sheep. They were found in a nest, lying in their original positions covered by presumably windblown sand. Small theropods, including *Velociraptor* and *Oviraptor*, were collected, and many turtles, lizards, numerous nests of dinosaur eggs, and mammals were unearthed. In sedimentological features the beds are apparently similar to the Djadokhta Formation of Mongolia.

Dinosaur Eggs

Dinosaur eggs from the Late Cretaceous are abundant in China. At first, only simple descriptions were published based on the outer structure of the eggshell. Recent studies, mainly by Zhao and collaborators, are based on observations of the microstructure of the eggshell. Microscopic analysis of eggshells from the Late Cretaceous of China indicate that at least 12–15

FIGURE 6 *Elongatoolithus* eggs from the Late Cretaceous of South China.

dinosaur species are represented by the eggs. We have observed eggshells from the Nanxiong Basin that are unusual, indicating eggs from the end of Cretaceous that did not hatch.

The dinosaurian eggs from the Wangshi Formation of the Late Cretaceous were studied by Chow in 1951. Thereafter, Young (1953) and Zhao (1979) restudied and reclassified two groups: Spheroolithid and Elongatoolithid. These eggs were distributed in the red clays of the middle-upper part of the Wangshi Formation (Fig. 6).

See also the following related entries:

ERENHOT DINOSAUR MUSEUM • INSTITUTE OF VERTEBRATE PALEONTOLOGY AND PALEOANTHROPOLOGY • LUFENG • MIDDLE ASIAN DINOSAURS • MONGOLIAN DINOSAURS • SINO-CANADIAN DINOSAUR PROJECT

References

Andrews, R. C. (1932). The new conquest of Central Asia. The Amer. Mus. Nat. Hist. of Central Asia. *American Museum of Nat. History, New York.* pp. 1–678.

Camp, C. C. (1935). Dinosaur remains from the Province of Szechuan. *Bull. Dept. Geol. Univ. Calif.* **23,** 467–471.

Currie, P. J. (1991). The Sino-Canadian Dinosaur Expeditions, 1986–1990. *Geotimes* **36 (4),** 18–21.

Currie, P. J., and Zhao, X. J. (1993). A new large theropod (Dinosauria, Theropoda) from the Jurassic of Xinjiang, People's Republic of China. *Canadian J. Earth Sci.* **30,** 2037–2081.

Dong, Z. M. (1973a). Cretaceous stratigraphy of Wuerho District, Dsungar (Zunggar) Basin. *Memoirs of the Institute of vertebrate Paleontology and Paleoanthropology Academia Sinica* **11,** 1–7.

Dong, Zhiming (1980). On the dinosaurian fauna and their stratigraphical distribution in China. *J. Stratigraphy* **4,** 24–38.

Dong, Z. M. (1992). Dinosaurian Faunas of China. 188 pp. Springer-Verlag, Berlin.

Dong, Z. M., Russell, D. A., and Currie, P. J. (1988). The Dinosaur Project, an international cooperation program on dinosaurs. *Vertebrata PalAsiatica* **26,** 235–240.

Dong, Z. M., Currie, P. J., and Russell, D. A. (1989). the 1988 field program of the Dinosaur Project. *Vertebrata PalAsiatica* **27 (3),** 293–295.

Dong, Z. M., Zhou, S. W., and Zhang, Y. H. (1983). The dinosaurian remains from Sichuan Basin, China. *Palaeontologia Sinica,* No. 162 (New Series C, No. 23), pp. 1–145., 43 plates. (In Chinese.)

Eberth, D. A., Russell, D. A., Braman, D. R., and Deino, A. L. (1993). The age of the dinosaur-bearing sediments at Tebch, Inner Mongolia, People's Republic of China. *Canadian. J. Earth Sci.* **30,** 2101–2106.

Hou, L.-H., Zhou, Z.-H., Martin, L. D., and Feduccia, A. (1995). A beaked bird from the Jurassic of China. *Nature* **377,** 616–618.

Lee, Gidong, *et al.* (1987). Eastern Asia in Cretaceous, new paleomagnetic data from south Korea and a new look at Chinese and Japanese data. *J. Geophys. Res.* **29,** 3580–3596.

Louderback, G. D. (1935). The stratigraphic relations of the Junghsien fossil dinosaur in Sichuan Red Bed of China. *Bull. Dept. Geol. Univ. Calif.* **23,** 459–466.

Osborn, H. F. (1924). Three new theropoda, *Protoceratops* Zone, central Mongolia. *Am. Mus. Nat. Hist. Novitates* **144,** 12 pp.

Peng, G. H. (1990). A new small ornithopod (*Agilisaurus louderbacki* gen. et sp. nov.) from Zigong, Sichuan, China. *Zigong Dinosaur Museum Newletter* **2,** 19–27 (in Chinese).

Riabinin, A. N. (1925). A restored skeleton of a huge Trachodon amurensis nov. sp. Izvestia. *Geol. Com.* XLIV (1) p. 1–12.

Riabinin, A. N., (1930). *Mandschurosaurus amurensis* nov. gen. sp., a hadrosaurian dinosaur from the Upper Cretaceous of Amur River. *Societe Paleon. de Russie,* Mem. S.2 p. 1–36. Leningrad.

Russell, D. A. (1993). The role of central Asia in dinosaurian biogeography. *Canadian J. Earth Sci.* **30,** 2002–2012.

Russell D. A., and Zheng, Z. (1993). A large mamenchisaurid from the Junggar Basin, Xinjiang, People's Republic of China. *Canadian J. Earth Sci.* **30,** 2082–2095.

Sereno, P. C., and Rao, C. G. (1992). Early evolution of avian flight and perching: new evidence from the Lower Cretaceous of China. *Science* **255,** 845–848.

Wiman, C. (1929). Die Kreide-dinosaurier aus Shantung. *Pal. Sin. C* **1,** 1–67.

Young, C. C. (1941). Gyposaurus sinenses (sp. nov.), a new Prosauropoda from the Upper Triassic Beds at Lufeng, Yunnan. *Bull. Geol. Soc. China.* **21,** 205–252.

Young, C. C. (1942). Yunnanosaurus huangi (gen. et sp. nov.), a new Prosauropoda from the red Beds at Lufeng, Yunnan. *Bull. Geol. Soc. China.* **22,** 63–104.

Young, C. C. (1951). The Lufeng Saurischian Fauna in China. *Pal. Sin. C* **13,** 1–96.

Zhao, X. J., and Currie, P. J. (1993). A large crested theropod from the Jurassic of Xinjiang, People's Republic of China. *Canadian J. Earth Sci.* **30,** 2027–2036.

Zhen, S., Mateer, N., and Lucas, S. G. (1985). The Mesozoic reptiles of China. Studies of Chinese fossil vertebrates. *Bull. Geol. Inst. Univ. Uppsala N.S.* **11,** 133–150.

Chinle Formation

J. MICHAEL PARRISH
Northern Illinois University
DeKalb, Illinois, USA

The Chinle Formation is an extensive sequence of sedimentary rocks that was deposited over much of what today is the Colorado Plateau during the Late Triassic (Carnian–Norian). Consisting primarily of a series of fluvial/floodplain deposits, the rivers that deposited Chinle sediments appear to have drained into a large lake in the middle of the basin, a location that corresponds to present-day southeastern Utah (Blakey and Gubitosa, 1983). Paleoclimatological studies suggest that Chinle climates were relatively humid and marked by a monsoonal weather pattern that produced striking seasonality (Dubiel *et al.*, 1991).

The Chinle Formation is abundantly fossiliferous, with extensive records of plants, vertebrates, and invertebrates. Plants include vast deposits of petrified wood referred to the form genus *Araucaryoxylon*, including logs tens of meters in length. Other plant records include the horsetail *Neocalamites* and abundant leaf records of ferns, seed ferns, and conifers (Ash, 1986). The invertebrate record primarily consists of nonmarine molluscs, conchostrachans (clamshrimp), and rare insect fossils. Floodplain deposits are dominated by amphibious vertebrates, notably the large-headed metoposaurid amphibians and the long-snouted phytosaurian archosaurs. In more upland environments, common taxa include the armored aetosaurs and (in local concentrations) the dicynodont synapsid *Placerias*, large carnivorous archosaurs such as *Postosuchus*, and early crocodylomorphs (Long and Padian, 1986). Dinosaurs known from the Chinle include the ornithischian tooth form genus *Revueltosaurus* (Padian, 1990), the ceratosaurian theropod *Coelophysis* (Padian, 1986; Colbert, 1989), and the herrerasaurid *Chindesaurus* (Long and Murry, 1995).

The basal units of the Chinle Formation include the Shinarump, a coarse conglomerate that is the major source for uranium ore in southeastern Utah, and the fluviolacustrine Monitor Butte Member. The main fossil-bearing parts of the Chinle Formation are contained in a section usually referred to as the Petrified Forest Member. This member is commonly subdivided into upper and lower units, separated by an extensive sheet sandstone, the Sonsela. The pollen and fossil vertebrate records are consistent with the lower unit being Carnian in age and the upper unit Norian (Litwin *et al*, 1991). Overlying the Petrified Forest Member are the Owl Rock Member, a thick series of lacustrine facies, and the fluviolacustrine Church Rock Member.

Lucas and Lucas (1989) have advocated merging the Chinle and DOCKUM FORMATIONS within the Chinle Group and proposed elevating many of the members to formation status. The details of this proposal have yet to be fully published, and as yet it has not been widely accepted by other workers. Others, such as Blakey and Gubitosa (1983) and Dubiel *et al.* (1991), maintain that the Chinle and Dockum formations were deposited in adjacent sedimentary basins at approximately the same time.

See also the following related entries:

ARCHOSAURIA • PETRIFIED FOREST • TRIASSIC PERIOD

References

Ash, S. L. (1986). Fossil plants and the Triassic Jurassic Boundary. In *The Beginning of the Age of Dinosaurs* (K. Padian, (Ed.), pp. 21–30. Columbia Univ. Press, New York.

Blakey, R. C., and Gubitosa, R. (1983). Late Triassic paleogeography and depositional history of the Chinle Formation, southern Utah and northern Arizona. In *Mesozoic Paleogeography of West Central United States* (M. W. Reynolds and E. D. Dolly, Eds.), Rocky Mountain Paleogeography Symposium 2, pp. 57–76. Rocky Mountain Section, Society of Economic Paleontologists and Minerologists.

Colbert, E. H. (1989). The Triassic dinosaur *Coelophysis*. *Bull. Museum. N. Arizona* **57**, 1–174.

Dubiel, R. F, Parrish, J. T., Parrish, J. M., and Good, S. C. (1991). The Pangean megamonsoon—Evidence from the Upper Triassic Chinle Formation, Colorado Plateau. *Palaios* **6**, 347–370.

Litwin, R. L., Traverse, A., and Ash, S. R. (1991). Preliminary palynological zonation of the Chinle Formation, Southwestern U.S.A., and its correlation to the Newark Supergroup (eastern U.S.A.). *Rev. Palynol. Paleobot.* **68**, 269–287.

Long, R. A., and Murry, P. A. (1995). Late Triassic (Carnian and Norian) tetrapods from the Southwestern United States. *New Mexico Museum Nat. History Sci. Bull.* **4,** 1–254.

Long, R. A., and Padian, K. (1986). Vertebrate biostratigraphy of the Late Triassic Chinle Formation, Petrified Forest Natural Park, Arizona: Preliminary results. In *The Beijing of the Age of Dinosaurs* (K. Padian, Ed.), pp. 161–170. Cambridge Univ. Press, New York.

Lucas, S. G., and Lucas, A. P. (1989). Revised Triassic stratigraphy in the Tucumcari Basin, east-central New Mexico. In *Dawn of the Age of Dinosaurs in the American Southwest.* (S. G. Lucas and A. P. Hunt, Eds.), pp. 150–170. New Mexico Museum of Natural History, Albuquerque.

Padian, K. (1986). On the type material of *Coelophysis* Cope (Saurischia: Theropoda) and a new specimen from the Petrified Forest of Arizona. *The Beginning of the Age of Dinosaurs* (K. Padian, Ed.), pp. 45–60. Cambridge Univ. Press, New York.

Padian, K. (1990). The ornithischian form genus *Revueltosaurus* from the Petrified Forest of Arizona (Late Triassic: Norian; Chinle Formation). *J. Vertebr. Paleontol.* **10,** 268–269.

Classification

see PHYLOGENETIC SYSTEM; PHYLOGENY OF DINOSAURS; SYSTEMATICS

Cleveland–Lloyd Dinosaur Quarry

JOSHUA B. SMITH
University of Pennsylvania
Philadelphia, Pennsylvania, USA

Located in the Brushy Basin Member of the Morrison Formation (Tithonian), the Cleveland–Lloyd Dinosaur Quarry represents one of the most impressive dinosaur mass accumulations known. The site, designated a national landmark in 1967, is located in Emery County in east-central Utah (Fig. 1). The date of the quarry's discovery is not known but the first elaborate excavations occurred from 1937 to 1939 under the direction of William Stokes of Princeton University (Stokes, 1986).

The quarry sits within a gray calcareous claystone, floodplain deposit capped by a micritic limestone and a volcanic ash (Bilbey, 1992, 1993). The preliminary interpretation of the paleoenvironment is a groundwater-fed wetland (Bilbey, 1992, 1993).

The Cleveland–Lloyd Quarry differs from many other mass assemblages (e.g., Derstler, 1995) in that the animals found do not appear to be allochthonous, water-transported carcasses. The remains found at the quarry are mostly disarticulated, and are commonly broken. Evidence of predation or scavenging is common at Cleveland–Lloyd, and it is one of the few places where carnosaur bones are marked (Molnar and Farlow, 1990). The bones are randomly oriented, and there is no evidence of a prevalent current direction.

One of the most striking features about the Cleveland–Lloyd fauna is its almost exclusively dinosaurian composition. Of these remains, 75% have been identified as the large theropod *Allosaurus fragilis*. Indeed, Madsen (1976) reported that *A. fragilis* material from at least 44 individuals had been recovered from the quarry by 1975. This is an uncommon occurrence because dinosaur predator / prey ratios are usually weighted towards the herbivores. Also, Dodson (1971) and Dodson *et al.* (1980) observed that in many dinosaur mass accumulations, theropods are greatly outnumbered (but see GHOST RANCH QUARRY).

The stratigraphy of the quarry has been well studied (e.g., Bilbey, 1992). The common belief is that it was a predator trap, and that it may indicate gregarious behavior in Allosaurus (Richmond and Morris, 1996).

Invertebrate and floral remains from the Cleveland–Lloyd Quarry are extremely rare. Madsen (1976) reported rare, disassociated charophytes (Plantae), several gyrogonites (from genera of charophyte reproductive structures), and as many as four genera of gastropods (Mollusca). The dinosaur faunal list compiled by Madsen (1976) includes the theropods *A. fragilis*), *Ceratosaurus* sp. (elements from ?one individual), *Stokesaurus clevelandi* (two individuals), *Marshosaurus bicentesimus* (two individuals); the sauropods *Camarasaurus cf. lentus* (two or three individuals) and elements belonging to at least one and per-

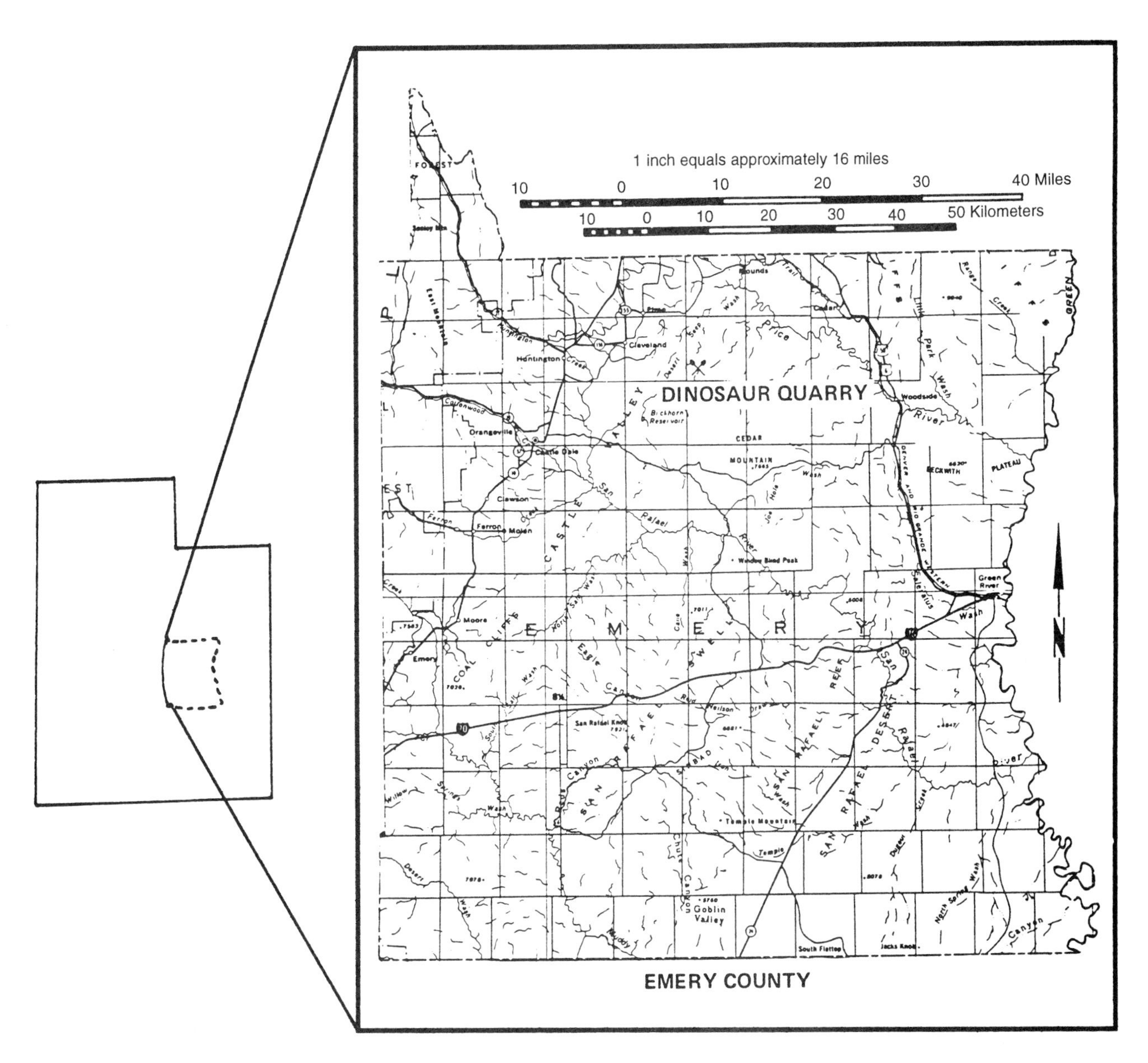

FIGURE 1 Locus map of the Cleveland–Lloyd Dinosaur Quarry in Emery County, Utah. (modified from Madsen, 1976).

haps several non-camarasaurid sauropods; the ornithischians *Camptosaurus cf. browni* (at least five individuals), *Stegosaurus cf. stenops* (at least two individuals), and *Stegosaurus* sp. (?-two or three individuals), and material related to one possible ankylosaur.

See also the following related entries:

Dinosaur National Monument • Morrison Formation • Museums and Displays

References

Bilbey, S. A. (1992). Stratigraphy and sedimentary petrology of the Upper Jurassic–Lower Cretaceous rocks at Cleveland–Lloyd dinosaur quarry with a comparison to the Dinosaur National Monument quarry, Utah, pp. 313. PhD dissertation, University of Utah, Provo.

Bilbey, S. A. (1993). Depositional environments in the Morrison Formation at the Cleveland–Lloyd Dinosaur Quarry, Utah. *J. Vertebr. Paleontol.* **13**(3), 26A.

Derstler, K. (1995). The Dragon's Grave—An *Edmontosaurus* bonebed containing theropod egg shells and juveniles, Lance Formation (Uppermost Cretaceous), Niobrara County, Wyoming. *J. Vertebr. Paleontol.* **15**(3), 26A.

Dodson, P. (1971). Sedimentology and taphonomy of the Oldman Formation (Campanian), Dinosaur Provincial Park, Canada. *Palaeogeogr. Palaeoclimatol. Palaeoecol.* **10,** 21–74.

Dodson, P., Behrensmeyer, A. K., Bakker, R. T., and McIntosh, J. S. (1980). Taphonomy and paleoecology of the dinosaur beds of the Jurassic Morrison Formation. *Paleobiology* **6**(2), 208–232.

Madsen, J. H., Jr. (1976). *Allosaurus fragilis*: A revised osteology. *Utah Geol. Survey Bull.* **109,** 163.

Miller, W. E., Horrocks, R. D., and Madsen, Jr., J. H. (1996). The Cleveland–Lloyd Dinosaur Quarry, Emery County, Utah: A U.S. Natural Landmark. *Brigham Young University Geological Studies* **41,** 3–24, four maps.

Molnar, R. E., and Farlow, J. O. (1990). Carnosaur paleobiology. In *The Dinosauria* (D. B. Weishampel, P. Dodson, and H. Osmolska, Eds.), pp. 210–224. Univ. of California Press, Berkeley.

Richmond, D. R., and Morris, T. H. (1996). The dinosaur death-trap of the Cleveland–Lloyd Quarry, Emery County, Utah. *Mus. N. Ariz. Bull.* **60,** 533–545.

Stokes, W. L. (1986). *The geology of Utah.* Utah Museum of Natural History Occasional Papers, No. 6.

Cleveland Museum of Natural History, Ohio, USA

see Museums and Displays

Cloverly Formation

W. Desmond Maxwell
New York College of Osteopathic Medicine
Old Westbury, New York, USA

The Lower Cretaceous Cloverly Formation (Aptian–Albian) is exposed in the Bighorn Basin of south-central Montana and northern Wyoming. It has an average thickness of approximately 90 m and consists of predominantly variegated claystones with channel-filling sandstones and conglomeratic sandstones (Moberly, 1962). The claystones represent soils resulting from fluvial overbank deposits and pyroclastic ash weathered in hot climates in seasonal swamps and lakes. The top of the Cloverly is marked by a transgressive marine facies, deposited by a southward-encroaching inland sea (Moberly, 1962; Young, 1970).

The terrestrial deposits of the Cloverly have yielded relatively few, but highly significant, dinosaurian remains. Most significantly perhaps are the remains of the raptorial, pack-hunting dromaeosaurid *Deinonychus antirrhopus* (Ostrom, 1969, 1990; Maxwell and Ostrom, 1995). Other important members of the dinosaurian fauna include the basal ornithopod *Tenontosaurus tilletti*, the nodosaurid *Sauropelta edwardsi*, and the small theropod *Microvenator celer* (Ostrom, 1970). Recent discoveries at a microsite include several varieties of dinosaurian eggshell and the only neonate dinosaurian remains known from any Lower Cretaceous deposit (Maxwell and Horner, 1994). Continued collecting at the site has produced new varieties of dinosaurian eggshell and additional remains of neonates, including the teeth of neonate dromaeosaurid and sauropod individuals.

The Cloverly fauna is a transitional one, mediating the sauropod- and stegosaur-dominated faunas of the Late Jurassic and the hadrosaur- and ceratopsian-dominated faunas of the Late Cretaceous. Further discoveries in the Cloverly and other Lower Cretaceous formations may help us to better understand the evolutionary mechanics of the Jurassic–Cretaceous transition in western North America.

See also the following related entries:

Cedar Mountain Formation • Cretaceous Period

References

Maxwell, W. D., and Horner, J. R. (1994). Neonate dinosaurian remains and dinosaurian eggshell from the Lower Cretaceous Cloverly Formation of south-central Montana. *J. Vertebr. Paleontol.* **14**(1), 143–146.

Maxwell, W. D., and Ostrom, J. H. (1995). Taphonomy and paleobiological implications of *Tenontosaurus–Deinonychus* associations. *J. Vertebr. Paleontol.* **15**(4), 707–712.

Moberly, R., Jr. (1962). Lower Cretaceous history of the Bighorn Basin, Wyoming. In *Wyoming Geological Association Guidebook, 17th Annual Field Conference* (R. L. Enyert and W. H. Curry, III, Eds.), pp. 94–101. Wyoming Geological Association.

Ostrom, J. H. (1969). Osteology of *Deinonychus antirrhopus,* an unusual theropod from the Lower Cretaceous of Montana. *Bull. Peabody Museum Nat. History* **30,** pp. 165.

Ostrom, J. H. (1970). Stratigraphy and paleontology of the Cloverly Formation (Lower Cretaceous) of the Bighorn Basin area, Wyoming and Montana. *Bull. Peabody Museum Nat. History* **35,** pp. 234.

Ostrom, J. H. (1990). The Dromaeosauridae. The *Dinosauria* (D. B. Weishampel, P. Dodson, and H. Osmolska, Eds.), pp. 269–279. Univ. of California Press, Berkeley.

Young, R. G. (1970). Lower Cretaceous of Wyoming and the southern Rockies. In *Wyoming Geological Association Guidebook, 22nd Annual Field Conference.* (R. L. Enyert, Ed.), pp. 147–160. Wyoming Geological Association.

Coelurosauria

John R. Hutchinson
Kevin Padian
University of California
Berkeley, California, USA

Coelurosauria ("hollow-tailed reptiles") (Fig. 1) was a name coined by F. von Huene (1914, 1920, 1926), based on the genus *Coelurus,* to include a variety of small, lightly built theropod dinosaurs. It was distinguished from Carnosauria, larger forms with great skulls and enormous teeth. Since that time both terms have been used in variously formal and informal senses, but it has often been thought that the large forms comprised a more or less natural group, the lineage beginning with Early Jurassic forms such as *Dilophosaurus* and extending through the Late Jurassic *Ceratosaurus, Megalosaurus,* and *Allosaurus* and through the Late Cretaceous tyrannosaurs (see Theropoda).

Despite occasional misgivings to the contrary, such as Ostrom's (1969) description of the maniraptoran *Deinonychus,* which hinted at the possible artificiality of the small coelurosaur—large carnosaur dichotomy, coelurosaurs have generally been considered an informal taxon of small theropods ranging from the Late Triassic through the Late Cretaceous. However, Gauthier (1986) redefined Coelurosauria in cladistic terms as a stem-based taxon comprising birds and all theropods closer to birds than to Carnosauria. Hence, Coelurosauria and Carnosauria were sister

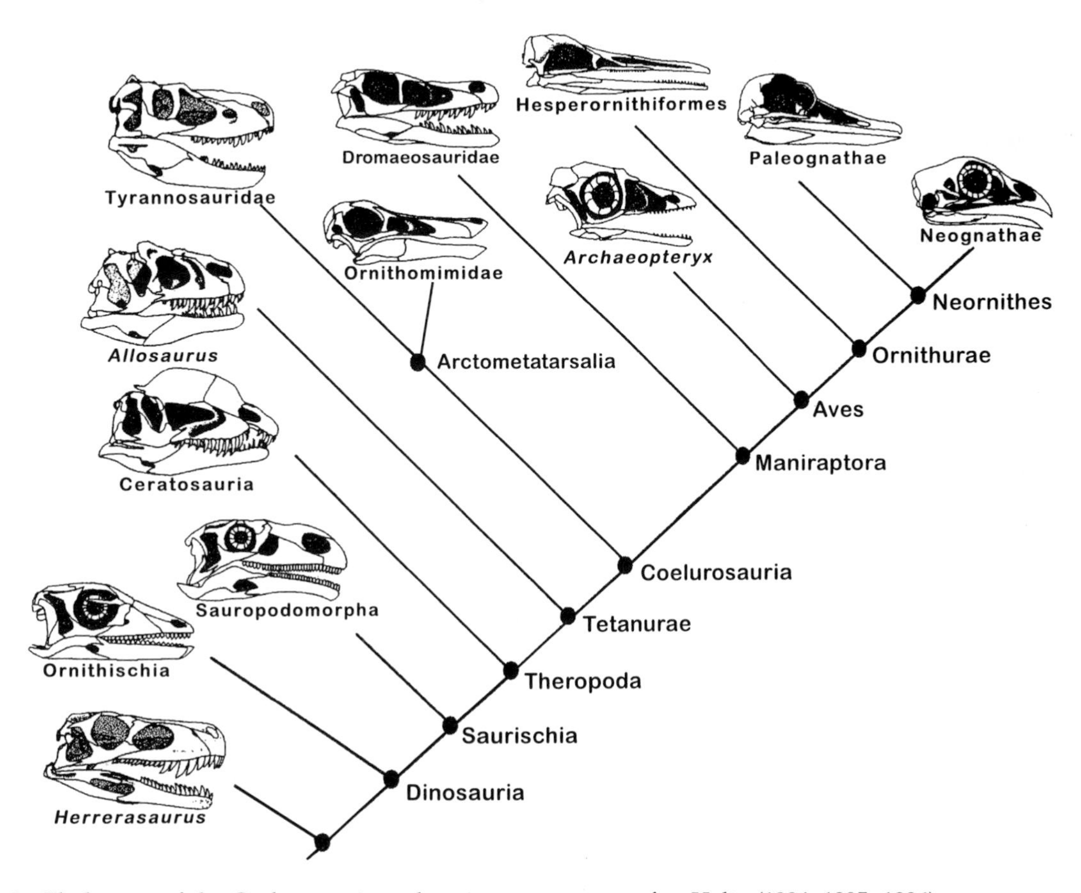

FIGURE 1 Phylogeny of the Coelurosauria and various outgroups; after Holtz (1994, 1995, 1996) and other sources, courtesy T. Rowe and L. Dingus.

taxa within what Gauthier (1986) called TETANURAE, a stem-based taxon. Holtz (1994) formalized AVETHEROPODA as the node-based taxon within Tetanurae comprising Coelurosauria and Carnosauria. Gauthier (1986) listed ORNITHOMIMIDAE, *Compsognathus, Ornitholestes, Coelurus, Microvenator, Saurornitholestes, Hulsanpes,* ELMISAURIDAE, Caenagnathidae, DEINONYCHOSAURIA, and AVIALAE (*Archaeopteryx* + Aves) as component taxa of Coelurosauria. To these can be added THERIZINOSAUROIDEA, now generally considered close to OVIRAPTORIDAE (Russell and Dong, 1993; Holtz, 1994, 1996a,b; Clark *et al.*, 1995).

Novas (1992) realized that tyrannosaurs did not belong in Carnosauria, and an analysis by Holtz (1994) indicated that they were the sister group to BULLATOSAURIA (TROODONTIDAE + ORNITHOMIMIDAE) within Coelurosauria. Hence tyrannosaurs, despite their great size, are not carnosaurs but rather coelurosaurs, as von Huene first recognized.

Gauthier's (1986) analysis did not recognize specific subgroups within Coelurosauria, except to link DEINONYCHOSAURIA with AVIALAE (*Archaeopteryx* + AVES in his formulation; Aves of most other workers). Gauthier's definition of Deinonychosauria (Colbert and Russell, 1969) included Troodontidae and Dromaeosauridae, but troodontids are now regarded as the sister taxon to ornithomimids; therefore, Deinonychosauria could be retained as the node corresponding to Holtz's (1994) unnamed "node 11." However, Holtz (1996b), in reanalyzing the nomenclature, called this node MANIRAPTORIFORMES.

Holtz's (1994, amended 1996b) analysis of the phylogenetic relationships among the major subclades of Theropoda found that within Coelurosauria, Dromaeosauridae and Avialae form a monophyletic group called MANIRAPTORA [following Gauthier's (1986) usage; Holtz 1996b *non* 1994]; a second monophyletic group (ARCTOMETATARSALIA) is formed by Ornithomimosauria + Troodontidae (Bullatosauria) + TYRANNOSAURIDAE; and OVIRAPTORIDAE, Caenagnathidae, and THERIZINOSAUROIDEA may form a third monophyletic group (designated OVIRAPTOROSAURIA by Russell and Dong, 1993). *Ornitholestes* and *Compsognathus* appear to be outgroups to all these taxa within Coelurosauria (Fig. 1). Recent analyses have also placed the enigmatic Late Cretaceous "carnosaur" *Dryptosaurus* as a basal coelurosaur (Denton, 1990; Holtz, 1996a), although this is strange given its late appearance. Other poorly known taxa, such as the recently discovered *Bagaraatan* (Osmólska, 1996) and *Deltadromeus* (Sereno *et al.*, 1996) and the long known but poorly understood *Coelurus*, may be other basal (i.e., nonmaniraptoriform) coelurosaurs, members of other less inclusive clades within Coelurosauria, or noncoelurosaurian tetanurines (see also Norman, 1990).

Coelurosauria is supported as a node by synapomorphies including an ischium reduced to two-thirds or less the length of the pubis, the loss of the ischial foot, an expanded circular orbit, a triangular obturator process on the ischium, an ascending process of the astragalus that is greater than one-fourth the length of the epipodium, and 15 or fewer caudal vertebrae bearing transverse processes (Holtz, 1994). Apart from *Compsognathus,* the other coelurosaurians share an ulna that is bowed posteriorly, a long and slender metacarpal III, a posterodorsal margin of the ilium that curves ventrally in lateral view, flexed cervical zygapophyses, a jugal expressed on the rim of the antorbital fenestra, and a first metacarpal that is one-third or less as long as the third (Holtz, 1994). Coelurosaurians apart from *Compsognathus* (Fig. 2) and *Ornitholestes* (Fig. 3), which Holtz (1996b) termed Maniraptoriformes, share a distally placed obturator process on the ischium, a third antorbital fenestra, and elongated anterior cervical zygapophyses (Holtz, 1994). Dromaeosauridae + Avialae form Maniraptora, and Troodontidae + Ornithomimosauria form Bullatosauria. *Avimimus* is removed from the analysis because it may be a chimera (Holtz, 1996a,b).

Phylogenetic analysis of the coelurosaurian groups must be regarded as substantially in flux. This is the result of several factors, including considerable homoplasy, the uncertain identity or association of some taxa (e.g., *Avimimus*), incomplete knowledge of some groups (e.g., Elmisauridae), differences in coding and polarizing characters and character states, and the programs and options used in computerized phylogenetic analysis. Nevertheless, considerable strides have been made by recent explorations and systematic analyses. The hypothesis that birds are coelurosaurs most closely related to dromaeosaurs has been sustained, ornithomimids and oviraptors have been shown to be not especially closely related within Coelurosauria, therizinosauroids appear to be coelurosaurs and not aberrant ornithischians or basal sauropodomorphs, and tyrannosaurs are closely related to ornithomimids and troodontids. It can be expected

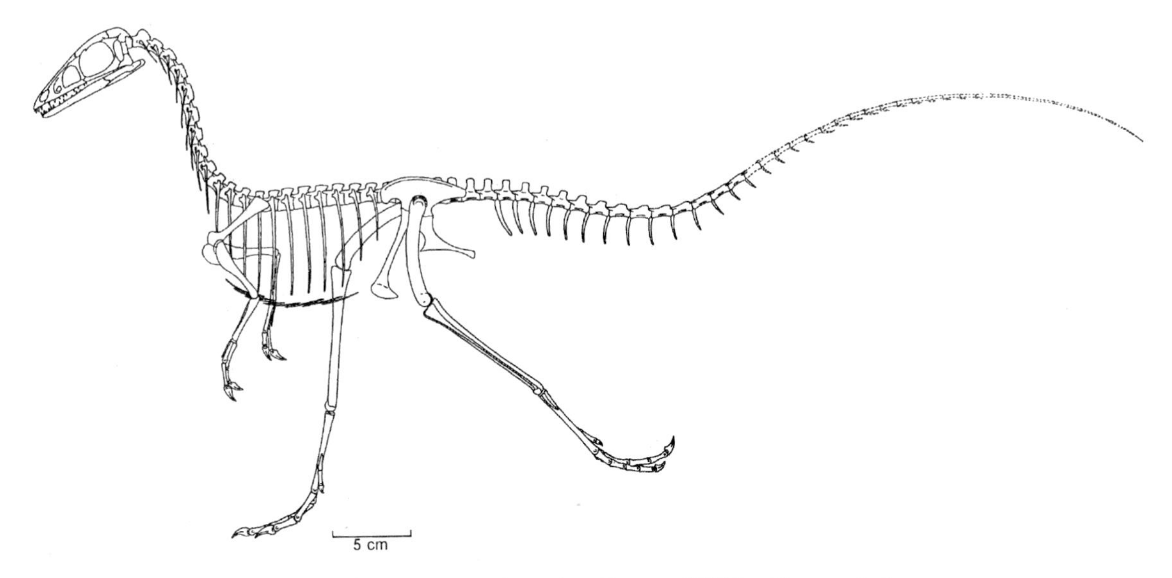

FIGURE 2 *Compsognathus,* after Ostrom (1978) with permission.

that future discoveries and analyses will change some details of the phylogeny as presented here.

Coelurosaurs, as small theropods, have traditionally been considered active, agile bipeds, feeding primarily on small tetrapods, sometimes including those of their own species. These dinosaurs show virtually all the features classically considered to demonstrate cursoriality (Coombs, 1978). The femur is shorter than the tibia and, as in all dinosaurs, is set off from the pelvic girdle by a rounded head placed at 90° to the femoral shaft. The knee is a hinge, as is the ankle, and the lower extremities (lower leg bones, metatarsals, and phalanges) are particularly elongated. The fibula is reduced and strap-like and does not move against the tibia; nor is there substantial movement between the astragalus and calcaneum nor between these proximal tarsals and the leg. The proximal tarsals and distal tarsals (which form a cap to the metatarsals) articulate to create a mesotarsal hinge joint. These features are characteristic not only of basal coelurosaurs but also of basal theropods, dinosaurs, and even ornithodirans (see BIPEDALITY; FUNCTIONAL MORPHOLOGY; ORNITHODIRA; ORNITHOSUCHIA). In this way, the femur's rotation is generally around a subhorizontal orientation, and during locomotion it is elevated and depressed through an angle of perhaps 30–45°. The tibia, meanwhile, swings in a wide anteroposterior arc, as do the bones of the foot, and it is distal to the knee that most of the excursion of the stride takes place. This is why, for example, ornithomimids are considered to have been ostrich-like, rapid runners, because their lower leg bones and metatarsals are so long. Conversely, *Deinonychus* has an almost surprisingly robust hindlimb with a low

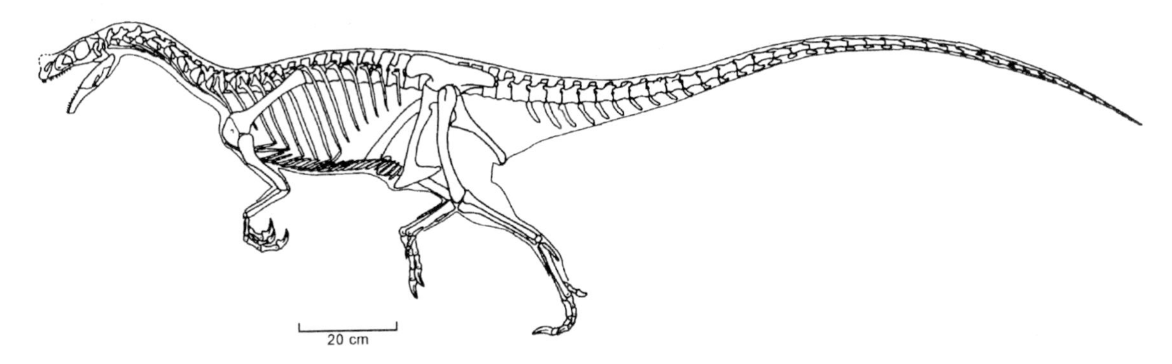

FIGURE 3 *Ornitholestes,* after Norman (1990).

tibia:femur ratio; Ostrom (1969) associated this anomaly with the function of the reverse-jointed second toe and hypertrophied claw in attacking prey rather than in maximizing running ability. Among ornithischian dinosaurs, only some basal forms, such as *Lesothosaurus* and *Heterodontosaurus,* are comparable to theropods in the cursorial ratios of their hindlimb elements. ORNITHISCHIANS took refuge, as did sauropodomorphs, not in their running ability but in large size, which made them more formidable prey and probably also facilitated the digestive processing of the plants that they ate. Several ornithischian clades also show evidence of group behavior, which would have made individual members more difficult targets (see BEHAVIOR). There is some evidence that some coelurosaurs, such as *Deinonychus,* foraged in packs (Ostrom and Maxwell, 1996), which may have given them some advantage over large prey.

Coelurosaurs are generally considered small theropods, and certainly most known Mesozoic coelurosaurs were small (under 2 m in length); most living coelurosaurs—i.e., birds (see BIRD ORIGINS)—are even smaller, seldom exceeding 1 m in length, and mostly 10–15 cm long. However, as shown previously, not all coelurosaurs were small. Whether one accepts *Eoraptor* and herrerasaurids as basal theropods, as dinosauriformes close to the origin of dinosaurs, or at some phylogenetic level in between (see HERRERASAURIDAE), it is clear that these forms are not large representatives of carnivorous dinosaurs (*Eoraptor* has a skull just over 10 cm long, and that of *Herrerasaurus* is about three times as long). Ceratosaurs, as currently understood, are small when they first appear in the fossil record (e.g., *Coelophysis* and *Syntarsus*), and the first well-known large forms are the Early Jurassic *Dilophosaurus* and the Late Jurassic *Ceratosaurus* (which, enigmatically, appears to be the sister taxon of the Cretaceous ABELISAURIDS as well as phylogenetically the most basal member of Ceratosauria; see Rowe and Gauthier, 1990). A great number of small problematic theropod remains, usually grouped in Coelurosauria but actually too indeterminate or basal to qualify, are known from the Late Triassic (Norman, 1990), but no very large forms are known. The true coelurosaurs, then, began at small size, as exemplified by their basal members *Compsognathus, Coelurus,* and *Ornitholestes,* all from the Late Jurassic.

However, large forms evolved several times within the lineage. Ornithomimids, such as *Harpymimus, Gallimimus,* and *Struthiomimus,* reached lengths of 3–5 m, with skulls up to 30 cm long (see ORNITHOMIMIDAE). The giant *Deinocheirus,* represented only by a shoulder girdle and forelimb, has a variety of apparently general theropod characters that make it difficult to classify but has the three subequal metacarpals characteristic of ornithomimids (see Norman, 1990, 292); if it is indeed a member of this group, or of Coelurosauria, it is certainly on the large end of the size range. Tyrannosaurids, now reclassified as coelurosaurs, are also among the largest, if not the largest known theropods. The skull of *Deinonychus* is approximately 30 cm long, so it must be regarded as at least a medium-sized carnivore, and *Utahraptor* must have been much larger. *Dryptosaurus,* whose precise systematic position is uncertain, is now regarded as a basal coelurosaur (Denton, 1990); it is very large, but it occurs in the latest Cretaceous, so it would probably be unwise to accord its size much influence in considering the state of basal coelurosaurs in the Late Jurassic, when the group is first known. Given the outgroup comparisons previously noted, *Dryptosaurus* and all the other large coelurosaurs are almost certainly secondarily large.

Finally, as noted previously, coelurosaurs are the only dinosaurian clade that did not become extinct by the end of the Cretaceous. Its surviving subclade, the birds, first appear in the Late Jurassic *(Archaeopteryx)* and so are known as far back in the fossil record as any group of coelurosaurs, as Gauthier (1986) noted. Intimations of their avian relationships have been provided dramatically with the discovery that at least some obviously nonavian coelurosaurs apparently had a type of feathered covering; this further suggests that the origin of feathers did not evolve directly for the purpose of flight, and whether thermoregulation, display, or a related function was the selective force for the evolution of feathers, the behavioral implication is clear that coelurosaurs were and are different from other, nonornithodiran reptiles, then and now (see AVES; BIRD ORIGINS; FEATHERED DINOSAURS).

See also the following related entries:

ARCTOMETATARSALIA • CARNOSAURIA • CERATOSAURIA • MANIRAPTORA • THEROPODA

References

Colbert, E. H., and Russell, D. A. (1969). The small Cretaceous dinosaur *Dromaeosaurus*. *Am. Museum Novitates* **2380,** 1–49.

Coombs, W. P., Jr. (1978). Theoretical aspects of cursorial adaptations in dinosaurs. *Q. Rev. Biol.* **53,** 393–418.

Denton, R. K., Jr. (1990). A revision of the theropod *Dryptosaurus (Laelaps) aquilunguis* (Cope 1869). *J. Vertebr. Paleontol.* **9**(Suppl. to No. 3), 20A.

Gauthier, J. A. (1986). Saurischian monophyly and the origin of birds. *Mem. California Acad. Sci.* **8,** 1–55.

Holtz, T. R., Jr. (1994). The phylogenetic position of the Tyrannosauridae: Implications for theropod systematics. *J. Paleontol.* **68,** 1100–1117.

Holtz, T. R., Jr. (1995). A new phylogeny of the Theropoda. *J. Vertebr. Paleontol.* **15**(Suppl. to No. 3), 35A.

Holtz, T. R., Jr. (1996a). Phylogenetic analysis of the nonavian Tetanurine dinosaurs (Saurischia, Theropoda). *J. Vertebr. Paleontol.* **16**(Suppl. to No. 3), 42A.

Holtz, T. R., Jr. (1996b). Phylogenetic taxonomy of the Coelurosauria (Dinosauria: Theropoda). *J. Paleontol.* **70,** 536–538.

Norman, D. B. (1990). Problematic Theropods: "Coelurosaurs". In *The Dinosauria* (D. B. Weishampel, P. Dodson, and H. Osmólska, Eds.), pp. 280–305. Univ. of California Press, Berkeley.

Novas, F. E. (1992). La evolución de los dinosaurios carnivóros. In *Los Dinosaurios y su Entorno Biotico. Actas II, Curso de Paleontológia en Cuenca* (J. L. Sanz and A. Buscalione, Eds.), pp. 123–163. Instituto "Juan de Valdes," Cuenca, Spain.

Osmólska, H. (1996). An unusual theropod dinosaur from the Late Cretaceous Nemegt Formation of Mongolia. *Acta Palaeontol. Polonica* **41,** 1–38.

Ostrom, J. H. (1969). Osteology of *Deinonychus antirrhopus*, an unusual theropod from the Lower Cretaceous of Montana. *Bull. Peabody Museum Nat. History* **30,** 1–165.

Ostrom, J. H. (1978). The osteology of *Compsognathus longipes*. *Zitteliana* **4,** 73–118.

Ostrom, J. H. (1990). Dromaeosauridae. In *The Dinosauria* (D. B. Weishampel, P. Dodson, and H. Osmólska, Eds.), pp. 269–279. Univ. of California Press, Berkeley.

Ostrom, J. H., and Maxwell, W. D. (1996). *J. Vertebr. Paleontol.* **16.**

Rowe, T. R., and Gauthier, J. A. (1990). Ceratosauria. In *The Dinosauria* (D. B. Weishampel, P. Dodson, and H. Osmólska, Eds.), pp. 151–168. Univ. of California Press, Berkeley.

Russell, D. A., and Dong, Z. (1993). The affinities of a new theropod from the Alxa Desert, Inner Mongolia, People's Republic of China. *Can. J. Earth Sci.* **30,** 2107–2127.

Sereno, P. C., Duthiel, D. B., Iarochene, M., Larsson, H. C. E., Lyon, G. H., Magwene, P. W., Sidor, C. A., Varricchio, D. J., and Wilson, J. A. (1996). Predatory dinosaurs from the Sahara and Late Cretaceous faunal differentiation. *Science* **272,** 986–991.

von Huene, F. (1914). Das naturliche System der Saurischia. *Zentralblatt Mineral. Geol. Paläontol. B* **1914,** 154–158.

von Huene, F. (1920). Bemerkungen zur Systematik und Stammesgeschichte einiger Reptilien. *Zeitschrift Indukt. Abstammungslehre Vererbungslehre* **24,** 162–166.

von Huene, F. (1926). The carnivorous Saurischia in the Jura and Cretaceous formations, principally in Europe. *Revista Museo de La Plata* **29,** 35–167.

Coevolution

see Plants and Dinosaurs

College of Eastern Utah Prehistoric Museum, Utah, USA

see Museums and Displays

Colombian Dinosaurs

Unidentified theropod dinosaurs have been excavated from the "Ortega" formation in Colombia.

see South American Dinosaurs

Coloniality

There is some evidence that some dinosaurs were at least periodically, and perhaps usually, colonial (i.e., that they congregated in large, cooperative groups). See Behavior entry for a fuller treatment of this issue and related inferences.

see Behavior

Color

Michael J. Ryan
Royal Tyrrell Museum of Palaeontology
Drumheller, Alberta, Canada

Anthony P. Russell
University of Calgary
Calgary, Alberta, Canada

The fossil record of life on earth consists of copious amounts of preserved hard tissues and relatively few records of soft tissues. For vertebrates, bones and teeth comprise by far the greatest component of the record, with soft tissues, including the integument, playing a minor role. In vertebrates the skeleton, whether it is relatively superficial (such as the bony scales of many fish) or deep (such as the long bones of the limbs and the ribs), is overlain by soft tissues and therefore is normally unpigmented. For vertebrates, then, we rely on superficial tissues, especially the SKIN, to reveal color and color patterns.

This, in itself, poses two major problems because soft tissues, including the skin, typically do not preserve well, and pigments that occur in biological tissues are generally not stable and thus do not normally withstand the rigours of fossilization. The geochemistry of fossilized pigments and their derivatives is poorly known and the material relatively rare. Almost all the information that we have on pigments of extinct organisms comes from mollusks, although evidence is also available for brachiopods, trilobites, crinoid echinoderms, and insects. Most animal pigments are soluble in water and thus are prone to rapid disappearance after death (Parsons and Brett, 1991). Melanins (pigments responsible for brown and black colors) are generally much less soluble, however, and thus have a better record of preservation. The more recent the fossil the better the chances of the preservation of any pigments. Although our evidence of pigments for mollusks is better than that for other groups and dates back to the Permian, most examples are known from remains no older than the Tertiary.

From what we do know of the preservation of pigmentation, it is dependent upon the original pigment composition (i.e., stability—most are unstable), the location in the body (hard parts vs soft parts), the mineralogy of the hard parts if the pigments are preserved there, the rapidity of burial, and the nature of the subsequent fossilization process.

If alteration of pigments occurs, color patterns (as opposed to true color preservation) may be evident. Thus, light/dark patterns may be evident, as in a specimen of the trilobite *Anomocare vittata* (Raymond, 1922) from the Ordovician, which was preserved with three transverse, alternating light/dark bands on its pygidium. Extensive evidence of color pattern preservation in fossil mollusks is documented by Hollingsworth and Barker (1991). The ink sacs of dibranchiate cephalopods are well known for Jurassic seriopods from the Holzmaden and Solhofen deposits of Germany (Boucot, 1990). Preserved color patterns are known for a variety of terrestrial fossil insects, including elaborate wing patterns with eye spots (Jarzembowski, 1984). For fossil vertebrates, a cryptic color pattern is preserved from an amphibian of the Upper Carboniferous (Lund, 1978), with chromatophores known from frogs from the Jurassic (Hecht, 1970) and the Eocene (Voight, 1935). Grande (1982) and others have described chromatophores for Eocene fish. The radius and ulna of an ornithomimid from the late Campanian of Alberta possesses short, black markings that are regularly spaced parallel and sub-parallel to the long axes of these bones (M. Ryan and A. Russell, personal observation). Their origin and association to the dinosaur are unknown.

A small, feathered compsognathid dinosaur from China (Late Jurassic–Early Cretaceous) exhibits a black discoloration on both sides of the preserved sclerotic ring. A bird feather from the Eocene Green River Shale of Colorado preserves a color pattern (Lambrecht, 1933). For the reasons outlined previously, color preservation in vertebrates is extremely rare. Holman and Sullivan (1981) described dark ocular markings on the turtle *Chrysemys* from the Miocene. Sullivan *et al.* (1988) report the preservation of the original color pattern of black spots against a reddish-brown matrix in the preserved keratinous epidermal scales of the early Paleocene turtle *Neurankylus*.

One of the most frequently asked questions by students of dinosaur paleontology is "what color were dinosaurs?" The simple answer is " we don't know" and probably never will because of the chemical composition of such pigments and their solubility, and because of the extreme unlikelihood that skin will be preserved at all, let alone with the colors intact. Such eventualities mean that speculation, based on what we know of pigmentation in living vertebrates, is our chief vehicle for understanding color and color patterns in dinosaurs.

Dinosaur artists from the first half of the century (e.g., Charles Knight and, to a lesser degree, Zdenek Burian and Rudolph Zallinger) typically painted dinosaurs in drab green, browns, or grays. These artists would have based their interpretations on modern-day analogs such as large mammals (elephants and hippopotamus) and reptiles [alligators, crocodiles, and large varanid lizards (e.g., the Komodo Dragon, *Varanus komodensis*)]. These animals usually have subdued colors over the majority of their bodies, although they often exhibit countershading and some spots of other colors. Recent artists (Mark Hallett, Greg Paul, William Stout, and many others) have expanded their pallet and presented us with brightly colored and patterned dinosaurs reflecting the relatively recent rethinking of dinosaurs as energetic, sociable animals. Modern vertebrates display a wide variety of color patterns in their integument (skin, scales, feathers, and hair). Based on preserved fossil skin impressions for dinosaurs (primarily large dinosaurs) of the Mesozoic, we know their skin was usually covered in non-bony scales of a variety of sizes, which abutted each other in a mosaic-like pattern. Modern reptiles, especially lizards and snakes, possess a wide variety of colors and color patterns. There is nothing, in theory, precluding dinosaurs from being similarly endowed. Colors and color patterns can serve a number of functions in modern reptiles including use in social interactions, thermoregulation, aposematism (color patterns that signal a noxious or otherwise dangerous quality of a prey to a potential predator, such as that found in the coral snake), and defense (cryptic or mimicking coloration or for suddenly revealed bright colors for startling potential predators) (see Pough, 1988, and references therein). Dinosaurs probably used coloration patterns in these ways as well, although which species possessed which qualities will probably never be known with certainty.

See also the following related entry:

SKIN

References

Boucot, A. J. (1990). *Evolutionary Paleobiology of Behaviour and Coevolution*, pp. 725. Elsevier, Amsterdam.

Carpenter, F. M. (1971). Adaptations among Paleozoic insects. *Proc. North Am. Paleontol. Congr.* **II,** 1236–1251.

Grande, L. (1982). A revision of the fossil genus *Knightia*, with a description of a new genus from the Green River Formation. *Am. Museum Novitiates*, **2731**.

Hecht, M. K. (1970). The morphology of *Eodiscoglossus*, a complete Jurassic frog. *Am. Museum Novitiates*, **2227**.

Hollingsworth, N. T. J., and Barker, M. J. (1991). Color pattern preservation in the fossil record: Taphonomy and diagenetic significance. In *The Processes of Fossilization* (S. K. Donovan, Ed.) pp. 105–119. Columbia Univ. Press, New York.

Holman, J. A., and Sullivan, R. M. (1981). A small herpetofauna from the type section of the Valentine Formation (Miocene: Barstovian), Cherry County, Nebraska. *J. Paleontol.* **55,** 138–144.

Jarzembowski, E. A. (1984). Early Cretaceous insects from southern England. *Mod. Geol.* **9,** 71–94.

Lambrecht, K. (1933). *Handbuch der Paläoornithologie*, pp. 1024. Borntraeger, Berlin.

Lund, R. (1978). Anatomy and relationships of the Family Phlegethontiidae (Amphibia, Aistopoda). *Ann. Carnegie Museum* **47,** 53–79.

Pough, F. H. (1988). Antipredator mechanisms in reptiles. *Biol. Reptilia* **16,** 1–152.

Raymond, P. E. (1922). A trilobite retaining color markings. *American Journal of Science* **4,** 461–464.

Sullivan, R. M., Lucas, S. G., Hunt, A. P., and Fritts, T. H. (1988). Color pattern on the selmacryptodiran turtle *Neurankylus* from the early Paleocene (Puercan) of the San Juan Basin, New Mexico. *Contrib. Sci., Nat. History Museum Los Angles County* **401**.

Voight, E. (1935). Die Erhaltung von Epithelzellen mit Zellkeren, von Chromatophoren und Corium in fossiler Froschhaut aus der mittel eozänen Braunkohle des Geiseltales. *Nova Acta Leopoldina*, **3,** 339–360.

Como Bluff

BRENT H. BREITHAUPT
University of Wyoming
Laramie, Wyoming, USA

One of the most renowned dinosaur collecting areas in the world is Como Bluff in the northern part of the Laramie Basin, in southeastern Wyoming. Como

Bluff is a roughly east–west trending anticline plunging to the west, with a gently dipping southern limb and a steeply dipping northern limb. Como Bluff is approximately 15 km long and 1.5 km wide. The southern limb (Como Ridge) of this breached anticline is capped by a highly resistant sandstone of the Cretaceous Cloverly Formation. Beautiful exposures of the Upper Jurassic Morrison Formation crop out beneath the Cloverly Formation cap rock on the north face of Como Ridge. The marine Jurassic Sundance Formation and some of the Triassic red beds are also exposed. The south face is a slightly vegetated, gentle slope of Cretaceous strata (e.g., Cloverly Formation, Mowry Shale, Muddy Sandstone, and Thermopolis Shale).

Como Bluff derives its name from the spring-fed Lake Como at the western end of Como Bluff. The lake was given this name by early surveyors who thought that this 1.5-km-long body of water resembled the famous Lake Como (Lago de Como) in the Italian Alps (Urbanek, 1988).

In 1877, Professor Othniel Charles Marsh at Yale University received a letter from Messrs. Harlow and Edwards, identified as Union Pacific Railroad workers in the Wyoming Territory. They reported finding numerous large bones of a "*Megatherium*" from "Tertiary Period" units near Laramie City, Wyoming Territory. They claimed to have found a 142-cm-long shoulder blade and a 25-cm-long vertebra. A few months after their first letter, Marsh received the bones in Connecticut. Marsh sent his field assistant, Samuel W. Williston, to check the site. Williston wrote back to Marsh of a long exposure containing dinosaur bones in an amount and diversity far exceeding any site that had yet been found.

As Marsh had suspected, the bones were from the same fossiliferous formation on which he had crews working in Colorado. Marsh transferred his crews to Wyoming Territory to begin collecting later that same year. Williston discovered that the two men who wrote to Marsh were the station agent and section foreman of the Union Pacific Railroad Como Station, approximately 90 km north of Laramie. Their full names, which they had not used to keep their find secret, were William Harlow Reed and William Edward Carlin. The value of prehistoric animal remains and the importance of keeping sites secret were the result of Marsh's already legendary confrontations with archrival fossil collector Edward Drinker Cope. Although Marsh's crews from Yale University did most of the collecting at Como Bluff from 1877 to 1889, Cope's field crews also opened quarries there at that time.

The Marsh–Cope rivalry extended into the field operations of the two rival camps (Ostrom and McIntosh, 1966; Colbert, 1968; Wilford, 1986). Reed and Carlin continued to work for the railroad but also helped Marsh's crews excavate fossil bones. Although Carlin eventually would work for Cope and then quit the fossil collecting business altogether, Reed was a devoted and loyal employee to Marsh. In fact, Reed became so enamored with discovering dinosaurs that he would eventually work full time as a fossil collector (Breithaupt, 1990). Some of the first published reconstructions of a sauropod (the skeleton of *Brontosaurus*, now known as *Apatosaurus*) were based on a fairly complete skeleton found by Reed at Como Bluff's Quarry 10 (Marsh, 1883, 1891). These famous reconstructions, in which *Camarasaurus* elements were used to replace various missing parts, including the head, dramatically affected views of this dinosaur for almost a century.

William Harlow Reed was born in Hartford, Connecticut, in 1848. In 1877, while working as a section foreman for the Union Pacific Railroad at Como Station, Wyoming Territory, he accidentally discovered large bones on the nearby ridge. These specimens launched him in a career in vertebrate paleontology that he would pursue for the next 38 years. Although frustrated by certain aspects of field-work and by lack of recognition as a field paleontologist, he was a diligent and loyal collector for Marsh. He gave this same dedication in later years to W. C. Knight at the University of Wyoming, W. J. Holland at the Carnegie Museum, and W. Granger at the American Museum of Natural History. Although not formally educated in the sciences, Reed's desire to learn, interest in natural phenomena, and association with the notable paleontologists of his time allowed him to gain a background in geology and paleontology. After more than 25 years of significant discoveries of dinosaurs, ichthyosaurs, plesiosaurs, pterosaurs, mammals, and cycads in Wyoming, Reed was given the position as curator of the museum and instructor in geology at the University of Wyoming in 1904. He held this position until his death in 1915.

In 1897, the American Museum of Natural History sent collecting crews to Como Bluff after Marsh and

Cope's crews had left the area. The American Museum crews, under the direction of Jacob Wortman and later Walter Granger (Professor Henry Fairfield Osborn's chief collectors) found some new dinosaur quarries, but most of the localities at Como Bluff were barren of fossils. Major collections were made for the American Museum from Como Bluff and the region around this famous site between 1897 and 1905 (Colbert, 1968; McIntosh, 1990). By the turn of the century, museum crews from the Carnegie Museum and the University of Wyoming were also making collections in the area. Although by the early 1900s most collecting in the Como Bluff area had ceased, museum crews continue to routinely excavate dinosaur bones and small, fossil vertebrate material from the region (Bakker, 1985, 1990).

Since the discovery of Como Bluff in 1877, thousands of vertebrate fossils have been recovered by institutions throughout the country. This extensive collecting has resulted in one of the best known terrestrial faunas in the world, ranging from bivalves to pterosaurs. The fossil vertebrate fauna from Como Bluff is dominated by large herbivorous dinosaurs. Five types of sauropods (e.g., *Apatosaurus, Diplodocus, Camarasaurus, Pleurocoelus,* and *Barosaurus)* have been identified from Como Bluff (Ostrom and McIntosh, 1966). Other herbivorous dinosaurs that inhabited the region during the Late Jurassic were the armored *Stegosaurus* and the bipedal *Camptosaurus, Laosaurus, Othnielia, Drinker,* and *Dryosaurus*. The carnivores *Allosaurus, Ceratosaurus, Ornitholestes,* and *Coelurus* were also present. Many smaller animals have been found fossilized at Como Bluff as well. These include lungfishes, frogs, salamanders, turtles, lizards, rhynchocephalians, crocodiles, pterosaurs, and a diverse group of early mammals. Although early studies interpreting the depositional environment of the Morrison Formation varied, current research suggests that the formation represents a complex fluviatile–lacustrine floodplain deposited in a seasonally wet–dry, warm-temperate environment (Dodson *et al.*, 1980).

See also the following related entries:

Bone Cabin Quarry • Morrison Formation

References

Bakker, R. T. (1985). *Dinosaur Heresies*, pp. 481. Morrow, New York.

Bakker, R. T. (1990). A new latest Jurassic vertebrate fauna from the highest levels of the Morrison Formation at Como Bluff, Wyoming. *Hunteria* **2,** 1–19.

Breithaupt, B. H. (1990). Biography of William Harlow Reed: The story of a frontier fossil collector. *Earth Sci. History* **9,** 6–13.

Colbert, E. H. (1968). *Men and Dinosaurs*, pp. 283. Dutton, New York.

Dodson, P., Behrensmeyer, A. K., Bakker, R. T., and McIntosh, J. S. (1980). Taphonomy and paleoecology of the dinosaur beds of the Jurassic Morrison Formation. *Paleobiology* **6,** 208–232.

Marsh, O. C. (1883). Principal characters of American Jurassic dinosaurs. Part VI: Restoration of *Brontosaurus*. *Am. J. Sci.* **26,** 81–85.

Marsh, O. C. (1891). Restoration of *Triceratops* and *Brontosaurus*. *Am. J. Sci.* **41,** 339–342.

McIntosh, J. S. (1990). The second Jurassic dinosaur rush. *Earth Sci. History* **9,** 22–27.

Ostrom, J. H., and McIntosh, J. S. (1966). *Marsh's Dinosaurs*, pp. 388. Yale Univ. Press, New Haven, CT.

Urbanek, M. (1988). *Wyoming Place Names*, pp. 233. Mountain Press, Missoula, MT.

Wilford, J. N. (1986). *The Riddle of the Dinosaur*, pp. 304. Knopf, New York.

Computers and Related Technology

Ralph E. Chapman
National Museum of Natural History
Smithsonian Institution
Washington, DC, USA

David B. Weishampel
Johns Hopkins University School of Medicine
Baltimore, Maryland, USA

Computers are in the process of revolutionizing the way we look at dinosaurs. Early uses of computers with dinosaurs date to the 1960's, concentrated in research applications (e.g., data analysis) and early attempts by some museums (e.g., the National Museum of Natural History of the United States) to store their collections data electronically. These applications became more and more prevalent, and progressively more sophisticated through the 1970's and

1980's. It is during the 1990s, however, that computers finally have expanded to start assisting paleontologists in the field, during the preparation process, and in making a great difference in how dinosaurs are being presented to the general public.

The application of computers and related technology to dinosaurs can be divided into five major areas: field work and specimen collection, specimen preparation, collections and data management, research, and exhibition. We will touch on these subjects in order.

Finding dinosaurs in the field still continues to be done mostly in the same way it has been done for more than a century: finding areas with the right rock from the right time and the right paleoenvironments, and walking the outcrops looking for exposed bones. For the most part this will continue to be the approach taken, but new technology promises to change much of the process of getting into the field, determining where you are once there, and even analyzing what has been found and what is still in the ground.

Formerly, it took much skill to use the available maps and keep track of where you were in the field; many mistakes were made and much of the old location data are inaccurate. This is changing because global positioning systems (GPS) using satellite technology can locate a position within a 100 m or less anywhere on earth and are being used by paleontologists during exploration. For example, a GPS system was used by the American Museum of Natural History in their new Gobi Desert expeditions (McKenna, 1992) to keep track of where they were in unmapped or badly mapped areas. These locality data have yet to be used extensively with geographic information systems (GIS), which combine geographic data with computer databases. In the future this will allow predictive mapping of potential outcrop areas, and the GIS will suggest where new prospecting should be done.

Small-scale geographic data (e.g., quarry maps of bone orientations and positions) were once taken using compasses and were plotted in field notebooks, but now are being taken automatically using electronic distance measurement devices with millimeter accuracy (see Jorstad and Clark, 1995, for work on paleohominid applications) and, in even smaller scale and with higher accuracy, using three-dimensional digitizers (e.g., see Jefferson, 1989, on Pleistocene Rancho La Brea material). Finally, technology is being developed to allow paleontologists to determine, in some cases, the nature of fossils buried in an area. One example is the application of geophysical diffraction tomography by Witten, *et al.* (1992), who tried to determine the extent of the material buried for a specimen of the sauropod *Seismosaurus*.

Computers and related technology have had only limited effect on specimen preparation but changes are well on the way. Standard X-rays have been used for years during specimen preparation (see Zangerl and Schultze, 1989), but more advanced imaging methods such as computed tomography (CT) are allowing, in some cases, significantly better indications of what fossil material is present in an unprepared specimen (see Clark and Morrison, 1994). In fact, the CT scanning of dinosaur eggs has become a standard operating procedure (e.g., Hirsch *et al.*, 1989), and many fetal dinosaur fossils are now being found. It may be common in the near future for preparators to have three-dimensional models of the specimens they are preparing as an aid to the process.

Specimen casting also will be changed by three-dimensional computer modeling. Fossils can be digitized using various scanning or related technologies and three-dimensional reconstructions made in wax, plastic, or some other medium for exhibition or study using methods of automated casting (also known as prototyping or, in some cases, stereolithography; see Burns, 1993) thus avoiding more destructive ways of casting specimens. These casts can also be varied in scale to enable them to be viewed at a more manageable scale; very large specimens can be reduced to allow more easy manipulation and very small specimens enlarged to allow them to be viewed without a microscope.

The use and development of collections materials has changed dramatically and will continue to change in the coming years due to computers. The initial transfer of collections data to an electronic format was done in the 1960's (e.g., the National Museum of Natural History of the United States), with many now using their third, fourth, or later generation of database software, and nearly all having some electronic storage. Where data used to be kept in card catalogs that took long search times to extract simple information, data are now available in large computer databases that can be searched for very complex information almost instantaneously. These databases are also being stored on-line and, in many cases, are available for searching on the Internet.

Another big change due to computer technology is the nature of data being made available; computer databases are not restricted to just text anymore. Image scanners are making it possible to store pictures of specimens as well as the text information that goes along with it. Furthermore, with the scanning technology being developed, three-dimensional computer images can now be stored and viewed from a variety of angles. A first attempt at storing such images was made by Rowe *et al.* (1993) for high-resolution CT scans of the skull of the cynodont *Thrinaxodon*. The CD-ROM released contains both the important descriptive literature and CT scan data that can be viewed from front-to-back, side-to-side, or top-to-bottom.

The use and storage of bibliographic data on dinosaurs is changing rapidly. The *Bibliography of Fossil Vertebrates* is being made available in electronic format and the next release of *A Bibliography of the Dinosauria* (Chure and McIntosh, 1989) will be made available in electronic format as well. Many dinosaur paleontologists have developed and maintained their own electronic bibliographic databases using their personal computers.

Computers have had a major impact on the types of research being done in the natural sciences. The relatively unquantitative approaches taken by scientists in the early days mostly have been replaced by quantitative ones as methods have become more and more rigorous. Computer-based studies of dinosaurs are still in their infancy, however, because most of the research done on them still proceeds mostly in a qualitative fashion. This is due, to a large part, to the difficulties involved in doing research on a group represented by relative few individuals that are often incomplete and fragmentary. However, there are a number of important exceptions and we will discuss them within four major areas of research: morphometrics, mathematical and/or statistical studies of variation used to solve taxonomic or evolutionary problems; phylogenetic analyses, the analysis of the relationships among taxa; functional morphology, studies of the biomechanics and locomotion of dinosaurs; and distributional analyses, studies of the distribution of dinosaurs through time and space.

Morphometrics is the quantitative analysis of shape. Before the availability of computers, paleontologists were limited to analyses of two or three variables, usually within the context of allometry, the study of size and its consequences. One classic study of this type is the analysis of bivariate (two-variable) allometry in groups of ceratopsian dinosaurs by Gray (1946). Another morphometric approach, the application of D'Arcy Thompson's (1942) transformation grids, could be done without computers and grids were generated for a number of dinosaur groups (e.g., Lull and Gray, 1949, for ceratopsians).

Computers allowed calculations to be done much faster than ever before, which opened the door for multivariate analyses, those using many variables simultaneously (e.g., Dodson, 1975, 1976; Chapman, *et al.*, 1981), as well as very sophisticated geometric methods of shape analysis (e.g., Chapman, 1990). These methods have led to a much better understanding of growth in dinosaurs, have allowed sexual dimorphs to be described in some cases (e.g., see Chapman, *et al.*, 1981, for the pachycephalosaurid *Stegoceras*), and have started to be used more within studies of phylogeny and functional morphology. The next step will be to do even more sophisticated analyses of shape in three-dimensions and use high-level computer graphics to show the results.

Phylogenetic Analyses try to determine relationships among the dinosaur taxa being studied. Here, we will concentrate on numerical cladistic analyses, which attempt to reconstruct these relationships on trees, called cladograms, using the principle of parsimony; looking for the shortest trees by minimizing the number of evolutionary steps needed to generate the tree and minimizing instances of convergence or parallelism.

Originally done by hand, most cladistic analyses of dinosaurs, as well as all other groups of organisms, are carried out on computers using programs such as *Phylogenetic Analysis Using Parsimony* (PAUP) (Swofford and Begle, 1993) and *MacClade* (Maddison and Maddison, 1992). These packages search for the shortest trees (cladograms) that account for the characters forming the database supplied by the researcher, while reducing the number of reversals and convergences in these characters. Computer programs are necessary because the number of possible trees increases exponentially as more taxa are studied.

Cladistic research on dinosaurs is burgeoning, and nearly every major group has been analyzed to some degree. Much of this work was begun in the 1980's, spearheaded by the research of Gauthier (1984, 1986), and it has become the standard for reconstructing phylogenetic relationships.

As more numerical cladistic analyses are done, we have begun to get a better understanding of how the different dinosaur groups are related and a much better understanding of the anatomy of dinosaurs and why they look the way they do. An additional way to use phylogenetic analysis in research is to follow the evolution of a single character (an anatomical feature) on a tree to see how it varies across a dinosaur taxon. Other ways are to superimpose geographic locations, ecological characteristics, or time on trees to see how they have influenced the history of dinosaurs. As computers get stronger, phylogenetic analyses will provide better information.

Functional morphology is the study of how organisms work. These studies are still relatively rare for dinosaurs but such analyses are now becoming more common. Functional analyses typically make use of architectural (e.g., Weishampel, 1993) and/or machine analogies to understand the evolution and operation of a particular anatomical structure. They commonly use physical models, graphical representations, mathematical computations, computer simulations, and thought experiences to analyze this anatomy. Examples include Alexander's (1989) analyses of dinosaur locomotion and Weishampel's (1981) study of the nasal systems of lambeosaurine hadrosaurids. Computers can help increase the level of sophistication possible in such studies, especially through the use of high-level computer graphics, and should provide a strong impetus for a great increase in the number of functional studies on dinosaurs.

To date, computer applications in dinosaur biomechanics have been limited to studies of feeding mechanisms and locomotion. For example, Weishampel (1984) has used a three-dimensional kinematics computer program (developed by engineers) to analyze a series of ornithopod skulls as chewing machines. Heinrich *et al.* (1993) studied locomotion in the Late Jurassic iguanodontian *Dryosaurus lettowvorbecki* by modeling the femur as a beam. Bone cross-sections provided indications of both strength and patterns of loading on the living dinosaurs. Studying juvenile and adult specimens allowed them to postulate changes in locomotory patterns with age for that species.

Distributional studies analyze the distribution of organisms through time and space. Dinosaur studies of this kind have been very limited so far because of the nature of the fossil record for dinosaurs, but a number of paleontologists are now actively studying distributional problems with some success. To date, studies using computer databases have been able to track dinosaur diversity through space and time (Weishampel and Norman, 1989; Dodson, 1990) as well as the rate of study of dinosaurs by paleontologists (Dodson and Dawson, 1991). The large compilation by Weishampel (1990) of dinosaur localities is making it possible to analyze dinosaur paleobiogeography quantitatively for the first time (R. Chapman and D. Weishampel, work in progress). Such studies, however, can only be done using computers because they involve the mathematical and statistical analysis of large matrices of data. Clearly, this is one area of research that will be expanding greatly because of computers and related technology.

The interface between the general public and dinosaurs is one area of great change because of computers. Computers make available a wide range of educational approaches for teaching people about dinosaurs, especially using CD-ROM and multi-media technology. More people are also gaining access to data about dinosaurs through the Internet using online computer databases.

One of the biggest effects will be in changing the ability of the public to visualize what dinosaurs looked like. Conventional approaches of reconstructing dinosaurs (e.g., Paul, 1987) are being supplemented by sophisticated three-dimensional computer graphics that use computer visualization technology (e.g., Nielson and Shriver, 1990) to help generate lifelike dinosaurs such as those seen in the film *Jurassic Park* (Shay and Duncan, 1993) and can support the development of more life-like robotic dinosaurs (e.g., Poor, 1991). The development and distribution of better systems for virtual reality will allow researchers and the public alike to tour a dinosaur's morphology and even view it from the inside [(see Fröhlich, *et al.*, (1995) and Stevens (1995) for discussions in the field of biology and medicine).

Once these approaches are developed, they will be used more in conjunction with the original fossil material within exhibitions. Most modern exhibits on dinosaurs include some computer technology and this will increase more with time. Clearly, computers can vastly improve how the general public is introduced to dinosaurs and increase their general knowledge.

See also the following related entries:

BIOGEOGRAPHY • BIOMETRICS • FUNCTIONAL MORPHOLOGY • PHYLOGENETIC SYSTEM

References

Alexander, R. M. (1989). *Dynamics of Dinosaurs and Other Extinct Giants*, pp. 167. Columbia Univ. Press, New York.

Burns, M. (1993). *Automated Fabrication. Improving Productivity in Manufacturing*, pp. 369. Prentice-Hall, Englewood Cliffs, NJ.

Chapman, R. E. (1990). Shape analysis in the study of dinosaur morphology. In *Dinosaur Systematics, Approaches and Perspectives* (K. Carpenter and P. J. Currie, Eds.), pp. 21–42. Cambridge Univ. Press, New York.

Chapman, R. E., Galton, P. M., Sepkoski, J. J., Jr., and Wall, W. P. (1981). A morphometric study of the cranium of the pachycephalosaurid dinosaur *Stegoceras*. *J. Paleontol.* **55**(3), 608–618.

Chure, D. J., and McIntosh, J. S. (1989). *A Bibliography of the Dinosauria (Exclusive of the Aves) 1677–1986*, Paleontology Series No. 1, pp. 226. Museum of Western Colorado.

Clark, S., and Morrison, I. (1994). CT scan of fossils. In *Vertebrate Paleontological Techniques, Volume 1* (P. Leiggi and P. May, Eds.), pp. 323–329. Cambridge Univ. Press, New York.

Dodson, P. (1975). Taxonomic implications of relative growth in lambeosaurid dinosaurs. *Systematic Zool.* **24**(1), 37–54.

Dodson, P. (1976). Quantitative aspects of relative growth and sexual dimorphism in *Protoceratops*. *J. Paleontol.* **50**(5), 929–940.

Dodson, P. (1990). Counting dinosaurs: How many kinds were there? *Proc. Natl. Acad. Sci. (USA)*, **87**, 7608–7612.

Dodson, P., and Dawson, S. D. (1991). Making the fossil record of dinosaurs. *Mod. Geol.* **16**(1/2), 3–15.

Fröhlich, B., Grunst, G., Krüger, W., and Wesche, G. (1995). The responsive workbench: A virtual working environment for physicians. *Comp. Biol. Med.* **25**(2), 301–308.

Gauthier, J. (1984). A cladistic analysis of the higher systematic categories of the Diapsida, pp. 564. PhD dissertation, Univ. of California, Berkeley.

Gauthier, J. (1986). Saurischian monophyly and the origin of birds. In *The Origin of Birds and the Evolution of Flight* (K. Padian, Ed.), Memoirs, No. 8, pp. 51–55. California Academy of Sciences.

Gray, S. W. (1946). Relative growth in a phylogenetic series and in an ontogenetic series of one of its members. *Am. J. Sci.* **244**, 792–807.

Heinrich, R. E., Ruff, C. B., and Weishampel, D. B. (1993). Femoral ontogeny and locomotor biomechanics of *Dryosaurus lettowvorbecki* (Dinosauria, Iguanodontia). *Zool. J. Linnean Soc.* **108**, 179–196.

Hirsch, K. F., Stadtman, K. L., Miller, W. F., and Madsen, J. H., Jr. (1989). Upper Jurassic dinosaur egg from Utah. *Science*, **243**, 1711–1713.

Jefferson, G. T. (1989). Digitized sonic location and computer imaging of Rancho La Brea specimens from the Page Museum salvage. *Curr. Res. Pleistocene*, **6**, 45–47.

Jorstad, T., and Clark, J. (1995). Mapping human origins on an ancient African landscape. *Prof. Surveyor*, **15**(4), 10–12.

Lull, R. S., and Gray, S. W. (1949). Growth patterns in the Ceratopsia. *Am. J. Sci.*, **247**, 492–503.

Maddison, W. P. and Maddison, D. R. (1992). *MacClade. Analysis of Phylogeny and Character Evolution, Version 3*, pp. 398. Sinauer, Sunderland, MA.

McKenna, P. C. (1992). GPS in the Gobi: Dinosaurs among the dunes. *GPS World*, **3**(6), 20–26.

Nielson, G. M., and Shriver, B. (Eds.). (1990). *Visualization in Scientific Computing*, pp. 282. IEEE Computer Society Press, Los Alamitos, CA.

Paul, G. S. (1987). The science and art of restoring the life appearance of dinosaurs and their relatives: a rigorous how-to guide. In *Dinosaurs Past and Present. Volume II* (S. J. Czerkas and E. E. Olson, Eds.), pp. 4–49. Los Angeles County Museum of Natural History/University of Washington Press, Seattle.

Poor, G. W. (1991). *The Illusion of Life: Lifelike Robotics*, pp. 96. Educational Learning Systems, San Diego, CA.

Rowe, T., Carlson, W., and Bottorf, W. (1993). *Thrinaxodon: Digital Atlas of the Skull*. Univ. of Texas Press, Austin.

Shay, D., and Duncan, J. (1993). *The Making of Jurassic Park, An Adventure 65 Million Years in the Making*, pp. 196. Ballantine Books, New York.

Stevens, J. E. (1995). The growing reality of virtual reality. *BioScience* **45**(7), 435–439.

Swofford, D. L., and Begle, D. P. (1993, March). *User's Manual for PAUP: Phylogenetic Analysis Using Parsimony*. Version 3.1, pp. 257. Laboratory of Molecular Systematics, Smithsonian Institution, Washington, DC.

Thompson, D'A. W. (1942). *On Growth and Form, The Complete Revised Edition*, 1992 ed., pp. 1116. Dover, New York.

Weishampel, D. B. (1981). Acoustic analysis of potential vocalization in lambeosaurine dinosaurs (Reptilia: Ornithischia). *Paleobiology* **7,** 252–261.

Weishampel, D. B. (1984). Evolution of jaw mechanisms in ornithopod dinosaurs. *Adv. Anat. Embryol. Cellular Biol.* **87,** 1–116.

Weishampel, D. B. (1990). Dinosaurian distribution. *The Dinosauria* (D. B. Weishampel, P. Dodson, and H. Osmolska, Eds.), pp. 63–139. Univ. of California Press, Berkeley.

Weishampel, D. B. (1993). Beams and machines: Modeling approaches to the analysis of skull form and function. *The Skull, Volume 3* (J. Hanken and B. K. Hall, Eds.), pp. 303–343. Univ. of Chicago Press, Chicago.

Weishampel, D. B., and Norman, D. B. (1989). Vertebrate herbivory in the Mesozoic: jaws, plants, and evolutionary metrics [Special paper]. *Geol. Soc. of Am.* **238,** 87–100.

Witten, A., Gillette, D. D, Sypniewski, J., and King, W. C. (1992). Geophysical diffraction tomography at a dinosaur site. *Geophysics* **57**(1), 187–195.

Zangerl, R., and Schultze, H.-P. (1989). X-radiographic techniques and applications. *Paleotechniques* (R. M. Feldmann, R. E. Chapman, and J. T. Hannibal, Eds.), The Paleontological Society, Spec. Publ. No. 4, pp. 165–178.

Connecticut River Valley

JOANNA WRIGHT
University of Bristol
Bristol, United Kingdom

Dinosaur tracks from the Connecticut Valley have been known for more than 150 years. They were first scientifically studied and described by Edward Hitchcock (1836, 1858). Hitchcock's fossil trackway collections are now held in the YALE PEABODY MUSEUM, Connecticut, and the Pratt Museum of Amherst College, Massachusetts. Most of the dinosaur tracks that have been found in the Connecticut Valley were collected in the 19th century. This is because at that time conditions in the area were ideal for fossil collection. Much of the economy was based on farming and the population was sparse by today's standards. The only suitable building stone in the area was the Turners Falls Sandstone and the Portland Formation: These are the sediments in which the footprints occur and many of the footprints were discovered in quarries. In the middle of the 1800's the Great Plains were discovered and exploited and many farms were abandoned as it became uneconomical to farm in New England. The abandoned farms were left to be reclaimed by the forest so that a large number of the quarry and farm footprint sites are now completely overgrown. In addition, dams were built at several points along the river, which had the effect of raising the water level along much of the river with the result that many of Hitchcock's footprint sites are now either submerged or only accessible by boat.

Previous Work

After Hitchcock, the only other person to review the whole of the track collection at Amherst College was Lull (1904, 1915), who attempted to sort out the morass of ichnogenera and ichnospecies names that was Hitchcock's legacy. From when he first started to study the fossil tracks Hitchcock had assigned them genus and species names in the Linnaean binomial system and he would change these as his perceptions of the trackmakers changed and he rarely acknowledged previous names in subsequent publications. Lull also tried to work out what (kinds of) animals made the different types of prints. Unfortunately, much of Lull's nomenclatural work served only to confuse the issue further because he tended to resurrect old and abandoned names.

In recent years the main workers on these prints have been Olsen and co-workers (Olsen and Baird, 1986; Olsen *et al.*, 1992), who also attempted to revise some of the tangled nomenclature. They demonstrated that if the lengths of specimens of the ichnogenera *Grallator*, *Anchisauripus*, and *Eubrontes* are plotted against their widths, all three types show a complete gradation in size and proportions and their species frequently overlap.

Geological Background

The track-bearing strata of the Connecticut Valley are contained in the Hartford and Deerfield Basins of Connecticut and Massachusetts, two of a series of rift basins along the east coast of North America that opened in the early Mesozoic in response to the NE–SW extension associated with the breakup of Pangaea. The sediments occur in a north–south elongate half graben extending over more than 160 km from the northern border of Massachusetts to Long Island Sound (Fig. 1). More than 4000 m of predominantly red, gray, and black clastic sediments and tholeiitic basalt was deposited in the basin during its approximately 35 million year existence during the Late Triassic and Early Jurassic (Fig. 2).

Strata of the Hartford and Deerfield basins constitute two major genetic sequences, a lower Late Triassic age fluvial and alluvial arkose and an upper Early Jurassic age lacustrine and alluvial siltstone to conglomerate with interbedded basalts very low in the sequence.

The basal strata, the New Haven and Sugarloaf Arkoses, are dominantly alluvial fan and braided stream redbeds laid down by rivers that flowed from the crystalline highlands in the east. Abundant calcretes suggest that the palaeoclimate was tropical and semi-arid with perhaps 100–500 mm of seasonal rain

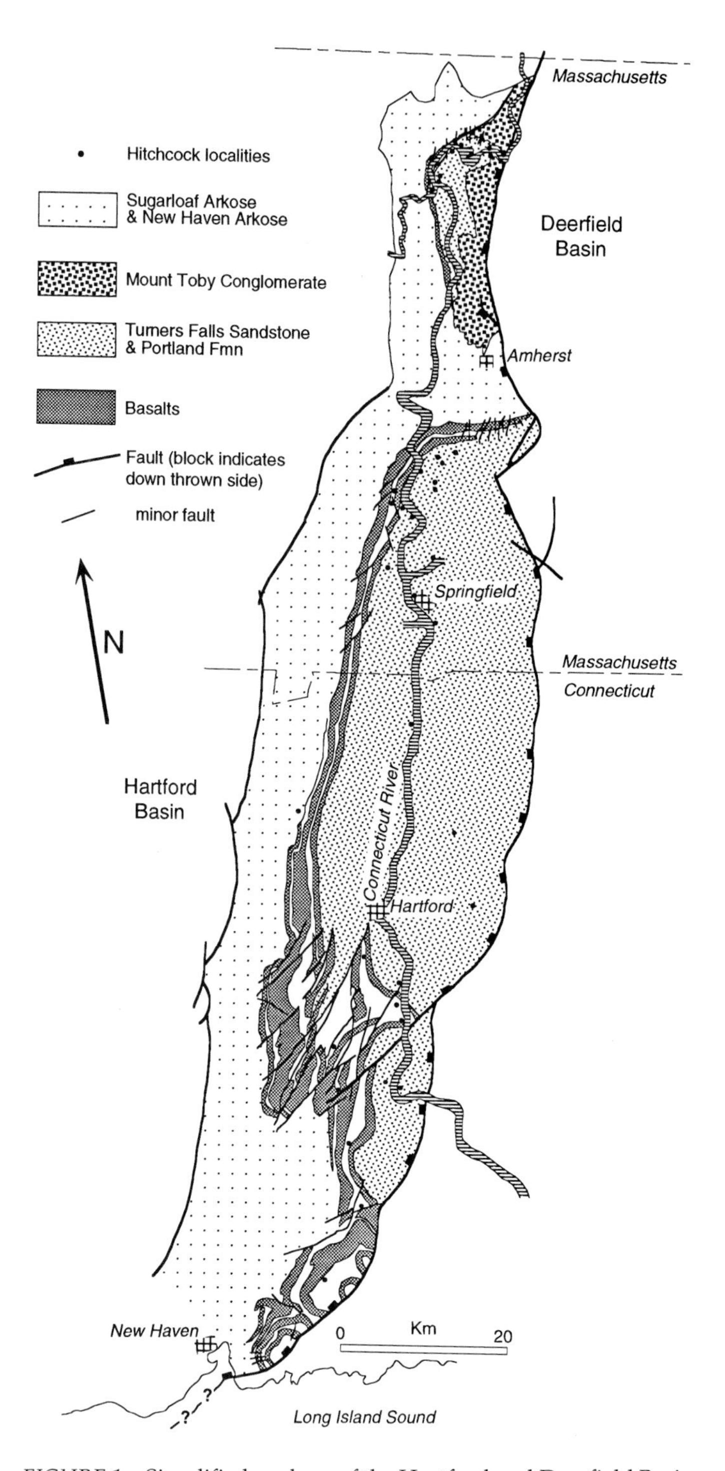

FIGURE 1 Simplified geology of the Hartford and Deerfield Basins.

and a long dry season. Only rare reptile remains have been recovered from these formations.

In the Deerfield Basin, the Mount Toby Conglomerate and Turners Falls Sandstone overlie the Sugarloaf Arkose and Deerfield Basalt; they largely reflect alluvial fan, allluvial plain floodplain, and local lacustrine depositional environments. In the Hartford Basin, the red and gray, fine-grained lacustrine deposits of the Shuttle Meadow and the East Berlin Formations are sandwiched between three tholeiitic basalt flows,

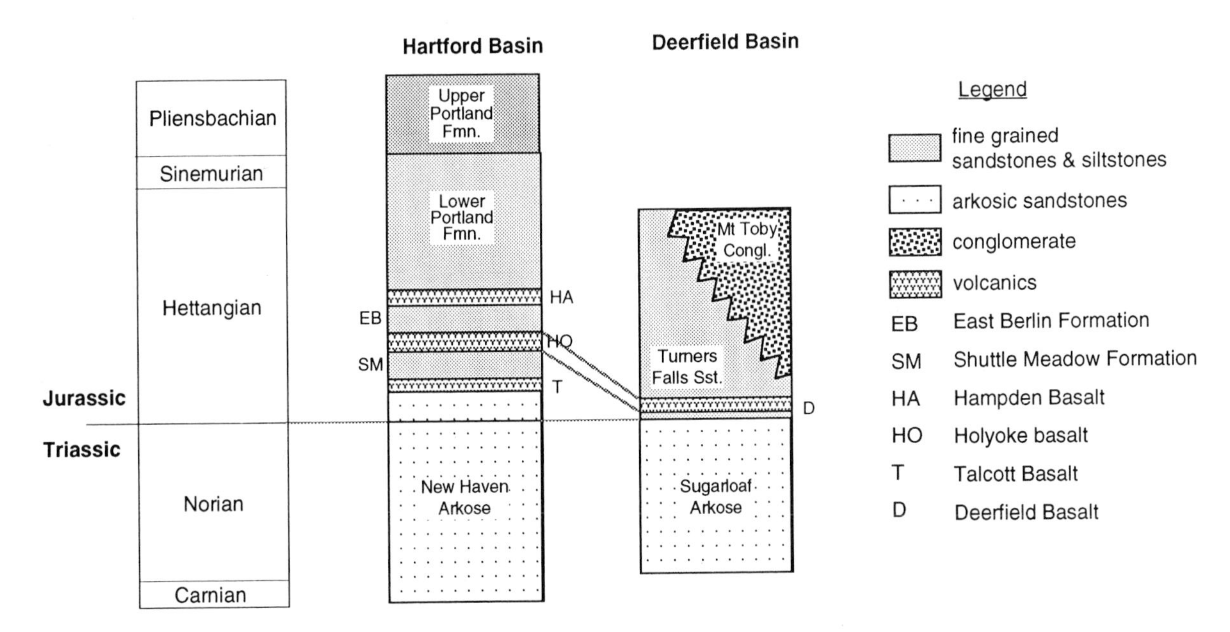

FIGURE 2 Correlation of the formations between the Hartford and Deerfield Basins.

the topmost of which is overlain by the Lower and then the Upper Portland Formation, which represent similar environments to that of the Turners Falls Sandstone and the Mount Toby Conglomerate, respectively. Olsen *et al.* (1992) have equated the Holyoke and Deerfield Basalts of the Hartford and Deerfield Basins, but the Mount Toby Conglomerate and Turners Falls Sandstone cannot currently be directly correlated with the sedimentary formations of the Hartford Basin. These sedimentary units, with the exception of the Mount Toby Conglomerate and the Upper Portland Formation, are the rocks in which most of the trackways occur. It is likely that the climate throughout this period was subtropical, monsoonal characterized by alternating episodes of high precipitation and aridity.

Faunal Diversity

Although the taxonomy of the footprint fauna has not recently been revised and can therefore not be taken as a reliable diversity indicator, the tracks can nevertheless be divided into several groups that represent the main types of animals present in the Connecticut Valley in the Early Jurassic (Fig. 3). *Grallator*, *Anchisauripus*, and *Eubrontes* represent small, medium, and large theropods, respectively. *Anomoepus* represents a small ornithopod (probably fabrosaurid). *Batrachopus* represents a small crocodilomorph. *Otozoum* has recently been identified as the tracks of a basal thyreophoran. Some tracks seem to have been produced by juvenile theropods. There are some indeterminate tridactyl footprints, most of which are probably theropod prints, and there are some indeterminate quadrupedal prints, perhaps made by an animal such as a sphenodontid reptile.

If all these kinds of prints, including invertebrate traces, are plotted as numbers of trackways (in the tracks collection of Amherst College) in a pie chart (Fig. 4), it can be seen that there is a very strange

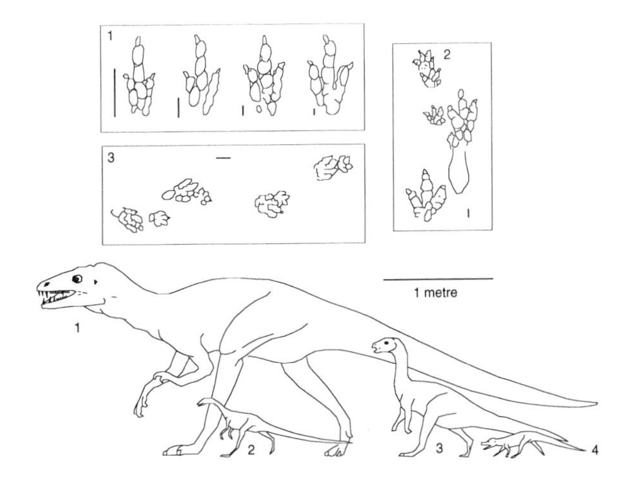

FIGURE 3 Some of the more common vertebrate tracks and their likely producers.

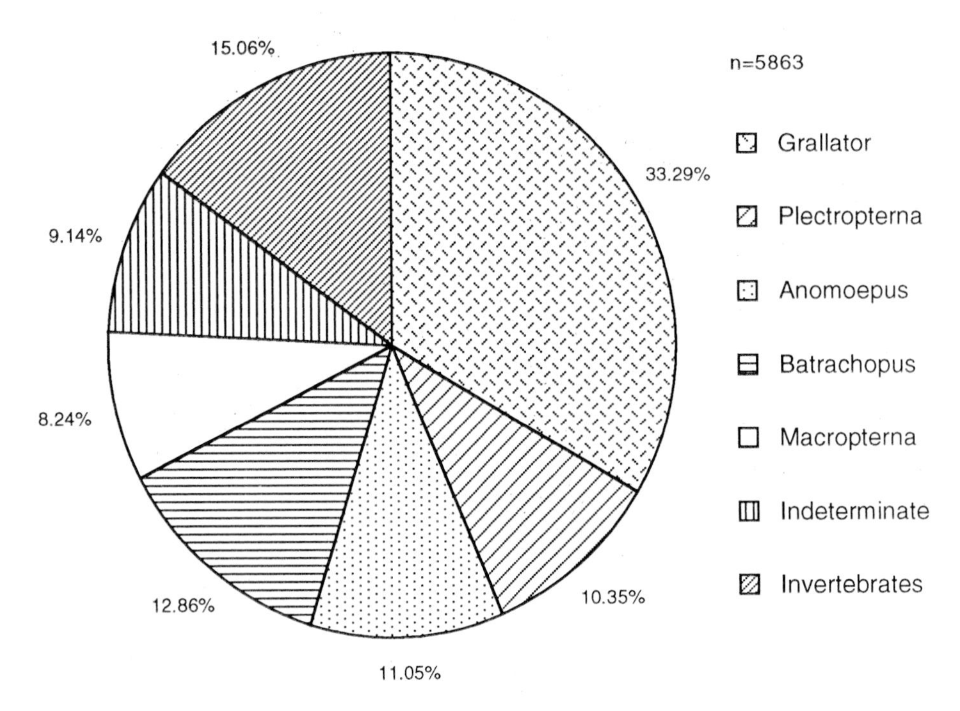

FIGURE 4 Pie chart showing the proportions of the different types of trackways found in the Valley.

faunal distribution pattern. One-third of the fauna is composed of theropods, 15% consists of invertebrates, and only 11% is made up of herbivores. This is obviously an unsustainable ecological situation. The maker of *Batrachopus* was probably an insectivore/carnivore and most of the indeterminate tridactyl prints were probably made by theropods—this would bring the proportion of carnivores to more than 50%. This is obviously not an accurate picture of the Early Jurassic fauna of the Connecticut Valley; therefore, what could be the explanation for the strong bias in the footprint fauna? There are several possible reasons. First, higher activity of the theropods—they moved around more so they made more footprints. Second, the theropods may have waited at strategic places, such as a watering hole, for their prey and thus spent a longer time in areas where their footprints were more likely to be preserved. Also, there is a possibility that our way of recognizing the makers of footprints is flawed and not all the footprints that are assigned to theropods were made by theropods. However, no successful method has been developed to test this hypothesis.

Invertebrates An unusual feature of the Connecticut Valley fauna is that it also contains very good invertebrate trackways; the sediments that preserve the invertebrate traces also preserve plant fossils. Many of the invertebrate trackmakers were hexapods; again the ichnotaxonomy is in great need of revision. The only person to have worked on the invertebrate traces is Edward Hitchcock; they have largely if not totally been ignored by subsequent workers, who preferred to concentrate on the larger, more spectacular dinosaur tracks and trackways [in Lull's (1953) book of 285 pages he devotes only 17 to the invertebrate traces].

Conclusions

The extensive trace fossil collections from the Connecticut Valley are a valuable resource. Both vertebrate and invertebrate traces are preserved and, in addition, plant fossils also occur in the same sediments; therefore, quite a complete picture of the fauna can be obtained. Much work still remains to be done, however, especially on the invertebrate traces, but there is, in addition, a great deal of scope for statistical analysis on the dinosaur footprints.

See also the following related entries:

FOOTPRINTS AND TRACKWAYS • JURASSIC PERIOD • NEWARK SUPERGROUP • TRIASSIC PERIOD

References

Hitchcock, E. (1836). Ornithichichnology. Description of the footmarks of birds (ornithichnites) on New Red Sandstone in Massachusetts. *Am. J. Sci.* **29,** 307–340.

Hitchcock, E. (1858). *Ichnology of New England: A Report of the Sandstone of the Connecticut Valley, Especially Its Fossil Footmarks*, pp. 232. White, Boston.

Lull, R. S. (1904). Fossil footprints of the Jura–Trias of North America. *Mem. Boston Soc. Nat. History* **5,** 461–557.

Lull, R. S. (1915). Triassic Life of the Connecticut Valley. *Connecticut Geol. Nat. History Surv. Bull.* **24.**

Olsen, P. E., and Baird, D. (1986). The ichnogenus *Atreipus* and its significance for Triassic biostratigraphy. In *The Beginning of the Age of Dinosaurs: Faunal Change across the Triassic–Jurassic Boundary* (K. Padian, Ed), pp. 61–87. Cambridge Univ. Press, Cambridge, UK.

Olsen, P. E., McDonald, N. G., Huber, P., and Cornet, B. (1992). Stratigraphy and paleoecology of the Deerfield Rift Basin (Triassic–Jurassic, Newark Supergroup), Massachusetts. In *Guidebook for Field Trips in the Connecticut Valley Region of Massachusetts and Adjacent States* (P. Robinson and J. B. Brady, Eds.), Vol. 2, pp. 488–535.

Connecticut State Museum of History, Connecticut, USA

see MUSEUMS AND DISPLAYS

Constructional Morphology

DAVID B. WEISHAMPEL
The Johns Hopkins University School of Medicine
Baltimore, Maryland, USA

The term "constructional morphology" as currently understood comes from an important evolutionary biological research program in Germany called Konstruktionsmorphologie. Founded by A. Seilacher at the University of Tübingen, constructional morphology treats biological form not just as a consequence of function and phylogeny, although these are admittedly necessary parts of the explanation of such form. However, because neither alone in isolation provides a sufficient understanding of form, nor do both together, constructional morphology adds a third factor in biological explanations of form—a *bautechnische* (roughly translated as architectural or fabricational) factor. Thus, biological form is understood only as the result of three interacting factors: ecological–adaptive, historic—phylogenetic, and *bautechnische.*

Traditional biology views form as the result of adaptation and phylogeny. *Bautechnik* recognizes that geometry, natural materials, and growth processes also regulate morphologic patterns. As a consequence, *bautechnische* factors are ahistorical elements that express biological possibilities and limits on evolutionary change that stem from the physical and chemical properties of available materials (strength and failure, elasticity, adhesion, and viscosity); cybernetic controls on development, maintenance, and repairs; and, finally, the geometry of pattern formation and self-organization. The particular materials, growth programs, and regulatory systems are naturally acquired as specific historical (evolutionary) events; hence the transfer of ahistorical factors to phylogenetic clades and lineages. Likewise, the ways in which these ahistorical factors affect reproductive/evolutionary fitness also shift such influences to the unique historical nexus of adaptation.

See also the following related entries:

BIOMECHANICS • BIOMINERALIZATION • CHEMICAL COMPOSITION OF DINOSAUR FOSSILS • FUNCTIONAL MORPHOLOGY • HISTOLOGY OF BONES AND TEETH • PHYLOGENETIC SYSTEM

References

Reif, W.-E., Thomas, R. D. K., and Fischer, M. S. (1985). Constructional morphology. *Acta Biotheor.* **34,** 233–248.

Seilacher, A. (1970). Arbeitskonzept zur Konstruktions–Morphologie. *Lethaia* **3,** 393–396.

Coprolites

KAREN CHIN
University of California
Santa Barbara, California, USA

Under exceptionally favorable conditions, ancient feces have been preserved as fossils called coprolites.

Because feces are largely composed of soft material, coprolites are usually much rarer than skeletal fossils. They can be locally abundant, however, and such concentrations contributed to early interest in these unusual formations. The first published report of fossil feces actually predates the earliest descriptions of dinosaur bones and was made without benefit of precedent when William Buckland (1823) compared some enigmatic white fossil lumps with fresh hyena feces and deduced a fecal origin. He later (Buckland, 1835) applied the term "coprolite" to fossilized feces.

Coprolites have been found on every continent, with the oldest known vertebrate specimens dating back to the Silurian (Gilmore, 1992). Although some Quaternary feces have been preserved through desiccation and resemble modern dried dung (e.g., Mead and Agenbroad, 1992), most coprolites have been substantially altered during fossilization. A small number of specimens have been preserved as carbonaceous compressions (e.g., Hill, 1976), but the overwhelming majority of coprolites are lithified.

Most fossil feces have been recognized by their familiar fecal shapes, but coprolites are highly variable fossils. This variation reflects differences in the animals that produced the feces, fluctuations in diet, and disparate diagenetic conditions. Numerous coprolite morphologies have been described, including spherical, cylindrical, fusiform, spiral, blocky, pancake-like, and amorphous forms. This range of shapes is similar to that found in modern animal droppings. The colors of coprolites also vary considerably: Different diagenetic regimes have resulted in brown, white, cream, orange, black, gray, and even bluish specimens.

As trace fossils, coprolites provide a record of animal activity and have the potential to supplement information obtained from skeletal fossils. Although coprolitic interpretations are complicated by diagenetic variability, well-preserved specimens can contain recognizable dietary inclusions that provide information on trophic interactions in ancient ecosystems. This information can be enhanced by some knowledge of the animal of origin.

Coprolite Provenance

The assignment of a coprolite to its source animal remains one of the more difficult problems of coprolite analysis. Unless a formed but unextruded fecal mass is found in the body cavity of an articulated specimen, its origin remains speculative. Certain factors, however, can help constrain a list of likely producers.

A spiral configuration is the only distinctive coprolite morphology that can be reliably associated with a type of source animal. Because the spiral valve that affects the egestion of coiled feces is absent in teleosts and tetrapods (Romer and Parsons, 1986), spiral coprolites are attributable to other taxa such as sharks, gars, or lungfish (Gilmore, 1992).

Spiral coprolites have been recovered from many Paleozoic and Mesozoic localities (Hantzschel *et al.*, 1968). The widespread distribution of these specimens indicates that feces deposited in aquatic environments have a high preservation potential. This suggests that many non-spiraled coprolites may have also been produced by aquatic animals or by terrestrial vertebrates that defecated in or near bodies of water. Although non-spiraled coprolites can have distinctive shapes and/or striations, such morphologies are found in feces produced by many different taxa (Thulborn, 1991; Hunt *et al.*, 1994). This necessitates the evaluation of non-morphological characters. Of prime consideration is the fact that the stratigraphic distribution of potential source animals must be consistent with the age, locality, and depositional environment of the coprolite itself. The co-occurrence of skeletal elements in the same sediments may pose strong arguments for associations between coprolites and source animals, but such associations remain speculative without additional evidence.

Coprolite size can provide information on possible animal producers, but interpretations of the significance of size must be made carefully. Although the quantity of egested feces is proportional to body size, direct correlations can be misleading. A small coprolite, for example, may have broken off a larger fecal mass. In addition, because some large extant animals produce quantities of small pelletoid feces, an isolated pelletoid coprolite could have been produced by a relatively large animal. Small animals, on the other hand, cannot produce large fecal masses. These considerations suggest that coprolite size should be primarily used to infer minimum sizes of possible producers. This criterion is particularly useful for identifying possible dinosaur coprolites.

Coprolite composition can also help constrain the number of likely producers by providing clues to the feeding strategies of source animals. Inclusions of bone fragments, teeth, fish scales, or mollusc shells, for example, provide evidence of carnivory. Unfortunately, recognizable dietary residues have often been destroyed by digestion and diagenesis. In such cases, carnivory can be implicated by a predominance of calcium phosphate. Bradley (1946) noted that most coprolites are phosphatic and that carnivore feces contain relatively high percentages of phosphorus. He suggested that carnivore feces are preferentially fossilized because of the availability of dietary calcium phosphate. This is consistent with the observation that permineralized coprolites containing substantial plant material are relatively rare and are almost invariably calcareous or siliceous.

Although it is difficult, if not impossible, to ascertain the origin of a coprolite, the concomitant analysis of stratigraphic occurrence, morphology, size, and composition can help characterize likely source animals. The informative value of these factors is variable, however. The identities of some coprolite producers may remain poorly resolved if coprolite specimens have very common features, whereas unusual or distinctive attributes might provide significant clues to coprolite provenance.

Dinosaur Coprolites

Many coprolites have been found in Mesozoic sediments, but few described specimens have been confidently attributed to dinosaurs. The identification of possible dinosaur coprolites is complicated by the fact that dinosaurs shared Mesozoic ecosystems with many other animals. Non-spiraled small or mid-sized phosphatic coprolites might be particularly difficult to identify because they could have been produced by a number of carnivorous vertebrates, including teleosts, turtles, crocodilians, or dinosaurs. Very large coprolites may be more reasonably ascribed to dinosaurs, but sizable bona fide coprolites are rare. The paucity of large specimens probably reflects preservational biases: Large fecal masses would have been highly susceptible to mechanical disruption, and large animals may have rarely defecated in depositional environments that were conducive to the fossilization of feces.

These considerations help account for the poor record of dinosaur coprolites. Moreover, some specimens previously interpreted as dinosaur feces must be re-evaluated, such as those from the Bernissart Iguanodon Quarry in Belgium. Bertrand (1903) initially attributed those coprolites to theropods, but a subsequent discussion (Abel, 1935) suggested that the feces could have been produced by crocodiles. In another report, Matley (1939) used size criteria to assign large Cretaceous coprolites from India to titanosaurs whose bones were found in the same sediments. The coprolites were probably not produced by herbivores, however, because the specimens are phosphatic and contain no traces of plant tissue. The large droppings are up to 10 cm wide and 17 cm long, and certainly implicate hefty source animals, but it is not clear if fecal masses produced by large carnivorous dinosaurs can be distinguished from those left by large crocodilians.

Still other purported dinosaur feces may not be coprolites at all. Large, bulbous, siliceous nodules commonly found in Jurassic deposits have sometimes been interpreted as dinosaur coprolites (Spendlove, 1979). These specimens, however, lack organic inclusions or other positive evidence that supports a fecal origin and may simply be inorganic concretions.

The origin of large Mesozoic plant-filled coprolites is less ambiguous, because few large herbivores coexisted with dinosaurs. One unusual grouping of more than 250 compressed pellets containing plant cuticle was found in Jurassic sediments in England (Hill, 1976). The individual pellets are small (8- to 18-mm diameter), but the total assemblage represents a sizable fecal mass that could have been produced by a herbivorous dinosaur. Much larger herbivore coprolites found in Montana are undoubtedly dinosaurian. These large (up to $24 \times 33 \times 34$ cm) blocky Cretaceous specimens contain conifer stem fragments. They lack a familiar coprolite shape, but a fecal origin has been corroborated by the presence of backfilled dung beetle burrows in the specimens (Chin and Gill, 1997).

The recognition of atypical coprolitic masses suggests that additional dinosaur droppings may be identified with more careful examination of Mesozoic sediments. Continued analyses of dinosaur feces will increase our understanding of dinosaur diets and their interactions with other organisms because coprolites can provide paleobiological information that is unavailable from skeletal fossils.

See also the following related entries:

Diet • Trace Fossils

References

Abel, O. (1935). *Vorzeitliche Lebensspuren*, pp. 644. Fischer, Jena, Germany.

Bertrand, C. E. (1903). Les Coprolithes de Bernissart. I. partie: Les Coprolithes qui ont été attribués aux Iguanodons. *Mem. Musee Royal d'histoire Nat. Belgique* **1**(4), 1–54.

Bradley, W. H. (1946). Coprolites from the Bridger Formation of Wyoming: Their composition and microorganisms. *Am. J. Sci.* **244,** 215–239.

Buckland, W. (1823). *Reliquiae Diluvianae*, pp. 303. Arno Press, New York [Reprint of the 1823 ed. published by Murray, London, 1978]

Buckland, W. (1835). On the discovery of coprolites, or fossil faeces, in the Lias at Lyme Regis, and in other formations. *Trans. Geol. Soc. London Ser. 2,* **3**(1), 223–236.

Chin, K., and Gill, B. D. (1997). Dinosaurs, dung beetles, and conifers: Participants in a Cretaceous food web. *Palaios*, in press.

Gilmore, B. G. (1992). Scroll coprolites from the Silurian of Ireland and the feeding of early vertebrates. *Palaeontology* **35**(2), 319–333.

Hantzschel, W., El-Baz, F., and Amstutz, G. C. (1968). *Coprolites, an Annotated Bibliography.* Memoir 108, pp. 132. Geological Society of America, CO.

Hill, C. R. (1976). Coprolites of *Ptiliophyllum* cuticles from the Middle Jurassic of North Yorkshire. *Bull. Br. Museum Nat. History Geol.* **27,** 289–294.

Hunt, A. P., Chin, K., and Lockley, M. G. (1994). The palaeobiology of vertebrate coprolites. In *The Palaeobiology of Trace Fossils* (S. K. Donovan, Ed.), pp. 221–240. John Wiley, Chichester, UK.

Matley, C. A. (1939). The coprolites of Pijdura, Central Provinces. *Rec. Geol. Surv. India* **74**(4), 530–534.

Mead, J. I., and Agenbroad, L. D. (1992). Isotope dating of Pleistocene dung deposits from the Colorado Plateau, Arizona and Utah. *Radiocarbon* **34**(1), 1–19.

Romer, A. S., and Parsons, T. S. (1986). *The Vertebrate Body*, pp. 679. Saunders College, Philadelphia.

Spendlove, E. (1979). Henry Mountain coprolites. *Rock Gem* **9,** 60–64.

Thulborn, R. A. (1991). Morphology, preservation and palaeobiological significance of dinosaur coprolites. *Palaeogeogr. Palaeoclimatol. Palaeoecol.* **83,** 341–366.

Corpus Christi Museum of Science and History, Texas, USA

see Museums and Displays

Cranbrook Institute of Science, Michigan, USA

see Museums and Displays

Cranial Comparative Anatomy

see Skulls, Comparative Anatomy

Craniofacial Air Sinus Systems

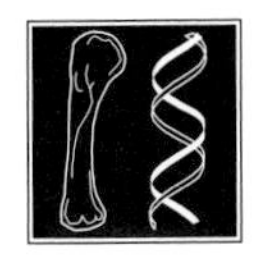

LAWRENCE M. WITMER
Ohio University
Athens, Ohio, USA

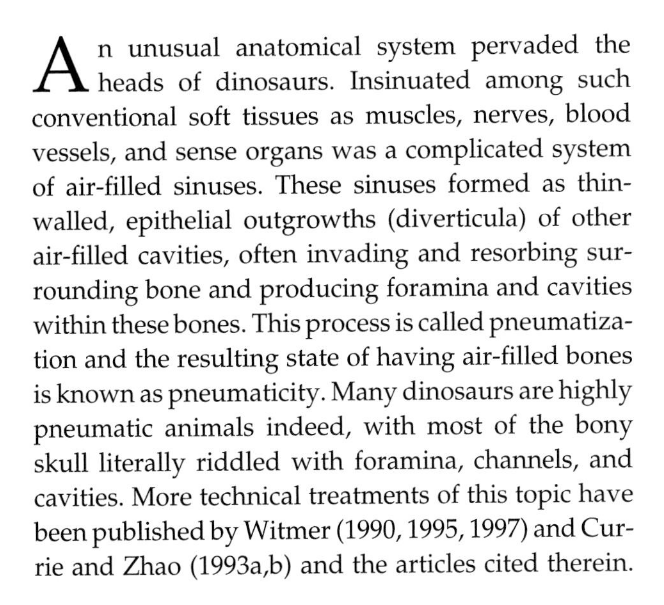

An unusual anatomical system pervaded the heads of dinosaurs. Insinuated among such conventional soft tissues as muscles, nerves, blood vessels, and sense organs was a complicated system of air-filled sinuses. These sinuses formed as thin-walled, epithelial outgrowths (diverticula) of other air-filled cavities, often invading and resorbing surrounding bone and producing foramina and cavities within these bones. This process is called pneumatization and the resulting state of having air-filled bones is known as pneumaticity. Many dinosaurs are highly pneumatic animals indeed, with most of the bony skull literally riddled with foramina, channels, and cavities. More technical treatments of this topic have been published by Witmer (1990, 1995, 1997) and Currie and Zhao (1993a,b) and the articles cited therein.

Pneumatic Systems in Dinosaurs

There are two well-known pneumatic systems in dinosaurs, one arising as outgrowths of the nasal cavity and the other as outgrowths of the tympanic (middle ear) cavity. Not just dinosaurs, but all archosaurs—living and extinct—have at least one main paranasal air sinus, known as the antorbital sinus, that forms as an outgrowth of the main nasal cavity (Fig. 1). The antorbital sinus produces a large cavity and opening in the side of the face known respectively as the antorbital cavity and antorbital fenestra. In many archosaurs, the antorbital sinus itself has subsidiary outgrowths that may pneumatize surrounding bones, producing so-called accessory cavities. In addition to the nearly ubiquitous antorbital sinus, a few kinds of dinosaurs have air sacs deriving from a different part of the nasal cavity, namely, the front-most portion known as the nasal vestibule. Such vestibular sinuses tend to pneumatize the bones surrounding the bony nostril (i.e., premaxilla and nasal). Humans and most other mammals have similar (but not homologous) paranasal sinuses; these are the sinuses that become congested when we have colds and that are involved in our "sinus headaches."

Paratympanic air sinuses are less common in archosaurs. In those non-dinosaurian taxa with paratympanic pneumaticity (such as crocodylomorphs and pterosaurs), the bones of the braincase are the ones that are usually invaded by air sacs. Among perhaps all archosaurs, tympanic recesses are best developed in theropod dinosaurs. As with paranasal sinuses, humans and most other mammals have a set of sinuses associated with the tympanic cavity, and particularly bad middle-ear infections may spread to our paratympanic air sinuses.

In addition to paranasal and paratympanic sinuses, there are other, more poorly known, systems pneumatizing the head skeleton. The first may simply represent diverticula from the cervical system of pulmonary air sacs that extend beyond the neck vertebrae into the occipital region of the skull. Some of the pneumatic cavities of certain theropod dinosaurs may result from these pulmonary diverticula. The second is the median pharyngeal system that forms as a midline outgrowth from the roof of the throat and invades the base of the skull in the region of the basisphenoid and basioccipital bones in many archosaurs. It is not always demonstrably of pneumatic origin in many archosaurs but is almost certainly so in theropod dinosaurs. Although the resulting "basisphenoid sinus" is often regarded as a derivative of a "median Eustachian tube," it is clearly distinct from the definitive auditory (Eustachian) tubes that, along with the tympanic cavities, have their embryological origins from the paired first pharyngeal (or branchial) pouches; whereas in some archosaurs the median system eventually connects up with the tympanic cavity, it does not always do so. Some have suggested that the median pharyngeal pneumatic system results from aeration of the embryonic hypophysial pouch (of Rathke)—a precursor of

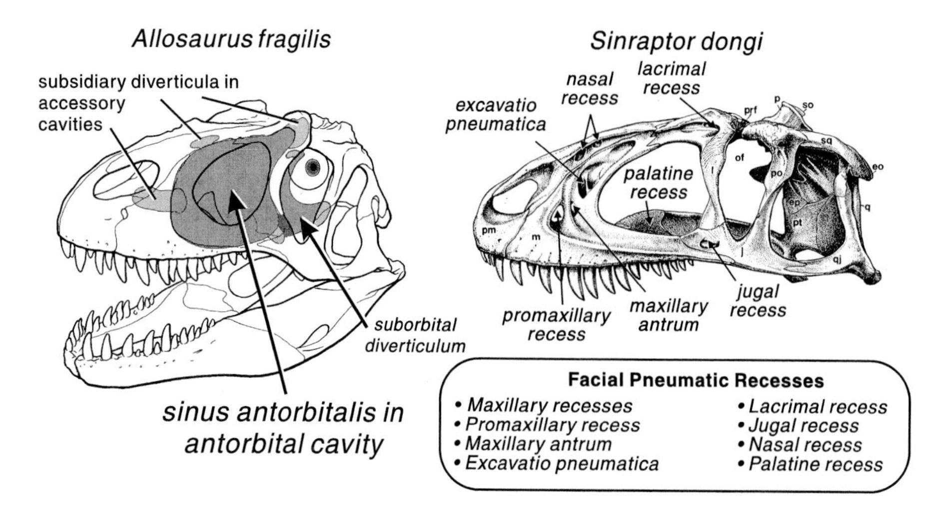

FIGURE 1 Paranasal pneumaticity in theropod dinosaurs. (Left) skull of *Allosaurus fragilis* in oblique view, showing the antorbital paranasal air sinus and some of its epithelial diverticula (modified from Witmer, 1997, with permission). (Right) skull of *Sinraptor dongi* in left lateral view with the major paranasal pneumatic accessory cavities labeled (modified from Currie and Zhao, 1994a, with permission). The accessory cavities result from pneumatization by the subsidiary diverticula of the antorbital sinus.

part of the pituitary gland—and this is an idea worthy of further investigation.

Ornithischian Dinosaurs

Paratympanic air sinuses are very uncommon in ornithischians. The middle ear sac was certainly present (as evidenced by, among other things, the discovery of columellae—the slender ear bones), but apparently it did not typically send out diverticula that invaded surrounding bones. A few ornithischians, however, such as the basal thyreophoran *Scelidosaurus* and the ornithopod *Hypsilophodon*, do seem to have some excavations of portions of the braincase that are best interpreted as being of pneumatic origin. In these forms, there is a fairly extensive cavity associated with the canal for the major artery supplying the brain, the cerebral (or internal) carotid artery. This cavity is directly behind and medially undercuts a curving ridge of bone—nearly ubiquitous in archosaurs—known as the otosphenoidal crest, which runs from the basipterygoid process to the paroccipital process and segregates the orbital contents in front from the middle ear contents behind. Although such a rostral tympanic recess is, as we will see, fairly common in theropod dinosaurs (as well as in a variety of other archosaurs), it is rather rare in ornithischians.

Paranasal pneumaticity, on the other hand, is present in probably all ornithischians in that, like all other archosaurs, they possessed an antorbital sinus. This antorbital air sinus was lodged in a cavity, the antorbital cavity, located in front of the orbit and bounded by primarily the maxilla and lacrimal, and sometimes also the jugal, palatine, and nasal. In fact, this description holds for most dinosaurs (indeed, most archosaurs). Pneumatic bony accessory cavities, produced by subsidiary diverticula of the antorbital sinus, are relatively rare in ornithischians, although they are present in a few taxa, such as the intramaxillary sinuses of *Protoceratops* and its relatives and the maxillary recesses of some basal thyreophorans. Higher thyreophorans, in particular ankylosaurid ankylosaurians, deserve special mention here in that within their highly armored skulls is a maze of pneumatic sinuses. The precise pattern and arrangement of ankylosaurid paranasal sinuses remain poorly known, and it is not entirely clear if some are accessory cavities of the main antorbital sinus, novel paranasal sinuses, or even diverticula of the nasal vestibule. Pneumatic evaginations of the nasal vestibule, however, are clearly expressed in lambeosaurine hadrosaurids, such as *Corythosaurus*. In lambeosaurines, the narial region is greatly enlarged, and the bones enclosing

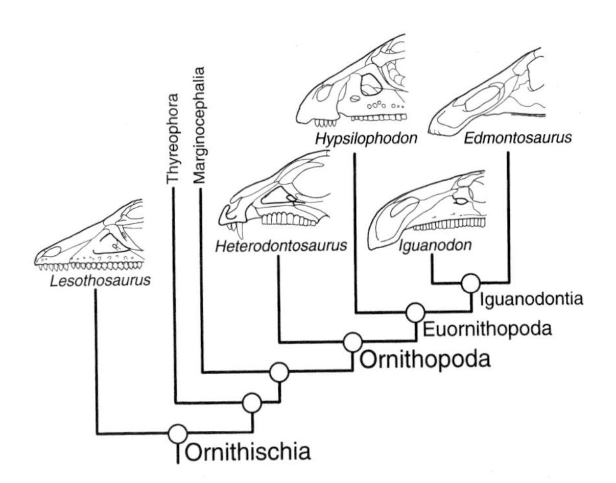

FIGURE 2 The evolving antorbital cavity of ornithischian dinosaurs, especially Ornithopoda. Most clades of ornithischians, such as Ornithopoda, show marked trends for reduction and enclosure of the antorbital cavity by laminae of the maxilla and lacrimal. Modified from Witmer (1997) and references cited therein with permission.

the nasal vestibule (the premaxilla and nasal) are folded into a complicated collection of passages and chambers, all of which are perched atop the remainder of the skull.

Perhaps the most remarkable aspect of skull pneumaticity in ornithischians is its recurrent trend for reduction. In other words, ornithischians tend to become relatively less pneumatic when comparing more advanced members of a clade with more basal members. For example (Fig. 2), the basal ornithischian *Lesothosaurus* has a more or less primitive—and hence, fairly large—antorbital cavity; it even has a small pneumatic accessory cavity associated with its palatine bone. In ornithopodan ornithischians, however, there is a general trend for reduction of the antorbital cavity and enclosure by lateral sheets of the maxilla and lacrimal bones such that the antorbital cavity becomes a relatively small and completely internalized space. The broad outlines of this trend can be viewed by making a phylogenetic march from a basal ornithopod such as *Heterodontosaurus* (which retains a relatively large antorbital cavity but also shows the beginnings of the lateral enclosure), through *Hypsilophodon* (which shows further lateral enclosure) and iguanodontians such as *Camptosaurus* and *Iguanodon* (which show further reduction and enclosure), to hadrosaurids (in which the antorbital cavity is relatively tiny and completely closed laterally). A similar phylogenetic trend can be identified in thyreophorans. These trends almost certainly relate to (especially in ornithopods) the expansion and elaboration of the feeding apparatus (in particular, the dentition and its bony supports). Thus, as the feeding apparatus expanded phylogenetically, the antorbital sinus and its bony cavity contracted.

Before concluding the discussion of ornithischian craniofacial pneumaticity, the supracranial cavities of ceratopsids such as *Triceratops* need to be considered. These cavities, formed by the folding of the frontal bones, are often referred to as "frontal sinuses" and are commonly thought to be of pneumatic origin. They are often compared to the frontal air sinuses of modern bovid mammals (cattle, sheep, etc.) because in both ceratopsids and bovids the sinuses form strutted chambers that extend up into the base of the horn cores. It is possible that the supracranial cavities of ceratopsids are indeed pneumatic, but the source of the air-filled diverticulum remains obscure. It is not yet clear how an outgrowth from either the nasal cavity or the tympanic cavity could reach the skull roof. Its mode of development is also unusual for a pneumatic recess. Thus, pending further research, the supracranial cavities of ceratopsids will remain functionally enigmatic.

Saurischian Dinosaurs

Saurischia includes neornithine birds, the most pneumatic of all known vertebrates, but not all saurischians display extensive craniofacial pneumaticity. In fact, other than the antorbital cavity itself, sauropodomorphs do not display many pneumatic features in their skulls, which is ironic because their axial skeletons are otherwise often marvels of pneumatization. Basal sauropodomorphs (i.e., prosauropods) have relatively primitive and simple antorbital cavities, although a subsidiary diverticulum of the antorbital sinus excavates a pneumatic accessory cavity in the nasal of *Plateosaurus*. The antorbital cavity of most sauropods is relatively reduced, being telescoped between the orbit behind and the greatly expanded nasal vestibule in front. As in ornithopods, the antorbital sinus appears to be, in a sense, "crowded out" by other structures. Paratympanic pneumaticity is also relatively poorly developed in sauropodomorphs. A few basal taxa, such as *Anchisaurus* and *Plateosaurus*, have, much as does *Hypsilophodon*, moderate development of a rostral tympanic recess (i.e., the cavity associated with the cerebral carotid artery and bounded by the otosphenoidal crest). Other paratympanic recesses appear to be virtually absent, although a few sauropods (e.g., *Camarasaurus*) have a deep excavation in the back surface of the quadrate bone that could be interpreted as pneumatic in nature.

In contrast to sauropodomorphs, theropod dinosaurs display the most extensive craniofacial pneumaticity of all archosaurs with perhaps the exception of some pterosaurs. Not only are the paranasal and paratympanic systems well developed in theropods but also the median pharyngeal system—for the first time in dinosaurs—takes on clearly pneumatic attributes. In virtually all theropods, the antorbital cavity is huge, occupying in some cases more than half of the total skull length; thus, the enclosed antorbital paranasal air sinus must have been very voluminous (Fig. 1). As in other archosaurs, the maxilla and lacrimal were the major bones housing the air sac, although the nasal, jugal, and palatine were also commonly involved. With the exception of the bizarre oviraptorosaurs, which have dramatically complicated pneumatic skulls, there is no evidence that the facial skeleton of theropods was pneumatized by any air sacs other than the antorbital sinus.

Nevertheless, one of the remarkable aspects of paranasal pneumaticity in theropods was the tendency for the antorbital sinus to send out subsidiary diverticula that penetrated into the adjacent facial bones, often producing large pneumatic accessory cavities (Fig. 1). Although many other archosaurs have more or less shallow pneumatic depressions on various facial bones, it is theropods alone that routinely exhibit a facial skeleton that is produced into an open series of hollowed struts and chambers. The most commonly observed accessory cavities are those in the maxilla, with the two most consistent maxillary sinuses being the promaxillary recess and the maxillary antrum. Both these accessory cavities are widely (but not universally) distributed among neotetanurans, and many (but again not all) ceratosaurians have a single maxillary sinus, which is probably homologous to the promaxillary recess of higher theropods. The lacrimal bone is also commonly pneumatized by a subsidiary diverticulum of the antorbital sinus. It is this lacrimal sinus that invades and hollows out the "horns" of theropods such as *Allosaurus* and *Ceratosaurus*. The nasal, jugal, and palatine bones are less frequently pneumatized, but in some cases these recesses can result in spectacular structures, such as the inflated nasal crest of *Monolophosaurus* or the puffed-up palatine bones of *Tyrannosaurus*.

The phylogenetic distribution of these paranasal pneumatic accessory cavities can be rather confusing. Some groups, such as tyrannosaurids, tend to be fairly consistent. Other groups, however, can be more variable. For example, among dromaeosaurids, *Deinonychus* has all the accessory cavities noted previously, but the closely-related form, *Velociraptor*, has only a few of them. Moreover, a few taxa, such as the basal tetanuran *Torvosaurus* and the aberrant maniraptoran *Erlikosaurus*, lack many or even all of the accessory cavities that phylogenetics indicates they "should" have. Nevertheless, despite these problems, one point that emerges is that the facial skeleton of theropods, as a group, is highly pneumatic; in fact, pneumaticity has been recorded in every bone of the facial skeleton except two (the prefrontal and vomer).

A final aspect of the paranasal air sinus system of theropods involves a diverticulum of the antorbital sinus that only very rarely pneumatizes bone. This air sac, the suborbital diverticulum, is almost always present in modern theropods (i.e., birds). It forms as an outgrowth of the back wall of the antorbital sinus and expands into the orbit where it often encircles the eyeball and interleaves with the jaw musculature. In at least a few nonavian theropods, there are good reasons to believe that a bird-like suborbital diverticulum was present (Fig. 1). The significance of this suborbital sac is that it provides a mechanism for actively ventilating the antorbital sinus (i.e., pumping air in and out). Movements of the lower jaw—such as in closing and opening the mouth—will set up positive and negative pressures in the suborbital sac because of its intimate relationship to the jaw muscles. These pressure changes are transferred to the antorbital sinus and, thus, like a bellows pump, air passes to and fro between the nasal cavity and antorbital sinus. This situation is unique: In other animals with paranasal sinuses, such as crocodilians and mammals, the sinuses are never actively ventilated but rather are stagnant, dead-air spaces. What role such a paranasal bellows pump plays in the physiology of birds and probably other theropods is still unknown.

As with the paranasal system, the paratympanic air sinuses are generally very diverse and extensive in theropods. Most of the sinuses invade the bones of the braincase and otic (ear) region, such as the prootic, opisthotic, basisphenoid, and basioccipital. Some of these recesses are obviously associated with the middle ear sac, but others, as mentioned previously, seem to result from median pneumatic outgrowths of the roof of the pharynx and perhaps even from extensions from the pulmonary air sacs in the neck. The pneumatic sinuses of the braincase will be briefly discussed using a hypothetical form (Fig. 3) because no known species has all the air cavities.

There are three fairly consistent pneumatic cavities that clearly derive from diverticula of the middle ear sac, and they all aptly bear the name "tympanic recess." Perhaps the most commonly encountered tympanic recess is the rostral or lateral tympanic recess noted previously in some other dinosaurs. This recess is located just behind the otosphenoidal crest in the area of the cerebral carotid artery foramen. In some theropods, such as *Syntarsus* or *Dilophosaurus*, this recess resembles that of other archosaurs in being a more or less simple but expanded cavity, whereas in others, such as some coelurosaurs, it becomes complicated and multichambered. For example, taxa such as *Deinonychus* and *Struthiomimus* have a discrete prootic recess within the rostral tympanic recess just ventral to the facial nerve foramen, and a varied group of tetanurans have a more ventral cavity, the subotic recess, that excavates the basal tubera. In some forms, such as ornithomimids, troodontids, and birds, the

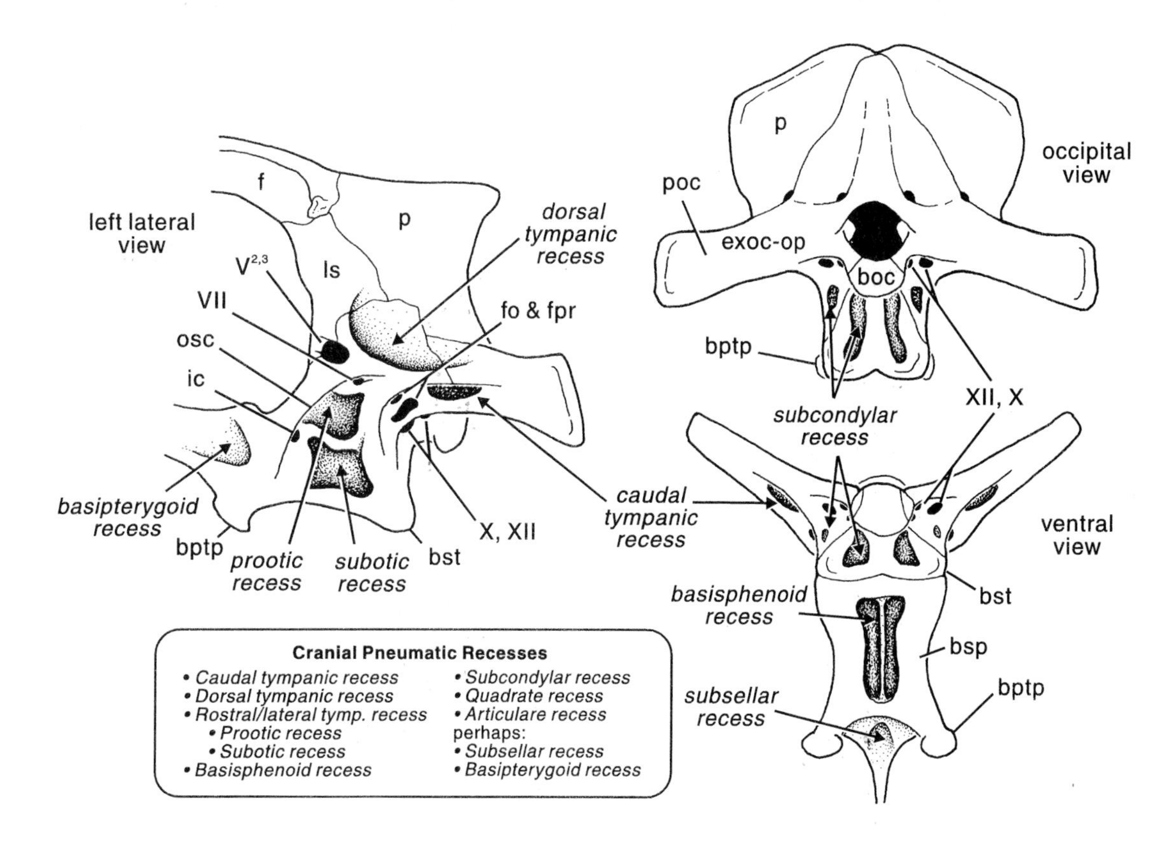

FIGURE 3 Braincase of a hypothetical theropod dinosaur showing the diversity of pneumatic recesses. Pneumatization from the middle ear sac produces the dorsal, caudal, and rostral tympanic recesses and the recesses within the quadrate and articular bones (not shown). The subcondylar recesses derive from either the middle ear sac or extensions from the pulmonary air sacs. Pneumatization from a median pharyngeal system produces the basisphenoid recess. The basipterygoid and subsellar recesses are not clearly pneumatic structures, and the source of the diverticulum, if present, is also uncertain. fo, fenestra ovalis (or vestibularis) [note: not foramen ovale, which is the maxillary n. foramen in mammals); fpr, fenestra pseudorotundum (or cochlearis); osc, otosphenoidal crest.

rostral tympanic recess and part of the main middle ear sac are covered laterally by a thin sheet of bone, the parasphenoid. The caudal tympanic recess is typically found in Coelurosauria (*Troodon* being an important exception) and involves a large air space within the paroccipital process that opens into the tympanic cavity via an oval foramen on the front of the base of the paroccipital process. The dorsal tympanic recess, a moderate to very deep depression on the dorsolateral surface of primarily the prootic bone, has a very patchy distribution but is found in ornithomimids, velociraptorine dromaeosaurids, and all known birds.

In a few groups of nonavian theropods, such as birds, tyrannosaurids, and at least some ornithomimids and troodontids, the quadrate and / or articular bones are also invaded by outgrowths of the middle ear sac. In some respects, such pneumaticity makes good sense in that the quadrate and articular bones have their embryological origins as parts of the first pharyngeal (= mandibular) arch; recall that the middle ear sac itself derives from the same pharyngeal arch. In fact, it is a mystery as to why more groups of theropods—especially bird-like forms such as *Deinonychus* that have extensive braincase pneumaticity—lack mandibular arch pneumaticity. Interestingly, crocodyliforms, many pterosaurs, and some other non-dinosaurian archosaurs have pneumatic quadrates and articulars.

The median pharyngeal system, a rudimentary blind pit or small foramen between the basioccipital and basisphenoid in other archosaurs, takes on the unmistakable appearance of an invasive air-filled cavity in theropod dinosaurs, in which it is usually referred to as a "basisphenoid sinus" (Fig. 3). Typically, the sinus is roughly pyramidal or conical, with its base being a large opening in the midline of the basicranium between the paired basal tubera and basipterygoid processes and its apex being directed dorsally toward the pituitary fossa. Even in basal forms such as *Coelophysis* and *Dilophosaurus*, the basisphenoid sinus is already somewhat expanded. However, in most tetanurans the basisphenoid sinus becomes a large expansive cavity, invading back into the basioccipital bone, sometimes even into the occipital condyle. Sometimes a median septum is preserved within the sinus, partially dividing it into left and right sides.

Paired openings on the occipital surface of the cranial base below the occipital condyle, known as the subcondylar recesses (Fig. 3), are not widely distributed in theropods but are well developed in tyrannosaurids and ornithomimids. These cavities are obviously of pneumatic origin in that they expand within the bone and are multichambered. What is uncertain is the source of the air-filled diverticulum. The diverticulum could indeed derive from the middle ear sac—that is, it is another tympanic recess. Certainly, the subcondylar recess is close enough to the tympanic cavity that it would not be a "far reach." Another idea, however, cannot be ruled out: The subcondylar recesses derive from diverticula of the cervical division of the lung air-sac system. The cervical air sacs pneumatize the backbone up to the second cervical vertebra in tyrannosaurids and ornithomimids and thus are very close to the occiput. Furthermore, they presumably interleaved with neck muscles that attached to the occiput in the vicinity of the subcondylar recesses. Choosing between a tympanic or pulmonary source is currently very difficult.

A couple of other cavities may be of pneumatic origin but are much more uncertain (Fig. 3). The basipterygoid recess is a depression within the lateral surface of the basipterygoid process within the orbit of theropods such as *Allosaurus*. If it is pneumatic, the source of the diverticulum is unclear. Instead, it may be a site of muscular attachment, perhaps for a palatal protractor (if indeed such muscles were present in nonavian theropods). The subsellar recess is a median ventral cavity located just in front of the basisphenoid recess at the base of the parasphenoid rostrum. As its name implies, it resembles the basisphenoid recess in being directed toward the pituitary fossa, but obviously both recesses could not result from a diverticulum tracking along the single embryonic hypophysial stalk (if in fact either did). Clearly, both the basipterygoid and subsellar recesses are structures that need a good deal more research.

Two more "problem sinuses" need to be considered. Large cavities within both the ectopterygoid and squamosal bones are clearly pneumatic. The source of the diverticula, however, remains unclear. The ectopterygoid recess is an almost ubiquitous feature of theropods. It usually takes the form of a simple, smooth-walled ventromedial "pocket" in the bone, although in some tyrannosaurids it becomes a multichambered affair. However, did the diverticulum come from the middle ear sac, the antorbital sinus, or from some unknown source (perhaps yet another pharyngeal outgrowth!)? All these ideas are possible but have problems. Likewise, the large recess within the squamosal bones of tyrannosaurids and ornithomimids could have been produced by subsidiary diverticula of either the antorbital sinus or the middle ear sac, and we currently have little strong evidence that allows us to make a reasoned choice.

In summary, theropod dinosaurs obviously exhibit an extraordinary—often bewildering—diversity of air-filled sinuses. Without question, pneumaticity is the single most important anatomical system shaping theropod skull morphology. Interestingly, however, it seems fraught with high levels of homoplasy (the evolutionary loss and/or convergent acquisition of a feature). Certainly, there are some very consistent, phylogenetically informative pneumatic characters, such as the presence of the two main maxillary sinuses in neotetanurans or the caudal tympanic recess in coelurosaurs. However, many of the pneumatic characters seem to have been evolved or been lost repeatedly—often yielding a morphology so similar that the only hint of homoplasy comes through phylogenetic analysis. The dorsal tympanic recess of coelurosaurs is a good, but by no means the only, example. This recess is found in Mesozoic birds, such as *Hesperornis* and *Archaeopteryx*, as well as the dromaeosaurids *Deinonychus* and *Velociraptor*. So far, so good, but it is absent in the dromaeosaurid *Dromaeosaurus*. It is also absent in the bullatosaur *Troodon* but is clearly present in the bullatosaur *Struthiomimus*. The addition of other taxa only complicates the picture further.

When the workings of pneumatic systems—that is, the soft tissue systems that literally produces the bony recesses—are understood, morphogenetic mechanisms for phylogenetic reversal and convergence become rather easy to envision. For example, in the diagram in (Fig. 4) showing a hypothetical ancestor–descendant sequence of species, the ancestral form has a relatively simple middle ear sac with no dorsal tympanic diverticulum. A descendant species might evolve some anatomical change in the conformation of the general region (in this case, a higher and shorter braincase) that, for some reason, permits evagination of a diverticulum, and this diverticulum excavates a pneumatic cavity on the prootic. A further

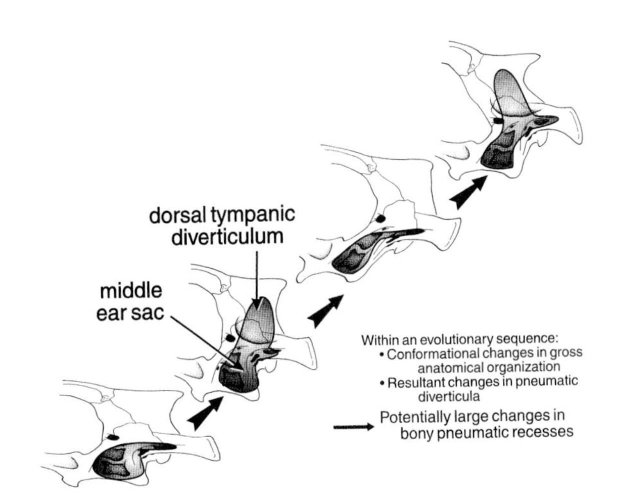

FIGURE 4 A transgression–regression model for homoplasy in pneumatic systems. Hypothetical evolutionary transformation series of theropod braincases depicting the changing status of a dorsal tympanic diverticulum of the middle ear sac and, hence, the variable presence or absence of the bony recess. As the anatomical conformation of the whole region changes, the possibility for a diverticulum to evaginate changes. Thus, the mechanism for pneumatic homoplasy is easily envisioned, with an apt analogy being the fluctuating sea levels associated with transgression and regression.

conformational change might prevent evagination of this diverticulum or cause it to shift its position, which would appear to us as a reversal. Subsequent changes could permit the diverticulum to evaginate once again, perhaps resulting in a bony recess that is identical to the second species, which then would represent convergence. The point is that these epithelial air sacs are highly labile structures, the position (or even the presence) of which is easily affected by surrounding anatomical structures. Thus, given this almost capricious ebb and flow of pneumatic diverticula, perhaps it is unreasonable even to expect these bony cavities to fall out neatly on a cladogram.

Functions of Sinuses

It is somewhat ironic that, despite air-filled sinuses being very prominent components of head anatomy in not just birds and other dinosaurs but also in mammals, the function of pneumaticity has remained obscure. Numerous ideas have been proposed over the years, including sinuses acting as shock absorbers, flotation devices, vocal resonators, thermal insulators, weight-reducing air bubbles, and the list goes on. Many of these ideas at first may seem absurd but still may apply to some animals in some cases. For example, the sinuses within the crests of lambeosaurine hadrosaurids are thought to have functioned as resonating chambers. However, these functions obviously cannot apply broadly across all animals that possess the sinus. In fact, because almost all the ideas that have been proposed fail because of limited applicability, the problem of the general function of pneumatic sinuses has been regarded as one of the great mysteries of craniofacial functional morphology.

It is significant that virtually all the suggestions hinge on the empty space within the sinuses being that which provides the suggested functional benefits. With regard to the lambeosaurine example discussed previously, the empty space would provide the chamber in which sound resonates. It now seems that it is this focus on the enclosed volume of air that has led us astray. Sinuses are not truly "empty," but rather they contain the air sac itself—that is, the thin epithelial balloon that lines the bony recess. It turns out that the epithelium—not the empty space—may be the key. Pneumatic epithelium (and its associated tissues) appear to have the intrinsic capacity to expand and to pneumatize bone. Therefore, these air sacs may simply be opportunistic pneumatizing machines, expanding as much as possible—but within certain limits. These limits are provided by the local biomechanical loading regimes (i.e., the stresses and strains within the bone substance). Bone is sensitive to and responsive to its local stress environment, such as the forces involved in chewing or biting or laying down new bone to maintain sufficiently strong and stiff structures. Thus, there are two competing tendencies, one involving expanding air sacs and the other involving mechanically mediated bone deposition. A compromise is struck ensuring that pneumatization does not actually jeopardize the strength of the whole structure. An interesting consequence of this "battle" is that mechanically "optimal" structures will result as a secondary and completely incidental by-product—in other words, there is no reason to invoke natural selection to act directly to produce structures of maximal strength but minimal materials. Put simply, architecturally "elegant" structures are automatic outcomes of these two intrinsic, but opposite, processes.

Thus, what is the function of the antorbital sinus or the diverticula of the middle ear sac? The answer is probably, at its base, no particular function whatsoever. That is not to say that any of these air sacs could not be secondarily pressed into service for some positive function, subsequently honed by natural selection. Such is likely the case for the resonating chambers of hadrosaurs and, additionally, aspects of the expanded tympanic cavities of theropod dinosaurs, which enhance hearing due to certain acoustic properties. However, at the same time, positive functions need not be sought for every pneumatic recess in every species that has them. Again, they are simply intrinsic properties of pneumatic systems.

This new view on the function of pneumaticity helps explain certain trends in various dinosaur groups. For example, the reduction of the antorbital cavity in ornithopod ornithischians that was discussed previously (Fig. 2) clearly reflects a situation in which bone deposition "won out" over pneumatically-induced resorption of bone. The evolution and refinement of chewing in ornithopods required extensive bony buttressing and reinforcement, and so expansion of the antorbital sinus was severely limited. On the other hand, skull pneumatization in theropods appears to have progressed with few constraints and was clearly opportunistic. Although carnivory may involve fairly large bite stresses (force per unit area), the stresses and strains that a theropod skull as a whole underwent were minimal in comparison with the repetitive masticatory forces to which chewing animals such as ornithopods were subject. Thus, in theropods, sinuses were generally free to expand. It is significant, however, that the remaining bony struts in theropod skulls are positioned in mechanically advantageous locations, giving the appearance of exquisite design. It is also worth mentioning that *Tyrannosaurus rex*, a theropod that secondarily became adapted for particularly hard biting, increased the dimensions of some bony buttresses and bars to resist the stresses, yet it maintained pneumaticity—and even expanded some sinuses—by pneumatizing the bones internally, yielding a skull composed of a series of hollow, often truly tubular bars and plates.

See also the following related entries:

BRAINCASE ANATOMY • POSTCRANIAL PNEUMATICITY • SKULL, COMPARATIVE ANATOMY

References

Currie, P. J., and Zhao, X.-J. (1993a). A new carnosaur (Dinosauria, Theropoda) from the Jurassic of Xinjiang, People's Republic of China. *Can. J. Earth Sci.* **30,** 2037–2081.

Currie, P. J., and Zhao, X.-J. (1994b). A new troodontid (Dinosauria, Theropoda) braincase from the Dinosaur Park Formation (Campanian) of Alberta. *Can. J. Earth Sci.* **30,** 2231–2247.

Witmer, L. M. (1990). The craniofacial air sac system of Mesozoic birds (Aves). *Zool. J. Linnean Soc.* **100,** 327–378.

Witmer, L. M. (1995). Homology of facial structures in extant archosaurs (birds and crocodilians), with special reference to paranasal pneumaticity and nasal conchae. *J. Morphol.* **225,** 269–327.

Witmer, L. M. (1997). The evolution of the antorbital cavity of archosaurs: A study in soft-tissue reconstruction in the fossil record with an analysis of the function of pneumaticity. *Mem. Soc. Vertebr. Paleontol (J. Vertebr. Paleontol.)* **17**(Suppl.), 1–75.

Crests

see ORNAMENTATION

Cretaceous Extinction

see EXTINCTION, CRETACEOUS

Cretaceous Period

EVA B. KOPPELHUS
Geological Survey of Denmark and Greenland
Copenhagen, Denmark

The Cretaceous is the last period of the Mesozoic. It lasted for approximately 80 million years, ending 65 million years ago. The name is derived from the Latin word "*creta*," which means "chalk," and refers to the thick beds of Cretaceous chalk that are characteristic of parts of Europe.

Once divided into three epochs, the Cretaceous of Europe is now divided into the Early and Late Cretaceous, which are further subdivided into 12 stages. These were all defined on the basis of strata in the Anglo–Paris–Belgian area. In North America, the period is also divided into Early (Comanchean) and Late (Gulfian) Cretaceous.

The earth's climate was generally warm in the Cretaceous. Early in the period, conditions were becoming more humid and seasonal, which favored explosive diversification of the floras and the animals that fed on them. By the end of the Cretaceous, global cooling led to a drop in diversity of plants and animals in the higher latitudes.

Throughout the Cretaceous, the continents drifted apart to approach their current positions. The Atlantic Ocean opened up, India reached Asia, and extensive inland seas subdivided the continental masses.

Cretaceous rocks are widely distributed and exposed, and it is not surprising that almost half of the known dinosaurs have been found at this level. Hadrosaurs and ceratopsians in particular are common in the Late Cretaceous of the Northern Hemisphere. It was a time of great diversification, and dinosaurs shared their world with many other vertebrate groups that have changed little since the Cretaceous. Placental mammals, birds, snakes, and many other animals established the bauplans that are familiar to us today. Insects and other invertebrates were taking on a more modern appearance. It was a time when the terrestrial floras changed from being entirely dominated by pteridophytes and gymnosperms in the early part of the Cretaceous to angiosperms in the Late Cretaceous. The floral assemblages from the Late Jurassic persisted into the Early Cretaceous in the lower latitudes, whereas angiosperms became more prominent in the higher latitudes. By the early Late Cretaceous, there was a dramatic change into assemblages dominated by angiosperms. Four floral provinces based on pollen have been recognized in the Cenomanian: northern Laurasia, southern Laurasia, northern Gondwana, and southern Gondwana (Brenner, 1976). By Santonian–Campanian times, angiosperm pollen assemblages have been used to divide Laurasia into a southern *Normapolles* province and a northern *Aquilapollenites* province (Batten, 1984). The Cretaceous is a most interesting period from an evolutionary point of view.

Toward the end of the period, however, things

changed dramatically. Ammonites and the great marine reptiles disappeared from the seas, pterosaurs disappeared from the air, and nonavian dinosaurs suffered a major extinction event.

See also the following related entries:

Extinction, Cretaceous • Geologic Time • Mesozoic Era • Jurassic Period • Triassic Period

References

Batten, D. J. (1984). Palynology, climate and the development of floral provinces in the northern hemisphere: A review. In *Fossils and Climate* (P. Brenchley, Ed.), pp. 127–164. Wiley, London.

Brenner, G. J. (1976). Middle Cretaceous floral provinces and early migration of angiosperms. In *Origin and Early Evolution of Angiosperms* (C. B. Beck, Ed.), pp. 23–47. Columbia Univ. Press, New York.

Crocodylia

James M. Clark
George Washington University
Washington, DC, USA

The term Crocodylia was originally coined to encompass living crocodiles, but as fossil forms were discovered, they were also included. Many of these, however, were outside the group formed by living crocodiles (crown group Crocodylia), although obviously related to them. T. H. Huxley separated the fossil and recent Crocodylia into three groups: Protosuchia, Mesosuchia, and Eusuchia, distinguished by the posterior extent of their secondary palate and other features. These categories are now recognized not as monophyletic units but as grades of organization, the first two being paraphyletic. Most investigators use the term Crocodylomorpha to include these taxa plus the closely related Sphenosuchidae, which themselves may be paraphyletic relatives of crocodiles but share with crocodiles an elongated radial–ulnar and other features. Some workers use the term Crocodylia to include "Protosuchia," "Mesosuchia," and "Eusuchia"; others prefer to confine the term to crown group Crocodylia (a subset of Eusuchia).

Because crocodilians and birds are each other's closest living relatives, the evolutionary history of each group's stem began at the same time, when they initially diverged from a common ancestor. The lineage leading to birds included dinosaurs and pterosaurs, whereas the lineage culminating in living crocodilians included a variety of extinct groups, such as phytosaurs, aetosaurs, rauisuchids, and a host of forms more or less similar to living crocodilians (see Archosauria). Two names are currently applied to the lineage leading to crocodilians; the inaptly named Pseudosuchia (stem based) is defined as crocodilians and all extinct taxa more closely related to them than to birds, whereas the Crurotarsi (node based) refers to the same group of taxa but is not based on the concept of representing the evolutionary limb leading to crocodilians (see Phylogenetic System). The crocodilian lineage first appears in the fossil record in the Middle Triassic, at about the same time the first fossils of the bird lineage (stem-based Ornithosuchia or node-based Ornithodira) occur. Although the 25 or so living species of crocodilians are all semiaquatic and have a somewhat sprawling gait, their distant ancestors held their hindlimbs erect and were probably terrestrial. At the other extreme, one extinct group, thalattosuchians (Early Jurassic–Early Cretaceous), included forms that were apparently committed to an aquatic lifestyle.

Many terrestrial groups were small, even smaller than the living dwarf crocodile (*Osteolaemus*), and two (*Candidodon* and *Chimaerasuchus*) developed mammal-like (cusped) teeth and may have chewed their food. The largest Pseudosuchian was *Deinosuchus*, a close relative of living crocodilians from the Late Cretaceous that may have reached 13 m (50 ft) in length. The first crown group crocodilians (i.e., those belonging to a living group of crocodilian) appear in the Late Cretaceous, and crocodilians apparently were little affected by the events surrounding the extinction of nonavian dinosaurs at the end of the Mesozoic.

See also the following related entry:

Archosauria

References

Brochu, C., and Poe, S. Crocodylia section of the Tree of Life web site. *http://phylogeny.arizona.edu/tree/phylogeny.html.*

Clark, J. M. (1994). Patterns of evolution in Mesozoic Crocodyliformes. In *In the Shadow of the Dinosaurs* (N. C. Fraser and H.-D. Sues, Eds.), pp. 84–97. Cambridge Univ. Press, Cambridge, UK.

Parrish, J. M. (1993). Phylogeny of the Crocodylotarsi, with reference to archosaurian and crurotarsan monophyly. *J. Vertebr. Paleontol.* **13**(3), 287–308.

Poe, S. (1997). Data set incongruence and the phylogeny of the Crocodylia. *Systematic Biol.*, in press.

Ross, C. A. (Ed.) (1989). *Crocodiles and Alligators.* Facts on File, New York.

Sereno, P. C. (1991). Basal archosaurs: Phylogenetic relationships and functional implications. *Soc. Vertebr. Paleontol. Mem.* **2**.

Wu, X.-C., and Sues, H.-D. (1996). Anatomy and phylogenetic relationships of *Chimaerasuchus paradoxus*, an unusual crocodyliform reptile from the Lower Cretaceous of Hubei, China. *J. Vertebr. Paleontol.* **16**(4), 688–702.

Wu, X.-C., Brinkman, D. B., and Lu, J.-C. (1994). A new species of *Shangungosuchus* from the Lower Cretaceous of Inner Mongolia (China), with comments on *S. chuhsienensis* Young, 1961 and the phylogenetic position of the genus. *J. Vertebr. Paleontol.* **14**(2), 210–229

Crystal Palace

William A. S. Sarjeant
University of Saskatchewan
Saskatoon, Saskatchewan, Canada

The creation of the world's earliest three-dimensional restorations of dinosaurs was an event of the mid-19th century and the consequence of a suggestion by Prince Albert, consort of Queen Victoria. The prince had been much involved in developing the Great Exhibition, held in London in 1851. Its principal feature was a highly innovative prefabricated structure of glass and iron, the Crystal Palace. When the exhibition closed, it was decided that the Crystal Palace should be dismantled and re-erected in park-like grounds at Sydenham, on the south side of London. The prince had been greatly intrigued by Richard Owen's accounts of antediluvian creatures, in particular the giant ground-sloth *Megatherium*; he suggested that life-sized models of these creatures might be displayed.

Sir Joseph Paxton, who had designed the Crystal Palace, was also given charge of planning the new park. He decided to include in its southwest corner a "geological illustration" of the British Isles in the form of structures that would represent all major stratigraphical horizons, together with the principal mineral reserves. David T. Ansted (1814–1880), a leading economic geologist, served as geological adviser, while Richard Owen supervised the construction of the models of the extinct creatures that would be superposed on this geological landscape. Because Owen had himself named the dinosaurs a decade earlier, he made sure that they would be featured. As Doyle (1994) reports,

> The "Geological Illustrations" were logically arranged in three separate but interconnected parts of the park, surrounding a tidal lake which acted as a reservoir for the great central fountains of the park. The first part of the exhibit was a cliffline "exposing" a series of natural strata, representing the older rocks of northern England, Wales and Scotland. The exhibit was three-dimensional and dynamic; a water course issued from a spring line at the base of the Carboniferous (Mountain) Limestone. The limestone cliff in turn had a three-quarter scale reconstruction of a lead mine through which observers could pass. There was an accurate representation of the Coal Measures of Clay Cross [Derbyshire], complete with a dipping and fracture coal seam and associated beds of ironstone, and an unconformity, a discordance between two rock sequences indicating tectonic upheaval and erosion. None of these features represented a compromise of scientific accuracy, and all were dynamic in their impact.
>
> The lake itself contained the younger formations, set in a series of small islands. Scientifically the exhibits make sense; a fault can be "mapped-in" to explain the relationship of the younger strata in the lake with the older strata in the cliffline. On the islands were represented the great reptiles of the Secondary (Mesozoic) Era and the mammals of the Tertiary (Cenozoic) Era, each sited on the geologically most appropriate strata getting successively younger to the northeast. (pp. 7–8)

The task of restoring the extinct beasts was given to Benjamin Waterhouse Hawkins (1807–1889), who had been Paxton's assistant superintendent. Hawkins was a skilled artist and sculptor; under Owen's supervision, first he made small-scale models, then clay models at presumed life-size. From these, the final

FIGURE 1 "The Secondary Island" in Crystal Palace Park, London (from Anonymous, 1893).

restorations were made, but the task was not simple. As Hawkins reported (1854),

> Some of these models contained 30 tons of clay, which had to be supported on four legs. . . . In the instance of the *Iguanodon*, it is not less than building a house upon four columns, as the quantities of material of which the standing *Iguanodon* is composed, consist of 4 iron columns 9 feet long by 7 inches diameter, 600 bricks, 650 5-inch half-round drain-tiles, 900 plain tiles, 38 casks of cement, 90 casks of broken stone, making a total of 640 bushels of artificial stone. (p. 448)

By present-day standards, the restorations fail in accuracy. The Carboniferous *Labyrinthodon* was, in consequence of a misattribution and misreading of New Red Sandstone *Chirotherium* footprints, reconstructed as a toad-like creature, while the mammal-like reptile *Dicynodon* was quite wrongly depicted—presumably on the basis of its almost toothless jaws—as a turtle-like creature with a carapace.

The dinosaurs especially suffered from Owen's misinterpretations. One must remember that the reconstructions were based only on partial skeletons and a presumed analogy with living reptiles. Nevertheless, to modern eyes, it is startling to see both *Megalosaurus* and *Iguanodon* depicted as extremely massive quadrupeds and the latter genus with a sharp, nasal horn. These errors were not unreasonable; rhinocerine lizards were well known to Owen (and indeed, several genera of rhinocerine dinosaurs were discovered subsequently), whereas there were no living parallels to what was, in truth, *Iguanodon*'s spike-like thumb. As for the concept of dinosaur bipedality, that had not even arisen: Wealden (Early Cretacous) dinosaur tracks, though already reported by Tagart (1846), were still thought to be those of birds.

The third dinosaur to be reconstructed, *Hylaeosaurus*, was correctly depicted as a quadruped. It is still poorly known, but is now considered to be an ankylosaur. The row of spikes, which Owen placed upon its back, is no longer acceptable; *Hylaeosaurus* seems instead to have had spikes directed laterally outward from its flanks.

Debus believes Owen had a particular motive in shaping these models (see Desmond, 1979), considering (Debus, 1993) that

> Owen enlisted dinosaurs in his crusade to defend paleontology from the progressive evolutionists, who claimed species naturally transmuted into more advanced forms. If dinosaurs were, as Hawkins constructed them, grandiose Mesozoic overlords, this would imply that there had been no progression. (p. 12)

Before the park was opened to the public, a New Year's dinner party was held on December 31, 1853. Its setting was unique because 12 of the 22 guests

FIGURE 2 The savants dining inside the unfinished *Iguanodon* (from *The Illustrated London News*, 1854).

actually dined inside the mould of the still-uncompleted *Iguanodon*, with the others at a table alongside. Spalding (1993) reports,

> Owen presided at the head of the table, which was appropriately located inside the head of the animal. Edward Forbes (1815–54; Professor of Natural History at the Museum of Practical Geology) wrote a song for the occasion, which the company sang with gusto.
>
> Thus the eminent cavorted, joked, and sang, and newspapers reported that their noise could be heard across the park. The humorous magazine *Punch* commented solemnly, in an article called "Fun in a Fossil," that "if it had been an earlier geological period they might perhaps have occupied the *Iguanodon*'s inside without having any dinner there." (pp. 66–67)

It is likely that the inspiration for this event came from Hawkins, whose father had attended a banquet hosted in 1802 by Charles Willson Peale inside the skeleton of a *Mastodon* discovered in New York State (Debus, 1993, p. 12). The park was officially opened on June 10, 1854, to an estimated 40,000 visitors. The dinosaurs—set on islands in the lake—excited much attention because they afforded the first demonstration to the public at large of how awesomely huge those creatures were. This may well have been the true beginning of "dinomania" (see Torrens, 1993, p. 277).

The exhibit established Hawkins's reputation and brought many further opportunities for illustrating extinct creatures. Soon his work was becoming even better known through his illustrations in books and on educational wall-charts. Eventually he was invited to the United States, where he studied vertebrate remains in several major museums, making casts of dinosaur bones in Philadelphia and being invited to develop a series of dinosaur restorations for a proposed "Palaeozoic Museum" in Central Park, New York. Unfortunately, this project fell victim to the machinations of politicians (Colbert, 1959; Desmond, 1974); also, none of the casts of American vertebrates prepared by Hawkins are known to survive (Debus,

1993, pp. 11, 18). However, 15 (of an original 17) mural paintings by Hawkins of the life of past geological epochs are still to be seen in Guyot Hall, Princeton University.

Although Sir Joseph Paxton's project for the geological illustrations was never quite carried to completion (Doyle and Robinson, 1993) and the Crystal Palace itself burned down in 1936, Waterhouse Hawkins's restorations still survive in the park in Sydenham. Somewhat oddly, they have been scheduled by the National Trust as "grade II listed buildings" and are protected against vandalism (McCarthy and Gilbert, 1994). They serve as a visible reminder of the earliest attempts at three-dimensional restoration of dinosaurs.

See also the following related entry:

HISTORY OF DINOSAUR DISCOVERIES

References

Anonymous. (1893). *Crystal Palace: Illustrated Guide to the Palace and Park.* Dickens and Evans, London.

Colbert, E. H. (1959). The Palaeozoic Museum in Central Park, Or the Museum that Never Was. *Curator* **2,** 137–150.

Debus, A. A. (1993). The first great paleo-artist: Benjamin Waterhouse Hawkins. *Earth Sci. News Bull. Earth Sci. Club Northern Illinois* **44**(September), 11–15; (October), 17–21.

Desmond, A. J. (1974). Central Park's fragile dinosaurs. *Nat. History* **63,** 64–71.

Desmond, A. J. (1979). Designing the dinosaur: Richard Owen's response to Robert Edmond Grant. *Isis* **70,** 224–234.

Doyle, P. (1994). Crystal Palace revisited. *Ethical Rec.* **99,** 7–8.

Doyle, P., and Robinson, E. (1993). The Victorian 'Geological Illustrations' of Crystal Palace Park. *Proc. Geol. Assoc.* **104,** 181–194.

Hawkins, B. W. (1854). On visual education as applied to geology. *J. Soc. Arts* **2,** 444–449.

Illustrated London News (1854). January 7, p. 22.

McCarthy, S. and Gilbert, M. (1994). *The Crystal Palace Dinosaurs . . . The Story of the World's First Prehistoric Sculptures.* (London: Crystal Palace Foundation), 99 p.

Spalding, D. A. E. (1993). *Dinosaur Hunters*, pp. ix, 310. Key Porter, Toronto. [Republ. by Prima, Rocklin, CA, 1995].

Tagart, E. (1846). On markings in the Hastings Sands near Hastings, supposed to be the footprints of birds. *Q. J. Geol. Soc. London* **2,** 262.

Cultural Impact of Dinosaurs

see POPULAR CULTURE, LITERATURE

Dakota Dinosaur Museum, North Dakota, USA

see Museums and Displays

Dallas Museum of Natural History, Texas, USA

see Museums and Displays

Dalton Wells Quarry

Brooks B. Britt
Museum of Western Colorado
Grand Junction, Colorado, USA

Kenneth L. Stadtman
Brigham Young University
Provo, Utah, USA

The Dalton Wells dinosaur quarry is an extraordinarily rich deposit of fossil bone—rich not only in the number of bones but also in the number of dinosaurian taxa represented. Most of the fauna is undescribed but is currently being studied by the authors. Stratigraphically, the quarry is at the base of the Cedar Mountain Formation and is located in east-central Utah, near the town of Moab. The site is particularly important to paleontologists because it provides a window into the little known world of Early Cretaceous dinosaurs in North America.

Casual collectors have known about this site for more than 50 years, and probably at least since the 1930s when a Civilian Conservation Corps camp was constructed less than 0.5 km from the bone-bearing layer. Mr. J. Leroy "Pop" Kay showed the site to James A. Jensen of Brigham Young University in the early 1960s when the site was thought to be within the Late Jurassic Morrison Formation. It was not until Lyn Ottinger discovered a partial, tooth-bearing iguanodontid maxilla that the significance of the site was recognized. Galton and Jensen (1979) made the maxilla the type of *Iguanodon ottingeri* and recognized that the fauna was Early Cretaceous in age.

The deposit is a bone bed approximately 0.3 km long, consisting mainly of disarticulated elements of many animals, including juveniles and adults. The bones of several individuals, however, occur in clusters. Articulated elements are rare but include, significantly, a sauropod cranium with three cervical vertebrae. The bone bed varies from the thickness of a single bone to about 0.75 m. The matrix is a silty mudstone with occasional fine-grained chert pebbles, and the bone-bearing horizon is virtually devoid of internal sedimentary structures. The bone bed is tentatively interpreted to have been deposited in a volcanic ash-choked stream. The rate of flow was such that even large sauropod vertebrae were often severely broken. Medium-bedded fluvial sandstones overlie the quarry unit, which in turn are capped by mudcracked, limy mudstone, preserving sauropod and ornithopod footprints made along the shore of a small, Early Cretaceous lake.

An analysis of the Dalton Wells fauna is still in the early stages, but with nearly 1000 prepared elements on hand six dinosaurian genera are recognized. Two genera from the locality have been described, the dromaeosaurid theropod *Utahraptor* (Kirkland *et al.*, 1993) and *Iguanodon ottingeri* (Galton and Jensen, 1975, 1979). Bones of an ornithomimid theropod have been recovered and are currently being described. Two sauropod genera are present, a titanosaurid and a possible camarasaurid. These represent, respectively, the earliest and latest occurrences of these clades in North America. The titanosaurid is recognized on the basis of strongly procoeleous caudal vertebrae, dorsal vertebrae with reclined neural spines, and ulnae with short but robust olecrenon processes. *Iguanodon ottingeri* is regarded as a *nomen dubium* by Norman and Weishampel (1991), but at least one iguanodontid genus is present in the fauna. It is characterized by tall neural spines similar to those of *Ouranosaurus*. The Dalton Wells ornithopod is large, with an estimated length of 8 meters. A small nodosaur is also present and appears to be the same genus as the undescribed nodosaur found at the nearby Gaston Quarry. The only nondinosaurian taxon is a turtle, represented by a single carapace fragment.

See also the following related entries:

CEDAR MOUNTAIN FORMATION • CRETACEOUS PERIOD

References

Galton, P. M., and Jensen, J. A. (1975). *Hypsilophodon* and *Iguanodon* from the Lower Cretaceous of North America. *Nature.*

Galton, P. M., and Jensen, J. A. (1979). Remains of ornithopod dinosaurs from the Lower Cretaceous of Colorado. *Brigham Young Univ. Geol. Stud.*, **25**, 1–10.

Kirkland, J. I., Burge, D., and Gaston, R. (1993). A large dromaeosaur (Theropoda) from the Lower Cretaceous of eastern Utah. *Hunteria* **2**(10), 1–16.

Norman, D. B., and Weishampel, D. B. (1991). Iguanodontidae and related ornithopods. In *The Dinosauria* (D. B. Weishampel, P. Dodson, and H. Osmolska, Eds.), pp. 510–533. Univ. of California Press, Berkeley.

Dashanpu

see ZIGONG MUSEUM

Dating Techniques

see BIOSTRATIGRAPHY; PALEOMAGNETIC CORRELATION; RADIOMETRIC DATING

Dayton Public Library Museum, Ohio, USA

see MUSEUMS AND DISPLAYS

Deccan Basalt

ASHOK SAHNI
Punjab University
Chandigarh, India

The Deccan Traps constitute one of the most extensive continental flood basalt provinces of the Phanerozoic and have now been radiometrically shown to lie at the Cretaceous–Tertiary boundary. The basaltic flows are intercalated at the base with thin, highly fossiliferous sedimentary horizons that have yielded a diverse biota including mammals, dinosaurs, lower microvertebrates, and freshwater flora. Along the western coast of India, the exposed thickness of the Deccan Traps exceeds 2 km, but it gradually thins to the east. The Deccan Traps have been subdivided into three subgroups composed of compound and sheet flows. Magnetostratigraphy suggests a N-R-N sequence for the composite flow, with the majority of flows showing reverse polarity, probably corresponding to the 29R Chron. Deccan volcanism is considered to be one possible cause of the mass extinctions at the end of the Cretaceous.

Deinonychosauria

Gauthier (1986) cladistically defined and diagnosed Colbert and Russell's (1969) term Deinonychosauria, including in it DROMAEOSAURIDAE and TROODONTIDAE. Gauthier's definition was taxon based and so requires a stem- or node-based definition in the PHYLOGENETIC SYSTEM. Because Troodontidae has been determined to be closer to Ornithomimidae than to Dromaeosauridae (Holtz, 1994), the latter taxon carries approximately the same information as Deinonychosauria. Accordingly, it is appropriate to redefine Deinonychosauria as a stem-based taxon comprising all maniraptorans closer to *Deinonychus* than to birds; its sister taxon is AVIALAE.

References

Colbert, E. H., and Russell, D. A. (1969). The small Cretaceous dinosaur *Dromaeosaurus*. *Am. Museum Novitates* **2380,** 1–49.

Gauthier, J. A. (1986). Saurischian monophyly and the origin of birds. *Mem. California Acad. Sci.* **8,** 1–55.

Holtz, T. R., Jr. (1994). The phylogenetic position of the Tyrannosauridae: Implications for theropod systematics. *J. Paleontol.* **68,** 1100–1117.

Denver Museum of Natural History

KENNETH CARPENTER
Denver Museum of Natural History
Denver, Colorado, USA

Located in Denver, Colorado, the Denver Museum of Natural History has had a long but sporadic history of collecting dinosaurs. The museum's research collection is modest, holding only approximately 300 dinosaur specimens, mostly from the Morrison and the Hell Creek formations. Dinosaurs on exhibit include articulated skeletons of the small dinosaur *Coelophysis*, *Stegosaurus*, and *Allosaurus* (Fig. 1), *Diplodocus*, the skull of *Brachiosaurus* from Colorado, five juvenile skeletons of the ornithopod *Othnielia*, and a skeleton of the hadrosaur *Edmontosaurus* that shows a partially healed injury from the attack of a *Tyrannosaurus*. In addition, there is a walk-through diorama showing two life-sized male *Stygimoloch* pachycephalosaurs fighting for a female. Current research by the museum is concentrated on the vertebrate fauna of the Morrison Formation, especially the dinosaurs, at Cañon City.

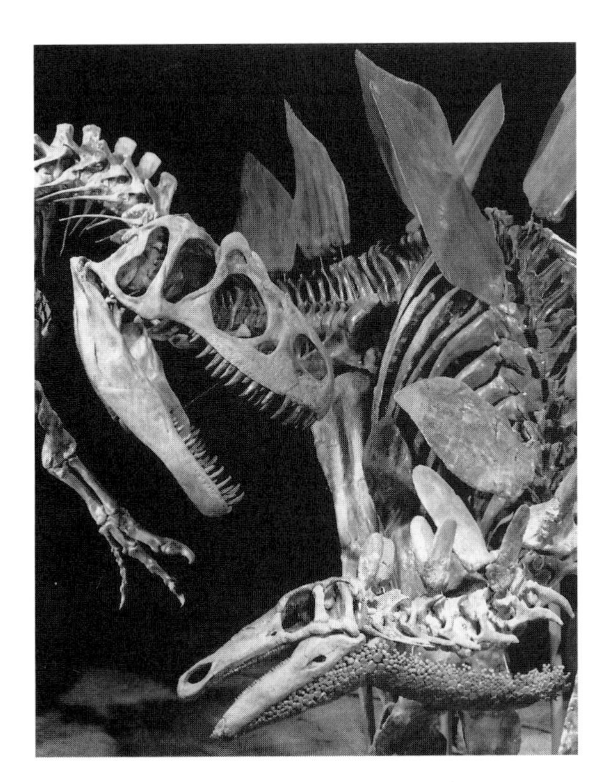

FIGURE 1 The skeletons of *Allosaurus* and *Stegosaurus* on exhibition at the Denver Museum of Natural History. Photo by Rick Wicker, courtesy of the Denver Museum of Natural History.

See also the following related entries:

CAÑON CITY • HELL CREEK FORMATION • MORRISON FORMATION

Department of Geology, University of Buffalo, New York, USA

see MUSEUMS AND DISPLAYS

Department of Geology, University of Texas at El Paso, Texas, USA

see MUSEUMS AND DISPLAYS

Devil's Canyon scovery Center, Fruita, Colorado, USA

see MUSEUMS AND DISPLAYS

Devil's Coulee Dinosaur Egg Historic Site

CLIVE COY
Royal Tyrrell Museum of Palaeontology
Drumheller, Alberta, Canada

Devil's Coulee is an unimposing pocket of Badlands that derives its name from a trident-shaped drainage system that ultimately feeds the Milk River in the southwestern corner of Alberta, Canada. In 1987, dinosaur embryos were discovered in eggs at Devil's Coulee by staff of the Royal Tyrrell Museum of Palaeontology. The embryonic skeletons found together in the nest were described as a new type of hadrosaurian dinosaur called *Hypacrosaurus stebingeri* (Horner and Currie, 1994). Devil's Coulee was the first dinosaur nesting site to be discovered in Canada and only the second in North America. Eggs and beautifully preserved embryonic skeletal material have now been found at numerous sites in Devil's Coulee. Six types of eggshell have been identified, although only two of these are represented by nests of eggs.

Although many dinosaur skeletons have been collected from Alberta during the past 100 years, they provide little evidence about the birth and development of dinosaurs. Study of the sediments of Devil's Coulee is assisting in our understanding of the environment in which these animals nested and providing clues about the season during which nesting occurred. Examination of the nests and nesting behavior of these dinosaurs may eventually provide insight into the yearly routines of these animals. Speculations can be made regarding the animals living in groups, possibly moving as herds. As modern birds do today, dinosaurs appeared to have nested in great numbers in close proximity to each other in an attempt to simply overwhelm any predators. Three-dimensional preservation of the eggs guides our understanding of dinosaur egg physiology and indicates that not all the eggs were fertile. The excellent skeletal remains at Devil's Coulee show a range of skeletons from embryos to nestlings. This sample of growth in a single species may indicate how often the dinosaurs reproduced, how long it took the young to develop, and what kinds of stresses (disease and predation) affected populations of duck-billed dinosaurs.

Devil's Coulee was purchased from the landowner by the government of Alberta and has been designated a protected historical resource, ensuring its safety for future generations. The Royal Tyrrell Museum has a continuing field excavation project at Devil's Coulee and has collected hundreds of specimens there, including teeth of half a dozen dinosaur taxa, two new mammals, a new bird, turtles, and amphibians.

See also the following related entries:

EGGS, EGGSHELLS, AND NESTS • JUDITH RIVER WEDGE

References

Horner, J. R., and Currie, P. J. (1994). Embryonic and neonatal morphology and ontogeny of a new species of *Hypacrosaurus* (Ornithischia, Lambeosauridae) from Montana and Alberta. In *Dinosaur Eggs and Babies* (K. Carpenter, K. F. Hirsch, and J. R. Horner, Eds.), pp. 312–336. Cambridge Univ. Press, Cambridge, UK.

Diet

MICHAEL J. RYAN
MATTHEW K. VICKARYOUS
Royal Tyrrell Museum of Palaeontology
Drumheller, Alberta, Canada

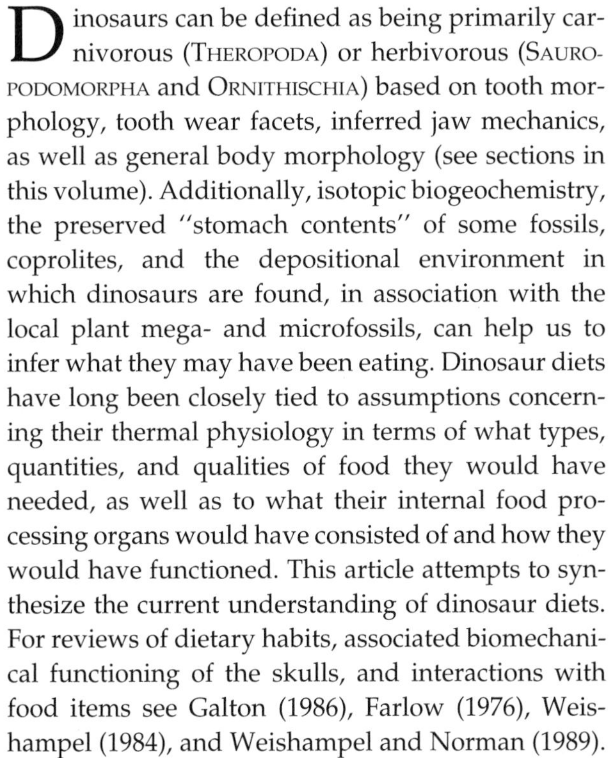

Dinosaurs can be defined as being primarily carnivorous (THEROPODA) or herbivorous (SAUROPODOMORPHA and ORNITHISCHIA) based on tooth morphology, tooth wear facets, inferred jaw mechanics, as well as general body morphology (see sections in this volume). Additionally, isotopic biogeochemistry, the preserved "stomach contents" of some fossils, coprolites, and the depositional environment in which dinosaurs are found, in association with the local plant mega- and microfossils, can help us to infer what they may have been eating. Dinosaur diets have long been closely tied to assumptions concerning their thermal physiology in terms of what types, quantities, and qualities of food they would have needed, as well as to what their internal food processing organs would have consisted of and how they would have functioned. This article attempts to synthesize the current understanding of dinosaur diets. For reviews of dietary habits, associated biomechanical functioning of the skulls, and interactions with food items see Galton (1986), Farlow (1976), Weishampel (1984), and Weishampel and Norman (1989).

Living carnivores and herbivores have selected a wide variety of food items to meet their dietary needs which occasionally seem inconsistent with their skeletal characteristics (e.g., herbivorous pandas). In addition to the demands of specific physiologies, diets can be at least partially determined by age, social organization, habitat preference, and the availability of food, all of which are not easily determined from the fossil record. Palaeontologists parsimoniously assume that the most common food item associated with a dinosaur will be its most probable dietary food. However, we know from living animals that some taxa can be very selective of their source of food and may even go for long periods of time without eating, even in the presence of edible items (e.g., many constrictor snakes and whales). Thus, all speculations about dinosaurian diets should be taken with caution.

Dinosaurs evolved from a carnivorous ornithosuchid ancestor sometime during the Early Triassic. By the time the first dinosaurs are recognized in the fossil record the two major dietary types appear to have been well established. *Eoraptor lunensis*, from the Ischigualasto Formation (Upper Triassic–middle Carnian), is perhaps the oldest known dinosaur and the basal member of the Theropoda. Its unique heterodont dentition has serrated, recurved crowns typical of theropods in the maxillae, whereas the lower teeth are leaf-shaped resembling those seen in basal sauropodomorphs (Sereno *et al.*, 1993). *Pisanosaurus mertii* (Bonaparte, 1976), from the same formation, is the oldest known ornithischian. Its closely packed teeth suggest herbivorous habits.

Theropoda

The generalized dinosaurian carnivore is a bipedal theropod with powerful legs designed for running, long forearms (reduced in tyrannosaurids) designed for some type of prey manipulation, and jaws with a wide gape and a large number of laterally compressed teeth. The teeth typically have serrations (denticles) on mesial and distal carinae. All carnosaurs appear to have evolved as active predators that could locate, track, and capture the appropriate

Hunting styles and the size of prey items that theropods utilized have long been debated in the literature and will continue to be, probably without resolution. Late Cretaceous tyrannosaurids were either the same size or larger than their prey items, whereas allosaurs from the Jurassic were in some cases 10 times smaller than adult sauropods. If they were limited to prey items of equal size or smaller, the tyrannosaurs could have made a healthy living off of hadrosaurs and ceratopsians, but the allosaurs either fed on immature sauropods, stegosaurs, or campotosaurs or developed a method to bring down the massive sauropods. Auffenberg (1981) has suggested that the jaws of *Allosaurus* functioned like the canines of sabretooth cats and that these dinosaurs used a "hit-and-run,'' possibly pack-style, attack that allowed them to inflict deep wounds and then quickly withdraw and wait for their prey to succumb to its wounds. Trackway evidence suggests that some theropods hunted prey (sauropods) in packs. Large carnivores, such as *Giganotosaurus* or *Tyrannosaurus rex*, would have been well suited for active predation but they may also have intimidated other carnivores out of their actively hunted prey and subsisted by scavenging. In addition to ''hit-and-run'' attacks, some theropods may have suffocated their prey by clamping their jaws onto their necks or practiced ambush-style attacks.

Most large theropods would have had powerful mandibular adductors providing for a strong bite. Erickson *et al.* (1996) reported on a *Triceratops* pelvis with multiple bite and puncture marks attributed to *T. rex*. Their research concluded that *T. rex* had the greatest bite force of any animal measured to date (up to 13,400 N) and suggested that *T. rex* had strong, impact-resistant teeth that could regularly puncture the bones of prey. Broken teeth in dinosaurs would not have caused long-term problems because dinosaurs demonstrate a pattern of continual replacement. Tooth wear on large theropod teeth and tooth marks on a variety of dinosaur bones (see Jacobson, this volume) indicates that these dinosaurs regularly tore the flesh off carcasses. Shed theropod teeth associated with skeletons and bone beds suggests that carnivorous dinosaurs were feeding at these sites. Tooth shape in *Carcharodontosaurus* and tooth and jaw shape in *Baryonyx* and *Spinosaurus* suggests that these dinosaurs may have been piscivorous.

Almost all theropods have hind limbs that are similar in shape and function. The femur is usually shorter than the tibia and designed for rapid running. The feet are functionally three-toed (except in therizinosaurids) and equipped with terminal claws that could have assisted in the killing and/or dismembering of prey. DROMAEOSAURS were small maniraptorans with a highly derived pedal digit II designed for hypertension and that terminated in an enlarged trenchant ungual. This claw seems to be designed to disembowel prey when the dromaeosaurid was either balanced on one foot or possibly leaping through the air. The forelimbs in CERATOSAURS, COELUROSAURS, and MANIRAPTORANS are strong and relatively long in order to assist in grappling prey. Maniraptorans probably possessed limited pronation/supination when the wrist was flexed or extended, allowing them at least to manipulate their food. One can visualize these theropods leaping at a large prey item, grasping hold with both hands, and slashing away with their raptorial claws. If these animals hunted in packs, as has been suggested by some authors (Ostrom, 1969), then no prey item may have been too large for them. Russell and Séguin (1982) have suggested that *Troodon formosus* may have been crepuscular, preying on small, nocturnal mammals or lizards.

Juvenile theropods probably preyed on different animals than they would as adults and may also have utilized different hunting strategies (e.g., pack hunting in juveniles vs solitary hunting in adults). The juveniles were more gracile than adults, and probably were fast, energetic hunters regardless of the hunting styles they may have adopted as adults. Invertebrates and small vertebrates may have constituted a large part of their diet (Farlow, 1976).

Some theropods (Ornithomimosauria and Oviraptorosauria) were secondarily edentulous; the jaws were covered in a horny beak (now known from an ornithomimid; P. Currie, personal communication, 1996). The jaws of ornithomimids are considered to have been relatively weak, suggesting that their diet consisted of soft items such as insects or small mammals. Oviraptorids have jaws that are powerfully built and connected to an akinetic skull, suggesting an extremely strong bite force. They probably fed on small lizards and mammals as well as insects and eggs (Fig. 1).

FIGURE 1 *Oviraptor mongoliensis* (© S. R. Bissette, 1996).

FIGURE 2 *Therizinosaurus*, a therizinosaurid from Asia (© S. R. Bissette, 1996).

Stomach contents are rarely preserved in theropods but are known for *Compsognathus* (Ostrom, 1978), containing the small lepidosaur *Bavarisaurus*, and small fragments of bone mesial to the gastralia of *Syntarus rhodesiensis* (Raath, 1969). A new, feathered compsognathid dinosaur from China also has a small lizard preserved in its stomach cavity (P. Currie, personal communication, 1996). Cannibalism has been reported for *Coelophysis* from the bone beds of Ghost Ranch, New Mexico.

The Therizinosauridae (= Segnosauridae) (Fig. 2) are a unique group of theropods with long arms and greatly elongated manual unguals. Various authors have suggested that these dinosaurs were piscivorous or even herbivorous.

Sauropodomorphs and Ornithischians

All sauropodomorphs and all ornithischians are believed to have been primarily or exclusively herbivorous. These dinosaurs can be divided broadly into two groups, gut processors and mouth processors. Gut processors are characterized by simple dentitions and are assumed to have had modified guts for the digestion of high-fiber, low-nutrition plants. Some appear to have had gastroliths to assist in the breakdown of plant matter. Fermentation in the gut would have further broken down the fibrous plant material and produced nutrients that subsequently were absorbed by the host. Mouth processors have modified dentitions and/or jaw structure/mechanics that allowed for active grinding or pulping of low-fiber, high-nutrition foods. Prosauropods and most ornithischians are believed to have possessed cheeks that allowed them to contain food in the mouth while they processed it.

When dinosaurs first appeared in the Middle to Late Triassic, the flora still contained some Late Paleozoic members (e.g., herbaceous lycopods, large arborescent and small herbaceous horsetails, and possibly some conifers such as *Brachyphyllum*) but was starting to take on the more gymnosperm-dominated aspect typical of late Mesozoic. These plants included a variety of ferns, true cycads, cycadeoids, ginkgos, conifers, and the seed fern order Caytoniales. These groups dominated the Jurassic and Early Cretaceous, varying in diversity and numbers depending on latitude and the local environment. By the Middle Jurassic such modern conifer families as Araucariaceae, Pinaceae, and Taxodiaxeae had appeared. Throughout the Jurassic and the Early Cretaceous, the sauropods were the dominant herbivores. Large groups of sauropods would have consumed immense quantities of plant matter and substantially impacted local environments. Some authors (e.g., Bakker, 1978) have suggested that by clearing large amounts of upperstory foliage, the underbrush was opened up to be exploited by the fast-growing, "weedy" angiosperms, which then quickly evolved and out-competed their gymnosperm relatives. By the Early Cretaceous (Valanginian) angiosperm pollen is present in the fossil record. The dramatic increase in angiosperm floras from the Barremian onward came at the expense of the cycads, cycadeoids, and the seed ferns. The evolution of the Late Cretaceous angiosperm-dominated forests at least parallels the evolution of the primary Late Cretaceous herbivores, and the Cerapoda (specifically the ornithopods and the ceratopsians) effectively replaced the sauropods in the Northern Hemisphere (see Basinger, this volume, for a complete review of Mesozoic floras).

The Prosauropoda first appeared in the Carnian and represent the first radiation of herbivorous dinosaurs. Prosauropods and the giant sauropods that replaced them in the Late Jurassic are considered to have been primarily gut processors, and gastroliths are known from *Massospondylus*, *Sellosaurus*, and *Seismosaurus*. The skulls are lightly built with a relatively weak jaw musculature. Prosauropod teeth are narrow, subconical, and closely spaced, showing little in the way of wear facets, indicating that the teeth did not regularly occlude. This suggests that they lived on relatively soft plants. A variety of sauropod families have teeth with distinctive wear patterns (see Fiorillo and Weishampel, TOOTH WEAR, this volume) indicating that different plant groups may have been utilized by different sauropods.

Prosauropods show a modest lengthening of the cervical vertebrae that would have allowed them to reach plants as high as 3 m, making them the dominant high-browsing animals of the Triassic. They may have fed on lycopsid fructifications (Weishampel, 1984). Likewise, the later sauropods exhibit extreme elongation of the neck and were designed to be high browsers. Some sauropods, such as the diplodocids *Apatosaurus* and *Diplodocus*, may have been able to achieve a tripodal stance using the long tail as a counterbalance to reach plant material more than 18 m above the ground (Bakker, 1971). Sauropod food groups may have included ferns, ginkgophytes, conifers, nilssonian fructifications, and possibly Czekanowskiales and Caytoniales. Late Cretaceous forms may have also fed on angiosperm fructifications (Weishampel, 1984).

The Ornithischia show a progressive increase in the complexity of the dentition and associated jaw mechanics from their first appearance in the Late Triassic until the end of the Cretaceous. All ornithischians appear to utilize some degree of oral processing. Basal ornithischians such as *Pisanosaurus* have closely packed teeth showing continuous wear, foreshadowing the dental batteries of more derived ornithischians. These early herbivores would have foraged on low-level ground cover. Their apparent lack of gastroliths suggests that these herbivores may have utilized food with lower fiber than the gut processors.

The Thyreophora includes ankylosaurs, stegosaurs, and their basal relatives *Scelidosaurus*, *Scutellosaurus*, and *Emausaurus*. Like the prosauropods, which they functionally replaced, and the sauropods, thyreophorans were mostly obligatory quadrupeds with simple spatulate dentition and were probably primarily gut processors. Other than in the basal *Scutellosaurus*, most thyreophorans show evidence of tooth wear from imprecise occlusion. Oral processing was likely limited to slicing and/or puncturing and crushing. Diet was probably restricted to relatively low-lying plants (e.g., 1 or 2 m above the ground). Most authors believe that these dinosaurs fed on nonabrasive, "soft" plants that may have included the fleshy components of bennettitalian, nilsonnialian (Jurassic), caytonialian inflorescences (Jurassic to Cretaceous), and angiosperm fructifications (Cretaceous) (Weishampel, 1984). Among the Marginocephalia, the pachycephalosaurs and, to a lesser degree, the psittacosaurs had dentitions very similar to those of thyreophorans and probably also relied on gut processing to a high degree. However, their bipedal lifestyle would have allowed them the ability to move more rapidly across the landscape, not unlike modern deer, perhaps in search of different, higher quality foods.

Ornithopods were the first group of herbivores to develop a transverse chewing stroke (side-to-side grinding), either by the slight rotation of the lower jaw as in heterodontosaurids or through the rotation of the upper jaw via pleurokinesis as in hypsilophodontids, iguanodontids, and hadrosaurids (Weishampel and Norman, 1989). This progression from simple isognathy (bilateral occlusion), seen in basal forms, to an increasingly more complex series of jaw mechanics and tooth/tooth row structures in the more advanced Iguanodontia allowed for more extensive processing of high-quality, low-fiber food. The basal ornithopods (Heterodontosauridae, Hypsilophodontidae, *Tenontosaurus*, and Dryosauridae) tended to be relatively small and were probably browsers of low (≤2 m and under) ground cover. Typically, the rostral portion of their premaxilla was narrow, edentulous, and covered by a cornified rhamphotheca suggesting that these dinosaurs selectively cropped their food. The more derived Euornithopoda (Hadrosauridae and Iguanodontidae) show a more advanced adaptation to herbivory. Their teeth are generally narrow and buccally lanceolate (maxillary) or leaf-shaped (dentary). Their diet was probably a mix of gymnosperms and the proliferating late Mesozoic angiosperms.

All members of the Hadrosauridae are characterized by having broad, edentulous, "duck-billed beaks" and a complex maxillary/dentary dentition organized into batteries of up to several hundred closely packed teeth. Dental batteries formed a single wear facet at the point of occlusion. Hadrosaur teeth differ from those of other ornithopods: they are taller than wide and have lancolate crowns bearing heavy enamel, required for efficiently grinding plant matter. Hadrosaurs appear to have been well suited to living off low-quality, high-fiber vegetation (Weishampel and Norman, 1989). Some authors have argued that some hadrosaurs invested in parental care of their young and may have brought food or provided regurgitate for nest-bound young.

Several hadrosaurs have been collected that have putative stomach contents (see Currie *et al.*, 1995) located in the thoracic cavity. The best studied consists of 1- to 4-cm-long sections of 5- to 7-year-old twigs from angiosperms and gymnosperms as well as seeds and seed pods. Although this material would be consistent with food processed through the jaws of a large hadrosaur, the authors could not rule out the possibility that this material was washed in after death.

All members of the Ceratopsia (Psittacosauridae, Protoceratopsidae, and Ceratopsidae) have a mouth that terminates in an edentulous parrot-like beak formed from the rostral and predentary bones. The surface of each would have been cornified and formed a sharp cutting surface. All ceratopsians have leaf-shaped teeth in both maxillae and dentaries. Psittacosaurs have a single tooth row of low, leaf-shaped teeth characterized by broad planar wear surfaces with self-sharpening cutting edges. Polished gastroliths have been associated with some skeletons, suggesting that gastric mills may have played some role in processing food.

Although the protoceratopsids have only a single replacement tooth present per position, the Ceratopsidae have a dental battery similar to, but less extensive than, that of hadrosaurs. Both protoceratopsids and ceratopsids have vertically inclined wear facets on the teeth, indicating that during chewing the power stroke was restricted to orthal slicing movements. Ceratopsians may have browsed at a level of approximately 2 m and under. Their sharp beaks appear to have been well adapted to cropping off small trees and even processing them whole. Their large guts may have housed large fermentary engines. Palynological evidence suggests that *Triceratops* may have subsisted on the small herbivorous plant, *Gunnera* (Rich, 1996). Isotopic (C^{13} and N^{15}) examinations of high-molecular-weight material isolated from fossils have recently been used to infer possible diets for some dinosaurs. Isotopic work by Ostrom *et al.* (1990) suggests that some ceratopsids may have been omnivorous.

Coprolites have been attributed to a variety of dinosaurs, but their use in determining dinosaur dietary behaviour has been limited to date. Coprolites attributed to herbivores can be used to infer dietary fiber content and food quality based on the presence or absence of various plant tissues. The presence of bone fragments, scales, or teeth can indicate carnivorous excrement.

A coprolite from the Maastrichtian Frenchman Formation of Saskatchewan with bone fragment inclusions from what appears to be a sub-adult ornithischian has been attributed to *T. rex* based in part on its large size.

See also the following related entries:

COPROLITES • GASTROLITHS • TEETH AND JAWS • TOOTH REPLACEMENT • TOOTH SERRATIONS IN CARNIVOROUS DINOSAURS • TOOTH WEAR

References

Auffenberg, W. (1981). *The Behavioral Ecology of the Komodo Monitor*, pp. 406. Univ. Presses of Florida, Gainesville.

Bakker, R. T. (1978). Dinosaur feeding behavior and the origin of flowering plants. *Nature* **274,** 661–663.

Bonaparte, J. F. (1976). *Pisanosaurus mertii* Casamiquela and the origin of the Ornithischia. *J. Paleontol.* **50,** 808–820.

Currie, P. J., Koppelhus, E. B., and Muhammad, A. F. (1995). "Stomach" contents of a hadrosaur from the Dinosaur Park Formation (Campanian, Upper Cretaceous) of Alberta, Canada. In *Sixth Symposium on Mesozoic Terrestrial Ecosystems and Biota, Short Papers* (A. Sun and Y. Wang, Ed.), pp. 111–114. Beijing, China.

Erickson, G. M., Van Kirk, S. D., Su, J., Levenston, M. C., Caler, W. E., and Carter, D. R. (1996). Bite-force estimation for *Tyrannosaurus rex* from tooth-marked bones. *Nature* **382,** 706–708.

Farlow, J. O. (1976). Speculations about the diet and foraging behavior of large carnivorous dinosaurs. *Am. Midland Nat.* **95,** 186–191.

Galton, P. M. (1986). Herbivorous adaptations of Late Triassic and Early Jurassic dinosaurs. In *The Beginning of the Age of Dinosaurs* (K. Padian, Ed.), pp. 203–221. Cambridge Univ. Press, New York.

Ostrom, J. H. (1969). Osteology of *Deinonychus antirrhopus*, an unusual theropod from the Lower Cretaceous of Montana. *Bull. Peabody Nat. History Museum* **30,** 1–165.

Ostrom, J. H. (1978). The osteology of *Compsognathus longipes* Wagner. *Zitteliana* **4,** 73–118.

Ostrom, P. H., Macko, S. A., Engel, M. H., Silfer, J. A., and Russell, D. A. (1990). Geochemical characterization of high molecular weight material isolated from Late Cretaceous fossils. *Org. Geochem.* **16,** 1139–1144.

Raath, M. A. (1969). A new coelosaurian dinosaur form the Forest Sandstone of Rhodesia. *Arnoldia* **4,** 1–25.

Rich, F. J. (1996). Palynological interpretation of matrix associated with a *Triceratops* burial site, Lance Creek Area, Wyoming. *Geol. Soc. Am.* **28**(4), 36. [Abstracts with programs]

Russell, D. A., and Séguin, R. (1982). Reconstruction of the small Cretaceous theropod *Stenonychosaurus inequalis* and a hypothetical dinosauroid. *Syllogeus* **37,** 1–43.

Sereno, P. C., Forster, C. A., Rogers, R. R., and Monetta, A. M. (1993). Primitive dinosaur skeleton from Argentina and the early evolution of Dinosauria. *Nature* **361,** 64–66.

Weishampel, D. B. (1984). Interactions between Mesozoic plants and vertebrates: Fructifications and seed predation. *Neues Jahrbuch Geol. Paläontol. Abhandlungen* **167,** 224–250.

Weishampel, D. B., and Norman, D. B. (1989). Vertebrate herbivory in the Mesozoic: Jaws, plants, and evolutionary metrics [Special paper]. *Geol. Soc. Am.* **238,** 87–100.

Dimorphism

see BEHAVIOR; VARIATION

Dinosaur Discovery Center, Cañon City, Colorado, USA

see MUSEUMS AND DISPLAYS

Dinosaur Eggs

see EGGS, EGGSHELLS, AND NESTS

Dinosaur Extinction

see EXTINCTION, CRETACEOUS; EXTINCTION, TRIASSIC

Dinosaur Footprints

see FOOTPRINTS AND TRACKWAYS; MEGATRACKSITES

Dinosaur Growth

see GROWTH AND EMBRYOLOGY

Dinosauria: Definition

KEVIN PADIAN
University of California
Berkeley, California, USA

The first use of the word "dinosaur" was in 1842, when the great British comparative anatomist and paleontologist, Richard Owen (Fig. 1), applied it to three partially known but impressively large fossil reptiles from the English countryside (see HISTORY: EARLY DISCOVERIES). The Dinosauria were inaugurated in a published version of a lecture on British fossil reptiles that Owen had given to the British Association for the Advancement of Science in Plymouth in August 1841. However, in the lecture, which went on for more than 2 hr, Owen evidently did not use the word Dinosauria because it was not reported in any account of the lecture (Torrens, 1992); its origin must be traced to the updated version that appeared in the report of the meeting, published in April 1842.

The exact date of the first use of the term Dinosauria is perhaps not as important as what it meant at the time and why Owen erected it. He based the Dinosauria on three previously named taxa: the large carnivore *Megalosaurus*, which Buckland had described in 1824; the ornithopod *Iguanodon*, the teeth of which were described by Mantell in 1822 and first recognized from a giant reptile in 1825; and the armored *Hylaeosaurus*, which Mantell had described in 1833 (Fig. 2). He did not include other material that could reasonably have been considered, such as *Palaeosaurus*, a tooth of uncertain origin (probably pseudosuchian), and *Thecodontosaurus*, known from a toothed lower jaw and perhaps some fragmentary material (now called "*Palaeosauriscus*"), both described by Riley and Stutchbury in 1836, as well as *Cetiosaurus*, which was described by Owen himself in 1842 (and at that time considered a kind of giant crocodile-like aquatic form with clawed, webbed feet and a swimming tail). He also did not consider any continental material, such as the nomenclatorially problematic taxon *Streptospondylus*, which at the time was based on both British and continental material (and which Owen considered a crocodile because the referred material included both crocodilian and dinosaurian remains); *Plateosaurus*, known from fragmentary remains described by Meyer in 1837; or *Poikilopleuron*, considered a megalosaurid by Eudes-Deslongchamps in 1838.

FIGURE 1 Portrait of Richard Owen, ca. 1858.

Owen erected the Dinosauria ("fearfully great lizards" as he translated the Greek) to receive these taxa because he recognized that they were completely distinct from other reptiles. They were large, but other fossil reptiles, including mosasaurs, plesiosaurs, ichthyosaurs, and some crocodiles, were also large; the dinosaurs, however, were terrestrial, not aquatic. Owen pointed to the five fused sacral vertebrae and the hips structured so that the animals demonstrably walked upright (Fig. 3; see PELVIS). These were not like living reptiles, and not only because of their size: They could not have sprawled. He also pointed to the height, breadth, and sculpturing of the dorsal neural arches; the two-headed ribs; the "broad and sometimes complicated coracoids and long and slender clavicles"; and the proportionally large but thin-walled limb bones that indicated terrestrial habits.

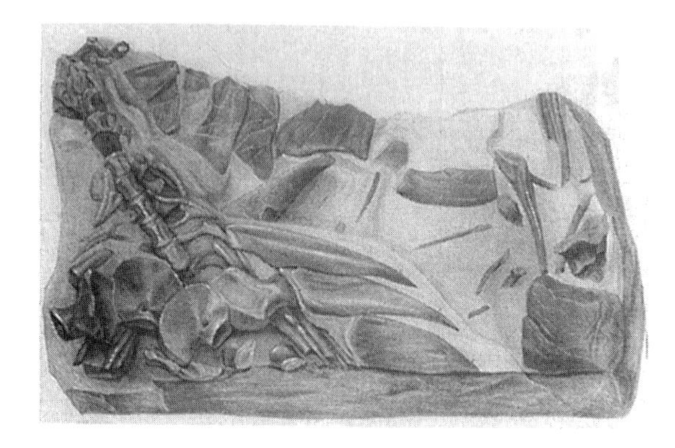

FIGURE 2 The jaw of *Megalosaurus*, part of the type specimen; Mantell's specimen of the Maidstone *Iguanodon*, on which he based his restoration; and the type specimen of *Hylaeosaurus* described by Mantell.

FIGURE 3 Owen's 1854 reconstruction of *Megalosaurus* as a quadruped. The lower jaw and other available skeletal bones are indicated.

Part of Owen's motivation in constructing this new taxon may not have been entirely taxonomic. In those early Victorian days, the concept of evolution had many meanings, and both materialistic and idealistic theories were proposed to explain it. The pattern of evolution, to many, meant continuous progress through time, from the earliest humble beginnings of life through the rise of the vertebrates to the ascent of mammals and their pinnacle, *Homo sapiens*. Owen did not accept progressivism because he knew from the fossil record that new forms continued to be produced, and life was not a ladder of ascending complexity. However, by showing that some extinct reptiles were more "advanced" structurally than living reptiles—that is, that they approached the mammalian and avian grades of organization—he could deny the validity of progressivism and its easy connection to materialistic transmutation, advocated by such scientists as Owen's rival at University College London, Robert Edmond Grant (Desmond, 1979).

Regardless of motivation, Owen's concept of the Dinosauria took hold, and the unveiling of Waterhouse Hawkins' statues of these dinosaurs at the CRYSTAL PALACE EXPOSITION in 1854 sealed their fixation in the minds of the public. Although almost no other new dinosaurs would be discovered in Britain until the 1970s, it will always be regarded as the birthplace of the dinosaurs. In the late 19th and early 20th centuries, more dinosaurs were discovered on the continent, including complete skeletons of *Iguanodon* in Belgium and of *Plateosaurus* in Germany. However, notably in the 1850s, the first dinosaurs in North America were discovered in New Jersey and soon after in Montana. By the 1880s, dinosaurs were well known from the western United States, and from both theropod and ornithopod remains it became clear that Owen was more correct than he had supposed: Many dinosaurs not only walked upright but also their short forelimbs proved that they walked bipedally—habits that had never been seen in a reptile (Desmond, 1975; see BIPEDALITY).

The concept of Dinosauria altered radically once again in 1887 and 1888 when Owen, now in his eighties and retired from the Natural History Museum that he had founded, was only a few years from his death. Harry Govier Seeley, a former student of Owen and an expert on both PTEROSAURS and anomodont dicynodont therapsids, surveyed the known skeletal material of dinosaurs and concluded that there were two consistent types of pelvic structure in this group. One type, which characterized a group that he named SAURISCHIA, was much like those of other reptiles, with a pubis directed mostly anteriorly and an ischium directed mostly posteriorly. The other type, which characterized a group that he named ORNITHISCHIA, superficially resembled the pelves of birds in having the shaft of the pubis retroverted to lie posteriorly next to the ischium; in most forms a new prong of the pubis developed anteriorly and outwardly, unlike the original anteroventrally and medially directed pubic shaft (see PELVIS). Seeley (1887, 1888) concluded that the Dinosauria was really composed of two separate groups, and so he erected Saurischia and Ornithischia as orders of the Reptilia. In retrospect, all that Seeley had done was to recognize the distinctness of Ornithischia and Saurischia based on the structure of the pelvis, vertebrae, braincase, and armor. In deconstructing the Dinosauria, he was neglecting the similarities in the vertebrae, pelvis, and hindlimb, among other structures, that Owen had noted in the animals on which he had originally based the taxon. However, Seeley could have been expected to treat those similarities as simply convergences, exigencies of large size and terrestrial living. He had used the same argument several years before in disputing T. H. Huxley's contention that birds were allied to dinosaurs (specifically theropods such as *Megalosaurus* and *Allosaurus*), asking why the similarities could not be merely convergences associated with bipedality, without indicating particularly close relationship (Desmond, 1982).

Seeley's question, which as Desmond (1982) notes probably "stemmed more from his love of morphological tabulation than any evolutionary imperative," nonetheless influenced paleontology for most of the ensuing century. The Dinosauria was no longer regarded universally as a natural group, and often the term "dinosaurs" was only informally used by paleontologists. Even within Saurischia, there were frequent doubts that Sauropoda and Theropoda had a particularly close relationship. Often, Saurischia was broken into three component groups, listed as if of equal rank: PROSAUROPODA, SAUROPODA, and THEROPODA. In all three editions of A. S. Romer's influential textbook *Vertebrate Paleontology* (1933, 1945, 1966), Dinosauria was not listed as a taxon, and Saurischia and Ornithischia were simply listed as orders of Reptilia. Until the 1970s, the question became largely a matter of individual judgment hinging largely on the taxonomic weight placed on various evolutionary similarities and differences among the various "dinosaurian" subgroups.

The 1970s brought the first stirrings of cladistics (phylogenetic systematics; see PHYLOGENETIC SYSTEM; SYSTEMATICS), a methodology that has since fundamentally changed not only the practice of taxonomy but also the approach to comparative biology as a whole. However, the initial paper that reconsidered the monophyly of Dinosauria was not primarily cladistic in its thrust but rather paleophysiological. R. T. Bakker and P. M. Galton (1974) argued that Dinosauria, including the birds, should be elevated to a new class of vertebrates, united by a suite of morphological features related to upright stance, bipedality, and a metabolic level elevated above those of typical reptiles. These arguments were based on the revival and general acceptance of the hypothesis that birds descended from theropod dinosaurs; on renewed comparative studies of the origins of dinosaurs augmented by the new discoveries of small Late Triassic archosaurs (*Lagosuchus* and *Lagerpeton*) with obvious cursorial capabilities and many similarities to early dinosaurs (see DINOSAUROMORPHA); and on considerable circumstantial evidence from bone histology, zoogeography, functional morphology, and inferences about behavior that suggested strongly that if dinosaurs were not exactly like living birds and mammals, they were more like them than like living reptiles (see PHYSIOLOGY). The debate raged for another decade, particularly regarding the last conclusion, and has flared episodically in various new forms since then (Thomas and Olson, 1980). However, if the taxonomic premise of Bakker and Galton's argument, that Dinosauria (including birds) should be considered a new class of vertebrates, was not generally accepted, the phylogenetic conclusion, that Dinosauria was monophyletic and included birds, almost universally was.

For most workers, the question of dinosaurian paleobiology must be separated from the question of their systematic identity. Phylogenetic systematics groups organisms into hierarchically arranged taxa, based on the distribution of shared derived characters, or synapomorphies (see SYSTEMATICS). In the 1980s this approach was first extensively applied to the groups hypothesized to comprise Dinosauria. The most thorough analysis, and most seminal from the standpoint of later work, was by Jacques Gauthier (1986), who, in the course of trying to decide this question, brought the levels of analysis below dinosaurs to Archosauria (the crown group formed by birds and crocodiles) and down to the level of Diapsida and its immediate outgroups (Gauthier, 1984). Gauthier determined that Dinosauria, including the monophyletic groups Ornithischia and Saurischia, was itself monophyletic, united by a suite of nine synapomorphies of the skull, shoulder, hand, hip, and hindlimb. Later workers have reanalyzed and modified this original listing, but the node has remained robust.

The question that ultimately arises from these historical considerations is the following: What, finally, are dinosaurs, and can this taxon have any stability? In this work, we follow the basic principles of the phylogenetic system developed in a series of papers by de Queiroz and Gauthier (1990, 1992, 1994)—as opposed to the traditional Linnean system and the attendant rules of nomenclature followed by the International Commission on Zoological Nomenclature. We omit traditional categories of hierarchical rank, such as order and family, and focus on the monophyly of taxa and their relationships to other taxa. The monophyly of a taxon depends on an adequate definition and an adequate diagnosis (see PHYLOGENETIC SYSTEM).

The diagnosis of a taxon is a matter of determining synapomorphies that apply to it; synapomorphies are hypotheses of homologous characters that obtain at particular hierarchical levels, but with increasing knowledge these may be shown to apply to more or less general hierarchical levels or they may be found to be homoplasies (convergences). Hence the stability of a taxon rests more on its definition than on its diagnosis. Definitions may be stem-based, node-based, taxon-based, or apomorphy-based (see PHYLOGENETIC SYSTEM); as the phylogenetic system has progressed, it turns out that the former two are far preferable to the latter two in the interests of stability. However, many taxa have been based cladistically on lists of taxa or on one or more presumed synapomorphies, and although their priority should be respected when at all possible, in some cases they need to be adjusted for uniformity and ease of use.

Gauthier (1986, p. 44) appears to have defined Dinosauria in a taxon-based sense as "Herrerasauridae*, Ornithischia, Saurischia, Sauropodomorpha, and Theropoda—including birds" (the asterisk denoted a metataxon, or a taxon with no synapomorphies of its own). However, from other contexts in the same work, including the diagnosis he gave of Dinosauria (1986, p. 45), it is clear that he meant to exclude Herrerasauridae per se. The diagnosis of Dinosauria begins by recognizing the Ornithischia, Sauropodomorpha, and Theropoda as monophyletic separately and as a group, and it ends by recognizing herrerasaurs and other ornithodirans as successively more remote outgroups (see also Gauthier, 1986, pp. 14–15). In either case, however, this is a taxon-based diagnosis. Recognizing this, Padian and May (1993) proposed "to define Ornithischia as all those dinosaurs closer to *Triceratops* than to birds, and Saurischia as all dinosaurs closer to birds than to Ornithischia. Dinosauria is defined as all descendants of the most recent common ancestor of birds and *Triceratops*." This made the taxa stem based and node based, respectively. Regrettably, it was only later that T. R. Holtz (personal communication) suggested that the first two described dinosaurs, *Megalosaurus* and *Iguanodon*, included in Owen's original Dinosauria, would have been more fitting end members than birds and *Triceratops*! As dinosaurian phylogeny is currently understood, this would have made no difference to the membership of the group, and it would have paid homage to Owen's foresight.

To be a dinosaur, then, according to current definition within the phylogenetic system, a given animal must be a member of the group descended from the most recent common ancestor of birds and *Triceratops*. The diagnosis of this group, and its membership, will change as we learn more about the included taxa and modify the distributions of synapomorphies accordingly. However, what cannot change in the phylogenetic system is the valid definition of Dinosauria.

See also the following related entries:

HISTORY OF DINOSAUR DISCOVERIES • ORNITHISCHIA • PHYLOGENY OF DINOSAURS • SAURISCHIA

References

Bakker, R. T., and Galton, P. M. (1974). Dinosaur monophyly and a new class of vertebrates. *Nature* **248,** 168–172.

de Queiroz, K., and Gauthier, J. (1990). Phylogeny as a central principle in taxonomy: Phylogenetic definitions of taxon names. *Systematic Zool.* **39,** 307–322.

de Queiroz, K., and Gauthier, J. (1992). Phylogenetic taxonomy. *Annu. Rev. Ecol. Systematics* **23,** 449–480.

de Queiroz, K., and Gauthier, J. (1994). Toward a phylogenetic system of biological nomenclature. *Trends Ecol. Evol.* **9,** 27–31.

Desmond, A. J. (1975). *The Hot-Blooded Dinosaurs: A Revolution in Palaeontology.* Blond & Briggs, London.

Desmond, A. J. (1979). Designing the dinosaur: Richard Owen's response to Robert Edmond Grant. *Isis* **70,** 224–234.

Desmond, A. J. (1982). *Archetypes and Ancestors: Palaeontology in Victorian London, 1850–1875.* Muller, London.

Gauthier, J. A. (1984). A cladistic analysis of the higher systematic categories of Diapsida. Unpublished Ph.D. dissertation, Department of Paleontology, University of California, Berkeley.

Gauthier, J. A. (1986). Saurischian monophyly and the origin of birds. *Mem. California Acad. Sci.* **8,** 1–55.

Owen, R. (1842). Report on British fossil reptiles, Part II. *Rep. Br. Assoc. Adv. Sci.* **1841** 60–294.

Padian, K., and May, C. L. (1993). The earliest dinosaurs. *New Mexico Museum Nat. History Sci. Bull.* **3,** 379–381.

Romer, A. S. (1933). *Vertebrate Paleontology,* 1st ed. Univ. of Chicago Press, Chicago.

Romer, A. S. (1945). *Vertebrate Paleontology,* 2nd ed. Univ. of Chicago Press, Chicago.

Romer, A. S. (1966). *Vertebrate Paleontology,* 3rd ed. Univ. of Chicago Press, Chicago.

Seeley, H. G. (1887). On the classification of the fossil animals commonly named Dinosauria. *Proc. R. Soc. London* **43**(206), 165–171.

Seeley, H. G. (1888). The classification of the Dinosauria. *Rep. Br. Assoc. Adv. Sci.* **1887,** 698–699.

Thomas, R. D. K., and Olson, E. C. (Eds.) (1980). *A Cold Look at the Warm-Blooded Dinosaurs.* Westview Press, Boulder, CO.

Torrens, H. (1992, April 4). When did the dinosaur get its name? *New Scientist,* 40–44.

Dinosaur Museum, Dorchester, United Kingdom

see Museums and Displays

Dinosaur National Monument

Daniel J. Chure
Dinosaur National Monument
Jensen, Utah, USA

Dinosaur National Monument is a unit of the U.S. National Park Service that straddles the Utah–Colorado border. It was established in 1915 to protect Upper Jurassic dinosaur fossils of the Morrison Formation, then being excavated by the Carnegie Museum. Some 350 tons of fossils were shipped back to the Carnegie from the discovery of the quarry in 1909 to the cessation of excavations in 1924. The remains of several hundred dinosaurs belonging to 10 genera have been found in the quarry, making it the most diverse Upper Jurassic dinosaur site in the world. Part of the quarry, with some 1500 dinosaur bones prepared in high relief and left *in situ,* is enclosed within the Quarry Visitor Center and can be viewed by the public. Recent backcountry excavations have uncovered a diverse Morrison flora and fauna including mammals, lizards, sphenodontians, turtles, and some of the earliest known frogs and salamanders.

See also the following related entries:

Judith River Wedge • Jurassic Period • Morrison Formation • Museums and Displays

Dinosauromorpha

Andrea B. Arcucci
Universidad Nacional de La Rioja
La Rioja, Argentina

Sereno (1991) defined Dinosauromorpha as "ornithodirans more closely related to the dinosaur-avian clade than to pterosaurs." Apart from dinosaurs themselves, this includes taxa informally known as "lagosuchids," small, long-limbed carnivorous archosaurs from the Middle Triassic of Argentina. They were originally described by Romer (1971) based on fragmentary specimens. Several other specimens recovered later allowed a more detailed description (Bonaparte, 1975; Arcucci, 1986). Recent reviews of

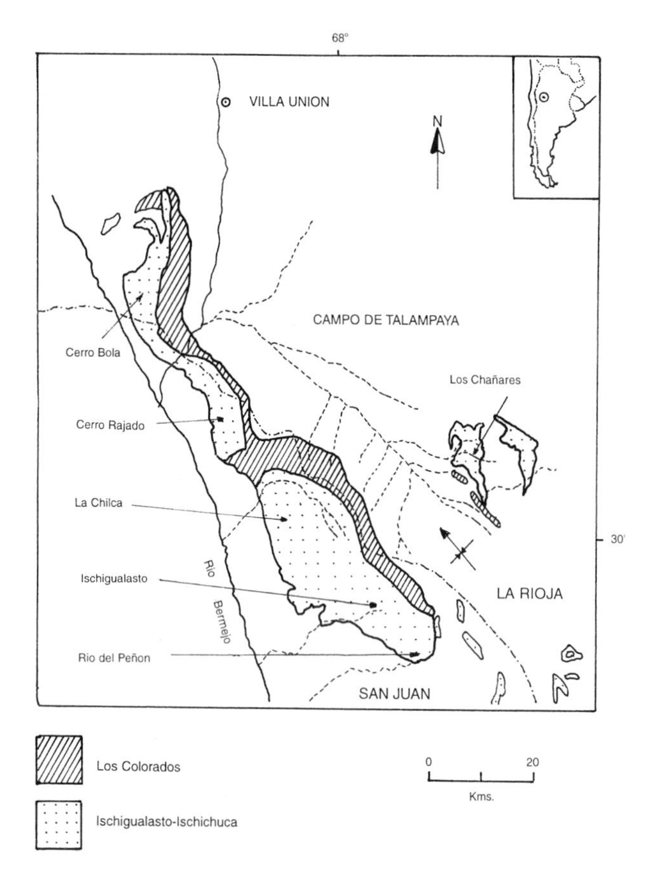

FIGURE 1 Location map showing the distribution of sites in the Triassic Ischigualasto–Villa Union basin in the north west of Argentina.

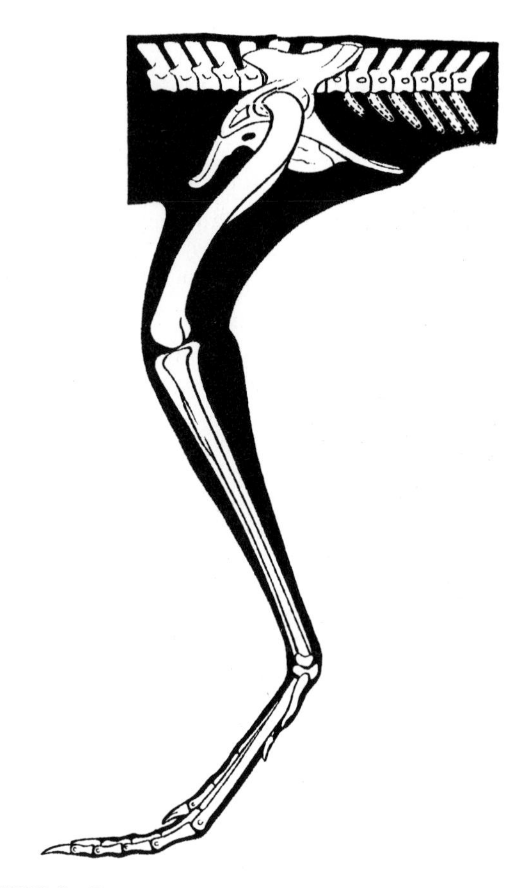

FIGURE 2 Reconstruction of *Lagerpeton* pelvis and hindlimb in lateral view. From Sereno and Arcucci (1994 a).

archosaurian phylogeny have found that these animals constitute the sister group of Dinosauria, showing several derived anatomical features that support their monophyly (Sereno and Arcucci, 1994a,b).

Marasuchus (*Lagosuchus*) and *Lagerpeton* are preserved in the same levels and were probably contemporary, taking part in one of the richest and most diverse paleofaunas recorded in the Middle Triassic worldwide. Although they are too fragmentary to evaluate their affinities, *Lewisuchus* and *Pseudolagosuchus*, which come from the same levels, probably belong to the same clade.

The beds of the Chañares Formation are extensively distributed in the Ischigualasto–Villa Union Basin in the southwest of the La Rioja Province, Argentina, near the Andes Range (Fig. 1). They consist of a relatively thin rock unit (about 60 m thick) divided into two members; the lower one yields the vertebrate fossils.

The paleoenvironment recorded from the Chañares Formation is an extensive floodplain, with thin layers of paleosols. The preservation of the bones is often very good, showing the smallest anatomical details. The skeletons are usually enclosed in strongly cemented nodules, without internal structure, that probably caused selective preservation reflected in the small and medium sizes of the animals recorded. There is a bias toward the preservation of certain body parts, such as the hindlimbs and vertebrae, apparently because they are lighter in structure than the rest of the skeleton, and not as a consequence of collection bias. The skull is only partially known in *Marasuchus* and *Lewisuchus*, and they share an overall profile of an elongated snout with numerous small serrated teeth. These dinosauromorph archosaurs consist of approximately 10 specimens and were apparently scarce in a large faunal sample of approximately 300 specimens of therapsids and other archosaurs recovered from the site. This proportion could correspond to the predator–prey ratio in the faunal assemblage.

Several genera of non-dinosaurian dinosauromorphs are known from these deposits. They are described here in order of their increasing proximity to dinosaurs. The position of *Eoraptor* and herrerasaurs is currently debated. They may be basal theropods, outside Dinosauria, or basal saurischians, depending on the analysis (see HERRERASAURIDAE; PHYLOGENY OF DINOSAURS).

Lagerpeton Romer, 1971 (Fig. 2)

Type species: *Lagerpeton chanarensis* Romer 1971

Diagnosis: Small archosaur with posterior dorsal vertebrae with anterodorsally inclined neural spines, iliac blade with sinuous dorsal margin, ischium with broad convex ventromedial flange and ventrally deep puboischiadic suture, proximal end of femur with flat anteromedial surface, astragalus with tongue-shaped posterior ascending process, pedal digit IV and metatarsal IV longer than pedal digit III and metatarsal III, respectively (in part from Sereno and Arcucci, 1994a) (Fig. 2).

Novas (1992) defined Dinosauriformes to include the common ancestor of *Lagosuchus* and dinosaurs and all its descendants.

Lagosuchus Romer, 1971

Type species: *Lagosuchus talampayensis* Romer 1971

Marasuchus Sereno and Arcucci, 1994

Type species: *Marasuchus lilloensis* (Romer 1971)

Diagnosis: Small archosaur with anterodorsally projected cervical neural spines, marked fossa ventral to the transverse process in the last cervical vertebrae and the first dorsals, mid-caudal vertebrae twice the length of the anterior caudals, and broad scapular blade (Fig. 3) (in part from Sereno and Arcucci, 1994b).

Pseudolagosuchus Arcucci, 1987

Type species: *Pseudolagosuchus major* Arcucci 1987

Diagnosis: Medium-sized archosaur with pubis longer than femur, elongated proximal caudal vertebrae?, and rounded process projected from the posterior face of the astragalus (in part from Arcucci, 1987).

Lewisuchus Romer, 1972

Type species: *Lewisuchus admixtus* Romer 1972

Diagnosis: Small archosaur with elongated cervical vertebrae, long and narrow scapular blade, small oval dermal scutes on the cervical, and dorsal neural spines (Fig. 4) (in part from Romer, 1972a).

These last two dinosauriforms are in revision at the moment, and although not much of the available

FIGURE 3 Reconstruction of *Marasuchus* (*Lagosuchus*); skull roof, hands, and gastralia restored. From Sereno and Arcucci (1994b).

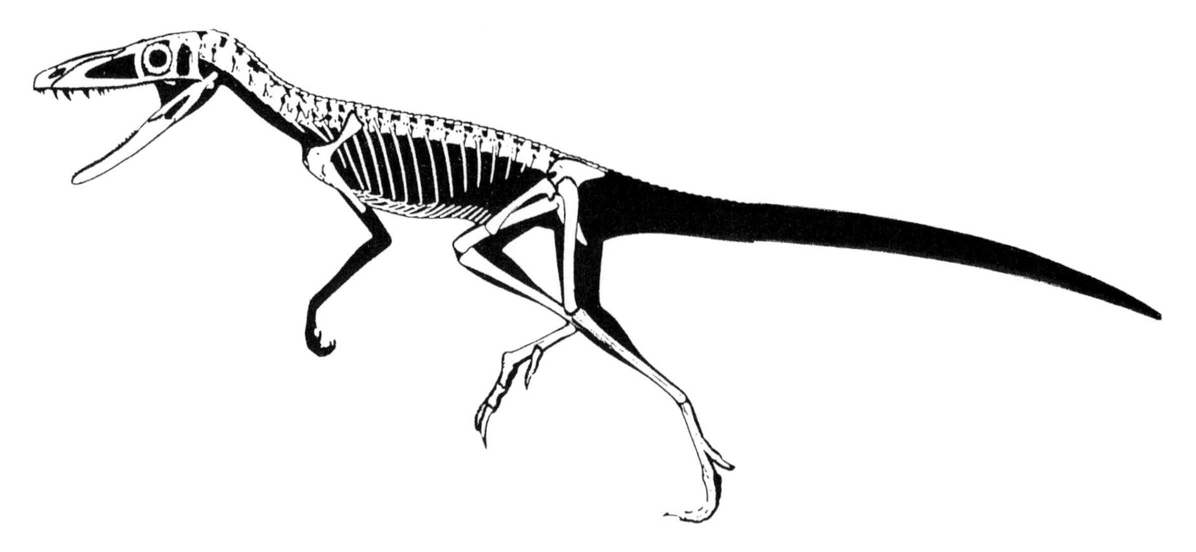

FIGURE 4 Reconstruction of *Lewisuchus*; skull roof, hands, and gastralia restored. Not to scale. Modified from Paul (1988).

material overlaps, some of it suggests that these could represent a single taxon.

The phylogenetic relationships of these animals are generally clear, but details are sketchy. *Lagerpeton chanarensis* is clearly associated with the dinosaurian radiation, but it is not possible to evaluate its precise relation to the dinosauriforms mentioned previously.

Its particular sacral and pedal specializations distinguish it from the rest of the archosaurian fauna from the Chañares Formation, but it shares a set of derived characters with Dinosauriformes and Dinosauria—for example, the transverse extension of the calcaneum, the acute corner of the astragalus, and the reduction of the articular facet for the fifth metatarsal (Fig. 5).

The Dinosauriformes (*Marasuchus, Lagosuchus, Pseudolagosuchus,* and probably *Lewisuchus*) share with dinosaurs several characteristic features such as the proportions of the forelimbs, the partly open acetabulae, the trochanteric shelf on the posterior side of the proximal part of the femur, and the parallelogram shape of the cervical vertebrae. These features strongly suggest that these Middle Triassic archosaurs are more closely related to Dinosauria than to *Lagerpeton* or the pterosaurs (Fig. 5). The fragmentary preservation of the available material of all these taxa keeps some phylogenetic relationships obscure, even after detailed analysis of the characters.

Functional Morphology

These reptiles are the first ones preserved in the fossil record that developed obligatory BIPEDALITY. They probably explored new ecological roles using locomotor capabilities not previously recorded in other tetrapods.

Although it is difficult to assign a specific locomotor gait to extinct taxa, despite the completeness of the available material, there is general consensus that *Lagerpeton* and *Marasuchus* were undoubtedly bipeds.

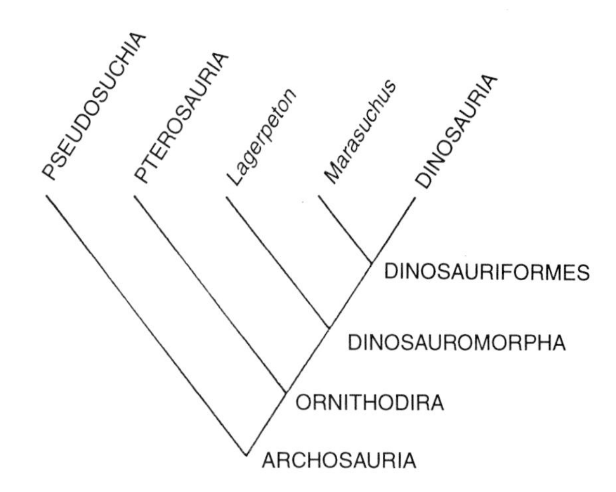

FIGURE 5 Cladogram depicting the phylogenetic relationships among the clades of basal archosaurs. From Sereno and Arcucci (1994a).

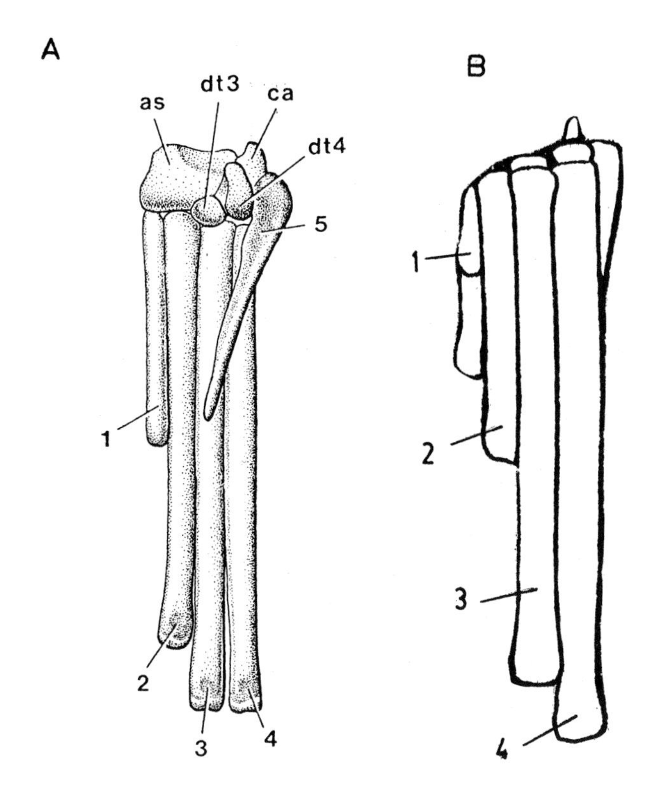

FIGURE 6 Dorsal view of the pes of (A) *Marasuchus* and (B) *Lagerpeton*, showing the differences among the tridactyl and didactyl pes. From Sereno and Arcucci (1994a).

Even though they show similar enlargement of the hindlimbs relative to the body, they present a completely different set of specializations in their pelvis and hindlimb. *Lagerpeton* has a very wide and short pelvis, relative to the limbs, and the pubis and ischium developed flat transverse surfaces, very much like the ones present in primitive archosaurs (Fig. 2). The astragalus and calcaneum are co-ossified in adults, unlike most known related archosaurs (except pterosaurs). The astragalus itself is unusual, presenting a posterior ascending process and lacking the anterior foramen. The last dorsal neural spines are inclined anteriorly, as in some saltatorial mammals (Arcucci, 1987). Finally, the functionally didactyl pres is also consistent with a saltatory gait, which could be similar to the living small ricochetal springhare *Pedetes* (Sereno and Arcucci, 1994a).

Marasuchus, on the other hand, has an elongated, rod-like pubis and ischium and a narrow pelvis in dorsal view. The dorsal neural spines incline posteriorly (Fig. 3). The pes is basically tridactyl, and the astragalus has an anterior ascending process, like those of dinosaurs, but not very developed. Its complete set of hindlimb functional features are the same as those in the other Dinosauriformes, including primitive dinosaurs, and appear for the first time in these reptiles (Fig. 6). Obviously, they represent evolutionary novelties that involved extensive changes in muscle attachments and the function of the hindlimb, but their meaning is difficult to reveal in the current state of our knowledge.

See also the following related entries:

ARCHOSAURIA • DINOSAURIA • ORNITHODIRA • ORNITHOSUCHIA • PHYLOGENY OF DINOSAURS • TRIASSIC PERIOD

References

Arcucci, A. B. (1986). Nuevos materiales y reinterpretacion de *Lagerpeton chañarensis* Romer (Thecodontia, *Lagerpetonidae* nov.) del Triasico medio de La Rioja, Argentina. *Ameghiniana* **23,** 233–242.

Arcucci, A. B. (1987). Un nuevo Lagosuchidae (Thecodontia–Pseudosuchia) de la fauna de Los Chanares (Edad Reptil Chañarense, Triasico medio), La Rioja, Argentina. *Ameghiniana* **24,** 89–94.

Bonaparte, J. F.(1975). Nuevos materiales de *Lagosuchus talampayensis* Romer (Thecodontia–Pseudosuchia) y su significado en el origen de los Saurischia. Chañarense inferior. Triasico medio de Argentina. *Acta Geol. Lilloana* **13,** 1–90.

Bonaparte, J. F. (1982). Faunal replacement in the Triassic of South America. *J. Vertebr. Paleontol.* **2,** 362–371.

Paul, G. (1988). *Predatory Dinosaurs of the World,* pp. 464. Simon & Schuster, New York.

Romer, A. S. (1971). The Chañares (Argentina) Triassic reptile fauna X. Two new but incompletely known long-limbed pseudosuchians. *Breviora* **378,** 1–10.

Romer, A. S. (1972a). The Chañares (Argentina) Triassic reptile fauna. XIV. *Lewisuchus admixtus* gen. et sp. nov. a further thecodont from the Chañares beds. *Breviora* **390,** 1–13.

Romer, A. S. (1972b). The Chañares (Argentina) Triassic reptile fauna. XV. Further remains of the thecodonts *Lagerpeton* and *Lagosuchus. Breviora* **394,** 1–7.

Sereno, P. C. (1991). Basal archosaurs: Phylogenetic relationships and functional morphology. *J. Vertebr. Paleontol.* **II** (Suppl.), 1–65.

Sereno, P. C., and Arcucci, A. B. (1994a). Dinosaur precursors from the Middle Triassic of Argentina: *Lagerpeton chañarensis. J. Vertebr. Paleontol.* **13,** 385–399.

Sereno, P. C., and Arcucci, A. B. (1994b). Dinosaur precursors from the Middle Triassic of Argentina: *Marasuchus lilloensis* gen. nov. *J. Vertebr. Paleontol.* **14,** 53–73.

Dinosaur Provincial Park

CLIVE COY
Royal Tyrrell Museum of Palaeontology
Drumheller, Alberta, Canada

Established in 1955, Dinosaur Provincial Park occupies an area of 73 square km along the Red Deer River near the center of southern Alberta, Canada. No single place of equal size on earth has produced as many individual skeletons of different dinosaurs or attracted so much research. In 1979, the park was designated a World Heritage Site by UNESCO in recognition of the exceptional abundance and diversity of dinosaur and other vertebrate fossils, the largest and most spectacular area of badlands found in Canada, and the endangered riparian habitat of plains cottonwood trees.

Meltwaters of retreating glaciers 12,000–14,000 years ago exposed 120 vertical meters of Upper Cretaceous sediments in the park. These sediments are divided into three distinct periods within the Judith River Group.

In 1909, a rancher from Alberta reported to Barnum Brown of the American Museum of Natural History that fossil bones like those on display at the museum were common on his ranch. Brown started working on the Red Deer River the following year, and by the summer of 1912 he had set up camp within current park boundaries. He continued working in the area until 1915. Brown's impressive collection included complete skeletons (with skulls) of *Centrosaurus, Corythosaurus, Gorgosaurus, Prosaurolophus,* and *Struthiomimus.* American fossil collector Charles H. Sternberg, and his three sons Charles M., Levi, and George, were hired by the Canadian Geological Survey (1911) to collect dinosaurs that would stay in Canada. The Sternbergs' first summer yielded the type specimen of *Gorgosaurus libratus,* two fine skulls of hadrosaurs, the horned dinosaurs *Chasmosaurus* (with skin impressions), *Centrosaurus,* and the type of the spike-frilled *Styracosaurus albertensis.* This friendly competition between Brown and the Sternbergs began a period of intense collection (1911–1917) that set the stage for more than 80 years of successful fossil collecting in the park. This has produced more than 250 articulated large skeletons representing 36 different species of dinosaur and another 84 species of vertebrates including fish, salamanders, frogs, turtles, lizards, crocodiles, pterosaurs, birds, and mammals. Dinosaur Provincial Park's rich bounty is housed in more than 30 institutions around the world.

The Royal Tyrrell Museum of Palaeontology has been conducting regular collection and research projects within the park since 1978. In 1987, the government of Alberta acknowledged the overwhelming significance of the park's fossiliferous deposits by building the permanent research station and field laboratory of the Royal Tyrrell Museum of Palaeontology in Dinosaur Provincial Park for use by researchers from around the world. The field station has a small but nice display of dinosaurs, including original specimens of *Centrosaurus* and *Daspletosaurus.*

See also the following related entries:

CRETACEOUS PERIOD • JUDITH RIVER WEDGE • ROYAL TYRRELL MUSEUM OF PALAEONTOLOGY

Dinosaur Ridge

MARTIN LOCKLEY
University of Colorado at Denver
Denver, Colorado, USA

Dinosaur Ridge, also known as the "Dakota Hogback," is a national natural landmark, that forms part of the elongate north–south ridge that comprises the easternmost range of the Rocky Mountain foothills, just west of Denver, Colorado. The segment named Dinosaur Ridge is situated between the town of Morrison in the south and the southern outskirts of Golden in the north and consists of eastward-dipping Jurassic Morrison Formation and Cretaceous (Dakota Group) strata.

Dinosaur Ridge is also the type section for the famous Late Jurassic Morrison Formation, which in 1877 began producing many well-known dinosaurs including *Allosaurus, Apatosaurus, Diplodocus,* and *Stegosaurus* (the Colorado State fossil). These discoveries sparked off a "dinosaur gold rush" and figured prominently in the famous "bone wars" between Edward D. Cope and Othniel C. Marsh.

Cretaceous Dakota Group strata at Dinosaur Ridge are replete with dinosaur tracks. Approximately 500 tracks, at as many as 10 different levels, represent the activity of more than 70 individuals. Most tracks are attributable to ornithopod dinosaurs, probably iguanodontids, that walked on all fours. These tracks have been assigned to the track type *Caririchnium* because of their similarity to tracks of the same name from Cretaceous strata in the Carir basin region of Brazil. There are also many unnamed footprints of slender-toed, theropod dinosaurs (probably coelurosaurs), and a few tracks of crocodiles.

Dinosaur Ridge is also considered a sister institution to Dinosaur Valley in western Colorado. Dinosaur Ridge is open to the public year-round and is furnished with interpretive signs and a visitors' center. Guidebooks and other documentation of the site are available (Lockley, 1990, 1991; Lockley and Hunt, 1995). Collecting is not permitted, but track replicas can be obtained from the visitors' center.

See also the following related entries:

FOOTPRINTS AND TRACKWAYS • MORRISON FORMATION

References

Lockley, M. G. (1990). *A Field Guide to Dinosaur Ridge*, pp. 29. Friends of Dinosaur Ridge and Univ. Colorado. at Denver, Denver.

Lockley, M. G. (1991). *Tracking Dinosaurs: A New Look at an Ancient World*, pp. 238. Cambridge Univ. Press, Cambridge, UK.

Lockley, M. G., and Hunt, A. P. (1995). *Dinosaur Tracks and Other Fossil Footprints of the Western United States*, pp. 338. Columbia Univ. Press, New York.

Dinosaur Society

DON LESSEM
Dinosaur Society
Waban, Massachusetts, USA

Distressed by the inadequacy of funding for dinosaur science and the poor quality of dinosaur-related education, author Don Lessem and paleontologists Dr. David Weishampel of Johns Hopkins University and Dr. Peter Dodson of the University of Pennsylvania founded a nonprofit organization, the Dinosaur Society, in 1991. Its purposes are to promote public education about dinosaur science and fund scientific research. Overseen by scientists on a voluntary basis, the Dinosaur Society is now a leading international funder of dinosaur research.

Since its inception, the society has published the monthly children's newspaper, *Dino Times*, which provides current and scientifically vetted findings on dinosaur science to a young audience. An adult quarterly, *The Dinosaur Report*, is also published.

A major source of financial support and educational value to the society since 1993 has been its traveling "Dinosaurs of Jurassic Park" exhibit, created by Research Casting International in cooperation with Amblin Entertainment and Universal Studios. Two editions of this exhibit have toured North America, Europe, and South America. Revenues from the exhibit have become the major source of income for the Dinosaur Society. Grants are awarded to applicants who submit proposals to a scientist review panel on a quarterly basis. Paleontological research, from the Arctic to Australia, and from Brazil to Montana, has been supported through the program. A special artists' grant program has also been established to encourage artists to assist scientists in the preparation of scientific illustrations.

The society is involved in many other programs to support dinosaur science, art, and education. It has created artists' displays and worked with artists to secure employment and gallery opportunities and has published their work in the society's own calendar. It has developed a teacher's kit and a list of recommended children's and adult dinosaur books, updated annually. The society hopes to remain a major force in furthering cooperative support for science and education among professional and amateur dinosaur workers and enthusiasts, young and old.

Dinosaur Valley

MARTIN LOCKLEY
University of Colorado at Denver
Denver, Colorado, USA

Dinosaur Valley, as used in Colorado, is an alternate name for the Grand Valley region in the greater Grand Junction–Fruita area. The term also refers more specifically to the main paleontological branch of the Museum of Western Colorado, situated on 4th and Main streets in Grand Junction and open to visitors year-round.

Within the greater Dinosaur Valley region there are a number of important dinosaur sites including Dinosaur Hill, the FRUITA PALEONTOLOGICAL AREA, RABBIT VALLEY, and Riggs Hill. Dinosaur Valley, Colorado, should not be confused with Dinosaur Valley State Park (in Texas) or other sites with the same or similar names in other parts of the world.

DINOSAUR RIDGE, in eastern Colorado, is considered a sister institution to Dinosaur Valley.

See also the following related entry:

MORRISON FORMATION

Dinoturbation

MARTIN LOCKLEY
University of Colorado at Denver
Denver, Colorado, USA

Dinoturbation is a term for trampling of soil or sedimentary substrates by dinosaurs. The word derives from the more general term "bioturbation," which refers to all manner of burrowing and disturbance of sedimentary substrates by plant roots, invertebrates, and vertebrates. A good example of bioturbation, Charles Darwin noted, is the constant recycling of soil through the gut of earthworms.

The term dinoturbation was first coined in 1980, in the same year that the term "Megabioturbation" was used to describe trampling caused by mammoths in Ice Age sediments. The term dinoturbation has caught on and been widely used in recent years, whereas the latter term has not. Dinoturbation is particularly common in late Mesozoic sediments in which the tracks of large gregarious herbivores (mainly sauropods and ornithopods) are abundant. At this time in earth history, trampling reached a peak (Lockley, 1991; Lockley and Hunt, 1995). It has even been suggested that this trampling stimulated the evolution of flowering plants, which are capable of rapid growth and regeneration. Certainly there is evidence that dinosaurs were destructive in trampling flora and fauna underfoot.

In theory, any dinosaur track is an example of dinoturbation, but the term is often taken to indicate extensive trampling. A "dinoturbation index" has been proposed (Lockley, 1991), based on the percentage of the surface covered by footprints, to define areas of light, moderate, and heavy trampling (respectively disturbing 0–33, 34–66, and 67–100% of the surface). Heavy dinoturbation can result from the biological activity of many animals, from the geological effects of long periods of exposure of the substrate to trackmakers, or from a combination of both factors. The term dinoturbation should not be used to describe trampling caused by vertebrates that are not dinosaurs.

See also the following related entries:

FOOTPRINTS AND TRACKWAYS • TAPHONOMY

References

Lockley, M. G. (1991). *Tracking Dinosaurs: A New Look at an Ancient World*, pp. 238. Cambridge Univ. Press, Cambridge, UK.

Lockley, M. G., and Hunt, A. P. (1995). *Dinosaur Tracks and Other Fossil Footprints of the Western United States*, pp. 338. Columbia Univ. Press, New York.

Disease

see PALEOPATHOLOGY

Display

see BEHAVIOR; ORNAMENTATION

Distribution and Diversity

PETER DODSON
University of Pennsylvania
Philadelphia, Pennsylvania, USA

We now know that dinosaurs lived on all seven continents. Antarctica was the last continent to produce dinosaur fossils, and the Early Jurassic theropod *Cryolophosaurus* was only described in 1994, whereas

Late Cretaceous ornithopod and ankylosaurian remains have been reported but not yet named. The distribution of dinosaurs throughout the Mesozoic is to an extent determined by the positions of the wandering continents. The earliest dinosaurs or dinosaur relatives currently known appear to be those of the Late TRIASSIC (Carnian) Ischigualasto Formation of Argentina and the Santa Maria Formation of Brazil. At this time, Pangea was substantially intact, and no major barriers impeded intercontinental dispersal. By the succeeding Norian stage of the Late Triassic, prosauropods had appeared in Germany, Greenland, and South Africa. Early Jurassic prosauropods reached China and the southwestern United States. By the end of the Cretaceous, approximately 163 million years after the first appearance of dinosaurs, the continents were close to their current positions. It is probable that some degree of faunal exchange was possible among South America, Antarctica, and Australia. For essentially the entire Mesozoic, Europe was an archipelago, and for significant lengths of time during the Jurassic and especially the Cretaceous, North America was flooded by epicontinental seas that tended to separate the eastern and western parts of the continent.

As a rule of thumb, dinosaurs that appeared early have a cosmopolitan distribution, whereas those appearing later are more restricted. PROSAUROPODS (Late Triassic) and SAUROPODS (Early Jurassic) are rather cosmopolitan, whereas CERATOPSIANS (late Early Cretaceous) are found only in eastern Asia and western North America (claims of solitary bone specimens from South America and Australia are unconvincing to some; but see POLAR DINOSAURS). HYPSILOPHODONTIDS (Middle Jurassic) are cosmopolitan, but LAMBEOSAURINE HADROSAURIDS are known only from western North America and eastern Asia. HADROSAURINE HADROSAURIDS (early Late Cretaceous) are predominantly of Laurasian distribution. Hadrosaurs reached South America (*Secernosaurus*), possibly by sweepstakes dispersal from North America. Many specimens have been collected, but few have been studied or described. STEGOSAURS (Middle Jurassic) are cosmopolitan, but show greater diversity in China than anywhere else. The sauropod family Titanosauridae has a broad distribution in the Late Cretaceous on southern continents. Titanosaurids also entered Europe (*Magyarosaurus* and *Hypselosaurus*) and North America (*Alamosaurus*), a continent from which sauropods had previously become extinct. The abelisaurids may be a Gondwanan family of THEROPODS of broad distribution (South America, India, or Europe?) perhaps related to Ceratosauria. It has been argued that rather than being typical of worldwide dinosaurs of the Late Cretaceous, dinosaurs of western North America and eastern Asia were isolated from the world dinosaur fauna and exhibited instead a high degree of endemism (e.g., tyrannosaurids, lambeosaurines, ceratopsians, and ankylosaurids).

Assessments of dinosaur diversity and of diversity trends through time are difficult matters. In the simplest form, a tabulation of described genera represents one estimation of dinosaur diversity. More than 600 genera of dinosaurs have been named, but of these only about 325 are currently (as of 1995) regarded as valid. However, the number has doubled since 1970 and may double again in another 25 years. Two further approaches have been taken to estimate dinosaur diversity. Dodson (1990) estimated the mean generic longevity of dinosaurs to be 7.7 million years (range: 5–10.5 million years). This suggests nearly 100% faunal turnover per geological stage. He estimated the maximum number of dinosaur genera living at one time (Campanian–Maastrichtian, latest Cretaceous) to be 100. Depending on several different models of diversity change through time, and integrating across intervals for which no fossils are known, this method yields estimates of dinosaur diversity on the order of 900–1200 genera. Russell (1994) estimated dinosaur diversity on the basis of the relationship between area of continental landmass and diversity. This method yields an estimate of 3300 genera. The two methods frame reasonable endmember estimates. One approach suggests that our knowledge of the fossil record of dinosaurs is currently nearly 30% complete; the other a more conservative 10% complete.

Did dinosaur diversity increase through time? If it is assumed that certain time intervals are relatively well sampled (e.g., Norian–Sinemurian, Late Triassic–Early Jurassic; Kimmeridgian–Tithonian, Late Jurassic; and Campanian–Maastrichtian, Late Cretaceous), relevant observations may be made. In a typical early dinosaur assemblage, there were one or two sauropodomorphs, one or two theropods, and possibly one ornithischian. In a typical Late Jurassic assemblage, there were three or more sauropods, two or more theropods, one or two stegosaurs, and one or two ornithopods. In a Late Cretaceous North American assemblage, there were three or more hadrosaurs, a basal ornithopod, a pachycephalosaur, two or more ankylosaurs, two or three ceratopsians, and three or more theropods. It thus appears that diversity increased through the Mesozoic. Two factors may be mentioned as contributing to increased diversity in the Late Cretaceous. One is the fractionation of continental landmasses compared with the earlier Mesozoic, leading to increased endemism. The other is the smaller body sizes of ornithopods compared to sauropods, with the attendant greater possibilities for niche specialization and consequent greater diversity at the community level.

See also the following related entries:

Biogeography • Migration • Plate Tectonics

References

Dodson, P. (1990). Counting dinosaurs: How many kinds were there? *Proc. National. Acad. Sci.* **87,** 7608–7612.

Dodson, P. and Dawson, S. D. (1991). Making the fossil record of dinosaurs. *Mod. Geol.* **16,** 3–15.

Russell, D. A. (1994). China and the lost worlds of the dinosaurian era. *Historical Biol.* **10,** 3–12.

Smith, A. G., Smith, D. G., and Funnell, B. M. (1994). *Atlas of Mesozoic and Cenozoic Coastlines*, pp. 99. Cambridge Univ. Press, Cambridge, UK.

Weishampel, D. B. (1990). Dinosaurian distributions. In *The Dinosauria* (D. B. Weishampel, P. Dodson, and H. Osmolska, Eds.), pp. 63–139. Univ. California Press, Berkeley.

Djadokhta Formation

Tom Jerzykiewicz
Geological Survey of Canada
Calgary, Alberta, Canada

The Djadokhta Formation of Campanian (Late Cretaceous) age is famous for yielding the first unquestionable finds of dinosaur eggs (discovered in 1922 by the third Asiatic Expedition led by Roy Chapman Andrews) and numerous exceptionally well-preserved dinosaur skeletons, most notably *Protoceratops andrewsi*, *Pinacosaurus grangeri*, and *Velociraptor mongoliensis* (Figs. 1–3). The type localities of the formation at Bayn Dzak, Ukhaa Tolgod, and Tugrig of pre-Altai Gobi (Mongolia) and correlative strata at Bayan Mandahu of the southern Gobi (Inner Mongolia, China) have been the subject of many later successful paleontological expeditions (most notably from the United States, Russia, Mongolia, Poland, and Canada).

The dominant lithology of the formation is poorly cemented reddish brown, fine-grained eolian sandstone. Beds of water-deposited sandstone, mudstone, and conglomerate are subordinate. Sedimentary facies indicate that the Djadokhta redbeds were deposited in semiarid, alluvial-to-eolian settings. The presence of mature caliche paleosols and wind-blown sediments suggests overall semiarid conditions of accumulation for the formation. Stratal geometry of wind-blown sediments indicates the presence of large straight-crested dune forms and smaller barchan and parabolic dunes. Some layers within the dunes are rich in trace fossils, suggesting that the dunes were organically rich and seasonally moist.

An assemblage of fossil vertebrates found in the Djadokhta Formation consists of ankylosaurs, ceratopsians, theropods, turtles, crocodiles, lizards, and mammals. The fossil vertebrates occur largely in association with fine-grained eolian sandstones and paleosols, and these have been interpreted as the remains of autochthonous 'faunal' components, adapted to living in semiarid environments. In contrast, rare and fragmentary specimens of large dinosaurs that occur in coarse-grained alluvial deposits have been interpreted as the remains of allochthonous faunal components. These rare components include the tyrannosaurid *Tarbosaurus*, another large, nontyrannosaurid theropod, a sauropod(s), and ornithopods.

The low diversity of the Djadokhta fossil assemblage and the overall small to medium size of its constituents indicate a relatively stressed paleoenvironment. Furthermore, the absence of fishes or any other undoubted aquatic organisms, together with the localized abundance of dermatemydid (terrestrial) turtles, further suggests a fully terrestrial vertebrate assemblage.

The environment of the Djadokhta dinosaurs developed during Late Cretaceous time as a result of block-faulting tectonic movements that affected central Asia and transformed large, Early Cretaceous perennial lakes into semiarid steppes located far from

FIGURE 1 Partly unearthed skeleton of "standing" *Protoceratops* in the Djadokhta Formation at Toogreeg. (a) Oblique view; (b) side view (photo from T. Jerzykiewicz, Polish–Mongolian Expedition of 1971).

the seashore. These semiarid steppes were partly covered by dune fields and drained by seasonal streams into intermittent ponds whose distribution and occurrence may have been controlled by plugged caliche horizons. Such a landscape, shaped to a large degree by eolian processes (including dust and sand storms), must have made living conditions rather difficult and supposedly contributed to an increased mortality of vertebrates during droughts.

The poses of some fossil dinosaurs in the eolian sandstones indicate that the animals died *in situ* and were not transported. A number of articulated *Protoceratops*, for instance, "stand" on their hindlimbs, with snouts pointing upward and forelimbs tucked in at their sides (Figs. 1a and 1b). Such a pose clearly indicates that the animals were encased within the surrounding sediment at the time of their deaths. It is likely that some of these creatures died while attempting to free themselves from a sandstorm deposit during or shortly after the storm event. In the exceptional co-occurrence of *Protoceratops* and *Velociraptor*—the so called "fighting dinosaurs" (Fig. 2), the theropod's right forelimb is "clenched" between the closed jaws of the prostrate CERATOPSIAN indicating rapid burial by eolian sands shortly after the animals came in contact with one another. Monospecific death assemblages of both *Protoceratops* and *Pinacosaurus* were also noted. In one case, a group of five medium-sized adult *Protoceratops* were found lying parallel aligned, side by side, and inclined about 20° from the horizontal, probably on a slope of a dune. All the specimens were lying on their bellies with their heads facing upslope. Yet another example consists of a group of juvenile *Pinacosaurus* that died *in situ*, probably as a result of burial during sandstorms (Fig. 3). The occurrence of large numbers of fossil vertebrate eggs in association with the eolian sandstone and semiarid paleosols provides additional evidence of a semiarid climate during deposition of the Djadokhta Formation.

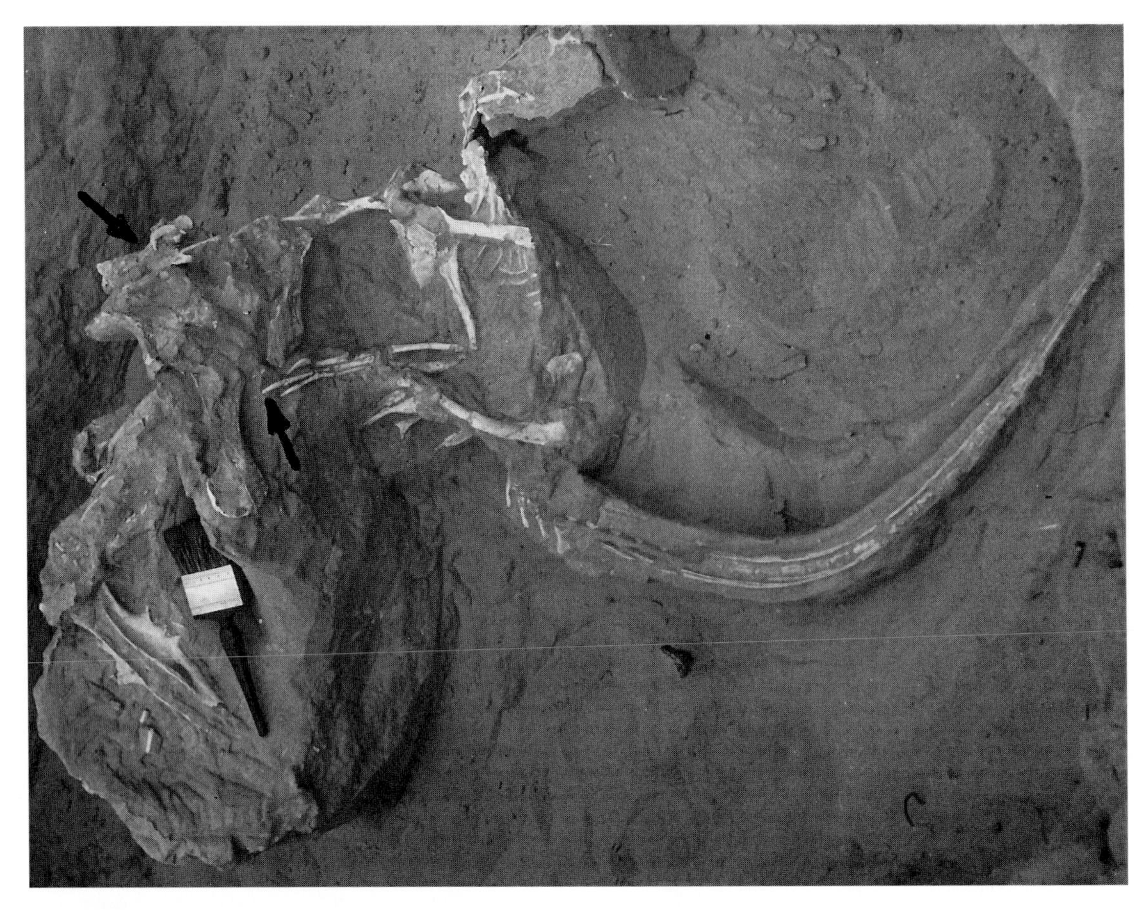

FIGURE 2 Partly unearthed skeletons of *Protoceratops* and *Velociraptor* at Toogreeg (pre-Altai Gobi). Note the claws (arrow) of the left hindfoot of the *Velociraptor* imbedded within the thorax of the *Protoceratops* and the claw (arrow) of the hand clasp the top of the skull of the presumed prey (photo from T. Jerzykiewicz, Polish–Mongolian Expedition of 1971).

FIGURE 3 Excavating the mass grave of juvenile *Pinacosaurus* at Bayan Mandahu. Inset shows one of the skulls recovered from the grave.

See also the following related entries:

American Museum of Natural History • Central Asiatic Expeditions • Chinese Dinosaurs • Mongolian Dinosaurs

References

Barsbold, R., and Perle, A. (1983). On taphonomy of a joint burial of juvenile dinosaurs and some aspects of their ecology. *Sovmestnaya Sovetsko–Mongolskaya Paleontologicheskaya Ekspeditsia. Trudy* **24,** 121–125. [In Russian].

Berkey, C. P., and Morris, F. K. (1927). *Geology of Mongolia, Natural History of Central Asia, Volume II,* pp. 475. American Museum of Natural History, New York.

Currie, P. J., and Peng, J. H. (1993). A juvenile specimen of *Saurornithoides mongoliensis* from the Djadokhta Formation (Upper Cretaceous) of northern China. *Can. J. Earth Sci.* **30,** 2224–2230.

Gradzinski, R., Kielan-Jaworowska, Z., and Maryanska, T. (1977). Upper Cretaceous Djadokhta, Barun Goyot and Nemegt formations of Mongolia, including remarks on previous subdivisions. *Acta Geol. Polonica* **27,** 281–318.

Granger, W., and Gregory, W. K. (1923). *Protoceratops andrewsi*, a preceratopsian dinosaur from Mongolia. *Am. Museum Novitates* **42,** 1–9.

Jerzykiewicz, T., Currie, P. J., Eberth, D. A, Johnston, P. A., Koster, E. H., and Zheng, J. J. (1993). Djadokhta Formation correlative strata in Chinese Inner Mongolia: An overview of the stratigraphy, sedimentary geology, and paleontology and comparisons with the type locality in the pre-Altai Gobi. *Can. J. Earth Sci.* **30,** 2180–2195.

Lefeld, J. (1971). Geology of the Djadokhta Formation at Bayn Dzak (Mongolia). *Palaeontol. Polonica* **25,** 101–127.

Tverdokhlebov, V. P., and Tsybin, Yu. I. (1974). Genezis verkhnemelovykh mestorozhdeniy dinozavrov Tugrikin-Us i Alagteg. *Sovmestnaya Sovetsko–Mongolskaya Paleontologicheskaya Ekspeditsia. Trudy* **1,** 314–319. [In Russian].

DNA

see Chemical Composition of Dinosaur Fossils; Computers and Related Technology; Genetics; Jurassic Park; Variation

Dockum Group

Phillip A. Murry
Tarleton State University
Stephenville, Texas, USA

Robert A. Long
Pleasanton, California, USA

The Dockum Group of western Texas consists of Upper Triassic-age redbeds, representing fluvial, lacustrine, and paleosol facies. Four Dockum units are recognized in Texas; in ascending order, these include an informal "Pre-Tecovas Horizon," Tecovas Formation, Trujillo Formation, and Cooper Canyon Formation. Dinosaur remains have been reported from all Dockum units in Texas, except the Trujillo Formation.

The poorly defined, mudstone-rich Pre-Tecovas Horizon is regarded as Middle Carnian age; localities of this age are in Howard, Borden, and Scurry counties within the southern portion of the Dockum depositional basin. Fossiliferous localities within the mudstone-dominated Tecovas Formation are found within the central Dockum basin of Crosby and Dickens counties, and in the lower mudstone units in the Palo Duro Canyon area (Randall County) and Canadian River Valley (Oldham and Potter counties) within the northern Dockum of Texas. The sandstone-rich Trujillo Formation is best exposed in the Palo Duro Canyon area, and fossiliferous units in the interbedded coarse and fine clastics of the Cooper Canyon Formation are known from Garza County, in the south-central portion of the Dockum Basin (Long and Murry, 1995).

Although thousands of reptilian individuals are represented in the Dockum collections, only a few are dinosaurian. Furthermore, most Dockum material reported as dinosaurs by previous workers has proven to be that of non-dinosaurian archosauromorphs. Cope (1893) named *Palaeoctonus dumblianus* and *P. orthodon* from isolated teeth from Palo Duro Canyon, and referred them to theropod dinosaurs. These teeth are actually phytosaurian.

Case (1922) referred a left femur (UMMP 3396) to the Dinosauria that in reality represents a robust aetosaurian, probably *Desmatosuchus*. He also referred a series of cervical and anterior dorsal vertebrae (UMMP 7507) found in Crosby County to *Coelophysis* (Case, 1922, 1927). In 1932, von Huene designated this specimen the holotype of a supposed saurischian dinosaur, *Spinosuchus caseanus*, and referred the braincase described by Case to this taxon. The braincase belongs to a rauisuchian (probably *Postosuchus*) and the vertebral column does not appear to be Saurischian. Relationships have been proposed among *Spinosuchus caseanus* and *Lotosaurus adentus* from the Middle Triassic of China (Fa-Kui, 1975) and *Ctenosauriscus koeneni* from the Middle Buntsandstein of Germany (Krebs, 1969). These three long-spined Triassic archosauromorphs are probably not closely related but are merely convergent in the development of dorsal sails. There is no evidence that these taxa are closely related to dinosaurs.

Caudal vertebrae (UMMP 7277 and UMMP 9805) and compressed recurved teeth from Crosby County, identified as theropod by Case (1927), cannot be referred with confidence to the Theropoda, although the teeth do appear to represent non-phytosaurian carnivorous archosauromorphs. Case (1932) also described and illustrated a large series of associated caudal vertebrae from Potter County that he referred to *Coelophysis* sp.; these are not referable to *Coelophysis* and appear to represent a rauisuchian. Elder (1987) reported "coelurosaur" material from Quarries 1, 3, and 3a at Otis Chalk, and Gregory (1972) listed *Coelophysis* from Howard, Borden, Crosby, Randall, and Potter counties. Although there are a number of lightly constructed bones in the TMM collections, there is no irrefutable evidence of dinosaurs or herrerasaurs in any of these collections other than the herrerasaurid *Chindesaurus*. Colbert (1961) referred a fragment of a right ilium (UMMP 11748) from Howard County to *Poposaurus* and concluded that *Poposaurus* represented a primitive theropod. However, subsequent workers have placed the poposaurs within the Rauisuchia.

Shuvosaurus inexpectatus is based on the remains of at least three individuals from the Post Quarry, including a well-preserved skull with lower jaw (TTU P9280). Chatterjee (1993) regarded *Shuvosaurus* as a new family of ORNITHOMIMOSAUR, based primarily on a comparison of skull characters including a pointed, hooked premaxillary beak, short preorbital region, participation of the nasal in the development of the maxillary fenestra, mandibular articulation of the quadrate by the lateral and medial condyle, and the presence of a secondary dental shelf at the jaw symphysis. We have not studied this skull in detail, and our conclusions are tentative. We believe that Chatterjee's diagnosis of *Shuvosaurus* fails to include it specifically within the Dinosauria. Probable remains of *S. inexpectatus* recovered in both New Mexico and Texas are closely associated with elements of *Chatterjeea elegans*, a rauisuchian based on postcrania (Long and Murry, 1995). These taxa closely match in size and preservation. The morphology of *Chatterjeea* has a number of convergent features with those of ornithomimosaurs, including attenuation of the vertebral column and limbs, development of a synsacrum, and the presence of an ornithomimosaur-like deltopectoral crest on the humerus. It is possible that all these specimens represent one taxon, in which case *Shuvosaurus* would have priority. Based on postcranial characters, it could not be referred to the Dinosauria. In this case, the range of *Shuvosaurus* would be at least from the Middle Carnian to Lower Norian.

Although many reports of dinosaurs in the Dockum Group are false, there are evidently both saurischians and ornithischians within the Upper Triassic deposits of Texas. Chatterjee (1986, p. 145) reported a juvenile specimen referable to *Coelophysis*, and examination of his collection verifies the presence of a small theropod dinosaur femur.

Chindesaurus bryansmalli is a 2- to 4-m-long HERRERASAURID dinosauromorph or dinosaur relative that is primarily based on a single specimen from the Lower Norian of Petrified Forest National Park. It is the most primitive North American dinosauromorph known. It is also among the oldest North American dinosauromorphs, as the proximal portion of a femur (TMM 31100-523) from Howard County is identical to that of the type specimen, and is believed to be of Middle Carnian age (Long and Murry, 1995). The Petrified Forest specimen is also the youngest known record of Herrerasauridae. Another Texas specimen of *Chindesaurus* is a right ilium (UMMP 8870) from Crosby County that Case (1927) tentatively referred to *Coelophysis*. The presence of a prominent groove along the ventral articular surface of the astragalus, the glutealform shape of its distal surface, and the apparent absence of a fibular facet on the astragalus differentiates *Chindesaurus* from *Herrerasaurus* (Long and Murry, 1995).

Chatterjee (1984) described *Technosaurus smalli* from the Cooper Canyon Formation near Post, Texas. According to Sereno (1991), at least a portion of the holotype (TTU P9021) displays features that are consistent with the "Prosauropoda," (basal sauropodomorpha) although exhibiting no clear sauropodomorph synapomorphies. However, Hunt and Lucas (1994) agree with Chatterjee that *Technosaurus* is an ornithischian, although the material could not be more specifically assigned.

Hunt and Lucas (1994) described isolated dentary and premaxillary teeth from Crosby County as *Tecovasaurus murryi* that they believe represents undoubted ornithischians. Other material probably referable to this species was collected by Murry (1982, 1986) from Crosby and Potter counties. Asymmetrically crowned teeth with large, compound denticles of this morphotype evidently show marked heterodonty. The range of dental variation in Triassic dinosaur teeth, convergence between non-dinosaurian archosauromorphs (especially rauisuchians) and dinosaurs, and convergence among dinosaurian taxa present problems in naming dinosaurs on the basis of isolated teeth. However, the presence of several dinosaur-like tooth morphotypes in the Texas collections may indicate that more, as yet undescribed, dinosaurs were present within the Upper Triassic Dockum Group of Texas.

See also the following related entries:

CHINLE FORMATION • TRIASSIC PERIOD

References

Case, E. C. (1922). New reptiles and stegocephalians from the Upper Triassic of Western Texas. *Carnegie Institution of Washington Publ.* **321,** 1–84.

Case, E. C. (1927). The vertebral column of *Coelophysis* Cope. *Univ. Michigan Museum Paleontol. Contrib.* **2**(10), 209–222.

Case, E. C. (1932). On the caudal region of *Coelophysis* sp. and on some new or little known forms from the Upper Triassic of Western Texas. *Univ. Michigan Museum Paleontol. Contrib.* **4**(3), 81–91.

Chatterjee, S. (1986). The Late Triassic Dockum vertebrates: their stratigraphic and paleobiogeographic significance. In *The Beginning of the Age of Dinosaurs. Faunal Change across the Triassic–Jurassic Boundary* (K. Padian, Ed.), pp. 139–150. Cambridge Univ. Press, Cambridge, UK.

Chatterjee, S. (1993). *Shuvosaurus*, a new theropod. *Nat. Geogr. Res. Exploration* **9**(3), 274–285.

Colbert, E. H. (1961). The Triassic reptile *Poposaurus. Fieldiana, Geol.* **14,** 59–78.

Cope, E. D. (1893). A preliminary report on the vertebrate paleontology of the Llano Estacado. Fourth Annual Report of the Geological Survey of Texas. pp. 11–87.

Elder, R. L. (1987). Taphonomy and paleoecology of the Dockum Group, Howard County, Texas. *J. Arizona–Nevada Acad. Sci.* **22,** 85–94.

Fa-Kui, Z. (1975). A new thecodont *Lotosaurus*, from Middle Triassic of Hunan (in Chinese). *Vertebr. Palasiatica* **13**(3), 144–147.

Gregory, J. T. (1972). Vertebrate faunas of the Dockum Group, Triassic, eastern New Mexico and west Texas. In *New Mexico Geological Society, Annual Field Conference Guidebook,* Vol. 23, pp. 120–123. New Mexico Geological Society.

Hunt, A. P., and Lucas, S. G. (1994). Ornithischian dinosaurs from the Upper Triassic of the United States. In *In the Shadow of the Dinosaurs. Early Mesozoic Tetrapods* (N. C. Fraser and H.-D. Sues, Eds.), pp. 227–241. Cambridge Univ. Press, Cambridge, UK.

Krebs, B. (1969). *Ctenosauriscus koeneni* (v. Huene), die Pseudosuchia und die Buntsandstein-Reptilien. *Eclogae Geol. Helvetiae* **62**(2), 697–714.

Long, R. A., and Murry, P. A. (1995). Late Triassic (Carnian and Norian) tetrapods from the southwestern United States. *New Mexico Museum Nat. History Bull.* **4,** 1–254.

Murry, P. A. (1982). Biostratigraphy and paleoecology of the Dockum Group, Triassic, of Texas, pp. 459. PhD dissertation, Southern Methodist University, Dallas, Texas.

Murry, P. A. (1986). Vertebrate paleontology of the Dockum Group, western Texas and eastern New Mexico. In *The Beginning of the Age of Dinosaurs. Faunal Change across the Triassic–Jurassic Boundary* (K. Padian, Ed.), pp. 109–137. Cambridge Univ. Press, Cambridge, UK.

Sereno, P. C. (1991). *Lesothosaurus*, "fabrosaurids," and the early evolution of Ornithischia. *J. Vertebr. Paleontol.* **11**(2), 168–197.

Dromaeosauridae

PHILIP J. CURRIE
Royal Tyrrell Museum of Palaeontology
Drumheller, Alberta, Canada

The first dromaeosaurids described were *Dromaeosaurus albertensis* (Matthew and Brown, 1922) and *Velociraptor mongoliensis* (Osborn, 1924), both from the Upper Cretaceous strata of the Northern Hemisphere. Because of the rarity of small theropod fossils, however, the significance of these animals was not fully understood until the discovery of *Deinonychus* in 1964. Since that time, dromaeosaurids have been a focal point for research on the interrelationships of theropods, the origin of birds, dinosaur physiology, dinosaur brain size, and dinosaur behavior.

Two subtaxa of dromaeosaurids are currently recognized. Velociraptorine dromaeosaurids include *Deinonychus* (Ostrom, 1969), *Saurornitholestes*, and *Velociraptor*. *Dromaeosaurus* is the only unquestionable dromaeosaurine dromaeosaurid, but the poorly known *Adasaurus* from Mongolia has also been referred to this genus. Giant dromaeosaurids from the Lower Cretaceous of the United States (*Utahraptor*), Japan, and Mongolia are poorly known and cannot be assigned to either subtaxon with confidence at this time.

Most dromaeosaurids were between 2 and 3 m in length. Their most conspicuous character is found in the second toe of the hindfoot, which bears a large raptorial claw. This claw is strongly recurved and was more than twice as long as the other claws on the foot. Because of its sharp point and knife-like lower edge, it was held off the ground in normal situations. Although dromaeosaurid footprints are unknown at present, the raised position of this claw can be seen in several articulated skeletons.

Many other dromaeosaurid apomorphies make this one of the better diagnosed theropod taxa. These include a pubis that is more bird-like than those of any other known dinosaur. The pubis faces down and backward, parallel to the ischium. The tail is unusual because there are long but delicate rods in the tail that extend anteriorly from the prezygopophyses and the haemal arches. They form a cable-like network that would have stiffened the tail without making it completely rigid. As in other theropods, inter-

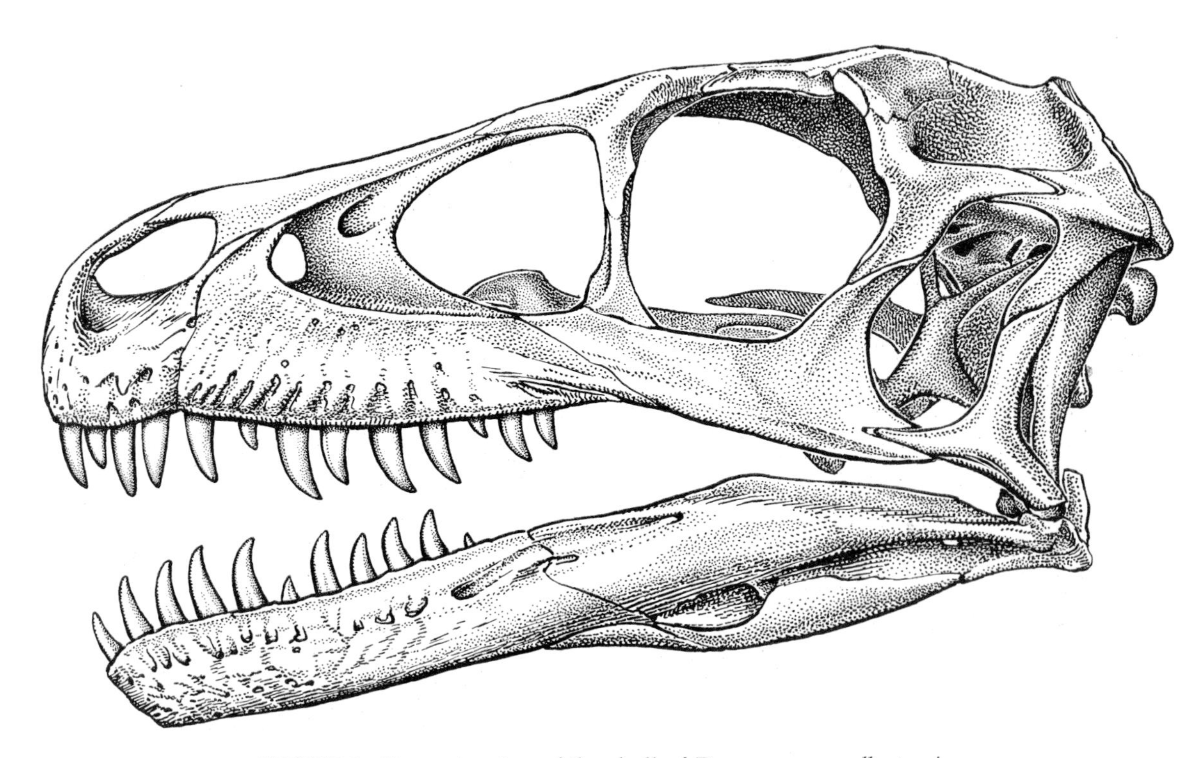

FIGURE 1 Reconstruction of the skull of *Dromaeosaurus albertensis*.

dental plates seem to have been present on the inside of the tooth rows, but they are fused together so that individual plates can no longer be distinguished. The jaw articulation is peculiar in that there is a tall, slender, vertical process behind the joint (Currie, 1995).

Velociraptorines and dromaeosaurines are easily distinguished from each other on the basis of differences in their teeth. The serrations on the front of a dromaeosaurine tooth are about the same size as the serrations on the back of the same tooth. In velociraptorines, the posterior denticles are much larger than the anterior ones. The teeth in the premaxilla of *Dromaeosaurus* are all about the same size, whereas the second tooth of this bone is the largest in velociraptorines. Dromaeosaurine skulls seem to have been more heavily constructed (Fig. 1).

Dromaeosaurids are often considered to be the most bird-like of the small theropods. The brain is relatively large, the lightly built skulls are often pneumatic, they have clavicles and ossified sternals (breastbones), their arms are relatively long, and the pubis is retroverted. Unlike most other Cretaceous theropods, the metatarsal bones (found in the flat of the foot of humans) are relatively short and unspecialized and are more similar to the metatarsals of early birds. Even the specialized raptorial claw has now been found in Cretaceous birds from Madagascar and Argentina. There are so many similarities between dromaeosaurids and birds that some have even speculated that dromaeosaurids were birds that lost their ability to fly. However, in some characters, such as the stiffened tail, dromaeosaurids are too specialized to have been good ancestors for birds. All known dromaeosaurids are Cretaceous in age, although they no doubt originated sometime in the Jurassic, and other maniraptoran taxa are known from the Late Jurassic (see BIRD ORIGINS).

The dromaeosaurid *Velociraptor* has become a well-known dinosaur thanks to its role in the book and movie called *Jurassic Park*. Depicted as relatively intelligent, vicious, warm-blooded, pack-hunting animals, dromaeosaurids have done much to change the public perception of dinosaurs as slow-witted, solitary, cold-blooded creatures. Like troodontids, dromaeosaurids had relatively large brains. Although this does not indicate that they were as intelligent as living birds and mammals, it does suggest that they had the same capabilities as some birds and mammals. The light, agile bodies, long fingers, and raptorial claws show that they were probably effective predators. This view is supported by a remarkable pair of skeletons from Mongolia. A *Velociraptor* and a *Protoceratops* were discovered together in Upper Cretaceous sediments, apparently locked in mortal combat. It appears that the predator attacked the herbivore and killed it. However, before the protoceratopsian died, it locked its jaws on the arm of the dromaeosaurid. It might have escaped if a sandstorm had not been in progress, and both animals were completely sealed in sand for 75 million years. Determining conclusively whether or not dinosaurs such as *Velociraptor* were warm-blooded has been elusive (see PHYSIOLOGY). However, the warm-blooded proponents gained some support in 1996 by the discovery in China of a pair of small 'feathered' theropods.

See also the following related entries:

COELUROSAURIA • MANIRAPTORA

References

Currie, P. J. (1995). New information on the anatomy and relationships of *Dromaeosaurus albertensis* (Dinosauria: Theropoda). *J. Vertebr. Paleontol.* **15,** 576–591.

Matthew, W. D., and Brown, B. (1922). The family Deinodontidae, with notice of a new genus from the Cretaceous of Alberta. *Am. Museum Nat. History Bull.* **46,** 367–385.

Osborn, H. F. (1924). Three new theropoda, *Protoceratops* Zone, central Mongolia. *Am. Museum Nat. History Novitates* **144,** 12.

Ostrom, J. H. (1969). Osteology of *Deinonychus antirrhopus*, an unusual theropod from the Lower Cretaceous of Montana. *Peabody Museum Nat. History Bull.* **30,** 165.

Dromaeosaurinae

see DROMAEOSAURIDAE

Drumheller Dinosaur and Fossil Museum

see MUSEUMS AND DISPLAYS

Dry Mesa Quarry

Brooks B. Britt
Museum of Western Colorado
Grand Junction, Colorado, USA

Brian D. Curtice
Brigham Young University
Provo, Utah, USA

The Dry Mesa Quarry (Uncompahgre) has yielded the most diverse dinosaurian fauna known from any single quarry in the Morrison Formation (Britt, 1991). All the common Morrison taxa such as *Camarasaurus*, *Diplodocus*, *Allosaurus*, *Stegosaurus*, and others are present, as are rare taxa such as *Ceratosaurus*, *Marshosaurus*, *Stokesosaurus*, and an unnamed nodosaurid (based on a scapulocoracoid). At least two small theropod taxa remain to be identified to the generic level. The quarry is the type locality of the robust megalosaurid-grade theropod *Torvosaurus* and the sauropods *Dystylosaurus*, *Supersaurus*, and *Ultrasauros* (= *Ultrasaurus* Jensen 1985). The validity of *Dystylosaurus* (Jensen, 1985) remains to be determined. Through the years, there has been much confusion over *Supersaurus* and *Ultrasaurus*. *Supersaurus*, an enormous diplodocid, is valid (B. Curtice *et al.*, manuscript in preparation) and still ranks as one of the largest animals known. It is now recognized that the 2.5-m-long scapulocoracoid that Jensen (1985) referred to *Ultrasauros*, once thought to be the largest dinosaur ever found, is referable to *Brachiosaurus*. Furthermore, the *Brachiosaurus* scapulocoracoid is slightly shorter than the largest one known from Tendaguru. The type specimen of *Ultrasauros* was not the scapulocoracoid but a large diplodocid dorsal vertebra, now referred to *Supersaurus*. Thus, *Ultrasauros* is a junior synonym of *Supersaurus* (Curtice *et al.*, manuscript in preparation).

The Dry Mesa site was discovered by Eddie and Vivian Jones, who reported their find to James A. Jensen of Brigham Young University (BYU). Jensen opened the quarry in the spring of 1972 and the quarry has been operated by BYU in nearly every field season since that time. More than 4000 elements have been recovered and the quarry face is more than 120 m in length. The quarry is positioned near the base of the Brushy Basin Member of the Morrison Formation. The bone-bearing lithosome is a fluvial deposit consisting of a broad, conglomeratic sandstone channel incised into overbank mudstone deposits. Bones were deposited at the base of trough crossbeds in the bottom of the river channel. Portions of bones not immediately buried in the sands were weathered before being completely buried by subsequent depositional events. Also, many of the bones exhibit varying degrees of pitting attributed to an as yet unknown biologic agent. Most of the skeletons were completely disarticulated, making it difficult to associate bones with particular taxa. In 1985, a partially articulated juvenile diplodocid was discovered. Before its discovery the few articulated bones consisted mainly of vertebral column segments. Richmond and Morris (1997) attributed the concentration of bones to a drought followed by a flash flood with a maximum flow velocity approaching 2 m/sec.

Although the quarry is most renowned for its dinosaurian fauna, its nondinosaurian composition is surprisingly diverse and remains to be studied in detail. The nondinosaurian fauna consists of fishes, including the lungfish *Ceratodus*, an amphibian, two turtle genera, a crocodilian, and a prototherian mammal (Prothero and Jensen, 1983). Many small vertebrates have been recovered from the matrix encasing the large dinosaur bones, including several amphibians, various reptiles, maniraptoran dinosaurs that may be either small dromaeosaurs or birds, and the type material of the pterodactyloid pterosaur *Mesadactylus ornithosphyos* (Jensen and Padian, 1989). As is the case with most Morrison Formation quarries, no plant fossils were preserved.

See also the following related entry:

Morrison Formation

References

Britt, B. B., (1991). Theropods of Dry Mesa Quarry (Morrison Formation, late Jurassic), Colorado, with emphasis on the osteology of Torvosaurus tanneri. *Brigham Young Univ. Geol. Stud.* **37,** 1–72.

Jensen, J. A., (1985). Three new sauropod dinosaurs from the Upper Jurassic of Colorado. *Great Basin Nat.* **45**(4), 697–709.

Jensen, J. A., and Padian, K. (1989). Small pterosaurs and dinosaurs from the Uncompahgre fauna (Brushy Basin

FIGURE 1 *Dryosaurus* with young (© Mark Schultz, 1996).

Member, Morrison Formation: ?Tithonian), Late Jurassic, western Colorado. *J. Paleontol.* **63**(3), 364–374.

Prothero, D. R., and Jensen, J. A. (1983). A mammalian humerus from the Upper Jurassic of Colorado. *Great Basin Nat.* **43**(4), 551–553.

Dryosauridae

Michael J. Ryan
Royal Tyrrell Museum of Palaeontology
Drumheller, Alberta, Canada

Dryosaurids (Fig. 1) were small, primitive, bipedal herbivorous ornithopods from the Upper Jurassic and Lower Cretaceous (Sues and Norman, 1990). Dryosaurids reached ~6.5 m in length, more than half of which was tail, and weighed ~70 kg. They closely resemble members of the Hypsilophodontidae and were originally placed within this family. Subsequent work recognized a number of unique features (see Sues and Norman, 1990) that placed dryosaurids in their own monophyletic family and made them a basal taxon within the Iguanodontia.

The Dryosauridae comprise four species with a wide distribution across both hemispheres. *Dryosaurus altus* is known from the Morrison Formation (late Kimmeridgian–early Tithonian) of Colorado, Wyoming, and Utah of the United States. *Dryosaurus lettowvorbecki* is known from the Tendaguru beds (late Kimmeridgian) of Mtwara, Tanzania. Despite the wide geographic separation, these two species appear to be very closely related and show only very minor differences (i.e., longer palpebrals in *D. altus* and a more pronounced olecranon process in *D. lettowvorbecki*). This close resemblance has been used to support the close ties of Laurasia and Gondwana into the latter part of the Jurassic (Galton, 1977).

Valdosaurus (Galton, 1977) is a poorly known descendant of *Dryosaurus* from the Lower Cretaceous. *Valdosaurus canaliculatus* (Galton, 1975) is known primarily from post-cranial material from the Wealden of the Isle of Wright and West Sussex, England, and the Bauxite of Cornet, Romania (Berriasian–Barremian). It is distinguished from *Dryosaurus* by the deeper, more prominent extensor groove on the distal end of the femur. *Valdosaurus nigeriensis* (Galton and Taquet, 1982) has been named from a single femur from the El Rhaz Formation of Agadez, Nigeria (late Aptian). Hooker *et al* (1991) have reported on ornithopod material from the Late Cretaceous of West Antarctica (Vega Island) with dryosaurid affinities.

Dryosaurids are relatively small with a short snout giving the skull a triangular appearance in lateral view. The premaxillae are edentulous and smooth and would have born a bony beak in life. The premaxillae do not enclose the external narial openings dorsally. A long dorsocaudal process of the premaxilla contacts the prefrontal and lacrimal separating the nasal from contact with the lacrimal and the maxilla (Sues and Norman, 1990). There is a small gap between the rostral margin of the maxilla and the premaxilla in both embryos and adults. In *D. altus* the palpebral is a long tapering element that extends completely across the orbit. In embryos, the palpebral extends to three-quarters of the orbit as in the adult *D. lettowvorbecki*. Of note is a rostral embayment (not present in the embryonic *D. altus*) on the quadrate similar to the foramen in the quadratojugal of *Hypsilophodon*. The dentary is narrow and straight margined with an elevated coronoid process. The predentary has a bilobed ventral process similar to that seen in ceratopsids. The thickly enameled buccal surfaces of the maxillary teeth and the corresponding lingual surfaces of the dentary teeth bear one prominent and a number of secondary ridges, the latter of which usually terminate at well-developed denticles along the tooth crown margin. These denticles are seen even in embryonic material (Scheetz, 1991). Dryosaurs would have been efficient processors of vegetable matter. The joint between the premaxilla and the maxilla functioned as a diagonal hinge allowing the skull to rotate slightly relative to the muzzle when the dinosaur chewed. This meant that as the jaws closed on food, with the teeth of the upper jaw rotated slightly outward increasing the shear force between food and teeth.

Postcranially, dryosaurs closely resemble hypsilophodontids possessing at least nine cervicals, 15(?) dorsals, six sacrals, and 29+ caudal vertebrae (complete tails are not known). The hind limbs are long and slender. The femur has anterior bowing seen in all small ornithopods. It possesses a distinct anterior intercondylar extensor groove on the distal head, a deep separation of the greater and lesser trochanters, and a very well-developed fourth trochanter with a pit at its base for insertion of the *M. caudifemoralis longus*. The manus is poorly known from incomplete material. The pedal formula is 2-3-4-5-0 with the presence of a vestigial metatarsal I. The structure of the legs would have made dryosaurs very fast runners—something they would have needed because they lacked any other defense against predators. Thulborn (1982) estimated their top speed to be in excess of 40 km/hr, making *Dryosaurus* one of the fastest ornithischian dinosaurs to be studied to date.

Dryosaurus is well known from bone bed material from both Africa and the United States. One site near Uravan, Colorado, has produced more than 2000 Dryosaurus bone fragments with a minimum of eight individuals ranging from 25 cm (hatchling) to 165 cm long. Carpenter (1994) has described an articulated baby *D. altus* skull from Dinosaur National Monument in Utah. Heinrich *et al.* (1993) examined the biomechanical strength of hatchling *D. lettowvorbecki* femora and determined that young dryosaurs were quadrupedal until at least 2.5 months of age, becoming functionally bipeds when they reached their next size class. Chinsamy (1995) examined the bone histology of *D. lettowvorbecki* and concluded that this dinosaur had a rapid, uninterrupted pattern of bone growth. Such growth is typical of mammals and birds that maintain a high, constant body temperature by endogenous means. However, like most reptiles, dryosaurs also appear to have had a pattern of indeterminate growth.

See also the following related entry:

Ornithopoda

References

Carpenter, K. (1994). Baby *Dryosaurus* from the Upper Jurassic Morrison Formation of Dinosaur National Monument. In *Dinosaur Eggs and Babies* (K. Carpenter, K. F. Hirsch, and J. R. Horner, Eds.), pp. 288–297. Cambridge Univ. Press, Cambridge, UK.

Chinsamy, A. (1995). Ontogenetic changes in the bone histology of the Late Jurassic Ornithopod *Dryosaurus lettowvorbecki*. *J. Vertebr. Paleontol.* **15,** 96–104.

Galton, P. M. (1977). The ornithopod dinosaur *Dryosaurus* and a Laurasia–Gondwanaland connection in the Upper Jurassic. *Nature* **268,** 230–232.

Galton, P. M., and Taquet, P. (1982). *Valdosaurus,* a hypsilophodontid dinosaur from the Lower Cretaceous of Europe and Africa. *Géobios* **15,** 147–159.

Heinrich, R. E., Ruff, C. B., and Weishampel, D. B. (1993). Femoral ontogeny and locomotor biomechanics of *Dryosaurus lettowvorbecki* (Dinosauria, Iguanodontia). *Zool. J. Linnean Soc.* **108,** 179–196.

Hooker, J. J., Milner, A. C., and Sequeira, S. E. K. (1991). An ornithopod dinosaur from the Late Cretaceous of West Antarctica. *Antarctic Res.* **3,** 331–332.

Scheetz, R. D. (1991). Progress report of juvenile and embryonic *Dryosaurus* remains from the Upper Jurassic Morrison Formation of Colorado. In *Guidebook for Dinosaur Quarries and Trackway Tour, Western Colorado and Eastern Utah* (W. R. Averett, Ed.), pp. 27–29. Grand Junction Geological Society, Grand Junction, CO.

Sues, H.-D., and Norman, D. (1990). Hypsilophodontidae, *Tenontosaurus,* Dryosauridae. In *The Dinosauria* (D. B. Weishampel, D. Dodson, and H. Osmolska, (Eds.), pp. 498–509. Univ. of California Press, Berkeley.

Thulborn, R. A. (1982). Speeds and gaits of dinosaurs. *Palaeogeogr. Palaeoclimatol. Palaeoecol.* **38,** 227–256.

Edmonton Group

David A. Eberth
Royal Tyrrell Museum of Palaeontology
Drumheller, Alberta, Canada

The Edmonton Group is an important dinosaur- and coal-bearing unit in the south-central portion of the Alberta Basin that ranges in age from latest Campanian to early Danian (Early Paleocene) and thus spans the Cretaceous–Paleogene boundary. It comprises a southeastward-thinning, largely nonmarine to shallow marine clastic wedge that is exposed in modern drainages throughout south-central Alberta. It conformably overlies and interfingers with marine shales of the late Campanian Bearpaw Formation and is overlain unconformably by sandstones of the Late Paleocene Paskapoo Formation. It has been a source of important dinosaur and other fossil vertebrate, invertebrate, and plant discoveries since the early part of the 20th century and, recently, has figured importantly in studies of terminal Cretaceous extinctions.

Stratigraphy

Gibson (1977) included the Horseshoe Canyon, Whitemud, Battle, and Scollard formations (ascending order) together in the Edmonton Group. This stratigraphic arrangement is widely accepted and is used here. The base of the group (Horseshoe Canyon Formation) is conformable on the underlying marine Bearpaw Formation. Whereas the Horseshoe Canyon–Whitemud and Whitemud–Battle formational contacts are regarded as conformable, a significant unconformity has been postulated for the Battle–Scollard contact (Russell, 1983) and Scollard–Paskapoo contact (Lerbekmo *et al.*, 1992).

Following Cant and Stockmal (1989), the wedge resulted from a major tectonic and basin-response event related to the accretion of the Pacific Rim–Chugach Terrane. In the terminology of Kaufmann (1977), the Edmonton Group records the R9 and T/R 10 cycles as well as the transgression that coincides with the onset of the Danian. In the sequence stratigraphic terminology of Haq *et al.* (1988), the Edmonton Group includes the uppermost portion of the Upper Zuni A supercycle (cycles 4.4 and 4.5) as well as the lower portion of the Tejas A supercycle (cycles 1.1 and 1.2). Although the base of the group is strongly diachronous west–east, paleomagnetic analyses of the dinosaur-bearing outcrops in the Red Deer River valley (Lerbekmo and Coulter, 1985) show that the lowest 40 m of these exposures is in the uppermost portion of the 33n magnetochron (latest Campanian), and that the uppermost 5 m of the group extends up into the 28r magnetochron. Radiometric data from bentonites reveal an age of approximately 72 or 73 Ma for the base of the exposed group in the Red Deer River valley (estimated from results published by Lerbekmo and Coulter, 1985). The top of the group occurs approximately 40 m above the Cretaceous–Paleogene (K–P) boundary and is estimated here at approximately 63.5 Ma. The K–P boundary has been recently dated at 64.70 [0.09 Ma (McWilliams and Baadsgaard, 1991)], whereas the Kneehills Tuff, a multistoried, silicified to bentonitic volcanic ash horizon in the Battle Formation, has yielded K–Ar dates that suggest an overall age of 66.5–67.0 Ma (inferred from Lerbekmo and Coulter, 1985).

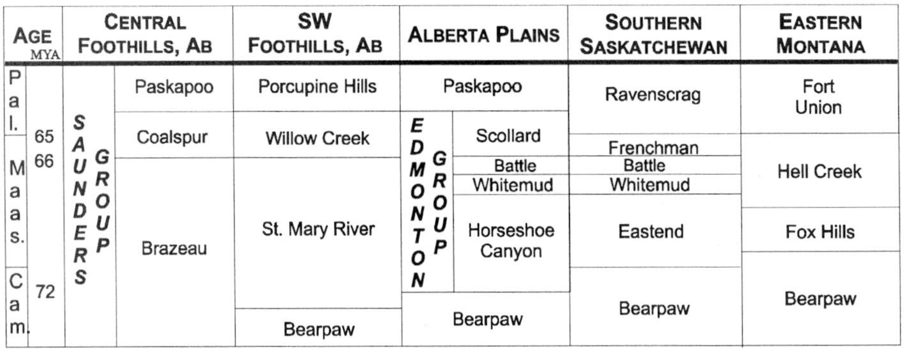

Age	MYA	Central Foothills, AB (Saunders Group)	SW Foothills, AB	Alberta Plains (Edmonton Group)	Southern Saskatchewan	Eastern Montana
Pal.		Paskapoo	Porcupine Hills	Paskapoo	Ravenscrag	Fort Union
	65	Coalspur	Willow Creek	Scollard	Frenchman	Hell Creek
Maas.	66	Brazeau	St. Mary River	Battle	Battle	Hell Creek
		Brazeau	St. Mary River	Whitemud	Whitemud	Hell Creek
		Brazeau	St. Mary River	Horseshoe Canyon	Eastend	Fox Hills
Cam.	72	Brazeau	Bearpaw	Bearpaw	Bearpaw	Bearpaw

FIGURE 1

Correlations (Fig. 1)

North of the North Saskatchewan River (at Edmonton) the Edmonton and Judith River clastic wedges become indistinguishable; there, the combined Campanian–Maastrichtian wedge is referred to as the Wapiti Formation. In the Foothills Belt of western Alberta, the Edmonton Group correlates with the St. Mary River and Willow Creek formations (southern Foothills) and with the middle to upper Brazeau and Coalspur formations (central Foothills). To the east into southern Saskatchewan, the Battle and Whitemud formations are still recognized but the Horseshoe Canyon and Scollard formations are equivalent to the Eastend and Frenchman formations, respectively. South-eastward into Montana, the group is equivalent to the Fox Hills and Hell Creek formations. Farther southward the Scollard is equivalent to the Lance and the lower part of the Tullock Formation.

Sediments and Paleoenvironments

Edmonton Group facies reflect a variety of depositional settings as well as tectonic and climatic influences along the Western Interior seaway during latest Campanian, Maastrichtian, and earliest Paleogene time. In general, coarse sediments range from litharenites to volcanic litharenites (Rahmani and Lerbekmo, 1975; Binda *et al.*, 1991). Claystones comprise bentonite and minor kaolinite.

The Horseshoe Canyon Formation (up to 230 m thick) forms an overall regressive unit comprising offlapping, stacked parasequences that record rapid changes in relative sea level (Ainsworth, 1994). A complex variety of depositional environments are present including offshore, shoreface, foreshore, barrier, estuarine, back-barrier, lagoonal, tidal flat, peat swamp, salt marsh, deltaic, distributary channel, oyster bank, meandering and straight fluvial channels, and overbank (Rahmani, 1988; Ainsworth, 1994). Incised valleys have been recently identified (Ainsworth and Walker, 1994; Eberth, 1996). The lower one-third of the formation displays a strong, nearshore marine influence and is depauperate in vertebrate fossils. The remaining two-thirds of the formation is dominated by coastal plain facies that yield the vast majority of vertebrate fossils.

The Whitemud Formation (up to 20 m thick) consists of light gray to white weathering bentonitic and kaolinitic sandstones, siltstones, and claystones. Its sediments were deposited on an extensive, very low-gradient alluvial plain that stretched across Alberta and western Saskatchewan. Depositional environments include meandering streams, swamps, ponds, and small lakes (Nambudiri and Binda, 1991).

The Battle Formation (up to 14 m thick) consists of dark gray to purplish black bentonites and bentonitic shales with thin interbedded, rooted, siliceous paleosols, erroneously referred to as tuffs. Like the Whitemud, its sediments are thought to have been deposited in very low-gradient alluvial plain to paludal environments during a period of frequent volcanic eruptions.

The Scollard Formation (up to 80 m thick) can be divided subequally into a lower, non-coaly interval and an upper, coaly interval (Eberth and O'Connell, 1995). The contact between the two units coincides approximately with the K–P boundary. The lower interval (latest Maastrichtian) comprises an alluvial succession of straight, sandy, paleochannel fills; levee and splay deposits; and a variety of rooted and mottled paleosols and floodbasin deposits. The upper (Paleogene) portion of the Scollard is dominated by meandering channel sediments deposited in environments characterized by a very high water table, extensive interchannel lakes, marshes, and peat swamps. Eberth and O'Connell (1995) proposed that the principal cause of the transition from straight to meandering channels upwards across the K–P boundary in southern Alberta was climatic, resulting from increased wetness, locally rising water tables, and changes in discharge characteristics and in-channel sediment load.

Along the Red Deer River, north of Drumheller, the K–P boundary has been accurately placed at the base of, or within the lowest few centimeters of, the thin, though locally extensive, subbituminous coal referred to as the Nevis, or No. #13 seam (Lerbekmo and St. Louis, 1986). This coal occurs approximately in the middle of the formation, at the lowest portion of the upper coaly unit. A kaolinitic clay layer is locally preserved within or at the base of the Nevis seam and has yielded an iridium abundance spike, an anomalous concentration of shocked quartz, and microdiamonds, and is now widely regarded as resulting from a bolide impact event.

Paleontology

The Edmonton Group yields an impressive and diverse assemblage of plant, invertebrate, and vertebrate fossils recording important aspects of the paleoecology of western Canada during the past 7 million years of the Cretaceous and the first 1 million years of the Paleogene (Tables I and II). Although the taxonomic literature is extensive, useful reviews can be found in a series of guidebooks published by staff

TABLE I Fossil Vertebrates from the Horseshoe Canyon Formation

- Chondrichthyes
  - Elasmobranchii
    - *Lamna*
  - Sclerorhynchidae
    - *Myledaphus bipartitus*
- Osteichthyes
  - Acipenseriformes
    - Acipenseridae
      - *Acipenser* sp.
      - *Diphyodus longirostris*
    - Polyodontidae
      - *Palaeopsephurus* sp.
  - Lepisosteiformes
    - Lepisosteidae
      - *Atractosteus* sp.
  - Amiiformes
    - Amiidae
      - *Cyclurus fragosa*
    - Aspidorhynchiformes
      - Aspidorhynchidae
        - *Belonostomus* sp.
- Amphibia
  - Anura
    - Undetermined gen. and sp.
  - Caudata
    - Undetermined gen. and sp.
- Reptilia
  - Chelonia
    - Trionychidae
      - *Aspideretes* sp.
    - Dermatemydidae
      - *Basilemys* sp.
  - Sauropterygia
    - Elasmosauridae
      - *Leurospondylus ultimus*
  - Archosauromorpha
    - *Champsosaurus albertensis*
  - Crocodylia
    - Crocodylidae
      - *Leidyosuchus* sp.
      - *Stangerochampsa*
  - Saurischia
    - Dromaeosauridae
      - cf. *Dromaeosaurus* sp.
      - *Saurornitholestes* sp.
    - Troodontidae
      - *Troodon formosus*
    - cf. Caenagnathidae
      - *Chirostenotes pergracilis*
    - Ornithomimidae
      - *Dromiceiomimus brevitertius*
      - *Ornithomimus edmontonensis*
      - *Ornithomimus currelli*
      - *Ornithomimus ingens*
      - *Struthiomimus altus*
    - Tyrannosauridae
      - *Albertosaurus sarcophagus*
      - *Daspletosaurus* cf. *D. torosus*
      - *Aublysodon* sp.
  - Ornithischia
    - Hypsilophodontidae
      - *Parksosaurus warreni*
    - Hadrosauridae
      - *Edmontosaurus annectens*
      - *Edmontosaurus regalis*
      - *Edmontosaurus edmontoni*
      - *Hypacrosaurus altispinus*
      - *Saurolophus osborni*
    - Pachycephalosauridae
      - *Stegoceras edmontonense*
      - *Stegoceras validus*
    - Nodosauridae
      - *Edmontonia longiceps*
      - *Panoplosaurus* sp.
    - Ankylosauridae
      - *Anodontosaurus lambei*
      - *Euoplocephalus tutus*
    - Ceratopsidae
      - *Anchiceratops ornatus*
      - *Anchiceratops longirostris*
      - *Arrhinoceratops brachyops*
      - *Pachyrhinosaurus canadensis*
- Mammalia
  - Multituberculata
    - Unidentified gen. and sp.
  - Marsupicarnivora
    - *Didelphodon coyi*

TABLE II Fossil Vertebrates from the Scollard Formation

- Chondrichthyes
  - Heterodontiformes
    - Palaeospinacidae
      - *Palaeospinax ejuncidus*
  - Elasmobranchii
    - Sclerorhynchidae
      - *Myledaphus bipartitus*
- Osteichthyes
  - Acipenseriformes
    - Acipenseridae
      - *Acipenser* sp.
  - Lepisosteiformes
    - Lepisosteidae
      - *Atractosteus* sp.
      - *Lepisosteus occidentalis*
  - Amiiformes
    - Amiidae
      - *Cyclurus fragosa*
      - *Amia* sp.
  - Aspidorhyncbiformes
    - Aspidorhynchidae
      - *Belonostomus* sp.
  - Elopiformes
    - Albulidae
      - *Paralbula* sp.
      - Unidentified gen. and sp.
  - Semionotiformes
    - Semionotidae
      - *Semionotus* sp.
      - Holostean sp. A
- Amphibia
  - Caudata
    - Scapherpetonidae
      - *Scapherpeton tectum*
      - *Urodela* sp.
  - Anura
    - *Anura* sp.
- Reptilia
  - Chelonia
    - Dermatemydidae
      - *Adocus* sp.
      - *Compsemys victa*
    - Trionychidae
      - *Aspideretes* sp.
    - Chelyridae
      - Unidentified gen and sp.
  - Archosauromorpha
    - Champsosauridae
      - *Champsosaurus* sp.
  - Crocodylia
    - Crocodylidae
      - *Leidyosuchus* sp.
  - Pterosauria
    - Pterodactyloidea
      - Unidentified gen. and sp.
  - Saurischia
    - Troodontidae
      - cf.*Troodon* sp.
    - Ornithomimidae
      - *Struthiomimus* sp.
      - Unidentified gen. and sp.
    - Tyrannosauridae
      - *Paronychodon* sp.
      - *Tyrannosaurus rex*
      - Unidentified gen. and sp.
  - Ornithischia
    - Hypsilophodontidae
      - cf. *Hypsilophodon* sp.
    - Thescelosauridae
      - *Parksosaurus* sp.
      - *Thescelosaurus neglectus*
      - *Thescelosaurus edmontonensis*
    - Pachycephalosauridae
      - ?*Pachycephalosaurus* sp.
      - *Stegoceras* sp.
    - Protoceratopsidae
      - *Leptoceratops gracilis*
    - Ceratopsidae
      - *Monoclonius* sp.
      - cf. *Torosaurus* sp.
      - *Triceratops albertensis*
    - Ankylosauridae
      - *Ankylosaurus magniventris*
- Aves
  - Unidentified gen. and sp.
- Mammalia
  - Multituberculata
    - Ectypodae
      - *Cimexomys priscus*
      - *Mesodma* cf. *florencae*
      - *Mesodma hensleighi*
      - *Mesodma thompsoni*
    - Ptilodontidae
      - *Cimolodon nitidus*
    - Cimolomyidae
      - *Cimolomys gracilis*
      - *Cimolomys trochuus*
  - Undetermined order
  - Deltatheridiidae
      - cf. *Deltatheroides cretacicus*
  - Marsupicarnivora
    - Didelphidae
      - *Alphadon marshi*
      - *Alphadon wilsoni*
      - *Alphadon rhaister*
    - Pediomyidae
      - *Pediomys elegans*
      - *Pediomys* cf. *florencae*
      - *Pediomys hatcherii*
      - *Pediomys krejcii*
    - Stagodontidae
      - *Didelphodon vorax*
      - *Eodelphis* sp.
  - Insectivora
    - Gypsonictopidae
      - *Gypsonictops hypoconus*
      - *Gypsonictops illuminatus*
    - Palaeoryctidae
      - *Cimolestes cerberoides*
      - *Cimolestes magnus*
      - *Cimolestes propalaeoryctes*
      - *Batodon tenuis*

at the ROYAL TYRRELL MUSEUM in Drumheller (e.g., Braman *et al.*, 1995) as well as papers by L. Russell (1964), D. Russell (1967), and Fox (1988).

It is the vertebrate faunas, especially the dinosaurs, that are the best known fossils from the Horseshoe Canyon Formation (Table I). Some of the earliest finds of significant remains in Alberta were made along the Red Deer River northwest of Drumheller, and since 1910 more than 100 complete or partial skeletons have been excavated. Hadrosaurs are the most numerous dinosaur preserved, comprising roughly 50% of the known articulated–associated remains. Ceratopsians and ornithomimids are present in subequal abundance and together comprise approximately 30% of known articulated–associated specimens. Tyrannosaurids make up only 5% of the

Predatory dinosaurs, such as *Allosaurus fragilis* from the Morrison Formation, are recorded in the western United States. Possible prey, like *Dryosaurus altus*, are identified by a coincident stratigraphic and geographic distribution and also occasionally by association of fossil remains. Illustration by Michael Skrepnick.

A sculpture depicting hatching *Protoceratops* at the Institute of Vertebrate Paleontology and Paleoanthropology in Beijing, China. Protoceratopsids first appear in Toogreeg and Alan Teg sediments in Mongolia and were last found in the Lance and Scollard Formations of Wyoming and Alberta, respectively. Photo by François Gohier.

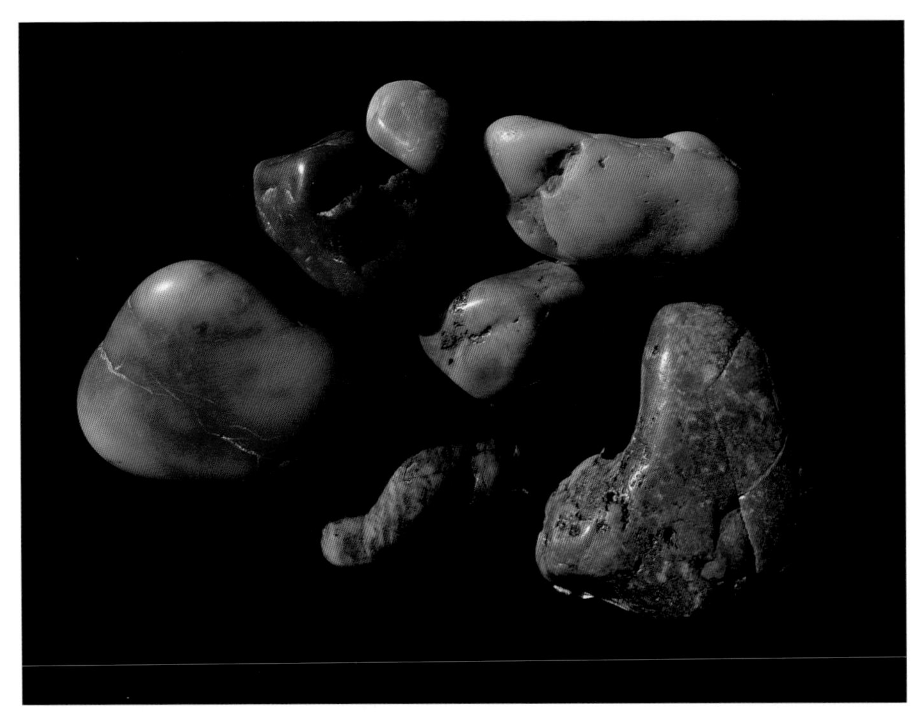

Clumps of smooth stones found in close association with particular dinosaur fossils, often inside the rib cage or apparent abdominal cavity, may be gastroliths (stomach stones). These stones might have functioned in the same way that small stones swallowed by some birds function; they aid in grinding up tough or fibrous plant food. Photo by François Gohier.

Theories of dinosaur extinction are numerous, but two extremes dominate: (1) extinction was a gradual, locality by locality, species by species disappearance; (2) a single, sudden, widespread event eradicated all dinosaurs extant at the close of the Cretaceous. Evidence of catastrophic events near the close of the Cretaceous does not constitute sufficient evidence of the latter theory. Illustration by Michael Rothman.

Group behaviors of dinosaurs cannot be observed directly, but comparative anatomy, the circumstances of preservation, footprints and trackways, and analogy suggest that complex behaviors may have been common. Depicted here is a lek of *Parasaurolophus*, but there is as yet no direct evidence for such lek structures among dinosaurs. Illustration by Larry Felder.

Dinosaur footprints and trackways are found widely. However, associating particular footprints with particular dinosaurs is not easy. These trackways have been used as evidence to support hypotheses of group behaviors, migration, locomotion patterns, and physiology. Photo by François Gohier.

Associations of particular dinosaurs with habitats are often speculative. *Iguanodon*, pictured here as a forest animal, may have thrived in many kinds of habitats, including the shores of a North American intercontinental sea. In fact, *Iguanodon* has also been recovered in sediments from Europe, Asia, Africa, and Australia. Illustration by Michael Rothman.

Lance Creek horizons are among the last Cretaceous formations to contain dinosaur remains. Rocks in the Cheyenne River Valley of Wyoming are assigned to the Lance Creek Formation. Photo by François Gohier

TABLE III
Stratigraphic Positions of Articulated Dinosaur Specimens below the Cretaceous–Paleogene Boundary [Base of the Nevis (No. 13) Seam][a]

Specimen	Position (m)
NMC 8861 cf. *Triceratops albertensis*	2.3
NMC 9542 cf. *Triceratops albertensis*	4.5
UA 9542 uncataloged cf. *Torosaurus sp.*	8.0
TMP 81.12.1 *Tyrannosaurus rex*	10.5
NMC 8862 *Triceratops albertensis*	33.2
NMC 8889 *Leptoceratops gracilis*	33.2
Uncollected large ceratopsian (NMC field records)	34.8
NMC 887, 8888 *Leptoceratops gracilis*	34.8
Uncollected large ceratopsian (NMC field records)	42.4
NMC 8880 *Ankylosaurus magniventris*	43.9
AMNH 5213 *Ankylosaurus magniventris*	45.4
NMC 8537 *Thescelosaurus neglectus*	48.5

[a] From Braman *et al.* (1995).

known assemblage. The most common taphonomic modes include mono- and multigeneric bone beds, vertebrate microfossil sites, and articulated to associated skeletons in overbank, channel thalweg, and point-bar deposits. Other common fossil types include macroplants (Bell, 1949, 1965; Aulenbach and Braman, 1991), nonmarine to marine invertebrates and trace fossils (Saunders, 1989), fish, amphibians, nondinosaurian reptiles, and mammals (Braman *et al.*, 1995).

Together, the Whitemud and Battle formations yield very little vertebrate fossil bone but are important sources of plants and coprolites (e.g., Broughton *et al.*, 1978; Binda *et al.*, 1991), especially in southern Saskatchewan. Plant microfossil, seed cuticle, and palynomorph assemblages have been examined at sites across Alberta and Saskatchewan and support the interpretation that together, the Whitemud and Battle form a chronostratigraphic datum in this region (Nambudiri and Binda, 1991). Sideritic vertebrate coprolites (attributed to dinosaurs) have been described by Broughton *et al.* (1978) and interpreted by Binda *et al.* (1991) as resulting from syndepositional methanic diagenesis, a process that may have been responsible for the dissolution of shells and bone in these same deposits. There are anecdotal reports of vertebrate microfossils from the Battle Formation in the Drumheller area.

The Scollard Formation (Table II) has yielded important dinosaur remains since Barnum Brown made the first collections in 1910. The fauna includes 20 different taxa of dinosaur and a large diversity of mammals (Russell, 1987). Dinosaur specimens are found throughout the lower one-half (latest Maastrichtian) of the Scollard, up to 2.3 m below the K–P boundary at the base of the Nevis seam (Table III). Articulated dinosaurs are rare and the fauna is known largely by vertebrate microfossil remains associated with channel lags. Hadrosaurs are not represented in collections from Alberta, but ankylosaurs are abundant, especially in the lowest 10 m of the exposures. Tyrannosaurids and ceratopsians dominate the vertebrate assemblage in the 20 m of section below the K–P boundary.

This dinosaur assemblage constitutes the most northerly known occurrence of the Lancian *Triceratops* fauna. An overview of the mammal fauna is presented by Fox (1988). In addition to the dinosaurs, a modest invertebrate assemblage comprising bivalves and gastropods is known as well as an extensive palynomorph assemblage (Braman *et al.*, 1995).

See also the following related entries:

Cretaceous Period • Extinction, Cretaceous • Hell Creek Formation • Lance Formation • Land Mammal Ages

References

Ainsworth, R. B. (1994). Marginal marine sedimentology and high resolution sequence analysis; Bearpaw–Horseshoe Canyon transition, Drumheller, Alberta, Canada. *Bull. Can. Petroleum Geol.* **42,** 26–54.

Ainsworth, R. B., and Walker, R. G. (1994). Control of estuarine valley-fill deposition by fluctuations of relative sea-level, Cretaceous Bearpaw–Horseshoe Canyon Transition, Drumheller, Alberta, Canada. In *Incised-Valley Systems: Origin and Sedimentary Sequences*, SEPM Spec. Publ. No. 51, pp. 159–174.

Aulenbach, K. R., and Braman, D. B. (1991). A chemical extraction technique for the recovery of silicified plant remains from ironstones. *Rev. Paleobot. Palynol.* **70,** 3–8.

Bell, W. A. (1949). Uppermost Cretaceous and Paleocene floras of western Alberta. Geological Survey of Canada, Bull. 11, pp. 231.

Bell, W. A. (1965). Upper Cretaceous and Paleocene plants of Western Canada. Geological Survey of Canada, Paper No. 65–35, pp. 46.

Binda, P. L., Nambudiri, E. M. V., Srivastava, S. K., Schmitz, M., Longinelli, A., and Iacumin, P. (1991). Stratigraphy, paleontology, and aspects of diagenesis of the Whitemud Formation (Maastrichtian) of Alberta and Saskatchewan. In *Sixth International Williston Basin Symposium* (J. E. Christopher and F. M. Haidl, (Eds.), Spec. Publ. No. 11, pp. 179–192. Saskatchewan Geological Society.

Braman, D. R., Johnston, P. A., and Haglund, W. M. (1995). Upper Cretaceous paleontology, stratigraphy and depositional environments at Dinosaur Provincial Park and Drumheller, Alberta. Canadian Paleontology Conference Field Trip Guidebook No. 4. Fifth Canadian Paleontology Conference, pp. 119. Geological Association of Canada.

Broughton, P. L., Simpson, F., and Whitaker, S. H. (1978). Late Cretaceous coprolites from western Canada. *J. Geol.* **89,** 741–749.

Cant, D. J., and Stockmal, G. S. (1989). The Alberta foreland basin: Relationship between stratigraphy and Cordilleran terrane-accretion events. *Can. J. Earth Sci.* **26,** 1964–1975.

Eberth, D. A. (1996). Origin and significance of mud-filled, incised valleys (Upper Cretaceous) in southern Alberta, Canada. *Sedimentology* **43,** 459–478.

Eberth, D. A., and O'Connell, S. C. (1995). Notes on changing paleoenvironments across the Cretaceous–Tertiary boundary (Scollard Formation) in the Red Deer River valley of southern Alberta. *Bull. Can. Petroleum Geol.* **43,** 44–53.

Fox, R. C. (1988). Late Cretaceous and Paleocene mammal localities of southern Alberta. Field Trip "A," Society of Vertebrate Paleontology 48th Annual Meeting. Occasional Paper of the Royal Tyrrell Museum of Palaeontology, No. 6, pp. 38.

Gibson, D. W. (1977). Upper Cretaceous and Tertiary coal-bearing strata in the Drumheller–Ardley region, Red Deer River Valley, Alberta. Geological Survey of Canada, Paper No. 76-35, pp. 41.

Haq, B. U., Hardenbol, J., and Vail, P. R. (1988). Mesozoic and Cenozoic chronostratigraphy and cycles of sea-level change. In *Sea-Level Changes: An Integrated Approach,* (C. K. Wilgus, B. S. Hastings, C. G. St. C. Kendall, H. W. Posamentier, C. A. Ross, and J. C. Van Wagoner, Eds.), Spec. Publ. No. 42, pp. 71–108. Society of Economic Paleontologists and Mineralogists.

Kauffman, E. G. (1977). Geological and biological overview: Western Interior Cretaceous Basin. *Mountain Geologist* **14,** 75–99.

Lerbekmo, J. F., Demchuk, T. D., Evans, M. E., and Hoye, G. S. (1992). Magnetostratigraphy, and biostratigraphy of the continental Paleocene of the Red Deer Valley, Alberta, Canada. *Bull. Can. Petroleum Geol.* **40,** 24–35.

Lerbekmo, J. F., and Coulter, K. C. (1985). Late Cretaceous to early Tertiary magnetostratigraphy of a continental sequence: Red Deer Valley, Alberta, Canada. *Can. J. Earth Sci.* **22,** 567–583.

Lerbekmo, J. F., and St. Louis, R. M. (1986). The terminal Cretaceous iridium anomoly in the Red Deer Valley, Alberta, Canada. *Can. J. Earth Sci.* **23,** 120–124.

McWilliams, M., and Baadsgaard, H. (1991). High resolution $^{40}Ar/^{39}Ar$ ages from Cretaceous–Tertiary boundary bentonites in western North America. Event Markers in Earth History, IGCP Project Nos. 216, 293, and 303, pp. 53. [Abstract]

Nambudiri, E. M. V., and Binda, P. L. (1991). Paleobotany, palynology and depositional environment of the Maastrichtian Whitemud Formation in Alberta and Saskatchewan, Canada. *Cretaceous Res.* **12,** 579–596.

Rahmani, R. A. (1988). Estuarine tidal channel and nearshore sedimentation of a Late Cretaceous epicontinental sea, Drumheller, Alberta, Canada. In *Tide-Influenced Sedimentary Environments and Facies* (P. L. de Boer, A van Gelder, and S. D. Nio, Eds.), pp. 433–471. Reidel, Dordrecht.

Rahmani R. A., and Lerbekmo, J. F. (1975). Heavy mineral analysis of Upper Cretaceous and Paleocene sandstones in Alberta and adjacent areas of Saskatchewan. In *The Cretaceous System in the Western Interior of North America* (W. G. E. Cladwell, Ed.), Spec. Paper No. 13, pp. 607–632. Geological Association of Canada.

Russell, L. S. (1983). Evidence for an unconformity at the Scollard–Battle contact, Upper Cretaceous strata, Alberta. *Can. J. Earth Sci.* **20,** 1219–1231.

Russell, D. A. (1967). A census of dinosaur specimens collected in western Canada. Natural History Papers, No. 36, pp. 13. National Museum of Canada.

Russell, L. S. (1964). Cretaceous non-marine faunas of northwestern North America. Life Sciences Contrib. No. 61, pp. 24. Royal Ontario Museum.

Russell, L. S. (1987). Biostratigraphy and palaeontology of the Scollard Formation, Late Cretaceous and Paleocene of Alberta. Life Sciences Contrib. No. 147, pp. 23. Royal Ontario Museum.

Saunders, T. (1989). Trace fossils and sedimentology of a Late Cretaceous progradational barrier island sequence: Bearpaw–Horseshoe Canyon Formation transition, Dorothy, Alberta, pp. 170. M.Sc. thesis, University of Alberta, Edmonton, Alberta, Canada.

Egg Mountain

John R. Horner
Montana State University
Bozeman, Montana, USA

Egg Mountain is the locality name for a site in the Willow Creek Anticline of western Montana, within the Upper Cretaceous (Campanian) Two Medicine Formation. Egg Mountain was discovered in 1979 by Princeton University undergraduate Fran Tannenbaum, while on an expedition under the direction of John Horner on the ranch of John and James Peebles. Miss Tannenbaum found a complete fossil egg sitting on top of a rock ledge, atop a small grass-covered hill. Excavation of the hill during the following 17 years has produced hundreds of eggs representing two dinosaurian taxa, as well as numerous skeletons of dinosaurs, lizards, and mammals. Egg Mountain is now owned by the Nature Conservancy.

When the Egg Mountain locality was first being worked, the skeletal elements were initially identified as belonging to the theropod dinosaur *Troodon* and thus many of the eggs were attributed to *Troodon*. During the following years it was discovered that the majority of skeletal elements actually belonged to a hypsilophodontid dinosaur that was eventually named *Orodromeus makelai* Horner and Weishampel 1988 to honor Bob Makela, who had accomplished the majority of the excavations and had named Egg Mountain. Subsequent discovery of an egg clutch from a nearby site named Egg Island contained the remains of embryonic skeletons initially identified as hypsilophodontid because this seemed to substantiate the identity of the eggs (Horner, 1987; Horner and Weishampel, 1988). In addition to numerous skeletons of *Orodromeus*, there were also discoveries of juvenile Troodon skeletons, varanid and teid lizards, and primitive mammals (Montellano, 1988).

In 1996, further preparation of the eggs containing the embryos revealed that the embryos were not representatives of *Orodromeus*, but rather of *Troodon* (Horner and Weishampel, 1996). Additional discoveries on Egg Mountain within the past few years have suggested the hypothesis that *Troodon* brooded its eggs similar to *Oviraptor*, constructed a sediment rim around its egg clutches, and nested in colonies (Horner, 1982; Varricchio *et al.*, 1997). The skeletons of *Orodromeus* might be food items brought to the nesting area by adults or, because a second type of egg is found in clutches on Egg Mountain (Horner, 1997), these may simply represent attritional mortality.

There are three separate, but distinct, nesting horizons within the 10-m thick rock unit comprising the Egg Mountain locality. The locality, interpreted as an island near the shores of an alkaline lake, covers less than 10,000 m^2 (Horner, 1987).

See also the following related entries:

Bastús Nesting Site • Behavior • Devil's Coulee • Eggs, Eggshells, and Nests

References

Horner, J. R. (1982). Evidence of colonial nesting and site fidelity among ornithischian dinosaurs. *Nature* **297,** 675–676.

Horner, J. R. (1987). Ecologic and behavioral implications derived from a dinosaur nesting site. In *Dinosaurs Past and Present, Vol. II* (S. J. Czerkas and E. C. Olson, Eds.), pp. 51–63. Univ. Washington Press, Seattle.

Horner, J. R. (1997). Rare preservation of an incompletely ossified embryo. *J. Vertebr. Paleontol.*, in press.

Horner, J. R., and Weishampel, D. B. (1988). A comparative embryological study of two ornithischian dinosaurs. *Nature* **332,** 256–257.

Horner, J. R., and Weishampel, D. B. (1996). Correction to: A comparative embryological study of two ornithischian dinosaurs (1988). *Nature* **383,** 103.

Montellano, M. (1988). *Alphadon halleyi* (Didelphidae, Marsupialia) from the Two Medicine Formation (Late Cretaceous, Judithian) of Montana. *J. Vertebr. Paleontol.* **8**(4), 378–382.

Varricchio, D. J., Jackson, F., Borlowski, J., and Horner, J. R. (1997). Nest and egg clutches for the theropod dinosaurs *Troodon formosus* and evolution of avian reproductive traits. *Nature* **385,** 247–250.

Eggs, Eggshells, and Nests

K. E. Mikhailov
Palaeontological Institute of the Russian Academy of Sciences
Moscow, Russia

Egg-laying is characteristic of all birds and most extant reptiles, with the exception of many secondarily viviparous snakes and lizards. All known species of birds, turtles, and crocodiles lay eggs with mineralized (calcite) shell. These eggs are hard-shelled (i.e., their shell is rigid and brittle), except in some chelonian taxa (sea turtles and some emydid turtles) that possess soft or pliable shells because the basic shell units are underdeveloped. The shell units are not well attached to one another, which permits the passage of water in and out of the egg. Some lizards, namely, geckos and dibamids, also lay hard-shelled eggs. In many other cases the soft shells of squamat eggs are encrusted with fine calcareous elements (globules, star-like structures, etc.). Soft-shelled eggs are unlikely to be fossilized for obvious taphonomic reasons.

Based on current understanding, many dinosaurs are known to have been oviparous, and some are known to have been viviparous. Correctly assigned dinosaurian eggs are now known from most of the major divisions of ornithischians and saurischians, including ornithopods (hadrosaurs and hypsilophodontids), ceratopsians (protoceratopsians), sauropods, and therizinosaurid ("segnosaur"), troodontid, oviraptorid, and perhaps dromaeosaurid theropods. No certain eggs are known from the stegosaurs and ankylosaurs. All known dinosaur eggs possessed hard shells with well-organized crystalline matter as in other hard-shelled vertebrate eggs.

The hard shells of reptilian and avian eggs consist of basic vertical shell units that start to grow from particular sites on the surface of the fibrous shell membrane. Mineralogically this consists of calcium carbonate, either in an aragonitic crystallographic form (turtles) or in calcitic form (all other reptiles and birds). In many cases the crystalline matter includes a network of organic matrix, and the fine organizations of both organic and mineral phases are in high accordance. The eggshell is pierced by numerous pore canals that enable gas exchange for the embryos. The morphology and ultra-microstructure of the basic shell units and the particular organization of the pore system are important characteristics in the designation and classification of fossil eggs, most of which belong to dinosaurs.

The general taxonomic assignment of fossil eggs is based on the stability of structural types of shells. All members of large systematic groups (such as turtles, crocodiles, and different divisions of birds and dinosaurs) exhibit their own unique eggshell structures. The only reliable argument for the final assignment of eggs to particular genera and species is the discovery of associated embryos and hatchlings, although the correlation of eggs with bones in some localities can be helpful for preliminary identification.

The large diversity of fossil egg remains, in particular those of dinosaurs, can only be described and systematically ordered using a special egg parataxonomy (as for footprints and other forms and organ taxa). The pertinent parataxonomic nomenclature initiated by the Chinese paleontologists is now generally accepted and coordinated with different structural classifications. The universal system of identifying fossil egg remains is referred to as Veterovata and includes three working categories, namely, oofamily, oogenus, and oospecies. Oofamilies are combined in larger structural groups correlated with the five basic types of eggshell structure known for vertebrates. Dinosaurian oofamilies are distributed amongst three of these types (Fig. 1). The oofamily and oogenus include the root "oolithus," derived from "oolithes" (meaning "stone egg"), which easily allows one to distinguish egg taxa from animal taxa in vertebrate fossil lists and catalogs. Such structural categories as eggshell morphotype, type of pore system, ornamentation type, egg shape type, and general range of eggshell thickness are used in the concise descriptions and diagnoses of ootaxa. Most recently known dino-

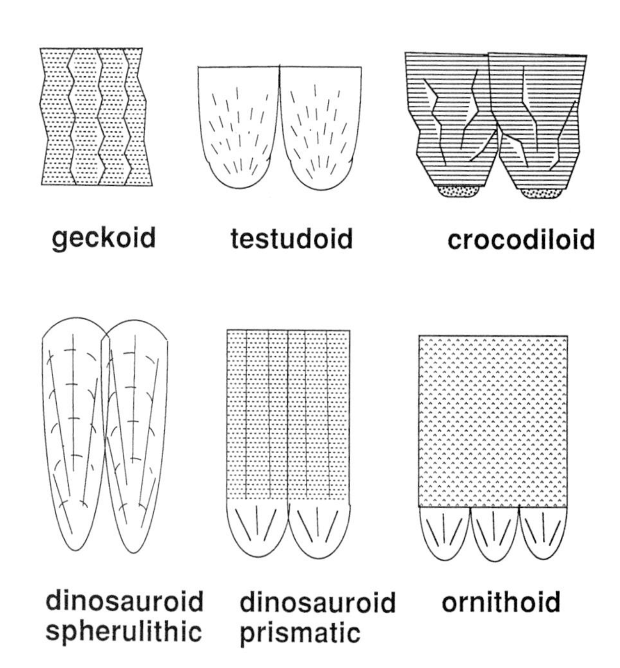

FIGURE 1 Shelled eggs are characteristic of amniotes. The microscopic structure of eggshells varies, and dinosaurs possess the lower three structures diagrammed above.

TABLE I Distribution of Eggshell Types and Oofamilies among Particular Dinosaur Lineages

Basic eggshell type	Oofamily	Taxonomic group
Testudoid		Chelonia
Geckoid		Gekkota
Crocodiloid		Crocodilia
	Spheroolithidae	Ornithopoda: hadrosaurs
	Ovaloolithidae	?Ornithopoda
Dinosauroid–spherulitic	Megaloolithidae	Sauropoda
	Faveoloolithidae	Sauropoda
	Dendroolithidae	"Segnosauria"
	Dictyoolithidae	?
Dinosauroid–prismatic	Prismatoolithidae	Ornithopoda: protoceratopsids, hypsilophodontids
Ornithoid	Elongatoolithidae	Theropoda
	Oblongoolithidae	?Theropoda or ?Aves
	Laevisoolithidae	Aves: Enantiornithine birds
	Gobioolithidae	Aves: Flying paleognath birds

saurian oofamilies can be positively coordinated with particular larger dinosaurian groups (Table I).

Three clear-cut basic types of eggshell structure can be distinguished for dinosaurs (Fig. 1): dinosauroid–spherulitic type (sauropods and therizinosaurs; hadrosaurines and possibly lambeosaurines), dinosauroid–prismatic type (protoceratopsians and hypsilophodonts), and ornithoid type (theropods). In the dinosauroid–spherulitic type, the basic shell unit roughly looks and grows like spherocrystal. Dinosauroid–prismatic shell displays spherocrystalline structure only in its lower (internal) one-half to one-fourth part, whereas the upper portion is prismatic in morphology with homogeneous calcite ultrastructure. The ornithoid type most strikingly differs from the others because the basic shell units are expressed as discrete items (mammillae) only in the lower one-sixth to one-third of the eggshell's thickness. Above (external to) this layer is a mass of biocrystalline material with a squamatic (= spongy) ultrastructure that comprises a single continuous layer. This structure of theropod eggshell is essentially similar to that of birds. The three remaining eggshell types are crocodiloid, testudoid, and geckoid, all of which have different states of eggshell ultra-microstructure than those of dinosaurs.

The known diversity of dinosaurian egg remains comprises eight oofamilies and 18 oogenera with more than 40 oospecies. Most of these are known from the Cretaceous of central Asia (China, Mongolia, Kyrgyzstan, eastern Kazakhstan, and Uzbekistan); some from the Cretaceous of southern Europe, India, and South America; and some from the Jurassic and Cretaceous of North America (Fig. 2, 3). Singular specimens have been found in the Cretaceous of

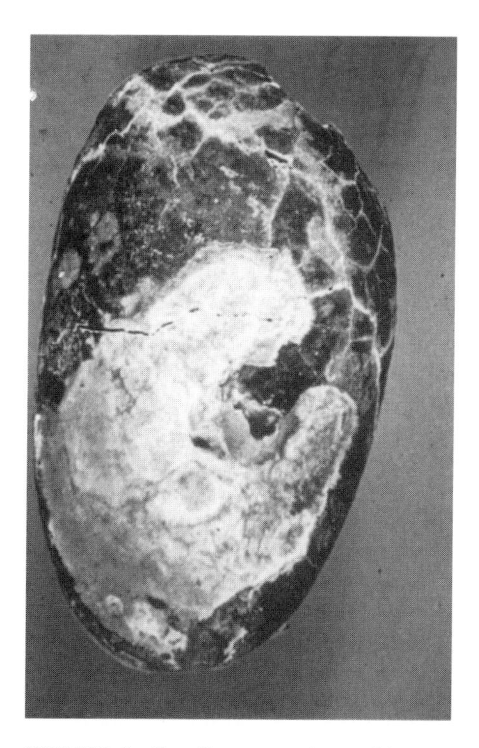

FIGURE 2 Fossil eggs such as this one are found in Ramos Arizpe County, west of the state of Coahuila, Mexico.

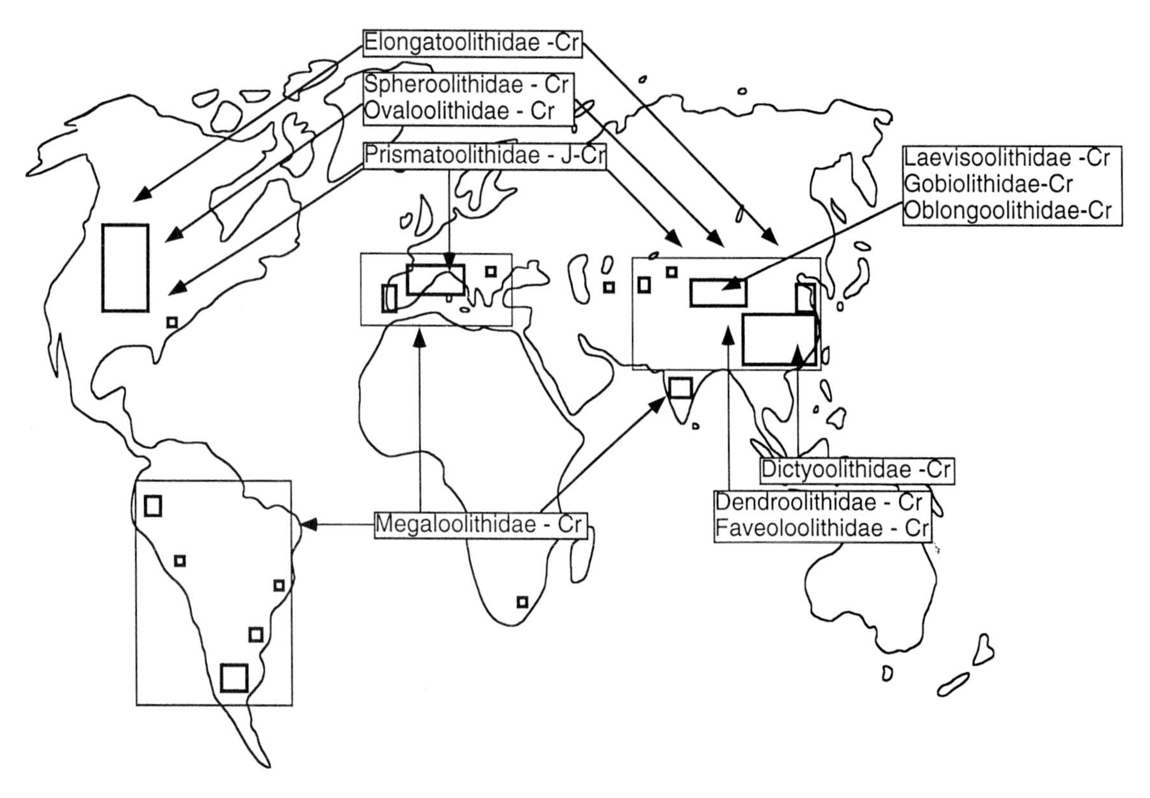

FIGURE 3 Eggshells of different structure and in different oofamilies have been unearthed around the world.

southern Africa. The oofamilies Elongatoolithidae, Prismatoolithidae, Spheroolithidae, Ovaloolithidae, and Megaloolithidae are the most widespread and diverse (6–12 oospecies); the others are restricted to Mongolia and China and are mostly monotypic. Additionally, two families (Laevisoolithidae and Gobioolithidae) of eggs of Mesozoic birds are known from the Upper Cretaceous of Mongolia. Many dinosaurian eggs and their shells still await formal description.

Known dinosaurian nests and eggs exhibit a surprising diversity of form. The eggs of sauropod dinosaurs (see parataxonomic correlation in Table I) that had been found in southern Europe and in the Gobi Desert of Mongolia imply that subsurface nest conditions were similar to those of sea turtles, with the eggs developing in a moist substrate not far from water. Those eggs are large and subspherical in shape and have a highly developed pore system (pore orifices occupy about half of the egg surface). In contrast, the eggs of hadrosaurs, hypsilophodonts, protoceratopsians, and theropods were laid in nests composed of soil and vegetation similar to those of extant alligators, some crocodiles, and even some birds (Australian megapodids). These nests seem to be rather primitive in arrangement in protoceratopsians and more sophisticated in theropods and hadrosaurs. The hadrosaurs and therizinosaurs produced subspherical eggs, whereas protoceratopsians, hypsilophodonts, and theropods laid strikingly different, elongate eggs that were often set in subvertical position and organized in two or three stacked circles. In many cases, the surfaces of dinosaur eggs are characteristically ornamented, although the functional significance of this feature is not yet clear.

See also the following related entries:

Bastús Nesting Site • Behavior • Devil's Coulee • Egg Mountain

References

Carpenter, K., Hirsch, K. F., and Horner, J. R. (1994). *Dinosaur Eggs and Babies*, pp. 372. Cambridge Univ. Press, Cambridge, UK.

Hirsch, K. F. (1989). Interpretations of Cretaceous and pre-Cretaceous eggs and eggshell fragments. In *Dino-*

saur Tracks and Traces (D. D. Gilette and M. G. Lockley, Eds.), pp. 89–97. Cambridge Univ. Press, Cambridge, UK.

Hirsch, K. F. (1994). The fossil record of vertebrate eggs. In *The Palaeoecology of Trace Fossils* (S. K. Donovan, Ed.), pp. 269–294. Wiley, Chichester, UK.

Horner, J. R. (1987). Ecologic and behavioral informations derived from a dinosaur nesting site. In *Dinosaur Past and Present, Vol. 2* (S. J. Czerkas and E. C. Olson, Eds.). Washington Univ. Press, Seattle.

Mikhailov, K. E. (1991). Classification of fossil eggshells of amniotic vertebrates. *Acta Palaeontol. Polonica* **36,** 193–238.

Mikhailov, K. E. (1997). Fossil and recent eggshells in amniotic vertebrates: Fine structure, comparative morphology and classification. *Spec. Papers Palaeontol.*, in press.

Mikhailov, K. E., Bray, E. S., and Hirsch, K. F. (1997). Parataxonomy of fossil egg remains (Veterovata): Basic principles and applications. *J. Vertebr. Paleontol.*, in press.

Zhao, Zi-kui. (1993). Structure, formation and evolutionary trends of dinosaurian eggshell. In *Structure, Formation and Evolution of Fossil Hard Tissues* (I. Kobayashi, H. Mutvei, and A. Sahni, Eds.), pp. 195–212. Tokai Univ. Press, Tokyo.

Egyptian Dinosaurs

see AFRICAN DINOSAURS

Elmisauridae

PHILIP J. CURRIE
Royal Tyrrell Museum of Palaeontology
Drumheller, Alberta, Canada

Elmisaurids are small, lightly built theropods of the Late Cretaceous of the Northern Hemisphere (Currie, 1990). Unfortunately, no cranial material has been identified for elmisaurids, which are best known from their feet and hands. The first specimens of *Elmisaurus* were recovered from the NEMEGT FORMATION of Mongolia (Osmólska, 1981) and were considered distinctive enough to erect a new family (the Elmisauridae). Currie and Russell (1988) noted similarities in morphological characters with *Chirostenotes* from North America and concluded that both *Elmisaurus* and *Chirostenotes* may be related to *Caenagnathus*, an oviraptorosaur. Recent preparation of a skeleton of *Chirostenotes* (Sues, 1994) has shown that it is probably synonymous with *Caenagnathus*.

Elmisaurus is a smaller animal than *Chirostenotes*, but its tarsometatarsus is co-ossified. This clue led to the discovery of elmisaurid material from Alberta (Currie, 1989). Small, toothless caenagnathid jaws from central Asia (Currie *et al.*, 1993) bearing the name *Caenagnathasia* may also be from elmisaurids. If better material confirms this suspicion, then Elmisauridae will become a junior synonym of Caenagnathidae. However, there are enough differences to suggest that making such a move at this time would be premature.

The first metacarpal of an elmisaurid is a straight, slender bone, intermediate in relative length between those of most theropods and those of ornithomimids. The first metacarpals are also straight and slender in *Chirostenotes* and *Microvenator* (Currie and Russell, 1988), whereas those of oviraptorids are shorter and stouter. Digit III is approximately 30% shorter than digit II, whereas the second finger of an oviraptorid is only slightly longer than the third. Whereas phalanx II-2 is the longest of the manual phalanges in *Elmisaurus* and caenagnathids, phalanx I-1 is longest in oviraptorids. Although elmisaurids, caenagnathids, and oviraptorids all have well-developed extensor ligament attachments on their manual unguals, only elmisaurids and caenagnathids have arctometatarsalian feet. Elmisaurids are unique among these three taxa in that the distal tarsals and the proximal ends of the metatarsals are co-ossified into a tarsometatarsus. The metatarsi of elmisaurids and caenagnathids are elongate. The elmisaurid metatarsus is strongly arched in section at midlength, with the third metatarsal deeply inset from the flexor surfaces of the second and fourth.

When more complete specimens are found, it is possible that elmisaurids will turn out to be caenagnathids. In the meantime, there are enough differences to maintain taxonomic separation. Even though it is clear that elmisaurids are closely related to caenagnathids, the lack of cranial material makes it impossible to determine whether they should be included in the Oviraptorosauria.

See also the following related entries:

OVIRAPTOROSAURIA • THEROPODA

References

Currie, P. J. (1989). The first records of *Elmisaurus* (Saurischia Theropoda) from North America. *Can. J. Earth Sci.* **26,** 1319–1324.

Currie, P. J. (1990). The Elmisauridae. In *The Dinosauria* (D. Weishampel, P. Dodson, and H. Osmólska, Eds.), pp. 245–248. Univ. of California Press, Berkeley.

Currie, P. J., and Russell, D. A. (1988). Osteology and relationships of *Chirostenotes pergracilis* (Saurischia, Theropoda) from the Judith River (Oldman) Formation of Alberta, Canada. *Can. J. Earth Sci.* **25,** 972–986.

Currie, P. J., Godfrey, S. J., and Nessov, L. (1993). New caenagnathid (Dinosauria: Theropoda) specimens from the Upper Cretaceous of North America and Asia. *Can. J. Earth Sci.* **30,** 2255–2272.

Osmólska, H. (1981). Coossified tarsometatarsi in theropod dinosaurs and their bearing on the problem of bird origins. *Palaeontol. Polonica* **42,** 79–95.

Sues, H.-D. (1994). New evidence concerning the phylogenetic position of *Chirostenotes* (Dinosauria: Theropoda). *J. Vertebr. Paleontol.* **14,** 48A.

Embryos

Embryos of dinosaurs have been found preserved within eggs in China, Mongolia, Canada, the United States, and South Africa. See GROWTH AND EMBRYOLOGY for a more complete reference.

see EGGS, EGGSHELLS, AND NESTS; GROWTH AND EMBRYOLOGY

Emery County Museum, Utah, USA

see MUSEUMS AND DISPLAYS

Enantiornithes

Enantiornithes is a group of Cretaceous birds distinguished by an unusual shoulder girdle and tarsometatarsal fusion. Not recognized as a group until 1981, they are now the most numerous and widespread group of Cretaceous birds, thanks to new discoveries in South America, China, North America, Australia, and Antarctica, accompanied by the recognition that considerable partially known material already in museum collections around the world pertains to enantiornithine taxa.

See also the following related entry:

AVES

Energetics and Thermal Biology

see PHYSIOLOGY

Environments of Dinosaur Preservation

Dinosaurs are preserved in rocks whose depositional circumstances vary from aolian (= sand dunes) sediments to intercontinental shallow sea sediments. Thus, dinosaurs probably inhabited most environments where large vertebrates now live, barring frigid areas of today's poles.

see PALEOCLIMATOLOGY; PALEOECOLOGY

Erenhot Dinosaur Museum

PHILIP J. CURRIE
Royal Tyrrell Museum of Palaeontology
Drumheller, Alberta, Canada

Erenhot is a small Chinese city on the border between Inner Mongolia and Mongolia. Because it is on the Beijing–Moscow trunk of the Trans-Siberian Railroad, many international visitors pass through it every year, little realizing its paleontological significance. It lies close to Iren Nor, the locality where the AMERICAN MUSEUM OF NATURAL HISTORY discovered the first central Asian dinosaurs in 1922. The type section of the Iren Dabasu Formation is located here. Believed to be Cenomanian–Campanian in age (Currie and Eberth, 1993), the limited exposures have proven to be amazingly rich in skeletons and bone beds. Large collections have been made by the CENTRAL ASIATIC, SINO–SOVIET, and SINO–CANADIAN expeditions and by smaller parties from the Beijing Natural History Museum, the Inner Mongolia Museum, and the INSTITUTE OF VERTEBRATE PALAEONTOLOGY & PALEOANTHROPOLOGY (Beijing). In 1989, after several years of planning and collection, a local museum opened in Erenhot. The Erenhot Dinosaur Museum displays mostly specimens from the Iren Dabasu Formation, including skeletons of *Bactrosaurus* and *Archaeornithomimus*, and isolated bones of hadrosaurs, sauropods, small theropods, and therizinosaurids.

See also the following related entries:

CHINESE DINOSAURS • MONGOLIAN DINOSAURS

Reference

Currie, P. J., and Eberth, D. A. (1993). Palaeontology, sedimentology and palaeoecology of the Iren Dabasu Formation (Upper Cretaceous), Inner Mongolia, People's Republic of China. *Cretaceous Res.* **14,** 127–144.

Espéraza

see MUSEÉ DES DINOSAURES

Euornithopoda

Euornithopoda was established by Sereno (1986) to include the ornithischian taxa heterodontosaurs, hypsilophodonts, iguanodonts, and hadrosaurs. The node is supported by several synapomorphies, including the ventral offset of the premaxillary tooth row relative to the maxillary tooth row, crescent-shaped paroccipital process, jaw joint lower than occlusal plane, and exclusion of the maxillary–nasal contact by the premaxilla. Sereno used the term Euornithopoda as a node-based sister taxon to MARGINOCEPHALIA within ORNITHISCHIA. Most authors have continued to use the term Ornithopoda for this taxon, although in Sereno's phylogeny the term Ornithopoda was confined to the Euornithopoda excluding heterodontosaurs.

See also the following related entry:

ORNITHOPODA

Reference

Sereno, P.C. (1986). Phylogeny of the bird-hipped dinosaurs (Order Ornithischia). *Natl. Geogr. Res.* **2,** 234–256.

Gryposaurus notabilis is an euornithopodan hadrosaur found in Cretaceous sediments that are exposed at Dinosaur Provincial Park, Alberta, Canada. (Photo by François Gohier.)

European Dinosaurs

Eric Buffetaut
Laboratoire de Paléontologie des Vertébrés,
Université Paris
Paris, France

The first dinosaur remains to be described scientifically were found in Europe at the beginning of the 19th century. Since then, however, European dinosaurs, which are often represented by incomplete specimens, have been somewhat overshadowed by finds of spectacular complete skeletons from other parts of the world, including North America, Africa, and Asia. Nevertheless, the European dinosaur record, which includes skeletal remains, footprints, and eggs, is probably the most complete in the world from a stratigraphic point of view, with relatively few gaps in a series of sites covering the time span from the Late Triassic to the end of the Cretaceous.

Many European countries (Fig. 1) have yielded remains belonging to various groups of dinosaurs and a stratigraphic rather than systematic or geographic presentation seems to be the most appropriate way to present the European record in a concise way. Only the main sites and assemblages have been included.

Late Triassic

In central and western Europe, the Late Triassic is largely represented by continental deposits, in which dinosaur bones and footprints sometimes occur in great abundance. Although there have been reports of dinosaur footprints in rocks older than the Late Triassic in several parts of Europe, including England and France, they appear to be erroneous or inconclusive. The European dinosaur record actually begins with small three-toed footprints from the Carnian of northern Bavaria and possibly from the Ladinian–Carnian boundary in the Swiss Alps. The earliest skeletal remains are Norian in age. The earliest relatively well-known European dinosaur is the prosauropod *Sellosaurus gracilis*, from the lower Stubensandstein of Württemberg, considered early Norian in age. Slightly more recent, probably middle Norian, levels of the Stubensandstein of the same region have yielded remains of the theropod *Halticosaurus*. Dinosaur bones first become abundant, however, in higher levels of the Late Triassic usually referred to the late Norian. Sites of that age include the famous German prosauropod localities in the Knollenmergel, such as Trossingen in Württemberg and Halberstadt in Sachsen-Anhalt, which have yielded many well-preserved skeletons of *Plateosaurus*, and various other sites in other parts of Germany, including the Grosse Gleichberg in Thuringia, where the theropod *Liliensternus* has been found. Outside Germany, *Plateosaurus* remains have been found in abundance in the upper Norian of Switzerland and in rocks of the same age in the French Jura mountains. Late Triassic dinosaur remains from Britain include the prosauropod *Thecodontosaurus* from the Magnesian Conglomerate near Bristol, which is Norian to Rhaetian in age. Dinosaur remains from the controversial "Rhaetian" stage, at the top of the Triassic, in various European countries (Britain, France, Germany, and Belgium) are usu-

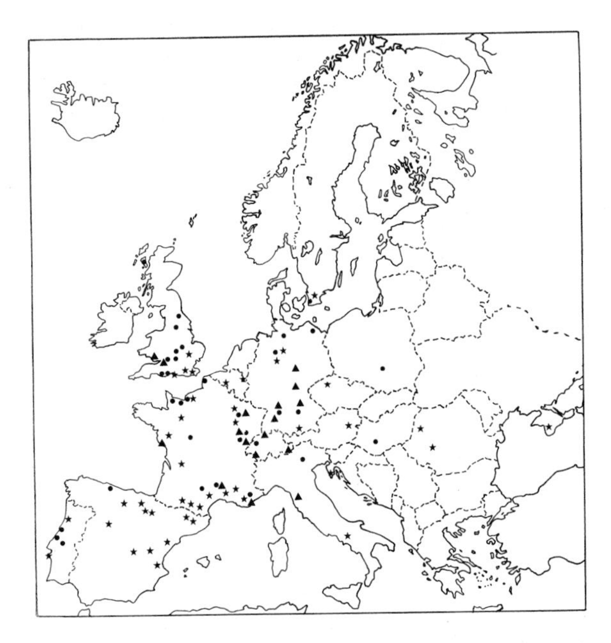

FIGURE 1 Map of Europe showing the main dinosaur localities. ▲, Triassic; ●, Jurassic; ★, Cretaceous.

ally very fragmentary elements from bone beds that are not easily identifiable; some can be referred to indeterminate prosauropods. An exception is the melanorosaurid prosauropod *Camelotia* from Somerset, England.

Norian dinosaur footprints, mostly attributable to theropods, are known from several sites in northern Bavaria, and also from northern Italy. Several footprint sites of Rhaetian age are known from France, the most important one being at Le Veillon (Vendée), on the Atlantic coast. There, hundreds of footprints have been found, most of them referable to theropods (*Grallator* and *Eubrontes*), although some have been attributed to ornithischians. "Rhaeto-Liassic" tridactyl footprints have also been reported from Scania in southern Sweden.

Early Jurassic

In Europe, the Early Jurassic is represented largely by marine deposits in which dinosaur remains are relatively infrequent. An incomplete theropod skeleton, recently redescribed as *Liliensternus airelensis*, comes from beds very close to the Triassic–Jurassic boundary, probably basal Hettangian in age, at Airel in Normandy (northwestern France). In England, more or less complete skeletons of the early thyreophoran *Scelidosaurus harrisonii* have been found since the 19th century in the marine Sinemurian of the Dorset coast. Very incomplete remains of another possible thyreophoran, *Lusitanosaurus*, were reported from the Sinemurian of Portugal. A third primitive thyreophoran, *Emausaurus ernsti*, was described on the basis of a skull and some postcranial elements from the Toarcian of Mecklenburg-Vorpommern in northern Germany. Hindlimb elements of a sauropod from the Toarcian Posidonienschiefer of Baden-Württemberg (southwestern Germany) have been described as *Ohmdenosaurus liasicus*. Besides these scattered skeletal remains from marine deposits, which also include a few isolated theropod bones or teeth from England, Scotland, France, and Germany, dinosaur footprints of Early Jurassic age occur in several European countries, often in calcareous rocks deposited in a beach or mudflat environment. Such trackways, mostly from theropods, are known from the Hettangian and Pliensbachian of the southern rim of the French Central Massif. Other Early Jurassic dinosaur footprints, most of them referable to theropods, have been reported from the Hettangian of Hungary and the Holy Cross Mountains of Poland. Among the early Liassic footprints from the Rovereto area in northern Italy, some have been referred to a very early sauropod.

Middle Jurassic

The Middle Jurassic dinosaurs of Europe are better known than those from the Lower Jurassic mainly because of a few comparatively rich localities in England and France. Bajocian forms are poorly known, although remains of the theropod *Megalosaurus* have been described from the Inferior Oolite of Dorset, England. Dinosaur remains have been found in abundance in several Bathonian formations in England, including the famous Stonesfield "Slate" which yielded the remains of *Megalosaurus bucklandi*, the first dinosaur to receive a proper generic name, in 1824. The Bathonian of England also contains remains of the large sauropod *Cetiosaurus oxoniensis* and stegosaurs. In France, the Bathonian of the Caen area in Normandy yielded remains of a large theropod, that was described as *Poekilopleuron bucklandi*, but may in fact belong to *Megalosaurus*. Theropod and sauropod teeth are also known from the marine Bathonian of Saint-Gaultier in central France.

The Callovian dinosaurs of Europe are relatively well known because of fairly numerous finds from the marine Lower Oxford Clay of England and its equivalents in Normandy. The English finds include theropods such as *Eustreptospondylus oxoniensis*, represented by a nearly complete skeleton, sauropods (*Cetiosauriscus* and *Ornithopsis*), ornithopods (*Callovosaurus*), stegosaurs (*Lexovisaurus*) and one of the oldest known ankylosaurs (*Sarcolestes*). The French record from Calvados in Normandy is somewhat less diverse but nevertheless includes theropods (with at least two forms, one of which seems to be referable to *Megalosaurus*; the status of *Piveteausaurus*, known from a braincase from Normandy, is uncertain), stegosaurs (with a partial skeleton of *Lexovisaurus*), and sauropods. Middle Jurassic dinosaur footprints have been reported from several sites in England.

Late Jurassic

The Late Jurassic European dinosaur record is good, with important localities in several countries. Oxfordian theropod remains are known from marine deposits in England (with *Metriacanthosaurus parkeri* from Dorset) and Normandy (with *Megalosaurus* bones from the Vaches Noires Cliffs of Calvados). Isolated theropod and sauropod teeth have been found in the Oxfordian fluvio-marine Sables de Glos of Normandy. One of the best sauropod skeletons ever found in Europe comes from the Oxfordian of Damparis (Jura, eastern France); it belongs to a brachiosaurid and was found associated with several theropod teeth, indicating *in situ* scavenging on the sauropod carcass during a phase of emersion of a carbonate platform. In Portugal, the Guimarota lignite mine, of Late Oxfordian or Kimmeridgian age, has yielded hypsilophodontid teeth and, recently, teeth referred to *Archaeopteryx*.

Kimmeridgian dinosaurs are known from the Kimmeridge clay of England. They include poorly known sauropods, stegosaurs (*Dacentrurus*), and ornithopods (including a fairly complete skeleton of *Camptosaurus prestwichii* from the vicinity of Oxford). The marine Kimmeridgian of the Normandy coast near Le Havre also contains dinosaur remains, including theropods, sauropods, and stegosaurs. In the Jura mountains of Switzerland, abundant sauropod tracks have been found in Kimmeridgian limestones, as well as an incomplete sauropod skeleton. Other sauropod footprints are known from rocks of the same age in Barkhausen in northern Germany. A few dinosaur bones have been reported from the marine Kimmeridgian of the Boulonnais, in northern France, but most of the remains from that area are from non-marine Tithonian rocks. They include bones and teeth of a large theropod, a large camarasaurid sauropod (*Neosodon*), and a small iguanodontid. The small theropod *Compsognathus* is known from two fairly complete skeletons from Tithonian lithographic limestones, in northern Bavaria and southeastern France.

Abundant dinosaur remains have been discovered in the Upper Jurassic (Kimmeridgian and Portlandian) of Portugal. The Kimmeridgian assemblage includes several sauropods, including a probable brachiosaurid, theropods, stegosaurs (*Dacentrurus*), an early ankylosaur (*Dracopelta*), and ornithopods. Footprints (of theropods and sauropods) are frequent in the Portlandian. The discovery of numerous dinosaur eggs in the Kimmeridgian of the Lourinha region is especially noteworthy. In Asturias (northern Spain), large footprints referable to sauropods have been reported from Kimmeridgian rocks.

Early Cretaceous

The most varied Early Cretaceous dinosaur assemblage in Europe comes from the non-marine Wealden beds of southern England (Sussex and Isle of Wight), which range in age from late Berriasian to early Aptian. The Wealden fauna includes several species of *Iguanodon*; other ornithopod genera such as *Hypsilophodon*, *Valdosaurus*, and *Vectisaurus*; the nodosaurids *Hylaeosaurus* and *Polacanthus*; the early pachycephalosaur *Yaverlandia*; several sauropods (including a brachiosaurid, a titanosaurid, and a possible diplodocid); and several theropods, among which is the probable spinosaurid *Baryonyx*. On the continent, the most famous Wealden dinosaur locality is undoubtedly the Bernissart coal mine in Belgium, where about 30 skeletons of *Iguanodon*, belonging to the species *I. bernissartensis* and *I. atherfieldensis*, were discovered in 1878. Dinosaurs also occur in the Wealden of northern Germany, where skeletal remains (including those of a theropod, an ornithopod, and the basal marginocephalian *Stenopelix*) are usually far less abundant than footprints. The latter include tracks of sauropods, ornithopods, and theropods from the region around Hanover (Bückeberg and Münchehagen). An exception is the Nehden locality in Westphalia, a karstic deposit of Aptian age, in which abundant remains of *I. bernissartensis* and *I. atherfieldensis* have been discovered.

In the eastern Paris Basin, a succession of alternating shallow marine and non-marine beds, ranging in age from Hauterivian to Aptian, has yielded well-dated *Iguanodon* remains that show a succession of species similar to that from the English Wealden. Remains of a brachiosaurid sauropod are also known from the Barremian of that area. In southern France, a few bones of an *Allosaurus*-like theropod have been described from the marine Valanginian of the department of Gard.

Early Cretaceous non-marine rocks in eastern Spain contain dinosaur assemblages. The mainly Barremian beds of the Galve area (Teruel Province) have yielded hypsilophodontids, theropods, and four forms of sauropods, including the camarasaurid *Aragosaurus*. From the early Aptian of the Morella region (Castellon Province), theropods, sauropods, ankylosaurs, and *Iguanodon* have been reported. Early Cretaceous ankylosaurs, iguanodontids, and theropods are also known from Burgos Province in north-central

Spain. In Cuenca Province in central Spain, the late Hauterivian to early Barremian lacustrine lithographic limestones at Las Hoyas have yielded a few dinosaur remains, including the early multitoothed ornithomimosaur *Pelecanimimus*. Abundant ornithopod and theropod footprints are known from the Lower Cretaceous of the La Rioja region in northern Spain. A few skeletal remains (of theropods, sauropods, and ornithopods) and tracks are also known from the Lower Cretaceous (Hauterivian and Aptian) of Portugal.

Among the few dinosaurs reported from Italy is the skeleton of a very small theropod from the Aptian lithographic limestones of Pietraroia (Benevento Province).

In eastern Europe, the most important Early Cretaceous dinosaur assemblage is that from the karstic bauxite deposits (Barremian to Aptian in age) of Cornet, in Transylvania (Romania). It mainly contains ornithopods (*Iguanodon*, *Valdosaurus*, and *Dryosaurus*), as well as a theropod.

On the northern outskirts of Europe, ornithopod and theropod footprints have been reported from the Barremian of Spitzbergen.

Albian dinosaurs are known from localities in England and France. The English material mainly comes from the Cambridge Greensand, mostly as disarticulated and reworked material. It includes sauropods, ornithopods (including a probable iguanodontid and a hadrosaurid), and ankylosaurs . In France, sauropod remains have been found in the Albian of the Pays de Caux and the Pays de Bray in Normandy, as well as in the "Gault" clay of the eastern Paris Basin. Dinosaurs from the phosphate-bearing Albian of northeastern France include the enigmatic theropod *Erectopus*. In southeastern France, a sauropod humerus was found in the Albian green sandstones of the Mont Ventoux.

Late Cretaceous

The European dinosaurs of the early part of the Late Cretaceous, up to the Campanian, are relatively poorly known. In England, the nodosaurid ankylosaur *Acanthopholis horridus* comes from the Cenomanian Chalk Marl, and an ornithopod has been reported from the Totternhoe Stone, also Cenomanian in age. A few isolated and fragmentary specimens of sauropods and theropods are known from the Cenomanian, Turonian, and Santonian of west-central France. A few dinosaur remains (belonging to a theropod and to the iguanodontid *Craspedodon*) have been collected from the Santonian of Lonzée in Belgium. Cenomanian footprints are known from Portugal (theropods and sauropods) and from Croatia (sauropods).

The European dinosaur record becomes tolerably good again in the Campanian. A few isolated specimens are known from marine rocks of that age in southern Sweden (theropod and ornithopod) and southwestern France (sauropod). One of the best early Campanian localities was found in lignite-bearing beds at Muthmannsdorf in Austria in the 19th century; it yielded a theropod, the ornithopod *Rhabdodon*, and the nodosaurid ankylosaur *Struthiosaurus*. In southern France, the Villeveyrac locality, of the same age, also contains *Rhabdodon*, an ankylosaur, and a small theropod. In southeastern France, an abelisaurid theropod, *Tarascosaurus*, has been reported from Campanian beds at Le Beausset (Var).

Maastrichtian dinosaurs are known from marine Chalk deposits in the Limburg region of Belgium and the Netherlands, near the city of Maastricht. They include a theropod of uncertain affinities (*Betasuchus*) and hadrosaurs. Hadrosaur remains have also been found in the marine upper Maastrichtian of the Bavarian Alps, and there is one hadrosaur record from the marine Maastrichtian of the Crimea in Ukraine. However, most of the Maastrichtian dinosaurs from Europe have been found in non-marine rocks in Transylvania, in southern France, in northern Spain, and in Portugal (some of the French-Iberian localities may actually be late Campanian in age). The Transylvanian assemblage is apparently late Maastrichtian in age. It includes *Rhabdodon*, the primitive hadrosaur *Telmatosaurus*, the nodosaurid ankylosaur *Struthiosaurus*, titanosaurid sauropods (*Magyarosaurus*), and poorly known theropods. Also worth mentioning is the occurrence of eggs associated with remains of hadrosaur embryos.

Whereas the Portuguese record is poor (with indeterminate theropods and ornithischians), there are good late Campanian to Maastrichtian dinosaur localities in various parts of Spain. In the Tremp basin of Catalonia, sauropods, hadrosaurs, and *Rhabdodon* occur in late Maastrichtian deposits. Footprints (including those of sauropods) and eggs have also been found. Other Late Cretaceous dinosaurs are known from various provinces of Spain. One of the richest sites is at Lano, near Vitoria in the Basque Country. It is of late Campanian to early Maastrichtian age and has yielded bones and teeth of abelisaurid theropods, titanosaurid sauropods, *Rhabdodon,* and the ankylosaur *Struthiosaurus.*

In southern France, dinosaur localities of early Maastrichtian age are known in great number in nonmarine formations extending from Provence in the east to the foothills of the Pyrenees in the west. The best known assemblages are from the Fox-Amphoux area in Provence and the upper valley of the Aude River. Skeletal remains indicate the occurrence of abelisaurid and dromaeosaurid theropods, armored titanosaurid sauropods, the ornithopod *Rhabdodon,* and an ankylosaur. Eggs are abundant in some areas, especially the Aix-en-Provence basin and the upper Aude valley. Although several types have been distinguished on the basis of shell microstructure, no clear associations with skeletal remains have yet been reported. Less numerous late Maastrichtian localities are also known, mainly from the Corbières region and the area of the Garonne valley. They contain a fauna dominated by hadrosaurs, accompanied by theropods and ankylosaurs. The southern French record thus suggests a faunal change, marked by a decline of titanosaurid sauropods and an expansion of hadrosaurs, during the Maastrichtian.

From this brief review, it appears that the European record covers most of dinosaur history in a remarkably complete way. In the Triassic and Jurassic, European dinosaur faunas showed clear resemblances to those of other continents, notably North America and Africa. In the Early Cretaceous, resemblances to North America were still marked (with such genera as *Iguanodon* and *Polacanthus* in common). In the Late Cretaceous, however, the European assemblages showed characteristics of their own, with taxa of "Gondwanan" affinities such as abelisaurid theropods and titanosaurid sauropods playing an important part together with endemic forms, whereas taxa of "Asiamerican" affinities were few. This of course reflects the changing paleogeographical history of the European continent during the Mesozoic, which alternatingly favored or limited faunal exchanges with North America, Asia, and Africa. The details of this complex biogeographical history still have to be worked out in detail.

See also the following related entries:

American Dinosaurs • Bastús Nesting Site • Canadian Dinosaurs • Crystal Palace • History of Dinosaur Discoveries • Musée des Dinosaures • Wealden Group

References

Buffetaut, E. (1995). *Dinosaures de France,* pp. 144. Editions du BRGM, Orléans, France.

Buffetaut, E., and Le Loeuff, J. (1991). Late Cretaceous dinosaur faunas of Europe: Some correlation problems. *Cretaceous Res.* **12,** 159–176.

Buffetaut, E., Cuny, G., and Le Loeuff, J. (1991). French dinosaurs: The best record in Europe? *Mod. Geol.* **16,** 17–42.

Bultynck, P. (1989). *Bernissart et les Iguanodons,* pp. 115. Institut Royal des Sciences Naturelles de Belgique, Bruxelles, Belgium.

Dantas, P. (1991). Dinossaurios de Portugal. *Gaia* **2,** 17–26.

Jaeger, M. (1986). *Die Dinosaurier der Schweiz und der Bundesrepublik Deutschland,* pp. 1–40. Schriften des Bodensee–Naturmuseums, Konstanz, Germany.

Jurcsak, T., and Kessler, E. (1991). The Lower Cretaceous paleofauna from Cornet, Bihor County, Romania and its importance. *Nymphaea* **21,** 5–32.

Leonardi, G., and Avanzini, M. (1994). Dinosauri in Italia. *Sci. Quaderni* **76,** 69–81.

Probst, E., and Windolf, R. (1993). *Dinosaurier in Deutschland,* pp. 316. C. Bertelsmann, München, Germany.

Sanz, J. L., Buscalioni, A. D., Moratalla, J. J., Francés, V., and Anton, M. (1990). Los reptiles mesozoicos del registro espanol. *Monogr. Museo Nacional Ciencias Nat,* 1–81.

Weishampel, D. B. (1990). Dinosaurian distribution. In *The Dinosauria* (D. B. Weishampel, P. Dodson, and H. Osmolska, Eds.), pp. 63–139. Univ. of California Press, Berkeley.

Weishampel, D. B., Grigorescu, D., and Norman, D. B. (1991). The dinosaurs of Transylvania. *Nat. Geogr. Res.* **7,** 196–215.

Everhart Museum, Pennsylvania, USA

see Museums and Displays

Evolution

J. David Archibald
San Diego State University
San Diego, California, USA

In everyday usage, evolution is used most often in the sense of change, but frequently there is the added connotation of progress. In biology there is no single definition for evolution, which is often qualified as organic or biological evolution. For the population biologist it might be expressed as a change in the frequency of certain gene types in a species or population with concomitant changes in phenotype (visible aspects of the organism). More generally we could define evolution as the change in a species or lineage of populations over some number of generations. One of the best definitions is that of Charles Darwin, who called it "descent with modification." This is the simplest and most eloquent definition because it embodies both elements that the majority of biologists feel are key in evolution—descent because there is continuity of life from one generation to the next through genetic inheritance, and modification because life demonstrably changes over time with changes in the genes. Note, however, that in none of these definitions or characterizations is there a clear explanation of a cause.

Charles Darwin (Fig. 1) is often and incorrectly portrayed as the scientist who "discovered" evolution; rather, he argued persuasively for the principal mechanism that seemed to explain most of what we see in nature. Darwin did not even care for the term evolution, preferring transmutation or descent with modification. The word evolution does not even appear in his famous book *On the Origin of Species by Means of Natural Selection* or in the *Preservation of Favoured Races in the Struggle for Life* that appeared in 1859. The word "evolved" does appear, however, but only once as the very last word in the text. It was the English philosopher, Herbert Spencer, who first used evolution in basically the sense we use it today. Previously, the word was usually used by a group of biologists who argued that all organisms began as a preformed individual within the sperm or egg of the parent. To them, evolution was the process of unfolding that occurred as the preformed individual (a homunculus in the case of humans) grew or unfolded from the sex cell. For Spencer, it became the unfolding of life in general.

It was also Spencer who coined the phrase "survival of the fittest" and the now largely discredited sociological theory of "social Darwinism." Although Darwin began to use the phrase in the fifth edition of *The Origin of Species*, it has had unfortunate connotations. As philosophers pointed out, it could be construed as a tautology; that is, a statement that is true simply because its terms so define it (all mothers are female). The fittest are by definition those that survive, thus the survival of the fittest becomes survival of the survivors. It might be more accurate to speak of the survival of the adequate because although Darwin later did use Spencer's phrase, he usually wrote about the variations that organisms possessed that could be advantageous or injurious to their survival. This was the essence of Darwin's incalculable contribution to evolutionary theory—that nature, just as the breeder of plants and animals, "selects" from among the available variations those that will be passed along to subsequent generations. This is Darwin's major contribution—natural selection. Natural selection can be defined as the differential contribution of heritable variation to the next generation.

Today, natural selection is still regarded as the major mechanism that drives the engine of evolutionary change. We now realize, however, that chance does play a very important role. A cataclysmic ecological event or even the piggybacking of one nonselected gene along with another that is under selective pressure could simply by chance determine which types of genes will survive.

From his time as naturalist on the naval surveying vessel, the *H. M. S. Beagle*, starting in 1831 through the publication of *The Origin of Species* in 1859, Darwin not only developed and honed his ideas on natural selection but also developed a general explanation of the process of evolution through his keen powers

FIGURE 1 Charles Darwin, 1809–1882.

of observation and deduction. The major points of Darwin's descent with modification (or evolution) by means of natural selection are as follows: (i) organisms within a species vary, (ii) some of these variations are inherited by offspring, (iii) more offspring are produced than can possibly survive, (iv) usually offspring with variations favored by the environment will survive, (v) surviving offspring will in turn usually leave more offspring with variations favored by environment, and (vi) over time variations favored by the environment will accumulate.

The two essential components of evolutionary thought that eluded Darwin are how the variations arise and how they are passed on to the next generation. Darwin's ideas on the origin of variations and how they are transmitted to the next generation hark back to the older ideas of the French scientist Lamarck, who argued in the early 19th century that influences of the environment can be passed along to the next generation. Much later Lamarck's ideas on evolution were somewhat incorrectly simplified to the phrase "inheritance of acquired characteristics." It was Darwin's contemporary, the Austrian monk Gregor Mendel, who began to reveal the secrets of genetics. It is ironic that as far as can be determined, Darwin did not know of Mendel's work, although Mendel did know of Darwin. Even if Darwin did know of Mendel, he clearly did not connect the importance of Mendel's findings to his own work on variation and natural selection.

Of the critics of Darwin's ideas on evolution, the only one of considerable interest here is Richard Owen, the originator in 1842 of the name and concept DINOSAURIA, and the describer in 1862 of the London specimen of the earliest bird, *Archaeopteryx lithographica*. Both the early ideas of dinosaurs and *Archaeopteryx* influenced and in turn were influenced by ideas on evolution. Early in their careers, Darwin and Owen were on quite cordial terms. Owen even described in 1840 fossil mammals from South America that Darwin brought back from his voyage on the *H. M. S. Beagle*. With the publication of *The Origin of Species* (Darwin, 1859), the relationship between Darwin and Owen became very strained. Often and incorrectly, but not surprisingly, Owen is portrayed as antievolutionist. In fact, it is clear that he thought change had occurred in the organic world. What was lacking in Owen's limited work on the subject, however, was a clearly articulated idea on the mechanism(s) of evolution. He did not totally reject Darwin's main theme of natural selection, but he ranked it as a minor factor.

The naming and defining of Dinosauria by Owen at a meeting in 1841 (published the following year) has been viewed as antievolutionary. Of course Darwin had not yet published his societally shattering work, but ideas about evolution (or transformation) were becoming more common in the scientific community. Owen's reconstructions of dinosaurs, later immortalized in the 1850s in statuary by Waterhouse Hawkins, were rather mammal like in that they all were portrayed in a quadrupedal position. This has been used to argue that Owen wished to show that modern reptiles are far less advanced than their dinosaurian precursors, a presumed blow against what we today would call progressive evolution. It is far more likely, however, that Owen was simply applying the best known taxonomic and anatomical techniques of the time to reconstruct his best estimation of what these creatures looked like. Any potential side benefits adding ammunition against the Darwin view of evolution would have been welcomed.

Turning to *Archaeopteryx*, Owen secured for the British Museum in 1862 what at the time was the best specimen of *A. lithographica*. It was hailed by some of Darwin's supporters as a missing link between reptiles and birds. It was also regarded by Darwin as an exception that proved the rule that the fossil record is very imperfect. Thus, the fossil intermediates that Darwin thought would be required if his theory was true should also be very uncommon. Owen did not regard *Archaeopteryx* as an intermediate but rather as a true bird that demonstrated his ideas of progress from the general to the more specific. As one of England's preeminent scientists of the time, jealousy over Darwin's success seems to have also prevented Owen from accepting natural selection as the single most important evolutionary mechanism.

Although evolutionary change as described by Darwin is eloquently simple, there are misconceptions about what evolution by natural selection is and is not. Two commonly held misconceptions about evolution are that it is progressive and that it has or shows purpose. These can be explored by starting with another common but incorrect view of evolution, the *scala naturae*. The "scale of nature" is traceable back to ancient Greek philosophers such as Aristotle who arranged life from humans, with the highest kind of soul, down (skipping some groups) through mammals, birds, reptiles, fishes, and finally plants at the bottom. This was the great and continuous chain of being that prevailed and continued to grow through the Middle Ages in Europe. This is also why, even today, we still commonly speak of "higher" and "lower," or "advanced" and "primitive" forms of life. Unconsciously we still view life as a linear hierarchy with humans at the pinnacle. In fact, all species are mosaics of characteristics that have changed little from the ancestor and other characteristics that are greatly changed. For example, both the "lowly" opossum and humans retain the ancestral condition of five digits on hands and feet; humans clearly evolved a much larger, more complex brain than an opossum; but opossums have reduced the number of tooth generations to essentially one whereas humans still must rely on the tooth fairy because we still retain two generations of teeth from our ancient ancestors.

Both the study of the pattern of the history of life and the study of the process of evolution emphatically show that life does not form a linear hierarchy. In fact, before evolution became widely accepted in the past century, some scientists who did not even subscribe to evolution showed that life is not a ladder, but rather a bush or tree; hence our metaphor of a phylogenetic tree to show the history of life. The great French comparative anatomist, paleontologist, and antievolutionist, Georges Cuvier (Fig. 2), argued that animal life could be arranged in four distinct plans (*gembranchements*). The four "branches" that Cuvier recognized were the vertebrates, the molluscs, the articulates (insects and worms), and the radiates (jellyfish, starfish, etc.). Except for the unnatural radiate grouping, we still recognize the other three groupings along with others within Animalia.

With the shift to evolutionary thinking in biology in the mid to late 19th century, it became clear that the different branches of organisms formed a bush of life taking its origin someplace in the deep recesses of time. We now know that many of these major branches of life show up in the fossil record by some 500 million years ago, not much after animals evolved hard supporting structures or skeletons that were

FIGURE 2 Georges Cuvier, 1769–1832. Photo courtesy of the National Library of Medicine.

easily preserved. Thus, our own major branch, the vertebrates, and that including insects, the arthropods, appeared within the same geological time frame rather than vertebrates showing up as the last and highest form of life. Humans are not more highly evolved than ants, simply differently evolved.

Even though there is clearly no ordering of life from the lowest to the highest species in a *scala naturae*, why can we not argue that within the branches of life the later appearing forms have progressed compared to older forms? In Darwinian terms, why can we not argue that because the species are still here, they are better adapted to their environment and thus are more progressive? The answer is that species have evolved and adapted (or became extinct) to conditions that prevailed at the time they were alive. The physical conditions, such as climate and topography, and the biological conditions, such as competition within species or food sources, are constantly shifting over geological time. So too must the species shift or become extinct. In fact, extinction is the unquestioned rule rather than the exception: Approximately between 90 and 99.9% of all species that have ever lived are extinct.

Progress can certainly be rejected as detectable in evolution. This does not mean that there are no detectable trends or directions, but these are determined only by looking backwards from the present. Also, our predictive powers are of only the broadest kind. Very broad predictions or directions in evolution are obvious if for no other reason than, like all other parts of the known universe, living organisms must obey physical laws. As early plant and animal life evolved, it invaded the land. For animals we know that this invasion was successfully accomplished a bare minimum of five times. Another big jump was to flight, which in vertebrates alone evolved a minimum of three times. Thus, we can detect major directions and even predict what the general physical requirements would be (lungs for breathing and wings for flight). There is another general direction in evolution—a general increase in complexity. This does not mean that all branches show increased complexity, but rather only that there is increased complexity in some branches. This occurs from the molecular level, with a general increase of genetic information over time in some groups, to the increase in larger structures such as the vertebrate brain, which has reached its greatest complexity in placental mammals. Although it might be tempting to view this increased complexity as progress, we must keep in mind that in only some societies is increased complexity equated with progress; in others an attainment of simplicity is progress. This strongly suggests that societal, value-based concepts such as progress are not applicable when examining questions in science—in this case evolution.

See also the following related entries:

GENETICS • HISTORY OF DINOSAUR DISCOVERIES • PHYLOGENETIC SYSTEM • SPECIATION • SPECIES • SYSTEMATICS

References

Brown, J. (1995). *Charles Darwin Voyaging*, pp. 605. Knopf, New York.

Darwin, C. (1859). *On the Origin of Species by Means of Natural Selection, or the Preservation of Favoured Races in the Struggle for Life*, pp. 502. John Murray, London.

Dennett, C. C. (1995). *Darwin's Dangerous Idea*, pp. 586. Simon & Schuster, New York.

Desmond, A., and Moore, J. (1991). *Darwin*, pp. 808. Warner Books, New York.

Futuyma, D. J. (1986). *Evolutionary Biology*, pp. 600. Sinauer, Sunderland, MA.

Gould, S. J. (1996). *Full House: The Spread of Excellence from Plato to Darwin*, pp. 244. Harmony, New York.

Minkoff, E. C. (1983). *Evolutionary Biology*, pp. 627. Addison-Wesley, Reading, MA.

Rupke, N. A. (1994). *Richard Owen Victorian Naturalist*, pp. 462. Yale Univ. Press, New Haven, CT.

Extinction

I. Extinction, Cretaceous

J. DAVID ARCHIBALD
San Diego State University
San Diego, California, USA

The phenomenon that we call extinction is arguably one of the most misunderstood biological events. Three common misperceptions contribute to this misunderstanding. First, extinction is thought to be rare, or at least may have been rare before the explosion of the human population. Second, extinction is viewed as a negative process, one that only brings destruction. Third, the simple disappearance of a species is thought to be the same as extinction.

Estimates of both the number of living (or extant) SPECIES and the rate at which they are being driven to extinction by humankind vary widely. Estimates for the numbers of species thought to grace the earth today range from a known number of 1.4 million to tens of millions (or more). Rates of present-day extinction are placed from as low as one per year to as high as one per day, with some even higher estimates. Within the constraints of the human timescale, extinction seems to occur at very low levels, even using the highest rates. So what if we lose one species per day if we have millions of species? If such levels continued, however, for only the extent of our Gregorian calendar—which is fast closing in on 2000 years—three-quarters of a million species would have disappeared. This would represent a loss of more than one-half of all known living species and would be very high even if the true number of extant species was closer to 10 million. How high a rate of extinction would this be compared to those in the geological past? It would rank within the top five, if not the top three, of the most significant episodes of extinction in the past 550 million years.

The process of extinction, however, was occurring well before the emergence of *Homo sapiens*, although we certainly have accelerated it to breakneck rates. Extinction has been ubiquitous during earth history since the origin of life approximately 3.5 billion years ago. All current estimates place the total percentage of species extinction throughout earth history at more than 90%, if not over 99%. This is a surprising figure for the uninitiated, but the surprise soon dissipates when we realize that evolution has constantly added new species throughout the past 3.5 billion years. Although there have been long intervals of increasing species diversity, life remained at a steady state for very long stretches of GEOLOGIC TIME. The only way a steady state of species numbers could be maintained, even while evolution is occurring, is for other species to become extinct. Thus, extinction, rather than being a rare and negative event in human time frames and sensibilities, is actually a very common and positive counterpoint to evolution. Without extinction, the vast majority of extant species, including *Homo sapiens*, could not and would not have arisen.

I have addressed the first two common misperceptions concerning extinction; first, that extinction is thought to be rare, and second, that extinction is viewed as a negative process. However, what of the third—that the simple disappearance of a species is thought to be the same as extinction? The disappearance of species, usually based on disappearance from the fossil record, often occurs because of vagaries of the fossil record. It must also be remembered that many species are only poorly represented in the fossil record, if at all, because they lack preservable hard parts such as a shell or skeleton. If they live in regions that usually do not promote fossilization, such as in the mountains or desert, lack of a fossil record is not by itself a good measure of whether extinction has occurred. Even if we have a good fossil record for a particular species, and it disappears locally, it may continue to survive elsewhere on the globe for a considerable length of time. For example, plants such as dawn redwoods, fish such as coelacanths, and small mammals such as rat opossums (small South American marsupials) were known as fossils before living representatives were discovered growing, swimming, and scampering about elsewhere.

Finally and probably most important for the disappearance of some species is the process of speciation. During speciation a parent species may give rise to a recognizably new species while the parent species remains relatively unchanged. In other instances the parent species disappears as it splits to form two or possibly even more new daughter species. This is pseudoextinction. Unlike true extinction, which occurs when all the individuals possessing a very similar but variable genome disappear, pseudoextinction occurs when the genomes of individuals in the two daughter species are sufficiently altered so that both are clearly set on new and different evolutionary paths. This most obviously has occurred when the resulting daughter species can no longer interbreed.

If we can establish with reasonable certainty that the true extinction of one or more species has occurred, the next step is to place the extinction(s) in the context of other extinctions. Were the extinctions in question normal (or background) or were they part of a mass extinction? The major difference between normal and mass extinctions is one of scale. Because extinctions have been sampled repeatedly throughout the better known past 550 million years of earth's history, it appears that the rate of extinction is roughly similar. Although it is difficult to provide a specific figure because of vagaries of the fossil record and differences between environments and different kinds of organisms, it can be safely said that normal extinction is below (often well below) 50%. There were intervals, however—five to be precise—during the past 550 million years that witnessed percentages of extinction well over 50%, in one case possibly reaching as high as 95%. This horrendously high level of extinction occurred at the end of the Permian Period approximately 250 million years ago. All five known mass extinctions, from oldest to youngest, were in (or at the end of) the Late Ordovician, Late Devonian, Late Permian, Late Triassic, and Late Cretaceous. There is an unresolved debate as to whether these five mass extinctions represent a separate class of extinction from normal extinctions or instead form a continuum with normal extinctions in both rate and cause.

The most recent mass extinction event near the end of the Late Cretaceous includes that of the dinosaurs, or more correctly the nonavian dinosaurs. The level of extinction for all species during the Cretaceous–Tertiary (or K–T) transition has been frequently given at approximately 75%, although there are no studies documenting this level of extinctions for species. For backboned animals or vertebrates only, the level of extinction hovers around 50% for species, but this is based on only one area, the Western Interior of North America.

Dinosaurs have come to represent not only the K–T mass extinction event for both scientists and the public but also extinction in general. Except for perhaps the hapless Dodo, nothing seems to epitomize extinction as much as dinosaurs. When we are dealing with large, relatively rare creatures such as dinosaurs, the problems of unraveling when, where, how much (magnitude), and how fast (rate) extinction occurred can be disheartening. Nevertheless, only after these questions have been addressed, even if not completely answered, can we turn to the question of possible causes.

The "common knowledge" is that dinosaurs became extinct at the same time everywhere on earth. If one is to believe not only the popular press but also some scientists, this common knowledge comes from a global record of dinosaur extinction at the K–T boundary. The idea that the dinosaurs disappeared from earth at the same time in the wink of an eye is not new; but the proposition in 1980 that an asteroid impact caused this very rapid decline and extinction of dinosaurs has given it new life. Some proponents of this theory have explicitly stated that these extinctions were essentially instantaneous around the world. Such explicit assertions about global records of dinosaurs are patently false. Most people are surprised to learn that the geographic coverage of dinosaur extinction is appallingly bad. The only region where we currently have a reasonably large sample of dinosaurs and contemporary vertebrates extending to near or at the K–T boundary (and have a fossil record of similar quality above this boundary) is in the Western Interior of North America, especially well known in the eastern part of Montana and into southern Canada. This region formed the eastern coast of a great inland sea that split North America in half in the Late Cretaceous. This was certainly an extensive region stretching for thousands of miles; nevertheless, it is a very limited region to use to explore questions of extinction on a global scale.

One study showed that of 26 dinosaur localities from near the end of the Cretaceous, 20 are from the Western Interior of North America. This represents more than 75% (20/26) of all the information we have about dinosaurs leading up to their extinction. The record of the last dinosaurs is biased not only in where the localities are found but also in the taxa and numbers of specimens. Of the 20 genera of dinosaurs known from the latest Cretaceous, 14 (70%) are known only from North America. Turning to numbers of specimens (those represented by articulated individuals composed of several bones up to a complete skeleton found in very close association), 95% of the 100 specimens known from the latest Cretaceous are from North America. These figures once more emphasize that the record of dinosaurs near the K–T boundary is almost exclusively North American.

In the next few years we may see a better global record of latest Cretaceous dinosaurs emerging. Especially promising are new finds in several sedimentary basins in China and localities in central South America. Until such time that we do have a more global record, arguments about the pace of dinosaur extinction on a global scale remain unsubstantiated speculation. For now, it must be emphasized that we simply have no record of dinosaurs that permits us to clearly show whether dinosaur extinction was catastrophically fast or glacially slow. Rather, the data we do have are more regional in scope and only permit us to examine questions of the magnitude and selectivity of these extinctions, but nothing of its pace.

Although the vertebrate record of the K–T boundary is almost exclusively limited to the Western Interior, the uppermost Cretaceous Hell Creek Formation in eastern Montana has yielded a taxonomically rich sample of 107 vertebrate species. The record includes 12 major vertebrate lineages—5 species of sharks and relatives, 15 of bony fishes, 8 of frogs and salamanders, 10 of multituberculate mammals, 6 of placental mammals, 11 of marsupial mammals, 17 of turtles, 10 of lizards, 1 of the unfamiliar champsosaurs, 5 of crocodilians, 10 of ornithischian dinosaurs, and 9 of saurischian dinosaurs (not including birds). Of these 107 vertebrate species or species, 49% (52 of 107) survived across the K–T boundary in the Western Interior. This is the minimum percentage survival of vertebrate species across the K–T boundary because some of the very rare species may have survived undetected. Twenty of the 107 are quite rare species, represented by fewer than 50 identifiable specimens out of 150,000 specimens estimated to have been recovered from the Hell Creek Formation. Although an accurate estimate is not possible, certainly some of these very rare species must have survived. The extreme and very improbable case would be if all 20 survived. This would be 67% (72 of 107) survival. This provides the extreme maximum percentage survival in the region. An educated guess, though no more than a guess, would be that no more than 60% of vertebrate species survived the K–T boundary.

When examined in greater detail, a very interesting pattern emerges within the record of survival and extinction for these Hell Creek Formation vertebrates. This is a pattern of disparity in species survival among the 12 major groups. Only 5 of the groups—sharks and relatives, marsupials, lizards, ornithischians, and saurischians—contributed to 75% of the extinction. What do sharks, lizards, marsupials, and the two lineages of dinosaurs have in common in these faunas, other than that each suffered at least 70% or more species extinction at the K–T boundary in western North America? If we are to understand causes of extinction at the K–T boundary we must

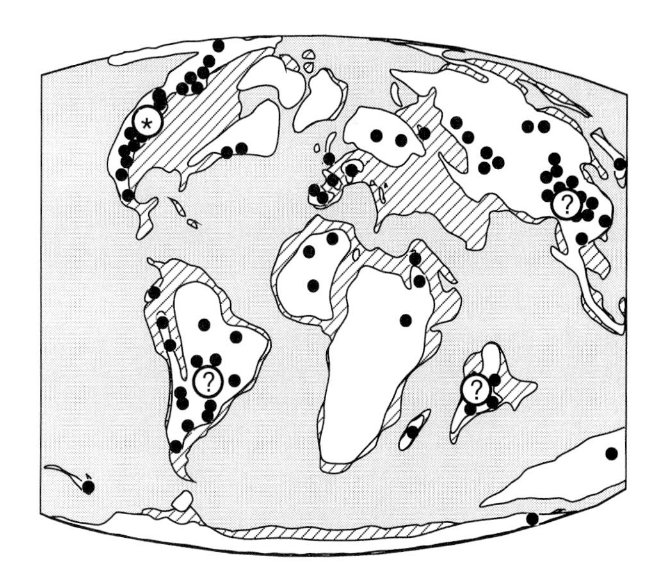

Late Cretaceous dinosaur bearing sites from around the world are reasonably common. These nonmarine lacustrine sites, however, rarely cross the Cretaceous/Tertiary boundary. Very few of the North American sites are known to span this important boundary (⊛). Localities that might span the boundary (?) are still being studied. Map by J. David Archibald (1996), from *Dinosaur Extinction and the End of an Era*. Columbia University Press. Reprinted by permission.

explain this disparate pattern of extinctions. Any theories about the cause(s) of extinction of dinosaurs and their contemporary vertebrates must be able to explain the previously discussed pattern. Any theories must explain why sharks, lizards, marsupials, ornithischians, and saurischians suffered very high levels of or even total extinction, whereas bony fishes, frogs and salamanders, multituberculate mammals, placental mammals, turtles, champsosaurs, and crocodilians suffered 50% or often much less extinction.

There are as many as 80 theories or variants of theories that have been proposed for dinosaur extinction. Many of these are either frivolous (Martians hunted the dinosaurs to extinction) or untestable (a plague spread through dinosaur populations). Three particularly important and testable hypotheses are the impact, volcanism, and marine regression theories. To these recently has been added another called the Pele theory that suggests that, among other things, the decrease in atmospheric oxygen would have had a detrimental affect on dinosaurs. We must await further information to test this last theory adequately. Only the impact and marine regression theories have been relatively thoroughly tested with the vertebrate fossil record by their respective proponents, although proponents of the volcanism theory suggest that many of the biotic responses to an impact would also be found with massive volcanism. One method to assess the efficacy of these theories is to examine them in the context of the known K–T vertebrate record, starting with the impact theory.

The original scientific paper in 1980 advocating the impact theory still offers the basic mechanism of how such an impact might cause extinction among both animals and plants, including vertebrates. The impact would create a dust cloud enveloping the globe for a few months to a year. Darkness would shroud the world as long as the dust remained in the atmosphere. Photosynthesis in the sea and on land ceased. As the plants died or became dormant, herbivores soon starved, followed by the carnivores.

Some of the best physical evidence of such an impact is the discovery of anomalously high levels of the rare earth element iridium at the K–T boundary and the probable remains of an impact crater. Although iridium is found on the earth, especially deep in the earth, high levels of iridium are associated with extraterrestrial events. Generally, the larger the impacting object, such as an asteroid, the greater the increase in signature elements such as iridium. The remains of a probable impact structure dubbed "Chicxulub," approximately 110 miles across at the northern tip of the Yucatan peninsula from at or near the K–T boundary, are strong evidence for an impact. Neither elevated levels of iridium nor an impact crater, however, are direct evidence for specific causes of extinction at the K–T boundary.

An incorrect assumption often made in testing the impact theory and its possible corollaries is that all major taxa show very high levels of extinction across the ecological spectrum on a global scale. As already discussed, for many organisms, but most notably dinosaurs and their contemporary vertebrates, there is no such global record at the species level. The impact-generated scenario of extremely high levels of catastrophic extinction across most environments is so broad spectrum and tries to explain so much that it is difficult to test. The burden of proof for sweeping, catastrophic extinction scenarios rests with the proposers of the theory. The various corollaries of the impact theory, such as a sudden cold snap, highly acidic rain, or global wildfires, are more easily tested using the known K–T vertebrate record.

A short, sharp decrease in temperature was not emphasized in the originally proposed hypothesis, but it soon became an important corollary of the impact theory. It is argued that if tremendous amounts of dust were injected into the atmosphere after a large impact, the darkness would not only suppress photosynthesis but also produce extremely cold temperatures. This hypothesized condition has become known as "impact winter." It is argued that following a large impact, ocean temperatures would decrease only a few degrees because of the huge heat capacity of the oceans, but on the continents, however, temperatures would be subfreezing from 45 day up to 6 months. The temperature would remain subfreezing for about twice the time of darkness caused by the dust.

If a suddenly induced, prolonged interval of subfreezing occurred in subtropical and tropical regions today, which vertebrates in this climate, which is similar to that of the latest Cretaceous, would be most affected? In general, ectothermic tetrapods would suffer most. Ectotherms, as the name suggests, heat or cool themselves using the environment. Endotherms such as mammals and birds generate their heat through metabolic activity. In endotherms approximately 80% of food consumption goes toward thermoregulation (regulation of body temperature).

Fishes, which are by and large ectothermic, would be generally less affected by a severe temperature drop.

Today, the northern limit of turtles and crocodilians is controlled by temperature. These animals cannot tolerate freezing, becoming sluggish or immobile at 10–15°C. Various amphibians and reptiles do inhabit areas with low winter temperatures or severe drought, but they have evolved methods of torpor (estivation and hibernation) to survive. These are the exceptions, however, because species diversity for ectothermic tetrapods is far higher in warmer climates. More important, we should not assume that Late Cretaceous ectothermic tetrapods living in subtropical to tropical climates such as in eastern Montana were capable of extended torpor. Torpor is most often preceded by decreases in ambient temperature, changes in light regimes, and decreases in food supply. The ectotherms in eastern Montana could not have anticipated a short, sharp decrease in temperature. This is true even if the impact had occurred during a Northern Hemisphere winter when temperatures would been slightly lower. We must remember that this was a subtropical to tropical setting, and thus the extended, subfreezing temperatures advocated by proponents of this corollary would have been devastating to ectotherms even during a terminal Cretaceous winter in Montana.

Except for a 70% decline in lizards, ectothermic tetrapods (frogs, salamanders, turtles, champsosaurs, and crocodilians) did very well across the K–T boundary. The corollary of a sudden temperature decrease simply does not fit with the vertebrate data at the K–T boundary. A latest (but not terminal) Cretaceous vertebrate fauna from northern Alaska strengthens the evidence that a hypothesized sudden temperature drop was not a likely cause of K–T boundary extinctions. Comparing the Late Cretaceous vertebrate faunas from Alaska and eastern Montana reveals a striking difference. Although the Alaskan fauna is decidedly smaller and with fewer species than that from eastern Montana, both have sharks, bony fishes, dinosaurs, and mammals. The Alaskan fauna, however, completely lacks amphibians, turtles, lizards, champsosaurs, and crocodilians. These taxa comprise 41 of 107 (38%) of the eastern Montana fauna. If even the fairly balmy temperature range of 2–8°C for Late Cretaceous Alaska was enough to exclude ectothermic tetrapods, a severe temperature drop to below subfreezing temperatures at the K–T boundary should have devastated the rich ectothermic tetrapod faunas at midlatitudes. These species flourished.

A second prominent corollary of the impact theory is highly acidic rain. The most commonly cited acids as products of an impact are nitric and sulfuric acid. It is argued that nitric acid would be produced by the combination of atmospheric nitrogen and oxygen as a result of the tremendous energy released by an impact. Sulfuric acid would be produced because large amounts of sulfur dioxide are vaporized from rock at the impact site. These acids would be precipitated in the form of rain. Estimates of the pH of these acid rains vary, but estimates reach as low as 0.0–1.5! It is suggested that global effects could have caused the pH of near-surface marine and fresh water to below 3. In today's environment, rain below a pH of 5.0 is considered unnaturally acidic. Rain as low as 2.4 has been recorded, but annual averages in areas affected by acid rain range from 3.8 to 4.4. Acid fogs and clouds from 2.1 to 2.2 pH have been recorded in southern California and have been known to bathe spruce–fir forests in North Carolina. The biological consequences of such low pH values vary from one vertebrate group to another but are always detrimental. Aquatic species (fish, amphibians, and some reptiles) are the first and most drastically affected, with those reproducing in water being the first to suffer. If pH becomes lower than approximately 3.0, adults often die. The effects on aquatic vertebrates across the K–T boundary would have been very bad if a pH of 3.0 was reached and truly horrendous if it hit 0.0 as suggested by some authors. Some advocates of K–T acid rain argue that the surrounding soils or bedrock would have buffered the aquatic systems, and they even suggest that limestone caves could have been important refugia for birds, mammals, amphibians, and small reptiles. The only problem with this scenario is that there were none of these kinds of buffering soils or bedrock or limestone caves in eastern Montana in the Latest Cretaceous. Based on what we know of our modern biota's reaction to acid rain, aquatic animals should have been devastated by acid rain at the K–T boundary. Of all the aquatic species, only sharks and their relatives show a drastic drop in eastern Montana. Thus, the likelihood of low pH rain is highly implausible.

A third corollary of the impact theory that receives various levels of support is global wildfire resulting from the aftermath of the impact. Soot and charcoal have been reported from several sites at the K–T boundary coincident with the enrichment of iridium noted earlier. It was argued that this pattern is unique and must come from the extremely rapid burning of vegetation equivalent to half of all the modern forests! Other scenarios argue that approximately 25% of the aboveground biomass burned at the end of the Cretaceous!

Such a global conflagration is really beyond our comprehension. In order to grasp the magnitude of this scenario, imagine one-quarter to one-half of all structures on the globe engulfed in flames within a matter of days or weeks. This still would be only a fraction of what is argued to have been burned at the K–T boundary. In such an apocalyptic global wildfire, much of the aboveground biomass all over the world would have been reduced to ashes. In fresh water, those plants and animals not boiled outright would have faced a rain of organic and inorganic matter unparalleled in human experience. These organisms would have literally choked on the debris or suffocated as oxygen was suddenly depleted with the tremendous influx of organic matter. The global wildfire scenario is so broad in its killing effects that it could not have been selective, but, as discussed earlier, the vertebrate pattern of extinction and survival is highly selective. Thus, it is no surprise that this scenario of equal opportunity losers does not show any significant agreement with the pattern of vertebrate extinction and survival at the K–T boundary.

Not only is there almost no fossil evidence supporting global wildfire but also the physical basis for such an event is suspect. It is argued that there is a global charcoal and soot layer that coincides with the K–T boundary, whose emplacement is measured in months. This also assumes that the sedimentary layer encasing the charcoal and soot was also deposited in only months. This is demonstrably not the case for at least one K–T section that continues to be cited in these studies—the Fish Clay of the Stevns Klint section on the coast of Denmark. The Fish Clay is a laterally discontinuous, complexly layered and burrowed clay reflecting the conditions at the time of its deposition. It is not the result of less than a year of deposition caused by an impact-induced global wildfire. Thus, carbon near the K–T boundary at Stevns Klint as well as in other sections is likely the result of much longer term accumulation during normal sedimentation.

When all of the corollaries of an impact of an asteroid or comet are compared to the pattern of extinction and survival for vertebrates at the K–T boundary in eastern Montana, there is relatively poor agreement. Without special pleading, these corollaries as currently proposed are unlikely causes of vertebrate extinction. This does not mean that all corollaries of an impact should be rejected, but it is imperative that those proposing the different corollaries separate those that are supported by the vertebrate fossil record from those that are not.

The next hypothesis is the volcanism theory. Although some proponents of the impact theory do not agree, many advocates of both theories feel that a number of the same physical events would have occurred at the K–T boundary if either extensive volcanism or an impact took place. They also say that the biological results would be similar. Given the previous discussion of how most of the corollaries of the impact theory do not test well against the vertebrate fossil record, the volcanism and impact theories are equally weak in their biological predictions. The major difference in these two theories is in their timing. Whereas the impact theory measures most of the cataclysmic effects in months or years, with physical effects possibly lingering for a few hundred or a few thousand years, the volcanism theory measures effects into the millions of years. The effects of many volcanic eruptions, such as that of Mt. St. Helens, linger for a few months or a few years. Many other episodes of volcanism are very prolonged. These are flood basalt eruptions. The best known in the United States is the 16-million-year-old Columbia River flows in the northwest. In the past 250 million years, arguably one of the biggest flood basalt eruptions occurred on the Indian subcontinent. This was occurring during (and is probably related to) the collision of the subcontinent with the remainder of Asia. Its most obvious manifestation today is the tallest mountain range in the world, the Himalayas. These flood basalts, known as the Deccan Traps, cover an immense part of both India and Pakistan. Individual flows in the sequence cover almost 4000 square miles with a volume exceeding 2400 cubic miles. Individual flows average 30–160 ft thick, sometimes reaching 500 ft. In western India the accumulations of lava flows is 7800 ft, or 1.5 miles thick. The flows originally may have covered almost 800,000 square miles with a volume possibly exceeding 350,000 cubic miles (an area the size of Alaska and Texas combined, save about 30,000 square miles, to a depth of more than 2000 ft). Based on radiometric dating, paleomagnetics, and vertebrate fossils, the bulk of the eruptions are centered around the K–T boundary, during a reversal in earth's magnetic poles known as 29R or 29 reversed. The number 29 represents the 29th reversal of the earth's magnetic field counting backwards from the present, which today has normal polarity by definition. The K–T boundary happens to fall in 29R.

What would the effect have been on the global biota if something the magnitude of the Deccan Trap erupted for tens of thousands of years or longer? One of the greatest effects would have been to increase and maintain a much higher level of particulate matter in the atmosphere. Whether it would have caused warming through a greenhouse effect, cooling because of less light, or simply prettier sunsets is not certain. The amount of CO_2 pumped into the atmosphere by the eruptions may have been a boon for green plants that require CO_2 for photosynthesis, but a reduction in light reaching the surface because of particulate matter may have canceled the effects of increased CO_2. The effects of added particulate matter might have prevailed for no other reason than that they would linger longer after eruptions ceased, whereas the release of CO_2 would have diminished much more rapidly after each eruption stopped.

If the latter scenario is correct, the longer term effects over a million years or more would be to push a global cooling. Most estimates suggest that regional if not the global climate cooled through the K–T transition. Because the time frame is moderately long, many species, especially smaller ones, on land or in the sea could adapt to changes, whereas larger species such as dinosaurs may not have been as fortunate. Although it probably was not a cause of extinction for most species, the cooling across the K–T boundary would have been an added stress.

A final long-term effect suggested for eruption of the Deccan Traps is reduced hatching success for eggs of herbivorous dinosaurs. Volcanic activity can release elements such as selenium that are highly toxic to developing embryos. Increased levels of selenium in the eggshells of dinosaurs are known from near the K–T boundary in southern France. Poisoning of eggs has also been reported from dinosaur eggs near the K–T boundary in Nanxiong Basin, southeastern China.

The final hypothesis that has been tested with the vertebrate fossil record is the marine regression/habitat fragmentation theory, or simply marine regression theory. Many areas of the terrestrial realm were repeatedly inundated by shallow epicontinental seas throughout geologic history. The term "epicontinental" refers to the occurrence of these very shallow seas upon the continental shelves and platforms rather than in deep ocean basins. Epicontinental seas reached depths of only 1500–2000 ft, very shallow compared to most large modern marine bodies. Epicontinental seas are almost nonexistent today, except for such bodies of water as Hudson Bay. It is known that during the Late Cretaceous, large areas of continents were submerged under warm, shallow epicontinental seas.

It became clear only recently just how dramatic the loss of these seas was leading up to the K–T boundary. There is absolutely no mistaking that the K–T loss of shallow seas (or increase in nonmarine area) is greater than at any time in the past 250 million years. The nonmarine area increased from 42 million m^2 to 53 million m^2—more than a 25% increase. This is the equivalent of adding the land area of all of Africa, the second largest continent today. The second largest increase in continental area in the past 250 million years occurred across the Triassic–Jurassic boundary. Like the K–T transition, this is also during one of the five universally recognized mass extinctions during the Phanerozoic or last 550 million years.

Some of the most dramatic additions of nonmarine areas at or near the K–T boundary occurred in North America. Near the end of the Cretaceous, maximum transgression divided North America into two continents. As regression continued until at or near the K–T boundary, coastal plains decreased in size and became fragmented; stream systems multiplied and lengthened; and as sea level fell, land connections were established or reestablished.

The driving force for these repeated inundations or transgressions of the lower-lying portions of continents is still not fully understood. The general consensus is that it is related to plate tectonics. It is thought that rises in sea level and inundations began as the motion of the plates increased. As this occurs, the margins along which the colliding plates converge are subducted or pushed downwards into the earth. This causes inundation by the seas upon shallow continental shelves and platforms.

Whatever the geophysical factors driving the process, the physical manifestations of marine regression, like impacts and volcanism, are important ultimate causes of extinction. Although these processes of marine transgression and regression were global in extent, a closer examination of North America is best because, as I have emphasized, this is where we have the vertebrate data at the K–T boundary. North America was split into two continents—a western continent (Laramidia) and an eastern continent (Appalachia)—by the Pierre Seaway for almost 40 million years during the Late Cretaceous. Most of our latest Cretaceous vertebrate fossils come from the east coast of Laramidia. The west coast of Appalachia as well as the eastern seaboard of Appalachia have also produced some specimens.

In the last few million years of the Cretaceous the Pierre Seaway began to regress from both Laramidia and Appalachia. At or just shortly before the K–T boundary, the seaway reached its nadir. Placement of the receding coastlines both north and south have not been well established, but we know the southern coastline reached well into Texas. There is no question that there was a dramatic reduction in coastal plains. This is exactly the kind of environment from which we are sampling the last of the Late Cretaceous vertebrate community. A common refrain is that because the total amount of land increased with the regression, dinosaurs should have had more, not less, area and more environments in which to live. We know with considerable certainty that dinosaurs did live in other environments such as the higher, drier Gobi Desert in Mongolia during part of the Late Cretaceous. Currently, however, the only well-known vertebrate communities that preserve dinosaurs at the K–T boundary are coastal. Thus, arguments about what dinosaurs and other vertebrates may or may not have done in other environments are moot. It is simply incorrect to say that the dinosaurs and other vertebrates may have survived elsewhere when we have little or no information about other environments.

The drastic reduction of coastal plains put tremendous pressure on some, especially large vertebrate species. Reduction of habitat, for example, in the Rift Valley system of East Africa, today first affects larger vertebrates, especially mammals. In the shrinking coastal plains of latest Cretaceous North America, the equivalent large vertebrates first affected were the dinosaurs. An additional problem, whether in East Africa today or the coastal plains of latest Cretaceous North America, is the fragmenting of the remaining habitat. This process, as a result of human activity, has become known as habitat fragmentation. In larger, undisturbed habitats, animals (and plants) can spread more freely from one area to another. If the habitat is fragmented, although the amount of habitat may not have been greatly altered, it will reduce the flow of species from one fragment to another.

For some species, even seemingly small barriers such as two-lane roads can be insurmountable. The results can be disastrous if viable populations cannot be maintained in the various fragments. Fragmentation can lead to extinctions. Barriers also arise in nature even among animals that would seem easily capable of dispersing. Although no doubt the result is very often extinction, we usually only see what survives in the form of differences between closely related species. Small arboreal primates in the rainforests of both South America and Africa form small fragmented groups that are isolated from each other often by rivers only tens of yards wide. Another example is the Kaibab squirrel on the North Rim of the Grand Canyon. Unlike its nearest relative, Abert's squirrel, which is found on the south side of the Grand Canyon and in the western United States and Mexico, the Kaibab squirrel is restricted to an area of only 20 $\times$ 40 miles. Fragmentation, in this case the development of the Grand Canyon, helped produce the differences, but the margin between this and oblivion for the Kaibab squirrel is slim.

The idea of habitat fragmentation not only extends to natural processes operating today but also to processes operating in the geological past. Although historical habitat fragmentation is not well understood by earth scientists, it is an all too real phenomenon among biologists studying the effects of human activity in modern rainforests and in urban settings. Declines of bird and mammal populations have been well documented in the city of San Diego as urban development divides and isolates habits in canyon areas. One would not expect that the natural equivalent of habitat fragmentation would be easily, if at all, preserved in the rock. The forcing factor for habitat fragmentation in the latest Cretaceous—marine regression—is a thoroughly documented fact during the waning years of the Late Cretaceous in North America. Globally, marine regression occurred within this same general time frame although how close in time it occurred in various regions is a matter of debate.

Theory predicts that large species would be the most severely affected by habitat fragmentation for the reasons discussed previously. During the K–T transition in eastern Montana, only 8 of 30 large species survived and these are partially or entirely aquatic (2 fishes, 1 turtle, 1 champsosaur, and 4 crocodilians), whereas all 22 large terrestrial species (and 1 aquatic species) became extinct (1 turtle, 1 lizard, 1 crocodilian, and 19 dinosaurs). Thus, predictions from habitat fragmentation fit the observed data very well.

As noted previously, two other major physical events occur with marine regression in addition to habitat fragmentation—stream systems multiply and lengthen and, as sea level falls, land connections are established or reestablished.

Only a few of the K–T boundary stream systems have been studied in detail in the Western Interior, and thus we do not know the exact drainage patterns for most latest Cretaceous and early Tertiary stream systems in the eastern part of Laramidia. Nevertheless, we are certain that as new land was added following marine regression in the early Tertiary, stream systems increased and lengthened. This process is another major corollary of marine regression. When freshwater habitats were bolstered following marine regression, most aquatic vertebrates did well, except those with close marine ties—sharks and some bony fishes. Such fishes may need to spend at least a portion of their life in a marine environment, in some instances to reproduce. The major group most likely to suffer would have been the sharks and their relatives. In fact, all five species of sharks disappeared. It is not clear, however, whether these disappearances from the Western Interior are actually extinctions at the K–T boundary or whether the shark species survived elsewhere in marine environments into the earliest Paleocene. The problem is that the definitively oldest marine sediments that postdate the K–T boundary in the Western Interior are no older than late early Paleocene in age. This means that there is a gap in marine sedimentation in the Western Interior of possibly 1 million years or more immediately after the K–T boundary. This pattern of disappearance and reappearance strongly suggests that as the Pierre Seaway regressed further and further away from eastern Montana, all sharks and relatives departed because connections to the sea became attenuated. New species of elasmobranchs did not occur in the area until a smaller transgression reached the Western Interior at or just before middle Paleocene times. This is known as the Cannonball Sea, which was a smaller seaway than the Pierre. The total disappearance of sharks and relatives is the only prediction that can be made with any certainty as a result of the loss of marine connections and the lengthening of stream systems. The increase of stream systems was a positive factor helping to mitigate other stresses that may have been put on the freshwater system.

New land areas were exposed as sea level lowered. In some cases this included the establishment or reestablishment of intercontinental connections. One such connection was the Bering land bridge joining western North America and eastern Asia. At various times during the Late Cretaceous this bridge appeared and then disappeared. This is suggested by similarities in parts of the Late Cretaceous vertebrate faunas in Asia and North America, especially the better studied turtles, dinosaurs, and mammals. Competition and extinction often result from biotic mixing, but predicting the fates of various taxonomic groups is usually not possible. An exception may have been the fate of marsupials in North America near the K–T boundary.

The oldest marsupials are known from approximately 100-million-year-old sites in western North America. By some 85 Mya, we know of about 10 species of marsupial. This rose and stayed at about 15 species from approximately 75 Mya until the K–T boundary approximately 65 Mya, when it plummeted to one species. These were all quite small mammals, from the size of a mouse up to a very well-fed opossum or raccoon. Their teeth were very much like those of extant opossums, with slicing crests and well-developed but relatively low cusps (compared to contemporary placental mammals) for poking holes in insect carapaces, seeds, or whatever they found. Most did not appear to be specialists on any particular food. With the reestablishment of the Bering land bridge (or at least closer islands) near the K–T boundary, a new wave of placental mammals appeared in western North America. These mammals (traditionally known as condylarths) were the very early relatives of modern ungulates and whales. Their appearance in North America coincides with the very rapid decline of marsupials near the K–T boundary. Within a million years of the K–T boundary, 30 species of these archaic ungulates are known in North America, and their numbers kept on rising. Our best guess now is that the lineage that gave rise to these mammals first appeared in middle Asia between approximately 80 and 85 million years ago and reached North America near the K–T boundary. What is of interest is that the archaic ungulate invaders had dentitions very similar to contemporary marsupials and presumably ate similar things. Its seems more than coincidence that marsupials did well in North America for approximately 20 million years only to almost disappear with the appearance of the ungulate clade. It is ironic that both marsupials and ungulates were joint invaders of South America very soon after the K–T boundary. Their dentitions were already beginning to show differentiation, with the marsupials headed toward carnivory and ungulates headed toward herbivory. It shows what a little cooperation can do.

These various physical events accompanying marine regression fit very well the pattern of extinction and survival described previously for the 107 vertebrate species from near the K–T boundary in eastern Montana. In fact, the patterns of extinction and survival for 11 of 12 of the major vertebrate groups agree very well with predictions from the marine regression theory discussed previously.

Marine regression, extraterrestrial impact, and massive volcanism are all major environmental events that occurred near or at K–T boundary. None of these physical events appears sufficient by itself to be crowned the sole cause of extinctions at the K–T boundary. The evidence outlined here overwhelmingly supports this view. What is less certain is whether all three were necessary for the pattern of extinctions we see at the K–T boundary. No one knows for sure. Both marine regression and an impact were apparently necessary to give us the pattern of turnover at the K–T boundary for all species, not just the vertebrates discussed in this essay. The role of volcanism or the effects of the Pele hypothesis are less certain. The effects of massive volcanism are formidable, but the purported biological effects have not been as closely explored as those of the other two events. The Pele hypothesis simply has not yet been properly studied and tested.

What is very clear is that at least three physical events—marine regression, extraterrestrial impact, and massive volcanism—did coincide near the K–T boundary, but this does not mean that all three events necessarily occurred simultaneously at this boundary. The greatest addition of nonmarine area in the past 250 million years brackets the Late Cretaceous mass extinctions. One of the largest known impact craters is thought to have been identified at very near the K–T boundary. Massive volcanism had been pouring out great quantities of lava sporadically for several million years during the K–T interval. This was clearly one of the most geologically complicated and biologically challenging episodes in earth history. Given these challenges it is surprising that more species did not succumb. It suggests that life has been more resilient than we dreamed, or have a right to hope for, given the stresses we are placing it under today.

See also the following related entries:

CRETACEOUS PERIOD • EVOLUTION • EXTINCTION, TRIASSIC

References

Archibald, J. D. (1996). *Dinosaur Extinction and the End of an Era: What the Fossils Say*, pp. 237. Columbia Univ. Press, New York.

Fastovsky, D. E., and Weishampel, D. B. (1996). *The Evolution and Extinction of the Dinosaurs*, pp. 460. Cambridge Univ. Press, Cambridge, UK.

MacLeod, N., and Keller, G. (Eds.) (1996). *The Cretaceous–Tertiary Mass Extinction: Biotic and Environmental Changes*, pp. 575. Norton, New York.

Raup, D. M. (1991). *Extinction: Bad Genes or Bad Luck*, pp. 210. Norton, New York.

II. Extinction, Triassic

MICHAEL J. BENTON
University of Bristol
Bristol, United Kingdom

The oldest dinosaurs date from the Carnian Stage (see HERRERASAURIDAE; ORIGIN OF DINOSAURS), and the group split into recognizable theropods, sauropodomorphs, and ornithischians during that time interval. However, dinosaurs were minor elements of the Carnian faunas, and they were generally modest sized. Dinosaurs became abundant, and they diversified further, during the subsequent Norian Stage.

Competition or Mass Extinction?

Until 1980, the origin of the dinosaurs was explained generally by a model of long-term competitive replacement. Inadequacies of stratigraphy, and of understanding of faunal compositions, suggested that dinosaurs had arisen some time during the Middle Triassic. The evidence for this consisted of a scattering of bony specimens from Europe and the Americas and of footprints largely from Europe. Some of the footprints were from Early Triassic sediments, and the picture of a long slow rise of the Dinosauria throughout the first half of the Triassic was widely accepted.

Competition Model The competitive model for Triassic faunal replacements (Bonaparte, 1982; Charig, 1984) proposed that the Early Triassic faunas, dominated by therapsid synapsids, gave way to Middle Triassic faunas dominated by rhynchosaurs and basal archosaurs, which were in turn replaced by Late Triassic dinosaur faunas. The evolutionary explanation for the successive replacements was that each group competed with its predecessors and rose to ascendancy because of some superior adaptations. The strongest argument for those superior adaptations was the change in archosaurian posture during the Triassic from the typical reptilian sprawling pattern to a fully erect posture in dinosaurs (Charig, 1972).

Mass Extinction Model The alternative mass extinction model (Benton, 1983, 1986a,b, 1994b) states that none of this evolutionary relay took place, but that the dinosaurs radiated into empty ecologic space after a mass extinction event had wiped out preexisting groups. The particular proposal is that dinosaur faunas in fact showed relatively little change until the second half of the Carnian Stage, when the rhynchosaurs and dicynodonts, as well as several groups of amphibians, basal archosaurs, and smaller reptiles, disappeared at, geologically speaking, the same time. The dinosaurs radiated only after the slate had been wiped clean of other moderate to large-sized herbivores.

Triassic Faunal Evolution

Pre-Carnian "Dinosaurs" Careful study of the supposed Early and Middle Triassic dinosaur remains showed that the bones were either definitely nondinosaurian or they were inadequate for identification (Benton, 1986b). Many so-called "dinosaur" remains from pre-Carnian rocks in Germany—named variously *Teratosaurus* or *Zanclodon*, for example—are teeth and vertebrae identifiable only as archosaurian. In addition, supposed dinosaur footprints from the Early and Middle Triassic of England turned out to be inorganic structures or clearly nondinosaurian (King and Benton, 1995), and the Middle Triassic material from France is equivocal.

Stratigraphy New studies of the stratigraphy of the Late Triassic and Early Jurassic dinosaur-bearing continental sediments (Olsen and Galton, 1977, 1984; Olsen and Sues, 1986; Hunt and Lucas, 1991; Benton, 1994a,b) have clarified the sequence of faunas (Fig. 1). The revisions have been based on new biostratigraphic evidence from palynomorphs and freshwater fishes, on rare ties with marine sequences dated by ammonoids, by sporadic magnetostratigraphic and radiometric tie points, and by comparisons of tetrapod faunas. The details of dating should improve considerably in the future as a result of further study of palynological evidence and of new data from magnetostratigraphy and scattered radiometric dates.

Continental Tetrapod Evolution The faunal succession during the Triassic (Fig. 2) suggests, for Gondwanaland and Laurasia, that there was a substantial mass extinction event during the second half of the Carnian Stage, perhaps 225 Ma ago at or near the Carnian–Norian (Crn–Nor) boundary. The oldest Triassic faunas, from several parts of the world, were dominated by the dicynodont *Lystrosaurus*, which represented more than 90% of all animals found. This astonishing monoculture, with parallels today only in agricultural situations, was probably the result of the vast end-Permian extinction event, which had cleared the typical Late Permian faunas of the dominant groups, such as dinocephalians, gorgonopsians, and pareiasaurs (see Reptiles).

During the Middle Triassic, faunas north and south became dominated by dicynodonts and/or rhynchosaurs as herbivores and cynodonts and basal archosaurs as carnivores. In no case did these archosaurs exceed 10% of the number of individuals in a fauna. This low representation was counter to the expectations of the classic competitive model for the rise of the basal archosaurs, and then for their progressive replacement by the dinosaurs.

Most Carnian continental tetrapod faunas were comparable in composition. Substantial Middle to Late Carnian tetrapod assemblages are known from Argentina, Brazil, the United States (New Mexico, Arizona, and Texas), Britain, Morocco, and India (Fig. 1). Typically, these faunas were still dominated by rhynchosaurs and/or dicynodonts as bulky herbivores, with basal archosaurs and/or cynodonts as carnivores. Many of them contain one or two specimens of modest-sized dinosaurs.

Tetrapod faunas of early Norian age from the

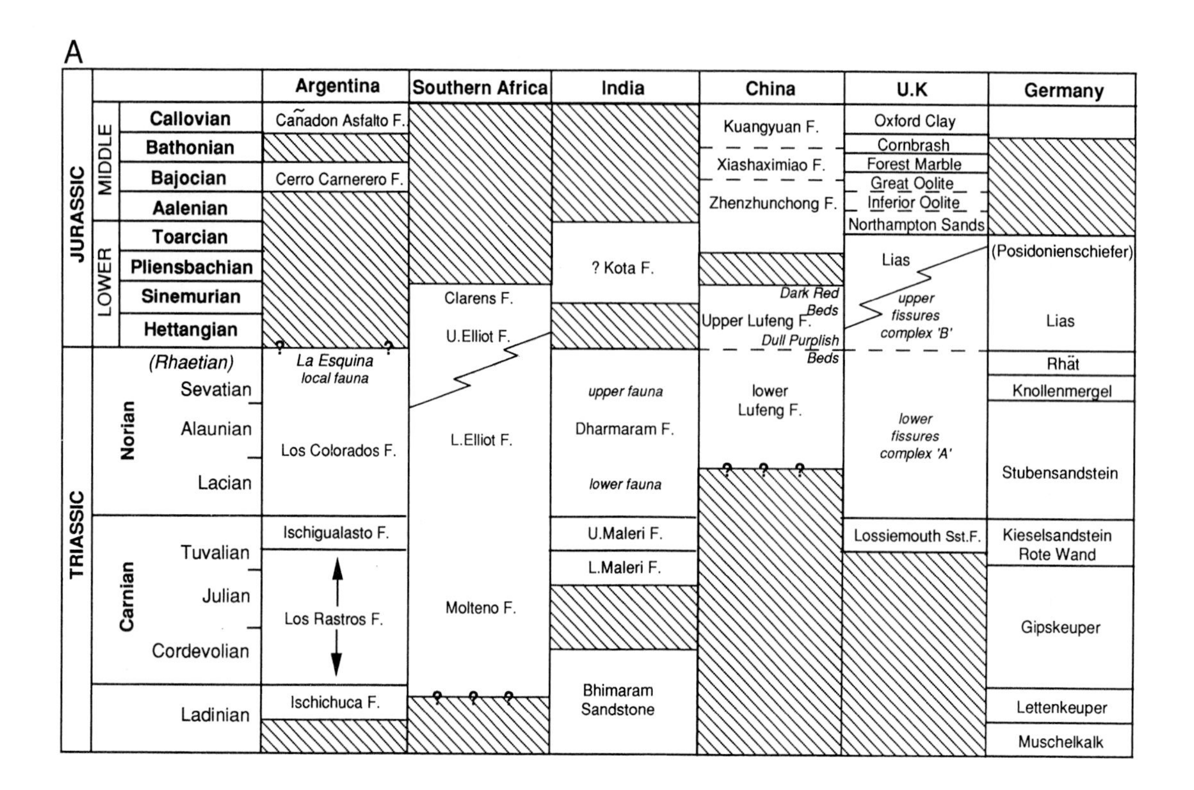

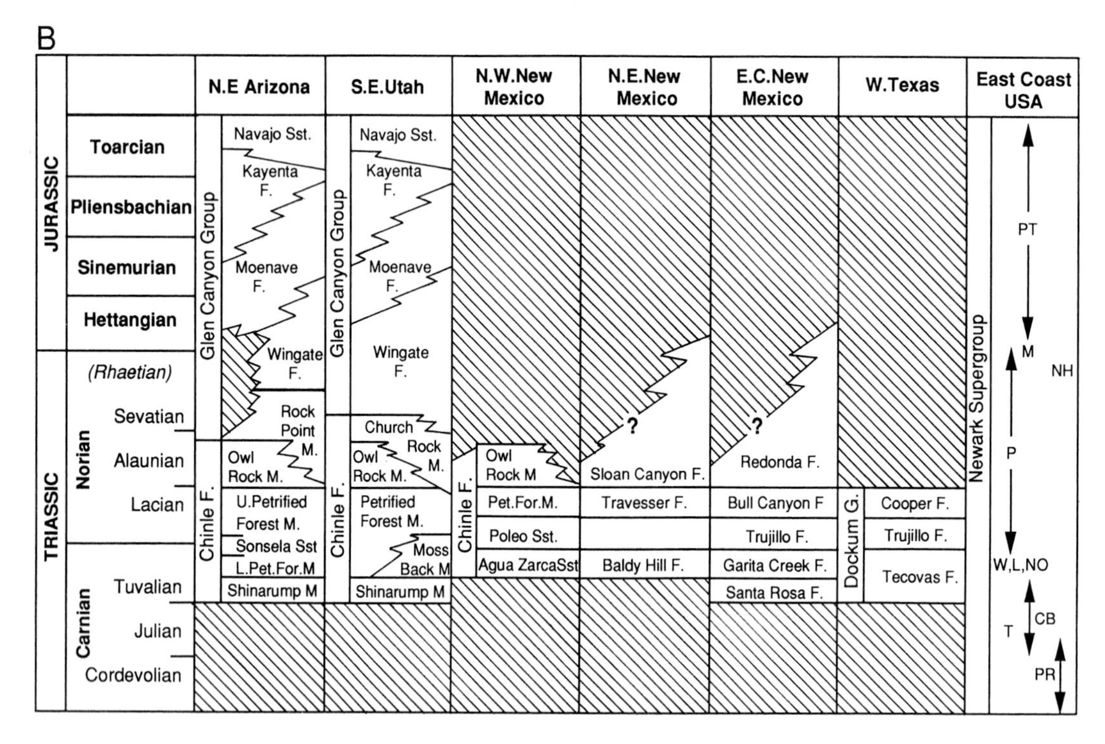

FIGURE 1 Stratigraphy of major Late Triassic and Early Jurassic continental tetrapod-bearing formations in Gondwanaland and Europe (A) and in North America (B). Abbreviations: CB, Cow Branch Formation; L, Lockatong Formation; M, McCoy Brook Formation; NH, New Haven Arkose; NO, New Oxford Formation; P, Passaic Formation; PR, Pekin Formation; PT, Portland Formation; T, Turkey Branch Formation; W, Wolfville Formation. (Based on various sources and summarized in Benton, 1994a.)

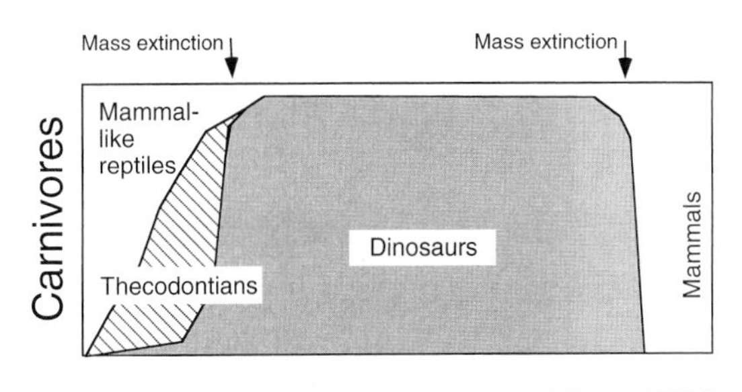

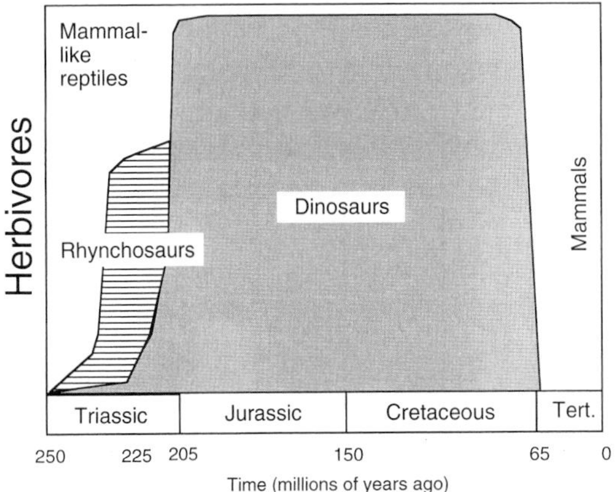

FIGURE 2 Relative abundances of different tetrapod groups through the Mesozoic. The patterns for carnivores (top) and herbivores (bottom) show evidence for a catastrophic loss of medium-sized to large animals in the late Carnian (225 Ma) and their replacement by dinosaurs. (Based on data in Benton, 1983.)

southwestern United States, Germany, and southern continents are quite different. Dinosaurs dominate some assemblages in abundance (consider the famous Ghost Ranch assemblage of *Coelophysis* in the upper Petrified Forest Member of the Chinle Formation and the rich assemblages of *Sellosaurus* and *Plateosaurus* from the German Stubensandstein) and to some extent in terms of diversity and size range. However, although some Norian assemblages are dominated by dinosaurs (25–60% of specimens), dinosaurs were not really diverse, with at most four or five genera at any time in the Stubensandstein. Sizes of Triassic prosauropods ranged up to 8 m for *Plateosaurus* and 10 m for *Riojasaurus*.

The Oldest Dinosaurs

The best known early fauna of dinosaurs comes from the Ischigualasto Formation of San Juan Province, Argentina. It consists of *Eoraptor lunensis* and *Herrerasaurus ischigualastensis* (variously regarded as basal theropods, basal saurischians, basal dinosaurs, or the closest relatives of dinosaurs though not to be members of Saurischia–Ornithischia) and the ornithischian *Pisanosaurus mertii* (Bonaparte, 1976; Sereno and Novas, 1992, 1994; Sereno *et al.*, 1993; Novas, 1994; Sereno, 1994) (see HERRERASAURIDAE; PHYLOGENY OF DINOSAURS). The dinosaurs were rare, however, representing only 5.7% of the fauna (13 of 228 specimens, according to Rogers *et al.*, 1993).

The Ischigualasto fauna was dominated by the rhynchosaur *Scaphonyx*, a pig-sized grubbing herbivore, and *Exaeretodon*, a medium-sized herbivorous cynodont. Other rarer elements include the bulky herbivorous dicynodont *Ischigualastia*, the basal archosaurs *Aetosauroides* (a herbivore), *Proterochampsa* (a piscivore), and *Saurosuchus* (a carnivore), as well as the dinosaurs (Fig. 3). Rogers *et al.* (1993) established a radiometric date of 228 Mya for the lower part of the Ischigualasto Formation, which, they argue, places it in the middle of the Carnian Stage.

Small dinosaur faunas are known elsewhere in

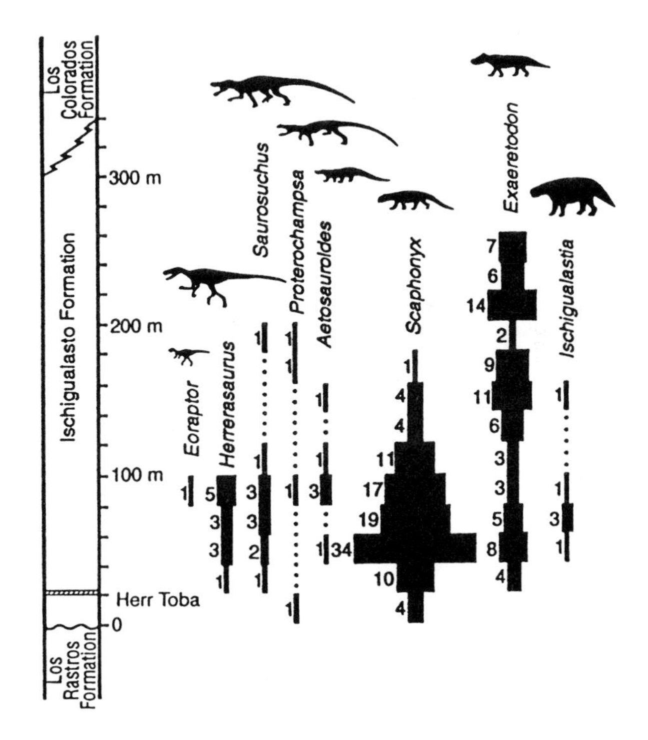

FIGURE 3 Range chart of reptiles through the thickness of the Ischigualasto Formation, Argentina, showing relative abundances, based on a collection of 228 specimens. Dinosaurs (*Eoraptor* and *Herrerasaurus*) occur rarely in the lower part of the formation. The most common animal is the rhynchosaur *Scaphonyx*, followed by the herbivorous cynodont synapsid *Exaeretodon*. The dicynodont *Ischigualastia* and the thecodontians *Saurosuchus*, *Proterochampsa*, and *Aetosauroides* are rare. The Herr Toba bentonite is dated 228 Ma. (Based on data in Rogers *et al.*, 1993.)

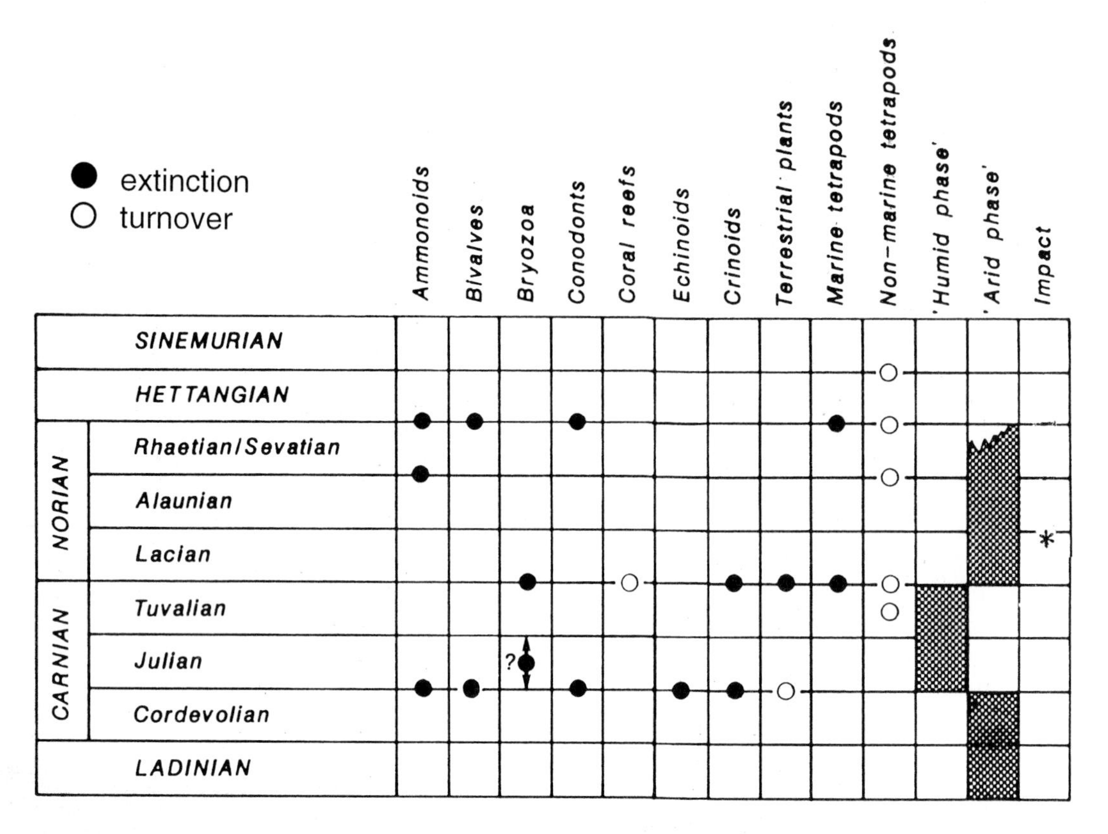

FIGURE 4 Major events in the Late Triassic, showing phases of mass extinction (●) and turnover (i.e., high extinction and origination rates; ○). Climatic changes and the Manicouagan impact are indicated. (Based on data in Simms and Ruffell, 1990, and Benton, 1991.)

rocks of middle to late Carnian age: in the Santa Maria Formation of Brazil (*Staurikosaurus*, a herrerasaur), the Maleri Formation of India (*Alwalkeria*), the lower part of the Petrified Forest Member, Arizona (*Coelophysis*), the Argana Formation of Morocco (*Azendohsaurus*), and perhaps the Lossiemouth Sandstone Formation of Scotland (*Saltopus*). In these faunas, dinosaurs represent 1–6% of specimens found. Although specimens are rare, the three main lines of dinosaurian evolution—the theropods, sauropodomorphs, and ornithischians—were already laid down during Carnian times and they had apparently diverged from a single common ancestor within only a few million years. The Carnian dinosaurs in all lineages were moderately sized animals, all lightweight bipeds less than 6 m long.

Late Triassic Mass Extinctions

During the Late Triassic, 20% or more of families of marine animals died out, scaling to some 50% of species, and this matches the severity of the K–T event (Sepkoski, 1990). Controversy revolves around three issues: (i) whether there was a single extinction event in the Late Triassic or more than one, (ii) whether the event(s) was catastrophic and caused by a major extraterrestrial impact, and (iii) whether the extinction(s) had anything to do with the rise of the dinosaurs.

The Triassic–Jurassic Boundary Event There is no question that there was a mass extinction at the end of the Triassic Period (Fig. 4), at the Triassic–Jurassic (Tr–J) boundary, 202 million years ago, when ammonoids and bivalves were decimated and when the conodonts finally disappeared (Sepkoski, 1990). On land too, several families of reptiles disappeared, particularly the last of the basal archosaurs and some nonmammalian synapsids (see REPTILES).

Recent work on earliest Jurassic vertebrate faunas has led to the claim (Olsen *et al.*, 1987, 1991; Hallam, 1990) that this Tr–J event was instrumental in triggering the radiation and huge success of the dinosaurs. In addition, a major impact crater site, the Manicouagan structure in Quebec, was identified as the smoking gun for a catastrophic extraterrestrial impact at the Tr–J boundary (Olsen *et al.*, 1987, 1991). Elevated levels of iridium were reported from a Tr–J boundary section in Austria (Badjukov *et al.*, 1987), and shocked quartz has been found at a Tr–J boundary section in Italy (Bice *et al.*, 1992). High levels of iridium in K–T boundary clays worldwide, and shocked quartz in many such sections, are, of course, taken as key evidence for a major impact (see EXTINCTION, CRETACEOUS).

The case for an impact at the Tr–J boundary is far from certain. Hallam (1990) failed to find the iridium anomaly in the Austrian section, and the nature of the lamellae in the shocked quartz was not adequate to rule out other explanations, such as a volcanic source for the material (Bice *et al.*, 1992). Furthermore, the Manicouagan impact structure was redated (Hodych and Dunning, 1992) away from the Tr–J boundary (202 Mya), well down in the Late Triassic, about 220 Mya. This redating means that the impact happened at a time when no mass extinctions were taking place.

The Crn–Nor Event A commonly expressed view (Olsen *et al.*, 1987, 1991; Hallam, 1990) has been that the postulated Crn–Nor extinction event (225 Mya) was restricted to nonmarine vertebrates and was a minor blip in the diversification of life compared to the Tr–J event. Recent studies of marine fossil records (Sepkoski, 1990; Simms and Ruffell, 1990) have indicated, however, that the foraminifera, ammonoids, bivalves, bryozoans, conodonts, reef corals, echinoids, and crinoids all showed global-scale extinctions either during the Carnian or at the Crn–Nor boundary (Fig. 4). Quantitative assessments of data on nonmarine tetrapods (Benton, 1986a,b, 1991, 1994a,b) have shown that the rate of loss of families was as great during the late Carnian interval as it was around the Tr–J boundary.

Causes Causes of the two Late Triassic extinction events are hard to determine. Neither corresponds to the Manicouagan crater, and evidence for impact is nonexistent for the Crn–Nor event and limited to some shocked quartz at the Tr–J boundary in Italy. Earthbound causes have been posited for both events.

Simms and Ruffell (1990) argued for a major climatic change across the tropical belt, from humid climates in the middle to late Carnian to arid climates in the Norian (Fig. 4). These might have caused major stresses for land plants and herbivorous tetrapods and led to the widespread extinction of previously hugely abundant dicynodonts and rhynchosaurs. Extinctions in the sea may have been caused by increased rainwater runoff into the sea and spread of fresh waters in previously marine basins.

The Tr–J event has been explained by a phase of oceanic anoxia in earliest Jurassic times (Hallam, 1990), an environmental crisis that would have affected mainly marine organisms. Effects on land animals would have been less severe, as seems to have been the case.

The current state of knowledge about mass extinctions is astonishingly limited, and yet it is much advanced over the information available 20 years ago. In comparison to the K–T event that may or may not have finished the dinosaurs (see EXTINCTION, CRETACEOUS), very little effort has been devoted to gaining an understanding of Late Triassic extinction events that may have kick-started the radiation of the clade Dinosauria.

See also the following related entries:

ARCHOSAURIA • EXTINCTION, CRETACEOUS • ISCHIGUALASTO FORMATION

References

Badjukov, D. D., Lobitzer, H., and Nazarov, M. A. (1987). Quartz grains with planar features in the Triassic–Jurassic boundary sediments from northern Limestone Alps, Austria. *Lunar Planet. Sci. Lett.* **18,** 38.

Benton, M. J. (1983). Dinosaur success in the Triassic: A noncompetitive ecological model. *Q. Rev. Biol.* **58,** 29–55.

Benton, M. J. (1986a). More than one event in the Late Triassic mass extinction. *Nature (London)* **321,** 857–861.

Benton, M. J. (1986b). The Late Triassic tetrapod extinction events. In *The Beginning of the Age of Dinosaurs: Faunal Change across the Triassic–Jurassic Boundary* (K. Padian, Ed.), pp. 303–320. Cambridge Univ. Press, Cambridge, UK.

Benton, M. J. (1991). What really happened in the Late Triassic? *Historical Biol.* **5,** 263–278.

Benton, M. J. (1994a). Late Triassic terrestrial vertebrate extinctions: Stratigraphic aspects and the record of the Germanic Basin. *Paleontol, Lombarda Nuova Ser. 2,* 19–38.

Benton, M. J. (1994b). Late Triassic to Middle Jurassic extinctions among continental tetrapods: Testing the pattern. In *In the Shadow of the Dinosaurs: Early Mesozoic Tetrapods* (N. C. Fraser and H.-D. Sues, Eds.), pp. 366–397. Cambridge Univ. Press, Cambridge, UK.

Bice, D. M., Newton, C. R., McCauley, S., Reiners, P. W., and McRoberts, C. A. (1992). Shocked quartz at the Triassic–Jurassic boundary in Italy. *Science* **255,** 443–446.

Bonaparte, J. F. (1976). *Pisanosaurus mertii* Casamiquela and the origin of the Ornithischia. *J. Paleontol.* **50,** 808–820.

Bonaparte, J. F. (1982). Faunal replacement in the Triassic of South America. *J. Vertebr. Paleontol.* **21,** 362–371.

Charig, A. J. (1972). The evolution of the archosaur pelvis and hind-limb: An explanation in functional terms. In *Studies in Vertebrate Evolution* (K. A. Joysey and T. S. Kemp, Ed.), pp. 121–155. Oliver & Boyd, Edinburgh, UK.

Charig, A. J. (1984). Competition between therapsids and archosaurs during the Triassic period: A review and synthesis of current theories. *Symp. Zool. Soc. London* **52,** 597–628.

Hallam, A. (1990). The end-Triassic mass extinction event. *Geol. Soc. Am. Spec. Paper* **247,** 577–583.

Hodych, J. P., and Dunning, G. R. (1992). Did the Manicouagan impact trigger end-of-Triassic mass extinction? *Geology* **20,** 51–54.

Hunt, A. P., and Lucas, S. G. (1991). The *Paleorhinus* Biochron and the correlation of the nonmarine Upper Triassic of Pangaea. *Palaeontology* **34,** 487–501.

King, M. J., and Benton, M. J. (1995). Dinosaurs in the Early and Mid Triassic, fact or fiction? The footprint evidence from Britain. Submitted for publication.

Novas, F. E. (1994). New information on the systematics and postcranial skeleton of *Herrerasaurus ischigualastensis* (Theropoda: Herrerasauridae) from the Ischigualasto Formation (Upper Triassic) of Argentina. *J. Vertebr. Paleontol.* **13,** 400–423.

Olsen, P. E., and Galton, P. M. (1977). Triassic–Jurassic extinctions: Are they real? *Science* **197,** 983–986.

Olsen, P. E., and Galton, P. M. (1984). A review of the reptile and amphibian assemblages from the Stormberg Group of Southern Africa, with special emphasis on the footprints and the age of the Stormberg. *Palaeontol. Africana* **25,** 87–110.

Olsen, P. E., and Sues, H.-D. (1986). Correlation of continental Late Triassic and Early Jurassic sediments, and patterns of the Triassic–Jurassic tetrapod transition. In *The Beginning of the Age of Dinosaurs* (K. Padian, Ed.), pp. 321–351. Cambridge Univ. Press, Cambridge, UK.

Olsen, P. E., Shubin, N. H., and Anders, M. H. (1987). New Early Jurassic tetrapod assemblages constrain Triassic–Jurassic tetrapod extinction event. *Science (New York)* **237,** 1025–1029.

Olsen, P. E., Fowell, S. J., and Cornet, B. (1991). The Triassic/Jurassic boundary in continental rocks of eastern North America; A progress report. *Geol. Soc. Am. Spec. Paper* **247,** 585–593.

Rogers, R. R., Swisher, C. C., III, Sereno, P. C., Forster, C. A., and Monetta, A. M. (1993). The Ischigualasto tetrapod assemblage (Late Triassic) and $^{40}Ar/^{39}Ar$ calibration of dinosaur origins. *Science* **260,** 794–797.

Sepkoski, J. J., Jr. (1990). The taxonomic structure of periodic extinction. *Geol. Soc. Am. Spec. Paper* **247,** 33–44.

Sereno, P. C. (1994). The pectoral girdle and forelimb of the basal theropod *Herrerasaurus ischigualastensis. J. Vertebr. Paleontol.* **13,** 425–450.

Sereno, P. C., and Novas, F. E. (1992). The complete skull and skeleton of an early dinosaur. *Science* **258,** 1137–1140.

Sereno, P. C., and Novas, F. E. (1994). The skull and neck of the basal theropod *Herrerasaurus ischigualastensis. J. Vertebr. Paleontol.* **13,** 451–476.

Sereno, P. C., Forster, C. A., Rogers, R. R., and Monetta, A. M. (1993). Primitive dinosaur skeleton from Argentina and the early evolution of Dinosauria. *Nature* **361,** 64–66.

Simms, M. J., and Ruffell, A. H. (1990). Climatic and biotic change in the late Triassic. *J. Geol. Soc. London* **147,** 321–327.

Extraterrestrial Impact Theory

see EXTINCTION, CRETACEOUS

Fabrosauridae

PENG GUANGZHAO
Zigong Dinosaur Museum
Sichuan, People's Republic of China

The Fabrosauridae was originally proposed by Galton (1972) as a group of basal ornithischians. It comprises some of the small, unarmored, and most primitive-looking ornithischians yet discovered—an early lineage in the ornithischian radiation.

Fabrosaurids are known from southern Africa in the Upper Triassic and Lower Jurassic and from China in the Middle and Upper Jurassic. Because they bear the generalized states for many anatomical characters of the Ornithischia, they have attracted considerable attention in phylogenetic studies. However, most genera previously referred to the Fabrosauridae are represented by fragmentary remains, sometimes no more than isolated teeth, and their affinities are debatable. In 1990, specimens of *Agilisaurus* from the Middle Jurassic of Zigong, Sichuan, China, provided new evidence for resolving the systematic problems of the Fabrosauridae.

The genus *Fabrosaurus*, type and only species *F. australis*, was established by Ginsburg (1964) on a partial right dentary with several teeth from the Upper Red Beds of Basutoland (Lesotho). Thulborn (1970, 1972) successively described the cranial and postcranial material of two more specimens from the same horizon as the type specimens in Lesotho, which he inferred were congeneric and conspecific with Ginsburg's specimens. He transferred *F. australis* from the Scelidosauridae to the Hypsilophodontidae.

Galton (1972) removed both Ginsburg's and Thulborn's specimens of *F. australis* from the Hypsilophodontidae to the newly created family Fabrosauridae. Subsequently, Galton (1978) removed Thulborn's specimens from *F. australis* and created a new genus as well as a new species for them, *Lesothosaurus diagnosticus*. However, Gow (1981) published a sharp criticism of Galton's procedure, considering the genus *Lesothosaurus* a "myth." Even so, the genus *Lesothosaurus* was adopted without comment by most authors. Sereno (1991) considered *F. australis* a *nomen dubium* and the Fabrosauridae invalid. He used *Lesothosaurus* rather than the Fabrosauridae in discussions of ornithischian phylogeny. However, by ICZN rules a family name does not become invalid with the loss of its eponymous genus, although these rules may not apply within the PHYLOGENETIC SYSTEM.

Thulborn (1992) opined that *F. australis* Ginsburg 1964 is a determinate species founded on diagnostic material, and that *Lesothosaurus diagnosticus* Galton 1978 is a subjective junior synonym. In his view, therefore, all fabrosaurid specimens so far described from the Elliot Formation and Clarens Formation of southern Africa are referable to the species *F. australis*, or at least to the genus *Fabrosaurus*.

Initially, Galton (1972) created the Fabrosauridae on the primitive character of marginally positioned cheek teeth. Later, Galton (1978) revised its diagnosis on the basis of approximately 40 characters, but the majority are plesiomorphic for Ornithischia rather than apomorphic for Fabrosauridae. Consequently, the inclusions of *Alcodon kuehnei* and *Trimucrodon cuneatus* from Portugal, *Tawasaurus minor* and *Xiaosaurus dashanpensis* from China, *Nanosaurus agilis*, *Revueltosaurus callenderi*, *Technosaurus smalli*, and *Scutellosaurus lawleri* from North America, and *Echinodon becklesii* from England have made this group a paraphyletic assemblage. Plesiomorphy, as commonly acknowledged, is phylogenetically uninformative; therefore, Sereno (1991) concluded that *A. kuehnei*, *T. cuneatus*, *X. dashanpensis*, *N. agilis*, and *R. callenderi* are ornithischians of unknown relation or *nomina dubia*, *E. becklesii* and *S. lawleri* are more closely related to other ornithischian subgroups, and *T. minor*

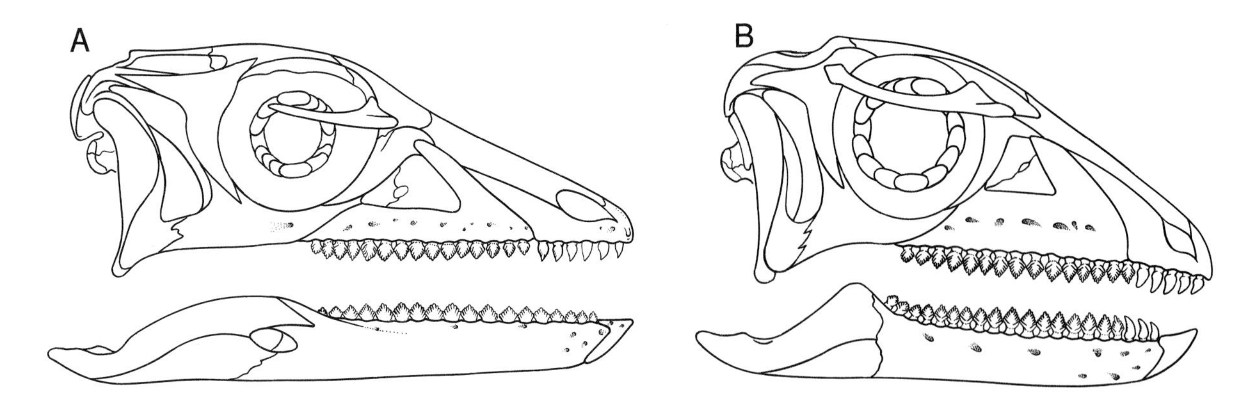

FIGURE 1 Reconstructions of skulls in lateral view. (A) *Fabrosaurus australis* (after Sereno, 1991); (B) *Agilisaurus louderbacki*.

and *T. smalli* are indeterminate Prosauropoda. This left *F. australis (= Lesothosaurus diagnostics)* as the only genus and species diagnosable on apomorphic features.

The recent discovery of a Middle Jurassic primitive ornithischian, *Agilisaurus louderbacki* Peng 1990, represented by a beautifully preserved skeleton from the Xiashaximiao Formation of Zigong, Sichuan Basin, China, has shed light on the systematic relationships of fabrosaurids. It is evident that Fabrosauridae is a monophyletic taxon diagnosed by the following unequivocal features:

1. lacrimal inserts into a narrow slot in the apex of the maxilla (Figs. 1A and 1B);
2. mandible with peculiarly salient finger-like retroarticular process (Figs. 1A and 1B);
3. especially short forelimb that is approximately 40% of the hindlimb in length;
4. ilium with a supra-acetabular flange over the anterior half of the acetabulum (Fig. 2);
5. posterior process of the ilium with a distinct brevis shelf that first turns medially and then downwards (Fig. 2);
6. ischium with a dorsal groove on the proximal shaft (Fig. 2);
7. pedal digit I reduced, with the splint-like shaft of metatarsal I and the ungual extending just beyond the end of the second metatarsal.

In addition to the characters listed previously, fabrosaurids may be unique in tooth morphology and structure. They differ from prosauropods in their less symmetrical tooth crown, with fewer erect denticles of unvarying size on the anterior and posterior edges. Unlike other primitive ornithischians, such as heterodontosaurids and hypsilophodontids, the teeth of fabrosaurids are thinly and uniformly enameled on either the buccal or lingual sides, each with a single round central vertical ridge.

Compared to *Agilisaurus, Fabrosaurus* is more primitive in having six premaxillary teeth, a flat maxilla, a lower coronoid eminence, a large external mandibular fenestra, and a very short prepubis. *Agilisauris* is clearly more derived. It differs from *Fabrosaurus* in being larger and more complex in structure, as well as in having the specializations of a well-developed rod-like palpebral, a longitudinal depression along the sutural line between the nasals, and a nutritive foramen on the femoral shaft.

Apart from the two genera mentioned previously, the Upper Jurassic ornithischian *Gongbusaurus* from

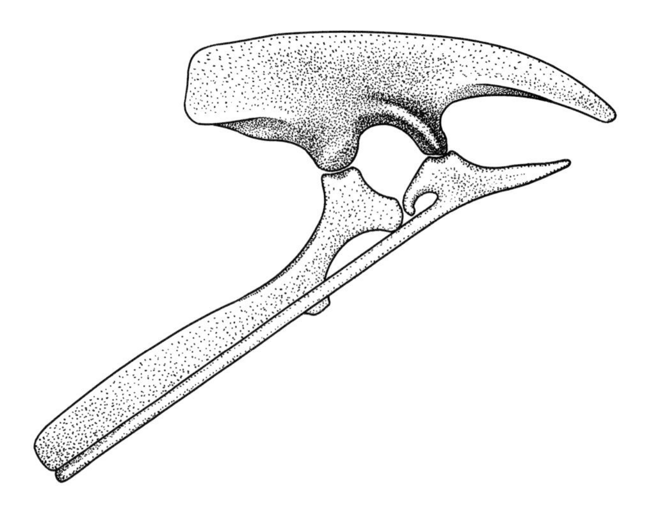

FIGURE 2 Reconstruction of the pelvic girdle of *Agilisaurus louderbacki* in lateral view.

China may be more closely related to fabrosaurids than to hypsilophodontids. The type species, *G. shiyii* Dong *et al.* 1983, from the Shangshaximiao Formation of Rongxian County in Sichuan Basin, is represented only by isolated teeth, but the included species *G. wucaiwanensis* Dong 1989, from the Shishugou Formation of Junggar Basin, Xinjiang, is represented by some fragmentary cranial and postcranial elements that show two apomorphic resemblances to fabrosaurids: the reduced pedal digit I with the splint-like metatarsal I and the ilium with a distinct brevis shelf. The most unequivocal feature in *Gongbusaurus* is that there are four distal tarsals. Although the other diagnostic characters of fabrosaurids are uncertain in *Gongbusaurus*, the long, deep posterior process of the ilium suggests that it is more closely related to *Agilisaurus* than to *Fabrosaurus*. The phylogenetic relationships of the three genera that constitute Fabrosauridae can be briefly indicated in Fig. 3.

The fabrosaurids are small (1- or 2-m body length). The skull is triangular, with large circular orbits on the sides that suggest that the eyes were huge and directed laterally. A long and rod-like palpebral (supraorbital) extends across the dorsal margin of the orbit. It might have been used as a support for the eyeball. Like all other ornithischians, fabrosaurids have a small predentary bone joined to the tips of the two mandibular rami. The first premaxillary tooth is set a little back from the tip of the premaxilla, and the general structure of this area suggests that it may have had a small horny beak. The lower beak fitted inside the upper beak and premaxillary teeth, forming a more effective cropping device.

In fabrosaurids, the leaf-like tooth crowns are mediolaterally compressed, with several coarse marginal denticles on either side of the apex and with thinly and uniformly enameled buccal and lingual sides. Single or paired, highly inclined wear facets are usually present. A single wear facet indicates that the tooth was worn directly against a single counterpart tooth in the opposing jaw; paired wear facets were produced when the tooth interlocked with two teeth in the opposing jaw. Therefore, the jaw action of fabrosaurids was strictly vertical. Such a jaw mechanism is the simplest among known ornithischians.

The forelimb of fabrosaurids was very much smaller than the hindlimb and terminated in a diminutive hand, whereas the hindlimb was distinctively elongated. The length of the forelimb is only approximately 40% of that of the hindlimb. Such a small forelimb could not have been used for locomotion, and it is obvious that fabrosaurids were bipedal. Adaptations for bipedalism are also evident in other skeletal features. The entire skeleton is very lightly built, with a largely fenestrated skull, a very short neck and trunk, and slender, hollow, and thin-walled limb bones, whereas the tail is much longer, occupying nearly half the total body length. Lightening of the skeleton implies weight reduction, which is most marked in front of the hips. The long tail presumably acted as a counterbalance for the weight of the body in front of the hips and as a compensating mechanism for shifts in the center of gravity.

In ornithischians, as in other dinosaurs and ornithodirans, the lengthening of the distal parts of the hindlimb is usually associated with rapid bipedal progression. The hindlimb of fabrosaurids is also somewhat unusual because the tibia is considerably longer than the femur, and metatarsal III exceeds half of the femur length. These higher hindlimb ratios suggest that fabrosaurids were adapted for bipedal, fast running.

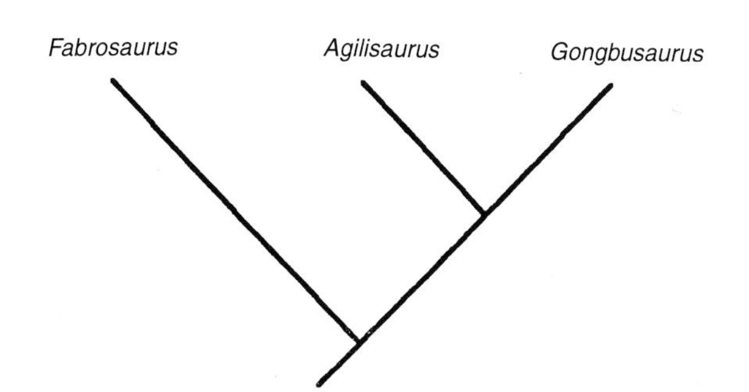

FIGURE 3 Phylogenetic relationships within the Fabrosauridae.

See also the following related entries:

HYPSILOPHODONTIDAE • ORNITHISCHIA

References

Colbert, E. H. (1981). A primitive ornithischian dinosaur from the Kayenta Formation of Arizona. *Museum Northern Arizona Bull.* **53,** 1–61.

Dong, Z. (1989). On a small ornithopod (*Gongbusaurus wucaiwanensis* sp. nov.) from Kelamaili, Junggar Basin, Xinjiang, China. *Vertebr. Palasiatica* **27,** 140–146. [In Chinese]

Dong, Z., and Tang, Z. (1983). Note on the new mid-

Jurassic ornithopod from Sichuan Basin, China. *Vertebr. Palasiatica* **21,** 168–172. [In Chinese]

Dong, Z., Zhou, S., and Zhang. Y. (1983). The dinosaurian remains from Sichuan Basin, China. *Palaeontol. Sinica New Ser. C* **23,** 1–145. [In Chinese]

Galton, P. M. (1972). Classification and evolution of ornithopod dinosaurs. *Nature* **239,** 464–466.

Galton, P. M. (1974). The ornithischian dinosaur *Hypsilophodon* from the Wealden of the Isle of Wight. *Bull. Br. Museum (Nat. History) Geol.* **25,** 1–152.

Galton, P. M. (1978). Fabrosauridae, the basal family of ornithischian dinosaurs (Reptilia: Ornithopoda). *Palaontol. Zeitschrift* **52,** 138–159.

Ginsburg, L. (1964). Decouverte d'un Scelidosaurien (Dinosaure ornithischien) dans le Trias superieur du Basutoland. *Comptes Rendus Acad. Sci. Paris* **258,** 2366–2368.

Gow, C. E. (1981). Taxonomy of the Fabrosauridae (Reptilia, Ornithischia) and the *Lesothosaurus* myth. *South Africa J. Sci.* **77,** 43.

Hunt, A. P. (1989). A new ornithischian dinosaur from the Bull Canyon Formation (Upper Triassic) of East Central New Mexico. In *The Dawn of the Age of Dinosaurs in the American Southwest* (S. G. Lucas and A. P. Hunt, Eds.), pp. 355–358. New Mexico Museum of Natural History, Albuquerque.

Marsh, O. C. (1877). Notice of some new vertebrate fossils. *Am. J. Sci.* **14,** 249–256.

Owen, R. (1861). On the fossil Reptilia of the Wealden and Purbeck Formations. Part V. *Palaeontogr. Soc.* **7,** 31–39. [Monograph]

Padian, K. (1990). The ornithischian form genus *Revueltosaurus* from the Petrified Forest of Arizona (Late Triassic: Norian; Chinle Formation). *J. Vertebr. Paleontol.* **10,** 268–269.

Peng, G. (1990). A new small ornithopod (*Agilisaurus louderbacki* gen. et sp. nov.) from Zigong, Sichuan, China. *Newaleters Zigong Dinosaur Museum* **2,** 19–27. [In Chinese]

Peng, G. (1992). Jurassic ornithopod *Agilisaurus louderbacki* (Ornithopoda: Fabrosauridae) from Zigong, Sichuan, China. *Vertebr. Palasiatica* **30,** 39–53. [In Chinese]

Santa Luca, A. P. (1984). Postcranial remains of Fabrosauridae (Reptilia: Ornithischia) from the Stormberg of southern Africa. *Palaeontol. Africana* **25,** 151–180.

Sereno, P. C. (1986). Phylogeny of the bird-hipped dinosaurs (Order Ornithischia). *Natl. Geogr. Res.* **2,** 234–256.

Sereno, P. C. (1991). *Lesothosaurus,* "fabrosaurids," and the early evolution of Ornithischia. *J. Vertebr. Paleontol.* **11,** 168–197.

Thulborn, R. A. (1970). The skull of *Fabrosaurus australis,* a Triassic ornithischian dinosaur. *Palaeontology* **13,** 414–432.

Thulborn, R. A. (1971). Origins and evolution of ornithischian dinosaurs. *Nature* **234,** 75–78.

Thulborn, R. A. (1972). The postcranial skeleton of the Triassic ornithischian dinosaur *Fabrosaurus australis. Palaeontology* **15,** 29–60.

Thulborn, R. A. (1973). Teeth of ornithischian dinosaurs from the Upper Jurassic of Portugal, with a description of a hypsilophodontid (*Phyllodon henkeli* gen. et sp. nov.) from the Guimarota Lignite. *Mem. Servicos Geol. Portugal* **22,** 89–134.

Thulborn, R. A. (1992). Taxonomic characters of *Fabrosaurus australis,* an ornithischian dinosaur from the Lower Jurassic of southern Africa. *Geobios* **25,** 283–292.

Weishampel, D. B., and Witmer, L. M. (1990). *Lesothosaurus, Pisanosaurus,* and *Technosaurus,* In *The Dinosauria* (D. B. Weishampel, P. Dodson, and H. Osmólska, Eds.), pp. 416–425. Univ. of California Press, Berkeley.

Farmington Museum, New Mexico, USA

see MUSEUMS AND DISPLAYS

Fatima

MARTIN LOCKLEY
University of Colorado at Denver
Denver, Colorado, USA

A Middle Jurassic dinosaur tracksite in the vicinity of Fatima, Portugal, reveals the two longest sauropod trackways in the world, measuring 142 and 147 m. The tracks are also distinctive because of the well-preserved impressions of the manus claws (digit I). Such claw impressions are not normally seen in sauropod trackways, and their position in life has been the subject of much debate (Lockley *et al.*, 1994).

See also the following related entries:

EUROPEAN DINOSAURS • FOOTPRINTS AND TRACKWAYS

Reference

Lockley, M. G., Santos, V. F., Meyer, C. A., and Hunt, A. P. (Eds.) (1994). *Aspects of Sauropod Biology.* Gaia: Geoscience Magazine of the National Natural History Museum, Lisbon, Portugal.

Faunas

see MESOZOIC FAUNAS

"Feathered" Dinosaurs

PHILIP J. CURRIE
Royal Tyrrell Museum of Palaeontology
Drumheller, Alberta, Canada

In 1996, several skeletons of a 1-m-long animal were found in Liaoning (People's Republic of China) that show feather-like structures covering the head, trunk, tail, arms, and legs. Named *Sinosauropteryx prima*, this animal is closely related to *Compsognathus* from the Upper Jurassic of Europe. The integumentary structures were simpler than true feathers, and each seems to be composed of a central rachis and branching barbs but lacks the aerodynamic quality of avian feathers. The longest "feathers" were about 3 cm in length, and it has been suggested that they were more suitable for insulation than they were as display structures. The discovery of these specimens has given additional support to the hypotheses that theropod dinosaurs were the direct ancestors of birds, and that some theropods were endothermic.

See also the following related entries:

BIRD ORIGINS • PHYSIOLOGY • SKIN

Feeding

see DIET; TEETH AND JAWS

Fernbank Museum of Natural History, Georgia, USA

see MUSEUMS AND DISPLAYS

Fernbank Science Center, Georgia, USA

see MUSEUMS AND DISPLAYS

Field Museum of Natural History, Illinois, USA

see MUSEUMS AND DISPLAYS

Flaming Cliffs

A commonly used reference for a famous Mongolian locality known as Bayn Dzak.

see BAYN DZAK

Floras, Mesozoic

see MESOZOIC FLORAS; PLANTS AND DINOSAURS

Footprints and Trackways

Martin Lockley
University of Colorado at Denver
Denver, Colorado, USA

The study of dinosaur footprints can be considered a branch of vertebrate paleontology, but it is also a branch of ichnology, the study of trace fossils. As such, dinosaur ichnology conforms to different procedures and conventions from those applied to the naming and description of dinosaur body fossils. Ichnotaxonomy, or the naming of trace fossils, allows ichnologists to give distinctive names (ichnogenera and ichnospecies) to footprints and other traces (see Coprolites; Trace Fossils). For example, the ichnospecies *Tyrannosauripus pillmorei* (referring to a tyrannosaur track discovered by the geologist Pillmore) was probably made by the species *Tyrannosaurus rex*, the only known animal of that type capable of making the track. Endings such as *-ipus, -opus, -podus,* or *-ichnium* indicate a trace fossil rather than the animal itself. In most cases, the species that made a particular track is not known, even though available evidence may narrow down the probable trackmaker to a more or less specific taxon with varying degrees of confidence.

The naming of tracks conforms to the same general principles as the naming of body fossils. That is, they are named on the basis of distinctive morphologies, with the designation of type specimens (holotypes), and must include descriptions and appropriate documentation. In practice there are clear morphological differences between tracks made by sauropods, theropods, and other major groups, such as ornithopods or ceratopsians. However, the difference between various theropod morphotypes may be more subtle. In addition, a significant percentage of tracks in any sample may be poorly preserved. Consequently, guidelines have evolved that recommend that tracks be named only when sufficient supplies of "well-preserved" material are available. Such guidelines recommend having sequences of tracks (trackways) available for study rather than isolated footprints (Sarjeant, 1989).

Unlike the traces of invertebrate animals that often represent behavior such as burrowing, traces of vertebrates usually reveal foot morphology and so reflect the anatomy of the trackmaker to some degree. For this reason, vertebrate tracks are often named after the inferred trackmaker as in the example of *Tyrannosauripus* or *Brontopodus* (meaning brontosaur track). There is no rule about the choice of such names, and many tracks are named after places, rock formations, people, and so on.

Edward Hitchcock (1793–1864) is considered the father of vertebrate ichnology. He described many dinosaur tracks from the Lower Jurassic of New England, but attributed most to birds or nondinosaurian reptiles. His most complete work (Hitchcock, 1858) contains many ichnogenus names (notably, *Grallator* and *Eubrontes*) still in use today. His work was revised by Richard Swan Lull in a series of papers (notably, Lull, 1953). Baron von Nopsca has been credited by some as the first European to produce a seminal work on vertebrate ichnology (Nopsca, 1923). Other notable encyclopedic and glossary contributions include Haubold (1971, 1984) and Leonardi (1987). The past decade has witnessed a renaissance in dinosaur ichnology, beginning with the publication of the proceedings of the First International Symposium on Dinosaur Tracks and Traces (Gillette and Lockley, 1989), two general books on dinosaur tracks (Thulborn, 1990; Lockley 1991), and regional studies (Leonardi, 1994; Lockley and Hunt, 1995).

Tracks or footprints (same meaning) are described on the basis of footprint morphology as one, two, three, four, or five toed (i.e., mono-, di-, tri-, tetra-, or pentadactyl, respectively). Hind footprints are referred to as pes or pedal impressions and front foot impressions as manus or manual impressions. Most quadrupedal dinosaurs had larger hind footprints than front footprints (= heteropody). Some dinosaurs, notably theropods and ornithopods, were bipedal. Trackmaker morphology and movement can be understood in part by noting whether the long axis of the foot rotates outward, points forward, or rotates inwards. Similarly, some dinosaurs have longer inner digits (I or II), a larger central digit (III), or longer outer digits (IV or V) referred to as entaxonic, mesaxonic, or ectaxonic, respectively. Trackmakers may have put their entire foot flat on the ground (plantigrade) or walked on their toes (digitigrade). All such features must be studied in relation to the overall pattern of the trackway (consecutive steps made by an individual trackmaker), in which one can measure step or pace (left to right or right to left), stride (left to left or right to right), and pace angulation (angle between lines of two consecutive steps). Different trackway patterns are characteristic of particular groups and reveal the stance of the trackmakers. For example, wide trackways indicate sprawling and narrow trackways indicate erect or upright stance.

Because tracks were made by living animals, they are useful for interpreting behavior. A basic feature of behavior is locomotion, and various formulae have been proposed to estimate dinosaur speed (μ) from trackways. The most widely used is $\mu = 0.25\ g^{0.5} \cdot \lambda^{1.67} h^{-1.17}$ proposed by Alexander (1976), where λ is stride length, h is hip height ($4 \times$ footprint length), and g is acceleration due to gravity. Other formulae are available (Thulborn, 1990), but none can calculate absolute speed owing to the fact that duration of step and stride (cadence) is not known. Thus, all calculated velocities are relative speed estimates. Trackways with increasing or decreasing step length may indicate acceleration or deceleration, whereas those with alternating long and short steps suggest limping. Other reports of unusual individual behavior, such as hopping or attacking, are not supported by the trackway evidence or are, at best, very controversial.

Parallel trackways of the same type on a single surface may indicate gregarious or herd behavior (Lockley, 1991, 1995). Such examples are particularly common among sauropod and ornithopod tracks in the Late Jurassic through Cretaceous and are characterized by regular spacing of trackways. Trackway orientations may also reflect ancient geography and are often recorded running parallel to ancient shorelines.

Trackways often occur in similar proportions in numerous track assemblages (*ichnocoenoses*) from particular sedimentary facies or rock formations. Such recurrent associations can be used to define particular ichnofacies that reflect the composition of animal communities in particular environments (Lockley *et al.*, 1994a). Current evidence suggests that tracks record distinctive animal communities in different settings, ranging from desert dunes to desert playas, humid swamps, or carbonate environments. Thus, tracks are useful for paleoecological census studies.

There is danger of misinterpreting tracks if students do not take adequate care to understand how they are preserved. For example, tracks can be transmitted downwards from the surface on which the foot came to rest, into one or more underlayers. Such transmitted tracks are known as ghostprints or underprints and are generally less distinct and somewhat larger than the true tracks in younger layers above. Although undertracks lack details of skin or pad impressions, they sometimes show well-preserved claw marks that are not seen in association with the true tracks. Some underlayers preserve only hind or front footprints from trackways of quadrupeds. Sauropod trackways that show only front footprints were for a long time interpreted as the result of swimming behavior but are now known to be underprints caused by the front feet having sunk deeper than the hind feet.

It is important to understand not only how tracks are preserved but also how they are made. The form of a track is more often than not a representation not only of the animal's foot anatomy but also of the kinematic pattern of its step cycle and the condition of the substrate when the track was made. All these features have to be considered in interpreting and taxonomizing trackways (Padian and Olsen, 1984).

Other aspects of understanding track preservation pertain to bias in the fossil record in general. In most cases, small tracks are rarer and less well preserved than large tracks. Tracks are also preferentially preserved along shorelines where substrates were wet. Thus, animals that habitually frequented dry areas are less well represented. However, because most major dinosaur groups are well represented in the Mesozoic track record, it can be inferred that the record is not excessively biased. The tracks of carnivorous dinosaurs (predators) are often more common than those of herbivores (prey) and relatively more common than the predator:prey ratios inferred from the bone record. Such inconsistencies have been attributed to higher activity levels among theropods, but other explanations are possible. For example, different predator:prey ratios may reflect the inclination of predators to patrol shorelines and hunt by sight. Predator:prey ratios based on trackway numbers should be corrected for estimated body size to give biomass estimates for ecological interpretations.

Comparison of footprint and bone records in various formations is also revealing for at least two reasons. First, tracks mainly provide evidence of terrestrial faunas, whereas some skeletal assemblages are dominated by the remains of aquatic vertebrates (fish, turtles, crocodiles, etc.). Such evidence establishes that tracks must be included in paleontological censuses in order to obtain as complete a picture as the fossil record will allow. Second, recent studies show that tracks are many orders of magnitude more abundant than bones in many formations. For example, in some deposits, hundreds or thousands of trackways have been recorded, whereas the skeletal record consists of only one or two or, in some cases, no individual remains. Such evidence gives us a new perspective on the importance of tracks by filling in many gaps in the fossil record (Lockley, 1996).

In recent decades ichnologists have recognized that the same track assemblages occur in rocks of the same or similar age over wide areas. Thus, tracks are useful for correlation or track stratigraphy referred to as palichnostratigraphy. It has become evident that dinosaur track correlations can be made from North America to Europe and Asia throughout much of the Mesozoic, but especially in the Late Triassic and Jurassic. In several cases the correlations can be made using two or three ichnogenera in formations or time intervals where no skeletal remains exist. Because tracks are abundant and trackmakers potentially mobile over wide areas, they fulfill the important criteria for use in stratigraphic correlation. Such correlations also reveal a prolonged record of faunal interchange between continents that are often shown as having no connections on paleogeographic maps.

Recent studies have often shown that tracks occur in exceptional abundance over wide areas in particular thin sedimentary units or even on single surfaces. Such extensive sites are referred to as megatracksites or even dinosaur freeways and can be measured on the order of thousands or even tens of thousands of square kilometers. Examples have been identified in both the Jurassic and Cretaceous of North America and Europe. Such megatracksites are associated with sediment aggradation during periods of rising sea level. Indeed, some studies indicate that both the track and bone records are more prolific and complete at times of elevated sea level. Such an abundant track record provides the raw data necessary for making track correlations. It should be emphasized, however, that megatracksites are essentially continuous track layers in a particular facies in a given region, whereas track correlations identify the same tracks (ichnogenera) in different regions that are often on different continents and in different facies.

Despite the fact that Edward Hitchcock first described dinosaur tracks (as those of birds) in 1836 before dinosaurs were known, dinosaur ichnology is still a relatively young and immature science. During the past decade, however, rapid progress has been made in the field, resulting in a significant new literature. Possibly the most important results pertain to the realization that tracks are incredibly abundant at literally thousands of sites on most continents. Such abundance provides a large database but also reveals the extent to which numerous formations have a better record of tracks than skeletal remains.

For historical reasons track ichnotaxonomy is somewhat confused, but recently progress has been made toward revising nomenclature. This important first step allows valid names to be applied in paleoecological and biostratigraphic studies, and so demonstrates important patterns of track distribution in time and space. It is also important to realize that ichnotaxonomy is a separate system from body fossil taxonomy, and that matching tracks with trackmakers can be done in some circumstances (in which both bones and tracks are known from a given formation or in which track morphologies are distinctive) but not in others (in which there are no bones to match the track record). Even so, it is desirable, though not essential, to have track names that reflect the morphology or affinity of the trackmaker. Such names must be applied judiciously, for example, only when trackmaker affinity can be deduced with high levels of confidence.

Many old track names are invalid because they are duplicative (junior synonyms) or are based on poorly preserved material (*nomina dubia*). As a result, compilations from old track literature (Weishampel *et al.*, 1990) can be misleading, as demonstrated by Lockley *et al.* (1994b). Such problems, however, are well within the abilities of ichnologists to solve. Through careful study of distinctive track assemblages, containing well-preserved material, we can hope that ichnotaxonomy will be progressively simplified and revised. Through comparison of the bone and track records, the affinity of trackmakers will become much clearer in most cases. Without doubt, the recent renaissance in dinosaur ichnology has brought the field into the paleontological mainstream. This integration of the track and skeletal records is taking place at a fundamental level, owing to the compelling evidence that the body fossil record is impoverished and less complete than it would be without track evidence and vice versa.

See also the following related entries:

Cabo Espichel • Cameros Basin Megatracksite • Carenque • Connecticut River Valley • Fatima • Khodja-Pil Ata • Las Hoyas • Megatracksites • Samcheonpo

References

Alexander, R. M. (1976). Estimates of speeds of dinosaurs. *Nature* **261,** 129–130.

Gillette, D. D., and Lockley, M. G. (Eds.) (1989). *Dinosaur Tracks and Traces,* pp. 454. Cambridge Univ. Press, Cambridge, UK.

Haubold, H. (1971). Ichnia Amphibiorium et Reptiliorum fossilium. In *Handbuch der Palaeoherpetologie, Teil 18* (O. Kuhn, Ed.), pp. 1–124. Fischer–Verlag, Stuttgard.

Haubold, H. (1984). *Saurierfährten.* Wittenberg Lutherstadt, Die Neue Brehm-Bucherei, pp. 232.

Hitchcock, E. (1858). *A Report on the Sandstone of the Connecticut Valley Especially Its Fossil Footmarks,* pp. 220. White, Boston. [Reprinted by Arno Press; in the Natural Sciences in America Series]

Leonardi, G. (Ed.) (1987). *Glossary and Manual of Tetrapod Footprint Palaeoichnology,* pp. 117. Departamento Nacional da Producão Mineral, Brasilia.

Leonardi, G. (1994). *Annotated Atlas of South American Tetrapod Footprints (Devonian to Holocene),* pp. 248. Companhia de Pesquisa de Recursos Minerais, Brasilia.

Lockley, M. G. (1991). *Tracking Dinosaurs: A New Look at an Ancient World,* pp. 238. Cambridge Univ. Press, Cambridge, UK.

Lockley, M. G. (1995). Track records. *Nat. History* **104,** 46–51.

Lockley, M. G. (1997). The paleoecological and paleoenvironmental utility of dinosaur tracks. In *Dinosaurs: A Sesquicentennial Celebration* (J. O. Farlow and M. Brett-Surman, Eds.), in press.

Lockley, M. G., and Hunt, A. P. (1995). *Dinosaur Tracks and Other Fossil Footprints of the Western United States,* pp. 338. Columbia Univ. Press, New York.

Lockley, M. G., Hunt, A. P., and Meyer, C. (1994a). Vertebrate tracks and the ichnofacies concept: Implications for paleoecology and palichnostratigraphy. In *The Paleobiology of Trace Fossils* (S. Donovan, Ed.), pp. 241–268. Wiley, New York.

Lockley, M. G., Santos, V. F., Meyer, C. A., and Hunt, A. P. (1994b). *Aspects of Sauropod Biology,* pp. 266. Gaia: Geosciencies Journal of the Natural History Museum of the University of Lisbon, Lisbon, Portugal.

Lull, R. S. (1953). Triassic life of the Connecticut Valley. State Geological and Natural History Survey of Connecticut Bull. 81, pp. 336.

Nopsca, F. (1923). Die familien der Reptilien. *Fortschr. Geol. Palaont. Heft* **2,** 1–210.

Padian, K., and Olsen, P. E. (1984). The track of *Pteraiehuus:* not pterosaurian, but crocodilian. *J. Paleontol.* **58,** 178–184.

Sarjeant, W. A. S. (1989). Ten paleoichnological commandments: A standardized procedure for the description of fossil vertebrate footprints. In *Dinosaur Tracks and Traces* (D. D. Gillette and M. G. Lockley, Eds.), pp. 369–370. Cambridge Univ. Press, Cambridge, UK.

Thulborn, R. A. (1990). *Dinosaur Tracks,* pp. 410. Chapman & Hall, London.

Weishampel, D. B., Dodson, P., and Osmólska, M. (1990). *The Dinosauria,* pp. 733. Univ. of California Press, Berkeley.

Forelimbs and Hands

Per Christiansen
Københavns Universitet
Copenhagen, Denmark

Plesiomorphically dinosaurs were bipedal animals (see Bipedality), and although most members of several lines later became obligatory quadrupedal, the quadrupedality probably evolved convergently. The earliest dinosaurs had long, slender forelimbs, about half as long as the hindlimbs, with a long, low deltopectoral crest, and rather small hands suited for manipulation, not progression. The relative forelimb length became reduced in most lines not adopting a quadrupedal posture, except maniraptorans (see Bird Origins). The earliest dinosaurs were all small animals and as some lines grew to immense sizes, the quadrupedal posture probably became a necessity. The mobility of the forelimbs of all dinosaurs was always somewhat parasagittal as in most larger extant mammals. SAUROPODS in particular appear to have had limbs that hardly allowed any lateromedial mobility at all, as in extant artiodactyls and perissodactyls and, of course, elephants (Fig. 1; Table I). The antebrachium of dinosaurs was usually morphologically quite similar to those of extant mammals or reptiles in overall view, but there was never a somewhat posteriorly directed, rounded humeral caput with a distinct neck as in most living mammals, and the proximal part of the humerus was lateromedially expanded, taking on a spoon-like shape. Other differences between dinosaurs and mammals were also apparent, but overall the functional anatomy of dinosaurian forelimbs is closer to mammals than reptiles. In quadrupedal dinosaurs the forelimbs appear never

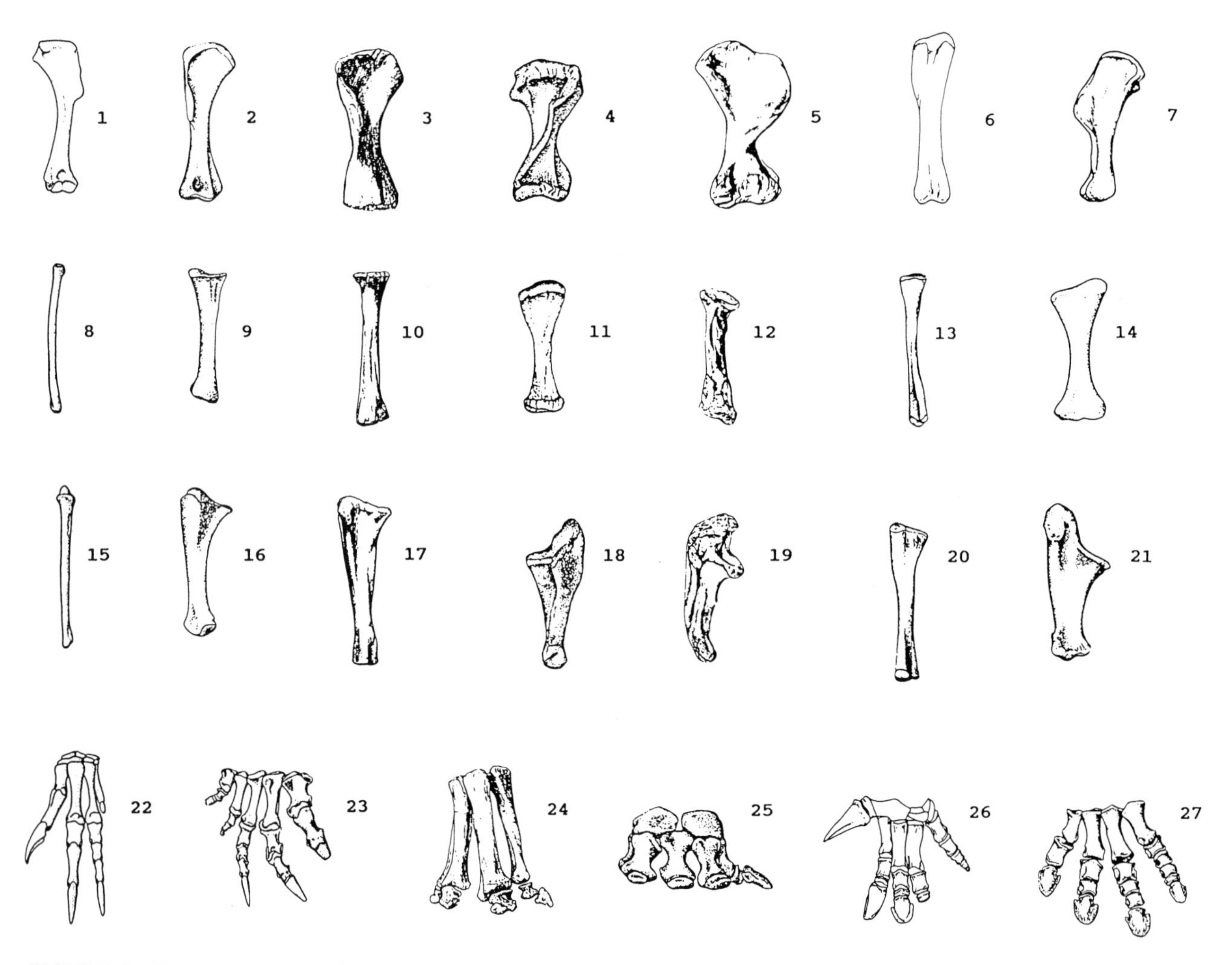

FIGURE 1 Comparative morphology of dinosaurian forelimbs. (1–7) Humeri (1–6, cranial view; 7, lateral view): 1, theropod *(Syntarsus)*; 2, prosauropod *(Plateosaurus)*; 3, sauropod *(Dicraeosaurus)*; 4, stegosaur *(Stegosaurus)*; 5, ankylosaur *(Euplocephalus)*; 6, ornithopod *(Camptosaurus)*; 7, ceratopsian *(Centrosaurus)*. (8–14) Radii (8, 10–12, cranial view; 9 and 14, lateral view; 13, medial view): 8, theropod *(Gallimimus)*; 9, prosauropod *(Plateosaurus)*; 10, sauropod *(Brachiosaurus)*; 11, stegosaur *(Stegosaurus)*; 12, ankylosaur *(Euplocephalus)*; 13, ornithopod *(Dryosaurus)*; 14, ceratopsian *(Chasmosaurus)*. (15–21) Ulnae (15, cranial view; 21, caudal view; 16, 18, and 19, lateral view; 17 and 20, medial view): 15, theropod *(Gallimimus)*; 16, prosauropod *(Plateosaurus)*; 17, sauropod *(Brachiosaurus)*; 18, stegosaur *(Stegosaurus)*; 19, ankylosaur *(Euplocephalus)*; 20, ornithopod *(Dryosaurus)*; 21, ceratopsian *(Triceratops)*. (22–27) Manus (cranial view): 22, theropod *(Syntarsus)*; 23, prosauropod *(Plateosaurus)*; 24, sauropod *(Brachiosaurus)*; 25, stegosaur *(Stegosaurus)*; 26, ornithopod (*Iguanodon*); 27, ceratopsian *(Centrosaurus)*.

TABLE I Forelimb Proportions in Some Dinosaurs Compared to Extant Mammals

Taxon	n^a	Size range of		Ratio
		Propodia (mm)	Epipodia (mm)	
Mammalia				
Carnivora				
Ursidae	7 (12)	142–422	108–385	0.76–0.94
Canidae	15 (24)	70–255	65–275	0.82–1.08
Artiodactyla				
Cervidae	11 (21)	116–385	115–416	0.88–1.08
Bovidae	32 (49)	90–409	83–361	0.87–1.34
Perissodactyla				
Rhinocerotidae	4 (8)	349–457	286–408	0.76–0.91
Proboscidea				
Elephantidae	2 (10)	475–1059	370–813	0.76–0.81
Dinosauria				
Prosauropoda				
Plateosauridae	2 (3)	335–420	187–230	0.51–0.58
Sauropoda				
Diplodocidae	3 (6)	750–1150	525–840	0.69–0.78
Camarasauridae	2 (4)	450–1004	298–702	0.64–0.71
Ceratopsia				
Ceratopsidae	5 (8)	541–826	316–396	0.48–0.64
Ornithopoda				
Hadrosauridae	6 (7)	47–607	52–657	0.92–1.16

[a] n is number of species and in parentheses are number of specimens.

to have sprawled as in modern reptiles but rather to have been mammalian in function, and the gait was largely parasagittal.

Prosauropoda

Archaic PROSAUROPODS, such as most Plateosauridae, Massospondylidae, Anchisauridae, and Yunnanosauridae, were at least facultatively bipedal, as indicated by their considerably longer and stronger hindlimbs. The Melanorosauridae and probably Blikanasauridae were quadrupedal because their appendicular anatomy approached that of sauropods. The humerus is the longest bone in the forelimb and the distal epiphysis is somewhat anteriorly and slightly medially angled compared to the long axis of the diaphysis, indicating elbow flexure and slight medial orientation of the epipodium. The deltopectoral crest is pronounced, indicating a rather strong upper arm and shoulder musculature, in accordance with the fairly large proximal part of the scapula. The epicondyles are moderately well developed, indicating quite powerful carpal flexors. The propodium : epipodium ratio is usually 0.5–0.7. The ulna has an anterior proximal depression for the radius and the two bones probably did not cross over much. The distal epiphysis is usually triangular and slightly angled, and the olecranon process is quite well developed. The proximal epiphysis of the radius is concave or saddle shaped as in many extant mammals, indicating fairly well-developed anteroposterior and less well-developed lateromedial mobility.

The carpus may have included as many as six elements (Young, 1941) but usually the proximal carpals are rarely preserved and largely unknown. Galton (1990) suggested that they were cartilaginous, which seems reasonable. Frequently found is the small, semicircular intermedium and three distal carpals, of which distal carpal I usually is the largest.

The metacarpus is distinctly asymmetrical with metacarpal I being the strongest and the rest decreasing in size numerically. Metacarpal IV and especially V are reduced. The phalangeal formula is usually 2-3-4-2-0. The articulating facets are usually of the condylar–cotylar type but are often somewhat ginglymoid. Digit I is massive and bears a large, somewhat recurved ungual, and the morphology of the hand indicates that slight supination was possible. Unguals are present only on digits I–III, with II and III much smaller and straighter than I. In the quadrupedal Melanorosauridae, the limb bones are more massive and longer compared to the hindlimb in other prosauropods, the epiphyses are less angled in relation to the diaphyses, and the asymmetry of the hand is less pronounced. However, no prosauropod appears to even have approached the highly apomorphic metacarpal anatomy of sauropods.

Sauropoda

The sauropod dinosaurs included the largest terrestrial animals of all time and thus faced problems of support of mass not experienced by most other dinosaurs. Despite the great diversity and longevity of the clade, their appendicular anatomy was rather conservative. All sauropods were obligatory quadrupeds with forelimbs primarily adapted for support of mass, and the limb bones appear never to have been hollow as in theropods. Sauropod locomotion has often been compared to that of elephants. A very elephantine trait is that all epiphyses of the three long bones of the forelimb are perpendicular to the long axes of their respective diaphyses, which lack curvature. This suggests that the pillar-like limb posture and graviportal mode of locomotion of recent elephants was also present in all sauropods. The forelimb bones were massive, more so than in modern elephants, and radius and ulna were both well developed. The propodial:epipodial proportions (0.55–0.79, highest in certain diplodocids) also point to a graviportal mode of locomotion, compared to 0.76–0.81 in recent elephants (Fig. 2). Archaic sauropods, such as the Euhelopodidae (*sensu* Upchurch, 1995), Camarasauridae, and paraphyletic Cetiosauridae, have forelimbs of moderate length compared to the hindlimbs. The Dicraeosauridae and Diplodocidae have shorter forelimbs. In the Brachiosauridae, the forelimbs are usually as long as or longer than the hindlimbs, mimicking the condition in proboscideans and giraffids. The deltopectoral crest is usually pronounced, indicating rather powerful propodial and scapular musculature. This is confirmed by the rather large size of the shoulder girdle, which is similar to those of elephants in this respect. The humeral epicondyles are modest, indicating modest manus flexors. The ulna is usually the longer of the two epipodial bones and has a more reduced olecranon process than in elephants, indicating less epipodial extensor power in accordance with a more limited capacity for forward propulsion. The epipodial bones probably crossed each other only

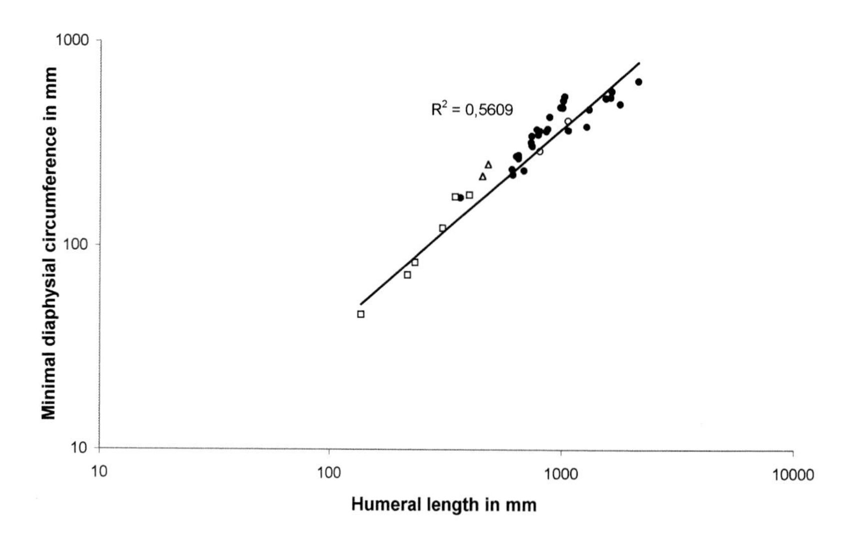

FIGURE 2 Propodial robusticity in sauropod dinosaurs (n = 29). ●, sauropods; ○, elephants; △, rhinoceroses; □, bovids.

modestly. The carpus was not elephantine; elephants have a double row of squarish or more rounded carpals where intercarpal movement occurs, whereas in sauropods there is only one row of block-like carpals (a large scapholunare and a smaller radial). This must have made the manus less flexible. Furthermore, the metacarpals are fully erect and form a semicircle, and the phalanges are reduced in number and size. Only digit I bears an ungual. Unlike those of prosauropods, all five metacarpals are stout and unreduced in size. The phalangeal formula is usually 2-1/2-1/2-1/2-1. Unlike elephants, sauropods did not have a posterior heel pad, and they were unique among very large terrestrial vertebrates in having an unguligrade forelimb posture. This stance probably made a pushoff against the ground impossible, unlike all extant mammals including elephants. The propulsive force must have come from propodial retractors exclusively. The single claw was probably not used in locomotion but possibly for grasping tree trunks during high browsing in a tripodal posture (Upchurch, 1994); hence its modesty in the giraffe-like Brachiosauridae.

Theropoda

Theropod forelimbs were always considerably shorter than the hindlimbs (except in Maniraptora) and unsuited for terrestrial progression. A general trend during theropod evolution (except Maniraptoriformes) was relative shortening of the forelimb and shortening of the hand compared to the arm. This trend reached its peak in tyrannosaurids, whereas the ORNITHOMIMOSAURIA, DROMAEOSAURIDAE, and OVIRAPTOROSAURIA reversed this trend. Proximally, the humerus was lateromedially expanded by the large medial tuberosity and prominent deltopectoral crest. This crest reached quite far down the diaphysis in *Herrerasaurus* and ornithisichians (Sereno, 1993) but was more proximal in theropods. In ornithomimids and tyrannosaurids, except *Tyrannosaurus rex*, the deltopectoral crest was markedly reduced. The humeral diaphysis was usually somewhat sigmoidal. The distal epiphysis is lateromedially expanded, often with a shallow olecranon fossa posteriorly, and cranially and slightly medially directed causing permanent elbow flexure and supination of the arm during flexion. The epicondyles were rather prominent, indicating quite powerful manual flexors. In ornithomimids, the diaphysis was straighter and the distal expansion and epicondyles were more modest. The antebrachium was always shorter than the propodium, the ulna being the longer of the two epipodial bones. The ulnohumeral ratio varied from 0.53–0.61 in tyrannosaurs, 0.7–0.9 in ornithomimids, to 0.95 in *Herrerasaurus* and theropod outgroups. The olecranon process was quite prominent, except in gracile ceratosaurs (Rowe and Gauthier, 1990). The ulnar diaphysis was cranially slightly convex and tapered distally. The radius was straighter, with a concave or saddle-shaped proximal epiphysis, and the radiohumeral ratio varied from 0.42–0.5 in tyrannosaurs, 0.76 in *Deinonychus*, to about 0.9 in *Herrerasaurus* and certain ornithomimids. The carpus is not well known in many taxa. In *Herrerasaurus* there are two proximal carpals, probably radial and ulnar, and five smaller distal carpals (Sereno, 1993). Ornithomimids have five carpals in all, whereas *Allosaurus* appears to have four—an intermedium, radial, and two unidentified carpals (Madsen, 1976). *Deinonychus* appears to have had only two ossified carpals (Ostrom, 1969), originally identified as radial and ulnar but recently interpreted as distal carpals I and II (fused) and III (see BIRD ORIGINS). The semilunate carpal also occurs in oviraptorosaurs, troodontids, and some less well-known taxa. In *Herrerasaurus* and theropod outgroups there are five metacarpals, the fifth being vestigial, and the phalangeal formula is 2-3-4-1-0. In all theropods metacarpal V is absent. Theropod unguals are always lateromedially compressed and recurved, but are straighter in ornithomimids. Ceratosaurs still have four metacarpals and a phalangeal formula of 2-3-4-1-X. However, tetanurans are tridactyl, and tyrannosaurs are didactyl with only a vestigial third metacarpal. Metacarpal I is short and stout, and metacarpal II is the longest. Phalanges are usually ginglymoid, especially in dromaeosaurs, and the unguals recurve with prominent flexor tubercles. The manus reaches its relative peak of development in dromaeosaurs, in which the hand is large and mobile and the phalanges, especially the large unguals, have prominent flexor tubercles. The manus is least developed in tyrannosaurs and certain abelisaurids.

Stegosauria

Stegosaurids were all obligatory quadrupeds with moderately long limbs and hindlimbs considerably longer than the forelimbs. The limb proportions were somewhat elephantine, indicating limited locomotory potential, and the radiohumeral ratio was approximately 0.67–0.81, compared to 0.76–0.81 in recent elephants. The humerus was stout and greatly expanded proximally and distally, and the diaphysis was much more robust than in elephants. Also, unlike elephants, the humeral muscle scars were large and the distal epiphysis was somewhat anteriorly directed compared to the long axis of the diaphysis, causing permanent elbow flexure. The deltopectoral crest is usually very pronounced, reaching one quarter or more down the diaphysis, and the medial tuberosity and epicondyles are pronounced, especially the medial epicondyle. Distally the posterior olecranon fossa was usually distinct but often rather shallow. The radius had moderately to greatly expanded proximal and distal epiphyses almost perpendicular to the long axis of the diaphysis. The proximal epiphysis was usually concave or more planar, whereas the distal epiphysis was inclined. The ulna is more robust and longer than the radius, and the olecranon process is large or very large, which indicates massive triceps musculature, causing the ulnohumeral ratio sometimes to exceed 1 (0.82–1.03). The ulnar diaphysis is somewhat triangular and tapers distally. There is a proximal anterior radial fossa. The manus is reminiscent of that of sauropods: The carpus consists of the block-like radial and ulnarulnar, both often subequal in size or the radial can be the larger. In juveniles of some species there appear to have been four carpals (Gilmore, 1914). There are five short and stout metacarpals that appear to be almost erect and form a semicircle as in sauropods. The phalanges are reduced in size and number, and only digit I appears to have borne a rather straight, lateromedially wide and hoof-like ungual. The phalangeal formula is usually 2-2-2-2-1.

Ankylosauria Ankylosaur forelimbs were shorter than the hindlimbs and stout. The humerus is short and massive with a short, thick diaphysis and great proximal and somewhat less distal expansion. The deltopectoral crest is large and is usually relatively larger in ankylosaurids than in nodosaurids. The epicondyles were well developed and the distal epiphysis was angled to the long axis of the diaphysis, causing permanent elbow flexure. The radius is the shorter of the two epipodial bones and has an ellipsoid or circular diaphysis. The proximal epiphysis is rounded and concave, and there are usually prominent proximal rugosities, perhaps for insertion of the interosseus. In ankylosaurids, the radiohumeral ratio is approximately 0.60. The ulna resembles that of stegosaurs in being proximally massive, in tapering distally, in being anteroposteriorly concave with a distinct proximal fossa for the radius, and in having a very prominent olecranon process. The ulnohumeral ratio is considerably higher than the radiohumeral ratio, generally approximately 0.85 or even more, due to the size of the olecranon process. The manus is not well known and most genera appear to have had a pentadactyl manus with short and stout phalanges, superficially resembling the hand of stegosaurs. The phalangeal formula appears to have been 2-3-3/4-2. There were dorsally flattened, hoof-like claws on digits II–IV.

Ornithopoda

The Ornithopoda was a large, long-lived, and successful dinosaurian clade (Fig. 3). Nonetheless, their appendicular anatomy did not vary much. All had considerably shorter forelimbs than hindlimbs, but most probably progressed quadrupedally at least some of the time. Ironically, the most archaic ornithopods, the Heterodontosauridae, had proportionally the longest forelimbs but do not appear to have used these in locomotion. The humerus is generally longer than the antebrachium in more archaic ornithopods, but in hadrosaurs the radiohumeral and ulnohumeral ratio often exceeds 1. The humeral diaphysis is slightly bowed or sigmoid, and the proximal and distal lateromedial expansions are moderate. The caput is anteroposteriorly thickened and distinct and is usually located more or less centrally. The anteriorly directed deltopectoral crest is strong, especially in heterodontosaurids and lambeosaurine hadrosaurs, in which it extends almost halfway down the diaphysis. The medial tuberosity is pronounced in heterodontosaurids but less well developed in iguanodontids and hadrosaurs. The distal epiphysis is cranially directed, more so in *Heterodontosaurus* than in iguanodontids or hadrosaurs. In well-preserved specimens the condyles are round and well developed. The olecranon fossa is weakly developed in *Heterodontosaurus* but is usually more pronounced in most iguanodontids and hadrosaurs. The radiohumeral ratio is 0.61 in *Camptosaurus*, approximately 0.7 in *Heterodontosaurus* (Weishampel and Witmer, 1990), and much higher in hadrosaurs (0.92–1.16). The proximal radial epiphysis is rounded and concave, the proximal and distal

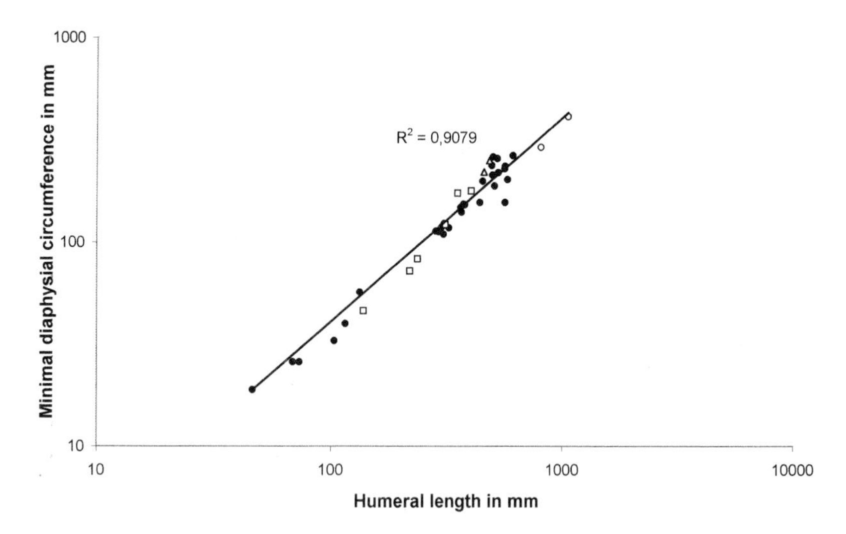

FIGURE 3 Propodial robusticity in ornithopod dinosaurs (n = 30). ●, ornithopods; ○, elephants; △, rhinoceroses; □, bovids.

expansions are modest to moderate, and the distal epiphysis is usually more squarish and planar. The diaphysis is straight and often rounded in iguanodontids and hadrosaurs but is more triangular in hypsilophodontids. The ulna is always longer than the radius, and the ulnohumeral ratio in *Camptosaurus* is 0.67–0.70. In *Heterodontosaurus* it is 0.80 (Weishampel and Witmer, 1990), and in hadrosaurs it is 1.02–1.21. The olecranon process is large in heterodontosaurids and most iguanodontids but more modest in hypsilophodontids and hadrosaurs. The diaphysis is relatively straight and tapers somewhat distally, although there is an expansion at the distal epiphysis. The distal epiphysis is usually slightly convex.

The carpus of *Heterodontosaurus* includes no fewer than nine bones, including the ulnar, radial, and pisiform in the proximal row of carpals, one medial carpal (tentatively identified as the centrale), and five distal carpals (Santa Luca, 1980). The carpals do not fuse to the metacarpals or to each other. The radial and ulnar were probably immovably attached to the antebrachium, and the large number of carpals must have provided the manus with substantial flexibility. In more derived conditions, the carpal number is reduced to three proximal carpals (of which the ulnar is the largest) and two smaller distal carpals in *Tenontosaurus* (Forster, 1990), three proximal carpals in *Dryosaurus* (the radial, intermedium, and ulnar, all approximately subequal in size), and only two maximal carpals in hadrosaurs (the radial and ulnar). The metacarpus includes five metacarpals in heterodontosaurids, hypsilophodontids, dryosaurids (Galton, 1981), and iguanodontids, but metacarpal I is absent in hadrosaurs. Unusual among dinosaurs, metacarpal I of *Heterodontosaurus* was directed medially, with the proximal lateral metacarpal condyle at right angles to the long axis of the bone and the lateral side longer than the medial side. In hypsilophodontids and iguanodontids, metacarpal V was directed laterally, and in iguanodontids metacarpal I was also directed medially. Anatomically, it appears different from the condition in *Heterodontosaurus*, suggesting convergence. The phalanges in *Heterodontosaurus* were quite long with ill-defined condylar–cotylar joints, and the phalangeal formula is 1-3-4-3-2. Digits I–III bore large, somewhat recurved unguals with large flexor tubercles (Santa Luca, 1980). In hypsilophodontids, iguanodontids, and hadrosaurs, the phalanges were shorter and stouter and the interphalangeal joints condylar–cotylar or ginglymoid, especially in the distal phalanges. The claws were hoof-like, except in *Iguanodon* and *Ouranosaurus*, in which digit I was transformed into a large, spike-like protuberance, probably a defensive weapon. Digits II–IV in hadrosaurs and iguanodontids (except *Camptosaurus)* were the longest and had their large metacarpals closely appressed to each other. These were probably enclosed in connective tissue to form a common foot. Digits I and V projected medially and laterally from this unit, respectively, and these animals frequently walked quadrupedally, as shown by trackways (e.g., Paul, 1987; Currie, 1989).

Ceratopsia

Among the CERATOPSIA, the small PSITTACOSAURIDAE were bipedal and the Protoceratopsidae were at least facultatively quadrupedal. The forelimb is considerably shorter than the hindlimbs in all ceratopsians. The humerus is gracile in small forms and becomes increasingly robust as linear dimensions increase. The humeral caput is located approximately centrally on the proximal part and extends somewhat caudally, indicating that the long axis of the bone was inclined posteriorly. The humerus is always longer than the antebrachium. The diaphysis is rather straight and approximately circular in psittacosaurids and protoceratopsids but has greater lateromedial than anteroposterior diameter in ceratopsids, unlike large modern herbivores such as large bovids or rhinoceroces (in which either the reverse is the case or the diaphysis is circular) (Fig. 4). The deltopectoral crest is quite large in psittacosaurids, larger still in protoceratopsids, very large in ceratopsids, and is truly gigantic in large chasmosaurines such as *Triceratops* and *Torosaurus*. The medial tuberosity is usually quite pronounced, especially among larger forms. The distal expansion is moderate in psittacosaurids and protoceratopsids but large in ceratopsids, in which the epicondyles are usually prominent. The distal condyles are rounded and well defined in uncrushed specimens and set at an angle compared to the long axis of the diaphysis, indicating permanent elbow flexure. The ulna is quite stout in smaller forms and massive in large ceratopsids. The diaphysis is somewhat triangular and tapers distally. The olecranon process is large in smaller forms and gigantic among large chasmosaurines. The radius is shorter than the ulna, rather straight, has moderate proximal and distal expansions in smaller taxa, and has large expansions in large taxa. The proximal epiphysis is usually somewhat medially inclined, whereas the distal epiphysis is more planar.

There has been dispute about the forelimb posture of ceratopsids for a long time. The morphology indicates that the elbow was directed posteriorly as in large modern herbivores, and trackways show an erect posture similar to large mammals, with manus prints only slightly wider than the hindlimb prints (Lockley and Hunt, 1995). The manus prints probably had to be slightly more lateral because the anterior ribcage was so wide. The carpus includes three unfused proximal carpals, identified as ulnar, intermedium, and radial, and a smaller distal carpal in psittacosaurids. Ceratopsids also have four carpals, whereas *Protoceratops* had five. Metacarpals I–III in psittacosaurs are well developed, but IV is reduced and V is vestigial. The phalangeal formula is 2-3-4-1-0. This is a problem because psittacosaurids are regarded as the sister group of the Neoceratopsia,

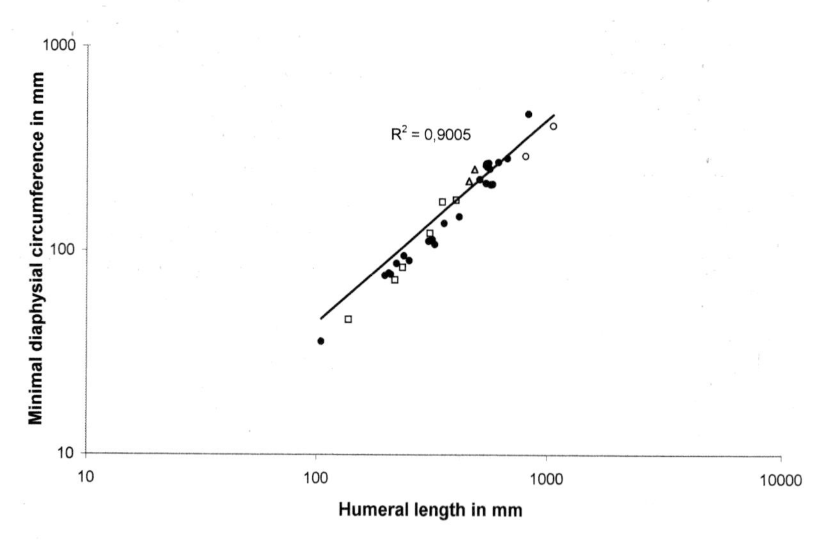

FIGURE 4 Propodial robusticity in ceratopsian dinosaurs ($n = 25$). ●, ceratopsians; ○, elephants; △, rhinoceroses; □, bovids.

which do not have these manus reductions, suggesting reversals will have to be accepted. In neoceratopsids, the hand is stout and wide, and metacarpals IV and V are somewhat reduced, but not to the same extent as in psittacosaurids. The metacarpals become progressively larger as linear dimensions increase. The phalanges are short and stout with cotylar–condylar articulating facets, and the terminal claws are broad, flat, and hoof-like in ceratopsids but are less pronounced in protoceratopsids. The phalangeal formula is 2-3-4-3-1/2.

References

Currie, P. J. (1989). Dinosaur footprints of Western Canada. In *Dinosaur Tracks and Traces* (D. Gillette and M. G. Lockley, Eds.), pp. 293–300. Cambridge Univ. Press, New York.

Forster, C. A. (1990). The postcranial skeleton of the ornithopod dinosaur *Tenontosaurus tilletti. J. Vertebr. Palaeontol.* **10**(3), 273–294.

Galton, P. M. (1981). *Dryosaurus*, a hypsilophodontid dinosaur from the Upper Jurassic of North America and Africa. Postcranial skeleton. *Palaeontol. Z.* **55**(3/4), 271–312.

Galton, P. M. (1990). Basal Sauropodomorpha–Prosauropoda. In *The Dinosauria* (D. B. Weishampel, P. Dodson, and H. Osmólska, Eds.), pp. 320–344. Univ. of California Press, Berkeley.

Gilmore, C. W. (1914). Osteology of the armoured Dinosauria in the United States National Museum, with special reference to the genus *Stegosaurus. Bull. U.S. Natl. Museum* **89,** 1–136.

Lockley, M. G., and Hunt, A. P. (1995). Ceratopsid tracks and associated ichnofauna from the Laramie Formation (Upper Cretaceous: Maastrichtian) of Colorado. *J. Vertebr. Palaeontol.* **15**(3), 592–614.

Madsen, J. H., Jr. (1976). *Allosaurus fragilis*: A revised osteology. *Bull. Utah Geol. Mineral. Survey* **109,** 1–163.

Ostrom, J. H. (1969). Osteology of *Deinonychus antirrhopus*, an unusual theropod from the Lower Cretaceous of Montana. *Peabody Museum Nat. History Bull.* **30,** 1–165.

Paul, G. S. (1987). The science and art of restoring the life appearance of dinosaurs and their relatives: A rigorous how-to guide. In *Dinosaurs Past and Present* (S. J. Czerkas and E. C. Olson, Eds.),Vol. II, pp. 5–49. Natural History Museum of Los Angeles County, Los Angeles.

Rowe, T., and Gauthier, J. (1990). Ceratosauria. In *The Dinosauria* (D. B. Weishampel, P. Dodson, and H. Osmólska, Eds.), pp. 151–168. Univ. of California Press, Berkeley.

Santa Luca, A. P. (1980). The postcranial skeleton of *Heterodontosaurus* tucki from the Stormberg of South Africa. *Ann. South Africa Museum* **79,** 159–211.

Sereno, P. C. (1986). Phylogeny of the bird-hipped dinosaurs (order Ornitischia). *Natl. Geogr. Soc. Res.* **2,** 234–256.

Sereno, P. C. (1993). The pectoral girdle and forelimb of the basal theropod *Herrerasaurus ischigualastensis. J. Vertebr. Palaeontol.* **13**(4), 425–450.

Upchurch, P. (1994). Manus claw function in sauropod dinosaurs. *Gaia* **10,** 161–171.

Upchurch, P. (1995). The evolutionary history of sauropod dinosaurs. *Philos. Trans. R. Soc. London Ser. B* **349**(1330), 365–390.

Weishampel, D. B., and Witmer, L. M. (1990). Heterodontosauridae. In *The Dinosauria* (D. B. Weishampel, P. Dodson, and H. Osmólska, Eds.), pp. 486–497. Univ. of California Press, Berkeley.

Young, C. C. (1941). A complete osteology of *Lufengosaurus huenei* (gen. et sp. nov.) from Lufeng, Yunnan, China. *Palaeontol. Sinica (N.S.) Ser. C* **7,** 1–53.

Fort Peck Power Project, Montana, USA

see Museums and Displays

Fort Worth Museum of Science and History, Texas, USA

see Museums and Displays

Fossilization and the Fossil Record

see Biomineralization; Chemical Composition of Dinosaur Fossils; Permineralization; Trace Fossils

Freilichtmuseum, Germany

see Museums and Displays

Frills

see Ornamentation

Fruita Paleontological Area

James I. Kirkland
Dinamation International Society
Fruita, Colorado, USA

Created in 1976, the Fruita Paleontological Area (FPA) was the first area set aside in the United States by the Bureau of Land Management as a specially protected management area based solely on its fossil resources. The FPA has been known locally for 100 years for its fossil dinosaur remains. In 1975, George Callison, Jim Clark, and Mark Norell, then of California State University at Long Beach, discovered significant occurrences of small vertebrate remains (Callison, 1987). Their discovery was followed by the discovery of the second known *Ceratosaurus* skeleton by Lance Ericksen of the Museum of Western Colorado. Over the ensuing years a diverse, well-preserved vertebrate fauna has been identified from the basal Brushy Basin Member of the Morrison Formation (Upper Jurassic, Kimmeridgian) at the FPA (Fig. 1).

Taking advantage of the remarkable three-dimensional outcrop at the FPA, the principles of high-resolution event stratigraphic methodologies have been used in the analysis of the periodic flood deposits (crevasse splays) to facilitate a detailed analysis of the interrelationships of individual fossil sites to each other and to specific local environments. These local environments include bank-controlled, low-sinuosity, gravelly river channel sandstones (ribbon sandstones), clay-dominated levee deposits, well-sorted sandstone proximal crevasse splay complexes, increasingly distal graded crevasse splays superimposed by carbonate-rich paleosoils, permanent abandoned channel ponds and / or springs, and temporary alkaline ponds characterized by barite nodules. Depositional base level was controlled by the distribution of active and abandoned channel–levee complexes (Kirkland *et al.*, 1990).

Within this framework, a taphonomic basis for the distribution of fossil remains can be discerned. Virtually all ribbon sandstones extend across the FPA, with flow to the north-northeast to the east-northeast, and preserve isolated dinosaur bones. One ribbon sandstone (ceratosaur channel) enters the area from the southeast curves due west before turning north and finally exiting the area to the northeast. The ceratosaur channel is very rich in disarticulated to at least semiarticulated dinosaur remains: *Ceratosaurus*, *Allosaurus*, *Camarasaurus*, *Apatosaurus*, *Stegosaurus*, and *Dryosaurus* have been identified to date. In addition, goniopholid crocodilians, turtles, and unionid clams occur in these sandstones, indicating permanent water flow.

Sediments indicating abandoned channel ponds and / or springs often overlie the ribbon sandstones. These environments are most readily identified by the abundance of carbonaceous plant materials preserved within them. They are also characterized by variable amounts of dinosaur bone, theropod teeth, and aquatic snails (mainly *Viviparous reesidei*). Isolated splays within these ponds sometimes preserve abundant unionid clams, which suggests that the abandoned channels were reoccupied during flooding events. At the top of the abandoned channel pond sequences, conchostracans become common with the loss of carbonaceous plant material. At one pond an algal limestone preserves several intact examples of the fish *Hullettia hawesi* with abundant silicified *Viviparous reesidei* and serves as the type locality for both taxa (Yen, 1952; Kirkland, 1997).

Proximal crevasse splay complexes are best developed at bends in the ceratosaur channel and are identified as being the source of several fossiliferous distal crevasse splays. In one case, theropod tracks were found in a proximal crevasse splay complex. Rooted gray calcareous splay slurries are associated with disarticulated and associated skeletons of microvertebrates, whereas more distal splays associated with oxidized, more mature soils lack vertebrate remains and are more strongly rooted. Better articulated microvertebrate remains have been found in association with laminated smectitic claystones identified as temporary alkaline pond deposits. These microvertebrates represent a mixed transported assemblage; the terrestrial elements tend to be better articulated, often with complete skulls and articulated jaws. The terrestrial taxa include one species of very small theropod dinosaur, one species of very small ornithischian di-

FIGURE 1 The Fruita Paleontological Area in Colorado is famous for Jurassic dinosaurs preserved in the Morrison Formation. James Kirkland, pictured above, has explored these sediments in search of dinosaur fossils. (Photo by François Gohier.)

nosaur (cf. *Echinodon*) (Callison and Quimby, 1984), common small terrestrial mesosuchian crocodilians, a small pterosaur, three species of rhynchocephalians, four species of lizards, the earliest snake?, and at least eight species of mammals (Callison, 1987). Small eggshell fragments have also been recovered (Hirsch, 1994). The aquatic elements include turtle, frog, lungfish, actinopterygians, crawfish, and the remains of caddisfly cases and were most likely transported from the river during flooding episodes. The alkaline ponds often preserve common conchostracans. Most of these taxa are new and are being described.

An accumulation of juvenile and hatchling-sized *Dryosaurus* bones and small egg fragments and associated small terrestrial crocodile bones is associated with a horizon of small calcareous nodules in a soil horizon. It has been interpreted as a site close to a *Dryosaurus* nesting site, where small mesosuchian crocodilians were preying on eggs and / or hatchlings (Kirkland, 1994).

This well-documented interplay of local environments and significant fossil localities makes the FPA an excellent natural laboratory for the study of Upper Jurassic faunas, floras, sedimentology, taphonomy, ecology, and climatology.

See also the following related entry:

MORRISON FORMATION

References

Callison, G. (1987). Fruita: A place for wee fossils: In *Paleontology and Geology of the Dinosaur Triangle* (W. R. Averett, Ed.), pp. 91–96. Museum of Western Colorado, Grand Junction, CO.

Callison, G., and Quimby, H. M. (1984). Tiny dinosaurs: Are they fully grown. *Journal of Vertebrate Paleontology* **3**(4), 200–209.

Hirsch, K. F. (1994). Upper Jurassic eggshells from the Western Interior of North America: In *Dinosaur Eggs and Babies* (K. Carpenter, K. F. Hirsch, and J. R. Horner, Eds.), pp. 137–150. Cambridge Univ. Press, Cambridge, UK.

Kirkland, J. I. (1994) Predation of dinosaur nests by terrestrial crocodilians. In *Dinosaur Eggs and Babies* (K. Carpenter, K. F. Hirsch, and J. R. Horner, Eds.), pp. 124–133. Cambridge Univ. Press, Cambridge, UK.

Kirkland, J. I. (1997). Morrison fishes. In The Morrison Formation: An interdisciplinary study (K. Carpenter, D. Chure, and J. I. Kirkland, Eds.), in press. Geological Society of America Special Paper.

Kirkland, J. I., Mantzios, C., Rasmussen, T. E., and Callison, G. (1990) Taphonomy and environments; Fruita Paleont. Resource Area, Upper Jurassic, Morrison Formation, W. Colorado: *J. Vertebr. Paleontol.* **10**(Suppl. to No. 3), 31A.

Yen, T.-C. (1952). Molluscan fauna of the Morrison Formation. United States Geological Survey Professional Paper No. 233-B, pp. 21–51.

Stage/ Age	Lithostratigraphic Units		Vertebrate Fauna	Dispositional Environment
Danian	OJO ALAMO SS			
Maastrichtian	KIRKLAND FRM	Naashoibito member	Alamo Wash local fauna	Inland flood plain
		De-na-zin member	lower Kirkland fauna	↓
		Farmington member		
		Hunter Wash member		
		Bisti member		Coastal flood plain
Campanian	FRUITLAND FRM		Fruitland fauna	Delta plain
	PICTURED CLIFFS SS			

FIGURE 1 Chronology, lithostratigraphy, vertebrate fauna, and general depositional environment of the late Campanian and early Maastrichtian of the San Juan Basin, New Mexico (modified from Lucas and Williamson, 1993).

Fruitland Formation

Michael J. Ryan
Royal Tyrrell Museum of Palaeontology
Drumheller, Alberta, Canada

The San Juan Basin of northeastern New Mexico encompasses strata of Late Cretaceous, Paleocene, and early Eocene age and contains rich vertebrate fossil assemblages. Excellent reviews of this region include Hunt and Lucas (1993) and Lucas and Williamson (1993). Throughout the Cretaceous, New Mexico was located at the western margin of an epicontinental seaway that stretched from the Gulf of Mexico to the Arctic Ocean. The late Campanian (Judithian) Fruitland Formation (Fig. 1) was named by Bauer (1916) and is at least 170 m thick. It is composed of a series of complexly interbedded channel sandstones, carbonaceous gray shales and siltstones, and coal beds that can reach thicknesses of 9 m and extend laterally over several kilometers. To the northwest, the formation was deposited as the landward facies of a marshy, delta complex that was poorly drained and subjected to frequent flooding. To the southeast, the formation displays a barrier shoreline signature. Paleochannels are lenticular, up to 6 m thick and represent high-sinuosity, mostly unbraided river channels with numerous vertebrate remains. Paleoflow was to the northeast at right angles to the shoreline that trended NW–SE. Unionid bivalves and nonmarine gastropods are common in the Fruitland indicative of the freshwater channel system of the local environment. Plant material is common throughout the Fossil Forest region of the formation with at least eight successive fossil forest horizons. These horizons preserve more than 400 *in situ* tree stumps and logs. In the north, the easily recognizable, persistent brown-colored ribbon sandstones mark the top of the formation, whereas in the south the stratigraphically highest, thick (>1 m) coal can be used as the formational contact (Hunt, 1992). The climate of the Fruitland was warm, humid, and seasonal based on the associated fossil leaf flora and fossil tree-ring evidence.

The upper Fruitland and lowermost Bisti Member of the overlying Kirtland Formation constitute one of two stratigraphic intervals containing vertebrate fossils in the Fruitland–Kirtland stratigraphic sequence, the other being the Naashoibito Member of the Kirtland. In the Fruitland Formation, fossils are equally common in both channel and interchannel environments. The Fruitland contains a diverse vertebrate assemblage including 10 chondrichthyans, 13 osteichthyans, two anurans, three urodeles, 10 chelonians, two teiid and two anguid lizards, one aniilid snake, five crocodilians, at least 15 dinosaurs, and more than three dozen mammals. This fauna comes primarily from the upper two-thirds of the formation in an outcrop belt between the San Juan River and the Kimbeto Wash. The dinosaurs are as follows: Saurischia: Tyrannosauridae (?*Albertosaurus libratus* and ?*Albertosaurus* sp., gen. et. sp. nov.), Ornithomimidae (cf. *Ornithomimus* sp. and Ornithomimidae indet.), Dromaeosauridae indet., Troodontidae indet., Titanosauridae (undescribed n. gen et sp.); Ornithischia: Hypsilophodontidae (?*Thescelosaurus* sp.), Lambeosauridae (*Parasaurolophus crystocristatus* and ?*Corythosaurus* sp.), Hadrosauridae indet., Hadrosauroidea footprints, Nodosauridae indet., Ankylosauridae indet., Pachycephalosauridae indet., Ceratopsidae [*Pentaceratops sternbergii*, Ceratopsidae indet. (= cf. *Chasmosaurus* Gilmore 1935)].

Microvertebrate localities are more common in the Fruitland than in the overlying Kirkland and account for almost all fish, amphibian, and lower reptiles recovered to date. Dinosaur specimens are largely made up of disarticulated and isolated material (limbs, ribs, and vertebrae) with partial skeletons and skulls being rare. Hydrodynamic sorting appears to be common. Theropods are poorly known from isolated material. The ankylosaurs are known from scutes. The hadrosaurs are known from partial skeletons with the Hadrosauridae known from a juvenile dentary (Wolberg, 1993). Eggshell fragments and nests (Wolberg, 1992) suggest that dinosaurian nesting may have occurred in the area.

See also the following related entries:

CRETACEOUS PERIOD • KIRTLAND FORMATION

References

Bauer, C. M. (1916). Contributions to the geology and paleontology of San Juan County, New Mexico. 1. Stratigraphy of a part of the Chaco River Valley. United States Geological Survey, Professional Paper No. 98P, pp. 271–278.

Gilmore, C. W. (1935). On the Reptilia of the Kirtland Formation of New Mexico, with descriptions of new species of fossil turtles. *Proc. United States Natl. Museum* **83,** 159–188.

Hunt, A. P., and Lucas, S. G. (1993). Cretaceous vertebrates of New Mexico. In *Vertebrate Paleontology in New Mexico* (S. G. Lucas and J. Zidek, Eds.), Bull. 2, pp. 77–91. New Mexico Museum of Natural History.

Lucas, S. G., and Williamson, T. E. (1993). Late Cretaceous to Early Eocene vertebrate biostratigraphy and biochronology of the San Juan Basin. In *Vertebrate Paleontology in New Mexico* (S. G. Lucas and J. Zidek, Eds.), Bull. 2, pp. 92–104. New Mexico Museum of Natural History.

Wolberg, D. L. (1992). Dinosaur nesting near the Cretaceous (Campanian–Maastrictian) marine shoreline; Evidence from the first infant dinosaur discovered in the San Juan Basin, of New Mexico. *Geol. Soc. Am.* **24**(7), 270. [Abstract]

Wolberg, D. L. (1993). A Juvenile hadrosaurid from New Mexico. *J. Vertebr. Palaeontol.* **13,** 367–369.

Fryxell Geology Museum, Illinois, USA

see MUSEUMS AND DISPLAYS

Fukui Prefectural Museum, Japan

see MUSEUMS AND DISPLAYS

Functional Morphology

Xiao-Chun Wu
Anthony P. Russell
University of Calgary
Calgary, Alberta, Canada

Function in a biological context may generally be defined as the use or action of structures (such as bones, dermal armor, muscles, etc.) of organisms. The study of functional morphology attempts to explain the diversity of structure and function exhibited by organisms by proposing theories of how structures work and how they have come to be. The limb skeletons of terrestrial tetrapods (such as dinosaurs) function to provide a rigid frame on which muscles can act. The actions of particular subsets of these muscles (which are controlled by the nervous system) on the elements of the legs enable these animals to effect locomotion. The wings of birds can be studied from the viewpoint of how they are involved in flight, but other attributes of their morphology and the basis of their design can be understood by realizing that such wings have evolved from the forelimbs of advanced theropod dinosaurs.

It has long been accepted that function and form (structure) are intimately related, despite little consensus on just what this relationship is. Much (current and previous) work on functional morphology is based on the assumption that form and function are intimately correlated and, thus, biological function can be deduced from a study of form or structure. Recently, however, some functional morphologists (such as Lauder, 1995) have postulated that form and function are related only at a very general level, and that they are often not tightly linked.

In contrast to neontology (the study of extant taxa), studies of the functional morphology of extinct organisms, and hence the Dinosauria, are often limited to deductions based on fossilized hard tissues (bones, teeth, dermal armor, eggs, etc.) or, less often, on traces of animal activities (impressions of external structures, trackways, etc.). However, functional reconstructions of dinosaur morphology can help us enrich our understanding of their paleobiology and, sometimes, reveal some of the subtleties of their evolutionary history (Weishampel, 1995). Methods currently employed in studies of dinosaur functional morphology may be referred to three categories: the nonhistorical approach, the historical (phylogenetic) approach, and the synthetic approach.

Nonhistorical Approach

The nonhistorical approach has a long history and thus numerically dominates studies of functional morphology in vertebrate paleontology. This method takes as its base aspects of biomechanics that are theoretically well substantiated and emphasizes analogies (with beams, machines, or extant animals as models or modern analogies, etc.) as universals from which to assess their functional relevance to extinct animals. Studies using mechanical analogies are inherently nonhistorical, whereas studies using extant animals as models or modern analogies usually do not consider the phylogenetic relationships among the taxa of interest. For any particular model to be applied, it has to be established on the basis of explicit, well-understood parameters that come from engineering, biology, or other related sources. There are many variants of the nonhistorical approach. Of them, the modeling method based on machine analogies is the most often used in studies of the functional morphology of extinct vertebrates. These are directed toward the understanding of the operation of a particular anatomical system. Other variants include experiments, graphic representation and mathematical computation, and computer-based simulations (see Weishampel, 1995). The following are examples of the most commonly used nonhistorical approaches in the study of the functional morphology of the Dinosauria.

Modeling Lever Devices The dinosaurian jaw, as it does in all vertebrates, operates as a class III lever, with the force acting at a point between the fulcrum (the articulation between the quadrate and articular) and the resistance (between the upper and lower den-

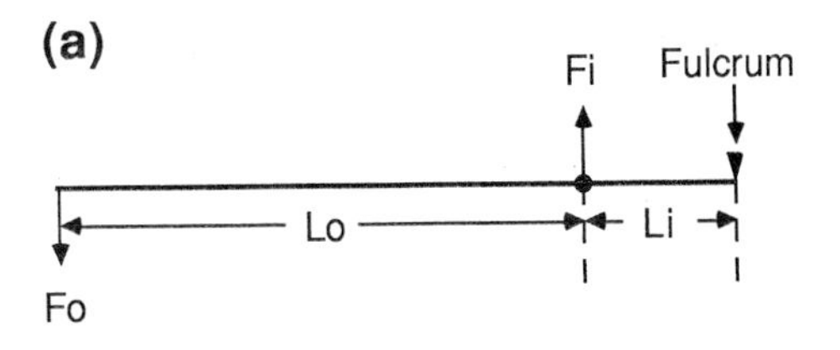

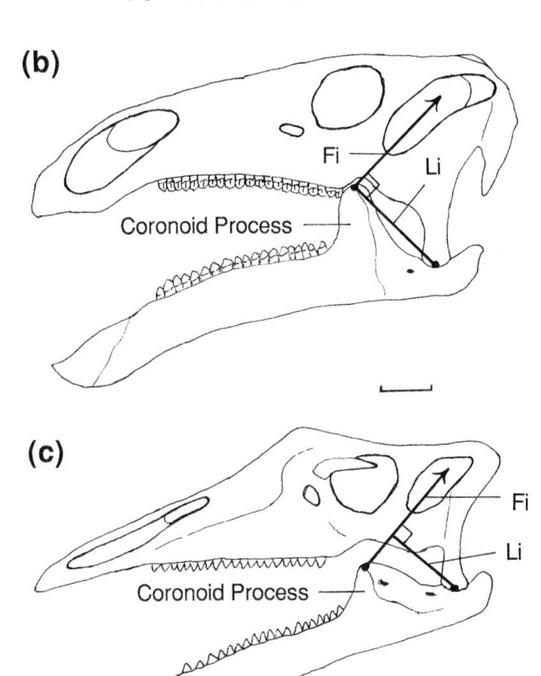

FIGURE 1 (a) Relationships among in-forces (F_i), out-forces (F_o), in-lever arm (L_i), and out-lever arm (L_o): $F_o L_o = F_i L_i$. From this, $F_o = F_i L_i / L_o$ (F_o increases when L_i increases and F_i remains the same). (b) Skull and mandible of *Iguanodon atherfieldensis* in lateral view. (c) Skull and mandible of *Ouranosaurus nigeriensis* in lateral view. Scale bars = 5 cm. [(b) and (c) modified from Norman and Weishampel, 1990, Figs. 25.1A and 25.2B, respectively]

titions). By the application of the principles of lever systems (Fig. 1a), one may postulate differences in the feeding mechanics of different dinosaurs.

Iguanodon (*I. atherfieldensis*) and *Ouranosaurus* (*O. nigeriensis*) are two ornithopod dinosaurs from the Early Cretaceous. Both of them were herbivorous. The lower jaw of these dinosaurs possesses a well-developed coronoid process posterior to the tooth row. This process is often considered to be the focus for the insertion of the major jaw-closing muscles and is taken to be the point at which the resultant force produced by the major jaw-closing muscles is applied (Figs. 1b and 1c). The vector of the resultant force can be estimated in terms of the position of the temporal fenestrae, the area from which the major jaw-closing muscles originated. In the cases of the two dinosaurs mentioned previously, the length of the in-lever arm (the acting moment of the force) is closely related to the position or height of the coronoid process; that is, the farther from the articulation or the taller the process is, the longer the in-lever arm. The longer the in-lever arm, the greater the force produced by a given muscle mass. It is very clear from the application of this type of lever analysis that *Iguanodon* could produce a more powerful bite than could *Ouranosaurus*, indicating that a diet of tough material could be accommodated more readily by the former. This hypothesis is also supported by tooth morphology. The teeth of *Iguanodon* are massive, with complicated closely packed crowns, in contrast to those in *Ouranosaurus*, which are lightly built, with simple crowns that do not overlap.

Modeling Circular Motion *Deinonychus* (*D. antirrhopus*) of the Early Cretaceous is a close relative of birds. The pes of *Deinonychus* is very peculiar in that its second digit is equipped with a massive and very strongly arched claw and ungual phalanx, with a curvature even greater than that seen in the raptorial-type manual unguals of other theropods (Fig. 2). The

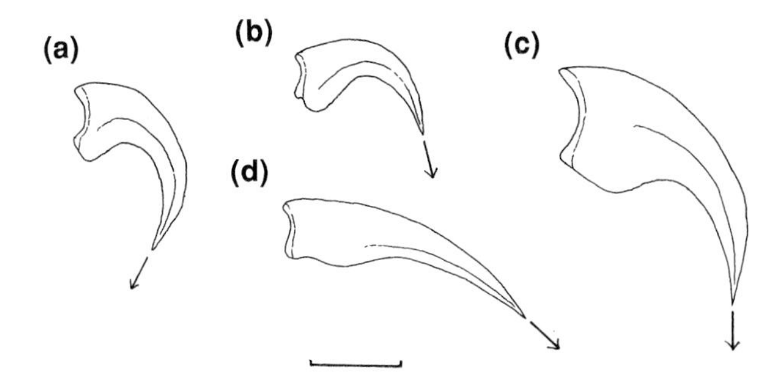

FIGURE 2 Comparison of the second pedal ungual of *Deinonychus* (a) with the raptorial type manual ungual of *Deinonychus* (b) and *Allosaurus* (c). The second ungual of the manus of *Ornithomimus* (d) provides a further comparison because it represents straight, nonraptorial ungual form. All are depicted with the chord of the articular facet arc oriented vertically. Scale bar = 5 cm (modified from Ostrom, 1969, Fig. 77).

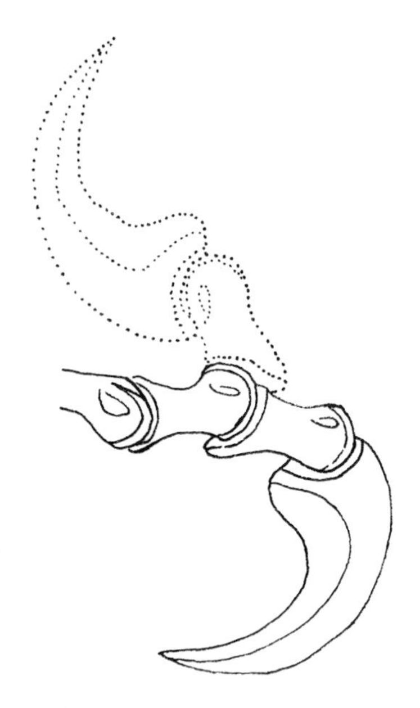

FIGURE 3 Medial outline of the second digit of the pes of *Deinonychus antirrhopus*, showing the maximum flexion (solid lines) and unusual degrees of extension (dashed lines) possible at the two distal joints (redrawn based on Ostrom, 1969, Fig. 76).

unusual morphology of the articulation between the penultimate phalanx and the ungual allows the claw of this digit to sweep an arc more than 90° (Fig. 3). On the basis of this, Ostrom (1969) proposed that the second digit of the pes in *Deinonychus* might have functioned as a tool for cutting or slashing prey rather than as a component used in locomotion, the primary role of the pes. This conclusion can be tested by functional analysis of the second digit, using circular motion as a mechanical model.

The ungual of the second digit of the pes can rotate by way of a diarthrotic joint with the penultimate phalanx. This kind of action is actually a circular motion (Fig. 4). The principle of circular motion indicates that the greatest force able to be produced by the ungual should act along the tangent line, being perpendicular to the radius of the acting circle at any given point of the circle. When compared to the manual unguals of other theropod dinosaurs (Figs. 4c and 4d), the ungual of the second digit of the pes of *Deinonychus* has significant mechanical advantage in producing force that could be used in cutting or slashing. This is indicated by the angle between the acting line of the ungual and the tangent line of the acting circle; that is, the smaller the angle, the stronger the force that can be produced by the ungual. Such an analysis indicates that a cutting or a slashing function was possible but cannot directly indicate whether the structure was actually used in this way. Analysis of the functional morphology of adjacent structures, such as the ankle, lower limb and femur, and their articulations, may, however, shed light on the feasibility of such a biological role.

Methods Using Graphic Representation and Mathematical Computation Investigations of the ontogenetic variation of patterns of locomotion in *Dryosaurus* (*D. lettowvorbecki*), a Late Jurassic iguanodontian dinosaur, provide an example of this method (Weishampel, 1995). This study was based on beam theory, which indicates that the cross-sectional morphology of a beam is altered when the loading acting on it changes. In this study the diaphysis of the femur was

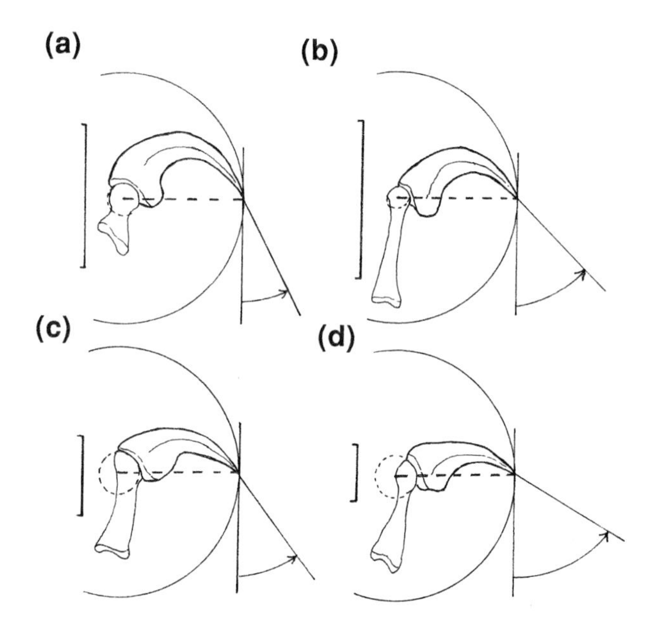

FIGURE 4 Comparison of ungual form and mechanics of the second pedal ungual of *Deinonychus* (a) with the second manual ungual of *Deinonychus* (b), the first manual ungual of *Allosaurus* (c), and the third manual ungual of *Ornitholestes* (d). Heavy dashed lines represent the radii of the acting circles of the distal ends of the unguals; dashed circles represent the projections of the curvature of the articular facets of the penultimate phalanges; angles measured in terms of the cutting edges of unguals and the tangent lines (perpendicular to radii) of acting circles at any given point. Scale bars = 10 cm (modified from Ostrom, 1969, Fig. 78).

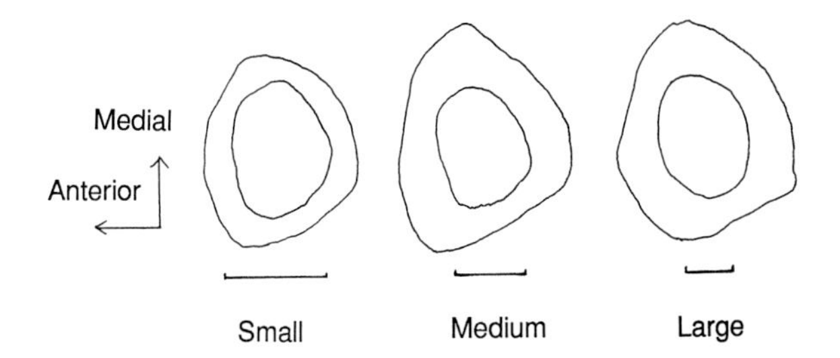

FIGURE 5 Cross-sectional shape of representative femora of three size classes of *Dryosaurus lettowvorbecki*. Scale bars = 1 cm (redrawn based on Weishampel, 1995, Fig. 3.3).

treated as a hollow beam. Geometric changes in cross section of the femur through growth may reflect, in turn, ontogenetic variations in the magnitude and orientation of mechanical loadings acting on it.

Relevant geometric data used for the study were obtained from the femoral cross sections of 27 femora of a growth series, represented by small-sized, medium-sized and large-sized individuals (Fig. 5). Several biomechanical calculations were derived from these data, including the cross-sectional area, cortical area, second moments of area (estimates of resistance to bending loads) about the X- and Y-axes (I_x, Y_x), maximum and minimum second moments (I_{max}, I_{min}), and the angle expressing the disposition of greatest bending strength relative to the mediolateral plane. These data were digitized and calculations were obtained using a computer program [a modified version of the program SLICE (Nagurka and Hayes, 1980)].

The results revealed that the cross-sectional area of cortical bone increases by 33% from small- to medium-sized femora, but that it then remains constant in large-sized femora. The cross-sectional shape of small-sized individuals is modified by an increase in the ratios of I_{max}/I_{min} (from 1.269 to 1.444) and I_x/I_y (from 1.208 to 1.307) in medium-sized femora, but by a decrease of the ratios of I_{max}/I_{min} (from 1.44 to 1.241) and I_x/I_y (from 1.307 to 1.016) in large-sized femora. These changes during ontogeny indicate a shift to a more mediolateral orientation of the greatest resistance to bending.

If the locomotor pattern did not change during the ontogeny of *Dryosaurus*, the cross-sectional properties of the femora would be predicted to change in a linear fashion in proportion to the increase in body weight. However, this is evidently not the case. The previously discussed results indicate that the observed biomechanical changes in cross-sectional morphology of the femur are apparently not directly related to the increase of body size. The actual effect that led to these changes may have resulted from a shift in the center of gravity as older individuals shifted from a quadrupedal to a bipedal mode of progression.

Information from other dinosaurs (*Orodromeus makelai* and *Maiasaura peeblesorum*) indicates that hatchling individuals had a relatively large head and a small tail. These body proportions result in the center of gravity lying far in advance of the hip (Fig. 6a). The negatively allometric growth of the head and the positively allometric growth of the tail necessarily result in a posterior shift of the center of gravity in adults (Fig. 6b). For a bipedal dinosaur, the farther the center of gravity is from the hip joint, the larger the moment acting about the femur and the greater the bending stresses are on that element. This mechanical constraint may have made habitual bipedality impossible for hatchlings of *D. lettowvorbecki*; that is, a reasonable locomotor pattern for the hatchlings would be quadrupedalism. Shifting from quadrupedal locomotion to bipedalism would double the force acting on the hindlimbs, which may be the primary causal factor associated with the significant and rapid increase of both the relative amount of bone and the maximum bending strength in medium-sized femora. (Further considerations of modifications of femoral biomechanics follow the small to

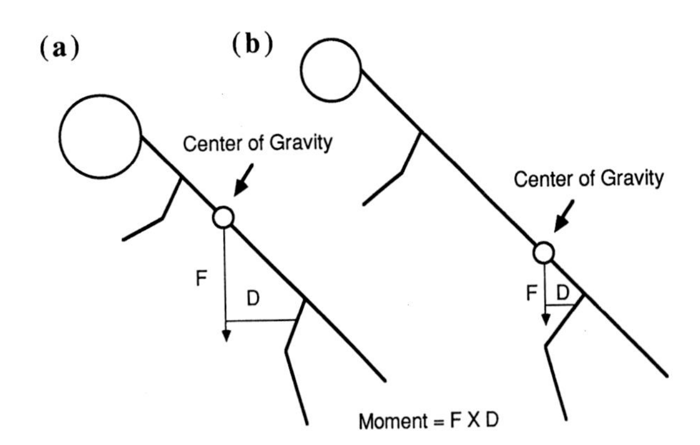

FIGURE 6 Schematic drawing of a hatchling (a) and a fully adult *Dryosaurus lettowvorbecki* (b), showing the biomechanical effect of the position of the center of gravity on the bending stress incurred by the femur (redrawn based on Weishampel, 1995, Fig. 3.4).

medium-sized transition, and these, as well as other aspects, are detailed in the original paper of Heinrich *et al.*, 1997.)

Methods Employed in the Estimation of Speeds of Dinosaurs Speeds of dinosaurs cannot be directly measured by the experiments used for extant animals. Speeds of dinosaurs, however, may be estimated from a combination of morphological features and the trackways that these animals left behind. To ensure that estimates of speeds are as accurate as possible, the type of gait (walk, trot, or run) being employed when the trackway was made must first be determined. Information gleaned from studies of various extant animals (Alexander, 1976) suggests that gait changes are accompanied by changes in relative stride length. Relative stride length may be defined as SL/H, where SL is the length of the animal's stride and H is the animal's height at the hip. Such observations reveal that the ratio of SL/H is less than 2.0 when animals are walking, but that it increases to greater than 2.0 when animals are trotting or galloping/running (SL/H is over 2.9). SL may be measured directly from trackways, and H can be estimated from the dimensions of the trackmaker's footprints. This assumes, of course, that the species responsible for making the trackway can be unequivocally identified. There are several methods for predicting the value of H, each making certain assumptions. The method outlined here is called the morphometric ratios approach (Alexander, 1976). This approach suggests that H can be calculated to be approximately four times the footprint length (FL) for a variety of dinosaurs, both bipeds and quadrupeds. Recently, this ratio (4 FL) has been modified; that is, for different dinosaurs of different sizes, different ratios should be used for the calculation of the speed (see Table I).

TABLE I
Supposed Relationships between Height at Hip Joint (H) and Length of Footprint (FL) in Different Bipedal Dinosaurs[a]

Small theropods	(FL < 25 cm): $H = 4.5$ FL
Large theropods	(FL > 25 cm): $H = 4.9$ FL
Small ornithopods	(FL < 25 cm): $H = 4.8$ FL
Large ornithopods	(FL > 25 cm): $H = 5.9$ FL
Small bipedal dinosaurs in general	(FL < 25 cm): $H = 4.6$ FL
Large bipedal dinosaurs in general	(FL > 25 cm): $H = 5.7$ FL

[a] Derived from Thulborn (1989).

Following gait determination, the speeds of dinosaurs can be calculated in terms of the relationship established between SL, H, and speed (V). For trackways with an SL/H value less than 2.0 (walking gait), the speed can be calculated as follows:

$$V = 0.25g^{0.5}SL^{1.67}H^{-1.17}, \quad (1)$$

where g represents gravitational acceleration, which equals 10 m/sec^2 (Alexander, 1976). For trackways with an SL/H value greater than 2.9 (running gait) calculations employing Eq. (2) are used:

$$V = [gH(SL/1.8H)^{2.56}]^{0.5}, \quad (2)$$

(Thulborn and Wade, 1984). Speeds of those dinosaurs that used a trotting gait, where the SL/H ratio varies from 2.0 to 2.9, may be calculated as the means of the two previously mentioned estimates (Thulborn, 1984).

As an example we present the analysis of two sets of trackways of the theropod *Anchisauripus* (*A. sp.*). These extend in opposite directions to each other (Fig. 7a). The length of a single footprint is about 0.17 m in both sets, suggesting that the trackmaker was a small theropod dinosaur (see Table I). Thus, the height at the hip joint is 4.5 times 0.17 m in this dinosaur. According to the ratio of SL/H, the left set of three trackways was made when the dinosaur was running ($SL/H = 2.98$), whereas the right set was made when the dinosaur was trotting ($SL/H = 2.24$). Therefore, the speed when the dinosaur was making the left set is calculated using Eq. (2): $V = [10 \text{ m} \times 0.765 \text{ m}(2.28 \text{ m}/1.8 \times 0.765)^{2.56}]^{0.5} = 5.274$ m/sec; that is, about 19 km/hr. The speed when the dinosaur made the right set should be the means of the results obtained using both Eqs. (1) and (2): about 11.4 km/hr.

Historical (Phylogenetic) Approach

Function in the evolutionary perspective has a historical context; that is, function, like morphological char-

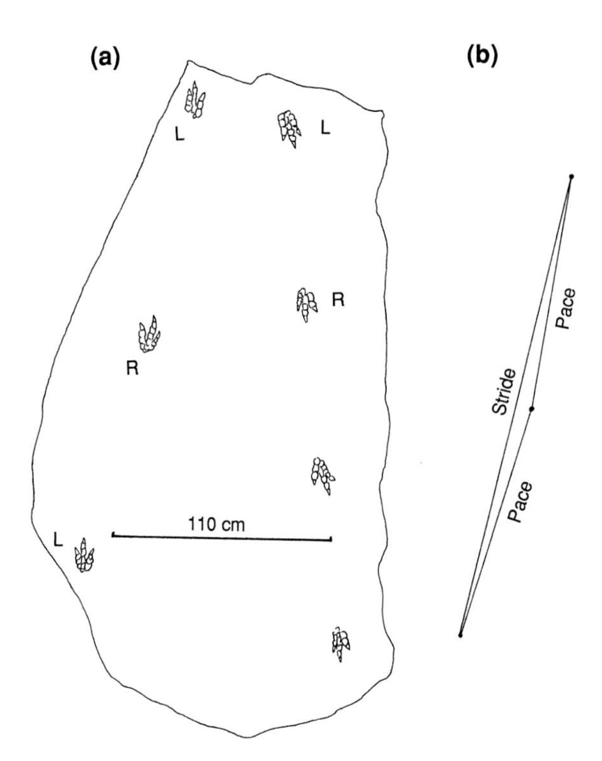

FIGURE 7 (a) Schematic drawing of *Anchisauripus* sp. trackways (modified from Padian and Olson, 1989, Fig. 24.4). (b) Diagram of a stride taken from the right trackway in a. Here the stride is represented as the distance between one placement and the next of the same foot. The pace is the distance between the placement of hindfeet from opposite sides of the body.

acters, has a phylogenetic distribution (plesiomorphic or apomorphic) in each member of a monophyletic group. A flapping function of the forelimbs in bird-like theropod dinosaurs or early birds (such as *Archaeopteryx*) may be the plesiomorphic state for flying, which may have given rise to the soaring ability of later birds (Padian, 1985). The phylogenetic approach is a recently developed method that does not make *a priori* gradistic assumptions regarding the functional predictions of extinct animals (Bryant and Russell, 1992; Witmer, 1995). This approach relies on related extant taxa to infer the function of a structure in fossil taxa. It emphasizes that the phylogenetic inference of the transfer of a function of a structure to a fossil taxon should be based on both the extant sister group of the fossil taxon and the second related extant outgroup, the fossil taxon being considered in a more inclusive monophyletic group (Fig. 8). Therefore, this approach has to begin with an explicit hypothesis of phylogenetic relationships of both the extant and fossil taxa of interest. Conceptually, the phylogenetic approach adopts the global method of outgroup comparison (Maddison *et al.*, 1984) in evaluating functional predictions for fossil taxa. Phylogenetic inference regarding a series of fossil taxa with similar relationships to extant taxa will necessarily be similar (Fig. 8c) because the transference of functional morphology from the extant taxa to any of the fossil taxa depends on the congruence or noncongruence of functional morphology in the two extant taxa in question.

When inferring functional attributes of fossil taxa using the phylogenetic approach there may be three

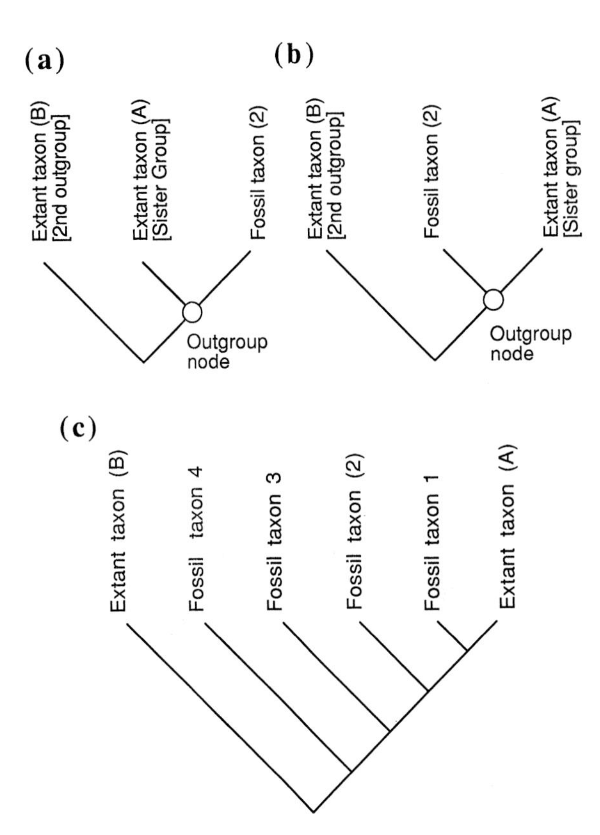

FIGURE 8 The historical (phylogenetic) approach to functional inference in fossil taxa. (a and b) The phylogenetic relationships of fossil taxon (2) with its two extant outgroups. These two depictions of relationship are identical, but the outgroup node is rotated through 180° between a and b. (c) A number of additional fossil taxa related to extant taxa (A and B) have been included in the pattern of relationships. Transference of functional inference from the extant taxa to any of the fossil taxa depends on the congruence or noncongruence of functional morphology in the two extant taxa in question. See Fig. 9 for an example of the principles involved.

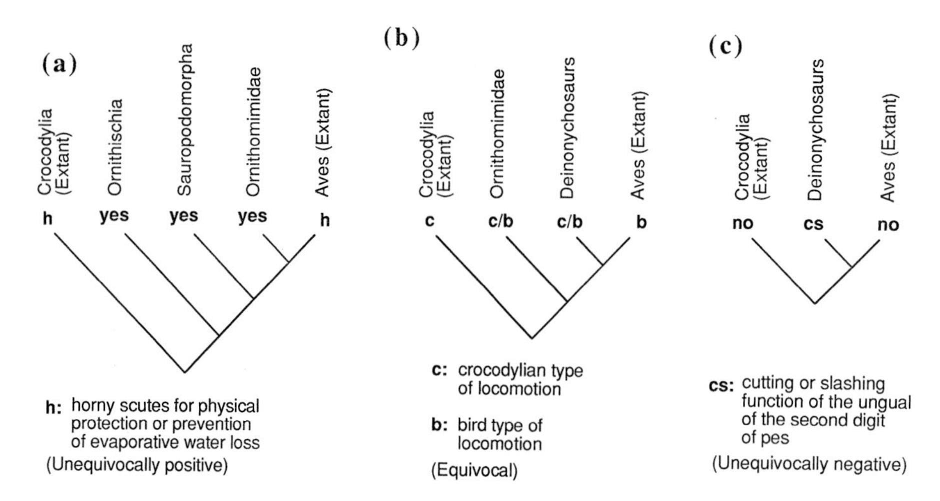

FIGURE 9 Phylogenetic inferences of function in fossil taxa on the basis of functional interpretation of two related extant taxa. (a) Both extant taxa possess the horny scute suspected to occur in fossil taxa, leading to an unequivocal assessment. (b) Two extant taxa conflict with each other in locomotor pattern, leading to an equivocal assessment for the fossil taxa. (c) Neither of two extant taxa has the suspected cutting or slashing function of the second digit of pes in the fossil taxon, leading to a negative assessment.

possible kinds of outcomes: (i) unequivocally positive, (ii) equivocal, or (iii) unequivocally negative inferences.

The unequivocally positive condition means that the function/structure postulation is in concordance with both the sister group and the second outgroup (Fig. 9a); for example, the phylogenetic inference of protective role (defense against attackers or against water loss from the body) of horny scutes (which cover the surface of the body) of dinosaurs is an unequivocal prediction because horny scutes are present on the surface of the body in both extant sister groups (birds, in which horny scutes occur in the distal area of the legs) and the second outgroup (the Crocodylia). The equivocal condition means that the postulation about function/structure is in conflict between the two extant groups (Fig. 9b). For example, the phylogenetic inference of either the bird type or crocodylian type of locomotor pattern (reflected by the function of the hindlimb and tail) for any nonavian theropod dinosaur is equivocal because the relevant function/structure differs between the two related extant groups. The unequivocal negative condition means that the postulation about function/structure is unique to the fossil taxon (Fig. 9c). For example, the function of the ungual of the second digit of the pes as a cutting or slashing tool is unique to the theropod dinosaur *Deinonychus* (see above) because no member of either the birds or the crocodylians possesses such a massive and greatly arched ungual of the second digit of the pes. It is very clear from the previous examples that only those functions/structures that are congruently present in the two related extant taxa can be unequivocally transferred by using the phylogenetic approach to the fossil taxon of interest (see SYSTEMATICS).

Synthetic Approach

The synthetic approach emphasizes the use of separate methods (phylogenetic, modeling, experimental, anatomical, etc.) to gather lines of evidence for producing a single, robust inference of the function of a structure in fossil taxa. The study of the functional evolution of the hindlimb and tail from primitive theropod dinosaurs to birds (Gatesy, 1990) is used here as an example of this approach.

The functional inferences about the hindlimb and tail in nonavian theropod dinosaurs are equivocal when the phylogenetic approach is applied alone (see above). Gatesy (1990), based on a combination of methods (phylogenetic, experimental, and anatomical), proposed a new hypothesis that suggests a transformational sequence of the hindlimb and tail, both structurally and functionally, within the Theropoda.

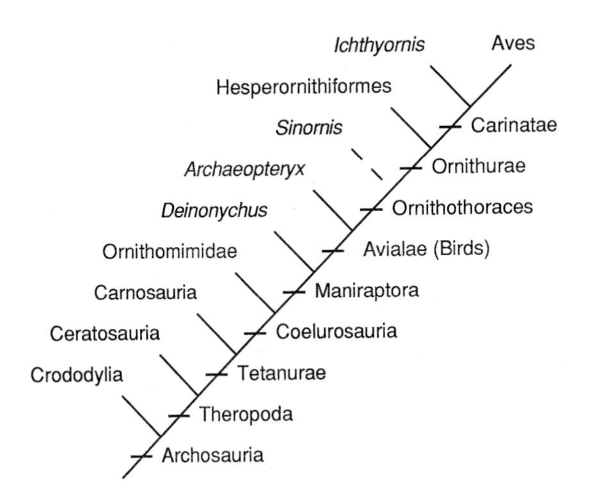

FIGURE 10 Simplified cladogram showing the phylogenetic relationships of fossil theropods with extant archosaurs (modified from Gatesy, 1995, Fig. 13.1).

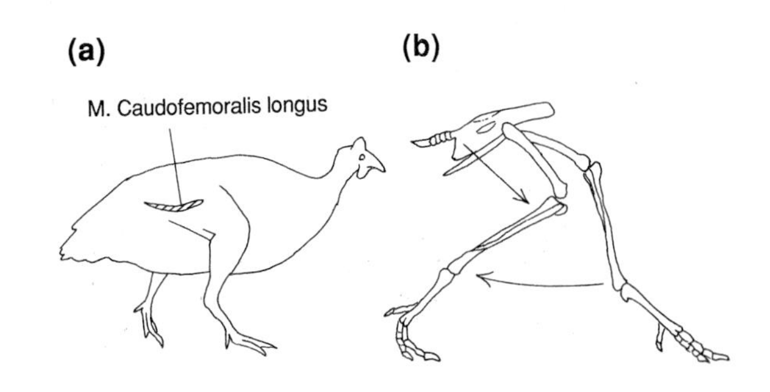

FIGURE 12 Terrestrial locomotion in guineafowl, *Numida*. (a) The M. caudofemoralis longus in guineafowl is a small, strap-like muscle that lacks the broad origin and prominent insertion of crocodylians. (b) Hindlimb position at the beginning and end of the propulsive phase at a slow run; the flexion of the knee (curved arrow) produced by the hamstrings (double-headed arrow) accounts for the majority of the foot displacement (modified from Gatesy, 1995, Fig. 13.5).

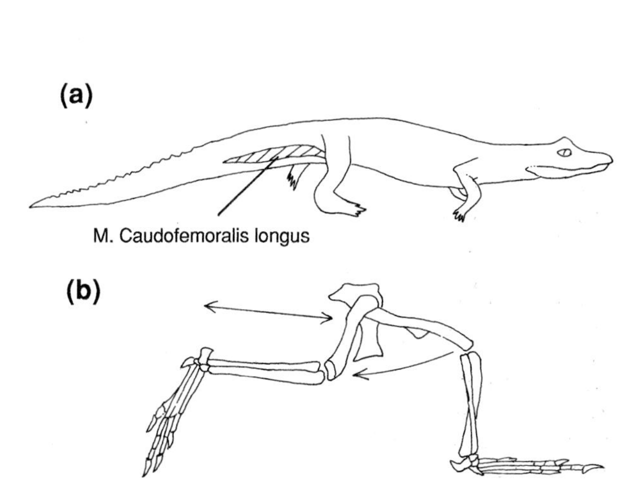

FIGURE 11 Terrestrial locomotion in the alligator, *Alligator*. (a) The M. caudofemoralis longus (M. CFL) in the alligator is a large, cylindrical muscle originating from tail base ventral to transverse processes of anterior caudal vertebrae and inserting on the fourth trochanter of the femur. (b) The position of hindlimb at the beginning and end of a propulsive phase; the majority of the foot displacement is produced by the large arc of femoral retraction (curved arrow) powered by the M. CFL (double-headed arrow) (modified from Gatesy, 1995, Fig. 13.4).

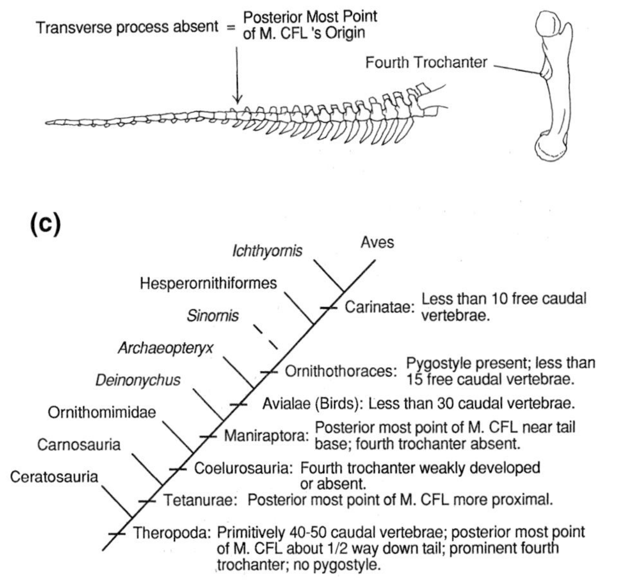

FIGURE 13 Skeletal indicators of the extent and size of the M. caudofemoralis longus (M. CFL). (a) The tail skeleton of the ornithomimid *Struthiomimus* illustrating the posteriormost point of the M. CFL's origin, which divides the tail into proximal (basal) and distal portions. (b) The medial view of the left femur of *Tyrannosaurus* showing the fourth trochanter, the insertion site of the M. CFL. (c) The distribution of characters relating to the M. CFL throughout theropod evolution (modified from Gatesy, 1995, Fig. 13.6).

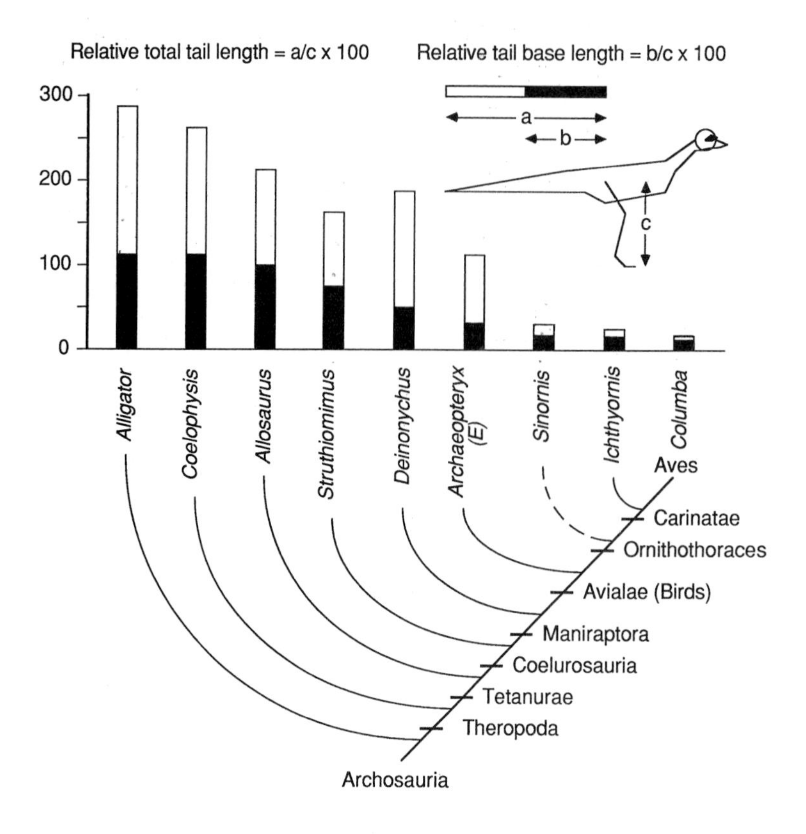

FIGURE 14 Plot of the relative tail length and relative tail base length throughout theropod evolution, as indicated by the ratio of tail length and tail base (anterior caudal vertebrae with transverse processes) length to hindlimb length (femur + tibia + metatarsal III) represented as a percentage for each taxon. Note the dramatic reduction in relative tail size and tail base length in the theropods most closely related to birds (modified from Gatesy, 1995, Fig. 13.7).

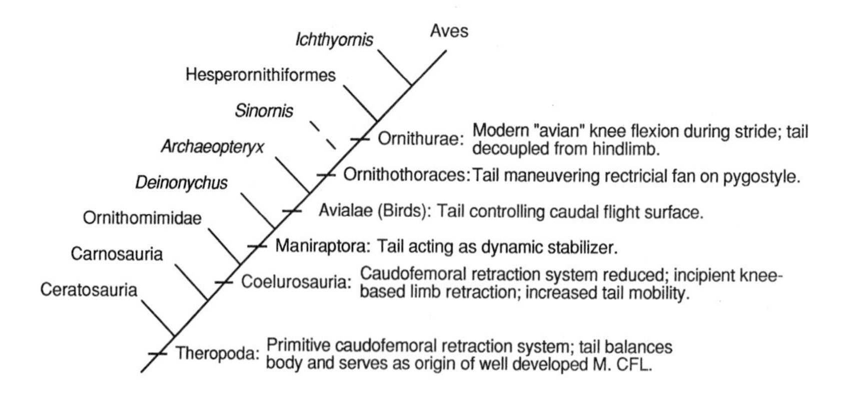

FIGURE 15 Proposed distribution of changes in the hindlimb and tail function throughout theropod evolution (modified from Gatesy, 1995, Fig. 13.8).

This result was obtained from a study that consisted of four steps: (i) adopting a commonly accepted cladogram (hypothesis) of phylogenetic relationships between fossil and extant taxa, (ii) analyzing relevant anatomical structures of extant taxa, (iii) assessing differences between extant taxa and their functional significance, and (iv) superimposing osteological correlates of an important muscular subsystem onto the cladogram to elucidate function in fossil taxa.

According to the adopted cladogram, two extant taxa, birds (the sister group) and the Crocodylia (the second outgroup), were chosen for the purpose of experiments and anatomical analyses (Fig. 10). Analyses of the anatomical structure of birds and crocodylians, based on dissection of the musculature of their representatives and kinematic information from X-ray film results and electromyographic investigations, indicate the importance of the tail and the M. caudofemoralis longus (CFL) in terrestrial locomotion in crocodylians. Differences in locomotor pattern are evident between the two extant taxa: During walking crocodylians primarily employ a caudofemoral retraction system to move the entire hindlimb about the hip joint (accomplished primarily by the powerful action of the M. CFL on the femur (Fig. 11); in contrast, birds do not have a long tail or a well-developed M. CFL and primarily use knee flexion to move the limb (Fig. 12).

In extant diapsids (crocodylians, birds, squamates, and sphenodontians) the M. CFL originates from just ventral to the transverse processes of the anterior caudal vertebrae (except for the first two) and inserts on the fourth trochanter of the femur, and the distalmost extent of the M. CFL's origin correlates well with the loss of the transverse processes of caudal vertebrae. Consequently, the size of the M. CFL in fossil theropods can be deduced from the total number (= length) of the anterior caudal vertebrae (basal tail) that bear transverse processes and from the robustness of the fourth trochanter of the femur. By superimposing modifications of the M. CFL's origin and insertion on the cladogram, it is revealed that the M. CFL has been continuously reduced in size throughout theropod evolution; that is, the size of the fourth trochanter and the length of the basal portion of the tail that could support the origin of the M. CFL concordantly decrease as the birds are approached (Fig. 13). In addition, the relative size of the tail and the relative length of the basal portion of the tail (for the origin of the M. CFL) are dramatically reduced in the theropod dinosaurs close to birds (Fig. 14).

It should be clear from anatomical features that primitive theropod dinosaurs mainly employed the caudofemoral retraction system for locomotion, as is reflected in the elongate basal portion of the tail and the pronounced fourth trochanter. With the progressive reduction of the relevant structures, the caudofemoral retraction system was gradually replaced by the knee-based limb retraction system in derived theropod dinosaurs and birds (Fig. 15). As for the function of the tail in bipedal theropods, it primitively functioned as a cantilever beam to balance the front of the body about the hip joint in early theropod dinosaurs that had a long tail. The tail functioned as a dynamic stabilizer in derived theropod dinosaurs in which it was dramatically reduced in size. With further structural changes, tail function underwent a series of modifications within birds (see Fig. 15).

See also the following related entries:

Biomechanics • Biometrics • Constructional Morphology

References

Alexander, R. McN. (1976). Estimates of speeds and dinosaurs. *Nature* **261,** 129–130.

Bryant, H. N., and Russell, A. P. (1992). The role of phylogenetic analysis in the inference of unpreserved attributes of extinct taxa. *Philos. Trans. R. Soc. London B* **337,** 405–418.

Gatesy, S. M. (1990). Caudofemoral musculature and the evolution of theropod locomotion. *Paleobiology* **16,** 170–186.

Gatesy, S. M. (1995). Functional evolution of the hindlimb and tail from basal theropods to birds. In *Functional Morphology in Vertebrate Paleontology* (J. J. Thomason, Ed.), pp. 206–219. Cambridge Univ. Press, Cambridge, UK.

Heinrich, R. E., Ruff, C. B., and Weishampel, D. B. (1997). Femoral ontogeny and locomotor biomechanics of *Dryosaurus lettowvorbecki* (Dinosauria: Iguanodontia). *Zool. J. Linnean* Soc., in press.

Lauder, G. V. (1995). On the inference of function from structure. In *Functional Morphology in Vertebrate Paleontology* (J. J. Thomason, Ed.), pp. 1–18. Cambridge Univ. Press, Cambridge, UK.

Maddison, W. P., Donoghue, M. J., and Maddison, D. R. (1984). Outgroup analysis and parsimony. *Systematic Zool.* **33,** 83–103.

Nagurka, M. L., and Hayes, W. C. (1980). An interactive graphics package for calculating cross-sectional properties of complex shapes. *J. Biomechanics* **13,** 419–451.

Norman, D. B., and Weishampel, D. B. (1990). Iguanodontidae and related Ornithopoda. In *The Dinosauria* (D. B. Weishampel, P. Dodson, and H. Osmólska, Eds.), pp. 510–533. Univ. of California Press, Berkeley.

Ostrom, J. H. (1969) Osteology of *Deinonychus antirrhopus,* an unusual theropod from the Lower Cretaceous of Montana. *Bull. Yale Peabody Museum Nat. History* **30,** 1–165.

Padian, K. (1985). The origins and aerodynamics of flight in extinct vertebrates. *Palaeontology* **28,** 423–438.

Padian, K., and Olsen, P. E. (1989). Ratite footprints and the stance and gait of Mesozoic theropods. In *Dinosaur Tracks and Traces* (D. D. Gillette and M. G. Lockley, Eds.), pp. 231–241. Cambridge Univ. Press, Cambridge, UK.

Thulborn, R. A. (1984). Preferred gaits of bipedal dinosaurs. *Alcheringa* **8,** 243–252.

Thulborn, R. A. (1989). *Dinosaur Tracks,* pp. 410. Chapman & Hall, London.

Thulborn, R. A., and Wade, M. (1984). Dinosaur trackways in the Queensland. *Mem. Queensland Museum* **21,** 413–517.

Weishampel, D. B. (1995). Fossil, function, and phylogeny. In *Functional Morphology in Vertebrate Paleontology* (J. J. Thomason, Ed.), pp. 34–54. Cambridge Univ. Press, Cambridge UK.

Witmer, L. M. (1995). The extant phylogenetic bracket and the importance of reconstructing soft tissues in fossils. In *Functional Morphology in Vertebrate Paleontology* (J. J. Thomason, Ed.), pp. 19–33. Cambridge Univ. Press, Cambridge, UK.

Garden Park Fossil Area, Cañon City, Colorado, USA

see MUSEUMS AND DISPLAYS

Gastralia

LEON CLAESSENS
Utrecht University
Utrecht, The Netherlands

Gastralia are dermal ossifications situated in the belly wall. They are present in Prosauropoda and Theropoda, and their presence has been suggested in the Sauropoda. Evidence for the existence of sauropod gastralia is not unequivocal, and it has been suggested that reported sauropod gastralia are in fact sternal ribs (Claessens, 1996). No gastralia are present in ornithischians.

The nomenclature pertaining to skeletal structures situated in the abdomen is in a considerable state of confusion. Endoskeletal cartilaginous or bone structures have often been confused with, or simply not distinguished from, the exoskeletal gastralia. Less desirable terms for the gastralia include abdominal ribs, parasternalia, and plastron. All of the gastralia together are often referred to as the abdominal cuirass or the abdominal basket.

The gastralia are of similar basic morphology in all prosauropods and theropods. Approximately 14–21, anteriorly pointing, V-shaped segments are present. Each segment generally consists of four rod-like, slender bones and a medial and a lateral gastralium on each side. In prosauropods, the lateral gastralia are most prominent, whereas in theropods the medial gastralia are generally the largest. The medial gastralia interlock on the ventral midline. They not only interlock transversely but also interlock longitudinally. Anteriorly or posteriorly, a fused, median, chevron-shaped segment may be present. Fused gastralia have sometimes been confused with furculae. However, gastralia and furculae are clearly distinguishable from one another. Many incorrect descriptions of dinosaur gastralia, based on pathologic bones, exist in the literature. A good description of theropod gastralia, which lacks only an understanding of how they interlock, is given by Lambe (1917).

In the lineage leading to dinosaurs, gastralia have experienced a reduction in the number of segments and the number of bones per segment. Ancient functions of the gastralia in protecting and supporting the belly lost their significance in dinosaurs. Perry (1983) suggested that the gastralia might have been passive respiratory structures, preventing a paradoxical shift of the viscera in the lung space during costal inspiration by mechanically stiffening the belly wall.

In contrast to the reduction in segments and number of bones per segment stands the development of an intricate interlocking mechanism and the regionalization and intensification of muscle or ligament scars in theropods. These developments argue for an active function of the gastralia. Muscle reconstruction points out that the gastralia can actively widen or narrow the theropod belly (Claessens, 1996). Possibly this is for a respiratory function. The gastralia may have functioned simply as an external analogue to the mammalian diaphragm. However, they may even have manipulated posteroventral diverticula of the lung.

See also the following related entries:

BIOMECHANICS • HISTOLOGY OF BONES AND TEETH • PHYSIOLOGY • SKELETAL STRUCTURES

References

Claessens, L. P. A. M. (1996). Dinosaur gastralia and their function in respiration. *J. Vertebr. Paleontol.* **16,** 28A.

Lambe, L. M. (1917). The Cretaceous carnivorous dinosaur *Gorgosaurus*. Geological Survey of Canada Memoir No. 100, pp. 1–84.

Perry, S. F. (1983). Reptilian lungs, functional anatomy and evolution. *Adv. Anat. Embryol. Cell Biol.* **79,** 81.

Gastroliths

PHILIP J. CURRIE
Royal Tyrrell Museum of Palaeontology
Drumheller, Alberta, Canada

Stones found within the digestive tract of living animals and within the abdominal cavity of fossil animals are often referred to as gastroliths. Generally, they are swallowed by some animals as a means of mechanically breaking down food. However, the presence of stones in the abdomen does not necessarily mean that they are gastroliths. For example, many swimming vertebrates swallow stones to increase their specific gravity and to lower their center of gravity (Currie, 1981). Gravel within the body cavities of the hadrosaur *Claosaurus* (Brown, 1907) and a tyrannosaur in the Paleontological Museum of the Mongolian Academy of Sciences was probably acquired postmortem during burial (see DIET).

To date, despite widespread rumors, consistent reports of stomach stones are only known for a few types of dinosaurs: prosauropods (Raath, 1974), sauropods, and psittacosaurids. In the case of sauropods, the ingested stones would have been held in a muscular gizzard somewhere along the digestive tract. Sauropods such as diplodocids and titanosaurids had thin, pencil-like teeth and only minimal capability of processing food in the mouth (Calvo, 1994). Leaves were probably stripped from branches by the comb-like teeth and swallowed without being chewed. These would have been ground up and broken down by the stones in the gizzard before passing further along the alimentary canal to be digested. One of the best examples of more than half a dozen reports of sauropod gastroliths was found in the abdominal region of the diplodocid sauropod *Seismosaurus*. The distribution of the stones was mapped and suggests that some were held in two pockets, one close to the front of the chest and the other in the region of the pelvis (Gillette, 1994). In all, more than 240 gastroliths were recovered from the excavation, the largest of which were approximately 10 cm across. The mechanical movement of the stones within the digestive tract and the chemical effects of the digestive acids give gastoliths a distinctive, high polish (Manley, 1991).

The presence of stones in the abdominal cavities of specimens of *Psittacosaurus* is more difficult to understand. Although not as highly adapted as in neoceratopsians, psittacosaurs nevertheless have relatively sophisticated dental batteries that must have done most of the food processing. The large number of psittacosaur specimens recovered from Lower Cretaceous lacustrine deposits of Asia suggests that these animals lived close to water. Perhaps the stones were swallowed primarily for ballast and had little to do with the digestive system. Overall, the only strong evidence suggesting that gastroliths were used to help break down food comes from diplododid (Gillette, 1994) and titanosaurid (Calvo, 1994) sauropods and from prosauropods (Raath, 1974).

See also the following related entries:

COPROLITES • DIET • PHYSIOLOGY • PLANTS AND DINOSAURS

References

Brown, B. (1907). Gastroliths. *Science* **25**(636), 392.

Calvo, J. O. (1994). Gastroliths in sauropod dinosaurs. *Gaia* **10,** 205–208.

Currie, P. J. (1981). *Hovasaurus boulei,* an aquatic eosuchian from the Upper Permian of Madagascar. *Palaeontol. Africana* **24,** 99–163.

Gillette, D. D. (1994). *Seismosaurus, the Earth Shaker*, pp. 205. Columbia Univ. Press, New York.

Manley, K. (1991). Two techniques for measuring surface polish as applied to gastroliths. *Ichnos* **1,** 313–316.

Raath, M. (1974). Further evidence of gastroliths in prosauropod dinosaurs. *Arnoldia* **7**(5), 1–5.

Genasauria

Genasauria was established by Sereno (1986) to represent the node linking THYREOPHORA + CERAPODA; that is, all ornithischians except *Lesothosaurus*. Several synapomorphies support this node, including the medial offset of the maxillary dentition (the inferred possession of "cheeks"), a spout-shaped mandibular symphysis, and other features. Presumably other fabrosaurids (whether Fabrosauridae is monophyletic is a matter of debate; see FABROSAURIDAE; PHYLOGENY OF DINOSAURS) would be excluded from Genasauria, as would *Pisanosaurus*, the earliest known ornithischian. Genasauria may be usefully redefined as a node-based taxon including the node-based taxa Thyreophora and Cerapoda and all descendants of their most recent common ancestor.

See also the following related entries:

FABROSAURIDAE • ORNITHISCHIA • PHYLOGENY OF DINOSAURS

Reference

Sereno, P. C. (1986). Phylogeny of the bird-hipped dinosaurs (Order Ornithischia). *Natl. Geogr. Res.* **2,** 234–256.

Genetics

J. MICHAEL PARRISH
Northern Illinois University
DeKalb, Illinois, USA

Despite the dramatic scenarios in Jurassic Park (Crichton, 1990), the concept of cloning dinosaurs from ancient DNA remains a fantasy and, barring some unforeseen technological breakthrough, seems a virtual impossibility. Nonetheless, some dramatic advances have occurred in our abilities to extract and amplify genetic material from ancient sources in the past several years.

The building blocks of the genetic code consist of nucleic acids. The double-stranded deoxyribonucleic acid (DNA), arranged in chromosomes, is the storage bank for genetic information in all living organisms.

The building blocks of DNA are individual nucleic acid residues that are linked by phosphodiester bonds to form long helical chains. Each nucleic acid link (nucleotide) is also attached to a complementary nucleotide on an opposing chain by hydrogen bonding, with the result that a double-stranded structure consisting of two helices of DNA is formed. Each set of three nucleotides codes for one of 21 amino acid residues, which are assembled through the activities of transcription and translation to form the proteins that perform the metabolic processes in cells.

The genetic code for a given organism is located in the cell nucleus, within massive, linear aggregations of DNA that are called chromosomes. Within a given species, there are as many combinations of genetic programs as there are individual organisms, each with its own unique complement of DNA sequences. During the process of sexual reproduction, genetic material from the two reproducing individuals is combined in the fertilized egg, resulting in the creation of a distinctive genetic program in the offspring made up of half of the genetic programs of each of the parents. During the process of duplication, various types of errors can occur in the reproduction of the genetic code, which produces new genetic sequences, or mutations. Most mutations are either harmful or harmless, but they are an important source of new genetic variability within a species.

The basis of the science of genetics lies in the expression of the genes on each half of the pairs of each chromosome that are represented in the organism. Gregor Mendel worked out the concept of dominance, which deals with the fact that most genes (sections on the chromosome that code for a particular feature of the organism) have two or more alternate states, or alleles, that differ in the extent to which they are expressed in the organism's appearance (phenotype). In the simplest case, a dominant allele will be expressed and a recessive allele suppressed when the two occur together. The recessive allele will be expressed only if two copies of it occur in the genome. Since Mendel's day, we have come to understand that the genetic expression of phenotype is a complex process involving multiple genes combining to produce some features and some genes controlling the expression of a number of different features. The key to understanding the genetics of a given organism still ultimately lies in how its genes are recorded as sequences of genetic code in the DNA of the chromosomes. Other than in birds, this information has been unavailable in dinosaurs because DNA was long thought to be extractable only from living or very recent tissue. In recent years, new investigations into the extraction of so-called ancient DNA have made some of the predictions in Jurassic Park closer to the realm of possibility.

The chief advance that made the detection of ancient DNA practical is the polymerase chain reaction (PCR), a technique that allows the rapid duplication of fragments of nucleic acids specified by primers containing the sequences at the start and end of the fragment to be analyzed (Pääbo *et al.*, 1989). PCR offers a rapid, easy method to generate multiple copies of what can be a very tiny fragment of nucleic acid. Once the multiple copies of the fragment are produced by PCR, they can be sequenced to determine how the sequence compares with those for the same region in related organisms. Before the advent of PCR, scientists tended to think that nucleic acids decayed as soon as their host organism died. However, attempts to amplify DNA samples from dried specimens of animals and plants showed that, under some conditions, nucleic acids are retained, in trace amounts, in the bodies of dead organisms.

DNA sequences were successfully amplified from Egyptian mummies, dried plant tissues, and the skins and bones of extinct organisms such as the Quagga (Pääbo, 1993). Still, most molecular biologists thought that DNA could, at best, be recovered from organisms a few thousand years old.

In 1990, Edward Golenberg and colleagues reported isolating DNA from a Miocene (17–20 my) Magnolia leaf, a finding that opened up the possibilities of finding DNA in fossil tissues. Since then, DNA has been reported in insects embedded in amber (Cano *et al.*, 1992; Desalle *et al.*, 1992) and from sabertoothed cats from the La Brea tar pits (Janszewski *et al.*, 1992).

In 1993, two groups, one at the Museum of the Rockies and the other at the California Polytechnic University, working on the problem of isolating DNA fragments from dinosaur bones had their projects popularized by an NSF press release prior to publication of results (Morell, 1993). These studies have continued, but no publications have yet come from these efforts. A recent paper (Woodward *et al.*, 1994) reported isolating DNA from a bone fragment of Cretaceous age, although the DNA sequence shows no clear affinities to any known groups of organisms.

The biggest challenge that faces those trying to establish that they have isolated ancient DNA lies in the difficulty of establishing that any DNA isolated from the sample indeed came from the fossil in question rather than from some kind of a contaminant. PCR will amplify any DNA sequence that matches the primers chosen, regardless of the organism of origin. Thus, trace amounts of bacterial, fungal, or human contaminants could readily be amplified from samples of fossilized tissues. In order to establish that DNA is from an organism of choice, its sequence has to be shown to closely match a similar sequence in a close modern relative. In the case of dinosaurs, DNA from birds and crocodilians are the molecules most appropriate for comparisons.

Even if DNA is extracted from dinosaur fossils, the likelihood is that it will be badly degraded and only preserved as short fragments. Thus, even if the technology existed, the notion of cloning dinosaurs from ancient DNA samples is well beyond our technological capabilities at this point. However, any DNA that could be isolated from dinosaurs might shed light on the questions of relationships among dinosaurs and between avian and nonavian dinosaurs and other groups such as crocodilians and lepidosaurs.

See also the following related entries:

Evolution • Jurassic Park

References

Cano, R. J., Poinar, H., and Poinar, G. O. (1993). Isolation and partial characterization of DNA from the bee *Proplebeia dominicana* (Apidae: Hymenoptera) in 25–40 million year old amber. *Med. Sci. Res.* **20,** 619–622.

Crichton, M. (1990). *Jurassic Park*, pp. 399. Knopf, New York.

Desalle, R., Gatesy, J., Wheeler, W., and Grimaldi, D. (1992). DNA sequences from a fossil termite in Oligo-Miocene amber and their phylogenetic implications. *Science* **257,** 1933–1936.

Golenberg, E. M., Giannasi, D. E., Clegg, M. T., Smiley, C. J., Durbin, M., Henderson, D., and Zurawski, G. (1990). Chloroplast DNA sequence from a Miocene Magnolia species. *Nature* **344,** 656–658.

Morell, V. (1993). Dino DNA: The hunt and the hype. *Science* **261,** 160–162.

Pääbo, S. (1993). Ancient DNA. *Sci. Am.* **269**(5), 86–92.

Pääbo, S., Higuchi, R., and Wilson, A. C. (1989). Ancient DNA and the polymerase chain reaction. *J. Biol. Chem.* **264,** 9709–9712.

Woodward, S. R., Weyland, N. J., and Bunnell, M. (1994). DNA sequence from Cretaceous Period bone fragments. *Science* **266,** 1229–1232.

Geological and Palaeontological Institute, Germany

see MUSEUMS AND DISPLAYS

Geologic Time

CARL C. SWISHER
Berkeley Geochronological Institute
Berkeley, California, USA

Dinosaurs first appeared in the fossil record in the early part of the Mesozoic Period approximately 225 million years ago, persisted for some 160 million years, and, except for birds, became extinct at the end of the Mesozoic, 65 million years ago. In comparison, mammals have been the dominant land vertebrates on the earth for only the past 65 million years or about one-third as long, whereas our own species has been around for only about 100,000 years, a small fraction of time that dinosaurs ruled. The amount of geologic time represented by the age of the Dinosaurs is difficult to comprehend. Although most of us are accustomed to speaking of tens and possibly thousands of years, when we speak of the difference between the ages of two different dinosaurs, for the most part, we speak of age differences in the millions of years.

Partly due to historical development and partly for ease in communication, ages of dinosaurs are usually given not as numerical ages but as belonging to a specifically named time or chronostratigraphic (*chrono* = time, *strata* = layer) unit. Names such as "Cretaceous Period" or "Campanian Age" are part of a hierarchical system that subdivides the geologic record of the earth into nested chronostratigraphic units. The names themselves reflect their historical development, most being derived from the names of geologic formations in Europe. During the 17th and 18th centuries, as geologists attempted to unravel complex stratigraphic relationships of economically valuable coal and mineral deposits, they turned their attention to the fossils found contained in the geologic strata. The nonrepetitive succession in the stratigraphic record of unique fossil aggregates was eventually accepted as a result of organic evolution. This provided geologists with a sequence of events that were unique in time and thus would permit the determination of the relationship of one event in geological history to another (Berry, 1968).

Although the subdivisions were originally descriptive rock units, they soon became interpretative units that were found to represent intervals of geologic time; hence the term chronostratigraphic. For the most part, the duration of the descriptive unit has been expanded to include other units that were found to contain fossils of similar appearance. The inherently vague boundaries between successive units have become precisely delimited with "golden spiked" boundary stratotypes. This history and transition from a descriptive rock unit to an interpretive time unit can be illustrated by the term Jurassic, the second period of the Mesozoic. The name Jurassic stems from geological observations made by Alexander von Humboldt during a geological excursion in southern France, western Switzerland, and northern Italy in 1795. Von Humboldt recognized that the Jura-Kalkstein, a massive formation in the Jura Mountains, was a distinctive rock unit of widespread geologic importance. Although the fossils from the Jura-Kalkstein were not described by Humboldt, it was precisely the fossils from the Jura-Kalkstein and their equivalents in England that provided the basis and testing ground for a new field of geologic study, that of using fossils to correlate and subdivide geologic time (Berry, 1968). A history in itself, the studies of primarily marine ammonite fossils from the Jura-Kalkstein and its correlatives by William Smith, Buckland, Conybeare, Phillips, von Buch, Brongniart, d'Orbigny, and Arkell, among many others, led to the foundations of the methods of biostratigraphy and biochronology and to the development of chronostratigraphic units such as the Jurassic, its name being derived from the original Jura-Kalkstein Formation (Berry, 1968; Harland *et al.*, 1990).

These studies eventually led to the recognition of 11 distinct subdivisions, or stages, for the Jurassic, based on fossil aggregates of ammonites. Finer divisions of the Mesozoic based on aggregates of ammonites were also proposed during this time by Oppel, who subdivided the Jurassic into a series of zones that were based on the overlapping stratigraphic ranges of fossil species. Oppel grouped his zones into a series of stages whose lowermost and uppermost stage boundaries coincided with the duration of the Jurassic (Berry, 1968). Following these principles, the Jurassic is now subdivided into 180 ammonite zones, which are in turn grouped to form the 11 stages (Gradstein *et al.*, 1995). Zones proved to be the building blocks and working units of biochronology; however, it is the stage that has proven most useful for global correlations. Zones appear to be greatly influenced by differences in latitude, bathymetry, and local paleoenvironmental conditions, as for example in comparing marine versus nonmarine environments. Similar subdivisions of the entire European geologic record led to a series of time stratigraphic units that serve as the basis by which we subdivide the entire Mesozoic today. The sequence of Mesozoic stages developed in Europe soon became known as the European Standard. The correlation of local zonations developed on rocks worldwide are correlated with the European Standard, which in turn provides a basis for the recognition and correlation of geologic time.

Today, we subdivide the Mesozoic into the TRIASSIC, JURASSIC, and CRETACEOUS PERIODS, which are in turn subdivided into 31 distinct stages or ages. Although these subdivisions separate Mesozoic time into discrete units, they tell us little about absolute time. One can address relative time using ammonite zonations and assuming constant evolutionary rates. Thus, the duration of each zone can be considered to reflect a similar amount of geologic time. However, the assumption of constant evolutionary rates is difficult to test by independent means. For the most part, true age information comes from the dating, by radioisotopic methods, of volcanic rocks found interbedded in the strata from which the fossil aggregate or chronostratigraphic unit is recognized. Dating methods such as K–Ar (potassium–argon) and U–Pb (uranium–lead), which are based on the radioactive decay of one isotope to another through time, are capable of resolving ages in the Mesozoic at best to approximately a tenth of a percent. Thus, for 200-million-year-old dinosaurs, the best age resolution we can expect at this point is plus or minus 200,000 years. Unfortunately, this uncertainty in age represents an immense amount of geologic time. For contrast, the amount of time represented by this uncertainty would encompass almost the entire evolutionary history of our own species.

Unfortunately, even this level of time resolution is rarely available in the Mesozoic. Dinosaur fossils that are associated with volcanic rocks suitable for radioisotopic dating are few and far between. An exception to this is the Late Cretaceous dinosaur faunas of the Western Interior of North America, where numerous bentonites or altered volcanic rocks occur that have yielded high-resolution radioisotopic dates. However, even here, most of the dated horizons occur in ammonite-bearing marine rocks. If we wish to apply these dates to a particular dinosaur, we must correlate the dated marine horizons through a series of biostratigraphic and paleomagnetic correlations into the nonmarine strata in which the dinosaur was discovered.

Except for the Late Cretaceous, the radioisotopic database available for the Mesozoic lacks the necessary precision to constrain the age of most stage boundaries. Dates for stage boundaries are derived primarily by interpolation methods. Currently, we use approximately 300 radioisotopic dates to calibrate the Mesozoic timescale. Unfortunately, different criteria in the selection of the radioisotopic dates, as well as different interpolation methods, have led to an array of Mesozoic timescales. The Cenozoic timescale has the unique ability to scale the magnetic polarity chrons according to geomagnetic anomaly profiles recorded on rocks from the Atlantic seafloor that in turn can be pinned to a selected set a radioisotopic dates. In contrast, the Mesozoic lacks a unifying interpolation concept because the geomagnetic anomaly profiles only extend back to the Late Jurassic (Callovian), whereas much of the Middle Cretaceous lacks a unique geomagnetic anomaly signature (for example, the interval from 125 to 84 Ma is predominantly of the same, normal geomagnetic polarity) (Gradstein *et al.*, 1995).

The current age calibration of the Mesozoic timescale is shown in Fig. 1; the precise numerical values

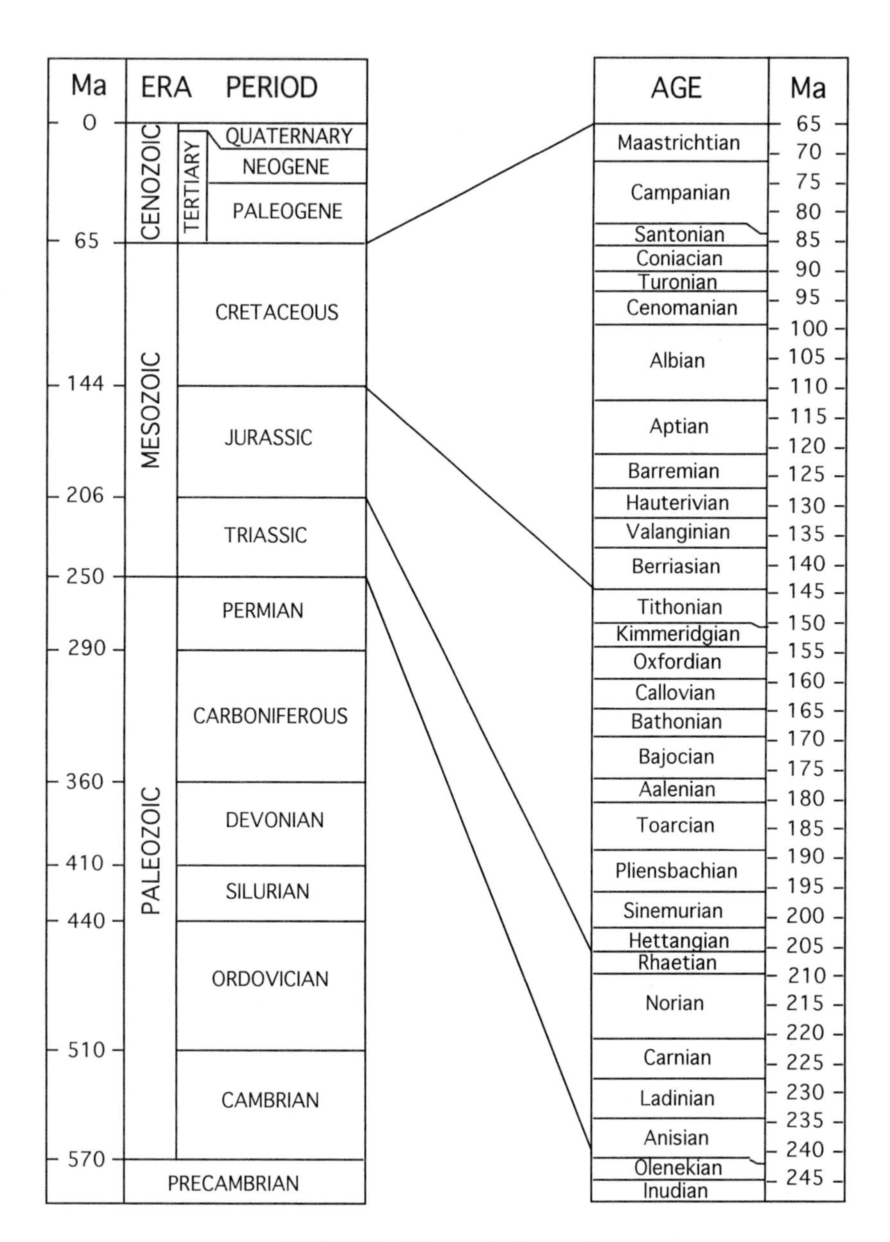

FIGURE 1 Mesozoic timescale.

can be found in Gradstein *et al.* (1995). The terminology and usage of the chronostratigraphic terms used here for the Mesozoic are found in Harland *et al.* (1990) and Gradstein *et al.* (1995). Often confusing to most of us is the dual terminology that accompanies most geologic timescales. Each subdivision of geologic time is given both as "time–rock units" and purely "time" units. Thus, in most timescales we commonly see the paired headings Era/Erathem, Period/System, Epoch/Series, Stage/Age, and even the relative terms early/lower and late/upper (Hedberg, 1976; Harland *et al.*, 1982). Recently, it has been argued that both of these headings are precisely equivalent and for the most part are chosen only in the context of a particular sentence. Thus, for example, we might speak of a fossil from the Upper Campanian JUDITH RIVER FORMATION being of Late Campanian Age. It is considered that these dual terms in geologic timescales are redundant and confusing. Thus, breaking from tradition, Harland *et al.* (1990) recommended that only the first term of each of the above couplets be used in chronostratigraphic scales.

Thus, Early Campanian Age, rather than Lower Campanian Stage is used. Late Campanian rocks or fossils would mean any rock or fossil that was formed or lived in Late Campanian time. Although this approach certainly simplifies terminology, it has not been followed by all workers and the reader should be aware of departures from this in some timescales. For example, the most recent timescale for the Mesozoic uses the term stage instead of age (Gradstein *et al.*, 1995).

The terminology and age calibration used in the Mesozoic timescale shown here (Fig. 1) is derived primarily from Gradstein *et al.* (1995). However, the usage of Harland *et al.* (1990) is followed for the Mesozoic and that of Berggren *et al.* (1995) for the Cenozoic, in the use of age rather than stage, although these two terms in the context of a timescale are considered equivalent. Here, the geologic timescale is divided into eras, periods, epochs , and ages, and associated relative subdivisions of early and late. Also noted here is the absence of epochs in the Mesozoic. For the Cenozoic, the well-known epochs, such as Paleocene, Eocene, Oligocene, Miocene, and Pliocene, are routinely used in discussing global correlations (Berggren *et al.*, 1996) but are not routinely used in the Mesozoic (Gradstein *et al.*, 1995). Epochs have been proposed for the Mesozoic and can be found in Harland *et al.* (1990). In many cases, these subdivisions are simply given letter / number designations, such as the K1 and K2, that subdivide the Cretaceous Period (Harland *et al.*, 1990).

For further details in the development and subdivisions of the Mesozoic and the geological timescale in general, the reader is referred to the references listed under References.

See also the following related entries:

BIOSTRATIGRAPHY • LAND MAMMAL AGES

References

Berggren, W. A., Kent. D. V., Swisher, C. C., III, and Aubry, M.-P. (1996). A revised Cenozoic geochronology and chronostratigraphy. In *Geochronology, Time Scales and Global Stratigraphic Correlation* (W. A. Berggren, D. V. Kent, M.-P. Aubry, and J. Hardenbol, Eds.), Spec. Publ. No. 54, pp. 129–212. SEPM.

Berry, W. B. N. (1968). *Growth of a Prehistoric Time Scale*, pp. 158. Freeman, San Francisco.

Gradstein, F. M., Agterberg, F. P., Ogg, J. G., Hardenbol, J., van Veen, P., Thierry, J., and Huang, Z. (1995). A Triassic, Jurassic, and Cretaceous time scale. In *Geochronology, Time Scales and Global Stratigraphic Correlation* (W. A. Berggren, D. V. Kent, M.-P. Aubry, and J. Hardenbol, Eds.), Spec. Publ. No. 54, pp. 95–126. SEPM.

Harland, W. B., Cox, A. V., Llewellyn, P. G., Pickton, C. A. G., Smith, A. G., and Walters, R. (1982). *A Geologic Time Scale*, pp. 263. Cambridge Univ. Press, Cambridge, UK.

Harland, W. B., Armstrong, R. L., Cox, A. V., Craig, L. E., Smith, A. G., and Smith, D. G. (1990). *A Geologic Time Scale 1989*, pp. 263. Cambridge Univ. Press, Cambridge, UK.

Hedberg, H. D. (1976). *International Stratigraphic Guide, A Guide to Stratigraphic Classification, Terminology, and Procedure*, pp. 200. Wiley, New York.

Geologische–Paläontologisches Institut, Switzerland

see MUSEUMS AND DISPLAYS

Geologisk Museum, Denmark

see MUSEUMS AND DISPLAYS

Geology Museum, Indian Statistical Institute, India

see MUSEUMS AND DISPLAYS

German East African Expedition

see TENDAGURU

Ghost Ranch

Clive Coy
Royal Tyrrell Museum of Palaeontology
Drumheller, Alberta, Canada

The Ghost Ranch *Coelophysis* quarry, named after a ranch located in the Chama River Valley, is located within the upper part of the Petrified Forest Member of the Upper Triassic Chinle Formation in northern New Mexico, United States. The first fossils of *Coelophysis* were collected in 1881 and described by Edward Drinker Cope in 1887 on the basis of fragmentary material. During the summer of 1947, Dr. Edwin Colbert led a field party from the American Museum of Natural History on what was intended to be a 2-week study of the sedimentology and fossil vertebrates of the Ghost Ranch area. This reconnaissance developed into one of the world's largest dinosaur excavations. Finds on the first day included a complete phytosaur skull *in situ*. On the morning of the next day, fragmentary bones of *Coelophysis* were discovered. Careful examination of the locality revealed the level the bones had weathered out from. It became apparent that this was an unusually rich deposit of bones, and further excavation exposed a remarkable accumulation of articulated skeletons, many of them complete, almost all of them *Coelophysis*. The evidence from the sedimentology and taphonomy of the quarry suggests that the *Coelophysis* carcasses were buried in a shallow stream channel that transported the remains a short distance before coming to rest. The relative completeness of the skeletons, and lack of damage to the bones, suggests that the bodies were not exposed before burial long enough for them to fully decay and disarticulate. The cause of death and burial history may be related, which suggests the animals may have been drowned by flood waters.

A considerable amount of overburden had to be removed to expose the initial quarry. The skeletons are deposited in a great profusion, interlaced among and on top of each other. In order to excavate a particularly good skeleton, it was often necessary to cut through another. The exacting task of collecting the skeletons was continued throughout that summer, 1948, 1949, 1951, and 1953. The quarry was reopened in the summer of 1981 by a joint crew from the Carnegie Museum, the New Mexico Museum of Natural History, the Museum of Northern Arizona, and the Peabody Museum of Yale University. In total, 28 blocks of *Coelophysis* material have been removed from the Ghost Ranch quarry and have gone to over a dozen institutions across North America. In 1989, Dr. Edwin Colbert published his long-awaited monograph on *Coelophysis*, a product of 42 years of dedicated research. In 1995, he published an account of the history of Ghost Ranch and the excavation of its quarry (see also H. L. Schwartz and D. D. Gillette, 1994).

See also the following related entries:

Ceratosauria • Chinle Formation • Triassic Period

References

Colbert, E. H. (1989). The Triassic dinosaur *Coelophysis*. Museum of Northern Arizona Bull. No. 57, pp. 1–160.

Colbert, E. H. (1995). *The Little Dinosaurs of Ghost Ranch*. New York: Columbia Univ. Press. 250 pp.

Schwartz, H. L., and Gillette, D. D. (1994). Geology and taphonomy of the *Coelophysis* Quarry, Ghost Ranch, New Mexico. *J. Paleontol.* **68,** 1118–1130.

Glen Canyon Group

Kevin Padian
University of California
Berkeley, California, USA

This set of Early Jurassic terrestrial formations outcrops in northern Arizona and parts of southeastern Utah and northwestern New Mexico. It extends over several hundred square kilometers and comprises the Wingate, Moenave, Kayenta, and Navajo formations (Fig. 1). The first two are predominantly reddish eolian sandstones, the Kayenta mainly alternates between a reddish "sandy facies" and a gray blue "silty facies," and the Navajo is composed of white and red sandstones that form a caprock over much of the cliffs and plateaus of the region. There are as yet no clear radiometric dates from the group, and most stratigraphic correlation has been based on vertebrate fossils, along with magnetostratigraphy and some pollen evidence (Clark and Fastovsky, 1986; Sues *et al.*, 1994).

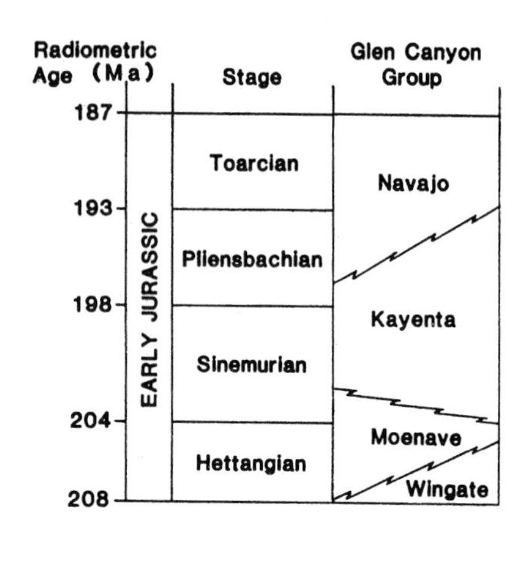

FIGURE 1 Stratigraphy and estimated age of the formations of the Glen Canyon Group, with a map of their exposures in Northern Arizona.

For some time, the Glen Canyon Group was considered of Late Triassic age because it is broadly coeval with the Newark Supergroup of eastern North America and also because several pieces of dermal armor from the Rock Head area had been attributed to aetosaurs, which are only known from the Late Triassic (see ARCHOSAURIA; EXTINCTION, TRIASSIC; PSEUDOSUCHIA). In the late 1970s, however, it became clear on the basis of footprints, pollen remains, and fishes that the basins of the Newark Supergroup contained deposits broadly spanning the Late Triassic and Early Jurassic, of which the latter were marked primarily by a loss of nondinosaurian, noncrocodilian archosaurian taxa. This forced a reconsideration of horizons worldwide that had been traditionally considered of Late Triassic age (Olsen and Galton, 1977, 1984). In the Glen Canyon Group, skeletal and footprint assemblages from the Moenave, Kayenta, and Navajo lacked typical Triassic marker taxa (the Wingate is relatively barren of fossils), and the scutes previously identified as aetosaurian were reidentified as being from *Scelidosaurus,* a thyreophoran dinosaur known from the southwest coast of England from deposits probably Sinemurian in age (Padian, 1989). Although it remains difficult to pin down absolute dates for specific strata within the Glen Canyon Group, it is now generally considered to be Early Jurassic in age, or almost entirely so (Fig. 1). It is still possible that the earliest strata encroach upon the Late Triassic, and that the uppermost Navajo Formation beds extend into the Middle Jurassic. All Glen Canyon strata lie above the J–0 unconformity supposed to separate Triassic and Jurassic rocks in the region, except the Rock Point Member of the Wingate Formation, which has been reassigned to the underlying Late Triassic Chinle Formation (Pipiringos and O'Sullivan, 1978).

Fossil vertebrates known from the Moenave Formation include the basal crocodilian *Protosuchus* and the crocodylomorph footprint *Batrachopus* (perhaps made locally by *Protosuchus,* although not diagnostic to the skeletal taxon), plus partial remains of fishes, turtles, and theropod dinosaurs (Olsen and Padian, 1986; Clark and Fastovsky, 1986; Sues *et al.*, 1994). From the Navajo Formation come the small ceratosaurian theropod dinosaur *Segisaurus,* various partial remains of basal sauropodomorph dinosaurs assigned to *Ammosaurus,* and a basal sauropodomorph trackway, *Navahopus,* as well as a partial specimen of a large tritylodontid cynodont. Remains of small crocodyliforms and of other footprints have also been reported (Clark and Fastovsky, 1986; Sues *et al.*, 1994).

The Kayenta Formation is by far the richest for vertebrate fossils, thanks primarily to explorations in the late 1970s and early 1980s led by Farish Jenkins and James M. Clark and indefatigable collection and processing of microvertebrate-bearing matrix by William R. Downs (Fig. 2). Currently known fossil verte-

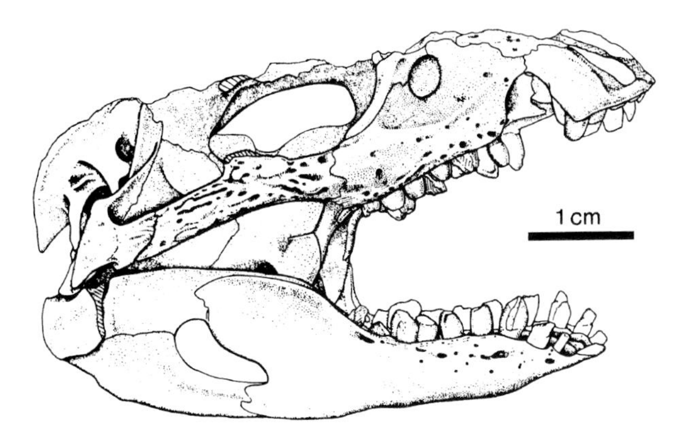

FIGURE 2 Partial skull and jaws of an unnamed *Edentosuchus*-like protosuchid crocodile in the University of California Museum of Paleontology, described by J. M. Clark (see Sues *et al.*, 1994).

brates include the ceratosaurian theropods *Dilophosaurus* and *Syntarsus,* numerous theropod footprints, the basal sauropodomorph *Massospondylus,* the basal thyreophorans *Scutellosaurus* and *Scelidosaurus,* and an undescribed heterodontosaurid; numerous skulls and skeletons of tritylodontid cynodont therapsids (including *Oligokyphus, Dinnebiton,* and *Kayentatherium),* the crocodilian *Eopneumatosuchus* (which may be the earliest known, terrestrial mesoeusuchian crocodilian), and several other new taxa (including a sphenosuchian crocodylomorph), plus numerous remains of turtles (*Kayentachelys*), sphenodontians, squamates, pterosaurs, and fishes (Clark and Fastovsky, 1986; Sues *et al.*, 1994). The vertebrate microfauna of the silty facies of the Kayenta, known from several localities in the "silty facies," includes hybodont sharks, frogs, caecilians (and perhaps salamanders), sphenodontians, protosuchid crocodyliforms, and other archosaurian and various remains not easily diagnosed (K. M. Curtis, 1989). Most remarkable, perhaps, are the recent discoveries of the earliest known morganucodont mammaliamorphs *(Dinnetherium)* and frogs on the continent and the earliest known caecilian amphibians anywhere (Sues *et al.*, 1994). In the space of a decade, the Kayenta Formation vertebrate fauna has become the most diverse that we have from the Early Jurassic in the world.

See also the following related entries:

Jurassic Period • Newark Supergroup

References

Clark, J. M., and Fastovsky, D. E. (1986). Vertebrate biostratigraphy of the Glen Canyon Group in northern Arizona. In *The Beginning of the Age of Dinosaurs: Faunal Change across the Triassic–Jurassic Boundary* (K. Padian, Ed.), pp. 285–301. Cambridge Univ. Press, New York.

Curtis, K. M. (1989). Unpublished M.Sc. thesis, Univ. of Calif.

Olsen, P. E., and Galton, P. M. (1977). Triassic–Jurassic tetrapod extinctions: Are they real? *Science* **197,** 983–986.

Olsen, P. E., and Galton, P. M. (1984). A review of the reptile and amphibian assemblages from the Stormberg of southern Africa, with special emphasis on the footprints and the age of the Stormberg. *Palaeontol. Africana* **25,** 87–110.

Olsen, P. E., and Padian, K. (1986). Earliest records of *Batrachopus* from the southwestern United States, and a revision of some Early Mesozoic crocodylomorph ichnogenera. In *The Beginning of the Age of Dinosaurs: Faunal Change across the Triassic–Jurassic Boundary* (K. Padian, Ed.), pp. 259–273. Cambridge Univ. Press, New York.

Padian, K. (1989). Presence of the dinosaur *Scelidosaurus* indicates Jurassic age for the Kayenta Formation (Glen Canyon Group, northern Arizona). *Geology* **17,** 438–441.

Pipiringos, G. N., and O'Sullivan, R. B. (1978). Principal unconformities in Triassic and Jurassic rocks, Western Interior, United States—A preliminary report. U. S. Geological Survey Profession Paper No. 1035-A, pp. 1–29.

Sues, H.-D., Clark, J. M., and Jenkins, F. A., Jr. (1994). A review of the Early Jurassic tetrapods from the Glen Canyon Group of the American Southwest. In *In the Shadow of the Dinosaurs: Early Mesozoic Tetrapods* (N. C. Fraser and H.-D. Sues, Eds.), pp. 284–294. Cambridge Univ. Press, New York.

Glen Canyon National Recreation Area, Arizona, USA

see Museums and Displays

Glen Rose Formation, Texas, USA

see Footprints and Trackways; Glen Rose, Texas

Glen Rose, Texas

Dale A. Winkler
Shuler Museum
Southern Methodist University
Dallas, Texas, USA

Near Glen Rose, Texas, well-preserved dinosaur tracks occur along the Paluxy River in the Early Cretaceous (Albian) Glen Rose Limestone. R. T. Bird undertook celebrated excavations of these tracks in the 1940s for the American Museum of Natural History (Bird, 1985). Tracks from the area that became Dinosaur Valley State Park (Farlow, 1987, 1993) were also distributed to other institutions including the University of Texas at Austin. Similar tracksites are abundant in central Texas. Tridactyl prints mostly attributed to large theropods are common (Shuler, 1917), but the most famous tracks are those of large sauropods. The tracks were made in soft but cohesive carbonate muds in shallow bays and estuaries along the ancient sea coast and were often filled with clastic sediments (sand and mud). Large three-toed prints are thought to have been made by the high-spined theropod *Acrocanthosaurus* and were named *Eubrontes? glenrosensis* by Shuler (1917) but referred to *Irenesauripus* by Langston (1974). Some blunt-toed tridactyl prints may have been made by the ornithopod *Iguanodon*. Sauropod prints were probably made by *Pleurocoelus* and these have been given the name *Brontopodus birdi* (Farlow *et al.*, 1989).

Tracks from near Glen Rose, and others in the Cretaceous of Texas, have been instrumental in developing new ideas about dinosaur behavior, including possible attacks by large theropods on the sauropods and herding behavior in sauropods.

See also the following related entries:

Behavior • Footprints and Trackways

References

Bird, R. T. (1985). *Bones for Barnum Brown* (V. T. Schreiber, Ed.), pp. 225. Texas Christian Univ. Press, Fort Worth, TX.

Farlow, J. O. (1987). *A guide to Lower Cretaceous dinosaur footprints and tracksites of the Paluxy River Valley, Somervell County, Texas.* South-Central Section, Geological Society of America, Field Trip Guidebook, Baylor University, Waco, TX.

Farlow, J. O. (1993). *The Dinosaurs of Dinosaur Valley State Park*, pp. 29. Texas Parks and Wildlife Press, Austin, TX.

Farlow, J. O., Pittman, J. G., and Hawthorne, J. M. (1989). *Brontopodus birdi*, Lower Cretaceous sauropod footprints from the U. S. Gulf Coastal Plain. In *Dinosaur Tracks and Traces* (D. D. Gillette and M. G. Lockley, Eds.), pp. 371–394. Cambridge Univ. Press, Cambridge, UK.

Hastings, R. J. (1987). New observations on Paluxy tracks confirm their dinosaurian origin. *J. Geol. Education* **35,** 4–15.

Kuban, G. J. (1989). Elongate dinosaur tracks. In *Dinosaur Tracks and Traces* (D. D. Gillette and M. G. Lockley, Eds.), pp. 57–72. Cambridge Univ. Press, Cambridge, UK.

Langston, W., Jr. (1974). Nonmammalian Comanchean tetrapods. *Geosci. Man* **8,** 77–102.

Shuler, E. W. (1917). Dinosaur tracks in the Glen Rose Limestone near Glen Rose, Texas. *Am. J. Sci.* **44,** 294–298.

Graduate Studies

Philip J. Currie
Royal Tyrrell Museum of Palaeontology
Drumheller, Alberta, Canada

Professional paleontologists normally only obtain full-time positions in museums and universities after they complete their graduate studies. It is possible to approach the study of vertebrate paleontology, including dinosaurs, by specializing in either biology or geology, but the selection of an appropriate university is often determined by the choice of a thesis supervisor. If an undergraduate student has a specific interest in dinosaurs, he or she should contact the experts in those fields before deciding to which graduate school to apply. Currently, the highest concentration of graduate schools that include research on dinosaurs are found in the United States [i.e., Brigham Young University, Columbia University, Harvard, Montana State University, Southern Methodist University, State University of New York at Stony Brook, Yale, and the universities of California (Berkeley), Chicago, and Pennsylvania], Europe (Cambridge, Oxford, Université Pierre et Marie Curie, Universität Bonn, and University of Bristol), and Canada (McGill University and the universities of Alberta, Calgary, and Toronto). Graduation with a doctoral degree is no guarantee that a student will obtain a job directly related to dinosaur paleontology, but the opportunities are greater today than ever before.

Group Behavior

On the basis of trackway information, circumstance of preservation, and other associations, some dinosaurs are thought to exhibit group behaviors such as herding and mass migration.

see BEHAVIOR; FOOTPRINTS AND TRACKWAYS; MIGRATION

Parallel trackways, like these from Jurassic sediments of the Morrison Formation, suggest group behavior among dinosaurs. Photo by François Gohier.

Growth and Embryology

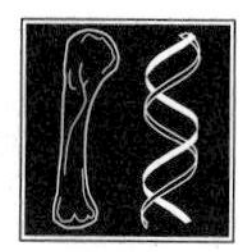

DAVID J. VARRICCHIO
Old Trail Museum
Choteau, Montana, USA

Growth

Growth involves not only increasing size but also important changes in bone shape and proportion (allometric growth), development of muscle attachment and joint surfaces, fusion of skeletal elements, and macro- and microscopic changes in bone tissue. Partial growth series exist for a variety of dinosaurs including *Coelophysis*, *Syntarsus*, *Allosaurus*, *Dryosaurus*, *Hypacrosaurus*, *Maiasaura*, *Stegosaurus*, *Psittacosaurus*, *Protoceratops*, and *Chasmosaurus*. Many dinosaurian growth trends (Table I) are ontogenetic, such as increased tooth counts, larger areas for muscle and ligament attachment, and more robust limb proportions, and likely reflect necessary shifts in the feeding and locomotor behavior of an individual as its body size increases. Nevertheless, some trends clearly vary among dinosaur species. Theropods, like crocodilians and unlike ornithischians, show little or no change in tooth counts (Madsen, 1976; Colbert, 1989). Due to changes in bone construction, the limb bones of some taxa become less robust with age.

Juvenile dinosaur skulls typically have short snouts, large orbits, and large brain cavities. During growth the snout elongates and the orbits and braincase become relatively smaller. Similar growth patterns occur in extant archosaurs (birds and crocodilians) and possibly result from the limited growth potential of the brain and eyes. Short, "cute" faces of juvenile dinosaurs may also have stimulated parenting behavior (see BEHAVIOR). Documentation of such ontogenetic trends can help evaluate the significance of characters used in phylogenetic analysis. For example, based on growth in *Dryosaurus*, Carpenter (1994) ascribed the differences between *Yandusaurus multidens* and *Y. hongheensis* to ontogeny.

Drastic changes in the ornithischian skull also result from the growth of crests, horns, and frills. Juveniles lack such features, and those of different species appear indistinguishable. Some named taxa, for example *Brachyceratops*, *Tetragonosaurus*, and *Procheneosaurus*, represent undiagnostic juveniles (Nopsca, 1933; Dodson, 1976; Sampson, 1993). Late development and possible dimorphism of these cranial ornaments suggests that they are important for sexual/social display and the products of sexual selection (Nopsca, 1933; Dodson, 1975a, 1976; Lehman, 1990; Sampson, 1993).

Dinosaur bones do not possess secondary centers of ossification, and most species lack the extensive fusion of bony elements seen in birds. Presumably, growth in most dinosaur species was indeterminate, continuing throughout the animal's life, and resulted in a broad range of adult sizes for a given species. Limb proportions may provide clues to adult size. In modern birds, adult limbs of small species differ from juvenile limbs of large species in both their proportions and their growth (Callison and Quimby, 1984). Testing whether this holds for dinosaur species will likely require a sample of independently verified adult dinosaur bones. With individual and sexual variation, size likely only approximates skeletal maturation (see Brinkman, 1988).

Skeletal fusion has also been used as an indicator of maturation. Elements that commonly fuse during ontogeny in dinosaurs include braincase elements, neural arches to vertebral centra, sacral vertebrae to sacral ribs, scapulae to coracoids, and pelvic elements. Some theropods show additional bird-like fusion of the clavicles and in their hands, legs, and feet. Recently, Brochu (1992) found in crocodilians that closure of neurocentral sutures proceeds in a caudal to cranial sequence. Sexual maturation probably occurs before closure of sutures on sacral vertebrae, whereas morphological maturation coincides with the closure of sutures on cervicals. No studies have yet addressed specifically the relationship between fusion and maturity for dinosaurs.

During growth, changes take place at the microscopic level within bone tissues. Generally, bone shows increasing organization and decreasing vascularity with age. For birds, mammals, and dinosaurs,

TABLE I
Some Ontogenetic Trends of Dinosaurs

A. Orbits diminish in relative size: *Coelophysis, Dryosaurus, Hypacrosaurus, Psittacosaurus, Protoceratops*

B. Frontals diminish in relative size: *Hypacrosaurus, Psittacosaurus, Protoceratops*

C. Tooth counts increase: Nodosaurs, hadrosaurs, *Psittacosaurus, Protoceratops, Chasmosaurus*

D. Face elongates: *Dryosaurus, Hypacrosaurus, Psittacosaurus, Protoceratops, Chasmosaurus*

E. Skull shortens and deepens: *Archaeopteryx*

F. Cranial crests develop: Lambeosaurs

G. Horns and frills develop: *Protoceratops, Chasmosaurus,* centrosaurines

H. Curved skull roof flattens: *Dryosaurus, Psittacosaurus*

I. Sagittal crest and jugal flange enlarge: *Psittacosaurus*

J. Lower jaw deepens and curves: *Protoceratops*

K. Neck elongates: *Coelophysis, Archaeopteryx*

L. Scapula becomes stouter: *Stegosaurus, Tyrannosaurus*

M. Humerus becomes more robust and delto-pectoral crest enlarges: *Protoceratops, Syntarsus, Troodon*

N. Olecranon process of ulna enlarges: *Stegosaurus, Syntarsus*

O. Femur head enlarges: *Stegosaurus*

P. Distal condyles of femur rotate medially: *Chasmosaurus*

Q. Lesser trochanter of femur lost: Nodosaurs, *Stegosaurus*

R. Trochanters and muscle insertion scars of limb bones become more pronounced and rugose: *Syntarsus, Stegosaurus, Chasmosaurus*

S. Limb elements become slightly less robust or do not change: *Archaeopteryx, Hypacrosaurus, Apatosaurus*

T. Limb elements become more robust: *Saurornithoides,* tyrannosaurids

U. Neural canal diameter diminishes in relative size: Hadrosaurs, *Camptosaurus*

Note. Data from Brown and Schlaikjer (1940), A–D, G, M; Sternberg (1955), C; Russell (1970), L. T.; Dodson (1975a, 1976), F, G; Coombs (1982), A, B, D, H, I; Galton (1982), L, N, O, R; Colbert (1989), A, K; Houck *et al.* (1990), E, K, S; Lehman (1990), C, G, P, R; Raath (1990), M, N, R; Currie and Peng (1993), T; Dong and Currie (1993), C, D, J; Sampson (1993), G; Carpenter (1994), A, D, H; Carpenter and McIntosh (1994), S; Chure *et al.* (1994), U; Horner and Currie (1994), A–D, F, S, U; and Jacobs *et al.* (1994), C, Q.

the histologic differences between juvenile and adult bone can be visible to the naked eye. Juvenile bone tends to have a rougher, more porous and possibly striated surface. Both microscopic (Nopsca, 1933; Chinsamy, 1990; Varricchio, 1993) and macroscopic (Ostrom, 1978; Callison and Quimby, 1984; Sampson, 1993) examinations of bone have proven valuable in determining the relative maturity and taxonomic significance of dinosaur specimens.

Growth Rates

Dinosaur growth rates have been estimated based on data for extant animals. Case (1978) assumed dinosaur ectothermy and predicted that *Protoceratops* and *Hypselosaurus* required approximately 20 and 62 years, respectively, to reach maturity. An endothermic assumption reduces these estimates by perhaps as much as a factor of 10. Dunham *et al.* (1989), assuming a reptilian resting metabolic rate, modeled *Maiasaura* as taking at least 5 and probably 10–12 years to reach maturity.

Bone microstructure reflects bone deposition rates and provides at least a qualitative record of growth speed (Ricqlès, 1980). Because of allometry and mechanical constraints, skeletal parts grow at different rates. Bone tissue then varies throughout a skeleton, within a single bone and with age. Histologic comparisons must incorporate these variations.

Dinosaurs commonly exhibit both fibrolamellar and zonal bone (Ricqlès, 1980; Reid, 1984, 1990). Though typical of mammals and birds, fibrolamellar bone also occurs in some fast-growing reptiles. The tissue represents rapid and continuous bone deposition. Zonal bone results from intermittent growth in which lines of arrested growth (LAGs) form "growth rings" separating periods of bone deposition at usually slow to moderate rates. Zonal bone is typical of, but not exclusive to, extant ectotherms. In extant nonavian diapsids, including crocodilians, these tissues represent annual fluctuations in growth and allow the aging of individuals. Possibly diapsids (or all tetrapods?) retained cyclic growth as a primitive feature. Furthermore, the cyclic growth of crocodilians is endogenous (Hutton, 1986) and some Mesozoic birds may also show LAGs (Chinsamy *et al.*, 1994). Zonal growth, if widespread among dinosaurs, would conservatively be interpreted as primitive and annual, but dinosaur outgroups must also be sampled

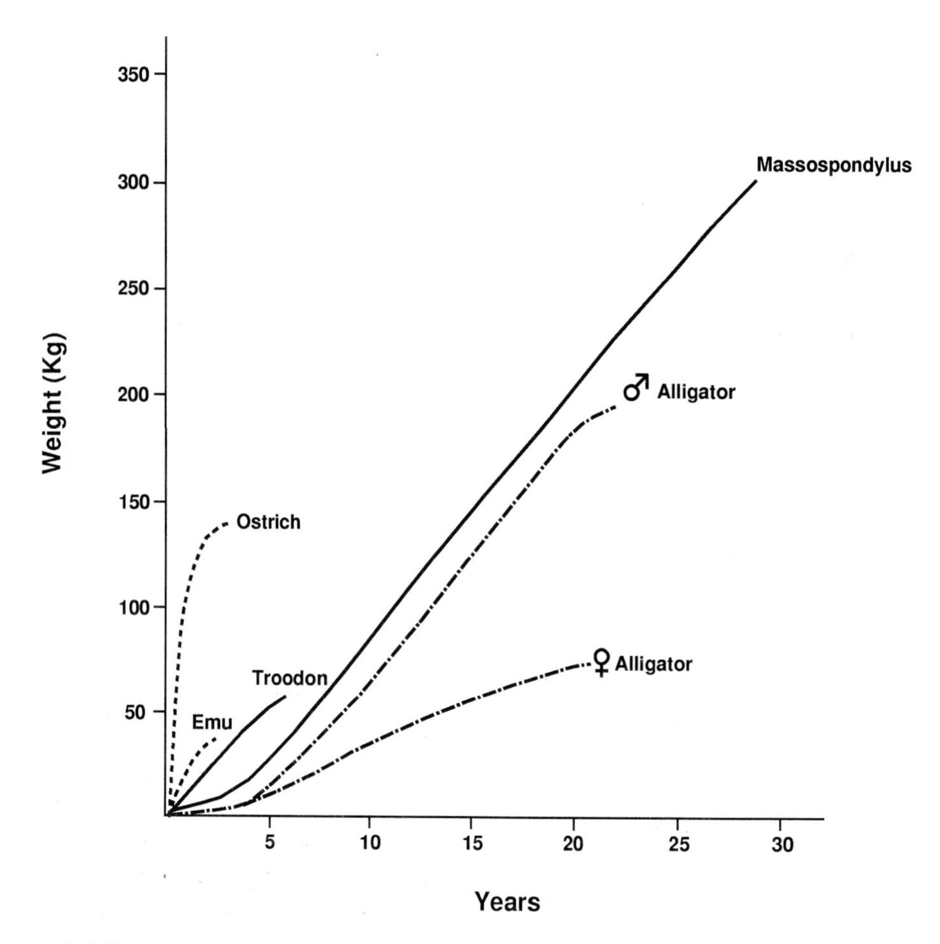

FIGURE 1 Growth curves for ostrich (Degen *et al.*, 1991), emu (Long, 1965), female and male American alligator (Chabrek and Joanen, 1979), *Massospondylus* (Chinsamy, 1993a), and *Troodon* (Varricchio, 1993). Adult dinosaur weights based on equations in Anderson *et al.* (1985). Dinosaur growth curves conjectural. Note that for all vertebrates, growth rates increase with increasing body size. A large reptile may grow as fast as a small bird or mammal. Fair, direct comparisons are those between animals of approximately equivalent adult size.

extensively. Only more histologic data from both fossil and extant vertebrates will allow testing of this.

Estimates of dinosaur growth, assuming the LAGs associated with zonal bone as annual, include the following: *Bothriospondylus*, 43 years to reach half size (Ricqlès, 1983); *Massospondylus*, 29 years to reach 300 kg (Chinsamy, 1993a); *Syntarsus*, 7 years to reach 20 kg (Chinsamy, 1990); and *Troodon*, 5 years to reach >50 kg (Varricchio, 1993). These rates exceed those of modern crocodilians but do not match the fast rates of birds (Fig. 1). *Troodon* grew as fast as some slowly maturing mammals of similar size (e. g., marsupials and primates). Despite a drastic difference in bone vascularity (Chinsamy, 1993b), *Massospondylus* and crocodilians may have had similar growth rates. This reflects either physiological differences or perhaps simply the negative allometry and slow relative growth of crocodilian limbs (Dodson, 1975b).

Zonal growth of dinosaurs could reflect nonannual cycles, environmental stress, egg-laying, or other unknown factors. The histology of many modern vertebrates remains undescribed, and independent testing of the dinosaur LAGs as annual is needed. Furthermore, given that the previously mentioned studies examined only a few skeletal elements, these growth estimates should be viewed as tentative. Nevertheless, a thorough histologic study of the small hypsilophodont *Orodromeus* also suggests only moderate growth rates (Ricqlès and Horner, 1997). In contrast, fibrolamellar tissue is common among many dinosaurs (Ricqlès, 1980; Reid, 1984). The bird-like growth plates of nestling *Maiasaura* likely allowed for rapid long bone growth (Barreto *et al.*, 1993) and support the idea that these altricial young grew quickly (Horner and Weishampel, 1988). Taphonomic evidence suggests persistent fast growth through at least the first year of life.

In at least three small theropods (*Syntarsus*, *Saurornitholestes*, and *Troodon*) poorly vascularized tissue with closely spaced LAGs marks the bone periphery. Late-stage growth in these forms was slow to determinate (Chinsamy, 1990; Reid, 1993; Varricchio, 1993).

Erickson (1992) demonstrated daily growth lines in modern archosaur teeth. Finding identical structures in dinosaurs, he estimated tooth growth and replacement rates. Hadrosaur teeth take approximately 200 days to develop and their tooth families (columns of functioning and developing teeth) shed a tooth every 60–90 days. In contrast, adult *Tyrannosaurus rex* shed a tooth every 2 or 3 years from a given socket.

Mass assemblages of trackways or bones can provide census-like data and, potentially, information on growth (Thulborn, 1990; Varricchio and Horner, 1993; Lockley, 1994). This requires a good taphonomic understanding of the locality, reconstruction of the population, and an age estimate of the cohorts present. Necessary assumptions include the following: tracks/bones represent one species, correlation between track/bone size and age, seasonally limited reproduction, and the assemblage samples the population accurately without preservational or behavioral biases. Trackway assemblages may have several advantages over bone beds (Lockley, 1994) because data are more easily obtained, trackways represent living populations (not death assemblages), behavior can be incorporated into interpretations, and trackway data are less likely to be time averaged. Although problematic, dinosaur taphonomic data, when combined with other sources (e.g., modeling or bone histology), may ultimately provide important growth information. Mammalian assemblages have already yielded this kind of data (Klein and Cruz-Uribe, 1984). Among dinosaurs, Thulborn (1990) recognized three size classes within the ornithopod ichnogenus *Wintonopus* at the Lark Quarry site. Insufficient data on the trackmaker prohibited testing if these size classes represent annual age classes. Size frequency data from bone beds suggest that juvenile *Maiasaura* grew as fast as large ungulates and reached 3 m in length, approximately 180 kg, in 1 year (Varricchio and Horner, 1993).

Embryology

Identifiable embryos found within eggs represent *Troodon*, *Hypacrosaurus*, and an oviraptorid (J. Horner, personal communication; Horner and Currie, 1994; Norell *et al.*, 1994). Additional eggs from the South African Triassic, Utah Jurassic, and the Mongolian and Montanan Cretaceous contain unrecognizable embryonic remains. Other specimens identified as embryonic on the basis of their association with eggshell, small size, bone texture, and degree of ossification include *Protoceratops*, *Bagaceratops*, *Breviceratops*, an unidentified Mongolian hadrosaur, *Maiasaura*, *Camptosaurus*, *Dryosaurus*, *Camarasaurus*, *Mussaurus*, therizinosaurs, and a dromaeosaurid (*Velociraptor*?). For most of these animals, embryonic remains consist of only one or a few specimens, providing only a limited view of dinosaur embryonic development.

Study of dinosaur nests and eggs can define the physiological environment of growth before hatching. For example, the nest material, egg placement, size, and abundance and distribution of eggshell pores reflect nest humidity, O_2 and CO_2 levels and the water–vapor exchange, and respiration rates of the embryos (Seymour, 1979; Grigorescu *et al.*, 1994).

Dinosaurs likely shared those developmental patterns found in both crocodilians and birds. Extant archosaurs have a cleidoic egg with an amnion, chorion, and allantois; macrolecithal zygote with cleavage completed while still in the female genital tract; disk-like early embryo (blastodisc); gastrulation through a primitive streak, though crocodilians also retain a blastopore; and similar development of the main embryonic tissues (neural tube, notochord, somites, and lateral plate), embryonic circulation, limb buds, and neural crest derivatives (Ferguson, 1985; Bellairs, 1991). Both crocodilian and bird embryos obtain much of their calcium from the eggshell, and interior resorption pits mark at least some dinosaur eggshell (Hirsch and Quinn, 1990). Late-stage embryos of modern archosaurs bear an epidermal caruncle or egg "tooth" made of keratin and lie with their head tucked down by their hindlimbs and tail (in crocodilians). *Troodon* and oviraptorid embryos curl in a similar fashion.

Crocodilian and avian development differ in several important features, including palate formation, stage at time of laying (crocodilians in mid-neurulation and birds in late blastula stage), the absence or presence of egg rotation during incubation, and temperature versus genetic sex determination. Furthermore, eggs of both crocodilians and megapode birds lack air cells. Their young switch quickly from embryonic to pulmonary respiration at hatching. Eggs of all other birds retain air cells and young make a slower transition to air breathing before hatching (O'Connor, 1984).

Ossification in birds usually begins about 36–61% of the way through incubation and follows a similar pattern. Limb elements begin to ossify first, followed by dermal bones (including furcula, pterygoid, and angular), girdles, ribs, remaining dermal bones, some endochondral skull elements (including ceratobranchials, exoccipitals, quadrate, basisphenoid, and basioccipital), vertebral arches and centra from anterior to posterior, remaining skull parts, carpals, and sternum and sternal ribs (Stark, 1989). Potentially, the degree of ossification could be used to distinguish the developmental stage of dinosaur embryos. Unfortunately, embryonic dinosaur bone (even late in development) remains thoroughly spongy (see Sochava, 1972). Not surprisingly, nearly all recognized dinosaur remains come from late-stage embryos.

The degree of ossification at hatching varies in modern archosaurs and correlates with behavior and growth (O'Connor, 1984). Poorly ossified altricial chicks remain nest-bound, rely heavily on parental care, and grow rapidly. More completely ossified precocial young of birds and crocodiles can move and feed on their own shortly after hatching, require less parental care, and grow more slowly. The degree of ossification observed in *Orodromeus* and oviraptorid embryos suggests that these dinosaurs hatched precocially. In contrast, bone development in *Maiasaura*, *Hypacrosaurus*, and possibly *Camptosaurus* may indicate altriciality (Horner and Weishampel, 1988; Horner and Currie, 1994; Chure *et al.*, 1994), although this has been questioned by some (Geist and Jones, 1996).

See also the following related entries:

Behavior • Eggs, Eggshells, and Nests • Heterochrony • Histology of Bones and Teeth • Physiology • Von Ebner Incremental Growth Lines

References

Barreto, C., Albrecht, R. M., Bjorling, D. E., Horner, J. R., and Wilsman, N. J. (1993). Evidence of the growth plate and the growth of long bones in juvenile dinosaurs. *Science* **262,** 2020–2023.

Bellairs, R. (1991). Overview of early stages of avian and reptilian development. In *Egg Incubation: Its Effects on Embryonic Development in Birds and Reptiles* (D. C. Deeming and M. W. J. Ferguson, Eds.), pp. 371–383. Cambridge Univ. Press, Cambridge, UK.

Brinkman, D. (1988). Size-independent criteria for estimating relative age in *Ophiacodon* and *Dimetrodon* (Reptilia, Pelycosauria) from the Admiral and Lower Belle Plains Formations of West-central Texas. *J. Vertebr. Paleontol.* **8,** 172–180.

Brochu, C. A. (1992). Ontogeny of the postcranium in crocodylomorph archosaurs. Unpublished M. A. thesis, University of Texas at Austin.

Callison, G., and Quimby, H. M. (1984). Tiny dinosaurs: Are they fully grown? *J. Vertebr. Paleontol.* **3,** 200–209.

Carpenter, K. (1994). Baby *Dryosaurus* from the Upper Jurassic Morrison Formation of Dinosaur National Monument. In *Dinosaur Eggs and Babies* (K. Carpenter, K. F. Hirsch, and J. R. Horner, Eds.), pp. 288–297. Cambridge Univ. Press, Cambridge, UK.

Case, T. J. (1978). Speculations on the growth rate and reproduction of some dinosaurs. *Paleobiology* **4,** 320–328.

Chinsamy, A. (1990). Physiological implications of the bone histology of *Syntarsus rhodesiensis* (Saurischia: Theropoda). *Palaeontol. Africana* **27,** 77–82.

Chinsamy, A. (1993a). Bone histology and growth trajectory of the prosauropod dinosaur *Massospondylus carinatus* Owen. *Mod. Geol.* **18,** 319–329.

Chinsamy, A. (1993b). Image analysis and the physiological implications of the vascularisation of femora in archosaurs. *Mod. Geol.* **19,** 101–108.

Chinsamy, A., Chiappe, L. M., and Dodson, P. (1994). Growth rings in Mesozoic birds. *Nature* **368,** 196–197.

Chure, D., Turner, C., and Peterson, F. (1994). An embryo of *Camptosaurus* from the Morrison Formation (Jurassic, Middle Tithonian) in Dinosaur National Monument, Utah. In *Dinosaur Eggs and Babies* (K. Carpenter, K. F. Hirsch, and J. R. Horner, Eds.), pp. 298–311. Cambridge Univ. Press, Cambridge, UK.

Colbert, E. H. (1989). The Triassic dinosaur *Coelophysis*. *Bull. Museum Northern Arizona* **57,** 1–174.

Dodson, P. (1975a). Taxonomic implications of relative growth in lambeosaurine hadrosaurs. *Systematic Zool.* **24,** 37–54.

Dodson, P. (1975b). Functional and ecological significance of relative growth in *Alligator*. *J. Zool. London* **175,** 315–355.

Dodson, P. (1976). Quantitative aspects of relative growth and sexual dimorphism in *Protoceratops. J. Paleontol.* **50,** 929–940.

Dunham, A. E., Overall, K. L., Porter, W. P., and Forster, C. A. (1989). Implications of ecological energetics and biophysical and developmental constraints for life-history variation in dinosaurs. In *Paleobiology of the Dinosaurs* (J. O. Farlow, Ed.), Spec. Paper No. 238, pp. 1–20. Geological Society of America.

Erickson, G. M. (1992). Verification of von Ebner incremental lines in extant and fossil archosaur dentine and tooth replacement rate assessment using counts of von Ebner lines. Unpublished M. Sc. thesis, Montana State University, Bozeman.

Ferguson, M. W. J. (1985). Reproductive biology and embryology of the crocodilians. *Biol. Reptilia* **14,** 329–491.

Galton, P. M. (1982). Juveniles of the stegosaurian dinosaur *Stegosaurus* from the Upper Jurassic of North America. *J. Vertebr. Paleontol.* **2,** 47–62.

Geist, N. R., and Jones, T. D. (1996). Juvenile skeletal structure and the reproductive habits of dinosaurs. *Science* **272,** 712–714.

Grigorescu, D., Weishampel, D., Norman, D., Seclamen, M., Rusu, M., Baltres, A., and Teodorescu, V. (1994). Late Maastrichtian dinosaur eggs from the Hateg Basin (Romania). In *Dinosaur Eggs and Babies* (K. Carpenter, K. F. Hirsch, and J. R. Horner, Eds.), pp. 75–87. Cambridge Univ. Press, Cambridge, UK.

Hirsch, K. F., and Quinn, B. (1990). Eggs and eggshell fragments from the Upper Cretaceous Two Medicine Formation of Montana. *J. Vertebr. Paleontol.* **10,** 491–511.

Horner, J. R., and Currie, P. J. (1994). Embryonic and neonatal morphology and ontogeny of a new species of Hypacrosaurus (Ornithischia, Lambeosauridae) from Montana and Alberta. In *Dinosaur Eggs and Babies* (K. Carpenter, K. F. Hirsch, and J. R. Horner, Eds.), pp. 312–336. Cambridge Univ. Press, Cambridge, UK.

Horner, J. R., and Weishampel, D. B. (1988). A comparative embryological study of two ornithischian dinosaurs. *Nature* **332,** 256–257.

Hutton, J. M. (1986). Age determination of living Nile crocodiles from the cortical stratification of bone. *Copeia* **1986**(2), 332–341.

Klein, R. G., and Cruz-Uribe, K. (1984). *The Analysis of Animal Bone from Archaeological Sites.* Univ. of Chicago Press, Chicago.

Lehman, T. M. (1990). The ceratopsian subfamily Chasmosaurinae: Sexual dimorphism and systematics. In *Dinosaur Systematics: Approaches and Perspectives* (K. Carpenter and P. J. Currie, Eds.), pp. 211–220. Cambridge Univ. Press, Cambridge, UK.

Lockley, M. G. (1994). Dinosaur ontogeny and population structure: Interpretations and speculations based on fossil footprints. In *Dinosaur Eggs and Babies* (K. Carpenter, K. F. Hirsch, and J. R. Horner, Eds.), pp. 347–365. Cambridge Univ. Press, Cambridge, UK.

Madsen, J. H. (1976). *Allosaurus fragilis*: A revised osteology. *Bull. Utah Geol. Miner. Survey* **109**.

Nopsca, F. (1933). On the histology of the ribs in immature and half-grown trachodont dinosaurs. *Proc. R. Zool. Soc. London* **1933,** 221–223.

Norell, M. A., Clark, J. M., Demberelyin, D., Rhinchen, B., Chiappe, L. M., Davidson, A. R., McKenna, M. C., Altangerel, P., and Novacek, M. J. (1994). A theropod dinosaur embryo and the affinities of the Flaming Cliffs dinosaur eggs. *Science* **266,** 779–782.

O'Connor, R. J. (1984). *The Growth and Development of Birds,* pp. 315. Wiley, Chichester, UK.

Ostrom, J. H. (1978). The osteology of *Compsognathus longipes* Wagner. *Zitteliana, Abhandlungen Bayerischen Staatssammlung Paläontol. historische Geol. (München)* **4,** 73–118.

Raath, M. A. (1990). Morphological variation in small theropods and its meaning in systematics: Evidence from *Syntarsus rhodesiensis.* In *Dinosaur Systematics: Approaches and Perspectives,* (K. Carpenter and P J. Currie, Eds.), pp. 91–106. Cambridge Univ. Press, Cambridge, UK.

Reid, R. E. H. (1984). The histology of dinosaurian bone, and its possible bearing on dinosaur physiology. *Symp. Zool. Soc. London* **52,** 629–662.

Reid, R. E. H. (1990). Zonal "growth rings" in dinosaurs. *Mod. Geol.* **15,** 19–48.

Reid, R. E. H. (1993). Apparent zonation and slowed late growth in a small Cretaceous theropod. *Mod. Geol.* **18,** 391–406.

Ricqlès, A. de (1980). Tissue structures of dinosaur bone: Functional significance and possible relation to dinosaur physiology. In *A Cold Look at the Warm-Blooded Dinosaurs* (D. K. Thomas and E. C. Olson, Eds.), pp. 104–140. Westview Press, Boulder, CO.

Ricqlès, A. de (1983). Cyclical growth in the long limb bones of a sauropod dinosaur. *Acta Palaeontol. Polonica* **28**(1/2), 225–232.

Ricqlès, A. de, and Horner, J. R. (1997). Ontogenetic histogenesis of the hypsilophodontid dinosaur *Orodromeus makeli.* Unpublished manuscript.

Sampson, S. D. (1993). Cranial ornamentation in ceratopsid dinosaurs: systematic, behavioral and evolutionary implications, pp. 299. Ph.D. dissertation, University of Toronto.

Seymour, R. S. (1979). Dinosaur eggs: Gas conductance through the shell, water loss during incubation and clutch size. *Paleobiology* **5,** 1–11.

Sochava, A. V. (1972). The skeleton of an embryo in a dinosaur egg. *Paleontol. J.* **6,** 527–531.

Stark, J. M. (1989). Zeitmuster der Ontogenesen bei nestflüchtenden und nesthockenden Vögeln. *Courier Fortschungsinstitut Senckenberg* **114,** 1–319.

Thulborn, T. (1990). *Dinosaur Tracks,* pp. 410. Chapman & Hall, London.

Varricchio, D. J. (1993). Bone microstructure of the Upper Cretaceous theropod dinosaur *Troodon formosus. J. Vertebr. Paleontol.* **13,** 99–104.

Varricchio, D. J., and Horner, J. R. (1993). Hadrosaurid and lambeosaurid bone beds from the Upper Cretaceous Two Medicine Formation of Montana: Taphonomic and biologic implications. *Can. J. Earth Sci.* **30,** 997–1006.

Growth Lines

KEVIN PADIAN
University of California
Berkeley, California, USA

The general phenomenon known as growth lines is actually a grouping of many different kinds of structures in the bones, teeth, scales, scutes, shells, and other hard parts of vertebrates and invertebrates. Even among tetrapods, growth lines are heterogeneous in form and causation. In dinosaurs, growth lines are found in the teeth and in the bones of the skull, jaws, and postcranial skeleton, including scutes. Hence, "growth lines" may be regarded as a somewhat heterogeneous subject.

The histological patterns in bones and teeth result from several complex and often interconnected causes (see HISTOLOGY OF BONES AND TEETH). Genetic and ontogenetic factors, which are specific to certain taxa, influence the local growth dynamics of tissues as well as the kinds of tissues that are deposited throughout ontogeny. These factors, in turn, stem from phylogenetic legacy, and so, to some extent, the type of tissues deposited are an inherited habit of evolutionary history characteristic of larger clades. Mechanical factors, related to size, age, use, and mechanical loading, affect the kind of tissue deposited and the rates at which it is deposited. Environmental factors related to climate, diet, and seasonal availability of nutrients, among other influences, may exert direct or mediated effects on the formation, resorption, and remodeling of bone, locally and systemically (de Ricqlès *et al.*, 1997).

Daily growth lines have recently been discovered in the teeth of dinosaurs (see VON EBNER INCREMENTAL GROWTH LINES). Such lines were first discovered in mammalian teeth in the late 1800s and were recently identified in crocodilians by multifluorochrome labeling (Erickson, 1996a). Because nearly identical lines are found in the dentine of the teeth of many dinosaurs, which like crocodiles have a common archosaurian background, they are concluded to have a common origin (Erickson, 1996b). These daily lines are endogenous; that is, part of the genetic program apparently not normally affected by seasonal or environmental conditions. The cone-in-cone structures that Marsh (1880) noted in the teeth of the Cretaceous bird *Hesperornis* probably reflect some type of periodic growth lamination. Other kinds of growth lines in teeth, however, reflect seasonal retardation or hiatus in the deposition of tissue. Lieberman (1993) discovered that growth lines in the cementum of sheep teeth are influenced by both the quality of feed and its texture. Cementum is formed in part by the modification of Sharpey's fibers, which are oriented with respect to mechanical stress; a diet of softer food changes the orientation of cementum structures, whereas feed with poorer nutrients produces a more robust growth line.

Growth lines are a common feature of most vertebrate bones, regardless of tissue type, size, or phylogenetic position (Castanet *et al.*, 1993). However, not all growth lines are the same in structure or causation. In most vertebrates, lines of arrested growth (LAGs or *LACs, lignes d'arrêt de croissance*) appear where the periosteum has temporarily stopped depositing bone. This can be indicated by a circumferential ring marking the interruption of the layer of cortical bone that is outermost at that time. LAGs may further be marked by the deposition of a ring of avascular or nearly avascular bone before normal tissue growth is resumed. A reversal line may form where cortical bone has been eroded during nondeposition.

In vertebrates primitively, bone growth is slow and bone tissue is not well vascularized. These two factors are known to be correlated: High growth rates are possible only with a strong supply of blood to the depositional tissues. Seasonal variations in ambient temperature may directly affect the rate and even temporary cessation of bone growth in vertebrates whose body temperature fluctuates markedly with ambient temperature (poikilothermic). Certain metabolic processes will not operate well or at all at lower temperatures. LAGs typically result. However, it does not necessarily follow that the presence of LAGs indicates either poikilothermy or an inability to metabolize because of environmental conditions. Growth lines are typically found in the jaws of mammals (de Buffrénil, 1982), and the long bones of hibernators such as polar bears deposit growth lines during this annual torpor. However, growth lines are not confined to teeth and jaws in mammals; they are found in the long bones of members of most mammalian orders regardless of size, reproductive cycle, diet, latitude, or realm (marine or terrestrial), principally in the periosteal bone that forms once the rapid growth of juveniles has subsided (Klezeval' and Kleinenberg, 1967). These growth lines are clearly environmentally induced in most of these taxa, indicating that such histological phenomena are not confined to ectotherms, and the number of annual lines in the long bones usually corresponds to those in the mandible.

LAGs in the bones of dinosaurs have been reported in many taxa [Reid, 1990, 1996; e.g., the basal sauropodomorph *Massospondylus* (Chinsamy 1990), the ceratosaurian theropod *Syntarsus* (Chinsamy 1993), the sauropod *Brachiosaurus* (Reid 1996), the hadrosaurs *Hypacrosaurus* and *Maiasaura* (Horner *et al.*, 1997), and enantiornithine birds (Chinsamy *et al.*, 1995)]. These LAGs are sometimes interpreted as indications that the animals in question were not metabolically able to deposit bone continually, due to environmental fluctuations. Reid (e.g., 1990) has suggested the possibility of a thermometabolic regime intermediate between those of living endotherms and ectotherms, but he has also noted that the presence of LAGs alone is not a comprehensive indicator of physiology. Clearly, a holistic approach is needed. What sets dinosaur bone apart from those of all other known reptiles (except pterosaurs; Padian *et al.*, 1997) is that the growth rings occur not in a typical matrix of slow-growing lamellar-zonal bone but in one of fast-growing fibrolamellar bone (see Histology of Bones and Teeth). When growth resumes after the LAG is deposited, it is of fibrolamellar bone (except in the most mature adults, in which it may become progressively avascular periosteal bone).

By analogy with living reptiles and amphibians, the growth lines seen in some dinosaurian long bones have been reasoned to reflect annual deposition (Chinsamy, 1990). This may well be true, although no independent evidence as yet supports the inference. If they were not annual, it would be difficult to point to another possible measure of periodicity. Because they were found in all examples of the femora (the only elements studied) that were sectioned, it can be ruled out that these LAGs occurred only in females undergoing the periodic calcium stress of egg-laying. In *Massospondylus*, the number of LAGs and their relative positions in the cortical diameter were correlated among a size-variable sample (ostensibly a growth series) to estimate the age of various individuals (Chinsamy, 1990). Because cortical bone is continually resorbed into the medullary cavity by osteoclasts, the long bones of dinosaurs are generally hollow, not mostly solid like those of most other reptiles and amphibians, in which the ontogenetic record of growth lines is more complete. In *Massospondylus* the distance between successive LAGs does not appear to be regular, nor does it vary regularly from one LAG to the next. If these rings are indeed annual, then either these dinosaurs must have deposited bone at irregular rates from year to year (perhaps in "good" and "bad" nutritional or environmental years) or the onset and duration of the environmental cues to produce an avascular lamination must have varied from one year to the next. Alternatively, the deposition could have been under endogenous control and no further causal pattern need be looked for, or a combination of factors may have obtained.

Not all growth lines are LAGs. An apparent hiatus in deposition can occur for other reasons. Cortical drift is a phenomenon in which change in shape of a bone during growth, caused by mechanical or other factors, forces concomitant changes on the inner architecture of a bone. As seen in cross section (Fig. 1), a bone may grow more rapidly on one side than another, and a zone of little or no growth may simultaneously form on another side, frequently forming a jagged line of resorption. Such zones of altered growth can be distinguished easily from LAGs if the full cross section of the bone is available. Another kind of growth line occurs when the type of bone tissue, or the orientation of the fibers, changes ontogenetically. Changes from one type of tissue deposition to another may be gradual but are often abrupt. Even a simple change in the orientation of fibers, such as can be seen from layer to layer in the "plywood" construction of some pterosaur bone (Padian *et al.*,

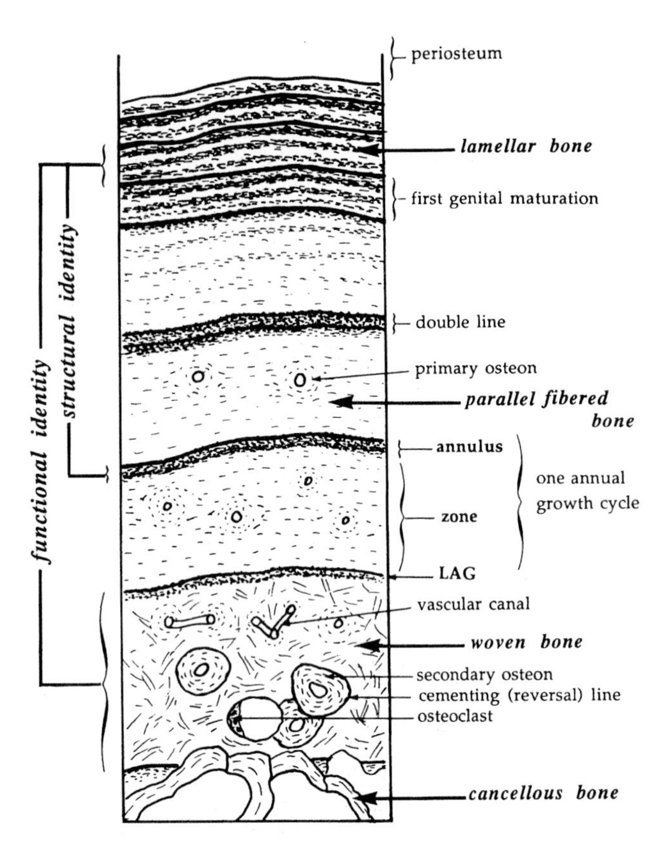

FIGURE 1 A generalized cross section of bone, with growth structures indicated (after Castanet *et al.*, 1993). Inferences about growth should not be made from partial sections like this one, nor from isolated bones, but rather from a comprehensive survey of skeletal tissues, preferably using an ontogenetic series.

1997), can appear to produce a "growth line," but it is easily seen under high microscopy that this is not a typical LAG and no avascular tissue or zone of nondeposition is produced (as with cortical drift). Rimblot-Baly *et al.* (1995, p. 9), studying the bone histology of the sauropod *Lapparentosaurus,* found a thin (4 mm) circular band of two or three rows of longitudinal canals in the humerus that they described as a "growth ring" (*anneau de croissance*) but not a LAG—a local expression of variation in the speed of radial growth.

There is considerable potential to examine the biological meaning of growth lines within paleoenvironmental contexts. For example, Chinsamy (1995) studied the femora of a growth series of the hypsilophodont *Dryosaurus* and found a complete absence of growth lines. In her previous studies, growth lines had always been found in other dinosaurian taxa, suggesting to her an intermediate thermometabolic level between endotherms and ectotherms or an inability to thermoregulate constantly throughout the year, given seasonal changes in the environment. Logically, the absence of growth lines in *Dryosaurus* might have suggested *(ceteris paribus)* a level of endothermy compatible with those of extant mammals and birds, but Chinsamy did not draw this conclusion. Recently, a paleopolar fauna of Australian dinosaurs, including various hypsilophodontids, ankylosaurs, and several theropods (see POLAR DINOSAURS), provided the opportunity to examine the histology of well-known lineages in an environment that must have had strong annual periodicity in sunlight and growing season for plants (Chinsamy *et al.*, 1996). However, the histological patterns of these bones appeared to be typical of those of respective taxa from nonpolar areas, and no histological similarities united the bones of the polar dinosaurs compared to those from elsewhere. Even the polar hypsilophodontids showed no trace of growth lines. This suggests that phylogenetic legacies and endogenous rhythms in the type and deposition of bone tissue may have been more important in these dinosaurs than were immediate, local environmental conditions. Hence, the presence or absence of LAGs in extinct taxa may by themselves say very little about thermometabolic regime or paleoenvironment.

Finally, it is worth noting that growth lines vary greatly in their presence and absence throughout the dinosaurian skeleton and are generally less common in long bones than elsewhere in the skeleton (see HISTOLOGY OF BONES AND TEETH). However, even their appearance in long bones is irregular and potentially deceptive. The number of rest lines found in various elements of the type specimen of the hadrosaur *Hypacrosaurus stebingeri* are as follows: femur, 10 or 11; tibia, 9; fibula, radius, 4 or 5; fibula, 4 or 5; rib, 10, ossified tendon, 10; metacarpal, none (Horner *et al.*, 1997). Reasons for this variation may in part be related to differential rates of deposition and absorption among bones of the skeleton (Castanet *et al.*, 1996). Variations in spacing and in Haversian remodeling correlate with the apparent number of rest lines, suggesting that not all bones are equally valuable indicators of age determination (Horner *et al.*, 1997), if the number of rest lines can be used in this context at all.

In summary, the term "growth lines" represents a mixed bag of histological phenomena that are the by-products of growth dynamics. Some are endogenous in origin, and others are affected by mechanical, ontogenetic, or environmental factors. Hence, caution should be observed in attributing physiological causal factors to the presence or distribution of growth lines in hard tissues of the skeleton.

See also the following related entries:

AGE DETERMINATION • PHYSIOLOGY • VON EBNER INCREMENTAL GROWTH LINES

References

Castanet, J., Francillon-Viellot, H., Meunier, F. J., and de Ricqlès, A. (1993). Bone and individual aging. In *Bone* (B. K. Hall, Ed.), Vol. 7B. CRC Press, Boca Raton, FL.

Castanet, J., Grandin, A., Abourachid, A., and de Ricqlès, A. (1996). Expression de la dynamique de croissance dans la structure de l'os périostique chez *Anas platyrhynchos. Comptes Rendus l'Acad. Sci. Paris Sci. Vie* **319,** 301–308.

Chinsamy, A. (1990). Physiological implications of the bone histology of *Syntarsus rhodesiensis* (Saurischia: Theropoda). *Palaeontol. Africana* **27,** 77–82.

Chinsamy, A. (1993). Bone histology and growth trajectory of the prosauropod dinosaur *Massospondylus carinatus* Owen. *Mod. Geol.* **18,** 319–329.

Chinsamy, A. (1995). Ontogenetic changes in the bone histology of the Late Jurassic ornithopod *Dryosaurus lettowvorbecki. J. Vertebr. Paleontol.* **15,** 96–104.

Chinsamy, A., Chiappe, L. M., and Dodson, P. (1995). Growth rings in Mesozoic birds. *Nature* **368,** 196–197.

Chinsamy, A., Rich, T., and Rich-Vickers, P. (1996). Bone histology of dinosaurs from Dinosaur Cove, Australia. *J. Vertebr. Paleontol.* **16**(Suppl. to No. 3), 28A.

de Buffrénil, V. (1982). Données préliminaires sur la présence de lignes d'arrêt de croissance périostiques dans la mandibule du marouin commun, *Phocoena phocoena* (L.), et leur utilisation comme indicature de l'âge. *Can. J. Zool.* **60,** 2557–2567.

de Ricqlès, A., Padian, K., and Horner, J. R. (1997). Comparative biology and the bone histology of extinct tetrapods: What does it tell us? Abstracts of the Fifth International Conference on Vertebrate Morphology, Bristol, July.

Erickson, G. M. (1996a). Daily deposition of dentine in juvenile *Alligator* and assessment of tooth replacement rates using incremental line counts. *J. Morphol.* **228,** 189–194.

Erickson, G. M. (1996b). Incremental lines of Von Ebner in dinosaurs and the assessment of tooth replacement rates using growth line counts. *Proc. Natl. Acad. Sci. USA* **93,** 14623–14627.

Horner, J. R., Padian, K., and de Ricqlès, A. (1997). Histological analysis of a dinosaur skeleton: Evidence of skeletal growth variation. Abstracts of the Fifth International Conference on Vertebrate Morphology, Bristol, July.

Klezeval', G. A., and Kleinenberg, S. E. (1967). *Age Determination of Mammals from Annual Layers in Teeth and Bones.* Akademiya Nauk SSSR, Institut Morpologii Zhivotnykh Im. A. N. Severtsova, Izdatal'stvo "Nauka," Moscow. [Israel Program for Scientific Translations, Jerusalem, 1969]

Lieberman, D. (1993). Life history variables preserved in dental cementum microstructure. *Science* **261,** 1162–1164.

Marsh, O. C. (1880). Odontornithes: A monograph on the extinct toothed birds of North America. *Rep. Geol. Exploration Fortieth Parallel* **VII,** 1–201.

Padian, K., de Ricqlès, A., and Horner, J. R. (1997). Pterosaur bone histology: Growth rates and mechanics. Abstracts of the Fifth International Conference on Vertebrate Morphology, Bristol, July.

Reid, R. (1990). Zonal "growth rings" in dinosaurs. *Mod. Geol.* **15,** 19–48.

Reid, R. (1996). Bone histology of the Cleveland–Lloyd dinosaurs and of dinosaurs in general, Part I: Introduction: Introduction to bone tissues. *Brigham Young Univ. Geol. Stud.* **41,** 25–71.

Rimblot-Baly, F., de Ricqlès, A., and Zylberberg, L. (1995). Analyse paléohistologique d'une série de croissance partielle chez *Lapparentosaurus madagascariensis* (Jurassique Moyen): Essai sur la dynamique de croissance d'un dinosaure sauropode. *Ann. Paléontol.* **81,** 49–86.

Gunma Prefectural Museum of History, Japan

see MUSEUMS AND DISPLAYS

Migrating hadrosaurs. Illustration by Michael Rothman.

Hadrosauridae

Catherine A. Forster
State University of New York at Stony Brook
Stony Brook, New York, USA

The Hadrosauridae, or "duck-billed" dinosaurs, are the most numerous, diverse, and widespread large-bodied herbivores of the Late Cretaceous. Hadrosaurids were large, facultative bipeds with body lengths up to 15 m and include approximately 27 genera, ranging in age from the late Cenomanian to late Maastrichtian. Hundreds of hadrosaurid specimens (from disarticulated elements to fully articulated skeletons) have been collected in North America, Central America, South America, Indochina, Europe, Russia, and Asia (Weishampel and Horner, 1990).

The first hadrosaurid material found was a single tooth and some worn bones from Montana and South Dakota. These scrappy pieces from the western frontier were described in 1856 by eminent Philadelphia paleontologist Joseph Leidy. The following year a partial hadrosaurid skeleton, the first nearly complete dinosaur discovered anywhere in the world, was unearthed in Dr. Leidy's own backyard—Haddonfield, New Jersey. This first good hadrosaurid specimen, named *Hadrosaurus foulkii* (Leidy, 1858), gave the world its first glimpse of a nearly complete dinosaur.

Hadrosaurids were one of the last major groups of dinosaurs to arise in the Mesozoic, with Cretaceous sediments recording their near simultaneous arrival worldwide. An Asian origin for hadrosaurids has been hypothesized (Milner and Norman, 1984), although other research suggests that their center of origin cannot yet be pinpointed (e.g., Brett-Surman, 1979). Regardless of where they began, once hadrosaurids arose they quickly diversified and spread worldwide, becoming the primary constituent of many herbivore faunas.

Hadrosaurid Origins and Relationships

Hadrosaurids are a monophyletic group of ornithopod dinosaurs, with all the diverse members of the group being derived from a common ancestor. Hadrosaurid monophyly has long been presumed (e.g., Lull and Wright, 1942; Ostrom, 1961), and recent cladistic character analyses have bolstered this hypothesis (e.g., Forster and Sereno, 1994, although see Weishampel and Horner, 1990). This common ancestry of hadrosaurids is revealed by the many unique shared characters, or synapomorphies, in hadrosaurid skeletons and skulls (Figs. 1 and 2). A few of these unique characters are the development of a complex dental battery, loss of the antorbital and surangular fenestrae, the absence of a free palpebral bone in the eye, increase in vertebral number to at least 12 cervicals and eight sacrals, and reduction of the number of wrist bones with loss of the first digit of the hand (e.g., Weishampel and Horner, 1990).

Nearly all known hadrosaurids can be placed in one of two subtaxa, each diagnosed by its own unique suites of characters. The two subtaxa are the Lambeosaurinae, or "hollow-crested" hadrosaurids, and the Hadrosaurinae, or "flat-headed" hadrosaurids. However, two very early, primitive hadrosaurids (*Telmatosaurus* and *Gilmoresaurus*) lie outside these two major groups (Weishampel and Horner, 1990; Weishampel, 1993). A third subgroup, the Saurolophinae, was once

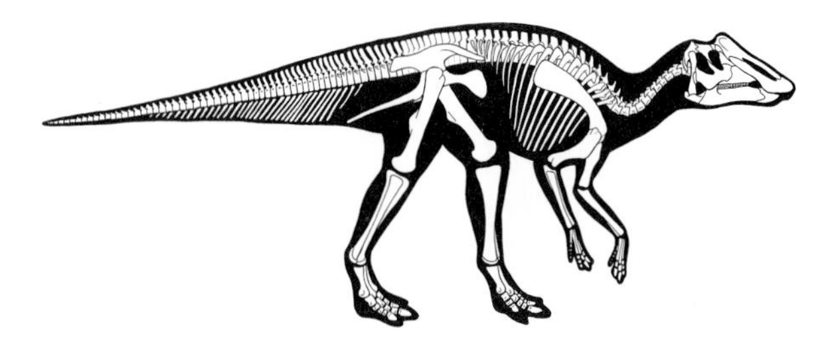

FIGURE 1 Skeleton with outline of *Prosaurolophus maximus,* a hadrosaurine hadrosaurid from the Campanian of North America. Based on ROM 787.

recognized (e.g., Lull and Wright, 1942; Ostrom, 1961) but has now been subsumed by the Hadrosaurinae (e.g., Brett-Surman, 1979; Weishampel and Horner, 1990).

Some of the unique characters that distinguish the Lambeosaurinae include completely separate nasal passages on the snout, short parietals, exclusion of the frontal from the orbit margin, an extensive contact

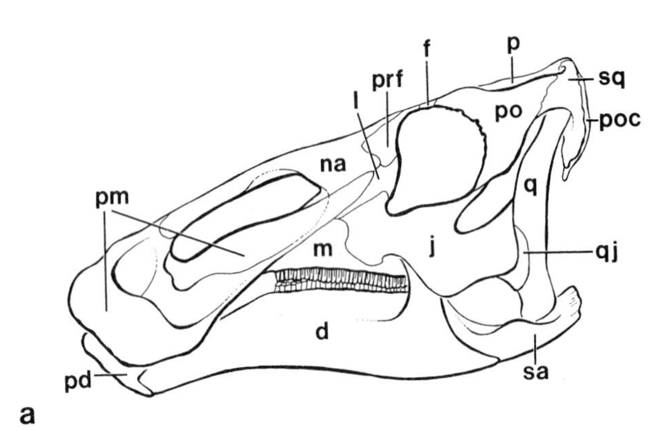

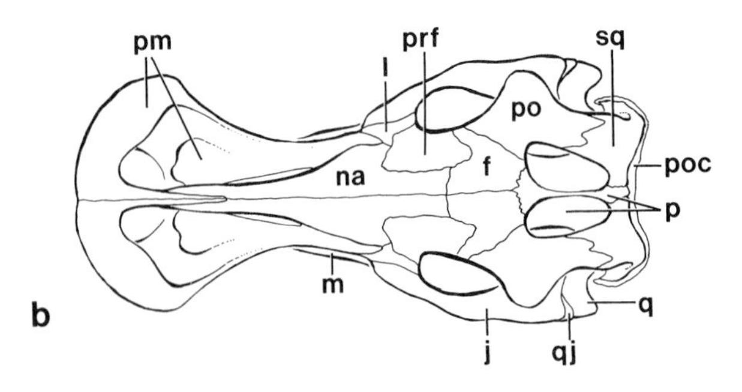

FIGURE 2 *Edmontosaurus regalis* (NMC 2288), a hadrosaurine hadrosaurid, in (a) dorsal and (b) left lateral view. Abbreviations: d, dentary; f, frontal; j, jugal; l, lacrimal; m, maxilla; na, nasal; p, parietal; pd, predentary; pm, premaxilla; po, postorbital; poc, paroccipital process; prf, prefrontal; q, quadrate; qj, quadratojugal; sa, surangular; sq, squamosal.

between squamosals, anteriorly bowed quadrates, and expansive, hollow cranial crests of various shapes that house complex nasal passages. The Hadrosaurinae are diagnosed in part by the development of depressions around the external nares (circumnarial depressions), the absence of marginal denticles on all teeth, a prominent anterior process on the maxilla, a narrow prepubic shaft, and a slender, tapering ischial shaft. Some hadrosaurines, such as *Prosaurolophus* (Fig. 1) and *Brachylophosaurus* (Fig. 3C), have low, solid cranial crests, whereas others, such as *Gryposaurus* (Fig. 3A) and *Edmontosaurus* (Fig. 2), lack any crest development.

More is known about aspects of life history in hadrosaurids than in any other group of dinosaurs. By examining their unique skeletal morphologies, as well as skin impressions, trackways, eggs and nesting sites, and information regarding growth series and sexual dimorphism, we are presented with an unparalleled opportunity to move beyond the bones and unveil the living hadrosaurid.

The skin that covered hadrosaurids has long since disappeared, but patches of skin impressions are occasionally recovered with specimens (e.g., Lull and Wright, 1942). Among the finest examples of skin impressions is a nearly complete, fossilized *Edmontosaurus* mummy (now at the American Museum of Natural History), which when unearthed "lay there with expanded ribs as in life, wrapped in the impressions of the skin whose beautiful patterns of octagonal plates marked the fine sandstone above the bones" (Sternberg, 1909, p. 275). Hadrosaurid skin was composed of a pavement of small, interlocking polygonal tubercles, or plates, arrayed in varying patterns around the body (e.g., Osborn, 1912). For instance, some skin patches show pavements of uniformly sized tubercles, whereas others have small background tubercles filling in spaces between larger ones. The tubercles never overlap and are quite thin—too thin and flimsy to offer hadrosaurids any protection against predators. Unfortunately, these impressions cannot reveal the color hadrosaurids were in life.

In addition to a pavement-like skin, impressions of a midline skin fold or frill have occasionally been found along the center of the back and tail. One well-preserved tail section shows a segmented midline frill composed of short, flexible, tab-like projections (Horner, 1984).

The entire front half of a hadrosaurid's mouth is

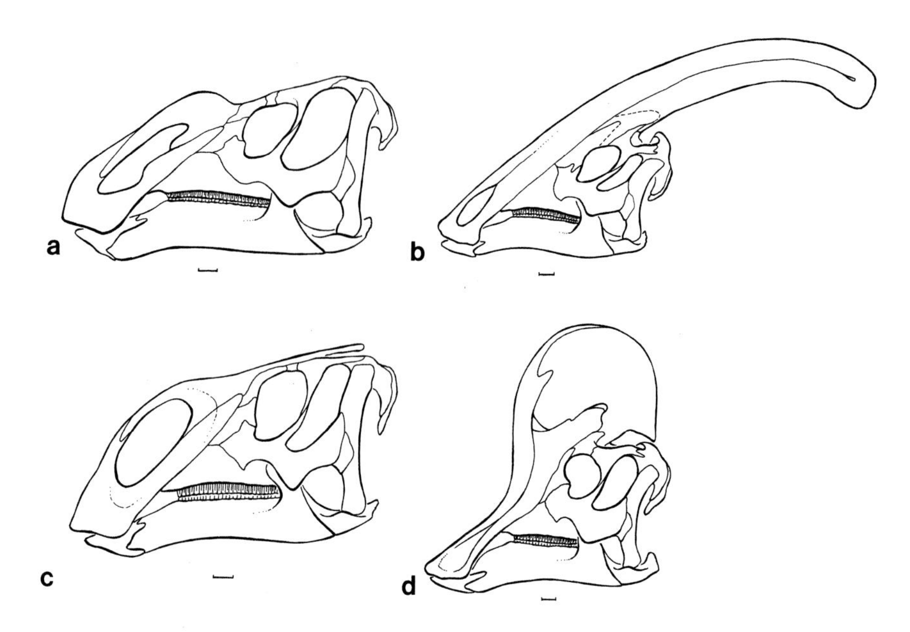

FIGURE 3 Hadrosaurid skulls in left lateral views, showing the basic differences between hadrosaurines (a, c) and lambeosaurines (b, d). Scale bars = 5 cm. a, *Gryposaurus* (NMC 2278); b, *Parasaurolophus* (ROM 768); c, *Brachylophosaurus* (NMC 8893) (note the flat, solid crest, derived entirely from the nasal bones, that projects horizontally over the posterior portion of the skull like a lid); d, *Corythosaurus* (ROM 871).

toothless. It flares outwards to varying degrees into a broad, flattish, semicircular "bill" (Fig. 2B), prompting early researchers to dub them the duck-billed dinosaurs. Along the sides of the mouth, however, hundreds of small teeth are tightly jammed together into molar-like "dental batteries."

"Mummified" specimens of both lambeosaurines (Sternberg, 1935) and hadrosaurines (Morris, 1970) show that a thick horny sheath, or rhamphotheca, was adhered to the front of the upper jaws. The rhamphotheca was tightly bound to the front of the premaxillary bones in the upper "bill" and extended out and downwards to broadly overlap the front of the lower jaw. A double row of bony projections, or denticles, studded the front margin of the premaxillae and probably helped bind the rhamphotheca tightly to the bone. No rhamphotheca has yet been found preserved on the lower jaw, although similar denticles on the predentary may indicate one was present. Although fossilized rhamphothecas have only been recovered in a small number of specimens, denticulate oral margins occur in all hadosaurid taxa and indicate their presence throughout the group. These rhamphothecas give hadrosaurids a much different appearance in life than the bone alone indicates and likely had a large effect on how they fed on plants.

Hypotheses for the use of this rhamphotheca range from stripping leaves and bark from trees to filtering soft vegetation from water. Morris (1970) concluded that the rhamphotheca was too weak to have cropped vegetation tougher than soft aquatic plants. However, Maryánska and Osmólska (1984) point out that recent mammalian herbivores efficiently crop fibrous grass with soft tongues and lips. Cows, for instance, lack upper teeth and crop grass by trapping it between their lower teeth and a horny pad on the upper jaw inside the lip. Hadrosaurs may have fed in a similar manner, using the overhanging rhamphotheca as a plant trap.

Teeth in hadrosaurids were restricted to the sides of the jaws and wedged tightly together (so tightly that no spaces remain between them) to form a dental battery. Each small, slender individual tooth was arranged in a specific position within the battery. Three teeth lined up from the inside to the outside of the jaw to form a single "tooth family." Numerous tooth families were then packed together, front to back, along the length of the jaw. Juvenile hadrosaurids had only 10 or 12 tooth families per jaw (producing 30–36 teeth across the surface of each dental battery), whereas adults average 45 tooth families (or 135 teeth on the surface of each dental battery). Because hadro-

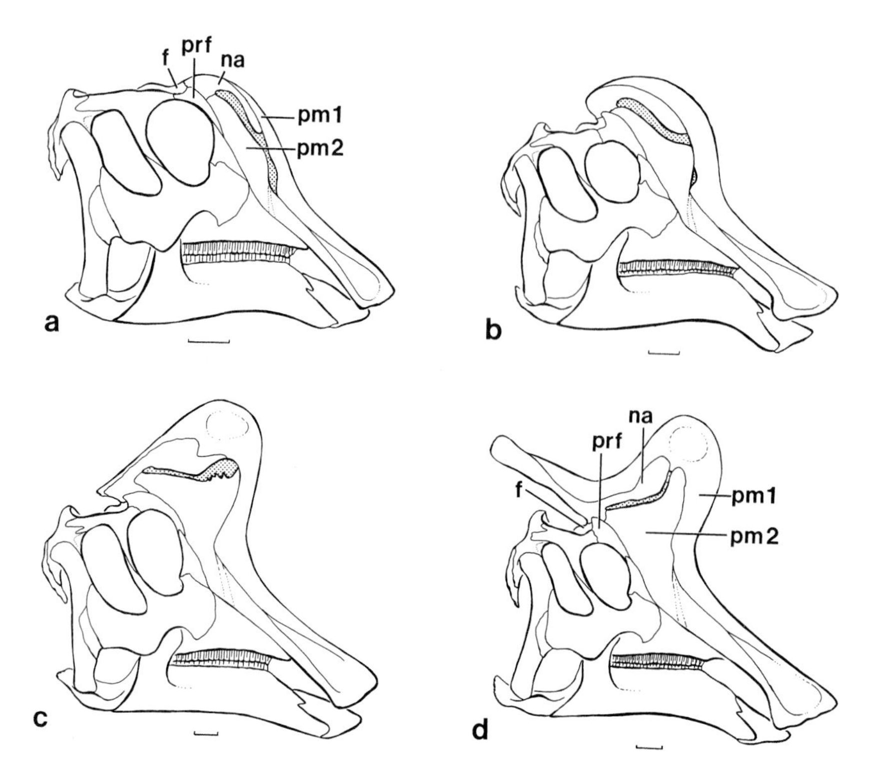

FIGURE 4 The progressive stages of crest growth in *Lambeosaurus* from juvenile through mature adult. The premaxillae, originally restricted to the area in front of the orbits, are pulled up and back with the nasals. Note, however, that their relationships to the surrounding elements remain unchanged. Also note the development of the "tail" on the posterior crest in the adult. Abbreviations: f, frontal; na, nasal; pm1, premaxilla 1; pm2, premaxilla 2; prf, prefrontal. Scale bars = 5 cm. a, ROM 758; b, NMC 8633; c, NMC 8703; d, NMC 2869.

saurids continually replaced their teeth throughout life, each tooth also had 2–4 replacement teeth lined up below it in the jaw, giving adult hadrosaurids more than 400 teeth in each quadrant of the jaw. Although each individual tooth was small and slender, this "strength in numbers" arrangement afforded hadrosaurids an effective, gapless, molar-like grinding machine through life.

With these efficient plant-processing dental batteries, what did hadrosaurids eat? Krausel (1922), in the only direct account of hadrosaurid diet, reported that the stomach contents of a fossilized *Edmontosaurus* mummy included conifer needles and twigs and unidentified "fruits and seeds." However, this report is questioned by Farlow (1987), who suspects the plant material was preserved alongside the specimen and does not represent stomach contents. Studies of jaw biomechanics in hadrosaurids do show they had a unique way of chewing, with both fore and aft and lateral jaw motions possible for grinding plant material. Muscular cheeks, whose boundaries are defined by indentations along the sides of the jaws, kept the plant material in the mouth while being ground between dental batteries (Galton, 1973). Based solely on their jaw mechanics and teeth, Weishampel (1984) concluded that hadrosaurs may have needed such efficient dental batteries to process fibrous plants of low nutritional value.

While examining the skeleton of *Hadrosaurus* well over a century ago, Leidy noted the great disparity between the lengths of its fore- and hindlimbs. He hypothesized that *Hadrosaurus* was bipedal, balancing itself upright on its hindlimbs and tail like a kangaroo, but added the caveat that *Hadrosaurus* may have also progressed in a quadrupedal fashion (Leidy, 1858). Leidy further speculated that *Hadrosaurus* may have been aquatic, a viewpoint echoed by later paleontologists who thought them water-bound and incapable of habitual terrestrial locomotion.

An upright, kangaroo-like posture for hadrosaurids was rejected by researchers who, after examining the articulations of the vertebral column and pelvis, realized that the back and tail were held horizontally rather than angled upwards from the sacrum (Galton, 1970). Still, the mode of hadrosaurid locomotion continued to be debated, with some researchers favoring bipedality (e.g., Ostrom, 1964) and others quadrupedality.

Brown (1912), after examining the *Edmontosaurus* mummy at the American Museum, thought it had "phalanges partly imbedded in skin." Osborn (1912) thought that this indicated a webbed hand and interpreted it as an aquatic adaptation for paddling about in water; this became one of the primary lines of evidence for researchers favoring an aquatic habit for hadrosaurids. However, it is likely that the "webbing" is actually digital pads that were displaced during mummification (Bakker, 1986). These digital pads would have cushioned the fingertips during quadrupedal locomotion, thus leaving hadrosaurids "paddleless." Furthermore, the metacarpals in hadrosaurids show scars along their lengths that indicate that they were tightly bound together in life—a poor arrangement for a paddle but an excellent adaptation for weight support on land.

Although digital pads and tightly bound metacarpals are adaptations for quadrupedal locomotion, the disparity in forelimb to hindlimb lengths that Leidy noted in 1858 remains outside the range of known obligatory quadrupeds. Hadrosaurid forelimbs are shorter and more lightly built than the long, robust hindlimbs, thus limiting stride length and limb strength during quadrupedal locomotion. In other words, a quadrupedal hadrosaurid could only have run as fast as its front legs would allow. Like their close relatives the iguanodontians, hadrosaurids have stiff, ossified tendons arranged in a trellis-like pattern down the dorsum of their back, sacrum, and tail. In a bipedal stance, the body and tail are cantilevered over the sacrum and hindlimbs. This creates large tensile forces across the top of the vertebral column that must be resisted to keep the animal from bending, or "hogging," across the sacrum. The arrangement of the tendons in a trellis-like pattern turns out to be an efficient system to counter these tensile forces. Although tendons are strong, ossification increases their tensile strength, allowing them to hold the vertebral column horizontal with greater ease and providing further evidence that hadrosaurids were well adapted for terrestrial locomotion (e.g., Ostrom, 1964; Galton, 1970).

The evidence for hadrosaurids as terrestrial animals is strong, but the choice between bipedalism and quadrupedalism seems ambiguous. Footprints attributed to hadrosaurs reveal that they walked both bipedally and quadrupedally (Lockley, 1994). Thulborn (1989, p. 48) thought that hadrosaurids were able to "switch between bipedal and quadrupedal gaits according to circumstances." When all evidence is taken together, hadrosaurids appear to have been primarily bipedal, particularly at higher speeds, but probably often walked on all four legs while walking and browsing on low vegetation (see BIPEDALITY; QUADRUPEDALITY).

The most striking feature of lambeosaurine hadrosaurids is their varying development of spectacular supracranial crests that are either fan shaped (*Lambeosaurus*, *Hypacrosaurus*, and *Corythosaurus*) or tubular (*Parasaurolophus*). These thin-walled crests house greatly distorted, convoluted, and hypertrophied nasal passages. In *Corythosaurus* (Fig. 3D), the nasal passages loop up and through the circular crest before passing down into the throat. In *Parasaurolophus* (Fig. 4B), the nasal passages pass backwards along the entire top of the tubular crest before doubling back to the throat. This entire route is more than 2 m long in some *Parasaurolophus* specimens. Additional chambers also develop inside lambeosaurine crests, resulting in a very complex internal structure.

In contrast to what is going on inside, the outer surface of the crest remains relatively uncomplicated. Lambeosaurine hatchlings lacked crests but as they grew the skull expanded upward and backward until it reached its adult form. The nasal passages, fairly straight in juveniles, were carried up and back inside the developing crest during growth. The changes in crest form from juvenile to adult specimens (Fig. 4) show (i) the crests are composed almost exclusively of the nasal and premaxillary bones, and (ii) although greatly distorted, these bones retain their original contacts with the bones around them (e.g., trace the contacts of the nasal bone with the prefrontal and frontal bones from Figs. 4A to 4D). Although the nasal and premaxilla become greatly hypertrophied with growth, their relationships to one another and to the surrounding bones do not change.

These enormous changes in cranial morphology from juvenile to adult confounded lambeosaurine systematics for many years. For example, juvenile specimens of *Corythosaurus* and *Lambeosaurus* were once considered different taxa from the adults. Called

Tetragonosaurus or *Procheneosaurus*, they were thought to be adults of small, crestless lambeosaurines. Later researchers, realizing that morphology within a taxa can differ depending on growth stage and sex, demonstrated that these were simply juvenile animals who had yet to grow their crests (e.g., Dodson, 1975).

The purpose of these flashy, hypertrophied nasal passages is not perfectly clear. Some researchers, thinking that hadrosaurids were aquatic, thought the crests either served as an air trap that facilitated feeding under water (e.g., Sternberg, 1935) or provided an attachment point for an elephant-like trunk that acted as a snorkel (Wilfarth, 1939). Ostrom (1961) thought that the enlarged supracranial crests may have served to heighten the sense of smell. Another hypothesis surmised that the crests were used in the same way a peacock would use his tail or an elk his antlers: as sexual display structures to compete for or attract mates (e.g., Dodson, 1975). Weishampel (1981a,b) thought the crests may have acted as resonance chambers for vocalization, perhaps also a tactic for attracting or competing for mates. Like an elk "bugling" during mating season, loudly trumpeting hadrosaurids may once have searched for mates across the Cretaceous landscape.

Some hadrosaurines also have supracranial crests. However, unlike the complex, hollow crests of lambeosaurines, hadrosaurines have small, solid crests that are simply thickened portions of the nasal and/or frontal bones. These crests extend only slightly above the skull roof and take on a variety of shapes in different taxa [e.g., *Prosaurolophus* (Fig. 1) and *Brachylophosaurus* (Fig. 3C)]. They never involve the nasal passages and appear to have evolved independent of the much larger crests of lambeosaurines.

In 1979, Jack Horner and Bob Makela discovered the jumbled bones of some tiny juvenile hadrosaurids in the TWO MEDICINE FORMATION of western Montana. This stunning discovery, along with subsequent discoveries of eggs and nests of this hadrosaurine (which was named *Maiasaura*), provides a remarkable glimpse into hadrosaurid reproduction and behavior. Recent discoveries of nests and babies of the lambeosaurine *Hypacrosaurus* correspond with those of *Maiasaura* (Horner and Currie, 1994), indicating a common pattern for reproductive behavior for lambeosaurines and perhaps for all hadrosaurids.

Maiasaura seems to have built large, circular earthen nesting mounds with concave centers. Up to 20 eggs per clutch were laid in a circular pattern inside these nests, which were then covered with vegetation whose heat of decomposition kept the eggs warm—like incubating eggs in a compost pile. Some living birds and crocodilians, the closest living relatives of dinosaurs, construct nests and incubate their eggs in the same manner. Each *Maiasaura* egg weighed about 1 kg, or a little less than 2% adult weight (Horner, 1982). Multiple nests have been found clustered together in colonies. There is also some evidence of "site fidelity," with the same nesting sites being used from season to season (Horner and Makela, 1979; Horner, 1982).

Some of these "fossilized" nesting mounds have been found containing remains of *Maiasaura* young. Examination of their bones shows two very interesting traits: (i) They were too large to be newly hatched and show wear on their teeth, indicating they had been eating plants for some time, and (ii) studies of their bone structure show that the ends of their bones were still poorly formed (Horner and Weishampel, 1988; Horner, 1994). These findings imply that *Maiasaura* young were "altricial," or nest-bound, for some period of time after hatching. Because they were eating plants while in the nest, food may have been carried to them by one or more *Maiasaura* adults. All this evidence implies that hadrosaurids nested in large colonies, returning year after year to lay their eggs in the same location. Instead of laying their eggs and leaving, they stayed first to protect the eggs from predators, then to feed and raise the nest-bound hatchlings. This type of parental BEHAVIOR, also found among some birds and crocodiles, is what prompted the name of this new hadrosaurine—*Maiasaura* is Greek for "good mother lizard" (Horner and Makela, 1979).

Hadrosaurids were a late starting but successful group of ornithopod dinosaurs that had spread nearly worldwide by the end of the age of dinosaurs. In addition to their well-preserved skeletons, we have an excellent idea of what they looked like, how they stood and walked, how and what they ate, and how they reproduced and grew. However, although they are united by many unique morphologies, and doubtlessly many behaviors, hadrosaurids remained a diverse group of dinosaurs, successful in both numbers and variety.

See also the following related entry:

ORNITHOPODA

References

Bakker, R. T. (1986). *The Dinosaur Heresies*, pp. 481. Kensington, New York.

Brett-Surman, M. K. (1979). Phylogeny and paleobiogeography of hadrosaurian dinosaurs. *Nature* **277**(5697), 560–562.

Brown, B. (1912). The osteology of the manus in the family Trachodontidae. *Bull. Am. Museum Nat. History* **31,** 105–108.

Dodson, P. (1975). Taxonomic implications of relative growth in lambeosaurine hadrosaurs. *Systematic Zool.* **24,** 37–54.

Farlow, J. O. (1987). Speculations about the diet and digestive physiology of herbivorous dinosaurs. *Paleobiology* **13,** 60–72.

Forster, C. A., and Sereno, P. C. (1994). Phylogeneny of the Hadrosauridae. *J. Vertebr. Paleontol.* **14**(Suppl.), 25A.

Galton, P. M. (1970). The posture of hadrosaurian dinosaurs. *J. Paleontol.* **44,** 464–473.

Galton, P. M. (1973). The cheeks of ornithischian dinosaurs. *Lethaia* **6,** 67–89.

Horner, J. R. (1982). Evidence of colonial nesting and "site fidelity" among ornithischian dinosaurs. *Nature* **297,** 675–676.

Horner, J. R. (1984). A "segmented" epidermal tail frill in a species of hadrosaurian dinosaur. *J. Paleontol.* **58,** 270–271.

Horner, J. R. (1994). Comparative taphonomy of some dinosaur and extant bird colonial nesting grounds. In *Dinosaur Eggs and Babies* (K. Carpenter, K. F. Hirsch, and J. R. Horner, Eds.), pp. 116–123. Cambridge Univ. Press, Cambridge, UK.

Horner, J. R., and Currie, P. J. (1994). Embryonic and neonatal morphology and ontogeny of a new species of *Hypacrosaurus* (Ornithischia, Lambeosauridae) from Montana and Alberta. In *Dinosaur Eggs and Babies* (K. Carpenter, K. F. Hirsch, and J. R. Horner, Eds.), pp. 312–336. Cambridge Univ. Press, Cambridge, UK.

Horner, J. R., and Makela, R. (1979). Nests of juveniles provides evidence of family structure among dinosaurs. *Nature* **282,** 296–298.

Horner, J. R., and Weishampel, D. B. (1988). A comparative embryological study of two ornithischian dinosaurs. *Nature* **332,** 256–257.

Krausel, R. (1922). Die Nahrung von Trachodon. *Paleontol. Zeitschrift* **4,** 80.

Leidy, J. (1858). *Proc. Acad. Nat. Sci. Philadelphia,* 215–218.

Lockley, M. A. (1994). Dinosaur ontogeny and population structure: Interpretations and speculations based on fossil footprints. In *Dinosaur Eggs and Babies* (K. Carpenter, K. F. Hirsch, and J. R. Horner, Eds.), pp. 347–365. Cambridge Univ. Press, Cambridge, UK.

Lull, R. S., and Wright, N. E. (1942). Hadrosaurian dinosaurs of North America. Geological Society of America Spec. Paper No. 40, pp. 1–242.

Maryánska, T., and Osmólska, H. (1984). Postcranial anatomy of *Saurolophus angustirostris* with comments on other hadrosaurs. *Paleontol. Polonica* **46,** 119–141.

Milner, A. R., and Norman, D. B. (1984). The biogeography of advanced ornithopod dinosaurs (Archosauria: Ornithischia)—A cladistic–vicariance model. In *Third Symposium on Mesozoic Terrestrial Ecosystems* (W.-E. Reif and F. Westphal, Eds.), pp. 145–150. Attempto Verlag, Tubingen, Germany.

Morris, W. J. (1970). Hadrosaurian dinosaur bills-morphology and function. *Los Angeles County Museum Contrib. Sci.* **193.**

Osborn, H. F. (1912). Integument of the iguanodont dinosaur *Trachodon. Mem. Am. Museum Nat. History* **1,** 33–54.

Ostrom, J. H. (1961). Cranial morphology of the hadrosaurian dinosaurs of North America. *Bull. Am. Museum Nat. History* **122**(2), 35–186.

Ostrom, J. H. (1964). A reconsideration of the paleoecology of hadrosaurian dinosaurs. *Am. J. Sci.* **262,** 975–997.

Sternberg, C. H. (1909). *The Life of a Fossil Hunter*, pp. 281. Indiana Univ. Press, Bloomington.

Sternberg, C. M. (1935). Hooded Hadrosaurs of the Belly River Series of the Upper Cretaceous. Canada Department of Mines Bulletin No. 77, Geological Series Vol. 52, pp. 1–37.

Thulborn, R. A. (1989). The gaits of dinosaurs. In *Dinosaur Tracks and Traces* (D. G. Gillette and M. G. Lockley, Eds.), pp. 39–50. Cambridge Univ. Press, Cambridge, UK.

Weishampel, D. B. (1981a). Acoustic analysis of potential vocalization in lambeosaurine dinosaurs (Reptilia: Ornithischia). *Paleobiology* **7,** 252–261.

Weishampel, D. B. (1981b). The nasal cavity of lambeosaurine hadrosaurids (Reptilia: Ornithischia): comparative anatomy and homologies. *J. Paleontol.* **55,** 1046–1057.

Weishampel, D. B. (1984). Interactions between Mesozoic plants and vertebrates: Fructifications and seed predation. *Neues Jahrbuch Geol. Palaeontol. Abhandlung* **167,** 224–250.

Weishampel, D. B. (1993). *Telmatosaurus transsylvanicus* from the late Cretaceous of Romania: The most basal hadrosaurid dinosaur. *Paleontology* **36,** 361–385.

Weishampel, D. B., and Horner, J. R. (1990). Hadrosauridae. In *The Dinosauria* (D. B. Weishampel, P. Dodson, and M. Osmolska, Eds.), pp. 534–561. Univ. of California Press, Berkeley.

Wilfarth, M. (1939). Die Nasienbasis der Lambeosaurinae. *Neuen Jahrbuch Miner. Geol. Palaontol. Abteilung B,* 24–39.

Hadrosaurinae

A clade of the Hadrosauridae along with the Lambeosaurinae.

see HADROSAURIDAE

Heimatmuseum, Germany

see MUSEUMS AND DISPLAYS

Hell Creek Flora

KIRK R. JOHNSON
Denver Museum of Natural History
Denver, Colorado, USA

The HELL CREEK FORMATION is an Upper Cretaceous (upper Maastrichtian) deposit that occurs in eastern Montana, northeastern Wyoming, and western North Dakota and South Dakota. It is bounded by middle Maastrichtian marine deposits below and lower Paleocene terrestrial deposits above. It is laterally equivalent to the LANCE FORMATION of Wyoming. The Hell Creek Formation is best known for its extensive and well-studied vertebrate fauna that includes the classic dinosaurs of the *Triceratops* zone or the so-called "Lancian" stage (Archibald, 1996). The Cretaceous–Tertiary (K–T) boundary occurs very near the upper contact of the Hell Creek Formation.

Until recently, the megaflora of the Hell Creek Formation has not received extensive attention or systematic description. The palynoflora of the Hell Creek Formation has been studied by a number of workers since the late 1960s and recently by Nichols (cited in Johnson *et al.*, 1989) and Hotton (1988). Although the Hell Creek Formation near the Fort Peck Reservoir in central Montana has produced most of the studied vertebrate remains and abundant palynomorphs, it has yielded few plant macrofossils. Instead, most Hell Creek plant fossils come from localities in the western part of the Dakotas where the Formation is approximately 110 m thick. Here the unit is overlain by the lower Paleocene Ludlow Member of the Fort Union Formation with the K–T boundary generally situated within 1 m above or below the formational contact (Johnson *et al.*, 1989; Johnson, 1992). Paleomagnetic analysis of the Hell Creek Formation near Marmarth, North Dakota, indicates that the upper 20 m of the formation lie in subchron C29R and the basal 90 m lie in subchron C30N. This analysis is in agreement with ammonite biostratigraphy in the underlying marine shales and suggests that the Hell Creek Formation represents the last 1.5–2 million years of the Cretaceous Period.

Extensive fieldwork in the Hell Creek Formation between 1983 and 1988 resulted in a collection of 7708 specimens from 57 localities near Marmarth, North Dakota, and smaller collections from 6 sites near Ekalaka, Montana, and four sites near Faith, South Dakota (Johnson, 1989). Subsequent collecting by the Denver Museum of Natural History between 1991 and 1995 near Marmarth, North Dakota, and Buffalo, South Dakota, has yielded specimens from 30 additional localities. A single good site (MPM 3317) was discovered by the Milwaukee Public Museum in McCone County, Montana, in the late 1980s. Thus, the megaflora of the Hell Creek is known from nearly 100 quarries in three states.

Composition of the Megaflora

The known megaflora (Johnson, 1989) consists of 190 plant morphotypes (1 bryophyte, 6 pteridophytes, 1 ginkgo, 1 cycadophyte, 9 conifers, and 172 angiosperms). Angiosperms represent about 90% of the flora both in number of species and in specimens collected. Important angiosperm families include the Lauraceae, Platanaceae, Magnoliaceae, Berberidaceae, and Arecaceae (palms), but a great number of the Hell Creek leaves have yet to be assigned to families. The majority of Hell Creek angiosperms belong to extinct genera. Although it is commonly reported in textbooks and popular articles, modern plant groups such as oaks, maples, willows, and grasses are not present in the Hell Creek. Conifers are represented by members of the Taxodiaceae, Araucariaceae, and Cheirolepidiaceae (extinct). Ginkgo is known from only a few localities. The sole cycad, *Nilssonia*, is present throughout the formation but is rare. Ferns are very uncommon.

Megafloral data were used to establish a zonation of the Hell Creek Formation in North Dakota (Johnson, 1989, 1992; Johnson and Hickey, 1990). This three-part zonation documents considerable floral change at the species level throughout the latest Cretaceous and a major extinction at the K–T boundary.

Composition of the Palynoflora

The palynoflora of the Hell Creek includes nearly 300 pollen and spore taxa, the vast majority of which cannot be assigned to extant families (Hotton, 1988). In terms of major plant groups, the compositions of the palynoflora and megaflora are in general agreement. In addition, the palynoflora documents a number of plants not seen in the megaflora, including *Sphagnum* mosses, the gnetalean plant *Ephedra*, and a variety of ferns including members of the Osmundaceae, Schizaeaceae, Polypodiaceae, and Gleicheniaceae.

Even though they were derived from the same source vegetation, there are differences between the palynofloral and megaflora records. In general, the entire Hell Creek Formation falls into a single zone, the *Wodehouseia spinata* Assemblage Zone, whereas the megaflora can be easily split into three zones. In addition, the earliest Paleocene megaflora of the overlying Fort Union Formation is almost entirely distinct from the megaflora of the Hell Creek, whereas the Fort Union palynoflora contains a number of taxa in common with the Hell Creek palynoflora. These differences can be attributed to three major causes: differential taxonomic resolution (leaf species generally represent botanical species, whereas pollen "species" may represent botanical genera or even families), variable preservation potential of different plant organs, and different taphonomic histories (pollen and spores are more abundant and can be transported much further than leaves).

Paleoecology

Angiosperms dominate the Hell Creek megaflora and were clearly the primary vegetation on the landscape. Ferns and fern allies are extremely rare as megafossils though more abundant in the palynoflora, which suggests that ferns were present but were primarily herbaceous. Conifers are locally abundant but rarely dominate. Fossil trunks and logs are present but extremely rare. Their near absence seems to be a function of not being preserved because deep-rooting horizons are abundant, and tree leaves are common. The few trunks that are preserved are usually carbonized rather than petrified. Trunk diameter is rarely larger than 30 cm. These observations suggest that the Hell Creek landscape was forested by small trees with trunks smaller than 30 cm and heights of generally less than 20 m. Analysis of Hell Creek paleosols indicates that soils were moderately to poorly drained and conditions were generally humid (Retallack, 1994). Coal seams are rare and thin in the Hell Creek Formation, suggesting that although the area was moist, it was rarely swampy.

The Hell Creek flora differs markedly from slightly older floras such as the early Maastrichtian Meeteetse flora, which is codominated by ferns, cycads, conifers, and angiosperms. By Hell Creek time, angiosperm-dominated woodlands appear to have displaced fern–cycad–palm meadows from the landscape of Wyoming, Montana, and the Dakotas.

Paleoclimate

Leaf margin analysis of dicotyledonous angiosperms from the Hell Creek Formation indicates that the climate ranged from temperate (mean annual temperature of 10°C) low in the formation to nearly tropical (mean annual temperature of 23°C) near the top of the formation (Johnson and Wilf, 1996). The abundance of palms and increase in species richness per locality in the upper portions of the formation also indicates that the warmest climates of Hell Creek time occurred during the final 300,000 years of the Cretaceous. Earliest Paleocene floras are as much as 13°C cooler than latest Cretaceous ones. A humid climate is supported by generally large leaf size, especially in the upper Hell Creek.

Significance to the K–T Boundary Debate

The Hell Creek flora represents the only well-studied latest Cretaceous megaflora that occurs in the same section as vertebrate fossils and the physical evidence (iridium anomaly and shocked minerals) for extraterrestrial impact. As a result, this flora is extraordinarily important for testing the abruptness and magnitude of the K–T extinction event. In general, the palynoflora is best suited for assessing the abruptness and regional extent of an extinction, whereas the megaflora, with its higher taxonomic resolution, is better suited for measuring the magnitude of the event. Detailed analyses of the megafloral (Johnson, 1989, 1992; Johnson *et al.*, 1989) and palynofloral (Hotton, 1988;

Johnson *et al.*, 1989) extinction patterns across the K–T boundary document an abrupt, widespread, high-magnitude floral extinction event at the K–T boundary (disappearance of 79% of megafloral taxa and 25–30% of palynoflora taxa). These floral extinctions are stratigraphically coincident (within 2 cm) with an iridium anomaly at six sites in Montana (Hotton, 1988) and one site in North Dakota (Johnson *et al.*, 1989). The magnitude of the extinctions and the precision of the microstratigraphy is a strong argument for a causal relationship between the asteroid impact and the K–T floral extinctions.

See also the following related entries:

Cretaceous Period • Edmonton Group • Extinction, Cretaceous • Lance Formation • Land Mammal Ages

References

Archibald, J. D. (1996). *Dinosaur Extinction and the End of an Era: What the Fossils Say*, pp. 237. Columbia Univ. Press, New York.

Hotton, C. L. (1988). Palynology of the Cretaceous–Tertiary boundary in central Montana, U.S.A., and its implications for extraterrestrial impact, pp. 732. Unpublished Ph. D. dissertation, University of California at Davis.

Johnson, K. R. (1989). High-resolution megafloral biostratigraphy spanning the Cretaceous–Tertiary boundary in the northern Great Plains, pp. 556. Unpublished Ph.D. dissertation, Yale University, New Haven, CT.

Johnson, K. R. (1992). Leaf–fossil evidence for extensive floral extinction at the Cretaceous–Tertiary boundary, North Dakota, USA. *Cretaceous Res.* **13,** 91–117.

Johnson, K. R. (1996). Description of seven common plant megafossils from the Hell Creek Formation (Late Cretaceous: late Maastrichtian), North Dakota, South Dakota, and Montana. *Proc. Denver Museum Nat. History Ser. 3* **3,** 1–48.

Johnson, K. R., and Hickey, L. J. (1990). Megafloral change across the Cretaceous/Tertiary boundary in the northern Great Plains and Rocky Mountains, USA. In *Global Catastrophes in Earth History: An Interdisciplinary Conference on Impacts, Volcanism, and Mass Mortality* (V. L. Sharpton and P. D. Ward, Eds.), Spec. Paper No. 247, pp. 433–444. Geological Society of America.

Johnson, K. R., Nichols, D. J., Attrep, M., Jr., and Orth, C. J. (1989). High-resolution leaf–fossil record spanning the Cretaceous/Tertiary boundary. *Nature* **340,** 708–711.

Hell Creek Formation

Donald F. Lofgren
Raymond Alf Museum
Claremont, California, USA

The Hell Creek Formation of Montana and the Dakotas represents deposition of fluvial nonmarine sediments along the eastern margin of an interior seaway that once stretched from what is now the Gulf of Mexico to the Arctic Ocean. Hell Creek Formation sediments are primarily sandstones, siltsones, and mudstones that represent channel and floodplain depositional settings.

In 1902, Barnum Brown from the American Museum of Natural History undertook the first paleontological survey of sedimentary rocks, which he would later name the Hell Creek Beds. Soon after beginning exploration, Brown discovered that the Hell Creek Formation contained a well-preserved dinosaur fauna that included *Triceratops*, "*Anatosaurus*," and the type specimen of *Tyrannosaurus rex*. Between 1902 and 1908, Brown excavated two partial skeletons of *T. rex* that were displayed at the Carnegie Museum and the American Museum, which brought him fame as a dinosaur paleontologist. Numerous types of dinosaurs have been collected from the Hell Creek Formation since the early 1900s and the formation is now widely recognized as one of the world's great dinosaur graveyards, containing a diverse assemblage of theropods, ornithopods, pachycephalosaurs, ankylosaurs, and ceratopsids.

The Hell Creek Formation is nearly devoid of lignites. In any one local area, the top of the formation is determined by the presence of the first significant lignite above the highest stratigraphic occurrence of dinosaur remains. In the past, these lignites were also used as a close approximation of the stratigraphic position of the Cretaceous–Tertiary boundary in any one local area. However, the onset of lignite formation varies from one area to the next and these lignites may be Late Cretaceous in some regions and earliest Paleocene at others. Thus, the Hell Creek Formation is almost entirely Late Cretaceous in age but the upper few meters may be Paleocene (based on the presence of Paleocene palynofloras). Abnormally high concentrations of the rare earth element Iridium have been found within some boundary lignites at certain sites and could represent the record of a bolide impact with earth that may have contributed to the extinction of dinosaurs.

A number of sandstone channel fills in the uppermost Hell Creek Formation yield isolated teeth and fragmentary skeletal remains of dinosaurs but also contain Paleocene palynofloras. These records of dinosaur remains in Paleocene channels have been interpreted as evidence of dinosaur survival into the Paleocene Epoch. However, Paleocene channels with dinosaur remains are deeply entrenched into Cretaceous sediments and the dinosaur remains, which are never articulated, may have been eroded and redeposited. Reworking would explain why dinosaur remains are absent in Paleocene floodplain sediments (a depositional setting in which reworking is highly unlikely) and present in Paleocene channel fills within the uppermost Hell Creek Formation. The hypothesis of Paleocene dinosaurs (other than birds) is still a matter of dispute among paleontologists.

See also the following related entries:

CRETACEOUS PERIOD • EXTINCTION, CRETACEOUS • HELL CREEK FLORA • LANCE FORMATION • MESOZOIC FLORAS • PLANTS AND DINOSAURS

References

Estes, R., Berberian, P., and Meszoely, C. A. M. (1969). Lower vertebrates from the late Cretaceous Hell Creek Formation, McCone County, Montana. *Breviora* **337,** 1–33.

Weishampel, D. B. (1990). Dinosaurian Distribution. In *The Dinosauria* (D. B. Weishampel, P. Dodson, and H. Osmólska, Eds.), pp. 63–139. University of California Press, Berkeley, California.

Hemodynamics

see PHYSIOLOGY

Herding

see BEHAVIOR; MIGRATION

Herrerasauridae

FERNANDO E. NOVAS
Museo Argentino de Ciencias Naturales
Buenos Aires, Argentina

Ischigualastian Age (Carnian) beds of South America have yielded remains of several taxa commonly considered the oldest known dinosaurs. Currently, the best represented early dinosaurs are three species of predatory Dinosauromorpha upon which current discussions of the origin and early diversification of dinosaurs are centered. These taxa are *Staurikosaurus pricei*, from the Santa Maria Formation of Brazil and *Herrerasaurus ischigualastensis* and *Eoraptor lunensis*, from the ISCHIGUALASTO FORMATION of Argentina (Fig. 1), all considered Carnian in age. The Santa Maria Formation has also yielded remains of *Spondylosoma*, the dinosaurian nature of which remains to be demonstrated. A brief account of the anatomy and phylogenetic relationships of the oldest carnivorous dinosaurs or near-dinosaurs is offered in this chapter. Other aspects concerned with the paleoecological circumstances associated with the emergence of dinosaurs are considered elsewhere (see EXTINCTION, TRIASSIC; SOUTH AMERICAN DINOSAURS).

Eoraptor

The basal predatory dinosaur *Eoraptor lunensis* is known from an almost complete skeleton of a carnivorous animal nearly 1 m long (Fig. 1A), found in the lower third of the Ischigualasto Formation (Sereno *et al.*, 1993). The discovery of *E. lunensis* increased our knowledge of dinosaurian groups present during the Carnian, representing a new species of carnivorous dinosaur that does not pertain to the herrerasaurid radiation: *Eoraptor* lacks the derived characters recognized for the Herrerasauridae (Novas, 1992, 1994).

On the basis of published information (Sereno *et al.*, 1993), *Eoraptor* is distinguished from other dinosaurs by the presence of a posterolateral process on the premaxilla and a heterodont dentition consisting of leaf-shaped premaxillary and anterior maxillary teeth. Interestingly, *Eoraptor* has a rather generalized anatomy for a dinosaur, lacking the derived skull characters diagnostic of THEROPODS, SAUROPODOMORPHS, and ORNITHISCHIANS. The same also applies

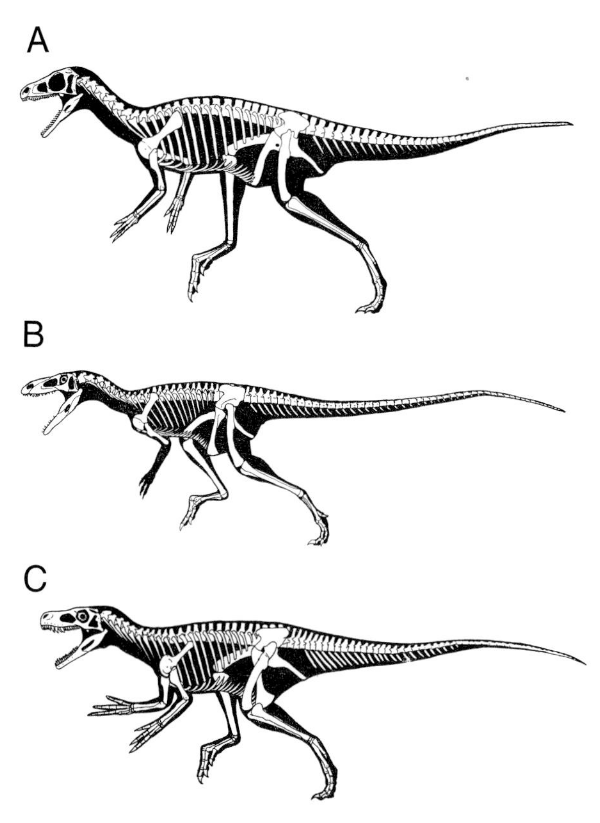

FIGURE 1 Skeletal reconstructions of (A) *Eoraptor lunensis*, (B) *Staurikosaurus pricei*, and (C) *Herrerasaurus ischigualastensis*. Modified from Paul (1988), Sereno *et al.* (1993), and Sereno (1994).

for the POSTCRANIAL SKELETON, although the presence of several derived characters support the consideration of *Eoraptor* as a saurischian theropod (Sereno *et al.*, 1993).

The vertebral column is made up of 24 presacrals, 3 sacrals, and a little more than 40 caudals. The neural spines of the mid- to posterior dorsals are axially expanded, a condition frequently found among basal dinosauriforms but not herrerasaurids, in which the neural spines are anteroposteriorly narrow (Novas, 1992). The scapular blade is broad, resembling that of *Marasuchus*, ornithischians, coelophysids, and sauropodomorphs. The forelimb, nearly one-half of the hindlimb in length, bears well-developed digits I, II, and III, but digits IV and V are strongly reduced, a condition that *Eoraptor* shares with *Herrerasaurus* and Tetanurae (but not with Ceratosauria). The manual unguals are not trenchant (Sereno *et al.*, 1993).

The pelvic girdle and hindlimb bones are rather primitive in appearance, excepting the presence of several derived features shared by Dinosauria (e.g., perforate acetabulum, brevis fossa on posteroventral ilium, and ischium with slender shaft; Novas, 1996). Sereno *et al.* (1993) concluded that the available specimen of *Eoraptor* represents an adult individual on the basis of the closure of sutures in the vertebral column and the partial fusion of the scapulocoracoid. However, fusion between neural arches and centra, as well as between pelvic and scapular girdle bones in herrerasaurids and basal sauropodomorphs, is not always correlated with large size (Novas, 1994). Moreover, Bonaparte (1996) has emphasized that the proportionally large orbital opening of *Eoraptor* may represent a juvenile trait.

Herrerasauridae

This is a clade of basal theropod dinosaurs known from Ischigualastian Age (Carnian) beds of South America (Reig, 1963; Colbert, 1970; Benedetto, 1973; Bonaparte, 1982) and Carnian and Norian beds from North America (Murry and Long, 1989; Long and Murry, 1995). South American herrerasaurids include *Staurikosaurus pricei* (Colbert, 1970; Galton, 1977) (Fig. 1B) from the Santa Maria Formation of southwest Brazil, and *Herrerasaurus ischigualastensis* (Reig, 1963) (Fig. 1C) from the Ischigualasto Formation, San Juan, Argentina. The North American representative is *Chindesaurus briansmalli* from the late Carnian–early Norian Chinle Formation (Murry and Long, 1989; Long and Murry, 1995).

Herrerasauridae was coined by Benedetto (1973) to include *H. ischigualastensis* and *S. pricei*. Recently, this clade was phylogenetically defined to encompass the common ancestor of the previously mentioned species plus all its descendants (Novas, 1992). Poorly known Triassic dinosaurs have sometimes been considered members of Herrerasauridae (e.g., Paul, 1988; Galton, 1985). These include *Alwalkeria maleriensis*, from the Maleri Formation (Carnian) of India (Chatterjee, 1987; Chatterjee and Creisler, 1994) and *Aliwalia rex* from the Elliot Formation (Norian) of South Africa (Galton, 1985). However, none of the synapomorphic traits diagnostic of the Herrerasauridae are documented in the previously mentioned dinosaurs, and so there is no valid reason to include them in the Herrerasauridae (Novas, 1992, 1994).

Herrerasaurids ranged from nearly 2 m (e.g., *Staurikosaurus)* to 4.5 m (e.g., *Chindesaurus* and *Herrerasaurus)* in length. They played the role of large pred-

FIGURE 2 A pair of *Herrerasaurus ischigualastensis* capturing the rhynchosaur *Scaphonyx sanjuanensis*. Illustration by J. Gonzalez.

ators, being surpassed in size among other Triassic predators only by the rauisuchians *Saurosuchus, Prestosuchus,* and *Postosuchus,* which ranged from 4 to 6 m in length (Bonaparte, 1981; Chatterjee, 1985). South American herrerasaurids evolved in faunas numerically dominated by rhynchosaurs and traversodonts, on which the herrerasaurids principally preyed (Fig. 2). Evidence for this is the discovery of a juvenile *Scaphonyx* (rhynchosaur) within the rib cage of one specimen of *Herrerasaurus.*

The anatomy of herrerasaurids is peculiar for its mixture of plesiomorphic and highly derived characters. For example, herrerasaurids have two sacral vertebrae associated with an enlarged pubic "foot" resembling that of Jurassic tetanurans (but not shared by ceratosaurs). This mélange of traits has created and still generates disagreements among paleontologists about their phylogenetic relationships. Herrerasaurids were variously considered theropods or saurischians of uncertain relationships (e.g., Reig, 1963; Benedetto, 1973), carnivorous sauropodomorphs (Galton, 1977; Cooper, 1981), and, recently, the sister group of the clade Saurischia + Ornithischia (Gauthier, 1986; Brinkman and Sues, 1987; Sereno and Novas, 1990; Novas, 1992; Holtz and Padian, 1995). However, many features of the skull, vertebral column, forelimb, and pelvic girdle anatomy suggest, to the contrary, that Herrerasauridae are saurischian theropods (Novas, 1994; Sereno and Novas, 1992, 1994) (Fig. 3).

Herrerasaurids have a long and low skull, subequal in length to the femur. The teeth are conical,

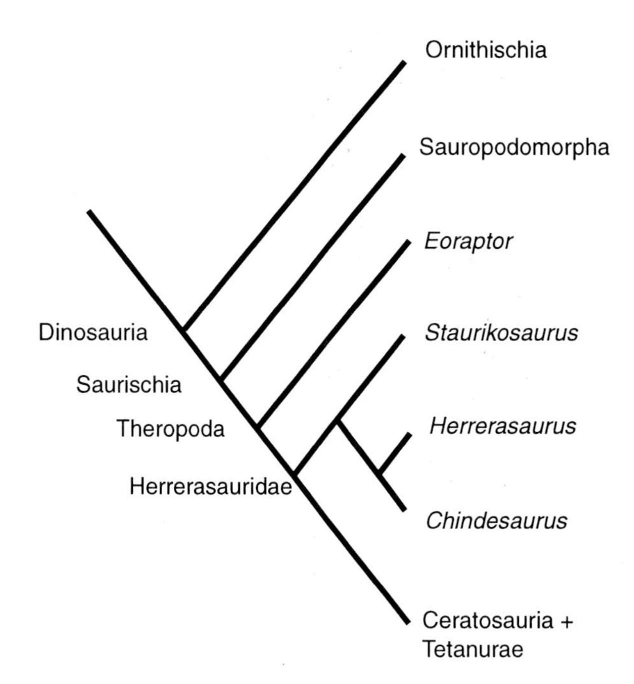

FIGURE 3 Cladogram depicting the phylogenetic relationships of *Eoraptor, Herrerasaurus, Staurikosaurus,* and *Chindesaurus,* as described in the text.

FIGURE 4 Reconstructed skull of *Herrerasaurus ischigualastensis* in left lateral view. From Sereno and Novas (1994) with permission.

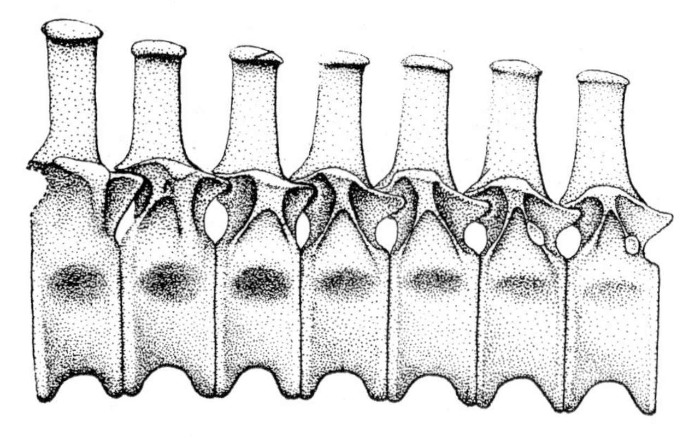

FIGURE 5 Posterior dorsal vertebrae (dorsals 9–15) of *Herrerasaurus ischigualastensis* in left lateral view. From Novas (1994) with permission. Scale bar = 25 mm.

trenchant, and serrated. *Herrerasaurus* has large "caniniform" teeth in a robust and dorsoventrally deep maxilla (Fig. 4; Novas, 1986). This set of characters, in conjunction with a deep jugal, suggests a powerful buccal apparatus. Potentially interesting is the presence in herrerasaurids of what appears to be an intramandibular joint, by which both the dentary and splenial slid over the postdentary bones (e.g., surangular and angular, respectively). This kind of mandibular articulation, also present in a somewhat different morphology in more derived theropods, allowed the toothed anterior segment of the jaw to flex around struggling prey, preventing escape from the mouth (Sereno and Novas, 1994).

The postcranial bones of *Herrerasaurus* are hollow. *Herrerasaurus* and *Staurikosaurus* probably had 24 presacral vertebrae (e.g., 9 cervicals + 15 dorsals), as in other ornithodiran archosaurs (e.g., *Marasuchus;* Bonaparte, 1975). Herrerasaurids are readily discernible from other ARCHOSAURS in that the dorsal vertebrae are axially shortened, not only the centra but also the transverse processes and neural spines (Fig. 5). The stout neural spines are tall and axially very short. In the anterior dorsal vertebrae the neural spines are subrectangular in cross section, but they are square in the posterior dorsals. Both *Herrerasaurus* and *Staurikosaurus* have accessory intervertebral articulations along the dorsal series, a feature only present in theropods and sauropodomorphs among ornithodirans (Gauthier, 1986; Novas, 1994).

Despite the derived characters of the presacral column, the sacrum is composed of only two sacrals (Fig. 6), the lowest number that occurs among dinosaurs. Originally, this character was interpreted in support

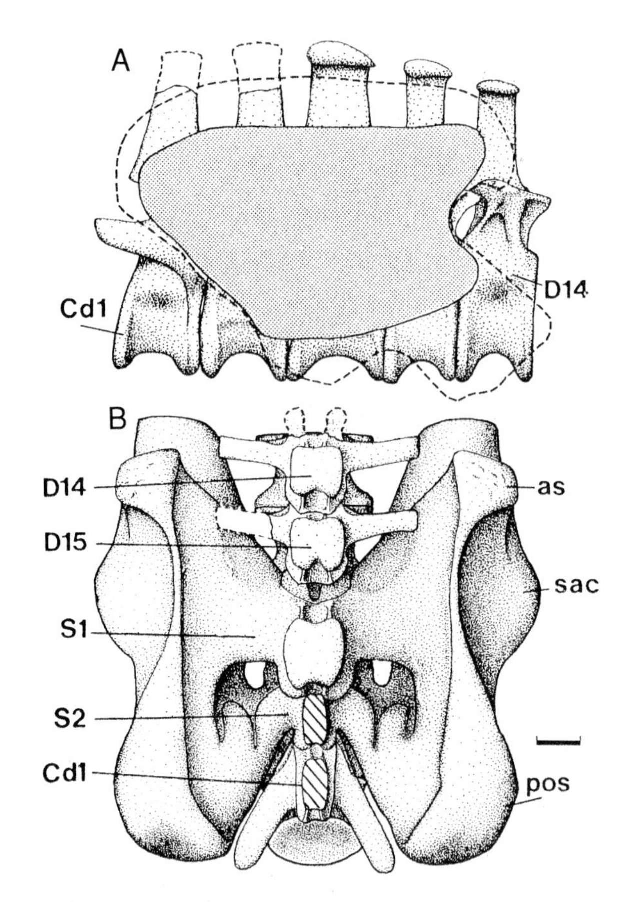

FIGURE 6 Dorsal vertebrae 14 and 15, sacrum, first caudal, and ilia of *Herrerasaurus ischigualastensis* (from Novas, 1994, with permission): (A) lateral and (B) dorsal views. (A) The area of attachment of sacral ribs 1 and 2 (stippled) on the right ilium (dashed line). Scale bar = 25 mm. as, anterior spine; Cd1, first caudal vertebra; D14 and D15, 14th and 15th dorsal vertebrae; pos, posterior spine; sac, supraacetabular crest; S1 and S2, first and second sacral vertebrae.

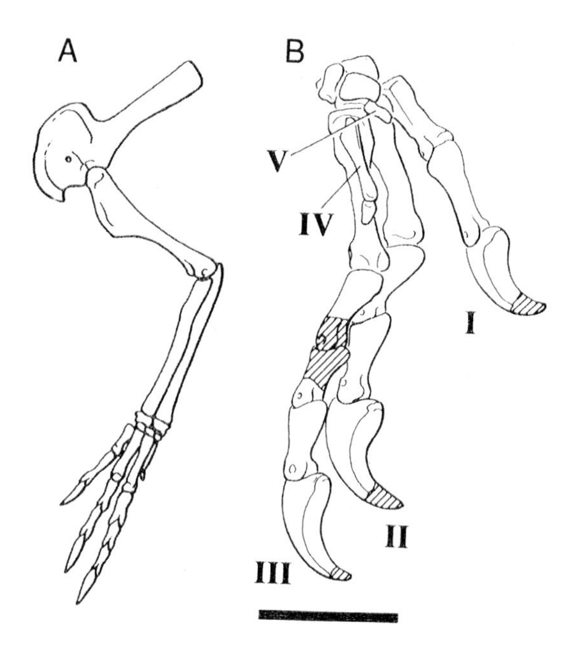

FIGURE 7 Left pectoral girdle and forelimb of *Herrerasaurus ischigualastensis* (A) (not to scale) and left manus in ventrolateral view (B) (scale bar = 5 cm). Modified from Sereno and Novas (1992) with permission. I–V, digits I –V.

of a position of Herrerasauridae outside the dinosaurs (Novas, 1992). However, newer cladistic analyses have supported herrerasaurids as theropods, and consequently the reduction in number of sacrals must be reinterpreted as an evolutionary reversal from dinosaur ancestors provided with three sacrals (Novas, 1994). Despite the reduced number of sacrals, the sacrum is widely attached to the ilia by very robust sacral ribs.

The total number of caudals is probably close to 50. The prezygapophyses become longer, so much so that behind caudal 35 the prezygapophyses overlap the posterior half of the preceding vertebra. This character, widely distributed among theropod dinosaurs, probably acted as a stabilization device for running and leaping.

The PECTORAL GIRDLE and forelimbs of herrerasaurids are also peculiar in several highly derived characters: For example, the scapular shaft (Fig. 7) is narrow and distally unexpanded, a condition resembling that of derived maniraptoran theropods such as dromaeosaurids and birds (Ostrom, 1976). The acromial process of the scapula is well developed, as in the Cretaceous tyrannosaurids (Lambe, 1917), suggesting an important development of the M. deltoides clavicularis, a humeral protractor and elevator (Nicholls and Russell, 1985).

Herrerasaurids were obligatory bipeds, a condition that is primitive to dinosaurs and apparently to ornithodirans in general. The forelimbs are less than half the length of the hindlimbs, but they bear an elongated manus (Fig. 7B; Sereno and Novas, 1992), a condition again resembling that of derived coelurosaurian theropods (Gauthier, 1986). In accordance with the predatory habits inferred from the jaws and teeth, the forelimbs have specialized features for prey capture. Metacarpals IV and V are splint-like and considerably more reduced than in any other Triassic dinosaur, including theropods (e.g., *Coelophysis*; Colbert, 1989), but digits I, II, and III bear elongate phalanges and large and trenchant claws adapted for grasping (Sereno and Novas, 1992).

The mosaic of derived and primitive features is even more impressive in the pelvic girdle and hindlimb (Fig. 8): The ilium is brachyiliac, the acetabulum remains bony and is perforated only by a relatively small and elliptical fenestra, and the brevis fossa is represented by a horizontal furrow in the posterolateral surface of the ilium, in contrast to the transversely wide fossa present in saurischians and ornithischians (Novas, 1992). The pubis, instead, looks highly derived, with its distal half anteroposteriorly expanded, ending distally in a robust pubic foot representing more than 25% of the pubic length.

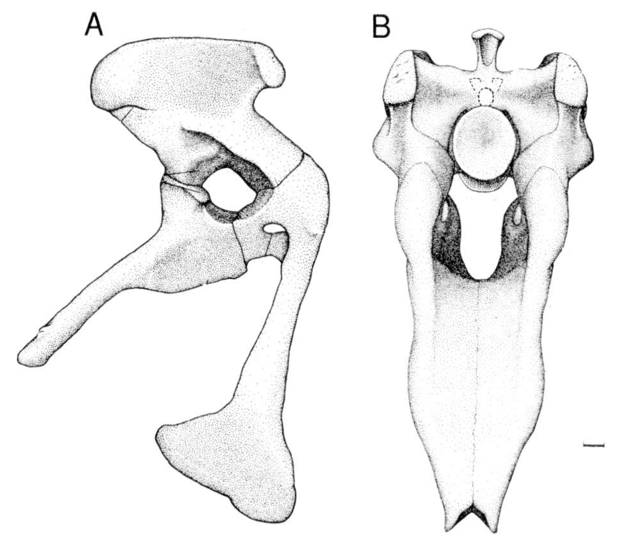

FIGURE 8 Pelvic girdle of *Herrerasaurus* in (A) lateral and (B) anterior views (scale bar = 25 cm).

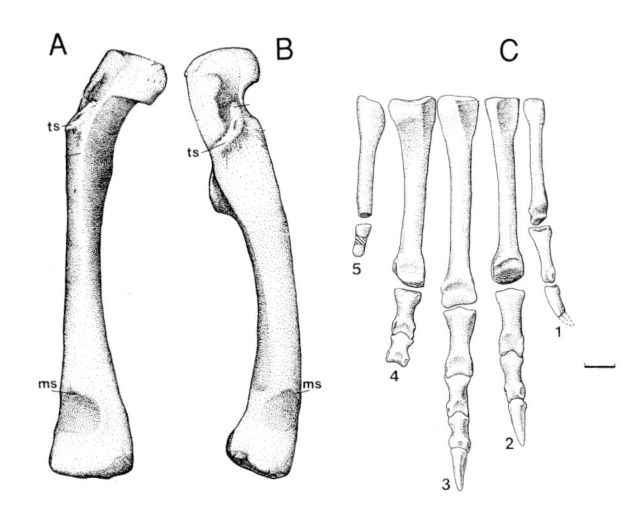

FIGURE 9 Hindlimb bones of *Herrerasaurus.* Right femur in (A) anterior and (B) lateral views; (C) right pes in anterior aspect. From Novas (1994) with permission. Scale bar = 25 mm. ms, muscle scar; ts, trochanteric shelf; 1–5, digits 1–5.

The hindlimb can be summarily described as an enlarged and more robust version of the hindlimb of the basal dinosauriform *Marasuchus* (Bonaparte, 1975; Sereno and Arcucci, 1994), especially in the presence of a trochanteric shelf on the lateral surface of the proximal femur, a modest cnemial crest on the proximal tibia, and a square distal articular surface of the tibia. Although the femur, tibia, and tarsus of the herrerasaurids show distinctive features of Dinosauria (e.g., Novas, 1996), the hindlimb bones lack derived characters that distinguish more advanced theropods (Gauthier, 1986). The tarsus of herrerasaurids resembles that of basal dinosauriforms, such as *Marasuchus* and *Pseudolagosuchus,* and may represent the kind of tarsus from which the remaining dinosaur ankle types evolved (Novas, 1989). The pes, known only in *Herrerasaurus* (Fig. 9), is composed of five metatarsals and has a phalangeal formula of 2-3-4-5-1. Digits 1–4 bear slightly curved, nonraptorial ungual phalanges, the longest and most recurved of which belongs to the second digit.

Herrerasaurus, Staurikosaurus, and *Chindesaurus:* General Comments

Herrerasaurus ischigualastensis (Reig, 1963) is one of the best known Triassic dinosaurs, and detailed anatomical information about it has become available in recent years (e.g., Novas, 1992, 1994; Sereno and Novas, 1992, 1994; Sereno, 1994). Nearly 10 specimens are known of *H. ischigualastensis,* all from the lower third of the Ischigualasto Formation, exposed at the "Hoyada de Ischigualasto," San Juan Province, Argentina (Stipanicic and Bonaparte, 1979). Two other species have been described from the same formation: *Ischisaurus cattoi* (Reig, 1963) and *Frenguellisaurus ischigualastensis* (Novas, 1986). However, their respective type specimens reveal the same synapomorphies recognized for *H. ischigualastensis,* and consequently they are considered junior synonyms of the latter taxon (Novas, 1994).

Among the autapomorphies of *Herrerasaurus,* the pubis exhibits the most remarkable ones (Fig. 8): It is proximally curved, resulting in an almost ventral and slightly posterior orientation of the distal end of the bone (Reig, 1963). This condition emerged convergently in late Mesozoic maniraptorans and segnosaurian theropods (Gauthier, 1986; Paul, 1988). Also, the distal end forms an extensive enlarged "foot," even more enlarged than in *Staurikosaurus.* Also curious is the existence of a definite and unusual subcircular muscle scar on the distal anterior surface of the femur (Fig. 9), a character only reported in *Herrerasaurus* among ornithodirans. This scar may represent peripheral areas of the origin of muscles related with protraction of the tibia.

Staurikosaurus pricei comes from the *Scaphonyx* assemblage zone (Barberena *et al.,* 1985) of the Santa Maria Formation, Ischigualastian Age. Contrary to recent claims (Brinkman and Sues, 1987; Sues, 1990), there is no evidence supporting the presence of *Staurikosaurus* in the Ischigualasto Formation (Novas, 1992, 1994). One potential autapomorphy of *S. pricei* is the presence of a distal bevel on the anterior margin of the pubis (Fig. 10), a feature apparently absent in other dinosaurs and immediate outgroups, in which the anterodistal surface is flat (e.g., *Marasuchus* and early sauropodomorphs) or slightly convex (e.g., *Herrerasaurus* and *Coelophysis*).

Chindesaurus briansmalli (Murry and Long, 1989) is a large predatory dinosaur (3 or 4 m long) that shares several features of Herrerasauridae, for example, two sacrals, brevis fossa absent, dorsal centra anteroposteriorly short and transversely compressed, and anterior iliac notch with a lateral vertical ridge. *Chindesaurus* shares with *Herrerasaurus* a transversely narrow pubic apron, in contrast with *Staurikosaurus* in which the pubis is transversely wider. In summary, *Chindesaurus* may be considered a later member of

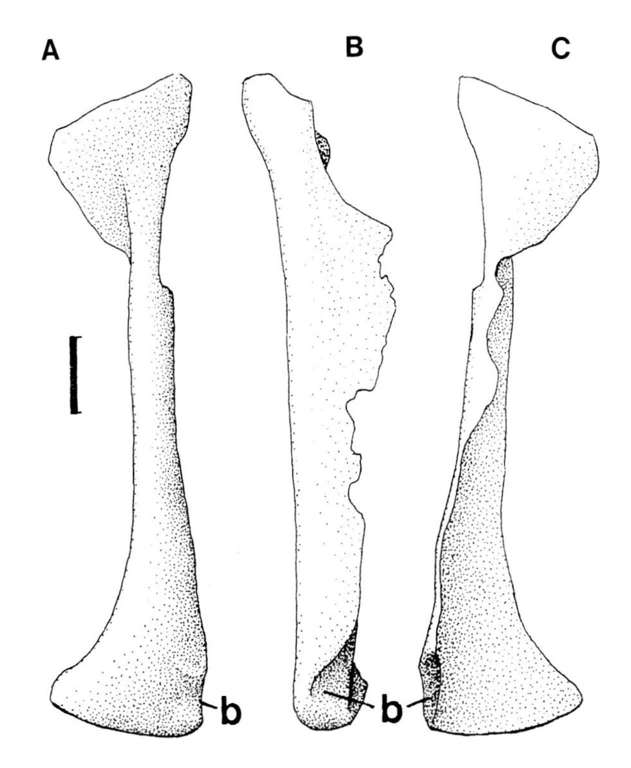

FIGURE 10 Right pubis of *Staurikosaurus* in (A) lateral, (B) anterior, and (C) medial views. From Novas (1994) with permission. Scale bar = 20 mm. b, anterodistal bevel.

the Herrerasauridae that survived into the early Norian in North America.

Phylogenetic Relationships of *Eoraptor* and Herrerasauridae

Eoraptor was described (Sereno *et al.*, 1993) as anatomically close to the predicted structure and size of the common dinosaurian ancestor, although the same authors listed 15 synapomorphic traits present in *Eoraptor* that are absent in the common dinosaurian ancestor and that support the placement of this taxon within Theropoda. Some contrary opinions have been raised against this interpretation (e.g., Padian and May, 1993), but an extensive cladistic analysis of this dinosaur is still lacking.

The position of the Herrerasauridae with respect to the remaining dinosaurs has been erratic since these dinosaurs were first described (Reig, 1963; Colbert, 1970). Herrerasaurids were often thought to be closely related to theropods or theropods themselves (e.g., Reig, 1963; Colbert, 1970; Benedetto, 1973). However, this interpretation was founded on plesiomorphic resemblances shared by herrerasaurids and theropods, for example, the "saurischian" kind of pelvis (e.g., pubis anteroventrally projected), carnivorous teeth, a large head, and a short neck.

Two alternative hypotheses have been recently proposed using cladistic methodology. One hypothesis supports Herrerasauridae as the sister group of the remaining dinosaurs [Ornithischia + (Sauropodomorpha + Theropoda)] (e.g., Gauthier, 1986; Brinkman and Sues, 1987; Benton, 1990; Sereno and Novas, 1990; Novas, 1992; Holtz and Padian, 1995). A second alternative depicts herrerasaurids nested within Saurischia as basal members of the Theropoda (Sereno and Novas, 1992, 1994; Sereno *et al.*, 1993; Novas, 1994). Such radically different results are explained by the fact that the latter hypothesis is supported by cranial and postcranial anatomical data of *Herrerasaurus* that only became available after recent discoveries (Sereno and Novas, 1992, 1994; Novas, 1994; Sereno, 1994).

Herrerasaurids are saurischians because they have the following derived traits also shared by theropods and sauropodomorphs: subnarial foramen present; jugal posterior process forked; jugal overlaps laterally onto lacrimal; epipophyses present on axis; axial postzygapophyses set lateral to prezygapophyses; hyposphene–hypantrum articulation in dorsal vertebrae; bases of metacarpals IV and V lie on the palmar surfaces of manual digits 3 and 4, respectively; digit 1 with proximal phalanx longer than its metacarpal; and ischium with a rod-like shaft.

Herrerasauridae have 12 theropod synapomorphies (e.g., cervical epipophyses prong-shaped; trochanteric shelf on the proximal femur; humerus nearly 50% of femoral length; metacarpals I–III with deep extensor pits; metacarpal IV strongly reduced, with fewer than two phalanges; metacarpal V strongly reduced and without phalanges; splenial posterior process spoon-shaped and horizontally placed below angular; angular anteriorly hooked; lacrimal exposed on skull roof; distal caudal prezygapophyses elongate; penultimate manual phalanges elongate; manual unguals of digit 2 and 3 enlarged, compressed, sharply pointed, strongly recurved, and with enlarged flexor tubercles; and pubis distally enlarged). In the sense used here, Theropoda includes not only Ceratosauria and Tetanurae but also Herrerasauridae and *Eoraptor*, as defended by Sereno *et al.* (1993) and Novas (1994).

In the context of the phylogenetic interpretation defended here, herrerasaurids apomorphically reversed some characters to the ancestral ornithodiran condition. Those features include the presence of two sacrals and the reduction of the brevis shelf to a slight ridge that laterally bounds a feebly developed brevis fossa. According to the hypothesis accepted here, the number of sacrals increased from two to three in the common dinosaur ancestor, but a vertebral segment was apomorphically lost in herrerasaurids. Similarly, the reduction of the brevis shelf to a slight ridge and the virtual absence of the brevis fossa are interpreted as apomorphic reversals uniting *Herrerasaurus*, *Staurikosaurus*, and *Chindesaurus*.

Conclusions

The basal theropods *Herrerasaurus*, *Staurikosaurus*, and *Eoraptor* shed light on early dinosaur diversification. First, the contemporaneous occurrence of these taxa with the ornithischian *Pisanosaurus* indicates that the main herbivorous and carnivorous lineages of dinosaurs were established during the middle Carnian. Second, if herrerasaurids and *Eoraptor* represent early theropods, then several lineages at the base of the phylogenetic tree of Dinosauria must have occurred before the deposition of the Santa Maria and Ischigualasto formations; that is, before early Carnian or late Ladinian times. Third, the Middle Triassic can be envisaged as an early stage in the history of dinosaurs characterized by a high degree of evolutionary activity, including morphological innovations leading to the origin of peculiar forms such as herrerasaurids. Also, during this stage the split into saurischians and ornithischians, and the origins of different subsets within each of these major clades, had taken place. Fourth, although early dinosaurs were relatively small forms (nearly 1 m long), they rapidly climbed the food pyramid to share the role of top predator with other large archosaurs as early as the middle Carnian. Fifth, accepting *Eoraptor* and Herrerasauridae as saurischian theropods, it follows that truly basal dinosaurs remain to be discovered.

See also the following related entries:

Dinosauria • Dinosauromorpha • Extinction, Triassic • Ornithodira • Phylogeny of Dinosaurs

References

Barberena, M., Araujo, D., and Lavina, E. (1985). Late Permian and Triassic tetrapods of Southern Brazil. *Natl. Geogr. Res.* **1,** 5–20.

Benedetto, J. L. (1973). Herrerasauridae, nueva familia de saurisquios trisicos. *Ameghiniana* **10,** 89–102.

Benton, M. J. (1990). Origin and interrelationships of dinosaurs. In *The Dinosauria* (D. B. Weishampel, P. Dodson, and H. Osmólska, Eds.), pp. 11–30. Univ. of California Press, Berkeley.

Bonaparte, J. F. (1975). Nuevos materiales de *Lagosuchus talampayensis* Romer (Thecodontia, Pseudosuchia) y su significado en el origen de los Saurischia. Chañarense inferior, Triásico medio de Argentina. *Acta Geol. Lilloana* **13,** 5–90.

Bonaparte, J. F. (1976). *Pisanosaurus mertii* Casamiquela and the origin of the Ornithischia. *J. Paleontol.* **50,** 808–820.

Bonaparte, J. F. (1981). Descripción de *Fasolasuchus tenax* y su significado en la sistemitica y evolución de los thecodontia. *Revista del Museo de Ciencias Nat. "Bernardino Rivadavia" Paleontol.* **3,** 55–101.

Bonaparte, J. F. (1982). Faunal replacement in the Triassic of South America. *J. Vertebr. Paleontol.* **2,** 362–371.

Bonaparte, J. F. (1996). *Dinosaurios de América del Sur*, pp. 174. Museo Argentino de Ciencias Naturales "B. Rivadavia.".

Brinkman, D., and Sues, H.-D. (1987). A staurikosaurid dinosaur from the Ischigualasto Formation (Upper Triassic) of Argentina and the relationships of the Staurikosauridae. *Palaeontology* **30**(3), 493–503.

Chatterjee, S. (1985). *Postosuchus*, a new thecodontian reptile from the Triassic of Texas and the origin of tyrannosaurs. *Philos. Trans. R. Soc. London B* **309,** 395–460.

Chatterjee, S. (1987). A new theropod dinosaur from India with remarks on the Gondwana–Laurasia connection in the Late Triassic. In *Gondwana Six: Stratigraphy, Sedimentology and Paleontology* (G. D. McKenzie, Ed.), Geophysical Monogr, No. 41, pp. 183–189. American Geophysical Union, Washington, DC.

Chatterjee, S., and Creisler, B. S. (1994). *Alwalkeria* (Theropoda) and *Morturneria* (Plesiosauria), new names for preoccupied *Walkeria* Chatterjee 1987 and *Turneria* Chatterjee and Small 1989. *J. Vertebr. Paleontol.* **14,** 142.

Colbert, E. (1970). A saurischian dinosaur from the Triassic of Brazil. *Am. Museum Novitates* **2405,** 1–39.

Colbert, E. (1989). The Triassic dinosaur *Coelophysis*. *Museum Northern Arizona Bull.* **57,** i–xv, 1–160.

Cooper, M. (1981). The prosauropod dinosaur *Massospondylus carinatus* Owen from Zimbabwe: Its biology, mode of life and phylogenetic significance. *Natl. Museum Rhodesia Occasional Papers* **6**(10), 689–840.

Galton, P. M. (1977). *Staurikosaurus pricei*, an early saurischian dinosaur from the Triassic of Brazil, with notes on the Herrerasauridae and Poposauridae. *Palaontol. Z.* **51,** 234–245.

Galton, P. M. (1985). The poposaurid thecodontian *Teratosaurus suevicus* v. Meyer, plus referred specimens mostly based on prosauropod dinosaurs, from the Middle Stubensandstein (Upper Triassic) of Nordwurttemberg. *Stuttgarter Beitrage Naturkunde Ser. B* **116,** 1–29.

Gauthier, J. A. (1986). Saurischian monophyly and the origin of birds. *Mem. California Acad. Sci.* **8,** 1–55.

Holtz, T. R., Jr., and Padian, K. (1995). Definition and diagnosis of Theropoda and related taxa. *J. Vertebr. Paleontol.* **15**(Suppl. to No. 3), 35A.

Lambe, L. M. (1917). The Cretaceous theropodous dinosaur *Gorgosaurus. Mem. Geol. Survey Can.* **100,** 1–84.

Long, R. A., and Murry, P. A. (1995). Late Triassic (Carnian and Norian) tetrapods from the southwestern United States. *New Mexico Museum Nat. History Sci. Bull.* **4,** 1–254.

Murry, P. A., and Long, R. A. (1989). Geology and paleontology of the Chinle Formation, Petrified Forest National Park and vicinity, Arizona and a discussion of vertebrate fossils of the southwestern Upper Triassic. In *Dawn of the Age of Dinosaurs in the American Southwest* (S. G. Lucas and A. P. Hunt, Eds.), pp. 29–64. New Mexico Museum of Natural History, Albuquerque.

Nicholls, E., and Russell, A. P. (1985). Structure and function of the pectoral girdle and forelimb of *Struthiomimus altus* (Theropoda: Ornithomimidae). *Palaeontology* **28,** 643–677.

Novas, F. E. (1986). Un probable terópodo (Saurischia) de la Formación Ischigualasto (Triásico superior), San Juan, Argentina. IV Congreso Argentino de Paleontología y Estratigrafía, Vol. 2, pp. 1–6 Mendoza, Argentina.

Novas, F. E. (1989). The tibia and tarsus in Herrerasauridae (Dinosauria, incertae sedis) and the origin and evolution of the dinosaurian tarsus. *J. Paleontol.* **63,** 677–690.

Novas, F. E. (1992). Phylogenetic relationships of basal dinosaurs, the Herrerasauridae. *Palaeontology* **35,** 51–62.

Novas, F. E. (1994). New information on the systematics and postcranial skeleton of *Herrerasaurus ischigualastensis* (Theropoda: Herrerasauridae) from the Ischigualasto Formation (Upper Triassic) of Argentina. *J. Vertebr. Paleontol.* **13,** 400–423.

Novas, F. E. (1996). Dinosaur monophyly. *J. Vertebr. Paleontol.* **16**(4), 723–741.

Ostrom, J. H. (1976). *Archaeopteryx* and the origin of birds. *Biol. J. Linnean Soc.* **8,** 91–182.

Padian, K., and May, C. L. (1993). The earliest dinosaurs. *New Mexico Museum Nat. History Sci. Bull.* **3,** 379–381.

Paul, G. (1988). *Predatory Dinosaurs of the World*, pp. 464. Simon & Schuster, New York.

Reig, O. A. (1963). La presencia de dinosaurios saurisquios en los "Estratos de Ischigualasto" (Mesotriásico superior) de las provincias de San Juan y La Rioja (República Argentina). *Ameghiniana* **3,** 3–20.

Sereno, P. (1994). The pectoral girdle and forelimb of the basal theropod *Herrerasaurus ischigualastensis. J. Vertebr. Paleontol.* **13,** 425–450.

Sereno, P. C., and Arcucci, A. B. (1994). Dinosaurian precursors from the Middle Triassic of Argentina: *Marasuchus lilloensis*, gen. nov. *J. Vertebr. Paleontol.* **14,** 53–73.

Sereno, P. C., and Novas, F. E. (1990). Dinosaur origins and the phylogenetic position of pterosaurs. *J. Vertebr. Paleontol.* **10,** 42A.

Sereno, P. C., and Novas, F. E. (1992). The complete skull and skeleton of an early dinosaur. *Science* **258,** 1137–1140.

Sereno, P. C., and Novas, F. E. (1994). The skull and neck of the basal theropod *Herrerasaurus ischigualastensis. J. Vertebr. Paleontol.* **13,** 451–476.

Sereno, P. C., Forster, C. A., Rogers, R. R., and Monetta, A. M. (1993). Primitive dinosaur skeleton from Argentina and the early evolution of Dinosauria. *Nature* **361,** 64–66.

Stipanicic, P., and Bonaparte, J. F. (1979). Cuenca triásica de Ischigualasto–Villa Unión (Provincias de La Rioja y San Juan). II Simposio de Geología Argentina, Vol. 1, pp. 523–575. Academia Nacional de Ciencias de Córdoba, Córdoba.

Sues, H.-D. (1990). Staurikosauridae and Herrerasauridae. In *The Dinosauria* (D. B. Weishampel, P. Dodson, and H. Osmólska, Eds.), pp. 143–147. Univ. of California Press, Berkeley.

Heterochrony

JOHN A. LONG
KENNETH J. MCNAMARA
Western Australian Museum
Perth, Western Australia

Heterochrony is the missing link between GENETICS and natural selection, providing the raw material upon which natural selection works (McNamara, 1997). It is the mechanism by which species evolve through changes in the timing or rate of development. Thus, if sexual maturity occurs earlier in the descendant than in the ancestor (progenesis) a juvenile morphology occurs in the descendant adult (paedomorphosis). Conversely, if onset of maturity is delayed (hypermorphosis), extension of rapid juvenile growth trajectories will result in attainment of a larger body size and morphologically more "developed" adults (peramorphosis). Similarly, paedomorphosis can occur by delaying the rate of growth (neoteny) or delaying the onset time of growth of a particular structure (postdisplacement); whereas peramorphosis can also occur by an increase in growth rate (acceleration) or initiating growth relatively earlier (predisplacement). These mechanisms have been reviewed in detail by McKinney and McNamara (1991). Most organisms evolved by a combination of peramorphic and paedomorphic processes. Surprisingly, little detailed work has been carried out on the role of heterochrony in dinosaur evolution. The only workers that have undertaken overviews of its importance are Weishampel and Horner (1994) and Long and McNamara (1995, 1997).

Tyrannosaurids

Tyrannosaurids differ from other theropods in the possession of several derived features, many of which are of heterochronic derivation, being related to changes in ontogenetic development, often peramorphic. Some of the observed ontogenetic changes in tyrannosaurids include closure and some fusion of cranial sutures; increase in serration count on teeth; skull deepens dorsoventrally with age and muzzle shortens; orbits become less rounded and the suborbital bar (where present) develops late in ontogeny (Fig. 1); metatarsals become more robust and thick; increase in relative length of presacral column; disproportionate growth of cervical neural spines; increase in relative size of limb girdles, especially in the size of the pubic boot; decrease in relative length of hindlimb but increase in width. Compared with skulls of juvenile theropods and of adults of smaller species, tyrannosaurids possessed a relatively larger head. This is probably a peramorphic feature, arising from allometric scaling with the large body size. The retention of a more slender skull in the smaller tyrannosaurids *Nanotyrannus* and *Maleevosaurus*, however, is probably an apaedomorphic feature. The peramorphic skull of *Tyrannosaurus* probably underlies the evolution of the massive hindlimbs in the larger forms. The very short forelimbs and manus, in which there are only two digits, however, are paedomorphic features. This trend is carried to extremes in the small

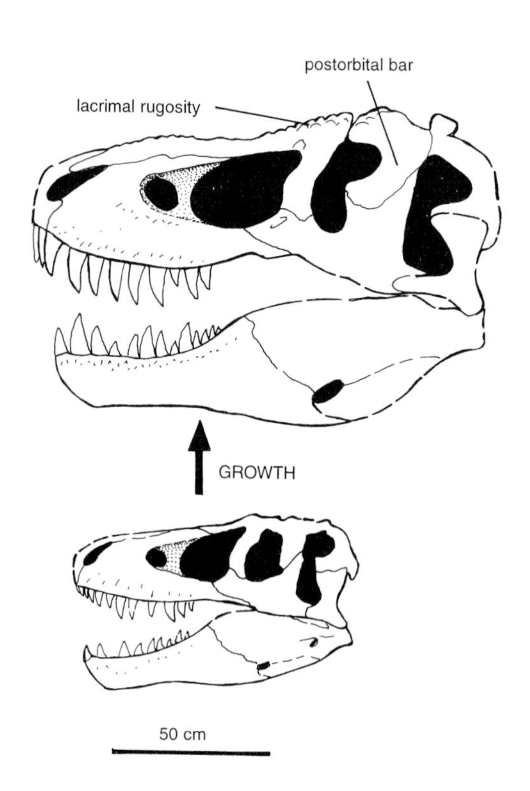

FIGURE 1 Juvenile and adult skulls of *Tyrannosaurus* (*Tarbosaurus*) *bataar* showing changes in general skull proportions and identifying significant areas of late-stage ontogenetic change.

avialian theropod *Mononykus*, which has only one large digit in each manus. This is dissociated heterochrony. It involves a combination of peramorphosis of body size, skull, and hindlimbs, with paedomorphic reduction in the forelimbs and manus. As such it may represent a developmental trade-off, with some features undergoing increased growth in descendants, with a compensatory, paedomorphic reduction in other areas. Reduction and modification of digits and reduced forearm size is a common trend in the evolution of large theropods, especially tyrannosaurids and some other derived maniraptorans, such as abelisaurids. Another important peramorphic trend occurs in the pelvis. During ontogeny the pelvis of *Albertosaurus libratus* undergoes major morphological changes, especially the pubis. Hatchling *Albertosaurus* possess a long, slender pubis with very weak development of the "boot." This structure increases greatly in size during ontogeny. Advanced adult tyrannosaurids, such as *Tyrannosaurus rex*, have a strongly expanded pubic boot with an anterior region larger than the posterior division. Analysis of tyrannosaurid bone structure suggests that their growth rate was relatively rapid. This suggests that peramorphic features in tyrannosaurids, even in large forms, may have been a function of delayed onset of maturation (hypermorphosis) combined with acceleration in growth of specific structures.

The Evolution of Birds from Theropods

Theropod dinosaurs are generally accepted as being the ancestral group from which birds arose. The evolution of birds illustrates another example of dissociated heterochrony. A number of features indicate that birds are, in some respects, paedomorphic theropod dinosaurs. Feathers may have been present on juvenile theropods. Support for this idea comes from the recent finding of *Sinosauropteryx prima*, a small feathered theropod dinosaur from the Cretaceous of China. Other paedomorphic features are retention of small body size, unfused bones, the large orbit, and tooth shape. The teeth of *Archaeopteryx* resemble the simple, peg-like teeth found in juvenile *Velociraptor*. Another important feature of birds, the enlargement of the forelimbs and manus, is a peramorphic effect arising from localized acceleration in growth of the bones. Dissociated heterochrony, again involving more specifically localized developmental trade-offs, occurred in the avialian theropod *Mononykus*. There is strong paedomorphic reduction in a number of the elements of the forelimbs, in particular the extreme reduction in the major and minor digits, and extremely short ulna and radius. This is in contrast to the presence of a massive carpometacarpus and robust claw of the alular digit.

Sauropodomorphs

Growth patterns in prosauropod dinosaurs have not been well documented, although comparisons can be made between very small juveniles, such as *Mussaurus*, with those of an adult from the same family, such as *Plateosaurus*. Ontogenetic differences include a great increase in snout length and corresponding increase in maxilla length anteriorly and in teeth numbers along the jaws; and more posterior development of the quadrate and enlargement of the ventral division of the temporal fenestra. Some of these trends are further developed in sauropods, in which the skull generally shows a continued enlargement of the maxilla anteriorly (Fig. 2). Moreover, a ventrally situated mandibular jaw joint with quadrate ventrally extended and very large ventral expansion of the temporal fenestra occurs in some groups (e.g., camarasaurids and brachiosaurids). Diplodocids and nemegtosaurids further continue the trend of maxilla expansion in having very deep maxillae, which extend further forward of the nares than in other sauropod groups (Fig. 2). This has had the effect of pushing the nares dorsally in some groups (e.g., diplodocids and brachiosaurids). These trends indicate peramorphic evolution of the skull, commencing with moderate ontogenetic change in prosauropods and extending to increased ontogenetic development throughout the lineages of camarasaurids, brachiosaurids, diplodocids, and nemegtosaurids. Juvenile material of the sauropod *Phuwiangosaurus sirindhornae* from Thailand shows that young sauropods possess a number of features that are otherwise found in the adults of much older sauropods, further indicating peramorphosis. The centra and pleurocoels of the juvenile vertebrae are very simple, unlike the more complex development seen in the adults. Primitive, Middle Jurassic, adult sauropods possessed a vertebral morphology similar to that displayed by these later juveniles. The femora of the juveniles of *Phuwiangosaurus* are more elongate than those in the adult, similar to the condition found in more primitive sauropod adults. In addition to such anatomical changes, the massive increase in size of sauropods and the greater elongation of the necks and tails from their prosauropod ancestors are clear evidence of either increased growth rate and/or longer period of growth. Sauropodomorphs were the largest land animals that ever lived. This was achieved through peramorphosis, perhaps arising from both hypermorphosis and acceleration.

Ornithopods

Ontogenetic changes mainly described in the ornithopods *Dryosaurus* and *Maiasaura* bone structure studies indicate that, as in tyrannosaurids, the young dinosaurs grew at a rapid rate. The following general trends in higher ornithopod ontogeny occurred: decrease in size of orbit; decrease in size of palpebral bones relative to orbit; increase in size (length and width) of premaxillary and nasal bones, eventually expanding to form "bill" in hadrosaurs; snout prolonged in large forms; prefrontal brow ridges in some forms (e.g., *Maisaura* and *Prosaurolophus*); new tooth

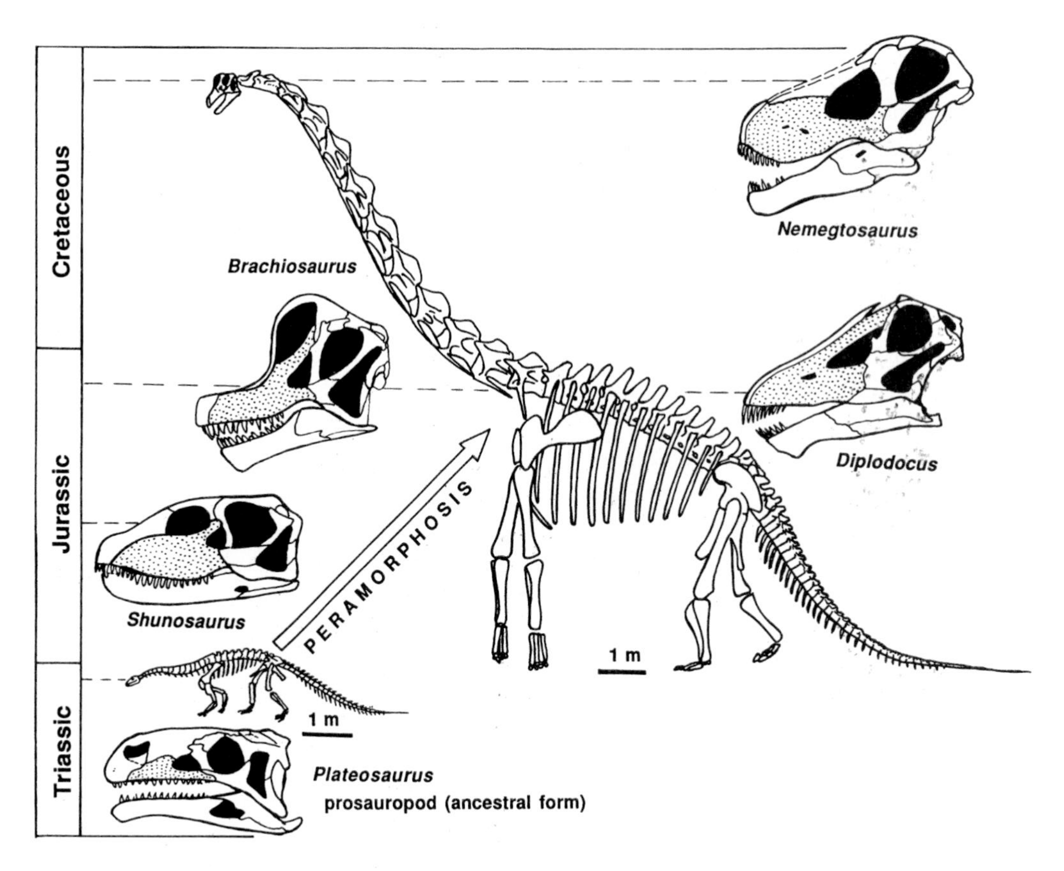

FIGURE 2 Comparisons between prosauropod skull and various sauropod skulls showing the peramorphic development of the maxilla (stippled). The massive size increase in sauropodomorphs from the Late Triassic through to the Late Jurassic is a peramorphic feature.

rows added to both lower and upper jaws; neural canal decrease in relative size in vertebrae; and neural spines increase in size. Major cranial differences between adult hypsilophodontids, *Dryosaurus*, and *Tenontosaurus* are the possession by *Tenontosaurus* of a longer snout, smaller orbit, straight quadrate bone, long and narrow external antorbital fenestra, the absence of premaxillary teeth (present only in hypsilophodontids within this clade), long curved retroarticular process on the mandible, short palpebral bone, and fenestra enclosed entirely by the quadratojugal. Although a few of these changes are paedomorphic, such as the short palpebral bones, relatively larger and elongate antorbital fenestra, and reduced digits and forearms, the most dramatic differences are peramorphic, such as the relatively straight quadrate bone, the longer snout length, development of the retroarticular process, and the relatively smaller orbits. The increase in growth to reach a large body size (up to 7.5 m) in *Tenontosaurus* may account for the reduced size of digits and forearms because these elements were probably dissociated from the overall peramorphic trends (Fig. 3), being development trade-offs similar to those in tyrannosaurids.

Premaxillary teeth occur in the most primitive stegosaurid genus *Huayangosaurus* as well as in basal thyreophorans such as *Scutellosaurus*, but were lost independently in advanced lineages of stegosaurids and in ankylosaurids. A similar paedomorphic loss of teeth within a monophyletic lineage occurred in bird evolution. Thus, in *Archaeopteryx* premaxillary teeth are present, but in the late Mesozoic Hesperornithiformes premaxillary teeth are absent but dentaries and maxillae are toothed. Complete loss of dentition occurs in later birds. The presence of premaxillary teeth is only known in small ornithischians (including

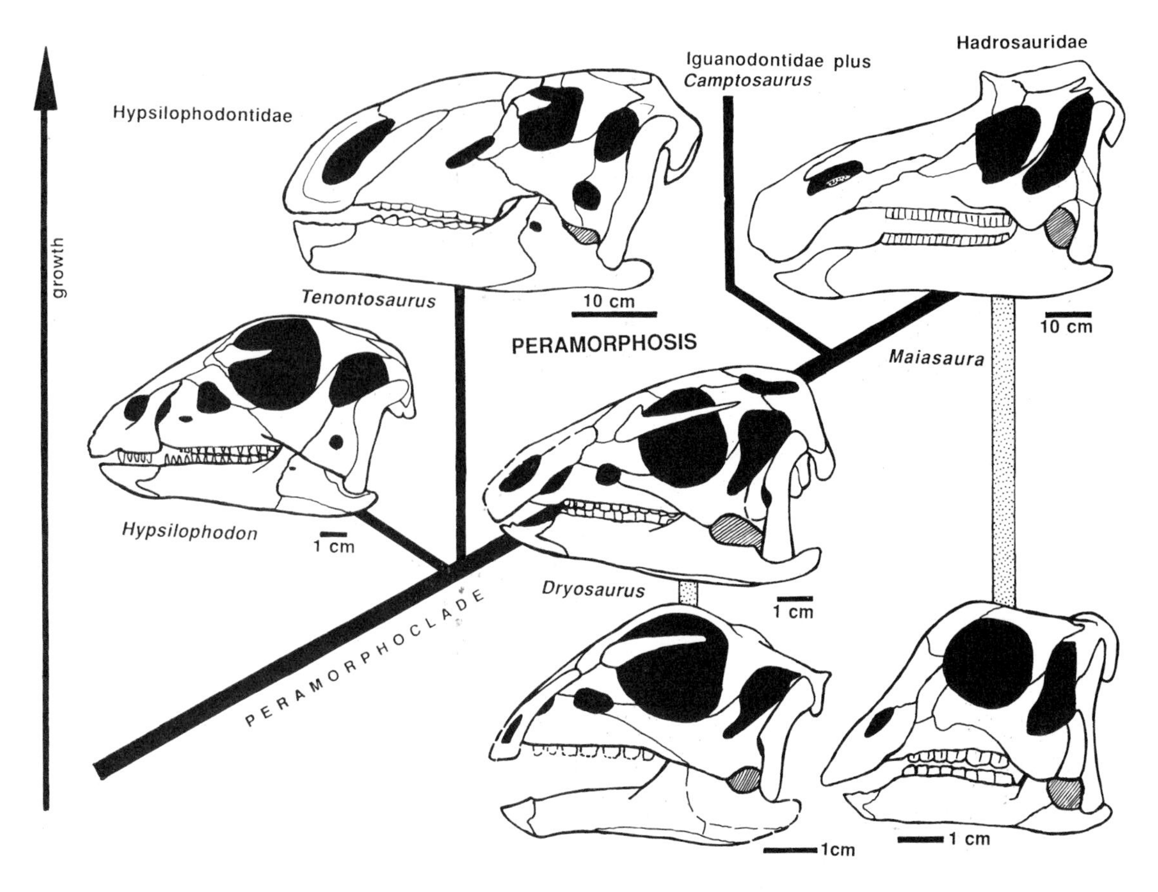

FIGURE 3 Growth series and effect of peramorphosis in euornithopod evolution.

basal taxa such as *Lesothosaurus*), and the loss of such teeth appears to be a feature of all large members of the group that exceed 4 m in length.

Ceratopsians

Ontogenetic development in ceratopsians has been described in *Bagaceratops*, *Leptoceratops*, *Protoceratops*, and *Breviceratops*. Most ontogenetic changes occurred in the scapula, femur, ilium, and tibia. Ontogenetic changes also occurred in the skull of neoceratopsians, including decrease in orbit diameter; increase in snout length (slight); slight increase in frill length, followed by subsequent shortening of the frill; widening of the frill; widening of the jugal and quadrate area; and development of the nasal horn (in *Protoceratops*). The observed evolutionary trends, in terms of both size increase from psittacosaurids to primitive and advanced neoceratopsians and increase in morphological complexity and extent of the frill and horn arrangements, indicate that peramorphosis has been a principal factor in ceratopsian evolution (Fig. 4). The postcranial skeleton is remarkably conservative in the two advanced groups, the Centrosaurinae and Chasmosaurinae. The major evolutionary trends are the flattening of the digits in the large ceratopsians to accommodate the greater weight, and increase in size of the olecranon process of the ulna in chasmosaurines. Most postcranial elements in *Chasmosaurus* grow nearly isometrically without major changes to overall morphology. The thickening and broadening of certain limb elements in large ceratopsids, such as *Triceratops*, would appear to be a peramorphic development for graviportal posture in response to the great increase in body weight.

Summary

Heterochrony has been a major factor in dinosaur evolution. Although, theoretically, paedomorphosis should be as frequent as peramorphosis, the dominant factor in dinosaur evolution was peramorphosis.

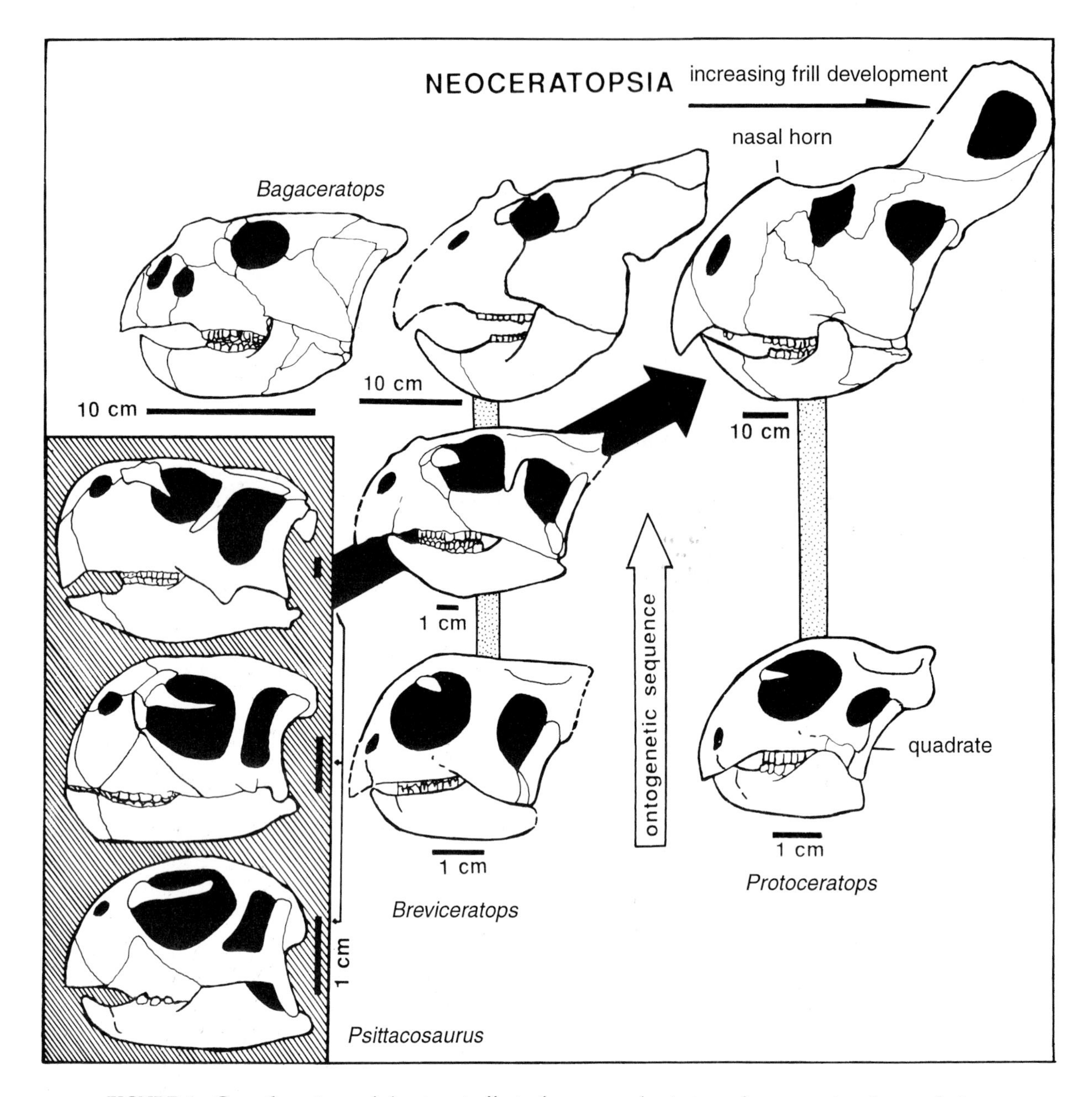

FIGURE 4 Growth series and dominant effect of peramorphosis in early neoceratopsian evolution.

Its dominance may well be due to the operation of more than one peramorphic process; hypermorphosis and acceleration. This domination may have been possible only because of developmental trade-offs, as shown by the mosaic of heterochronic features found in many groups. The same effect contributed to the evolution of birds from dinosaurs, but paedomorphic features were as important as peramorphic features in this adaptive breakthrough that resulted in dinosaurs taking to the air.

See also the following related entries:

EVOLUTION • GROWTH AND EMBRYOLOGY

References

Long, J. A., and McNamara, K. J. (1995). Heterochrony in dinosaur evolution. In *Evolutionary Change and Heterochrony* (K. J. McNamara, Ed.), pp. 151–168. Wiley, Chichester, UK.

Long, J. A., and McNamara, K. J. (1997). Heterochrony: The key to dinosaur evolution. In *Dino Fest 2* (D. Wolberg, Ed.). Journal of Paleontology memoir.

McKinney, M. L., and McNamara, K. J. (1991). *Heterochrony: The Evolution of Ontogeny*, pp. 420. Plenum, New York.

McNamara, K. J. (1997). *Shapes of Time*. Johns Hopkins Univ. Press, Baltimore.

Weishampel, D. B., and Horner, J. R. (1994). Life history syndromes, heterochrony, and the evolution of Dinosauria. In *Dinosaur Eggs and Babies* (K. Carpenter, K. F. Hirsch, and J. R. Horner, Eds), pp. 227–243. Cambridge Univ. Press, Cambridge, UK.

Heterodontosauridae

JOSHUA B. SMITH
University of Pennsylvania
Philadelphia, Pennsylvania, USA

Heterodontosauridae was erected by Romer (1966) to acknowledge shared characteristics between the small ornithopods *Heterodontosaurus* Crompton and Charig 1962 and *Lycorhinus* Haughton 1924. Heterodontosaurids are small (1 or 2 m long), bipedal, herbivorous ORNITHISCHIANS known only (except for some dubious material from North America) from the Early Jurassic of South Africa (Fig. 1). They are considered the most primitive members of the Ornithopoda (Weishampel and Witmer, 1990).

Lycorhinus skull material, described by Haughton (1924), was the first appearance of these small herbivores in the literature. *Heterodontosaurus* was discovered during a 1961–1962 expedition and was named by Harvard paleontologist A. W. Crompton (then of the SOUTH AFRICAN MUSEUM) and A. J. Charig of the British Museum. *Abrictosaurus consors* was added to the Heterodontosauridae by J. Hopson of the University of Chicago in 1975.

Of the three genera that comprise the Heterodontosauridae, two are based on incomplete or incompletely described material. The three genera are represented each by a single species: *Heterodontosaurus tucki* is known from two well-preserved skulls and a single articulated skeleton (Fig. 2), whereas material referred to *A. consors* and *Lycorhinus angustidens* consist, respectively, of two skulls (one associated with a partial skeleton) and isolated dentary and maxillae elements.

According to Weishampel and Witmer (1990), the Heterodontosauridae can be diagnosed on the basis of high-crowned cheek teeth with chisel-shaped crowns, denticles on the uppermost third of the crown, and the presence of large, canine-like teeth in both the premaxilla and the dentary (Fig. 3). The cheek teeth denticles are restricted to the apical third of the tooth crowns. This and the presence of the large, caniniform teeth characterize the Heterodontosauridae as a monophyletic group. Precise diagnosis of the characteristics and position of the taxon is difficult, however, given that two of the three genera are known only from incomplete, poorly preserved specimens.

Only a small amount of research has been done on the Heterodontosauridae compared to other ornithischian taxa. This is probably a result of the small sample size and the fragmentary condition of most of the known material. As a result, we know precious little about the paleobiology of these early ornithischians. The high, chisel-shaped crowns of heterodontosaurid cheek teeth suggest that they were herbivores. This is substantiated by the rostral region of the premaxilla of *H. tucki*, which is edentulous and probably bore a rhampotheca (a tough, horny covering) in life that could have facilitated the cropping of vegetation. Also of note concerning the skulls of *Heterodontosaurus* and *Lycorhinus* is the presence of the large, caniniform teeth (Crompton and Charig, 1962; Hop-

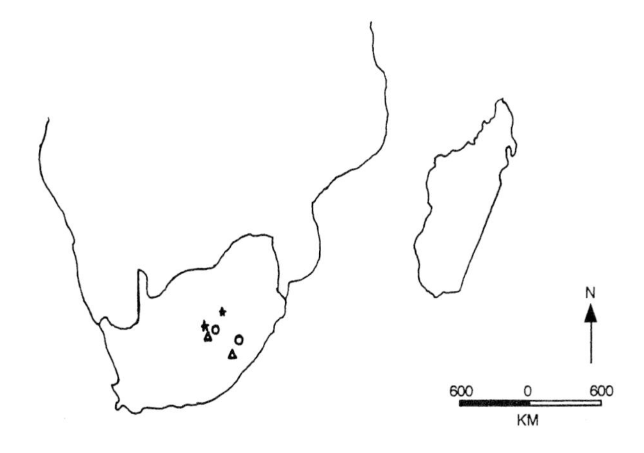

FIGURE 1 Locus map of South Africa showing sites where *Lycorhinus angustidens* (*), *Abrictosaurus consors* (△), and *Heterodontosaurus tucki* (○) material have been found (modified from Weishampel and Witmer, 1990).

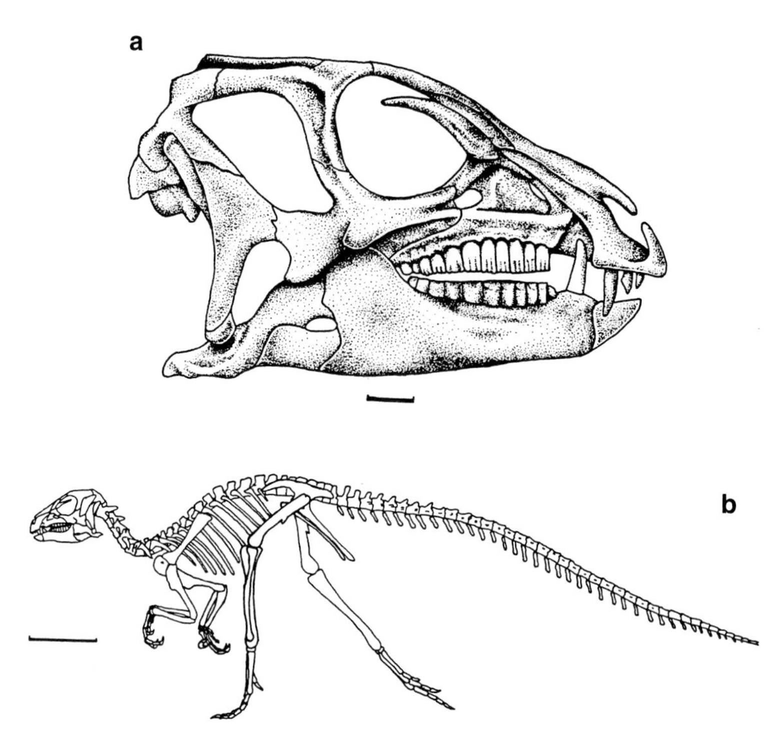

FIGURE 2 (a) Lateral view of the skull of *Heterodontosaurus tucki*. Scale bar = 10 mm (after Weishampel and Witmer, 1990). (b) Reconstruction of the skeleton of *H. tucki*. Scale bar = 10 cm (after Weishampel and Witmer, 1990).

son, 1975) (Figs. 2 and 3). Although these teeth may have been used for some aspect of feeding, some researchers (Thulborn, 1974; Molnar, 1977) have suggested that they were used for intraspecific ritual or combat. All known members of the Heterodontosauridae have these large caniniform teeth, and although they might seem to be an odd thing to find on a herbivore, there are extant deer taxa with similar teeth.

It is generally agreed (Charig and Crompton, 1974; Hopson, 1975; Santa Luca *et al.*, 1976; Santa Luca, 1980) that *H. tucki* and *L. angustidens* were at least habitual bipeds (see BIPEDALITY). In the one articulated skeleton of *H. tucki*, nine cervical (neck) vertebrae are preserved in articulation (minus the atlas and the front half of the centrum of the last cervical, which are fragmentary or not preserved at all), which are arranged to give the neck a sigmoidal-shaped curvature common in bipeds. The articulated skeleton of *H. tucki* also allows an accurate reconstruction of the limbs and the trunk. The forelimb proportions do not suggest a quadrupedal stance for the animal, although the articulations of the cervical and dorsal (trunk) vertebrae suggest that the trunk of the animal was held horizontal to the ground (Weishampel and Witmer, 1990) (Fig. 2b).

Ossified tendons connect the neural spines of the last few dorsal vertebrae but do not extend along the caudal vertebrae. Thus, the spine anterior to the pelvis was strengthened while the tail remained flexible, perhaps counterbalancing the anterior portion of the animal. Paul (1987) reconstructed *Heterodontosaurus* as a quadrupedal galloper, but Weishampel and Witmer (1990) point out that the relationships among the bones forming the shoulder girdle would inhibit movement enough to make rapid quadrupedal movement unlikely. Also, the ungual phalanges of the front feet were laterally compressed, claw like, and apparently unsuitable for locomotion. The ungual phalanges of ornithischians that are hypothesized to have walked quadrupedally (e.g., iguanodontids and hadrosaurids) tend to be extended laterally into hooflike structures for better weight distribution. It seems unlikely, based on the material that has been referred to the heterodontosaurids, that they spent much time in quadrupedal stances (see QUADRUPEDALITY).

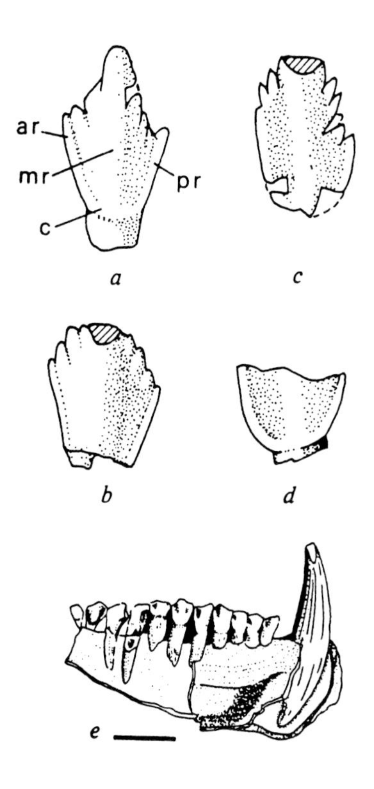

FIGURE 3 Detail of the dentition of heterodontosaurids. (a–d) 3rd, 4th, 5th, and 11th cheek teeth, respectively, of *Abrictosaurus consors* (after Hopson, 1975). (e) Medial view of the left dentary of *Lycorhinus angustidens*, showing articulation and detail of caniniform tooth. Scale bar = 10 mm (after Hopson, 1975).

The forelimb and manus morphology of *Heterodontosaurus* raises some interesting questions about their diet. The teeth are certainly not the teeth of carnivores, although their morphology does not preclude insectivorous behavior. Size would restrict these dinosaurs to the lower understory levels of vegetation, but Weishampel and Witmer (1990) suggest that the configuration of the front claws of *Heterodontosaurus*, which appear to be well suited to digging, might allow the animal to pursue insects in their burrows just as easily as it would facilitate digging roots and tubers.

In the hindlimb of *Heterodontosaurus*, the tibia is fused to the fibula. The calcaneum and the astragalus are completely fused to the fibula and the tibia as well as to each other. Weishampel and Witmer (1990) called this the tibiofibulotarsus. Weishampel and Witmer (1990) and Santa Luca (1980) used the presence of the tibiofibulotarsus to reinforce their suggestions that heterodontosaurids could move, at least some of the time, in a rapid, bipedal fashion.

Most of the material that is confidently assigned to the Heterodontosauridae has been recovered from the Early Jurassic Elliot and Clarens Formations in the Stormberg Series of east-central South Africa. These Hettangian–?Sinemurian-aged units are characterized by floodplain siltstones and eolian (wind-deposited) sandstones with some minor fluvial sandstones. Anderson and Anderson (1970) and Visser (1984) concluded that these rocks were deposited as an alternating sequence of floodplain and eolian deposits. This may indicate that there were localized wet/dry cyclical climate changes in the Stormberg Series during the Early Jurassic, much like those hypothesized to have occurred in rocks of similar age and character in the southwestern United States and the NEWARK SUPERGROUP of eastern North America (Olsen *et al.*, 1989).

The rocks of the Stormberg Series are highly oxidized red beds (again similar to the sedimentary fill of the Newark Supergroup rift basins) that do not preserve a great deal of organic matter, and the material that exists is generally incomplete or fragmentary. Even so, the fauna that has been recovered from the Stormberg is quite rich. Among the finds have been early mammals, cynodonts, sphenosuchians, crocodilians, and the prosauropod *Massospondylus carinatus* (Weishampel and Witmer, 1990).

See also the following related entry:

ORNITHOPODA

References

Anderson, H. M., and Anderson, J. M. (1970). A preliminary review of the biostratigraphy of the uppermost Permian, Triassic and Jurassic of Gondwanaland. *Palaeontol. Africana* **13**(Suppl.), 1–22.

Broom, R. (1904). On the dinosaurs of the Stormberg, South Africa. *Ann. South African Museum Ser. 5* **1,** 445–447.

Charig, A. J., and Crompton, A. W. (1974). The alleged synonomy of *Lycorhinus* and *Heterodontosaurus. Ann. South African Museum* **64,** 167–189.

Crompton, A. W., and Charig, A. J. (1962). A new ornithischian from the Upper Triassic of South Africa. *Nature* **196,** 1074–1077.

Gow, C. E. (1975). A new heterodontosaurid from the Red Beds of South Africa showing clear evidence of tooth replacement. *Zool. J. Linnean Soc. London* **57**, 335–339.

Haughton, S. H. (1924). The fauna and stratigraphy of the Stormberg series. *Ann. South African Museum* **12**, 323–497.

Hopson, J. A. (1975). On the generic separation of the ornithischian dinosaurs *Lycorhinus* and *Heterodontosaurus* from the Stormberg Series (Upper Triassic) of South Africa. *South African J. Sci.* **71**, 302–305.

Molnar, R. E. (1977). Analogies in the evolution of display structures in ornithopods and ungulates. *Evolutionary Theor.* **3**, 165–190.

Olsen, P. E., and Baird, D. (1986). The ichnogenus *Atreipus* and its significance for Triassic biostratigraphy. In *The Beginning of the Age of Dinosaurs: Faunal Change across the Triassic–Jurassic Boundary* (K. Padian, Ed.), pp. 61–87. Univ. of Cambridge Press, New York.

Olsen, P. E., Schlische, R. W., and Gore, P. J. W. (1989). Tectonic, depositional, and paleoecological history of Early Mesozoic rift basins, eastern North America. American Geophysical Union Field Trip Guidebook T351, pp. 174.

Paul, G. S. (1987). The science and art of restoring the life appearance of dinosaurs and their relatives. In *Dinosaurs Past and Present, Vol. II* (S. J. Czerkas and E. C. Olson, Eds.), pp. 171–176. Natural History Museum of Los Angeles County, Los Angeles.

Romer, A. S. (1966). *Vertebrate Paleontology*, pp. 468. Univ. of Chicago Press, Chicago.

Santa Luca, A. P. (1980). The postcranial skeleton of *Heterodontosaurus tucki* (Reptilia, Ornithischia) from the Stormberg of South Africa. *Ann. South African Museum* **79**(7), 159–211.

Santa Luca, A. P., Crompton, A. W., and Charig, A. J. (1976). A complete skeleton of the Late Triassic ornithischian *Heterodontosaurus tucki*. *Nature* **264**, 324–328.

Thulborn, R. A. (1974). A new heterodontosaurid dinosaur from the Upper Triassic red beds of Lesotho. *Zool. J. Linnean Soc. London* **55**, 151–175.

Visser, J. N. J. (1984). A review of the Stormberg Group and Drakensberg Volcanics in Southern Africa. *Palaeontol. Africana* **25**, 5–27.

Weishampel, D. B., and Witmer, L. M. (1990). Heterodontosauridae. In *The Dinosauria* (D. B. Weishampel, P. Dodson, and H. Osmólska, Eds.), pp. 486–497. Univ. of California Press, Berkeley.

Himeji Museum of Natural History, Japan

see Museums and Displays

Hindlimbs and Feet

Per Christiansen
Københavns Universitet
Copenhagen, Denmark

Plesiomorphically dinosaurs were bipedal animals (see Bipedality) progressing exclusively on their long, slender hindlimbs (Fig. 1; Table I). The early dinosaurs displayed a number of distinctive traits found in all later lines, such as the distinct tibial cnemial crest and a lateral process on the astragalus for immobile attachment to the calcaneum, although this is markedly reduced in certain later groups. The presence of an ascending process on the astragalus is also a distinct dinosaurian hallmark, becoming very tall and bird-like in tetenuran theropods (including birds). Actually the process may represent a separate ossification center and should rightly be called the pretibiale because a true astragalar ascending process is a synapomorphy of carinate birds. All dinosaurs and archaic birds, such as the extant ratites, have a pretibiale anterodistally on the tibia, but the process usually fuses completely to the astragalus very early in ontogeny and thus resembles an ascending astragalar process. However, the term ascending process is almost without exception always used to describe the pretibiale of dinosaurs and archaic birds, so it is also used here. Although many dinosaurian lines later adopted quadrupedality, this probably always occurred convergently and the hindlimbs remained longer than the forelimbs in all dinosaurs, except brachiosaurine sauropods and birds. The femur always had a medially offset caput that fit into the acetabulum, subequal distal condyles, and an epipodium aligned in the same longitudinal plane as the long axis of the femoral diaphysis, producing an erect posture similar to that of most extant mammals. The action of the hindlimbs seems to have been restricted to a fore and aft motion, resulting in a parasagittal gait as in most living Carnivora, all Artiodactyla, Perissodactyla, Proboscidea, and Avialae, but unlike most Insectivora or Rodentia and completely different from those of all reptiles, extant or extinct.

Prosauropoda

Both the bipedal and quadrupedal prosauropods had strong hindlimbs compared to the forelimbs. The fem-

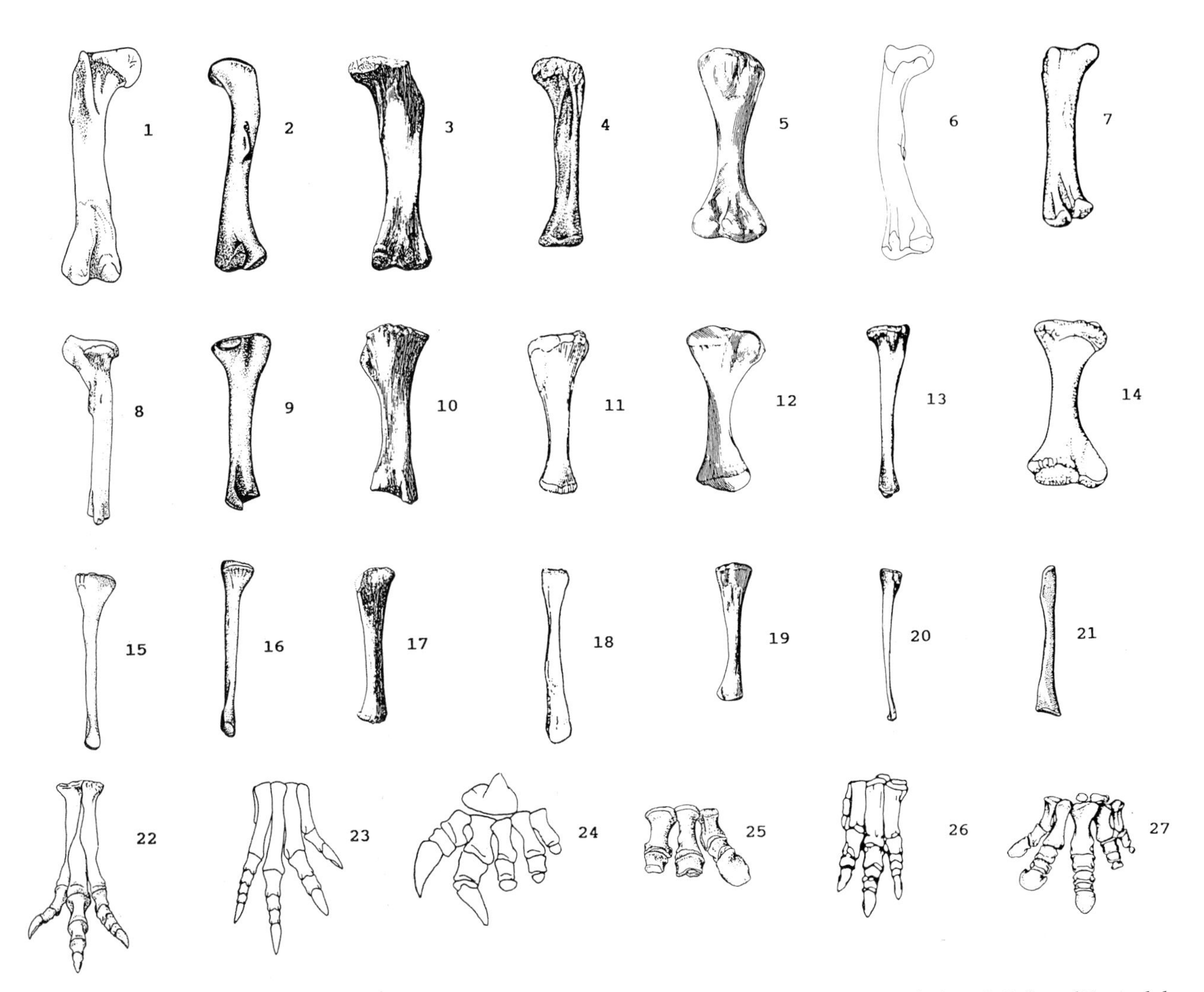

FIGURE 1 Comparative morphology of dinosaurian hindlimbs. (1–7) Femorae (1 and 4, cranial view; 2, 3, 5, and 7, caudal view): 1, theropod *(Tyrannosaurus);* 2, prosauropod *(Plateosaurus);* 3, sauropod *(Brachiosaurus);* 4, stegosaur *(Stegosaurus);* 5, ankylosaur *(Euplocephalus);* 6, ornithopod *(Dryosaurus);* 7, ceratopsian *(Styracosaurus).* (8–14) Tibiae (14, cranial view; 9, 10, and 12, lateral view; 8, 11, and 13, medial view): 8, theropod *(Allosaurus);* 9, prosauropod *(Plateosaurus);* 10, sauropod *(Barosaurus);* 11, stegosaur *(Stegosaurus);* 12, ankylosaur *(Euplocephalus);* 13, ornithopod *(Dryosaurus);* 14, ceratopsian *(Triceratops).* (15–21) Fibulae (15, 16, 18–21, lateral view; 17, medial view): 15, theropod *(Allosaurus);* 16, prosauropod *(Plateosaurus);* 17, sauropod *(Barosaurus);* 18, stegosaur *(Stegosaurus);* 19, ankylosaur *(Euplocephalus);* 20, ornithopod *(Dryosaurus);* 21, ceratopsian *(Centrosaurus).* (22–27) Pes (cranial view): 22, theropod *(Tyrannosaurus);* 23, prosauropod *(Blikanasaurus);* 24, sauropod *(Apatosaurus);* 25, stegosaur *(Stegosaurus);* 26, ornithopod (*Iguanodon*); 27, ceratopsian *(Centrosaurus).*

oral diaphysis is sigmoidal even in the large quadrupedal elanorosauridae and the caput lacks a distinct neck. The fourth trochanter is large and more proximally placed in small taxa (Galton, 1990a), and the caudal transverse processes persist for some distance from the pelvis, underlining the importance of propodial propulsion in the hindlimb. The epicondyles are fairly well developed and the cnemial crest quite large. The tibia is longer and considerably more massive than the fibula. It has slightly angled proximal and distal epiphyses, and in accordance with the femoral morphology, the hindlimb appears to have had permanent flexure, even in large quadrupedal taxa. The femur is always longer than the epipodium, and the epipodium : propodium ratio is 0.7–0.95, which is consistent with medium-sized to large medioportal extant mammals. Even the melanorosaurids appear to have slightly exceeded the sauropods in relative

TABLE I Hindlimb Proportions in Some Dinosaurs Compared to Extant Mammals

Taxon	n[a]	Size range of Propodia (mm)	Size range of Epipodia (mm)	Ratio
Mammalia				
Carnivora				
Ursidae	7 (12)	160–505	120–367	0.72–0.78
Canidae	15 (24)	71–275	81–302	0.92–1.19
Artiodactyla				
Cervidae	11 (21)	152–443	180–490	1.05–1.24
Bovidae	32 (49)	118–481	127–465	0.95–1.29
Perissodactyla				
Rhinocerotidae	4 (8)	402–548	281–392	0.66–0.73
Proboscidea				
Elephantidae	2 (10)	620–1172	325–771	0.52–0.66
Dinosauria				
Prosauropoda				
Plateosauridae	2 (3)	550–740	350–560	0.64–0.76
Sauropoda				
Diplodocidae	4 (8)	1135–1790	790–1123	0.59–0.70
Camarasauridae	2 (5)	555–1351	350–867	0.58–0.81
Ceratopsia				
Ceratopsidae	5 (9)	695–1068	476–668	0.63–0.81
Ornithopoda				
Hadrosauridae	8 (9)	80–1297	77–1008	0.80–1.07
Theropoda				
Allosauroidea	3 (3)	768–884	747–830	0.85–1.08
Ornithomimidae	5 (5)	439–519	444–608	1.01–1.19
Tyrannosauridae	2 (6)	654–1289	689–1199	0.91–1.09

[a] n is number of species and in parentheses are number of specimens.

locomotory potential. A small ascending process of the astragalus fits into a cranial depression distally on the tibia, and the astragalar body is rectangular and somewhat dorsoventrally flattened. The calcaneum is much smaller. Together they form a rather convex distal epiphysis for the metatarsals, indicating greater tarsal flexibility than in sauropods. The foot is short and broad unlike cursorial animals. Metatarsals II–IV are well developed in all prosauropods, whereas metatarsal I is slightly reduced but still prominent: prosauropods appear to have carried mass with all four digits. Metatarsal V is vestigial. The phalangeal formula is usually 2-3-4-5-0/1. Unguals are present on digits I–IV but are not as recurved as in the manus.

Sauropoda

Sauropod hindlimbs were longer than the forelimbs, except in the Brachiosauridae. All three major leg bones were well developed as in modern elephants, unlike in more cursorial mammals, in which the fibula is often greatly reduced (Fig. 2). The bones were more massive than in recent elephants; the diaphyses were straight and appear never to have had a hollow medullary cavity as in theropods. As in the forelimb, the epiphyses are perpendicular to the long axes of their respective diaphyses, strongly suggesting that the hindlimbs were pillar-like and the mode of progression graviportal as in recent elephants. The hindlimb proportions were also elephantine, with the epipodium making up 0.58–0.81 of propodial length compared to 0.52–0.66 in recent elephants, which is considerably shorter than most medioportal to cursorial mammals. The femoral caput is massive and rarely protruding much above the greater trochanter. A distinct neck is absent. The fourth trochanter is usually only moderately developed, and the caudal transverse processes persist only a fairly short distance

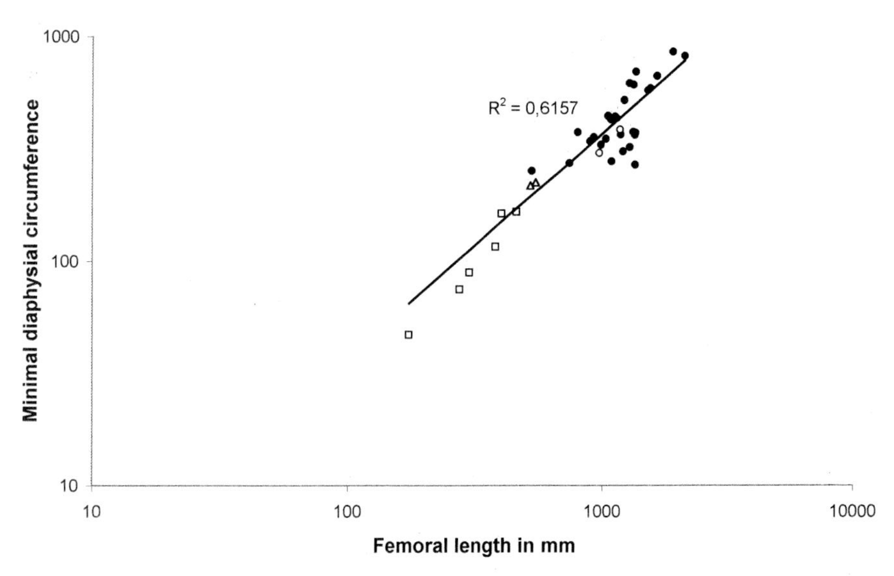

FIGURE 2 Propodial robusticity in sauropod dinosaurs (n = 28). ●, sauropods; ○, elephants; △, rhinoceroses; □, bovids.

from the sacrum, indicating moderately developed propodial retractor musculature. In accordance with this, the brevis shelf is also rather modest. The femoral epicondyles are modest, and the intercondylar fossa is usually rather pronounced indicating modest phalangeal flexors in accordance with the morphology of the foot. The cnemial crest is relatively stronger than that in recent elephants, indicating a more powerful gastrocnemius. The fibula is as long as or longer than the tibia, as in recent elephants, but unlike most other mammals in which the tibia is longer. McIntosh (1990) suggested that the long fibula compensated for the modesty of the calcaneum, which is absent in the Diplodocidae. The astragalus is the largest tarsal bone and is somewhat convex dorsally, indicating increased anteroposterior mobility compared to the carpus. The metatarsals are inclined, and trackways confirm the presence of a posterior heel pad as in recent elephants. The phalangeal formula is usually 2-3-3/4-1/2-1. Most sauropods had unguals on digits I–III, but some even had a fourth (Jensen, 1988; McIntosh *et al.*, 1992). The powerful pedal flexor musculature probably provided a relatively strong pushoff against the ground, compensating for the lack of this from the forelimbs.

Theropoda

Theropods were all obligatory bipeds with long, slender hindlimbs, well-developed limb muscle scars, and large ilia for support of massive epipodial flexors. The femoral diaphysis was circular or ellipsoidal and slightly sigmoidal, and the distal epiphysis was angled compared to the long axis of the diaphysis, causing permanent knee flexure, even in gigantic tyrannosaurs. This trait suggests a relatively high locomotory behavior, even in large theropods. As among modern medioportal to cursorial animals, theropod hindlimb bones grew increasingly robust and the limb posture progressively more upright as the linear dimensions increased, thus reducing the transverse stress on the diaphysis during locomotion (Fig. 3). The femoral head was connected to the diaphysis through a constricted neck, which is more pronounced among smaller taxa. The greater trochanter was usually less prominent than the lesser trochanter, the latter being very prominent among large taxa such as allosaurs and tyrannosaurs. The fourth trochanter was prominent among more archaic taxa with prominent proximal caudal transverse processes but became progressively reduced in derived tetanurans, in which the transition point in the tail moved cranially (e.g., ornithomimids, oviraptorids, dromaeosaurs,

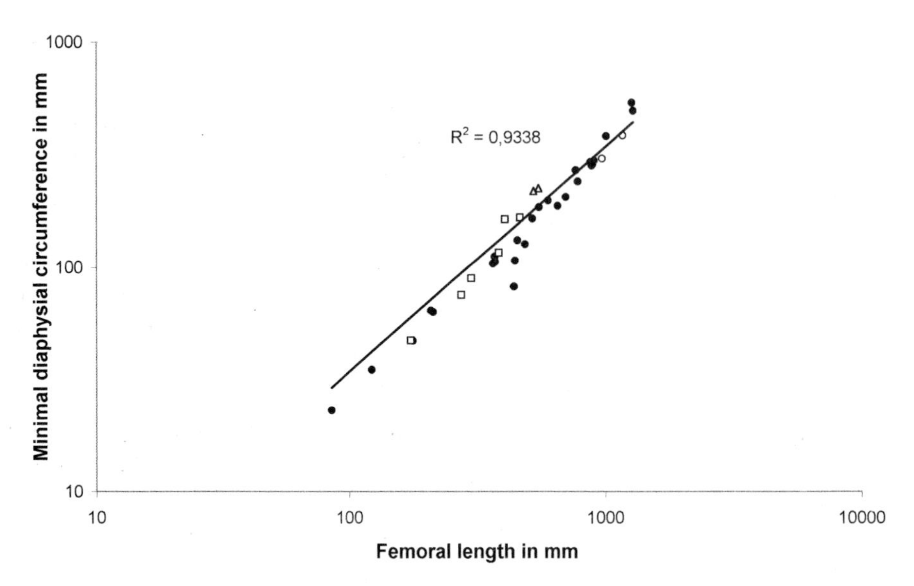

FIGURE 3 Propodial robusticity in theropod dinosaurs (n = 26). ●, theropods; ○, elephants; △, rhinoceroses; □, bovids.

and tyrannosaurids). This indicates evolution from a more reptilian mode of progression with extensive hip flexion, powered mainly by the caudifemoralis longus, to a more bird-like mode of progression with increasingly extensive knee flexion (Gatesy, 1990). The distal epiphysis is moderately expanded with well-defined condyles and usually fairly prominent epicondyles. The tibia is the longest epipodial bone and the astragalus is usually fused to the tibia distally in adult specimens, forming a tibiotarsus. It is thus included in the functional length of the epipodium. The functional epipodial length is very large in certain small to medium-sized taxa (e.g., epipodial : propodial ratio of 1.1–1.2 in ornithomimids, 1.26 in *Microvenator,* and 1.32 in *Saurornithoides*; the latter two are as high as in highly cursorial extant artiodactyls and perissodactyls). Among the large tyrannosaurids it is 1.05–1.10 in *Albertosaurus* and 0.91–0.95 in *Tyrannosaurus rex*, exceeding the ratios in large extant bovids and more archaic large theropods such as *Allosaurus* (0.85) or *Sinraptor* (0.93). The cnemial crest is prominent, especially among larger taxa, indicating powerful gastrocnemius and iliotibialis musculature, the latter being confirmed by the dolichoiliac pelvis, and in later taxa there is a proximolateral fibular crest. Distally the tibia bears a cranial depression for the astragalar ascending process. The fibula is always shorter and considerably thinner than the tibia with prominent proximal and modest distal anteroposterior expansions, and it always reaches the tarsus, unlike the situation in many extant birds. In the tarsus, usually the large astragalus makes up all the medial condyle and parts or most of the lateral condyle; the smaller calcaneum makes up the rest; and two to four distal tarsals, sometimes fused to the metatarsus, especially in ceratosaurs, form a bird-like tarsometatarsus. The ascending process of the astragalus is very low in primitive theropods but becomes progressively larger in more advanced taxa. More distant relatives such as *Herrerasaurus* were functionally tetradactyl, probably also supporting mass on metatarsal I, although metatarsal III was the longest as in all dinosauromorphs. The reduction of metatarsal V was not as pronounced and it still bore a phalanx. The phalangeal formula of *Herrerasaurus* is 2-3-4-5-1 (Novas, 1989). All theropods, with the exception of the Dromaeosauridae and Troodontidae, were functionally tridactyl with a vestigial fifth metatarsal, a reduced first metatarsal that no longer reached the tarsus and in some cases was reversed, and a phalangeal formula of 2-3-4-5-0. In large taxa, the metatarsals and phalanges are robust. The Ornithomimosauria lost the first and fifth metatarsals altogether, except in the most archaic taxa. In the Dromaeosauridae the pes is highly apomorphic with deeply ginglymoid phalangeal articulating facets, especially on digit II, and the foot is functionally didactyl because digit II is short, robust, and highly mobile dorsoventrally, with large flexor tubercles and a very large, lateromedially compressed, and strongly recurved sickle-shaped ungual with an enormous flexor tubercle. This highly raptorial design was probably for killing large prey in a slashing fashion. The Troodontidae shared this morphology superficially, but the second ungual was smaller and straighter, the flexor tubercles and ginglymoid facets were less well developed, and the metatarsus was more slender and probably unable to withstand the considerable transverse stress imposed on it while delivering the powerful slashing kicks that dromaeosaurs probably employed when subduing prey. The presence of an arctometatarsus, a strongly compressed proximal third metatarsal, in tyrannosaurs, ornithomimosaurs, elmisaurs, and troodontids, but not dromaeosaurs, suggests that the unusual morphology of the second digit and claw in troodontids and dromaeosaurids evolved convergently.

Stegosauria

The hindlimbs of stegosaurids are always considerably longer than the forelimbs, and analyses of the center of gravity in stegosaurids suggest that they supported about two-thirds of their mass on the hindlimbs. The femur resembles that of many sauropods in having a long, straight diaphysis, lateromedially wider than anteroposteriorly, and moderate distal expansion. The caput is not set on a neck and the lesser and fourth trochanters are usually rather modestly developed. The distal epiphysis is almost perpendicular to the long axis of the diaphysis and the condyles are rugose and not as distinct as in mammals resembling sauropods. The hindlimb thus appears to have been elephantine in posture. The lateral epicondyle is usually considerably less well developed than the medial epicondyle. The epipodium is much shorter than the propodium and the tibiofemoral ratio is approximately 0.54–0.62, closely resembling the ratios for recent elephants. The tibia is much more massive than the fibula; it has a straight diaphysis and carries all the body mass. The proximal epiphysis is perpendicular to the long axis of the diaphysis but the distal epiphysis is transversely angled, being lower medially, and has an anterior fossa for the ascending astragalar process. The cnemial crest is usually large and extends a considerable distance down the diaphysis, indicating much stronger pedal flexors than those of extant elephants. The fibula is slender and straight and slightly shorter than the tibia. There appear to have been only two proximal tarsals, although cartilaginous distal tarsals may have been present (Galton, 1990b). The astragalus is much larger than the calcaneum and in adults both are usually fused to the distal end of the tibia and fibula, respectively, and to each other. They usually form a large articulating face instead of distinct condyles. Metatarsals II–IV were well developed, metatarsal V was vestigial or absent, and metatarsal I appears always to have been absent. The proximal epiphyses of metatarsals II–IV are slightly inclined posteriorly, indicating a semidigitigrade pes. The phalanges are short and stout and a large flat, hoof-like ungual is present on digit II. The phalangeal formula is *X*-2-2/3-2/3-0/*X*.

Ankylosauria

The femur is the longest element in the hindlimb of ankylosaurs, which have a long, straight, and stout diaphysis, with a considerably greater lateromedial than anteroposterior diameter. This resembles the condition in recent elephants and rhinoceroses but is unlike that in other large ungulates. The head is stout and usually occupies the normal medial position in nodosaurids but is displaced more laterally in ankylosaurids, and there is no neck. The greater and lesser trochanters are usually modest, whereas the fourth trochanter is large. The distal epiphysis is posteriorly angled compared to the long axis of the diaphysis, providing permanent knee flexure. The epicondyles are large. The tibia is stout with a large cnemial crest and great distal lateromedial expansion. It is approximately 80–87% of femoral length, which is higher than in extant elephants or rhinoceroses. As in sauropods the proximal and distal epiphyses are somewhat rugose, indicating cartilaginous covering. The fibula is slender and shorter than the tibia, and all anteroposterior dimensions exceed the lateromedial ones, as in most modern mammals. Proximally, there is usually a great lateromedial expansion and a smaller distal expansion, and the diaphysis is twisted about its long axis. The pes is poorly known in most specimens but is usually short and stout, and the metatarsals are proximally free of each other. The astragalus and calcaneum are sometimes almost subequal in size, or the astragalus is the larger element. The calcaneum sometimes appears to be absent or completely

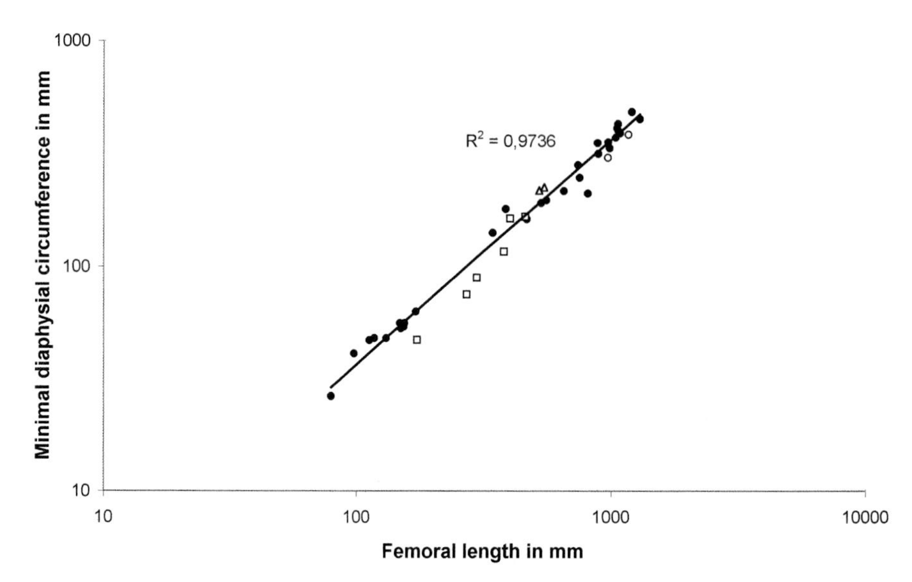

FIGURE 4 Propodial robusticity in ornithopod dinosaurs (n = 30). ●, ornithopods; ○, elephants; △, rhinoceroses; □, bovids.

fused to the astragalus. In adult specimens, the astragalus probably always fuses to the tibia and the suture is very indistinct (Coombs, 1986). Metatarsals I and V appear to be absent and metatarsals II–IV are long and robust. The phalanges are short, stout, and block-like, resembling phalanges from stegosaurids or ceratopsids. The claws are dorsally flattened and hoof-like. The phalangeal formula is *X*-3-4-5-*X*.

Ornithopoda

The femur is long, and the diaphysis is rather straight in Heterodontosauridae and Hadrosauridae but is distinctly sigmoidal in lateral or medial view in the Hypsilophodontidae, Dryosauridae, and Iguanodontia. Smaller taxa are gracile, whereas larger taxa are considerably more robust (Fig. 4). The caput is medially directed and set off from the diaphysis by a distinct neck. The greater trochanter is modest in heterodontosaurids but somewhat more developed in later taxa, making up much of the proximolateral part of the bone. The lesser trochanter is separated from the greater trochanter by a distinct cleft and is prominent and somewhat wing-like in hypsilophodontids, dryosaurids, iguanodontids, and most hadrosaurs, superficially resembling the condition in certain large theropods, but in heterodontosaurids it is less distinct and confluent with the greater trochanter. The fourth trochanter is a well-developed, spike-like, medioventrally projecting process in heterodontosaurids, hypsilophodontids, and dryosaurids, but it is more prosauropod-like and less ventrally projecting in iguanodontids and hadrosaurs. The distal epiphysis is only moderately expanded and along with the curvature of the diaphysis in some, and somewhat caudally directed epiphysis in all taxa, indicates permanent knee flexure, even in large forms such as *Edmontosaurus*. As in modern animals, the hindlimb posture appears to become progressively more upright as linear dimensions increase, although this is less apparent than among modern animals. The flexor sulcus and extensor groove are usually prominent, especially in hadrosaurs in which the latter can form an enclosed tube anterodistally. In heterodontosaurids and hypsilophodontids these features are indistinct. The tibia is straight and, like the femur, becomes progressively more robust as size increases. The lateromedial proximal expansion is generally rather well developed, especially in hadrosaurs, and the proximal epiphysis is planar or with two lightly concave cotyles. The cnemial crest is usually well developed and slightly laterally directed cranially, although it is more modest in certain hypsilophodontids. Distally there is usually a lateromedial expansion, and the distal part of the tibia is twisted somewhat laterally about the long axis of the diaphysis.

The astragalus is immovably attached to the tibia

in adults, increasing the functional length of the tibia. In *Heterodontosaurus* the astragalus is fused to the tibia and calcaneum. The hindlimbs are quite gracile and the tibiofemoral ratio in *Camptosaurus* is 0.95–1, and 0.88–0.98 in hadrosaurs, which are similar to extant *Bos, Bison,* and *Syncerus*. Even in giant *Edmontosaurus,* the tibia still amounts to approximately 80% of the femoral length, significantly more than in elephants. The fibula is slightly shorter and thinner than the tibia with moderate proximal and distal lateromedial expansions. The distal epiphysis is usually planar or slightly convex. Unusual among ornithopods, the fibula is fused to the tibia distally in *Heterodontosaurus* (Santa Luca, 1980). The astragalus is much larger than the calcaneum and extends craniodistally on the tibia. The ascending process is always rather low and wide. There are three distal tarsals in *Heterodontosaurus,* which are fused to the proximal tarsals: two in *Tenontosaurus* (Forster, 1990) and *Dryosaurus* (Galton, 1981) and one in hadrosaurs (Weishampel and Horner, 1990). All ornithopods were digitigrade. Archaic iguanodontids such as *Camptosaurus* possessed five metatarsals but metatarsal V was vestigial and never bore a phalanx. In all ornithopods, metatarsal III is the longest. Heterodontosaurids, dryosaurids, and hypsilophodontids may have supported mass on digits I–IV, although digit I was considerably shorter. Iguanodonts and hadrosaurs only supported mass on digits II–IV, based on evidence from trackways. The second to fourth metatarsals are much stronger and longer than the first and are closely appressed proximally to provide strength and stability to the foot. *Ouranosaurus* and hadrosaurs have slightly asymmetrical proximal intermetatarsal articulating facets, causing the distal metatarsals to diverge away from each other. The phalanges were moderately long, with condylar–cotylar or ginglymoid articulating facets, and the phalangeal formula was 2-3-4-5-X. Most iguanodontids had only a vestigial first metatarsal, and in all hadrosaurs except *Claosaurus* it was absent. *Claosaurus* is also unusual in being the only ornithopod in which gastroliths have been reported (Brown, 1907). The claws of archaic ornithopods, such as heterodontosaurids and hypsilophodontids, were still slightly recurved and ungual-like, but iguanodontids and hadrosaurs had distinctly hoof-like claws on digits II–IV.

Ceratopsia

The hindlimbs of all ceratopsians were considerably longer than the forelimbs. In the Protoceratopsidae, which appear to have been at least facultatively quadrupedal, the discrepancy is so large and the hindlimbs so cursorially proportioned compared to the forelimbs that it seems likely that these animals progressed bipedally during fast locomotion (Fig. 5). In

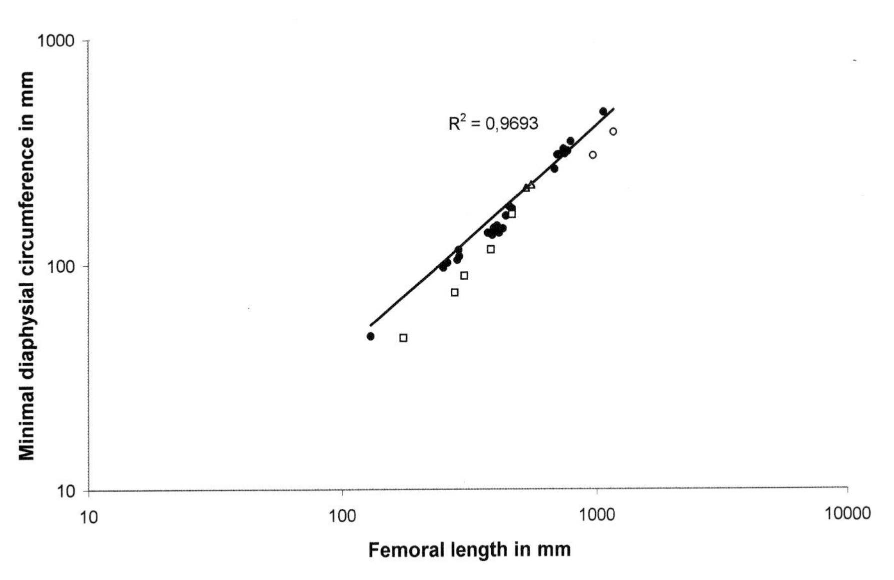

FIGURE 5 Propodial robusticity in ceratopsian dinosaurs (n = 26). ●, ceratopsians; ○, elephants; △, rhinoceroses; □, bovids.

psittacosaurids and protoceratopsids, the tibiofemoral ratio is approximately 1.12–1.20, but in ceratopsids the femur is always longer, with the exception of certain juvenile specimens. In large forms, the tibiofemoral ratio is 0.66–0.76, compared to 0.66–0.74 in extant rhinoceroses. Only gigantic *Triceratops* approaches elephantine proportions with a tibiofemoral ratio of approximately 0.63. As is most large modern herbivores, the diaphysis is relatively straight with a slight curvature distally, thus inclining the distal epiphysis posteriorly in relation to the long axis of the diaphysis and causing permanent knee flexure. The lateromedial diameter of the diaphysis always considerably exceeds the anteroposterior diameter in ceratopsids, more so than in most living megaherbivores, except elephants. However, part of this is often due to crushing. In psittacosaurids and protoceratopsids, the diaphysis is more circular. The greater trochanter is well developed and the lesser trochanter distinct in psittacosaurids and protoceratopsids, but less so in ceratopsids. The fourth trochanter is large in psittacosaurids and protoceratopsids, being crest-like in the latter and resembling that of archaic ornithopods in the former, but is relatively less pronounced in ceratopsids. The extensor groove is distinct but usually rather shallow and the flexor sulcus prominent, in accordance with the presumably strong pedal musculature. The distal epiphysis is moderately expanded and the condyles are well developed and rounded in uncrushed specimens. The tibia is stout and straight, with large proximal and distal expansions, and the cnemial crest is rather modest in psittacosaurids, quite large in protoceratopsids, and large to very large in ceratopsids. The astragalus forms all the distal medial condyle and often the medial part of the lateral condyle, the lateral part formed by the smaller calcaneum. There are two distal tarsals, although Lull (1933) reported three in *Chasmosaurus*. The pes is digitigrade and always considerably larger than the manus, even in large quadrupedal forms. Mass was probably carried only on metatarsals II–IV in psittacosaurids because metatarsal I is considerably shorter and metacarpal V vestigial, and the pes resembles an ornithopod pes. The phalangeal formula is 2-3-4-5-0 and the terminal claws are rather straight and lateromedially compressed, taking on an ungual-like appearance. In protoceratopsids and ceratopsids, metatarsals I–IV probably carried mass, and metatarsal V was vestigial. In protoceratopsids, the metatarsals were closely appressed to each other proximally, whereas the ceratopsid foot was more elephantine. The phalanges were less elongate than in psittacosaurids and stout in ceratopsids. The phalangeal formula is 2-3-4-5-0. The claws were broad, flat, and hoof-like, more so in ceratopsids than in protoceratopsids.

References

Brown, B. (1907). Gastroliths. *Science* **25**(636), 392.

Coombs, W. P., Jr. (1986). A juvenile dinosaur referable to the genus *Euplocephalus* (Reptilia, Ornitischia). *J. Vertebr. Palaeontol.* **6**(2), 162–173.

Forster, C. A. (1990). The postcranial skeleton of the ornithopod dinosaur *Tenontosaurus tilletti*. *J. Vertebr. Palaeontol.* **10**(3), 273–294.

Galton, P. M. (1981). *Dryosaurus*, a hypsilophodontid dinosaur from the Upper Jurassic of North America and Africa. Postcranial skeleton. *Palaeontol. Z.* **55**(3/4), 271–312.

Galton, P. M. (1990a). Basal Sauropodomorpha–Prosauropoda. In *The Dinosauria* (D. B. Weishampel, P. Dodson, and H. Osmólska, Eds.), pp. 320–344. Univ. of California Press, Berkeley.

Galton, P. M. (1990b). Stegosauria. In *The Dinosauria* (D. B. Weishampel, P. Dodson, and H. Osmólska, Eds.), pp. 435–455. Univ. of California Press, Berkeley.

Gatesy, S. M. (1990). Caudofemoral musculature and the evolution of theropod locomotion. *Palaeobiology* **16**(2), 170–186.

Jensen, J. A. (1988). A fourth new sauropod dinosaur from the Upper Jurassic of the Colorado Plateau and sauropod bipedalism. *Great Basin Nat.* **48**(2), 121–145.

Lull, R. S. (1933). A revision of the Ceratopsia or horned dinosaurs. *Peabody Museum Nat. History Bull.* **3,** 1–175.

McIntosh, J. S. (1990). Sauropoda. In *The Dinosauria* (D. B. Weishampel, P. Dodson, and H. Osmólska, Eds.), pp. 345–401. Univ. of California Press, Berkeley.

McIntosh, J. S., Coombs, W. P., Jr., and Russell, D. A. (1992). A new diplodocid sauropod (Dinosauria) from Wyoming, USA. *J. Vertebr. Palaeontol.* **12**(2), 158–167.

Novas, F. E. (1989). The tibia and tarsus in Herrerasauridae (Dinosauria, *incertae sedis*) and the origin of the dinosaurian tarsus. *J. Palaeontol.* **63**(5), 677–690.

Santa Luca, A. P. (1980). The postcranial skeleton of *Heterodontosaurus tucki* from the Stormberg of South Africa. *Ann. South African Museum* **79,** 159–211.

Weishampel, D. B., and Horner, J. R. (1990). Hadrosauridae. In *The Dinosauria* (D. B. Weishampel, P. Dodson, and H. Osmólska, Eds.), pp. 534–561. Univ. of California Press, Berkeley.

Histology of Bones and Teeth

Robin E. H. Reid
Chesham Bois
Amersham, United Kingdom

Bones and teeth are built from mixtures of organic and mineral matter, of which the mineral part does not decay after death, although it may later suffer diagenetic alteration. Organic contents are usually mostly lost in fossilization, but biochemical traces (see Chemical Composition of Dinosaur Fossils) and inert material may persist.

The main mineral constituent of bones and teeth is calcium phosphate, in the form of hydroxyapatite microcrystals (see Biomineralization), although variants with, for example, some strontium or fluorine content may also occur. Bones may be built wholly from tissues classed as bone or also have parts formed from cartilage, which then is itself partly mineralized. Soft cartilage is normally lost after death, but the mineralized form persists unless abraded. Teeth are built partly from tissues called enamel and dentine (which are not classed as bone) but usually also from a form of bone called cementum. Some reptiles (e.g., ichthyosaurs) can have a further form of bone (pulp bone) inside their teeth. The normal organic constituents of bone are living cells (osteocytes), bundled collagen fibers, and a proteoglycan (acid mucopolysaccharide) content, whereas cartilage has living cells (chondrocytes) in a proteoglycan (chondroitin sulfate) matrix in which fibrils of collagen or elastin may also occur. Dentine resembles bone chemically, but enamel contains noncollagenous proteins.

Many fossil bones and teeth retain their original phosphate content and mineral microstructure, although others are more or less recrystallized or show molecule by molecule replacements. In the last case, microstructure may again be well preserved, but original polarization patterns may be lost. Internal spaces in bones (e.g., marrow cavities) may be partly or completely filled with sediment that has entered through vascular or pneumatic foramina or through access due to damage before burial; but both large spaces and smaller ones (e.g., those occupied by osteocytes) may show only diagenetic infilling, by, for example, calcite, gypsum, pyrite, barites, or chalcedony. Diffusion of bone phosphate into sediment may also occur. The pulp cavities of teeth are usually matrix filled.

Bones, Bone, and Cartilage

Dinosaurian bones can be divided generally into bones from the skeleton proper and bones formed in the skin and known as osteoderms. Skeletal bones are divisible into membrane bones, formed directly from fibrous tissue, and cartilage bones, in which bone replaces cartilage during growth. The limb bones and vertebrae, for instance, are cartilage bones and often retain traces of their origin by having smooth "articular" surfaces formed from calcified cartilage. In life, this underlays the true articular cartilage, now lost.

Most bone tissues are laid down by soft tissues called the periosteum and the endosteum, which coat external and internal bone surfaces, respectively. Bone is first formed under them as unmineralized osteoid, with bundled collagen fibers built from numerous individual fibrils. Mineralization is controlled by cells called osteoblasts, which become transformed into osteocytes by bone enclosing them as it grows. Thereafter, their main function is involvement in molecular exchanges between bone and body fluids via the bloodstream. Osteocytes occupy small spaces called lacunae, from which finely branching cellular processes radiate along minute tunnels called canaliculi. Most bone in the skeleton proper is bone of this type, described as osteoblastic bone. Some bones, however, show local developments of tissues known as metaplastic bone, formed from cartilage or fibrous tissues (tendons and ligaments) in the absence of osteoblasts through other types of cells (chondrocytes and fibrocytes) assuming their mineralizing function. This type of bone can also occur in osteoderms. Cells included in it become osteocytes in effect, but the lacunae they occupy do not emit canaliculi.

The matrix of cartilage is formed partly by cells termed chondroblasts, located in a covering perichondrium. Like osteoblasts, some become enclosed in the matrix they form, becoming chondrocytes; but, unlike osteocytes, they retain the ability to form new matrix and to undergo division. The latter ability is especially important in parts in which cartilage is replaced by bone described as endochondral bone (see Bone Tissues, p. 331), where replacement is preceded by multiplication and hypertrophy of chondrocytes and by calcification of the matrix between them. This calcified cartilage is not classified as a bone tissue.

Growth, Remodeling, and Reconstruction

Besides bone accretion, the growth of bones commonly involves progressive external remodeling by bone resorption, and may also involve internal reconstruction in which resorbed bone is replaced by new bone. Internal reconstruction is possible because bone is normally alive throughout an animal's life. Many dinosaurs had bones that were extensively reconstructed.

Before looking at bone tissues described below, it is useful to have a picture of how a typical limb bone grows. In embryos, such bones are first detectable as tracts of dense mesenchymal tissue, in which some of the cells become chondroblasts and secrete a rod of cartilage. The remaining cells become a surrounding perichondrium, containing cells that can give rise to more chondroblasts or cells with other functions. Later, cartilage at the center of the growing rod calcifies and is then invaded by tissue of perichondrial origin containing cartilage-resorbing cells called chondroclasts. Osteoblasts appear in this tissue, which then becomes the endosteum, and begin to deposit bone on the walls of the spaces excavated by chondroclasts. This process spreads toward the ends of the rod, to which cartilage soon becomes restricted. The bone formed is typically cancellous (spongy) and is termed endochondral because of its origin within cartilage. Meanwhile, osteoblasts appear in the external perichondrium, which then becomes the periosteum, and begin to deposit a sheath of periosteal bone. Growth in length proceeds by terminal formation of new cartilage, replaced progressively by endochondral bone, and growth in thickness proceeds by external accretion of periosteal bone.

By hatching or birth, most such bones have assumed a form showing a central shaft (diaphysis), subterminal expansions (metaphyses), and articulating ends (epiphyses). Growth proceeds basically as before; but because bone, unlike cartilage, cannot grow interstitially, diaphyses can only grow in length through a process involving the external resorption of bone in the metaphyses. This remodeling is carried out by bone-resorbing cells known as osteoclasts, which appear in the metaphyseal periosteum, and may trigger other internal changes (see Compacted Coarse Cancellous Bone, p. 335). In other forms of external remodeling, limb bones or flat bones can change their curvature by resorption occurring on one side while accretion occurs on the other, with parts "drifting" laterally away from their original positions. Similarly, the neural canals of vertebrae can enlarge and migrate dorsally during growth, with bone being resorbed dorsally and laterally but deposited ventrally.

Growing bones may also be remodeled internally by osteoclasts in the endosteum, which resorb bone to produce medullary (marrow) cavities in limb bones, and the air-filled cavities in pneumatic bones. A marrow cavity may first appear during early development and then spread both longitudinally and radially, until only periosteal bone surrounds it in the shaft. This bone is then itself resorbed from within as growth continues, until medullary expansion ceases. The formation of pneumatic cavities is similar, except that they are also invaded by tissues of external origin. In internal reconstruction, bone resorption by endosteal osteoclasts is followed by renewed deposition by osteoblasts, producing tissues of various types (see Haversian Bone and Endosteal Lining Bone, p. 333, 334). Structures formed in this manner are bounded by sharp resorption lines, which mark the centrifugal limit of bone resorption. For example, expansion of a medullary cavity may be followed by the formation of a secondary lining, formed from new bone that grows centripetally. The "internal circumferential lamellae" of traditional medical nomenclature are formed in this manner. The change from resorption to deposition is called reversal, and resorption lines are hence also called reversal lines. Resumed deposition may also be marked by a thin initial deposit of apparently nonfibrous cement, on which the term cementing line is based.

Bone Tissues: Fibrillar Organization

Tissue classification at the microstructural level is based on the thickness and arrangement of collagen fiber bundles, using nomenclature of Weidenreich (1930) as anglicized by Pritchard (1956) or variant systems (e.g., Smith, 1960; Pritchard, 1972). In the version used here (Pritchard, 1956) three main types are distinguished:

1. coarsely bundled bone, with parallel-fibered and woven-fibered varieties;
2. finely bundled bone, with parallel-fibered and lamellar varieties; and
3. finely and coarsely bundled bone, with mixed contents.

The main types are also described more simply as parallel-fibered bone, woven bone, and lamellar bone, to which the term nonlamellar is sometimes added. Bone tissues laid down externally (periosteally) may be coarsely or finely bundled, but those formed internally are normally finely bundled. In osteoblastic tissues, variations are broadly related to growth rates, with woven bone formed fastest and lamellar bone slowest. The most highly organized tissue is lamellar bone, in which fiber bundles, for example, 2–4 μ in diameter, are arranged in regular layers, or lamellae, and follow crossing directions in successive layers. The osteocytes are usually widely spaced, flattened, or elongated in the plane of lamellation and sometimes also arranged in regular layers. At the other extreme, woven bone has fiber bundles, for example, 30 μ thick, interwoven irregularly and globular or irregularly shaped osteocytes arranged without order. Other tissues show various intermediate conditions.

Although microstructure does depend on fiber patterns, this limited nomenclature gives an oversimplified picture of the range of variations seen in practice, and also involves other problems. Intermediates between nominal types occur, and named varieties can also have variants. Woven bone (s. s.), for instance, can grade into stratified tissues or bone in which fiber bundles show radial and plumose arrangements; also, several different arrangements of fiber bundles are known from lamellar bone. Stratified variants of woven bone (s. s.) can appear to be parallel fibered only if examined in transverse and longitudinal sections. Pritchard (1972) grouped such tissues with true parallel-fibered coarsely bundled bone as bundle bone; however, although woven bone is always osteoblastic, some coarsely bundled parallel-fibered tissues are metaplastic. The terms parallel fibered and woven fibered can also mislead if read literally because they refer to the arrangement of fiber bundles and not to that of individual collagen fibers.

Fossil bone commonly shows collagen loss, requiring use of accessory characters to identify fibrillar types. The form, size, and arrangement of osteocyte lacunae can be used to identify woven bone; but they are not diagnostic in other cases, except in a form of lamellar bone in which canaliculi are arranged at right angles to lamellae. Lamellar bone can often be recognized by use of crossed polarizers, when it shows a characteristic pattern of fine light and dark banding, plus a four-armed "axial cross" when it forms concentric structures. In theory (Enlow, 1969; de Ricqlès, 1975), parallel-fibered bone should show uniform transitions from transition to extinction as the stage is rotated, whereas woven bone should be dark in all positions. However, stratified tissues that are not parallel fibered can show a false "parallel-fibered" extinction pattern, and both patterns can occur in altered tissues of any type.

Bone Tissues: Further Classification

The further classification and nomenclature of bone tissues is based on various aspects of microstructure, gross structure, mode of origin, location, and function used in various combinations. As seen previously, bone can be osteoblastic or metaplastic and, when osteoblastic, periosteal or endosteal. In the convention used here (Seitz, 1907; Gross, 1934), it is primary if formed during growth and secondary if formed during internal reconstruction after preexisting bone has been resorbed. It is cellular if included cells (osteocytes) are present and acellular if they are absent. If dense, it is compact bone (or compacta), and if spongy it is cancellous bone (or spongiosa). When compact, it is vascular if traversed by blood capillaries, which occupy channels called vascular (or Haversian) canals, and it is avascular if these are absent. If compact bone is deposited centripetally within internal spaces, it is said to form osteons and is called osteonic bone. Vascular canals and osteons are classified as primary or secondary in the same way as bone tissues.

Most dinosaurian bones are like those of other terrestrial vertebrates, although a few show features only known from dinosaurs. Many bones from the skeleton proper are formed entirely from osteoblastic bone, but some contain minor amounts of metaplastic bone. External surfaces are normally formed from compact bone, except when articular or at sutures. Lengthwise growth of long bones follows the pattern seen in turtles and crocodiles, without formation of the separately calcified epiphyses seen in lizards and mammals. Some bones show only compact bone surrounding internal cancellous bone; but others contain a medullary (marrow) cavity, and supposed pneumatic spaces occur in saurischian vertebrae and ribs. Thin flat bones and partitions between internal spaces may be formed from compact bone only. Periosteal compact bone is often like that of large fast-growing mammals and birds (e.g., cattle and ostriches) but can also be like that of normal reptiles or change from mammal like to typically reptilian during growth. Secondary internal reconstruction is often extensive in both compact and cancellous bone, but it can also be limited or lacking. External remodeling also occurred, but less extensively than in mammals.

Bone tissues seen in dinosaurs can be summarized briefly as including all the types usual in large mammals, ratite birds, and crocodiles, plus one apparently unique tissue. Further details are as follows:

Periosteal Bone (Sensu Lato) For convenience, the term periosteal bone is used for both bone of wholly periosteal origin (periosteal bone s. s.) and tissues formed in part from osteonic bone laid down in primary vascular spaces. It is typically compact bone, except in early ontogeny and in some marine forms (e.g., whales and ichthyosaurs). Four main compact types can be distinguished, although intergrades and mixtures also occur:

1. uniform bone, with no evident textural variations;
2. lamellated bone, showing small-scale cyclical variations in fibrillar and mineral microstructure;
3. fibrolamellar bone, in which a finely cancellous initial framework built from woven bone is compacted by internal deposition of lamellar or fine parallel-fibered osteonic bone; and
4. zonal bone (Fig. 1), in which one of the above types forms a series of major growth rings, separated by (i) simple resting lines, marking pauses in growth; (ii) thin layers of dense or lamellated bone, termed annuli and marking slowed growth; or (iii) both together.

Fibrolamellar bone (Fig. 2) can be divided further into reticular, parallel-osteoned, laminar, and radiate types according to whether the osteons reticulate irregularly or form parallel cylinders or circumferential or radial plates. The additional term accretionary bone is convenient for distinct superficial deposits, formed after normal growth has ceased. In other usages, lamellated bone with resting lines between lamellae was called pseudolamellar by Enlow (1969); but this term has also been used by de Ricqlès (1975) as meaning parallel-fibered bone. De Ricqlès (1974) groups lamellated and zonal tissues together as lamellar–zonal bones; but lamellation and zonation are different phenomena, which can also occur independently, and some zonal bone has zones built from fibrolamellar bone. He also divides laminar bone in the above sense into plexiform and laminar types, according to the presence or absence of radial vascular canals; however, they can appear to be absent

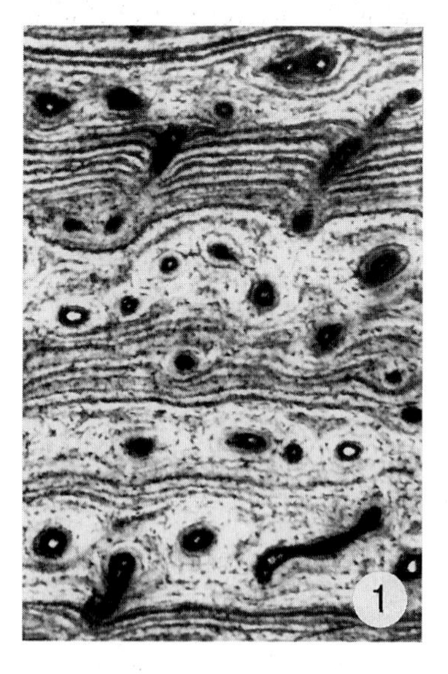

FIGURE 1 Zonal bone, with lamellated bone forming annuli between successive nonlamellated zones. The large round perforations are cross sections of vascular canals. *Allosaurus*, local development in a humerus; ×64.

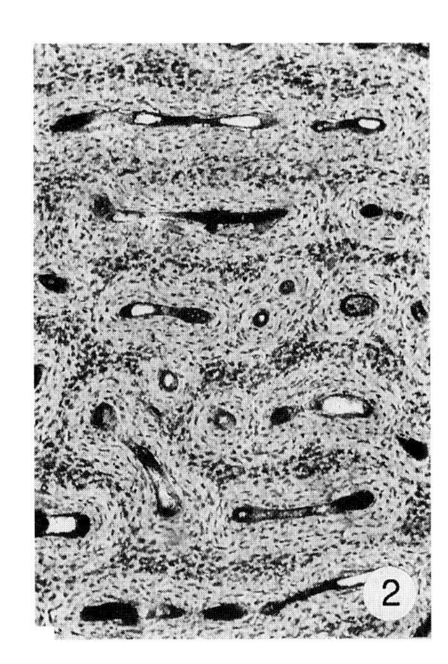

FIGURE 2 Fibrolamellar bone, in a form intermediate between laminar and parallel osteoned. Bone of the periosteal framework is seen as dark trabeculae in which mineral-filled osteocyte lacunae (small dark bodies) are larger than those in the lighter osteonic bone. Some lacunae in the primary osteons show concentric arrangement around vascular canals. Sauropod *indet.* (*"Cetiosaurus"*) limb bones; ×48.

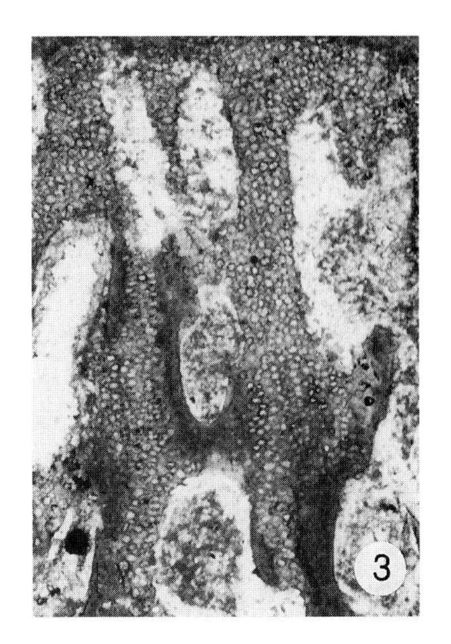

FIGURE 3 Endochondral ossification, as seen at the proximal end of a *Camptosaurus* humerus. In the lower part, early endochondral bone is seen as a dark tissue coating the walls of cavities excavated through calcified cartilage by advancing marrow processes. The cartilage is identified by the presence of numerous small globular chondrocyte lacunae, giving a foam-like appearance. ×64.

when present because they pass through the plane of transverse sections obliquely.

Periosteal bone from dinosaurs is commonly fibrolamellar bone and then usually parallel osteoned or laminar although sometimes reticular; however, zonal bone can also occur in limb bones and bones other than limb bones. Some show fibrolamellar variants that lack a true woven framework or both a woven framework and an osteon system. Accretionary bone is rare but is known from some genera (e.g., *Brachiosaurus* and *Troodon*). Very young dinosaurs may show fibrolamellar bone with imperfectly formed osteons or simply a finely cancellous tissue built from woven bone.

Endochondral Bone Endochondral bone is cancellous endosteal bone formed during growth by a process involving the destructive replacement of cartilage (Fig. 3). After cartilage cells (chondrocytes) in the growth zones have multiplied and hypertrophied, the matrix around them calcifies and the cells in it die. This calcified cartilage is next attacked from below by marrow cells termed chondroclasts, which excavate cavities on whose walls the first endochondral bone is deposited. Removal of the rest of the cartilage plus further bone deposition then produces a pattern of bone trabeculae with marrow-filled interspaces. In dinosaurs the process can often be seen in "frozen" form beneath articular surfaces and may also be seen at sutures or at the tips of neural spines.

Haversian Bone Haversian bone is osteonic endosteal bone that replaces preexisting bone tissues in the course of internal (or Haversian) reconstruction and occurs in both compact and cancellous forms (Fig. 4). The term has commonly been applied to the compact form only; but both result from variants of one process and they can intergrade completely. When compact Haversian bone replaces primary compact bone, some primary vascular canals are first enlarged by osteoclasts to form resorption spaces, in which osteoblasts then deposit new bone centripetally to form cylindrical secondary osteons with central secondary canals. The bone deposited is usually lamellar bone but can be fine parallel-fibered bone initially. The cross-sections of the secondary osteons are traditionally called "Haversian systems," and primary bone left between them is called interstitial bone. If the

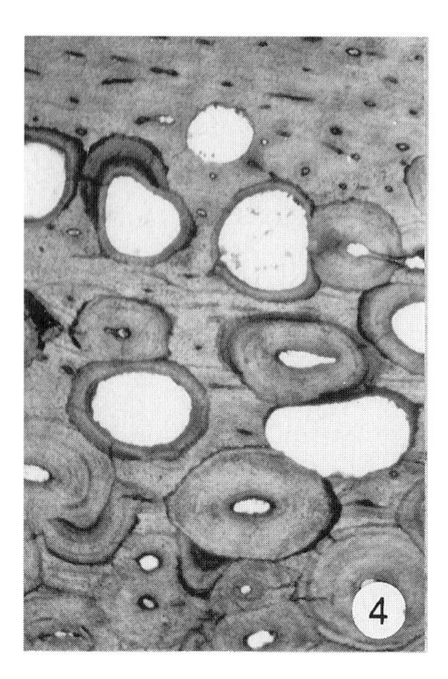

FIGURE 4 The start of Haversian reconstruction, showing primary bone (top), resorption spaces, and partly and completely formed "Haversian systems" (cross sections of compact secondary osteons). In the middle part, primary bone is still seen as interstitial bone, but the lowest part shows only a dense Haversian tissue. *Allosaurus,* lachrymal; ×64.

process continues repeatedly, the ultimate product is dense Haversian bone, built entirely from the last-formed secondary osteons and segments of partly resorbed earlier ones. Replacement of compact bone by cancellous Haversian bone is similar, but the resorption spaces formed are larger and only receive thin linings of new bone (Fig. 5). Again, repeated reconstruction produces a wholly secondary tissue in which trabeculae are said to show brecciate structure if formed in part from remnants of partly resorbed osteons. Replacement of endochondral bone begins with local resorption of primary trabeculae and can again lead to their replacement by a wholly secondary tissue.

Both types of Haversian replacement are often extensive in dinosaurs, and replacement of primary bone can be virtually total. However, it can also be limited or even lacking in some bones. The reason for this is not known. In limb bones, compact Haversian bone may be limited to metaphyses or localized under muscle scars (Reid, 1984).

Endosteal Lining Bone The term lining bone is convenient for secondary endosteal bone laid down on the walls of internal cavities in hollow bones, which may be marrow-filled or air-filled. It is related to Haversian tissues, being similarly formed in spaces produced by internal bone resorption, but has various special features and functions. It commonly forms compact bone only, but it can also form local outgrowths, which may branch and unite to form cancellous bone. It is also commonly avascular and built from lamellar bone, but it can also be vascular and nonlamellar. In marrow-filled limb bones, it may only be formed after active growth has ceased or formed and resorbed repeatedly during pauses in medullary expansion. It can extend around medullary cavities continuously, to form "internal circumferential lamellae," or be formed on one side only in a bone that is "drifting" toward the other side and then compensates for bone lost by external resorption (Fig. 6). In mammals (Enlow, 1969), external bone resorption can expose it over large parts of diaphyseal surfaces, but this is not usual in reptiles. In pneumatic structures, it typically forms thin avascular linings but can also grow out to form transverse struts in flight bones, as in pterosaurs and birds.

In dinosaurs, lining bone may be present or absent in hollow limb bones and is a constant feature of the cavernous ("pneumatic") vertebrae and ribs of saurischians. It is typically compact, thinly developed, and avascular, but may sometimes thicken locally into vascular bone in cavernous structures (Reid, 1997).

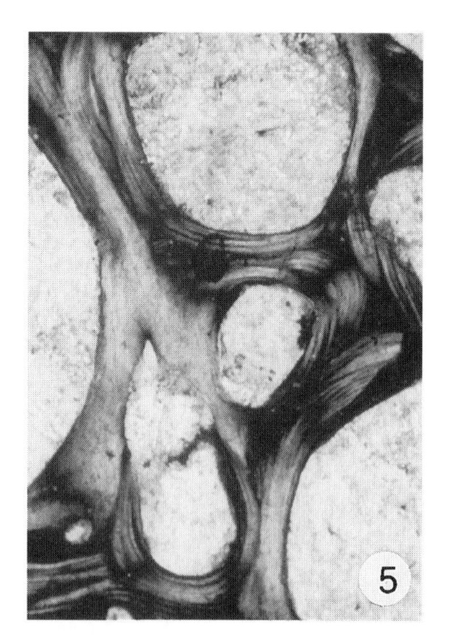

FIGURE 5 Haversian secondary cancellous bone, as seen with crossed polarizers, with brecciate structure at middle right and bottom left. *Allosaurus,* ischium; ×64.

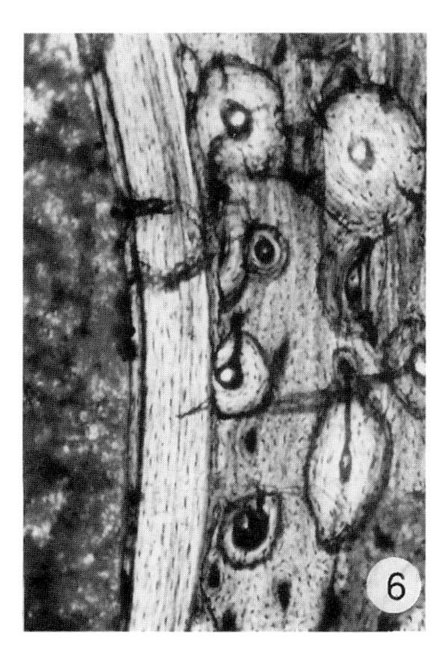

FIGURE 6 Lining bone (left of center) forming "internal circumferential lamellae" around a medullary cavity, seen filled with dark matrix at left. Bone to the right of the lining bone is periosteal bone, in part replaced by "Haversian systems." Their junction is a sharp reversal line, marking the limit of medullary expansion. Small theropod *indet.*, radius; ×64.

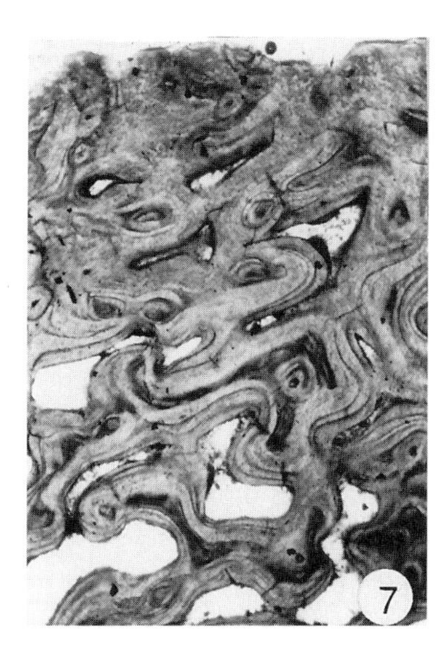

FIGURE 7 Compacted coarse cancellous bone, formed under an external resorption surface seen at top. The "swirling" pattern is a characteristic of this tissue. *Allosaurus*, ischium; ×21.

Compacted Coarse Cancellous Bone In certain situations, bone that is normally cancellous is converted into compact bone by bone accretion on the surfaces of trabeculae (Fig. 7). Cancellous bone of any type may be affected. The most common trigger of compaction is external bone resorption, when this would otherwise lead to the exposure of cancellous bone. However, it can also be triggered by internal cavitation in cavernous bones or in other ways. It is generally most developed in mammals, in which compacted bone may form large parts of diaphyseal surfaces, and less developed in reptiles. In dinosaurs it mainly occurs in metaphyses, in remodeled parts of pelvic bones, and near articular parts of cavernous bones, although sometimes it also occurs in alveolar bone (see Alveolar Bone) or in toe bones (Reid, 1997).

Alveolar Bone As in basal archosauromorphs and crocodiles, the teeth of dinosaurs were formed in deep tooth-bearing grooves, within which individual sockets were built by special alveolar bone tissue that lined the grooves and extended across them between teeth (Reid, 1997). This was formed initially as cancellous bone built from woven bone, but was sometimes compacted locally by accretion of slower growing tissues (Fig. 8). It was formed and resorbed repeatedly in relation to tooth replacement and the lateral expansion of the grooves during growth. For the latter reason, its junction with their walls is typically a reversal line, except in apparently old specimens that show

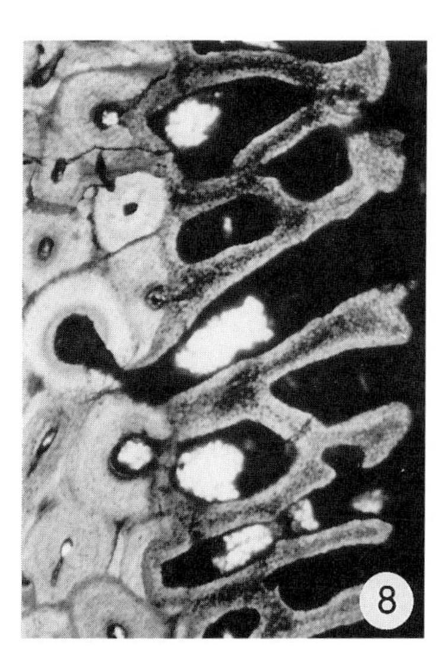

FIGURE 8 Cancellous alveolar bone (center and right) growing out from the wall of a tooth-bearing groove, formed by the dense Haversian bone seen at left. Although not sharp, their junction can be recognized as a reversal (resorption) line by the way some "Haversian systems" along it are truncated. The nonlamellar nature of the alveolar bone is also seen. *Camarasaurus*, dentary; ×32.

Haversian bone spreading into it. In theropods, upgrowths of alveolar bone form the interdental plates seen on the lingual (medial) sides of the teeth, although periosteal bone may also spread thinly onto them. In hadrosaurs the upper parts of teeth are in lateral contact, but alveolar bone still extends around their bases.

Metaplastic Bone This term (Haines and Mohouiddin, 1968) is applied to bone formed by nonpathological ossification (or metaplasia) of tissues that are not normally ossified in the absence of typical osteoblasts. It may arise from cartilage or from fibrous tissue forming tendons and ligaments, in the latter case mainly where these are attached to bones although sometimes also within tendons. Cells in the tissues affected (chondrocytes and fibrocytes) become in effect osteocytes if they persist; but the lacunae they occupy do not emit canaliculi. However, lack of canaliculi does not mark fossil bone as metaplastic because their loss in fossilization is common.

Metaplastic bone formed from cartilage is not known with certainty from dinosaurs; but bone ossified from the interspinous ligament is common in saurischians, occurring under the ligament scars seen on neural spines (Fig. 9) (Reid, 1997). Intratendinous ossification at the origins or insertions of muscles is also known but uncommon.

FIGURE 9 Densely fibrous bone produced by metaplastic ossification of an interspinous ligament during growth. *Haplocanthosaurus*, caudal vertebra, horizontal section through neural spine; ×64.

Special Cases Four kinds of dinosaurian bones have histological features requiring special mention.

Pachycephalosaur Skull Caps Bone forming the skull cap of *Stegoceras* contains numerous closely spaced radial vascular canals enclosed by primary osteons formed within a framework of woven bone (Reid, 1997). This tissue has the structure of fibrolamellar bone, although it is not a normal member of the fibrolamellar complex. Its high vascularity is more suggestive of a heat exchanging function than of head-butting habits.

Cavernous Bones In saurischians the centra of presacral vertebrae are commonly cavernous, with a plexus of internal cavities up to several centimeters wide between anastomosing plates of compact bone. The cavities arise by internal cavitation during growth and open to the exterior via lateral foramina. When stoutly developed (e.g., 3- to 5-mm thick) the bony plates typically show a contrast between coring and lining tissues, of which the former may be of several kinds (Reid, 1997). Near lateral surfaces, the coring bone may be relict periosteal or Haversian bone, of cortical origin, whereas near articular surfaces it is typically compacted endochondral bone. In deeper parts, however, these tissues are commonly replaced by Haversian bone, which may spread outward into the lining bone. When thinner, the plates may instead be formed entirely from lining bone, apart from patches of Haversian bone at plate junctions. Histologically identical bone occurs in the pneumatic parts of elephant skulls, supporting the view (e.g., Janensch, 1947) that such vertebrae were also pneumatic.

Tail Rods and "Ossified Tendons" The special tail rods of the theropod *Deinonychus* look like ossified tendons but seem to have arisen through tendon sheaths, developing a periosteal function (Ostrom, 1976). Unattached rods bracing the vertebral column in ornithischians are commonly called "ossified tendons," suggesting metaplastic structures, but usually consisting mainly or wholly of Haversian bone (Broili, 1922; Moodie, 1928), which is osteoblastic and can only have been formed by an endosteum. In *Iguanodon* (Reid, 1997) the outer parts may show a highly vascular tissue in which osteocyte lacunae emit normal canaliculi. This again must be an osteoblastic tissue and implies an external periosteum.

Osteoderms Dinosaurian osteoderms have been little studied, but periosteal, Haversian, and metaplastic tissues are known from them. At one extreme, the dorsal plates of Stegosaurs are formed mainly from periosteal and Haversian bone, though with a strongly fibrous tissue that could be metaplastic in the basal parts (de Buffrénil *et al.*, 1986). At the other extreme, flat dorsal plates of *Polacanthus* were built entirely by metaplastic ossification of large interwoven fiber bundles, like those seen in crocodile hides. Between these extremes various mixtures can occur (Reid, 1997).

Bone Pathology

Histological abnormalities in bone can result from various factors, including faulty or excessive ossification (osteomalacia and osteosclerosis), excess bone resorption during aging (osteoporosis), bone cancers, and reactions to injuries, inflammatory conditions, bacterial or viral infections, and cancerous metastases. Some of their effects can be recognized in dinosaurs; however, except in obvious cases (e.g., osteoporosis, healed fractures, and arthritic states), diagnosis can be difficult because of the absence of soft parts. Most involve aberrant developments of periosteal, endochondral, or Haversian bone; but reactive growths (e.g., Campbell, 1966) and ossification of soft tissues may also occur.

Bone and Physiology

Because they now occur mainly in endotherms, both Haversian and fibrolamellar bone have been claimed to show that dinosaurs had high metabolic rates (e.g., Bakker, 1972; de Ricqlès, 1974); but their value as evidence can be doubted on various grounds. They are not seen, for instance, in many small mammals and birds (e.g., sparrow and shrews), although these have the highest metabolic rates among modern endotherms. In contrast, Ferguson (1984) found Haversian reconstruction extensive in scutes of American alligators more than 6–8 years old, despite their low metabolic rates, and Reid (1987) found dense Haversian bone from a tortoise. Unlike modern reptiles, dinosaurs could form fibrolamellar bone continuously, as mammals and birds can; but its periodic formation in some turtles and crocodiles was recognized by Enlow (1969) before the start of the "hot-blooded" controversy. In young American alligators, which form it in early years, the mass-specific resting metabolic rate of a 7-kg specimen was found by Coulson and Hernandez (1983) to be only 1/12 of that of a comparable dog. It was also formed continuously by early therapsids that show no other signs of being endotherms (Kemp, 1982). Furthermore, although many dinosaurs had fibrolamellar periosteal bone (de Ricqlès, 1980; Table 1), some had bone with zonal "growth rings" like those of modern reptiles. Their occurrence, once doubted by de Ricqlès (1980), is now known from a wide range of dinosaurs (Reid, 1990), although mainly from bones other than limb bones. Bone with and without growth rings can then occur in different parts of one skeleton. Also, when bones grew asymmetrically, zonation could also be restricted to the parts where growth was slowest, despite all parts presumably having grown at the same metabolic rate. The bearing of bone on dinosaurian metabolic levels is hence best seen as uncertain; it cannot, in any case, throw light on how temperature was regulated.

The ability of dinosaurs to sustain rapid growth to huge sizes, implied by nonzonal fibrolamellar bone in, for example, *Brachiosaurus* (Gross, 1934), is, however, plain evidence of possession of cardiovascular and hemal systems able to support such growth, which would not have been possible otherwise. Furthermore, histological agreement between saurischian cavernous bones and modern pneumatic ones (Reid, 1997) supports the anatomical argument (e.g., Janensch, 1947) for their having an air sac system, in turn suggesting that saurischians at least had aerobic activity metabolism.

Teeth: General Features, Formation, and Histology

Dinosaurian teeth are basically of the hollow-cone type seen in crocodiles, with various modifications reflecting different diets and feeding styles. They were located in separate sockets, except in forms with tooth batteries, and were presumably held in place by a periodontal ligament. Tooth replacement occurred throughout life, so tooth germs were presumably budded repeatedly from a persistently active dental lamina.

Teeth are built primarily from acellular tissues called dentine and enamel, to which a special form of bone called cementum may be added. Dentine resembles bone chemically and in its level of mineral content (ca. 70%); but enamel is less than 3% organic and its protein content is noncollagenous, consisting of glycoproteins called enamelins and amelogenins. The former predominate in lower vertebrates (e.g., sharks) and the latter in tetrapods (Slavkin *et al.*, 1984). Odontogenesis in dinosaurs would generally have followed the pattern seen in crocodiles (e.g., Westergaard and Ferguson, 1987), which parallels that seen in mammals apart from being repetitive. In early development, linear invaginations of epithelial tissue along what will be the tooth-bearing surfaces grow inward to form dental laminae, from the deeper parts of which structures termed enamel organs are next produced at intervals. These each develop an indented and mesenchyme-filled base, forming a dental papilla on which a tooth will be formed within the organ. Epithelial tissues forming the outsides of the organs and covering their dental papillae internally are called enamel epithelia, classed as outer and inner, respectively, and the internal space between them is filled by a tissue called the stellate reticulum. Mesenchymal cells along the contact with the inner enamel epithelium differentiate into odontoblasts that begin to lay down dentine, retreating from it as it forms but leaving cellular processes in fine radial spaces called dentine tubules. Cells on the inside of the inner enamel epithelium are next transformed into ameloblasts and start laying down enamel, which forms first at the top of the tooth and spreads downwards. If cementum is formed, it is added on the outside of the basal parts later.

Dentine and enamel are often well preserved in dinosaurian teeth, and fine details of the contact between them can sometimes be recognized. Dentine may show both major and minor growth rings, of which the former can give an indication of how long the teeth remained in the jaws if they are assumed to be annual. Johnston (1979) claimed examples from tyrannosaurs and hadrosaurs to be evidence of ectothermy; but this does not follow in animals whose periosteal bone is typically nonzonal.

See also the following related entries:

Biomineralization • Chemical Composition of Dinosaur Fossils • Growth and Embryology • Growth Lines • Permineralization • Physiology

References

Bakker, R. T. (1972). Anatomical and ecological evidence of endothermy in dinosaurs. *Nature* **238,** 81–85.

Broili, F. (1992). Über den feineren Bau der "verknöcherten Sehnen" (verknöcherten Muskeln) von *Trachodon. Anatomischer Anzeiger* **55,** 465–475.

Campbell, J. G. (1966). A dinosaur bone lesion resembling avian osteoporosis with some remarks on the development of the lesions. *J. R. Microsc. Soc.* **85,** 163–174.

Coulson, R. A., and Hernandez, T. (1983). Alligator metabolism: Studies on chemical reactions *in vivo. Comp. Biochem. Physiol.* **74,** i–iii, 1–182.

de Buffrénil, V., Farlow, J. O., and de Ricqlès, A. J. (1986). Growth and function of *Stegosaurus* plates: Evidence from bone histology. *Paleobiology* **12,** 459–473.

de Ricqlès, A. J. (1974). Evolution of endothermy: Histological evidence. *Evolutionary Theor.* **1,** 51–80.

de Ricqlès, A. J. (1975). Recherches paléohistologiques sur les os longs des Tétrapodes, VII, sur la classification, la signification fonctionelle et l'histoire des tissue osseux des tetrapodes: Premiere partie, Structures. *Ann. Paleontol. (Vertebr.)* **61,** 49–129.

de Ricqlès, A. J. (1980). Tissue structure of dinosaur bone: Functional significance and possible relationship to dinosaur physiology. In *A Cold Look at the Warm-Blooded Dinosaurs* (R. D. K. Thomas and C. Olson, Eds.), pp. 103–139. American Association for the Advancement of Science, Selected Symposium 28, pp. 514. Westview, Boulder, CO.

Enlow, D. H. (1969). The bone of reptiles. In *Biology of the Reptilia* (C. Gans and A. de A. Bellairs, Eds.), pp. 45–80. Academic Press, London.

Ferguson, M. W. J. (1984). Craniofacial development in *Alligator mississipiensis.* In *The Structure, Development and Evolution of Reptiles* (M. W. J. Ferguson, Ed.), pp. 223–273. Zoological Society of London Symposia 52, pp. 697. Academic Press, London.

Gross, W. (1934). Die Typen den mikroskopischen Knochenbaues bei fossiler Stegocephalen und Reptilian. *Zeitschrift Anat.* **103,** 731–764.

Haines, R. W., and Mohouiddin, A. (1968). Metaplastic bone. *J. Anat.* **103,** 527–538.

Janensch, W. (1947). Pneumatizität bei Wirbeln von Sauropoden und anderen Saurischiern. *Palaeontogr. Suppl.* **7,** 1–125.

Johnston, P. A. (1979). Growth rings in dinosaur teeth. *Nature* **278,** 635–636.

Kemp, T. S. (1982). *Mammal-like Reptiles and the Origin of Mammals,* pp. 363. Academic Press, London.

Moodie, R. L. (1928). The histological nature of ossified tendons found in dinosaurs. *Am. Museum Novitates* **311,** 1–15.

Ostrom, J. H. (1969). Osteology of *Deinonychus antirrhopus*, an unusual theropod from the Lower Cretaceous of Montana. *Peabody Museum Nat. History Bull.* **30,** 1–165.

Pritchard, J. J. (1956). General anatomy and physiology of bone. In *The Biochemistry and Physiology of Bone* (G. H. Bourne, Ed.), pp. 1–125. Academic Press, New York.

Pritchard, J. J. (1972). General histology of bone. In *The Biochemistry and Physiology of Bone 2nd Edition* (G. H. Bourne, Ed.), pp. 1–20. Academic Press, New York.

Reid, R. E. H. (1984). The histology of dinosaurian bone, and its possible bearing on dinosaurian physiology. In *The Structure, Development and Evolution of Reptiles* (M. W. J. Ferguson, Ed.), pp. 629–663. Zoological Society of London Symposia 52, pp. 697. Academic Press, London.

Reid, R. E. H. (1987). Bone and dinosaurian "endothermy." *Mod. Geol.* **11,** 133–154.

Reid, R. E. H. (1990). Zonal "growth rings" in dinosaurs. *Mod. Geol.* **15,** 19–48.

Reid, R. E. H. (1997). Bone histology of the Cleveland–Lloyd Dinosaurs, and of Dinosaurs in General. Part I: Introduction; Introduction to Bone tissues. *Brigham Young Univ. Geol. Studies,* in press.

Seitz, A. L. L. (1907). Vergleichende Studien über den mikroskopischen Knochenbau fossiler und rezenter Reptilien, und dessen Bedeutung für das Wachstum und Umbildung des Knochengewebes in allegemeinen. *Abhandlungen der kaiserlichen Leopold–Carolignischen deutschen Akademie der Naturforscher, Nova Acta* **87,** 230–370.

Slavkin, H. C., Zeichner-David, M., Snead, M. L., Graham, E. E., Samuel, N., and Ferguson, M. W. J. (1984). Amelogenesis in Reptilia: evolutionary aspects of enamel gene products. In *The Structure, Development and Evolution of Reptiles* (M. W. J. Ferguson, Ed.), pp. 275–304. Zoological Society of London Symposia, 52, pp. 697. Academic Press, London.

Smith, J. W. (1960). Collagen fibre patterns in mammalian bone. *J. Anat.* **94,** 329–344.

Weidenreich, F. (1930). Das Knochengewebe. In *Handbuch der mikroskopischen Anatomie des Menschen* (W. H. M. von Mollendorff, Ed.). pp. 391–520. Springer-Verlag, Berlin.

Historical Museum of Hokkaido, Japan

see Museums and Displays

History of Dinosaur Discoveries

I. Early Discoveries

William A. S. Sarjeant
University of Saskatchewan
Saskatoon, Saskatchewan, Canada

Dragons form a persistent component of legend, as do giant humans. In theory, such legends might well be based on dinosaur remains. In practice, in every instance investigated so far, the bones attributed to such monsters have proved to be those of giant mammals (most often mammoths or mastodonts).

It is possible that dinosaur bones were noticed early and did form a basis for some other legendary creatures. The French–Canadian traveler Jean-Baptiste L'Heureux, while living among the Piegans in Alberta, Canada, was shown such bones at a date before 1871—long before the first scientific discovery of dinosaurs in that area—and told they were the remains of "the grandfather of the buffalo" (Spalding, 1993). In general, though, the huge bones of these extinct creatures seem to have been ignored either because, being petrified, they were not recognized as skeletal remains or because they were just too big to be noticed.

In contrast, the fossil footprints of the dinosaurs and their archosaurian progenitors were certainly noticed. The bushmen of Namibia, themselves efficient trackers, not only observed such footprints but also envisaged a bipedal trackmaker very much like a dinosaur.

Tracks preserved in the Lower Cretaceous sandstones of Paraíba, Brazil, were certainly noticed by the local Amerindians because they were incorporated into a design involving other symbols of unknown significance, carved beside the footprints into the rock surface.

Dinosaur tracks in the red sandstones of the Rhine valley may well have given rise to the legend of the hero Siegfried's slaying of the dragon, a rather undersized dragon according to early illustrations! (Fig. 1). Tridactyl tracks in similar sandstones of the Connecticut valley were thought to be those of the raven that failed to return to Noah's ark, and a dinosaur footprint built into the porch wall of a church in Cheshire, England, was improbably designated "The Devil's Toenail"!

Dinosaur eggs—sometimes intact but more often fragmentary—occur in the "Flaming Cliffs" of the Gobi Desert and were certainly known to ancient people there because shell fragments have been found rearranged into geometrical shapes. However, it is doubtful whether their character was recognized.

Even after the remains of dinosaurs came to be noticed, their nature was long misunderstood. The first published record of a dinosaur bone was in Robert Plot's *Natural History of Oxfordshire* (1677); the dinosaur bone had been found in Cornwall, England, by Sir Thomas Pennyston. It comprised what Plot termed the "*capita Femoris inferiora*" (Fig. 2). He recognized it as being "a real *Bone*, now petrified" but supposed, from its size, that it must be "the Bone of some Elephant, brought hither during the Government of the Romans in Britain."

The bone was refigured in 1763 by Richard Brookes, in his *Natural History of Waters, Earths, Stones, Fossils and Minerals*. His illustration, inverted from Plot's, was captioned "*Scrotum humanum*," surely in reference to the appearance of the condyles. Because this was arguably a binomen and postdated the publication of Linnaeus's *Systema Naturae* (10th ed., 1758), which is accepted as the effective starting point of zoological nomenclature, it has been mischievously argued that the earliest dinosaur generic name was *Scrotum* and the earliest specific name *S. humanum*! However, the binomen was merely descriptive; it was not intended to establish any taxon and has never been so used. The bone was probably part of a megalosaur femur, but because it is lost, that cannot be confirmed.

Two further dinosaur bone discoveries had been made in England during the intervening years. Specimen A1 in the geological collection of John Wood-

FIGURE 1 The dragon of the Drachenfels (from Kircher, 1665).

ward, of which a posthumous catalog was published in 1728, consists of portions of the shank of a dinosaur limb bone (though not, of course, so named). It survives in the Woodwardian Museum, Cambridge, and is probably the earliest discovered dinosaur bone that is still identifiable. That museum also contains part of a scapula of *Megalosaurus*, presented to it in 1784 by one Dr. Watson; no report of this was published. A further donation to the museum, made in 1809 and again from Oxfordshire, was identified by H. G. Seeley in 1869 as the caudal centrum of a cetiosaur; thus, it was the earliest discovered saurischian bone.

Furthermore, much earlier finds had been made in the Stonesfield State of Oxfordshire by Joshua Platt in 1755: three vertebrae "of enormous size," destined never to be described or illustrated, and an incomplete left femur, weighing 200 lb (more than 90 km), 29 in. (74 cm) long, and of width varying from 4 to 8 in. (10.2–20.4 cm). Platt's letter reporting his discoveries, accompanied by an excellent illustration of the femur (Fig. 3), was published in 1758 in the *Philosophical Transactions* of the Royal Society of London. However, though part of Platt's collection survives in the BM(NH), the dinosaur bones are lost.

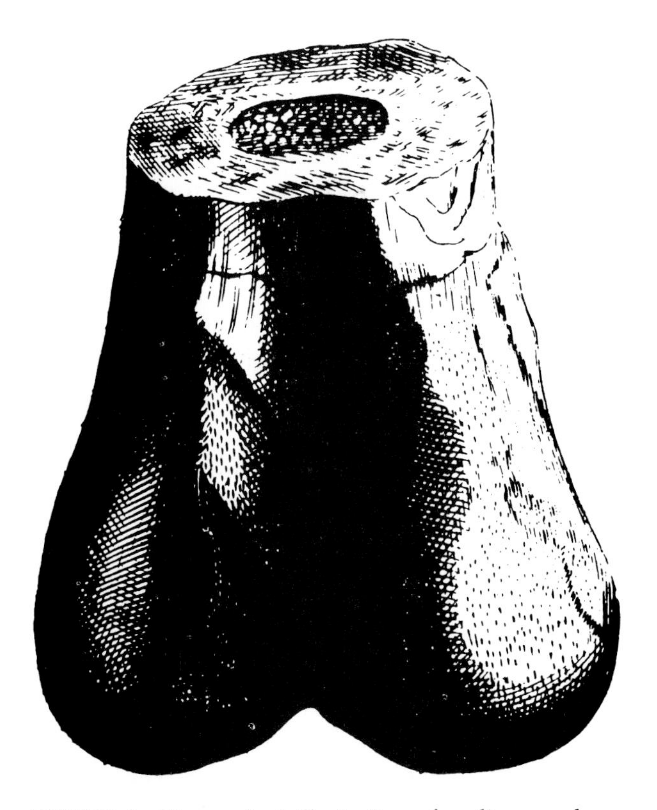

FIGURE 2 The earliest illustration of a dinosaur bone (from Plot, 1677).

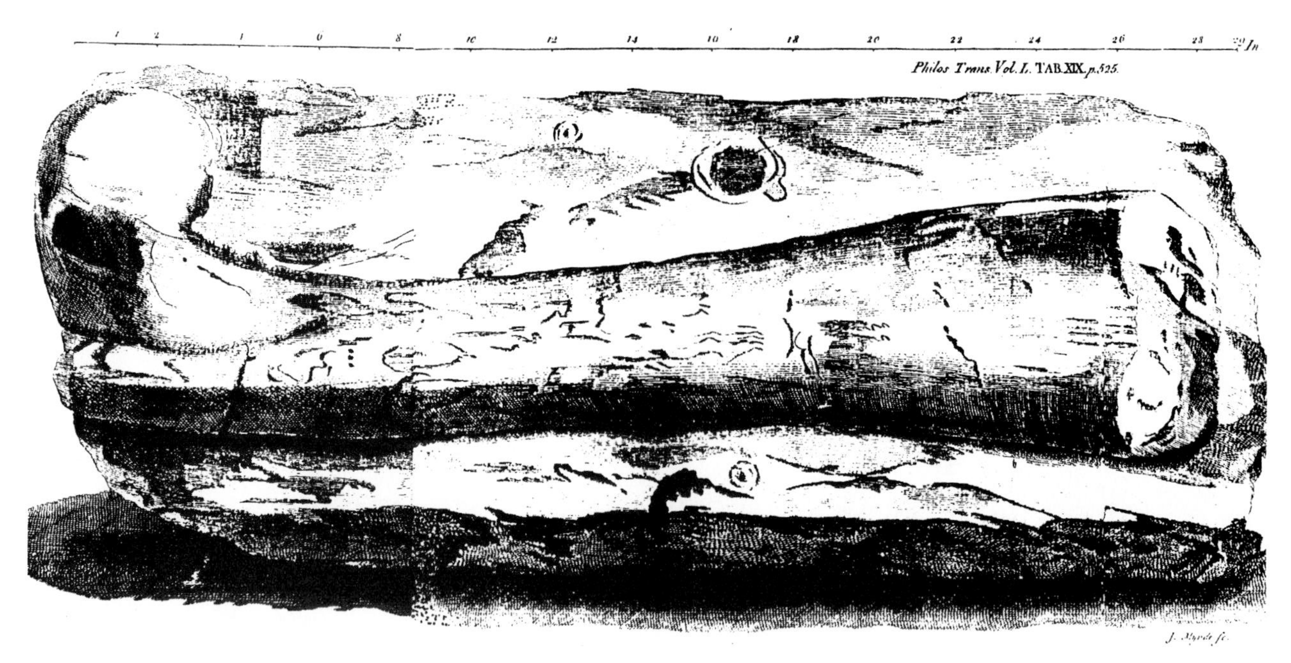

FIGURE 3 An incomplete saurischian femur from Oxfordshire (drawing by J. Mynde for J. Platt, 1758).

In France, the first finds may have been made before 1776, when the Abbé Dicquemare reported some vertebrae and a femur from the Jurassic strata of Normandy; however, this is unconfirmed because the bones were not illustrated and are now lost. Fossils collected there by another cleric, the Abbé Bachelet, were sent to the great anatomist Georges Cuvier for identification. Although noting certain unusual features, he thought they were vertebrae of crocodiles and it was only much later that their dinosaurian character was perceived.

The earliest discovery of dinosaur remains in America appears to have taken place in 1787, when Casper Wistar and Timothy Matlack reported the finding of a large bone, said to have been a "thigh-bone," in Late Cretaceous deposits of Gloucester County, New Jersey. The bone was never figured and is also lost; it was perhaps that of a hadrosaur. Subsequently, during his famous exploring expedition with Meriwether Lewis, William Clark discovered an enormous bone on the south bank of the Yellowstone River, close to the present town of Billings, Montana. Clark thought it to be the rib of a large fish but, because it was 3 ft (almost 1 m) in length "tho' a part of the end appears to have been broken off," it must have been one of the dinosaur bones for which that region is renowned.

The Wealden (Early Cretaceous) strata of England, later to be so important in the history of dinosaur study, may have yielded bones earliest to Thomas Webster at some date between 1812 and 1816. However, his only published mention of his finds, in 1829, reported them merely as "large saurian bones." Thus, it was not until well into the 19th century that these extinct vertebrates came clearly to be recognized.

In the history of the scientific description of dinosaurs, there can be no question that William Buckland (1784–1856) has priority. That this has passed unperceived results from the fact that Buckland, a polymath with very many social, societal, and scientific concerns, did not adequately document his own activities and was often dilatory in completing his scientific papers. It is only from the writings of others that we can confidently assign him that priority.

Some time before 1818 Buckland obtained, from the Middle Jurassic Stonesfield Slate of Stonesfield, Oxfordshire, the teeth and part of the lower jaw of what he recognized as a gigantic reptile (Fig. 4). The bones were seen by Cuvier during a visit to Oxford that year and were mentioned in correspondence between Buckland and Cuvier's Irish assistant, Joseph Pentland, in 1820. They were first named in 1822, by the physician and geologist James Parkinson in his *Outlines of Oryctology*, as *Megalosaurus* (Greek *megalos*,

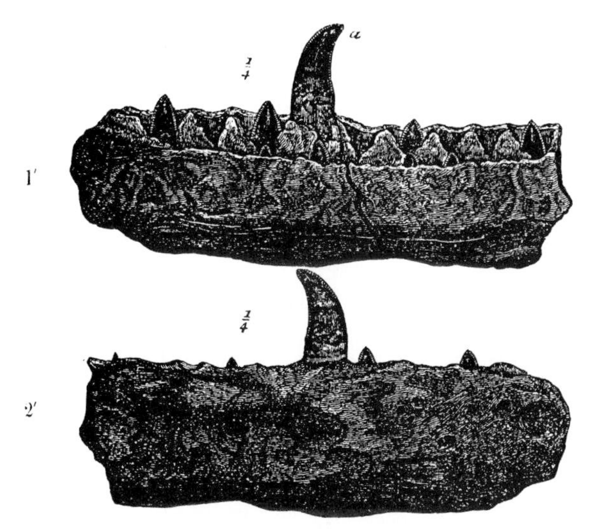

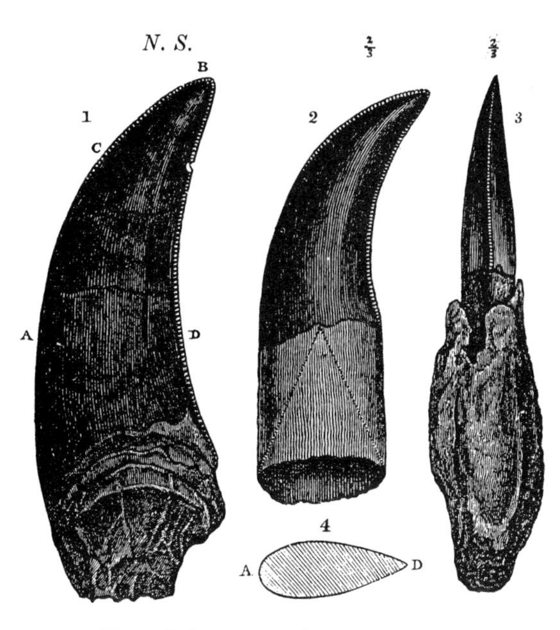

FIGURE 4 Right lower jaw and teeth of *Megalosaurus* from the Stonesfield Slate of Oxfordshire (after Buckland, 1836, plate 23).

great *sauros*, a lizard: here latinized). Parkinson noted that "the animal must, in some instances, have attained a length of forty feet [12 m] and stood eight feet [2.4 m] high," expressing also the hope that "a description may shortly be given to the public."

Further bones of "the Stonesfield reptile" had been presented to the Geological Society of London in 1822 by James Vetch. Gideon Algernon Mantell (1790–1852) mentioned the bones as being those of a "gigantic crocodile" in his *Fossils of the South Downs* (1822). It was probably the stimulus of these other works that caused Buckland at last to begin the serious description and illustration of his finds. The results were presented to the Geological Society in February 1824 and published in the society's *Transactions* later that year. Shortly afterwards, Cuvier virtually reproduced Buckland's description in his *Recherches sur les Ossements Fossiles*, using drawings sent to him by the British geologist. Mantell's own contributions to dinosaur discovery have been a focus for romantic stories, in particular of the finding of the first teeth by his wife Mary Ann in a heap of roadmetal while her surgeon husband was visiting a patient. However, Mantell's own journal does not report this incident and his later writings are, at best, ambiguous concerning the circumstances of the discoveries. All that can be said with certainty is that, by November 1821, Mantell's collection contained at least six dinosaur teeth. It is likely, as Dean (1993) speculates, that those teeth were purchased from quarrymen working the Wealden ironstones around Cuckfield, Sussex.

Mantell displayed a series of Wealden fossils, including the teeth, to the Geological Society in June 1822; but his idea—that they might be those of a gigantic herbivorous reptile—received little support. Consequently, he sent one of the teeth, and some of the bones, to Paris for Cuvier to examine. The great anatomist's response was disappointing; he considered the teeth to be those of a rhinoceros and the bones to be those of some unknown large mammal.

Only during a visit to the Hunterian Museum of the Royal College of Surgeons, early in 1824, did Mantell receive the crucial clue when another visitor, Samuel Stutchbury, pointed out to him how closely the Wealden teeth resembled those of the living iguana, albeit very much larger. Mantell promptly wrote again to Cuvier and this time received a gratifyingly positive response; yes, Cuvier now thought that these were indeed the teeth of a gigantic herbivorous reptile. In February 1825—almost exactly one year after Buckland's lecture on *Megalosaurus*—Mantell lectured to the Royal Society of London on his "newly discovered fossil reptile," which he named *Iguanodon* ("iguana tooth") (Fig. 5). Though not the first dinosaur to be named, it was certainly the first herbivorous dinosaur.

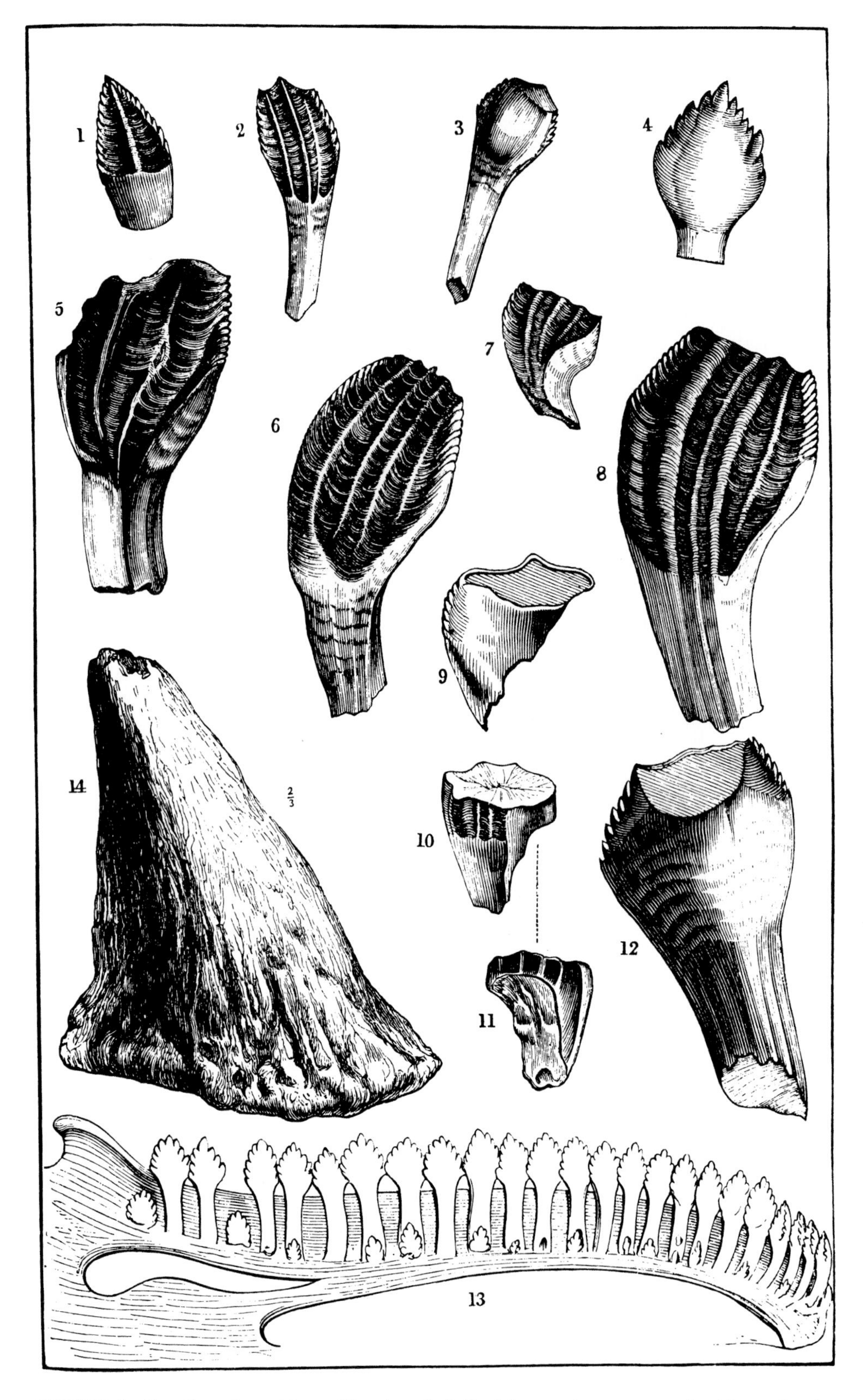

FIGURE 5 Fossil teeth and horn-like thumb (14) of *Iguanodon*, compared with teeth of a living iguana (1–4), and a restoration of an iguana jaw (13) (after Buckland, 1836, plate 24).

Interestingly, it is likely that Buckland had found *Iguanodon* bones at almost the same time as Mantell's acquisitions around Christmas 1821 from Wealden strata of the Isle of Wight. However, he misinterpreted them as "cetacean" until Mantell's work had alerted him to their true character. By the time of his 1824 paper, he also acquired for the Ashmolean Museum a series of bones from Oxfordshire that were even larger and more spectacular. Some of these had been collected by the paleontologist Hugh Strickland; others by John Kingdon (or Kingdom) who reported his finds to the Geological Society; and yet others by Buckland himself. Although initially referring to them as cetacean, Buckland perceived early that they belonged to "some yet undescribed reptile of enormous size, larger than the iguanodon." However, though he wrote that "I am collecting scattered fragments . . . in hope ere long of being able to make of its history," he never did so, thus missing the opportunity of describing the first sauropod dinosaur. That privilege was left to Richard Owen (1804–1892), whose eventual description of *Cetiosaurus* ("monstrous reptile") in 1841 also used bones from Mantell's collection and from localities in Northamptonshire, Buckinghamshire, and Yorkshire.

The description of this gigantic reptile formed a part of a lengthy "Report on British Fossil Reptiles," presented by Owen to the British Association for the Advancement of Science at its Plymouth meeting in 1841. It is from that meeting that the name "dinosaur" has been considered to date, its sesquicentennial being consequently celebrated in 1991. However, as Torrens (1992, 1993) pointed out, Owen did not employ that name at Plymouth. Instead, he introduced it into the manuscript of his lecture prior to its publication in 1842, writing of the "distinct tribe or suborder of Saurian Reptiles, for which I would propose the name Dinosauria." The celebration was thus one year too soon!

Between 1825 and Owen's naming of the dinosaurs, there had been many further discoveries. Mantell had continued to acquire fossil bones, most notably a block of arenaceous limestone from a quarry near Maidstone, Kent, displaying the partial skeleton of *Iguanodon*. It was this find, facetiously called his "mantel-piece," that permitted him to make the reconstruction upon which the Sydenham restoration was to be based. Mantell also discovered and described from the Sussex Wealden the articulated anterior portion of the skeleton of an armored reptile that, in 1832, he christened *Hylaeosaurus* ("woodland lizard"), as well as other bones that he believed wrongly to be those of a megalosaur.

The record of dinosaurs had been extended backward in time by the description by Samuel Stutchbury, working in association with S. H. Riley on bones from the Triassic "Magnesian Conglomerate" of the Bristol region (1836). Three different saurians were reported, two of which [*Thecodontosaurus* ("sheath-toothed reptile") and *Palaeosaurus* ("ancient reptile")] were accepted as being dinosaurs (though the nature of the latter genus is now considered dubious). Shortly afterwards, the extensive remains of another Triassic dinosaur, *Plateosaurus*, were reported by Hermann von Meyer from Germany (1837). A tooth described from Jurassic strata of southern Russia by A. Zborzewski (1834) was given the name *Macrodontophion* ("big tooth producer"); the tooth was perhaps that of a carnosaur. In France, Jacques-Armand Eudes-Deslongchamps (1838) had found a fragmentary dinosaur skeleton in the Middle Jurassic strata of Normandy. He perceived resemblances to *Megalosaurus*; however, he thought the differences sufficient to merit giving it a new generic name *Poekilopleuron* ("mottled-rib").

Throughout the first 30 years of dinosaur discovery, it was assumed that these great reptiles were quadrupeds. When life-size restorations of them were made by Benjamin Waterhouse Hawkins, under the supervision of Owen, they were reconstructed as massive animals walking upon all fours (Fig. 6). *Iguanodon* was depicted as horned, like the rhinoceros to which Cuvier had once compared its teeth; only later was it recognized that the supposed "horn" was in fact its spike-like thumb. Prior to the public display in 1854, some of London's savants dined inside the *Iguanodon*, with Owen sitting proudly within its head (see Crystal Palace).

Seventeen years after Owen's talk, discoveries in the United States forced a revision of the concepts of dinosaurs. Large bones were found in Cretaceous marls of New Jersey by a farmer, John Hopkins, who brought them to the attention of his friend William P. Foulke of Philadelphia. Foulke succeeded in excavating a quantity of these bones and passed them for study to Joseph Leidy (1823–1891), who was profes-

FIGURE 6 *Iguanodon* and *Megalosaurus*, portrayed as quadrupeds battling to the death (drawing by Rieu in L. Figuier, 1866).

sor of anatomy in that city. By the end of the year, Leidy had presented an account of them to the Philadelphia Academy of Sciences (1858). They were those of a herbivorous dinosaur with a remarkable duck bill; Leidy placed them in a new genus—*Hadrosaurus* ("bulky reptile"). Moreover, noting the great disproportion between the fore and hind parts of the skeleton, he concluded that "this great extinct herbivorous reptile may have been in the habit of sustaining itself, kangaroo-like, in an erect position on its back extremities and tail." From that interpretation to the perception of bipedality, not only in *Hadrosaurus* but also in such other disparate genera as *Iguanodon* and *Megalosaurus*, was but a small step. This was confirmed by the belated recognition that tracks in early Mesozoic red sandstones of New England and Europe were not of birds, as hitherto supposed, but of dinosaurs or their bipedal progenitors.

With the expansion of the study of dinosaur remains in North America and in Europe, this branch of paleontology moved forward into what is here styled "The first golden period."

References

Buckland, W. (1836). *Bridgewater Treatise VI*, vol. 2.

Buffetaut, F., Cuny, G., and Le Loeuff, J. (1993). The discovery of French dinosaurs. *Mod. Geol.* (Halstead Memorial Vol.) **18**(2), 161–182.

Colbert, E. H. (1968). *Men and Dinosaurs*. Dutton, New York.

Dean, D. R. (1993). Gideon Mantell and the discovery of *Iguanodon*. *Mod. Geol.* (Halstead Memorial Vol.) **18**(2), 209–219.

Delair, J. B., and Sarjeant, W. A. S. (1975). The earliest discoveries of dinosaurs. *Isis* **66**, 5–25.

Figuier, L. (1866). *The World Before the Deluge*.

Halstead, L. B., and Sarjeant, W. A. S. (1993). *Scrotum humanum* Brookes—The earliest name for a dinosaur? *Mod. Geol.* (Halstead Memorial Vol.) **18**(2), 221–224.

Kircher, A. (1665). *Mundus Subterraneous*, Vol. 8. Amsterdam.

Platt, J. (1758). *Philos. Trans. R. Soc.* **50,** plate XIX.

Plot, R. (1677). *Natural History of Oxfordshire*.

Sarjeant, W. A. S. (1987). The study of fossil vertebrate footprints. A short history and selective bibliography.

In *Glossary and Manual of Tetrapod Footprint Palaeoichnology* (G. Leonardi, Ed.), pp. 1–19. República Federativa do Brasil, Departamento Nacional da Produçáo Mineral, Brasília.

Spalding, D. A. E. (1993). *Dinosaur Hunters*. Key Porter Books, Toronto.

Taquet, P. (1983). Cuvier–Buckland–Mantell et les dinosaures. In *Actes du Symposium Paléontologique* (E. Buffetaut, J. M. Mazin, and E. Salmon, Eds.), pp. 475–494. G. Cuvier, Montbéliard.

Torrens, H. S. (1992). When did the dinosaur get its name? *New Scientist* **134,** 40–44.

Torrens, H. S. (1993). The dinosaurs and dinomania over 150 years. *Mod. Geol.* (Halstead Memorial Vol.) **18**(2), 257–286.

II. First Golden Period in the USA

Brent H. Breithaupt
University of Wyoming
Laramie, Wyoming, USA

Although the remains of dinosaurs were known from the Old World centuries before those in western North America, most specimens were fragmentary in nature. These scrappy, initial finds resulted in "creative" interpretations and reconstructions of prehistoric beasts. Fairly complete skeletal material of dinosaurs did not come to light until the latter part of the 1800s, when major discoveries were made in the western part of the United States. This material allowed scientists and the public to gain an appreciation of the diversity, complexity, and uniqueness of dinosaurs. Although articulated skeletons and extensive bone beds are currently known from this area of North America, the first discoveries of dinosaurs in the West were also fragmented, isolated material.

Early Discoveries

The first discoveries of dinosaur remains in the Rocky Mountain West undoubtedly were made by the Native Americans that occupied this area of the country. These people probably noticed the petrified bones now known to occur commonly in this region but may not have understood why these remains were present or their importance. Various tribal beliefs regarding the occurrences of fossil material are yet to be investigated, although stories of ancient giants (Betts, 1871) and huge, burrowing serpents (Colbert, 1968) are known to have been used to explain the presence of some large, prehistoric animal remains in the West.

Dinosaur remains were most likely first described from the Rocky Mountain West in 1806, when a large "fish" bone was discovered by William Clark of the Lewis and Clark expedition in an area of Cretaceous deposits in present-day Montana. Unfortunately, the specimen and other records concerning it are not available, but the description of its size and location suggest a late Mesozoic dinosaur bone. In the latter half of the 19th century, the eminent scientist Ferdinand Vandiveer Hayden (referred to by some Native Americans as the "man who picks up stones running") collected many of the first fossils in the West associated with the Geological and Geographical Survey of the Territories. In 1855 Hayden collected teeth from Cretaceous badlands in present-day Montana. The following year, Joseph Leidy (1856) described these dinosaur fossils and assigned the "lacertilian-like" teeth to three new genera (i.e., *Palaeoscincus*, *Troodon*, and *Trachodon*). From another tooth type he designated the name *Deinodon*, a relative of the carnivorous dinosaur *Megalosaurus*. Also in 1856, Leidy named the dinosaur *Thespesius* (now possibly *Edmontosaurus*) based on two caudal vertebrae and a first phalanx given to him by Hayden from the "Great Lignite" Formation of Nebraska Territory (now South Dakota).

In 1859 John Strong Newberry (geologist on Captain John N. Macomb's survey in present-day Utah) discovered the remains of a giant, herbivorous dinosaur. Edward Drinker Cope (1877) described this new Jurassic reptile, *Dystrophaeus* (now possibly *Camarasaurus*). F. V. Hayden (1868; 1869) wrote of the occurrences of huge Cretaceous saurian fossils in various spots across Wyoming Territory, primarily along the line of the Union Pacific railroad. In 1868 Othniel Charles Marsh traveled the length of the transcontinental railroad and is reputed to have been shown a large bone at Como Station in southeastern Wyoming Territory. Although Marsh said that he recognized the rocks in this area as being Jurassic in age, his main interest at Como Station in 1868 was to collect the little axolotl-like salamanders that lived in nearby Lake Como. Hayden obtained a bone identified as a "petrified horse hoof" from Middle Park, Colorado, in 1869. This specimen was later described by Leidy

(1873) as the vertebra of the carnivorous dinosaur *Poicilopleuron* (now known as *Allosaurus*). These early discoveries of dinosaurs in the West were only the "tip of the iceberg" of what would later be found.

Major Discoveries: 1870s to the Turn of the Century

In 1872 Drs. F. B. Meek and H. M. Bannister, while working for Hayden's Geological Survey of the Territories, discovered dinosaurian remains in the latest Cretaceous rocks of western Wyoming near Black Butte Station. E. D. Cope (1872) (Fig. 1) assigned this partial skeleton to a new genus and species of dinosaur, *Agathaumas sylvestris*. *Agathaumas sylvestris* was one of the largest, best preserved, and most complete dinosaurs known at that time. Cope's excitement with the discovery of these bones was manifested in his name meaning "marvelous saurian of the forest." Although its relationship to other dinosaurs was unclear in 1872, the postcranial remains resemble later finds of *Triceratops*.

In the race to name new fossil animals in the late 1800s, noted vertebrate paleontologists Othniel Charles Marsh (Fig. 2) and Edward Drinker Cope were the primary protagonists and fierce adversaries.

FIGURE 2 Othniel Charles Marsh, 1831–1899. Reproduced with permission from *Science* (1889).

FIGURE 1 Edward Drinker Cope, 1840–1897. Reproduced with permission from *Science* (1897).

Marsh and Cope ardently strove to scientifically outdo one another by naming and describing new taxa of fossil animals. The animosity that each felt for the other was evident not only in their writings but also projected into the field crews as well. As a consequence of their rivalry and the rich vertebrate fossil deposits of the West, a tremendous surge in the scientific knowledge of ancient animals resulted. The opening of the West associated with the railroad and the development of cities in that region allowed paleontologists to travel to areas of vast outcrops of Mesozoic strata to hunt for dinosaur remains. No longer did scientists have to rely on haphazard finds of specimens, but rather they were able to visit areas of paleontological potential themselves to discover the remains of ancient organisms new to science.

Spectacular discoveries of dinosaurs were made by Marsh, Cope, and their field crews from the Upper Jurassic Morrison Formation in 1877 in both Wyoming (i.e., Como Bluff) and Colorado (i.e., Garden Park and near Morrison) (Ostrom and McIntosh, 1966). Without question, the discovery of dinosaurs in the Morrison Formation had a tremendous impact on the science of paleontology. Some of the first major quarries of dinosaur remains found anywhere were made in this unit. These localities were exceptional not only because of the number of specimens found but also because of the exceptional preservation and diversity of animals uncovered. Because some of the first nearly complete dinosaur skeletons were found at these sites, these discoveries had a tremendous influence on the understanding of ancient reptiles and the general evolution of life. As the result of these discoveries, dinosaurs emerged from an eclectic array of natural curiosities as a distinct race of extinct animals. Discoveries of dinosaur remains from western North America subsequently led to important explorations for dinosaurs worldwide. In addition, dinosaurs uncovered in the Mesozoic units of the West influenced both the philosophy and the architecture of museums throughout the world.

Between 1877 and 1910 some of the most notable paleontologists (e.g., Marsh, Cope, Wortman, Granger, Lull, Matthew, Osborn, Brown, Williston, Sternberg, Gilmore, Riggs, Hatcher, Douglass, and Reed), many from eastern institutions, traveled west to collect dinosaur remains at some of the premier Mesozoic sites (e.g., Como Bluff, Garden Park, Morrison, Bone Cabin Quarry, Sheep Creek, Dinosaur National Monument, Judith River Formation badlands, and type Lance Formation area) anywhere in the world. Many of these people and places pivotal to our current understanding of dinosaurs are discussed in detail elsewhere in this book. Knowledge of Mesozoic organisms and the evolution of life was forever changed by the work in the West in the late 1800s and early 1900s. Organized paleontological field expeditions were amassed to travel west to indulge in the "smorgasbord" of fossils present. Trainloads of fossil specimens were collected and sent back east. Although Cope and Marsh (who initiated much of this work) carried their bitter "fossil feud" to their deaths, by the 1890s the "bone wars" were essentially over in the West and other institutions had begun collecting there (McIntosh, 1990). Work at various sites within fossiliferous Mesozoic strata continued into the 20th century. Many North American museums began their collections of dinosaurs in the late 1800s or early 1900s in this region of North America. Dinosaur specimens from the Rocky Mountain West are currently on display in many prominent, eastern museums (e.g., NATIONAL MUSEUM, AMERICAN MUSEUM OF NATIONAL HISTORY, CARNEGIE MUSEUM, Field Museum of National History, and YALE PEABODY MUSEUM) as well as around the world.

Conclusion

Dinosaurs have been enormously popular for many years. They are used commercially to market a wide variety of products (e.g., gasoline, pasta, candy, books, cereals, and movies). Public recognition of these beasts is the result of more than 100 years of study. Many of the most popular dinosaurs (e.g., *"Brontosaurus," Allosaurus, Stegosaurus, Camarasaurus, Tyrannosaurus, Triceratops,* and *Diplodocus*) are known from discoveries made initially during the so-called "Golden Years" of dinosaur collecting in the Rocky Mountain West. The foundations of many modern paleontological concepts and field techniques are based on the work done at that time. Fossil discoveries made during the late 1800s and early 1900s bolstered the emergence of vertebrate paleontology worldwide. Today, fossils collected during the Golden Years highlight exhibit and research collections in museums around the world. Following the important discoveries made in the 19th century, institutions around the country are still making significant finds in the Mesozoic strata of the West. Continued work within these units is adding to our understanding of the climate, depositional regimes, and organisms of the Mesozoic Era.

See also the following related entries:

CANADIAN DINOSAURS • CENTRAL ASIATIC EXPEDITIONS

References

Betts, C. (1871, October). The Yale College Expedition of 1870. *Harper's New Monthly Magazine,* 663–671.

Colbert, E. H. (1968). *Men and Dinosaurs: The Search in the Field and Laboratory,* pp. 283. Dutton, New York.

Cope, E. D. (1872). On the existence of Dinosauria in the Transition Beds of Wyoming. *Proc. Am. Philos. Soc.* **12,** 481–483.

Cope, E. D. (1877). On a Dinosaurian from the Trias of Utah. *Proc. Am. Philos. Soc.* **16,** 579–587.

Hayden, F. V. (1868). Second annual report of the United States Geological Survey of the Territories, embracing Wyoming. In *1873, 1st, 2nd, and 3rd Annual Reports of the United States Geological Survey of the Territories for the Years 1867, 1868, 1869* (F. V. Hayden, Geologist-in-Charge), pp. 67–102. Department of the Interior, Government Printing Office, Washington, DC.

Hayden, F. V. (1869), Third annual report of the United States Geological Survey of the Territories, embracing Colorado and New Mexico. In *1873, 1st, 2nd, and 3rd Annual Reports of the United States Geological Survey of the Territories for the Years 1867, 1868, 1869* (F. V. Hayden, Geologist-in-Charge), pp. 103–251. Department of the Interior, Government Printing Office, Washington, DC.

Leidy, J. (1856). Notices of the remains of extinct reptiles and fishes, discovered by Dr. F. V. Hayden in the badlands of the Judith River, Nebraska Territory. *Proc. Acad. Nat. Sci.* **1856,** 72–73.

Leidy, J. (1873). Contributions to the extinct vertebrate fauna of the western territories. In *Report of the U.S. Geological Survey of the Territories* (F. V. Hayden, Geologist-in-Charge). Department of the Interior, Government Printing Office, Washington, DC.

McIntosh, J. S. (1990). The second Jurassic dinosaur rush. *Earth Sci. History* **9,** 22–27.

Ostrom, J. H., and McIntosh, J. S. (1966). *Marsh's dinosaurs,* pp. 388. Yale Univ. Press, New Haven, CT.

III. Quiet Times

Eric Buffetaut
Laboratoire de Paléontologie des Vertébrés
Université Paris
Paris, France

The period between 1930, which marks the end of the Central Asiatic Expeditions of the American Museum of Natural History, and 1970, which can be considered the beginning of the so-called "dinosaur renaissance," is one of "quiet times" mainly in terms of concepts: During this period there was little change in the image of dinosaurs in the minds of scientists and public alike. Franz Nopcsa, one of the most imaginative dinosaur specialists of his generation, committed suicide in 1933, at a time when he was beginning to explore interesting new approaches such as paleohistology and the geographical distribution of dinosaurs in terms of Alfred Wegener's continental drift. Many years elapsed until these innovative approaches were taken up again. Dinosaurs were often considered an oddity in the history of life—a bizarre offshoot that, however spectacular, had relatively little evolutionary importance. Although some experts, such as Friedrich von Huene in Germany, C. M. Sternberg in Canada, and Charles W. Gilmore in the United States, published valuable and extensive contributions on dinosaurs during the 1930s, for many paleontologists of that time emphasis was on the mechanisms of evolution or on the great evolutionary innovations that led from "lower" to "higher" vertebrates, and eventually culminated in man—and dinosaurs seemed to play little part in that dramatic development. This attitude continued well into the 1960s, at least in some countries.

Nevertheless, many important dinosaur discoveries were made during that period and were to pave the way for later conceptual developments. During the 1930s, fieldwork was to some extent limited by political unrest in many countries and by lack of funds caused by the Great Depression. However, the severe economic crisis and the resulting unemployment also made cheap labor available for large-scale excavations at some important sites. A good example from Germany is provided by Reinhold Seemann's excavations for the Stuttgart Museum at the Late Triassic Trossingen locality in 1932, which yielded a large number of *Plateosaurus* remains. American examples can be found in Roland T. Bird's work for the American Museum of Natural History, including his excavations at the Late Jurassic (Morrison Formation) Howe Quarry in Wyoming (1934), which yielded a huge accumulation of sauropod bones, and his pioneering work, begun in 1938, on the theropod and sauropod footprint sites of the Lower Cretaceous of Texas. Chance discoveries also led to more limited excavations at particular sites, some of which are of special interest. For example, in 1934 the Oxfordian Damparis quarry of the Jura region of eastern France yielded one of the most complete skeletons of a sauropod (an early brachiosaurid) ever found in Europe, which was described by Albert François de Lapparent in 1943.

Outside the Western world, the 1930s were also marked by the emergence of studies on dinosaurs by local paleontologists in eastern Asia. In 1935, C. C. Young (Yang Zhongjian) described an ankylosaur from Ningxia Province in northwestern China. In 1938, excavations were begun in the Lufeng Beds (Late Triassic to Early Jurassic) of Yunnan in southern China and resulted in the discovery of a large number of prosauropod remains, described during the 1940s by Young. In 1936, the Japanese paleontologist T. Nagao described *Nipponosaurus*, a hadrosaur from southern Sakhalin Island (then a part of Japan). Farther south, in Indochina, the first dinosaur bones (of ornithopods and sauropods) from Southeast Asia were found in the Cretaceous of Laos during the late 1930s by the French geologist J. H. Hoffet.

World War II put an end to many paleontological projects, including A. F. de Lapparent's excavations at the Fox–Amphoux dinosaur locality in the Upper Cretaceous of Provence, which had begun in 1939. This did not prevent him, however, from reviewing the Late Cretaceous dinosaurs from southern France in an extensive monograph published in 1947. The war also caused the destruction by air raids of several valuable dinosaur collections in European cities, including Stromer's collection of Cretaceous dinosaurs from Egypt in Munich and the Jurassic and Cretaceous specimens from Normandy kept in Caen and Le Havre. Fortunately, the dinosaurs collected from the Upper Jurassic of Tendaguru in East Africa before 1914, which were kept in Berlin, largely escaped destruction, and Werner Janensch's work on that remarkable collection went on after the war until the 1960s.

Among the major efforts in dinosaur paleontology undertaken in the immediate postwar years were the excavations carried out in the Upper Triassic of New Mexico at Ghost Ranch by Edwin H. Colbert and associates from the American Museum of Natural History, which resulted in the discovery of many skeletons of the early theropod *Coelophysis bauri*. Colbert's activity as a popularizer, both through several books and through the opening of a new dinosaur hall at the American Museum, was important in keeping the American public interested in dinosaurs at a time when relatively few paleontologists were working in that group. In 1958, the opening of the permanent visitor center at Dinosaur National Monument, the former "Carnegie Quarry," in Utah was another expression of a desire to respond to enduring public interest in dinosaurs.

Just after the war, in 1946, a first Russian expedition under I. Efremov was sent to the Gobi Desert of Mongolia. It was followed by others in 1948 and 1949 and resulted in the discovery of a large amount of dinosaur material from the Upper Cretaceous, including various forms from the Nemegt Formation that had not been found by the earlier American expeditions, such as the large tyrannosaurid *Tarbosaurus*.

The 1950s also saw the last efforts of colonial vertebrate paleontology when French paleontologists discovered important dinosaur localities in various parts of the crumbling French colonial empire. R. Lavocat's work centered on the Cretaceous of Morocco (with the mysterious sauropod *Rebbachisaurus*) and Madagascar, whereas A. F. de Lapparent explored the Cretaceous dinosaur localities of the central and southern Sahara, thus revealing the great paleontological potential of that vast desert.

In China, an active local school of vertebrate paleontology developed during the 1950s after the creation of the Institute of Vertebrate Paleontology and Paleoanthropology in Beijing. Important fieldwork was done in the early 1950s by C. C. Young and associates on the Cretaceous dinosaur localities of Shandong Province, where the pioneering researches of H. C. Tan and Otto Zdansky in the 1920s had resulted in Wiman's description of the sauropod *Euhelopus zdanskyi* and the hadrosaur *Tanius sinensis* in 1929. Young's 1958 memoir on the Shandong dinosaurs included descriptions of the unusual (and perhaps chimeric) Late Cretaceous hadrosaur *Tsintaosaurus spinorhinus*, with its cranial spike, and of Early Cretaceous psittacosaurs.

In Europe, the 1950s and 1960s saw a renewal of interest in the dinosaur eggs from the Upper Cretaceous of southern France, which had been reported as early as the mid-19th century by Pouech and Matheron. R. Dughi and F. Sirugue, from the Aix-en-Provence Museum, collected intensively in Provence and attempted to distinguish egg types on the basis of shell microstructure. During that period, one of the most active dinosaur specialists in Europe was A. F. de Lapparent, who, besides his work in the Sahara, described important Late Jurassic material (including theropods, sauropods, and stegosaurs) from Portugal with R. Zbyszewski in 1957. Lapparent also contributed to the renaissance of studies on dinosaur ichnology when, in 1967, he described with C. Montenat the remarkable terminal Triassic to basal Jurassic site at Le Veillon, on the Atlantic coast of France, where thousands of footprints were recorded. He was also the first to report the occurrence of "polar" dinosaurs when he described *Iguanodon* footprints from the Lower Cretaceous of Spitzbergen in 1962.

Large-scale expeditions in search of dinosaurs were started again in the 1960s. Among them were several French expeditions to the southern Sahara, in the Republic of Niger, where uranium prospecting had led to finds of well-preserved vertebrates of Early Cretaceous age. One of the major results of these expeditions was the discovery by Philippe Taquet of nearly complete skeletons of a new iguanodontid, *Ouranosaurus nigeriensis*.

In central Asia, work was resumed in Mongolia when POLISH–MONGOLIAN EXPEDITIONS (in 1963, 1964, 1965, and 1971) under the leadership of Zofia Kielan-Jaworowska again explored the Late Cretaceous formations of the Gobi Desert, with remarkable success. Among the most spectacular finds were well-preserved ankylosaurs, hadrosaurs, tyrannosaurs, ornithomimosaurs, oviraptorosaurs, pachycephalosaurs, protoceratopsids, and sauropods, plus hitherto unknown forms of mysterious affinities such as the huge long-armed theropod *Deinocheirus*. The careful descriptions of these finds in a series of large monographs by several Polish paleontologists were to increase considerably our knowledge of the Late Cretaceous dinosaurs of Asia.

This renewed field activity was accompanied during the 1960s by the beginnings of a reassessment of many long-held conceptions about dinosaurs, which often turned out to be poorly founded. Coming after the pioneering work of Nopcsa and Heidsieck in the 1930s, and the research of Enlow and Brown in the 1950s, large-scale investigations of bone microstructure in a variety of fossil reptiles, including dinosaurs, were launched by Armand de Ricqlès in Paris in the late 1960s. This work contributed to the reappraisal of dinosaur physiology that was to culminate in the debate about "hot-blooded dinosaurs" during the following decade. Using different approaches, John Ostrom in New Haven, Connecticut, was one of the main actors in this reassessment of dinosaur habits and mode of life when, in the 1960s, he challenged the long-held belief that hadrosaurs were predominantly aquatic animals and when he suggested on the basis of footprint evidence that some dinosaurs were gregarious. One of his major contributions to this debate was directly based on field discoveries. In 1969 he described a new theropod found in Montana in 1964 as *Deinonychus antirrhopus* and interpreted it as a fast and agile predator, quite different from the then common image of dinosaurs as sluggish animals.

The quiet times between 1930 and 1970 were not as quiet as a superficial comparison with more "exciting" times before and after might suggest. Despite often adverse economic and political conditions, much fieldwork was done and many new dinosaurs were discovered in parts of the world that had previously been little visited by paleontologists. There may have been some stagnation in the image paleontologists had of dinosaurs, which were often represented as sluggish giants that spent a large part of their time wallowing in swamps. Nevertheless, during the 1960s this image began to crumble, and the use of various new approaches led to the reassessments that culminated in the debates of the 1970s and 1980s and in the current tremendous development of every field of dinosaur research.

References

Bird, R. T. (1985). *Bones for Barnum Brown*. Texas Christian Univ. Press, Fort Worth.

Buffetaut, E. (1995). *Dinosaures de France*. Éditions du BRGM, Orléans, France.

Colbert, E. H. (1968). *Men and Dinosaurs: The Search in Field and Laboratory*. Dutton, New York.

Dong, Z. (1992). *Dinosaurian Faunas of China*. China Ocean Press / Springer-Verlag, Beijing / Berlin.

Kielan-Jaworowska, Z. (1969). *Hunting for Dinosaurs*. MIT Press, Cambridge, MA.

Probst, E., and Windolf, R. (1993). *Dinosaurier in Deutschland*. Bertelsmann, Munich.

Rojdestvenski, A. (1960). *Chasse aux dinosaures dans le désert de Gobi*. Arthème Fayard, Paris.

Spalding, D. A. (1993). *Dinosaur Hunters*. Key Porter Books, Toronto.

Taquet, P. (1995). *L'Empreinte des Dinosaures*. Éditions Odile Jacob, Paris.

IV. Research Today

LOUIE PSIHOYOS
Boulder, Colorado, USA

Two hundred years ago, all the dinosaur specimens known to science could have fit in your pocket. Fifty years later, when paleontologist Sir Richard Owen coined the name dinosaur, all the specimens known to the world might have filled your kitchen cupboards. (Queen Victoria knighted Sir Richard Owen in 1884 as a K.C.B., or Knight Commander of the Order of the Bath. This was a kind of lifetime-achievement award for general public service. He retired that same year, only after he was sure that there was enough public funding for the British Museum.)

Then, shortly after the Civil War in America was over, everything changed and a new era of dinosaur collecting began.

It is hard not to be nostalgic about the earliest days of collecting dinosaurs, back before the West was truly conquered and every town did not have a similar avenue of fast-food stores, gas stations, motels, and shopping malls. Back then, men such as Edward Cope, O. C. Marsh, and Charles Sternberg were roaming around virgin territory, and virtually every fossil they picked up represented some exotic new addition to science. Fossils were shipped back to East Coast museums by the boxcar, as a handful of collectors rushed to log these bizarre new creatures into the science books.

Today there are hundreds of authorities on dinosaurs, thousands of books on the subject, and museums around the world are bulging with specimens. Indeed, all the world seems like it must have been explored by now and there might not be anything new left to be discovered. It is getting harder and harder to find badlands not covered with human footprints.

Until recently, part of the appeal of fossil collecting in the Gobi Desert was its remoteness. On the 1995 AMERICAN MUSEUM OF NATURAL HISTORY expedition, I recall a surreal vision of a busload of Japanese tourists being dropped off near our remote campsite in the Flaming Cliffs to snap a few pictures and pile back in to the bus, leaving most of us on the expedition dumbfounded and shaking our heads. When J. B. Shackelford, the cameraman for Roy Chapman Andrews of the famed CENTRAL ASIATIC EXPEDITIONS, discovered this particular area in 1922, the first creature he found was a *Protoceratops*. It was sitting on a small precipice and he simply plucked off the skull, like it was a vase on a pedestal. Today it is rare to make the effort to collect a *Protoceratops* because it is so commonly known. They are still quite literally all over the Flaming Cliffs.

So why bother collecting in the Gobi? What can be discovered after a 70-year parade of collectors combing the area? What can now be found that has not already been found? Plenty. The area that became the American Museum of Natural History's Xanadu in the Gobi was passed over by several decades of explorers, and UKHAA TOLGOD, Mongolian for "brown hills," is now one of the most productive, best preserved Cretaceous dinosaur sites in the world.

Like scientists or doctors, paleontologists these days are realizing that they cannot keep up with their whole field as generalists, so they are headed toward specializing. Also, because they are concentrating their expertise on a more narrow avenue of research, the whole field is expanding at a breathtaking pace.

The crew I accompanied in the Gobi, headed by Michael Novacek and Mark Norell of the American Museum of Natural History and Demberelyn Dashzeveg from the Mongolian Academy of Sciences, were interested primarily in small, rare, bird-like theropods such as *Mononykus*, *Oviraptor*, and *Velociraptor*. The crucial question they were trying to answer was "How are dinosaurs related to modern birds?" On their quest they ignored the virtual cemetery of beautiful ankylosaurs that surrounded us. Although ankylosaurs are interesting to some, for these scientists they simply did not answer the questions they were asking, and the excavation of those specimens would have diluted the efforts and resources of their mission (see CENTRAL ASIATIC EXPEDITIONS).

A few years ago when Paul Sereno of the University of Chicago was applying cladistics, the new accepted method of animal classification, to dinosaurs, he realized that to really understand the origins of dinosaurs he would have to find some complete specimens. One of the regions bearing early dinosaurs of the Late Triassic was in South America, an area that had been thoroughly researched by some of the best paleontologists for 40 years but with little to show for their effort except some scraps. Undaunted, Sereno and crew not only found the missing links, but the specimens they found were nearly complete, giving the field a new window into the beginning of dinosaurs (see ORIGIN OF DINOSAURS).

The first dinosaur specimens found in America were actually dinosaur trackways, first cited by the clergyman and professor Edward Hitchcock in the early 1800s in the CONNECTICUT RIVER VALLEY. At this time, the term dinosaur had not even been coined, nor had enough specimens yet been found to determine what they looked like. However, Hitchcock noted that the huge footprints he saw had a mysteriously bird-like quality, an insight that now seems prophetic.

Dinosaur tracks were largely dismissed by paleontologists as inconsequential curiosities until recently. James Farlow of Indiana, Father Guiseppe Leonardi of Venice, and Martin Lockley of Colorado have specialized in tracking dinosaurs and have led the field in identifying thousands of new sites worldwide. More important, the new kinds of questions generated by their work and previous findings have spawned a whole new valid field of research. They have found that the footprints can reveal behavior in a way that the bones cannot. Trackways can be used to determine gait, speed, and posture and to analyze the question of whether dinosaurs traveled alone or in herds. Also, although there are only a few thousand mounted dinosaurs worldwide, there are millions of dinosaur tracks. By agreeing on a general classification for footprints and systematically mapping and identifying tracks made by thousands of individuals, Leonardi, Lockley, and Farlow are undertaking the creation of a worldwide dinosaur census.

John McIntosh is a retired physics professor whose main passion has always been sauropods, those long-necked, long-tailed creatures with a ridiculously small head. Retirement from physics has unleashed Jack to pursue his sauropod research as never before. Now that he is ostensibly retired, he reports that he is actually busier than ever and that he is now in the process of publishing several papers describing new sauropods from all around the world.

Jack Horner, on the other hand, hates sauropods. He is a curator at the MUSEUM OF THE ROCKIES in Bozeman, Montana. In his view sauropods take too long to dig up, take up too much storage space, and are too heavy to pick up and study. However, when he does find a sauropod, he still at least feels inclined to collect it for the museum. Horner likes small dinosaurs—the smaller the better. In fact he prefers baby dinosaurs, the kind that you can put in a coffee can and take back to your desk and study. He is an expert on baby dinosaurs. He has found hundreds of eggs and dozens of nests. He has mapped whole nesting grounds, and he has come up with some very important new data on how dinosaurs grew up. Horner has used his research to study bone growth and parenting behavior—areas of dinosaur research that were previously thought to be incomprehensible. He thinks that there are probably baby dinosaurs in every formation in the world, and that they have gone unnoticed because paleontologists have had the wrong search image. They are predisposed to looking for large fossils. He has proved this theory to the delight and embarrassment of his colleagues on their own prospecting grounds. Jack probably would know even more about baby dinosaurs and dinosaur parenting behavior, if he could just stop finding sauropods.

Bob Bakker has been concentrating on COMO BLUFF, Wyoming—a region he has steadfastly worked for more than 20 years. (Como Bluff was first worked by O. C. Marsh's crew more than a century ago.) Bakker is known the world over as the gadfly of paleontology because of his radical ideas about warm-blooded dinosaurs. In 1996 Bakker reported finding the best specimen to date of an *Apatosaurus* skull, as well as the first documented case of sauropod gastralia, delicate stomach cage bones. Some museums have already begun remounting their dinosaurs to reflect this discovery.

China, once regarded as a paleo-backwoods for research and discovery, is now proving to be a whole new untapped resource for large, previously unknown gaps in the fossil record. Philip Currie of the ROYAL TYRRELL MUSEUM OF PALAEONTOLOGY in Alberta, Dale Russell, who was Curator at the Canadian Museum of Nature, and Dong Zhiming of the Institute of Vertebrate Paleontology and Paleoanthropology led the Canada–China Dinosaur Project, a 5-year joint expedition to search for Chinese and Canadian fossils. Their effort produced the largest traveling exposition of dinosaurs ever assembled, and now they can hardly keep pace with the general reporting of discoveries flooding out of China.

All over the world—in Antarctica, the Arctic, Africa, North America, South America, China, and Australia—whole new regions are being tapped, gaps in the record are being filled, and intermediate forms of life are being discovered. All of this gives a new resolution to the earth's history.

Advances in medical research have also aided paleontological work. Powerful new software with CT scanning has enabled researchers to nondestructively probe the internal structures of delicate fossils and map the matrix-filled nerve paths and brain cases. Fossil DNA recovery, once the stuff of science fiction, may prove to be a new frontier as pioneering researchers from China to Scotland and the United States push the envelope in this new field. New acid preparation techniques, developed by Kevin Aulenback of the Tyrrell Museum, have allowed the viewing of three-dimensional Cretaceous plants, the food of most dinosaurs, with stunning resolution never before imagined.

In fact, everything from fossil pollen to dinosaur dung is now being scrutinized with the arsenal of new techniques, forcing the past to reveal its secrets. Nearly everything that is found in the field has the potential to uncover new information as never before imagined. Also, the plethora of new material available to research does not stop at fossils coming from the field. There is enough undescribed material and unprepared material currently on museum shelves around the world to keep a future crop of paleontologists asking questions for several generations.

Today, a whole new generation of dinosaur researchers is coming up fast on the heels of the senior researchers mentioned previously. Nurtured, tutored, and field tested, the graduate students of these guiding lights are asking new questions, probing with new techniques, and prodding the fossil record to reveal more than ever was thought possible. The field is expanding rapidly and the specific examples cited in this article are only a few instances of the range of research being done on dinosaurs. New dinosaur museums are now being built and nearly every existing museum in the world has revamped its exhibits or is expanding them to reflect the new findings. In fact, right now, as the millennium approaches, is the golden age of dinosaur discovery. More likely, it is just the beginning.

References

Horner, J. R., and Gorman, J. (1988). *Digging Dinosaurs.* Workman Publishing, New York.

Jacobs, L. L. (1993). *Quest for the African Dinosaurs.* Villard Books, New york.

Novacek, M. J. (1996). *Dinosaurs of the Flaming Cliffs.* Knopf, New York.

Psihoyos, L. (1994). *Hunting Dinosaurs.* Random House, New York.

Horns

see BEHAVIOR; ORNAMENTATION; SKELETAL STRUCTURES; SKULL, COMPARATIVE ANATOMY

Houston Museum of Natural Science, Texas, USA

see MUSEUMS AND DISPLAYS

Howe Quarry

BRENT H. BREITHAUPT
University of Wyoming
Laramie, Wyoming, USA

While collecting in the Lower Cretaceous rocks of the Big Horn Basin for New York's AMERICAN MUSEUM OF NATURAL HISTORY in 1932, Barnum Brown was notified of the existence of large bones on the ranch of Mr. Barker Howe. A preliminary reconnaissance of the site revealed an area that promised to have an extremely rich quarry. Brown returned to this area just off of the west flank of the Big Horn Mountains north of Shell, Wyoming, in 1934 with a field crew (financed by the Sinclair Refining Company). Although only two sauropod skeletons were initially uncovered, later that summer a veritable herd had been discovered. Skeletal remains were crisscrossed and interlocked in a confusing, almost inextricable, manner in a clay unit beneath a relatively thick sandstone layer. For two months no bones were collected and the "log jam" of skeletal parts was exposed, mapped, and viewed by thousands of visitors from the United States and Europe. After an ingenious, highly accurate quarry map was made, the specimens were painstakingly removed from the quarry. Approximately six months after the quarry was opened in 1934, the last box of fossils was loaded for New York.

The Howe Quarry produced an enormous amount of Late Jurassic dinosaur remains. The quarry, measuring 14 × 20 m, contained one of highest single concentrations of Jurassic dinosaur bones ever found. More than 4000 bones were uncovered representing at least 20 animals. The quarry was dominated by sauropods such as *Barosaurus* (the head and neck of a rare juvenile from this quarry are currently on display in New York), *Diplodocus*, *Apatosaurus*, and *Camarasaurus*. In addition to these beasts, remains of *Camptosaurus* and teeth of *Allosaurus* were also found. The remains in the quarry are generally disarticulated, but a number of bones show some association and run vertically through the strata representing an accumulation of desiccated carcass parts that were washed into a small depression. Footprints, skin impressions, and gastroliths were also found in the Howe Quarry. More than 30 metric tons of bones were collected from the Howe Quarry. Packed in 144 large cases, these specimens filled a large boxcar almost to the roof (Brown, 1935; Norell *et al.*, 1995).

In the early 1990s, other collectors resumed work in this area of Wyoming. The Howe Quarry was expanded and additional material uncovered. In 1991 a nearly complete (95%), partially articulated, subadult *Allosaurus* (with at least 14 pathologic bones) was uncovered just north of the Howe Quarry (Breithaupt, 1996). This specimen, nicknamed "Big Al," is one of the most complete skeletons of *Allosaurus* ever found and indicates the potential for important new discoveries to be made in areas of historic exploration.

See also the following related entries:

History of Dinosaur Discoveries • Jurassic Period • Morrison Formation • Skin

References

Breithaupt, B. H. (1996). The discovery of a nearly complete *Allosaurus* from the Jurassic Morrison Formation, eastern Bighorn Basin, Wyoming. In *Forty-Seventh Annual Field Conference Guidebook* (C. E. Bowen, S. C. Kirkwood, and T. S. Miller, Eds.), pp. 309–313. Wyoming Geological Association, Casper.

Brown, B. (1935). Sinclair dinosaur expedition, 1934. *Nat. History* **36,** 3–13.

Colbert, E. H. (1968). *Men and Dinosaurs*, pp. 283. Dutton, New York.

Dodson, P., Behrensmeyer, A. K., Bakker, R. T., and McIntosh, J. S. (1980). Taphonomy and paleoecology of the dinosaur beds of the Jurassic Morrison Formation. *Paleobiology* **6,** 208–232.

Norell, M. A., Gaffney, E. S., and Dingus, L. (1995). *Discovering Dinosaurs in the American Museum of Natural History*, pp. 204. Knopf, New York.

Hunterian Museum, United Kingdom

see Museums and Displays

Hyabashibara Museum of Natural History, Japan

see Museums and Displays

Hyogo Museum, Japan

see Museums and Displays

Hypsilophodontidae

Hans-Dieter Sues
Royal Ontario Museum
Toronto, Ontario, Canada

The Hypsilophodontidae are a diverse and widely distributed group of ornithopod dinosaurs with a fossil record ranging in time from the Middle Jurassic to the end of the Cretaceous (170–65 mybp). Fossils of hypsilophodontid ornithopods are known from Europe, western North America, China, Antarctica, and Australia (Sues and Norman, 1990). They may be defined as all ornithopods closer to *Hypsilophodon* than to *Iguanodon*.

Traditional evolutionary scenarios interpreted the Hypsilophodontidae as a basal stem group from which several lineages of larger (iguanodontian) ornithopods independently descended (Galton, 1974). However, several recent cladistic analyses of ornithopod interrelationships have clearly supported the monophyly of both Hypsilophodontidae and Iguanodontia (Sereno, 1986; Weishampel and Heinrich, 1992), although the monophyly of the latter is doubted (Norman and Weishampel, 1990).

Hypsilophodontids are facultatively if not obligately bipedal, small to medium-sized (1–4 m long) ornithopod dinosaurs. Diagnostic features for this group include the presence of a cingulum on the dentary teeth, the rod-like shape of the prepubic process, and the ossification of the sternal segments of the more anterior dorsal ribs (Sues and Norman, 1990; Weishampel and Heinrich, 1992). Hypsilophodontids retain a number of primitive features such as the presence of premaxillary teeth.

Thescelosaurus neglectus from the Upper Cretaceous (Maastrichtian) of Alberta and Saskatchewan (Canada) and of Montana, South Dakota, and Wyoming is considered the basal member of the hypsilophodontid clade although it is the latest to appear in the fossil record. It is also the largest known hypsilophodontid.

All other known hypsilophodontid taxa appear to

differ from *Thescelosaurus* in the possession of two derived features of the skull: the frontals are longer anteroposteriorly than wide transversely, and an angle of about 35° is formed between the base and the long axis of the braincase (Weishampel and Heinrich, 1992).

Yandusaurus hongheensis and possibly synonymous taxa, such as *Agilisaurus multidens* from the Middle Jurassic Shaximiao Formation of Sichuan (China) and *Othnielia rex* from the Upper Jurassic (Kimmeridgian–Tithonian) Morrison Formation of Colorado, Utah, and Wyoming, share the reduction of the quadratojugal and the distinct curvature of the pubic shaft. The still poorly known *Drinker nisti* from the Morrison Formation of Wyoming is a small form apparently closely related to *Othnielia*.

Hypsilophodon, from the Wealden (Lower Cretaceous: Barremian–Aptian) of England and Spain, and possibly from the Lower Cretaceous of Texas, is by far the best documented member of the Hypsilophodontidae. Superbly preserved material of *H. foxii* from the Wealden of the Isle of Wight documents the entire skeleton, although most of the specimens probably represent immature individuals. *Hypsilophodon foxii* appears to be most closely related to *Parksosaurus warreni* from the Upper Cretaceous (Maastrichtian) Horseshoe Canyon Formation of Alberta, with which it shares the exclusion of the jugal from the margin of the antorbital fenestra (Fig. 1).

The possession of a small bony boss on the outer surface of the jugal may ally *Zephyrosaurus schaffi* from the Lower Cretaceous (Aptian–Albian) Cloverly Formation of Montana with *Orodromeus makelai* from the Upper Cretaceous (Campanian) Two Medicine Formation of Montana.

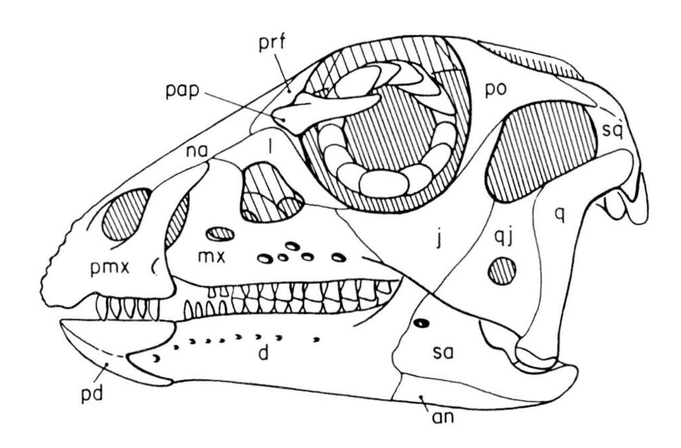

FIGURE 1 Skull of *Hypsilophodon foxii* (from P. M. Galton, 1974).

Three hypsilophodontid genera have been described from the Lower Cretaceous (Aptian–Albian) of Australia: *Atlascopcosaurus loadsi* and *Leaellynasaura amicagraphica* from the Otway Group of Victoria and *Fulgurotherium australe* (initially misidentified as a small theropod) from the Griman Creek Formation of New South Wales. Currently, all these forms are represented only by fragmentary remains and thus cannot be readily placed in any of the aforementioned groupings. However, the fossils from Victoria are noteworthy for their occurrence within the Antarctic circle during Early Cretaceous times.

Due to an incorrect early interpretation of the first toe (hallux) as being reversed, *Hypsilophodon* was long considered an arboreal dinosaur similar in habits to the present-day tree kangaroos. However, the proportions of the hind legs, especially the distinct elongation of the tibia and third metatarsal, clearly indicate that *Hypsilophodon* and its relatives were, in fact, fast bipedal runners (Galton, 1974). The long tail, which was partially stiffened by ossified tendons, would have functioned not only for counterbalancing but also as a dynamic stabilizer during fast running (Fig. 2). The teeth and inferred mode of jaw function, with its emphasis on shearing, are consistent with plant-eating habits. Judging from their size and stature, hypsilophodontids probably fed on vegetation within 1 m of the ground. The skull had a remarkable system of movable joints linking the tooth-bearing bones of the snout with those of the rest of the skull. That system allowed the upper jaws to rotate outward during mastication (pleurokinesis; Norman, 1984). As the mandible was brought upward its teeth came in contact with the upper teeth on both sides of the snout. At a certain point, further raising of the lower jaws would force the teeth of the upper jaws apart. This rotated the upper teeth against the opposing lower teeth in a grinding motion. The "cheek" teeth are typically heavily worn with obliquely inclined wear facets. The front end of the narrow snout was covered by a small horny beak, which, along with a similar beak covering the predentary bone of the mandible, must have formed an effective cropping device. The dentary and maxillary teeth were distinctly inset from the side of the face and may indicate that hypsilophodontids had muscular cheeks to retain fodder in the mouth during chewing.

Numerous nests with clutches of eggs of *Orodromeus makelai* have been discovered in the Campanian-age Two Medicine Formation of northwestern Mon-

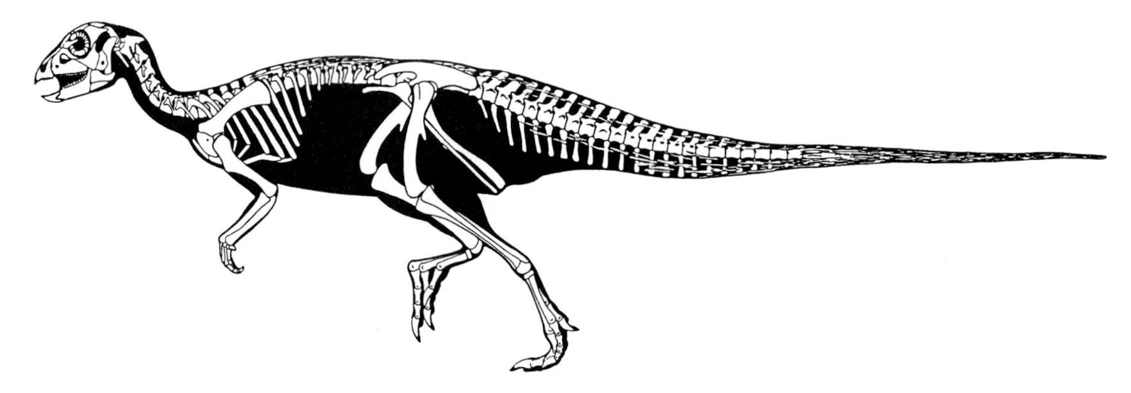

FIGURE 2 Skeletal reconstruction and body silhouette of *Hypsilophodon foxii* (from P. M. Galton, 1974).

tana. The limb bones of embryonic skeletons found in some of those eggs have well-formed joint surfaces and indicate that the hatchlings apparently left the nest soon after hatching to fend for themselves (precocial behavior). This is unlike the behavior inferred for the hatchlings of the large hadrosaur *Maiasaura peeblesorum* from the same formation and region, which apparently stayed in the nest for some time after hatching and must have depended on parental support (altricial behavior) (Horner and Weishampel, 1988).

See also the following related entries:

FABROSAURIDAE • ORNITHOPODA

References

Galton, P. M. (1974). The ornithischian dinosaur *Hypsilophodon* from the Wealden of the Isle of Wight. *Bull. Br. Museum Nat. History Geol.* **25,** 1–152.

Horner, J. R., and Weishampel, D. B. (1988). A comparative embryological study of two ornithischian dinosaurs. *Nature* **332,** 256–257.

Norman, D. B. (1984). On the cranial morphology and evolution of ornithopod dinosaurs. In *The Structure, Development and Evolution of Reptiles* (M. W. J. Ferguson, Ed.), Symposium of the Zoological Society of London, Vol. 52, pp. 521–547. Academic Press, London.

Norman, D. B., and Weishampel, D. B. (1990). Iguanodontidae and related ornithopods. In *The Dinosauria* (D. B. Weishampel, P. Dodson, and H. Osmólska, Eds.), pp. 510–533. Univ. of California Press, Berkeley.

Sereno, P. C. (1986). Phylogeny of the bird-hipped dinosaurs (Order Ornithischia). *Natl. Geogr. Res.* **2,** 234–256.

Sues, H.-D., and Norman, D. B. (1990). Hypsilophodontidae, *Tenontosaurus*, and Dryosauridae. In *The Dinosauria* (D. B. Weishampel, P. Dodson, and H. Osmólska, Eds.), pp. 498–509. Univ. of California Press, Berkeley.

Weishampel, D. B., and Heinrich, R. E. (1992). Systematics of Hypsilophodontidae and basal Iguanodontia (Dinosauria: Ornithopoda). *Historical Biol.* **6,** 159–184.

Iguanodontidae

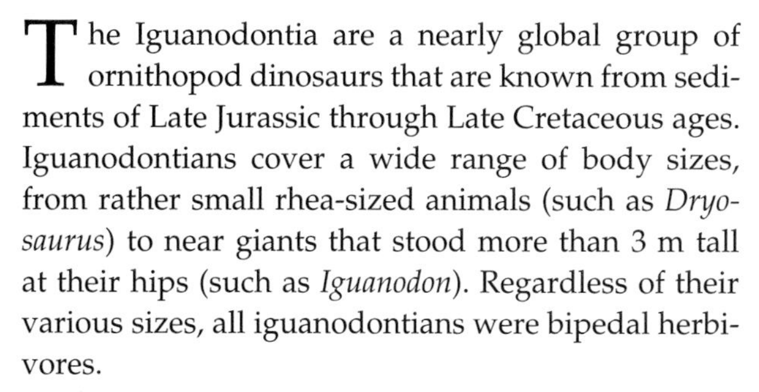

Catherine A. Forster
State University of New York
Stony Brook, New York, USA

The Iguanodontia are a nearly global group of ornithopod dinosaurs that are known from sediments of Late Jurassic through Late Cretaceous ages. Iguanodontians cover a wide range of body sizes, from rather small rhea-sized animals (such as *Dryosaurus*) to near giants that stood more than 3 m tall at their hips (such as *Iguanodon*). Regardless of their various sizes, all iguanodontians were bipedal herbivores.

The Iguanodontia are also important for historical reasons: One of its members, *Iguanodon mantelli*, is one of the three founding constituents of Dinosauria (Owen, 1842). Iguanodontians also gave rise to the major group of ornithopods of the Late Cretaceous: the Hadrosauridae, or "duck-billed" dinosaurs.

Iguanodon and Early Dinosaur History

While strolling through the Sussex countryside in England around 1822, Mary Ann Mantell discovered some unusual, large, fossilized teeth. After examining these enigmatic teeth, her physician husband, Dr. Gideon Mantell, named them *Iguanodon* ("iguana tooth") because of their superficial resemblance to those of a living iguana (Mantell, 1825). This was only the second dinosaur ever named, predated only by the theropod Megalosaurus the year before (Buckland, 1824). When Sir Richard Owen proposed the name "Dinosauria" in 1841, *Iguanodon* was one of the charter members.

For many years, very little was known of *Iguanodon* outside of a few scraps of bone and teeth. In fact, so little was known of any dinosaur that reconstructions of these animals were more fancy than fact. Mantell imagined *Iguanodon* to be a giant lizard up to 100 ft long, with sprawling limbs and an elongate body (Benton, 1989). The first widely known dinosaur reconstructions, of *Iguanodon* and *Megalosaurus*, were erected for the Crystal Palace exhibition center in Sydenham Park, London, in 1854. Richard Owen, with sculptor Waterhouse Hawkins, reconstructed *Iguanodon* as a heavy-set, scaly quadruped with a large spike on its nose like that of a rhinoceros. These bizarre but wonderful monsters invoked the first public sensation over dinosaurs (Norman, 1985).

In 1857, a partial hadrosaur skeleton, *Hadrosaurus foulki*, was dug up near Haddonfield, New Jersey. Finally, the world got its first good glimpse of how dinosaurs may have really looked—but many questions were still left unanswered (Leidy, 1858). Then, 20 years later, another *Iguanodon* discovery broke the scientific, and the dinosaur, world wide open.

On February 28, 1878, two coal miners uncovered what they excitedly thought was a large tree trunk filled with gold, deep in an underground mine called the Sainte-Barbe Pit near Bernissart in southwestern Belgium. After close examination, it was concluded that the treasure trove was actually a fossil bone filled with pyrite, or "fool's gold." However, this was just the tip of the iceberg. A few days later, the Sainte-Barbe Pit started to produce one of the greatest dinosaur discoveries of all times—more than 30 complete, articulated skeletons of *Iguanodon* (Casier, 1978; Norman, 1980; Bultynck, 1989; Martin and Bultynck, 1990). These were the first complete dinosaur skeletons ever discovered and still remain one of the greatest accumulations of a single taxon of dinosaur. For paleontologists, it turned out to be a treasure trove indeed.

It took 3 years to remove more than 130 tons of these specimens from the coal mine. Brought to the Royal Institute of Natural Sciences in Brussels, the *Iguanodon* specimens were prepared from the coaly rock and preserved (Martin and Bultynck, 1990). Dr. G. A. Boulenger (as cited in Van Beneden, 1881) named the dinosaur *Iguanodon bernissartensis.* These discoveries from the Belgian coal mine showed the world what a complete dinosaur really looked like— and confirmed Joseph Leidy's hypothesis (based on *H. foulki*) that these animals were bipeds rather than heavy, rhino-like quadrupeds.

Today, most of these specimens of *I. bernissartensis* are on display at the Royal Institute of Natural Sciences in Brussels, Belgium. Many of them are standing, mounted side by side within an enormous glass cage. Numerous others have been left in their original poses, lying on their sides as found entombed in the coal mine. This astounding array of *Iguanodon* skeletons remains the most impressive display of a single taxon of dinosaur anywhere in the world.

Relationships among Iguanodontians

Among the many *I. bernissartensis* skeletons found at Bernissart was a complete skeleton of a smaller, more delicate animal—*Iguanodon atherfieldensis.* Excellent partial specimens of *I. atherfieldensis* have also been found in England (Norman, 1986). A third species has also been named from the United States— *I. lakotaensis.* In addition to *Iguanodon,* many other iguanodontians have been studied and named, including *Camptosaurus* (Late Jurassic; United States and England); *Dryosaurus* (Late Jurassic, United States and Tanzania), *Kangnasaurus* (age unknown; South Africa), *Muttaburrasaurus* (Early Cretaceous; Australia); *Ouranosaurus* (Early Cretaceous; Niger), *Probactrosaurus* (Early Cretaceous; China), *Rhabdodon* (Late Cretaceous; Romania), and *Tenontosaurus* (Early Cretaceous, United States) (Norman and Weishampel, 1990; Sues and Norman, 1990). Other undescribed iguanodontians have also been reported from the southern United States, South Africa, Niger, and Argentina.

Iguanodontians do not form a united, natural ("monophyletic") group; rather, they form an assemblage (a "paraphyletic" group) of genera that spans the long gap between hypsilophodontids and hadrosaurids (Milner and Norman, 1984; Norman, 1984; Sereno, 1984, 1986; Forster, 1990; Norman and Weishampel, 1990). Numerous morphological changes can be observed through this "span" of iguanodontians, with primitive taxa, such as *Tenontosaurus* and *Dryosaurus,* looking much like hypsilophodontids, and derived taxa, such as *Probactrosaurus,* closely resembling hadrosaurids.

Many of these changes along the span to hadrosaurids are easy to observe. For instance, basal iguanodontians (such as *Tenontosaurus*) have ossified (bony) tendons arranged in long bundles down the back and out along the tail vertebrae. These tendons parallel the vertebral column. This arrangement of tendons is identical to that seen in hypsilophodontids. However, more derived iguanodontians (such as *Iguanodon*) have ossified tendons arranged in a crisscrossing trellis-like configuration down the back and tail. This tendon arrangement is identical to that seen later in hadrosaurids. Many other morphological changes can also be observed: Teeth are lost in the front of the mouth, and the cheek teeth begin to pack closer and closer together into the beginnings of a dental battery; the foot is reduced from four toes to only three; the wrist bones fuse together into one solid piece; the thumb becomes a large spike that fuses with the wrist; and the sternal plates (breast bones) change from kidney shaped to hatchet shaped.

By tracing these accumulations of new features through the iguanodontians, we can arrange them in a sequence from hypsilophodontids to hadrosaurs. For instance, *Tenontosaurus* still has parallel bundles of ossified tendons on its back, kidney-shaped sternals, four toes on each foot, and unfused wrist bones. *Camptosaurus* still has a kidney-shaped sternal plate and four toes on each foot but has fused wrist bones and trellis-like ossified tendons. *Iguanodon* has hatchet-shaped sternals, three toes on each foot, fused wrist bones, and trellis-like tendons. Therefore, we know that *Tenontosaurus* is more basal than *Camptosaurus,* which in turn is more basal than *Iguanodon.* Hadrosaurids arose from an unidentified iguanodontian ancestor, but where and when this happened is not known.

Summary

Iguanodontians were a very successful and varied group of ornithopods. Ranging in size from small to gigantic, they are known from nearly every continent, both north and south, and have sometimes been found in great abundance. Some iguanodontians seem to have been very wide ranging; for instance, *Dryosaurus* is known from both the United States and Tanzania, and *Camptosaurus* is known from the United States and England. Although primarily bipedal, iguanodontians also appear to have been capable of walking on all four legs, perhaps to browse on long growing plants.

See also the following related entries:

HADROSAURIDAE • ORNITHOPODA

References

Benton, M. (1989). *On the Trail of the Dinosaurs*, pp. 143. Crescent Books, New York.

Buckland, W. (1824). Notice on *Megalosaurus* or great fossil lizard of Stonesfield. *Trans. Geol. Soc. London* **21,** 390–397.

Bultynck, P. (1989). *Bernissart et les Iguanodons*, pp. 115. L'Institut Royal des Sciences Naturelles de Belgique, Bruxelles.

Casier, E. (1978). *Les Iguanodons de Bernissart*, Editions du Patrimoine, pp. 116. L'Institut Royal des Sciences Naturelles de Belgique, Bruxelles.

Forster, C. A. (1990). The postcranial skeleton of the ornithopod dinosaur *Tenontosaurus tilletti*. *J. Vertebr. Paleontol.* **10,** 273–294.

Leidy, J. (1858). *Proc. Acad. Nat. Sci. Philadelphia*, 215–218.

Mantell, G. A. (1825). Notice on the *Iguanodon*, a newly discovered fossil reptile, from the sandstone of Tilgate Forest, in Sussex. *Philos. Trans. R. Soc. London* **115,** 179–186.

Martin, F., and Bultynck, P. (1990). *The Iguanodons of Bernissart*, pp. 51. Institut Royal des Sciences Naturelles de Belgique, Brussels.

Milner, A. R., and Norman, D. B. (1984). The biogeography of advanced ornithopod dinosaurs (Archosauria: Ornithischia)—A cladistic–vicariance model. In *Third Symposium on Mesozoic Terrestrial Ecosystems* (W.-E. Reif and F. Westphal, Eds.), pp. 145–150. ATTEMPTO-Verlag, Tubingen, Germany.

Norman, D. B. (1980). *On the Ornithischian Dinosaur Iguanodon bernissartensis from the Lower Cretaceous of Bernissart (Belgium)*, Memoire No. 178. L'Institut Royal des Sciences Naturelles de Belgique, Bruxelles.

Norman, D. B. (1984). A systematic reappraisal of the reptile order Ornithischia. In *Third Symposium on Mesozoic Terrestrial Ecosystems* (W.-E. Reif and F. Westphal, Eds.), pp. 157–162. ATTEMPTO-Verlag, Tubingen, Germany.

Norman, D. B. (1985). *The Illustrated Encyclopedia of Dinosaurs*, pp. 208. Salamander Books, London.

Norman, D. B. (1986). On the anatomy of *Iguanodon atherfieldensis* (Ornithischia: Ornithopoda). Bulletin de L'Institut Royal des Sciences Naturelles de Belgique, Sciences de la Terre, Vol. 56.

Norman, D. B. and Weishampel, D. B. (1990). Iguanodontidae and related Ornithopods. In *The Dinosauria* (D. B. Weishampel, P. Dodson, and M. Osmólska, Eds.), pp. 510–533. Univ. of California Press, Berkeley.

Owen, R. (1842). Report on British fossil reptiles. Pt. II. *Rep. Br. Assoc. Adv. Sci.* **11,** 60–204.

Sereno, P. C. (1984). The phylogeny of the Ornithischia: A reappraisal. In *Third Symposium on Mesozoic Terrestrial Ecosystems* (W.-E. Reif and F. Westphal, Eds.), pp. 219–226. ATTEMPTO-Verlag, Tubingen, Germany.

Sereno, P. C. (1986). Phylogeny of the bird-hipped dinosaurs (Order Ornithischia). *Natl. Geogr. Res.* **2**(2), 234–256.

Sues, H-D., and Norman, D. B. (1990). Hypsilophodontidae, *Tenontosaurus*, Dryosauridae. In *The Dinosauria* (D. B. Weishampel, P. Dodson, and M. Osmólska, Eds.), pp. 499–509. Univ. of California Press, Berkeley.

Van Beneden, P. J. (1881). Sur l'arc pelvien chez les dinosauriens de Bernissart. *Bulletin Academie Royal Belge. Ch. Sci.* **3,** 600–608.

Indian Dinosaurs

ASHOK SAHNI
Punjab University
Chandigarh, India

Fragmentary bones, later to be identified as dinosaurs, were first described by Captain (later Major–General) Sleeman in 1828 from the now classic sections of Jabalpur Cantonment (Matley, 1921) in India. The first systematic description of Indian dinosaurs was carried out by Lydekker (1877), who erected the new sauropod taxon *Titanosaurus indicus* on the basis of two caudal vertebrae and an imperfectly preserved femur. Later, Matley laid the foundations of the study of Indian dinosaurs with the collection and description of several taxa from central India (Matley, 1921) and south India (Matley, 1929), culminating in a monographic description of all known material at the time (von Huene and Matley, 1933). The next major phase of investigations started in the early 1960s when scientists from the Geological Studies Unit, Indian Statistical Institute, Calcutta, began to explore the Pranhita–Godavari Basin (Jain *et al.*, 1979). In the past decade, the contributions of the Geological Survey of India and Panjab and Delhi Universities have been largely instrumental in reviving interest in Indian dinosaurs (Sahni *et al.*, 1994).

The occurrence of Indian dinosaurs is currently confined to central and southern India, and the record extends to all three Mesozoic periods. The Triassic and Jurassic dinosaurs are known from the Pranhita–Godavari Gondwana Rift Basin, where a complete sedimentary sequence is found ranging from the Upper Permian to the Lower Cretaceous (Dasgupta, 1993). Cretaceous dinosaur localities are much more widespread and occur in two contrasting geologic settings: one in the basal Deccan volcano–sedimentary sequences (Sahni *et al.*, 1994) and the other in a regressive phase (Ariyalur Formation) toward the top of a paralic sequence yielding abundant shallow marine benthic invertebrates (Yadagiri and Ayyasami, 1979). Indian Cretaceous dinosaurs are represented by a diversity of fossils comprising cranial and postcranial material, BRAINCASES, nests and eggshell fragments, coprolites, and FOOTPRINTS (Fig. 1).

The Triassic record includes the small coelurosaur *Walkeria maleriensis*, which represents one of the oldest dinosaurs known from Asia. It was reported by

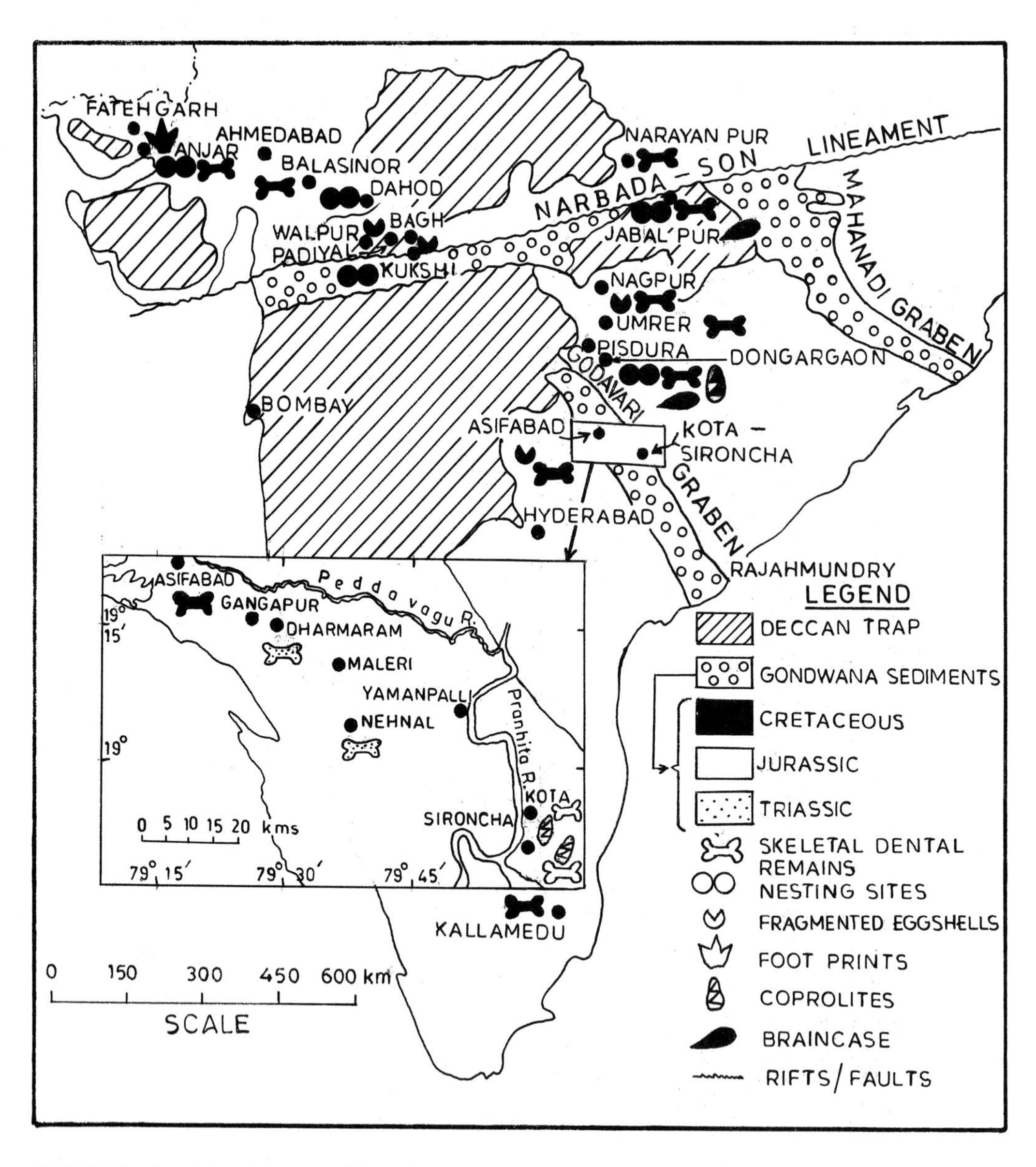

FIGURE 1 Spatial and temporal distribution of Indian dinosaurs based on skeletal material, traces, and tracks.

Bayan Dzak in Mongolia, also known as the Flaming Cliffs, is among the more famous of dinosaur localities. Made legendary by the American Museum of Natural History expeditions led by Roy Chapman Andrews, many of the most glorious discoveries revealing much about dinosaur biology were made while scientists were prospecting these formations. Photo by François Gohier.

Displays of dinosaurs in museums fascinate children and adults of all cultures. The Dinosaur Hall of the Museum of Natural History in Ulaan Baatar (pictured above) displays the rich fossil record unearthed from the Flaming Cliffs and elsewhere in Mongolia. Photo by François Gohier.

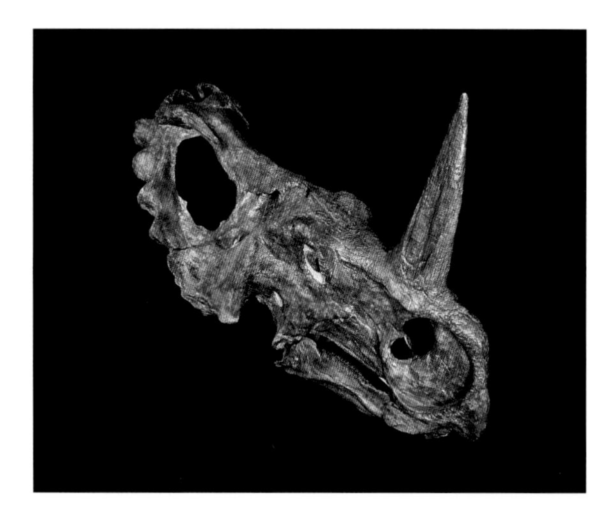

Many interesting speculations about ceratopsians pertain to the function of the cranial casques and horns. For example, *Centrosaurus apertus* has only a single horn and moderately developed casques. Photo by François Gohier.

In contrast, *Arrhinoceratops,* (below) has a much better developed casque and two horns, one above each eye. Although it seems likely that these oddities relate to real differences in the behavior of these dinosaurs, there is no direct or indirect evidence compelling enough to reject or support any single explanation. Photo by François Gohier.

One of the earlier bits of evidence used to support Wegner's theory of continental drift was the distribution of fossils. The dinosaur fossils of Antarctica suggest that this now frigid continent once occupied a more equable global position that supported animals and plants of tropical, subtropical, and temperate nature. Illustration by William Stout.

Coelophysis bauri (left: the foot, leg, and some vertebrae are shown) was a small to medium dinosaur, whereas *Camarasaurus* (shown below) was considerably larger. Both were a part of the Saurischia, dramatically illustrating the great range of variation among dinosaurs. Photo by François Gohier; illustration by Michael Skrepnick.

Hadrosaurs were plant-eating dinosaurs, as this tooth battery from *Edmontosaurus annectens* shows. Such dental structures are analogous to the structures found in large mammal plant eaters living today, such as elephants and horses. Photo by François Gohier.

Disagreement swirls around *Tyrannosaurus*. Some scientists are convinced *Tyrannosaurus* and its near relatives were active, agile predators. Others are equally sure that this "tyrannical lizard" was a lumbering scavenger. The answer may be that both are partly true. Illustration by Michael Rothman.

Dinosaur Provincial Park in Alberta, Canada, has been and continues to be a rich source of dinosaur bones, such as this hadrosaur being exhumed. Photo by François Gohier.

Chatterjee (1987) from the Late Triassic Maleri Formation at Nennal, Andhra Pradesh. *Walkeria* appears to be a lightly built, predaceous, bipedal form with a functionally tridactyl bird-like foot. The age of the Maleri Formation, which contains at least four generically similar taxa to the Dockum fauna of Texas, has been established on the basis of associated vertebrates, including metoposaurid and chigutisaurid temnospondyls, rhynchosaurid and phytosaurid reptiles, and traversodontid cynodonts (Dasgupta, 1993).

Dinosaurs have also been discovered in the overlying Dharmaram Formation (Kutty, 1969; Jain and Roychowdhury, 1987). The Dharmaram dinosaurs housed in the Museum of the Indian Statistical Institute are still to be systematically described. The limited information on these Rhaetian–Norian forms (Jain and Roychowdhury, 1987) suggests the presence of at least two prosauropods, one representing a large plateosaurid and the other a small anchisaurid.

The best known Jurassic dinosaur from India is *Barapasaurus tagorei* (Jain *et al.*, 1975, 1979) from the Kota Formation (Early Jurassic) of the Pranhita–Godavari Basin. *Barapasaurus*, on display at the Museum of the Indian Statistical Institute, is a large sauropod with rather slender limbs (Fig. 2). Its spoon-shaped teeth bear coarse denticles on their sides. The cervical and anterior dorsal centra are opisthocoelous, whereas all others are nearly platycoelous (Jain *et al.*, 1975). The sacrum has four co-ossified vertebrae with "waisted" amphiplatyan centra. The girdles are typically saurischian in nature. Some interesting aspects of *Barapasaurus* have been highlighted by Jain *et al.* (1979). These include early attempts at gigantism, bone reduction and economy in the vertebrae, and peculiar modifications in the region of the neural canal with the development of interlamellar hollows between the infradiapophysial laminae.

The presence of a possible ornithischian in the Kota

FIGURE 2 Mounted skeleton of *Barapasaurus tagorei* at the Museum of the Indian Statistical Institute, Calcutta (reprinted with permission).

FIGURE 3 Sauropod nest from the Lameta Formation, Balasinor, Kheda District, Gujarat, showing circular outlined cross sections with thick shells.

Formation has been postulated on the basis of personal observation by E. H. Colbert in 1977 based on material collected by the Geological Survey of India (Jain, 1980). An Early Jurassic age for the Kota Formation has been established on the basis of an associated and fairly diverse vertebrate assemblage that includes pholidophorid teleosteans, coelacanthid crossopterygians, the cetiosaurid, a campylognathoid pterosaur, a teleosaurid crocodylian, and kuhneotheriid and amphidontid symmetrodonts (Dasgupta, 1993).

The Cretaceous dinosaurs of India are the most diverse and best known. The fauna is dominated by saurischians (sauropods and theropods) and a few doubtful ornithischians. The sauropods are characterized by at least two well-defined genera, *Titanosaurus* and *Antarctosaurus*. Another sauropod, cf. *Laplatasaurus madagascarensis* (von Huene and Matley, 1933), from the Pisdura locality, needs confirmation as to its affinities with the corresponding taxa from South America and Madagascar. The theropods are more diverse and can be grouped into about five taxa. The large carnivores are represented by *Indosuchus raptorius* (Abelisauridae) and *Indosaurus matleyi* (Abelisauridae), whereas the smaller-sized predaceous carnivores are exemplified by *Compsosuchus, Jubbulporia, Laevisuchus, Dryptosauroides, Coeluroides*, and *Ornithomimoides*.

At least three Cretaceous ornithischian genera have been named: *Lametasaurus* and *Brachypodosaurus* (Matley, 1923; Chakravarti, 1935) and *Dravidosaurus* (Yadagiri and Ayyasami, 1979). The taxonomic status of all three taxa is controversial (Chakravarti, 1935; Buffetaut, 1987) and until diagnostic material is forthcoming, the occurrence of stegosaurids and large ornithischians should be considered doubtful.

The age of the Lameta dinosaurs has been the subject of considerable recent work. Based on their microfossil assemblages and bio- and magnetostratigraphy (Sahni and Bajpai, 1988; Buffetaut, 1987; Courtillot *et al.*, 1986), the LAMETA FORMATION is considered to be largely Maastrichtian in age and coeval with the beds at Dongargaon and Pisdura (Jain and Sahni, 1983; Mohabey *et al.*, 1993).

In recent years, spectacular finds of dinosaur (sauropod) nesting sites have been documented from the Lameta Formation occurring extensively across central peninsular India. To date, hundreds of nesting grounds and hatcheries have been reported (Mohabey, 1990). The EGGS are large and spherical (Fig. 3) and range in size from 14 to more than 20 cm. These have tentatively been assigned to the Titanosauridae. On the basis of microstructural characteristics, the eggs have been grouped into four or five morphostructural types (Sahni *et al.*, 1994). Of general interest is the fact that one of the thick-shelled morphotypes from the Indian localities is similar to eggs attributed to the titanosaurid *Hypselosaurus* from Aix-en-Provence, France (Vianey-Liaud *et al.*, 1987). Thin ornithischian eggshell fragments assignable either to small dinosaurs or to birds have also been described from the intertrappeans of Kutch (Bajpai *et al.*, 1993).

Oxygen isotope analysis of the sauropod eggshells from the Lameta Formation from widespread localities (Sarkar *et al.*, 1991) suggests that the dinosaurs imbibed water from small rivers and ponds. On the other hand, the carbon isotope values indicate that they were eating plants that follow the C_3 photosynthetic pathway, such as small palms and conifers.

Large COPROLITES have been attributed to dinosaurs because of their size, morphology, and association with dinosaurian skeletal material from the Jurassic Kota Formation and from the Lameta Formation of the Pisdura area (Jain, 1983; Matley, 1939). During the collection of dinosaur fossil bones from the Kota Formation, Jain (1983) mentions the occurrence of a large number of coprolites—rounded, oval, or elliptical in shape and marked by dessication cracks. Matley (1939) reported the occurrence from Pisdura of numerous coprolites, up to 170 mm in length, that he regarded as belonging to the titanosaurid sauropods.

Currently, two braincases of sauropod dinosaurs have been recorded from India. The first of these constitutes part of the type material of the sauropod

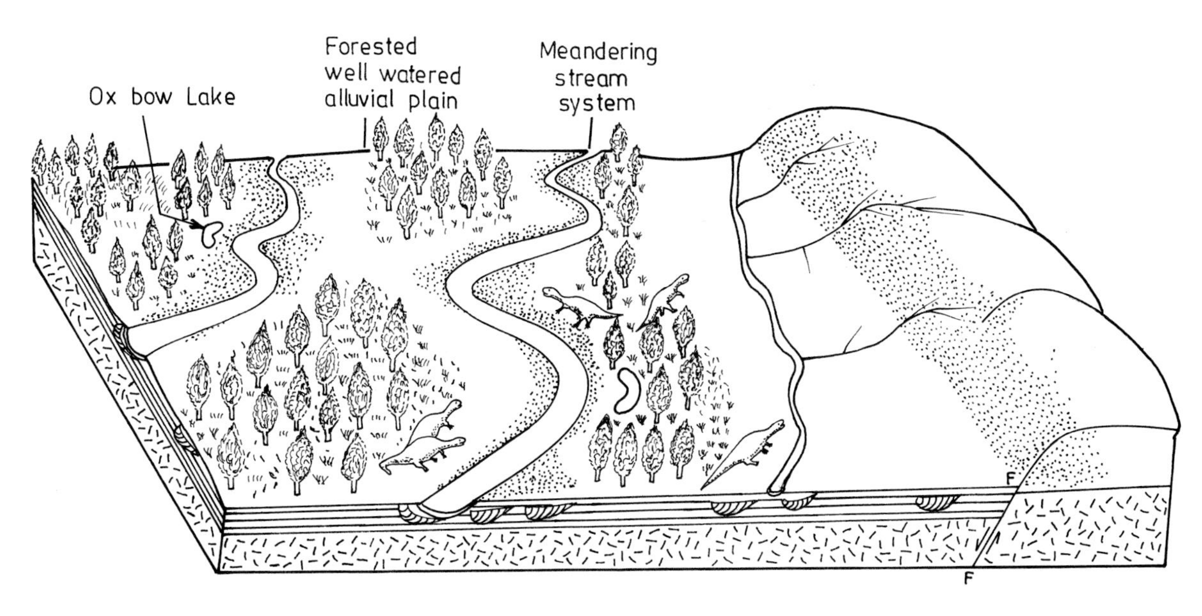

FIGURE 4 Schematic diagram of Late Triassic paleoenvironments: well-watered landscape with meandering rivers derived from fault-truncated basins to the east.

taxon *Antarctosaurus septentrionalis* (von Huene and Matley, 1933), whereas the second comprises a specimen from the Dongargaon locality described by Berman and Jain (1982) from the Lameta Formation.

The Triassic and Jurassic dinosaur localities are situated in the intracratonic rift basin of the Pranhita–Godavari Rift Basin. *Walkeria maleriensis* occurs in the Late Triassic Maleri Formation, which consists of red clays, fine to medium sandstones, and peloidal calcarenites and calcirudites (Dasgupta, 1993). This early dinosaur is found in association with the lungfish *Ceratodus*, labyrinthodont amphibians, rhynchosaurs, the eosuchian *Malerisaurus*, and therapsids. The landscape appears to be seasonally well watered with a preponderance of water bodies (Fig. 4) in which unionids and a variety of fishes and semiaquatic tetrapods lived (Jain, 1980).

The Kota Formation consists of fluvial sediments with a prominent highly fossiliferous limestone band representing a playa-type condition (Maulik and Rudra, 1986). The limestone shows desiccation cracks and was probably laid down in a moderately hot climate under mainly sediment–depauperate conditions. Jain (1980) has suggested that *Barapasaurus* was terrestrial in habit but probably fed on nearby aquatic vegetation (Fig. 5).

The environments of the Indian Lameta dinosaurs indicate the existence of a semiarid alluvial plain undergoing active pedogenesis (Fig. 6). Massive and nodular calcretes form the dominant lithology for the nesting sites and occur as "limestone" outcropping in discontinuous patches across central peninsular India. The sandy carbonate horizons show desiccation cracks, honeycomb calcretes, brecciation, and pebbles derived from sheetwash events (Tandon *et al.*, 1990; Sahni *et al.*, 1994). The dinosaurian skeletal material is either found in coarse conglomerates as in the Kheda District of Gujarat or found as lag deposits or, more commonly, as reworked, fragmented, and weathered bones (Mohabey *et al.*, 1993).

The relationships of Indian dinosaurs along with those of associated freshwater biotas have been used to assess the contiguity or isolation of the Indian Plate in the context of other Gondwana landmasses (Sahni and Bajpai, 1988). In general terms, the Indian dinosaurs have cosmopolitan affinities. According to Chatterjee (1987) and Jain (1990), the Late Triassic *Walkeria* is similar to *Procompsognathus* of Germany, *Coelophysis* of North America, and *Syntarsus* from Zimbabwe and North America. In the absence of coeval, comparable material from the Lower Jurassic from other parts of the world, it is more difficult to evaluate the relationships of the Kota *Barapasaurus*. It was probably a specialized offshoot of the central sauropod evolutionary line.

In one of the latest assessments of the affinities of Indian Cretaceous dinosaurs, Buffetaut (1987) has commented on the inadequacies in our knowledge

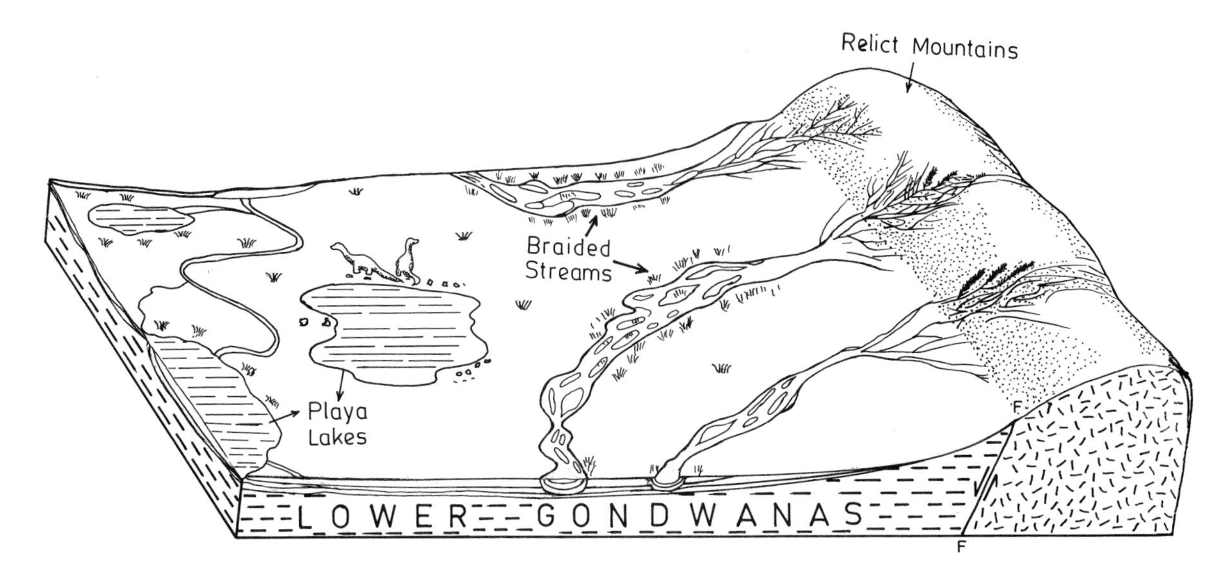

FIGURE 5 Schematic diagram of Lower Jurassic Kota Formation: braided river system with coarse sediments proximal to provenance with fine limey clays and muds toward the distal margins. Playa lakes that are subaerially exposed become common.

of Gondwana dinosaurs. Based on the meager evidence at hand, the Indian sauropods are close to those known from Argentina. To the extent that titanosaurids are also known from southern France and Cretaceous sauropod eggshell types from both regions are similar, affinities of the Indian forms to those of southern Europe cannot be ruled out. However, relationships to the Madagascan taxon, *Laplatasaurus madagascarensis*, must await the discovery of better comparative material (Buffetaut, 1987).

Despite a long history of study, much still has to be learned about Indian dinosaurs. Many of the previously described forms, in particular the sauropod taxa, are based on inadequate material and are in need of systematic revision. Fairly complete dinosaurian skeletal material, barring a few notable exceptions, is generally lacking. The presence and nature of Indian ornithischians needs to be better documented because reliance cannot be placed on earlier descriptions. Until the taxonomic status of Indian dinosaurs

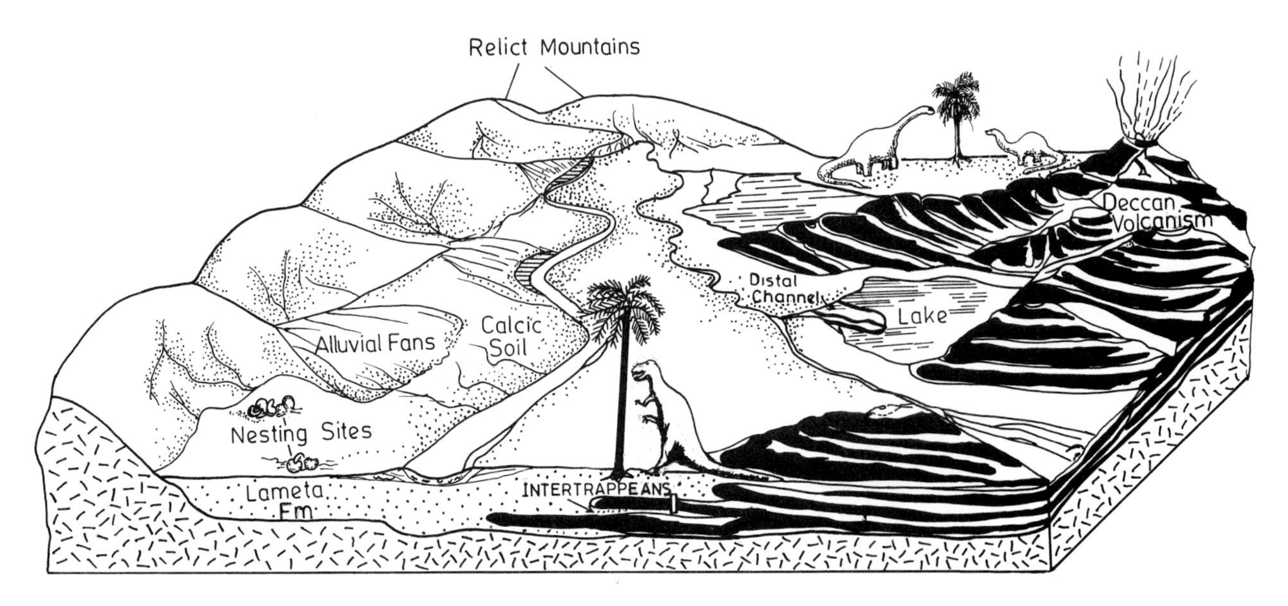

FIGURE 6 Schematic diagram of Uppermost Cretaceous Lameta Formation: semiarid alluvial plain with well-developed alluvial fan systems and active pedogenesis forming massive calcretes. Sheetwash activity is frequent.

is better known, it is difficult to comment on their affinities and paleobiogeographic relationships. Finally, the Lameta dinosaur hatcheries provide an exciting area for studying nesting behavior and preferences.

See also the following related entries:

AFRICAN DINOSAURS • CHINESE DINOSAURS • DECCAN BASALT • MONGOLIAN DINOSAURS

References

Bajpai, S., Sahni, A., and Srinivasan, S. (1993). Ornithoid eggshells from Deccan intertrappean beds near Anjar (Kachchh), western India. *Curr. Sci. Bangalore* **64**(1), 42–45.

Berman, S. D., and Jain, S. L. (1982). The braincase of a small sauropod dinosaur (Reptilia: Saurischia) from the Upper Cretaceous Lameta Group, Central India, with a review of Lameta Group Localities. *Ann. Carnegie Museum Pittsburgh* **51,** 406–422.

Buffetaut, E. (1987). On the age of the dinosaur fauna from the Lameta Formation (Upper Cretaceous) of Central India. *Newslett. Stratigr.* **18,** 1–6.

Chakravarti, D. K. (1935). Is *Lametasaurus indicus* an armored dinosaur? *Am. J. Sci.* **230,** 138–141.

Chatterjee, S. (1987). A new theropod dinosaur from India with remarks on the Gondwana–Laurasia connection in the Late Triassic. In *Geophysical Monograph 41* (G. D. Mckenzie, Ed.), pp. 183–189. International Gondwana Symposium 1985, Columbus, OH.

Courtillot, V., Besse, J., Vendamme, D., Montigny, R. J., Jaeger, J. J., and Capetta, H. (1986). Deccan flood basalts at the Cretaceous–Tertiary boundary? *Earth Planetary Sci. Lett.* **80,** 361–374.

Dasgupta, K. (1993). Some contributions to the stratigraphy of the Yerrapalli Formation, Pranhita–Godavari Valley, Deccan, India. *J. Geol. Soc. India* **42,** 223–230.

Ghevariya, Z. G., and Srikarni, C. (1990). Dinosaur foot tracks from Cretaceous rocks of Kachchh Gujarat, India. In *Cretaceous Event Stratigraphy and the Correlation of the Nonmarine Strata* (A. Sahni and A. Jolly, Eds.), Seminar cum Workshop Contribution Vols. IGCP 216 and 245, pp. 94–98. Panjab University, Chandigarh, India.

Jain, S. L. (1980). The continental Lower Jurassic fauna from the Kota Formation, India, aspects of vertebrate history. In *Aspects of Vertebrate History; Essays in Honor of E. H. Colbert* (L. L. Jacobs, Ed.), pp. 99–123. Museum of Northern Arizona Press.

Jain, S. L. (1983). Spirally coiled coprolites from the Upper Triassic Maleri Formation, India. *Palaeontology* **26,** 813–829.

Jain, S. L. (1990). An Upper Triassic vertebrate assemblage from Central India. *Bull. Indian Geol. Assoc. Chandigarh* **23**(2), 67–84.

Jain, S. L., and Roychowdhury, T. (1987). Fossil vertebrates from the Pranhita–Godavari Valley (India) and their stratigraphic correlation. In *Geophysical Monograph 41* (G. D. Mckenzie, Ed.), pp. 219–228. International Gondwana Symposium, Columbus, OH.

Jain, S. L., and Sahni, A. (1983). Some Upper Cretaceous vertebrates from Central India and their palaeogeographic implications. In *Cretaceous of India* (H. K. Maheshwari, Ed.), pp. 67–84. Indian Association of Palynostratigraphers, Lucknow, India.

Jain, S. L., Kutty, T. S., Roychowdhury, T., and Chatterjee, S. (1975). The sauropod dinosaur from the Lower Jurassic Kota Formation of India. *Proc. R. Soc. London A* **188,** 221–228.

Jain, S. L., Kutty, T. S., Roychowdhury, T., and Chatterjee, S. (1979). Some characteristics of *Barapasaurus tagorei*, a sauropod dinosaur from the Lower Jurassic of Deccan, India. *Proc. 4th Int. Gondwana Symp. (1977) Calcutta* **1**(3), 204–216.

Kutty, T. S. (1969). Some contribution to the stratigraphy of the Upper Gondwana Formations of India. *J. Geol. Soc. India* **10**(1), 33–48.

Lydekker, R. (1877). Notice of new and other vertebrata from Indian Tertiary and Secondary Rocks. *Rec. Geol. Survey India* **10,** 30–43.

Matley, C. A. (1921). On the stratigraphy, fossils and geological relationships of the Lameta beds of Jubbulpore. *Rec. Geol. Survey India* **53,** 142–164.

Matley, C. A. (1923). Note on an armoured dinosaur from the Lameta beds of Jubbulpore. *Rec. Geol. Survey India* **55**(2), 105–109.

Matley, C. A. (1929). The Cretaceous dinosaurs of the Trichinopoly District, and the rocks associated with them. *Rec. Geol. Survey India* **61**(4), 337–349.

Matley, C. A. (1939). The coprolites of Pijdura, central provinces. *Rec. Geol. Survey India* **74,** 535–547.

Maulik, P. K., and Rudra, D. K. (1986). Trace fossils from the freshwater Kota Limestone of the Pranhita–Godavari Valley, south central India. *Bull. Geol. Mines Metallurgical Soc. India* **54,** 114–123.

Mohabey, D. M. (1990). Dinosaur eggs from Lameta Formation of western and central India: their occurrence, and nesting behaviour. In *Cretaceous Event Stratigraphy and the Correlation of the Nonmarine Strata* (A. Sahni and A. Jolly, Eds.), Seminar cum Workshop Contribution Vols. IGCP 216 and 245. Panjab University, Chandigarh, India.

Mohabey, D. M., Udhoji, S. G., and Verma, K. K. (1993). Palaeontological and sedimentological observations on nonmarine Lameta Formation (Upper Cretaceous) of Maharashtra, India: Their palaeoecological and palaeoenvironmental significance. *Palaeogeogr. Palaeoclimatol. Palaeoecol.* **105,** 83–94.

Sahni, A., and Bajpai, S. (1988). Cretaceous–Tertiary boundary events: The fossil vertebrate, palaeomagnetic and radiometric evidence from peninsular India. *J. Geol. Soc. India* **32,** 382–396.

Sahni, A., Tandon, S. K., Jolly, A., Bajpai, S., Sood, A., and Srinivasan, S. (1994). Upper Cretaceous dinosaur eggs and nesting sites from the Deccan volcano–sedimentary province of peninsular India. In *Dinosaur Eggs and Babies* (K. Carpenter, K. Hirsute, and J. Horner, Eds.), pp. 204–226. Cambridge Univ. Press, Cambridge, UK.

Sarkar, A., Bhattarcharya, S. K., and Mohabey, D. M. (1991). Stable isotope analysis of dinosaur eggshells, palaeoenvironmental implications. *Geology* **19,** 1068–1071.

Tandon, S. K., Verma, V. K., Jhingran, V., Sood, A., Kumar, S., Kohli, R. P., and Mittal, S. (1990). The Lameta beds of Jabalpur, central India. Deposits of fluvial and pedogenically modified semi-arid fan-palustrine flat systems. In *Cretaceous Event Stratigraphy and the Correlation of the Nonmarine Strata* (A. Sahni and A. Jolly, Eds.), Seminar cum Workshop Contribution Vols. IGCP 216 and 245. Panjab University, Chandigarh, India.

Vianey-Liaud, M., Jain, S. L., and Sahni, A. (1987). Dinosaur eggshells (Saurischia) from the Late Cretaceous intertrappeans and Lameta Formation (Deccan, India). *J. Vertebr. Paleontol.* **7,** 408–424.

von Huene, F., and Matley, C. A. (1933). The Cretaceous saurischia and ornithischia of the Central Provinces of India. *Mem. Geol. Survey India Palaeontol. Indica* **21,** 1–72.

Yadagiri, P., and Ayyasami, K. (1979). A new Stegosaurian dinosaur from Upper Cretaceous sediments of South India. *J. Geol. Soc. India* **20**(2), 521–530.

Injuries

see PALEOPATHOLOGY

Institute and Museum for Geology and Paleontology, University of Tubingen, Germany

see MUSEUMS AND DISPLAYS

Institute de Paléontologie, Muséum National d'Histoire Naturelle, Paris

The building housing the Institute de Paléontologie, in the rue Buffon adjacent to the Jardin des Plantes, includes the classic collection of fossil vertebrates on which Baron Georges Cuvier founded his theories of comparative anatomy, the vanished worlds of past ages, and the first scientific demonstration of extinction. On exhibit in the gallery of paleontology are several casts and original specimens of dinosaurs, and the collections house historically important fossil as well as recent vertebrate materials, mainly from France and Africa. Among these specimens is a great variety of ornithischian and saurischian dinosaurs.

Triassic dinosaurs from France include postcranial bones of *Plateosaurus engelhardti* (Norian) from the *marnes irisées* of Saint Lothain in the Jura region and postcranial bones of a prosauropod from the Upper Triassic of Alzon (Gard). From Morocco comes the important early prosauropod *Azendohsaurus laaroussii,* represented by a dentary and teeth, from the Argana Formation (Carnian), collected by J. M. Dutuit in 1965. There are also postcranial bones of the prosauropod *Euskelosaurus* from the Lower Elliot Formation of Maphutseng, Lesotho. Early Jurassic dinosaurs are represented only by a lower jaw of the fabrosaurid *Lesothosaurus* (*"Fabrosaurus australis"*) from the Upper Elliot Formation of Maphutseng, Lesotho. The Lesotho dinosaurs were collected by F. and P. Ellenberger, L. Ginsburg, J. Fabre, and B. Battail in 1955, 1956, and 1959.

There are several important late Middle and Late Jurassic dinosaur specimens from France. The theropod *Poekilopleuron bucklandii* is now known only from a cast of the bones described by Eudes-Deslongchamps in 1838 (see MEGALOSAURUS); the original bones were destroyed in Caen during the World War II. These come from the Calcaire de Caen (middle Bathonian) in Normandy. The braincase of the theropod *Piveteausaurus divesensis* comes from the *marnes de Dives* (late Callovian) of Calvados, Normandy. Part of a sauropod skeleton from the late Oxfordian described as *"Bothriospondylus madagascariensis"* comes from Damparis (Jura). Some vertebrae of a theropod from the Kimmeridgian of Honfleur, initially known as the "Gavial de Honfleur," were the first French dinosaur bones to be described, although Cuvier initially believed the specimen was an unusual crocodile. It was later named *"Streptospondylus cuvieri."* Recently collected and fully known is the almost complete skeleton of the coelurosaur *Compsognathus corallestris* from the Portlandian of Canjuers (Var). Middle Jurassic sauropods are also represented by postcranial material of *Cetiosaurus mogrebiensis* from the Bathonian of El Mers, in the High Atlas mountains of Morocco, and by *Bothriospondylus madagascariensis* and *Lapparentosarus madagascariensis* from Kamoro (Isalo III: Bathonian), Madagascar, collected by Ginsburg in 1962. The Moroccan specimens were collected by Lapparent in 1940 and 1941, and a complete skeleton was collected by Monbaron and Taquet in 1980.

French Cretaceous dinosaurs include the theropod *Genusaurus sisteronis*, known from part of a pelvis and of a posterior limb, collected by Taquet in 1986 from the middle Albian of Bevons (Alpes de Haute Provence), and a hadrosaurid dentary from the *calcaire Nankin* (Maastrichtian) of Saint Martory (Haute Garonne). There are also materials of Titanosaurinae referred to "*Hypselosaurus*," of the hypsilophodont *Rhabdodon priscus*, and of a nodosaurid referred to "*Struthiosaurus*," as well as theropod teeth and bones from the Maastrichtian of Fox Amphoux (Var), collected by Lapparent in 1939. Eggs are also known from Maastrichtian deposits of the Aix Basin (Bouches du Rhone).

African Cretaceous dinosaurs are well represented in the museum's collections. They include vertebrae of the sauropod *Rebbachisaurus garasbae* and teeth and isolated bones of the giant carnosaur *Carcharodontosarus saharicus* from Kem-Kem, Morocco (Albian), collected by Lavocat in 1950. From the Aptian of Gadoufaoua, Niger, collected by Taquet from 1966 to 1972, come several dinosaurs including the iguanodontid *Ouranosaurus nigeriensis*, represented by a cast (the type specimen is in the National Museum of Niger, Niamey, with a second original specimen in Venice), a femur of the hypsilophodont *Valdosaurus nigeriensis*, and two premaxillaries of the theropod *Spinosaurus*. Expeditions to the Berivotro (Campanian) beds of Madagascar collected a dentary of the theropod *Majungasaurus crenatissimus*, a braincase of what was initially described (now doubted) as a pachycephalosaur, *Majungatholus atopus*, and vertebrae of the sauropod *Titanosaurus madagascariensis*. The dinosaur collections are completed by several specimens from the United States, including a skull of *Triceratops horridus* from the Lance Formation of Wyoming and part of a skeleton of the hadrosaur *Edmontosaurus (Anatosaurus) annectens* from Niobrara County, Kansas, both purchased from C. H. Sternberg in 1911. On exhibit in the galleries are casts of the dinosaurs *Allosaurus fragilis*, *Diplodocus carnegiei*, *Iguanodon bernissartensis*, *Tyrannosaurus rex*, and *Tarbosaurus bataar*.

Institute of Vertebrate Paleontology and Paleoanthropology

Dong Zhiming
Institute of Vertebrate Paleontology and Paleoanthropology
Chinese Academy of Sciences
Beijing, People's Republic of China

Paleovertebrate research in China has taken its place in the mainstream of world paleontology. The center of this work is the Institute of Vertebrate Paleontology and Paleoanthropology (IVPP) of the Chinese Academy of Sciences. The scope of scientific research at the IVPP includes paleovertebrates (including paleoanthropology) and related fields of biology and geology.

The IVPP grew out of the Cenozoic Research Laboratory founded in 1929 as a subsidiary of the Geological Survey of China. In that year, the Chinese scientist Pei Wen-zhong found the first cranium of "Peking Man," which firmly helped to establish the institute. The IVPP became an independent research laboratory in 1953, was named the Institute of Vertebrate Paleontology in 1957, and received its current name in 1960. The founder of the IVPP was Professor Yang Zhongjian (1897–1979), or C. C. Young as he is better known in the Western world.

There are currently three departments—Paleoichthyology and Paleoherpetology, Paleomammalogy, and Paleoanthropology. Mesozoic dinosaurs and birds and related stratigraphic and paleoecological studies come under the domain of the first department. In addition, the IVPP runs a paleozoological museum in Beijing and the palaeoanthropological museum and research center at Zhoukoudian (the original Peking Man site). The latter was added to the World Heritage List by UNESCO in 1987.

There are currently 230 staff and workers in the IVPP, among them 96 researchers and 40 technicians (preparation, casting, moulding, scientific illustration, and photography). Staff edit and publish three periodicals. *Vertebrata PalAsiatica* was first published in English in 1957. Since 1961, this journal has been published mainly in Chinese with foreign language abstracts (usually English and Russian). It should be noted that this journal was not published between

1967 and 1972. The other two journals are *Acta Anthroplogica Sinica* and *Fossils*, a popular publication initiated in 1973. There are also two monograph series, *Paleontological Sinica* and *Memoirs of the IVPP*.

In 1993, the IVPP moved into a new building, which houses offices, a three-story public exhibition hall, and storage for the institution's 200,000 cataloged specimens of vertebrates. Dinosaurs on display include *Lufengosaurus*, *Monolophosaurus*, *Psittacosaurus*, *Protoceratops*, and *Tuojiangosaurus*. Many specimens from the IVPP can be seen worldwide in traveling exhibitions. The library has more than 65,000 books and reprints, many acquired from the personal libraries of the late Drs. C. C. Young, W. C. Pei, Davidson Black, W. G. Kuhne, and F. Weidenreich.

The IVPP trains students and is interested in international cooperation and scientific exchange. Recently, there have been joint projects to study dinosaurs in cooperation with institutions in Belgium, Canada, Japan, and Mongolia. I hope that more colleagues and friends will come and visit us at the IVPP.

See also the following related entries:

CHINESE DINOSAURS • ERENHOT DINOSAUR MUSEUM • LUFENG • MONGOLIAN DINOSAURS • MUSEUMS AND DISPLAYS

Instituto Miguel Lillo (Facultad de Ciencias Naturales), Argentina

see MUSEUMS AND DISPLAYS

Institut Royal des Sciences Naturelles de Belgique, Belgium

see MUSEUMS AND DISPLAYS

Integument

see COLOR; "FEATHERED" DINOSAURS; ORNAMENTATION; SKELETAL STRUCTURES; SKIN

Intelligence

DALE A. RUSSELL
North Carolina State University
Raleigh, North Carolina, USA

The behavioral complexity (intelligence) of Mesozoic dinosaurs cannot be observed and of necessity must be assessed by examining the relative size of the brain. However, nervous tissues of dinosaurs have not been recognized in the fossil condition, and structures of the central nervous system are only identifiable through the bony structures that once contained them. Thus, some evidence relating to the size and shape of the olfactory nerves, bulb and tracts, cerebral hemispheres, midbrain, optic nerves, pituitary gland, medulla, and major cranial nerves in dinosaurs has been obtained by examining the fossil bone that once surrounded the brain.

As early as 1896, O. C. Marsh presented excellent illustrations and descriptions of endocasts of the brain cavity in seven species of dinosaurs. Calculating simple ratios between brain weight as indicated by endocast volumes and body weight estimated from skeletons, Marsh concluded that the brain was relatively smaller in several varieties of dinosaurs than in living crocodilians, and that the brain–body ratio in *Stegosaurus* was the smallest observed in any land vertebrate. The proverbial stupidity of dinosaurs thereby became rooted in the public mind. Because Marsh also discovered an enlargement of the neural canal in the sacral region of some dinosaurs, including *Stegosaurus*, much speculation has been focused on the possibility that dinosaurs had a "second brain" capable of generating "afterthoughts." However, subsequent studies indicate that this enlargement was probably filled with nonnervous tissues.

Jerison, in his fundamental work *Evolution of the Brain and Intelligence* (1973), found that brain weight in living animals is approximately proportional to the two-thirds power of body weight. Thus, because dinosaurs were on the average much larger than living animals, their brains only appear to be unusually small. In fact, brain–body proportions in 10 species of dinosaurs, when corrected for body size effects, were comparable to those of living reptiles. The stigma of surpassing stupidity was removed from dinosaurs. Indeed, at approximately the same time, Russell (1969, 1972) observed that the brain–body proportions in two small Late Cretaceous theropods from Canada were comparable to those of ratite birds.

Consideration of relative brain size in dinosaurs then passed from observation to more controversial matters of inference. In the debate on metabolic rates in dinosaurs, Hopson (1977, 1980) compared all available data on brain–body proportions in dinosaurs (including 12 species) to those in living crocodilians (in which the brain is approximately 1/25th that of an average mammal of similar body weight). He found that the brain was relatively large in small theropods, the skeletal adaptations of which suggested unusual locomotor speed and agility. According to Hopson, these were the only dinosaurs that could possibly have possessed metabolic rates approaching those of birds. In assessing the relative size of the brain in a small ornithopod, Rich and Rich (1989) found values in excess of the dinosaurian average.

If, as Marsh (1896) observed, the brains of dinosaurs were small relative to those of living birds and mammals, then it might be suspected that the size of the brain had gradually increased in terrestrial vertebrates through the intervening time. Russell (1982, 1983) postulated that, due to interorganismal competition, relative brain size increased exponentially during the Phanerozoic, a trend that would be in conflict with a current view that evolution is a random process driven by changes in the physical environment. However, it accords with Hopson's observation that the size of the dinosaur brain is approximately intermediate between that of reptiles and birds and with the independently derived conclusions of Wyles *et al.* (1983) that the evolution of brain in vertebrates is essentially an autocatalytic (exponential) process. Drawing on analogies between the anatomy of small, bipedal theropods and bipedal hominids, as well as trends in the evolution of brain size in birds (which are living dinosaurs) and mammals, Russell and Séguin (1982) carried out a "thought experiment." They postulated that, had the dinosaurs not become extinct, some might have assumed a human-like body form through autocatalytic brain growth and convergent selection since the end of Mesozoic time. The resulting model, termed a "dinosauroid," has attracted both criticism and praise.

The diversity of terrestrial plants and vertebrates remained relatively constant through much of Mesozoic time. Noting recent biogeographical work that suggests that successful animal immigrants are often characterized by relatively greater brain size and often arise in regions of relatively high biodiversity, Russell (1997a,b) suggests that neurological flexibility is linked to biodiversity. The latter in turn has long been known to be affected by physical environmental conditions. Therefore, the apparent stability in brain size during early and mid-Mesozoic time may have been a result of stagnating biodiversity under gradually worsening physical environmental conditions (e.g., declining atmospheric carbon dioxide levels and increasing aridity). A brief but profound excursion in physical environmental conditions evidently resulted from the impact of a comet at the end of the Mesozoic Era and produced a planetwide biological catastrophe in the course of which the dinosaurs became extinct. However, biodiversity rapidly returned to preextinction levels, and the general trend toward increasing brain size continued in surviving birds and mammals without a major interruption.

Much work remains to be done on the dinosaurian brain. The morphology of conduits for nerves and blood vessels through the braincase wall, as well as that of auditory and pneumatic structures associated with the braincase, has proven to be useful in assessing dinosaurian relationships. However, estimates of brain size are available for less than 5% of the known genera of dinosaurs. Although some small theropods (troodontids and ornithomimids) are known to have possessed a ratite-sized brain, numerical evaluations of brain–body proportions are available for only one species (*Troodon formosus*). The brain was also relatively large, by dinosaurian standards, in several other groups of small theropods (including some dromaeosaurids, oviraptorosaurs, and avimimids), but no quantitative evaluations of brain–body proportions have so far been attempted.

A significant part of the reason for the relatively little attention that has been given to the dinosaurian brain is due to the difficulties involved in obtaining accurate data. Braincase material is neither abundant nor well preserved, and the brain cavity is often crushed or filled with resistant matrix. The brain seldom filled its cavity to the extent of leaving surface impressions on the surrounding bony surfaces so that estimates of its original volume are often not well constrained. Nearly all of the body weights used in estimating brain–body proportions are derived from models of dinosaurs, where the results can vary widely according to the anatomical interpretations of the artist. However, Anderson *et al.* (1985) have devised a quantitative means of estimating weight from measurements of the supporting limb bones, and Paul (1988) has reassessed weights using carefully constructed models. It is hoped that information on the nature of the brain in dinosaurs will rapidly increase in the future.

See also the following related entries:

BEHAVIOR • PALEONEUROLOGY

References

Anderson, J. F., Hall-Martin, A., and Russell, D. A. (1985). Long-bone circumference and weight in mammals, birds and dinosaurs. *J. Zool. London* **207,** 53–61.

Hopson, J. A. (1977). Relative brain size and behavior in archosaurian reptiles. *Ann. Rev. Ecol. Systematics* **8,** 429–448.

Hopson, J. A. (1980). Relative brain size in dinosaurs: Implications for dinosaurian endothermy. *Am. Assoc. Adv. Sci Selected Symp.* **28,** 287–310.

Jerison, H. J. (1973). *Evolution of the Brain and Intelligence,* pp. xiv, 482. Academic Press, New York.

Marsh, O. C. (1896). The dinosaurs of North America. U. S. Geological Survey, 16th Annual Report, 1894–95, pp. 133–244.

Rich, T. H. V., and Rich, P. V. (1989). Polar dinosaurs and biotas of the Early Cretaceous of southeastern Australia. *Natl. Geogr. Res.* **5,** 15–53.

Russell, D. A. (1982). Quasi-exponential evolution: Implications for intelligent extraterrestrial life. Twenty-fourth Plenary Meeting, Committee on Space Research, Ottawa, Canada, May 16–June 2, 1982, pp. 517. [Abstract]

Russell, D. A. (1983). Exponential evolution: Implications for intelligent extraterrestrial life. *Adv. Space Res.* **3,** 95–103.

Russell, D. A. (1994). Dinosaurs within the history of multicellular organisms. University of Houston, Lunar and Planetary Institute Technical Report: Workshop on new developments regarding the K/T event and other catastrophes in earth history, pp. 100–101.

Russell, D. A. (1997a). *The Place of Dinosaurs in the History of Life.* Indiana Univ. Press/Paleontological Society, Bloomington.

Russell, D. A. (1997b). The significance of the extinction of the dinosaurs. Geological Society of America, Spec. Paper.

Russell, D. A. and Séguin, R. (1982). Reconstruction of the small Cretaceous theropod *Stenonychosaurus inequalis* and a hypothetical dinosauroid. *Syllogeus* **37,** pp. 43.

Wyles, J. S., Kunkel, J. G., and Wilson, A. C. (1983). Birds, behavior and anatomical evolution. *Proc. Natl. Acad. Sci. USA* **80,** 4394–4397.

Ischigualasto Formation

RAYMOND R. ROGERS
Cornell College
Mount Vernon, Iowa, USA

The earliest skeletal records of dinosaurs are preserved in rocks of Late Triassic age (Carnian) on several continents, including North America (Chinle Formation), South America (Ischigualasto and Santa Maria formations), India (Maleri Formation), and Africa (Timesgadiouine Formation). Some of the oldest and most complete skeletons of early dinosaurs yet discovered have been found within the Ischigualasto Formation of northwestern Argentina. Dinosaurs recovered from Ischigualasto sediments include the ornithischian *Pisanosaurus* and the primitive (and occasionally controversial) theropods *Herrerasaurus* and *Eoraptor* (Sereno and Novas, 1992; Sereno *et al.*, 1993; Holtz, 1995; Holtz and Padian, 1995). Other tetrapods in the Ischigualasto vertebrate assemblage include nondinosaurian archosaurs (e.g., *Saurosuchus* and *Proterochampsa*), herbivorous and carnivorous cynodonts, dicynodonts, rhynchosaurs, and temnospondyl amphibians.

Geologic Setting

Badland exposures of the Ischigualasto Formation crop out in the Ischigualasto–Ville Unión Basin of northwestern Argentina (Figs. 1 and 2). This basin is one of several small rift basins that formed along the western margin of South America before the breakup of Pangea. The Ischigualasto Formation is included within the Agua de la Peña Group, which is a thick succession of continental Triassic strata that accumulated during the Middle and Late Triassic. Beds of the Ischigualasto Formation rest upon sediments of the Los Rastros Formation (this contact is characterized locally by marked angular discordance) and pass up into spectacular red and orange cliffs of the Los Colorados Formation. Three-toed (tridactyl) footprints and trackways presumably attributable to theropod dinosaurs have been discovered within the underlying Los Rastros Formation (Forster *et al.*, 1995), and the prosauropod *Riojasaurus* was discovered within the overlying Los Colorados Formation. The Chañares Formation, which preserves dinosaurian precursors such as *Lagosuchus* (*Marasuchus* of Sereno and Arcucci, 1994) and mammalian precursors such

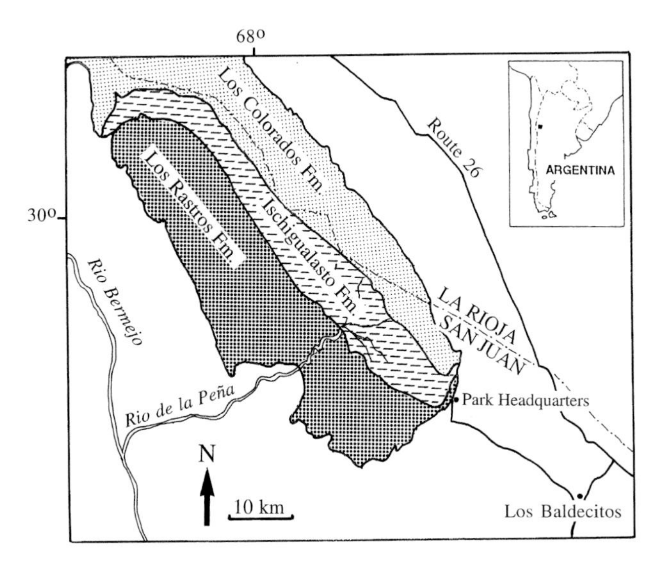

FIGURE 1 Location of the Ischigualasto–Ville Unión Basin in San Juan and La Rioja provinces, northwestern Argentina (inset). The basin preserves several thousand meters of continental Triassic deposits, including the tetrapod-bearing Ischigualasto, Los Rastros, Los Colorados, and Chañares formations.

as *Probainognathus* and *Massetognathus*, is also included within the Aqua de la Peña Group.

Deposits of the Ischigualasto Formation consist of medium- to coarse-grained conglomeratic sandstones, siltstones, and silty mudstones. Sandstone beds display trough cross-bedding indicative of current activity within shallow streams. Siltstone and mudstone beds show abundant evidence of soil formation, including root traces and color banding. The paleoenvironmental setting of the Ischigualasto Formation has been reconstructed as both a lush, lake-dominated setting characterized by a humid climate (Robinson, 1971) and a predominantly fluvial (river-dominated) setting that experienced seasonal variations in water availability (Bossi, 1970). Recent work within the formation supports the latter view: Sedimentological data and the scarcity of freshwater vertebrates and invertebrates are consistent with a seasonal, water-limited fluvial setting (Rogers *et al.*, 1992, 1993).

FIGURE 2 Expansive badlands of the Ischigualasto Formation stretching across the Ischigualasto–Ville Unión Basin.

Several thin volcanic ash horizons (bentonites) are interbedded within the Ischigualasto section. Radioisotopic dating of one of these bentonite beds indicates that deposition of the Ischigualasto Formation began approximately 228 million years ago (Rogers *et al.*, 1993). This age places the formation and its vertebrate fossils firmly within the Carnian Stage.

Taphonomy

Tetrapod fossils in the Ischigualasto Formation are typically preserved as isolated carcasses or disarticulated skeletal elements, most often in fine-grained overbank deposits (Fig. 3). Concentrated deposits of fossils such as bone beds and vertebrate microsites are conspicuously rare. Most Ischigualasto tetrapods occur in the lower two-thirds of the formation, where the rhynchosaur *Scaphonyx* and the cynodont *Exaeretodon* are most abundant. Isolated skulls, particularly those of *Exaeretodon*, are extremely common (one horizon alone preserves at least 15 isolated crania of this animal). The relative scarcity of fossils in the upper third of the formation is somewhat puzzling because neither sedimentary deposits nor preservational styles differ significantly from those that characterize the richly fossiliferous lower part of the section.

Among Ischigualasto dinosaurs, the skeletal remains of *Herrerasaurus* are most abundant. *Eoraptor* and *Pisanosaurus* are known from relatively few specimens. As a whole, dinosaur fossils comprise only a small fraction (~6%) of the total tetrapod sample recovered from the formation, but theropod dinosaurs account for approximately 40% (percentage based on minimum number of individuals) of the known terrestrial carnivores (Rogers *et al.*, 1993).

FIGURE 3 *In situ* skeleton of *Eoraptor* (Sereno *et al.*, 1993), a primitive theropod dinosaur recovered from the Ischigualasto Formation in 1991. Arrow points to skull and attached neck vertebrae; hindlimbs and tail are situated above scale card.

Summary

The Carnian Ischigualasto Formation preserves a rich fossil record that spans a critical time in dinosaurian evolution. Based on the analysis of Ischigualasto sediments, we can conclude that early dinosaurs inhabited fluvial and floodplain settings comparable to their Jurassic and Cretaceous counterparts. Like many of the later dinosaurs, Ischigualasto dinosaurs may have also been adapted to fairly dry, possibly even seasonal conditions (at least with respect to the availability of water). The composition of the Ischigualasto tetrapod assemblage further suggests that dinosaurs as a whole were rather minor components of vertebrate faunas during the Carnian, at least within the Ischigualasto–Ville Unión Basin.

Radioisotopic dates derived from a volcanic ash bed within the Ischigualasto Formation indicate that by approximately 228 million years ago the major dinosaurian lineages (Ornithischia, Sauropodamorpha, and Theropoda) were already established. It apparently took another 10 million years or so before dinosaurs would assume the role of dominant large-bodied tetrapods in most terrestrial ecosystems. The eventual success of the dinosaurs may be linked to one or more mass extinction events that occurred at the end of the Triassic Period (Benton, 1983).

See also the following related entries:

South American Dinosaurs • Triassic Period

References

Benton, M. J. (1983). Dinosaur success in the Triassic: A noncompetitive ecological model. *Q. Rev. Biol.* **58,** 29–55.

Bossi, G. E. (1970). Asociaciones mineralogicas de las arcillas en la Cuenca de Ischigualasto–Ischichuca parte II: Perfiles de la Hoyada de Ischigualasto. *Acta Geol. Lilloana* **XI,** 75–100.

Forster, C. A., Arcucci, A. B., Marsicano, C. A., Abdala, F., and May, C. L. (1995). New vertebrate material from the Los Rastros Formation (Middle Triassic), La Rioja Province, northwestern Argentina. *J. Vertebr. Paleontol.* **15**(Suppl. to No. 3), 29A.

Holtz, T. R. (1995). A new phylogeny of the theropoda. *J. Vertebr. Paleontol.* **15**(Suppl. to No. 3), 35A.

Holtz, T. R., and Padian, K. (1995). Definition and diagnosis of theropoda and related taxa. *J. Vertebr. Paleontol.* **15**(Suppl. to No. 3), 35A.

Robinson, P. L. (1971). A problem of faunal replacement on Permo–Triassic continents. *Palaeontology* **14,** 131–153.

Rogers, R. R., Forster, C. A., May, C. L., Monetta, A., and Sereno, P. C. (1992). Paleoenvironment and taphonomy of the dinosaur-bearing Ischigualasto Formation (Upper Triassic, Argentina). Fifth North American Paleontological Convention, Abstracts and Program, pp. 249.

Rogers, R. R., Swisher, C. C., III, Sereno, P. C., Monetta, A. M., Forster, C. A., and Martínez, R. N. (1993). The Ischigualasto tetrapod assemblage (Late Triassic, Argentina) and $^{40}Ar/^{39}Ar$ dating of dinosaur origins. *Science* **260,** 794–797.

Sereno, P. C., and Arcucci, A. B. (1994). Dinosaurian precursors from the Middle Triassic of Argentina: *Marasuchuc lilloensis*, gen. nov. *J. Vertebr. Paleontol.* **14,** 53–73.

Sereno, P. C., and Novas, F. E. (1992). The complete skeleton of an early dinosaur. *Science* **258,** 1137–1140.

Sereno, P. C., Forster, C. A., Rogers, R. R., and Monetta, A. M. (1993). Primitive dinosaur skeleton from Argentina and the early evolution of the Dinosauria. *Nature* **361,** 64–66.

Israeli Dinosaurs

see African Dinosaurs

Iwaki Museum of Coal and Fossils, Japan

see Museums and Displays

Japanese Dinosaurs

YOICHI AZUMA
Fukui Prefectural Museum
Fukui City, Japan

YUKIMITSU TOMIDA
National Science Museum
Shinjyuku, Tokyo, Japan

Research History

A hadrosaur found in southern Sakhalin in 1935 and described as *Nipponosaurus sachalinensis* (Nagao, 1936) was the first dinosaur ever found in Japan. However, Sakhalin became a part of the USSR after World War II, and no further dinosaur fossils were found in the current Japanese territory for more than 30 years. In 1978, a fragment of a sauropod humerus was found in the Upper Cretaceous of Iwate Prefecture; this became the first dinosaur fossil found within the current Japanese territory (Hasegawa *et al.*, 1991). Since then, a number of dinosaur body and footprint fossils have been found in the Cretaceous at various localities in Japan (Fig. 1). These discoveries were made in part by children and amateur collectors, but in recent years prefectural museums, universities, and local education boards began to organize expeditions, and dinosaur fossil material has been accumulated with accelerated speed.

The Tetori Group, which ranges from the Late Jurassic to the Early Cretaceous of central Japan, has produced most of the dinosaur fossils found in Japan to date. For about the past 10 years the Fukui Prefectural Museum, Gifu Prefectural Museum, and the Shiramine-mura Education Board have been organizing expeditions to the Cretaceous part of the Tetori Group and have collected many dinosaur specimens. A small expedition to the Cretaceous Mifune Group, distributed in west-central Kyushu, produced some theropods and ornithischians and further finds of dinosaur fossils seem promising.

Dinosaur Fossils and Their Horizons

Dinosaur fossils are from 14 different localities or areas in Japan. They are explained briefly from north to south.

Iwate Pref., Iwaizumi-cho; Miyako Group (Late Aptian to Early Albian) A fragmentary humerus of a sauropod was found. Its Japanese name is "Moshiryu." It was once questionably identified as *Mamenchisaurus* sp. (Diplodocidae) (Hasegawa *et al.*, 1991). However, the recent discovery of *M. sinocanadorum* from Xinjiang, China, revealed that *Mamenchisaurus* is not a member of Diplodocidae (Russell and Zheng, 1993). We believe that the fragmentary nature of the specimen does not hold any characters to identify it at a generic level or probably at the family level either. It is probably best identified as Sauropoda indet.

Fukushima Pref., Iwaki City and Hirono-cho; Futaba Group (Coniacian to Santonian) A hadrosaur tooth, an ornithischian cervical vertebra, a theropod tibia (Hasegawa *et al.*, 1987), and a sauropod tooth have been found.

Gumma Pref., Nakazato-mura; Sebayashi Fm. (Aptian) A theropod cervical vertebra (Hasegawa *et al.*, 1984) and a number of dinosaur footprints (Matsukawa and Obata, 1985) have been reported. The "footprints" are about 50 imprints on a bedding plane with ripple marks. It is said that they include four different kinds of dinosaurs. However, the preservation is so poor and the outline of each imprint is so unclear that the possibility remains that they may not be dinosaur footprints.

Toyama Pref., Oyama-cho; Tetori Group, Itoshiro Subgroup (Valanginian) A theropod and a sauropod footprint have been reported (Goto, 1992).

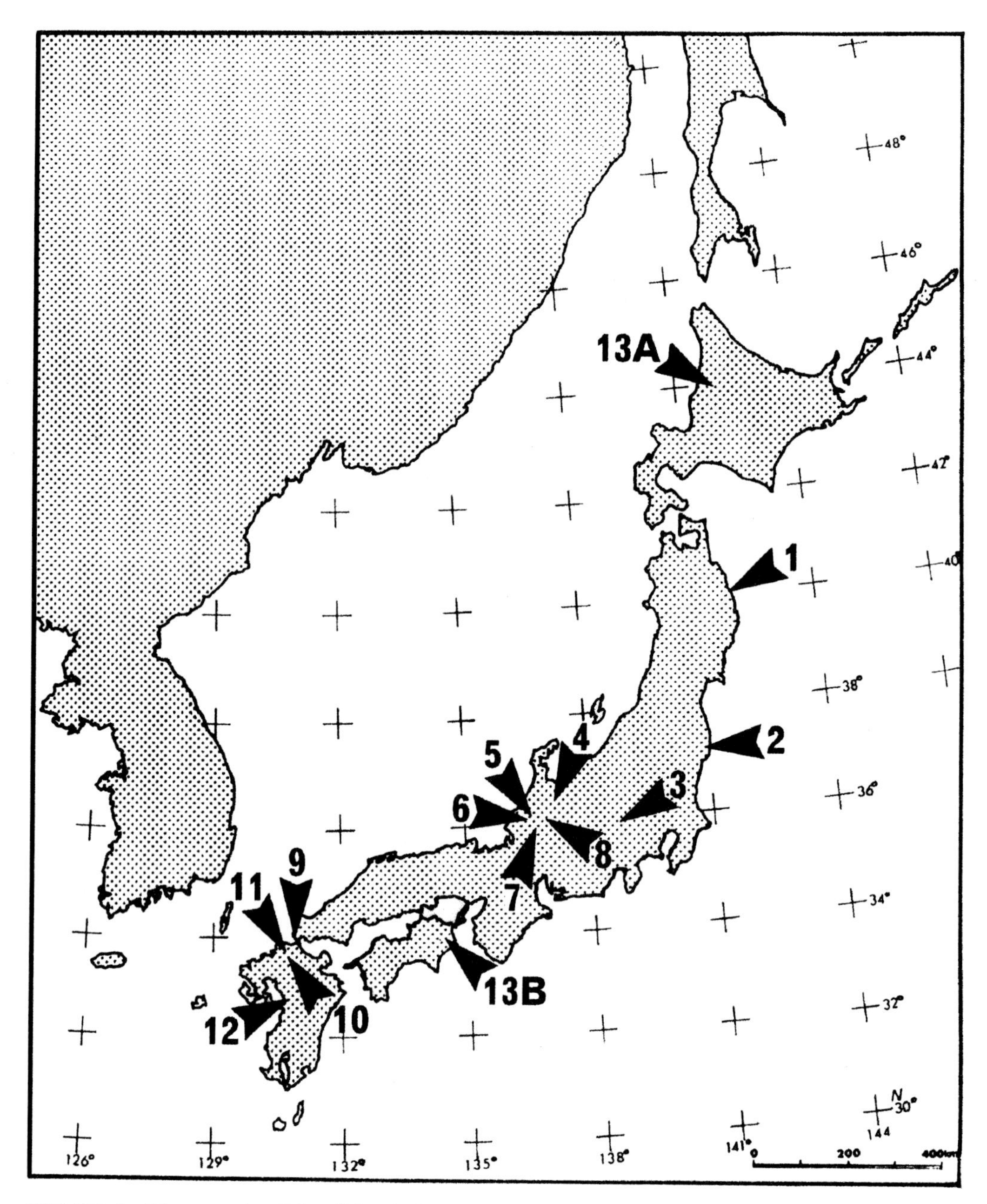

FIGURE 1 Dinosaur fossil localities in Japan. 1, Iwate Pref., Iwaizumi-cho; 2, Fukushima Pref., Iwaki City and Hirono-cho; 3, Gumma Pref., Nakazato-mura; 4, Toyama Pref., Oyama-cho; 5, Ishikawa Pref., Shiramine-mura; 6, Fukui Pref., Katsuyama City; 7, Fukui Pref., Izumi-mura; 8, Gifu Pref., Shirakawa- and Shokawa-mura; 9, Yamaguchi Pref., Shimonoseki City; 10, Fukuoka Pref., Miyata-cho; 11, Fukuoka Pref., Kitakyushu City; 12, Kumamoto Pref., Mifune-cho; 13, (A)Hokkaido, Obira-cho and (B) Tokushima Pref., Katsu-ura-cho.

Ishikawa Pref., Shiramine-mura; Tetori Group, Itoshiro Subgroup (Valanginian) and Akaiwa Subgroup (Barremian to Aptian) Two theropod teeth and several ornithischian teeth have been reported (Azuma, 1991). Two footprints each of an iguanodont and a theropod have also been found (Azuma and Takeyama, 1991; Azuma, 1993). These co-occur with a number of ganoid scales and turtle shell fragments.

Fukui Pref., Katsuyama City; Tetori Group, Akaiwa Subgroup (Barremian to Aptian) A dromaeosaurid (claws, several phalanges, humerous, ulna, femur, tibia, metatarsus, astragalus, maxilla, dentary, and isolated teeth) and an iguanodontid (premaxilla, maxilla, dentaries, angular, quadrate, parietal, several isolated teeth, cervical and dorsal vertebrae, synsacrum, caudal vertebrae, phalanges, etc.) are known. The dromaeosaurid is an extremely large species of the family and is unusual among Early Cretaceous coelurosaurs. The iguanodontid may be reconstructed to be about 5 m in length, and some of its skull characters are considered in between *Iguanodon* and *Probactrosaurus*. More than 300 bones and teeth from taxa other than the previously mentioned two species have been excavated, including teeth and vertebrae of theropods, spoon-shaped teeth of a sauropod, and teeth and bones of a few ornithischian taxa. In addition to the dinosaurs, numerous ganoid scales and turtle shell fragments and a nearly complete skeleton of a crocodile have been found. All these fossils were collected from a sandstone bed 1.5–2 m thick.

Several horizons about 1.5 m above the dinosaur-bearing sandstone contain a number of footprints of dinosaurs and other vertebrates, including birds (Azuma, 1993). Dinosaur footprints include theropods, sauropods, and ornithopods, and some of them are continuous trackways. Ornithopod footprints are mainly iguanodontids, and they range from 18 to 70 cm in length, probably representing different growth stages. Furthermore, these iguanodont footprints include both two-legged and four-legged gaits. The largest theropod footprint is 68.5 cm in length, indicating the presence of a large carnosaur.

Fukui Pref., Izumi-mura; Tetori Group, Itoshiro Subgroup (Valanginian) and Akaiwa Subgroup (Barremian to Aptian) Several footprints of theropods and ornithopods, as well as a number of bird footprints, have been found (Azuma *et al.*, 1992; Azuma and Takeyama, 1991; Tomida and Azuma, 1992).

Gifu Pref., Shirakawa and Shokawa-mura; Tetori Group, Itoshiro Subgroup (Valanginian) Iguanodontid footprints preserved on a ripple-marked sandstone bed (Kunimitsu *et al.*, 1990), and teeth of theropods, a hypsilophodontid, and another ornithopod have been found (Gifu Pref. Dinosaur Research Group, 1992a,b; Hasegawa *et al.*, 1990).

Yamaguchi Pref., Shimonoseki City; Toyonishi Group, Kiyosu-e Fm. (Latest Jurassic or Earliest Cretaceous) Footprints of an iguanodontid and a theropod were recently reported to occur on a sandstone bed (Okazaki, 1994), but we believe that both sets of footprints are iguanodontids. The geologic age of the fossil-bearing bed is uncertain; most likely it is earliest Cretaceous.

Fukuoka Pref., Miyata-cho; Kammon Group, Sengoku Fm. (Neocomian) An incomplete tooth of a large theropod has been found. A new genus and species of Megalosauridae, *Wakinosaurus satoi*, was proposed based on this poorly preserved specimen (Okazaki, 1992a). However, because no diagnostic characters taxonomically distinguish the specimen from other theropods at the genus and family levels, this proposed name should be abandoned. It is best identified as Theropoda indet.

Fukuoka Pref., Kitakyushu City; Kammon Group. Wakino Subgroup (Neocomian) Several teeth of a small theropod, a sauropod tooth, and two ceratopsian-like teeth have been reported (Okazaki, 1992b), co-occurring with fragmentary bones of turtles and crocodiles and ganoid scales.

Kumamoto Pref., Mifune-cho; Mifune Group (Cenomanian) Discoveries of a tooth (Hasegawa *et al.*, 1992) and a metatarsal, tibia, and phalanx of possibly the same individual of an allosaurid-like theropod (Tamura *et al.*, 1991) have been reported, but both identifications are questionable. Turtles, crocodiles, a pterosaur bone, and a jaw of an eutherian mammal also co-occurred.

Other Localities Discoveries of a hadrosaur femur and ischium from the Ezo Group (Coniacian) of Obira-cho, northwestern Hokkaido (Fig. 1, 13A) and an iguanodontid tooth fragment from the Tazukawa Formation of Monobe Group (late Hauterivian to Barremian) of Katsu-ura-cho Tokushima Pref. (Fig. 1, 13B) have been announced in the newspapers.

Summary

Japanese dinosaur fossils, except in a very few cases, are mostly isolated teeth and bones. Therefore, taxonomic studies are very diffcult, and current identifications in the literature include a number of doubts and questions. The dinosaur localities other than the Tetori Group (Fig. 1. 4–8), Kammon Group (Fig. 1, 10 and 11), and Mifune Group (Fig. 1, 12) are located within the Outer Belt of the Japanese tectonic divisions. The localities in the Outer Belt are of shallow marine sediments, and the dinosaur fossils are allochthonous. Therefore, there is not much hope for more dinosaur fossils from those localities. On the other hand, the Tetori, Kammon, and Mifune groups are terrestrial sediments, and dinosaur fossils are more or less autochthonous. Thus, more dinosaur material can be expected from further excavations. Also, dinosaur fossils from those groups are all from the Early Cretaceous, and it is expected that in the near future they will provide information on the dinosaurs and paleoenvironments of the Early Cretaceous, which is sometimes called the "twilight zone" of dinosaur fossils.

The Tetori Group is subdivided into three subgroups (Kuzuryu, Itoshiro, and Akaiwa in ascending order) and shows a characteristic transition of sedimentary environments from marine to freshwater through brackish water environments (Maeda, 1961). The Kuzuryu Subgroup has yielded only a skeleton of a land lizard, *Tedorosaurus asuwaensis* (Shikama, 1969). Dinosaur fossils have been found from the upper part of the Itoshiro Subgroup and the Akaiwa Subgroup. They include theropods, sauropods, and ornithopods, and important members include a dromaeosaurid and an iguanodontid. Co-occurring taxa include crocodiles, turtles, ganoid fishes, and birds. The dinosaur fauna of the Tetori Group is Early Cretaceous in age, and it represents a fauna that lived in a wet lowland area in a warm, humid climate located on the Asian continental edge. It also represents the dinosaur fauna of the Far East, and its significance should be emphasized.

See also the following related entry:

CHINESE DINOSAURS

References

Azuma, Y., (1991). Early Cretaceous dinosaur fauna from the Tetori Group Central Japan—Research of dinosaurs from the Tetori Group (1). In *Professor Shizuka Miura Memorial Volume*, pp. 55–69. Fukui Univ., Fukui, Japan. [In Japanese with English abstract]

Azuma, Y. (1993). The Early Cretaceous dinosaur ichnofauna and its paleoenvironmental development in the Tetori Group, Japan. Unpublished doctoral dissertation, pp. 155. University of Tokyo, Tokyo.

Azuma, Y., and Takeyama, K. (1991). Dinosaur footprints from the Tetori Group, central Japan—Research of dinosaurs from the Tetori Group (4). *Bull. Fukui Prefectural Museum* **4,** 33–51. [In Japanese with English abstract]

Azuma, Y., Sugimori, T., Yamada, K., Kojima, T., and Takeyama, K. (1992). Two dinosaur footprints from the Tetori Group of Izumi Village, Fukui Prefecture, central Japan. *Bull. Jpn Sea Res. Inst. Kanazawa Univ.* **24,** 19–33. [In Japanese with English abstract]

Gifu Prefecture Dinosaur Research Group (1992a). The Tetori Group in Oshirakawa area, Shirakawa-mura, Gifu Prefecture, Japan. *Bull. Gifu Prefectural Museum* **13,** 1–8. [In Japanese]

Gifu Prefecture Dinosaur Research Group (1992b). The Tetori Group in Ogamigo area, Shokawa-mura, Gifu Prefecture, Japan. *Bull. Gifu Prefectural Museum* **13,** 9–16. [In Japanese]

Goto, M. (1992). Dinosaur footprint from Ooyama Town, Toyama Prefecture. Abstracts of the 99th Annual Meeting of the Geological Society of Japan, p. 41. [In Japanese]

Hasegawa, Y., Kase, T., and Nakajima, S. (1984). A large vertebrate fossil from the Sanchu Graben. Abstracts of the 91st Annual Meeting of the Geological Society of Japan, p. 219. [In Japanese]

Hasegawa, Y., Kouda, Y., Watanabe, T., Oshida, K., Takizawa, A., and Suzuki, S. (1987). Dinosaur fossils from the Futaba Group, Hirono, Fukushima Prefecture. Abstracts of the 136th Regular Meeting of the Palaeontological Society of Japan, p. 4. [In Japanese]

Hasegawa, Y., Okura, M., and Manabe, M. (1990). Smaller dinosaur, Hypsilophodon tooth from Gifu Prefecture. Abstracts of the 139th Regular Meeting of the Palaeontological Society of Japan, p. 36. [In Japanese]

Hasegawa, Y., Manabe, M., Hanai, T., Kase, T., and Oji, T. (1991). A diplodocid dinosaur from the Early Cretaceous Miyako Group of Japan. *Bull. Natl. Sci. Museum Tokyo Ser. C* **17,** 1–9.

Hasegawa, Y., Murata, M., Wasada, K., and Manabe, M. (1992). The first carnosaur (Saurischia; Theropoda) from Japan; A tooth from the Cenomanian Mifume Group of kyushu. *Sci. Rep. Yokohama Natl. Univ. Ser. 2* **39,** 41–49.

Kunimitsu, M., Shikano, K., Sugiyama, M., and Hasegawa, Y. (1990). Dinosaur footprints from the Tetori Group, Shirakawa Village, Gifu Prefecture. Abstracts of the 1990 Annual Meeting of the Palaeontological Society of Japan, p. 101. [In Japanese]

Maeda, S. (1961). On the geological history of the Mesozoic Tetori Group in Japan. *J. College Arts Sci. Chiba Univ.* **3,** 369–426. [In Japanese with English abstract]

Matsukawa, M.. and Obata, I. (1985). Dinosaur footprints and other indentation in the Cretaceous Sebayashi Formation, Sebayashi, Japan. *Bull. Natl. Sci. Museum Ser. C* **11,** 9–36.

Nagao, T. (1936). *Nipponosaurus sachalinensis*—A new genus and species of trachodont dinosaur from Japanese Saghalien. *J. Faculty Sci. Hokkaido Imperial Univ. Ser. IV* **2,** 187–220.

Okazaki, Y. (1992a). A new genus and species of carnivorous dinosaur from the Lower Cretaceous Kwanmon Group, northern Kyusyu. *Bull. Kitakyushu Museum Nat. History* **11,** 87–90.

Okazaki, Y. (1992b). Dinosaur fossil locality newly found from the Kwanmon Group. Abstracts of the 1992 Annual Meeting of the Palaeontological Society of Japan, p. 15. [In Japanese]

Okazaki, Y. (1994). Footprints of Jurassic dinosaur from the coast at Yoshimo, Yamaguchi Prefecture. Abstracts of the 143rd Regular Meeting of the Palaeontological Society of Japan, p. 48. [In Japanese]

Russell, D. A., and Zheng, Z. (1993). A large mamenchisaurid from the Junggar Basin, Xinjiang, People's Republic of China. *Can. J. Earth Sci.* **30,** 2082–2095.

Shikama, T. (1969). On a Jurassic reptile from Miyamacho, Fukui Prefecture, Japan. *Sci. Rep. Yokohama Natl. Univ. Section 2* **15,** 25–34.

Tamura, M., Okazaki, Y., and Ikegami, N. (1991). Occurrence of carnosaurian and herbivorous dinosaurs from upper formation of Mifune Group, Japan. *Mem. Faculty Education Kumamoto Univ. Nat. Sci.* **40,** 31–45. [In Japanese with English abstract]

Tomida, Y., and Azuma, Y. (1992). The earliest known bird footprints from the lower most Cretaceous, Tetori Group, Japan. *J. Vertebr. Paleontol.* **12,** 56A. [Abstracts]

Jaws

see SKULL, COMPARATIVE ANATOMY; TEETH AND JAWS

Judith River Wedge

DAVID A. EBERTH
Royal Tyrrell Museum of Palaeontology
Drumheller, Alberta, Canada

The Judith River Formation and Judith River Group are important dinosaur-bearing units in the northern portion of the Western Interior of the United States and southern portion of the Western Canada Sedimentary Basin, respectively. These units comprise the down-dip portion of a multilobed, eastward-thinning, nonmarine to shallow marine clastic wedge (herein termed the Judith River wedge) that is exposed discontinuously in river drainages from central Alberta and west-central Saskatchewan (south of 54°) through north- and south-central Montana. In the subsurface, the wedge forms a nearly continuous unit across this same region interfingering with eastward-thickening marine shales of the Pakowki, Lea Park, and Claggett formations and Pierre Shale (below) and the Bearpaw Formation (above). Following Cant and Stockmal (1989) the wedge can be viewed as having been deposited as a result of major tectonic and basin-response events related to the accretion of the Insular Superterrane along the northwestern margin of North America during the Campanian.

Stratigraphy

The western Canadian portion of the wedge, the Judith River Group, comprises in ascending order the Foremost, Oldman, and Dinosaur Park formations. In north-central Montana the wedge is referred to as the Judith River Formation and is the lithostratigraphic equivalent of the Oldman Formation of southern Alberta. In the terminology of Kaufman (1977) the Judith River wedge records the R8 and T9 cycles. In the sequence stratigraphic terminology of Haq *et al.* (1988) it records the lower of the lower Zuni A supercycle (cycles 4.1–4.3). In the southern Alberta plains, the base of the Judith River Group has been correlated with the base of the 33n polarity chron and the *Baculites asperiformis* biozone (Lerbekmo, 1989). In this same area, the top has been correlated with the *Baculites compressus* biozone (see Eberth *et al.*, 1990). Single crystal laser fusion $^{40}Ar/^{39}Ar$ analyses of sanidine grains from suites of bentonites from both outcrop and core suggest that the base of the Judith River Group in southeastern Alberta is slightly older than 79 Ma, whereas the top of the unit is approximately 74 Ma (Eberth and Deino, 1992). Whereas the top of the Judith River Formation/Group is widely regarded as a chronostratigraphic datum east to west across Alberta (the Lethbridge coal zone/Bearpaw contact), the base is known to be strongly diachronous, becoming progressively older to the west and comprising an offlapping succession of parasequences (Hamblin, 1995). In north-central Montana the base of the Judith River Formation correlates with the base of the Oldman Formation in southeastern Alberta and has been dated at approximately 78 Ma (Eberth and Deino, 1992). In north-central Montana the top of the Judith River Formation has been dated at approximately 75 Ma (Rogers, 1994). Thus, exposures of the Judith River Formation in north-central Montana can be regarded as a more basinward equivalent of the Judith River Group in Canada. Farther south into Wyoming the Mesaverde Formation is the name given to correlative clastics of the Judith River wedge. Using multistep marine biozone correlations with European stratotype sections, Lillegraven and McKenna (1986) concluded that nonmarine mammal assemblages from the Mesaverde Formation (and thus, other vertebrate assemblages from the Judith River wedge) are late Campanian to early Maastrichtian in age.

Stratigraphic Nomenclature

USA The terms Judith Group and Judith River Group were first applied to the nonmarine sediments cropping out along the Judith and Missouri rivers in north-central Montana by Hayden (1871) and Meek (1876), respectively. Stanton and Hatcher (1905) were the first to successfully correlate, on a lithostratigraphic basis, the "Judith River beds" of Montana with Dawson's Belly River series along the Milk River in southern Alberta. Bowen (1915) first applied the term Judith River Formation to Stanton and Hatcher's Judith River beds and confirmed the Late Cretaceous age of the unit throughout Montana and into the southern plains of Alberta. The terminology in Montana has remained unchanged since 1915.

Canada The Campanian nonmarine clastic wedge of the southern Alberta Plains was first referred to as the Belly River series by Dawson (1883), who inadvertently included within it the underlying Pakowki marine shales and sediments that are today included in the Milk River Group. Since Dawson's time, a confusing and complex history of stratigraphic nomenclature for these beds has ensued (see McLean, 1971, for details) with sporadic attempts to resolve the problem. Russell and Landes (1940) assigned a portion of these sediments (between 49° and 50° latitude) to two units, the Foremost and Oldman formations. McLean (1971) considered the Foremost and Oldman formations difficult to distinguish and argued against their usage, assigning these clastics instead to the Judith River Formation. His decision to apply the name Judith River, rather than Belly River, to these sediments was based on priority of the former name and its consistent usage in the western United States for almost 100 years. Eberth and Hamblin (1993) divided McLean's Judith River Formation into the Foremost, Oldman, and Dinosaur Park formations and elevated the Judith River Formation to group status. Although this system has gained popularity among outcrop geologists, usage within western Canada remains inconsistent. Current practice within the resource industry is to refer to the entire wedge as the Belly River Group. Outcrop geologists and paleontologists prefer the terms Judith River Group, Judith River Formation, and Oldman Formation for fossiliferous exposures in the southern Alberta Plains.

Correlations

In Canada, the Judith River Group correlates westward with the Belly River Group in the southern Foothills (Jerzykiewicz and Norris, 1994) and the lower Brazeau Formation in the central Foothills (Fig. 1). North of approximately 54° latitude, the Judith River Group and Edmonton Group clastic wedges become indistinguishable; there the combined Campanian–Maastrichtian wedge is referred to as the Wapiti Formation. Eastward into Saskatchewan, the lowest portion of the Judith River Group, the Foremost Formation, is represented by isolated parasequences nested within marine shales of the Pakowki (Lea

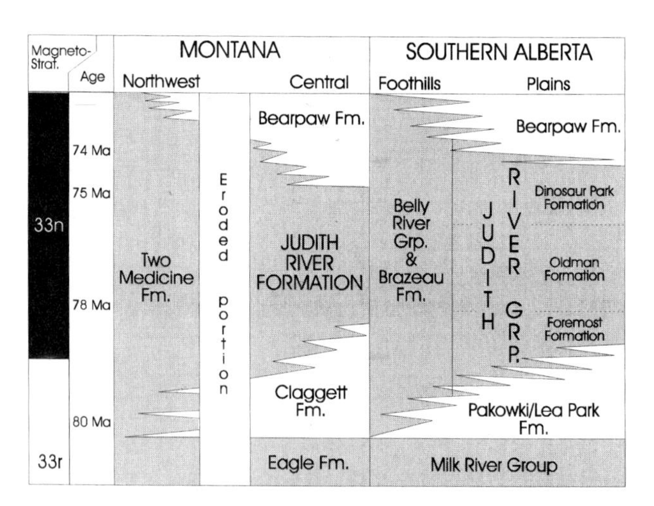

FIGURE 1 Stratigraphic correlation chart of the Judith River wedge in southern Alberta and Montana.

Park) Formation (Kwasniowski, 1993). Notable among these is the Ribstone Creek Member, an important hydrocarbon reservoir. The Oldman Formation is an important water reservoir throughout southwestern Saskatchewan and extends basinward in the subsurface as far as Regina, Saskatchewan. The Judith River Formation of north-central Montana correlates lithostratigraphically, sedimentologically, petrologically, and petrographically with the Oldman Formation of southern Alberta. With the exception of exposures of the Judith River Formation along the Milk River near Havre, along Kennedy Coulee south of the International border, and in Wheatland County, Montana (Fiorillo, 1991), there are no successions similar to Canada's Foremost or Dinosaur Park formations. In Montana, the Judith River Formation correlates westward with the more time-inclusive Two Medicine Formation (Fig. 1; Lorenz, 1981; Rogers, 1994). Farther south into Wyoming, correlative nonmarine strata are included in the more time-inclusive Mesaverde Formation (e.g., Lillegraven and McKenna, 1986).

Sediments and Paleoenvironments

The Judith River wedge comprises a variety of sediment and facies types that reflect a variety of sources, areas, depositional settings, and climatic influences along the Western Interior seaway during the Campanian. In general, the sediments range from litharenites to volcanic litharenites (e.g., Eberth and Hamblin, 1993; Rogers, 1995). In Canada, the Foremost and Dinosaur Park formations are arranged in offlapping and onlapping parasequence sets, respectively (Hamblin and Abrahamson, 1993; Hamblin, 1994). Offshore, shoreface, foreshore, barrier, estuarine, back-barrier, lagoonal, swamp, marsh, deltaic, high-sinuosity fluvial channel, and floodplain facies are all represented in varying abundance in these successions and have been described in numerous local and site-specific reports (see Eberth and Hamblin, 1993, for the most recent review). The Oldman Formation of Canada and the Judith River Formation of central Montana west of approximately 109° longitude comprise alluvial sediments deposited by a system of low-sinuosity rivers (Eberth and Hamblin, 1993; Noad, 1993). In Montana, east of 109° longitude, the Judith River Formation exhibits abundant nearshore to shoreline facies arranged in a lower succession of offlapping progradational parasequences and an upper succession of vertically aggraded, onlapping parasequences (Rogers, 1995).

Paleontology and Paleoecology

The paleontologic resources of the Judith River wedge have been recognized since the 1870s. Although the great abundance and broad diversity of dinosaurs from the Judith River wedge are world renowned, these beds also yield an impressive and important array of nondinosaurian vertebrates (Table 1) and invertebrates, micro- and macroplants, and traces (see Eberth and Brinkman, 1994). Whereas the literature on the systematics and taxonomy of fossils from the Judith River wedge is vast, there have been relatively few paleoecologic studies (Dodson, 1971; Beland and Russell, 1978, 1979; Currie and Dodson, 1984; Koster, 1987; Wood *et al.*, 1988; Fiorillo, 1991). Recent paleoecologic studies have been carried out using refined stratigraphic and sedimentologic data (e.g., Brinkman, 1990; Eberth, 1990; Rogers, 1995; Eberth and Brinkman, 1996).

Within the three formations of the Judith River Group in Canada, dinosaur fossils are most abundant in the Dinosaur Park Formation, a prolific fossil-producing unit that has yielded almost 300 species of Campanian age plants and animals. Currently, approximately 40 species of dinosaur are known from the wedge, and although taxonomic questions still remain, new discoveries continue to be made almost yearly. The Dinosaur Park Formation is unique among Campanian age nonmarine strata throughout the world in the abundance of articulated and associated dinosaur remains that it contains. More than 300 articulated to associated finds have been made from the small stretch of badlands (75 km^2) along the Red Deer River in and around Dinosaur Provincial Park alone. Eberth (1994) attributed this abundance to a unique combination of tectonic activity and its influences on the depositional parameters within an estuarine setting. Fiorillo (1991) described an unusually rich vertebrate assemblage from Careless Creek, Montana, and attributed fossil abundance there to a paleologjam within a fluvial setting.

Although yielding far fewer articulated vertebrate specimens than the Dinosaur Park Formation, the Oldman Formation of Canada and its lithostratigraphic equivalent, the Judith River Formation in

TABLE I Fossil Vertebrate Assemblage from the Judith River Wedge

- Chondrichthyes
  - Holocephali
    - Chimaeriformes
      - Rhiochimaeridae
        - *Elasmodus* cf. *greenoughi*
  - Elasmobranchii
    - Selachii
      - Hybodontidae
        - *Hybodus montanensis*
    - Euselachii
      - Odontaspidae
        - *Odontaspis sangiunei*
      - Cretoxyrhinidae
        - *Plicatolamna arcuata*
      - Synechodontidae
        - *Eucrossorhinus*
      - Carchariidae
        - *Odontaspis*
      - Rhinobatidae
        - *Protoplatyrhina*
      - Sclerorhynchidae
        - *Ischyrhiza* cf. *avonicola*
        - *Ischyrhiza mira*
        - *Myledaphus bipartitus*
- Osteichthyes
  - Acipenseriformes
    - Acipenseridae
      - *Acipenser albertensis*
    - Polyodontidae
      - Gen. et sp. indet.
  - Aspidorhynchiformes
    - Aspidorhynchidae
      - *Belonostomus longirostris*
  - Lepisosteiformes
    - Lepisosteidae
      - *Atractosteus occidentalis*
  - Amiiformes
    - Amiidae
      - *Kindleia*
      - Gen. et sp. indet.
  - Semionotiformes
    - incertae sedis
      - Gen. et sp. indet. (Holostean A)
      - Gen. et sp. indet. (Holostean B)
  - Elopiformes
    - Elopidae
      - *Paratarpon apogerontus*
    - Albulidae
      - *Coriops amnicolus*
  - ?Elopiformes
    - Phylodontidae
      - *Paralbula casei*
  - Osteoglossomorpha
    - Hyodontidae
      - Gen. et sp. indet.
  - Salmoniformes
    - Esocideae
      - Gen. et sp. indet.
  - Acanthopterygii
    - incertae sedis
      - Gen. et sp. indet.
- Amphibia
  - Salientia
    - Ascaphidae
      - Gen. et sp. nov.
    - Discoglossidae
      - cf. *Scotiophyrne*
    - Pelobatidae
      - *Eopelobates*
  - Caudata
    - Scapherpetontidae
      - *Scapherpeton tectum*
      - cf. *Lisserpeton bairdi*
    - Amphiumidae
      - *Proamphiuma*
    - Batrachosauroididae
      - *Opisthotriton kayi*
      - *Prodesmodon copei*
    - Sirenidae
      - *Habrosaurus dilatus*
  - Allocaudata
    - Albanerpetontidae
      - *Albanerpeton* sp.
- Reptilia
  - Chelonia
    - Baenidae
      - *Neurankylus eximus*
      - *Boremys pulchra*
      - *Plesiobaena antiqua*
    - Macrobaenidae
      - *"Clemmys"* cf. *bachmani*
    - Dermatemydidae
      - *Adocus lineolatus*
      - *Basilemys variolosa*
    - Trionychidae
      - *Aspideretes foveatus*
      - *Aspideretes alleni*
    - Chelydridae
      - Gen. et sp. indet.
  - Sauropterygia
    - Polycotylidae
      - Gen. et sp. indet.
  - Archosauromorpha
    - Champsosauridae
      - *Champsosaurus natator*
  - Sauria
    - Teiidae
      - *Chamops*
      - *Leptochamops denticulatus*
    - Scincidae
      - *Sauriscus*
    - Anguidae
      - *Pancelosaurus*
    - Xenosauridae
      - Gen. et sp. indet.
    - Necrosauridae
      - *Parasaniwa*
      - *Paraderma*
      - *Colpodontosaurus*
    - Varanidae
      - *Palaeonsaniwa canadensis*
    - incertae sedis
      - Gen. et sp. nov.
  - Ophidia
    - Aneliidae
      - *Coniophis*
  - Crocodylia
    - Alligatoridae
      - *Albertochampsa langstoni*
      - *Brachychampsa*
    - Crocodylidae
      - *Leidyosuchus canadensis*
  - Pterosauria
    - Pterodactyloidea
      - *Quetzalcoatlus* sp.
      - Gen. et sp. indet.
  - Saurischia
    - Segnosauridae
      - cf. *Erlikosaurus*
    - Ornithomimidae
      - *Dromiceiomimus samueli*
      - *Ornithomimus edmontonicus*
      - *Struthiomimus altus*
    - Troodontidae
      - *Troodon formosus*
      - *Stenonychosaurus inequalis*
      - *Pectinodon bakkeri*
    - Dromaeosauridae
      - *Dromaeosaurus albertensis*
      - *Saurornitholestes langstoni*
    - cf. Dromaeosauridae
      - *Paronychodon lacustris*
    - Caenagnathidae
      - *Caenagnathus collinsi*
      - *Caenagnathus sternbergi*
    - cf. Caenagnathidae
      - *Chirostenotes pergracilis*
      - *Macrophalangia canadensis*
    - Elmisauridae
      - *Elmisaurus elegans*
    - Tyrannosauridae
      - *Albertosaurus libratus*
      - *Daspletosaurus torosus*
      - *gracile tyrannosaur*
      - *Aublysodon* sp.
  - Ornithischia
    - Hadrosauridae
      - *Brachylophosaurus canadensis*

continues

Continued

Gryposaurus notabilis
Kritosaurus incurvimanus
Prosaurolophus maximus
Corythosaurus casuaris
Lambeosaurus lambei
Lambeosaurus magnicristatus
Parasaurolophus walkeri
Thescelosauridae
Thescelosaurus cf. *neglectus*
Pachycephalosauridae
Gravitholus albertae
Ornatotholus browni
Pachycephalosaurus sp.
Stegoceras validus
Full-domed form
Nodosauridae
Panoplosaurus mirus
Edmontonia longiceps
Ankylosauridae
Euoplocephalus tutus
Protoceratopsidae
cf. Leptoceratops
Ceratopsidae
Anchiceratops sp.
Centrosaurus apertus
Chasmosaurus belli
Chasmosaurus kaiseni
Eoceratops canadensis
Monoclonius lowei
Styracosaurus albertensis
Aves
cf. Neognathidae
Apatornithidae
Apatornis sp.
Mammalia
Multituberculata
Neoplagiaulacidae
Cimexomys judithae
Mesodma primaeva
Cimolodontidae
Cimolodon electus
Cimolodon similis
Cimolomyidae
Cimolomys clarki
Meniscoessus major
Gen. et sp. indet.
Theria incertae sedis
Deltatheridiidae
cf. *Deltatheroides*
Marsupialia
Didelphidae
Alphadon praesagus
Alphadon halleyi
Alphadon sp.
Pediomyidae
Pediomys clemensi
Pediomys sp.
Stagodontidae
Boreodon matutinus
Eodelphis cutleri
Eodelphis browni
Eutheria / Insectivora
Leptictoidea
Gypsonictops lewisi
cf. Nyctitheriidae
Paranyctoides sternbergi
Palaeoryctidae
Cimolestes sp.
Gen. et sp. Indet.

Montana, have yielded hundreds of vertebrate "microfossil" localities that consist of the disarticulated and concentrated physicochemical remains of vertebrates (including dinosaurs). Important paleoecological studies based on these vertebrate microfossil assemblages have been published by Sahni (1972), Brinkman (1990), Rogers (1995); and Eberth and Brinkman (1996). Rogers (1995) explained some of these occurrences in the context of a sequence stratigraphic analysis.

Whereas vertebrates comprise the most numerous and dramatic component of the Judith River wedge's fossil assemblage, oysters are common in the lowermost and uppermost Judith River beds, and nonmarine mollusks are frequently encountered within the Oldman and Judith River formations. Widespread penecontemporaneous dissolution of aragonite appears to have been responsible for removing the latter faunal component from the Foremost and Dinosaur Park Formations.

High-quality macroplant fossils are uncommon within the Judith River Group and are best represented by coal beds and silicified tree trunks. The latter commonly occur as allochthonous components of channel sandstones, especially in the Dinosaur Park Formation. Carbonaceous leaf and leaf-impression fossils occur locally but are rare. In contrast, the combined palynologic and microplant assemblage currently comprises at least 170 taxa (see Eberth and Brinkman, 1994).

See also the following related entries:

CANADIAN DINOSAURS • CRETACEOUS PERIOD • DEVIL'S COULEE • EGG MOUNTAIN • TWO MEDICINE FORMATION

References

Beland, P., and Russell, D. A. (1978). Paleoecology of Dinosaur Provincial Park (Cretaceous), Alberta, interpreted from the distribution of articulated vertebrate remains. *Can. J. Earth Sci.* **15,** 1012–1024.

Beland, P., and Russell, D. A. (1979). Ectothermy in dinosaurs: Paleoecological evidence from Dinosaur Provincial Park, Alberta. *Can. J. Earth Sci.* **16,** 250–255.

Bowen, C. F. (1915). The stratigraphy of the Montana Group with special reference to the position and age of the Judith River Formation in north-central Montana.

United States Geological Survey, Professional Paper 90-I, pp. 95–154.

Brinkman, D. B. (1990). Paleoecology of the Judith River Formation (Campanian) of Dinosaur Provincial Park, Alberta, Canada: Evidence from vertebrate microfossil localities. *Palaeogeogr. Palaeoclimatol. Palaeoecol.* **78,** 37–54.

Cant, D. J., and Stockmal, G. S. (1989). The Alberta foreland basin: Relationship between stratigraphy and Cordilleran terrane-accretion events. *Can. J. Earth Sci.* **26,** 1964–1975.

Currie, P. J., and Dodson, P. (1984). Mass death of a herd of ceratopsian dinosaurs. In *Third Symposium of Mesozoic Terrestrial Ecosystems, Short Papers* (W. E. Reif and F. Westphal, Eds.), pp. 61–66. Attempto Verlag, Tubigen, Germany.

Dawson, G. M. (1883). Preliminary report on the geology of the Bow and Belly River region, North-west Territory, with special reference to coal deposits. Geological and Natural History Survey and Museum of Canada, Report of Progress, 1880–82, Pt. B, pp. 1–23.

Dodson, P. (1971). Sedimentology and taphonomy of the Oldman Formation (Campanian), Dinosaur Provincial Park, Alberta (Canada). *Palaeogeogr. Palaeoclimatol. Palaeoecol.* **10,** 21–74.

Eberth, D. A. (1990). Stratigraphy and sedimentology of vertebrate microfossil sites in the uppermost Judith River Formation (Campanian), Dinosaur Provincial Park, Alberta, Canada. *Palaeogeogr. Palaeoclimatol. Palaeoecol.* **78,** 1–36.

Eberth, D. A. (1994). Tectonics and taphonomy: Why does Alberta have so many dinosaurs? *Reservoir (Can. Soc. Petroleum Geol.)* **21,** 2–3. [Abstract]

Eberth, D. A., and Brinkman, D. B. (1994). Sequence stratigraphy, depositional environments and paleoecology of the Judith River Group (Upper Cretaceous), southern Alberta. Field Trip Guidebook PR-3, Canadian Society of Petroleum Geologists, Annual Meeting, Calgary, Alberta, pp. 69.

Eberth, D. A., and Brinkman, D. B. (1996). Paleoecology of an estuarine, incised-valley fill in the Dinosaur Park Formation (Judith River Group) of southern Alberta, Canada. *Palaios* **11.**

Eberth, D. A., and Deino, A. L. (1992). A geochronology of the non-marine Judith River Formation of southern Alberta; SEPM 1992 Theme Meeting, Mesozoic of the Western Interior, Ft. Collins, Abstracts with Programs, p. 24.

Eberth, D. A., and Hamblin, A. P. (1993). Tectonic, stratigraphic, and sedimentologic significance of a regional discontinuity in the upper Judith River Formation (Belly River Wedge) of southern Alberta, Saskatchewan, and northern Montana. *Can. J. Earth Sci.* **30,** 174–200.

Eberth, D. A., Braman, D. B., and Tokaryk, T. T. (1990). Stratigraphy, sedimentology and vertebrate paleontology of the Judith River Formation (Campanian) near Muddy Lake, west-central Saskatchewan. *Bull. Can. Petroleum Geol.* **38,** 387–406.

Fiorillo, A. R. (1991). Taphonomy and depositional setting of Careless Creek Quarry (Judith River Formation), Wheatland County, Montana, USA. *Palaeogeogr. Palaeoclimatol. Palaeoecol.* **81,** 281–311.

Hamblin, A. P. (1994). The Dinosaur Park Formation of the Upper Cretaceous Judith River (Belly River) Group, subsurface of southern Alberta. Geological Survey of Canada, Open File Report No. 2831, pp. 17.

Hamblin, A. P. (1995). Stratigraphic architecture of the Campanian Belly River (Judith River) Group, southern Alberta Plains, surface and subsurface. Geological Survey of Canada, Open File Report No. 3058, pp 33–37.

Hamblin, A. P., and Abrahamson, B. (1993). Offlapping progradational cycles and gas pool distribution in the upper Cretaceous 'Basal Belly River' sandstones, Judith River Group, southern and central Alberta. Geological Survey of Canada, Open File Report No. 2672, pp. 19.

Haq, B. U., Hardenbol, J., and Vail, P. R. (1988). Mesozoic and Cenozoic chronostratigraphy and cycles of sea-level change. In *Sea-Level Changes: An Integrated Approach* (C. K. Wilgus, B. S. Hastings, C. G. St. C. Kendall, H. W. Posamentier, C. A. Ross, and J. C. Van Wagoner, Eds.), Spec. Publ. 42, pp. 71–108. Society of Economic Paleontologists and Mineralogists.

Hayden, F. V. (1871). Geology of the Missouri Valley. Preliminary report of the United States Geological Survey of Wyoming and portions of contiguous territories (a second annual report of progress), pp. 85–188.

Jerzykiewicz, T., and Norris, D. K. (1994), Stratigraphy, structure and syntectonic sedimentation of the Campanian 'Belly River' clastic wedge in the southern Canadian Cordillera. *Cretaceous Res.* **15,** 367–399.

Kauffman, E. G. (1977). Geological and biological overview: Western Interior Cretaceous Basin. *Mountain Geol.* **14,** 75–99.

Koster, E. H. (1987). Vertebrate taphonomy applied to the analysis of ancient fluvial systems. In *Recent Developments in Fluvial Sedimentology* (F. G. Ethridge and R. M. Flores, Eds.), Spec. Publ. 39, pp. 159–168. Society of Economic Paleontologists and Mineralogists.

Kwasniowski, S. M. (1993). Effects of sea level variation on the Upper Cretaceous (Campanian), Foremost Formation in southeastern Alberta and southwestern Saskatchewan. Geological Association of Canada, Annual Meeting, Edmonton; Abstracts with Programs, p. A-55.

Lerbekmo, J. F. (1989). The stratigraphic position of the 33–33r (Campanian) polarity chron boundary in southeastern Alberta. *Bull. Can. Petroleum Geol.* **37,** 43–47.

Lillegraven, J. A., and McKenna, M. C. (1986). Fossil mammals from the "Mesaverde" Formation (Late Cretaceous, Judithian) of the Bighorn and Wind River basins, Wyoming, with definitions of Late Cretaceous North American land-mammals "ages." *Am. Mus. Novitates* **2840,** 1–68.

Lorenz, J. C. (1981). Sedimentary and tectonic history of the Two Medicine Formation, Late Cretaceous (Campanian), northwestern Montana. Unpublished Ph.D. thesis, pp. 215. Department of Geology, Princeton University, Princeton, NJ.

McLean, J. R. (1971). Stratigraphy of the Upper Cretaceous Judith River Formation in the Canadian Great Plains. Saskatchewan Research Council, Geology Division, Report No. 11, pp. 96.

Meek, F. B. (1876). A report on the invertebrate Cretaceous and Tertiary fossils of the upper Missouri country. United States Geological Survey of the Territories, Vol. 9.

Noad, J. J. (1993). An analysis of changes in fluvial architecture through the upper Oldman member, Judith River Formation of the Upper Cretaceous, southern Alberta, Canada. Unpublished M. Sc. thesis. Birbeck College, University of London, London.

Rogers, R. R. (1994). Nature and origin of through-going discontinuities in nonmarine foreland basin strata, Upper Cretaceous, Montana: Implications for sequence analysis. *Geology* **22,** 1119–1122.

Rogers, R. R. (1995). Unpublished Ph.D. thesis. University of Chicago, Chicago.

Russell, L. S., and Landes, R. W. (1940). Geology of the southern Alberta plains. Geological Survey of Canada, Memoir No. 221, pp. 223.

Sahni, A. (1972). The vertebrate fauna of the Judith River Formation, Montana. *Bull. Am. Museum Nat. History* **147,** 321–412.

Stanton, T. W., and Hatcher, J. B. (1905). Geology and paleontology of the Judith River beds. United States Geological Survey, Bull. No. 257.

Wood, J. M., Thomas, R. G., and Visser, J. (1988). Fluvial processes and vertebrate taphonomy: The Upper Cretaceous Judith River Formation, south-central Dinosaur Provincial Park, Alberta, Canada. *Palaeogeogr. Palaeoclimatol. Palaeoecol.* **66,** 127–143.

Jurassic Park

Mary Schweitzer
Montana State University
Bozeman, Montana, USA

Don Lessem
Dinosaur Society
Waban, Massachusetts, USA

Based on the best-selling book by Michael Crichton, the film *Jurassic Park* was released in 1993. The film promoted the theory of dinosaurs as the ancestors of birds and helped reverse stereotypical impressions of dinosaurs as dim-witted, swamp-bound sluggards. Among the press and general public of all ages, it engineered a fresh interest in dinosaurs and dinosaur science, with broad current and future repercussions for the future of science education, support of paleontology, and commercial development of dinosaur products. Through the Dinosaur Society, the film's director Stephen Spielberg provided support for dinosaur research at the Institute for Vertebrate Paleontology and Paleoanthropology (IVPP) in Beijing. The IVPP's senior dinosaur paleontologist, Dong Zhiming, named a Jurassic ankylosaur *Tianchisaurus nedegoaperferkimorum* in honor of the cast. The tongue-twisting species name was an acronym for the first letters of the last name of the film's performers. Most substantively, Mr. Spielberg authorized the Dinosaur Society to construct an educational exhibit, "The Dinosaurs of Jurassic Park," to generate funding for dinosaur research.

The best science fiction is said to walk the line between the current and the unrealized, and this is certainly the case with *Jurassic Park*, a work that seems to be only a few steps ahead of the headlines of scientific journals.

One of the first premises of the movie is that DNA, the "molecule of life" that codes the genetic composition of living organisms, can be preserved for 65–100 million years. There have been reports of DNA recovered from extinct organisms, such as the woolly mammoth or the horse-like quagga, but these specimens are not truly "old" in a geological sense. DNA has been reported to have been recovered from specimens hundreds of millions of years old, but all of these reports still await independent verification. Whether DNA can be preserved for millions of years is still an issue very much open to debate and has by no means been accepted by the general scientific community.

In *Jurassic Park*, a Cretaceous insect that bit a dinosaur became trapped in amber and preserved along with its blood meal. Mosquitoes and other biting insects have indeed been identified in amber that dates to the Cretaceous. It is not unreasonable to suggest that some of these may have fed on dinosaurs, although this would not have been their only choice of meal. The closest living relatives of the dinosaurs, the birds and crocodilians, have red blood cells that (unlike mammals) retain their nuclei with a full complement of DNA. Thus, it is reasonable to assume that if such a blood meal were preserved, DNA would be preserved as well. However, from at least 65 million years later, could we tell if the DNA belonged to a dinosaur, bird, or other closely related animal? Distinguishing which dinosaur the blood belonged to would be almost impossible. Comparing the amber-derived DNA with dinosaurs' closest living relatives would not necessarily shed much light on the issue because these modern descendants have had at least 65 million years to accumulate molecular changes after the dinosaurs died out.

There are other difficulties in bringing an animal to life from bits of ancient DNA. In part this has to do with the degraded state of DNA recovered from any ancient tissue, even those only a few hundred years old. Human DNA contains approximately 108 base pairs, or "letters" of the DNA alphabet. About the maximum length of DNA we have been able to verifiably retrieve from ancient tissues is 102 base pairs, several orders of magnitude less than is contained in any living vertebrate—indeed at least an order of magnitude smaller than the simplest bacterium. Additionally, this ancient DNA contains base modifications and other chemical changes that make the enzymes we use to replicate it inefficient and prone to error. However, suppose that we could somehow overcome these huge stumbling blocks in our quest to bring back the dinosaurs. The *Jurassic Park* scientists "filled in" missing pieces of DNA with frog DNA to reach the number of bases needed for a complete organism. This solved the problem of overcoming an all-female population because in certain circumstances, frogs may literally change sexes if the population is overwhelmingly one sex. However, frogs are very distant from dinosaurs from a phylogenetic standpoint. Their developmental stages vary greatly from those of "higher" vertebrates, and using frog DNA would cause far greater problems than it would solve. In our quest to regenerate the dinosaurs, we could use phylogenetically closer DNA, taken from modern archosaurs (crocodiles and birds). This is not currently feasible, given that the entire genome of any living vertebrate has not yet been sequenced, but it is conceivable that we could do this in the not-too-distant future. Then we would have a complete set of DNA blueprints, supposedly to code for a dinosaur (see GENETICS).

Species are identified, in part, by chromosome number. There is probably a wide range of possibilities for dinosaur chromosome numbers based on number comparisons with living taxa, and we have no way of knowing how many chromosomes a *T. rex* or a *Triceratops* possessed. Having the correct chromosome number is vital to the successful development of an animal, and even an extra little bit of chromosome has a very deleterious, often lethal effect. Not only would it be necessary to know the chromosome number for each dinosaur but also we would need to know which genes belong on each chromosome, and in which order these genes are arrayed. Very often in development, the controlling genes are arranged linearly on the chromosome, in the order in which they are turned on and off. Disruption of this order results in severe, and again, often lethal deformities. However, if these obstacles could all be overcome, and we could have a complete, ordered, whole blueprint of a dinosaur in the right order with the right chromosome number, we would still have some big problems.

Currently, we can "clone" an organism in one of two ways. First, we can induce a fertilized egg to split and then develop independently as two individuals. This process is similar to that which gives rise to identical twins. Alternatively, we can remove a fertilized nucleus from one cell and insert it into an anucleate cell from a similar or identical organism and then place this cell into the proper environment for development. This latter method would have to be used with our dinosaur genetic material. However, with dinosaurs, we have no cells in which to insert our manufactured blueprint. It is not just the DNA that is crucial to correct development. Hormonal and environmental cues act upon the developing cells in order to provide signals for the production of developmental proteins and physiological changes. Sometimes this environment is strongly temperature dependent, but we do not know the temperatures at which dinosaurs incubated their EGGS. Again, we would only have living relatives as a guide, and maybe this educated guesswork would be sufficient for some dinosaur taxa, but certainly not all of them.

However, suppose that all these problems could be overcome. Suppose that we could find enough DNA to ensure that it was really dinosaurian; that we could piece together the vast majority of DNA that was missing or degraded or damaged from pieces of bird and crocodile DNA; that we could learn how many chromosomes a tyrannosaur, a *Triceratops*, an *Apatosaurus*, or a *Gallimimus* possessed; that we could correctly guess the placement of every single base pair with the correct number of intervening, noncoding sequences, on each chromosome; that we could find the appropriate "host" cell to receive this genetic cocktail; and that we could find the right combination of temperatures, light, hormones, and other protein signals to ensure proper development, to the point that we could produce a living dinosaur embryo. The odds against this occurring are astronomical, but even if it did, we would still be a very long way from Jurassic Park (or, more accurately, Cretaceous Park). The dinosaur's genetic programming would be designed to deal with the environment, the foods, the enzymes, the diseases, and the parasites of the world as it was at least 65 million years ago. If it was a plant eater, the enzymes produced in its digestive tract would be designed to break down plants with which it had coevolved. Sixty-five million years of changes in plant proteins, cell walls, and toxins would have occurred in the plants descended from those on which the dinosaur fed. Most of the plants that made up its DIET are long extinct.

One could make the argument that if one could overcome the obstacles to resurrect a dinosaur, one could certainly resurrect the plants that made up its diet, but this will not solve this issue completely. The interactions of microorganisms that inhabit the digestive tract of all living herbivores and aid in digestion are incredibly complex, sometimes species specific. If these microorganisms infested the gut of a dinosaur they did not coevolve with, the chances are very great that it would result in severe or lethal diseases. Dinosaur dependence on lysine supplied by their caretakers would be the least of their dietary problems. If the resurrected dinosaur was a carnivore, these problems would be intensified. It is unlikely that they would possess the enzymes that would allow them to subsist on a diet of cow or sheep, let alone lawyers. The problems of living on a diet to which one is not adapted was illustrated when the Europeans placed Native Americans on reservations and fed them a diet based on wheat flour instead of the native grains and roots that had been their staple for the 10,000 or 12,000 years they had been isolated from their ancestral Asian populations. The Native Americans sickened, and many died as a result. This would be multiplied many times over in the dinosaurs, in which isolation from new food sources would be effectively millions, not thousands of years.

Therefore, manufacturing dinosaurs to be lysine dependent would probably not be a necessity for controlling these animals and preventing their escape. However, even this premise is in error. Lysine is defined as an essential amino acid. No animal can produce it, and all must obtain it from foods they eat. Fortunately, this amino acid can be easily obtained by including plant material in a diet. Because the tropical island on which Jurassic Park was established was well covered with vegetation, obtaining lysine would not be any more of a problem for escaping dinosaurs than it would be for cows, if they had the enzymes to digest the plants.

Although *Jurassic Park* is an imaginative scenario, future research in diverse fields such as paleontology, ecology, embryology, genetics, geology, molecular biology, and preservation processes, to name a few, may bring us ever closer to understanding the world of the dinosaurs. Imagination in the form of *Jurassic Park* lets us visualize the world of the dinosaurs. The years of research and the real science behind the scenes gave these visionaries their starting point. For the critical viewer, a successful blending of science, art, and fiction has opened a window into a world long gone.

See also the following related entry:

POPULAR CULTURE, LITERATURE

Jurassic Period

PETER DODSON
University of Pennsylvania
Philadelphia, Pennsylvania, USA

The Jurassic Period is the middle of the three periods of the Mesozoic Era, and extends from approximately 208 to 146 Ma. The earth's climate was generally warm in the Jurassic, with subtropical climates perhaps as far north as 60° N latitude, but seasonal aridity is also a striking attribute. Although rifting began in the Triassic, significant breakup of Pangea was not evident until the Middle Jurassic, with the opening of the North Atlantic. The Gondwanan continents

(South America, Africa, Australia, India, and Antarctica) remained in substantial contact throughout the Jurassic. In the Late Jurassic, Laurasia was somewhat fragmented, inasmuch as much of Europe was an archipelago. Nonetheless, this archipelago seemingly offered stepping stones allowing limited faunal exchange between Africa and North America. Gymnosperms were the dominant large trees, including diverse conifers and ginkgoes, and bennettitaleans were dominant small trees and shrubs. Ferns were the dominant herbaceous plants. The last nonmammalian carnivores lived in the Jurassic and the first mammals and birds appeared. Frogs, salamanders, turtles, lizards, and crocodilians either appeared or radiated during the Jurassic. Pterosaurs radiated into long-tailed and short-tailed, toothed and toothless types. In the Jurassic, dinosaurs are found on all seven continents, recently having been discovered in Antarctica. During the Jurassic, dinosaurs became dominant terrestrial herbivores and carnivores, with sauropods attaining gigantic size. Significant Early Jurassic dinosaur assemblages of low diversity are known from southern Africa, India, China, and southwestern United States; more or less isolated finds come from Europe. Important Middle Jurassic assemblages come from China, Madagascar, Europe, and Argentina. Important Late Jurassic assemblages come from China, western United States, Europe, and Tanzania. Sauropods were typically the most abundant dinosaurs of their communities, with stegosaurs also being important. The last prosauropods are widespread in Early Jurassic assemblages. Small and medium-sized ornithopods were subordinant in Jurassic dinosaur communities. Ankylosaurians were rare elements. Theropods include a variety of small and medium-sized ceratosaurs, megalosaurs, and allosaurids and diverse small maniraptorans.

See also the following related entries:

Cretaceous Period • Geologic Time • Mesozoic Era • Mesozoic Floras • Triassic Period

References

Behrensmeyer, A. K., Damuth, J. D., DiMichele, W. A., Potts, R., Sues, H.-D., and Wing, S. L. (Eds.) (1992). *Terrestrial Ecosystems Through Time*, pp. 568. Univ. of Chicago Press, Chicago.

Weishampel, D. B., Dodson, P., and Osmolska, H. (Eds.) (1990). *The Dinosauria*, pp. 733. Univ. of California Press, Berkeley.

Kagoshima Prefectural Museum, Japan

see MUSEUMS AND DISPLAYS

Kanagawa Prefecture Living Planet and Earth Museum, Japan

see MUSEUMS AND DISPLAYS

Kasaoka City Museum, Japan

see MUSEUMS AND DISPLAYS

Kayenta Formation

see GLEN CANYON GROUP

Kazakhstan Dinosaurs

see MIDDLE ASIAN DINOSAURS

Kenyan Dinosaurs

see AFRICAN DINOSAURS

Keuper Formation

HARTMUT HAUBOLD
Martin Luther Universität
Halle, Germany

The Upper Triassic (uppermost Ladinian to Norian) continental sediments of the Germanic Basin in central Europe, mainly in Germany, are referred to as the Keuper Formation, or simply the Keuper. The environment was in the process of becoming semiarid, although the dinosaur-bearing deposits originated in fluvial channel systems. These horizons are dominated by sandstones, mudstones, and paleosols. Important dinosaur occurrences of the middle Keuper are known from the lower to upper Stubensandstein, from the Knollenmergel, and from the Feuerletten of middle Norian age. Prosauropods include *Sellosaurus gracilis, Thecodontosaurus? hermannianus*, and the very common *Plateosaurus engelhardti*. The rarer theropods are *Halticosaurus* and *Procompsognathus* of the middle Stubensandstein of Pfaffenhofen in Wuerttemberg and *Liliensternus* in the Knollenmergel in Thuringia. Mass accumulations of nearly monospecific content are excavated from the so-called *Plateosaurus* bone beds near Trossingen in Wuerttemberg, near Halberstadt in Sachsen-Anhalt, near Frick in Switzerland (Sander, 1992), and near Weissenburg in Bavaria (Wellnhofer, 1994). Taphonomic studies formerly interpreted these as of eolian origin—*Plateosaurus* herds dying in desert storms. Recent sedimentological studies relate the Trossingen sites to mud flows and mudhole traps, and most other localities, to fluvial accumulations. On the eastern margin of the Germanic Basin, the *Plateosaurus*–conglomerate is widespread in Franconia (Bavaria). Upper Keuper, or Rhaetic, deposits in western Europe, eastern France, and England have produced the remains of prosauropods and theropods: *Camelotia, Thecodontosaurus,* and *Liliensternus.* Dinosaur records earlier than mid-Norian from the Keuper are very fragmentary and their identification is questionable.

See also the following related entries:

European Dinosaurs • Triassic Period

References

Sander, P. M. (1992). The Norian *Plateosaurus* bonebeds of central Europe and their taphonomy. *Palaeogeogr. Palaeoclimatol. Palaeoecol.* **93,** 255–299.

Wellnhofer, P. (1994). Prosauropod dinosaurs from the Feuerletten (middle Norian) of Ellingen near Weissenburg in Bavaria. *Rev. Paleobiol.,* Spec. Vol. 7 (1993), 263–271.

Khodja-Pil Ata

Martin Lockley
University of Colorado at Denver
Denver, Colorado, USA

A Late Jurassic dinosaur tracksite at the village of Khodja-Pil Ata in eastern Turkmenistan reveals one of the largest dinosaur footprint-bearing surfaces in the world, measuring about 300 × 100 m (= 30,000 m^2). The tracks are those of two types of carnivorous dinosaurs that have been assigned the names "*Megalosauropus*" and "*?Therangospodus.*" The longest trackway can be followed for 311 m, making it by far the longest recorded trackway on earth. Preliminary research indicates that the same two track types are found in rocks of the same age in Europe and North America.

See also the following related entries:

Carenque • Fatima • Footprints and Trackways • Middle Asian Dinosaurs

Kingman Museum of Natural History, Michigan, USA

see Museums and Displays

Kirtland Formation

Michael J. Ryan
Royal Tyrrell Museum of Palaeontology
Drumheller, Alberta, Canada

The Maastrichtian age Kirtland Formation (see Fig. 1 under Fruitland Formation) in the San Juan Basin of northwestern New Mexico was originally named by Bauer (1916) as the Kirtland Shale. It erosionally overlies the late Campanian (Judithian) Fruitland Formation and is separated from the overlying Ojo Alamo Sandstone by a disconformity. Excellent reviews of this formation include Hunt and Lucas (1992, 1993). Throughout the Cretaceous, New Mexico was located at the western margin of an epicontinental seaway that stretched from the Gulf of Mexico to the Arctic Ocean. The upper Campanian Fruitland Formation represents a variety of alluvial plain environments. As the inland sea regressed toward the end of the Cretaceous, the fluvial floodplains prograded, depositing almost 600 m of mudstone, siltstone, shale, sandstone, and conglomerate. Vertebrate fossils are found throughout almost the entire column. The lenticular paleochannels in the lower Kirtland Formation represent moderately well-drained overbank environments with low to intermediate velocity, highly braided channels. Paleoflow in the southwest of the formation was to the northeast, whereas in the north it was to the southeast, suggesting a concentric drainage pattern (Hunt and Lucas, 1992).

The one upper and four lower members of this formation contain separate dinosaur assemblages with at least 12 and 9 dinosaurs, respectively. In ascending order, the lower early Maastrichtian (Edmontonian) fauna occurs in the Bisti, Hunter Wash, Farmington Sandstone, and De-na-zin members. The fauna consists of two chondrichthyans, four osteichthyans, 14 chelonians, four crocodylians, three mammals, and at least 9 dinosaurs. The dinosaurs are as follows: Saurischia: Ornithomimidae (cf. *Struthiomimus* sp.) Dromaeosauridae indet., Tyrannosauridae (*Albertosaurus* sp. and *Aublysodon* cf. *A. mirandus*); Ornithischia: Nodosauridae (?*Euplocephalus* sp. indet.), Ankylosauria indet., Ceratopsidae (*Pentaceratops sternbergii*), Hadrosauridae [*Anasazisaurus horneri* (= *Kritosaurus navajovius* Horner 1992)], and Lambeosauridae (*Parasaurolophus* sp.). This part of the Kirtland preserves a few rare partial skeletons and skulls, with *Aublysodon* known from at least two incomplete skeletons (Archer and Babiarz, 1992). The lower Kirtland Formation is intermediate in diversity and composition between the coastal plain fauna of the Fruitland Formation and the inland vertebrate upper Kirtland fauna.

The late Maastrichtian (Lancian) upper fauna is restricted to the uppermost Naashoibito Member and is present from the basal conglomerate to within 3 m below the overlying Ojo Alamo Sandstone. The distinctive purple mudstones of the Naashoibito Member are separated from the underlying four members of the Kirtland in a nonconforming manner. This member represents deposition by low-sinuosity meandering and braided streams associated with well-drained floodplains. Well-preserved fossils are confined to meandering stream channels. Vertebrate remains from this member are known as the Alamo Wash local fauna and refer to material collected from the Alamo Wash region. This fauna consists of two osteichthyans, 10 chelonians, two crocodylians, five mammals, and at least 13 dinosaurs. These include the following: Saurischia: Tyrannosauridae (?*Albertosaurus* sp. and cf. *Tyrannosaurus* sp.), Ornithomimidae indet., Dromaeosauridae indet., ?Troodontidae indet., and Titanosauridae (*Alamosaurus sanjuanensis*); Ornithischia: Hadrosauridae [*Naashoibitosaurus ostromi* (= *Kritosaurus navajovius* Horner 1992 and *Edmontosaurus saskatchewanensis* Hunt and Lucas 1992)], Lambeosauridae (*Parasaurolophus tubicen*), Ankylosauridae indet., Nodosauridae (?*Panoplosaurus* sp. and Nodosauridae indet.), and Ceratopsidae (*Torosaurus* cf. *T. latus* and Ceratopsidae indet.). Of note is the *Pentaceratops* (Lehman, 1993). The consensus is that the juvenile frill fragment from the Naashoibito Member previously referred to *Pentaceratops* cannot be confidently referred to this species. Hence, *Pentaceratops*, although endemic to the San Juan Basin, is not present in the Kirtland and is known only from a single species in the Fruitland Formation.

The Alamo Wash fauna is dominated by turtles, hadrosaurs, and ceratopsians. This reflects both the scarcity of microvertebrate localities in the Kirtland and the fact that the Kirtland was better drained than the Fruitland and would not have been as favorable to fish and herptiles. The dinosaurs found here comprise largely disarticulated and isolated material, with partial skeletons and skulls being rare. *Tyrannosaurus* is identified on the basis of a single isolated tooth and ?*Albertosaurus* from a single metatarsal. *Parasaurolophus* represents one of the youngest occurrences of this dinosaur. *Naashoibitosaurus ostromi* n. gen et n. sp. (Hunt and Lucas, 1993) is known from a partial skull and skeleton and possesses a *Gryposaurus*-like nasal arch but differs from *Gryposaurus* in other nasal features. Ankylosaurs are known from scutes with the nodosaurid material resembling the dorsal scutes of *Edmontonia*.

See also the following related entries:

Cretaceous Period • Edmonton Group • Fruitland Formation • Hell Creek Formation • Lance Formation • Two Medicine Formation

References

Archer, B., and Babiarz, J. P. (1992). Another tyrannosaurid dinosaur from the Cretaceous of northwest New Mexico. *J. Paleontol.* **66,** 690–691.

Bauer, C. M. (1916). Contributions to the geology and paleontology of San Juan County, New Mexico. 1. Stratigraphy of a part of the Chaco River Valley. United States Geological Survey, Professional Paper No. 98P, pp. 271–278.

Horner, J. R. (1992). Cranial morphology of *Prosaurolophus* (Ornithischia: Hadrosauridae) with descriptions of two new hadrosaurid species and an evaluation of hadrosaurid phylogenetic relationships. Museum of the Rockies, Occasional Paper No. 2, pp. 110.

Hunt, A. P., and Lucas, S. G. (1992). Stratigraphy, paleontology and the age of the Fruitland and Kirtland formations (Upper Cretaceous), San Juan Basin, New Mexico. In *San Juan Basin IV* (S. J. Lucas, B. S. Kues, T. E. Williamson, and A. P. Hunt, Eds.), Guidebook No. 43, pp. 217–239. New Mexico Geology Society.

Hunt, A. P., and Lucas, S. G. (1993). Cretaceous vertebrates of New Mexico. In *Vertebrate Paleontology in New Mexico* (S. G. Lucas and J. Zidek, Eds.), Bull. 2, pp. 77–91. New Mexico Museum of Natural History.

Lehman, T. M. (1993). New data on the ceratopsian dinosaur *Pentaceratops sternbergii* Osborn from New Mexico. *J. Paleontol.* **67,** 279–288.

Kitakyushu Museum of Natural History, Japan

see MUSEUMS AND DISPLAYS

Kyoto Municipal Science Center for Youth, Japan

see MUSEUMS AND DISPLAYS

Kyrgyzstan Dinosaurs

see MIDDLE ASIAN DINOSAURS

"Kritosaurus" australis is a hadrosaurine from the Cretaceous of Patagonia, unearthed from the Los Alamitos Formation, Rio Negro, Argentina. Photo by François Gohier.

Laboratoire de Paleontologie, Faculté des Sciences, France

see MUSEUMS AND DISPLAYS

Lambeosaurinae

A clade of the Hadrosauridae.

see HADROSAURIDAE

Lameta Formation

ASHOK SAHNI
Punjab University
Chandigarh, India

The Lameta Formation is a thin but extensive horizon that underlies the basal flows of the Deccan volcanic sedimentary sequence. The type section was designated by Medlicott in 1860, but it was left to Matley (1921) to work out in detail the excellent sequence exposed at Lameta Ghat on the banks of the Narbada river near Jabalpur. Since 1928, when the first dinosaur bones were noticed by a British army officer, numerous dinosaur fossils have been excavated and described. The most productive localities are in the Balasinor, Jabalpur, and Pisdura regions, traversing a linear E–W stretch of approximately 700 km. Lameta sediments represent deposits of a semiarid alluvial plain undergoing active pedogenesis. On the basis of its stratigraphic position, vertebrate assemblages, and associated microfossils (ostracods, charophytes, and pollens), the age of the Lameta Formation is now considered to be Maastrichtian.

See also the following related entries:

CRETACEOUS PERIOD • DECCAN BASALT • INDIAN DINOSAURS

Lambeosaurines like *Parasaurolophus* have been unearthed from Late Cretaceous sediments of North America (Judith River Formation and Fruitland Formation) and Asia. Photo by François Gohier.

Lance Formation

BRENT H. BREITHAUPT
University of Wyoming
Laramie, Wyoming, USA

The most prolific unit in Wyoming for the occurrence of Late Cretaceous vertebrate fossils is the Lance Formation. It is dominated by rocks of a regressive sequence consisting of nonmarine, coastal floodplain sandstones, mudstones, and marls, with marginal marine sandstones and shales in its lower parts. The latest Cretaceous environment in Wyoming was a warm temperate to subtropical, seasonal floodplain on the west coast of the eastward-regressing intracratonic seaway. The Lance Formation encompasses a fairly short period of geologic time (approximately 1.5 million years) at the end of the Maastrichtian. It reaches more than 750 m in thickness and is found throughout Wyoming. It was separated from strata of equivalent age by the developing Rocky Mountains, which divided the region into various restricted depositional basins. Because of the mammalian fauna found in the Lance Formation, this unit has been assigned to the Lancian "age." Much work has been and will continue to be done on this vertebrate fossil-rich formation in Wyoming. Although vertebrate fossils are known from 11 counties in Wyoming (Breithaupt, 1985), the most abundant and diverse faunas in the Lance Formation are from Niobrara (Hatcher *et al.*, 1907; Estes, 1964; Clemens, 1964, 1966, 1973; Derstler, 1994) and Sweetwater (Breithaupt, 1982) counties.

The Lance Formation contains one of the best known Late Cretaceous vertebrate faunas. The diverse fauna contains various cartilaginous and bony fishes, frogs, salamanders, champsosaurs, turtles, lizards, snakes, crocodiles, pterosaurs, mammals, birds, and some of the best known Cretaceous dinosaurs (e.g., *Triceratops, Torosaurus, Tyrannosaurus, Edmontosaurus, Pachycephalosaurus, Ankylosaurus, Edmontonia, Thescelosaurus, Troodon, Dromaeosaurus,* and *Ornithomimus*). Abundant remains of small vertebrates and ceratopsian dinosaurs were collected for O. C. Marsh by J. B. Hatcher during the years 1889–1894 (Hatcher *et al.*, 1907).

Named for a small drainage (Lance Creek) in the eastern part of Wyoming, the Lance Formation is best known for the exposures found in that region. However, the 1872 discovery of a partial skeleton of a dinosaur from the western part of the state by Drs. F. B. Meek and H. M. Bannister (while working for the Hayden Geological Survey of the Territories) was the first indication of the paleontological importance of this unit (Breithaupt, 1982, 1994). E. D. Cope (1872) collected and described the material and named a new species of dinosaur *Agathaumas sylvestris*. The presence of dinosaur remains in this unit were helpful in resolving the debate concerning the age of units of this type in the Rocky Mountain Region at the time (Knowlton, 1922). A Cretaceous, not Tertiary, age was strongly suggested for the Lance Formation (called the Laramie Formation in the 1870s) by this dinosaur discovery. This interpretation disputed earlier fossil floral studies regarding the age.

Agathaumas sylvestris was one of the first ceratopsian dinosaurs discovered and was considered in the early 1870s to be one of the largest animals ever to walk the earth. Currently, *Agathaumas* is thought to be a form of *Triceratops* (the most common horned dinosaur to have inhabited Wyoming) and sauropods hold the record for terrestrial vertebrate size. O. C. Marsh (1889) defined the genus *Triceratops* on material he had originally called *Ceratops horridus* from the Lance Formation of Niobrara County, Wyoming. The type area of the Lance Formation in east-central Wyoming has produced hundreds of *Triceratops* fossils, including at least 100 skulls (Derstler, 1994).

Since the discovery of *Agathaumas*, literally thousands of Late Cretaceous dinosaur remains have been recovered from the Lance Formation. Fossil vertebrate material ranging from important microscopic elements to extensive bone beds (with nearly complete, sometimes articulated dinosaur skeletons) are known. Spectacular specimens like the dinosaur "mummies" (hadrosaur skeletons surrounded by skin impressions) have also been found in the Lance Formation (Lull and Wright, 1942). The first partial skeleton of *Tyrannosaurus rex* (originally named *Dynamosaurus imperiosus*) was found in this unit in 1900.

Today, dinosaur fossils from Wyoming's Lance Formation highlight exhibit and research collections in museums throughout the world. Even after more than a century of work, important vertebrate paleontological discoveries are still being made from the Lance Formation in Wyoming.

See also the following related entries:

CRETACEOUS PERIOD • EDMONTON GROUP • HELL CREEK FORMATION

References

Breithaupt, B. H. (1982). Paleontology and paleoecology of the Lance Formation (Maastrichtian), east flank of Rock Springs Uplift, Sweetwater County, Wyoming. *Contrib. Geol.* **21,** 123–151.

Breithaupt, B. H. (1985). Nonmammalian vertebrate faunas from the Late Cretaceous of Wyoming. In *Thirty-Sixth Annual Field Conference Guidebook* (G. E. Nelson, Ed.), pp. 159–175. Wyoming Geological Association, Casper.

Breithaupt, B. H. (1994). The first dinosaur discovered in Wyoming. In *Forty-Fourth Annual Field Conference Guidebook* (G. E. Nelson, Ed.), pp. 15–23. Wyoming Geological Association, Casper.

Clemens, W. A. (1964). Fossil mammals of the type Lance Formation, Wyoming. Part I. Introduction and multituberculata. *Univ. California Publ. Geol. Sci.* **48,** 1–105.

Clemens, W. A. (1966). Fossil mammals of the type Lance Formation, Wyoming. Part II. Marsupialia. *Univ. California Publ. Geol. Sci.* **62,** 1–122.

Clemens, W. A. (1973). Fossil mammals of the type Lance Formation, Wyoming. Part III. Eutheria and summary. *Univ. California Publ. Geol. Sci.* **94,** 1–102.

Cope, E. D. (1872). On the existence of Dinosauria in the Transition Beds of Wyoming. *Proc. Am. Philos. Soc.* **12,** 481–483.

Derstler, K. (1994). Dinosaurs of the Lance Formation in eastern Wyoming. In *Forty-Fourth Annual Field Conference Guidebook* (G. E. Nelson, Ed.), pp. 127–146. Wyoming Geological Association, Casper.

Estes, R. (1964). Fossil vertebrates from the Late Cretaceous Lance Formation, eastern Wyoming. *Univ. California Publ. Geol. Sci.* **49,** 1–180.

Hatcher, J. B., Marsh, O. C., and Lull, R. S. (1907). The Ceratopsia. United States Geological Survey Monograph No. 49, pp. 300. United States Government Printing Office, Washington, DC.

Knowlton, F. H. (1922). The Laramie Flora of the Denver Basin. United States Geological Survey Professional Paper No. 130, pp. 175. United States Government Printing Office, Washington, DC.

Lull, R. S., and Wright, N. E. (1942). Hadrosaurian dinosaurs of North America. Geological Society of America Spec. Paper No. 40, pp. 1–242.

Marsh, O. C. (1889). Notice of gigantic horned Dinosauria from the Cretaceous. *Am. J. Sci.* **38,** 173–175.

Landesmuseum für Naturkund, Germany

see MUSEUMS AND DISPLAYS

Land-Mammal Ages

SPENCER G. LUCAS
New Mexico Museum of Natural History and Science
Albuquerque, New Mexico, USA

Wood *et al.* (1941) introduced the term land-mammal "age" (LMA) to refer to intervals of Tertiary time characterized by distinctive mammalian fossil assemblages from western North America. Currently, land-mammal "ages" have been introduced to encompass Cenozoic time intervals on most of the world's continents (Savage and Russell, 1983). For Mesozoic time, land-mammal "ages" have been proposed only for the Late Cretaceous of western North America. However, more broadly based land-vertebrate "ages" (or faunachrons) have been introduced for parts of the Mesozoic record of Asia, South America, and North America.

Biochronological Units

An aggregation of mammalian fossils in a body of sedimentary rock (strata) is an assemblage zone. Vertebrate paleontologists have long referred to such an assemblage zone as a fauna or local fauna. The geologic time equivalent to a fauna is a LMA. This means that LMAs are biochronological units—intervals of geologic time recognized by distinctive fossils (Tedford, 1970).

LMAs are not the ages of formal stratigraphic terminology. Such ages are intervals of geologic time equivalent to stages, which are bodies of strata. Each age is thus based directly on strata in a specific area; these strata are the stratotype of the stage age. However, LMAs lack such stratotypes, so to call them ages is to use that term in a different sense than is usual, which is why the term "age" should be placed in quotation marks.

Among vertebrates, not only mammals can be used to recognize intervals of geologic time. In the Mesozoic, dinosaurs, turtles, and some other kinds of vertebrates can be biochronologically useful. For this reason, some workers use the term land-vertebrate "age" (LVA). Because LMAs and LVAs are not formal ages in stratigraphy, Lucas (1993a) introduced the term faunachron to refer to the time equivalent to the duration of a fauna. He argued that the more precise term land-vertebrate faunachron should replace LMA and LVA.

Definition and Name

To define a LMA or LVA, a distinctive assemblage of vertebrate fossils must be identified. The name of the LMA or LVA is a geographic name taken from the place where (or very close to where) the vertebrate fossils were collected. Most LMA and LVA names have been taken from the rock formation in which the fossils are found. The rock formation name is based on a place name. However, using the rock formation name can cause confusion because it implies that the LMA or LVA refers to the duration of deposition of the formation and not just to the duration of the vertebrate fossil assemblage, which is often much shorter. It is less confusing to choose another place name for the LMA or LVA. For example, the Late Triassic Ischigualastian LVA of Argentina (Bonaparte, 1966) is named for the Ischigualasto Formation, although the Ischigualastian vertebrates do not occur throughout the Ischigualasto Formation. In contrast, the Late Triassic Adamanian LVA of western North America (Lucas and Hunt, 1993) is named after Adamana where the fossils occur and not after the Petrified Forest Formation, which contains the fossils.

Rationale

Phanerozoic (the past 570 million years) time is largely divided into intervals based on the stratigraphic ranges of marine invertebrates and microfossils (e.g., Harland *et al.*, 1990). Most correlations (determinations of equivalence of age or stratigraphic position) are at the stage-age level. For example, the last stage of the Cretaceous is the Maastrichtian, named after a town in Holland. The stratotype of the Maastrichtian is marine strata that contain ammonites (extinct cephalopod) and microfossils by which the Maastrichtian is readily recognized and correlated to marine strata elsewhere.

The last Mesozoic dinosaurs lived during the Maastrichtian, but their fossils almost always occur in rocks deposited by rivers and lakes that lack marine fossils. These rocks cannot be easily correlated to Maastrichtian marine strata. This leads to imprecision in correlation, especially when trying to assign a precise age within the Maastrichtian (which is about 5 million years long) to dinosaur fossils.

Rather than trying to correlate dinosaurs imprecisely with the Maastrichtian, time intervals based on dinosaur and associated nonmarine fossils can produce more precise correlations. LMAs and LVAs for nonmarine fossil assemblages can provide this precision. The Lancian LVA is the age of late Maastrichtian dinosaurs and other vertebrate fossils in western North America. Correlation of Lancian vertebrates can be made with more precision and certainty than can their correlation to the Maastrichtian. LMAs and LVAs thus provide a framework for correlation independent of marine fossils that can be very useful in biochronologically organizing nonmarine vertebrate fossil assemblages.

Problems

Some workers would rather not create the new concepts and terminology necessary to define LMAs and LVAs. They prefer to use the marine ages of the Mesozoic, no matter how imprecise correlation of terrestrial vertebrate fossil assemblages is to these ages.

Precise correlations using LMAs and LVAs are not always easy and can be problematic. Nevertheless, these problems are inherent to all correlation based on fossils and do not detract from the utility of LMAs and LVAs.

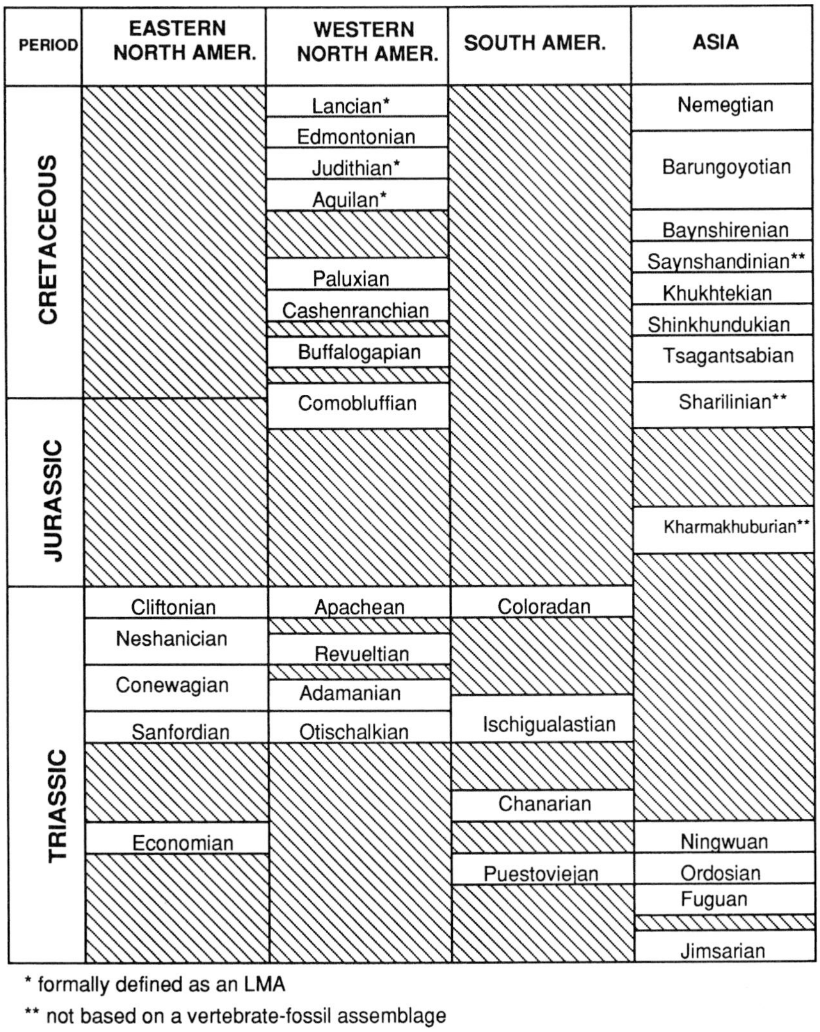

PERIOD	EASTERN NORTH AMER.	WESTERN NORTH AMER.	SOUTH AMER.	ASIA
CRETACEOUS		Lancian* Edmontonian Judithian* Aquilan* Paluxian Cashenranchian Buffalogapian Comobluffian		Nemegtian Barungoyotian Baynshirenian Saynshandinian** Khukhtekian Shinkhundukian Tsagantsabian Sharilinian**
JURASSIC				Kharmakhuburian**
TRIASSIC	Cliftonian Neshanician Conewagian Sanfordian Economian	Apachean Revueltian Adamanian Otischalkian	Coloradan Ischigualastian Chanarian Puestoviejan	Ningwuan Ordosian Fuguan Jimsarian

* formally defined as an LMA

** not based on a vertebrate-fossil assemblage

FIGURE 1 Only a small number of LVAS and LMAS (*) have been defined for the Mesozoic.

Mesozoic LVAs

Dorf (1942) and Russell (1964, 1975) first defined five LVAs for the Cretaceous of western North America. Lillegraven and McKenna (1986) redefined four of these as LMAs (Fig. 1). These four Late Cretaceous LMAs are the only formally defined Mesozoic LMAs.

The biochronological organization of the fossil record of Mesozoic vertebrates is still in its infancy. Other than the Cretaceous LMAs previously discussed, LVAs have only been proposed for the Triassic of China (Lucas, 1993b) and Argentina (Bonaparte, 1966), the Late Triassic of western (Lucas and Hunt, 1993) and eastern (Huber *et al.*, 1993) North America, the Late Jurassic–Early Cretaceous of western North America (Lucas, 1993c), and the Late Jurassic–Cretaceous of Mongolia (Jerzykiewicz and Russell, 1991) (Fig. 1). Russell (1993) proposed marine vertebrate ages for the Cretaceous of western North America.

Existing LVAs provide a good reflection of the level of temporal resolution possible using terrestrial vertebrates to subdivide geological time. Once an LVA is established, it challenges biochronologists to undertake further research to correlate and subdivide (if possible) the LVA. Cenozoic LMAs have so challenged biochronologists that a system of very precise and far-reaching correlations can now be undertaken using Cenozoic mammal fossils. Such correlations also can be achieved with Mesozoic vertebrates once LVAs are defined.

See also the following related entries:

Biostratigraphy • Geologic Time • Paleomagnetic Correlation • Radiometric Dating

References

Bonaparte, J. F. (1966). Cronologia de algunas formaciones triasicas de Argentina. Basad en restos de tetrapodos. *Asoc. Geol. Argentina Revista* **21,** 20–38.

Dorf, E. (1942). Upper Cretaceous floras of the Rocky Mountain region. II: Flora of the Lance Formation at its type locality, Niobrara County, Wyoming. Carnegie Institution of Washington Publ. No. 508, pp. 79–159.

Harland, W. B., Armstrong, R. L., Cox, A. V., Craig, L. E., Smith, A. G., and Smith, D. G. (1990). *A Geologic Time Scale 1989*, pp. 263. Cambridge Univ. Press, Cambridge, UK.

Huber, P., Lucas, S. G., and Hunt, A. P. (1993). Vertebrate biochronology of the Newark Supergroup Triassic, eastern North America. *New Mexico Museum Nat. History Sci. Bull.* **3,** 179–186.

Jerzykiewicz, T., and Russell, D. A. (1991). Late Mesozoic stratigraphy and vertebrates of the Gobi basin. *Cretaceous Res.* **12,** 345–377.

Lillegraven, J. A., and McKenna, M. C. (1986). Fossil mammals from the "Mesaverde" Formation (Late Cretaceous, Judithian) of the Bighorn and Wind River basins, Wyoming, with definitions of Late Cretaceous North American land-mammal "ages." *Am. Museum Novitates* **2840,** 1–68.

Lucas, S. G. (1991). Dinosaurs and Mesozoic biochronology. *Mod. Geol.* **16,** 127–138.

Lucas, S. G. (1993a). The Chinle Group: Revised stratigraphy and biochronology of Upper Triassic nonmarine strata in the western United States. *Museum Northern Arizona Bull.* **59,** 27–50.

Lucas, S. G. (1993b). Vertebrate biochronology of the Triassic of China. *New Mexico Museum Nat. History Sci. Bull.* **3,** 301–306.

Lucas, S. G. (1993c). Vertebrate biochronology of the Jurassic–Cretaceous boundary, North American Western Interior. *Mod. Geol.* **18,** 371–390.

Lucas, S. G., and Hunt, A. P. (1993). Tetrapod biochronology of the Chinle Group (Upper Triassic), western United States. *New Mexico Museum Nat. History Sci. Bull.* **3,** 327–329.

Russell, D. A. (1993). Vertebrates in the Cretaceous Western Interior sea. Geological Association of Canada Spec. Paper No. 39, pp. 665–680.

Russell, L. S. (1964). Cretaceous non-marine faunas of northwestern North America. *R. Ontario Museum Life Sci. Contrib.* **61,** 1–24.

Russell, L. S. (1975). Mammalian faunal succession in the Cretaceous System of western North America. Geological Association of Canada Spec. Paper No. 13, pp. 137–161.

Savage, D. E., and Russell, D. E. (1983). *Mammalian Paleofaunas of the World*, pp. 432. Addison-Wesley , Reading, MA.

Tedford, R. H. (1970). Principles and practices of mammalian geochronology in North America. *Proc. North Am. Paleontol. Convention* **1,** 666–703.

Wood, H. E., II, Chaney, R. W., Clark, J., Colbert, E. H., Jepsen, G. L., Reeside, J. B., Jr., and Stock, C. (1941). Nomenclature and correlation of the North American continental Tertiary. *Geol. Soc. Am. Bull.* **52,** 1–48.

Lanier Museum of Natural History, Georgia, USA

see Museums and Displays

Laotian Dinosaurs

see Southeast Asian Dinosaurs

Lark Quarry Environmental Park, Australia

see Museums and Displays

Las Hoyas

Bernardino P. Pérez-Moreno
José L. Sanz
Universidad Autónoma de Madrid
Madrid, Spain

The Las Hoyas *Konservat-lagerstätte* is located at the Serranía de Cuenca (placed at the Iberian Mountain Ranges), province of Cuenca, Spain. It is Barremian (Lower Cretaceous) in age, belonging to the Calizas de la Huérguina Formation, which is lacustrine and

FIGURE 1 *Pelecanimimus polyodon* Pérez-Moreno *et al.* 1994. LH-7777B (counterpart) without preparation. The material from Las Hoyas fossil site is usually preserved as a part and a counterpart, with the specimen roughly sagittally split.

palustrine in origin (Fregenal-Martínez, 1991; Fregenal-Martínez and Melíndez, 1993). Las Hoyas has yielded thousands of fossil remains of plants (e.g., cycads, ferns, conifers, and primitive angiosperms) and invertebrates (e.g., insects, crustaceans, and vertebrates) (Sanz *et al.*, 1988, 1990; Sanz and Buscalioni, 1994). Among the latter, articulated skeletons of fish, salamanders, frogs, lizards, crocodiles, nonavian dinosaurs, and birds are valuable specimens to clarify the evolutionary history of these groups during the Lower Cretaceous.

Nonavian dinosaur material from Las Hoyas has been relatively scarce up to now. The laminated limestones have provided an isolated theropod tooth. Fragmentary sauropod vertebrae have been found in massive limestone facies. Lateral detritic facies to the lacustrine limestones of Las Hoyas have yielded some *Iguanodon* bones in Buenache de la Sierra (Francís and Sanz, 1989).

The most significant nonavian dinosaur material from Las Hoyas was discovered in 1993: the anterior half of an ornithomimosaur skeleton (Fig. 1). The preserved bones include the skull (with the hyoid), complete cervical and almost complete dorsal vertebral series, ribs, pectoral girdle, a large sternum, and forelimbs. Some impressions that can be clearly observed behind and below the skull and around the neck, pectoral girdle, right humerus, and elbow probably correspond to integumentary structures. The name proposed for this ornithomimosaur is *Pelecanimimus polyodon* Pérez-Moreno *et al.* 1994, and its diagnosis is as follows:

> Small ornithomimosaur (2–2.5 m long). Skull with long and shallow snout (maximum length about 4.5 times the maximum height). About 220 teeth: 7 premaxillary, about 30 maxillary, and about 75 in the dentary. Heterodont. Maxillary teeth larger than the dentary teeth. Teeth unserrated, with a constriction between the crown and the root. There are no complete interdental plates. The maxilla has teeth only in its anterior third and a sharp edge in the posterior zone. The rostral half of the lower jaw has a straight ventral edge. Ulna and radius are tightly adhered distally. Metacarpal ratio is 0.81 : 1 : 0.98 (Pérez-Moreno *et al.*, 1994).

The most important traits that place *Pelecanimimus* inside Ornithomimosauria are related to the forelimb skeleton, especially the derived condition of the hand for such an ancient ornithomimosaur (Pérez-Moreno and Sanz, 1995).

One of the most striking differences between *Pelecanimimus* and other ornithomimosaurs is the presence of about 220 tiny teeth. The high number of teeth, along with its morphology, has allowed a new hypothesis of the evolutionary process toward the toothless condition in Ornithomimosauria, which is different from the process usually envisaged up to now, which suggested a progressive reduction in the number of teeth from the primitive tetanurine condition (up to 80 teeth with tall blade-like crowns). The phylogenetic hypothesis suggests an alternative evolutionary process based on a functional analysis of the character "increasing number of teeth." The elevated number of teeth would be an adaptation for cutting and ripping as long as the space between adjacent teeth was preserved (the condition present in the node BULLATOSAURIA Holtz 1994), while it would have the effect of working as a beak if spaces were filled by more teeth (in the node Ornithomimosauria). The adaptation to a cut-and-rip function therefore becomes an exaptation with a slicing effect, eventually leading to the cutting edge seen in most ornithomimosaurs. The tendency to substitute the dentition by a beak is shown by the posterior–anterior replacement of teeth by a cutting border on the maxilla of *Pelecanimimus*.

The avian fossil record of the Lower Cretaceous is relatively scarce. Nevertheless, the knowledge of the avian evolutionary history immediately after the *Archaeopteryx* times has been increased in recent years (Wellnhofer, 1994; Chiappe, 1995; see AVES and BIRD ORIGINS). The outcrops that have generated most new information are placed in China (Sereno and Rao, 1992; Zhou *et al.*, 1992) and Spain (Sanz *et al.*, 1988, 1995, 1996; Sanz and Bonaparte, 1992; Sanz and Buscalioni, 1992, 1994).

The current avian record from Las Hoyas includes four published skeletal remains and about half a dozen isolated feathers. The first bird skeleton was discovered about 10 years ago and named *Iberomesornis romerali* Sanz and Bonaparte 1992 (Fig. 2). The specimen is relatively small (the femur is about 15 mm long) and lacks the skull, anterior cervical vertebrae, and hands. *Iberomesornis* presents a singular combination of primitive and derived character states. The sacropelvic elements and hindlimbs are, in general, primitive. Both proximal and distal tarsal bones are free, neither fused to each other nor to the tibia or to the metatarsals. The metatarsals show no trace of fusion. *Iberomesornis* has a primitive number of sacral vertebrae (5; like *Archaeopteryx*), although the number of dorsal vertebrae (11) seems to be intermediate between the primitive condition present in *Archaeopteryx* (13 or 14) and that of modern birds (4–6). *Iberomesornis* shares three main derived character states with the clade ENANTIORNITHES + Ornithurae: the presence of free caudal vertebrae plus a pygostyle and avian furcula and coracoids. The pygostyle is relatively large, formed by about 10–15 vertebrae. The furcula has a low clavicular angle and a conspicuous styloid furcular process. The coracoid is

FIGURE 2 *Iberomesornis romerali* Sanz and Bonaparte 1992. LH-22. The specimen is exposed in left lateral view. Scale in millimeters.

strut like with a broad expansion for contact with the sternum.

The holotype of *Concornis lacustris* Sanz and Buscalioni 1992 is fairly complete (Fig. 3), lacking the skull and the neck. The specimen lies in ventral aspect with its limbs partially flexed. The sternum is deeply notched and keeled in its posterior region. The forelimbs retain a primitive character, the presence of claws, but they have proportions similar to those of modern birds. The vertebral centra are amphicoelous. The pubes form a distal symphysis. The tibiotarsus is fused just proximally. *Concornis* shares with the Enantiornithes 12 synapomorphies. Among the most significant enantiornithine traits present in *Concornis* are a humerus with a prominent bicipital crest, a coracoid with convex lateral margin, dorsal vertebral centra with strong lateral grooves, medial condyle of the distal end of the tibiotarsus bulbous, metatarsal IV significantly smaller than metatarsals II and III, and laterally excavated furcula.

The third avian specimen is a poorly preserved isolated foot, identified as cf. *Iberomesornis* sp. (Sanz

FIGURE 3 *Concornis lacustris* Sanz and Buscalioni 1992. LH-2814. The specimen is exposed in ventral view. Scale in millimeters.

and Buscalioni, 1994). The fourth one is a new taxon, *Eoalulavis hoyasi* Sanz *et al.* 1996 (Fig. 4). The specimen is composed of the posterior cervical and complete dorsal vertebral series, the thoracic girdle, both forelimbs, and part of the sacrum, pelvic girdle, and proximal end of the hindlimbs. *Eoalulavis* belongs to the Enantiornithes, as does *Concornis*, although its singular sternal morphology, along with other autopomorphies in the vertebrae and furcula, has allowed us to propose for it a new taxon. The sternum is unique even considering DINOSAURIA as a whole: It is lanceolate, with a foot-like caudal expansion, a faint carina, and a deep rostral cleft, in which the extremely long hypocleidium probably fit. This bony sternum could be the remnant of a larger cartilaginous structure (pointing toward the existence of a widely unossified large sternum in early birds, including *Archaeopteryx*). Besides the sternum, the most unusual features of *Eoalulavis* are preserved thanks to the excellent preservational properties of the Las Hoyas fossil site. The first one is the digestive contents found within a limonitic mass inside the thoracic cavity. Inside this limonitic mass, some organic particles have been identified as crustacean exoskeletal elements, providing the oldest known direct evidence of trophic habits in Aves. The second one is the preservation of the alula in natural position, the oldest known alula in the fossil record. This structure is fundamental in modern birds for low-speed flight (including take off

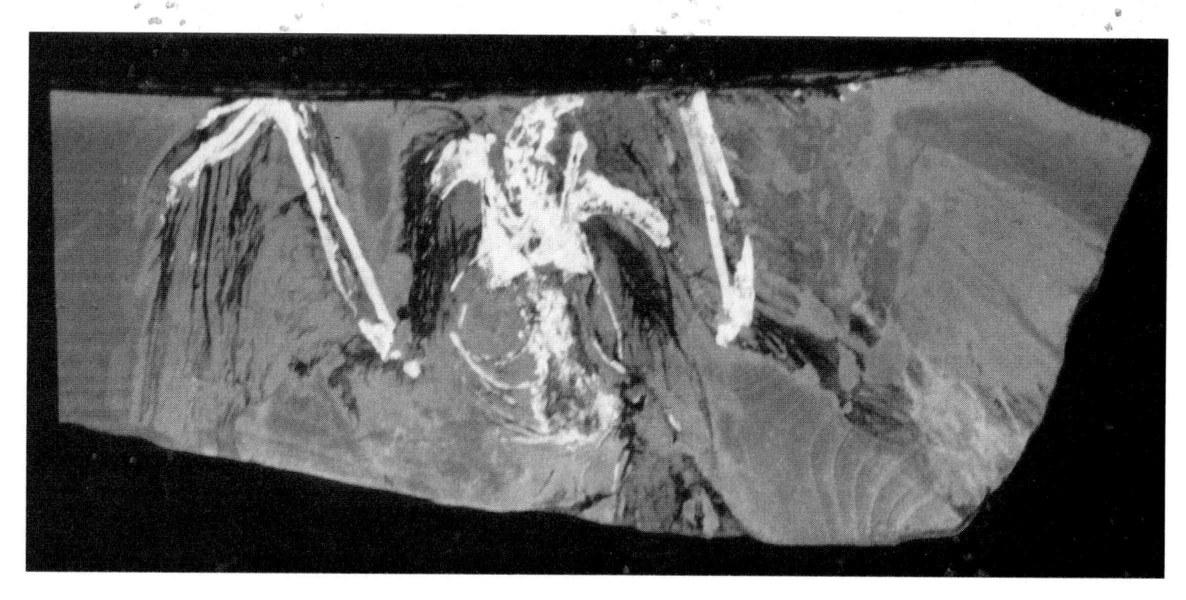

FIGURE 4 *Eoalulavis hoyasi* Sanz *et al.* 1996. Ultraviolet induced fluorescence photograph of LH-13500B (counterpart) before preparation. Feathers are visible in black clearly attached to the alular digit, major digit, forearm, and around the body.

and landing) and maneuverability. Its absence in the well-preserved wings of the Eichstätt and Berlin specimens of *Archaeopteryx* suggests the existence of aerodynamic constraints during low-speed flights and maneuvers for the famous Urvögel. This is consistent with the primitive characteristics of its flight apparatus, which lacks most of the structures usually correlated with an enhanced flying ability. The basic modern flight apparatus (strut-like coracoids, pygostyle, etc.) appears in the basalmost ornithothoracine birds such as *Iberomesornis*, implying important differences between the flying capabilities of these birds and *Archaeopteryx*. The presence of an alula in *Eoalulavis* suggests that, as early as 115 million years ago, birds possessed a modern structural complex in order to perform low-speed flight and precise maneuverability control.

See also the following related entries:

Aves • Cretaceous Period • European Dinosaurs

References

Chiappe, L. M. (1995). The first 85 million years of avian evolution. *Nature* **378,** 349–355.

Francís, V., and Sanz, J. L. (1989). Restos de dinosaurios del Cretácico inferior de Buenache de la Sierra (Cuenca). In *La Fauna del pasado en Cuenca. Ayuntamiento de Cuenca*, pp. 125–144.

Fregenal-Martínez, M. A. (1991). El sistema lacustre de Las Hoyas (Cretácico inferior, Serranía de Cuenca): Estratigrafía y sedimentología, pp. 226. Masters thesis, Universidad Complutense, Madrid.

Fregenal-Martínez, M. A., and Melíndez, N. (1993). Sedimentología y evolución paleogeográfica de la cubeta de Las Hoyas (Cretácico inferior, Serranía de Cuenca). *Cuad. Geol. Ibérica* **17,** 231–256.

Holtz, T. R., Jr. (1994). The phylogenetic position of the Tyrannosauridae: Implications for theropod systematics. *J. Paleontol.* **68**(5), 1100–1117.

Pérez-Moreno, B. P., and Sanz, J. L. (1995). The hand of *Pelecanimimus polyodon*. A preliminary report. II International Symposium on Lithographic Limestones, Extended abstracts, pp. 115–117.

Pérez-Moreno, B. P., Sanz, J. L., Buscalioni, A. D., Moratalla, J. J., Ortega, F., and Rasskin-Gutman, D. (1994). A unique multitoothed ornithomimosaur from the Lower Cretaceous of Spain. *Nature* **370,** 363–367.

Sanz, J. L., and Bonaparte, J. F. (1992). A new order of birds (class Aves) from the Early Cretaceous of Spain. In *Papers in Avian Paleontology. Honoring Pierce Brodkorb* (K. E. Campbell, Ed.), Science Ser., Vol. 36, pp. 39–49. Natural History Museum of Los Angeles County, Los Angeles.

Sanz, J. L., and Buscalioni, A. D. (1992). A new bird from the Early Cretaceous of Las Hoyas, Spain, and the early radiation of birds. *Palaeontology* **35**(4), 829–845.

Sanz, J. L., and Buscalioni, A. D. (1994). An isolated bird foot from the Barremian (Lower Cretaceous) of Las Hoyas (Cuenca, Spain). *Geobios Mem. Spec.* **16,** 213–217.

Sanz, J. L., Bonaparte, J. F., and Lacasa, A. (1988). Unusual Lower Cretaceous birds from Spain. *Nature* **331,** 433–435.

Sanz, J. L., Díguez, C., Fregenal-Martínez, M. A., Martínez-Delclís, X., Melíndez, N., and Poyato-Ariza, F. J. (1990). El yacimiento de fósiles del Cretácico inferior de Las Hoyas, Provincia de Cuenca (España). *Comun. Reunión Tafonomía Fosilización*, 337–355.

Sanz, J. L., Chiappe, L. M., and Buscalioni, A. D. (1995). The osteology of *Concornis lacustris* (Aves: Enantiornithes) from the Lower Cretaceous of Spain and a reexamination of its phylogenetic relationships. *Am. Museum Novitates* **3133**.

Sanz, J. L., Chiappe, L. M., Pérez-Moreno, B. P., Buscalioni, A. D., Moratalla, J. J., Ortega, F., and Poyato-Ariza, F. J. (1996). A new Lower Cretaceous bird from Spain: Implications for the evolution of flight. *Nature*.

Sereno, P. C., and Rao, C. (1992). Early evolution of avian flight and perching: New evidence from the Lower Cretaceous of China. *Science* **255,** 845–848.

Wellnhofer, P. (1994). New data on the origin and early evolution of birds. *C. R. Acad. Sci. Paris t. 319, Sér. II*, 299–308.

Zhou, Z., Jin, F., and Zhang, J. (1992). Preliminary report on a Mesozoic bird from Liaoning, China. *Chinese Sci. Bull.* **37**(16), 1365–1368.

Laurasiatic Dinosaurs

see American Dinosaurs; Asian Dinosaurs; Canadian Dinosaurs; Chinese Dinosaurs; European Dinosaurs; Mexican Dinosaurs

Leicestershire Museum, United Kingdom

see Museums and Displays

Legislation Protecting Dinosaur Fossils

VINCENT L. SANTUCCI
Slippery Rock University
Slippery Rock, Pennsylvania, USA

The science of PALEONTOLOGY and the role of the paleontologist continues to evolve. As recent discoveries shed more light and pose new questions, the direction of paleontological research and the methodologies employed also change. Growth within the science has created the opportunity for paleontologists to expand new directions in the management of fossils resources.

A new generation of paleontologists is emerging. This new breed is concerned with the issues associated with the management and protection of fossils as nonrenewable resources. These issues have surfaced as the result of changes within the social and political environment regarding fossils as well as changes in legislation and regulations associated with paleontological resources.

Legislation and practices vary greatly around the world. Historically, in socially developed countries much or all land was controlled by its rulers, whether local or national. In underdeveloped or unsettled lands there were no such laws, and fossils were collected by both academic paleontologists and commercial interests. Much of this material became prized specimens of museums in North America and Europe. Increasingly, legislation has been established by recently developed countries to restrict or prohibit the removal or export of fossil material. This is particularly true in countries such as Tanzania, Ethiopia, and many other South American and African countries, in addition to Canadian provinces. However, there is still considerable variation. For example, in Germany, the state of Bavaria does not control the excavation or sale of fossils from privately owned land; but in the state of Baden-Württemberg, all fossils belong to the state, whether they are found on public or privately owned land, and the owners of private land are paid a nominal finder's fee in exchange for their fossils. In the United States, permission to collect fossils must be granted by the owner in the case of privately owned land. To collect on lands belonging to American Indian nations, permission must be obtained from the Department of the Interior, the Bureau of Indian Affairs, and all tribal entities and local landowners.

The laws developed for the management and protection of fossils in the United States are complex and confusing. Paleontological laws and regulations exist at both the federal and state levels. It is important to recognize that fossil laws can vary considerably from state to state and even between the federal land managing agencies. Often the legislation intended for the protection of fossils is too specific or too vague, presenting problems in both interpretation and enforcement. Additionally, various groups of fossils are often addressed independently within regulations. For example, the collection of limited amounts of petrified wood may be permitted in an area where fossil bones and teeth cannot be collected. Currently, there is no legislation that is specific for dinosaurs. Dinosaurs are grouped with all other vertebrate fossils.

An examination of the federal laws pertaining to fossils suggests a need for more specific and comprehensive laws to provide for the adequate management and protection of fossils on federal lands. A review of some of the key federal paleontological resource legislation is provided below. In addition, the regulations established within the various federal land managing agencies is briefly outlined. It is important to recognize that laws and regulations are changing rapidly and the information provided may soon be obsolete.

The Antiquities Act (16 USC 431–433) was established in 1906 to protect "objects of antiquity" from unauthorized collection or vandalism. This act required that permits be issued for the excavation of any antiquities found on public land and the Antiquities Permit was utilized by many federal land managing agencies until the 1980s.

In 1974, the Antiquities Act was challenged in the case United States v. Diaz (Ninth Circuit Court of Appeals). The phrase "object of antiquity" was determined to be unconstitutionally vague and this court decision established a judicial precedent. As the result of the Diaz case, federal land managing agencies discontinued to recognize the Antiquities Act as the statutory authority for regulating fossil resources.

The Archeological Resources Protection Act (ARPA) (Public Law 95-341) was established in 1979 to protect archeological resources on federal lands.

Protection of paleontological resources under ARPA is limited to only those fossil "specimens found in an archeological context."

The National Park Service (NPS) was established under the Organic Act of 1916 (Public Law 64-235). The mission of the NPS defined in the Organic Act includes "to conserve the scenery and the natural and historic objects . . . by such means as will leave them unimpaired for the enjoyment of future generations." Paleontological resources are regulated under 36 CFR (Code of Federal Regulations) Section 2.1 (a)(iii). Any disturbance, injury, or removal of nonfossilized or fossilized paleontological specimens without a permit can be subjected to a fine up to $5000 or imprisonment not exceeding 6 months or both.

The Bureau of Land Management (BLM) regulates paleontological resources under the Federal Land Policy and Management Act of 1976 (FLPMA) (Public Law 94-579). FLPMA enables the BLM to manage fossils through the concepts of multiple use and sustained yield of natural resources. BLM regulates paleontological resources and issues collecting permits through the state offices. Unauthorized collection of fossils on BLM lands is punishable by fines up to $1000 or imprisonment of up to 1 year or both.

The U.S. Forest Service areas are managed through broad authority under the National Forest Management Act of 1976 (Public Law 94-588) and the Forest Service General Provisions. Regulations are established that prohibit "excavating, damaging, or removing any vertebrate fossil or removing any paleontological resources for commercial purposes without a special use authorization." The Forest Service allows the collection of invertebrate and paleobotanical fossils for noncommercial purposes without a permit. Under 16 U.S. Code, Section 551 and 36 CFR, Section 261.9 fines up to $500 and or imprisonment of not more than 6 months or both can be levied for illegal collection of fossils on Forest Service lands.

The use of permits is essential for the sound management of paleontological along with other resources. Agencies manage a diversity of resources in multiple use areas. Endangered or sensitive resources may occur in areas that overlap paleontological resources. The permit serves as an educational tool by providing the permittee with information on other resource concerns, park regulations, and acceptable practices.

New legislation and changes in agency regulations regarding the management and protection of paleontological resources can be promulgated at any time. Site resource managers should be prepared to make researchers and collectors aware of these changes through the permit process and other means available. It remains incumbent upon anyone interested in collecting dinosaurs and other fossils that land ownership is known and appropriate permission is obtained prior to fossil excavation.

In 1995, a lobbying organization named SAFE (Save America's Fossils for Everyone) was established by the Society of Vertebrate Paleontology, with the purpose of establishing responsible legislation regarding the protection and preservation of fossils on federally owned public land. This is the first known instance of a professional scientific society taking such a step in the interest of preserving the paleontological heritage of the United States, and it signals the urgency felt by a large portion of the profession to educate legislators and the public on this issue.

Leonard Hall, University of North Dakota, North Dakota, USA

see MUSEUMS AND DISPLAYS

Lesotho Dinosaurs

Dinosaur fossils are found in Late Triassic and Early Jurassic sediments of the Lower and Upper Elliot Formations of Lesotho.

see AFRICAN DINOSAURS

Libyan Dinosaurs

see AFRICAN DINOSAURS

Life History

David B. Weishampel
Johns Hopkins University School of Medicine
Baltimore, Maryland, USA

Dinosaurs epitomize large size among terrestrial animals, but few studies have attempted to explore the ontogenetic and phylogenetic relationship of how such size is achieved and its consequences for their general biology; morphology, reproduction, and other aspects of life history.

The evolution of life histories is seen as changes in ecology, BEHAVIOR, and/or PHYSIOLOGY that increase or maximize the fitness of organisms through reproduction. These changes can include shifts in brood size, relative maturity and size of young, relative age distribution of reproductive effort, and interactions of reproductive effort with adult mortality.

The two end members of the continuum of life histories are known as r- and K-strategies. Those organisms that are exposed principally to density-independent selection, with relatively little negative feedback on growth rate by dwindling resources, are said to be exposed to r-selection, which often is a consequence of large, frequent, and unpredictable environmental fluctuations; frequent catastrophe mortality; superabundant resources; and lack of crowding life history traits related to r-selection (i.e., found in r-strategists), including early and rapid production of large numbers of rapidly developing young, high fecundity, early maturation, short life span, limited parental care, rapid (i.e., precocial) development, and a greater proportion of available resources committed to reproduction. Mosquitoes and oysters, with their abundant reproductive efforts and lack of parental care, are commonly viewed as r-strategists.

In contrast, so-called K-strategists are predominantly exposed to density-dependent selection, in which one kind of organism increases at the expense of another. Such K-selection, in crowded, stable, and benign environments, results in low reproductive effort with late maturation, longer life, and a tendency to invest a great deal of parental care in small broods of late-maturing (i.e., altricial) offspring. Humans, with their low reproductive output and extensive parental care, are one of the best examples of K-strategists.

What about dinosaurs? The best of the life history information comes from EUORNITHOPOD dinosaurs—from *Orodromeus, Rhabdodon, Telmatosaurus, Maiasaura,* and *Hypacrosaurus.* Among SAURISCHIANS, aspects of life histories are known only in *Oviraptor, Troodon,* and, of course, extant birds (for which the best life history data exist among recent vertebrates!). Based on a phylogenetic analysis of life history features conducted by D. B. Weishampel (Johns Hopkins University) and J. R. Horner (Museum of the Rockies), *Orodromeus* appears to retain the primitive condition among euornithopods of laying a reasonably large number of eggs and having relatively large, skeletally mature, precocial hatchlings. In contrast, HADROSAURIDS (i.e., *Telmatosaurus, Maiasaura,* and *Hypacrosaurus*) evolved the new condition of relatively larger adults producing—and caring for—skeletally immature hatchlings. On this basis, hadrosaurids are regarded as K-strategists investing in altricial young.

Among nonavian saurischians, on the other hand, life history data are extremely rare. Much of the evidence from nesting sites and skeletal morphology of juvenile and embryonic material suggests that hatchlings of the MANIRAPTORANS *Troodon* and *Oviraptor* may have been precocial, although whether this represents the retention of the primitive r-selected condition or a newly evolved strategy within the theropod clade remains to be seen.

See also the following related entries:

Eggs, Eggshells, and Nests • Growth and Embryology • Heterochrony

Reference

Weishampel, D. B., and Horner, J. R. (1994). Life history syndromes, heterochrony, and the evolution of Dinosauria. In *Dinosaur Eggs and Babies* (K. Carpenter, K. F. Hirsch, and J. R. Horner, Eds.), pp. 229–243. Cambridge Univ. Press, New York.

Life Sciences Museum, Pierce College, California, USA

see Museums and Displays

Life Span

see AGE DETERMINATION; GROWTH AND EMBRYOLOGY; GROWTH LINES; LIFE HISTORY

Linnaean System of Classification

see PHYLOGENETIC SYSTEM

Lippisches Landesmuseum, Germany

see MUSEUMS AND DISPLAYS

Locomotion

see BIPEDALITY; FOOTPRINTS AND TRACKWAYS; QUADRUPEDALITY

Lommiswil

MARTIN LOCKLEY
University of Colorado at Denver
Denver, Colorado

The Lommiswil dinosaur tracksite, situated in the classic Jura Mountains near Solothurn, Switzerland, is one of the largest dinosaur tracksites in Europe. It reveals many trackways of large, Upper Jurassic sauropods, including one that can be followed for 90 m. Study of this site, and others in the region, led to the recognition of the first known megatracksite in Europe. The site is open to the public.

See also the following related entries:

EUROPEAN DINOSAURS • FOOTPRINTS AND TRACKWAYS • MEGATRACKSITES

Reference

Meyer, C. A. (1993). A sauropod dinosaur megatracksite from the late Jurassic of northern Switzerland. *Ichnos* **3**, 29–38.

Long Necks of Sauropods

EBERHARD FREY
Staatliches Museum für Naturkünde
Karlsruhe, Germany

JOHN MARTIN
Leicestershire Museum
Leicester, United Kingdom

Flexible or Rigid?

One of the best known features of sauropods, the large terrestrial herbivores of the Mesozoic, is their long neck. Life reconstructions regularly show them with the neck held almost vertically, or in a flexed "S" curve, enabling the animals to browse high in the trees or to fight (e.g., Bakker, 1986; Paul, 1987). Other paleontologists (e.g., Coombs, 1975; Martin, 1987; Dodson, 1990) have argued that there were rather severe mechanical limits to the flexibility of sauropods' necks as a result of their complex intervertebral articulation systems. Part of the problem is that some reconstructions of sauropod necks are speculative and based, intuitively, on recent "analogous" long-necked animals. The principles of BIOMECHANICS apply to sauropods' necks (as they do to all "biostructures") and, when the fossil record is good enough, can be used to help resolve the dilemma.

Braced Beams

An unbraced beam—a simple girder, for example—supported at one end like a crane jib has its upper

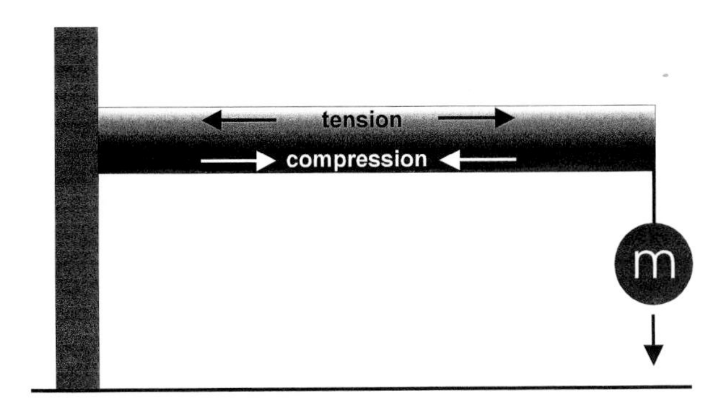

FIGURE 1 A simple, unbraced beam supported at one end only. The effect of gravity on the unsupported structure is to put the upper part under tension while compressing the lower part.

part stressed by tensile forces while the lower part is under compression. Most engineering beams are, in fact, braced to reduce these stresses. A flexible, segmented beam cannot be unbraced, however: Because each segment needs to be free to move against its neighbors, the whole length of the structure must be braced above, below, or both.

Sauropod Necks Sauropod necks comprised several segments held against each other but free to move in a controlled way, and so were braced. Because they were a compromise between the mechanical optimum (rigid and stable beam) and a living construction (segmented, with controlled flexibility), the observed situation is that the possible bracing systems were more or less combined so that mainly dorsally braced, mainly ventrally braced, or dual systems are known in the Sauropoda. The dorsal, tensile bracing elements were unmineralized connective tissues such as tendons, muscles, and ligaments, whereas the ventral, incompressible elements were rigid or elastic tissues, such as bone or cartilage. Muscles, in addition to applying power, provided hydraulic stabilization for the other structures. The bracing systems seen in sauropod necks are not those of engineering masts; sauropod necks are structures with a "normal" orientation less than 90° (above or below horizontal); "down" (the direction of gravitational force on the construction) is always anatomically ventral and indicated by the incompressible bracing system. Not even in *Brachiosaurus* was the neck held straight up under normal conditions: There is no biomechanical evidence in the fossil record for swan-necked sauropods.

Classes of Sauropod Necks

Dorsally Braced Dorsally braced sauropod neck constructions appear to have been the least common. A beam braced dorsally by flexible, tensile elements allows great mobility but stability is reduced; in effect the vertebral segments are simply slung beneath the bracing "cables." Bone and ligament locks at the intervertebral joints and powerful, lateral tendinomuscular systems are necessary to control mobility. These, and the massive suspension ligaments that (especially in the nuchal area) must bridge many segments, imply large mass and thus limit the length and number of segments in the neck.

Examples of dorsally braced systems are seen in *Dicraeosaurus* and *Apatosaurus*. Both had neural spines at least three times the height of the vertebral centrum. In *Dicraeosaurus* the neural spines were deeply forked. Bifurcation had the effect of increasing the volume of suspension ligaments, which presumably ran in the space between the paired neural spines from the hips to the back of the skull. In the neck these formed a powerful nuchal ligament with additional lateral branches inserting on the outer faces of the neural spines. Unlike sauropods with ventrally braced necks (see below), the cervical ribs of these animals did not extend into adjacent vertebral segments; because ventral bracing was minimal, lateral flexibility was enhanced.

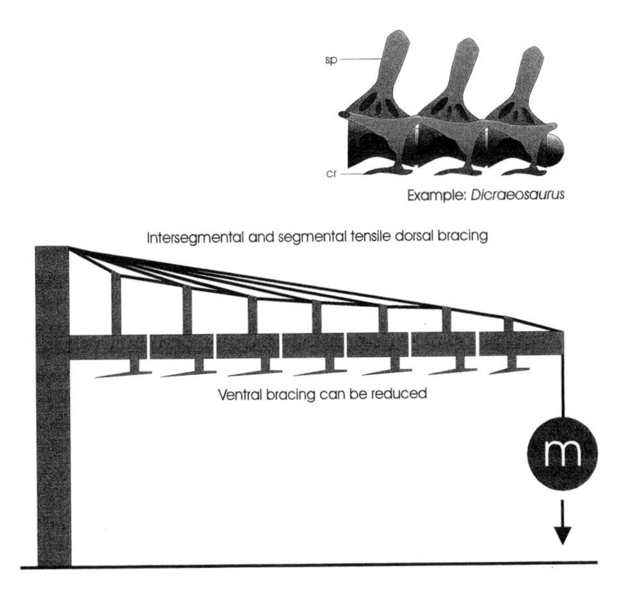

FIGURE 2 A segmented beam with mainly dorsal bracing and (top) three schematic neck vertebrae of *Dicraeosaurus*, with very high neural spines and short cervical ribs.

Apatosaurus was a special case of dorsal bracing. Its neural spines were less tall but very widely divided for a massive nuchal ligament whose action was kept close to the vertebral centra. The very large, wide transverse processes and the stout but short cervical ribs (not extending into adjacent segments) provided long lever arms for the huge lateral musculature necessary to give stability in this otherwise weakly braced construction. *Apatosaurus* must have had a uniquely powerful but flexible neck among sauropods.

Ventrally Braced In contrast to dorsally braced systems, a segmented beam braced ventrally by incompressible elements is inherently inflexible and the mass of stabilizing structures can be reduced. The bracing system can only transmit compressive load if it is structurally continuous. In sauropod necks the ventral bracing system was provided by the longitudinally extended cervical ribs. The ribs were bound together by connective tissue and muscles, a construction also seen in the ventral bracing system of crocodilians (Frey, 1988). The larger the overlap of the cervical ribs, the more segments combined as functional units and the more rigid the neck became.

The smaller mass associated with this style of ventral bracing permits an increase in the length and number of segments and consequently in the length of the beam. Therefore, ventrally braced systems are characteristic of very long-necked sauropods, particularly the Chinese taxa *Mamenchisaurus*, *Euhelopus*, and *Omeisaurus*. These had cervical ribs considerably in excess of the length of one segment (more than two in *Euhelopus* and possibly three in *Mamenchisaurus*). The elongated ribs formed overlapping bundles of up to four in the mid part of the neck. The neural spines were undivided and low—less than the height of the vertebral centrum—and the centra themselves were long and low. The neck of *Brachiosaurus* is not completely known (the crucial posterior neural arches are missing) but, with its long cervical ribs (perhaps two segments long) and undivided neural spines the front part at least seems to have been primarily ventrally braced. All these sauropods' necks were inherently inflexible structures except, presumably, for mobile occipital and shoulder areas.

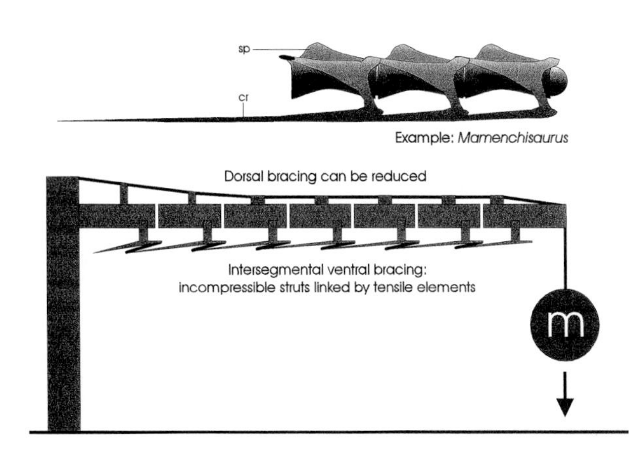

FIGURE 3 A segmented beam with mainly ventral bracing and (top) three schematic neck vertebrae of *Mamenchisaurus*, with low neural spines and bundles of elongated cervical ribs.

A special case of ventrally braced neck construction is seen in *Camarasaurus*, in which, although the neural spines were undivided until segment 11 of the 12 and the cervical ribs extended over three segments to brace the neck strongly, the vertebrae were relatively wide and short, with high neural arches and widely separated zygapophyses. *Camarasaurus* had a very powerful but inflexible neck.

Dual Systems All other sauropods in which the cervical vertebrae are well known had necks with both bracing systems combined. Dorsal bracing is indicated by neural spines whose height was about the same as that of the centra—the spines were sometimes divided (e.g., *Diplodocus*) in the posterior segments—and ventral bracing was supplied by cervical ribs somewhat longer than one segment, giving a moderate overlap between adjacent vertebrae. Dual constructions include *Diplodocus*, *Cetiosaurus*, and *Haplocanthosaurus*, all of which had moderately long, relatively inflexible necks.

Taxonomy of Necks

Recent cladistic analyses of sauropods (e.g., Upchurch, 1994) are not inconsistent with the relationships suggested by neck constructional types. Dual (combined) constructions occur in *Cetiosaurus*, *Brachiosaurus*, and *Haplocanthosaurus*, and also, interestingly, among the prosauropods in which the condition of the neck is known, such as *Plateosaurus* (von Huene, 1907/1908). Ventrally braced constructions are clustered among the Chinese Euhelopidae, whereas dorsally braced systems occur, perhaps independently, among derived taxa such as *Dicraeosaurus* and *Apatosaurus*. Only the systematic position of *Diplodocus* (with its dual neck bracing) close to *Apatosaurus*, and of *Camarasaurus* close to *Brachiosaurus* and *Haplocanthosaurus*, are problematic.

See also the following related entries:

SAURISCHIA • SAUROPODA

References

Alexander, J. McN. (1985). Mechanics of posture and gait of some large dinosaurs. *Zool. J. Linnean Soc. London* **83,** 1–25.

Bakker, R. T. (1986). *Dinosaur Heresies*, pp. 481. Morrow, New York.

Coombs, W. (1975). Sauropod habits and habitats. *Palaeogeogr. Palaeoclimatol. Palaeoecol.* **17,** 1–33.

Dodson, P. (1990). Sauropod paleoecology. In *The Dinosauria* (D. B. Weishampel, P. Dodson, and H. Osmólska, Eds.), pp. 402–407. Univ. of California Press, Berkeley.

Frey, E. (1988). Das Tragsystem der Krokodile-eine biomechanische und phylogenetische Analyse. *Stuttgarter Beitrage Naturkünde* **426,** pp. 60.

Jensen, J. A. (1985). Three new sauropod dinosaurs from the Upper Jurassic of Colorado. *Great Basin Nat.* **45,** 697–709.

Martin, J. (1987). Mobility and feeding of *Cetiosaurus* (Saurischia: Sauropoda)—Why the long neck? In *4th Symposium of Terrestrial Mesozoic Ecosystems, Short Papers* (P. J. Currie and E. H. Koster, Eds.), pp. 154–159. Tyrrell Museum of Paleontology, Drumheller, Alberta, Canada.

Paul, G. S. (1987). The science and art of restoring the life appearance of dinosaurs and their relatives. In *Dinosaurs Past and Present: Vol. II* (S. J. Czerkas and E. C. Olson, Eds.), pp. 5–49. Natural History Museum of Los Angeles County, Los Angeles.

Upchurch, P. (1994). Manus claw function in sauropod dinosaurs. *Gaia* **10,** 161–171.

von Huene, F. (1907/1908). Die Dinosaurier der Europäischen Triasformation mit Berücksichtigung der aussereuropäischen Vorkommnisse. *Geol. Palaeont. Abhandl. Suppl.* **1,** 1–419.

Los Rastros

see ISCHIGUALASTO FORMATION

Louisiana State University Museum of Geoscience, Louisiana, USA

see MUSEUMS AND DISPLAYS

Lufeng

DONG ZHIMING
Institute of Vertebrate Paleontology and Paleoanthropology
Chinese Academy of Sciences
Beijing, People's Republic of China

The Lufeng, a small inland basin, is situated 102 km northwest of Kunming City, the capital of Yunnan Province, China. The early Mesozoic strata of the Lufeng Basin have traditionally been divided into the Lower and Upper Lufeng Formations. They consist of a mosaic of lacustrine, fluvial, and overbank deposits. The Lower Lufeng Formation has yielded very abundant vertebrate fossils, whereas only isolated fish spines, teeth, and scales and fragmentary turtle shells are known from the Upper Lufeng Formation. The early stratigraphic studies of the Lufeng Basin by Bien (1940, 1941) were summarized by Sun *et al.* (1985). The Lower Lufeng Formation is divisible into two members, lower "dull purplish beds" and upper "dark red beds" (Fig. 1).

Since 1938, many dinosaur remains have been discovered in the Lufeng Basin, which has been visited and surveyed by many Chinese and foreign paleontologists. The Lufeng Basin is an important source of information for our understanding of Triassic–Jurassic tetrapod assemblages. To date, seven genera and 12 species of dinosaurs have been reported from this famous Upper Triassic and Lower Jurassic fossil site (Sun *et al.*, 1985). They include prosauropods (*Lufengosaurus, Yunnanosaurus,* and *Anchisaurus*), theropods (*Dilophosaurus, Lukousaurus,* and *Sinosaurus*), an ornithopod (*Diachungosaurus*), and the stegosaur *Tatisaurus* (Young, 1951, 1982a,b; Simmons, 1965; Sun *et al.*, 1985; Dong, 1992). Among the tritylodonts, most can be referred to *Bienotherium, Yunnanodon,* and *Oli-*

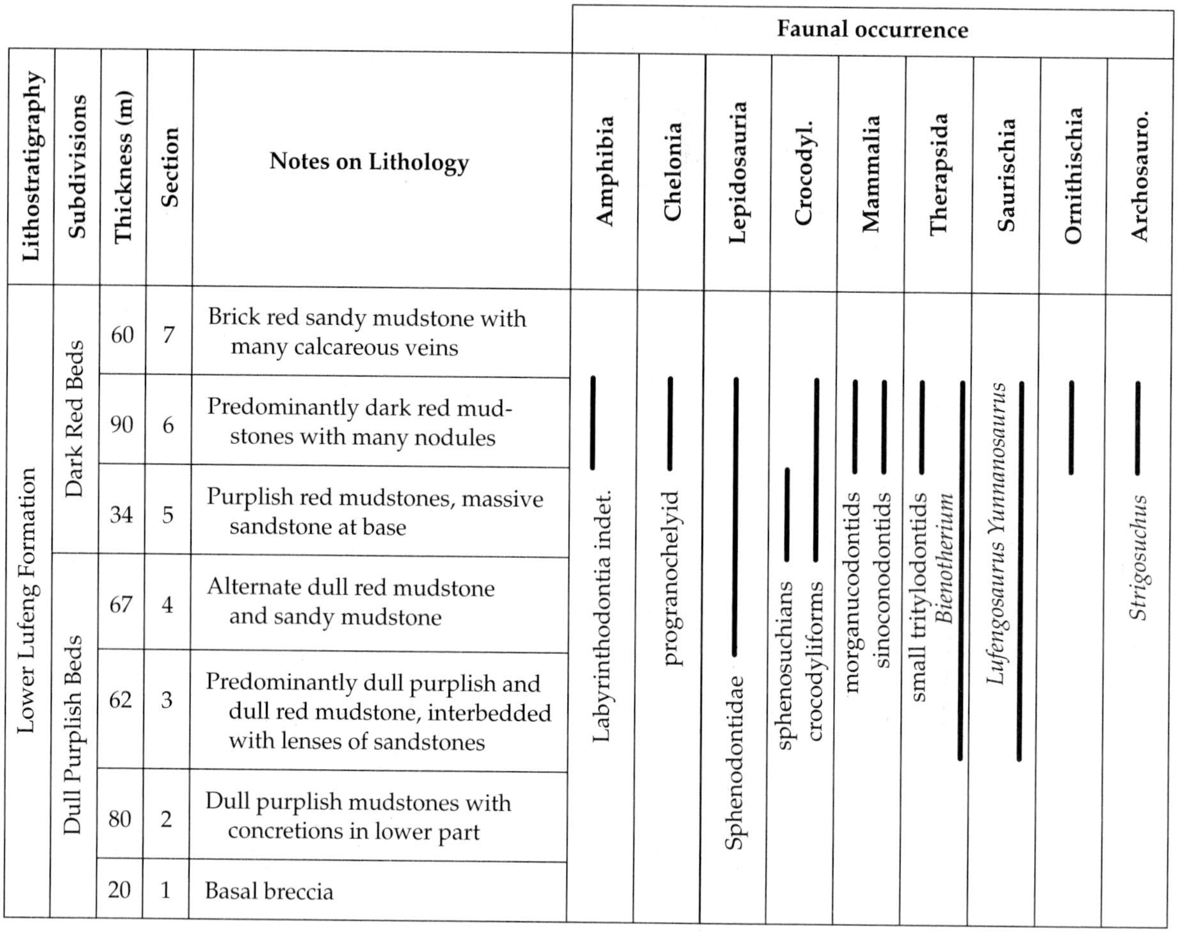

Lithostratigraphy	Subdivisions	Thickness (m)	Section	Notes on Lithology
Lower Lufeng Formation	Dark Red Beds	60	7	Brick red sandy mudstone with many calcareous veins
		90	6	Predominantly dark red mudstones with many nodules
		34	5	Purplish red mudstones, massive sandstone at base
	Dull Purplish Beds	67	4	Alternate dull red mudstone and sandy mudstone
		62	3	Predominantly dull purplish and dull red mudstone, interbedded with lenses of sandstones
		80	2	Dull purplish mudstones with concretions in lower part
		20	1	Basal breccia

Faunal occurrence	
Amphibia	Labyrinthodontia indet.
Chelonia	progranochelyid
Lepidosauria	Sphenodontidae
Crocodyl.	sphenosuchians; crocodyliforms
Mammalia	morganucodontids; sinoconodontids
Therapsida	small tritylodontids; *Bienotherium*
Saurischia	*Lufengosaurus Yunnanosaurus*
Ornithischia	
Archosauro.	*Strigosuchus*

FIGURE 1 Stratigraphic distribution of vertebrates in the Lower Lufeng Formation (modified from Luo and Wu, 1994).

gokyphus. Mammals consisted of two genera (*Sinoconodon* and *Morganucodon*) and an unnamed new taxon (Luo and Wu, 1994).

See also the following related entries:

Chinese Dinosaurs • Erenhot Dinosaur Museum • Extinction, Triassic • Jurassic Period • Triassic Period

References

Bien, M. N. (1941). "Red Beds" of Yunnan. *Bull. Geol. Soc. China* **21,** 159–198.

Dong, Z. M. (1992). *Dinosaurian Faunas of China*, pp. 1–188. Springer-Verlag, Berlin/New York.

Luo, Z., and Wu, X.-C. (1994). The small tetrapods of the Lower Lufeng Formation, Yunnan, China. *In the Shadow of the Dinosaurs, Early Mesozoic Tetrapods* (Fraser and Sues, Eds.), pp. 251–270. Cambridge Univ. Press, Cambridge, UK.

Simmons, D. J. (1965). The non-therapsid reptiles of the Lufeng Basin, Yunnan, China. *Fieldiana Geol.* **15,** 1–93.

Sun, A. L., Cui, G., Li, Y., and Wu, X. C. (1985). A verified list of Lufeng saurischian fauna. *Vertebr. Palasiatica* **23,** 1–12.

Wu, X. C., and Chatterjee, S. (1993). *Dibothrosuchus elaphros*, A crocodylomorph from the Lower Jurassic of China and the phylogeny of the Sphenosuchia. *J. Vertebr. Paleontol.* **13**(1), 58–89.

Young, C. C. (1951). The Lufeng saurischian fauna in China. *Palaeontol. Sinica New Ser. C* **13,** 1–96.

Young, C. C. (1982a). A new ornithopod from Lufeng, Yunnan. In *Selected Works of Yang Zhung-jian*, pp. 29–35. Science Press, Beijing.

Young, C. C. (1982b). On a new genus of dinosaur from Lufeng, Yunnan. In *Selected Works of Yang Zhung-jian*, pp. 38–42. Science Press, Beijing.

Malagasy Dinosaurs

see African Dinosaurs

Malawi Dinosaurs

see African Dinosaurs

Mali Dinosaurs

see African Dinosaurs

Maniraptora

Kevin Padian
University of California
Berkeley, California, USA

Maniraptora (Greek: *manus,* hand; *raptus,* to seize) was coined by Gauthier (1986) to emphasize that "Avialae and Deinonychosauria (or at least dromaeosaurs)" shared a common ancestor exclusive of Ornithomimidae. Hence it is a stem-based taxon, and Gauthier listed a series of synapomorphies that diagnosed it. However, some confusion has been generated in that it has also been regarded as a node-based taxon including "Avialae and Deinonychosauria" and all descendants of their most recent common ancestor, or alternatively as a node-based taxon including all descendants of the first coelurosaur to possess the "maniraptoran condition" (see below). The latter group would encompass virtually all coelurosaurs except *Compsognathus,* including *Ornitholestes,* Dromaeosauridae, Aves, Oviraptoridae, Elmisauridae, *Avimimus,* Tyrannosauridae, Troodontidae, and Ornithomimosauria (Holtz, 1994, pp. 1105, 1107). This was rationalized because the eponymous group's "maniraptoran condition" and other synapomorphies were hypothesized to have been present in the common ancestor of these taxa and secondarily reversed in Ornithomimidae and other taxa. However, Holtz (1996) recognized that his previous work had changed the concept of Maniraptora from a stem-based to a less stable apomorphy-based taxon and had also included the group (Ornithomimidae) explicitly regarded as the sister taxon of the Maniraptora as originally conceived. Accordingly, he amended the definition to "all theropods closer to birds than to ornithomimids," corresponding to the stem containing node 12 of his cladogram (1994, Fig. 4). Currently, this would comprise only Dromaeosauridae and Avialae (Fig. 1).

However, some questions about the composition of Maniraptora remain. Gauthier's (1986) conception of Deinonychosauria included Dromaeosauridae and Troodontidae, with some reservation about the membership of Troodontidae. Troodontids are now considered close to ornithomimids, which leaves Dromaeosauridae (see Bullatosauria). However, there may be some question about whether *Dromaeosaurus, Deinonychus,* and *Velociraptor,* and possibly *Hulsanpes, Adasaurus,* and *Saurornitholestes* (Ostrom, 1990) actually form a natural group. *Velociraptor* and *Deinonychus* appear to be very similar. *Deinonychus,* traditionally considered a member of Dromaeosauridae, has been more or less interchangeably used with Dromaeosauridae in phylogenetic definitions. This may become problematic if these taxa are not at least sister groups. Furthermore, Gauthier (1986) defined Avialae as a node-based taxon including *Archaeopteryx* and Neornithes (crown group birds). Because this node

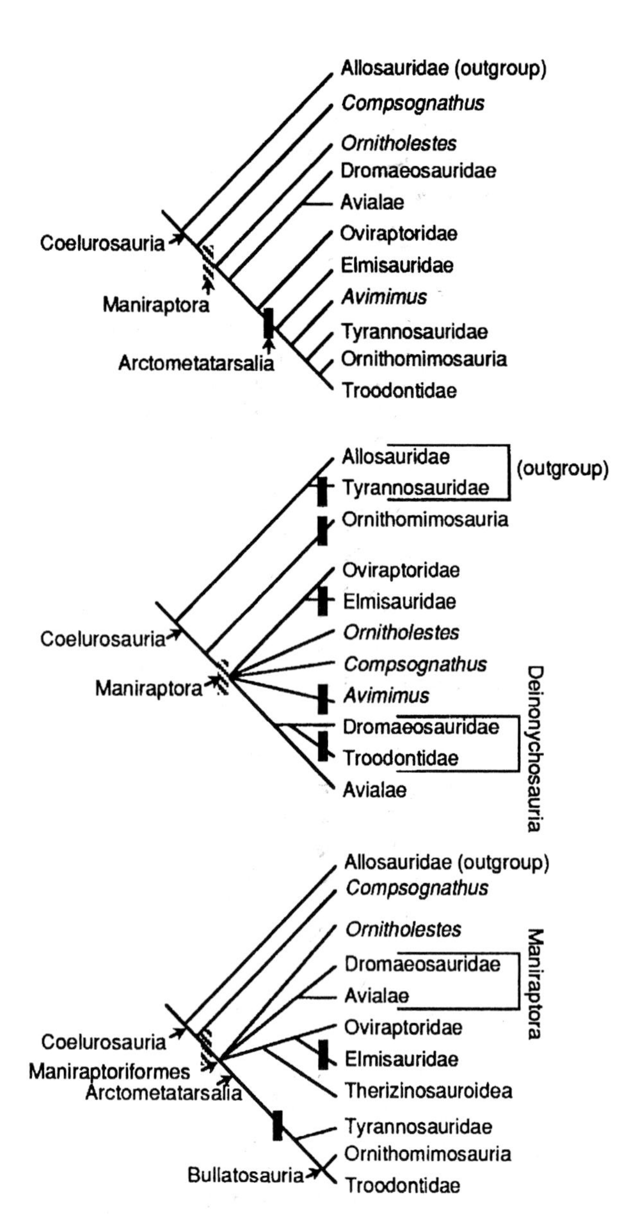

FIGURE 1 Definition and phylogenetic relationships of the Maniraptora (from Holtz, 1996). (Top) Phylogeny proposed by Holtz (1994), defining Maniraptora to include Ornithomimosauria and relatives; this contradicted the stem-based definition of Gauthier (1986) (center), which specifically excluded them as a sister taxon to Maniraptora. (Bottom) Holtz's reformulation of both the definition and the phylogeny of these coelurosaurs, as noted in text. Solid bar is the acquisition of the proximally pinched "arctometatarsalian" condition of the foot; striped bar is the acquisition of the "maniraptoriform" forelimb and manus. The stem leading to Dromaeosauridae and Avialae comprises Maniraptora.

has traditionally been called Aves, in this work the term Avialae is recognized as the stem comprising Neornithes and all maniraptorans closer to them than to *Deinonychus* (see AVIALAE). Deinonychosauria is recognized as the stem-based sister taxon to Avialae, defined as *Deinonychus* and all maniraptorans closer to it than to birds.

Holtz (1996) further established the node-based taxon Maniraptoriformes, to replace his (1994) use of Maniraptora, as the most recent common ancestor of *Ornithomimus* and birds and all descendants of that common ancestor. It comprises four stem groups: the ARCTOMETATARSALIA, Maniraptora, OVIRAPTOROSAURIA, and *Ornitholestes* (see MANIRAPTORIFORMES).

The maniraptoran condition, which now pertains to Maniraptoriformes, refers to features of the forelimb discussed by Gauthier (1986, pp. 32–33, among other characters diagnosing his Maniraptora), including a forelimb elongated to nearly 75% (or more) of the presacral vertebral column, the hand longer than the foot, the ulna bowed posteriorly, a semilunate carpal, and a bowed, thin metacarpal III. The semilunate ("half-moon-shaped") carpal (Fig. 2) is almost certainly a fusion of distal carpals I and II because in basal theropods distal carpal I overlaps the base of digit II, and in forms with a semilunate carpal distal carpal II is never separately ossified, although distal carpal III may be present. Ostrom (1969, 1990) identified this bone as the radial, but this is unlikely because the radial, a proximal carpal, caps the radius distally and does not articulate with the metacarpals but with the distal carpals, which in turn articulate with the metacarpals.

The maniraptoran condition of the forelimb is recognized not only for the phylogenetic unity of its characters but also for the functional significance of its structure. The elongated forelimb and hand, clearly not used in terrestrial locomotion, are interpreted to have provided a greater reach in predation. The semilunate carpal allowed the wrist to flex in a dorsoventral plane, as in other theropods, but also in a mediolateral plane with a substantial excursion (Ostrom, 1969); thus, as the animal reached forward, the hands could also swing upward and outward. Flexion of the fingers and claws would complete a grasping or slicing stroke. Indications that this behavior, as predicted by Ostrom (1969) for *Deinonychus*, was effective were given only a few years later in 1971, with the discovery at Tugrig by the POLISH–

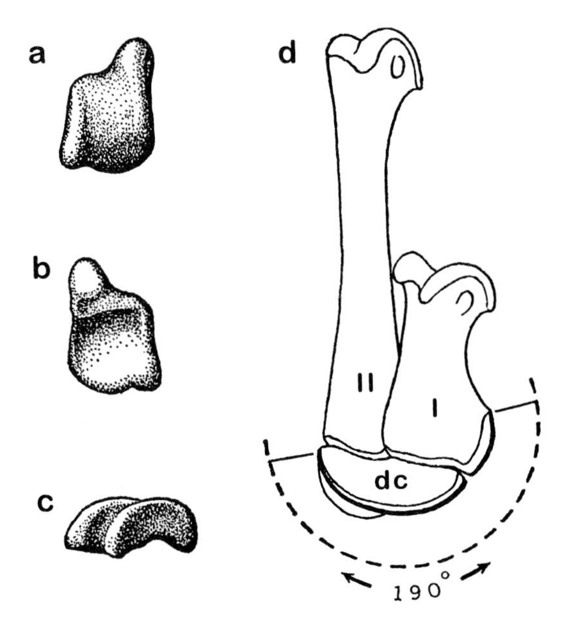

FIGURE 2 Semilunate carpal of Maniraptoriformes (after Ostrom, 1969) for *Deinonychus.* (a–c) Right semilunate carpal, seen in proximal (a), distal (b), and ventral (c) views; (d) left semilunate carpal (dc) in articulation with metacarpals I and II, showing range of lateral flexion.

MONGOLIAN EXPEDITIONS of the famous "fighting dinosaurs" of Mongolia. This tableau, frozen in sandstone, depicts a *Velociraptor* with one hand grabbing the frill of a *Protoceratops* and the other lodged in its jaws; its foot, armed with the enlarged trenchant claw of the second digit, appears to have been lodged in the throat region of the beaked plant eater (Unwin *et al.*, 1995).

A further significance of the maniraptoriform forelimb, especially in its condition in Maniraptora, is that the mediolateral movement of the wrist is functionally identical and phylogenetically homologous to the excursion of the wrist in birds during the flight stroke (Gauthier and Padian, 1985). The hand twists about an axis that is inclined with respect to the axis of the normal "hinge joint" of the wrist. This movement provides the component of thrust to the flight stroke, the essential difference between flying and merely gliding, as well as the initiation of the wing's recovery stroke at the wrist (Fig. 3). If birds evolved from terrestrial ancestors, they would not have had to proceed through a fully arboreal, gliding stage providing they had a flight stroke capable of delivering thrust as well as lift (see BIRD ORIGINS).

Other synapomorphies of Maniraptora include a reduced or absent prefrontal bone, prominent axial epipophyses, hypapophyses in the cervicothoracic region, a "transition point" in the tail that begins between caudals 7 and 11, a subrectangular coracoid, a ventral curve to the posterodorsal border of the ilium, a pubic peduncle of the ilium that is longer ventrally than the ischiadic peduncle, a retroverted pubis, a pubic foot with an anterior reduction (convergent in other coelurosaurs), an ischium reduced to two-thirds of the length of the pubis or less (this is reversed in birds after *Mononykus)* with a distally placed obturator process on the ischium, a near confluence of the anterior trochanter with the proximal end of the femur, the extreme reduction or absence of the fourth trochanter on the femur, and the slight elongation of pedal digit IV to bring it nearly subequal to digit III (Gauthier, 1986). As Gauthier pointed out, some of these characters are not known in all coelurosaurs,

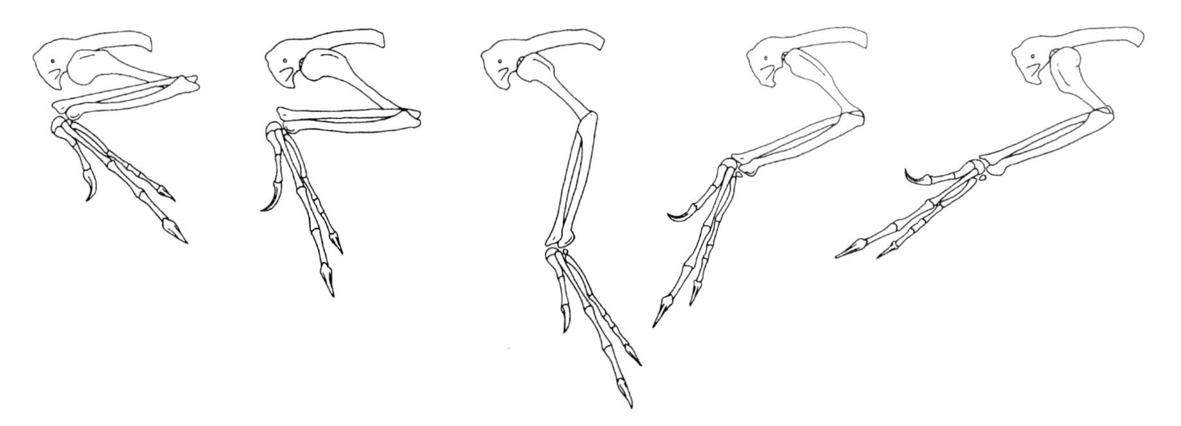

FIGURE 3 The predatory stroke of the forelimb in *Deinonychus,* similar to the avian flight stroke. Right shoulder girdle and forelimb in lateral view (after Gauthier and Padian, 1985, with permission).

so some may apply to more general levels within the group or they may have evolved convergently in other taxa.

Novas and Puerta (1997) have described *Unenlagia comahuensis*, a maniraptoran coelurosaur. They hypothesize that this taxon is the sister taxon of Avialae.

See also the following related entries:

AVES • BIRD ORIGINS • DROMAEOSAURIDAE

References

Gauthier, J. A. (1986). Saurischian monophyly and the origin of birds. *Mem. California Acad. Sci.* **8**, 1–55.

Gauthier, J. A., and Padian, K. (1985). Phylogenetic, functional, and aerodynamic analyses of the origin of birds and their flight. In *The Beginnings of Birds* (M. K. Hecht, *et al.*, Eds.), pp. 185–197. Freunde des Jura-Museums, Eichstatt, West Germany.

Holtz, T. R., Jr. (1994). The phylogenetic position of the Tyrannosauridae: Implications for theropod systematics. *J. Paleontol.* **68**, 1100–1117.

Holtz, T. R., Jr. (1996). Phylogenetic taxonomy of the Coelurosauria (Dinosauria: Theropoda). *J. Paleontol.* **70**, 536–538.

Novas, F. E., and Puerta, P. F. (1997). New evidence concerning avian origins from the Late Cretaceous of Patagonia. *Nature* **387**, 390–392.

Ostrom, J. H. (1969). Osteology of *Deinonychus antirrhopus*, an unusual theropod from the Lower Cretaceous of Montana. *Bull. Peabody Museum Nat. History* **30**, 1–165.

Ostrom, J. H. (1990). Dromaeosauridae. In *The Dinosauria* (D. B. Weishampel, P. Dodson, and H. Osmólska, Eds.), pp. 269–279. Univ. of California Press, Berkeley.

Unwin, D. M., Perle, A., and Trueman, C. (1995). *Protoceratops* and *Velociraptor* preserved in association: Evidence for predatory behavior in dromaeosaurid dinosaurs. *J. Vertebr. Paleontol.* **15**(Suppl. to No. 3), 57A–58A.

Maniraptoriformes

KEVIN PADIAN
University of California
Berkeley, California, USA

Holtz (1996) established the node-based taxon Maniraptoriformes as the most recent common ancestor of *Ornithomimus* and birds and all descendants of that common ancestor. It comprises four stem groups: the ARCTOMETATARSALIA, MANIRAPTORA, OVIRAPTOROSAURIA, and *Ornithrolestes* (emend. Holtz, 1994, following Gauthier, 1986). It corresponds to Holtz's (1994, Fig. 4) "node 11," distinguished by a distally placed obturator foramen on the ischium, a tertiary antorbital fenestra, and elongated anterior cervical zygapophyses. Principal taxa included in this group are Maniraptora (DROMAEOSAURIDAE + AVES) and Arctometatarsalia (TYRANNOSAURIDAE, ORNITHOMIMIDAE, and TROODONTIDAE). The OVIRAPTORIDAE, ELMISAURIDAE, and THERIZINOSAURIDAE, which comprise the clade OVIRAPTOROSAURIA, are also included (Holtz, 1994, Fig. 4; 1996, Figs. 2 and 3). *Ornitholestes* also belongs to Maniraptoriformes on the basis of the common acquisition of the "maniraptoriform" forelimb and hand (Holtz, 1994, node 10); however, Holtz (1996, Figs. 2 and 3) collapsed nodes 10 and 11, implying that Maniraptoriformes is diagnosed not only by the three features of node 11 cited previously but also by the maniraptoriform hand that characterizes Maniraptoriformes, as defined in his (1996) text, plus *Ornitholestes.* The compositions of these taxa are somewhat unstable, but the definition stands; hence, all coelurosaurs that may prove to be descended from the most recent common ancestor of *Ornithomimus* and birds are Maniraptoriformes.

References

Gauthier, J. A. (1986). Saurischian monophyly and the origin of birds. *Mem. California Acad, Sci.* **8**, 1–55.

Holtz, T. R., Jr. (1994). The phylogenetic position of the Tyrannosauridae: Implications for theropod systematics. *J. Paleontol.* **68**, 1100–1117.

Holtz, T. R., Jr. (1996). Phylogenetic taxonomy of the Coelurosauria (Dinosauria: Theropoda). *J. Paleontol.* **70**, 536–538.

Manitoba Museum of Man and Nature, Canada

see MUSEUMS AND DISPLAYS

Manus, Comparative Anatomy

see FORELIMBS AND HANDS

Marginocephalia

Marginocephalia (see CERAPODA, Fig. 1) was established by Sereno (1986) to represent Pachycephalosauria + Ceratopsia. It was based on four synapomorphies, including the narrow parietal shelf, the posterior squamosal shelf, the short posterior premaxillary palate, and short postpubic process and loss of the pubic symphysis. Sereno regarded Marginocephalia as the sister taxon to EUORNITHOPODA (Ornithopoda of other authors; see PHYLOGENY OF DINOSAURS) within Cerapoda. An alternate view (see NEOCERATOPSIA; PHYLOGENY OF DINOSAURS) is that the origin of Marginocephalia is within Ornithopoda, perhaps among hypsilophodonts; if so, Cerapoda would be a problematic taxon. Marginocephalia can be redefined as a node-based taxon comprising the two node-based taxa Ceratopsia and Pachycephalosauria and all descendants of their most recent common ancestor. The most basal form known is *Stenopelix*, a fragmentary postcranial skeleton with broad hips reminiscent of both pachycephalosaurs and ceratopsians, but clearly not a member of either group (Dodson 1990); it is known from the Berriasian (Early Cretaceous) of Germany.

See also the following related entries:

CERAPODA • CERATOPSIA • ORNITHISCHIA • ORNITHOPODA • PACHYCEPHALOSAURIA • PHYLOGENY OF DINOSAURS

References

Dodson, P. 1990. Marginocephalia. pp. 562–563 in *The Dinosauria* (D. B. Weishampel, P. Dodson, and H. Osmólska, eds.). University of California Press, Berkeley.

Sereno, P. C. (1986). Phylogeny of the bird-hipped dinosaurs (Order Ornithischia). *Natl. Geogr. Res.* **2,** 234–256.

Megalosaurus

KEVIN PADIAN
University of California
Berkeley, California, USA

Megalosaurus was the first dinosaur to be described, though it was not first described as a dinosaur for the simple reason that dinosaurs were not then known (review in Swinton, 1955). It was one of the three forms (and was both the only theropod and the only saurischian form) on which Owen (1842) erected the Dinosauria (along with the ornithischians *Iguanodon* and *Hylaeosaurus;* see DINOSAURIA, DEFINITION). Ironically, 175 years after its baptism, *Megalosaurus* remains largely enigmatic and confusing.

Dean William Buckland (1824) of Oxford described the first known remains of the taxon (see HISTORY, EARLY DISCOVERIES); these consisted of a lower jaw (Fig. 1), several vertebrae and partial ribs, most of the pelvis, and the right femur of a specimen collected from the Stonesfield Slate (Middle Jurassic) of Stonesfield, Oxfordshire. The name *Megalosaurus* had first been given to the specimen when it was mentioned in Parkinson's *An Introduction to the Study of Fossil Organic Remains* (1822). In that same year Gideon Mantell, in *The Fossils of the South Downs*, first mentioned the teeth, discovered by his wife in Early Cretaceous outcrops in Sussex, that eventually became part of the basis for the genus *Iguanodon* in 1825. *Megalosaurus*, or "great reptile," was not given a specific epithet by either Parkinson or Buckland; that was eventually done by Hermann von Meyer (1832), who named it *M. bucklandi.* Owen eventually produced a reconstruction (Fig. 2) that pictured the animal as a quadruped; the bipedality of theropods would not be known until Cope and Marsh described remains of *Dryptosaurus (Laelaps)* from New Jersey in 1866 and 1877 (Desmond, 1975). Owen's reconstruction was made the basis of a statue by Waterhouse Hawkins in the CRYSTAL PALACE Exposition near London.

At the point at which it was first described, 160 years ago, the concept of *Megalosaurus* reached the peak of its nomenclatorial clarity. Since then, remains

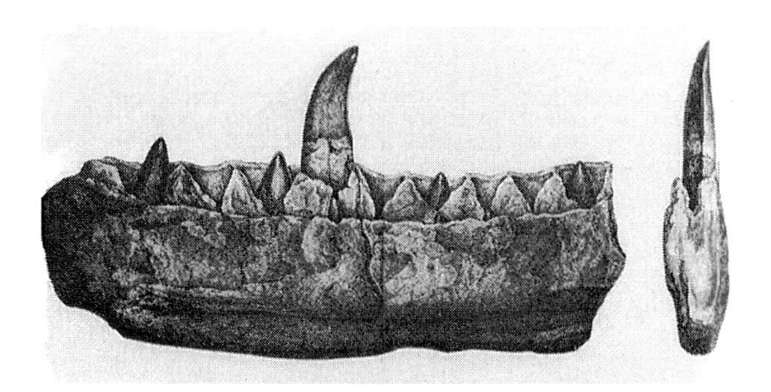

FIGURE 1 Lower jaw of *Megalosaurus*, the type specimen described by Buckland (1824).

FIGURE 2 Owen's reconstruction of *Megalosaurus,* with the available skeletal material shown.

from all over the world, ranging from the Early Jurassic through the latest Cretaceous, have been referred to the genus primarily because it is so incomplete that in its general characters it resembles nearly any large theropod. At least two dozen species, mostly based on incomplete or indeterminate material, have been assigned to *Megalosaurus,* and many of these were eventually separated to form the basis for other valid or equally questionable taxa, including *Magnosaurus, Poikilopleuron, Eustreptospondylus,* and *Dilophosaurus,* to name only a few (reviewed in Steel, 1970; Molnar, 1990; Molnar *et al.*, 1990). It has been mixed on at least two occasions with British thyreophoran remains: *"Nuthetes" destructor,* described by Owen from the middle Purbeck of Dorset, was a chimera of "megalosaurid" bones and teeth and thyreophoran dermal ossicles (see Steel, 1970, p. 34), and the type specimen of the thyreophoran *Scelidosaurus harrisonii* [first mentioned by Owen in an entry titled "Palaeontology" in the *Encyclopedia Britannica* (1859), then in his *Palaeontology* (1860, 2nd ed.), and finally described in 1861] consisted of the partial femur of a large megalosaurid theropod, to which Owen (1861) added first a large megalosaurid knee joint and then much more complete skull and associated skeletal material (Owen, 1863) that were ornithischian (historical review in Newman, 1968, who advocated reestablishing the name *Scelidosaurus* on the basis of the skull and skeleton of the juvenile thyreophoran). The name, however irregularly, has been applied to the thyreophoran material ever since. These vignettes illustrate only part of the tremendously confusing history of *Megalosaurus.*

However, in detail the type material has some more particular characters. These were reviewed by Waldman (1974), based on comments by A. D. Walker. The maxilla and dentary have 12 or 13 teeth with anteriorly and posteriorly positioned carinae (not oblique). The dentary is straight and has a symphysial facet. The vertebrae are short and the is scapula large, with an anterior expansion of the middle part of the blade, and the humerus is stout. The pubis has an extensive symphysis and a small distal thickening, with no foot. The ischium is downwardly curved posteriorly. The femur is massive and the lesser trochanter is placed well down its shaft; the tibia is approximately 83% of the femur in length.

These characters are sufficient to tell that *Megalosaurus* is not a COELUROSAUR or a CARNOSAUR. It is similar in many respects to *Dilophosaurus,* including its tooth count, the somewhat anteriorly expanded scapula, the stout humerus, the form of the pelvic girdle, and the placement of the lesser trochanter; however, some of these resemblances may be plesiomorphic or a function of large size. Nevertheless, they indicate that Welles had reason to assign his Early Jurassic theropod to *Megalosaurus*—before the dual cranial crests on the former, once discovered, suggested to him the distinct name *Dilophosaurus* for this material (see CERATOSAURIA).

Megalosaurus has been a magnet for the association of other taxa, such as *Poikilopleuron* (destroyed during World War II), *Eustreptospondylus,* and at one point or another several other genera that are now placed elsewhere, including *Dilophosaurus.* These taxa, grouped variously into Megalosauridae, Megalosauria, or other permutations of the generic name, have yet to demonstrate a unity that can be diagnosed by a unique suite of characters; hence, suprageneric taxa based on *Megalosaurus* are to date questionable. However, further study of the materials in question is needed in order to determine whether the oldest dinosaurian generic name can be vindicated by a unique diagnosis and a clear sense of phylogenetic position and stratigraphic range.

References

Buckland, W. (1824). Notice on *Megalosaurus* or great fossil lizard of Stonesfield. *Trans. Geol. Soc. London* **21,** 390–397.

Desmond, A. J. (1975). *The Hot-Blooded Dinosaurs: A Revolution in Palaeontology.* Blond & Briggs, London.

Molnar, R. (1990). Problematic theropoda: "Carnosaurus." In *The Dinosauria* (D. B. Weishampel, P. Dodson, and H. Osmólska, Eds.), pp. 306–316. Univ. of California Press, Berkeley.

Molnar, R., Kurzanov, S. M., and Dong, Z. (1990). Carnosauria. In *The Dinosauria* (D. B. Weishampel, P. Dodson, and H. Osmólska, Eds.), pp. 169–209. Univ. of California Press, Berkeley.

Newman, B. (1968). The Jurassic dinosaur *Scelidosaurus harrisoni*, Owen. *Palaeontology* **11,** 40–43.

Owen, R. (1842). Report on British fossil reptiles, Part II. *Rep. Br. Assoc. Adv. Sci.* **1841,** 60–294.

Owen, R. (1861). A monograph of a fossil dinosaur (*Scelidosaurus harrisonii,* Owen) of the Lower Lias. Monograph of the Fossil Reptilia of the Liassic Formations, Part I (1859), pp. 1–14, Plates I–VI.

Owen, R. (1863). *Scelidosaurus harrisonii,* continued. Monograph of the Fossil Reptilia of the Liassic Formations, Part II (1860), pp. 1–26, Plates I–XI.

Steel, R. (1970). *Saurischia. Handbuch der Paläoherpetologie, Teil 14* (O. Kuhn, Ed.). Fischer-Verlag, Stuttgart.

Swinton, W. E. (1955). *Megalosaurus,* the Oxford dinosaur. *Adv. Sci.* **12,** 130–134.

von Meyer, H. (1832). *Paläologica zur Geschichte der Erde.* Frankfurt am Main.

Waldman, M. (1974). Megalosaurids from the Bajocian (Middle Jurassic) of Dorset. *Palaeontology* **17,** 325–339.

Welles, S. P. (1984). *Dilophosaurus wetherilli* (Dinosauria, Theropoda): Osteology and comparisons. *Palaeontographica* **185A,** 85–180.

Megatracksites

MARTIN LOCKLEY
University of Colorado at Denver
Denver, Colorado, USA

Megatracksites are regionally extensive single surfaces, or thin layers of strata, that yield abundant tracks. Hence, the term has a specific stratigraphic meaning. Known examples from the Jurassic and Cretaceous of North America and Europe are on the order of hundreds to tens of thousands of square kilometers. All known examples are associated with marine coastal plain environments, and within each megatracksite the same track types are consistently found throughout (see also DINOSAUR FOOTPRINTS).

The term megatracksite should not be applied indiscriminately to any geographic area where tracks are abundant, unless the criteria of a specific surface or thin unit of strata are met.

See also the following related entries:

FOOTPRINTS AND TRACKWAYS • LOMMISWIL

References

Lockley, M. G. (1991). *Tracking Dinosaurs: A New Look at an Ancient World*, pp. 238. Cambridge Univ. Press, Cambridge, UK.

Lockley, M. G., with Hunt, A. P. (1997). *Dinosaur Tracks and Other Fossil Footprints of the Western United States*, in press. Columbia Univ. Press, New York.

Memphis Pink Palace Museum, Tennessee, USA

see MUSEUMS AND DISPLAYS

Mesa Southwest Museum, Arizona, USA

see MUSEUMS AND DISPLAYS

Mesozoic Era

KEVIN PADIAN
University of California
Berkeley, California, USA

The Mesozoic Era, often informally known among terrestrial fossil vertebrate workers as the "Age of Dinosaurs," is one of three eras (Paleozoic, Mesozoic, and Cenozoic) in the Phanerozoic Eon (approximately 535 mya to present). The Mesozoic Era comprises three periods (TRIASSIC, JURASSIC, and CRETACEOUS). Each period is divided into stages. (These divisions, with their estimated ages in millions of years, are reproduced on the inside covers of this book.) The stages have mostly been established on the basis of marine deposits, which are often difficult to correlate with terrestrial sediments without (i) marine–terrestrial interfingering of sediments, (ii) occurrence of terrestrial index fossils—usually palynomorphs (fossil pollen) or diagnostic terrestrial vertebrates, or (iii) radiometric dates (see GEOLOGIC TIME; RADIOMETRIC DATING). Dinosaurian diversity through the Mesozoic Era appears to grow steadily once corrections are accounted for (Dodson, 1990): Dinosaur-bearing deposits are mostly from lowland terrestrial exposures, which will not have been preserved during times of marine encroachment or regression, usually resulting in erosion (Padian and Clemens, 1985). Upland environments generally have a poor fossil record. Surprisingly, arid environments may preserve animals well if they can be buried quickly and preserved from decomposition and exposure; interdune deposits can be especially productive (e.g., GLEN CANYON GROUP; DJADOKHTA FORMATION). The tectonic history of the earth has determined through time the factors contributing to the deposition, preservation, and exposure of dinosaur-bearing rocks, and the present distribution of these fossils cannot be understood without the framework of tectonic history.

The Triassic (250–206 mya) comprises the Early Triassic (Scythian Stage), Middle Triassic (Anisian and Ladinian stages), and Late Triassic (Carnian and Norian stages; in some regions a Rhaetian Stage is recognized). The earliest members of Saurischia and Ornithischia are known from the Carnian Stage, from both skeletal remains and footprints; apparently earlier records have been mistaken or cannot be verified (King and Benton, 1996). This is also when the first members of other ORNITHODIRAN groups are known (see DINOSAUROMORPHA; ORIGIN OF DINOSAURS). Ornithischian dinosaurs are very rare and fragmentary in the Late Triassic, but a few basal sauropodomorphs *(Plateosaurus)* and theropods *(Coelophysis)* are known. However, Middle Triassic terrestrial vertebrates are generally poorly known because terrestrial Middle Triassic outcrops are not extensive anywhere in the world.

The Jurassic (206–144 mya) comprises the Early Jurassic (Hettangian, Sinemurian, Pliensbachian, and Toarcian stages), the Middle Jurassic (Aalenian, Bajocian, Bathonian, and Callovian stages), and the Late Jurassic (Oxfordian, Kimmeridgian, and Tithonian stages). In the Early Jurassic the first thyreophorans, ornithopods, and sauropods are known. The Middle Jurassic is not well exposed around the world because this was a time of great marine transgressions onto the continental surfaces; nonetheless, in recent years the first stegosaurs, a variety of sauropods, and ceratosaurian theropods have been found, particularly in China. It is in the Late Jurassic that the first great flowering of dinosaurian life is known, mostly from the MORRISON FORMATION of North America, the TENDAGURU beds of Africa, and coeval formations in China. Stegosaurs, ankylosaurs, hypsilophodontids, carnosaurs, coelurosaurs, and a variety of sauropods seem to appear suddenly; however, the relatively poorly known Middle Jurassic record may be contributing to this apparently sudden effect. The first birds appear in the Late Jurassic.

The Cretaceous (144–66.5 mya) comprises the Early Cretaceous (Berriasian, Valanginian, Hauterivian, Barremian, Aptian, and Albian stages; the first four are collectively known as the Neocomian) and the Late Cretaceous (Cenomanian, Turonian, Coniacian, Santonian, Campanian, and Maastrichtian stages). There is no formally recognized "Middle Cretaceous." In the early Early Cretaceous, dinosaur fossil records are well exposed in China, Europe, Africa, and in parts of North and South America and Australia. Iguanodontids, abelisauroid, and various coelurosaurian theropods, and perhaps the first poorly known members of some Late Cretaceous dinosaur groups are known from these times. The Late Cretaceous represents the final Mesozoic flowering of dinosaurs: ankylosaurids, hadrosaurs, pachycephalosaurs, ceratopsians, titanosaurs, abelisaurids, and a broad variety of coelurosaurs, including the first great radiation of birds (mostly Enantiornithes: see AVES). Much of the Cretaceous diversification of dinosaurs is traced to the diversification of plants, particularly angiosperms, although climatic conditions and other factors must also be considered (see MESOZOIC FLORAS; PLANTS AND DINOSAURS). Dinosaurian diversity since the Late Triassic experienced rapid change, with few genera outlasting a few million years and most are much more ephemeral (Padian and Clemens, 1985; Dodson, 1990). Hence, turnover at the generic level was high, amounting to nearly 100% from one geologic formation to the next. The end-Cretaceous extinction, then, was not a question of an increase in extinction rates but one of a drop in origination rates not easily explicable by catastrophic agents (see EXTINCTION).

Mesozoic climates were more equable than those today (see PALEOCLIMATOLOGY), and temperatures were generally warmer than they are today in temperate and polar regions—although the Mesozoic terrestrial world was not exactly the benign "hothouse" often pictured in traditional artists' reconstructions. There were no polar icecaps, so seasonality was less pronounced over most of the globe. Climates were warm during the early Mesozoic; a drying phase is widely thought to have occurred in the latest Triassic and Early Jurassic, and a cooling trend settled in during the Late Jurassic and Early Cretaceous, returning to warmer temperatures in the Late Cretaceous. Some cooling occurred in the Paleocene and Eocene, when climates became wetter again, after the Mesozoic.

Much of Mesozoic climatic equability can be traced to the configuration of the continents (see TECTONICS), which were collected together as the supercontinent Pangaea at the beginning of the Mesozoic. Beginning in the Late Triassic, the Atlantic Ocean began to "unzip" the Laurasian continents, north to south, and the Tethys seaway, separating the northern and southern continents, had begun to widen by the Middle Jurassic. Still the continents were in more or less a single landmass for long afterward, and extensive provinciality of terrestrial faunas is not obvious until the Early Cretaceous, at which point it generally increases through the rest of the Mesozoic. Separation of India and Australia–Antarctica from Africa and South America is not evident until the Early Cretaceous, and biotic exchange may have persisted across the Cape of Good Hope until well into the Tertiary. Africa was well separated from the other southern continents by the mid-Cretaceous, and India was an island by that time, heading northward toward the southern edge of Asia. The Laurasian continents appeared to continue biotic exchange through the Mesozoic, but the faunas are not cosmopolitan; seemingly they were interrupted either by marine excursions (eastern and western North America; North America–Europe) or by ecological barriers (North America–Asia) at least from time to time. For example, congeneric hadrosaurs *(Saurolophus)* are known in both North America and Asia in the Maastrichtian; different genera of closely related pachycephlosaurs are known from the Campanian and Maastrichtian of both continents; basal ceratopsians (*Psittacosaurus*) are known from Asia but not North America, protoceratopsids are known from both continents, but Ceratopsidae are restricted to the Western Hemisphere.

The Mesozoic Era may be considered the Age of Dinosaurs, but birds, which are living dinosaurs, number some 10,000 species today, having increased in diversity throughout the Tertiary, which has traditionally been considered the "Age of Mammals," although there are only approximately 6000 mammal species today. In this context, as Gauthier (1986) pointed out, perhaps the Age of Dinosaurs did not end with the Mesozoic Era. Nevertheless, the relative species diversity must be put into the context that birds are the only group of living dinosaurs and, taxonomically and ecologically diverse though they may be, mammals outdo them in ecological diversity today and have done so since the Mesozoic Era ended.

See also the following related entries:

CRETACEOUS PERIOD • GEOLOGIC TIME • JURASSIC PERIOD • RADIOMETRIC DATING • TRIASSIC PERIOD

References

Dodson, P. (1990). Counting dinosaurs: How many kinds were there? *Proc. Natl. Acad. Sci. USA* **87,** 7608–7612.

Gauthier, J. A. (1986). Saurischian monophyly and the origin of birds. *Mem. California Acad. Sci.* **8,** 1–55.

King, M. J., and Benton, M. J. (1996). Dinosaurs in the Early and Mid Triassic?—The footprint evidence from Britain. *Palaeogeogr. Palaeoclimatol. Palaeoecol.* **122,** 213–225.

Padian, K., and Clemens, W. A. (1985). Terrestrial vertebrate diversity: Episodes and insights. In *Phanerozoic Diversity Patterns* (J. W. Valentine, Ed.), pp. 41–96. Princeton Univ. Press, Princeton, NJ.

Mesozoic Faunas

HANS-DIETER SUES
Royal Ontario Museum
Toronto, Ontario, Canada

The Mesozoic Era (245–65 mybp) is often informally referred to as the "Age of Reptiles" because it was during that interval of geological time that the great diversification of diapsid REPTILES took place on land, in the seas, and in the air. At the beginning of the Mesozoic, nonmammalian therapsids ("mammal-like reptiles") were still the most common land-living tetrapods, much as they were during the preceding Permian Period, but therapsid-dominated communities soon gave way to communities characterized by an abundance of diapsids, especially archosaurs.

Even after more than a century of intensive worldwide paleontological exploration, much remains to be learned about the history of the faunas of Mesozoic land-dwelling vertebrates. However, the broader outlines of the historical development of Mesozoic continental ecosystems are now slowly emerging (for a review see Wing and Sues, 1992).

The Triassic period (245–200 mybp) is named for the threefold standard sequence of sedimentary rocks characteristic of that time interval in central Europe. There, the Lower Triassic (Buntsandstein) is represented by continental strata deposited under apparently semiarid to arid conditions and containing few vertebrate fossils. The Middle Triassic (Muschelkalk) comprises mainly shallow-water marine carbonates. The Upper Triassic (Keuper) forms an extensive sequence of mostly continental strata with locally abundant tetrapod remains. This mixture of sedimentary rocks forms a major obstacle to direct correlation with Triassic deposits in other regions of the world, and thus intercontinental correlations continue to be the subject of much debate.

Most of the principal groups of present-day land vertebrates, including crocodilians, mammals, squamate reptiles (lizards and snakes), and turtles, or their closest relatives, made their first appearance in the fossil record during the Triassic. Furthermore, dinosaurs and pterosaurs originated during the latter part of that period. The supercontinent Pangaea existed during the entire length of the Triassic, and there apparently were few, if any, significant barriers for the dispersal of land animals across this giant landmass.

The Middle and Late Triassic witnessed the rapid evolutionary radiation of archosaurian reptiles and a more or less concomitant decline in the diversity of therapsids. Middle Triassic archosaurian diversity comprised an abundance of crurotarsal archosaurs and a few lightly built, bipedal forms, such as *Lagerpeton* and *Marasuchus*, which had mesotarsal ankle joints and were closely related to dinosaurs (see DINOSAUROMORPHA). Continental strata of Middle Triassic age are best documented from western Argentina, southern Brazil, and Tanzania. The oldest known ornithischian (*Pisanosaurus*) and two of the earliest ?saurischian dinosaurs (*Eoraptor* and *Herrerasaurus*) are from the Upper Triassic (Carnian) Ischigualasto Formation (228 mybp) of northwestern Argentina, indicating that the divergence of the two main lineages of dinosaurs occurred even earlier in the Triassic. The subsequent evolutionary diversification of dinosaurs apparently took place rapidly and, by the end of the Triassic Period, these reptiles were already very common worldwide. Only a few lineages of therapsids persisted, comprising forms approaching and finally reaching a mammalian level of structural organization toward the end of the Triassic.

The Jurassic Period (200–145 mybp), named for the Jura Mountains of Switzerland, represented an interval of widespread seas. Many of the characteristic sedimentary rocks of that period in England and Germany are marine in origin, and this makes precise correlation with continental strata elsewhere very difficult.

Early Jurassic assemblages of continental tetrapods differed from the preceding Late Triassic ones principally in the absence of several lineages of nondinosaurian archosaurs that, along with a number of other groups of reptiles and temnospondyl amphibians, had become extinct at the end of the Triassic. Both the dinosaurs and the small land-living vertebrates (including early precursors of mammals) were surprisingly cosmopolitan in distribution, even though the breakup of the supercontinent Pangaea was already under way. Dinosaurian communities comprised various sauropodomorphs and both small and large ceratosaurian theropods such as the crested *Dilophosaurus*. Middle Jurassic continental strata are still poorly known, but recent discoveries in Argentina and China indicate a considerable variety of dinosaurs, especially sauropods and basal tetanuran theropods.

The well-sampled Late Jurassic vertebrate communities from the western United States and East Africa (Tanzania) were characterized by the great abundance and diversity of dinosaurs. Especially noteworthy is the occurrence of a variety of very large sauropods such as *Brachiosaurus* and *Diplodocus*. The American and East African assemblages shared a number of dinosaurian taxa in common, which indicates the persistence of a land connection through Europe to Gondwana. Apparently, Late Jurassic dinosaurian assemblages from Sichuan, China (the exact geological age of which remains uncertain), appear to be quite distinct in their specific faunal composition; it is currently unclear whether this difference is due to geographic isolation or geological age. In addition to the radiation of the two major dinosaurian lineages, the Jurassic Period is noteworthy for the diversification of the principal "modern" groups of small tetrapods including lizards, frogs and salamanders, mammals and various closely related groups, and turtles. Furthermore, the first undisputed birds, descendants of small theropod dinosaurs, appeared in the Late Jurassic.

The Cretaceous, named for its extensive chalk deposits, represented the longest period of the Mesozoic (145–65 mybp) and witnessed the acme of dinosaurian evolution. At the end of that period, all dinosaurian lineages except birds, along with scores of other terrestrial and marine organisms, became extinct. The causes of this major biotic crisis form the subject of continuing scientific debate (see EXTINCTION).

The Cretaceous record of continental vertebrates is highly biased toward the last two stages of the period, the Campanian and Maastrichtian. Early Cretaceous tetrapod assemblages are still inadequately known. In communities from the Lower Cretaceous of Europe and western North America, large ornithopod dinosaurs such as *Iguanodon* were more common and more diverse than in Late Jurassic communities, but sauropods appear to have been much less abundant.

Late Cretaceous tetrapod communities are best documented by a series of assemblages from western North America and Mongolia. The latter regions formed parts of a landmass, which is sometimes referred to as Asiamerica and was separated from eastern North America and adjoining regions of Europe by a large midcontinental seaway. The assemblages from western North America represented, for the most part, lowland floodplain environments. They were dominated by the duck-billed hadrosaurs and the horned ceratopsians, large herbivores that, at least in some taxa, apparently foraged in great herds. Some paleontologists have proposed a coevolutionary scenario that links the radiation of these dinosaurian plant eaters to the rise and rapid diversification of flowering plants (angiosperms). The main predatory dinosaurs in Late Cretaceous Asiamerica were tetanuran theropods, especially the large Tyrannosauridae and the "sickle-toed" Dromaeosauridae. Late Cretaceous vertebrate communities from Mongolia and adjacent regions of China appear to represent drier, apparently at least seasonally arid environments. Although they shared many dinosaurs and other vertebrates in common with western North America, these Asian faunas also include a number of possibly endemic taxa of dinosaurs and other animals and lack advanced ceratopsians. Late Cretaceous "Gondwanan" tetrapod communities from Argentina, Brazil, India, and Madagascar were characterized by the abundance of titanosaurid sauropods, the rarity or absence of hadrosaurs, and the apparent absence of ceratopsians. Theropod dinosaurs were represented in Gondwana mainly by the Abelisauridae.

Although public attention focuses on the diversity of dinosaurs, the Cretaceous Period is also noteworthy for the first appearance in the fossil record, and initial evolutionary radiation, of both marsupial and placental mammals, the stem groups of modern birds, and many lineages of present-day amphibians and reptiles. These groups survived the terminal Cretaceous extinctions and became key constituents of continental ecosystems during the Tertiary Period.

See also the following related entries:

MESOZOIC ERA • MESOZOIC FLORAS

References

Weishampel, D. B., Dodson, P., and Osmólska, H. (Eds.) (1990). *The Dinosauria.* Univ. of California Press, Berkeley.

Wing, S. L., and Sues, H.-D. (1992). Mesozoic and early Cenozoic terrestrial ecosystems. In *Terrestrial Ecosystems through Time: Evolutionary Paleoecology of Terrestrial Plants and Animals* (A. K. Behrensmeyer, J. Damuth, W. A. DiMichele, R. Potts, H.-D. Sues, and S. L. Wing, Eds.), pp. 326–416. Univ. of Chicago Press, Chicago.

Mesozoic Floras

James F. Basinger
University of Saskatchewan
Saskatoon, Saskatchewan, Canada

The dinosaurs experienced a world much different than our own. In addition to all the environmental consequences of a warmer global climate, the vegetation clothing the landscape was of fundamentally different character to that we see around us.

Perhaps that most significant difference is the dominion of the flowering plants on much of today's land surface. This group, which numbers in the hundreds of thousands of species, evolved rather late in the Mesozoic. It is difficult to conceive of a world without them, and yet we must re-create this vegetation of the past in order to comprehend the world of the dinosaurs. Virtually all land ecosystems depend on energy stored by vascular plant photosynthesis. Large dinosaurian herbivores may have been the most conspicuous consumers of plants, but we commonly forget that the flesh consumed by the carnosaur was merely this energy taking a different form.

In our survey of Mesozoic vegetation, it is useful first to consider the evolutionary history of the major groups of land plants as told by the fossil record. It will then be possible to step back and briefly view Mesozoic environments.

The transition from the Paleozoic to the Mesozoic, the Permian–Triassic boundary, is marked by the most severe extinction of marine organisms in the history of life. The vast majority of species living in the Permian seas did not survive into the Triassic (see Erwin, 1993). Land animals, including reptiles, were not devastated to the same degree; neither were land plants. Nevertheless, global environmental changes occurring at this time strongly influenced the evolution of all land organisms.

The continents of the Paleozoic were somewhat distributed about the earth. Oceans surrounded these continents and commonly transgressed well inland to form great epicontinental seas. As the Paleozoic came to a close, motion of the earth's crustal plates slowly drove the continents together so that they coalesced into a vast supercontinent called Pangaea. Pangaea was so large that its interior developed a severely continental, dry climate. To make matters worse, ocean levels fell and the continents became almost completely emergent. This further dried out the land and probably resulted in disastrous loss of habitat for marine life.

The drying out of the land surface occurred gradually, over a period of approximately 40 million years, from the mid-Permian through much of the Triassic. The earth's land surface did not all become dry at the same time, but moist regions became gradually more restricted. Also, not all land surfaces became dry; the southern polar regions remained moist, although cool, and supported coal forests. Nevertheless, the vast coal swamps of the Carboniferous and Permian disappeared, and with them went the arborescent lycopsids (giant club mosses or scale trees) and sphenopsids (giant horsetails) and most of the archaic seed plants (cordaitalean conifers and early pteridosperms) and archaic ferns. The Triassic was a time of reorganization of land communities and the rise of the major groups of conifers, cycads, cycadeoids, ginkgophytes, later pteridosperms, and ferns that characterized global vegetation of the Jurassic and much of the Cretaceous.

The following overview of Mesozoic plant evolution is of necessity superficial because in diversity alone plants far exceeded the dinosaurs, the subject of the rest of this substantial encyclopedia. The balloon diagram (Fig. 1) presents a summary of the evolution of the major groups of land plants of the Mesozoic. Many groups of Paleozoic plants that became extinct near the Paleozoic–Mesozoic boundary have been omitted, and many minor groups have either been merged with related forms or entirely left off for sake of simplicity. Additional information on the plant fossil record can be found in the excellent paleobotanical textbooks by W. N. Stewart and G. W. Rothwell (1993) and T. N. Taylor and E. L. Taylor (1993).

Early land plants generally seem to fall naturally into two groups, the lycopsids and everything else. In the modern world, lycopsids occupy a variety of

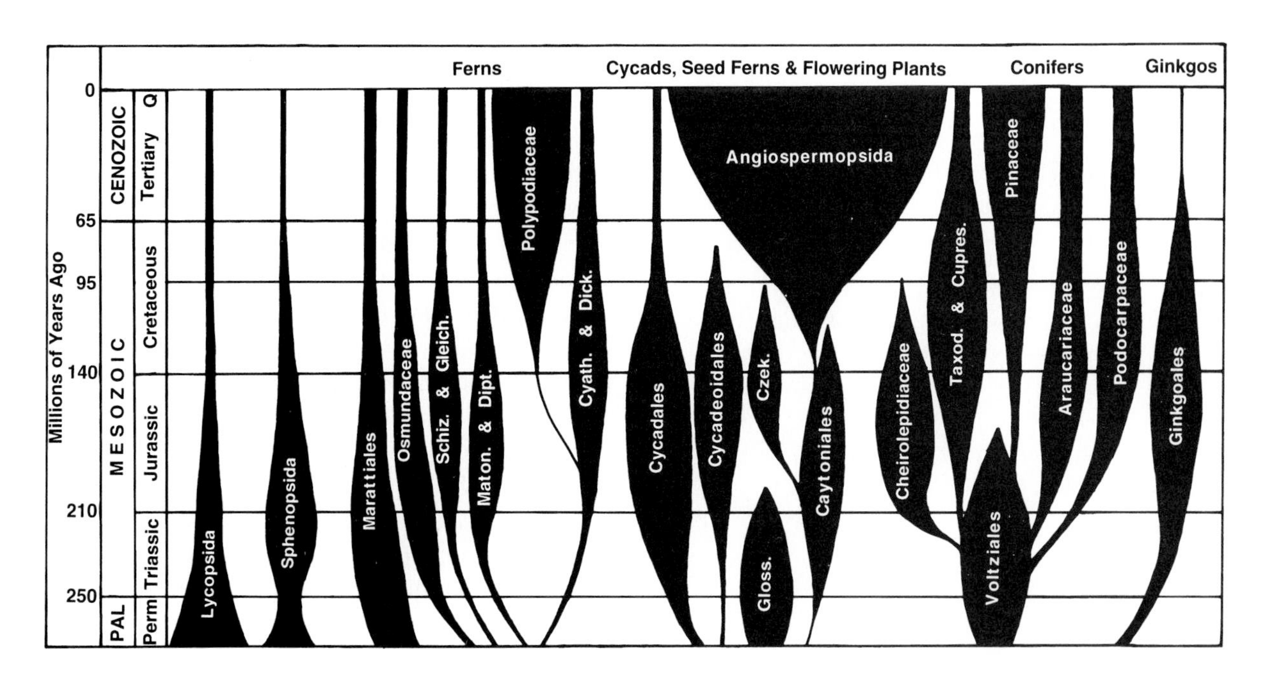

FIGURE 1 Mesozoic plant evolution. The width of each balloon is meant to generally represent both diversity and abundance so that the importance and success of each group can be followed through time and compared with other groups. Many groups of lesser significance to the Mesozoic have not been considered or have been included in related groups. PAL, Paleozoic; Perm, Permian; Schiz. & Gleich., Schizaeaceae and Gleicheniaceae; Maton. & Dipt., Matoniaceae and Dipteridaceae; Cyath. & Dick., Cyatheaceae and Dicksoniaceae; Czek., Czekanowskiales; Gloss., Glossopteridales; Taxod. & Cupres., Taxodiaceae and Cupressaceae.

inconspicuous roles as forest, prairie, and aquatic plants herbs (nonwoody plants). *Phylloglossum* and *Lycopodium* (ground pines), *Selaginella* (club mosses), and *Isoetes* (quillworts) alone survive. The herbaceous lineages of the ground pines and club mosses can be traced well back into the Paleozoic and probably have changed little in appearance and habit throughout the Mesozoic and Cenozoic.

The quillworts, on the other hand, are the last of a long line of primarily woody plants that included the great lepidodenrids (scale trees) of the Carboniferous coal swamps. The scale trees died out in the Permian, and by the end of the Triassic only a handful of dwarf isoetaleans remained, persisting in shallow water habitats.

The evolution of sphenopsids is similar to that of the lycopsids. All sphenopsids possess the distinctive feature of whorled leaves and branches, as does the single surviving genus *Equisetum* (horsetails). Large calamite trees of the Carboniferous occupied significant roles in the coal swamps but are gone before the Mesozoic as the equatorial swamp habitats disappeared. A variety of unusual sphenopsids, all herbs to low shrubs, including several genera of horsetails, existed during the Triassic and Jurassic. It is possible that these plants were important colonizers of open habitats, areas typically now occupied by flowering plants. By the end of the Mesozoic, only *Equisetum* remained.

Ferns (Pteropsida) experienced a great turnover from the Permian to Triassic. Few modern groups of ferns can be traced beyond the Triassic, and few of the Carboniferous ferns can be linked closely to Mesozoic forms (see Tidwell and Ash, 1994) (Fig. 2). Nevertheless, ferns of one type or another have occupied major roles in the forest understory and as colonizers of open habitats since the Paleozoic and continue to do so today.

The fern family Marattiaceae can be traced back into the Paleozoic with confidence. Typically with squat, thick stems and massive leaves, they are among the most spectacular plants. In the Carboniferous, *Psaronius* was abundant in the coal swamps. From the Permian through to the Jurassic, members of this

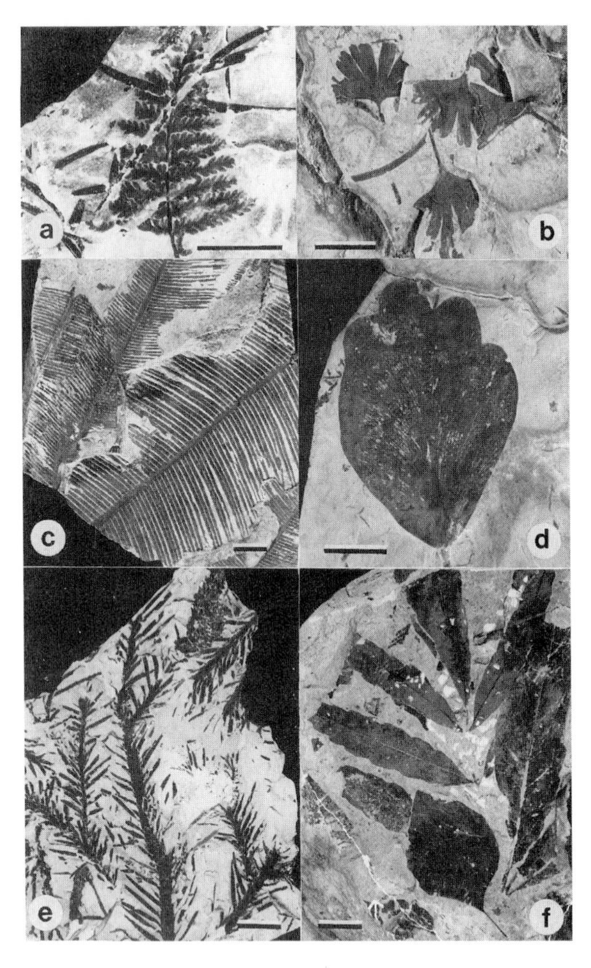

FIGURE 2 Some examples of important groups of plants during the Mesozoic. Fossils in a–e are from the Early Cretaceous of western Canada. Fossil in f is from the Late Cretaceous of western Canada. Scale bars = 2 cm. (a) A common fern of the genus *Coniopteris*; (b) fan-shaped leaves of *Ginkgo*; (c) pinnately divided leaves of the cycad *Ptilophyllum*; (d) a leaflet of a caytonialian seed fern called *Sagenopteris*; (e) the taxodiaceous conifer *Elatocladus*; (f) this pinnately compound angiosperm leaf has been referred to as *Cupanites*, the circular leaf ("*Zyzyphus*"); however, relationships to living angiosperms are unknown.

family were diverse and ranged widely and can be found as fossils on most continents. They probably were very important in the forest understory. During the Jurassic they declined in importance, perhaps as a consequence of competition from the numerous other groups of ferns evolving at this time. Today, the family is a small group scattered throughout the tropics.

The Osmundaceae (cinnamon and royal ferns) is a primitive family of small ferns that can be traced back to the Permian and is allied with some of the Carboniferous ferns. The fronds may be borne on a small squat stem or a creeping rhizome and are commonly dimorphic; that is, they produce sterile leafy portions and fertile portions. When crushed, the fertile portion produces a reddish brown powder of spores and sporangia that looks like ground cinnamon. The Osmundaceae was probably very important ground cover in the Triassic and Jurassic, giving way to other groups of ferns in the Cretaceous. It persists today as very widely distributed ferns of moist woodlands and marshes.

Most other families of the filicalean ferns (including Schizaeaceae, Gleicheniaceae, Matoniaceae, Dipteridaceae, Dicksoniaceae, Cyatheaceae, and the extinct Temskyaceae) appear to have originated in the Mesozoic from poorly defined Permian–Triassic ferns. These families become diverse during the Jurassic and Cretaceous. Expansion of the role of these families during that time may have been at the expense of the more archaic marattiaceous and osmundaceous ferns as well as the sphenopsids. It is also likely that the return to moister climates worldwide during the Early to Middle Jurassic contributed to a vast expansion of habitat suitable for ferns.

Although ferns are unable to produce woody secondary tissue, some members of the Dicksoniaceae and Cyatheaceae have been able to achieve tree height by development of a massive and armored primary stem. Today, tree ferns of these families are important in many tropical and subtropical forests, even becoming part of the canopy. The abundance of large frond fragments in the fossil record of the Jurassic and Cretaceous indicates that tree ferns were abundant, possibly as forest margin and understory shrubs to small trees. In the modern world this habitat is most likely to be occupied by angiosperms. Decline of many of these families in the Late Cretaceous through the Tertiary is most likely attributable to the diversification of angiosperms. The Temskyaceae was driven to extinction by the end of the Cretaceous, and the Matoniaceae was reduced from a worldwide tropical distribution to only two genera surviving in Malaysia and Indonesia.

The Polypodiaceae, the most widespread and abundant of all modern groups of ferns and the group with which we are most familiar, has perhaps the poorest Mesozoic record. Diversification of this family appears to have begun in the Cretaceous and continues to the present day. In fact, the tremendous success of the Polypodiaceae parallels that of the angiosperms. Whereas the rise of the angiosperms doomed many groups of plants, the Polypodiaceae apparently prospered with them.

The seed ferns (pteridosperms) and ferns of the Paleozoic are excellent examples of convergent evolution. Members of both groups adapted to swamp and wetland habitats and in doing so came to resemble one another so closely that it may be impossible to distinguish fern from seed plant on the basis of fronds alone. These early seed ferns became extinct along with the scale trees and calamites at the close of the Paleozoic but were survived by innovative seed plants that likely played a key part in the evolution of angiosperms.

The earliest known of these "later pteridosperms" are the glossopterids of the Southern Hemisphere. *Glossopteris* has been found on all fragments of the southern supercontinent of Gondwana and was strong evidence for Wegener's theory of continental drift. This group included large trees that dominated much of the southern mid- to high latitudes. The glossopterids diminish to extinction in the Triassic as the cycads and conifers diversify.

The Caytoniales includes an odd assortment of trees and shrubs of Late Triassic to Early Cretaceous age, some from the southern mid- to high latitudes and others from the north. *Sagenopteris*, a name applied to fossil leaves of the *Caytonia* family, is common in Jurassic to Early Cretaceous mixed forests of the northern mid to high latitudes. They may have had a similar appearance, and perhaps role, in these dominantly coniferous forests as broad-leaved hardwoods have in our modern mixed forests. There is some indication that the northernmost seed ferns were deciduous.

The Czekanowskiales is another enigmatic group of trees or shrubs that inhabited the northern high latitudes of the Jurassic and Early Cretaceous. Like the Caytoniales, they were inclined to experimentation with the morphology of their reproductive organs. This, and the coincidence of the extinction of the later pteridosperms with the appearance of early flowering plants, has not gone unnoticed by those searching for the origins of the angiosperms.

Ginkgo biloba, the maidenhair tree of the East, is the sole living member of a lineage that extends back into the Paleozoic and was present in essentially its current form when the dinosaurs roamed. The maidenhair tree is truly a living fossil. We should consider ourselves fortunate to still possess this magnificent plant because not only was it decimated throughout its former range by Tertiary global climatic cooling but also it is believed now extinct in its native eastern China through human land use and habitat loss. It survives only in cultivation.

Ginkgophytes reached their zenith in the Jurassic and Cretaceous, at one time spanning the globe from pole to pole. Diversity was never great but included a number of plants with deeply dissected leaves in addition to those with the fan-shaped leaves typical of the living species. Ginkgos are especially conspicuous in high-latitude fossil floras, where they apparently filled a niche now occupied by broad-leaved deciduous hardwoods. Species of the northern deciduous forests persisted into the Tertiary as ginkgos elsewhere became extinct, probably through competition with angiosperms. During the Cenozoic, the range of ginkgos dwindled from circumpolar to a handful of sites in China. Ironically, it has by human intervention once more achieved worldwide distribution.

Only 10 genera of cycads exist now, scattered around the globe in tropical to subtropical climates. They typically have thick, short, sparsely or unbranched stems with a crown of pinnately divided leaves. Their large seeds are borne in cones that may be gigantic, weighing many kilograms. The primitive appearance is not deceiving because they are the remnants of a once diverse group that was probably second only to the conifers in many Mesozoic forests.

The origins of the cycads can be traced as far as the Carboniferous, but it is not until the Triassic that they assume a major role. Although they become diverse, there is little departure from the morphology typified by the few living relicts. Their short stature and slow growth would ill-suit them to the forest canopy, but they may have been successful understory plants and colonizers of open areas. Their stout stems and tough leaves are ideally suited to drought tolerance, and they were prominent in equatorial arid to semiarid regions. Their global distribution included even the high latitudes, where they may have been deciduous and frost tolerant, features not found in any living species. Cycads decline through the Cretaceous in direct contrast to the rise of the angiosperms. By the end of the Cretaceous, cycads were probably near their current level of diversity, and today they continue to tenaciously resist extinction.

That cycads were so common throughout much of the Mesozoic, and within reach of herbivorous dinosaurs, begs us to question their component of dinosaur diets. Leaves of most cycads appear to have been stiff and resistant to casual browsing. Nevertheless, the horny beaks and tooth batteries of ornithischians may have been more than a match for them. It may be that the sauropods were less well equipped to deal with most cycads. The growing tip of most cycads is deeply hidden by an armor of leaf bases and stiff bracts, perhaps as protection against predation. Unlike conifers, cycads are generally unable to branch and die if the shoot tip is destroyed.

Had dinosaurs depended on cycads in the Jurassic and Early Cretaceous, they must have had little difficulty in altering their diet to other plants because dinosaurs continued to enjoy their rule in the Late Cretaceous long after cycads had become scarce and largely replaced by angiosperms.

The cycadeoids paralleled the cycads in time, space, and appearance. Arising in the Triassic, they became abundant and worldwide in the Jurassic and Early Cretaceous. Although many became more highly branched and shrubby, others resembled cycads so closely that a microscopic examination is necessary to differentiate them. Their reproductive structures, however, differ vastly from the simple unisexual cones of cycads. Many had bisexual cones, an uncommon feature outside of the angiosperms, although the resemblance to flowers is purely superficial. Profound differences in reproductive organs between cycads and cycadeoids indicate an extremely remote relationship.

Similar vegetative appearance, and the common occurrence of both cycads and cycadeoids in the same deposits, indicates that the two groups occupied similar niches and may have had interchangeable roles in many environments. Both decline in the Late Cretaceous, but the cycadeoids failed to survive and became extinct well before the end of the Cretaceous.

Conifers arose in the Carboniferous but did not become common until the beginning of Mesozoic (see Miller, 1977; Beck, 1988). As a group, they were well suited to the extensive dry land habitats of the Triassic and soon came to dominate the forest canopies throughout the world. Permian and Triassic conifers belong to the Voltziales, the "transition conifers," named so because they are perceived as a group transitional between early seed plants and living groups of conifers. All modern families of conifers, including the Araucariaceae, Podocarpaceae, Taxodiaceae, Cupressaceae, and Pinaceae, are first recognizable in the Late Triassic to Middle Jurassic and replace the Voltziales before the end of the Jurassic.

The voltzialean conifers would have looked much like many modern conifers: tall and highly branched, with needle-like leaves. Seed cones show considerable diversity but are all variations on the theme found in most modern conifers, in which seeds are borne on the surfaces of overlapping scales that are assembled into cones. Petrified woods of these trees are quite similar to woods of a number of living conifers, particularly members of the Araucariaceae and Taxodiaceae. The name *Araucarioxylon* has been used for many of these fossil woods, indicating their resemblance to wood of the modern *Araucaria* (e.g., Norfolk Island pine and monkey puzzle). Petrified conifer woods of the Triassic and Jurassic, including huge tree trunks found in so-called petrified forests of the western United States, are generally the remains of voltzialeans. As may be deduced, the distinction between voltzialeans and modern conifers is blurry.

Various members of the Voltziales were found on all continents. Aridity and global warmth during the Triassic to Early Jurassic appears, however, to have created an effective barrier to migration across the low, equatorial latitudes, tending to isolate groups adapted to the more humid mid- to high latitudes. This isolation was increased as Pangaea began to break apart, creating an oceanic barrier between the northern and southern continents. Many conifer families evolving in the Jurassic were more or less confined to either the northern or Southern Hemisphere, failing to effectively cross the equatorial barriers.

The Pinaceae (pine family) is entirely of the Northern Hemisphere. With a record extending well back into the Jurassic, it seems to have been most common in the mid latitudes of the Mesozoic. A number of now extinct genera flourished in the Cretaceous, but only a few of the living genera, such as pines and golden larch, can be traced this far. Although the pollen record of the family indicates that many of the living genera may have been present in the Cretaceous, most do not enter the macrofossil record until the Tertiary. This has led to speculation that many evolved in alpine regions during the Late Cretaceous and early Tertiary and did not become widespread until onset of Tertiary global climatic deterioration. Most members of the Pinaceae are now restricted to montane, boreal, and northern mixed forests of the mid- to high latitudes. The Pinaceae has been mistakenly portrayed as an archaic gymnospermous group evolutionarily eclipsed by the angiosperms. Although some conifers fit this description, the Pinaceae underwent diversification in parallel with the angiosperms, although on a more modest scale, during the Cretaceous and Tertiary. It is now diverse, widespread, and highly successful in a wide range of habitats and dominates vast tracts of boreal forest.

The Araucariaceae is now entirely restricted to the Southern Hemisphere. The Norfolk Island pines and monkey puzzles are familiar house and garden trees. Despite numerous reports of wood and leaves attributed to this family from the Northern Hemisphere, only a few are credible. It seems that most of these remains belong to other families, and that the Araucariaceae made few inroads into the Northern Hemisphere in the Jurassic but was extirpated there in the Cretaceous.

The Podocarpaceae is also a widely distributed Southern Hemispheric family. Members of this family are commonly called "yews" and "pines," although there is no relationship to these Northern Hemispheric trees. The fossil record of the podocarps in the Northern Hemispheric record is limited to a few arguable macrofossil reports and a record of pollen grains that more probably belong to the Pinaceae. Its southern record, on the other hand, is rich and reveals it to have been a common feature of southern forests throughout the Jurassic and Cretaceous. It continues to occupy a variety of habitats, from tropical rainforest to alpine shrub, throughout much of the Southern Hemisphere.

The Taxodiaceae (redwood family) today consists of a handful of relicts of a once prominent group spanning both hemispheres. Nine genera, most of each with only a single surviving species, are limited to small areas of North America and eastern Asia; only a single genus survives in Australia. Included are the bald cypress (*Taxodium*), dawn redwood (*Metasequoia*), and the largest living trees, the redwood and giant sequoia (*Sequoia* and *Sequoiadendron*). During the Jurassic and Cretaceous, however, members of this family were major constituents of mid- and high-latitude forests, and they are common fossils in dinosaur-bearing rocks throughout the Northern Hemisphere. As a group they grow rapidly and tend to form large trees, ideally suiting them to canopy domination in moist forests. They may well have formed a part of the diet of high browsers. From limited evidence of their presence in coal deposits, members of this family may well have commonly formed swamp forests, thereby contributing significantly to the formation of Jurassic and Cretaceous coal reserves. Some, such as dawn redwood, bald cypress, and Chinese cypress (*Glyptostrobus*), evolved deciduousness and became abundant in the high northern deciduous forests. The family appears to have been decimated first by angiosperm competitors, then by Tertiary climatic deterioration.

The Cupressaceae (cypress or cedar family) is the only conifer family that successfully diversified in both hemispheres. Branches of these plants typically bear scale-like leaves arranged in whorls or opposite pairs and are known from rocks as old as Late Triassic. Although widely distributed, cedars are rarely abundant in fossil assemblages and do not seem to have occupied the prominent position of other conifers. They are closely related to the redwoods and have been included with them in Fig. 1.

The Cheirolepidiaceae is the only major family of Jurassic and Cretaceous conifers that failed to survive into the modern world. It was a diverse and enormously successful group that was united by its production of the peculiar *Classopolis* pollen. Leaves were needle-like to scale-like, and branches may resemble those of other families such as the Taxodiaceae and Cupressaceae. Many members of this family appear to have been well adapted to arid and semiarid conditions, and an abundance of *Classopolis* pollen is commonly interpreted as indicative of hot and dry climate. The overwhelming abundance of these plants throughout the equatorial regions during much of the Jurassic and Cretaceous, and the typical absence there of other types of conifers such as the Pinaceae, reveals strong latitudinal zonation of vegetation that was controlled by moisture.

From available evidence, the angiosperms originated in the Early Cretaceous but did not become particularly common or abundant until the Late Cretaceous (see Beck, 1976; Friss *et al.*, 1987). There are no confirmed reports of any flowering plants from pre-Cretaceous rocks. Angiosperm fossils reveal a slow increase in diversity and abundance up to about the Lower–Upper Cretaceous boundary and then an explosive radiation that resulted in their common occurrence in fossil floras worldwide. Early angiosperms appear to have included shrubs and herbs that occupied an assortment of marginal habitats, possibly as weedy opportunists. The efficiency of their reproductive process, which included the development of the flower, was revolutionary. In concert with their exploitation of various pollinators, particularly insects, the angiosperms invaded virtually every possible habitat. They appear first in the low latitudes; by the end of the Lower Cretaceous they were found in both north and south polar regions.

Although some angiosperm families, such as the Platanaceae (sycamores and plane trees), are conspicuous and abundant in Late Cretaceous floras, most modern families are not recognizable. Unfortunately, great numbers of early reports on Cretaceous fossil plants list numerous modern genera and families of flowering plants. Most of these assignments are now considered inappropriate; the fossils actually representing extinct families, even orders, of angiosperms. Nevertheless, recent advances in the study of Cretaceous fossil flowers is revealing early members of some familiar families in the Cretaceous record as well as an assortment of unusual forms that seem to have left no descendants. This is currently one of the most exciting areas of paleobotanical research (Friis *et al.*, 1987).

During the Late Cretaceous, angiosperms overwhelmed the equatorial regions and eventually entirely displaced the cheirolepidiaceous conifers, the cycadeoids, and, to a large extent, the cycads and many equatorial ferns. Unfortunately, the macrofossil record from low latitudes is poor, so we have little appreciation for the nature of the vegetation. The pollen record indicates the presence of a diversity of angiosperms of unknown affinity as well as an abundance of palms.

In the mid- to high latitudes of both hemispheres, the angiosperms flourished but did not entirely displace the gymnosperms. In the Northern Hemisphere the Taxodiaceae maintained a firm grip on many of the polar floras until mid-Tertiary climatic decline. The Pinaceae became diverse in the Tertiary and continues to be tremendously successful, particularly in boreal temperate regions. In the Southern Hemisphere, the podocarps and araucarian conifers are still significant forest constituents.

In considering the evolutionary implications of the angiosperms, it is important to keep in mind that many of the herbaceous flowering plants that we now associate with open areas and forest understory, including the entire grass family, the sedges, the aster family, and many others, are unknown from the Cretaceous. Grasslands did not exist until the mid-Tertiary. Open areas and semiarid regions would most likely have been covered with angiospermous shrubs, ferns, cycads, and other shrubby gymnosperms. Grazing was not a dinosaurian pastime.

For the most part, the transition from Paleozoic to Mesozoic floras occurred during the Late Permian to Middle Triassic as a result of profound changes to land environments, including widespread aridity. Throughout the Mesozoic, the equatorial regions appear to have been generally drier and the polar regions moister. The fossil record from the equatorial regions is, however, extremely poor. Most of our evidence from low latitudes comes from the pollen record of marine sediments. Because we do not know which plants produced many of the types of pollen grains, this is most unsatisfactory. The equatorial regions have been very difficult to explore, in part because too much vegetation obscures the rocks, a problem that the human population may soon remedy. More extensive work in these areas is of critical importance to our understanding of plant evolution.

There is currently much debate concerning the presence or absence of permanent ice anywhere on the earth, except at very high elevation. There is some indication of ice-related sediments at very high latitudes, but this is at odds with the record of fossil plants in the polar regions and with evidence for warm polar oceans throughout the Mesozoic. Antarctica was near or on the pole at that time, but its rock record is now concealed by ice, which is a great inconvenience. Much more data from the polar regions may help to answer this question. Our limited knowledge of plant diversity in the very high latitudes could also benefit from more extensive investigation of the Arctic and Antarctic, which are areas that are difficult, and expensive, to explore.

The Mesozoic may be conveniently divided into four packages of time for our purposes. Most of the Triassic is a time of floral change as the Paleozoic component declines and the Mesozoic floras are established. The Lower to Middle Jurassic finds climates becoming more humid in general; modern families of seed plants and ferns replace more archaic or transitional forms. From the Late Jurassic to Early Cretaceous, the *Classopolis* equatorial zone expands, signaling warmer global climates and perhaps a broader equatorial arid belt. Angiosperms evolve but have little impact on floras. The Late Cretaceous sees the onslaught of the angiosperms.

As a result of continuity, or at least close proximity, of land masses during much of the Mesozoic, plant dispersal was limited primarily by latitudinal temperature and moisture gradients. As a result, vegetation was rather uniform throughout climatic zones. The considerable floristic differences that now exist between continents straddling the same climatic zones appear to be a primarily Tertiary phenomenon.

The following is meant to be an imprecise view of global vegetation, drawing from the previous discussion of the evolution and distribution of major groups of plants, and with liberal interpretation. A recent English translation of V. A. Vakhrameev's book, *Jurassic and Cretaceous Floras and Climates of the Earth* (1991), is a rich source of data on this topic. Three paleogeographic maps (Fig. 3) depict global vegetation patterns in the Early–Middle Jurassic, Late Jurassic, and Late Cretaceous.

During the Triassic, the arid equatorial interior of Pangaea was most likely host to drought-tolerant cycads and cycadeoids, with archaic ferns and voltzialean conifers occupying suitable habitats. In the southern mid- to high latitudes, for example, Australia and southern Africa, the glossopterid pteridosperms declined during this period, and caytonialean pteridosperms (especially *Dicroidium*), cycads, ginkgophytes, and voltzialean conifers became more important. North of the equatorial belt were cycads, voltzialean conifers, various ginkgos, and ferns.

During Early to Middle Jurassic times, the equatorial regions became dominated by cheirolepidiaceous conifers, cycads, and cycadeoids—assemblages commonly interpreted as indicative of hot and dry conditions. Aridity appears to have been a feature of the equatorial belt throughout much of the Mesozoic.

South of this belt, in Australia, southern Africa, India, southern South America, and Antarctica, existed a diverse flora of ferns, cycads, cycadeoids, ginkgophytes, pteridosperms, and araucarian and podocarp conifers. There is no evidence of a south polar vegetation zone at this time.

To the north, the midlatitude belt of northern North America and central and southern Europe and Asia experienced warm, mainly subhumid conditions and hosted forests of conifers, particularly Taxodiaceae and Cheirolepidiaceae. The Yorkshire Jurassic Flora, superbly described by T. M. Harris (1961, 1964, 1969, 1974, 1979), finely represents this northern midlatitude vegetation zone. A tremendous diversity of cycads, cycadeoids, ginkgos, pteridosperms (especially *Sagenopteris*), ferns, and shrubby sphenopsids completes the picture of a rich, lush vegetation. Central and southern North America appear to have been more subequatorial at that time and experienced a hot, semiarid to subhumid climate, with perhaps an open forest and scrub vegetation.

In the north polar region, including northernmost North America and Europe and much of Siberia, floras were much lower in diversity. Here, the climate was humid, with moderate temperatures; even in the polar regions there was no severe frost, and forests could grow at the North Pole. However, continuous light in the summer and continuous darkness in the winter above the Arctic Circle creates seasonality so that plants had to become dormant even though the temperature was above freezing. It seems that relatively few plants could adapt to these conditions. One of the mechanisms evolved to deal with winter dormancy is deciduousness, a feature commonly found among trees in the polar regions. This zone typically includes land above approximately 60°N and in the Jurassic hosted deciduous ginkgos, the enigmatic deciduous conifer *Podozamites*, a number of archaic Pinaceae, the extinct deciduous pteridosperm *Czekanowskia*, ferns, and a few cycads.

Global climate apparently warmed through to the end of the Jurassic, with the result that the Cheirolepidiaceae–*Classopollis* equatorial arid zone expanded into the midlatitudes. This seems to have had the effect of pinching the old midlatitude warm and subhumid climatic belt against the north polar zone.

North polar floras apparently changed little during this time, remaining dominated by early Pinaceae and deciduous ginkgos, *Podozamites*, and *Czekanowskia*. Northward movement of North America brought moderate and humid conditions to the northern half of the continent, and at the Jurassic–Cretaceous boundary thick coals developed in the northwestern United States and western Canada. The fossil floras of this area are similar to those of the circumpolar Siberian region.

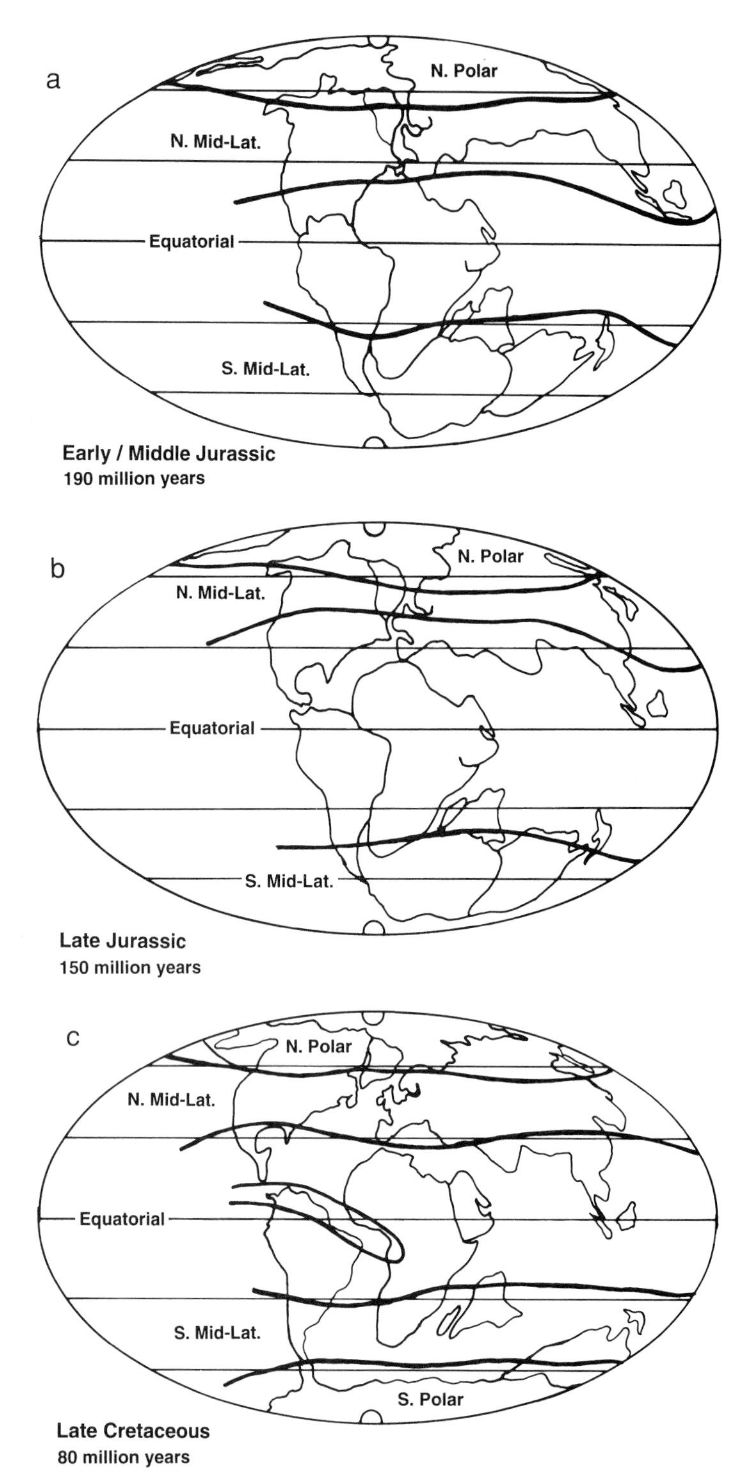

FIGURE 3 Paleogeographic maps for (a) Early–Middle Jurassic, (b) Late Jurassic, and (c) Late Cretaceous. The equatorial vegetation zone was hot and dry, although later in the Cretaceous a narrow central humid belt formed as the South Atlantic opened (the narrow belt drawn on the equator, northern South America, and West Africa). The north midlatitude zone (N. Mid-Lat.) was warm and subhumid to semiarid. The south midlatitude zone (S. Mid-Lat.) was warm and subhumid to humid. The north polar zone (N. Polar) was moderate and humid. The south polar zone (S. Polar) formed in the Late Cretaceous and was moderate and humid. Maps have been adapted from Vakhrameev (1991) and Smith *et al.* (1994).

The Early Cretaceous was a time of amelioration of global climate and a spread of more humid climate equatorward. The equatorial regions remained hot and dry, but the midlatitudes became more humid. Throughout central and northern North America and Siberia, coal deposits indicate that climate was humid. The flora of this region, above about paleolatitude 50°N, retained its Jurassic character in general. However, through the Early Cretaceous there is a considerable increase in the importance of the Taxodiaceae and a corresponding decrease in Czekanowskiales. Many of the taxodiaceous conifers seem to have adopted a deciduous habit and went on to play a major role in northern polar forests.

The floras of the subhumid to semiarid northern midlatitudes of the Early Cretaceous were much richer than the polar regions, with forests of various conifers, including principally Cheirolepidiaceae, Pinaceae, and Taxodiaceae. Cycads, cycadeoids (especially the barrel-shaped trunks of *Cycadeoidea*), and ferns were abundant.

Equatorial regions were apparently still primarily arid from Late Jurassic to Early Cretaceous times because they remained dominated by Cheirolepidiaceae, cycads, and cycadeoids. With the continuing breakup of the supercontinents, and the opening of seaways between continents, the low latitudes appear to develop a mosaic of arid to humid environments. Southern midlatitudes and polar regions experienced subhumid to humid climates. Dominant forest trees included primarily araucarian and podocarp conifers, with varying numbers of Cheirolepidiaceae, ginkgos, and a few other conifers. Ferns were diverse and abundant. Cycads and cycadeoids were also diverse and many show considerable similarity to Northern Hemispheric forms, reminding us of their very wide distribution. Primary differences between Northern and Southern Hemispheric floras of this time lay in the dominant canopy trees: The Pinaceae, Taxodiaceae, and Czekanowskiaceae were largely restricted to the Northern Hemisphere and the Araucariaceae and Podocarpaceae to the Southern Hemisphere. As noted previously, it appears that the floras of the very high southern latitudes did not differ much from those of the midlatitudes, unlike in the north, where a distinct north polar vegetation developed. This could be due to a lack of understanding of very high southern latitude vegetation because Antarctica was furthest south and is poorly explored for fossils. Also, perhaps southern plants were generally more tolerant of polar seasonality and were less constrained by latitude. It may also be that Antarctica had not yet positioned itself over the South Pole so that a true polar climate had not yet developed.

During the Early Cretaceous, angiosperms make their ominous appearance in the pollen and macrofloral records of the equatorial regions then spread to the higher latitudes toward the end of the Early Cretaceous.

A transformation of global vegetation occurred beginning in the Albian (latest Early Cretaceous) and continuing in the Late Cretaceous. From humble beginnings, the angiosperms evolved to overwhelm almost all terrestrial environments. Most of the known Cretaceous angiosperms were woody plants. Although many would have been large, canopy trees, angiosperms were especially important as shrubs. The role of angiosperms as herbs at this time is poorly known. Given the apparently poor representation of herbaceous angiosperms in both the macroflora and the microflora until the Tertiary, ground cover in open areas and under canopies may have consisted of mostly ferns.

The decline or extinction of numerous groups of ferns and gymnosperms appear to have been a direct consequence of habitat loss to the angiosperms. The most devastating effect was on the cycads, cycadeoids, and Cheirolepidiaceae of the equatorial regions. Most of our knowledge of this region comes from the pollen record, and unfortunately the affinities of these pollen grains to living angiosperms is generally unknown. The abundance of palm pollen in equatorial deposits has led some to refer to this as the Late Cretaceous palm zone.

In the northern midlatitude zone, including central and southern North America, southern Greenland, most of Europe, and southern and eastern Asia, climate varied from subhumid to semiarid. Angiosperms were commonly abundant and apparently were dominant forest-forming trees in many areas, although conifers of the Taxodiaceae and Pinaceae are still numerous in many floras. The shrub component of the vegetation consisted of a mix of angiosperms, cycads, and ferns.

In the Northern Hemisphere, two floristic zones become established in the Late Cretaceous, recognized on the basis of their pollen floras. The *Normapolles* zone of the midlatitudes, particularly in the Atlantic region, was probably like the modern moist subtropics and tropics. The north polar regions were distinguished by the *Aquillapollenites* complex, with mixed vegetation most like that of modern midlatitudes. The modern affinities of these two pollen groups are unknown. Shifts in the boundaries between these two zones reveal climatic fluctuation during the Late Cretaceous.

In the north polar regions, including much of Canada, Alaska, and northern Asia, a diversity of angiospermous trees and shrubs invaded the northern forests but did not entirely displace the deciduous taxodiaceous conifers and ginkgos of the region. These gymnosperms were prominent members of the northern deciduous floras until the mid-Tertiary. The Czekanowskiaceae, formerly widespread in the north, was apparently driven to extinction by first the northern expansion of the Taxodiaceae during the Cretaceous and then the arrival of the angiosperms. The angiosperms of the northern forests were largely of unknown modern affinity but may have played a role in the evolution of many taxa of the Arcto–Tertiary circumpolar deciduous flora. The Pinaceae, now such an important part of northern forests, is not as conspicuous in the Cretaceous macrofossil record as it is in the pollen record. This may indicate that the Pinaceae was an important part of high-altitude floras before becoming widespread in high-latitude floras.

Southern Hemisphere floras were similarly enriched in angiosperms in the Late Cretaceous, although there were pronounced differences in the types of angiosperms present compared to those in the Northern Hemisphere. Araucarian and podocarp conifers continued to be common. By the end of the Cretaceous, the southern polar regions had developed a distinct vegetation dominated by podocarps and angiosperms, notably including Nothofagaceae (southern beeches), a prominent southern high-latitude and -altitude family.

Although never approaching the coolness of the modern earth, the global climates of the Mesozoic did vary considerably. Temperatures rose during the Late Cretaceous to a maximum in the Campanian, then fell gradually during the last few million years of the Mesozoic. On a local level, the effects of this change can be seen in floristic changes and in the southward shift in the humid, coal-forming zone in western North America from above 60°N to below 50°N. Nevertheless, the magnitude of climatic deterioration is not substantially greater than that at other times in the Mesozoic, and there is no evidence that even the polar regions experienced severe frost at any time during this interval. Although rates of plant extinction appear to increase toward the boundary, and continue at a relatively high level following the boundary, the plant record does not indicate a crisis as a consequence of global climatic change.

Global climate warmed again to a maximum several million years after the boundary. The Cretaceous–Tertiary boundary, then, roughly coincides with the climatic minimum between the Late Cretaceous and early Tertiary climatic optima. The boundary rather precisely coincides with the deposition of a worldwide iridium-rich fallout layer that was a consequence of a massive asteroid impact on the Yucatan Peninsula of Mexico. Detailed study of this boundary layer in western North America reveals instantaneous devastation of forests over huge areas, with recovery taking centuries. In some places a tremendous number of fern spores in the few millimeters above the impact layer indicates that the landscape was clothed by ferns for many years until forests reestablished themselves. Interestingly, there is no correspondingly abrupt extinction of plants because most types return shortly above the boundary. The terrestrial record at the boundary is poor elsewhere in the world but seems to indicate a less spectacular end to the Cretaceous on other continents than we see in North America. The role of this impact in the terminal Cretaceous extinction event is currently widely debated and is one of the most fascinating controversies in paleontology today.

The cumulative changes that we see occurring across the Cretaceous–Tertiary boundary create a transition from the Mesozoic floras to the Cenozoic. Unlike the Cretaceous, early Tertiary fossil plants are commonly recognizable as members of living families, and there was a rapid modernization of the global flora during the first 20 million years of the Tertiary. Perhaps this is a suitable boundary, then, for the plants as well as the dinosaurs, because it rightly recognizes the advent of the modern flora as well as fauna. Also, it properly places early angiosperms among the Mesozoic floras as an integral part of the world of the ruling reptiles.

See also the following related entries:

MESOZOIC ERA • MESOZOIC FAUNAS

References

Beck, C. B. (Ed.) (1976). *The Origin and Early Evolution of Angiosperms*, pp. 341. Columbia Univ. Press, New York.

Beck, C. B. (Ed.) (1988). *The Origin and Evolution of Gymnosperms*, pp. 504. Columbia Univ. Press, New York.

Erwin, D. H. (1993). *The Great Paleozoic Crisis: Life and Death in the Permian*, pp. 327. Columbia Univ. Press, New York.

Friis, E. M., Chaloner, W. G., and Crane, P. R. (Eds.) (1987). *The Origins of Angiosperms and Their Biological Consequences*, pp. 358. Cambridge Univ. Press, Cambridge, UK.

Harris, T. M. (1961). *The Yorkshire Jurassic Flora. I. Thallophyta–Pteridophyta*, pp. 212. British Museum (Natural History), London.

Harris, T. M. (1964). *The Yorkshire Jurassic Flora. II. Caytoniales, Cycadales and Pteridosperms*, pp. 191. British Museum (Natural History), London.

Harris, T. M. (1969). *The Yorkshire Jurassic Flora. III. Bennettitales*, pp. 186. British Museum (Natural History), London.

Harris, T. M. (with J. Miller and W. Millington) (1974). *The Yorkshire Jurassic Flora. IV Czekanowskiales, Ginkgoales*, pp. 150. British Museum (Natural History), London.

Harris, T. M. (1979). *The Yorkshire Jurassic Flora. V Coniferales*, pp. 167. British Museum (Natural History), London.

Miller, C. N. (1977). Mesozoic conifers. *Bot. Rev.* **43,** 218–280.

Stewart, W. N., and Rothwell, G. W. (1993). *Paleobotany and the Evolution of Plants*, 2nd ed., pp. 521. Cambridge Univ. Press, New York.

Smith, A. G., Smith, D. G., and Funnell, B. M. (1994). *Atlas of Mesozoic and Cenozoic Coastlines*, pp. 99. Cambridge Univ. Press, Cambridge, UK.

Taylor, T. N., and Taylor, E. L. (1993). *The Biology and Evolution of Fossil Plants*, pp. 982. Prentice Hall, New York.

Tidwell, W. D., and Ash, S. R. (1994). A review of selected Triassic to Early Cretaceous ferns. *J. Plant Res.* **107,** 417–442.

Vakhrameev, V. A. (1991). *Jurassic and Cretaceous Floras and Climates of the Earth* (J. F. Litvinov, Trans.; N. F. Hughes, Ed.). pp. 318. Cambridge Univ. Press, Cambridge, UK.

Mexican Dinosaurs

RENE HERNANDEZ-RIVERA
Instituto de Geología UNAM
Del. Coyoacan, Mexico

The first recorded material pertaining to the dinosaurs of Mexico was published by Dr. W. Janensch in 1926. He reported the discovery of ceratopsian bones collected in the "Soledad beds" in the state of Coahuila in north-central Mexico (Fig. 1).

Lull and Wright (1942) wrote that in the upper part of the Snake Ridge Formation large dinosaur bones (and teeth) were found at locality No. 49 "in the gap between the Mustenas and Magallanes Peak" in the state of Sonora in northwestern Mexico. The large hadrosaur material was submitted to Barnum Brown for identification, and he considered it a new species but not definable. It was, at that time, the most southerly hadrosaurian occurrence on record. From the teeth he inferred the horizon to be Campanian or Maastrichtian.

In 1954, Wann Langston, Jr., and Millis H. Oakes of the University of California Museum of Paleontology reported hadrosaur bones at Punta San Isidro in the state of Baja California. The material consists of foot bones of two specimens similar in size to those of the genus *Kritosaurus*.

Between June and September 1959, a team of graduate students from the University of Louisiana, under the direction of Clarence O. Durham and Grover E. Murray, did fieldwork in part of the Parras Basin and reported dinosaur remains in association with marine molluscs. Later the locality was reexamined by a team from the University of Texas. They collected remains of at least four dinosaurs. One type was identified by J. A. Wilson and E. H. Colbert as the ceratopsid *Monoclonius*. Other specimens were hadrosaurid (Murray *et al.*, 1960).

During the summer of 1966 a research team from the Los Angeles County Museum of Natural History, under the direction of William Morris, worked with the permission of the Mexican government in cooperation with the Geological Institute of the National Autonomous University of Mexico (IGLUNAM). The team prospected for vertebrate faunas in the K–T boundary beds near El Rosario in Baja California.

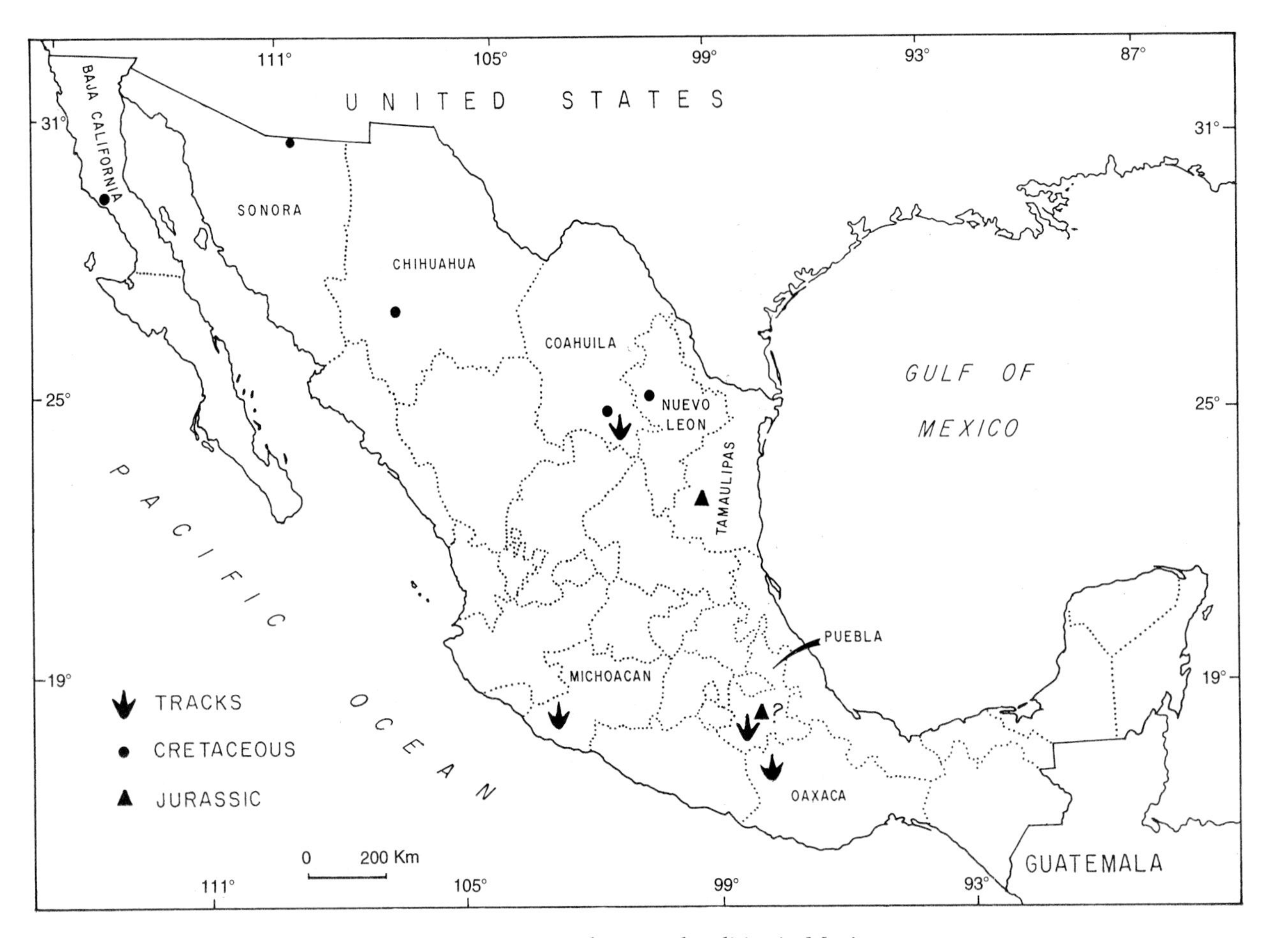

FIGURE 1 Major dinosaur localities in Mexico.

They found many dinosaur remains, mainly crested hadrosaurs. Other groups discovered included tyrannosaurs, dromaeosaurs, and ankylosaurs. At that time, the locality was considered one of the richest areas on the North American Pacific margin. The most abundant form was a big crested hadrosaur, referred first to the genus *Hypacrosaurus altispinus* (Morris, 1967) but in 1972 new skull fragments appeared and demonstrated that the specimen was more closely allied to the genus *Lambeosaurus* (Morris, 1972). More fieldwork in 1970, by the same group of researchers, was done south of the "Arroyo el Rosario," near the village of the same name. Harley J. Garbani discovered a theropod dinosaur in a locality designed by Kilmer as "La Bocana Roja" Formation (Kilmer, 1963). Ralph Molnar studied the specimen and proposed it as a new genus and species of theropod, *Labocania anomala*. The dinosaurs from the El Gallo Formation (Campanian) of Baja California include a tyrannosaurid (cf. *Albertosaurus*), an ornithomimid, *Troodon formosus*, *Saurornitholestes sp.*, Dromaeosauridae indet., ?*Lambeosaurus laticaudus*, ?*Lambeosaurus* sp., hadrosaurine indet., Ankylosauria indet., Nodosauridae, cf. *Euoplocephalus*, and Ceratopsia indet. La Bocana Roja Formation dinosaurs (Molnar, 1974) are probably Campanian in age and include *Labocania anomala*.

In 1980 a team of the IGLUNAM, organized by Shelton Applegate, Luis Espinosa, and Victor Torres and under the direction of Dr. Ismael Ferrusquia-Villafranca, prospected in the Cretaceous beds in the state of Coahuila looking for Mesozoic mammals. In the city of Torreon they met Dr. Luis Maeda, who had in his collection large fragmented dinosaur bones collected in the Ejido Presa San Antonio, Parras County. They visited the locality and met Mr. Ramon Lopez, who showed them many areas with exposed bones. Later they met Mr. Jose Rojas, an amateur collector, in Saltillo, the capital city of Coahuila. He had collected many dinosaur bones near Ejido Rincon

FIGURE 2 Dinosaur bones in "Dinosaur Hill," southwest of the state of Coahuila, Mexico.

FIGURE 3 Hadrosaur skin impression found in "Dinosaur Hill," Ejido de Rincon Colorado, state of Coahuila.

Colorado, General Cepeda County, 47 km to the west of Saltillo. One of the specimens was an almost complete hadrosaur skeleton with skin impressions, and another was a ceratopsian. Mr. Rojas donated that material to the IGLUNAM. In 1985 a team from the Royal Ontario Museum, organized by Dr. Christopher McGowan, Kevin Seymour, Andrew Leitch, and Brian Iwama, did fieldwork in the Ejido Presa San Antonio. They collected partial skeletons of two hadrosaurs. Other material reported belongs to theropods. The list of dinosaur material reported from the Cerro del Pueblo Formation (Campanian) of the state of Coahuila (A. Leitch and K. Seymour, personal communication; Murray *et al.*, 1960) includes a tyrannosaurid, an ornithomimid, a dromaeosaurid, a hadrosaurine, an ankylosaurid, and a ceratopsid (probably centrosaurine). The late Maastrichtian Soledad Beds (Janensch, 1926; Lull, 1933) have produced an indeterminate ceratopsid.

In the fall of 1987, the IGLUNAM supported the present author's initiation of formal studies of Cretaceous dinosaurs. Sixty percent of one specimen was collected in 1988 in the Ejido Presa San Antonio, and the specimen was prepared and mounted in the Geological Museum. The skeleton was identified tentatively as a *Kritosaurus*, 7 m long, with a pathological character in its left hand. In the same quarry where this material was excavated came parts of two lambeosaurines, identified mainly by the ischia. The project to study Mexican dinosaurs gained support from the state government of Coahuila, which created the Paleontological Commission of the SEPC and the Dinamation International Society. In 1993, in cooperation with the IGLUNAM, the Dinamation International Society started a very ambitious project to collect and study Cretaceous dinosaurs from the state of Coahuila. The first fieldwork was in Ejido Rincon Colorado on a small hill called Cerro de la Virgen, but now renamed Dinosaur Hill because of the great number of specimens found. The specific work area is now known as Dinosaur Valley, with a surface near 40 km^2 (Fig. 2). The material is well preserved and includes skin impressions and articulated skeletons. The most abundant specimens are hadrosaurines and lambeosaurines, followed by ceratopsids. We have also found theropod teeth and bones (Fig. 3). The list of taxa identified from Coahuila as of the end of 1995 included *Albertosaurus*, an indeterminate ornithomimid, cf. *Dromaeosaurus*, a possible titanosaurid, cf. *Kritosaurus*, cf. *Lambeosaurus*, an ankylosaurid, *Centrosaurus*, and *Chasmosaurus*. Other vertebrates col-

FIGURE 4 Coprolites are plentiful in the fossil localities in the state of Coahuila, Mexico.

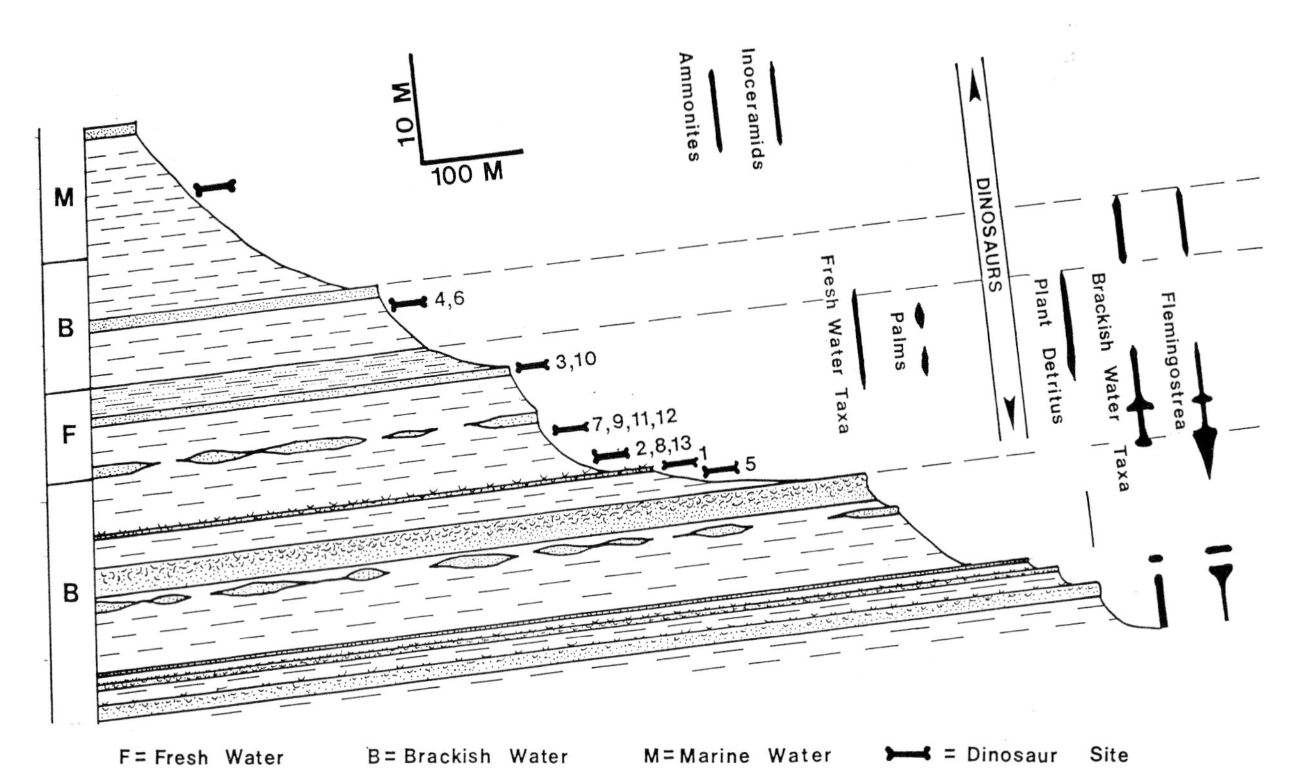

FIGURE 5 A schematic section of "Dinosaur Hill" showing the great diversity and wealth of fossil material, both animal and plant.

lected include fishes (sharks and the sawfish *Schizorhiza*), crocodylomorphs, and turtles. Most of the bones are associated with or covered by invertebrates. Coprolites are also abundant (Fig. 4). The paleoecology of this area consisted of deltaic sediments deposited in a lagoon of brackish water that had connections to the sea (Fig. 5).

Jurassic dinosaur bones are known from only two places in Mexico. One is in Huizachal Canyon, 25 km southwest of Ciudad Victoria, the capital city of Tamaulipas, where Dr. James Clark found therapsids, including tritylodontids, in 1982. Later we discovered a small ornithischian dinosaur here, which is represented by eight isolated teeth. There are isolated theropod teeth and two theropod specimens, including the back of the skull. Some of the larger bones may be from sauropods (Fig. 1). The other locality is near the town of San Felipe Ameyaltepec in the state of Puebla, but the material is very fragmentary, and the only identifiable bone seems to be from a sauropod (Fig. 1).

Dinosaur footprints from Mexico have been reported by Ferrusquia-Villafranca *et al.* (1995) from three localities. The first locality, in the state of Oaxaca, has footprints assignable to one theropod and two sauropod families of Middle Jurassic age. The second locality, in the state of Michoacan, has two

FIGURE 6 (A) A fossil egg found in Ramos Arizpe County, west of the state of Coahuila, Mexico. Length of the specimen is 10 cm. (B, inset) CAT scan of the same egg; the embryo is visible at the right lower edge of the image.

theropod and two ornithopod taxa represented, which are of Late Jurassic age. Finally, in the state of Puebla, one ornithopod and one sauropod taxa of Maastrichtian age have been identified from their footprints (Fig. 5).

In the summer of 1995, members of the Commission of Paleontology SEPC discovered Maastricthtian footprints near Saltillo, Coahuila, and assigned them to the Ornithopoda and Theropoda. In the fall of 1994, a fossil egg was found by Prof. Yolanda De Leon in Ramos Arizpe County. Studies of computerized tomography have shown the presence of an embryo inside the egg, the identification of which is still under study (Fig. 6).

See also the following related entries:

AMERICAN DINOSAURS • CANADIAN DINOSAURS • SOUTH AMERICAN DINOSAURS

References

Clark, J. M., Montellano, J. M., Hopson, J. A., Hernandez, R., and Fastovski, D. E. (1994). An Early or Middle Jurassic tetrapod assemblage from the La Boca Formation, northeastern Mexico. In *In the Shadow of the Dinosaurs, Early Mesozoic Tetrapods* (N. C. Fraser and H.-D. Sues, Eds.). Cambridge Univ. Press, Cambridge, UK.

Ferrusquia-Villafranca, I., Jimenes-Hidalgo, E., and Bravo-Cuevas, V. (1995). Jurassic and Cretaceous dinosaur footprints from Mexico: Additions and revision. *J. Vertebr. Paleontol.* **15**(Suppl. to No. 3), 28A.

Janensch, W. (1926). Dinosaurier-Reste aus Mexico. *Centralbl. Min. Geol. Pal.* **1926**(B), 192–197.

Langston, W., and Oakes, M. H. (1954). Hadrosaurs in Baja California. *Bull. Geol. Soc. Am.* **65,** 1344. [Abstract]

Lull, R. S., and Wright, N. E. (1942). Hadrosaurian dinosaurs of North America. Geological Society of America Spec. Paper No. 40, pp. xii, 242.

Molnar, R. E. (1974). A distinctive theropod dinosaur from the Upper Cretaceous of Baja California (Mexico). *J. Paleontol.* **48**(5), 1009–1017.

Morris, W. J. (1967). Baja California: Late Cretaceous dinosaurs. *Science* **155,** 1539–1541.

Morris, W. J. (1971). Mesozoic and Tertiary vertebrates in Baja California. *Natl. Geogr. Soc. Res. Rep.* **1965,** 195–198.

Morris, W. J. (1972). A giant hadrosaurian dinosaur from Baja California. *J. Paleontol.* **46**(5), 777–779.

Morris, W. J. (1973). A review of Pacific Coast hadrosaurs. *J. Paleontol.* **47**(3), 551–556.

Morris, W. J. (1981). A new species of hadrosaurian dinosaur from the Upper Cretaceous of Baja California ?*Lambeosaurus laticaudus*. *J. Paleontol.* **55**(2), 453–462.

Murray, G. E., Boyd, D. R., Durham, C. O., Forde, R. H., Lawrence, R. M., Lewis, P. D., Martin, K. G., Weidie, A. E., Wilbert, W. P., and Wolleben, J. A. (1960). Stratigraphy of Difunta Group, Parras Basin, states of Coahuila & Nuevo Leon, Mexico. *XXI Int. Geol. Congr. Pt.* 5, 82–96.

Weishampel, D. B., Dodson, P., and Osmólska, H. (Eds.) (1990). *The Dinosauria*, pp. 733. Univ. of California Press, Berkeley.

Michigan State University Museum, Michigan, USA

see MUSEUMS AND DISPLAYS

Microvertebrate Sites

JUDITH A. SCHIEBOUT
Louisiana State University
Baton Rouge, Louisiana, USA

Microvertebrate sites are fossil sites in which the animal remains are recovered by screening bulk samples of sedimentary rock. Usually, the fossils recovered are disarticulated and/or fragmentary. Microvertebrate sites often yield the remains of the smallest animals in a fauna (Fig. 1), but they are not restricted to remains of small animals. Large animals such as big dinosaurs may be represented by small hard parts, such as teeth, or by fragments of large bones. A tooth of a young ANKYLOSAUR recovered by screening was the first ankylosaur found in west Texas (Standhardt, 1986) (Fig. 2).

If screening produces important fossils, the area from which the sediment was derived is considered a microvertebrate site even if no process of concentration of fossils took place. However, usually some process of concentration of fossils occurred either before burial or during exposure by erosion. Many microvertebrate sites represent sedimentological concentrations of small vertebrate remains, such as at the bases of river channels. Channels and other erosional features such as small gullies on floodplains concentrate small objects such as pebbles, loose teeth and bones,

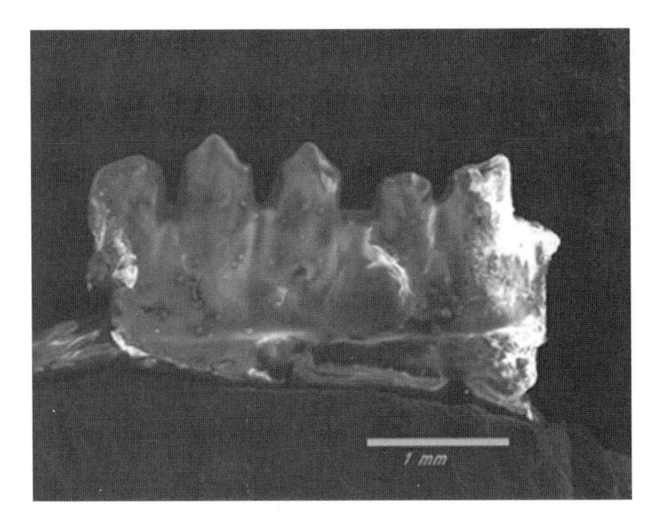

FIGURE 1 Cretaceous lizard jaw recovered by screening, Late Cretaceous of Big Bend National Park Texas. Scientific research in the park requires special permits. For comparison, a penny is approximately 19 mm in diameter.

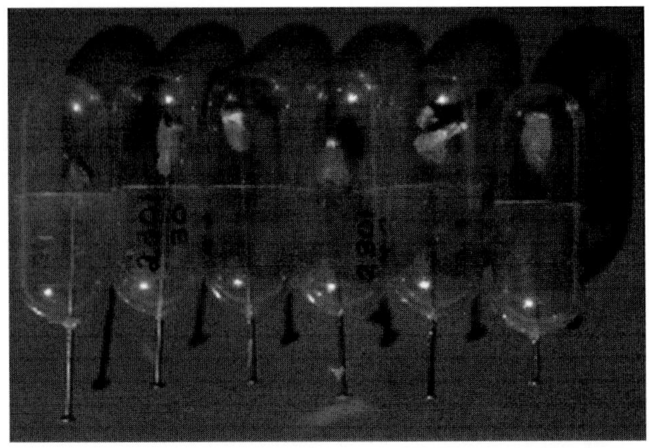

FIGURE 3 Microvertebrate specimens in the vertebrate paleontological collection of the LSU Museum of Natural Science. Pins have been driven partly through the ends of numbered gelatin capsules and the specimens affixed to the pinpoints with wax to ensure that they are protected from possible damage.

and soil-formed nodules. The fossil remains move as sedimentary particles (Voorhies, 1969; Dodson, 1973; Korth, 1979). Because a vertebrate skeleton is composed of components of a variety of sizes, shapes, and hardness, fossils recovered from such sites commonly have undergone disarticulation, breakage, and wear from transport. Often, teeth or jaw fragments are the only identifiable fossils. Teeth are the densest and most resistant parts of most vertebrate skeletons (Fig. 3). They can be of considerable taxonomic and paleoecological interest because they are the animals' main method of processing food. Tooth form and arrangement in the jaw can provide much information about animals' diet.

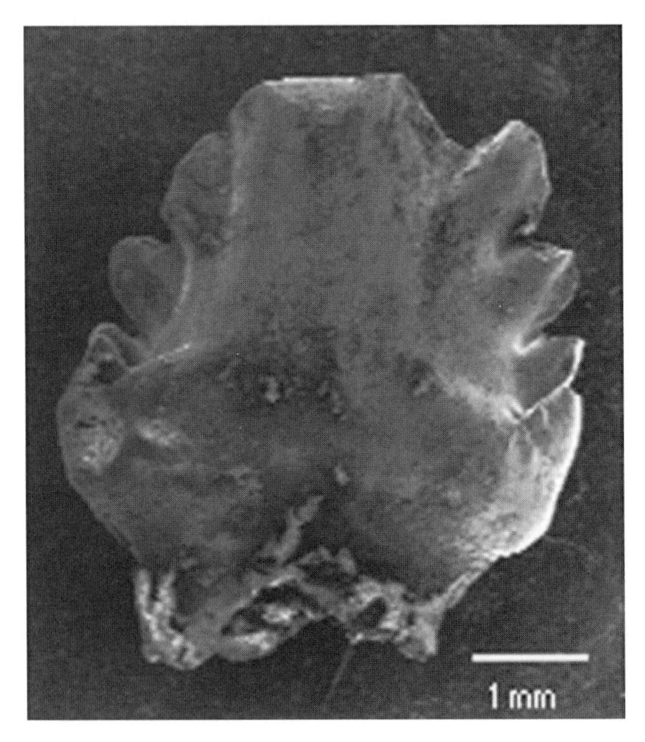

FIGURE 2 Ankylosaur tooth recovered by screening, Late Cretaceous of Big Bend National Park Texas.

Taphonomy yields insight on how microvertebrate sites form. Taphonomy, first defined by Efremov (1940), is the study of what happens to animal remains between their death and their study by scientists (Behrensmeyer and Hill, 1980; Shipman, 1981; Lyman, 1994). Areas of accumulation of sedimentary rock usually receive a spotty influx of dead animals because any surface at a given time usually does not contain carcasses and would not, even if a catastrophe wiped out all vertebrates. If an animal dies on a floodplain or in a marsh, it may lie unburied until the next flood, suffering changes wrought by scavengers and weathering. That flood may winnow the animal remains from soils and concentrate them in low spots or may move them into channels where they will eventually be concentrated in the deposits formed at channel bases. Reworking of fossils also takes place as rivers erode older deposits, winnowing out animal remains and concentrating them with more recent bones and teeth in the channel bases. Weathering, abrasion, and passing through the digestive tract of predators or scavengers before being incorporated

FIGURE 4 Julia Sankey indicates a fossil-bearing conglomerate, Late Cretaceous of Big Bend National Park, which has yielded dinosaur, shark, fish, frog, mammal, and lizard remains, when treated with acetic acid and screened.

into sediments (Mellett, 1974) may alter the remains. Transport may result in animals being buried in environments very different from those in which they lived. For example, dinosaur remains may be found with remains of marine vertebrates such as sharks in material concentrated in channels in marsh or tidal deposits (Fig. 4).

Sometimes quarry sites that have yielded large animals also serve as microvertebrate sites, yielding important fossils when sediments surrounding the large fossils are screened. Carcasses of large forms may have been responsible for the accumulation of the remains of small animals that fed upon the carcasses or took shelter among the bones.

Microvertebrate sites are discovered in several ways. When small fossils are embedded in soft material such as poorly lithified mudstone, they tend to be concentrated on the surface by rain and wind erosion. Paleontologists can find rock rich in small fossils by noticing such surface concentrates (Fig. 5). The fossils in microvertebrate sites are coarser than clay or fine sand size; therefore, they will usually be associated with pebbles or soil-formed nodules, which can be sought in searching badlands that expose floodplain mudstones. If pebbles were scarce in a depositional environment, fossils, such as the teeth and bones of small animals, may make up a large percentage of the hard objects in their size range. In this case, modern ants may prove helpful to paleontologists by gathering fossils to concentrate near their nest openings to armor the entrances. An example of a locality found using this technique is Bug Creek Anthills in Montana (Sloan, 1965). In areas where carbonate nodules were common in the ancient soils—for example, in southern areas such as Big Bend in Texas—these nodules are handy for ants to use. In such localities, the trick of examining anthills to find microvertebrate concentrations for screening does not work.

Methods of recovering specimens from microvertebrate sites vary. The main uniting circumstance is that the specimens are scattered through large volumes of sediment and are too small for either recovery by quarrying in the field or by jacketing and removal to the laboratory. Microvertebrate sites are usually an exception to the rule of careful and delicate collecting involved in preparing vertebrates, in which matrix is painstakingly picked away from the fossils. If sediments at a microsite are poorly consolidated,

FIGURE 5 Barbara Standhardt and Frank Haas search the ground surface of a Big Bend Late Cretaceous microvertebrate site for the richest area to bag for return to the laboratory.

FIGURE 6 Volunteer Margaret Steele shovels sediment for bagging at a Late Cretaceous microvertebrate site in Big Bend.

FIGURE 8 Paul Seifert wet screens in a washtub at Castolon, Big Bend National Park.

they may simply be shoveled into bags (Fig. 6) or loaded by earthmoving equipment into a truck (McKenna *et al.*, 1994) for transport to a screening location. Well-consolidated and cemented rock may need to be broken up with a hammer or jackhammer. Dry screening in the field can sometimes be used, but usually the rock matrix must be transported (Fig. 7) to water or even to a laboratory for wet screening, possibly preceded by chemical treatment to disaggregate the rock. Acids such as acetic acid may be used to dissolve carbonate cement and/or nodules and concretions to free the coarse particles, including the fossils, from a strongly cemented rock. Fossil bones and teeth are composed of the mineral apatite and are little affected by some weak acids such as acetic,

FIGURE 7 A helicopter carries bags of sediment for screening from a remote part of Big Bend National Park.

the acid found in vinegar. Amounts of rock screened may be high and the yield of fossils very small. A test washing to determine whether a site is productive and should be a focus of years of study may be as large as a quarter ton as in Big Bend. This seems like an enormous amount, but a ton of floodplain mudstone from the Late Cretaceous of Big Bend would fit under a folding card table.

The history of bulk screening began with Claude Hibbard of the University of Michigan, who developed wooden boxes with screen bottoms and two partially screened sides that could be placed in a stream and agitated to remove mud by a person standing in the water (Hibbard, 1949). McKenna (1994) gives a summary of practices in screening microvertebrate sites. Many young paleontologists have spent part of their scientific apprenticeship standing in streams or, in arid parts of western North America, in cattle tanks or ponds, or even sitting beside washtubs, bending-over versions of the wooden screen boxes first developed by Hibbard (Figs. 8, 9, 10).

FIGURE 9 LSU screening camp at Castolon in Big Bend National Park. Material from the screens is being dried on black plastic sheets.

In the past several decades, microvertebrate screening and sorting has been used in dinosaur-bearing horizons throughout the Mesozoic Era. Remains of tiny dinosaurs, typically the shed tooth crowns of juveniles, have been recovered in this way. Loose dinosaur teeth vary in identifiability, but small theropod teeth are often diagnostic at the genus or species level (Currie, et al. 1990; Fiorillo and Currie, 1994). Remains of other small vertebrates are more frequently recovered, and they have increased both our knowledge of the taxonomic diversity of non-dinosaurian vertebrates throughout the MESOZOIC ERA, and our understanding of the kinds of faunas, and hence the environments, in which dinosaurs lived (a very few examples include Sloan and Van Valen 1965; Estes 1964; Jacobs and Murry 1980; Bryant 1989; Rowe et al. 1992; Kaye and Padian 1994; Lofgren 1995; Eberth and Brinkman 1997).

FIGURE 10 An LSU field crew wet screens in the Rio Grande River, Big Bend National Park.

See also the following related entries:

BIOSTRATIGRAPHY • TAPHONOMY

References

Behrensmeyer, A. K., and Hill, A. (Eds.) (1980). *Fossils in the Making*. Univ. of Chicago Press, Chicago.

Bryant, L. J. (1989). Non-dinosaurian lower vertebrates across the Cretaceous-Tertiary boundary in northeastern Montana. *Univ. Of Calif. Publications, Geological Sciences* **134,** 1–107.

Currie, P. J., Rigby, Jr., J. K., and Sloan, R. E. (1990). Theropod teeth from the Judith River Formation of southern Alberta, Canada. pp. 107–125 in K. Carpenter and P. J. Currie (eds.), *Dinosaur Systematics: Perspectives and Approaches.* Cambridge University Press, New York.

Dodson, P. (1973). The significance of small bones in paleoecological interpretation. *Univ. Wyoming Contrib. Geol.* **12,** 15–19.

Dooley, A. C., Jr., Schiebout, J. A., and Dooley, B. S. An inexpensive, environmentally friendly method for bulk acid processing of carbonate rocks to obtain microvertebrates. In *Vertebrate Paleontological Techniques, Vol. 2* (P. Leiggi and P. May (Eds.), submitted for publication. Cambridge Univ. Press, New York.

Eberth, D. A., and Brinkman, D. B. (1997). Paleoecology of an estuarine, incised-valley fill in the Dinosaur Park Formation (Judith River Group, Upper Cretaceous) of Southern Alberta, Canada. *Palaios* **12**(1), 43–58.

Efremov, J. A. (1940). Taphonomy: New branch of paleontology. *Pan-Am. Geol.* **74**(2), 81–93.

Estes, R. (1964). Fossil vertebrates from the Late Cretaceous Lance Formation, eastern Wyoming. *Univ. of Calif. Publications, Geological Sciences* **49,** 1–187.

Fiorillo, A. R., and Currie, P. J. (1994). Theropod teeth from the Judith River Formation (Upper Cretaceous) of south-central Montana. *Journal of Vertebrate Paleontology* **14**(1), 74–80.

Hibbard, C. W. (1949). Techniques of collecting microvertebrate fossils. *Contrib. Museum Paleontol. Univ. Michigan* **3**(2), 7–19.

Jacobs, L. L., and Murry, P. A. (1980). The vertebrate community of the Triassic Chinle Formation near St. Johns, Arizona. pp. 55–72 in L. L. Jacobs (ed.), *Aspects of Vertebrate History.* Museum of Northern Arizona Press, Flagstaff.

Kaye, F. T., and Padian, K. (1994). Microvertebrates from the Placerias Quarry: a window on Late Triassic vertebrate diversity in the American southwest. pp. 171–196 in N. C. Fraser and H.-D. Sues (eds.), In *the Shadow of the Dinosaurs: Early Mesozoic Tetrapods.* Cambridge University Press, New York.

Korth, W. W. (1979). Taphonomy of microvertebrate fossil assemblages. *Ann. Carnegie Museum* **48**(15), 235–285.

Lofgren, D. L. (1995). The Bug Creek problem and the Cretaceous-Tertiary transition at McGuire Creek. *Univ. of Calif. Publications, Geological Sciences* **140,** 1–200.

Lyman, R. L. (1994). *Vertebrate Taphonomy.* Cambridge Univ. Press, New York.

McKenna, M. C., Bleefeld, A. R., and Mellett, J. S. (1994). Microvertebrate collecting: Large-scale wet sieving for fossil microvertebrates in the field. In *Vertebrate Paleontological Techniques, Vol. 1* (P. Leiggi and P. May, Eds.), pp. 93–111. Cambridge Univ. Press, New York.

Mellett, J. S. (1974). Scatological origin of microvertebrate fossil accumulations. *Science* **185,** 349–350.

Rowe, T., Cifelli, R. L., Lehman, T. M., and Weil, A. (1992). The Campanian Terlingua Local Fauna, with a summary of other vertebrates from the Aguja Formation, Trans-Pecos Texas. *Journal of Vertebrate Paleontology* **12**(4), 472–493.

Shipman, P. (1981). *Life History of a Fossil: An Introduction to Taphonomy and Paleoecology.* Harvard Univ. Press, Cambridge, MA.

Sloan, R. E., and Van Valen, L. (1965). Late Cretaceous mammals from Montana. *Science* **148,** 220–227.

Sloan, R. L. (1965). Cretaceous mammals from Montana. *Science* **148**(3667), 220–227.

Standhardt, B. R. (1986). Vertebrate paleontology of the Cretaceous/Tertiary transition of Big Bend National Park, Texas, pp. 299. Unpublished. Ph.D. dissertation, Louisiana State University, Baton Rouge.

Voorhies, M. R. (1969). Taphonomy and population dynamics of an early Pliocene vertebrate fauna, Knox County, Nebraska. *Univ. Wyoming Contrib. Geol. Spec. Papers* **1,** 1–69.

Middle Asian Dinosaurs

MICHAEL J. RYAN
Royal Tyrrell Museum of Palaeontology
Drumheller, Alberta, Canada

Middle Asia is a distinct region recognized by geographers to encompass the republics of Kazakhstan, Kyrgystan, Tajikistan, Turkmenistan, and Uzbekistan. Research on dinosaurs from this region began in 1882 when Russian geologist G. D. Romanovsky discovered dinosaur tracks along the Yagnob River. In the 1910s, Cretaceous dinosaurs (predominantly HADROSAURS and THEROPODS) were discovered in the Kyzyl Kum Desert and subsequently studied by A. N. Riabinin (1939). Middle Asian localities were examined by I. A. Efremov (1944), and subsequent fieldwork and research was done by N. N. Verzilin and A. K. Rozhdestvensky (1975). The most recent work in this area was carried out by the late Lev A. Nesov and his students. Research on Middle Asian dinosaurs would greatly benefit from thorough translations of Nesov's extensive research papers.

Middle Asian place and formation names have varied from paper to paper over the years, depending on the transliteration. Whenever possible, the formation names used here follow previously published English conventions. Formations listed in parentheses have been translated from Russian without reference to known standardized English versions. The information listed is derived primarily from Suslov (1987) and Nesov (1989, 1995) and does not include information on eggshell or footprint localities (see Nesov, 1995, for these). Bones have been found in Turkmenistan, but they have not been identified.

Republic of Uzbekistan

Karakalpakian Autonomous Republic of Uzbekistan Khodzhakul Fm, lower or middle Albian: Segnosauria indet.; upper Albian: cf. Coeluridae, *Ceratosaurus* (Ceratosauria), "*Laelaps*" sp. cf. *L. explanatus* (Theropoda), Sauropoda (Diplodocidae or Titanosauridae), Shamosaurinae (Ankylosauridae), *Gilmoreosaurus*(?) *atavus* (Hadrosauridae), cf. Hypsilophodontidae, *Kulceratops kulensis* (Protoceratopsidae?)

"Sultanbobin" Fm, upper Aptian: Sauropoda

Khodzhakul Fm, lower Cenomanian: Deinonychosauria, Pectinodon, *Troodon asiamericanus*, "*Laelaps*" sp. (cf. "*L*". *explanatus*), Segnosauria indet., Sauropoda (Diplodocidae or Titanosauridae), Shamosaurinae, *Gilmoreosaurus* sp., "*Gadrosaurus*" sp. and other Hadrosauridae, *Asiaceratops salsopaludalus* (Protoceratopsidae)

"Beshtuybin" Fm, lower Turonian: vertebrae

Kyzyl Kum Desert of Uzbekistan "Uchkuduk" Fm, lower Turonian: Theropoda, Hadrosauridae

Bissekty Fm, upper Turonian, *Alectrosaurus* sp., cf. *Aublysodon* sp. (Tyrannosauridae), *Dryptosaurus* sp. or cf. *Hypsibema*? sp., Coeluroidea, Deinonychosauria, *Caenagnathasia martinsoni* (Caenagnathidae), Oviraptoridae, *Archaeornithomimus*(?) *bissektensis* (Ornithomimidae), Dromaeosauridae, *Asiamericana asiatica* (Saurodontidae), *Euronychodon asiaticus*, cf. *Richardoestesia* sp. (Theropoda indet.), Segnosauria indet., Sauropoda (cf. Diplodocidae or cf. Titanosauridae), Ankylosauridae, Hadrosauridae (*Cionodon*? *kysylkumense* and *Gilmoreosaurus arkhangelskyi*), *Turanoceratops tardabilis* (Ceratopsidae)

Beleutin Fm, upper Turonian: *Itemirus medullaris* (Theropoda incertae sedis)

"Kyunyur" Fm, Upper Cretaceous: Theropoda, Ankylosauridae, *Gilmoreosaurus* (Hadrosauridae)

Other Localities in Uzbekistan ?Lower "Balaban" Fm (Upper Jurassic) and Shurab (middle Jurassic) Fm: Camarasauridae, Diplodocidae

"Hodgiabad" Fm, Lower Cretaceous: vertebrae

"Kuebikin" Fm, upper Albian/Cenomanian: *Alectrosaurus* sp.

Zhirkindeck Fm, lower Turonian: *Alectrosaurus*?, Sauropoda?, *Gilmoreosaurus* sp. (cf. *G. arkhangelskyi*)

"Zheirantuy" Fm, lower Turonian: Theropoda

"Syuksyuk" Fm, Santonian: Hadrosauridae

"Palvantash" Fm, Santonian: bones

"Getyumtau" Fm, lower Campanian: Theropoda, Hadrosauridae

Republic of Kazakhstan

Neocomian Sands, Lower Cretaceous: *Embasaurus minax* (Megalosauridae?)

"Chanak" Fm, Lower Cretaceous: bones

"Buralkenyuntuz" or "Koturbulak" Fm, upper Turonian: Deinonycheridae, Ankylosauridae, Hadrosauridae

"Koturbulak" Fm, Cenomanian: bones

"Buralkenyuntuz" Fm, upper Turonian: bones

Lower part of Dabrazin Fm and "Syuksyuk" Fm, Santonian: Theropoda, Sauropoda, *Procheneosaurus convincens* (Hadrosauridae), *Asiamericana*

"Syuksyuk" Fm, Santonian: Theropoda, Hadrosauridae

Bostobe Fm, Santonian: *Alectrosaurus*, Dromaeosauridae, Caenagnathidae, Segnosauria, Sauropoda, Stegosauria, *Bactrosaurus*, *Jaxartosaurus*, *Aralosaurus tuberiferus*, *Arstanosaurus akkurganensis* (Hadrosauridae), Ceratopsidae

Zhuravlevo Fm, upper Santonian/lower Campanian: *Hadrosaurus* sp.

"Syuksyuk" Fm, upper Santonian or Campanian–lower Masstrichtian: Coelosauridae

?Fm, ?Santonian/Campanian: *Tarbosaurus* aff. *bataar*, Ceratopsidae

"Ayat" and Zhuravlevo fms, lower Campanian: Hadrosauridae

Lower Dabrazin Fm, middle Campanian: cf. *Troodon* sp.

Upper "Syuksyuk" Fm, Upper Cretaceous: vertebrae

"Syuksyuk" Fm, Upper Cretaceous: Tyrannosauridae, *Antarctosaurus*?, *Velociraptor*, *Aralosaurus tuberiferus*, Deinodontidae, ?Titanosauridae, Stegosauria, Nodosauridae, Ceratopsia, Hadrosauridae

"Koturbulak" Fm, Upper Cretaceous: bones

"Severozaisan" Fm, Upper Cretaceous?: vertebrae

Republic of Kyrgyzstan

"Koktuy" Fm, Lower Jurassic: bones

"Balaban" Fm, Middle Jurassic: "Coelurosauria", Megalosauridae or Ceratosauria, Sauropoda; upper Middle Jurassic: Megalosauridae or Ceratosauria, other Theropoda, Cetiosauridae, Camarasauridae; Middle or Upper Jurassic: Theropod, ?*Apatosaurus*

"Alamyushik" Fm, Lower Cretaceous: Theropoda

"Hodgiabad" Fm, ?middle Aptian: Theropod, Ornithopoda

"Tokubay" Fm, upper Aptian–lower Cenomanian: bones

"Kurshab" Fm, upper Albian–lower Turonian: "*Trachodon*"

"Sharihan" Fm, Cenomanian: Carnosauria femur, Theropoda, Sauropoda, Hadrosauridae

Yalovach Fm, lower Santonian: Tyrannosauridae, Sauropoda, Hadrosauridae

Republic of Tajikistan

Yalovach Fm, Upper Cretaceous: cf. *Alectrosaurus* sp., *Velociraptor* cf. *mongoliensis* (Dromaeosauridae), *Ornithomimus* cf. *asiaticus*, Ornithomimidae indet., Caenagnathidae?, *Antarctosaurus*? *jaxarticus* (Titanosauridae), *Bactrosaurus prynadai*, *Jaxartosaurus aralensis*, cf. Hypsilophodontidae, Nodosauridae indet., Ceratopsidae? indet.

Yalovach Fm, lower Santonian: Theropoda

Lower part of Palvantash Fm, upper Santonian: Hadrosauridae

References

Efremov, I. A. (1944). The dinosaur horizon of Middle Asia and some aspects of the stratigraphy. *Izvestia Akademii Nauk SSSR Ser. Geol.* **3,** 40–58. [In Russian]

Nesov, L. A. (1989). In *Theoretical and Applied Aspects of Modern Paleontology* (T. N. Bogdanova and L. I. Khozatsky, Eds.), pp. 144. Proceedings of the XXXIII Session of the All-Union Paleontological Society, Leningrad. [In Russian]

Nesov, L. A. (1995). *Dinosaurs of Northern Eurasia: New Data ABOUT Assemblages, Ecology and Palaeogeography,* pp. 156. Univ. of Saint Petersburg, Saint Petersburg. [In Russian]

Riabinin, A. N. (1939). The Upper Cretaceous vertebrate fauna from the Upper Cretaceous of south Kazakhstan. I. Reptilia. Pt 1. Ornithischia. *Nauchno-issledoviia Geol. Inst. Trudy* **118,** 1–40. [In Russian]

Rozhdestvensky, A. K. (1975). The study of dinosaurs in Asia. *J. Palaeontol. Soc. India* **20,** 102–119.

Suslov, U. V. (1987). The Upper Cretaceous dinosaurs of Prichimkention Chuli. *Materials History Fauna Flora Kazakhstan* **9,** 23–32. [In Russian]

Migration

Gregory S. Paul
Baltimore, Maryland, USA

To migrate long distances on land requires three characteristics: sufficiently large size, long striding legs, and sufficient aerobic power. Mice, for example, cannot traverse long distances, nor can large snakes and turtles. Migrating on land is harder than migrating in the sea or in the air because the energy cost of moving a given distance is highest on land and least in water (swimming is up to a dozen times more efficient than walking). Also, sea creatures can take advantage of ocean currents. Many marine fish, reptiles, and mammals migrate very long distances, sometimes across entire oceans. Insects, bats, and birds also migrate great distances, in extreme cases from pole to pole. In comparison, land migrations are rather modest. The greatest terrestrial migrators are polar mammals, who are under strong pressure to seek out the best seasonal conditions. Caribou trek up to 5000 km a year, but this movement is confined to a rather small area of 400 km across. Polar bears move up to 2500 km a year, partly in the water. In lower latitudes strong wet–dry seasons can inspire seasonal movements. The annual migrations of saiga and Serengeti gnu cover up to 1000 km round-trip. The long migrations often attributed to plains bison and African elephants before European disruption are controversial. It is notable that caribou and gnu are specialized for migrating because they have the most energy-efficient legs known. In herding animals the size of the juveniles may be a limiting factor; migrating ungulate calves weigh at least 15 kg. Carnivores do not migrate with their herbivore prey; rather, they exploit herbivores as they move through their territory. The latter fact exposes the point that migrations are dangerous. The tendency of land animals to migrate in mass aggregations at speeds above 2 or 3 km/hr reduces their vulnerability to predation but increases the possibility of mass disasters at overcrowded water crossings. Migrations are energy-taxing and arduous journeys that are undertaken only if there are compelling advantages to doing so. This may be why few mammals migrate, and no reptiles do so.

Migrations in the fossil record are difficult to identify but are suggested by the following taphonomic characteristics. Migrating animals often take advantage of the clear run provided by shorelines (Cohen *et al.*, 1993). Therefore, mass, unidirectional trackways paralleling a shoreline are suggestive of migratory movements, especially if they were made at 2 or 3 km/hr or more. Mass aggregations can be killed by volcanic events, fast-acting epidemics, or suffer high mortality at river crossings. Mass death assemblages that are not attributable to gradual accumulation (as in droughts) are therefore strongly suggestive of migratory habits (Coombs, 1990). If a dinosaur species was limited in distribution, it probably did not migrate very far. A wider distribution is compatible with seasonal migrations but may simply indicate dispersed habitation, as per the circumpolar distribution of wolves and moose.

Almost all the larger dinosaurs had the long striding legs needed to migrate moderate distances (Fig. 1; the big-clawed, fat-bellied THERIZINOSAURS and short-limbed ANKYLOSAURS may have been exceptions). The 2–10 km/hr speeds recorded by the majority of dinosaur trackways are high enough for long-distance movements (Paul, 1994). No dinosaur had the ex-

FIGURE 1 The great size of hadrosaurs may have enhanced their migratory abilities, but they lacked the extremely energy-efficient legs that make gracile caribou the longest ranging modern land animals. Figures drawn to same scale.

tremely gracile, highly energy-efficient legs of specialized long-range migrators.

The traditional view of dinosaurs as REPTILES, many semiaquatic, tended to suppress speculation that dinosaurs migrated. An exception was Huene, who suggested that bone beds of Late Triassic prosauropods were death assemblages that recorded mass migrations across a central European desert. This scenario has been challenged by recent work indicating that the bone beds accumulated gradually (Coombs, 1990).

The giant SAUROPODS of the Late Jurassic Morrison Formation have been considered potential migrators because they combined great energy needs with a seasonally dry habitat and because they may have roamed to the bordering highlands in search of large cobbles suitable for GASTROLITHS. These speculations are plausible, but there is little direct fossil evidence that sauropods migrated. Parallel sauropod Jura–Cretaceous trackways following shorelines hint at migratory behavior.

The best evidence for migrating dinosaurs are the extensive bone beds of Late Cretaceous HADROSAURS and CERATOPSIDS recently discovered in western North America from Texas to Alaska, along what used to be the western, north–south running coast of an ancient interior seaway (Hotton, 1980; Horner and Gorman, 1988; Currie, 1989). It is difficult to explain the accumulations as gradual; instead, they appear to be the result of sudden mass deaths, attributable to vulcanism or drownings at river crossings. Numbers of individuals involved in each event range from the hundreds to the tens of thousands. Age range in the bone beds is from moderately sized (and therefore mobile) juveniles to adults (the implication is that smaller juveniles either did not migrate or did so independently). The provincial organization of western coastal dinosaur faunas into nearshore, shore, and interior, as well as distinct north–south populations (Lehman, 1996), suggests that the straight-line range of any migrations was limited in most or all cases.

The last point contradicts suggestions that Alaskan dinosaurs moved far south to avoid the polar winter. In the Late Cretaceous northern Alaska was as little as 10° from the north paleopole, and the Arctic Circle may have been near or in northern Alberta (Fig. 2). It has been argued that the high-latitude dinosaurs were migrating north–south in order to escape winter conditions that, although much milder than today, included extended dark, cool and perhaps freezing temperatures, and floral dormancy near the poles. Similar journeys have been suggested for Australian dinosaurs living near the south paleopole of the time. In this view dinosaurs moved toward the pole to enjoy the abundant floral growth of long summer days and then to lower latitudes in order to find winter food or to seek the warmth needed by dinosaurs with reptile-like energetics. In the latter scenario, Hotton (1980) portrayed high-latitude dino-

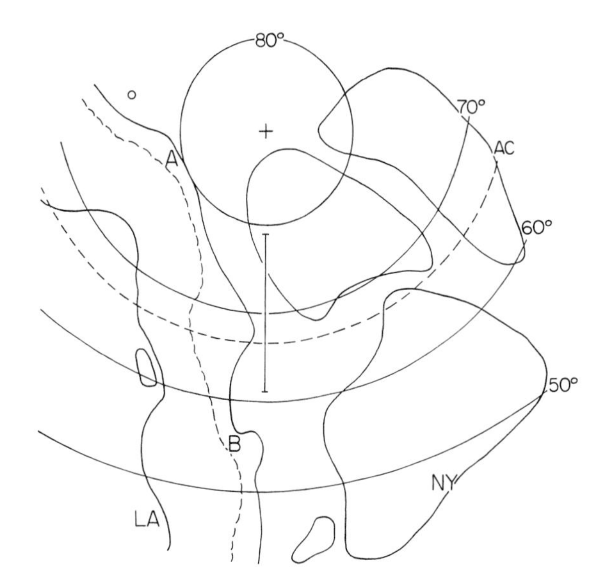

FIGURE 2 (A) North slope and (B) Montana–Wyoming indicate the possible north–south zone of migration of Late Cretaceous dinosaurs along the northwestern coast of the Niobrara seaway (cartographic distortion causes the coastal plain to appear straight rather than properly curved). The paleomagnetic pole is indicated by the open circle; the calculated spin pole by the cross. The future locations of New York and Los Angeles are marked. Scale bar = 2000 km.

saurs as "happy wanderers" that moved up to 6000 km a year. To move from northern Alaska far enough south to warm reptilian bodies in the winter may have required even further movements, up to 9000 km round-trip (equal to a New York to Los Angeles round-trip). However, such long migrations may have posed insurmountable energy problems. The cost for a juvenile hadrosaur or ceratopsid of 500 kg to move 6000–9000 km was about 900,000–1,400,000 kcal. The annual active energy budget of the same juvenile reptilian dinosaurs would have been only 800,000–1,500,000 kcal, too low to power such a long trip. Nor could reptilian dinosaurs have produced enough sustainable aerobic exercise power to move at the 2 or 3 km/hr speeds typical of migrating land animals. If dinosaurs had mammal-like energetics, then they would have had the power needed for very long migrations, at least in principle. The great size of many dinosaurs also facilitated long movements. In this view the consensus that some large dinosaurs migrated is supported.

However, scenarios that portray dinosaurs as greater migrators than mammals may be excessive, and it is probable that the great majority of dinosaurs did not migrate at all. The absence of specialized, unusually energy-efficient limbs may have limited those large dinosaurs that did migrate to about 500 km straight-line one way each year. In this case, a minority of herbivorous dinosaurs near and far from the poles migrated in order to seek out the best intraregional conditions, and to avoid overexploiting the food resources in one place, rather than to span part of a continent to find the best seasonal climatic regime (Paul, 1994). In this scenario improvements of high-latitude winter conditions were at most modest and alleviated but did not solve major thermal and food problems.

See also the following related entry:

BEHAVIOR

References

Cohen, A. S., Halfpenny, J., Lockley, M., and Michel, E. (1993). Modern vertebrate tracks from Lake Manyara, Tanzania and their paleobiological implications. *Paleobiology* **19,** 433–458.

Coombs, W. P. (1990). Behavior patterns of dinosaurs. In *The Dinosauria* (D. B. Weishampel, P. Dodson, and H. Osmólska, Eds.), pp. 32–42. Univ. of California Press, Berkeley.

Currie, P. J. (1989). Long-distance dinosaurs. *Nat. History* **98**(6), 60–65.

Horner, J. R., and Gorman, J. (1988). *Digging Dinosaurs.* Workman, New York.

Hotton, N. (1980). An alternative to dinosaur endothermy: The happy wanderers. In *A Cold Look at the Warm-Blooded Dinosaurs* (R. D. K. Thomas and E. C. Olson, Eds.), pp. 311–350. AAAS, Washington, DC.

Lehman, T. (1996). Late Campanian dinosaur biogeography in the western interior of North America. In *Dinofest International Symposium* (D. L. Wolberg and E. Stump, Eds.), p. 73. Arizona State University, Tempe.

Paul, G. S. (1994). Physiology and migration of North Slope dinosaurs. In *1992 Proceedings of the International Conference on Arctic Margins* (D. K. Thurston and K. Fujita, Eds.), pp. 405–408. United States Department of the Interior, Anchorage.

Mill Canyon Dinosaur Trail, Utah, USA

see MUSEUMS AND DISPLAYS

Milwaukee Public Museum, Wisconsin, USA

see MUSEUMS AND DISPLAYS

Mineral Research and Exploration Institute, Turkey

see MUSEUMS AND DISPLAYS

Mongolian Dinosaurs

R. Barsbold
Academy of Sciences of Mongolia
Ulaan Bataar, Mongolia

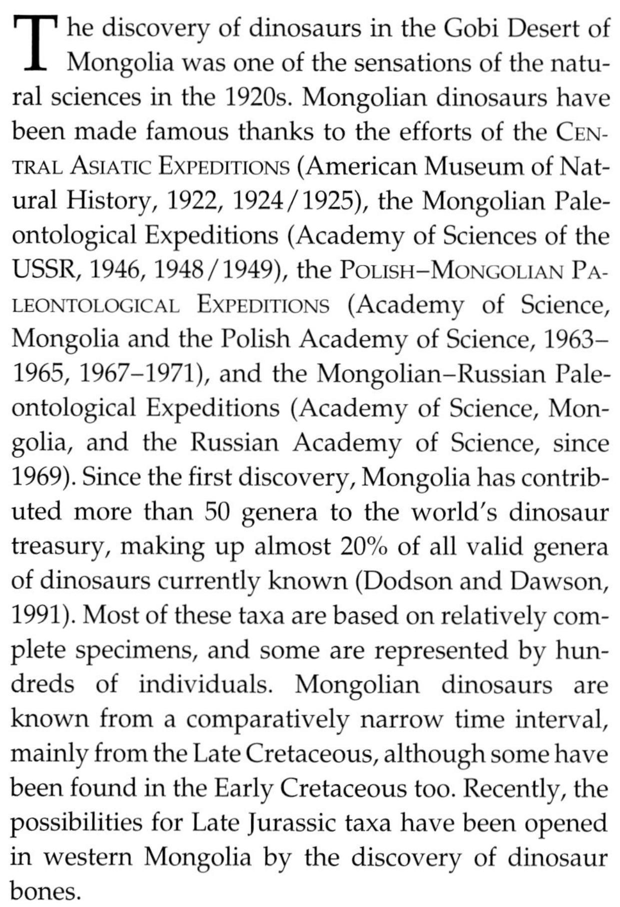

The discovery of dinosaurs in the Gobi Desert of Mongolia was one of the sensations of the natural sciences in the 1920s. Mongolian dinosaurs have been made famous thanks to the efforts of the Central Asiatic Expeditions (American Museum of Natural History, 1922, 1924/1925), the Mongolian Paleontological Expeditions (Academy of Sciences of the USSR, 1946, 1948/1949), the Polish–Mongolian Paleontological Expeditions (Academy of Science, Mongolia and the Polish Academy of Science, 1963–1965, 1967–1971), and the Mongolian–Russian Paleontological Expeditions (Academy of Science, Mongolia, and the Russian Academy of Science, since 1969). Since the first discovery, Mongolia has contributed more than 50 genera to the world's dinosaur treasury, making up almost 20% of all valid genera of dinosaurs currently known (Dodson and Dawson, 1991). Most of these taxa are based on relatively complete specimens, and some are represented by hundreds of individuals. Mongolian dinosaurs are known from a comparatively narrow time interval, mainly from the Late Cretaceous, although some have been found in the Early Cretaceous too. Recently, the possibilities for Late Jurassic taxa have been opened in western Mongolia by the discovery of dinosaur bones.

The nonmarine Cretaceous, which produces most dinosaurs of Mongolia, is divided into stages with successive dinosaur communities (Jerzykiewicz and Russell, 1991). These are as follows: Tsagan–Tsav (Berriasian–Valanginean; dinosaurs include an indefinite psittacosaurid), Shine–Khuduk (Hauterivian–Barremian; *Harpymimus*, psittacosaurs, and *"Iguanodon"*), Barun–Bayan (Albian–Cenomanian; nests of large round eggs), Bayan–Shire (Cenomanian–Turonian; therizinosaurs, ornithomimids, and indefinite sauropods), Djadokhta (lower Senonian; dromaeosaurids, *Saurornithoides*, oviraptorids, *Pinacosaurus*, hadrosaur babies, and protoceratopsids), Barun–Goyot (upper Senonian; eggs, oviraptorids, ankylosaurs, pachycephalosaurs, and *Bagaceratops*), and Nemegt (upper Senonian; oviraptorids, *Therizinosaurus, Deinocheirus, Mononykus*, tyrannosaurids, hadrosaurs, *Saichania* and other ankylosaurs, pachycephalosaurs).

Late Cretaceous theropod dinosaurs are better known from Mongolia than anywhere else. Perhaps the most famous of these is *Velociraptor mongoliensis*, which was 1.5 m long. It had a low skull, a tail stiffened by rod-like extentions of the prezygapophyses and hemal arches, a semilunate (pulley-like) carpal, an opisthopubic pelvis, and a functionally didactylous foot in which the second pedal digit was raised above the ground to support a strongly enlarged ungual phalanx. It was one of the Mongolian dinosaurs discovered in 1923 and was one of the first to show evidence of the connections between dinosaur faunas of central Asia and North America. One specimen is well known as part of the "Fighting Dinosaurs" discovery, in which the predator's sharp claws were found thrust into the tissues of the head and abdomen of a *Protoceratops*.

Other Mongolian dromaeosaurids include *Adasaurus mongoliensis*, which had a deep skull and a reduced pedal ungual on the second digit, and *Hulsanpes perlei*, which had a long and slender metatarsus.

Gallimimus bullatus was on average 3.5 m long, although one young specimen is 1.2 m long and the type specimen was almost 8 m. This was the first ornithomimid to show a parasphenoid capsule, a synapomorphy shared with troodontids and other ornithomimids. Other Mongolian ornithomimosaurs include *Anserimimus planinychus*, a 3.5-m-long form with flattened manual unguals; *Harpymimus okladnikovi*, a similarly sized species with reduced teeth in the front of the mandible and a short metacarpal I; and *Garudimimus brevipes*, which retained its first pedal digit. Mongolian ornithomimids are more diverse than their North American relatives.

Troodontids as a whole had the largest relative brain sizes of all known dinosaurs. The metatarsus was elongated, although most of the weight was carried by metatarsal IV. Like dromaeosaurids, the second pedal digit was specialized to carry the claw off the ground, although the ungual was not as well developed. *Saurornithoides mongoliensis* from the Djadokhta beds was approximately 2 m long, whereas the younger *S. junior* was larger and had more teeth. Two other troodontids, *Borogovia gracilicrus* and *Tochisaurus nemegtensis*, are only known from partial skeletons.

Oviraptorids are diverse, and include *oviraptor, conchoraptor, ingenia*, and a genus that will be described shortly. Nests and embryos are known for these animals, which appear to have brooded their eggs like birds.

Elmisaurus rarus was a small, gracile animal with the third metatarsal pinched proximally between metatarsals II and IV. Related forms have been found in North America.

The relationships of some Mongolian carnivorous dinosaurs are difficult to determine. These include *Deinocheirus mirificus*, which has 2.4-m-long forelimbs. The metacarpals and fingers are subequal in length and have massive, curved ungual phalanges. It is generally assigned to the Ornithomimosauria because of resemblance of manual and forelimb structure, but there are some significant differences in the outlines of the forelimb bones, and unfortunately nothing else is known of this dinosaur. The functional significance of the gigantic forelimbs is also unknown. Rozhdestvensky (1970) supposed that the "terrible hand" dinosaur fed on termites and used its massive claws to break open termite nests.

Therizinosaurus cheloniformis had 2-m-long front limbs with a three-fingered hand. There is a pulley-like semilunate carpal in the wrist, and the first metacarpal is short. The laterally compressed unguals are gigantic, and the one on the first finger can be 0.7 m long. The claws on the second and third digits were shorter. The first remains of *Therizinosaurus* were thought to belong to a "turtle-like lizard" (Maleev, 1954) and were later assigned to chelonians. Recently, however, it has been included with *Segnosaurus, Erlikosaurus*, and others within the Therizinosauroidea (Russell and Dong, 1993).

Erlikosaurus andrewsi is a medium-sized form with a well-preserved skull. The jaws are toothless at the front, the external nares are greatly elongated, a secondary palate is well developed, the basicranium and ear region is enlarged and pneumatized, the pelvis is opisthopubic, the ilium is broad and short with a flaring anterior flange, and four toes of each foot contacted the ground. *Segnosaurus galbinensis* and *Enigmosaurus mongoliensis* are two other therizinosaurs. Collectively these are one of the most unusual types of dinosaurs discovered within the past two decades. At first they were conditionally considered as aberrant theropods but were subsequently designated as Late Cretaceous relicts of the "prosauropod–ornithischian transition" (Paul, 1984) or as the sister group of sauropodomorphs (Barsbold and Maryanska, 1990). Recently, therizinosaurs have been unambiguously shown to be theropods (Russell and Dong, 1993; Clark *et al.*, 1994). Clearly, the morphology of group is very peculiar.

Avimimus portentosus is a small (less than 1 m) animal with some bird-like characters, including a proximally fused carpometacarpus and ilia that are inclined medially. The ulna is interpreted by Kurzanov (1981, 1982) as showing evidence of feather attachments.

Mononykus olecranus is a similar sized animal that has biconvex posterior dorsal vertebrae and keeled posterior synsacral vertebrae. The forelimb is very short and robust, with a prominent deltopectoral crest, a large and massive olecranon, a flat carpometacarpus, and a functionally monodactylous manus with a robust ungual phalanx. The pelvis is opisthopubic, and the third metatarsal is proximally reduced. The first fragmentary remains of *Mononykus* were found by the Central Asiatic Expeditions but remained unrecognized in storage until more complete material had been found from Bugin–Tsav. One of the most interesting finds in recent times, *Mononykus* has been classified as a basal bird (Perle *et al.*, 1994). Although it displays many avian features, it has also retained many theropod characters and is one of the best examples showing the complexity of the theropod–bird transition. The forelimb specializations are comparable to those of some fossorial tetrapods, but the shortness of the front limb argues against this.

Tyrannosaurids are characteristic of Late Cretaceous faunas of the Northern Hemisphere. Closely related to *Tyrannosaurus, Tarbosaurus bataar* is the 12- to 13-m-long Mongolian representative of this group (Maleev, 1974). It is very common at some sites in the Nemegt beds. *Alectrosaurus olseni* and *Alioramus remotus* are the other two tyrannosaurids known from Mongolia.

Several sauropods have been described from the Upper Cretaceous strata of Mongolia. *Nemegtosaurus mongoliensis* has a lightly built but elongate skull with slender, peg-like teeth at the front of the jaws. The nasal does not contact the maxilla, but the squamosal reaches the quadratojugal. *Quaesitosaurus orientalis* has a rather broad snout, and the short squamosal does not contact the quadratojugal. The third genus, *Opisthocoelicaudia skarzynskii*, was 12 or 13 m long. The dorsal vertebrae are strongly opisthocoelous, bearing prominent ball-and-socket articulations. The tail is very short for a sauropod and has only approximately 35 vertebrae. Of the three sauropod species, the first two are represented by incomplete skulls of diplodocid character, whereas *Opisthocoelicaudia* lacks the skull and neck but has a camarasaurid skeleton.

Psittacosaurus mongoliensis was approximately 1.5 m long. The snout is short and tapers anteriorly into a tall, narrow, parrot-like beak. The external naris is small and is positioned high on the snout, whereas the antorbital fossa and fenestra are both absent. The lightly built postcranial skeleton has a manual fourth digit with only small terminal phalanx. Other species have been described from successive Lower Cretaceous horizons.

Protoceratopsian dinosaurs are the most frequently recovered Mongolian dinosaurs. *Protoceratops andrewsi* is 2 m long. Its skull tapers anteriorly into a narrow edentulous beak, although the premaxilla retains two teeth. There is an antorbital fossa. The parietosquamosal frill extends behind the skull, exceeding about half of its total length in mature animals. The most common and well-known Mongolian dinosaur, more than a hundred specimens of *P. andrewsi* have been found since its first discovery in 1922. Babies were recognized first for this dinosaur, and recently at Tugrigin–Shire a flock of 15 hatchlings was recovered. All of the tiny skeletons lay closely packed with their bodies bent but similarily oriented. No doubt this regularity in disposition of the babies will provide information about the behavior and paleoecology of *P. andrewsi*. The abundance of specimens has allowed several researchers to attempt to analyze ontogenetic growth and sexual dimorphism. *Protoceratops* is one of the "fighting dinosaurs" and, even though it was the victim, it holds in its mouth the right arm of the predator and was undoubtedly the cause of its death.

There are several other Mongolian protoceratopsians. *Microceratops gobiensis* was a small, lightly built, cursorial form, with a short but fenestrated frill. *Bagaceratops rozhdestvensky* has a short snout and frill, an additional antorbital fenestra, and a prominent nasal horn core. *Breviceratops kozlowskii* has a short and narrow snout, a large, deep antorbital fossa, and a short, flattened nasal horn. The largest protoceratopsian is *Udanoceratops tschizhovi*, with a rostral to quadrate length of 65–70 cm and a probable total length of more than 3 m. The rostral bone is tall and narrow, there is no premaxillary dentition, and the nasals do not form a horn core.

Several pachycephalosaurids have been described on the basis of Mongolian specimens. *Homalocephale calathocercos* was 1.5 m long and had a flat, table-like, ornamented skull roof. The almost round supratemporal fenestrae are broadly separated. *Goyocephale lattimorei* was an animal of similar size, but the supratemporal fenestrae are longitudinally oval, and the bridge between them is narrow. The medial part of the skull roof is ornamented. *Tylocephale gilmorei* was also about the same size, but the skull has a strongly elevated dome that incorporates the postorbitals and supraorbitals. Because the parietosquamosal shelf is narrow, supratemporal fenestrae are probably absent. *Prenocephale prenes* was also approximately 1.5 m long, but the skull has a high dome that incorporates the prefrontals, supraorbitals, postorbitals, and parts of the squamosals. The parietosquamosal shelf is not developed, and the supratemporal fenestrae are closed. Pachycephalosaurs were perhaps a rare component of Mongolian dinosaur communities, and their fossils were identified later than those of most other Mongolian dinosaur taxa.

"Iguanodon" orientalis was an 8- to 10-m-long iguanodontid. The skull has an unusually large external naris, its margins formed mostly by the highly arched nasal. This animal has a heavily constructed postcranial skeleton with powerfully built limbs and an enlarged conical pollex on the hand. The generic assignment of this species is conditional pending a more complete study.

Only two hadrosaurs have been described so far. *Saurolophus angustirostris* was a huge animal, up to 14 m long. Like the North American species, the nasals extend posterodorsally above the orbit to form a solid crest, which is supported from behind by upraised frontal buttresses. This is one of the most common Mongolian dinosaurs from the latest Cretaceous beds, and there are many fine specimens that even include abundant skin impressions. *Barsboldia sicinskii* is a lambeosaurine with very long and club-shaped neural spines in the dorsal vertebrae.

Mongolia has one of the best fossil records for ankylosaurid ankylosaurs. *Pinacosaurus grangeri* was the first Mongolian species described on the basis of adequate material. Its skull is rather short and broad, with anteriorly positioned nares. Each of these is subdivided by a premaxillary septum that separates a dorsal foramen leading to the respiratory passage from a ventral one that is connected to a premaxillary sinus. The occipital condyle is oriented posteroventrally, and the postcranial skeleton is rather light with slender limb bones. *Talarurus plicatospineus* is approximately 5 or 6 m long and has an elongate but relatively small skull. The occipital condyle is oriented posteroventrally, the exoccipital is high and perpendicular to the skull roof, and the occiput has shifted slightly backward beyond skull roof. As in *Pinacosaurus*, the maxillary teeth have cingulum-like thickenings, cut externally by furrows. The postcranial skeleton is rather heavily built, and the limb bones are massive. Originally described as *Talarurus disparoserratus*, Tumanova (1987) has redescribed the type material as *Maleevus disparoserratus*. *Saichania chulsanensis* was approximately 7 m long. The skull is short and broad, and each terminally situated large nostril is subdivided by a septum into a large, oval foramen leading into the air passage and a more ventral second opening that connects with a premaxillary sinus. The occipital condyle is oriented ventrally, the exoccipital is low, and the postcranial skeleton is heavily built with rather massive limb bones. *Tarchia gigantea* is another large ankylosaurid, but it has a high, short exoccipital that is perpendicular to the skull roof. The occipital condyle is oriented posteroventrally, and the occiput has shifted slightly behind the skull roof. *Amtosaurus magnus* is a poorly known form based on an incomplete skull. *Shamosaurus scutatus* is a large ankylosaurid in which the skull roof is completely obscured by fusion of the osteoderms. The postorbital osteoderms are not horn-like, osteoderms do not close the quadrate cotylus, the anterior part of the snout is narrow, the occipital condyle is oriented ventrally, and the quadrate and paroccipital process are fused. *Shamosaurus* is generally considered to be the earliest and most primitive of the ankylosaurids.

This quick survey of Mongolian dinosaurs shows how rich the faunas were in central Asia during Cretaceous times. Field programs have been intensified in recent years, and new species of dinosaurs continue to be recovered. However, the recovery of data and specimens is also making great contributions to our understanding of the behavior, ecology, growth, and physiology of dinosaurs.

See also the following related entries:

Central Asiatic Expeditions • Chinese Dinosaurs • Erenhot Dinosaur Museum • Middle Asian Dinosaurs • Polish–Mongolian Paleontological Expeditions • Soviet–Mongolian Paleontological Expeditions

References

Barsbold, R., and Maryanska, T. (1990). Segnosauria. In *The Dinosauria* (D. B. Weishampel, P. Dodson, and H. Osmólska, Eds.), pp. 408–415. Univ. of California Press, Berkeley.

Clark, J. M., Perle, A., and Norell, M. A. (1994). The skull of *Erlicosaurus andrewsi*, a Late Cretaceous "Segnosaur" (Theropoda: Therizinosauridae) from Mongolia. *Am. Museum Novitates* **3115**, 1–39.

Dodson, P., and Dawson, S. D. (1991). Making the fossil record of dinosaurs. *Modern Geol.* **16**, 3–15.

Jerzykiewicz, T., and Russell, D. A. (1991). Late Mesozoic stratigraphy and vertebrates of the Gobi Basin. *Cretaceous Res.* **12**, 345–377.

Maleev, E. A. (1954). New turtle-like reptile in Mongolia. *Priroda* **1954**, 106–108. [In Russian]

Maleev, E. A. (1974). Gigantic carnosaurs of the family Tyrannosauridae. *Joint Soviet–Mongolian Paleontol. Expedition Trans.* **1**, 132–191. [In Russian]

Paul, G. S. (1984). The segnosaurian dinosaurs: Relics of the prosauropod–ornithischian transition? *J. Vertebr. Paleontol.* **4**, 507–515.

Perle, A., Chiappe, L. M., Barsbold, R., Clark, J. M., and Norell, M. (1994). *Am. Museum Novitates* **3105**, 1–29.

Rozhdestvensky, A. K. (1970). Giant claws of enigmatic Mesozoic reptiles. *Paleontol. J.* **1970**, 117–125.

Russell, D. A., and Dong, Z. M. (1993). The affinities of a new theropod from the Alxa Desert, Inner Mongolia, China. *Can. J. Earth Sci.* **30**, 2107–2127.

Tumanova, T. A. (1987). The armored dinosaurs of Mongolia. *Joint Soviet–Mongolian Paleontol. Expedition Trans.* **32**, 1–80 [In Russian].

Mongolian Museum of Natural History, Mongolia

see Museums and Displays

Moroccan Dinosaurs

see African Dinosaurs

Morphology, Functional

see Biomechanics; Constructional Morphology; Functional Morphology

Morphometrics

see Biometrics

Morris Museum, New Jersey, USA

see Museums and Displays

Morrison Formation

Kenneth Carpenter
Denver Museum of Natural History
Denver, Colorado, USA

The Morrison Formation is a widespread deposit occupying more than 1.5 million km^2 of the western United States. The formation is best known for its spectacular skeletons of the giant sauropods, such as *Diplodocus* and *Apatosaurus*. The formation was deposited on a vast, broad, low plain following the retreat of the Sundance Sea. In the southwestern outcrops, the Morrison contains many eolian sandstones, indicating a hot, arid sand desert (Carpenter *et al.*, 1997). Northwards, the sediments show an increase in fluvial, or river, activity, with long, ribbon-like sandstone bodies marking the paths of ancient meandering rivers. Adjacent to these sandstones are thick deposits of mudstone, formed by the deposition of mud from floods that spilled out onto the ancient floodplain. Thin limestone beds scattered among the mudstone mark the location of small lakes and ponds. Finally, the northern exposures of the Morrison in Montana contain coal, suggesting a much wetter and swampier environment than that to the south.

It is in the sandstones and mudstones that most of the dinosaur carcasses were buried. Today, the bones of various dinosaurs can be seen at Dinosaur National Monument preserved where they lay 145 million years ago. However, dinosaurs are not the only fossils that have been found in the Morrison Formation. Fishes, frogs, salamanders, lizards, crocodiles, pterosaurs, dinosaur eggs, and shrew- to rat-sized mammals are also known, mostly from fragmentary remains.

Recent radiometric dating shows that the Morrison ranges between 155 and 148 million years. Thus, the Morrison is entirely Late Jurassic in age (Kowallis *et al.*, 1997).

See also the following related entries:

Bone Cabin Quarry • Cleveland–Lloyd Dinosaur Quarry • Como Bluff • Dinosaur National Monument • Dry Mesa Quarry • Fruita Paleontological Arena • Howe Quarry • Jurassic Period

References

Carpenter, K., Chure, D., and Kirkland, J. (Eds.) (1997). The Morrison Formation: An interdisciplinary study. *Mod. Geol.*, in press.

Kowallis, B. K., Christiansen, E. H., Deino, A. L., Peterson, F., Turner, C. E., Kunk, M. J., and Obradovich, J. D. (1997). The age of the Morrison Formation. In The Morrison Formation: An interdisciplinary study (K. Carpenter, D. Chure, and J. Kirkland, Eds.). *Mod. Geol.*, in press.

Morrison Natural History Museum, Colorado, USA

see Museums and Displays

Musculature

J. Michael Parrish
Northern Illinois University
DeKalb, Illinois, USA

Because musculature is rarely fossilized, our information about muscle patterns in nonavian dinosaurs is almost entirely inferential. Muscles have been restored in a wide variety of dinosaurs, and this process is a necessary prerequisite for realistic flesh restorations of dinosaurs and other extinct vertebrates (Paul, 1987). The restoration process requires several different kinds of information:

1. the arrangements and innervations of muscles within extant archosaur groups (crocodilians and birds);
2. the extent to which these patterns are conservative between crocodilians and birds;
3. the extent to which attachments of muscles to bones are visible in crocodilians and birds and the extent to which similar attachment sites are visible in the dinosaur being studied.

Muscles can be restored with some confidence when they are conservative in arrangement between crocodilians and birds. Such conservative muscles can be considered to have been inherited in both extant groups from a common archosaur ancestor. Because birds are descended from dinosaurs, it is a reasonable assumption that the same pattern present in crocodilians and birds would also be present in dinosaurs, although the possibility always exists that the pattern was modified independently in one or more lines of dinosaurs other than the line leading directly to birds.

A common practice has been to restore muscles in dinosaurs based on the arrangement of muscles in one modern relative, usually a bird (e.g., Galton, 1969; Russell, 1972). This approach is probably sensible if the goal is to determine what the dinosaur's body looked like with flesh on the bones. It becomes more problematical if the goal is some type of functional interpretation of the dinosaur based on the restoration. In such cases, much more attention needs to be paid to the extent to which muscle attachments can be determined on the dinosaur's bones and how informative those attachment sites are with respect to the mechanical arrangements and relative sizes of the muscle(s) in question. Comparative reconstructions of jaw and limb muscles in a variety of dinosaurs are provided by Norman (1985) and Paul (1987).

Many studies have focused on restorations of the jaw musculature in dinosaurs. Lull (1908) and Haas (1955, 1963, 1969) restored the muscles of mastication in a variety of herbivorous dinosaurs. Carnivorous dinosaur cranial musculature has been restored less frequently (Adams, 1919; Raath, 1977; Molnar, 1993). As is the case in mammals, herbivorous dinosaurs had well-developed muscles for holding the jaws closed and for side to side (and sometimes forward and backward) movement of the jaws for grinding plant material (Norman and Weishampel, 1985). Carnivorous dinosaurs, on the other hand, had well-developed musculature for rapid, powerful, scissor-like closing of the jaws. Sites of obvious muscle attachment include the retroarticular process at the rear of the mandible, the coronoid process, a raised eminence on the mandible just anterior to the jaw joint that is developed as a raised tongue in ornithischians, and the descending flange of the pterygoid bone of the palate.

Forelimb musculature has been reconstructed in theropods (Ostrom, 1972; Nicholls and Russell, 1985; Smith and Carpenter, 1990), ankylosaurs (Coombs, 1978), and ceratopsians (Russell, 1935). Osteological and inferred muscular patterns are quite different in bipedal dinosaurs, such as *Iguanodon* (Norman, 1980) and *Deinonychus* (Ostrom, 1974), than they are in quadrupeds such as ankylosaurs and ceratopsians.

The muscles of the pelvic limb have been the object of many studies, notably the classic work of Osborn and Gregory (1918), Osborn (1919), Romer (1923, 1927), and subsequent studies by L. S. Russell (1935) on ceratopsians, D. A. Russell (1972) on *Dromiceiomimus*, Coombs (1979) on ankylosaurs, Tarsitano (1983) on *Tyrannosaurus rex*, and Gatesy (1991) on theropods. The insertions of several muscles are clearly visible on most dinosaur hindlimbs, notably the fourth trochanter on the posterior surface of the femur (insertion of the bellies of the caudifemoralis) and the iliofibularis trochanter on the anterior surface of the fibula. The wide variety of pelvic arrangements in dinosaurs has been linked to changes in musculature, although the origins of various muscles from that structure are mostly difficult to infer. Walker (1977) suggested that the retroverted pubis in ornithischians and birds was made possible by the shift of the origins of the anterior parts of the puboischiofemoralis internus from the pubis to the transverse processes of the posterior dorsal vertebrae early in archosaur evolution.

Some interesting functional problems can be addressed by looking at evidence for musculature in dinosaurs; these generally involve one or at most a few muscles that have readily identifiable attachment points that appear to be clearly homologous between the fossil and the recent relatives. A good example of this approach is Gatesy's (1991) analysis of the caudifemoralis longus muscle in archosaurs. He compared the condition in crocodilians, in which the muscle is the chief retractor of the hindlimb, with that in birds, in which the muscle, when present, is small and does not significantly retract the femur. Instead, most retraction of the hindlimb occurs through knee flexion. In most dinosaurs, the crocodilian pattern, with a long tail and broad transverse processes on the proximal caudal vertebrae for caudifemoralis longus origin, is observed. Coelurosaurian theropods have reduced tails that led Gatesy to suggest that the transition to the avian hindlimb orientation and pattern of movement may have occurred in the maniraptoran line prior to the origin of birds.

Another way that muscle attachment patterns can be useful in inferring function in dinosaurs lies in the relative position of indicators of muscle insertion on the limb bones. For example, the fourth trochanter is much more distal in its location on the femur of *Brachiosaurus* than it is on the femur of *Deinonychus*. A proximal location of a muscle insertion maximizes the movement that that muscle can produce around the fulcrum of the joint(s) in which it produces movement (the hip joint in this case), whereas a more distal location, all other factors being equal, will produce less movement around the joint per unit muscle contraction. Of course, another way to increase the mechanical advantage of a muscle is to increase the relative length of the bone; animals that are fast runners generally have much longer legs than slower animals of the same size.

Variations in the relative size of a muscle attachment may, in some cases, relate to the relative mass of the muscle in question. For instance, in modern mammals, the deltopectoral crest (insertion for the deltoideus, the major extensor of the humerus) is much more massive in fossorial mammals than in cursorial mammals of comparable size.

See also the following related entries:

Biometrics • Bipedality • Forelimbs and Hands • Functional Morphology • Hindlimbs and Feet • Pectoral Girdle • Pelvis, Comparative Anatomy • Quadrupedality • Skeletal Structures

References

Adams, L. A. (1919). A memoir on the phylogeny of the jaw muscles in recent and fossil vertebrates. *Ann. N. Y. Acad. Sci.* **28,** 51–66.

Coombs, W. P. (1978). Forelimb muscles of the Ankylosauria (Reptilia: Ornithischia). *J. Paleontol.* **52,** 642–657.

Coombs, W. P. (1979). Osteology and myology of the hindlimb in the Ankylosauria. *J. Paleontol.* **53,** 666–684.

Galton, P. M. (1969). The pelvic musculature of the dinosaur Hypsilophodon (Reptilia: Ornithischia). *Postilla* **131,** 1–64.

Gatesy, S. M. (1991). Caudifemoral musculature and the evolution of theropod locomotion. *Paleobiology* **16,** 170–186.

Gregory, W. K. (1919). The pelvis of dinosaurs; A study of the relations between muscular stresses and skeletal form. *Copeia* **69,** 18–20.

Gregory, W. K., and Camp, C. L. (1918). Studies in comparative osteology and myology. Part III. *Bull. Am. Museum Nat. History* **38,** 447–563.

Haas, G. (1955). The jaw musculature in *Protoceratops* and in other ceratopsians. *Am. Museum Novitates* **1729,** 1–24.

Hass, G. (1963). A proposed reconstruction of the jaw musculature of *Diplodocus. Ann. Carnegie Museum* **36,** 139–157.

Haas, G. (1969). On the jaw musculature of *Ankylosaurus. Am. Museum Novitates* **2399,** 1–11.

Lull, R. S. (1908). The cranial musculature and the origin of the frill in ceratopsian dinosaurs. *Am. J. Sci.* **25**(4), 387–399.

Molnar, R. (1993). The cranial morphology of *Tyrannosaurus rex. Paleontographica A* **217,** 137.

Nicholls, E. L., and Russell, A. P. (1985). Structure and function of the pectoral girdle and forelimb of *Struthiomimus altus* (Theropoda: Ornithomimidae). *Palaeontology* **28,** 643–677.

Norman, D. (1985). *The Illustrated Encyclopedia of Dinosaurs*, pp. 208. Crescent Books, New York.

Norman, D., and Weishampel, D. B. (1985). Ornithopod feeding mechanisms: Their bearing on the evolution of herbivory. *Am. Nat.* **126,** 151–164.

Ostrom, J. H. (1964). A functional analysis of jaw mechanics in the dinosaur *Triceratops. Postilla* **88,** 1–35.

Ostrom, J. H. (1974). The pectoral girdle and forelimb function of *Deinonychus* (Reptilia: Saurischia): A correction. *Postilla* **165,** 1–11.

Paul, G. S. (1987). The science and art of restoring the life appearance of dinosaurs and their relatives. In *Dinosaurs Past and Present Vol. II* (S. J. Czerkas and E. C. Olson, Eds.), pp. 5–49. Natural History Museum, Los Angeles County, Los Angeles.

Raath, M. A. (1977). The anatomy of the Triassic theropod *Syntarsus rhodesiensis* (Saurischia: Podokesauridae) and a consideration of its biology, pp. 230. Ph.D. dissertation, Rhodes University.

Romer, A. S. (1923). The pelvic musculature of saurischian dinosaurs. *Bull. Am. Museum Nat. History* **48,** 605–617.

Romer, A. S. (1927). The pelvic musculature of ornithischian dinosaurs. *Acta Zool.* **8,** 225–275.

Russell, D. A. (1972). Ostrich dinosaurs from the Late Cretaceous of Western Canada. *Can. J. Earth Sci.* **9,** 375–402.

Russell, L. S. (1935). Musculature and function in the Ceratopsia. *Bull. Natl. Museum Can.* **77,** 39–48.

Smith, M. B., and Carpenter, K. (1990). Forelimb mechanics of *Tyrannosaurus rex. J. Vertebr. Paleontol.* **9,** 43A.

Tarsitano, S. (1983). Stance and gait in theropod dinosaurs. *Acta Paleontol. Polonica* **28,** 251–264.

Walker, A. D. (1977). Evolution of the pelvis in birds and dinosaurs. In *Problems in Vertebrate Evolution* (S. M. Andrews, R. S. Miles, and A. D. Walker, Eds.), pp. 319–357. Academic Press, London.

Musée des Dinosaures, Espéraza, Aude, France

JEAN LE LOEUFF
Musée des Dinosaures
Espéraza, France

The Musée des Dinosaures opened in 1992 in Espéraza, 50 km south of Carcassonne (southern France). The museum is managed by the Association DINOSAURIA, a nonprofit organization composed of both professional and amateur paleontologists; it is now led by Dr. Eric Buffetaut (Paris). The collections of the Musée des Dinosaures consist mainly of Late Cretaceous dinosaur bones excavated from southern France since 1989 and of casts of some of the best specimens preserved in French private collections. This Late Cretaceous (Campanian–Maastrichtian) European fauna is very different from the contemporaneous North American and Asian faunas and is closer to the "Gondwanan" (from South America, Africa, India, and Madagascar) Cretaceous faunas. The best represented dinosaur in the collections of the museum is the new titanosaurid sauropod *Ampelosaurus atacis*, a large armored plant-eating animal. Ornithopod dinosaurs are represented by the conservative *Rhabdodon priscus* and *Rhabdodon septimanicus* and recently discovered hadrosaurids of late Maastrichtian age. Theropod dinosaurs consist of a small, unnamed dromaeosaurid of north Tethysian affinities and larger south Tethysian abelisaurid-like dinosaurs. A few ankylosaur bones are also housed in the collections. The Upper Aude Valley is rich in dinosaur egg localities, and the museum has a fairly good collection of EGGS AND NESTS representing different paleoenvironments. The richest dinosaur egg site at Rennes-le-Château has just been acquired by the museum and will be excavated in the near future. The richest dinosaur bone locality of the area is at Campagne-sur-Aude (the type locality of *Ampelosaurus atacis*) and is now accessible to visitors. A museum project is to create a "Dinosaur Valley" in the dinosaur-rich area between Couiza and Quillan, including the construction of a great paleontological museum and the development of educational programs at two dinosaur localities and a dinosaur park.

Other dinosaurs in the museum include a few Triassic plateosaurids (casts from Eastern France and Thailand), many casts of the Late Jurassic Thai sauropod *Phuwiangosaurus sirintornae*, and Jurassic camarasaurids and theropods from western and northern France (including casts of the teeth of *Neosodon praecursor*). Cretaceous dinosaurs include newly discovered sauropods from the Albian of Normandy, casts of *Rhabdodon robustus*, *Telmatosaurus transylvanicus*, and *Magyarosaurus dacus* from Romania, and recently collected bones of *Spinosaurus aegyptiacus*, *Aegyptosaurus*, *Rebbachisaurus garasbae*, and *Carcharodontosaurus saharicus* from the Cenomanian of Morocco. A partial skeleton of *Psittacosaurus meileyingensis* from China is also on exhibit. Casts of isolated bones or skulls of other "classic" dinosaurs (*Allosaurus*, *Iguano-*

don, Triceratops, and *Tyrannosaurus*) are also exhibited at the museum.

Museum fieldwork includes approximately 3 months each year excavating in southern France and expeditions to Thailand, Morocco, and other regions. The museum is associated with a team of the Centre National de la Recherche Scientifique in Paris, directed by Dr. Eric Buffetaut.

See also the following related entry:

European Dinosaurs

Musée National du Niger, Niger

see Museums and Displays

Museo Argentino de Ciencias Naturales, Argentina

see Museums and Displays

Museo Civico di Storia Naturale de Milano, Italy

see Museums and Displays

Museo Civico di Storia Naturale di Venezia, Italy

see Museums and Displays

Museo Civico Naturales, Mexico

see Museums and Displays

Museo de Ciencias Naturales de la Universidad Nacional del Comahue, Argentina

see Museums and Displays

Museo de la Plata, Argentina

see Museums and Displays

Museo dell'Instituto di Paleontologia, Italy

see Museums and Displays

Museum d'Histoire Naturelle, Switzerland

see Museums and Displays

Museo di Scienze Naturali "E. Caffi," Italy

see Museums and Displays

Museum Hauf, Germany

see Museums and Displays

Museum Nacional, Brazil

see Museums and Displays

Museum of Comparative Zoology, Harvard University

The Museum of Comparative Zoology (MCZ) was founded in 1859 by Louis Agassiz, the great Swiss natural historian who was provided a professorship and funding for a museum by Harvard University officials after his successful series of Lowell lectures in Boston in 1845 (Lurie, 1960; Winsor, 1991). VERTEBRATE PALEONTOLOGY at the MCZ began early with Louis Agassiz's noted collection of fossil fishes and was augmented by dinosaur and other FOOTPRINTS from the Connecticut River valley and other remains during the directorship of his son Alexander. The original building between Oxford and Divinity Streets was enlarged and expanded over the years, and plans of the contemporary structure in 1889 show preparation, collection, and lecture rooms dedicated to fossil vertebrates. It was not, however, until Alfred S. Romer moved to Harvard from the University of Chicago in 1934 that the MCZ began to amass in earnest its collection of dinosaurs. Romer's expeditions to the Permian of Texas had brought back a great many pelycosaurian synapsids (see REPTILES), and he began to work in the Triassic of Argentina as early as 1958 in search of early dinosaurs and mammals (Romer and Jensen, 1966). The series of expeditions that continued through 1968 succeeded in providing a previously unknown chapter in the history of terrestrial vertebrates, including some of the earliest dinosauriforms and the immediate forerunners of dinosaurs brought back in the spectacularly successful field season of 1964–1965. These included the ornithodirans *Lagosuchus* (?= *Marasuchus*), *Lagerpeton*, and *Lewisuchus*, as well as numerous pseudosuchians and other archosaurs described by Romer (e.g., 1971, 1972a,b) and colleagues in a series of papers in the MCZ series *Breviora* between 1966 and 1972. A fine skeleton of the herrerasaurid *Staurikosaurus*, from the Santa Maria Formation of Brazil, was described by Colbert (1970) and Galton (1977), and another partial skeleton from the Ischigualasto Formation of Argentina, collected by Romer in 1958, was described by Brinkman and Sues (1987).

Expeditions led by F. A. Jenkins, Jr. to the CLOVERLY FORMATION in 1972 and 1974 yielded specimens of the ornithopod *Tenontosaurus* and the theropod *Deinonychus*, as well as a suite of Early Cretaceous mammals that included triconodonts and *Gobiconodon ostromi*. In the late 1970s and early 1980s, work in the Kayenta and associated formations of the GLEN CANYON GROUP of Arizona collected what has become the most diverse vertebrate fauna that we have from the Early Jurassic in the world. The MCZ collections excavated the ceratosaurian theropod *Syntarsus*, the basal sauropodomorph *Massospondylus*, and an undescribed heterodontosaurid, numerous skulls and skeletons of tritylodontid cynodont therapsids, and the crocodilian *Eopneumatosuchus*, plus numerous remains of turtles, sphenodontians, squamates, pterosaurs, and fishes. The earliest known morganucodont mammaliamorphs and frogs on the continent, and the earliest known caecilian amphibians, have been of particular systematic and functional value in understanding the early evolutionary histories of these groups (the fauna is reviewed in Sues *et al.*, 1994). A diverse vertebrate microfauna was also collected and screenwashed from the Kayenta; in addition to hybodont sharks, it generally reflects the macrofauna in the presence of frogs, caecilians and perhaps salamanders, sphenodontians, protosuchid crocodyliforms, and other archosaurian and various remains.

Recent expeditions by the MCZ (1988–1995) to the Fleming Fjord region (Late Triassic: Norian) of Greenland have yielded mostly as yet undescribed remains of basal sauropodomorphs, theropods and theropod footprints (Gatesy and Middleton, 1996), plus other archosaurs and associated fauna that includes aetosaurs, amphibians, and a pterosaur (Jenkins *et al.*, 1994), as well as an unusual haramiyid mammal with intact dentition. Other expeditions to the Triassic of Morocco have collected typical Late Triassic faunas of phytosaurs and metoposaurs.

References

Brinkman, D. B., and Sues, H.-D. (1987). A staurikosaurid dinosaur from the Upper Triassic Ischigualasto Formation of Argentina and the relationships of the Staurikosauridae. *Palaeontology* **30,** 493–503.

Colbert, E. H. (1970). A saurischian dinosaur from the Triassic of Brazil. *Am. Museum Novitates* **2405,** 1–39.

Galton, P. M. (1977). On *Staurikosaurus pricei*, an early saurischian dinosaur from Brazil, with notes on the Herrerasauridae and Poposauridae. *Paläontol. Z.* **51,** 234–245.

Gatesy, S. M., and Middleton, K. M. (1996). Sinking dinosaurs: Sub-surface preservation of foot kinematics in Greenlandic theropods. *J. Vertebr. Paleontol.* **16**(Suppl. to No. 3), 37A.

Jenkins, F. A., Jr., Shubin, N. H., Amaral, W. W., Gatesy, S. M., Schaff, C. R., Clemmensen, L. B., Downs, W. R., Davidson, A. R., Bonde, N., and Osbeck, F. (1994). Late Triassic continental vertebrates and depositional environments of the Fleming Fjord Formation, Jameson Land, East Greenland. *Meddeleleser Grønland Geosci.* **32,** 1–25.

Lurie, E. (1960). *Louis Agassiz: A life in science.* Univ. of Chicago Press, Chicago.

Romer, A. S. (1971). The Chañares (Argentina) Triassic fauna. X. Two new and incompletely known long-limbed pseudosuchians. *Breviora* **378,** 1–10.

Romer, A. S. (1972a). The Chañares (Argentina) Triassic fauna. XV. Further remains of the thecodonts *Lagerpeton* and *Lagosuchus*. *Breviora* **394,** 1–7.

Romer, A. S. (1972b). The Chañares (Argentina) Triassic fauna. XVI. Thecodont classification. *Breviora* **395,** 1–24.

Romer, A. S., and Jensen, J. A. (1966). The Chañares (Argentina) Triassic fauna. II. Sketch of the geology of the Rio Chañarense–Rio Gualo region. *Breviora* **252,** 1–20.

Sues, H.-D., Clark, J. M., and Jenkins, F. A., Jr. (1994). A review of the Early Jurassic tetrapods from the Glen Canyon Group of the American Southwest. In *In the Shadow of the Dinosaurs: Early Mesozoic Tetrapods* (N. C. Fraser and H.-D. Sues, Eds.), pp. 284–294. Cambridge Univ. Press, New York.

Winsor, M. P. (1991). *Reading the Shape of Nature: Comparative Zoology at the Agassiz Museum.*

Museum of Earth Science, Brigham Young University

David K. Smith
Pima Community College
Tucson, Arizona, USA

One of the most extensive collections of Jurassic vertebrates in the world is housed at the Earth Science Museum of Brigham Young University (BYU). Most of this material consists of dinosaur remains from the Upper Jurassic Morrison Formation, Dry Mesa Quarry, and Cleveland–Lloyd Quarry. Isolated finds include nearly complete, articulated *Allosaurus* crania from the Hinkle Site in east-central Utah and the Red Seeps Site on the San Rafael Swell. A nearly complete, articulated *Ceratosaurus* cranium, also from the San Rafael Swell, is currently being described. A hadrosaur specimen with skin impressions was recently recovered from the Book Cliffs in eastern Utah.

A large theropod pedal phalanx, the first element from what would be Dry Mesa Dinosaur Quarry on the Uncompaghre Uplift of western Colorado, was found by Vivian Jones. It was turned over to James Jensen and Kenneth Stadtman of BYU in 1972. With the support of the National Forest Service, an expedition was mounted to locate more of the specimen. The following summer, this expedition found an extensive bone bed on Dry Mesa. As more material was uncovered, the large theropod was referred to as a new species, *Torvosaurus tanneri*. Other dinosaur material from this site include *Allosaurus, Apatosaurus, Camarasaurus, Supersaurus*, and *Brachiosaurus*. The diversity of material found at Dry Mesa probably meets or exceeds that of many of the other major Morrison quarries. Collection continues from this site at the present time, and the probability of new discoveries remains high.

Work began at the Cleveland–Lloyd Dinosaur Quarry in Emery County, Utah, in 1927 and has continued intermittently to the present (Madsen, 1976). Expeditions under William Stokes from Princeton University worked the quarry from 1939 to 1941. In 1960 and 1965, an expedition from the University of Utah cooperative dinosaur project under Stokes and James Madsen continued quarrying dinosaur remains at this site. The most common species present is *Allosaurus fragilis*. A nearly complete allometric series is present from this one theropod, although few of the individuals reach the size attained by allosaurs from Como Bluff and Dry Mesa. Other dinosaurs present at this site include *Camptosaurus, Marshosaurus, Stokesosaurus*, and *Camarasaurus*. Much of this material is now housed at the BYU Earth Science Museum.

Under the direction of Kenneth Stadtman, the Earth Science Museum is increasing its collection of Lower Cretaceous vertebrates. The Dalton Wells Quarry, in the Cedar Mountain Formation near Moab, Utah, is currently being worked by Brigham Young University and the Museum of Western Colorado. Sauropod and nodosaurid material has been found here. Elements referred to the dromaeosaurid *Utahraptor* are also housed in the Earth Science Museum.

Preparation and research on the dinosaur collections continues at a high rate at the Earth Science Museum. With the support of Dr. Wade Miller, recent projects have included quarry TAPHONOMY, as well as dinosaur SYSTEMATICS, HISTOLOGY, and morphometrics.

See also the following related entries:

CLEVELAND–LLOYD DINOSAUR QUARRY • DRY MESA QUARRY • MORRISON FORMATION

Reference

Madsen, J. (1976). *Allosaurus fragilis.* A revised osteology. *Utah Geol. Miner. Survey Bull.* **109,** 1–163.

Museum of Earth Sciences, France

see MUSEUMS AND DISPLAYS

Museum of Earth Sciences, Morocco

see MUSEUMS AND DISPLAYS

Museum of Geology, South Dakota, USA

see MUSEUMS AND DISPLAYS

Museum of Isle of Wight Geology, United Kingdom

see MUSEUMS AND DISPLAYS

Museum of Natural History, Wisconsin, USA

see MUSEUMS AND DISPLAYS

Museum of Natural Sciences Saskatchewan, Canada

see MUSEUMS AND DISPLAYS

Museum of Northern Arizona, Arizona, USA

see MUSEUMS AND DISPLAYS

Museum of Paleontology, California, USA

see UNIVERSITY OF CALIFORNIA MUSEUM OF PALEONTOLOGY

Museum of Science and History, Florida, USA

see MUSEUMS AND DISPLAYS

Museum of Texas Tech University, Texas, USA

see MUSEUMS AND DISPLAYS

Museum of the Rockies

MARY SCHWEITZER
Montana State University
Bozeman, Montana, USA

The Museum of the Rockies is an extension of the campus of Montana State University in Bozeman, Montana. Following its central theme of "Montana: One Place Through Time," the museum has exhibits depicting the pioneering history of the region, Native American populations and their history, archaeological excavations, geological history and diversity, and, of course, vertebrate paleontology. The focus of vertebrate paleontological research and field exploration at the Museum of the Rockies is the evolution of Mesozoic terrestrial ecosystems within the region. Field exploration includes sites in the Chugwater Formation (Triassic), several MORRISON FORMATION exposures (Jurassic), and the CLOVERLY–Kootenai Formations (Early Cretaceous). Exploration and excavations continue in the Two Medicine–Judith River complex, representing the Late Cretaceous, and the St. Mary–HELL CREEK complex, of the latest Cretaceous. Data from field site studies are used to elucidate patterns of evolutionary change and include taphonomic interpretations and paleoecological studies. The large monospecific dinosaur bone beds found at some of these sites allow for statistical analyses of populations of hadrosaurs and ceratopsians. These data can shed light on the existence of sexual dimorphism within species, allometric growth patterns, and predator–prey relationships. Insight into dinosaurian BEHAVIOR has been developed from these dig sites, including herding and nesting behavior.

In addition to active field studies, research in the VERTEBRATE PALEONTOLOGY labs at the Museum of the Rockies includes ontogenetic and phylogenetic studies using various imaging techniques. Histological studies help illuminate physiological questions such as growth rates, inferred from bone growth series. One goal of the histology lab is the development of a database of histological samples that will eventually be accessible to researchers over the World Wide Web. Biomechanical studies include strength studies, range of motion analyses, and speed and locomotion studies of various dinosaur taxa. Paleopathologies, in the form of both healed fractures and disease processes, are under study in our labs and may shed light on the habits and lifestyle of *Allosaurus fragilis*. With the completion of a new research wing, an extant–extinct molecular facility will be part of the evolutionary studies carried out at the museum. Molecular data will be included in phylogenetic studies of extant taxa, and retrieval of molecular components from fossil specimens of varying ages will shed light not only on the phylogenies of these taxa but also on the nature and processes of degradation of the molecules themselves.

One goal of the Museum of the Rockies is to use the most up-to-date research available to bring the dinosaurs "to life." The vertebrate paleontology exhibits within the Dinosaur Hall at Museum of the Rockies are open to the general public. The displays are based on information obtained from Egg Mountain and include full-sized, fleshed-out *Maiasaura* guarding their nests and hatchlings. Also on display are nesting grounds of *Orodromeus makelai*, a small ornithopod found in the same region. Pterosaurs and champsosaurs complete the ecosystem portrayed here. The exhibits also include a cast of the skull of the museum's own *Tyrannosaurus rex* as well as the hindlimb and forelimb of the same specimen. This is the first complete forelimb of *T. rex* ever discovered.

See also the following related entries:

CRETACEOUS PERIOD • EGG MOUNTAIN • JUDITH RIVER WEDGE • MUSEUMS AND DISPLAYS • TWO MEDICINE FORMATION

Museum of Victoria, Australia

see MUSEUMS AND DISPLAYS

Museum of Western Colorado, Colorado, USA

see MUSEUMS AND DISPLAYS

Museums and Displays

Daniel J. Chure
Dinosaur National Monument
Jensen, Utah, USA

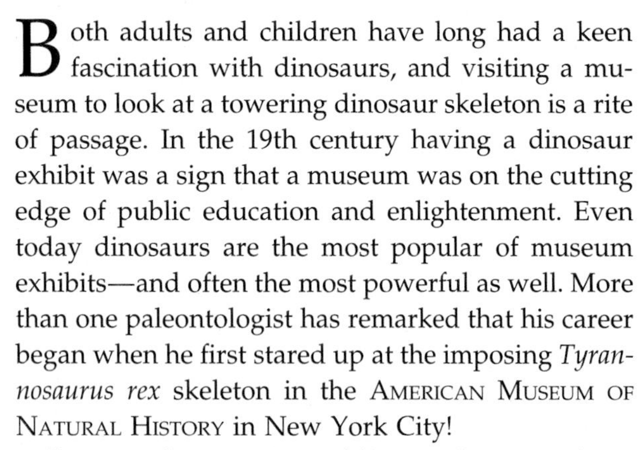

Both adults and children have long had a keen fascination with dinosaurs, and visiting a museum to look at a towering dinosaur skeleton is a rite of passage. In the 19th century having a dinosaur exhibit was a sign that a museum was on the cutting edge of public education and enlightenment. Even today dinosaurs are the most popular of museum exhibits—and often the most powerful as well. More than one paleontologist has remarked that his career began when he first stared up at the imposing *Tyrannosaurus rex* skeleton in the American Museum of Natural History in New York City!

Recent advances in molding and casting have made it possible for most museums to have at least a dinosaur skull if not a skeleton on display, and the number of dinosaur exhibits increases each year. In addition to traditional museum displays, fossils are now being exhibited and interpreted "in the ground." Sometimes this is skeletal material (such as at Dinosaur National Monument), but increasingly footprint localities are being opened to the public. Thus, you have an unprecedented ability to view dinosaurs from around the world and to see firsthand their spectacular, if sometimes confusing, diversity.

The following list is worldwide in scope and includes all sites known to the author where the public can see some type of dinosaur fossils—original bones, tracks, or replicas. It does not include exhibits that are primarily concrete monstrosities, robotic models, or temporary and traveling exhibits. The only exception to these criteria are Waterhouse Hawkins' Crystal Palace models, which are of great significance historically. Visiting hours may vary seasonally (or with fluctuating budgets); a mailing address is provided when possible to help plan your visit.

North America

Canada

Alberta

Dinosaur Provincial Park, Box 60, Patricia, Alberta TOJ 2KO

Drumheller Dinosaur and Fossil Museum, Box 2135, Drumheller, Alberta TOJ OYO

Provincial Museum of Alberta, 12845 102 Avenue, Edmonton, Alberta T5N OM6

Royal Tyrrell Museum of Palaeontology, Box 7500, Drumheller, Alberta TOJ OYO

Manitoba

Manitoba Museum of Man and Nature, 190 Rupert Avenue, Winnipeg, Manitoba R3B ON2

Ontario

Canadian Museum of Nature, Box 3443, Station D., Ottawa, Ontario K1P 6P4

Royal Ontario Museum, 100 Queen's Park, Toronto, Ontario M5S 2C6

Quebec

Redpath Museum, McGill University, 859 Sherbrooke Street West, Montreal, Quebec H3A 2K6

Saskatchewan

Museum of Natural Sciences, University of Saskatchewan, Saskatoon, Saskatchewan S7N OWO

Royal Saskatchewan Museum of Natural History, College Avenue and Albert Street, 2445 Albert St., Regina, Saskatchewan S4P 3V7

United States

Alabama

Anniston Museum of Natural History, Box 1587, Anniston, AL 36202

Department of Geology, University of Alabama, Tuscaloosa, AL

Red Mountain Museum, 1421 22nd Street South, Birmingham, AL 35205

Alaska

Dorothy C. Page Museum, 323 Main Street, Wasilla, AK 99687

University of Alaska Museum, 907 Yukon Drive, Fairbanks, AK 99775

Arizona

Arizona Museum of Science and Technology, 147 East Adams, Phoenix, AZ 85004

Glen Canyon National Recreation Area, Visitor Center, c/o Glen Canyon National Recreation Area, Box 1507, Page, AZ 86040

Mesa Southwest Museum, 53 North MacDonald Street, Mesa, AZ 85201

Museum of Northern Arizona, 3001 North Fort Valley Road, Route 4, Box 720, Flagstaff, AZ 86001

Petrified Forest National Park, Box 2217, Petrified Forest National Park, AZ 86028

Arkansas

County Seat and Courthouse Tracks, 421 North Main, Nashville, AR 71852

Arkansas Geological Commission, 3815 West Roosevelt Road, Little Rock, AR 72204

Arkansas Museum of Science and History, MacArthur Park, Little Rock, AR 72202

University Museum, Room 202, University of Arkansas, Garland Street, Fayetteville, AR 72701

California

California Academy of Sciences, Golden Gate Park, San Francisco, CA 94118

Life Sciences Museum, Pierce College, Woodland, CA

Museum of Paleontology, University of California, Valley Life Science Building, Berkeley, CA 94720

Natural History Museum of Los Angeles County, 900 Exposition Boulevard, Los Angeles, CA 90007

Raymond Alf Museum, 1175 West Base Line Road, Claremont, CA 91711

San Bernardino County Museum, 2024 Orange Tree Lane, Redlands, CA 92374

San Diego Natural History Museum, Balboa Park, Box 1390, San Diego, CA 91112

Colorado

Denver Museum of Natural History, 2001 Boulevard, Denver, CO 80205

Devil's Canyon Discovery Center, Box 307, Fruita, CO 81521

Dick's Rock Museum, 490 Moraine Route, Estes Park, CO 80517

Dinosaur Discovery Center, Box 313, Canon City, CO 81215-0313

Dinosaur Hill, c/o Museum of Western Colorado, Box 20000-5020, Grand Junction, CO 81502

Dinosaur Ridge, c/o Morrison Natural History Museum, Box 564, Morrison, CO 80456

Garden Park Fossil Area, c/o Garden Park Paleontology Society, Box 313, Canon City, CO 81215

Morrison Natural History Museum, P.O. Box 564, Morrison, CO 80456

Museum of Western Colorado, Box 20000-5020, Grand Junction, CO 81502

Purgatoire Valley, near La Junta, c/o National Forest Service, 1321 East 3rd, La Junta, CO 81050

Rabbit Valley Trail Through Time, c/o Museum of Western Colorado, Box 20000-5020, Grand Junction, CO 81502

Riggs Hill, c/o Museum of Western Colorado, Box 20000-5020, Grand Junction, CO 81502

University of Colorado Museum, Henderson Building, Campus Box 218, Boulder, CO

Connecticut

Connecticut State Museum of History, University of Connecticut, Route 195, Storrs, CT 06269-3023

Dinosaur State Park, West Street, Rocky Hill, CT 06067

Powder Hill Dinosaur Park, c/o Chamber of Commerce, 393 Main Street, Middlefield, CT 06455

Science Center, Church Street, Weslyan University, Middletown, CT 06459

Yale–Peabody Museum of Natural History, Whitney Avenue and S Street, New Haven, CT 06511

Florida

Museum of Science and History, 1025 Museum Circle, Jacksonville, FL 32207-9854

Georgia

Fernbank Museum of Natural History, 767 Clifton Road, NE, Atlanta, GA 30307

Fernbank Science Center, 156 Heaton Park Drive, Atlanta, GA 30307

Lanier Museum of Natural History, 2601 Buford Dam Road, Buford, GA 30518

Weinman Mineral Museum, P.O. Box 1255, Cartersville, GA 30120

Idaho

Idaho State Museum of Natural History, Box 8096, Idaho State University, Pocatello, ID 83209

Illinois

Field Museum of Natural History, Lake Shore Drive at Roosevelt Road, Chicago, IL 60605

Fryxell Geology Museum, 820 38th Street, Rock Island, IL 61201

Geology Museum, Western Illinois University, Macomb, IL 61455

Indiana

Children's Museum of Indianapolis, Box 3000, Indianapolis, IN 46208

Joseph Moore Museum, Earlham College, Richmond, IN

Kansas

Department of Geology, University of Wichita, Wichita, KS

Ottawa County Historical Museum, 110 South Concord, Minneapolis, KS 67467

Sternberg Museum, Fort Hays State University, Campus Drive, Hays, KS 67601

University of Kansas Museum of Natural History, Dyche Hall, 14th and Jayhawk Boulevard, Lawrence, KS 66045

Louisiana

Audubon Institute—Pathways to the Past, 6500 Magazine Street, New Orleans, LA 70118

LSU Museum of Geoscience, 109 Howe–Russell Geoscience Complex, Louisiana State University, Baton Rouge, LA 70803

Massachusetts

Barton Cove Footprint Quarry, c/o Greenfield County Chamber of Commerce, 395 Main Street, Greenfield, MA 01301

C. Nash Dinosaur Museum, Route 116, Amherst Road, South Hadley, MA 01075

Dinosaur Footprints Reservation, c/o the Trustees of Reservations, Western Regional Office, Box 792, Stockbridge, MA 01261

Museum of Comparative Zoology, Harvard University, 26 Oxford Street, Cambridge, MA 02138

New England Paleontological Society, Barre, MA

Pratt Museum of Natural History, Amherst College, Amherst, MA 01002

Springfield Science Museum, 236 State Street, Springfield, MA 01103

Wistariahurst Museum, 238 Cabot Street, Holyoke, MA 01041

Michigan

Cranbrook Institute of Science, 500 Lone Pine Road, Box 801, Bloomfield Hills, MI 48013

Kingman Museum of Natural History, 175 Limits, Battle Creek, MI 49017

Michigan State University Museum, West Circle Drive, East Lansing, MI 48824

University of Michigan Exhibit Museum, Alexander G. Ruthven Museum Building, 1109 Geddes Avenue, Ann Arbor, MI 48109

Minnesota

Science Museum of Minnesota, 30 East Tenth Street, Saint Paul, MN 55101

Mississippi

Dunn–Seiler Museum, Mississippi State University, Box 5167, Mississippi State University, MS 39762

Missouri

Department of Geology, Washington University, St. Louis, MO

Montana

Carter County Museum, 100 Main Street, Ekalaka, MT 59324

Fort Peck Power Project, U.S. Army Corps of Engineers, Box 208, Fort Peck, MT 59223

Garfield County Museum, Box 325, Jordan, MT 59337

Museum of the Rockies, 600 West Kagy Boulevard, Bozeman, MT 59717

Old Trail Museum, Choteau, MT 59422

Pine Butte Swamp Preserve, HC 58, Box 34B, Choteau, MT 59442

Upper Musselshell Museum, 11 South Central, Harlowton, MT 59036

Nebraska

University of Nebraska State Museum, Morrill Hall, Lincoln, NE 68588

Nevada

Las Vegas Museum of Natural History, 900 Las Vegas Boulevard North, Las Vegas, NV 89101

New Jersey

Morris Museum, 6 Normandy Heights Road, Morristown, NJ 07960

New Jersey State Museum, CN-530, 205 West State Street, Trenton, NJ 08625

Princeton Natural History Museum, Guyot Hall, Washington Street, Princeton, NJ 08544

Rutgers University Geology Museum, Geology Hall c.a.c., Rutgers University, New Brunswick, NJ 08903

New Mexico

Clayton Lake State Park, R.R. Box 20, Seneca, NM 88437

Farmington Museum, 302 North Orchard, Farmington, NM 87401

Geology Museum, University of New Mexico, Albuquerque, NM 87125

New Mexico Museum of Natural History, 1801 Mountain Road NW, Albuquerque, NM 87104

Ruth Hall Museum of Paleontology, Ghost Ranch Conference Center, Abiquiu, NM 87510

New York

American Museum of Natural History, Central Park West and 79th Street, New York, NY 10024

Buffalo Museum of Science, 1020 Humboldt Parkway, Buffalo, NY 14211

Department of Geology, University of Buffalo, Buffalo, NY 14211

Long Island Natural History Museum, Long Island University

New York State Museum / New York State Geological Survey, State Education Center, Albany, NY 12230

North Carolina

Natural Science Center, 4301 Lawndale Drive, Greensboro, NC 27401

North Carolina Museum of Life and Science, 433 Murray Avenue, Durham, NC 27705

North Carolina State Museum of Natural Science, Box 27647, Raleigh, NC 27647

North Dakota

Dakota Dinosaur Museum, Dickenson, ND 58601

Joachim Regional Museum, Visitor's Center, 314 Third Avenue, West, Dickinson, ND 58601

Leonard Hall, University of North Dakota, Box 8358, Grand Forks, ND 58202

Ohio

Cleveland Museum of Natural History, One Wade Oval Drive, University Circle, Cleveland, OH 44106

Dayton Public Library Museum, 2600 Deweese Parkway, Dayton, OH

McKinley Museum of History, 800 McKinley Monument Drive NW, Canton, Ohio 44708-4800

Oklahoma

Dinosaur Quarries, c/o Bonnie Heppard, Box 36, Kentoon, OK 73946

Dinosaur Track Site, Cimarron Heritage Center, Box 1146, Boise City, OK 73933

Oklahoma Museum of Natural History, 1335 Asp Avenue, Norman, OK 73019

Pennsylvania

Academy of Natural Sciences, 1900 Ben Franklin Parkway, Philadelphia, PA 19103

Carnegie Museum of Natural History, 4400 Forbes Avenue, Pittsburgh, PA 15213

Earth and Mineral Science Museum, Pennsylvania State University, 112 Steidle Building, University Park, PA 16802

Everhart Museum, Nay Aug Park, Scranton, PA 18510

State Museum of Pennsylvania, Harrisburg, PA 17108-1026

Wagner Free Institute of Science, 17th St. and Montgomery Avenue, Philadelphia, PA 19121

South Dakota

Black Hills Museum of Natural History, 217 Main Street, Hill City, SD 57745

Museum of Geology, South Dakota School of Mines and Technology, 501 East Joseph Street, Rapid City, SD 57701

Tennessee

Frank H. McClung Museum, 1327 Circle Park Drive, University of Tennessee, Knoxville, TN 37919

Memphis Pink Palace Museum, 3050 Central Avenue, Memphis, TN 38111

Texas

Annie Riggs Museum, Fort Stockton Historical Society, 301 South Main, Fort Stockton, TX 79735
Brazosport Museum of Natural Science, 400 College Drive, Lake Jackson, TX 77566
Corpus Christi Museum of Science and History, 1900 North Chaparral, Corpus Christi, TX 78401
Dallas Museum of Natural History, Box 150349, Dallas, TX 75315
Department of Geology, University of Texas at El Paso, El Paso, TX 79968
Dinosaur Flats, 4831 FM 2673, Canyon Lake, TX 75960
Dinosaur Gardens, Box 98, Moscow, TX 75960
Dinosaur Tracks, c/o Chamber of Commerce, Box 126, Hondo, TX 78043
Dinosaur Valley State Park, Box 396, Glen Rose, TX 76043
Fort Worth Museum of Science and History, 1501 Montgomery Street, Fort Worth, TX 76107
Houston Museum of Natural Science, One Hermann Circle Drive, Houston, TX 77030
Museum of Texas Tech University, 4th and Indiana Avenue, Lubbock, TX 79409
Panhandle–Plains Historical Museum, Western Texas State University, Box 967, 2401 Fourth Avenue, Canyon, TX 79016
Petroleum Museum, 1500 I-20 West, Midland, TX 79701
R. A. Vines Environmental Science Center, 8856 Westview Drive, Houston, TX 77055
Shuler Museum of Paleontology, Southern Methodist University, Department of Geological Science, Dallas, TX 75275-0395
Strecker Museum, South Fourth and Speight, Box 97154, Waco, TX 76798
Texas Memorial Museum, University of Texas at Austin, 2400 Trinity, Austin, TX 78705
Witte Museum of History and Science, 3801 Broadway, San Antonio, TX 78209

Utah

Brigham Young University Earth Science Museum, 1683 North Canyon Road, Provo, UT 84602
Cleveland–Lloyd Dinosaur Quarry, c/o Bureau of Land Management, P.O. Drawer A.B., 900 North 700 East, Price, UT 84501
College of Eastern Utah Prehistoric Museum, 155 East Main Street, Price, UT 84501
Dan O'Laurie Museum, 118 East Center St., Moab, UT 84532
Dead Horse Point State Park, Box 609, Moab, UT 84532
Dinosaur Museum, 754 South 200 West, Blanding, UT 83511
Dinosaur National Monument, Box 128, Jensen, UT
George S. Eccles Dinosaur Park, 1544 East Park Boulevard, Ogden, UT 84401
Glen Canyon National Recreation Area, near Bullfrog, c/o Glen Canyon National Recreation Area, Box 1507, Page, AZ 86040
Mill Canyon Dinosaur Trail, c/o Bureau of Land Management, Grand Resource Area, 885 South Sand Flats Road, Moab, UT 84532
Museum of the San Rafel, 95 North 100 East, Castle Dale, UT 84513
Potash Road Dinosaur Tracks, c/o Bureau of Land Management, Grand Resource Area, 885 South Sand Flats Road, Moab, UT 84532
Red Fleet Reservoir State Park, Steinaker Lake North 4335, Vernal, UT 84078
Sauropod Tracksite, c/o Bureau of Land Management, Grand Resource Area, 885 South Sand Flats Road, Moab, UT 84532
Utah Field House of Natural History State Park, 235 East Main Street, Vernal, UT 84078
Utah Museum of Natural History, University of Utah, Salt Lake City, UT 84112
Warner Valley Tracksite, c/o Bureau of Land Management, Dixie Resource Area, 255 North Bluff Street, Saint George, UT 84770
Weber State University Museum of Natural Science, 3750 Harrison Boulevard, Ogden, UT 84408

Vermont

Montshire Museum of Science, Box 770, Montshire Road, Norwich, VT 05055

Virginia

Museum of Culpeper History, 140 East Davis Street, Culpeper, VA 22701
Virginia Living Museum, 524 Clyde Morris Boulevard, Newport News, VA 23601
Virginia Museum of Natural History, 1001 Douglas Avenue, Martinsville, VA 24112
Virginia Polytechnic Institute and State University, Blacksburg, VA

Washington

Burke Museum, University of Washington, DB-10, Seattle, WA 98185
Pacific Science Center, 200 Second Avenue North, Seattle, WA 98102
Washington State University, Pullman, WA

Washington, DC

National Museum of Natural History, Smithsonian Institution, Tenth Street and Constitution Avenue NW, Washington, DC 20560

Wisconsin

Milwaukee Public Museum, 800 West Wells Street, Milwaukee, WI 53233
Museum of Natural History, University of Wisconsin, Stevens Point, WI
University of Wisconsin Geology Museum, 1215 West Dayton Street, Madison, WI 53706

Wyoming

Greybull Museum, 325 Greybull Street, Greybull, WY 82462
Sheridan College Geology Museum, 3059 Coffeen Avenue, Sheridan, WY 82801
Tate Museum, Casper Community College, Casper, WY 82062
Thermopolis Dinosaur Museum, Thermopolis, WY
University of Wyoming Geological Museum, Laramie, WY 82071
Western Wyoming State College, Rock Springs, WY
Wyoming Dinosaur Center, Rt. 3, Box 209, Thermopolis, WY 82443

Mexico

Museo Civico Naturales, Villahermosa, Tabasco
Natural History Museum, Mexico City

South America

Argentina

Instituto Miguel Lillo (Falcultad de Ciencias Naturales), Miguel Lillo 205, AR 4000 San Miguel de Tucuman
Museo Argentino de Ciencias Naturales, "Bernardino Rivadavia," Av. Angel Gallardo 470, Buenos Aires 1405
Museo "Carmen Funes," Plaza Huicul, Neuguen
Museo de Ciencias Naturales de la Universidad Nacional del Comahue, Buenos Aires 1400, Neuquen
Museo de La Plata, University La Plata, Paseo del Bosque s/n 1900, La Plata

Brazil

City of Araraquara, Ouro District, Sao Paulo (tracks in sidewalks)
City of Franca, Sao Paulo (tracks in sidewalks)
City of Rifaina, Sao Paulo (tracks in sidewalks)
Museum Nacional, 20492 Quinto da Boa Vista, Sao Critovao, Rio de Janeiro 20940

Uruguay

Museo Nacional de Historia Naturales, Casilla de Correo 399, 1100 Montivedeo

Europe

Austria

Naturhistorisches Museum Wien, Vienna

Belgium

Bernissart Museum, Bernissart, Hainaut
Institut Royal des Sciences Naturelles de Belgique, Rue 29, B-1040, Bruxelles

Denmark

Geologisk Museum, Copenhagen
Museum von Gram, Gram

France

Laboratoire de Paleontologie, Faculté des Sciences, place Eugene-Bataillon, Montpelier 34095, Cedex 05
Musée des Dinosaures, Espéraza
Muséum Guimet d'Histoire Naturelle (Musée de Lyon), 28 Boulevard des Belges, 13004 Marseille
Muséum National d'Histoire Naturelle, Institute de Paleontologie, 8 rue Buffon, 75005 Paris
Muséum of Earth Sciences, Nancy
Natural History Museum, Aix-en-Provence
Natural History Museum, Le Havre
Natural History Museum, Nantes

Germany

Bavarian State Collection for Paleontology and Historical Geology, Richard-Wagner-Strasse 10/2, 8000 Munich 2
Beldheim Castle, Hildburghausen

Freilichtmuseum, near Bad rehurg-Loccum, Munchenhagen
Geological and Palaeontological Institute, University of Munster, Pferdegasse 3, D4400, Munster
Heimatmuseum, Trossingen
Institute and Museum for Geology and Paleontology, University of Tübingen, Sigwartstrasse 10, 7400 Tübingen 1
Jura Museum, Willibaldsburg, Eichstätt 8078
Landesmuseum für Naturkund, Münster
Lippisches Landesmuseum, Detmold
Museum am Scholerberg, Onasbruck
Museum des Geologisch–Palaontologischen Institutes der Universität, Münster
Museum für Naturkunde der Humboldt Universität zu Berlin, Paläontologisches Museum, Unter der Linden 6, 108 Berlin
Museum Hauff, Holzmaden
Museum Heineanum, Halberstadt
Museum im Institute für Geologie und Paläontologie der Universität, Gottingen
Niedersachisches Landesmuseum, Hannover
Schaumburgisches Heimatmuseum, Eulenberg, Rintlen
Senckenberg NaturMuseum, Forschunginstitut Senckenberg, Senckenberganlage 25, 6000 Frankfurt am Main 1
Staatliches Museum für Naturkunde, Rosenstein 1, D-7000, Stuttgart 1
Stadtmuseum, Brilon
Westphalisches Landesmuseum fur Naturkunde, Münster

Hungary

Temeszettudomanyi Muzeum, Fold-es Oslenytar, Muzeum korut 14-16, H-1088 Budapest

Italy

G. Capellini Museum, Bologna
Museo Civico di Storia Naturale de Milano, Corso Venezia 55, 20121 Milano
Museo Civico di Storia Naturale di Venezia, S. Croce 1730, 30125 Venice
Museo dell'Instituto di Paleontologia, Universita de Modena, Via Universita, 4, 41100 Modena
Museo di Scienze Naturali "E. Caffi," 74100 Bergamo

Poland

Paleobiology Institute, Academy of Sciences, Al Zwirki I Wigury 93, 02-089 Warsaw

Portugal

Servicos Geologicos de Portugal, Rua Academia das Ciencias, Lisbon

Russia

Central Geological and Prospecting Museum, St. Petersburg
Orlov Museum of Paleontology, Palaeontological Institute, Academy of Sciences, Profsoyuznaya 113, Moscow 117321

Spain

La Rioja Province, north-central Spain, footprints under protective structures
Museo Nacional de Ciencias Naturales, Jose Gutierrez, Abscal 2, Madrid 28006
Ribadesella area, northern Spain, tracks exposed in coastal cliffs and exposures

Sweden

Paleontological Museum, Uppsala University, Box 256, 751 05 Uppsala

Switzerland

Geologische–Paläontologisches Institüt, Universität Basel, Bernoullistrasse 32, CH 4056, Basel
Museum d'Histoire Naturelle, Route de Malagnou, Case Postale 434, 1211 Genève 6

Turkey

Mineral Research and Exploration Institute, Ankara, Turkey

United Kingdom

Birmingham Museum, Dept. of Natural History, Chamberlin Square, Birmingham B3 3DH
British Museum (Natural History), Cromwell Road, London SW7 5BD
City of Liverpool Museum, Liverpool, England
Crystal Palace Park, Syndenham, London SE20
The Dinosaur Museum, Icen Way, Dorchester, Dorset DT1 1EW
Hunterian Museum, The University, Glasgow G12 8QQ

The Leicestershire Museum, 96 New Walk, Leicester LE1 6TD
Museum of Isle of Wight Geology, Sandown Library, Isle of Wight PO36 8AF
Royal Scottish Museum, Chambers St., Edinburgh EH1 1JF
Sedgwick Museum, Cambridge University, Downing St., Cambridge CB2 3EQ
University Museum, Parks Road, Oxford OX1 3PW

Africa

Morocco

Museum of Earth Sciences, Universite Mohamed V, Avenue Ibn Batouta, BP No. 1014, Rabat

Niger

Musee National du Niger, B.P. 248, Niamey

South Africa

Albany Museum, Grahamston
Bernard Price Institute of Palaeontology, University of Witwatersrand, Jan Smuts Avenue, Johannesburg 2001
National Museum, Bloemfontein
South African Museum, Box 61, Capetown 8000

Zimbabwe

Museum of Natural History, Bulawayo
National Museum of Zimbabwe, Harare

Asia

India

Geology Museum, Indian Statistical Institute, 203 Barrackpore Trunk Road, Calcutta 700-035

Japan

Energy Land, Shirahama
Gunma Prefectural Museum of History, Maebashi, Gunma
Himeji Museum of Natural History, Himeji
Historical Museum of Hokkaido, Sapporo, Hokkaido
Hokkaido Centennial Office, Hokkaido
Hyabashibara Museum of Natural History, 2-3 Shimoishi 1-Chome, Okayama Shi 700
Hyogo Museum, Hyogo
Ibaraki Prefectural Museum of Natural History, 700 Ozaki, Iwai City, Ibaraki Prefecture
Institute for Breeding Research, Tokyo University of Agriculture, Tokyo
Iwaki Museum of Coal and Fossils, Iwaki, Fukuishima
Kagoshima Prefectural Museum, Kagoshima, Kagoshima
Kanagawa Prefecture Living Planet and Earth Museum, 499 Iryuda Odawara City, Kanagawa Ken 250
Kasaoka City Museum, Kasaoka
Kitakyushu Museum of Natural History, Kitakyushu, Fukuoka
Kyoto Municipal Science Center for Youth, Kyoto
Nagashima Spa, Nagashima
Nakasoto Dinosaur Center, Gunma Prefecture
National Science Museum, Ueno Park, Tokyo
Natural History Museum, Tokai University, Shimizu, Shikuoka
Niigata Prefectural Natural Science Museum, Niigata, Niigata
Osaka Museum of Natural History, Osaka
Saito Ho-on Kai Museum of Natural History, Sendai, Miyagi
Takikawa Museum of Art and Natural History, Takikawa, Hokkaido
Tochigi Prefectural Museum, Utsunomiya, Tochigi
Toyama Science Museum, Toyama
Toyohashi Museum of Natural History, Toyohashi, Aichi

Mongolian People's Republic

Mongolian Museum of Natural History, Ulan-Bataar 46

People's Republic of China

Beijing Natural History Museum; Beijing
Beipei Museum, Beipei, Sichuan Province
Chengdu College of Geology Museum, Chengdu
Erenhot Dinosaur Museum, Erenhot, Inner Mongolia
Institute of Vertebrate Paleontology and Paleoanthropology Museum, Box 643, Beijing
Lufeng Dinosaur Museum, Lufeng, Yunnan
Natural History Museum of Chongqing, Chongqing, Sichuan
Prosauropod skeletons, *in situ*, near villages of Heilongtan–Zhangjiawa–Dawa, Lufeng, Yunnan
Zigong Dinosaur Museum, Dashanpu, Sichuan

South Korea

National Science Museum of Korea, Seoul

Samcheanpo, southern coast of South Korea (footprints)

Taiwan

National Museum of Natural Science, Taichung

Australia

Australian Museum, Box A285, Sidney, New South Wales 2000

Lark Quarry Environmental Park, 120 km southwest of Winton, central west Queensland (tracks under protective structure), c/o Queensland National Parks and Wildlife Service, Queensland

Museum of Victoria, 285-321 Russell St., Melbourne, Victoria 3000

National Dinosaur Museum, Young, New South Wales

Queensland Museum, Gregory Terrace, Fortitude Valley, Queensland 4006

South Australian Museum, North Terrace, Adelaide, South Australia 5000

Western Australian Museum, Perth, Western Australia

New Zealand

Canterbury Museum, Rolleston Ave., Christchurch 1

References

The following books provide more detailed information on many of the exhibits listed previously, although none is complete.

Cohen, D., and Cohen, S. (1992). *Where to Find Dinosaurs Today*, pp. 209. Puffin Unicorn Books, New York.

Costa, V. (1994). *Dinosaur Safari Guide, Tracking North America's Prehistoric Past*, pp. 259. Voyageur Press, Stillwater, MN.

Halls, K. M. (1996). *Dino-Trekking, the Ultimate Dinosaur Lover's Travel Guide*, pp. 206. Wiley, New York.

Lambert, D. (1990). *The Dinosaur Data Book*, pp. 320. Avon Press, New York.

Lockley, M. G. (1991). *Tracking Dinosaurs, a New Look at an Ancient World*, pp. 238. Cambridge Univ. Press, Cambridge, UK.

Lockley, M. G., and Hunt, A. P. (1995). *Dinosaur Tracks and Other Fossil Footprints of the Western United States*, pp. 338. Columbia Univ. Press, New York.

Norman, D. (1985). *The Illustrated Encyclopedia of Dinosaurs*, pp. 208. Crescent Books, New York.

Probst, E., and Windolf, R. (1993). *Dinosaurier in Deutschland*, pp. 316. C. Bertelsmann, Munchen, Germany.

Skwara, T. (1992a). *Old Bones and Serpent's Stones, a Guide to Interpreted Fossil Localities in Canada and the US. Vol.1. The Eastern States*. McDonald & Woodward.

Skwara, T. (1992b). *Old Bones and Serpent's Stones, a Guide to Interpreted Fossil Localities in Canada and the US. Vol. 2. The Western States*. McDonald & Woodward.

Stokes, W. L. (1988). *Dinosaur Tour Book*, pp. 64. Starstone, Salt Lake City, UT.

Museums at Blackhawk, California, USA

see MUSEUMS AND DISPLAYS

Museum von Gram, Denmark

see MUSEUMS AND DISPLAYS

Museum Ziemi, Poland

see MUSEUMS AND DISPLAYS

Nagashima Spa, Japan

see Museums and Displays

Nakasota Dinosaur Center, Japan

see Museums and Displays

Namibian Dinosaurs

see African Dinosaurs

National Dinosaur Museum, Australia

see Museums and Displays

National Museum, Bloemfontein, South Africa

Anusuya Chinsamy
South African Museum
Cape Town, South Africa

Skeletal elements of the basal sauropods *Euskelosaurus* and *Massospondylus* are well represented in the collections of the National Museum, which was founded in 1877 in Bloemfontein, South Africa. The only known skull of *Euskelosaurus* is housed in this museum, which also has some fabrosaurid skeletal material. The *Euskelosaurus* skull and some skeletal remains of *Lesothosaurus* are on display in the Palaeontology Hall at this museum.

See also the following related entries:

African Dinosaurs • Albany Museum • Bernard Price Institute for Palaeontological Research • South African Museum

National Museum of Natural History, Washington, DC, USA

see Museums and Displays

National Museum of Natural Science, Taiwan

see Museums and Displays

National Museum of Zimbabwe, Harare

see Museums and Displays

National Science Museum, Japan

see Museums and Displays

National Science Museum of Korea, South Korea

see Museums and Displays

Natural History Museum, Aix-en-Provence, France

see Museums and Displays

Natural History Museum of Chongqing, People's Republic of China

see Museums and Displays

Natural History Museum, Le Havre, France

see Museums and Displays

Natural History Museum, London

Paul G. Davis
National Science Museum
Tokyo, Japan

The Natural History Museum (formerly known as the British Museum of Natural History—and also more affectionately as "The BM") has its origins with Sir Hans Sloane (1660–1753). Sloane was a "gentleman collector" who gathered and amassed a sizable collection of "natural curiosities," "antiquities," scientific instruments, and books during his lifetime. Upon his death, he bequeathed this legacy to the nation (it must be added that, with a rather sharp piece of legal work, his will secured the necessary money from the government to do this).

The first British Museum opened in June 1753, in Montagu House, Bloomsbury, London. Further collections were acquired either by donation or by purchase from other gentleman collectors. Important paleontological collections gained in this way include those of W. Smith (the "Father of Geology"), C. D. E. König, the Sowerby family, G. Mantell (famed for his association with *Iguanodon*), W. Gilbertson, and T. Davidson. The museum is also the repository for the material collected by Charles Darwin on his Beagle voyage.

The growth of the natural history collections led to a tripartite division into botany, zoology, and mineralogy and geology in 1837.

Collections continued to grow during the Victorian era with many specimens from the British Empire being shipped back by expatriate naturalists. Further specimens came from the Geological Survey, which transferred all of its foreign material in 1880, and the Geological Society of London, which transferred all of its collections in 1911.

In 1845, the museum moved to a new location in Bloomsbury. By 1856, the natural history collections had surpassed all others and a new building was sought. Sir Richard Owen (famous for establishing the name Dinosauria) played an important and instrumental role in the design and implementation of this new building. The impressive, purpose-built, Romanesque building was moved into in 1881 at its present location in South Kensington, London. Along with this move came a restructuring in organization and the museum became the British Museum (Natural History).

In 1956, the geology department changed its name to Paleontology. The British Museum (Natural History) became independent of the British Museum in 1963 and eventually changed its name to the Natural History Museum. A new, purpose-built building (the east wing) was added in 1976 to hold the Department of Palaeontology and its collections. In 1986, The Natural History Museum incorporated the former Geological Museum, which had been constructed on an adjacent site in 1935. A new gallery linking the two buildings was opened in 1988.

The collections of the Natural History Museum are organized according to zoological nomenclature. The total number of specimens is in excess of 9 million. There are more than 30,000 specimens of fossil REPTILES and amphibians. Type and figured specimens include the dinosaurs described by Sir Richard Owen, Gideon Mantell, Harry G. Seeley, Thomas H. Huxley, and others. The fossil bird collection includes the London specimen of *Archaeopteryx lithographica* (see AVES).

In 1991 the museum revamped its dinosaur displays, producing an excellent modern display exhibiting much of the current knowledge and controversies of dinosaur PALEONTOLOGY. It achieved a difficult balance between skeletal mounts and interactive displays.

References

Cocks, R., and Cleevely, R. (1992). The Natural History Museum, London, Palaeontological Collections. *Geol. Today* (July/August), 151–153.

Stearn, W. T. (1981). The Natural History Museum at South Kensington. London: Heinemann.

Whitehead, P., and Keates, C. (1981). *The British Museum of Natural History*, pp. 128. Summerfield Press/Philip Wilson, London.

Natural History Museum of Los Angeles County, California, USA

see MUSEUMS AND DISPLAYS

Natural History Museum, Mexico City, Mexico

see MUSEUMS AND DISPLAYS

Natural History Museum, Nantes, France

see MUSEUMS AND DISPLAYS

Natural History Museum, Tokai University, Japan

see MUSEUMS AND DISPLAYS

Naturhistorisches Museum, Wien, Austria

see MUSEUMS AND DISPLAYS

Nemegt Formation

HALSZKA OSMÓLSKA
Polska Akademia Nauk
Warsaw, Poland

The Nemegt Formation (previously referred to as the "Upper Nemegt Beds") includes the youngest Late Cretaceous sediments known in the Gobi Basin of Mongolia. This rock unit was discovered in 1948 by the Paleontological Expeditions of the USSR Academy of Sciences and it was then referred to as the "fossiliferous series" or "subaqueous deltaic-channel deposits." Because no radiometric data are available for these beds and there are no marine sequences in the region, the absolute age of the Nemegt Formation cannot be independently determined and its relative age is debatable. The latter has been estimated by various authors as late Campanian through early Maastrichtian, early Maastrichtian, or middle Maastrichtian. The Nemegt locality was chosen as the type locality of the Nemegt Formation. Other than the type locality, the formation is exposed in numerous other parts of the Gobi Altai, Trans-Altai Gobi, and eastern Gobi (e.g., Altan Ula I-IV, Tsagan Khushu, Khermeen Tsav II, Nogon Tsav, Bugin Tsav, and Gurlin Tsav). The Nemegt Formation conformably overlies the Barun Goyot Formation, but an erosional hiatus separates it from overlying middle Paleocene strata. The total thickness of the Nemegt Formation is estimated at about 400 m.

The Nemegt Formation is vertically and laterally variable and is dominated by poorly cemented, yellowish to gray brown, rarely red sandstone. Intercalations of sandy mudstones, well-cemented sandstones, and intraformational conglomerates or gravels are numerous but thin. Fining-upward, 2- to 9-m-thick sequences of sandstones, considered to be deposited on alluvial plains by meandering rivers, are distinctive for the entire formation. Channel deposits predominate over overbank facies. Evidence suggests a warm, subhumid climate with seasonal rains.

In the channel deposits of the Nemegt Formation, fossilized indeterminate wood of Araucariaceae, gastropods, and pelecypods are common. In the overbank facies, well-differentiated charophyte oogonia and ostracods abound. Aquatic and amphibious vertebrates are represented by unidentified fish vertebrae, turtles, and rare crocodiles. The absence of terrestrial microvertebrates, such as lizards and mammals, is striking, whereas megavertebrates, represented by large and medium-sized dinosaurs, are abundant and well differentiated. The dinosaur remains mostly occur in the channel-bar sediments. The complete articulation of many dinosaur skeletons suggests burial at or near the death sites. Complete dinosaur EGGS are rare, but eggshells are sporadically found.

The assemblage of dinosaurs found in the Nemegt Formation contains theropods, sauropods, hadrosaurids, pachycephalosaurs, and ankylosaurs. A distinctive character of this fauna is the unusually high diversity of theropods, which includes 14 monospecific genera belonging to at least six families. Among these theropods, *Tarbosaurus bataar* and *Gallimimus bullatus* are represented by 10 or more specimens, whereas there are only single specimens for most other species. Among herbivorous dinosaurs, the large hadrosaurid *Saurolophus angustirostris* is as common as *T. baatar*, whereas other species are rare. Another striking feature of the dinosaur assemblage of the Nemegt Formation is the lack of neoceratopsians, although their primitive representatives (Protoceratopsidae) occurred in the older Barun Goyot and Djadokhta formations. In contrast, the advanced neoceratopsians of the Ceratopsidae are common in contemporaneous North American strata.

Dinosaur species found in the Nemegt Formation include the theropods *Adasaurus mongoliensis, Gallimimus bullatus, Alioramus remotus, Maleevosaurus novojilovi, Anserimimus planinychus, Oviraptor mongoliensis, Avimimus portentosus, Saurornithoides junior, Bagaraatan ostromi, Tarbosaurus bataar, Borogovia gracilicrus, Therizinosaurus cheloniformis, Deinocheirus mirificus, Tochisaurus nemegtensis,* and *Elmisaurus rarus*. Herbivores include the sauropods *Nemegtosaurus mongoliensis* and *Opisthocoelicaudia skarzynskii*, the ornithopods *Saurolophus angustirostris* and *Barsboldia sicinskii*, the pachycephalosaurs *Homalocephale calathocercos* and *Prenocephale prenes*, and the ankylosaur *Tarchia gigantea*. Dinosaur eggs and eggshells have been identified as *Spheroolithus* sp., *Laevisoolithus sochavai, Subtiliolithus microtuberculatum* shells, elongatoolithid shells, and smooth ?protoceratopsid shells.

See also the following related entries:

BARUN GOYOT FORMATION • DJADOKHTA FORMATION • MONGOLIAN DINOSAURS • POLISH–MONGOLIAN PALEONTOLOGICAL EXPEDITIONS

References

Efremov, I. A. (1950). Taphonomy and geological chronicle. *Proc. Paleontol. Inst. USSR Acad. Sci.* **24,** 1–176. [In Russian]

Gradzinski, R. (1970). Sedimentation of dinosaur-bearing Upper Cretaceous deposits of the Nemegt Basin, Gobi Desert. *Palaeontol. Polonica* **21,** 147–229.

Gradzinski, R., Kielan-Jaworowska, Z., and Maryanska, T. (1977). Upper Cretaceous Djadokhta, Barun Goyot and Nemegt formations of Mongolia, including remarks on previous subdivisions. *Acta Geol. Polonica* **27,** 281–318.

Jerzykiewicz, T., and Russell, D. A. (1991). Late Mesozoic stratigraphy and vertebrates of the Gobi Basin. *Cretaceous Res.* **12,** 345–377.

Osmólska, H. (1980). The Late Cretaceous vertebrate assemblages of the Gobi Desert, Mongolia. *Mém. Soc. Géol. France* **139,** 145–150.

Neoceratopsia

PETER DODSON
University of Pennsylvania
Philadelphia, Pennsylvania, USA

The Neoceratopsia are commonly known as the horned dinosaurs, the last major group of dinosaurs to appear in the fossil record and among the most diverse. The Neoceratopsia comprise the Protoceratopsidae and the Ceratopsidae, taxa primarily known from Late Cretaceous deposits of eastern Asia and western North America. The term neoceratopsian ("new horned dinosaur") designates all ceratopsians more derived than the Psittacosauridae. Horned dinosaurs were four-legged ornithischians that ranged in size from barely a meter in length to giants 8 m long or longer. All are characterized by skulls that were large compared to their bodies. The largest had greater skulls than any other terrestrial animals that ever lived, measuring more than 2.5 m in total length in large specimens of both *Torosaurus* and *Pentaceratops* (Fig. 1). Neoceratopsian skulls usually include a prominent cranial frill, various patterns of horns on the face (ceratopsian meaning "horn face"), and additional ornaments such as facial and jugal bosses and scallops, bumps, and spikes on their frills.

Neoceratopsians have an excellent fossil record, one of the very best of all groups of dinosaurs. There are 22 genera and 30 valid species currently described. Most taxa are known from many specimens. More than half are known from essentially complete skeletons, and others are based on excellent skulls. There are some 30 specimens per genus, more than four times the average for all dinosaurs. *Psittacosaurus, Protoceratops,* and *Triceratops* are three of the 10 most abundant dinosaurs in museum collections.

It is hypothesized that ceratopsians share a common ancestry with the PACHYCEPHALOSAURIA (dome-headed dinosaurs), a common character being the

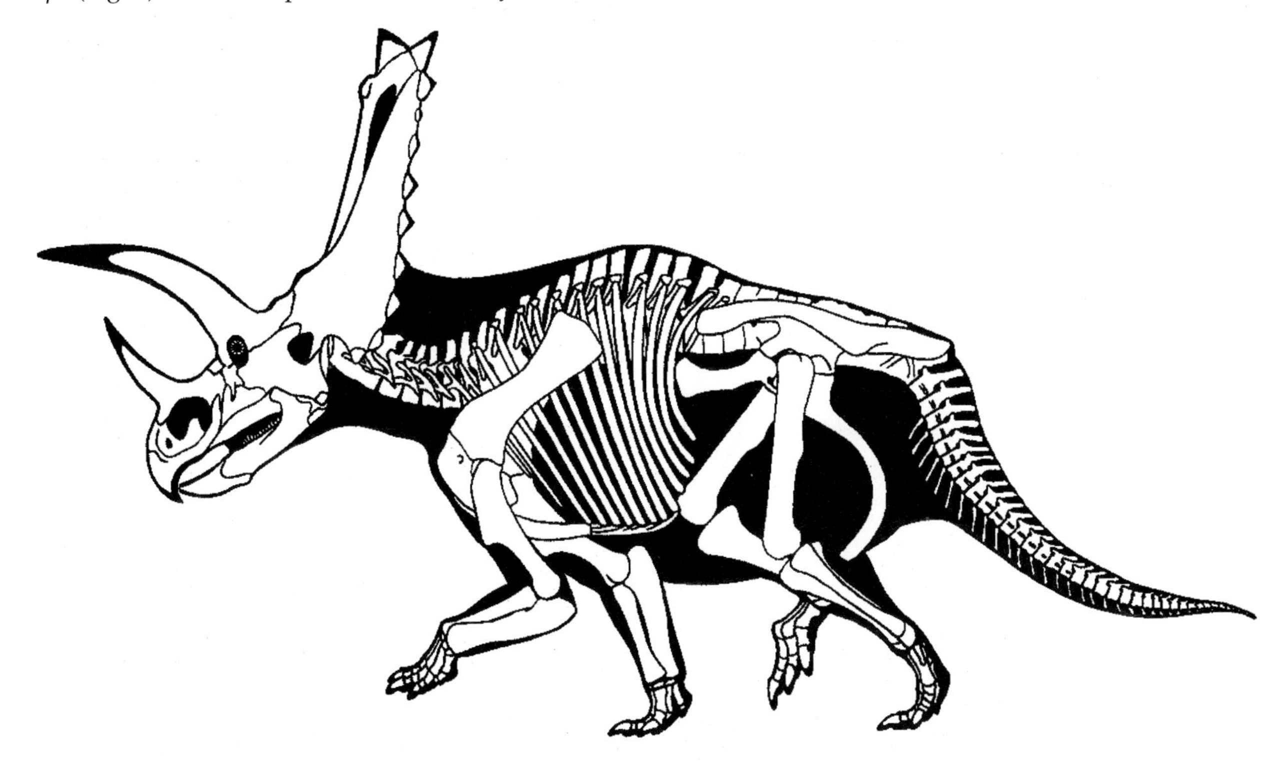

FIGURE 1 The skeleton of *Pentaceratops*, a chasmosaurine ceratopsid, by Gregory S. Paul (from Dodson, Peter. *The Horned Dinosaurs*. Copyright © 1996 by Princeton University Press. Reprinted by permission of Princeton University Press).

presence of a frill, formed by the squamosal and parietal bones, that overhangs the back of the skull (see MARGINOCEPHALIA). Both groups lack the obturator process, a bony tab on the ischium that supports the pubis. In this respect, marginocephalians resemble basal ornithischians rather than ornithopods. Otherwise, it would seem that ceratopsians and pachycephalosaurs might have arisen from generalized small ornithopods at about the level of *Hypsilophodon*. Ornithopod-like bipedalism is persistent in pachycephalosaurs and in basal ceratopsians (*Psittacosaurus*), but within neoceratopsians, increased size of the skull quickly rendered bipedalism impractical.

Derived characters by which the CERATOPSIA are distinguished from all other dinosaurs include a special bone, the rostral, situated in front of the premaxilla. The rostral bone takes the form of a narrow, sharply keeled beak that articulates with a lower beak formed by the predentary bone. Another distinctive character is the presence of laterally flaring jugal bones in the cheek region that may form distinct ventral or ventrolateral processes. All ceratopsians have at least the beginnings of a crest at the back of the skull.

Additional characters identify the Neoceratopsia as a monophyletic taxon within the Ceratopsia (Fig. 2). Many represent an exaggeration of characters incipient in basal ceratopsians (psittacosaurs). Some of these include increase in the size of the head relative to the body, development of sharp points at the tips of the upper and lower beaks, increased prominence of the parietosquamosal frill, development of a forward slope at the back of the skull as the quadrate bone slants forward, fusion of the first four vertebrae of the neck for support of the enlarged head, and gentle downward curvature of the ischium.

Traditionally neoceratopsians are divided into two groups, the Protoceratopsidae and the Ceratopsidae. The Protoceratopsidae are basal neoceratopsians, but whether they form a monophyletic clade or merely a grade is controversial; the consensus is that they are a grade. These are small animals 1–2.5 m in length. They are rare in North America (*Leptoceratops* and *Montanoceratops*) and sometimes very common in Mongolia (*Protoceratops*) (Fig. 3). *Leptoceratops*, described in 1914 from a partial specimen found in Alberta, is regarded as the most generalized of all protoceratopsids, with a very small crest and no nasal horn. Its true nature was not recognized until three more or less complete skeletons were described by C. M. Sternberg in 1951. When *Protoceratops* was described from Mongolia in 1923, it seemed terribly

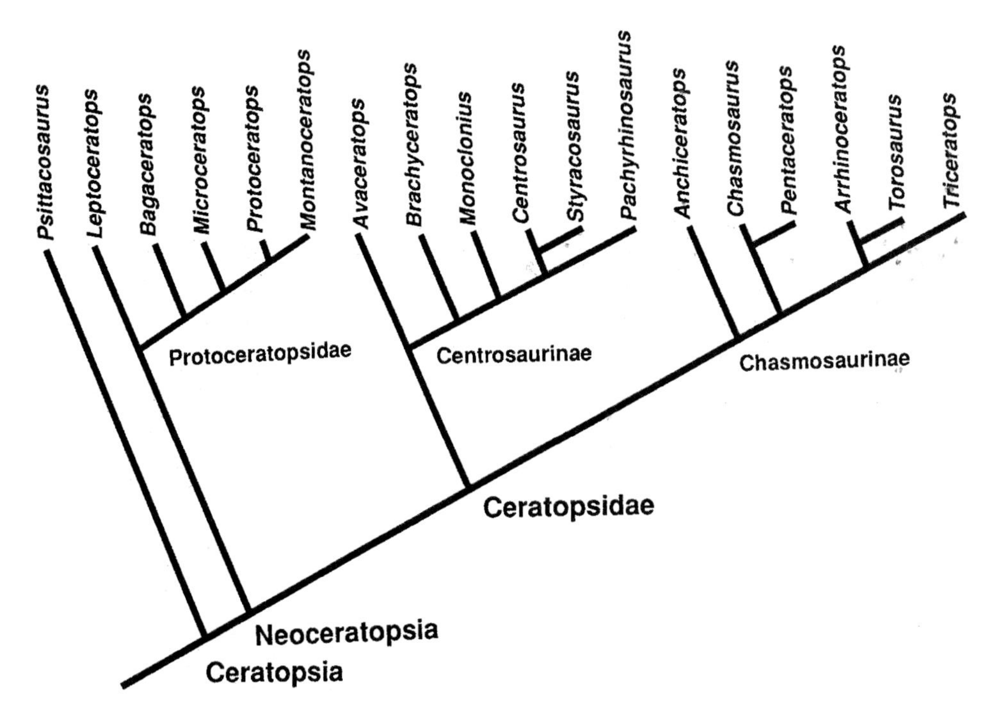

FIGURE 2 Phylogeny of the Ceratopsia, after Dodson and Currie (1990).

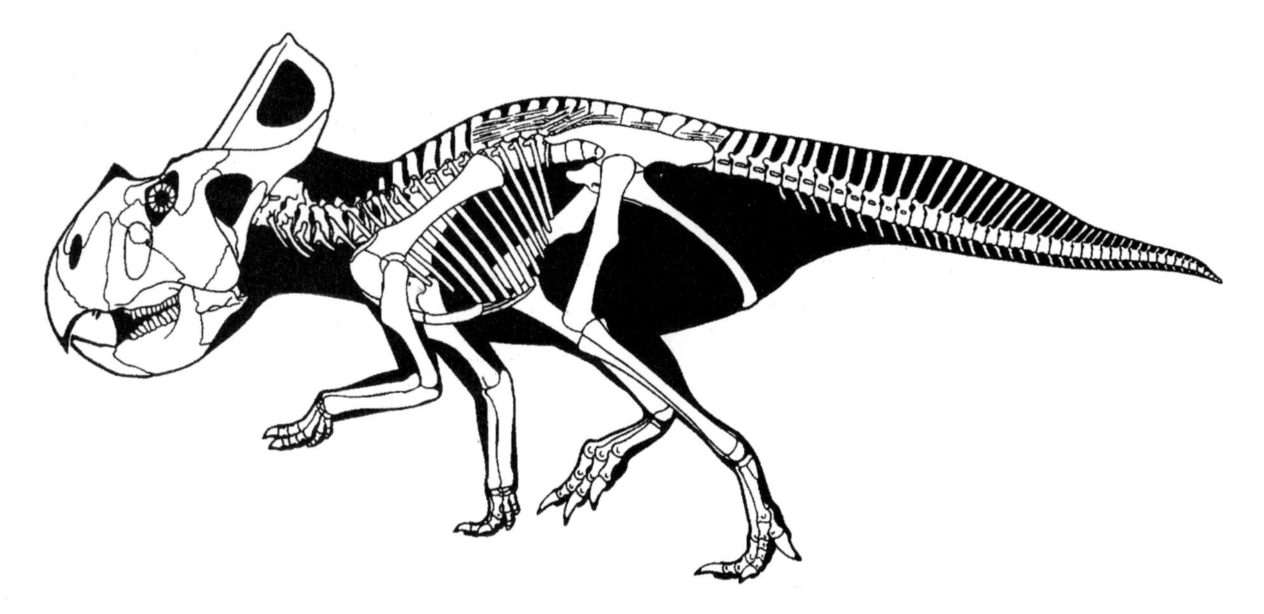

FIGURE 3 The skeleton of *Protoceratops*, by Gregory S. Paul (from Dodson, Peter. *The Horned Dinosaurs.* Copyright © 1996 by Princeton University Press. Reprinted by permission of Princeton University Press).

primitive, but now it is recognized as highly derived within a radiation of basal forms, a number of which have been described from Asia since 1975 (*Bagaceratops, Breviceratops,* and *Udanoceratops*). No protoceratopsid has any hint of horns over the eye, and none has a fully developed nose horn. The nose horn may be absent (*Leptoceratops*), incipient (*Protoceratops*), small but distinct (*Bagaceratops* and *Montanoceratops*), or unknown. The frill varies from small and without openings (*Leptoceratops*) to large and fully fenestrated (*Protoceratops*). In all cases, the frill lacks additional ornamentation. *Protoceratops* and *Breviceratops* retain premaxillary teeth; all others have lost them.

All neoceratopsians retain the generalized five-fingered forefoot; there are four digits on the hindfoot. In protoceratopsids there are six to eight fused sacral vertebrae. The ungual bones at the tips of the fore- and hindfeet are either claw shaped or tapering with blunt tips. In protoceratopsids, the femur is always shorter than the tibia.

No embryonic specimens of *Protoceratops* have been confirmed (i.e., found within an egg), but very small (possible hatchling) skulls with a basal skull length of 5 cm are known from both Mongolia and China. By comparison, maximum adult basal length is 37 cm and total skull length is 50 cm. In specimens as small as half the maximum adult size, an apparent sexual dimorphism in prominence of the frill and face becomes evident. No other protoceratopsids are known from large enough samples to confirm sexual dimorphism. Numerous EGGS AND NESTS have been attributed to *Protoceratops*, but recently many of these have been shown to belong to *Oviraptor*, previously thought to have been the predator of *Protoceratops* eggs.

The Ceratopsidae constitute a very highly cohesive group of medium- to large-sized ceratopsians from Late Cretaceous deposits of western North America. The monophyly of this group is confirmed by numerous derived characters. Skulls are large, and prominent horns are found over either the nose or the eyes, but usually not both (except *Pachyrhinosaurus*). The frill formed by the parietal and squamosal bones typically is about as long as or longer than the rest of the skull. The frill may be additionally ornamented with scallops, spikes, or hooks. Parietal fenestrae may be exaggerated (*Chasmosaurus* and *Pentaceratops*) or absent (*Avaceratops* and *Triceratops*). The opening for the nostril is large. The sharp beak is toothless, as in most protoceratopsids. Ceratopsids had dental batteries consisting of numerous interlocking teeth in which the roots of each tooth are split by the tooth beneath it in the replacement series. Also unique to ceratopsids is the vertical orientation of the plane of contact between upper and lower tooth batteries, resulting in a peculiar slicing action. The occipital condyle by which the back of the skull contacts the vertebral column is spherical and in some cases as large as a small grapefruit (up to 115 mm in diameter).

There are 10 fused sacral vertebrae in ceratopsids. The femur is always longer than the tibia, and the ungual phalanges are blunt, rounded, and hoof-like.

The Ceratopsidae are divided into two subtaxa, the Centrosaurinae and the Chasmosaurinae, that represent a genuine split, neither being more derived than the other. Both subtaxa appear fully developed in late Campanian deposits (e.g., Judith River Formation of Montana and Alberta). The common ancestor of the two groups presumably is to be sought in prelate Campanian strata not yet discovered.

The Centrosaurinae are also known as the short-frilled or short-squamosaled ceratopsids (Fig. 4). In these, the squamosal is always significantly shorter than the parietal, usually less than 50 cm long. These horned dinosaurs are of moderate size, typically 5 m long (extremes: *Avaceratops*, 4 m; *Pachyrhinosaurus*, 7 m) and more restricted both in time and distribution than the chasmosaurines. Typical skulls were up to 1.5 m in length. The face was relatively short and high. Centrosaurines emphasize the nasal horn, which may be erect or may curve forward or backward. *Styracosaurus* had a nasal horn 50 cm tall. The horns above the eyes are generally either nonexistent or modestly developed. Typical centrosaurines with lavishly ornamented parietals are *Centrosaurus* and *Styracosaurus*. *Pachyrhinosaurus*, the latest occurring centrosaurine and the most widespread geographically (Alberta to Alaska), is the most unusual in that it lacked facial horns altogether, possessing instead a very thick, textured boss over the nose and eyes. New centrosaurines (*Einiosaurus* and *Achelousaurus*) recently discovered in Montana possibly bridge the gap between *Styracosaurus* and *Pachyrhinosaurus*. The status of *Monoclonius*, the first validly named ceratopsian, with a thin, unornamented frill, is controversial. By one interpretation, it simply represents the generalized juvenile form before the definitive adult characters appeared. In the other interpretation, it is a legitimate paedomorphic form that maintained juvenile characters as a reproductive adult of large size. The status of *Avaceratops* and *Brachyceratops*, acknowledged juveniles, is similarly controversial but not necessarily invalidated because of their juvenility.

The Chasmosaurinae, the long-frilled or long-squamosaled ceratopsids, are distributed from Mexico and Texas to Alaska and survived to the end of the age of dinosaurs, albeit at reduced diversity (*Triceratops* and *Torosaurus*). At least three different representatives attained skull sizes of 2–2.5 m in length, corresponding to body lengths of 8 m (*Triceratops, Torosaurus,* and *Pentaceratops*). The face was relatively long and low. These dinosaurs emphasized paired horns over the eyes, with modest development of the nasal horn. The frill was longer than the basal length of the skull, and the squamosal extends the full length of the frill. The squamosal ranged from 1 to 1.5 m in length, with a distinctive bend in the middle to accommodate elevation of the frill. The parietal did not have elaborate spikes or hornlets. The position of *Triceratops* within the Ceratopsidae

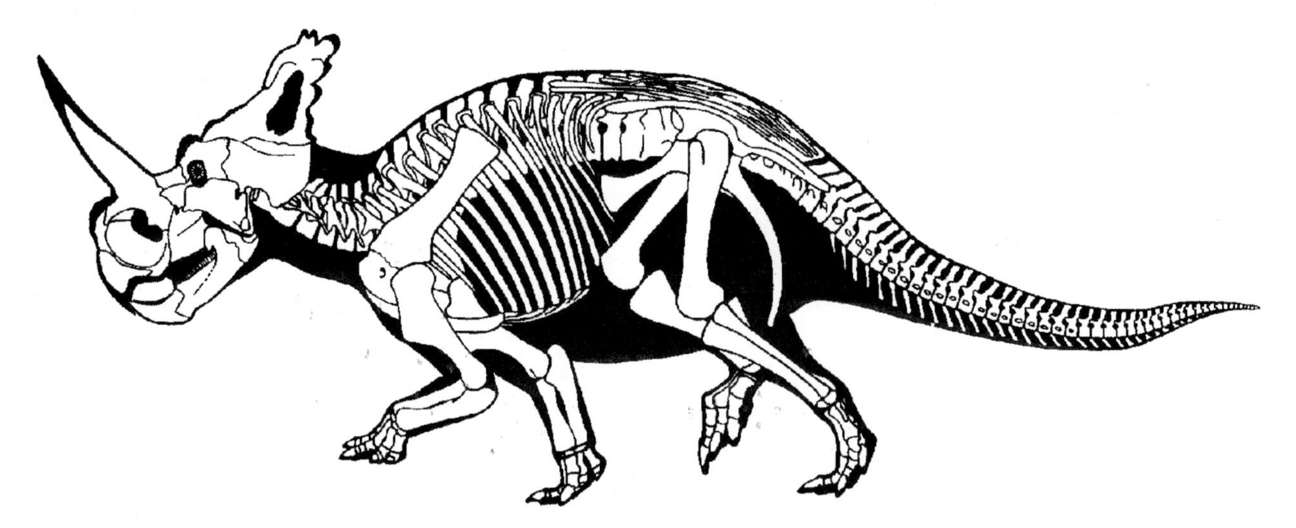

FIGURE 4 The skeleton of *Centrosaurus*, a centrosaurine ceratopsid, by Gregory S. Paul (from Dodson, Peter. *The Horned Dinosaurs*. Copyright © 1996 by Princeton University Press. Reprinted by permission of Princeton University Press).

has been controversial because it has a secondarily shortened frill. However, the configuration of the horns, face, and squamosal all demonstrate that it is a genuine chasmosaurine, despite its peculiarities, which also include a solid frill lacking parietal fenestrae.

The unique features of neoceratopsians are frills and horns, both of which invite interpretation in terms of behavioral biology (Fig. 5). The bony cranial frill may have been a visual display structure. Neoceratopsians are readily interpreted as abundant, gregarious animals, many of which may have been herding animals. The abundance of many species in the fossil record, and the tendency of some them, particularly the centrosaurines *Centrosaurus, Styracosaurus, Einiosaurus, Achelousaurus,* and *Pachyrhinosaurus* and the chasmosaurine *Chasmosaurus,* to form single-species bone beds are consistent with herding behavior, especially when a size range of individuals is represented. A behavioral model for horns and frills suggests that centrosaurines, with short frills and dominant unpaired nasal horns, had reduced display value in the frill. Comparatively dangerous combat between rival males is hypothesized because of the difficulty of safely coupling opposing nasal horns. By contrast, chasmosaurines had long frills with enhanced display value for ritualized interaction between rivals, and the dominant paired horns over the eyes could have been easily coupled when rivals fought.

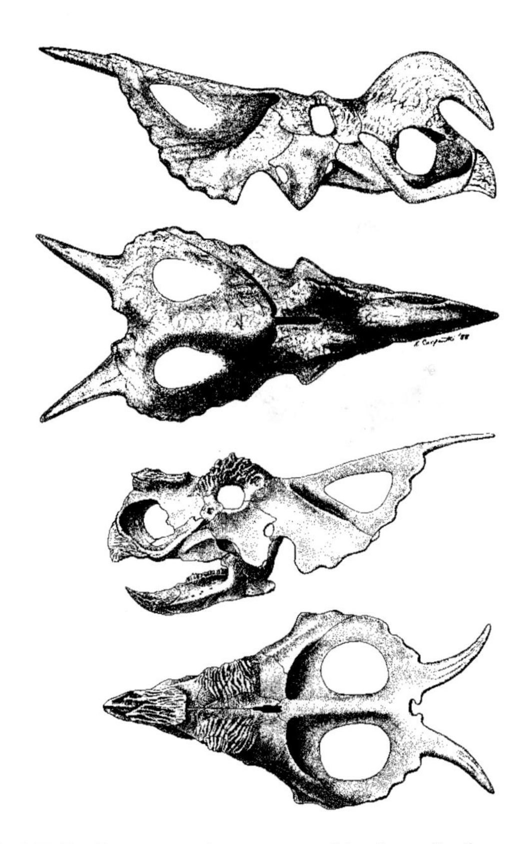

FIGURE 5 Centrosaurine ceratopsids show further variation in the nose horn and frills. (Above) Lateral and top views of *Einiosaurus procurvicornis;* (below) lateral and top views of *Achelousaurus horneri.* (Drawings by Ken Carpenter and Robert Walters in Dodson, 1996, and after Kris Ellingsen, from Sampson, 1995).

The case for the frill as a sexual display organ is enhanced by the bimodal distribution of measurements of the frill in *Protoceratops.* In presumed male *Protoceratops*, the frill is wider and higher, and more conspicuous; correlated with these features of the frill are a stronger arching of the nasal bones as the "protonose horn" and a stronger flaring of the jugal bones in the cheek region. In contrast, putative females have a frill that is lower and narrower, less arching of the nasals and less flaring of the jugals. In addition to *Protoceratops*, sexual dimorphism has been inferred in *Centrosaurus, Chasmosaurus,* and *Triceratops;* reasonable sample sizes exist for all of these. Indeed, it is almost an a priori expectation that animals with such conspicuous structures as frills and horns would, by analogy with deer and horned ruminants, show sexual differences. Ceratopsian frills also may have provided frameworks for elongated jaw muscles, but for several reasons this seems less attractive as a general explanation in addition to the probable sexual display function of the frill. Elongation of muscles in itself does not lead to increased strength, and placing such muscles on an exposed external surface makes them subject to puncture by opposing horns or jaws. The surfaces of protoceratopsid frills are smooth, but many ceratopsids had a grooved texture that was incompatible with muscle attachment.

Neoceratopsians were herbivores, but ceratopsids developed an unusual vertical shearing mechanism in the jaws. Dental batteries chopped tough, fibrous food but did not grind it. The teeth were probably subject to high rates of wear due to edge-to-edge contact of upper and lower teeth. Fibrous cycads and palms have been suggested as suitable food sources for ceratopsids, but in the northern half of western North America (i.e., Wyoming and north), where ceratopsids are found in greatest abundance, neither group of plants was common. Cycads were now in steep decline, and palms were just appearing, spreading from south to north. Palms were available to *Pentaceratops* in New Mexico but probably not to

Centrosaurus in Alberta. Angiosperm trees in the Late Cretaceous were primarily less than 10 cm in diameter, easily knocked over by horned dinosaurs; perhaps they formed a readily obtainable source of fibrous food.

The posture of ceratopsids has been controversial. The HINDLIMBS were clearly upright, with a slight outward orientation. The front limbs were initially mounted with the humerus horizontal, in extreme push-up position. Recently, an esthetic preference has led to artistic reconstructions of ceratopsids with fully upright posture and running with the unrestrained freedom of ungulates. However, the architecture of the shoulder and of the humerus genuinely constrain the posture and movement of the forelimb, making a mammalian carnivore or ungulate posture unlikely. A prominent coracoid restricts forward movement of the humerus, and prominent internal processes on the head of the humerus constrain medial movement (adduction) of that bone. Possible ceratopsid trackways suggest that the forelimbs were not really wide-tracked. If this inference is correct, some bend at the elbow must have compensated for the apparent abduction of the humerus. In any case, it is not easy to transform ceratopsids into high-grade runners.

Protoceratopsids are found in both western North America and in central Asia. The greatest diversity of protoceratopsids is in Asia, but the most generalized (*Leptoceratops*) and the most derived (*Montanoceratops*) representatives are found in North America. Protoceratopsids have been claimed from both South America (*Notoceratops*) and Australia (unnamed). The former is a single jaw lacking teeth, and the latter is an ulna. Both claims are dubious on biogeographic grounds due to the isolated positions of the respective continents and the lack of corroborating evidence. Each specimen is more plausibly interpreted as a local endemic, for which there is ample evidence, particularly in South America.

As currently known, ceratopsids are strictly North American. Given the importance of protoceratopsids in Asia, ceratopsids should be expected there. Indeed, claims have been made on the basis of fragmentary materials (*Turanoceratops*), but none of these has yet been convincing. If large, horn-bearing neoceratopsians are found in Asia, the question will arise as to whether these represent an endemic radiation of forms more closely related to each other and to an Asian protoceratopsid such as *Protoceratops* than to a North American ceratopsid; or will such Asian neoceratopsians actually prove to nest phylogenetically within the Ceratopsidae as currently understood? The former seems more likely.

See also the following related entries:

BEHAVIOR • DIET • GROWTH AND EMBRYOLOGY • ORNAMENTATION • PSITTACOSAURIDAE • QUADRUPEDALITY

References

Clark, J. M. (1995). An egg thief exonerated. *Nat. History* **104**(6): 56–57.

Dodson, P. (1990). On the status of the ceratopsids *Monoclonius* and *Centrosaurus*. In *Dinosaur Systematics—Approaches and Perspectives* (K. Carpenter and P. J. Currie, Eds.), pp. 231–243. Cambridge Univ. Press, New York.

Dodson, P. (1993). Comparative craniology of the Ceratopsia. *Am. J. Sci.* **293A,** 200–234.

Dodson, P. (1996). *The Horned Dinosaurs*. Princeton Univ. Press, Princeton, NJ.

Dodson, P., and Currie, P. J. (1990). Neoceratopsia. In *The Dinosauria* (D. B. Weishampel, P. Dodson, and H. Osmólska, Eds.), pp. 593–618. Univ. of California Press, Los Angeles.

Dong, Z., and Currie, P. J. (1993). Protoceratopsian embryos from Inner Mongolia, People's Republic of China. *Can. J. Earth Sci.* **30,** 2248–2254.

Farlow, J. O., and Dodson, P. (1975). The behavioral significance of frill and horn morphology in ceratopsian dinosaurs. *Evolution* **29,** 353–361.

Johnson, R. E., and Ostrom, J. H. (1994). The forelimb of *Torosaurus* and an analysis of the posture and gait of ceratopsians. In *Functional Morphology in Vertebrate Paleontology* (J. Thomason, Ed.), pp. 205–218. Cambridge Univ. Press, New York.

Lehman, T. M. (1990). The ceratopsian subfamily Chasmosaurinae: Sexual dimorphism and systematics. In *Dinosaur Systematics—Approaches and Perspectives* (K. Carpenter, and P. J. Currie, Eds.), pp. 211–229. Cambridge Univ. Press, New York.

Norell, M. A., Clark, J. M., Dashveg, D., Barsbold, R., Chiappe, L. M., Davidson, A. R., McKenna, M. C., Perle, A., and Novacek, M. J. (1994). A theropod dinosaur embryo and the affinities of the Flaming Cliffs dinosaur eggs. *Science* **266,** 779–782.

Sampson, S. D. (1995). Horns, herds and hierarchies. *Nat. History* **104**(6), 36–40.

Sereno, P. C. (1986). Phylogeny of the bird-hipped dinosaurs. *Natl. Geogr. Res.* **2,** 234–236.

Neotetanurae

This node-based taxon was established by Sereno *et al.* (1994) to define "allosauroids and coelurosaurs," which share a more recent common ancestry than either does with Torvosauroidea (which includes *Afrovenator,* Torvosauridae, and Spinosauridae). Eleven synapomorphies diagnose the taxon, including a posteriorly facing retroarcticular surface of the articular, L-shaped posterior chevrons, a furcula, and the presence of a preacetabular fossa on the ilium. This taxon is the same as AVETHEROPODA (Holtz, 1994), which precedes it in published priority.

References

Holtz, T. R., Jr. (1994). The phylogenetic position of the Tyrannosauridae: Implications for theropod systematics. *J. Paleontol.* **68,** 1100–1117.

Sereno, P. C., Wilson, J. A., Larsson, H. C. E., Dutheil, D. B., and Sues, H.-D. (1994). Early Cretaceous dinosaurs from the Sahara. *Science* **265,** 267–271.

Newark Supergroup

HANS-DIETER SUES
Royal Ontario Museum
Toronto, Ontario, Canada

The early Mesozoic Newark Supergroup in eastern North America comprises several thousand meters of Triassic fluvial and lacustrine sedimentary rocks as well as Jurassic sedimentary deposits interbedded with flows of basalt lavas (Olsen *et al.*, 1989). These strata were deposited in 13 or so major rift basins that developed during a lengthy episode of crustal extension and thinning preceding the breakup of the supercontinent Pangaea and the opening of the northern Atlantic Ocean in the Jurassic. Although traditionally regarded as essentially unfossiliferous apart from footprints, renewed exploration of the sedimentary rocks of the Newark Supergroup in recent decades has resulted in the discovery of several well-dated assemblages of reptiles and other vertebrates ranging in age from the Middle Triassic (Anisian) to the Early Jurassic (Toarcian). In addition, some of the strata have long been renowned for the great abundance and diversity of footprints and trackways of reptiles, including dinosaurs. Much of the geochronological dating for the Newark Supergroup is based on the distribution of characteristic types of pollen and spores and(or) on radiometric dates for the interbedded lava flows. These dates require verification and further refinement by magnetostratigraphy.

The oldest tetrapod assemblage, from the Fundy basin in Nova Scotia, appears to be early Middle Triassic (Anisian) in age. The rare skeletal remains are very fragmentary and frequently unidentifiable but include a partial skull of a trematosaurid amphibian that proved useful for dating the fossiliferous strata. A much more diverse tetrapod assemblage from the Richmond basin in Virginia is dominated by gomphodont cynodonts and may be slightly older than other Late Triassic tetrapod communities from North America. Diverse assemblages of late Carnian tetrapods from the Wolfville Formation of the Fundy basin, the middle Pekin Formation of the Deep River basin in North Carolina, and the New Oxford Formation of the Gettysburg basin in Pennsylvania share a number of taxa with those from the CHINLE Group and DOCKUM FORMATION of the American southwest. These communities all are characterized by the presence of phytosaurs and(or) metoposaurid amphibians. They are also noteworthy for including the oldest records in eastern North America of both ornithischian and prosauropod dinosaurs, although the currently known fossils mainly comprise isolated teeth. To date, only a few skeletal remains for Norian age reptiles are known from the Hartford and Newark basins of the northeastern United States and from the Fundy basin.

The lower portion of the McCoy Brook Formation of the Fundy basin has yielded vast numbers of isolated bones and teeth as well as a few partial skulls and skeletons of various tetrapods, including dinosaurs of earliest Jurassic age. The tetrapods are similar to, and in some cases identical with, those from the upper Stormberg Group of southern Africa. Although the fossil assemblages represent a diversity of paleoenvironments, they all share the absence of key groups such as procolophonid parareptiles and phytosaurian archosaurs that characterized Late Triassic continental biotas in North America and neighboring regions. On the other hand, most of the tetrapod lineages documented from the McCoy Brook Formation

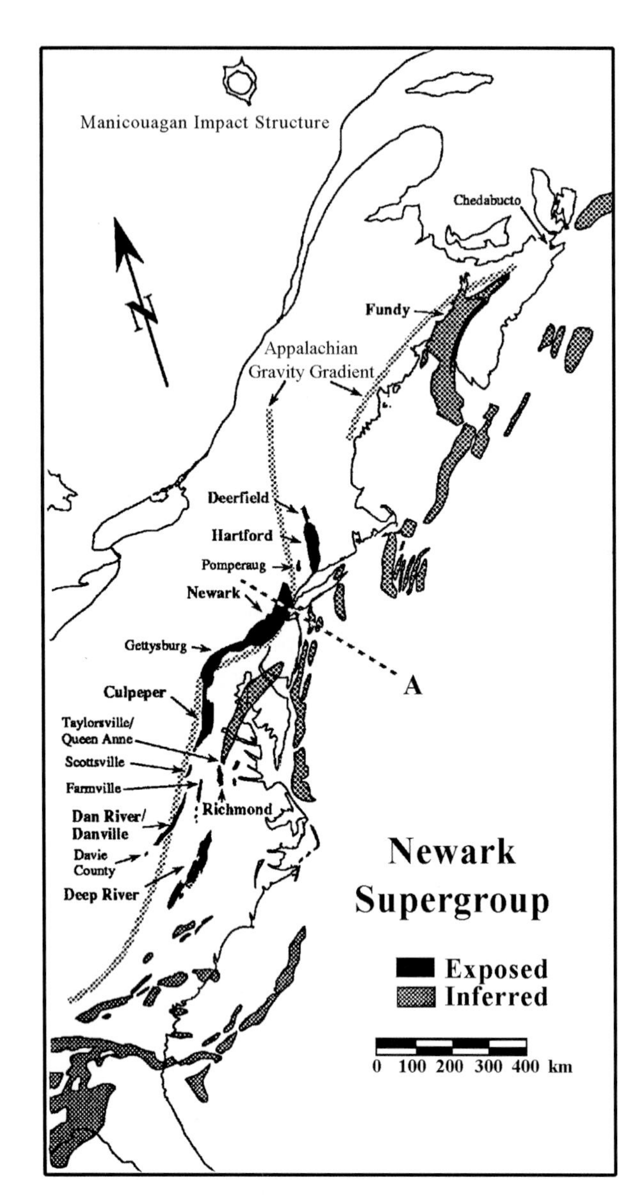

FIGURE 1 Exposed and inferred rocks of the basins of the Newark Supergroup. From Olsen *et al.*, 1989.

are already present in Late Triassic communities. This picture is consistent with hypotheses of a major extinction event near or at the Triassic–Jurassic boundary. The extensive record of pollen and spores from the sedimentary rocks of the Newark Supergroup (and elsewhere) also indicates major changes among land plants near or at the Triassic–Jurassic boundary. The dinosaurs from the McCoy Brook Formation are documented by small ornithischians with teeth similar to those of *Lesothosaurus* from southern Africa and by partial skeletons of as yet unidentified prosauropods. Trackways indicate the presence of both large and small theropod dinosaurs.

Slightly younger sedimentary rocks of the Portland Formation in the Hartford basin of Connecticut have yielded skeletons of the medium-sized prosauropod dinosaurs *Ammosaurus* and *Anchisaurus* and bones of unidentifiable small, lightly built theropods ("*Podokesaurus*"). Based on associated pollen and spores, these Early Jurassic dinosaurs are Pliensbachian or possibly Toarcian in age.

Although much less abundant than those from the well-known Triassic and Early Jurassic strata of the American Southwest, the early Mesozoic tetrapods from the Newark Supergroup are of great importance because of the considerable temporal range of the fossil record from the Middle Triassic well into the Jurassic and the diversity of paleoenvironments recorded in the sedimentary rocks.

See also the following related entries:

CONNECTICUT RIVER VALLEY • EXTINCTION, TRIASSIC • ROCKY HILL DINOSAUR STATE PARK • TRIASSIC PERIOD

Reference

Olsen, P. E., Schlische, R. W., and Gore, P. J. W. (Eds.) (1989). Tectonic, depositional, and paleoecological history of early Mesozoic rift basins, Eastern North America. 28th International Geological Congress, Field Trip Guidebook No. T351. American Geophysical Union, Washington, DC.

New Jersey State Museum, New Jersey, USA

see MUSEUMS AND DISPLAYS

New Mexico Museum of Natural History, New Mexico, USA

see MUSEUMS AND DISPLAYS

Origin of Dinosaurs

Kevin Padian
University of California
Berkeley, California, USA

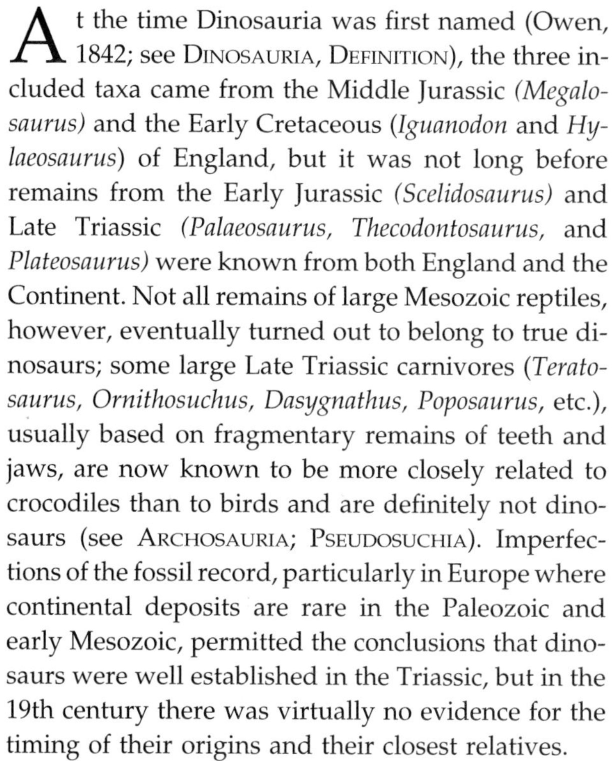

At the time Dinosauria was first named (Owen, 1842; see Dinosauria, Definition), the three included taxa came from the Middle Jurassic *(Megalosaurus)* and the Early Cretaceous (*Iguanodon* and *Hylaeosaurus*) of England, but it was not long before remains from the Early Jurassic *(Scelidosaurus)* and Late Triassic *(Palaeosaurus, Thecodontosaurus,* and *Plateosaurus)* were known from both England and the Continent. Not all remains of large Mesozoic reptiles, however, eventually turned out to belong to true dinosaurs; some large Late Triassic carnivores (*Teratosaurus, Ornithosuchus, Dasygnathus, Poposaurus,* etc.), usually based on fragmentary remains of teeth and jaws, are now known to be more closely related to crocodiles than to birds and are definitely not dinosaurs (see Archosauria; Pseudosuchia). Imperfections of the fossil record, particularly in Europe where continental deposits are rare in the Paleozoic and early Mesozoic, permitted the conclusions that dinosaurs were well established in the Triassic, but in the 19th century there was virtually no evidence for the timing of their origins and their closest relatives.

Toward the latter years of the 19th century and the early decades of the 20th century, various groups of archosaurs became better known, until a classification emerged that recognized five suborders of the Order Archosauria: Thecodontia, Crocodylia, Pterosauria, Ornithischia, and Saurischia (see Archosauria; Pseudosuchia; Thecodontia). Thecodontia was supposed to be the basal group from which all the others, as well as the birds, evolved (see Bird Origins). In the late 1970s and the 1980s, consensus began to form that these "thecodontians" were not a natural group but a wastebasket taxon of various early archosaurs; moreover, most taxa placed in the thecodontian taxon Pseudosuchia were either closer to dinosaurs or closer to crocodiles. Gauthier (1986) formalized this viewpoint by defining, in the phylogenetic system, Pseudosuchia as all archosaurs closer to crocodiles than to birds, and Ornithosuchia as all archosaurs closer to birds than to crocodiles (Fig. 1).

What is particularly interesting from an historical point of view about the question of the origin of dinosaurs is that until the 1970s there were virtually no archosaurs known that were closer to dinosaurs than to crocodiles. A few Triassic archosaurs, such as *Ornithosuchus* and *Scleromochlus,* known since the late 1800s or early 1900s, made reasonable candidates for dinosaurian ancestors. Walker (1964) recognized the similarity of *Ornithosuchus* to large carnivorous theropod dinosaurs and thoughtfully discussed a series of Late Triassic archosaurs that might have been close to dinosaurian ancestry; however, his recognition of *Ornithosuchus* as a carnivorous dinosaur has not been upheld by further analyses, and it is now considered closer to crocodiles (Sereno, 1991b). *Scleromochlus* was advanced by von Huene (1914) as a possible ancestor of pterosaurs, but he also saw it as close to dinosaurs (1956), as did Romer (1966). A Late Triassic (Norian) contemporary of *Scleromochlus,* also from the Elgin Sandstones of Scotland, was a small, long-legged form called *Saltopus,* and both authors advanced it as a possible dinosaur; however, *Saltopus* is so poorly preserved that it is little more than "a stain on a rock," as its curator, A. J. Charig, once

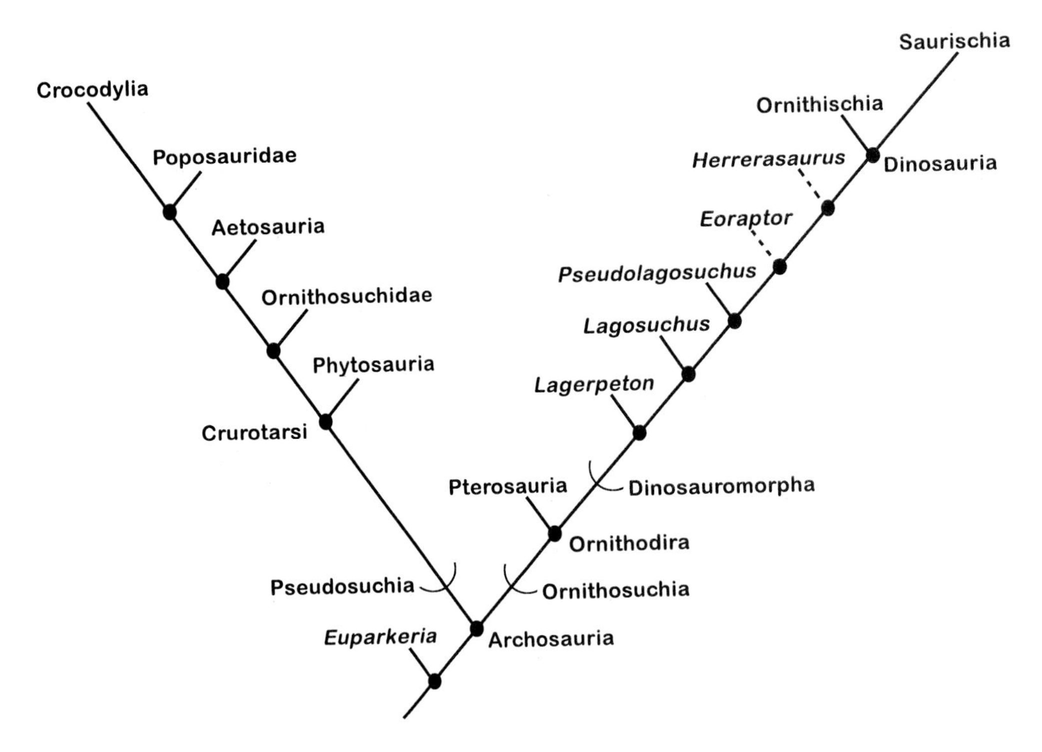

FIGURE 1 Phylogeny of the immediate out-groups of Dinosauria. For an alternative arrangement see Herrerasauridae.

wryly noted (personal communication; however, see Benton and Walker, 1985). *Scleromochlus* has recently been linked by phylogenetic analysis to the PTEROSAURS (see PTEROSAUROMORPHA), and although pterosaurs themselves are the closest major group to dinosaurs, they are so transformed in so many aspects of their skeletons that they provide little information about the immediate origins of dinosaurs.

The expeditions of Harvard University to the Late Triassic (Carnian) beds of the Rio Chañarense and Rio Guala regions of Argentina (1958–1968), led by A. S. Romer and J. A. Jensen, brought back the first fauna of nondinosaurian archosaurs that helped to fill the gap between basal archosaurs and dinosaurs (see MUSEUM OF COMPARATIVE ZOOLOGY, HARVARD UNIVERSITY). Small, slender, long-limbed forms known mostly from hindlimb remains were given the names *Lagosuchus* ("rabbit-crocodile") and *Lagerpeton* ("rabbit-reptile") by Romer (1971, 1972; see Dinosauromorpha). *Lewisuchus*, a third form, was poorly understood and originally thought to be close to crocodiles. More material of *Lagosuchus* was eventually prepared and studied by José Bonaparte (1975) in Argentina, and he persuasively argued for the intermediate status of *Lagosuchus* between typical thecodontians and dinosaurs, citing features of the vertebral column, shoulder girdle, pelvis, and hindlimb. Bonaparte was right: In nearly every respect, apart from a few individual features, the incompletely known *Lagosuchus* had the features that would have been expected in a dinosaurian ancestor. This, however, countered the very influential view of Charig *et al.* (1965) that dinosaurs had descended from large quadrupedal archosaurians somewhat like Erythrosuchidae or the smaller *Euparkeria* (Ewer, 1965; Fig. 2), which was long used as a model for a generalized "thecodontian" from which dinosaurs might have evolved.

Lagerpeton and *Lagosuchus* (Fig. 3) shared important features with dinosaurs (and pterosaurs) (Gauthier, 1986; Sereno and Arcucci, 1990; Sereno, 1991a). Although the femur is rather sigmoid in *Lagerpeton*, much like a crocodile, in the other ORNITHODIRA it is straighter, with a well-offset head that fits into a shallow acetabulum with some development of a supraacetabular ridge. The femur is long and slender, but the tibia is longer, and the fibula is reduced so that virtually no rotation of the lower leg takes place at the knee (see ARCHOSAURIA). The ankle is mesotarsal, so it works like a hinge, and the foot

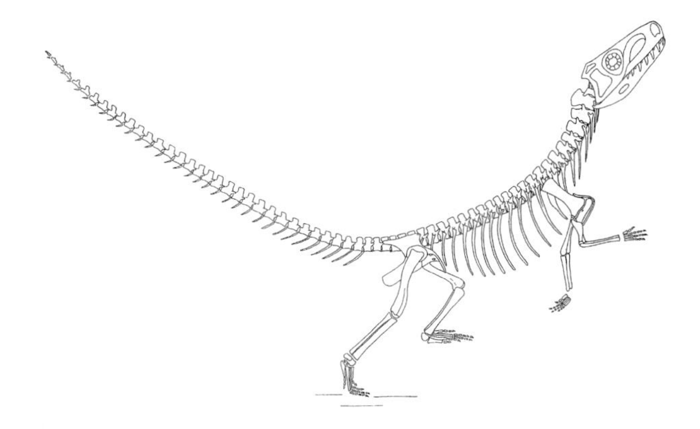

FIGURE 2 *Euparkeria,* an archosauromorph closely related to basal archosaurs (after Ewer, 1965).

bones (metatarsals) are elongated and closely appressed. The foot is strange in *Lagerpeton:* The first metatarsal and its phalanges are quite short, the second toe is about twice as long as the first, the third is about a third as long as the second, and the fourth is longer still, whereas the fifth is reduced to a splint-like metatarsal with no phalanges. This is a decidedly unusual configuration, which Sereno and Arcucci (1993) likened to a saltatory mammal, also because the spines of its posterior dorsal vertebrae incline anteriorly instead of posteriorly.

Sereno and Arcucci (1994) also restudied *Lagosuchus* and concluded that the type specimen of *L. talampayensis* was too poorly preserved to refer material to it with confidence; moreover, the proportions of the forelimb and hindlimb were different from those in other, better preserved specimens referred to the species. They decided to restrict *L. talampayensis* to the relatively poorly preserved type specimen and erected a new genus, *Marasuchus lilloensis,* for reception of the referred material as well as the type material of *Lagosuchus lilloensis* Romer, which they concluded was not specifically different. In all particulars except the proportions of the forelimb, however, the type material of *L. talampeyensis* agrees with other *Lagosuchus* material, and whether or not the association of this forelimb with the type specimen is firmly established, *Lagosuchus* and *Marasuchus* are in other respects phylogenetically indistinguishable. The hindlimb is generally similar to that of *Lagerpeton* and other ornithodirans but more like dinosaurs in having an offset femoral head and a straighter shaft. The acetabulum was reported by Bonaparte (1975) as partly open, but this may be an artifact of preservation. The foot is also more dinosaurian than that of *Lagerpeton* because the asymmetry of the digits is not as pronounced, and the third toe is the longest. Additionally, the vertebral column is well regionalized, with discrete cervical, anterior dorsal, and posterior dorsal vertebrae, as contrasted with those of Pseudosuchia (Bonaparte, 1975; Sereno and Arcucci, 1994). These features suggest a bipedal animal with more or less horizontal back and arched neck, running rapidly on hindlimbs that had elongated tibiae and metatarsals for improved cursorial abilities.

These animals, however, were not the closest to dinosaurs discovered in the Late Triassic of Brazil and Argentina. In 1942 von Huene had recovered teeth and postcranial material of a dinosaurian or near-dinosaurian form that he called *Spondylosoma;* once thought to be a "primitive saurischian," its status is now uncertain. Reig (1963) described the "saurischian dinosaurs" *Herrerasaurus* and *Ischisaurus,* and Colbert (1970) described *Staurikosaurus* from these deposits (see also Galton, 1977; HERRERASAURIDAE). To most investigators, however, these were not true saurischian dinosaurs; rather, they were primitive, but very dinosaurian, and were placed in Saurischia because the latter group of dinosaurs generally appeared to retain many characters of more basal archosaurs (Gauthier, 1986). Phylogenetic analysis of Dinosauria (Gauthier, 1986) supported this general contention by placing Herrerasauridae and related forms outside Dinosauria (i.e., Ornithischia + Saurischia; see DINOSAURIA: DEFINITION). Other analyses (Novas, 1992, 1993; Sereno, 1993; Sereno and Novas, 1992, 1993) have placed Herrerasauridae and the more recently discovered *Eoraptor* (Sereno *et al.*, 1993) as basal Theropoda, although a position outside Theropoda, as either basal saurischians or as the closest dinosaurian sister group, has also been found (Holtz, 1995; Holtz and Padian, 1995).

Whatever the particular outcome of this controversy, it has little effect on the timing of diversification of Dinosauria (Benton, 1990; Hunt, 1991; Sereno *et al.*, 1993). Herrerasaurids and the basal ornithischian *Pisanosaurus* are known from the Ischigualasto Formation (latest Ladinian or early Carnian), which implies that even if herrerasaurids are not theropods or not even saurischians, the Saurischia must have appeared and begun to diversify by that time (Sereno *et al.*, 1993). Herrerasaurid material, usually represented by isolated bones or incomplete skeletons, has

FIGURE 3 *Lagosuchus (Marasuchus)* (from Sereno, 1991).

been recovered from a number of localities in the American southwest (Hunt, 1994); perhaps the latest known occurrence of herrerasaurids, although it was announced at the time of its discovery as the "world's oldest dinosaur," is the type specimen of *Chindesaurus* from the Norian beds of the Petrified Forest National Park, Arizona (Long and Murry, 1995). Once herrerasaurids are better known, it may be that considerable Late Triassic material from Europe and Africa, now assigned to other taxa, may pertain. For now, it does not appear that herrerasaurs survived the Triassic.

Apart from isolated ornithischian teeth in eastern and southwestern North America, parts of the Texas material called "*Technosaurus*" that are not sauropodomorph (Sereno, 1991a), and some fabrosaurid material from the Lower Elliott Formation of southern Africa, ornithischians are relatively inconspicuous in the Late Triassic (Hunt and Lucas, 1994).

Sauropodomorphs are similarly elusive. Perhaps the earliest known is the fragmentary ornithischian *Azendohsaurus* from Morocco (see PROSAUROPODA), possibly of middle Carnian age; the Carnian–Norian Lower Elliott Formation of southern Africa has also produced *Melanorosaurus, Euskelosaurus,* and *Blikanasaurus*. From the Norian of South America come *Mussaurus, Riojasaurus,* and *Coloradisaurus* (= *Coloradia),* and *Sellosaurus, Plateosaurus,* and *Thecodontosaurus* are known from Europe (Galton, 1990). North America seems devoid of sauropodomorphs until the Early Jurassic (*Anchisaurus, Ammosaurus,* and *Massospondylus*); some isolated teeth from the Late Triassic have been referred to the group, but they are more probably ornithischian (Hunt and Lucas, 1994).

Late Triassic theropods are a mixed lot; apart from the candidacy of *Eoraptor* and herrerasaurids as theropods, only *Coelophysis* is really well preserved, and it is represented by hundreds of specimens from the Ghost Ranch Quarry of the Chinle Formation and elsewhere (?late Carnian–Norian; Colbert, 1989; Padian, 1986; but see Sullivan *et al.*, 1996). Rowe and Gauthier (1990) recognized *Coelophysis* as a member of Ceratosauria and relatively derived for an early member of the group. According to their analysis, the only two relatively well-preserved Triassic theropods, *Coelophysis* and *Liliensternus* (late Norian: Knollenmergel, Germany), are ceratosaurs, the sister taxon to TETANURAE (whose first members appear in the Middle Jurassic, e.g., *Monolophosaurus).* More poorly preserved theropods, of which the best known are probably *Procompsognathus* (Norian, Germany; Ostrom, 1981; Sereno and Wild, 1992) and *Walkeria* (Norian, Maleri Formation, India; Chatterjee, 1987), include many possible or doubtful forms whose principal synapomorphy appears to be poor or incomplete preservation (Norman, 1990). Material referred to theropod dinosaurs by Lucas *et al.* (1992), from the Placerias Quarry of Arizona, is of Carnian age. These authors (see also Hunt and Lucas, 1994; Hunt, 1991) provide useful reviews of early dinosaur occurrences.

It appears, then, that dinosaurs, including herrerasaurs or not, are first known from the early Carnian on the basis of both bones and footprints (King and Benton, 1996). Their origins appear to be from small, lightly built, cursorial bipeds of which much is not known, and nothing at all would be known if it were not for the South American fossil record.

It has often been wondered why dinosaurs, who kept a low profile during the Late Triassic, survived the end-Triassic extinctions and went on to dominate terrestrial faunas as no vertebrate group has before or since (see EXTINCTION, TRIASSIC). Competitive models, opportunistic models, and null hypotheses have all been advanced and discussed (Benton, 1983, 1996), but ecological models are even harder to test in the past than in the present. It is obvious how different dinosaurs were from the various pseudosuchians (of which only crocodiles survived the Triassic) and pterosaurs (which flourished until the very end of the Cretaceous) that arose with them in the Late Triassic; it is less clear why they persisted over the other ornithodirans, represented by *Lagosuchus* and *Lagerpeton*, that are really not so dissimilar to them. This ornithodiran lineage, in evolving obligate bipedality (they could not sprawl), may have been committed to a more active way of life. Their arms were freed from locomotory needs, and their hands and fingers were prehensile. Certainly the bone histology of pterosaurs and dinosaurs indicates rapid growth, and perhaps high metabolic levels (at least higher sustained levels than seen in reptiles today), whereas this was decidedly not the case for any Triassic pseudosuchians. However, the concept of "different strokes for different folks" must be considered. Pseudosuchians were a varied lot in the Triassic, ranging from amphibious, crocodile-like forms to large macropredaceous carnivores to armored plant eaters. They were clearly not in competition with the small, early dinosaurs. If comparison to Late Cretaceous terrestrial vertebrate extinctions is any indication (see EXTINCTIONS), at that time the warm-blooded mammals and birds generally survived, as did the cold-blooded turtles, crocodiles, champsosaurs, lizards and snakes, and amphibians. Improved posture and gait, and perhaps metabolism, may have prepared dinosaurs for new ways of life that could be exploited as the Mesozoic Era unfolded; but these are still, and may remain, matters of conjecture.

See also the following related entries:

DINOSAURIA • EXTINCTION, TRIASSIC • PHYLOGENY OF DINOSAURS

References

Benton, M. J. (1983). Dinosaur success in the Triassic: A noncompetitive ecological model. *Q. Rev. Biol.* **58,** 29–55.

Benton, M. J. (1990). Origin and interrelationships of dinosaurs. In *The Dinosauria* (D. B. Weishampel, P. Dodson, and H. Osmólska, Eds.), pp. 11–30. Univ. of California Press, Berkeley.

Benton, M. J. (1996). On the nonprevalence of competitive replacement in the evolution of tetrapods. In *Evolutionary Paleobiology* (D. Jablonski, D. H. Erwin, and J. H. Lipps, Eds.), pp. 185–210. Univ. of Chicago Press, Chicago.

Benton, M. J., and Walker, A. D. (1985). Palaeoecology, taphonomy, and dating of Permo–Triassic reptiles from Elgin, northeast Scotland. *Palaeontology* **28,** 207–234.

Bonaparte, J. F. (1975). Nuevos materiales de *Lagosuchus talampayensis* Romer (Thecodontia–Pseudosuchia) y su significado en el origen de los Saurischia. Chañarense inferior, Triásico medio de Argentina. *Acta Geol. Lilloana* **13,** 5–90.

Charig, A. J., Attridge, J., and Crompton, A. W. (1965). On the origin of the sauropods and the classification of the Saurischia. *Proc. Linnean Soc. London* **176,** 197–221.

Chatterjee, S. (1987). A new theropod dinosaur from India with remarks on the Gondwana–Laurasia connection in the Late Triassic. *Geophys. Monogr.* **41,** 183–189.

Colbert, E. H. (1970). A saurischian dinosaur from the Triassic of Brazil. *Am. Museum Novitates* **2405,** 1–39.

Colbert, E. H. (1989). The Triassic dinosaur *Coelophysis. Bull. Museum Northern Arizona* **57,** 1–174.

Ewer, R. F. (1965). The anatomy of the thecodont reptile *Euparkeria capensis* Broom. *Philos. Trans. R. Soc. London B* **248,** 379–435.

Galton, P. M. (1977). On *Staurikosaurus pricei*, an early saurischian dinosaur from Brazil, with notes on the Herrerasauridae and Poposauridae. *Paläontol. Z.* **51,** 234–245.

Galton, P. M. (1990). Basal Sauropodomorpha–Prosauropods. In *The Dinosauria* (D. B. Weishampel, P. Dodson, and H. Osmólska, Eds.), pp. 320–344. Univ. of California Press, Berkeley.

Gauthier, J. A. (1986). Saurischian monophyly and the origin of birds. *Mem. California Acad. Sci.* **8,** 1–55.

Holtz, T. R., Jr. (1995). A new phylogeny of the Theropoda. *J. Vertebr. Paleontol.* **15**(Suppl. to No. 3), 35A.

Holtz, T. R., Jr., and Padian, K. (1995). Definition and diagnosis of Theropoda and related taxa. *J. Vertebr. Paleontol.* **15**(Suppl. to No. 3), 35A.

Hunt, A. P. (1991). The early diversification pattern of dinosaurs in the Late Triassic. *Mod. Geol.* **16,** 43–60.

Hunt, A. P. (1994). Vertebrate paleontology and biostratigraphy of the Bull Canyon Formation (Chinle Group:

Norian), east-central New Mexico, with revisions of the families Metoposauridae (Amphibia: Temnospondyli) and Parasuchidae (Reptilia: Archosauria). Ph.D. dissertation, University of New Mexico, Albuquerque.

Hunt, A. P., and Lucas, S. G. (1994). Ornithischian dinosaurs from the Upper Triassic of the United States. In *In the Shadow of the Dinosaurs: Early Mesozoic Tetrapods* (N. C. Fraser and H.-D. Sues, Eds.), pp. 225–241. Cambridge Univ. Press, New York.

King, M. J., and Benton, M. J. (1996). Dinosaurs in the Early and Mid Triassic?—The footprint evidence from Britain. *Palaeogeogr. Palaeoclimatol. Palaeoecol.* **122,** 213–225.

Long, R. A., and Murry, P. (1995). Late Triassic (Carnian and Norian) tetrapods from the southwestern United States. *New Mexico Museum Nat. History Sci. Bull.* **4,** 1–254.

Lucas, S. G., Hunt, A. P., and Long, R. A. (1992). The oldest dinosaurs. *Naturwissenschaften* **79,** 171–172.

Norman, D. B. (1990). Problematic Theropoda—"Coelurosaurs." In *The Dinosauria* (D. B. Weishampel, P. Dodson, and H. Osmólska, Eds.), pp. 280–305. Univ. of California Press, Berkeley.

Novas, F. E. (1992). Phylogenetic relationships of the basal dinosaurs, the Herrerasauridae. *Palaeontology* **35,** 51–62.

Novas, F. E. (1993). New information on the systematics and postcranial morphology of *Herrerasaurus ischigualastensis* (Theropoda: Herrerasauridae) from the Ischigualasto Formation (Late Triassic) of Argentina. *J. Vertebr. Paleontol.* **13,** 400–424.

Ostrom, J. H. (1981). *Procompsognathus:* Theropod or thecodont? *Palaeontographica A* **175,** 179–195.

Owen, R. (1842). Report on British fossil reptiles, Part II. *Rep. Br. Assoc. Adv. Sci.* **1841,** 60–294.

Padian, K. (1986). On the type material of *Coelophysis* Cope (Saurischia: Theropoda) and a new specimen from the Petrified Forest of Arizona (Lae Triassic: Chinle Formation). In *The Beginning of the Age of Dinosaurs: Faunal Change across the Triassic–Jurassic Boundary* (K. Padian, Ed.), pp. 45–60. Cambridge Univ. Press, New York.

Reig, O. A. (1963). La presencia de dinosaurios saurisquios en los "Estratos de Ischigualasto" (Mesotriasico superior) de las provencias de San Juan y La Rioja (Republico Argentina). *Ameghiniana* **3,** 3–20.

Romer, A. S. (1966). *Vertebrate Paleontology,* 3rd ed. Univ. of Chicago Press, Chicago.

Romer, A. S. (1971). The Chañares (Argentina) Triassic fauna. X. Two new and incompletely known long-limbed pseudosuchians. *Breviora* **378,** 1–10.

Romer, A. S. (1972). The Chañares (Argentina) Triassic fauna. XV. Further remains of the thecodonts *Lagerpeton* and *Lagosuchus. Breviora* **394,** 1–7.

Sereno, P. C. (1991a). *Lesothosaurus,* "fabrosaurids," and the early evolution of Ornithischia. *J. Vertebr. Paleontol.* **11,** 168–197.

Sereno, P. C. (1991b). Basal archosaurs: Phylogenetic relationships and functional implications. *J. Vertebr. Paleontol.* **11**(Suppl. 4), 1–53.

Sereno, P. C. (1993). The pectoral girdle and forelimb of the basal theropod *Herrerasaurus ischigualastensis. J. Vertebr. Paleontol.* **13,** 425–450.

Sereno, P. C. (1995). Theropoda: Early evolution and major patterns of diversification. *J. Vertebr. Paleontol.* **15**(Suppl. to No. 3), 52A–53A.

Sereno, P. C., and Arcucci, A. B. (1990). The monophyly of crurotarsal archosaurs and the origin of bird and crocodile ankle joints. *Neues Jahr. Geol. Paläontol. Abhandlungen* **180,** 21–52.

Sereno, P. C., and Arcucci, A. B. (1993). Dinosaurian precursors from the Middle Triassic of Argentina: *Lagerpeton chanarensis. J. Vertebr. Paleontol.* **13,** 385–399.

Sereno, P. C., and Arcucci, A. B. (1994). Dinosaurian precursors from the Middle Triassic of Argentina: *Marasuchus lilloensis,* gen. nov. *J. Vertebr. Paleontol.* **14,** 53–73.

Sereno, P. C., and Novas, F. E. (1992). The complete skull and skeleton of an early dinosaur. *Science* **258,** 1137–1140.

Sereno, P. C., and Novas, F. E. (1993). The skull and neck of the basal theropod *Herrerasaurus ischigualastensis. J. Vertebr. Paleontol.* **13,** 451–476.

Sereno, P. C., and Wild, R. (1992). *Procompsognathus:* Theropod, "thecodont," or both? *J. Vertebr. Paleontol.* **12,** 435–458.

Sereno, P. C., Forster, C. A., Rogers, R. R., and Monetta, A. M. (1993). Primitive dinosaur skeleton from Argentina and the early evolution of Dinosauria. *Nature* **361,** 64–66.

Sullivan, R. M., Lucas, S. G., Heckart, A., and Hunt, A. P. (1996). The type locality of *Coelophysis,* a Late Triassic dinosaur from north-central New Mexico (USA). *Paläontol. Z.* **70,** 245–255.

von Huene, F. (1914). Beiträge zur Kenntnis der Archosaurier. *Geol. Paläontol. Abhandlungen N.F.* **13**(1), 1–53.

von Huene, F. (1942). *Die fossilen Reptilien des südamerikanischen Gondwanalandes. Ergebnisse der Sauriergrabungen in Südbrasilien 1928/29.* Beck'sche Verlagsbuchhandlung, Munich.

von Huene, F. (1956). *Paläontologie und Phylogenie der niederen Tetrapoden.* Fischer-Verlag, Jena.

Walker, A. D. (1964). Triassic reptiles from the Elgin area: *Ornithosuchus* and the origin of carnosaurs. *Philos. Trans. R. Soc. London B* **248,** 53–134.

Orlov Museum of Paleontology

L. P. TATARINOV
Academy of Sciences
Moscow, Russia

The Orlov Museum of Paleontology in Moscow is a part of the Palaeontological Institute of the Russian Academy of Sciences. It was founded in 1937. In 1988 the museum was placed in a new building erected under a specific architectural design with four corner towers. Besides the museum, this building houses the administration of the institute, vertebrate and some other laboratories, the conference hall, and other operations. Mounted in the courtyard are statues of some fossil animals, including dinosaurs.

The interior of the museum is decorated with many pieces of ornamental art, e.g., animal bas-relief, ceramic compositions, forged copper compositions, and frescoes. The pivotal one is the vertically oriented monumental ceramic panorama, the "Tree of Life," placed in one of the towers. The floor and ceiling of this tower are made of numerous mirrors, repeatedly reflecting the whole composition.

The total space of the building is 10,000 m^2, and the exhibition space is 4200 m^2. Approximately 10,000 specimens are exhibited. There are five main exhibition halls: the introductory one, the Paleozoic invertebrates, the Paleozoic vertebrates, the Mesozoic, and the Cenozoic. Dinosaurs are exhibited in the Mesozoic hall.

The Mesozoic hall is divided into main space and upper gallery. The hall is decorated with monumental color murals about Mesozoic life, and the gallery is decorated with bas-relief. The largest specimen in the hall is the cast of *Diplodocus* presented by the United States, but most of the exposition comprises specimens from Mongolia and Middle Asia. Among these are skeletons of the iguanodontid *Probactrosaurus*, the hadrosaurs *Corythosaurus*, *Arstanosaurus*, and *Saurolophus*, the ankylosaurs *Pinacosaurus* and *Talarurus*, ceratopsians including *Psittacosaurus*, *Protoceratops*, and the theropod *Tarbosaurus*. In glass cases many additional materials are exposed, including skulls of the sauropod *Quaesitosaurus*, the theropods *Velocipator* and *Tarbosaurus*, and a forelimb of the segnosaur *Therezinosaurus*, the ornithopods *Iguanodon* and *Saurolophus*, the ankylosaurians *Saichania*, *Shamosaurus*, and *Pinacosaurus*, and the ceratopsians *Proceratops* and *Bagaceratops*. EGGS of various dinosaurs, including one with embryonic bones, FOOTPRINTS from the Upper Jurassic of Tadjikistan, and other materials are all on display. Besides dinosaurs, the Mesozoic hall exhibits some marine invertebrates, insects, plants, fishes, birds, marine reptiles, crocodiles, turtles, and the pterosaurians *Sordes* and *Phobeter*.

The main collections of the Palaeontological Institute number approximately 120,000 specimens (invertebrates + vertebrates + plants) placed in storage.

See also the following related entries:

CHINESE DINOSAURS • MIDDLE ASIAN DINOSAURS • MONGOLIAN DINOSAURS

Reference

Palaeontological Institute Palaeontological Museum named after academician Yu. A. Orlov. Tokio: Academy of Sciences of Russia, 1992, pp. 134.

Ornamentation

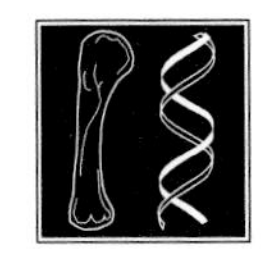

Matthew K. Vickaryous
Michael J. Ryan
Royal Tyrrell Museum of Palaeontology
Drumheller, Alberta, Canada

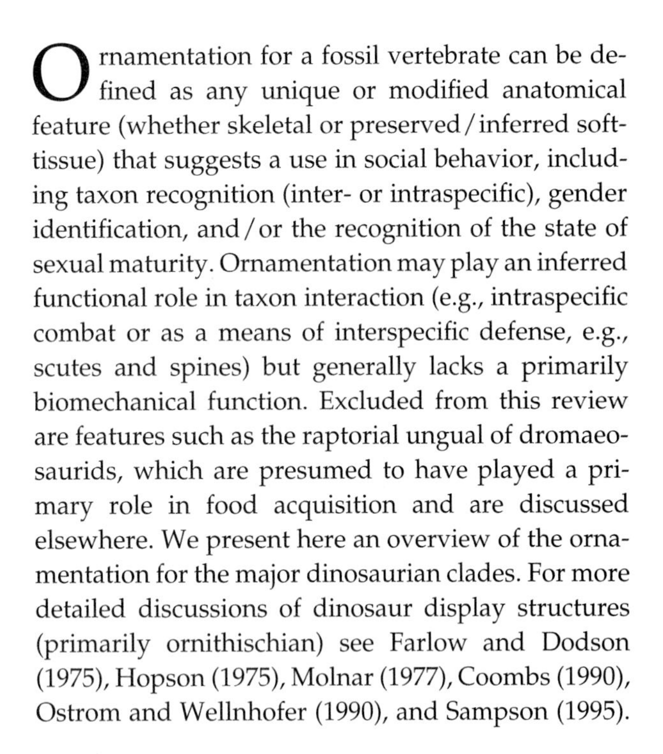

Ornamentation for a fossil vertebrate can be defined as any unique or modified anatomical feature (whether skeletal or preserved/inferred soft-tissue) that suggests a use in social behavior, including taxon recognition (inter- or intraspecific), gender identification, and/or the recognition of the state of sexual maturity. Ornamentation may play an inferred functional role in taxon interaction (e.g., intraspecific combat or as a means of interspecific defense, e.g., scutes and spines) but generally lacks a primarily biomechanical function. Excluded from this review are features such as the raptorial ungual of dromaeosaurids, which are presumed to have played a primary role in food acquisition and are discussed elsewhere. We present here an overview of the ornamentation for the major dinosaurian clades. For more detailed discussions of dinosaur display structures (primarily ornithischian) see Farlow and Dodson (1975), Hopson (1975), Molnar (1977), Coombs (1990), Ostrom and Wellnhofer (1990), and Sampson (1995).

Theropoda

Saurischians are not usually regarded as exemplifying ornamentation, but both sauropods and theropods exhibit a variety of unusual features. Some of the best documented theropod cranial features include the paired blade-like nasolacrimal crests of *Dilophosaurus wetherilli* and *Syntarsus kayentakatae* and the slender nasal horn of *Ceratosaurus nasicornis*. More common, however, are the lacrimal and postorbital rugosities found in *Ceratosaurus* and the Allosauridae, Sinraptoridae, and Tyrannosauridae. Nasal rugosities have been documented in most large theropod groups (Abelisauridae, Allosauridae, Sinraptoridae, and Tyrannosauridae), with the most prominent outgrowths being displayed by the tyrannosaurid *Alioramus remotus* (Fig. 1). This theropod has a double row of five vertical blades, continuous with the lacrimal and postorbital rugosities. Some of the more bizarre cranial features demonstrated by large theropods include the robust frontal horns of *Carnotaurus sastresi*, the trenchant lacrimal crest of *Cryolophosaurus ellioti* (Hammer and Hickerson, 1994), and the elaborate midline crest of *Monolophosaurus jianji* (Zhao and Currie, 1993) (Fig. 2). This crest, formed by the fusion of the premaxillae, nasals, lacrimals, and frontals, is noteworthy in that it is pneumatized, with connections to the antorbital fossae. Other rostral crests include the shallow nasal crest of *Baryonyx walkeri*, the delicate premaxillary nasal crest of *Ornitholestes hermanni*, and the highly pneumatized crests of *Oviraptor mongoliensis* (premaxillae, nasals, frontals, and parietals) and *Oviraptor philoceratops* (premaxillae, nasals, and frontals). Cranial ornamentation in theropods has been suggested to have been used in intraspecific combat (e.g., head-to-head interactions) but the excavated bone seen in rugose lacrimals and the delicate nature of midline crests argues against this.

Most theropods appear to lack postcranial ornamentation. When present, the most familiar example is that of elongate neural spines. Elongate spinous processes are considered to be polyphyletic and are found in a variety of poorly known theropods (*Acrocanthosaurus atokensis, Altispinax dunkeri, B. walkeri, Becklespinax altispinax*, and *Spinosaurus aegyptiacus*). Theropod dermal ossifications are only known for *C. nasicornis*. A small Late Jurassic (?Early Cretaceous) compsognathid-like theropod from China as been reported to have feathers.

Sauropoda

Sauropods are poorly known cranially with preserved skulls appearing to lack any form of traditional ornamentation. Postcranially, several groups demonstrate some degree of skeletal modification. The Dicraeosauridae are characterized in part by elongate neural spines that are strongly bifurcated presacrally (up to the mid-dorsals). The spinous processes may have given rise to a low midline vertical sail. Osteo-

FIGURE 1 *Alioramus remotus* (© D. L. Sloan, 1996).

derms, including relatively large (>10 cm) plate-like scutes and smaller (<10 cm) amorphous ossicles, have been well documented for the Titanosauridae. Titanosaurid armor appears to be associated with the dorsal and lateral sides of the body and can be readily differentiated from that of ankylosaurs based on gross morphology. The euhelopodids *Shunosaurus lii* and *Omeisaurus tianfuensis* have evolved a tail club, a unique polyphyletic feature previously considered restricted to the Ankylosauridae. The incipient clubs are fusiform in shape and appear to be derived from the hyperossification and fusion of three to five distal caudal vertebrae. Skin impressions are covered elsewhere in this volume, but here it is noteworthy to mention the nonossified midline *Iguana*-like spines apparently present in some diplodocids. These spines likely functioned in recognition displays, although they may have aided in passive defense by strengthening the hide. Other sauropod features, such as trenchant manual unguals, elongate necks (e.g., *Mamenchisaurus*), and/or tails (e.g., *Diplodocus*), are not herein considered ornamentation.

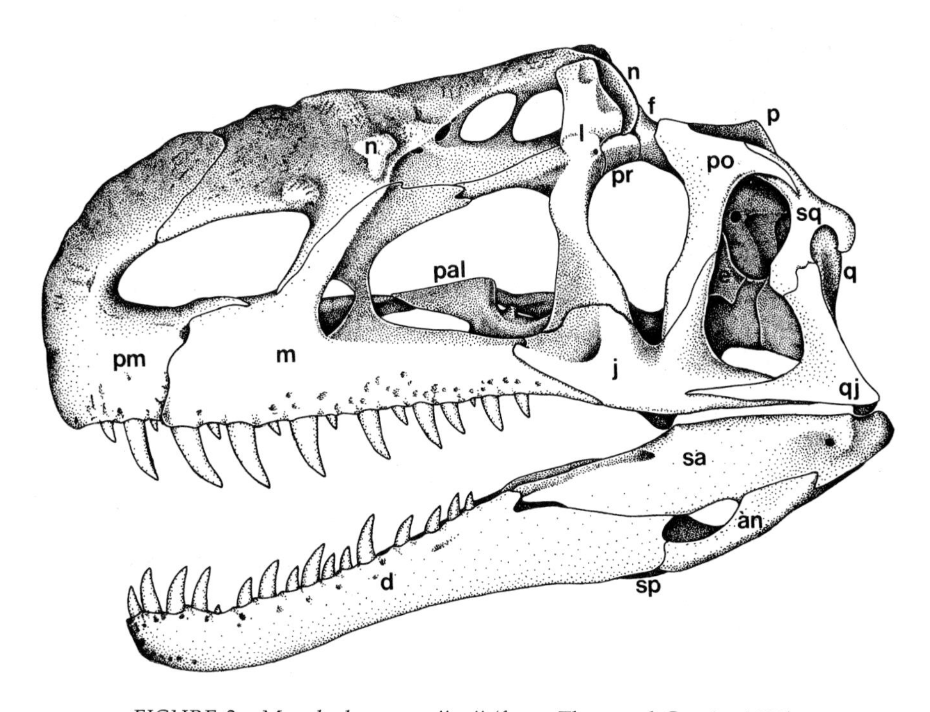

FIGURE 2 *Monolophosaurus jianji* (from Zhao and Currie, 1993).

Ornithopoda

The Ornithopoda represents a large and diverse group of ornithischians, with many taxa exhibiting diagnostic forms of ornamentation. The Heterodontosauridae are primitive members, diagnosed in part by the presence of caniniform teeth in the premaxillae and dentaries. Caniniform teeth may have functioned in intraspecific (and possibly interspecific) combat, much like modern suids. Heterodontids also have a laterally projecting bony boss on the jugal.

Some members of the Iguanodontidae (*Iguanodon* and *Ouranosaurus*) are noted for possessing a spike-like manual digit I that potentially may have been used as a display device. The iguanodontids *"Iguanodon" orientalis* and *Muttaburrasaurus langdoni* both have enlarged external nares, formed as a result of dorsomedial expansion of the premaxillae and nasals. Cranially, *Ouranosaurus nigeriensis* has developed a small pair of nasal protuberances, but it is the elongated neural spines of the axial skeleton that make this dinosaur distinctive.

Skeletal ornamentation in the Hadrosauridae is almost exclusively restricted to the cranium. In the Lambeosaurinae, the skull is dominated by a large crest formed from the expansion and elaboration of the premaxillae and nasals (Weishampel and Horner, 1990). Well-known examples include the plate-like crest of *Corythosaurus,* the hatchet-like crest of *Lambeosaurus,* and the caudally oriented, tube-like crest of *Parasaurolophus*. Each of these crests contain convoluted nasal passages and diverticula that have been postulated as resonating chambers. *Hypacrosaurus altispinus* has elongated dorsal neural spines in addition to the nasopremaxillary crest. Other hadrosaurs are generally more conservative in their cranial adornment, although some taxa from the Saurolophinae possessed a solid spine-like nasal crest. Members of the Gryposaurinae, Saurolophinae, as well as *Brachylophosaurus* typically demonstrate enlarged external nares and nasal crests, which may have potentially been used for intraspecific combat (e.g., broadside shoving) (Weishampel and Horner, 1990). In "flat-headed" hadrosaurs (e.g., *Edmontosaurus*), circumnarial excavations potentially held inflatable diverticula for use in auditory and visual signaling. Diverticula may also have been present in gyposaurines, saurolophines, and *Brachylophosaurus*. *Tsintaosaurus spinorhinus* appears to possess a spine-like nasal crest that is oriented anterodorsally (as opposed to posterodorsally as in gryposaurines), although its reality has been disputed. The crests in some species have been recognized as being sexually dimorphic (Dodson, 1975). Some hadrosaurs (e.g., *Anatosaurus annectus* and *Corythosaurus casuarius*) have been found with *in situ* midsagittal integumentary folds running the length of the body. This fold potentially acted in display, similar in fashion to the nonosseous spines of some diplodocids.

Thyreophora

Thyreophorans are the archetypal armored dinosaurs, characterized by the presence of parasagittal rows of bony ossifications. Primitive members, including forms such as *Scelidosaurus harrisoni, Scutellosaurus lawеri*, and *Emausaurus ernsti*, demonstrate a dermal armor made up of various small scutes and ossicles (Coombs *et al.*, 1990). These scutes have a diverse morphology, which appears to be related in part to their relative position along the body.

The Stegosauria can be distinguished on the basis of rows of vertically oriented (although perhaps dorsolaterally displaced) plates or spines dorsal to the vertebral column. New evidence from Colorado indicates that *Stegosaurus* did indeed have two rows of alternating plates, and that the distal caudal spines were oriented posterolaterally (Carpenter and Small, 1993), resolving a long-standing paleontological controversy. In the primitive stegosaur *Huayangosaurus taibaii* the skull has small postorbital horns. Postcranially, it demonstrates a lateral row of keeled scutes and parascapular spines. Other stegosaurs are known to also have had parascapular spines, as well as gular armor, dermal ossicles, and small keeled scutes in the sacral–femoral areas. The midline plates may have had a thermoregulatory function.

Ankylosaurs demonstrate the most extreme examples of dermal armor in the Dinosauria. In addition to the mosaic of dermal plates on the skull, the oval to rectangular scutes (with or without keels or ridges), and the amorphous ossicles, ankylosaurs may also exhibit cervical rings, gular armor, and bony eyelids [e.g., *Euoplocephalus tutus* and *Pawpawsaurus campbelli* (Lee, 1996)]. The various taxa of the Ankylosauria can be in part diagnosed on the basis of armor morphology. The Nodosauridae (e.g., *Edmontonia rugosidens*) are typified by "solid" (nonexcavated) scutes and conical spines, directed laterally and anterolaterally. The presence of pyramidal squamosal and jugal horns, supraorbital wedges of armor, excavated scutes, and tail clubs are indicative of the Ankylosauridae. Within the ankylosaurids, the Polacanthinae (e.g., *Polacanthus foxii*) retain the lateral spines of nodosaurids in addition to the development of dorsally oriented spines and sacral armor (rosettes of fused ossicles and scutes co-ossified to the ilia) (Carpenter *et al.*, 1996). The unnamed group that contains the Shamosaurinae (*Shamosaurus scutatus*) and the Ankylosaurinae (e.g., *E. tutus*) (Fig. 3) demonstrates a reduction/loss of the dorsal and lateral spines (Carpenter *et al.*, 1996). In shamosaurines, the scutes making up the cranial armor become indistinct, whereas in ankylosaurines the scutes remain discernible rostrally. The presence of armor (scutes, spines, and tail clubs) appears to be related to ontogeny. Subadult *Pinacosaurus grangeri* fail to demonstrate body armor, tail clubs, and cranially bound dermal plates. As a result, *Pinacosaurus* is the only ankylosaur for which the sutures of the skull are known dorsally.

The tails of stegosaurs (spines), ankylosaurids (clubs), and probably some euhelopodid sauropods (clubs) likely functioned in active defense. It has also been proposed that the entire tail could have acted in intraspecific combat and as display features. The hypothesis of ankylosaurid tail clubs functioning in mimicry is rejected based on lack of evidence. In rare instances, the plates and spines of stegosaurs (e.g., *Stegosaurus stenops*) and the dermal armor of ankylosaurs (e.g., *Edmontonia rugosidens*) are preserved *in situ*, providing for derivations of arrangement.

Pachycephalosauria

The hypertrophied skulls of pachycephalosaurs are unique among the Dinosauria. Most pachycephalosaurs are best (or exclusively) known from cranial material. The frontoparietal region generally exhibits the greatest degree of thickening forming a high, rounded dome, but the nasals, palpebrals, postorbitals, and squamosals may also thicken and/or become ornamented. Minor contributions to cranial ornamentation may also be demonstrated by the premaxillae, maxillae, and jugals. Ossified nodes and spines are frequently found on the rostral and squamosal regions. An extreme example of pachycephalosaur cranial ornamentation is demonstrated by *Stygimoloch spinifer*. Brown and Schlaikjer (1943) suspected that the domes of *Stegoceras validum* may show sexual dimorphism.

Ceratopsia

Ceratopsians are the group of dinosaurs whose members demonstrate the greatest degree of cranial ornamentation. The Psittacosauridae possess the flaring jugals characteristic of all ceratopsians although they lack the extreme caudal extension of the parietal/squamosal frill seen in most members of the Neoceratopsia. Among the Neoceratopsians, the parietosquamosal frill is variably developed, reaching its extreme form in *Torosaurus latus*, which has the largest skull of any terrestrial vertebrate (2.4 m). In the Protoceratopsidae, the frill is only moderately developed in most taxa with the exceptions being *Protoceratops andrewsi* and *Montanoceratops cerorhynchus*. Variation

FIGURE 3 *Euoplocephalus tutus* (© S. R. Bissette, 1996).

FIGURE 4 *Styracosaurus albertensis* (illustration by D. L. Sloan, © RTMP, 1996).

in the size and shape of the frill of *P. andrewsi* has been used to infer sexual dimorphism in this species (Dodson, 1976). The midline and posterior margin of the frill is formed from a single fused parietal. Enlarged, elongated squamosals form the lateral margins of the frill. In adult centrosaurines, epoccipitals fuse to the margins of parietals and grow with the underlying bone to develop into spikes [*Achelousaurus horneri*, *Styracosaurus albertensis* (Fig. 4), and *Einosaurus procurvicornus*], hooks (*Centrosaurus apertus*), or recurved spikes (*Pachyrhinosaurus* n. sp.). Epoccipitals also fuse to the parietals of chasmosaurines and the squamosals of both subfamilies but do not undergo significant modification here. The frill can be highly fenestrated (e.g., *Chamsmosaurus*) or solid (e.g., *Triceratops*) and would have been covered by a thick dermis. The frill would have been structurally too weak to act as an effective shield during predatory attacks but it could have acted as an excellent, possibly colorful, display device. Typically, centrosaurines have large nasal horns and very small brow horns on the postorbitals, whereas chasmosaurines have very large brow horns and reduced nasal horns. Both structures would have been covered in a horny sheath and could have made impressive offensive weapons. The nasal horns in centrosaurines are outgrowths of the nasal bones, which fused with maturity. The nasal horns of chasmosaurines appear to be separate ossifications (epinasals). The small pyramidal brow horns of subadult centrosaurines are variably modified in adult specimens with the horn frequently (but not always) undergoing a form of resorption, leaving in its place a low, pitted structure. There is no evidence for a separate ossification forming the brow horn in centrosaurines. *Pachyrhinosaurus* n. sp. undergoes one of the most bizarre ontogenetic transformations of the skull seen in any vertebrate. In juveniles, the nasal is a low, thin blade; the postorbitals have small, low horns; and the frill is thin and unadorned. As an adult, the nasals developed into a large sprawling boss of bone that covered the face; the brow horns became large, shallow, rugose pits;

and the frill thickened to support laterally extending, caudal spines and a unique midline "unicorn-like" spike(s). It has been suggested that the large nasal boss supported a keratinous horn in life.

See also the following related entries:

BEHAVIOR • CERATOPSIA • HADROSAURIDAE • PACHYCEPHALIA • SKELETAL STRUCTURES

References

Brown, B., and Schlaikjer, E. M. (1943). A study of the troodont dinosaurs with the description of a new genus and four new species. *Bull. Am. Museum Nat. History* **82,** 121–149.

Carpenter, K., and Small, B. (1993). New evidence for plate arrangement in *Stegosaurus stenops* (Dinosauria). *J. Vertebr. Paleontol.* (Suppl. to No. 3) **13,** 28A.

Carpenter, K., Kirkland, J., Miles, C., Cloward, K., and Burge, D. (1996). Evolutionary significance of new ankylosaurs (Dinosauria) from the Upper Jurassic and Lower Cretaceous, Western Interior. *J. Vertebr. Paleontol.* **16**(Suppl. to No. 3), 25A.

Coombs, W. P., Jr. (1990). Behaviour patterns of dinosaurs. In *The Dinosauria* (D. B. Weishampel, P. Dodson, and H. Osmólska, Eds.), pp. 32–42. Univ. of California Press, Berkeley.

Coombs, W. P., Jr., Weishampel, D. B., and Witmer, L. M. (1990). Basal Thyreophora. In *The Dinosauria* (D. B. Weishampel, P. Dodson, and H. Osmólska, Eds.), pp. 427–434. Univ. of California Press, Berkeley.

Dodson, P. (1975). Taxonomic implications of the relative growth in lambeosaurine hadrosaurs. *Systematic Zool.* **24,** 37–54.

Dodson, P. (1976). Quantitative aspects of relative growth and sexual dimorphism in *Protoceratops. J. Paleontol.* **50,** 929–940.

Farlow, J. O., and Dodson, P. (1975). The behavioral significance of frill and horn morphology in ceratopsian dinosaurs. *Evolution* **29,** 353–361.

Hammer, W. R., and Hickerson, W. J. (1994). A crested theropod dinosaur from Antarctica. *Science* **264,** 828–830.

Hopson, J. A. (1975). The evolution of cranial display structures in hadrosaurian dinosaurs. *Paleobiology* **1,** 21–43.

Lee, Y.-N. (1996). A new nodosaurid ankylosaur (Dinosauria: Ornithischia) from the Paw Paw Formation (Late Albian) of Texas. *J. Vertebr. Paleontol.* **16,** 232–245.

Molnar, R. E. (1977). Analogies in the evolution of display structures in ornithopods and ungulates. *Evolutionary Theor.* **3,** 165–190.

Ostrom, J. H., and Wellnhofer, P. (1990). *Triceratops*: An example of flawed systematics. In *Dinosaur Systematics: Approaches and Perspectives* (K. Carpenter and P. J. Currie, Eds.), pp. 245–254. Cambridge Univ. Press, Cambridge, UK.

Sampson, S. D. (1995). Horns, herds, and hierarchies. *Nat. History,* **104**(6), 36–40.

Weishampel, D. B., and Horner, J. R. (1990). Hadrosauridae. In *The Dinosauria* (D. B. Weishampel, P. Dodson, and H. Osmólska, Eds.), pp. 534–561. Univ. of California Press, Berkeley.

Zhao, X.-J., and Currie, P. J. (1993). A large crested theropod from the Jurassic of Xinjiang, People's Republic of China. *Can. J. Earth Sci.* **30,** 2027–2036.

Ornithischia

KEVIN PADIAN
University of California
Berkeley, California, USA

Ornithischia was established by Seeley in lectures presented in 1887 (published in 1888); he divided Owen's Dinosauria into two groups based on the form of the pelvis, the braincase, the vertebrae, and the armor (see DINOSAURIA: DEFINITION). Since that time they have universally been regarded as a natural group. Both Sereno (1986) and Gauthier (1986) regarded them as monophyletic groups by cladistic analysis. Ornithischia and Saurischia are sister taxa, stem groups of Dinosauria; hence any member of Dinosauria must belong to one of these two groups, following its establishment by Owen (1842). *Iguanodon* and *Hylaeosaurus,* the two earliest described ornithischians, were explicitly numbered in this group by Owen.

Within Ornithischia, a number of major groups have traditionally been recognized. These include the STEGOSAURIA, ANKYLOSAURIA, CERATOPSIA, PACHYCEPHALOSAURIA, and several groups lumped within or variously separated from a group called ORNITHOPODA, including FABROSAURIDAE, HETERODONTOSAURIDAE, HYPSILOPHODONTIDAE, IGUANODONTIDAE, and HADROSAURIDAE. At times, Ceratopsia and Pachycephalosauria have also been included in Ornithopoda or considered derivatives of Ornithopoda.

Ornithischia can be defined as all Dinosauria closer to *Triceratops* than to birds (Padian and May, 1993). Sereno (1986) established a thorough cladistic diagnosis of the group based on 32 synapomorphies of the teeth, skull, and postcrania. Salient features include leaf-shaped teeth with triangular crowns and constricted roots, ossified palpebral bone, predentary bone, roughened edentulous tip of the premaxilla (and loss of at least one pair of teeth), reduction of antorbital fossa and fenestra, five or more sacral vertebrae, posteroventrally directed pubis, pendant fourth trochanter of femur, pedal digit V reduced to a splint with no phalanges, and ossified epaxial tendons (Fig. 1).

Sereno (1986) further divided the Ornithischia into major groups (Fig. 2). The basal taxon, apart from the poorly known *Pisanosaurus* from the Late Triassic of Argentina and *Technosaurus* from the Late Triassic of Texas, is *Lesothosaurus* from the Early Jurassic of South Africa, and the closely related forms *Fabrosaurus* (also from the Early Jurassic of South Africa) and *Agilosaurus* from the Middle Jurassic of Zigong, Sichuan, China), sometimes collectively placed (with other forms) in the Fabrosauridae. The monophyly of the Fabrosauridae is sometimes questioned because their unity appears to depend on plesiomorphic or questionable characters, and many forms are very poorly preserved. Sereno separated *Lesothosaurus* and other basal forms from the GENASAURIA, forms with "cheeks," comprising THYREOPHORA and CERAPODA. Thyreophora in turn includes the Stegosauria and Ankylosauria and their relatives. Cerapoda includes the EUORNITHOPODA (Ornithopoda of most authors) + MARGINOCEPHALIA (Pachycephalosauria and Ceratopsia and their relatives). Ornithopoda includes Heterodontosauridae, Hypsilophodontidae, Iguanodontidae, and Hadrosauridae, the first three of which are often not regarded as monophyletic but as grades of ornithopod evolution. Some authors hypothesize that Marginocephalia did not evolve from a common ancestor with basal ornithopods in Cerapoda but rather evolved from within Ornithopoda, perhaps somewhere within the general hypsilophodontid grade (see PHYLOGENY OF DINO-

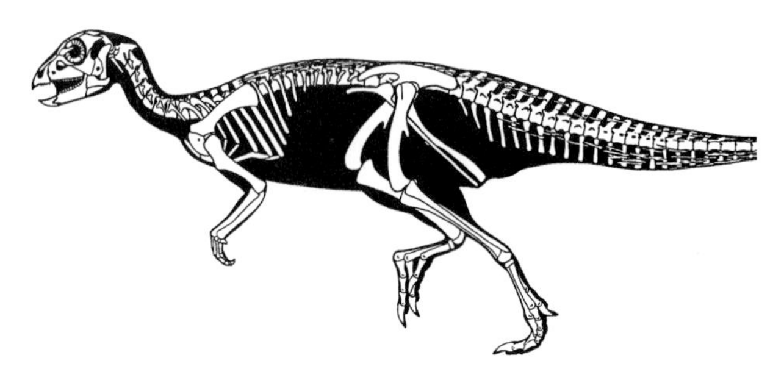

FIGURE 1 Reconstruction of *Hypsilophodon foxii* by Gregory S. Paul, showing the basic ornithischian synapomorphies described in the text.

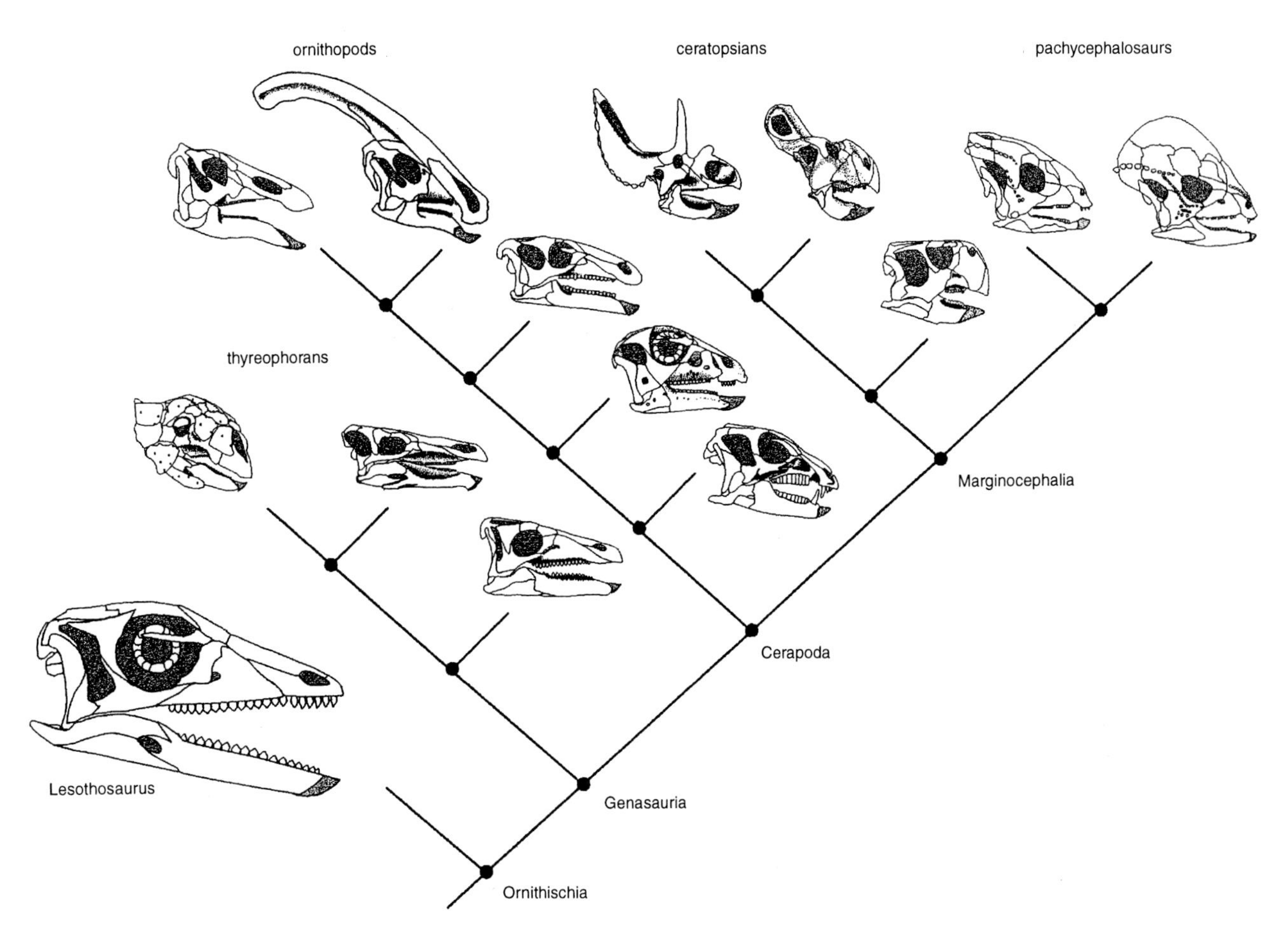

FIGURE 2 The major groups of Ornithischia, as node defined by Sereno (1986); phylogeny courtesy of T. Rowe and L. Dingus.

SAURS). This idea is based on the form of the preacetabular portion of the ilium, the prepubic prong of the pubis, the form of the teeth, and other features. Benton (1990, pp. 26–28) reviews some of the recent assessments of ornithischian interrelationships.

Given that *Pisanosaurus,* however incompletely preserved, is an ornithischian, the origin of the group can be traced back to at least as early as the late Carnian. Isolated teeth from the eastern United States and northern Africa also indicate the Carnian presence of the group; but inasmuch as both theropodan and sauropodomorphan saurischian dinosaurs are recorded in the Carnian, an ornithischian divergence would have been assured by that time. What is interesting about the ornithischian fossil record is that no well-preserved ornithischian skeletons appear until the Early Jurassic (e.g., *Lesothosaurus, Heterodontosaurus, Scutellosaurus,* and *Scelidosaurus*), leaving a stratigraphic gap of approximately 15 million years or more. Unfortunately, Late Triassic forms are so poorly preserved that we know little about the early evolution of ornithischian characters and the initial radiation of the group.

It is assumed that all members of the Ornithischia were herbivores (see DIET; TEETH AND JAWS) and that they were initially bipedal, with several groups independently evolving facultative (iguanodontids and hadrosaurs) or full (Neoceratopsia and all but the most basal Thyreophora) quadrupedal habits (see BIPEDALITY; QUADRUPEDALITY). The radiation of ornithischian groups seems to have begun by the Early Jurassic with the first appearance of basal thyreophorans and the evolution of stegosaurs (and presumably ankylosaurs) by the Middle Jurassic. Stegosaurs seem not to have survived the Early Cretaceous but ankylosaurs flourished in the Late Cretaceous; neither group, however, seems to have been very diverse, unless their typical habitats biased against their preservation in the fossil record. On the ornithopodan side, the stratigraphic record seems to show an expected gradal divergence of heterodontosaurids in the Early Jurassic, hypsilophodontids in the Late Jurassic, iguanodontids in the Early Cretaceous, and hadrosaurs in the Late Cretaceous, although each of these had broader stratigraphic ranges. Marginocephalians may be represented by *Stenopelix,* an incompletely known form not clearly assignable to either Pachycephalosauria or Ceratopsia, in the earliest Cretaceous of Europe. Pachycephalosauria are exclusively Late Cretaceous, with one report from the Early Cretaceous *(Yaverlandia),* and Ceratopsia are known from the latest Early Cretaceous *(Psittacosaurus)* through the end of the Maastrichtian. The Late Cretaceous witnessed a great radiation of hadrosaurs and neoceratopsians, which evolved strikingly similar dental batteries (see TEETH AND JAWS) evidently used in the oral processing of tough plant material, as well as success of the simpler-toothed pachycephalosaurs and ankylosaurids. The radiation of the hadrosaurs and neoceratopsians is often linked to the success of the early angiosperms (see MESOZOIC FLORAS; PLANTS AND DINOSAURS).

Ecologically, ornithischian dinosaurs, along with the herbivorous sauropodomorphs, comprised the first major guild of plant-eating tetrapods on land. There were, of course, earlier large herbivores, including the half-ton pareiasaurs (distant relatives of the turtles in Permian times) and the dicynodonts (two-tusked but otherwise toothless relatives of the mammals), ranging from small forms up to about the same size as pareiasaurs in the Permian and Triassic (see REPTILIA). However, the ornithischians took the cutting, triangular, crenelated, plant-slicing tooth form similar to that of pareiasaurs (and living iguanas) and used it as the starting point of an adaptive radiation. In all lineages (although independently several times), the front teeth were lost and the predentary bone, a neomorph that formed the lower beak, perched on the tip of the dentaries. With the progressive reduction and loss of the upper front teeth, a horny beak was an effective plant gatherer. This took the form of a sharp, often hooked, parrot-like point in ceratopsians or a broad, flared, duck-billed shape in hadrosaurs.

Meanwhile, the cheek teeth tended to become progressively separated from the front of the jaw by a wide diastema, especially in ornithopods. Stegosaurs and ankylosaurs kept the very simple row of denticled, chisel-shaped teeth, arranged *en echelon* (a dozen or so larger ones in nodosaurids, ranging up to three dozen smaller ones in ankylosaurids, whereas stegosaurs have two dozen or more, with slightly less in more derived forms), with a strong, off-center keel on the labial side of each tooth. Ornithopods tended to combine the tooth row anteroposteriorly into a more or less single grinding surface, with enamel that was stronger on the labial side of the upper teeth and on the lingual side of the lower teeth. Occlusion continually wore the teeth into beveled surfaces with sharp cutting edges, and in iguanodontids and hadrosaurs a complex dental battery of several simultaneously erupted rows of replacement teeth evolved (see TEETH AND JAWS; TOOTH WEAR). This dental battery effectively created a food processing surface of several hundred teeth per quadrant—almost a thousand in the skull at once (see HADROSAURIDAE). These teeth were replaced continually, as in all vertebrates except mammals (which produce only two sets), probably at intervals of half a year to a year, although precise data remain to be collected for all forms (see TEETH; VON EBNER GROWTH LINES). Aiding mastication was a particular type of kinesis in the maxillary region and the lower jaws, which probably guarded against strains of chewing tough or unusually hard material by absorbing shock, and also formed a principal component of the mastication cycle (see HADROSAURIDAE; IGUANODONTIDAE; Norman, 1984; Weishampel, 1984; Norman and Weishampel, 1985). A similar type of dental battery was independently evolved in the Neoceratopsidae.

If the dental specializations of some ornithischians accounted for much of their success, especially in the Cretaceous, it does not fully account for their diversity. One of the things that makes ornithischians so interesting is the panorama of spikes, plates, scutes, clubs, crests, horns, shields, and domes that are otherwise unmatched among animals—even the flashy horns and antlers of artiodactyls and other hoofed mammals of the Cenozoic Era. Some of this ornamentation, such as the scutes of thyreophorans, may have begun as a means of protection against the teeth and claws of predators. Ankylosaurs obviously took this to an extreme, evolving bands of armor in several forms, including scutes, spikes, and consolidated pelvic shields of dermal bone. Ankylosaurids also added a stiffened posterior tail with a club of fused scutes, obviously for defense. However, stegosaurs lost the lateral dermal armor and reduced their scutes to a single row along each side of the backbone (plus a shoulder spike), primitively with small pentagonal plates in the front half and conical spikes in back but progressing in some forms (the classic *Stegosaurus*) to large plates along the entire column but for two pairs of spikes at the end of the tail. Thermoregulation is only one of many functions proposed for these plates (see PHYSIOLOGY; STEGOSAURIA), but any proposed function must explain how it evolved, what was the phylogenetic context of its appearance and use, and why related forms lacked it. Why, for example, would *Stegosaurus* be particularly in need of these flashy plates for thermoregulation when they are not nearly as showy in related forms that are nearly as large? Was the thermoregulatory function, which has been reasonably supported by both histological and bioengineering studies, the original impetus for the evolution of the plates or was this an exaptive function?

The same argument might be made for the frills and horns of ceratopsians and the domes of pachycephalosaurs. Ceratopsian frills do not seem to have played an important role in jaw mechanics, as once thought, and the often hypothesized protective function accounts poorly for the wide fenestration seen almost unexceptionally in the frills (*Triceratops* is the sole counterexample). The frills are also frequently ornamented with accessory spikes, scutes, and hooks that loom large in frontal display but (with perhaps the exception of the spikes of *Styracosaurus*) would scarcely have deterred predators. Protection, defense, and aggression might more reasonably be ascribed to the nose horns of centrosaurines and the eye horns of chasmosaurines, but there is little hard evidence to determine whether these were commonly used weapons against potential predators or potential rivals, or perhaps most frequently, simply warning devices. Certainly it is difficult to account for the roughened nasal boss of the derived centrosaurine, *Pachyrhinosaurus*, in terms of aggression or defense, and it appears to be part of an unusual continuum of horn modification in this lineage (Sampson, 1995).

As for pachycephalosaurs, a function in inter- and intraspecific head-butting has been generally accepted, or at least considered reasonable, by the professional community and popular audiences for several decades. However, the picture may be more complex for the group's range. The basal forms appear to have flat skulls that are not greatly thickened. More derived forms have much thicker, kneecap-shaped domes of columnar bone, secondarily closed antorbital and temporal fenestrae, and knobs and spikes in rows behind the eyes and on the back of the skull. Still other derived forms have the cranial dome laterally compressed into a sagittal crest, with nodules and spikes protruding horizontally backward from the skull. Few of these can be considered compatible with the straight-on, high-impact head-butting seen in mountain sheep today. A kneecap is not an effective surface to avoid the torque of high-energy impacts that are even slightly off center, and a sagittal crest is inconceivable as a contact surface for this type of behavior. Flank-butting, however, is a simpler behavior that could have been accommodated by the flat skulls of basal pachycephalosaurs (which might also have engaged in head-shoving) as well as the kneecap-shaped domes of more derived forms. However, it does not seem conceivable that the sagittal crest and posterior spikes of *Stygimoloch* and other forms could have been used in any function other than display. Only histological, functional, and phylogenetic analyses of these structures and their contemporary behavioral analogs are likely to test hypotheses about this complex picture of pachycephalosaurian skull function with necessary rigor.

The patterns seen in all the ornithischians previously discussed underscore what appears to have been a basic, reiterated theme in ornithischian evolution: Display, probably as much for specific recognition and mating as for intra- and interspecific aggression and protection, was the most reasonable explanation for much of the diversity of skeletal oddities in ornithischians. Much of the species diversity in neoceratopsians and hadrosaurs is determinable almost entirely on the basis of cranial features, with postcranial structures showing far less variation and little diagnostic value to the species level. Although sexual dimorphism cannot clearly be established for every one of these ornithischian lineages (see VARIATION), it has been shown both in clearly display-oriented forms (e.g., lambeosaurines; Dodson, 1975) and in the more cryptic ceratosaurian theropod dinosaurs (e.g., Raath, 1991). Natural selection was obviously important in many aspects of ornithischian feeding and locomotory structures and functions, but sexual selection, so often an antagonist of natural selection, was equally clearly a force determining the diversity of a great many ornithischian lineages throughout the Jurassic and Cretaceous. Dinosaurs evolved at such a high rate that their turnover is virtually 100% from one geologic formation to the succeeding one (Padian and Clemens, 1985; but see Dodson, 1990). Their extinction at the end of the Cretaceous seems not to have been from an acceleration of their normal extinction rates (which were already about as high as they could be) but rather from a drop in their origination rate of new species. This drop cannot be ascribed to any kind of gradual change in behavior, inasmuch as so many of the latest Cretaceous lineages (pachycephalosaurs, ceratopsians, and hadrosaurs) seem to have built their species diversity largely on the basis of sexual selection, which was not obviously diminished in their latest forms. It is far more likely that gradual environmental changes led to their decline, as documented by their disappearances from the record in the last several million years of the Cretaceous (Archibald, 1996).

See also the following related entries:

DINOSAURIA • PHYLOGENY OF DINOSAURS • SAURISCHIA

References

Archibald, J. D. (1996). *Dinosaur Extinction and the End of an Era: What the Fossils Say*. Columbia Univ. Press, New York.

Benton, M. J. (1990). Origin and interrelationships of dinosaurs. In *The Dinosauria* (D. B. Weishampel, P. Dodson, and H. Osmólska, Eds.), pp. 11–30. Univ. of California Press, Berkeley.

Dodson, P. (1975). Taxonomic implications of relative growth in lambeosaurine hadrosaurs. *Systematic Zool.* **24,** 37–54.

Dodson, P. (1990). Counting dinosaurs: How many kinds were there? *Proc. Natl. Acad. Sci. USA* **87,** 7608–7612.

Gauthier, J. A. (1986). Saurischian monophyly and the origin of birds. *Mem. California Acad. Sci.* **8,** 1–55.

Norman, D. B. (1984). On the cranial morphology and evolution of ornithopod dinosaurs. In *The Structure, Development and Evolution of Reptiles* (M. W. J. Ferguson, Ed.), pp. 521–547. Oxford Univ. Press, Oxford.

Norman, D. B., and Weishampel, D. B. (1985). Ornithopod feeding mechanisms: Their bearing on the evolution of herbivory. *Am. Nat.* **126,** 151–164.

Owen, R. (1842). Report on British fossil reptiles, Part II. *Rep. Br. Assoc. Adv. Sci.* **1841,** 60–294.

Padian, K., and May, C. L. (1993). The earliest dinosaurs. *New Mexico Museum Nat. History Sci. Bull.* **3,** 379–381.

Raath, M. (1991). Morphological variation in small theropods and its meaning in systematics: Evidence from *Syntarsus rhodesiensis*. In *Dinosaur Systematics: Approaches and Perspectives* (K. Carpenter and P. J. Currie, Eds.), pp. 91–106. Cambridge Univ. Press, New York.

Sampson, S. (1995). Two new horned dinosaurs from the Upper Cretaceous Two Medicine Formation of Montana; with a phylogenetic analysis of the Centrosaurinae (Ornithischia: Ceratopsidae). *J. Vertebr. Paleontol.* **15,** 743–760.

Sereno, P. C. (1986). Phylogeny of the bird-hipped dinosaurs (Order Ornithischia). *Natl. Geogr. Res.* **2,** 234–256.

Weishampel, D. B. (1984). Evolution of jaw mechanisms in ornithopod dinosaurs. *Adv. Anat. Embryol. Cell Biol.* **87,** 1–110.

Ornithodira

This node group within Ornithosuchia was established by Gauthier (1986) to include *Lagosuchus*, Pterosauria, Herrerasauridae, Ornithischia, Sauropodomorpha, and Theropoda, and by implication all the descendants of their most recent common ancestor, including birds. It was diagnosed by a suite of features not found in outgroups, including Ornithosuchidae (later removed to Pseudosuchia), *Euparkeria* (later removed to outside Archosauria), and Pseudosuchia. This included the loss of the postfrontal, interclavicle, and parasagittal osteoderms; the reduction of the primitive form of the clavicle; the posteroventrally facing glenoid fossa of the pectoral girdle; the reduction of the coracoid; the reduction of manual digits IV and V; the tibia and fibula longer than the femur (reversed in large forms); the mesotarsal ankle; elongated, closely appressed metatarsals; digitigrade pes; and other features. With some reformulation of the characters and of the levels of synapomorphies, the node has been generally regarded as a strong one (see ORNITHOSUCHIA).

Ornithodirans were primitively small, lightly built, bipedal forms with long limbs, shorter trunks than those in outgroup taxa, forelimbs freed for grasping and other functions, and hindlimbs that promoted upright stance and parasagittal gait (Gauthier, 1986;

Sereno, 1986; Gatesy, 1990; see ARCHOSAURIA; BIPEDALITY; DINOSAUROMORPHA; ORIGIN OF DINOSAURS). These features are generally thought to have contributed to the success of dinosaurs, pterosaurs, and their relatives, but it is difficult to test alternative models of competitive replacement and environmental opportunism (Benton, 1983) that may have contributed to their success.

See also the following related entries:

ARCHOSAURIA • DINOSAURIA • DINOSAUROMORPHA

References

Benton, M. J. (1983). Dinosaur success in the Triassic: A noncompetitive ecological model. *Q. Rev. Biol.* **58,** 29–55.

Gatesy, S. M. (1990). Caudofemoralis musculature and the evolution of theropod locomotion. *Paleobiology* **16,** 170–186.

Gauthier, J. A. (1986). Saurischian monophyly and the origin of birds. *Mem. California Acad. Sci.* **8,** 1–55.

Sereno, P. C. (1986). Phylogeny of the bird-hipped dinosaurs (Order Ornithischia). *Natl. Geogr. Res.* **2,** 234–256.

Ornithomimosauria

HALSZKA OSMÓLSKA
Polska Akademia Nauk
Warsaw, Poland

Ornithomimosauria (Barsbold, 1976) were predominantly Cretaceous coelurosaurian dinosaurs from Eurasia and North America, although some debatable ornithomimosaur remains have been also reported from the Late Jurassic of Africa and the Late Triassic of North America. Ornithomimosaurs were lightly built cursorial theropods (Fig. 1), up to 5 m long, with long, slender hindlimbs, relatively long, three-fingered forelimbs, long neck, and small skull. Ornithomimosaurians are defined as all bullatosaurs closer to *Ornithomimus* than to *Troodon* (Holtz, 1994). Diagnostic features include the enlarged beak-like premaxilla; the secondary palate formed by the premaxilla and maxilla; reduced lower temporal fenestra; lightly constructed humerus with reduced deltopectoral crest; long hand with subequal metacarpals and digits; less trenchant unguals; metatarsus elongated, about 80% tibial length; and short, stout pedal digits (Gauthier, 1986).

The ornithomimosaur skull (Figs. 2a–2c) is distinctive due to its large eye sockets, long, shallow snout, shortened temporal region, and a bony capsule of an unknown functional meaning present at the base of braincase; jaws are toothless in all but the basal forms and were covered in life by a horny beak. The antorbital fossa is very extended, with two or three antorbital fenestrae, the posteriormost of which is the largest, although it is shorter than the eye socket. The nasal is very long and the nostrils are elongated and placed close to the tip of the snout; no horn-like, nasal projections have ever been found in any ornithomimosaur. The jugal arch is very shallow. The temporal region of the skull is short and the lower temporal fenestra is reduced in size; the quadrate is strongly oblique in advanced ornithomimosaurs, which results in a forward shift of the mandibular articulation. The upper temporal fenestra is small and the upper temporal arcade, which bounds it from the side, is straight and shifted medially. The braincase roof is convex, with posteroventrally sloping, short parietals and relatively long, posteriorly wide frontals; most of the dorsal margin of the eye socket is formed of the frontal. The brain cavity is spacious and the brain size was probably close to that of an ostrich of comparable body size. The large eyeball was protected by a ring of bony plates.

The vertebral column includes 10 cervical, 13 dorsal, 5 or 6 sacral, and up to 40 caudal vertebrae. Most of the cervicals are elongated and have very low neural spines and ribs fused to the vertebrae. The spines are higher on the dorsals and sacrals and gradually decrease in height toward the end of tail. Each dorsal, except the last one, bears a pair of free ribs. In the proximal half of the tail, vertebrae bear transverse processes; along the distal half of the tail, the vertebrae have very long articular processes that stiffened that area of the tail.

The shoulder girdle has a relatively large coracoid (Fig. 3a), usually provided with a long, pointed posteroventral process and with a conspicuous tubercle for attachment of the biceps muscle. Clavicles have never been found in any ornithomimosaur and the shoulder girdle was probably mobile. The sternum is known only in one species.

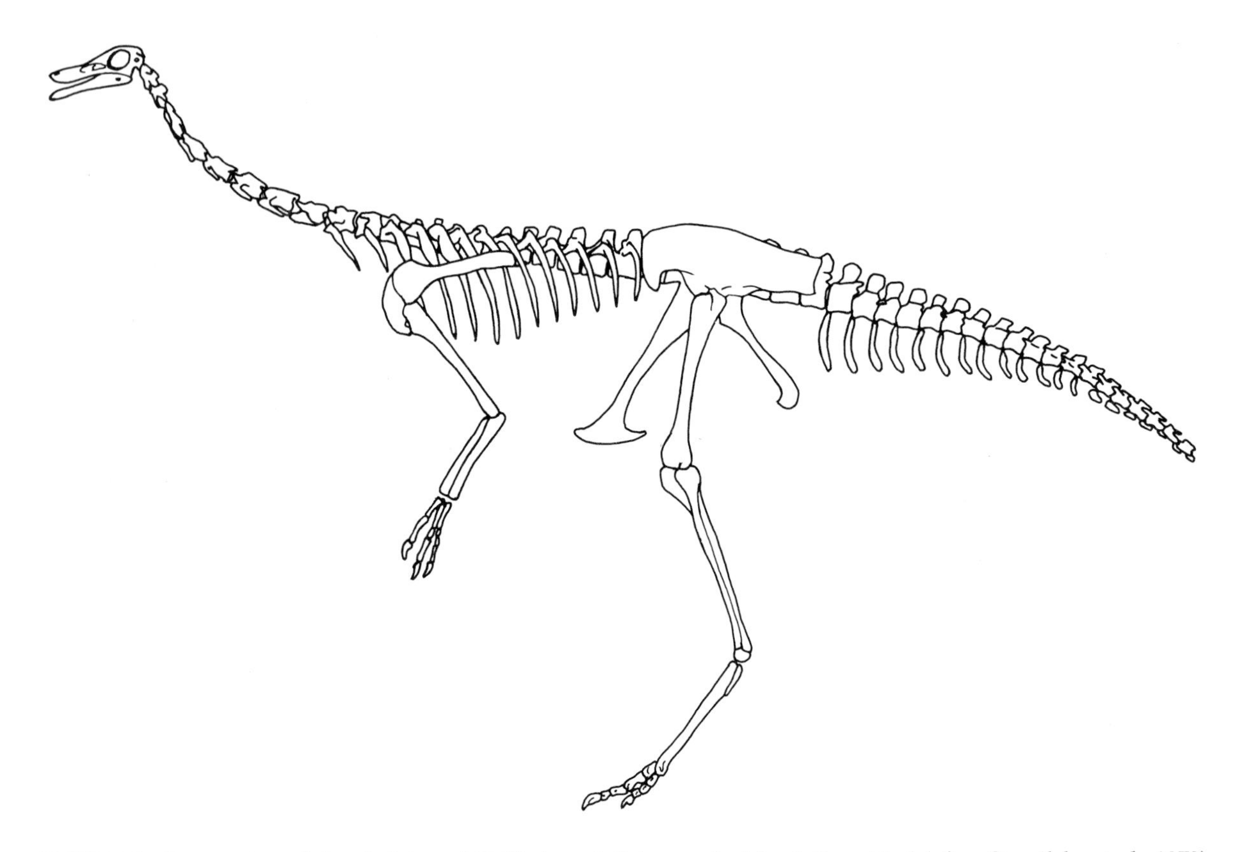

FIGURE 1 Restoration of the skeleton of *Gallimimus bullatus*; end of the tail omitted (after Osmólska *et al.*, 1972).

In all adequately known ornithomimosaurs, the forelimb is about half as long as the hindlimb. The humerus (Fig. 3b) is long, with a straight, usually twisted shaft and a slightly expanded, highly placed deltopectoral crest for attachment of the brachial musculature. The ulna and radius tightly adhere to each other distally and presumably could not move separately. The wrist bones are flat, reduced, and unsuited for any extensive mediolateral movements of the hand. Except in *Gallimimus bullatus*, the hand (Figs. 3c–3g) constitutes more than a third of the total forelimb length and has three fingers; fingers are subequal in length in all species except *Harpymimus okladnikovi*, in which the first finger is shorter. Claws of the hand are very variable in shape, from being weakly recurved to almost straight. They are usually lateromedially compressed, except in *Anserimimus planinychus* (Fig. 3h).

The pelvis (Fig. 4a) is propubic with long ilia and long, slender pubes and ischia; the distal ends of the pubes are anteroposteriorly extended, boot shaped, and fused; and the ischia are also fused distally.

The hindlimb is long and slender and the femur is usually 10–20% shorter than tibia (Figs. 4b–4f). The pes is tridactyl with relatively short toes, and the first toe is entirely lacking in all species except *Garudimimus brevipes*. The metatarsus is strongly elongated, in most species equaling about 80% of the femur length; the medial metatarsal is constricted proximally. The pedal unguals are pointed and subtriangular, with sharp edges developed as "spurs" on each side; the ventral surface of the unguals is flat and, instead of the flexor tubercle, it bears a semicircular depression.

Relationships of the Ornithomimosauria with other theropods have been difficult to establish. According to Currie (1985), the presence of the parasphenoid capsule, which also occurs in the basicranium of the Troodontidae, is evidence of the common origin of these two coelurosaurian groups. Pérez-Moreno *et al.* (1994) and Holtz (1994) supported this hypothesis (Fig. 5). Barsbold and Osmólska (1990) were reluctant to accept a close ornithomimosaur–troodontid relationship. They argued that ornithomimosaurs lack the characteristic wrist and metatarsus structures of

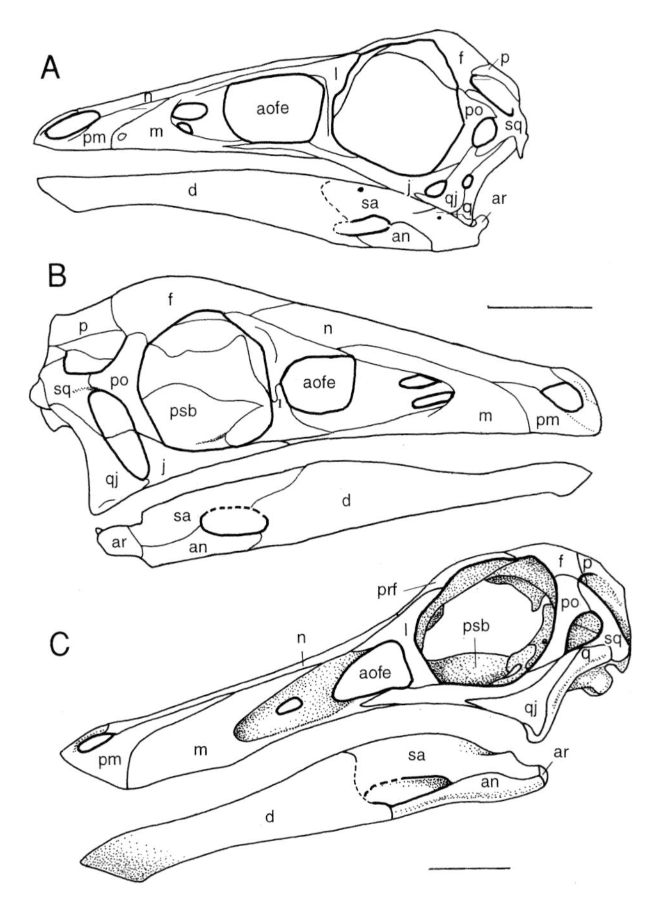

FIGURE 2 Skull and mandible (lateral view) in (A) *G. bullatus*, (B) *Garudimimus brevipes*, and (C) *Dromiceiomimus brevitertius*. A, original; B, after Osmólska *et al.* (1972); C, after Russell (1972). Scale = 5 cm. psb, parasphenoid bulbous structure.

troodontids, which make the latter group close to members of another theropod clade, Maniraptora. However, the wrist of ornithomimosaurs is clearly autapomorphic.

The oldest unquestionable ornithomimosaur species are the Lower Cretaceous (Aptian–Albian) *Harpymimus okladnikovi* Barsbold and Perle 1984 from Mongolia and *Pelecanimimus polyodon* Pérez-Moreno *et al.* 1994 (upper Hauterivian–lower Barremian) from Spain. Both species are primitively toothed forms, but their dentition is (respectively) either rudimentary or strongly specialized. The former species still displays a more primitive structure of the manus, with the first metacarpal shorter than the other two. The toothless *Shuvosaurus inexpectatus* Chatterjee 1993, reported as a possible ornithomimosaur from the Late Triassic of Texas, in some characters of its skull parallels the most advanced ornithomimosaurs. However, it differs significantly from ornithomimosaurs (e.g., in the lack of the parasphenoid capsule and the unusual anterior extent of the antorbital fossa), and until it is more extensively studied, its ornithomimosaur relationships remain unproved. Similarly, the Late Jurassic Tanzanian *Elaphrosaurus bambergi* Janensch 1920 is too poorly known to be considered an unquestionable ornithomimosaur, although it bears some ornithomimosaur-like features in its humerus and metatarsus.

The radiation of the Ornithomimosauria presumably occurred in the Late Cretaceous because they are most differentiated in Campanian–Maastrichtian strata, displaying a diversity of manus specializations reflected in variable forms of manual claws. These strata have also provided the most numerous ornithomimosaur remains.

Ornithomimosauria, as understood here, include nine genera (Fig. 5): the North American *Ornithomi-*

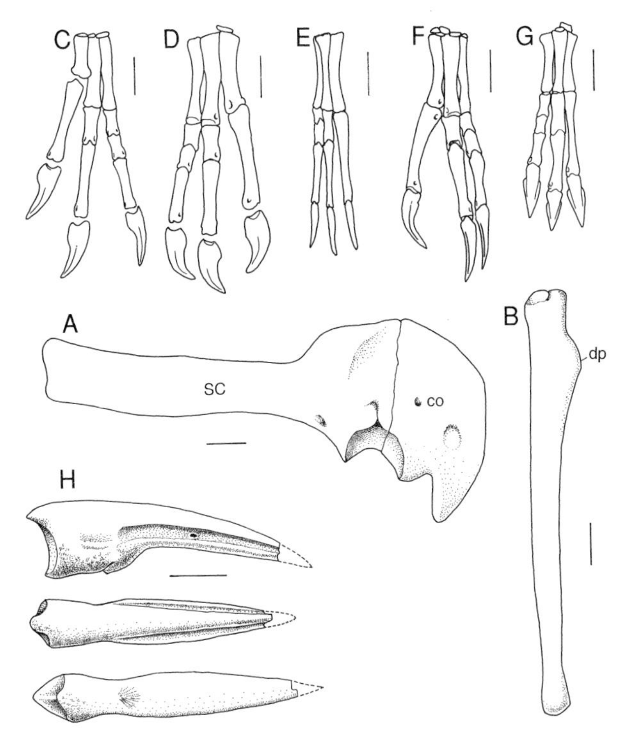

FIGURE 3 Shoulder girdle (A, lateral view) and humerus (B, lateral view) in *G. bullatus*; manus (extensor side) in (C) *Harpymimus okladnikovi*, (D) *G. bullatus*, (E) *Ornithomimus edmontonensis*, (F) *Struthiomimus altus*, (G) *Anserimimus planinychus*, and (H) manual claw in *A. planinychus* (lateral, extensor, and flexor sides). A, B, D, after Osmólska *et al.* (1972); F, after Osborn (1917). A–G, scale = 5 cm; H, scale = 2 cm.

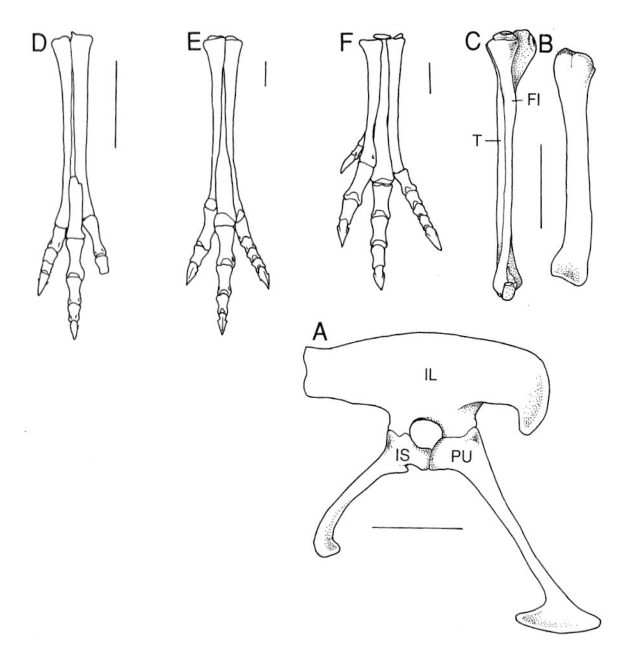

FIGURE 4 Pelvis (A, lateral view), femur (B, lateral view), crus with proximal tarsals (C, lateral view), and pedes (extensor side) in (D) *G. bullatus*, (E) *S. altus*, and (F) *G. brevipes*. A–D after Osmólska *et al.* (1972). Scale = 5 cm.

mus Marsh 1890, *Struthiomimus* Osborn 1917, and *Dromiceiomimus* Russell 1972 (all from the Late Cretaceous); the Asian *Archaeornithomimus* Russell 1972, *Gallimimus* Osmólska *et al.* 1972, *Garudimimus* Barsbold 1981, and *Anserimimus* Barsbold 1988 (all from the Late Cretaceous); the Asian *Harpymimus* Barsbold and Perle 1984 and the European *Pelecanimimus* Pérez-Moreno *et al.* 1994 are Early Cretaceous forms. These genera include 11 valid species. The gigantic,

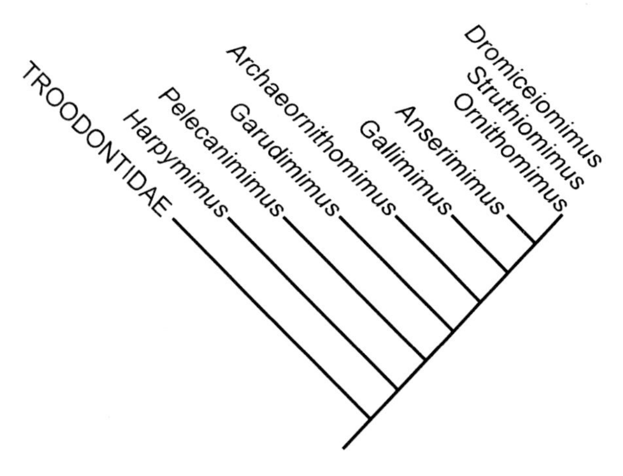

FIGURE 5 Supposed interrelationships of ornithomimosaur genera.

but very fragmentary, Late Cretaceous Asian *Deinocheirus* Osmólska and Roniewicz 1970 is also included in the Ornithomimosauria because of its subequal metacarpals and phalanges and other features.

Barsbold and Osmólska (1990) grouped the ornithomimosaur genera in three subtaxa: Ornithomimidae Marsh 1890, Harpymimidae Barsbold and Perle 1984, and Garudimimidae Barsbold 1981. To the contrary, Smith and Galton (1990) considered them to represent a single taxon: Ornithomimidae.

There is no consensus as to the feeding habits of the ornithomimosaurs. Some authors considered them carnivores, feeding on small vertebrates and invertebrates (Russell, 1972; Osmólska *et al.*, 1972; Barsbold and Osmólska, 1990), whereas others (Nicholls and Russell, 1985) argued for their herbivory. The long crus and metatarsus suggest that most ornithomimosaurs were able to make long strides and were among the fleetest theropods so far known; their speed was estimated (Thulborn, 1982) at 35–60 km / hr. Barsbold (1983) considered that some of them were moderately fast, wading animals.

See also the following related entries:

ARCTOMETATARSALIA • AVIALAE • BULLATOSAURIA • COELUROSAURIA • TROODONTIDAE

References

Barsbold, R. (1976). On the evolution and systematics of the late Mesozoic carnivorous dinosaurs. *Trans. Joint Soviet–Mongolian Paleontol. Expeditions* **3,** 68–75. [In Russian]

Barsbold, R. (1981). The toothless carnivorous dinosaurs of Mongolia. Sovm. Sov.–Mong. Paleontol. Eksped. *Trudy* **15,** 28–39 [In Russian].

Barsbold, R. (1983). The carnivorous dinosaurs from the Cretaceous of Mongolia. *Trans. Joint Soviet–Mongolian Paleontol. Expeditions* **19,** 1–117. [In Russian]

Barsbold, R. (1988). A new late Cretaceous ornithomimid of Mongolia. *Paleontol. J.* **1,** 122–125. [In Russian]

Barsbold, R., and Osmólska, H. (1990). Ornithomimosauria. In *The Dinosauria* (D. B. Weishampel, P. Dodson, and H. Osmólska, Eds.), pp. 225–244. Univ. of California Press, Berkeley.

Barsbold, R., and Perle, A. (1984). On the first new find of a primitive ornithomimosaur. *Paleontol. J.* **2,** 121–123. [In Russian]

Chatterjee, S. (1993). *Shuvosaurus,* a new theropod. *Natl. Geogr. Res. Exploration* **9**(3), 274–285.

Currie, P. J. (1985). Cranial anatomy of *Stenonychus inequalis* (Saurischia, Theropoda) and its bearing on the origin of birds. *Can. J. Earth Sci.* **22,** 1643–1658.

Gauthier, J. A. (1986). Saurischian monophyly and the origin of birds. *California Acad. Sci. Mem.* **8,** 1–55.

Holtz, T. R. (1994). The phylogenetic position of the Tyrannosauridae: Implications for theropod systematics. *J. Paleontol.* **68,** 1100–1117.

Janensch, W. (1920). Über *Elaphrosaurus bambergi* und die Megalosaurier aus den Tendaguru–Schichten Deutsch–Ostafricas. *Sitzungsberichte Gesellschaft Naturforschender Freunde Berlin* **8,** 225–235.

Nicholls, E. L., and Russell, A. P. (1985). Structure and function of the pectoral girdle and forelimb of *Struthiomimus altus* (Theropoda: Ornithomimidae). *Palaeontology* **28,** 643–677.

Osmólska, H., Roniewicz, E., and Barsbold, R. (1972). A new dinosaur, *Gallimimus bullatus* n. gen., n. sp. (Ornithomimidae) from the Upper Cretaceous of Mongolia. *Palaeontol. Polonica* **27,** 103–143.

Pérez-Moreno, B. P., Sanz, J. L., Buscalioni, A. D., Moratalla, J. J., Ortega, F., and Rasskin-Gutman, D. (1994). A unique multitoothed ornithomimosaur dinosaur from the Lower Cretaceous of Spain. *Nature* **30,** 363–367.

Russell, D. A. (1972). Ostrich dinosaurs from the Late Cretaceous of western Canada. *Can. J. Earth Sci.* **9,** 375–402.

Smith, D., and Galton, P. (1990). Osteology of *Archaeornithomimus asiaticus* (Upper Cretaceous, Iren Dabasu Formation, People's Republic of China). *J. Vertebr. Paleontol.* **10,** 255–265.

Thulborn, R. A. (1982). Speeds and gaits of dinosaurs. *Palaeogeogr. Palaeoclimatol. Palaeoecol.* **38,** 227–256.

Ornithopoda

Ornithopoda, or "bird-foot," has traditionally applied to ORNITHISCHIAN dinosaurs with three-toed (bird-like) feet; however, all ornithischians have three-toed feet and, in general, the term has referred to bipedal forms. Hence, it has not usually included Thyreophora (Stegosauria and Ankylosauria) or Ceratopsia. In current phylogenetic parlance, this concept of Ornithopoda would be paraphyletic because ornithischians (in fact, dinosaurs in general) were primitively bipedal (see BIPEDALITY). Sereno (1986) used the term EUORNITHOPODA as a node-based sister taxon to Marginocephalia within Ornithischia. Euornithopoda includes heterodontosaurs, hypsilophodonts, iguanodonts, and hadrosaurs and is supported by several synapomorphies, including the ventral offset of the premaxillary tooth row relative to the maxillary tooth row, crescent-shaped paroccipital process, jaw joint lower than occlusal plane, and exclusion of the maxillary nasal contact by the premaxilla. Most authors have continued to use the term Ornithopoda for this taxon, although in Sereno's phylogeny the term Ornithopoda was confined to the Euornithopoda excluding heterodontosaurs. Ornithopoda in the general sense excluded Marginocephalia, regarding the two taxa as sister groups; however, an alternate view (see CERAPODA; NEOCERATOPSIA; PHYLOGENY OF DINOSAURS) is that the origin of Marginocephalia is within Ornithopoda, perhaps among hypsilophodonts; if so, Cerapoda would be a problematic taxon.

See also the following related entries:

HADROSAURIDAE • HETERODONTISAURIDAE • HYPSILOPHODONTIDAE • IGUANODONTIDAE • MARGINOCEPHALIA • ORNITHISCHIA • PHYLOGENY OF DINOSAURS

Reference

Sereno, P. C. (1986). Phylogeny of the bird-hipped dinosaurs (Order Ornithischia). *Natl. Geogr. Res.* **2,** 234–256.

Ornithosuchia

The stem-based clade Ornithosuchia was established by Gauthier (1986) to comprise birds and all Archosauria closer to birds than to crocodiles. Its sister taxon was the Pseudosuchia, redefined to comprise crocodiles and all archosaurs closer to crocodiles than to birds (see ARCHOSAURIA; PSEUDOSUCHIA). As originally constituted, Ornithosuchia included *Euparkeria* (questionably), Ornithosuchidae, *Lagosuchus*, Pterosauria (including *Scleromochlus*), Herrerasauridae, Ornithischia, Sauropodomorpha, and Theropoda (including birds) (see also BIRD ORIGINS; DINOSAURIA: DEFINITION; DINOSAUROMORPHA; PHYLOGENY OF DINOSAURS; PTEROSAURIA; PTEROSAUROMORPHA).

Subsequent work determined that *Euparkeria* was outside Archosauria, as suspected; Ornithosuchidae (here including *Ornithosuchus, Riojasuchus, Venaticosuchus,* and all descendants of their common ancestor; Sereno, 1991) was found to belong to Pseudosuchia (Sereno and Arcucci, 1990); and *Lagosuchus* was divided into two genera, *Lagosuchus* and *Marasuchus* (Sereno and Arcucci, 1994). Also, this listing did not include probable members known only from incomplete remains, such as *Lagerpeton, Pseudolagosuchus,* and *Lewisuchus* (see DINOSAUROMORPHA).

As currently constituted, the node ORNITHODIRA (Gauthier, 1986) includes, in order of increasing proximity to dinosaurs, Pterosauromorpha, *Lagerpeton, Lagosuchus/Marasuchus, Pseudolagosuchus/Lewisuchus, Eoraptor, Staurikosaurus,* and *Herrerasaurus* (Holtz and Padian, 1995). Alternatively, if the latter three taxa are regarded as basal Theropoda (Sereno *et al.*, 1993), they are included within Dinosauria (Fig. 1).

Before the establishment of Ornithosuchia, the term Pseudosuchia was widely used for archosaurs that were not dinosaurs, pterosaurs, or crocodiles, but Pseudosuchia was generally considered the ancestral source of the last three. With Gauthier's redefinition of these two major archosaurian stem groups, most archosaurs not belonging to those three terminal clades fell out on the pseudosuchian side of Archosauria. Given the later work outlined previously, leading to the present composition of Ornithosuchia, it is intriguing that every one of these taxa, with the exception of Pterosauromorpha and Dinosauria, was unknown before the 1960s. That decade witnessed Reig's (1963) description of *Herrerasaurus* and the incomplete *Ischisaurus*, those of *Staurikosaurus* by Colbert (1970) and Galton (1977), and the menagerie of early Late Triassic forms such as *Lagerpeton, Lagosuchus, Lewisuchus,* and others discovered in Argentina and described by A. S. Romer in the late 1960s and early 1970s (see MUSEUM OF COMPARATIVE ZOOLOGY). The first three, plus fragmentary material earlier described by von Huene (1942) as *Spondylosoma,* were traditionally considered primitive Saurischia. Therefore, our knowledge of immediate dinosaurian forerunners and relatives has come almost entirely from the work of Romer and colleagues who have followed up his work since 1972.

With the removal of *Euparkeria* and Ornithosuchidae from Ornithosuchia, the composition of the last is the same as that of Ornithodira, at least at present, but it is possible that basal ornithosuchians that are not ornithodirans may be discovered; hence the utility of separate stem- and node-based definitions.

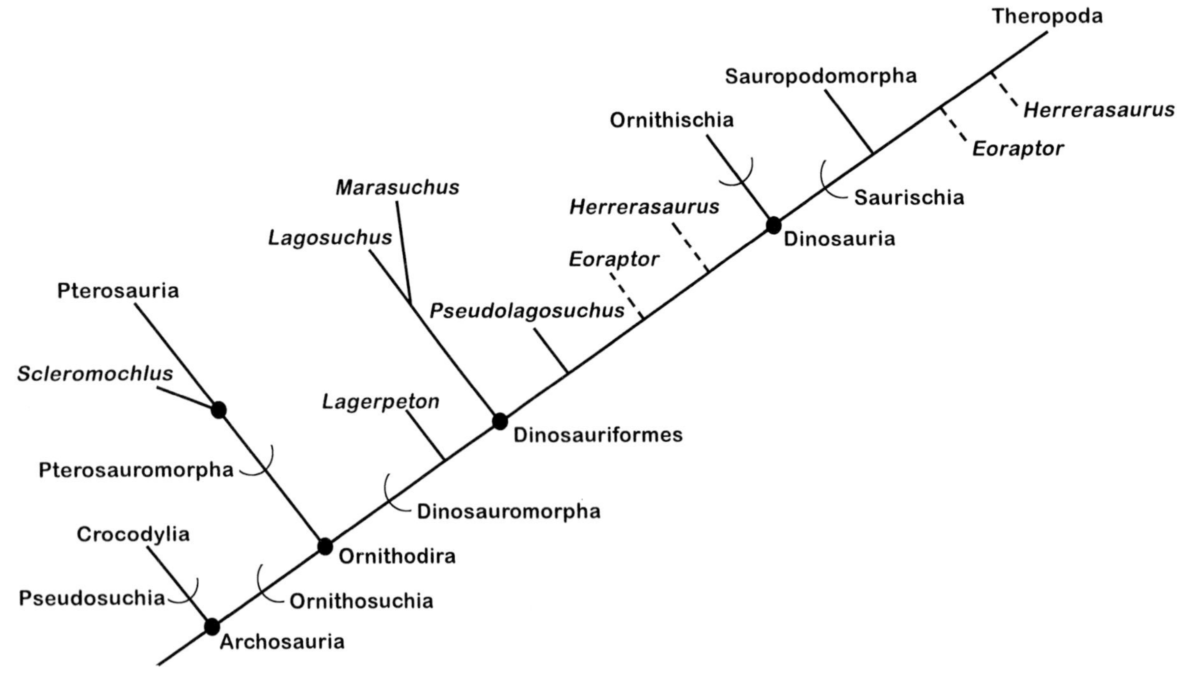

FIGURE 1 Phylogeny of the Ornithosuchia.

References

Colbert, E. H. (1970). A saurischian dinosaur from the Triassic of Brazil. *Am. Museum Novitates* **2405,** 1–39.

Galton, P. M. (1977). On *Staurikosaurus pricei*, an early saurischian dinosaur from Brazil, with notes on the Herrerasauridae and Poposauridae. *Paläontol. Z.* **51,** 234–245.

Gauthier, J. A. (1986). Saurischian monophyly and the origin of birds. *Mem. California Acad. Sci.* **8,** 1–55.

Holtz, T. R., Jr., and Padian, K. (1995). Definition and diagnosis of Theropoda and related taxa. *J. Vertebr. Paleontol.* **15**(Suppl. to No. 3), 35A.

Reig, O. A. (1963). La presencia de dinosaurios saurisquios en los "Estratos de Ischigualasto" (Mesotriasico superior) de las provencias de San Juan y La Rioja (Republico Argentina). *Ameghiniana* **3,** 3–20.

Sereno, P. C. (1991). Basal archosaurs: Phylogenetic relationships and functional implications. *J. Vertebr. Paleontol.* **11**(Suppl. 4), 1–53.

Sereno, P. C., and Arcucci, A. B. (1990). The monophyly of crurotarsal archosaurs and the origin of bird and crocodile ankle joints. *Neues Jahr. Geol. Paläontol. Abhandlungen* **180,** 21–52.

Sereno, P. C., and Arcucci, A. B. (1994). Dinosaurian precursors from the Middle Triassic of Argentina: *Marasuchus lilloensis*, gen. nov. *J. Vertebr. Paleontol.* **14,** 53–73.

Sereno, P. C., Forster, C. A., Rogers, R. R., and Monetta, A. M. (1993). Primitive dinosaur skeleton from Argentina and the early evolution of Dinosauria. *Nature* **361,** 64–66.

von Huene, F. (1942). *Die fossilen Reptilien des südamerikanischen Gondwanalandes. Ergebnisse der Sauriergrabungen in Südbrasilien 1928/29.* Beck'sche Verlagsbuchhandlung, Munich.

Osaka Museum of Natural History, Japan

see Museums and Displays

Ottawa County Historical Museum, Kansas, USA

see Museums and Displays

Oviraptorosauria

R. Barsbold
Academy of Sciences of Mongolia
Ulaan Baatar, Mongolia

The Oviraptorosauria is a Late Cretaceous group of toothless, carnivorous dinosaurs, mostly of small size, with lightly built skeletons that display most common theropod characters. The group includes two Mongolian taxa—Oviraptoridae Barsbold 1976 and Ingeniidae Barsbold 1983 and the poorly known North American Caenagnathidae (Sternberg, 1940). The Mongolian taxa include the skulls and postcrania of several species, some of which are superbly preserved. Two skeletons of caenagnathids have been found recently but have not been described yet.

The following characters are considered as diagnostic for the group: skull deep, strongly fenestrated; snout short; toothless jaws form a broad base for horny beak; mandible with medial process on articular; external mandibular fossa large; anterodorsal margin of mandible arched; and dentary with two long posterior processes (Fig. 1). A definition of the group is difficult because the systematic relationships among some incompletely known coelurosaurs remain problematic. Oviraptorosaurs appear to be weakly aligned with elmisaurids and perhaps therizinosaurs (Holtz, 1994, 1996), but more analysis is needed.

Oviraptor philoceratops was described by Osborn in 1924 and, consequently, is the best known oviraptorid. It is a 2.5-m-long animal with a deep skull, large orbits, and an unusually large, rectangular infratemporal fenestra. It has a short snout with a deep, elongate external naris that is positioned high. The tall premaxilla forms the anterior part of snout and has three dorsal processes. The medial one forms the longitudinally oriented, posterodorsally sloping and strongly pneumatized crest, whereas two lateral processes border the external naris. The crest is laterally compressed and is positioned above the snout in front of the orbits. The anterior margin of crest has two subparallel and elongate accessory openings of unknown function. The back of the crest is made up of the nasal bones and part of the frontals. In life, the crest was probably covered by a horny sheath as it

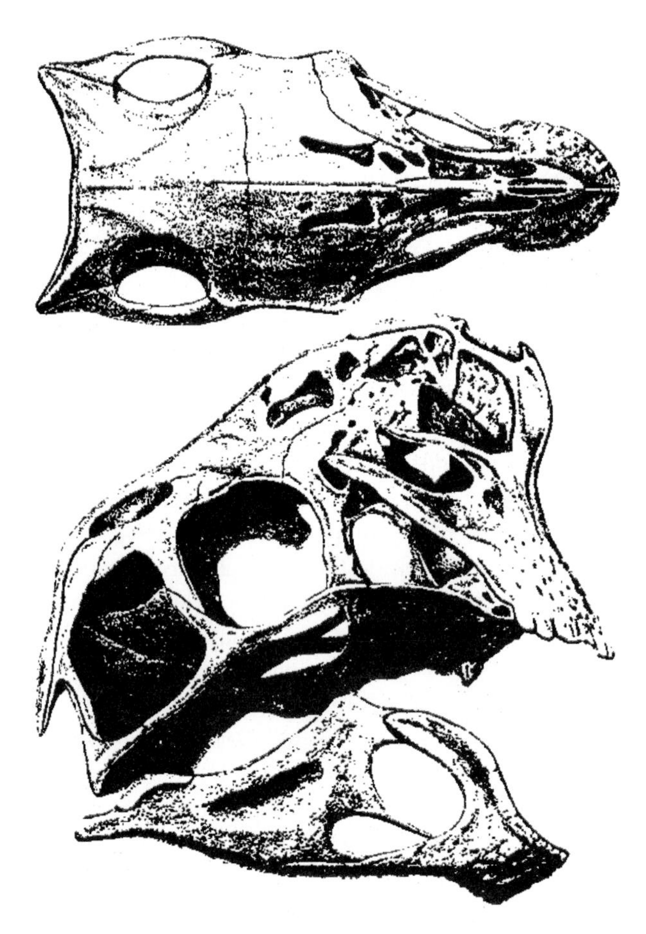

FIGURE 1 Skull of *Oviraptor philoceratops*. From G. S. Paul (1988), *Predatory Dinosaurs of the World* (Simon and Shuster) with permission.

is in cassowaries. The antorbital fossa is large. The maxilla has a ventrally convex occlusal margin. Along with the premaxilla, it forms the secondary palate, which bears a pair of parallel, convex ridges. Near the midline, the palatal portion of the maxilla has a sharp, tooth-like projection that contacts its opposite fellow. The anterior part of the fused vomers is squeezed between the projections. The palatine is reduced. The ventral surface of the pterygoid has a longitudinal trough, which is bordered laterally by an overhanging ridge that extends forward as a robust bar on the ectopterygoid and palatine. The triradiate jugal has slender processes. The large external mandibular fenestra is incompletely subdivided into two parts by a thin, anteriorly directed process of the surangular. At the back of the lower jaw, the distinctive articular is flattened, round in shape, and bears a longitudinally convex ridge that divides the articular surface into two subcircular, horizontal regions, the medial one of which is usually larger. Anteriorly, the fused mandibles form a deep, spoon-like base for the beak. The dorsal process of dentary is highly arched where it meets the surangular, and the coronoid process is prominent. There are 10 cervical, 13 dorsal, 6 sacral, and about 40 caudal vertebrae as in most theropods. However, with the exception of the distal part of the tail, all centra are pleurocoelous. Also, despite the number of vertebrae, the tail was short and would not have reached the ground. The front margin of the acromion process of the scapula is thick and flat to form a subtriangular articular surface for the clavicle. The clavicles are robust and are fused medially into a furcula. On the humerus, the deltopectoral crest is well developed. The wrist has a pulley-like semilunate carpal. There were three fingers, the first of which was supported by a short metacarpal. The claw-bearing unguals were laterally compressed and curved. Like most theropods, the pelvis is propubic. In contrast with other oviraptorids, the ilium is a relatively high bone. The pubis is posteriorly concave and has a small distal foot. The ischium has a large obturator process at midlength along the shaft. The greater trochanter of the femur is massive, and is well separated from the femoral neck, whereas the fourth trochanter is weak. The metatarsus is only moderately elongate, and the third metatarsal is not pinched as it is in caenagnathids, troodontids, and other Cretaceous theropods. There are four toes, of which the first is short and the other three are weight bearing. The foot unguals are rather compressed dorsoventrally and are weakly curved. *Oviraptor* specimens have been recovered from Shabarakh-Usu (also known as the Flaming Cliffs, Bayan-Dzag, and Bayn Dzak), Tugrigin-Shire, and Zamyn-Khond in Mongolia, and from Bayan Mandahu in Inner Mongolia.

Rinchenia mongoliensis was about the same size as *Oviraptor*, but the skull has a higher dome-like crest that incorporates the parietals in addition to the premaxillae, nasals, and frontals. The postcranium is more lightly built, and the cervical vertebrae are rather low in comparison with those of *Oviraptor*. Its remains were found at Altan-Ula.

Conchoraptor gracilis (Barsbold, 1981) was only about 1.3 m long. The skull has neither a crest nor a dome, the ilium is shallow, the second and third fingers are slender and subequal in size, and the claws were narrow and slightly curved. Specimens of *Conchoraptor* were found at Herem.

The Ingeniidae includes *Ingenia yanshini* Barsbold 1981, which was about 1.3 m long. The skull lacks a crest and the skeleton is rather robust. In the hand, the digits are short and subequal in length, and the first finger is robust with a large, massive, and strongly curved claw. The second and third digits were successively thinner, and their claw-bearing unguals are only slightly curved. The ilium is shallow. *Ingenia* remains are found at Bugin-Tsav in Mongolia.

Oviraptorosaurs include the more poorly known Caenagnathidae, specimens of which consist mostly of lower jaws that were initially assigned to birds (Sternberg, 1940; Cracraft, 1971). Osmólska (1976) recognized similarities with *Oviraptor* and correctly referred the Caenagnathidae to the Theropoda. Mandibles of oviraptors and caenagnathids are similar in the peculiar structure of the articular, in their toothlessness, in the size of the external mandibular fenestra, and in general outline. However, an oviraptorid jaw has a relatively larger external mandibular fenestra, a more highly arched dentary, a more prominent coronoid process, and a more pronounced concave outline posterior to the beak. Postcranially, oviraptorid and caenagnathid skeletons are similar in having pleurocoels in the anterior caudal vertebrae, but differ in that the third metatarsal of caenagnathids is constricted proximally (Fig. 2).

Two species of *Caenagnathus* have been described from Alberta based on lower jaw elements. Discovery of an unprepared caenagnathid skeleton in a 60-year-old museum collection has recently revealed that postcranial skeletons previously identified as *Chirostenotes* and *Macrophalangia* are caenagnathids. An undescribed caenagnathid genus from the Maastrichtian of Alberta, Saskatchewan, and the Dakotas has a flat, spade-like termination to the lower jaw.

The only other caenagnathid currently known is *Caenagnathasia martinsoni* Currie, Godfrey, and Nessov 1993. This was a very small animal that was described on the basis of several lower jaws from Uzbekistan.

The first *Oviraptor* was found by the Central Asiatic Expeditions at Shabarakh Usu. The specimen was lying on top of a nest of fossil eggs that were initially assigned to *Protoceratops andrewsi*. The identification of the eggs as protoceratopsian was based on the fact that most specimens, including babies, recovered from the site are *Protoceratops*. The association of the theropod with the eggs has been expressed in the name of this toothless dinosaur—*Oviraptor philoceratops*—which means "predator that loves to eat the

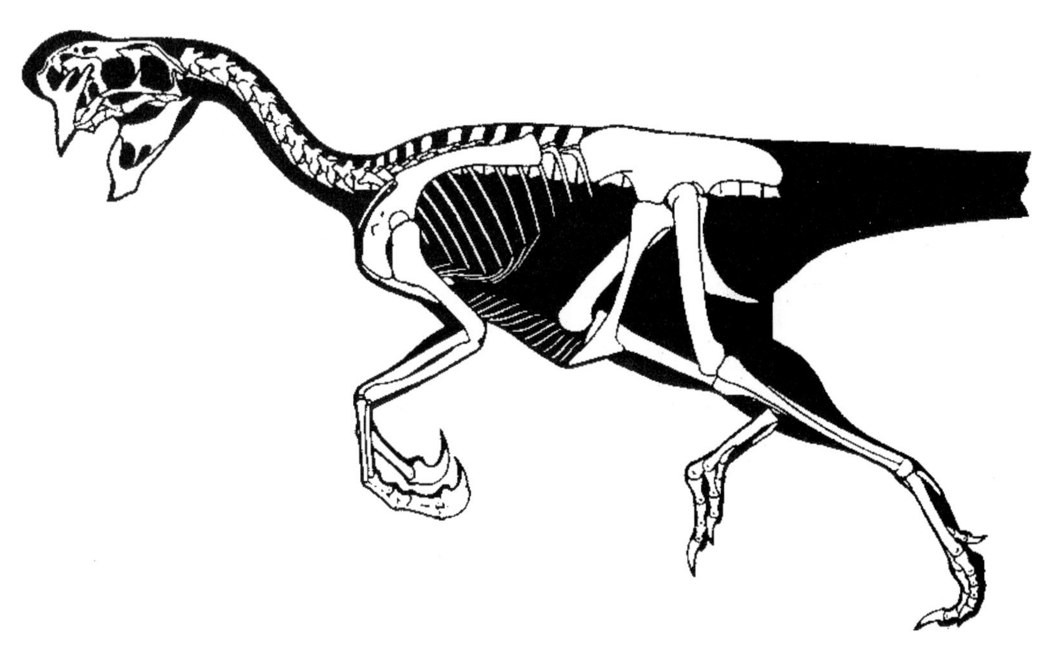

FIGURE 2 Skeleton of *Oviraptor philoceratops*. From G. S. Paul (1988), *Predatory Dinosaurs of the World* (Simon and Shuster) with permission.

eggs of horned dinosaurs." Much later, in 1994, at UKHAA-TOLGOD in the southwest of Mongolia, a skeleton of *Oviraptor* was found lying on a nest of eggs. Moreover, in 1990 a similar find was made in Inner Mongolia, northern China (Dong and Currie, 1996). Meanwhile, eggs previously identified as *Protoceratops* were reassigned on the basis of another lucky find—an embryo of *Oviraptor* (Norell *et al.*, 1994).

It has become clear that oviraptorids found lying on eggs were not feeding on them, as was inferred for a long time, but were either protecting their own nests or died in the process of laying their eggs. These unique finds provide rare information on parental care and other aspects of oviraptorosaur life. It is difficult to understand why representatives of only one restricted theropod group have been found in the fossil record on their nests. Taphonomic processes of the paleoenvironments of oviraptorosaurs played a major role in the burial and preservation of these unique samples. Dong and Currie (1996) attribute the burial of the nests to sandstorms. Indeed, bone-containing deposits at the Flaming Cliffs and Ukhaa-Tolgod in Mongolia and at BAYAN MANDAHU in China (Jerzykiewicz *et al.*, 1993) clearly show that sand dunes were widespread during some intervals of the Late Cretaceous.

A Note Added by the Editors

Oviraptorosauria was established by Barsbold (1976b) to emphasize the distinctness of his newly erected Oviraptoridae (1976a). Osmólska (1976) included the Caenagnathidae, known only from isolated mandibles of two named species of *Caenagnathus,* in the group. The systematic position of Caenagnathidae has been controversial, but recently a senior synonymy with Elmisauridae has also been suggested (Currie and Russell, 1988). Holtz (1996) recognized Oviraptoridae and Elmisauridae as sister taxa, so either position of the Caenagnathidae would reflect similar results. Holtz (1996) also linked Therizinosauridae to these two taxa. The original establishment of Oviraptorosauria is no different in content from Oviraptoridae; and because *Caenagnathus* is now recognized as a junior synonym of *Chirostenotes*, the expansion of Oviraptoridae to Oviraptorosauria provides little information. Russell and Dong (1993) designated Oviraptorosauria to comprise Oviraptoridae, Elmisauridae, and Therizinosauroidea. In the phylogenetic system, Oviraptorosauria may be defined to include Oviraptoridae and all taxa closer to *Oviraptor* than to birds. Currently, this group may include Oviraptoridae, Caenagnathidae (= Elmisauridae), and Therizinosauridae.

See also the following related entries:

CANADIAN DINOSAURS • CENTRAL ASIATIC EXPEDITIONS • ELMISAURIDAE • GROWTH AND EMBRYOLOGY • MONGOLIAN DINOSAURS

References

Barsbold, R. (1976a). On a new Late Cretaceous family of small theropods (Oviraptoridae fam. n.) of Mongolia. *Doklady Akad. Nauk. SSSR* **226,** 685–688. [In Russian]

Barsbold, R. (1976b). On the evolution and systematics of the late Mesozoic dinosaurs. *Sovm. Soviet–Mongolian Paleontol. Eksped. Trudy* **3,** 76–92. [In Russian with English summary]

Barsbold, R. (1981). Toothless carnivorous dinosaurs of Mongolia. *Trans. Joint Soviet–Mongolian Paleontol. Expedition* **15,** 28–39. [In Russian]

Barsbold, R. (1983). Carnivorous dinosaurs from the Cretaceous of Mongolia. *Trans. Joint Soviet–Mongolian Paleontol. Expedition* **19,** 5–120. [In Russian]

Cracraft, J. (1971). Caenagnathiformes: Cretaceous birds convergent in jaw mechanism to dicynodont reptiles. *J. Paleontol.* **45,** 805–809.

Currie, P. J., and Russell, D. A. (1988). Osteology and relationships of *Chirostenotes pergracilis* (Saurischia, Theropoda) from the Judith River (Oldman Formation) of Alberta, Canada. *Can. J. Earth Sci.* **25,** 972–986.

Currie, P. J., Godfrey, S. J., and Nessov, L. (1993). New caenagnathid (Dinosauria: Theropoda) specimens from the Upper Cretaceous of North America and Asia. *Can. J. Earth Sci.* **30,** 2255–2272.

Dong, Z. M., and Currie, P. J. (1996). On the discovery of an oviraptorid skeleton on a nest of eggs at Bayan Mandahu, Inner Mongolia, People's Republic of China. *Can. J. Earth Sci.* **33.**

Holtz, T. R., Jr. (1994). The phylogenetic position of the Tyrannosauridae: Implications for theropod systematics. *J. Paleontol.* **68,** 1100–1117.

Holtz, T. R., Jr. (1996). Phylogenetic taxonomy of the Coelurosauria (Dinosauria: Theropoda). *J. Paleontol.* **70,** 536–538.

Jerzykiewicz, T., Currie, P. J., Eberth, D. A., Johnston, P. A., Koster, E. H., and Zheng, J. J. (1993). Djadokhta Formation correlative strata in Chinese Inner Mongolia: An overview of the stratigraphy, sedimentary geology and paleontology and comparisons with the type locality in the pre-Altai Gobi. *Can. J. Earth Sci.* **30,** 2180–2195.

Norell, M. A., Clark, J. M., Dashzeveg, D., Barsbold, R., Chiappe, L. M., Davidson, A. R., McKenna, M. C., Perle, A., and Novacek, M. J. (1994). A theropod dinosaur embryo and the affinities of the Flaming Cliffs dinosaur eggs. *Science* **256,** 779–782.

Osborn, H. F. (1924). Three new Theropoda, *Protoceratops* zone, central Mongolia. *Am. Museum Novitates* **144,** 1–12.

Osmólska, H. (1976). New light on the skull anatomy and systematic position of *Oviraptor. Nature (London)* **262,** 683–684.

Russell, D. A., and Dong, Z. (1993). The affinities of a new theropod from the Alxa Desert, Inner Mongolia, People's Republic of China. *Can. J. Earth Sci.* **30,** 2107–2127.

Sternberg, R. M. (1940). A toothless bird from the Cretaceous of Alberta. *J. Paleontol.* **14,** 81–85.

Oxford Clay

Joshua B. Smith
University of Pennsylvania
Philadelphia, Pennsylvania, USA

The Oxford Clay, or Oxford Clay Formation (Cox *et al.*, 1992), one of the classic formations in British stratigraphy and paleontology, has been studied as a lithologic entity since the early 19th century (Buckland as cited in Phillips, 1818). This highly fossiliferous marine deposit, which crops out extensively in central England (Fig. 1), was named on the basis of excellent exposures near the historic city of Oxford, in Oxfordshire, England. The unit is well known for its abundant ammonites, cephalopods, bivalves, gastropods, and marine reptiles, especially ichthyosaurs (Duff, 1975; McGowan, 1978).

The Oxford Clay Formation is composed primarily of argillaceous marine mudstones of Callovian (163–169 Ma) and Oxfordian (156–163 Ma) ages. It is generally subdivided into three units, which had traditionally been termed the lower, middle, and upper Oxford Clay. These have recently been formally renamed by Cox *et al.* (1992) as the Peterborough, Stewartby, and Weymouth Members.

The Callovian-aged Peterborough Member consists of a brownish gray, organic-rich mudstone dominated by aragonitic ammonites and bivalves with basal silty *Gryphaea*-rich beds (Duff, 1975; Cox *et al.*, 1992). Its lithology is interpreted by Duff (1975) as representing a marine transgression. This rise in sea level during the Callovian represents a change from the lagoon- and estuarian-dominated environments common in England during the Bathonian (169–176 Ma).

The Stewartby Member (Callovian) is dominated by pale to medium gray, silty, calcareous, blocky mudstones and *Gryphaea*-rich calcareous siltstones (Cox *et al.*, 1992). Less fossiliferous than the Peterborough Member, the Stewartby Member is interpreted as a deeper-water, low-energy marine environment with a low sedimentation rate (Hudson and Palframan, 1969).

The Oxfordian Weymouth Member is composed of blocky, calcareous mudstones and occasional dark gray carbonaceous layers and siltstones (Cox *et al.*, 1992). Fossils are more common in the Weymouth than in the Stewartby Member; the assemblage is dominated by *Gryphaea* and ammonite faunas (Wright, 1986; Cox *et al.*, 1992). Wright (1986) concluded that the lithology of the Weymouth Member indicates moderately deep, calm marine conditions and reflects the maximum extent of the regional transgression that began during the Callovian.

Though the Oxford Clay Formation is dominated by marine faunas, including its famed specimens of ichthyosaurs and ammonites, it is not devoid of dinosaur remains. They have been found in four counties in both the Peterborough (P) and Stewartby Members (S) (Fig. 1):

1. Northamptonshire (P): *Lexovisaurus durobrivensis* (including *Omosaurus vetustus* partim) (Stegosauria: Stegosauridae)
2. Dorset (P): *Cetiosaurus* sp. (Sauropoda: Cetiosauridae); *Lexovisaurus durobrivensis* (Stegosauria: Stegosauridae)
3. Cambridgeshire (S): *Cetiosauricus stewarti* (Sauropoda: Diplodocidae); Hypsilophodontid indet. (Ornithopoda); *Camptosaurus leedsi* (Ornithopoda: ?Iguanodontid indet.); *Lexovisaurus durobrivensis* (Stegosauria: Stegosauridae); *Sarcolestes leedsi* (Ankylosauria: Nodosauridae)

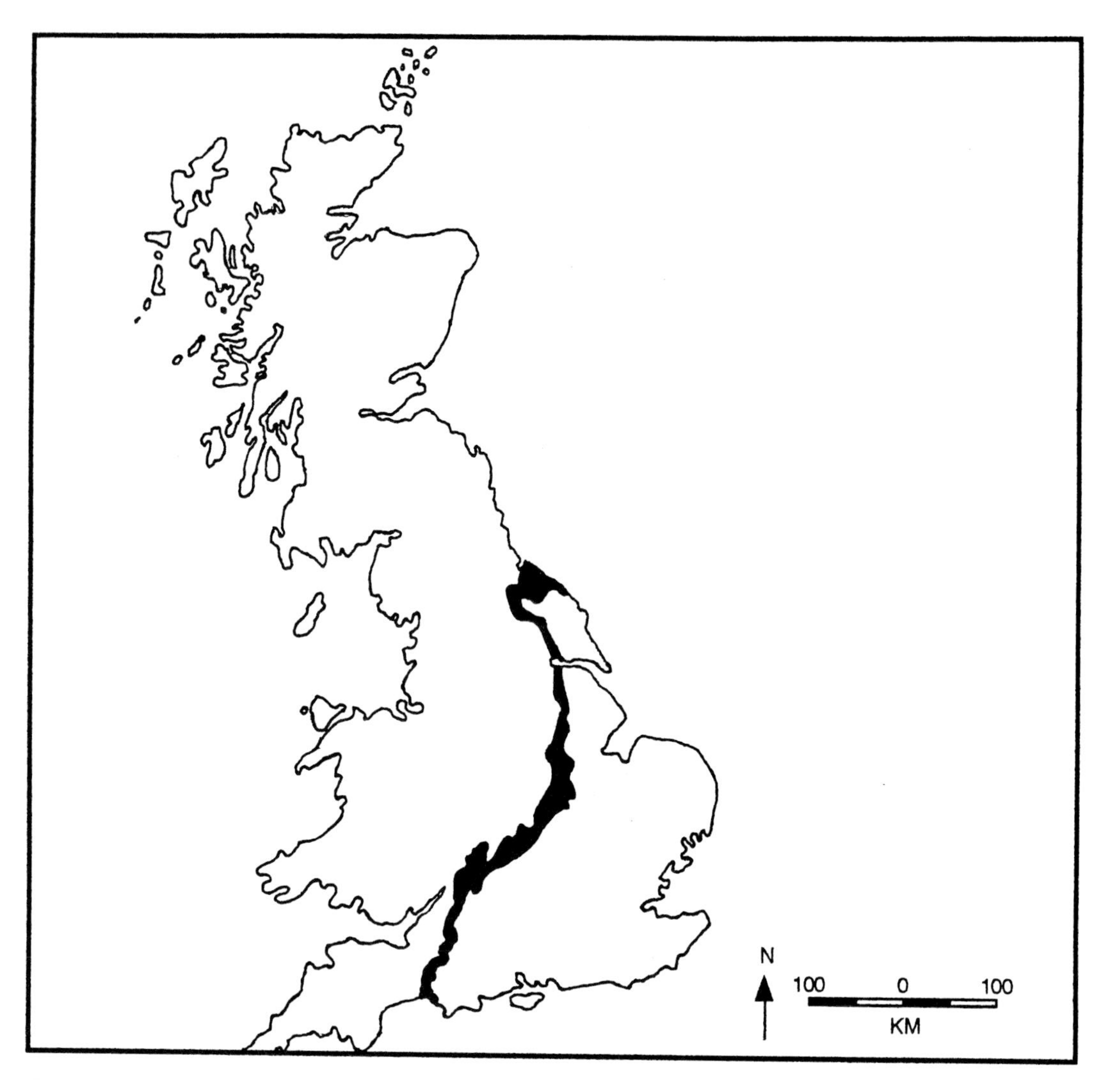

FIGURE 1 Locus map of the outcrop pattern of the Oxford Clay Formation (black) in southern Great Britain (modified from Benton and Spencer, 1995).

4. Oxfordshire (S): *Eustreptospondylus oxoniensis* (Theropoda: ?Carnosauria); *Omosaurus vetustus* (Stegosauria: Stegosaurid indent.)

See also the following related entries:

EUROPEAN DINOSAURS • HISTORY OF DINOSAUR DISCOVERIES • NATURAL HISTORY MUSEUM, LONDON

References

Benton, M. J., and Spencer, P. S. (1995). *Fossil Reptiles of Great Britain*, pp. 386. Chapman & Hall, London.

Cox, B. M., Hudson, J. D., and Martill, D. M. (1992). Lithostratigraphic nomenclature of the Oxford Clay (Jurassic). *Proc. Geol. Assoc.* **103,** 343–345.

Duff, K. L. (1975). Palaeoecology of a bituminous shale—The lower Oxford Clay of central England. *Palaeontology* **18**(3), 443–482.

Hudson, J. D., and Palframan, D. F. B. (1969). The ecology and preservation of the Oxford Clay fauna at Woodham, Buckinghamshire. *Q. J. Geol. Soc. London* **124,** 387–418.

McGowan, C. (1978). Further evidence for the wide geographical distribution of ichthyosaur taxa (Reptilia: Ichthyosauria). *J. Paleontol.* **52,** 1155–1162.

Wright, J. K. (1986). The Upper Oxford Clay at Furzy Cliff, Dorset: stratigraphy, paleoenvironment and ammonite fauna. *Proc. Geol. Assoc.* **97,** 221–228.

Pachycephalosauria

Hans-Dieter Sues
Royal Ontario Museum
Toronto, Ontario, Canada

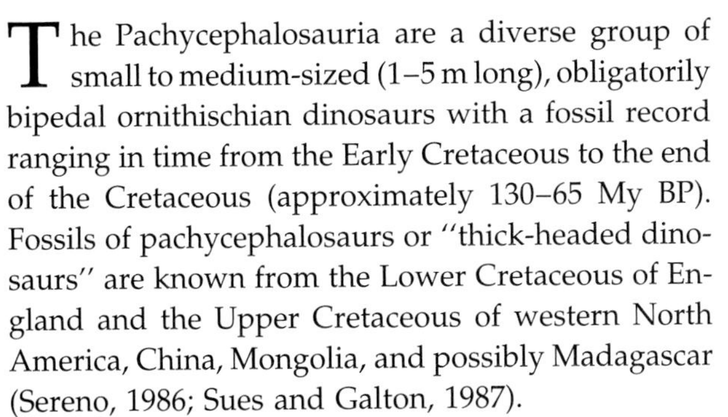

The Pachycephalosauria are a diverse group of small to medium-sized (1–5 m long), obligatorily bipedal ornithischian dinosaurs with a fossil record ranging in time from the Early Cretaceous to the end of the Cretaceous (approximately 130–65 My BP). Fossils of pachycephalosaurs or "thick-headed dinosaurs" are known from the Lower Cretaceous of England and the Upper Cretaceous of western North America, China, Mongolia, and possibly Madagascar (Sereno, 1986; Sues and Galton, 1987).

Sereno (1986) defined Pachycephalosauria, along with Ceratopsia, as the two major nodes within Marginocephalia. In *Pachycephalosaurus,* the frontal and parietal bones of the skull roof are thickened dorsoventrally. Numerous other uniquely derived characters in the skull and postcranial skeleton clearly support the monophyly of the Pachycephalosauria (Maryańska and Osmólska, 1974; Sereno, 1986; Sues and Galton, 1987). The external surfaces of the skull bones are, for the most part, strongly ornamented. The postorbital and squamosal typically bear distinct bony tubercles. The anterior and medial walls of the orbits are fully ossified. The jugal contacts the quadrate ventral to the quadratojugal. The basicranial region of the skull is foreshortened anteroposteriorly. The squamosal is broadly expanded on the deep occiput. The dorsal vertebrae have a distinctive double ridge-in-groove articulation between the zygapophyses of successive vertebrae. The sacral and proximal caudal vertebrae bear long ribs. A "basket" of multiple rows of intertwining ossified tendons surrounds the more distal portion of the tail. The slender scapula is much longer than the humerus. The forelimb is short relative to the hindlimb. The ilium has a horizontal, anteriorly broad preacetabular process and a medial flange on the postacetabular process. The small pubis is excluded from participation in the acetabulum. The pelvic girdle is unusually broad transversely compared to those of other bipedal dinosaurs.

Pachycephalosaurs retain several primitive features. The teeth of the heterodont dentition are quite small. The margins of the mediolaterally compressed, somewhat leaf-shaped crowns of the maxillary and posterior dentary teeth bear distinct denticles. The premaxilla retains teeth. Also, the ischium lacks an obturator process.

The relationships of pachycephalosaurs to other ornithischian dinosaurs remain uncertain. The most plausible hypothesis was advanced by Sereno (1986), who argued that pachycephalosaurs are most closely related to the Ceratopsia and therefore united the two groups in a clade Marginocephalia. The principal diagnostic feature for this grouping is the presence of a parietosquamosal shelf along the posterior edge of the skull roof, which overhangs the occiput. However, there are still few other diagnostic characters to support the monophyly of Marginocephalia.

The geologically oldest taxon generally referred to the Pachycephalosauria is *Yaverlandia bitholus* from the Wealden (Lower Cretaceous: Barremian) of the Isle of Wight, England (Galton, 1971). Each of the fused frontals is slightly domed and has a somewhat sculptured dorsal surface.

Based on the structure of the skull roof, most authors distinguish two groups among Late Cretaceous pachycephalosaurs—the "flat-headed" Homalocephalidae and the "dome-headed" Pachycephalosauridae. The interrelationships of pachycephalosaurs remain poorly understood because most taxa are known only from fragmentary cranial remains, and populational, ontogenetic, and sexual variation are poorly understood.

The Homalocephalidae are currently known only from the Upper Cretaceous of China and Mongolia. The skull roof is rather evenly thickened and has a flat dorsal surface, which is covered by numerous pits. The supratemporal fenestrae are relatively large. The poorly known *Wannanosaurus yansiensis* from the ?Upper Cretaceous of Anhui (China) may be the least-derived member of this group because of the large size of the supratemporal fenestrae and apparent absence of distinct bony tubercles along the posterior margin of the squamosal (Sereno, 1986). *Homalocephale calathocercos* from the Nemegt Formation (Maastrichtian) of Mongolia is the best known representative of the Homalocephalidae (Maryańska and Osmólska, 1974).

Goyocephale lattimorei from apparently slightly older Upper Cretaceous beds in Mongolia differs from *Homalocephale* mainly in the possession of fewer sacral vertebrae and in the proportionately larger size of the supratemporal fenestrae.

In the Pachycephalosauridae, the frontals and parietals are greatly thickened and fused into a single dome-like structure composed of solid bone. *Ornatotholus browni* from the Judith River Group (Campanian) of Alberta has distinctly thickened frontals and parietals, but the dome is low and divided by a shallow depression into frontal and parietal portions. The most common pachycephalosaurid in Campanian-age continental strata of western North America is *Stegoceras validum*, which retains a shelf-like structure around the frontoparietal dome (Sues and Galton, 1987). In the most derived pachycephalosaurids, such as *Pachycephalosaurus wyomingensis* from the Upper Cretaceous (Campanian–Maastrichtian) of western North America and *Prenocephale prenes* from the Nemegt Formation of Mongolia (Maryańska and Osmólska, 1974), the frontoparietal is extremely thick, and the postorbital, prefrontal, and squamosal become almost entirely integrated into the dome. As a result, the supratemporal fenestrae became completely closed in adult individuals. In one specimen of *P. wyomingensis*, the frontoparietal dome of a 64-cm-long skull reaches a maximum thickness of more than 22 cm. *Stygimoloch spinifer* from the Upper Cretaceous (Maastrichtian) Hell Creek Formation of the western United States has a proportionately deeper but transversely narrower frontoparietal dome surrounded by large knob-like bony projections. The tubercles on the posterolateral corner of the squamosal form massive horn cores.

The structure of the dentition indicates that pachycephalosaurs were herbivores. Jaw movements were

FIGURE 1 Pachycephalosaurs had domed heads and thickened skull bones, which suggests to some researchers that they engaged in the sort of head-butting behavior depicted above, though head-to-flank pushing may have been more likely. Illustration by Bob Bakker, from L. Psihoyos (1994) *Hunting Dinosaurs*, with permission.

presumably restricted to simple vertical (orthal) motion.

Based primarily on the structure of the skull, it is now generally thought that the thickened skull roof was used as a battering ram for fighting with conspecifics, much as in present-day bighorn sheep (Galton, 1971). However, the domes of the Pachycephalosauridae have a transversely convex and often smooth surface and thus seem more suitable for head-to-flank butting rather than head-to-head butting (Fig. 1). The flat-headed *Homalocephale* and its relatives may have employed their thickened but flat skull roof for head-to-head pushing, much as in present-day marine iguanas (*Amblyrhynchus*). The frontoparietal domes of *Stegoceras* and other pachycephalosaurids show considerable variation in shape and size, which has led to the creation of a plethora of taxa but presumably merely reflects ontogenetic and(or) sexual differences. Skull caps of *S. validum* show two major types in terms of the relative size and thickness of the domes; the thicker and relatively larger domes have been identified as representing males.

See also the following related entries:

Marginocephalia • Ornithopoda

References

Galton, P. M. (1971). A primitive dome-headed dinosaur (Ornithischia: Pachycephalosauridae) from the Lower Cretaceous of England and the function of the dome of pachycephalosaurids. *J. Paleontol.* **45,** 40–47.

Maryańska, T., and Osmólska, H. (1974). Pachycephalosauria, a new suborder of ornithischian dinosaurs. *Palaeontol. Polonica* **30,** 45–102.

Sereno, P. C. (1986). Phylogeny of the bird-hipped dinosaurs (Order Ornithischia). *Natl. Geogr. Res.* **2,** 234–256.

Sues, H.-D., and Galton, P. M. (1987). Anatomy and classification of the North American Pachycephalosauria (Dinosauria: Ornithischia). *Palaeontographica A* **198,** 1–40.

Palaeontological Institute, Academy of Sciences, Russia

see Museums and Displays; Orlov Museum of Paleontology

Paleobiogeography

see Biogeography

Paleobiology Institute, Academy of Sciences, Poland

see Museums and Displays

Paleoclimatology

Hartmut Haubold
Martin Luther Universität
Halle, Germany

Understanding the biology of dinosaurs in some aspects depends on and is influenced by the climatic model of the Mesozoic age. Traditionally, the Mesozoic is thought of as a period of long-term equable warm climates when climatic zonation was less differentiated than it is today (Colbert, 1964). The large body size and physiology of successful land animals such as dinosaurs are consistent with this paleoclimatic model if it is assumed that they were ectothermic giants following Cope's rule of the relationship between body volume and temperature. A different view regarding paleoclimate started with the development of an endothermic interpretation of dinosaur physiology by Ostrom (1969) and Bakker (1971). The possibility that dinosaurs were homeothermic, or even endothermic, would have given them a relatively high degree of climatic independence. Bakker argued that dinosaurs probably achieved homeothermy in a warm, arid climatic regime as an adaptive reaction to high-temperature stress during the early Mesozoic. A decade earlier, Lapparent (1960) reported the discovery of polar dinosaurs in the form of *Iguanodon*-like footprints from the Lower Cretaceous of Svalbard. This first record of dinosaurs at high latitudes (about 70° north) has been used as evidence in favor of equable warm Mesozoic climate, dinosaurian ectothermy, and dinosaurian endothermy.

Since the 1980s, the climate model of the Mesozoic

has been profoundly modified by geological data. The evidence suggests significant climatic variation during the MESOZOIC ERA, which can now be climatically subdivided into three main phases: (i) a warm Late Permian to mid-Jurassic period, (ii) a cooler mid-Jurassic to Early Cretaceous phase, and (iii) a warm time lasting from the Late Cretaceous until the early Tertiary (Frakes *et al.*, 1992). Shorter term (third-order sea level cyclicity with periods from 0.5 or 1 up to 10 million years in duration) climatic fluctuations have been documented within each of these major phases by Haq *et al.*(1987). Kemper (1987) demonstrated relatively long warm periods with high sea levels, separated by cold periods during the Cretaceous. These papers suggest the following:

1. Dinosaurs, as dominant Mesozoic land animals, were optimally adapted to changing climates. This is consistent with the view held long before Ostrom (1969) that dinosaur distribution was somewhat independent of climate, and that it is of doubtful value as simple indicators of climatic conditions.
2. The polar records of dinosaurs correlate with times of high sea level (warm periods).
3. The pattern of sea level fluctuations and correlated sedimentation rates were controlling factors of frequency of dinosaur fossilization during the Mesozoic (Haubold, 1990). Their best fossil record is in Late Jurassic, Barremian to Cenomanian, and Campanian to Maastrichtian times. During colder times, characterized by low sea levels, they have a reduced fossil record. Sea levels controlled both dinosaur distribution and our knowledge of their distribution through the fossil record. Consequently, reconstructing any kind of patterns using dinosaurs and other terrestrial vertebrates must be an oversimplified model.

Large and small species of dinosaurs lived at high latitudes during warm phases. This shows that they were physiologically capable of invading extreme regions. Distribution data and paleoclimatological models suggest that dinosaurs were highly adaptive animals that dispersed widely under any varying climatic conditions. Dispersal into high latitudes may have been facilitated by annual and seasonal migrations, although in some cases permanent polar existence was possible for some taxa (e.g., see Molnar and Wiffen, 1994, for polar occurrences in the Late Cretaceous of New Zealand).

The modified model of Mesozoic climates and our current knowledge of dinosaur distribution, diversity, and physiology show that dinosaurs at each geological stage must have developed very different kinds of climatic specializations and adaptations, just as mammals and birds have today. In itself, the climate does not appear to have limited the dispersal of dinosaurs. Land plants are more directly controlled by climatic conditions and, in turn, were probably a more direct influence on the distribution and diversity of herbivorous dinosaurs from the equator to the poles.

See also the following related entries:

PHYSIOLOGY • PLATE TECTONICS • POLAR DINOSAURS • TROPHIC GROUPS

References

Bakker, R. T. (1971). Dinosaur physiology and the origin of mammals. *Evolution* **25,** 636–658.

Colbert, E. H. (1964). Climatic zonation and terrestrial faunas. In *Problems in Palaeoclimatology* (A. E. M. Narin, Ed.), pp. 617–639. Interscience, London.

Frakes, L. A., Francis, J. E., and Syktus, J. I. (1992). *Climate Modes of the Phanerozoic,* pp. 274. Cambridge Univ. Press, Cambridge, UK.

Haq, B. U., Hardenbol, J., and Vail, P. R. (1987). Chronology of fluctuating sea levels during the Mesozoic. *Historical Biol.* **4,** 75–106.

Kemper, E. (1987). Das Klima der Kreidezeit. *Geol. Jahrbuch A* **86,** 5–185.

Lapparent, A. F. de (1960): Decouverte de traces de pas de dinosauriens dans le Cretace du Spitsberg. *Comptes Rendus 1 Acad. Sci. Paris* **251,** 1399–1400.

Molnar, R. E., and Wiffen, J. (1994). A Late Cretaceous polar dinosaur fauna from New Zealand. *Cretaceous Res.* **15,** 698–706.

Ostrom, J. H. (1969). Terrestrial vertebrates as indicators of Mesozoic climates *Proc. North Am. Paleontol. Convention* **D,** 347–376.

Paleoecology

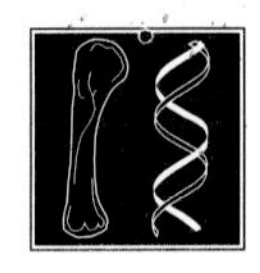

PETER DODSON
University of Pennsylvania
Philadelphia, Pennsylvania, USA

Paleoecology is, in general, the study of the environments, interactions, trophic relationships, and behavior of extinct organisms. Dinosaur paleoecology is a subject too broad to be reviewed exhaustively in this format; rather, the sources for understanding the ecology of dinosaurs, both as individual species and as members of communities, will be outlined and selected examples highlighted. In a general way, the sources are physical (sedimentary environment, climate, and paleogeography) and biological (faunal associations, abundance data, plant associations, anatomy, and physiology). Some of these matters are considered in more detail in other entries.

The fossil record of dinosaurs is biased toward lowland sedimentary environments near the sea. Probably most dinosaur fossils come from within 200 km of the sea. Isolated specimens are found in marine deposits (Alabama, Kansas, Montana, New Jersey, Alberta, Saskatchewan, England, and France) or in hypersaline lagoons (France, Spain, and Germany). In other cases, dinosaurs are associated with marine fossils, suggesting nearby shorelines (Mexico, Tanzania, and Texas). Genuine upland environments have been identified (TWO MEDICINE FORMATION and St. Mary River Formation, Late Cretaceous, Montana), as have situations far removed from marine influence (MORRISON FORMATION, Late Jurassic, western United States; Late Cretaceous, Gobi Desert, Mongolia). Burial in windblown sand has been recorded in the Late Cretaceous of China and Mongolia. Typical terrestrial environments for the preservation of dinosaurs are fluviatile sediments including channel sands, point bar deposits, channel fills, and reduced or oxidized overbank deposits.

Climates were generally much warmer and more equable in the Mesozoic than today, although the Late Triassic and Late Cretaceous were probably not quite as warm as the Jurassic and Early Cretaceous. Recently, both northern and southern polar dinosaurs have been discovered, including Early Jurassic *Cryolophosaurus* from Antarctica, Early Cretaceous *Leaellynasaura* from Australia, and Late Cretaceous *Pachyrhinosaurus* and *Edmontosaurus* from the North Slope of Alaska. It is not clear that any of them would ever have experienced ice or snow. For *Pachyrhinosaurus*, the buffering effect of the ocean and adjacent interior seaway would have ameliorated the climate, but it is likely that the interior of Alaska did receive snow. All polar dinosaurs would have undergone seasonal darkness, opening the possibility of either migration or hibernation during periods of darkness, comparative cool, and cessation of plant growth. *Leaellynasaura* was small enough to hibernate, although we have no evidence that it did so. The paleogeographic configuration apparently did not permit migration away from darkness. *Pachyrhinosaurus* and *Edmontosaurus* were too large for hibernation to have been likely, and seasonal long-distance migration down the western shore of the interior seaway was likely. *Pachyrhinosaurus* and *Edmontosaurus* fossils are known from both ends of the hypothesized migration track, although this does not confirm that they actually did migrate.

Habitats in which dinosaurs are preserved typically show evidence of a mesic, humid climate. Channel deposits, chemically reduced sediments, and evidence of plants indicate this. In the Late Jurassic Morrison Formation, general oxidation of sediments indicated by bright red and purple colors, scarcity of channel deposits, and rarity of plants together suggest at least a seasonally dry climate. The local abundance of very large-bodied herbivores indicates that plants and rainfall were at least periodically sufficient to support large herbivores. In the Late Cretaceous of central Asia, indications of general dryness and aridity are widespread. These include red colors, dune sands, and alkali lake deposits. The body sizes of dinosaurs from the DJADOKHTA and BARUN GOYOT formations are small. The dinosaur fauna includes protoceratopsids, small ankylosaurids, pachycephalosaurids, and abundant small theropods. By contrast, the NEMEGT FORMATION, the youngest of the Mongolian Late Cretaceous beds, consists of pale sediments, fluviatile sands, plant remains, and large-bodied dinosaurs including hadrosaurs, ankylosaurids, sauropods, and the predator *Tarbosaurus*. These observations strongly suggest environmental control of faunal distribution. Lush conditions favor large body size. Small body size is also observed in island faunas. For example, the hadrosaur *Telmatosaurus*, the nodosaur *Struthiosaurus*, and the titanosaur *Magyarosaurus* from the Late Cretaceous "Hateg Island" (Romania) in the Tethys Sea of Europe are among the smallest members of their respective families.

In North America, faunal transects may be made from lowland formations such as the Judith River Formation of Montana and Alberta to upland deposits such as the Two Medicine Formation. Generally, there is higher diversity in the lowlands, as in the Judith River of Alberta, than in the uplands (Two Medicine Formation). It is also striking that although the separation of the upland from the lowland faunas is only a matter of a few hundred kilometers, there is almost no mixing of faunas. In the upland Two Medicine Formation, caliche deposits indicate seasonal dryness; drought is interpreted as a major cause of death for several major bone beds here.

In the best studied situations—that is, those with large numbers of specimens collected over a period of years—one can obtain a consistent idea of community diversity of dinosaurs. The Late Jurassic Morrison Formation provides such an example. This formation is widespread across the western United States. Dinosaurs are generally rare, but from New Mexico and Oklahoma to Montana and South Dakota they are locally abundant in bone beds of up to 10,000 elements, sometimes including reasonably complete specimens. Three sauropods (*Camarasaurus*, *Apatosaurus*, and *Diplodocus*), *Stegosaurus*, and the large theropod predator *Allosaurus* are nearly ubiquitous. The medium-sized ornithopod *Camptosaurus* is common, and the smaller ornithopods, *Dryosaurus* and *Othnielia*, apparently less so. Two other sauropods are locally common, and several others are very rare, as are all nonallosaurid theropods. Morrison dinosaur diversity is moderate. There is a strikingly unbalanced Morrison fauna at the CLEVELAND–LLOYD QUARRY in Utah, in which 80% of the fauna consists of specimens of *Allosaurus* of small to large size. This deposit is usually interpreted as a predator trap.

In the well-studied Early Cretaceous CLOVERLY FORMATION of Wyoming and Montana, diversity is very low. There is only a single common dinosaur, the ornithopod *Tenontosaurus*. There is a ubiquitous but much less common small predator, *Deinonychus*. Uncommon is the nodosaurid *Sauropelta*. Rare elements include a tiny theropod and a sauropod.

In the Late Cretaceous Judith River of Alberta, common faunal elements based on the repeated occurrence of isolated skeletons include the lambeosaurines *Corythosaurus* and *Lambeosaurus*, the hadrosaurines *Prosaurolophus* and *Gryposaurus*; the centrosaurine *Centrosaurus*, the chasmosaurine *Chasmosaurus*, the ankylosaurid *Euoplocephalus*, the nodosaurid *Panoplosaurus*, and the theropods *Albertosaurus* and *Daspletosaurus*. Other faunal members are less common, but for some of these, complementary evidence from bone beds reinforces their importance. These include the hadrosaurine *Brachylophosaurus*, the centrosaurines *Styracosaurus* and *Monoclonius*, the ornithomimids *Ornithomimus*, *Struthiomimus*, and *Dromeiciomimus*, and the pachycephalosaur *Stegoceras*. Still other faunal elements are extremely rare, apart from tooth specimens: the lambeosaurine *Parasaurolophus* and a variety of small theropods including *Dromaeosaurus*, *Caenagnathus*, *Elmisaurus*, *Chirostenotes*, *Saurornitholestes*, and *Troodon*. The JUDITH RIVER GROUP at DINOSAUR PROVINCIAL PARK is an example of a very high diversity dinosaur assemblage; no known community anywhere demonstrates a comparable diversity. Outcrop area in Alberta is much smaller than outcrop area of the HELL CREEK FORMATION of Montana, the Morrison Formation of the western United States, or the Late Cretaceous formations of Mongolia. An anomaly of dinosaur faunas from Mongolia is the diversity and comparative abundance of small and large theropods, whose diversity exceeds that of herbivorous dinosaurs at all stratigraphic levels sampled.

Mass accumulations of single species (monotypic bone beds) suggest group behavior, particularly when a wide size range of specimens is included. Among theropods, there is a mass accumulation of Late Triassic *Coelophysis* at GHOST RANCH, New Mexico, where specimens of all sizes are preserved in articulation. There are bone beds with many specimens of *Syntarsus*, both *S. rhodesiensis* from Zimbabwe and South Africa and *S. kayentakatae* from Arizona. Both *Coelophysis* and *Syntarsus* show gracile and robust forms that are taken to represent males and females. Cranial crests and ornaments first appear in

Dilophosaurus (Early Jurassic, Arizona), and are also seen in *Cryolophosaurus* (Early Jurassic, Antarctica) and *Monolophosaurus* (Middle Jurassic, China). These cranial ornaments are presumed sexual display organs, but none of the crested theropods have sample sizes that allow determination of dimorphic distribution of these features. Cranial display organs make sense in social, gregarious animals, where other members of the species are available to receive and interpret messages of sexual status. In the Jurassic, monotypic concentrations of dinosaurs are rare; mass accumulations in North America, Africa, and Asia often represent attritional, polytypic assemblages. In the Late Cretaceous, single-species bone beds again become important. Although monotypic bone beds involving hadrosaurines (*Maiasaura* and *Edmontosaurus*) are known, single-species bone beds of horned dinosaurs are seemingly much more common. Lambeosaurine hadrosaurs had sexually dimorphic cranial crests, but rarely contribute to monotypic bone beds. The elaborate patterns of horns and frills in horned dinosaurs strongly suggest gregarious habits. Horned dinosaurs present interesting contrasts among themselves. Certain species, including *Psittacosaurus, Protoceratops,* and *Triceratops,* are among the most abundant of all dinosaurs, but none of these dinosaurs is prone to occur in monotypic bone beds. They may have had dense but loosely organized populations without having moved in herds. By contrast, monotypic bone beds of other species are well known, especially *Centrosaurus, Styracosaurus, Pachyrhinosaurus, Chasmosaurus, Einiosaurus,* and *Achelousaurus.* These ceratopsids, most of which are centrosaurine or short-frilled ceratopsids, may well have moved in single-species herds that included young animals.

In the Judith River Group of Alberta, where both hadrosaurs and ceratopsids are abundant, the former are preserved as complete skeletons more frequently than the latter, which are preserved as isolated skulls or partial skeletons. This may indicate partial ecological segregation, including somewhat more thoroughly terrestrial habits in ceratopsids. Nonetheless, the ceratopsid *Anchiceratops* is found both in brackish water deltaic deposits and in freshwater fluviatile deposits of the Horseshoe Canyon Formation of Alberta, and *Chasmosaurus* is found in interdistributary swamp deposits in Texas; therefore, it is difficult to generalize on the preferences of ceratopsids. By contrast, protoceratopsids, which are rare in North America and common in Asia, are always found in well-drained upland deposits.

An important dimension of the assessment of dinosaur paleoecology stems from the anatomy of the animals themselves. For example, the recognition that the anatomical construction of sauropods resembles that of terrestrial, elephantine animals rather than aquatic, hippopotamus-like animals led to the successful recasting of sauropods as adept, terrestrial animals. Indeed, all dinosaurs are fundamentally long-legged terrestrial animals, although relative leg length and locomotor performance vary. The posture of large theropods has been rearranged to cantilever over the hips, with the vertebral column rendered horizontal, the head brought down to a lower (and more menacing) level, and the tail elevated. Similarly, ornithopods, which have a shallower articular arc on the acetabular surface of the ilium than do theropods, have been repositioned downward, and slow-moving iguanodontids are now regarded as secondary quadrupeds. Arguments about locomotor performance in ceratopsids center on controversies about forelimb anatomy and posture. Interpretations based on apparent lack of mobility of the scapulocoracoid and morphology of shoulder and elbow joint surfaces is allegedly contradicted by trackways suggesting narrow carriage of quadrupedal herbivores.

Little specific information is available about food sources of dinosaurs. Stomach contents are limited to a few cases (*Coelophysis* with ?cannibalized young, *Compsognathus* with an ingested lizard, and *Edmontosaurus* with conifer needles). Among predators, there is a general relationship between predator size and prey size. Small theropods probably ate nondinosaurian prey, including mammals, nondinosaurian reptiles, amphibians, and even insects. Toothless types such as oviraptorids and ornithomimosaurs give rise to particularly fertile speculation that includes molluscs, insects, grubs, fruit, and even fibrous plant material. In the Jurassic and in the Cretaceous of Gondwana, sauropods often outweighed the largest predators by a factor of 10. Because lions pose no threats to elephants, rhinoceros, or hippopotamus today, it is hard to understand how large, healthy sauropods could have been threatened by contemporary predators. Rather, ornithopods, young or weakened sauropods, and carrion seem more likely food sources. Very large size in sauropods might be viewed as a defensive strategy against predation. It is striking that in the Cretaceous of North America, where sauropods were largely absent, the large predators were on a par with their potential prey. *Tyrannosaurus* is one of the largest predators of all time. However, even its abilities were limited, and encounters with healthy, alert *Triceratops* or *Torosaurus* must have been of uncertain outcome and chancy at best. Some argue that *Tyrannosaurus* was actually a scavenger; it almost certainly included carrion, when available, in its diet (see DIET; TEETH AND JAWS).

The recognition that hadrosaurs were not weak-toothed, as was mistakenly believed in the 19th century, but rather possessed strong teeth in dental batteries and an unusual mechanism of jaw kinesis led to a renewed appreciation of the sophistication of these animals as high-grade herbivores. The role of stomach stones in digestion has long been controversial. Ornithischians with dental batteries (hadrosaurs and ceratopsids) certainly lacked them, but they have been reliably reported in the basal ceratopsian *Psittacosaurus* and in the sauropodomorphs *Massospondylus, Vulcanodon,* and *Seismosaurus*. They could conceivably have been present in all sauropods, but there are few documented associations. In the best documented example, that of *Seismosaurus*, the stomach stones are great in number (ca. 250) but small in size (modal size about 5 cm) with a total volume of only a few liters. A few generalizations are available, one of which is that for most of the Mesozoic, flowering plants did not constitute an important food source, and it is speculated that energy flow through Jurassic herbivorous communities dependent on ferns, cycads, and conifers may have been much slower than it is currently. Dominant herbivores in the Late Cretaceous of Laurasia potentially fed at lower levels than did Jurassic communities, and flowering plants of weedy habits in riparian habitats conceivably were an important food source but one that was geographically limited on the landscape. The established coniferous forest and fern and sphenophyte ground cover persisted (see MESOZOIC FLORAS).

The concept of dinosaur PHYSIOLOGY significantly affects our understanding of dinosaur ecology. Without reviewing all the arguments involved, it is fair to present two models as end members of the spectrum of possibilities. One is the reptilian model, suggesting that dinosaurs were strictly bradymetabolic, ectothermic poikilotherms, and the other is that they were avian- or mammalian-grade tachymetabolic, homeothermic endotherms. There are several general points to make. Most of the Mesozoic was much warmer and more equable than today. Dinosaurs enjoyed the benefits of warm bodies and thermally elevated metabolic rates most of the time. Large body size, which is a distinctive thermoregulatory strategy on its own, conferred substantial thermal stability on dinosaurs (inertial homeothermy). The internal thermal response time of a cold-blooded sauropod to external change in environmental temperature was measured in weeks. Thus, dinosaurs more than 1–3 metric tons had warm, essentially constant body temperatures. Dinosaurs were phylogenetically diverse: There is no reason to think that all dinosaurs had the same physiology. There is also no reason to think that the reptilian model and the avian/mammalian model exhaust all the possibilities. One idea is that dinosaurs may have had high metabolic rates during early growth and then had more economical low metabolic rates as adults (heterometabolism).

The leatherback turtle, *Dermochelys*, has been advanced as a potential physiological intermediate, with a metabolic rate higher than that of typical reptiles but lower than that of typical birds or mammals. This reptile is a long-distance migrator and maintains warm body temperatures in cold arctic waters, a thermally hostile regime. The term gigantothermy has been used to describe the physiological state predicated on slightly elevated metabolic rate combined with large body size, the use of peripheral tissues as insulation, and active control of blood flow to retain or to disperse heat from the body as appropriate. Most dinosaurs were excellent candidates for gigantothermy (see PHYSIOLOGY).

In terms of biophysical ecology, physiological models suggest that small to medium-sized dinosaurs of less than 1000 kg could probably have tolerated elevated metabolic rates without danger of overheating, but very large dinosaurs such as sauropods with high metabolic rates would have been at grave risk of overheating without the use of water as a heat sink. Dinosaurs of 3000 kg, such as hadrosaurs, could have done well with intermediate metabolic rates but high metabolic rates might have posed problems if water and shade were not available. None of this is to say that any dinosaur had an avian or mammalian metabolic rate. It is conceivable that none did.

As far as is known, all dinosaurs were egg-layers, but details are known for very few. Dinosaur EGGS, once considered very rare following early discoveries in France and Mongolia, have recently been discovered in abundance in China, India, and North America. Additional discoveries have come from South America and Europe. Embryos within eggs constitute the only acceptable evidence for linking a species of dinosaur with a particular egg type. Embryos within eggs are known for *Orodromeus, Oviraptor,* and *Hypacrosaurus.* Clutches of embryos and/or neonates are known for *Maiasaura* as well. Claims of maternal care of hatchling dinosaurs are based on circumstantial evidence, although not inherently implausible.

Growth series are now known for many species of dinosaurs. Few data yet bear directly on growth rates. Evidence from bone histology suggests that the small theropod *Syntarsus* grew to an adult size of 25 kg in 8 years, whereas the 250-kg prosauropod *Massospondylus* achieved maximum size in 14 years. It is possible that elevated growth rates were achieved by ornithopods. The histologies of *Dryosaurus, Orodromeus,* and the "Proctor Lake ornithopod" all suggest rapid, uninterrupted growth. Lacking growth rings, it is difficult to assign absolute values to growth rates. Sexual dimorphism becomes evident when *Protoceratops* is about half maximum adult size and in lambeosaurine hadrosaurs that are about two-thirds maximum adult size. These figures suggest a reptilian growth pattern. It has been suggested that no matter what the growth rate and adult size, all dinosaurs may have reached reproductive age by age 20; otherwise, the survival rate of offspring would have to have been unreasonably high.

See also the following related entries:

DIET • PALEOCLIMATOLOGY • PLATE TECTONICS • TAPHONOMY • TROPHIC GROUPS

References

Behrensmeyer, A. K., Damuth, J. D., DiMichele, W. A., Potts, R., Sues, H.-D., and Wing, S. L. (Eds.) (1992). *Terrestrial Ecosystems through Time*, pp. 568. Univ. of Chicago Press, Chicago.

Carpenter, K., Hirsch, K. F., and Horner, J. R. (Eds.) (1994). *Dinosaur Eggs and Babies*, pp. 372. Cambridge Univ. Press, Cambridge, UK.

Chinsamy, A., and Dodson, P. (1995). Inside a dinosaur bone. *Am. Scientist* **83,** 174–180.

Clemens, W. A., and Nelms, L. G. (1993). Paleoecological implications of Alaskan terrestrial vertebrate fauna in latest Cretaceous times at high paleolatitudes. *Geology* **21,** 503–506.

Currie, P. J. (Ed.) (1993). Results from the Sino–Canadian Dinosaur Project. *Can. J. Earth Sci.* **30,** 1997–2272.

Dodson, P. (1975). Taxonomic implications of relative growth in lambeosaurine hadrosaurs. *Systematic Zool.* **24,** 37–54.

Dodson, P., Behrensmeyer, A. K., Bakker, R. T., and McIntosh, J. S. (1980). Taphonomy and paleoecology of the dinosaurs beds of the Upper Jurassic Morrison Formation. *Paleobiology* **6,** 208–232.

Horner, J. R. (1984). Three ecologically distinct vertebrate faunal communities from the Late Cretaceous Two Medicine Formation of Montana, with discussion of evolutionary pressures induced by interior seaway fluctuations. Montana Geological Society 1984 Field Conference, pp. 299–303.

Rich, T. H. V., and Rich, P. V. (1989). Polar dinosaurs and biotas of the Early Cretaceous of southeastern Australia. *Natl. Geogr. Res.* **5,** 15–53.

Rogers, R. R. (1990). Taphonomy of three dinosaur bone beds in the Upper Cretaceous Two Medicine Formation of northwestern Montana: Evidence for drought-related mortality. *Palaios* **5,** 394–413.

Spotila, J. R., O'Connor, M. P., Dodson, P. and Paladino, F. V. (1991). Hot and cold running dinosaurs: Body size, metabolism and migration. *Mod. Geol.* **16,** 203–227.

Varricchio, D. J. (1995).Taphonomy of Jack's birthday site, a diverse dinosaur bonebed from the Upper Cretaceous Two Medicine Formation of Montana. *Palaeogeogr. Palaeoclimatol. Palaeoecol.* **114,** 297–323.

Varricchio, D. J., and Horner, J. R. (1993). Hadrosaurid and lambeosaurid bone beds from the Upper Cretaceous Two Medicine Formation of Montana: Taphonomic and biologic implications. *Can. J. Earth Sci.* **30,** 997–1006.

Weishampel, D. B., and Norman, D. B. (1989). Vertebrate herbivory in the Mesozoic; Jaws, plants, and evolutionary metrics. *Geol. Soc. Am. Spec. Papers* **238,** 87–100.

Weishampel, D. B., Dodson, P., and Osmólska, H. (Eds.) (1990). *The Dinosauria,* pp. 733. Univ. of California Press, Berkeley.

Weishampel, D. B., Grigorescu, D., and Norman, D. B. (1991). The dinosaurs of Transylvania. *Natl. Geogr. Res.* **7,** 196–215.

Paleohistology

see HISTOLOGY OF BONES AND TEETH

Paleomagnetic Correlation

J. F. Lerbekmo
University of Alberta
Edmonton, Alberta, Canada

The earth's magnetic field is probably produced by convection currents in the liquid outer part of the earth's core, which is believed to be composed mostly of nickel and iron. The field produced is largely dipolar due to the spinning of the earth and has lines of force analogous to those that would be produced if a large bar magnet were placed at the center of the earth (Fig. 1). However, the best model of this field would have the bar magnet inclined slightly to the spin axis or geographic poles. However, because the field is likely produced by a number of convection cells, it is not steady in direction or intensity. The magnetic pole migrates continuously around the geographic pole in irregular loops, producing what is called the "secular variation." In these loops, the magnetic pole may drift up to some 20° from the geographic pole, but when averaged out over a period of a few thousand years, the mean position of the magnetic pole coincides closely with the geographic pole.

A phenomenon of the earth's magnetic field that makes it useful for time correlations is that it episodically reverses polarity. That is, the north magnetic pole moves to the position of the south magnetic pole and vice versa. This is called a geomagnetic polarity reversal, and it happens at very irregular intervals ranging from a few tens of thousands of years to a few tens of millions of years. During a reversal the main dipole component of the field apparently decreases in intensity, making the field less stable and allowing it to rebuild with the opposite polarity. The present configuration, in which the "north-seeking" end of a compass needle points toward the north magnetic pole, has lasted for the past approximately 700,000 years and is therefore called "normal" polarity. When the same end of the compass needle would point toward the south magnetic pole, the field is said to be "reversed." The time required for such a change to take place is usually on the order of a few thousand years. In terms of geologic time, this is essentially instantaneous.

Because a polarity reversal is a global phenomenon, if it can be identified in the rock column it represents a worldwide time horizon. Fortunately, a very magnetic iron oxide mineral, magnetite (and a solid solution series, titanomagnetite), occurs in small quantities in almost all rocks. It originates by crystallizing out of molten material such as that which produces lavas and volcanic ashes. As the crystallized magnetite cools through a temperature called the "Curie point" (575°C), a strong magnetization is locked in the crystal that is aligned with the geomagnetic field at that location. This magnetization will not be released unless the crystal is again heated to this temperature or subjected to a strong electric field. Therefore, a cooled lava has preserved in its magnetite the direction of the earth's field at the time of its crystallization. When these and other previously molten rocks, such as granites, weather at the earth's surface, the more chemically resistant minerals, including magnetite, are washed away as sediment. As this sediment settles out of a stream or a standing body of water, the magnetite crystals orient themselves along the lines of force of the earth's magnetic field, like small compass needles. After settling, they are trapped in that orientation by compaction from the weight of sediment deposited on top. Because the lines of force (Fig. 1) are steeply inclined in the high latitudes, the direction of magnetism exhibited by the

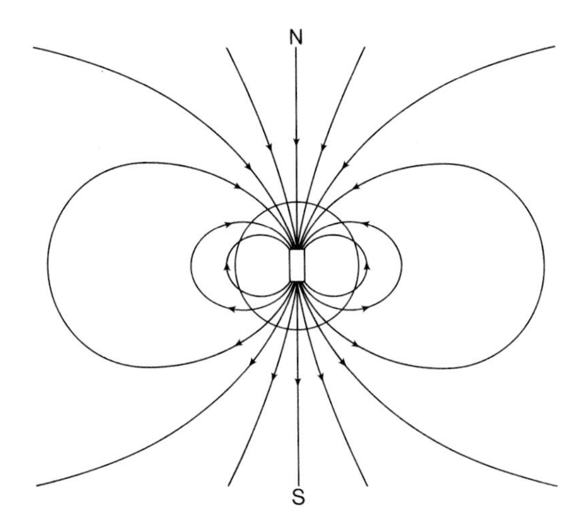

FIGURE 1 Illustration of the time-averaged major dipole component of the earth's magnetic field as modeled by a bar magnet at the earth's center. Arrows along lines of force indicate direction and inclination of a free-swinging compass needle at various latitudes for "normal" polarity. Arrows would be reversed for "reversed" polarity.

magnetite grains will also be steeply inclined at these latitudes. On the other hand, the lines of force are nearly horizontal at low latitudes. Therefore, the steepness or inclination of the magnetic direction (and whether up or down) is of most importance at high latitudes in determining whether the polarity is normal or reversed, whereas the compass direction (declination), whether north or south, is most important at low latitudes. If a sample of the compacted sediment or sedimentary rock (or lava) is later collected, and the orientation of the sample in the field is marked on it, it can be taken back to the laboratory and the direction of magnetization in the rock (as preserved by the magnetite grains) measured with a sensitive magnetometer. Thus, the approximate position of the earth's magnetic pole at the time of deposition of the rock can be determined. If a sequence of rocks is systematically sampled, any polarity changes that took place during their deposition can be located. Sampling can be done by collecting a block of rock (hand sample) and cutting it into appropriate sizes in the laboratory or by taking an oriented core, usually with a diamond drill.

Since the early 1960s, numerous exposures of strata of many ages have been studied in this way. Because a sequence of normal and reversed polarity intervals (magnetozones) of one age may look much like another of different age, the approximate age of the rocks must be known from the fossils they contain or from absolute dating (most commonly by radiometric dating of suitable minerals in a volcanic ash). In this way, it has been possible to build up a polarity timescale in which the ages of the normal and reversed periods (called magnetochrons) have been approximately determined by interpolating between a number of well-determined absolute ages. The magnetochrons have been numbered from 1 (present) backward to 34 (Early Cretaceous). For the Jurassic Period the numbering starts over, prefixed by the letter M (for example, see Harland *et al.*, 1990). Therefore, if specific magnetozones can be identified in a sequence of rocks, the polarity boundaries offer precise time correlation of any fossils present—for example, dinosaurs.

As more detailed studies are done, more subdivisions of the chrons (subchrons) are being identified, providing more polarity boundary time horizons and better time resolution. For example, currently controversy rages over the degree of synchroneity of extinctions, including the dinosaurs, at the Cretaceous–Tertiary (K–T) boundary. Alvarez *et al.* (1980) showed that the extinction of Cretaceous marine planktonic foraminifera in Italy coincided with an anomalously high concentration of the element iridium, which is very rare at the surface of the earth but relatively abundant in meteorites. They thus attributed the K–T boundary extinctions in general to the impact of a large (10- to 15-km diameter) meteorite, making them essentially instantaneous. This anomalous iridium concentration has since proven to be a global feature and has come to be accepted as the best marker of the K–T boundary. Recently, a prime candidate for a large K–T boundary impact site has been recognized under the northern coast of the Yucatan peninsula in Mexico (Hildebrand *et al.*, 1991; Sharpton *et al.*, 1993). However, with the iridium anomaly as a time marker, paleontologists studying plant and animal extinctions, both on land and in the sea, have noted that extinctions were taking place both before and after the iridium concentration (Keller, 1989; Sweet and Braman, 1992) and have therefore contested the synchroneity of the extinctions. It has been pointed out that rapid sea level and climatic changes were also going on at K–T boundary time, providing stresses that probably led to many of the extinctions. Paleontologists have also noted irregularities in the extinction levels of some organisms, suggesting time gaps in the sedimentary record in some K–T boundary sections, producing only an apparent sudden extinction (MacLeod and Keller, 1991).

As an aid to resolving the controversy over synchroneity at the K–T boundary, Lerbekmo *et al.* (1995) have identified a short normal polarity subchron extending across the boundary within chron 29r in continental sedimentary rocks in seven K–T boundary sequences in Alberta and Saskatchewan in western Canada and in Montana in the United States. This subchron appears to have lasted only from some 25,000 years before the K–T boundary (as defined by the iridium anomaly) through a similar time period after the boundary. This study is now being extended to other sections, including marine sequences that are believed to be continuous. The presence or absence of this subchron, especially in marine rocks, will help determine whether or not some seemingly sudden extinctions at the K–T boundary are only apparently sudden because of missing sedimentary record at the boundary.

See also the following related entries:

BIOSTRATIGRAPHY • GEOLOGIC TIME • PLATE TECTONICS • RADIOMETRIC DATING

References

Alvarez, L. W., Alvarez, W., Asaro, F.. and Michel, H. V. (1980). Extraterrestrial cause for the Cretaceous–Tertiary extinction. *Science* **208,** 1095–1108.

Harland, W. B., Armstrong, R. L., Cox, A. V., Craig, L. E., Smith, A. G., and Smith, D. G. (1990). *A Geologic Time Scale 1989*, pp. 263. Cambridge Univ. Press, Cambridge, UK.

Hildebrand, A. R., Penfield, G. T., Kring, D. A., Pilkington, M., Camargo-Zanoguera, A., Jacobsen, S. B., and Boynton, W. V. (1991). Chicxulub crater: A possible Cretaceous/Tertiary boundary impact crater on the Yucatan Peninsula, Mexico. *Geology* **19,** 867–871.

Keller, G. (1989). Extended period of extinctions across the Cretaceous/Tertiary boundary in planktonic foraminifera of continental-shelf sections: Implications for impact and volcanism theories. *Geol. Soc. Am. Bull.* **101,** 1408–1419.

Lerbekmo, J. F., Sweet, A. R., and Duke, M. J. M. (1995). A normal polarity subchron which embraces the K–T boundary: A measure of sedimentary continuity across the boundary and synchroneity of boundary events. *Geol. Soc. Am, Spec. Paper*, in press.

Macleod, N., and Keller, G. (1991). Hiatus distributions and mass extinctions at the Cretaceous/Tertiary boundary. *Geology* **19,** 497–501.

Sharpton, V. L., Burke, K., Camargo-Zanoguera, A., Hall, S. A., Lee, S., Marin, L. E., Suárez-Reynoso, G., Quezada-Muñeton, J. M., Spudis, P. D., and Urrutia-Fucugauchi, J. (1993). Chicxulub multiring impact basin: Size and other characteristics derived from gravity analysis. *Science* **261,** 1564-1567.

Sweet, A. R., and Braman, D. R. (1992). The K–T boundary and contiguous strata in western Canada: Interactions between paleoenvironments and palynological assemblages. *Cretaceous Res.* **13,** 31–79.

Paleoneurology

EMILY A. BUCHHOLTZ
Wellesley College
Wellesley, Massachusetts, USA

The study of the nervous system of dinosaurs is constrained by the nature of neural tissue, which is soft and readily decays after death. However, the central nervous system (brain and spinal cord) is supported and protected in life by the bony skull and vertebrae, which preserve evidence of the size and structure of neural tissues to a variable extent. This structure is very conservative among all vertebrates in general outline, and yet flexible in detail. Modes of dominant sensory input as well as the ability of an animal to process and respond to this input can be inferred from details of the structure and relative size of sense organs and corresponding integrative areas of the brain. Additionally, nerves enter and leave the skull through foramina whose pattern and location is often taxon specific and thus a source of phylogenetic data. Information concerning the nerves of the peripheral nervous system, which travel almost exclusively through soft tissue, is unavailable to the paleontologist.

The relationship of brain size to body size has been documented among living vertebrates of all classes. Brain size is negatively allometric relative to body size, following the power function $E = kPx$, where E is brain size, P is body size, and k is a constant equal to the y-intercept on a log–log plot (Jerison, 1973; Hopson, 1977). The slope (x) of the regression line defined by this function varies modestly among vertebrate classes (and also among authors), but is very close to 0.67. Y-intercepts vary by a power of 10. Birds and mammals have brains approximately 10 times as large as those of bony fish, amphibians, and reptiles of the same body size. These relationships allow prediction of expected brain size for a vertebrate of a given body size and class. Jerison (1973) defined the "Encephalization Quotient" (EQ) as the value of E_i/E_e, where E_i is the measured and E_e the expected brain size of an animal of a given body size. An EQ of 1.0 falls on the regression line and indicates a brain size equal to that predicted on the basis of body size. In general, the larger the encephalization quotient, the more plastic and less stereotypical the behavior of the animal. This behavioral flexibility is an approximation of "intelligence."

Evaluation of dinosaur brain size requires estimation of both brain size and body size. Brain sizes are typically estimated by measurement of water displacement of endocranial casts or by "double integration" (Jerison, 1973). Resulting volumes may be converted to mass by multiplying by specific gravity. All estimates of brain size are complicated by the variable relationship between braincase size and brain size in living reptiles and, presumably, in dinosaurs. Although some dinosaur endocasts show fine structure indicating that the brain was nearly equal in volume to the braincase, there is little evidence of such an intimate relationship for most endocasts. Hopson (1977), following Jerison (1973), estimated that the brain filled 50% of the braincase except for specimens (sauropods and troodontids) that had indications of more extensive fill. Dinosaur body sizes are typically estimated on the basis of scale models (Alexander, 1989). Model volume is determined by water displacement and scaled up to life size; mass may again be determined by a conversion using specific gravity. Recent models typically portray dinosaurs more sleekly than older reconstructions, with a resulting decrease in predicted body mass, an increase in the ratio of brain mass to body mass, and an increase in estimated encephalization quotient.

Brain and body sizes calculated with the methodological constraints mentioned previously (Jerison, 1973) indicate that, in general, dinosaur relative brain size was similar to that of living bony fish, amphibians, and reptiles. These data suggest a modal behavioral flexibility or intelligence for dinosaurs similar to that of these animals. Hopson (1977) analyzed differences in relative brain size among dinosaur taxa and found a range of EQ values from 0.2 to 5.8, more than a 25-fold variation. He suggested an association between large relative brain size and high activity levels and degree of agility during locomotion, as predicted from limb anatomy and display structures. Large relative brain size is also associated to at least some extent with small absolute body size, carnivory, and bipedalism. The EQ of *Troodon* (*Stenonychosaurus*), recently recalculated at 6.48 by Currie (1993), is the largest known among dinosaurs. It falls within the range of those of living birds and suggests a behavioral complexity similar to that of these dinosaur descendants. Sauropods, quadrupedal herbivores that used immense size instead of speed for defense, have the lowest dinosaur EQs, even when the full volume of their braincases is used to estimate brain size. Despite the very low EQs of sauropods, it is *Stegosaurus* that has been most disparaged in respect to relative brain size in the popular press. As pointed out by Edinger (1961), these comments reflect popular exaggeration of estimates based on incompletely known brain casts of individuals of completely unknown body size.

Relative brain size has been used by various authors as a predictor of metabolic regime in extinct taxa. Living vertebrates with highly encephalized brains are almost exclusively endotherms, and both brain size and metabolic rate scale with body mass. The possibility that a causal link exists between relative brain size and metabolic rate has been cautiously suggested by several authors as reviewed by Farlow (1990), although a mechanism of action has not been recognized. McNab and Eisenberg (1989) convincingly demonstrated that the mass-independent variation in brain size is not statistically correlated with the mass-independent variation in basal metabolic rate in mammals. Barring contradictory evidence from living reptiles, it appears that brain size does not currently offer evidence of dinosaur metabolic rates or support for theories of "hot-blooded" dinosaurs.

Prediction of sensory modalities employed by various dinosaur subgroups is possible based on well-known principles of brain organization that compartmentalize referral areas of olfactory, visual, and auditory input to the olfactory lobes, optic lobes, and cerebellum. Although subject to the same constraints of preservation and reconstruction mentioned previously, it is clear that some dinosaurs (such as the herbivorous pachycephalosaurs) depended much more heavily on olfaction, and others (such as the theropods) on vision, for dominant sensory input. Such inferences from brain anatomy are supported in exceptional cases by osteological evidence of the size and structure of the sense organs themselves. Examples include the sclerotic rings that supported enlarged eyes and the osseous labyrinths that housed the semicircular canals. Currie (1993) noted the large relative size of the cerebellum (especially the floccular lobe) in small theropods and suggested that these genera may have had a refined sense of balance.

Analogous to the relationship between the braincase and the brain, the neural canal supports can be used to predict the size of the enclosed spinal cord. Variations in spinal cord cross-sectional area in living taxa reflect the number of incoming (sensory), integrative, and outgoing (motor) neurons at a given level. Enlargements are most marked at pectoral and pelvic levels, where nerves supply the limbs and reflect both limb size and density of innervation. When controlled for limb size, relative size of spinal cord enlargements allows prediction of neural density and limb use in dinosaurs. Limbs designed for manipulation (e.g., the forelimbs of *Deinonychus*) received particularly dense innervation. However, the tiny forelimbs of *Tyrannosaurus* and other large theropods received only marginal neural supplies, suggesting biologically insignificant function(s) commensurate with their small size (Giffin, 1995).

Enlargements several times larger than the braincase occur in the neural canals of sacral vertebrae of stegosaur and sauropod dinosaurs (Janensch, 1939). Such enlargements are not consistent in size or segmental location with those predicted on the basis of limb or tail use and certainly may not be used to posit the presence of a second, posterior "brain" in dinosaurs (Edinger, 1961). Somewhat more likely is the presence of a glycogen body similar to that housed within the sacrum of birds and responsible for their marked endosacral enlargements (Giffin, 1991).

Spinal cord anatomy may also be used as a tool in determining vertebral homologies and limb placement among extinct taxa. Enlargements of the canal at limb levels allow reconstruction of limb-level neural plexuses, estimation of limb position even when girdles have no vertebral connections, and estimation of transitions between different vertebral series.

See also the following related entries:

BEHAVIOR • BRAINCASE ANATOMY • INTELLIGENCE

References

Alexander, R. McN. (1989). *Dynamics of Dinosaurs and Other Extinct Giants*, pp. 167. Columbia Univ. Press, New York.

Currie, P. J. (1993). A new troodontid (Dinosauria, Theropoda) braincase from the Dinosaur Park Formation (Campanian) of Alberta. *Can. J. Earth Sci.* **30,** 2231–2247.

Edinger, T. (1961). Anthropocentric misconceptions in paleoneurology. *Proc. Rudolf Virchow Med. Soc. New York* **19,** 56–107.

Farlow, J. O. (1990). Dinosaur energetics and thermal biology. In *The Dinosauria* (D. B. Weishampel, P. Dodson, and H. Osmólska, Eds.), pp. 43–55. Univ. of California Press, Berkeley.

Giffin, E. B. (1991). Endosacral enlargements in dinosaurs. *Mod. Geol.* **16,** 101–112.

Giffin, E. B. (1995). Postcranial paleoneurology of the Diapsida. *J. Zool. (London)* **235,** 389–410.

Hopson, J. A. (1977). Relative brain size and behavior in archosaurian reptiles. *Annu. Rev. Ecol. Systematics* **8,** 429–448.

Hopson, J. A. (1979). Paleoneurology. In *Biology of the Reptilia, Volume 9* (C. Gans, Ed.), pp. 39–146. Academic Press, London.

Janensch, W. (1939). Der sakrale Neuralkanal einiger Sauropoden und anderer Dinosaurier. *Paläontol. Zeitschrift* **21,** 171–194.

Jerison, H. J. (1973). *Evolution of the Brain and Intelligence*, pp. 482. Academic Press, New York.

McNab, B. K., and Eisenberg, J. F. (1989). Brain size and its relation to the rate of metabolism in mammals. *Am. Nat.* **133**(2), 157–167.

Paleontological Department of the Natural History Museum, Ulaan Baatar

see PALEONTOLOGICAL MUSEUM OF THE MONGOLIAN ACADEMY OF SCIENCES

Paleontological Institute, Russian Academy of Sciences

see ORLOV MUSEUM OF PALEONTOLOGY

Paleontological Museum of the Mongolian Academy of Sciences, Ulaan Baatar

PHILIP J. CURRIE
Royal Tyrrell Museum of Palaeontology
Drumheller, Alberta, Canada

The museum in Ulaan Baatar, Mongolia, has one of the most outstanding collections and displays of dinosaurs in the world. Now housed in the Central State Museum in the downtown area, the Paleontological Museum will one day be housed in its own facility. It traces its origins to 1954 when one of the *Tarbosaurus* skeletons collected by the Russian expeditions to the Nemegt Basin was returned from Moscow. Three Mongolian expeditions headed by the geographer O. Namnandorzh collected another *Tar-*

bosaurus skeleton and other specimens in 1964, 1967, and 1968. Rinchen Barsbold was a student when the POLISH–MONGOLIAN PALEONTOLOGICAL EXPEDITIONS started in 1963, but by the end of the decade he had been hired as the paleontologist in Ulaan Baatar. In addition to collections made in collaboration with the Poles and Russians (see SOVIET–MONGOLIAN PALEONTOLOGICAL EXPEDITIONS), specimens have been collected by Barsbold and Mongolian colleagues in many internationally sponsored expeditions. These have produced one of the richest collections of dinosaurs anywhere. The displays are constantly changing as new specimens are acquired and others are rotated into traveling exhibitions that go to Europe, Japan, and the United States. Nevertheless, the two halls are more or less exclusively devoted to MONGOLIAN DINOSAURS of the Cretaceous. At any time the visitor can expect to see between two and four mounted skeletons of *Tarbosaurus*; one or more oviraptorid skeletons; numerous nests of eggs; sauropod limb bones; skulls of *Conchoraptor, Homalocephale, Nemegtosaurus, Oviraptor, Pinacosaurus, Protoceratops, Saichania, Saurornithoides,* and *Tarbosaurus* ; and skeletons of hadrosaur and protoceratopsian embryos and hatchlings, *Gallimimus, Protoceratops, Psittacosaurus, Saichania,* and *Saurolophus*. The most spectacular display is undoubtedly the famous "fighting dinosaurs," the skeletons of a *Protoceratops* and a *Velociraptor* that died while locked in mortal combat.

Paleontological Museum, Uppsala University, Sweden

see MUSEUMS AND DISPLAYS

Paleontology

PAUL G. DAVIS
National Science Museum
Tokyo, Japan

Paleontology is simply the study of fossils. Fossils (from the Latin *fossilis,* meaning "dug up") are the remains of once living animals. Fossil remains include skeletons (any hard tissue, e.g., shell, bone, chitin, etc.), tracks, impressions, trails, borings, and natural casts. A fossil is also often defined as older than the last glaciation (i.e., older than 10,000 years). Younger remains are termed subfossils. Paleontology (in Europe, Canada, and other parts of the world it is spelled palaeontology) is a scientific discipline that is at the junction of the biological and geological sciences. Because of its position, it encompasses many subdisciplines.

These subdisciplines are usually subdisciplines themselves of either geology or biology, and to denote that they are applied to fossils the prefix *paleo-* is added. For example, the major subdisciplines of paleontology are as follows:

Paleobiology: The study of the biology of ancient living organisms.

Paleoecology: The study of the ecology of ancient living organisms.

Paleobiogeography: The study of the distribution of fossils.

Taxonomy: The naming and biological relationships of fossils.

Taphonomy: The study of the processes of the formation of fossils (fossilization).

Paleopathology

DARREN H. TANKE
Royal Tyrrell Museum of Palaeontology
Drumheller, Alberta, Canada

BRUCE M. ROTHSCHILD
Arthritis Center of Northeast Ohio
Youngstown, Ohio, USA

Paleopathology is the study of ancient disease and injuries (trauma) in the fossil to subfossil record. The field of paleopathology brings together researchers of differing backgrounds, allowing for comprehensive interdisciplinary studies. Most studies have concentrated on pathology of early humans, although pathologies of other vertebrates are also known as far back as the Permian. In the past, paleopathological studies have undergone several periods of popularity and are currently enjoying a revival in interest in the professional arena. This is at least in part in response to rigorous collecting activities and technological advances in the medical field [for example, CT scan and magnetic resonance imaging (MRI) technologies]. Now diagnosis and interpretation of pathological conditions can be reinterpreted and understood with a much higher degree of confidence. Earlier workers had small numbers of comparative specimens. This, coupled with a preliminary and limited knowledge of diseases and their effects on bones, often resulted in misdiagnoses. For example, the chronic and erroneous misuse of the term "arthritis" to describe dinosaur pathology has only recently been addressed (Rothschild, 1990), but usage of the term still continues to appear in modern scientific papers and popular works.

Dinosaur pathology has been a minor and relatively neglected area of study, and the whole field is badly in need of a review and rigorous overhaul. Less than 1% of currently existing diseases leave any traces on bones. Despite medical technology breakthroughs, and interdisciplinary studies in conjunction with paleontologists, veterinarians, biologists, and medical doctors, our understanding of the whole picture of dinosaur paleopathology is somewhat limited. However, bones, ichnites (footprints), eggs, skin impressions, and insects in amber all provide additional tantalizing information and circumstantial evidence regarding the illnesses and maladies of dinosaurs. Although bones tell us mostly how dinosaurs appeared, our steadily increasing knowledge of dinosaur pathology will give us a better and more complete appreciation of how these remarkable animals lived, behaved, suffered, and died from day to day.

Our awareness of dinosaur pathology has actually been known longer than dinosaurs themselves. Whereas the name Dinosauria was not established until 1841, osseous pathological conditions (vertebral fusions) were first documented and figured by Eudes-Deslongchamps in 1838 in the theropod dinosaur *Poikilopleuron*. The discovery of Upper Jurassic and Late Cretaceous dinosaurs in North America late in the 19th century revealed dinosaurian pathologies and stimulated interest on a small scale. Until the 1970s, dinosaur injuries have been described mostly as casual observations. The work was mostly descriptive in nature, with no real attempt to determine etiology (cause) or quantify data.

Although numerous pathological dinosaur specimens are known and more are collected annually, relatively few have been described. Those that were documented more than 20 years ago are suitable for reappraisal using the latest medical technologies and data. Therefore, this report is largely based on unpublished findings.

True tooth decay or dental caries have not been documented in dinosaurs. There are several good reasons for this. Because tooth replacement was a continuous and rapid process, worn-out teeth would have been ejected from the mouth long before dental caries developed. DIET and tooth crown morphology also probably played roles in prevention of tooth decay. The low, rounded bunodont dentitions of extant suids, ursids, and hominids are more prone to tooth decay than any of the tooth morphologies possessed by dinosaurs.

Tooth fractures are known mostly in large theropods such as Late Cretaceous tyrannosaurids, in which 10–75% of the distal crowns of some specimens were snapped off lengthways and/or transversely. The roots in many cases were evidently retained in the mouth for extended periods of time, often developing subsequent heavy and/or abnormal wear facets (Farlow and Brinkman, 1994). Premortem tooth fracture or damage is usually related to feeding behavior, in which teeth contact the bone of the prey during feeding. Malocclusion is a possible explanation for large theropod teeth that bear wear facets on opposing sides (see TEETH AND JAWS).

Occasionally, one finds large theropod teeth bearing serration marks from another tooth. Although some of these may have been caused by malocclusion, anomalous orientation in some suggests an external source is responsible, such as a bite from an aggressive conspecific (Tanke and Currie, 1997, and manuscript in preparation).

Oral abscesses are known from two hadrosaur jaws. Both were collected in Dinosaur Provincial Park (Alberta, Canada) at the end of the 19th century. Curiously, despite the subsequent collection and observation of literally thousands of hadrosaur jaws, these remain the only two examples known.

Healed or healing bone fractures are one of the most common types of bone pathologies found in extant wild animals. It is therefore not surprising that this is also true for dinosaurs. A healing fracture is immediately recognizable by the presence of a swollen area or callus at the fracture site. Constant usage or movement of the affected element and infection sometimes resulted in an improper reknitting of the bone segments and a false joint or pseudoarthrosis was formed. Bone fracture repair in dinosaurs apparently followed the same general sequence of events for bone repair in extant vertebrates, although it has not been determined whether the processes were closer to those of extant reptiles, birds, or mammals. Study in this area might shed light on dinosaurian PHYSIOLOGY. Fractures are the single most common pathological condition identified in dinosaurs. In the Late Cretaceous deposits of DINOSAUR PROVINCIAL PARK, Alberta, Canada, more than 100 examples of healed or healing fractures are found each field season. Only fully active animals regularly fracture their bones today. By virtue of the propensity of dinosaur bone fractures, we can say that some dinosaur families also had active lifestyles. In some dinosaurs, such as dromaeosaurids and ornithomimids, evidence of healed fractures is rarely seen. This may reflect a taphonomic bias, or it might indicate that traumatic fractures were not conducive to long-term survival. Extensive collecting programs in "monospecific" or low-diversity bone beds allow us the opportunity to understand the relative frequency of fractures among single dinosaur populations or herds. In short-frilled horned dinosaurs, at least, it can be stated with a high degree of confidence that evidence of healed or healing bone fractures is rare. This finding is not in accordance with past popular interpretations of horned dinosaurs as aggressive animals (Tanke, 1989).

Repetition of similar fractures affecting the same bones or regions of skeletons requires explanation. It may be indicative of a repeated behavior or stressful activity. For example, numerous examples of crushed and healing distal caudal vertebrae in Late Cretaceous hadrosaurs from Alberta may have been caused by accidental intraspecific trampling injuries within large herds. The most commonly observed fractured and healed skeletal elements in dinosaurs are ribs, fibulae, phalanges, caudal neural spines, and caudal centra, although other bones can be affected too. Massive skull and unilateral jaw fractures with total healing have been seen in ceratopsians. Bilateral fracturing of the dentaries of a *Brachylophosaurus* and with good realignment and near perfect healing has been observed. However, the teeth immediately above the fracture site were misaligned, indicating tooth germ trauma had occurred. Severe uni- or bilateral fractures of the pelvis (affecting the ischium) in hadrosaurs and iguanodonts are occasionally seen.

The absence of healed or healing fractures in certain bones is easily explained. A dinosaur suffering from a broken neck, back, or major limb bone (e.g., femur) would likely die from its injuries and thus not have a chance to heal and potentially contribute its pathologic bones to the fossil record.

The most common form of arthritis in humans, osteoarthritis, was extremely rare in dinosaurs. Although osteoarthritis was once widely considered common in dinosaurs, this is not so. Dinosaur arthritis is really a spinal deformity called "spondylosis deformans." The major sign of osteoarthritis is osteophytes or bone spurs. Such spurs often form on dinosaur vertebrae, but they do not occur in limb bone joints; therefore, they do not qualify as osteoarthritis (Rothschild, 1990). The only known examples of true osteoarthritis are in the ankles of 2 of the 39 specimens of *Iguanodon* from Bernissart, Belgium.

Recognition of tumors and cancer in dinosaurs is problematic. Bone abnormalities have previously been mistaken for tumors. Moodie (1916) erroneously interpreted a caudal fusion in a sauropod dinosaur as a hemangioma. Such nonpathologic fusions actually affect a quarter to half of all sauropod tails studied and are actually examples of diffuse idiopathic skeletal hyperostosis (DISH). Recent widespread newspaper reports of bone cancer in an *Allosaurus* humerus await confirmation.

Opportunities to identify specific infections await the availability of specific antisera (immunofluorescence) or DNA probe techniques. An infection is recognized by the reactive new bone (often filigree in appearance), draining sinuses, disorganized internal trabecular structure, or characteristic periosteal reaction. Osteomyelitic infections can be visually spectacular, with large quantities of extraneous and irregular rugose bone being deposited on the infected area. Infections are often secondarily related to trauma or serious compound bone fractures, in which injury to overlying tissues can leave the wound open to infection. It is not clear if such infections were actually rare or if affected dinosaurs failed to survive the infections. In fact, specimens showing rampant osteomyelitis at the time of death suggest the latter. A long-standing osteomyelytic infection was found in an Early Cretaceous hypsilophodont tibia collected from the southern coast of Australia. This region was relatively close to the South Pole at that time, which raises the question whether dinosaur bones healed in colder climates at the same rate as those from warmer regions.

Congenital disorders of the skeleton are rarely seen. A fetal *Hypacrosaurus* from Devil's Coulee in southern Alberta had a wedge-shaped dorsal hemivertebrae. A nestling *Maiasaura* was found to have had a severely twisted metatarsal. An asymmetrical occipital condyle of a hadrosaur was found in Dinosaur Provincial Park. It would have been interesting to see how the rest of the cranium was commensurately affected. A round, smooth-rimmed hole completely perforating a dorsal rib head was seen in a *Psittacosaurus* skeleton. Supernumerary serration rows have been observed on tyrannosaurid teeth and probably reflect genetic variation (Erickson, 1995).

Another commonly observed pathology in dinosaurs is the ossification of spinal longitudinal ligaments (DISH). It is often confused with arthritis. DISH is not actually a disease but acts to strengthen the back at a structurally weak or stressed point. Ossified ligaments (tendons) in ceratopsians, ankylosaurs, hadrosaurs, iguanodonts, and pachycephalosaurs serve a similar function. Fusion of two to four caudal vertebrae in sauropods (Moodie, 1916; Rothschild and Berman, 1991) may have served an even more important function. In addition to helping keep the tail in the air, it may have been integral to the mating act, maintaining the cloaca (in the female) in a position more accessible for mating. Co-ossification of vertebrae occurs occasionally in one form of arthritis called spondyloarthropathy. When fusions occur, the centra involved may even mimic a section of bamboo in appearance. Similar fusion is present in the anterior cervical vertebrae of ceratopsian dinosaurs, although this is a developmental phenomenon to assist in the support of the massive head. Vertebral fusions may also develop from injuries, infection, or congenital defects. Infections produce significant distortion of bony architecture, which is easily recognized on gross examination, in cross-sections, or in X-ray photographs.

Multilayering of eggshell has been observed occasionally in dinosaur eggs and is attributed to the egg being retained or recirculated in the oviduct of a stressed individual. Such eggs are frequently found in extant snakes, turtles, and birds, but the reason for this condition is not always clear. Thinning eggshell of sauropods (cf. *Hypselosaurus*) from the uppermost Cretaceous deposits of France has been implicated in dinosaur extinction, but the hypothesis needs a vigorous test. Footprints exhibiting pathological conditions are also known but have only been reported recently (Lockley *et al.*, 1994). Curiously, many of the trackways are referable to large carnivorous dinosaurs that show missing or curled digits or close appression of two toes. A trackway of the latter even indicates that the individual was limping, which is circumstantial evidence that dinosaurs suffered (as expected) some degree of pain from their injuries. Toe injuries either reflect direct trauma or congenital defects. Amputation or sloughing off of individual toes could be expected if the injured areas became seriously infected. Such amputations have been observed in extant lizards. Foot pathologies in larger bipedal dinosaurs may be related to more active lifestyles or simply reflect fragility of the narrow, protruding digits. Certainly, the more massive elephantine-like foot shapes possessed by a large number of plant-eating dinosaurs (sauropods, hadrosaurs, ankylosaurs, and ceratopsians) would have been less susceptible to traumatic injuries.

All living animals are affected by a variety of internal and external parasites. Frozen Pleistocene mammals have been found with well-preserved, taxonomically diverse internal parasites. Amber (fossilized tree resin) reveals that a wide variety of essentially modern-appearing insect parasites or disease vectors were already present by at least the Early Cretaceous. Dinosaurs undoubtedly served as hosts to a wide variety of parasitic organisms, although currently there is no direct fossil evidence for these. Fossilized dinosaur skin impressions show that hadrosaurs, ceratopsians, and other dinosaurs had rough, beaded skin that would have provided many nooks and crannies in which ectoparasites could lodge. Borkent (1995) speculated that several species of blood-sucking midges could have fed on dinosaurs, biting around the eyes of hadrosaurs or the underside of vascularized ceratopsian neck frills. Circular lesions or rimless crater-like depressions on the dorsal surfaces of some ankylosaur scutes resemble bone damage caused by ectoparasites in extant and fossil turtle shells and might provide a clue to possible dinosaur parasites.

Long-term observation of dinosaur bones in the field or in large museum collections can reveal patterns of osteopathy that affected specific dinosaur taxa or specific regions of the body. Frequency of similar injuries to a specific body region among a large specimen sample requires explanation. Animals today engage in a variety of activities, some of which are more likely to result in bone injuries. Therefore, behavior plays a strong role in the development of certain pathological conditions. Among fossil mammalian faunas, the recovery of fossils showing serious injuries that had healed has been used as evidence of "compassion," whereby healthy members of a species fed and protected an injured individual. Such interpretations may not always be correct, especially if one tries to extrapolate this behavior to dinosaurs. Any given population of vertebrates today will have its share of seriously injured or deformed individuals, which in most cases are not cared for by their fellows. Such animals may only be simply tolerated. Seriously injured individuals of some dinosaurs, such as hadrosaurs or ceratopsians, may have been ignored by others within the herd. Such an individual could have remained anonymous among a herd of many thousands, giving it the time needed to repair bone and tissue damage.

Injuries found in dinosaurs from failed attacks are often attributed to violent inter- and intraspecific interactions. Injuries are rarely found. Relatively few animals are fortunate enough to escape after having been caught and injured. Fewer of these would ever be preserved as fossils. The myth of "many horn and frill injuries" in ceratopsians is overstated and can be traced to a couple of injured specimens that have often been mentioned, mostly in the popular literature. Battles between *Tyrannosaurus* and *Triceratops* that caused injuries to both are also undocumented. Although such scenarios probably did occur, most theropods probably preferred prey that was less dangerous to catch and subdue.

Paleopathology is entering a period of exponential growth. Although technical and popular publications addressing dinosaur paleopathology number fewer than 500, only in the past decade have dinosaur pathologies really gained the attention they deserve. Earlier studies treated them more as curiosity items. Systematic screening of specimens in museum and in the field and the introduction of paleoepidemiologic techniques permit the development of testable hypotheses. This is leading to an enhanced understanding of dinosaur lifestyles, physiology, and habitats. The study of dinosaur paleopathology has even been used to understand and treat human osteopathy. Although semantic confusion led to the widespread and erroneous belief that osteoarthritis was common in dinosaurs, investigation led to the recognition of the importance of joint stability (rather than weight of the individual) as a major determinant of disease risk. This in turn spawned new ideas and approaches to the treatment of the human disease. The use of X-rays, CT scans, MRI, and histologic, biochemical, DNA, and immunologic analyses further expands these opportunities.

See also the following related entries:

BEHAVIOR • EGGS, EGGSHELLS, AND NESTS

References

Borkent, A. (1995). *Biting midges in the Cretaceous amber of North America* (Diptera: Ceratopogonidae). pp. 237. Packhuys, Leiden.

Erickson, G. M. (1995). Split carinae on Tyrannosaurid teeth and their implications in their development. *J. Vertebr. Paleontol.* **15**(2), 268–274.

Eudes-Deslongchamps, J. A. (1838). Memoire sur le Poekilopleuron bucklandi, Grand Saurien Fossile, Intermediare Entre les Crocodiles et les Lezards. *Mem. Soc. Linn. Normandie* **VI,** 37–146.

Farlow, J. O., and Brinkman, D. L. (1994). Wear surfaces on the teeth of Tyrannosaurs. In *Dinofest* (G. D. Rosenberg and D. L. Wolberg, Eds.), Spec. Publ. No. 7, pp. 165–175. Paleontological Society, Knoxville, TN.

Lockley, M. G., Hunt, A. P., Moratalla, J., and Matsukawa, M. (1994). Limping dinosaurs? Trackway evidence for abnormal gaits. *Ichnos* **3,** 193–202.

Moodie, R. L. (1916). Two caudal vertebrae of a Sauropodous dinosaur exhibiting a pathological lesion. *Am. J. Sci.* **191**(4), 530–531.

Rothschild, B. M. (1990). Radiologic assessment of osteoarthritis in dinosaurs. *Ann. Carnegie Museum* **59,** 295–301.

Rothschild, B. M., and Berman, D. (1991). Fusion of caudal vertebrae in Late Jurassic sauropods. *J. Vertebr. Paleontol.* **11,** 29–36.

Rothschild, B. M., and Martin, L. D. (1993). *Paleopathology—Disease in the Fossil Record*, pp. 386. CRC Press, Boca Raton, FL.

Rothschild, B. M., and Tanke, D. H. (1992). Paleopathology of vertebrates: Insights to lifestyle and health in the geological record. *Geosci. Can.* **19**(2), 73–82.

Tanke, D. H. (1989). K/U Centrosaurine (Ornithischia: Ceratopsidae) paleopathologies and behavioral implications. *J. Vertebr. Paleontol.* **9**(3), 41A.

Tanke, D. H. (1997). The rarity of paleopathologies in "short-frilled" Ceratopsians (Reptilia: Ornithischia: Centrosaurinae): Evidence for non-aggressive intraspecific behavior. In *Paleopathology* (B. M. Rothschild and S. Shelton, Eds.), in press. Univ. of Texas Museum of Natural History, Austin.

Tanke, D. H., and Currie, P. J. (1995). Intraspecific fighting behavior inferred from toothmark trauma on skulls and teeth of large Carnosaurs (Dinosauria). *J. Vertebr. Paleontol* **15**(3), 55A.

Paleophysiology

see PHYSIOLOGY

Paleotemperature

see PALEOCLIMATOLOGY

Parental Care

see BEHAVIOR

Pathology

see PALEOPATHOLOGY

Pectoral Girdle

KEVIN PADIAN
University of California
Berkeley, California, USA

The bones of the pectoral (shoulder) girdle (*sensu stricto*) in dinosaurs include two endochondral elements, the scapula and coracoid, and a dermal element, the clavicle. Endochondral elements are preformed in cartilage and replaced by bone during ontogeny, whereas dermal bones form directly as bone. For convenience the sternum will also be discussed briefly here, although it is most often associated with the ribs (see POSTCRANIAL AXIAL SKELETON).

The elements of the pectoral girdle were originally attached to the skull, and most of them were dermal elements (cleithrum, supracleithrum, upper and lower postcleithra, posttemporal, and clavicle). They formed a half circle behind the operculum (Fig. 1). The cleithrum formed the attachment for the scapulocoracoid arch, which supported the forelimb. Through the evolution of tetrapods, the dermal elements were gradually reduced and lost, and the clavicles came to rest anterior and medial to the scapulocoracoid, articulating near the glenoid (shoulder socket). This is essentially the situation in basal archosaurs, which also retain the interclavicles of basal tetrapods that persist in PSEUDOSUCHIANS but are lost in ORNITHOSUCHIANS.

The pectoral girdle was originally attached to the skull roof via the posttemporal, but once tetrapods came onto land they developed a discrete atlas–axis complex, the beginning of the neck, and the need to move the head relative to the body effectively broke the connection between skull and shoulder girdle. The shoulder girdle, unlike the pelvic girdle, was not originally attached to the vertebral column, and it has moved free of bony attachments to the axial skeleton in all tetrapods ever since, with a few exceptions (all turtles and some derived pterosaurs and birds in which the posterior ends of the scapulae attach to a fused series of dorsal vertebral spines). This lack of a set axial "anchor" for the shoulder girdle has led to differing interpretations of the orientation of the dinosaurian shoulder girdle (e.g., dromaeosaurids and birds; see Ostrom, 1976; Martin, 1983) and the potential functions of the forelimb (e.g., the question of ceratopsid "galloping"; see Dodson, 1996, for a review).

In dinosaurs (Fig. 2) the scapula is somewhat straplike, usually three times or more as long as its average width, and expanded slightly to meet the diameter of the coracoid at that end. However, the scapula (Fig. 3) is unusually narrow and short in herrerasaurids, which have been variously regarded as basal theropods, basal saurischians, or dinosaurian outgroups (see HERRERASAURIDAE). The scapula is also expanded dorsally in most sauropodomorphs and in some ceratosaurian theropods, and in most sauropods and also in stegosaurs the proximal end is also expanded anteriorly. Nodosaurid ankylosaurs and some cera-

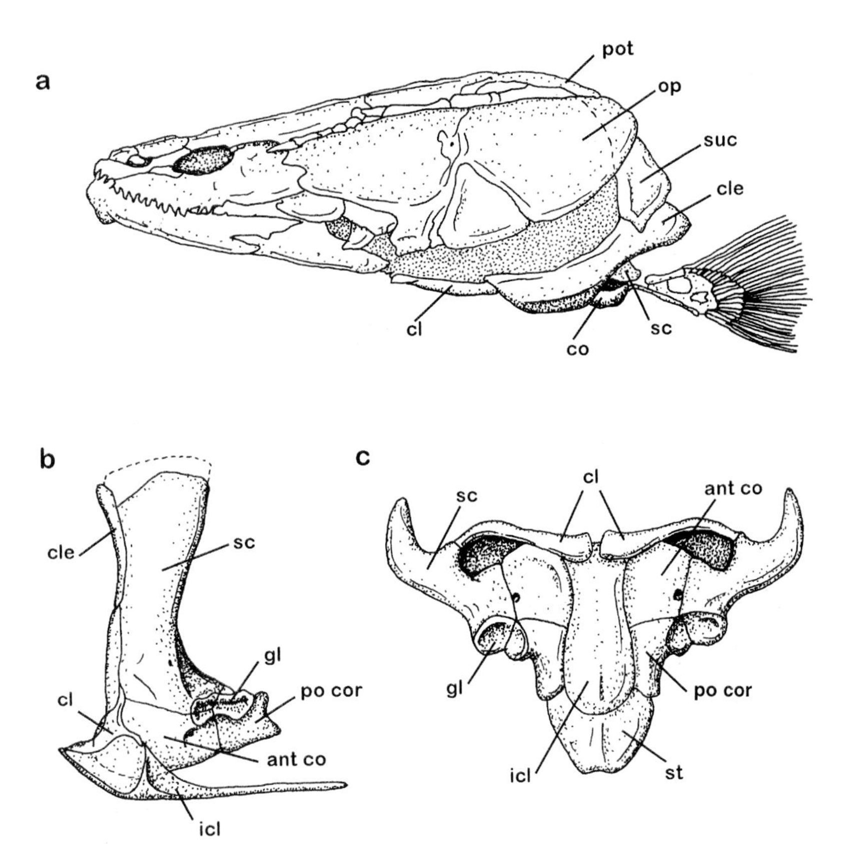

FIGURE 1 Evolution of the vertebrate shoulder girdle. In the bony fish (a; after Gregory) and other nontetrapods, the pectoral girdle is attached to the postopercular part of the skull at the posterior skull roof. In a typical early tetrapod (b, the synapsid *Dimetrodon;* after Romer), the shoulder girdle is separated from the skull, and its principal supportive elements are endochondral. In the lepidosaur *Sphenodon* (c), the cleithrum is lost, although the coracoid's two centers of ossification are retained. Compare to further evolution in Figs. 2–4. ant co, anterior coracoid ossification; cl, clavicle; cle, cleithrum; co, coracoid; gl, glenoid fossa; icl, interclavicle; op, opercular; po cor, posterior coracoid ossification; pot, posttemporal; sc, scapula; st, sternum; suc, supracleithrum. Not to scale.

topsids have a pronounced spine running longitudinally on the lateral face of the scapula. Presumably this spine separated muscle masses, much as in mammals, that protracted and retracted the forelimbs. The scapula is generally fused to the coracoid in adults; the scapulocoracoid in some forms has a discernible acromion process for the articulation of the clavicles, when present.

The coracoid is generally subcircular in dinosaurs but may take on a more rounded rectangular (e.g., ankylosaurids and elmisaurids) or triangular (*Deinonychus)* form. A coracoid foramen, which carries the supracoracoid nerve and associated blood vessels, is situated slightly anteroventral to the glenoid fossa, which marks the junction between scapula and coracoid. The glenoid fossa is a biplanar structure with an open articulation of 120° or greater; its articular surface is slightly concave on its face, and convexly rounded on its edges, which often form a pronounced lip. The glenoid appears to have been covered by a cartilaginous surface in life, and it apparently permitted a wide degree of movement. This conclusion is supported by the corresponding morphology of the head of the dinosaur humerus, which tends to be a biconvex, elongated surface curving anteromedially to meet the proximal end of the deltopectoral crest.

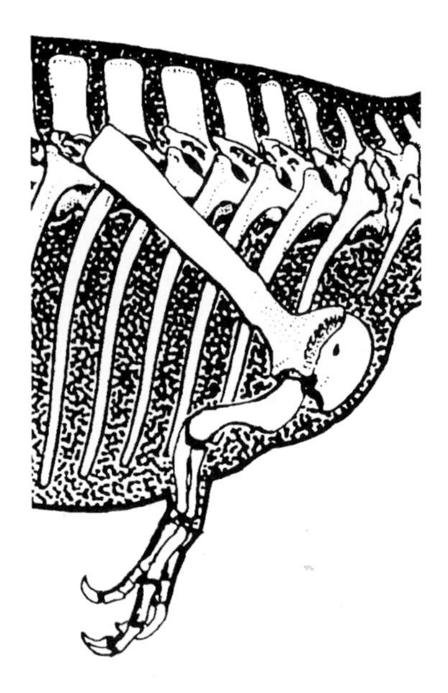

FIGURE 2 Position of the shoulder girdle with respect to the rib cage in a typical dinosaur (*Sinraptor;* after Currie and Zhao, 1993).

The posteroventral corner of the coracoid may be somewhat pronounced for articulation with the bony sternum, which is known in some forms (e.g., ankylosaurids); in ornithomimids, elmisaurids, oviraptorids, and dromaeosaurids (although not *Deinonychus)* the coracoid is narrower dorsoventrally, and the posteroventral end of the coracoid is attenuated. This form often indicates contact with the sternum, which has been well preserved in oviraptorids (Barsbold, 1983).

The distribution of clavicles in dinosaurs (Fig. 4) was recently reviewed in an excellent paper by Bryant and Russell (1993). Clavicles are known in a number of saurischian dinosaurs, including the sauropodomorph taxa Massospondylidae *(Massospondylus),* Plateosauridae *(Plateosaurus),* Cetiosauridae *(Shunosaurus),* and Diplodocidae *(Diplodocus)* (Galton, 1990; McIntosh, 1990). The interclavicle identified in the basal sauropodomorph *Massospondylus* (Cooper, 1981) is short and rod-like and is better interpreted as a clavicle (Sereno, 1991); in general form it resembles the clavicles reported in the basal ceratopsians *Psittacosaurus* (Sereno, 1990), *Protoceratops* (Brown and Schlaikjer, 1940), and *Leptoceratops* (Sternberg, 1951), the only ornithischians in which clavicles have been reported. Among theropod saurischians, the basal ceratosaurs *Coelophysis* and *Segisaurus* (Camp, 1936) have rib-like clavicles, curved at both ends with a gentle bend in the shaft. The pectoral ends are flattened and so can be easily distinguished from ribs, with which they are often confused and therefore overlooked. Clavicles have also been reported in the abelisaurid ceratosaur *Carnotaurus* (Bonaparte *et al.*, 1990).

In an increasingly broad variety of TETANURAE, including *Allosaurus* and an unnamed allosaurid (Chure and Madsen, 1996), Troodontidae (Russell and Dong, 1993), oviraptorids (Barsbold, 1983), tyrannosaurids (*Albertosaurus, Daspletosaurus, Gorgosaurus;* Makovicky and Currie, 1996), and possibly dromaeosaurids (*Velociraptor*; Ostrom, 1976, 1990; but see Barsbold, 1983), the presence of a furcula has now been clearly demonstrated, as it has always been recognized in the maniraptoran *Archaeopteryx* and all other birds. A furcula, the "wishbone" in birds, has always been assumed to represent fused clavicles because it forms the foramen triosseum at its junction with the

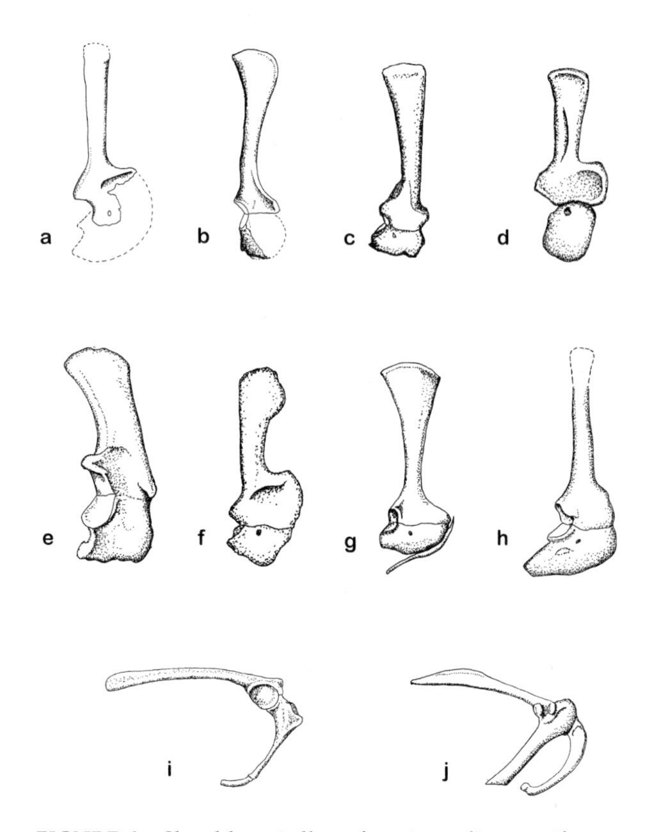

FIGURE 3 Shoulder girdles of various dinosauriformes: (a) *Herrerasaurus,* (b) *Heterodontosaurus,* (c) *Iguanodon,* (d) *Stegosaurus,* (e) *Euoplocephalus,* (f) *Brachiosaurus,* (g) *Coelophysis,* (h) *Ornithomimus,* (i) *Archaeopteryx,* and (j) the eagle *Aquila.* Not to scale.

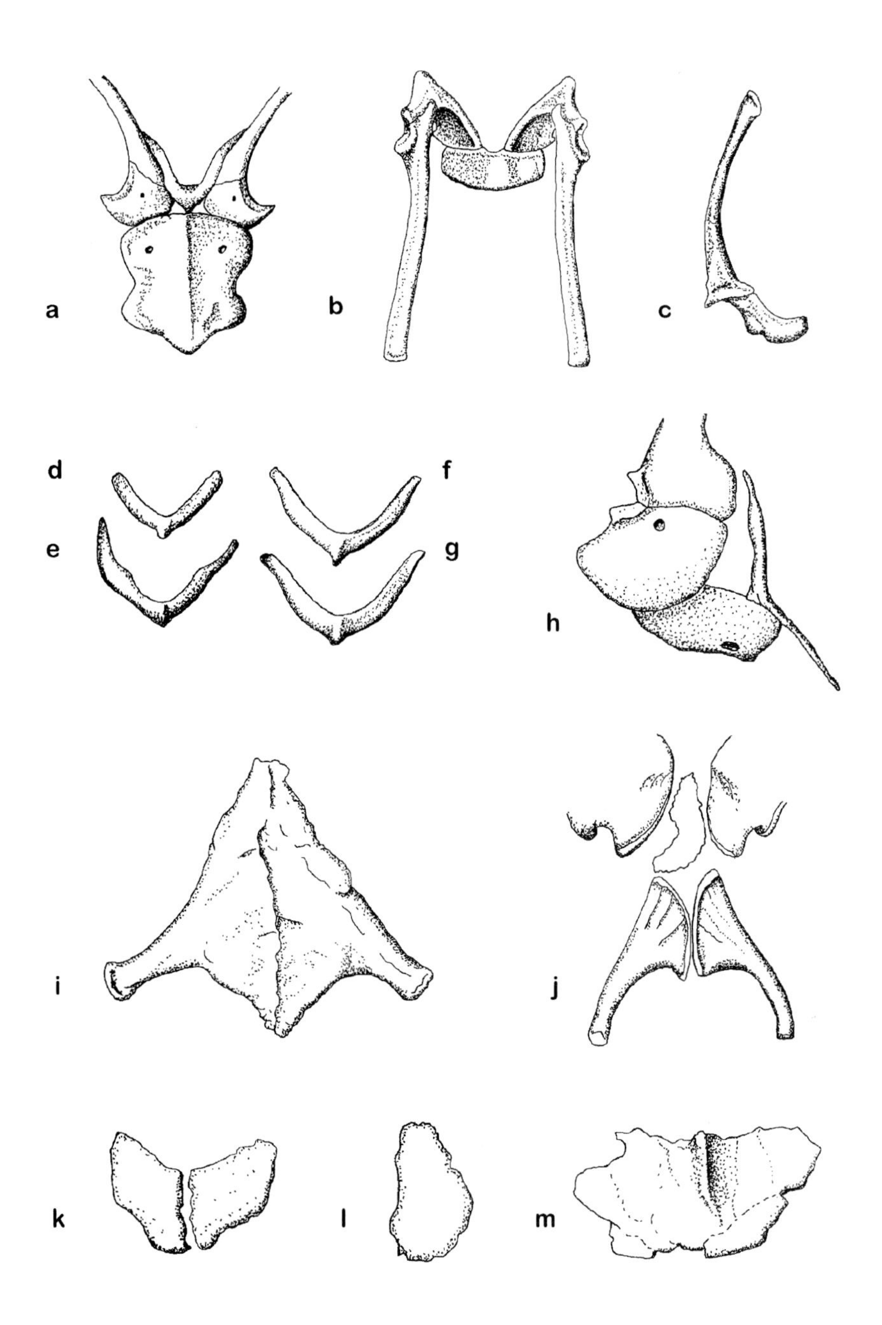

FIGURE 4 Sterna and associated structures of dinosaur pectoral girdles. (a) sternum, scapulocoracoid, and furcula of the oviraptorid *Ingenia* (after Barsbold, 1983), in anterior (ventral) view. (b) Scapulocoracoids and sternum of *Archaeopteryx* (after Wellnhofer, 1993). (c) Right scapulocoracoid of *Oviraptor* in anterior (cranial) view. (d–g) Furculae of *Oviraptor* (d and e) and *Ingenia* (f and g). (h) Right scapulocoracoid, left coracoid, and furcula of a new allosauroid (after Chure and Madsen, 1996), in right anterior view. (i–m) Sterna of *Euoplocephalus* (i), *Iguanodon* (j), *Plateosaurus* (k), *Diplodocus* (l), and *Sinraptor* (m). Not to scale.

scapula and coracoid, and at the midline it articulates directly, ligamentously or membraneously, with the ventral keel of the scapula. These connections are like those in other archosaurs (e.g., crocodiles), including birds (in more basal saurians the clavicles may contact the interclavicle instead of the sternum, which is deep to the interclavicle).

It is still a question of some interest whether the clavicles of all dinosaurs are homologous, whether those of basal theropods are homologous to the furcula reported in many tetanurans, and whether this structure is homologous to the furcula of birds. Bryant and Russell (1993) maintained that the identification of clavicles in a number of dinosaurs was uncertain because the elements are not preserved in the vast majority of genera; hence hypotheses of (i) the presence but nonpreservation or nondiscovery, (ii) the variable presence in cartilage or bone, therefore inconsistent discovery, (iii) repeated loss of an element synapomorphic to Dinosauria, and (iv) repeated neomorphic formation of an element may be equally parsimonious explanations. Chure and Madsen (1996) provide clear criteria for distinguishing between a furcula and median gastralia, allaying some concerns about identification raised by Bryant and Russell. Some of the V-shaped median elements identified by Gilmore (1920) as gastralia and by Madsen (1976) as an anteroposterior series of gastralia are actually an ontogenetic series of furculae; this had led earlier authors to multiply the number of gastral elements and so widen the body unnecessarily (Chure and Madsen, 1996).

At this point it seems most likely that the furcula is a synapomorphy of Tetanurae because it is found in both carnosaurs (*Allosaurus* and a related form) and various coelurosaurs. The morphology is variously V-shaped or slightly U-shaped, but its form and connections are apparently identical. Apparently, no theropods outside Tetanurae had a furcula, but the morphology and position of the isolated clavicles of several ceratosaurs suggest that these elements fused and became more flattened, stout, and boomerang-shaped in Tetanurae. Not to accept this would mean accepting the proposition that the clavicles of basal theropods were completely lost and the furcula was acquired as a complete neomorph, inasmuch as the two structures have not been found together in the same animal, unless one has been misidentified. Birds are members of Tetanurae, and *Archaeopteryx* has a boomerang-shaped furcula just like those of other Tetanurae; therefore, this homology seems beyond reasonable doubt (see BIRD ORIGINS).

A sternum, formed of two plates (typically fused in adults), has been reported in stegosaurs, a variety of ornithopods, ceratopsians, as well as in many sauropodomorphs, in the carnosaur *Sinraptor* (Currie and Zhao, 1993), Dromaeosaurids, oviraptorids, ?TROODONTIDS, and recently in *Archaeopteryx* and Enantiornithine birds (Fig. 4). In stegosaurs it is poorly ossified, whereas in ankylosaurids it consists of two fused plates, united to form a rhomboidal structure, with a pair of posterolateral projections. A sternal plate has been reported in *Heterodontosaurus*, but is not known in hypsilophodontids. Iguanodontids and hadrosaurs appear to have posterior projections on their paired, kidney-shaped sternal plates much as ankylosaurids do. This reniform sternal shape is also preserved in *Psittacosaurus*, which has posterior scalloping for articulation with the ribs, and in various neoceratopsians. Long, narrow sternals are known in several pachycephalosaurs. In sauropodomorphs (the plateosaurid *Lufengosaurus* and many cetiosaurids, diplodocids, and titanosaurids) the sternal plates are separate and ellipsoid; the anterior end is thicker than the posterior end, which is rugose for the reception of sternal ribs. In oviraptorids (*Oviraptor* and *Ingenia*) the sternals are plate-like and sometimes fused, with an anterolateral groove for reception of the coracoid, as in birds and pterosaurs. *Archaeopteryx* appears to have an anteroposteriorly short, rectangular sternum, judging from the one specimen in which it is partly visible (Wellnhofer, 1993).

Because the sternum is usually cartilaginous in reptiles, its failure to be preserved in the fossil record can only be regarded as a lack of information and not necessarily a phylogenetic loss. Also, because ontogenetic series of fossil reptiles are so incomplete, we do not know if the interclavicle has been lost or simply fused with the sternum in many extinct forms. Because the first is preformed in cartilage and the second is not, separate centers of ossification could be expected; however, it is not unusual for bony elements to become mixtures of dermal and endochondral bone (Romer, 1956, p. 50), and the furcula of the quail *Coturnix* was reported to be just such a bone (Lansdown, 1968; but see Russell and Joffe, 1985).

In brief, the construction of the dinosaurian shoulder girdle reflects a phylogenetic heritage in which a move from quadrupedal forms to apparently smaller, bipedal forms removed the selective force of locomotion from the evolution of the forelimb, which was now freed for grasping, manipulation, and other functions (see BIPEDALITY; DINOSAUROMORPHA; FORELIMBS AND HANDS; ORIGIN OF DINOSAURS; QUADRUPEDALITY). The simplification of the shoulder girdle and loss of most dermal girdle elements may have been related to this. The typical basal pectoral form is a strap-like scapula slightly constricted at midlength, fused in adults to a subcircular coracoid, with a posterolaterally and slightly ventrally facing glenoid fossa. The scapula became very thin and strap-like, and the coracoid elongated to meet the sternum (in some coelurosaurian groups), and the clavicles apparently retained by basal theropods from archosaurian ancestors were modified into the furcula that is still seen in birds. Meanwhile, with the return to quadrupedality in many large herbivorous dinosaurs (both sauropods and ornithischians), the shoulder girdle became more robust, often enlarging transversely and developing spines for the accommodation of the origins of muscles used in forelimb locomotion. Unlike the basal archosaurs from which ornithodirans evolved, however, the movement of the forelimb in dinosaurian quadrupeds was more parasagittal (fore and aft) than in crocodiles and other reptiles, as FOOTPRINTS AND TRACKWAYS make clear. There is still much to be learned about the mobility, function, and kinematics of the shoulder girdle and forelimb of dinosaurs, especially in those secondarily returning to quadrupedality.

See also the following related entries:

BIPEDALITY • FORELIMBS AND HANDS • PELVIS, COMPARATIVE ANATOMY • QUADRUPEDALITY

References

Barsbold, R. (1983). Carnivorous dinosaurs from the Cretaceous of Mongolia. *Trudy Sovmestnaya Sovetsko–Mongol'skaya Paleontol. Ekspeditsiya* **19,** 1–117. [In Russian]

Bonaparte, J. F., Novas, F. E., and Coria, R. A. (1990). *Carnotaurus sastrei* Bonaparte, the horned, lightly built carnosaur from the middle Cretaceous of Patagonia. *Contrib. Sci. Nat. History Museum Los Angeles County* **416,** 1–42.

Brown, B., and Schlaikjer, E. M. (1940). The structure and relationships of *Protoceratops*. *Ann. N. Y. Acad. Sci.* **40,** 133–266.

Bryant, H. N., and Russell, A. P. (1993). The occurrence of clavicles within Dinosauria: Implications for the homology of the avian furcula and the utility of negative evidence. *J. Vertebr. Paleontol.* **13,** 171–184.

Camp, C. L. (1936). A new type of small bipedal dinosaur from the Navajo Sandstone of Arizona. *Univ. California Publ. Geol. Sci.* **24,** 39–56.

Chure, D. J., and Madsen, J. H. (1996). On the presence of furculae in some non-maniraptoran theropods. *J. Vertebr. Paleontol.* **16,** 573–577.

Cooper, M. R. (1981). The prosauropod dinosaur *Massospondylus carinatus* Owen from Zimbabwe: Its biology, mode of life and phylogenetic significance. *Occasional Papers Natl. Museums Monuments Rhodesia Ser. B Nat. Sci.* **6,** 689–840.

Currie, P. J., and Zhao, X.-J. (1993). A new carnosaur (Dinosauria, Theropoda) from the Jurassic of Xinjiang, People's Republic of China. *Can. J. Earth Sci.* **30,** 2037–2081.

Dodson, P. (1996). *The Horned Dinosaurs: A Natural History*. Princeton Univ. Press, Princeton, NJ.

Galton, P. M. (1990). Basal Sauropodomorpha—Prosauropoda. In *The Dinosauria* (D. B. Weishampel, P. Dodson, and H. Osmólska, Eds.), pp. 320–344. Univ. of California Press, Berkeley.

Gilmore, C. W. (1920). Osteology of the carnivorous dinosauria in the United States National Museum, with special reference to the genera *Antrodemus (Allosaurus)* and *Ceratosaurus*. *Bull. U.S. Natl. Museum* **110,** 1–159.

Lansdown, A. B. G. (1968). The origin and early development of the clavicle in the quail (*Coturnix c. japonica*). *J. Zool. London* **156,** 307–312.

Madsen, J. H. (1976). *Allosaurus fragilis*: A revised osteology. *Utah Geol. Miner. Survey Bull.* **109,** 1–163.

Makovicky, P. J., and Currie, P. J. (1996). Discovery of a furcula in tyrannosaurid theropods. *J. Vertebr. Paleontol.* **16**(Suppl. to No. 3), 50A.

Martin, L. D. (1983). The origin of birds and of avian flight. *Curr. Ornithol.* **1,** 105–129.

McIntosh, J. S. (1990). Sauropoda. In *The Dinosauria* (D. B. Weishampel, P. Dodson, and H. Osmólska, Eds.), pp. 345–401. Univ. of California Press, Berkeley.

Ostrom, J. H. (1976). *Archaeopteryx* and the origin of birds. *Biol. J. Linnean Soc.* **8,** 91–182.

Ostrom, J. H. (1990). Dromaeosauridae. In *The Dinosauria* (D. B. Weishampel, P. Dodson, and H. Osmólska, Eds.), pp. 269–279. Univ of California Press, Berkeley.

Romer, A. S. (1956). *Osteology of the Reptiles*. Univ. of Chicago Press, Chicago.

Russell, A. P., and Joffe, D. J. (1985). The early development of the quail (*Coturnix c. japonica*) furcula reconsidered. *J. Zool. London A* **206,** 69–81.

Russell, D. A, and Dong, Z. (1993). A nearly complete skeleton of a new troodont dinosaur from the Early Cretaceous of the Ordos Basin, Inner Mongolia, People's Republic of China. *Can. J. Earth Sci.* **30,** 2163–2173.

Sereno, P. C. (1990). Psittacosauridae. In *The Dinosauria* (D. B. Weishampel, P. Dodson, and H. Osmólska, Eds.), pp. 579–592. Univ of California Press, Berkeley.

Sereno, P. C. (1991). Basal archosaurs: Phylogenetic relationships and functional implications. *J. Vertebr. Paleontol.* **11**(Suppl. 4), 1–53.

Sternberg, C. M. (1951). Complete skeleton of *Leptoceratops gracilis* Brown from the Upper Edmonton member on Red Deer River, Alberta. *Bull. Natl. Museum Can.* **123,** 225–255.

Wellnhofer, P. (1993). Das siebte Exemplar von *Archaeopteryx* aus den Solnhofener Schichten. *Archaeopteryx* **11,** 1–48.

Pelvis, Comparative Anatomy

Diego Rasskin-Gutman
National Museum of Natural History, Smithsonian Institution
Washington, DC, USA

The pelvic girdle is a bony structure positioned between the sacral vertebrae and the hindlimbs that is composed of three paired endochondral bones—the ilium, pubis, and ischium—that together form an articular socket for the hindlimb called the acetabulum. The ventral opening between the pubis and the ischium is known as the obturator fenestra. In lateral view, the triradiate design shows three rod-like radii. The ilium extends dorsally and the pubis and the ischium are positioned ventrally, with the pubis anterior and the ischium posterior (see Fig. 1). The open acetabulum is more or less diagnostic of the pelvis of the Dinosauria. The primitive condition of having a closed or partly open acetabulum is found in its ornithodiran outgroups.

The two main groups of dinosaurs, Ornithischia and Saurischia, can be distinguished because of the different spatial disposition of the elements of the pelvis. The Saurischia ("reptile-hipped") are mainly characterized by the plesiomorphic condition of having the pubic shaft pointing forward (propubic). The Ornithischia ("bird-hipped") have the original pubic shaft pointing backward (opisthopubic). No name has been proposed for a totally downward (vertical) position of the pubis, as in some coelurosaurs (e.g., the Dromaeosauridae). A suitable name, however, would be ortho- or mesopubic, indicating the intermediate position between the pro- and opisthopubic conditions.

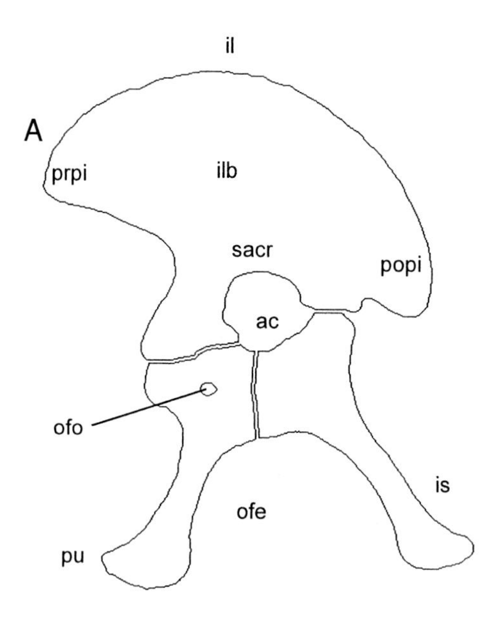

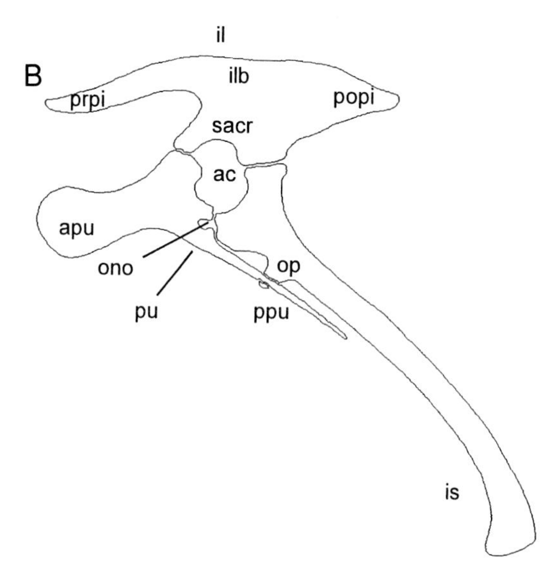

FIGURE 1 Some characters in a schematic lateral view, left side of dinosaur pelvis. (A) Saurichia and (B) Ornithischia; drawn to the same size. ac, acetabulum; apu, anterior ramus of the pubis (prepubis); il, ilium; ilb, iliac blade; is, ischium; ofe, obturatum fenestra; ofo, obturator foramen; ono, obturator notch; op, obturator process; popi, postacetabular process of the ilium; prpi, preacetabular process of the ilium; pu, pubis; sacr, supra-acetabular ridge.

Despite the etymology of the term Ornithischia, one must bear in mind that birds did not evolve from ornithischian dinosaurs, but rather from coelurosaurian theropods closely related to the Dromaeosauridae (Gauthier, 1986), whose public position is almost certainly evolutionarily intermediate between typical saurischians and birds. The opisthopubic condition of the pelvis in both ornithischians and birds is considered an evolutionary convergence. On the other hand, the Therizinosauria, a strange group from the Cretaceous of China and Mongolia, shows a clear opisthopubic condition in the pelvic girdle, although this was evidently acquired independently of dromaeosaurs and birds.

The pelvis is connected to other skeletal structures: the vertebral column, the hindlimbs, and, in some groups, the posterior gastralia. The hindlimb and the pelvis are joined between the femoral head and the acetabulum. The medial face of the iliac blade contacts the vertebral ribs of the sacral region, or sacrum, of the column. Some dorsal vertebrae and some proximal caudals may also participate in the sacrum. The number of vertebrae taking part in the sacral complex varies from 3 sacrals (e.g., the prosauropod *Plateosaurus)* to more than 10 (presacrals, sacrals, and caudals) in the Ankylosauria and in extant birds. These vertebrae are often fused, contributing to the strength and the structural support of the whole pelvic region. The functional relationship between the pubis and the gastralia (also known as ventral ribs, although their origins are dermal and not endochondral) has yet to be studied. Gastralia are found in saurischians, including primitive birds, but not derived birds, and are totally absent in the Ornithischia.

The pelvic girdle has a dual role in locomotion. First, it supports the weight of the individual via the hindlimbs. Second, it connects the hindlimbs to the rest of the organism through the acetabulum. As a result of its position, the pelvic girdle has also secondarily taken on protective functions. Along with the sacral region of the vertebral column it forms a bony ring that surrounds the posterior coelomic cavities and the respiratory, digestive, and urogenital systems. The formation of the pelvic ring is especially distinctive in Sauropodomorpha, in which the three bones of the hip are, in general, sturdier than in the other Saurischians and are twisted and medially fused forming a characteristic "apron-like" sheet retained in many respects from dinosaurian outgroups. Theropods have markedly rod-like pubes that usually contact each other only on their distal ends. In derived birds, however, the symphyses are lost.

Dinosaur hips have several distinctive features compared to other archosaurs: (i) a deep, open acetabulum; (ii) a well-developed preacetabular process of the iliac blade; (iii) an iliac supra-acetabular crest or ridge; and (iv) elongated pubis and ischium. This consistent pattern has been associated with a phylogenetic trend toward an upright posture. The deep open acetabulum is associated with the parasagittal position of the hindlimb. Thus, the femoral head in dinosaurs points inwardly at a right angle to the femoral shaft so that it fits inside the socket provided by the acetabulum. A stout iliac supra-acetabular crest or ridge helps to transmit the weight of the individual to the ground. The well-developed preacetabular process is associated with an increase in the surface area for muscular attachments that are involved in femoral protraction (M. iliofemoralis).

Most characters found in the pelvic girdle are related to the shape and proportions of the bone processes. Some, like the preacetabular and postacetabular processes of the ilium, allow the pelvic musculature to be attached efficiently to this structural support. Other processes are involved in the joints of the three bones. Each pelvic bone has processes contacting the other two bones. The medial face of the iliac blade has sutural contacts for the sacral ribs. Muscles attached to the pelvic bones in tetrapods are involved in both locomotory and nonlocomotory functions. However, mostly the former have been studied in dinosaurs. Important locomotory muscles with attachment origin in the pelvic girdle are the iliofemoralis, ischiofemoralis, caudofemoralis brevis, puboischiofemoralis, and iliotibialis. Openings or fenestrae in the pelvic bones are the obturator foramen and pubic fenestra in the pubis and the ischiadic foramen in the ischium. Other important characters found in the pelvic girdle are the presence or absence of a medial contact (symphysis) between the pubes and/or the ischia, the presence or absence of an anterior and/or posterior (brevis) fossa for muscle attachment in the ilium, the presence or absence of the antitrochanter on the lateral face of the ilium, the presence or absence of an obturator process on the ischium, the relative position of the pubis (forward, downward, or backward), the presence or absence of an anterior ramus on the pubis,

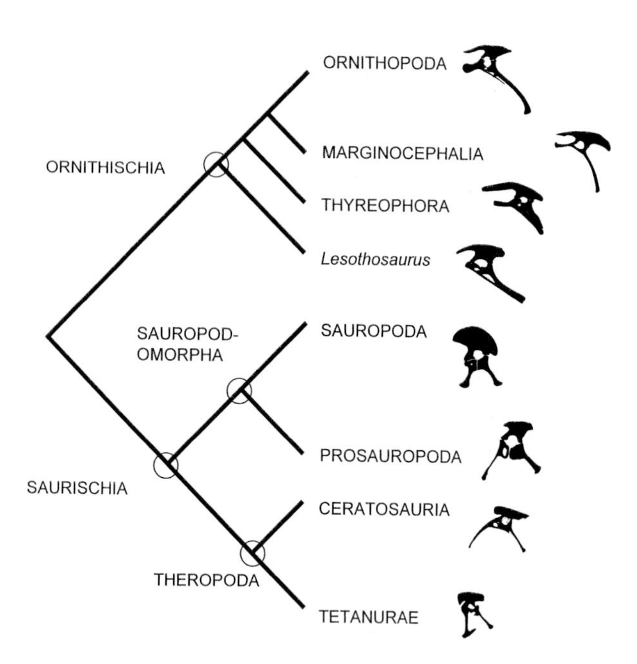

FIGURE 2 General consensus of the phylogenetic relationships among the Dinosaurian groups discussed in this chapter. The drawings are in lateral view, left side; not to scale.

the presence or absence of a pubic foot, and the mediolateral apron-like form of the pubis. Figure 2 shows the main groups of dinosaurs with some pelvic girdles.

The Theropoda pelvic girdle is primitively propubic, although several coelurosaurian groups (e.g., Maniraptora) show a more downward (vertical) position of the pubic shaft. This condition has led some authors to refer to the dromeosaur pelvis as being opisthopubic. However, controversy remains over the exact orientation of the pubis in this clade; some authors reconstruct it as being truly opisthopubic, whereas others position it more vertically (mesopubic). It was clearly variable because birds are fully opisthopubic. The preacetabular blade in theropods is long, a condition that has been termed dolichoiliac. A pronounced brevis fossa (Gauthier, 1986) on the caudal part of the ilium projects internally behind the acetabulum, presumably for the attachment of the M. caudofemoralis brevis. This fossa is broad in ceratosaurs but narrow in tetanurans.

In Ceratosauria, all three paired bones are fused in adults. In this group, the pubis retains a plate-like shape proximally, whereas it is rod-like distally. This proximal plate-like portion of the pubis is rarely preserved and fuses proximally to the ischium (forming a puboischiadic plate, as in archosaurs primitively). Two fenestrae are present in this plate: the obturator foramen, small and located immediately ventral to the acetabulum, and a larger pubic fenestra, found only in Ceratosauria, located immediately ventral to the obturator foramen. The rod-like portion of the pubis is sometimes anteriorly convex.

In Tetanurae, the ischium shows an ischiadic foramen and an obturator process. The pubis has a distal expansion called the pubic foot or boot that points backward. The ilium is compressed mediolaterally and has an anterior fossa, presumably for the insertion of the M. cupedicus. In Ornithomimosauria, the preacetabular process is curved downward, culminating in a sharp end.

The Sauropodomorpha pelvic girdle is propubic and brachyiliac in the shape of the iliac blade (as opposed to the dolichoiliac condition of Theropoda). This condition is characterized by a short postacetabular and preacetabular processes. The pubis of this clade presents two very important features: (i) a strong symphysis that fuses both pubes, which are proximally twisted, and (ii) the presence of a large obturator foramen, immediately behind the acetabular portion. The pelvis of the prosauropoda has a distinctive apron-like distal part of the pubis. In Sauropoda, the pubic process of the ilium is longer than the ischiadic one, and the preacetabular process is markedly convex. These features are also primitive for dinosaurs, being found in their outgroups.

The Ornithischia have several synapomorphic characters in their hips. The most significant one is the true opisthopubic condition, in which the pubis has rotated to lie parallel to the ischium shaft. There is also a well-developed prepubic process or anterior ramus of the pubis that, in primitive forms such as *Lesothosaurus*, is small, whereas in more derived forms it is conspicuous, pointing forward, upward, and outward and almost as large as the posterior ramus (in many forms, e.g., *Iguanodon*, the prepubis is larger than the posterior ramus). The ilium shows a lateral expansion of the ischiadic process, a long and thin preacetabular process, and a deep caudal process. The pubis shows an obturator notch rather than a foramen as in Sauropodomorpha. The medial symphyses are restricted to the distal ends in both the pubis and the ischium.

In the Stegosauria, proximal bony sheets of both pubis and ischium partially occlude the acetabulum. The ischium ends sharply. The other group of the Thyreophora, the Ankylosauria, have sacral and caudal ribs fused to the medial face of the iliac blade. The pubis is reduced and often is not part of the acetabulum. The preacetabular process of the ilium is expanded and the acetabulum is closed secondarily.

In the PACHYCEPHALOSAURIA, the ischium shows a long proximal pubic process that, rather than contacting the pubis, contacts the pubic process of the ilium, thus closing the acetabulum. As a consequence, the pubis is not part of the acetabulum. This bone has a reduced posterior ramus, while showing a conspicuous prepubis. This ramus is flattened dorsoventrally, as is the preacetabular process of the ilium. The CERATOPSIA have an iliac process called the antitrochanter. This lateral projection of the supra-acetabular margin of the ilium is very distinctive and is shared by hadrosaurs. Overall, the Ceratopsian pelvis is broad transversely, with stout sacral ribs fused to the inner surface of the iliac blade.

The most important features of the pelvis of the ORNITHOPODA, in relation to the remaining Ornithischia, is the enlargement of the anterior ramus of the pubis and the presence of a tabular obturator process on the ischiadic shaft. However, primitive ornithopods, such as the Heterodontosauridae, lack these two features.

Trends in the evolution of the dinosaurian pelvis can be summarized as follows. In the immediate outgroups of dinosaurs the pelvis is basically triradiate, as it is in tetrapods plesiomorphically. The ilium is short, with very little extension anterior to the acetabulum and only a moderate postacetabular extension. Two sacral vertebrae were the condition in dinosaurian outgroups except PTEROSAURIA, which had at least 3 primitively. HERRERASAURS retain the condition of 2 sacrals, although a third vertebra (the last dorsal) is incipiently sacralized. Five sacrals were soon found in theropods, sauropodomorphs, and various ornithischian lineages, which could incorporate up to 6 or 8 (pachycephalosaurs), 9 (neoceratopsians), or 10 (ankylosaurs). The ilium generally elongated to accommodate these vertebrae, and the ventral side of its postacetabular portion took on a deep, inverted U-shape to house the powerful caudofemoralis muscle. The ilium is also rather tall in sauropods and in tyrannosaurs, perhaps as a functional correlate of size. It is laterally everted and recurved as a "hood" over the hip region in ankylosaurs.

The ischium, generally the most conservative of the three bones, primitively points posteroventrally at an angle of about 45° to the horizontal. In dinosaurs plesiomorphically it is about the length of the femur, as is the pubis, but it shortens to about two-thirds the pubic length in maniraptoran coelurosaurs (then elongates again in birds), becomes thin and rod-like in hadrosaurines, attains a pronounced anteroposterior bow in marginocephalians and a posterior bend in some sauropods, and is very much reduced in nodosaurid ankylosaurs. The ischium is reduced to a tiny splint fused to the pubis in alvarezsaurids such as *Mononykus* (see AVES).

The pubis is the most variable of the three bones. Plesiomorphically it pointed down and forward at an angle of about 45° to the horizontal, and the left and right halves were connected by a broad apron. This is the plesiomorphic condition for archosaurs, as seen, for example, in the basal archosauromorph *Euparkeria,* and it was also clearly retained by basal sauropodomorphs; the apron became narrower but was still basally retained in Theropoda, particularly in Ceratosauria. The distal end of the pubis in theropods was originally rounded and knob-like, as in basal sauropodomorphs, and the ischium was slightly more robust than the pubis; however, in Tetanurae a blade-like posterior process, sometimes called the "foot" or "boot," appeared on the distal end of the pubis, and a corresponding anterior extension is present in ornithomimids, allosaurs, and tyrannosaurs, among other coelurosaurs. In maniraptorans, as noted previously, the pubis appears to be retroverted, but the degree of retroversion is not always clear for many taxa because the bones are not preserved in normal articulation. The degree of retroversion may have varied among closely related taxa, including *Deinonychus*, *Velociraptor*, and *Archaeopteryx*. Within Aves the pubis is fully retroverted, and within Ornithurae it is usually essentially horizontal (see AVES). Retroversion appears to have occurred independently in therezinosaurids, which are closer to ornithomimids than to maniraptorans (see COELUROSAURIA).

The pubis in ornithischians is fully retroverted, even in basal forms such as *Pisanosaurus* and *Lesothosaurus*. Plesiomorphically the original pubic prong lies parallel to the ischium, contacted by the obturator process, and the two bones are nearly of the same length. A secondary process of the pubis extends not downward and inward, like the original pubis, but slightly upward and outward. This forward process of the pubis is rudimentary at first and does not protrude anteriorly as far as the preacetabular process of the ilium. Both processes elongate at approximately the same rate in more derived forms, with salient exceptions: Ankylosaurs reduce the pubis almost completely, and in ornithopods beginning with hypsilophodontids the anterior pubic process is longer than the preacetabular iliac process. In iguanodontids and hadrosaurs the anterior end of the pubic process flares into a spatulate shape. Meanwhile, the original prong of the pubis retains its more or less original form in stegosaurs, is lost entirely in ankylosaurs, and shrinks to a rod about two-thirds of the ischial length in iguanodontids and to half its length or less in hadrosaurs; it extends only slightly behind the acetabulum in pachycephalosaurs and basal ceratopsians, convergently in some lambeosaurine hadrosaurs, and does not extend posteriorly past the acetabulum in most ceratopsids and some pachycephalosaurs.

The acetabulum is primitively a closed structure in archosaurs, and this condition is retained by pterosaurs, *Lagerpeton*, and other basal ornithodirans. It has been described as open or partly open in *Lagosuchus*, a form very close to dinosaurs, but this has also been described as artefactual. The pelvis is partly open in herrerasaurs and dinosaurs plesiomorphically, and the supra-acetabular ridge is more pronounced, especially in ceratosaurian theropods. The acetabulum becomes fully open independently within theropods, sauropodomorphs, and ornithischians, although it is secondarily closed in ankylosaurs and incipiently so in stegosaurs. There is no strict delineation between "open" and "partly open" except as a matter of degree, usually referring to the extent of the emarginated inner wall of the acetabulum and the contacts between the pelvic bones.

See also the following related entries:

BIPEDALITY • HINDLIMBS AND FEET • QUADRUPEDALITY

References

Charig, A. J. (1972). The evolution of the archosaur pelvis and hind-limb: An explanation in functional terms. In *Studies in Vertebrate Evolution* (K. A. Joysey and T. S. Kemp, Eds.), pp. 121–155. Winchester, New York.

Galton, P. M. (1969). The pelvic musculature of the dinosaur *Hypsilophodon* (Reptilia: Ornithischia). *Postilla* **131,** 1–64.

Gatesy, S. M. (1990). Caudofemoral musculature and the evolution of theropod locomotion. *Paleobiology* **16**(2), 170–186.

Gauthier, J. A. (1986). Saurischian monophyly and the origin of birds. *Mem. Calif. Acad. Sci.* **8,** 1–55.

Parrish, J. M. (1986). Locomotor adaptations in the hindlimb and pelvis of the Thecodontia. *Hunteria* **1**(2), 2–35.

Rasskin-Gutman, D., and Buscalioni, A. D. (1996): Affine transformation as a model of virtual form change for generating morphospaces. In *Advances in Morphometrics* (L. F. Marcus, M. Corti, A. Loy, D. Slice, and G. Naylor, Eds.), pp. 169–178. Plenum, New York.

Romer, A. S. (1923). Crocodilian pelvic muscles and their avian and reptilian homologues. *Bull. Am. Museum Nat. History* **48,** 533–552.

Walker, A. D. (1977). Evolution of the pelvis in birds and dinosaurs. In *Problems in Vertebrate Evolution* (S. M. Andrews, R. S. Miles, and A. D. Walker, Eds.), pp. 319–357. Academic Press, New York.

Weishampel, D., Dodson, P., and Osmólska, H. (Eds.) (1990). *The Dinosauria.* Univ. of California Press, Berkeley.

Permineralization

CLIVE TRUEMAN
University of Bristol
Bristol, United Kingdom

Fresh bone contains approximately 24–26% organics and has a porosity of approximately 40% (Wang and Cerling, 1994), whereas fossil bones can be broadly characterized by a loss of organic material within the bone and infilling of the pore spaces with diagenetic minerals.

The loss of organics occurs principally when reduced organic matter (such as fatty acids, marrow tissue, blood cells, etc) is oxidized either biologically or abiologically. The loss of this material creates spaces within the vascular network of the bone, allowing pore waters to flow through the bone. The flow of these fluids subjects the bone to chemical conditions different from those found in the body, where the bone mineral was held in equilibrium. These nonequilibrium conditions cause chemical changes in the bone mineral such as a loss of structural carbonate and increased crystallinity. These changes to the apatite lattice are the principal chemical differences between fresh and fossil bones, actual replacement of bone mineral being rather rare.

The pore waters that flow through the bone are also enriched in various ions, obtained by weathering of rocks and soils. When these waters reach the chemical fronts encountered by the changing bone material, and oxidizing organic matter, they exchange many of their dissolved ions. Many of the trace elements carried by these waters are incorporated directly into the bone apatite, where they may be used to give information concerning the environment of death and burial, but many of the major elements are precipitated within the vascular spaces, forming the infilling or permineralizing minerals.

Common permineralizing minerals are calcite and silica. Calcite ($CaCO_3$) is the most common permineralizing mineral, principally due to the ease of transport of Ca^{2+} and CO_3^- ions in solution. Weakly acidic waters may transport both of these, and a mild chemical front will stimulate their precipitation as calcite. Silica (SiO_4) is also a very common permineralizing phase but requires a lower pH for transport. Silica is, however, the most common constituent of all rocks on earth, and therefore forms a very important phase in all pore waters. Commonly, waters that contain little or no calcite will mineralize with silica; however, the two phases are not mutually exclusive.

Other common permineralizing mineral phases include pyrite (FeS_2). This is very common in anoxic conditions. Other more unusual phases may include barite ($BaSO_4$), gypsum ($CaSO_4 \cdot 2H_2O$), opal (a hydrated variety of silica), siderite (Fe_2CO_3), and malachite ($CuCO_3 \cdot OH$).

Mineralization reduces porosity and hence forms an effective seal to further pore water transport; therefore, frequently only one or two phases of minerals are seen. However, when a bone has been transported through many environments in a short space of time, many different phases of infilling minerals may be present, and by careful observation of the relationships of these minerals the taphonomic pathway of that bone may be deduced. Jane Clarke has used this feature of permineralization to deduce the complex taphonomic history of a reworked *Iguanodon* vertebra, with the changing chemical conditions of burial producing different suites of authigenic minerals (Clarke, 1991; Clarke and Barker, 1993). This data may be presented in two ways—as a "minstrat" diagram showing the order of mineral precipitation or as diagenetic pathways displayed on Eh/pH diagrams. Different chemical conditions will occur within individual bones remaining in the same environment, however, simply due to differential rates of pore water flow, organic decomposition, and local microenvironments. Therefore, a varied mineral suite may be displayed by bones with a seemingly simple taphonomic history.

Permineralizing phases may give information concerning the environment of fossilization but are not as informative as the wide suite of trace elements included in bone apatite during the early diagenetic passage of pore waters.

See also the following related entries:

BIOMINERALIZATION • CHEMICAL COMPOSITION OF DINOSAUR FOSSILS

References

Clarke, J. B. (1991). Diagenetic and compactional history of an *Iguanadon* vertebra from the Vectis Formation of the Isle of Wight. *Proc. Isle of Wight Nat. History Archaeol. Soc.* **10,** 149–158.

Clarke, J. B., and Barker, M. J. (1993). Diagenesis in *Iguanodon* bones from the Wealden Group, Isle of Wight, southern England. *Kaupia Darmstadtler Beitrage Naturgeschichte* **2,** 55–65.

Wang, Y., and Cerling, T. E. (1994). A model of fossil tooth and bone diagenesis: Implications for paleodiet reconstruction from stable isotopes. *Palaeogeogr. Palaeoclimatol. Palaeoecol.* **107,** 281–289.

Peruvian Dinosaurs

see SOUTH AMERICAN DINOSAURS

Pes

see HINDLIMBS AND FEET

Petrifaction

see PERMINERALIZATION

Petrified Forest

J. Michael Parrish
Northern Illinois University
DeKalb, Illinois, USA

Petrified Forest National Park, located in east-central Arizona, consists of a series of spectacular exposures of the Upper Triassic Chinle Formation including the scenic badlands of the Painted Desert. As the name suggests, this area was originally noted for its extensive exposures of giant, fossilized logs, which occur in locally abundant assemblages throughout the park and the adjoining region. The logs were known to the native Americans of the region, and the Piutes apparently recognized them as fossilized wood (Ash, 1972). Collections of the petrified wood were made by Army expeditions as early as the 1850s, and the first geological studies of the region date from that decade as well. With enthusiastic collectors threatening the fossil resources of the region, Lester Ward of the U. S. Geological Survey protected the region in 1899 and it was subsequently designated a national monument in 1906, then expanded and made a national park in 1962 (Ash, 1972).

Although known largely for its fossil wood, the Petrified Forest is also the site of many significant vertebrate fossil occurrences, mostly from the fossiliferous Petrified Forest Member of the Chinle Formation. Significant collections from the park are now housed at a number of institutions around the country, most notably the American Museum of Natural History and the University of California Museum of Paleontology, whose collections include a few fossils found by John Muir in the early 1900s.

Chinle sediments in the park region comprise fluvial systems and associated floodplain and pond deposits. The lower exposures are dominated by meandering streams, and the fossil soils in the floodplain sediments indicate that the entire system was extremely wet. This is also supported by the abundance of aquatic taxa, notably phytosaurs and metoposaurid amphibians. Later exposures consist of braided streams and floodplain sediments indicating somewhat drier conditions. It is from these sediments that the greatest abundance of fully terrestrial vertebrate taxa are found (Parrish, 1993). At its uppermost extent, the Park's Chinle exposures are made up by the Owl Rock Member, which represents a record of a widespread lake present late in Chinle history (Dubiel, 1993).

The Triassic faunas of Petrified Forest National Park are dominated by phytosaurian archosaurs, the armored archosaurian aetosaurs, and by the flat-headed metoposaurid amphibians. Rarer elements of the fauna include the hippo-like dicynodont *Placerias*, long-legged cursorial crocodylomorphs, and the large, predatory poposaurid archosaurs. At least three distinct types of early dinosaurs have been found in the park. A specimen of *Coelophysis* was discovered in the Petrified Forest in 1982 (Padian, 1986). The same locality also produced a number of teeth of the herbivorous ?ornithischian *Revueltosaurus* (Padian, 1990). In 1983, a partial skeleton of the herrerasaurid dinosauromorph *Chindesaurus* (Long and Murry, 1995) was discovered in the Painted Desert.

Because of its extensive exposures of Chinle rocks, badlands topography, and abundant records of fossil floras and faunas, Petrified Forest National Park remains one of the most fruitful places to study the ecosystems of the Late Triassic.

See also the following related entries:

Chinle Formation • Dockum Group • Extinction, Triassic • Museums and Displays • Newark Supergroup • Triassic Period

References

Ash, S. R. (1972). The search for plant fossils in the Chinle Formation. *Museum Northern Arizona Bull.* **47,** 45–58.

Dubiel, R. F. (1993). Depositional setting of the Owl Rock Member of the Chinle Formation, Petrified Forest National Park and Vicinity, Arizona. *New Mexico Museum Nat. History Bull.* **3,** 117–122.

Long, R. A., and Murry, P. A. (1995). Late Triassic (Carnian and Norian) tetrapods from the Southwestern United States. *New Mexico Museum Nat. History Sci. Bull.* **4,** 1–254.

Long, R. A., and Padian, K. (1986). Vertebrate biostratigraphy of the Late Triassic Chinle Formation, Petrified Forest National Park, Arizona: Preliminary results. In *The Beginning of the Age of Dinosaurs* (K. Padian, Ed.), pp. 161–169. Cambridge Univ. Press, New York.

Padian, K. (1986). On the type material of *Coelophysis* Cope (Saurischia: Theropoda), and a new specimen from the Petrified Forest of Arizona (Late Triassic: Chinle Formation). In *The Beginning of the Age of Dinosaurs* (K. Padian, Ed.), pp. 45–60. Cambridge Univ. Press, New York.

Padian, K. (1990). The ornithischian form genus *Revueltosaurus* from the Petrified Forest of Arizona (Late Triassic: Norian; Chinle Formation). *J. Vertebr. Paleontol.* **10,** 268–269.

Parrish, J. M. (1993). Distribution and taxonomic composition of fossil vertebrate accumulations in the Upper Triassic Chinle Formation, Petrified Forest National Park. *New Mexico Museum Nat. History Bull.* **3,** 393–396.

Phalanges

see FORELIMBS AND HANDS; HINDLIMBS AND FEET

Phylogenetic System

KEVIN PADIAN
University of California
Berkeley, California, USA

Phylogenetic systematics and phylogenetic taxonomy obviate the utility of the Linnaean system, and many concepts of nomenclature and priority require reconsideration as a result (see SYSTEMATICS). De Queiroz and Gauthier (1990, 1992, 1994; see also Rowe, 1987), in a series of papers, have begun to lay out the basic framework of such reconsiderations. Some basic concepts may be summarized.

First, taxa in the phylogenetic system are defined by their ancestry in several different ways. A node-based taxon is the clade stemming from the most recent common ancestor of two other taxa. For example, Dinosauria has been defined as all descendants of the most recent common ancestor of *Triceratops* and birds (Padian and May, 1993). A stem-based taxon includes all entities that share a more recent common ancestor with one recognized taxon than with another. For example, Saurischia includes birds and all dinosaurs that are closer to birds than to *Triceratops*, and Ornithischia includes *Triceratops* and all dinosaurs that are closer to it than to birds (Padian and May, 1993). These are the most stable kinds of definitions because they rely on ancestry, which logically cannot be affected by changes in group membership or diagnoses. Other kinds of definitions include the apomorphy-based definition, in which a clade is defined by the first ancestor to possess a particular synapomorphy, and the taxon-based definition, comprising a given list of taxa, presumably united by their most recent common ancestor. However, the apomorphy-based definition can change easily if the apomorphy is found to pertain to a more general level (e.g., if feathers are given to define birds but later turn up in nonavian theropod dinosaurs), or if, on the basis of the distribution of other synapomorphies, the apomorphy turns out to be a homoplasy (e.g., if a taxon of theropod dinosaurs is based on large size, and it turns out that large size evolved several times in different lineages). The taxon-based definition may be the most unstable: For example, the original membership of ORNITHOSUCHIA (Gauthier, 1986, p. 43) included *Euparkeria* and Ornithosuchidae, two taxa that later turned out to be outside ARCHOSAURIA and members of PSEUDOSUCHIA, respectively (Fig. 1). Had Ornithosuchia and Pseudosuchia been taxon-based instead of stem-based taxa within the node-based Archosauria (Gauthier, 1986, p. 42), tremendous confusion would have resulted.

Second, whereas taxa are defined by ancestry, they are diagnosed by synapomorphy (Rowe, 1987). As noted previously, every synapomorphy at one level is a plesiomorphy at a less inclusive level, so the hierarchy of nested sets of synapomorphies is what ultimately tests our concepts of ancestry and thereby the validity of taxonomic groups. It is important to use the dual criteria of definition and diagnosis because, as knowledge increases, the level at which any given synapomorphy applies is subject to change. This will not affect the validity of the taxon because it is defined by ancestry; however, if it eventuates that no synapomorphies characterize a given taxon, then it will be subsumed by the next most general level or will become a metataxon (Gauthier, 1986). However, this depends on the way that the taxon was defined. For example, with the removal of *Euparkeria* and Ornithosuchidae from Ornithosuchia, some workers considered that Ornithosuchia and ORNITHODIRA were redundant names for the same taxon.

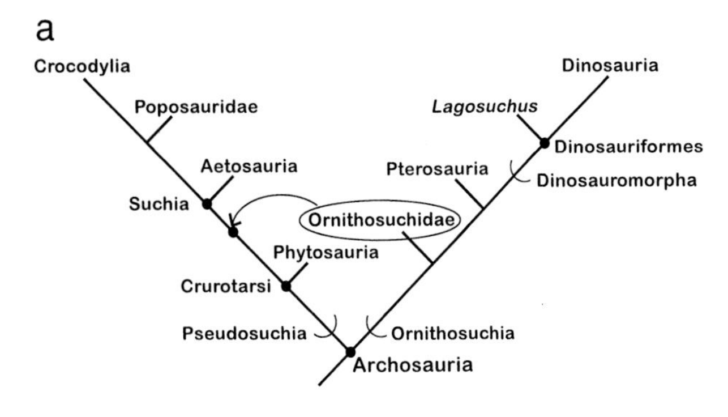

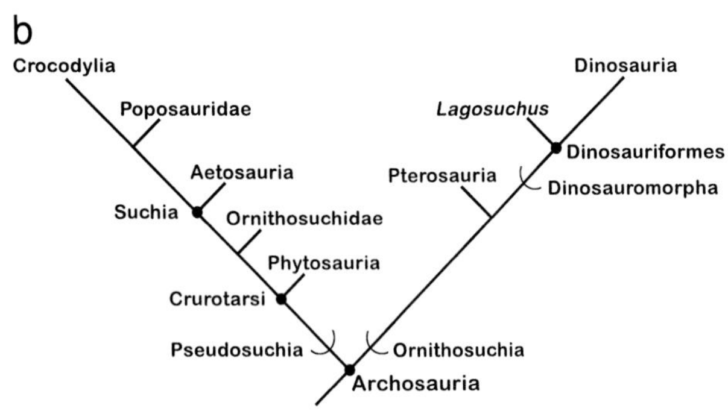

FIGURE 1 The removal of *Euparkeria* and Ornithosuchidae from Ornithosuchia (a) did not affect the definition of the latter taxon, although it did affect its composition and diagnosis (b) because it was diagnosed by other synapomorphies.

This is not true: Ornithosuchia is a stem-based taxon, whereas Ornithodira is a node-based taxon (Gauthier, 1986). It would still be possible to find a new animal that was closer to birds than to crocodiles, yet not possessing all the synapomorphies of Ornithodira. If this animal, for example, had some but not all ornithodiran synapomorphies, its position within Ornithosuchia outside Ornithodira would be clear, and Ornithodira itself would not be affected in either definition or ancestry. However, the synapomorphies shared by the newly discovered taxon would now apply to a more general level. These eventualities are predicted and accommodated by the phylogenetic system.

Other Uses of Phylogenetic Systematics

In addition to constructing phylogenies, cladograms can provide tests of various kinds of evolutionary hypotheses and can provide insight into long-standing evolutionary problems by approaching them in a different way (Padian *et al.*, 1994). Increasingly, cladistic analysis has been used in conjunction with hypotheses of vicariance biogeography based on understanding of tectonically driven changes in the positions and configurations of continental and oceanic masses through time (see BIOGEOGRAPHY).

Phylogeny and Stratigraphy Fine-scaled biostratigraphic work depends on fine-scaled taxonomy. For this reason, the use of monophyletic groups is important because paraphyletic taxa can suggest both artificially short and long ranges. Cladistic analysis can show that if two groups are sister taxa, even though the fossil record of one is more ancient, then the other's must be equally ancient. These "ghost lineages" (Norell, 1992) provide us with expectations of what we should be finding in the fossil record, given appropriate facies and circumstances of preservation. The next logical step is to compare the sequence of phylogenetic splitting with the sequential record of preservation in the geologic column in order to get a quantification of how complete the record is among groups (Norell and Novacek, 1992a,b), and this work shows striking differences in completeness among the records of dinosaurian taxa. Finally, studies of the diversity of groups through time depend on the use of monophletic taxa. The most obvious examples from the vertebrate record include the difference in the evolutionary "relays" of dominance in diversity through time when the "mammal-like reptiles" are recognized not as reptiles but as synapsids (thus pulsing synapsid diversity in the Early and Late Permian, Early Triassic, and Cenozoic as opposed to the pulses of reptilian diversity in the Late Permian and Late Triassic through Cretaceous), and the recognition that birds are dinosaurs (thus implying that dinosaurs are not only not extinct but are also the most diverse clade of terrestrial tetrapods in the Cenozoic, thereby adding another "pulse" of reptilian diversity).

Molecular Paleontology It is now generally understood that changes in biomolecules do not occur at constant rates, although the rates may be quite regular in certain molecules in certain groups. Biochemical evolutionary clocks must be calibrated by external sources. Although occasionally claims are made to calibrate the separation of evolutionary lineages by "geological events," such as the opening of the Atlantic or the separation of Australia from other continents, in reality these "events" are usually far more imprecise and of greater duration in time than calibration against appearances in the fossil record.

The use of monophyletic taxa is of paramount importance in calibrating molecular clocks, but this is often not enough. For example, in order to determine the rates of change in some proteins in birds, their split from crocodiles, their closest extant sister taxon, is often used. The first known bird appeared approximately 150 mya (Late Jurassic) and the first known crocodilian (crocodylomorph, actually) approximately 210 mya (Late Triassic). Reasoning from the necessary presence of ghost lineages, we would have to expect earlier birds or bird relatives as far back as 210 mya, and in fact other dinosaurs and ornithodirans are indeed known as far back as the Late Triassic. However, the bird–crocodile split is even earlier because these two taxa are end members of the Pseudosuchia and Ornithosuchia, respectively, within Archosauria. Hence, the most ancient member of any of these last three groups will automatically establish the minimum age of the other two. Because the earliest known undisputed archosaur appears in the Middle Triassic, approximately 235 mya, the calibration of this split now increases by 25 my, or nearly 12%. The fossil record of turtles extends only to the Late Triassic, but cladistic analysis suggests that their sister taxa split from the other extant reptiles (Diapsida) in the Pennsylvanian—a 50% difference in stratigraphic range.

Evolution of Major Adaptations A phylogeny based on a suite of characters from all parts of the skeleton can test hypotheses about the process and pattern of adaptations seen in a given lineage. Hypotheses about the evolution of a function should be generated and supported based on evidence independent of the phylogeny, of course (Padian, 1994). However, phylogenies can test whether such functional hypotheses of evolution are congruent with the pattern of structural change using all the evidence of the skeleton. One may ask, for example, whether the ceratopsian frill first evolved for defense. In Neoceratopsia, the frills of the basal members and indeed most members of the clade are broadly fenestrated, whereas the frill is entire in *Triceratops*. Whether *Triceratops* used its entire frill as a shield is another question, but it cannot be said that the ceratopsian frill evolved for defense because the entire frill is not basal for the group.

See also the following related entries:

DINOSAURIA • PHYLOGENY OF DINOSAURS • SYSTEMATICS

References

de Queiroz, K., and Gauthier, J. (1990). Phylogeny as a central principle in taxonomy: Phylogenetic definitions of taxon names. *Systematic Zool.* **39,** 307–322.

de Queiroz, K., and Gauthier, J. (1992). Phylogenetic taxonomy. *Annu. Rev. Ecol. Systematics* **23,** 449–480.

de Queiroz, K., and Gauthier, J. (1994). Toward a phylogenetic system of biological nomenclature. *Trends Ecol. Evol.* **9,** 27–31.

Gauthier, J. A. (1986). Saurischian monophyly and the origin of birds. *Mem. California Acad. Sci.* **8,** 1–55.

Norell, M. A. (1992). Taxic origin and temporal diversity: The effect of phylogeny. In *Phylogeny and Extinction* (M. J. Novacek and Q. D. Wheeler, Eds.), pp. 89–118. Columbia Univ. Press, New York.

Norell, M. A., and Novacek, M. J. (1992a). The fossil record: Comparing cladistic and paleontologic evidence for vertebrate history. *Science* **255,** 1690–1693.

Norell, M. A., and Novacek, M. J. (1992b). Congruence between superpositional and phylogenetic patterns: Comparing cladistic patterns with fossil records. *Cladistics* **8,** 319–338.

Padian, K. (1994). Form vs. function: The evolution of a dialectic. In *Functional Morphology and Vertebrate Paleontology* (J. J. Thomason, Ed.), pp. 264–277. Cambridge Univ. Press, New York.

Padian, K., and May, C. L. (1993). The earliest dinosaurs. *New Mexico Museum Nat. History Sci. Bull.* **3,** 379–381.

Padian, K., Lindberg, D. R., and Polly, P. D. (1994). Cladistics and the fossil record: The uses of history. *Annu. Rev. Earth Planet. Sci.* **22,** 63–91.

Rowe, T. R. (1987). Definition and diagnosis in the phylogenetic system. *Systematic Zool.* **36,** 208–211.

Phylogeny of Dinosaurs

KEVIN PADIAN
University of California
Berkeley, California, USA

Since the early 1970s, the acceptance and use of phylogenetic systematics, or cladistics, has grown to a virtual hegemony. VERTEBRATE PALEONTOLOGY was one of the earliest disciplines to incorporate cladistic analyses of phylogenetic relationships, and analyses of major dinosaurian subgroups were well in place by the mid-1980s, with phylogenies of subgroups proliferating to the present.

There are still some unanswered questions about the monophyly and relationships of some major dinosaur lineages, as well as of individual taxa and small groups. The following summary indicates consensus as well as continuing controversy (Fig. 1), which is discussed further in entries of individual taxa so noted. More extensive discussions of diagnoses and synapomorphies of individual taxa may also be found under their separate entries.

Dinosauria

The limits of Dinosauria are determined by their definition in the phylogenetic system. The two stem groups of Dinosauria are ORNITHISCHIA and SAURISCHIA; any taxon that does not belong to one of these two groups is not a dinosaur. Over the years, considerable controversy has surrounded the question of whether herrerasaurs are dinosaurs. The answer is yes if they share all the synapomorphies of Ornithischia + Saurischia. However, workers have yet to agree on a definitive list of synapomorphies for Dinosauria (e.g., Gauthier, 1986; Benton, 1990; Novas, 1991; Sereno, 1992, 1993; Sereno *et al.*, 1993; Sereno and Novas, 1992; Holtz and Padian, MS.). Many characters are ambiguous or subject to interpretation, and some are not known in the immediate outgroups of dinosaurs, such as *Pseudolagosuchus, Lewisuchus, Lagosuchus (Marasuchus), Lagerpeton*, and pterosaurs, or they may be autoapomorphically transformed in certain taxa (e.g., the forelimbs of pterosaurs or the apomorphic reduction of the phalangeal formula in *Eoraptor* and *Herrerasaurus*). Clearly a great number of characters indicate that *Eoraptor* and *Herrerasaurus* are the closest known taxa to true dinosaurs, if they are not dinosaurs themselves (see DINOSAURIA: DEFINITION; HERRERASAURIDAE; ORIGIN OF DINOSAURS; ORNITHOSUCHIA).

Ornithischia + Saurischia appear to share the following characters. (Note that these hypotheses of synapomorphy do not indicate what all members of these taxa share, but what their most recent common ancestor had; obviously all characters are subject to modification as evolution proceeds—by definition.) The manus is asymmetrical, with its fourth and fifth manual digits reduced but still retaining a phalangeal formula of 2-3-4-3-2. Claws are present only on the first three digits. There are three fully incorporated sacral vertebrae, and there is a true U-shaped brevis fossa on the ventral surface of the ilium behind the acetabulum. *Eoraptor* and *Herrerasaurus* do not share these features, but they share many others with Dinosauria, including a semiperforate acetabulum with supra-acetabular buttress, the loss of the postfrontal (which participates in the supratemporal opening), reduced fourth and fifth digits on the manus, and a well-developed ascending process of the astragalus situated on the anterior face of the tibia.

Ornithischia

This group is united by many features, including the toothless premaxillary tip, the predentary bone, reduced antorbital fenestra, at least slight buccal emargination of the jaw bones, the leaf-shaped teeth, jaw joint lower than the tooth row, five sacral vertebrae, and the opisthopubic pubis (Sereno, 1986). The basal members of this group include the poorly known *Pisanosaurus* (Late Triassic, Argentina) and *Technosaurus* (Late Triassic, Texas), as well as the better known *Lesothosaurus* (Early Jurassic, South Africa) (Weishampel and Witmer, 1990). The other ornithischians are united into a node-based group called GENASAURIA (Sereno, 1986) based on their possession of fully emarginated jaw bones, which suggest the presence of "cheeks" or muscular pouches lateral to

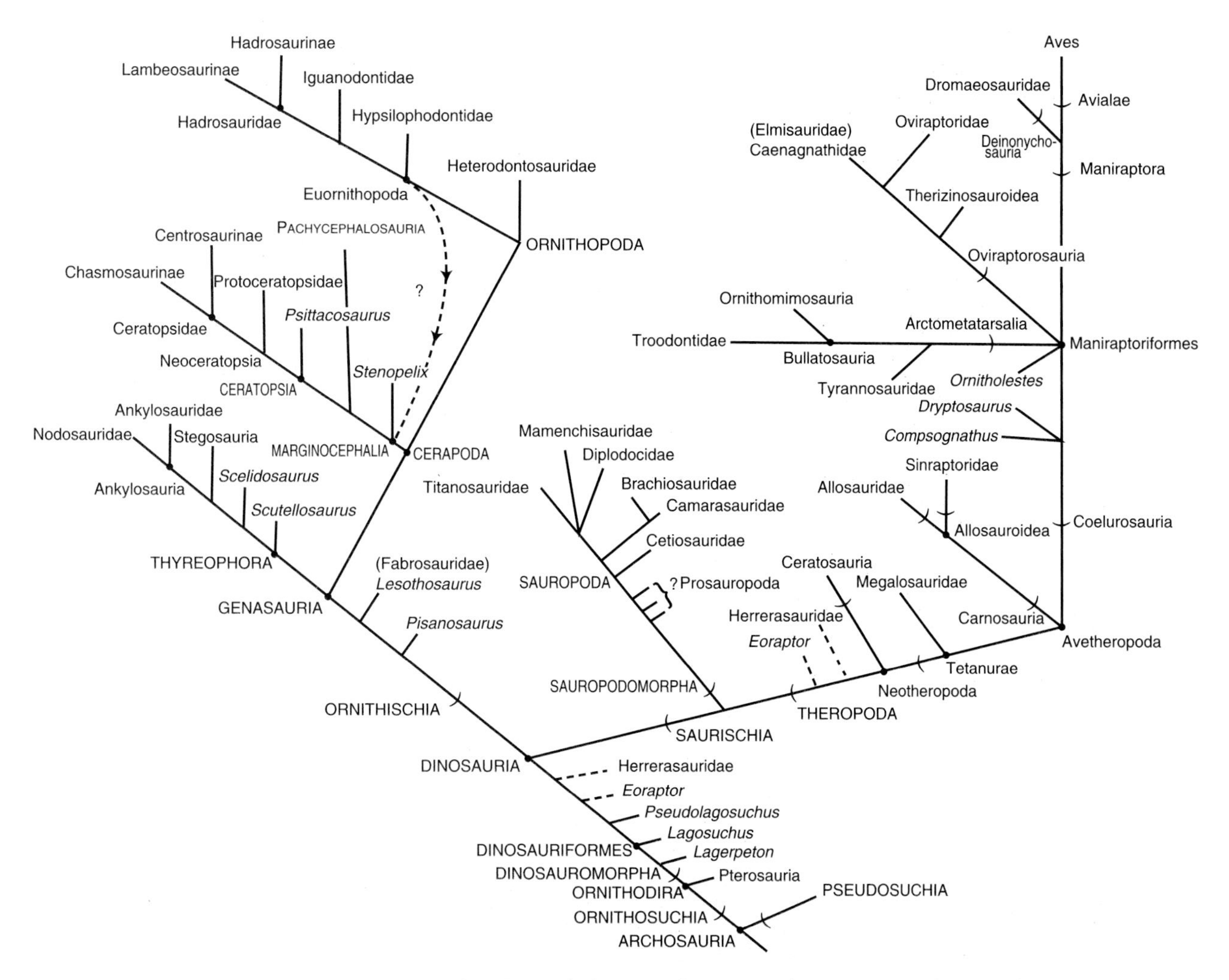

FIGURE 1 Phylogeny of Dinosauria

the mouth for storing and manipulating food while it is being chewed, as well as a spout-shaped symphysis on the lower jaw and several other features. Genasauria is in turn divided into two node-based groups: THYREOPHORA, the stegosaurs and ankylosaurs and their relatives, and CERAPODA, the remaining ornithischians (Sereno, 1986).

Thyreophora

This group is distinguished by keeled scutes of dermal armor in parasagittal rows along the body. The most basal well-known member of the group appears to be *Scutellosaurus* (Early Jurassic, Arizona), a small (skull length about 5 cm), apparently bipedal form with gracile forelimbs and hindlimbs, in which the tibia is slightly longer than the femur. Small scutes of various sizes and shapes appear to have covered the body and proximal tail. *Scelidosaurus* (Early Jurassic, England and Arizona) was considerably larger (skull length >20 cm) and presumably quadrupedal, though its tibia was also slightly longer than the femur, usually associated with a tradition of cursoriality and often bipedality. *Scelidosaurus* is considered closer to the remaining thyreophorans than is *Scutellosaurus* because it is larger and its jaws have more of the typical S-shaped curvature seen in stegosaurs and ankylosaurs, among other features (Sereno, 1986; Coombs *et al.*, 1990).

The STEGOSAURIA and ANKYLOSAURIA are the two largest clades of Thyreophora, both node defined (Sereno, 1986). Together, these two groups share two lateral supraorbital bones at the dorsal margin of the orbit, a rectangular scapula, long preacetabular and short postacetabular processes of the ilium, short metapodials, the loss of the first pedal digit, and other features (Sereno, 1986). Stegosaurs reduced the dermal armor to a single row of plates or spikes, often considerably enlarged, arranged in either an opposite or alternating pattern along the backbone, depending on the genus (Galton, 1990a); pairs of thick spikes appear to have been the basal condition, and a single, long shoulder spike is also found in most forms. Stegosaurs also have small heads and short, stocky forelimbs, with long straight hindlimbs in which the femur is considerably longer than the tibia. The unguals are hoofed. Stegosaurs are known from about a dozen genera, ranging from the Middle Jurassic of China and Europe through the Late Jurassic of China, North America, Europe, and Africa and the Early Cretaceous of China and Europe. Two Late Cretaceous forms from India have been questioned but otherwise are the only known post-Albian records.

Ankylosaurs are generally divided into two node-based clades, the Nodosauridae and the Ankylosauridae (Sereno, 1986; Coombs and Maryańska, 1990). Both have short broad skulls covered with bony osteoderms that obscure the antorbital and temporal openings. The skull is very flat and low, and there is a complete bony palate separating the processes of respiration and alimentation. The body is broad and squat, and the limbs are short. Ankylosaurs, unlike stegosaurs, retained a full complement of dermal armor and elaborated it into rows of several, sometimes alternating shapes, including keeled or spined plates, polygonal pustules, spikes, and symmetrical ovals or rounded rectangles. Ankylosaurids additionally have a "tail club" formed of coalesced vertebrae and dermal scutes, absent in nodosaurids, and a pronounced fore–aft arch to the skull, which is broader than it is long and bears posterolateral "horns" formed of dermal scutes. Ankylosaurs are mostly Cretaceous forms: ankylosaurids are only known from the Late Cretaceous (North America and Asia), whereas nodosaurids range throughout the Cretaceous (Europe, North America, Australia, possibly South America, and Antarctica), with one form (*Dracopelta*) assigned from the Late Jurassic and another (*Sarcolestes*) from the Middle Jurassic of Europe. Nodosaurids are generally less specialized than ankylosaurids, which could create the impression that the latter may have evolved from the former sometime in the Cretaceous. Ankylosaurs and stegosaurs thus have approximately equally ancient first known appearances in the fossil record, suggesting an approximate divergence time by the Middle Jurassic.

Cerapoda

The remaining ornithischians are assigned to this node-based group, which is distinguished by a broad diastema between the premaxillary and maxillary teeth, asymmetrical enamel in the cheek teeth (thicker on the buccal side of the upper teeth and thicker on the lingual side of the lower teeth), no more than five premaxillary teeth, and a fully open acetabulum, among other features (Sereno, 1986). Two principal node-defined taxa comprise the Cerapoda: the ORNITHOPODA [Euornithopoda of Sereno (1986); generally termed Ornithopoda by Weishampel and others in Dodson *et al.* (1990) as well as in most other treatments] and the MARGINOCEPHALIA (Sereno, 1986). The Ornithopoda as a group appear to show a distinct set of trends toward larger size, loss of premaxillary teeth, robusticity of cheek teeth, closure of antorbital and mandibular fenestrae, elaboration of the forward prong and reduction of the backward prong of the pubis, graviportal posture, and even quadrupedal stance, including hoof-like unguals on both fore and hind digits. This set of trends seems to follow a gradation from HETERODONTOSAURS through HYPSILOPHODONTS through IGUANODONTS to the noncrested and crested HADROSAURS. However, it is also possible to find characters that unite several known forms into apparently monophyletic clusters for which the gradational names can be retained (Sereno, 1986; Norman, 1984; Sues and Norman, 1990; Norman and Weishampel, 1990; Weishampel and Horner, 1990; Horner, 1990), although experts disagree on this. Heterodontosaurs are mostly known from the Early Jurassic, though some persist later in that period. Hypsilophodonts are mostly known from the Late Jurassic and Early Cretaceous, with a couple of forms persisting to the latest Cretaceous. Iguanodonts, which few experts regard as monophyletic, first appear in the Late Jurassic and the forms closer to hadrosaurs appear in the Early Cretaceous; some members of the group appear to persist to the latest Cretaceous. Finally, the hadrosaurs are strictly Late Cretaceous. Thus, the known stratigraphic relationships of ornithopods appear to parallel closely their phylogenetic relationships.

The Marginocephalia comprise the node-based PACHYCEPHALOSAURIA and the CERATOPSIA, united by the posterior skull shelf formed by the parietals and squamosals. Some authors (Sereno, 1986) regard this group as a separate stem from basal cerapodan stock, and others (Norman, 1984) regard some characters (e.g., the form of the pubis and the maxillary diastema) as indicating a closer relationship within Ornithopoda, perhaps branching off among the hypsilophodont nodes. Again, discovery of more complete basal taxa may inform this question further. Because marginocephalians are strictly Cretaceous—in fact, nearly all but the apparent basal form *Stenopelix* are restricted to the Late Cretaceous—there would be a considerable stratigraphic disjunction in the separation between the ornithopod (earliest Jurassic) and marginocephalian (late Early Cretaceous) lineages were they to be regarded as sister taxa (Dodson, 1990).

Pachycephalosaurs are distinguished by their flat, thickened skull roofs, ornamented with small tubercles continuing along the posterior sides and back of the skull. Through the history of the group the skull roofs thicken further, forming a patellar dome with clusters of spikes and nodules around its sides. In some forms these spikes become quite long, directed posteriorly, and the dome can develop a sagittal crest; the antorbital and temporal openings become closed (Sereno, 1986; Maryańska, 1990). Ceratopsians parallel ornithopods in many evolutionary ways: They tend to become larger, with a reduction of front teeth and the development of great dental batteries in the cheek teeth; the cranial ornamentation becomes more elaborate and separates two of the most derived subtaxa; and postcranially, as size increases, quadrupedality replaces bipedality and the unguals become more hoof like. The basal forms, the psittacosaurs, have a parrot-like beak and the distinctive rostral bone that characterizes all ceratopsians. These were apparently bipedal forms for the most part, but the protoceratopsids, which do not appear to be a monophyletic group (Sereno, 1986), grew to 3 m or more in length and were mostly quadrupedal; they first show the characteristic posterior skull frill of the rest of the ceratopsians, which is centrally open in all forms except *Triceratops*. The ceratopsids proper can be divided into the centrosaurs, which sported large nasal horns and small orbital horns, and the chasmosaurs, which had small nasal horns and large orbital horns (Dodson and Currie, 1990).

Saurischia

This second great branch of Dinosauria is distinguished by long posterior cervical vertebrae, axial postzygapophyses lateral to prezygapophyses, and fully developed hyposphene–hypantrum articulations in the trunk vertebrae. Digit II is the longest in the manus, both metacarpals IV and V are palmar to digits III and IV, and the "saurischian pollex" (Gauthier, 1986) is more robust than the other digits, bearing an enlarged ungual, with the first metacarpal no more than half the length of the second and the first with offset distal condyles that direct the thumb medially. There is ambiguity about many other proposed synapomorphies (Gauthier, 1986; Sereno and Novas, 1992; Novas, 1993; Sereno *et al.*, 1993), but the monophyly of Saurischia is still well supported (Holtz and Padian, MS.).

Saurischia is separated into two stem-based groups, SAUROPODOMORPHA and THEROPODA (Gauthier, 1986; Padian and May, 1993), both ranging from the Late Triassic to the Late Cretaceous (except birds, which are members of Theropoda and still survive). Sauropodomorphs have small, high-snouted skulls, long necks, and stout, strong forelimbs with robust first toe and claw. The tibia is shorter than the femur, and the trunk between the shoulder and hip girdles is at least as long as the forelimb (Gauthier, 1986). Sauropodomorpha has been separated into PROSAUROPODA and SAUROPODA, but the monophyly of the former (e.g., Galton, 1990b) has been strongly questioned (e.g., Gauthier, 1986), and most (though not all) of its hypothesized synapomorphies can be equally viewed as transitional characters to the larger, longer necked, more ponderous sauropods. Under this view, the "thecodontosaurids," plateosaurids, and melanorosaurids can be seen as successively closer to sauropods, but more work is clearly needed. All "prosauropod" forms are restricted to the Late Triassic and Early Jurassic.

Sauropoda

The large size of this group is accompanied by even longer necks and relatively smaller skulls (compared to the bulk of the animal) than in "prosauropods," and the tails are also elongated. The nares are large and dorsally placed, the vertebrae are lightened by pleurocoels and their spines and processes become arch-like and laminar, and the number of phalanges is reduced in the manus and pes. There is, however, a fifth metatarsal, which is severely reduced in all "prosauropods." Some major groups within Sauropoda have been traditionally identified; they include vulcanodontids, cetiosaurs, camarasaurs, diplodocids, brachiosaurs, and titanosaurs, but the interrelationships of these groups remain problematic. It is not clear if any of them (except titanosaurs) are monophyletic, and McIntosh (1990) feels that there is too much homoplasy and not enough associated skulls and skeletons known for most taxa to make cladistic analysis meaningful. The Early Jurassic vulcanodontids and the Middle to Late Jurassic cetiosaurs appear to be outside the remaining taxa; sauropods persisted in some parts of the world until the latest Cretaceous, but most lineages had become extinct before then.

Theropoda

The only carnivorous dinosaurs, theropods seem to have retained this ancestral habit among dinosaurs from their ornithodiran predecessors, as witnessed by the distinctive shape of the claws and teeth. Among the synapomorphies of theropods, the lacrimal is broadly exposed on the skull roof, the vomers are fused anteriorly, the ectopterygoid is expanded with a ventral fossa, the presacral vertebrae are pleurocoelous, there are five sacral vertebrae, a "transition point" (Gauthier, 1986) appears approximately halfway down the tail, distal carpal 1 overlaps metacarpals I and II, the penultimate manual phalanges are longer than the others, the preacetabular portion of the ilium is enlarged, there is a pronounced brevis fossa on the ilium, the femur is convexly bowed and not sigmoid, and the fibula is straplike and attached to a crest on the lateral side of the tibia.

Gauthier (1986) separated Theropoda into an initial dichotomy of CERATOSAURIA (Rowe and Gauthier, 1990), which includes *Coelophysis, Syntarsus,* and virtually all Late Triassic and Early Jurassic theropods, and TETANURAE, a group that bears more derived characters such as the reduction of the hand to three fingers, a large, posteriorly placed maxillary fenestra, a more anterior caudal transition point, a straplike scapula, a manus at least two-thirds as long as the humerus plus forearm, and other characters (Gauthier, 1986). These animals are virtually all post-Middle Jurassic forms. Tetanurae was in turn divided into redefined CARNOSAURIA and COELUROSAURIA. Many characters uniting Carnosauria were suspected by Gauthier (1986) of being correlates of large size, and this was borne out by later analyses by Novas (1992) and Holtz (1994), who showed that tyrannosaurs were actually coelurosaurs, a group thought to comprise only relatively smaller forms. The Carnosauria still has an objective definition as theropods closer to *Allosaurus* than to birds, and the Coelurosauria are closer to birds than to allosaurs (Holtz and Padian, 1997). The systematics of both groups are largely problematic, and the nomenclature has been complicated by conflicting uses of the same terms (reviews in Holtz, 1994, 1996; Holtz and Padian, MS.). Within Coelurosauria, the MANIRAPTORA includes birds and all coelurosaurs closer to birds than to ornithomimids, and ARCTOMETATARSALIA are all coelurosaurs closer to ornithomimids than to birds. The DROMAEOSAURIDS appear to be the closest taxon to *Archaeopteryx* and the birds (AVES). Particularly in regard to Saurischia, the use of nomenclature, definitions, and group memberships needs to be standardized and synapomorphies hypothesized for these reorganized taxa before their phylogenetic relationships can be fully clarified. Nevertheless, the advances briefly summarized here may be regarded as the signal achievement of dinosaurian study within the past two decades.

See also the following related entries:

DINOSAURIA • PHYLOGENETIC SYSTEM • SYSTEMATICS

References

Benton, M. J. (1990). Origin and interrelationships of dinosaurs. In *The Dinosauria* (D. B. Weishampel, P. Dodson, and H. Osmólska, Eds.), pp. 11–29. Univ. of California Press, Berkeley.

Coombs, W. P, Jr., and Maryańska, T. (1990). Ankylosauria. In *The Dinosauria* (D. B. Weishampel, P. Dodson, and H. Osmólska, Eds.), pp. 456–483. Univ. of California Press, Berkeley.

Coombs, W. P., Jr., Weishampel, D. B., and Witmer, L. M. (1990). Basal thyreophora. In *The Dinosauria* (D. B. Weishampel, P. Dodson, and H. Osmólska, Eds.), pp. 427–434. Univ. of California Press, Berkeley.

Dodson, P. (1990). Marginocephalia. In *The Dinosauria* (D. B. Weishampel, P. Dodson, and H. Osmólska, Eds.), pp. 562–563. Univ. of California Press, Berkeley.

Dodson, P., and Currie, P. J. (1990). Neoceratopsia. In *The Dinosauria* (D. B. Weishampel, P. Dodson, and H.

Osmólska, Eds.), pp. 593–618. Univ. of California Press, Berkeley.

Dodson, P., Weishampel, D. B., and Osmólska, H. (Eds.) (1990). *The Dinosauria.* Univ. of California Press, Berkeley.

Galton, P. M. (1990a). Stegosauria. In *The Dinosauria* (D. B. Weishampel, P. Dodson, and H. Osmólska, Eds.), pp. 435–455. Univ. of California Press, Berkeley.

Galton, P. M. (1990b). Basal Sauropodomorpha—Prosauropods. In *The Dinosauria* (D. B. Weishampel, P. Dodson, and H. Osmólska, Eds.), pp. 320–344. Univ. of California Press, Berkeley.

Gauthier, J. A. (1986). Saurischian monophyly and the origin of birds. *Mem. California Acad. Sci.* **8,** 1–55.

Holtz, T. R., Jr. (1994). The phylogenetic position of the Tyrannosauridae: Implications for theropod systematics. *J. Paleontol.* **68,** 1100–1117.

Holtz, T. R., Jr. (1996). Phylogenetic taxonomy of the Coelurosauria (Dinosauria: Theropoda). *J. Paleontol.* **70,** 536–538.

Holtz, T. R., Jr., and Padian, K. (MS.). Definition and diagnosis of the carnivorous dinosaurs (Dinosauria: Theropoda) and related taxa. Manuscript in preparation.

Horner, J. R. (1991). Evidence of diphyletic origination of the hadrosaurian (Reptilia: Ornithischia) dinosaurs. In *Dinosaur Systematics: Perspectives and Approaches* (K. Carpenter and P. J. Currie, Eds.). pp 179–187. Cambridge Univ. Press, Cambridge, UK.

Maryańska, T. (1990). Pachycephalosauria. In *The Dinosauria* (D. B. Weishampel, P. Dodson, and H. Osmólska, Eds.), pp. 564–577. Univ. of California Press, Berkeley.

McIntosh, J. S. (1990). Sauropoda. In *The Dinosauria* (D. B Weishampel, P. Dodson, and H. Osmólska, Eds.), pp. 345–401. Univ. of California Press, Berkeley.

Norman, D. B. (1984). A systematic appraisal of the reptile order Ornithischia. In *Third Symposium on Mesozoic Terrestrial Ecosystems: Short Papers* (W.-E. Reif and F. Westphal, Eds.), pp. 157–162. Attempto-Verlag, Tübingen, Germany.

Norman, D. B., and Weishampel, D. B. (1990). Iguanodontidae and related Ornithopoda. In *The Dinosauria* (D. B. Weishampel, P. Dodson, and H. Osmólska, Eds.), pp. 510–533. Univ. of California Press, Berkeley.

Novas, F. E. (1992). Evolucion de los dinosaurios carnivoros. In *Los Dinosaurios y su Entorno Biótico. Actas II, Curso de Paleontología en Cuenca* (J. L. Sanz and A. Buscalioni, Eds.), pp. 123–163. Instituto "Juan de Valdés," Ayuntamiento de Cuenca.

Novas, F. E. (1993). New information on the systematics and postcranial morphology of *Herrerasaurus ischigualastensis* (Theropoda: Herrerasauridae) from the Ischigualasto Formation (Late Triassic) of Argentina. *J. Vertebr. Paleontol.* **13,** 400–424.

Padian, K., and May, C. L. (1993). The earliest dinosaurs. In *The Nonmarine Triassic* (S. G. Lucas and M. Morales, Eds.), Bull. No. 3, pp. 379–381. New Mexico Museum of Natural History and Science.

Rowe, T. R., and Gauthier, J. A. (1990). Ceratosauria. In *The Dinosauria* (D. B. Weishampel, P. Dodson, and H. Osmólska, Eds.), pp. 151–168. Univ. of California Press, Berkeley.

Sereno, P. C. (1986). Phylogeny of the bird-hipped dinosaurs (Order Ornithischia). *Natl. Geogr. Res.* **2,** 234–256.

Sereno, P. C. (1992). Complete skull and skeleton of an early dinosaur. *Science* **257,** 1137–1140.

Sereno, P. C. (1993). The pectoral girdle and forelimb of the basal theropod *Herrerasaurus ischigualastensis. J. Vertebr. Paleontol.* **13,** 425–450.

Sereno, P. C, and Novas, F. E. (1992). The complete skull and skeleton of an early dinosaur. *Science* **258,** 1137–1140.

Sereno, P. C., Forster, C. A., Rogers, R. R., and Monetta, A. M. (1993). Primitive dinosaur skeleton from Argentina and the early evolution of Dinosauria. *Nature* **361,** 64–66.

Sues, H.-D., and Norman, D. B. (1990). Hypsilophodontidae, *Tenontosaurus,* and Dryosauridae. In *The Dinosauria* (D. B. Weishampel, P. Dodson, and H. Osmólska, Eds.), pp. 498–509. Univ. of California Press, Berkeley.

Weishampel, D. B., and Horner, J. R. (1990). Hadrosauridae. In *The Dinosauria* (D. B. Weishampel, P. Dodson, and H. Osmólska, Eds.), pp. 534–561. Univ. of California Press, Berkeley.

Weishampel, D. B., and Witmer, L. M. (1990). Heterodontosauridae. In *The Dinosauria* (D. B. Weishampel, P. Dodson, and H. Osmólska, Eds.), pp. 486–497. Univ. of California Press, Berkeley.

Physiology

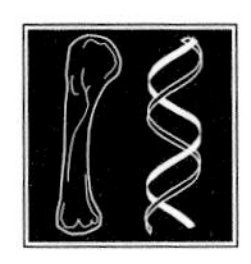

Kevin Padian
University of California
Berkeley, California, USA

The physiology of dinosaurs, except birds, cannot be measured directly because they are all extinct. Inferences about the physiology of Mesozoic dinosaurs and any other extinct animals must be made on the basis of several principles that also apply to function and behavior. One is systematic: Living relatives of extinct organisms have certain structures by which they behave or perform in certain ways, and so the same is reasonably inferred for their extinct relatives on the basis of structural evidence. For example, tooth form, jaw gape, body size, and limb proportions of living Felidae (cats, lions, etc.) present a spectrum of morphologies by which the hunting and feeding behavior of specific extinct felids can reasonably be retrodicted. Another principle is analogous: Rhinoceros have nasal horns that they use to gore when threatened; centrosaurine ceratopsid dinosaurs had nasal horns, and presumably they did the same thing. However, both principles are ontologically problematic; there is nothing to say that an extinct lineage behaved in a way not exactly represented by any extant form, and most structures may function in more than one way.

In attempting to assess nonskeletal features of extinct forms, it is useful to consider first the "extant phylogenetic bracket" (Witmer, 1995) of their closest relatives and second the phylogenetic sequence of the acquisition of traits by which specific adaptations are recognized (Greene, 1986; Padian, 1987, 1991; Weishampel, 1995). For example, warm-bloodedness in birds can be directly measured, as can the high exercise metabolism associated with flight. However, the evolutionary sequence of these features, along with others such as the development of feathers (presumably from fringed scales, but see Bird Origins), one-way respiration, and the pneumatized skeleton, can only be understood with reference to a complete phylogeny of birds and their closest nonavian relatives based on other unrelated characters. In this comprehensive approach, which is a hallmark of comparative biology (Brooks and McLennan, 1991), it is seldom that one characteristic is diagnostic of a functional or physiological complex, and one feature seldom tells the entire evolutionary story. Consequently, the holistic approach of comparative biology, considering both taxonomic association and analogical similarities in a phylogenetic framework that incorporates all lines of evidence, is most likely to provide meaningful information (Padian, 1991).

The question of dinosaurian physiology is an old one; in fact, it was integral to their creation. When Richard Owen named the Dinosauria in 1842, he knew that their great size, terrestrial habits, and upright stance made them anything but typical reptiles, and he allowed that in these features they approached the mammalian grade of organization. However, Owen was still a typologist, and the fact that dinosaurs were reptiles and not mammals by the essential characters of their skeletons allowed him to conclude that the features he saw in the dinosaurian skeleton were simply adaptive, admitting no overt conclusions about elevated physiology. Birds and mammals were members of Haemothermia by virtue of their insulatory pelage (feathers and fur); reptiles were not warm-blooded, and could not be (Desmond, 1975, 1979).

As many authors have noted, through the 1960s dinosaurs were considered physiologically typical reptiles, although much larger than their extant relatives. Narratives of dinosaurian life emphasized the warm, benign hothouse climate of the Mesozoic Era, the shallow swamps that helped the sauropods to support their otherwise impossible weights, and the torpor of the giant carnosaur sleeping off his last meal under a tree, until hunger awoke him to start the cycle anew. In the late 1960s and 1970s, however, this orthodoxy was challenged, initially by J. H. Ostrom and R. T. Bakker. Ostrom's (1970) argument questioned the assumption that dinosaurs must have been restricted to warm, moist climates; as he noted, although paleolatitudinal positions of the continents differed from today's, dinosaurs still had a broader north–south range than do extant reptiles; on the other hand, seasonality was lower, equability higher, and there were no polar icecaps. Nevertheless, even assessments of the zoogeography of extant reptiles are often mistakenly typecast: Today's reptiles live not only in warm, moist areas but also in hot dry ones and (in the case of *Sphenodon* or the marine iguana) in cold and moist or even cold and wet environments. Bakker's (1968, 1971) arguments were more far-reaching and were more directly responsible for approximately a decade of debate that culminated in the publication of a book of perspectives on the question (Thomas and Olson, 1980). Many of the ideas raised by Bakker, his critics, and others are discussed below.

Although the debate of the 1970s was ostensibly about the physiology of dinosaurs, most of the arguments were framed not in terms of cellular metabolism and enzyme regulation but in terms of BEHAVIOR. Most discussions investigated whether dinosaurs were relatively active or inactive, whether they foraged all day or mostly rested, whether they could run, the likelihood that they could maintain high activity levels, the role of display and agonistic behavior in intra- and interspecific interactions, and so on. Of course, most of these have physiological capability at their base, but only in one respect, discussed below, was this even indirectly approached.

The definition of the question of dinosaur physiology in the 1970s was often centered on the catchphrase "warm-bloodedness." In discussions, several very different thermometabolic phenomena were often conflated. "Warm-blooded" and "cold-blooded" are relative terms that refer to an animal's normal body temperature at rest: relatively high or relatively low. "Homeothermic" and "poikilothermic" refer to whether the body temperature is relatively constant or varies according to the ambient temperature. "Endothermic" and "ectothermic" refer to whether the major heat source for metabolic activity is internal cellular catalysis or external to the body. "Heterothermic" is an intermediate position in which the animal profits from both sources. These terms are themselves, to be sure, only descriptors of the gross effects of cellular processes, but they demonstrate how far the debates about dinosaurian physiology were necessarily removed from actual physiological data and parameters. A few of the lines of evidence invoked in these debates are summarized briefly in the following sections; zoogeography was already mentioned [these topics are covered extensively in Thomas and Olson (1980); see also Bakker (1986, 1987) and Farlow (1990) for an excellent review of physiology and Coombs (1990) on behavior; see also BEHAVIOR].

Posture and Gait

Because dinosaurs were obliged to stand upright, which would require more metabolic energy than sprawling, this must have required an active aerobic metabolism comparable to that of mammals. Furthermore, the speeds of dinosaurs as measured in some trackways could have been quite high, suggesting the ability to sustain high metabolic levels. These arguments are plausible, although there are other possibilities. Passive locking mechanisms for joints while standing at rest (as in horses) might have been possible, although there are no instances reported or argued for dinosaurs, and cold-blooded animals are typically capable of bursts of speed, which could have been recorded in trackways.

Brain Size

Dinosaurs had a considerable range of relative brain size (see BRAINCASE ANATOMY; INTELLIGENCE; PALEONEUROLOGY). The Encephalization Quotient (EQ) is a measure of relative brain to body size compared to a crocodile of similar (often extrapolated) size. With crocodiles set at 1.00, the EQs of dinosaurs varied from 0.20 to 5 or higher, and the highest figures, approaching those of birds and mammals, pertained to some hadrosaurs and especially coelurosaurs, the group from which birds evolved and of which they are members (see BIRD ORIGINS). In theory, a large brain would require a strong blood supply carrying abundant oxygen, implying high respiratory rates associated with high metabolism. Although plausible, the range of dinosaur brain sizes still ranks well below those of mammals and birds, and even low-EQ birds, such as ratites, are not particularly intelligent; nor is there a necessary connection between brain size and physiology. Finally, we cannot be sure that the brain filled the braincase entirely in dinosaurs.

Bird Origins

The argument that birds evolved from small carnivorous dinosaurs (see BIRD ORIGINS) prompted inferences that the dinosaurs from which they evolved may also have been warm-blooded. It is not directly possible to confirm that either the earliest birds or their nonavian ancestors were warm-blooded. Recently, the discovery of lines of arrested growth in the bones of some Cretaceous birds has prompted the conclusion that these birds were not fully endothermic (Chinsamy *et al.*, 1994); however, the presence of growth lines alone indicates little about thermophysiological regime (see GROWTH LINES). Ruben (1991) has also argued that birds could have evolved flight under an anaerobic metabolic regime, but this has been questioned (e.g., Speakman, 1993), and in any event the issue is not whether this is theoretically possible but what specific evidence of early avian evolution can test alternative scenarios.

Prey–Predator Ratios

Warm-blooded predators have much higher caloric needs than cold-blooded predators because they have to sustain a constantly higher aerobic metabolism. Dinosaurian faunas were the first in which most of the large vertebrates counted in censuses were herbivores, not carnivores, and they share this feature with mammals, implying that only a relatively few macropredators could be supported by the community. However, there are some problems with these inferences, reasonable though they may be. Collections of fossil vertebrates may not be comprehensive, and collecting bias depends on many factors. Fossil assemblages cannot be known to reflect the proportions or compositions of living populations. Moreover, it can only be an assumption that the densities of predator populations are in all cases necessarily limited by the densities of prey populations. Finally, even if these questions were removed, the metabolic inference could apply only to the predator species and not to the prey species.

Gregariousness

It is certain that at least some dinosaur species traveled in large groups. Whether or not the famous GHOST RANCH assemblage of the small Triassic theropod *Coelophysis*, numbering in the hundreds of skeletons, is an instantaneous mass kill, the preponderance of *Coelophysis* in the assemblage, compared to any other taxon, is so high that it is highly implausible that these skeletons accumulated one by one or in small numbers. Similar mass death assemblages are known through the Late Cretaceous, involving hadrosaurs and ceratopsians, among other taxa. It has been argued that gregarious behavior, also supported by myriad trackways and megatracksites (see FOOTPRINTS AND TRACKWAYS), implies high metabolism, but this can easily be countered by the presence of the same phenomenon in ants, bees, and various fishes, among many taxa that are not warm-blooded.

Parental Care and Social Behavior

Social behavior has been inferred from a variety of evidence in many lineages of dinosaurs (see BEHAVIOR), including trackways, sexual dimorphism, and skeletal ornamentation ostensibly related to display that may have repelled rivals and attracted mates. Parental care is inferred from the apparently helpless condition of many dinosaurs at the time of hatching, based on the paucity of embryonic bone and the preponderance of cartilage found in some dinosaur embryos (see GROWTH AND EMBRYOLOGY). Although these features are certainly characteristic of warm-blooded vertebrates, they are also both characteristic of cichlid fishes and many other cold-blooded taxa, including insects. Still, among living tetrapods only birds and mammals show these features.

Histology of Bones

This is the most direct line of preserved paleontological evidence about cellular physiology because bone cells and tissues are direct records of metabolic activity. Furthermore, all bones are by no means the same in these features. However, patterns are not always simple or easily interpreted. Dinosaurs (including birds), like pterosaurs and some therapsid synapsids (including mammals), predominantly deposited fibrolamellar bone throughout most tissues of the skeleton; this bone tissue reflects high growth rates in living forms, and it seems generally accepted that high growth rates were the norm in Mesozoic dinosaurs (see Farlow, 1990). Whereas other tetrapods, including cold-blooded ones, can deposit fibrolamellar bone under certain conditions (e.g., when raised under optimal conditions by people), this is almost never able to be sustained throughout life and is usually confined to juvenile stages, in which growth is highest among all tetrapods (see HISTOLOGY OF BONES AND TEETH). Growth lines, as indicated previously, are also present in the bones of many dinosaurs, but these by themselves are poor indicators of thermometabolic regime. Some dinosaurs, like some mammals, have them and some do not; they vary in presence, number, and form from bone to bone in the same skeleton, and they can form for reasons related to phylogenetic legacy, environment, diet, ontogeny, or mechanical factors. Their presence has been inter-

The Crystal Forest, part of the Petrified Forest National Park, is in the Triassic Chinle Formation. The fossilized trunks of the once majestic forest litter a Northern Arizona landscape much different today from what it must have been when this rocky forest was vital. Photo by François Gohier.

Nanotyrannus, from the Hell Creek Formation and in the same lineage as *Tyrannosaurus*, possesses characteristics of carnivorous dinosaurs and also some specialized characteristics. For example, it is possible that some tyrannosaurids, including *Nanotyrannus*, had overlapping visual fields. Illustration by Larry Felder.

Triceratops are common dinosaur fossils from the Late Cretaceous of North America. The habitat as pictured above might have been like the African savannas of today. Illustration by Michael Skrepnick.

The end of the Cretaceous Period is represented by the clay layer at the bottom of the pen in the photo above. The Red Deer River Valley, north of Drumheller, Alberta, Canada, is just one of many places where distinct and often very narrow sediments mark the end of dinosaurs. Photo by François Gohier

Stratigraphic sequences like those revealed at Ghost Ranch in New Mexico attest to the length of Earth's natural history. In the foreground are the red Triassic rocks of the Chinle Formation. In the distance, lying atop the Chinle, are Jurassic cliffs of the Entrada Formation. The slopes above these cliffs are also Jurassic rocks but of the more recent Morrison Formation. Finally, the upper cliffs are Dakota sandstones of the Cretaceous. These sediments represent part of the span of time over which dinosaurs existed. Photo by François Gohier.

Hypsilophodon is a member of the Ornithischia, one of the two major lineages of dinosaurs. Illustration by Michael Rothman.

The active predators skulking in the distance belong to the genus *Troodon*. The small dinosaur in the foreground might have been their quarry. Illustration by Donna Braginetz.

preted to indicate a physiological regime "intermediate" between those of reptiles on the one hand and mammals and birds on the other hand, but it is not clear what metabolic features are intermediate or how this supposition would be tested. It is probably conservative to say that Mesozoic dinosaurs were different in some respects from birds and mammals, but probably more like them in factors related to growth and tissue deposition regimes than like other reptiles.

Large Size

Because most Mesozoic dinosaurs exceeded the mass of the largest living terrestrial animals by up to an order of magnitude or more, it is thought that their great bulk conferred an inertial homeothermy. That is, their internal (core) temperatures would have been quite stable, and perhaps (though not necessarily) elevated above those of typical reptiles, because their problem may have been to shed heat rather than to gain it. This would explain, perhaps, why they deposited bone tissue of the type and growth rate common to mammals. However, both small species of dinosaurs and juveniles of large species deposit the same kind of bone, so the special explanation of large size does not seem to fit. Some herpetologists have regarded dinosaurs as "good reptiles," albeit large ones, maintaining that their metabolic regimes are consistent with what can be extrapolated from large living reptiles such as the leatherback turtle (e.g., Paladino and Spotila, 1994). These turtles use vascular countercurrent heat exchangers, an insulating subepidermal fatty layer, and large body size to maintain body temperatures well above the ambient temperature; this strategy is called "gigantothermy." However, there is no evidence that dinosaurs used any of these mechanisms. Leatherbacks have low-capacity and low-pressure respirocirculatory systems, and they swim in an environment that requires far less energy to travel long distances. They generate considerable heat within the body core by swimming, and they are well insulated by internal fat (Paul and Leahy, 1994). Also, as noted previously, dinosaurs did not grow, stand, move, live, or feed like leatherback turtles, so the analogy is questionable at best.

Oxygen Isotope Ratios

The ratio of ^{18}O to ^{16}O in bone phosphate depends on body temperature and the $^{18}O/^{16}O$ ratio of body water at the time of bone deposition. Arguments have been made that the ratios are similar throughout the skeletons of dinosaurs, indicating that their body temperature was relatively constant throughout (Barrick, 1994; Barrick and Showers, 1994). However, the data have not convinced scientists who regard the variation as too high and uncontrolled; moreover, many think that the method is not measuring the original isotopes of body water but those of the sediments since deposition. This method is still somewhat new and requires further investigation before being accepted as a line of evidence one way or another.

Nasal Turbinates

Small scroll-like bones in the nasal passages of living mammals and birds help to conserve heat and moisture as the animals breathe. These turbinates have been found in several cynodont relatives of mammals (Hillenius, 1995) and in some fossil birds, although none more basal so far than the Late Cretaceous *Hesperornis* (Ruben, 1995). Some workers regard the presence or absence of these bones as a litmus test for endothermy, arguing that water and heat loss would be too great without them to sustain high metabolic rates; others suggest that drinking extra water and other behavioral strategies could compensate for the difference. Ruben *et al.* (1996) estimated that the cross-sectional area of the nasal passages of three Mesozoic dinosaurs was too small to accommodate nasal turbinates, casting doubt on their physiological abilities. However, it is difficult to estimate the extent of soft tissues in wholly extinct clades, and there is as yet unpublished evidence that some dinosaurs may have had turbinate-like structures. Two other issues have yet to be explored. First, the absence of certain derived structures related to adaptations does not mean that the adaptation itself was absent. For example, *Archaeopteryx* lacked fused wrist and hand bones, a deeply keeled sternum, an acrocoracoid process on the shoulder girdle, and many other features, but doubts that it could fly—although not as well as living birds—have never been seriously received. Finally, it is unlikely that any single feature will emerge as definitive in establishing an entire physiological regime of extinct animals. However, nasal turbinates and other related structures may emerge as important correlates of metabolic and behavioral strategies in fossil animals once a more complete survey of their distribution can complement a more complete understanding of their evolution and function in living animals.

In summary, the physiology of extinct dinosaurs

can only be perceived dimly through a glass clouded by a remote separation from their living biology and their environments. Several conclusions are instrumental to guiding future research. First, warm-bloodedness is not an "either/or" proposition but a complex of physiological characteristics, each of which has a sliding scale. Second, no single feature is likely to tell the tale of any metabolic adaptation or performance. Third, Mesozoic dinosaurs varied in adult size from the chicken-sized *Compsognathus* to the gigantic *Brachiosaurus* and other sauropods. These animals could no more have had similar metabolic strategies than a bat and a whale or an elephant and a cheetah. As a corollary, little about dinosaurian physiology can be understood without reference to their phylogeny and evolution. Finally, it is likely that dinosaurs were not exactly like any living animals in most aspects of their biology. When all the lines of evidence mentioned previously are taken in totality, it is clear that dinosaurs were not like any living reptiles, and in many respects they were more similar to birds and mammals in some respects than to typical living reptiles. However, the issue is problematic, and conclusions to the contrary can be found in thoughtful essays by Farlow *et al.* (1995), Ruben (1995), and Paladino and Spotila (1994).

References

Bakker, R. T. (1968). The superiority of dinosaurs. *Discovery* **3,** 11–22.

Bakker, R. T. (1971). Dinosaur physiology and the origin of mammals. *Evolution* **25,** 636–658.

Bakker, R. T. (1986). *The Dinosaur Heresies.* Morrow, New York.

Bakker, R. T. (1987). The return of the dancing dinosaurs. In *Dinosaurs Past and Present, Vol. I* (S. M. Czerkas and E. C. Olson, Eds.), pp. 38–69. Univ. of Washington Press, Seattle.

Barrick, R. E. (1994). Thermal physiology of the Dinosauria: Evidence from oxygen isotopes in bone phosphate. *Paleontol. Soc. Spec. Publ.* **7,** 243–254.

Barrick, R. E., and Showers, W. J. (1994). Thermophysiology of *Tyrannosaurus rex*: Evidence from oxygen isotopes. *Science* **265,** 222–224.

Brooks, D. R., and McLennan, D. A. (1991). *Phylogeny, Ecology, and Behavior: A Research Program in Comparative Biology.* Univ. of Chicago Press, Chicago.

Chinsamy, A., Chiappe, L. M., and Dodson, P. (1994). Growth rings in Mesozoic birds. *Nature* **368,** 196–197.

Coombs, W. P. (1990). Dinosaur paleobiology: Behavior patterns of dinosaurs. In *The Dinosauria* (D. B. Weishampel, P. Dodson, and H. Osmólska, Eds.), pp. 32–42. Univ. of California Press, Berkeley.

Desmond, A. J. (1975). *The Hot-Blooded Dinosaurs: A Revolution in Palaeontology.* Blond & Briggs, London.

Desmond, A. J. (1979). Designing the dinosaur: Richard Owen's response to Robert Edmond Grant. *Isis* **70,** 224–234.

Farlow, J. O. (1990). Dinosaur paleobiology: Dinosaur energetics and thermal biology. In *The Dinosauria* (D. B. Weishampel, P. Dodson, and H. Osmólska, Eds.), pp. 43–55. Univ. of California Press, Berkeley.

Farlow, J. O., Dodson, P., and Chinsamy, A. (1995). Dinosaur biology. *Annu. Rev. Ecol. Systematics* **26,** 445–471.

Greene, H. W. (1986). Diet and arboreality in the emerald monitor, *Varanus prisinus*, with comments on the study of adaptation. *Fieldiana (Zool.) N.S.* **31,** 1–12.

Hillenius, W. J. (1995). Turbinates in therapsids: Evidence for Late Permian origins of mammalian endothermy. *Evolution* **48,** 207–229.

Ostrom, J. H. (1970). Terrestrial vertebrates as indicators of Mesozoic climates. *Proc. North Am. Paleontol. Convention* **D,** 347–376.

Padian, K. (1987). A comparative phylogenetic and functional approach to the origin of vertebrate flight. In *Recent Advances in the Study of Bats* (B. Fenton, P. A. Racey, and J. M. V. Rayner, Eds.), pp. 3–22. Cambridge Univ. Press, Cambridge, UK.

Padian, K. (1991). Pterosaurs: Were they functional birds or functional bats? In *Biomechanics and Evolution* (J. M. V. Rayner and R. J. Wootton, Eds.), pp. 145–160. Cambridge Univ. Press, Cambridge, UK.

Paladino, F. V., and Spotila, J. S. (1994). Does the physiology of large living reptiles provide insights into the evolution of endothermy and paleophysiology of extinct dinosaurs? *Paleontol. Soc. Spec. Publ.* **7,** 261–274.

Paul, G. S., and Leahy, G. D. (1994). Terramegathermy in the time of the titans: Restoring the metabolics of colossal dinosaurs. *Paleontol. Soc. Spec. Publ.* **7,** 177–198.

Ruben, J. (1991). Reptilian physiology and the flight capacity of *Archaeopteryx. Evolution* **45,** 1–17.

Ruben, J. (1995). The evolution of endothermy in mammals and birds: From physiology to fossils. *Annu. Rev. Physiol.* **57,** 69–95.

Ruben, J., Hillenius, W. J., Geist, N. R., Leitch, A., Jones, T. D., Currie, P. J., Horner, J. R., and Espe, G. III (1996). The metabolic status of some Late Cretaceous dinosaurs. *Science* **273,** 1204–1207.

Speakman, J. R. (1993). Flight capabilities in *Archaeopteryx. Evolution* **47,** 336–340.

Thomas, R. D. K., and Olson, E. C. (Eds.) (1980). *A Cold Look at the Warm-Blooded Dinosaurs.* Westview Press, Boulder, CO.

Weishampel, D. B. (1995). Fossils, function, and phylogeny. In *Functional Morphology in Vertebrate Paleontology* (J. J. Thomason, Ed.), pp. 34–54. Cambridge Univ. Press, New York.

Witmer, L. M. (1995). The extant phylogenetic bracket and the importance of reconstructing soft tissues in fossils. In *Functional Morphology in Vertebrate Paleontology* (J. J. Thomason, Ed.), pp. 19–33. Cambridge Univ. Press, New York.

Plants and Dinosaurs

Bruce H. Tiffney
University of California
Santa Barbara, California, USA

Plants form an important but little appreciated component of the world of the dinosaurs. They are the source of energy for the entire ecosystem and influence at least three aspects of herbivore, and thus carnivore, biology. First, plants cannot move. Hence, physical features (e.g., climate and soil) determine the distribution and density of plants. This in turn determines the global distribution of herbivores. Second, the quality of plants as an herbivore food resource varies depending on their initial chemical composition, the ease of their digestion, and their ability to tolerate the disturbance attendant upon herbivory. Thus, some plants are rich in sugars and oils, whereas others are not; some are composed of easily broken cell walls, whereas others have tough cell walls rich in digestion-inhibiting chemicals; some resprout in days to weeks, whereas some resprout in months. These factors influence the size and numbers of herbivores. Third, plants create the three-dimensional environment, influencing the evolutionary options open to vertebrates. An open environment permits large animals; a closed environment favors small ones.

Herbivory is not a one-way street. Herbivores are part of the selective environment of plants, influencing evolution by their choice of plant materials to consume, the quantities they consume, and the frequency of consumption. Additionally, dinosaurs must have played a role in the dispersal of some Mesozoic plants; for example, the large, flesh-covered, colorful seeds of cycads were most likely disseminated by dinosaurs.

From available evidence, the history of plant–dinosaur interaction may be broken into three stages. In the first, extending from the inception of the dinosaurs through roughly the mid-Cretaceous, dinosaurs coevolved with a gymnosperm-dominated vegetation. The second stage commenced in the mid-Cretaceous with the initial radiation of the angiosperms and corresponded with a radiation of ornithischian herbivores. The third stage followed the K–T boundary, at which time angiosperms apparently developed closed (densely forested) communities for the first time, creating a three-dimensional environment favorable to the radiation of avian dinosaurs. The formulation and testing of hypotheses about the dynamic factors involved in these differences is in its infancy, and a clear consensus has not emerged.

The first stage in this history involved several clades of "primitive" seed plants (commonly termed gymnosperms) including cycads, cycadeoids, seed ferns, and conifers, together with an understory of ferns and allies. Although these gymnosperms embraced a diversity of biological solutions, they all were woody and ranged from shrubs to trees in stature. They also commonly possessed armored trunks and thick, resistant foliage. By analogy to living relatives, it is probable that many possessed chemicals that were difficult to digest or that inhibited digestion. Also by analogy, some may have recovered from damage rather slowly. In summary, they formed an apparently low-quality (by contrast to living angiosperms) source of food. This was probably particularly true of those communities at lower paleolatitudes that were more drought adapted and probably more open. Higher paleolatitude communities may have been cooler and wetter and thus possibly more productive. The dominant herbivorous dinosaurs found in association with the lower latitude communities were the giant, high-feeding sauropods, although smaller forms were also present. One hypothesis is that the generally large stature and posited low food quality of the available gymnosperms influenced the size and morphology of the herbivores because the lower quality food necessitated the consumption of larger quantities of food and its retention for longer times. However, other factors may also have entered into the equation, including the digestive physiology of large dinosaurs. It has also been suggested that the large sauropods fed on "fern prairies," which would have provided a higher quality and more disturbance-tolerant food source. It is noteworthy that there is some correlation between the occurrence of sauropod herbivores and gymnosperms in the later Cretaceous, suggesting an association between the two.

The second stage in the history was ushered in by the radiation of the angiosperms or flowering plants. Angiosperms represent an advanced clade of gymnosperms that is characterized by several features. They are much more "weedy" and disturbance-tolerant than gymnosperms, and unlike any other group of seed plants, they possess a pronounced ability to reproduce vegetatively by horizontal stems or by root sprouting. This allows them to dominate an area by vegetative rather than sexual reproduction. Additionally, most angiosperms possess structurally weak foliage, often not as rich in digestion-inhibiting chemicals as observed in gymnosperms. The angiosperms first appeared in the later portion of the Early Cretaceous and diversified rapidly through the Late Cretaceous. Although they became taxonomically dominant (numbers of species) very quickly, there is some debate when they actually came to physically dominate the vegetation. There is reasonable evidence that closed angiosperm communities like those of the present day first appeared in the Tertiary, and that Cretaceous angiosperms may have formed an open community by comparison. Their diversification was closely paralleled by a radiation of lower feeding ornithischian herbivores, dominated by the hadrosaurs and ceratopsians. This radiation is notable in two respects. First, from current data, more than 50% of all dinosaur taxonomic diversity stems from this radiation, which occurred in perhaps the last 40 million years of the dinosaur's 160-million-year history (excluding birds). Second, anecdotal evidence suggests that the individuals in each species were very numerous because mass kill sites involve hundreds to perhaps thousands of animals. This is in contrast to the much lower numbers of individual dinosaurs found in currently known mass kill sites of the pre-mid-Cretaceous. The ornithischian radiation and its specific and individual diversity could be related to the rapid growth, disturbance tolerance, and relatively higher food quality provided by the angiosperms. However, the possibility exists that moderating global climates in the Cretaceous may also have allowed a greater density of ferns and their allies and/or other gymnosperms, increasing the general food base. Changes in dinosaur biology (e.g., evolving dentition in hadrosaurs) may also have allowed more efficient use of plant resources. The patterns of radiation clearly suggest (contrary to one early hypothesis) that the radiation of the angiosperms did not cause the extinction of the dinosaurs by poisoning them with a new food source.

One aspect of this transition deserves further note. The sauropod–gymnosperm association involves very large herbivores. Large herbivores tend both to inhabit open environments and to create and maintain them by the intensity of their herbivory. Indeed, it is possible that the extreme disturbance created by dinosaurian herbivory may have provided a selective environment that favored the evolution and radiation of angiosperms. Because many gymnosperms were large trees, one can envision the common occurrence of open "parklands" in the Jurassic and Early Cretaceous, with low shrubs between trees and perhaps ferns in moister areas, permitting the passage of large herbivores. The radiation of angiosperms in the mid- and Late Cretaceous may have initiated a process of infilling of these open communities—a process that culminated with the appearance of closed-canopy angiosperm communities in the early Tertiary. Ornithischians may have maintained a disturbed selective environment for the weedy angiosperms at the expense of gymnosperms, but it is also possible that, if angiosperm communities became increasingly more dense in the later Cretaceous, they altered the environment in which the ornithischian herbivores functioned. The balance between the appearance of a possibly increasingly closed vegetation and the ability of these large herbivores to "mow it back," and thus maintain space for themselves, presents an interesting puzzle for which the pieces are dimly perceived.

Whether the appearance of closed communities in the early Tertiary reflects the loss of large herbivores or whether it involved additional factors (e.g., climatic changes), it clearly had a major influence on the dynamic of plant–animal interactions. Most of the Mesozoic was dominated by large, disturbance-creating herbivores that were predators upon whole plants and that operated in an open environment. Even the average Late Cretaceous dinosaur herbivores, although smaller than their predecessors, were far larger than the average living herbivores. The Cretaceous–Tertiary transition in herbivore size was sudden and involved a drop in herbivore average size of several orders of magnitude. A resulting feature is that Tertiary and living herbivores (including birds) generally fit within the three-dimensional plant community and feed on the organs of plants, apparently allowing the development of more complicated plant–animal interactions.

See also the following related entries:

COPROLITES • DIET • HELL CREEK FLORA • MESOZOIC FLORAS

References

Coe, M. J., Dilcher, D. L., Farlow, J. O., Jarzen, D. M., and Russell, D. A. (1987). Dinosaurs and land plants. In *The Origins of Angiosperms and Their Biological Consequences* (E. M. Friis, W. G. Chaloner, and P. R. Crane, Eds.), pp. 225–258. Cambridge Univ. Press, Cambridge, UK.

Farlow, J. O. (1987). Speculations about the diet and digestive physiology of herbivorous dinosaurs. *Paleobiology* **13,** 60–72.

Krassilov, V. A. (1981). Changes of Mesozoic vegetation and the extinction of dinosaurs. *Palaeogeogr. Palaeoclimatol. Palaeoecol.* **34,** 207–224.

Tiffney, B. H. (1992). The role of vertebrate herbivory in the evolution of land plants. *Palaeobotanist* **41,** 87–97.

Vakhrameev, V. A. (1991). *Jurassic and Cretaceous Floras and the Climates of the Earth.* Cambridge Univ. Press, Cambridge, UK.

Weishampel, D. B., and Norman, D. B. (1989). Vertebrate herbivory in the Mesozoic; Jaws, plants and evolutionary metrics. In *Paleobiology of the Dinosaurs* (J. O. Farlow, Ed.), Spec. Paper 238, pp. 87–100. Geological Society of America, Denver.

Wing, S. L., and Tiffney, B. H. (1987). The reciprocal interaction of angiosperm evolution and tetrapod herbivory. *Rev. Palaeobot. Palynol.* **50,** 179–210.

Plate Tectonics

ELLA HOCH
Geological Museum
Copenhagen, Denmark

The concept of earth as the center of the world and a firm basement for created life changed throughout Western history under the influence of practical experience, scientific research, and a free exchange of knowledge. Analytical observation methods were reborn in Europe with the Renaissance, and freedom of thought gained acceptance in most Western societies during subsequent centuries. Ideas proliferated with the invention and spread of printing.

A stage toward comprehension of planetary evolution was the recognition of large-scale vertical movements of the terrestrial crust. To this also belongs the concept of a sunken Atlantis, originally expounded by Plato. It still was in vogue around 1596 when a great European geographer and classical scholar, Abraham Ortelius, in the third edition of his *Thesaurus Geographicus* (a dictionary of classical toponyms)with caution reasoned that America had been torn away from Europe and Africa by earthquakes and flood, and accordingly will seem to be elongated toward the west. Vestiges of the rupture, Ortelius noted, can be seen on a map of the world by careful study of the coasts of said continents where they face each other. This is the oldest known statement on continental drift, a conceptual forerunner of plate tectonics.

Continental drift was defended by Alfred Wegener, who perished at the age of 50 on his third expedition in Greenland in 1930. He was by no means the first since Ortelius to remark on the symmetry of South America's and Africa's Atlantic margins. Some, such as the English Francis Bacon (in *Novum Organum*, 1620) and the French Comte de Buffon (in *Theorie de la Terre*, 1749), had plainly noted it; others, among whom the American Frank Bursley Taylor is the more well known, had explained it by horizontal displacements of the continents. Taylor's (1910) argument, on a sliding of the continents from the northern and southern hemispheres toward the equator and a spreading of North–South America to one side and Europe–Africa to the other side of the fixed Mid-Atlantic Ridge, rested on Eduard Suess's interpretation of the mountain plan of Asia (Taylor, 1910, p.226) in *Das Antlitz der Erde* (Vienna, F. Tempsky, 1885–1909 (3 vols.)). This latter, translated as *The Face of the Earth*, was an important work of its time but is now long outdated.

Alfred Wegener, born in Berlin and educated primarily as a climatologist, collected evidence from an array of scientific fields, such as paleontology, geology, and physics, into an hypothesis of drifting of the relatively lightweight continents on the denser basement. The basic *sima* (after its high content of silicon and magnesium) was the basaltic ocean floor believed to continue under the continental rocks, the *sial* (high content of silicon and aluminum). Wegener published his arguments in a now famous book, *Die Entstehung der Kontinente und Ozeane* (Braunschweig,

F. Vieweg), in 1915 and revised them according to new knowledge in editions in 1920, 1922, and 1929. The latter two were translated into English (among other languages) as *The Origin of Continents and Oceans* (1924, 1966).

Wegener's hypothesis was strongly criticized, if not rejected, during his lifetime and a generation thereafter by the geophysical community, who knew of no feasible mechanism for a natural displacement of the continents over the surface of the planet Earth. Some paleontologists welcomed continental drift as an explanation to otherwise incomprehensible paleobiogeographic patterns. However, the majority of the paleontologists repudiated it and stuck to the traditional concept of intercontinental land bridges exposed or submerged according to their needs. Geologists in general, on the other hand, had rather favorable opinions of the hypothesis. So did, for instance, Professor Daly from Harvard University, who in 1923 concluded a critical review of the Taylor–Wegener hypothesis by the statement that geologists have good reason to retain the root idea embodied in these writings. Wegener also enjoyed support from the two remarkable geologists Arthur Holmes and Alexandre L. du Toit.

Wegener's introduction to Greenland, with the Denmark Expedition of 1906–1908, was as a meteorologist. However, a later object of his Greenland studies was the possible drift of Greenland (as part of the North American continent) relative to Europe. He compared new longitudinal observations with those made by Sabine in East Greenland in 1823 and others obtained on the Germania Expedition in 1870 and calculated an annual relative separation rate of several meters. This result, obtained under rather primitive conditions, is now known to be conspicuously wrong. It was, however, the best possible incentive for continued work along the initiated line. Aided by satellites and other technically and materially sophisticated equipment, we can observe that the Greenland–Norwegian Sea grows laterally by 3–5 cm / year. Modern scientific substantiation of Wegener's idea began in the second half of the 1950s. Together with names of important contributors, the state of plate tectonic art in the mid-1960s is reviewed by B. C. King (1967) in the Introduction to *The Origin of Continents and Oceans*.

Plate tectonics acknowledges the dichotomy between the relatively lightweight continents and the relatively heavy ocean floors, which may have existed as far back as 3.6 Ga, in the Archaean (Nisbet, 1987). The 20- to 70-km-thick continental crust is mainly intrusive granitoids (which may be generated by plumes and above subduction zones) and gneissic material overlain by sediments and volcanic rocks. The 4- to 10-km-thick oceanic crust originated by extrusion in lithosphere rifts of hot, molten basalt solidifying as dykes covered by lava (pillow lava, if extruded below water). The lithosphere is the outer, colder, more brittle part of the earth, including the crust, with a surface temperature of slightly above 0°C. Inside the lithosphere, which may have a thermal base from 40–50 km (below oceans) to 400–500 km (below continents) (Nisbet, 1987), is the asthenosphere, which is warmer—due to radioactive processes in the earth's interior—and more ductile. A widely accepted plate tectonic working hypothesis states that convection currents in the asthenosphere move rafts or plates of the lithosphere horizontally at varying speeds, from a few to up to 16–20 cm / year (Meert *et al.*, 1993). Time and location of lithospheric rifting depend on asthenospheric convection with rising limbs of convection cells located under the spreading centers (the rifts).

Eruption from lithosphere rifts add material to the crust, and lithosphere plates grow away from the rifts. Topographical highs of the ocean floor along the rifts, such as the Mid-Atlantic Ridge, are explained by the heat and accumulation of lava near the rifts. Due to cooling, thus growing density, the basaltic ocean floor sinks slowly during lateral dislocation. Ocean depth above ridges may be 0–1000 m and above basaltic floor at great distances from the ridges 4–5000 m.

Drifting lithosphere plates, whether ocean floor, continent, or both in one plate (e.g., the western Atlantic Ocean floor + the Americas), may move (i) apart, with constructive margins being formed at the lithosphere rifts (e.g., the Atlantic Ocean floor); (ii) along each other, with conservative margins interfering in transform faults where accumulated stress is intermittently released in earthquakes (e.g., San Andreas Fault); or (iii) toward each other, with convergent or destructive margins. Continent–continent collision typically creates ranges of folded crust (e.g., the Alps) or may result in continental underthrusting (Indian crust beneath the Tethyan Himalaya; Zhao *et al.*, 1993); continent–oceanic plate collision can lead to oceanic plate sliding over the continental margin onto the continent (e.g., the Semail Ophiolite, Oman), but most often results in subduction of the oceanic plate beneath the continent (e.g., Pacific plates under the Americas). Subducted oceanic plates, it is believed, melt at depth, and melt products rise in the lithosphere, part of them to become extruded in volcanic ranges that run parallel to the subduction zone, such as the Rocky Mountains and Cordilleran Ranges on the American continents. Subduction under an oceanic plate gives rise to a volcanic island arc above the subducted plate. Due to these recycling processes, no ocean floor exists older than the Jurassic. A hot spot, an area of high volcanic activity, is explained as rising rock melt, a plume, possibly initially generated in the asthenosphere. Its location is stationary, and a plate moving over it is marked by a chain of progressively older volcanoes corresponding to periodic plume activity (e.g., the Hawaiian–Emperor Chain). Erosion, transport, and deposition of rock material by wind, water, organisms, or other agents continually reshape the surface of the continents, which react isostatically by subsidence of the loaded parts and rise of the eroded parts. Typically, continental margins are subaqueous (the shelf), as are many basins on the continents, some being open to the oceans. Shelf seas and intracontinental seaways, although relatively shallow (below 200–500 m), make as serious barriers to terrestrial life as do oceans. Continental sedimentary basins *sensu lato* trap animal and plant remains and may develop into fossil reservoirs depending on geochemical environment.

The distribution of fossils reflects plate tectonic events. *Mesosaurus*, a small reptile, in Lower Permian rocks on opposite sides of the South Atlantic Ocean suggested a split to Wegener. Known dinosaurs based on less than 3000 whole or partial skeletons from a space of time around 160 million years support the notion of an early Mesozoic Pangea, a supercontinent including all continental land. Late Triassic prosauropods were some of the Pangean dinosaurs. Fossil prosauropods are now found in China, Europe, Greenland, South America, etc., testifying to post-Triassic rifting of the Pangean lithosphere.

Several clades of mid-Mesozoic dinosaurs in central Asia (Siberia and China) are different from dinosaurs elsewhere, and biogeographic isolation of central Asia from Neopangea (Pangea without central Asia) is suggested (Russell, 1993). To the south, central Asia was limited by the Tethys Ocean. Toward Europe, an intracontinental seaway may have constituted a faunal barrier.

The Cretaceous Tethys Ocean separated Gondwana, the southern continent, from Laurasia, the northern continent. Neo-Atlantic rifting, initiated in the Triassic, finally split South America from Africa during the Cretaceous and North America from Europe during the Tertiary. In the Late Cretaceous, a longitudinal Western Interior Seaway divided the North American part of Laurasia into western North America, with intermittent land connection to central Asia, and eastern North America, intermittently connected via Greenland to Europe.

See also the following related entries:

BIOGEOGRAPHY • DISTRIBUTION AND DIVERSITY • GEOLOGIC TIME • MIGRATION

References

Daly, R. A. (1923). A critical review of the Taylor–Wegener hypothesis. *J. Washington Acad. Sci.* **13,** 447–448.

King, B. C. (1967). Introduction. In *The Origin of Continents and Oceans* (A. Wegener, Ed.). Methuen, London.

Meert, J. G., Van der Voo, R., Powell, C. McA., Li, Z.-X., McElhinny, M. W., Chen, Z., and Symons, D. T. A. (1993). A plate-tectonic speed limit? *Nature* **363,** 216–217.

Nisbet, E. G. (1987). *The Young Earth. An Introduction to Archaean Geology*, pp. 402. Allen & Unwin, Boston.

Russell, D. A. (1993). The role of central Asia in dinosaurian biogeography. *Can. J. Earth Sci.* **30,** 2002–2012.

Taylor, F. B. (1910). Bearing of the Tertiary mountain belt on the origin of the earth's plan. *Bull. Geol. Soc. Am.* **21,** 179–226.

Zhao, W., Nelson, K. D., and Project INDEPTH Team (1993). Deep seismic reflection evidence for continental underthrusting beneath southern Tibet. *Nature* **366,** 557–559.

Pneumaticity

see CRANIOFACIAL AIR SINUS SYSTEMS; POSTCRANIAL PNEUMATICITY

Polar Dinosaurs

THOMAS H. RICH
Museum of Victoria
Victoria, Australia

ROLAND A. GANGLOFF
University of Alaska Museum
Fairbanks, Alaska, USA

WILLIAM R. HAMMER
Augustana College
Rock Island, Illinois, USA

Until 15 years ago, polar habitats were hardly considered fertile ground for dinosaur research. However, studies completed since 1985 strongly suggest that they may have served as both a refuge and birthplace for some dinosaur groups. (The identification of polar dinosaurs refers to dinosaurs that lived within the polar circles of their time, not necessarily within the current polar circles.) In the Northern Hemisphere, eight areas ranging from the Late Jurassic(?) to the Late Cretaceous have produced dinosaurs that lived within the paleo-Arctic Circle. In the Southern Hemisphere, two areas within the Early Cretaceous paleo-Antarctic Circle have yielded dinosaurs, and three others are known from the ?Early Jurassic to Late Cretaceous that fall just outside it. These occurrences are summarized in Tables I and II. Most of the material from both hemispheres was collected during the past 20 years, and much of it remains to be fully described (Fig. 1).

North Polar Dinosaurs

Dinosaur remains and associated faunas have been described from the Late Jurassic–Early Cretaceous and the Late Cretaceous of Russia, the Late Jurassic and late Early Cretaceous through latest Cretaceous of Alaska, the latest Cretaceous (Maastrichtian) of Canada, and the Early Cretaceous of Spitzbergen (Table I). Clemens and Nelms (1993) inferred some of the adaptations to a polar environment of the fauna recovered from the Kogosukruk Tongue of the Prince Creek Formation, Colville River, Alaska (Late Cretaceous: Campanian–Maastrichtian). They pointed out that its dinosaurs and mammals closely resemble those of contemporaneous sites in Wyoming, Montana, and Alberta, but there is a striking difference in the paucity of known terrestrial ectothermic tetrapods: frogs, salamanders, turtles, champsosaurs, lizards, and snakes, and crocodilians are completely absent. Hypsilophodontids are unknown from this region so far, though *Thescelosaurus*, an oddly primitive form, is found in the Late Cretaceous of the northern United States, and hypsilophodontids are common in the Australian polar assemblages (see below). A small *Albertosaurus* is the only evidence of a top carnivore in the assemblage, though this implies that larger ones were present, and sample size is not sufficient to assess the complete faunal diversity with confidence. The tyrannosaurids are known primarily from teeth that are 40–50% of the adult size of southern conspecifics. Nevertheless, the absence of the typical ectotherms found so abundantly in southern faunas appears to be real and raises questions about the ability of dinosaurs to tolerate climatic conditions that evidently could not be tolerated by other reptiles and amphibians. The possibility of smaller body size in the tyrannosaurids, combined with the abundance of hadrosaur juveniles and even rare hatchlings (Carpenter and Alf, 1994), may reflect year-round adaptations to high paleoaltitudes during the North American Cretaceous.

South Polar Dinosaurs

Antarctic dinosaurs come from Antarctica proper as well as from paleo-Antarctic sites in Australia and New Zealand (Table II).

The first Antarctic dinosaur fossils were not dis-

TABLE I Northern Hemisphere

Asia, Russia, Sakha, Kempendyay site, Yakutia, Teheteh River, 62°31′ N, 119°23′ E
Rock unit: Suntar Series
Age: Late Jurassic–Early Cretaceous?
Paleolatitude: 63–70° N
Reference: Rozhdestvensky (1973)
Fauna: Stegosauria: *cf. Stegosaurus;* Sauropoda: Camarasauridae; Carnosauria: Allosauridae; small theropod

Asia, Russia, Koryak River, a tributary of the Kakanaut River near Lake Pekulneyskoje, 62°58′ N, 176° E
Rock unit: Kakanaut Formation
Age: Late Cretaceous (middle Maastrichtian)
Paleolatitude: 65–72° N
Paleoenvironment: Volcanogenic continental sandstones and other terrigenuous deposits
Reference: Nessov and Golovneva (1990)
Fauna: Hadrosauridae; Troodontidae: *Troodon cf. T. formosus;* large theropod; Aves?

North America, United States of America, Alaskan Peninsula near Black Lake, 56°29′ N, 158°39′ W
Rock unit: Naknek Formation
Age: Late Jurassic (Kimmeridgian–Tithonian)
Paleolatitude: 66° N
Paleoenvironment: Fluvial
Reference: Conyers (1978)
Fauna: Theropod footprints

North America, Alaska, Colville River, north-central North Slope. Various localities ranging from 69°58′ to 70°05′ N, 151°38′ to 151°31′ W
Rock unit: Kogosukruk Tongue of the Prince Creek Formation
Age: Late Cretaceous (late Campanian–early Maastrichtian). Radiometrically dated between 68 and 73 myBP utilizing K-Ar on glass shards and sanidine
Paleolatitude: 65–85° N
Paleoenvironment: Coastal flood plain
Reference: Clemens and Nelms (1993).
Fauna: Chondrichthyes; Osteichthyes; Ceratopsia: *Pachyrhinosaurus cf. P. canadensis, ?Anchiceratops sp., ?Brachyceratops sp.;* Tyrannosauridae: small *Albertosaurus sp.;* Troodontidae: *Troodon sp.;* Dromaeosauridae: *?Saurornitholestes;* Hypsilophodontidae: *cf. Thescelosaurus;* Multituberculata: *Cimolodon nitidus;* Marsupialia: Near *Pediomys cf. P. elegans;* Placentalia: *Gypsonictops sp.*

North America, United States of America, Alaska, western North Slope. Various sites ranging from 69°7′ N, 153°17′ W to 69°22′ N, 161°27′ W
Rock units: Corwin, Chandler, and Ninuluk formations
Age: Early Cretaceous (Albian) to Late Cretaceous (Cenomanian)
Paleolatitude: 65–75° N
Paleoenvironment: Fluvial, delta plain
Reference: Gangloff (1994)
Fauna: Testudines: ?Dermatemydidae; Hadrosauridae: footprints and skin impression; Theropoda: tooth

North America, south-central Alaska, Talkeetna Mountains, 61°56′ N, 147°39′ W and 61°52′ N, 147°21′ W
Rock unit: Matanuska Formation
Age: Late Cretaceous (Turonian and late Campanian–early Maastrichtian)
Paleolatitude: 68–70° N
Paleoenvironment: Shallow marine to prodelta
Reference: Gangloff (1995)
Fauna: Hadrosauridae; Ankylosauridae: *Edmontonia sp.*

North America, Canada, District of Franklin, Bylot Island, 73°30′ N, 79° W
Rock unit: "Kanguk Formation"
Age: Late Cretaceous (Maastrichtian)
Paleolatitude: 68–70° N

continues

Continued

Paleoenvironment: Shallow marine to prodelta
Reference: Russell (1988)
Fauna: Chondrichthyes; Hadrosauridae: Lambeosaurinae; theropod; plesiosaur; mosasaur; Aves: *Hesperornis sp.*

North America, Canada, Northwest Territories, District of Mackenzie, East Little Bear River, 64°33′ N, 125°50′ W
Rock unit: Summit Creek Formation
Age: Late Cretaceous (Maastrichtian)
Paleolatitude: 65–78° N
Paleoenvironment: Fluvial, floodplain
Reference: Russell (1984)
Fauna: Ceratopsia

North America, Canada, Yukon Territory, Peel River drainage, 66°N, 135°W
Rock unit: Bonnet Plume Formation
Age: Late Cretaceous (Maastrichtian)
Paleolatitude: 68–80° N
Paleoenvironment: Fluvial, floodplain, paludal
Reference: Russell (1984)
Fauna: Hadrosauridae

Svalbard (= Spitzbergen), southern Heerland at 77°40′ N, 18°50′ E and West Spitzbergen at 78° N, 15° E
Rock unit: Helvetiafjellet Formation
Age: Early Cretaceous
Paleolatitude: 67–69° N
Paleoenvironment: Fluvial, floodplain
Reference: Edwards *et al.* (1978)
Fauna: (Footprints only): Iguanodontidae; Carnosauria

covered until 1986 on James Ross Island off the coast of the Antarctic Peninsula (Olivero *et al.*, 1986; Fig. 2). This find, of a Late Cretaceous ankylosaur, was followed 3 years later by the collection of a partial skeleton of a Late Cretaceous hypsilophodont from nearby Vega Island (Fig. 2) by the British Antarctic Survey (Hooker *et al.*, 1991). These two specimens represent the only Cretaceous dinosaurs known from Antarctica.

During the 1990–1991 austral summer, Early Jurassic dinosaurs were discovered on Mt. Kirkpatrick in the Transantarctic Mountains, approximately 650 km from the geographic South Pole near the Beardmore Glacier (Hammer and Hickerson, 1994; Fig. 2). The Jurassic locality produced dinosaur fossils belonging to at least three different theropods and a prosauropod. These animals were found associated with fragmentary remains of a tritylodont (synapsid) and a pterosaur.

There are several reasons why Antarctic dinosaurs were discovered so late and are relatively rare. First, 98% of the surface area of the continent is ice covered, leaving relatively few places to look. The inaccessibility of many areas has additionally delayed exploration of the Mesozoic sediments that are exposed. Second, rocks of Late Triassic, Jurassic, and Cretaceous age, the time when dinosaurs existed, are rare in areas where rocks are exposed. The youngest rocks in the Transantarctic Mountains, the largest region where exploration is possible, are early Middle Jurassic in age, and they only occur on the highest peaks. Although some Cretaceous exposures exist in the Antarctic Peninsula region, they are primarily marine deposits. Because dinosaurs were terrestrial they are rarely preserved in marine rocks. The two individual Cretaceous dinosaurs from the Antarctic both represent unusual occurrences where dinosaur skeletons were washed offshore after death and deposited in marine sediments.

The Jurassic dinosaurs from Antarctica occur at an elevation of more than 4000 m in a tuffaceous siltstone (river bank deposit) high in the Falla Formation. All but 3 of the more than 100 bones collected from this site during a single 8-week field season were concentrated in one small area of exposure approximately 1 m thick and 5 m wide. The Falla Formation overlies

TABLE II Southern Hemisphere

Antarctica, Mt. Kirkpatrick, Beardmore Glacier area, Transantarctic Mountains, 84° S, 165° E
Rock unit: Falla Formation
Age: Early Jurassic
Paleolatitude: 61–70° S
Paleoenvironment: Foreland basin flood plain
Reference: Hammer and Hickerson (1994)
Fauna: ?Allosauridae: *Cryolophosaurus ellioti;* small theropod; prosauropod; tritylodontid; pterosaur

Antarctica, Antarctic Peninsula, James Ross Island, 63.7° S, 57.5° W
Rock unit: Santa Marta Formation (Marambio Group)
Age: Late Cretaceous (Campanian)
Paleolatitude: 65° S
Paleoenvironment: Shallow marine
Reference: Olivero *et al.* (1991)
Fauna: Ankylosauria indet.

Antarctica, Antarctic Peninsula, Vega Island, 63.6° S, 57.5° W
Rock unit: Cape Lamb Member of the Lopez de Bertodano Formation
Age: Late Cretaceous (Campanian–Maastrichtian)
Paleolatitude: 65° S
Paleoenvironment: Shallow marine
Reference: Milner and Hooker (1992)
Fauna: Hypsilophodontidae indet.

Australia, Victoria, numerous sites centered on 38°45′ S, 143°30′ E and 38°40′ S, 145°40′ E
Rock units: Strzelecki and Otway Groups
Age: Early Cretaceous (Aptian–Albian)
Paleolatitude: 66.8–77.8° S
Paleoenvironment: Rift valley flood plain
Reference: Rich and Vickers-Rich (1994)
Fauna: Ceratodontidae: Ceratodus avus, Ceratodus nargun, Ceratodus sp., Coccolepis woodwardi, Wadeichthys oxyops, Koonwarria manifrons; Leptolepididae: *Leptolepis koonwarri;* Labyrinthodont: *Koolasuchus cleelandi;* Hypsilophodontidae: *Fulguratherium australe, Leaellynasaura amicagraphica, Atlascopcosaurus loadsi;* Victorian hypsilophodontid type 1; Victorian hypsilophodontid type 2; Victorian hypsilophodontid type 3; Ankylosauria: *cf. Minmi;* Protoceratopsidae: aff. *Leptoceratops;* Ornithomimidae: *Timimus hermani;* Oviraptorosaur?; Allosauridae: *Allosaurus sp.;* Testudines: *Chelycarapookus arcuatus;* Cryptodira; Plesiosaur; Pterosaur; Aves; Crocodilia

Australia, southeastern Queensland, Taloona Station near Roma, 26°05′ S, 149°03′ E
Rock unit: Walloon Coal Measures (Injune Creek Group)
Age: Early Jurassic (?Bajocian)
Paleolatitude: 56–65°
Paleoenvironment: Intracratonic flood plain
Reference: Longman (1927)
Fauna: Sauropoda: *Rhoetosaurus brownei*

New Zealand, North Island, Mangahouanga Stream, 39° S, 176°45′ E
Rock unit: Maungataniwha Member of the Tahora Formation
Age: Late Cretaceous (Campanian)
Paleolatitude: 66° S
Paleoenvironment: Nearshore marine
Reference: Molnar and Wiffen (1994)
Fauna: Dryosauridae; Theropod; Ankylosauria; Sauropoda?; Testudines; Plesiosauroidea; Mosasauridae; Pterodactyloidea

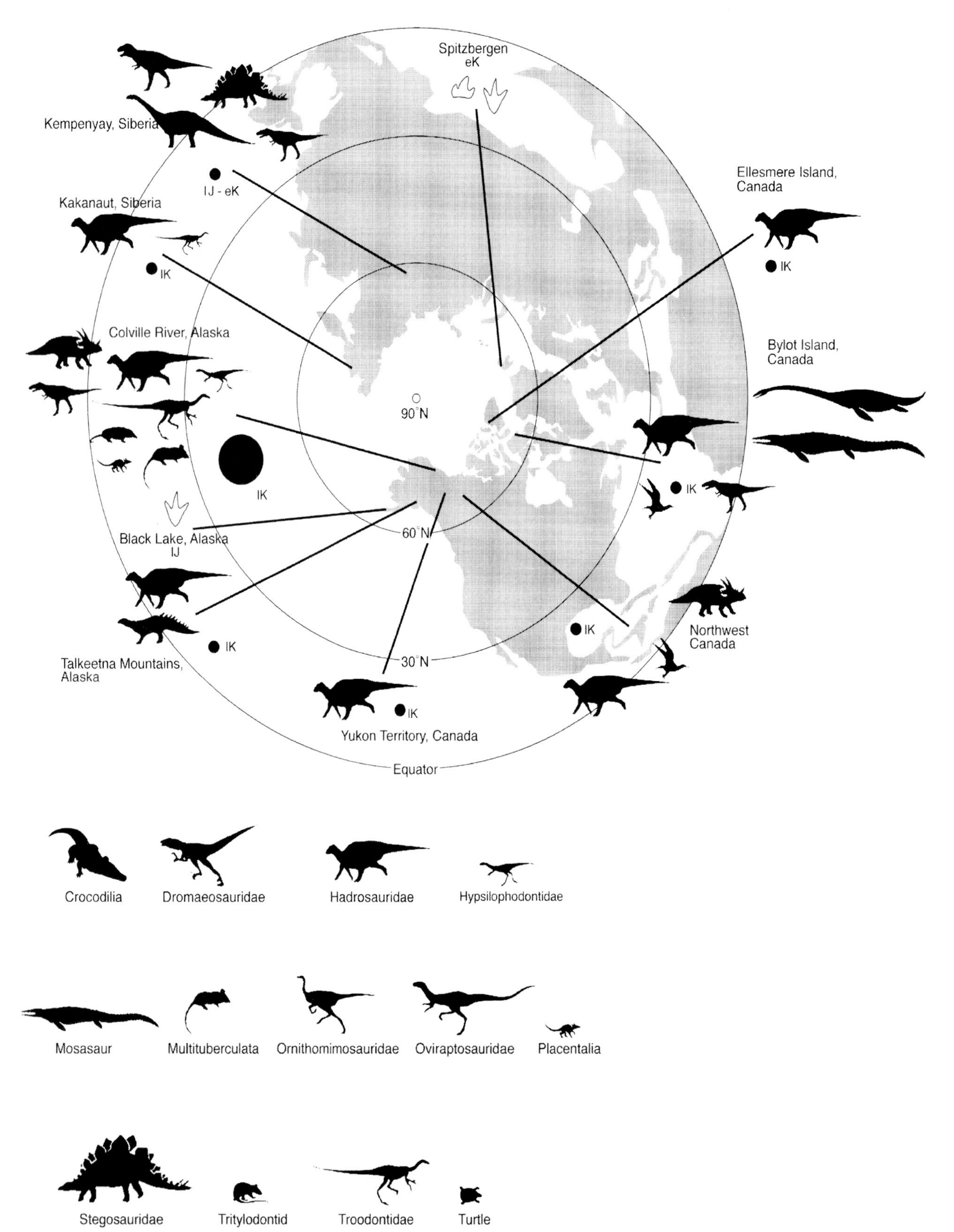

FIGURE 1 Distribution of polar dinosaurs. Dinosaurs from polar latitudes in either hemisphere are not common. Most such sites have been discovered since 1975. None have been fully studied. Of all these sites, that on the Colville River, Alaska, has the greatest potential to yield a detailed picture about these animals. For more than 200 km on the left bank of that river there are beautifully exposed outcrops that range in age from Albian to Maastrichtian. Fossil bones have been found at many places along that river. By contrast, the potential of sites shown as having less than 100 bones is much more problematical. The apparent greater diversity of the southeastern Australian dinosaur assemblage compared

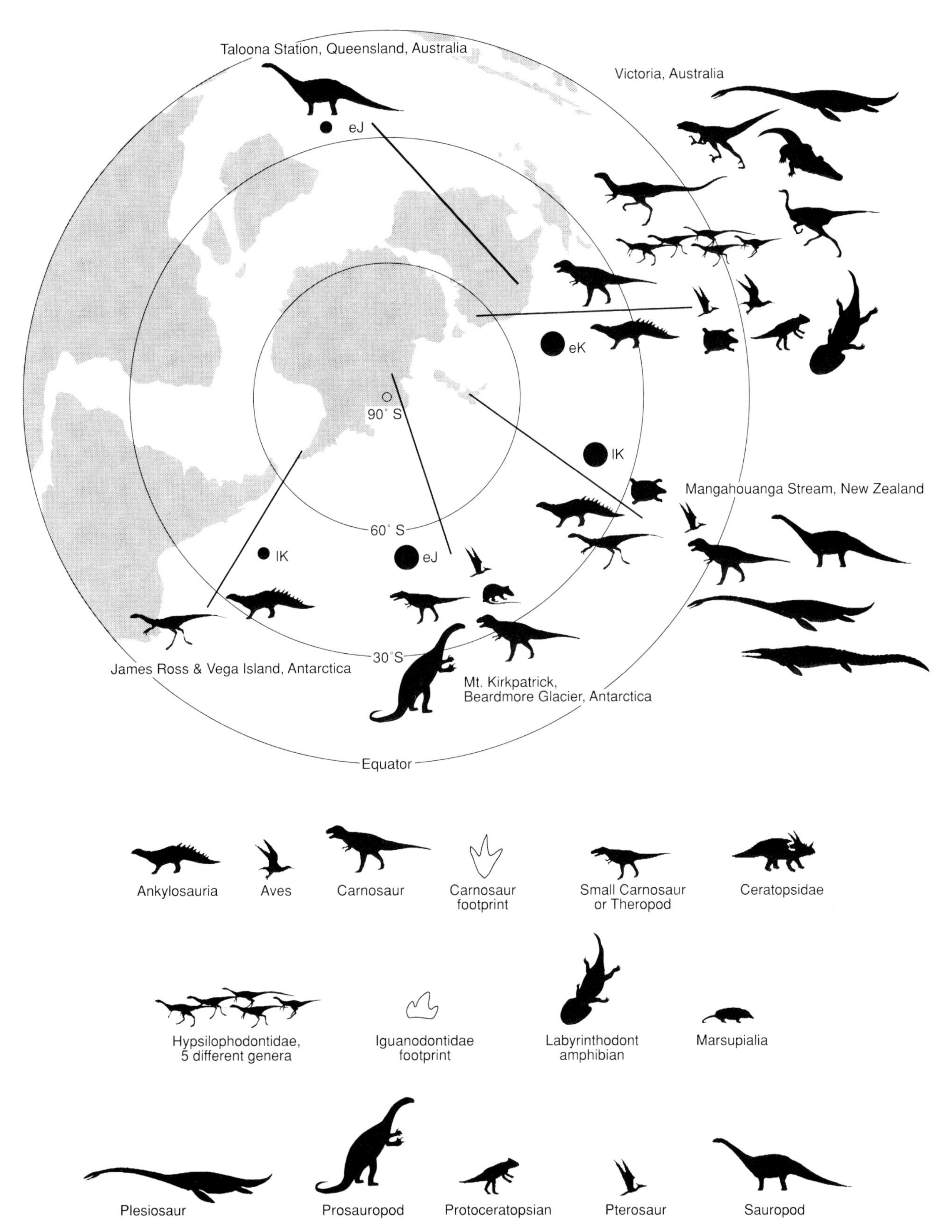

with elsewhere is probably due in part to the fact that more effort has been expended in that area. The base map is taken from Smith *et al.* (1981) for the early Late Cretaceous (earliest Cenomanian), 100 million years ago. Some dinosaur sites significantly different in age were formed at paleolatitudes quite different from the positions shown because of continental movement between the time of accumulation of the fossils and the age of the base map. Geological age of site or sites: eJ, Early Jurassic, 178–208 Ma; lJ, Late Jurassic, 145–157 Ma: eK, Early Cretaceous, 97–145 Ma; lK, Late Cretaceous, 65–97 Ma. Quantity of fossil material: ●, greater than 1000 bones; ●, 100–1000 bones; •, less than 100 bones.

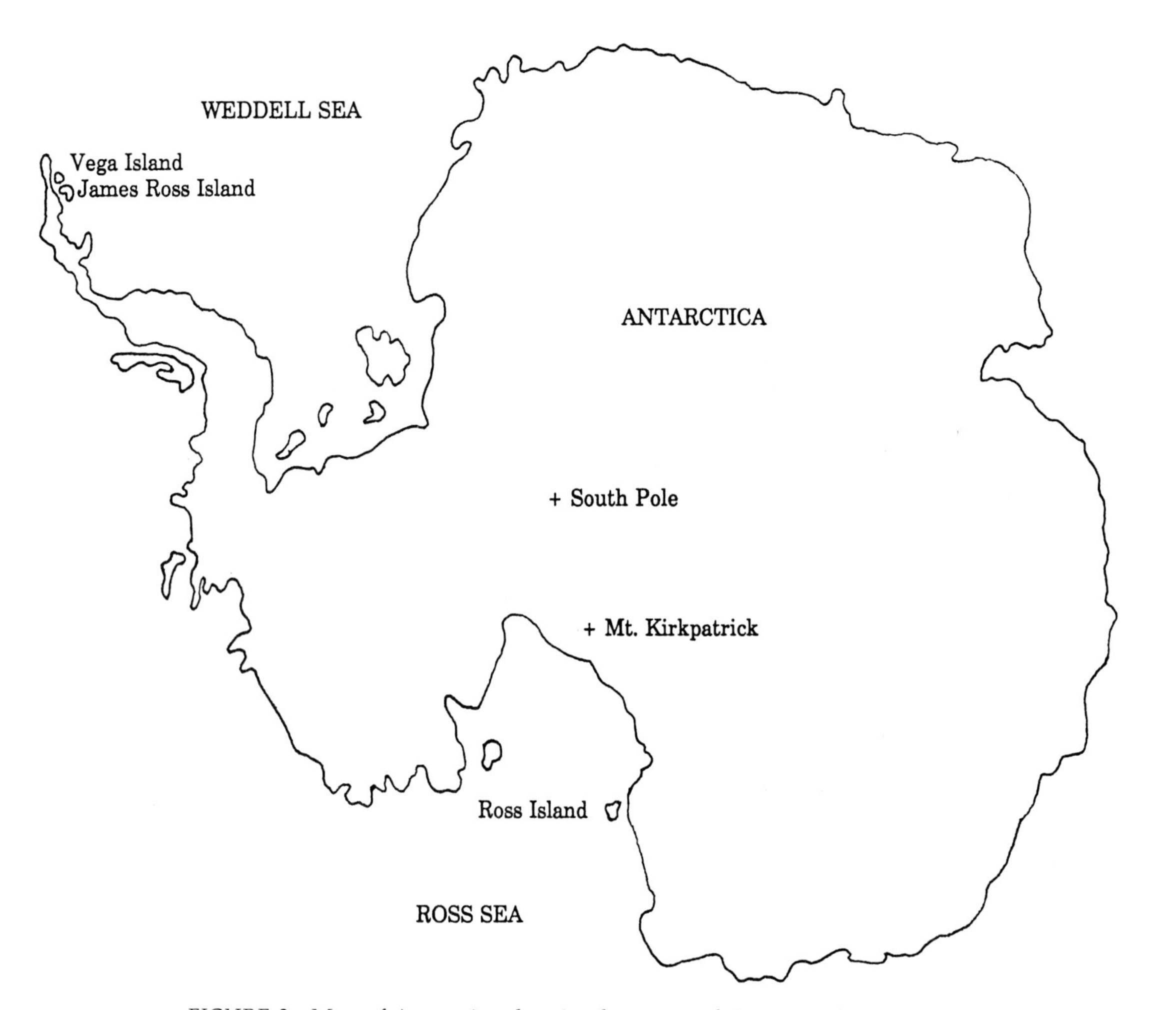

FIGURE 2 Map of Antarctica showing locations of Antarctic dinosaur sites.

the Triassic Fremouw Formation in the southern portions of the Transantarctic Mountains. The Fremouw Formation has yielded faunas of Early to early Middle Triassic age that consist mainly of synapsids and temnospondyls (amphibians) that lived prior to the first appearance of the dinosaurs (Hammer, 1990).

Igneous rocks that have been dated at 177 million years (early Middle Jurassic) intrude the upper portion of the Falla Formation where the dinosaurs occur, indicating they lived before that time. The presence of a large tritylodont and a large plateosaurid prosauropod indicate more precisely an Early Jurassic (Pleinsbachian–Toarcian) age for the fauna from the Falla Formation. Similar tritylodonts from South Africa (*Tritylodon maximus*) and China (*Bienotheroides*) are restricted to the Early Jurassic, and plateosaurid prosauropods from other continents do not occur later than the Early Jurassic.

Most of the Jurassic dinosaur specimens from Antarctica belong to a new genus and species of theropod, *Cryolophosaurus ellioti* ("frozen crested reptile"; Hammer and Hickerson, 1994). *Cryolophosaurus* was a large carnivore (skull length 65 cm, body length approx 7 or 8 m) with a unique crested skull (Fig. 3). The skull is high and narrow and the nasals are high ridges that run along the front of the skull and merge into the highly furrowed crest that sits just above the orbits (Fig. 4). The crest consists mainly of the lacrimal bone and extends the entire width of the skull between the orbits. Two horns lie immediately adjacent to the crest on either side. The crest apparently functioned as a display feature during certain types of social behavior such as mating. It is thin, its highly furrowed surface is decorative, and it would have been ineffective as a weapon.

Other theropods, particularly *Dilophosaurus* and

FIGURE 3 Reconstruction of the theropod *Cryolophosaurus* from the Jurassic of Antarctica.

Monolophosaurus, have cranial display crests. However, in the case of both of these animals the display crests are most obvious from a lateral (side) view. The crest of *Cryolophosaurus*, on the other hand, is most visible from a frontal view. This probably indicates behavioral differences between *Cryolophosaurus* and the other crested theropods, assuming the crests in all three were visual display features.

A second unusual characteristic of the *Cryolophosaurus* skull can be seen in the postorbital region, where the lateral temporal opening is actually split into two openings by processes of the postorbital and squamosal bones. Only a few tyrannosaurid specimens show a similar fusion among all the theropods. Aside from the unique features mentioned, the skull of *Cryolophosaurus* is similar to later Jurassic tetanuran theropods such as *Allosaurus* and *Monolophosaurus*.

Elements of the postcranial skeleton of *Cryolophosaurus* collected include a femur, pubis, ilium, ischium, tibiotarsus, tibia, fibula, two metatarsals, and numerous vertebrae. The femur, tibiotarsus, and metatarsals show primitive theropod characteristics that are also found in early ceratosaurs such as *Dilophosaurus*. However, the pelvic features are more derived, again similar to the later Jurassic theropods such as *Monolophosaurus*.

Cryolophosaurus thus shows a mixture of characters, some of which link it to the more derived theropods of the later Jurassic and some of which are more primitive, like the earlier Jurassic ceratosaur *Dilophosaurus*. It appears most likely that *Cryolophosaurus* is related to the later Jurassic tetanuran theropods (probably the allosaurids) but because it is much older it is not surprising that it retains some primitive features.

Other Jurassic theropod taxa from Antarctica are represented only by teeth. Some of the bones of *Cryolophosaurus* show evidence of scavenging and at least two different scavenging theropods are represented by broken teeth found near gnawed bones.

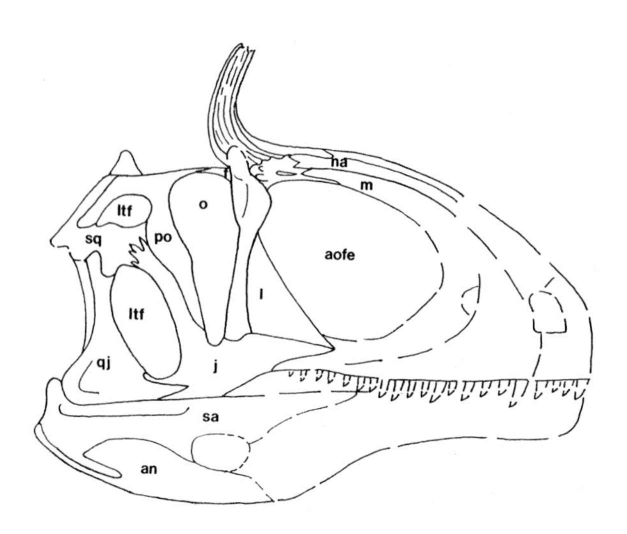

FIGURE 4 Lateral view drawing of the skull of *Cryolophosaurus*; an, angular; aofe, antorbital fenestra; f, frontal; j, jugal; l, lacrimal, ltf, lateral temporal fenestra; m, maxilla; na, nasal; o, orbit; po, postorbital; qj, quadratojugal; sa, surangular; sq, squamosal.

Elements of the prosauropod identified from the Mt. Kirkpatrick site include a partial articulated foot that includes the astragalus and four metatarsals (Fig. 5), the distal end of a femur, and, possibly, a few cervical vertebrae. The Antarctic animal is among the largest of the prosauropods, and the size and features of the foot indicate it is related to the plateosaurids such as *Plateosaurus* from the Late Triassic of Germany and *Lufengosaurus* from the Early Jurassic of China.

The James Ross Island ankylosaur came from the Gamma Member of the Santa Marta Formation (Olivero *et al.*, 1986). This Campanian (Late Cretaceous) marine unit also contains gastropods, bivalves, ammonites, and plesiosaurs (marine reptiles). The one incomplete specimen of a probable nodosaurid ankylosaur collected includes skull and other skeletal fragments and armor plates.

The Antarctic hypsilophodont was collected from a marine deep-shelf silty mudstone in the Late Cretaceous Lopez de Bertodano Formation on Vega Island. The specimen includes disarticulated elements of the skull, cervical, dorsal, and sacral vertebrae; portions of both pectoral girdles and humeri; and parts of the pelvis (ilium and ischium) (Hooker *et al.*, 1991). The length of the animal, estimated to be 4 or 5 m, makes it one of the largest of the hypsilophodonts (normal size range 2 or 3 m). Preliminary investigations indicate that although this animal shows features consistent with other hypsilophodonts, it also has some unique features, particularly in the dentition. The humeri and pelvis have some characteristics similar to *Dryosaurus* and *Valdosaurus*.

Australian dinosaurs are known from the Early Jurassic through the Late Cretaceous. From the Early or Middle Jurassic of southeastern Queensland, Australia, has come a partial skeleton of one of the earliest sauropods, *Rhoetosaurus brownii* Longman 1927.

A single astragalus suggests that the well-known form *Allosaurus* may have persisted into the Aptian of southeastern Australia after having become extinct elsewhere at the end of the Jurassic.

On the basis of one or a few bones, the presence of four groups previously known in the Late Cretaceous of the Northern Hemisphere has been suggested in the Early Cretaceous of southeastern Aus-

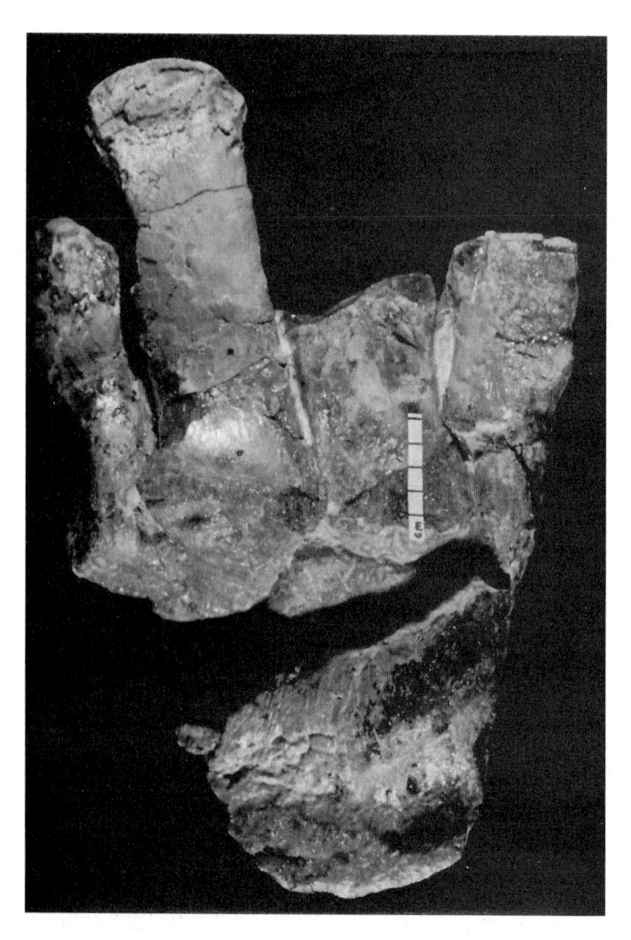

FIGURE 5 Dorsal (top) view of Antarctic prosauropod foot, showing astragalus and metatarsals I–IV.

tralia. An ulna from the Aptian bears an uncanny resemblance to that of the Maastrichtian protoceratopsian *Leptoceratops gracilis* from Alberta, Canada. This suggests that Polar Gondwana may have been the place of origin for the neoceratopsians because they are known nowhere else prior to the Late Cretaceous (see NEOCERATOPSIA).

Ornithomimosaurs are represented in the Albian of southeastern Australia by femora and vertebrae distinct enough to base a new genus and species, *Timimus hermani* (Rich and Vickers-Rich, 1994) on them. Together with the Late Jurassic *Elaphrosaurus bambergi* from Tendaguru, Tanzania, this material suggests a presence for this group on the Gondwana continents prior to the Late Cretaceous, when ornithomimosaurs are best known in the Northern Hemisphere. The recent publication of *Shuvosaurus inexpectatus* Chatterjee 1993 from the Late Triassic of Texas implies that ornithomimosaurs might have had a much longer history in the Northern Hemisphere than previously suspected (but see ORNITHOMIMOSAURIA).

Oviraptorosaurs, previously represented exclusively in the Late Cretaceous of the Northern Hemisphere, appear to have been present in the Albian of southeastern Australia based on a partial surangular and a vertebra (Currie *et al.*, 1996).

More than 15 isolated teeth of dromaeosaurids suggest the presence of another typically Late Cretaceous group of the Northern Hemisphere in the Aptian and Albian of southeastern Australia (Currie *et al.*, 1996).

Among polar dinosaurs from the Southern Hemisphere, only one feature of one individual has yet been interpreted as an adaptation to life in a high latitude environment. This may reflect more the fact that very little is known about these animals and less published than that they rarely displayed marked differences between themselves and their lower latitude contemporaries. The feature in question is the enlargement of the optic lobes of the brain of the hypsilophodontid *Leaellynasaura amicagraphica* (Rich & Rich, 1989) from the Aptian of southeastern Australia in comparison to the same structure on hypsilophodontids from lower latitudes. Hypertrophy of this structure formed the basis for the suggestion that this animal had enhanced ability to see under the low light conditions that would have prevailed during the prolonged periods of continuous darkness each winter.

Hypsilophodontid dinosaurs are generally rare in most dinosaur assemblages. Even where specimens are relatively common as on the Isle of Wight, their taxonomic diversity is not great. Southeastern Australia is a marked exception to that generality. At least six species in five or six genera occur there, just over half the total dinosaurs recognized in that region to date. P. Currie (personal communication) has suggested that hypsilophodontids may have been primarily an upland group at lower latitudes (hence their general rarity there) and are better represented in polar southeastern Australia because of its cooler conditions.

In this regard, it may be noteworthy that the dinosaur known from the Late Cretaceous of Vega Island, Antarctic Peninsula, is a hypsilophodontid or dryosaurid (Milner and Hooker, 1992), and that of the four found at Mangahouanga Stream, New Zealand, one is a probable dryosaurid, a family closely related to hypsilophodontids (Molnar and Wiffen, 1994).

This explanation for the apparent preponderance of these groups at high latitude may well be true. However, in the much more fossiliferous Liscomb bone bed on the Colville River, Alaska, hypsilophodontids are unknown despite the extensive collecting carried out there (Clemens and Nelms, 1993, personal communication). The age of the site is late Campanian to early Maastrichtian (radiometric dates = 68–74 Ma), and by that time there may not have been the variety of hypsilophodontids that there was earlier in the Cretaceous. Several teeth and an ungual of a hypsilophodontid have been collected in older (Campanian) channel deposits upriver from the Liscomb bone bed site.

Today, there are no birds or terrestrial mammalian families restricted to the polar regions. As far as the available record has been analyzed, the same can be said of the polar dinosaurs. Although some new genera and species have been recognized among them, they all belong to families also known at lower latitudes.

However, given the fragmentary nature of much of the evidence for the existence of various groups, this high degree of overlap with low-latitude dinosaur assemblages could in part be an artifact. If one had only a single tooth of the extinct South American litopterns and that group was otherwise unknown, depending on the species, the most parsimonious familial identification might be Equidae. In that case one would be parsimonious but one would be wrong. The recent identification of protoceratopsians in the Aptian of Australia (Rich and Vickers-Rich, 1994) could be a parallel case. On the basis of a single bone, rather than propose an entirely new group of vertebrates that would share an uncanny resemblance in the form of the ulna to protoceratopsians but could in some other as yet unknown way be distinct from them, the specimen was allocated to this known taxon, which extended its record not only to another continent but also backwards in time by at least 15 million years. At our current state of knowledge, that identification is quite plausible but could eventually prove to be fundamentally in error if that Australian "protoceratopsian" proves to be a radically different animal when it is better known.

There is no evidence such as tillites to suggest that continental ice sheets existed during the Mesozoic at high latitudes. However, it has been inferred on the basis of the occurrence of dropstones as large as 3 m across in fine-grained marine sediments deposited during the Late Jurassic and Early Cretaceous that winter seasonal ice did form at high paleolatitudes in both hemispheres (Frakes *et al.*, 1992).

On the North Slope of Alaska, studies assessing the mean annual paleotemperature have been carried out on the sediments producing the dinosaurs. Paleobotanical evidence based on leaf margin and stomata structure together with the overall composition of the flora have been taken to suggest a mean annual temperature of 2–8°C.

In southeastern Australia, similar modes of analysis produced conflicting results. There, the paleobotanical evidence suggested a mean annual temperature of 10°C (Parrish *et al.*, 1991), whereas an oxygen isotope estimate is -2 ± 5°C (Gregory *et al.*, 1989), which is the difference between Chicago and Anchorage today. Although the biological implications of these two estimates are quite different, they are concordant in that the paleoclimate was far from tropical.

No matter what the paleotemperature was, polar dinosaurs would have had to adapt to prolonged periods of annual darkness each winter. Although year-round residency has not been proven in either hemisphere, several lines of evidence strongly suggest this possibility. The suggestion has been made that in the geological past the earth's rotational axis might have been significantly closer to being oriented perpendicular to the plane of the ecliptic, thus reducing the length of continuous darkness each winter at high latitudes (e.g., Douglas and Williams, 1982). However, the criticism of Laplace (1829) against the earth's obliquity having shifted more than a few degrees from its present orientation as it does over a period of approximately 41,000 years has never been refuted.

The presence of abundant and diverse assemblages of polar dinosaurs has raised several important questions regarding the behavior, physiology, and extinction of all dinosaurs. Polar dinosaurs are an indisputable fact. They occupied high latitudes in both hemispheres for more than 45 million years. Establishing whether they were year-round residents or long-distance seasonal migrants is directly related to their thermoregulation and energetics. This question also affects extinction theories that depend on short-term weather effects such as a bolide-induced "nuclear winter." Currently, polar dinosaurs have produced more questions than answers.

The work on the polar dinosaurs is just beginning. Just how distinctive they were from their lower latitude contemporaries is unknown. Beyond a doubt, the known area with the greatest potential for producing further knowledge about polar dinosaurs in either hemisphere is the 200 km of outcrops of terrestrial sediments on the left bank of the Colville River, Alaska. All of the Late Cretaceous is represented in those deposits with fossil vertebrates now known from sites of many different ages. The Falla Formation of Antarctica seems the most promising area in the Southern Hemisphere in terms of future potential. Clearly, to find significantly more information about Gondwana polar dinosaurs outside the area of the Falla Formation, new areas such as the Cretaceous coalfields of New Zealand need to be investigated (Rich, 1975).

See also the following related entries:

CANADIAN DINOSAURS • MIGRATION • PLATE TECTONICS

References

Carpenter, K., and Alf, K. (1994). Global distribution of dinosaur eggs, nests, and babies. In *Dinosaur Eggs and Babies* (K. Carpenter, K. F. Hirsch, and J. R. Horner, Eds.), pp. 15–30. Cambridge Univ. Press, Cambridge, UK.

Chatterjee, S. (1993). *Shuvosaurus*, a new theropod. *Res. Exploration* **9,** 274–285.

Clemens, W. A., and Nelms, L. G. (1993). Paleoecological implications of Alaskan terrestrial vertebrate fauna in latest Cretaceous time at high paleolatitudes. *Geology* **21,** 503–506.

Conyers, L. (1978). Letters, notes, and comments. *Alaska* **XLIV,** 30.

Currie, P. J., Vickers-Rich, P., and Rich, T. H. (1996). Possible oviraptorosaur (Theropoda, Dinosauria) specimens from the Early Cretaceous Otway Group of Dinosaur Cove, Australia. *Alcheringa* **20,** 73–79.

Douglas, J. G., and Williams, G. E. (1982). Southern polar forests: The Early Cretaceous floras of Victoria and their palaeoclimatic significance. *Palaeogeogr. Palaeoclimatol. Palaeoecol.* **39,** 171–185.

Edwards, M. B., Edwards, R., and Colbert, E. H. (1978). Carnosaurian footprints in the Lower Cretaceous of eastern Spitsbergen. *J. Paleontol.* **52,** 940–941.

Frakes, L. A., Francis, J. E., and Syktus, J. I. (1992). *Climate Modes of the Phanerozoic.* Cambridge Univ. Press, New York.

Gangloff, R. A. (1994). The record of Cretaceous dinosaurs in Alaska an overview. Proceedings of the 1992 International Conference on Arctic Margins, Alaska Geological Society, pp. 399–404.

Gangloff, R. A. (1995). *Edmontonia* sp., the first record of an ankylosaur from Alaska. *J. Vertebr. Paleontol.* **15,** 195–200.

Gregory, R. T., Douthitt, C. B., Duddy, I. R., Rich, P. V., and Rich, T. H. (1989). Oxygen isotopic composition of carbonate concretions from the lower Cretaceous of Victoria, Australia: Implications for the evolution of meteoric waters on the Australian continent in a paleopolar environment. *Earth Planetary Sci. Lett.* **92,** 27–42.

Hammer, W. R. (1990). Triassic terrestrial vertebrate faunas of Antarctica. In *Antarctic Paleobiology: Its Role in the Reconstruction of Gondwanaland* (T. N. Taylor and E. L. Taylor, Eds.), pp. 42–50. Springer-Verlag, New York.

Hammer, W. R., and Hickerson, W. J. (1994). A crested theropod dinosaur from Antarctica. *Science* **264,** 828–830.

Hooker, J. J., Milner, A. C., and Sequeira, S. E. K. (1991). An ornithopod dinosaur from the Late Cretaceous of West Antarctica. *Antarctic Sci.* **3**(3), 331–332.

Laplace, P. S. (1829). *Mecanique Celeste,* Vol. II.

Longman, H. A. (1927). The giant dinosaur *Rhoetosaurus brownei. Mem. Queensland Museum* **9,** 1–18.

Milner, A. C., and Hooker, J. J. (1992). An ornithopod dinosaur from the upper Cretaceous of the Antarctic Peninsula. *J. Vertebr. Paleontol.* **12,** 44A.

Molnar, R. E., and Wiffen, J. (1994). A Late Cretaceous polar dinosaur fauna from New Zealand. *Cretaceous Res.* **15,** 689–706.

Nessov, L. A., and Golovneva, L. B. (1990). History of the flora, vertebrates and climate in the late Senonian of the north-eastern Koriak. In *Continental Cretaceous of the USSR. Collection of Research Papers,* Project 245, pp. 191–212. IUGS, Vladivostok, Russia. [In Russian]

Olivero, E., Scasso, R., and Rinaldi, C. (1986). Revision of the Marambio Group, James Ross Island, Antarctica. *Contrib. Antarctic Inst. Argentina* **331,** 1–30.

Olivero, E. B., Gasparini, Z., Rinaldi, C. A., and Scasso, R. (1991). In *Geological Evolution of Antarctica* (M. R. S. Thomson, J. A. Crame, and J. W. Thomson, Eds.), pp. 617–622. Cambridge Univ. Press, New York.

Parrish, J. M., Spicer, R. A., Douglas, J. G., Rich, T. H., and Vickers-Rich, P. (1991). Continental climate near the Albian South Pole and comparison with climate near the North Pole. *Geol. Soc. Am.* **23,** A302. [Abstracts with Programs]

Rich, T. H. (1975). Potential pre-Pleistocene fossil tetrapod sites in New Zealand. *Mauri Ora* **3,** 45–54.

Rich, T. H., and Rich, P. V. (1989). Polar dinosaurs and biotas of the early Cretaceous of southeastern Australia. *Natl. Geogr. Res.* **5,** 15–53.

Rich, T. H., and Vickers-Rich, P. (1994). Neoceratopsians & Ornithomimosaurs: Dinosaurs of Gondwana origin? *Res. Exploration* **10,** 129–131.

Rozhdestvensky, A. K. (1973). The study of Cretaceous reptiles in Russia. *Paleontologicheskii Zhurnal.* **1973,** 90–99. [In Russian]

Russell, D. A. (1984). A check list of the families and genera of North American dinosaurs. Syllogeus Series No. 53. National Museums of Canada.

Russell, D. A. (1988). A check list of North American marine Cretaceous vertebrates including fresh water fishes. Occasional Paper of the Tyrrell Museum of Palaeontology, No. 4.

Smith, A. B., Hurley, A. M., and Briden, J. C. (1981). *Phanerozoic Paleocontinental World Maps.* Cambridge Univ. Press, New York.

Polish–Mongolian Paleontological Expeditions

Halszka Osmólska
Polska Akademia Nauk
Warsaw, Poland

Eight Polish–Mongolian Paleontological Expeditions between 1963 and 1971 worked in Mongolia where they explored the Upper Cretaceous and Tertiary deposits in the southern Mongolian Gobi Desert and western Mongolia (Kielan-Jaworowska and Dovchin, 1969; Kielan-Jaworowska and Barsbold, 1972; Gradziński *et al.*, 1977). The expeditions were a cooperative scientific program between the Polish and Mongolian Academies of Sciences and were organized by the Institute of Paleozoology (now the Institute of Paleobiology) in Warsaw and the Biological Research Institute, and subsequently the Geological Institute, in Ulaan Bataar. The main aim of the expeditions was to collect Cretaceous and Tertiary vertebrates along with data on their biological and physical environments. Organization and scientific leadership of the expeditions was undertaken by Zofia Kielan-Jaworowska.

The Polish–Mongolian expeditions to Mongolia were preceded by the expeditions of the American Museum of Natural History Expeditions from 1922 to 1930 (known as the Third CENTRAL ASIATIC EXPEDITION) and by the expeditions of the Institute of Paleontology of the USSR Academy of Sciences (1946–1949). Because the localities were all in uninhabited regions, sometimes hundreds of kilometers from the nearest settlements, the expeditions had to be completely self-sufficient. Three of the trucks, the equipment, petroleum, and most of food were brought from Poland. As worked out in the agreement between the Polish and Mongolian Academies of Sciences, most of the dinosaurs found remained in Mongolia, whereas the majority of mammal and lizard remains went to Poland.

The first (1963) expedition (including 11 Polish and Mongolian members) was a reconnaissance to reexamine localities previously prospected by the American and Soviet expeditions and to search for new outcrops with the potential of yielding Cretaceous and Tertiary vertebrates.

In 1964 the expedition consisted of 16 persons. June and July were devoted to exploration of the late Campanian or middle Maastrichtian dinosaur-bearing beds of the Nemegt Formation at the localities of Altan Ula, Nemegt, and Tsagan Khushu and the Paleocene deposits at Naran Bulak, all within the Nemegt Basin in the southern Gobi. In August, one group searched for early or middle Campanian mammals, lizards, and dinosaurs in the outcrops of Djadokhta Formation at Bayn Dzak in the pre-Altai Gobi, at the foot of the Flaming Cliffs (discovered by the American Expeditions), and inspected the neighboring Paleocene outcrops at Khashaat. The other group visited some known mammal-bearing Tertiary sites south of the Gobi Altai range, discovering a few new sites on its way.

The 1965 expedition (23 participants) was one of the largest. Most of the time the expedition team worked in two groups. Nine people spent a month at Bayn Dzak, discovering some new sites with small vertebrates and significantly augmenting the mammal and lizard collection. Dinosaur skeletons, including *Protceratops* and *Pinacosaurus* juveniles, numerous dinosaur eggs and nests, tortoises, and a previously unknown, small crocodile (*Gobiosuchus*) were also found. The second, larger group (14 people) worked in the Nemegt Basin, mainly at Altan Ula IV and Nemegt, but also visited Altan Ula III and Tsagan Khushu localities. Numerous skeletons of dinosaurs, tortoises, and crocodiles were found, as well as fish remains, ostracodes, tree trunks, and charophyte oogonia. Except for numerous *Tarbosaurus* and *Saurolophus* skeletons previously found by the Soviet Expeditions, the dinosaur skeletons collected represented new taxa of ornithomimids, sauropods, and pachycephalosaurids. In the middle of July, the Bayn Dzak group joined the Nemegt team for several days and then headed toward the Lakes Valley in the west of Mongolia to look at Tertiary deposits.

After completion of the expedition of 1965, fieldwork in Mongolia was suspended for 1 year. In 1967 and each of the successive three years, a group of five Polish and Mongolian paleontologists stayed for several weeks in the Bayn Dzak region to search for small Late Cretaceous vertebrates but no excavations were undertaken.

The 1970 expedition was also a larger one and included 15 participants. At the beginning, the entire team worked at Bayn Dzak, enlarging the mammal and lizard collection but also investigating outcrops in the vicinity. When the camp at Bayn Dzak was closed, the expedition headed south to the Nemegt Basin. The main camp was set up at Nemegt (Northern Sayr). The outcrops at Altan Ula II and III and the Shiregin Gashun (= Shiregeen Gashoon) Basin north of the Nemegt range were also visited. In addition to several *Tarbosaurus* and *Saurolophus* skeletons, oviraptorid and pachycephalosaurid skulls and ornithomimid and ankylosaurid skeletons were discovered.

During the second half of the stay in Nemegt, work concentrated mainly on the red-colored deposits of the Campanian Barun Goyot Formation underlying the yellowish beds of the Nemegt Formation. They cropped out in several places, including Khulsan, a locality discovered by the expedition, some 7 km east of the Nemegt locality. The red beds of this region, considered unfossiliferous until then, proved rich beyond expectation and yielded many dinosaurs, including oviraptorids and droameosaurids, dinosaur eggs, and numerous mammal and lizard skulls.

The team of the 1971 expedition consisted of 15 persons. The first 6 weeks were spent at Khulsan. In addition to numerous mammals, lizards, and eggs, the skull of a new pachycephalosaurid and a skull with partial postcranium of an unknown ankylosaurid were found. The next 6 weeks were spent in Altan Ula IV prospecting the Nemegt Formation outcrops. Several trips were made to Shiregin Gashun and Khermeen Tsav (both discovered by the Soviet–Mongolian expeditions). An almost complete skeleton of *Microceratops* was discovered in older, presumably early Late Cretaceous deposits of the Shiregin Gashun section. The main aim of the Khermeen Tsav trips was to search for small mammals and lizards in the red beds of a lower part of the Upper Cretaceous section, later referred to as the red beds of Khermeen Tsav I and II, which are considered roughly coeval with the Barun Goyot Formation. The collection of mammals and lizards was spectacular, but only a few dinosaurs were recovered, although they were interesting because they included some new ankylosaurid, protoceratopsid, and oviraptorid taxa.

In August, the expedition proceeded to Toogreeg, a locality situated approximately 30 km west of Bayn Dzak. During the 2-week stay in Toogreeg, the most famous of the Gobi Desert specimens, "the fighting dinosaurs" (*Protoceratops* and *Velociraptor*) were found by Dr. A. Sulimski in the beds coeval with the Djadokhta Formation.

Collections assembled during the eight Polish–Mongolian expeditions include several thousand specimens of dinosaurs, lizards, turtles, crocodilians, and mammals, as well as numerous dinosaur (and other vertebrate) eggs and nests. Microfaunal and microflora samples produced ostracods, diplopods, and charophyte oogonia.

The dinosaurs identified so far include 24 species (Table I), 15 of which were new, representing 24 genera. Some of dinosaur groups (pachycephalosaurids, ornithomimids, and elmisaurids) were found in this region by the Polish–Mongolian expeditions for the first time. Many are represented by exceptionally well-preserved, complete postcranial skeletons and/or skulls.

The lizard collection includes about 300 specimens,

TABLE I Dinosaur Taxa Found during Polish–Mongolian Palaeontological Expeditions

Saurischia	Ornithischia
Sauropoda	Ornithopoda
Opisthocoelicaudia skarzynskii Borsuk-Bialynicka 1977	*Saurolophus angustirostris* Rozhdestvensky 1952
Nemegtosaurus mongoliensis Osmólska 1981	*Barsboldia sicinskii* Mayańska and Osmólska 1981
Theropoda	Pachycephalosauria
Galimimus bullatus Osmólska, Roniewicz, and Barsbold 1972	*Homalocephale calathocercos* Maryańska and Osmólska 1974
Deinocheirus mirificus Osmólska and Roniewicz 1969	*Prenocephale prenes* Maryańska and Osmólska 1974
Velociraptor mongoliensis Osborn 1924	*Tylocephale gilmorei* Maryańska and Osmólska 1974
Elmisaurus rarus Osmólska 1981	Ceratopsia
Avimimus portentosus Kurzanov 1981	*Microceratops gobiensis* Bohlin 1953
Protoceratops andrewsi Granger and Gregory 1923	*Ingenia yanshini* Barsbold 1981
Conchoraptor gracilis Barsbold 1986	*Bagaceratops rozhdestvenskyi* Maryańska and Osmólska 1975
Archaeornithoides deinosauriscus Elzanowski and Wellnhofer 1992	*Breviceratops kozlowski* (Maryańska and Osmólska 1975
Tarbosaurus bataar Maleev 1955	Ankylosauria
Pinacosaurus grangeri Gilmore 1933	*Saichania chulsanensis* Maryańska 1977
Hulsanpes perlei Osmólska 1982	*Tarchia gigantea* Maleev 1956
Borogovia gracilicrus Osmólska 1987	
Anserimimus planinychus Barsbold 1988	

which represent 32 species of 29 genera; 23 of these species and 21 genera were new. The collection of turtles is less abundant and less diverse but includes about 8 turtle species (3 new). Three crocodylian species (1 new genus and species) were also recovered. The Late Cretaceous mammals embrace about 170 specimens, the majority of which are more or less complete skulls, sometimes associated with postcrania. To date, it is one of the most significant collections of Cretaceous mammals in the world. Fourteen mammal species belonging to 14 monotypic genera were described, 10 of which were new. Significant Tertiary mammals were also recovered by the expeditions.

Expedition narratives and descriptions of specimens collected by the Polish–Mongolian expeditions were published mostly in 10 volumes of the *Results of the Polish–Mongolian Palaeontological Expeditions*, issued between 1969 and 1984 in *Palaeontologia Polonica.* Additional dinosaur materials collected by the expeditions are still studied.

See also the following related entries:

Barun Goyot Formation • Bayn Dzak • Central Asiatic Expeditions • Mongolian Dinosaurs • Nemegt Formation

References

Gradziński, R., Kielan-Jaworowska, Z., and Maryańska, T. (1977). Upper Cretaceous Djadokhta, Barun Goyot and Nemegt formations of Mongolia, including remarks on previous subdivisions. *Acta Geol. Polonica* **27,** 281–318.

Kielan-Jaworowska, Z., and Dovchin, N. (1969). Narrative of the Polish–Mongolian Palaeontological Expeditions 1963–1965. *Palaeontol. Polonica* **19,** 7–30.

Kielan-Jaworowska, Z., and Barsbold, R. (1972). Narrative of the Polish–Mongolian Palaeontological Expeditions 1967–1971. *Palaeontol. Polonica* **27,** 5–13.

Popular Culture, Literature

Donald F. Glut
Burbank, California, USA

Although dinosaurs have become strongly rooted in popular culture, it is as living creatures rather than fossils that they are most appealing and known to the general public. Some of the earliest appearances of "live" dinosaurs in popular culture are found in the pages of literature. Such literary interpretations stemmed from the paleontologist's ability to extrapolate from fossil evidence in order to imagine these long-extinct organisms as live animals.

Living dinosaurs have inhabited popular fiction for nearly a century and a half—almost as long as they have been known to science. Charles Dickens, in the opening passage of *Bleak House*, was, it seems, the first author to fantasize a live dinosaur ("a Megalosaurus, forty feet long or so, waddling like an elephantine lizard up Holborn Hill") in a published story. Dickens' otherwise nonfanciful tale first appeared in 1853, 31 years after James Parkinson introduced the name *Megalosaurus* (see History of Dinosaur Discoveries) to the paleontological lexicon.

Nineteenth-century stories featuring dinosaurs (as well as other Mesozoic reptiles) were rare. In his 1864 fantastic novel *Voyage au centre de la Terre* (*Journey to the Center of the Earth*), author Jules Verne explored the already old scenario of a so-called "hollow earth" populated by fantastic denizens. Among Verne's subterranean inhabitants were various creatures including ichthyosaurs and plesiosaurs (nondinosaurian Mesozoic reptiles) that were otherwise extinct upon the surface world.

The plot basis of living dinosaurs surviving into the modern world, usually in some remote or isolated place segregated from the ravages of time, has become a staple in the subdivision of literature known today as "science fiction." In most examples of this literary subgenre, explorers, scientists, big-game hunters, or other modern-day characters discover a lost land, only to fall prey to its numerous perils. The prototype and template for such stories was *The Lost World* (1912), one of the tales centering upon bombastic scientist Professor Challenger, written by Sir Arthur Conan Doyle. In this fiction, then-known dinosaurs, such as *Megalosaurus* and *Iguanodon*, as well as other fauna, live in a tropical paradise located atop an isolated South American plateau.

From the century's teen years through early 1950s, most tales involving "lost worlds" and surviving dinosaurs were published in cheaply produced "pulp magazines" that satisfied the public's hunger for entertainment. Among the authors in this arena was the prolific Edgar Rice Burroughs, who explored lost world territories in numerous fantastic novels. Two years after the Doyle story, Burroughs gave new life to the old hollow earth idea with *At the Earth's Core,* first serialized in the pulp pages of *All-Story Weekly*. Burroughs' version of the world within our world was called Pellucidar. It teemed with numerous life forms long extinct on the surface, including the "Mahars," a ruling race of very large, intelligent pterosaurs resembling the genus *Rhamphorhynchus*.

Pellucidar was a popular fantasy world, and Burroughs continued to transport readers to its various exotic realms in a series of sequels: *Pellucidar* (1915), *Tanar of Pellucidar* (1929), *Tarzan at the Earth's Core* (1930), *Back to the Stone Age* (1935), *Land of Terror* (1944), and *Savage Pellucidar* (short-story collection first published as an anthology in 1963).

In *Tarzan at the Earth's Core*, Burroughs' most famous hero encountered many prehistoric creatures including a *Stegosaurus* (called a "dyrodor" by Pellucidarians). The animal broke various scientific laws by launching itself off a cliff side, moving its dorsal plates into a horizontal position to function as glider wings. Another dinosaur met by Tarzan in the story was a *Triceratops* (known in Pellucidar as a "Gyor"). This was not the first time Tarzan had seen a living *Triceratops*, or at least an evolved variation on it. In *Tarzan the Terrible* (1921), such animals, now equipped with carnivore's teeth, talons, and dorsal plates, inhabited Burroughs' lost land of Pal-ul-don (where they were called "gryfs").

Arguably the most inventive of Burroughs's lost worlds was the mountainous prehistoric island of Caprona (also called Caspak), featured in three rather short novels (*The Land That Time Forgot*, *The People That Time Forgot*, and *Out of Time's Abyss*) that originally appeared in 1918 in *The Blue Book* pulp periodical. In this trilogy, Burroughs enhanced the basic lost world concept by introducing the science-fictional idea of a single being progressing through a gamut of evolutionary stages—egg, fish, amphibian, reptile, mammal, and inevitably human—as it moved across the island in a south-to-north direction.

Most dinosaur-related pulp fiction was not as well conceived (nor well written) as that of Burroughs. These stories were usually unpolished, often lurid tales that have not endured beyond their original pulp-magazine appearances. Some of them, such as *Lords of the Underworld* (1941), *The Lost Warship* (1943), *Dinosaur Destroyer* (1949), all appearing in *Amazing Stories* and employing the science-fictional plot device of time travel to bring modern humans back to ancient times, perpetuated the popular misconception that dinosaurs and early man coexisted. In *King of the Dinosaurs* (in *Fantastic Adventures*, 1945), one of the more outlandish examples of this brand of fiction, Tarzan-like hero Toka discovers a lost land where dinosaurs (here called "big snakes") with human-level intelligence play baseball using humans for balls.

Some authors, whose writing careers started in pulps, made the transition into more respectable publications. *The Greatest Adventure*, written by Dr. Eric Temple Bell (pseudonym John Taine), was first published in 1945 in the pulp *Famous Fantastic Mysteries* and has since been reprinted in various book editions. Not just another lost world variation, this novel featured dinosaurs that fell oddly outside the criteria of any known scientific classification. Finally, they were revealed to be the products of extraterrestrial aliens' genetic manipulations.

Ray Bradbury, a former pulp magazine writer who went on to become one of the most respected authors in the fantasy and science fiction genres, expressed his interest in dinosaurs in a number of short stories. In "The Beast from 20,000 Fathoms" (1951, reprinted as "The Fog Horn"), a huge prehistoric reptile falls in love with the baleful sound emanating from a lighthouse (Fig. 1); the classic "A Sound of Thunder" (1952) reveals the evolutionary consequences of stepping on a butterfly during a time trip back to the Cretaceous Period; and "The Prehistoric Producer" (1962) combines Bradbury's love of dinosaurs with that of fantastic movies. These stories, as well as new pieces, were eventually collected and published in the anthology *Dinosaur Tales* (1983).

L. Sprague de Camp authored "A Gun for Dinosaur," which first saw print in a 1956 issue of *Galaxy Science Fiction* magazine. Again utilizing the concept of time travel, with wealthy hunters journeying back to the Mesozoic to "bag" big theropods, this short story, like Bradbury's, has become a science fiction classic. The author's interest in dinosaurs, however, transcended science fiction when, in 1968, de Camp, with coauthor wife Catherine Crook de Camp, produced the nonfiction book *The Day of the Dinosaur*.

In recent years, the concept of live dinosaurs has evolved dramatically from their pulp magazine predecessors (wherein the animals were generally portrayed as lumbering monsters stalking humans or engaged "in mortal combat"). David Gerrold, in his novel *Deathbeast* (1978), incorporated contemporary ideas (mostly championed by paleontologist Robert T. Bakker) relating to possible dinosaurian endo-

FIGURE 1 "McDunn fumbled with the switch. But even as he flicked it on, the monster was rearing up." From "The Fog Horn," by Ray Bradbury. Illustration by Alma Roussy, © 1982 by Academic Press Canada.

thermy (see PHYSIOLOGY). Harry Harrison's *West of Eden* (1984) and its two sequels expanded the evolved dinosaur or "dinosauroid" speculative exercise introduced 2 years earlier by paleontologists Dale A. Russell and R. Seguin (see INTELLIGENCE).

Michael Crichton's *Jurassic Park* (1990), in which dinosaurs were cloned using DNA extracted from insects trapped in Mesozoic amber, became a sweeping best-seller, resulting in the highest grossing motion picture made to date of its release and creating a worldwide dinosaur craze that escalated the public's fascination and love for dinosaurs to new plateaus. With the issuance of *Jurassic Park,* and its 1995 sequel *The Lost World*, more stories involving dinosaurs are enjoying publication, including anthologies (e.g., *Dinosaur Fantastic*, 1993, edited by M. Resnick and M. H. Greenberg) of new and reprinted tales centering on various aspects of these extinct animals.

See also the following related entry:

JURASSIC PARK

References

Ash, B. (Ed.) (1977). *The Visual Encyclopedia of Science Fiction*, pp. 352. Pan Books, London.

Camp, L. Sprague de, and Crook de Camp, C. (1968). *The Day of the Dinosaur* pp. xiii, 319. Doubleday, New York.

Glut, D. F. (1980). *The Dinosaur Scrapbook*, pp. 320. Citadel Press (Lyle Stuart), Secaucus, NJ.

Lupoff, R. A. (1968). *Edgar Rice Burroughs: Master of Adventure* (Rev. ed.), pp. 317. Ace Books, New York.

Parkinson, J. (1822). *Outlines of Orcyctology: An Introduction to the Study of Fossil Organic Remains, Especially Those Found in the British Strata*, pp. viii, 350. London.

Rottensteiner, F. (1975) *The Science Fiction Book: An Illustrated History*, pp. 160. New American Library, New York.

Russell, D. A., and Seguin, R. (1982). Reconstruction of the small Cretaceous theropod *Stenonychosaurus inequalis* and a hypothetical dinosauroid: *Syllogeus* **37**.

Portuguese Dinosaurs

see EUROPEAN DINOSAURS; FOOTPRINTS AND TRACKWAYS

Postcranial Axial Skeleton, Comparative Anatomy

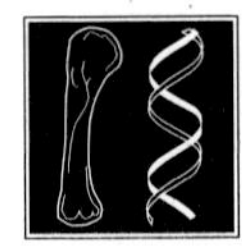

PETER MAKOVICKY
University of Copenhagen
Copenhagen, Denmark

Historically, the axial skeleton (composed most prominently of vertebrae and ribs) of most major groups of dinosaurs has received little attention, with the notable exception of that of the SAUROPODA. In the latter group, the predominance of vertebral material, as well as the elaboration of vertebral structures in advanced taxa, forced early workers to appreciate the importance of vertebral characters in phylogeny reconstruction. This understanding has not yet been fully extended to the other major groups, although recent works have improved the general acceptance of focusing on vertebrae.

Dinosaur ancestors were small, bipedal archosaurs with a lightly constructed axial skeleton. The dinosaurian outgroups have a vertebral composition of nine cervicals, 15 dorsals, two sacrals, and approximately 50 caudals. This composition is plesiomorphic for DINOSAURIA as a whole, but apomorphic changes from this condition occurred early in the evolutionary history of both the SAURISCHIA and ORNITHISCHIA. This comparative study attempts to compare the anatomy of various groups against a phylogenetic background and to draw attention to several general functional aspects (see ORIGIN OF DINOSAURS).

All saurischian dinosaurs are characterized by possessing presacral vertebrae in which a pair of laminar struts support each transverse process from below (Figs. 1 and 2). The transverse process and its struts delimit three fossae between them, which are termed the infraprezygapophyseal, infradiapophyseal, and infrapostzygapohyseal in anteroposterior order (Fig. 1). Within Saurischia, the groups Theropoda and Sauropoda have independently developed pneumatic diverticula that invade the neural arch through the fossae (Britt, 1993). The presence of hyposphene–hypantrum articulations between dorsals is a further derived feature uniting Saurischia (Figs. 1 and 2).

For theropods, the primitive vertebral count appears to be nine cervicals, 15 dorsals, two sacrals, and more than 40 caudals. There is some ambiguity with regard to the number of sacrals in *Herrerasaurus*, a discussion that ultimately is important to the question of whether the possession of three sacrals is a synapomorphy uniting all dinosaurs. There is some question whether HERRERASAURIDAE falls within Dinosauria (see PHYLOGENY OF DINOSAURS). All theropods, but not Herrerasauridae, have incorporated one of the dorsal elements into the neck. In birds that are more derived than *Archaeopteryx*, this process has progressed even further and the dorsal count is reduced to 11 or fewer.

The cervical vertebrae are parallelogram-shaped in lateral view, and the neck has a sigmoid curvature. The anterior cervical elements arch in a dorsally con-

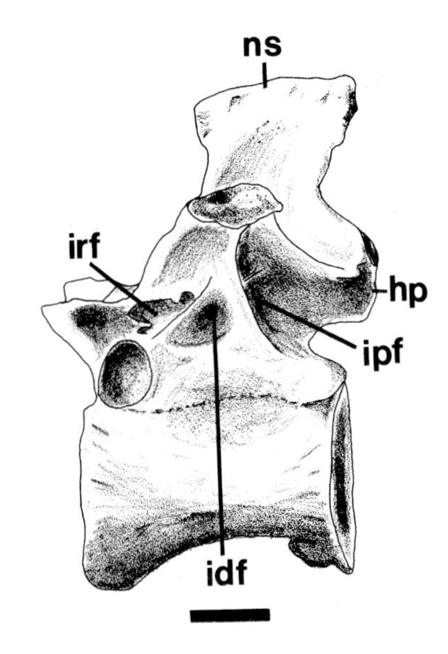

FIGURE 1 Middorsal vertebra of cf. *Ornithomimus* (RTMP 93.62.1) in lateral view. hp, hyposphene; idf, infradiapophyseal fossa; ipf, infra-postzygapophyseal fossa; irf, infra-prezygapophyseal fossa; ns, neural spine. Scale bar = 2 cm.

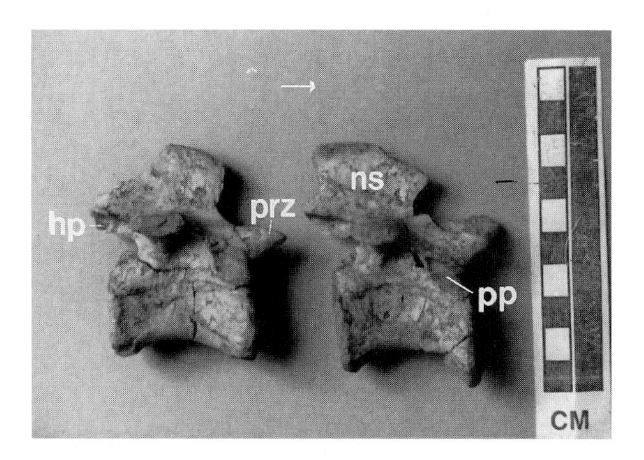

FIGURE 2 Two dorsal vertebrae of *Massospondylus*. Note the hyposphene and lateral fossac of the arch. hp hyposphene; ns, neural spine; pp, parapophysis; prz, prezygapophysis. The small arrow points cranially.

vex curve and the posterior cervicals form a dorsally concave arc. Primitively, the centra are rectangular in cross section and elongate, but in larger theropods, such as torvosaurids, allosauroids, and tyrannosaurids, the vertebrae become subcircular in cross section and relatively short axially. Megalosaurs, torvosaurids, and allosauroids developed opisthocoelous intercentral articulations between the cervicals. These changes are best viewed as modifications in response to enhanced mechanical stress on the vertebral column in connection with increased body mass and especially skull size. The great shortening of the vertebral centra is accompanied by a concomitant increase in the sigmoid curvature of the neck, especially in tyrannosaurids. Theropods are characterized by having pneumatic cervicals and sometimes dorsals. Various characters related to this pneumatization are of taxonomic importance.

Although they are highly diverse, theropod neural arches have a large number of features in common. The neural spine of the axis is larger than on the following cervicals, and its relative size is related to the proportional size of the skull. Generally, in theropods the size of the neural spine decreases progressively in the posterior direction, and in some lightly built ceratosaurs and coelurosaurs, including most birds, the posterior cervicals appear practically devoid of neural spines. Instead, the muscles of the *spinalis* system originate from a long nuchal tendon. The cervical diapophyses are ventrally directed and short, for which reason the neural arch fossae are often reduced.

The dorsal centra are shorter than those of the cervicals. On each of the first three dorsals, there is a ventrally situated insertion for the M. longus colli ventralis, and in, for example, troodontids, dromaeosaurids, and some birds, the insertion forms a prominent hypapophysis. The gradual migration of the parapophysis onto the arch occurs over a section of the vertebral column between the 11th and 16th presacrals in tetanurines and is restricted to the first three or four dorsals in MANIRAPTORA. In all theropods, height and breadth of the neural spines increases throughout the back in the direction of the sacrum, and larger forms possess highly sculptured ligament scars. The zygapophyses are small and closely spaced. Auxiliary hypantrum–hyposphene articulations are usually present posterior to the second or third dorsal (Fig. 1).

The sacrum comprises two vertebrae in *Herrerasaurus*, four in *Dilophosaurus*, and five or more elements in all other theropods. Incorporation of additional sacral vertebrae has occurred independently in several theropod lineages (Holtz, 1994). Both sacral ribs and transverse processes participate in the contact between the sacrum and ilia. In many ceratosaurs, the distal ends of adjacent sacral transverse processes fuse and close off fontanelles between them.

The tail consists of fewer than 50 vertebrae in all theropods, and in COELUROSAURIA there are no more than 40 caudal elements. The centra are short and taller than wide proximally, but become elongate and wider than tall distally. The anteriormost caudals have relatively large neural spines and transverse processes for attachment of the levator caudae and caudifemoralis longus musculature, respectively. These apophyses diminish in size distally, and the level at which they disappear, the transition point, is phylogenetically significant. A distal caudal of a theropod is characterized by elongate prezygapophyses that wrap tightly around the neural arch of the preceding element. In dromaeosaurids, the prezygapophyses are hyperelongated and may exceed central length by a factor of 12. On the other hand, some taxa, such as oviraptorosaurs, therizinosaurs, and *Coelurus*, have secondarily reduced the prezygapophyses of the distal caudals. Ribs are most probably absent in theropod caudal vertebrae.

All theropods, but not *Herrerasaurus*, have lost the

atlantal rib, and loss of the axial rib may be diagnostic for Ceratosauria (Gauthier, 1986). The poor preservation of most ceratosaur specimens precludes absolute confirmation of this character, however. In tetanurans, the axial rib is double headed, as are the remaining postaxial cervical ribs of all theropods. Primitively, the shaft of a cervical rib is longer than the centrum with which it articulates. In some coelurosaurian taxa, such as DROMAEOSAURIDAE, TROODONTIDAE, ORNITHOMIMOSAURIA, and ornithoracine birds, the shafts are greatly shortened, however, and do not exceed central length. Although Gauthier (1986) suggested that fusion between cervical vertebrae and ribs represents a coelurosaurian synapomorphy, this feature is also seen in some ceratosaurs and its phylogenetic significance remains unclear because of a patchy and inconsistent distribution.

The capitula and tubercula of the anterior dorsal ribs are widely separated, but this subsides posteriorly and the last presacral usually bears a single-headed rib. A pneumatic space occurs between the two attachment points in some taxa and may extend part way down the rib shaft. Distally, the end of a rib articulates with a gastral element. The gastral cuirasse generally consists of two elements on either side of the midline. Midline fusion is not uncommon, however, and is seen in tyrannosaurids, troodontids, and allosaurids. The total number of gastral arches usually exceeds the total number of dorsal rib pairs, and one specimen of *Ornithomimus,* for example, has 19 gastral arches for 12 pairs of dorsal ribs.

Chevrons are present throughout the theropodan tail, with the exception of the first three or four most proximal caudals. Proximally, they are long, rod-shaped, and slightly recurved in the posterior direction. Distally, the chevrons become reduced in height and axially elongate. They add to the overall stiffening of this part of the vertebral column. In many taxa the chevron pedicels are fused proximally but become separate in more distal elements.

Primitively, sauropodomorphs have a vertebral composition of paired proatlantal arches, 10 cervical, 15 dorsal, three sacral, and 50 caudal vertebrae. These counts appear largely unchanged throughout prosauropods, but in sauropods there is a general increase in the number of cervicals, with a concomitant reduction in the number of dorsal vertebrae. One or two additional sacral elements are also incorporated in a number of sauropod taxa, but this character was probably acquired independently in several lineages.

In the majority of prosauropods, the cervical centrum is elongate, rectangular in section, and devoid of invasive pneumatics. The intercentral articulations are moderately angled, and the neck was held in a shallow, sinuous curve. Intercentral articulations are generally platycoelous or amphiplatyan. The anteriormost centra are keeled and all presacral centra are moderately constricted. The neural arch bears a low, elongate neural spine, as well as short zygapophyses and transverse processes, and displays a number of features that are primitive for all dinosaurs. Central length increases up to the eighth cervical in *Plateosaurus,* whereafter it decreases. Neural spines increase gradually in size toward the dorsal series.

The anterior dorsals are laterally constricted and keeled. Central length remains subequal throughout the length of the dorsal series (Fig. 2). The arches are low and the distance between the neural canal and zygapophyses is small. The parapophysis reaches the neurocentral suture in the third dorsal and has fully migrated onto the neural arch by the fifth dorsal (Fig. 2). Laminae extending from the various apophyses toward the spine are not present in prosauropods, with the exception of advanced forms such as *Patagosaurus* (Bonaparte, 1986). In *Plateosaurus* and *Massospondylus,* the dorsal neural spines are low, axially broad, and have a slightly convex dorsal edge (Fig. 2). The sacrum is composed of three elements in most taxa, although incorporation of an additional caudosacral is known in some Jurassic forms. The sacrals contact the ilia through ribs as well as transverse processes, but a sacricostal yoke reminiscent of sauropods is not present.

The tail varies in length among taxa, with an evolutionary trend to reduce the number of caudals from the 50 elements that are plesiomorphic for Dinosauria. Neural spines and transverse processes throughout the majority of caudal elements, combined with the lack of zygapophyseal elongation, suggest that the tail was relatively muscular and mobile throughout its length.

Sauropods have opisthocoelous cervical vertebrae, and all but the most basal taxa have variously developed invasive pneumatics (Fig. 3). The cervical neural arches are some of the most variable, and therefore diagnostic, parts of the sauropod skeleton (Fig. 3). The prezygapophyses are generally large, widely sep-

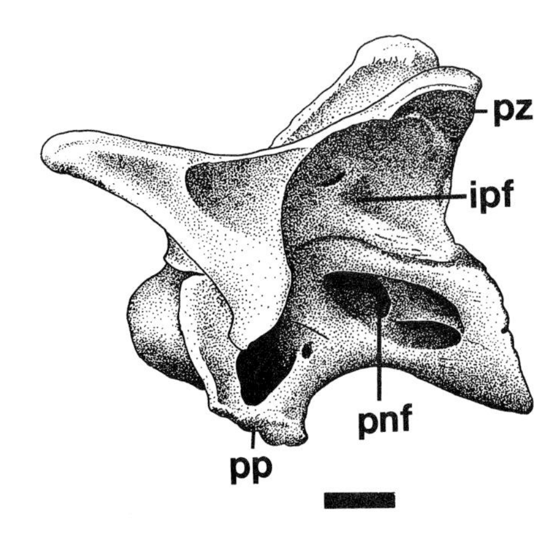

FIGURE 3 Anterior cervical vertebra of *Diplodocus* in lateral view. Note the extensive degree of pneumatization and pronounced anterior articular condyle. ipf, infra-postzygapophyseal fossa; pnf, pneumatic fossa; pp, parapophysis; pz, postzygapophysis. Scale bar = 5 cm. Redrawn from Ostrom and McIntosh (1966).

arated, and devoid of auxiliary articulations. The neural spine is low and, in most taxa, connected to the pre- and postzygapophyses by paired supra-prezygapophyseal and supra-postzygapophyseal laminae. In taxa that do not possess bifurcated neural spines, a ligament scar for attachment of the interspinous ligaments is broad and well developed and extends down between the zygapophyses. Bifurcated neural spines occur in a number of taxa, including the Diplodocidae, Mamenchisauridae, *Dicraeosaurus*, and *Opisthocoelicaudia*, and this character apparently originated independently several times within the Sauropoda. The space between the two divided halves of the neural spine was occupied by a long nuchal tendon originating within the dorsal or sacral series and inserting onto the axis. A typical sauropod cervical has a pendant transverse process that together with the ventral extent of the parapophysis displaces the cervical rib well below the level of the centrum (Fig. 3).

The dorsal centra are relatively elongate and remarkably lightly constructed for such heavy animals. The parapophyses migrate onto the arch progressively over a section of three or four anterior dorsals. The dorsal migration continues onto the transverse process throughout the dorsal series, and the parapophysis merges with the diapophyseal facet in the last presacral element. The main structures of the arches are reduced to a series of thin laminae that are intricately arranged to brace each other. The neural spine is tetraradiate in most taxa and can additionally be supported by two pairs of laminae originating on the zygapophyses. The transverse processes are likewise composed of a large number of laminae. The two infradiapophyseal laminae and the three related fossae common to saurischians are subdivided in various ways in different taxa. Other characters typical of sauropod dorsals are the presence of large hyposphene–hypantrum articulations, a marked elevation of the zygapophyses above the neural canal, and a progressive shortening of the transverse processes in the direction of the sacrum. A few sauropod taxa have hyperelongated neural spines, which must have formed a dorsal sail in these animals.

The sauropod sacrum comprised from four to six vertebrae: The exact number varies ontogenetically as well as phylogenetically. The sacral centra are more massive in appearance than the dorsals, although they are still pneumatic. The neural spines of the three original sacrals are fused to each other in most taxa, whereas the first and last sacrals have freestanding spines. The sacral ribs fuse to one another proximally and distally, thereby isolating a series of fontanelles. Distally, the three central ribs fused to form a huge plate, or sacricostal yoke, that sutures with the ilium and also with other pelvic elements along the dorsal rim of the acetabulum.

Caudal centra are cylindrical to oval in section and are apneumatic in most taxa. Diplodocids, however, have anterior caudals invaded by lateral pneumatic fossae and may also possess ventral excavations. The nature of the intercentral articulations between anterior caudals is important in sauropod taxonomy. Cetiosaurids and diplodocids have slightly procoelous anterior caudals, and pronounced procoely is found in the Cretaceous Titanosauridae. *Opisthocoelicaudia*, in contrast, has deeply opisthocoelic anterior caudals. Further distally, the intercentral articulations adopt an amphiplatyan or platycoelic morphology in all sauropods. The neural spines are tall, single, and devoid of supportive laminae. The transverse processes are short and support a plate-like rib in the anterior dorsals. Distally, the centra tend to become proportionally more elongate as the neural spines and ribs become reduced in size. Ribs are generally absent beyond the 15th caudal. Fusion of a series of caudal vertebrae in the midsection of the tail has been noted in a number of specimens. A moderate elongation of the prezygapophyses is present in the distal caudals.

Some taxa such as diplodocids have an increased number of caudals, of which the most distal ones form a whip-like end on the tail.

A ventral cuirasse of GASTRALIA is present in prosauropods. Gastralia are rarely reported in sauropods but are present in at least one specimen. A sauropodomorph cervical rib is long, slender, and exceeds the length of the vertebra that bears it. In sauropods, the anterior dorsal ribs display only slight lateral curvature to allow for support of the body via scapular suspension. The posterior dorsal ribs are far more arched to accommodate the huge viscera. Sauropods often have pneumatized ribs. The morphology of the sacral and caudal ribs has been mentioned.

The location of the first chevron varies between sauropodomorph taxa but is found within the first five caudals. The chevrons are rod-like anteriorly and extend posteroventrally at an oblique angle. This morphology is present throughout the tail in prosauropods, where the chevrons are very elongate. In advanced sauropods, the distal chevrons are axially elongate, and each bears a pair of long processes both anteriorly and posteriorly.

Vertebral synapomorphies uniting all known ornthischians are the presence of ossified tendons and a minimum of five sacral vertebrae (Sereno, 1986). Furthermore, Ornithischia is characterized by the complete loss of gastralia. Ancestral ornithischians are small bipedal creatures, but the radiation into a number of highly specialized lineages brought about fundamental changes in the structure of the vertebral column.

A number of predominantly Triassic basal ornithischians, such as *Pisanosaurus* and *Technosaurus,* have been described, but *Lesothosaurus diagnosticus* is the only taxon represented by more than fragmentary remains. This Early Jurassic species is regarded as the sister group of all other ornithischians (Sereno, 1986). The total vertebral count is unknown, but current phylogenetic ideas imply a count of nine cervicals, 13 dorsals, five sacrals, and approximately 50 caudals. The cervical centra are pinched laterally and bear a ventral keel. The cervical neural spine is represented by a low ridge and the diapophysis is extremely short. The dorsals are faintly keeled and have platycoelous articulations. Each dorsal neural spine is rectangular in lateral view. The postzygapophysis fully overhangs the posterior end of the centrum, whereas the distal tip of the prezygapophysis is level with the anterior intercentral articulation. The transverse processes are directed horizontally. *Lesothosaurus* has five sacral vertebrae, which are not co-ossified. The anterior caudal vertebrae are characterized by having tall, thin neural spines and ribs. The latter disappear in the distal caudals, anterior to the last vertebra that bears a neural spine.

The ribs are relatively unspecialized. In the dorsal ribs, the lateral edge is more acute than the internal one. The distance between capitulum and tuberculum decreases posteriorly concurrently with the migration of the parapophyses of the dorsal vertebrae. Chevrons are unknown for this taxon (Thulborn, 1972).

The armored ornithischian taxa that are united in the THYREOPHORA display a number of peculiar specializations of the vertebral column. These are to a high degree related to their reversal to obligate quadrupedality and a robust build.

There are nine cervicals in stegosaurs, and proatlantal elements have not been reported. These have probably been lost in Thyreophora. Fusion between the neurapophyses and atlas intercentrum is seen in larger specimens of some stegosaur genera such as *Stegosaurus* and *Huayangosaurus.* The axis is keeled and surmounted by a low neural spine, and fusion between the odontoid and axis centrum is seen in some specimens. Stegosaur cervicals are diagnostic in the possession of a low, roughened ridge that extends horizontally along the length of the centrum. The neural arch bears a low neural spine that gradually lengthens toward the dorsal series. The parapophyses contact the arch on the 10th presacral element, and together with an increasingly steep angling of the transverse processes mark the transition to the back.

Stegosaur dorsal vertebrae are distinct in having extremely elongated neural arch pedicels (Gilmore, 1914). In some taxa, the neural canal also expands to fill much of the space between the pedicels (Fig. 4), whereas in others the diameter of the canal remains small. The arches have elongate transverse processes set at an angle of up to 70° from the horizontal, and the tips of the transverse processes of a posterior dorsal are nearly as tall as the neural spine (Fig. 4). A fusion of the prezygapophyses at the midline forms a U-shaped trough. The fused postzygapophyses are shaped as a corresponding convexity. The dorsal ribs are double headed, with a large distance between the tuberculum and capitulum to accommodate for the length and angle of the transverse processes. Proxi-

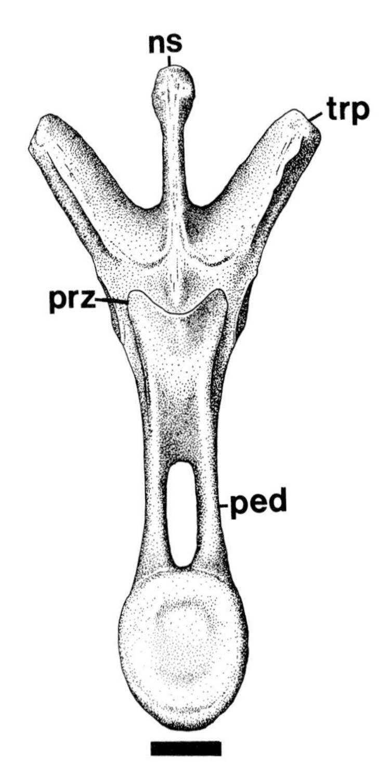

FIGURE 4 Anterior dorsal of *Stegosaurus* in anterior view. The elongate neural arch pedicels and steeply inclined transverse processes are characteristic features of stegosaurs. ns, neural spine; ped, neural arch pedicel; prz, prezygapophysis; trp, transverse process. Scale bar = 5 cm. Redrawn from Ostrom and McIntosh (1966).

mally, each rib is T-shaped in section, but this gradually subsides into an oval cross-section toward the distal end. Stegosaurs have uniquely lost the ossified tendons present in other ornithischians.

Stegosaurs are more primitive than many other ornithischians in having no more than five sacrals, although a few large specimens apparently added an element from the front of the tail. The degree of co-ossification is far less developed than in, for instance, the ornithopods, but the neural spines are fused into a continuous plate in some taxa. The most distinct feature of the sacral vertebrae is the hypertrophied expansion of the neural canal. Although the expansion was in part due to the enlarged motor nerve supply to the hindlimb musculature, its full extent has proven difficult to explain. This feature was formerly interpreted as an accommodation for a second brain, which was to coordinate the movements of the posterior half of the body, but this claim has been refuted. The sacral ribs are robust and form a sacricostal yoke as in many other ornithischians. The ribs are ankylosed to the ventral border of the elongated transverse processes, and the latter may fuse into a continuous plate as in *Kentrosaurus*.

The anterior caudal centra are short and massive and are slightly broader than tall. Each bears a slender and elongate neural spine that leans posteriorly. The top of the spine is transversely expanded, and in more distal elements this expansion is dorsally bifurcated. A short, massive rib is present on the anteriormost caudals but becomes reduced posteriorly. Caudal ribs disappear before neural spines, which are present well into the tail. Large and rugose chevron facets are present on most of the caudals, and the first chevron is found behind the third caudal. The chevrons are short, robust, and moderately straight. In more proximal chevrons, the facets are united at the midline, but this contact is lost in more distal elements.

Ankylosaurs possess seven or eight cervical vertebrae. The neurapophyses and atlas intercentrum co-ossify firmly, and in some ankylosaurid genera the atlas and axis are also fused, similar to neoceratopsians. The postaxial cervicals have amphiplatyan to amphicoelous centra that are wider than tall. Their neural spines are often expanded in the transverse plane and subtriangular in anterior view, although interspecific variation may be pronounced. Each cervical transverse process is robust and extends horizontally. Ankylosaurs are derived in the possession of a proportionally large neural canal in the cervical series. The zygapophyses of each cervical extend well beyond the intercentral articulations, suggesting that the centra were separated by a thick cartilaginous disc. Unlike the other cervicals, the atlas bears a single-headed rib. All other cervical ribs are double headed, and in the posterior half of the neck the ribs may fuse serially.

The dorsal series is characterized by migration of the parapophyses toward the neural arch as well as by a laterodorsal reorientation of the transverse processes. The neural spines are axially elongate and rectangular in profile. Although most ankylosaurs have amphiplatyan to amphicoelous vertebrae, some nodosaurid genera have a small knob on the anterior intercentral articulation that fits into a concomitant depression on the preceding vertebra (Coombs, 1978). This feature has been interpreted as a remnant of the embryonic notochord. The last three to six dorsals co-ossify to form a sacral rod. This region is characterized by having the neural spines fused into lamina that join the neural spines of the sacrals. The inclination of the transverse processes subsides toward the sacrum. In each posterior dorsal, the parapophysis and diapophyseal facet become confluent and form a long rugose tract extending along the whole length of the transverse process. As in stegosaurs, the prezygapophyses may fuse along the midline to form a U-shaped trough, with the postzygapophyses forming a rod that fits into this aperture. The anterior two or three ribs are short and one or more pairs may contact the coracoid. A posterior dorsal rib, in contrast, forms a semicircle with a large diameter. The angle between the capitulum and tuberculum is very small. When articulated, the ribs form a large and extremely wide, barrel-shaped rib cage. The ribs of the last few dorsals extend laterally to contact the enlarged preacetabular blade of the ilium.

The ankylosaur sacrum consists of five or six elements, of which the caudosacral often remains unfused. The neural spines of the anterior sacrals coossify into a lamina, and the zygapophyses are also fused. As in stegosaurs, the neural canal is markedly expanded in the anterior half of the sacrum. The sacral transverse processes are elongated and are fused with the sacral ribs along the full length of their ventral faces. The lengths of the transverse processes and sacral ribs diminish toward the tail so that the preacetabular portion of each ilium is much further removed from the midline than the postacetabular blade.

The tail comprises between 20 and 40 vertebrae, with the lower number present in the Ankylosauridae. Proximally, the centra are circular in section but become progressively more square to hexagonal in section toward the distal end. Chevron facets are well developed on ankylosaurid caudals. The neural spine is shaped as a parallelogram with a moderate posterior inclination. The ribs disappear anterior to the 15th caudal. In the distal two-thirds of the tail in ankylosaurids, the prezygapophyses become elongate and wrap around a posteriorly directed wedge formed by the postzygapophyses of the preceding caudal. This morphology is analogous to that seen in the tails of theropods and served a similar purpose of stiffening the tail, which bears a large club at the distal end. In nodosaurids, in which a tail club is absent, the caudals do not display the derived modifications seen in the ankylosaurids. Chevrons are present throughout the tail. They are V-shaped anteriorly but become elongate and low distally. In ankylosaurids, the distal chevrons overlap one another, thus adding to the stiffness of the tail. This feature is not present in the tails of nodosaurs, in which the chevrons meet but do not overlap. Thick bundles of ossified tendons are present in the region of the tail club in ankylosaurids.

The ornithopods had an extensive adaptive radiation in the Cretaceous, a fact that is clearly demonstrated by the diversity of vertebral morphologies in the various ornithopod lineages. Hypsilophodont vertebral columns are generally primitive and differ only marginally from the anatomy seen in *Lesothosaurus.* The iguanodontian lineage is characterized by the progressive addition of neomorphic vertebrae and by the arrangement of the ossified tendons into a lattice-like mesh along the neural spines of the posterior dorsals, sacrals, and anterior caudals. Heterodontosaurs also appear to retain the primitive condition but differ from the other two lineages in the reduction of the number of dorsals to 12.

All ornithopods have paired preatlantal elements that articulate between the occiput and neurapophyses. In *Hypsilophodon* there are 9 cervicals, but the number increases throughout the Iguanodontia, reaching a maximum of 15 in lambeosaurine hadrosaurs. In most ornithopods, each cervical centrum is lateroventrally pinched and bears a broad ventral keel. In the hypsilophodonts, the centra are platycoelic, whereas derived iguanodontians, such as *Camptosaurus* and *Iguanodon*, have opisthocoelic intercentral articulations. The parapophysis straddles the neurocentral suture throughout the cervical series of hypsilophodontids, much as in *Lesothosaurus.* In other euornithopod taxa, the first contact between the parapophysis and neural arch occurs in the region of transition between neck and back.

The axis bears an elongate neural spine with an extensive anterior projection above the neural canal. Posterior to the axis, the anterior cervical vertebrae of all euornithopods are practically devoid of a neural spine (Fig. 5), although this structure becomes more prominent in the last few cervicals. In hypsilophodonts, the zygapophyses and transverse processes of a cervical neural arch are short as in *Lesothosaurus.* Iguanodontians show a derived morphology of the cervical neural arches, where each prezygapophysis is displaced far from the axial midline and originates from the dorsal face of the transverse process. A semi-

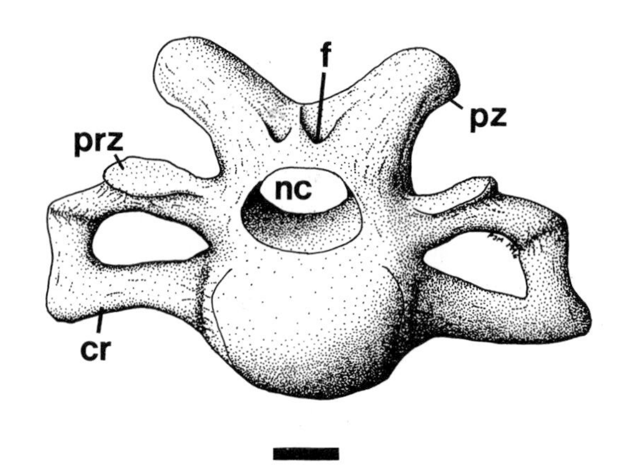

FIGURE 5 Anterior cervical of hadrosaur in anterior view. Note the convex articular condyle, the laterally displaced prezygapophyses, and fused cervical ribs. cr, cervical rib; f, foramen or semicircular pit; nc, neural canal; prz, prezygapophysis; pz, postzygapophysis. Scale bar = 2 cm.

circular pit is present on either side of the rudimentary neural spine in hadrosaurs and closely related taxa (Fig. 5). The postzygapophyses are massive and overhang the posterior end of the centrum markedly.

The dorsal centra of hypsilophodonts and heterodontosaurs are spool shaped, in contrast to those of iguanodonts, which are more tall than wide. All ornithopods have amphiplatyan to platycoelic dorsal vertebrae. The parapophysis is found on the neural arch throughout the dorsal series and migrates toward the distal end of the transverse process in posterior dorsals. In hadrosaurs, the transverse processes are angled strongly up and back in the posterior dorsal vertebrae. They are triangular in transverse section with the ventral apex formed by a thick, supportive infradiapophyseal wall or lamina. The neural spines are low and thin in hypsilophodonts but are tall in the larger iguanodontians. This development is most pronounced in lambeosaurine hadrosaurs, in which the spine height may exceed the height of the centrum by a factor of five. The zygapophyses are small in hypsilophodonts but are larger and fused along the midline in hadrosaurs and other iguanodontians. There are between 15 and 20 dorsals in Ornithopoda.

The development of the sacrum varies widely in ornithopods. Hypsilophodonts have a plesiomorphic count of five vertebrae. Galton (1974) noted an interesting dimorphism in *Hypsilophodon*, in which some specimens have five sacrals and others six. The morphology of the first sacral rib is markedly different between the two sacrum types. Iguanodonts incorporate a much larger number of vertebrae into the sacrum. This culminates in the hadrosaurs, in which up to 12 vertebrae may be co-ossified in the adult sacrum. Ankylosis is pronounced in hadrosaurs and related taxa, whereas the sacra of hypsilophodonts are more weakly co-ossified. The sacral ribs of the anterior elements fuse to form a sacricostal yoke that sutures with the ilium and constitutes part of the pubic articulation.

There are 45 to 70 caudals in ornithopods, with the higher values present in hadrosaurs. The caudal centra of *Hypsilophodon* are spool shaped and bear large chevron facets at either end. The neural spine of each caudal leans posteriorly, but the caudal rib is laterally directed and horizontal. In iguanodontians, the caudal ribs are directed posteriorly. Each chevron facet is formed as a large, ventrally sloping, beveled area bearing two round depressions. Chevrons are present throughout most of the tail but are absent between the first few caudals as well as in the most distal portion of the tail. A diminutive, flattened chevron is present between the first two caudals of *Hypsilophodon*, however. The first one or two chevrons are short and flattened, but the remaining ones are long and extend posteroventrally.

All ornithopods have short, robust cervical ribs, and all but the atlantal rib are double headed. The last few cervical ribs are longer and more robust than the preceding ones. The shafts of the anterior dorsal ribs are markedly curved, and their distal ends bear rugosities that represent the attachment points of the cartilaginous sternal ribs. Toward the sacrum, the ribs become shorter and straighter. The capitulum and tuberculum approach each other progressively, and the last one or two dorsals bear a single-headed rib. It seems apparent that caudal ribs are present, and the contribution of the neural arch to the transverse process of a caudal vertebra may be quite limited. Ossified tendons are present through the posterior part of the back, sacrum, and anterior half of the tail in ornithopods. Hypsilophodonts have the simple axial type of tendons that extend in bundles between the transverse processes. Iguanodontians, on the other hand, have a derived condition, in which the ossified tendons are arranged in a rhomboidal lattice along the vertebral column.

Psittacosaurs are generally rather conservative in their vertebral morphology, but neoceratopsians display a large number of autapomorphies, many of which are functionally related to the greatly increased size of the skull. The atlantal centrum, axis, and third cervical are fused in all neoceratopsians to form a syncervical, and the cup for reception of the occipital condyle is deep. The atlantal intercentrum is fused to the anterior end of the anterior cervical complex, a fact that led to some confusion about the total number of elements in the syncervical (Hatcher *et al.*, 1907).

Nine cervicals are present throughout the Ceratopsia, a figure that appears to be primitive for Ornithischia. The cervical centra are short and roughly circular in shape, being slightly taller than wide. The neural spines and transverse processes are shorter than in the dorsals but generally much more prominent than the equivalent structures of other ornithischian taxa. Both types of processes lengthen gradually in the posterior direction, and the transition between the cervical and dorsal series is not pronounced in ceratopsians. Hatcher *et al.* (1907) placed the beginning of the dorsal series at the 10th presacral based on the dorsal angling and expanded, robust morphology of the transverse process.

The parapophysis does not reach the neural arch until the third dorsal. The dorsal neural spines gradually increase in size toward the sacrum, and they are moderately inclined posteriorly. The parapophysis shifts onto the transverse process and is located near the distal end in the last dorsals. The transverse processes lengthen up to the 10th dorsal, whereafter the trend is reversed. The zygapophyses are large and closely spaced, displaying a transverse inclination that subsides as they approach the sacrum. Ossified tendons run forward from the sacrum to the third dorsal or so, attaching to the individual vertebrae near the tips of the neural spines.

In psittacosaurs, there are only 6 sacral vertebrae, but the number increases to 7 or 8 in protoceratopsians and to 10 in neoceratopsians. The sacrum forms a shallow, dorsally convex arc. Each sacral centrum is excavated ventrally by a broad furrow. The neural spines of sacrals 2–5 stand vertically and are fused to form a plate. These sacral vertebrae are further characterized by the distal fusion of their sacral ribs, such that an acetabular bar or sacricostal yoke is formed. In contrast to all other sacrals, these elements show clear separation between the iliac sutural contacts of the diapophyses and sacral ribs. In the posterior five elements, which are of caudal origin, the neural spines lean posteriorly.

There are about 45 caudal vertebrae in all ceratopsians, although individual variation with differences of up to 10 vertebrae is known. The proximal caudals are short and bear moderately tall, posteriorly inclined neural spines. Caudal ribs are present, overlapping the neurocentral suture. In most taxa, the ribs disappear in the midcaudal region, close to the 20th caudal vertebra. Each centrum bears a pair of large chevron facets at either end, and each facet is formed as a depression. Chevrons are present in all but the most distal caudals. They are long and cylindrical anteriorly and are directed sharply posteroventrally. Ossified tendons are present in the anterior caudals of all ceratopsian taxa.

The postcranial skeleton of pachycephalosaurs is poorly known, despite their wide distribution across the globe. Cervical vertebrae are unknown, except for an undescribed specimen of *Pachycephalosaurus*. The dorsals are amphiplatyan and are longer than wide. The arches bear a rectangular neural spine with a slight posterior inclination. In all save the last dorsal vertebra, the transverse process is dorsolaterally directed. The prezygapophyses of the dorsals bear a longitudinal ridge, and a corresponding groove is visible on the postzygapophyses. This ridge and groove articulation served to prevent transverse movement between adjacent vertebrae and is an autapomorphy of the Pachycephalosauria. The last dorsal bears a horizontal transverse process. In the Mongolian pachycephalosaur *Homalocephale*, the sacrum consists of six vertebrae (Maryańska and Osmólska, 1974). These are low and elongate, and the middle four sacrals are keeled. The sacral ribs are sutured to the ends of their respective transverse processes and do not form a true sacricostal yoke. These conditions are plesiomorphic compared to those of neoceratopsians.

The centrum of each of the proximal caudals is rounded and lacks chevron facets. The intercentral articulations are angled, and when articulated would have given the tail a downward curvature. From the fifth caudal backwards, the centra develop prominent chevron facets, and the midcaudal and distal vertebrae bear a ventral sulcus. Each anterior caudal neural arch has a long and slender neural spine. Large ribs are fused to each of the anterior caudals, but the ribs disappear around caudal 9. The neural spines become strongly reduced and posteriorly inclined toward the distal end of the tail. Chevrons are present posterior to the fifth caudal. They are Y-shaped, without fusion of the articulations, and have an oval cross section of the shaft. A complicated mesh of ossified tendons, consisting of six differing rows, forms a stiffening basket around the caudal vertebral column of pachycephalosaurs.

Although space prevents a detailed discussion, the adaptive significance of the anatomy of the vertebral column will now be analyzed for the various major groups of dinosaurs.

Primitively, archosaurs possess a proatlas, usually seen as paired, triangular, and flat ossifications that articulate with the occipital condyle and the atlantal neurapophyses. Several dinosaurian lineages appear to have lost the proatlas. These include most theropods among the Saurischia and the ceratopsians and thyreophorans within Ornithischia. The primary function of the proatlas was to stiffen and thus strengthen the occipital joint. Such an element would have been disadvantageous in the two ornithischian groups by limiting mobility of the head, which was already restricted in its range of movement by the short and stiff neck. Because a highly mobile neck and skull were instrumental in the feeding behavior of theropods, the proatlas appears to have been lost early in the evolution of that group.

The number of cervicals, as well as the proportional length of the neck, varies widely in dinosaurs. The proportional length of the neck is highly dependent on the size of the skull. In forms with relatively light crania, such as many coelurosaurs, prosauropods, and sauropods, the neck is often elongate. The only parameter controlling variation in the relative length of the neck in nonavian theropods, prosauropods, basal ornithischians, and basal ornithopods is the length of the cervical centra. In other groups, however, the number of cervicals is not fixed and thus plays an important role in alterations in the length of the neck. Among sauropods, for instance, the incorporation of anterior thoracic vertebrae into the neck, along with formation of neomorph cervicals in some taxa, appears to be the main mechanism in the progressive development of longer necks. An opposite trend is seen in certain ornithischians, which are characterized by having large and massive skulls. Ankylosaurs have only seven or eight cervical vertebrae with shortened centra, whereby the moment arm between the skull and front limbs is shortened. In addition to this reduction, some ankylosaurs fuse anterior cervical elements into a syncervical, which provides a more robust articulation for the heavy skull. The latter feature is even more prominent in neoceratopsians. Both of these groups are almost strictly quadrupedal, and evolution of a heavy skull is correlated with the use of the front limbs to provide support for the cranial end of the body. In contrast to these taxa, ornithopods show incomplete or no fusion between the odontoid and axis, which indicates far less mechanical stress exerted on the neck by the weight and movements of the skull. Although hadrosaurs, like some sauropods, have evolved a number of neomorph cervicals, other neomorph vertebrae are present throughout all parts of the vertebral column. For hadrosaurs, the formation of neomorph elements may therefore be related to an overall increase in size rather than to an allometric development of a given part of the body.

Traditionally, the transition between the neck and back is defined as the position of the first vertebra that bears ribs that contact the sternum. This definition is rarely applicable to dinosaur specimens because they are generally either incomplete or disarticulated. Cervicodorsal transitions are therefore deduced by the presence of other characters. These include migration of the parapophysis onto the neural arch, a reorientation of the diapophysis to a laterodorsal direction, and an increase in the height of the neural spine. The shift in the articular facets for the rib indicates a change to the thoracic rib morphology with a wide separation between the capitulum and tuberculum. In addition to these more general features, theropods have the cervicodorsal transition marked by distinct hypapophyses in the first three dorsals. In most saurischians the cervicodorsal transition is distinct, whereas, for example, stegosaurs and ceratopsians display a more gradual transition. In quadrupedal ornithischians, the neck is held almost horizontally away from the trunk and below the level of the back. This added mechanical strain is reflected in the larger size of the neural spines, the transverse processes of the cervical series, and the more gradational changes in size of these apophyses between neck and back.

The posterior dorsal series of most dinosaurs is rigid, with limited rotation and flexion between adjacent vertebrae. In ornithischians, the ossified tendons are responsible for this inflexibility, whereas the development of accessory hypantrum–hyposphene articular facets serves the same purpose in Saurischia. The ridge and groove articulations of the posterior dorsal zygapophyses in pachycephalosaurs represent an analogous development to the saurischian hypantrum–hyposphene. A modern analogy is seen in the lumbars of even-toed ungulates such as the Cervidae, in which each prezygapophysis forms a semicircular articulation that wraps tightly around a postzygapophysis of the preceding vertebra. Although stegosaurs have lost the ossified tendons, and these are reduced in other thyreophorans, the posterior dorsals fuse to form a sacral rod, with the same result of forming a rigid lumbar region. Inflexibility of the lumbar region helps brace the pelvic girdle to the trunk. The mechanical effect of this bracing is that locomotory force generated by each hindlimb is transmitted to the body as a whole rather than to ipselateral vectorial components, as in a semierect lacertilian type of locomotion.

The progressive incorporation of additional dorsal and caudal vertebrae into the sacrum seen in most dinosaurian lineages is also a result of the need to brace the pelvic girdle. To some degree, however, it may also be a passive result of evolution of a larger ilium in conjunction with increased cursoriality. Addition of sacral elements appears to have low phylogenetic consistency throughout Dinosauria and evolved independently in many lineages. Many large dinosaurs fuse some of the sacral ribs to form a sacricostal yoke. This yoke sutures to the ilium in an area from the pubic peduncle to the ischiadic peduncle and provides a medial extension of the acetabulum for direct transmission of force between the vertebral column and the hindlimbs.

The tail served two related functions in dinosaurs. The proximal portion served as an origin for the main muscles powering hindlimb retraction, and all nonavian dinosaurs possessed a coupled tail–hindlimb type of locomotion. The tail was also a counterbalance to the trunk and head in the cantilevered body plan of bipedal dinosaurs. Most dinosaurs show a dorsally concave, downward flexure of the anteriormost portion of the tail, whereafter the tail straightens out toward the horizontal. Such a lowering of the center of gravity (the tail) relative to the point of support (the sacrum) enhances the physical stability of the system. However, high stability leads to a decrease in maneuverability, and for high-speed turns the tail was probably raised and thrown to one side or the other. In most bipedal taxa, the tail is stiffened. In ornithischians, this phenomenon is due to the presence of ossified tendons in the proximal part. These may be arranged in a quite complex fashion, as seen in ornithopods and pachycephalosaurs. Theropods differ by having a mobile proximal section of the tail, but the distal end was stiffened by progressive elongation of the prezygapophyses to overlap preceding element(s). The number of proximal caudals becomes gradually reduced in higher theropods, and the stiff distal part becomes proportionately longer. This development culminates in the highly specialized tail of the dromaeosaurids, which was probably used to stabilize acrobatic leaps in conjunction with the use of their raptorial toe unguals. In general, caudal structure suggests that theropods were more maneuverable than ornithischians.

The stabilizing function of the tail lost importance in most groups that reverted to a strictly quadrupedal locomotion, such as neoceratopsians and thyreophorans. Not surprisingly, liberation of tail movement from overall stability prompted some taxa within these quadrupedal groups to develop secondary caudal functions. Both stegosaurs and ankylosaurids used their tails for agonistic behavior: the former with a distal armament of spikes, the latter with massive tail clubs. Various sauropods also show similar adaptations, such as the tail club of *Omeisaurus* or the whiplash of diplodocids.

See also the following related entries:

FORELIMBS AND HANDS • HINDLIMBS AND FEET

References

Bonaparte, J. F. (1986). The early radiation and phylogenetic relationships of the Jurassic sauropod dinosaurs, based on vertebral anatomy. In *The Beginning of the Age of Dinosaurs* (K. Padian, Ed.), pp. 247–258. Cambridge Univ. Press, Cambridge, UK.

Britt, B. B. (1993). Pneumatic postcranial bones in dinosaurs and other archosaurs. Unpublished Ph.D. thesis, University of Calgary.

Coombs, W. P. , Jr. (1978). The families of the ornithischian dinosaur order Ankylosauria. *Palaeontology* **21**(1),143–170.

Galton, P. M. (1974). The ornithischian dinosaur *Hypsilophodon* from the Wealden of the Isle of Wight. *Bull. Br. Museum (Nat. History) Geol.* **25**(1), 1–152.

Gauthier, J. (1986). Saurischian monophyly and the origin of birds. *Mem. California Acad. Sci.* **8,** 1–55.

Gilmore, C. W. (1914). Osteology of the armored Dinosauria in the U.S. National Museum, with special reference to the genus *Stegosaurus*. U.S. National Museum Bull. No. 89.

Hatcher, J. B., Marsh, O. C., and Lull, R. S. (1907). The Ceratopsia. *U.S. Geol. Survey Monogr.* **XLIX,** 1–300.

Holtz, T. R., Jr. (1994). The phylogenetic position of the Tyrannosauridae: Implications for theropod systematics. *J. Paleontol.* **68**(5), 1100–1117.

Maryańska, T., and Osmólska, H. (1974). Pachycephalosauria, a new order of ornithischian dinosaurs. *Palaeontol. Polonica* **30,** 45–102.

Ostrom, J. H., and McIntosh, J. S. (1966). *Marsh's Dinosaurs, the Collections from Como Bluff*, pp. 388. Yale Univ. Press, New Haven, CT.

Sereno, P. C. (1986). Phylogeny of the bird-hipped dinosaurs (Order Ornithischia). *Natl. Geogr. Res.* **2**(2), 234–256.

Thulborn, R. A. (1972). The post-cranial skeleton of the Triassic ornithischian dinosaur *Fabrosaurus australis*. *Palaeontology* **15**(1), 29–60.

Postcranial Pneumaticity

Brooks B. Britt
Museum of Western Colorado
Grand Junction, Colorado, USA

Descriptions of sauropods and theropods frequently note the presence of "pleurocoels" (large fossae or foramina) in cervical and dorsal vertebrae. Until recently, however, few paleontologists were aware of the origin of the fossae and foramina that are traces of air-filled diverticula and are osteological traces of a derived respiratory system (Britt, 1993). Birds are the only extant taxon with pneumatic postcranial bones and they serve as a model for the study of fossil pneumatic bones.

The presence of postcranial pneumatic bones among fossil, nonavian vertebrates was first recognized in pterosaurs by Von Meyer (1837), who cited the presence of pneumatic bones as evidence that pterosaurs could fly. The pneumatic nature of dinosaurian bones was first recognized by Sir Richard Owen, the British comparative anatomist. Owen (1856) astutely noted that the three fossae on the side of the vertebral neural arch of the theropod *Altispinax* "probably receiv[ed] parts of the lungs in the living animal." Later, Seeley (1870) described two sizable but lightly constructed vertebral centra with large external foramina and internal chambers as a new genus and species, *Ornithopsis hulkei*. He noted similarities to pterosaur vertebrae but concluded that they were from an animal closely related to the Dinosauria. The pneumatic characters of the vertebrae led Seeley to surmise that *Ornithopsis* had an avian-grade respiratory system. [*Ornithopsis* is now known to be a brachiosaurid sauropod (McIntosh, 1990; Blows, 1995]. With the discovery of well-preserved sauropods in the late 1870s the cavernous interiors and large external fossae of vertebral centra were commonly interpreted to be of pneumatic origin. At the turn of the century, the pneumatic hypothesis fell out of favor. This shift appears to have been due to Henry Farfield Osborn (1904), who opposed the theory in a popular publication. The German paleontologist Janensch (1947), too, was skeptical of the pneumatic hypothesis but his study of sauropod vertebrae convinced him of their pneumatic nature. Unfortunately, his research went largely unnoticed and few workers recognized the pneumatic nature of saurischian vertebrae.

Avian Respiratory System

The pneumatic nature of avian postcranial bones has long been known but their relationship to the respiratory system was first noted by Hunter (1774), who dramatically demonstrated the connection by severing the humerus of a waterfowl and tying off the trachea to show that air "passed to and from the lungs by the canal in this bone." The bones are pneumatized by diverticula (elongate, balloon-like extensions) of air sacs. Bird lungs consist of two major components: the lung core and its attached air sacs. Compared to other vertebrates the respiratory core of the lung is extremely small in relation to body size. The core consists of a dense network of subparallel tubes, termed parabronchi, through which air flows essentially unidirectionally in an anterior direction. Gaseous exchange occurs only in the lung core. A network of blood vessels surrounds the parabronchi, resulting in a dense organ quite unlike the lightly constructed lung of other extant vertebrates. The diameters of the parabronchi are so small that if the inner walls of parabronchi touch, surface tension will not allow reinflation and the bird suffocates. For this reason, pulmonary cavity volume is tightly constrained by the rib cage and connective tissues. The other component of the lungs are the air sacs, which are thin-walled, nearly avascular, muscleless, balloon-like appendages of the lung core. Gaseous exchange does not occur in the air sacs. There are generally nine major air sacs in birds that function primarily as passive bellows. Volume changes of the air sacs are induced primarily by raising and lowering the sternum. The pneumatic diverticulae extend from some of the air sacs into various areas of the body, including intermuscular areas, and alongside and into bones. These diverticula are composed of mesenchymal cells that can actively attack bone, especially around nutrient foramina. Usually the postcranial bones are not pneumatized until the individual is nearly fully grown. For example, in *Gallus*, the domestic chicken, the cervical vertebrae are not pneumatically invaded until approximately Day 35 and the dorsal vertebrae after 63 days (Hogg, 1984). The marrow and fat that normally occupies the interior of a bone is absorbed by the invading diverticula (Müller, 1907; Bremer, 1940). The walls of pneumatic bones are thinner than those of apneumatic bones, in part due to absorption of some of the spongiosa and inner layers of compact bone (Müller, 1907). Bones that are typically pneumatized in birds include cervical, dorsal, and sacral vertebrae, and cranial dorsal ribs, humeri, coracoids, scapulae, ilia, and femora. The postcranial bones of strong swimmers and divers, such as the common loon, *Gavia immer*, however, frequently lack pneumatic postcranial elements.

Pneumatic Features

Pneumatic diverticulae, being soft tissue, are not fossilized. Tracks and traces of these diverticulae, however, are often preserved on the exterior of bones. Also, it is these traces that permit the recognition of bones that were pneumatic in life. The most obvious pneumatic features are the large foramina or fossae on the sides of vertebral centra. In the literature these features are typically termed pleurocoels (= side cavity), a relatively inaccurate term that should be abandoned in favor of the more accurate anatomical terms fossae and foramina. Less obvious but more widely distributed are the pneumatic features of the neural arch. Generally, there are five pairs of major pneumatic foramina / fossae on the neural arch. The most prominent features are the three major fossae on each side of the neural arch. These are located below the transverse process and are primarily defined by two laminae, the infradiapophysial laminae, that ventrally buttress the transverse process. In anterior and / or posterior view, a pair of pneumatic foramina or fossae are often present dorsolateral to the neural canal. Accessory pneumatic foramina and fossae are also common and diagnostic in some taxa. Pneumatic diverticulae leave minor but diagnostic canals and fossae on the neural arch.

The pneumatic features described in the preceding paragraph are external, but pneumatization may also involve the interior of bones. The interiors of most fossilized saurischian vertebrae and long bones contain sizable cavities, but alone these spaces do not indicate a bone was pneumatic. The key to recognizing internally pneumatized bones is the presence of a large, pneumatic foramen leading to interior chambers. These large foramina represent the point of invasion by a pneumatic diverticulum. Bones lacking the large foramina were not pneumatic but were simply filled with marrow and fatty tissues. Pneumatic vertebrae are divided into two structural types: camerate and camellate. Camerate vertebrae are characterized by relatively thick walls and several large internal chambers. This is the underived condition and is typically found in sauropods and nontetanuran theropods. Camellate vertebrae are characterized by thin walls and supernumerary, small internal chambers. Typically, the transverse process and neural spine of camellate vertebrae appear swollen or "inflated." Camellate vertebrae are present in nearly all tetanuran theropods, including extant birds.

Nonavian Pneumaticity

Among nonavian archosaurs, only saurischians and pterosaurs have pneumatic postcranial bones. The SAURISCHIA is composed of SAUROPODOMORPHA (PROSAUROPODA + SAUROPODA) and THEROPODA. Among nonavian saurischians postcranial pneumatization is limited to vertebrae and ribs. In the Prosauropoda, vertebral pneumatization is largely restricted to the external surfaces of the neural arches, with minor expressions on the centra of proximal cervical vertebrae. Internal pneumatization is limited or absent. On the neural arches pneumatization is expressed mainly as three lateral fossa, all ventral to the transverse process.

Saurópods exhibit the most spectacular examples

of superficial pneumatization, particularly on the neural arches, which are composed of a network of delicate laminae. On the neural arches, bony buttresses separating adjacent pneumatic diverticulae were reduced to thin laminae by remodeling induced by the diverticulae. The cross-shaped transverse section diagnostic of sauropod dorsal vertebrae spines is a function of four large pneumatic grooves. These pneumatic grooves reduce the spine to four laminae, anterior and posterior ligament scars, and lateral laminae. Sauropod centra typically bear a single large fossa or foramen on each side—the so-called pleurocoeleous condition. With some exceptions, such as *Brachiosaurus* and *Euhelopus*, the vertebrae of sauropods are of the camerate type. In most sauropods all postatlantal precaudal vertebrae are pneumatic and the caudal vertebrae are apneumatic. The diplodocids are exceptional in that the proximal caudal vertebrae are also pneumatic. In early sauropods, such as the *Barapasaurus* and *Vulcanodon*, the dorsal vertebral centra lack internal chambers but the centra bear broad fossae. The cervical ribs of most sauropods are pneumatic and in some taxa, such as *Brachiosaurus*, the anterior dorsal ribs are also pneumatic (Janensch, 1947).

The differences between the camerate and camellate types of pneumatic vertebrae are best exemplified in the Theropoda. The vertebrae of nontetanuran theropods are characterized by robust neural arches with internally apneumatic processes or processes with only the proximal portions being hollow. The standard complement of external pneumatic fossae and foramina are present. Like most sauropods, the transverse processes are composed entirely of thin laminae. The centra are composed of several large chambers connected to the exterior via large foramina or of simple, broad fossae limited to the exterior of the centrum. In less derived theropods, such as *Syntarsus* and *Coelophysis*, the neural arches of most cervical and dorsal vertebrae are pneumatic, and pneumatization of the centra is limited to the proximal cervical vertebrae. Thus, most nontetanuran theropod vertebrae are of the camerate type. This contrasts sharply with the tetanuran theropods, which are characterized by camellate vertebrae. This character is a new one supporting the monophyly of Tetanurae. The neural arches are generally highly pneumatized, both internally and externally. In coelurosaurs, such as tyrannosaurids and ornithomimids, external pneumatic fossae are connected to the interior via pneumatic foramina. These foramina connect with numerous, small chambers that fill the core of the neural arch, transverse processes, and the spine. External pneumatic fossae are generally absent on the centra, and the pneumatic foramina are relatively small but connect via a pneumatic canal leading to the interior of the centrum. The cervical ribs are pneumatized, as are cranial dorsal ribs in rare cases. In tetanurans in general, all postatlantal cervical vertebrae and the cranial two or three dorsal vertebra centra are pneumatic. As is the case with nontetanuran theropods, pneumatization of the neural arches is more extensive, with pneumatic fossae and/or foramina on nearly all presacral vertebrae. In some coelurosaurs, cranial sacral vertebrae centra are also pneumatic, and in oviraptorids and caenagnathids even the caudal vertebrae are pneumatic. It should be noted that the basal avian *Archaeopteryx* has pneumatic vertebrae and ribs. In the London specimen a large pneumatic foramen is present on an anterior dorsal vertebra. Based on the distribution of pneumatic vertebrae in other theropods, it was predicted that the cervical vertebrae were also pneumatic (Britt, 1993). A pneumatic fossae is present on an anterior dorsal rib and the cervical ribs are hollow, suggesting that they were pneumatic, a hypothesis supported by the fact the cervical ribs of nearly all theropods are pneumatic. As with nonavian theropods, *Archaeopteryx* lacks pneumatic limb bones and other nonaxial bones. In other words, the pneumatic characters of *Archaeopteryx* are no more derived than those of nonavian theropods.

Function and Significance

Among the hypothesized advantages of postcranial pneumatic bones in birds, mass reduction is the most widely accepted. The mass reduction comes not only from the displacement of marrow and fatty tissues with air but also from the accompanying thinning of bone that occurs during the process of pneumatization. Postcranial pneumatization in birds is generally thought to be an adaptation to flight. The fact that homologous features are present in the ancestral theropods, however, negates the flight adaptation hypothesis and suggests that postcranial pneumatization is an exaptation to flight. The advantage of mass reduction in the long necks of sauropods is easy to understand. The benefit to theropods is not readily apparent but any reduction of mass would reduce energy requirements and provide a slight advantage in acceleration and deceleration. It has been suggested that air sacs play a role in thermoregulation in birds (Salt and Zeuthen, 1960), but Menaum and Richards (1975) concluded that such a role was unlikely. The main hope for understanding the roles of postcranial pneumatization in fossil saurischians lies in a better understanding of their role in modern birds, a topic that has been little studied.

The presence of pneumatic postcranial bones in saurischians suggests the lungs had well-developed air sacs with elongate diverticulae that encompassed and invaded the presacral vertebrae. This suggests that saurischian lungs had evolved to a stage that facilitated the development of the modern avian lung as had been hypothesized by Perry (1989). The development of camellate pneumatic vertebrae in tetanuran theropods may mark the development of an even more avian-like lung. The presence of advanced, partitioned lungs in saurischians and in theropods in particular could be interpreted to indicate the attainment of an endothermic metabolism. Nevertheless, the presence of an efficient respiratory system in itself does not require endothermy because an increase in respiration efficiency is advantageous to an organism of any metabolic grade.

The presence of pneumatic postcranial bones in the sister taxa PTEROSAURIA and DINOSAURIA (= ORNITHODIRA) suggests that their common ancestor shared this evolutionary novelty. Therefore, it follows that a search among less derived archosaurs for this character will help paleontologists identify the most plausible sister taxon to Ornithodira. The hypothesis that among the Theropoda camellate-type bones are an apomorphy of TETANURAE allows a test of Gauthier's (1986) phylogenetic hypothesis independent of characters he utilized. This test supports the monophyly of the Tetanurae (Britt, 1993). Finally, the presence of pneumatic vertebrae and ribs in theropods is further evidence that birds are theropods.

See also the following related entries:

BIRD ORIGINS • CRANIOFACIAL AIR SINUS SYSTEMS • GASTRALIA

References

Blows, W. T. (1995). The Early Cretaceous brachiosaurid dinosaurs *Ornithopsis* and *Eucamerotus* from the Isle of Wight, England. *Paleontology* **38**(1), 187–197.

Bremer, J. L. (1940). The pneumatization of the head of the common fowl. *Journal of Morphology* **67**, 143–157.

Britt, B. B. (1993). Pneumatic postcranial bones in dinosaurs and other archosaurs, unpublished Ph.D. thesis, University of Calgary, Geology. pp. 383.

Gauthier, J. A. (1986). Saurischian monophyly and the origin of birds. In *Origin of Birds and Evolution of Flight* (K. Padian, Ed.), pp. 1–55. California Academy of Sciences, San Francisco.

Hogg, D. A. (1984). The development of pneumatisation in the postcranial skeleton of the domestic fowl. *J. Anat.* **139,** 105–113.

Hunter, J. (1774). An account of certain receptacles of air, in birds, which communicate with the lungs, and are lodged both among the fleshy parts and in the hollow bones of those animals. *R. Soc. London Philos. Trans.* **64,** 205–213.

Janensch, W. (1947). Pneumatizitat bei wirbeln von sauropoden und anderen saurischiern. *Palaeontographica* **3**(Suppl. 7), 1–25.

McIntosh, J. S. (1990). Sauropoda. In *The Dinosauria* (D. B. Weishampel, P. Dodson, and H. Osmólska, Eds.), pp. 345–401. Univer. of California Press, Berkeley.

Müller, B. (1907). The air sacs of the pigeon. *Smithsonian Miscellaneous Collections* **50,** 365–414.

Osborn, H. F. (1904). Manus, sacrum and caudals of Sauropoda. *Am. Museum Nat. History Bull.* **20,** 181–190.

Owen, R. (1856). Monograph on the fossil reptilia of the Wealden and Purbeck Formations—Part III, Dinosauria (*Megalosaurus*). *Palaeontogr. Soc. Monogr.* **9,** 1–26.

Perry, S. F. (1989). Mainstreams in the evolution of vertebrate respiratory structures. In *Form and Function in Birds, Volume 4* (A. S. King and J. McLelland, Eds.), pp. 1–67. Academic Press, London.

Seeley, H. G. (1870). On *Ornithopsis*, a gigantic animal of the pterodactyle kind from the Wealden. *Ann. Magazine Nat. History (Ser. 4)* **5,** 279–283.

Von Meyer, H. (1837). Die Bayreuthen Petrefakten-Sammlungen. *Neues Jahrbuch Mineral. Geognosie Geol. Petrefakten-Kunde* **1837,** 314–316.

Posture and Gait

see BIPEDALITY; FOOTPRINTS AND TRACKWAYS; QUADRUPEDALITY

Potash Road Dinosaur Tracks, Utah, USA

see MUSEUMS AND DISPLAYS

Powder Hill Dinosaur Park, Connecticut, USA

see Museums and Displays

Pratt Museum of Natural History, Massachusetts, USA

see Museums and Displays

Princeton Natural History Museum, New Jersey, USA

see Museums and Displays

Problems with the Fossil Record

Ralph E. Molnar
Queensland Museum
Queensland, Australia

The problem with the fossil record is that it does not tell us everything we would like to know. In other words, the fossil record is not a completely transparent window on the past, and information about fossil organisms and their lives has been lost. This bias arises from two additive components (Table I). First, the record of fossils that exist in rocks is biased. Second, our study of it has produced a biased result. These two components of the fossil record have been termed the preservational and observational records, respectively (McKinney, 1991).

Preservational Record

The bias of the preservational record arises from two main causes: Nowhere has sedimentation been continuous throughout geologic time, and not all past organisms are preserved as fossils. Several factors conspire to bias the preservational record. Fossils may not be found because of limited areas of exposure of fossiliferous rock (Raup, 1976). They are more plentifully exposed, and hence more easily found, where the surface area of the rock is increased by erosion into badlands, mountains, etc. Only rarely have dinosaurian fossils been recovered as the result of human activity serving the same purpose, such as from wells (as *Erectopus*) or mines (as *Iguanodon*).

Accessibility is relevant to getting into the record as well as to recovering fossils from it. Fossils are preserved only where sediments are deposited. On the continents, areas of deposition are limited. Olshevsky (1991) very roughly estimated them as 10% of the continental area: Even at twice that much this is still a significant limitation. Dinosaurs must either die in the area of deposition or their carcasses or bones must be transported there. Large bones are not transported as far as small ones, so this hydrody-

TABLE I
Major Source of Bias in the Fossil Record

Preservational record
Accessibility
Area of outcrop
Amount of topographic relief
Completeness of record, geological factors
Area of deposition
Groundwater chemistry in region of deposition
Sorting during transport
Stratigraphic completeness
Subsequent destruction (i.e., age)
Completeness of record, biological factors
Population size
Habitat (in relation to areas of deposition)
Body size
Robustness of skeletal elements
Proportion of time spent away from areas of deposition
Observational record
Scientific
Interest in displays
Fossil size (ease of locating large specimens)
Number of workers
Historical bias (of Dodson, 1990)
Geographic bias
Other cultural
Interest in prehistory
Sample size

namic sorting may result in assemblages of only small bones from small animals: Apparently this has happened at Lightning Ridge in Australia.

Potential fossils may be destroyed during diagenesis. Acidic groundwaters tend to dissolve bone, and regions with such waters will lose bony fossils, although plant material may be preserved. Once buried, replacement of the bone by minerals, with subsequent cement and crystal growth, can destroy not only the internal structure but also the whole bone. This can happen after burial or during weathering.

However, the most important aspect is the degree to which sediments were continuously deposited, the stratigraphic completeness. This has been discussed in detail by van Andel (1981) and Schindel (1982). Both agree that most units have accumulated sporadically and hence represent more time during which nothing was deposited than during which sediments were laid down. Pre-Pliocene sediments record only from 2 to 45% of the time between their initial and final deposition (Schindel, 1982).

Subsequent erosion, metamorphism, or subduction often destroys fossils. The older the deposit, the longer the period over which these have acted, so age is significant. This bias can be shown dramatically for dinosaurs (Padian and Clemens, 1985). Approximately 40% of dinosaur genera are known from the final 6.25% (10 million years) of their Mesozoic time range (Dodson, 1990). For certain groups this effect is more marked: 52% of well-known large theropods lived in this final 10 million years (Molnar *et al.*, 1990) and they include all but three of the taxa known from complete skeletons.

Biological factors that control which organisms become incorporated into the record include population biology, ecology, and anatomy. Carnivorous populations are smaller than those of herbivores for reasons of thermodynamics and energy loss. Thus, carnivores, especially top carnivores, are rare as fossils, even though they may be preserved in the same relative proportions as herbivores. Other species with small populations are likewise rare in the record.

It has been argued (Olshevsky, 1991) that areas of deposition do not necessarily support highly diverse assemblages of large vertebrates. Today, lowland rainforest and coastal swamps are not home to diverse groups of large mammals, although floodplains often are, and this may have been the case for dinosaurs as well.

Obviously the size and robustness of the skeleton are important. Small bones break up more easily than large ones, so small vertebrates are less likely to be represented, other things being equal. Flying animals, especially birds, are less likely to be preserved than those living on land, which in turn are less likely to be preserved than those living in the water directly over areas of deposition. Flying creatures may also be underrepresented because weight limitations imply hollow bones. Hollow bones are not substantially weaker under the stresses imposed during life but break up more readily under the presumably much greater stresses of transportation and burial. Furthermore, flying creatures may range widely and, even if living near areas of deposition, spend some of their time away from such areas, thus reducing their chances of being preserved.

Observational Record

The second suite of biases concerns those arising from problems with the collecting and study of fossils. Dinosaurs make spectacular museum displays; therefore, large, well-preserved, and complete material has always been sought over smaller or more fragmentary specimens. In addition, large fossils are more easily seen—even if not more easily collected—than small ones.

The number of dinosaur workers, reflecting times during which such work is fashionable, is another obvious bias (Sheehan, 1977). In addition, Dodson (1990) has documented a probably related historical bias: There have been about five times as many workers since 1970 as before.

Sometimes there are geographical biases, and work is limited by national boundaries. Dinosaurian fossils have been excavated in the western United States and Canada for about a century, but only recently have the deposits in Mexico been explored and studied.

Fashionable research problems to some extent reflect cultural interests and biases, the cultural *Weltanschauung.* Scientific curiosity spread with European colonization, and an interest in the prehistoric past spread with it. Cultures in which dinosaurian research has flourished have either been in Europe or European colonies (e.g., North and South America) or those with a strong sense of, and philosophic interest in, time and history (e.g., China, India, and Japan). Milieus in which an interest in the ancient past is not politically correct, such as fundamentalist Christian and Muslim societies that attach no value to times before Christ or Muhammad, produce no paleontology. However, cultures change: Spanish-speaking cultures that traditionally showed little interest in dinosaurs are now doing so, at least in Argentina, Mexico, and Spain.

Another source of bias, not attributable to either of these major groups, is simply sample size. Reportedly, even large sample sizes can produce a 25% difference in the species represented (Koch, 1987; Lewin, 1987).

The ideal method of discovering how incomplete the fossil record is would be to compare the number of fossils of some group (or groups) of organisms at some given period of time with the numbers of those organisms (or groups) alive at that time. This seems to be impossible. Therefore, we are left with making an estimate of the completeness of the record, at best an uncertain endeavor. Attempts have recently been made by Dodson (1990) and Olshevsky (1991). Both estimated how many dinosaurian genera existed. Dodson suggested 1050 ± 150 and Olshevsky approximately 6000. The discrepancy is due to the different assumptions used and Olshevsky's belief that there were very many more small dinosaurian taxa than now recognized. Although apparently large, the discrepancy is within one order of magnitude, a difference often regarded as insignificant by theoreticians in the physical sciences, and so we should perhaps not make too much of it.

They then compared the estimated numbers of dinosaurian genera with numbers of genera reliably identified from fossils to gain some notion of the completeness of the record. Dodson (1990) suggests that 25% of dinosaurian genera have been recovered, whereas Olshevsky (1991) thinks this is only 4 or 5%, biased toward the larger ones. How accurate are these estimates? We do not know, although presumably they are not so far off that, for example, 90% of dinosaurs are already known.

Another method for estimating the completeness of the record would involve genera known or inferred to have been present, but not represented as fossils. For example, *Camptosaurus* is known from the Kimmeridgian Morrison Formation and the Barremian Lakota Formation of the United States but not from intervening rocks. Nonetheless, if it is correctly identified, it lived somewhere during the time between the Kimmeridgian and Barremian. So far a survey of such taxa has not been carried out, no doubt partly due to disagreement over the accuracy of taxonomic identifications.

The so-called "ghost taxa" are identified by comparing the fossil record with a phylogenetic analysis. For example, ceratosaurs are known from the Late Triassic but their sister group, tetanurans, only appears in the Jurassic. If these are both valid monophyletic groups, then there must have been at least one still unknown, Late Triassic tetanuran—a ghost taxon. No survey of the proportion of ghost taxa to known lineages for any given time interval has been made, but this would provide another way of estimating the completeness of the record.

Statistical methods of estimating the completeness have also been proposed. These provide confidence intervals for the origin and disappearance of fossil taxa. These figures may then be compared with known fossils to obtain an estimate in much the same way as for missing or ghost taxa described previously. Some details, and references, may be found in Marshall (1990).

The fossil record does not tell us all we wish to know, but we wish to know many things. For some questions the record is a better source of information than for others (Padian and Clemens, 1985). Although the representation of taxonomic diversity has received much attention (all of this entry so far), the record is incomplete in other ways. Anatomical incompleteness is so obvious that we may fail to consider it at all. For dinosaurs, little of the soft tissues is ever preserved, but there are skin impressions (e.g., Osborn, 1912; Czerkas, 1992) and endocasts, from which the form of the brain and associated structures can reliably be ascertained (Hopson, 1979).

Even BEHAVIOR, seemingly ephemeral, is not always lost, as is strikingly demonstrated by Boucot (1990). TRACKWAYS are particularly useful for preserving a record of past behavior (e.g., Thulborn and Wade, 1984), as is PALEOPATHOLOGY. Some ecological inferences are also reliable, although the ability to support opposing conclusions with the same data is shown by comparing Bakker (1972) with Farlow (1978). Taphonomic studies can reveal whether dinosaurs lived in the area where they were preserved or were transported there as carcasses (e.g., Dodson, 1971). Statistical studies can elucidate whether similar specimens represent different sexes (e.g., Kurzanov, 1972) or parts of a growth series of one species (e.g., Dodson, 1975).

Such studies are inferential, however, and not based on direct observation. They provide only a reasonable degree of plausibility, not certainty. Tudge's (1985) clever essay showed that there are aspects, possibly many, about fossil organisms about which we simply cannot be certain.

The bias of the fossil record, both preservational and observational, permits only a partial view of the past. The significance of what we do not see is controversial but important. Because allopatric speciation was thought to involve small population sizes and rapid evolutionary rates, this was long used as an excuse for the absence (or alleged absence) of transitional forms from the fossil record. Gould and Eldredge reinterpreted this as being due not to bias but accurately representing evolution (by punctuated equilibrium). Further discoveries, such as those by Sheldon (1987), indicate that without careful work the record may not be sufficiently transparent to test hypotheses such as punctuated equilibrium.

In some topics of interest, such as the Cretaceous–Tertiary extinctions, the transparency of the record can be very significant. Signor and Lipps (1982) showed that the rate of extinction, whether sudden or gradual, can be irretrievably masked by the bias of the record. Likewise, times of origin and extinction allow estimates of evolutionary rates, but the accuracy of the estimates depends on how accurately these times are recorded (Marshall, 1990).

As pointed out by McKinney (1991), the fossil record is not a book but a filter. It filters the information we receive from the past, and in doing so loses information. McKinney maintains that it also filters noise. This permits us to perceive large-scale patterns more easily. Furthermore, much of the information that went into the record was redundant and so we can, if careful, infer much that has been lost.

So far only one method for testing the reliability of the record, as opposed to its completeness, has been proposed. Norell and Novacek (1992) compared the occurrence of fossil taxa of dinosaurs and mammals with their positions on cladograms. A properly constructed cladogram is regarded (at least by cladists) as giving a reliable sequence of the appearance of related groups. They found that for some groups, e.g., pachycephalosaurs and synapsids, the record provides an accurate source of such information. However, for others, such as protoceratopsians and hadrosaurs, much information is missing. Curiously, a large sample size did not guarantee accuracy, nor did a small one guarantee inaccurate information.

See also the following related entries:

BEHAVIOR • DISTRIBUTION AND DIVERSITY • EVOLUTION • PLATE TECTONICS • TAPHONOMY

References

Bakker, R. T. (1972). Anatomical and ecological evidence of endothermy in dinosaurs. *Nature* **229,** 172–174.

Boucot, A. J. (1990). *Evolutionary Paleobiology of Behavior and Coevolution*, pp. 725. Elsevier, New York.

Czerkas, S. A. (1992). Discovery of dermal spines reveals a new look for sauropod dinosaurs. *Geology* **20,** 1068–1070.

Dodson, P. (1971). Sedimentology and taphonomy of the Oldman Formation (Campanian), Dinosaur Provincial Park, Alberta (Canada). *Palaeogeogr. Palaeoclimatol. Palaeoecol.* **10,** 21–74.

Dodson, P. (1975). Taxonomic implications of relative growth in lambeosaurine hadrosaurs. *Systematic Zool.* **24,** 37–54.

Dodson, P. (1990). Counting dinosaurs: How many kinds were there? *Proc. Natl. Acad. Sci. USA* **87,** 7608–7612.

Farlow, J. O. (1976). A consideration of the trophic dynamics of a late Cretaceous large-dinosaur community (Oldman Formation). *Ecology* **57,** 841–857.

Hopson, J. A. (1979). Paleoneurology. In *Biology of the Reptilia, Neurology A* (C. Gans, R. G. Northcutt, and P. Ulinski, Eds.), pp. 39–146. Academic Press, London.

Koch, C. F. (1987). Prediction of sample size effects on the measured temporal and geographic distribution patterns of species. *Paleobiology* **13,** 100–107.

Kurzanov, S. M. (1972). O polovom dimorfizme prototseratopov. *Paleontologicheskii Zhurnal* **1972,** 104–112.

Lewin, R. (1987). Statistical traps lurk in the fossil record. *Science* **236,** 521–522.

Marshall, C. R. (1990). Confidence intervals on stratigraphic ranges. *Paleobiology* **16,** 1–10.

McKinney, M. L. (1991). Completeness of the fossil record: An overview. In *The Processes of Fossilization* (S. K. Donovan, Ed.), pp. 66–83. Belhaven, London.

Molnar, R. E., Kurzanov, S. M., and Dong, Z. (1990). Carnosauria. In *The Dinosauria* (D. B. Weishampel, P. Dodson, and H. Osmólska, Eds.), pp. 733. Univ. of California Press, Berkeley.

Norell, M. A., and Novacek, M. J. (1992). Congruence between superpositional and phylogenetic patterns: comparing cladistic patterns with fossil records. *Cladistics* **8,** 319–337.

Olshevsky, G. (1991). A revision of the parainfraclass Archosauria Cope, 1869, excluding the advanced Crocodylia. *Mesozoic Meanderings* **2,** 1–267.

Osborn, H. F. (1912). Integument of the iguanodont dinosaur *Trachodon. Mem. Am. Museum Nat. History* **1,** 33–54.

Padian, K., and Clemens, W. A. (1985). Terrestrial vertebrate diversity: Episodes and insights. In *Phanerozoic Diversity Patterns* (J. W. Valentine, Ed.), pp. 41–96. Princeton Univ. Press, Princeton, NJ.

Raup, D. M. (1976). Species diversity in the Phanerozoic: An interpretation. *Paleobiology* **2,** 289–297.

Schindel, D. E. (1982). The gaps in the fossil record. *Nature* **297,** 282–284.

Sheehan, P. M. (1977). A reflection of labor by systematists? *Paleobiology* **3,** 325–328.

Sheldon, P. R. (1987). Parallel gradualistic evolution of Ordovician trilobites. *Nature* **330,** 561–563.

Signor, P. W., III, and Lipps, J. H. (1982). Sampling bias, gradual extinction patterns and catastrophes in the fossil record. *Geol. Soc. Am. Spec. Paper* **190,** 291–296.

Thulborn, R. A., and Wade, M. (1984). Dinosaur trackways in the Winton Formation (mid-Cretaceous) of Queensland. *Mem. Queensland Museum* **21,** 413–517.

Tudge, C. (1985). Bones of contention. *New Scientist* [April 4] 26–27.

van Andel, T. (1981). Consider the incompleteness of the geological record. *Nature* **294,** 297–298.

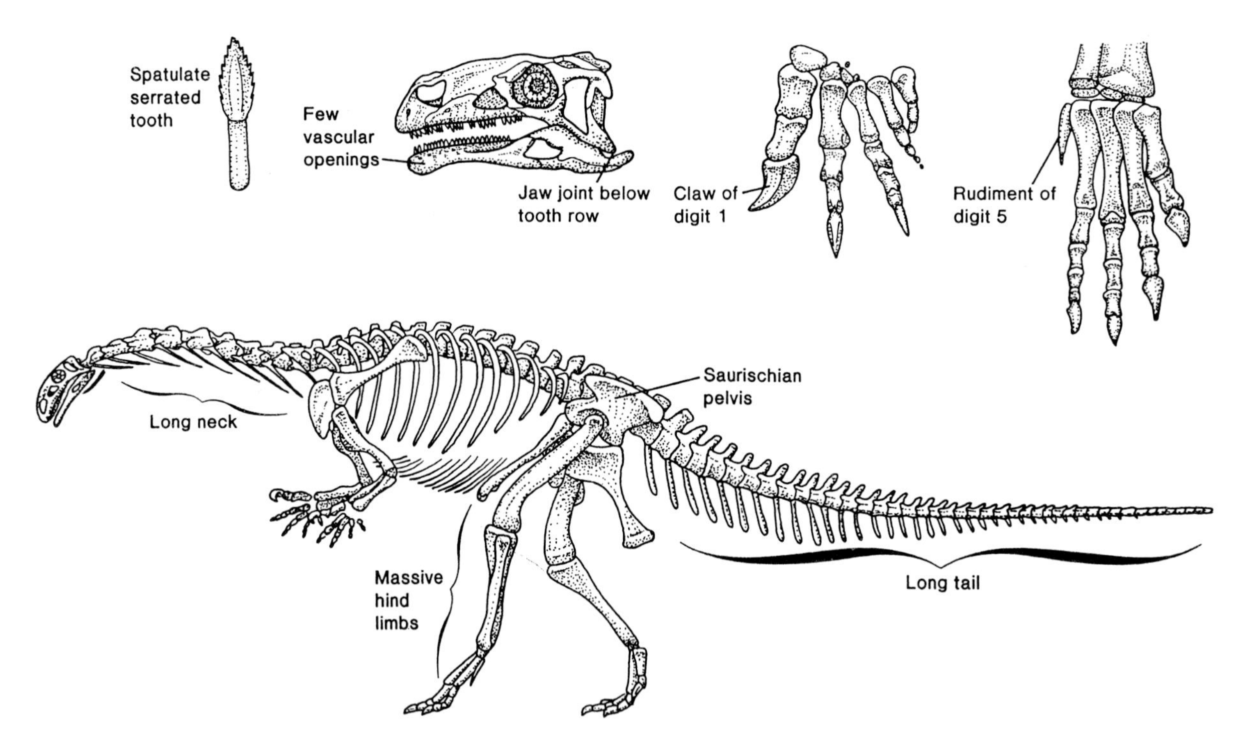

The skeleton of 6–8-meter-long *Plateosaurus* from the Upper Triassic of western Europe shows many characteristic features of prosauropods. Illustration by Phil Sims, from S. Lucas (1994) *Dinosaurs* with permission of the McGraw-Hill Companies.

Prosauropoda

Paul Upchurch
Cambridge University
Cambridge, United Kingdom

Prosauropods often receive only the most cursory of treatments in many dinosaur texts. This is partly the result of our ignorance of many aspects of prosauropod evolution and paleoecology. These "basal sauropodomorphs" first appear in the middle Carnian (~230 Mya) and disappear at the end of the Toarcian (~178 Mya; Fig. 1). During this 50-million-year period prosauropods were usually the most numerous large terrestrial animals. Their remains have now been found on all continents, including Antarctica (Hammer and Hickerson, 1994). Prosauropods also appear to represent the first tetrapod group specialized for high browsing (Bakker, 1978). Anyone interested in the first dinosaur radiation must give considerable attention to the prosauropods. However, questions remain about their monophyly and relationships (see Sauropodomorpha).

The first prosauropod remains were discovered in 1818 during the sinking of a well at East Window, Connecticut (Galton, 1976). Although initially interpreted as human bones (Smith, 1820), they were soon recognized as those of a reptile and are now considered to have belonged to *Anchisaurus.* More prosauropod material soon followed with the discoveries of *Thecodontosaurus* in England (Riley and Stutchbury, 1836, 1840), *Plateosaurus* in Germany (von Meyer, 1837), and *Massospondylus* in South Africa (Owen, 1854). The current century has seen more material recovered from Europe, North America, and Africa, as well as new taxa from Asia (Young, 1941, 1942, 1951) and Argentina (Bonaparte, 1972, 1978). There are now 16 valid prosauropod genera (many of which contain just a single species), plus several recently discovered Chinese forms that require further study (Dong, 1992).

Prosauropods have traditionally been divided into three families: Thecodontosauridae, Plateosauridae ("Anchisauridae"), and Melanorosauridae (Romer, 1956; Charig *et al.,* 1965). The Thecodontosauridae (Figs. 1, 2A, 2B, and 3A) are represented by small (body length = 2 m), possibly bipedal forms, such as *Thecodontosaurus.* Plateosaurids (Figs. 1, 2C and 2D, and 3B) are larger animals (3–6 m long) and include *Plateosaurus* itself and *Lufengosaurus* from China. The Melanorosauridae (Figs. 1 and 3C) is the least well known of these "traditional" families. It includes large (8–11 m long) quadrupedal forms such as *Riojasaurus* from the Upper Triassic of Argentina (Bonaparte, 1972). *Chinshakiangosaurus* is a possible melanorosaurid from the Lower Jurassic of China (Dong, 1992). If this rather fragmentary specimen has been correctly identified, it represents the largest known prosauropod, apparently measuring 12 or 13 m in length. In recent years, there has been a tendency to add further families to the classification of prosauropods. The Blikanasauridae (Galton and van Heerden, 1985) contains *Blikanasaurus* from the Upper Triassic of South Africa. Only the distal end of the hindlimb, including tarsal elements and most of the pes, is known from this genus. Nevertheless, *Blikanasaurus* appears to represent a heavily built quadrupedal form distinct from melanorosaurids. Other families, such as the Massospondylidae (von Huene, 1914), Yunnanosauridae (Young, 1942), and Anchisauridae (Marsh, 1885), were accepted by Galton (1990) in a comprehensive review of prosauropods. For the purposes of discussion, however, a threefold division of the prosauropods (thecodontosaurids, plateosaurids, and melanorosaurids) will be employed here (Fig. 1).

Melanorosaurids have frequently been regarded as likely ancestors of the Sauropoda (Romer, 1956; Charig *et al.,* 1965). This view is based on rather superficial similarities between these two groups, such as their large body size, quadrupedal stance, and straight femoral shafts. Several authors have recently suggested that prosauropods possess a suite of derived characters that are absent in sauropods (Sereno, 1989; Galton, 1990; Upchurch, 1993; Gauffre, 1994). These prosauropod autapomorphies include the presence of a keratinous beak and fleshy cheek (which also occur in ornithischians and some theropods), a relatively tall and dorsally directed maxillary as-

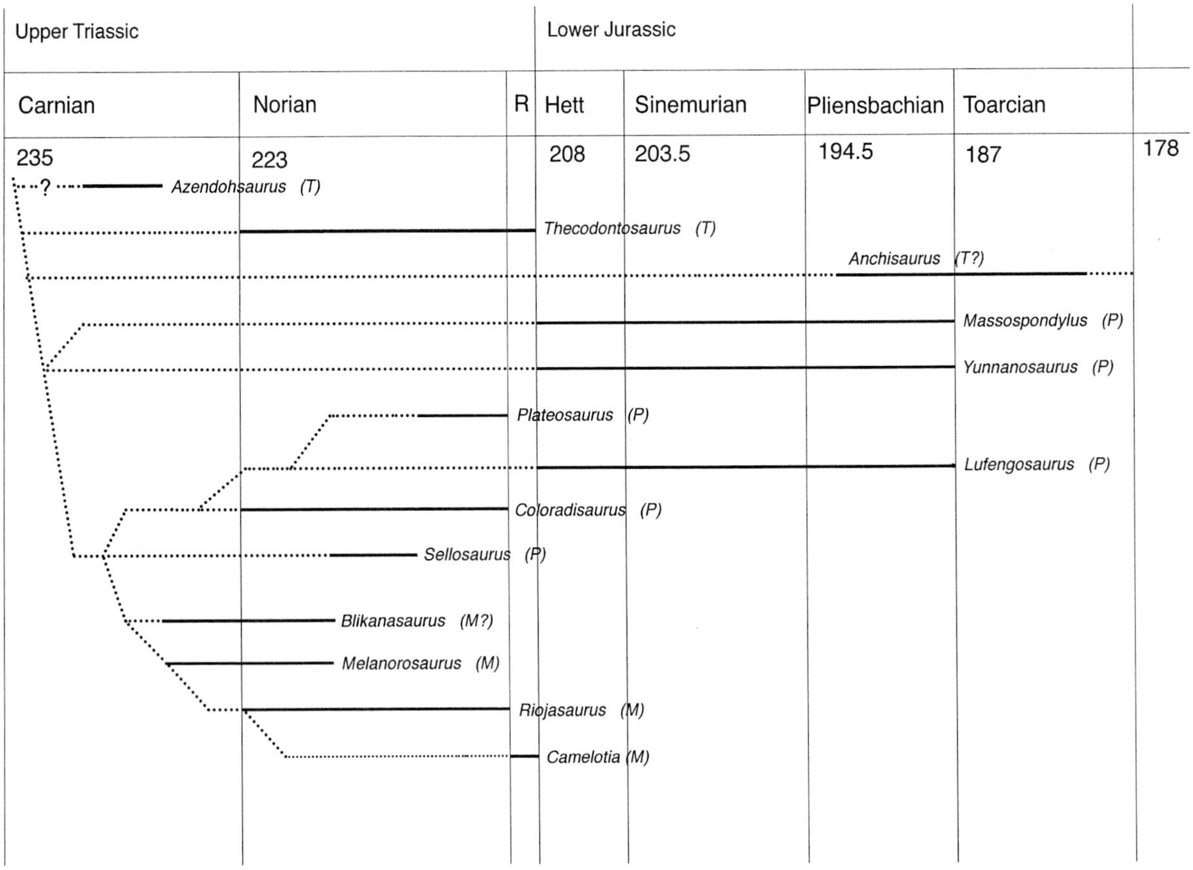

Upper Triassic | Lower Jurassic

Carnian	Norian	R	Hett	Sinemurian	Pliensbachian	Toarcian	
235	223		208	203.5	194.5	187	178

? Azendohsaurus (T)
Thecodontosaurus (T)
Anchisaurus (T?)
Massospondylus (P)
Yunnanosaurus (P)
Plateosaurus (P)
Lufengosaurus (P)
Coloradisaurus (P)
Sellosaurus (P)
Blikanasaurus (M?)
Melanorosaurus (M)
Riojasaurus (M)
Camelotia (M)

FIGURE 1 Prosauropod phylogeny, based on the cladogram and stratigraphic ranges presented in Galton (1990) and the stratigraphic dates listed in Harland *et al.* (1990). The numbers along the horizontal axis are in millions of years before the present. Hett, Hettangian; R, Rhaetian. M, P, and T indicate members of the Melanorosauridae, Plateosauridae, and Thecodontosauridae, respectively.

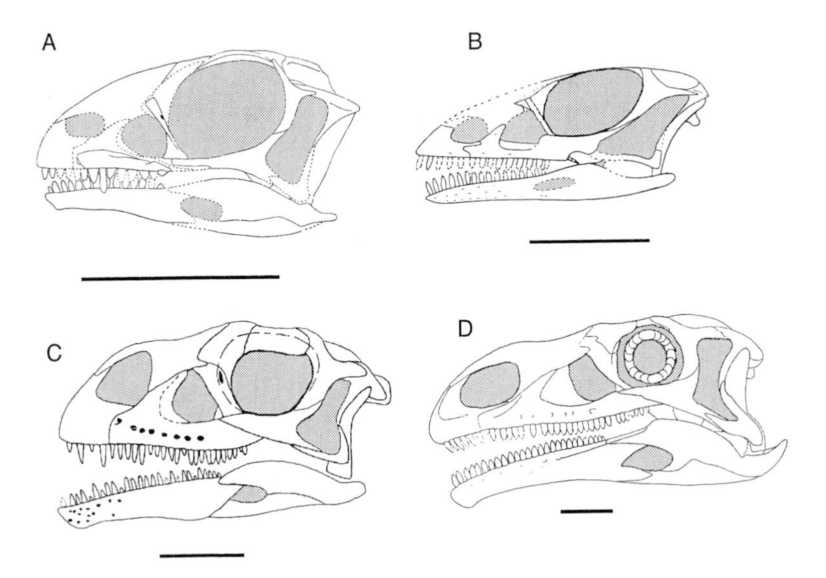

FIGURE 2 Prosauropod skulls in left lateral view. A, *Thecodontosaurus* (after Kermack, 1984); B, *Anchisaurus* (after Galton, 1976); C, *Massospondylus* (after Gow *et al.*, 1990): D, *Plateosaurus* (after Galton, 1984). Scale bars = 50 mm.

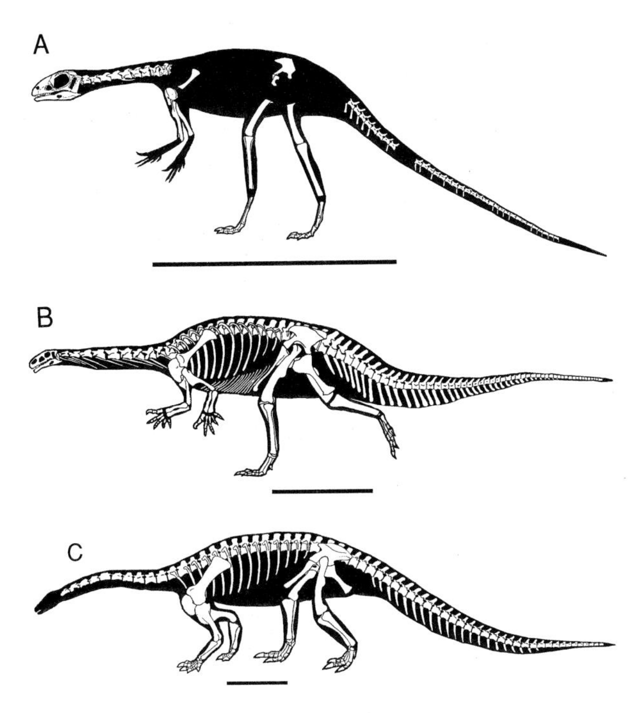

FIGURE 3 Skeletal reconstructions of representative prosauropods, in left lateral view. A, *Thecodontosaurus* (modified from Kermack, 1984); B, *Plateosaurus* (modified from Weishampel and Westphal, 1986), C, *Riojasaurus* (modified from Bonaparte, 1972). Scale bars = 20 cm (A) and 100 cm (B, C).

cending process, the structure of the manus (see below), the characteristically triangular cross section through the distal end of the ischium, and the reduction of the fifth metatarsal (which has also occurred in other dinosaurs, except sauropods). Sauropods and prosauropods are, therefore, commonly regarded as monophyletic sister groups. A monophyletic Prosauropoda, then, can be defined as Thecodontosauridae, Plateosauridae (Anchisauridae), Melanosauridae, and all Sauropodomorpha closer to them than to Sauropoda. The "trend" toward increased body size and quadrupedality within the prosauropods is interpreted as resulting from convergent evolution.

There have been only a very small number of cladistic analyses of prosauropod evolutionary relationships (Gauthier, 1986; Galton, 1990; Upchurch, 1993), all of which have produced rather different results. The phylogeny shown in Fig. 1 is based on Galton (1990) because this analysis provides the most detail. Several factors hamper the study of prosauropod evolution. Prosauropods tend to have rather similar postcranial anatomy, apart from differences in absolute size and relative proportions. Thus, most of the "useful" characters are found in the skull. Thecodontosaurid and plateosaurid skulls are well known, but few data are currently available for the crania of melanorosaurids. A skull belonging to *Riojasaurus* has been discovered but has not yet been described (Galton, 1985a). Currently, therefore, the evolutionary relationships of melanorosaurids must be based on only a small number of postcranial characters. The second problem arises from the interpretation of the significance of certain characters. In all three of the analyses listed previously, *Anchisaurus* and *Thecodontosaurus* were found to represent successively more distant sister taxa to the remaining prosauropods (Fig. 1). These thecodontosaurids are best represented by remains of small, possibly juvenile or subadult individuals. *Massospondylus* and *Sellosaurus* include specimens of juvenile and adult individuals (Cooper, 1981; Gow *et al.*, 1990; Galton and Bakker, 1985). The juveniles often possess "primitive" characters (and therefore more closely resemble thecodontosaurids), whereas the adults possess more "advanced" plateosaurid characters. The position of *Thecodontosaurus* and *Anchisaurus* in Fig. 1 may reflect ontogeny rather than phylogeny. This problem is illustrated by "*Efraasia*," which was considered to be a primitive prosauropod by Gauthier (1986) but is now simply regarded as a juvenile of the more advanced prosauropod *Sellosaurus* (Galton and Bakker, 1985). Studies of prosauropod ontogeny have rarely been attempted despite the fact that several genera (*Massospondylus, Mussaurus, Plateosaurus,* and *Sellosaurus)* are known from numerous individuals from different growth stages. Such analyses may eventually suggest that shifts in ontogenetic timing (see HETEROCHRONY) have played an important role in sauropodomorph evolution, as proposed by Bonaparte and Vince (1979).

The PALEOECOLOGY of prosauropods has been a subject for considerable debate, with diet being especially controversial. It has been suggested that prosauropods were carnivores (Owen, 1854; Cooper, 1981), omnivores (Kermack, 1984), or herbivores (Galton, 1976, 1985b). The history of prosauropod taxonomy is partly responsible for the "carnivore" hypothesis. Prosauropods have frequently been interpreted as theropods (and vice versa) or have been regarded as the ancestors of both sauropods and theropods. For example, Marsh (1896) placed *Anchisaurus* within the

Theropoda; von Huene (1914) and others (see Upurch, 1993) viewed prosauropods as primitive members of the Saurischia that contained lineages ancestral to the sauropods and carnosaurs; and recently, some authors (Colbert, 1970; Cooper, 1981) have suggested that *Herrerasaurus* was a primitive prosauropod, whereas some regard it as a basal theropod (Sereno and Novas, 1992) or outside Dinosauria altogether (Holtz and Padian, 1995).

The most detailed case for prosauropod carnivory was put forward by Cooper (1981), who provided arguments in favor of prosauropod carnivory and against herbivory. For example, he noted the presence of large, laterally compressed, serrated teeth, recurved "carnivorous" teeth in association with many prosauropod skeletons, and a grasping "raptorial" manus armed with large claws (especially digit I). Against prosauropod herbivory, Cooper lists the absence of wear facets on the teeth, a dentition and jaw apparatus suitable for cutting rather than chewing, and a habitat that was arid and therefore lacked soft vegetation and fresh water. Galton (1984, 1985b) countered Cooper's arguments and suggested that prosauropods were herbivorous. The recurved serrated teeth found in association with prosauropod remains are most plausibly explained as teeth shed by theropods or other carnivorous archosaurs while they scavenged the prosauropod carcass. Most dinosaurs, including ornithischians and early sauropods, possess serrated teeth. In prosauropods, as in other herbivorous dinosaurs, the teeth have a few large serrations, whereas in theropods the serrations are finer and more numerous. The raptorial hand could have been used for a number of functions, most of which are compatible with herbivory (see below). Many of Cooper's other points rest on the assumption that prosauropods should show characteristics similar to those found in mammalian herbivores. However, unlike mammals (and derived ornithischians such as hadrosaurs and ceratopsids), most reptilian herbivores do not grind or chew their food. Thus, the absence of wear facets on prosauropod teeth (except in *Yunnanosaurus)* indicates little oral processing of the food, not necessarily carnivory. Galton (1984, 1985b) noted that many prosauropods have a jaw joint that is offset ventral to the dentary tooth row. This offset increases the leverage of the adductor musculature and also causes the upper and lower tooth rows to be more nearly parallel during the onset of occlusion. This feature, coupled with the shape and placement of the teeth, would allow the prosauropod jaws to slice effectively through even tough plant material. The lateral surface of the prosauropod dentary is marked by a ridge that runs forward and slightly downward from the region of the coronoid eminence. Galton (1984) interprets this ridge, and the small number of large vascular foramina on the lateral surface of the dentaries and maxillae, as evidence for the presence of a fleshy cheek. Such a structure would help to retain the chopped plant material within the mouth before swallowing. Discoveries of concentrations of small stones in the rib cages of several prosauropods, including *Massospondylus* (Bond, 1955; Raath, 1974) and *Sellosaurus* (von Huene, 1932; Galton, 1973), indicate that they possessed a "gastric mill." The absence of wear on the teeth is consistent with the possession of stomach stones because most of the processing of the plant material into smaller particles probably occurred after swallowing. Taylor (1993) noted that gastroliths are found in carnivores (sea lions, plesiosaurs, and crocodiles), insectivores (pangolins and aardwolf), and herbivores (gallid birds, sauropods, and some ornithischians), suggesting no clear correlation with DIET.

However, the function of stomach stones may differ between aquatic and terrestrial animals. Aquatic forms that employ an "underwater flight" style of locomotion also usually possess GASTROLITHS, which are thought to be important for buoyancy control. In terrestrial animals, however, buoyancy is irrelevant and it seems likely that gastroliths are involved in food processing. Taylor states that the only terrestrial animals with a gastric mill are herbivores or insectivores (gastroliths are never found in true carnivores). Given that prosauropods are generally regarded as having been terrestrial, the presence of gastroliths makes a carnivorous diet very unlikely. Finally, prosauropod remains are usually very numerous and make up a high proportion of individuals from the deposits in which they are found. For example, Benton (1983) calculates that 75% of the individuals from the Knollenmergel deposits (late Norian) of Germany are *Plateosaurus*, whereas *Lufengosaurus* and *Yunnanosaurus* make up 82% of the specimens from the Lower LUFENG Formation of China. In today's terrestrial environments, a large portion of the biomass is formed by herbivorous animals (a majority of carnivores would not be ecologically stable for more than a very short time period). Thus, unless some extremely unusual taphonomic bias has been at work, the abundance of prosauropod remains also argues for a largely herbivorous diet.

The prosauropod manus, with its unusual structure, has played an important role in studies of the paleoecology of this group. As in dinosauromorphs plesiomorphically, the manus possesses five digits, with the inner three (I–III) being more robust and generally better developed than the outer two (IV and V). In all prosauropods, except *Thecodontosaurus*, metacarpal I is particularly robust. Metacarpals II–IV have also increased in relative robustness in all prosauropods except *Thecodontosaurus* and *Anchisaurus* (Galton, 1990). The typical phalangeal formula is 2-3-4-3-2, but the number of phalanges on the outer two digits is a variable (Galton and Cluver, 1976). Cooper (1981) suggested that the manus of *Massospondylus* was unsuitable for bearing any weight during quadrupedal locomotion. He noted that metacarpals IV and V lie "stacked" vertically beneath metacarpal III, forming the palmar region of the manus into a "tube"-like structure. Most authors, however, have reconstructed the prosauropod manus with the five metacarpals lying parallel and in approximately the same plane, making a weight-bearing function more plausible (Bonaparte, 1972; Galton, 1976; Galton and Cluver, 1976). The structure of the first manual digit is particularly unusual in prosauropods. The distal end of metacarpal I is very asymmetrical, possessing an enlarged lateral portion of the ginglymus. The distal articular region extends well onto the dorsal surface of the metacarpal. The proximal and distal ends of the first phalanx are rotated (about the long axis of the bone) by approximately 45° relative to each other. The second phalanx is a large, laterally compressed ungual. The modifications found in metacarpal I and phalanx I-1 allow this large claw to swing through an extremely wide arc. When the first digit is flexed, the claw swings downwards and lateroposteriorly so that it is folded across the palm. Cooper (1981) questioned the amount of possible flexion in *Massospondylus*, noting the presence of a ventral lappet at the proximal end of phalanx I-1. However, taking into account the true length of the claw in life (where a keratinous sheath would also have been present), Cooper admits that the tip of the claw could have contacted the palmar surface. When the first digit is extended, the claw swings forward and medially. Rotation of the claw against the distal end of phalanx I-1, combined with the expansion of the distal ginglymus of metacarpal I, allows hyperextension. In this hyperextended position, the claw is lifted away from the substrate, with its point projecting medially. Large claws are also present on digits II and III, but these do not appear to show the same range of movement as that in digit I. Suggestions for the function of the prosauropod manus have ranged from grasping small prey and tearing open carcasses (Cooper, 1981) to weight-bearing during locomotion (Thulborn, 1990). Given that prosauropods seem to have been mainly herbivorous, the more carnivorous functions of the manus are unlikely. The large sharp claws may, nevertheless, have formed a useful weapon for defense against predators and/or intraspecific combat. The hypothesis that prosauropods walked (at least some of the time) on "all-fours" is made more feasible by their ability to lift the large claw on digit I away from the substrate. This possibility has now received support from trackway evidence (see below). The capacity of the prosauropod manus to grasp could have been useful during herbivorous feeding. The hand may have held a branch steady while the animal fed. Prosauropods, as high browsers, could have increased their feeding range by rearing up into a bipedal stance. The grasping manus may have been used to hold onto the trunk or branches of a tree in order to increase the stability of the animal while in this stance. These various possible functions are not mutually exclusive. Indeed, the concept of a multifunctional manus is not without precedent among the dinosaurs. Norman (1980) notes that the manus of *Iguanodon* possesses a large thumb spike (for defense), hoof-like unguals on digits II–IV (related to weight-bearing during QUADRUPEDALITY), and a slender fifth digit that probably allowed grasping.

There has been some debate over the habitual gaits of prosauropods. In general, the thecodontosaurids have been regarded as the most bipedal forms, with the larger plateosaurids and melanorosaurids considered to be facultative bipeds and obligate quadrupeds, respectively (Galton and Cluver, 1976). This view is largely based on comparison of the lengths of the forelimbs and trunk region relative to the length of the hindlimb. Forms with longer forelimbs might be expected to be more quadrupedal. However, most dinosaurs, including prosauropods, have FORELIMB/HINDLIMB ratios of less than 0.60 (only sauropods have values above 0.65). Animals with a relatively long trunk region will have a center of gravity further from the acetabulum, making it more difficult to be bipedal. Galton (1976) notes that bipedal dinosaurs such as hadrosaurs have hindlimb/trunk ratios of 1.1 or higher, whereas quadrupedal dinosaurs such as sauropods and ceratopsians have values of 0.9 or less. Most prosauropods have values in the intermediate range (1.1–0.9). Galton (1976) noted that some of the smaller prosauropods, such as *Anchisaurus,* had longer trunks than larger, supposedly more quadrupedal forms (e.g., *Massospondylus* and *Plateosaurus).* Thus, the study of ratios does not seem to provide a very clear picture of the gaits of prosauropods. Cooper (1981) took a different approach to the question of prosauropod gaits and concluded that the "anchisaurids" (plateosaurids) were probably habitually bipedal. This view was partly based on the structure of the manus in *Massospondylus.* The large claw would have hampered quadrupedal locomotion and the "tubular" arrangement of the metacarpals would have led to dislocation if any weight had been placed on the manus (but see above). He also lists 12 characters that, according to Romer (1966), were essential to a bipedal animal. These include (1) a long balancing tail, sheathing powerful hindlimb musculature; (2) a strong sacrum with more than two vertebrae; (3) a long ilium with the acetabulum high up near the backbone; (4) an inturned femoral head and open acetabulum; (5) a prominent fourth trochanter; (6) a long tibia and reduced fibula; (7) reduction of the tarsal bones so that the astragalus and calcaneum are the only large elements; (8) hindfoot extending straight forward with the center toe the longest and most developed; (9) the outer two toes frequently modified and reduced; (10) a triradiate pelvis; (11) forelimbs considerably shorter and much weaker than the hindlimbs; and (12) some tendency toward loss of fingers in advanced bipeds. Cooper states that, except for character 3, every one of these characters is present in *Massospondylus* and other anchisaurid (plateosaurid) prosauropods. This impressive list does not, however, provide strong evidence for bipedality. For example, characters 1–4, 7, 9, and 10 are present in the quadrupedal sauropods, and some of these, plus character 12, are present in the smaller, less graviportal quadrupedal mammals. Other characters, such as 6 and 8, are not well developed in prosauropods. Thus, the hindfoot is not tridactyl with a noticeably longer third digit; rather, it is functionally tetradactyl (Thulborn, 1990) with digits III and IV of similar length.

Evidence from trackways provides some insight into prosauropod locomotion. For example, the ichnogenera *Navahopus* (from the Lower Jurassic Navaho Sandstone of Arizona), *Tetrasauropus,* and *Pseudotetrasauropus* (from the Upper Triassic of South Africa) appear to be trackways made by prosauropods walking quadrupedally (Baird, 1980; Thulborn, 1990). The manus prints show at least four digits. In some cases, a large claw on digit I can be seen projecting medially. This characteristic claw impression is usually only present in the deeper tracks (Thulborn, 1990), as would be expected if the claw was carried in its hyperextended position during locomotion. The pes prints are also tetradactyl, with toes of nearly equal length. Prosauropod and certain basal archosaurian pedes are similar in general structure and would be expected to leave very similar tracks (Thulborn, 1990). Thus, true prosauropod trackways can only be recognized when these animals walked quadrupedally (leaving the characteristic manus prints). Therefore, *Navahopus, Tetrasauropus,* and *Pseudotetrasauropus* do provide evidence in support of the view that at least some prosauropods were capable of quadrupedal locomotion. However, the absence of bipedal prosauropod tracks cannot be interpreted as evidence that prosauropods never walked bipedally. Bipedal prosauropod tracks may exist but might never be recognized.

In summary, the small thecodontosaurid prosauropods were probably bipedal. The larger plateosaurids may have used a quadrupedal gait when moving slowly (perhaps while moving between feeding sites) and a bipedal gait for faster locomotion (perhaps during predator avoidance). Melanorosaurids, like sauropods, were probably habitually quadrupedal, only rearing up into a bipedal stance during high browsing and mating.

Prosauropods are occasionally depicted as living in rather harsh desert environments. This is not without justification because the Upper Elliot Formation of South Africa (which has produced *Massospondylus)* appears to represent arid or semiarid conditions (Cooper, 1981) and *Navahopus* indicates a prosauropod walking quadrupedally up the side of a sand dune (Baird, 1980). Nevertheless, Galton (1984) noted that even the Elliot Formation has produced fish and crocodiles that indicate the presence of rivers or other areas of fresh water. Indeed, according to Galton (1990), the majority of prosauropod material has been found in mudstones, siltstones, or marls (this is even true for *Massospondylus*). For example, Dong (1992) notes that the Fengjiahe Formation of Yunnan, which has produced the typical "*Lufengosaurus* fauna," represents fluviolacustrine deposits. Like extant mammalian herbivores, prosauropods may have lived in a variety of environments with some forms inhabiting arid areas. In general, however, it seems that most prosauropods lived in rather mesic conditions.

Current information provides only a glimpse into prosauropod LIFE HISTORY and social BEHAVIOR. At least some prosauropods were oviparous. A possible nest of *Massospondylus* EGGS was reported by Grine and Kitching (1987). The nest contained six eggs, each approximately 65 mm long and 55 mm in diameter. Bonaparte and Vince (1979) described the discovery of a prosauropod nest from the Upper Triassic of Argentina. This nest, which was assigned to the new genus *Mussaurus*, consisted of numerous eggshell fragments and the remains of five hatchlings. Moratalla and Powell (1994) suggest that *Mussaurus* young may have been altricial, i.e., the young remained in the nest after hatching and were looked after by one or both parents.

Despite our ignorance of most aspects of prosauropod ontogeny, a few "trends" during growth can be detected. Apart from changes in relative proportions, larger (and presumably older) individuals have more numerous teeth and more robust metapodials. Juvenile prosauropods, like many "adult" sauropods, show a marked increase in the size of the teeth at the anterior end of the upper jaw. The fact that large numbers of prosauropod skeletons have been found together may indicate that these animals lived in herds, but there are currently no trackway data to support this (Thulborn, 1990). Prosauropods lack the crests, frills, and other display structures common in many theropods and ornithischians. Nevertheless, male and female prosauropods may have differed in overall size and robustness. Weishampel and Chapman (1990), for example, performed a morphometric analysis on the femora of *Plateosaurus*. Their results suggest the presence of two morphs distinguished by the robustness of the proximal and distal ends of the femur.

No prosauropod remains have been found in deposits younger than the Toarcian (Galton, 1990), and it is assumed that this lineage of sauropodomorphs became extinct at the end of the Lower Jurassic. The reasons for the disappearance of the Prosauropoda are not well understood. Galton (1976) suggested that the masticatory apparatus of ornithischian dinosaurs was superior to that of prosauropods, and that competition between these groups may have driven the latter to extinction. Prosauropods probably faced competition for plant material at both lower feeding levels (ornithischians) and during high browsing (sauropods). However, prosauropods had presumably been competing with ornithischians and sauropods for approximately 30 million years; in that time they achieved a global distribution and were usually the most abundant large terrestrial animals. Competition, at least by itself, does not seem to be an adequate explanation for the disappearance of the prosauropods. Whatever the cause of their extinction, there can be little doubt that prosauropods represent an early dinosaur radiation of major importance.

See also the following related entries:

SAURISCHIA • SAUROPODA • SAUROPODOMORPHA

References

Baird, D. (1980). A prosauropod dinosaur trackway from the Navaho Sandstone (Lower Jurassic) of Arizona. In *Aspects of Vertebrate History. Essays in Honour of Edwin Harris Colbert* (L. L. Jacobs, Ed.), pp. 219–230. Museum of Northern Arizona Press, Flagstaff.

Bakker, R. T. (1978). Dinosaur feeding behaviour and the origin of flowering plants. *Nature* **274,** 661–663.

Benton, M. J. (1983). Dinosaur success in the Triassic: A noncompetitive ecological model. *Q. Rev. Biol.* **58,** 29–55.

Bonaparte, J. F. (1972). Los tetrápodos del sector superior de la Formación Los Colorados, La Rioja, Argentina (Triásico Superior). I Parte. *Opera Lilloana* **22,** 1–183.

Bonaparte, J. F. (1978). *Coloradia brevis* n. gen. et n. sp. (Saurischia, Prosauropoda), dinosaurio Plateosauridae de la Formación Los Colorados Triásico superior de La Rioja, Argentina. *Ameghiniana* **15,** 327–332.

Bonaparte, J. F., and Vince, M. (1979). El hallazgo del primer nido de Dinosaurios Triásicos (Saurischia, Prosauropoda), Triásico superior de Patagonia, Argentina. *Ameghiniana* **16,** 173–182.

Bond, G. (1955). A note on dinosaur remains from the Forest Sandstone (Upper Karoo). *Arnoldia* **2,** 795–800.

Charig, A. J., Attridge, J., and Crompton, A. W. (1965). On the origin of the sauropods and the classification of the Saurischia. *Proc. Linnean Soc. London* **176,** 197–221.

Colbert, E. H. (1970). A saurischian dinosaur from the Triassic of Brazil. *Am. Museum Novitates* **2405,** 1–39.

Cooper, M. R. (1981). The prosauropod dinosaur *Massospondylus carinatus* Owen from Zimbabwe: Its biology, mode of life and phylogenetic significance. *Occasional Papers Nat. Museum Rhodes B Nat. Sci.* **6,** 689–840.

Dong, Z. M. (1992). *Dinosaurian Faunas of China.* Springer-Verlag, Berlin.

Galton, P. M. (1973). On the anatomy and relationships of *Efraasia diagnostica* (v. Huene) n. gen., a prosauropod dinosaur (Reptilia: Saurischia) from the Upper Triassic of Germany. *Palaeontol. Z.* **47,** 229–255.

Galton, P. M. (1976). Prosauropod dinosaurs of North America. *Postilla* **169,** 1–98.

Galton, P. M. (1984). Cranial anatomy of the prosauropod dinosaur *Plateosaurus* from the Knollenmergel (Middle Keuper, Upper Triassic) of Germany. *Geol. Palaeontol.* **18,** 139–171.

Galton, P. M. (1985a). Notes on the Melanorosauridae, a family of large prosauropod dinosaurs (Saurischia, Sauropodomorpha). *Géobios* **19,** 671–676.

Galton, P. M. (1985b). Diet of prosauropod dinosaurs from the Late Triassic and Early Jurassic. *Lethaia* **18,** 105–123.

Galton, P. M. (1990). Basal Sauropodomorpha–Prosauropods. In *The Dinosauria* (D. B. Weishampel, P. Dodson, and H. Osmólska), pp. 320–344. Univ. of California Press, Berkeley.

Galton, P. M., and Bakker, R. T. (1985). Cranial anatomy of the prosauropod dinosaur '*Efraasia diagnostica,*' a juvenile individual of *Sellosaurus gracilis* from the Upper Triassic of Nordwittenberg, West Germany. *Stuttgarter beitrage Naturkunde Ser. B (Geol. Palaontol.)* **117,** 1–15.

Galton, P. M., and Cluver, M. A. (1976). *Anchisaurus capensis* and a revision of the Anchisauridae (Reptilia, Saurischia). *Ann. South Africa Museum* **69,** 121–159.

Galton, P. M., and van Heerden, J. (1985). Partial hindlimb of *Blikanasaurus cromptoni* n. gen. and n. sp. representing a new family of prosauropod dinosaurs from the Upper Triassic of South Africa. *Géobios* **18,** 509–516.

Gauffre, F.-X. (1994). The prosauropod dinosaur *Azendohsaurus laaroussii* from the Upper Triassic of Morocco. *Palaeontology* **36**(4), 897–908.

Gauthier, J. (1986). Saurischian monophyly and the origin of birds. *Mem. California Acad. Sci.* **8,** 1–55.

Gow, C. E., Kitching, J. W., and Raath, M. A. (1990). Skulls of the prosauropod dinosaur *Massospondylus carinatus* Owen, in the collections of the Bernard Price Institute for palaeontological research. *Palaeontol. Africa* **27,** 45–58.

Grine, F. E., and Kitching, J. W. (1987). Scanning electron microscopy of early dinosaur egg shell structure: A comparison with other rigid sauropod eggs. *Scanning Microsc.* **1**(2), 615–630.

Hammer, W. R., and Hickerson, W. J. (1994). A crested theropod dinosaur from Antarctica. *Science* **264,** 828–830.

Harland, W. B., Armstrong, R. L., Cox, A. V., Craig, L. E., Smith, A. G., and Smith, D. G. (1990). *A Geologic Time Scale.* Cambridge Univ. Press, Cambridge, UK.

Holtz, T. R., Jr., and Padian, K. (1995). Definition and diagnosis of Theropoda and related taxa. *J. Vert. Paleontol.* **15** (suppl. to no. 3), 35A.

Kermack, D. (1984). New prosauropod material from South Wales. *Zool. J. Linnean Soc. London* **82,** 101–117.

Marsh, O. C. (1885). Names of extinct reptiles. *Am. J. Sci. Ser. 3* **29,** 169.

Marsh, O. C. (1896). The dinosaurs of North America. Sixteenth Annual Report of the U. S. Geological Survey, Pt. I.

Moratalla, J. J., and Powell, J. E. (1994). Dinosaur nesting patterns. In *Dinosaur Eggs and Babies* (K. Carpenter, K. F. Hirsch, and J. R. Horner, Eds.), pp. 37–46. Cambridge Univ. Press, Cambridge, UK.

Norman, D. B. (1980). On the ornithischian dinosaur *Iguanodon bernissartensis* from the Lower Cretaceous of Bernissart (Belgium). *Inst. R. Sci. Nat. Belgique* **178,** 1–103.

Owen, R. (1854). *Descriptive Catalogue of the Fossil Organic Remains of Reptilia Contained in the Museum of the Royal College of Surgeons of England.* British Museum (Natural History), London.

Raath, M. A. (1974). Fossil vertebrate studies in Rhodesia: Further evidence of gastroliths in prosauropod dinosaurs. *Arnoldia* **7**(5), 1–7.

Riley, H., and Stutchbury, S. (1836). A description of various fossil remains of three distinct saurian animals discovered in the autumn of 1834, in the Magnesian conglomerates of Durdham Down, near Bristol. *Proc. Geol. Soc. London* **2,** 397–399.

Riley, H., and Stutchbury, S. (1840). A description of various fossil remains of three distinct saurian animals discovered in the autumn of 1834, in the Magnesian conglomerates of Durdham Down, near Bristol. *Trans. Geol. Soc. London* **5**(2), 349–357.

Romer, A. S. (1956). *Osteology of the Reptiles.* Univ. of Chicago Press, Chicago.

Romer, A. S. (1966). *Vertebrate Paleontology,* 3rd ed., pp. 468. Univ. of Chicago Press, Chicago.

Sereno, P. C. (1989). Prosauropod monophyly and basal sauropodomorph phylogeny. *J. Vertebr. Paleontol.* **9**(Suppl.), 38A.

Sereno, P. C., and Novas, F. E. (1992). The complete skull and skeleton of an early dinosaur. *Science* **258,** 1137–1140.

Smith, N. (1820). Fossil bones found in red sandstones. *Am. J. Sci.* **2,** 146–147.

Taylor, M. A. (1993). Stomach stones for feeding or buoyancy? The occurrence and function of gastroliths in marine tetrapods. *Philos. Trans. R. Soc. London B* **341,** 163–175.

Thulborn, R. A. (1990). *Dinosaur Tracks,* pp. 410. Chapman & Hall, London.

Upchurch, P. (1993). The anatomy, phylogeny and systematics of sauropod dinosaurs. Unpublished Ph.D. dissertation, University of Cambridge, Cambridge, UK.

von Huene, F. (1914). Saurischia et Ornithischia triadica ("Dinosauria" triadica). *Fossilium Catalogus 1 Animalia* **4,** 1–21.

von Huene, F. (1932). Die fossile Reptile-Ordnung Saurischia, ihre Entwicklung und Geschichte. *Monogr. Geol. Palaeontol.* (Pt. I and II, Ser. I) **4,** 1–361.

von Meyer, H. (1837). Mitteilung an Prof. Bronn *(Plateosaurus engelhardti). N. J. Mineral. Geol. Palaeontol.* **1837,** 317.

Weishampel, D. B., and Chapman, R. E. (1990). Morphometric study of *Plateosaurus* from Trossingen (Baden-Württemberg, Federal Republic of Germany). In *Dinosaur Systematics: Approaches and Perspectives* (K. Carpenter and P. J. Currie, Eds.), pp. 43–52. Cambridge Univ. Press, Cambridge, UK.

Weishampel, D. B., and Westphal, F. (1986). Die Plateosaurier von Trossingen. *Austellungskat. Univ. Tübingen* **19,** 1–27.

Young, C. C. (1941). A complete osteology of *Lufengosaurus huenei* Young (gen. et sp. nov.). *Palaeontol. Sinica Ser. C* **7,** 1–53.

Young, C. C. (1942). *Yunnanosaurus huangi* (gen. et sp. nov.) a new Prosauropoda from the Red Beds of Lufeng, Yunnan. *Bull. Geol. Soc. China* **22,** 63–104.

Young, C. C. (1951). The Lufeng saurischian fauna in China. *Palaeontol. Sinica Ser C* **13,** 1–96.

Protoceratopsidae

see NEOCERATOPSIA

Provincial Museum of Alberta, Canada

see MUSEUMS AND DISPLAYS

Pseudofossils

PAUL G. DAVIS
National Science Museum
Tokyo, Japan

A pseudofossil is an object that is not a fossil but looks like one. The term pseudofossil covers a series of organic, inorganic, and man-made structures that resemble fossils. Because the term pseudofossil covers such a wide range of structures, paleontologists often subdivide it into four more precise terms.

Problematicum (plural problematica) can also be termed dubiofossil or problematic fossil. These pseudofossils include a named group into which geological structures are placed if their affinity is uncertain but their origin is considered to be organic (Allaby and Allaby, 1990). It can be defined more simply as those specimens in which a question exists as to their organic or inorganic nature (Hofmann, 1971). Also, if the identification of a true fossil is problematic, it can be considered to be a pseudofossil. For example Robert Plot's 1677 description and figure of *Scrotum humanum* in his book *The Natural History of Oxfordshire* is the first record of a dinosaur. It is probably the distal end of a *Megalosaurus* femur, but because of its incorrect assignment (dinosaurs were unrecognized at that time), it also becomes a "Problematicum pseudofossil." Allaby and Allaby (1990) noted that a problematic pseudofossil is removed from the Problematica when more certain evidence of relationship is discovered. Fortunately for science, Plot's specimen was lost, otherwise the dinosaur genus *Megalosaurus* could have been suppressed (as a junior synonym) and would have been replaced with the rather unattractive name *Scrotum*! There are many other examples of Problematica and Wills and Sepkoski (1993) give a comprehensive listing.

Sedimentary pseudofossils include many sedimentary rock structures that are mistaken for fossils. These are usually created by the transport and deposition of the sediments and sedimentary rocks. For example, a sedimentary structure called "cone-in-cone," which is formed by pockets of pressurized water in mudrocks, can resemble the pygidia of trilobites. Also, drag marks (or tool marks) are created when a pebble or other debris is dragged and scraped across a soft sediment surface by current activity.

These have been mistaken for feeding or locomotion traces of animals; they are correctly referred to as pseudo-trace fossils.

Pseudofossils can also be created during or after lithification (the process of turning sediments into rock). Dendrites are iron or manganese oxides that are deposited on bedding or fracture surfaces during the weathering of the rock. As their name suggests (*dendros* = tree in Greek), the pattern they form can be mistaken for plant fossils. Sedimentary pseudofossils are the most numerous of the four categories and the reader is directed to Gutstadt (1979) and Häntzschel (1975) for a more complete listing.

Igneous and metamorphic (including tectonic) structures have been mistaken for fossils. For example, moss agates (a variety of quartz that forms in the "frozen" gas bubbles of solidified lava) are often mistaken for plant fossils. Also mistaken for plant fossils are slickensides, which are the (polished and often lined) surfaces of a fault plane. Most examples of this grouping of pseudofossils occur in pre-Cambrian rocks because here real fossils are often cryptic and difficult to recognize and the rocks are usually metamorphosed (further complicating the distinction between pseudofossils and true fossils). Hofmann (1971) includes many examples of this grouping of pseudofossils.

Man-made pseudofossils are termed artifacts. Whether the artifact was manufactured deliberately or not is of no consequence to their classification. Therefore, casts and replicas of fossils are in the strictest sense pseudofossils, although these are rarely used to deliberately mislead people. Perhaps the most famous example of an artifact pseudofossil is Piltdown Man, which misled the leading paleoanthropologists of the 1920s for several decades (see Gee, 1996, for the latest interpretation of this palaeontological whodunit).

See also the following related entry:
CHEMICAL COMPOSITION OF DINOSAUR FOSSILS

References

Allaby, A., and Allaby, M. (Eds.) (1990). *The Concise Oxford Dictionary of Earth Science*, pp. 410. OUP, Oxford.

Gee, H. (1996). Box of bones 'clinches' identity of Piltdown palaeontology hoaxer. *Nature* **381,** 261–262.

Gutstadt, A. M. (1979). Pseudofossil. In *The Encyclopedia of Paleontology* (R. W. Fairbridge and D. Jablonski, Eds.). Dowden, Hutchinson, & Ross, Stroudsburg.

Häntzschel, W. (1975). Trace fossils and Problematica. In *Treatise on Invertebrate Paleontology, Pt. W, Miscellanea* (C. Teichert, C., Ed.) Suppl. 1. GSA / Univ. Kansas Press, Lawrence.

Hofmann, H. J. (1971). Precambrian fossils, pseudofossils and problematica in Canada. *Geol. Surv. Can. Bull.* **189,** 146.

Wills, M. A., and Sepkoski, J. J., Jr. (1993). Problematica. In *The Fossil Record 2* (M. J. Benton, Ed.). Chapman & Hall, London.

Pseudosuchia

KEVIN PADIAN
University of California
Berkeley, California, USA

The term Pseudosuchia, or "false crocodiles," was established by Zittel (1887–1890) as a suborder of the Order CROCODYLIA, to include the armored aetosaurs *Aetosaurus, Typothorax,* and *Dyoplax.* It accompanied the Parasuchia, a term founded by Huxley in 1859 to receive *Stagonolepis* and *Belodon,* which he regarded as early crocodiles. Of these, the former (which Agassiz, in describing the original specimen, a skin impression, took for a ganoid fish) eventually turned out to be an aetosaur, whereas the latter is a phytosaur. Von Huene eventually separated the Pseudosuchia and Parasuchia (Phytosauria) from the crocodiles and regarded them as separate branches of the Archosauria. They came to be included traditionally in the Order Thecodontia, along with the Proterosuchia and Erythrosuchia (Krebs, 1976, provides an historical summary; see also ARCHOSAURIA).

Knowledge of the archosaurs referred to Pseudosuchia expanded with discoveries in the Late Triassic Elgin Sandstones of Scotland in the late 1800s and early 1900s (*Ornithosuchus, Erpetosuchus, Stagonolepis,* and *Scleromochlus*), the Early Triassic deposits of South Africa and Tanganyika in the early 1900s (*Euparkeria* and *Parringtonia*), and von Huene's work in the late Middle Triassic Santa Maria beds of Brazil (*Rauisuchus, Prestosuchus,* and *Procerosuchus*) in the 1930s. Meanwhile, a wide diversity of phytosaurs, aetosaurs, and other forms were being discovered in the southwestern United States, Europe, and elsewhere. In the 1950s and 1960s, further work in South America by Reig, Casamiquela, and Bonaparte resulted in the discoveries of the ornithosuchid *Riojasuchus,* the rauisuchian *Saurosuchus,* and several aetosaurs. Expeditions by Harvard University and Argentinian scientists in the 1950s and 1960s, led by A. S. Romer and colleagues, resulted in the discoveries of *Chanaresuchus* and *Gracilisuchus* as well as many important ornithodirans (see MUSEUM OF COMPARATIVE ZOOLOGY).

The Archosauria came to comprise Thecodontia, Crocodylia, Pterosauria, and Ornithischia and Saurischia (the latter two were generally regarded only informally as Dinosauria; see DINOSAURIA, DEFINITION). Thecodontia was considered the "stem group" of Archosauria, from which the other groups independently evolved (see Charig, 1976a,b). Birds were also widely considered to have evolved from indeterminate archosaurs (see BIRD ORIGINS).

With the advent of the cladistic system, Gauthier (1984, 1986) redefined Archosauria as a crown group comprising crocodiles and birds and all the descendants of their most recent common ancestor (see ARCHOSAURIA). As a result, the Thecodontia as traditionally constituted was divided into two major monophyletic clades but (as always understood) did not constitute a monophyletic group itself; consequently, its utility in the phylogenetic system is superseded by the term Archosauria, and "Thecodontia" is no longer in use.

Most of the taxa traditionally included in the pseudosuchians are more closely related to crocodiles than to birds. Gauthier (1986), recognizing this, named this stem Pseudosuchia, even though (ironically) it would include crocodiles themselves, who would then be members of a larger taxon called "false crocodiles." The sister taxon to Pseudosuchia is all archosaurs closer to birds than to crocodiles. Gauthier named this stem ORNITHOSUCHIA. As a matter of further irony, the Ornithosuchidae *(Ornithosuchus* and its relatives) were soon discovered to belong to Pseudosuchia (Sereno and Arcucci, 1990), but this has no effect on the defining name. Sereno and Arcucci (1990) and Sereno (1991) regarded the Ornithosuchia and Pseudosuchia as invalid for this reason, but the names depend on definition, not composition (see PHYLOGENETIC SYSTEM; SYSTEMATICS). They named the node within Pseudosuchia uniting Parasuchia, Ornithosuchidae, and Suchia (aetosaurs, Rauisuchia, Poposauridae, and Crocodylomorpha) the Crurotarsi. Unfortunately, Sereno (1991, p. 26) defined Crurotarsi as a node but later (p. 42) defined it as a stem. Padian and May (1993) regarded Crurotarsi as an invalid junior synonym of Pseudosuchia, which is true according to Sereno's stem-based definition but incorrect according to his prior definition. Hence, Pseudosuchia is a stem, and Crurotarsi is a valid node within Pseudosuchia. By the same token, Ornithosuchia is a valid stem, and ORNITHODIRA is a node within Ornithosuchia (Fig. 1).

As Sereno (1991, p. 41) notes, since the late 1970s consensus has developed that archosaurs could be broadly split into two clades on the basis of different ankle joints, called crurotarsal and mesotarsal. The crurotarsal ankle, found in Pseudosuchia, is distinguished by a strong concavoconvex articulation between the astragalus and calcaneum (the proximal tarsals). In this articulation the astragalar surface is concave in Ornithosuchidae and convex in the other forms. The calcaneum rotates along with the distal tarsals and the rest of the foot as a unit against the unit formed by the astragalus and leg. This configuration produced the ability for both hinge-like movement and posterolateral excursion that may be seen in the crocodile's "high walk" and sprawling posture, respectively (Parrish, 1986, 1987, 1993; see ARCHOSAURIA). In the mesotarsal ankle, found in Ornithosuchia, the astragalus and calcaneum articulate immovably with each other and with the tibia and fibula; a "hinge" between the proximal tarsals and the distal tarsals, which form a functional unit with the rest of the foot, produces a parasagittal gait and the inability to walk as sprawling reptiles do. This configuration is most easily seen in living birds (see ARCHOSAURIA; BIPEDALITY; HINDLIMBS AND FEET). To divide archosaurs into only two ankle types is a structural oversimplification (Sereno, 1991), but anatomically and functionally the Pseudosuchia have a unity in the ankle and other characters, and a very different and unique ankle type characterizes the Ornithosuchia.

The Pseudosuchia were the most successful group of Triassic REPTILES. Their separation from Ornithosuchia appears to have occurred at least by Middle Triassic (Anisian) times because the earliest pseudosuchians are known from Anisian *(Ticinosuchus)* or Ladinian *(Rauisuchus)* strata, and the earliest ornitho-

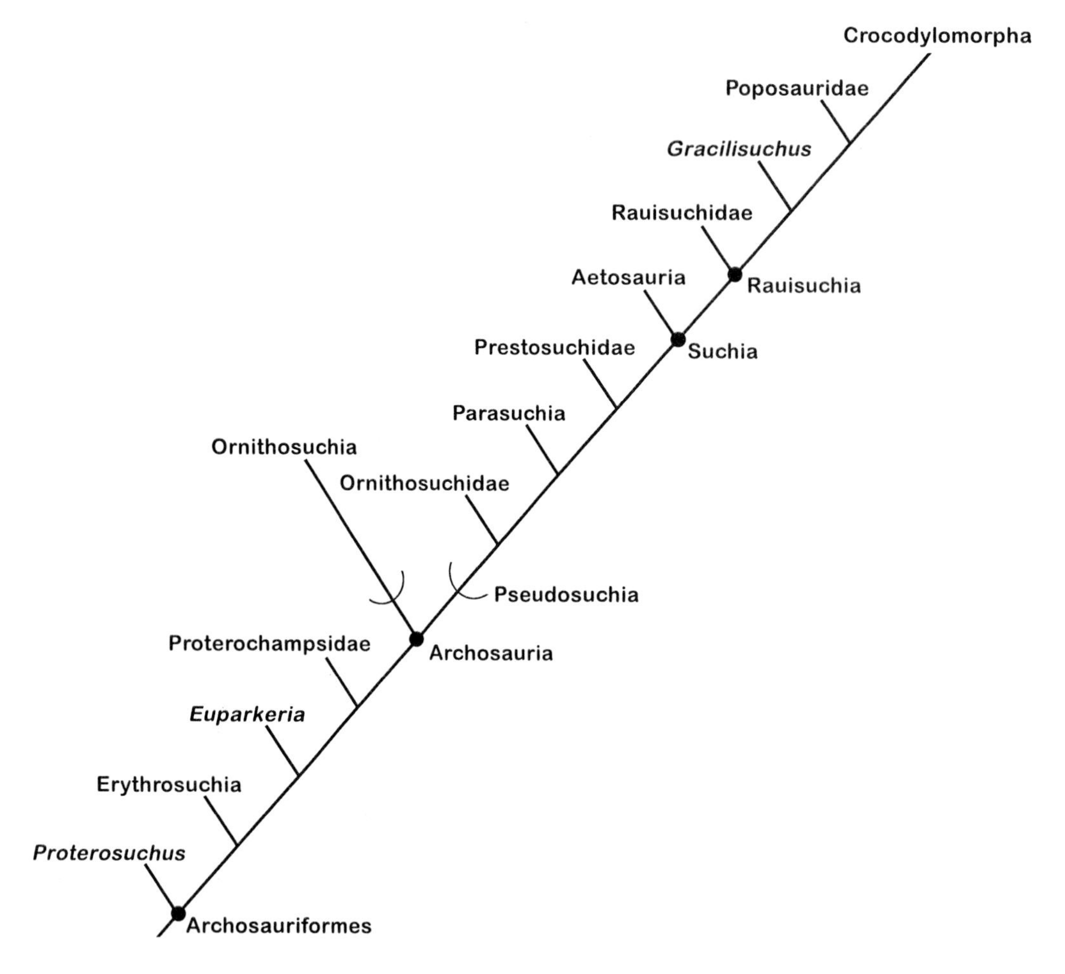

FIGURE 1 Phylogeny of the Pseudosuchia, largely after Parrish (1993).

suchians can be traced to the early Carnian. The pseudosuchian radiation included amphibious carnivores (phytosaurs and derived crocodiles), terrestrial carnivores (rauisuchians, poposaurs, ornithosuchids, and basal crocodylomorphs), and terrestrial herbivores (aetosaurs). They filled a rather different set of adaptive zones than did dinosaurs: They were aquatic whereas no dinosaurs were, they did not radiate into the giant carnivorous mode as some dinosaurian carnivores did (although some phytosaurs were as long as 13 m), and their herbivorous radiation was limited to the aetosaurs, with their chisel-shaped teeth and toothless, pig-like snouts. They appear to have been very effective, mostly quadrupedal terrestrial carnivores. Basal crocodylomorphs, the sphenosuchians such as *Terrestrisuchus, Hesperosuchus*, and related forms, were lightly built and gracile and probably bipedal. Because bipedality does not seem to have been common or primitive in the Pseudosuchia, the situation in sphenosuchians is probably derived and may not have been the primitive state for Crocodyliformes (see Parrish, 1987; Benton and Clark, 1988). By the Early Jurassic, the protosuchid Crocodyliformes appear to have taken on the amphibious way of life and functional morphology of the skeleton generally seen in extant crocodiles. All pseudosuchian groups except Crocodylomorpha (sphenosuchians + Crocodyliformes) became extinct by the end of the Triassic, although most extended into the Norian Stage. Latest Triassic exposures are not known well enough worldwide to determine the tempo and mode of these extinctions (see EXTINCTION, TRIASSIC).

See also the following related entries:

ARCHOSAURIA • ORNITHOSUCHIA

References

Benton, M. J., and Clark, J. M. (1988). Archosaur phylogeny and the relationships of the Crocodylia. In *The Phylogeny and Classification of the Tetrapods, Volume I* (M. J. Benton, Ed.), pp. 295–338. Clarendon, Oxford.

Charig, A. J. (1976a). Introduction: Subclass Archosauria. In *Handbuch der Paläoherpetologie, Teil 13: Thecodontia* (O. Kuhn, Ed.), pp. 1–6. Fischer-Verlag, Stuttgart.

Charig, A. J. (1976b). Order Thecodontia. In *Handbuch der Paläoherpetologie, Teil 13: Thecodontia* (O. Kuhn, Ed.), pp. 7–10. Fischer-Verlag, Stuttgart.

Gauthier, J. A. (1984). A cladistic analysis of the higher systematic categories of Diapsida. Unpublished Ph.D. dissertation, Department of Paleontology, University of California, Berkeley.

Gauthier, J. A. (1986). Saurischian monophyly and the origin of birds. *Mem. California Acad. Sci.* **8,** 1–55.

Krebs, B. (1976). Pseudosuchia. In *Handbuch der Paläoherpetologie, Teil 13: Thecodontia* (O. Kuhn, Ed.), pp. 40–98. Fischer-Verlag, Stuttgart.

Padian, K., and May, C. L. (1993). The earliest dinosaurs. *New Mexico Museum Nat. History Sci. Bull.* **3,** 379–381.

Parrish, J. M. (1986). Locomotor adaptations in the hindlimb and pelvis of the Thecodontia. *Hunteria* **1,** 1–35.

Parrish, J. M. (1987). The origin of crocodilian locomotion. *Paleobiology* **13,** 396–414.

Parrish, J. M. (1993). Phylogeny of the Crocodylotarsi and a consideration of archosaurian and crurotarsan monophyly. *J. Vertebr. Paleontol.* **13,** 287–308.

Sereno, P. C. (1991). Basal archosaurs: Phylogenetic relationships and functional implications. *J. Vertebr. Paleontol.* **11**(Suppl. 4), 1–53.

Sereno, P. C., and Arcucci, A. B. (1990). The monophyly of crurotarsal archosaurs and the origin of bird and crocodile ankle joints. *Neues Jahr. Geol. Paläontol. Abhandlungen* **180,** 21–52.

Zittel, K. A. von (1887–1890). *Handbuch der Paläontologie, Part 1: Paläozoologie, 3.* Oldenbourg, Munich.

Psittacosauridae

PAUL SERENO
University of Chicago
Chicago, Illinois, USA

Psittacosaurids, or parrot-beaked dinosaurs, comprise a small group of herbivorous ornithischians, known only from Lower Cretaceous sediments of Asia. Not exceeding 2 m in length, the psittacosaurid postcranium is remarkably primitive compared to other ornithischians, with hindlimb proportions of a facultative biped. The skull, in contrast, is highly modified, with a tall, parrot-like snout, which is proportionately shorter than the snout of any other ornithischian. Psittacosaurids are the earliest ceratopsians, a group that would flourish in the Late Cretaceous in Asia and western North America.

Discovery

Early Work Despite a rich fossil record, psittacosaurids have received scant attention in the paleontologic literature (Sereno 1990, 1997). Psittacosaurid remains were first uncovered in the Mongolian People's Republic in 1922 by the Third Asiatic Expedition of the AMERICAN MUSEUM OF NATURAL HISTORY. Based on two nearly complete skeletons, Osborn (1923) erected two taxa, *Psittacosaurus mongoliensis* and *Protiguanodon mongoliense*. These are now regarded as pertaining to the single species *Psittacosaurus mongoliensis.* Some fine juvenile material collected by the American Museum expeditions remained unstudied until recently (Coombs, 1982) and documents the body proportions of hatchlings. Several additional species of *Psittacosaurus* described since then from the same regions are now also referred to *P. mongoliensis,* the most common and widespread of psittacosaurids.

Well-preserved specimens of a smaller species were discovered in China during the 1950s, described as *Psittacosaurus sinensis* and *Psittacosaurus youngi.* Now referred to the single species *P. sinensis,* this psittacosaur is also quite widespread geographically within northern China.

Recent Work Four additional species of psittacosaurids have been named in recent years from fossils found in northern and western China. Two of these, *P. meileyingensis* and *P. xinjiangensis*, are quite distinct (Sereno and Zhao, 1988; Sereno *et al.*, 1988). The other two, *P. neimongoliensis* and *P. ordosensis* (Russell and Zhao, 1996), are very close in form to *P. mongoliensis* and *P. sinensis,* respectively. Finally, as yet unpublished psittacosaur remains have been found recently in lake deposits in northern China from levels below those that have yielded all previous psittacosaurid fossils. This earliest Cretaceous psittacosaurid is a small-bodied species that retains two small subconical premaxillary teeth. In most respects, it looks quite similar to other species of *Psittacosaurus* and may eventually be recognized as the seventh species of this genus.

Psittacosaurid Features

The preorbital portion of the skull in psittacosaurids is less than 40% of total skull length, shorter than in any other ornithischians. Other diagnostic psittacosaurid features include the elevated position of the external naris on the snout. The nasal bone extends as narrow processes ventral to the external nares, a very unusual configuration. The broad lateral surface of the snout is formed by the premaxilla, another unusual configuration that separates the maxilla and external naris by a wide margin. The sutural pattern on the side of the snout is very characteristic, with the premaxilla, maxilla, lacrimal, and jugal sutures converging to a point. In addition, the snout wall is solid, with no development of an antorbital fenestra or fossa (depression).

The postcranial skeleton is generally very similar to that of a primitive ornithopod, such as HYPSILOPHODON. The most unusual and diagnostic features of psittacosaurids involve the manus. Psittacosaurids have a very asymmetrical manus, with only three functional digits. Most other ornithischians retain all five manual digits. In psittacosaurids, the manual unguals are expanded and slightly hoof-shaped.

Evolution

Psittacosaurids are primitive CERATOPSIANS. The front of the snout is formed by the median rostral bone, found only in ceratopsians. This bone forms a cap over the snout and supports most of the upper horny bill. Psittacosaurids share a number of additional derived characters (synapomorphies) with other more advanced ceratopsians (neoceratopsians), all of which are located in the skull. The shape of the ceratopsian skull is unusual and also characterizes psittacosaurids. In dorsal view, the jugal projects laterally well beyond the external margin of the orbit. This lateral jugal projection is developed as a pyramidal horn in several species, most prominently in *P. sinensis.*

Other cranial features linking psittacosaurids with other ceratopsians include the posterior shelf that overhangs the occiput. Made principally of the parietal bone in ceratopsians, this shelf is later elaborated as a posterodorsally angling shield, or frill, in ceratopsids. The ventral process of the ceratopsian predentary, the bone supporting the lower bill, is broad and expands in transverse width toward its proximal end. The broad proportions of this process also occurs in other ceratopsians, which firmly brace the anterior junction of the lower jaws.

Distribution and Temporal Range

Psittacosaurids are currently known from central Asia, broadly defined to include eastern and western China, Mongolia, and southern Siberia. A dentary from Thailand has been referred to *Psittacosaurus* (Buffetaut and Suteethorn, 1992), but better material is needed to reliably confirm its presence there. Similarly, supposed psittacosaur material from Japan (Manabe and Hasegawa, 1991) awaits the discovery of more complete specimens.

Psittacosaurids are known mainly from beds believed to be mid-Cretaceous in age (Aptian–Albian). This age is supported by associated invertebrate, freshwater fish (*Lycoptera*), turtles (*Hangiemys* and *Mongolemys*), dinosaurs (*Iguanodon* and *Shamosaurus*), and mammals (*Prokennalestes* and *Prozalambdalestes*). The new undescribed primitive psittacosaurid with premaxillary teeth from northern China (Liaoning) is clearly older than other species because it is derived from a formation beneath that that yielded *P. mongoliensis* and *P. meileyingensis* in the same region. Although primitive avian fossils (*Confuciusornis*) found near the new primitive psittacosaurid were claimed to be latest Jurassic in age, it is more likely that the basal psittacosaurid is earliest Cretaceous (Barriasian–Valanginian) in age, or about 140 million years old. If so, psittacosaurids would have a temporal range of at least 30 million years, and over this lengthy interval exhibit remarkably little change.

Functional Morphology

The psittacosaurid dentition has self-sharpening cutting edges and is suited for mastication of plant material. Even the crowns of small hatchlings are truncated by wear. Polished GASTROLITHS, often exceeding 50 in number, are associated with several psittacosaurid skeletons and were used in the breakdown of plant materials. Thus, psittacosaurids maintained the primitive ornithischian pattern of tooth-to-tooth cropping and gastrolith grinding in the breakdown of plant matter. Later neoceratopsians would abandon

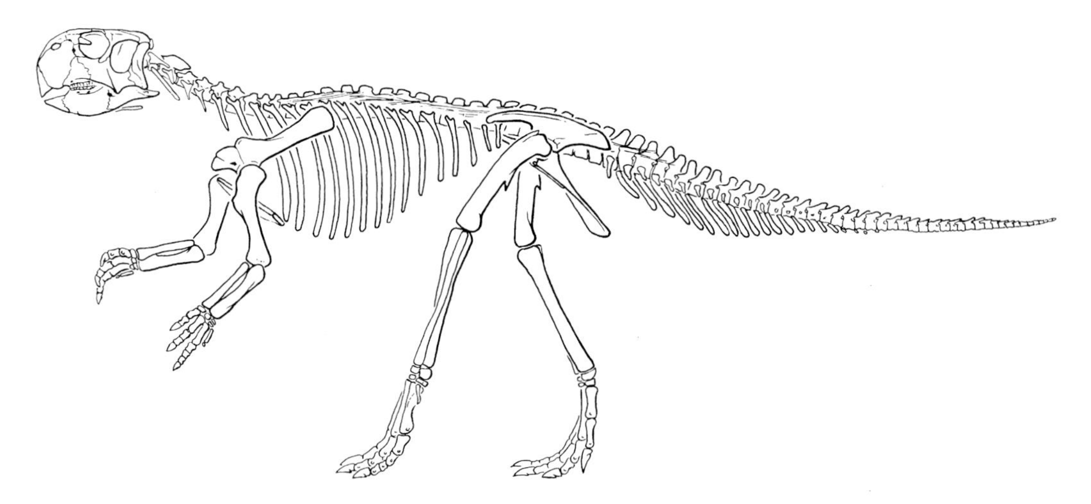

FIGURE 1 Skeletal reconstruction of an adult *Psittacosaurus mongoliensis*, a psittacosaurid that grew from 5-in. long hatchlings to a length of approximately 6 ft when mature. *Psittacosaurus mongoliensis* is the best known and most common psittacosaurid, represented by approximately 100 partial and complete skeletons.

gastroliths and rely entirely on enlarged tooth batteries.

If limb length is estimated by summing the lengths of respective fore- and hindlimb bones, the length of the forelimb in *P. mongoliensis* is approximately 58% that of the hindlimb, somewhat longer than that in *Hypsilophodon* (51%). Proportions within the hindlimb in *Psittacosaurus* are comparable to those in *Hypsilophodon*. Given the proportions of the fore- and hindlimbs, it is reasonable to suppose that psittacosaurs were facultatively bipedal and would operate normally on two legs. The somewhat lengthened forelimbs and more hoof-shaped unguals on the manus, however, indicate that psittacosaurids did use their forelimbs for locomotion, perhaps at the slowest speeds.

See also the following related entries:

BIPEDALITY • CERATOPSIA • NEOCERATOPSIA

References

Buffetaut, E., and Suteethorn, V. (1992). A new species of the ornithischian dinosaur *Psittacosaurus* from the Early Cretaceous of Thailand. *Palaeontology* **35,** 801–812.

Coombs, W. P. (1982). Juvenile specimens of the ornithischian dinosaur *Psittacosaurus*. *Palaeontology* **25,** 89–107.

Manabe, M., and Hasegawa, Y. (1991). The Cretaceous dinosaur fauna of Japan. *Paleontol. Contrib. Univ. Oslo* **364,** 41–42.

Osborn, H. F. (1923). Two Lower Cretaceous dinosaurs from Mongolia. *Am. Museum Novitates* **95,** 1–10.

Russell, D. A., and Zhao, Z.-J. (1996). New psittacosaur occurrences in Inner Mongolia. *Can. J. Earth Sci.* **33,** 637–648.

Sereno, P. C. (1990). Psittacosauridae. In *The Dinosauria* (D. B. Weishampel, P. Dodson, and H. Osmólska, Eds.), pp. 579–592. Univ. of California Press, Berkeley.

Sereno, P. C. (1997). Systematics, evolution and polar wanderings of margin-headed dinosaurs (Ornithischia: Marginocephalia). In *The Age of Dinosaurs in Russia and Mongolia* (M. Benton, E. Kurochkin, M. Shishkin, and D. Unwin, Eds.), in press. Cambridge Univ. Press, Cambridge, UK.

Sereno, P. C., and Zhao, Z.-J. (1988). *Psittacosaurus xinjiangensis* (Ornithischia: Ceratopsia), a new psittacosaur from the Lower Cretaceous of northwestern China. *J. Vertebr. Paleontol.* **8,** 353–365.

Pterosauria

KEVIN PADIAN
University of California
Berkeley, California, USA

First known as early as the 1780s from the Late Jurassic SOLNHOFEN limestones of Bavaria, pterosaurs, the "winged reptiles" of the MESOZOIC ERA, have been regarded as biological oxymorons ever since. That they flew, using wings of skin stretched from the body and supported by the forelimb and a tremendously elongated outer finger, is generally accepted, but nearly every other idea about their paleobiology has been contested at one time or another. Currently, their known stratigraphic range is from the latest Triassic (Norian) through the latest Cretaceous (Maastrichtian), and they have now been found on every continent. They are known to range from sparrow-sized to giant forms with wingspans exceeding 10 or 11 m *(Quetzalcoatlus),* and as far as the available record indicates, they seem generally to have occupied many adaptive zones that the birds took over during the Tertiary, after coexisting with them for approximately 85 million years.

Pterosaurs, like birds and bats, used a down-and-forward stroke in flight that creates a ring-shaped vortex wake that provides the forward thrust component of flight (Padian, 1983, Padian and Rayner, 1993). Pterosaurs had expanded, keeled breastbones like those of birds, and the coracoids braced the shoulder girdle to the sternum (Fig. 1). The humerus also had an expanded, proximally concentrated deltopectoral crest for insertion of the flight muscles. Even the earliest known pterosaurs had a fully developed flight apparatus; but the largest pterosaurs, like the largest birds, undoubtedly spent most of their time soaring (Padian, 1987).

Sereno (1991) defined Pterosauria as a list of commonly accepted taxa and all the descendants of their most recent common ancestor; hence it may be regarded as a node-based group (see PTEROSAUROMORPHA). According to cladistic analyses by Padian, Gauthier, and Sereno (see Sereno, 1991), pterosaurs are the closest major sister group to dinosaurs within the ornithodiran archosaurs, and the small, poorly preserved Late Triassic (Carnian) form *Scleromochlus* is their closest known sister group. *Scleromochlus* (see PTEROSAUROMORPHA) has a large skull and long limbs in which the humerus is longer than the scapula and the forearm is longer still. The bowed femur is exceeded in length by the lower leg, in which the fibula is greatly reduced and fused to the tibia. The ankle is mesotarsal and there are four elongated, closely appressed metatarsals, plus an aberrant, somewhat reduced fifth metatarsal. These features are only found otherwise in pterosaurs, but most other bones of the pterosaurian skeleton are so modified for flight that it is difficult to establish many skeletal comparisons to other ornithodirans.

Pterosaurs have been traditionally divided into the Rhamphorhynchoidea (long tailed, with moderately long metacarpals and a long fifth toe of two tapering, curved phalanges) and the Pterodactyloidea (short tailed, with elongated metacarpals and a fifth toe reduced to only a nubbin of the metatarsal); the former group ranged from the Late Triassic to the Late Jurassic and the latter from the Late Jurassic to the Late Cretaceous (Wellnhofer, 1991). The "Rhamphorhynchoidea" has now been generally abandoned as a taxon because it is not monophyletic: The Pterodactyloidea evolved from this general group of basal pterosaurs, and *Rhamphorhynchus* itself is one of its closest known relatives (Figs. 2 and 3).

According to most cladistic analyses, pterosaurs shared a close common ancestor with dinosaurs and their closest relatives, such as *Lagosuchus/Marasuchus, Lagerpeton,* and *Pseudolagosuchus.* All these animals are small and presumably agile, cursorial, and digitigrade, and their hindlimbs are characterized by well-offset femoral heads, bowed femoral shafts, a long

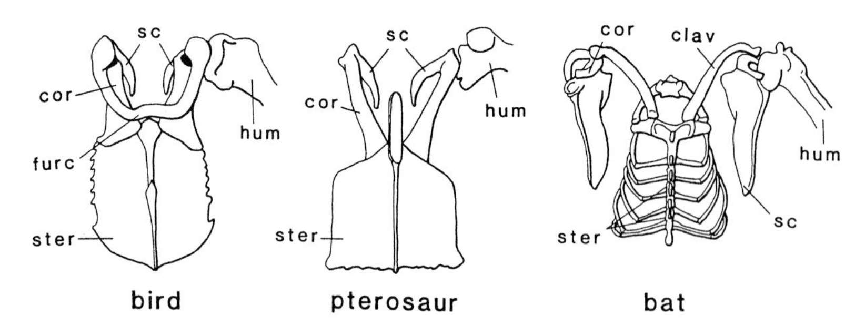

FIGURE 1 Pectoral girdles of a bird, bat, and pterosaur, indicating the primary structures used in flight.

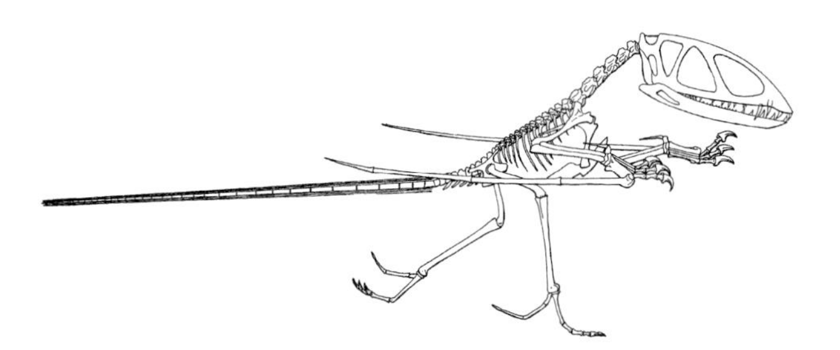

FIGURE 2 The basal pterosaur *Dimorphodon* (up to 1.5 m wingspan), restored as a bipedal cursor.

tibia and reduced fibula, a mesotarsal ankle, and elongated metatarsals. These features have suggested erect stance and parasagittal gait in all these animals; therefore, presumably pterosaurs were similarly abled by both structural–functional analogy and by the homology of common descent (Padian, 1983). However, this view has been frequently questioned by authors who prefer a quadrupedal pose and often a more sprawling gait for many or most pterosaurs (Wellnhofer, 1988, 1991; Unwin, 1989).

The wings of pterosaurs were invested with a system of long, fine, possibly collagenous or keratinous stiffening fibers that radiated through the wing following the structural lines of the bird's feather shafts and the bat's fingers (Fig. 4). They are never found bent or folded and were obviously of aerodynamic importance, forming with the skin membrane and other tissues a composite structure that was both flexible and airworthy (Padian and Rayner, 1993). In other parts of the epidermis, similar "fibers," shorter and more twisted in appearance, have often been reported, as well as other epidermal hair-like structures (Padian and Rayner, 1993, Unwin and Bakhurina, 1994), but these are not like the wing's structural fibers. In the few taxa in which wing outlines have been preserved, different aspect ratios are suggested. Many specimens of *Rhamphorhynchus* have a narrow wing that could scarcely have reached posterior to the caudal region, whereas at least one specimen of *Pterodactylus* seems to have wing membranes attaching to its femora, and more than one specimen of *Sordes* appear to have the wing connected to the ankle, with a broad interfemoral membrane that is free of the tail. The problem is that, once out of the main wing, the characteristic structural fibers are not found, so it is difficult to tell what is integumental fiber, what may have been accessory lift membranes, and what may simply be soft parts of a flattened and decayed Mesozoic "roadkill." It is possible that the extent of the membranous attachment varied in pterosaurs; but the functional morphology of the hindlimb does not vary discernibly in any pterosaur, so it is difficult to see that it would have interfered with bird-like terrestrial locomotion.

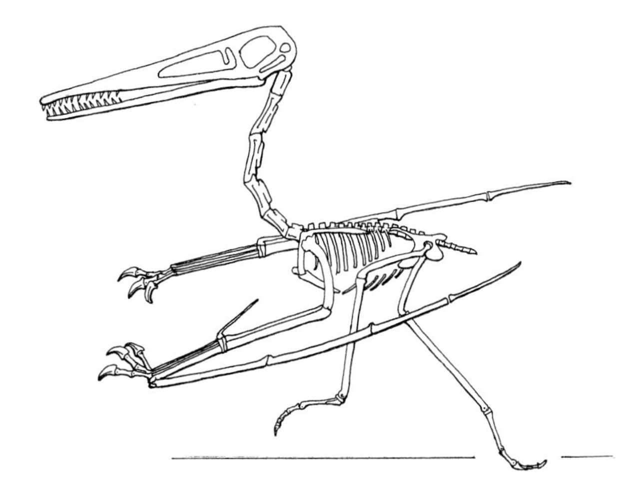

FIGURE 3 The pterodactyloid *Pterodactylus* (approximately 0.5 m wingspan).

Pterosaurs are often pictured as terrestrial quadrupeds, but if their hindlimbs worked like those of birds and other dinosaurs there would seem to have been no need for the forelimbs to be used in walking. However, the prephalangeal elements were so long, especially in the pterodactyloids, that it would have been easy to reach the ground for facultative QUADRUPEDALITY. Quadrupedal footprints have been assigned to pterosaurs, but these have been reassigned to crocodiles on the basis of the functional morphology of the pterosaur limbs (Padian, 1983) and the resemblance of many features of the tracks in question to those of experimentally produced caiman tracks (Padian and Olsen, 1984). However, new tracks frequently surface that are proposed to be those of pterosaurs, and the debates continue (Wellnhofer, 1991; Lockley *et al.*, 1995; Mazin *et al.*, 1995).

Pterosaurs had large skulls that were often elongated, and most appear to have trapped fish, insects, and other small prey. Some skulls were adorned with crests that were probably sexually dimorphic, correlated with adult size and pelvic width (Bennett, 1993). The bones of the skeleton were hollow, and the bone

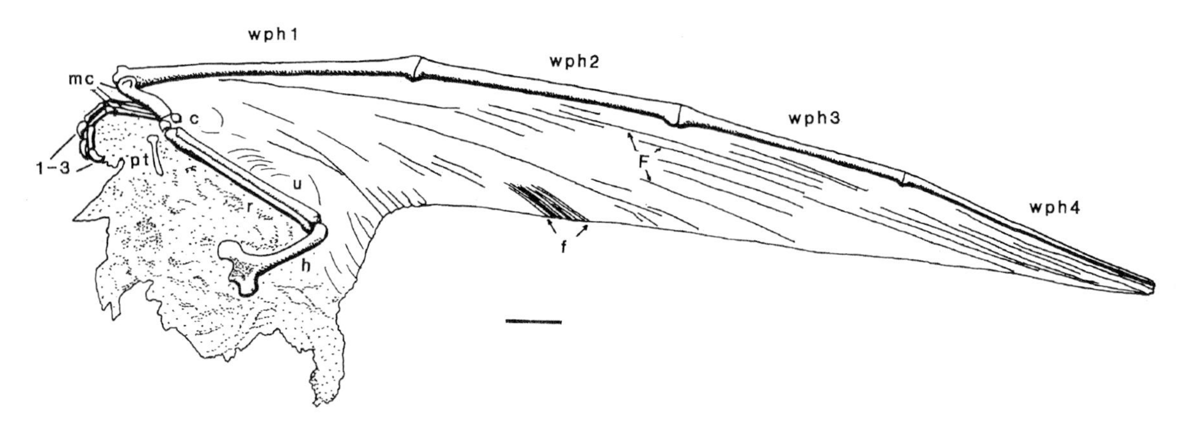

FIGURE 4 Diagrammatic representation of the pterosaur wing, based on a specimen of *Rhamphorhynchus* in the Bavarian State Collections (see Padian and Rayner, 1993).

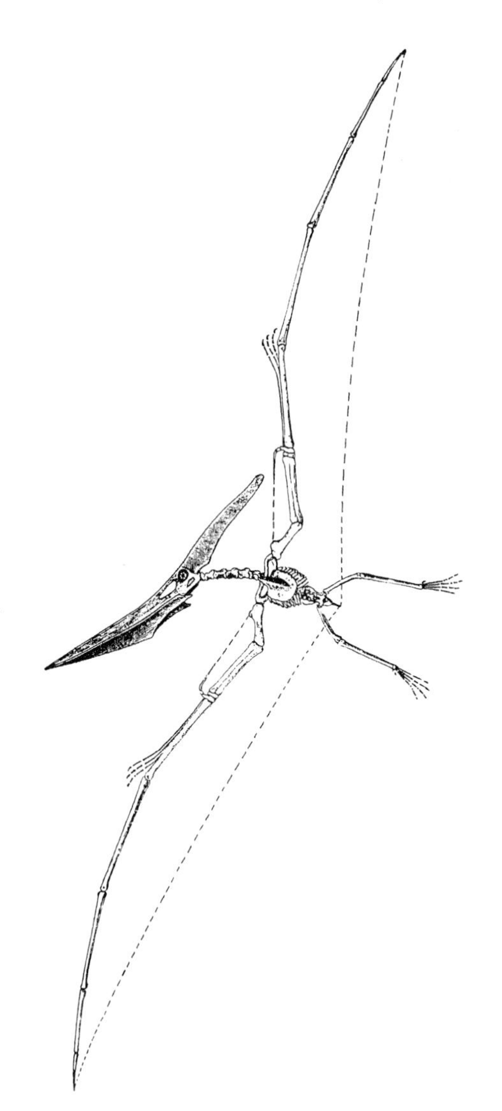

FIGURE 5 *Pteranodon*, a Late Cretaceous pterodactyloid with a wingspan of up to 7 m, as restored by G. F. Eaton in 1910.

walls were thinner than in any other tetrapod. Many of the long bones were pneumatic, expanding the respiratory tissue internally to cool the blood from the exertions of flight. The bones are primarily of fibrolamellar tissue that evidently grew very rapidly, with few or no lines of interrupted growth; there is extensive Haversian remodeling of much of this bone, and the long bones are often internally braced by a network of narrow struts, much as in birds. Their large forelimbs often make the hindlimbs seem small and spindly by comparison, but the hindlimbs are proportioned in accordance with the torso. The head and neck are lightly constructed, and the hollow bones of the forelimb have broad diameters because diameter is the principal variable on which depends the second moment of inertia, which correlates with resistance to the torsion forces encountered in flight. Hence, pterosaurs are something of a front-loaded optical illusion (Fig. 5).

If the indications of their biology are correct, pterosaurs could not have flown as soon as they hatched and must have been cared for while they grew rapidly to fledging size. Some evidence indicates that they lived in large terrestrial colonies (Bell and Padian, 1995). By the end of the Maastrichtian their diversity had apparently dwindled to little over a few species in a subclade, the Azhdarchidae, that included both the giant *Quetzalcoatlus* and smaller forms. The record is too sparse at the species level to provide any indication of catastrophic extinction or rapid decline, although through the Late Cretaceous the other pterodactyloid subclades disappeared from the record one by one (Wellnhofer, 1991). Whatever their ultimate fate, pterosaurs were the first vertebrate fliers, and their geological time span of success is only now being matched by the birds.

See also the following related entries:

ARCHOSAURIA • ORNITHODIRA • ORNITHOSUCHIA

References

Bell, C. M., and Padian, K. (1995). Pterosaur fossils from the Cretaceous of Chile: Evidence for a pterosaur colony on an inland desert plain. *Geol. Magazine* **132,** 31–38.

Bennett, S. C. (1993). The ontogeny of *Pteranodon* and other pterosaurs. *Paleobiology* **19,** 92–106.

Lockley, M. G., Logue, T. J., Moratella, J. J., Hunt, A. P., Schultz, R. J., and Robinson, J. W. (1995). The fossil trackway *Pteraichnus* is pterosaurian, not crocodilian: Implications for the global distribution of pterosaur tracks. *Ichnos* **4,** 7–20.

Mazin, J.-M., Hantzpergue, P., Lafaurie, G., and Vignaud, P. (1995). Des pistes de ptérosaures dans le tithonien de Crayssac (Quercy, France). *Comptes Rendus Acad. Sci. Paris* **321**(IIa), 417–424.

Padian, K. (1983). A functional analysis of flying and walking in pterosaurs. *Paleobiology* **9**(3), 218–239.

Padian, K. (1987). A comparative phylogenetic and functional approach to the origin of vertebrate flight. *In Recent Advances in the Study of Bats* (B. Fenton, P. A. Racey, and J. M. V. Rayner, Eds.), pp. 3–22. Cambridge Univ. Press, Cambridge, UK.

Padian, K., and Olsen, P. E. (1984). The track of *Pteraichnus*: Not pterosaurian, but crocodilian. *J. Paleontol.* **58,** 178–184.

Padian, K., and Rayner, J. M. V. (1993). The wings of pterosaurs. *Am. J. Sci.* **293A,** 91–166.

Sereno, P. C. (1991). Basal archosaurs: Phylogenetic relationships and functional implications. *Soc. Vertebr. Paleontol. Mem.* **2,** 1–53.

Unwin, D. M. (1989). A predictive method for the identification of vertebrate ichnites and its application to pterosaur tracks. In *Dinosaur Tracks and Traces* (D. D. Gillette and M. G. Lockley, Eds.), pp. 259–274. Cambridge Univ. Press, New York.

Unwin, D. M., and Bakhurina, N. N. (1994). *Sordes pilosus* and the nature of the pterosaur flight apparatus. *Nature* **371,** 62–64.

Wellnhofer, P. (1988). Terrestrial locomotion in pterosaurs. *Historical Biol.* **1,** 3–16.

Wellnhofer, P. (1991). *The Illustrated Encyclopedia of Pterosaurs.* Crescent, New York.

Pterosauromorpha

KEVIN PADIAN
University of California
Berkeley, California, USA

Pterosauromorpha may be defined as the node group PTEROSAURIA and all ornithodiran archosaurs closer to them than to dinosaurs. This concept is a stem-defined taxon coordinate to DINOSAUROMORPHA, which is all ornithodirans closer to dinosaurs than to pterosaurs.

Because Sereno (1991) node-defined Pterosauria as a list of inclusive taxa and all the descendants of their most recent common ancestor, the concept of Pterosauromorpha is practically as well as theoretically useful: Several known taxa, as well as potential discoveries, may have some or all of the diagnostic features of known pterosaurs but cannot be considered members of Pterosauria.

For example, an animal with the flight apparatus of known pterosaurian genera, but more basal than these forms, would belong to Pterosauromorpha but not Pterosauria. The same is true for forms that share synapomorphies of pterosaurs but did not fly: The Late Triassic *Scleromochlus*, from the Norian of Scotland (Fig. 1), has been considered as the closest known relative of pterosaurs by von Huene (1914), Padian (1984), Gauthier (1986), Padian *et al.* (1995), and with reservations by Sereno (1991). It shares with them a skull that is nearly the volume of the torso; a forearm longer than humerus, both of which are longer than the scapula; a tibia longer than femur with a splint-like fibula; and four elongated, closely appressed metatarsals with a fifth angled posterolaterally. It is a small form (skull length approximately 3 cm), preserved only as impressions of several skeletons and partial skeletons, in fragments of a medium-coarse red sandstone whose grain size is large enough to be mistaken for some fine morphological details.

Hence, there is unlikely to be consensus about its anatomy and systematic position. The forelimb is considerably shorter than the hindlimb and lightly built, like those of ornithodirans plesiomorphically, so it is thought to have been a good bipedal runner, although hopping and climbing have also been proposed functions. Its locomotory features suggest an origin of pterosaurs from common ornithodiran ancestors with

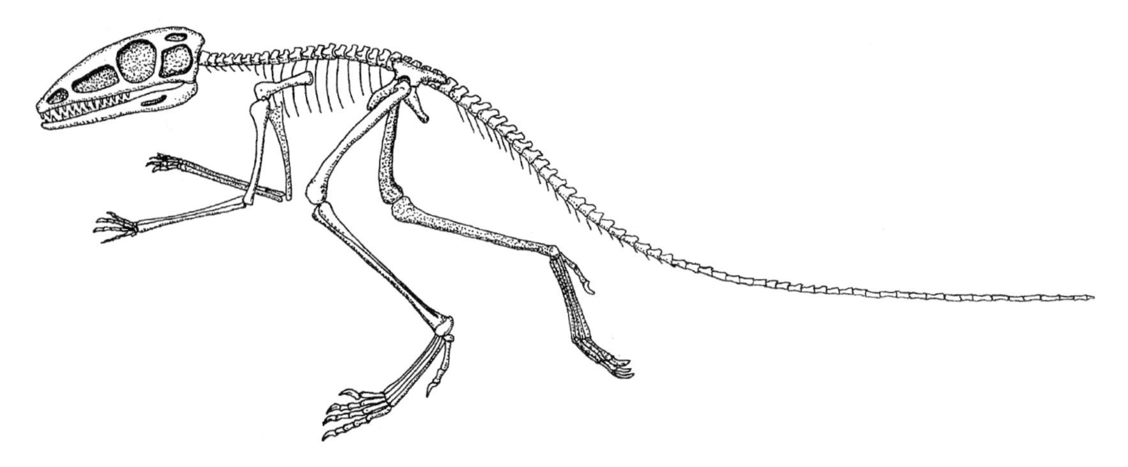

FIGURE 1 *Scleromochlus taylori*, as reconstructed to reflect the view of recent authors who have regarded its hindlimb structures as adapted for digitigrade bipedality.

dinosaurs that were small, terrestrial, bipedal, and cursorial (Padian, 1984, 1991; Sereno, 1991).

References

Gauthier, J. A. (1986). Saurischian monophyly and the origin of birds. *Mem. California Acad. Sci.* **8**, 1–55.

Padian, K. (1984). The origin of pterosaurs. In *Symposium on Mesozoic Terrestrial Ecosystems: Short papers* (W.-E. Reif and F. Westphal, Eds.), pp. 163–168 Third. Attempto, Tubingen, Germany.

Padian, K. (1991). Pterosaurs: Were they functional birds or functional bats? In *Biomechanics and Evolution* (J. M. V. Rayner and R. J. Wootton, Eds.), pp. 145–160. Cambridge Univ. Press, Cambridge, UK.

Padian, K., Gauthier, J. A., and Fraser, N. C. (1995). *Scleromochlus taylori* and the early evolution of Pterosaurs. *J. Vertebr. Paleontol.* **15**(Suppl. to No. 3), 47A.

Sereno, P. C. (1991). Basal archosaurs: Phylogenetic relationships and functional implications. *J. Vertebr. Paleontol.* **11**(Suppl. 4), 1–53.

von Huene, F. (1914). Beiträge zur Geschichte der Archosaurier. *Geol. Paläontol. Abhandlungen N.F.* **13**, 1–53.

Purgatoire

Martin Lockley
University of Colorado at Denver
Denver, Colorado, USA

The Purgatoire Valley dinosaur tracksite, situated in southeastern Colorado, has become one of the most famous in the world. It was first reported in 1935 and qualifies as the first report of sauropod tracks, predating discoveries in 1938–1939 at what is now Dinosaur Valley State Park in Texas. The Purgatoire Valley tracks are found in the famous Morrison Formation and comprise the largest mapped site in North America, consisting of more than 1300 footprints representing the activity of no fewer than 100 animals.

Parallel trackways of subadult sauropods have been illustrated in many publications (e.g., Lockley, 1991; Lockley and Hunt, 1995) as a classic example of evidence of gregarious behavior. The site also reveals evidence that dinosaurs killed clams by trampling them. Located in a relatively remote region on Forest Service land that is designated as the Picketwire Canyonlands, it is open to the public, although access by private motor vehicles is not permitted.

See also the following related entries:

Behavior • Footprints and Trackways • Migration

References

Lockley, M. G. (1991). *Tracking Dinosaurs: A New Look at an Ancient World*, pp. 238. Cambridge Univ. Press, Cambridge, UK.

Lockley, M. G., and Hunt, A. P. (1995). *Dinosaur Tracks and Other Fossil Footprints of the Western United States*, pp. 338. Columbia Univ. Press, New York.

Quadrupedality

KEVIN PADIAN
University of California
Berkeley, California, USA

Quadrupedality, the habit of walking on four legs, is secondary in dinosaurs; their common ancestor and their out-groups, down to the level of ORNITHODIRA, were primitively bipedal (see ARCHOSAURIA; BIPEDALITY). This is known through a combination of phylogenetic and functional analyses. Functional analysis gives us reasonable evidence that animals were bipedal or quadrupedal based on several features including (i) their overall size, (ii) the relative size and proportional differences between their forelimbs and hindlimbs, (iii) the morphology of the forelimb and manus (e.g., the form of the digits and metapodials), (iv) estimations of the body's center of gravity with respect to the pelvic girdle, and (v) articulations of the bones of the vertebral column, girdles, and limbs that can confirm or refute the ability to move easily on four legs (Alexander, 1985). Phylogenetic analysis of the taxa under study arranges them in an evolutionary sequence, and the functional analyses may then be used to study the evolutionary pattern of change in posture, gait, and other features (see Padian, 1987, for a review of methodology).

An example of this method may be seen among the THYREOPHORA. Because their bodies are so squat and heavily armored and their legs are so short, ANKYLOSAURS have almost universally been considered quadrupedal. STEGOSAURS have also, in the main, been considered quadrupedal, partly because they are large and also because their plates and spikes were thought to be of most defensive use in a four-legged posture. However, bipedal RECONSTRUCTIONS of stegosaurs are occasionally encountered (note reconstructions in Czerkas, 1987); they are usually pictured standing on their hindlimbs with their forelimbs leaning against a bush or tree, in the act of feeding. *Scelidosaurus*, a smaller (approximately 3 m total length) Early Jurassic form, has often been considered a basal ankylosaurian; it has usually been considered quadrupedal because its limb bones are relatively stocky and the forelimb–hindlimb disparity is not large; however, its tibia is longer than its femur, which is often taken for an indication of cursoriality in a bipedal mode (Coombs, 1978). The final taxon to be considered here is *Scutellosaurus*, a small armored form from the Early Jurassic (Colbert, 1981) with short forelimbs and long hindlimbs, initially identified as a FABROSAURID (basal ornithischian).

Phylogenetic analysis (Fig. 1) has shown that ankylosaurs and stegosaurs are sister taxa, with *Scelidosaurus* and *Scutellosaurus* as successively more remote out-groups to these (Sereno, 1986). Hence, *Scutellosaurus* is not a fabrosaurid and *Scelidosaurus* is not a basal ankylosaur; these animals are all thyreophorans (see also PHYLOGENY OF DINOSAURS). Using what is understood of the posture and gait of each taxon, in the context of this phylogenetic sequence it can be inferred that the basal thyreophorans were bipedal (*Scutellosaurus*) and then perhaps quadrupedal or facultatively bipedal (*Scelidosaurus* and some basal stegosaurs) to fully quadrupedal (ankylosaurs and most stegosaurs) (Fig. 2).

Using similar methods of analysis, testing, and inference, it has been determined that quadrupedality evolved in dinosaurs at least four separate times. Within ORNITHISCHIA, as noted previously, Thyreophora adopted quadrupedality from their original bipedal habit. CERATOPSIANS did the same: The basal forms (PSITTACOSAURIDAE) were bipedal, "protoceratopsids" were perhaps facultatively bipedal but are mostly pictured as quadrupedal (e.g., Tereshchenko, 1996); and ceratopsids were clearly quadrupedal. Among ORNITHOPODS, HADROSAURS are often considered to have been at least facultatively quadrupedal,

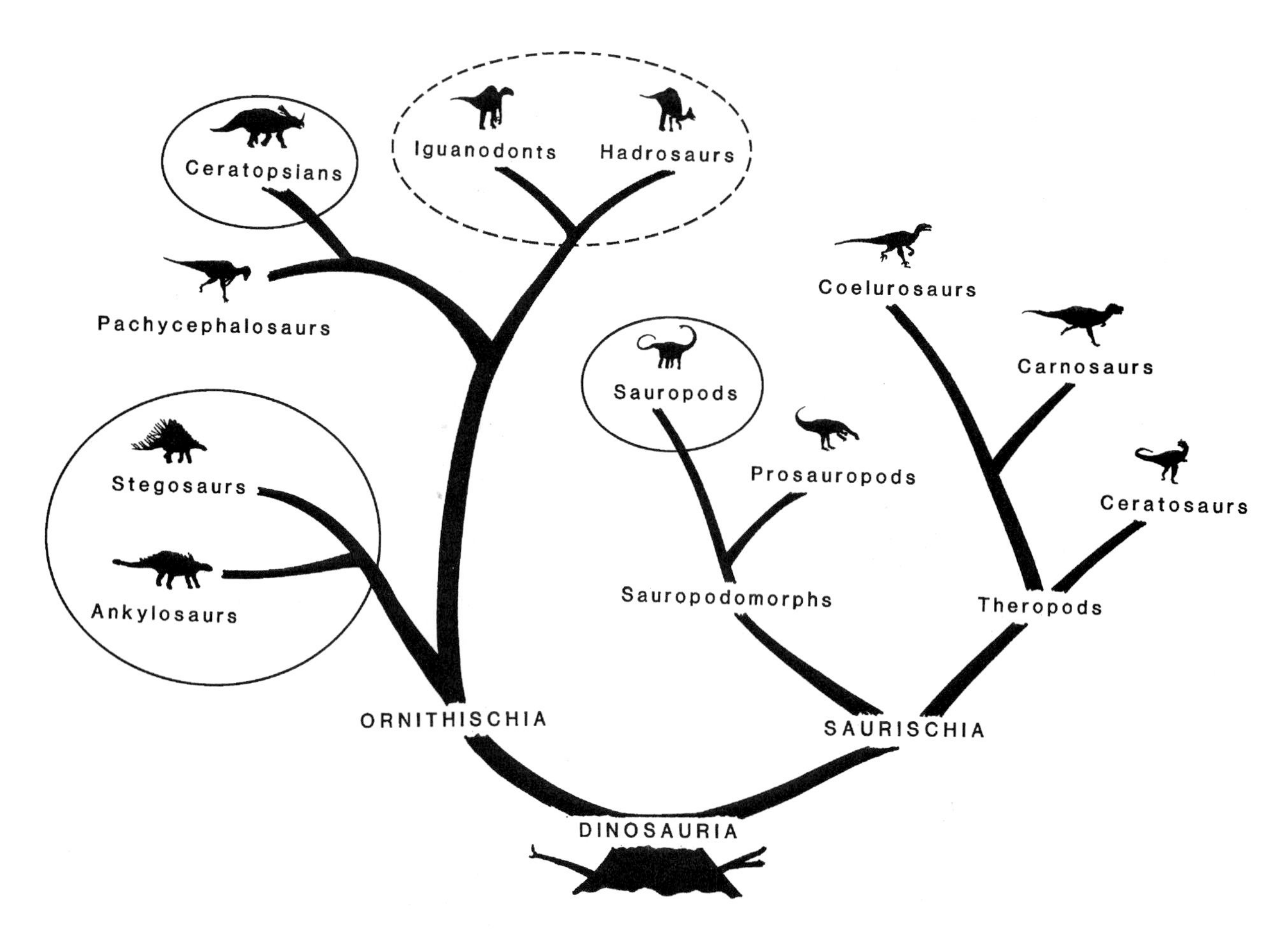

FIGURE 1 Phylogeny of Dinosauria, indicating independently derived quadrupedal (circles) and facultatively quadrupedal (broken circles) lineages.

and sometimes entirely so, because of their great size, the horizontal disposition of the backbone, and the flattened, spatulate shape of the unguals, not only on the three main pedal digits but also on digits II and III of the manus (IV is short and button shaped, in contrast to the claw-like shape of these unguals in the bipedal HYPSILOPHODONTIDS). IGUANODONTS also share these features and have been pictured in both bipedal and quadrupedal modes. However, bipedality has also been vigorously argued (Galton, 1970; Maryanska and Osmólska, 1983; Weishampel, 1995).

Within Saurischia, the view adopted by most contemporary authors is that SAUROPODA evolved from a paraphyletic PROSAUROPODA (e.g., Gauthier, 1986), although several authors (e.g., Sereno, Galton, Gauffre, and Upchurch) have supported the monophyly of Prosauropoda (see also PHYLOGENY OF DINOSAURS). All known sauropods are quadrupedal because of their ponderous size, and it appears that the more highly derived prosauropods (e.g., melanorosaurids and blikanosaurids) were mainly or fully quadrupedal, whereas the smaller, more basal forms may have habitually shifted between two and four legs. If the former phylogenetic view is taken, then it is likely that quadrupedality evolved only once in SAUROPODOMORPHA; if the latter view is taken, then quadrupedality was independently acquired in sauropods and in derived prosauropods.

Quadrupedal dinosaurs, of course, retained the erect posture and parasagittal gait of their bipedal ancestors; this can be established on the basis of the well-offset femoral heads and the persistence of the mesotarsal ankle joint, among other features (see BIPEDALITY); even the largest sauropods appear to have walked on their toes, which were supported by a hoof-like horny pad much as elephants have, judg-

FIGURE 2 Quadrupeds and bipeds have evolved within different dinosaur lineages. Pictured above are reconstructions of both locomotory types on display at the Carnegie Museum of Natural History. Photo by François Gohier.

ing from their footprints (see FOOTPRINTS AND TRACKWAYS). In varied dinosaurian taxa, the habit of quadrupedality or bipedality may have varied depending on the size of the species, the age of the individual, or the particular feeding or locomotory activity of the moment; these categories are not dichotomous in an evolutionary or functional sense.

The shift in basal ornithodirans from a quadrupedal to a bipedal stance involved a reorganization of posture and gait from sprawling to erect and parasagittal (see ARCHOSAURIA; BIPEDALITY). Another evolutionary problem, only partly explored, is what happened to the forelimb and shoulder girdle in the course of returning to a quadrupedal stance in the dinosaurian lineages that adopted it. Most discussion to date has concerned the CERATOPSIA. Reconstructions of these animals since the late 1800s have alternately depicted the forelimbs as sprawling or tucked in mammalian fashion parallel to the body, elbows flexed backward. Proponents of both views have maintained that the alternative position is unlikely or impossible. Footprints assigned to ceratopsians from the Late Cretaceous appear to support the former idea, but many questions remain. Some workers have even maintained that the shoulder girdle was mobile enough to facilitate galloping (e.g., Bakker, 1987), but this proposal has largely been met with skepticism (see review of these issues in Dodson, 1996). Little has been done to study the transition to quadrupedality in thyreophorans or in large ornithopods. As for the Sauropodomorpha, the first sauropods already have columnar limbs similar to the hindlimbs, and the manus appears to have adapted well to the habit, keeping the metapodials off the ground and moving in a parasagittal fashion (see FOOTPRINTS AND TRACKWAYS). If "prosauropods" are viewed as transitional to sauropods, details of the functional transition must be sought among the phylogenetic nodes of basal sauropodomorphs. If Prosauropoda is monophyletic, then the anchisaurid–plateosaurid–melanorosaurid sequence can be studied in its own right, and the origin of quadrupedality in the sudden appearance of Sauropoda in the Early Jurassic must remain mysterious.

See also the following related entries:

BIPEDALITY • FORELIMBS AND HANDS • HINDLIMBS AND FEET • PECTORAL GIRDLE • PELVIS, COMPARATIVE ANATOMY

References

Alexander, R. McN. (1985). Mechanics of posture and gait of some large dinosaurs. *Zool. J. Linnean Soc.* **83,** 1–25.

Bakker, R. T. (1987). The return of the dancing dinosaurs. In *Dinosaurs Past and Present, Vol. I* (E. C. Olson and S. M. Czerkas, Eds.), pp. 38–69. Univ. of Washington Press, Seattle.

Colbert, E. H. (1981). A primitive ornithischian dinosaur from the Kayenta Formation of Arizona. *Museum Northern Arizona Bull.* **53,** 1–61.

Coombs, W. P. (1978). Theoretical aspects of cursorial adaptations in dinosaurs. *Q. Rev. Biol.* **53,** 393–418.

Czerkas, S. A. (1987). A reevaluation of the plate arrangement on *Stegosaurus stenops.* In *Dinosaurs Past and Present, Vol. II* (E. C. Olson and S. M. Czerkas, Eds.), pp. 83–99. Univ. of Washington Press, Seattle.

Dodson, P. (1996). *The Horned Dinosaurs: A Natural History.* Princeton Univ. Press, Princeton, NJ.

Galton, P. M. (1970). The posture of hadrosaurian dinosaurs. *J. Paleontol.* **44,** 464–473.

Gauthier, J. A. (1986). Saurischian monophyly and the origin of birds. *Mem. California Acad. Sci.* **8,** 1–55.

Maryańska, T., and Osmólska, H. (1983). Some implications of hadrosaurian postcranial anatomy. *Acta Palaeontol. Polonica* **28,** 205–207.

Padian, K. (1987). A comparative phylogenetic and functional approach to the origin of vertebrate flight. In *Recent Advances in the Study of Bats* (B. Fenton, P. A. Racey, and J. M. V. Rayner, Eds.), pp. 3–22. Cambridge Univ. Press, Cambridge, UK.

Sereno, P. C. (1986). Phylogeny of the bird-hipped dinosaurs (Order Ornithischia). *Natl. Geogr. Res.* **2,** 234–256.

Tereshchenko, V. S. (1996). A reconstruction of the locomotion of *Protoceratops. Paleontol. J.* **30,** 232–245.

Weishampel, D. B. (1995). Fossils, function, and phylogeny. In *Functional Morphology in Vertebrate Paleontology* (J. J. Thomason, ed.), pp. 34–54. Cambridge Univ. Press, New York.

Queensland Museum, Australia

see MUSEUMS AND DISPLAYS

Radiometric Dating

D. A. EBERTH
Royal Tyrrell Museum of Palaeontology
Drumheller, Alberta, Canada

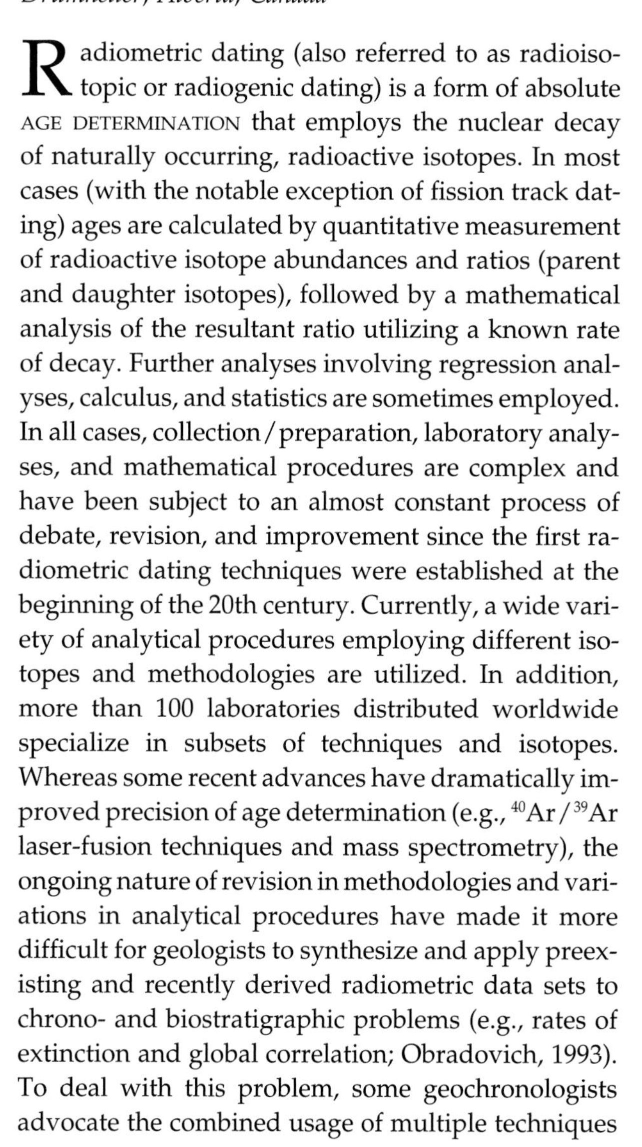

Radiometric dating (also referred to as radioisotopic or radiogenic dating) is a form of absolute AGE DETERMINATION that employs the nuclear decay of naturally occurring, radioactive isotopes. In most cases (with the notable exception of fission track dating) ages are calculated by quantitative measurement of radioactive isotope abundances and ratios (parent and daughter isotopes), followed by a mathematical analysis of the resultant ratio utilizing a known rate of decay. Further analyses involving regression analyses, calculus, and statistics are sometimes employed. In all cases, collection/preparation, laboratory analyses, and mathematical procedures are complex and have been subject to an almost constant process of debate, revision, and improvement since the first radiometric dating techniques were established at the beginning of the 20th century. Currently, a wide variety of analytical procedures employing different isotopes and methodologies are utilized. In addition, more than 100 laboratories distributed worldwide specialize in subsets of techniques and isotopes. Whereas some recent advances have dramatically improved precision of age determination (e.g., $^{40}Ar/^{39}Ar$ laser-fusion techniques and mass spectrometry), the ongoing nature of revision in methodologies and variations in analytical procedures have made it more difficult for geologists to synthesize and apply preexisting and recently derived radiometric data sets to chrono- and biostratigraphic problems (e.g., rates of extinction and global correlation; Obradovich, 1993). To deal with this problem, some geochronologists advocate the combined usage of multiple techniques in an effort to establish concordance among analyses and ensure their long-term validity (e.g., Baadsgaard *et al.*, 1988), whereas others reanalyze previously collected and dated samples using the newer methodologies (e.g., Obradovich, 1993). Unfortunately, although these approaches clearly yield more reliable results, they are neither economically feasible nor time efficient for the average earth scientist. Thus, although the precision of modern analyses has increased, there remains widespread difficulty in integrating old and new data sets to confidently determine and assess "big-picture" chronostratigraphic patterns.

Dinosaur-bearing successions typically comprise alluvial clastics. Age determination of the fossils in these successions is typically achieved by radiometric dating of associated pyroclastics (e.g., Thomas *et al.*, 1990), other igneous rocks such as basalts (e.g., Eberth *et al.*, 1993), and in rare cases authigenic clays such as glauconite. In this way a bracketed age range, or minimum/maximum ages, are ascribed to the fossils that fall above and below these igneous rocks. Where datable minerals and rocks are not present, other forms of stratigraphic correlation (e.g., biostratigraphy, event stratigraphy, and sequence stratigraphy) are used to link the fossiliferous beds to a datable succession (e.g., Eberth *et al.*, 1990). Each radiometric dating technique works best on certain minerals and over a set age range. In this context, the techniques that have been most reliably applied to dating dinosaur-bearing strata include K-Ar, $^{40}Ar/^{39}Ar$, Rb-S, and U-Pb. These techniques are described below and some published applications are cited.

K-Ar Dating

Until the onset of the 1990s K-Ar dating was the mostly widely employed methodology for dating sedimentary successions. Thus, in general, the majority of references to radiometrically dated sedimentary successions, biozone and sequence frameworks, as well as timescales that predate the mid-1990s are drawn largely from K-Ar databases (e.g., Obradovich and Cobban, 1975; Haq *et al.*, 1988; Harland *et al.*, 1989). Beginning in the late 1980s advances in $^{40}Ar/^{39}Ar$ techniques (see below) resulted in ages of much greater precision than the standard K-Ar method, and K-Ar ages and collated K-Ar-based timescales began to be superseded by those derived from $^{40}Ar/^{39}Ar$ analyses (cf., Eberth *et al.*, 1992; Thomas *et al.*, 1990). Ironically, standard K-Ar dating (as well as other standard techniques) is still necessary to calibrate $^{40}Ar/^{39}Ar$ dating, and thus it remains an important component of modern age determinations. K-Ar dating covers an age range greater than 3 Ma and employs naturally occurring radioactive ^{40}K and its daughter product ^{40}Ar in potassium-bearing minerals. Potassium is abundant throughout the lithosphere and the technique is thus applicable to a wide variety of geological environments. Datable potassium-bearing rocks that are most commonly found in stratigraphic association with fossil-bearing successions include pyroclastics, basalts, and glauconites. Precision, when measured as 2 standard deviations (95%), is typically 2–4%.

$^{40}Ar/^{39}Ar$

Beginning in the late 1980s the $^{40}Ar/^{39}Ar$ method has rapidly increased in popularity due to its greater precision and has now supplanted K-Ar dating as the standard radiometric dating tool (see above). It is now widely used in studies involving dinosaurs and Mesozoic events and BIOSTRATIGRAPHY (e.g., Goodwin and Deino, 1989; Thomas *et al.*, 1990; Eberth and Deino, 1992; Rogers *et al.*, 1993; Swisher *et al.*, 1993). $^{40}Ar/^{39}Ar$ dating is a derivative of K-Ar dating and is useful over the same range of geologic time (greater than 3 Ma) and for most of the same materials. The greater precision of the $^{40}Ar/^{39}Ar$ method derives from the fact that samples are analyzed only for isotope ratios rather than ratios and isotope abundances. Because separate analyses are not conducted to determine K and Ar isotope abundances (as is the case for standard K-Ar dating), analytical error is reduced. However, the process is more involved in necessitating that samples be irradiated alongside a standard of known age (determined by K-Ar or other standard dating techniques). Isotopic ratios are determined using either laser-fusion or step-heating techniques, the former requiring only very small samples and in some cases only single mineral grains. The fact that, in some cases, only small samples are needed implies that precise ages can also be derived from core samples (Eberth and Deino, 1992). In general, then, this method is a combination of relative and absolute dating, is costlier, although more precise than standard K-Ar dating, and involves considerably more time (for neutron irradiation and cooldown) than standard K-Ar dating.

In the Western Interior Basin, the technique has been applied to feldspars (sanidines and plagioclase) and biotites derived from fresh and altered pyroclastics (bentonites and tonsteins being the most common) as well as whole rock basalts interbedded or associated with clastic strata. Whereas standard K-Ar ages are typically reported from the Mesozoic with uncertainties (2 standard deviations) of ±2–4% or greater, ages derived from $^{40}Ar/^{39}Ar$ analyses are commonly reported with uncertainties of less than ±0.5%. Furthermore, statistical treatment of ages derived from replicate analyses (from small volume, single samples) allows additional refinement of the ages [e.g., Swisher *et al.* (1993) report standard errors of approximately 0.1% for samples collected at the Cretaceous–Paleogene boundary in Montana]. In theory, such levels of reputed precision make the $^{40}Ar/^{39}Ar$ method well suited to help determine rates of change in Mesozoic physical and ecological systems and to test aspects of evolution/extinction hypotheses. $^{40}Ar/^{39}Ar$ dating has been applied to the question of the global correlation of the Cretaceous–Paleogene boundary, as a means of calibrating biostratigraphic and sequence stratigraphic frameworks (e.g., Obradovich, 1993), and for hundreds of site-specific studies.

Rb-Sr

Regarded as a time-consuming method (Geyh and Schleicher, 1990), Rb-Sr covers an age range greater than 10 Ma and is used to date both whole rock and minerals. Although used predominantly in dating igneous rocks, limited success has been achieved in applying it to authigenic clay minerals. Its application to sedimentary rock successions and associated fossiliferous deposits is typically limited to phenocrysts in fresh pyroclastics and their altered products (bentonites). The method typically involves the calculation of an isochron, which is in turn derived from analysis of four or more cogenetic samples. The isochron methodology is necessary to correct for the presence of Sr in the nascent rock and/or mineral and its contribution to the amount of decay product in the sample. Precision (2 standard deviations) is typically 1–5%. Baadsgaard *et al.* (1988) have successfully used Rb-Sr in combination with other techniques to date the Cretaceous–Paleogene boundary in the northern portion of the Western Interior Basin. Their published Rb-Sr results from Alberta (63.9 ± 0.6 Ma), Saskatchewan (64.5 ± 0.4 Ma), and Montana (63.7 ± 0.6 Ma) are close to one another, but slightly younger than the $^{40}Ar/^{39}Ar$ age of 65.16 ± 0.04 Ma published for the Cretaceous–Paleogene boundary in Montana by Swisher *et al.* (1993).

U-Pb Dating

U-Pb dating is a standard procedure that employs uranium- and thorium-bearing minerals in igneous and metamorphic rocks and sometimes whole rock. In dinosaur-bearing successions U-Pb dating has almost exclusively been applied to zircons (derived from associated igneous rocks). As in the Rb-Sr dating method, ratios are usually assessed by using isochrons and/or concordia diagrams, thus necessitating several cogenetic fractions for each age determination. It can be applied to U-Th-bearing igneous rocks greater than 1 Ma and its precision can be as good as 1%. Baadsgaard and Lerbekmo (1982) and Baadsgaard *et al.* (1988) successfully employed U-Pb dating to assess the age of the Cretaceous–Tertiary boundary in Montana. Their results were very close to those that they had generated using Rb-Sr dating.

See also the following related entries:

BIOSTRATIGRAPHY • GEOLOGIC TIME • PALEOMAGNETIC CORRELATION

References

Baadsgaard, H., and Lerbekmo, J. F. (1982). Rb-Sr and U-Pb dating of bentonites. *Can. J. Earth Sci.* **20,** 1282–1290.

Baadsgaard, H., Lerbekmo, J. F., and McDougall, I. (1988). A radiometric age for the Cretaceous–Tertiary boundary based upon K-Ar, Rb-Sr, and U-Pb ages of bentonites from Alberta, Saskatchewan, and Montana. *Can. J. Earth Sci.* **25,** 1088–1097.

Eberth, D. A., and Deino, A. L. (1992). Geochronology of the non-marine Judith River Formation of southern Alberta. Abstracts of the Society of Economic Paleontologists and Mineralogists, 1992, Theme Meeting: Mesozoic of the Western Interior, pp. 24–25.

Eberth, D. A., Braman, D. R., and Tokaryk, T. T. (1990). Stratigraphy, sedimentology and vertebrate paleontology of the Judith River Formation (Campanian) near Muddy Lake, west-central Saskatchewan. *Bull. Can. Petroleum Geol.* **38,** 387–406.

Eberth, D. A., Thomas, R. G., and Deino, A. L. (1992). Preliminary K-Ar dates from bentonites in the Judith River and Bearpaw Formations (Upper Cretaceous) of Dinosaur Provincial Park, southern Alberta, Canada. In *Aspects of Nonmarine Cretaceous Geology* (N. J. Mateer and P. J. Chen, Eds.), 296–304. China Ocean Press, Beijing.

Eberth, D. A., Russell, D. A., Braman, D. R., and Deino, A. L. (1993). The age of the dinosaur-bearing sediments at Tebsch, Inner Mongolia, People's Republic of China. *Can. J. Earth Sci.* **30,** 2101–2106.

Geyh, M. A., and Schleicher, H. (1990). *Absolute Age Determination: Physical and Chemical Dating Methods and Their Application*, pp. 503. Springer-Verlag, Berlin/New York.

Goodwin, M. B., and Deino, A. L. (1989). The first radiometric ages from the Judith River Formation (Upper Cretaceous), Hill County, Montana. *Can. J. Earth Sci.* **26,** 1384–1391.

Haq, B. U., Hardenbol, J., and Vail, P. R. (1988). Mesozoic and Cenozoic chronostratigraphy and cycles of sea-level change. In *Sea-Level Changes: An Integrated Approach* (C. K. Wilgus, B. S. Hastings, C. G. St. C. Kendall, H. W. Posamentier, C. A. Ross, and J. C. Van Wagoner, Eds.), Society of Economic Paleontologists and Mineralogists Spec. Publ. No. 42, pp. 71–108.

Harland, W. B., Armstrong, R. L., Cox, A. V., Craig, L. E., Smith, A. G., and Smith, D. G. (1989). *A Geologic Time Scale 1989*, pp. 236. Cambridge Univ. Press, Cambridge, UK.

Obradovich, J. D. (1993). A Cretaceous time scale. In *Evolution of the Western Interior Basin* (W. G. E. Caldwell and E. G. Kauffman, Eds.), Geological Association of Canada Spec. Paper No. 39, pp. 379–396.

Obradovich, J. D., and Cobban, W. A. (1975). A time-scale for the Late Cretaceous of the Western Interior of North America. In *The Cretaceous System in the Western Interior of North America* (W. G. E. Caldwell, Ed.), Geological Association of Canada Spec. Paper No. 11, pp. 31–54.

Rogers, R. R., Swisher, C. C., III, and Horner, J. R. (1993). ^{40}Ar/^{39}Ar age and correlation of the nonmarine Two Medicine Formation (Upper Cretaceous), northwestern Montana, U.S.A. *Can. J. Earth Sci.* **30,** 1066–1075.

Swisher, C. C., III, Dingus, L., and Butler, R. F. (1993). ^{40}Ar/^{39}Ar dating and magnetostratigraphic correlation of the terrestrial Cretaceous–Paleogene boundary and Puercan Mammal Age, Hell Creek–Tullock formations, eastern Montana. *Can. J. Earth Sci.* **30,** 1981–1996.

Thomas, R. G., Eberth, D. A., Deino, A. L., and Robinson, D. (1990). Composition, radioisotopic ages, and potential significance of an altered volcanic ash (bentonite) from the Upper Cretaceous Judith River Formation, Dinosaur Provincial Park, southern Alberta, Canada. *Cretaceous Res.* **11,** 125–162.

Raptors

PHILIP J. CURRIE
Royal Tyrrell Museum of Palaeontology
Drumheller, Alberta, Canada

The term "raptor" is correctly applied to various types of birds of prey, including hawks, eagles, owls, and vultures. However, since the release of the movie *Jurassic Park* in 1993, it has become increasingly common for people to refer informally to small predatory dinosaurs as raptors. This started as an abbreviation of the name *Velociraptor* (Fig. 1), but has been applied to other, related forms of the Dromaeosauridae, or even to any of the small maniraptoran THEROPODS. The word "raptor" is derived from a Latin word that means thief, robber, or predator and has been incorporated into the names of other theropods (i.e., *Oviraptor, Sinraptor*) that would not normally be referred to as raptors.

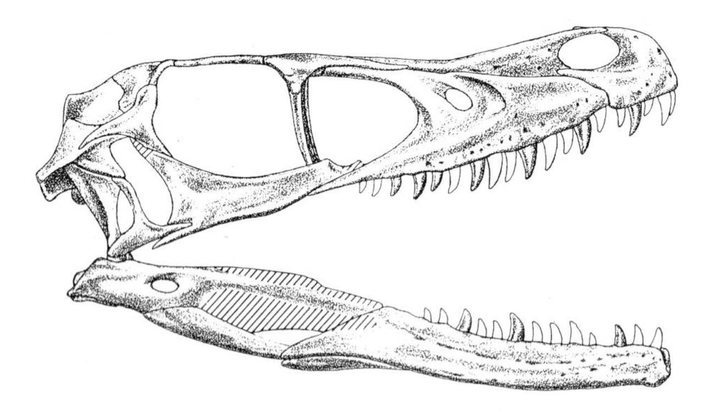

FIGURE 1 *Velociraptor mongoliensis* is one of the "raptors" made famous by *Jurassic Park*. Illustration by Gregory Paul.

See also the following related entry:

DROMAEOSAURIDAE

Reconstruction and Restoration

SYLVIA J. CZERKAS
The Dinosaur Museum
Monticello, Utah, USA

The reconstruction of fossilized skeletal remains is the process of filling in the missing parts of individual bones and/or the assembly of entire and partial skeletons. When reconstructing an individual bone, the bone is sculpted or drawn as close as possible to the original shape and size it would have had in life. In the reconstruction of full or partial skeletons, the individual skeletal elements are assembled into the natural positions that they would have held when the animal was alive. Technical drawings and three-dimensional methods using the actual fossil remains, or cast replicas of the fossils, can be employed in a skeletal reconstruction.

The scientific process of reconstruction involves studying the elements of the skeleton and comparing them with any other existing specimens of the same species and also comparing them with all other similar species. Detailed measurements of the bones are calculated, and then based on comparative anatomy and deduction, all the missing parts are filled in. When a partial or entire skeleton is assembled, the comparative anatomy of both extinct and living vertebrates is considered. The joints are articulated so that they do not extend beyond what is mechanically possible and so that they are held in natural and believable positions (Fig. 1). An important part of presenting lifelike poses for exhibit and study is the incorporation of the possible behavioral aspects and locomotor capabilities of the animal. Consideration

FIGURE 1 Skeletal reconstruction of a nestling *Maiasaura* based on fossil material in the collection of the Museum of the Rockies. Length is 85 cm. Reconstruction by Stephen Czerkas (© 1984).

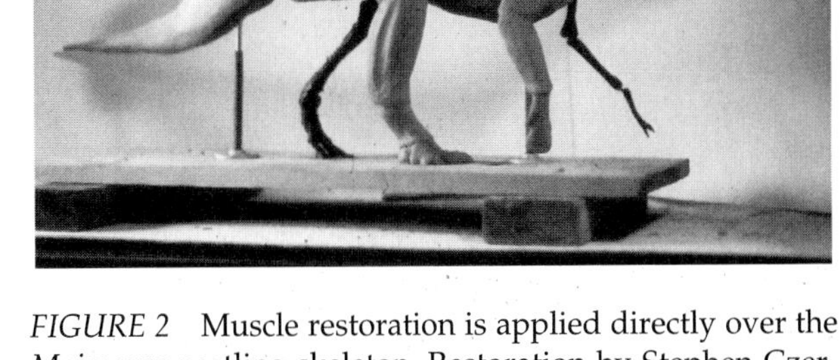

FIGURE 2 Muscle restoration is applied directly over the *Maiasaura* nestling skeleton. Restoration by Stephen Czerkas (© 1984).

of the attitudes and poses of contemporary animals can also be an aid in skeletal reconstructions. This process of skeletal reconstruction is necessary before an accurate recreation of the musculature and then a life restoration can be made.

The restoration of the life appearance of extinct animals consists of putting muscles and then skin onto a skeletal reconstruction (Figs. 2 and 3). This is done in order to visualize what the animal would have looked like when it was alive. It can then be placed into a suitable environment (Fig. 4).

The life restoration can be in the form of either an illustration or a sculpture, but in both cases the starting point is building the muscle structure onto the skeleton. When creating a life illustration, a detailed drawing of the skeleton is needed to start so that the muscles can overlay the skeleton. It is also desirable to have replica study casts of the bones to work from. For life-size sculptures, replicas of the bones should be incorporated into the work whenever possible so that the muscles can then be directly modeled with clay in three dimensions (Figs. 5 and 6). In scale models, the use of both detailed drawings and study casts are valuable for achieving accurate measurements (Figs. 7 and 8). Muscle scars on the bones are studied for attachment points and indications of muscle mass and strength. In combination with comparative anatomy studies of similar animals, this allows the restoration of the musculature of extinct animals.

After the muscles are restored, the skin detail is

FIGURE 3 Skin detail is applied over the muscle restoration of the *Maiasaura* nestling. Restoration by Stephen Czerkas (© 1984).

FIGURE 4 The completed life-size sculpture of the nestling *Maiasaura*. Sculptural restoration by Stephen Czerkas (© 1984).

FIGURE 5 Musculature is applied over replicas of the fossil bones on three life-size *Deinonychus* sculptures, which are 3.3 m in length. Restoration by Stephen Czerkas (© 1989).

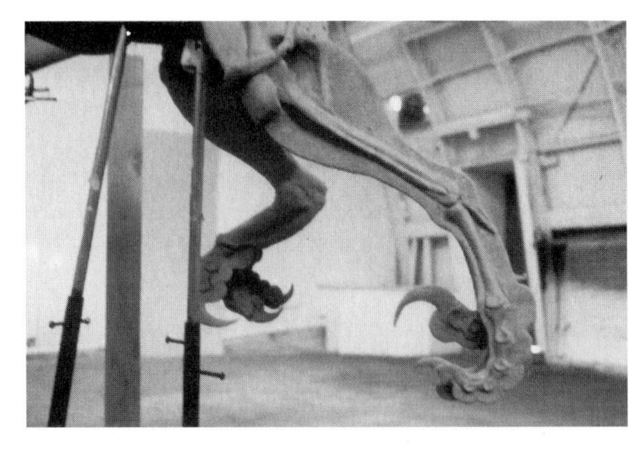

FIGURE 6 *Deinonychus* leg shows how the muscle mass relates to the underlying skeleton. Restoration by Stephen Czerkas (© 1989).

placed over them. Much is known about the SKIN detail of different types of dinosaurs and other extinct animals. Each has its own distinctive patterns that should be incorporated into the life restoration.

The external coloration of the hide of the animal should be based on all the known aspects of the animal's environment and behavior. Color can have a multiplicity of functions from thermal regulation to species recognition and camouflage (Fig. 9). One function of color in REPTILES is the absorption and elimination of heat. Observations of modern reptiles show that many can change hue throughout the day and night as they adjust to heat conditions around them. Another function of color is camouflage for both predatory and prey animals to blend into their environment. The SIZE of the animal should be considered, because generally the larger the animal, the more subtle the coloration. The observation of the living animal kingdom can provide clues to the function of color and can aid in choosing color and patterns for extinct animals. Because color of extinct animals is speculative, overstatements should be avoided so as not to detract from what information is more precisely known.

The study of living animals can provide other information about animals of the past. Form and function are incontrovertibly linked so that living animals can reveal much about the shape and activities of extinct animals. Areas of study involving BIOMECHANICS, locomotion, and the BEHAVIOR and psychol-

FIGURE 7 A detailed drawing of a *Stegosaurus* skeleton. Reconstruction by Stephen Czerkas (© 1986).

FIGURE 8 *Stegosaurus* 1/10-scale model is 66 cm in length. Restoration by Stephen Czerkas (© 1986).

FIGURE 9 *Deinonychus* sculpture shown completed with final coloration and in an environment. Restoration by Stephen Czerkas (© 1989).

ogy of living animals should be incorporated into life restorations of extinct animals.

The complete fossil record should be considered in a restoration for what it can tell about herd or pack behavior and what other types of animals that may have been found associated with it. The paleobotany and geology should be studied for what might be revealed about the environment of the animals.

The purpose of a restoration is to bring the animal to life through the use of scientific evidence and artistry. It is always important to have the scientific evidence dictate the content. Once the animal is given a visual image that is presented to the public, it is difficult to change incorrect information. Artists and their scientific advisers have a responsibility to present as clearly as possible all known aspects of an extinct animal.

See also the following related entries:

BIOMECHANICS • BIOMETRICS • COLOR • FUNCTIONAL MORPHOLOGY • MUSCULATURE • MUSEUMS AND DISPLAYS • SIZE • SIZE AND SCALING • SKIN

Red Fleet Reservoir State Park, Utah, USA

see MUSEUMS AND DISPLAYS

Red Mountain Museum, Alabama, USA

see MUSEUMS AND DISPLAYS

Redpath Museum, McGill University, Canada

see MUSEUMS AND DISPLAYS

Relative Size

see SIZE AND SCALING

Reproduction

see BEHAVIOR; REPRODUCTIVE BEHAVIOR AND RATES

Reproductive Behavior and Rates

Gregory S. Paul
Baltimore, Maryland, USA

Despite the finding of dinosaur eggs in the year *On the Origin of Species* was published, and the famed Gobi Desert finds in the 1920s, little scientific attention was paid to dinosaur reproduction until the discovery of dinosaur nesting complexes starting in the 1970s. Since then our knowledge of dinosaur reproduction has expanded enormously, although it remains fragmentary (Coombs, 1989, 1990; Carpenter *et al.*, 1994). Past views that dinosaur reproduction was "reptilian" in nature and included little parental care have been replaced to a certain extent by an "avian–mammalian" view that may in some cases overemphasize their parenting skills.

The sexual aspect of dinosaur reproduction is preserved via the presence of sexual dimorphism and potential display and combat characters. A repeated pattern of two morphs, gracile and robust, has been recognized in many theropods. It has been argued that the gracile morphs are males, in part because of the configuration of bones at the base of the tail. Females are larger than males in a number of tetrapods (including birds of prey and whales), and species with robust females tend to be nonsocial and frequently predaceous. Herd-forming herbivores in which males fight for territory and/or females often exhibit male size–strength superiority, and this may have been true of most plant-eating dinosaurs. Many dinosaurs exhibited potential sexual skeletal features, some of which may be developed dimorphically between the sexes. The features include cranial tusks, hornlets, large horns, crests, and nasal diverticula, as well as erect body armor plates, spines, and tail clubs. Well-developed caniniform teeth of heterodontosaurs have been both suggested and denied as a male character. Hadrosaur head crests appear to have been better developed in one sex, probably males, in at least some examples. Ceratopsid crests and especially horns also exhibit signs of dimorphic development. Visual displays to establish male dominance, increase species identity, and enhance breeding success are possible functions of prominent dental, cranial, and skeletal features. The hollow crests and nasal diverticula of hadrosaurs may have served similar purposes as sound-generating auditory display features. The crests of lambeosaurine dinosaurs and some theropods appear too delicate for physical use. The more solidly constructed nasal ridges of some large ornithopods, tail clubs of ankylosaurids and a few sauropods, lateral body spikes of armored dinosaurs, and especially the horns of large ceratopsids were potential weapons for establishing breeding dominance among males. Possible horn wounds in ceratopsid skulls may record combat of this nature. Sharp teeth, sharp beaks, and large claws could also be used for the same purposes. Even kicking may be an important way of establishing breeding rights, and broken ribs in predaceous and herbivorous dinosaurs may record such combat.

Intraspecific weaponry often evolves in order to minimize the potential for injury and death. The interlocking nature of large ceratopsid brow and crest horns, the nasal boss of pachyrhinosaurs, the flat-topped or horn-rimmed heads of some pachycephalosaurs, and the low, ridge-shaped nasal crests of some large ornithopods appear to have been adapted to minimize damage. However, this principle should not be exaggerated. Serious injury and mortality are often significant among extant large mammals fighting over breeding rights: Male lions and hippos are notable examples of this phenomenon. The same may have been the case among some dinosaurs, especially ceratopsids with large parrot-like beaks and long nasal horns. The full development of potential sexual display features in hadrosaurs (head crests) and ceratopsians (horns and head frills) suggests that these dinosaurs at least did not become sexually active until nearly fully grown—an attribute reminiscent of birds and mammals rather than of reptiles, which often become fertile when they are a fraction of adult size.

Little attention has been paid to the mating posture

of dinosaurs, which poses interesting problems in terms of the great size of many examples, and the erect dorsal armor of a few species. All dinosaurs had large tails and bore most of their mass on their hindlimbs (whose bones are always stronger than those of the forelimbs), so it was probably easier for even the most gigantic males to bipedally mount females than it may appear. In living diapsid reptiles and birds the male copulatory organs are internal except when needed, and the same was probably true of dinosaurs.

Our understanding of dinosaur nesting and parenting behavior is limited by a lack of data for the great majority of skeletal taxa and by the uncertain identity of most dinosaur egg nests (the origins of many of the long-known famous Cretaceous nests in France and central Asia are either in dispute or have only recently been resolved). What is known of dinosaurian nesting and parenting suggests that it was highly diverse, perhaps more so than within any extant major tetrapod groups (Table I). The reproductive flexibility of dinosaurs can be attributed to their combination of oviparity (which allowed the production of few or many young), often rapid growth (allowed care of young to be nil or to be provided until fully grown within a reasonable period of time), the absence of lactation (which allowed parental feeding to range from nil to substantial), and high locomotory capacity (which allowed but did not require extensive social-parenting activities, including long-range foraging trips by parents feeding nestlings). In these attributes dinosaurs were arguably most like their larger avian relatives. No dinosaur is known to have been viviparous in the manner of mammals and some reptiles. The well-calcified shells of dinosaur eggs may not have been compatible with live birth.

Many if not all dinosaurs were like most reptiles and large birds in being oviparous *r*-strategists (producing large numbers of eggs). A notable feature of many dinosaur nests is the careful placement of the eggs both individually and relative to one another in distinctive patterns (see EGGS, EGGSHELLS, AND NESTS). The organization is greater than that observed in most extant egg-laying amniotes. It has been argued that dinosaurs used their hands to make the final arrangement of the eggs before burying them. It has been countered that the eggs were deposited directly into their final positions. Among herbivores, nest building and egg laying may have been timed so that emergence of hatchlings coincided with maximal seasonal floral growth, whether this was spring at higher latitudes or the beginning of the wet season at lower latitudes. Predatory dinosaurs may have timed the same event to coincide with maximum prey availability. Some herbivorous dinosaurs nested in colonies. Optimal local soil conditions could have encouraged egg-burying dinosaurs to nest in the same location. In addition, colonial nesting reduced but did not eliminate predation—whether by large theropods attacking large adults or by small predators hunting adults, juveniles, and eggs—by providing group protection and/or by overwhelming the feeding capacity of the local predator population. The nests of some small dinosaurs regularly appear to contain more eggs than could be laid by one female. This implies communal nesting, which is practiced by some large birds.

After egg deposition, dinosaurian care of eggs and young may have ranged from none at one extreme to the feeding and guarding of nestlings at the other. Eggs broadcast in simple hole nests were the most likely to be abandoned after burial: mound and open nests were the most likely to be guarded. The incubation temperatures of fermenting mound nests may or may not have been regulated by adults in a manner similar to megapode fowl, adding or removing nest material as needed. Recent and remarkable discoveries have included the discovery of small advanced theropods lying directly atop their eggs. The nests and eggs show a combination of the reptilian and the avian (Norell et al., 1995; Varricchio et al., 1997). The eggs are larger than expected in reptiles, but were deposited in pairs, and they were partly buried, which indicates a degree of environmental heating. The eggs were also partly exposed, and the position of some of the adults—with the forelimbs carefully draped over eggs—was like that of brooding ratites (Fig. 1). Although brooding may include shading exposed eggs, the ultimate reason to leave eggs exposed is to incubate them with body heat. The near avian brooding of these theropods appears to have been very different from the brooding observed in some snakes (Table I). Indeed, because the narrow body and slender arms of brooding theropods could not entirely cover the eggs, it is possible that some form of insulating pelage more completely covered the eggs.

The long snouts and large teeth of young tyrannosaurs are unusual for dinosaurs and suggest that they

TABLE I Nesting and Parenting Behavior in Cenozoic Amniotes and Dinosaurs

Reptiles

General—Oviparous and viviparous. Because eggs can be small relative to females, reproductive output can be low to very high. Eggs usually buried in soil. Because low aerobic exercise abilities prevent extensive social interaction and parental foraging, young are precocial and abandoned upon deposition. Slow growth also hinders parenting over extended periods.

Python—Broods exposed eggs with sinuous body warmed by muscular contractions.

Crocodilians—Females of some examples build and maintain fermenting vegetation mound nest incubators. Nests usually guarded, to the point adults may lay atop nest (but eggs are not exposed and brooded). May assist hatchlings. Young often gregarious, sometimes herded and strongly defended by parent, which can socialize with young because they are energy efficient swimmers.

Mammals

Monotremes oviparous; the rest viviparous. Some small examples nest, often in dens. Reproductive output low to high in small examples, large size of calves limits large examples to low output. High aerobic exercise capacity and rapid growth allows extensive parenting over most of juvenile period. All juveniles nurse, rendering them highly dependent on parents. Juveniles altricial to precocial.

Birds

General—Oviparous. Reproductive output low. Nests from simple to complex usually constructed, sometimes in colonies. Because parents have high aerobic metabolisms and are warm-bodied, eggs are usually exposed and brooded, and juveniles receive extensive care including feeding in or near nest. Rapid growth facilitates parental care.

Juvenile altricial to precocial. Megapode fowl build and maintain fermenting vegetation mound nest incubators. Juveniles precocial and independent upon hatching.

Ratites (living and recent)—Reproductive output low in island examples, high in continental examples. Multiple females may deposit eggs in one nest. Precocial young leave nest soon after hatching and are largely self-feeding although minor parental assistance may occur. Gregarious young are herded and protected by adults; sometimes many broods are combined into a large crèche under care of one adult pair.

Dinosaurs

General—Probably all oviparous. Reproductive output high, with some low rate examples possible. Growth moderately to very rapid.

Simple hole nests (many unassigned eggs)—Nests may be isolated or in groups. Either abandoned immediately or guarded. In some cases semi-intact hatched eggs suggest young left nest soon after hatching.

Tyrannosaur juveniles—long snouts of juveniles suggest independence.

Hypsilophodont and other dinosaur nests with eggs laid vertically and in complex patterns—Careful placement of eggs in vertical position suggests burial rather than brooding. Nests often in colonies, spacing of nests by one adult body length indicates nests were guarded. In some cases semi-intact hatched eggs suggest young left next soon after hatching. In some cases the presence of fast-growing, short-snouted juveniles with well-ossified limb joints near nest suggests that precocial chicks remained in nesting colony for an extended period, perhaps cared for and fed by parents.

Small theropod nests—Ratite-like posture of adult *Oviraptor* skeletons atop exposed eggs strongly suggests incubation via brooding. Trampled eggshells and damaged bones of small vertebrates in a nest tentatively assigned to small theropods suggest altricial nestlings fed by parents.

Hadrosaur mound nest colonies, bone beds, and trackways—Eggs incubated in fermenting vegetation mounds, built and probably maintained by parents in large colonies; spacing of nests by one adult body length suggests the latter. Trampled eggshells, and the presence of short-snouted juveniles in nest, suggest altricial chicks remained in open-pit nest and were fed by parents. Trackways and skeletal associations suggest that postnestling juveniles were gregarious and perhaps independent. Calves joined large herds when about one-half adult dimensions.

Ceratopsid bone beds—Sudden death assemblages suggest juveniles moved with adults when one-fifth adult dimensions.

Sauropod trackways—Young too small to join adult herds until they were one-third adult dimensions (about 1 ton).

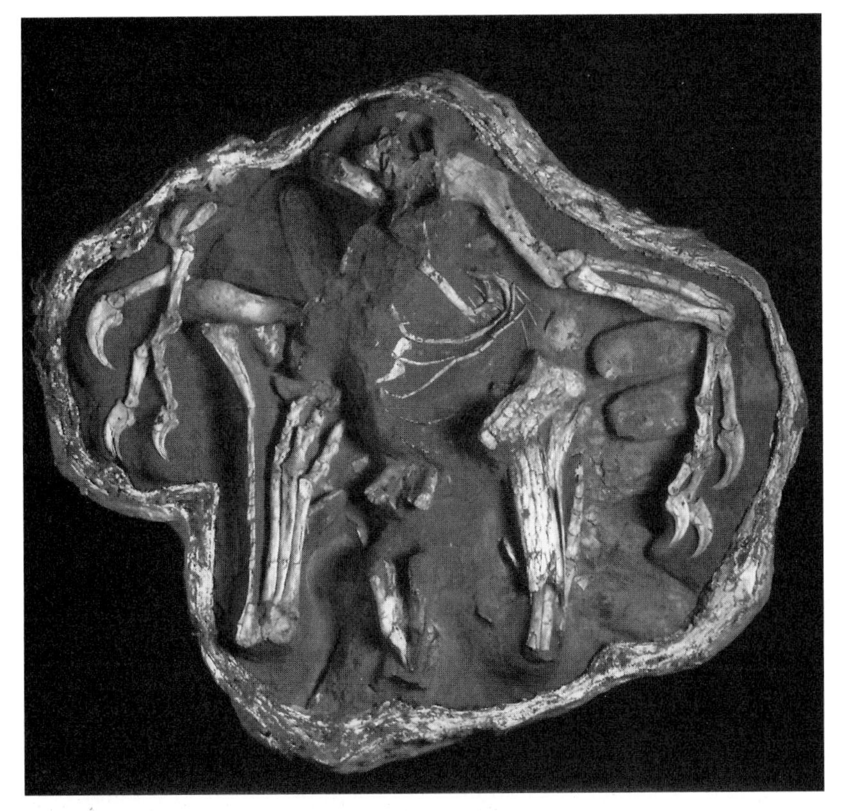

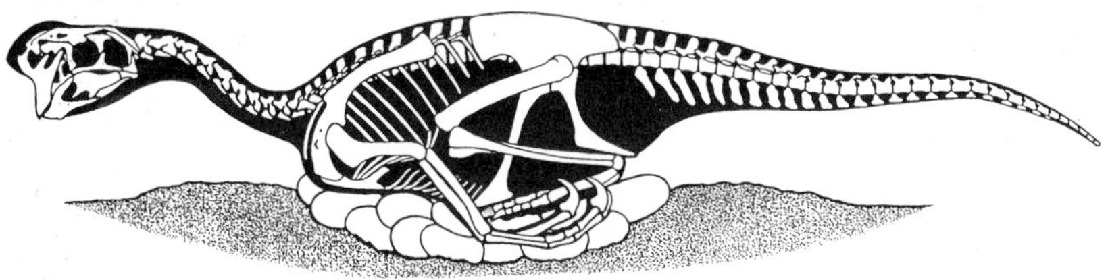

FIGURE 1 (Top) The preserved nest of an *Oviraptor* lying atop a spiral of eggs, viewed from above, with left side of the animal on the left. (Bottom) A restoration of the dinosaur brooding the eggs. Note that in both images the deep pelvis fits into the hollow circle formed by the eggs, and that the arms are draped over the eggs. No living reptile broods its eggs in this manner (contrary to Geist and Jones, 1996). Images courtesy M. Norell, American Museum of Natural History.

may have been independent hunters immediately or soon after hatching. As for those dinosaurs that cared for their offspring, the production of large numbers of young would have tended to reduce the parental attention received by each baby dinosaur. Also, most dinosaurs had small, simple brains, so their parent–offspring relationships were probably more stereotyped and limited than that observed in larger brained tetrapods. Therefore, most dinosaurs probably did not provide the intense one-on-one parenting observed in mammals and most birds. The short snouts characteristic of most if not all herbivorous and some predaceous dinosaurs may have acted as visual cues to incite parental attention and care (Horner and Gorman, 1988). If any adults fed their young, they may have done so with regurgitants, which may have been an important means of transferring the gut bacteria needed to digest fodder from one generation to the next. It is not known whether any small dinosaurs brooded their nestlings, but it is plausible considering the evidence for egg brooding. Arguments that the tiny hatchlings of some small

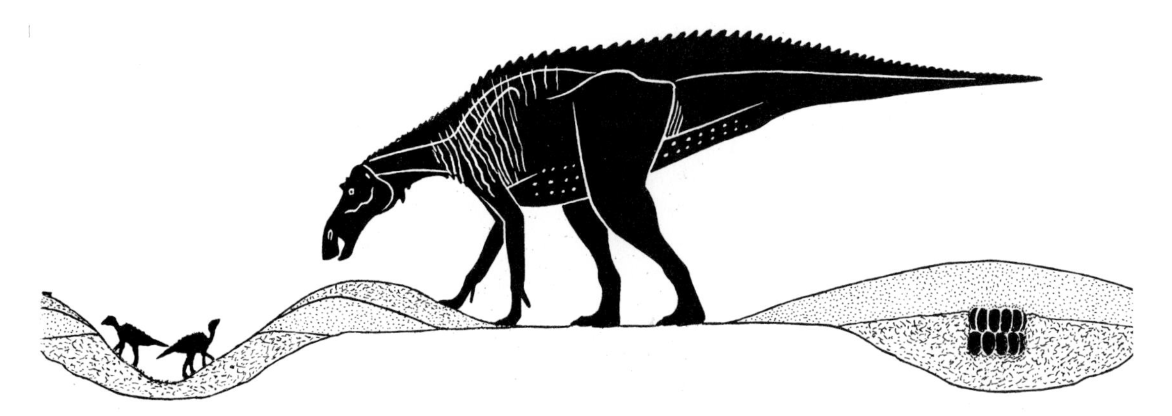

FIGURE 2 Restored nesting behavior in the 2.5-ton hadrosaur *Maiasaura*. Eggs were incubated in large, fermenting vegetation mound nests, spaced one adult's body length from other nests in large colonies. As eggs hatched, the nest was converted into a pit in which the nestlings lived.

dinosaurs were too small to be parented are countered by the extreme size disparity between newly hatched and adult ratites. Some of the largest-brained small theropods may have practiced the most comprehensive parenting among dinosaurs—but note that some hole nest eggs, probably abandoned immediately upon deposition, have been tentatively attributed to troodontids. Large dinosaurs were too heavy to brood their offspring. The extreme size disparity between adults and young of the largest dinosaurs may have rendered parental care impractical. The young juveniles of large dinosaurs were probably too small to join herds of adults, where they could have been trampled. Postnestling juveniles that were not independent were probably under the guardianship of only their parents, or perhaps in a multibrood crèche under the care of one pair of adults, until they were large enough to join herds. There is some evidence that independent and parented juveniles were gregarious. Suggestions that trackway patterns show juvenile sauropods within a protective ring of adults have been discounted, but it remains possible that the calves received active or passive protection by associating with larger individuals.

Hadrosaurs built mound nests in colonies, and the hatchlings weighed a few kilograms (Fig. 2; Horner and Gorman, 1988; Horner, 1996). Geist and Jones (1996) argued that the ossified pelves of baby hadrosaurs indicate well-developed locomotory abilities associated with precocial habits, but Horner (1996) countered that the leg bone shafts were so poorly ossified that the babies should have been immobile and therefore altricial. Trampling of the eggshells into small bits also suggests that the nests were inhabited for an extended period, as does the finding of individuals a few times larger than the hatchlings in or near the nests. Usually, fully immobile nestlings are limited to more isolated predator-free areas. It is possible that hadrosaur nestlings were semialtricial, with just enough locomotary ability to flee the nest if a predator penetrated the parental defenses. Some juvenile crocodilians, which are fully precocial and can wait for prey to move into their range, remain near their nest for months or years. However, the bone microstructure of hadrosaur chicks indicates that they grew much more rapidly than reptiles, which take years to grow the amount observed in hadrosaur nestlings. Such fast-growing herbivores would be under pressure to range far and wide if they fed themselves. The most logical explanation of why young hadrosaurs remained in and/or near their nests is because their parents fed them there. This would have been highly advantageous to the chicks; at no cost to themselves they received large quantities of food that dramatically boosted the pace of growth over that which can be achieved without parental feeding. If, for example, parent-fed nestlings grew about 15 kg in 1 or 2 months, then their growth rates matched those of the fastest growing birds (ostrich chicks take about 70 days to grow a similar amount). In this view, parental feedings of altricial nestlings were a means by which giant adults boosted the growth rates of tiny juveniles, until they were large enough to move in the company of adults without being trampled. Hadrosaur nestlings apparently lived exposed to the elements in open-pit nests. Temperatures may have ranged from low during cool rain storms to very high under sunlight. Bird nestlings that live in similarly harsh nesting colonies, and are not shielded by their parents, have well-developed thermoregulatory controls. The same may have been true of hadrosaur nestlings, a possibility supported by bone isotope analysis (Barrick and Showers, 1995). The browsing pressure placed on the flora surrounding a nesting colony by hadrosaurs foraging for food for their charges may have been very high, and the parents may have had to range out a dozen or more kilometers each day toward the end of the nesting period. Lack of sufficient food due to drought and/or overpopulation, thermal stress due to extreme weather, flooding, and disease vectors were common potential sources of mass mortality that could wipe out most of a year's production of offspring in a hadrosaur nesting colony.

Some dinosaurs may have been no more parental than other reptiles, but the level of egg care and parenting exhibited in some dinosaurs appears to have been above that practiced by any living reptile, including crocodilians. Conversely, dinosaurian parenting was probably not at the very sophisticated and intense levels practiced by most mammals. The descendants of terrestrial dinosaurs, birds, have both retained and modified a set of reproductive strategies patterned after their ancestral group.

Because gigantic dinosaurs laid small eggs, they had the highest adult to juvenile mass ratios observed among tetrapods—many tens of thousands in the case of the largest sauropods (the ratio is about 30 in elephants). Such extreme initial–final size ratios required very fast juvenile growth in order to reach sexual maturity within two or three decades. Animals must start breeding within that time frame in order to ensure that enough juveniles survive to reproduce. The size superiority of dinosaurs vis-à-vis land mammals cannot be attributed to the indeterminate adult growth of the former. Termination of growth and life is nearly coincident in bull elephants, and incremental growth cannot account for the size disparity of 20–100 tons (like other animals, dinosaurs could not live and grow for more than 100–150 years because death becomes a statistical certainty). Consider a 100-ton sauropod that started reproducing at one-third adult mass at age 20 or 30. Its peak growth rate would approach or exceed 5 kg per day, a rate otherwise observed only in whales. Even adult growth would have been about 2 kg per day, a rate far in excess of that of which living terrestrial reptiles are capable.

The reproductive strategies of large dinosaurs, birds, and mammals share the characteristic of rapid growth, which boosts rates of population expansion. The reproductive strategies of large dinosaurs and large birds differs from that of large mammals in two key regards: lactation or its absence, and rate of reproduction (Table II). Regarding the latter, large dinosaurs were not slow-breeding *K*-strategists like mammals of equal size. Instead, they were fast-breeding *r*-strategists with annual reproductive outputs similar to those of many smaller reptiles and mammals and large birds (Fig. 3). The annual and lifetime

TABLE II
Observed and Predicted Population Dynamics for Large Land Tetrapods

***K*-strategist megamammals**—Birth rates low. Growth fast; generational turnover fairly rapid. Juvenile mortality moderate. Adult populations of megamammals must be high in order to raise the few highly dependent, nursing young they produce. If adult populations drop too low, not enough juveniles survive to mature and slowly produce enough new young to reestablish the population. Because the available resource base must be divided among a large number of adults, the size of individuals is limited to about 20 tons. Populations skewed toward adults; adult population densities relatively low.

***r*-Strategist reptiles**—Egg deposition rates high. Growth slow; generational turnover slow. Juvenile mortality high; nonnursing young fully or partly independent of adults. Growth is too slow to take full advantage of the size potential associated with rapid reproduction, so adult masses limited to 1 ton. Populations skewed toward adults; adult population densities very high.

***r*-Strategist megadinosaurs**—Egg deposition rates high. Growth fast; so generational turnover probably fairly rapid. Juvenile mortality probably high; nonnursing young fully or partly independent of adults. Combination of *r*-strategy reproduction and independent young allowed whale-sized adults because even if small adult populations were lost, only a small population of nonnursing juveniles needed to reach sexual maturity and start repopulating the habitat. Populations skewed toward juveniles; adult population densities somewhat lower than those of megamammals of similar size, extremely low over 30 tons.

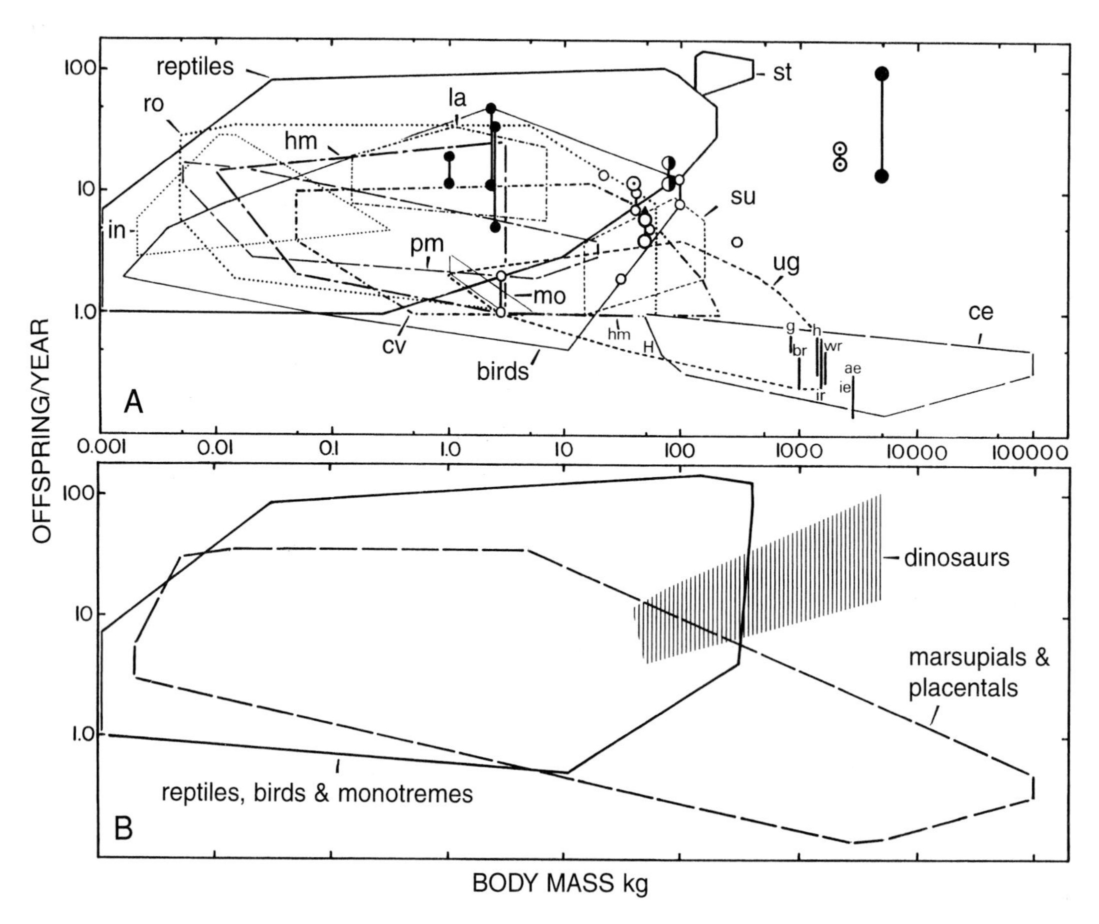

FIGURE 3 (A) Annual reproductive potential as a function of the body mass of breeding females in living and recent amniotes and dinosaurs. (B) The same data reorganized to compare oviparous versus viviparous taxa and dinosaurs. Reproductive potential is total egg or newborn production. Most living groups are enclosed in least-area polygons; nonpredaceous marsupials from 3 to 40 kg fall upon the single line indicated; marine turtles (st), megapodes (small solid circles), continental ratites (rhea, emu, cassowary, and ostrich; small open circles), island ratites (kiwi and moas; small circles with dots), elephants (African, ae; Asian, ie), giraffes (g), black rhinos (br), hippos (h), Indian rhinos (ir), white rhinos (wr), humans (H). The major mammalian groups plotted are cetaceans (ce), ungulates (ug), suids (su), carnivores (cv), rodents (ro), lagomorphs (la), insectivores (in), herbivorous marsupials (hm), predaceous marsupials (pm), and monotremes (mo). Dinosaurs plotted are the sauropod *Hypselosaurus* (tentative; large solid circles), hadrosaurs *Maiasaura* and *Hypacrosaurus* (large bold circles with dots), hypsilophodonts (large circle with dot), and the theropods *Oviraptor* (large half-solid circles) and *Troodon* (tentative; large open circles). From Paul in Carpenter *et al.* (1994).

reproductive output of a sauropod was at least 50 times higher than that of an elephant. Large ornithischians outbred rhinos of similar size by almost as much. Also, female dinosaurs could breed every year or more often, whereas extended gestation periods force big female mammals to space births by at least 2 years. Juvenile mortality was presumably much higher in dinosaurs than in mammals. Also, compared to smaller birds and mammals with similar annual reproductive rates, slower generational turnover limited large dinosaur reproductive output. Even after these factors are taken into account, the potential population growth rates of large dinosaurs under optimal realistic conditions were probably many times higher than the 6–12% expansion rates observed in large mammals.

In ecological terms, large dinosaurs may have been "weed species" with rates of population recovery and dispersal well above those seen in large mammals (Janis and Carrano, 1992; Farlow, 1993; Paul as cited in Carpenter *et al.*, 1994). Because periodic loss of the adult portion of the population had to be tolerable, it was possible for species to evolve oversized adults whose populations were too low to ensure their stability—the extreme scarcity of extremely large, 50- to 150-ton sauropods suggests that this view is correct. A single large sauropod was the rough equivalent—in terms of energy intake, population density, and reproductive output—of an entire elephant herd. As a consequence, dinosaurs should have been able to achieve far larger adult masses than mammals living on similar resource bases. For example, South America supported 50- to 100-ton sauropods in the Cretaceous compared to 5-ton proboscideans in the Neogene. High recovery and dispersal rates also rendered large dinosaurs highly resistant to extinction, which may explain the long-term stability of most major large dinosaur groups. In comparison, slow-breeding large mammals are much more sensitive to disruption, and most major terrestrial groups (uintatheres, arsinoitheres, brontotheres, and indricotheres) have been geologically short-lived. The greater resistance of fast-breeding large dinosaurs to extinction vis-à-vis similar-sized mammals complicates attempts to explain their loss at the K–T boundary. It has been suggested that temperature fluctuations at the K–T boundary catastrophically distorted male–female sex ratios in dinosaurs because the ratio may have been dependent on the temperature of the incubating eggs. The closest reptilian relatives of dinosaurs, crocodilians, have temperature-dependent ratios, but the more closely related birds have genetically determined sex ratios. It is therefore possible that most or all dinosaurs had the latter system, and this hypothesis of dinosaur extinction cannot be falsified or supported.

See also the following related entries:

BEHAVIOR • EGG MOUNTAIN • EGGS, EGGSHELLS, AND NESTS • GROWTH AND EMBRYOLOGY

References

Barrick, R. E., and Showers, W. J. (1995). Oxygen isotope variability in juvenile dinosaurs (*Hypacrosaurus*): Evidence for thermoregulation. *Paleobiology* **21,** 552–560.

Carpenter, K., Hirsch, K. F., and Horner, J. R. (1994). *Dinosaur Eggs and Babies.* Cambridge Univ. Press, Cambridge, UK.

Coombs, W. P. (1989). Modern analogs for dinosaur nesting and parental behavior. *Geol. Soc. Am. Spec. Paper* **238,** 21–53.

Coombs, W. P. (1990). Behavior patterns of dinosaurs. In *The Dinosauria* (D. B. Weishampel, P. Dodson, and H. Osmólska, Eds.), pp. 32–42. Univ. of California Press, Berkeley.

Farlow, J. O. (1993). On the rareness of big, fierce animals: Speculations about the body sizes, population densities, and geographic ranges of predatory mammals and large carnivorous dinosaurs. *Am. J. Sci.* **239,** 167–199.

Geist, R., and Jones, T. D. (1996). Juvenile skeletal structure and the reproductive habits of dinosaurs. *Science* **272,** 712–714.

Horner, J. R. (1996). New evidence for post-eclosion parental attention in *Maiasaura peeblesorum. J. Vertebr. Paleontol.* **16**(Suppl. to No. 3), 42A.

Horner, J. R., and Gorman, J. (1988). *Digging Dinosaurs.* Workman, New York.

Janis, C. M., and Carrano, M. (1992). Scaling of reproductive turnover in archosaurs and mammals: Why are large terrestrial mammals so rare? *Acta Zool. Fennica* **28,** 201–206.

Norell, M. A., *et al.* (1995). A nesting dinosaur. *Nature* **378,** 774–776.

Varricchio, D. J. et al. (1997). Nest and egg clutches of the dinosaur *Troodon formosus* and the evolution of avian reproductive traits. *Nature* **385,** 247–250.

Reptiles

MICHAEL J. BENTON
University of Bristol
Bristol, United Kingdom

Origin of the Reptiles

The Reptilia, as traditionally understood, is a paraphyletic group because it excludes two major descendant clades, the birds (see AVES) and the mammals. However, cladistic terminology has reemphasized the use of Amniota to include traditional (paraphyletic) reptiles, plus birds and mammals. Reptilia, as defined, is restricted to living reptiles (including birds) and all amniotes closer to them than to mammals. Many hundreds of species of fossil reptiles are known, and these indicate the former existence of a great diversity of major extinct lineages (Carroll, 1987; Benton, 1997).

The date of origin of the Amniota (= Reptilia + Mammalia and their relatives) is currently debated. A very early amniote, *Westlothiana lizziae*, has been reported from the Early Carboniferous (350 Myr) of Scotland, but the affinities of this creature are disputed (Smithson *et al.*, 1994). It could perhaps be a reptile, as suggested by some characters of the back of the skull, but other evidence suggests it may be a microsaur, a group of superficially reptile-like amphibians (Spencer, 1994).

If *Westlothiana* is not a reptile, the next oldest amniote is *Hylonomus* (Carroll and Baird, 1972) from the mid-Carboniferous (310 Myr) Cumberland Group of Nova Scotia (Fig. 1). *Hylonomus* is represented by several specimens preserved in the unusual location of hollow tree trunks, preserved in life position. The sedimentary evidence suggests that forests of tall seedferns were felled by major floods, leaving the stumps to rot in a mass of surrounding debris. The small reptiles crept into the hollow stumps in search of insect prey and then became trapped themselves. *Hylonomus* possesses an astragalus and calcaneum in the ankle as well as a tall skull shape. Most Carboniferous amphibians had very low broad skulls, similar to those of modern frogs, but reptiles typically have narrower and taller skulls, which allows longer and more adaptable jaw muscles to be located inside the back portion of the skull.

Amniote Evolution

Amniote evolution during the late Carboniferous shows evidence for a series of major phylogenetic splits that laid the basis for the major amniote groups: the Anapsida, Diapsida, and Synapsida. These three are readily distinguished by differences in posterior skull architecture: Anapsids (= "no hole") have no openings behind the orbit (eye socket), diapsids (= "two hole") have two temporal openings, and synapsids (= "same hole") have only a lower temporal opening (Fig. 2). A cladogram showing postulated relationships of the major amniote groups is shown in Fig. 3. The anapsid condition is primitive for amniotes because it is also found in amphibians and the other nonamniote outgroups to amniotes.

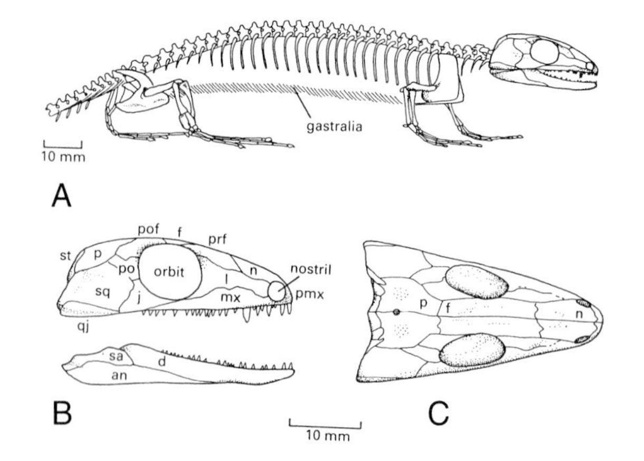

FIGURE 1 One of the oldest reptiles, *Paleothyris*, relative of *Hylonomus*, from the mid-Carboniferous of Nova Scotia; skeleton (A) and skull in side view (B) and from above (C) (modified from Carroll and Baird, 1972).

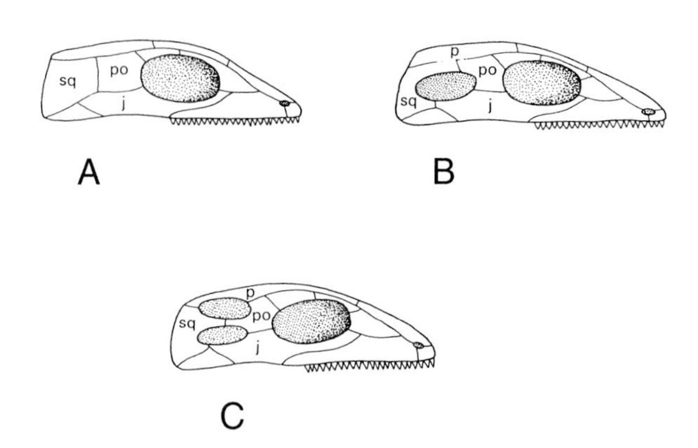

FIGURE 2 The three main skull patterns found among amniotes; the anapsid (A), synapsid (B), and diapsid (C) (modified from Benton, 1990).

Anapsida The phylogenetic position of the early Anapsida is currently debated, with many studies suggesting several separate cladistic branches of reptiles with the primitive anapsid skull pattern during the Carboniferous and Permian. The chief idea (Gauthier *et al.*, 1988) was that there was a sidebranch, termed the parareptiles, that consisted variably of the marine Early Permian mesosaurs, the superficially lizard-like Late Permian millerettids, the small Permo–Triassic procolophonids, and the large Late Permian herbivorous pareiasaurs. This parareptile clade was seen as distinct from the major modern anapsid clade, the turtles, which arose in the Late Triassic.

Recent studies of the parareptiles, on the other hand (Spencer, 1994; Laurin and Reisz, 1995; Lee, 1995), show that these fall along the lineage leading to turtles (Fig. 2), and hence a major evolutionary gap is filled. The clade Anapsida, which also includes the captorhinids such as *Hylonomus*, and relatives of the

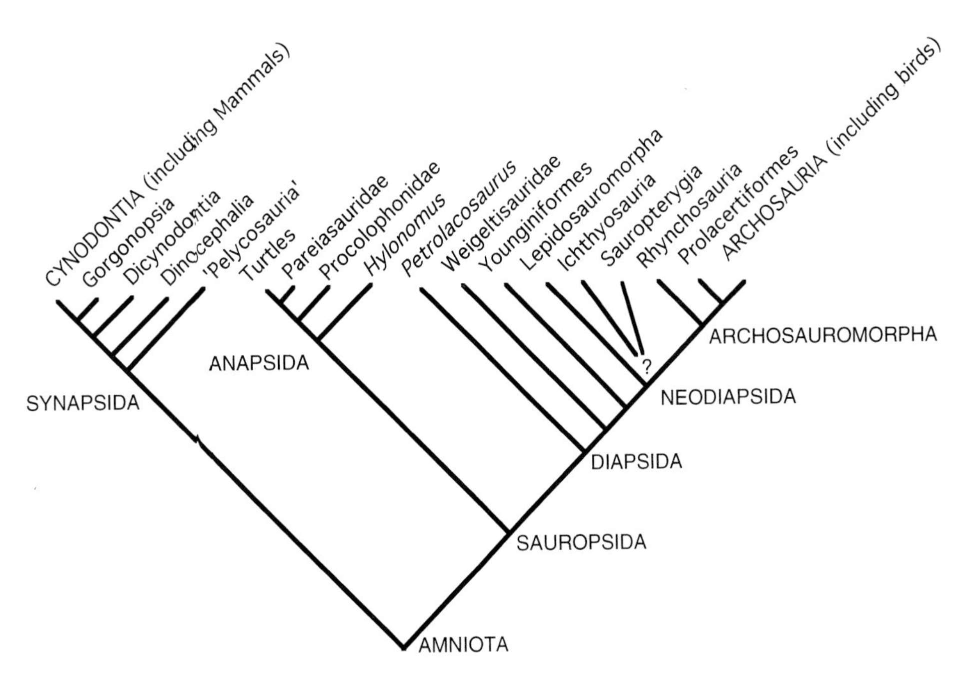

FIGURE 3 Cladogram showing relationships of the major amniote groups (modified from Benton, 1990, and Spencer, 1994).

Permo–Carboniferous, is diagnosed by several characters of the skull and vertebrae (Spencer, 1994).

Synapsida The Synapsida, consisting of the mammals today and their ancestors of the Carboniferous to Jurassic, have long been recognized as a distinctive amniote clade. They are characterized by features of the skull (Spencer, 1994). The phylogeny of synapsid amniotes has been partially worked out (Kemp, 1982; Hopson and Barghusen, 1986; King, 1988).

The oldest synapsid is *Archaeothyris*, a "pelycosaur" (basal synapsid) from the Morien Group of Nova Scotia of late Carboniferous (300 Myr) age. The group flourished particularly during the Permian and Triassic, when it included many of the dominant land-living tetrapods. The pelycosaurs were particularly important during the Early Permian, and they include the famous sail-backed animals such as *Dimetrodon* and *Edaphosaurus* (Fig. 4A).

The synapsid clade Therapsida, which originated in the mid-Permian, consists of the dicynodonts plus cynodonts (including mammals). It included a wide array of small to large herbivores and carnivores. The dinocephalians, represented by meat-eating and plant-eating titanosuchians and plant-eating tapinocephalians, were restricted to the Late Permian. Some of the herbivores, such as *Moschops* and *Estemmenosuchus*, were bulky animals, as large as rhinoceroses but with larger bodies and shorter legs. The herbivorous dicynodonts were more uniform in appearance and habits. They generally had only two teeth, the tusk-like canines, and sharp-edged jaw bones that were used to cut their plant food. The dicynodonts were preyed upon by the saber-toothed gorgonopsians, whose massive canine teeth and wide gape were adaptations to piercing the thick skin of their prey.

At the end of the Permian, the dinocephalians, gorgonopsians, and the pareiasaurs (large anapsid herbivores) died out, probably as a result of the major environmental crisis at that time which precipitated the biggest mass extinction of all time in the seas. The dicynodonts survived into the Triassic and radiated again (Fig. 4B). One other seemingly minor therapsid group survived the Permo–Triassic extinction event—the cynodonts. These modest-sized carnivores arose in the latest Permian and radiated during the Triassic, producing a number of carnivorous (Fig. 4C) and herbivorous lineages. Many of the cynodonts may have been endothermic ('warm blooded'), and one lineage evolved into the true mammals in the Late Triassic or Early Jurassic.

The class Mammalia does not have a universally

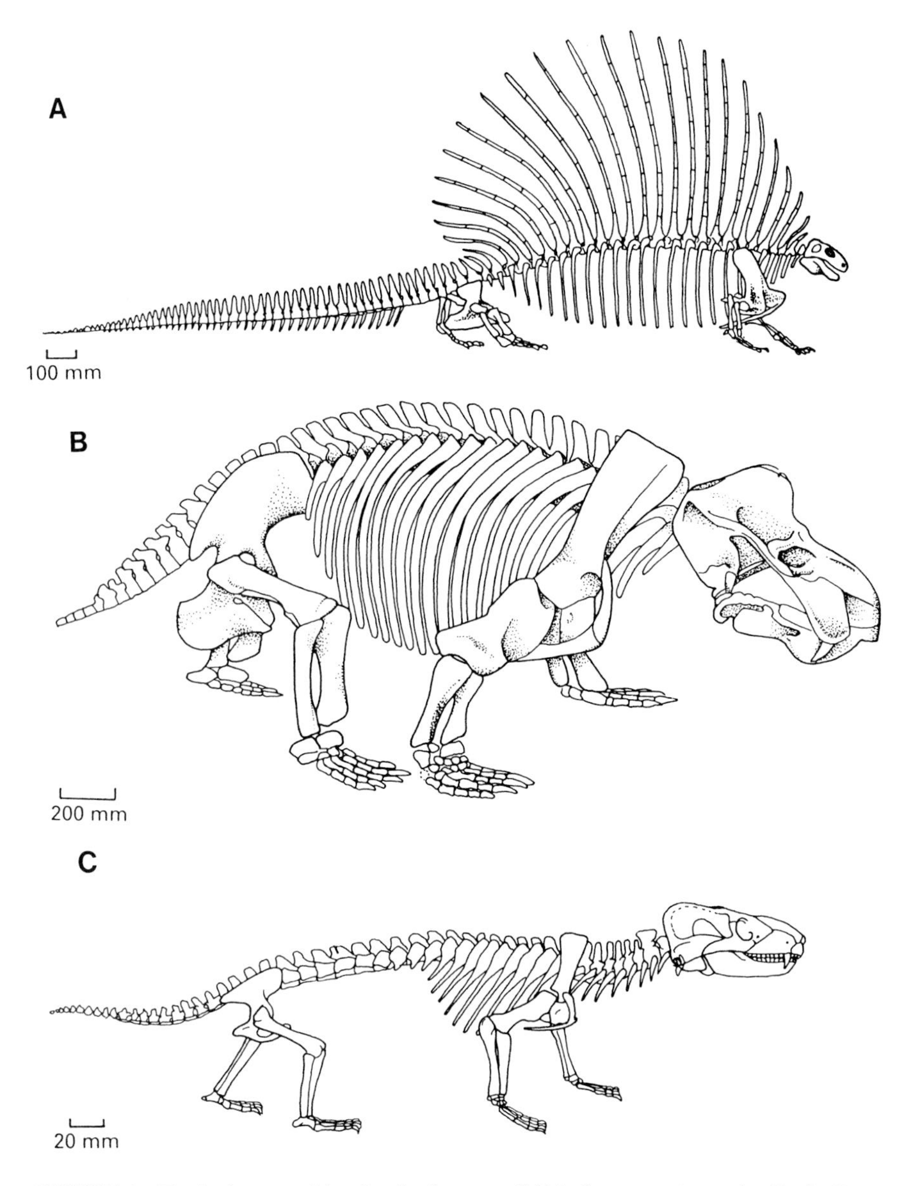

FIGURE 4 Typical synapsids: the "pelycosaur" *Edaphosaurus* from the Early Permian of Texas (A), the dicynodont *Kannemeyeria* from the Early Triassic of South Africa (B), and the cynodont *Thrinaxodon* from the Early Triassic of South Africa (C). A, modified from Romer and Price (1940); B, after Pearson (1924); C, modified from Jenkins (1971).

accepted definition. Some would restrict it to crown-group mammals (Gauthier *et al.*, 1988) and others to include fossil groups with a particular set of "mammalian" features. It is diagnosed by the possession of a single bone, the dentary, in the lower jaw (reptiles typically have seven: the dentary, splenial, coronoid, angular, prearticular, surangular, and articular) and by the possession of three middle ear ossicles (hammer, anvil, and stirrup or malleus, incus, and stapes) compared to only one (the stapes) in the primitive condition. Mammals are also characterized by hair, but fossil evidence for hair is limited and it appears to occur in some nonmammalian cynodonts.

Diapsida The third major amniote clade, the Diapsida, is represented today by the crocodiles, birds, lizards, and snakes. Diapsids of the past include dinosaurs and pterosaurs, as well as, probably, the marine

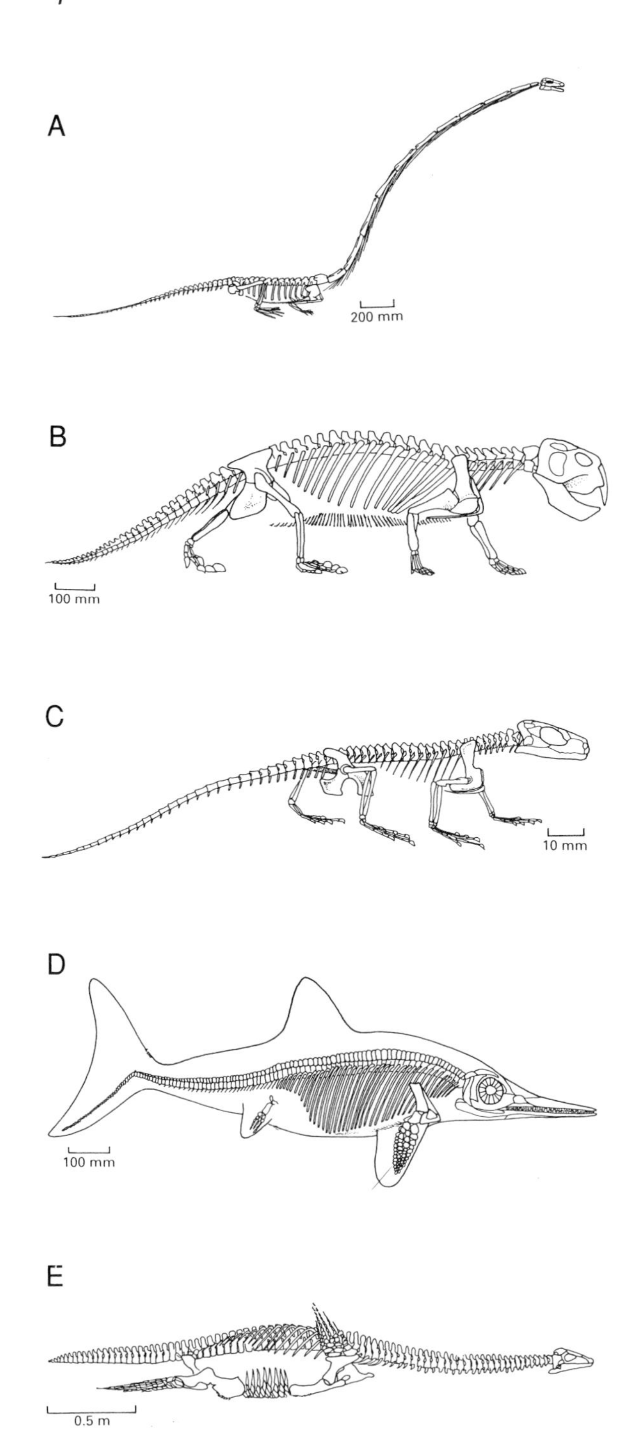

FIGURE 5 Typical diapsid reptiles: the prolacertiform *Tanystropheus* from the Middle Triassic of Switzerland (A), the rhynchosaur *Hyperodapedon* from the Late Triassic of Scotland (B), the sphenodontid *Planocephalosaurus* from the Late Triassic of England (C), the ichthyosaur *Ichthyosaurus* from the Early Jurassic of England (D), and the plesiosaur *Cryptocleidus* from the Late Jurassic of England (E). A, modified from Wild (1973); B, modified from Benton (1983); C, modified from Fraser and Walkden (1984); D, after Stromer (1912); E, modified from Brown (1981).

ichthyosaurs and plesiosaurs. The Diapsida arose during the late Carboniferous, their first representative being *Petrolacosaurus* from Kansas. In addition to their pair of temporal openings, diapsids are characterized by a suborbital foramen, an opening in the palate behind the internal naris.

The fossil record of diapsids is poorly known until the Late Permian, when a number of distinctive groups arose. The prolacertiforms had elongate necks, the younginiforms, largely from Africa, were moderate-sized terrestrial and freshwater animals, and the weigeltisaurs were specialized gliders. Of these, only the prolacertiforms survived the Permo–Triassic mass extinction event, and they radiated during the Triassic, including the bizarre *Tanystropheus* from the Middle Triassic of central Europe, a 3-m animal that had a neck up to 2 m long (Fig. 5A).

The prolacertiforms belong to a broader clade called the Archosauromorpha, all of which have nostrils close to the midline of the snout, closely articulating astragalus and calcaneum in the ankle, and other features (Benton, 1985; Evans, 1988; Laurin, 1991). New archosauromorph groups became established in the Triassic including the rhynchosaurs (Fig. 5B), bulky plant eaters with hooked snouts and broad tooth-bearing plates, and the archosaurs, the group that includes the dinosaurs.

Also during the Triassic another major diapsid group, the lepidosauromorphs, became established. These reptiles, generally small in size and with a tympanic recess on the quadrate, a specialized ankle, and other features (Benton, 1985; Evans, 1988; Laurin, 1991), gave rise to the sphenodontids (Fig. 5C) in the Late Triassic, which are ancestors of the modern tuatara, *Sphenodon*, of New Zealand. Close relatives, the lizards, are first known from the Middle Jurassic and the snakes from the Early Cretaceous.

The marine ichthyosaurs and sauropterygians have a "euryapsid" skull pattern, in which there is only one (upper) temporal opening. They may be phylogenetically close to the lepidosauromorphs. Their skull pattern probably arose by loss of the lower temporal bar below the lower temporal opening. The ichthyosaurs (Fig. 5D) were dolphin-like fish eaters that lived throughout the Mesozoic and fed on fishes and cephalopods. The sauropterygians had a similar time span and dietary range, and they were represented by the long-neck nothosaurs and the mollusc-eating placodonts in the Triassic and by the familiar plesiosaurs (Fig. 5E) and pliosaurs during the Jurassic and Cretaceous.

The archosaurs arose in the latest Permian and radiated, mainly as carnivores, during the Early and Middle Triassic. New carnivorous and herbivorous groups of archosaurs arose in the Late Triassic, including the dinosaurs, pterosaurs, and crocodylomorphs. These forms became diverse and abundant in the last 20 Myr of the Late Triassic after a profound extinction event 225 Myr ago, in the late Carnian Stage (see EXTINCTION, TRIASSIC) when the rhynchosaurs, dicynodonts, and others died out.

Living Reptiles

Reptiles today—turtles, crocodiles, lizards and snakes, and birds—are an odd assemblage of animals, often seen as primitive and unsuccessful in comparison to birds and mammals. This view is based on a misunderstanding of reptilian physiology and behavior and of reptilian history. Extant reptiles (except birds) are ectothermic ("cold blooded"), which tends to restrict their distribution to warm climates, but many lizards and snakes are adapted to life in arid conditions where birds and mammals cannot survive as well. Reptiles have efficient water conservation mechanisms, and they require approximately one-tenth the amount of food of a mammal of equivalent weight because they do not control their body temperatures by internal physiological means. The reptiles are a natural group if they include birds, and it is meaningless to compare modern reptiles only with dominant amniote groups of the past such as various synapsids, reptiles, dinosaurs, pterosaurs, or ichthyosaurs. However, living birds and other reptiles outnumber mammals in species diversity, so it would seem misleading to call the time since the Cretaceous the "Age of Mammals" (Gauthier, 1986; Gauthier *et al.*, 1988).

See also the following related entries:

ARCHOSAURIA • CROCODYLIA • DINOSAUROMORPHA • PHYLOGENETIC SYSTEM • PHYLOGENY OF DINOSAURS

References

Benton, M. J. (1983). The Triassic reptile *Hyperodapedon* from Elgin: Functional morphology and relationships. *Philos. Trans. R. Soc. London Ser. B* **302,** 605–720.

Benton, M. J. (1985). Classification and phylogeny of the diapsid reptiles. *Zool. J. Linnean Soc.* **84,** 97–164.

Benton, M. J. (1997). *Vertebrate Palaeontology, Biology and Evolution*, second edition, pp. 456. Chapman & Hall, London.

Brown, D. S. (1981). The English Upper Jurassic Plesiosauroidea (Reptilia) and a review of the phylogeny and classification of the Plesiosauria. *Bull. Br. Museum (Nat. History) Geol. Ser.* **35,** 253–347.

Carroll, R. L. (1987). *Vertebrate Paleontology and Evolution*, pp. 698. Freeman, San Francisco.

Carroll, R. L., and Baird, D. (1972). Carboniferous stem-reptiles of the family Romeriidae. *Bull. Museum Comp. Zool.* **143,** 321–363.

Evans, S. E. (1988). The early history and relationships of the Diapsida. In *The Phylogeny and Classification of the Tetrapods* (M. J. Benton, Ed.), Systematics Association Spec. Vol. 35A, pp. 221–260. Clarendon Press, Oxford.

Fraser, N. C., and Walkden, G. M. (1984). The postcranial skeleton of the Upper Triassic sphenodontid *Planocephalosaurus robinsonae*. *Palaeontology* **27,** 575–595.

Gauthier, J. A. (1986). Saurischian monophyly and the origin of birds. *Mem. California Acad. Sci.* **8,** 1–55.

Gauthier, J. A., Kluge, A. G., and Rowe, T. (1988). The early evolution of the Amniota. In *The Phylogeny and Classification of the Tetrapods. Volume 1. Amphibians, Reptiles, Birds* (M. J. Benton, Ed.), Systematics Association Spec. Vol. 35A, pp. 103–155. Clarendon Press, Oxford.

Hopson, J. A., and Barghusen, H. R. (1986). An analysis of therapsid relationships. In *The Ecology and Biology of Mammal-like Reptiles* (N. Hotton, III, P. D. MacLean, J. J. Roth, and E. C. Roth, Eds.), pp. 83–106. Smithsonian Institution Press, Washington, DC.

Jenkins, F. A., Jr. (1971). The postcranial skeleton of African cynodonts. *Bull. Peabody Museum Nat. History* **36,** 1–216.

Kemp, T. S. (1982). *Mammal-like Reptiles and the Origin of Mammals*, pp. 363. Academic Press, London.

King, G. M. (1988). Anomodontia. *Handbuch der Paläoherpetologie* **17C,** 1–174.

Laurin, M. (1991). The osteology of a Lower Permian eosuchian from Texas and a review of diapsid phylogeny. *Zool. J. Linnean Soc.* **101,** 59–95.

Laurin, M., and Reisz, R. R. (1995). A reevaluation of early amniote phylogeny. *Zool. J. Linnean Soc.* **113,** 165–223.

Lee, M. S. Y. (1993). The origin of the turtle body plan: bridging a famous morphological gap. *Science* **261,** 1716–1721.

Lee, M. S. Y. (1995). Historical burden in systematics and the interrelationships of the parareptiles. *Biol. Rev.* **70,** 459–547.

Pearson, H. S. (1924). A dicynodont reptile reconstructed. *Proc. Zool. Soc. London* **1924,** 827–855.

Reisz, R. R., and Laurin, M. (1991). *Owenetta* and the origin of turtles. *Nature (London)* **349,** 324–326.

Romer, A. S., and Price, L. I. (1940). Review of the Pelycosauria. *Spec. Paper Geol. Soc. Am.* **28,** 1–538.

Smithson, T. R., Carroll, R. L., Panchen, A. L., and Andrews, S. M. (1994). *Westlothiana lizziae* from the Viséan of East Kirkton, West Lothian, Scotland and the amniote stem. *Trans. R. Soc. Edinburgh Earth Sci.* **84,** 383–412.

Spencer, P. S. (1994). The early interrelationships and morphology of Amniota. Ph.D. thesis, University of Bristol, Bristol, UK.

Stromer, E. v. (1912). *Lehrbuch der Paläozoologie.* Treuber, Leipzig, Germany.

Wild, R. (1973). Die Triasfauna der Tessiner Kalkalpen. XXIII. *Tanystropheus longobardicus* (BASSANI)(Neue Ergebnisse). *Abhandlungen Schweizerische Paläontol. Gesellschaft* **95,** 1–162.

Riggs Hill Museum of Western Colorado, Colorado, USA

see MUSEUMS AND DISPLAYS

Rocky Hill Dinosaur Park

JOANNA L. WRIGHT
University of Bristol
Bristol, United Kingdom

The dinosaur tracks at Rocky Hill, Connecticut, were first uncovered in 1966 during excavations for the foundations of a state building (McDonald and Olsen, 1989). A temporary building was erected to protect the tracks but this was later destroyed in a storm. A permanent modern museum, opened in 1996, displays approximately 500 tracks. The exposures here have revealed more than 2000 reptile tracks; most of them have been reburied to preserve them for study in the future. The trackway surface at Rocky Hill is in the East Berlin Formation of the Hartford Basin. The tracks are preserved in a dark gray mudstone that was deposited along the margins of a large lake of fluctuating water level (Coombs, 1980). Tracks of THEROPODS (*Grallator, Anchisauripus,* and *Eubrontes*) and small crocodylomorphs (*Batrachopus*) have been identified at this site (Ostrom and Quarrier, 1968). By far the most common tracks at the Rocky Hill site are those of large theropods (*Eubrontes*) of a similar size to the near contemporary *Dilophosaurus* from Arizona.

Dinosaur Behavior

Some tracks of an unusual morphology have evoked some discussion (Fig. 1). They consist of a small semicircular claw imprint followed by a broad shallow depression made by the first phalangeal pad of digit III (Coombs, 1980). This pad print lies in front of the traces produced by the claws of digits II and IV, which each consist of a depression or pit connected by a shallow groove to a second pit at the rear of the track. Some prints have small mounds of sediment behind the hindmost pit of the lateral traces. Coombs (1980) suggested that these tracks were made by partially submerged theropods using their hindfeet to push themselves along in the water. These "swimming" tracks are of variable depth and were made by medium to large theropods. It is considered that the unusual morphology of these tracks cannot be explained by undertracking because they show very clear claw marks. The stride length and hypothesized posture of the trackmaker (Fig. 2) indicates that the water depth at the time of the formation of these

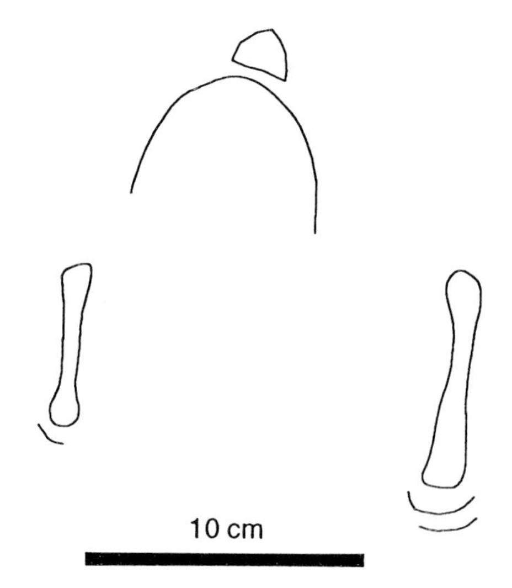

FIGURE 1 Sketch of a theropod "swimming" track.

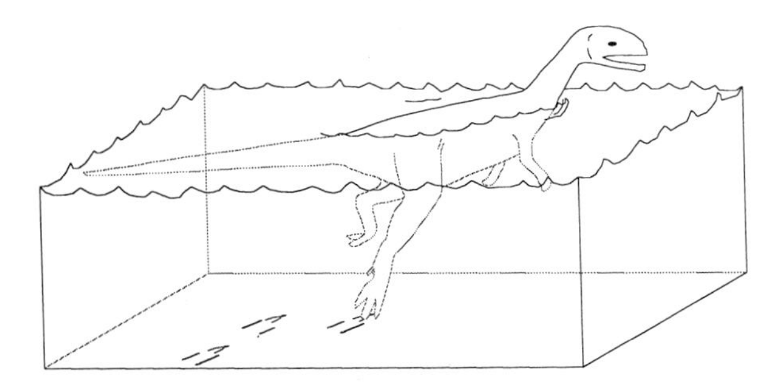

FIGURE 2 Reconstruction of the posture of the "swimming" track producer

tracks was 1.5–2.5 m. The Mount Tom tracksite near Holyoke, Massachusetts, shows several parallel trackways running in the same direction (Ostrom, 1972). Of 22 *Eubrontes* trackways, 19 run subparallel to one another (within a 30° arc). This statistically unlikely distribution has been interpreted as evidence for gregarious behavior in these large theropod dinosaurs (Ostrom, 1972). The tracks at Rocky Hill, however, are disposed in random orientations, and this, together with the preservational variation of the tracks, suggests that the Rocky Hill tracks were imprinted over a longer period of time (Ostrom, 1972).

See also the following related entries:

BEHAVIOR • CONNECTICUT RIVER VALLEY • FOOTPRINTS AND TRACKWAYS • JURASSIC PERIOD • NEWARK SUPERGROUP

References

Coombs, W. P., Jr. (1980). Swimming ability of carnivorous dinosaurs. *Science* **207,** 1198–1200.

McDonald, N. G., and Olsen, P. E. (1989). In *Tectonic, depositional and paleoecological history of Early Mesozoic rift basins, Eastern North America* (P. Olsen and Gore, Eds.), 28th International Geological Congress, Field Trip Guidebook No. T351, pp. 114–115.

Ostrom, J. H. (1972). Were some dinosaurs gregarious? *Palaeogeogr. Palaeoclimatol. Palaeoecol.* **11,** 287–301.

Ostrom, J. H., and Quarrier, S. S. (1968). The Rocky Hill dinosaurs. In *Guidebook for field trips in Connecticut: New England Intercollegiate Geological Conference* (P. M. Orville, Ed.), 60th meeting, Connecticut Geological and Natural History Survey Guidebook No. 2, pp. 1–12.

Royal Ontario Museum

HANS-DIETER SUES
Royal Ontario Museum
Toronto, Ontario, Canada

The Late Cretaceous sedimentary rocks of the badlands in southern Alberta have yielded numerous excellently preserved skeletal remains of a great diversity of dinosaurs. The Red Deer and South Saskatchewan rivers have carved deep valleys in the prairies to expose sandstones, mudstones, and coal seams that were formed between approximately 83 and 65 million years ago. During that time, forelands in the shadows of the rising Rocky Mountains bordered a large inland sea that covered much of the North American midcontinent including the conterminous prairie provinces of Canada. Abundant vegetation flourished in a setting of rivers draining eastward into that sea and provided the basis for rich continental ecosystems supporting a great diversity of dinosaurs.

Although the first bones of Late Cretaceous dinosaurs in western Canada were discovered in the 1870s by George M. Dawson, son of the pioneering Canadian geologist, Sir William Dawson, it was not until the early 20th century that Canadian institutions became interested in collecting dinosaurian remains on a large scale. Starting in 1910, Barnum Brown, working for the AMERICAN MUSEUM OF NATURAL HISTORY in New York, explored the Red Deer River region using a specially built boat. By 1912 he had amassed a spectacular collection of well-preserved dinosaurian skeletons. Upon learning of Brown's success, the authorities of the Geological Survey of Canada became concerned about the exodus of too many dinosaurian fossils to the United States. They decided to retain the services of another veteran American dinosaur hunter, Charles Hazelius Sternberg (1850–1943). From 1912 to 1915, rival field crews headed by Brown and Sternberg collected many magnificent skeletons in the same general region of southern Alberta. Sternberg's sons, Charles Mortram (1885–1981), George Friar (1883–1969), and Levi (1894–1976), continued to collect dinosaurs in Alberta for a number of Canadian institutions after Brown's crews has ceased work there.

The splendid collection of Late Cretaceous dinosaurs from Alberta housed in the Royal Ontario Museum has always been one of principal attractions for visitors of that institution. It is eloquent testimony to the determination and efforts of two men, William Arthur Parks (1868–1936) and Levi Sternberg. After an early research career focusing on the study of Paleozoic invertebrates, Parks, professor of geology at the University of Toronto and director of the Royal Ontario Museum of Palaeontology since its opening in 1914, became interested in the dinosaurs from western Canada. An early attempt by one of his staff members to collect such material in 1912 yielded only fragmentary remains. Parks began to look for financial resources to undertake fieldwork aimed at the recovery of dinosaurian skeletons for display in the new museum and eventually was successful. During the first field season in 1918, Parks and his assistant, Robert Wilson, skillfully collected the excellently preserved skeleton of the duck-billed dinosaur *"Kritosaurus" incurvimanus*. Parks soon came to realize that he needed an experienced preparator for this specimen and recruited Levi Sternberg as chief preparator of fossil vertebrates in 1919. When first exhibited at the museum in the spring of 1920, the skeleton of *"Kritosaurus"* attracted much public attention and encouraged financial support for continuing exploration. Wilson left the museum in 1920 and was replaced with Sternberg's brother-in-law, Gustav Eric Lindblad (1897–1962). Continuing with few exceptions until 1935, annual expeditions to the region of the Red Deer River recovered many excellent specimens of dinosaurs. The first two campaigns were headed by Parks himself; Sternberg and Lindblad led all later field parties. Each season produced new and scientifically important specimens. The most spectacular discovery was the skull and partial skeleton of the distinctively crested hadrosaur *Parasaurolophus walkeri*, found during the 1920 campaign and named by Parks in 1923 after the museum's principal benefactor, Sir Edmund Walker. From 1920 until his retirement and death in 1936, Parks described many new taxa of dinosaurs based on finds made during the museum's annual fieldwork.

In 1954, Levi Sternberg, accompanied by Gordon A. Edmund (b. 1924), who became a curator of vertebrate paleontology at the Royal Ontario Museum that year, made his last collection of vertebrate fossils from the Upper Cretaceous of southern Alberta. The most noteworthy result of that campaign was the recovery of four hadrosaur skulls.

Loris S. Russell (b. 1904) became assistant director of the Royal Ontario Museum of Palaeontology in 1937. Although he published several important papers on Cretaceous dinosaurs from Alberta, his principal research interests were Late Cretaceous and early Tertiary mammals and stratigraphic problems. Russell left the Royal Ontario Museum for a senior position at the National Museum of Canada in Ottawa in 1950 but returned to the Royal Ontario Museum as head of the Life Sciences Division in 1963. For many years, even long after his retirement in 1971, Russell continued fieldwork in the Late Cretaceous strata of southern Alberta and Saskatchewan.

The Royal Ontario Museum of Palaeontology and Zoology became integrated into the present-day Royal Ontario Museum in 1957, and that institution became independent from the University of Toronto in 1968. Research on dinosaurs ceased for the most part. In 1986, a field party under the leadership of Gordon Edmund and Christopher McGowan (b. 1942) explored and collected dinosaurian remains from Late Cretaceous strata in Coahuila Province, Mexico, but this material is very fragmentary and has never been studied in detail. In 1992, Hans-Dieter Sues (b. 1956) joined the museum as a curator and commenced work on a number of projects concerning dinosaurs from Canada and elsewhere.

See also the following related entries:

CANADIAN DINOSAURS • JUDITH RIVER WEDGE

Royal Saskatchewan Museum, Canada

see MUSEUMS AND DISPLAYS

Royal Scottish Museum, Scotland

see MUSEUMS AND DISPLAYS

Royal Tyrrell Museum of Palaeontology

BRUCE G. NAYLOR
Royal Tyrrell Museum of Palaeontology
Drumheller, Alberta, Canada

The Royal Tyrrell Museum of Palaeontology (Fig. 1) opened its doors to the public in the fall of 1985. Situated in the midst of the dinosaur-bearing beds of the latest Cretaceous Horseshoe Canyon Formation, the museum's dinosaur hall is home to 43 mounted skeletons, including original specimens of *Albertosaurus, Brachylophosaurus, Centrosaurus, Chasmosaurus, Coelophysis, Corythosaurus, Daspletosaurus, Edmontosaurus, Gorgosaurus, Gryposaurus, Hypacrosaurus, Lambeosaurus, Montanoceratops, Ornithomimus, Prosaurolophus, Struthiomimus,* and *Tyrannosaurus*.

The museum now houses more than 93,000 cataloged dinosaur specimens, 29,000 from DINOSAUR PROVINCIAL PARK and more than 2000 from the DEVIL'S COULEE Nest Site. Satellite operations at the park and Devil's Coulee continue to uncover new specimens on a yearly basis—a complete ornithomimid with an uncrushed skull and the almost complete skeleton of a juvenile albertosaur are the most exciting discoveries of the past few years.

Multidisciplinary research programs concentrate on the Late Cretaceous of the Western Interior, with curators working together and with visiting scientists from around the world. Fieldwork has not been limited to Alberta, with concentrated and ongoing work in China and other areas outside of the province.

Rutgers University Geology Museum, New Jersey, USA

see MUSEUMS AND DISPLAYS

FIGURE 1 The Royal Tyrrell Museum, a leading center of dinosaur research, attracts thousands of visitors and scientists from around the world annually.

Saito Ho-on Kai Museum of Natural History, Japan

see MUSEUMS AND DISPLAYS

Samcheonpo

MARTIN LOCKLEY
University of Colorado at Denver
Denver, Colorado, USA

Situated on the southern coast of South Korea, the dinosaur tracksites of the Samcheonpo region comprise one of the richest concentrations of dinosaur footprints known anywhere. More than 500 trackways have been recorded from more than 200 discrete stratigraphic levels. The tracks are Cretaceous in age and consist of large numbers of sauropod and ornithopod footprints. Sauropod and ornithopod tracks rarely occur at the same stratigraphic levels, suggesting that two different trackmaking groups visited the area at different times. The sauropod track sample is one of the largest known anywhere and includes a large number of tracks attributed to small (juvenile) individuals. The ornithopod track sample is large, with many levels providing parallel trackway evidence of gregarious behavior. Bird tracks are also abundant in the region.

See also the following related entry:

FOOTPRINTS AND TRACKWAYS

Reference

Lockley, M. G., Santos, V. F., Meyer, C. A., and Hunt, A. P. (Eds.) (1994). *Aspects of Sauropod Biology*. Gaia: Geoscience Magazine of the National Natural History Museum, Lisbon, Portugal.

San Bernardino County Museum, California, USA

see MUSEUMS AND DISPLAYS

San Diego Natural History Museum, California, USA

see MUSEUMS AND DISPLAYS

Saurischia

KEVIN PADIAN
University of California
Berkeley, California, USA

Saurischia (Fig. 1) was established by Seeley in lectures presented in 1887 (published in 1888); he divided Owen's Dinosauria into two groups based on the form of the pelvis, the braincase, the vertebrae, and the armor (see DINOSAURIA, DEFINITION). Although for most of the 20th century, the idea that Saurischia was a natural group was doubted by many workers, Gauthier (1986) established the monophyly of Saurischia by cladistic analysis. Saurischia and Ornithischia are sister taxa, stem groups of Dinosauria; hence any member of Dinosauria must belong to one of these two groups, following its establishment by Owen (1842). *MEGALOSAURUS*, the earliest described saurischian, was explicitly included in this group by Owen. In the PHYLOGENETIC SYSTEM, Saurischia can be defined as all Dinosauria closer to birds than to *Triceratops* (Padian and May, 1993).

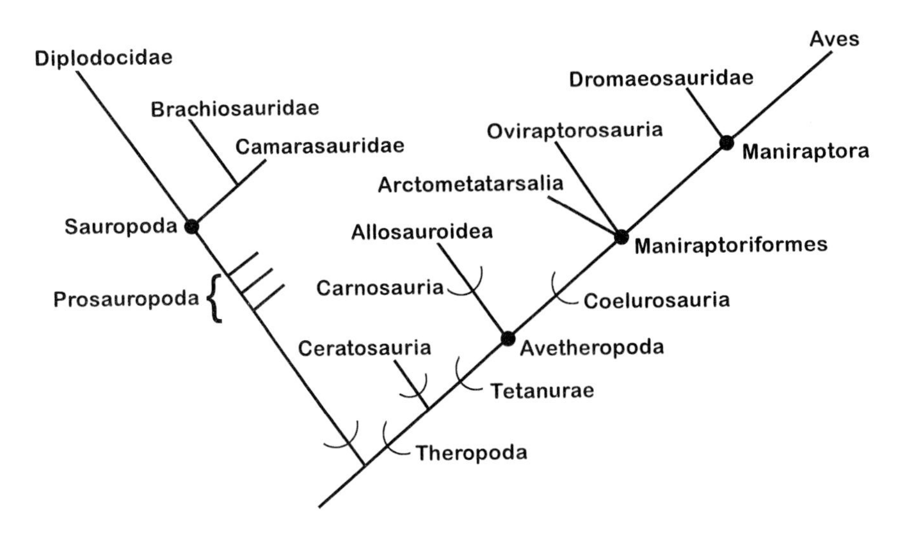

FIGURE 1 Phylogeny of Saurischia.

Most dinosaurian students of the early 20th century, if pressed, probably would not have strongly advocated the monophyly of Saurischia as they would have for Ornithischia. It was generally recognized that saurischians retained many features of more "primitive" reptiles (e.g., Romer, 1933), and although grouping them together was perhaps more than a simple matter of convenience, their association based on the retention of primitive characters was not unacceptable, as it is within the phylogenetic system. THEROPODS and SAUROPODOMORPHS, while retaining Owen's basic dinosaurian features of pelvis and hindlimb stance, were clearly very different, as their skulls, dentitions, body plans, and functional anatomy showed. Charig *et al.* (1965), in a very influential paper, were among those who concluded that theropods and saurpodomorphs had descended from independent archosaurian lineages; another very influential paper (Bakker and Galton, 1974) maintained that "prosauropod" dinosaurs were a primitive grade from which both ornithischians and other saurischians evolved. These conclusions, however, depended heavily on the confusion of associated materials in assemblages as well as on the taxonomic inclusion of nondinosaurs within dinosaurian groups on the basis of primitive or convergent features. For example, Charig *et al.* (1965) recognized that the serrated, recurved, carnivorous teeth found with postcranial bones otherwise identical to those of prosauropods did not belong together; hence there was no further need to advocate a lineage of carnivorous prosauropods from which other saurischians might have evolved. Also, until the 1980s it was generally considered that forms such as *Teratosaurus, Zanclodon, Poposaurus,* and *Ornithosuchus* were theropod "carnosaurs" based on their macropredaceous features of jaws and teeth; however, recent cladistic analyses have found that all these valid taxa, plus other presumed carnosaurs from the Triassic, are in fact pseudosuchians (Ornithosuchidae, Rauisuchidae, etc.).

Gauthier (1986, pp. 15–18) found a suite of features characteristic of Saurischia. Contact is reduced or lost between the maxillary process of the premaxilla and the nasal. The temporal MUSCULATURE extends onto the frontal bones rather than just the parietals. The posterior cervicals are elongated, so the neck appears sensibly longer than in ornithischians or dinosaurian out-groups (Fig. 2). The axial postzygapophyses are set lateral to the prezygapophyses rather than equally distant from the central midline. Epipophyses are present on the anterior cervical postzygapophyses, and the trunk vertebrae bear hyposphene–hypantrum accessory intervertebral articulations. The manus is elongated to more than 45% of the humerus plus forearm. The manus is also markedly asymmetrical: In other dinosaurs and their outgroups the third finger is the longest, but in saurischians it is invariably the second; the fourth and fifth fingers are reduced and the bases of their metacarpals line on the palmar surfaces of the third and fourth

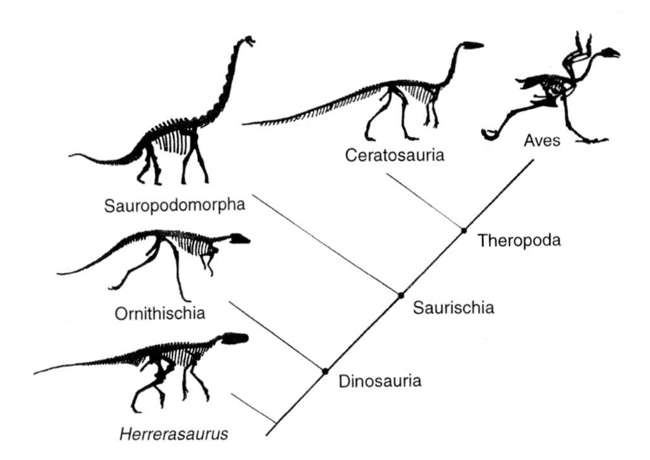

FIGURE 2 Saurischians have elongated necks, sometimes formed by adding cervicals (e.g., in sauropods and derived birds), but originally by elongating the individual cervical vertebrae, particularly the posterior ones (courtesy T. Rowe and L. Dingus).

digits, respectively (Fig. 3). The pollex (thumb) in saurischians is also unique: Its metacarpal is short, only about half the length of the second metacarpal, and its distal ends bear two condyles that are strongly asymmetrical, having the effect of setting off the digit at a more pronounced angle from the rest of the hand. This digit is also robust, with a large claw, and its first phalanx is longer than any other in the hand. These features are not found in ornithischians nor in herrerasaurs or in other ornithodirans.

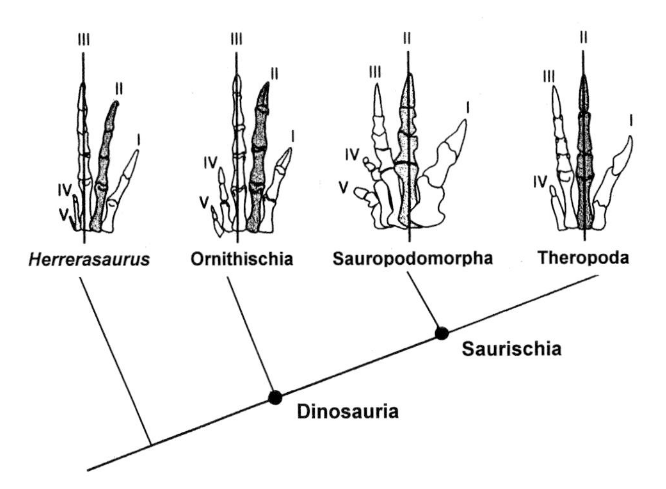

FIGURE 3 The second digit is the longest in every saurischian dinosaur, including living birds, but in virtually no other animals (courtesy T. Rowe and L. Dingus).

Because ornithischians are relatively derived, and saurischians appear to retain many generalized features of archosaurs, they have served as an unintentional repository for numerous taxa that are outside Dinosauria. Gauthier (1986, p. 15) noted that saurischians have traditionally been considered "primitive dinosaurs" with respect to ornithischians; hence, Galton (1977) described *Herrerasaurus* as a "primitive saurischian," even though it was described as "primitive" compared to all other dinosaurs. The same could be said of the treatment of *Staurikosaurus* by Colbert (1970) and Galton (1977), of *Herrerasaurus* and *Ischisaurus* by Reig (1963), of both *Herrerasaurus* and *Staurikosaurus* by Benedetto (1973), and of *Spondylosoma* by von Huene (1942) (see HERRERASAURIDAE; STAURIKOSAURIDAE). Very primitive dinosaurs or near-dinosaurs were placed in Saurischia by default because these forms lacked obvious diagnostic ornithischian features. In Gauthier's analysis, however, these forms are outside Dinosauria (see also Holtz, 1995; Holtz and Padian, 1995), although some later analyses (e.g., Novas, 1992, 1993; Sereno and Novas, 1992, 1993; Sereno, 1993) have placed them as basal theropods (see HERRERASAURIDAE), along with *Eoraptor* (Sereno *et al.*, 1993) (see PHYLOGENY OF DINOSAURS).

Within Saurischia, two stem taxa are recognized: SAUROPODOMORPHA, which includes PROSAUROPODA and SAUROPODA, and THEROPODA, the carnivorous dinosaurs (Gauthier, 1986). Prosauropods were given their name because they have traditionally appeared to be the forerunners of sauropods. The subgroups Anchisauridae (Thecodontosauridae), Plateosauridae, and Melanorosauridae (sometimes including Blikanosauridae; see SAUROPODOMORPHA; PROSAUROPODA) were considered to form a grade of increasing size, longer necks, and tendency toward quadrupedality that culminated in the appearance of the Sauropoda, which were already large, ponderous quadrupeds from their first appearance. Not all workers have accepted this picture. Charig *et al.* (1965) maintained that the melanorosaurids appeared much too early in time to be directly ancestral to sauropods, which are and were then first known from the Early Jurassic. They presumed instead that prosauropods and sauropods had descended independently from quadrupedal archosaurs ("thecodontians") of the Triassic, and the prosauropod lineage had basally acquired bipedality. This corresponded to the basic conclusion of these authors that dinosaurs in general were secondarily bipedal, having descended from lineages related to basal archosaurians such as *Euparkeria* and erythrosuchians of the Early Triassic. Subsequent analyses of dinosaur origins have not sustained this general view (see ARCHOSAURIA; DINOSAUROMORPHA; ORIGIN OF DINOSAURS; PHYLOGENY OF DINOSAURS), although the paper by Charig *et al.* stimulated much discussion and thought at the time and was influential for decades afterward.

The monophyly of Sauropodomorpha is not currently in doubt and has been sustained by a series of characters (Gauthier, 1986; see SAUROPODOMORPHA). Several recent analyses by Sereno (1989), Galton (1990), Upchurch (1993), and Gauffre (1994) have sustained the monophyly of Prosauropoda, and the supporting evidence is discussed in the appropriate entries. In other analyses, this supporting evidence largely consists of plesiomorphic characters. Sereno's character analysis has not been published; Galton's hypothesized synapomorphies are mostly plesiomorphies or states that can be seen equally as transitional between basal dinosauriformes and sauropods (a reversal in reduction of the fifth pedal digit is required); characters listed by Upchurch and Gauffre have yet to be reviewed by other workers (see PROSAUROPODA). Upchurch (see SAUROPODOMORPHA) notes that the acceptance of prosauropod monophyly raises several questions, including why there should be as yet no Triassic record of Sauropoda, why a convergent trend developed in prosauropod lineages and in sauropods by the Early Jurassic, and why prosauropods disappeared in the Early Jurassic, just as the very similar sauropods were coming onto the scene. Sauropods may have been competitively superior in the contemporary ecological milieu; unfortunately our record of Early Jurassic sauropods is not terribly rich and, as noted, nothing is known of them in the Triassic, so the answers to these intriguing questions are still shrouded in the mists of time.

Regardless of their origins, Sauropoda soon became the largest animals ever to walk on land, with estimates of adult size ranging from 30 to 50 and even 80 tons, by some accounts. The first described sauropods were the incomplete remains of Middle and Late Jurassic cetiosaurids from England, initially taken by Owen for giant, whale-like crocodiles (this is why he did not include them in Dinosauria in 1842; see DINOSAURIA: DEFINITION). The association with water accompanied sauropods from that time onward, even though it was soon clear that their limbs were columnar and the carcasses of the cetiosaurids that had first been described from marine sediments had washed in (see Desmond, 1975, for an historical review). Still, it was difficult for many to accept that these reptiles were capable of supporting their great bulks on land, and so they were usually pictured in marshes and swamps, long necks bent over to feed on aquatic vegetation (for which, one presumes, the long necks would not have been necessary). Some workers inferred reasonably that the dorsally situated nares on the skull would have been effective snorkeling devices for submerged animals avoiding predators; however, elementary physics demonstrates the impossibility of breathing under such circumstances (Colbert, 1952). Bakker (1971), reviving a well-supported but politically suppressed argument of Riggs (1903), argued that sauropods were much more like elephants than like hippos, citing the columnar limbs and the comparative angles of their joints, the digitigrade posture, shortened digits, and hoof-like tracks. A fully terrestrial view of sauropods accords well with the paleoecology of the deposits in which they are found (Dodson *et al.*, 1980).

Sauropod relationships are extremely problematic. McIntosh (1990; see SAUROPODA) notes the inevitability of extensive character convergence in nearly all anatomical areas of the skeleton. Some workers have perforce relied on girdle and limb characters for systematic indications; others have used vertebral characters. The central difficulty is that so few skulls are preserved, and association of skulls with known skeletons is not always easy. Vulcanodontid sauropods are the earliest known (Early Jurassic), and they have the combination of characters that would be expected: long, but not extraordinarily long necks and a generalized pelvis with columnar limbs. Cetiosaurs (primarily Middle and Late Jurassic) are not well known from skull material but there are ample records of (unfortunately incomplete) postcranial remains. The Late Jurassic is when great sauropod faunas are first known, including camarasaurids, diplodocids, and brachiosaurids. All these taxa are known from both the Morrison Formation and the Tendaguru beds. It appears that a diplodocid lineage evolved in China perhaps by the Middle Jurassic, but certainly by the Late Jurassic, and continued through the Late Cretaceous (e.g., *Mamenchisaurus* and *Nemegtosaurus*), but their relationships to other sauropods are not fully clear. Finally, in the Cretaceous, titanosaurids are the dominant sauropods; they are mostly Gondwanan forms but some reached southern North America by the Late Cretaceous (see KIRTLAND FORMATION; FRUITLAND FORMATION). Otherwise, sauropods had mostly disappeared from North America by the end of the Early Cretaceous, but the reasons are unknown.

Theropods, although the only carnivorous group of dinosaurs, are the most diverse clade. As noted previously, theropods retain the primitive meat-eating habit, and many of the skeletal correlates, of their nondinosaurian ancestors (see DINOSAUROMORPHA; HERRERASAURIDAE; ORNITHODIRA; PHYLOGENY OF DINOSAURS; THEROPODA). Herrerasaurids and *Eoraptor* are placed as basal theropods by some authors (see HERRERASAURIDAE), but others reserve judgment because these taxa lack many synapomorphies of both theropods and saurischians, such as the long neck, the elongated second digit, and the failure of the first metatarsal to contact the ankle (Fig. 4; see PHYLOGENY OF DINOSAURS). According to Gauthier (1986) and Rowe and Gauthier (1990), Theropoda can be divided into two basal stem groups: CERATOSAURIA and TETANURAE. Ceratosaurs are known from the Late Triassic (*Coelophysis* and *Liliensternus*) through the Early Jurassic (*Dilophosaurus, Segisaurus, Sarcosaurus,* and *Syntarsus*) and Late Jurassic (*Ceratosaurus*), and they now include the ABELISAURIDAE, which roamed southern continents throughout the Cretaceous. The stem taxon Tetanurae, according to Gauthier (1986), comprises two principal stems, CARNOSAURIA and COELUROSAURIA, which Holtz (1994) united at the node NEOTETANURAE. Carnosauria comprises all taxa closer to *Allosaurus* than to birds. These are the classic Late Jurassic and Early Cretaceous large carnivores. Coelurosauria comprises all taxa closer to birds than to *Allosaurus,* and there is a wide diversity of these Late Jurassic and Cretaceous forms, including TROODONTIDS, ELMISAURIDS, ORNITHOMIMIDS, DROMAEOSAURIDS, THERIZINOSAURIDS, OVIRAPTOROSAURIA, TYRANNOSAURIDS, and AVES. As far as current knowledge permits, the Cretaceous radiations of theropods seem to have been very different on different continents (Holtz, 1994, 1996; Sereno, 1995). Coelurosaurs seem to have dominated Laurasian continents, especially in North America and Asia. Abelisaurids, the surviving ceratosaurs (Neoceratosauria), spread throughout South America and Africa and into Europe; carnosaurs survived into the Early Cretaceous of Africa and are generally known from taxa in South America and Asia as well, but *Giganotosaurus* (Late Cretaceous, Patagonia) is the latest Cretaceous record of the group. Aves, which first appeared in the Late Jurassic, is the most successful theropod taxon, having radiated over all continents during the Cretaceous (especially Enantiornithes: see AVES), surviving to the present day.

The stark ecological differentiation between the

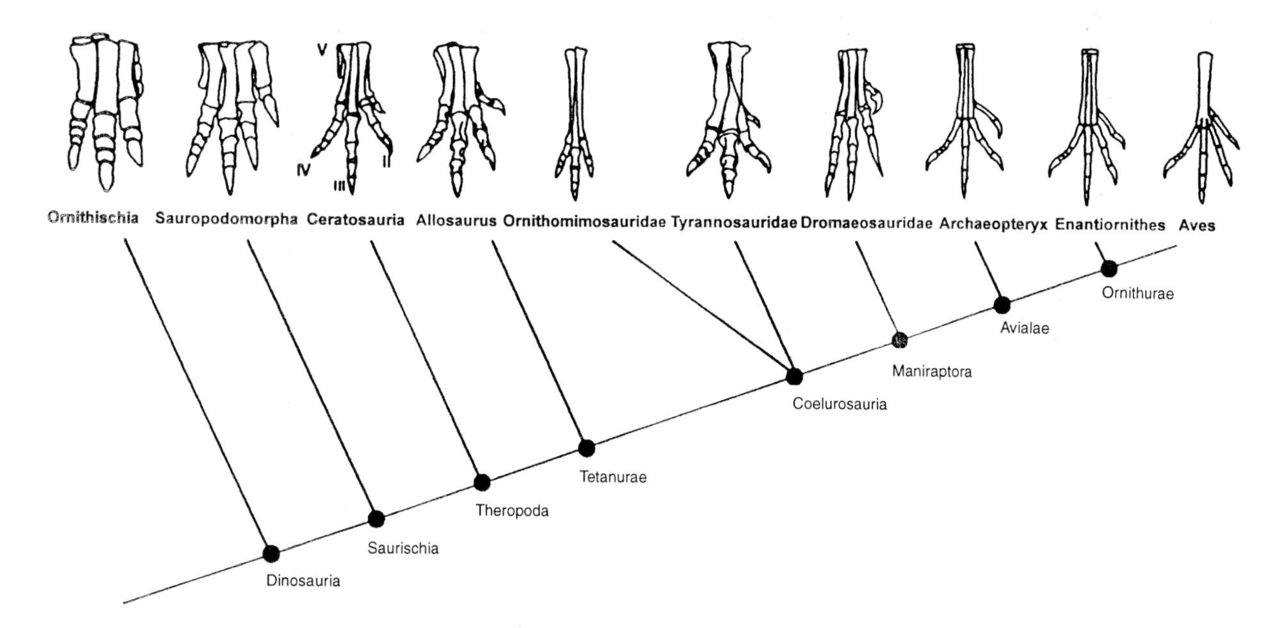

FIGURE 4 The foot of saurischian dinosaurs and their outgroups (courtesy T. Rowe and L. Dingus).

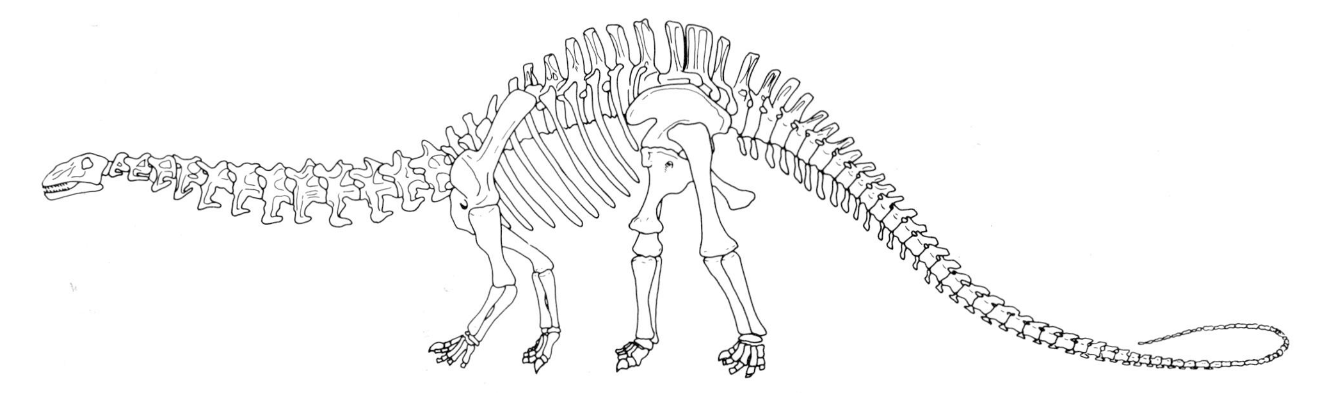

FIGURE 5 *Apatosaurus* is a sauropod whose remains have been excavated from the Morrison Formation of Colorado, Wyoming, and Oklahoma. The quadrupedal body plan of these behemoths is in contrast to theropod saurischians, whose derived members are small, lightly built bipeds.

two major lineages of Saurischia seems to have been in place by the Carnian Stage of the Late Triassic and continued through the Cretaceous. Sauropodomorphs tended to large size by the Early and Middle Jurassic, and by the Late Jurassic included the hugest beasts ever to roam the earth (Fig. 5). Their long necks enabled them to browse in the highest trees, and at least in diplodocids a series of canoe-shaped hemal arches midway along the tail has been invoked as an aid to using the tail as a prop for rearing up on the hindlimbs in order to feed even higher in the trees. Brachiosaurs had no such structures but their forelimbs were longer than their hindlimbs, and their necks were longer than their tails. Some criticism has been directed at a recently mounted exhibit in the American Museum of Natural History showing a female *Barosaurus* rearing up on hindlimbs to defend her offspring against marauding allosaurs. A similar bipedal action was portrayed for *Brachiosaurus* in the film *Jurassic Park*. It is not possible to know how typical or plausible such poses were, but it is certain that males had to mount females from behind in order to procreate, so at least some bipedal activity, with concomitant stress on the hindlimbs, was necessary and regular.

Theropods, meanwhile, became the most diverse group of dinosaurs, at least in terms of species diversity, with well over 100 valid genera currently accepted and perhaps another 25–50 questionable or dubious forms named from Cretaceous deposits—not counting Aves, of which at least another two dozen Mesozoic genera are now known (see AVES) and thousands more from the Tertiary and recent. Nonavian theropods were themselves ecologically diverse. One tends to think of small lizard and insect chasers, on the one hand, and macropredaceous terrors, on the other hand, but the situation was more complex. *Baryonyx* may have been a fish eater; ornithomimids and oviraptors may have been omnivorous or even frugivorous as the opportunity presented itself. The giant, long-armed *Deinocheirus* was at one time conjectured to have been an anteater. At least one specimen of *Coelophysis* is found with the remains of a juvenile in its stomach. Large carnivores, meanwhile, evolved in several lineages of theropods, taking over from the mid-sized rauisuchid, ornithosuchid, and herrerasaurid carnivores of the Late Triassic. Ceratosaurs achieved large size in the Early Jurassic *Dilophosaurus*, the Late Jurassic *Ceratosaurus*, and the Cretaceous abelisaurids of the southern continents. True carnosaurs, related to *Allosaurus*, also proliferated in the Late Jurassic and Early Cretaceous and at least one lineage, represented by *Giganotosaurus*, seems to have survived through the Late Cretaceous in Patagonia. Meanwhile, Late Cretaceous tyrannosaurids independently evolved large size from coelurosaurian ancestors; they may have originated in Asia and migrated to North America. Dinosaurs represented the first ecological guild of terrestrial animals to be dominated by herbivorous types (ornithischians and sauropodomorphs) rather than by meat eaters mainly eating each other, with few plant eaters. Nevertheless, carnivores were omnipresent and ecologically diverse, much as they are in mammalian communities today.

See also the following related entries:

DINOSAURIA • ORNITHISCHIA

References

Bakker, R. T. (1971). The ecology of the brontosaurs. *Nature* **229,** 172–174.

Bakker, R. T., and Galton, P. M. (1974). Dinosaur monophyly and a new class of vertebrates. *Nature* **248,** 168–172.

Benedetto, J. L. (1973). Herrerasauridae, nueva familia de saurisquios triásicos. *Ameghiniana* **10,** 89–102.

Charig, A. J., Attridge, J., and Crompton, A. W. (1965). On the origin of the sauropods and the classification of the Saurischia. *Proc. Linnean Soc. London* **176,** 197–221.

Colbert, E. H. (1952). Breathing habits of the sauropod dinosaurs. *Ann. Magazine Nat. History* **5**(12), 708–710.

Colbert, E. H. (1970). A saurischian dinosaur from the Triassic of Brazil. *Am. Museum Novitates* **2405,** 1–39.

Desmond, A. J. (1975). *The Hot-Blooded Dinosaurs: A Revolution in Palaeontology.* Blond & Briggs, London.

Dodson, P., Behrensmeyer, A. K., Bakker, R. T., and McIntosh, J. S. (1980). Taphonomy and paleoecology of the dinosaur beds of the Jurassic Morrison Formation. *Paleobiology* **6,** 208–232.

Galton, P. M. (1977). *Staurikosaurus pricei*, an early saurischian dinosaur from the Triassic of Brazil, with notes on the Herrerasauridae and Poposauridae. *Paläontol. Z.* **51,** 234–245.

Galton, P. M. (1990). Basal Sauropodomorpha—Prosauropods. In *The Dinosauria* (D. B. Weishampel, P. Dodson, and H. Osmólska, Eds.), pp. 320–344. Univ. of California Press, Berkeley.

Gauffre, F. X. (1994). The prosauropod dinosaur *Azendohsaurus laaroussii* from the Upper Triassic of Morocco. *Palaeontology* **36,** 897–908.

Gauthier, J. A. (1986). Saurischian monophyly and the origin of birds. *Mem. California Acad. Sci.* **8,** 1–55.

Holtz, T. R., Jr. (1994). The phylogenetic position of the Tyrannosauridae: Implications for theropod systematics. *J. Paleontol.* **68,** 1100–1117.

Holtz, T. R., Jr. (1995). A new phylogeny of the Theropoda. *J. Vertebr. Paleontol.* **15**(Suppl. to No. 3), 35A.

Holtz, T. R., Jr. (1996). Phylogenetic taxonomy of the Coelurosauria (Dinosauria: Theropoda). *J. Paleontol.* **70,** 536–538.

Holtz, T. R., Jr., and Padian, K. (1995). Definition and diagnosis of Theropoda and related taxa. *J. Vertebr. Paleontol.* **15**(Suppl. to No. 3), 35A.

McIntosh, J. S. (1990). Sauropoda. In *The Dinosauria* (D. B. Weishampel, P. Dodson, and H. Osmólska, Eds.), pp. 345–401. Univ. of California Press, Berkeley.

Novas, F. E. (1992). Phylogenetic relationships of the basal dinosaurs, the Herrerasauridae. *Palaeontology* **35,** 51–62.

Novas, F. E. (1993). New information on the systematics and postcranial morphology of *Herrerasaurus ischigualastensis* (Theropoda: Herrerasauridae) from the Ischigualasto Formation (Late Triassic) of Argentina. *J. Vertebr. Paleontol.* **13,** 400–424.

Owen, R. (1842). Report on British fossil reptiles, Part II. *Rep. Br. Assoc. Adv. Sci.* **1841,** 60–294.

Padian, K., and May, C. L. (1993). The earliest dinosaurs. *New Mexico Museum Nat. History Sci. Bull.* **3,** 379–381.

Reig, O. A. (1963). La presencia de dinosaurios saurisquios en los "Estratos de Ischigualasto" (Mesotriasico superior) de las Provincias de San Juan y La Rioja (Republica Argentina). *Ameghiniana* **3,** 3–20.

Riggs, E. S. (1903). *Brachiosaurus altithorax*, the largest known dinosaur. *Am. J. Sci.* **15**(4), 299–306.

Romer, A. S. (1933). *Vertebrate Paleontology.* Univ. of Chicago Press, Chicago.

Rowe, T. R., and Gauthier, J. A. (1990). Ceratosauria. In *The Dinosauria* (D. B. Weishampel, P. Dodson, and H. Osmólska, Eds.), pp. 151–168. Univ. of California Press, Berkeley.

Seeley, H. G. (1887) [1888]. On the classification of the fossil animals commonly named Dinosauria. *Proc. R. Soc. London* **43**(206), 165–171.

Seeley, H. G. (1888). The classification of the Dinosauria. *Rep. Br. Assoc. Adv. Sci.* **1887,** 698–699.

Sereno, P. C. (1989). Prosauropod monophyly and basal sauropodomorph phylogeny. *J. Vertebr. Paleontol.* **9**(Suppl. to No. 3), 38A.

Sereno, P. C. (1993). The pectoral girdle and forelimb of the basal theropod *Herrerasaurus ischigualastensis. J. Vertebr. Paleontol.* **13,** 425–450.

Sereno, P. C. (1995). Theropoda: Early evolution and major patterns of diversification. *J. Vertebr. Paleontol.* **15**(Suppl. to No. 3), 52A–53A.

Sereno, P. C., and Novas, F. E. (1992). The complete skull and skeleton of an early dinosaur. *Science* **258,** 1137–1140.

Sereno, P. C., and Novas, F. E. (1993). The skull and neck of the basal theropod *Herrerasaurus ischigualastensis. J. Vertebr. Paleontol.* **13,** 451–476.

Sereno, P. C., Forster, C. A., Rogers, R. R., and Monetta, A. M. (1993). Primitive dinosaur skeleton from Argentina and the early evolution of Dinosauria. *Nature* **361,** 64–66.

Upchurch, P. (1993). The anatomy, phylogeny and systematics of sauropod dinosaurs. Unpublished Ph.D. thesis, University of Cambridge, Cambridge, UK.

von Huene, F. (1942). *Die fossilen Reptilien des südamerikanischen Gondwanalandes. Ergebnisse der Sauriergrabungen in Südbrasilien 1928/29.* Beck'sche Verlagsbuchhandlung, Munich.

Sauropoda

John S. McIntosh
Middletown, Connecticut, USA

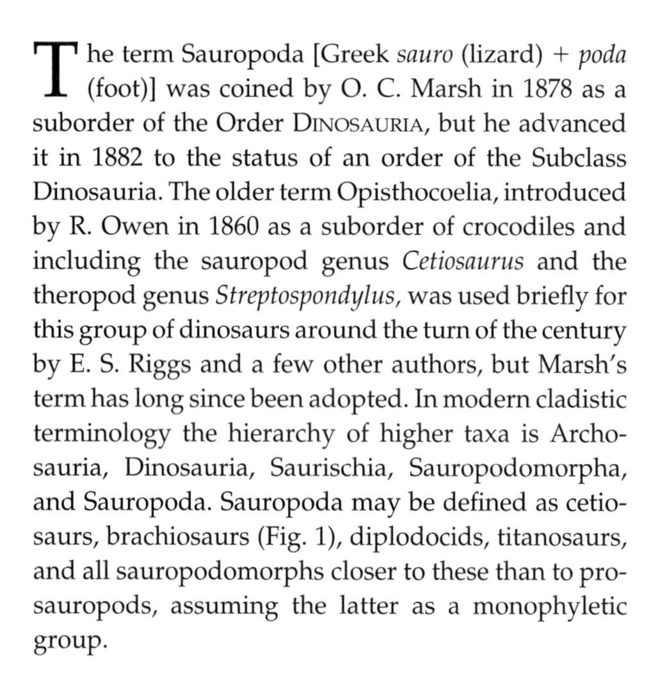

The term Sauropoda [Greek *sauro* (lizard) + *poda* (foot)] was coined by O. C. Marsh in 1878 as a suborder of the Order Dinosauria, but he advanced it in 1882 to the status of an order of the Subclass Dinosauria. The older term Opisthocoelia, introduced by R. Owen in 1860 as a suborder of crocodiles and including the sauropod genus *Cetiosaurus* and the theropod genus *Streptospondylus*, was used briefly for this group of dinosaurs around the turn of the century by E. S. Riggs and a few other authors, but Marsh's term has long since been adopted. In modern cladistic terminology the hierarchy of higher taxa is Archosauria, Dinosauria, Saurischia, Sauropodomorpha, and Sauropoda. Sauropoda may be defined as cetiosaurs, brachiosaurs (Fig. 1), diplodocids, titanosaurs, and all sauropodomorphs closer to these than to prosauropods, assuming the latter as a monophyletic group.

Anatomy

The sauropods were, for the most part, very large quadrupedal animals including the largest terrestrial beasts that ever lived. The head was small, the neck long, and the tail long. The articulation between the neck and the skull was a very weak one. Thus, there was a tendency for the skull to snap off and drift downstream in the rivers in which the dead animals had fallen. Consequently, for many years the skulls in most of the otherwise well-established taxa were unknown. That situation is being gradually rectified, but there are still many forms in which the skull is unknown, including all the members of one very large group, the Titanosauridae.

Superficially, the sauropod skulls may be divided into two types. The first (basal) of these have large, spoon-shaped teeth and nares (nostrils) that open outward. Those of the other type are long snouted with peg-like teeth confined to the front part of the jaws and nares that open upward on the top of the skull. Of the five pairs of major openings in the sauropod skull, the nares, subcircular orbits (containing a sclerotic ring), and lateral temporal fenestrae face outward and are the largest. The supratemporal fenestrae are smaller and open upward, whereas the anteriormost antorbital fenestrae are the smallest. Most genera lack mandibular fenestrae except for a small circular opening in the surangular. The palate is highly vaulted.

The primal number of presacral vertebrae in the Sauropodomorpha is 25. There are two primary sacrals to which are often added one or more from each of the dorsal and caudal series. The typical prosauropod *Plateosaurus* has 10 neck or cervical vertebrae, 15 trunk or dorsals and three sacrals (a caudosacral having been added), and approximately 46 tail or caudal vertebrae. The trend in the sauropods is for the "cervicalization" of the presacral column, i.e., the modification of anterior dorsals into cervicals, with the addition of as many as 4 or 5 additional cervicals in some of the Chinese forms. The presacral openings in the sides of the centra (pleurocoels) are simple in more basal forms but complex in the neck region of many derived forms and very deep in the dorsal region of some. The neural arches and spines, particularly in more derived forms, are greatly reduced to a set of struts and laminae providing maximum strength for minimum weight—a true triumph of engineering. The sacrum in most adult genera consists of five functioning sacrals, including one dorsosacral and two caudosacrals. In the Cretaceous titanosaurids and in very old individuals of an occasional Jurassic form, a second dorsosacral is incorporated. The tail is long, with about 45 vertebrae in basal forms and more than 80 in some of the diplodocids, the last 40 or so being reduced to simple rods of bone that form a whiplash. The scapula has an expanded lower (anterior) end to which, in adults, a large coracoid usually fuses. In several genera, what are supposed to be slender rod-like clavicles have been reported. Recently, one genus, *Apatosaurus*, has been shown to possess a full set of gastralia. The pelvic bones are large plates, usually not fused to one another. The ilium has developed a large anterior lobe, which tends to differentiate the sauropods from the prosauropods.

FIGURE 1 *Brachiosaurus* are among the largest dinosaurs, with necks and forelimbs which are longer than their tails and hindlimbs. Recovered from the Morrison Formation and Tendaguru rocks of North America and Africa, respectively, these sauropods may be related to camarasaurs (from Paul, 1988).

The limb bones tend to have reduced central cavities. The humeri are longer than the radii and ulnae, and the femur is longer than the tibia and fibula. In most forms the hindlimbs are longer than the forelimbs, but in some brachiosaurids this is reversed. The wrist or carpus is reduced to three elements in basal forms, to two elements in most later forms, and to a single element in at least one genus (*Apatosaurus*). The manus has five digits with reduced phalanges on the outer four. Usually, only digit I possesses a claw, but there is some evidence that in some of the Cretaceous forms even this claw is absent. The ankle or tarsus is reduced to a large astragalus, strongly interlocking with the tibia, and a much smaller globular calcaneum, which disappears in the diplodocids. The pes has five digits with claws on digits I, II, and III.

Systematics

Six to eight or nine sauropod subtaxa are currently recognized. The most basal, the Vulcanodontidae, is represented by *Vulcanodon* from the Triassic–Jurassic boundary in Zimbabwe. It is incompletely known and might possibly be an advanced prosauropod. The skull and vertebrae of the neck and trunk are unknown, as is the very diagnostic anterior part of the ilium. The girdle bones resemble those of both the sauropods and prosauropods, whereas the limbs appear to be more nearly sauropod, except that the femur retains an inner trochanter, a feature absent in all later sauropods.

The Cetiosauridae is a generalized group of largely Middle Jurassic sauropods, in most of which the skull is unknown. The teeth are of the broad type. The vertebral spines in the neck and trunk region are simple (uncleft). The tail is of only moderate length and the hindlimbs are only moderately longer than the fore. Typical representatives are *Cetiosaurus* in England, *Patagosaurus* in Argentina, *Barapasaurus* in India, and *Bellusaurus* in China. *Haplocanthosaurus* is a survivor from the Upper Jurassic of North America. It retained a large number of trunk vertebrae (14). A primitive Middle Jurassic Chinese form, *Shunosaurus*, is known from a number of complete skeletons with skulls. Several are mounted in Zigong. This animal possessed a number of small spatulate teeth and retained an opening in the lower jaw. It had 12 vertebrae in the neck, 13 in the trunk, 4 in the sacrum, and 44 in the tail. The middle chevron bones are of the forked type typical of the Diplodocidae to be discussed below. A bony club somewhat reminiscent of that found in the ankylosaurs was present on the end of the tail. A contemporary, somewhat heavier, related Chinese sauropod was *Datousaurus*. Reference of the two Chinese forms to a separate taxa from the other cetiosaurids is not unlikely but should be delayed until more is known of the latter genera.

A related but more advanced group is the Brachiosauridae. It is characterized by large teeth, greatly elongated neck vertebrae with very long cervical ribs, and deep pleurocoels (side cavities) in the dorsal vertebrae. The tail was relatively short. Most characteristic were the greatly elongated forelimbs. In the typical Upper Jurassic *Brachiosaurus* from Tanzania and Colorado, the forelimbs were longer than the hindlimbs. The skull of this animal was large, with a vaulted nasal bone. There were 13 vertebrae in the neck, 11 or 12 in the trunk, and 5 in the sacrum. The metacarpals were especially elongated, and the claw on digit I of the manus was greatly reduced. *Brachiosaurus* was one of the heaviest land animals that ever lived. Colbert estimated its weight to be 86 tons, but D. Russell calculates this as too high. Mounted skeletons are to be found in Berlin and Chicago. A similar animal with slightly shorter forelimbs, found in the Middle Jurassic of Madagascar, has been referred to the ill-defined British genus *Bothriospondylus*. To this animal may belong a large amount of juvenile material named *Lapparentosaurus*, also from Madagascar. *Pleurocoelus*, from the Lower Cretaceous of Maryland and Texas, also has greatly hollowed-out vertebrae, but more slender teeth, and appears to be a smaller member of this group.

The Camarasauridae resembles the Brachiosauridae in that the neck and trunk vertebrae were very hollow and the cervical ribs very long. On the other hand, these vertebrae were more specialized in that the vertebral spines in much of the column possessed a deep U-shaped cleft. The hindlimbs were longer than the forelimbs. *Camarasaurus* was the most common North American sauropod. It was found in the Upper Jurassic of Colorado, Wyoming, Utah, and other Morrison Formation locales. Its skull was short with a "bulldog-like" muzzle and large spoon-shaped teeth. There were 12 vertebrae in both the neck and trunk, 5 in the sacrum, and 53 in the tail. Mounted skeletons are found in New Haven, Pittsburgh (where there is a 5-m-long, nearly complete juvenile skeleton), Washington, DC, Drumheller, Alberta, Canada, and Albuquerque. The European representative of the group is the very similar *Aragosaurus*. *Euhelopus*, from the Upper Jurassic of China, had a skull resembling that of *Camarasaurus*, with similar teeth. The vertebral spines are only slightly cleft, and the neck and trunk vertebrae have increased to 17 and 14, respectively. This has led some experts to refer it to the Camarasauridae and others to the Chinese group Mamenchisauridae, which will be discussed below. A partial skeleton is mounted in Uppsala, Sweden. Another controversial form is *Opisthocoelicaudia* from the Upper Cretaceous of Mongolia. This animal possessed the deeply U-shaped cleft spines but had a very short tail with vertebrae unlike those of any other sauropod. The centra articulated with one another with a ball on the front and a cup on the back (opisthocoelous condition). The all-important head and neck are unknown. Until these parts are recovered, *Opisthocoelicaudia* may be referred to the Camarasauridae with a query.

The Diplodocidae have been found almost exclusively in rocks of Upper Jurassic age. These include two of the best known dinosaurs, the North American *Diplodocus* and *Apatosaurus* (also known by its junior synonym, *Brontosaurus*). These animals had skulls with protruding muzzles and nostrils that opened upward on the top of the head. The very characteristic teeth were small and peg shaped and were confined to the front of the jaws. The neck was long (15 vertebrae), the trunk short (10 vertebrae), and the tail very long with 80+ vertebrae ending in a whiplash. The vertebral spines in much of the neck and trunk possessed a deep V-shaped cleft. An important character of the family involves the shape of the chevron bones in the mid-tail region. Instead of being directed downward, these elements developed fore and aft extensions so as to lie almost horizontally, a feature most highly developed in *Diplodocus* itself. The forelimbs were the shortest of those in any sauropod family. *Apatosaurus* differed from *Diplodocus* in being much more robust. Adults of both genera could exceed 23 m (75 ft) in length, but Colbert has estimated their weights at 30–35 and 11 or 12 tons, respectively. A contemporary of these animals in North America, also found in East Africa, was *Barosaurus*. This animal closely resembled *Diplodocus* but had a longer neck and shorter trunk. Incomplete remains of three truly gigantic animals from the same beds in North America have been named *Supersaurus*, *Seismosaurus*, and *Amphicoelias*. Adults may have attained lengths up to or exceeding 38 m (125 ft). All three are of the very slender type and have not been absolutely distinguished from *Diplodocus* and *Barosaurus* (see DALTON WELLS QUARRY). A basal diplodocid, *Cetiosauriscus*, is known from a partial skeleton found in England. Mounted skeletons of *Diplodocus* are found in Pittsburgh, Washington, DC, Denver, Houston, and Frankfurt, Germany. Casts of the Pittsburgh specimen, *Diplodocus carnegii*, were sent to London, Paris, Berlin, Madrid, Bologna, Vienna, St. Petersburg, Mexico City, and La Plata, Argentina. Mounts of *Apatosaurus* are found in New Haven, New York, Pittsburgh, Chicago, and Laramie, Wyoming. A skeleton of *Barosaurus* has recently been mounted in New York in a controversial pose—rearing up on its hindlegs to protect a juvenile from a marauding *Allosaurus*. Finally, the partial skeleton of *Cetiosauriscus* is mounted in London.

An offshoot of the typical diplodocids is a subtaxon, the Dicraeosaurinae. (Bonaparte and collaborators consider these animals to belong to a separate taxon, Dicraeosauridae.) *Dicraeosaurus* from the Upper Jurassic of Tanzania has a diplodocid-like skull, limb, and girdle bones. The 12 neck and 12 trunk vertebrae are conservative in number but are distinguished by the extraordinary development of the vertebral spines in the neck and trunk. These were very high and deeply cleft. This trend reached its maximum development in *Amargasaurus*, from the Lower Cretaceous of Argentina. In this animal the spines made up three-quarters of the total height of some vertebrae. Mounted skeletons of *Dicraeosaurus* are found in Berlin and of *Amargasaurus* in Buenos Aires.

A more distantly related taxon is the Mamenchisauridae from the Jurassic of China. The Upper Jurassic *Mamenchisaurus* was a truly bizarre creature with an enormously long neck of 19 elongated cervical vertebrae. The cleft of some of the spines in the shoulder region and the typical forked diplodocid chevrons led to its early inclusion in the Diplodocidae. However, the recent discovery of a partial skull with broad camarasaur-like teeth and an associated neck that appears to belong to *Mamenchisaurus* indicates that this animal apparently diverged from the diplodocid line before the development of the slender peg teeth. The Middle Jurassic Chinese genus *Omeisaurus* also had an inordinately long neck with 17 vertebrae and may be related to *Mamenchisaurus*. It had broad teeth and lacked the cleft vertebral spines, but it had diplodocid forked chevrons. Mounted skeletons of *Mamenchisaurus* and *Omeisaurus* are found in Beijing and in Zigong and Chongqing, respectively.

Finally, two isolated skulls from the Upper Cretaceous of Mongolia appear to be of the general diplodocid type. They have been named *Nemegtosaurus* and *Quaesitosaurus*. Their true relationships must await future skeletal discoveries.

The Titanosauridae were the last survivors of the Sauropoda. Numerous remains of this group are found throughout the world in rocks of Cretaceous, especially Upper Cretaceous, age. However, it remains the most enigmatic taxon because no even partly complete skull or articulated skeleton has ever been found. The presence of very small slender teeth led to the association of the titanosaurids with the Diplodocidae for some years, but it is now virtually certain that this feature evolved independently in the two taxa. The spines of the vertebrae of the neck and back were simple, not bifid. A second dorsal vertebra had been incorporated into the sacrum to give a total of six. Perhaps the most characteristic anatomical feature of this family is seen in the tail vertebrae. The bodies (centra) of these vertebrae articulated by having a deep cup on the front and a large ball on the back (procoelous condition). The chevrons were simple, not forked. Another very significant feature was the development of several types of dermal armor ranging from small globular ossicles to moderate-sized plates. The titanosaurids ranged in size from relatively small to truly gigantic, rivaling the huge diplodocids mentioned previously. Fragmentary and disarticulated remains are found in abundance in India, France, Romania, parts of Africa, and especially in South America, where numerous genera have been named. Among the best known are the moderately sized *Saltasaurus*, *Titanosaurus*, and *Neuquenosaurus*; the large *Antarctosaurus* and *Argyrosaurus*; and some very large unnamed animals. The earliest member of the taxon is often taken to be *Janenschia* (incorrectly referred to as *Tornieria*) from Tanzania. The recently discovered and not yet fully described *Malawisaurus* from the Lower Cretaceous of Malawi promises to throw much light on the origins of the group. The titanosaurids died out in the upper part of the Lower Cretaceous period in North America but were apparently reintroduced from South America late in the Upper Cretaceous. However, they only got as far north as Utah before the final dinosaurian extinction. Ironically, some of the best articulated titanosaur material is that of the large North American genus *Alamosaurus*. Composite mounted skeletons of titanosaurids are found in several museums in Argentina, namely, in Buenos Aires (*Saltasaurus*) and La Plata (*Neuquenosaurus*).

Based on vertebral differences, Bonaparte and collaborators have separated two Lower Cretaceous Argentinian sauropods, *Andesaurus* and *Argentinosaurus*, from the other titanosaurs as a distinct subtaxon Andesaurinae. Recently, they have raised it to a separate taxon. Full understanding of these relationships must await the discovery of more material and fuller phylogenetic analyses. Significant at this time is the gigantic size of *Argentinosaurus*.

Paleobiology

At one time the sauropods were thought to be semiaquatic animals dwelling in shallow seas and lakes, but most experts now agree that they were essentially terrestrial, venturing occasionally into rivers. As clearly shown by the teeth, they were herbivorous. The recent discovery of gastroliths in the gut of *Seismosaurus* confirms, as some have suspected for years, that these stones were used to aid in digestion. The animals walked in an upright pose, not (as Hay and Tornier had earlier supposed) in a crocodile-like sprawl. Bakker has argued that they were endothermic (warm blooded), but most writers regard them as homeothermic, i.e., their huge bulk permitted heat to be retained so that the practical benefits of endothermy could be enjoyed without the presence of a "biological thermostat" (see PHYSIOLOGY). From a study of the footprints, Alexander has estimated their normal walking speed at about 1 m per second (see BIOMECHANICS; FOOTPRINTS AND TRACKWAYS; SIZE AND SCALING).

See also the following related entries:

PROSAUROPODA • SAURISCHIA • SAUROPODAMORPHA

References

Bonaparte, J. F. (1986). The early radiation and phylogenetic relationships of the Jurassic sauropod dinosaurs, based on vertebral anatomy. In *The Beginnings of the Age of Dinosaurs* (K. Padian, Ed.), pp. 247–258. Cambridge Univ. Press, Cambridge, UK.

McIntosh, J. S. (1990). Sauropoda. In *The Dinosauria* (D. B. Weishampel, P. Dodson, and H. Osmólska, Eds.), pp. 345–401. Univ. of California Press, Berkeley.

A complete set of references is found in:

Chure, D. J., and McIntosh, J. S. (1989). A bibliography of the Dinosauria (exclusive of the Aves) 1677–1986. Museum of Western Colorado Paleontology Series No. 1, pp. 226.

Sauropod Necks

see LONG NECKS OF SAUROPODS

Sauropodomorpha

P. UPCHURCH
Cambridge University
Cambridge, United Kingdom

Sauropodomorpha may be defined as Sauropoda + Prosauropoda and all saurischians closer to them than to birds. Sauropodomorpha and Theropoda are thus the two stem taxa (sister groups) within Saurischia. An overview of sauropodomorph biology and evolutionary history is complicated by questions about its "constituent" taxa and their evolutionary relationships. In addition, the morphological variation presented by sauropodomorphs makes generalizations difficult. "Sauropodomorpha" [first coined by von Huene (1932)] contains the "herbivorous" saurischians Prosauropoda and Sauropoda. Prosauropods have been found in Upper Triassic and Lower Jurassic deposits and are known from every continent. They range from 2-m-long bipedal forms, such as *Thecodontosaurus,* to the 10-m-long quadrupedal *Riojasaurus.* Sauropods first appear at the Triassic–Jurassic boundary and survive to the end of the Cretaceous. They may have attained a virtually global distribution as early as the Lower/Middle Jurassic (Upchurch, 1997). THERIZINOSAURS (Segnosaurs) were originally interpreted as unusual theropods (Perle, 1979; Barsbold and Perle, 1980) but were later included with reservations within the Sauropodomorpha (Gauthier, 1986; Sereno, 1989). They are medium to large (up to 5 m long), bipedal forms from the Middle and Upper Cretaceous of Asia. Therizinosaurids resemble prosauropods in their general body form, but they also display a number of ornithischian and theropod characters. Recently, more complete material, belonging to a Middle Cretaceous therizinosaurid (*Alxasaurus*), was described by Russell and Dong (1993). The cladistic analysis performed by these authors places the therizinosaurids within the tetanuran theropods. Moreover, prosauropods and sauropods share several derived states (such as the ascending process of the astragalus "keying" into the distal end of the tibia) that are absent in therizinosaurids. Although the exact relationships of therizinosaurids remain controversial, the view that they belong to the Sauropodomorpha is no longer strongly advocated (see THERIZINOSAURIA).

True sauropodomorphs can be diagnosed by a number of derived characters, including lanceolate tooth crowns, anteroposterior elongation of the cervical vertebrae, long trunk region relative to length of hindlimb, and the structure of the astragalus–tibia contact mentioned previously (Gauthier, 1986; Upchurch, 1993, and references therein). The relatively LONG NECKS of prosauropods and particularly sauropods suggest that sauropodomorphs usually exploited high-browsing niches. Like basal ornithischians, sauropodomorphs have rather simple dentitions, and in most cases the jaw actions were also simple (Barrett and Upchurch, 1997). There seems to have been relatively little oral processing of plant material (except, perhaps, in brachiosaurid and camarasaurid sauropods), which is consistent with the presence of GASTROLITHS ("stomach stones") in both prosauropods and sauropods. The large size of many sauropodomorphs may reflect the need for slow fermentation and/or bacterially assisted breakdown of the tough, low-nutrient content plants available during the Upper Triassic and Jurassic.

The apparent "trend" toward increased body size and QUADRUPEDALITY within the Prosauropoda has led to suggestions that prosauropods were ancestral to sauropods (von Huene, 1932; Romer, 1956; Bonaparte, 1986). This view has received some support from cladistic analysis (Gauthier, 1986) (see Sauropoda). Bonaparte (1986) reported an Upper Triassic "melanorosaurid prosauropod" that possesses rather sauropod-like presacral vertebrae. Whether this animal provides support for prosauropod paraphyly or is actually the earliest known sauropod depends on the rest of the skeleton, which is currently undescribed. Recent work, however, has provided a suite of prosauropod specializations absent in sauropods (Sereno, 1989; Galton, 1990; Upchurch, 1993; Gauffre, 1994). Prosauropods and sauropods can therefore be regarded, not without controversy, as monophyletic sister groups.

The acceptance of a monophyletic Prosauropoda raises several issues. First, the existence of Upper Triassic sauropods must be inferred, despite the total absence of direct fossil evidence from deposits of this age. The earliest prosauropod is *Azendohsaurus* from the middle Carnian (~229 Mya) of Morocco (Gauffre, 1994), whereas the earliest known sauropod is *Vulcanodon* (~208 Mya). Thus, the first 15–20 million years (at least) of sauropod evolution is currently unknown (Upchurch, 1997). Second, the trend toward a "sauropod-like" condition within the prosauropods must be interpreted as convergent EVOLUTION. Finally, the EXTINCTION of the prosauropods at the end of the Lower Jurassic cannot be interpreted as a gradual transformation into the Sauropoda. Prosauropods and sauropods were apparently similar in their ecology and overlapped in geographic and stratigraphic distribution. Why prosauropods became extinct, while sauropods went on to flourish throughout the Jurassic and much of the Cretaceous, is not understood. Perhaps the answer to this mystery lies, in part at least, in the longer necks and even greater body size of sauropods. Clearly, there is still much to be learned about sauropodomorph paleobiology and evolution.

See also the following related entries:

PROSAUROPODA • SAURISCHIA • SAUROPODA • THEROPODA

References

Barrett, P. M., and Upchurch, P. (1995). Sauropod feeding mechanisms: Their bearing on palaeoecology. 6th Symposium on Mesozoic Terrestrial Ecosystems, Beijing.

Barsbold, R., and Perle, A. (1980). Segnosauria, a new infraorder of carnivorous dinosaurs. *Acta Palaeontol. Polonica* **25,** 185–195.

Bonaparte, J. F. (1986). The early radiation and phylogenetic relationships of Jurassic sauropod dinosaurs, based on vertebral anatomy. In *The Beginning of the Age of Dinosaurs* (K. Padian, Ed.), pp. 247–258. Cambridge Univ. Press, Cambridge, UK.

Galton, P. M. (1990). Basal Sauropodomorpha–Prosauropods. In *The Dinosauria* (D. B. Weishampel, P. Dodson, and H. Osmólska), pp. 320–344. Univ. of California Press, Berkeley.

Gauffre, F.-X. (1994). The prosauropod dinosaur *Azendohsaurus laaroussii* from the Upper Triassic of Morocco. *Palaeontology* **36**(4), 897–908.

Gauthier, J. (1986). Saurischian monophyly and the origin of birds. *Mem. California Acad. Sci.* **8,** 1–55.

Perle, A. (1979). Segnosauridae—A new family of theropods from the Late Cretaceous of Mongolia. *Sovm. Soviet–Mongolean Paleontol. Eksped. Trudy* **8,** 45–55.

Romer, A. S. (1956). *Osteology of the Reptiles.* Univ. of Chicago Press, Chicago.

Russell, D. A., and Dong, Z. (1993).The affinities of a new theropod from the Alxa desert, Inner Mongolia, People's Republic of China. *Can. J. Earth Sci.* **30**(10/11), 2107–2127.

Sereno, P. C. (1989). Prosauropod monophyly and basal sauropodomorph phylogeny. *J. Vertr. Paleontol.* **9**(Suppl.), 38A.

Upchurch, P. (1993). The anatomy, phylogeny and systematics of sauropod dinosaurs, pp. 489. Unpublished Ph.D. dissertation, University of Cambridge, Cambridge, UK.

Upchurch, P. (1997). The evolutionary history of sauropod dinosaurs. *Philos. Trans. R. Soc. London B,* in press.

von Huene, F. (1932). Die fossile Reptile–Ordnung Saurischia ihre Entwicklung und Geschichte. *Monogr. Geol. Palaeontol.* **4**(Pts. I and II, Ser. I), 1–361.

Schaumburgisches Heimautmuseum, Germany

see MUSEUMS AND DISPLAYS

Sedgwick Museum, Cambridge University, United Kingdom

see MUSEUMS AND DISPLAYS

Segnosauria

Segnosauria are properly known as Therizinosauria and appear to be aberrant members of the MANIRAPTORA.

see THERIZINOSAURIA

Senckenberg Nature Museum, Germany

see MUSEUMS AND DISPLAYS

Servicos Geologicos de Portugal, Portugal

see MUSEUMS AND DISPLAYS

Sexual Behavior

see REPRODUCTIVE BEHAVIOR AND RATES

Sexual Dimorphism

see VARIATION

Sino–Canadian Dinosaur Project

PHILIP J. CURRIE
Royal Tyrrell Museum of Palaeontology
Drumheller, Alberta, Canada

The Canada–China Dinosaur Project (CCDP) was conceived in 1982 while the ROYAL TYRRELL MUSEUM OF PALAEONTOLOGY was being planned. The agreement between the Tyrrell Museum (Drumheller), the Canadian Museum of Nature (Ottawa), and the INSTITUTE OF VERTEBRATE PALEONTOLOGY AND ANTHROPOLOGY (Beijing) was signed in 1985 and initiated the following year with joint expeditions to the Junggar Basin of northwestern China, to the badlands of Alberta, and to the Canadian Arctic islands. In 1987, work focused on the Middle and Late Jurassic of the Junggar Basin. The expedition numbered 45 people, and specimens discovered included the holotypes of *Sinraptor dongi* and *Mamenchisaurus sinocanadorum*. In Canada, DEVIL'S COULEE was discovered, producing the first hadrosaur embryos. Expeditions in 1988 were spread across northern China and recovered the spectacular troodontid *Sinornithoides* in the Ordos Basin and the first pod of *Pinacosaurus* juveniles at Bayan Mandahu. The Junggar Basin was the major focus in China the following year, although a CCDP expedition to the Canadian arctic recovered dinosaur bones on Bylot Island. By 1990, the major effort was put into BAYAN MANDAHU, where significant discoveries were made almost daily. These included the second group of juvenile ankylosaurs and an *Oviraptor* sitting on a nest of eggs. The last joint expeditions were in 1991, when Chinese scientists and technicians joined their Canadian colleagues in Dinosaur Park and at a new footprint site at Grande Cache in Alberta. The field expeditions collected more than 60 tons of fossils in China alone, and two volumes of scientific papers have already been published (Currie, 1993, 1996).

The expeditions attracted a great deal of attention and were well covered by the media, popular books (Grady, 1993), and films. A major traveling exhibition (variously referred to as the Dinosaur Project, the Greatest Show Unearthed, and Dinosaur World Tour) was assembled using Chinese and Canadian dinosaurs that had never before been seen by the public. This exhibition has already been shown in four Canadian cities, Osaka (Japan), Singapore, and Sydney (Australia), and will remain on the road until at least the end of the century.

See also the following related entries:

CANADIAN DINOSAURS • CENTRAL ASIATIC EXPEDITIONS • CHINESE DINOSAURS • MONGOLIAN DINOSAURS • SINO–SOVIET EXPEDITIONS • SINO–SWEDISH EXPEDITIONS

References

Currie, P. J. (1991). The Sino-Canadian Dinosaur Expeditions, 1986–1990. *Geotimes* **36**(4), 18–21.

Currie, P. J. (Ed.) (1993). Results from the Sino–Canadian Dinosaur Project. *Can. J. Earth Sci.* **30**(10/11), 1997–2272.

Currie, P. J. (Ed.) (1996). Results from the Sino–Canadian Dinosaur Project, Volume 2. *Can. J. Earth Sci.* **33**(4), 511–648.

Grady, W. (1993). *The Dinosaur Project*, pp. 261. McFarlane, Ross & Walters, Toronto.

Sino–Soviet Expeditions

PHILIP J. CURRIE
Royal Tyrrell Museum of Palaeontology
Drumheller, Alberta, Canada

In 1959 and 1960, the Paleontological Institute in Moscow and the INSTITUTE OF VERTEBRATE PALEONTOLOGY AND PALEOANTHROPOLOGY ran two expeditions into Inner Mongolia. The field parties numbered more than 60 people, not including local support, using up to 16 vehicles and two "Stalin 100" bulldozers. Under the leadership of Chow Minchen (Zhou Mingzhen) and Anatole K. Rozhdestvensky, dinosaurs were collected at Erenhot in 1959, relying on information from the Central Asiatic expeditions (1922–1930) of the American Museum of Natural History. New sites in the Alashan desert were worked in 1960, and Lower and Upper Cretaceous dinosaurs were recovered (including *Chilantaisaurus* and *Probactrosaurus*), some of which are now on display in the ORLOV MUSEUM OF PALEONTOLOGY in Moscow. The enormous collections were split between Beijing and Moscow, although most of them remain unprepared. Joint fieldwork was supposed to last for 5 years, eventually pushing west into Kazakhstan. Unfortunately, the political and economic relationship between China and the USSR disintegrated, and the Russians were forced to pull out of the project in the middle of the second season. The Chinese scientists and technicians continued the planned expeditions alone for several years, but funding became scarce as the government paid back loans to Moscow, and the Cultural Revolution terminated all expeditions in 1965.

See also the following related entries:

Central Asiatic Expeditions • Chinese Dinosaurs • Mongolian Dinosaurs

References

Lavas, J. R. 1994. Dragons from the Dunes, the Search for Dinosaurs in the Gobi Desert. Published by the author, Panmure, New Zealand. 138 pp.

Sino–Swedish Expeditions

Niall J. Mateer
University of California
Berkeley, California, USA

The Swedish involvement in the exploration of China began in 1914 with the appointment of J. Gunnar Andersson as advisor to the Chinese government on mining. During his travels as a mining inspector, he found abundant vertebrate and plant fossils, mostly in Cenozoic strata in eastern and central China. In 1921, a paleontology student, Otto Zdansky, traveled to China to assist Andersson in the collection of fossils and to send them to Uppsala, Sweden, for detailed study. The sauropod *Euhelopus zdanskyi* Wiman was excavated by Zdansky in Shandong province in 1923. In 1927, Zdansky was replaced by Birger Bohlin. This activity was funded by Axel Lagrelius and overseen by the professor of geology in Uppsala, Carl Wiman. Much of this collection, including *Euhelopus*, remains in Uppsala.

The Swedish explorer, Sven Hedin planned a major scientific expedition to map and document the geology, geography, and natural and human history of this region.

The Hedin expedition was executed on a large scale involving many people, rarely together at the same time, and lasted from 1927 to 1935. It became known as the "wandering university." More than 60 volumes now comprise the *Reports from the Scientific Expedition to the North-Western Provinces of China under the Leadership of Dr. Sven Hedin*, which cover a broad range of scientific topics. Bohlin was the paleontologist on the expedition and published 14 volumes in the *Reports* series, only 1 of which pertained to dinosaurs (*Fossil Reptiles from Mongolia and Kansu*, 1953); the other volumes were principally on Tertiary plants and on Paleozoic fossils. In this monograph, Bohlin erected a number of new genera and species, mostly on fragmentary material, which include the dinosaurs *Sauroplites scutiger, Troodon bexelli, Microceratops gobiensis, M. sulcidens, Peishansaurus philemys,* and *Heishansaurus pachycephalus*. Bohlin found these fossils at locations in southern Nei-Mongol and Gansu provinces.

In 1935 the political situation in western China, and in much of China generally, made further expeditions untenable, and the two decades of Sino–Swedish collaboration ceased. The Swedes Andersson, Bohlin, Hedin, Zdansky, and others played an important role in training a first generation of Chinese geologists and paleontologists.

See also the following related entries:

Chinese Dinosaurs • Mongolian Dinosaurs

Size

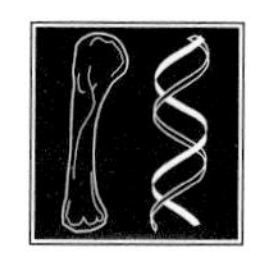

EDWIN H. COLBERT
Museum of Northern Arizona
Flagstaff, Arizona, USA

Perhaps the predominant perception of the dinosaurs, by paleontologists and by the general public alike, has to do with gigantic size. Not all the dinosaurs were giants—giants being defined according to a suggestion put forth by Owen-Smith (1992) as land-living vertebrates having weights (or probable weights) in excess of 1000 km or 1 metric ton. As for the weights of giant dinosaurs, these have been separately determined by Colbert (1962) and by Alexander (1989) by obtaining the volumes of accurate scale models and, on the assumption that weights were essentially equivalent to or slightly less than the density of water, by extrapolating up the values for the models to natural size. Weights have also been estimated by using the circumference of limb bones (Anderson *et al.*, 1985).

The fossil record demonstrates that the dinosaurs show a decided evolutionary trend toward giants, as defined by Owen-Smith (1992). Among the major subdivisions of the dinosaurs, namely, the saurischian taxa THEROPODA and SAUROPODOMORPHA and the ornithischian taxa ORNITHOPODA, STEGOSAURIA, ANKYLOSAURIA, and CERATOPSIA, all contain genera of gigantic size.

In the Pangean world of Mesozoic times, characterized by tropical and subtropical climates, by ample botanical food supplies, and by broad avenues for the worldwide distributions of active tetrapods, there must have been some cogent evolutionary advantage for so many of the dinosaurian lineages to have evolved into giant size. Basically, it probably was a matter of the conservation of energy—a result of the ratio of body surface to mass. In large reptiles the overriding influence of large body mass to stabilize the physiology of the animals is of paramount importance, as will be discussed.

Another advantage enjoyed by the giant dinosaurs was freedom to a large degree from predation. We see this attribute as very important in the lives of modern elephants, animals that can go where they please with the ingrained knowledge that they will not be prevented from carrying on their daily activities by the predators around them. (This statement ignores, of course, the activities of modern man, armed with powerful weapons and spreading across the land to preempt habitats hitherto reserved for nonprimate occupants.)

Of course there were disadvantages for the giant dinosaurs as well. One was their immense weight, which increased by the cube of linear dimensions as these reptiles increased in size, to be supported by the strength of the bones that increased only by the square of the linear dimensions of their cross sections. Obviously, there must have been huge stresses on bone and muscle in the giant dinosaurs, as detailed by Alexander (1989), while at the same time there probably were severe restraints on the behavior of the giants. One sees this latter consideration illustrated by the behavior of modern elephants, which under certain conditions must be extremely cautious in their movements. How much more so would have been the case with animals that often may have outweighed the largest elephants by a factor of 10! Consequently the skeleton in the large dinosaurs, especially in the gigantic sauropods, shows adaptations for massive strength, especially in the limbs, associated with interlocking truss-like structures in the vertebrae, the latter to afford strength combined with light weight. It is significant that the huge sauropods, as well as many other large dinosaurs, were obligatory quadrupeds, even though their ancestors were bipedal.

Because of the large size of many dinosaurs, these reptiles had to cope with other problems besides those of bodily strength. Thus, there was the problem of body temperature. Were the dinosaurs ectothermic or "cold-blooded" animals like modern amphibians and reptiles, or were they endothermic or "warm-blooded" animals like modern birds and mammals? An international debate, often expressed with considerable emotion, has raged around this subject for the past two decades, with no solution acceptable to all participants in the discussions (Thomas and Olson, 1980) (see PHYSIOLOGY).

More than half a century ago Colbert *et al.* (1946) performed temperature experiments with alligators in Florida and established that the internal core temperature of these reptiles was directly proportional to body mass. Thus, the largest alligators accumulated and lost body heat much more slowly than did the smaller individuals. By extrapolation it was concluded that the large dinosaurs, although probably basically ectothermic, were characterized by body temperatures that were relatively stable through each 24-hr time period. Consequently the very large dinosaurs probably were something between ectothermic and endothermic: They may be characterized as having been inertial homeotherms, having some of the attributes of endothermy without entailing the costs, particularly the ingestion of large amounts of food that are necessary for the endothermic mode of life. They must have been reptiles of a very special kind.

In this respect a cautionary note has been sounded by Ostrom (1980), who has pointed out that the physiology and lifestyles of the smaller dinosaurs may have been significantly different from what has been invoked for the giants.

A difficult problem attendant upon huge size is that of overheating. Elephants dump excess body heat by radiation through the huge ears. The large dinosaurs probably controlled overheating largely through behavior, as do extant reptiles—by going into the shade or into water to cool off. Some dinosaurs, for example, the stegosaurs, possessed large plates on the back that probably served as temperature-controlling devices (Farlow *et al.*, 1976), but not all stegosaurs were so equipped, and neither were their relatives and ancestors.

A problem concerning the lifestyles of the dinosaurs has to do with food intake, especially among the gigantic brontosaurs. How did such enormous reptiles manage to gather enough food during each 24-hr period to survive? Haas (1963) suggested that these dinosaurs may have fed upon freshwater molluscs, which provided much more energy per gram than plant food. However, most students agree that the sauropods, as well as all of the ornithopod dinosaurs, were herbivorous, feeding upon bulky plants that in many instances were not particularly nutritious. The relatively small, weak jaws and the restricted number of small, often very weak teeth in the sauropods pose a dilemma in this regard. Such jaws and teeth afford evidence that prompted Weaver (1983) to call the sauropod dinosaur an "improbable endotherm."

As for the sauropods, there is the problem of the height of the brain above the heart when these dinosaurs carried the head in an elevated position. The giraffe, whose heart is more than 2 m below the brain, has solved the problem of blood supply by the development of extremely high blood pressure (and a rete mirabile at the base of the skull to keep the brain from bursting) and very thick elastic arterial walls. How much greater must have been the problem for a large sauropod, in which the brain might have been 10 m or more above the heart when the head was elevated! It would seem that blood pressures may have been extremely high.

As for the dinosaurian brain, to judge from endocranial casts it was a characteristic reptilian brain, structurally primitive and extraordinarily small compared to the size of the individual. In fact, when compared to the relative brain size of crocodiles enlarged to similar sizes, dinosaur brains ranged from 20 to 400% or more of expected volume: Brain size scales to only the two-thirds power of body mass (see BEHAVIOR, BRAINCASE ANATOMY). A few of the smaller theropod dinosaurs had relatively large brains, and of these, the brain of *Troodon* has been characterized by Hopson (1980) as falling "unequivocally within the range of living endotherms." However, for most of the dinosaurs the brain, regardless of size, would nonetheless seem to have enabled these reptiles to employ rather complex behavior patterns, as dinosaurian behavior may be interpreted from footprints and trackways, from nests, and from other fossil evidence. In this regard it is useful to look at the Komodo dragon or ora (*Varanus komodoensis*), a very large Indonesian lizard that ambushes and kills large mammals and behaves in ways that might not be expected in an ectothermic reptile (Auffenberg, 1981).

In essence, the evolutionary history of the dinosaurs is a story of varied archosaurian reptiles adapted to numerous environments and to remarkably varied lifestyles. Although the dinosaurs had their beginnings as comparatively small forms, with the transition from Triassic to Jurassic times the evolutionary trend toward growth in size was very strong, if not dominant. These were the tetrapod vertebrates that, small and large but especially large, ruled the continents during Jurassic and Cretaceous times.

See also the following related entries:

AGE DETERMINATION • BIOMECHANICS • GROWTH AND EMBRYOLOGY • LIFE HISTORY • SIZE AND SCALING

References

Alexander, R. McN. (1989). *Dynamics of Dinosaurs and Other Extinct Giants*, pp. vii and 167. Columbia Univ. Press, New York.

Anderson, J. F., Hall-Martin, A., and Russell, D. A. (1985). Long bone circumference and weight in mammals, birds, and dinosaurs. *J. Zool. (A)* **207,** 53–61.

Auffenberg, W. (1981). *The Behavioral Ecology of the Komodo Monitor*, pp. x and 405. Univ. of Florida, Gainesville.

Colbert, E. H. (1962). The weights of dinosaurs. *Am. Museum Novitates* **2076,** 1–16.

Colbert, E. H., Cowles, R. B., and Bogert, C. M. (1946). Temperature tolerances in the American alligator and their bearing on the habits, evolution, and extinction of the dinosaurs. *Am. Museum Nat. History Bull.* **86,** 327–374.

Farlow, J. O., Thompson, C. V., and Rosner, D. E. (1976). Plates of *Stegosaurus*: Forced convection heat loss fins? *Science* **192,** 1123–1125.

Haas, G. (1963). A proposed reconstruction of the jaw musculature of *Diplodocus*. *Ann. Carnegie Museum* **36,** 139–157.

Hopson, J. A. (1980). Relative brain size in dinosaurs. Implications for dinosaurian endothermy. In *A Cold Look at Warm-Blooded Dinosaurs* (R. D. K. Thomas and E. C. Olson, Eds.), American Association for the Advancement of Science Symposium Series, pp. 287–310. Westview, Boulder, CO.

Ostrom, J. H. (1980). The evidence for endothermy in dinosaurs. In *A Cold Look at Warm-Blooded Dinosaurs* (R. D. K. Thomas and E. C. Olson, Eds.), American Association for the Advancement of Science Symposium Series, pp. 15–54. Westview, Boulder, CO.

Owen-Smith, R. N. (1992). *Megaherbivores. The Influence of Very Large Body Size on Ecology*. Cambridge Univ. Press, Cambridge, UK.

Thomas, R. D. K., and Olson, E. C. (Eds.) (1980). *A Cold Look at Warm-blooded Dinosaurs*, pp. 514. Westview, Boulder, CO.

Weaver, J. C. (1983). The improbable endotherm: The energetics of the Sauropod dinosaur *Brachiosaurus*. *Paleobiology* **9,** 173–182.

Size and Scaling

R. McNEILL ALEXANDER
University of Leeds
Leeds, United Kingdom

The dinosaurs catch our imagination because some of them were extremely large. Among the largest of which we have reasonable complete fossil skeletons, *Diplodocus* was about 25 m long (as long as a tennis court) and *Brachiosaurus* was 13 m high (tall enough to look over the top of a four-story building). Length and height, however, are not necessarily the most revealing indicators of size: A python may be longer than an elephant and a giraffe is taller than an elephant but most of us would regard the elephant, the heaviest of the three, as the largest. We would like to know how heavy large dinosaurs were in life, but we cannot find that out by weighing fossils.

We can only estimate the masses of dinosaurs. One method requires a scale model of the living animal, scaled to the dimensions of the skeleton and fleshed out to the likely shape of the intact animal (good models are sold by many museums). The model's volume is measured. The simplest way of doing this is to dunk it in a jar that is brim-full of water and measure the volume of water that overflows, but there is a more accurate method that depends on Archimedes' Principle (Fig. 1). Once the volume of the model is known, the volume of the dinosaur can be calculated by scaling up, and the mass can be calculated by multiplying by the animal's likely density. Modern reptiles, for example, crocodiles, have almost the same density as water—1.00 kg per cubic meter—and it seems reasonable to assume the same for dinosaurs.

An alternative method depends on measuring the circumferences of the humerus and femur (the upper arm bone and thigh bone). Heavier animals have thicker leg bones. A formula based on measurements of extant mammals has been used to calculate dinosaur masses from the circumferences of their bones.

Both methods are liable to error. Model making is uncertain because it depends on the restorer's judgment of how plump or skinny the animal was—different modelers will make models of quite different volumes. Bone measurements are also unreliable:

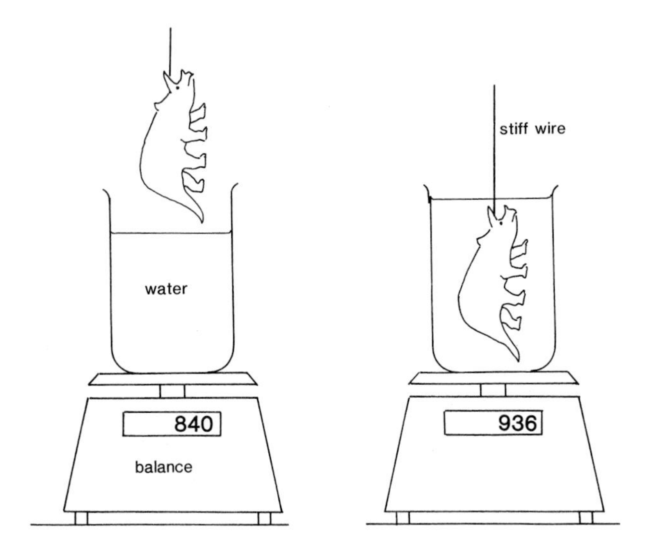

FIGURE 1 Weighing a dinosaur; in this example *Triceratops.* With a container of water on its pan, the balance registers 840 g. The scale model is then held submerged in the water (but not touching the container) and the balance registers 936 g. By Archimedes' Principle, the extra 96 g is the mass of a volume of water, equal to the volume of the model: It represents the upward push of the water on the model. The model was made to 1/40 scale so we can get the mass of a volume of water equal to the actual dinosaur by multiplying by 40 × 40 × 40 = 64,000. (The actual dinosaur was 40 times as long, 40 times as wide, and 40 times as high as the model). The result, about 6 million g or 6 tons represents the mass of the dinosaur if dinosaurs, like living reptiles, had about the same density as water.

For example, a hippopotamus will have thicker bones than a giraffe of the same body mass. However, the methods do give rough indications, some of which are shown in Table I. They tell us that the largest dinosaurs were many times heavier than any extant land animal but not as heavy as the biggest whales.

Only dinosaurs of which we have reasonably complete skeletons are included in Table I. Others, such as *Supersaurus* and *Argentinosaurus*, known from only a few bones each, are thought by some paleontologists to have been even heavier.

The sizes of the biggest dinosaurs raise many questions. One that has often been asked is how could such large animals have supported their enormous weight? To appreciate this problem, imagine two dinosaurs of identical shape, one twice as long as the other. It is twice as long, twice as wide, and twice as high as the other so it has eight times the volume (2 × 2 × 2 = 8) and is eight times as heavy. Its leg bones have twice the diameter of the bones of the smaller dinosaur, so they have four (2 × 2) times the cross-sectional area and are four times as strong. They are only four times as strong but have to support eight times the weight. Thus, the bigger an animal is, the harder it is for its legs to support it.

That argument depends on the assumption that different-sized animals are exactly the same shape. The assumption is to some degree false. For example, among living animals, gazelles and buffaloes are close relatives, but a buffalo has relatively shorter, thicker legs than a gazelle would if it were scaled up to the same body mass. Similar size effects can be seen in dinosaurs. For example, *Tyrannosaurus* has relatively shorter, thicker leg bones than a scaled-up *Compsognathus* would have.

We can express this more quantitatively. If differently sized animals were geometrically similar (that is, if small ones were exact scale models of large ones) the lengths and diameters of all their parts would be proportional to (body mass)$^{0.33}$; that is, to the cube root of body mass. Among antelopes and their relatives leg bone lengths are about proportional to (body mass)$^{0.25}$ (therefore, larger antelopes have relatively shorter bones) and bone diameters to (body mass)$^{0.38}$ (larger antelopes have relatively thicker bones). These changes in proportion are more marked in antelopes

TABLE I
Masses of Dinosaurs[a]

	Tons
Tyrannosaurus rex	5–8
Allosaurus fragilis	1–2
Struthiomimus currellii	0.1
Brachiosaurus brancai	30–90
Apatosaurus louisae	30–40
Diplodocus carnegiei	6–20
Iguanodon bernissartensis	5
Stegosaurus ungulatus	2–3
Triceratops prorsus	6–9
Masses of modern mammals	
African elephant, *Loxodonta africana*	3–5
White rhinoceros, *Ceratotherium simum*	1.6–2.5
Blue Whale, *Balaenoptera musculus*	80–150

[a] Estimated as explained in the text. This table gives a rough indication of the ranges of masses calculated by different paleontologists.

than in some other mammal groups, but even in antelopes they are not enough for bone strength to keep pace with body weight. The skeletons of large antelopes are weaker, relative to body weight, than those of small ones. The same is true of dinosaurs, but we cannot be as precise about the changes of dinosaur bone dimensions with body mass because dinosaurs cannot be weighed accurately.

Strenuous activities need strong legs. For example, peak forces on human feet are about equal to body weight in walking, 2.5 times body weight in jogging, 3.5 times body weight in sprinting, and even more in jumping. The more athletic an animal is, the stronger its legs must be. Buffaloes, whose legs are weaker relative to body weight than those of gazelles, cannot gallop as fast or jump as far as gazelles. Elephants, the largest land animals today, cannot jump or gallop and can manage only a moderately slow run.

An attempt has been made to judge how athletic large dinosaurs may have been by using measurements of their leg bones to calculate "strength indicators." These tell us how strong the bones were compared to the animal's body weight. An animal whose strength indicators are low is too weak to be a good athlete. Elephants have low strength indicators: Values between 7 and 11 units for different bones were measured on the carcass of a moderate-sized (2.5-ton) female African elephant. Very similar values (6–14 units) were calculated for leg bones of a 35-ton *Apatosaurus.* Although this dinosaur was many times heavier than the elephant, its leg bones were strong enough for it to be as active as an elephant. It would have had no trouble supporting its weight on land and could probably have managed a slow run. The femur (thigh bone) of an 8-ton *Tyrannosaurus* had a strength indicator of 9 units, showing that it too could not have been fast or athletic. However, leg bones of a *Triceratops* of about the same mass had strength indicators of 12–21 units, indicating more athletic ability: Perhaps it could gallop. For comparison, values for a 0.5-ton African buffalo were rather higher (21–27 units) and values for small mammals would be much higher.

Large size not only affects athleticism but also has other physical consequences. For example, the higher the head is above the heart, the greater blood pressure has to be. For a *Brachiosaurus* to pump blood to its brain, 8 m above its heart, it would need a much higher blood pressure even than the tallest giraffe (see PHYSIOLOGY). Also, large objects heat and cool more slowly than small ones. Even if large dinosaurs had reptile-like metabolism, their body temperatures would have been almost constant, day and night.

Other consequences of size are less closely related to physics. The complexity of the tasks that a brain can undertake depends on the number of nerve cells it contains. However, cells are about the same size (approximately 1/100 mm diameter) in animals of all sizes. This suggests that a lion with a cat-sized brain would have as many brain cells as a cat and be capable of cat-like behavior. Actually, lions have bigger brains than cats: 225 g in a 110-kg lion compared to 28 g in a 2.4-kg cat. However, the lion brain is relatively smaller than that of the cat: 0.2% of body mass compared to 1.1%. As a general rule, the masses of related animal's brains are proportional not to body mass but to (body mass)$^{0.65}$.

Dinosaur brain sizes have been estimated from the volumes of the cavities in skulls. Reptiles today have brains only one-tenth the mass of brains of typical mammals of equal body mass, but even so dinosaur brains may seem remarkably small. The largest known, that of *Tyrannosaurus,* could have had a mass of only 200 g—0.003% of body mass. In contrast, a 200-kg alligator had a 14-g brain—0.007% of body mass. The difference in percentage may seem large, but it is what would be expected as a consequence of the size difference. Large dinosaurs had brains of about the sizes we would expect in extant reptiles of equal mass.

Similar considerations apply to eyes. The effectiveness of an eye depends on its absolute size, not on its size relative to the animal's body. Larger animals have relatively smaller eyes (0.04% of body mass in a lion compared to 0.4% in a cat). It is apparent from the sizes of eye sockets in skulls that the same applied to dinosaurs.

See also the following related entries:

AGE DETERMINATION • BIOMECHANICS • GROWTH AND EMBRYOLOGY • LIFE HISTORY • SIZE

References

Alexander, R. McN. (1989). *Dynamics of Dinosaurs and Other Extinct Giants.* Columbia Univ. Press, New York.

McGowan, C. (1994). *Diatoms to Dinosaurs. The Size and Scale of Living Things.* Island Press, Washington, DC.

Skeletal Chronology

see AGE DETERMINATION; GROWTH LINES

Skeletal Structures

KEVIN PADIAN
University of California
Berkeley, California, USA

The most commonly preserved skeletal structures of dinosaurs are bones and teeth because their chemical components are most durable of all the parts of the body (see CHEMICAL COMPOSITION OF DINOSAUR FOSSILS; HISTOLOGY OF BONES AND TEETH; ORNAMENTATION; TEETH AND JAWS). The bones of the skeleton include those of the skull itself (see SKULL, COMPARATIVE ANATOMY), the axial skeleton, the PECTORAL GIRDLE and the FORELIMB AND HAND, the pelvic girdle and the HINDLIMB AND FOOT, and various bones of dermal origin that take the form of Gastralia, scutes, spikes, plates, horns, and other ossicles (see ORNAMENTATION). SKIN and skin impressions are known for a variety of dinosaurs, mostly in the Late Cretaceous, and there are instances of naturally mummified remains that preserve traces of some MUSCULATURE and other soft parts. True "horn" (keratin) is almost never preserved except as impressions, for example, of the horny "beak" or rhamphotheca that characterized ornithomimids and ornithischians, much as in extant birds; and these structures are most frequently inferred from the roughened, well-vascularized surface of the jaw bones. Horny sheaths on the claws, or their impressions, are recorded in several dinosaurs, including *Archaeopteryx*, the first known bird. Horny pads are not directly preserved in sauropod dinosaurs but are inferred from the form of their FOOTPRINTS AND TRACKWAYS, which suggest an elephantine, wedge-shaped horny pad that makes the shape of the foot rounder and the forms of individual digits indistinct, excepting the claws of the inner digits. Other soft parts are more difficult to preserve and consequently to reconstruct. The shape and size of the pelvic canal, the positioning of gastroliths (stomach stones), when present, and the attitude of the fossilized carcass after death (often distorted by postmortem contractions of tendons) can provide some idea of the extent of certain soft parts, as can the rugosities on bones that indicate muscular attachments (see MUSCULATURE). The internal BRAINCASE ANATOMY preserves the general shape of the brain, but caution must be exercised because it is not known whether the brain in dinosaurs completely filled the cavity as it does in birds and mammals but does not in other tetrapods (see INTELLIGENCE; PALEONEUROLOGY). Other organs of the nervous system include the brainstem, spinal cord, and afferent and efferent nerves indicating the extent of neural control and dedication to local organs (see PALEONEUROLOGY). The standard reference for the skeletal anatomy of dinosaurs and other reptiles remains Romer's (1956) *Osteology of the Reptiles*. Papers in Czerkas and Olson (1987), especially those by Paul, Czerkas, Hallett, and Bakker, deal with inferences of soft parts that can be drawn from fossil anatomy and functional morphology; the classic work in the field of the reconstruction of soft part anatomy of extinct animals is summarized by Charles R. Knight (1947).

See also the following related entries:

GASTRALIA • HISTOLOGY OF BONES AND TEETH

References

Czerkas, S. M., and Olson, E. C. (Eds.) (1987). *Dinosaurs Past and Present*, 2 vols. Univ. of Washington Press, Seattle.

Knight, C. R. (1947). *Animal Drawing: Anatomy and Action for Artists*. McGraw-Hill, New York.

Romer, A. S. (1956). *Osteology of the Reptiles*. Univ. of Chicago Press, Chicago.

Skin

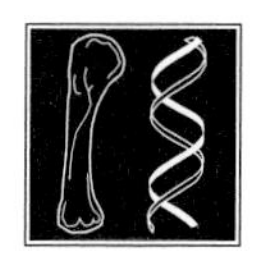

STEPHEN A. CZERKAS
The Dinosaur Museum
Monticello, Utah, USA

Although skeletal remains can be used to reconstruct an approximation of a dinosaur's body shape, only fossilized skin impressions can provide tangible evidence to show what the animal looked like in life. It is not appropriate to depict dinosaurs with scaleless, smooth skins like those of cetaceans or elephants. Except the feathered *Archaeopteryx* and the other birds among the Dinosauria, until recently all known examples of dinosaur skin have had reptilian scales (but see FEATHERED DINOSAURS).

Typical dinosaur skin consists of tuberculate scales. They are not imbricated as seen in many forms of lizards and snakes. Analogous forms of tuberculate squamation can be seen on extant chameleons. The size of individual tubercles varies in different parts of the body. A basic ground plan consisting of the smallest scales can usually be found on the underside of the animal. Larger scales can be arranged in clusters or broad areas found in increasing abundance on the animal's outer surface. When adequately preserved, ornate patterns can often be discerned. Apparently, most if not all dinosaurs were much more decorative than is commonly appreciated. Bony ossicles, or scutes, are well known among some kinds of dinosaurs, but similar protuberances are now known to have been present without necessitating an ossified substructure.

Several examples of skin patterns are well known from different dinosaur genera. As a result, various kinds of dinosaurs can be identified or portrayed as readily distinguishable from each other. Also, features that would not be readily indicated by skeletal features can be revealed. In fact, what dinosaurs really looked like, and even how they behaved, can often be misinterpreted or remain obscure until skin impressions reveal the true form and function of the animal.

Duck-billed hadrosaurs and lambeosaurs are the most common dinosaurs found with preserved skin impressions. It is unclear if this represents a preservational bias from physical environmental conditions or other factors, including the relatively large number of specimens available. It has not been uncommon for skin impressions to go unnoticed or even be sacrificed in order to reveal the underlying bones of seemingly unique specimens. This practice is becoming increasingly corrected thanks to a better appreciation of what fossil skin represents.

There is a great difficulty in locating fossil skin impressions when excavating or preparing a specimen from its surrounding matrix. Skin impressions may or may not be preserved along with the bones. If they are, the surface layer can easily be cut through by even the most experienced professional. Often if not for the serendipitous natural separation caused by a fortunate blow of the hammer, fossil skin impressions would go unnoticed.

Many preservational conditions can preserve skin impressions. Like fossil bones, the process of preservation begins when the surrounding matrix covers the animal's remains. Unlike bones, however, which are infilled by minerals during the fossilization process, skin impressions are usually a natural mold of the surface of the animal's skin by the surrounding matrix. This may happen at any stage of the animal's desiccation, and the resulting fossil may reflect anything from the full true form to a dried and twisted mummy. In extremely rare cases, it is even possible for more than just the surface texture of the skin to be preserved. Depending on the preservational conditions, the actual epidermis has been known to become mineralized. Also, although rare, even the subcutaneous tissues of muscle, ligaments, or tendons can be preserved if the conditions are right.

Preserved skin and tissues may be regarded as true body fossils like conventional skeletal remains. However, skin can also form trace fossils in some instances. FOOTPRINTS are regarded as trace fossils, and yet in a few rare examples made by hadrosaurs, the preservation is so fine that even the texture of the scales remains clearly visible.

The first discovery of dinosaur skin impressions

FIGURE 1 Fragments of sauropod skin preserved in positive form on the right, and the corresponding negative mold made by overlying matrix on the left. Scale bar = 10 cm.

FIGURE 2 Medium-sized dermal spine of a sauropod (see Fig. 4b). Scale bar = 10 cm.

was made by S. H. Beckles in 1852. It consisted of only a small patch approximately 21 × 20 cm in area, but it clearly showed hexagonal scales ranging in size from 1 to 3 cm in diameter, radiating in a rosette pattern. This discovery occurred so early during the history of paleontology that the animal it belonged to (a cetiosaurid sauropod) was not recognized as a dinosaur but rather as a giant crocodilian. Perhaps because of its supposed crocodilian affinities, the prominent rounded scales were uncontroversial. Today, in retrospect, this first piece of dinosaur skin has a greater significance in demonstrating that the long-necked sauropods were indeed covered in a scaly hide.

As with most dinosaur restorations, those of sauropods often neglect the available details made possible from fossil skin impressions. Instead, there is often a circular rationale based on realistic portrayals that instill the belief that sauropods had a smooth, wrinkled hide like that of an elephant. The underlying effect from this misperception further exaggerates other popular beliefs that if dinosaurs look so mammalian, why should they not be considered endothermic? Certainly, the theory of dinosaurs being warm blooded will remain controversial. However, if dinosaurs were accurately portrayed according to their fossilized skin impressions, one could hardly dismiss the biological affinities of their reptilian appearance. In fact, recent discoveries of sauropod skin not only reaffirm the noticeably scaly nature of their hide but also demonstrate that they looked far more reptilian than previously suspected (Figs. 1, 2, and 3). The presence of dermal spines, much like those on the back of iguanas or crocodilian tails, are now known to have existed on sauropods. The exact patterns of the dermal spines remain unknown, but indications suggest that the spines were aligned on top of the midline of the animal's tail, and possibly on the back and neck as well. Additional spines, shorter and more blunt, also appear to have decorated the sides of these

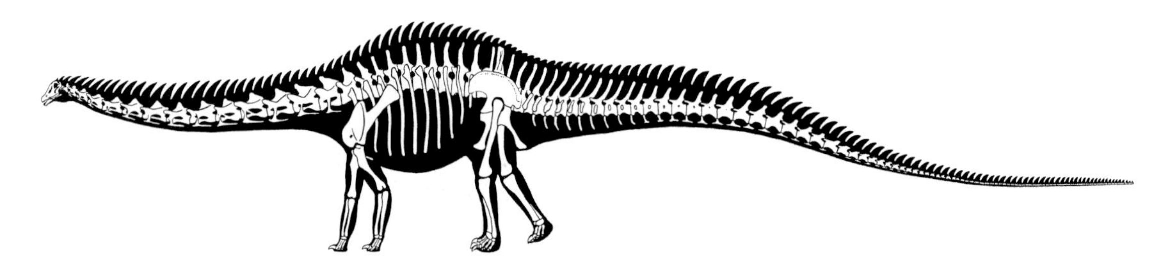

FIGURE 3 Skeletal reconstruction of a diplodocid sauropod with body outline illustrating dermal spines. Total length is approximately 14 m. Illustration © 1992 Stephen Czerkas.

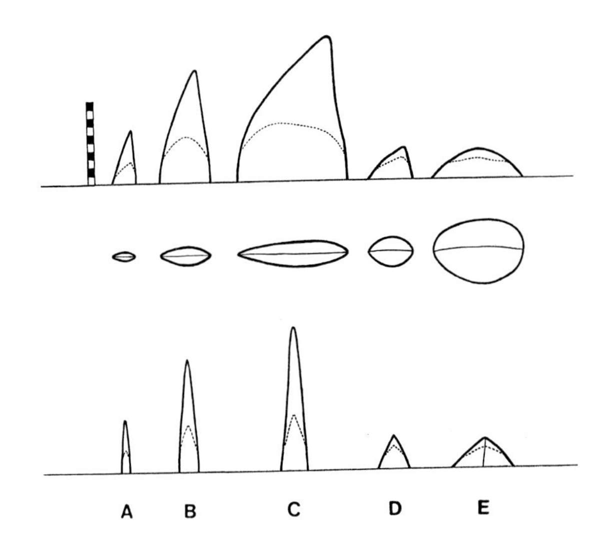

FIGURE 4 Outlines of sauropod dermal spines in (top) left side view, (middle) top view, and (bottom) front or back view. A, B, and C are narrow and pointed spines from the midline of the animal. D and E are wide, blunt spines from the sides of the animal. Broken lines represent lower portions of the spines that are preserved dimensionally, in the round. Scale bar = 10 cm.

animals (Fig. 4). The dermal spines are known from an undescribed sauropod resembling *Barosaurus* and *Diplodocus.* Whether or not such spines occurred on any or all other sauropods remains uncertain. One may anticipate that sauropods will continue to be depicted without the proper scalation or dermal spines. However, no physical evidence substantiates the traditional image of smooth-skinned sauropods.

The dermal spines of diplodocid sauropods are relatively large and, despite their size, they lack bony cores. However, further evidence supporting a scaly reptilian hide can be demonstrated by the presence of bony scutes associated with some Cretaceous sauropods, such as the titanosaur *Saltasaurus.* The discovery of dermal spines on sauropods provides evidence of what these dinosaurs looked like in life. It helps place in perspective the difficulty of making life restorations based on limited evidence. Also, it suggests that without incorporating the evidence available from fossil skin impressions, there is a potentially high degree of error in the interpretation of all dinosaurs, not just sauropods.

The abundant examples of duckbill dinosaur skin (Figs. 5 and 6) provide the most comprehensive evidence for interpreting what these particular animals looked like. Still, misinterpretations can creep in. On the famous "mummified" specimen collected by the Sternbergs in 1908, there are large folds of skin above the cervical vertebrae. The original description of this area above the neck was regarded as an ornamental "frill" of loosely folded skin. Consequently, almost all extended areas of skin located above the vertebrae have been regarded as pertaining to a frill. Contradicting the ornamental frill interpretation is the fact that the complete upper margin is rarely found intact, and when it is, it has a series of dermal spines. These spines are similar in construction to those now found on sauropods in that they do not have basal osteoderms, despite their considerable size. They appear to be composed of purely keratinous, hypertrophied tubercles. These spines are definitely known to be located over the back and tail of hadrosaurs, and to maintain the original description of a loosely folded

FIGURE 5 Ornamental section of duckbill skin with large, low conical scales from the underside of the pelvic region.

FIGURE 6 Detail of negative cast of duckbill skin with large, low conical scale. Conical scale is approximately 3 cm in length.

FIGURE 7 Skeletal reconstruction of a duckbill, *Corythosaurus,* showing typical dermal spines over a deep, muscular neck with nuchal ligaments.

frill is now unwarranted. Instead, it is more likely that the dermal spines continued along the crest of the neck, which was much deeper and more heavily muscled than previously supposed (Fig. 7). Considering the highly desiccated condition of the 1908 mummy, the segmented folds over the neck are apparently caused by the dried nuchal ligaments.

Also well preserved but poorly understood are the hands of duckbill dinosaurs. Several examples are now known. The best of these are from the Sternbergs' 1908 mummy in the American Museum of Natural History in New York and Sternberg's 1910 mummy in the Senckenberg Museum in Frankfurt, Germany. The right hand of the 1908 specimen was described first as being paddle-like in that all of the digits were connected within an integumentary web. Because iguanodonts and duckbill dinosaurs were thought to be bipedal and possibly aquatic in their habits, the possibility of the forelimbs being used as a regular part of terrestrial progression was dismissed, and the supposed aquatic adaptation of the animals' hands has been largely accepted. Curiously, the right hand of the Senckenberg specimen was similarly preserved and more complete. It showed an even more paddle-like web of skin extending over and beyond the bones of the fingers. The aquatic modification appeared to be unequivocal.

However, the left hand of the Senckenberg specimen does not support such a conclusion. This left hand is the best preserved in retaining its natural shape. Comparison of both flattened right hands with the more rounded left hand shows that the flattened hands were distorted by desiccation. The more fully fleshed-out hand indicates that it must have been used in supporting the body and was a highly functional participant during locomotion. Most of the weight was probably thrust upon the digitigrade third and fourth fingers. It is unknown but likely that the inner digit was equipped with a nail or hoof. The one next to it is the largest and appears to have had a prominent hoof, whereas the digit next to that definitely does not. Also, the last digit on the outside of the hand appears to have had a compressed nail. The outer finger is not a totally free structure. The manner in which the outer finger is enclosed in a separate web of skin does not readily support the grasping capabilities suggested for *Iguanodon* but does ironically have the potential of being a paddle-like structure, especially when the finger is directed outward at a nearly right angle. This position is often preserved among *Iguanodon* and duckbills. Unlike the Senckenberg specimen, the outer finger on the left hand of the 1908 mummy is extended outward and is connected with a broad web of skin that affirms the paddle-like adaptation and aquatic capabilities for these dinosaurs. Therefore, skin impressions have provided evidence, not revealed by skeletal remains, that suggests an extensive range of both terrestrial and aquatic behavior.

The skeletal structure of the tails of iguanodonts suggests that the tail was deep and narrow, making it well adapted for swimming. There are skin impressions known that extend well above the vertebrae, possibly to further emphasize the aquatic ability of the animal. Although ornamental clusters with enlarged conical scales are possible on the upper region, the tails are usually uniform in having large, polygonal scales a centimeter or more in size covering the tail.

Several beaks of duckbill dinosaurs and an *Iguanodon* have been preserved. They vary in size, but all show a crenulated tooth-like outer edge. As Mantell suggested as early as the 1850s, it is likely that iguanodonts had cheeks, but still no skin impressions have been found to substantiate this. An undescribed specimen of *Maiasaura* appears to have a dewlap, or throat wattle, which one can only imagine must have been used for display.

The skin of ceratopsians is known from only a few examples. The best of these are from *Centrosaurus* and

FIGURE 8 Skin fragment from over the upper thigh of the ceratopsian, *Chasmosaurus*. Note the large, round, flattened scales.

Chasmosaurus. Both are similar in having large, round, flattened scales surrounded by prominent polygonal tubercles arranged in rosette patterns (Figs. 8 and 9). Coincidentally, the areas preserved are both from over the upper right thigh and adjacent region of the torso. Compared to iguanodonts, the scalation of ceratopsians is larger overall and represents a thicker, protective hide. On a young *Triceratops* skull, a carbonaceous powdery layer 1 or 2 cm thick was found surrounding a horn core and was probably the remains of the outer sheath. A small fragment of skin is known from the ankle of a *Psittacosaurus*, although the indistinct nodes overlying the lower jaw of another individual may be an artifact. There is a *Protoceratops* skeleton in which the skin is poorly preserved over parts of the skull, but it is notable in having the desiccated and sunken eyelids intact within the orbit.

FIGURE 9 Life-size restoration of a *Styracosaurus* with typical ceratopsian scalation. Total length is approximately 6 m.

Other than an unpublished report from China, no actual skin impressions are as yet known for stegosaurians, although bony ossicles and the characteristic plates and spikes represent ossified dermal structures of the skin. It appears that *Stegosaurus* was covered by bony ossicles not only in the area of the throat but also elsewhere on the body, over the thigh and tail as well. The bony plates were aligned in a paired alternating pattern along the back instead of in matching pairs. The only alternating dermal structures found on extant reptiles, such as *Cyclura*, are caused when a single row of dermal spines grow large enough to crowd against each other so as to overlap.

Ankylosaurs and nodosaurs have skin patterns distinct from those of stegosaurs. The paired plates that are often portrayed vertically on top of nodosaurs were incorrectly influenced by comparison with stegosaurs. The triangular, spike-like plates of these heavily armored dinosaurs were quite different in construction from those of stegosaurs, especially at their bases. Totally unlike stegosaurs, these pointed, armored plates are positioned horizontally in pairs, predominantly over the shoulder region and tail.

Patterns are highly variable among ankylosaurs and nodosaurs. Scutes can be keeled, conical, or flattened. Smaller bony nodules surround the larger scutes adding some flexibility to a heavily armored hide. The bony armor is arranged in transverse bands across the backs of the animals. The bands acted like hinges and may have allowed the animal to fold into a defensive posture. This never developed to the extent seen in armadillos. Nodosaurs can be readily discerned from ankylosaurs in not having the characteristic bony tail club.

Skin impressions from small, nonvolant theropods are almost unknown. Speculations based on *Archaeopteryx* and the theory of dinosaurian ancestry of birds suggest that a plumage-like covering may have ex-

FIGURE 10 Life-size restoration of the horned theropod, *Carnotaurus*. Total length of the head is approximately 60 cm.

isted on some bird-like theropods. There may now be some evidence of this (see FEATHERED DINOSAURS).

A recently discovered ornithomimid, *Pelecanimimus*, does have patches of possible integumentary structures preserved below the skull, along the bottom of the neck, around the right humerus and elbow, and in a small patch behind the skull. Apparently, neither definite scales nor insulating structures can be discerned. The texture is described as being composed of a primary system of subparallel fibers arranged perpendicular to the bone surface and a less conspicuous secondary system parallel to it. Preservational factors may have obscured the original details of the actual skin. However, it is significant in suggesting an outline of the throat that may indicate a gular pouch.

A single row of bony ossicles is preserved over the neural spines of the neck and tail of a *Ceratosaurus*. Also, three small patches of tyrannosaurid skin impressions are known. They suggest a scaly hide consisting of typical dinosaurian tubercles. However, skin impressions from *Carnotaurus* provide the most information as to what some large theropods looked like (Fig. 10). The skin was preserved on the right side of the animal. The largest patch was found near the base of the tail. It revealed a basic plan of small tubercles approximately 5 mm in diameter, surrounding low, conical studs 4 or 5 cm in diameter that were arranged in irregular rows approximately 8–10 cm apart (Fig. 11). This ornamentation extended along the sides of the tail with increasingly larger studs near the top. The conical studs have no bony cores, and in that respect they are similar to those of sauropods and iguanodonts. Smaller patches of skin with conical studs were found over the ribs, scapula, neck, and skull.

No skin impressions of dinosaurs preserve the actual color of any animal. The highly variable patterns of different-sized scales, such as are seen on hadrosaurs, have led to the speculation that different colors would have accentuated the patterns. However, this cannot be verified because pigmentation is not restricted or necessarily regulated by the size of scales. The skin colors of dinosaurs may have played important functions in their lives. Patterns of protective camouflage, coloration for sexual display, and controllable changes in tone for thermoregulation are all possible though not provable. Although not preserving the actual colors, it is possible that indications of a color pattern could be preserved as mineralized tonal differences on fossil dinosaur skin. This appears to have occurred on a mosasaur specimen indicating

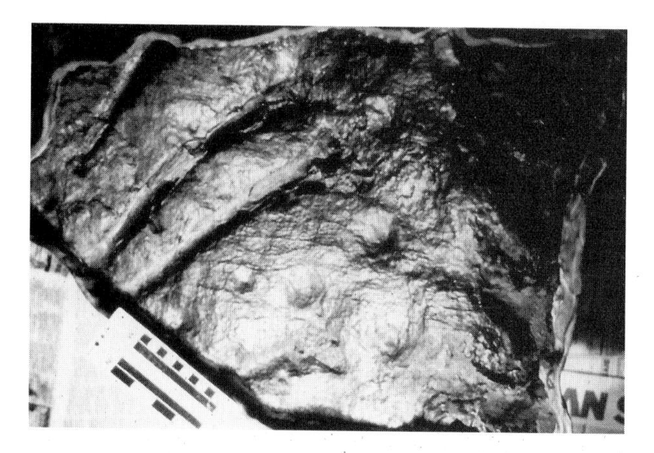

FIGURE 11 Negative mold of a large skin section from the tail of the theropod, *Carnotaurus*. Note the crater-like depressions that represent the distinctive conical spines that were found over the sides of the animal. Scale bar = 10 cm.

a banding pattern and on a turtle carapace in which a carbonized impregnation of the shell apparently reflects the original color pattern.

Skin impressions provide the only verifiable evidence of the external features of dinosaurs. As has been more readily apparent in recent years, additional discoveries and a more thorough appreciation of fossil skin impressions will no doubt continue to change and enhance the accuracy of how dinosaurs are portrayed.

See also the following related entries:

COLOR • FEATHERED DINOSAURS • ORNAMENTATION • SKELETAL STRUCTURES

References

Bonaparte, J. F., Novas, F. E., and Coria, R. A. (1990). *Carnotaurus sastrei* Bonaparte, the horned, lightly built Carnosaur from the Middle Cretaceous of Patagonia. *Natural History Museum of Los Angeles County, Contributions in Science* No. 416, pp. 1–42.

Czerkas, S. A. (1994). The history and interpretation of sauropod skin impressions. In *Aspects of Sauropod Paleobiology* (M. G. Lockley, V. F. dos Santos, C. A. Meyer, and A. P. Hunt, Eds.), Gaia No. 10. (Lisbon, Portugal).

Lull, R. S., and Wright, N. E. (1942). Hadrosaurian dinosaurs of North America. *Geol. Soc. Am. Spec. Papers* **40,** pp. 272.

Osborn, H. F. (1912). Integument of the iguanodont dinosaur *Trachodon. Mem. Am. Museum Nat. History* **II,** 33–54.

Sternberg, C. M. (1925). Integument of *Chasmosaurus belli. Can. Field Nat.* **XXXIX,** 108–110.

Skull, Comparative Anatomy

A complete description of the comparative anatomy of the skull of dinosaurs is beyond the scope of this work. However, briefly, the vertebrate skull is often divided into parts composed of groups of separate bones. These bones are, in turn, characterized by the developmental processes and patterns that lead to their formation. The brain and most of the sensory organs are housed in the cranium, which is divided into parts. The basicranium is composed mostly of endochondral bone—bone preformed in cartilage that later becomes ossified, whereas the dermatocranium is composed of both membrane and dermal bone. The dermatocranium in dinosaurs is sometimes expanded into various craniofacial air sinuses, cranial crests, ridges, or frills. The function of these structures is not definitely known, nor is it definitely known whether or not dinosaur skulls were sexually dimorphic, although the latter seems likely. The mandible or lower jaw is composed of both dermal and endochondral bone and, in some dinosaurs, is hinged usually between the dentary (the tooth-bearing bone) and the postdentary elements. There is also a hinge found in some dinosaurs between the anterior and posterior dermatocranial bones. The mobility and function of these hinges has been the subject of speculation.

See also the following related entries:

BRAINCASE ANATOMY • CRANIOFACIAL AIR SINUS SYSTEMS • ORNAMENTATION • PALEONEUROLOGY • TEETH AND JAWS • VARIATION

References

de Beer, G. R. (1985). *The Development of the Vertebrate Skull*, pp. 554, 143 plates. Univ. of Chicago Press, Chicago.

Hanken, J., and Hall, B. K. (Eds.) (1993). *The Skull*, 3 vols. Univ. of Chicago Press, Chicago.

Smithsonian Institution

M. K. BRETT-SURMAN
George Washington University
Washington, DC, USA

The Department of Paleobiology of the National Museum of Natural History, Smithsonian Institution is the repository for more than 40,000,000 plant, invertebrate, and vertebrate fossils. These are housed in more than 450 separate collections. The Dinosaur Collections include more than 1500 specimens ranging from isolated teeth to complete skulls with skeletons. These collections can be divided into three core collections, with some overlap: The Marsh and Gilmore Collections, Type Collection, and exhibits.

The Marsh Collection was acquired at the beginning of the 20th century and comprises those specimens collected under the auspices of O. C. Marsh using primarily government funds. Some of the world's most famous Jurassic dinosaurs, such as

Ceratosaurus, *Stegosaurus*, *Allosaurus*, and *Diplodocus*, are represented by complete skulls and / or skeletons. These were collected mostly from COMO BLUFF in Albany County, Wyoming and Garden Park, Fremont County, Colorado. There is also a large collection of *Triceratops* skulls collected for Marsh from Lance Creek, Wyoming, by the legendary field man John Bell Hatcher.

The Gilmore Collection was made during the first third of the 20th century and consists of mostly Cretaceous dinosaurs from New Mexico and Montana. It includes the first mass assemblage of late juvenile dinosaurs found in North America (TWO MEDICINE FORMATION). Other collections include the largest assemblage of dinosaurs from the east coast of North America ("Arundel Formation"), plus the only titanosaur from North America (*Alamosaurus*), and one of the best collections of pathologic dinosaur bones.

The Type Collection consists of the following dinosaurian taxa (those marked by an asterisk are on display): "*Agathaumas milo*" Cope, *Alamosaurus sanjuanensis* Gilmore, *Allosaurus medius* Marsh, "*Antrodemus valens*" Leidy, "*Arrhinoceratops*" *utahensis* Gilmore, **Brachyceratops montanus* Gilmore, **Camptosaurus* "*browni*" Gilmore, *Camptosaurus* "*depressus*" Gilmore, **Camptosaurus* "*nanus*" Marsh, "*Ceratops montanus*" Marsh, **Ceratosaurus nasicornis* Marsh, *Diceratops hatcheri* Lull in Hatcher, **Edmontosaurus annectens* Marsh, "*Coelurus gracilis*" Marsh, "*Creosaurus potens*" Lull, "*Dystrophaeus viaemalae*" Cope (the Smithsonian's first Type dinosaur, collected in 1859), "*Hadrosaurus paucidens*" Marsh, "*Hadrosaurus tripos*" Cope (no longer considered a dinosaur), "*Hoplitosaurus*" *marshi* (Lucas), "*Hypsibema crassicauda*" Cope, "*Labrosaurus ferox*" Marsh, "*Morosaurus*" *agilis* Cope, "*Neosaurus missouriensis*" (Gilmore) (originally named as a sauropod but now considered Hadrosauridae indet.), "*Ornithomimus affinis*" Gilmore, "*Ornithomimus sedens*" Marsh, "*Ornithomimus tenuis*" Marsh, *Pachycephalosaurus wyomingensis* (Gilmore), *Edmontonia rugosidens* (Gilmore), *Pleurocoelus altus* Marsh, *Pleurocoelus nanus* Marsh, "*Polyonax mortuarius*" Cope, "*Priconodon crassus*" Marsh, **Stegosaurus stenops* Marsh, *Stegosaurus* "*sulcatus*" Marsh, *Stegosaurus* "*ungulatus*" Marsh, *Styracosaurus ovatus* Gilmore, *Texasetes pleurohalio* Coombs, **Thescelosaurus neglectus* Gilmore, "*Thespesius occidentalis*" Leidy, *Triceratops* "*alticornis*" (Marsh), *Triceratops* "*calicornis*" Marsh, **Triceratops* "*elatus*" Marsh, *Triceratops* "*galeus*" Marsh, *Triceratops* "*obtusus*" Marsh, and *Triceratops* "*sulcatus*" Marsh.

Those exhibit specimens that are at least 50% complete are *Albertosaurus*, *Allosaurus*, *Brachyceratops*, *Camarasaurus*, *Camptosaurus*, *Ceratosaurus*, *Corythosaurus*, *Diplodocus*, *Edmontosaurus*, *Heterodontosaurus* (cast), *Maiasaura* (cast of a baby), *Stegosaurus*, *Thescelosaurus*, and *Triceratops*.

Exhibits of nonprimary-type material include skulls and skeletons of *Allosaurus*, *Stegosaurus*, *Camarasaurus*, *Diplodocus*, *Heterodontosaurus*, *Albertosaurus*, *Triceratops*, *Tyrannosaurus* (cast of the AMERICAN MUSEUM OF NATURAL HISTORY skull), *Corythosaurus* ("mummified" tail), and *Centrosaurus* (skull).

Reference

Gilmore, C. W. (1941). A history of the division of vertebrate paleontology in the U.S. National Museum. *Proc. U.S. Natl. Museum* **90**(3109), 308–377.

Solnhofen Formation

HARTMUT HAUBOLD
Martin Luther Universität
Halle, Germany

In the Upper Jurassic lithographic limestones of Solnhofen, only nine specimens of only two kinds of dinosaurs have been found: *Archaeopteryx* and *Compsognathus*. Both have been considered as the smallest known dinosaurs and are of particular interest concerning the understanding of dinosaur and bird evolution. The exceptional preservation in the lithographic limestones led to recovery of the first known Mesozoic bird feather. Stratigraphically, the Solnhofen Formation has been assigned to the lower part of the lower Tithonian. Its occurrence is restricted to the southern Franconian Alb of Bavaria. The evenly layered limestones are intercalated with fine-layered marls that originated in depressions between algal–sponge reefs in lagoon-like environments positioned on the northern rim of the Tethys. The carbonate accumulations and fossil assemblage indicate warm, semiarid climatic conditions (Viohl, 1985). *Compsognathus* and *Archaeopteryx* lived on islands along the shoreline of a slowly emerging central German swell to the north. Within the fossil record of the formation there is no evidence of trees or cliffs, but bushland vegetation alternated with open plains. In addition to *Compsognathus* and *Archaeopteryx*, pterosaurs and lizards have been recovered from the Solnhofen limestones. The rareness of dinosaurs in this intensively sampled deposit is surprising.

See also the following related entries:

BIRD ORIGINS • EUROPEAN DINOSAURS

References

Barthel, W., Morris, S. C., and Swinburne, N. C. (1990). *Solnhofen*, 2nd ed. Cambridge Univ. Press, New York.

Viohl, G. (1985). Geology of the Solnhofen Lithographic Limestone and the habitat of *Archaeopteryx*. In *The Beginnings of Birds* (M. K. Hecht, J. H. Ostrom, G. Viohl, and P. Wellnhofer, Eds.), pp. 31–44. Eichstatt. Freude des Jura-Museums Eichstätt, Willibaldsburg.

South African Museum

ANUSUYA CHINSAMY
South African Museum
Cape Town, South Africa

The South African Museum was established in 1825, making it the oldest museum in South Africa and one of the oldest outside Europe. The type specimens of the basal ornithopods *Heterodontosaurus tucki* and *Lycorhinus angustidens*, as well as that of the Malawian sauropod *Torniera dixeya* and those of the prosauropods *Euskelosaurus browni* and *Melanorosaurus readi*, are located at the South African Museum in Cape Town. Several specimens representative of the South African dinosaur fauna are in the collection, notably, postcranial elements of the prosauropods *Euskelosaurus* and *Blikanasaurus*. Cranial and postcranial elements of *Massospondylus* are well represented at the South African Museum, including specimens from Tanzania and Lesotho. Some South African sauropod material (including *Algoasaurus bauri*), as well as fabrosaurid (indet.) and dryosaurid (= *Kangnasaurus coetzeei*) skeletal elements, is also present.

A Late Cretaceous South African theropod (possibly avian) specimen has been recovered for the museum. Postcranial specimens of the sauropods *Janenschia robusta*, *Brachiosaurus* sp, and *Barosaurus africanus*, as well as the stegosaur *Kentrosaurus aethiopus*, all from the Tendaguru beds of Tanzania, can also be seen in Cape Town.

See also the following related entries:

AFRICAN DINOSAURS • ALBANY MUSEUM • BERNARD PRICE INSTITUTE FOR PALEONTOLOGICAL RESEARCH • NATIONAL MUSEUM, BLOEMFONTEIN • SAUROPODA

South American Dinosaurs

Fernando E. Novas
Museo Argentino de Ciencias Naturales
Buenos Aires, Argentina

Dinosaurs from South America are known from Triassic, Jurassic, and Cretaceous rocks (Fig. 1). Currently, nearly 50 dinosaur species (diagnosable on the basis of derived osteological features) have been identified in this continent (Table I). Representatives of Saurischia and Ornithischia were documented in South America as well as immediate dinosaur forerunners (Fig. 2). Saurischians are better known than ornithischians, the latter currently being represented by only 10 species (Table I). Chronologically, dinosaurs are known from Carnian, Norian, Bajocian, Callovian, Hauterivian, and Aptian through Maastrichtian times (Fig. 3 and Table I).

Dinosaur remains have been found in several countries of South America (Weishampel, 1990), although discoveries in Argentina have shed the most relevant and informative fossil evidence for the understanding of many important aspects in dinosaur evolution (e.g., their origins in Triassic times and the effects of the fragmentation of Gondwana over the evolution of dinosaur communities).

Triassic

The Ischigualasto-Villa Unión Basin of northwest Argentina (Stipanicic and Bonaparte, 1979) is filled by an almost continuous sequence of sediments of Early, Middle, and Late Triassic ages (Fig. 1). The formations identified in this basin yielded remains of several archosaurs of principal interest in the early evolution of dinosaurs. Especially relevant are the Los Chañares (Ladinian) and Ischigualasto (Carnian) formations (Bonaparte, 1982). From Los Chañares Formation were discovered skeletons of *Lagosuchus talampeyensis, Marasuchus lilloensis,* and *Pseudolagosuchus major* (Bonaparte, 1975; Arcucci, 1987), the proximate sister taxa of Dinosauria (Fig. 2; Novas, 1992a, 1997a; Sereno and Arcucci, 1994). These slender archosaurs are among the first tetrapods that acquired bipedal and digitigrade postures. The earliest known saurischians and ornithischians have been recorded in the Ischigualasto Formation of northwest Argentina and Santa Maria Formation of southeast Brazil (Bonaparte, 1982). The basal radiation of the Theropoda is represented by *Eoraptor lunensis* and Herrerasauridae (= *Staurikosaurus pricei* + *Herrerasaurus ischigualastensis*), which constitute successive sister taxa of the Ceratosauria + Tetanurae clade (Novas, 1992a, 1994a; Sereno and Novas, 1992, 1994; Sereno *et al.*, 1993). As noted elsewhere (see Herrerasauridae; Phylogeny of Dinosaurs; Theropoda), the systematic position of these animals is still controversial. Also present is *Pisanosaurus mertii* (Casamiquela, 1967), a basal ornithischian with a mixture of derived and plesiomorphic features (Bonaparte, 1976; Novas, 1989; Sereno, 1991; Weishampel and Witmer, 1990).

The documented morphological diversity of the South American Carnian dinosaurs suggests that an extensive evolutionary radiation occurred before the Late Triassic. Following the method described by Norell (1993), the minimum age for the origin of the Dinosauria can be predicted as early as the Ladinian on the basis of the presence of immediate sister taxa of Dinosauria in the Ladinian Los Chañares Formation (Fig. 3).

The evolutionary novelties evolved in the common dinosaurian ancestor (mainly involving pelvic girdle and hindlimb anatomy) apparently did not trigger an immediate numerical dominance but probably promoted extensive morphological disparity (Novas, 1997a). The early radiation of dinosauriforms (the clade formed by dinosaurs and their immediate forerunners; Novas, 1992a) was characterized by a sustained increase in body size, from 0.50 m in Mesotriassic dinosauriforms (e.g., *M. lilloensis*) up to 6 m in Carnian dinosaurs (e.g., *H. ischigualastensis*; Novas, 1994a,b). This increase in body size also involved a shifting from insectivorous toward megapredatory feeding habits and was accompanied by an increase

	Neuquén Basin (Argentina)	San Jorge Basin (Argentina)	Nordpatagonian Masif (Argentina)	NW Subandean Basin (Argentina)	Magallanes Basin (Argentina)	Araripe Basin (Brasil)	Ischigualasto Villa Unión Basin (Arg.)	Deseado Masif (Arg.)	Paraná Basin (Brazil)
Maastrichtian	Allen	Laguna Palacios	Los Alamitos	Lecho	Chorrillo				
Campanian									
Santonian	Neuquén Gr. Río Colorado	Bajo Barreal							
Coniacian	Neuquén Gr. Río Colorado	Bajo Barreal							
Turonian	Neuquén Gr. Río Neuquén	Bajo Barreal			Mata Amarilla				
Cenomanian	Neuquén Gr. Río Limay								
Albian	Rayoso	Castillo / La Colonia				Santana			
Aptian		Mata siete / Gorro Frigio				Santana			
Barremian									
Hauterivian									
Valanginian									
Berriasian									
Malm									
Callovian		Cañadón Asfalto							
Bathonian									
Bajocian		Cerro Carnerero							
Aalenian									
Lias									
Rhaetian									
Norian							Los Colorados	El tranquilo	
Carnian							Ischigualasto		Santa María
Ladinian							Los Chañares		Santa María

FIGURE 1 Tentative stratigraphic correlation among dinosaur-bearing formations in Argentina, based on regional geology, sequential geology, and vertebrate fossil content. Data gathered from Ardolino and Delpino (1987), Barcat *et al.* (1989), Bonaparte (1982, 1986a), Bonaparte *et al.* (1987, 1990), Kellner (1997), Legarreta and Gulisano (1989), Powell (1986), and Riccardi and Rolleri (1980).

in numerical abundance in Carnian times. In the early Late Triassic, herrerasaurids entered the terrestrial biotas as large, highly predatory forms, sharing with rauisuchid archosaurs the role of superpredators. In the Ischigualasto Formation herrerasaurids are numerically more abundant than other contemporary terrestrial carnivores (e.g., *Saurosuchus galilei, Venaticosuchus rusconii,* and *Chiniquodon theotonicus*). However, there are no clear patterns of competitive exclusion among these lineages, mainly because herrerasaurids, ornithosuchids, and rauisuchids survived until the Norian (Bonaparte, 1972, 1982; Sereno *et al.*, 1993).

In consequence, the available Triassic record of Argentina does not entirely support the interpretation that dinosaur dominance was reached in an empty ecospace after several nondinosaurian groups (e.g., rhynchosaurs; dicynodont, traversodont, and chiniquodont synapsids) became extinct (contra Benton, 1988). On the contrary, dinosauriforms coevolved during the Middle to early Late Triassic with the previously mentioned nondinosauriform tetrapods, amplifying the range of body size, species diversity, and feeding habits. In summary, the ecological diversification of the dinosauriforms documented in the Carnian can be seen as part of an uninterrupted coevolutionary process inaugurated in the Ladinian (Novas, 1994b).

The small sized (~1 m long) ornithischian *P. mertii* is known from only one specimen, suggesting that it constituted a minor component in the Triassic biotas of West Gondwana (Bonaparte, 1976, 1982). The Ischigualasto Formation has not currently yielded any sauropodomorph bones, in sharp contrast to the overlaying Los Colorados Formation (Norian), in which sauropodomorphs are numerically important (Bonaparte, 1972, 1982). However, the minimum age for the origin of the Sauropodomorpha can be predicted

TABLE I Dinosaur Taxa Recorded in Argentina[a]

Basal Ornithischia
Pisanosaurus mertii (Ischigualasto Fm.; Carnian; San Juan)
Ankylosauridae
N. gen. et sp. (Allen Fm.; early Maastrichtian; Neuquén)
Stegosauridae
Stegosauridae indet. (La Amarga Fm; Hauterivian; Neuquén)
Dryomorpha
Gasparinisaura cincosaltensis (Río Colorado Fm., Anacleto Mb.; Coniacian–Santonian; Neuquén)
Loncosaurus argentinus (Mata Amarilla Fm.; Santonian–early Campanian; Santa Cruz)
N. gen et sp. (Bajo Barreal Fm.; ?Cenomanian; Chubut)
Hadrosauridae
Secernosaurus koerneri (?Laguna Palacios Fm.; Maastrichtian; Chubut)
"Kritosaurus" australia (Los Alamitos Fm.; Campanian–early Maastrichtian; Río Negro).
Lambeosaurinae indet. (Allen Fm.; early Maastrichtian; Río Negro)
Ceratopsia?
Notoceratops bonarelli (?Laguna Palacios Fm.; Maastrichtian; Chubut)
Abelisauria
Velocisaurus unicus (Río Colorado Fm.; Coniacian–Santonian; Neuquén)
Ligabueino andesi (La Amarga Fm.; Hauterivian; Neuquén)
Noasaurus leali (Lecho Fm.; Maastrichtian; Salta)
Xenotarsosaurus bonapartei (Bajo Barreal Fm.; ?Cenomanian; Chubut)
Abelisaurus comahuensis (Allen Fm.; Maastrichtian; Río Negro)
Carnotaurus sastrei (La Colonia Fm.; Albian?, Chubut)
N. gen. et sp. (Río Limay Fm.; Cenomanian; Neuquén)
Basal Tetanurae
Piatnitzkysaurus floresi (Cañadán Asfalto Fm.; Callovian; Chubut)
Giganotosaurus carolini (Río Limay Fm.; Cenomanian; Neuquén)
Avialae
Alvarezsaurus calvoi (Río Colorado Fm.; Coniacian–Santonian; Neuquén)
Patagonykus puertai (Río Neuquén Fm.; Turonian; Neuquén)
Neuquenornis volans (Río Colorado Fm.; Coniacian–Santonian; Neuquén)
Yungavolucris brevipedalis (Lecho Fm.; Maastrichtian; Salta)
Soroavisaurus (Lecho Fm.; Maastrichtian; Salta)
Lectavis bretincola (Lecho Fm.; Maastrichtian; Salta)
Patagopteryx deferriasi (Río Colorado Fm.; Coniacian–Santonian; Neuquén)
Basal Sauropodomorpha
Riojasaurus incertus (Los Colorados Fm.; Norian; La Rioja)
Coloradisaurus brevis (Los Colorados Fm.; Norian; La Rioja)
Mussaurus patagonicus (El Tranquilo Fm.; Norian; Santa Cruz)
Basal Sauropoda
Amigdalodon patagonicus (Cerro Carnerero Fm.; Bajocian; Chubut)
Volkheimeria chubutensis (Cañadán Asfalto Fm.; Callovian; Chubut)
Patagosaurus fariasi (Cañadán Asfalto Fm.; Callovian; Chubut)
Diplodocid-related sauropods
Limaysaurus tessonei (Río Limay Fm.; Cenomanian; Neuquén)
Gen. et sp nov. (Rayoso Fm.; Albian; Neuquén)
Gen. et sp. indet. (Bajo Barreal Fm.; ?Cenomanian; Chubut)
Dicraeosauridae
Amargasaurus cazaui (La Amarga Fm.; Hauterivian; Neuqu)
Titanosauria
Chubutisaurus insignis (Gorro Frigio Fm.; Aptian; Chubut)
Andesaurus delgadoi (Río Limay Fm.; Cenomanian; Neuquén)

(*continues*)

Continued

Epachtosaurus sciuttoi (Bajo Barreal Fm.; ?Cenomanian; Chubut)
Argentinosaurus huinculensis (Río Limay Fm.; Cenomanian; Neuquén)
Argyrosaurus superbus (Bajo Barreal Fm.; ?Cenomanian; Chubut)
Antarctosaurus wichmanianus (Río Colorado Fm.; Coniacian–Santonian; Río Negro).
"Antarctosaurus" giganteus (?Río Neuquén Fm.; Turonian; Neuquén)
Titanosaurus araukanicus (Allen Fm.; Maastrichtian; Río Negro)
Neuquensaurus australis (Allen Fm.; Maastrichtian; Río Negro)
Aeolosaurus rionegrinus (Allen Fm. and Los Alamitos Fm.; late Campanian–early Maastrichtian; Río Negro)
Saltasaurus loricatus (Lecho Fm.; Maastrichtian; Salta)

[a] Only taxa known by skeletal material are listed. Stratigraphic and geographic provenances given are restricted to the holotypes. Dismissed from this list are referred specimens represented by poorly known materials. See Leonardi (1989) and Casamiquela (1964) for dinosaur ichnotaxa. This list updates and amends that given by Weishampel (1990).

as early as the Carnian on the basis of the presence of their sister taxon (e.g., Theropoda) in the Ischigualasto Formation (Sereno and Novas, 1992; Fig. 3). Late Triassic dinosaurs are represented in South America by the melanorosaurid *Riojasaurus incertus* and the plateosaurids *Coloradisaurus brevis* and *Mussaurus patagonicus* (Bonaparte, 1972, 1979, 1982; Bonaparte and Vince, 1979).

The "explosive evolution" (e.g., rapid increase in numerical abundance and body size) manifested by "prosauropod" dinosaurs during the Norian was interpreted as the result of opportunistic evolution after the extinction of several nondinosauriform tetrapods (e.g., rhynchosaurs and traversodonts; Benton, 1988). However, a tacit competitive scenario is entailed in this interpretation: If sauropodomorphs were present in the Carnian, as predicted previously, their rarity or virtual absence in the Ischigualasto Formation can be explained as a consequence of the presence of successful competitors (rhynchosaurs and traversodonts).

Jurassic

The record of Jurassic dinosaurs in South America is scarce. The early evolution of Sauropoda is represented by the Bajocian *Amygdalodon patagonicus* and the Callovian *Volkheimeria chubutensis* and *Patagosaurus fariasi*, from Patagonia (Bonaparte, 1986a). These taxa are interpreted as successive sister taxa of more derived sauropods (Bonaparte, 1986a). Theropoda is represented by the Callovian basal tetanurine *Piatnitzkysaurus floresi*, which resembles the European *Eustreptospondylus cuvieri* (Bonaparte, 1986a). Scarce remains of a presumed ornithischian have been documented in the Jurassic La Quinta Formation of Venezuela (Russell *et al.*, 1992). Probably the maker of the footprints named *Delatorrichnus goyenechei* (Casamiquela, 1964) was a small quadrupedal ornithischian. During the Jurassic, Patagonia had positive hydrologic balance which favored the development of lake systems (Uliana and Biddle, 1988), around which large sauropods and theropod dinosaurs prospered. Dinosaur taxa from Patagonia (Bonaparte, 1986a) resemble nearly contemporary relatives from other parts of the world (e.g., England and India), suggesting the presence of a relatively uniform dinosaur fauna throughout the world. However, local environmental conditions in South America are represented by the Oxfordian La Matilde Formation, in which small dinosaurs (e.g., *D. goyenechei*, *Sarmientichnus scagliai*, and *Wildeichnus navesi*) were documented (Casamiquela, 1964). A similar dinosaur assemblage comes from the Late Jurassic–Early Cretaceous Botucatu Formation, southeast Brazil (Leonardi, 1989), deposited under desert conditions and also disturbed by widespread volcanic activity (Soares, 1981; Uliana and Biddle, 1988). These disturbed environmental conditions in central South America prevailed from Kimmeridgian through Neocomian times, approximately 27 ma (Uliana and Biddle, 1988). This paleodesert probably controlled dinosaurian DISTRIBUTIONS, creating a filter for intracontinental dispersion of animals and plants.

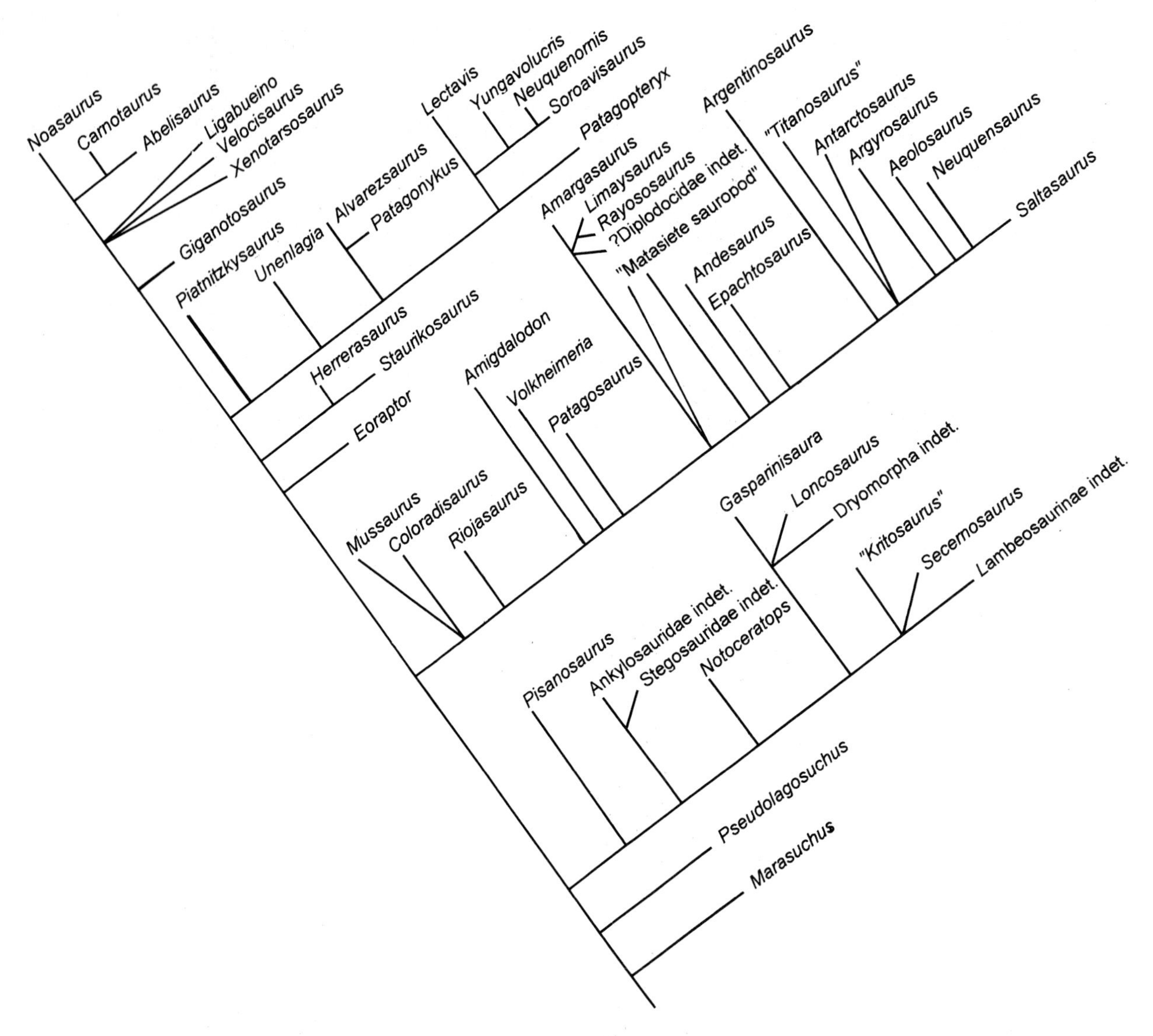

FIGURE 2 Cladogram depicting the phylogenetic relationships of dinosaur taxa recorded in South America. The relationships of South American taxa with relatives of other continents are indicated. Data gathered from Bonaparte (1986a, 1991a,b, 1994), Bonaparte *et al.* (1990), Bonaparte and Coria (1993), Calvo and Salgado (1997), Chiappe (1995), Coria and Salgado (1993b, 1995, 1997a), Novas (1991, 1992a,b, 1994a, 1997a,b,c), Powell (1986, 1987), Salgado and Bonaparte (1991), Salgado and Martínez (1993), Salgado and Calvo (1997), Salgado *et al.* (1997b), Sereno (1991), Sereno and Arcucci (1994), Sereno *et al.* (1993), Sereno and Novas (1992).

Cretaceous

During the Cretaceous South America was inhabited by a wide variety of dinosaurian clades. Ornithischians are recorded in this continent mostly from Upper Cretaceous rocks, although their presence in the Early Cretaceous is confirmed by ornithopod footprints from Brazil and Chile (Leonardi, 1989) and skeletal remains in Patagonia (Bonaparte, 1994). Ornithischians appear to be less diversified than those from the Cretaceous of North America, although the record of representatives of Stegosauria, Ankylosauria, Ornithopoda, and probably Ceratopsia suggests that a hidden diversity of ornithischians remains to be discovered. Ornithischians more frequently discovered in South America are basal iguanodontians (e.g., Dryomorpha) of small to medium size (Coria and Salgado, 1993a, 1997a; R. Martínez, personal communication). Theropoda is represented by the neo-

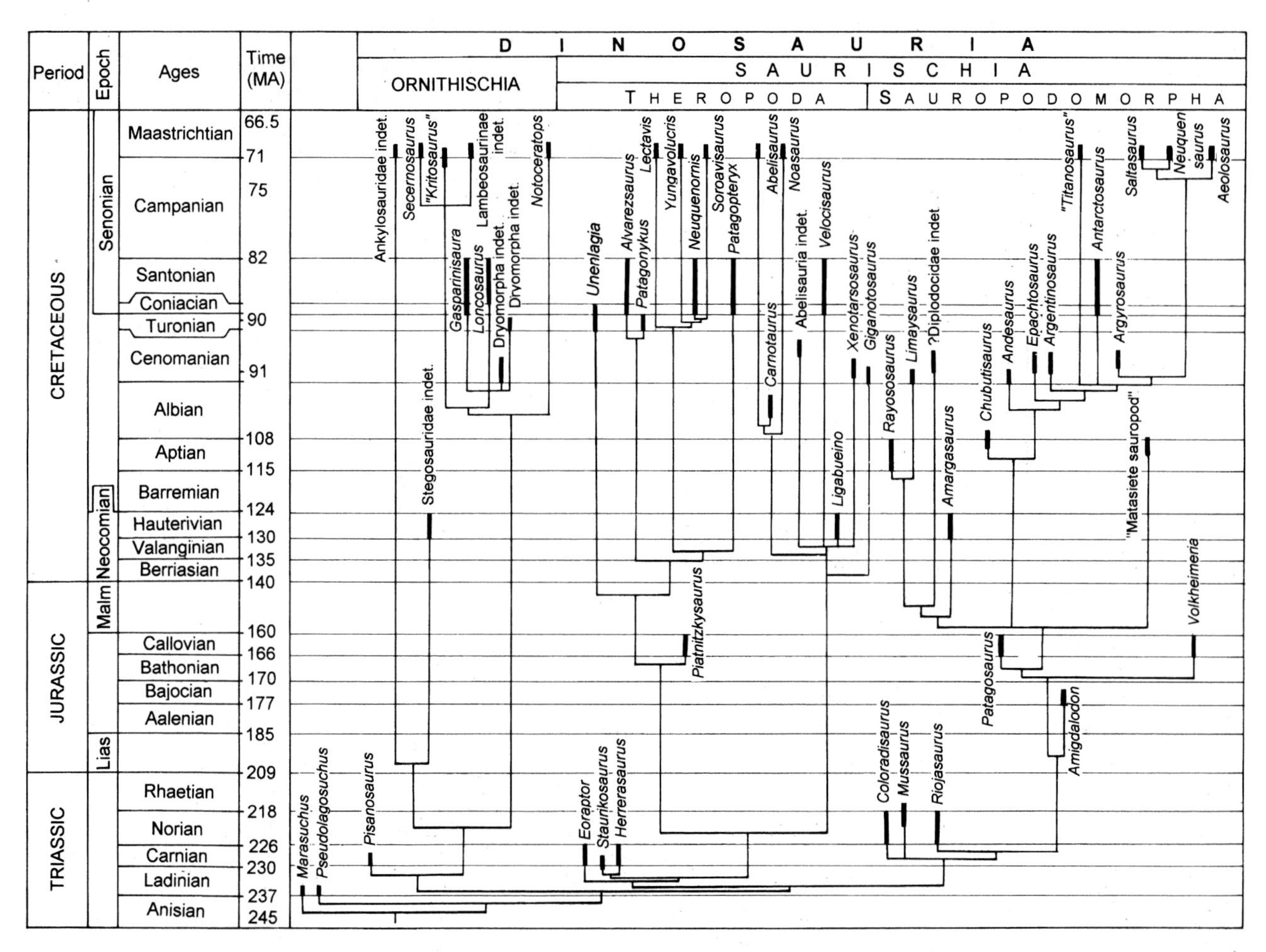

FIGURE 3 Phylogenetic diagram depicting the relationships and chronostratigraphic distribution of dinosaur taxa recorded in South America on the basis of data depicted in Figs. 1 and 2.

ceratosaurian Abelisauria (Novas, 1992b), as well as tetanurans of different pedigree, including early birds (Chiappe, 1995). Abelisaurs in particular underwent a significant evolutionary radiation during the Cretaceous in South America (Bonaparte, 1991a). Abelisaurs were active predators, some of them no larger than a chicken (e.g., *Velocisaurus unicus* and *Ligabueino andesi*; Bonaparte, 1991b, 1994), but other taxa reached 10 m in length (e.g., Abelisauridae; Bonaparte and Novas, 1985; Bonaparte *et al.*, 1990). Sauropod dinosaurs constituted the dominant group of megaherbivores in the South American Cretaceous landscape (Bonaparte and Kielan-Jaworowska, 1987). However, Leonardi (1989) pointed out that ichnological associations yield a sauropod to ornithischian ratio of 1:2, in contrast to the ratio of skeletal findings on the continent. Sauropods attained a high diversity of forms not recorded in the Cretaceous of other continents. Two main sauropod clades evolved in South America during the Cretaceous (Calvo and Salgado, 1997): the Hauterivian through Cenomanian diplodocid-related forms (including *Amargasaurus cazaui* and *Limaysaurus tessonei*; Salgado and Bonaparte, 1991; Calvo and Salgado, 1997), and Titanosauridae and their kin (Bonaparte and Coria, 1993; Salgado and Calvo, 1997) recorded from Albian (e.g., *Chubutisaurus insignis*; Salgado, 1993) through Maastrichtian times (Powell, 1986).

The following major intervals can be recognized for the Cretaceous of South America on the basis of different faunal assemblages:

1. Hauterivian interval: The local fauna of La Amarga Formation (northwest Patagonia) yielded remains of the oldest known abelisaur (*L. andesi*; Bonaparte, 1994, 1997), a stegosaur (Bonaparte 1994, 1997),

and the dicraeosaurid sauropod *A. cazaui* (Salgado and Bonaparte, 1991; Salgado and Calvo, 1992).

2. Aptian–Albian interval: The Aptian–Albian interval is characterized by the presence of very large sauropods (e.g., *Chubutisaurus insignis* from the Gorro Frigio Formation and indeterminate sauropod from the Matasiete Formation; Salgado, 1993; Martínez *et al.*, 1989a). The presence of large theropods in this interval is documented by fragmentary teeth (Del Corro, 1974), nearly half the size of those of *Tyrannosaurus rex*. During the Aptian–Albian, abelisaurs evolved toward bizarre forms such as the horned *Carnotaurus sastrei* (it must be noted, however, that the age of the La Colonia Formation, which yielded *C. sastrei*, was variously considered to be Albian and recently as Maastrichtian; Bonaparte *et al.*, 1990; Ardolino and Delpino, 1987). Fragmentary evidence from Brazil suggests the possible presence of spinosaurid tetanurines (Rauhut, 1994) in South America during the Albian (Kellner, 1997).

3. Cenomanian through Santonian interval: This interval of the Cretaceous, spanning ~14 ma, involves the time of deposition of the Neuquén Group (Legarreta and Gulisano, 1989) and presumed equivalent formations of other sedimentary basins (e.g., San Jorge Basin; Barcat *et al.*, 1989; Fig. 1). It is characterized by the explosive evolution of the titanosaur clade, including truly gigantic forms (e.g., *Argentinosaurus huinculensis*, "*Antarctosaurus*" *giganteus*, *Argyrosaurus superbus*; von Huene, 1928; Powell, 1986; Bonaparte and Coria, 1993), along with diplodocid-related sauropods (Sciutto and Martínez, 1994), some of them closely related to the African *Rebbachisaurus garasbae* (Bonaparte, 1997; Calvo and Salgado, 1997).

   The "Neuquenian fauna" also includes small to medium-sized basal iguanodontians such as the Santonian *Loncosaurus argentinus*, *Gasparinisaura cincosaltensis*, an indeterminate tiny dryosaurid from the Turonian (Novas, 1997b), as well as yet undescribed Cenomanian ornithopods from the Bajo Barreal Formation (R. Martínez, personal communication). Footprints from the Cenomanian Río Limay Formation reveal the presence of several ornithischian (e.g., ornithopod) taxa (Calvo, 1991).

   The evolution of the Theropoda was also prolific at this time. Ceratosaurians are known by the turkey-sized *Velocisaurus unicus* (Bonaparte, 1991b) and by larger indeterminate taxa resembling *Carnotaurus sastrei* (Coria *et al.*, 1991; Coria and Salgado, 1993). The Cenomanian basal tetanurine *Giganotosaurus carolini* (Coria and Salgado, 1995) is one of the largest known theropods (6–8 tons), which attained macropredaceous habits independently from the Laurasian *T. rex*. The other large group of tetanurines known from South America are Late Cretaceous birds (Avialae *sensu* Gauthier, 1986). Avialans are represented on this continent by several taxa more derived than *Archaeopteryx lithographica* (e.g., Alvarezsauridae, Enantiornithes, and *Patagopteryx deferriasi*; Bonaparte, 1991b; Chiappe, 1995; Novas, 1997b,c). They were small to medium-sized animals (0.60–2.00 m long) that fed on small food items, probably insects and fruits. Interestingly, the coeval alvarezsaurids and *P. deferriasi* secondarily reversed to nonvolant, cursorial habits. Enantiornithine birds are also recorded in the Santonian beds of the Neuquén Group (Chiappe, 1995).

4. Maastrichtian interval: On the basis of the available record, an extinction event seems to have occurred during the Campanian, coincident with a severe orogenic phase in South America (Zambrano, 1981). After that, a major Maastrichtian transgressive phase took place (Zambrano, 1981; Uliana and Biddle, 1988), and wide lacustrine deposits developed in Patagonia. A new faunistic assemblage came on the scene, corresponding to the latest Campanian to early Maastrichtian Alamitian fauna (Bonaparte, 1987). During the Alamitian, basal iguanodontians as well as the gigantic sauropods characteristic of the previous "Neuquenian fauna" became extinct, as did the diplodocid-related forms. New ornithischian clades are recorded during the Alamitian (e.g., "*Kritosaurus*" *australis*, Lambeosaurine indet., Ankylosauridae, and Ceratopsidae?; von Huene, 1928; Bonaparte, 1987; Powell, 1987; Salgado and Coria, 1997), which probably arrived from North America (Bonaparte, 1986b). Derived titanosaurs (e.g., *Neuquensaurus australis* and "*Titanosaurus*" *araukanikus*; Powell, 1986) were smaller in size than those of earlier times, although titanosaurs of large size are known from supposedly Maastrichtian beds of southern Patagonia (Chorrillo Formation; Powell, 1986). Interestingly, the previously mentioned replacement of herbivorous dinosaurs is coincident

with the diversification of the angiosperm tree *Nothofagus*, which occurred during late Campanian–Maastrichtian in the southern South America–Antarctic Peninsula region (Romero, 1993). Titanosaur–hadrosaur communities of Alamitian age lived in lacustrine environments, related to a major Maastrichtian transgressive phase over most of South America (Uliana and Biddle, 1988). Alamitian age theropods of large size (nearly 10 m long) belong to the ceratosaurian subclade Abelisauridae (*Abelisaurus comahuensis*; Bonaparte and Novas, 1985), although Maastrichtian abelisaurs also include the turkey-sized *Noasaurus leali* (Bonaparte, 1991a), which evolved a raptorial sickle claw in pedal digit II, independent of dromaeosaurs and troodontids (Novas, 1992b).

Unfortunately, to date dinosaur remains from the latest Maastrichtian have not been recorded.

Paleobiogeographic Evolution of South American Dinosaurs during the Cretaceous

South American dinosaur faunas are in contrast with those from Asia and North America mainly because in the latter two continents both tetanurans and ornithischians are quite diverse, and sauropods constitute a minor component (Bonaparte and Kielan-Jaworowska, 1987). These sharp differences in faunal composition can be interpreted as the result of the breakup of Laurasia and Gondwana in the course of the Cretaceous, resulting in the evolution of endemic dinosaur taxa for each continental mass (Bonaparte, 1986b).

Although most of the evidence supporting the isolation of Gondwanan from Laurasian tetrapod faunas comes from Late Cretaceous rocks, the presence in the Early Cretaceous of South America of several taxa not currently recorded in Neocomian formations from the Northern Hemisphere (e.g., Cloverly Formation; Russell, 1993) suggests that the isolation between Laurasia and Gondwana was already complete during the Neocomian. Such taxa include the previously mentioned dinosaurs from the Hauterivian La Amarga Formation (e.g., dicraeosaurids, basal abelisaurs, and stegosaurs; Bonaparte, 1994; Salgado and Bonaparte, 1991). However, the differences between southern South America (e.g., Patagonia) and Laurasian continents are probably due not only to the physical separation mentioned previously but also to more or less effective barriers, intrinsic from West Gondwana (e.g., extensive deserts in the Paraña and Chaco basins in northeast South America; Soares, 1981; Zambrano, 1981; Uliana and Biddle, 1988).

The Early and "middle" Cretaceous dinosaur faunas from South America correspond to the interval (e.g., Neocomian through Albian) during which this continent was geographically connected with Africa. As a result, some dinosaurian taxa are shared in common by both continents (e.g., stegosaurs, dicraeosaurids, and basal diplodocids closely related to *R. garasbae*; Salgado and Bonaparte, 1991; Salgado and Calvo, 1992; Calvo and Salgado, 1997; McIntosh, 1990). From Aptian through Santonian times (an interval of nearly of 33 my; Haq and Van Eysinga, 1994), South America progressively increased its geographic isolation from other continents (Scotese and Golonka, 1992). Complete isolation occurred at some point between 106 my (Albian) and 84 my (Campanian), with the full development of a marine barrier between the two continents (Pitman *et al.*, 1993). Bizarre tetanurans from the Cenomanian of northern Africa (e.g., *Carcharodontosaurus saharicus, Bahariasaurus ingens,* and *Spinosaurus aegyptiacus*; Rauhaut, 1994) suggest that Patagonia and that region of Africa were partially isolated during the beginning of the Late Cretaceous. However, spinosaurid-like teeth recently found in the Albian of Brazil (Kellner, 1997) suggest that the north portions of both South America and Africa retained faunistic affinities, corresponding to the south–north direction of opening of the South Atlantic (Uliana and Biddle, 1988; Pitman *et al.*, 1993).

On the basis of the available vertebrate fossil record and according to current paleogeographic reconstructions (Scotese and Golonka, 1992), South America renewed contact with North America presumably during Campanian to Maastrichtian times, when intense orogeny caused continentalization of Caribbean regions (Bonaparte, 1986b; Zambrano, 1981; Pitman *et al.*, 1993). Some clades interpreted to have evolved in South America, or more generally Gondwana, reached Asia during the Late Cretaceous (e.g., titanosaurids and alvarezsaurids; Bonaparte, 1986b; Novas, 1997b). The presence of indisputable titanosaurids in the Maastrichtian of North America (e.g., *Alamosaurus sanjuanensis*), as well as the record of Laurasian taxa in the Late Campanian–

Maastrichtian of South America (e.g., Hadrosaurinae and Lambeosaurinae; Bonaparte, 1987; Powell, 1987), can be explained as the result of faunal interchange (Bonaparte, 1986b).

See also the following related entries:

African Dinosaurs • American Dinosaurs • Canadian Dinosaurs • Mexican Dinosaurs

References

Arcucci, A. (1987). Un nuevo Lagosuchidae (Thecodontia–Pseudosuchia) de la Fauna de Los Chañares (Edad Reptil Chaêarense, Triásico medio), La Rioja, Argentina. *Ameghiniana* **24**(1/2), 89–94.

Ardolino, A., and Delpino, D. (1987). Senoniano (continental-marino), comarca nordpatagónica, Provincia del Chubut, Argentina. *X Congr. Geol. Argentino Tucumán* **3,** 193–196.

Barcat, C., Cortiñas, J., Nevistic, V., and Zucchi, H. (1989). Cuenca Golfo San Jorge. In *Cuencas Sedimentarias Argentinas* (G. Chebli and L. Spalletti, Eds.), Serie de Correlación Geológica 6, pp. 319–345. Univ. Nacional de Tucumán, Tucumán, Argentina.

Benton, M. J. (1988). The origin of the dinosaurs. *Mod. Geol.* **13,** 41–56.

Bonaparte, J. F. (1972). Los tetrápodos del sector superior de la Formación Los Colorados, La Rioja, Argentina (Triásico superior). Parte I. *Opera Lilloana* **XXII,** 1–183.

Bonaparte, J. F. (1975). Nuevos materiales de *Lagosuchus talampayensis* Romer (Thecodontia, Pseudosuchia) y su significado en el origen de los Saurischia. Chañarense inferior, Triásico medio de Argentina. *Acta Geol. Lilloana* **13,** 5–90.

Bonaparte, J. F. (1976). *Pisanosaurus mertii* Casamiquela and the origin of the Ornithischia. *J. Paleontol.* **50,** 808–820.

Bonaparte, J. F. (1979). *Coloradia brevis* n. gen et n. sp. (Saurischia–Prosauropoda) dinosaurio plateosaurido de la Formación Los Colorados, Triásico Superior de La Rioja, Argentina. *Ameghiniana* **15**(3/4), 327–332.

Bonaparte, J. F. (1982). Faunal replacement in the Triassic of South America. *J. Vertebr. Paleontol.* **2,** 362–371.

Bonaparte, J. F. (1986a). Les dinosaures (carnosaures, allosaurides, sauropodes, cetiosaurides) du Jurassique Moyen de Cerro Cóndor (Chubut, Argentine). *Ann. Paleontol.* (Vertèb–*Invertèbr*) **72**(3/4), 247–386.

Bonaparte, J. F. (1986b). History of the terrestrial Cretaceous vertebrates of Gondwana. *IV Congr. Argentino Paleontol. Biostratigrafía Actas* **2,** 63–95.

Bonaparte, J. F. (1987). The Late Cretaceous fauna of Los Alamitos, Patagonia, Argentina. *Revista Museo Argentino Ciencias Nat. "Bernardino Rivadavia" Paleontol.* **3**(3), 103–179.

Bonaparte, J. F. (1991a). The Gondwanan theropod families Abelisauridae and Noasauridae. *Historical Biol.* **5,** 1–25.

Bonaparte, J. F. (1991b). Los vertebrados fósiles de la Formación Río Colorado, de la ciudad de Neuquén y cercanías, Cretácico Superior, Argentina. *Revista Museo Argentino Ciencias Nat. "Bernardino Rivadavia" Paleontol.* **4,** 17–123.

Bonaparte, J. F. (1994). Patagonia: The world's smallest dinosaur. A new discovery. *Ligabue Magazine* **24,** 136–137.

Bonaparte, J. F., and Coria, R. A. (1993). Un nuevo y gigantesco saurópodo titanosaurio de la Formación Río Limay (Albiano–Cenomaniano) de la provincia del Neuquén, Argentina. *Ameghiniana* **30**(3), 271–282.

Bonaparte, J. F., and Kielan-Jawarowska, Z. (1987). Late Cretaceous dinosaur and mammal faunas of Laurasia and Gondwana. In *Fourth Symposium on Mesozoic Terrestrial Ecosystems, Short Papers* (P. J. Currie and E. H. Koster, Eds.), pp. 24–29. Tyrrell Museum of Palaeontology, Drumheller, Alberta, Canada.

Bonaparte, J. F., and Novas, F. E. (1985). *Abelisaurus comahuensis*, n. g. et n. sp., Carnosauria del Cretácico Tardío de Patagonia. *Ameghiniana* **21**(2–4), 259–265.

Bonaparte, J. F., and Vince, M. (1979). El hallazgo del primer nido de dinosaurios triásicos de Patagonia, Argentina. *Ameghiniana* **16,** 173–182.

Bonaparte, J. F., Novas, F. E., and Coria, R. A. (1990). *Carnotaurus sastrei* Bonaparte, the horned, lightly built carnosaur from the middle Cretaceous of Patagonia. *Contrib. Sci. Nat. History Museum Los Angeles County* **416,** 1–42.

Calvo, J. (1991). Huellas de dinosaurios en la Formación Río Limay (Albiano–Cenomaniano), Picón Leuf, Prov. del Neuquén, República Argentina (Ornithischia, Saurischia: Sauropoda–Theropoda). *Ameghiniana* **28**(3/4), 241–258.

Calvo, J., and Bonaparte, J. (1991). *Andesaurus delgadoi* gen. et sp. nov. (Saurischia: Sauropoda), dinosaurio Titanosauridae del la Formación Río Limay (Albiano–Cenomaniano), Neuquén, Argentina. *Ameghiniana* **28**(3/4), 303–310.

Calvo, J., and Salgado, L. (1997). *Limaysaurus tessonei*, a New Sauropoda of the Albian–Cenomanian of Argentina: New Evidence on the Origin of the Diplodocidae, GAIA, in press.

Casamiquela, R. (1964). *Estudios icnológicos. Problemas y métodos de la icnología con aplicación al estudio de pisadas mesozoicos (Reptilia, Mammalia) de la Patagonia*, pp. 229. Ministerio de Asuntos Sociales, Provincia de Río Negro.

Casamiquela, R. M. (1967). Un nuevo dinosaurio ornitisquio triásico (*Pisanosaurus mertii*; Ornithopoda) de la Formación Ischigualasto, Argentina. *Ameghiniana* **5,** 47–64.

Chiappe, L. M. (1995). The first 85 million years of avian evolution. *Nature* **378,** 349–355.

Coria, R. A., and Salgado, L. (1993a). Un probable Dryomorpha basal (Ornithischia–Ornithopoda) de la Formación Río Colorado, Provincia de Río Negro. *Ameghiniana* **30**(1), 103.

Coria, R. A., and Salgado, L. (1993b). Un probable nuevo neoceratosauria Novas, 1989 (Theropoda) del Miembro Huincul, Formación Río Limay (Cretácico presenoniano) de Neuquén. *Ameghiniana* **30**(3), 327.

Coria, R. A., and Salgado, L. (1995). A new giant carnivorous dinosaur from the Cretaceous of Patagonia. *Nature* **258.**

Coria, R. A., and Salgado, L. (1997a). A basal iguanodontian (Ornithischia: Ornithopoda) from the Late Cretaceous of South America. *J. Vertebr. Paleontol.*, in press.

Coria, R. A., and Salgado, L. (1997b). "*Loncosaurus argentinus*" Ameghino, 1898 (Ornithischia: Ornithopoda): A revised description with comments on its phylogenetic relationships. *Ameghiniana,* in press.

Coria, R. A., Salgado, L., and Calvo, J. (1991). Primeros restos de dinosaurios Theropoda del Miembro Huincul, Formación Río Limay (Cretácico tardío presenoniano), Neuquén, Argentina. *Ameghiniana* **28**(3 / 4), 405–406.

Del Corro (1974). Un nuevo megalosaurio (carnosaurio) del Cretácico de Chubut (Argentina). *Comun. Museo Argentino Ciencias Nat. "Bernardino Rivadavia" Paleontol.* **1,** 37–44.

Giménez, O. (1993). Estudio preliminar del miembro anterior de los saurópodos titanosáuridos. *Ameghiniana* **30**(2),154.

Haq, B. U., and Van Eysinga, F. W. B. (1994). *Geological Time Table,* 4th ed. Elsevier, Amsterdam.

Kellner, (1997). Remarks on Brazilian dinosaurs. *Queensland Museum Mem.*, in press.

Legarreta, L., and Gulisano, C. (1989). Análisis estratigráfico secuencial de la Cuenca Neuquina (Triásico Superior–Terciario Inferior). In *Cuencas Sedimentarias Argentinas* (G. Chebli and L. Spalletti, Eds.), Serie de Correlación Geológica No. 6, pp. 221–243. Univ. Nacional de Tucumán, Tucumán, Argentina.

Leonardi, G. (1989). Inventory and statistics of the South American dinosaurian ichnofauna and its paleobiological interpretation. In *Dinosaur Tracks and Traces* (D. D. Gillette and M. G. Lockley, Eds.), pp. 165–178. Cambridge Univ. Press, Cambridge, UK.

Martínez, R., Giménez, O., Rodriguez, J., and Luna, M. (1989a). *Hallazgo de restos de saurópodos en cañadán Las Horquetas, Formación Matasiete (Aptiano), Chubut*, pp. 49–51. VI Jornadas Argentinas de Paleontología de Vertebrados, San Juan.

Martínez, R., Giménez, O., Rodriguez, J., and Luna, M. (1989b). Un titanosaurio articulado del género *Epachtosaurus,* de la Formación Bajo Barreal, Cretácico del Chubut. *Ameghiniana* **26**(3 / 4), 246.

McIntosh, J. (1990). Sauropoda. In *The Dinosauria* (D. B. Weishampel, P. Dodson, and H. Osmólska, Eds.), pp. 345–401. Univ. of California Press, Berkeley.

Norell, M. (1993). Tree-based approaches to understanding history: Comments on ranks, rules, and the quality of the fossil record. *Am. J. Sci. A* **293,** 407–417.

Novas, F. E. (1989). The tibia and tarsus in Herrerasauridae (Dinosauria, incertae sedis) and the origin and evolution of the dinosaurian tarsus. *J. Paleontol.* **63,** 677–690.

Novas, F. E. (1991). Relaciones filogenéticas de los dinosaurios terópodos ceratosaurios. *Ameghiniana* **28**(3 / 4), 410.

Novas, F. E. (1992a). Phylogenetic relationships of basal dinosaurs, the Herrerasauridae. *Palaeontology* **35,** 51–62.

Novas, F. E. (1992b). La evolución de los dinosaurios carnívoros. In *Los dinosaurios y su entorno bíotica. Actas II Curso de Paleontología en Cuenca* (J. L. Sanz and A. Buscalioni, Eds.), pp. 125–163. Instituto "Juan de Valdés," Ayuntamiento de Cuenca, Spain.

Novas, F. E. (1994a). New information on the systematics and postcranial skeleton of *Herrerasaurus ischigualastensis* (Theropoda: Herrerasauridae) from the Ischigualasto Formation (Upper Triassic) of Argentina. *J. Vertebr. Paleontol.* **13,** 400–423.

Novas, F. E. (1994b). Origen de los dinosaurios. *Investigación Ciencia* **217,** 52–59.

Novas, F. E. (1997a). Dinosaur monophyly. *Journal of Vertebrate Paleontology,* in press.

Novas, F. E. (1997b). Anatomy of *Patagonykus puertai* n. gen. n. sp. (Theropoda, Maniraptora, Alvarezsauridae), from the Late Cretaceous of Patagonia. *Journal of Vertebrate Paleontology,* in press.

Novas, F. E. (1997c). Alvarezsauridae, Cretaceous maniraptorans from Patagonia and Mongolia. *Queensland Museum Mem.*, in press.

Pitman, W. C., III, Cande, S., LaBreque, J, and Pindell, J. (1993). Fragmentation of Gondwana: The separation of Africa from South America. In *Biological Relationships between Africa and South America* (P. Goldblatt, Ed.), pp. 15–34. Yale Univ. Press, New Haven, CT.

Powell, J. E. (1986). Revisión de los titanosáuridos de América del Sur. Argentina. Unpublished doctoral thesis, Universidad Nacional de Tucumán, Tucumán, Argentina.

Powell, J. E. (1987). Hallazgo de un dinosaurio hadrosaurido (Ornithischia, Ornithopoda) en la Formacián Allen (Cretácico superior) de Salitral Moreno, Provincia de Río Negro, Argentina. *X Congr. Geol. Argentino San Miguel de Tucumán* **3,** 149–152.

Rauhut, O. (1994). Zur systematischen Stellung der afrikanischen Theropoden *Carcharodontosaurus* Stromer 1931 und *Bahariasaurus* Stromer 1934. *Berliner Geowissenchaften Abhandlungen* **16,** 357–375.

Riccardi, A., and Rolleri, E. (1980). Cordillera patagónica austral. II Simposio de Geología Argentina Vol. 2, pp. 1173–1306. Academia Nacional de Ciencias de Córdoba, Córdoba.

Romero, E. (1993). Cretaceous floras of South America. In *Biological Relationships between Africa and South America* (P. Goldblatt, Ed.), pp. 15–34. Yale Univ. Press, New Haven, CT.

Russell, D. A. (1993). The role of Central Asia in dinosaurian biogeography. *Can. J. Earth Sci.* **30**(10/11), 2002–2012.

Russell, D. E., Rivas, O. O., Battail, B., and Russell, D. A. (1992). Déco uverte de vertébrés fossiles dans la Formation de La Quinta, Jurassique du Vénézuéla Occidental. *C. R. Acad. Sci. Paris* **314**(2), 1247–1252.

Salgado, L. (1993). Comments on *Chubutisaurus insignis* Del Corro (Saurischia, Sauropoda). *Ameghiniana* **30**(3), 265–270.

Salgado, L. (1997). *Pellegrinisaurus powelli,* nov. gen. et sp. (Sauropoda, Titanosauridae) from the Upper Cretaceous of Lago Pellegrini, Northwestern Patagonia, Argentina. *Ameghiniana,* in press.

Salgado, L., and Bonaparte, J. F. (1991). Un nuevo saurópodo Dicraeosauridae, *Amargasaurus cazaui* gen. et sp. nov., de la Formación La Amarga, Neocomiano de la Provincia del Neuquén, Argentina. *Ameghiniana* **28**(3/4), 333–346.

Salgado, L., and Calvo, J. (1992). Cranial osteology of *Amargasaurus cazaui* Salgado and Bonaparte (Sauropoda, Dicraeosauridae) from the Neocomian of Patagonia. *Ameghiniana* **29**(4), 337–346.

Salgado, L., and Calvo, J. (1993). Report of a sauropod with amphiplatyan mid-caudal vertebrae from the Late Cretaceous of Neuquén province (Argentina). *Ameghiniana* **30**(2), 215–218.

Salgado, L., and Calvo, J. (1997). Origin and evolution of titanosaurid sauropods. II: The cranial evidence. *Ameghiniana,* in press.

Salgado, L., and Coria, R. (1993a). El género *Aeolosaurus* (Sauropoda, Titanosauridae) en la Formación Allen (Campaniano–Maastrichtiano) de la Provincia de Río Negro, Argentina. *Ameghiniana* **30**(2), 119–128.

Salgado, L., and Coria, R. (1993b). Consideraciones sobre las relaciones filogenéticas de *Opisthocoelicaudia skarzynskii* (Sauropoda) del Cretácico superior de Mongolia. *Ameghiniana* **30**(3), 339.

Salgado, L., and Coria, R. (1997). First evidence of an armoured ornithischian dinosaur in the Late Cretaceous of North Patagonia, Argentina. *Ameghiniana,* in press.

Salgado, L., and Martínez, R. (1993). Relaciones filogenéticas de los titanosáuridos basales *Andesaurus delgadoi* y *Epachtosaurus* sp. *Ameghiniana* **30**(3), 339–340.

Salgado, L., Calvo, J., and Coria, R. (1991). Estudio preliminar de dinosaurios saurópodos de la Formación Río Limay (Cretácico presenoniano, Provincia de Neuquén). *Ameghiniana* **28**(1/2), 134.

Salgado, L., Coria, R., and Calvo, J. (1997a). Presencia del género *Aeolosaurus* (Sauropoda, Titanosauridae) en la Formación Los Alamitos, Cretácico superior de la Provincia de Río Negro. *Paulacoutiana,* in press.

Salgado, L., Coria, R., and Calvo, J. (1997b). Evolution of titanosaurid sauropods. I: Phylogenetic analysis based on the postcranial evidence. *Ameghiniana,* in press.

Sciutto, J., and Martínez, R. (1994). Un nuevo yacimiento fosilífero d e la Formación Bajo Barreal (Cretácico tardío) y su fauna de saurópodos. *Nat. Patagónica Ciencias de la Tierra* **2,** 27–47.

Scotese, C. R., and Golonka, J. (1992). PALEOMAP Paleogeographic Atlas, PALEOMAP Progress Report No. 20, Department of Geology, University of Texas at Arlington.

Sereno, P. C. (1991). *Lesothosaurus,* "fabrosaurids," and the early evolution of Ornithischia. *J. Vertebr. Paleontol.* **11,** 168–197.

Sereno, P. C., and Arcucci, A. B. (1994). Dinosaurian precursors from the Middle Triassic of Argentina: *Marasuchus lilloensis,* gen. nov. *J. Vertebr. Paleontol.* **14,** 53–73.

Sereno, P. C., and Novas, F. E. (1992). The complete skull and skeleton of an early dinosaur. *Science* **258,** 1137–1140.

Sereno, P. C., and Novas, F. E. (1994). The skull and neck of the basal theropod *Herrerasaurus ischigualastensis. J. Vertebr. Paleontol.* **13,** 451–476.

Sereno, P. C., Forster, C. A., Rogers, R. R., and Monetta, A. M. (1993). Primitive dinosaur skeleton from Argentina and the early evolution of Dinosauria. *Nature* **361,** 64–66.

Soares, P. C. (1981). Estratigrafia das forma es Jurásico–Cretáceas na bácia do Paraná–Brasil. In *Cuencas Sedimentarias del Jurásico y Cretácico de América del Sur* (W. Volkheimer and E. Musacchio, Eds.), pp. 271–304. Buenos Aires.

Stipanicic, P., and Bonaparte, J. F. (1979). Cuenca triásica de Ischigualasto-Villa Unión (Provincias de La Rioja y San Juan). II Simposio de Geología Argentina, Vol. 1, pp. 523–575. Academia Nacional de Ciencias de Córdoba, Córdoba.

Uliana, M. A., and Biddle, K. T. (1988). Mesozoic–Cenozoic paleogeographic and geodynamic evolution of southern South America. *Revista Brasileira Geociencias* **18**(2), 172–190.

von Huene, F. (1928). Los saurisquios y ornitisquios del Cretácico argentino. *Anal. Museo de La Plata.*

Weishampel, D. B. (1990). Dinosaurian distribution. In *The Dinosauria* (D. B. Weishampel, P. Dodson, and H. Osmólska, Eds.), pp. 63–139. Univ. of California Press, Berkeley.

Weishampel, D. B., and Witmer, L. M. (1990). *Lesothosaurus, Pisanosaurus,* and *Technosaurus*. In *The Dinosauria* (D. B. Weishampel, P. Dodson, and H. Osmólska, Eds.), pp. 416–425. Univ. of California Press, Berkeley.

Zambrano, J. J. (1981). Distribución y evolución de las cuencas sedimentarias en el continente sudamericano durante el Jurásico y el Cretácico. In *Cuencas Sedimentarias del Jurásico y Cretácico de América del Sur* (W. Volkheimer and E. Musacchio, Eds.), pp. 9–44. Buenos Aires.

South Australian Museum, Australia

see MUSEUMS AND DISPLAYS

Southeast Asian Dinosaurs

ERIC BUFFETAUT
Laboratoire de Paléontologie des Vertébrés
University of Paris
Paris, France

Although dinosaur remains were found in Laos by the French geologist J. H. Hoffet as early as the 1930s, few discoveries were made until the 1980s, when a systematic search for dinosaur localities was started in the thick continental formations of northeastern Thailand by a Thai–French group. Although many gaps remain in our knowledge of Southeast Asian dinosaurs, both skeletal elements and footprints have been reported and sites are now known in the Upper Triassic, Jurassic, and Cretaceous. The best record is from Thailand, but Laos has also yielded promising results. The Thai specimens come from a thick series of nonmarine formations, known as the Khorat Group, which forms most of the Khorat Plateau of northeastern Thailand and ranges in age from Late Triassic to Early Cretaceous. The Laotian finds are from rocks that are lateral equivalents of the top of the Khorat Group, on the eastern bank of the Mekong River.

Thailand

The oldest known dinosaur specimen currently known from Southeast Asia is a PELVIS fragment, consisting of the fused distal ends of the ischia, of a prosauropod found in the Nam Phong Formation, of Late Triassic age, in Khon Kaen Province, Thailand. Although this material is too fragmentary to warrant an accurate identification, it indicates a large, heavily built animal that seems different from most of the abundant PROSAUROPODS found in the not too distant (but probably slightly younger) localities of the Lufeng area in Yunnan (southern China).

The next evidence of dinosaurs from the Khorat Plateau is in the form of trackways on the surface of a sandstone bed belonging to the Phra Wihan Formation at Hin Lat Pa Chad, in Phu Wiang National Park, Khon Kaen Province. The Phra Wihan Formation was long considered Middle to Late Jurassic in age, but recent palynological evidence suggests that it may in fact belong to the basal Cretaceous. Be that as it may, 10 trackways have been found at Hin Lat Pa Chad. Most of the FOOTPRINTS are quite small, with pes lengths usually under 10 cm, and must have been made by small dinosaurs not taller than 50 cm at the hip. Some of the footprints show three long, slender, and pointed toes and were probably left by theropods. Other tracks are less easy to interpret, with broader toes and what seem to be small tridactyl manus impressions lateral to the pes impressions. They resemble trackways from other parts of the world that have been referred to small ORNITHISCHIANS, although this interpretation has been challenged. Preliminary estimates based on the Hin Lat Pa Chad trackways suggest low speeds of 1 or 2 km per hour. The general picture suggested by the site is that of a lakeshore with various small dinosaurs walking slowly along the water's edge.

Most of the dinosaur remains hitherto found in Thailand are from the red clays and sandstones of the Sao Khua Formation, which overlies the Phra Wihan Formation. Although the Sao Khua Formation was long held to be Late Jurassic in age, the basal Cretaceous age of the Phra Wihan Formation shows that it must in fact be placed in the Early Cretaceous. Dinosaur localities have been found in the Sao Khua Formation at many places on the Khorat Plateau, the main sites being in Phu Wiang National Park and at Phu Pha Ngo and Wat Sakawan near the city of Kalasin. The most common dinosaurs in the Sao Khua Formation are SAUROPODS. A partial articulated skeleton was found at Phu Wiang and has been made the type of a new taxon, *Phuwiangosaurus sirindhornae*. Several more complete skeletons have since been discovered at Wat Sakawan. This middle-sized sauropod (about 15 m long), with its transversely broad cervical vertebrae with bifurcated neural spines, is quite different from the Chinese genera *Euhelopus*, *Mamenchisaurus*, and *Omeisaurus*. It shows some resemblances to *Camarasaurus*, but several differences apparently preclude its inclusion in the Camarasauridae, and it has not yet been placed in another subtaxon. Interestingly, many bones of juvenile sauropods referable to *P. sirindhornae*, some of them quite small, have been found in the Sao Khua Formation at Phu Wiang. They provide useful information about the characters of young sauropods, which in some parts of their anatomy seem to have been more reminiscent of basal sauropods than of adults of their own species.

THEROPODS are represented in the Sao Khua Formation by several forms. A few very small bones, apparently from adults, have been referred to a *Compsognathus*-like animal. Peculiar isolated TEETH, showing very limited lateral compression, unserrated carinae, and ridged enamel, apparently belong to an unusual theropod, possibly a spinosaurid, which has been named *Siamosaurus suteehorni*. More "normal," compressed, and serrated theropod teeth are also common in the Sao Khua Formation, and an incomplete skeleton of a large theropod recently found at Phu Wiang has been described as *Siamotyrannus isanensis*, the earliest and most primitive known tyrannosaurid. In addition, one of the Phu Wiang localities has yielded fairly abundant vertebrae and limb bones of what is clearly an early but fairly advanced small ornithomimosaur.

The only evidence of dinosaurs from the Phu Phan Formation, which overlies the Sao Khua Formation and is Early Cretaceous in age, consists of a set of three-toed footprints on a sandstone slab at Phu Luang, in Loei Province. They were made by several fairly large (about 1.80 m high at the hip) theropods that may have been moving in a group at a speed of about 8 km per hour.

The geologically youngest dinosaur remains hitherto found in Thailand are from the Khok Kruat Formation, at the top of the Khorat Group, which is referable to the late Early Cretaceous (Aptian–Albian) on paleontological evidence. The dinosaurs from the Khok Kruat Formation include theropods, represented by compressed, serrated teeth and very fragmentary skeletal elements, and a representative of the primitive ceratopsian *Psittacosaurus*. Jaws with teeth of this small ornithischian have been found near the city of Chaiyaphum and described as a new species, *Pisttacosaurus sattayaraki*. The occurrence of *Psittacosaurus* in Thailand shows that this genus, far from being restricted to Mongolia, Siberia, and the northern part of China, was also present in southern Asia in the Early Cretaceous. This is not surprising in view of the fact that the Indochina Block, which includes the Khorat Plateau of northeastern Thailand, has been in contact with mainland Asia since at least the Triassic, and probably much earlier.

Laos

Dinosaur remains were reported from the Cretaceous of the area around Muong Phalane, in southern Laos, as early as 1936 by the French geologist J. H. Hoffet. Hoffet briefly described his finds of various disarticulated bones in 1942 and 1944, under the difficult circumstances of the Japanese occupation of Indochina. He reported the occurrence of a sauropod, which he considered a titanosaurid ("*Titanosaurus falloti*"), and HADROSAURS, which he referred to the genus *Mandschurosaurus*, originally described from Manchuria by Riabinin as a new species, *Mandschurosaurus laosensis*. According to Hoffet, the dinosaur-bearing beds at Muong Phalane were Late Cretaceous, probably Campanian, in age. After Hoffet's death and the end of French rule in Indochina, most of the dinosaur material he had collected was apparently lost.

A reappraisal of Hoffet's discoveries, based on his descriptions and figures, leads to different conclusions from those he had reached. The sauropod material he described (mostly poorly preserved limb bones and a single amphicoelous caudal vertebra) is not reminiscent of titanosaurids and should be considered *incertae sedis*. As to the ornithischian material, it included, among other postcranial elements, well-preserved ilia that exhibit characters found in IGUANODONTIDS and primitive hadrosaurids. An accurate identification is difficult, but there is no special reason to refer the Lao material to *Mandschurosaurus*, all the more so that this taxon is now considered a *nomen dubium*. The age of the dinosaur assemblage from Muong Phalane is certainly older than Hoffet assumed. Correlations with the previously mentioned Khok Kruat Formation of the Khorat Plateau, on the other bank of the Mekong, seem possible, and a late Early Cretaceous to, at most, basal Late Cretaceous age is likely for the Muong Phalane dinosaurs. This seems to be confirmed by new dinosaur finds by a French–Laotian team, including remains of sauropod and ornithopods said to be iguanodontids in the Muong Phalane area, although no detailed description has yet been published. Ornithopod footprints have also been reported briefly from the same area of southern Laos.

Although our knowledge of southeast Asian dinosaurs is still far from complete, there is no longer any doubt about the great paleontological potential of the thick series of Mesozoic continental rocks in northeastern Thailand and Laos. Nonmarine Mesozoic rocks also occur in other parts of Thailand (notably the southern peninsula, where Jurassic vertebrates, but no dinosaurs, have been found), as well as in Malaysia, and their exploration may well lead to further dinosaur discoveries in the future, although the dense vegetation that covers these regions makes prospecting difficult. Meanwhile, further work on localities already known should also shed more light on the evolution and biogeographic history of dinosaurs in that part of Asia. Although some elements of the Southeast Asian dinosaur assemblages, such as the Thai *Psittacosaurus*, have close relatives in other regions of the Asian continent, other forms, such as the sauropod *Phuwiangosaurus*, are unexpectedly different from their counterparts from China. The discovery of the early tyrannosaur *Siamotyrannus* and of the early but already advanced ornithomimosaur from Phu Wiang shows that finds from Southeast Asia may lead to reappraisals of the evolutionary history of some important groups of dinosaurs.

See also the following related entries:

CHINESE DINOSAURS • INDIAN DINOSAURS

References

Buffetaut, E. (1991). On the age of the Cretaceous dinosaur-bearing beds of southern Laos. *Newslett. Stratigraphy* **24,** 59–73.

Buffetaut, E., and Suteethorn, V. (1993). The dinosaurs of Thailand. *J. Southeast Asian Earth Sci.* **8,** 77–82.

Buffetaut, E., Suteethorn, V., and Tong, H. (1996). The earliest known tyrannosaur from the Lower Cretaceous of Thailand. *Nature* **381,** 689–691.

Martin, V., Buffetaut, E., and Suteethorn, V. (1994). A new genus of sauropod dinosaur from the Sao Khua Formation (Late Jurassic or Early Cretaceous) of northeastern Thailand. *Comptes Rendus l'Acad. Sci. Paris II* **319,** 1085–1092.

Soviet–Mongolian Paleontological Expeditions

CLIVE COY
Royal Tyrrell Museum of Palaeontology
Drumheller, Alberta, Canada

Russian explorer and geologist Vladimir A. Obruchev described the first fossils from the Gobi Desert of Mongolia in 1892. However, in the 1920s Outer Mongolia came under Soviet influence and within reach of Soviet paleontologists. This may have contributed to the inability of the CENTRAL ASIATIC EXPEDITIONS of the AMERICAN MUSEUM OF NATURAL HISTORY to gain permission to venture beyond the borders of Inner Mongolia after 1925.

In 1941, a decade after the Americans had departed, the Committee of Scientific Affairs of the Mongolian People's Republic proposed a joint Soviet–Mongolian paleontological exploration of Outer Mongolia to the USSR Academy of Sciences. In June of that year, before the expedition could be organized, the Nazi invasion of the Soviet Union and ensuing war postponed all plans for joint explorations.

With the cessation of hostilities in August 1945, negotiations between Mongolia and the Soviet Union concerning joint paleontological exploration resumed. In 1946, the Soviet Paleontological Institute prepared a reconnaissance team to send into the Gobi under the scientific auspices of the Soviet Academy of Sciences, with I. Orlov as advisor. This initial team was led by I. A. Efremov, with A. Rozhdestvensky as second in command and six additional scientists and technicians.

Whereas the Central Asiatic Expeditions had used light touring cars to probe the remote Gobi Desert, the Soviets traveled in large, modern heavy-duty trucks and open-sided all-terrain field cars. Gone was the limiting reliance on camel caravans for resupplying an extended expedition into the vast wastes of the Gobi. The Soviet crew spent 2 months in the summer of 1946 making two traverses: One was in the southeastern Gobi, where rich deposits of large and medium-sized dinosaur skeletons were discovered in the NEMEGT Valley. On the basis of knowledge gained from this preliminary trip, plans were formulated for more extensive expeditions to be sent in 1948 and 1949.

The expeditions of 1948 and 1949 were well outfitted and composed of 19 scientific staff members, eight laborers and technicians, and six drivers. The fleet of large trucks and fast scout cars enabled Efremov and his team to penetrate more deeply into the Gobi than the American Expeditions could have done.

The first excavations of 1948 were opened up in the southeastern Gobi at a locality called Sayn Shand. Here, several skeletons of the new ankylosaurid *Talarurus plicatospineous* and fragmentary material of *Talarurus disparoserratus* were recovered from the Upper Cretaceous strata. The expedition then turned westward to Bayn Dzak, where the Central Asiatic Expeditions had made dramatic discoveries in the 1920s. As the Americans had found a quarter century before, the Soviets found abundant *Protoceratops* skeletons and dinosaur eggs. Additional finds included a well-preserved large ankylosaurid, *Syrmosaurus vimicaudus* (a junior synonym of *Pinacosaurus grangeri*).

As the Soviet team moved south toward the border with China, they came upon a broad depression surrounded by an expanse of trackless, shifting desert. The Nemegt Valley extends 180 km east to west, and up to 70 km north to south. Within the confines of this isolated area the crew located numerous Cretaceous "dinosaur cemeteries" scattered across the valley floor. These deposits included hadrosaurs, carnosaurs, sauropods, ankylosaurs, and ornithomimids. Among the inhospitable gorges of red sandstone the expedition found seven skeletons of *Tarbosaurus bataar*, a new large theropod closely related to the North American genus *Tyrannosaurus*.

At a Nemegt locality near Altan Ula, the team discovered a dense burial of large hadrosaurs (>12 m long and 7.7 m tall), some with skin impressions, which Rozhdestvensky (1960) described as *Saurolophus angustirostris*. So productive was the locality that it was christened "Dragon's Tomb." Certainly the most puzzling animal recovered from the Nemegt by the Soviet–Mongolian expeditions was *Therizinosaurus cheloniformis*, originally thought to be a gigantic turtle, represented by unguals varying from 30 to more than 60 cm in length. Therizinosauridae were eventually recognized as similar to the dinosaurs known as segnosaurs, and these are now regarded as coelurosaurian theropods.

The expedition of 1949 retraced the route taken in 1948 and also explored further west to the border with Sinkiang, China. At a locality known as the Valley of the Big Lakes (Altan Teli), a deposit of Pliocene mammals was found and excavated. A 1.5-m-thick layer contained the remains of thousands of animals that may have drowned in a flash flooding event. The 1949 season ended with a return to the Nemegt Valley and continued excavation at sites found in 1948 and others in the southeastern Gobi.

In total the Soviet–Mongolian expeditions collected 460 wooden crates totaling 120 tons of specimens, most new to science. A large percentage of this material was recovered from the Nemegt Valley, helping to establish it as one of the world's most important localities of Late Cretaceous dinosaurs, alongside the Red Deer River of Alberta and the Hell Creek area of northern Montana. Today, material from these expeditions is on display in Ulan Bator, People's Republic of Mongolia, and at the Academy of Sciences' Paleontological Museum in Moscow.

From 1969 to 1989, the Soviet–Mongolian Paleontological Expeditions continued, exploring almost every corner of the country, collecting fantastic dinosaur skeletons on display in Ulaan Baatar and Moscow, and producing a significant series of research papers. With the dissolution of Soviet influence in Mongolia, the work is now being continued by new teams from all over the world.

See also the following related entries:

BARUN GOYOT FORMATION • BAYN DZAK • DJADOKHTA FORMATION • MONGOLIAN DINOSAURS

References

Lavas, J. R. (1993). *Dragons from the Dunes. The Search for Dinosaurs in the Gobi Desert*, pp. 138. Privately published by the author, Aukland, New Zealand.

Rozhdestvensky, A. (1960). *Chasse aux Dinosaures dans le Désert de Gobi* (In the Footsteps of Dinosaurs in the Gobi Desert), pp. 301. Librairie Arthème Fayard, Paris.

Speciation

J. David Archibald
San Diego State University
San Diego, California, USA

A broad definition of speciation is that it is the formation of new species from a preexisting species. Speciation can be further subdivided into three components. First, how many new species are recognizable after speciation? Second, how rapidly does the speciation occur? Third, what is the mode of speciation?

The first subdivision or component of speciation is whether the process does or does not result in the splitting into two or more new species. Speciation that does not result in the splitting into two or more lineages is known by such terms as anagenesis, straight-line, or phyletic evolution. It is envisioned that through this process, by natural selection and chance, a single species becomes demonstrably modified over time. Thus, we have evolution of a single species lineage but no additions to the number of species. If a single species splits so that there are now two species where there originally was one, the process of cladogenesis or "clade birth" has occurred. Among biologists, anagenesis sometimes is not viewed as true speciation but only as the modification of an existing lineage. To these biologists, only cladogenesis, which produces an additional lineage, is true speciation. Whether anagenesis is or is not called a form of true speciation, evolutionary change within a single lineage has been argued to occur by many paleontologists, including some working on dinosaurs. One such study found no evolutionary change in four dinosaur lineages through most of the 5-million-year deposition of the Judith River Formation in western North America during the Late Cretaceous; however, with the transgression (landward expansion) of shallow seas and the deposition of the Two Medicine Formation, anagenetic evolution occurred in all four lineages (Horner *et al.*, 1992).

If the evolutionary relationships of either fossil or recent species have been properly analyzed, we can argue that a speciation event may have been anagenetic or cladogenetic. However, it is only when we add the time dimension of a very good fossil record that our assessments of cladogenesis and anagenesis can be more thoroughly evaluated. Unfortunately, in the case of dinosaurs, the record is seldom good enough to discern patterns of speciation. The above example of anagenesis is an exception.

The second component of speciation—namely, how rapidly it occurs—is also nearly impossible to assess using dinosaurs. There is a spectrum of opinion varying from the view that speciation is a very gradual process to the view that it is extremely rapid, essentially instantaneous within our ability to perceive it in the fossil record. These two views, as to the rate of evolution and speciation, are called gradualism and punctuated equilibrium. In the extreme version of gradualism, species (or actually the individuals in the species) are varying constantly over time so that if we could take fossil samples at small intervals we would see continual change in the species. If the species never splits to form additional species we have continual, gradual anagenetic change. If splitting or cladogenesis does occur, the resultant species continue to gradually diverge from one another over time. In the extreme punctuated equilibrium model, no major changes occur in a given species over time except possibly some random shifts back and forth in some minor aspect of form. The vast majority of change is concentrated at the time a species splits to form two or more species. These new species then settle down and change little over time until the next cladogenetic event. The term punctuated equilibrium comes from this process of very rapid or punctuated change during speciation with little change or stasis during the equilibrium times between speciations.

Charles Darwin's name has been invoked in support of gradualism, with reasonably good justification. Darwin did emphasize a gradual process of the accumulation of changes in species over time. He used the word gradual or a derivative many times in *On the Origin of Species*, first published in 1859. However, the terms of scientific question, and the meanings of some words, have changed over time. In his *Journal of Researches* (1839), Darwin described the earthquake at Coquimbo, Chile, as a "gradual change," looking downslope at the step terraces formed by previous earthquakes ("gradual" comes from Latin *gradus* or "step"). Darwin did, however, realize that stasis of some species over long periods of time is also quite common. The sole figure in *On the Origin of Species* illustrates a hypothetical PHYLOGENY showing the trajectory and fates of 10 species of one genus over 10^{14} generations. Two species are shown radiating and diverging in a gradual manner (which Darwin identified as a gradual process) to yield 14 descendant lineages. The other 9 original species are shown as static through time with only 1 eventually surviving to yield a descendant in the 10^{14} generation. It is striking that most species in this diagram do not change (i.e., there is stasis). It is very clear that Darwin intended this in his diagram. In referring to what we now call stasis, Darwin noted (1859, p. 121) that "the other nine species . . . of our original genus may for a long period continue transmitting unaltered descendants." Thus, although Darwin did emphasize gradual change, he also did recognize in several places in his book that species may for some time remain static, or as he noted, "unaltered."

Testing gradualism and punctuated equilibrium requires a very good to excellent fossil record, which unfortunately excludes dinosaurs from use in the study of these competing views of the tempo of speciation. Studies of gradualism and punctuated equilibrium use fossil organisms that are abundant and for which we have a relatively continuous record: for example, marine and freshwater molluscs, microscopic marine creatures, and small mammals. No consensus has been reached as to which, if either, of these views of speciation is most reflective of evolution as a whole. It may turn out that both are valid depending on the organism under study.

The third component of speciation is the mode. With some exceptions, mode of speciation is not a question that can be easily (if at all) addressed using the fossil record. This is especially true for larger, rarer species such as dinosaurs. Nevertheless, it is a subject that impinges on our understanding of the evolution of all species. A variety of names have been used to describe modes of speciation. Arguably the three most commonly used are allopatry, sympatry, and peripatry. The first modes, allopatry and sympatry, deal mostly with the issue of the geography of speciation. In classic allopatry, a parent species becomes geographically split into two or more genetically isolated populations of relatively similar sizes. The form of separation can range from continental-sized separations such as during continental drift or mountain building, down to the scale of a newly formed stream dividing populations of the same species on either side. In the vast majority of cases, when a species splits into subpopulations one or more of these becomes extinct, gene flow (individuals or their propagules move around) is maintained at lower levels, or the subpopulations later reunite. In extremely rare cases, however, the original subpopulations had slightly different genetic compositions, which over time are further amplified by natural selection in slightly differing environments (see EVOLUTION). In these rare cases, if and when the subpopulations later come in contact, they are no longer capable of interbreeding to produce offspring that in turn can survive and produce fertile offspring. This process of geographically dividing one species to form new ones is allopatry.

Sympatric speciation occurs even when the species does not become divided geographically. This mode of speciation is arguably not as common as allopatry, but there appear to be some documented cases involving plants and insects. In sympatry, it is envisioned that although organisms of the same species may live in close proximity to each other, certain males and females preferentially choose to mate with other. If the differences that resulted in this preferential mating are inheritable, over a number of generations these differing populations could be reproductively driven apart. Sympatry might also arise if one or more individuals simultaneously show mutations that could lead to reproductive isolation within a larger population. This scenario is even more likely if the mutated individual is self-fertilizing or can reproduce by budding to form new individuals. These cases are more common in plants than animals.

Finally, peripatry, which can take place either allopatrically or sympatrically (but the former is more likely), occurs when a very small subset of an existing species is isolated. An obvious case of peripatry occurs when one or a few organisms of a given species reach very distant, isolated areas (such as oceanic islands like Hawaii or the Galapagos) where they are clearly isolated from the parent populations. Although extinction for such small, peripheral isolates is far more common, on extremely rare occasions the small subset that differs from the parent population might survive and thrive, forming a new species.

We cannot be certain which if any of these modes characterized dinosaurs, but by analogy with modern large mammals, allopatry seems the most likely.

See also the following related entries:

EVOLUTION • GENETICS • SPECIES

References

Barrett, P. H., Weinshank, D. J., and Gottleber, T. T. (Eds.) (1981). *A Concordance to Darwin's Origin of Species*, 1st ed., pp. 834. Cornell Univ. Press, Ithaca, NY.

Darwin, C. (1859). *On the Origin of Species by Means of Natural Selection, or the Preservation of Favoured Races in the Struggle for Life*, pp. 502. John Murray, London.

Horner, J. R., Varricchio, D. J., and Goodwin, M. (1992). Marine transgressions and the evolution of Cretaceous dinosaurs. *Nature* **358,** 59–61.

Species

J. DAVID ARCHIBALD
San Diego State University
San Diego, California, USA

The formal naming of any animal, including those of dinosaurs, is governed by a set of published rules called the International Code of Zoological Nomenclature. The smallest unit or taxonomic group recognized by the code is the species group. Although the species group may include several higher level (superspecies) or lower level (subspecies) groupings, the most important in this group is the species.

All taxonomic units in the Linnean system (genera, families, orders, etc.) have names that always begin with a capital letter (except species), are Latin or Greek in origin (or at least given a Latinized ending), and are not used more than once (except species). An example of a higher taxon following these three rules is Dinosauria, which is capitalized, was formed of two Latin words, "terrible" and "lizard," by Richard Owen (1841), and although the species referred to Dinosauria have changed over time, the same name has not been applied to other major groups.

Although similar in some ways to other higher taxonomic names, species names are also quite different. They are formed and written in a manner unlike any other higher taxonomic units such as genera, families, orders, etc. Species names are always binomials, that is, composed of two names; the first is the generic name (e.g., *Triceratops*), which always begins with a capital letter, and the second is the specific or trivial name, which always begins with a lowercase letter (e.g., *horridus*). The species name is often shortened to the initial of the generic name plus the trivial name, if there is no potential for confusion as to what species is intended. Thus, *Tyrannosaurus rex* is often shortened to *T. rex*. Both the species name, such as *Tyrannosaurus rex*, and the generic name, such as *Tyrannosaurus*, are always set apart from the remaining text in a different typeface, which in the above cases is italics. As with other higher taxonomic names, the generic portion of a species name may only be used once, but the trivial name can be used any number of times. For example, the trivial name "gracilis" has been applied to (among others) the prosauropods *Dromicosaurus gracilis* and *Sellosaurus gracilis*, the sauropod *Barosaurus gracilis*, and to the theropods *Coelurus gracilis* and *Conchoraptor gracilis*.

In order for a species name to be recognized it must be published in some print format that is widely distributed, it must be represented by a holotype or type specimen kept in a generally recognized public repository, it must have a proper diagnosis that distinguishes it from other similar species, and it must be properly illustrated.

Even when all the proper steps are taken, problems can arise, some of which are discussed below. The major problem at the species level is that it may be determined that the same species has been given two names or that the type specimen upon which the original species was based is so poorly preserved (or lost) that later comparisons of new or existing material cannot be made properly. Both are very common problems in dinosaur SYSTEMATICS, mostly because of the fragmentary nature of many of these creatures' remains. When two or more species names are given to the same material or a study reveals that a number of different specimens given different names are best regarded as the same species, all later names are treated as junior synonyms of the earliest published name. This is providing that the earliest name fulfills the criteria outlined in the previously.

These problems are not unique to the species level; they also occur at higher levels. Because of the lack of a proper understanding of the role of variation in species and the competitive rush to name new ones, many dinosaur species named in the 19th century have turned out to be junior synonyms. For example, the well-known ceratopsid *Triceratops horridus*, named by O. C. Marsh in 1889, is thought to have as many as 10 junior synonyms, 6 of which were named by Marsh in the late 19th century and 4 more named by various authors in the first half of the 20th century.

In addition to the obvious differences in appearance between species names and those of higher taxa, species are sometimes viewed as fundamentally different from higher taxa: A species is the only level within which evolution is occurring. To others, species are simply the smallest taxonomic units that can be recognized but are not fundamentally different from higher taxa. This duality of opinion has resulted in a number of different species concepts, of which the three major ones are noted here—biological, evolutionary, and phylogenetic species.

Biological species are "groups of actually or potentially interbreeding natural populations which are reproductively isolated from other such groups" (Mayr, 1942, p. 120). This concept emphasizes reproductive and genetic isolation of populations rather than phenotypic (or visible) differences in morphology, behavior, etc. of organisms. This has been the dominant species concept for almost 50 years, although it does have some shortcomings. First, it is readily applicable to only sexually reproducing species. Second, it cannot be readily applied to fossil organisms, especially larger, more poorly represented animals such as dinosaurs. Even among dinosaurs, however, there are the relatively rare cases of mass, almost instantaneous accumulations of individuals that clearly represent a snapshot of a population of one species of dinosaur from one place and one time (*Centrosaurus* and *Pachyrhinosaurus* from Alberta, Canada, *Maiasaura* from Montana, and *Edmontosaurus* from Alaska). Such mass death assemblages are similar to censusing all the adults, juveniles, and young of some large grazing mammal from the Serengeti Plains of modern-day Africa. Third, the ability to interbreed is what is termed an ancestral retention in systematics. In systematic studies of taxa at all levels, including species, the only nonarbitrary and biological cogent taxonomy reflects evolutionary history, but ancestral retentions may not be indicative of evolutionary relationship. For example, the possession of a heart in all vertebrates is a holdover from before the origin of vertebrates that tells us little about relationships between any major groups of vertebrates; also, the presence of hair in primates (our order) is not indicative of close relationships between any primates because it is an ancestral retention found in most mammals. Similarly, two populations that are genetically divergent might still be able to interbreed. Interbreeding is retained from their common ancestry. However, one of the populations may share a more recent common ancestry with a third population with which it no longer can interbreed because of some physical or biological barrier.

An evolutionary species is "a lineage (an ancestral-descendant sequence of populations) evolving separately from others and with its own unitary evolutionary role and tendencies" (Simpson, 1961, p. 153). This concept is an attempt to expand the biological species concept backward in time so that it has evolutionary depth. It answers some of the shortcomings of the biological species concept but introduces another. If we could trace backwards in time from a living species, we would find an unbroken lineage leading to the origin of life. This is simply because life is a continuum through time. The fossil record is such, however, that we would never be able to trace a single species backwards in time. Thus, because of the breaks in the fossil record and possibly because of the quite rapid formation of new species (see SPECIATION), we have interruptions in the lineages. These interruptions permit successive groupings of similarly looking organisms to be placed into what have been called chrono- or paleospecies. As long as the breaks are present or the formation of species is more rapid than our ability to detect it, few problems arise. When the fossil record becomes denser, the rather arbitrary division into chrono- or paleospecies also becomes apparent. Fortunately, for dinosaur paleontology this is not a problem because the record of dinosaurs is never very dense or complete over time. This also means that dinosaurs make poor subjects for elucidating species concepts and speciation.

A phylogenetic species is "the smallest diagnosable cluster of individual organisms within which there is a parental pattern of ancestry and descent" (Cracraft, 1983, p. 169). This concept emphasizes the evolutionary nature of species, but overcomes the theoretical issue inherent in the evolutionary concept of eventually tracing any living species back to the origin of life. Unlike the biological species concept, it is less limited in application by mode of reproduction and thus is more broadly applicable. Implicit in the phylogenetic species concept is that species are not fundamentally different from other higher taxa; rather, taxonomic units (e.g., orders, families, genera, etc.) are simply arbitrary constructs to permit us to group successively larger groups of organisms. A species is simply the smallest such unit. Within the above definition, "diagnosable" refers to any characters that help us recognize the species in question relative to other species, but most important are those characters that are unique to that particular cluster of organisms as determined by a phylogenetic analysis (see SYSTEMATICS). Such unique characters arguably permit us to recognize clusters of organisms that resemble each other because of evolutionary descent. Thus, this concept unites the ideas of genetic and evolutionary continuity with our ability to detect evolutionary changes by examining changes in characters within organisms. This union of theory and practicality has made this a more widely accepted species concept. It is equally applicable to a study of ferns or of dinosaurs.

Two of the criticisms of this concept are that it will force us to name every small, slightly variable population of organisms and that, although applicable to very fast evolving lineages, more slowly evolving lineages are not so easily diagnosed. If true, the first criticism might be valid, but when such small, variable populations are examined closely they more often than not lack diagnosable, uniquely derived characters. There has not been an explosion of new species as a result. The second criticism is mostly driven by arguments as to whether evolution in general and speciation in particular represent a very slow, gradual process or a very rapid one (see SPECIATION). If it is rapid, there would be little problem in diagnosing clusters of individual organisms, but if it were very slow there would be no nonarbitrary way to group organisms. Whether this is really an issue remains a matter of some debate.

No matter what species concept is used, there remain obstacles to naming and comparing species of fossil organisms. These obstacles, which are especially keen when dealing with large, fragmentary, and relatively rare animals such as dinosaurs, are geographic and geological age differences, individual variation, sexual dimorphism and ontogenetic change, and comparing species using noncomparable parts.

Over geological time, evolution occurs and species change. Fortunately, some of the changes are preserved in the fossil record and we can thus detect them. If, however, changes cannot be detected in species from lower to higher strata (or layered rock), then either there has been no discernible evolution or we are simply looking at a portion of the animal that did not change. Stasis or lack of change over time is common in the fossil record, especially among some groups of organism. Although we cannot make any hard and fast rules, dinosaurs seem to change more slowly than their contemporaries among the very small mammals but faster than the turtles and fishes found in the same rocks. This means that a particular dinosaur species may range over a number of intervals of recognizable geological time.

This is also true for geographic ranges of dinosaurs. Although genera or families of dinosaurs on occasion occur on more than one continent, species are seldom so far ranging. Species of dinosaurs are seldom recognizable over areas that encompass more than the area of one or two of the larger U.S. states or Canadian provinces.

Understanding that individual variation is important in evolution is one of Charles Darwin's major contributions to evolutionary thought (see EVOLUTION). Variations among individuals are important because they are the raw material, so to speak, on which evolution operates. Individual variations versus variations between species can be especially difficult to sort when examining fossil assemblages. This has given rise in the study of both living and fossil organisms to morphometric studies that statistically analyze the variations found between individuals in the same species and between different species. From compilations based on both living and fossil species, we have a good idea which parts of the morphology vary within and between species. Dinosaurs are no exception. For example, differences in the presence or absence, form, and size of the crest among hadrosaurids have been important in determining differences in species as well as higher taxa. Analysis of variations in the size and form of the frill has served a similar function among ceratopsids.

Both sexual dimorphism (differences in the sexes) and ontogenetic (life history) changes have confounded attempts at recognizing the proper number of species of animals, including dinosaurs, present in a fossil assemblage. Very commonly in animals there are considerable differences between the sexes beyond the obvious differences in reproductive organs. In living mammals we see this in the possession of horns or antlers more often in males and in the brighter coloration in male birds and lizards. Sexual dimorphism has been suggested in at least seven major lineages of dinosaurs and thus it may be an important factor in assessing the number of species present in any given fossil assemblage. If not cognizant of these possible differences, males and females of some fossil species could be named as different species.

One way to test whether observed VARIATIONS are caused by sex differences is to examine proportions of the kinds of morphology found in the sample. If they are in a ratio of 50:50, there is a good change that the differences are the result of sex differences. If the ratios are not 50:50, however, this unfortunately does not mean these different fossil forms are not males and females. This is because very often sex ratios in species can be decidedly skewed, especially with an overabundance of females.

Differences are also apparent as individuals grow. In mammals there is a definite cycle of TOOTH REPLACEMENT of up to two generations. In dinosaurs, there is continuous tooth replacement. Thus, tooth cycles and wear are not reliable assessments of age in dinosaurs. A characteristic of growth in both mammals and some reptiles is the co-ossification or fusion of bones (notably in the skull) upon reaching adult size. In some reptiles, growth continues at a very reduced rate as the animal grows older so that many of the bones in the skull remain unfused. Also, in both mammals and reptiles various parts of the animal are proportioned differently in adults and juveniles. Good examples are the eyes or the orbits in the skull that housed the eyes. These are disproportionately larger in younger individuals.

Two dinosaurian examples of possible sexual dimorphism or ontogenetic changes are found in the South African ornithischian family, HETERODONTOSAURIDAE, and among HADROSAURIDS. Some individuals in species referred to Heterodontosauridae possess canine-like teeth at the front of the jaw, whereas others do not. In small deer such as the Musk Deer (*Moschus* sp.), the males lack antlers but have large canines, which are much smaller in females. The differences within heterodontosaurids could be because different species are represented; the differences could be the result of sex differences as in Musk Deer; or those specimens lacking canines could be younger animals. All three possibilities have been suggested. Among some species of hadrosaurs, some individuals are thought to be males because they possess larger, more elaborated crests on their heads. These could have been used in display or possibly even ritualized combat, although severe blows would have damaged the structures because they are relatively lightly constructed with internal spaces.

Finally, it is not unusual for species (dinosaurs and others) to be named using different parts of the animal. This has lead to the same species having a number of names based on different parts of the animal's skeleton or general confusion as to whether different parts represent the same or different species. An example of the latter case involves the Late Cretaceous species *Troodon formosus*. In 1856, Joseph Leidy named this species based on a single, small recurved tooth with serrations. He thought it belonged to some kind of meat-eating reptile. In 1902, Lawrence Lambe reported finding similar teeth and also some quite thick skull fragments that he named *Stegoceras validus*. In 1924, Charles Gilmore described a thickened skull and partial skeleton of a dinosaur that resembled Lambe's 1902 find, but because the teeth near the front of the upper jaw in this thick-headed skull resembled the isolated teeth referred to *T. formosus*, he thought it best to refer to his material as *Troodon validus* rather than *Stegoceras validus*. As more specimens were recovered and described, it became clear that *T. validus* is best regarded as *S. validus*. *Stegoceras* belongs to the group of dinosaurs referred to as ORNITHISCHIANS, specifically the bone-headed dinosaurs or PACHYCEPHALOSAURS. The small, recurved teeth with serrations belong to the small, meat-eating saurischian group known as TROODONTIDS, including *Troodon*. Thus, the skull parts and teeth do not even belong to the same group of dinosaur, let alone the same species. This shows the danger of naming species on fragmentary or noncomparable parts.

See also the following related entries:

EVOLUTION • GENETICS • PHYLOGENETIC SYSTEM • SPECIATION

References

Cracraft, J. (1983). Species concepts and speciation analysis. *Curr. Ornithol.* **1,** 159–187.

Mayr, E. (1942). *Systematics and the Origin of Species,* pp. 334. Columbia Univ. Press, New York.

Norman, D. (1985). *The Illustrated Encyclopedia of Dinosaurs,* pp. 208. Crescent, New York.

Rose, K. D., and Bown, T. (1993). Species concepts and species recognition in Eocene primates. In *Species, Species Concepts, and Primate Evolution* (W. H. Kimbel and L. B. Martin, Eds.), pp. 299–330. Plenum, New York.

Simpson, G. G. (1961). *Principles of Animal Taxonomy,* pp. 247. Columbia Univ. Press, New York.

Weishampel, D. B., Dodson, P., and Osmólska, H. (1990) *The Dinosauria,* pp. 733. Univ. of California Press, Berkeley.

Spines

see ORNAMENTATION

Spinosauridae and Baronychidae

ANGELA MILNER
Natural History Museum
London, United Kingdom

The Spinosauroidea is a group of highly derived, long-snouted Cretaceous tetanurans from the Barremian to Cenomanian in Europe, North Africa, and perhaps Brazil. The group is characterized by the following apomorphies: elongation of the jaws, especially in the prenarial region; a moderately to well-developed terminal "spoon-shaped" expansion or "rosette" in both upper and lower jaws; upturned symphysial end of the dentary with a constricted region immediately posterior to it; premaxillary tooth count increased to seven; and teeth with reduced labiolingual compression, slightly recurved or straight crowns, and finely serrated (*Baryonyx*) or unserrated carinae (*Spinosaurus*) (Charig and Milner, 1997). As defined by Charig and Milner (1997), the Spinosauroidea includes two families defined on characters of jaws and teeth only: Baryonychidae Charig and Milner 1986 for the Barremian *Baryonyx* Charig and Milner 1986, from Surrey, England, and *Baryonyx* sp. from the Aptian of Niger, originally described as spinosaurid indet. by Taquet (1984); and Spinosauridae Stromer 1915 for *Spinosaurus* Stromer 1915, and provisionally *Irritator* Martill *et al.* 1996 and *Angaturama* Kellner and Campos 1996, both from the Aptian of Brazil. A cladistic analysis of the spinosauroids by Charig and Milner (1997) placed them within the Tetanurae as the sister group to the Neotetanurae, and with *Megalosaurus* and *Torvosaurus,* respectively, as progressively more distant out-groups to the combined Spinosauroidea + Neotetanurae clade. The relationship between *Baryonyx, Megalosaurus,* and *Torvosaurus* is weakly supported because of the fragmentary nature of the material and missing data.

Spinosaurus aegyptiacus is known only from fragmentary remains, including a dentary and distinctive long-spined dorsal vertebrae, of an individual at least 12 m long from the Cenomanian of Egypt (Stromer, 1915). Unfortunately, that material was destroyed in Munich during World War II. Similar skull material from the basal Cenomanian of Morocco is now under study (Milner, manuscript in preparation), and a dentary fragment and cervical vertebra have been described as *Spinosaurus maroccanus* by Russell (1996).

Baryonyx is the best known spinosauroid, based on a material discovered in the WEALDEN of Surrey in 1983. It comprises a single skeleton, about 60% complete. Its distinguishing characters include the prenarial extension of the snout into a spatulate rostrum; an increase in the number of the dentary teeth, which are twice as numerous per unit length of jaw as the opposing maxillary teeth; an unusually robust forelimb with a broad humerus but with epipodials only half the humerus length; and at least one pair of huge manual talons, probably from the pollex. The neck was straight without the usual theropod curvature, and the dorsal vertebrae were short spined, in contrast to *Spinosaurus.* Lack of fusion between elements of the skull and between the neural arches and centra suggests that the animal was immature despite its estimated length of approximately 10 m.

Baryonyx has an unusually long and low skull with the naris set well back from the end of the snout and an S-shaped maxillary tooth row, conferred by the expanded terminal rosette. Those features together with the high dentary tooth count, and the possession of a pair of unusually large manual unguals, have been interpreted as adaptations to piscivory and perhaps specialist scavenging (Charig and Milner, 1990, 1997). The discovery of acid-etched scales and teeth of a contemporary Wealden fish, *Lepidotes*, inside the rib cage of the holotype corroborates that hypothesis. All the spinosauroid remains come from floodplain, lake, or near-shore marine sediments together with abundant fish fossils. *Baryonyx* may represent a clade of specialized large piscivores that inhabited river, lake, and coastal margins.

See also the following related entries:

TETANURAE • THEROPODA

References

Charig, A. J., and Milner, A. C. (1986). *Baryonyx*, a remarkable new theropod dinosaur. *Nature (London)* **324,** 359–361.

Charig, A. J., and Milner, A. C. (1990). The systematic position of *Baryonyx walkeri*, in the light of Gauthier's reclassification of the Theropoda. In *Dinosaur Systematics: Approaches and Perspectives* (K. Carpenter and P. J. Currie, Eds), pp. 127–140. Cambridge Univ. Press, Cambridge, UK.

Charig, A. J., and Milner, A. C. (1997). *Baryonyx walkeri*, the fish-eating dinosaur from the Wealden of Surrey. *Bull. Nat. History Museum Geol.* **53**(1).

Kellner, A. W. A., and Campos, D. de A. (1996). First Early Cretaceous theropod dinosaur from Brazil with comments on Spinosauridae. *Neues Jahrbuch Geol. Paläontol. Abhandlungen Stuttgart* **199**(2), 151–166.

Martill, D. M., Cruickshank, A. R. I., Frey, E., Small, P. G., and Clarke, M. (1996). A new crested maniraptoran dinosaur from the Santana Formation (Lower Cretaceous) of Brazil. *J. Geol. Soc. London* **153**(1), 5–8.

Russell, D. A. (1996). Isolated dinosaur bones from the Middle Cretaceous of the Tafilalt, Morocco. *Museum Natl. d'Histoire Nat. (Paris) Bull. Ser. 4* **18**(Section. C, Nos. 2–3), 349–402.

Stromer, E. (1915). Ergebnisse der Forschungensreisen Prof. E. Stromers in den Wüsten Ägyptens. II. Wirbeltier-Reste der Baharîye-Stufe (unterstes Cenoman). 3. Das Original des Theropodes *Spinosaurus aegyptiacus* nov. Gen., nov. Spec. *Abhandlungen Königlich Bayerischen Akad. Wissenschaften Mathematisch–Physikalische Klasse Munich* **28**(Band 3), 1–28.

Taquet, P. (1984). Une curieuse spécialisation du crâne de certains Dinosaures carnivores du Crétacé: Le museau long et étroit des Spinosauridés. *Comptes Rendus Hebdomadaires Séances Acad. Sci. Paris* **299**(2–5), 217–222.

Springfield Science Museum, Massachusetts, USA

see MUSEUMS AND DISPLAYS

State Museum, Brilon, Germany

see MUSEUMS AND DISPLAYS

State Museum for Natural History, Stuttgart

The state collections of Baden-Württemberg, Germany, are housed in a magnificent new museum, opened in the mid-1980s, in a large park on the north side of the city of Stuttgart called Rosenstein. This park is a part of what was formerly a hunting preserve for the king of Baden-Württemberg, and his residence there is now the original museum building. When collections and research facilities became too large for this structure in the 1960s, the paleontological unit was moved to quarters at nearby Ludwigsburg; a new research facility including collection storage, plus the magnificent new exhibit hall, were constructed in the 1980s. The paleontological collections, among the best in Europe, include a great variety of vertebrates from the Early Jurassic Holzmaden limestones of the region, highlighted by numerous ichthyosaurs, plesiosaurs, pterosaurs, and spectacular invertebrates. Late Triassic dinosaurs and other vertebrates are well represented and number among them several mounted skeletons and much additional material of the sauropodomorphs *Plateosaurus, Sellosaurus,* and *Ohmdenosaurus,* the early theropod *Procompsognathus,* and the early turtle *Proganochelys,* as well as placodonts, fishes, metoposaurs, and other giant amphibians.

See also the following related entry:

MUSEUMS AND DISPLAYS

State Museum of Pennsylvania, Pennsylvania, USA

see MUSEUMS AND DISPLAYS

Staurikosauridae

This taxon was erected by Galton (1977) to emphasize the distinction between its eponymous and only genus, *Staurikosaurus*, and *Herrerasaurus*, with which it had been placed in the Herrerasauridae (Benedetto, 1973). Most other authors have followed Benedetto's lead (see Novas, 1992; HERRERASAURIDAE), but other analyses have found Herrerasauridae to be paraphyletic, with *Herrerasaurus* closer to DINOSAURIA. In any event, Staurikosauridae communicates no further phylogenetic information than that it contains a single genus; it is generally regarded as subsumed within Herrerasauridae.

See also the following related entry:

HERRERASAURIDAE

References

Benedetto, J. L. (1973). Herrerasauridae, nueva familia de saurisquios triásicos. *Ameghiniana* **10,** 89–102.

Galton, P. M. (1977). *Staurikosaurus pricei*, an early saurischian dinosaur from the Triassic of Brazil, with notes on the Herrerasauridae and Poposauridae. *Paläontol. Z.* **51,** 234–245.

Novas, F. E. (1992). Phylogenetic relationships of the basal dinosaurs, the Herrerasauridae. *Palaeontology* **35,** 51–62.

Stegosauria

PETER M. GALTON
University of Bridgeport
Bridgeport, Connecticut, USA

Stegosauria may be defined as all thyreophoran ornithischians closer to *Stegosaurus* than to *Ankylosaurus*. They are diagnosed by several features (Sereno, 1986), including the presence of a shoulder spike and the reduction of dermal armor to two parasagittal rows of anterior plates and posterior spikes. Originally all the armored herbivorous dinosaurs were referred to the Stegosauria, but now the group is restricted to medium to large (up to 9 m in total body length) quadrupedal forms with an extensive system of erect plates and spines along the middle of the back and tail (for details see Dong, 1990; Galton, 1990, Sereno and Dong, 1992). The earliest possible record consists of fragmentary remains from the Middle Jurassic (lower Bathonian) of England (Galton, 1985), but almost complete skeletons with skulls are known from the Bathonian–Callovian of the People's Republic of China. Stegosaurs are best represented in the Upper Jurassic by excellent articulated skeletal material from Africa, Asia, Europe, and North America. The latest record is a partial skeleton from the Upper Cretaceous (Coniacian) of India. This was described by Yadagiri and Ayyasami (1979), who also reported still undescribed stegosaur material from the latest Cretaceous of India. However, these are doubted by many workers (see INDIAN DINOSAURS).

The dorsal vertebrae are tall due to the height of the pedicels of the neural arch, the part immediately above the neural canal that bears the rib-supporting transverse processes. These are directed upwards by as much as 50–60° from the horizontal in middorsal vertebrae; this angle decreases to about 25–40° in anterior and posterior dorsals. The sacrum consists of five or six fused vertebrae. In several genera there is sexual dimorphism, with one form bearing an additional slender first sacral rib. The neural canal is considerably enlarged to accommodate the nerve plexus for the very large hindlimb and perhaps a glycogen storage organ, but contrary to a common misconception, this did not constitute a "sacral brain." A comparable dilation of the neural canal of the anterior dorsals provided room for the plexus to the massive forelimb. Articulated remains of stegosaurs show no trace of ossified tendons; therefore, given their ubiquitous occurrence in all other groups of ornithischians, this absence probably represents a secondary loss in stegosaurs. The posterior caudal vertebrae are stout in order to support the tail spines.

The inferior part of the scapula forms a broad plate and the humerus is short but massive with expanded ends. The manus is quite elephant-like, with one or two large, block-like proximal carpals, but distal carpals have not been preserved. As in most other ornithischians, the pelvic girdle is tetraradiate, but the prepubic process is proportionally long. The femur is slender in side view, with a straight shaft of nearly uniform width. The tibia is short and massive, with the distal end often fused to the slender fibula, the massive astragalus, and the small calcaneum. Metatarsal I and its digit are absent but Metatarsal II, III, and IV are short and massive.

Apart from a pair of shoulder spines, the osteoderms of stegosaurs are situated above the vertebral column, rather than being over the back as a whole, and are angled upward and slightly outward. Viewed from the side, the osteoderms form a series that grade from short, erect plates anteriorly to longer, posterodorsally angled spines posteriorly. The series ends with a proportionally narrow pair of spines that continue beyond the last tail vertebra. In most stegosaurs all the osteoderms are paired, so there is a left and right representative of each type of plate and spine. In *Stegosaurus* the armor consists of a series of 17 erect and thin plates of varying sizes, plus two pairs of tail spines (Fig. 1). No 2 plates have exactly the same shape or size, so they have been reconstructed as a single median row or as two rows of staggered alternates (see CAÑON CITY).

Huayangosaurus is the sister taxon to all other stegosaurs, which have lost the premaxillary teeth (seven in *Huayangosaurus*); fused the dorsal margins of the sacral ribs to form a nearly solid plate between the ilia, further increased the length of the prepubic process; lost the row of scutes along either side of the trunk; and elongated the femur, which is at least 1.5 times longer than the humerus (cf., 1 : 1 in *Huayangosaurus*) (Sereno and Dong 1992). However, the systematic relationships of the other stegosaurs are unclear. Stegosaurs are relatively rare as fossils; probably fewer than a dozen forms are valid (Galton, 1990).

Stegosaurs were probably important low-level browsers (maximum height approximately 1 m), with a locomotory habit comparable to that of elephants (i.e., graviportal). As in most other ornithischians, the sides of the mouth region were bordered by structures functionally equivalent to the cheeks of mammals; therefore, while the soft foliage and fructifications were being chewed, they did not fall out the sides of the mouth.

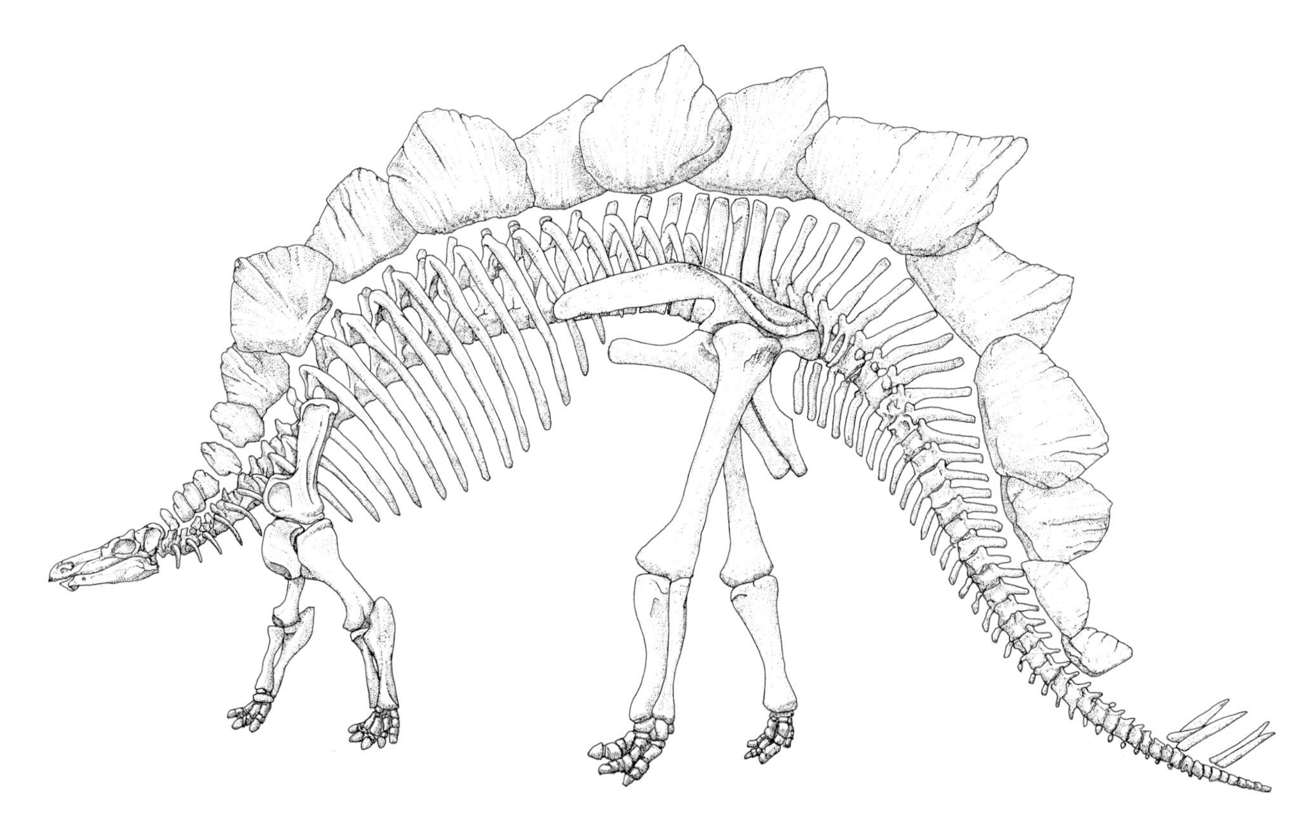

FIGURE 1 *Stegosaurus* are recovered from Morrison Formation rocks of Colorado and Wyoming. Related dinosaurs are found in rocks of Europe, southern Africa, India, and China.

The pattern of plates and spines is characteristic for each species and it was probably important for intraspecific recognition. The HISTOLOGY of the plates of *Stegosaurus* provides clues to their functions (Buffrénil *et al.*, 1986). The plates could not have functioned as armor because they do not consist of thick compact bone and, because the bases of the plates were firmly embedded in the skin, they could not be suddenly unfolded to act as a deterrent against attacking animals. However, the plates were probably used for sexual display and perhaps for temperature regulation, though it is not clear why related forms lacked them. In an alternating arrangement, the plates would have functioned well as a forced convection fin to dissipate heat or as heat absorbers (Farlow *et al.*, 1976). The plates formed a scaffolding for the support of a richly vascularized skin that would have acted as an efficient heat exchange structure. A linear arrangement of the plates would indicate that radiation and convection were equally important as modes of heat transfer. A heat-absorbing role for the plates makes sense if *Stegosaurus* was an ectotherm, whereas heat loss by radiation or forced convection would be useful if *Stegosaurus* was ectothermic or to any degree endothermic (Buffrénil *et al.*, 1986).

See also the following related entries:

ANKYLOSAURIA • THYREOPHORA

References

Buffrénil, V. de, Farlow, J. O., and Ricqlès, A. de (1986). Growth and function of *Stegosaurus* plates. Evidence from bone histology. *Paleobiology* **12,** 459–473.

Dong, Z. (1990). Stegosaurs of Asia. In *Dinosaur Systematics. Approaches and Perspectives* (K. Carpenter and P. J. Currie, Eds.), pp. 255–268. Cambridge Univ. Press, Cambridge, UK.

Farlow, J. O., Thompson, C. V., and Rosner, D. E. (1976). Plates of *Stegosaurus*: Forced convection heat loss fins? *Science* **192,** 1123–1125.

Galton, P. M. (1985). British plated dinosaurs (Ornithischia, Stegosauria). *J. Vertebr. Paleontol.* **5,** 211–254.

Galton, P. M. (1990). Stegosauria. In *The Dinosauria* (D. B. Weishampel, P. Dodson, and H. Osmólska, Eds.), pp. 435–455. Univ. of California Press, Berkeley.

Sereno, P. C. (1986). Phylogeny of the bird-hipped dinosaurs (order Ornithischian). *Natl. Geogr. Res.* **2,** 234–296.

Sereno, P. C., and Dong, Z. (1992). The skull of the basal stegosaur *Huayangosaurus taibaii* and a cladistic analysis of Stegosauria. *J. Vertebr. Paleontol.* **12,** 318–343.

Yadagiri, P., and Ayyasami, K. (1979). A new stegosaurian dinosaur from Upper Cretaceous sediments of south India. *J. Geol. Soc. India* **20,** 251–530.

Sternberg Museum, Kansas, USA

see MUSEUMS AND DISPLAYS

Sudanese Dinosaurs

see AFRICAN DINOSAURS

Systematic Philosophies

see PHYLOGENETIC SYSTEM; SYSTEMATICS

Systematics

Xiao-Chun Wu
Anthony P. Russell
University of Calgary
Calgary, Alberta, Canada

The prime goal of biological systematics is to establish a general reference system for comparative biology (Brooks, 1981). This is achieved by the recognition and classification (naming sets within a hierarchical arrangement of taxa) of organisms. In general, there are three contemporary approaches to systematics: phylogenetic systematics (usually called cladistics), evolutionary (traditional) systematics, and phenetics (generally called numerical taxonomy). For dinosaur systematists, evolutionary systematics and phenetics are currently practiced, but phylogenetic systematics has recently become increasingly popular due to its philosophical rigor and is used most commonly. It will thus be introduced here.

Phylogenetic systematics was formally founded by the German entomologist Willi Hennig (1950, 1966), although in many ways its basic tenets had long been a feature of systematic practice in general. Recently, this approach has been considerably improved, both theoretically and methodologically, by a number of its proponents (e.g., Eldredge and Cracraft, 1980; Wiley, 1981, 1987; Maddison *et al.*, 1984; Wiley *et al.*, 1991).

Philosophical Foundations

Philosophically, phylogenetic systematics adopts the hypotheticodeductive method of Popper (1959) for scientific reasoning. The principles of this method are formulated to give adequate logical justification for proposing a scientific hypothesis, which must be structured in such a way that it contains testable implications; that is, it must predict something that may be actually or potentially observed (Wiley, 1981). In this approach, manifestly false hypotheses are eliminated or rejected, and as yet unfalsified alternative hypotheses are said to be more highly corroborated. The application of the hypotheticodeductive method to phylogenetic systematics is mainly based on two biological axioms (principles): (i) Evolution occurs such that species have certain phylogenetic (genealogical) relationships to each other and (ii) phylogenetic relationships may be reconstructed using characters of organisms (Wiley, 1981). In other words, life has evolved and the diversity of organisms has been produced by speciation with character modification—new intrinsic features arising from time to time and being inherited by descendants. Consequently, species are phylogenetically (genealogically) related to each other and their relationships may be recovered by analysis of their characters. This is true not only for species but also for supraspecific taxa, if each is hypothesized to have originated as a single species. For example, a hypothesis suggests that hadrosaurs are more closely related to tyrannosaurs than they are to ankylosaurs. This hypothesis is logically structured under the two biological principles. Characters (character modifications) should be able to help test the hypothesis. One would predict that hadrosaurs and tyrannosaurs would share one or more unique characters not shared by ankylosaurs, thus indicating that they are more closely related to each other than either is to ankylosaurs. Conversely, one might falsify this hypothesis with other unique characters that are shared by hadrosaurs and ankylosaurs to the exclusion of tyrannosaurs. If several hypotheses compete against each other in an attempt to explain the same data, parsimony (the principle of economy of argument) is used to select the hypothesis that can explain the data in the most economical way. For the purposes of phylogenetic systematics, the most parsimonious or simplest hypothesis is the preferred one because it uses the fewest *ad hoc* statements (assumptions) to explain the full array of available data. For any three taxa (1, 2, and 3) with three characters (*Xx*, *Yy*, and *Zz*) there are three possible depictions of relationships based on the assumption of a dichotomous branching pattern (Figs. 1a–1c). These three taxa may have the following distribution of ancestral (primitive) character states (capital letters) and derived (novel) character states (lowercase letters): Taxon 1 bears characters

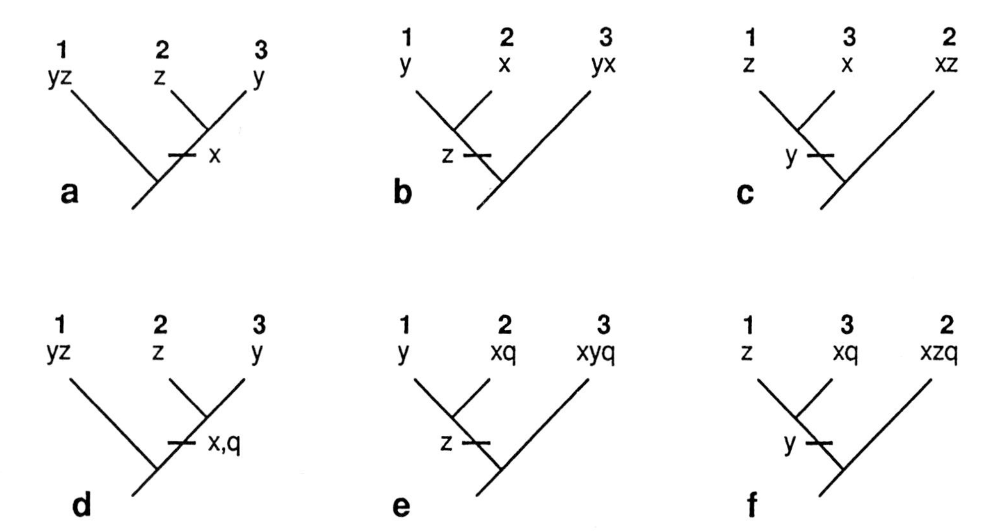

FIGURE 1 Decisions concerning the hypotheses of relationships among any three taxa based on character or character state distribution, and the application of the principle of parsimony. For any three taxa there are three possible depictions of relationships based on the assumption of a dichotomous branching pattern (see text for details).

Xyz; taxon 2 bears *xYz;* and taxon 3 bears *xyZ*. Thus, taxa 2 and 3 share derived state *x*, taxa 1 and 2 share derived state *z*, and taxa 1 and 3 share derived state *y*. The character state transformations required to explain the distribution of character states in the three taxa in question are the same (five) for each of the three cladograms. Thus, the patterns in the three cladograms are equally economical (parsimonious) hypotheses of relationships among the three taxa and none of them can provide a better explanation of the data than the others. If we examine a fourth character with states *Q* (ancestral) and *q* (derived), this may help in the resolution of the problem. As indicated in Figs. 1d–1f, there are now the following combinations of characters for the three taxa: *XyzQ* for taxon 1, *xYzq* for taxon 2, and *xyZq* for taxon 3. In this situation the derived state *q* is shared by taxa 2 and 3 (Fig. 1d). For character state *q*, it requires only one evolutionary event to explain it in Fig. 1d, whereas two independent derivations of *q* are called for in the other two (Figs. 1e and 1f). Consequently, the total number of character state transformations are fewest in Fig. 1d, which represents the most parsimonious hypothesis of relationships among the three taxa and is accepted as the preferred summary of the data.

The hypotheticodeductive method is consistently employed in phylogenetic systematics in order to reconstruct the history of the life.

Basic Ideas of Phylogenetic Systematics

Phylogenetic systematics attempts to establish phylogenetic (genealogical) relationships among groups of organisms and produces classifications that are entirely consistent with those relationships. Phylogenetic systematics posits that because some character transformations (modifications) occur earlier than others in evolution, the reconstruction of phylogenetic relationships among organisms must rely on a nested set of evolutionary resemblances; in other words, similarities among organisms are hierarchically ordered as the expected outcome of the evolutionary process itself (Eldredge and Cracraft, 1980). Phylogenetic systematists contend that ancestors never obtain a set of evolutionary novelties (character modifications) unique to themselves and thus cannot be defined and recognized. For this reason, phylogenetic systematics concentrates on the search for monophyletic groups. Phylogenetic relationships among organisms may be expressed by a sequence of nested sets or a hierarchical pattern of hypothesized monophyletic groups that are linked by sister groups (or sister group relationships rather than ancestor–descendant relationships, as has been considered in evolutionary systematics) and are generally depicted in so-called cladograms(branching diagrams) (Fig. 2).

A monophyletic taxon or group in phylogenetic

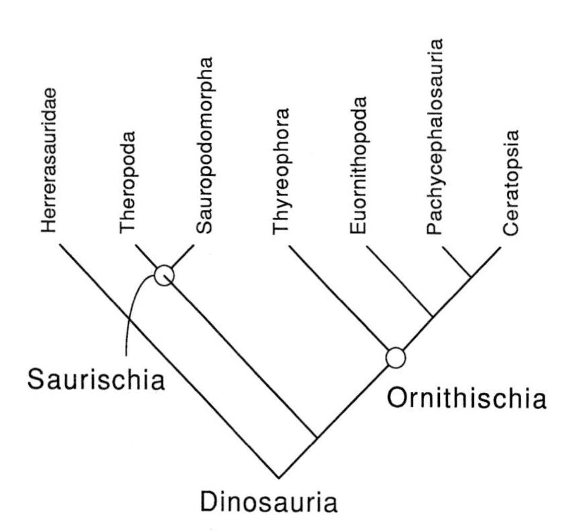

FIGURE 2 A cladogram of the major groups of the Dinosauria (modified from Benton, 1990, Fig. 1.5). Each level (node) of the hierarchy (branch point) is defined by one or more evolutionary novelties.

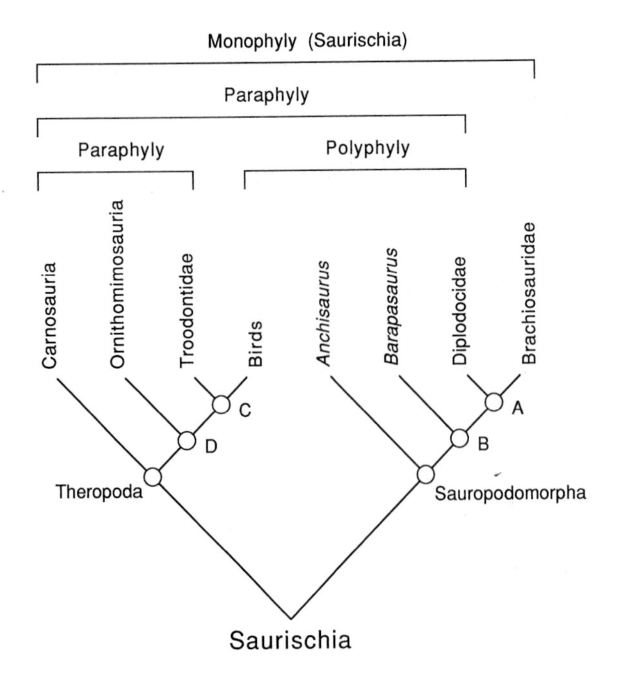

FIGURE 3 Concepts of monophyly, paraphyly, and polyphyly in phylogenetic systematics. The cladogram for the major groups of the Saurischia is modified from Fig. 1.5 of Benton (1990). Within the Saurischia, the groups Theropoda, Sauropodomorpha, and A–D are monophyletic. Of these groups, the Theropoda and Sauropodomorpha share the same hierarchical level, groups B and D share the same level lower in the hierarchy, and groups A and C share a same level at the lowest point of the hierarchy.

systematics means a natural group of species, which comprises an ancestral species and all its descendants. In a cladogram this is referred to as a clade (Fig. 3). Members of a monophyletic taxon or group share a set of common ancestral relationships not shared with any other species placed outside the taxon or group. In other words, a monophyletic taxon is an evolutionary entity representing a part of evolutionary history. The ORNITHISCHIA (Fig. 2) is, for instance, a monophyletic group of the DINOSAURIA. Its members are phylogenetically linked to each other by sharing a set of evolutionary similarities, such as the presence of the predentary bone and the opisthopubic pelvis.

The sister group of any monophyletic group is that taxon that is phylogenetically the most closely related to the monophyletic group in question. For example, the SAURISCHIA is the sister group of the Ornithischia (Fig. 4). The ancestor of a monophyletic group cannot be its sister group because the ancestor is a member of the monophyletic group.

Phylogenetic systematics emphasizes evolutionary similarities (derived or descendant characters or character states) in the establishment of phylogenetic relationships among the groups of organisms (Fig. 1), in contrast to the use of overall (general) similarities, as has been the common practice for evolutionary (traditional) systematics. The establishment of a monophyletic group is accomplished by the recognition

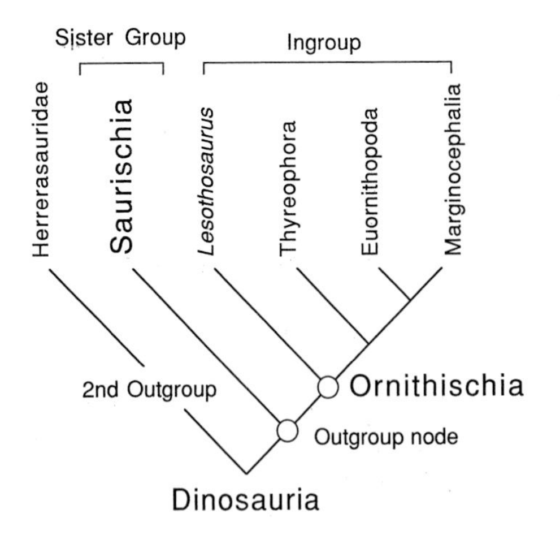

FIGURE 4 Concepts of the in-group, sister group, and out-group. A sister group is the immediate relative (the first out-group) of the in-group. The cladogram for the major groups of the Ornithischia is derived from Fig. 1.5 of Benton (1990).

of the descendant (derived) characters (or states) that are shared by the members of the group. The members of a group (or taxon) may share a number of similar features, but only those that are derived relative to the sister group of that group demonstrate close relationships. For example, all dinosaurs share many features, but CERATOPSIANS, PACHYCEPHALOSAURIANS, and the other ornithischians are united in having a set of derived characters, such as the presence of a predentary bone and an opisthopubic pelvis. These features suggest that these dinosaurs are more closely related to each other than any of them is to other dinosaurs because of the simplest assumption that these features arose once and not on multiple occasions. The common possession of these special characteristics thus allows us to hypothesize certain patterns of relationship (Fig. 2).

In phylogenetic systematics, a derived (descendant) character (or state) is called an apomorphy or an apomorphic character (or state); a primitive (ancestral) character (or state) is called a plesiomorphy or a plesiomorphic character (or state); a primitive character (or state) shared by two or more taxa is called a symplesiomorphy or a symplesiomorphic character (or state); a character (or state) unique to a taxon is called an autapomorphy or an autapomorphic character (or state); and a derived character (or state) shared by two or more taxa is called a synapomorphy or a synapomorphic character (or state). All these terms must be determined in relation to a particular level of a particular hierarchical pattern of hypothesized monophyletic groups. The presence of the predentary bone and opisthopubic pelvis are two synapomorphies of the Ornithischia within the Dinosauria, but these two characters become two symplesiomorphies of any subordinate group, such as the Pachycephalosauria (see Fig. 2), within the Ornithischia (less inclusive taxon). On the other hand, these two characters are uniquely derived for the Ornithischia; hence, they are two autapomorphies of the taxon. An autapomorphy could be a synapomorphy of a monophyletic group at a given taxonomic level, but a synapomorphy is not necessarily an autapomorphy. The absence of the gastralia (abdominal ribs) is one of the synapomorphies of the Ornithischia, but it is not unique to the taxon because the lack of the gastralia occurs independently in most sauropodomorph dinosaurs of the Saurischia—the sister group of the Ornithischia. An autapomorphy is of no value in establishing phylogenetic relationships of a group with other groups but is instead a defining characteristic. The presence of the predentary bone and the opisthopubic pelvis do not help determine the relationships of the Ornithischia to other dinosaurs that lack these characters or to determine relationships within the Ornithischia, but they play a major role in the diagnosis of this taxon.

Phylogenetic systematics considers taxa or groups that cannot be diagnosed by one or more synapomorphies to be either paraphyletic or polyphyletic. Such taxa or groups do not represent natural entities produced by evolution. In phylogenetic systematics, a paraphyletic taxon or group is an artificial taxon that includes an ancestral species, but excludes some of its descendants, and could be defined only by both the presence or absence of synapomorphies. In other words, a paraphyletic group does not have derived characters unique to itself; its members, together with the excluded descendants, share the characters derived relative to the sister group, but its members lack the derived character unique to those excluded descendants. For example, the THEROPODA in the traditional classification of dinosaurs (Romer, 1956) is a paraphylytic group because it excludes the avian (bird) descendants evolved from the common ancestor (Fig. 3). The members of the Theropoda share many synapomorphies with birds but lack the synapomorphies of birds and have no derived characters exclusive to themselves. A polyphyletic taxon or group is an artificial taxon that does not include the common ancestor and all its descendants. For example, the group including birds, *Anchisaurus*, *Barapasaurus*, and the Diplodocidae in Fig. 3 is polyphyletic.

In phylogenetic systematics synapomorphies subsume the concept of symplesiomorphies (see above). A synapomorphy is a shared derived character (similarity) inherited from a preexisting character of the immediate common ancestor; a symplesiomorphy is a shared primitive character (similarity) inherited from a preexisting character of the ancestor more distant from the immediate common ancestor. Viewed in this way, phylogenetic systematics simply considers synapomorphies (and symplesiomorphies) as homologies and recognizes that homology cannot be empirically determined before the establishment of the phylogenetic relationships of the groups under study. Every statement of homology is a hypothesis that has to be subjected to testing. Whether or not a similar character in two or more taxa is homologous depends on its congruence with other hypothesized homologous characters for diagnosing the phylogenetic relationships among the taxa in question. For example, one could hypothesize that the mandibular condyle sitting below the tooth row is a homologous character of the Dinosauria according to its morphology, function, and position. However, this hypothesis does not match the distribution of many other derived characters, on the basis of which an alternative hypothesis is proposed that suggests that the mandibular condyle sitting below the tooth row has been independently evolved in the Ornithischia and in some sauropodomorph dinosaurs of the Saurischia (Benton, 1990). This kind of character is termed a homoplasy; that is, a homoplasy is a character of similar appearance evolved independently by convergence (or parallel) in two or more taxa that are not closely related to each another.

In phylogenetic systematics the reconstruction of phylogenetic relationships among groups of organisms relies on the establishment of monophyletic taxa, and the recognition of a monophyletic taxon depends on the search for synapomorphies (derived homologies). Consequently, the determination of whether or not a character (or state) is apomorphic or plesiomorphic plays a key role in phylogenetic systematics and is termed character polarization. Phylogenetic systematics considers characters that consist only of two states (plesiomorphic and apomorphic) to be binary characters (binary transformation series). If characters comprise a plesiomorphic state and two or more apomorphic states, these characters are termed multistate characters (multistate transformation series).

The most commonly used approach for determining character state polarity is the method of out-group comparison. An outgroup is a group that is not included in the group under study. The latter is called the in-group (Fig. 4). The most important outgroup is the sister group of the group in question. A sister group or an outgroup can be a species or any monophyletic supraspecific taxon. A recent hypothesis (Sereno and Arcucci, 1993) suggests that the sister group of the Dinosauria is the species *Marasuchus lilloensis* from the Middle Triassic of Argentina.

The method of outgroup comparison may be summarized as follows: If a character state found in a monophyletic group is also found in its sister group, then this character state is taken to be plesiomorphic. We have, for example, two binary characters: (i) the presence of the predentary bone versus the absence of that bone and (ii) the well-developed ascending process of the astragalus versus the low ascending process of that bone in the Ornithischia. The predentary bone (character i) does not occur in the sister group Saurischia, but the well-developed ascending process of the astragalus (character ii) does. Thus, the presence of the predentary bone is taken to be an apomorphic character state and the well-developed ascending process of the astragalus is considered to be a plesiomorphic character state for the Ornithischia. The tentative character state polarity of the two characters can only be confirmed, however, if the hypothesized state of each character for the sister group is also present in more remote outgroups (the second outgroup) of the Ornithischia. That is, if the state of a character (such as the absence of the predentary bone) that occurs in the Saurischia is also characteristic of the HERRERASAURIDAE (Fig. 4), then the alternative state (such as the presence of the predentary bone) of the character present in the Ornithischia would be unequivocally determined to be apomorphic.

In the practice of phylogenetic systematics, actual polarization of a character can be somewhat more complicated than the previous simple example, especially if one does not know the exact sister group but only a set of possible sister groups, if the sister group is a monophyletic group possessing both plesiomorphic and apomorphic states of characters, if the in-group is not monophyletic, or if a given character evolved in the sister group independently. For methods that deal with these uncertainties, readers are encouraged to read the article of Maddison *et al.* (1984), in which some formal rules concerning argumentation about character polarities are discussed in detail. In addition, for details about polarizing multistate characters and ordering character states, readers are referred to Wiley *et al.* (1991).

Reconstruction of Phylogenies

For the reconstruction of phylogenies one needs first to compile a data matrix, which consists of the mem-

TABLE I Characters and Character States Used in a Phylogenetic Study of the Chasmosaurinae[a]

Character	State
1. Flange along anterior margin of external naris	Partial (0); complete (1)
2. Supraorbital horn core curvature	Anterior (0); posterior (1)
3. Dimensions of posterior margin of parietal fenestra	Transverse width less (0) or more (1) than twice antero-posterior depth
4. Posterior frill transverse width	Less (0) or more (1) than twice skull width across orbits
5. Supraorbital horn core length	Long (0); short (1)
6. Jugal–squamosal contact below laterotemporal fenestra	Absent (0); present (1)
7. Premaxillary posterolateral process	Evenly tapered with tip of maxilla nasal suture (0); abruptly narrowed with tip isolated on nasal (1)
8. Parietal fenestra diameter	Less (0) or more (1) than preorbital length
9. Parietal posteromedian embayment	Weak or absent (0); marked (1)

[a] From Forster *et al.* (1993). 0 denotes the plesiomorphic state; 1 denotes the apomorphic state.

ber taxa of the ingroup, at least the sister group, descriptions of the selected characters, and the character state distribution of each character for all of the taxa in question. Before compiling a data matrix, one needs to code each state of each character. A character state code is usually a numerical or alphabetical symbol that represents a particular character state (Tables I and II). By convention, the plesiomorphic state is coded 0 and the apomorphic state is coded 1 (or 2, 3, and so on for multistate characters). By the method of outgroup comparison, one can determine the polarities of the characters and score a suitable character state for each character for each taxon. If the condition of a character is unknown for a certain taxon, owing to lack of preservation, then a "?" state is scored for that taxon. When a data matrix is completed, it is ready for analysis.

There exist several computer algorithms for finding the preferred (most parsimonious) hypothesis of phylogenetic relationships among the taxa included in the data matrix. The computer algorithm Phylogenetic Analysis Using Parsimony (PAUP; Swofford, 1993) is widely used in studies of phylogenetic relationships among groups of the Dinosauria. For the basic procedures of reconstructing phylogenies, we have used the study, in a slightly modified form, of the phylogenetic relationships of a new ceratopsid dinosaur as an example (Forster *et al.*, 1993).

Table I outlines the descriptions of the nine characters and character states used in this study. All nine characters are binary, consisting of a plesiomorphic state and a single apomorphic state. By means of outgroup comparison, the polarity of each character is determined for each taxon. In Table II, the ceratopsid dinosaur *Chasmosaurus mariscalensis*, a new species of the genus *Chasmosaurus* of the subfamily Chasmosaurinae, has been included. The member taxa of the

TABLE II
Character State Distributions of the Nine Binary Characters Presented in Table I, Which Vary among *Pentaceratops sternbergii* and the Three Species of *Chasmosaurus*[a]

	Character (out-group)	
Taxa	12345	6789
Other chasmosaurines	00000	0000
Ingroup		
Chasmosaurus mariscalensis	111?0	0111
Chasmosaurus belli	11111	1111
Chasmosaurus russelli	11111	1111
Pentaceratops sternbergii	00000	0111

[a] For establishing polarity and coding the characters, the taxon "Other chasmosaurines" (a monophyletic group) was chosen as the sister group (closest out-group) of the in-group that includes *Pentaceratops* and *Chasmosaurus*. 0 indicates the plesiomorphic state, 1 indicates the apomorphic state, and ? indicates that a character is unknown for a taxon owing to lack of preservation. Data are from the appendix provided by Forster *et al.* (1993).

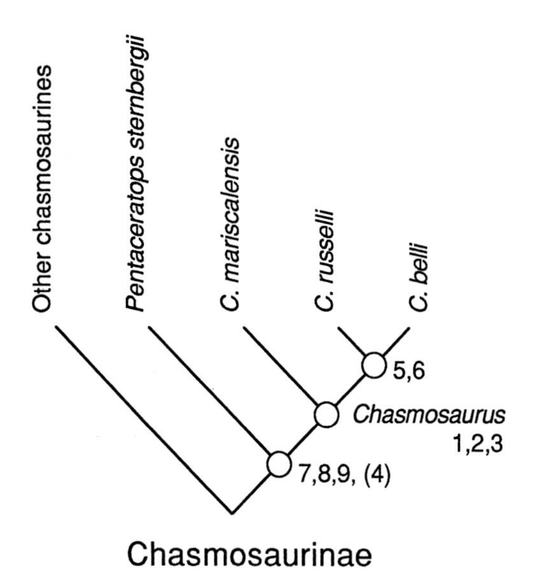

FIGURE 5 The most parsimonious cladogram, produced by PAUP (Version 3.1.1), for the data matrix presented in Tables 1 and 2. This depicts phylogenetic relationships among the chasmosaurines. "Other chasmosaurines" is a monophyletic taxon and was chosen as the sister group of the in-group. The latter includes *Pentaceratops* and *Chasmosaurus*.

ingroup include four species representing the genera *Chasmosaurus* and *Pentaceratops*. The most appropriate outgroup is the sister group of the ingroup in question. In this case, it has been taken to be the monophyletic taxon including all other chasmosaurines. Character state distributions are listed for all five taxa. The data matrix was analyzed using the branch-and-bound algorithm of PAUP, Version 3.1.1 (Swofford, 1993). The analysis yielded a most parsimonious cladogram, which depicts a hierarchical pattern of the relationships among the four in-group taxa (Fig. 5). *Chasmosaurus russelli* and *C. belli* are more closely related to each other than either is to the other taxa and thus form a monophyletic subgroup. This hypothesis is supported by their sharing of two synapomorphies (characters 5 and 6). The closest relative of *C. russelli* + *C. belli* is *C. mariscalensis*, the new ceratopsid, because it shares with the former cluster three synapomorphies (characters 1–3). In other words, the genus *Chasmosaurus* is a monophyletic taxon that can be diagnosed by characters 1–3. Furthermore, the genera *Pentaceratops* (represented by *P. sternbergii*) and *Chasmosaurus* form a monophyletic subgroup at a higher level within the Chasmosaurinae. Three synapomorphies (characters 7–9) shared by them support this hypothesis. Character 4 is an equivocal synapomorphy for the genus *Chasmosaurus* due to its uncertainty in the new species *C. mariscalensis*.

Phylogenetic Trees

Phylogenetic trees are branching diagrams delineating the hypothesized genealogical ties and sequence of historical events linking individuals, populations, species, or supraspecific taxa of organisms. Some phylogenetic systematists (such as Eldredge and Cracraft, 1980) do not think of cladograms as phylogenetic trees but suggest instead that they subsume the logical structure of a set of trees. However, many others (such as Wiley, 1981; Wiley *et al.*, 1991) consider cladograms to be phylogenetic trees. They are aware that because ancestral species cannot be identified, such trees taken from cladograms are incomplete and they can portray only sister group relationships and a relative time axis. Furthermore, phylogenetic trees are not facts but hypotheses that suggest probable patterns of evolutionary events and may change with our increasing understanding of organisms. In a real sense, cladograms are merely a graphic depiction of the most parsimonious summary of character state distribution. Phylogenetic trees can be converted into classifications with a minimum of required conventions (see below).

Phylogenetic Classification

Biological classifications are systems of words that are used to organize the diversity of life and/or to reflect our estimate of nature's own organization of life (Wiley, 1981). In phylogenetic systematics, a classification must be consistent with the phylogeny on which it is based and must be capable of expressing the sister group relationships among the groups classified. In other words, anyone other than the original investigator is, from the set of words of a phylogenetic classification, able to read the exact relationships among the taxa included and reconstruct the topology of the source cladogram. For historical consistency, phylogenetic systematics recommends that one should make every effort to alter preexisting classifications as little as possible. There are a number of

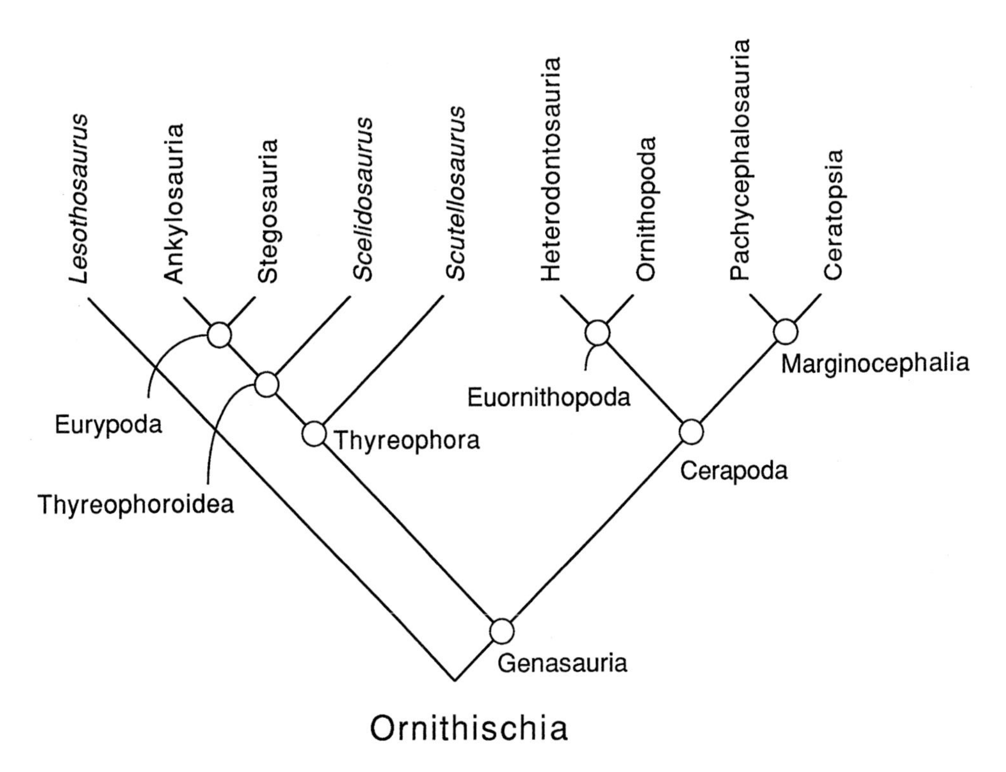

FIGURE 6 A cladogram portraying relationships among major groups of the Ornithischia (modified from Sereno, 1986, Figs. 2 and 3).

methods, with their own conventions, for reconstructing phylogenetic classifications. The method introduced below is currently the most commonly used in phylogenetic systematics.

In accordance with the structure of phylogenetic trees, one can consistently assign sister groups in the classification the same rank and thus name every hypothesized monophyletic taxon. In such a classification, rank within a given part denotes relative time of origin. Here, we present an example taken from the phylogenetic study of ornithischian dinosaurs by Sereno (1986). Figure 6 is the resulting cladogram showing the relationships among the major groups of the Ornithischia. According to the pattern of phylogenetic relationships among the Ornithischia, one can classify the included taxa hierarchically as is indicated in Figs. 3 or 4 (two alternative methods are presented).

See also the following related entries:

DINOSAURIA • PHYLOGENETIC SYSTEM

References

Benton, M. J. (1990). Origins and interrelationships of Dinosaurs. In *The Dinosauria* (D. B. Weishampel, P. Dodson, and H. Osmólska, Eds.), pp. 11–30. Univ. of California Press, Berkeley.

Brooks, D. R. (1981). Hennig's parasitological method: A proposed solution. *Systematic Zool.* **30,** 229–249.

Eldredge, N., and Cracraft, J. (1980). *Phylogenetic Patterns and the Evolutionary Process,* pp. 349. Columbia Univ. Press, New York.

Forster, C. A., Sereno, P. C., Evans, T. W., and Row, T. (1993). A complete skull of *Chasmosaurus mariscalensis* (Dinosauria: Ceratopsidae) from the Aguja Formation (Late Campanian) of west Texas. *J. Vertebr. Paleontol.* **13**(2), 161–170.

Hennig, W. (1950). *Grundzüge einer Theorie der Phylogenetischen Systematik.* Deutcher Zentralverlag, Berlin.

Hennig, W. (1966). *Phylogenetic Systematics,* pp. 263. Univ. of Illinois Press, Urbana.

Maddison, W. P., Donoghue, M. J., and Maddison, D. R. (1984). Outgroup analysis and parsimony. *Systematic Zool.* **33**(1), 83–103.

Popper, K. R. (1959). *The Logic of Scientific Discovery.* Harper Torchbooks, New York.

Romer, A. S. (1956). *Osteology of the Reptiles*, pp. 772. Univ. of Chicago Press, Chicago.

Sereno, P. C. (1986). Phylogeny of the bird-hipped dinosaurs (Order ornithischia). *Natl. Geogr. Res.* **2,** 234–256.

Sereno, P. C., and Arcucci, A. B. (1993). Dinosurian precursors from the Middle Triassic of Argentina: *Lagerpeton chanarensis. J. Vertebr. Paleontol.* **13,** 385–399.

Swofford, D. L. (1993). Phylogenetic Analysis Using Parsimony, Version 3.1.1. Illinois Natural History Survey, Chicago.

Wiley, E. O. (1981). *Phylogenetics. The Principles and Practice of Phylogenetic Systematics*, pp. 439. Wiley, New York.

Wiley, E. O. (1987). The evolutionary basis for phylogenetic classification. In *Systematics* (P. Hovenkamp, E. Gittenberger, E. Hennipman, R. de Jong, M. C. Roose, R. Sluys, and M. Zandee, Eds.), pp. 55–64. Utrecht Univ. Press, Utrecht, The Netherlands.

Wiley, E. O., Siegel-Causey, D., Brooks, O. R., and Funk, V. A. (1991). The compleat cladist. A primer of phylogenetic procedures. The University of Kansas Museum of Natural History Spec. Publ. 19.

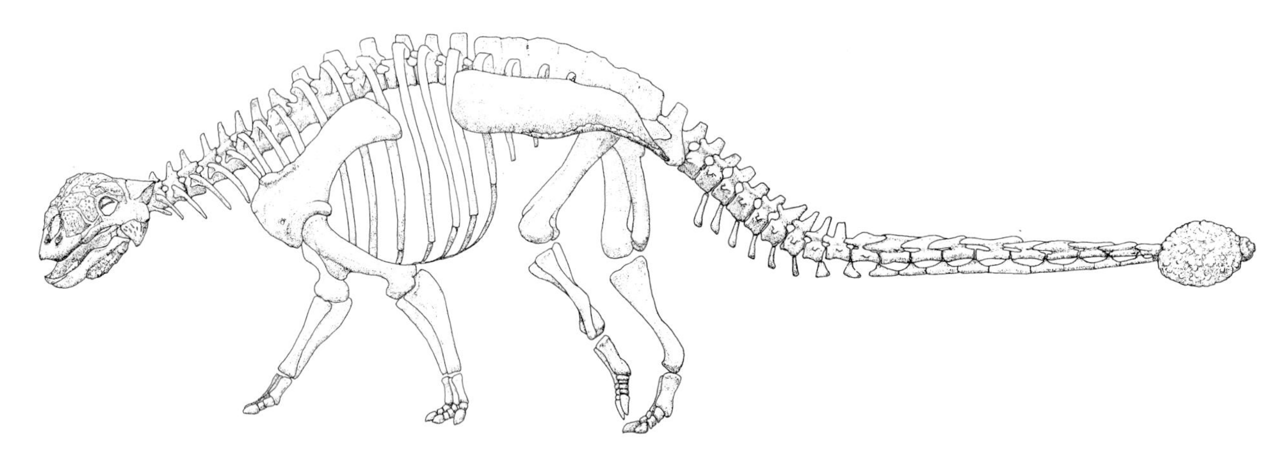

Only ankylosaurids possess tail clubs. These clubs are formed by bones, embedded in the skin, becoming enlarged and fusing to each other and, eventually, to the vertebrae of the tail.

Tail Clubs

Tail clubs, like those pictured on the facing page, are found only in ankylosaurs.

see ANKYLOSAURIA; ORNAMENTATION

Tajikistan Dinosaurs

see MIDDLE ASIAN DINOSAURS

Takikawa Museum of Art and Natural History, Japan

see MUSEUMS AND DISPLAYS

Tanzanian Dinosaurs

see AFRICAN DINOSAURS; TENDAGURU

Taphonomy

ANTHONY R. FIORILLO
Dallas Museum of Natural History
Dallas, Texas, USA

Taphonomy is the study of the pathways an organism passes through on its way to the fossil record. Although the term was coined in 1940 by the Russian paleontologist I. A. Efremov, and the science has flourished since, paleontologists have recognized the principles of taphonomy for far longer. For example, by correlating the orientation of brachiopod valves in the Paleozoic rocks of New York with other paleocurrent indicators found in the units, James Hall in 1843 understood the significance of the orientation of brachiopod valves as an independent means for determining paleocurrents. Similarly, William Buckland in 1823 studied the tooth-marked bones left by modern hyenas and compared the patterns to fossil bones in an effort to determine behavioral patterns of certain extinct carnivores. Modern taphonomy combines observational and experimental data to provide interdisciplinary insight into paleobiological, sedimentological, and archaeological problems.

Traditionally, taphonomic studies have focused on the biases or information loss as organisms are incorporated into the fossil record. These studies compare living communities with death assemblages to assess postmortem information loss.

Recently, workers have also focused their efforts on the information gain through the study of taphonomy. The recognition of taphofacies and comparative taphonomy has provided a framework that utilizes the differing survivorship of various skeletal elements within and between taxa within depositional contexts. This framework provides the means for bracketing the ancient environmental conditions under which the fossil deposit formed. Workers addressing evolutionary questions can appreciate the effects of time averaging on the frequency of species morphotypes. In biostratigraphic studies taphonomy plays a role in understanding the distribution of fossils through the rock record and the resulting influence on range endpoints of taxa. Understanding the processes of bone modification has helped archaeologists differentiate between culturally and nonculturally derived bone assemblages, resulting in a clearer picture of behavior of early humans. Bone modification studies provide clues to the predepositional or no-depositional history of fossil localities.

Much of the effort spent in vertebrate taphonomy

has been focused on faunas from the Tertiary or those faunas in an archaeological context. Studies of Mesozoic faunas, though few in number, are growing and proving to be insightful. For example, detailed studies of some ceratopsian bone beds show that these animals likely lived in herds (e.g., Currie and Dodson, 1984; Ryan, 1992). These animals, then, may have displayed courtship and defense BEHAVIORS similar to those of modern herding animals. Another study compared bones from dinosaur and Tertiary and modern mammal localities (Fiorillo, 1991a). This study showed the paucity of tooth-marked bone in the Mesozoic assemblages compared to assemblages from the Cenozoic. These data suggest Mesozoic predators did not utilize the bone-chewing niche that the later Cenozoic mammalian carnivores did. In still another taphonomic study, Rogers (1990) examined three Cretaceous dinosaur-bearing bone beds and suggested that periodic drought played a significant role in the formation of Mesozoic bone assemblages.

The growing body of taphonomic work focused throughout the vertebrate fossil record has illustrated many of the processes affecting the formation of bone assemblages. As these studies continue to grow in number, a more complete picture of the origin of the vertebrate fossil record and its application to geological, biological, and archaeological questions will become more clear.

Taphonomy is the key to better understanding ancient environments, but to interpret a fossil locality precisely, a precise taphonomic methodology is needed. Just as there is a proper way to describe a fossil, there is a proper way to describe a fossil assemblage. With respect to vertebrate taphonomy, studying parameters of a bone assemblage, such as the spatial distribution of fossils, the degree of skeletal articulation, the orientation of fossil material, and bone modification features, provides the means for determining the biological and sedimentological processes that influenced the formation and preservation of a particular vertebrate fossil assemblage.

The three-dimensional, or spatial, distribution of fossils within a single stratigraphic unit is typically heterogeneous in terrestrial vertebrate fossil localities. Concentrations of bones along the basal part of a fluvial unit with a sharp base can be used to infer a catastrophic origin (e.g., a flood) for the bone assemblage. These assemblages represent very short periods of time, ranging from days to years. In contrast, a distribution of fossils through a large stratigraphic interval can suggest a slower rate of accumulation and may represent a time ranging from tens to thousands of years. An approximation of the time represented by a fossil assemblage is critical for interpreting potential ecological relationships.

The degree of skeletal articulation at a given site provides a key to measuring the relative degree of postmortem disturbance within a fossil assemblage. For example, there is a continuum from an articulated skeleton to a sample of isolated, unassociated elements with each preservational mode suggesting something different about the history of burial. An assemblage of articulated skeletal material suggests relatively quick burial, probably with little carcass transport, before processes such as scavenging, weathering, or fluvial activity could disassociate or destroy skeletal elements. An assemblage of isolated bones represents a relatively longer period of time during which much of the carcass is removed and bones disassociate.

Another means of measuring postmortem disturbance is through a bone census. The different parts of a vertebrate skeleton behave differently within a stream environment. Lighter, less dense elements tend to be washed away sooner than the heavier, more dense elements. Subsets of the vertebrate skeleton that behave similarly in the stream environment are called Voorhies groups. Voorhies group I consists of the least dense elements, such as vertebrae and ribs, whereas Voorhies group III is made up of the most dense elements, such as skulls and jaws. Voorhies group II elements are the midrange density elements such as limb bones. Outside of the channel environment, on the landscape, mammalian predators have proven to be a factor in differential bone preservation by selectively removing parts of the skeletons of prey animals. Therefore, a bone census at a fossil site provides a means to evaluate what is missing from a fossil locality and provides insight into the mechanism for removal.

In assemblages of disarticulated bones, the orientation of the long bones helps determine the effects of fluvial activity at the site during its formation. The significance of fossil orientations has been discussed through the years. Several idealized distributions have been put forth, with each having a dramatically different origin. A uniform three-dimensional distribution of fossils means that some bones were buried

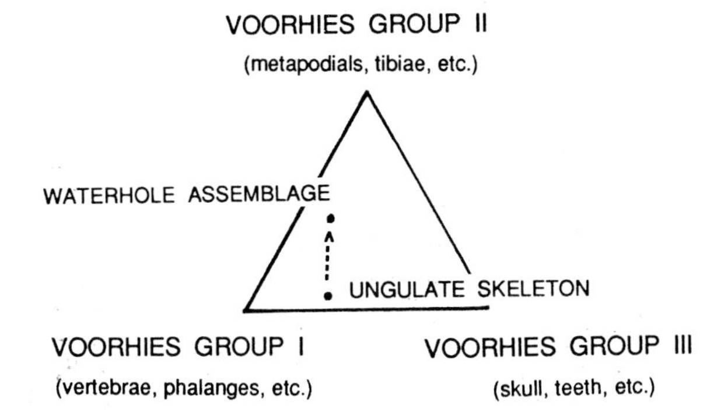

FIGURE 1 Ternary plot comparing the skeletal element representation at a Miocene fossil site in western Nebraska with the idealized skeletal element distribution of the most common animal from the site (data from Fiorillo, 1988). Notice that this graphical portrayal of data easily highlights the paucity of lighter, less dense skeletal elements at the fossil site with respect to the expected distribution.

in unstable orientations. Such a burial position can be the result of deposition within a thick, viscous medium such as a mudflow, or the unstable positions may be secondarily produced by a biological process such as trampling on a soft substrate. A preferred distribution, that is, similarly aligned linear bones, within a horizontal plane suggests a fluvial overprint due to directional flow. In contrast, a nonpreferred distribution within a horizontal plane suggests fluvial currents, if present, were not sufficient to align bones at the site.

Another measure of the fluvial overprint on a vertebrate assemblage is to estimate the hydraulic compatibility of the bones with the surrounding sediment. Based on studies of settling velocities of bones and spheroidal quartz grains, one can quantitatively determine if the stream currents responsible for the sediment at a fossil site were of sufficient strength to transport the bones to the locality.

Bone modification studies provide details of the postmortem history of individual bones, which may include biological influences as well as physical influences. Bone modification features have been defined as those features that are the result of any postmortem, prediagenetic process, such as trampling, weathering, or scavenging, that alters the morphology of once-living bone. Typically, bone modification excludes pathological features, or diagenetic features, such as mineral recrystallization in bone cavities or the effects of strain. Generally, these features can be divided into two groups—those that are the result of fluvial processes, such as abrasion, and those that result from exposure on the land surface. These land surface features can be further subdivided into those resulting from biological activity, such as trampling and scavenging, and those resulting from physical processes, such as weathering.

In addition to the descriptive analysis of taphonomic parameters, there are various graphic means employed to describe and compare fossil assemblages. Two of the more useful ones are shown in Figs. 1 and 2.

The three points of the triangle in Fig. 1 represent each of the three Voorhies groups. Plotted on the diagram are two data points; one is an idealized ungulate skeleton, whereas the other is that of the relative abundances of bones from a Tertiary mammal locality in Nebraska (Fiorillo, 1988). This simple plot shows that the fossil site is skewed toward the more dense skeletal elements and away from Voorhies group I, inferring that some selective removal of less dense elements took place at the site during its formation.

Figure 2, referred to as a series of taphograms (Behrensmeyer, 1991), is a multivariate comparison of taphonomic parameters between sites. Three sites with similar sample sizes but from three different depositional environments are compared. This graphic means of comparison immediately reveals certain similarities and differences between sites. For example, the site with the lowest taxonomic diversity is the pond assemblage, and the site with the highest diversity is the stream assemblage. The skeletons in the pond deposit only showed a tendency to be disarticulated; therefore, there is a low taphonomic overprint on this assemblage of bones. In contrast, the waterhole assemblage shows evidence of a heavy taphonomic overprint, particularly with respect to biological factors such as predation and trampling. The high taxonomic diversity of the stream deposit assemblage, compared to the other two environments, suggests that streams tend to mix faunal elements from several environments, thus preserving representatives from different ecosystems within one deposit.

Graphic methods such as these facilitate the comparison of fossil assemblages from various depositional environments. They provide the reader with a quick and easy means to detect similarities and differences between sites.

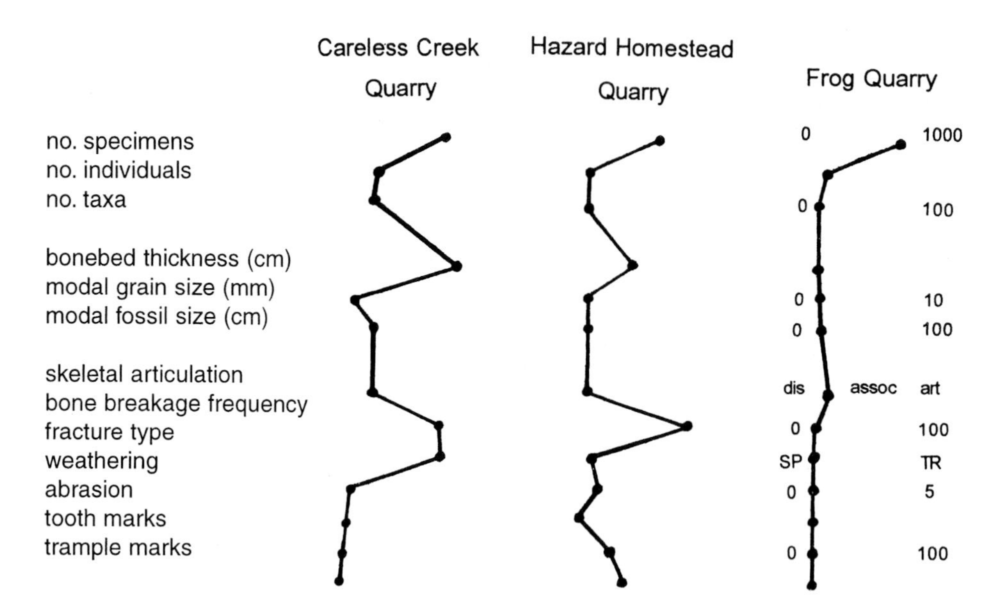

FIGURE 2 A series of taphograms illustrating the differences in various taphonomic parameters between three fossil assemblages. Data for Careless Creek Quarry, Hazard Homestead Quarry, and Frog Quarry are from Fiorillo (1988, 1991b) and Henrici and Fiorillo (1993), respectively. The scales for each taphogram are shown in the Frog Quarry taphogram. An unmarked scale refers to the preceding marked scale. dis, disarticulated; assoc, associated; art, articulated; SP, spiral; TR, transverse. This method of graphically portraying data allows one to easily visualize the basic taphonomic differences and similarities between fossil assemblages.

Taphonomy is a tool to bridge the gap in understanding between what is preserved at a fossil locality and how reliably the site reflects the ancient ecosystem within which it formed. Careful description of various taphonomic parameters provides the basis for a more thorough understanding of the various biological and sedimentological processes contributing to the formation of particular fossil assemblages.

See also the following related entries:

BIOSTRATIGRAPHY • GEOLOGICAL TIME • PALEOECOLOGY

References

Behrensmeyer, A. K. (1991). Terrestrial vertebrate accumulations. In *Taphonomy: Releasing the Data Locked in the Fossil Record* (P. A. Allison and D. E. G. Briggs, Eds.), pp. 291–335. Plenum, New York.

Currie, P. J., and Dodson, P. (1984). Mass death of a herd of ceratopsian dinosaurs. In *Third Symposium on Mesozoic Terrestrial Ecosystems* (W. E. Reif and F. Westphal, Eds.), pp. 61–66. Attempto–Verlag, Tubingen, Germany.

Fiorillo, A. R. (1988). Taphonomy of Hazard Homestead Quarry (Ogallala Group), Hitchcock County, Nebraska. *Contrib. Geol. Univ. Wyoming* **26,** 57–97.

Fiorillo, A. R. (1991a). Prey bone utilization by predatory dinosaurs. *Palaeogeogr. Palaeoclimatol. Palaeoecol.* **88,** 157–166.

Fiorillo, A. R. (1991b). Taphonomy and depositional setting of Careless Creek Quarry (Judith River Formation), Wheatland County, Montana, U. S. A. *Palaeogeogr. Palaeoclimatol. Palaeoecol.* **81,** 281–311.

Henrici, A. C., and Fiorillo, A. R. (1993). Catastrophic death assemblage of Chelomophrynus bayi (Anura, Rhinophyrnidae) from the middle Eocene Wagon Bed Formation of central Wyoming. *J. Paleontol.* **67,** 1016–1026.

Rogers, R. R. (1990). Taphonomy of three monospecific dinosaur bone beds in the Upper Cretaceous Two Medicine Formation, Northwestern Montana: Evidence for mass mortality related to episodic drought. *Palaios* **5,** 394–413.

Ryan, M. J. (1992). The taphonomy of a Centrosaurus (Reptilia: Ornithischia) bone bed (Campanian), Dinosaur Provincial Park, Alberta, Canada. M. Sc. thesis, University of Calgary, Calgary.

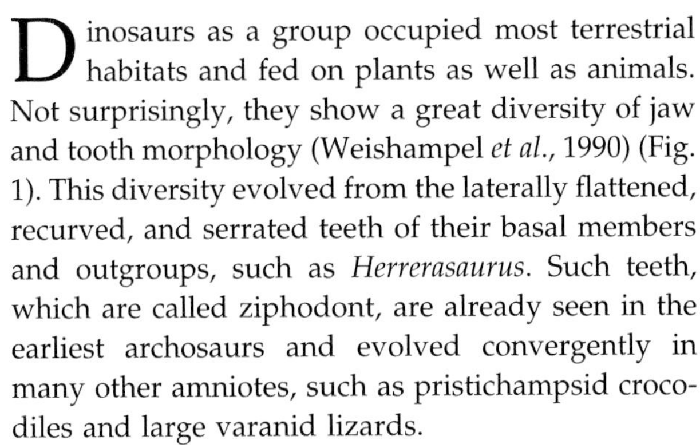

Teeth and Jaws

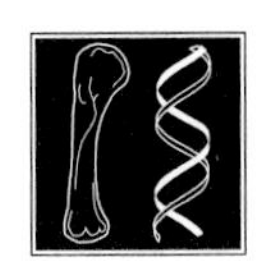

P. MARTIN SANDER
Universität Bonn
Bonn, Germany

Dinosaurs as a group occupied most terrestrial habitats and fed on plants as well as animals. Not surprisingly, they show a great diversity of jaw and tooth morphology (Weishampel *et al.*, 1990) (Fig. 1). This diversity evolved from the laterally flattened, recurved, and serrated teeth of their basal members and outgroups, such as *Herrerasaurus*. Such teeth, which are called ziphodont, are already seen in the earliest archosaurs and evolved convergently in many other amniotes, such as pristichampsid crocodiles and large varanid lizards.

Amniote teeth primarily consist of two tissues, enamel and dentine. Enamel is the outer hard and shiny but brittle covering that is supported by a tougher but softer core made of dentin, a tissue somewhat similar to bone. The outside of the enamel layer occasionally is covered by a third tissue, cementum. Cementum is rare in reptiles but important in some mammal teeth (for a comprehensive treatment see Peyer, 1968).

As opposed to mammals, teeth in dinosaurs and nearly all other vertebrates are replaced many times during the life of the animal. In tooth loss, the root is resorbed as the old crown is pressured by the incoming replacement tooth. Finds of isolated rootless teeth represent such spent teeth. Rootless carnivore teeth are particularly frequent finds with the carcasses of herbivorous dinosaurs.

In some groups of dinosaurs, rapid tooth replacement is combined with heavy wear on the teeth. Because most wear is caused by contact between the upper and lower teeth of a dentition, wear patterns are of great interest for understanding jaw movement. Microscopic traces of wear on the tooth surface (microwear) differ depending on the kind of food ingested and can be used to reconstruct diet.

A common feature in vertebrate skulls is the ability of certain parts of the skull to move against each other during the feeding process. This phenomenon is called kinesis and primarily functions to improve the angle of attack of the jaw muscles and the dentition. However, most dinosaurs lack skull kinesis. Exceptions are certain carnosaurs, ornithopods, and birds (as avian dinosaurs).

The primitive tetrapod feeding process consists of tearing or biting off the food and swallowing the chunks without further manipulation in the mouth. This method is not very efficient for fast nutrient extraction, the prerequisite for a fast metabolism. Food reduction (i.e., mechanical breakdown) combined with thorough mixing with digestive fluid, on the other hand, decreases digestion time in the stomach, thus increasing the speed of nutrient release. This process can take place in the oral cavity, as in some dinosaurs and most mammals, or in a special chamber of the stomach (gizzard) as in some other dinosaurs and in birds.

Although orientational terms vary in the more specialized literature, labial is the side of a tooth facing the outside, whereas lingual (or buccal) is the inward direction, facing the tongue. Anterior (or cranial) is toward the tip of the snout and posterior (caudal) is toward the jaw joint. Apical is toward the tooth tip and basal is toward the root. Serrations are the line of fine bumps called denticles along the cutting edge of some teeth.

Carnivores

Because the out-groups of dinosaurs were carnivores with recurved, serrated teeth (see DINOSAUROMORPHA), the theropod condition is generally plesiomorphic (for overviews, see Currie *et al.*, 1990; Farlow *et al.*, 1991). The cross section of their ziphodont teeth is lenticular, and the cutting edges generally bear well-developed serrations consisting of closely spaced denticles. Looking straight down at the denticles, they have a squarish or oval outline. The denticles are separated from each other by a slot that cuts meat fibers. Because ziphodont teeth are adapted to cutting and slicing flesh (Abler, 1992), they are a reliable indicator of a carnivorous diet.

A general distinction is conveniently made be-

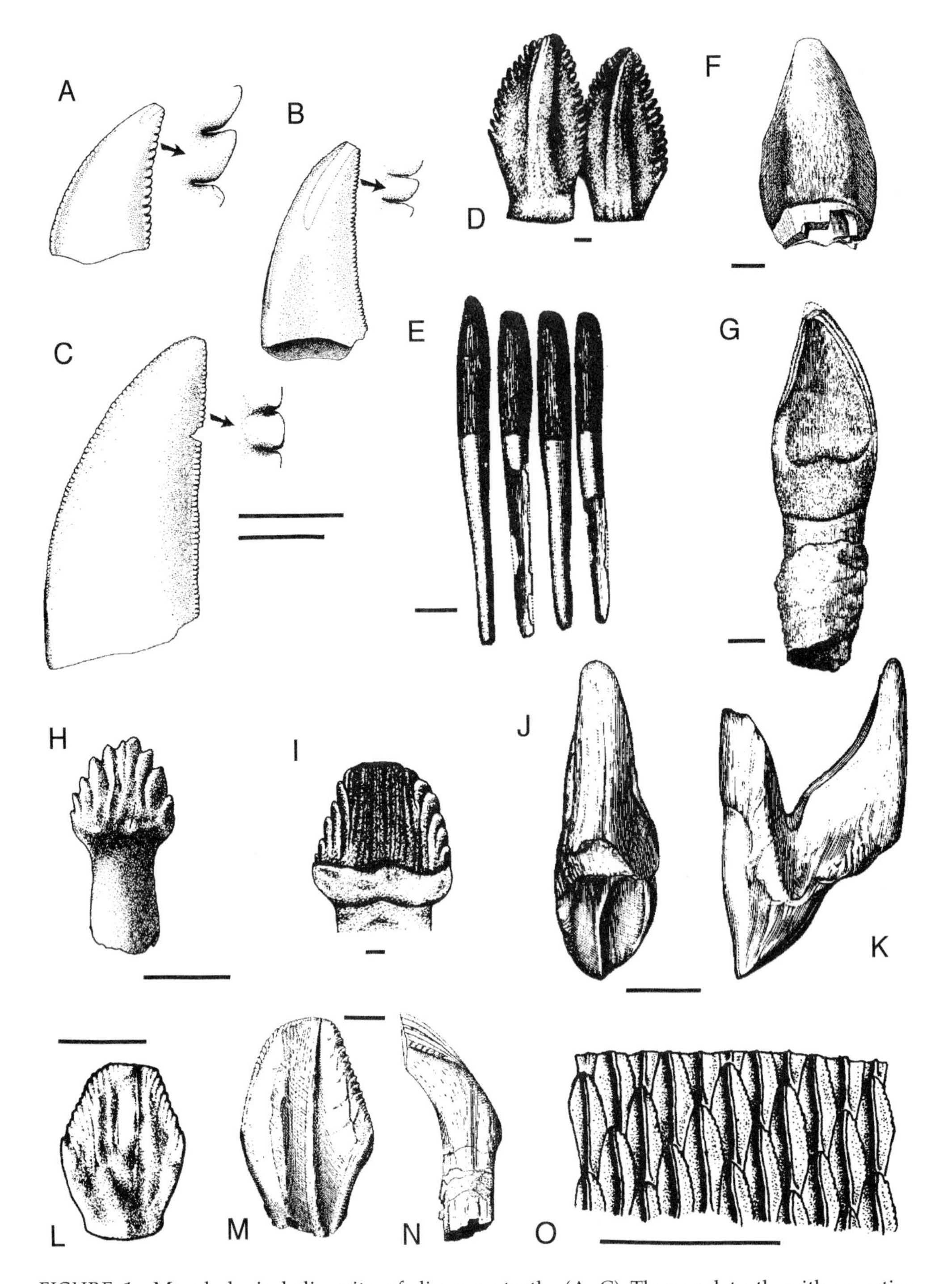

FIGURE 1 Morphological diversity of dinosaur teeth. (A–C) Theropod teeth with serrations enlarged (from Fiorillo and Currie, 1994, with permission); (A) *Troodon*, (B) *Saurornitholestes*, (C) plesiomorphic large theropod; (D) the prosauropod *Plateosaurus* (from Galton, 1990a, with permission); (E) the sauropod *Diplodocus* (from Marsh, 1884, with permission); (F, G) the brachiosaurid sauropod *Oplosaurus* in (F) labial view (from Swinton, 1973, with permission) and (G) lingual view (from Glut, 1982, with permission); (H) the ankylosaur *Edmontonia* (from Coombs and Maryańska, 1990, with permission); (I) the stegosaur *Stegosaurus* (from Galton, 1990b, with permission); (J, K) a maxillary tooth of the ceratopsian *Triceratops* in (J) labial view and (K) anteroposterior view (both from Marsh, 1891, with permission); (L) dentary tooth of the ornithopod *Dryosaurus* in lingual view (from Weishampel and Heinrich, 1992, with permission); (M, N) dentary tooth of the ornithopod *Iguanodon* in (M) lingual view and (N) anteroposterior view (from Swinton, 1973, with permission); (O) part of mandibular tooth battery of the hadrosaur *Lambeosaurus* in lingual view (from Weishampel and Horner, 1990, with permission). Small scale bar in A–C = 5 mm; large scale bar in A–C = 1 mm. Scale bar in D = 1 mm. Scale bars in E–K and in M and N = 1 cm. Scale bar in L = 5 mm. Scale bar in O = 5 cm.

tween large theropods, primarily tyrannosaurids, some ceratopsians, and carnosaurs, and a plethora of small forms of different systematic affinity (most cretursaurs and ceratosaurs) that, nevertheless, have rather similar dentitions.

CARNOSAURS and other large theropods are characterized by large skulls with relatively few (between approximately 16 and 22 per tooth row) but large teeth. In most large theropods the teeth were the primary killing tools, supported by strong jaw and neck muscles. The largest teeth are characteristically not found at the front of the snout but more toward the middle, increasing flesh-cutting efficiency. Also because of their killing function, teeth of large theropods are stout in comparison to those of smaller forms. Among the large forms, the tyrannosaurs show an additional modification from the plesiomorphic pattern: The front teeth have a D-shaped cross section with both carinae oriented posteriorly. These teeth are sometimes termed incisiform teeth, and they are particularly useful for seizing and cutting tissue.

SPINOSAURS and the strange (perhaps related) *Baryonyx* from the Lower Cretaceous are also large theropods. Their teeth, however, are only slightly compressed laterally, with poorly developed cutting edges and weak or no serrations. *Baryonyx* has 32 teeth per lower jaw ramus but only approximately 16 teeth (which are larger) in the upper jaw. Based on tooth shape and other evidence, a diet of fish has been suggested for these forms.

In small theropods, the teeth generally are more numerous and of the plesiomorphic dinosaurian type. Small theropods either took relatively small prey, including carrion (plesiomorphic forms), or had evolved other killing tools such as strongly clawed hands or a single large claw in the hindfoot (dromaeosaurs). Identification of small theropod teeth to genus and species is not easy because most are recurved and serrated. Some taxa show a distinctive morphology (e.g., *Troodon*, with its rather coarse serration, which is somewhat reminiscent of prosauropods; Currie *et al.*, 1990).

At least two lineages of small theropods, ornithomimosaurs, and oviraptorosaurs greatly reduced or lost their teeth and evolved a bony beak that was covered by a horny sheath. One such keratinous sheath was recently discovered in an ornithomimosaur (P. Currie, personal communication). Although seemingly aberrant, tooth reduction combined with beak evolution is not uncommon among dinosaurs nor in amniotes as a whole (e.g., in turtles, large pterosaurs, and dicynodonts). The lack of teeth makes the dietary preference of these theropods rather difficult to assess. A beak is a rather universal tool suitable for a wide variety of food (compare diets of birds and turtles). Thus, the anatomy and biomechanics of the jaw may offer the primary clues. ORNITHOMIMOSAURS have a small head with an elongate snout on a long and slender neck that seems best suited for picking up small prey, such as small vertebrates and insects, as well as plant matter. OVIRAPTOROSAURIDS, on the other hand, have short jaws capable of great biting forces, as suggested by the short lever arm and the massive construction of the jaws. Great biting forces are useful for feeding on hard-shelled prey, such as molluscs and arthropods, but are also seen in seed-eating birds. Feeding on eggs, as suggested by the name of the group, also may have been facilitated by such a jaw anatomy, but there is no direct evidence and no modern analog.

THERIZINOSAURS ("segnosaurs") are only known from a few finds, and there is little variation among their dentitions. The teeth are small to very small, of even size, and numerous. The tip of the upper jaw is edentulous and may have been covered by a horny sheath. Somewhat similar to prosauropod teeth, the teeth have an expanded crown, giving them a leaf-like shape that is enhanced by a coarse serration. As in prosauropods, dental morphology together with other anatomical features point to a diet of plants.

Herbivores

Sauropodomorpha Basal sauropodomorphs, or prosauropods, as exemplified by the very well-known *Plateosaurus* from the Norian of central Europe, had relatively small but elongate skulls with slender lower jaws that carried numerous closely spaced teeth of approximately equal size. The teeth appear relatively small in comparison to the skull size. Their crowns are anteroposteriorly expanded from the root and have a very characteristic leaf-like shape. It is laterally flattened and the edge is covered by coarse denticles. These denticles superficially resemble the serrations on theropod teeth but differ in their round cross section, lack of a slot, and their apical inclination. The individual teeth overlap each other slightly in an *en echelon* fashion, with the posterior edge of one tooth covering the buccal side of the anterior edge of the succeeding tooth.

Although the superficial resemblance to serrated theropod teeth has in the past led to at least one proposal that prosauropods were carnivorous, it is now generally agreed that they fed primarily on plant matter. This is indicated by the large number but small and even size of the teeth, as well as by the fact that the jaw joint lies in many prosauropods significantly below the line of the tooth row. The tooth shape is rather similar to that of recent herbivorous lizards and nearly all ornithiscians. In addition, the small skull and large body cavity of prosauropods also suggest herbivory.

All sauropods unquestionably were herbivores. This is underscored by the fact that they took the small heads and large body cavities to extremes unseen before or since. For a long time it has been recognized that there are two tooth types among sauropods. One consists of stout, spatulate teeth that are curved lingually and may retain vestigial denticles along the edges. The other tooth type is generally characterized as slender and pencil or peg shaped, in which the crown is not expanded more widely than the root. The spatulate form occurs in camarasaurids and brachiosaurids, with their broad muzzles, whereas the pencil-shaped form is typical for diplodocids and titanosaurs, with their more elongate and rounded muzzles. Intermediate tooth forms are known, particularly in early forms. However, it must be noted that our knowledge of sauropod cranial anatomy is limited because complete skulls are known for only a few forms (see SAUROPODA).

The spatulate tooth type is associated with more robust jaws and a long tooth row with upright teeth. The pencil-like teeth, on the other hand, are concentrated in the anterior fourth of the jaw and are directed somewhat forward (procumbent). The tooth types also differ in the amount of wear (heavy in the spatulate teeth but slight in the peg-like teeth). Clearly, brachiosaurids and camarasaurids had stronger dentitions able to cope with tougher plant matter than did forms with pencil-like teeth.

Tooth morphology is only part of the story of sauropod food processing, however. It has long been observed (but rarely properly documented) that well-rounded pebbles with a high polish are associated with some sauropod skeletons. These pebbles represent stomach stones or gastroliths that presumably were lodged in the walls of the muscular gizzard, forming a so-called gastric mill that was the primary organ of mechanical food breakdown. This gastric mill, which is also seen in extant birds, obviated the need for extensive oral processing of the food.

Ornithischia Dental and jaw evolution within the Ornithischia is well documented by the fossil record, but this cannot be said about the transition from the plesiomorphic ziphodont tooth form to those of early ornithischians. Ornithischia as a group are characterized by an exclusively herbivorous DIET (a derived trait) as well as by a particular anatomical feature of the lower jaw, the predentary. This neomorphic, unpaired bone is attached to the front of the dentaries and lacks teeth. In conjunction with the predentary, ornithischians show a strong phylogenetic tendency to modify the anterior part of the jaws into a beak. All but the most primitive members of the group have progressively lost their premaxillary and anterior dentary teeth, and a horny sheath covers this region instead. However, this beak is not synapomorphic for a particular taxon but has evolved convergently at least three times: in the advanced thyreophorans, in the advanced ornithopods, and in the CERATOPSIA.

In analogy to the muzzle shape of modern herbivores, beak morphology in ornithischian dinosaurs can be used to interpret diet. Narrow, pointed beaks are suggestive of selective feeding on high-nutrient foods such as young shoots, buds, and fruits. Wide, flat beaks probably indicate bulk feeding on low-nutrient foods such as fern, cycad, and angiosperm leaves and conifer needles.

Plesiomorphic ornithischian teeth, as exemplified by *Lesothosaurus*, are differentiated into simple conical teeth in the premaxilla and laterally compressed teeth in the maxilla and dentary. The triangular tooth crown of the latter is much expanded over the root and lined by distinct denticles. On the buccal side of the upper teeth and the lingual side of the lower teeth, the denticles continue as ridges to the base of the crown. This morphological inversion between upper and lower cheek teeth is characteristic of many ornithischian dinosaurs.

Among the THYREOPHORA, basal forms such as *Scutellosaurus* retain a generalized ornithischian feeding apparatus. Stegosaurs and ankylosaurs, on the other hand, have well-developed beaks combined with small and relatively weak teeth. In addition, the tooth row has shifted medially, away from the margins of the dentary and premaxilla. The resulting labial shelf on the jaw bones is indicative of fleshy cheeks that prevented the food from falling out of the mouth.

The snout of stegosaurs is long and low, with a fairly narrow muzzle. The teeth have long roots and expanded crowns. The expansion is greatest near the base of the crown, forming a shelf surrounding the crown called a cingulum. Above the cingulum, the crown is somewhat laterally flattened and is divided by numerous vertical ridges. The individual ridges have a semicircular cross section and terminate in large, upright denticles where they intersect the cutting edges.

Ankylosaur teeth are similar in morphology to those of stegosaurs but are relatively weaker and smaller. The snout is much shorter and the muzzle is wider. The strong beak seems to have been the primary food-gathering organ in these animals. Although tooth wear does occur, neither stegosaurs nor ankylosaurs show regular wear patterns. This indicates the lack of precise occlusion.

Ornithopods best exemplify the trends described previously for ornithischians in general. In addition to beak evolution, there are two important modifications related to increased oral processing of food: the evolution of cheeks and of secondary occlusal surfaces.

Basal ornithopods, such as heterodontosaurs and hypsilophodonts, still retain premaxillary teeth in combination with a beak. As in basal ornithischians, the premaxillary teeth differ in shape from those of the cheek region. Heterodontosaurs derive their name from this and a pair of large fangs of the presumptive males.

In hypsilophodonts, the maxillary and posterior dentary teeth are leaf shaped, laterally flattened, and lingually curved. Only one side of the teeth bears enamel (lingual in the lower teeth and labial in the upper teeth). The enameled face of the crown is covered by several sharp vertical ridges, one of which is more prominent than the others. This main ridge divides the enamel-bearing face into two unequal parts, resulting in an asymmetrical crown.

Tooth shape is best observed in isolated replacement teeth because functional teeth in a complete dentition were rapidly cut down by wear. In fact, the maxillary and posterior dentary teeth are evenly worn down to one level by contact with the opposing tooth row. Together, the teeth thus form a secondary combined occlusal surface of the kind observed in hypsodont (high-crowned) mammals. This secondary occlusal surface ("dental belting") is a hallmark of ornithischian dinosaurs (Weishampel, 1984; Norman and Weishampel, 1985). It must have greatly increased food processing efficiency by providing a large grinding surface to reduce plant matter to small particles. This oral processing also required fleshy cheeks to retain food in one oral cavity during processing; this is indicated by labial shelves on the upper and lower jaws.

In conjunction with the evolution of a secondary occlusal surface, there are modifications in the skull to facilitate the use of these surfaces for grinding. In hypsilophodontids and more derived ornithopods, this is achieved by a type of skull kinesis called pleurokinesis. It consisted of a rotation of the maxillary bone around its long axis (Weishampel, 1984; Norman and Weishampel, 1985), although the angle of rotation was only a few degrees. Heterodontosaurids similarly were able to slightly rotate the lower jaws around their long axes. The larger Late Jurassic ornithopods such as *Iguanodon* basically continued along this evolutionary trend. Improvements to the secondary occlusal surface were mainly made through modifications of the tooth replacement pattern. Tooth replacement became more continuous in each tooth position, always providing a functional tooth and thus an occlusal surface without gaps. Iguanodontid teeth show a stronger asymmetry than more plesiomorphic ornithopod teeth, and the main ridge has increased in prominence at the expense of the accessory ones. The snout is elongate and the muzzle wide and rounded.

A further evolutionary increase in the rate of tooth replacement as well as the number of tooth positions led to the so-called dental batteries of the most advanced ornithopods, the hadrosaurs (Weishampel, 1984; Norman and Weishampel, 1985). As a result of this change, the individual teeth in the tooth batteries became closely packed and are much smaller than those in iguanodontids. The tooth crown morphology is simplified as well. The enameled side of the crown is diamond shaped, with a single central longitudinal ridge, and is set at an angle of approximately 60° to the root. Up to five teeth are stacked in one tooth position: Up to three were functional at any one time.

The resulting occlusal surface is thus subdivided by the hard enamel ridges, as in hypsodont mammals (e.g., horses). The dental batteries are separated from the wide beak by a gap called a diastema that characterizes nearly all ornithiscians but is particularly marked here. This diastema is also characteristic of hypsodont herbivorous mammals.

MARGINOCEPHALIANS infrequently evolved some dental specializations also seen in other groups. PACHYCEPHALOSAURS have a narrow snout with a heterodont dentition consisting of 15–20 teeth per side and jaw. The anterior dentary and premaxillary teeth are caniniform, i.e., larger than the cheek teeth, conical, and recurved. The cheek teeth are small and plesiomorphic in shape and are somewhat similar to those of stegosaurs and ankylosaurs. Cheek tooth wear differs between species, from confluent occlusal surfaces to wear restricted to distinctive facets.

Ceratopsia, including the PSITTACOSAURS, also have very highly developed feeding apparatus, including dental batteries in the most derived forms, the Ceratopsidae. However, the snout of ceratopsians differs from those of ornithopods in its narrowness. All Ceratopsia had well-developed horny beaks supported by predentary and rostral bones. This rostral bone, which is a synapomorphy of the Ceratopsia, is an unpaired neomorphic element at the tip of the upper jaw much like the predentary in the lower jaw.

The premaxilla is completely toothless, and a well-developed diastema separates the beak from the cheek teeth. Neoceratopsian teeth have a laterally flattened crown with enamel only on one side. This side has sharp vertical ridges and coarse serrations. Crown morphology is thus somewhat reminiscent of ornithopod teeth but with one important difference: The secondary occlusal surface is not oblique, but vertical. The teeth are closely set into tooth batteries much as in hadrosaurs. A unique feature of these teeth is their two-pronged roots, which interlock with the underlying teeth.

The vertical orientation of the secondary occlusal surface is related to the vertical cutting action of the rigid jaws, which presumably served to chop up the foodstuff finely rather than grind it up as in hadrosaurs. Oral food processing by chopping action was supported by a fleshy cheek, as in all advanced ornithischians. Although gastroliths have been reported in psittacosaurs, their function is rather enigmatic in light of the presumed oral processing in the group.

In secondarily aquatic tetrapods, such as plesiosaurs and crocodiles, stomach stones are common and serve as ballast to counteract the buoyancy of the lung. Although psittacosaurs are generally associated with lake deposits, their skeleton lacks any indication of aquatic adaptation. Swallowing stones to control buoyancy is thus not corroborated by anatomical evidence.

Enamel Microstructure

Enamel is the most highly mineralized tissue in the body, with a mineral content of more than 95% (bone, 60–70%). The mineral found in bone as well as in enamel is hydroxyapatite, a phosphatic mineral. The hydroxyapatite occurs in the form of micrometer-sized crystallites that may be arranged into higher-order structures.

Because it is highly mineralized, enamel is not a living tissue and cannot be modified by the animal after it is deposited by the ameloblasts, the cells making up the enamel organ. This means that enamel structure is under direct genetic control (although the form of individual dentitions may vary because of environmental and ontogenetic factors) and thus may be of great value in systematic studies. Enamel depostion is appositional, i.e., from the dentine (enamel dentine junction) outwards. Again, because it is highly mineralized, enamel is usually very well preserved, so fossil and recent specimens can be easily compared. These features of enamel have led to considerable work on mammalian enamel structure (Koenigswald and Sander, 1997). In mammals, the hydroxyapatite crystallites are combined into so-called prisms or rods that can be observed through a light microscope. Mammalian enamel microstructure is determined by two major constraints: phylogeny and biomechanics, mainly protection against fracturing. Reptilian (including dinosaurian) enamel generally lacks enamel prism (Sander, 1997) and is best studied with the scanning electron microscope. These differences presumably explain why relatively little work has been done on nonmammalian enamel (Fig. 2).

A survey of selected common dinosaur taxa indicates great microstructural variety in the group, in accordance with the great variability in dental morphology and function. Some dinosaurs (various small theropods and *Plateosaurus*) have a very thin enamel layer (5–20 μm) with a very simple structure con-

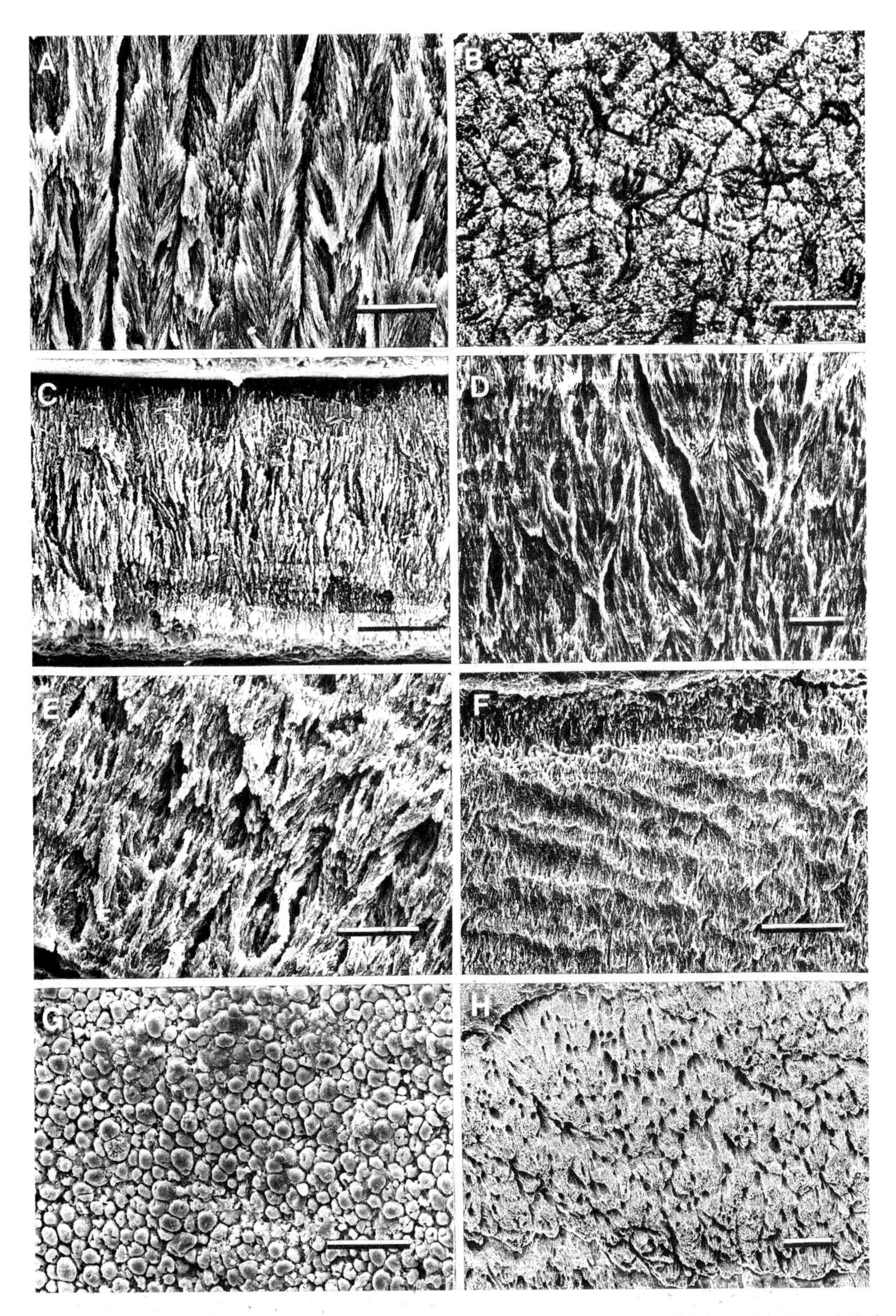

FIGURE 2 SEM photographs of dinosaurian tooth enamel microstructure. In all frames but B and G the outer enamel surface is at or beyond the top. (A, B) An indeterminate tyrannosaurid in (A) longitudinal section and (B) tangential section. Note the high degree of structural organization into columnar units. (C) The prosauropod *Plateosaurus*, longitudinal section. The thin enamel largely consists of parallel enamel crystallites. (D) The sauropod *Diplodocus*, cross section. The columnar units are less well defined than in the tyrannosaurid. (E) The ankylosaur *Palaeoscincus*, longitudinal section. The degree of organization is intermediate between that of *Plateosaurus* and *Diplodocus*. (F) The advanced ornithopod *Iguanodon*, longitudinal section. A thick layer of wavy enamel is covered by a thin outer layer of another kind of wavy enamel. (G) This unetched surface (not a section!) of *Anatosaurus* enamel shows the peculiar micromorphology common in hadrosaurs. (H) Indeterminate ceratopsid cut obliquely from the inside of the tooth. The enamel consists of columnar units intimately associated with enamel tubules (holes). Scale bars in A–E = 10 μm. Scale bars in F–H = 30 μm.

sisting of crystallites parallel to each other and normal to the enamel dentine junction. This enamel structure is plesiomorphic at the level of Amniota. Such a simple enamel structure is correlated with small teeth that in some cases were not of great functional sophistication (stegosaurs and ankylosaurs).

The large theropods investigated also have relatively thin enamel but with a much more highly organized structure. In *Tyrannosaurus*, the enamel consists of very regular columnar units. Columnar units are structural units bounded by planes of converging enamel crystallites. The units usually have a polygonal cross section. The enamel of the only sauropod investigated, *Diplodocus*, is also characterized by columnar units.

Both ornithischian groups that evolved tooth batteries, advanced ornithopods and advanced ceratopsians, have rather unusual features in their enamel microstructure. Both neoceratopsians and hadrosaurs (e.g., *Anatosaurus*) have a complex micromorphology on their enamel surface that gives the teeth of these groups their commonly observed dull appearance. The significance of this micromorphology is unclear. Derived ornithopods (e.g., *Iguanodon* and hadrosaurs) are characterized by an autapomorphic enamel type called wavy enamel. Wavy enamel consists of crystallites in a staggered and whorled arrangement. This enamel type may have evolved to reduce abrasion and so could have helped to maintain sharp cutting edges in the high-wear dentitions of derived ornithopods. Wavy enamel is one of the few known instances in which reptilian enamel microstructure is of systematic value. Neoceratopsian enamel, on the other hand, is dominated by columnar units.

Comparing dinosaur enamel with that of other nonmammalian amniotes, it becomes clear that the development of the enamel surface is more responsible for the enamel microstructure in prismless enamel than are functional or phylogenetic constraints. Many reptile teeth have an adaptive surface morphology with wrinkles, ridges, and keels. Surprisingly, these are only expressed in the enamel and not in the underlying dentine. Exceptions to this are the serrations of archosaur teeth, including those of phytosaurs, dinosaurs, and crocodiles. Because enamel grows by apposition, the development of the enamel surface morphology can be traced through growth lines. If the developmental program calls for an enamel surface with wrinkles or ridges, the internal structure of the enamel may be a mere by-product. This makes sense in view of the fact that reptilian teeth, including dinosaur teeth, are replaced frequently. A lesser premium than in mammals is placed on the protection of the individual tooth, but each successive tooth must look the same. The only case in which functional constraints on dinosaurian enamel microstructure were probably present is the wavy enamel of derived ornithopods.

It must be emphasized, however, that the study of dinosaur enamel microstructure is still very much in its infancy because few taxa have been studied and the functional significance of particular enamel types is unclear.

Adaptations and Their Systematic Occurrence

This review of dinosaur jaw and tooth morphology and function may have made it apparent that certain evolutionary trends and adaptations can be observed repeatedly. Most important is the evolution of herbivory, which happened at least twice in dinosaurs (in the Sauropodomorpha and in the Ornithischia). In fact, all dinosaurs but the Theropoda were herbivorous.

Herbivory leads to dental specialization not only in dinosaurs but also in all other groups of tetrapods. Whereas sauropodomorphs had relatively simple teeth only suited for cropping plant matter, ornithischians early on evolved a beak for this task. The dentition in conjunction with fleshy cheeks then served to process the food in the oral cavity. The degree of oral processing varied considerably. Two advanced groups, the hadrosaurs and the neoceratopsians, independently evolved a very high level of oral processing through secondary occlusal surfaces. Hadrosaurs as well as neoceratopsians thus show great functional convergence to each other as well as to several groups of mammalian herbivores.

However, the dinosaurian solution to the problem of high rates of tooth wear caused by oral processing is fascinatingly different from that of mammals. Mammals compensate for heavy wear by ever-growing cheek teeth, which evolved heterochronically through the loss of the roots. In hadrosaurs and neoceratopsians, on the other hand, the same demand is served by greatly increased tooth replacement rates and a dense packing of the teeth into tooth batteries. Dinosaur tooth and jaw morphology thus are extremely varied but can easily be understood as a combination of phylogenetic heritage and dietary specialization.

See also the following related entries:

Diet • Skeletal Structures • Skull, Comparative Anatomy • Tooth Marks • Tooth Replacement • Tooth Serrations • Tooth Wear • Von Ebner Growth Lines

References

Abler, W. L. (1992). The serrated teeth of tyrannosaurid dinosaurs, and biting structures in other animals. *Paleobiology* **18,** 161–183.

Currie, P. J., Rigby, J. K., and Sloan, R. E. (1990). Theropod teeth from the Judith River Formation of southern Alberta, Canada. In *Dinosaur Systematics—Approaches and Perspectives* (K. Carpenter and P. J. Currie, Eds.), pp. 107–125. Cambridge Univ. Press, Cambridge, UK.

Farlow, J. O., Brinkman, D. L., Abler, W. L., and Currie, P. J. (1991). Size, shape, and serration density of theropod dinosaur lateral teeth. *Mod. Geol.* **16,** 161–198.

Koenigswald, W. v., and Sander, P. M. (Eds.) (1997). *Tooth Enamel Microstructure*, pp. 280. Balkema, Rotterdam.

Norman, D. B., and Weishampel, D. B. (1985). Ornithopod feeding mechanisms: Their bearing on the evolution of herbivory. *Am. Nat.* **126,** 151–164.

Peyer, B. (1968). *Comparative Odontology*, pp. 347. Chicago Univ. Press, Chicago.

Sander, P. M. (1997). Prismless enamel in amniotes: Terminology, function, and evolution. In *Development, Function and Evolution of Teeth* (M. F. Teaford, M. W. J. Ferguson, and M. M. Smith, Eds.). Cambridge Univ. Press, New York.

Weishampel, D. B. (1984). The evolution of jaw mechanics in ornithopod dinosaurs (Reptilia: Ornithischia). *Adv. Anat. Embryol. Cell Biol.* **87,** 1–110.

Weishampel, D. B., Dodson, P., and Osmólska, H. (Eds.) (1990). *The Dinosauria*, pp. 733. Univ. of California Press, Berkeley.

Temeszettudomanyi Muzeum, Hungary

see Museums and Displays

Tendaguru

Gerhard Maier
Calgary, Alberta, Canada

Tendaguru, Africa's remarkable Jurassic fossil site, is located approximately 75 km (40 miles) northwest of Lindi, Tanzania. Engineer W. Bernhard Sattler recognized the significance of enormous bones he encountered when prospecting for commercial garnet deposits in what was then a colony of Germany. His reports reached government officials in Europe, who alerted paleontologist Eberhard Fraas of Stuttgart. Fraas spent a dysentery-wracked week in 1907 confirming the presence of a rich accumulation of dinosaur remains. Unable to investigate further, Fraas contacted an old friend, Wilhelm von Branca. Branca, director of Berlin's Natural History Museum, rapidly mobilized the resources of his museum to raise private funding, assemble equipment, and select expedition leaders.

In 1909 Werner Janensch and Edwin Hennig left Lindi and marched inland for 4 days. They hired approximately 180 locals to clear brush, establish a camp, and remove tons of overburden. Boheti bin Amrani was the expedition's indispensable overseer of the African workforce. The goals of the expedition were to collect dinosaur skeletons and determine the relative age and stratigraphic sequence of the deposits. Monsoon rains interrupted work, forcing the two Berlin scientists north in January 1910. They traveled along the Central Railway, gathering invertebrate fossils and describing the local geology. The second field season commenced in April 1910. A village soon formed around the base of Tendaguru Hill as the crew of 420 Africans were joined by their families. Plaster and clay-encased bones were regularly carried to the coast on the heads and backs of porters. In Lindi they were crated and loaded onto Arab sailboats. Transferred to steamships, the bone harvest reached Berlin weeks later where it soon filled every available space in the museum. Again the Germans headed north to avoid the rains, resuming their efforts at Tendaguru in 1911 with a staff of 480. They were joined by Hans von Staff, a geomorphologist from Berlin. Additional sites were opened and eventually more than 100 localities were investigated within an area covering 30 square km. Finds ranged from isolated bones to articulated skeletons and extensive bone beds. Returning to Berlin after a 2 1/2-year effort, Janensch and Hennig concentrated on overseeing the preparation activities and publishing their results. Funds were made available for a final season in 1912, which was led by Berlin vulcanologist Hans Reck. Reck and his wife Ina supervised 150 laborers.

In early 1913 the German Tendaguru Expedition struck camp for the last time. Approximately 225,000 kg of fossils had been manhandled to Lindi in more than 5000 bearer loads. More than 850 wooden crates had been shipped to Berlin. The colony was engulfed in a protracted guerrilla campaign during World War I, then mandated to Great Britain. Aware of the potential of Tendaguru, the British Museum of Natural History launched the British Museum East African Expedition in 1924. An experienced Canadian fossil collector, William Cutler, was placed in command. His assistant was Louis Leakey, later to gain renown as a paleoanthropologist. One hundred men reopened the quarries that year. Leakey returned to his studies at Cambridge and Cutler carried on until his tragic death from malaria in the summer of 1925. It was difficult to obtain financial backing in postwar Britain, but by 1925 F. W. H. Migeod, an African traveler, was appointed as Cutler's replacement. Forty locals excavated from the wet season of 1925 through to the end of 1926. A substantial haul was packed out, but neither of the two Europeans at the site had any experience collecting fossils. Migeod was succeeded by geologist John Parkinson for the 1927–1928 season.

Parkinson carefully examined the strata and found himself in disagreement with several of the assertions made by the Germans. Migeod returned from England in 1929. The final season in 1930 included a trip to neighboring Malawi. Work at Tendaguru ceased in January 1931. Several hundred crates had been transported to London, including the partial skeleton of a large sauropod dinosaur. The British expeditions had excavated a substantial number of bones, but little was ever described scientifically or placed on public display. Funding had remained modest throughout the years. German efforts, seriously hampered by postwar hyperinflation, produced impressive results. Geologically the Tendaguru Beds were named and their age and stratigraphy clarified. These Upper Jurassic beds consisted of three horizons of terrestrial marls alternating with marine sandstones. It is believed that rivers periodically carried sediments and fossils into a lagoon or estuary, where they were subsequently buried under marine deposits. In addition, the geomorphology and paleontology of the southern coast and the north-central strata was studied and described. Extant insects and mammals had been collected and even prehistoric lithic artifacts were found and described.

A multidisciplinary group examined thousands of invertebrate, paleobotanical, and geological specimens. Dozens of invertebrate genera and species were established: cephalopods, corals, bivalves, brachiopods, and gastropods. A diverse vertebrate fauna was described: five dinosaur families, crocodiles, bony fish, sharks, pterosaurs, and a mammal. Skeletons were prepared, studied, and mounted: the stegosaur *Kentrosaurus* (1924), the theropod *Elaphrosaurus* (1936), the sauropod *Dicraeosaurus* (1931), the magnificent sauropod *Brachiosaurus* (1937), and finally the hypsilophodont *Dryosaurus* (1961). The terrestrial fauna was dominated by sauropods (*Brachiosaurus, Barosaurus, Dicraeosaurus, Tornieria,* and *Janenschia*) but also included large numbers of stegosaurs and ornithopods, as well as carnosaurs.

The dinosaurian fauna is broadly comparable to that of another classic region, the Morrison Formation of the Western United States. The original expedition members continued publishing until the early 1960s and even today specimens are being reexamined. Though Tendaguru had been deserted since the 1930s, the locality has been revisited. In 1977 and 1978, Dr. Dale Russell of the Canadian Museum of Nature attempted to reopen the site. Dr. Christa Werner of the Technical University of Berlin returned to assess current conditions in 1994. Both groups were accompanied by Tanzanian scientists.

See also the following related entries:

African Dinosaurs • History of Dinosaur Discoveries • Jurassic Period • Morrison Formation

References

Hennig, E. (1912). *Am Tendaguru, Leben und Wirken einer deutschen Forschungsexpedition.*

Parkinson, J. (1931). *The Dinosaur in East Africa.*

Tetanurae

John R. Hutchinson
Kevin Padian
University of California
Berkeley, California, USA

The clades Tetanurae and Ceratosauria were established by Gauthier (1986) as two stem-based sister taxa comprising Theropoda. Tetanurae was defined in a stem-based sense to include birds and all other theropods more closely related to birds than to Ceratosauria. These were listed as Carnosauria, *Compsognathus*, *Ornitholestes*, *Coelurus*, *Microvenator*, *Saurornitholestes*, *Hulsanpes*, Elmisauridae, Caenagnathidae, Ornithomimidae, Deinonychosauria, and Avialae. All these taxa except Carnosauria are members of the Coelurosauria, and together Carnosauria and Coelurosauria are the stem-defined sister taxa comprising Avetheropoda (Holtz, 1994, 1996). Some taxa, including the Torvosauridae, Spinosauridae, *Afrovenator*, and poorly understood "megalosaurs" such as *Piatnitzkysaurus*, *Eustreptospondylus*, *Poikilopleuron*, and *Megalosaurus*, appear to fall outside the Avetheropoda but are still basal members of the Tetanurae (Holtz, 1994, 1995). The term "megalosaur," which dates back to William Buckland's 1824 description of *Megalosaurus*, is widely considered an unnatural assemblage of incompletely known or poorly preserved medium-sized to large theropods of the Early to Late Jurassic and Early Cretaceous (see Megalosaurus). New work on these enigmatic early tetanurans is in progress, focusing on the removal of indeterminate remains and taxa that clearly belong to other clades and on the definition and diagnosis of a monophyletic Megalosauridae, if this is indeed possible.

Tetanurae is diagnosed by numerous synapomorphies, including increased pneumaticity in the skull, a manus reduced to only three digits, a reduced fibula that is clasped by the tibia, an anteriorly placed horizontal groove on the astragalar condyles, an upper (maxillary) tooth row restricted to anterior to the eye socket, and an obturator notch on the ischium (Gauthier, 1986; Holtz, 1994; Sereno *et al.*, 1994). The name Tetanurae ("stiff tails") was devised by Gauthier (1986) in reference to the transition point in the theropod tail that seems to mark the posteriormost extent of the caudofemoralis longus muscle and evidences a shift anteriorly during theropod evolution (especially within the Tetanurae), correlated with a putative reduction in the flexibility of the posterior region of the tail (Gatesy, 1990). This trend toward the reduced importance of the tail as part of the theropod locomotory module has considerable implications for the evolution of theropod locomotion and the origin of flight in Aves (Gatesy and Dial, 1996; see Functional Morphology).

Recent discoveries of basal tetanurine theropods have shown some surprising diversity, especially Middle Jurassic to Early Cretaceous forms such as *Sinraptor*, *Afrovenator*, *Baryonyx*, and *Monolophosaurus*. These discoveries have also shed light on the early history of the Tetanurae because many of these taxa appear to represent intermediate evolutionary side branches between the Ceratosauria and Avetheropoda lineages (Sereno *et al.*, 1994, 1996). The Spinosauridae, currently known by *Spinosaurus* and related material from North Africa and *Baryonyx* from England plus *Angaturama* and *Irritator* from Brazil (Kellner and Campos, 1996), seems to be one such side branch that persisted long enough to become fairly cosmopolitan in distribution (Milner, 1996). The Torvosauridae, represented by *Torvosaurus*, is another side branch known from the Late Jurassic Dry Mesa Quarry in Utah. According to Sereno *et al.* (1994, 1996), along with the recently described Early Cretaceous *Afrovenator* from the Sahara Desert, the torvosaurids and spinosaurids form the node-based taxon Torvosauroidea, a group of nonneotetanurine theropods. However, resolution of this region of the theropod family tree is complicated by the fragmentary condition or poorly known status of some of the apparently basal taxa such as the megalosaurs.

See also the following related entries:

Carnosauria • Ceratosauria • Coelurosauria • Theropoda

References

Gatesy, S. M. 1990. Caudofemoral musculature and the evolution of theropod locomotion. *Paleobiology* **16**.

Gatesy, S. M., and Dial, K. P. (1996). Locomotory modules and the evolution of flight. *Evolution* **50,** 331–340.

Gauthier, J. A. (1986). Saurischian monophyly and the origin of birds. *Mem. California Acad. Sci.* **8,** 1–55.

Holtz, T. R., Jr. (1994). The phylogenetic position of the Tyrannosauridae: Implications for theropod systematics. *J. Paleontol.* **68,** 1100–1117.

Holtz, T. R., Jr. (1995). A new phylogeny of the Theropoda. *J. Verebr. Paleontol.* **15**(Suppl. to No. 3), 35A.

Holtz, T. R., Jr. (1996). Phylogenetic taxonomy of the Coelurosauria (Dinosauria: Theropoda). *J. Paleontol.* **70,** 536–538.

Kellner, A. W., and Campos, D. de A. (1996). First Early Cretaceous theropod dinosaur from Brazil with comments on Spinosauridae. *Neues Jahr. Geol. Paläontol. Abhandlungen* **199,** 151–166.

Milner, A. C. (1996). Morphology, relationships and ecology of spinosaurs, aberrant long-snouted Cretaceous theropods. *J. Vertebr. Paleontol.* **16**(Suppl. to No. 3), 53A.

Sereno, P. C., Wilson, J. A., Larsson, H. C. E., Dutheil, D. B., and Sues, H.-D. (1994). Early Cretaceous dinosaurs from the Sahara. *Science* **266,** 267–271.

Sereno, P. C., Duthiel, D. B., Iarochene, M., Larsson, H. C. E., Lyon, G. H., Magwene, P. W., Sidor, C. A., Varricchio, D. J., and Wilson, J. A. (1996). Predatory dinosaurs from the Sahara and Late Cretaceous faunal differentiation. *Science* **272,** 986–991.

Thailand Dinosaurs

see SOUTHEAST ASIAN DINOSAURS

Thecodontia

The etymology of *thecodont* and *thecodontian* means "socket-tooth" and refers to the setting of the teeth in deep alveoli or "sockets," in which they are replaced by the next generation of tooth through the resorption of the root and the shedding of the crown; a new crown replaces the old by growing in beneath and often slightly behind the former. This feature is shared by all ARCHOSAURIA primitively and their outgroups among Archosauromorpha, including Proterosuchia, Erythrosuchia, Proterochampsia, and other taxa. Lizards and snakes have a pleurodont dentition in which the teeth are attached only to the inner wall of the jawbone, and sphenodontids have an acrodont dentition in which the teeth sit on top of the jawbone.

The taxonomic concept of Thecodontia has a long and, as usual with categories of this sort, tortured history that was reviewed concisely by Charig (1976), obviating most discussion here (see also PSEUDOSUCHIA). Richard Owen established the group over a period of years, beginning in 1841, to recognize reptiles with their teeth set in sockets. This originally included *Palaeosaurus, Thecodontosaurus,* and *Cladeiodon,* all (as it eventually turned out) dinosaurs or questionable dinosaurs, but certainly what Cope (1869) later called Archosauria. Owen (1842, 1859) added some forms that were not dinosaurs, such as the phytosaur *Belodon,* the archosauromorph protorosaur *Protorosaurus,* and the synapsid pelycosaur *Bathygnathus.* Thecodontia was soon clearly perceived as an unnatural assemblage. The taxon was variously used or ignored by workers until about the beginning of the 20th century; in use, it was generally regarded as a subgroup of Archosauria. In most synoptic classifications, the Order Archosauria was divided into the suborders Thecodontia, CROCODYLIA, PTEROSAURIA, ORNITHISCHIA, and SAURISCHIA, and Thecodontia was considered the "stem group" from which the others (and, usually, the birds) arose (see BIRD ORIGINS). The analysis of its paraphyly and the separation of "thecodontians" [often confusingly called "thecodonts," as Charig (1976) points out] into monophyletic groups either closer to crocodiles or to birds among Archosauria are detailed separately (see ARCHOSAURIA; PSEUDOSUCHIA; ORNITHOSUCHIA; PHYLOGENETIC SYSTEM; SYSTEMATICS). "Thecodontia" is now regarded as a paraphyletic, wastebasket term and is not used by phylogenetic systematists because its content and meaning are synonymous with those of Archosauria. The term "thecodont" still refers to the dental configuration of socketed teeth.

References

Charig, A. J. (1976). Thecodontia. In *Handbuch der Paläoherpetologie, Teil 13* (O. Kuhn, Ed.), pp. 7–10. Fischer-Verlag, Stuttgart.

Cope, E. D. (1869). Synopsis of the extinct Batrachia, Reptilia and Aves of North America. *Trans. Am. Philos. Soc. N.S.* **14,** 1–252.

Owen, R. (1841). *Odontography.* Balliere, London.

Owen, R. (1842). Report on British fossil reptiles, Part 2. *Rep. Br. Assoc. Adv. Sci.* **11**(1841), 60–204.

Owen, R. (1859). Palaeontology. In *Encyclopedia Britannica,* 8th ed., Vol. 17, pp. 91–176. Black, Edinburgh.

Therizinosauria

Dale A. Russell
North Carolina State University
Raleigh, North Carolina, USA

Until recently, little was known about therizinosaurs (segnosaurs). They were regarded as a strange group of Asian dinosaurs of which the appearance was essentially unknown, the relationships were dubious in the extreme, and the ecology and evolution was a topic of nearly unconstrained speculation. Until 1990, only seven skeletal fragments had been deascribed, scattered across more than 30 million years of the Cretaceous record of central Asia. A more complete understanding of therizinosaurs came with the recovery in 1988 of associated skeletal parts preserved in lake sediments of Middle Cretaceous age in the Alashan (Alxa) Desert of northern China.

Prior to this discovery, huge isolated claws of therizinosaurs had been referred to an otherwise unknown giant turtle, an isolated forelimb was inferred to represent unknown theropods with very long arms, and cranial and pelvic elements were considered to belong to late surviving descendants of early Mesozoic prosauropods. However, the Chinese materials demonstrated that therizinosaurs were clearly aberrant theropods, and that isolated skeletal parts from Cretaceous strata in China and Mongolia actually belonged to a single group of dinosaurs. They were apparently restricted to Asia from Middle (Albian, 112–97 Ma) through Late Cretaceous (Maastrichtian, 74–65 Ma) time, although fragmentary North American records await confirmation. Skeletal restorations are available for *Alxasaurus* and *Therizinosaurus*.

Details of the bones of the wrist and the structure of the large inner toe indicate that therizinosaurs were TETANURANS or "stiff-tailed" THEROPODS. Additional skeletal attributes suggest that, within tetanurans, therizinosaurs may have been rather closely related to OVIRAPTOROSAURS, ORNITHOMIMIDS, and TROODONTIDS (Russell and Dong, 1993; Clark *et al.*, 1994).

Therizinosaurs include six described genera (*Alxasaurus, Enigmosaurus, Erlikosaurus, Nashiungosaurus, Segnosaurus,* and *Therizinosaurus*) of bipedal thero-

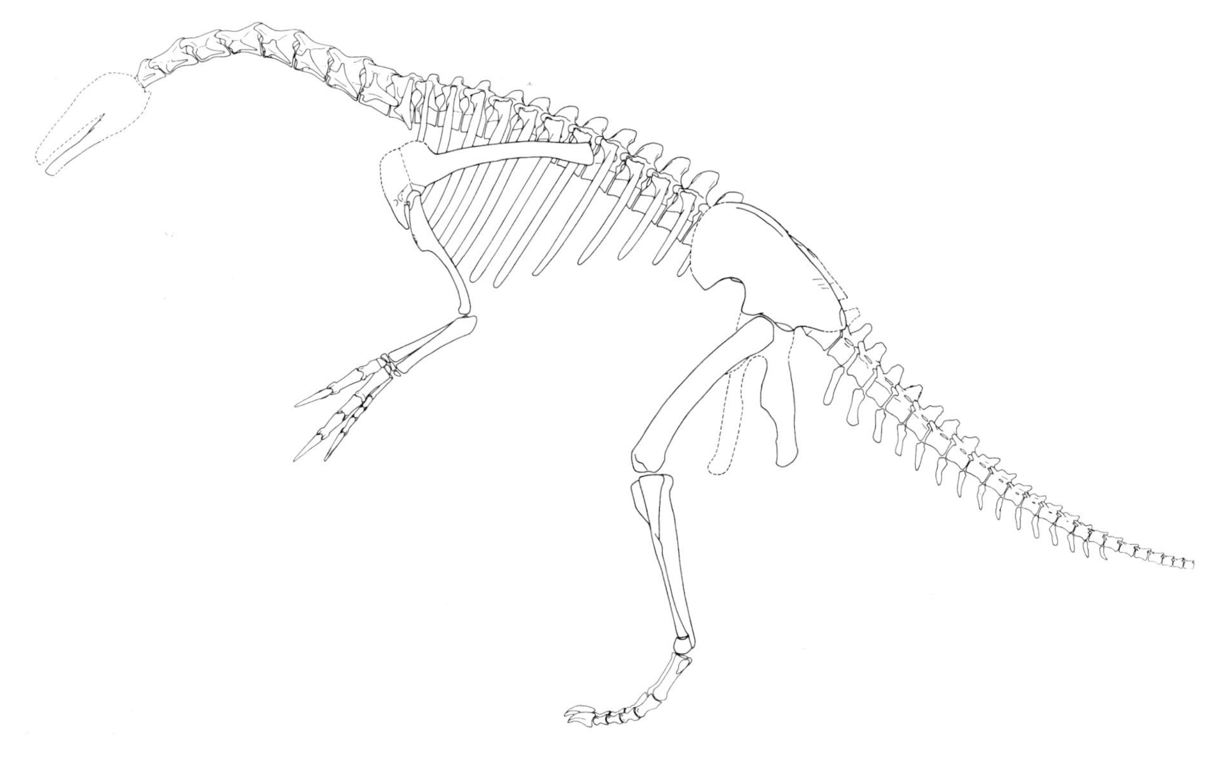

FIGURE 1 *Alxasaurus* is a therizinosaurian dinosaur unearthed from lacustrine rocks of Albian age. This tetanuran was found in the Alxa Desert of Inner Mongolia.

pods with unusually small heads, long necks, and abbreviated tails (Fig. 1). Therezinosaurs may be defined as the aforementioned taxa and all others closer to them than to oviraptorosaurs, ornithomimids, and troodontids. They are diagnosed by a number of unusual features. The teeth are small, and the anterior part of the muzzle was modified into a bill-like structure in which teeth were entirely lacking. Both the BRAINCASE and vertebrae were penetrated by air-filled cavities. The FORELIMBS were elongated and ended in large claws that were either recurved or nearly straight in various genera. The upper part of the PELVIS flared broadly toward the front, and its lower elements were inclined backwards. The foot was short and broad. Therizinosaurs are estimated to have ranged from nearly 4 to 8.5 m in length and approximately 160 kg to nearly 2 metric tons in weight.

Therizinosaurs were atypical theropods in that the SKULLS were small, weakly armed with teeth, and carried at the end of a long, lightly constructed neck. Head/body proportions differed greatly from those in CARNOSAURS, in which the skull was large and powerfully built, and the teeth were well-suited for penetrating large prey. The neck is reminiscent of that of sauropods, and in the larger forms the body, though bipedal, is ponderously constructed. It seems likely that therizinosaurs constituted a second saurischian experiment in herbivory (after that of sauropodomorphs). Indeed, the general form of their skeletons converges remarkably on that of PROSAUROPODS. Their remains are typically found in sediments deposited in river or lake environments where the animals had presumably been browsing among riparian bushes and trees.

It has been suggested (Russell and Russell, 1993) that therizinosaurs "sat" upon their pelvis, using their long arms and flexible necks to bring fronds within reach of their snouts. They thus may have paralleled certain herbivorous mammals in using their forelimbs to manipulate vegetation. Their general similarity to chalicotheres ("clawed horses") is remarkable, and similarities can also be seen in gorillas and extinct ground sloths. These dinosaurs and mammals are evidence of adaptive convergence in dinosaurian and mammalian radiations, although imperfectly expressed because of their widely differing ancestors at different times in earth history.

See also the following related entries:

MANIRAPTORA • OVIRAPTOROSAURIA

References

Clark, J. M., Perle, A., and Norell, M. A. (1994). The skull of *Erlicosaurus andrewsi*, a Late Cretaceous "segnosaur" (Theropoda: Therizinosauridae) from Mongolia. *Am. Museum Novitates*, in press.

Russell, D. A., and Dong, Z. M. (1993). The affinities of a new theropod from the Alxa Desert, Inner Mongolia, People's Republic of China. *Can. J. Earth Sci.* **30,** 2107–2127.

Russell, D. A., and Russell, D. E.. (1993). Mammal–dinosaur convergence. *Natl. Geogr. Res. Exploration* **9,** 70–79.

Thermopolis Dinosaur Museum, Wyoming, USA

see MUSEUMS AND DISPLAYS

Thermoregulation

see PHYSIOLOGY

Theropoda

Philip J. Currie
Royal Tyrrell Museum of Palaeontology
Drumheller, Alberta, Canada

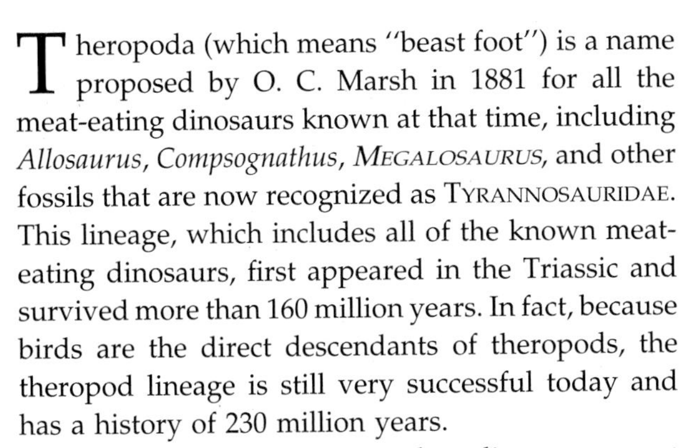

Theropoda (which means "beast foot") is a name proposed by O. C. Marsh in 1881 for all the meat-eating dinosaurs known at that time, including *Allosaurus*, *Compsognathus*, MEGALOSAURUS, and other fossils that are now recognized as TYRANNOSAURIDAE. This lineage, which includes all of the known meat-eating dinosaurs, first appeared in the Triassic and survived more than 160 million years. In fact, because birds are the direct descendants of theropods, the theropod lineage is still very successful today and has a history of 230 million years.

Theropods tend to be rare but diverse at most localities. Two exceptions are the GHOST RANCH *Coelophysis* bone bed and the CLEVELAND–LLOYD QUARRY (which is dominated by *Allosaurus*). Theropods usually make up less than 20% of the fossils recovered at any site. Although rare in numbers, theropods are very diverse. About 40% of approximately 300 genera of dinosaurs currently recognized as valid are theropods, which also comprise approximately 50% of the recognized families of dinosaurs. This compares well with mammals, in which approximately half of the recognized families are carnivores, insectivores, piscivores, and/or omnivores. Because of their relative rarity but high diversity, theropod interrelationships are not as well understood as those of most taxa of herbivorous dinosaurs, and theropod taxonomy is more volatile and susceptible to change.

The clades Theropoda and Sauropodomorpha form two stem-based sister taxa that make up the Saurischia. Theropoda can be defined in a stem-based sense to include birds and all other theropods more closely related to birds than to Sauropodomorpha, such as *Plateosaurus* and *Diplodocus*.

Usually theropods were slender, long-legged, bipedal animals that were capable of moving faster than contemporary herbivorous dinosaurs (Fig. 1). Most have blade-like teeth with serrated ridges (see TOOTH SERRATIONS). The claws, especially on the hand, are usually recurved and end in sharp points. Unlike most ornithischian dinosaurs, theropods always have hollow limb bones. They also show a tendency toward having pneumatic bones (cranial and axial) in the front part of the body (see CRANIOFACIAL AIR SINUS SYSTEMS; POSTCRANIAL PNEUMATICITY). In some theropods (OVIRAPTOROSAURIA), pneumatic bones are found as far back as the middle of the tail, and in birds most of the bones of the limbs and limb girdles can also be air-filled. None of these characters are unique to theropods, even though they are significant aspects of theropod appearance. Even the elaborate pneumatic systems shared by nonavian theropods and birds are similar to those of pterosaurs and sauropods, suggesting pneumaticity might be a primitive character that was secondarily lost in ornithischians and prosauropods. The absence of such pneumatism in *Lagosuchus* and other dinosaurian out-groups suggests that it developed convergently, however.

Unique characters that define Theropoda are found in the skull and postcranial skeleton. The lacrimal extends onto the top of the skull, there is a well-developed intramandibular joint, prominent processes (epipophyses) are found above the postzygopophyses of cervical vertebrae, the prezygapophyses of caudal vertebrae are elongate, there are at least five sacral vertebrae and a transition point in the tail, the scapular blade is strap-like, the humerus is less than half the length of the femur, the manus is elongate although the outer two fingers are reduced or lost, there are distinctive ligament attachment pits on the distal extensor surfaces of the metacarpals, the penultimate phalanges of the manus are elongate, the ilium has an enlarged preacetabular region, the distal end of the pubis is expanded, there is a shelf-like ridge of bone near the head of the femur for the attachment of muscles, and the fibula is strap-like and attached to a crest on the side of the tibia.

Specimens discovered in Upper Triassic (Carnian) rocks of Argentina include some of the oldest and most primitive dinosaurs that are currently known. One of these animals, *Herrerasaurus*, was for a long time considered too primitive to be classified as either

FIGURE 1 In some phylogenetic hypotheses, *Eoraptor* is thought to be near the root of the theropod clade (illustration by M. Skrepnick). See the endpapers of this book for a hypothesis of dinosaur phylogeny.

saurischian or ornithischian. However, the recovery of better specimens in recent years suggests that this animal has a moveable joint in the middle of the lower jaw (Sereno and Novas, 1993) like all but a few theropods that have secondarily fused this articulation.

Members of the HERRERASAURIDAE are also considered as basal theropods by some authors because of the elongate prezygapophyses in the distal part of the tail. However, because herrerasaurids have only two sacral vertebrae, short necks, a manus in which the third finger is longest, and a well-developed metatarsal V, some researchers consider them either as outside DINOSAURIA or as basal saurischians (Gauthier, 1986; Holtz and Padian, 1995).

Another 1-m-long animal found in the same rocks as *Herrerasaurus* is anatomically closer to what would be considered the ancestral morphotype of both saurischian and ornithischian dinosaurs. *Eoraptor* lacks the intramandibular joint and probably is not a theropod.

Theropoda can be divided into two basal stem groups. CERATOSAURIA are known from Upper Triassic to Upper Jurassic rocks and may include the Cretaceous ABELISAURIDAE of South America and Africa. Characteristic forms are *Coelophysis* and *Syntarsus*. *Dilophosaurus* was a related genus that had attained a relatively large size by Early Jurassic times. These three animals have been united with *Ceratosaurus* of the Late Jurassic into the Ceratosauria, although the union is based on relatively few derived features. The second stem taxon, TETANURAE, includes two major branches known as the CARNOSAURIA and COELUROSAURIA. Distinctive tetanuran characters include the presence of a large opening (maxillary fenestra) in front of the antorbital fenestra, the presence of a fur-

FIGURE 2 *Ornithomimus* have been excavated from Maastrichtian and Campanian age sediments of the Denver, Kaiparowits, Judith River, and Horseshoe Canyon Formations of Alberta, Colorado, and Utah (illustration by D. Braginetz).

cula, and extensive contact between the first and second metacarpals. Other characters that were used to set up the Tetanurae, such as the loss of the fourth finger and the presence of an obturator process on the ischium, are not as universal as they were once thought to have been (Currie and Zhao, 1993).

The carnosaurs were once united primarily by large size, but recently it has become widely accepted that some of the largest theropods (tyrannosaurids) are in fact overgrown coelurosaurs as proposed more than 70 years ago (von Huene, 1926), whereas large Abelisaurids and *Ceratosaurus* seem to represent more primitive taxa. Even though many genera have been removed from the Carnosauria, the clade still includes well-preserved Jurassic forms such as *Allosaurus* and the Sinraptoridae (Currie and Zhao, 1993). Currently, it is not clear whether the Late Cretaceous Carcharodontosaurids are late-surviving carnosaurs (Sereno *et al.*, 1996) or abelisaurids.

Coelurosauria is made up of all taxa closer to birds than to *Allosaurus* and includes a wide diversity of mostly Cretaceous forms, including DROMAEOSAURIDAE, ELMISAURIDAE, ORNITHOMIMOSAURIA, OVIRAPTOROSAURIA, THERIZINOSAURIA, TROODONTIDAE, TYRANNOSAURIDAE, and AVES. Elmisauridae may be a junior synonym of Caenagnathidae, and segnosaurs are included in the Therizinosauria. Gauthier (1986) redefined the Coelurosauria to include ornithomimids and a clade he referred to as the MANIRAPTORA. The latter has been redefined by Holtz (1996) to include only all theropods closer to birds than to ornithomimids. He went on to establish the taxon Maniraptoriformes, defined as the most recent common ancestor of *Ornithomimus* and birds and all descendants of that common ancestor (Fig. 2). Maniraptoriformes are characterized by a pulley-like wrist joint that allowed the hand to be folded back against the body. This characteristic was secondarily reduced or lost in ornithomimids and tyrannosaurids. Recently discovered compsognathids (see FEATHERED DINOSAURS) have well-preserved hands that suggest that this character was more widespread among coelurosaurs. Although they dominated northern continents, new evidence suggests that nonavian coelurosaurs had spread to the southern continents by the end of the era (Novas, 1996). Birds, the most successful theropod taxon, first appeared in the Late Jurassic and had attained a worldwide distribution before the nonavian theropods disappeared at the end of the Cretaceous.

There has been considerable speculation about which families of theropods are most closely related to birds (see BIRD ORIGINS). The most bird-like theropods are mostly Late Cretaceous forms, although this probably reflects preservational biases because small theropods are very poorly known in Jurassic rocks. Most Cretaceous theropod families clearly had long independent histories before they appear in the fossil record. Avimimids, oviraptorids, and ornithomimids were the most bird-like in appearance because of the convergent evolution of toothless, beaked skulls. Dromaeosaurids are currently considered to be the most likely avian ancestral stock, but troodontids show many bird-like characters as well. Both families are represented by well-preserved skeletons from Lower Cretaceous rocks, and Late Jurassic dromaeosaurid and troodontid teeth have also been reported.

Within the Theropoda, certain evolutionary trends can be observed over the course of their Mesozoic history. In the skull, pneumatization of the snout becomes more pronounced in advanced theropods, and accessory antorbital openings appear in front of the antorbital fenestra. By the Cretaceous, most theropods seem to have had a degree of stereoscopic vision with the eyes facing more forward than they had been in earlier forms. The brains had become relatively larger (see INTELLIGENCE; PALEONEUROLOGY), especially in the smaller theropods. The epipophyses in cervical vertebrae tended to be lost in more advanced theropod lineages as the neck took on a stronger curvature and the vertebrae became regionally more distinct from each other. The fingers tended to become more elongate in coelurosaurs, and there is a strong tendency for further reduction of the number of fingers. Many of the Jurassic theropods had three functional fingers (Fig. 3) with only a vestige of the fourth, and the fourth is completely lost in allosaurids. Tyrannosaurids independently reduced the number of fingers to two, and *Mononykus* retained only the first finger. In the hips, the distal expansion of the pubis became an enlarged boot in many of the theropod families, and the obturator foramen of the ischium opens up into a notch proximal to the obturator process. The limbs of theropods became progressively

FIGURE 3 Carnosaurs like *Monolophosaurus* may have been agile predators or lumbering scavengers (illustration by M. Skrepnick).

longer over time, especially the tibia, fibula, and the metatarsals. The lengthening of leg elements and changes in proportions reflect faster running abilities. As theropods became faster, they needed more control and better shock absorption in their feet. The ascending process of the astragalus became much higher, extending as much as 25% of the length of the tibia. The lower end of the fibula and the calcaneum had a tendency to become smaller, and the latter element was lost as a distinct element in some families. Theropods had three toes (second to fourth) that functioned in support, although therizinosaurs had secondarily enlarged the first metatarsal to support an enlarged first digit, whereas dromaeosaurids and troodontids became functionally didactylous. The upper part of the third metatarsal became narrow and splint-like in most of the Late Cretaceous theropods, and metatarsals II and IV contacted each other at the top (Holtz, 1994). The lower end of the third metatarsal would have contacted the ground first when a theropod was running. Through a system of bony contacts and elastic ligaments, the shock of impact was transferred to, and dissipated through, the adjacent metatarsals, thereby reducing the chance of foot injuries (Wilson and Currie, 1985). In elmisaurids, avimimids, and birds, the heads of the three main metatarsals fused with the distal tarsals to form a unified tarsometatarsal bone.

Theropods had a universal distribution, and specimens have been found from the North Slope of Alaska to Antarctica. Nevertheless, the ecological preferences of theropods are poorly understood, mostly because of the rarity of specimens in the vast majority of the species.

Wherever herbivorous dinosaurs are found, theropods are also recovered. The environmental extremes were the coastal lowlands of Late Cretaceous western North America (see Judith River Wedge) and the deserts of central Asia. The Djadokhta Formation (Upper Cretaceous) of Mongolia and equivalent beds in China represent dry, stressful environments that were suitable for only a few species of dinosaurs. Although hundreds of skeletons have been collected at these sites, the dinosaurian diversity is low, and none of the animals was more than 4 m long. Other than a few isolated teeth and bones that appear to have been washed in from other environments or reworked from older sediments, tyrannosaurids are not found in the Djadokhta. However, small theropods, including dromaeosaurids (Fig. 4), oviraptorids, and troodontids, seem to have been relatively common in these stressed environments (Dashzeveg *et al.*, 1995).

Theropod genera were less restrained than ornithischian genera by climate and vegetation and seem to have had wider geographic ranges, as in modern Carnivora (Farlow, 1993). Dinosaur Provincial Park and Devil's Coulee represent two synchronous but distinct habitats separated by approximately 300 km. There was a more consistent source of water at Dinosaur Park 75 million years ago that produced a well-watered, well-vegetated ecosystem. Devil's Coulee was farther from the coastline and was undoubtedly drier on at least a seasonal basis. Reflecting differences in the food resources, the composition of herbivores was different, with the hadrosaurs *Corythosaurus* and *Lambeosaurus* dominating in Dinosaur Provincial Park and *Hypacrosaurus* in Devil's Coulee. Ankylosaurs also seem to have been rarer in Dinosaur

FIGURE 4 *Utahraptor* possessed, like all dromaeosaurs, a large hyperextended trenchant ungual. Such a structure must have rendered dromaeosaurs especially threatening to prey (illustration by D. Braginetz).

Provincial Park. However, evidence suggests that the carnivores of the two sites are the same taxa—*Gorgosaurus libratus*, *Saurornitholestes langstoni*, and *Troodon formosus* being the most common ones.

Isolated small theropod bones and teeth are more common than those of large theropods in most Cretaceous localities, but smaller species tend to be rarer as articulated skeletons. It is difficult to determine how large the populations of small theropods might have been in relation to prey species because nondinosaurian animals would have comprised a large part of their diets. Ornithomimids may also have been omnivorous and are indeed very common at some sites. Immature large theropods probably competed with small species of theropods, which could explain the virtual absence of small theropods at most known dinosaur localities of the Jurassic. However, it is also possible that in some large theropod species, the juveniles associated with mature animals in packs or family groups, and therefore would not have been competing directly with the small species of theropods. There is some evidence to suggest that tyrannosaurids maintained family groups, so it is not surprising that the greatest diversity of small theropods is found in paleoenvironments where tyrannosaurids were the large predators.

The behavior of carnivorous dinosaurs was unquestionably as diverse as that of modern mammalian carnivores. Insight into theropod behavior is provided by a few rather exceptional finds. At Ghost Ranch, larger specimens of *Coelophysis* have been recovered with smaller specimens inside the rib cage and were presumably eating the juveniles. Such cannibalism is widespread in animals of all kinds, so it is not surprising that it was practiced by at least some theropods.

There is some controversy concerning whether large theropods, especially *Tyrannosaurus rex*, were active hunters or scavengers. The hunter/scavenger debate is somewhat meaningless because most carnivores are both. Hyenas are highly efficient scavengers that nevertheless hunt and kill their own prey whenever possible. On the other hand, even lions and tigers will scavenge carcasses that they come across. Theropods probably also would have been opportunists, but it is unlikely that they would have found enough food if they relied solely on this lifestyle. Theropods almost universally have limb proportions that exceed those of the assumed prey animals and were presumably faster animals. This relationship would not have been maintained if the big theropods were all scavengers. There is at least one example of a theropod (*Velociraptor*) that attacked a living herbivore (*Protoceratops*) in Cretaceous Mongolia (see Dromaeosauridae).

Evidence of nesting behavior is known for *Oviraptor* (Norell *et al.*, 1995; Dong and Currie, 1996) and *Troodon* (Varricchio *et al.*, 1996), thanks to the fortuitous recovery of adult skeletons associated with nests of eggs.

Full and partial skeletons of at least 1000 individuals of *Coelophysis* at Ghost Ranch in New Mexico (Schwartz and Gillette, 1994) strongly suggest that this dinosaur was gregarious. Such large accumulations of theropods do not appear to have been uncommon during Late Triassic and Early Jurassic times. There is a similar accumulation of skeletons in Zimbabwe of the closely related *Syntarsus*. A Lower Jurassic trackway site at Holyoke (Massachusetts) shows evidence of 20 theropods moving as a group (see Connecticut River Valley). Possibly, these theropods formed large packs only for short periods of time for breeding or migration. In Upper Jurassic rocks of the Cleveland–Lloyd Quarry in Utah, the remains of as many as 50 individuals of *Allosaurus* were found in a single bone bed (Miller *et al.*, 1996), and it has been suggested that the site may have been a predator trap. Lower Cretaceous footprint sites in British Columbia suggest that small theropods were moving in packs of a half dozen or so individuals, whereas large theropods were moving in smaller groups of 2 or 3 animals. In Montana, the associated skeletal remains of *Deinonychus* and *Tenontosaurus* have been used to show packing behavior in dromaeosaurids (Maxwell and Ostrom, 1995). A bone bed of articulated *Albertosaurus* remains from Alberta, and the association of a large *Tyrannosaurus* skeleton in South Dakota with the remains of several smaller individuals, suggests that tyrannosaurids were also packing animals.

See also the following related entries:

Ceratosauria • Herrerasauridae • Saurischia • Sauropodomorpha • Tetanurae

References

Currie, P. J., and Zhao, X. J. (1993). A new large theropod (Dinosauria, Theropoda) from the Jurassic of Xinjiang, People's Republic of China. *Can. J. Earth Sci.* **30,** 2037–2081.

Dashzeveg, D., Novacek, M. J., Norell, M. A., Clark, J. M., Chiappe, L. M., Davidson, A., McKenna, M. C., Dingus, L., Swisher, C., and Perle, A. (1995). Extraordinary preservation in a new vertebrate assemblage from the Late Cretaceous of Mongolia. *Nature* **374,** 446–447.

Dong Z. M., and Currie, P. J. (1996). On the discovery of an oviraptorid skeleton on a nest of eggs at Bayan Mandahu, Inner Mongolia, People's Republic of China. *Can. J. Earth Sci.* **33,** 631–636.

Farlow, J. O. (1993). On the rareness of big, fierce animals: Speculations about the body sizes, population densities, and geographic ranges of predatory mammals and large carnivorous dinosaurs. *Am. J. Sci.* **293A,** 167–199.

Gauthier, J. (1986). Saurischian monophyly and the origin of birds. In *The Origin of Birds and the Evolution of Flight* (K. Padian, Ed.), pp. 1–55. California Academy of Sciences, San Francisco.

Holtz, T. R., Jr. (1994). The phylogenetic position of the Tyrannosauridae: Implications for theropod systematics. *J. Paleontol.* **68,** 1100–1117.

Holtz, T. R., Jr. (1996). Phylogenetic taxonomy of the Coelurosauria (Dinosauria: Theropoda). *J. Paleontol.* **70,** 536–538.

Holtz, T. R., Jr., and Padian, K. (1995). Definition and diagnosis of Theropoda and related taxa. *J. Vertebr. Paleontol.* **15**(Suppl. to No. 3), 35A.

Marsh, O. C. (1881). Classification of the Dinosauria. *Am. J. Sci. Ser. 3* **23,** 81–86.

Maxwell, W. D., and Ostrom, J. H. (1995). Taphonomy and paleobiological implications of *Tenontosaurus–Deinonychus* associations. *J. Vertebr. Paleontol.* **15,** 707–712.

Miller, W. E., Horrocks, R. D., and Madsen, J. H., Jr. (1996). The Cleveland–Lloyd Dinosaur Quarry, Emery County, Utah: A U.S. Natural Landmark (including history and quarry map). *Brigham Young Univ. Geol. Stud.* **41,** 3–24.

Norell, M. A., Clark, J. M., Chiappe, L. M., and Dashzeveg, D. (1995). A nesting dinosaur. *Nature* **378,** 774–776.

Novas, F. E. (1996). New theropods from the Late Cretaceous of Patagonia. *J. Vertebr. Paleontol.* **16,** 56A.

Schwartz, H. L., and Gillette, D. D. (1994). Geology and taphonomy of the *Coelophysis* quarry, Upper Triassic Chinle Formation, Ghost Ranch, New Mexico.

Sereno, P. C., and Novas, F. E. (1993). The skull and neck of the basal theropod *Herrerasaurus ischigualastensis*. *J. Vertebr. Paleontol.* **13,** 451–476.

Sereno, P. C., Dutheil, D. B., Iarochene, S. M., Larsson, H. C. E., Lyon, H. G., Magwene, P. M., Sidor, C. A., Varricchio, D. J., and Wilson, J. A. (1996). Late Cretaceous dinosaurs from the Sahara. *Science* **272,** 986–991.

von Huene, F. (1926). The carnivorous Saurischia in the Jura and Cretaceous formations principally in Europe. *Museo de La Plata Revista* **29,** 35–167.

Wilson, M. C., and Currie, P. J. (1985). *Stenonychosaurus inequalis* (Saurischia: Theropoda) from the Judith River (Oldman) Formation of Alberta: New findings on metatarsal structure. *Can. J. Earth Sci.* **22,** 1813–1817.

Thyreophora

Kenneth Carpenter
Denver Museum of Natural History
Denver, Colorado, USA

The Thyreophora consists of the armored dinosaurs, hence the name, which means "shield bearer." The group may be defined as stegosaurs, ankylosaurs, and all taxa closer to them than to Cerapoda. Diagnostic features include rows of scutes along the backbone and the lateral body surface. Thyreophora consists of the stegosaurs, ankylosaurs, and several basal forms from the Lower Jurassic, including *Scutellosaurus* from Arizona, *Emausaurus* from Germany, and *Scelidosaurus* from England. The armor consists of flat, keeled plates, erect, thin plates, or long spines. This armor is usually arranged along the neck, back, and tail where it would offer the greatest protection or where it would be most visible in display BEHAVIOR. In addition, some armor is also present along the sides of the body and on the skull of the ankylosaurs and some primitive thyreophorans.

Scutellosaurus is the most basal form, having a skeleton otherwise similar to that of the basal ornithischian, *Lesothosaurus*. *Scutellosaurus* was bipedal, balancing its body over the hindlegs with a very long tail. *Scelidosaurus*, on the other hand, was more ponderous, walking quadrupedally on stout limbs. The skeleton of *Emausaurus* is unfortunately not very well known but may have been more similar to that of

Scelidosaurus than to *Scutellosaurus*. All three of these basal thyreophorans had small, primitive, triangular teeth indicating a DIET of plants.

See also the following related entries:

ANKYLOSAURIA • STEGOSAURIA

References

Coombs, W., and Maryańska, T. (1990). Ankylosauria. In *The Dinosauria* (D. B. Weishampel, P. Dodson, and H. Osmólska, Eds.), pp. 456–483. Univ. of California Press, Berkeley.

Coombs, W. P., Jr., Weishampel, D. B., and Witmar, L. M. (1990). Basal Thyreophora. In *The Dinosauria* (D. B. Weishampel, P. Dodson, and H. Osmólska, Eds.), pp. 427–434. Univ. of California Press, Berkeley.

Sereno, P. C. (1986). Phylogeny of the bird-hipped dinosaurs (Order Ornithischia). *Natl. Geogr. Res.* **2**, 234–256.

Tochigi Prefectural Museum, Japan

see MUSEUMS AND DISPLAYS

Tooth Marks

AASE R. JACOBSEN
Royal Tyrrell Museum of Palaeontology
Drumheller, Alberta, Canada

Tooth marks (bite traces) are important indications of predator/prey interaction in recent and ancient environments, and they provide information on bone consumption, feeding behavior, and feeding strategy of the taxa involved. Tooth-marked dinosaur bones seem more common than previously thought. A study of 1000 dinosaur bones from the Dinosaur Park Formation (Upper Cretaceous) (Jacobsen, 1995) showed up to 14% tooth-marked hadrosaur bones from isolated occurrences. A much lower frequency was found on ceratopsian bones (4% tooth-marked bones) and tyrannosaurid bones (2% tooth-marked bones), which corresponds with Fiorillo (1991), who reported 2–5% tooth-marked dinosaur bones. The low frequency might be due to the fact that these bones originate from bone beds and articulated skeletons, or it might show a prey preference of theropods on hadrosaurs. Other factors, such as preservation, collecting, and preparation biases, can also influence these frequencies.

It is possible to correlate the denticle size and shape of theropod teeth with the morphology of serration marks, and rows of parallel tooth marks can be correlated with the intertooth spacing of theropod teeth. Using these methods the feeding theropod has been identified on a variety of prey bones, and tyrannosaurid tooth marks have been identified on bones from hadrosaurids, ceratopsids, *Saurornitholestes*, and other tyrannosaurids (Jacobsen, 1995). *Troodon* tooth marks have been found on a ceratopsian bone, and *Saurornitholestes* tooth marks have been found on an ornithomimid caudal (Jacobsen, 1995). A *Triceratops* sp. pelvis has tooth marks attributable to *Tyrannosaurus rex* (Erickson and Olson, 1996), and a pterosaur bone had tooth marks and a *Saurornitholestes* tooth imbedded in the bone (Currie and Jacobsen, 1995).

Theropod tooth marks appear as linear grooves, parallel marks, or puncture marks, and they are found on every part of a skeleton and randomly distributed and oriented on the bone. The frequency, distribution, and location on bones indicate that theropods did not chew and consume bones for nutrition by chewing on distal ends of bones, like what is known for mammalian carnivores (Erickson and Olson, 1996; Fiorillo, 1991; Jacobsen, 1995). Instead, they probably bit and ripped off meat during feeding and had accidental tooth/bone contacts resulting in tooth-marked bones.

See also the following related entries:

BEHAVIOR • PALEOPATHOLOGY • TOOTH REPLACEMENT • TOOTH SERRATIONS • TOOTH WEAR

References

Currie, P. J., and Jacobsen, A. R. (1995). An azhdarchid pterosaur eaten by a velociraptorine theropod. *Can. J. Earth Sci.* **32**, 922–925.

Erickson, G. M., and Olson, K. H. (1996). Bite marks attributable to *Tyrannosaurus rex:* Preliminary description and implications. *J. Vertebr. Paleontol.* **16,** 175–178.

Fiorillo, A. R. (1991). Prey bone utilization by predatory dinosaurs. *Palaeogeogr. Palaeoclimatol. Palaeoecol.* **88,** 157–166.

Jacobsen, A. R. (1995). Predatory behaviour of carnivorous dinosaurs: Ecological interpretations based on tooth marked dinosaur bones and wear patterns of theropod teeth. Unpublished M.Sc. thesis, University of Copenhagen, Denmark.

Tooth Replacement Patterns

GREGORY M. ERICKSON
University of California
Berkeley, California, USA

The teeth of dinosaurs were continually replaced throughout life. This process allowed the establishment of successively larger teeth through ontogeny and the replacement of worn and broken teeth with sharper successors. Most dinosaurs had one or two replacement teeth developing at any one time for each tooth position, the plesiomorphic condition for the Dinosauria (Edmund, 1960). An average tooth functioned approximately 300 days in smaller dinosaurs with this condition (such as *Deinonychus antirrhopus*), whereas tooth turnover in giant taxa (such as *Tyrannosaurus rex*) may have taken up to 800 days (Erickson, 1996). In some herbivorous taxa (e.g., ceratopsids and hadrosaurids) the numbers of replacement teeth increased to as many as six per tooth position. These increases allowed more rapid rates of tooth formation and replacement and reflect shifts in diet to abrasive foodstuffs requiring oral processing (Erickson, 1996). Mean tooth replacement rates ranged from 50 to 100 days for derived members of such herbivorous lineages (e.g., *Triceratops horridus, Maiasaura peeblesorum,* and *Edmontosaurus regalis*; Erickson, 1996). In most dinosaurs the sequence of tooth turnover consisted of (i) the resorption of the root of a functioning tooth, (ii) the subsequent lateral migration of the next successive replacement tooth into the resorption void, (iii) the occlusal migration of the replacement tooth into the pulp cavity of the functioning tooth, (iv) the partition of the functional tooth crown, and finally, (v) the eruption of the replacement tooth into the functional position.

In the late 1800s it was noted that the patterns of tooth loss in some extant reptiles were not random because consecutive teeth along the tooth row were generally out of phase with one another in stage of development, and the simultaneous loss of adjacent teeth was somehow prevented (e.g., Röse, 1893). Presumably this patterning ensured that localized toothlessness due to tooth turnover did not hinder an animal's capacity to feed. Edmund (1960) demonstrated that such patterns were pervasive throughout the Reptilia, including extinct taxa such as dinosaurs. How such patterns of replacement are maintained has been a topic of debate for decades (see review by Edmund, 1960).

One of the competing hypotheses to explain reptilian dental patterning is the "tooth wave" theory proposed by Edmund (1960, 1962). In this mechanism the teeth are intitiated (and consequently replaced) in "waves" (the stimulus for which Edmund suggested is "chemical") that progress unidirectionally down along the tooth row and affect alternate tooth positions. The waves are temporally spaced so that consecutive functional teeth are in opposite stages of development, thus avoiding the disadvantages of simultaneous loss. Although this theory seems to explain the dental patterns seen in adult reptiles, it falls short in predicting the patterns observed during embryogenesis. For example, reptilian teeth are not progressively initiated unidirectionally down the length of the tooth row, and the first teeth formed are rarely from the first or last position in the tooth row (Osborn, 1970; Westergaard and Ferguson, 1986). This has led many to abandon the tooth wave theory and to search for a different explanation.

Mechanisms invoking more localized control of tooth formation and replacement have recently gained favor with many morphologists [e.g., the "tooth family theory" of Osborn (1971) and the "tooth position theory" of Westergaard (1983)]. In these mechanisms a zone of inhibition (the nature of which, like Edmund's "wave stimulus," remains a mystery) surrounds each developing tooth and prevents the simultaneous formation of teeth in immediately adjacent tooth positions by suppressing the development of their respective replacement teeth. As the teeth move occlusally, so do the zones of inhibition, thereby allowing the replacement teeth and the teeth in adjacent tooth positions to begin to form. In Westergaard's (1983) model, jaw growth can also create uninhibited regions suitable for tooth initiation. Weishampel (1991) has demonstrated that such local control mechanisms can theoretically produce nearly any of the dental patterns observed in extant and fossil reptiles, including dinosaurs, and that they can account for the purported waves of tooth replacement.

See also the following related entries:

DIET • TEETH AND JAWS • TOOTH MARKS • TOOTH REPLACEMENT RATES • TOOTH SERRATIONS • TOOTH WEAR

References

Edmund, A. G. (1960). Tooth replacement phenomenon in the lower vertebrates. Life Sciences Division of the Royal Ontario Museum, University of Toronto, Contribution No. 52, pp. 1–190.

Edmund, A. G. (1962). Sequence and rate of tooth replacement in the Crocodilia. Life Sciences Division of the Royal Ontario Museum, University of Toronto, Contribution No. 56, pp. 1–42.

Erickson, G. M. (1996). Incremental lines of von Ebner in dinosaurs and the assessment of tooth replacement rates using growth line counts. *Proc. Natl. Acad. Sci. USA* **93,** 14623–14627.

Osborn, J. W. (1970). New approach to Zahnreihen. *Nature* **225,** 343–346.

Osborn, J. W. (1971). The ontogeny of tooth succession in *Lacerta vivipara* Jacquin (1787). *Proc. R. Soc. London Ser. B* **179,** 261–289.

Röse, C. (1893). Über die Zahnentwicklung der Krockodile. *Morphol. Arbeiten* **3,** 195–228.

Weishampel, D. B. (1991). A theoretical morphological approach to tooth replacement in lower vertebrates. In *Constructional Morphology and Evolution* (N. Schmidt-Kittler and K. Vogel, Eds.), pp. 295–310. Springer-Verlag, Berlin.

Westergaard, B. (1983). A new detailed model for mammalian dentitional evolution. *Z. Zool. Systematik Evolutionsforschung* **21,** 68–78.

Westergaard, B., and Ferguson, M. W. J. (1986). Development of the dentition in *Alligator mississippiensis*: Early embryonic development in the lower jaw. *J. Zool. Ser. A* **210,** 575–597.

Tooth Replacement Rates

see VON EBNER INCREMENTAL GROWTH LINES

Tooth Serrations in Carnivorous Dinosaurs

WILLIAM L. ABLER
Oak Park, Illinois, USA

The leaf-shaped, crenelated teeth of basal ORNITHISCHIAN dinosaurs *(Lesothosaurus,* most thyreophorans, and *Heterodontosaurus)* and basal SAUROPODOMORPHS *(Massospondylus)* broadly resemble those of extant herbivorous lizards. It was for this reason that Mantell named what turned out to be the first described dinosaur *Iguanodon* ("iguana-tooth") in 1822, based on a suggestion of similarity by Samuel Stutchbury, who had just returned from seeing living iguanas in South America. Two years afterward, Buckland first described the very different serrated teeth of *Megalosaurus,* which he recognized as a carnivore. Owen (1860) noticed that the serrations on the teeth of carnivorous dinosaurs tilt slightly toward the point of the tooth, and Lambe (1917) noticed that tyrannosaurid teeth are mostly in "pristine" condition, not chipped, cracked, or worn. He concluded from this that tyrannosaurs ate mostly soft food, although the prevailing view has been that serrated theropod teeth are "razor-sharp" and function like a serrated steak knife. Both views are erroneous, as mechanical analysis shows. The physical construction of serrated teeth is well suited for carnivory, and microdamage sustained by the teeth records many of the masticatory and other movements of the teeth during feeding.

Serrated teeth appear to be plesiomorphic for THEROPODS because they are found throughout both CERATOSAURS and TETANURANS. Although both ornithischians and sauropodomorphs have modified their diets, and hence their feeding apparatus, toward herbivory, *Eoraptor* and HERRERASAURS have serrated teeth, suggesting that the habit was primitive for dinosaurs. The teeth of more distantly related dinosaur

FIGURE 1 Whole tyrannosaurid tooth from the Judith River Formation (Late Cretaceous). Note serrations on anterior (convex) and posterior (concave) edges. Serrations of the posterior row continue onto the surface of the tooth, forming the "tails" (caudae) described in the text (see also Erickson, 1995).

outgroups [*Pseudolagosuchus, Lagosuchus (Marasuchus),* and *Lagerpeton*] are unknown (see DINOSAUROMORPHA), but flattened, serrated teeth also occur in some PSEUDOSUCHIANS (some parts of the jaws in some phytosaurs and in rauisuchians), suggesting that this tooth form commonly arose in carnivorous lineages. The cross-sectional shape of such teeth varies from an elongated, laterally flattened ellipse (most coelurosaurs) to a D-shape (phytosaurs and some rauisuchians), to nearly circular (tyrannosaurs). The cross-section may change from the base of the tooth toward the tip. Serrations appear on keel-like structures on the anterior and posterior faces of the teeth, with some variation (Figs. 1 and 2). The gross shape of teeth rarely seems to be diagnostic to the species level, although the relative shape and number of serrations may help to distinguish the teeth of some taxa (Farlow *et al.*, 1991).

Serrations are not simple or uniform structures. Rows of serrations may sometimes bifurcate (Erickson, 1995). Along a keel, individual serrations may show a "tail" ("cauda") that continues from the serration onto the surface of the tooth, either perpendicular to its long axis or angled toward its base (Fig. 2). Serrations of both anterior and posterior rows ordinarily have the shape of a simple cube, but those of the posterior row may be saddle shaped. The internal anatomy of serrations is also not uniform, and their

FIGURE 2 Rare example of pathological tyrannosaurid tooth (courtesy Royal Tyrrell Museum of Palaeontology) with multiple rows of serrations (see also Erickson, 1995). The cause of the pathology is not known. Serrations run parallel to the long axis of the tooth and are of normal size and shape. Each row trends toward either the anterior or posterior edge of the tooth. It is perhaps surprising that a pathologically produced structure would have such normal morphology.

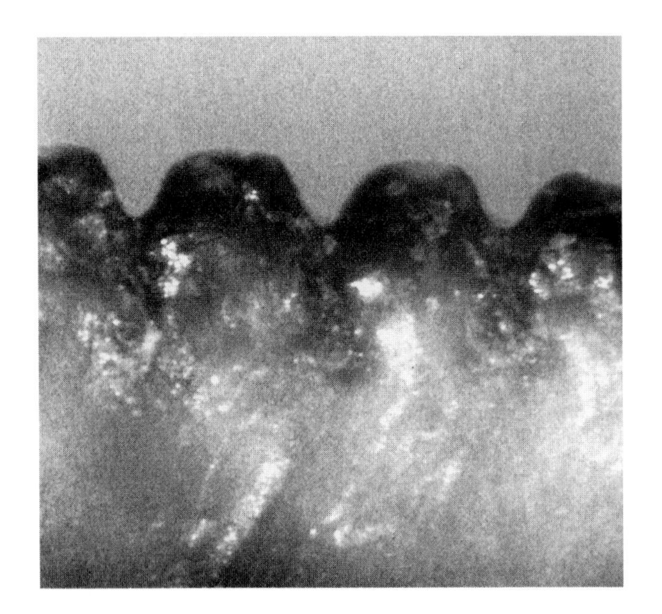

FIGURE 3 Naturally split tyrannosaurid tooth reveals that the outer edge of each serration consists of an enamel "cap" (operculum), and that the junction between each pair of serrations has a flask-shaped ampulla at the bottom; the histology of these structures may be diagnostic to some taxonomic levels (Abler, 1992).

complexity indicates constructional patterns that may be diagnostic for some taxa (Fig. 3). However, the presence or absence of serrations varies enough within lineages and even within species that their features may be considered independently evolved to some extent. Unfortunately, little is known of the genetic and developmental factors that influence the formation and pattern of serrations.

Experimental studies of serrated teeth and tooth models (Abler, 1992) have produced some results counterintuitive to traditional expectations about tooth function in theropods. Contrary to shark teeth (*Carcharodon megalodon*), which cut meat like a hacksaw blade, tyrannosaurid teeth cut meat more like a dull, smooth blade. Other properties of tyrannosaurid teeth are unlike those of any knife. Minute surface scratches run in all directions, suggesting nonorthal movement either by the predator or the prey (Fig. 4). Narrow slots (diaphyses) forming the junction between neighboring serrations act as tiny frictional vises that grip and hold the tiny fibers of tendon that are uniformly distributed in meat. These fibers, and the ribbons of meat attached to them, could have given the tyrannosaurs an infectious bite if left to rot in the jaw (Fig. 5). The hollow pockets between neighboring serrations (Fig. 4) afford maximum exposure to the diaphyses and also trap grease so effectively that it cannot be removed completely, even by specially made steel spoons. This suggests an accessory function of promoting infectiousness of the bite.

If the inference is valid that the particular structure of these serrations promoted infection, then the further inference would be supported that tyrannosaurs were active predators, inasmuch as it would have been useless to infect carrion in this way. The vulnerability of the big tyrranosaurs to injury from falling (Farlow *et al.*, 1995) suggests that they might have

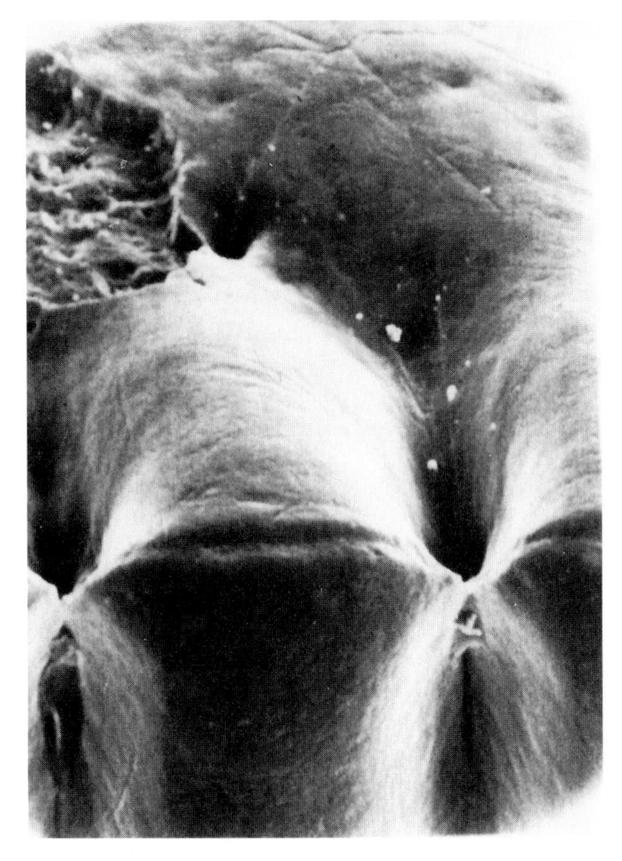

FIGURE 4 Apical view of neighboring serrations on a tyrannosaurid tooth. Note that the serrations are thicker mediolaterally (top to bottom in photomicrograph) than from tip to base of tooth (left to right in photomicrograph). A thin groove runs along the apices of the serrations. Note the narrow junction (diaphysis) between serrations, which grips and holds meat fibers, as shown in Fig. 5. Pockets between serrations, seen to either side (above and below in photomicrograph) of the diaphyses, are the cellae that trap and hold grease and meat debris, possibly facilitating an infectious bite. A few wear scratches are visible on the tooth surface. SEM photo by W. F. Simpson, Field Museum of Natural History, Chicago.

FIGURE 5 Tyrannosaur tooth serrations after experimentally cutting meat. Note the ribbons of meat fiber streaming from the serrations, where they have been trapped and held by high-pressure slots (diaphyses) between serrations. This view of the tyrannosaurid is perhaps as terrifying as any, inasmuch as the trapped meat fibers, if left to rot, would have guaranteed an infectious bite.

hunted from ambush rather than by pursuit. This would not, of course, exclude carrion feeding as a second dietary strategy; in fact, carrion would have been a good source of infectious bacteria, as it is for the Komodo Monitor (Auffenberg, 1981). Taking the analogy further, it is possible that tyrannosaurs need not always have killed prey by subduing it, but rather by infecting it with superficial bites and allowing the course of disease to weaken the prey to the point where it could be dispatched. Despite the occasional broken tyrannosaur tooth and tooth mark in dinosaur bone, tooth-to-bone contact is not commonly encountered (Fiorillo, 1991), but there are salient exceptions indicating a puncture-and-drag strategy potentially applicable either to killing or to feeding (Erickson *et al.*, 1996). Tooth-to-tooth occlusion is known for some theropods, but because all the teeth are incisive in shape, it is clear that they did not chew their food but swallowed it in chunks, detaching it from the carcass by strength rather than by the sharpness of the teeth.

See also the following related entries:

DIET • TEETH AND JAWS • THEROPODA • TOOTH WEAR

References

Abler, W. L. (1992). The serrated teeth of tyrannosaurid dinosaurs, and biting structures in other animals. *Paleobiology* **18,** 161–183.

Auffenberg, W. (1981). *The Behavioral Ecology of the Komodo Monitor.* Univ. of Florida Press, Gainesville.

Erickson, G. M. (1995). Split carinae on tyrannosaurid teeth and implications of their development. *J. Vertebr. Paleontol.* **15,** 268–274.

Erickson, G. M., Van Kirk, S. D., Su, J., Levenston, M. E., Caler, W. E., and Carter, D. R. (1996). Bite-force estimation for *Tyrannosaurus rex* from tooth-marked bones. *Nature* **382,** 706–708.

Farlow, J. O., Brinkman, D. L., Abler, W. L., and Currie, P. J. (1991). Size, shape, and serration density of theropod dinosaur lateral teeth. *Mod. Geol.* **16,** 161–198.

Farlow, J. O., Smith, M. B., and Robinson, J. M. (1995). Body mass, bone "strength indicator" and the cursorial ability of Tyrranosaurus rex. *J. Vert. Paleo.* **15,** 713–725.

Fiorillo, A. R. (1991). Prey bone utilization by predatory dinosaurs. *Palaeogeogr. Palaeoclimatol. Palaeoecol.* **88,** 157–166.

Lambe, L. (1917). The Cretaceous theropodous dinosaur *Gorgosaurus. Mem. Geol. Survey Can.* **100,** 1–84.

Owen, R. (1860). *Palaeontology,* 2nd ed. Black, Edinburgh.

Tooth Wear

ANTHONY R. FIORILLO
Dallas Museum of Natural History
Dallas, Texas, USA

DAVID B. WEISHAMPEL
Johns Hopkins University
Baltimore, Maryland, USA

Based on comparative studies with living animals it has been demonstrated that microwear patterns on teeth provide direct evidence of the qualitative differences in diet between animals. Studies of microwear on the teeth of fossil animals can be used to determine patterns of food use within ancient ecosystems. Despite its widespread use in other paleontological fields, most notably paleoanthropology, the study of tooth wear and its implications about feeding in dinosaurs has been pursued in only a few cases.

Preliminary work has been started on certain dinosaur groups. Examination of wear facets for features is usually done with a scanning electron microscope at 100–600×. Common wear features found on both mammalian and magnifications of dinosaurian teeth are pits and scratches (or striae). Some authors have defined pits as having length : width ratios of less than 2 : 1, whereas others have used a ratio of 4 : 1. Pits are generally associated with consumption of hard food items. The degree of development of scratching—that is, the length, depth, and density of the striae—can be used to infer relative food hardness consumed between animals with only scratches on their wear facets. In modern animals, microwear patterns can change with seasonal variation in diet.

A. R. Fiorillo (Dallas Museum of Natural History) has been studying microwear patterns on the teeth of some sauropod dinosaurs. His work has shown that a complex system of food resource partitioning can be demonstrated for *Camarasaurus* and *Diplodocus,* the two most common synchronic sauropod dinosaurs from the Upper Jurassic Morrison Formation of western North America. The patterns of microwear show that, in general, *Camarasaurus* ate coarser foods than did *Diplodocus.* However, the juveniles of *Camarasaurus* have patterns of microwear that are very similar to *Diplodocus,* suggesting that there was a period of dietary overlap between the young of *Camarasaurus* and the young and adults of *Diplodocus.*

Because wear features are ultimately produced by the movement between the upper and lower teeth, linear microwear features also provide a record of the pattern of chewing during which food is subdivided. This part of the masticatory cycle is called the power stroke. This relationship between striation, orientation, and power stroke movement has been known for some time in mammals, but because extant nonmammalian tetrapods lack a power stroke, this approach to studying "extinct" chewing has mostly been limited to fossil mammals and not extended to those dinosaurs that chewed.

There are other aspects of tooth wear that can be used to determine power stroke direction. In particular, the form of the enamel–dentine interface (the junction between the dentine, the softer inner tissue of the tooth, and the enamel, the harder outer coating) provides such information. When the power stroke direction moves from the dentine region across the harder enamel, the dentine will be more worn than the enamel. This produces a step between the two tissues. Conversely, if the power stroke moves from enamel across the dentine, the enamel protects the inner dentine and the interface will be flush.

In 1983 and 1984, D. B. Weishampel (Johns Hopkins University) used this connection between tooth wear and jaw mechanics to study chewing in ornithopod dinosaurs. At that time there was a variety of hypotheses about ornithopod jaw mechanisms but few attempts to test any of them. To do so, Weishampel treated ornithopod skulls as multilink chewing machines designed to subdivide plant matter. Using computer modeling, it was possible to construct and compare a great many hypothetical jaw systems that ranged from relatively simple to complex intracranial mobility.

Technically, each ornithopod skull was divided into sites of potential mobility and a given suite of moveable elements was considered a particular jaw mechanics model. With the sites of mobility located in three-dimensional space, each model was manipulated in cyberspace to simulate chewing.

The critical test for each of the models was its ability to predict the power stroke direction during the chewing cycle and the kind of tooth wear (wear striation direction and enamel–dentine configuration) that would have been created between occluded teeth. Because tooth wear provides a precise signature of jaw movement during the power stroke, these predictions from a modeled mechanism were then tested against tooth wear on actual ornithopod dentitions.

Tooth wear parameters were best predicted by two jaw mechanics models. The first model has a hingelike articulation between the maxilla and associated elements against the premaxilla and skull roof. Found only in euornithopods, this mechanism, termed pleurokinesis by D. B. Norman (Cambridge University), provides a transverse power stroke through the lateral rotation of the maxillae in concert with the continued upward movement of the lower jaws. The other ornithopod jaw mechanism, found in heterodontosaurids, demonstrates transverse chewing through a slight rotational mobility of the lower jaws about their long axes that is achieved by modified symphyseal joints at the front of the lower jaws.

Fiorillo, working with isolated teeth, noticed that the scratch marks on the teeth of both *Diplodocus* and *Camarasaurus* shows preferred orientations. This suggests that these sauropods had similar translational jaw movements during mastication. This finding is inconsistent with the expected nonpreferred distribution of striae orientations resulting from the previously inferred compressive bite- or nip-feeding strategy proposed for *Diplodocus.*

These studies in dinosaur microwear are few in

number but they highlight the potential contribution that more such studies can make to understanding aspects of dinosaur paleobiology.

See also the following related entries:

DIET • TOOTH REPLACEMENT PATTERNS • TOOTH SERRATIONS IN CARNIVOROUS DINOSAURS

References

Fiorillo, A. R. (1991). Dental microwear on the teeth of *Camarasaurus* and *Diplodocus:* Implications for sauropod paleoecology. In *Fifth Symposium on Mesozoic Terrestrial Ecosystems and Biota* (Z. Kielan-Jaworowska, N. Heintz, and H. A. Nakrem, Eds.), pp. 23–24. Univ. Oslo, Oslo.

Fiorillo, A. R. (1995). Enamel microstructure in *Diplodocus, Camarasaurus,* and *Brachiosaurus* (Dinosauria: Sauropoda) and its lack of influence on resource partitioning by sauropods in the Late Jurassic. In *Sixth Symposium on Mesozoic Terrestrial Ecosystems and Biota* (A. Sun and Y. Wang, Eds.), pp. 147–149. China Ocean Press, Beijing.

Fiorillo, A. R. (1997). Dental microwear patterns from the sauropod dinosaurs *Camarasaurus* and *Diplodocus:* Evidence for resource partitioning in the Late Jurassic of North America. *Historical Biol.,* in press.

Norman, D. B., and Weishampel, D. B. (1985). Ornithopod feeding mechanisms: Their bearing on the evolution of herbivory. *Am. Nat.* **126,** 151–164.

Weishampel, D. B. (1984). Evolution of jaw mechanisms in ornithopod dinosaurs. *Adv. Anat. Embryol. Cell Biol.* **87,** 1–110.

Weishampel, D. B., and Norman, D. B. (1989). The evolution of occlusion and jaw mechanics in Late Paleozoic and Mesozoic herbivores. *Geol. Soc. Am. Spec. Paper* **238,** 87–100.

Toyama Science Museum, Japan

see MUSEUMS AND DISPLAYS

Toyohashi Museum of Natural History, Japan

see MUSEUMS AND DISPLAYS

Trace Fossils

KEVIN PADIAN
University of California
Berkeley, California, USA

Fossils are generally divided into actual body remains (altered or unaltered; see SKELETAL STRUCTURES) and trace fossils, which are considered to include any records of an organism's behavior (in a broad sense) that are preserved in the fossil record.

The most common trace fossils encountered in the dinosaur fossil record are FOOTPRINTS AND TRACKWAYS. These may be found as individual prints, as single trackways of an individual, and as trackway assemblages that may extend over many square kilometers (megatracksites). Tracks provide many kinds of information. If the foot anatomy is sufficiently preserved, some idea of the general identity of the trackmaker may be had by comparison with the manus and pedes of skeletal remains. For example, both the crocodyliform *Protosuchus* and the crocodylomorph *Batrachopus* are found in the Early Jurassic Moenave Formation of Arizona (Olsen and Padian, 1986). *Protosuchus* may have made these *Batrachopus* tracks. On the other hand, the trackway form that we call *Batrachopus* may also have been made by other crocodylomorphs, not only *Protosuchus* (the foot anatomy and the step cycle recorded by the trackway may not necessarily have been particular to that genus), so the taxonomy of skeletal remains and the parataxonomy of trackways are kept separate.

Whether the trackmaker can be generally identified or only guessed at, trackways also provide indications of the kinematics of the step cycle—how the animal walked. This helps establish some correlates of posture and gait that are necessarily incompletely known from skeletal anatomy. For example, trackways of theropod dinosaurs are always narrow (this in fact applies to all dinosaurs: The feet were held under the body), and from their earliest occurrences in the Late Triassic, usually one foot is placed almost directly in front of the other, with the toes turned slightly inward. This did not change through the MESOZOIC ERA despite the great size evolved by some theropods, including tyrannosaurids. Nor did it change in the evolution of birds from small theropods despite the reduction of the tail, the hyperdevelopment of the forelimbs, and the transition to arboreal life. Bird footprints still have the same general form as those of Mesozoic dinosaurs, suggesting that although major patterns of musculature and skeletal anatomy have changed greatly over 200 million years, the basic particulars of the step cycle have not (Padian and Olsen, 1989; Gatesy, 1990).

Tracks may also reveal a considerable amount of information about the conditions of the substrate at the time the track was made. An important caveat to remember is that the exposed surface of a trackway may not be the surface actually trod upon by the animal: It could be a lower layer of pushed-down, compacted, and distorted sediment—an "underprint." A true surface track, however, can be assessed by the clarity of the foot impressions, and consideration of the degree to which the various digits, phalanges, and pad impressions are preserved is fundamental to understanding the substrate condition. For example, evidence of slumping, trailing, dragging, and slit-like digit impressions usually indicate high water content. Usually in these cases details of the anatomy are unclear, and although some idea of the kinematics of the step cycle may be had, assessments of the identity of the trackmakers can be widely divergent (Padian and Olsen, 1984; Lockley *et al.*, 1995).

Teeth and jaws may also leave trace fossils. Predatory dinosaurs were not reluctant to use their teeth on prey, and tooth marks can be found on bones, revealing something about feeding behavior. The famous mount in the American Museum of Natural History, immortalized by the famous Charles Knight painting of an *Allosaurus* skeleton crouched over the backbone of a sauropod dinosaur, was inspired by the tooth marks on the vertebrae. Recently, a ceratopsian pelvis with puncture-and-drag markings testified to the trenchant and rapacious feeding strategy of a large tyrannosaurid (Erickson and Olson, 1996).

Skin impressions have been known and associated with dinosaurs for almost a century. When directly preserved with diagnostic skeletal remains, skin impressions can clarify understanding of what the exterior of the animals actually looked like, and they are now known for a surprising diversity of forms (see SKIN). Most surprising, perhaps, is the recent revelation of specimens of an Early Cretaceous theropod allied to *Compsognathus,* preserved with the remains of what appear to be rudimentary feathers along its dorsal midline (see FEATHERED DINOSAURS). Feathers and feather impressions are already known from the Late Jurassic (e.g., *Archaeopteryx)* and Early Cretaceous; in fact, it was the feathers of *Archaeopteryx* that showed scientists that this was a true bird, however ancient and dinosaurian it was in many of its skeletal features.

Eggs and eggshells are an increasingly important source of information about dinosaur nesting, breeding, and parental BEHAVIOR (see Carpenter *et al.*, 1994), particularly given the recent increase in discoveries of eggs, nests, and embryos and inferences that have been drawn about the social structure and behavioral patterns of dinosaurs. The parataxonomy of eggs and eggshells has developed based on the size and shape of the egg, as well as on the shell microstructure, now much better known as a result of scanning electron microscopy. The shell microstructure appears to be diagnostic of particular taxa, and in all cases more similar to the microstructure of bird eggs than to those of any other vertebrates. Still, only when diagnostic material of embryonic bones is discovered inside the eggshells can a true taxonomic assessment be made (see EGG MOUNTAIN; EGGS, EGGSHELLS AND NESTS; UKHAA TOLGOD).

Coprolites, the fossilized remains of feces, have received increased attention in recent years. Although seldom can an individual coprolite be traced to a specific taxon or skeletal individual, coprolites can be good paleoenvironmental indicators: Ponds or muddy bottoms may accumulate great numbers of them, as in the Late Triassic Placerias Quarry of Arizona. These remains may be screenwashed to recover the broken and acid-leached bones of small vertebrates that passed through the guts of larger carnivores (Jacobs and Murry, 1980), or they can be thin-sectioned and disaggregated to yield pollen remains that indicate the plants that lived in the region and that perhaps were consumed by herbivores.

In addition to these types of remains, trace fossils can include burrows (lungfishes and some therapsids), gastroliths or stomach stones (swallowed and used alternatively for food processing in dinosaurs or ballast in plesiosaurs and housed in two different internal organs; see Gillette, 1995), and any other structures or records of behavior produced by organisms, such as the egg cases of sharks or the cocoons of insects—all biogenic in origin.

The giants of dinosaur heritage are the sauropods. *Diplodocus carnegii* was among these giants but certainly not the largest. Some scientists consider sauropods sprawling, sluggish,water-bound behemoths, whereas others consider them active, gracile, upright, and alert (as pictured above). Illustration by Michael Skrepnick.

Near Cañon City, Utah, above the red Permian Cliffs just beyond the trees are Jurassic rocks of the Morrison Formation. Well known for the giant sauropods excavated from these rocks, the Morrison is among the more widely distributed exposed outcrops of dinosaur-bearing sediments. Photo by François Gohier.

The ornithomimids were active theropod dinosaurs far along the lineage leading to birds. Although they look much like today's ostriches, these features are convergent (acquired independently) from those of living birds. Illustration by Donna Braginetz

Brent Breithaupt, a curator at the University of Wyoming Geological Museum, is shown standing in front of an *Apatosaurus* skeleton mounted for public display. Photo by François Gohier.

The function of the plates of *Stegosaurus* is not established. Defense from predators, radiative heat dissipation, and competitive displays are possibilities. *Stegosaurus stenops*, shown above, exhibits one pattern of plate arrangement. Related stegosaurids possess plates and spikes of different sizes and arrangements. Illustration by Michael Skrepnick.

Albertosaurus libratus (right) attacking *Lambeosaurus lambei* may have used an ambushing strategy of hunting, just as many predators do today. Waterholes have always been good places for ambush. Illustration by Michael Skrepnick.

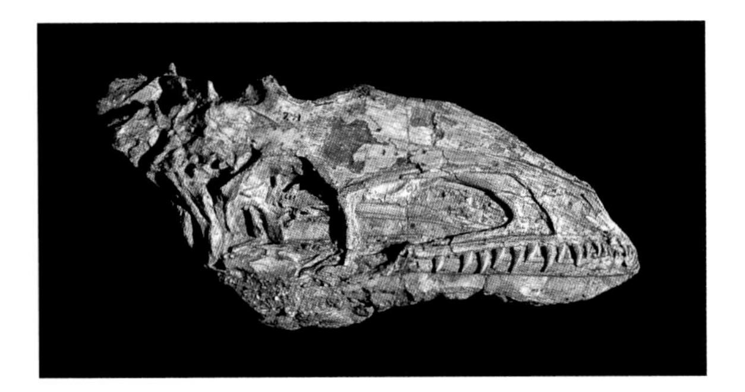

Coelophysis bauri was a medium-sized dinosaur. This 180-mm (about 7 in.) skull was found in the Triassic sediments of the Chinle Formation, Ghost Ranch, New Mexico. Photo by François Gohier.

Corythosaurus (right) is a hadrosaur from the Judith River Formation of Montana of the late Cretaceous that lived inland along lower coastal plains. Illustration by Donna Braginetz.

Dinosaur prospecting is hard, backbreaking work, often done miles from the amenities of civilization. The usual day involves rising at dawn and working all day hunched over in the baking sun, finding nothing or being disappointed with what is found. The exhilarating glory of one good discovery sustains prospectors sometimes for years of disappointing futility. The Mygatt-Moore Quarry in the Jurassic Morrison Formation in Rabbit Valley, Colorado, is typical in its isolation and spartan living conditions. Photo by François Gohier.

See also the following related entries:

COPROLITES • EGGS, EGGSHELLS, AND NESTS • FOOTPRINTS AND TRACKWAYS • GASTROLITHS • MEGATRACKSITES • SKIN

References

Carpenter, K., Hirsch, K. F., and Horner, J. R. (Eds.) (1994). *Dinosaur Eggs and Babies*. Cambridge Univ. Press, New York.

Erickson, G. M., and Olson, K. H. (1996). Bite marks attributable to *Tyrannosaurus rex*: Preliminary description and implications. *J. Vertebr. Paleontol.* **16,** 175–178.

Gatesy, S. M. (1990). Caudofemoralis musculature and the evolution of theropod locomotion. *Paleobiology* **16,** 170–186.

Gillette, D. D. (1995). *The Earth-Shaker Seismosaurus*. Columbia Univ. Press, New York.

Jacobs, L. L., and Murry, P. E. (1980). The vertebrate community of the Triassic Chinle Formation near St. Johns, Arizona. In *Aspects of Vertebrate History: Essays in Honor of Edwin Harris Colbert* (L. L. Jacobs, Ed.), pp. 55–72. Museum of Northern Arizona Press, Flagstaff.

Lockley, M. G., Logue, T. J., Moratalla, J. J., Hunt, A. P., Schultz, R. J., and Robinson, J. W. (1995). The fossil trackway *Pteraichnus* is pterosaurian, not crocodilian: Implications for the global distribution of pterosaur tracks. *Ichnos* **4,** 7–20.

Olsen, P. E., and Padian, K. (1986). Earliest records of *Batrachopus* from the Southwest U.S., and a revision of some early Mesozoic crocodilomorph ichnogenera. In *The Beginning of the Age of Dinosaurs: Faunal Change across the Triassic–Jurassic Boundary* (K. Padian, Ed.), pp. 259–274. Cambridge Univ. Press, Cambridge, UK.

Padian, K., and Olsen, P. E. (1984). The track of *Pteraichnus*: Not pterosaurian, but crocodilian. *J. Paleontol.* **58,** 178–184.

Padian, K., and Olsen, P. E. (1989). Ratite footprints and the stance and gait of Mesozoic theropods. In *Proceedings of the First International Conference on Dinosaur Tracks and Traces* (D. D. Gillette and M. Lockley, Eds.), pp. 231–241. Cambridge Univ. Press, New York.

Tracks

see FOOTPRINTS AND TRACKWAYS

Triassic Extinction Event

see EXTINCTION, TRIASSIC

Triassic Period

J. MICHAEL PARRISH
Northern Illinois University
DeKalb, Illinois, USA

The Triassic Period (245–208 my) comprises the first third of the Mesozoic Era, the Age of Dinosaurs, although skeletal fossils of dinosaurs proper are not known until more than halfway through the period. During the Triassic, the continents were, with the exception of a few isolated blocks, grouped together as the supercontinent Pangaea, which sat nearly symmetrically on the equator during most of the period. Triassic climates were warm and humid, with strong, monsoonal weather patterns providing marked seasonality (Parrish and Curtis, 1982).

Triassic vertebrates were dominated by two different clades of amniotes. In the early part of the Triassic, the greatest diversity and abundance of terrestrial vertebrates was found in the Synapsida, the group that includes mammals and their close relatives. Nonmammalian synapsids were extremely abundant during the Early and Middle Triassic. True mammals appeared in the Late Triassic, at approximately the same time that the earliest dinosaurs appear. The other large group of terrestrial vertebrates during the Triassic is the Archosauromorpha, consisting of dinosaurs, crocodiles, pterosaurs, and a number of fossil groups. Archosauromorphs appear in the Late Permian, and several lineages, notably the large, carnivorous erythrosuchids and the beaked rhynchosaurs, are important parts of Early and Middle Triassic faunas. For the last half of the Triassic, the dominant terrestrial vertebrates are mostly archosaurs, a derived subset of the Archosauromorpha that excludes basal groups such as the rhynchosaurs. The Late Triassic archosaurs include long-snouted, crocodile-like phytosaurs; the armored, herbivorous aetosaurs; and a variety of carnivorous forms, from small, long-legged sphenosuchians to giant rauisuchids, some (such as *Fasolasuchus* from the Ischigualasto Formation of Argentina) with skulls well over 1 m in length. Near the beginning of the Late Triassic, the first skeletal fossils of dinosauriforms and perhaps true dinosaurs appear: *Staurikosaurus* from Brazil (Colbert, 1970) and *Herrerasaurus* (Reig, 1963), *Eoraptor* (Sereno *et al.*, 1993), and *Pisanosaurus* (Casamiquela, 1967) from Argentina. Dinosaurs remain relatively rare elements of most terrestrial faunas during the Carnian stage of the Late Triassic. However, during the succeeding Norian stage, the herbivorous prosauropod dinosaurs (notably the genus *Plateosaurus*) suddenly become the dominant volumetric component of terrestrial faunas, particularly in Europe. The success of prosauropods at this time has been linked to their large size and long necks, which allowed them to feed in the upper parts of the forest canopy, a niche previously unavailable to large herbivores. Another locally abundant Norian group comprises the initial radiation of ceratosaurian theropods, which are best known from the spectacular death assemblage of the cassowary-sized *Coelophysis* from New Mexico. Another important Mesozoic archosaur group appears in the Norian—the flying pterosaurs.

Faunal turnover occurred continually throughout the Triassic—a phenomenon that may be linked to the climatic changes that accompanied the northward movement of Pangaea. An extinction at the Carnian–Norian boundary resulted in a turnover among archosaurs and extinction of many herbivorous groups such as the rhynchosaurs and the dicynodont synapsids (Benton, 1986; Olsen and Sues, 1986). The end of the Triassic is marked by an extinction that results in the disappearance of the archosaurs other than dinosaurs, crocodylomorphs, and pterosaurs.

See also the following related entries:

Cretaceous Period • Extinction, Triassic • Jurassic Period • Mesozoic Era

References

Benton, M. J. (1986). More than one event in the late Triassic mass extinction. *Nature* **321,** 857–861.

Casamiquela, R. M. (1967). Un nuovo dinosaurio ornitisquio Triasico (*Pisanosaurus mertii*; Ornithopoda) de la Formacion Ischigualasto, Argentina. *Ameghiana* **5**(2), 47–64.

Colbert, E. H. (1970). A saurischian dinosaur from the Triassic of Brazil. *Am. Museum Novitates* **2405,** 1–39.

Olsen, P. E., and Sues, H.-D. (1986). Correlation of continental Late Triassic and Early Jurassic sediments, and patterns of Triassic–Jurassic tetrapod transition. In *The Beginning of the Age of Dinosaurs* (K. Padian, Ed.), pp. 321–351. Cambridge Univ. Press, New York.

Parrish, J. T., and Curtis, R. L. (1982). Atmospheric circulation, upwelling, and organic-rich rocks in the Mesozoic and Cenozoic eras. *Palaeogeogr. Palaeoclimatol. Palaeoecol.* **40,** 31–66.

Sereno. P. C., Forster, C. A., Rogers, R. R., and Monetta, A. M. (1993). Primitive dinosaur skeleton from Argentina and the early evolution of Dinosauria. *Nature* **361,** 64–66.

Troodontidae

David J. Varricchio
Old Trail Museum
Choteau, Montana, USA

Troodontids, known primarily from the Cretaceous of Asia and North America (Table I), are small long-legged theropods with large brains, highly pneumatic skulls, many teeth, and retractable second-toe claws. Ferdinand Hayden collected the first troodontid specimen, a single tooth, in 1855 along the Judith River in the Nebraska Territory, now central Montana. The next year Joseph Leidy described this find, and *Troodon formosus* survives somewhat tenuously as the oldest dinosaur name still recognized as valid from the Western Hemisphere. The true nature of troodontids remained obscure for more than 100 years. This resulted primarily from a lack of specimens, though Charles Gilmore compounded the problem in 1924 when he synonymized *Stegoceras* and *Troodon* and established the family Troodontidae for thick-domed ornithischian dinosaurs. Both genus and family names were commonly applied to pachycephalosaurs for the next 25 years until the theropod nature of the Troodontidae could be established (Russell, 1948). Although Henry Farfield Osborn named *Saurornithoides mongoliensis* as early as 1924, a fuller picture of troodontid morphology only began to emerge approximately 25 years ago with the description of new *T. formosus* (= *Stenonychosaurus inequalis*) material from Alberta (Russell, 1969). Subsequent discoveries from Alberta (Currie, 1985, 1987; Wilson and Currie, 1985; Currie and Zhao, 1993) and new specimens and species from the Cretaceous of Mongolia (Barsbold, 1974; Barsbold *et al.*, 1987; Osmólska, 1987; Kurzanov and Osmólska, 1991; Currie and Peng, 1993; Russell and Dong, 1993) firmly established the Troodontidae as a well-defined group sharing unique features of the skull, dentition, and foot. The following description is based on the previously cited references, undescribed *Troodon* material from Montana, and the review of Osmólska and Barsbold (1990). Troodontidae may be defined as *Troodon, Sinornithoides, Saurornithoides, Borogovia,* and all coelurosaurs closer to them than to ornithomimids, oviraptorisaurs, or other well-defined taxa. Possible shared derived features (synapomorphies) of the Troodontidae include high number of maxillary and dentary teeth; teeth with a constriction between root and crown and large, hooked denticles with rounded pits between their bases; anteriorly split splenial; sharp sagittal crest; lateral depression for middle ear and a complex sinus system; pneumatic quadrate; distal caudals with neural groove rather than spine; lateral knob-like trochanter on the femur; fused calcaneum and astragalus; tall ascending process on astragalus; metatarsus with a laterally compressed metatarsal II, a long and strongly pinched metatarsal III, and a stout metatarsal IV; and a modified second toe allowing retraction of an enlarged ungual.

The long, low troodontid skull ends in a narrow U-shaped snout and lacks the bony protuberances of other theropods (Fig. 1). The skull has a slight arch: The skull roof slopes upwards to the posterior margin of the orbit, then bends and slopes downwards to the snout tip. Large orbits face laterally (*Sinornithoides*) to anterolaterally (*Troodon*) and indicate good, possibly stereoscopic vision. A narrow sagittal crest, similar to those found in some large theropods, marks the posterior roof of the skull. A large brain cavity housed a brain comparable in size to those of modern ratite birds (e.g., ostrich and emu) and its internal dimensions suggest a well-developed sense of balance but a poorer sense of smell in comparison to other theropods, such as dromaeosaurids and tyrannosaurids.

"Lateral depressions" lie on each side of the braincase posteriorly (Fig. 1) and several passageways penetrate the bones of this region. These circular depressions contained the middle ear, the air-filled space between the tympanic membrane and inner ear, and an elaborate craniofacial sinus system similar to that found in extant birds. Unusual effects of this sinus system include a bulbous parasphenoid capsule, found elsewhere only in the ornithomimosaurs, and a bird-like pneumatic (air-filled) quadrate. A large and well-developed middle ear likely gave troodontids excellent hearing and possibly enhanced low-

TABLE I Species of Troodontidae[a]

		MONGOLIA	CHINA	NORTH AMERICA
Late Cretaceous Forms	Maastrichtian	***Borogovia gracilicrus*** **Osmólska, 1987** ***Tochisaurus nemegtensis*** **Kurzanov and Osmólska, 1991** ***Saurornithoides junior*** **Barsbold, 1974**		***Troodon formosus*** **Leidy, 1856**
	Campanian	***Saurornithoides mongoliensis*** **Osborn, 1924**		
Early Cretaceous Forms	Unknown Stage	**indeterminate troodontid** **Barsbold, et al., 1987**	***Sinornithoides youngi*** **Russell and Dong, 1993**	

[a]Questionable troodontids, known only from fragmentary material and not shown, include: *Koparion douglassi*, tooth, Upper Jurassic, Utah (Chure, 1994); *Ornithodesmus cluniculus*, sacrum, Lower Cretaceous, England (Howse and Milner, 1993); and *Bradycneme draculae and Heptasteornis andrewsi*, partial tibiotarsi, Upper Cretaceous, Romania (Paul, 1988). *Troodon formosus* includes *Polyodontosaurus grandis*, *Stenonychosaurus inequalis*, and *Pectinodon bakkeri*.

frequency sound detection. The sinus system would have lightened the skull and potentially, through side-to-side connections, improved sound localization.

Troodontids possess large skull openings, the antorbital and maxillary fenestrae (Fig. 1). The spaces defined by each connect through the posterior portion of the maxilla. Another complex sinus system likely filled these spaces (Witmer, 1987) and possibly extended into some of the bones of the palate. The maxilla forms part of the floor of the nasal openings, an unusual feature among small theropods, and also bears a long, broad shelf that contributed to the palate. A complete troodontid palate is unknown and how this bony shelf may have affected troodontid respiration and physiology is not currently understood. Both the upper and lower jaws of troodontids, like those of all other theropods, have interdental plates. These were previously thought to be absent as in all birds except *Archaeopteryx*.

Troodontid teeth are both unusual and critical to troodontid taxonomy. The group includes three taxa named and described solely on teeth: *T. formosus*; *Pectinodon bakkeri*, considered a possible junior synonym of *T. formosus* (Currie, 1987); and *Koparion douglassi*, potentially the oldest troodontid (Chure, 1994). Because *T. formosus* is the type species of *Troodon*, the type genus of the family Troodontidae, the utility of the name Troodontidae depends directly on the diagnostic worth of the *T. formosus* type specimen at the species level. Current usage follows Currie (1987), who accepted the validity of *T. formosus*. Although some recent finds may challenge this usage, caution is necessary. Troodontid taxonomy has already had a rough history and future finds, such as skulls with well-preserved dentitions, or microscopic and chemical analyses may verify Currie's 1987 interpretation.

Tooth counts range from a low of approximately 96 in *S. mongoliensis* to a high of 122 in *Troodon*. Among theropods only *Coelophysis, Syntarsus*, and perhaps *Baryonyx* (at least in the lower jaw) possess similar numbers of teeth, and only the unusual ornithomimosaur *Pelecanimimus* has more (Perez-Moreno *et al.*, 1994). Tooth size and shape vary according to jaw position: More anterior teeth are smaller, subcylindrical, and somewhat incisor-like. More posterior teeth

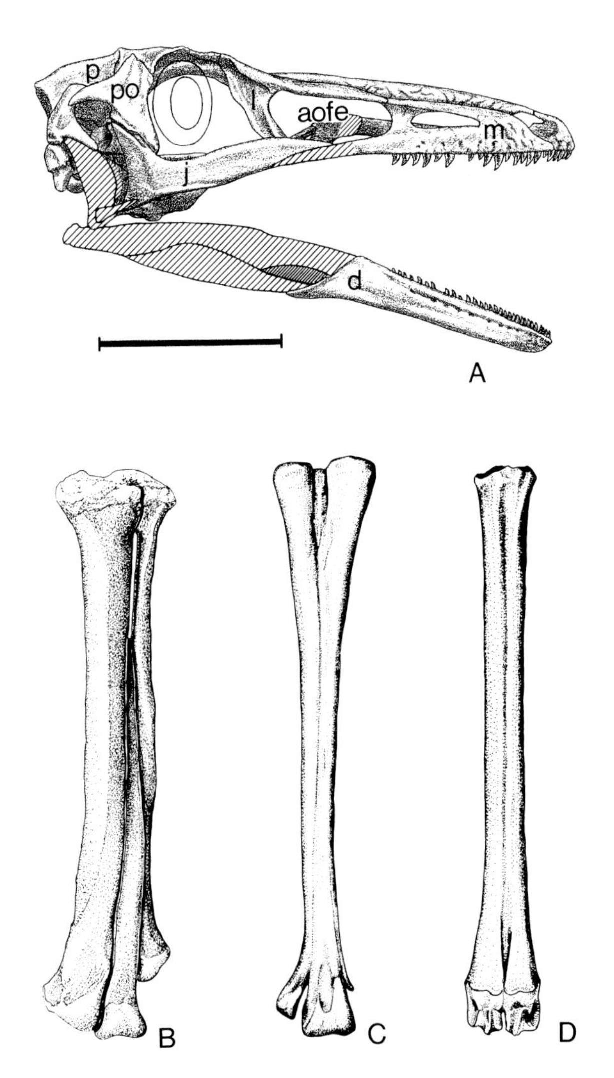

FIGURE 1 (A) Skull of *Saurornithoides junior*; narrow opening in maxilla is maxillary fenestra. Scale bar = 10 cm. (B, C, and D) Metatarsals of functionally two-toed vertebrates showing convergences of metatarsus structure. B, right *Troodon*; C, right juvenile ostrich; D, left mule deer. B and C show metatarsals II, III, and IV; D shows metatarsals II and III. Drawn in anterior view and to the same length of metatarsal III. Illustrations by Greg Paul (A), Elaine Nissen (B), and Frankie Jackson (C and D).

are larger, laterally compressed, sharply tapered, and recurved. Maxillary teeth tend to be larger than those of the dentary. Although all teeth bear posterior serrations, the presence and size of anterior serrations vary both among taxa and by tooth position. All Cretaceous troodontid teeth have large denticles, strongly curved toward the tooth tip. Despite the smaller overall tooth size, the basal length of *Troodon* denticles roughly equals those of tyrannosaurids [Farlow *et al.*, 1991 (see Figs. 4 and 5)]. Distinct round pits mark the grooves between adjacent denticles, and all troodontid teeth have a constriction between root and crown, a feature also found in early birds. *Koparion* differs significantly from Cretaceous troodontids in having blunt denticles and an overall squat tooth shape; only future finds can verify its inclusion in the group.

The troodontid dentition differs markedly from those of contemporary theropods, but its function is unclear. High tooth counts combined with denticle shape provide troodontids with an extraordinary high number of sharply pointed cusps. The numerous cusps may have served as an efficient saw for cutting flesh but not bone (Abler, 1992), for slicing plant material (Holtz *et al.*, 1994), or in the capture of small prey. Some mammalian carnivores that specialize on insects, such as the bat-eared fox, have increased tooth and/or cusp counts (Ewer, 1973). Insectivory, therefore, may be another possibility for troodontids.

Although troodontid material is incomplete, elongate cervical vertebrate suggest that the neck was long. Centra of all cervical vertebrae and the first one or two dorsals bear pleurocoels, openings for an air sac system. Troodontid sacra consist of six vertebrae with fused centra, neural spines, and arches. Caudal vertebrae all have a longitudinal ventral groove and become relatively more elongate down the tail. Both transverse processes and neural spines diminish posteriorly and are rudimentary to absent beyond the tenth caudal, or approximately a quarter of the way down the tail. A shallow groove replaces the neural spine on the distal caudals. Chevrons initially project downward but flatten by the tenth caudal. Elongate anterior articulating processes on the vertebrae, as well as flattened chevrons, probably aided in stiffening the tail. Pockets for the attachment of interspinous ligaments—well-developed on troodontid cervical, dorsal, and anterior caudal vertebrae—are reduced to circular pits just above the spinal canal on distal caudals, a feature that may be unique to troodontids.

Like those of *Deinonychus* and *Archaeopteryx*, the strap-like scapula of troodontids shows little expansion for the acromial process, whereas the coracoid has a prominent biceps tubercle. Thulborn [1984 (see Fig. 4E)] interpreted an isolated V-shaped bone of *Troodon*, originally described as a ventral rib, as a furcula; that is, fused clavicles. This is most likely fused gastralia, however. In contrast, the slender rod-like clavicles in the beautifully articulated *Sinornithoides* specimen lie separately. Humerus curvature and deltopectoral crest size vary with species and individual size. The ulna bows and a semilunate carpal allowed mediolateral movement of the sharply clawed, three-fingered hand. Several taxa of small theropods and birds share these features.

The troodontid PELVIS is poorly known. The pubic shaft points anteroventrally, making the pelvis propubic, as in nearly all theropods, and may bear a distal foot. Unlike most other theropods, the foot extends farther anteriorly than posteriorly. The strongly curved femur possesses four proximal trochanters in large individuals: a lesser and a greater, which extend to the proximal end of the femur and are separated only by a shallow notch; a lower lateral knob (posterior trochanter of Currie and Peng, 1993); and a posterior fourth. Dromaeosaurids, *Archaeopteryx,* and perhaps birds have lesser and greater trochanters of similar form and placement, and most theropods primitively retain a fourth trochanter. Tibiae of all species are long and slender. The strongly reduced calcaneum fuses to the astragalus and bears no facet for receiving the reduced fibula. The exceptionally long ascending process of the astragalus extends over 25% of the tibia shaft.

Metatarsals fit closely together and are reminiscent of the cannon bone of ungulates such as deer and antelope (Fig. 1). Proximally, metatarsal III pinches between metatarsals II and IV in a fashion similar to but more extreme than that in ornithomimosaurs and tyrannosaurids. Whereas metatarsal II is laterally compressed and narrow, metatarsal IV is stout. Modified phalanges of the second toe allowed retraction of an enlarged and (except in *Borogovia)* recurved ungual. The claw borne by this toe may have served as an offensive weapon. Articulated specimens show that the second toe was normally retracted, protecting the claw from wear and making the foot functionally two toed (didactyl). Although dromaeosaurids possess a similar modified second toe, the ungual is notably more recurved and laterally compressed, and the metatarsus is much stouter.

Complete HINDLIMBS are known only for *Sinornithoides* and *Troodon,* and femur length and tibia-astragalus/femur and metatarsus/tibia-astragalus length ratios are respectively: 140 mm, 141 and 56%; and 322 mm, 115 and 58%. [Note that previous estimates for *Troodon* of Varricchio (1993) are wrong.] The elongate didactyl foot and hindlimb of troodontids probably made them some of the faster or at least more efficient runners among dinosaurs. Histologic examination of a small sample of *Troodon* hindlimbs (Varricchio, 1993) reveals an abundance of fibrolamellar tissue, a type of well-vascularized and fast-growing bone. This and other tissues suggest high activity levels and attainment of adult size (>50 kg) in approximately 5 years.

Origin of and interrelationships within the troodontids are poorly understood. Good specimens come only from the Cretaceous of Asia and North America, and most are far from complete. Early Cretaceous forms exhibit typical troodontid features; therefore, presumably there should be older relatives. Jurassic and Early Cretaceous finds, such as *Koparion* and *Ornithodesmus cluniculus,* may provide clues to the clade's origin, but these specimens are fragmentary and of dubious affinity. Nevertheless, remaining troodontids form a well-defined group.

The Troodontidae have been variously placed within the Theropoda (Fig. 2). Both dromaeosaurids and troodontids share a specialized second toe, perhaps the result of common ancestry (Ostrom, 1969). Other shared features may not be unique to the two taxa and specific features of the second toe differ. Troodontids, ornithomimosaurs, and perhaps tyrannosaurids may represent a more closely related group (Thulborn, 1984; Holtz, 1994) because all display similar modifications of the metatarsus. Additionally, ornithomimosaurs and troodontids share a long, low skull arched behind the orbit, a bulbous parasphenoid capsule, more elongate cervical vertebrae with low neural spines, and an elongate tibia.

A new multitoothed ornithomimosaur, *Pelecanimimus,* could explain the disparity between toothless ornithomimosaurs and toothy troodontids (Perez-Moreno *et al.*, 1994). Troodontids may share several derived features with the enigmatic "segnosaurs" such as *Erlikosaurus andrewsi* (Clark *et al.*, 1994). Several features of troodontids also suggest a close relationship with birds: a complex middle ear region and associated sinus system, a pneumatic quadrate, and constriction between tooth root and crown. However, known troodontids possess some derived features, for example, the modified metatarsus, which is not found in the most primitive birds, and troodontids occur too late in the Cretaceous to be bird ancestors. Convergences between various theropods and birds could be widespread, and it will take time and additional specimens to sort them out from synapomorphies.

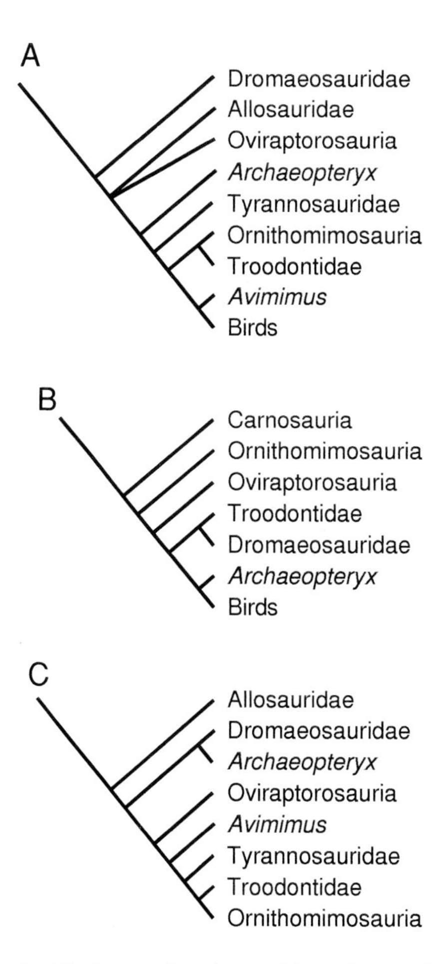

FIGURE 2 Phylogenetic relationships of some theropods and birds as proposed by Thulborn (1984) (A), Gauthier (1986) (B), and Holtz (1994) (C). Note the variability of the troodontids' position relative to dromaeosaurids, ornithomimosaurs, *Archaeopteryx,* and birds. Currently, this represents the most recent state of consensus.

Troodontids apparently inhabited a variety of environments ranging from arid to wetter, more seasonal climates of Mongolia and from the warm temperate of New Mexico through the cool temperate of Alaska's North Slope (Clemens and Nelms, 1993). The occurrences of *Troodon* in Alberta suggest a preference for drier, more inland areas (Russell, 1969; Brinkman, 1990). The only multi-individual assemblage accumulated on the shores of a small lake and may reflect an ecologic preference or group behavior (Varricchio, 1995).

In 1993, an adult *Troodon* was found lying directly on a clutch of at least eight eggs. These eggs conform in size, shape, and both macro- and microscopic eggshell features to eggs previously assigned on the basis of embryonic remains to the hypsilophodont *Orodromeus* (Horner and Weishampel, 1988). Subsequent reexamination of these embryonic remains shows them to be *Troodon* and not *Orodromeus* (Varrichio *et al.*, 1997). Thus, previously published discussions on *Orodromeus* nesting behavior (Horner, 1982, 1984, 1987) now apply to *Troodon* and eggs tentatively assigned to *Troodon* (Hirsch and Quinn, 1990) represent an unknown taxon.

Troodon clutches contain up to 24 tightly spaced, elongate eggs laid on end. A low earthen mound encircles the clutch. Like those of *Oviraptor, Troodon* eggs seemingly occur in pairs. The egg/adult size ratio, eggshell structure, and egg pairs suggest *Troodon* had an avian egg-laying style but with two functional ovaries and oviducts. Thus, rather than retaining an entire clutch and laying it en masse like crocodilians, *Troodon,* and at least some other theropods, would have produced two eggs nearly simultaneously, laying pairs at one or multiple-day intervals. Large clutches would require days or perhaps weeks to be completed. The adult *Troodon* and earthen mound associated with clutches may reflect incubation behavior or protection of the eggs during the long, egg-laying preincubation period.

See also the following related entries:

BULLATOSAURIA • COELUROSAURIA • INTELLIGENCE • ORNITHOMIMOSAURIA

References

Abler, W. L. (1992). The serrated teeth of tyrannosaurid dinosaurs, and biting structures in other animals. *Paleobiology* **18,** 161–183.

Barsbold, R. (1974). Saurornithoididae, a new family of small theropod dinosaurs from central Asia and North America. *Palaeontol. Polonica* **30,** 5–22.

Barsbold, R., Osmólska, H., and Kurzanov, S. M. (1987). On a new troodontid (Dinosauria, Theropoda) from the early Cretaceous of Mongolia. *Palaeontol. Polonica* **32,** 121–132.

Brinkman, D. B. (1990). Paleoecology of the Judith River Formation (Campanian) of Dinosaur Provincial Park, Alberta, Canada: Evidence from vertebrate microfossils localities. *Palaeogeogr. Palaeoclimatol. Palaeoecol.* **78,** 37–54.

Chure, D. J. (1994). *Koparion douglassi,* a new dinosaur from the Morrison Formation (Upper Jurassic) of Dinosaur National Monument: The oldest troodontid (Theropoda: Maniraptora). *Brigham Young Univ. Geol. Stud.* **40,** 11–15.

Clark, J. M., Altangerel, P., and Norrell, M. A. (1994). The skull of *Erlicosaurus andrewsi,* a Late Cretaceous "Segnosaur" (Theropoda: Therizinosauridae) from Mongolia. *Am. Museum Novitates* **3115**.

Clemens, W. A., and Nelms, L. G. (1993). Paleoecological implications of Alaskan terrestrial vertebrate fauna in latest Cretaceous time at high paleolatitudes. *Geology* **21**, 503–506.

Currie, P. J. (1985). Cranial anatomy of *Stenonychosaurus inequalis* (Saurischia, Theropoda) and its bearing on the origin of birds. *Can. J. Earth Sci.* **22**, 1643–1658.

Currie, P. J. (1987). Bird-like characteristics of the jaws and teeth of troodontid theropods (Dinosauria, Saurischia). *J. Vertebr. Paleontol.* **7**, 72–81.

Currie, P. J., and Peng, J.-H. (1993). A juvenile specimen of *Saurornithoides mongoliensis* from the Upper Cretaceous of northern China. *Can. J. Earth Sci.* **30**, 2224–2230.

Currie, P. J., and Zhao, X.-J. (1993). A new troodontid (Dinosauria, Theropoda) braincase from the Dinosaur Park Formation (Campanian) of Alberta. *Can. J. Earth Sci.* **30**, 2231–2247.

Ewer, R. F. (1973). *The Carnivores*, pp. 494. Cornell Univ. Press, Ithaca, NY.

Farlow, J. O., Brinkman, D. L., Abler, W. L., and Currie, P. J. (1991). Size, shape and serration of theropod dinosaur lateral teeth. *Mod. Geol.* **16**, 161–198.

Gauthier, J. (1986). Saurischian monophyly and the origin of birds. *Mem. California Acad. Sci.* **8**, 1–55.

Hirsch, K. F., and Quinn, B. (1990). Eggs and eggshell fragments from the Upper Cretaceous Two Medicine Formation of Montana. *J. Vertebr. Paleontol.* **10**, 491–511.

Holtz, T. R. (1994). The phylogenetic position of the Tyrannosauridae: Implications for theropod systematics. *J. Paleontol.* **68**, 1100–1117.

Holtz, T. R., Brinkman, D. L., and Chandler, C. L. (1994). Denticle morphometrics and a possibly different life habit for the theropod dinosaur *Troodon*. *J. Vertebr. Paleontol.* **14**(Suppl. 3), 30A.

Horner, J. R. (1982). Evidence for colonial nesting and "site fidelity" among ornithischian dinosaurs. *Nature* **297**, 675–676.

Horner, J. R. (1984). The nesting behavior of dinosaurs. *Sci. Am.* **250**, 130–137.

Horner, J. R. (1987). Ecological and behavioral implications derived from a dinosaur nesting site. In *Dinosaurs Past and Present, Volume II* (S. J. Czerkas and E. C. Olson, Eds.), pp. 51–63. Natural History Museum of Los Angeles County, Los Angeles.

Horner, J. R., and Weishampel, D. (1988). A comparative embryological study of two ornithischian dinosaurs. *Nature* **332**, 256–257.

Howse, S. C. B., and Milner, A. R. (1993). *Ornithodesmus*—A maniraptoran theropod dinosaur from the Lower Cretaceous of the Isle of Wight, England. *J. Palaeontol.* **36**, 425–437.

Kurzanov, S. M., and Osmólska, H. (1991). *Tochisaurus nemegtensis* gen. et sp. n., a new troodontid (Dinosauria, Theropoda) from Mongolia. *Palaeontol. Polonica* **36**, 69–76.

Osmólska, H. (1987). *Borogovia gracilis* gen. et sp. n., a new troodontid dinosaur from the Late Cretaceous of Mongolia. *Acta Palaeontol. Polonica* **32**, 133–150.

Osmólska, H., and Barsbold, R. (1990). Troodontidae. In *The Dinosauria* (D. B. Weishampel, P. Dodson, and H. Osmólska, Eds.), pp. 259–268. Univ. of California Press, Berkeley.

Ostrom, J. H. (1969). Osteology of *Deinonychus antirrhopus*, an unusual theropod from the Lower Cretaceous of Montana. *Peabody Museum Nat. History Bull.* **30**, 1–165.

Paul, G. S. (1988). *Predatory Dinosaurs of the World*. Simon & Schuster, New York.

Perez-Moreno, B. P., Sanz, J. L., Buscalioni, A. D., Moratalla, J. J., Ortega, F., and Rasskin-Gutman, D. (1994). A unique multi-toothed ornithomimosaur dinosaur from the Lower Cretaceous of Spain. *Nature* **370**, 363–367.

Russell, D. A. (1969). A new specimen of *Stenonychosaurus* from the Oldman Formation (Cretaceous) of Alberta. *Can. J. Earth Sci.* **6**, 595–612.

Russell, D. A., and Dong, Z.-M. (1993). A nearly complete skeleton of a new troodontid dinosaur from the Early Cretaceous of the Ordos Basin, Inner Mongolia, People's Republic of China. *Can. J. Earth Sci.* **30**, 2163–2173.

Russell, L. S. (1948). The dentary of *Troodon*, a genus of theropod dinosaurs. *J. Paleontol.* **22**, 625–629.

Thulborn, R. (1984). The avian relationships of *Archaeopteryx*, and the origin of birds. *Zool. J. Linnean Soc.* **82**, 119–158.

Varricchio, D. J. (1993). Bone microstructure of the Upper Cretaceous theropod dinosaur *Troodon formosus*. *J. Vertebr. Paleontol.* **13**, 99–104.

Varricchio, D. J. (1995). Taphonomy of Jack's Birthday Site, a diverse dinosaur bonebed from the Upper Cretaceous Two Medicine Formation of Montana. *Palaeogeogr. Palaeoclimatol. Palaeoecol.* **114**, 297–323.

Varricchio, D. J., Jackson, F., Borkowski, J. J., and Horner, J. R. (1997). Nest and egg clutches of the dinosaur *Troodon formosus* and the evolution of avian reproductive traits. *Nature* **385**, 247–250.

Wilson, M. C., and Currie, P. J. (1985). *Stenonychosaurus inequalis* (Saurischia: Theropoda) from the Judith River (Oldman) Formation of Alberta: New findings on metatarsal structure. *Can. J. Earth Sci.* **22**, 1813–1817.

Witmer, L. M. (1987). The nature of the antorbital fossa of archosaurs: Shifting the null hypothesis. In *Fourth Symposium on Mesozoic Terrestrial Ecosystems, Short Papers* (P. J. Currie and E. H. Koster, Eds.), Occasional Papers No. 3, pp. 234–239. Tyrrell Museum of Paleontology, Drumheller, Alberta, Canada.

Trophic Groups

James O. Farlow
Indiana University–Purdue University
at Fort Wayne
Fort Wayne, Indiana, USA

Most organisms on this planet ultimately depend on the sun as the source of energy for biological processes. Solar thermonuclear reactions result in the profligate radiation of electromagnetic energy in all directions away from our star. A tiny fraction of this energy is intercepted by the earth, and a small portion of that is fixed as organic material by photosynthesis. Some of this primary production is consumed by herbivores, some of whom in turn fall prey to carnivores. Herbivores and carnivores thus represent two important trophic (feeding) levels in the food chains of biological communities.

An animal is a biochemical machine that uses energy from its food to power the activities and processes necessary for life. Among these processes is the synthesis of new biological tissues (biomass), which are assembled using nutrients from the food as constituents. Such newly synthesized materials are manifest as increases in body size during the growth of young animals, in the accumulation of fatty storage tissues as a hedge against lean times, and in reproduction, which generates new individuals of the species. These processes of biological production are necessary if an individual animal's genes are to be passed on to the next generation. Enough individuals of the species must be successful in these productive endeavors if the species itself is to survive. The acquisition of food is consequently one of the most important activities of an animal's life. This was as true for dinosaurs during the Mesozoic Era as it is for living animals.

The adaptive radiation of dinosaurs resulted in the evolution of a host of species in both the herbivore and carnivore trophic levels. Identifying the food habits of many dinosaurs is fairly straightforward. The sophisticated grinding dentitions of hadrosaurids and their close relatives (Weishampel, 1984; Norman and Weishampel, 1985; Bakker, 1986) and the slicing dentitions of ceratopsids (Ostrom, 1966; Bakker, 1986) are reasonably interpreted as adaptations for processing coarse, fibrous vegetation. Sauropod teeth do not seem well constructed for grinding or slicing fodder, but look adequate to the job of nipping, plucking, or stripping foliage for further processing in the gut (Farlow, 1987b; Dodson, 1990; Gillette, 1994). The sharp, often serrate-edged teeth of most theropods (Currie *et al.*, 1990; Farlow *et al.*, 1991; Abler, 1992) constitute unambiguous evidence for carnivory in these reptiles.

Morphological evidence of food habits in some dinosaur groups, however, is more equivocal. Oviraptorosaurs, ornithomimosaurs, prosauropods, and ankylosaurs are forms whose diets have been the subject of some controversy (Barsbold and Osmólska, 1990; Barsbold *et al.*, 1990; Coombs and Maryanska, 1990; Galton, 1990); their dentitions (or lack thereof) do not provide unambiguous information about their diets. Furthermore, it is conceivable that many dinosaur taxa were omnivores that derived most of their nutrition from plants but occasionally supplemented their diets with animal matter (Paul, 1991).

Morphology is not the only source of evidence about dinosaur diets, however. Taphonomic associations of the bones or teeth of predatory dinosaurs with the remains of their presumed prey (Ostrom, 1969; Buffetaut and Suteethorn, 1989), co-occurrences of trackways of carnivores and herbivores that do not appear coincidental (Farlow, 1987a), bite marks in bone (Molnar and Farlow, 1990), preserved gut contents (Ostrom, 1978; Colbert, 1989), and coprolites (Thulborn, 1991; Chin, 1990; Chin *et al.*, 1991; Chin and Gill, 1993) provide circumstantial or even *prima facie* testimony about dinosaur food habits.

Perhaps the most novel evidence about dinosaur diets comes from stable isotope geochemistry. High-molecular-weight organic materials extracted from teeth and bones of tyrannosaurids and dromaeosaurids from the Late Cretaceous Judith River Formation have relatively high δ^{15} nitrogen (δ^{15}N) values, whereas the same ratios for ceratopsids, nodosaurids, *Stegoceras*, and hadrosaurids are lower (Ostrom *et al.*, 1993). In living animals, the δ^{15}N ratio increases from herbivores to carnivores in food chains, and so the results of Ostrom *et al.* (1993) are consistent with inferences about the diets of Judith River dinosaurs based on skeletal morphology. Interestingly, the δ^{15}N values obtained for ornithomimids suggest that these dinosaurs were more likely carnivores than herbivores. In contrast, the δ^{15}N values of nodosaurs are consistent with herbivory. Bocherens *et al.* (1994) reported that the strontium content of enamel and dentine is lower for Judith River theropod than for ceratopsian and hadrosaur teeth, again consistent with the inferred trophic positions of these dinosaurs and with strontium levels observed in modern herbivorous and carnivorous animals.

As biochemical machines, animals are inefficient at converting the food they eat into new animal biomass. A portion of the ingested food (generally greater in herbivores than in carnivores) is completely indigestible and exits the animal's body as feces (potential coprolites). Heat is released by the breakdown of chemical bonds of food during digestion and also by the subsequent synthesis of new biomass; although this heat may be used in thermoregulation by the animal, it nonetheless represents food energy that cannot be converted into new biomass. Excreted wastes of the animal constitute yet another energy loss.

Because of the inherent inefficiency in the conversion of ingested energy into new biomass, the total amount of biomass synthesized at one trophic level is considerably less than the amount produced at the preceding trophic level. In food chains in which plants are eaten by large-bodied herbivores, which in turn are eaten by large-bodied carnivores, the total number of herbivorous animals will be greater than the total number of carnivores. There are many more zebras and antelopes than lions and hyenas on the plains of East Africa and many more deer than wolves and cougars in western North America.

The relative abundance of herbivore and carnivore skeletal specimens in fossil vertebrate assemblages usually shows this same pattern (Farlow, 1990; Schult and Farlow, 1992). The ratio of the number of specimens of herbivores to carnivores (including dinosaurs in both trophic groups) in the upper unit of the CHINLE FORMATION of the PETRIFIED FOREST (Arizona) is approximately 3 herbivores : 1 carnivore (Parrish, 1993); in the Late Triassic ISCHIGUALASTO FORMATION of South America, the ratio is approximately 7 : 1 (Rogers *et al.*, 1993). For the Late Jurassic MORRISON FORMATION of the western United States, the ratio of herbivorous to carnivorous dinosaur specimens is about 5 : 1 (Coe *et al.*, 1987); because the average size of Morrison herbivorous dinosaurs is much greater than that of carnivorous dinosaurs, the reconstructed herbivore : carnivore biomass ratio is greater than the ratio based on counts of individual specimens (Farlow, 1990).

Although herbivores generally do outnumber carnivores in dinosaur skeletal assemblages, the herbivore : carnivore specimen ratio may not exactly reflect the herbivore : carnivore ratio of the living fauna. Skeletal assemblages accumulate in a variety of ways, each of which is subject to its own sampling biases. For example, the herbivore : carnivore ratio for large dinosaurs (ankylosaurs, hadrosaurs, ceratopsians, and tyrannosaurs) of the Late Cretaceous JUDITH RIVER FORMATION of Alberta can be as low as 6 : 1 or as high as 100 : 1, depending on whether the count is made of all collected specimens, "articulated" specimens, surface-collected "microfaunal" scrap, or screen-washed skeletal and tooth debris (Schult and Farlow, 1992; Farlow, 1993). Furthermore, most such specimen counts are made over the entire stratigraphic range of the formation. Although this may give a picture of the relative numbers of herbivores and carnivores averaged over the interval of time during which the formation's sediments accumulated, it may not accurately reflect the number of herbivores and carnivores that existed at any particular time during that interval (Behrensmeyer *et al.*, 1992).

Some fossil assemblages, such as "low-diversity" bone beds, do represent the numbers of individuals that died over a very short interval of time. However, even these assemblages have their shortcomings. Low-diversity bone beds are dominated by a single species, having formed during or after the mass deaths of herding animals (Wood *et al.*, 1988; Varricchio and Horner, 1993), and so do not constitute a representative sample of the entire dinosaur fauna.

Counts of trackways provide an entirely different way of estimating the number of herbivorous and carnivorous dinosaurs in Mesozoic formations (Lockley, 1991; Lockley and Hunt, 1994). However, these are subject to the same problems of time averaging, or of biased sampling, as are skeletal assemblages. Although in some formations the herbivore : carnivore ratio estimated from trackways is approximately the same as that estimated from skeletal assemblages, it is not unusual for dinosaur footprint assemblages in some geologic settings to be dominated by the tracks of carnivores (Lockley, 1991; Schult and Farlow, 1992—assuming that the trackmakers have been identified correctly) to an extent unlikely to have been typical of the living fauna. This may reflect greater activity on the part of theropods than herbivorous dinosaurs (Schult and Farlow, 1992).

Yet another problem arises if one wants to do more than tabulate the number of dinosaurs of different trophic groups in a formation. If one is interested in reconstructing the dynamics of predator–prey feeding interactions (Farlow, 1976) or in using the relative numbers of carnivores and herbivores to determine dinosaur metabolic rates (Bakker, 1986), it is not enough to know which dinosaurs were herbivores and which were carnivores. It is also necessary to know which dinosaur species actually fed upon which other species and to have some idea of the totality of feeding interactions in which the various dinosaur species were involved. There is as yet no way of obtaining this kind of information; therefore, counts of herbivore and carnivore skeletal specimens or trackways do not shed much light on the contentious question of whether dinosaurs were endotherms or ectotherms (Farlow, 1990). Although the identification of trophic groups in dinosaur faunas and the attempt to determine the relative abundances of members of those groups are important aspects of the paleoecological interpretation of rock units in which dinosaur fossils occur, we may never be able to reconstruct the food webs of Mesozoic terrestrial communities with as much resolution as is (at least potentially) possible for modern vertebrate communities.

See also the following related entries:

DIET • PALEOCLIMATOLOGY • PALEOECOLOGY

References

Abler, W. L. (1992). The serrated teeth of tyrannosaurid dinosaurs and biting structures in other animals. *Paleobiology* **18,** 161–183.

Bakker, R. T. (1986). *The Dinosaur Heresies: New Theories Unlocking the Mystery of the Dinosaurs and Their Extinction*. pp. 482. Morrow, New York.

Barsbold, R., and Osmólska, H. (1990). Ornithomimosauria. In *The Dinosauria* (D. B. Weishampel, P. Dodson, and H. Osmólska, Eds.), pp. 225–248. Univ. of California Press, Berkeley.

Barsbold, R., Maryańska, T., and Osmólska, H. (1990). Oviraptorosauria. In *The Dinosauria* (D. B. Weishampel, P. Dodson, and H. Osmólska, Eds.), pp. 249–258. Univ. of California Press, Berkeley.

Behrensmeyer, A. K., Damuth, J. D., DiMichele, W. A., Potts, R., Sues, H.-D., and Wing, S. L. (Eds.) (1992). *Terrestrial Ecosystems through Time: Evolutionary Paleoecology of Terrestrial Plants and Animals*, pp. 568. Univ. of Chicago Press, Chicago.

Bocherens, H., Brinkman, D. B., Dauphin, Y., and Mariotti, A. (1994). Microstructural and geochemical investigations on Late Cretaceous archosaur teeth from Alberta, Canada. *Can. J. Earth Sci.* **31,** 783–792.

Buffetaut, E., and Suteethorn, V. (1989). A sauropod skeleton associated with theropod teeth in the Upper Jurassic of Thailand: Remarks on the taphonomic and paleoecological significance of such associations. *Palaeogeogr. Palaeoclimatol. Palaeoecol.* **73,** 77–83.

Chin, K. (1990). Possible herbivorous dinosaur coprolites from the Two Medicine Formation (Late Cretaceous) of Montana. *J. Vertebr. Paleontol.* **9**(3), 17A.

Chin, K., and Gill, B. D. (1993). Evidence for dinosaur/insect interactions in the Cretaceous (Campanian) Two Medicine Formation of Montana. *J. Vertebr. Paleontol.* **13**(3), 29A.

Chin, K., Brassell, S., and Harmon, R. J. (1991). Biogeochemical and petrographic analysis of a presumed dinosaurian coprolite from the Upper Cretaceous Two Medicine Formation, Montana. *J. Vertebr. Paleontol.* **11**(3), 22A.

Coe, M. J., Dilcher, D. L., Farlow, J. O., Jarzen, D. M., and Russell, D. A. (1987). Dinosaurs and land plants. In *The Origins of Angiosperms and Their Biological Consequences* (E. M. Friis, W. G. Chaloner, and P. R. Crane, Eds.), pp. 225–258. Cambridge Univ. Press, Cambridge, UK.

Colbert, E. H. (1989). The Triassic dinosaur *Coelophysis*. *Museum Northern Arizona Bull.* **57.**

Coombs, W. P., Jr., and Maryańska, T. (1990). Ankylosauria. In *The Dinosauria* (D. B. Weishampel, P. Dodson, and H. Osmólska, Eds.), pp. 456–483. Univ. of California Press, Berkeley.

Currie, P. J., Rigby, J. K., Jr., and Sloan, R. E. (1990). Theropod teeth from the Judith River Formation of southern Alberta, Canada. In *Dinosaur Systematics: Approaches and Perspectives* (K. Carpenter and P. J. Currie, Eds.), pp. 107–125. Cambridge Univ. Press, Cambridge, UK.

Dodson, P. (1990). Sauropod paleoecology. In *The Dinosauria* (D. B. Weishampel, P. Dodson, and H. Osmólska, Eds.), pp. 402–407. Univ. of California Press, Berkeley.

Farlow, J. O. (1976). A consideration of the trophic dynamics of a Late Cretaceous large-dinosaur community (Oldman Formation). *Ecology* **57,** 841–857.

Farlow, J. O. (1987a). *Lower Cretaceous Dinosaur Tracks, Paluxy River Valley, Texas*, pp. 50. South Central Section, Geological Society of America, Baylor University, Waco, TX.

Farlow, J. O. (1987b). Speculations about the diet and digestive physiology of herbivorous dinosaurs. *Paleobiology* **13,** 60–72.

Farlow, J. O. (1990). Dinosaur energetics and thermal biology. In *The Dinosauria* (D. B. Weishampel, P. Dodson, and H. Osmólska, Eds.), pp. 43–55. Univ. of California Press, Berkeley.

Farlow, J. O. (1993). On the rareness of big, fierce animals: Speculations about the body sizes, population densities, and geographic ranges of predatory mammals and large carnivorous dinosaurs. *Am. J. Sci.* **293A,** 167–199.

Farlow, J. O.,. Brinkman, D. L., Abler, W. L., and Currie, P. J. (1991). Size, shape, and serration density of theropod dinosaur lateral teeth. *Mod. Geol.* **16,** 161–198.

Galton, P. M. (1990). Basal Sauropodomorpha—Prosauropoda. In *The Dinosauria* (D. B. Weishampel, P. Dodson, and H. Osmólska, Eds.), pp. 320–344. Univ. of California Press, Berkeley.

Gillette, D. D. (1994). *Seismosaurus the Earth Shaker*, pp. 205. Columbia Univ. Press, New York.

Lockley, M. G. (1991). *Tracking Dinosaurs: A New Look at an Ancient World*, pp. 238. Cambridge Univ. Press, Cambridge, UK.

Lockley, M., and Hunt, A. (1994). *Fossil Footprints of the Dinosaur Ridge Area*, pp. 52. Friends of Dinosaur Ridge/Univ. of Colorado at Denver, Denver.

Molnar, R. E., and Farlow, J. O. (1990). Carnosaur paleobiology. In *The Dinosauria* (D. B. Weishampel, P. Dodson, and H. Osmólska, Eds.), pp. 210–224. Univ. of California Press, Berkeley.

Norman, D. B., and Weishampel, D. B. (1985). Ornithopod feeding mechanisms: Their bearing on the evolution of herbivory. *Am. Nat.* **126,** 151–164.

Ostrom, J. H. (1966). Functional morphology and evolution of the ceratopsian dinosaurs. *Evolution* **20,** 290–308.

Ostrom, J. H. (1969). Osteology of *Deinonychus antirrhopus*, an unusual theropod from the Lower Cretaceous of Montana. *Peabody Museum Nat. History Bull.* **30,** 1–165.

Ostrom, J. H. (1978). The osteology of *Compsognathus longipes*. *Zitteliana* **4,** 73–118.

Ostrom, P., Macko, S. A., Engel, M. H., and Russell, D. A. (1993). Assessment of trophic structure of Cretaceous communities based on stable nitrogen isotope analyses. *Geology* **21,** 491–494.

Parrish, J. M. (1993). Distribution and taxonomic composition of fossil vertebrate accumulations in the Upper Triassic Chinle Formation, Petrified Forest National Park. In *The Nonmarine Triassic* (S. G. Lucas and M. Morales, Eds.), Bull. No. 3, pp. 393–396. New Mexico Museum of Natural History and Science, Albuquerque.

Paul, G. S. (1991). The many myths, some old, some new, of dinosaurology. *Mod. Geol.* **16,** 69–99.

Rogers, R. R., Swisher, C. C, III, Sereno, P. C., Monetta, A. M., Forster, C. A., and Martínez, R. N. (1993). The Ischigualasto tetrapod assemblage (Late Triassic, Argentina) and ^{40}Ar/^{39}Ar dating of dinosaur origins. *Science* **260,** 794–797.

Schult, M. F., and Farlow, J. O. (1992). Vertebrate trace fossils. In *Trace Fossils* (C. G. Maples and R. R. West, Eds.), Paleontological Society Short Courses in Paleontology No. 5, pp. 34–63. Univ. of Tennessee, Knoxville.

Thulborn, R. A. (1991). Morphology, preservation and palaeobiological significance of dinosaur coprolites. *Palaeogeogr. Palaeoclimatol. Palaeoecol.* **83,** 341–366.

Varricchio, D. J., and Horner, J. R. (1993). Hadrosaurid and lambeosaurid bone beds from the Upper Cretaceous Two Medicine Formation of Montana: Taphonomic and biologic implications. *Can. J. Earth Sci.* **30,** 997–1006.

Weishampel, D. B. (1984). Evolution of jaw mechanisms in ornithopod dinosaurs. *Adv. Anat. Embryol. Cell Biol.* **87,** 1–110.

Wood, J. M., Thomas, R. G., and Visser, J. (1988). Fluvial processes and vertebrate taphonomy: The Upper Cretaceous Judith River Formation, south-central Dinosaur Provincial Park, Alberta, Canada. *Palaeogeogr. Palaeoclimatol. Palaeoecol.* **66,** 127–143.

Trossingen

David B. Weishampel
Johns Hopkins University School of Medicine
Baltimore, Maryland, USA

Perhaps the most famous bone bed ever discovered in Germany is in southern Baden-Württemberg, just outside the small town of Trossingen. Here, beginning in 1911, workers first from Stuttgart, then from Tübingen, and again from Stuttgart began uncovering in a red sandy clay a jumble of bones pertaining to one of the early members of SAUROPODOMORPHA called *Plateosaurus*.

In fact, there are two fossil vertebrate localities at Trossingen: the lower mill (unteren Mühle), where the Stubensandstein outcrops and from which *Teratosaurus trossingensis* (= *Sellosaurus gracilis*) was recovered, and the upper mill (oberen Mühle), where the Knollenmergel is found and from which abundant *Plateosaurus* and rare turtle material (*Proganochelys*) have been collected. The oberen Mühle has proved the more important. It was worked three different time during the early part of this century. The first specimen (a plateosaur fifth metatarsal) was collected in 1906 by a young schoolboy; it was subsequently given to his teacher, then placed in the care of E. Fraas of the königlichen Naturalienkabinett Stuttgart (now the Staatliches Museum für Naturkunde Stuttgart). In 1911 and 1912, Fraas and workers from the Naturalienkabinett uncovered approximately 12 individual plateosaurs.

Fraas was followed at Trossingen by F. von Huene of the Institut und Museum für Geologie and Paläontologie in Tubingen in 1921–1923. During these three summers, von Huene extracted in abundance some of the first complete and near-complete plateosaur remains from Trossingen.

The collections that von Huene made at Trossingen featured prominently in his work on the early evolution of saurischian dinosaurs. In the late 1910s and early 1930s, he revised all that was known about these dinosaurs, in doing so assigning all the Trossingen material to eight different species distributed within two genera. In addition, von Huene was one of the first dinosaur paleontologists to discuss the taphonomic significance of dinosaur bone beds, suggesting that herds of migrating *Plateosaurus* traveled across Europe and that Trossingen represents where some of these animals succumbed to the hardship of travel.

The Staatliches Museum für Naturkunde in Stuttgart returned to Trossingen one final time in 1932. Under the supervision of R. Seeman, workers enlarged the quarry and collected approximately 50–60 new individuals, many of them complete from head to tail. The work of 1932 ceased shortly after a wall of the quarry collapsed on October 14, killing one worker and injuring several others.

No further quarrying has been done in Trossingen. The site is currently protected by the state of Baden-Württemberg as a conservation area.

Interest in the taphonomy of the Trossingen bone bed and in the systematics of the plateosaurs from this locality resumed in the 1980s and 1990s. In his work on the systematics of basal sauropodomorphs, P. M. Galton (University of Bridgeport) revised the taxonomy of the specimens from Trossingen, and in so doing synonymizing all material with *Plateosaurus engelhardti*. Interested in the same question, but using a morphometric approach, D. B. Weishampel (Johns Hopkins University) and R. E. Chapman (U.S. National Museum of Natural History) similarly found the existence of a single, perhaps sexually dimorphic species.

Originally thought to be a prosauropod graveyard indicating death by drought by von Huene, P. M. Sander (University of Bonn) ascribed the abundance of plateosaurs to miring: Large animals such as *Plateosaurus* may have been drawn to drink from shallow ponds that occupied the floodplains of the region (the reconstructed depositional environment of the Knollenmergel at Trossingen); once these animals stepped onto the viscous mud of the pond bottoms, they became trapped in the ooze and thereafter died, perhaps with theropods scavenging on these abundant carcasses.

See also the following related entries:

European Dinosaurs • Extinction, Triassic • Prosauropoda • Triassic Period

References

Galton, P. M. (1985). Cranial anatomy of the prosauropod dinosaur *Plateosaurus* from the Knollenmergel (Middle Keuper, Upper Triassic) of Germany. II. All the cranial material and details of soft-part anatomy. *Geol. Palaeontol.* **19,** 119–159.

Sander, P. M. (1992). The Norian *Plateosaurus* bonebeds of central Europe and their taphonomy. *Palaeogeogr. Palaeoclimatol. Palaeoecol.* **93,** 255–299.

Weishampel, D. B., and Chapman, R. E. (1990). Morphometric study of *Plateosaurus* from Trossingen (Baden-Württemberg, Federal Republic of Germany). In *Dinosaur Systematics: Approaches and Perspectives* (K. Carpenter and P. J. Currie, Eds.), pp. 43–51. Cambridge Univ. Press, Cambridge, UK.

Tugrik, Tugrig, Tugreeg, Tugreek, Tugrigan Shir

see Djadokhta Formation

Two Medicine Formation

RAYMOND R. ROGERS
Cornell College
Mt. Vernon, Iowa, USA

The Upper Cretaceous Two Medicine Formation of Montana has yielded a respectable assortment of Late Cretaceous dinosaurs but is undeniably best known for its remarkable preservation of dinosaur EGGS (and the occasional dinosaur embryo), nests, and colonial nesting grounds (Horner, 1982, 1994; Horner and Makela, 1979; Horner and Weishampel, 1988; Horner and Currie, 1994). Charles W. Gilmore (collecting for the United States National Museum of Natural History) and Barnum Brown (collecting for the AMERICAN MUSEUM OF NATURAL HISTORY) led successful expeditions into the dinosaur-bearing strata of the Two Medicine Formation during the early years of this century. Recent paleontological inquiries have been directed by Jack Horner of the MUSEUM OF THE ROCKIES.

The Two Medicine Formation is exposed in the high plains to the east of Glacier National Park in northwestern Montana (Fig. 1). A variety of sedimentary rocks comprise the formation, and the outcrop belt of these ancient sediments consists of widespread badland exposures that are ideal both for paleontological prospecting and for geological investigations (Fig. 2). Through the analysis of Two Medicine sediments we have been able to ascertain the ancient environmental preferences of some of Montana's best known dinosaurs. Our ability to reconstruct the habitat preferences and lifestyles of Two Medicine dinosaurs and their contemporaries has also been refined through the study of how and why fossils are preserved within the formation (TAPHONOMY).

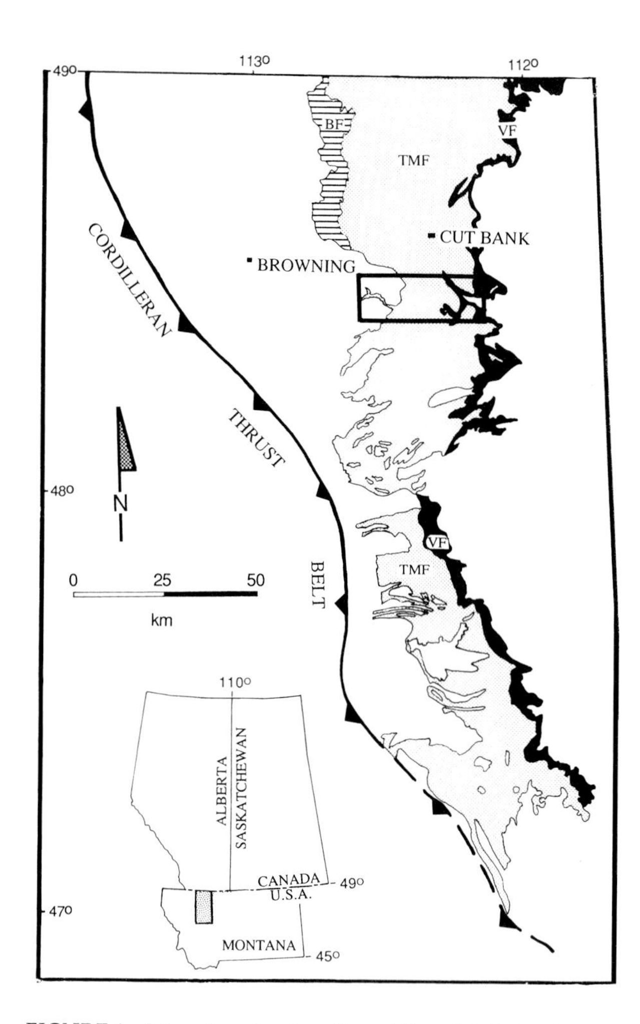

FIGURE 1 Map showing location of Two Medicine Formation in northwestern Montana.

Geologic Setting

Age and Correlation Deposition of the Two Medicine Formation commenced with the regression of the Colorado Sea and continued until maximum transgression of the Bearpaw Sea (Gill and Cobban, 1973). Correlation with marine ammonite zones suggests that the formation is probably Campanian in age (using Campanian Stage boundaries of Harland *et al.*, 1990). Radioisotopic dating of volcanic ash horizons (bentonites) interbedded within the formation also yields a Campanian age (Rogers *et al.*, 1993). On the basis of $^{40}Ar/^{39}Ar$ analyses, deposition of the Two Medicine Formation commenced approximately 83 million years ago and continued until approximately 74 million years ago.

The Two Medicine Formation consists of nonmarine sediments that accumulated between the Cordilleran mountain belt and the oscillating western shoreline of the Cretaceous Interior Seaway. Nonmarine and nearshore marine equivalents within central Montana include the Eagle and Judith River formations, which thin eastward into the fully marine Claggett and Bearpaw shales. Erosion over the Sweetgrass arch has isolated exposures of the Two Medicine Formation from correlative sedimentary deposits in central and eastern Montana. Correlative rock units in western Canada include the Belly River, Bearpaw,

FIGURE 2 Widespread badland exposures typical of the Two Medicine Formation.

Milk River, and Pakowki formations and the Judith River Group (Eberth and Hamblin, 1993). The recently named (and dinosaur-laden) Dinosaur Park Formation (Eberth and Hamblin, 1993), which is exposed throughout much of DINOSAUR PROVINCIAL PARK, Alberta, correlates with upper reaches of the Two Medicine Formation (Rogers *et al.*, 1993).

Sedimentology The Two Medicine Formation consists of approximately 560 m of terrestrial sediments that were derived from mountainous regions and volcanic centers to the west and south of present-day exposures. Rare sandstone beds deposited in shallow marine environments occur low in the section. The formation as a whole is dominated by gray green claystone and siltstone, with lesser amounts of fine- to medium-grained gray sandstone (Fig. 3). Claystone and siltstone beds are generally thin (<1 m) and tabular in geometry and often show evidence of soil formation (e.g., root traces, color banding, and zones of coalesced carbonate nodules). Sandstone beds range from less than 1 to 10 m in thickness, although most are less than 4 m thick. Thin deposits that include claystone and carbonate pebbles, fossil wood, and clam shell fragments and rare dinosaur bones are common above the basal contacts of sandstone beds. In most cases these basal lag deposits are overlain by cross-bedding indicative of current activity. The upper surfaces of sandstone beds typically preserve root traces and burrows made by small invertebrates (arthropods?).

Paleoenvironmental Reconstruction Rocks of the Two Medicine Formation represent the deposits of a variety of marginal marine and nonmarine environments that bordered the western margin of the Western Interior Seaway. Streams draining the Cordilleran highlands flowed across the Two Medicine alluvial plain toward the sea. These streams were relatively shallow, and they were not very sinuous. Calcareous soils developed between stream channels, and these soils supported a variety of Cretaceous vegetation. Shallow floodbasin ponds and lakes populated by small freshwater invertebrates (clams and snails) and aquatic vertebrates (primarily turtles and champsosaurs) were common. Volcanic ashfalls blanketed the Two Medicine alluvial plain on numerous occasions: These ash beds (bentonites) presumably record explosive volcanic events in the Elkhorn Mountains, which is a nearby volcanic center that was active during deposition of the Two Medicine Formation. As a result of these volcanic episodes, dinosaurs on the Two Medicine alluvial plain may have on occasion experienced hardships ranging from ash-entombed food supplies to rather instantaneous mass death.

Characteristics of sediments, fossil soils (paleosols), and fossil plants (Crabtree, 1987) suggest that the Two Medicine paleoclimate was seasonal (at least with respect to water availability). There is also evidence that Campanian droughts may have led to pulses of dinosaur mortality in the Two Medicine ecosystem (Rogers, 1990).

FIGURE 3 Fine-grained floodplain deposits such as these comprise the bulk of the Two Medicine Formation and yield most of the dinosaur fossils. Arrow points to a resistant sandstone ledge that was deposited in a laterally accreting river channel.

Taphonomy

Fossil vertebrates have been collected from the Two Medicine Formation since the early years of the 20th century (Gilmore, 1914, 1917), and a diverse vertebrate assemblage dominated by dinosaurs has been documented (Horner, 1989). The great wealth of paleontological data retrieved from the Two Medicine Formation has led to the description of numerous dinosaur type specimens (Horner and Makela, 1979; Horner and Weishampel, 1988; Horner, 1992; Sampson, 1995) and to a host of inferences concerning various aspects of dinosaurian EVOLUTION, PALEOECOLOGY, and BEHAVIOR (Horner and Makela, 1979; Horner, 1982; Rogers, 1990; Horner *et al.*, 1992; Sampson, 1995).

From a taphonomic perspective, the richly fossiliferous Two Medicine Formation is characterized by a variety of vertebrate skeletal concentrations superimposed on a background of dispersed skeletal material. These skeletal concentrations range from dinosaur nest sites and bone beds to diverse concentrations of microvertebrate skeletal debris.

Nest Sites The Two Medicine Formation is world renowned for its preservation of dinosaur nests and nesting horizons (Fig. 4). Nest sites that preserve both hatched eggs and unhatched eggs with intact embryos have been found (e.g., Horner and Currie, 1994). Clutches often appear to be largely undisturbed, and dinosaur eggs are in many cases preserved in orderly configurations. The disarticulated bones of both juvenile and embryonic dinosaurs are regularly found scattered upon nesting horizons. Dense concentrations of juvenile bones that presumably represent the remains of hatchling cohorts are occasionally found within and around nest sites (Fig. 5).

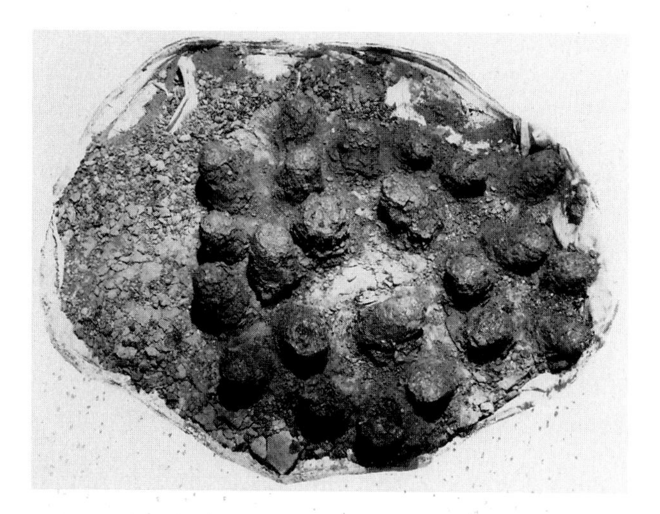

FIGURE 4 Egg clutch of a *Troodon* recovered from the Egg Mountain field area (courtesy of the Museum of the Rockies).

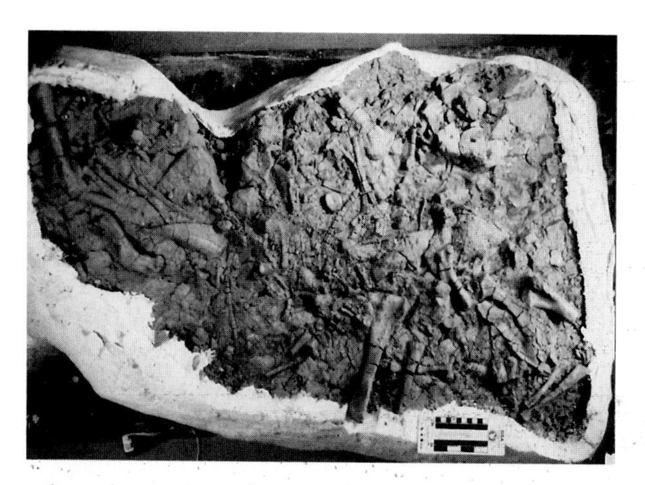

FIGURE 5 Collected block Two Medicine bone bed (Museum of the Rockies No. 548) that preserves the skeletal remains of numerous baby hypacrosaurs. This particular locality presumably represents the demise of a cohort of hatchling dinosaurs.

The sedimentology of most nesting horizons in the Two Medicine Formation indicates burial in floodplain sediments, presumably as a result of overbank flood events. The exquisite preservation of many nest sites suggests that at least some Two Medicine nesting grounds may have been localized in floodplain environs proximal to active stream channels, where the magnitudes of overbank flood events would have been more likely to entomb nests and associated juvenile skeletal debris. At least one nesting horizon in the Two Medicine Formation may represent a land surface blanketed by a volcanic ashfall (Horner, 1994). Nest sites preserved in the classic Egg Mountain locality near Choteau, Montana, are stratigraphically associated with thin limestone beds that accumulated in shallow lakes.

Bone Beds The vast majority of dinosaur bone beds known from the Two Medicine Formation yield multi-individual bone assemblages dominated by the skeletal remains of one species of dinosaur (Gilmore, 1917; Rogers, 1990). Eight low-diversity bone beds that preserve concentrations of either hadrosaurs or ceratopsians have been excavated (Fig. 6). In contrast,

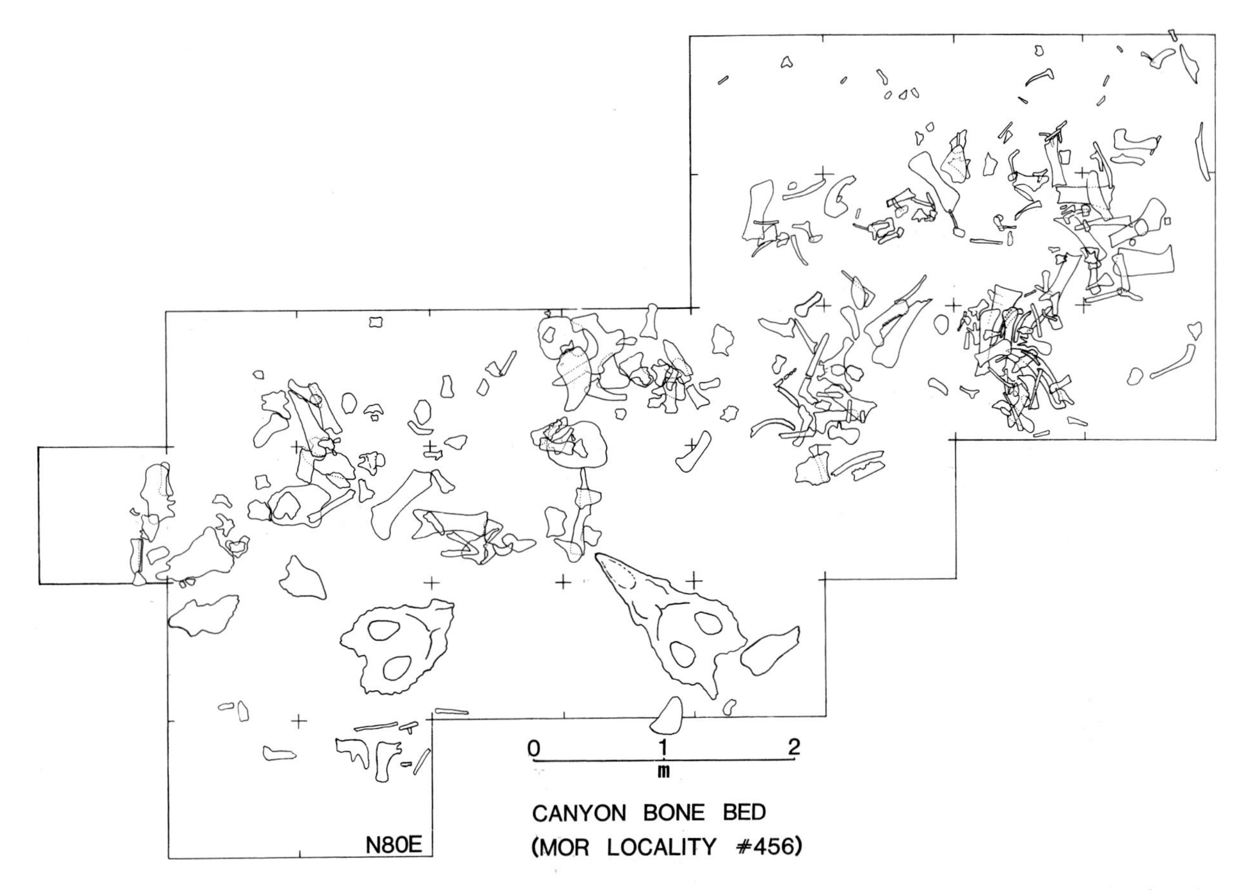

FIGURE 6 Map of a dinosaur bone bed in the Two Medicine Formation (modified from Rogers, 1990). This site (Canyon Bonebed; Museum of the Rockies No. 456) preserves the disarticulated remains of at least seven ceratopsian dinosaurs (*Einiosaurus procurvicornis*).

only one dinosaur bone bed that preserves a diverse array of dinosaur taxa has been exhumed and studied (Varricchio, 1995).

All bone beds currently known within the Two Medicine Formation occur in fine-grained sediments that were deposited in ancient floodplain or abandoned stream channel settings (in modern environments oxbow lakes form in abandoned stream channels). The skeletal remains of dinosaurs found in Two Medicine bone beds are almost always disarticulated (Fig. 6). The bones themselves are generally in very good condition, and most show minimal effects of surficial weathering (e.g., surface cracking or flaking). Taphonomic attributes of fossil bones (e.g., degrees of fragmentation, weathering, and abrasion) tend to show minimal variation within any given bone bed. However, the degree of bone breakage can vary between bone beds: Some localities preserve extremely delicate elements such as cranial bones in pristine condition, whereas others yield abundant fractured and partial elements. Scratch marks (indicative of trampling in abrasive sediments) and rare tooth marks (indicative of scavenging) have been documented at a few sites (Rogers, 1990). Borings in bones that may be attributable to carrion beetles occur in one *Prosaurolophus* bone bed (Rogers, 1992).

The taphonomy of the known Two Medicine bone beds suggests some type of recurrent dinosaur mass mortality in the Campanian Two Medicine ecosystem. Low-diversity concentrations of hadrosaurian and ceratopsian dinosaurs (herds?) apparently fell victim again and again to some type of killing agent. Proposed killing agents in the Two Medicine ecosystem include volcanic ashfalls (Hooker, 1987), drought (Rogers, 1990), and cyanobacterial toxicosis (botulism) (Varricchio, 1995).

Microvertebrate Concentrations Microvertebrate assemblages (microsites) are concentrations of dissociated teeth, scales, scutes, and / or small dense bones. They can form in a variety of settings, and they are most often attributed to either fluvial processes (hydrodynamic sorting and concentration) or predation (scatological concentrations). Microvertebrate assemblages typically preserve a diverse array of species and are thus frequently analyzed by paleontologists in order to reconstruct paleoecological associations.

Concentrations of vertebrate microfossils are relatively rare within the Two Medicine Formation. The few MICROVERTEBRATE SITES currently known preserve concentrations of small skeletal elements (e.g., teeth, scales, vertebrae, and phalanges) derived from a variety of aquatic and terrestrial vertebrate taxa, along with freshwater invertebrates (clams and snails). Plans are being developed to systematically screen wash and sample these important localities. By doing so, we will obtain a more comprehensive picture of the Two Medicine vertebrate community.

Patterns of Preservation The fossil remains of dinosaurs and their contemporaries are not distributed evenly within the sediments of the Two Medicine Formation. The lower half of the formation preserves widely scattered dinosaur bones and teeth along with rare bones, teeth, scutes, and scales of freshwater fish, turtles, crocodilians, and champsosaurs. In contrast, the upper half of the Two Medicine Formation is remarkably fossiliferous, and it is within the upper 200–250 m of the formation that most of the dinosaur bone beds and dinosaur nesting sites described previously occur. Variations in the large-scale stratigraphic distribution of vertebrate fossils within the Two Medicine Formation are best explained by changing depositional conditions (e.g., changing rates of sediment accumulation) and fossil preservation potentials associated with the onset of the Bearpaw transgressive phase (Rogers, 1994, 1995).

Conclusion

The Upper Cretaceous Two Medicine Formation of northwestern Montana ranks among the world's most productive dinosaur-bearing formations. Paleontologists have scoured the Two Medicine badlands since the early years of this century, and they have discovered and described a wide array of dinosaurian taxa. However, what truly sets the Two Medicine Formation apart from most other dinosaur-bearing formations is its snapshot-like record of dinosaur ecology and behavior. The numerous nest sites and bone beds preserved within the formation provide important clues that pertain to various aspects of dinosaurian nesting habits and herd structure. Taphonomic and geologic data retrieved from the formation also provide insights into the inner workings of a Late Cretaceous ecosystem. Current research along these lines includes a study of dinosaur COPROLITES, which are preserved in abundance within the sediments of the Two Medicine Formation (Chin, 1995). In the foreseeable future, we may be able to supplement our reconstructions of dinosaur behavior and ecology with a menu of dinosaurian dietary preferences.

See also the following related entries:

CRETACEOUS PERIOD • EGG MOUNTAIN • JUDITH RIVER WEDGE • WILLOW CREEK ANTICLINE

References

Chin, K. (1995). The paleobiological implications of herbivorous dinosaur coprolite fabrics from the Two Medicine Formation, Montana. *J. Vertebr. Paleontol.* **15**(Suppl. to No. 3), 23A.

Crabtree, D. R. (1987). Angiosperms of the northern Rocky Mountains: Albian to Campanian (Cretaceous) megafossil floras: *Ann. Missouri Botanical Garden* **74,** 707–747.

Eberth, D. A., and Hamblin, A. P. (1993). Tectonic, stratigraphic, and sedimentologic significance of a regional discontinuity in the upper Judith River Group (Belly River wedge) of southern Alberta, Saskatchewan, and northern Montana. *Can. J. Earth Sci.* **30,** 174–200.

Gill, J. R., and Cobban, W. A. (1973). Stratigraphy and geologic history of the Montana Group and equivalent rocks, Montana, Wyoming, and North and South Dakota. United States Geological Survey Professional Paper No. 776, pp. 37.

Gilmore, C. W. (1914). A new ceratopsian dinosaur from the Upper Cretaceous of Montana, with a note on *Hypacrosaurus. Smithsonian Miscellaneous Collections* **43,** 1–10.

Gilmore, C. W. (1917). *Brachyceratops*, a ceratopsian dinosaur from the Two Medicine Formation of Montana, with notes on associated fossil reptiles. United States Geological Survey Professional Paper No. 103, pp. 1–45.

Harland, W. B., Armstrong, R. L., Cox, A. V., Craig, L. E., Smith, A. G., and Smith, D. G. (1990). *A Geologic Time Scale 1989*, pp. 264. Cambridge Univ. Press, Cambridge, UK.

Hooker, J. S. (1987). Late Cretaceous ashfall and the demise of a hadrosaurian "herd." Geological Society of America, Rocky Mountain Section, Abstracts with Programs, No. 19, pp. 284.

Horner, J. R. (1982). Evidence of colonial nesting and site fidelity among ornithischian dinosaurs. *Nature* **297,** 675–676.

Horner, J. R. (1989). The Mesozoic terrestrial ecosystems of Montana. 1989 Montana Geological Society Field Conference, Montana Centennial, pp. 153–162.

Horner, J. R. (1992). Cranial morphology of *Prosaurolophus* (Ornithischia: Hadrosauridae) with descriptions of two new hadrosaurid species and an evaluation of hadrosaurid phylogenetic relationships. Museum of the Rockies Occasional Paper No. 2, pp. 119.

Horner, J. R. (1994). Comparative taphonomy of some dinosaur and extant bird colonial nesting grounds. In *Dinosaur Eggs and Babies* (K. Carpenter, K. F. Hirsch, and J. R. Horner, Eds.), pp. 116–123. Cambridge Univ. Press, Cambridge, UK.

Horner, J. R., and Currie, P. J. (1994). Embryonic and neonatal morphology and ontogeny of a new species of *Hypacrosaurus* (Ornithischia, Lambeosauridae) from Montana and Alberta. In *Dinosaur Eggs and Babies* (K. Carpenter, K. F. Hirsch, and J. R. Horner, Eds.), pp. 312–336. Cambridge Univ. Press, Cambridge, UK.

Horner, J. R., and Makela, R. (1979). Nest of juveniles provides evidence of family structure among dinosaurs. *Nature* **282,** 96–298.

Horner, J. R., and Weishampel, D. B. (1988). A comparative embryological study of two ornithischian dinosaurs. *Nature* **332,** 256–257.

Horner, J. R., Varricchio, D. J., and Goodwin, M. B. (1992). Marine transgressions and the evolution of Cretaceous dinosaurs. *Nature* **358,** 59–61.

Rogers, R. R. (1990). Taphonomy of three dinosaur bonebeds in the Upper Cretaceous Two Medicine Formation of northwestern Montana: Evidence for drought-related mortality. *Palaios* **5,** 394–413.

Rogers, R. R. (1992). Non-marine borings in dinosaur bones from the Upper Cretaceous Two Medicine Formation, northwestern Montana. *J. Vertebr. Paleontol.* **12,** 528–531.

Rogers, R. R. (1994). Nature and origin of through-going discontinuities in nonmarine foreland basin strata, Upper Cretaceous, Montana: Implications for sequence analysis. *Geology* **22,** 1119–1122.

Rogers, R. R. (1995). Sequence stratigraphy and vertebrate taphonomy of the Upper Cretaceous Two Medicine and Judith River formations, Montana. Unpublished Ph.D. thesis, University of Chicago, Chicago.

Rogers, R. R., Swisher, C. C., and Horner, J. R. (1993). $^{40}Ar/^{39}Ar$ age and correlation of the non-marine Two Medicine Formation (Upper Cretaceous), northwestern Montana. *Can. J. Earth Sci.* **30,** 1066–1075.

Ross, C. P., Andrews, D. A., and Witkind, I. J. (1955). Geologic map of Montana. United States Geological Survey, Map MR-411.

Sampson, S. D. (1995). Two new horned dinosaurs from the Upper Cretaceous Two Medicine Formation of Montana; with a phylogenetic analysis of the Centrosaurinae (Ornithischia: Ceratopsidae). *J. Vertebr. Paleontol.* **15,** 743–760.

Varricchio, D. J. (1995). Taphonomy of Jack's Birthday Site, a diverse dinosaur bonebed from the Upper Cretaceous Two Medicine Formation of Montana. *Palaeogeogr. Palaeoclimatol. Palaeoecol.* **114,** 297–323.

Tyrannosauridae

Kenneth Carpenter
Denver Museum of Natural History
Denver, Colorado, USA

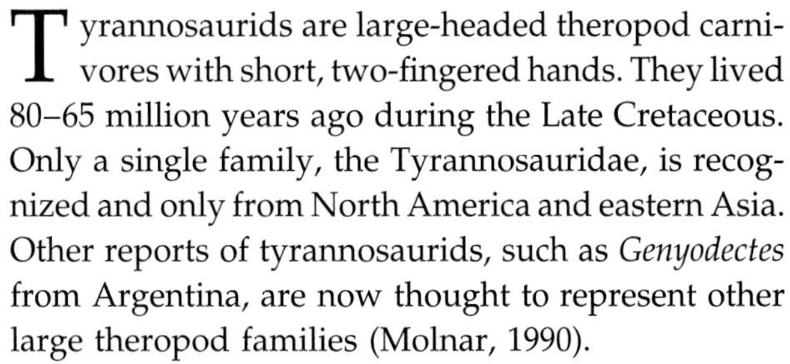

Tyrannosaurids are large-headed theropod carnivores with short, two-fingered hands. They lived 80–65 million years ago during the Late Cretaceous. Only a single family, the Tyrannosauridae, is recognized and only from North America and eastern Asia. Other reports of tyrannosaurids, such as *Genyodectes* from Argentina, are now thought to represent other large theropod families (Molnar, 1990).

Many of the adaptations seen in the tyrannosaurids are centered around their role as top predator in the Late Cretaceous ecosystem. As with most predators, emphasis is on the mouth as the killing and feeding organ. This adaptation follows a long trend in theropod evolution in which the size of the mouth, hence the size of the bite, is increased, enlarging the skull. However, a larger skull cannot be supported on a long, slender neck, so the neck and body are concomitantly shortened. These changes may be seen by comparing the skull length to the length of the presacral vertebrae. In the early, primitive *Coelophysis*, the skull is 24% of the presacral vertebral length; in *Allosaurus*, an intermediate sized theropod, it is 40%; and in *Tyrannosaurus* it is 47%.

The tyrannosaurid SKULL is a laterally compressed, rectangular box when viewed from the side (Fig. 1). The boxy look is due to the deep snout, which resists the strains placed on it during the bite. In contrast, the snout is long and tapering and the teeth are small in the primitive *Coelophysis*. In this animal the power of the bite was proportionally less than that in the tyrannosaurids. The TEETH in tyrannosaurids are large, serrated blades arranged along curved jaws so that when the mouth is closed almost all the teeth engage the prey at the same time. With *Coelophysis*, the teeth are arranged along straight-edged jaws and the teeth engage from back to front as the mouth is closed. *Tyrannosaurus* has also gone one step further than most tyrannosaurids by strengthening the teeth. This was accomplished by making the teeth thicker so that in cross-section they are almost as wide from side to side as long from front to back. Even with this adaptation, broken teeth are found mingled among the fossilized carcasses of their prey.

In allosaurids, kinesis allows the skull to "deform" or absorb stress at joints between various skull bones. This allows the skull to accommodate larger bites. In tyrannosaurids, the skull was apparently akinetic or nearly so. Again, having a more rigid skull may reflect an adaptation for a more powerful bite because energy is not lost or dissipated by "deforming" the skull. Some intramandibular movement was probably possible between the dentary and the surangular and angular. Movement may have also been possible between the two dentaries at the symphysis because the bones are loosely joined by cartilage.

The orbit is large in tyrannosaurids, as are both the olfactory and optic nerve tracts of the endocast (Osborn, 1912a). Thus, as with most predators, both smell and vision must have been very keen. What is less certain is if tyrannosaurids had stereoptic vision (depth perception). This stereoptical development is best seen in *Tyrannosaurus* in which the snout is narrow in front of the orbit, allowing the eyes to see forward. The overlapping fields of vision would allow the predator to judge distance to the prey. This overlapping vision was apparently not as well developed in the more primitive tyrannosaurid *Gorgosaurus* because the snout is not constricted in front of the orbits. However, it is possible that the eyeballs extended beyond the rim of the orbits, allowing some degree of overlapping vision; unfortunately, we will probably never know if this was the case.

The upper surface of the nasals is very rough, indicating that some sort of structure may have been present, such as a horn-like protuberance (Baird, personal communication). In *Gorgosaurus* and *Albertosaurus*, a "horn" was also present just in front of the eyes. This horn was developed from the lacrimal, just above the lacrimal fenestra. Interestingly, *Tyrannosaurus* does not have much of a lacrimal horn, although the surface of the lacrimal is very rough like the surface of the nasals. Perhaps their horn was

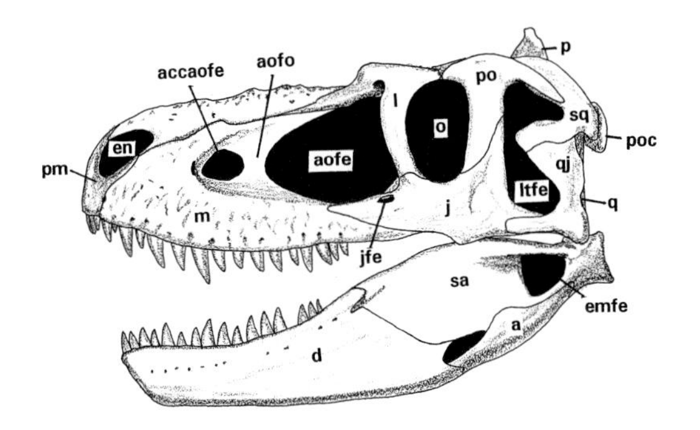

FIGURE 1 Skull of the tyrannosaurid, *Gorgosaurus libratus.*

formed from a cuticle-like material. The purpose of horns in front of the eyes and possibly on the snout is peculiar in a carnivore. Most probably the structures were for display, either sexual or agonistic.

The postcrania of tyrannosaurids indicate that the body was carried bipedally, with the tail counterbalancing the body over the hindlegs (Fig. 2). The cervical vertebrae are short and wide. The neural spines are not the tall blades seen in more primitive theropods but rather are short and wide for the attachment of powerful muscles and ligaments. The bulldog-like neck would have supported the large, heavy head with strong muscles for pulling flesh off the carcass. The body of tyrannosaurids is deep to accommodate the internal organs in a foreshortened body. Rods of bone, called GASTRALIA, support and protect the internal organs. They extend across the abdominal cavity from the coracoids to the pubis.

The FORELIMBS are short and could not have been used in locomotion. Their purpose is the subject of debate. Some paleontologists, such as Jack Horner, argue that the arms are too short to have any use (Horner and Lessem, 1993). I, on the other hand, argue that the well-developed muscle scars indicate large, powerful muscles on the arms. Such well-muscled arms must have had some function, possibly to hold a struggling hadrosaur prey against the body prior to killing it (Smith and Carpenter, 1990; Carpenter and Smith, 1995). The hand has only two fingers rather than the functional three that characterize all other theropods. The phalanges of digit I have a twist in them so that the claw angles away from digit II. This feature may have ensured that the struggling prey could not easily slip off of the claws. Instead, once engaged the claws were locked into the flesh.

The hindlegs are large and powerful for carrying the body. The legs were most probably flexed so that the femur projected somewhat forward in a bird-like manner. This flexure compensated for the greater mass of the body in front of the PELVIS. The feet, as in most theropods, are raptor-like, ending in three sharp claws. These talons could hold the carcass down while the neck pulled the head back, ripping flesh from the carcass. Whether tyrannosaurids could run very fast or not is debated among paleontologists. Some, such as Edwin Colbert (1983), argue that the large size of tyrannosaurids, especially *Tyrannosaurus*, precludes running. Instead, he argues that they had a fast elephantine walk. Others, such as Greg

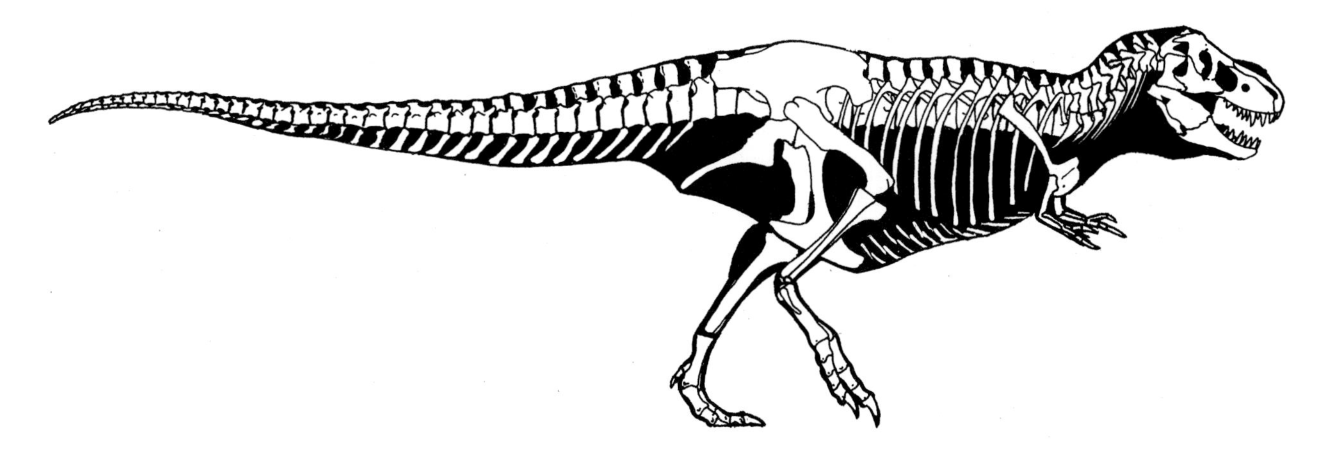

FIGURE 2 Skeletal reconstruction of the tyrannosaurid, *Tyrannosaurus rex.*

Paul (1988), argue that the bird-like hindleg indicates much faster locomotion, perhaps up to 70 km/hr.

Certainly, changes in the upper foot and ankle compared with that of *Allosaurus* would seem to indicate a more stable foot for running. The ascending process of the astragalus is very tall and firmly attached to the tibia with cartilage. This ensures that the ankle remains steady over uneven ground. In addition, the central metatarsal is constricted on its proximal end to a long slender splint. This splint is surrounded by the outer metatarsals, restricting the amount of movement between the three bones. This adaptation, as well as the sturdy ankle, would certainly keep the foot from twisting on uneven ground; therefore, perhaps tyrannosaurids did run, but how fast no one knows.

Traces of SKIN impression are known for a specimen of *Gorgosaurus libratus* (Day, personal communication). These show small rounded or hexagonal scales on the tail. The pattern on the rest of the body is unknown, but if other dinosaur skin impressions (e.g., hadrosaurs; Osborn, 1912b) are any indication, then the rest of the body was probably covered in hexagonal scales of different sizes. Impressions of skin around a badly weathered skull of *Tyrannosaurus (= Tarbosaurus) bataar* in Mongolia showed the presence of a wattle or bag of skin under the jaws (Mikhailov, personal communication). Perhaps this bag of skin enabled large chunks of prey to be swallowed.

It is also possible that the skin was a brightly colored dewlap similar to that seen in some lizards today.

The origin of the family Tyrannosauridae is obscure because so few theropod remains are known from the early and middle part of the Late Cretaceous. They first appeared 80 million years ago, and persisted for another 15 million years only to vanish in the great dinosaur extinction at the end of the Mesozoic. Traditionally, it was thought that tyrannosaurids evolved from allosaurids of the Jurassic (e.g., Colbert, 1983; Paul, 1988). However, recent cladistic analysis indicates that, in fact, tyrannosaurids are giant coelurosaurs more closely related (i.e., a "sister group") to the ornithomimids than they are to allosaurids (Holtz, 1994).

See also the following related entries:

ARCTOMETATARSALIA • CARNOSAURIA • COELUROSAURIA • THEROPODA

References

Carpenter, K., and Smith, M. B. (1995). Osteology and functional morphology of the forelimb in Tyrannosauridae as compared with other theropods (Dinosauria). *J. Vertebr. Paleontol.* **15**(Suppl. to No. 3), 21A.

Colbert, E. H. (1983). *Dinosaurs: An Illustrated History*, pp. 224. Hammond, Maplewood, NJ.

Horner, J. R., and Lessem, D. (1993). *The Complete T. Rex*, pp. 239. Simon & Schuster, New York.

Molnar, R. (1990). Problematic Theropoda: "carnosaurs." In *The Dinosauria* (D. B. Weishampel, P. Dodson, and H. Osmólska, Eds.), pp. 306–317. Univ. of California Press, Berkeley.

Osborn, H. F. (1912a). Crania of *Tyrannosaurus* and *Allosaurus*. *Mem. Am. Museum Nat. History* **1**, 1–30.

Osborn, H. F. (1912b). Integument of the iguanodont dinosaur *Trachodon*. *Mem. Am. Museum Nat. History* **1**, 33–54.

Paul, G. (1988). *Predatory Dinosaurs of the World*, pp. 464. Simon & Schuster, New York.

Smith, M. B., and Carpenter, K. (1990). Forelimb biomechanics *of Tyrannosaurus rex*. *J. Vertebr. Paleontol.* **10**(Suppl. to No. 3), 43A.

Tyrrell Museum of Palaeontology

see ROYAL TYRRELL MUSEUM OF PALAEONTOLOGY

Ukhaa Tolgod

MARK A. NORELL
American Museum of Natural History
New York, New York, USA

In 1993, during the third year of the Mongolian Academy of Sciences–American Museum of Natural History joint expedition to the Gobi Desert, an extremely rich Late Cretaceous locality was discovered in the NEMEGT Basin (Dashzeveg *et al.*, 1995). The locality was named Ukhaa Tolgod, roughly translating to "small brown hills." By Gobi standards the aerial expanse of Ukhaa Tolgod is very small, measuring only approximately 50 square km, with the main exposures being only a fraction of that. However, because of its prolific richness and remarkable preservation, it is one of the most important Mesozoic localities yet discovered (Novacek, 1996; Norell and Novacek, 1996).

The vertebrate fauna is diverse and especially rich in small vertebrates (Dashzeveg *et al.*, 1995; Novacek *et al.*, 1994, 1996). More than 400 specimens of 8 described taxa of mammals plus several more undescribed forms have been recovered as of 1996. Multituberculates (at least 5 described taxa) are most common, but several therians are represented by amazing specimens. Lizards are more common than mammals (more than 500 specimens and at least 20 taxa; see Gao and Norell, 1996) and turtles are occasionally encountered. Although ubiquitous (probably nearly 100 individuals), *Saichania* type ANKYLOSAURS and *Protoceratops andrewsi* are the only ornithischian dinosaurs yet recovered. THEROPODS are much more diverse (Norell *et al.*, 1996). Two birds, *Mononykus* (more than 10 specimens) and *Gobipteryx*, have been recovered. Large theropods are represented only by teeth; two distinct types have been found. DROMAEOSAURS are known from adults and babies (Norell *et al.*, 1994). These appear to be *Velociraptor* or *Velociraptor*-like forms and are known from excellent cranial material. A single undescribed TROODONTID is known from two fragmentary specimens (including an excellent BRAINCASE). Spectacular material of at least 3 oviraptorid taxa is known from Ukhaa Tolgod. These include large and small forms, adults and juveniles, and even embryos and adults sitting on nests (Norell *et al.*, 1994, 1995: Norell and Clark, 1997). Some of these remains clarified the misconception that the most common type of eggs found in Djadokhta-like sediments are those of *Protoceratops*—instead showing definitively that they can be referred to oviraptors (Norell *et al.*, 1994, Clark, 1995).

Lithologically, Ukhaa Tolgod resembles other typical Djadokhta-like deposits. Red sandstones with massive cross-beds are intermixed with structureless sands and fluvial and pond deposits (D. Loope *et al.*, manuscript in preparation). The absolute age of Ukhaa Tolgod is hard to determine because no sediments suitable for radiometric dating have been discovered. Faunally, Ukhaa Tolgod contains a mix of taxa found in Djadokhta and BARUN GOYOT rock units; consequently, it is viewed as roughly equivalent in age with these other localities, pending additional evidence.

See also the following related entries:

AMERICAN MUSEUM OF NATURAL HISTORY • BAYAN MANDAHU • BAYN DZAK • CENTRAL ASIATIC EXPEDITIONS • DJADOKHTA FORMATION • MONGOLIAN DINOSAURS

References

Clark, J. M. (1995). An egg thief exonerated. *Nat. History* **June,** 56–58.

Dashzeveg, D., Novacek, M. J., Norell, M. A., Clark, J. M., Chiappe, L. M., Davidson, A., McKenna, M. C., Dingus, L., Swisher, C., and Perle, A. (1995). Unusual preservation in a new vertebrate assemblage from the Late Cretaceous of Mongolia. *Nature* **374,** 446–449.

Gao, K., and Norell, M. A. (1996). A rich lizard assemblage from Ukhaa Tolgod, Gobi Desert, Mongolia. *Vertebr. Paleontol.* **16**(Suppl. to No. 3), 36A.

Loope, D., Dingus, L. R., and Swisher, C., III (1997). Life and death on Late Cretaceous Dunes, Nemegt Basin, Mongolia. Submitted for publication.

Norell, M. A., and Novacek, M. J. (1996). Expedition Mongolia: A window on the past. *Explorer's J.* **74**(2), 24–26.

Norell, M. A., Clark, J. M., Dashzeveg, D., Barsbold, R., Chiappe, L. M., Davidson, A. R., McKenna, M. C., Perle, A., and Novacek, M. J. (1994). A theropod dinosaur embryo and the affinities of the Flaming Cliffs Dinosaur eggs. *Science* **266,** 779–782.

Norell, M. A.., Clark, J. M., Chiappe, L. M., and Dashzeveg, D. (1995). A nesting dinosaur. *Nature* **378,** 774–776.

Norell, M. A., Clark, J. M., and Chiappe, L. M. (1996). Djadoktha series theropods: A summary review. Dinofest International Symposium Program and Abstracts.

Novacek, M. J. (1996). *Dinosaurs of the Flaming Cliffs.* Anchor, New York.

Novacek, M. J., Dashzeveg, D., and McKenna, M. C. (1994). Late Cretaceous mammals from Ukhaa Tolgod, Mongolia. *J. Vertebr. Paleontol.* **14**(Suppl. to No. 3), 40A.

Novacek, M. J., Norell, M. A., Dingus, L., and Dashzeveg, D. (1996). Dinosaurs, mammals, birds, and lizards from the Late Cretaceous Ukhaa Tolgod fauna, Mongolia. *J. Vertebr. Paleontol.* **16**(Suppl. to No. 3), 56A.

University Museum, Oxford, United Kingdom

see MUSEUMS AND DISPLAYS

University Museum, University of Arkansas, Arkansas, USA

see MUSEUMS AND DISPLAYS

University of Alaska Museum, Alaska, USA

see MUSEUMS AND DISPLAYS

University of California Museum of Paleontology

The University of California Museum of Paleontology (UCMP) is located in newly designed, state-of-the-art facilities in the renovated Valley Life Sciences Building on the campus of the University of California, Berkeley. The UCMP fossil collections rank among the six largest in the Western Hemisphere and are the largest of any university museum in the hemisphere. The museum's collections were founded upon material originally collected by the Whitney survey in the 1870s; by act of the California Legislature, the museum is the official repository of the state's fossil collections, which also include fossil vertebrate material from all over the western United States as well as Europe, Australia, Africa, and Asia. Much of the dinosaurian and other lower vertebrate material from these continents was collected by Charles Camp during a trip around the world that he made in the late 1930s (e.g., Camp, 1935).

Vertebrate paleontology had early beginnings at the museum; some of its first specimens, from the Triassic deposits near the PETRIFIED FOREST in Arizona, were deposited by John Muir. John C. Merriam, who worked widely on fossil vertebrates from thalattosaurs to mammals, was the first director of the Museum of Paleontology.

Charles M. Camp, who joined the museum in 1920, began his career by collecting fossil vertebrates from the Triassic beds of the CHINLE FORMATION in and around the Petrified Forest in Arizona and nearby New Mexico, spurred by a summons from the museum's great benefactor, Miss Annie Alexander. Camp thought that much of what he collected from the southwestern Triassic, in addition to remains of phytosaurs, aetosaurs, and metoposaurian amphibians, was dinosaurian, but much of this material has proved to be of poposaurian pseudosuchians or of indeterminate archosaurian origin. The famous Placerias Quarry near St. Johns, Arizona, collected by Camp and crews in the early 1930s, was a bonanza of remains of its eponymous dicynodont, but many other archosaurs are present, including several bones of indeterminate basal dinosaurs or dinosauriformes. From the Navajo Formation (Early Jurassic) of northern Arizona, Camp described the small ceratosaurian *Segisaurus* in 1936, the first theropod in which clavicles were identified. The large ceratosaurian *Dilophosaurus* was collected by S. P. Welles from the Navajo lands of Arizona in the 1940s; the known material consists of two skeletons, one as yet undescribed but different in some particulars from the type skeleton (Welles, 1984). A field party led by K. Padian and R. A. Long collected a skeleton of *Coelophysis* from the Upper Chinle Formation of the Petrified Forest National Park (PFNP) in 1983 (Padian, 1986). A herrerasaurid, *Chindesaurus,* was collected from the PFNP by Long and a field crew in 1984, and it reposes in the PFNP collections.

Other theropods in the UCMP collections include material of *Allosaurus* (a cast including original material) and *Marshosaurus* from the CLEVELAND–LLOYD QUARRY of the MORRISON FORMATION, and extensive (mostly isolated) material of theropods from the Late Cretaceous JUDITH RIVER and HELL CREEK formations of Montana, collected by expeditions under the guidance of W. A. Clemens and M. B. Goodwin. This includes material of *Tyrannosaurus rex,* ornithomimids, and dromaeosaurids, as well as *"Pectinodon"* troodontid teeth; among Aves, there are the first alvarezsaurid bones known from North America, a relatively complete skeleton of the enantionithiform *Avisaurus,* and remains of hesperornithiform birds. Sauropodomorph remains are few, represented notably by *Ammosaurus* from the Navajo Formation (Early Jurassic) of Arizona.

Ornithischian dinosaurs are also well represented in the UCMP. Examples of the basal ornithischian tooth genus *Revueltosaurus* were collected from the Petrified Forest National Park (Padian, 1990). Among thyreophorans, bones of an incomplete skeleton of *Scutellosaurus* and scutes of *Scelidosaurus* were collected from the Kayenta Formation of Arizona by a field party led by J. M. Clark and D. E. Fastovsky in 1981. These dinosaurs, in addition to *Dilophosaurus* and material collected by crews from the MUSEUM OF COMPARATIVE ZOOLOGY, were instrumental in establishing the Early Jurassic age of the GLEN CANYON GROUP (Padian, 1989). However, the most extensive holdings of ornithischians come from the Late Cretaceous Judith River and Hell Creek formations of Montana. These include incomplete skeletal remains of the questionably basal hypsilophodontid *Thescelosaurus;* a fully preserved skull of *Edmontosaurus annectens* and other hadrosaurs, including an undescribed skull of *Parasaurolophus* from Utah; an ontogenetic series of *Edmontosaurus* collected from the North Slope of Alaska (in trust from the University of Alaska); extensive ceratopsian material including skulls of *Triceratops,* plus skull and postcranial remains of the pachycephalosaurs *Stegoceras, Ornatotholus, Pachycephalosaurus, Stygimoloch,* and additional new and indeterminate material (Goodwin, 1990). Among an extensive collection of mostly marine vertebrates (plesiosaurs, mosasaurs, etc.) from the Panoche Hills of the Central Valley of California is a skull and relatively complete skeleton of a hadrosaurine allied to *Saurolophus,* although badly damaged by gypsum.

References

Camp, C. L. (1935). Dinosaur remains from the province of Szechuan, China. *Univ. California Publ. Geol. Sci.* **23,** 467–472.

Camp, C. L. (1936). A new type of small bipedal dinosaur from the Navajo Sandstone of Arizona. *Univ. California Publ. Geol. Sci.* **24,** 39–56.

Goodwin, M. B. (1990). Morphometric landmarks of pachycephalosaurid cranial material from the Judith River Formation of northcentral Montana. In *Dinosaur Systematics: Approaches and Perspectives* (K. Carpenter and P. J. Currie, Eds.), pp. 189–201. Cambridge Univ. Press, New York.

Padian, K. (1986). On the type material of *Coelophysis* (Saurischia: Theropoda), and a new specimen from the Petrified Forest of Arizona (Late Triassic: Chinle Formation). In *The Beginning of the Age of Dinosaurs: Faunal Change across the Triassic–Jurassic Boundary* (K. Padian, Ed.), pp. 45–60. Cambridge Univ. Press, Cambridge, UK.

Padian, K. (1989). Presence of the dinosaur *Scelidosaurus* indicates Jurassic age for the Kayenta Formation (Glen Canyon Group, northern Arizona). *Geology* **17,** 438–441.

Padian, K. (1990). The ornithischian form genus *Revueltosaurus* from the Petrified Forest of Arizona (Late Triassic: Norian; Chinle Formation). *J. Vertebr. Paleontol.* **10,** 268–269.

Welles, S. P. (1984). *Dilophosaurus wetherilli* (Dinosauria, Theropoda): Osteology and comparisons. *Paläontographica A* **185,** 85–180.

University of Colorado Museum, Colorado, USA

see Museums and Displays

University of Kansas Museum of Natural History, Kansas, USA

see Museums and Displays

University of Nebraska State Museum, Nebraska, USA

see Museums and Displays

University of Wisconsin Geology Museum, Wisconsin, USA

see Museums and Displays

University of Wyoming Geological Museum, Wyoming, USA

see Museums and Displays

Uruguayan Dinosaurs

see South American Dinosaurs

Utah Museum of Natural History, Utah, USA

see Museums and Displays

Uzbekistan Dinosaurs

see Middle Asian Dinosaurs

Variation

Scott D. Sampson
New York College of Osteopathic Medicine
Old Westbury, New York, USA

Michael J. Ryan
Royal Tyrrell Museum of Palaeontology
Drumheller, Alberta, Canada

Variation is the *sine qua non* of biological EVOLUTION, providing the raw materials for sorting mechanisms such as natural selection and genetic drift. An understanding of variation (explicit or assumed) is essential for addressing virtually all paleontological questions. For example, investigation of variation within species is required to examine biological issues such as growth and sexual dimorphism. Establishing parameters of within-species variation is also prerequisite to assessing the taxonomic and phylogenetic significance of morphological characters. Clearly, we cannot observe directly the BEHAVIOR of extinct organisms, and thus cannot apply criteria such as reproductive isolation when defining fossil species. By necessity, then, paleontologists must grapple with delimiting species strictly on morphological grounds (i.e., "morphospecies").

Like snowflakes, no two bones (fossil or recent) are exactly alike. One can always detect at least minor morphological differences or variations. The central question for paleontologists is the following: Does the perceived variation reflect taxonomic diversity (phylogeny) or variation within a species? That is, does the sample contain two (or more) species or only one? If the latter, is the variation due to growth stages (ontogeny), individual differences, sexual dimorphism, geographic or stratigraphic variation, or some combination of these? Without some confidence in the "reality" of the taxon or taxa under study, all further inferences must be considered suspect. This contribution addresses individual variation among dinosaurs in the broadest sense, including sexual dimorphism and ontogenetic changes as well as individual variation proper.

Dinosaur investigators (and most other vertebrate paleontologists) attempting to assess variation are faced with at least three major obstacles:

1. Poor-quality specimens: A large proportion of all dinosaur specimens are incompletely preserved, fragmentary, and/or distorted by postmortem processes.
2. Lack of soft tissue information: Bones are but a single component of an animal's anatomy, and numerous investigations of living fauna demonstrate that soft tissue structures are often more important in recognizing species and comprehending variation generally.
3. Small samples: According to a recent tally (Dodson, 1990), approximately 50% of the 350 or so recognized species of dinosaurs are based on single specimens, and complete skulls and skeletons are known for only 20% of all dinosaurs. Even the best represented taxa are known from statistically small samples, thereby making quantitative assessments of variation difficult or impossible.

Despite these inherent pitfalls, studies conducted during the past few decades have dramatically increased our understanding of dinosaur variation, elucidating old issues and posing new questions, in fields ranging from taxonomy and phylogenetics to FUNCTIONAL MORPHOLOGY and paleobiology. Although the first and second problems discussed previously have remained relatively intractable, the third has been addressed by increased collecting efforts and, particularly, study of mass death assemblages. In addition, new analytical techniques have provided numerous insights. Rather than an exhaustive survey, we present here an overview of the current knowledge of dinosaur variation, highlighting specific illustrative studies.

Analyzing Variation

Variation can be subdivided into two types of characters or data: qualitative (discrete) and quantitative (metrical or morphometric). Paleontologists have traditionally relied on discrete characters and to a great extent this remains true today among dinosaur workers largely due to problems in sample size and preservation. Important advances have been made, however, in the methodology applied to discrete character data. The advent of PHYLOGENETIC SYSTEMATICS, or cladistics, has provided a rigorous, testable methodology for tracking character evolution as well as developing hypotheses of historical relationships. Thus, it is possible to determine with greater confidence which characters are useful in resolving phylogenetic relationships and which represent independent evolutionary development in distinct lineages (i.e., convergence and homoplasy). Several monophyletic dinosaur groups, or clades, have now been subject to rigorous cladistic analysis, including larger scale clades [Saurischia (Gauthier, 1986), Ornithischia (Sereno, 1986), and Dinosauria (Benton, 1990)] and smaller groups [Hypsilophodontidae (Weishampel and Heinrich, 1992), Chasmosaurinae (Forster, 1990), and Centrosaurinae (Sampson, 1997)]. A number of cladistic studies of dinosaurs are currently in progress.

Morphometric approaches to analyzing variation in dinosaurs have taken several forms. Ratios of unrelated linear measurements (e.g., femur length/tibia length) have commonly been used to discriminate taxa. Allometric, bivariate, and multivariate statistical methods [especially principal components analysis (PCA)] have been extremely beneficial in sorting out the validity of congeneric taxa from the plethora of dinosaurs named since the turn of the century as well as in determining the existence of sexual dimorphism [*Protoceratops* (Dodson, 1975), lambeosaurids (Dodson, 1976), *Stegoceras* (Chapman *et al.*, 1981), *Chasmosaurus* (Lehman, 1990), *Plateosaurus* (Weishampel and Chapman, 1990), and *Triceratops* (Forster, 1990)].

One such group of morphometric methods employs bony landmark data (homologous points) in an attempt to analyze shape variation, either through application of multivariate techniques (e.g., PCA; Forster, 1990) or through geometric-based methods such as resistant–fit theta rho analysis (Chapman, 1990; Chapman and Brett-Surman, 1990). Landmark-based methods can be valuable in determining patterns of variation and morphological change, and they have the potential to provide information relating to phylogenetic and functional problems. However, the determination of homologous landmarks, particularly on skulls, can be problematic due to confounding factors such as postmortem crushing and variation in the contacts of skull elements. Chapman (1990) also notes that shape analysis does not provide a substitute for alternative metrical approaches to phylogenetic analysis, but adds that these methods may be helpful in determining which characters carry a strong phylogenetic signal.

Dinosaurs and Bone Beds

In the late 19th and early 20th centuries, paleontologists tended toward a strict typological view of species, establishing new taxa based on minimal morphological differences. Dinosaur hunters during this period, often employed by large museums, searched primarily for complete, display-quality specimens. In recent decades, paleontologists have adopted a more ecological approach to the subject of species and variation, using studies of modern animals to establish approximate ranges of taxonomic diversity within ecosystems. Concomitantly, dinosaur collectors have turned their attention to mass death assemblages, or "bone beds," which often contain the remains of multiple individuals from a single species. Whereas earlier fieldworkers often bypassed bone beds because the jumbled remains were not conducive to museum displays, recent studies have preferentially sought out these mass death localities because of their potential to elucidate patterns of variation.

The information potential of a single, complete, articulated dinosaur skeleton is far less than that of a mass death assemblage of that taxon containing the disarticulated remains of numerous individuals. Low-diversity sites are particularly valuable because they can provide relatively large samples from one geographic locality and one stratigraphic horizon, thereby permitting some confidence that the vast majority of the fossils represent a single species and perhaps a single population. These assemblages often include hundreds of individuals covering a wide range of variation, including adults, subadults, juveniles, and very likely males and females. Proposed killing agents responsible for these mass accumulations include drought (Rogers, 1990), drowning in flooded rivers (Currie and Dodson, 1984; Ryan, 1992), and volcanism (Hooker, 1987).

Mass death assemblages of dinosaurs have been discovered virtually worldwide and represent an ever-increasing number of taxa. Theropod examples include *Coelophysis* (Colbert, 1989, 1990), *Syntarsus* (Raath, 1990), and *Allosaurus* (Madsen, 1976). Prosauropods and sauropods are commonly found in low-diversity mass accumulations and include such taxa as *Lufengosaurus* (Young, 1951) and *Plateosaurus* (Weishampel and Westphal, 1986; Galton, 1986) among prosauropods and *Diplodocus* and *Camarasaurus* among sauropods (Dodson *et al.*, 1980). Ornithischian mass death assemblages are overwhelmingly dominated by various ornithopods and ceratopsids, including *Iguanodon* (Norman, 1980, 1986), *Maiasaura* (Horner and Gorman, 1988), *Prosaurolophus* (Rogers, 1990), *Centrosaurus* (Currie and Dodson, 1984; Ryan, 1992), *Styracosaurus* (Wood *et al.*, 1988), *Pachyrhinosaurus* (Langston, 1975), *Einiosaurus* (Rogers, 1990; Sampson, 1997), and *Chasmosaurus* (Lehman, 1989).

Ontogeny and Phylogeny

With several notable exceptions (e.g., *Protoceratops* from Mongolia), dinosaur finds prior to the 1970s were thought to be restricted to adult specimens. Since that time, there have been numerous important discoveries, including many mass death assemblages, encompassing the full range of growth stages, from embryos and hatchlings to juveniles, subadults, and adults. This new information has radically altered our conception of dinosaurs and allowed paleontologists to address topics that were previously inaccessible: for example, ontogenetic changes in gross morphology (Dodson, 1975; Raath, 1990; Sampson *et al.*, 1997), ontogenetic changes in bone histology (Ricqlès, 1980; Horner and Gorman, 1988; Chinsamy, 1992), nesting patterns and parental care (Horner and Makela, 1979; Horner, 1984), locomotor changes during growth (Norman, 1980; Heinrich *et al.*, 1993), and general life history strategies (Dunham *et al.*, 1989; Weishampel and Horner, 1994).

It is now possible to examine patterns of growth within dinosaur species and then compare these patterns across species. For example, Horner and Weishampel (Horner, 1984; Horner and Weishampel, 1988; Weishampel and Horner, 1994) contrast hatchling skeletons of hypsilophodontids such as *Orodromeus* and *Dryosaurus*, which exhibit relatively advanced degrees of ossification of the limb bones (particularly the distal femur), with those of hadrosaurs such as *Maiasaura* and *Hypacrosaurus*, in which the limb bones are relatively poorly ossified. The authors postulate that this differential in skeletal development reflects varying life history strategies, precociality in hypsilophodontids and altriciality in hadrosaurs, with concomitant differences in behaviors such as parental care.

Ontogenetic patterns have also been identified in later stages of growth in dinosaurs. Dodson (1975), in a biometric study of the skulls of lambeosaurine hadrosaurs, calculated that the premaxillary–maxillary crests did not become apparent until about 35% maximum body size was achieved and did not develop fully until late in maturity. Raath (1990) analyzed variation in a bone bed sample of *Syntarsus* and distinguished two adult morphs (see Sexual Dimorphism and Individual Variants) that could be differentiated only after the animals approached adult size. Similarly, Sampson (1995; Sampson *et al.*, 1997) described delayed expression of putative secondary sexual characters (horns and frills) in centrosaurine ceratopsids and compared this to similar ontogenetic patterns of delayed expression in extant taxa. Late development, especially in male secondary sexual characters, has been demonstrated in numerous extant animals, usually in association with social hierarchies. Prolonged or delayed growth produces a predictable pattern of visual signals permitting animals of varying ages and strengths to assess each other without the need for physical conflict. More important, prolonged growth may allow the attainment of larger size and more elaborate features, thereby increasing the potential for success in mate competition.

In addition to establishing patterns of within-species variation, ontogenetic information has been influential in taxonomic reassessments of some dinosaur groups. Specifically, several studies have questioned the validity of certain taxa within a clade, suggesting that some named taxa may actually represent juveniles or subadults of closely related forms. The pioneering study of this kind was the previously mentioned study of Dodson (1975), which examined variation in lambeosaurine hadrosaurs from the Campanian of Alberta. Dodson argued for the validity of only two genera and 3 species compared to the three genera and 12 species in use. He postulated that *Procheneosaurus*, rather than being a small adult hadrosaur, represents juveniles of *Corythosaurus* and *Lambeosaurus*. Similarly, Sampson *et al.* (1997) demonstrated that juvenile and subadult horned dinosaurs are often similar across taxa; they conclude that two taxa of centrosaurine ceratopsids, *Brachyceratops montanenis* and *Monoclonius crassus*, are actually composed of juveniles and subadults of better established forms such as *Centrosaurus* and *Einiosaurus*.

Sexual Dimorphism and Individual Variants

Sexual dimorphism, or variation between the sexes of a species, is a common phenomenon in living vertebrates and has been postulated for a relatively large number of dinosaur taxa, including several examples of both saurischians and orthnithischians. Sexual dimorphism in living animals is typically associated with body size and/or secondary sexual characters such as horns, antlers, frills, pelage, and plumage. The same pattern apparently holds true for dinosaurs.

Hadrosaurian (duck-billed) dinosaurs are extremely common in the Late Cretaceous deposits of North America and Asia, dominating assemblages in many terrestrial formations. Dodson (1975), in the aforementioned bivariate study of 36 lambeosaurine hadrosaur skulls found as isolated specimens in Alberta, argued for sexual dimorphism in crest shape, with males having the larger, "showier" form.

Perhaps because of the relatively large samples (many derived from mass death assemblages), and the presence of varied and well-developed cranial ornaments, sexual dimorphism has often been postulated in horned dinosaurs (Fig. 1). Lehman (1989, 1990) used data derived from a paucispecific bone bed to argue for the presence of dimorphic orientations of supraorbital horn cores in *Chasmosaurus* and other members of the subfamily Chasmosaurinae. Females are thought to have anterolaterally directed horns, whereas males are characterized by more erect, dorsally directed horns. An undescribed species of *Pachyrhinosaurus*, known from an extensive bone bed sample, includes two forms of nasal bosses ($N = 13$), one concave and the other convex (Sampson *et al.*, 1997). A much smaller sample of *Einiosaurus procurvicornis* preserves two varieties of adult nasal horn cores ($N = 5$), one erect and the other strongly procurved (Sampson, 1997).

Caution is required, however, in any inference of sexual dimorphism in ceratopsids and in dinosaurs generally. The same characters likely to express dimorphism—that is, secondary sexual characters—are also the most likely to vary among individuals generally, as witnessed in numerous living animals. It is not surprising, therefore, that ceratopsid horns and frills are also subject to extreme ontogenetic and individual variation (Sampson, 1995; Sampson *et al.*, 1997). In one bone bed sample dominated by the remains of *Centrosaurus* (Ryan, 1992), nasal horn specimens vary in orientation from forward curved to erect to recurved, with intermediate gradations. Were this a smaller or biased sample, preserving only two morphs (e.g., forward and backward oriented horns), one might (erroneously) infer sexual dimorphism. Similarly, supraorbital horn cores in the same sample of *Centrosaurus* are extremely variable, ranging from small pointed horns to pitted masses to true pits. Thus, in the absence of much larger samples, any conclusions regarding sexual dimorphism in ceratopsid dinosaurs must remain speculative.

Perhaps the largest collection of articulated skeletons known for any single dinosaur taxon is that for the ceratopsian *Protoceratops andrewsi*. Literally dozens of isolated specimens of this medium-sized ceratopsian have been recovered from Late Cretaceous deposits in Mongolia. Following a detailed osteological study by Brown and Schlaikjer (1940), Kurzanov (1972) and Dodson (1976) argued on morphometric grounds for the existence of sexual dimorphism in *Protoceratops*. Dodson (1976) examined 24 skulls using principal components analysis and found several characters to be dimorphic. In particular, the posterior frill, despite considerable ontogenetic and individual variation, occurs in two adult morphs, one more upright than the other (thought to be male and female, respectively).

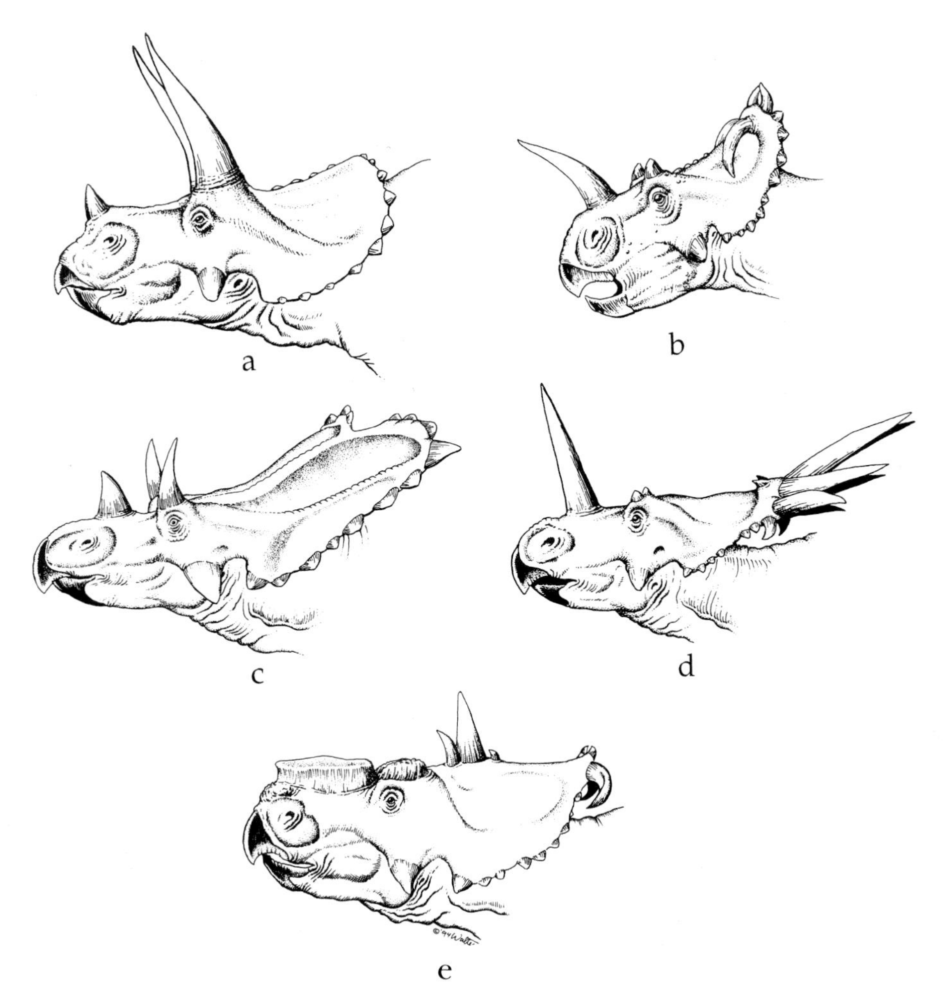

FIGURE 1 Interspecific variation is especially apparent in these ceratopsids, which variously illustrate horns, crests, and frills. (a) *Triceratops*; (b) *Centrosaurus*; (c) *Chasmosaurus*; (d) *Styracosaurus*; (e) *Pachyrhinosaurus*. By Gregory S. Paul (from Dodson, Peter. *The Horned Dinosaurs.* Copyright © 1996 by Princeton University Press. Reprinted by permission of Princeton University Press).

Investigators have noted potential dimorphism in several theropod dinosaurs, including *Syntarsus* (Raath, 1990), *Coelophysis* (Colbert, 1990), and *Tyrannosaurus* (Carpenter, 1990; Larson, 1992). Bone bed deposits have once again proven valuable in the demonstration of variation. The North American Upper Triassic form, *Coelophysis*, is known from hundreds of specimens. Colbert (1990) cautiously speculates that two adult skeletons from one locality may represent sexual dimorphs. One specimen, thought by Colbert to represent a male, has a larger skull, longer neck, shortened forelimbs, and fused sacral vertebral spines, whereas the other individual lacks these characters. Raath (1990) is more confident in asserting the presence of sexual dimorphism in a mass assemblage of the small theropod *Syntarsus*, a close relative of *Coelophysis* known from the Early Jurassic of Zimbabwe. Morphometric analysis of postcranial elements from a sample of more than 30 individuals of *Syntarsus* yields a bimodal distribution in supposed adults, with one morph characterized by more robust limb elements and better developed muscle scars. In contrast to most studies of this type, Raath postulates that the more robust form is female, citing precedents known from extant raptorial birds. Carpenter (1990) came to a similar conclusion in a study of variation in *Tyrannosaurus*, noting the existence of robust and gracile individuals and claiming the more robust individuals to be female. This contention is supported by the fact that the angle between the ischia and tail vertebrae is greater in the more robust form, perhaps in response to egg-laying requirements. Larson (1992) concurs with this view, arguing that the more robust individuals of *Tyrannosaurus* appear to lack a hemal arch in the proximal caudal vertebra that is present in gracile specimens, presumably a further accommodation for the passage of eggs. An equivalent dimorphic condition exists in modern crocodilians.

Conclusion: The Future of Dinosaur Variation Studies

Despite the large number of recent, often innovative studies, our understanding of dinosaur variation remains nascent. Although skeletal remains will continue to restrict paleontologists to a limited subset of morphology and variation, much can be done and great advances are virtually inevitable as more fossils are recovered, described, and analyzed. The study of variation is truly a matter of numbers; the more individuals considered, the greater the chances of determining the nature of morphological differences, whether intra- or interspecific. To date, the impact of morphometric analyses and quantitative data has been relatively limited in dinosaur PALEONTOLOGY. In general, morphometric and statistical analyses are reasonable options only when sample sizes are sufficiently large (minimum number of individuals > 20). However, samples of this size and larger are increasingly becoming available and are awaiting analysis, and it is likely that morphometric studies will become increasingly important, particularly as new analytical tools are developed.

Undoubtedly, mass death assemblages will play a major role in illuminating the nature of variation in dinosaurs. Paucispecific (low-diversity) bone beds provide unique opportunities to examine large samples of potentially single species, thereby facilitating investigation of patterns relating to ontogeny, sexual dimorphism, and individual differences. Particularly desirable are more complete ontogenetic series for a wide range of taxa, which would be amenable to a variety of analytical approaches. Histological study of ontogenetic series may prove useful in calibrating relative age estimations and thereby determining rates of absolute growth (Weishampel and Horner, 1994). More analyses of cross-sectional area will lead to greater comprehension of functional morphological changes associated with growth. Finally, continued study of ontogenetic changes in gross morphology will lead to a better understanding of various aspects of dinosaur paleobiology, from parental care and resource use to life history strategies and social organization.

See also the following related entries:

GENETICS • GROWTH AND EMBRYOLOGY • PROBLEMS WITH THE FOSSIL RECORD • SPECIATION • SPECIES • SYSTEMATICS • TAPHONOMY

References

Benton, M. J. (1990). Origin and interrelationships of dinosaurs. In *The Dinosauria* (D. B. Weishampel, P. Dodson, and H. Osmólska, Eds.), pp. 11–30. Univ. of California Press, Berkeley.

Brown, B., and Schlaikjer, E. M. (1940). The structure and relationships of *Protoceratops. Ann. N. Y. Acad. Sci.* **40,** 133–266.

Carpenter, K. (1990). Variation in *Tyrannosaurus rex.* In *Dinosaur Systematics: Approaches and Perspectives* (K. Carpenter and P. J. Currie, Eds.), pp. 141–145. Cambridge Univ. Press, Cambridge, UK.

Chapman, R. E. (1990). Shape analysis in the study of dinosaur morphology. In *Dinosaur Systematics: Approaches and Perspectives* (K. Carpenter and P. J. Currie, Eds.), pp. 21–42. Cambridge Univ. Press, Cambridge, UK.

Chapman, R. E., and Brett-Surman, M. K. (1990). Morphometric observations on hadrosaurid ornithopods. In *Dinosaur Systematics: Approaches and Perspectives* (K. Carpenter and P. J. Currie, Eds.), pp. 163–177. Cambridge Univ. Press, Cambridge, UK.

Chapman, R. E., Galton, P. M., Sepkoski, J. J., Jr., and Wall, W. P. (1981). A morphometric study of the cranium of the pachycephalosaurid dinosaur *Stegoceras. J. Paleontol.* **55**(3), 608–616.

Chinsamy, A. (1992). Ontogenetic growth of the dinosaurs *Massospondylus carinatus* and *Syntarsus rhodesiensis. J. Vertebr. Paleontol.* **12,** 23A. [Abstract]

Colbert, E. H. (1989). The Triassic dinosaur *Coelophysis. Museum Northern Arizona Bull.* **57,** i–xv, 1–160.

Chinsamy, A. (1990). Variation in *Coelophysis bauri.* In *Dinosaur Systematics: Approaches and Perspectives* (K. Carpenter and P. J. Currie, Eds.), pp. 81–90. Cambridge Univ. Press, Cambridge, UK.

Currie, P. J., and Dodson, P. (1984). Mass death of a herd of ceratopsian dinosaurs. In *Third Symposium on Mesozoic Terrestrial Ecosystems, Short Papers* (W.-E. Reif and F. Westphal, Eds.), pp. 61–66. Verlag, Tubingen, Germany.

Dodson, P. (1975). Taxonomic implications of relative growth in lambeosaurine hadrosaurs. *Systematic Zool.* **24**(1), 37–54.

Dodson, P. (1976). Quantitative aspects of relative growth and sexual dimorphism in *Protoceratops. J. Paleontol.* **50,** 929–940.

Dodson, P. (1990). On the status of the ceratopsids *Monoclonius* and *Centrosaurus.* In *Dinosaur Systematics: Ap-*

proaches and Perspectives (K. Carpenter and P. J. Currie, Eds.), pp. 211–229. Cambridge Univ. Press, Cambridge, UK.

Dodson, P., Behrensmeyer, A. K., Bakker, R. T., and MacIntosh, J. S. (1980). Taphonomy and paleoecology of the Upper Jurassic Morrison Formation. *Paleobiology* **6,** 208–232.

Dunham, A. E., Overall, K. L., Porter, W. P., and Forster, C. (1989). Implications of ecological energetics and biophysical developmental constraints for life-history variation in dinosaurs. *Geol. Soc. Am. Spec. Paper* **238,** 1–19.

Forster, C. A. (1990). The cranial morphology of *Triceratops*, and a preliminary phylogeny of the Ceratopsia. Doctoral thesis, University of Pennsylvania, Philadelphia.

Galton, P. M. (1986). Prosauropod dinosaur *Plateosaurus* (= *Gressylosaurus*) (Saurischia: Sauropodomorpha) from the Upper Triassic of Switzerland. *Geol. Palaeontol.* **20,** 167–183.

Gauthier, J. (1986). Saurischian monophyly and the origin of birds. *Mem. California Acad. Sci.* **8,** 1–55.

Heinrich, R. E., Ruff, C. B., and Weishampel, D. B. (1993). Femoral ontogeny and locomotor mechanics of *Dryosaurus lettowvorbecki* (Dinosauria, Iguanodontia). *Zool. J. Linnean Soc.* **108,** 179–196.

Hooker, J. S. (1987). Late Cretaceous ashfall and the demise of a hadrosaurian "herd." Geological Society of America, Rocky Mountain Section, Abstracts with Program, No. 19, pp. 284.

Horner, J. R. (1984). The nesting behavior of dinosaurs. *Sci. Am.* **250**(4), 130–137.

Horner, J. R., and Gorman, J. (1988). *Digging Dinosaurs.* Workman, New York.

Horner, J. R., and Makela, R. (1979). Nest of juveniles provides evidence of family structure among dinosaurs. *Nature* **282**(5736), 296–298.

Horner, J. R., and Weishampel, D. (1988). A comparative embryological study of two ornithischian dinosaurs. *Nature* **332,** 256–257.

Kurzanov, S. M. (1972). Sexual dimorphism in protoceratopsians. *Palaeontol. J.* **1972,** 91–97.

Langston, W. L., Jr. (1975). The ceratopsian dinosaurs and associated lower vertebrates from the St. Mary River Formation (Maestrichtian) at Scabby Butte, southern Alberta. *Can. J. Earth Sci,* **12,** 1576–1608.

Langston, W. L., Jr., Currie, P. J., and Tanke, D. H.. (1997). A new species of *Pachyrhinosaurus* from the Late Cretaceous of Alberta. Manuscript in preparation.

Larson, P. (1992). Sexual dimorphism in the abundant Upper Cretaceous theropod *Tyrannosaurus rex. J. Vertebr. Paleontol.* **12**(Suppl. to No. 3), 38A. [Abstract]

Lehman, T. M. (1989). *Chasmosaurus mariscalensis*, sp. nov., a new ceratopsian dinosaur from Texas. *J. Vertebr. Paleontol.* **9**(2), 137–162.

Lehman, T. M. (1990). The ceratopsian subfamily Chasmosaurinae: Sexual dimorphism and systematics. In *Proceedings of the Dinosaur Systematics: Approaches and Perspectives* (K. Carpenter and P. J. Currie, Eds.), pp. 211–229. Cambridge Univ. Press, Cambridge, UK.

Madsen, J. H. (1976). *Allosaurus fragilis*: A revised osteology. *Utah Geol. Mineral. Survey Bull.* **1091,** 1–163.

Norman, D. B. (1980). On the ornithischian dinosaur *Iguanodon bernissartensis* from the Lower Cretaceous of Bernissart (Belgium). *Inst. R. Sci. Nat. Belgique Mem.* **178,** 1–103.

Norman, D. B. (1986). On the anatomy of *Iguanodon atherfieldensis* (Ornithischia: Ornithopoda). *Bull. Inst. R. Sci. Nat. Belgique Sci. Terre* **56,** 281–372.

Raath, M. A. (1990). Morphological variation in small theropods and its meaning in systematics: Evidence from *Syntarsus rhodensis*. In *Dinosaur Systematics: Approaches and Perspectives* (K. Carpenter and P. J. Currie, Eds.), pp. 91–105. Cambridge Univ. Press, Cambridge, UK.

Ricqlès, A. J. de (1980). Tissue structures of dinosaur bone: Functional significance and possible relation to dinosaur physiology. In *A Cold Look at the Warm-Blooded Dinosaurs* (R. D. K. Thomas and E. C. Olson, Eds.), pp. 103–139. Westview Press, Boulder, CO.

Rogers, R. R. (1990). Taphonomy of three dinosaur bone beds in the Upper Cretaceous Two Medicine Formation of northwestern Montana: Evidence for drought-related mortality. *Palaios* **5,** 394–413.

Ryan, M. (1992). The taphonomy of a *Centrosaurus* (Reptilia: Ornithischia) bone bed (Campanian), Dinosaur Provincial Park, Alberta, Canada. Masters thesis, University of Calgary, Alberta, Canada.

Sampson, S. D. (1995). Horns, herds, and hierarchies. *Nat. History* **194**(6), 36–40.

Sampson, S. D. (1997). Two new horned dinosaurs from the Upper Cretaceous Two Medicine Formation of Montana; with a phylogenetic analysis of the Centrosaurinae (Ornithischia: Ceratopsidae). *J. Vertebr. Paleontol.*, in press.

Sampson, S. D., Ryan, M. J., and Tanke, D. H. (1997). The ontogeny of centrosaurine dinosaurs (Ornithischia: Ceratopsidae), with new information from mass death assemblages. Manuscript in preparation.

Sereno, P. C. (1986). Phylogeny of bird-hipped dinosaurs (order Ornithischia). *Natl. Geogr. Soc. Res.* **2,** 234–256.

Weishampel, D. B., and Chapman, R. E. (1990). Morphometric study of *Plateosaurus* from Trossingen (Baden-Wurttenberg, Federal Republic of Germany). In *Dinosaur Systematics: Approaches and Perspectives* (K. Carpenter and P. J. Currie, Eds.), pp. 43–51. Cambridge Univ. Press, Cambridge, UK.

Weishampel, D. B., and Heinrich, R. E. (1992). Systematics of Hypsilophodontidae and basal Iguanodontia (Dinosauria: Ornithopoda). *Historical Biol.* **6,** 159–184.

Weishampel, D. B., and Horner, J. R. (1994). Life history syndromes, heterochrony, and the evolution of Dinosauria. In *Dinosaur Eggs and Babies* (K. Carpenter, K. F. Hirsch, and J. R. Horner, Eds.), pp. 229–243. Cambridge Univ. Press, Cambridge, UK.

Weishampel, D. B., and Westphal, F. (1986). Die Plateosaurier von Trossingen. *Ausstellungskat Univ. Tübingen* **19,** 1–27.

Wood, J. M., Thomas, R. G., and Visser, J. (1988). Fluvial processes and vertebrate taphonomy: The Upper Cretaceous Judith River Formation, south-central Dinosaur Provincial Park, Alberta, Canada. *Palaeogeogr. Palaeoclimatol. Palaeoecol.* **66,** 127–143.

Young, C.-C. (1951). The Lufeng saurischian fauna in China. *Paleontol. Sinica Ser. C* **13,** 1–96.

Velociraptorinae

see DROMAEOSAURIDAE

Venezuelan Dinosaurs

see SOUTH AMERICAN DINOSAURS

Vertebrae, Comparative Anatomy

see POSTCRANIAL AXIAL SKELETON

Vertebrata

MICHAEL J. BENTON
University of Bristol
Bristol, United Kingdom

Dinosaurs are vertebrates, as indicated by their possession of bones. The vertebrates include all living and fossil backboned animals: fishes, amphibians, reptiles, birds, and mammals and all descendants of their most recent common ancestor (Carroll, 1987; Benton, 1990). The Vertebrata is the major division of the Chordata, a group that also includes the small, and rather obscure, groups of the tunicates and lancelets such as amphioxus. All chordates have a flexible stiffening rod in their backs termed the notochord. They also have a concentration of sense organs at the head end and a tail, which is identified as the part of the body extending behind the anus. These characters set chordates apart from all other animals, such as clams, sea urchins, corals, crabs, and the like, which are usually collectively known as invertebrates.

Vertebrates all have an identifiable SKULL of some sort, a toughened box that surrounds the brain and protects it. In most vertebrates, the skull and other elements of the skeleton are made from bone. Bone is a mix of living tissue and mineral tissue, the living materials consisting of the tough protein collagen as well as blood vessels, fat, and some muscular tissues on the surface, set in a crystalline matrix of apatite (calcium phosphate).

The oldest specimens of vertebrates date from the Cambrian and Ordovician periods (see GEOLOGIC TIME), approximately 500 my ago, and these are partial specimens of small jawless forms. The first jawed vertebrates arose during the Silurian Period, and the first land vertebrates, tetrapods, crept onto land during the Devonian Period. The first fully land-living animals, the oldest amniotes, arose during the Carboniferous, and they laid EGGS with tough shells (see REPTILES). The mammals first appeared during the Latest Triassic or Early Jurassic, and birds (see AVES) arose, probably during the Jurassic Period, from dinosaurs.

See also the following related entry:

VERTEBRATE PALEONTOLOGY

References

Benton, M. J. (1990). *Vertebrate Palaeontology*, pp. 377. Chapman & Hall, London.

Carroll, R. L. (1987). *Vertebrate Paleontology and Evolution*, pp. 698. Freeman, San Francisco.

Vertebrate Paleontology

Paul G. Davis
National Science Museum
Tokyo, Japan

Vertebrate paleontology is the study of fossil vertebrates, including dinosaurs. Vertebrates (more formally termed Vertebrata) are a subphylum of the phylum Chordata (the "chordates"). Chordates are united in the possession of a rod of flexible tissue (the notochord) that is medially situated and traverses each animal from front to back. The Vertebrata share the advanced character of a notochord (mostly incorporated into the spinal column) that is protected within either the cartilage or the bone that makes up the skull and vertebral column.

Vertebrata includes all fish, amphibians, reptiles, birds, and mammals. Many of these Linnaean taxonomic groups are no longer valid in terms of cladistic natural (or monophyletic) groups, but most biologists and paleontologists still use them informally. For example, it would become tedious to use the term "amniotes not including mammals and birds" when in simplistic terms a "reptile" is meant.

See also the following related entries:

Paleontology • Vertebrata

Virginia Living Museum, Virginia, USA

see Museums and Displays

Virginia Museum of Natural History, Virginia, USA

see Museums and Displays

Von Ebner Incremental Growth Lines

Gregory M. Erickson
University of California
Berkeley, California, USA

Incremental growth lines can be observed in the dentine of thin-sectioned mammalian teeth using light microscopy (Owen, 1840–1945). The smallest of these structural laminations are known as incremental lines of von Ebner and have a spacing interval <20 μm

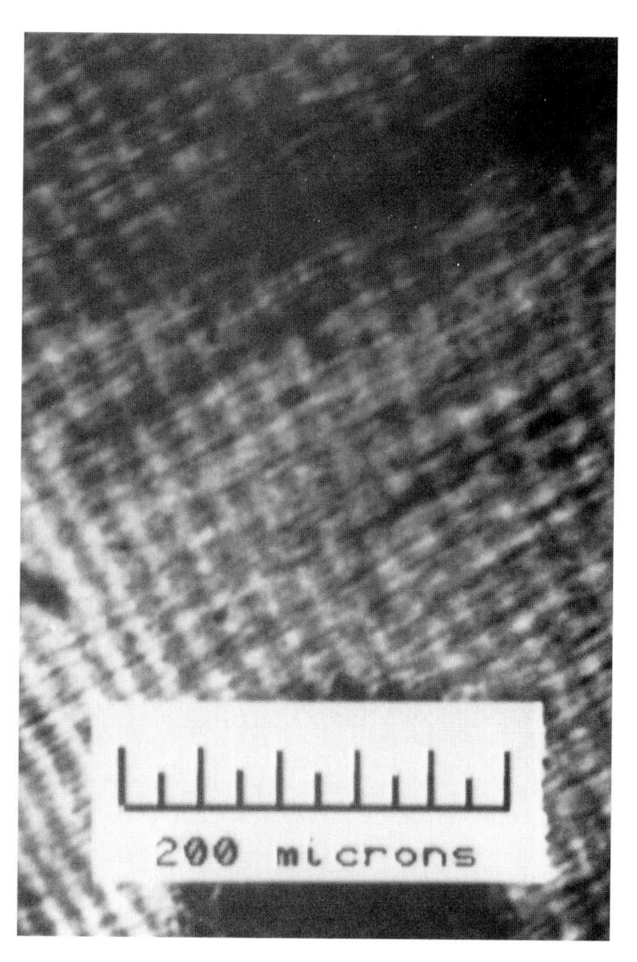

FIGURE 1 Dinosaur dentine exhibiting incremental lines of von Ebner. The growth lines extend from the lower right to upper left corners of the plate. Reprinted from G. M. Erickson, Incremental lines of Von Ebner in dinosaurs and the assessment of tooth replacement rates using growth line counts. *Proc. Nat. Acad. Sci.* **93,** 14623–14627. Copyright (1996) National Academy of Sciences, USA.

(i.e., <2/1000th of a meter). In most taxa these lines reflect daily fluctuations in mineral deposition due to the influences of endogenous biorhythms. Recently, von Ebner incremental lines were observed in the dentine of dinosaurs (Erickson, 1996; Fig. 1). Because rates of deposition of dentine had not been studied in reptiles before this, it was not clear whether the lines reflected daily growth, as in most mammals. A study of tooth formation in American alligators and common caiman using periodic chemical labeling indicated that the von Ebner lines were deposited daily (Erickson, 1997). Because crocodilians are closely related to dinosaurs phylogenetically (both share an archosaurian ancestry) and both groups have comparable dentitions morphologically, it is likely that the dinosaur incremental lines also mark daily growth. Thus, in theory, total counts of growth lines within dinosaur and crocodile teeth reveal tooth formation rates. (Because archosaurian reptiles shed their teeth throughout life, it is not possible to determine the age of individuals using growth line counts.) By subtracting the total number of incremental lines in a functional tooth (i.e., a tooth that has erupted and is being used in feeding) from those found in its immediate successor (i.e., a developing replacement tooth that will erupt following the partition of the functional tooth), tooth replacement rate can be ascertained (Erickson, 1997). Adult tooth replacement rates in dinosaurs appear to have ranged from ~50 to 800 days depending on the species. Larger toothed taxa exhibited the slowest rates of replacement, whereas smaller toothed species and those with tooth batteries (i.e., ceratopsians and hadrosaurs with up to seven teeth formed simultaneously per tooth position) showed the fastest rates. Dinosaur tooth replacement rates slowed through ontogeny (i.e., development), as has been demonstrated in extant reptiles.

See also the following related entries:

Growth and Embryology • Histology of Bones and Teeth • Tooth Replacement

References

Erickson, G. M. (1996). Incremental lines of von Ebner in dinosaurs and the assessment of tooth replacement rates using growth line counts. Proceedings of the National Academy of Science, USA **93,** 14623–14627.

Erickson, G. M. (1996). Daily deposition of incremental lines in alligator dentine and the assessment of tooth replacement rates using incremental line counts. *J. Morphol.* **228,** 189–194.

Owen, R. (1840–1845). *Odontography*, pp. 655. Ballière, London.

Washington State University Museum, Washington, USA

see Museums and Displays

Wealden Group

D. B. Norman
University of Cambridge
Cambridge, United Kingdom

The Wealden Group of England consists of fluvial, lacustrine, and lagoonal deposits that crop out across a broad area of southern and eastern England in two adjacent subbasins: Kent, Surrey, and Sussex (the geographic area known as the Weald, due south of London) and on the southern shores of the Isle of Wight and adjacent county of Dorset, along the central south coast of mainland England (geographically Wessex).

The Wealden of the Weald subbasin encompasses the Berriasian–Barremian stratigraphic stages and can be divided approximately into lower and upper divisions: The lower is arenaceous-influenced Hastings Beds and the upper is predominantly argillaceous Weald Clay. The Hastings Beds (~400 m) show a degree of cyclicity in deposition, and each cycle shows an onset of clays and siltstones, which coarsen upwards into cross-bedded sandstones, the latter frequently pocked by lenses of bone-rich gravel. Each cycle is completed by a return to silt-dominated deposits with horsetail stands (*Equisetites* sp.) and a final return to argillaceous rocks that form the base of the next cycle. This set of sedimentological regimes was probably laid down in lacustrine/lagoonal mudplain environments in which salinity was controlled by the relative rates of freshwater runoff and evaporation (Allen, 1981). Soil horizons, *in situ* plants, and footprints all testify to the shallow water conditions throughout much of the time zone. The Weald Clay (450 m) represents predominantly mudplain deposition, with occasional pulses of coarser material. Toward the top of the sequence increasingly brackish conditions prevail, marking the onset of the widespread marine transgression of the mid-Cretaceous and the deposition of the lower Greensand (Aptian and Albian) across much of southern and eastern England.

In the Wessex subbasin, the Wealden deposits are Barremian in age and comprise the Wessex Formation (formerly the Wealden Marls) and the overlying Vectis Formation (formerly the Wealden Shales). The Wessex Formation (reaching a maximum thickness of 530 m on the Isle of Wight) is a varicolored sequence (predominantly red) of mudstones and sandstones. Sedimentologically the Wessex Formation was deposited as alluvium by a meandering river system (Stewart, 1983). The Vectis Formation comprises gray mudstones and siltstones and, although well represented on the Isle of Wight, is largely absent in Dorset outcrops. The environment would appear to have been predominantly lagoonal and subject to increasing salinity and storm influence toward the top of the section (Stewart *et al.*, 1991).

Among the first accounts of the Wealden District was a report on "form'd stones found at Hunton in Kent" published in the *Philosophical Transactions of the Royal Society* in 1684. The general anticlinal nature of the surface rocks of the area was known to John Farey, who in 1806 produced, in the form of a diagram, a geological section across the Weald. This probably formed the foundations for the William Smith maps published in 1815. Gideon Mantell's monographic review of the geology of southeast England (1822) established the freshwater origin of many of the rocks, and his subsequent work (1822–1851) resulted in the discovery of a number of new and important fossil remains—notably those of the dinosaurs *Megalosaurus, Iguanodon, Hylaeosaurus, Pelorosaurus*, and *Regnosaurus* (all but the first of which he named). Since Mantell's time, considerable attention has been paid to the Wealden District as a whole through systematic geological studies, borehole examination, and paleontological investigations.

From a dinosaurian standpoint, the Wealden proved to be a relatively rich hunting ground, with regular discoveries from quarries throughout much of the last century. New material has continued to be recovered regularly from the shore cliffs in this (following the decline of local quarrying industry, or its replacement by heavy powered equipment such as draglines, which militate against new discoveries).

The dinosaur fauna includes a diversity of forms. *Calamospondylus oweni* is a small theropod of questionable status, based on fragmentary remains from the Barremian of the Isle of Wight. "*Megalosaurus dunkeri*" is a large THEROPOD of dubious status, based on isolated teeth and bones of general Wealden age from localities across the Wealden District. A large, well-preserved partial skeleton of an *Allosaurus*-like theropod *Neovenator salerii* has been excavated from the Isle of Wight (Barremian) in recent years. It is hoped that this will help to resolve many of the taxonomic problems concerning the identification of large theropod remains in the Wealden. A distinctively high-spined large theropod (*Becklespinax* sp.) is based solely on a row of three articulated dorsal vertebrae collected from Battle (Valanginian), near Hastings. *Baryonyx walkeri* is a unique large-clawed, large theropod with an unusually flattened, crocodile-like snout. It is based on a partial skeleton recovered from the Weald Clay (Barremian) at Smokejacks Pit, Ockley, Surrey.

Wealden SAUROPODS include *Cetiosaurus conybeari*, based on a row of four associated anterior caudal vertebrae [collected from "Mantell's quarry" at Whiteman's Green (Valanginian)], one distal caudal, and an isolated chevron of distinctly sauropod (brachiosaur?) appearance. *Pelorosaurus conybeari* is an isolated large humerus collected from Mantell's quarry. It has no proven association with *C. conybeari*, but both could be reasonably supposed to come from a brachiosaur-like sauropod. W. T. Blows (1995) reviewed the Barremian sauropod remains and concluded that there were at least two distinctive types of brachiosaur: *Ornithopsis*, based on an extremely poorly preserved cervical centrum, and *Eucamerotus*, based on a series of five better preserved dorsal vertebrae. In addition, a partial skeleton of a small sauropod has been recently recovered from the Isle of Wight, which may well help to clarify what is currently a very confused situation. *Pleurocoelus valdensis* is based on isolated teeth found across the Wealden District but is an extremely dubious sauropod taxon.

Ornithischia of the Wealden include the ORNITHOPOD *Hypsilophodon foxii*. There are numerous beautifully preserved skeletons of this small (2-m-long), cursorial ornithopod from the Isle of Wight. This is one of the best known of the Wealden dinosaurs, largely as a result of the monographic studies of Peter Galton. *Valdosaurus canaliculatus* was recognized as a distinct taxon in the 1970s by Peter Galton and appears to be a medium-sized ornithopod (femoral size suggests that adults may have been up to 5 m long). This dinosaur seems to be most similar to the Jurassic ornithopod *Dryosaurus* in both postcranial and dental morphology. *Iguanodon anglicus* (nominus typus) is based on the original type material (teeth alone) from Whiteman's Green, near Cuckfield (Valanginian), and was used to establish the genus *Iguanodon*. *Iguanodon hollingtonensis* is a gracile, long-spined species of *Iguanodon* based on a number of specimens collected from the lower part of the Wealden succession (Norman, 1987). It appears to have attained a body length on the order of 6 m. *Iguanodon dawsoni* was a contemporary (Valanginian) of *I. hollingtonensis* but is of more robust form, with short dorsal neural spines and a distinctive form to its ilium. It is known from two partial skeletons. *Iguanodon atherfieldensis* is one of the "classic" forms of *Iguanodon*. It is a small species of more gracile general build than *I. hollingtonensis*, with relatively long neural spines. Only found in the Barremian at the top of the Wealden succession and in the base of the overlying Greensand Formation in Kent, it is known from several well-preserved skeletons from the Isle of Wight and from the Weald Clay of Surrey. *Iguanodon bernissartensis*, a contemporary of *I. atherfieldensis*, was larger and more robust and was first described from the Barremian of Bernissart, Belgium. Similar skeletons were subsequently discovered in the Wealden of the Isle of Wight and the Weald Clay of Surrey (Smokejacks Pit).

The PACHYCEPHALOSAUR *Yaverlandia bitholus* is based on an isolated skull roof, which resembles that of *Hypsilophodon* but shows more pronounced thickening of the frontal bones. There is an incipient dome equivalent to that seen in later pachycephalosaurs. The fact that to date no other material attributable to this taxon has been discovered would normally be used as a note of extreme caution, but experience with the theropod *Baryonyx* has shown that even large animals can evade discovery in the Wealden through at least 150 years of intensive collecting. *Regnosaurus northamptoni* is a long-standing enigma represented by only a small section of jaw with only the roots of broken teeth visible. Barrett and Upchurch (1995) assigned it to the Stegosauria. The ankylosaur *Hylaeosaurus armatus* was first described from a partial skeleton by Mantell in 1833. This taxon has remained enigmatic largely because it has never been adequately prepared from the arenaceous block in which it is preserved. Isolated elements suggest that this was a nodosaurid ankylosaur characteristic of the Valanginian. The skeleton of *Polacanthus foxii* was discovered in the late 1870s on the Isle of Wight and has recently been restudied. More material has also been discovered. Although many attempts have been made to synonymize *Hylaeosaurus* (front end of an ankylosaur) with *Polacanthus* (back end of an ankylosaur), it seems likely that they will remain distinct taxa and may well reflect the Upper/Lower Wealden faunal split that is seen in the other dinosaurian taxa.

No trace of ceratopsian material has been discovered to date in the Wealden of England, despite the German Wealden *Stenopelix*, which seems to be basal marginocephalian.

The Wealden succession exposed in southern England encompasses much of the Lower Cretaceous (Berriasian–Barremian). The sedimentary environment was fairly consistently estuarine/lacustrine/lagoonal and forms a depositional setting capable of sampling the terrestrial and semiaquatic fauna of geographic region across this time zone. The outcrops, though much restricted compared to the previous century, still produce new material on a regular basis (sporadic in inland quarries and regularly on coastal sites) and will continue to do so provided that coastal management and sea defense work do not materially affect the rate of erosion of existing exposures. The dinosaur fauna of the Wealden is dominated by ornithopods (*Iguanodon*, *Hypsilophodon*, and *Valdosaurus*) with comparatively rare nodosaurid ankylosaurs (*Polacanthus* and *Hylaeosaurus*), brachiosaurid, and diplodocid sauropods and "erratics" such as a pachycephalosaur (*Yaverlandia*) and a STEGOSAUR (*Regnosaurus*). It is also predicted that rare remains of psittacosaur-like CERATOPSIANS will be discovered at some time in the future in the uppermost Wealden beds. Carnivores are, as is typical of the majority of dinosaur bearing sequences, relatively rare and currently quite poorly documented, with the notable exception of the rather bizarre *Baryonyx* and the recently discovered allosaur-like theropod *Neovenator*. The Wealden will always be of considerable importance to dinosaur studies simply because of the historical burden placed on it by the early workers (most notably Gideon Mantell, Richard Owen, Reverend William Fox, John Hulke, and Harry Seeley); however, the discoveries of the past decade (excellent material of sauropods, ornithopods, and ankylosaurs, as well as the spectacular *Baryonyx* and the new large theropod) suggest that the Wealden will continue to be important for prospecting and study in the future as well.

See also the following related entries:

CRETACEOUS PERIOD • EUROPEAN DINOSAURS • HISTORY OF DINOSAUR DISCOVERIES

References

Allen, P. (1981). Pursuit of Wealden models. *J. Geol. Soc. London* **138,** 379–405.

Barrett, P. M., and Upchurch, P. (1995). *Regnosaurus northamptoni*, a stegosaurian dinosaur from the Lower Cretaceous of southern England. *Geol. Magazine* **132,** 213–222.

Blows, W. T. (1995). The early Cretaceous brachiosaurid dinosaurs *Ornithopsis* and *Eucamerotus* from the Isle of Wight. *Palaeontology* **38,** 187–198.

Mantell, G. A. (1822). *The Fossils of the South Downs; or, Illustrations of the Geology of Sussex*, pp. 327. Lupton Relfe, London.

Norman, D. B. (1987). Wealden dinosaur biostratigraphy. In *Fourth Symposium on Mesozoic Terrestrial Ecosystems, Short Papers* (P. J. Currie and E. H. Koster, Eds.), pp. 161–166. Tyrrell Museum of Palaeontology, Drumheller, Alberta, Canada.

Stewart, D. J. (1983). Possible suspended load channel deposits from the Wealden Group (Lower Cretaceous) of southern England. *Spec. Publ. Int. Assoc. Sedimentol.* **6,** 369–384.

Stewart, D. J., Ruffel, A., Wach, G., and Goldring, R. (1991). Lagoonal sedimentation and fluctuation salinities in the Vectis Formation (Wealden Group, Lower Cretaceous) of the Isle of Wight. *Sedimentary Geol.* **72,** 117–134.

Weber State University Museum of Natural Science, Utah, USA

see Museums and Displays

Weinman Mineral Museum, Georgia, USA

see Museums and Displays

Willow Creek Anticline

John R. Horner
Museum of the Rockies
Montana State University
Bozeman, Montana, USA

The 50-m-thick Willow Creek Anticline of western Montana is the 1.5-km^2 area that initially yielded evidence of dinosaur social behaviors. Discovered originally in 1977 by Marion Brandevold and her family (Horner and Gorman, 1988), the area has been worked from 1978 until the present by researchers from Princeton University and the Museum of the Rockies, Montana State University, led by John Horner. The anticline area has produced an immense amount of data concerning dinosaur social behaviors and dinosaur ecology.

From the base of the anticline, in greenish mudstone sediments interpreted as overbank deposits are the remains of several nests of the hadrosaur *Maiasaura peeblesorum* Horner and Makela 1979. These and other nests, found on what are interpreted as nesting horizons (Horner, 1982), yield the information from which hypotheses have originated concerning parental care among some dinosaurs (Horner, 1984). Above the overbank mudstones unit is a dark gray mudstone containing a huge accumulation of skeletal elements attributed to *Maiasaura* (Lorenz and Gavin, 1984). This assemblage, representing numerous growth stages (Varricchio and Horner, 1993), has been interpreted as a group of maiasaurs killed catastrophically by volcanic ash (Hooker, 1987).

Resting conformably over the bone bed horizon is a lacustrine member that contains lake sediments, islands, and shoreline facies. The lake beds have yielded pterosaur remains (Padian, 1984), and islands such as Egg Mountain have yielded numerous skeletons and eggs. Above the lacustrine member is a sequence of greenish mudstone layers, each of which produce *Maiasaura* nesting sites; some of these include eggs with embryos and partial clutches.

The dinosaur fauna of the Willow Creek Anticline includes *M. peeblesorum, Orodromeus makelai, Troodon* cf. *formosus, Saurornitholestes* sp., *Daspletosaurus* sp., an ornithomimid, and a nodosaurid.

See also the following related entry:

Judith River Wedge

References

Hooker, J. S. (1987). Late Cretaceous ashfall and the demise of a hadrosaurian "herd." Geological Society of America, Rocky Mountain Section, Abstracts with Programs, No. 19, pp. 284.

Horner, J. R. (1982). Evidence of colonial nesting and "site fidelity" among ornithischian dinosaurs. *Nature* **297,** 675–676.

Horner, J. R. (1984). The nesting behavior of dinosaurs. *Sci. Am.* **250**(4), 130–137.

Horner, J. R., and Gorman, J. (1988). *Digging Dinosaurs,* pp. 210. Workman, New York.

Horner, J. R., and Makela, R. (1979). Nest of juveniles provides evidence of family structure among dinosaurs. *Nature* **282,** 296–298.

Lorenz, J. C., and Gavin, W. (1984). Geology of the Two Medicine Formation and the sedimentology of a dinosaur nesting ground. In *Montana Geological Society, 1984 Field Conference, Northwestern Montana, Guidebook* (J. D. McBane and P. B. Garrison, Eds.), pp. 175–186. Montana Geological Society, Billings, MT.

Padian, K. (1984). A large pterodactyoid pterosaur from the Two Medicine Formation (Campanian) of Montana. *J. Vertebr. Paleontol.* **4**(4), 516–524.

Varricchio, D. J., and Horner, J. R. (1993). Hadrosaurid and lambeosaurid bone beds from the Upper Cretaceous Two Medicine Formation of Montana: Taphonomic and biologic implications. *Can. J. Earth Sci.* **30,** 997–1006.

Yale Peabody Museum

Mary Ann Turner
Yale University
New Haven, Connecticut, USA

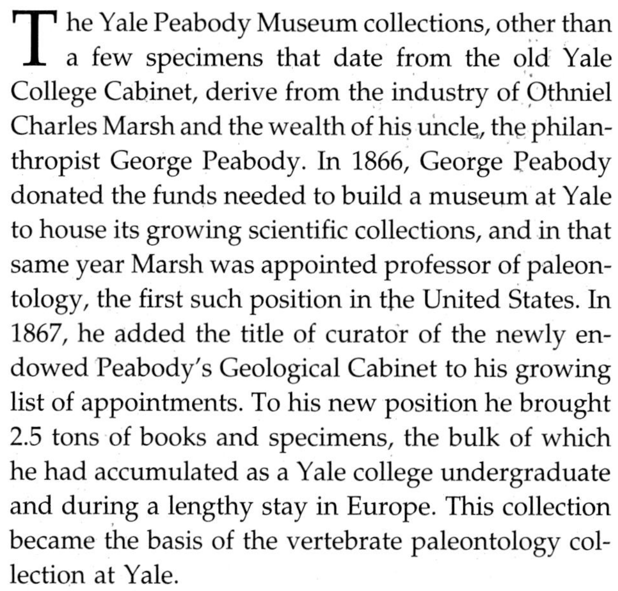

The Yale Peabody Museum collections, other than a few specimens that date from the old Yale College Cabinet, derive from the industry of Othniel Charles Marsh and the wealth of his uncle, the philanthropist George Peabody. In 1866, George Peabody donated the funds needed to build a museum at Yale to house its growing scientific collections, and in that same year Marsh was appointed professor of paleontology, the first such position in the United States. In 1867, he added the title of curator of the newly endowed Peabody's Geological Cabinet to his growing list of appointments. To his new position he brought 2.5 tons of books and specimens, the bulk of which he had accumulated as a Yale college undergraduate and during a lengthy stay in Europe. This collection became the basis of the vertebrate paleontology collection at Yale.

During a lifetime at Yale, Marsh amassed an impressive collection of vertebrate fossils. Although ultimately he, and the Peabody Museum, would become famous for his collection of dinosaurs from the American West, Marsh's earliest collections were of crocodiles, mosasaurs, turtles, and scraps of dinosaurs from the Cretaceous of New Jersey. The Yale College Scientific Expeditions of 1870–1873, under his leadership, explored and collected throughout the American West. Although the collections made by these expeditions are historically and scientifically important, most of the material collected was mammalian. However, Marsh and his students did make large collections of Cretaceous pterosaurs, mosasaurs, turtles, fishes, and birds from the Niobrara Formation of western Kansas. Included in this collection was the sole dinosaur skeleton that he described as the holotype of the ornithopod *Claosaurus agilis*.

After 1874, Marsh began to rely almost exclusively on "professional" collectors. Some of them began as local residents who found bones on their own and contacted Marsh; others were trained people sent to the West by Marsh. In April, 1877, Marsh learned of the discovery of dinosaur bones near Morrison, Colorado, from a local schoolteacher, Arthur Lakes. Marsh sent Benjamin F. Mudge to Morrison to help Lakes, but within a few months sent him to Garden Park, Colorado, where bones had also been discovered. At Garden Park he was joined by Samuel Wendell Williston. Within a short period of time, however, Marsh abandoned the quarry and sent Williston back to Morrison to work with Lakes. When bones were first reported from Como Bluff, Wyoming, later that year, Williston and Lakes were sent there. Marsh had collectors working in the Jurassic Como Bluff quarries collecting dinosaurs until 1889. One of these quarries, Quarry 9, yielded Marsh's collection of Jurassic mammals as well as the types of the species for which the dinosaurian genera *Camptosaurus*, *Brontosaurus*, and *Coelurus* were created. A nearby Cretaceous quarry yielded the skeleton of the type of *Nodosaurus textilis*. People such as William Reed and William Carlin, Arthur Lakes, Samuel Wendell Williston and his brother Frank, Arthur Lakes, Fred Brown, and others all collected for Marsh at Como over the years (see History: Early Discoveries; North American Dinosaurs).

Some years later John Bell Hatcher, a Yale graduate who had collected mammals for Marsh in the 1880s, was sent to Wyoming to acquire a skull that apparently was the source of horn cores previously sent to Marsh. Hatcher acquired the skull and, over the next 4 years, an additional 32 Cretaceous ceratopsian skulls. According to his biographers, Schuchert and LeVene (1940), Marsh named 344 new species and 161 new genera of fossil vertebrates. His genera include *Brontosaurus, Apatosaurus, Stegosaurus,* and *Triceratops.* In 1883 he was appointed United States Vertebrate Paleontologist, a position he held until his death in 1899. His work and collections were praised by both Charles Darwin and Thomas Huxley as affording some of the most crucial evidence supporting the theory of evolution.

Marsh did not mount original material because he believed that it should be available for research. After his death in 1899, Charles Beecher, his successor as head of the museum, arranged to have mounted a skeleton of the hadrosaur *Edmontosaurus annectens.* The skeleton, one of two collected by Hatcher in 1891, was mounted in 1901 and is one of the first two dinosaur skeletons mounted in North America. The mount was removed in 1914, before the old museum was torn down, and was later placed in the current museum. Also in 1901 (Yale's bicentennial), Beecher had the PELVIS and HINDLIMBS of the holotype of *Brontosaurus excelsus* mounted. The complete skeleton was later mounted in the current museum. Marsh's successor was Richard Swann Lull, who served the division in one curatorial capacity or another from 1906 until his death in 1957. He also served as director of the Peabody Museum from 1922 to 1936. It was under his tenure that the *Stegosaurus* skeleton was mounted, the museum moved into its current building, the Great Hall of Dinosaurs was opened, and the Marsh collection was made available for research. He and his students, M. R. Thorpe, E. L. Troxell, and G. G. Simpson, to name a few, inherited the task of identifying and publishing many of the neglected specimens collected earlier by Marsh and his colleagues. Lull's interest in the fossil footprints of the East Coast "Triassic" redbeds, begun years before while teaching at the Massachusetts Agricultural College (now the University of Massachusetts) in Amherst, culminated in the publication of the *Triassic Life of the Connecticut Valley* in 1915 and a massive revision of the same book in 1953. Lull also published studies of ceratopsian and hadrosaurian dinosaurs.

The vertebrate paleontologists who followed Lull at Yale were primarily interested in mammals. The arrival of J. T. Gregory in 1946 heralded the return to the study of lower vertebrates at Yale. Gregory held the position of curator for nearly 15 years. Although much of his tenure was spent heading up renovations to the exhibits and buildings, he also led a number of expeditions, making substantial collections of Triassic vertebrates from New Mexico, Permian fissure-fill material from Ft. Sill, Oklahoma, and dinosaur material from Hanksville, Wyoming.

John H. Ostrom took over the responsibility for the lower vertebrate collections in 1961. Ostrom's research and writings on the functional morphology of dinosaurs and the evolution of birds served as the main impetus behind the "dinosaur renaissance" of the past 20 years (see PHYSIOLOGY). His work on the Early Cretaceous fauna of Wyoming and Montana resulted in the discovery of a number of new dinosaurs, including *Tenontosaurus tilletti, Sauropelta edwardsi,* and, of course, *Deinonychus antirrhopus.* During the late 1970s, exploration of the NEWARK SUPERGROUP of eastern North America by graduate students Paul Olsen, Amy McCune, and others, under the supervision of Keith S. Thomson, collected thousands of freshwater fishes, reptiles, and invertebrates and many footprints from this important period at the beginning of the Age of Dinosaurs, but no Triassic–Jurassic dinosaur remains were discovered.

In 1985, the Peabody Museum acquired the fossil vertebrate collection of Princeton University. Princeton's early collecting efforts had focused on mammals from North and South America. However, in 1957, Donald Baird arrived at Princeton "to help with the program of renovation of the fossil vertebrate museum," and like so many of his predecessors, he chose to remain there. Through his field collecting efforts, Princeton became a major repository of fossil lower vertebrates from the Linton Coal Mines in Ohio and the NEWARK SUPERGROUP. John R. Horner arrived in the mid-1970s and held the position of assistant curator until 1982, when he left to join the staff of the MUSEUM OF THE ROCKIES. In the fall of 1978, Horner returned from a field season in Montana with the skull of a hadrosaur and bags of bones representing the partial skeletons of 15 "baby" dinosaurs from a nest. These remains belonged to an undescribed genus of dinosaur that was later named *Maiasaura*, the "good mother lizard." His continuing work on baby dinosaurs and dinosaur behavior has been instrumental in fleshing out dinosaurs so that they have become much more than simply an odd assortment of old bones.

See also the following related entries:

BEHAVIOR • EGG MOUNTAIN • GROWTH AND EMBRYOLOGY

References

Baird, D. (1989). Medial Cretaceous carnivorous dinosaur and footprints from New Jersey. *Mosasaur* **4,** 53–63.

Berman, D. S., and McIntosh, J. S. (1978). Skull and relationships of the upper Jurassic sauropod *Apatosaurus* (Reptilia, Saurischia). *Bull. Carnegie Museum Nat. History* **8,** 1–35.

Galton, P. M. (1976). Prosauropod dinosaurs (Reptilia: Saurischia) of North America. *Peabody Museum Nat. History Postilla* **169,** 1–98.

Hatcher, J. B., Marsh, O. C., and Lull, R. S. (1907). The Ceratopsia. United States Geological Survey Monograph No. XLIX, pp. 300.

Horner, J. R. (1983). Cranial osteology and morphology of the type specimen of *Maiasaura peeblesorum* (Ornithischia: Hadrosauridae), with discussion of its phylogenetic position. *J. Vertebr. Paleontol.* **31**(1), 29–38.

Hunt, A. P., and Lucas, S. G. (1994). Ornithischian dinosaurs from the Upper Triassic of the United States. In *In the Shadow of the Dinosaurs: Early Mesozoic Tetrapods* (N. C. Fraser and H.-D. Sues, Eds.). Cambridge Univ. Press, Cambridge, UK.

Lull, R. S. (1915). Triassic life of the Connecticut Valley. *Connecticut Geol. Nat. History Survey Bull.* **24**.

Lull, R. S. (1919). The sauropod dinosaur *Barosaurus* Marsh. *Mem. Connecticut Acad. Arts Sci.* **VI,** 1–42.

Lull, R. S. (1921). The Cretaceous armored dinosaur *Nodosaurus textilis* Marsh. *Am. J. Sci.* **5**(i), 97–126.

Lull, R. S., and Wright, N. E. (1942). Hadrosaurian dinosaurs of North America. *Geol. Soc. Am. Spec. Paper* **40**.

Marsh, O. C. (1880). Odontornithes: A monograph on the extinct toothed birds of North America. *Rep. Geol. Exploration Fortieth Parallel* **VII,** 1–201.

Marsh, O. C. (1896). The dinosaurs of North America. Sixteenth Annual Report of the United States Geological Survey 1894–1895, Part 1, pp. 133–244.

Ostrom, J. H. (1966). Functional morphology and evolution of the ceratopsian dinosaurs. *Evolution* **20,** 290–308.

Ostrom, J. H. (1969). Osteology of *Deinonychus antirrhopus,* an unusual theropod from the lower Cretaceous of Montana. *Yale Univ. Peabody Museum Nat. History Bull.* **30,** 1–165.

Ostrom, J. H. (1970). Stratigraphy and paleontology of the Cloverly Formation (lower Cretaceous) of the Bighorn Basin area, Wyoming and Montana. *Yale Univ. Peabody Museum Nat. History Bull.* **35,** 1–234.

Ostrom, J. H., and McIntosh, J. S. (1966). *Marsh's Dinosaurs. The Collection from Como Bluff,* pp. 388. Yale Univ. Press, New Haven, CT.

Ostrom, J. H., and Wellnhofer, P. (1986). The Munich specimen of *Triceratops* with a revision of the genus. *Zitteliana* **14,** 111–158.

Schuchert, C., and LeVene, C. M. (1940). *O. C. Marsh, Pioneer in Paleontology,* pp. 541. Yale Univ. Press, New Haven, CT.

Troxell, E. L. (1925a). The Bridger crocodiles. *Am. J. Sci.* **5**(ix), 29–72.

Troxell, E. L. (1925b). *Thoracosaurus,* a Cretaceous crocodile. *Am. J. Sci.* **5**(x), 229–231.

Zhucheng Dinosaur Museum, People's Republic of China

see MUSEUMS AND DISPLAYS

Zigong Dinosaur Museum

DONG ZHIMING
Institute of Vertebrate Paleontology and Paleoanthropology, Academia Sinica
Beijing, People's Republic of China

Zigong, a historically and culturally important city of China, is situated in southwest Sichuan Province, approximately 240 km southwest of Chengdu City, the capital of Sichuan Province. It is widely known as the "Salt City" of China. From the beginning of the Han Dynasty (202 B.C.–220 A.D.), Zigong people have extracted salty water from a well to produce pure salt. Dashanpu, a small town, is situated 11 km northeast of the center of Zigong City. The dinosaur quarry, near the eastern part of Dashanpu 1.2 km from the central district, is called Wuanian Deng hill, beside the Neijiang–Zigong highway. The area is characterized by low rolling hills, with an elevation of 50–70 m. The dinosaurian quarry is 320 m above sea level and the floor of Wuanian Deng hill. The exposed sediments are the Middle Jurassic deposits of the Lower Shaximiao Formation (Fig. 1).

Discovery and Collection of Dashanpu Dinosaur Quarry

The first fossils were discovered by local geologists from the second Geological Exploration Party of the Sichuan Provincial Bureau of Geology and Mineral Resources while working in the Dashanpu area in 1972. They found a caudal vertebra of a dinosaur on the cliff beside the road that crosses the Neijiang–Leshe highway at Dashanpu. The first excavation was jointly organized by the INSTITUTE OF VERTEBRATE PALEONTOLOGY AND PALEOANTHROPOLOGY (IVPP) and the Chongqing Museum in 1977. A large, incomplete skeleton of a sauropod was collected at Dashanpu Quarry. This specimen was named *Shunosaurus lii* by the author and colleagues (Dong *et al.*, 1983).

In 1979, the Dashanpu site was opened by the Southwest Oil / Gas Company as a building park. The author and Zhou Shiwu of the Chongqing Museum of Natural History led an expedition to the Sichuan Basin seeking dinosaur localities. On December 17, they arrived at the Dashanpu site and found a great number of dinosaurian remains exposed *in situ.* The Dashanpu dinosaur quarry was excavated by a jointly organized team of IVPP, Chongqing Museum, and Zigong Salt History Museum. The author led this excavation at Dashanpu Quarry from 1979 to 1981. During that time, more than 40 tons of dinosaur fossils were excavated. Approximately 8000 bones have been collected, many of them large and completely preserved. Some occurred as isolated bones and some were articulated in skeletons. These included several type specimens of SAUROPODS (*Shunosaurus* and *Datousaurus*), THEROPODS (*Gasosaurus*), and STEGOSAURS (*Huayangosaurus*) that are housed at IVPP and the Chongqing Museum.

From July 1981 to May 1982, the quarry was excavated by a Sichuan expedition team sent by the Scientific and Cultural Bureau of Sichuan Province. This excavation was led by He Xinlu of the Chengdu College of Geology and Fang Qiren of the Chongqing Museum. They conducted a large-scale scientific excavation for 1 year. The fossils are housed at the Zigong Dinosaur Museum. Fossils from this expedition, reported by He and colleagues, included small

FIGURE 1 The Zigong Dinosaur Museum has been built over the famous dinosaurian quarry of Dashanpu.

ornithopods and the large sauropod *Omeisaurus* (He and Zhou, 1980, 1984).

During 1984, while the Dinosaur Museum of Zigong was under construction, the quarry was cleared by the staff of the Zigong Dinosaur Museum. More than 500 m^2 were excavated. The Zigong Dinosaur Museum collection included a complete skull of a labyrinthodont, *Sinobrachyops placenticephalus,* and a nearly complete skull of a tritylodontid, *Bienotheroides zigongensis.* The former is the youngest element of a labyrinthodont found in Asia.

The material collected from the Dashanpu Quarry includes complete skeletons of sauropods, carnosaurs, stegosaurs, ornithopods, pterosaurs, amphibians, and fishes. They represent more than 100 individual animals and include at least 12 reptilian forms, including 6 different kinds of dinosaurs. There are plant-eating, long-necked sauropods more than 20 m long, a fierce meat-eating theropod only 1.4 m long, an ornithopod dinosaur, and the world's most perfectly preserved primitive stegosaur discovered to date. Also found within the assemblage was a pterosaur, the first to be discovered from the Middle Jurassic of China, and a plesiosaur, a long-necked aquatic REPTILE that lived in rivers and lakes.

Dashanpu Quarry has proven to be one of the richest and most important dinosaur localities in China and the world. The excavated site covers 2800 sq m, and work is still going on. It is estimated that the distribution of fossils could cover an area of more than 20,000 sq m. The site is remarkable, both for its great number of dinosaur fossils from the Middle Jurassic and for its great diversity of species. Many of these finds have helped to fill in gaps in the evolution of dinosaurs (Fig. 2).

Strata of Dashanpu Dinosaur Site

The Mesozoic deposits of Zigong area are continental red bed sequences, composed of 1700–2000 m of sandstone and mudstones. They lie unconformably on the coal-bearing formation of the Late Triassic and appear to represent fluvial, lacustrine, and floodplain environments. Many geologists have worked this area (Xie *et al.*, 1983). This outcrop of Mesozoic rocks contains fossil vertebrates in four sequences: the Zhenchuchong Formation, the Ziliujing Formation of the Lower Jurassic, the Lower Shaximiao Formation of the Middle Jurassic, and the Upper Shaximiao Formation of the Upper Jurassic.

The Dashanpu dinosaur fossils come from a large

FIGURE 2 Collecting dinosaurian fossils from the Dashanpu quarry (1980) before the Museum was built. The quarry is still open, producing fossils, and is accessible to both resident and visiting scientists.

FIGURE 3 Display of *Shunosaurus* in the Zigong Dinosaur Museum.

TABLE I
Vertebrate Fauna from Dashanpu Site

Chondrichthyes
Hybodontidae
Chondrichthyes: Hybodontidae indet.
Chondrichthyes: Semionotidae: *Lepidotes* sp.
Dipnoi: *Ceratodus zigongensis*
Osteichthyes
Semionotiformes
Semionotidae
Lepidotes sp.
Dipnoi
Ceratodontidae
Ceratodus zigongensis
Amphibia
Temnospondylia
Brachyopidae
Sinobrachyops placenticephalus
Reptilia
Chelonia
Chengyuchelyidae
Chengyuchelys baenoides
C. zigongensis
C. cf. zigongensis
C. dashanpuensis
Crocodylia
Hsisosuchidae
Hsisosuchidae indet.
Sauropterygia
Pliosauridae
Pliosauridae indet.
Pterosauria
Rhamphorhynchidae
Angustinaripterus longicephalus
Saurischia
Theropoda
Megalosauridae
Gasosaurus constructus
Sauropodomorpha
Camarasauridae
Shunosaurinae
Protognathosaurus oxyodon
Shunosaurus lii
Cetiosaurinae
Datousaurus bashanensis
Mamenchisauridae
Omeisaurus tianfuensis
Ornithischia
Ornithopoda
Fabrosauridae
Agilisaurus louderbacki
Xiaosaurus dashanpensis
Stegosauria
Huayangosauridae
Huayangosaurus taibaii
Therapsida
Tritylodontidae
Polistodon chuannanensis
Bienotheroides zigongensis

geological lens of gray–green sandstones and mudstones located in the middle part (about 60 m over the basal part) of the Lower Shaximiao Formation.

Zigong Dinosaur Museum

The Zigong Dinosaur Museum (Fig. 1) was built on the site and opened in the spring of 1987. It is the first specialized dinosaur museum to be established in Asia. Zigong Dinosaur Museum covers a total area of approximately 24,600 m^2. The main building, approximately 6000 m^2, comprises a basement and first and second display floors. The burial site is preserved in the basement of the museum, which is not accessible to the public. Visitors are only permitted to watch the burial scene from the corridor. The exhibition halls, approximately 3600 m^2, display nearly 10 complete skeletons of various kinds of dinosaurs found on the site and a series of associated animal fossils (Fig. 3). An outstanding feature of this magnificent museum is that it has been erected over the stratum containing the dinosaur skeletons. In the opinion of those who have worked at the quarry, the rich fossiliferous bed is far from mined out and probably extends to the east, west, and north of the current quarry. We estimate that the distribution is more than 17,000 m^2. That is, only a small part of the dinosaur cemetery, approximately 2800 m^2, has been excavated.

Systematics

The list of vertebrate fauna from Dashanpu site is shown in Table I.

See also the following related entries:

CHINESE DINOSAURS • MUSEUMS AND DISPLAYS

References

Dong, Z. (1979). On the dinosaurian faunas and their stratigraphic distribution in Sichuan Basin, Second Congress of Stratigraphy, China, Beijing.

Dong, Z. (1980). A new dinosaurian site—Dashanpu. *Fossil* **1980.**

Dong, Z. (1985). A Middle Jurassic Labyrinthodont (*Sinobrachyops placenticephalus* gen. et sp. nov.) from Dashanpu, Zigong, Sichuan. *Vertebr. PalAsiatica* **23**(4), 301–306.

Dong, Z. (1993). Origin and phylogeny of Stegosaurs. *Gakken Mook* **4,** 104–109. [In Japanese]

Dong, Z., and Tang, Z. (1983). Note on a new Mid-Jurassic ornithopod from Sichuan Basin, China. *Vertebr. PalAsiatica* **XXI**(2), 168–171.

Dong, Z., and Tang, Z. (1984). Note on a new Mid-Jurassic sauropod (*Datousaurus bashanensis* gen. et sp. nov.) from Sichuan Basin, China. *Vertebr. PalAsiatica* **22**(1), 69–75.

Dong, Z., and Tang, Z. (1985). Note on a new Mid-Jurassic carnosaur (*Gasosaurus constructus* gen. et sp. nov.) from Sichuan Basin, China. *Vertebr. PalAsiatica* **23**(1), 77–83.

Dong, Z., Tang, Z., and Zhou, S. (1982). Note on the new Mid-Jurassic stegosaur from Sichuan Basin, China. *Vertebr. Palasiatica* **XX**(1), 63–157.

Dong, Z., Zhou, S., and Zhang, Y. (1983). The dinosaurian remains from Sichuan Basin, China. *Paleontol. Sinica* **162**(New Ser. C, No. 23), 19–25.

Dong, Z., Peng, G., and Huang, D. (1989). The discovery of the bony tail club of sauropods. *Vertebr. PalAsiatica* **27**(3), 219–224.

Fang, Q. (1987). A new species of Middle Jurassic turtles from Sichuan. *Acta Herpetol. Sinica* **6**(1), 65–69.

He, X. L., and Cai, K. (1983). A new species of *Yandusaurus* (hypsilophodont dinosaur) from Middle Jurassic of Dashanpu, Zigong, Sichuan. *J. Chengdo College Geol. Suppl.* **1,** 5–14.

He, X., and Cai, K. (1984a). The tritylodont remains from Dashanpu, Zigong, Sichuan. *J. Chengdo College Geol. Suppl.* **2,** 32–45.

He, X., and Cei, K. (1984b). The ornithopod dinosaurs. In *The Middle Jurassic Dinosaurian Fauna from Dashanpu, Zigong, Sichuan,* Vol. 1, pp. 1–71. Sichuan Scientific and Technological Publishing House, Chengdo.

He, X. L., Yon, D., and Su, C. (1983). A new pterosaur from the Middle Jurassic of Dashonpu, Zigong, Sichuan. *J. Chengdo College Geol. Suppl.* **1,** 33–39.

He, X., *et al.* (1984). *Omeisaurus tianfuensis*—A new species of *Omeisaurus* from Dashanpu, Zigong, Sichuan. *J. Chengdo College Geol. Suppl.* **2,** 13–31.

He, X., Li, K., and Cai, K. (1988). The sauropod dinosaurs (2). In *The Middle Jurassic Dinosaurian Fauna from Dashanpu, Zigong, Sichuan,* Vol. 4, pp. 1–145. Sichuan Scientific and Technological Publishing House, Chengdo.

Sun, A. (1986). New material of *Bienotheroides* (Tritylodont reptile) from Shaximiao Formation of Sichuan. *Vertebr. PalAsiatica* **24**(3), 165–170.

Xia, W., and Li, X. (1988). The burial environment of dinosaurs and characteristics of lithofacies and paleogeography. In *The Middle Jurassic Dinosaur Fauna from Dashanpu, Zigong, Sichuan,* Vol. 5, pp. 1–112. Sichuan Scientific and Technological Publishing House, Chengdo.

Xia, W., *et al.* (1983). Primary study of composition and microstructure of the dinosaurian skeleton in Lower Shaximiao Formation (Middle Jurassic) at Dashanpu, Zigong, Sichuan Province. *J. Chengdu College Geol. Suppl.* **1,** 34–40.

Xia, W., Li, X., and Yi, Z. (1984) The burial environment of dinosaur fauna in Lower Shaximiao Formation of Middle Jurassic at Dashanpu, Zigong, Sichuan. *J. Chengdo College Geol. Suppl.* **2,** 46–59.

Yan, D., Cai, K., and Pi, X. (1983). Primary report on the dinosaurian expeditions of Dashanpu, Zigong, Sichuan. *J. Chengdo College Geol. Suppl.* **1,** 4–14.

Yeh, H.-k. (1982). Middle Jurassic turtles from Sichuan, SW China. *Vertebrata PalAsiatica* **20**(4), 282–289.

Yeh, H.-k. (1990). Fossil turtles from Dashanpu, Zigong, Sichuan. *Vertebr. PalAsiatica* **28**(4), 305–311.

Zhang, Y. (1988). The sauropod dinosaur (l). In *The Middle Jurassic Dinosaurian Fauna from Dashanpu, Zigong, Sichuan,* Vol. 3, 1–84. Sichuan Scientific and Technological Publishing House, Chengdo.

Zhang, Y., Yen, D., and Peng, G. (1984). New material of *Shunosaurus* from Middle Jurassic of Dashanpu, Zigong, Sichuan. *J. Chengdo College Geol. Suppl.* **2,** 1–12.

Zhou, S. W. (1983). A nearly complete skeleton of stegosaur from Middle Jurassic of Dashanpu, Zigong, Sichuan. *J. Chengdo College Geol. Suppl.* **1,** 14–26.

Zhou, W. (1984). Stegosaurs. In *The Middle Jurassic Dinosaurian Fauna from Dashanpu, Zigong, Sichuan,* Vol. 2, pp. 1–52. Sichuan Scientific and Technological Publishing House, Chengdo.

RESOURCES

DINOSAUR GENERA

This alphabetized classification list has been compiled by George Olshevsky. It comprises all dinosaur generic names known to have been published anywhere, with or without formal scientific descriptions. The list includes a total of 806 dinosaur genera in all. It excludes most birds after *Archaeopteryx*.

The names printed in ***bold italic type*** are generally recognized as scientifically valid; other names are vernacular, preoccupied, or otherwise scientifically invalid. Many of the listed genera are synonyms of other genera. The names appear here regardless of their validity or synonymy, but each name is associated in some way, formally or informally, with real material. The names of dinosaur footprints and egg genera, or of fictitious dinosaurs, are not included. The author(s) and year of publication are provided for each name, to facilitate locating original references. Most are found in the *Bibliography of Fossil Vertebrates*.

NOTES:

[*nomen nudum*] = name lacking a description and/or a type specimen
vide = name attributed to the first indicated author(s) by the second
/ = name preoccupied by the second indicated author(s)
* = genus not presently considered to be dinosaurian

Aachenosaurus Smets, 1888*
Abelisaurus Bonaparte & Novas, 1985
Abrictosaurus Hopson, 1975
Abrosaurus Ou, 1986 *vide* Zhang & Li, 1996
Acanthopholis Huxley, 1867
Achelousaurus Sampson, 1995
Acracanthus Langston, 1947 *vide* Czaplewski, Cifelli & Langston, 1994
Acrocanthosaurus Stovall & Langston, 1950
Actiosaurus Sauvage, 1882*
Adasaurus Barsbold, 1983
Aegyptosaurus Stromer, 1932
Aeolosaurus J. Powell, 1988
Aepisaurus Gervais, 1852
Aetonyx Broom, 1911
Afrovenator Sereno, Wilson, Larsson, Dutheil & Sues, 1994
Agathaumas Cope, 1872
Aggiosaurus Ambayrac, 1913*
Agilisaurus Peng, 1990
Agrosaurus Seeley, 1891
Alamosaurus Gilmore, 1922
Albertosaurus Osborn, 1905
Albisaurus Fritsch, 1905*
Alectrosaurus Gilmore, 1933
Algoasaurus Broom, 1904
Alioramus Kurzanov, 1976
Aliwalia Galton, 1985
Allosaurus Marsh, 1877
Alocodon Thulborn, 1973
Altispinax von Huene, 1923
Alvarezsaurus Bonaparte, 1991
Alwalkeria Chatterjee & Creisler, 1994
Alxasaurus Russell & Dong, 1994 (not 1993)
Amargasaurus Salgado & Bonaparte, 1991
Ammosaurus Marsh, 1891
Ampelosaurus Le Loeuff, 1995
Amphicoelias Cope, 1877
Amphisaurus Marsh, 1882/Barkas, 1870
Amtosaurus Kurzanov & Tumanova, 1978
Amurosaurus Bolotsky & Kurzanov *vide* Nessov, 1995 [*nomen nudum*]
Amygdalodon Cabrera, 1947
Anasazisaurus Hunt & Lucas, 1993
Anatosaurus Lull & Wright, 1942
Anatotitan Brett-Surman *vide* Chapman & Brett-Surman, 1990
Anchiceratops Brown, 1914
Anchisaurus Marsh, 1885
Andesaurus Calvo & Bonaparte, 1991
Angaturama Kellner & Campos, 1996
Ankistrodon Huxley, 1865*
Ankylosaurus Brown, 1908

Anodontosaurus C. M. Sternberg, 1929
Anoplosaurus Seeley, 1879
Anserimimus Barsbold, 1988
Antarctosaurus von Huene, 1927
Anthodon Owen, 1876*
Antrodemus Leidy, 1870
Apatodon Marsh, 1877
Apatosaurus Marsh, 1877
Aragosaurus Sanz, Buscalioni, Casanovas & Santafe, 1987
Aralosaurus Rozhdestvensky, 1968
Archaeoceratops Dong & Azuma, 1996 [*nomen nudum*]
Archaeopteryx von Meyer, 1861
Archaeornis Petronievics *vide* Petronievics & Woodward, 1917
Archaeornithoides Elzanowski & Wellnhofer, 1992
Archaeornithomimus D. A. Russell, 1972
Arctosaurus Adams, 1875*
Argentinosaurus Bonaparte & Coria, 1993
Argyrosaurus Lydekker, 1893
Aristosaurus van Hoepen, 1920
Aristosuchus Seeley, 1887
"*Arkanosaurus*" Sattler, 1983 [*nomen nudum*]
Arrhinoceratops Parks, 1925
Arstanosaurus Suslov *vide* Suslov & Shilin, 1982
Asiaceratops Nessov & Kaznyshkina *vide* Nessov, Kaznyshkina & Cherepanov, 1989
Asiamericana Nessov, 1995
Asiatosaurus Osborn, 1924
Astrodon Johnston, 1859
Astrodonius Kuhn, 1961
Atlantosaurus Marsh, 1877
Atlascopcosaurus T. Rich & P. Rich, 1989
Aublysodon Leidy, 1868
Austrosaurus Longman, 1933
Avaceratops Dodson, 1986
Avalonia Seeley, 1898 / Walcott, 1889*
Avalonianus Kuhn, 1961*
Avimimus Kurzanov, 1981
Avipes von Huene, 1932
Avisaurus Brett-Surman & Paul, 1985*
Azendohsaurus Dutuit, 1972
Bactrosaurus Gilmore, 1933
Bagaceratops Maryanska & Osmólska, 1975
Bagaraatan Osmólska, 1996
Bahariasaurus Stromer, 1934
Barapasaurus Jain, Kutty, Roy-Chowdhury & Chatterjee, 1975
Barosaurus Marsh, 1890
Barsboldia Maryanska & Osmólska, 1981
Baryonyx Charig & Milner, 1986
Basutodon von Huene, 1932*
Bathygnathus Leidy, 1854*
Becklespinax Olshevsky, 1991
Bellusaurus Dong, 1990
Belodon von Meyer, 1842*
Betasuchus von Huene, 1932
Bihariosaurus Marinescu, 1989
Blikanasaurus Galton & van Heerden, 1985
Borogovia Osmólska, 1987
Bothriospondylus Owen, 1875
Brachiosaurus Riggs, 1903
Brachyceratops Gilmore, 1914
Brachylophosaurus C. M. Sternberg, 1953
Brachypodosaurus Chakravarti, 1934
Brachyrophus Cope, 1878
Brachytaenius von Meyer, 1842*
Bradycneme Harrison & C. A. Walker, 1975
Brasileosaurus von Huene, 1931*
Breviceratops Kurzanov, 1990
Brontoraptor Redman, 1995 [*nomen nudum*]
Brontosaurus Marsh, 1879
Bruhathkayosaurus Yadagiri & Ayyasami, 1987
Bugenasaura Galton, 1995
Caenagnathasia Currie, Godfrey & Nessov, 1994 (not 1993)
Caenagnathus R. M. Sternberg, 1940
Calamosaurus Lydekker, 1891
Calamospondylus Fox, 1866
Calamospondylus Lydekker, 1889 / Fox, 1866
Callovosaurus Galton, 1980
Camarasaurus Cope, 1877
Camelotia Galton, 1985
Camptonotus Marsh, 1879 / Uhler, 1864
Camptosaurus Marsh, 1885
Campylodon von Huene, 1929 / Cuvier & Valenciennes, 1832
Campylodoniscus Kuhn, 1961
Carcharodontosaurus Stromer, 1931
Cardiodon Owen, 1841
Carnosaurus von Huene, 1929 [*nomen nudum*]
Carnotaurus Bonaparte, 1985
Cathetosaurus Jensen, 1988
Caudocoelus von Huene, 1932
Caulodon Cope, 1877
Centemodon Lea, 1856*
Centrosaurus Lambe, 1904 / Fitzinger, 1843
Ceratops Marsh, 1888 / Rafinesque, 1815
Ceratosaurus Marsh, 1884
Cetiosauriscus von Huene, 1927
Cetiosaurus Owen, 1841
Changtusaurus Zhao, 1983 [*nomen nudum*]
Chaoyoungosaurus Zhao, 1983 [*nomen nudum*]
Chasmosaurus Lambe, 1914
Chassternbergia Bakker, 1988
Cheneosaurus Lambe, 1917
Chialingosaurus Young, 1959
Chiayusaurus Bohlin, 1953
Chienkosaurus Young, 1942*
Chihuahuasaurus Ratkevich *vide* [Anonymous] 1997 [*nomen nudum*]
Chilantaisaurus Hu, 1964
Chindesaurus Long & Murry, 1995
Chingkankousaurus Young, 1958

Chinshakiangosaurus Yeh, 1975 [*nomen nudum*]
Chirostenotes Gilmore, 1924
Chondrosteosaurus Owen, 1876
Chuandongocoelurus He, 1984
Chubutisaurus del Corro, 1974
Chungkingosaurus Dong, Zhou & Zhang, 1983
Cionodon Cope, 1874
Cladeiodon Owen, 1841*
Claorhynchus Cope, 1892
Claosaurus Marsh, 1890
Clarencea Brink, 1959*
Clasmodosaurus Ameghino, 1898
Clepsysaurus Lea, 1851*
Clevelanotyrannus Bakker, Williams & Currie *vide* Currie, 1987 [*nomen nudum*]
Coelophysis Cope, 1889
Coelosaurus Leidy, 1865/[Anonymous, but known to be Owen] 1854
Coeluroides von Huene & Matley, 1933
Coelurosauravus Piveteau, 1926*
Coelurosaurus von Huene, 1929 [*nomen nudum*]
Coelurus Marsh, 1879
Colonosaurus Marsh, 1872*
Coloradia Bonaparte, 1978/Blake, 1863
Coloradisaurus Lambert, 1983
Compsognathus Wagner, 1861
Compsosuchus von Huene & Matley, 1933
Conchoraptor Barsbold, 1986
Corythosaurus Brown, 1914
Craspedodon Dollo, 1883
Crataeomus Seeley, 1881
Craterosaurus Seeley, 1874
Creosaurus Marsh, 1878
Cryolophosaurus Hammer & Hickerson, 1994
Cryptodraco Lydekker, 1889
Cryptosaurus Seeley, 1869
Cumnoria Seeley, 1888
Cylindricodon Owen *vide* Ingles & Sawyer, 1979/Jaeger, 1828
Cystosaurus Geoffroy Saint-Hilaire, 1833*
Dacentrurus Lucas, 1902
Dachungosaurus Chao, 1985 [*nomen nudum*]
Dakosaurus Quenstedt, 1856*
Damalasaurus Zhao, 1983 [*nomen nudum*]
Dandakosaurus Yadagiri, 1982
Danubiosaurus Bunzel, 1871
Daptosaurus Brown *vide* Chure & McIntosh, 1989
Daspletosaurus D. A. Russell, 1970
Dasygnathoides Kuhn, 1961*
Dasygnathus Huxley, 1877/MacLeay, 1819*
Datousaurus Dong & Tang, 1984
Deinocheirus Osmólska & Roniewicz, 1970
Deinodon Leidy, 1856
Deinonychus Ostrom, 1969
Deltadromeus Sereno, Duthiel, Iarochene, Larsson, Lyon, Magwene, Sidor, Varricchio & Wilson, 1996
Denversaurus Bakker, 1988
Dianchungosaurus Young, 1982
Diceratops Hatcher *vide* Lull, 1905
Diclonius Cope, 1876
Dicraeosaurus Janensch, 1914
Didanodon Osborn, 1902
Dilophosaurus Welles, 1970
Dimodosaurus Pidancet & Chopard, 1862
Dinodocus Owen, 1884
Dinosaurus Rütimeyer, 1856/Fischer de Waldheim, 1847
Dinotyrannus Olshevsky *vide* Olshevsky, Ford & Yamamoto, 1995
Diplodocus Marsh, 1878
Diplotomodon Leidy, 1868
Diracodon Marsh, 1881
Dolichosuchus von Huene, 1932
Doratodon Seeley, 1881*
Doryphorosaurus Nopcsa, 1916
Dracopelta Galton, 1980
Dravidosaurus Yadagiri & Ayyasami, 1979*
Drinker Bakker, Galton, Siegwarth & Filla, 1990
Dromaeosaurus Matthew & Brown, 1922
Dromiceiomimus D. A. Russell, 1972
Dromicosaurus van Hoepen, 1920
Dryosaurus Marsh, 1894
Dryptosauroides von Huene & Matley, 1933
Dryptosaurus Marsh, 1877
Dynamosaurus Osborn, 1905
Dyoplosaurus Parks, 1924
Dysalotosaurus Virchow, 1919
Dysganus Cope, 1876
Dyslocosaurus McIntosh, Coombs & D. A. Russell, 1992
Dystrophaeus Cope, 1877
Dystylosaurus Jensen, 1985
Echinodon Owen, 1861
Edmarka Bakker, Kralis, Siegwarth & Filla, 1992
Edmontonia C. M. Sternberg, 1928
Edmontosaurus Lambe, 1917
Efraasia Galton, 1973
Einiosaurus Sampson, 1995
Elaphrosaurus Janensch, 1920
Elmisaurus Osmólska, 1981
Elopteryx Andrews, 1913
Elosaurus Peterson & Gilmore, 1902
Emausaurus Haubold, 1991
Embasaurus Riabinin, 1931
Enigmosaurus Barsbold & Perle, 1983
Eoceratops Lambe, 1915
Eoraptor Sereno, Forster, Rogers & Monetta, 1993
Epachthosaurus J. Powell, 1990
Epanterias Cope, 1878
Ephoenosaurus [Anonymous] 1839 [*nomen nudum*]*
Epicampodon Lydekker, 1885*
Erectopus von Huene, 1922
Erlikosaurus Perle *vide* Barsbold & Perle, 1980
Euacanthus Owen *vide* Tennyson, 1897 [*nomen nudum*]
Eucamerotus Hulke, 1872
Eucentrosaurus Chure & McIntosh, 1989

Eucercosaurus Seeley, 1879
Eucnemesaurus van Hoepen, 1920
Euhelopus Romer, 1956
Euoplocephalus Lambe, 1910
Eupodosaurus Boulenger, 1891*
Eureodon Brown *vide* Olshevsky, 1991
Euronychodon Telles-Antunes & Sigogneau-Russell, 1991
Euskelosaurus Huxley, 1866
Eustreptospondylus Walker, 1964
Fabrosaurus Ginsburg, 1964
Fenestrosaurus Osborn, 1924 [*nomen nudum*]
Frenguellisaurus Novas, 1986
"Fukuisaurus" Lambert, 1990 [*nomen nudum*]
Fulengia Carroll & Galton, 1977
Fulgurotherium von Huene, 1932
"Futabasaurus" Lambert, 1990 [*nomen nudum*]
Gadolosaurus Saito, 1979 [*nomen nudum*]
Galesaurus Owen, 1859*
Gallimimus Osmólska, Roniewicz & Barsbold, 1972
Galtonia Huber, Lucas & Hunt, 1993
Garudimimus Barsbold, 1981
Gasosaurus Dong, 1985
Gasparinisaura Coria & Salgado, 1996
Gastonia Bakker, 1994 [*nomen nudum*; in the 1995 volume of *World Book Science Year*]
Genusaurus Accarie, Beaudoin, Dejax, Fries, Michard & Taquet, 1995
Genyodectes Woodward, 1901
Geranosaurus Broom, 1911
Giganotosaurus Coria & Salgado, 1995
Gigantosaurus Seeley, 1869
Gigantosaurus E. Fraas, 1908 / Seeley, 1869
Gigantoscelus van Hoepen, 1916
Gigantspinosaurus [Anonymous] 1993 [*nomen nudum*]
Gilmoreosaurus Brett-Surman, 1979
Giraffatitan Paul, 1988
Gobipteryx Elzanowski, 1974*
Gobisaurus Spinar, Currie & Sovak, 1994 [*nomen nudum*]
Gongbusaurus Dong, Zhou & Zhang, 1983
Gorgosaurus Lambe, 1914
Goyocephale Perle, Maryanska & Osmólska, 1982
Gracilisuchus Romer, 1972*
Gravisaurus Norman, 1989 [*nomen nudum*]
Gravitholus Wall & Galton, 1979
Gresslyosaurus Rütimeyer, 1857
Griphornis Owen *vide* Woodward, 1862
Griphosaurus Wagner, 1861
Gryponyx Broom, 1911
Gryposaurus Lambe, 1914
Gwyneddosaurus Bock, 1945*
Gyposaurus Broom, 1911
"Hadrosauravus" Lambert, 1990 [*nomen nudum*]
Hadrosaurus Leidy, 1859
Hallopus Marsh, 1881*
Halticosaurus von Huene, 1908
Haplocanthosaurus Hatcher, 1903
Haplocanthus Hatcher, 1903 / Agassiz, 1845
Harpymimus Barsbold & Perle, 1984
Hecatasaurus Brown, 1910
Heishansaurus Bohlin, 1953
Helopus Wiman, 1929 / Wagler, 1832
Heptasteornis Harrison & C. A. Walker, 1975
Herbstosaurus Casamiquela, 1974*
Herrerasaurus Reig, 1963
Heterodontosaurus Crompton & Charig, 1962
Heterosaurus Cornuel, 1850
Hierosaurus Wieland, 1909
Hikanodon Keferstein, 1834
"Hironosaurus" Hisa, 1988 [*nomen nudum*]
"Hisanohamasaurus" Lambert, 1990 [*nomen nudum*]
Homalocephale Maryanska & Osmólska, 1974
Honghesaurus [Anonymous] 1981 [*nomen nudum*]
Hoplitosaurus Lucas, 1902
Hoplosaurus Seeley, 1881
Hortalotarsus Seeley, 1894
Huayangosaurus Dong, Tang & Zhou, 1982
Hudiesaurus Dong & Azuma, 1996 [*nomen nudum*]
Hulsanpes Osmólska, 1982
Hylaeosaurus Mantell, 1833
Hypacrosaurus Brown, 1913
Hypselorhachis Charig, 1967 [*nomen nudum*]*
Hypselosaurus Matheron, 1869
Hypsibema Cope, 1869
Hypsilophodon Huxley, 1869
Hypsirophus Cope, 1878
Iguanodon Mantell, 1825
Iguanosaurus [Anonymous] 1824 [*nomen nudum*]
Iliosuchus von Huene, 1932
Indosaurus von Huene & Matley, 1933
Indosuchus von Huene & Matley, 1933
Ingenia Barsbold, 1981
Inosaurus de Lapparent, 1960
Irritator Martill, Cruickshank, Frey, Small & Clarke, 1996
Ischisaurus Reig, 1963
Ischyrosaurus Hulke, 1874 *vide* Lydekker, 1888 / Cope, 1869
Itemirus Kurzanov, 1976
Iuticosaurus Le Loeuff, 1993
Jainosaurus Hunt, Lockley, Lucas & Meyer, 1995 (not 1994)
Janenschia Wild, 1991
Jaxartosaurus Riabinin, 1937
Jenghizkhan Olshevsky *vide* Olshevsky, Ford & Yamamoto, 1995
Jensenosaurus Olshevsky vide B. D. Curtice, Stadtman & L. J. Curtice, 1996 [*nomen nudum*]
"Jiangjunmiaosaurus" [Anonymous] 1987 [*nomen nudum*]
Jingshanosaurus Zhang & Yang, 1995
Jubbulpuria von Huene & Matley, 1933
Jurapteryx Howgate, 1985
"Jurassosaurus" Dong *vide* Holden, 1992 [*nomen nudum*]
"Kagasaurus" Hisa, 1988 [*nomen nudum*]
Kaijiangosaurus He, 1984
Kakuru Molnar & Pledge, 1980
Kangnasaurus Haughton, 1915

"Katsuyamasaurus" Lambert, 1990 [*nomen nudum*]
Kelmayisaurus Dong, 1973
Kentrosaurus Hennig, 1915
Kentrurosaurus Hennig, 1916
"Kitadanisaurus" Lambert, 1990 [*nomen nudum*]
Klamelisaurus Zhao, 1993
Koparion Chure, 1994
Koreanosaurus Kim, 1979 [*nomen nudum*]
Kotasaurus Yadagiri, 1988
Kritosaurus Brown, 1910
Kulceratops Nessov, 1995
Kunmingosaurus Chao, 1985
Kuszholia Nessov, 1992*
Labocania Molnar, 1974
Labrosaurus Marsh, 1879
Laelaps Cope, 1866 / Koch, 1839
Laevisuchus von Huene & Matley, 1933
Lagerpeton Romer, 1971*
Lagosuchus Romer, 1971*
Lambeosaurus Parks, 1923
Lametasaurus Matley, 1923
Lanasaurus Gow, 1975
Lancangosaurus Dong, Zhou & Zhang, 1983 [*nomen nudum*]
Lancanjiangosaurus Zhao, 1983 [*nomen nudum*]
Laornis Marsh, 1870
Laosaurus Marsh, 1878
Laplatasaurus von Huene, 1927
Lapparentosaurus Bonaparte, 1986
Leaellynasaura T. Rich & P. Rich, 1989
Leipsanosaurus Nopcsa, 1918
Leptoceratops Brown, 1914
Leptospondylus Owen, 1854
Lesothosaurus Galton, 1978
Lexovisaurus Hoffstetter, 1957
Libycosaurus Bonarelli, 1947*
Ligabueino Bonaparte, 1996
Likhoelesaurus Ellenberger, 1972 [*nomen nudum*]
Liliensternus Welles, 1984
Limnosaurus Nopcsa, 1899 / Marsh, 1872
Lisboasaurus Seiffert, 1973*
Loncosaurus Ameghino, 1898
Longisquama Sharov, 1970*
Longosaurus Welles, 1984
Lophorhothon Langston, 1960
Loricosaurus von Huene, 1929
Lucianosaurus Hunt & Lucas, 1994
Lufengocephalus Young, 1974
Lufengosaurus Young, 1941
Lukousaurus Young, 1948
Lusitanosaurus de Lapparent & Zbyszewski, 1957
Lycorhinus Haughton, 1924
Macelognathus Marsh, 1884*
Macrodontophion Zborzewski, 1834
Macrophalangia C. M. Sternberg, 1932
Macroscelosaurus Muenster *vide* von Meyer, 1847 [*nomen nudum*]*
Macrurosaurus Seeley, 1876
"Madsenius" Lambert, 1990 [*nomen nudum*]
Magnosaurus von Huene, 1932
Magulodon Kranz, 1996 [*nomen nudum*]
Magyarosaurus von Huene, 1932
Maiasaura Horner & Makela, 1979
Majungasaurus Lavocat, 1955
Majungatholus Sues & Taquet, 1979
Malawisaurus Jacobs, Winkler, Downs & Gomani, 1993
Maleevosaurus Carpenter, 1992
Maleevus Tumanova, 1987
Mamenchisaurus Young, 1954
Mandschurosaurus Riabinin, 1930
Manospondylus Cope, 1892
Marasuchus Sereno & Arcucci, 1994*
Marmarospondylus Owen, 1875
Marshosaurus Madsen, 1976
Massospondylus Owen, 1854
Megacervixosaurus Zhao, 1983 [*nomen nudum*]
Megadactylus Hitchcock, 1865 / Fitzinger, 1843
"Megadontosaurus" Brown *vide* Ostrom, 1970
Megalosaurus Buckland, 1824
Melanorosaurus Haughton, 1924
Metriacanthosaurus Walker, 1964
Microceratops Bohlin, 1953
Microcoelus Lydekker, 1893
Microdontosaurus Zhao, 1983 [*nomen nudum*]
Microhadrosaurus Dong, 1979
Micropachycephalosaurus Dong, 1978
Microsaurops Kuhn, 1963
Microvenator Ostrom, 1970
"Mifunesaurus" Hisa, 1985 [*nomen nudum*]
Minmi Molnar, 1980
Mochlodon Seeley, 1881
Mongolosaurus Gilmore, 1933
Monkonosaurus Zhao *vide* Dong, 1990
Monoclonius Cope, 1876
Monolophosaurus Zhao & Currie, 1994 (not 1993)
Mononychus Perle, Norell, Chiappe & Clark, 1993 / Schueppel, 1824
Mononykus Perle, Norell, Chiappe & Clark, 1993
Montanoceratops C. M. Sternberg, 1951
Morinosaurus Sauvage, 1874
Morosaurus Marsh, 1878
"Moshisaurus" Hisa, 1985 [*nomen nudum*]
Mussaurus Bonaparte & Vince, 1979
Muttaburrasaurus Bartholomai & Molnar, 1981
Mymoorapelta Kirkland & Carpenter, 1994
Naashoibitosaurus Hunt & Lucas, 1993
Nanosaurus Marsh, 1877
Nanotyrannus Bakker, Williams & Currie, 1988
Nanshiungosaurus Dong, 1979
Nectosaurus Versluys, 1910 / Merriam, 1905
Nedcolbertia [Anonymous] 1996 [*nomen nudum*]
Nemegtosaurus Nowinski, 1971
Neosaurus Gilmore *vide* Gilmore & Stewart, 1945 / Nopcsa, 1923
Neosodon de la Moussaye, 1885

Neovenator Hutt, Martill & Barker, 1996
Neuquensaurus J. Powell, 1992
Ngexisaurus Zhao, 1983 [*nomen nudum*]
Niobrarasaurus Carpenter, Dilkes & Weishampel, 1995
Nipponosaurus Nagao, 1936
Noasaurus Bonaparte & J. Powell, 1980
Nodosaurus Marsh, 1889
Notoceratops Tapia, 1918
Nurosaurus Dong, 1992 [*nomen nudum*; later spelled *Nuoerosaurus*]
Nuthetes Owen, 1854*
Nyasasaurus Charig, 1967 [*nomen nudum*]*
Ohmdenosaurus Wild, 1978
Oligosaurus Seeley, 1881
Omeisaurus Young, 1939
Omosaurus Owen, 1875 / Leidy, 1856
Onychosaurus Nopcsa, 1902
Opisthocoelicaudia Borsuk-Bialynicka, 1977
Oplosaurus Gervais, 1852
Orinosaurus Lydekker, 1889
Ornatotholus Galton & Sues, 1983
Ornithodesmus Seeley, 1887
Ornithoides Osborn, 1924 [*nomen nudum*]
Ornitholestes Osborn, 1903
Ornithomerus Seeley, 1881
Ornithomimoides von Huene & Matley, 1933
Ornithomimus Marsh, 1890
Ornithopsis Seeley, 1870
Ornithosuchus Newton, 1894*
Ornithotarsus Cope, 1869
Orodromeus Horner & Weishampel, 1988
Orosaurus Huxley, 1867 / Peters, 1862
Orthogoniosaurus Das-Gupta, 1931
Orthomerus Seeley, 1883
Oshanosaurus Chao, 1985 [*nomen nudum*]
Othnielia Galton, 1977
Ouranosaurus Taquet, 1976
Oviraptor Osborn, 1924
Ovoraptor Osborn, 1924 [*nomen nudum*]
Pachycephalosaurus Brown & Schlaikjer, 1943
Pachyrhinosaurus C. M. Sternberg, 1950
Pachysauriscus Kuhn, 1959
Pachysaurops von Huene, 1961
Pachysaurus von Huene, 1908 / Fitzinger, 1843
Pachyspondylus Owen, 1854
Palaeoctonus Cope, 1877*
Palaeopteryx Jensen, 1981
Palaeosauriscus Kuhn, 1959*
Palaeosaurus Riley & Stutchbury, 1836 / Geoffroy Saint-Hilaire, 1833*
Palaeoscincus Leidy, 1856
Panoplosaurus Lambe, 1919
Paraiguanodon Brown *vide* Olshevsky, 1991 [*nomen nudum*]
Paranthodon Nopcsa, 1929
Pararhabdodon Casanovas-Cladellas, Santafe-Llopis & Isidro-Llorens, 1993
Parasaurolophus Parks, 1922
Pareiasaurus Owen, 1876*
Parksosaurus C. M. Sternberg, 1937
Paronychodon Cope, 1876
Parrosaurus Gilmore, 1945
Parvicursor Karhu & Rautian, 1996
Patagonykus Novas, 1996
Patagosaurus Bonaparte, 1979
Patricosaurus Seeley, 1887
Pawpawsaurus Lee, 1996
Pectinodon Carpenter, 1982
Peishansaurus Bohlin, 1953
Pekinosaurus Hunt & Lucas, 1994
Pelecanimimus Perez-Moreno, Sanz, Buscalioni, Moratalla, Ortega & Rasskin-Gutman, 1994
Pellegrinisaurus Salgado, 1996
Pelorosaurus Mantell, 1850
Peltosaurus Brown *vide* Chure & McIntosh, 1989 (see also Glut, 1972) / Cope, 1873
Pentaceratops Osborn, 1923
Phaedrolosaurus Dong, 1973
Phuwiangosaurus Martin, Buffetaut & Suteethorn, 1994
Phyllodon Thulborn, 1973
Piatnitzkysaurus Bonaparte, 1979
Picrodon Seeley, 1898*
Pinacosaurus Gilmore, 1933
Pisanosaurus Casamiquela, 1967
Piveteausaurus Taquet & Welles, 1977
Plateosauravus von Huene, 1932
Plateosaurus von Meyer, 1837
Pleurocoelus Marsh, 1888
Pleuropeltus Seeley, 1881
Pneumatoarthrus Cope, 1870*
Podokesaurus Talbot, 1911
Poekilopleuron Eudes-Deslongchamps, 1838
Polacanthoides Nopcsa, 1928
Polacanthus Owen *vide* [Anonymous] 1865
Polyodontosaurus Gilmore, 1932
Polyonax Cope, 1874
Poposaurus Mehl, 1915*
Prenocephale Maryanska & Osmólska, 1974
Priconodon Marsh, 1888
Priodontognathus Seeley, 1875
Probactrosaurus Rozhdestvensky, 1966
Proceratops Lull, 1906
Proceratosaurus von Huene, 1926
Procerosaurus von Huene, 1902*
Procerosaurus Fritsch, 1905 / von Huene, 1902
Procheneosaurus Matthew, 1920
Procompsognathus E. Fraas, 1913
Prodeinodon Osborn, 1924
Proiguanodon van den Broeck, 1900 [*nomen nudum*]
Prosaurolophus Brown, 1916
Protarchaeopteryx Ji Q. & Ji S., 1997
Protiguanodon Osborn, 1923
Protoavis Chatterjee, 1991
Protoceratops Granger & Gregory, 1923
Protognathosaurus Olshevsky, 1991

Protognathus Zhang, 1988 / Basilewsky, 1950
Protorosaurus von Meyer, 1830*
Protorosaurus Lambe, 1914 / von Meyer, 1830
Protrachodon Nopcsa, 1923 [*nomen nudum*]
Pseudolagosuchus Arcucci, 1987
Psittacosaurus Osborn, 1923
Pteropelyx Cope, 1889
Pterospondylus Jaekel, 1913
Qinlingosaurus Xue, Zhang & Bi, 1996
Quaesitosaurus Bannikov & Kurzanov *vide* Kurzanov & Bannikov, 1983
Rachitrema Sauvage, 1882*
Rapator von Huene, 1932
Rayososaurus Bonaparte, 1996
Rebbachisaurus Lavocat, 1954
Regnosaurus Mantell, 1848
Revueltosaurus Hunt, 1989
Rhabdodon Matheron, 1869
Rhadinosaurus Seeley, 1881
Rhodanosaurus Nopcsa, 1929
Rhoetosaurus Longman, 1925
Rhopalodon Fischer de Waldheim, 1841*
Ricardoestesia Currie, Rigby & Sloan, 1990
Rileya von Huene, 1902 / Howard, 1888
Rileyasuchus Kuhn, 1961
Rioarribasaurus Hunt & Lucas, 1991
Riojasaurus Bonaparte, 1969
Riojasuchus Bonaparte, 1969*
Roccosaurus van Heerden *vide* Anderson & Cruickshank, 1978 [*nomen nudum*]
Saichania Maryanska, 1977
Saltasaurus Bonaparte & J. Powell, 1980
Saltopus von Huene, 1910*
"*Sanchusaurus*" Hisa, 1985 [*nomen nudum*]
Sangonghesaurus Zhao, 1983 [*nomen nudum*]
Sanpasaurus Young, 1944
Sarcolestes Lydekker, 1893
Sarcosaurus Andrews, 1921
Sauraechmodon Falconer, 1861
Saurolophus Brown, 1912
Sauropelta Ostrom, 1970
Saurophaganax Chure, 1995
Saurophagus Stovall *vide* Ray, 1941 / Swainson 1831
Sauroplites Bohlin, 1953
Saurornithoides Osborn, 1924
Saurornitholestes Sues, 1978
Scaphonyx Woodward, 1908*
Scelidosaurus Owen, 1859
Scleromochlus Woodward, 1907*
Scolosaurus Nopcsa, 1928
Scrotum Brookes, 1763
Scutellosaurus Colbert, 1981
Secernosaurus Brett-Surman, 1979
Segisaurus Camp, 1936
Segnosaurus Perle, 1979
Seismosaurus Gillette, 1991
Sellosaurus von Huene, 1908
Shamosaurus Tumanova, 1983
Shanshanosaurus Dong, 1977
Shantungosaurus Hu, 1973
Shanyangosaurus Xue, Zhang & Bi, 1996
Shunosaurus Dong, Zhou & Zhang, 1983
Shuvosaurus Chatterjee, 1993*
Siamosaurus Buffetaut & Ingavat, 1986
Siamotyrannus Buffetaut, Suteethorn & Tong, 1996
Sigilmassasaurus D. A. Russell, 1996
Silvisaurus Eaton, 1960
Sinocoelurus Young, 1942
Sinornithoides D. A. Russell & Dong, 1994 (not 1993)
Sinosauropteryx Ji Q. & Ji S., 1996
Sinosaurus Young, 1948
Sinraptor Currie & Zhao, 1994 (not 1993)
Smilodon Plieninger, 1846 / Lund, 1842*
Sonorasaurus [Anonymous] 1995 [*nomen nudum*]
Sphenospondylus Seeley, 1882
Spinosaurus Stromer, 1915
Spinosuchus von Huene, 1932*
Spondylosoma von Huene, 1942
Squalodon Grateloup, 1840*
Staurikosaurus Colbert, 1970
Stegoceras Lambe, 1902
Stegopelta Williston, 1905
Stegosaurides Bohlin, 1953
Stegosaurus Marsh, 1877
Stenonychosaurus Sternberg, 1932
Stenopelix von Meyer, 1857
Stenotholus Giffin, Gabriel & Johnson, 1988
Stephanosaurus Lambe, 1914
Stereocephalus Lambe, 1902 / Lynch Arribalzaga, 1884
Stereosaurus Seeley, 1869 [*nomen nudum*]*
Sterrholophus Marsh, 1891
Stokesosaurus Madsen, 1974
Strenusaurus Bonaparte, 1969
Streptospondylus von Meyer, 1830
Struthiomimus Osborn, 1916
Struthiosaurus Bunzel, 1870
Stygimoloch Galton & Sues, 1983
Stygivenator Olshevsky *vide* Olshevsky, Ford & Yamamoto, 1995
Styracosaurus Lambe, 1913
Succinodon von Huene, 1941*
Suchoprion Cope, 1877*
"*Sugiyamasaurus*" Lambert, 1990 [*nomen nudum*]
Supersaurus Jensen, 1985
Symphyrophus Cope, 1878
Syngonosaurus Seeley, 1879
Syntarsus Raath, 1969
Syrmosaurus Maleev, 1952
Szechuanosaurus Young, 1942
Talarurus Maleev, 1952
Tanius Wiman, 1929
Tanystropheus von Meyer, 1855*
Tanystrosuchus Kuhn, 1963
Tapinocephalus Owen, 1876*

Tarascosaurus Le Loeuff & Buffetaut, 1991
Tarbosaurus Maleev, 1955
Tarchia Maryanska, 1977
Tatisaurus Simmons, 1965
Taveirosaurus Telles-Antunes & Sigogneau-Russell, 1991
Tawasaurus Young, 1982
Technosaurus Chatterjee, 1984
Tecovasaurus Hunt & Lucas, 1994
Teinurosaurus Nopcsa, 1928 emend. 1929
Teleocrater Charig, 1956 [*nomen nudum*]*
Telmatosaurus Nopcsa, 1903
Tenantosaurus Brown *vide* Chure & McIntosh, 1989
Tenontosaurus Ostrom, 1970
Teratosaurus von Meyer, 1861*
Termatosaurus von Meyer & Plieninger, 1844*
Tetragonosaurus Parks, 1931
Texasetes Coombs, 1995
Thecocoelurus von Huene, 1923
Thecodontosaurus Riley & Stutchbury, 1836
Thecospondylus Seeley, 1882
Therizinosaurus Maleev, 1954
Therosaurus Fitzinger, 1843
Thescelosaurus Gilmore, 1913
Thespesius Leidy, 1856
Thotobolosaurus Ellenberger, 1972 [*nomen nudum*]
Tianchisaurus Dong, 1993
Tianchungosaurus Zhao, 1983 [*nomen nudum*]
Tichosteus Cope, 1877
Tienshanosaurus Young, 1937
Timimus Rich & Vickers-Rich, 1994
Titanosaurus Lydekker, 1877
Titanosaurus Marsh, 1877/Lydekker, 1877
Tochisaurus Kurzanov & Osmólska, 1991
Tomodon Leidy, 1865/Dumeril 1853
Tonouchisaurus Barsbold, 1994 [*nomen nudum*]
Tornieria Sternfeld, 1911
Torosaurus Marsh, 1891
Torvosaurus Galton & Jensen, 1979
Trachodon Leidy, 1856
Trialestes Bonaparte, 1982*
Triassolestes Reig, 1963/Tillyard, 1918*
Tribelesodon Bassani, 1886*
Triceratops Marsh, 1889
Trimucrodon Thulborn, 1973
Troodon Leidy, 1856
Tsagantegia Tumanova, 1993
Tsintaosaurus Young, 1958
Tugulusaurus Dong, 1973
Tuojiangosaurus Dong, Li, Zhou & Zhang, 1977
Turanoceratops Nessov & Kaznyshkina *vide* Nessov, Kaznyshkina & Cherepanov, 1989
Tylocephale Maryanska & Osmólska, 1974
Tylosteus Leidy, 1872
Tyrannosaurus Osborn, 1905
Tyreophorus von Huene, 1929 [*nomen nudum*]
Udanoceratops Kurzanov, 1992
Ugrosaurus Cobabe & Fastovsky, 1987
Uintasaurus Holland, 1919
Ultrasauros Jensen *vide* Olshevsky, 1991
Ultrasaurus Kim, 1983
Ultrasaurus Jensen, 1985/Kim, 1983
Umarsaurus Maryanska & Osmólska, 1981 *vide* Olshevsky, 1992 [*nomen nudum*]
Unicerosaurus Armstrong, 1987 [*nomen nudum*]*
Unquillosaurus J. Powell, 1979
Utahraptor Kirkland, Burge & Gaston, 1993
Valdoraptor Olshevsky, 1991
Valdosaurus Galton, 1977
Vectensia Delair, 1982 [*nomen nudum*]
Vectisaurus Hulke, 1879
Velocipes von Huene, 1932
Velociraptor Osborn, 1924
Velocisaurus Bonaparte, 1991
Venaticosuchus Bonaparte, 1971*
Volkheimeria Bonaparte, 1979
Vulcanodon Raath, 1972
Wakinosaurus Okazaki, 1992
Walgettosuchus von Huene, 1932
Walkeria Chatterjee, 1987/Fleming, 1823
Wannanosaurus Hou, 1977
Wuerhosaurus Dong, 1973
Wyleyia Harrison & C. A. Walker, 1973
Wyomingraptor [Anonymous] 1997 [*nomen nudum*]
Xenotarsosaurus Martinez, Gimenez, Rodriguez & Bochatey, 1986
Xiaosaurus Dong & Tang, 1983
Xuanhanosaurus Dong, 1984
Xuanhuasaurus Chao, 1985 [*nomen nudum*]
Yaleosaurus von Huene, 1932
Yandusaurus He, 1979
Yangchuanosaurus Dong, Chang, Li & Zhou, 1978
Yaverlandia Galton, 1971
Yezosaurus Obata & Muramoto, 1977 [*nomen nudum*]*
Yimenosaurus Zhang, 1993
Yingshanosaurus Zhou, 1984 [*nomen nudum*]
Yubasaurus He, 1975? [*nomen nudum?*]
Yunnanosaurus Young, 1942
Zanclodon Plieninger, 1846*
Zapsalis Cope, 1876
Zatomus Cope, 1871*
Zephyrosaurus Sues, 1980
Zigongosaurus Hou, Chao & Chu, 1976
Zizhongosaurus Dong, Zhou & Zhang, 1983

Further Reading

Alexander, R. McN. (1989). *Dynamics of Dinosaurs and Other Extinct Giants.* 167 pp. Columbia University Press, New York.
A biomechanician's eye-view of size, function, and locomotion.

Alvarez, W. (1997). *T. rex and the Crater of Doom.* Princeton University Press.
A view of the end-Cretaceous extinctions by one of the scientists most involved in the discovery of the iridium layer.

Archibald, J. D. (1996). *Dinosaur Extinction and the End of an Era: What the Fossils Say.* 237 pp. Columbia University Press, New York.
An essay on the Cretaceous/Tertiary boundary and a detailed exposition of extinctions that occurred at this critical moment in history. The focus is on the evidence and its limitations. Readable discourse with respect to scientific methods, this long argument is engaging in part because of the use of the first person throughout. It is a good look at the workings of the scientific mind.

Bakker, R. T. (1986). *The Dinosaur Heresies.* 481 pp. William Morrow and Company, Inc., New York.

Benton, M. (1996). *The Penguin Historical Atlas of the Dinosaurs.* 144 pp. Penguin Books, London.
Well illustrated with many color maps, photographs, and diagrams, this thin paperback reviews important aspects of dinosaur science. Topics covered include general science issues such as evolution and geological time, as well as esoterica such as the dwarf dinosaurs of Hungary.

Bonaparte, J. F. (1978). *El Mesozoico de America del Sur y sus tetrapods.* Opera Lilloana, **26:** 1–596.

Bonaparte, J. F. (1996). *Dinosaurios de America del Sur.* 174 pp. Museo Argentino de Ciencias Naturales "Bernardino Rivadaria," Buenos Aires.

Carpenter, K., and Currie, P. J. (1990). *Dinosaur Systematics.* 318 pp. Cambridge University Press, Cambridge, UK.

Carpenter, K., Hirsch, K. F., and Horner, J. R. (1994). *Dinosaur Eggs and Babies.* 372 pp. Cambridge University Press, Cambridge, UK.

Carroll, R. L. (1988). *Vertebrate Paleontology and Evolution.* 698 pp. W. H. Freeman and Company, New York.

Colbert, E. H. (1984). *The Great Dinosaur Hunters and Their Discoveries.* 283 pp. Dover Publications, Inc.
Portraits of noted dinosaur field researchers, written by an author who himself ranks high in that category.

Colbert, E. H. (1995). *The Little Dinosaurs of Ghost Ranch.* 250 pp. Columbia University Press, New York.

Colbert, E. H. (1992). *William Diller Matthew–Paleontologist: The Splendid Drama Observed.* 275 pp. Columbia University Press, New York.
W. D. Matthew was one of the American Museum of Natural History's most prominent paleontologists and a participant in in the Central Asiatic Expeditions of the AMNH. This biography was prepared by E. H. Colbert, his son-in-law and himself a famous paleontologist.

Costa, V. (1994). *The Dinosaur Safari Guide: Tracking North America's Prehistoric Past.* 260 pp. Voyageur Press, Stillwater, OK.
Filled with maps, directions, telephone numbers, and highlights of dinosaur exhibits, this is a pocket book for the peripatetic dino-tourist. It is not directed to the serious student of paleontology, but could still be useful to anyone who wants to visit the past while traveling today.

Czerkas, S. J. and Olson, E. C. (1987). *Dinosaurs Past and Present, Vol. I* (161 pp.) *Vol. II* (149 pp.) Natural History Museum of Los Angeles County, Los Angeles, CA.
Interactions between science and art in dinosaurian paleontology.

Desmond, A. (1976). *The Hot-Blooded Dinosaurs.* 238 pp. The Dial Press/James Wade, New York.
Historically and scientifically informed study of the status of this debate from the 1870s to the 1970s.

Desmond, A. (1982). *Archetypes and Ancestors.* 287 pp. Blond and Briggs, London.
Victorian paleontological and evolutionary controversies, including those surrounding the first known dinosaurs.

Dodson, P. (1996). *The Horned Dinosaurs: A Natural History.* 346 pp. Princeton University Press, Princeton, NJ.
Horned dinosaurs, or ceratopsids, are among the most familiar of all dinosaurs. From descriptions of variation and the latest research to the history of ceratopsid excavations, this is a must for both scientists and students.

Fastovsky, D. E., and Weishampel, D. B. (1996). *The Evolution and Extinction of the Dinosaurs.* 461 pp. Cambridge University Press, Cambridge, UK.
An overview of dinosaur science, presenting a description of how professional paleontologists study dinosaurs but written at a level for the generalist or student, with a number of informal sidebar sections.

Fraser, N. C. and Sues, M.-D. (1994). *In the Shadow of the Dinosaurs.* 435 pp. Cambridge University Press, Cambridge, UK.

Gallagher, W. B. (1997). *When Dinosaurs Roamed New Jersey.* 176 pp. Rutgers University Press, New Brunswick, NJ.

Gillette, D. D. (1994). *Seismosaurus: The Earth Shaker.* 205 pp. Columbia University Press, New York.
Anyone interested in a firsthand and personal account of the excavation of the largest land animal known will enjoy this book. Mark Hallett's wonderful illustrations also make this book worth having.

Gillette, D. D. and Lockley, M. G. (1989). *Dinosaur Tracks and Traces.* 454 pp. Cambridge University Press.

Horner, J. R. and Gorman, J. (1988). *Digging Dinosaurs.* 210 pp. Workman Publishing, New York.

Horner, J. R. and Lessem, D. (1993). *The Complete T. rex.* 238 pp. Simon and Schuster, New York.

Jacobs, L. (1993). *Quest for the African Dinosaurs.* 314 pp. Villard Books, New York.

Johnson, K. R., and Stucky, R. K. (1995). *Prehistoric Journey: A History of Life on Earth.* 144 pp. Denver Museum of Natural History / Roberts Rinehart Publishers, Boulder, CO.
This is a well-illustrated book that highlights selected episodes of the natural history of life.

Lockley, M. (1995). *Dinosaur Tracks and Other Fossil Footprints of the Western United States.* 338 pp. Columbia University Press, New York.

Lucas, S. (1994). *Dinosaurs: The Textbook.* 290 pp. Wm. C. Brown Publishers, Dubuque, IA.

McGowan, C. (1991). *Dinosaurs, Spitfires and Sea Dragons.* 365 pp. Harvard University Press, Cambridge, MA.

Norell, M. A., Gaffney, E. S., and Dingus, L. (1995). *Discovering Dinosaurs.* 204 pp. Nevraumont Publishing Company, Inc., New York.

Norman, D. (1994). *Dinosaur!* 288 pp. Macmillan, New York.
A reprinted paperback edition of Norman's earlier 1991 book based upon a four-part television series. Well illustrated with color photographs, color illustrations, and numerous black-and-white line drawings. This authoritative treatment is as good a read as it is affordable.

Novacek, M. (1996). *Dinosaurs of the Flaming Cliffs.* 267 pp. Doubleday, New York.
This wonderful book is great reading. The intimacy of insight and the frank enthusiasm and affection for a life in paleontology make this account of discovery a gem.

Ostrom, J.H. and McIntosh, J. S. (1966). *Marsh's Dinosaurs: The Collections from Como Bluff.* 388 pp. Yale University Press, New Haven.
Scientific and historical explanations of O. C. Marsh's greatest dinosaurs.

Padian, K. [ed.] (1986). *The Beginning of the Age of Dinosaurs.* 378 pp. Cambridge University Press, Cambridge, UK.

Padian, K. [ed.] (1986). *The Origin of Birds and the Evolution of Flight.* Memoirs, California Academy of Sciences, number 8: 152 pp.

Padian, K., and Chure, D. J. [eds.] (1989). *The Age of Dinosaurs, Short Courses in Paleontology (2).* 210 pp. The Paleontological Society.

Preston, D. J. (1986). *Dinosaurs in the Attic.* 309 pp. Ballantine Books, New York.
Collected historical vignettes of the American Museum of Natural History; only three or four chapters deal specifically with paleontological subjects. Nevertheless, this well-written book is a most entertaining glimpse of the idiosyncratic people who have contributed to the growth of a leading research facility.

Russell, D. A. (1977). *A Vanished World, Natural History Series No. 4.* 142 pp. National Museum of Natural Sciences, National Museums of Canada, Ottawa.

Russell, D. A. (1989). *An Odyssey in Time: The Dinosaurs of North America.* 238 pp. University of Toronto Press.

Sarjeant, W. A. S. [ed.] (1995). *Vertebrate Fossils and the Evolution of Scientific Concepts.* 622 pp. Gordon and Breach Publishers, New York.

Sun Ailing (1989). *Before Dinosaurs: Land Vertebrates of China 200 Million Years Ago.* 109 pp. China Ocean Press.

Taquet, P. (1994). *L'Empreinte des Dinosaures.* 363 pp. Editions Odile Jacob, Paris.

Thulborn, T. (1990). *Dinosaur Tracks.* 410 pp. Chapman & Hall, London.

Wallace, J. (1994). *The American Museum of Natural History's Book of Dinosaurs and Other Ancient Creatures.* 144 pp. Simon & Schuster, New York.
This well-written and nicely produced coffee-table type book commemorates the opening of the AMNH's newly renovated vertebrate paleontology exhibit halls. Much of the book is devoted to the history of paleontological exploration, especially the central role that the AMNH has had in this history.

Weishampel, D. B., Dodson, P., and Osmólska, H. (1990). *The Dinosauria.* 733 pp. University of California Press, Berkeley, CA.
A remarkable book filled with detailed information.

Wellnhofer, P. (1991). *Illustrated Encyclopedia of Pterosaurs.* 192 pp. Crescent Books, New York.

Wilford, J. N. (1985). *The Riddle of the Dinosaur.* 304 pp. Alfred A. Knopf, New York.

CHRONOLOGY

1555 In *L'histoire de la nature des oyseaux (Natural History of Birds)*, **Pierre Belon** classifies 200 species and describes homologies in the skeletons of all vertebrates.

1596 European geographer and classical scholar **Abraham Ortelius,** in the third edition of his *Thesaurus Geographicus*, reasons that America was torn away from Europe and Africa by earthquakes and flood. Evidence of the rupture, he suggests, can be seen on a map of the world by careful study of the coasts of those continents where they face each other; this constitutes the oldest known statement on continental drift.

1650s **James Ussher,** Bishop of Armagh, calculates the Biblical creation of Earth to have occurred in 4004 B.C.

1669 Danish geologist and anatomist **Nicolaus Steno (Niels Stensen)** publishes *Prodromus,* a treatise on geology that sets forth the principles of original horizontality, superposition, and lateral continuity; Steno correctly describes the process by which living things become fossils.

1677 In *Natural History of Oxfordshire,* **Robert Plot** publishes a figure of the end of a large bone, which he labels the "Enigmatic Thigh Bone." The bone is believed to be the first mention of dinosaur remains. Plot interprets the remains as those of a giant form of a modern animal, rather than those of a dinosaur.

1686 **John Ray** publishes the first of three long volumes describing and classifying 18,600 plant species, laying important groundwork later used by Linnaeus.

1693 **Ray** follows up his classification of plants with a book logically classifying animals according to their hoofs, toes, and teeth. He correctly identifies whales as mammals.

1728 A posthumously published catalogue of the geological collection of **John Woodward** lists as Specimen A1, portions of the shank of a dinosaur limb-bone (though not so named). The specimen, probably the earliest discovered, still identifiable dinosaur bone, survives in the Woodwardian Museum, Cambridge.

1735 **Carolus Linnaeus** publishes *Systema Naturae (System of nature)*, introducing the system of binomial classification for plants still used today, which he extends in later volumes to include animals.

1749 **George-Louis Leclerc de Buffon** advances a theory of evolution in his first of 44 volumes of *Natural History*. He also suggests that the earth was formed by a collision of a huge comet with the sun some 75,000 years ago.

1751 In his *Philosophia botanica*, **Linnaeus** rejects any ideas of evolution.

1755 **Joshua Platt** finds, in the Stonesfield Slate of Oxfordshire, three vertebrae "of enormous size" as well as an incomplete left femur weighing 200 lbs. He reports his discoveries in a letter published in **1758** in the *Philosophical Transactions* of the Royal Society of London.

1763 **Richard Brookes,** in his *Natural History of Waters, Earths, Stones, Fossils and Minerals,* refigures Plot's 1677 bone. His illustration, inverted from Plot's, is captioned "*Scrotum humanum*" because the appearance of the condyles suggests a resemblance to the scrotum of a male mammal.

1770 Quarry workers discover the 4-foot-long skull and jaws of *Mosasaurus* in chalk mines near Maastricht, Holland.

1776? The French **Abbé Dicquemare** reports the discovery of some vertebrae and a femur from the Jurassic strata of Normandy. The discovery is unconfirmed, as the bones were not illustrated and are now lost. The **Abbé Bachelet,** another cleric, sends fossils from this area to anatomist **Georges Cuvier** for identification; Cuvier initially believes the remains to be vertebrae of crocodiles but later realizes that they are something quite different.

1785 The theory of uniformitarianism, or the very gradual and regular evolution of the earth's structure, is introduced in *Theory of the Earth*, by **James Hutton**. According to his ideas, the earth is so old that it is almost impossible to determine when it began. This theory holds that the processes which operate on the earth today are essentially the same as those which have operated in the past.

1787 In the earliest discovery of dinosaur remains in America, **Casper Wistar** and **Timothy Matlack** report the finding of a large "thighbone" in Late Cretaceous deposits of Gloucester County, New Jersey. The bone, most likely belonging to a hadrosaur, was never illustrated and is now lost.

1795 After French emperor Napoleon's army captures Maastricht, the jaw of *Mosasaurus* discovered by quarry workers in 1770 is confiscated and sent to Paris for study by **Georges Cuvier**. He identifies it as a "giant lizard."

1795 During a geological excursion in southern France, western Switzerland, and northern Italy, **Alexander von Humboldt** recognizes that the Jura-Kalkstein, a massive formation in the Jura Mountains, is a distinctive rock unit of widespread geologic importance; the name "Jurassic" stems from these observations.

1796 **Cuvier** observes that there is no living lizard with a jaw as large as that of *Mosasaurus,* thus indicating that earlier species have become extinct.

1798 English economist **Thomas Malthus** publishes *Essay on the Principle of Population,* suggesting that population tends to increase more rapidly than food supplies. Later evolutionary theorists were influenced by his description of overpopulation as a natural tendency unless checked by negative environmental conditions.

1801 **Jean Baptiste Lamarck,** later regarded as the founder of invertebrate paleontology, is the first to classify invertebrates in *Systéme des animaux sans vertèbres (System for animals without vertebras),* in which he also includes the beginnings of his distinctive theory of evolution.

1802 **Pliny Moody**, a farm boy in the Connecticut River Valley, discovers the first identifiable dinosaur tracks. These are described as the tracks of "Noah's Raven."

1806 A very large bone is discovered by **William Clark** of the famed Lewis and Clark expedition, in an area of Cretaceous deposits near the present town of Billings, Montana. Though identified then as the rib of a huge fish, the description of its size and location suggests a Late Mesozoic dinosaur bone.

1809 In *Philosophie zoologique,* **Lamarck** states his beliefs that animals evolved from simpler forms, that unused body parts degenerate or new body parts develop as needed for survival, and that such acquired characteristics can be inherited.

1811? At about the age of 12, **Mary Anning** finds some remains in the Lyme Regis cliffs on the southern shores of England, at first thought to be those of a crocodile, but later identified as the first remains of an ichthyosaur, a common marine reptile of the Mesozoic. She then becomes a respected fossil collector in her adult years.

1812–1816 "Large saurian bones" are discovered by **Thomas Webster** in the Wealden (Early Cretaceous) strata of England, an area that later becomes very important in the history of dinosaur study. Webster reports the finding in **1829.**

1815 British civil engineer **William Smith** produces a complete geological map of England and Wales, the first geological map to be published; he uses the fossil sequence for correlation of geological strata and demonstrates the principle of faunal succession.

1817 **Georges Cuvier,** a pioneer in comparative anatomy, originates a system of zoological classification based on the idea that the anatomical structures of all organs of the body are functionally related.

1822 For the first time, a dinosaur fossil is correctly identified as coming from a large, reptilian animal distinct from modern species; earlier discoveries were identified as giant humans, dragons, larger forms of living animals, or other such creatures. According to some histories, **Mary Ann Mantell** of Sussex, England, was the person who actually found the large, fossilized teeth and recognized them as unique. Other accounts dispute this and credit the discovery to her husband, **Dr. Gideon Mantell**, a physician and geologist. The discovery takes place in the Wealden strata of the Isle of Wight and is subsequently announced by Gideon Mantell in *The Fossils of South Downs.*

1822 In *An Introduction to the Study of Fossil Organic Remains,* physician and geologist **James Parkinson** describes a tooth of what he names *Megalosaurus* (from the Greek *megalos,* "great" and *sauros,* "lizard"); this, like *Iguanodon,* was discovered in England.

1822–1824 A number of bones are discovered at Oxfordshire by paleontologist **Hugh Strickland,** by **John Kingdon** (or **Kingdom**), and by **William Buckland.** Though he initially referred to these bones as "cetacean," Buckland was later alerted to their true character by Dr. Mantell's work.

1824 **Samuel Stutchbury,** while visiting the Hunterian Museum of the Royal College of Surgeons, points out to another visitor, Gideon Mantell, how closely the noted Wealden teeth resemble those of a living iguana, only on a larger scale. Georges Cuvier at first considered the teeth to be those of a rhinoceros but later determined that they must be the teeth of a gigantic herbivorous reptile.

1824 **William Buckland,** a professor at Oxford University, describes *Megalosaurus* as the "Great Fossil Lizard of Stonesfield." This proves to be the first scientific description of a dinosaur.

1825 *Iguanodon,* "iguana tooth," is named by **Gideon Mantell** in a lecture presented to the Royal Society of London. *Iguanodon* becomes the second dinosaur–and the first herbivorous dinosaur–ever named.

1830 Scottish geologist **Charles Lyell** publishes the first in his landmark three-volume *Principles of Geology,* in which he describes the uniformitarian theory that is still generally accepted today. He divides the geological system into three groups which he names Eocene, Miocene, Pliocene.

1832 **Dr. Gideon Mantell** describes *Hylaeosaurus,* another giant "lizard" found in Wealden.

1834 A tooth described from Jurassic strata of southern Russia by **A. Zborzewski** is given the name *Macrodontophion,* "big tooth producer"; the tooth is perhaps that of a carnosaur.

1836 **Samuel Stutchbury,** working with **S. H. Riley,** describes bones from the Triassic "Magnesian Conglomerate" of the Bristol region. Three different saurians are reported, two of which (*Thecodontosaurus,* "sheath-toothed reptile," and *Palaeosaurus,* "ancient reptile") are accepted as being dinosaurs (though the nature of the latter genus is now considered dubious).

1837 The extensive remains of another Triassic dinosaur, *Plateosaurus,* are reported by **Hermann von Meyer** from Germany.

1838 **Jacques-Amand Eudes-Deslongchamps** finds a fragmentary dinosaur skeleton in the Middle Jurassic strata of Normandy, France. He perceives resemblances to *Megalosaurus,* but thinks the differences warrant his giving it a new generic name *Poekilopleuron,* "mottled-rib."

1838 **Matthias Jakob Schleiden** discovers that all living plant tissue is composed of cells and cell products. The following year, **Theodor Schwann** announces that the same is also true for animal tissue. Their discoveries are known as the Schleiden–Schwann cell theory.

1841 The British anatomist **Richard Owen** recognizes that *Iguanodon* and *Megalosaurus* are more like mammals than lizards, in having an erect posture with limbs held beneath the body. He describes the first sauropod dinosaur, *Cetiosaurus,* "monstrous reptile," utilizing bones from Mantell's collection and from several other localities in Britain.

1842 **Owen** introduces the name "Dinosauria" (from the Greek words *deinos,* "terrible" or "dreadful" and *sauros,* "lizard") into the manuscript of his 1841 Plymouth lecture prior to its publication, writing of the "distinct tribe or suborder of Saurian Reptiles, for which I would propose the name Dinosauria."

1844 The last surviving great auk, a flightless diving bird of the northern oceans, is killed, thus demonstrating the fact of extinction.

1848? Massachusetts clergyman and professor **Edward Hitchcock** describes the unusual footprints known as the tracks of "Noah's Raven" that were found earlier in the Connecticut River Valley of New England. He notes that the huge footprints have a mysterious bird-like quality.

1854 Artist and sculptor **Benjamin Waterhouse Hawkins,** in collaboration with Richard Owen, constructs a lifesize restoration of *Iguanodon;* a group of London's eminent thinkers dine inside the hollow body of the *Iguanodon* prior to its public display.

1855 **Ferdinand Vandiveer Hayden** leads a geological expedition to the Cretaceous badlands of Montana and discovers the first identified dinosaur remains in the western United States. Teeth collected by the Hayden Survey are described by paleontologist **Joseph Leidy** in **1856.**

1855 The first identifiable American discovery of dinosaur skeletal remains (*Hadrosaurus*) is made in Haddonfield, New Jersey.

1856 Workers near the Neander River in Germany discover the first-known skeleton of a Neanderthal man in a limestone cave; this is an important first step in determining the evolution of the human species.

1858 **Edward Hitchcock,** considered by many to be the founder of vertebrate ichnology, publishes *A Report on the Sandstone of the Connecticut Valley Especially Its Fossil Footmarks,* his most complete work, containing many ichnogenus names still in use today.

1858 In a report to the Philadelphia Academy of Sciences, **Joseph Leidy** describes the first hadrosaur, *Hadrosaurus foulkii,* collected from a sand quarry in New Jersey. He is the first to show a bipedal posture for dinosaurs.

1859 **Charles Darwin** publishes *On the Origin of Species,* a book outlining his theory of evolution (descent with modification) by natural selection. It causes enormous controversy and becomes the foundation of a completely new approach to biology.

1860s? At some time earlier than 1871, long before the first scientific discovery of dinosaurs in the area of Alberta, Canada, the French-Canadian traveler **Jean-Baptiste L'Heureux,** while living among the Piegan Indians, is shown some bones that are characterized as remains of "the grandfather of the buffalo."

1862 **Richard Owen** describes the London specimen of the earliest bird, *Archaeopteryx lithographica,* hailed by some of Darwin's supporters as a "missing link" between reptiles and birds.

1865 After years of research with thousands of garden pea plants, the Austrian monk **Gregor Mendel** publishes his landmark theories of genetics in an obscure publication. His ideas remain generally unknown for more than thirty years.

1866, 1877 **Edward Drinker Cope** and **Othniel Charles Marsh** describe remains of *Dryptosaurus* (*Laelaps*) from New Jersey; these remains indicate the bipedal gait of theropods.

1868, 1869 **F. V. Hayden** writes of the occurrences of huge Cretaceous saurian fossils in various spots across Wyoming Territory, primarily along the line of the Union Pacific railroad.

1868 **Marsh** travels the length of the recently established transcontinental railroad. At Como Station in southeastern Wyoming Territory, he is mainly interested in collecting the small axolotl-like salamanders from nearby Lake Como.

late 1860s **Thomas H. Huxley,** a leading proponent of Charles Darwin's theory of evolution, argues the relationship of dinosaurs to birds, comparing *Archaeopteryx* and *Compsognathus* and citing 35 different characteristics to support his theory.

1871 **Darwin** publishes *The Descent of Man,* in which he applies his principles of evolution to the human race.

1872 **Drs. F. B. Meek** and **H. M. Bannister,** while working for Hayden's Geological Survey of the Territories, discover dinosaurian remains in the latest Cretaceous rocks of western Wyoming, near Black Butte Station. **Cope** then assigns this partial skeleton to a new genus and species of dinosaur, *Agathaumas sylvestris* ("marvelous saurian of the forest"). *A. sylvestris* is one of the largest and best-preserved dinosaurs known at that time.

1873 **Leidy** describes as the vertebra of the carnivorous dinosaur *Poicilopleuron* (now known as *Allosaurus*) a bone obtained from Middle Park, Colorado, in **1869,** identified by **Hayden** as a "petrified horse hoof."

1876 In *The Geographical Distribution of Animals,* English naturalist **Alfred Wallace** establishes biogeographical support for the evolutionary theory.

1877 The "Great Dinosaur Rush" begins in the American West. **Marsh** and **Cope** are the primary protagonists and fierce adversaries. As a consequence of their rivalry and the rich vertebrate fossil deposits of the West, there is a great surge forward in the knowledge of ancient animals.

1877 **Cope** describes a giant herbivorous dinosaur whose remains were discovered in Utah by geologist **John Strong Newberry** in **1859,** calling this new Jurassic reptile *Dystrophaeus* (now possibly *Camarasaurus*).

1877 Both **Marsh** and **Cope** learn about large bones at Morrison Colorado; **O. W. Lucas** discovers dinosaur bones at Cañon City, Colorado, for Cope; **William Edward Carlin** and **William Harlow Reed,** a pair of Union Pacific Railroad workers, open the dinosaur locality at Como Bluff, Wyoming, assisting Marsh with its excavation.

1877–1889 Numerous genera of ancient vertebrates, such as *Diplodocus, Apatosaurus, Triceratops,* and *Stegosaurus,* are identified by Cope and Marsh during their notorious "Bone Wars."

1877 **Henry Fairfield Osborn** and some friends go to Wyoming on a lark to collect fossils, unknowingly beginning, for Osborn, what will become a highly distinguished career in paleontology.

1877 While working for the Geological Survey of India, the English naturalist and geologist **Richard Lydekker** creates the first systematic description of the dinosaurs of the Indian subcontinent. He establishes a new sauropod taxon, *Titanosaurus indicus.*

1878 The discovery of the Bernissart iguanodons in a coal mine in southwestern Belgium marks the beginning of **Louis Dollo's** life-long study of the Bernissart fossils; the Sainte-Barbe pit eventually yields more than 30 complete, articulated skeletons of *Iguanodon,* the first complete dinosaur skeletons ever discovered. The name *Iguanodon bernissartensis* is provided by **Dr. G. A. Boulenger** in 1881.

1882 **Walther Flemming** publishes his discovery of chromosomes and mitosis in *Cell Substance, Nucleus, and Cell Division.* This discovery fits in neatly with Mendel's research into genetics, though that was largely overlooked and unknown at this time.

1884 With the "Great Dinosaur Rush" of the western U. S. now extending north to Canada, **Joseph Burr Tyrrell** discovers extensive beds of Cretaceous dinosaur remains along the cliffs of the Red Deer River in southern Alberta, Canada. The Red Deer River region subsequently proves to be one of the most prolific dinosaur sites of the world. (The Tyrrell Museum in Drumheller, Alberta, noted for its spectacular displays of dinosaurs, is named to honor Joseph Burr Tyrrell.)

1891 **H. F. Osborn** joins the staff of the American Museum of Natural History (AMNH), eventually becoming its Director and building the museum into one of the leading natural history institutions in the world.

1895 In Pittsburgh, the steel magnate and philanthropist **Andrew Carnegie** builds a museum that bears his name, Carnegie Museum.

1896 Following up on the work of Wilhelm Roentgen, discoverer of X rays, **Henri Becquerel** discovers radioactivity emitted by a uranium compound. This later leads to the use of radioactivity levels to date ancient materials.

1897 The first major AMNH expedition to the West is launched. It goes initially to Como Bluff, Wyoming, and shortly thereafter discovers a bonanza dinosaur deposit at nearby Bone Cabin.

1898 **Andrew Carnegie** begins to send field parties to the Western states to find suitable dinosaurs for display in his museum.

1900 **Gregor Mendel's** work with genetics is rediscovered and recognized as significant by three biologists working independently, Hugo de Vries of the Netherlands, Carl Correns of Germany, and Erich von Tschermak of Austria. They finally give publicity and credence to Mendel's theories.

1902 **Barnum Brown,** of the American Museum of Natural History, discovers the first *Tyrannosaurus rex* at the Hell Creek Formation, Montana; **between 1902 and 1908** he excavates two partial *T. rex* skeletons, which are displayed at the Carnegie and American Museums, winning fame as a dinosaur paleontologist.

1907 Engineers prospecting for garnet deposits discover a vast Jurassic fossil site at Tendaguru Hill in southern Africa. **Werner Janensch** and **Edwin Hennig** lead the German Tendaguru Expedition from **1909** to **1913.** One important result is the discovery of *Brachiosaurus.*

1907 **Thomas Hunt Morgan** begins genetic experiments with chromosome-simple fruit flies that prove the function of chromosomes, demonstrate the role of mutation in heredity, and illustrate sex-linked characteristics.

1908 **Charles Hazelius Sternberg** and his sons **Charles M., George,** and **Levi** discover, in the Lance Creek area of Wyoming, the mummified skeleton of *Edmontosaurus annectens,* later purchased by Henry Fairfield Osborn for the American Museum of Natural History. The Sternberg family eventually compiles large collections of dinosaurs for most of the major museums of the world.

1909 **Earl Douglass,** one of Andrew Carnegie's dinosaur prospectors, discovers *Barosaurus* fossils near Vernal, Utah. This locality eventually becomes Dinosaur National Monument, which opens to the public in **1958.**

1911 The English geologist and geophysicist **Arthur Holmes** first uses radioactive decay as a means of dating rocks.

1915 **Alfred Lothar Wegener,** German climatologist and geologist, sets forth the theory of continental drift.

1917 **Charles Sternberg,** under the sponsorship of the Canadian Geological Society, discovers in the Red Deer River region of Canada the most complete carnivorous dinosaur skeleton known from North America (*Albertosaurus libratus*).

1921 **Edgar Dacqué** initiates phylogenetically oriented paleontology.

1922 **Roy Chapman Andrews** makes his first visit to Central Asia, followed by a second in **1923** and a third in **1925.** These legendary Central Asiatic expeditions make a number of spectacular discoveries in the rich fossil fields of Mongolia, including the first dinosaur nests, at Bayn Dzak.

1922 **J. B. Shackelford,** the cameraman for **Roy Chapman Andrews,** discovers the remains of a *Protoceratops* in the Flaming Cliffs area of the Gobi Desert. The following year, paleontologist **George Olson** discovers the remains of an *Oviraptor* associated with the nest of a *Protoceratops.*

1923 **Baron Franz von Nopsca** produces a seminal work on vertebrate paleontology, *Die familien der Reptilien.* He explores interesting new approaches, such as paleohistology and the geographical distribution of dinosaurs in terms of Wegener's theory of continental drift.

1929 Deoxyribononucleic acid (DNA), the genetic material of most organisms, is discovered by **Phoebus Levene.**

1934 The excavations of **Roland T. Bird** for the American Museum of Natural History, including his excavations at the Late Jurassic (Morrison Formation) Howe Quarry in Wyoming, yield an enormous accumulation of sauropod bones.

1934 The Oxfordian Damparis quarry of the Jura region of eastern France yields one of the most complete skeletons of a sauropod (an early brachiosaurid) ever found in Europe, described by **Albert-François de Lapparent** in **1943.**

1935 At this time or in the years immediately following, the first dinosaur bones of ornithopods and sauropods from southeast Asia are found in the Cretaceous of Laos by the French geologist **J. H. Hoffet.**

1936 The Japanese paleontologist **T. Nagao** describes *Nipponosaurus,* a hadrosaur from southern Sakhalin island, north of Japan in the Sea of Okhotsk.

1937 **Theodor Dobzhansky** publishes *Genetics and the Origin of the Species,* finally linking Darwin's evolutionism and Mendel's theory of mutation in genetics.

1938 **Roland T. Bird** begins his pioneering work on the theropod and sauropod footprint sites of the Lower Cretaceous of Glen Rose, Texas.

1938 Excavations are begun in the Lufeng Beds (Late Triassic to Early Jurassic) of Yunnan in southern China, resulting in the discovery of a large number of prosauropod remains, described during the **1940s** by **C. C. Young.**

1939 European dinosaur specialist **A. F. de Lapparent** begins excavations at the Fox-Amphoux dinosaur locality in the Upper Cretaceous of Provence, but his work is interrupted by World War II. He reviews the Late Cretaceous dinosaurs from southern France in an extensive monograph published in **1947.**

1944 Based on experiments with pneumococci, **Oswald T. Avery** determines that DNA is material carrying genetic properties in nearly all living organisms.

1946 A Russian expedition under **I. Efremov** is sent to the Gobi Desert of Mongolia, followed by others in **1948** and **1949,** and resulting in the discovery of a large amount of dinosaur material from the Upper Cretaceous, including various forms such as the large tyrannosaurid *Tarbosaurus* from the Nemegt Formation, which had not been found by the earlier American expeditions.

1947 **Willard Frank Libby** introduces a method of dating archaeological objects by determining the concentration of radioactive carbon-14 in the item.

1947 At Ghost Ranch, New Mexico, fossil collectors led by **Edwin H. Colbert** discover an assemblage of skeletons of *Coelophysis bauri,* one of the earliest of all dinosaurs.

1950s **de Lapparent** explores the Cretaceous dinosaur localities of the central and southern Sahara, revealing the great paleontological potential of that vast desert.

1953 The discovery of the Great Global Rift in the Atlantic Ocean by **Maurice Ewing** and **Bruce Charles Heezen** establishes the study of plate tectonics and revolutionizes geology.

1953 **Francis H. Crick** and **James D. Watson** suggest the double helix model for DNA, giving a logical explanation for how genetic traits are transferred.

1953 The Institute of Vertebrate Paleontology of the Chinese Academy of Sciences, a center for paleo-vertebrate research in China, founded by **Professor Yang Zhongjian (C. C. Young),** is established in Beijing.

1957 **de Lapparent** describes with **R. Zbyszewski** various important Late Jurassic materials (including theropods, sauropods, and stegosaurs) from Portugal.

1958–1968 **Alfred S. Romer's** expeditions to the Permian of Texas provide a previously unknown chapter in the history of terrestrial vertebrates, including some important early mammal relatives.

1960s In expeditions to the southern Sahara, **Philippe Taquet** discovers nearly complete skeletons of a new iguanodontid, *Ouranosaurus nigeriensis.*

1960s John Ostrom challenges the long-held belief that hadrosaurs were predominantly aquatic animals; he also suggests on the basis of footprint evidence that some dinosaurs were gregarious.

1962 de Lapparent reports the discovery of polar dinosaurs in the form of *Iguanodon*-like footprints from the Lower Cretaceous of Spitzbergen, the first record of dinosaurs at high latitudes.

1963, 1964, 1965, and **1971** Polish-Mongolian expeditions under the leadership of **Zofia Kielan-Jaworowska** again explore the Late Cretaceous formations of the Gobi Desert, finding many well-preserved forms of known dinosaurs, as well as previously unknown forms.

1967 de Lapparent describes with **C. Montenat** the remarkable terminal Triassic to basal Jurassic site at Le Veillon, on the Atlantic coast of France, where thousands of footprints are recorded, contributing to the renaissance of studies on dinosaur ichnology.

late 1960s Large-scale investigations of bone microstructure in a variety of fossil reptiles, including dinosaurs, are launched by **Armand de Ricqlès** in Paris. This work contributes to the reappraisal of dinosaur physiology that later culminates in the debate about "hot-blooded dinosaurs" during the following decade.

1969 Ostrom describes a new theropod found in Montana in **1964** as *Deinonychus antirrhopus* "terrible claw," and interprets it as a fast and agile predator, quite different from the then-popular image of dinosaurs as sluggish animals, also noting anatomical similarities between this specimen and the primitive bird *Archaeopteryx*.

1970s Robert T. Bakker and **Peter Galton** publish a paper arguing that the order Aves (birds) should be included in the class Dinosauria, along with the dinosaur orders Ornithischia and Saurischia.

1979 Jack Horner and **Bob Makela** discover the jumbled bones of some tiny juvenile hadrosaurids (*Maiasaura*) in the Two Medicine Formation of western Montana, providing a remarkable glimpse into hadrosaurid reproduction and behavior.

1980 Physicist **Luis Alvarez** and his geologist son **Walter** publish a scientific paper advocating the extraterrestrial impact theory of Cretaceous extinction. They propose that the disappearance of nonavian dinosaurs from the fossil record was caused by the sudden collision of a massive asteroid with the Earth, throwing dust into the atmosphere and causing an abrupt climatic change.

1984 Paul Olsen and **Peter Galton,** building on recent work by **Bobb Schaeffer, Bruce Cornet,** and **Nick McDonald,** show that half of the "Late Triassic" beds of the Newark Supergroup are actually Early Jurassic in age. They extended their correlation worldwide to deposits in western North America, China, Europe, Southern Africa, and India, effectively adding an Early Jurassic dinosaur record where there had been almost nothing.

1986 Jacques Gauthier creates the first well-accepted, detailed phylogeny of the diapsids. Gauthier's research substantiates the longstanding proposal that birds are actually theropod dinosaurs, uniting *Archaeopteryx* with modern birds, as Ostrom had proposed in the 1970s.

1986 Paul Sereno produces the first well-accepted cladistic analysis of ornithischian dinosaurs, complementing Gauthier's work on saurischians.

1986 The Canada–China Dinosaur Project begins. This five-year joint expedition in search of Chinese and Canadian fossils makes important discoveries at, among other sites, Devil's Coulee, Canada (**1987**), and Bayan Mandahu, China.

late 1980s Paul Sereno, in collaboration with Argentine colleagues **José Bonaparte, Fernando Novas, Andrea Arcucci,** and others, recovers many early Late Triassic fossils of dinosaurs and dinosaur relatives, notably *Herrerasaurus* and *Eoraptor,* continuing the work of **Romer's** (**late 1950s–early 1960s**) South American expeditions.

1990 Edward Golenberg and colleagues isolate DNA from a Miocene Magnolia leaf, opening up the possibilities of finding DNA in fossil tissues.

1990 Publication of the novel ***Jurassic Park*** (and its subsequent release as a film in **1993**) sparks great popular interest in dinosaurs and dinosaur science.

1993 The flightless bird *Mononykus* ("one claw") is described by **Perle Altangerel** with **Mark Norell, Luis Chiappe,** and **James Clark** of the AMNH.

1995 Scientists discover in the hills of the Gobi Desert the fossil of a 9-foot-long beaked carnivorous dinosaur, *Oviraptor*, nesting on its eggs like a brooding bird. The eggs had previously been attributed to *Protoceratops*.

1996 Chinese paleontologists discover the remains of *Sinosauropteryx prima,* a meter-long dinosaur, in Liaoning, showing featherlike structures covering the head, trunk, tail, arms, and legs. The discovery lends support to the hypothesis that theropod dinosaurs were the direct ancestors of birds. Another feathered dinosaur, *Confuciusornis sanctus,* "holy Confucius bird," is discovered in the same area, as is the feathered dinosaur *Protarchaeopteryx*.

1996 Argentine paleontologist **Fernando Novas** recovers the fossil of a 90-million year old very birdlike nonavian dinosaur, *Unenlagia* ("half bird") *comahuensis* in northwest Patagonia.

GLOSSARY

The following glossary defines approximately 600 specialized terms that are used in the articles in the *Encyclopedia of Dinosaurs.* This glossary is intended for the general reader who may not be familiar with the technical terminology of dinosaur research. It includes terms in such fields as vertebrate zoology, anatomy, physiology, genetics, evolution, behavior, ecology, paleontology, systematics, and geology.

The glossary thus includes definitions of terms that are used to describe dinosaurs, but not the names of the dinosaurs themselves, such as Ankylosauria and Stegosauria. For reference to these items, see the Table of Contents and the Index.

abduction the movement of a body part away from the midline axis of the body. Compare ADDUCTION.

absolute age the age of an object as established by some precise dating method, such as radiometric dating. Compare RELATIVE AGE.

acceleration an evolutionary change in which the descendant species has a faster growth rate than the ancestral species.

accretion **1.** in anatomy, a process in which the growth or building up of some structure or mass occurs by the accumulation of fresh layers of material on top of existing layers. **2.** a similar process in geology; e.g., the gradual buildup of land on a shoreline by wave action.

acetabulum *plural,* **acetabula.** a cuplike socket in the bones that form the hips and pelvis, into which the head of the femur fits.

acidosis a metabolic disturbance in which the normal acid–base balance of the body is altered and the blood or tissue becomes excessively acidic.

acrodont **1.** describing a form of tooth arrangement in which the teeth sit on top of the jawbone and are often fused to it. **2.** an animal having this arrangement of teeth. Thus, **acrodontia.** Compare THECODONT, PLEURODONT.

acromion a bony process in the outer end of the scapula (shoulder blade).

adaptation **1.** the general ability of a species to undergo evolutionary change in response to its environment. **2.** a particular developmental, behavioral, anatomical, or physiological change in a population of organisms, in response to environmental conditions.

adduction the movement of a body part toward the midline axis of the body. Compare ABDUCTION.

aeolian relating to or caused by the action of wind; dispersed by the wind.

affinity a close relationship or common ancestry of organisms, suggested by resemblances in their anatomy, behavior, or other features.

agonistic behavior the interaction of members of the same species in various behavioral patterns in response to social conflict; e.g, attack, retreat.

allele one of two or more alternate forms of a gene occupying the same chromosome location.

allochthonous not originating at its present site; formed elsewhere. Compare AUTOCHTHONOUS.

allometry **1.** the relative growth rates of the different parts of a given organism. **2.** the study or measurement of the size or growth rate of organisms. Also, **allometrics.**

altricial describing a species in which the young are relatively undeveloped when born. Compare PRECOCIAL.

altricial behavior the fact or activity of caring for newly born offspring by their parent or parents. Also, PARENTAL BEHAVIOR.

alula *plural,* **alulae.** **1.** the portion of a bird's wing that corresponds to the human thumb; this structure is fundamental in modern birds for low-speed flight and maneuverability. **2.** a group of small feathers at this site. Thus, **alular.**

alveoli *singular,* **alveolus.** in anatomy, a cavity, pit, or socket; e.g., the numerous tiny air cells in the lungs in which the exchange of oxygen and carbon dioxide takes place, or the bony sockets in the jaw in which the roots of the teeth are held in place. Thus, **alveolar.**

ammonite **1.** a type of ammonoid fossil that has a thick, strongly ornamented shell with complex lines of juncture. **2.** see AMMONOID.

ammonoid a widespread, diverse group of mollusks (Ammonoidea), noted for their large, distinctive chambered shell. Though now extinct, ammonoids occurred in vast numbers in ancient seas within a distinct time span; thus they are important in fossil study because their presence (or absence) can be used to date the rocks in which they are found (or not found).

amnion a fluid-filled sac enveloping the embryo or fetus during development.

amniote one of the Amniota, land-dwelling vertebrates having an amnion (see above); amniotes thus include modern reptiles, birds, and mammals.

amphibious living, or capable of living, both on land and in water.

anaerobic without oxygen; not requiring or involving the presence of oxygen.

anagenesis an evolutionary change in which modified forms replace one another in continuous succession without branching into new taxa; descent without modification and the formation of new species. Thus, **anagenetic.**

anapsid one of the Anapsida, a group of amniotes including the oldest known forms, distinguished by a skull having a complete covering of dermal bones, with no openings (fenestrae) in the side of the skull behind the orbit (eye socket); e.g., turtles.

angiosperm any of the Angiospermae, the flowering plants; a plant that bears its seeds in a closed seed vessel.

anticline an upward fold of stratified rock in which the sides slope down and away from the crest; the oldest rocks are in the center, and the youngest ones on the outside.

antorbital relating to or located in the area in front of the orbit; i.e., in front of the eye socket.

antorbital fenestra a distinctive opening in the skull in front of the eye socket, characteristic of the archosaurs.

apatite a phosphate mineral that is a major constituent of sedimentary phosphate rocks and also of the bones and teeth of vertebrates.

apomorph a character in an organism that is derived from, but that is no longer the same as, an ancestral character. Thus, **apomorphy, apomorphic.** Compare PLESIOMORPH.

aposematism the fact of having distinctive coloration (or another such feature) that acts as a deterrent or warning to potential predators.

a priori Latin for "from (something) previous;" based on what is generally known; said of a conclusion about a specific instance that derives from a knowledge of the relevant general facts or conditions.

aquatic living in the water.

arboreal living exclusively or mostly in trees, shrubs, and bushes.

articulate 1. in anatomy, to form a joint; join. 2. of fossilized bones, to remain joined as they would have been in life. Thus, **articular, articulation.**

assemblage a large group of fossils or other items found in the same location and regarded as being from the same time period, though not necessarily as being positioned in a manner that reflects their arrangement in life.

assemblage zone an aggregation of fossils in a body of sedimentary rock.

astragalus the tarsal bone that joins with the tibia and calcaneum to form the ankle joint. Thus, **astragalar.**

autapomorphic describing an apomorphic or derived character unique in a particular species or other taxonomic group.

authigene a rock or mineral found at the site where it was formed. Thus, **authigenic.**

autochthonous found at the place where it originated; located at its original site. Compare ALLOCHTHONOUS.

avian 1. describing or relating to birds (Aves). 2. resembling birds, or in the manner of birds.

basal 1. situated at or forming the base. 2. specifically, placed at the base or "trunk" of a phylogenetic tree. 3. in general, being the earliest or ancestral form of a lineage.

basal group a group that is outside a more derived clade.

basicranium the base of the skull.

battery a distinctive tooth pattern in which a large number of small, slender individual teeth are wedged very tightly together along the length of the jaw, with multiple teeth stacked in one tooth position.

binomial nomenclature the traditional system for giving scientific names to organisms, developed by Linnaeus, in which the name of an organism is made up of two words, the genus name and the species name; e.g., *Oviraptor mongoliensis.*

bioapatite apatite (a phosphate mineral) that has an organic source.

biochronology the study of the age of the earth based on the relative dating of rocks and geological events, by the use of fossil evidence.

bioclimatology the study of the way in which climate relates to and affects the activities and characteristics of plants and animals.

biodiversity the taxonomic or ecological variety of species at a given time or in a given area.

biodynamics the study of the factors that affect or cause the metabolism of organisms. Thus, **biodynamic.**

biogenic 1. resulting from the actions of living organisms. 2. necessary for life and life processes.

biogeochemistry the study of the interrelationship between plant and animal life and the chemical makeup of the earth.

biogeography the study of the location and distribution of animal and plant life in the earth's environment. Thus, **biogeographer.**

biomass 1. the total estimated weight or body mass of all the organisms in a given habitat or site. 2. the complete body tissue of one individual animal.

biomechanics 1. the science or activity of examining organisms in the context of mechanical laws and principles. 2. specifically, an analysis of the motion of the body of a given organism.

biometrics the statistical study of issues or events in biology; a mathematical analysis of biological data. Also, **biometry.**

biomineralization the process in which animal or plant material becomes converted to mineral material.

biorhythms the collective array of internal conditions and processes (e.g., temperature, diet) that affect the growth and metabolism of an organism.

biostratigraphy the study of the distribution of fossils in distinct layers of rock.

bioturbation the disturbance of the soil surface or subsurface by living organisms; e.g., the extension of plant roots or the burrowing of moles, gophers, and the like.

biozone 1. the time period in which a given species has existed or did exist. 2. an area that includes all the strata deposited during the period of time that a given taxon lived.

bipedality the fact or ability of walking on two legs rather than four.

bivariate describing a variable condition that occurs simultaneously with another variable; usually depicted in a graph.

body fossil a fossil consisting of an actual body part or parts of an organism, as opposed to a mark left by the activity of the organism.

bonebed or **bone bed** a sedimentary layer having a significant concentration of fossil bones, bone fragments, and other such organic remains.

boss a raised ridge or rounded body part; e.g., such a part on the front central area of an animal's facial horns, above the eyes and nose.

bp or **BP** an abbreviation for "before present;" i.e., before the present time.

braincase the portion of the skull that encloses and protects the brain.

bryophyte a member of the Bryophyta, the large group of plants lacking vascular tissues and true roots, stems, and leaves; i.e., mosses, liverworts, and hornworts.

calcaneum the bone that forms the heel or rear part of the hindfoot. Thus, **calcaneal.**

caliche 1. a crust of salt or lime on the surface of materials such as bone or stone after they have been buried or exposed to moisture for an extended time. 2. an opaque concentration of calcium carbonate and other minerals found in layers at or near the surface of stony soils in arid or semiarid regions.

cancellous bone spongy bone; a type of bone having tissues that are not closely packed. Compare COMPACT BONE.

carapace a tough, shieldlike layer of material covering all or part of the dorsal surface of an animal.

carnivore an animal whose diet consists exclusively or mainly of the flesh of other animals. Thus, **carnivorous.**

carotid either of the two main arteries carrying blood from the heart to the head.

carpal 1. describing or relating to the wrist, or a structure analogous to the wrist. 2. the bone of the wrist.

catastrophism 1. the historic view, now generally rejected, that the observable differences in form and distribution of fossils were caused by cataclysms that no longer occur, followed by new periods of creation. Compare UNIFORMITARIANISM. 2. the current view that certain mass extinctions were the result of such cataclysmic events as the impact of very large meteorites.

CAT scan computer(ized) axial tomography scan, the technique of using a computer to process data from a tomograph, in order to display a reconstruction of an organism's body in cross section. See also TOMOGRAPHY.

caudal in anatomy, describing or relating to the tail.

cementum a thin layer of calcified tissue that covers the enamel layer at the root of a tooth, aiding in holding the tooth in place in its socket.

centrum *plural,* **centra.** in anatomy, the body of a vertebra as opposed to the arches.

cephalopod one of a group of marine invertebrate animals (Cephalopoda) including the squid, octopus, and nautilus.

cervical in anatomy, describing or relating to the neck or a neck-shaped region.

character any distinctive feature or trait of an organism, or any difference among organisms, that can be used to construct a classification or to estimate phylogeny.

character state any of the range of expressions or conditions of a particular character.

chorion the outermost fetal membrane lying outside the embryo itself.

chronostratigraphy 1. the study of geologic history based on an analysis of the age of distinctive rock layers (or fossil aggregations) and their time sequence. 2. a set of descriptive time units derived from such a study. Thus, **chronostratigraphic.**

clade a branch of the evolutionary tree; a taxon (group of organisms) made up of a common ancestor and all of its related descendants. See also MONOPHYLETIC.

cladistics a method of classifying organisms in which hypotheses about evolutionary relationships are the basis for classification, and the criterion for establishing groups of organisms is the recency of common ancestry, based on the identification of shared, derived characters. Also, **cladism.**

cladogenesis the process of developing clades; the formation of new lineages.

cladogram a branching diagram of the distributions of synapomorphies among taxa, and hence the degree of relation among species, based on shared characters derived from a common ancestor.

clavicle the bone joining the breastbone to the shoulder blade; the collarbone.

climatology the study of climate, especially its effects on living organisms.

clone 1. a group of genetically identical cells derived from a single parent cell. 2. a sequence of DNA material from one organism, artificially inserted into another organism. 3. in popular use, an organism that has been artificially brought to life by the extraction of a cell or cells from another living (or dead) organism. 4. to produce a new, genetically identical organism in this manner.

cold-blooded a popular term describing an organism whose internal body temperature is relatively lower than that of a warm-blooded animal.

collagen a gelatinous protein that is present in all multicellular organisms, especially in connective tissue. Thus, **collagenous.**

coloniality a behavior pattern in which a large number of individuals of a single species congregate in the same limited area, as for breeding; observed in modern birds and mammals and also indicated in some dinosaurs.

compact bone a type of bone having dense, closely packed tissue, with little interstitial space. Compare CANCELLOUS BONE.

competition the simultaneous use of a limited resource by two or more species, resulting in conflicting efforts by those species for continued survival.

computerized or **computed (axial) tomography** see CAT SCAN.

condyle in anatomy, a rounded projection at the end of a bone.

congeneric describing or relating to species that belong to the same genus.

conservative **1.** tending to remain unchanged. **2.** specifically, showing a relatively low rate or extent of evolutionary change; similar to an ancestral group.

conspecific describing or relating to individuals that belong to the same species.

continental drift the theory or phenomenon that the formation of the present continents resulted from the breakup and displacement of a single huge land mass over geologic time. See PLATE TECTONICS.

convergent evolution an evolutionary change that produces similar characters in two or more distantly related forms, as a result of their common, but separate, adaptation to similar environmental conditions. Also, **convergence.**

coprolite a fossil composed of ancient animal fecal (waste) matter.

coracoid a bone that extends from the upper ventral surface of the scapula to the sternum in reptiles. In mammals, this bone is reduced to a bony process on the scapula.

coronoid a membrane bone on the upper side of the lower jaw, nearer to the back of the jaw and lacking teeth.

cosmopolitan having a worldwide or very wide distribution.

craniology the study of the physical structure of the cranium.

cranium the skull with the exception of the lower jaw, especially the portion of the skull that encloses and protects the brain. Thus, **cranial.**

crown group a group of organisms consisting of the last common ancestor of the group and all of the living descendants; e.g., Archosauria has been defined as a crown group including two extant groups of archosaurs, birds and crocodiles, plus their common fossil relatives.

crus the lower leg; the leg from the knee to the foot.

cryptic **1.** not obvious or clear from the available evidence; uncertain or ambiguous. **2.** of an organism, marked or colored in such a way as to blend in with the surrounding environment.

crystallography the scientific study of the formation and properties of crystals (i.e., solids that have a regularly repeating arrangement of atoms).

CT scan see CAT SCAN.

cursorial **1.** of limbs or digits, adapted for running. **2.** of an organism, able to move rapidly over the ground; having the behavior pattern of running. Compare GRAVIPORTAL.

cycad any of the Cycades, an order of gynosperms (cone-bearing plants), typically having broad, unbranched stems with a crown of large, pinnate leaves at the end.

cynodont "dog-toothed," as applied to mammals and some of their closest fossil relatives.

dactyl a finger or toe; a digit.

Darwinian of or relating to the research and theories of Charles Darwin (English naturalist, 1809-1882), especially the proposition that the presence of many diverse species on earth is the result of the survival of those individual organisms within a given species whose traits are more likely to ensure reproductive success, and the consequent dying out of those individuals with less favorable traits. This reinforcement of favorable traits leads to divergence within the species and ultimately to the origin of new species. Thus, **Darwinism.** See also NATURAL SELECTION.

daughter species in the process of species development, a recognizably new species that evolves from a parent species.

defensive behavior any activity engaged in by an animal to avoid falling prey to a predator.

density-dependent describing a pattern in which the rate of a biological phenomenon (e.g., reproduction) is affected by population density, either the number of individuals of an organism's own species or those of other species, or both. For example, the birth rate in a density-dependent group will tend to decrease as the surrounding population expands, and tend to increase as it declines.

density-independent describing a pattern in which the rate of a biological phenomenon (e.g., reproduction) is independent of the population density of the environment.

dental battery see BATTERY.

dentary the lower jawbone; this bone often bears the teeth of the lower jaw.

denticle a small tooth or toothlike structure.

dentin or **dentine** a hard, bonelike tissue that is the main substance of teeth.

dentition the type, number, or arrangement of the teeth.

derived **1.** in general, not original; coming later. **2.** describing a character or distinguishing feature altered from an earlier form of a species or other taxonomic group. Compare PRIMITIVE.

dermal relating to the dermis or to the skin in general.

dermis the inner layer of the skin, beneath the epidermis.

diagenesis the various processes affecting deposits and organic remains after deposition and before recovery.

diaphysis the shaft of a long bone; the portion of a bone between the extremities (epiphyses).

diapsid one of the Diapsida, a group of reptiles distinguished by the presence of two openings in the side of the skull behind the orbit (eye socket); contrasted with the synapsids (one such skull opening) and the anapsids (no opening); the group includes pterosaurs, dinosaurs, marine reptiles, crocodiles, birds, lizards and snakes.

diffractometry the study of the structure of a crystal by observing the changes in amplitude or phase of an X-ray beam or other energy waves penetrating its structure.

digit a finger, toe, or other such homologous structure.

digitigrade walking (or running) with the weight supported along the length of the toes or digits, while the hind part of the foot is more or less raised off the ground; the pattern of many four-legged animals; e.g., the modern canines and felines. Compare PLANTIGRADE, UNGULIGRADE.

dimorphism the existence of two distinct genetically determined forms of the same species, such as distinct male and female forms or distinct young and mature forms. Thus, **dimorph, dimorphous.**

dinosauroid a name proposed for a theoretical modern dinosaur with a human-like body form, based on a model of extrapolated or projected evolution since the end of Mesozoic time.

dinoturbation the trampling of the soil surface or subsurface by dinosaurs, especially by large herbivorous dinosaurs. See also BIOTURBATION.

directional selection a process of natural selection in which variation is selected for, resulting in a character change in one direction in the overall genetic makeup of a population.

display any body feature or pattern of behavior that acts as a physical signal or indicator to others, either of the same species or other species.

disruptive selection a process of natural selection in which extremes are selected for, while intermediate examples are selected against, thus resulting in divergent subpopulations. Compare STABILIZING SELECTION.

distal 1. in anatomy, far away from the center or the attached end; outer. **2.** far apart; widely spaced.

distribution the total geographic range in which a given species or group occurs.

distributional analysis the scientific study of the distribution of a given category of organisms through time and space.

diurnal of an organism, primarily or exclusively active in the daytime.

diversity 1. the fact of being diverse; existing in a variety of forms or types; e.g., a wide variety of different species. **2.** a measurement of the number of species and their population size within a given community.

diverticulum *plural,* **diverticula.** an enclosed sac that can either occur normally or be caused by the herniation of the mucous membrane through a weakness or defect in the muscular wall of a tubular organ; e.g., the intestine.

DNA deoxyribonucleic acid, the complex chemical substance within the cells of a living organism that contains or forms the genetic (hereditary) information for the organism.

dominance 1. the fact of a particular species having a more significant effect on the ecological conditions of its environment than any of the other species present there, because of its greater population, body size, aggressiveness, or other such factors. **2.** the fact of an individual animal within a species group having higher status than others of the group in choice of food, mates, living space, and so on, with others in the group deferring to this individual. **3.** the fact of one alternate form of a gene tending to mask the expression of the other corresponding form. Thus, **dominant.**

dorsal relating to or the back or upper side of an animal or body part.

ecomorphology the study of the relationship between the morphology (form) of an organism and its environment and ecology.

ecosystem a local biological community and its pattern of interaction with its environment.

ectopic out of place; in an abnormal position or location.

ectothermy the condition of acquiring most heat energy for metabolic process from the external environment; e.g., by basking. Thus, **ectotherm, ectothermic.** Compare ENDOTHERMY.

egestion the process of eliminating waste material from the digestive tract as feces.

embryo the stage of a vertebrate in early development, before birth or hatching. Thus, **embryonic.**

embryology the scientific study of embryos; the study of the development of an organism from a fertilized egg.

enamel a form of calcium phosphate found in teeth.

endemic relating to a native species or population that occurs in a very restricted range. Thus, **endemism.**

endogenous occurring within the body; relating to or involving internal processes.

endoskeleton the cartilaginous and bony skeleton, excluding any part of the skeleton that is of dermal origin. Compare EXOSKELETON.

endosteum a fibrous connective tissue that coats the internal surface of bones and that has the potential to form new bone material.

endothermy the ability to regulate body temperature through the body's own metabolism. Thus, **endotherm, endothermic.** Compare ECTOTHERMY.

enigmatic not providing definite evidence or conclusions; not clearly defined; uncertain; puzzling.

eolian another spelling of AEOLIAN: caused by wind.

epicondyle a protrusion on the surface of a bone, above its condyle (rounded projection).

epidermis the outer layer of skin, external to the dermis.

epiphysis the end portion of a long bone.

epipodium the region of the rear part of the foot, or the bones of this area. Thus, **epipodial.**

epithelial relating to or composing the epithelium.

epithelium a type of animal tissue consisting of cells that are tightly packed together with little material between them; the tissue covering the outer surface of the body and also certain internal surfaces of body cavities.

erosion a combination of various processes in which the materials of the earth's surface are loosened, dissolved, or worn away, and then are transported from one place to another by natural agents.

escarpment a cliff or steep slope produced by erosion or faulting, typically separating two level or less sloped areas.

evagination the process of turning inside out in relation to the normal position.

evolution the processes and patterns of change through time; in living organisms, descent with modification.

exogenous occurring outside the body; relating to or involving external processes.

exoskeleton hard outer supporting structures other than bone or cartilage, produced by the epidermis; e.g., scales, armor plates, teeth, hoofs, or nails, or feathers. Compare ENDOSKELETON.

extant in existence; currently living; not extinct.

extinct not in existence as a group; no longer living.

extinction the death of every member of a species or taxonomic group without any descendants, causing the worldwide disappearance of the group. See also TRUE EXTINCTION, PSEUDOEXTINCTION.

facies 1. a part of a rock whose age or fossil content is distinct from the rest of the formation. 2. the features reflecting the exact environmental conditions that existed when a rock was formed or deposited.

facultative having the ability to live and adapt to various conditions, while not being restricted to those conditions or mode of life; e.g., a **facultative quadruped** is an animal that normally moves or stands on two limbs but can also do so on four limbs.

fauna 1. the animal life of a particular area or a particular time period. 2. animal life in general; animals, as opposed to plants (flora).

faunal extinction see MASS EXTINCTION.

femur *plural,* **femora.** the long, large bone of the upper leg that forms a joint with the hip bone above and the tibia and patella below.

fenestra *plural,* **fenestrae**. in anatomy, a window-like break or opening in a surface, especially one in or between bones, as in the skull or jaw of certain vertebrates. Thus, **fenestral, fenestrated, fenestration.**

fibrolamellar bone bone material resulting from a pattern of development that involves continuous growth at a rapid rate. Thus, **fibrolamellar growth.** Compare ZONAL BONE.

fibula a bone of the lower and outer part of the leg that joins with the tibia and femur above and the tibia and talus below.

flora 1. the plant life of a particular area or a particular time period. 2. plant life in general; plants, as opposed to animals (fauna).

fluvial 1. relating to or produced by rivers. 2. situated or living near a lake, or in a lake.

food chain the related feeding patterns of a series of organisms. A typical food chain begins with green plants that derive their energy from sunlight, then continues with organisms that eat these plants, then other organisms that consume the plant-eating organisms, then decomposers that break down the dead bodies of all those organisms in such a way that they can be used as soil nutrients by plants, and so on, as the chain continues.

food pyramid a visualization of a food chain as a pyramid, with a large number of individual organisms on the lowest layer, a smaller but still numerous group on the next layer, and so on in successively smaller layers up to the top of the pyramid; e.g., an ecosystem with a large number of plants, which are eaten by numerous plant-eating animals, which in turn are eaten by relatively few meat-eating animals.

foramen *plural,* **foramina.** in anatomy, a small opening, hole, or pore between two adjacent structures, especially in a bone or in cartilage; smaller than a fenestra.

fossa *plural,* **fossae.** in anatomy, a trench, depression, or channel.

fossil a part or record of behavior of an organism that has been preserved from past time, usually in sedimentary rock.

fossiliferous fossil-bearing; having or yielding fossils.

fossilization the complex process that transforms the remains of a whole or partial organism into a solid artifact, preserving, sometimes down to molecular detail, the anatomy of the original organism.

fossil record the cumulative evidence about organisms and conditions of the past that is provided by fossils.

fossorial 1. of limbs or digits, adapted for digging or burrowing. 2. of an organism, having the behavior pattern of digging or burrowing.

frill in certain dinosaurs (i.e., neoceratopsians), a distinctive, often large plate of bone extending upward and outward at the back of the skull, formed by the parietal and squamosal bones; it may have functioned as a sexual display structure.

functional morphology the scientific study of the movements and pattern of locomotion of an organism, especially in relation to its form or structure.

fusiform in anatomy, tapered at the ends; spindle-shaped.

fusion in anatomy, the act or fact of being joined together, either naturally or abnormally; e.g., adjacent body parts. Thus, **fused.**

gait an animal's distinctive way of running or moving; i.e., on two legs or four, at a rapid or slow pace, and so on.

gastralia *singular,* **gastralium.** in certain dinosaurs, formations of bone situated in the wall of the abdomen, possibly having a role in breathing.

gastrolith a pebble or small stone that is swallowed by an animal and kept in the digestive tract; gastroliths act as an aid to digesting food and this function has also been proposed for some fossil animals.

genera see GENUS.

genome the complete set of genes within a cell or an individual of a given species.

genotype the genetic makeup of an individual organism, especially as contrasted with its external appearance. Compare PHENOTYPE.

genus *plural,* **genera.** a rank of classification in the taxonomic hierarchy, ranking above a species and thus including several closely related species.

geologic time 1. the period of time from the formation of the earth to the beginning of recorded history; prehistoric time. 2. a very long span of time extending over millions of years.

ghost lineage an assumed fossil record of a lineage not yet found, but reasonably inferred to exist, on the basis of the presence of a related group; e.g., sister groups have a unique common ancestor and thus even if one group displays a more ancient record than the other, the second can be assumed to be equally ancient.

girdle in anatomy, a curved or circular structure, especially one that encircles another structure, such as the pectoral or pelvic girdle.

Gondwana the supercontinent or land mass that fragmented millions of years ago to form the present continents of the Southern Hemisphere. Also, **Gondwanaland.**

gracile gracefully slender; having a slim and graceful form or build.

graviportal describing a land animal that is large and heavily built and thus relatively slow-moving. Compare CURSORIAL.

gregarious associating with others of the same species; living in groups rather than in isolation.

group behavior **1.** any association of two or more animals of the same species that is considered to be more than accidental or random proximity. **2.** specifically, the assemblage and interaction for a period of time of a significant number of members of the same species, as for purposes of breeding, predation, or protection.

gymnosperm any of the Gymnospermae (Pinophyta), the cone-bearing plants; a plant that bears its seeds in cones rather than in an ovary or fruit as do the flowering plants (angiosperms).

habitat the usual or typical location and environment in which an organism lives.

hallux the first digit of the foot; the big toe or another analogous structure.

hatchling a newly hatched organism; a dinosaur or other organism that has just hatched from an egg.

Haversian bone a type of secondary bone, either compact or cancellous (spongy), that replaces primary bone; it forms a series of distinctive vascular canals (**Haversian canals**); among modern vertebrates, the compact form of Haversian bone is found only in endotherms (i.e., birds, mammals) and thus its presence in certain dinosaurs suggests that they grew in similar ways. Thus, **Haversian growth, Haversian system.** (From Clopton *Havers*, 1650-1702, English physician noted for his research in bone structure.)

head-butting a distinctive behavior pattern in which two members of the same species, typically two males contesting for dominance of a group, repeatedly collide head to head with each other; observed in some modern mammals, e.g., bighorn sheep, and also speculated to have occurred in certain dinosaurs.

herbivore an organism whose diet consists exclusively or mainly of plants. Thus, **herbivorous.**

herding the behavior pattern of forming large groups or herds; typically used to refer to larger plant-eating animals.

heterochrony the condition of having a different onset and cessation of growth, or a different rate of growth for a given feature, relative to the onset, cessation, or rate of development of this same feature in an ancestor. Thus, **heterochronous, heterochronic.**

heterodonty the possession of a variety of tooth types, as in many modern mammals. Thus, **heterodontous.**

heterometabolism the condition of having significant variation in its metabolic rate; e.g., a higher rate during early growth and a lower rate as an adult. Thus, **heterometabolic.**

histology the study of the structure and function of cells and tissues. Thus, **histological.**

holotype an individual specimen that serves as the basis for naming and describing a new species; i.e., the "legal" or official name bearer for this species.

homeothermy the fact of maintaining a relatively constant body temperature regardless of changes in the environmental temperature. Thus, **homeotherm, homeothermic.**

homologous similar by reason of common ancestry, as a comparable organ or structure in two different species of organisms.

homology **1.** a condition of similarity in structure and development of organs or other parts of the body, by reason of common ancestry; the parts may or may not have similar functions. **2.** a condition in which two similar characters in different species of organisms are derived from a single character in their common ancestor.

homoplasy **1.** a structural or developmental resemblance present in two or more species of organisms but not a result of their common ancestry. **2.** in cladistics, a character present in at least two clades that is absent in the common ancestor of the two clades. Thus, **homoplasia, homoplastic.**

horizon a distinct level or layer in the geologic column, indicating a particular interval of geologic time.

humeral relating to the upper arm or forelimb.

humerus the bone of the upper arm or forelimb.

hyper- a prefix meaning "more than" or "greater than," or "excessively."

hyperextension the unusual or excessive extension of a joint, limb, or other body part. Thus, **hyperextended.**

hypermorphosis an evolutionary change in which sexual maturity occurs later in the descendant species than in the ancestor.

hypersaline highly salty; especially, having a higher salt content than ordinary seawater. Thus, **hypersalinity.**

hypertrophy the unusual or excessive enlargement or expansion of an organ or other body part. Thus, **hypertrophied.**

ichno- a prefix meaning "footprint" or "track."

ichnofauna the general array of evidence of fossilized animal tracks and other such activity in an area.

ichnofossil a fossilized mark or remnant formed in soft sediment by the movement of an animal; e.g., a footprint.

ichnogenus a genus identified on the basis of evidence of footprints or other visible signs of activity.

ichnology the study of footprints and other marks preserved as fossil evidence of animal and plant activities. Also, PALEOICHNOLOGY.

ichnorecord the collective body of evidence provided by fossilized animal tracks and other such activity in an area.

ichnospecies a species identified on the basis of evidence of footprints or other visible signs of activity.

ichnotaxonomy the classification of organisms on the basis of evidence of tracks and other such movements.

ilium the dorsal bone of the pelvic girdle (hip girdle).

index fossil a distinct, identifiable organism that is abundant and widespread in fossilized form, but that is confined to a particular span of geologic time and thus can be utilized to geologically date the rocks in which it occurs.

ingroup in a study of a group of organisms (taxa), usually presumed to have a single common ancestor, the particular group that is under study, in comparison with other closely related groups (the outgroup or groups).

inorganic not organic; not composed of living or formerly living beings.

insectivore an organism whose diet consists exclusively or mainly of insects. Thus, **insectivorous.**

in situ Latin for "in place;" in the original site; e.g., a fossil found *in situ* is regarded as being located at the site of the death of the organism, as opposed to another site to which it may have been moved by some subsequent force or activity.

integument the skin or covering of an organism or body part.

integumentary relating to or composed of skin; involving or affecting the skin.

intercalation the presence of a body of rock interbedded or interlayered with another body of different rock.

invertebrate any of those animals that do not have an internal skeleton or endoskeleton.

ischium the ventral bone of the pelvic girdle (hip girdle).

jugal **1.** in anatomy, describing a part that connects other parts in the manner of an oxen yoke. **2.** relating to the cheekbone or a structure analogous to this.

juvenility or **juvenality** the condition of being a young or immature animal.

keratin the fibrous protein matter that is the main constituent of hair, nails, horn, and other such epidermal structures in vertebrates. Thus, **keratinous.**

kinematics the study of the course or pattern of motion; e.g., an animal's motion.

K-selection a pattern of reproduction of a given species, generally characterized by late maturation, an extended life span, and significant parental involvement in the rearing of a limited number of slowly developing young; e.g., humans. Also, **K-strategy, K strategists.** Compare R-SELECTION.

K-T boundary **1.** the transition in geologic time from the end of the Cretaceous (K) to the beginning of the Tertiary (T), approximately 65 million years ago. **2.** a site or area showing evidence of rocks and other materials from both periods.

K-T extinction a phenomenon occurring at the end of the Cretaceous era, i.e., about 65 million years ago, involving the mass extinction of numerous species including many dinosaurs; regarded as a continuum of related events but not established as having a single overriding cause.

lacrimal relating to or producing lacrimal fluid (tears) from the eyes.

lacrimal bone one of two paired bones located in the medial wall of the orbit (eye socket).

lacuna *plural,* **lacunae.** **1.** an interior opening or space. **2.** specifically, the minute internal cavities found within bone.

lacustrine **1.** relating to or produced by lakes. **2.** situated or living near a river, or in a river.

LAGs lines of arrested growth, a pattern of development in an organism in which there are pauses in bone deposition and an associated slower rate of growth.

Lamarckian according to the principles of Jean Baptiste Lamarck (1744-1829, French naturalist); specifically, favoring the theory (now generally discredited) that changes in the characteristics of an organism in response to environmental effects during its lifetime can be passed on to its offspring.

lamella *plural,* **lamellae.** in anatomy, a thin sheet or plate of tissue, as is found in compact or **lamellar bone.** Thus, **lamellation.**

Laurasia the hypothetical supercontinent or land mass that fragmented millions of years ago to form the present continents of the Northern Hemisphere, including North America, Eurasia, and Greenland, but not India.

lineage a continuous line of descent from a particular ancestor; an evolutionary sequence over time.

lines of arrested growth see LAGs.

Linnaean according to the principles of Carolus Linnaeus (Latinized name of Carl von Linné, 1707-1778, Swedish botanist); specifically, employing the traditional system of classification of organisms established by Linnaeus, based on similarity of form and structure.

locomotion the ability of an organism to move from place to place, or the manner in which this is done. Thus, **locomotory.**

ma an abbreviation for "million (years) ago."

macro- a prefix meaning "very large" or "on a very large scale."

macroevolution the patterns and processes of evolution above the level of populations and species.

macrofossil a fossil that is large enough to be studied directly, without the aid of a microscope.

macroscopic **1.** visible to the naked eye; not requiring the aid of a microscope. **2.** involving a large scale or a large area of study.

malocclusion a condition in which the teeth do not close or align properly in a bite.

mammilla *plural,* **mammillae.** a nipple or nipplelike structure.

mandible **1.** the bone or bones forming the lower jaw in vertebrates. **2.** the lower bill in birds.

mandibular located in or relating to the mandible or jaw.

manus *plural,* **manus.** the hand or forefoot.

marine relating to or living in the ocean.

mass extinction the death of every member of a number of diverse animal groups due to global ecological circumstances that suggest a common or related cause. Also, FAUNAL EXTINCTION.

matrix *plural,* **matrices.** the rock or other unified natural material in which a fossil, crystal, or other element is embedded.

maxilla *plural,* **maxillae.** the bone or bones forming the upper jaw, often bearing teeth.

maxillary located in or relating to the maxilla or jaw.

medioportal describing a land animal that is characterized as moving at a moderate rate, intermediate between graviportal (slow-moving) and cursorial (running) types. Compare CURSORIAL.

megafauna **1.** larger forms of animal life, or the largest forms within a given community. **2.** specifically, animal life that is visible to the naked eye as opposed to being visible only through a microscope.

megaflora 1. large, or relatively large, plant life. 2. specifically, plant life that is visible to the naked eye.

megatracksite a term for a site in which many different dinosaur tracks are present.

megavertebrate a very large form of vertebrate life; e.g., modern elephants and whales, certain dinosaurs.

metabolism the sum of physical and chemical processes by which a living organism maintains life. Thus, **metabolic.**

metacarpal 1. located in or relating to the metacarpus (hand). 2. one of the bones of the metacarpus; usually there is one metacarpal for each digit.

metacarpus the portion of the hand (or an analogous structure) between the wrist and the fingers.

metatarsal 1. located in or relating to the metatarsus. 2. one of the bones of the metatarsus; usually there is one metatarsal for each digit.

metatarsus the area of the foot (or hindfoot) between the instep and the toes.

micro- a prefix meaning "very small" or "on a small scale."

microevolution the process of evolution within populations of organisms, such as changes in gene frequency or chromosome number.

microfauna 1. very small forms of animal life, or the smallest forms within a given community. 2. specifically, animal life that is not visible to the naked eye and can only be seen through a microscope.

microflora 1. very small forms of plant life. 2. specifically, plant life that is not visible to the naked eye and can only be seen through a microscope.

microfossil a fossil whose typical form is microscopic in size.

microplate a relatively small plate (rigid unit of the earth's crust).

microsite 1. a very precise and specific location. 2. an assemblage of microscopic fossils.

microstratigraphy the study of the stratigraphy (arrangement of layers of rock) of a very precise area.

microvertebrate a very small or relatively small vertebrate, especially one that cannot be seen with the naked eye.

migration 1. any movement of an animal from one location to another relatively distant location. 2. specifically, a behavior pattern (**migratory behavior**) involving the regular or recurring movement from one location to another by a group of animals of a given species, as a response to changes in temperature, weather, or food availability, or other environmental conditions; characteristic of various modern species, e.g., caribou, and possibly of some large dinosaurs.

migratory relating to or characterized by migration.

mineralization 1. the fact or process of organic (animal or plant) material being transformed to inorganic (mineral) material. 2. the process in which the mineral constituent of bone (e.g., calcium phosphate) is formed from other, softer tissue.

mineralogy the scientific study of the physical and chemical properties of minerals.

modern living at the present time, or during recent historic time; extant.

modification change, especially change in the form or appearance of an organism.

mollusk or **mollusc** a member of the Mollusca, bilaterally symmetrical invertebrates including clams, snails, octopuses, squid, and others; soft-bodied organisms, typically with a calcium carbonate shell.

monophyletic describing a group of organisms sharing a single common ancestor and including all the descendants of this common ancestor. Thus, **monophyly.**

monospecific occurring in or forming a single species.

morph 1. the form or structure of a given organism or species. 2. see MORPHOTYPE.

morphogenesis the development of the form and structure of an organism during the life span of that individual. Thus, **morphogenetic.**

morphological relating to or based on the study of morphology (the form of an organism).

morphology 1. the scientific study of the form and structure of organisms, especially their external form. 2. the form and structure of a given organism, considered as a whole.

morphometrics the scientific measurement or analysis of the shape or form of organisms.

morphotype one example of the differentiation in form or structure of a population that exhibits such differences. Also, MORPH.

mummy the natural preservation of an ancient animal specimen so as to include not only material that would ordinarily survive over time (e.g., bone) but also tissue that would ordinarily have disappeared (e.g., skin), thus giving the specimen a somewhat lifelike appearance. Thus, **mummified, mummification.**

musculature the characteristic arrangement or system of muscles of a given organism.

my, mya, myr an abbreviation for "million years (ago)."

mybp an abbreviation for "million years before present."

nares *singular,* **naris.** the openings of the nasal cavity; the nostrils. Thus, **narial.**

natural selection the process in evolution identified by Charles Darwin, according to which the following life processes continue to take place over the expanse of time: (1) organisms within a species vary; (2) some of these variations are inherited by offspring; (3) more offspring are produced than can possibly survive for the full lifespan of the species, given competition for limited resources; (4) usually offspring with variations favored by the environment will survive; (5) surviving offspring will in turn usually leave more offspring with variations favored by the environment; and (6) over time, these variations favored by the environment will tend to accumulate in subsequent generations.

neo- a prefix meaning "new" or "newer."

neonate a newly born organism.

neoteny 1. the fact of retaining some juvenile characteristics after reaching sexual maturity. 2. an evolutionary change in which the descendant species has a slower growth rate than the ancestral species.

nesting behavior the activity of forming a receptacle or shelter to hold and hatch eggs and to rear newborn offspring; found in modern birds and reptiles and also described in dinosaurs.

niche the unique position or status occupied by a particular species within a larger ecological community, characterized by the physical area it inhabits and by its role within the community.

nocturnal of an organism, primarily or exclusively active at night.

node-based describing a taxonomic group that is defined as the descendants of the most recent common ancestor of two other groups; e.g., Dinosauria has been defined as all the descendants of the common ancestor of *Triceratops* and birds. Also, **node-defined.** Compare STEM-BASED.

nomenclature a name, or a system of naming. See also BINOMIAL NOMENCLATURE.

nomen dubium Latin for "dubious name;" a name originally or historically proposed for a given organism or group, but subsequently considered inaccurate or inadequate for any of various reasons; e.g., a poor or incomplete description.

nomen novum Latin for "new name;" a name proposed as the replacement for an existing name.

nomen nudum Latin for "naked name;" a name that is not regarded as valid because of inadequate or lost specimens on which the description is based.

nonavian **1.** not applying to or found in birds. **2.** applying to the Dinosauria other than birds.

obligate or **obligatory** restricted or limited to a single mode of behavior or to a specific environmental condition; e.g., an **obligate quadruped** walks only on four legs, never on two.

occipital **1.** relating to the lower back of the skull. **2.** the compound bone forming this area.

occiput the lower rear portion of the skull.

occlusion the contact pattern of the bite; the manner in which the teeth close together.

olecranon a projection on the proximal end of the ulna (larger bone of the forearm or forelimb).

olfactory relating to the nose or the sense of smell, or to nerves or other body parts involved with smelling.

ontogeny the developmental history of an individual organism. Thus, **ontogenous, ontogenetic.**

oo- a prefix meaning "egg."

oocyte a developing egg cell.

oogonium *plural,* **oogonia.** a female germ cell that gives rise to an oocyte (developing egg cell).

oology the scientific study of eggs.

oospecies, oogenus, oofamily an identification of organisms by species (genus, family) according to the classification of their eggs or fossilized egg remains.

orbit the eye socket; the bony cavity in the skull in which the eyeball is located. Thus, **orbital.**

organic relating to living beings or material from living beings.

organism an individual living being; any individual living animal, plant, or other life form.

ornamentation a visible body feature of an organism that functions primarily in social behavior rather than in the life processes of the individual organism; e.g., distinctive coloration that serves to identify this individual as male or female to others of the same species.

ornith- a prefix meaning "bird" or "birdlike."

ornithology the scientific study of birds.

osseous relating to or resembling bone; bony.

ossicle **1.** a small bone or bonelike structure. **2.** one of the small bones of the middle ear.

ossification the process of forming bone.

osteo- a prefix meaning "bone" or "relating to bones."

osteoblast an immature bone cell that actively participates in bone formation. Thus, **osteoblastic.**

osteocyte a bone cell.

osteoderm a bone or bonelike structure embedded in the skin.

osteolith a fossil bone that has become completely mineralized.

osteology the scientific study of bones.

osteopathy or **osteopathology** disease or injury of the bones.

outgroup in a classification of groups of organisms (taxa), the sister group (first outgroup) or another related group (second outgroup, etc.) that is compared to the particular group under study (the ingroup).

overburden the upper part of a sedimentary layer that compresses and consolidates the underlying material.

oviparous producing eggs; bearing offspring that develop and hatch in an egg outside the maternal body; e.g., modern birds and reptiles. Thus, **oviparity.** Compare VIVIPAROUS.

pack hunting a form of behavior involving the cooperation of a group of individuals of the same species in hunting prey; e.g., modern wolves.

paedogenesis **1.** the production of offspring by an immature form of an organism. **2.** see PROGENESIS.

paedomorphosis an evolutionary change in which the adults of a descendant species retain some of the characteristics of juvenile forms of the ancestral species. Thus, **paedomorphic.** Compare PERAMORPHOSIS.

palatine **1.** relating to or located in the palate (roof of the mouth). **2.** either of two bones that form the hard palate along with the maxillae.

paleo- or **palaeo-** a prefix meaning "past" or "ancient."

paleobiogeography the study of the geographic distribution of animals and plants in the geologic past.

paleobiology **1.** the study of the living forms and life processes of the geologic past. **2.** specifically, the study of existing fossil organisms.

paleobotany the study of plants and their relationships to other organisms and to each other in the geologic past.

paleochannel a remnant of a stream channel in older rock that has been filled with or buried by the sediment of younger overlying rocks.

paleoclimate the climate during a certain interval of geologic time.

paleoclimatology the study of climate in a designated interval of the geologic past.

paleoecology the study of the relationships between organisms and their environments in the geologic past, with special attention to fossil communities.

paleoenvironment the environmental conditions of the geologic past.

paleogeography the study of the distribution and form of the earth's landmasses in the geologic past.

paleohominid a human ancestor living in the ancient past.

paleoichnology see ICHNOLOGY.

paleomagnetics the study of the earth's magnetic field through geologic time.

paleoneurology the study of the nervous system of organisms from the geologic past.

paleontology the scientific study of the life forms and activities of the geologic past, especially through the analysis of plant and animal fossils.

paleopathology the study of the effects of disease, injury, and other traumas on the organisms of the geologic past.

paleosol an ancient soil that was present or was formed in the geologic past.

paleotemperature the prevailing temperature during a certain interval of geologic time.

palynoflora a collective term for the fossilized remains of spores and pollen grains.

palynology the study of fossilized remains of spores and pollen grains.

palynomorph a microfossil formed mainly from fossilized pollen and spores, but also typically containing some animal material.

Pangea or **Pangaea** a presumed original continent, consisting of all the land area of the earth that later divided into the supercontinents of Laurasia and Gondwana, from which the present continents derived.

paraphyletic relating to a taxonomic group that includes a hypothetical common ancestor and some, but not all, the descendants of this ancestor. Thus, **paraphyly.**

parental behavior or **parenting** see ALTRICIAL BEHAVIOR.

parent species in the process of species development, an existing species that gives rise to one or more recognizably new species (daughter species) and that itself may then continue relatively unchanged or may disappear.

parietal **1.** relating to the outer wall of an organ or cavity. **2.** either of a pair of bones (sometimes fused into a single bone) forming the sides and top of the skull. **3.** relating to these bones or to the skull in general.

parsimony **1.** the scientific or philosophical principle that an uncertain phenomenon or condition can be described by the simplest, most direct, and most obvious explanation that is available, even though other, more complex explanations could also be possible. **2.** specifically, in describing evolutionary relationships, the principle that a description involving fewer steps or changes should be tested before one involving more. Thus, **parsimonious.**

patella the flat, triangular bone covering the front of the knee, or a structure analogous to the knee; the kneecap. Thus, **patellar.**

PCR polymerase chain reaction, a laboratory technique for the repeated copying of a sequence of DNA material, using specialized heat-stable enzymes of bacteria that exist at extremely high temperatures; the technique results in a large quantity of the DNAsequence and it thus permits the detection and retrieval of a specific sequence that was only a tiny fragment of the original DNA material.

pectoral relating to the chest or shoulder.

pectoral girdle the bones of the shoulders to which the arms or forelimbs are attached.

pelage the fur, hair, or other surface covering of animal.

pelvic girdle the bones of the hips.

peramorphosis an evolutionary change in which the juveniles of a descendant species display some of the characteristics of adult forms of the ancestral species. Thus, **peramorphic.** Compare PAEDOMORPHOSIS.

periosteum a fibrous connective tissue that covers bones and that has the potential to form new bone material. Thus, **periosteal.**

permineralization a process of fossil formation in which additional mineral material is deposited in the pore spaces of originally hard animal parts.

pes *plural,* **pedes.** the foot or hind foot.

phalanges *singular,* **phalanx.** the bones of the fingers or the toes, or of other structures analogous to these. Thus, **phalangeal.**

phenotype the physical appearance and structure of an individual organism, resulting from the interaction of its genetic constitution with the conditions of its environment. Compare GENOTYPE.

photoperiodic affected by or involving daily or seasonal changes in the available light.

photoperiodism physiological and behavioral responses of an organism to daily or seasonal changes in the available light. Also, **photoperiodicity.**

phyletic extinction see PSEUDOEXTINCTION.

phylogenetic relating to the evolutionary relationships within and between groups of organisms.

phylogenetic tree a branching diagram that displays evolutionary relationships among and between organisms by the metaphor of a "tree" in which branches near the top of the tree diverged more recently than those near the base of the tree.

phylogeny evolutionary relationships and history; the sequence of events that make up the evolutionary past, often depicted by a branching treelike diagram. Also, **phylogenetics, phylogenesis.**

physiology the sum of the physical processes that occur within a living organism in order to maintain life. Thus, **physiological.**

plantigrade standing or walking with the **plantar** surface (the sole of the foot) in contact with the ground; e.g., modern bears, humans. Compare DIGITIGRADE.

plate tectonics the theory or principle that the earth's crust is divided into a relatively small number of large, rigid plates that move independently in relation to one another, thus causing significant deformation and other activity along the plate margins.

plesiomorph a character that is present in an ancestral form and also retained in a descendant or descendants. Thus, **plesiomorphic, plesiomorphy.** Compare APOMORPH.

pleurodont **1.** describing a form of tooth arrangement in which the teeth are attached to the inner wall of the jawbone rather than being set in sockets or fused to the top of the jaw. **2.** an animal having such an arrangement of teeth. Thus, **pleurodontia.** Compare THECODONT, ACRODONT.

pneumatic **1.** relating to or involved in the passage of air. **2.** of a body part, conveying air or allowing the passage of air. Thus, **pneumaticity, pneumatization.**

poikilothermy the fact of having a body temperature that will vary significantly according to changes in the temperature of the surrounding environment. Thus, **poikilotherm, poikilothermic.** Compare HOMEOTHERMY.

polymerase chain reaction see PCR.

polymorph one of the two or more forms (morphs) of a group exhibiting polymorphism (see below).

polymorphic occurring in various forms; relating to or displaying polymorphism. Also, **polymorphous.**

polymorphism **1.** the fact of having significant differences in body form or structure among the individual organisms within a single species or other taxonomic group, occurring when different forms of a gene appear in the same population. **2.** the fact of an individual organism's having significantly different forms at different stages of its life cycle.

polyphyletic describing an associated group of organisms that do not share a single, exclusive common ancestor. Thus, **polyphyly.**

post- in anatomy, closer to the rear or tail than some other comparable part.

postdisplacement an evolutionary change in which a descendant species has a delayed onset of development or growth in certain structures, in comparison with the ancestral species. Compare PREDISPLACEMENT.

postmortem after death; occurring or present subsequent to the death of an organism.

pre- in anatomy, closer to the front or head than some other comparable part.

precocial describing a species in which the young are relatively advanced in development when born. Compare ALTRICIAL.

precursor **1.** in general, something that is present at an earlier time than another thing. **2.** an earlier form from which a later form is descended.

predator an animal that feeds by hunting, killing, and eating other, typically smaller, animals.

predatory behavior the activity of a predator; the fact of feeding on live prey. Also, **predation.**

predentary **1.** relating to or located at the front of the lower jaw. **2.** a bone at the front of the lower jaw, characteristic of ornithischians.

predisplacement an evolutionary change in which a descendant species has an earlier onset of development or growth in certain structures, in comparison with the ancestral species. Compare POSTDISPLACEMENT.

prehensile of a body part, used in gripping or grasping; e.g., the hand or an analogous structure, or the tail.

premortem before death; occurring or existing prior to death.

preparator a person whose work or skill is preparing dinosaur models or reconstructions for display.

primary bone bone that is formed during the growth of the organism. Compare SECONDARY BONE.

primitive **1.** in general, earlier and less developed. **2.** describing a character or feature found in the common ancestor of a species or other taxonomic group.

process a projection or outgrowth of bone.

progenesis an evolutionary change in which sexual maturity occurs earlier in the descendant species than in the ancestor.

propodium the front region of the foot, or the bones of the front of the foot. Thus, **propodial.**

proto- a prefix meaning "earliest" or "first."

proximal in anatomy, toward the center or the attached end; inner; closer.

pseudoextinction a condition in which all the individuals of a given taxon disappear, but this fact is accompanied by the appearance of one or more new taxa directly descended from this species. Dinosaurs, if assumed to include modern birds, are an example of pseudoextinction rather than true extinction. Also, PHYLETIC EXTINCTION.

pseudofossil an object that has the appearance of a fossil but actually is not one, either a natural object mistaken for a fossil or a man-made object meant to resemble a fossil.

pubic relating to or located in the region of the pubis.

pubis the bone forming the front of the pelvic girdle (hip girdle).

pulmonary relating to or involving the lungs.

quadrupedality the fact or ability of walking on four legs rather than two.

radiation **1.** the process by which a group of species diverge from a common ancestral form, resulting in an overall increase in biological diversity. **2.** the occurrence of this phenomenon over a relatively short time.

radiocarbon dating a technique of estimating the age of ancient organic materials (e.g., bone or shell) by measuring the loss over time of radioactive carbon (carbon-14), which has a precisely known rate of decay; the absorption of radiocarbon in living tissues ceases with death and thus the amount remaining gives an indication of the time elapsed since the death of the organism.

radiometric dating a method of dating that involves the measurement of the decay in certain naturally occurring radioactive isotopes (atoms of the same element that differ in atomic weight) that decay at a constant known rate; e.g., measurement of the change of potassium-40 to argon-40. Also, **radioisotopic dating, radiogenic dating.**

radius the shorter, medial bone of the forearm or forelimb.

ramus either of the two branches of the lower jaws.

rank the position of a given level of classification in relation to the levels above and below; e.g., genus is a more general category than species but less inclusive than family.

raptor **1.** a modern bird of prey, such as a falcon, hawk, or eagle. **2.** a dinosaur described as having carnivorous, predatory behavior similar to that of such birds.

ratite **1.** having a flat breastbone. **2.** one of a group of birds characterized by a flat breastbone; e.g., the modern ostrich or kiwi.

recessive the fact of one alternate form of a gene tending to be masked by the expression of the other corresponding form (the dominant form).

reconstruction **1.** the process of preparing complete individual bones, or an entire or partial skeleton, of an ancient, extinct animal by utilizing the available fossil material from this animal and, as necessary, filling in the missing elements with other materials. **2.** a bone or skeleton assembled in this manner. See also RESTORATION. **3.** a pattern of internal bone growth in which new bone replaces preexisting bone tissues.

relative age a statement of the approximate age of an object or feature in comparison with some other object or feature, rather than in terms of its age in years. Compare ABSOLUTE AGE.

respiration the process of breathing; the exchange of gases between a living organism and its surrounding atmosphere.

restoration **1.** the process of creating a complete individual model of an ancient animal by placing materials that simulate muscles and skin over a skeletal reconstruction of the animal. **2.** a sculpture or illustration prepared in this manner.

retroversion in anatomy, the tipping or turning backward of an organ or other body part. Thus, **retroverted.**

rhamphotheca the thick, horny covering of a bird's or turtle's beak, or a comparable structure at the front of the upper jaw in some dinosaurs (e.g., hadrosaurids).

robust **1.** strongly built or formed; sturdy and powerful. **2.** of a method of analysis or investigation, likely to produce an accurate inference or likely to resist falsification; verified by past results.

rostral **1.** relating to or located at the rostrum (beak or snout). **2.** a bone at the tip of the snout.

rostrum the beak or snout of an animal.

r-selection a pattern of reproduction of a given species, generally characterized by early maturation and large production of great numbers of rapidly developing young, a short life span, and negligible parental involvement in the rearing of young; e.g., mosquitoes. Also, **r-strategy, r-strategists.** Compare K-SELECTION.

rugose having a rough surface.

ruminant **1.** feeding by a distinctive process that involves taking in (plant) food quickly, partly digesting it, and then bringing it back up to the mouth and chewing it more thoroughly at a later time. **2.** an animal that feeds in this manner; e.g., modern cattle and sheep.

sacral **1.** located in or relating to the sacrum. **2.** one of the bones of the sacrum.

sacrum the group of fused vertebrae forming the attachment of the pelvis to the spinal column.

sagittal **1.** relating to the imaginary midline of the body dividing it vertically into two symmetrical halves. **2.** relating to the line of junction of the two halves of the skull.

saurian or **sauroid** **1.** a lizard. **2.** relating to or resembling a lizard.

scansorial able to climb; having the behavior pattern of climbing.

scapula *plural,* **scapulae.** the shoulder blade.

scavenger an animal that feeds on the flesh of dead animals, but that typically eats the flesh of prey which has died naturally or has been killed by others, as opposed to actively hunting and killing live prey itself.

scute a bony plate or knob under the skin.

secondary bone bone that is formed during internal reconstruction after preexisting bone tissue has been dissolved and reconstituted. Compare PRIMARY BONE.

sedimentology **1.** the study of sediments and sedimentary rocks. **2.** the sediments found in a particular area.

sediments fragments of organic or inorganic material that are carried or deposited by wind, water, or ice, and that then are accumulated in unconsolidated layers on the surface of the earth.

selection the principle that organisms possessing a certain hereditary characteristic will tend to reproduce at a more successful rate than others of the same population lacking this feature, and will thus become a higher percentage of the population in successive generations.

septum *plural,* **septa.** in anatomy, a dividing wall or partition.

sexual behavior all activity that promotes or leads to reproduction.

sexual dimorphism the fact of having distinct male and female forms of the same species, especially visible differences in coloration, body shape, size, and so on.

sister group one of the two corresponding branches in an evolutionary tree or diagram, resulting from the splitting of a single line of descent; one of two groups more closely related to each other than to any other group under consideration. Also, **sister clade.**

sister species one of a pair of species arising from a single event of speciation; a species that is the closest relative of another species.

skeletogenesis the growth or formation of the skeleton.

social behavior any association of two or more members of a given species for a period of time, in circumstances other than the usual female-male interaction for reproduction; e.g., nesting, herding, hunting in packs.

spatulate in anatomy, shaped like a spoon or spatula.

specialized of a character or feature, greatly modified from the original ancestral state, in response to specific environmental conditions. Thus, **specialization.**

speciation the formation of new species; the process by which a distinct population develops as a result of evolutionary processes and forces.

species **1.** the fundamental level of classification used in the systematic identification of living things; the category that provides, along with the next higher level of genus, the scientific name of a plant or animal. **2.** a distinct group of organisms that is so classified by any of various standards, typically on the basis of having a common ancestry and the ability to reproduce freely within the group, but not outside it.

sphenoid a large, wedge-shaped bone at the base of the skull. Thus, **sphenoidal.**

squamosal a bone at the rear side of the skull, associated with the suspension of the jaw from the skull.

squamous covered with or composed of scales; scaly. Also, **squamatic.**

stabilizing selection a process of natural selection in which genetic variation is selected against, resulting in a population from which peripheral variants are eliminated, thus maintaining the existing state of adaptation. Compare DISRUPTIVE SELECTION.

stem-based describing a taxonomic group that is defined as all those entities that share a more recent common ancestor with one group than another; e.g., Saurischia includes birds and all those dinosaurs that are closer to birds than they are to *Triceratops*. Also, **stem-defined.** Compare NODE-BASED.

sternal relating to or involving the chest (breastbone).

sternum the long, flat bone at the center of the chest; the breastbone.

stratigraphy the study and description of the manner in which rocks or other material form strata (see below) on the surface of the earth.

stratum *plural,* **strata.** a distinct, observable segment of rock or other material forming a uniform or similar layer on the surface of the earth, visibly separated from other layers above and below it.

subfamily a classification of organisms that is more precise than a family but more inclusive than a genus; a subdivision of a family.

subgenus *plural,* **subgenera.** a classification of organisms that is more precise than a genus but more inclusive than a species; a subdivision of a genus.

subspecies 1. the taxonomic rank immediately below a species; a subdivision of a species. **2.** a distinct group of organisms so classified, typically isolated geographically from other populations of the same species but potentially able to interbreed with them; also often displaying certain minor differences in form or appearance.

subtaxon *plural,* **subtaxa.** a subdivision of a taxon.

suite a group or array of characteristics associated with a given organism or species.

super- a prefix meaning "above" or "greater."

supercontinent an earlier and larger land mass composed of currently existing continents.

supra- in anatomy, a prefix meaning "over" or "above;" e.g., the *supracranial* region is the upper surface of the cranium; a *supratemporal* feature is located above the temporal bone.

symphysis 1. in anatomy, a fusion point between two structures. **2.** specifically, a joint between two bones in which the joint cavity is filled by fibrous cartilage.

symplesiomorph a plesiomorph (character present in an ancestral form, and also in descendant forms) that is shared by two or more groups, thus indicating common ancestry for these groups. Thus, **symplesiomorphic, symplesiomorphy.**

syn- a prefix meaning "associated; together" or "fused; united."

synapomorph an apomorph (character derived from, but no longer the same as, an earlier character) that is shared by two or more groups. Thus, **synapomorphic, synapomorphy.**

synapsid one of the Synapsida, a group of amniotes including mammals and their relatives closer to them than to reptiles; identified by a single opening in the skull behind the orbit (eye socket).

synchronicity the fact of two (or more) events or conditions occurring at the same time. Also, **synchrony, synchronousness.**

systematics the branch of science that involves classifying and naming organisms according to certain established, consistent principles.

talus 1. the heel. **2.** slumped rock at the base of a hill.

taphogram a visual record of the changes that have affected an assemblage of animal remains between death and the present fossil record.

taphonomy 1. the scientific study of the changes that have affected animal remains in the period between the death of the animal and the discovery and examination of the animal as part of the fossil record; the changes may include biological and geological effects as well as physical influences; e.g., scavenging, trampling, or other interference by living organisms including humans. **2.** the process by which an assemblage of fossils is formed. Thus, **taphonomic.**

tarsus 1. the region in which the leg and the foot join; the ankle joint. **2.** the bones forming this joint. Thus, **tarsal.**

taxa the plural of TAXON.

taxon *plural,* **taxa. 1.** any group of organisms considered to be sufficiently distinct from other groups to be classified together. **2.** specifically, a group of organisms classified together on the basis of sharing a single common ancestor.

taxonomic relating to taxonomy, or according to the principles of taxonomy.

taxonomy 1. the scientific theories and techniques that are involved in classifying organisms into groups according to an established set of principles. **2.** specifically, the classification of organisms into groups on the basis of common ancestry.

temporal 1. a bone on either side of the skull, forming part of its lateral surface. **2.** of or relating to this region of the skull.

terrestrial 1. living on land as opposed to living in water. **2.** living on the ground as opposed to living in trees.

territoriality a behavior pattern (**territorial behavior**) in which an organism, or group of same-species organisms, will live or spend extended time within one generally defined area, which is defended against intrusion by others of the same species.

testudoid or **testudinal** relating to or resembling a tortoise or turtle.

tetrapod 1. an animal having four legs or limbs. **2.** amphibians and amniotes.

thanatocoenosis an assemblage of fossils that formed after the death of the organisms but that does not necessarily indicate a corresponding behavioral association of these organisms in life. Also, **thanatocoenose.**

thecodont 1. describing a form of tooth arrangement in which the teeth are set in sockets. 2. an obsolete name for an animal having such an arrangement of teeth. Thus, **thecodontia.** Compare ACRODONT, PLEURODONT.

therapsid one of the Therapsida, a clade of the Synapsida. See SYNAPSID.

thermoregulation 1. the various processes by which the body of an organism maintains its internal temperature. 2. the ability to maintain a relatively constant internal body temperature (i.e, to **thermoregulate**).

tibia the inner and larger bone of the hind leg.

tomography the recording of internal body images by means of an X-ray device (**tomograph**) that moves an X-ray source in one direction as the film is moved in the opposite direction, to show a predetermined feature of interest in detail while blurring the detail of other features.

tooth battery see BATTERY.

trace any mark or other evidence left by the activity of an animal; e.g., a footprint or bite mark.

trace fossil 1. fossil evidence of the movement or activity of an organism; e.g., a surviving footprint or track, as opposed to an actual body fossil. 2. see INDEX FOSSIL.

tracksite a location of fossil footprints.

trackway a location of fossil footprints showing a pattern of movement along a certain course.

tridactyl or **tridactylous** having three digits on the hand or foot.

trilobite a member of the Trilobita, an extinct group of marine arthropods having a three-lobed body; trilobites are an important feature of the fossil record, especially the Cambrian.

trochanter either of two bony processes at the upper end of the femur.

trophic relating to food or to the feeding process. Thus, **trophism.**

true extinction the disappearance of all the individuals possessing a very similar but variable genome, without the appearance of any subsequent daughter species. Compare PSEUDOEXTINCTION.

turbinate 1. having an inverted, scroll-like shape. 2. a structure having this shape; e.g., the spongy bones of the nasal passages in living mammals and birds.

tympanic relating to, involving, or affecting the ear or eardrum.

type 1. the single individual specimen with which a scientific name is associated. 2. the taxon that determines the name of the higher taxon to which it belongs. Thus, **type genus, type species.**

type locality the geographic site at which a type specimen or type species was located and collected.

type section the original exposed location that has been used to designate a certain stratigraphic unit.

type series a collection of specimens that are selected to serve as the basis for describing the group (taxon).

type specimen an original or ideal specimen that is designated as the model example of a given species or other group of organisms, and that is used as the basis for describing the group.

ubiquitous found everywhere, or found widely.

ulna the larger bone of the forearm or forelimb.

ungual 1. a nail, claw, or hoof. 2. relating to or describing a nail, claw, or hoof.

ungulate 1. a hoofed animal. 2. relating to or describing hoofed animals.

unguligrade standing or walking on hoofs or on the tips of the digits.

uniformitarianism 1. the presumption that the laws of Nature have not changed or altered through time; in effect, a denial of the supernatural. 2. the view that the significant geological and biological changes of the past were caused by the same physical forces observable in the present. Compare CATASTROPHISM.

vascular relating to or involving the blood vessels.

vascularization the formation or development of the blood vessels.

ventral relating to the underside or lower surface of an animal or body part.

vertebra *plural,* **vertebrae.** one of the bones of the spinal column.

vertebrate any of those animals having a backbone.

vesicle in anatomy, a small cavity, sac, or air space. Thus, **vesicular.**

vestibular 1. in anatomy, relating to or designating an opening at the entrance to a canal. 2. relating to or involving the inner ear.

viviparous producing live young; bearing offspring that develop within the maternal body and then are born alive; e.g., most modern mammals. Compare OVIPAROUS.

vocalization the production of sounds by an organism.

warm-blooded a popular term describing an organism whose internal body temperature is typically higher than the external temperature, often maintained at a constant level.

ziphodont a type of tooth having an inwardly curved, serrated shape; found in some dinosaurs, lizards, and crocodiles.

zonal bone bone material resulting from a pattern of development that involves slow to moderate growth at intermittent periods. Thus, **zonal growth.** Compare FIBROLAMELLAR BONE.

INDEX

A

Index

B

C

D

F

G

H

M

N

O

P

Q

T

U

V

W

X

Y

Z

ISBN 0-12-226810-5